Keys to Neotropical and Antarctic Fauna

Thorp and Covich's Freshwater Invertebrates

A Global Series of Books on the Identification,
Ecology, and General Biology of Inland Water Invertebrates
by Experts from Around the World

Fourth Edition
Series Editor: James H. Thorp

Volume I: Ecology and General Biology
Edited by James H. Thorp and D. Christopher Rogers
Published 2015

Volume II: Keys to Nearctic Fauna
Edited by James H. Thorp and D. Christopher Rogers
Published 2016

Volume III: Keys to Neotropical Hexapoda
Edited by Neusa Hamada, James H. Thorp, and D. Christopher Rogers
Published 2018

Volume IV: Keys to Palaearctic Fauna
Edited by D. Christopher Rogers and James H. Thorp
Published 2019

Volume V: Keys to Neotropical and Antarctic Fauna
Edited by Cristina Damborenea, D. Christopher Rogers and James H. Thorp
Published 2020

Volumes in Development
Keys to Australasian Fauna
Keys to Afrotropical Fauna

Planned Future Volumes
Keys to Oriental and Oceanian Fauna
Keys to Oriental and Oceanian Hexapoda
Keys to Nearctic Hexapoda
Keys to Palaearctic Hexapoda

Related Publications
Field Guide to Freshwater Invertebrates of North America
James H. Thorp and D. Christopher Rogers
First Edition (2011)

Keys to Neotropical and Antarctic Fauna

Thorp and Covich's Freshwater Invertebrates – Volume V

Fourth Edition

Edited by

Cristina Damborenea
Consejo Nacional de Investigaciones Científicas y Técnicas,
Museo de La Plata, Facultad Ciencias Naturales y Museo,
Universidad Nacional de La Plata, Argentina

D. Christopher Rogers
Kansas Biological Survey and, The Biodiversity Institute
University of Kansas, Lawrence, KS, USA

James H. Thorp
Kansas Biological Survey and
Department of Ecology and Evolutionary Biology
University of Kansas, Lawrence, KS, USA

ELSEVIER

ACADEMIC PRESS
An imprint of Elsevier

Academic Press is an imprint of Elsevier
125 London Wall, London EC2Y 5AS, United Kingdom
525 B Street, Suite 1650, San Diego, CA 92101, United States
50 Hampshire Street, 5th Floor, Cambridge, MA 02139, United States
The Boulevard, Langford Lane, Kidlington, Oxford OX5 1GB, United Kingdom

Notices

Knowledge and best practice in this field are constantly changing. As new research and experience broaden our understanding, changes in research methods, professional practices, or medical treatment may become necessary.

Practitioners and researchers must always rely on their own experience and knowledge in evaluating and using any information, methods, compounds, or experiments described herein. In using such information or methods they should be mindful of their own safety and the safety of others, including parties for whom they have a professional responsibility.

To the fullest extent of the law, neither the Publisher nor the authors, contributors, or editors, assume any liability for any injury and/or damage to persons or property as a matter of products liability, negligence or otherwise, or from any use or operation of any methods, products, instructions, or ideas contained in the material herein.

Library of Congress Cataloging-in-Publication Data

A catalog record for this book is available from the Library of Congress

British Library Cataloguing-in-Publication Data

A catalogue record for this book is available from the British Library

ISBN: 978-0-12-804225-0

For information on all Academic Press publications visit our website at https://www.elsevier.com/books-and-journals

Publisher: Candice Janco
Acquisitions Editor: Louisa Munro
Editorial Project Manager: Emily Thomson
Production Project Manager: Paul Prasad Chandramohan
Cover Designer: Greg Harris

Typeset by TNQ Technologies

Last digit is the print number: 10 9 8 7 6 5 4 3 2

Dedications from the Editors

To Laura, Cecilia, and Santiago, my wonderful children, and to Raul for their comprehension.
María Cristina Damborenea

To our authors who made this book possible.
D. Christopher Rogers

To our wonderful children, Sara and Zachary, for loving the natural world and understanding their father's obsession with it.
James H. Thorp

Contents

Contributors to Volume V

Roberto G. Alves [Chapter 12: Phylum Annelida] Departamento Zoologia, Universidade Federal de Juiz de Fora, Juiz de Fora, MG, Brazil; email: gama.alves@ufjf.edu.br

Monica Ammon Fernandez [Chapter 11: Phylum Mollusca] Laboratório de Malacologia / Laboratório de Referência Nacional para Esquistossomose - Malacologia, Instituto Oswaldo Cruz, Fundação Oswaldo Cruz, Rio de Janeiro, Brazil; email: ammon@ioc.fiocruz.br

Sonia Barbosa dos Santos [Chapter 11: Phylum Mollusca] Laboratório de Malacologia Límnica e Terrestre, Departamento de Zoologia, Universidade do Estado do Rio de Janeiro (UERJ), Brazil; email: gundlachia@yahoo.com.br

Gary M. Barker [Chapter 11: Phylum Mollusca] Landcare Research, Lincoln, Canterbury, New Zealand; email: barkerg@landcareresearch.co.nz

Maria Isabel Berning [Chapter 11: Phylum Mollusca] Bavarian State Collection of Zoology (ZSM), Section Mollusca, Münich, Germany; email: gomez.berning@googlemail.com

Felipe Bezerra Ribeiro [Chapter 23: Class Malacostraca] Departamento de Zoologia (Laboratório de Carcinologia), Instituto de Biociências, Universidade Federal do Rio Grande do Sul, Brazil; email: fbribeiro.bio@gmail.com

Matthew G. Bolek [Chapter 10: Phylum Nematomorpha] Department of Integrative Biology, Oklahoma State University, Stillwater, Oklahoma, USA; email: bolek@okstate.edu

Georgina Bond-Buckup [Chapter 23: Class Malacostraca] Independent Researcher, retired from the Universidade Federal do Rio Grande do Sul, Porto Alegre, RS, Brazil; email: gbond_buckup@yahoo.com.br

Francisco Brusa [Chapter 5: Phylum Platyhelminthes] División Zoología Invertebrados, Museo de La Plata, FCNyM, UNLP / CONICET, Argentina; email: fbrusa@fcnym.unlp.edu.ar

Alessandra Bueno [Chapter 23: Class Malacostraca] Departamento de Biologia, Universidade Federal de Lavras, Brazil; email: aapbueno@yahoo.com.br

Luciana Burdman [Chapter 2: Protozoa] Institute of Environmental Research and Engineering, University of San Martin, CONICET, Buenos Aires, Argentina; email: lburdman@unsam.edu.ar

Ludimila Calheira [Chapter 3: Phylum Porifera] Departamento de Zoologia, Universidade Federal de Pernambuco, Recife, Brazil; email: calheiralaurindo@gmail.com

Ana Isabel Camacho [Chapter 23: Class Malacostraca] Museo Nacional de Ciencias Naturales (CSIC), Departamento de Biodiversidad y Biología Evolutiva, Madrid, Spain; email: mcnac22@mncn.csic.es

Aline Carvalho de Mattos [Chapter 11: Phylum Mollusca] Laboratório de Malacologia, / Laboratório de Referência Nacional para Esquistossomose - Malacologia, Instituto Oswaldo Cruz, Fundação Oswaldo Cruz, Rio de Janeiro, Brazil; email: amattos@ioc.fiocruz.br

Silvana Carvalho Thiengo [Chapter 11: Phylum Mollusca] Laboratório de Malacologia / Laboratório de Referência Nacional para Esquistossomose - Malacologia, Instituto Oswaldo Cruz, Fundação Oswaldo Cruz, Rio de Janeiro, Brazil; email: sthiengo@ioc.fiocruz.br

María Cristina Claps [Chapter 2: Protozoa] Instituto de Limnología "Dr. R. A. Ringuelet", La Plata, Buenos Aires, Argentina; email: claps@ilpla.edu.ar

Rosa Graciela Cohen [Chapter 19: Class Branchiopoda] Departamento de Biodiversidad y Biología Experimental, Facultad de Ciencias Exactas y Naturales, Universidad de Buenos Aires, Argentina; email: cohen@bg.fcen.uba.ar

Gonzalo A. Collado [Chapter 11: Phylum Mollusca] Facultad de Ciencias, Departamento de Ciencias Básicas, Universidad del Bío-Bío, Chillán, Chile; email: collado.gonzalo@gmail.com

Yanis Cruz-Quintana [Chapter 21: Class Copepoda] Escuela de Acuicultura y Pesquería, Facultad de Ciencias Veterinarias, Universidad Técnica de Manabí, Ecuador; email: cqyanis@gmail.com

Maria Gabriela Cuezzo [Chapter 11: Phylum Mollusca] Instituto de Biodiversidad Neotropical (CONICET-UNT), Universidad Nacional de Tucumán, Tucumán, Argentina; email: gcuezzo@webmail.unt.edu.ar

Cristina Damborenea [Chapter 1: Introduction, Chapter 2: Protozoa, Chapter 5: Phylum Platyhelminthes, Chapter 11: Phylum Mollusca, Chapter 12: Phylum Annelida] División Zoología Invertebrados, Museo de La Plata, FCNyM, UNLP / CONICET, Argentina; email: cdambor@fcnym.unlp.edu.ar

Paul De Ley [Chapter 9: Phylum Nematoda] Department of Nematology, University of California Riverside, Riverside, CA, USA; email: ude.rcu@yeled.luap

Cristina de Villalobos [Chapter 10: Phylum Nematomorpha] Museo de Ciencias Naturales, FCNyM, UNLP, Argentina; email: villalo@fcnym.unlp.edu.ar

María Irene Deserti [Chapter 4: Phylum Cnidaria] Instituto de Investigaciones Marinas y Costeras, Universidad Nacional de Mar del Plata - CONICET, Mar del Plata, Buenos Aires, Argentina; email: mdeserti@mdp.edu.ar

Edinaldo Nelson dos Santos-Silva [Chapter 21: Class Copepoda] Instituto de Pesquisas da Amazônia, Manaus, AM, Brazil; email: nelson@inpa.gov.br

Marcelo E. Doucet [Chapter 9: Phylum Nematoda] Instituto de Diversidad y Ecología Animal (CONICET-UNC) and Centro de Zoología Aplicada, Universidad Nacional de Córdoba, Córdoba, Argentina; email: mdoucet@efn.uncor.edu

Maria Cristina Dreher Mansur [Chapter 11: Phylum Mollusca] Grupo de Pesquisa do CNPq, Biodiversidade de Moluscos Continentais, Museu de Ciências Naturais, Porto Alegre, RS, Brazil; email: mcrismansur@gmail.com

Elizangela Feitosa da Silva [Chapter 11: Phylum Mollusca] Laboratório de Referência Nacional para Esquistossomose, Malacologia, Instituto Oswaldo Cruz, Fundação Oswaldo Cruz, Rio de Janeiro, Brazil; email: efeitosa@ioc.fiocruz.br

Leonardo Fernández [Chapter 2: Protozoa] CIRENYS, Universidad Bernardo O'Higgins, Santiago, Chile; email: limnoleo@gmail.com

Nelson Ferretti [Chapter 16: Phylum Arthropoda: Introduction and Arachnida] Instituto de Ciencias Biológicas y Biomédicas del Sur (INBIO-SUR-CONICET-UNS), Bahıa Blanca, Buenos Aires, Argentina; email: nferretti@conicet.gov.ar

Diego Fontaneto [Chapter 8: Phylum Rotifera] Water Research Institute (IRSA), National Research Council (CNR), Verbania Pallanza, Italy; email: d.fontaneto@ise.cnr.it

Natalia A. Fredes [Chapter 16: Phylum Arthropoda: Introduction and Arachnida] Departamento de Biología, Universidad Nacional de Mar del Plata, Argentina; email: fredesnaty@gmail.com

Stuart R. Gelder [Chapter 12: Phylum Annelida] Department of Science and Math, University of Maine at Presque Isle, Presque Isle, Maine 04769, USA; email: stuart.gelder@umpi.edu

João Gil [Chapter 11: Phylum Annelida] Centre d'Estudis Avançats de Blanes, CEAB-CSIC, Blanes, Catalunya, Spain; email: gil@ceab.csic.es

Christopher J. Glasby [Chapter 12: Phylum Annelida] Museum & Art Gallery of the Northern Territory, Darwin, Australia; email: Chris.glasby@nt.gov.au

Samuel Gómez [Chapter 21: Class Copepoda] Instituto de Ciencias del Mar y Limnología, UNAM. Unidad Académica Mazatlán, Mazatlán, Mexico; email: samuelgomez67@gmail.com

Exequiel R. González [Chapter 23: Class Malacostraca] Universidad Santo Tomás, Santiago, Chile; email: egonzalezb@santotomas.cl

Priscila A. Grohmann [Chapter 4: Phylum Cnidaria] Instituto de Biologia, Departamento de Zoologia, Universidade Federal do Rio de Janeiro, Rio de Janeiro, Brazil; email: grohmann@biologia.ufrj.br

Jürgen Guerrero-Kommritz [Chapter 23: Class Malacostraca] Development of Applied Biology, Fundabas Foundation, Bogotá, Colombia; email: greledone@hotmail.com

Roberto Guidetti [Chapter 15: Phylum Tardigrada] Department of Life Sciences, University of Modena and Reggio Emilia, Modena, Italy; email: roberto.guidetti@unimore.it

Martha A. Gutiérrez-Aguirre [Chapter 21: Class Copepoda] Universidad de Quintana Roo, Cozumel, Mexico; email: margutierrez@uqroo.edu.mx

Diego Eduardo Gutiérrez Grégoric [Chapter 11: Phylum Mollusca] División Zoología Invertebrados, Museo de La Plata, FCNyM, UNLP / CONICET, Argentina; email: dieguty@yahoo.com.ar

Neusa Hamada [Chapter. 17: Phylum Arthropoda: Introduction to Crustacea and Hexapoda] Coordenação de Biodiversidade, Instituto Nacional de Pesquisas da Amazônia-INPA, Manaus, AM, Brazil; email: nhamada@inpa.gov.br

Ben Hanelt [Chapter 10: Phylum Nematomorpha] Department of Biology, University of New Mexico, Albuquerque, NM, USA; email: bhanelt@unm.edu

Brenda J. Hann [Chapter 19: Branchiopoda] Department of Biological Sciences, University of Manitoba, Canada; email: hann@cc.umanitoba.ca

Janet Higuti [Chapter 20: Ostracoda] State University of Maringá, Centre for Research on Limnology, Ichthyology and Aquaculture, Graduate Programme in Ecology of Inland Water Ecosystems, Maringá, Paraná, Brazil; email: janethiguti@gmail.com

Cristián Ituarte [Chapter 11: Phylum Mollusca] División Zoología Invertebrados, Museo Argentino de Ciencias Naturales, Buenos Aires, Argentina; email: ituarte@macn.gov.ar

Carlos G. Jara [Chapter 23: Class Malacostraca] Instituto de Ciencias Marinas y Limnológicas, Universidad Austral de Chile, Valdivia, Chile; email: cjara@uaChapter.cl

Susana B. José de Paggi [Chapter 8: Phylum Rotifera] Instituto de Limnologia, Santa Fe, Argentina; email: sjdepaggi@gmail.com

Łukasz Kaczmarek [Chapter 15: Phylum Tardigrada] Department of Animal Taxonomy and Ecology, Adam Mickiewicz University in Poznań, Poznań, Poland; email: kaczmar@amu.edu.pl

Anush Kosakyan [Chapter 2: Protozoa] Institute of Parasitology, Biology Centre of the Czech Academy of Sciences, České Budějovice, Czech Republic / Institute of Biosciences, Department of Zoology, University of Sao Paulo, Brazil; email: anna.kosakyan@gmail.com

Valentyna Krashevska [Chapter 2: Protozoa] J.F. Blumenbach Institute of Zoology and Anthropology, Georg August University Göttingen, Göttingen, Germany; email: valentyna.krashevska@biologie.uni-goettingen.de

Gabriela C. Küppers [Chapter 2: Protozoa] División Invertebrados, Museo Argentino de Ciencias Naturales "Bernardino Rivadavia", Buenos Aires, Argentina; email: gabrielakuppers@gmail.com

Enrique M. Lara [Chapter 2: Protozoa] Laboratory of Soil Biodiversity, University of Neuchâtel, Neuchâtel, Switzerland / Real Jardín Botánico de Madrid, CSIC, Spain; email: enrique.lara@rjb.csic.es

Paola Lax [Chapter 9: Phylum Nematoda] Instituto de Diversidad y Ecología Animal (CONICET-UNC) and Centro de Zoología Aplicada, Universidad Nacional de Córdoba, 5000 Córdoba, Argentina; email: laxpaola@gmail.com

Ana Maria Leal-Zanchet [Chapter 5: Phylum Platyhelminthes] Instituto de Pesquisas de Planárias and Programa de Pós-Graduação em Biologia, Universidade do Vale do Rio dos Sinos, UNISINOS, São Leopoldo, Rio Grande do Sul, Brazil; email: zanchet@unisinos.br

Luiz Eduardo Macedo de Lacerda [Chapter 11: Phylum Mollusca] Laboratório de Malacologia Límnica e Terrestre, Departamento de Zoologia, Universidade do Estado do Rio de Janeiro (UERJ), Brazil; email: lacerdauerjbio@yahoo.com.br

Célio Magalhães [Chapter 23: Class Malacostraca] Instituto Nacional de Pesquisas da Amazônia, Manaus, AM, Brazil; email: celiomagalhaes.inpa@gmail.com

Daniel Mansur Pimpão [Chapter 11: Phylum Mollusca] Instituto Brasileiro do Meio Ambiente e dos Recursos Naturais Renováveis (IBAMA), GO, Brazil; email: danielpimpao@yahoo.com.br

Fernando L. Mantelatto [Chapter 23: Class Malacostraca] Laboratório de Bioecologia e Sistemática de Crustáceos (LBSC), Departamento of Biologia, Faculdade de Filosofia, Ciências e Letras de Ribeirão Preto (FFCLRP), Universidade de São Paulo (USP), Ribeirão Preto, SP, Brazil; email: flmantel@usp.br

Mercedes R. Marchese [Chapter 12: Phylum Annelida] Instituto Nacional de Limnología (CONICET-UNL), Santa Fe, Argentina; email: mercedes.marchese@gmail.com

María Cristina Marinone [Chapter 8: Phylum Rotifera] Departamento de Biodiversidad y Biología Experimental, Facultad de Ciencias Exactas y Naturales, Universidad de Buenos Aires, Argentina; email: marinone@bg.fcen.uba.ar

Taisa M. Marques [Chapter 21: Class Copepoda] Laboratório de Ecologia Molecular e Parasitologia Evolutiva, Universidade Federal do, Paraná, Curitiba, Paraná, Brazil; email: taisaorsi@gmail.com

Koen Martens [Chapter 20: Class Ostracoda] Royal Belgian Institute of Natural Sciences (RBINS), Brussels, Belgium; Department of Biology, University of Ghent, Gent, Belgium; email: darwinula@gmail.com

Daniel Martin [Chapter 11: Phylum Annelida] Centre d'Estudis Avançats de Blanes, CEAB-CSIC, Catalunya, Spain; email: dani@ceab.csic.es

Pablo A. Martínez [Chapter 16: Phylum Arthropoda: Introduction and Arachnida] Departamento de Biología, Universidad Nacional de Mar del Plata, Argentina; email: pamartin@mdp.edu.ar

Sandra McInnes [Chapter 15: Phylum Tardigrada] British Antarctic Survey, Natural Environment Research Council, Cambridge, UK; email: s.mcinnes@bas.ac.uk

Nancy F. Mercado-Salas [Chapter 21: Class Copepoda] Centrum für Naturkunde, CeNak/Center of Natural History Hamburg, Universität Hamburg, Zoological Museum, Hamburg, Germany; Martin-Luther-King-Platz 3, 20146 Hamburg, Germany; email: nancy.mercado.salas@uni-hamburg.de

Igor Christo Miyahira [Chapter 11: Phylum Mollusca] Universidade Estadual do Rio de Janeiro, Rio de Janeiro, Brazil; email: icmiyahira@yahoo.com.bre

Diane R. Nelson [Chapter 15: Phylum Tardigrada] Department of Biological Sciences, East Tennessee State University, USA; email: nelsond@etsu.edu

Carolina Noreña [Chapter 5: Phylum Platyhelminthes] Departamento Biodiversidad y Biología Evolutiva, Museo Nacional de Ciencias Naturales (CSIC), Madrid, Spain; email: mcnnj92@mncn.csic.es

Thiago Nunes Antoniazzi [Chapter 11: Phylum Mollusca] Universidade Federal do Rio Grande do Sul, Porto Alegre, Brazil; email: thiago.antoniazzi@hotmail.com

Alejandro Oceguera-Figueroa [Chapter 12: Phylum Annelida] Laboratorio de Helmintología, Instituto de Biología, Universidad Nacional Autónoma de México; email: aoceguera@ib.unam.mx

Janine Oliveira Arruda [Chapter 11: Phylum Mollusca] Grupo de Pesquisa do CNPq, Biodiversidade de Moluscos Continentais, Museu de Ciências Naturais, Porto Alegre, Rio Grande do Sul, Brazil; email: arrudajo@gmail.com

Maria Cristina Orellana Liebbe [Chapter 13: Phylum Ectoprocta (Bryozoa)] Facultad de Ciencias, Universidad Católica de la Santísima Concepción, Concepción, Chile; email: mcorella@ucsc.cl

Ximena Maria Constanza Ovando [Chapter 11: Phylum Mollusca] Laboratório de Malacologia Límnica e Terrestre, Departamento de Zoologia, Universidade do Estado do Rio de Janeiro (UERJ), Brazil; email: velliger08@gmail.com

Manuel Pedraza [Chapter 23: Class Malacostraca] Programa de Pós-Graduação em Sistemática, Taxonomia Animal e Biodiversidade, Museu de Zoologia da Universidade de São Paulo, Brazil; email: manupedrazam@gmail.com

Carlos Pedraza-Lara [Chapter 23 Class Malacostraca] Licenciatura en Ciencia Forense, Facultad de Medicina, Universidad Nacional Autónoma de México, Ciudad Universitaria, Distrito Federal, Mexico; email: pedrazal@gmail.com

Almir Rogério Pepato [Chapter 16: Phylum Arthropoda: Introduction and Arachnida] Departamento de Zoologia, Universidade Federal de Minas Gerais, Belo Horizonte, Brazil; email: apepato@gmail.com

Marcela Peralta [Chapter 23: Class Malacostraca] Instituto de Invertebrados, Fundación Miguel Lillo, San Miguel de Tucumán, Argentina; email: maperalta@csnat.unt.edu.ar

Gilmar Perbiche-Neves [Chapter 21: Class Copepoda] Departamento de Hidrobiologia, Universidade Federal de São Carlos, SP, Brazil; email: gilmarperbi Chaptere83@gmail.com

Ulisses Pinheiro [Chapter 3: Phylum Porifera] Departmento de Zoologia Universidade Federal de Pernambuco, Recife, Brazil; email: uspinheiro@hotmail.com

Fabio Bettini Pitombo [Chapter 18: Class Cirripedia] Departamento de Biologia Marinha, Universidade Federal Fluminense, Niterói, RJ, Brazil; email: fpitombo@gmail.com

Jean-Pierre Pointier [Chapter 11: Phylum Mollusca] Laboratoire Ecosystèmes Aquatiques Tropicaux et Méditerranéens, Université de Perpignan, France; email: pointier@univ-perp.fr

Gabriel Pompozzi [Chapter 16: Phylum Arthropoda: Introduction and Arachnida] Sección Entomología, Facultad de Ciencias, Universidad de la República, Montevideo, Uruguay; email: gabrielpompozzi@gmail.com

Daniel Previattelli [Chapter 21: Class Copepoda] Departamento de Zoologia, Instituto de Biociências, Universidade de São Paulo, São Paulo, SP, Brazil; email: dpreviattelli@gmail.com

W. Wayne Price [Chapter 23: Class Malacostraca] Department of Biology, University of Tampa, Tampa, Florida, USA; email: wprice@ut.edu

Thiago Quintão Araújo [Chapter 7: Phylum Gastrotricha] Laboratório de Evolução de Organismos Meiofaunais, Departamento de Biologia Animal, Instituto de Biologia, Universidade Estadual de Campinas, Campinas, São Paulo, Brazil; email: araujotq@gmail.com

Lorena Rebecchi [Chapter 15: Phylum Tardigrada] Department of Life Sciences, University of Modena and Reggio Emilia, Italy; email: lorena.rebecchi@unimore.it

Carlos E.F. da Rocha [Chapter 21: Class Copepoda] Departamento de Zoologia, Instituto de Biociências, Universidade de São Paulo, São Paulo, SP, Brazil; email: cefrocha@usp.br

D. Christopher Rogers [Chapter 1: Introduction, Chapter 16: Phylum Arthropoda: Introduction and Arachnida, Chapter 17: Phylum Arthropoda: Introduction to Crustacea and Hexapoda, Chapter 19: Class Branchiopoda, Chapter 23: Class Malacostraca] Kansas Biological Survey and, The Biodiversity Institute, University of Kansas, Lawrence, KS, USA; email: branchiopod@gmail.com

Erich Harry Rudolph Latorre [Chapter 23: Class Malacostraca] Departamento de Ciencias Biológicas y Biodiversidad, Universidad de Los Lagos, Osorno, Chile; email: erudolph@ulagos.cl

Ana María Santana-Piñeros [Chapter 21 Class Copepoda] Facultad de Ciencias Veterinarias, Universidad Técnica de Manabí, Ecuador; email: anasantana4@gmail.com

Sandro Santos [Chapter 23: Class Malacostraca] Departamento de Ecologia e Evolução, Universidade Federal de Santa Maria, Santa Maria, RS, Brazil; email: sandro.santos30@gmail.com

Andreas Schmidt-Rhaesa [Chapter 10: Phylum Nematomorpha] Zoological Museum, University Hamburg, Hamburg, Germany; email: andreas.schmidt-rhaesa@uni-hamburg.de

André Rinaldo Senna Garraffoni [Chapter 7: Phylum Gastrotricha] Laboratório de Evolução de Organismos Meiofaunais, Departamento de Biologia Animal, Instituto de Biologia, Universidade Estadual de Campinas, São Paulo, Brazil; email: arsg@unicamp.br

Ferry Siemensma [Chapter 2: Protozoa] Julianaweg 10, 1241VW Kortenhoef, The Netherlands; email: ferry@arcella.nl

Pedro Henrique da Silva Conceição [Chapter 16: Phylum Arthropoda: Introduction and Arachnida] Departamento de Zoologia, Universidade Federal de Minas Gerais, Belo Horizonte, Brazil; email: pedro.katekyo@gmail.com

Thiago da Silva Paiva [Chapter 2: Protozoa] Laboratório de Protistologia, Departamento de Zoologia, Instituto de Biologia, CCS, Universidade Federal do Rio de Janeiro, RJ, Brazil; email: tpaiva@biologia.ufrj.br

Sérgio N. Stampar [Chapter 4: Phylum Cnidaria] Departamento de Ciências Biológicas, Universidade Estadual Paulista (UNESP), FCL / Campus de Assis, São Paulo, Brazil; email: stampar@assis.unesp.br

Malin Strand [Chapter 6: Phylum Nemertea] Swedish Species Information Centre, Swedish University of Agricultural Sciences, Uppsala, Sweden; email: malin.strand@slu.se

Eduardo Suárez-Morales [Chapter 21: Class Copepoda, Chapter 22: Class Branchiura] El Colegio de la Frontera Sur, Chetumal, Mexico; email: esuarez@ecosur.mx

Per Sundberg [Chapter 6: Phylum Nemertea] Department of Marine Sciences, University of Gothenburg, Gothenburg, Sweden; email: per.sundberg@marine.gu.se

James H. Thorp [Chapter 1: Introduction; Chapter 17: Phylum Arthropoda: Introduction to Crustacea and Hexapoda] Kansas Biological Survey and Department of Ecology and Evolutionary Biology, University of Kansas, USA; email: thorp@ku.edu

Tarmo Timm [Chapter 12: Phylum Annelida] Estonian University of Life Sciences, Centre for Limnology, Tartumaa, Estonia; email: Tarmo.Timm@emu.ee

Antonio Alejandro Vázquez [Chapter 11: Phylum Mollusca] Laboratory of Malacology, Institute of Tropical Medicine Pedro Kourí, Havana, Cuba; email: antonivp@ipk.sld.cu

Robert Wallace [Chapter 8: Phylum Rotifera] Department of Biology, Ripon College, Ripon, WI, USA; email: WallaceR@ripon.edu

Timothy S. Wood [Chapter 13: Phylum Bryozoa, Chapter 14: Phylum Entoprocta] Department of Biological Sciences, Wright State University, Dayton, Ohio, USA; email: tim.wood@wright.edu

Fernanda Zanca [Chapter 10: Phylum Nematomorpha] Centro de Estudios Parasitológicos y de Vectores (CEPAVE, CONICET-UNLP), La Plata, Argentina; email: fmzanca@fcnym.unlp.edu.ar

About the Editors

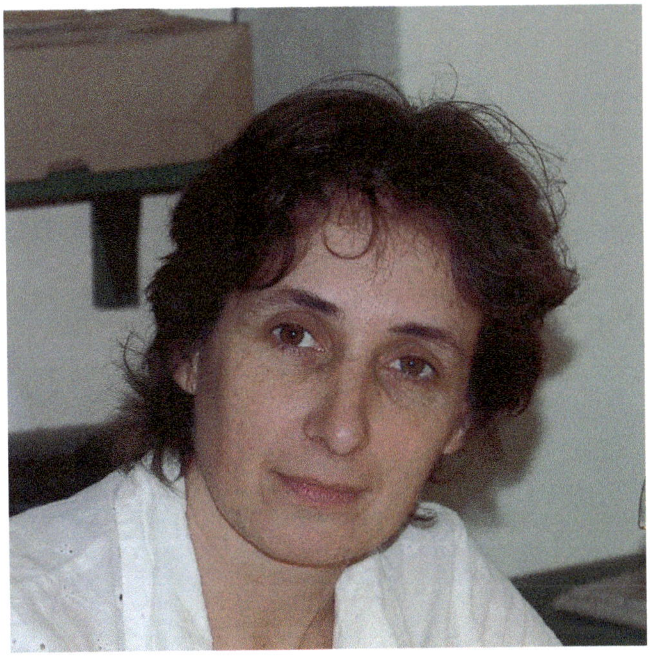

Dr. M. Cristina Damborenea is a research zoologist at the Consejo Nacional de Investigaciones Cientificas & Técnicas (Argentina), Curator of Invertebrate Collections of Museo de La Plata (Argentina), and Professor of Invertebrate Zoology at Facultad de Ciencias Naturales y Museo, La Plata University (Argentina). She not only specializes in free living Platyhelminthes but also studies other noninsect groups of aquatic invertebrates as well as aquatic invasive species in South America. Cristina Damborenea created and leads a research group in free living Platyhelminthes of the Neotropical region. She has numerous peer-reviewed publications focused on the taxonomy and ecology of invertebrates, invasive mollusks, scientific field guides, and studies of the dissemination of scientific knowledge.

Dr. D. Christopher Rogers is a research zoologist at the University of Kansas with the Kansas Biological Survey and is affiliated with the Biodiversity Institute, with numerous research projects all over the world. He received his PhD degree from the University of New England in Armidale, NSW, Australia. Christopher specializes in freshwater and terrestrial crustaceans (particularly Branchiopoda and Malacostraca) and the invertebrate fauna of seasonally astatic wetlands on a global scale. He has more than 150 peer-reviewed publications in crustacean taxonomy and invertebrate ecology, as well as published popular and scientific field guides and identification manuals to freshwater invertebrates. Christopher is an Associate Editor for the Journal of Crustacean Biology and a founding member of the Southwest Association of Freshwater Invertebrate Taxonomists. He has been involved in aquatic invertebrate conservation efforts all over the world.

Dr. James H. Thorp has been a Professor in the Department of Ecology and Evolutionary Biology at the University of Kansas (Lawrence, KS, USA) and a Senior Scientist in the Kansas Biological Survey since 2001. Prior to returning to his alma mater, Professor Thorp was a Distinguished Professor and Dean at Clarkson University, Department Chair and Professor at the University of Louisville, Associate Professor and Director of the Calder Ecology Center of Fordham University, Visiting Associate Professor at Cornell, and Research Ecologist at the University of Georgia's Savannah River Ecology Laboratory. He received his Baccalaureate from the University of Kansas (KU) and both Masters and PhD degrees from North Carolina State. Those degrees focused on zoology, ecology, and marine biology, with an emphasis on the ecology of freshwater and marine invertebrates. Dr. Thorp has been on the editorial board of three freshwater journals and is a former President of the International Society for River Science. He teaches freshwater, marine, and invertebrate courses at KU, and his Master and Doctoral graduate students work on various aspects of the ecology of communities through macrosystems in rivers, reservoirs, and wetlands. Professor Thorp's research interests and background are highly diverse and span the gamut from organismal biology to community, ecosystem, and macrosystem ecology. He works on both fundamental and applied research topics using descriptive, experimental, and modeling approaches in the field and lab. While his research emphasizes aquatic invertebrates, he also studies fish ecology, especially as related to food webs. He has published more than 130 refereed journal articles, books, and chapters, including three single-volume editions of *Ecology and Classification of North American Freshwater Invertebrates* (edited by J.H. Thorp and A.P. Covich) and five volumes in the current fourth edition of *Thorp and Covich's Freshwater Invertebrates.*

Preface to the Fourth Edition

Those readers familiar with the first three editions of our invertebrate book (*Ecology and Classification of North American Freshwater Invertebrates*, edited by J.H. Thorp and A.P. Covich) will note that the fourth edition has expanded from a North American focus to worldwide coverage of inland water invertebrates. We gave our book series on inland water invertebrates the name *Thorp and Covich's Freshwater Invertebrates* to: (a) associate present with past editions, unite current volumes, and link to future editions; (b) establish a connection between the ecological and general biology coverage in Volume I with the taxonomic keys in the remaining volumes; and (c) give credit to Professor Alan Covich for his work on the first three editions. For the sake of brevity, we refer to the current edition as T&C V. Whether the fifth edition of T&C will ever appear is certainly problematic, but who knows? At present, we are considering producing up to 11 volumes in the fourth edition.

While I am the sole series editor at this point, Christopher has been a major and highly valued partner in developing ideas for the fourth edition and has been a coeditor on all five volumes now in print (senior editor on Volume IV). He will also likely play a major role in many of the remaining volumes because of his diverse and global knowledge of freshwater invertebrates, especially in the area of taxonomy. In the next decade, we hope to publish 4–6 more volumes, with an Australasian volume likely to appear in 2021. We are still seeking experts in fields of invertebrate taxonomy for various zoogeographic regions to serve as highly dependable coeditors, especially those who both work and live in the zoogeographic regions covered by the various future volumes.

Our concept for T&C IV included producing one book (Volume I, published in 2015) with six chapters on general environmental issues applicable to many invertebrates, followed by 35 chapters devoted to individual taxa at various levels (order to phylum, or even multiple phyla in the case of the protozoa). Volume I was designed both as an independent book on ecology and general biology of the various freshwater invertebrate taxa and as a companion volume for users of the keys in the regional taxonomic volumes, thereby reducing the amount of information duplicated in the taxonomic volumes. All taxonomic volumes to be published over the next decade will contain both keys for identifying invertebrates in specific zoogeographic regions and descriptions of detailed anatomical features needed to employ those keys.

While the vast majority of authors in volumes of T&C editions I–III were from the United States or Canada, we attempted in T&C IV to attract authors from many additional countries in other continents. Although we largely succeeded in this goal, we expect the fifth edition of T&C—if it is ever published—to continue increasing the proportion of authors from outside North America as our books become better known internationally.

Our goals for T&C IV are to improve the state of taxonomic and ecological knowledge of inland water invertebrates, to help protect our aquatic biodiversity, and to encourage more students to devote their careers to working with these fascinating organisms. These goals are especially important because the verified and probable future losses of species in wetlands, ponds, lakes, creeks, and rivers around the globe exceed those in most terrestrial habitats.

James H. Thorp

Preface to Volume V

This is the fifth volume of the fourth edition of *Thorp and Covich's Freshwater Invertebrates* (T&C IV). Information on the ecology and general biology of the groups can be found in Volume I (*Ecology and General Biology*, edited by Thorp and Rogers, 2015), the companion text for the current and all remaining books in this series. All taxonomic volumes (other than those focused exclusively on Hexapoda) are expected to consist of an introductory chapter, a chapter on protozoa (multiple kingdoms), and multiple chapters on individual phyla from Cnidaria to Arthropoda. Some of the chapters are very small (e.g., Chapter 14 on Entoprocta), whereas others are very large (especially Chapter 23 on Malacostraca).

A typical chapter includes a short introduction, a brief discussion of limits to identification of taxa in that chapter, important information on terminology and morphology that is needed to use the keys, techniques for preparing and preserving material for identification (also covered in Volume I), the taxonomic keys, and a few references. In large chapters, such as those on Mollusca (11), Annelida (12), and Arthropoda (16–23), different individuals have contributed separate sections, and thus there are multiple sections on introduction through keys and references. While this may confuse some readers, it has allowed us to gain contributions from an increased number of experts around the world.

The multilevel keys are formatted to enable users to work easily at the level of their taxonomic expertise and the needs of their project. For that reason, we separated keys by major taxonomic divisions. For example, a student in a college course might work through one or more of the initial crustacean keys to determine the family in which a freshwater shrimp belongs. In contrast, someone working on an environmental monitoring project might need to identify a snail or crab to genus or even species, and thus would use the relevant, detailed keys that require more background experience.

We have asked authors to include only taxa that are recognized internationally by publication in reputable scientific journals that follow the International Code of Zoological Nomenclature. Thus, no taxa that have merely been proposed should have been included even if they have been identified by the world's expert on that group. "Common" species are not designated in the keys themselves because a common species in one area may not be common in another, and this designation can lead to overly frequent and false identifications. Authors have been encouraged to end the keys at the point where further identification without genetic analysis is not practical or when it is clear that too many of the extant fauna have yet to be described in scientific publications.

Users of these keys need to realize that taxonomy is a growing and vibrant field in which new taxa are being described and previously accepted relationships reevaluated. For some users, this volume may be sufficient for their needs, but for others, a companion text listing known species in a smaller geographic region may also be helpful.

This edition is strongly focused on species found in fresh through saline inland waters, with a nonexclusive emphasis on surface waters, thereby reflecting the bias of existing scientific literature. Again, most estuarine and parasitic species are not covered in this book, but we do discuss species whose life cycle includes a free-living stage (e.g., Nematomorpha) and species that live in hard freshwaters through to brackish waters even though they may be normally associated with estuarine or marine habitats in some parts of their life cycles (e.g., some shrimp and crabs).

It is our hope that scientists and students from around the world will benefit from this volume, especially those studying organisms in the Neotropics. Suggestions for improving future volumes are welcome.

Editors
M. Cristina Damborenea
D. Christopher Rogers
James H. Thorp

Acknowledgments for Volume V

Many people contributed to this volume in addition to the chapter authors and those acknowledged in individual chapters. We greatly appreciate all our colleagues who have contributed information, figures, or reviews to Volume V and also thank those who provided similar services for the earlier editions, upon which the present book partially relies. We are again grateful to the highly competent people at Academic Press/Elsevier who helped in many aspects of the book's production from the original concept to the final marketing. In particular, we appreciate our association with Elsevier editors including Emily Thomson and Louisa Hutchins and earlier Laura Kelleher, as well as the entire global production teams for Elsevier, especially those led by Paul Prasad Chandramohan.

Cristina Damborenea
D. Christopher Rogers
James H. Thorp

Chapter 1

Introduction[1]

James H. Thorp
Kansas Biological Survey, Lawrence, KS, United States, Department of Ecology and Evolutionary Biology, University of Kansas, Lawrence, KS, United States

D. Christopher Rogers
Kansas Biological Survey, Lawrence, KS, United States, The Biodiversity Institute, University of Kansas, Lawrence, KS, United States

Cristina Damborenea
División Zoología Invertebrados, Museo de La Plata, FCNyM-UNLP, Buenos Aires, Argentina. CONICET

Chapter Outline

INTRODUCTION TO THIS VOLUME AND CHAPTER 1

The first volume of *Thorp and Covich's Freshwater Invertebrates* (Thorp & Rogers, 2015) provided global coverage of the ecology, general biology, phylogeny, and collection techniques for inland water invertebrates. The focus of all volumes in our series after Volume I has been or will be on taxonomic keys for individual bioregions of the world. The taxonomic keys provide users with means to work at their level of need and expertise. Our authors have tried to limit the specificity of the keys to those taxa where comprehensive knowledge is available for most taxa in a particular group or where identification is only possible for certain life stages. It is our hope that these books demonstrate the limits of our knowledge on the biodiversity for different regions and taxa, and thus will inspire future studies in those areas.

In addition to the first five volumes currently in print (Thorp & Rogers, 2015, 2016; Hamada et al., 2018; Rogers & Thorp, 2019, and the current volume), a sixth volume should appear in 2021 (Keys to Australasian Fauna; Ahyong et al., 2021) and work has recently commenced on a seventh volume (Keys to Afrotropical Fauna; Day et al., 2024?). The general invertebrates and the Hexapoda of the Oriental/Oceana bioregions are expected to be covered in two future volumes, along with more detailed coverage of Hexapoda in the Nearctic and Palaearctic bioregions.

The Neotropical region as defined in this volume may have faunal overlap with the Nearctic fauna in the Baja Peninsula at the Tropic of Cancer, in the Mexican Lowlands, and through the southern Florida Peninsula. For specimens collected adjacent to these regions that do not closely correspond with the keys presented herein, please consult Volume II of this book series. Note also that the current volume treats insects only to order because greater detail appeared in Vol. III: Keys to Neotropical Hexapoda (Hamada et al., 2018).

We consider this series of 10–12 volumes to be a globally expanded fourth edition of the earlier *Ecology and Classification of North American Freshwater Invertebrates* (edited by Thorp & Covich in 1991, 2001, and 2010). While some labs may have multiple copies of the "Keys to Fauna" in their region, we also recommend that they have

1 This chapter was written to be a useful starting point for the previous, current, and future taxonomic volumes in all zoogeographic regions. Consequently, there will be only moderate differences in this chapter among various volumes.

Thorp and Covich's Freshwater Invertebrates. https://doi.org/10.1016/B978-0-12-804225-0.00001-0

at least one copy of Volume I to obtain useful background information on each invertebrate group, including global coverage of the ecology, general biology, phylogeny, and collection techniques for inland water invertebrates.

The current chapter is organized into an introduction, a description of the taxonomic focus of this particular volume, a section explaining the organization of most taxonomic chapters, and a key to larger taxonomic groups. This chapter's key is designed to help the reader locate the most pertinent chapter (a task probably useful mostly for students and relatively novice taxonomists) and begin identifying organisms in their samples. Readers will note that chapters within and among volumes vary in the specificity of their taxonomic keys. This reflects both the likely percent of the fauna that has been named and how easily taxa can be separated by alpha taxonomic methods and associated keys. While there are occasional references to genetic separation of taxa, the keys are primarily based on microscope identifications.

NATURE AND FAUNA OF THE NEOTROPICS AND ANTARCTICA

As is clear from the title, this volume covers inland water invertebrates of both the Neotropics and Antarctica, even though the former bioregion will likely represent over 99.9% of the taxa identified in this volume. Antarctica needed to be covered in one of our volumes, however, and we chose to put it here because many inland water species on Antarctica and surrounding islands occur in the southern portion of the Neotropical region, with a few also present in the southern tip of Africa and Australia. We have asked authors to include any species in their keys that have been identified from Antarctica and to indicate their general location. A map of the general region covered by Antarctica and nearby polar-type environments appears as Fig. 1.1.

For some readers, especially those from temperate zones, "neotropical" may suggest a warm climate. It is important to recognize, however, that the Neotropical bioregion stretches from the middle of Mexico (warm) through the equatorial countries (often hot) to the frigid areas of Tierra del Fuego (Argentina) and the islands of Cape Horn (Chile). It also includes both low and high altitude sites which vary greatly in their temperature range. Moreover, some regions are hot or cold deserts and others range up to very humid environments. The freshwater invertebrate fauna of these diverse regions will vary accordingly. Many of these diverse habitats are described in Chapter 2 of Volume I.

Fauna of Antarctica and Nearby Islands

Within the general Antarctic region covered in this chapter are a variety of mostly small islands (Fig. 1.1), some of which have more reported species than on the frozen continent. For example, continental Antarctica lacks permanent colonization of insects, whereas their diversity on islands range from 2 to 200 species, most of which are Diptera (flies) and Coleoptera (beetles) (Chown & Convey, 2016). Within aquatic systems, zooplankton species (e.g., rotifers and copepods) are abundant relative to most other aquatic invertebrates (other than protozoa). While additional taxa continue to be found, scientists face the problem of determining whether they are native species or newly arrived immigrants (e.g., Hughes & Convey, 2012). Table 1.1 lists the known composition of higher taxa of inland water invertebrates reported from the mainland and/or islands near Antarctica and cold Southern Ocean. This list is undoubtedly incomplete, but it will give the reader an idea of which chapters and keys to examine in this volume in search of specific taxa in these southern polar habitats and some of the scientific literature on these cold water invertebrates.

COMPONENTS OF TAXONOMIC CHAPTERS

This volume is an identification manual to the inland water invertebrates of the Neotropical and Antarctic regions where we present information needed to diagnose and determine these organisms to various taxonomic levels. Other information concerning ecology, morphology, physiology, phylogeny, and both collecting and culturing techniques can be found in Volume I of this series. All remaining chapters in the current volume are limited to a single phylum or class, except for Chapter 2's coverage of multiple phyla of unicellular protists. Chapter 2 is designed for readers who only need general information about protists. We have attempted to include the following five sections in those chapters: (1) a brief overview of the higher level taxon; (2) a description of identification limitations for each taxon; (3) details of pertinent terminology and morphology; (4) information on preparing and preserving specimens for identification; and (5) taxonomic keys (separated by level of identification). A restricted number of especially pertinent references are given in each chapter following appropriate taxonomic sections. Readers can find a much more extensive list of references to their group in Volume I (Chapters 3 and 7–41) along with more details on collecting, preparation, and preserving major taxa. Figures in each chapter are limited to those needed for effective use of the keys. For additional anatomical information, including figures, see the relevant chapter in Volume I.

HOW TO USE THIS VOLUME

There is an old maxim that says "keys are written by people who do not need them for people who cannot use

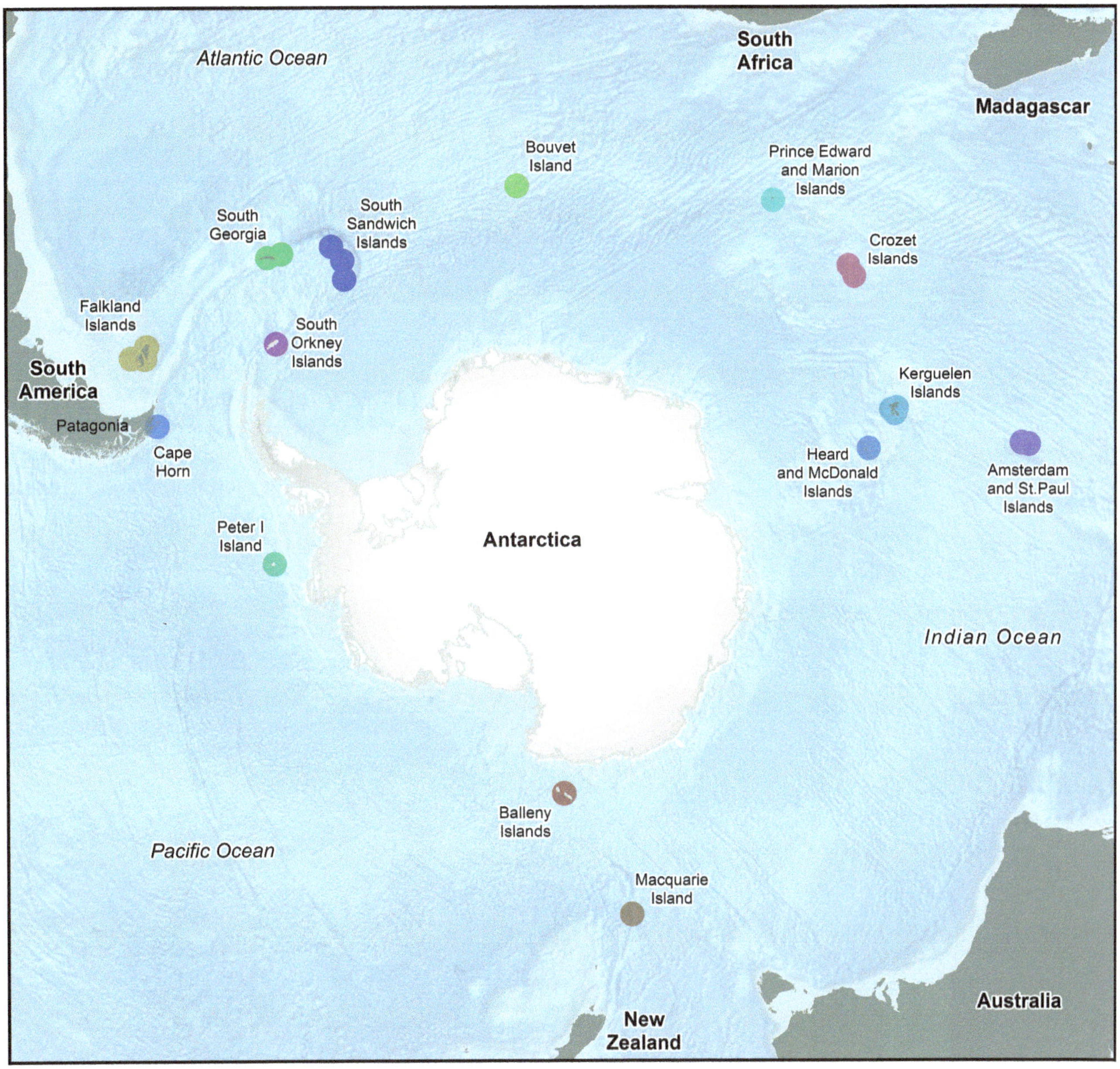

FIGURE 1.1 Map of Antarctica and surrounding islands and continents. The map shows the approximate locations of a number of islands where collection of inland water invertebrates have taken place. *Map based on a variety of sources and produced by Nicholas Kotlinksi.*

them." We have made every effort to make these keys as user-friendly as publication limitations would permit.

Each chapter includes a basic introduction to the morphology and terminology used in diagnosing the taxa of that section. Limitations to the current state of taxonomic knowledge are also presented so that the reader may gauge the reliability of the information presented. Only the established, peer reviewed scientific literature is used to define the taxonomic categories and epithets included. All names, as far as we are aware, conform to the International Code of Zoological Nomenclature. All nomina and taxonomic arrangements used, as well as the rejection of old names, was based on peer reviewed scientific literature. Names from unpublished manuscripts, dissertations, "in house" designations, or records that have not been validated are not acceptable. Provisional names and species designated "taxon 1" or "species 1" were not used unless they were previously recognized and accepted in the peer reviewed scientific literature (Richards & Rogers, 2011). No new species descriptions or previously unpublished taxonomic arrangements are presented.

The keys are dichotomous (no triplets or quadruplets are used) and are hierarchical. Thus, for a given group, the first keys are to the highest taxonomic category. The second set of keys is to the next level, the third set to the level below that one, and so on, down to the lowest justifiable taxonomic level based on current knowledge of that group. This level is different for different groups depending upon the

TABLE 1.1 Currently Reported Invertebrate Taxa Found on Antarctica and Surrounding Islands

Taxon	Reported Location	Comments and Recent References
General invertebrates	ISL	Downie et al. (2000); Dartnall (2003, 2005); Barber-James (2007); Coulson (2007); Dartnall & Smith (2012); Velasco-Castrillon et al. (2014)
Kingdom Protista	ANT, ISL	Lab cultures from water collected at freshwater sites. Tong et al. (1997); Xu et al. (2016)
Phylum Platyhelminthes	ISL	Ball & Hay (1977); Dartnall (1995, 2005); Dartnall et al. (2005); Dartnall & Smith (2012)
Phylum Gastrotricha	ISL	Dartnall (1995, 2005); Downie et al. (2000); Dartnall & Smith (2012)
Phylum Rotifera	ANT, ISL	Low diversity in Antarctica, but group is relatively abundant compared to other taxa on southern ocean islands. Dartnall (1995, 2005); Janiec (1996); Downie et al. (2000); Izaguirre et al. (2003); Dartnall et al. (2005); Camacho (2006); Segers (2008); Dartnall & Smith (2012); Fontaneto et al. (2015)
Phylum Nematoda	ANT, ISL	Dartnall (1995, 2005); Janiec (1996); Downie et al. (2000); Dartnall et al. (2005); Andrássy & Gibson (2007); Dartnall & Smith (2012); Raymond et al. (2013)
Phylum Annelida	ANT, ISL	Downie et al. (2000); Dartnall (2005); Dartnall et al. (2005); Rodriguez & Rico (2008); Dartnall & Smith (2012)
Phylum Tardigrada	ANT, ISL	Includes limnoterrestrial species. Janiec (1996); Dartnall (1995, 2005); McInnes & Pugh (1998); McInnes et al. (2001); Downie et al. (2000); Dartnall & Smith (2012)
Phylum Arthropoda		
Subphylum Chelicerata	ISL	Pugh & Dartnall (1994); Dartnall (1995, 2005); Dartnall et al. (2005); Dartnall & Smith (2012)
Subphylum Crustacea		
Class Hexapoda		
Subclass Entognatha	Questionable	Ice-free areas on ANT, but may really be accidental occurrences of terrestrial species. Dartnall & Smith (2012)
Subclass Insecta	ISL ANT	Mostly midges and diving beetles; none reported from Antarctica. Arnold & Convey (1998); Jones et al. (2003); Dartnall et al. (2005); Chown & Marshall (2008); Chown & Convey (2016); Allegrucci et al. (2012)
Class Branchiopoda	ANT, ISL	Mostly Cladocera but one anostracan. Janiec (1996); Dartnall (1995); Pugh et al. (2002); Dartnall et al. (2005); Camacho (2006); Dartnall & Hollwedel (2007); Hawes (2009); Dartnall & Smith (2012)
Class Maxillopoda	ANT ISL	Bayly (1994), Bayly et al. (2003) Dartnall (1995, 2005); Janiec (1996); Downie et al. (2000); Pugh et al. (2002); Izaguirre et al. (2003); Camacho (2006); Dartnall & Smith (2012)
Class Ostracoda	ISL	Pugh et al. (2002); Dartnall (2005); Dartnall & Smith (2012)
Class Malacostraca		Pugh et al. (2010)
Order Isopoda	ISL	Pugh et al. (2002); Dartnall et al. (2005)
Order Amphipoda	ISL	Pugh et al. (2002)

Note: This list is almost certainly incomplete and will need to be updated as more inland water habitats are explored in this area. Climate change may eventually lead to additional colonization opportunities by freshwater invertebrates. This action and ongoing possibilities for aerial colonization from other continents of some taxa may eventually increase the freshwater animal diversity on the continent and surrounding islands. See full citation in the chapter references.
ANT, presence on continental Antarctica and adjacent islands; *ISL*, islands distant from the Antarctic continent but within the Southern Ocean.

state of resolution in the scientific literature. Organisms not identifiable beyond a particular taxonomic level are left at that level.

Properly prepared keys typically employ specific, primary, diagnostic characters. Older keys often use different characters than the more recent keys. This shift in primary characters results from systematists and taxonomists testing the importance of characters. The ultimate goal of the systematist is to have the taxonomic characters used in our keys converge as closely as possible with biological reality. To a nontaxonomist, the resulting process may erroneously seem merely to be "lumping and splitting" taxonomic groups rather than the result of employing the scientific method to reveal natural relationships.

Surprisingly, many users do not know how to interpret a dichotomous key, making the fundamental assumption that a correct identification answer is always present in the key. This assumption generally takes one of the following three forms:

1. *All species are identifiable using a given key.* Many new species have yet to be described, let alone discovered. Generalized geographic ranges are provided for most taxa presented herein, yet species ranges shrink, swell, and change elevation constantly, particularly as weather and climate patterns shift. Species disperse, colonize, and suffer stochastic local extinctions. In addition to these natural processes, some species are introduced intentionally or accidentally by humans, and sometimes their establishment allows other species to invade as well. Furthermore, some regions are better studied than others: many parts of the world have not been explored, nor the fauna described. There are still plenty of new taxa to encounter.
2. *All variation is accounted for in the key.* As stated above, identification keys use specific, primary, diagnostic characters. Problems in identification are compounded by taxa that: (a) have different character states at different times; (b) only have diagnostic characters at certain life stages or in certain genders; and/or (c) have severely truncated morphology (often due to lack of sexual selection) and lack morphological characters to separate the species. Furthermore, new variation within taxa is continually developing, and thus one cannot assume that species are immutable or develop tools predicting those changes.
3. *The key is a sufficient identification tool in and of itself.* A key is just a tool. The fact that one has a bolt that needs removing and a wrench of the correct size does not mean that the bolt can be loosened. Similarly, identification keys are tools to aid in taxon identification. They are primarily tools to eliminate incorrect taxa from the range of possible choices, narrowing the field to the names that may be applicable. Keys are the process of elimination. The possibility that the specimen to be identified is new, a hybrid, anomalous, or a recent invasive colonist is always a possible answer. This is fundamental to using any identification key.

Once one arrives at a name or group of possible names for a specimen in hand, the specimen should then be compared against descriptions, distribution maps, and figures of that and other taxa in that group. The descriptions, figures, and maps are other tools to be used in identification. Direct comparison of the specimen at hand with identified museum material or using molecular comparisons is also sometimes necessary for a correct identification.

In many chapters you will discover that many or sometimes most of the species in a given taxon have not been identified or that most of the identifications are from one country or region. In those cases it is extremely important that you do not select a species name from the key unless all the descriptions fit that specimen. Instead, you should only select the next higher taxon (e.g., genus instead of species or subfamily instead of genus). To do otherwise will introduce avoidable errors into your project and possibly into the scientific literature.

Species are not immutable, fixed in location and form. They change constantly and will continue to do so, confounding keys and any other identification method, such as trait tables, character matrices, or even genetic analyses. This is why biology is far behind physics in the development of unified theories: biology is far more complex than physics, as it involves more mutable interacting parts and processes.

KEY TO KINGDOMS AND PHYLA IN THIS VOLUME

A major change in the identification keys for our fourth edition has been to include multiple keys per chapter that generally start with a class level key and proceed to finer and finer divisions. These allow users to work at their levels of interest, need, and skill without having to wade through extraneous taxa not in the direct line to the taxon of interest.

The following key was derived in part from Chapter 1 in Volume I of the fourth edition. It is meant to allow you to move to the next level of keys, which will be in individual chapters. The term "kingdom" and the specific names associated with it (i.e., Animalia and Protista) are used here loosely to designate common groupings of phyla, but these have highly questionable taxonomic significance.

1 Multicellular, heterotrophic organisms as individuals or colonies (sometimes with symbiotic autotrophs) .... kingdom Animalia 2

1′ Unicellular (or acellular) organisms present as individuals or colonies with nuclei irregularly arranged; heterotrophic and/or autotrophic; multiple phyla within the autotrophic protozoa phyla .... kingdom Protista [Chapter 2]

2(1) Radially symmetric or asymmetric organisms living individually or in colonies 3

2′ Individuals bilaterally symmetric 4

3(2) Surface not porous; oral tentacles always present around a closeable mouth; colonial or single, mostly single polyp forms (primarily hydra) or rarely medusoid form (freshwater jellyfish; adults with a single central body cavity opening to the exterior and surrounded by cellular endoderm, acellular mesoglea, and cellular ectoderm .... phylum Cnidaria [Chapter 4]

3′ Surface porous; colonial; tentacles absent; no closable orifices; without discrete organs; cellular-level (or incipient tissue-level) construction; variable, nondistinct colony shapes, including encrusting, rounded, or digitiform growth forms; skeleton of individual siliceous spicules and a collagen matrix; internal water canal system; may contain symbiotic algae; the sponges ..... phylum Porifera [Chapter 3]

4(2) Oral region with numerous tentacles or cilia distributed around the mouth; organism never with eversible jaws and never vermiform as adult 5

4 Oral region with two or no tentacles, or tentacles behind the mouth 7

5(4) Oral region with tentacles, organisms in gelatinoids or branching colonies 6

5′ Oral region ringed with cilia, muscular pharynx (mastax) with complex set of jaws; single free swimming, or semi-sessile living singly or in small colonies; wheel animals, or rotifers .... phylum Rotifera [Chapter 8]

6(5) Oral tentacles (the lophophore) in a "U" or "horseshoe" shape around mouth; anus opens outside of lophophore; colonial animals, often in massive colonies attached to hard surfaces; true bryozoans .... phylum Ectoprocta (Bryozoa) [Chapter 13]

6′ Both mouth and anus open within the oral tentacles; individual (noncolonial) animals with a calyx containing a single whorl of 8–16 ciliated tentacles .... phylum Entoprocta [Chapter 14]

7(4) Not with the combination of characteristics described below 8

7′ Small (50–800 μm), spindle- or tenpin-shaped, ventrally flattened with a more or less distinct head bearing sensory cilia; cuticle usually ornamented with spines or scales of various shapes; posterior of body often formed into a furca with distal adhesive tubes; gastrotrichs .... phylum Gastrotricha [Chapter 7]

8(7) Anterior mouth and posterior anus present 9

8′ Flattened or cylindrical, acoelomate worms with only one, ventral digestive tract opening; sometimes with evident head; turbellarian flatworms (commonly called planaria, a nonspecific and usually incorrect name) .... phylum Platyhelminthes [Chapter 5]

9(8) Vermiform or not, eversible oral proboscis not present, although eversible jaws or other mouthparts may occur 10

9′ Long, flattened, unsegmented worms with an eversible proboscis; ribbon worms .... phylum Nemertea [Chapter 6]

10(9) Body not enclosed in a single, spiraled shell or in a hinged, bivalved shell; or if a bivalved shell is present, then animal has jointed legs 11

10′ Soft-bodied coelomates whose viscera is covered (in freshwater species) by a single or dual (hinged), hard calcareous shell; with a ventral muscular foot; fleshy mantle covers internal organs; snails, clams, and mussels .... phylum Mollusca [Chapter 11]

11(10) Segmented legs absent in all life stages; if jaws are present, then body with at least 15 segments 12

11′ Adults and most larval stages with legs; if larvae without legs or prolegs (some insects), then cephalic region with paired mandibles, or eversible head, always with 14 or fewer body segments 14

12(11) Organism vermiform, not segmented 13

12′ Organism vermiform or not, body segmented .... phylum Annelida [Chapter 12]

13(12) Body cylindrical, usually tapering at both ends; cuticle without cilia, often with striations, punctuations, minute bristles, etc.; 1 cm long (except family Mermithidae, < 6 cm); nematodes, roundworms .... phylum Nematoda [Chapter 9]

13′ Body with anterior tip normally obtusely rounded or blunt, posterior tip may be bi- or trilobed; cuticle opaque to dark brown or black, and epicuticle usually crisscrossed by minute grooves; length several cm to 1 m, width 0.25–3 mm; only adults with free-living stage; hairworms or horsehair worms .... phylum Nematomorpha [Chapter 10]

14(11) Four pairs of clawed, nonjointed legs; water bears .... phylum Tardigrada [Chapter 15]

14′ Adults and most larvae with jointed legs, or legs lacking, or more or less than four pairs........ phylum Arthropoda [Chapter 16][2]

2 Chapter 16 includes a key that will lead to subsequent chapters covering this phylum.

REFERENCES

Ahyong, S., D.C. Rogers & J.H. Thorp (eds.). 2021. Keys to Australasian Fauna. Volume VI in: Thorp and Covich's Freshwater Invertebrates, Fourth Edition. Academic Press, Elsevier, Boston, MA. (in press).

Allegrucci, G., G. Carchin, P. Convey & V. Sbordoni. 2012. Evolutionary geographic relationships among orthocladine chironomid midges from maritime Antarctic and sub-Antarctic islands. Biological Journal of the Linnean Society 106:258–274.

Andrássy, I. & J.A.E. Gibson. 2007. Nematodes from saline and freshwater lakes of the Vestfold Hills, East Antarctica, including the description of *Hypodontolaimus antarcticus* sp. n. Polar Biology 30: 669–678.

Arnold, R.J. & P. Convey. 1998. The life history of the diving beetle, *Lancetes angusticollis* (Curtis) (Coleoptera: Dytiscidae), on sub-Antarctic South Georgia. Polar Biology 20: 153–160.

Ball, I.R. & D.A. Hay. 1977. The taxonomy and ecology of new monocelid flatworm from Macquarie Island (Platyhelminthes, Turbellaria). Bijdragen tot de Dierkunde 47: 205–214.

Barber-James, H. 2007. Freshwater invertebrate fauna of the Tristan da Cunha Islands (South Atlantic Ocean), with new records for Inaccessible and Nightingale Islands. Transactions of the Royal Society of South Africa 62: 24–36.

Bayly, I.A.E. 1994. *Gladioferens* Henry (Copepoda: Calanoidea) discovered in Antarctica: *G. antarcticus* sp. nov. described from a lake in the Bunger Hills. Polar Biology 14: 253–259.

Bayly, I.A.E., J.A.E. Gibson, B. Wagner & K.M. Swadling. 2003. Taxonomy, ecology and zoogeography of two East Antarctic freshwater calanoid copepod species: *Boeckella poppei* and *Gladioferens antarcticus*. Antarctic Science 15: 439–448.

Camacho, A. 2006. Planktonic microbial assemblages and the potential effects of metazooplankton predation on the food web of lakes from the maritime Antarctica and sub-Antarctic islands. Reviews in Environmental Science and Technology 5: 167–185.

Chown, S.L. & P. Convey. 2016. Antarctic entomology. Annual Review of Entomology 61: 119–137.

Chown, S.L. & D.J. Marshall. 2008. Appendix VII. Terrestrial invertebrates of the Prince Edward Islands. in: S.L. Chown & P.W. Froneman (eds.). The Prince Edward Islands: Land–Sea Interactions in a Changing Ecosystem. Sun Press, Stellenbosch: 400–409.

Coulson, J.C. 2007. Terrestrial and Freshwater Invertebrate Fauna of the High Arctic Archipelago of Svalbard. Zootaxa 1448: 41–58.

Dartnall, H. 1995. The rotifers of Heard Island: preliminary survey with notes on other freshwater groups. Papers and Proceedings of the Royal of Tasmania 129: 7–15.

Dartnall, H.J.G. 2003. Additions to the freshwater fauna of the Heard Island. Paper and Proceedings of the Royal Society of Tasmania 137: 75–79.

Dartnall, H.J.G. 2005. Freshwater invertebrates of subantarctic South Georgia. Journal of Natural History 39: 3321–3342.

Dartnall, H.J.G. & W. Hollwedel. 2007. A limnological reconnaissance of the Falkland Islands; with particular reference to the waterfleas (Arthropoda: Anomopoda). Journal of Natural History 41: 1259–1300.

Dartnall, H.J.G., W. Hollwedel & J.C. de Paggi. 2005. The freshwater fauna of Macquarie Island, including a redescription of the endemic water-flea *Daphnia gelida* (Brady) (Anomopoda: Crustacea). Polar Biology 28: 922–939.

Dartnall, H.J.G. & V.R. Smith. 2012. Freshwater invertebrates of sub-Antarctic Marion Island. African Zoology 47: 203–215.

Downie, R.H., P. Convey, S.J. McInnes & P.J.A. Pugh. 2000. The non-marine invertebrate fauna of Deception Island (Maritime Antarctic): a baseline for a comprehensive biodiversity database. Polar Record 36: 297–304.

Fontaneto, D., N. Iakovenko & W.H. De Smet. 2015. Diversity gradients of rotifer species richness in Antarctica. Hydrobiologia 761: 235–248.

Hamada, N., J.H. Thorp & D.C. Rogers (eds.). 2018. Keys to Neotropical Hexapoda. Volume IV in: Thorp and Covich's Freshwater Invertebrates, Fourth Edition. Academic Press, Elsevier, Boston, MA.

Hawes, T.C. 2009. Origins and dispersal of the Antarctic fairy shrimp. Antarctic Science 21: 477–482.

Hughes, K.A. & P. Convey. 2012. Determining the native/non-native status of newly discovered terrestrial and freshwater species in Antarctica – Current knowledge, methodology and management action. Journal of Environmental Management 93: 52–66.

Izaguirre, I., L. Allende & M.C. Marinone. 2003. Comparative study of the planktonic communities of three lakes of contrasting trophic status at Hope Bay (Antarctic Peninsula). Journal of Plankton Research 25: 1079–1097.

Janiec, K. 1996. The comparison of freshwater invertebrates of Spitsbergen (Arctic) and King George Island (Antarctic). Polish Polar Research 17: 173–202.

Jones, A.G., S.L. Chown, T.J. Webb & K.J. Gaston. 2003. The free-living pterygote insects of Gough Island, South Atlantic Ocean. Systematics and Biodiversity 1: 213–273.

McInnes, S.J., S.L. Chown, H.G.J. Dartnall & P.J.A. Pugh. 2001. *Milnesium* cfr. *tardigradum* (Milnesiidae, Apochela, Tardigrada): A monitor of high altitude meiofauna on Sub-Antarctic Marion Island. Zoologischer Anzeiger – A Journal of Comparative Zoology 240: 461–465.

McInnes, S.J. & P.J.A. Pugh. 1998. Biogeography of limno-terrestrial Tardigrada, with particular reference to the Antarctic fauna. Journal of Biogeography 25: 31–36.

Pugh, P.J.A. & H.J.G. Dartnall. 1994. The Acari of fresh- and brackish water habitats in the Antarctic and sub-Antarctic regions. Polar Biology 14: 401–404.

Pugh, P.J.A., H J.G. Dartnall & S.J. McInnes. 2002. The non-marine Crustacea of Antarctica and the Islands of the Southern Ocean: biodiversity and biogeography. Journal of Natural History 36: 1047–1103.

Pugh, P.J.A., H.G.J. Dartnall & S.J. McInnes. 2010. The non-marine Crustacea of Antarctica and the Islands of the Southern Ocean: biodiversity and biogeography. Journal of Natural History 36: 1047–1103.

Raymond, M.R., D.A. Wharton & C.J. Marshall. 2013. Nematodes from the Victoria Land coast, Antarctica and comparisons with cultured *Panagrolaimus davidi*. Antarctic Science 26: 15–22.

Richards, A.B. & D.C. Rogers. 2011. Southwest Association of Freshwater Invertebrate Taxonomists (SAFIT) list of freshwater macroinvertebrate taxa from California and adjacent states including standard taxonomic effort levels. 266 p..

Rodriguez, P. & E. Rico. 2008. A new freshwater oligochaete species (Clitellata: Enchytraeidae) from Livingston Island, Antarctica. Polar Biology 31: 1267–1279.

Rogers, D.C. & J.H. Thorp (eds.). 2019. Keys to Palaearctic Fauna. Volume III in: Thorp and Covich's Freshwater Invertebrates, Fourth Edition. Academic Press, Elsevier, Boston, MA. 920 p.

Segers, H. 2008. Global diversity of rotifers (Rotifera) in freshwater. Hydrobiologia 595: 49–59.

Thorp, J.H. & D.C. Rogers (eds.). 2015. Ecology and General Biology. Volume I in: Thorp and Covich's Freshwater Invertebrates, Fourth Edition. Academic Press, Elsevier, Boston, MA. 1118 p.

Thorp, J.H. & D.C. Rogers (eds.). 2016. Keys to Nearctic Fauna. Volume II in: Thorp and Covich's Freshwater Invertebrates, Fourth Edition. Academic Press, Elsevier, Boston, MA. 740 p.

Tong, S., N. Vors & D.J. Patterson. 1997. Heterotrophic fagellates, centrohelid heliozoa and filose amoebae from marine and freshwater sites in the Antarctic. Polar Biology 18: 91–106.

Velasco-Castrillon, A., J.A.E. Gibson & M.I. Stevens. 2014. A review of current Antarctic limno-terrestrial microfauna. Polar Biology 37: 1517–1531.

Xu, Y., C. Shao, X. Fan, A. Warren, K.A.S. Al-Rasheid, W. Song & N. Wilbert. 2016. New contributions to the biodiversity of ciliates (Protozoa, Ciliophora) from Antarctica, including a description of *Gastronauta multistriata* nov. spec. Polar Biology 39: 1439–1453.

Chapter 2

Protozoa

Gabriela C. Küppers, Anush Kosakyan, Ferry Siemensma, María Cristina Claps, Thiago da Silva Paiva, Leonardo D. Fernández, Luciana Burdman, Valentyna Krashevska, Enrique Lara and Cristina Damborenea

Chapter Outline

Thorp and Covich's Freshwater Invertebrates. https://doi.org/10.1016/B978-0-12-804225-0.00002-2

Introduction to Free-Living Protozoa

Leonardo D. Fernández

Centro de Investigación en Recursos Naturales y Sustentabilidad (CIRENYS), Universidad Bernardo O'Higgins, Santiago, Chile

Cristina Damborenea

CONICET, División Zoología Invertebrados, Museo de La Plata, La Plata, Argentina

INTRODUCTION

Classification of unicellular eukaryotes has been historically under review. In 1677, Antony van Leeuwenhoek first described these organisms under the name of "animalcules." Almost a century later, unicellular eukaryotes were grouped together under the name of "Infusoria," a term possibly coined by Martin F. Ledermüller in 1763 (Kudo, 1931). They were later included within Protozoa (i.e., "early animals"), a taxon originally created as a class (Goldfuss, 1820) and then elevated to phylum (von Siebold, 1854) and finally to kingdom (Owen, 1860). Around the same time of the creation of the kingdom Protozoa, unicellular eukaryotes were alternately included within the kingdom Protoctista (i.e., first created beings) to formally include mixotrophic representatives and multicellular organisms that did not behave like "true animals" or "zoa" (Hogg, 1860). Six years later, the kingdom Protista ("the first of all, primordial") was created to include all those organisms that were neither animals nor plants, such as bacteria and unicellular eukaryotes (Haeckel, 1866). From then, until the end of the 20th century, kingdoms Protozoa, Protoctista, and Protista were redefined, revalidated, and/or invalidated by several researchers. However, in the light of current molecular phylogenetic evidence, these kingdoms are no longer accepted by most modern protistologists (Hausmann et al., 2003; Adl et al., 2005, 2012; O'Malley et al., 2012; Geisen et al., 2017). Only some classic terms, such as "protist," are still retained to informally describe eukaryotes with a unicellular level of organization, without cell differentiation into tissues (Adl et al., 2005). The term "protist" is also increasingly being used by protistologists working in freshwater unicellular eukaryotes from the Neotropical region (e.g., Fernández et al., 2015; Schiaffino et al., 2016).

Unicellular eukaryotes have received special attention in the past decades. Recent advances wrought by molecular phylogenetic studies introduced new perspectives in the knowledge of these organisms. Terms such as algae (phototrophic protists) and protozoa (most heterotrophic protists, other than those that closely resemble fungi, predominantly nonfilamentous heterotrophic species) lost their boundaries and have no taxonomic value. The term protist describes eukaryotes with a unicellular level of organization, without cell differentiation into tissues. Alternate vegetative morphology restricted to sexual reproduction, quiescent, or resistant stages, such as cysts, might occur (Adl et al., 2005). Free-living protozoa are unicellular and heterotrophic and may also be mixotrophic (occurrence of phagotrophy and phototrophy in the same organism). They usually feed on bacteria, algae, and even zooplankton, depending on their size.

Free-living protozoa play a vital ecological role in aquatic environment. Because of their high sensitivity to environmental changes, they have been used as bioindicators and biomonitors of pollution. Hydrological cycles, increased nutrients, physical and chemical changes, and phytoplankton blooms in water bodies interfere in the dynamics of this community affecting the trophic network.

This chapter, as in the other volumes of the series, provides a synthesis on the diversity, ecology, and distribution of free-living protozoa found in freshwater Neotropical ecosystems, such as ciliates, flagellates, and sarcodines. The two latter groups consist of a number of largely unrelated evolutionary lineages.

LIMITATIONS

Free-living protozoa are diverse and abundant in freshwater ecosystems. The study of their diversity, ecology, and geographical distribution is incipient compared with that of multicellular organisms, particularly in the Neotropical region (Fernández et al., 2016; Schiaffino et al., 2016). Even though they play pivotal roles in natural ecosystems (Geisen et al., 2018), their study has contributed little to the knowledge of ecology and evolutionary biology, mainly because research on protozoa exhibits some limitations (Fenchel, 1987). For instance, most free-living protozoa exhibit microscopic sizes, making their morphological identification and counting difficult. Often, species identification requires training in specific microscopy and staining techniques, skills that not many researchers manage to date. They also suffer from a significant Linnean shortfall, due to the global deficit of taxonomists dedicated to the study of the different groups that make up Protozoa. In addition, there is a comparatively poor knowledge on their systematics, mainly due to: (1) lack of adequate methods for the fixation and long-term preservation; (2) absence of type specimens for most species; (3) lack of morphological features for species delimitation; (4) inadequate species descriptions; and (5) great areas undersampled (e.g., the Neotropics) (Esteban et al., 2015).

According to the aforementioned background, the keys presented herein reach different levels of identification, according to the groups.

TERMINOLOGY AND MORPHOLOGY

Most protozoans are microscopic, usually 5–250 μm long. Some species have the size range of bacteria and rarely exceeding 6–7 mm. The morphology is varied, e.g., single independent cells that may or may not be motile, filamentous species, and colonial (linked by a common stalk, or sheath, or cytoplasmic extensions) and multinucleate forms.

They possess different locomotory structures: cilia, flagella, or pseudopodia (fingerlike cytoplasmic extensions). Another relevant characteristic is the membrane that allows the exchange with the environment. This membrane may be present alone (naked forms), but many protozoa have a thickened membrane and form a semirigid or rigid pellicle. They also can secrete scales or a shell-like test or can accumulate external particles.

The types of nutrition are varied. The boundaries between autotrophs and heterotrophs are not strict: some heterotrophs retain their prey's plastids; other heterotrophs form symbioses with photosynthetic species; and many photosynthetic species are also phagotrophic and osmotrophic.

KEY TO PROTOZOA

Major Functional Groups

1	Primary locomotion and feeding organelles are cilia or flagellae	2
1′	Locomotion organelles are pseudopodia, but cilia may or may not be present (Figs. 2.1–11).	**Ameboid protozoa [p. 13]**
2(1)	Simple or compound cilia, at least in one part of the life cycle, infraciliature always present (Figs. 2.12–23)	**Ciliophora [p. 43]**
2′	With flagella (Figs. 2.24, 25)	**Flagellated protozoa [p. 75]**

REFERENCES

Adl, S.M., A.G.B. Simpson, M.A. Farmer, R.A. Andersen, O.R. Anderson, J.R. Barta, S.S. Bowser, G. Brugerolle, R.A. Fensome, S. Fredericq, T.Y. James, S. Karpov, P. Kugrens, J. Krug, C.E. Lane, L.A. Lewis, J. Lodge, D.H. Lynn, D.G. Mann, R.M. Mccourt, L. Mendoza, O. Moestrup, S.E. Mozley-Standridge, T.A. Nerad, C.A. Shearer, A.V. Smirnov, F.W. Spiegel & M.F.J.R. Taylor. 2005. The new higher level classification of eukaryotes with emphasis on the taxonomy of protists. Journal of Eukaryotic Microbiology 52: 399–451.

Adl, S.M., A.G.B. Simpson, C.E. Lane, J. Lukes, D. Bass, S.S. Bowser, M.W. Brown, F. Burki, M. Dunthorn, V. Hampl, A. Heiss, M. Hoppenrath, E, Lara, L. Le Gall, D.H. Lynn, H. McManus, E.A.D. Mitchell, S.E. Mozley-Stanridge, L.W. Parfrey, J. Pawlowski, S. Reuckert, L. Shadwick, C.L. Schoch, A. Smirnov & F.W. Spiegel. 2012. The revised classification on eukaryotes. Journal of Eukaryotic Microbiology 59: 429–493.

Esteban, G.F., B.J. Finlay & A. Warren. 2015. Free-living Protozoa. Pages 113–132 *in:* J. Thorp & D.C. Rogers (eds.), Ecology and General Biology, Thorp and Covich's Freshwater Invertebrates (Fourth edition). Academic Press, Elsevier, Boston, MA.

Fenchel, T. 1987. Ecology of Protozoa: The biology of free-living phagotrophic protists. Science Technical Publishers, Madison. 197 pp.

Fernández L.D., B. Fournier, R. Rivera, E. Lara, E.A.D. Mitchell & C.E. Hernández. 2016. Water–energy balance, past ecological perturbations and evolutionary constraints shape the latitudinal diversity gradient of soil testate amoebae in southwestern South America. Global Ecology and Biogeography 25: 1216–1227.

Fernández, L.D., E. Lara & E.A.D. Mitchell. 2015. Checklist diversity and distribution of testate amoebae in Chile. European Journal of Protistology 51: 409–424.

Geisen, S., E.A.D. Mitchell, S. Adl, M. Bonkowski, M. Dunthorn, F. Ekelund, L.D. Fernández, A. Jousset, V. Krashevska, D. Singe, F.W. Spiegel, J. Walochnik & E. Lara. 2018. Soil protists: A fertile frontier in soil biology research. FEMS Microbiology Reviews 42: 293–323.

Geisen, S., E.A.D. Mitchell, D.M. Wilkinson, S. Adl, M. Bonkowski, M.W. Brown, A.M. Fiore-Donno, T.J. Heger, V.E.J. Jassey, V. Krashevska, D.J.G. Lahr, K. Marcisz, M. Mulot, R. Payne, D. Singer, O.R. Anderson, D.J. Charman, F. Ekelund, B.S. Griffiths, R. Rønn, A. Smirnov, D. Bass, L. Belbahri, C. Berney, Q. Blandenier, A. Chatzinotas, M. Clarholm, M. Dunthorn, A. Feest, L.D. Fernández, W. Foissner, B. Fournier, E. Gentekaki, M. Hajek, J. Helder, A. Jousset, R. Koller, S. Kumar, A. La Terza, M. Lamentowicz, Y. Mazei, S.S. Santos, C.V.W. Seppey, F.W. Spiegel, J. Walochnik, A. Winding & E. Lara. 2017. Soil protistology rebooted: 30 fundamental questions to start with. Soil Biology and Biochemistry 111: 94–103.

Goldfuss, G. A. 1820. Handbuch der Zoologie. I. Abtheilung. Johann Leonhard Schrag, Nürnberg.

Haeckel, E. 1866. Generelle morphologie der organismen. Vol. II. Georg Reimer, Berlin.

Hausmann, K., N. Hülsmann & R. Radek. 2003. Protistology (Third edition). Schweizerbart'sche, Stuttgart.

Hogg J. 1860. On the distinctions of a plant and an animal and on a Fourth Kingdom of Nature. Edinburgh New Philosophical Journal 12:216–225.

Kudo R.R. 1931. Handbook of Protozoology. C.C. Thomas, Springfield, IL; Baltimore, MD. 476 pp.

O'Malley, M.A., A.G.B. Simpson & A.J. Roger. 2012. The other eukaryotes in light of evolutionary protistology. Biology and Philosophy 28: 299–330.

Owen, R. 1860. Palaeontology. Adam and Charles Black, Edinburgh.

Schiaffino, M.R., E. Lara, L.D. Fernández, V. Balagué, D. Singer, C.C.W. Seppey, R. Massana & I. Izaguirre. 2016. Microbial eukaryote communities exhibit robust biogeographical patterns along a gradient of Patagonian and Antarctic lakes. Environmental Microbiology 18: 5249–5264.

von Siebold, C.T. 1854. Anatomy of the Invertebrata. Gould and Lincoln, Boston.

KEY TO AMOEBOID PROTOZOA

1	Pseudopodia blunt, or filiform, sometimes branching or anastomosing (Figs. 2.1–11)	**Amoebae [p. 20]**
1′	Pseudopodia long and slender, supported by microtubules (axopodia) that are able to contract rapidly, radiating three dimensionally from the cell; filopodia; with or without skeletal elements (Fig. 2.25 E, F)	Centrohelida and heliozoan-like

Amoebae

Anush Kosakyan[a]
Institute of Parasitology, Biology Centre, Czech Academy of Sciences, 37005, Ceske Budejovice, Czech Republic.

Ferry Siemensma[a]
Kortenhoef, the Netherlands

Leonardo D. Fernández
Centro de Investigación en Recursos Naturales y Sustentabilidad (CIRENYS), Universidad Bernardo O'Higgins, Santiago, Chile

Luciana Burdman
Institute of Environmental Research and Engineering, University of San Martin, CONICET, Buenos Aires, Argentina

Valentyna Krashevska
J.F. Blumenbach Institute of Zoology and Anthropology, Georg August University Göttingen, Göttingen, Germany

Enrique Lara
Laboratory of Soil Biodiversity, University of Neuchâtel, Neuchâtel, Switzerland; Real Jardín Botánico de Madrid, CSIC, Spain

INTRODUCTION

Amoeboid protists comprise a large and very diverse group with over 15,000 described species (Adl et al., 2005). Most are free-living organisms that can be found in nearly any moist or aquatic habitat, aerobic or anaerobic. However, some are endosymbiotic, including some human pathogens. The ancient term Rhizopoda was used to designate those unicellular eukaryotes that move and feed by means of cytoplasmic streaming or by means of pseudopodia, temporary cytoplasmic extensions (Tekle et al., 2008); the name "amoeba" is derived from an ancient Greek word, which means "changing." Organisms with these characteristics can be found in all branches of the eukaryotic tree, thus rendering the taxon polyphyletic and therefore invalid (Pawlowski & Burki, 2009).

Before the advent and rise of molecular systematics, amoebae were traditionally divided into shelled (Testacea) and naked amoebae (Gymnamoebae) (Hoogenraad & De Groot, 1940). Since the application of phylogenetic approaches to protist taxonomy, this traditional view of amoeboid systematics has changed tremendously and led to a still ongoing revision of amoeboid classification (Tekle et al., 2008; Adl et al., 2012; Sierra et al., 2013; Kang et al., 2017). The largest number of amoeboid organisms can be found within the supergroups Amoebozoa and Rhizaria (Adl et al., 2012). Amoebozoa includes many of the naked amoebae (e.g., *Amoeba proteus*), testate lobose amoebae (Arcellinida, e.g., *Arcella*), and pelobionts (e.g., *Pelomyxa, Entamoeba histolytica*) as well as cellular (Dictyostelida), acellular (e.g., Myxogastrida), and protosteloid slime molds (Tekle et al., 2008). Rhizaria includes foraminiferans, radiolarians, as well as a huge array of filamentous amoebae. It also includes several groups of testate (filose) amoebae, the Euglyphida being the best known (e.g., *Euglypha, Trinema*).

The origin of the first organisms resembling testate amoebae dates back to the evolutionary diversification of eukaryotes in the Neoproterozoic (Porter et al., 2003; Lahr et al., 2019) and represents perhaps the oldest indisputable traces of eukaryotic life known to date (Berney & Pawlowski, 2006; Parfrey et al., 2011). Modern genera have, however, not been found before the Cretaceous period (Schmidt et al., 2010). These organisms are abundant, diversified, and widespread in soils, freshwater sediments, and water column and also marginally in marine systems (Golemansky, 1970). Their diversity has been estimated to about 2,000 morphospecies, with approximately two-thirds being affiliated to Arcellinida (Meisterfeld, 2002). Recent developments in molecular barcoding and electron microscopy are showing that this number is a clear underestimation, as traditional morphospecies have been widely split into genetically and ecologically different units (Heger et al., 2011b; Kosakyan et al., 2012, 2013; Chatelain et al., 2013; Singer et al., 2015, 2018). Diversity estimates may, therefore, need to be multiplied by several orders of magnitude (Kosakyan et al., 2016a). These organisms undergo traditional cell division, although there are examples of complex life cycles with naked (i.e., nontestate) stages; sex is supposed to occur frequently (Iudina & Sukhanova, 2000; Lahr et al., 2011). They feed on small items, bacteria, and small algae for the smallest species, although *Cryptodifflugia oviformis* is able to perform a kind of pack hunting, and dozens of these amoebae are able to kill nematodes much larger in size and shape (Geisen et al., 2015). Larger species feed commonly on small animals such as rotifers

[a] Anush Kosakyan and Ferry Siemensma contributed equally to this section.

PROTOZOA

and nematodes but also on the dead organic matter (Yeates & Foissner, 1995; Gilbert et al., 2003). Several arcellinid species (such as *Cucurbitella* spp and some "*Netzelia* spp" previously known as *Difflugia*, see these taxonomic changes in Kosakyan et al., 2016a) are known to have a planktonic phase in which the organisms are floating in the water column, hunting for diatoms whose frustules will be incorporated in the amoebae's tests (Meisterfeld, 1991). This behavior is widespread in family Hyalospheniidae, which are known for incorporating scales from their prey (mainly Euglyphida testate amoebae) in their own test, thus recycling silica (Lahr et al., 2015). Some species can form symbiotic relationships with green algae, most often in oligotrophic environments such as peat bogs (Gomaa et al., 2014; Lara & Gomaa, 2017). Others (genus *Paulinella*) have developed a primary endosymbiosis with a cyanobacterium, which resembles the first associations that eventually led to the rise of Archaeplastida (Marin et al., 2005).

Most testate amoebae research has been focused on bioindication issues, as many of these organisms have narrow ecological requirements. Their empty shells may remain intact in sediments and peats for long periods, and thus the composition of their past communities can be inferred. These shells have been extensively used as palaeoecological proxies for the reconstruction of past environmental conditions (Lamentowicz Ł. et al., 2008; Lamentowicz M. et al., 2008; Lamentowicz & Obremska, 2010; Payne et al., 2012; Swindles et al., 2014). In comparison with other commonly used palaeoecological markers such as pollen, they have a shorter generation time, which implies that testate amoeba communities will react faster to environmental changes (Mitchell et al., 2008). However, testate amoebae gained the attention of researchers also in other fields such as biomonitoring (Heger et al., 2012; Delaine et al., 2016), ecotoxicology (Nguyen-Viet et al., 2008; Meyer et al., 2015; Amacker et al., 2018), and forensic sciences (Szelecz et al., 2014; Seppey et al., 2015).

Testate amoebae have been in the center of a debate in the early 2000s concerning microbial biogeography. While some researchers claimed that all microorganisms had a worldwide distribution (Finlay, 2002), testate amoebae served as examples of microbes with a limited geographical distribution (Wilkinson, 2001). Indeed, hyalospheniid genera like *Apodera, Alocodera,* and *Certesella* have never been found northward of a desert belt surrounding the Tropic of Cancer (Smith et al., 2008; Heger et al., 2011a). Beyond the debate on cosmopolitanism, patterns of distribution have been evidenced in testate amoebae. For instance, it has been shown that common large-scale correlations between testate amoeba diversity and present-day environmental factors related to temperature (Lara et al., 2016), water availability (Fournier et al., 2016; Krashevska et al., 2012), or both (Fernández et al., 2016) might stem from an ancestral adaptation of these protists to tropical-like conditions (i.e., the need for warm, humid climates is a phylogenetically conserved trait in testate amoebae) (Fernández et al., 2016). Although diversity, evolution, ecology, and biogeography of testate amoebae are currently a dynamic field of research (Kosakyan et al., 2016a), our knowledge about these organisms is still very limited. These organisms are known to science since the beginning of the 19th century (Leclerc, 1816; Ehrenberg, 1838); however, most fundamental questions regarding their biology remained unanswered. Most studies on these microorganisms have been developed in the Holarctic region. Data from the Southern Hemisphere are still rare and sparse, indicating a limited knowledge on these protists along a considerable fraction of the terrestrial surface of the planet (Fernández et al., 2015).

Here, we present the naked and testate amoebae diversity of the Neotropical Region and give an identification key to the genus level.

LIMITATIONS

Southern Hemisphere amoebae have been by far less studied than their counterparts from the Northern Hemisphere. Most of the valid **naked amoebae** species were described from Europe and North America, and a few are known from the Mideast, India, Australia, or Japan. Surprisingly, one relatively well-studied region is Antarctica, whereas the amoeba biogeography in most regions of Asia, South America, and Africa is virtually unknown. Only two studies (Friz, 1992) so far reported naked amoeba from Amazon River and Mexican Gulf. Therefore, we can hardly provide any record from Neotropics. Yet, relying on personal investigations in some parts of the Neotropics (Peru, Brazil, Chile, Argentina), we assume that almost all known species can be found there. It is very difficult distinguishing most species with a sufficient level of confidence, and TEM studies are required for that; the microsystematics of Gymnamoebae is based on EM features. This requires a huge amount of time, trained specialist, and special facilities. However, as a general rule, nonspecialists will try to identify an amoeba using light microscopy. Therefore, we have limited the naked amoeba keys mainly to light microscopical level.

Compared with naked amoebae, **testate amoebae** are better recorded in Neotropics. The southern part of South America corresponding to Chile, Brazil, Argentina, and Uruguay (i.e., "Southern Cone") represents probably the best surveyed region in the Southern Hemisphere. Central America and other South American countries remain until now almost uncharted territories. Even within Brazil, large regions have never been surveyed for testate amoebae due to their inaccessibility, especially in the Amazonas Basin. The following publications are noteworthy to mention for testate amoebae records in Neotropics (Wailes, 1913; Jung,

1942; Bonnet, 1966; Fernández et al., 2015 for Chile; Vucetich, 1974, 1975; Vucetich & Lopretto, 1995 for Argentina; Lansac-Toha et al., 2014; Gomes e Souza, 2008 for Brazil; Sigala et al., 2016 for Mexico; Haman & Kohl, 1994 for Peru; Daday, 1905 for Paraguay; Decloitre, 1963 for Venezuela, Krashevska et al., 2007, 2010 for Ecuador).

Regardless of their origin, the identification of testate amoebae species can sometimes be difficult. Indeed, often species can be determined only through a thorough examination of small ultrastructural details (Kosakyan et al., 2013), whereas others are prone to a high phenotypic plasticity (Mulot et al., 2017). As a result, the taxonomy is leaky.

Moreover, working on testate amoebae presents several difficulties: most species are impossible to maintain in pure culture, as they feed exclusively on other eukaryotes. Molecular barcoding or phylogeny based on isolated cells of these large protists is difficult, since their shells host most of the time many symbionts and undigested preys. Because of this, the molecular data set is very limited with the exception of certain groups such as Hyalospheniidae (Kosakyan et al., 2016b).

TERMINOLOGY AND MORPHOLOGY

Identification of amoebae can be challenging both in naked and in testate amoebae, whereas for both groups, different morphological characters must be considered for taxon identification.

Naked amoebae are single-celled organisms, consisting of cytoplasm and one or more nuclei (Fig. 2.1 A, B). The morphology of the nucleus, its size, and number can be useful differential characters. Two principal types of nuclei are found in amoebae: vesicular nuclei with one, usually central nucleolus, sometimes with a few additional very small nucleoli, and ovular nuclei with several to many small nucleoli. The main mass of the cytoplasm, called granuloplasm, contains optically visible granules, crystals (Figs. 2.1 J, 2 B, C), and other inclusions such as lipid globules and glycogen bodies. The most obvious inclusions found in mainly larger amoebae are crystals (Fig. 2.1 J). Large amoebae like *A. proteus* and *Polychaos dubium* are usually filled with numerous crystals, which are contained in vacuoles. Most species of amoeba contain bipyramidal crystals of orthorhombic form. Only freshwater amoebae produce crystals, but crystals are absent in *Pelomyxa* and some other species. Crystals shape is usually very characteristic for a given species, but appearance and shape of crystals are influenced by culture conditions and type of prey.

Some amoebae contain bacteria and small green algae inside their cytoplasm. These organisms have a symbiotic relationship with their host and are called endosymbionts (Fig. 2.3 D). Host cells acquire their endosymbionts by two different mechanisms, first by inheriting when cells divide or second by acquisition from the surrounding environments with a highly selective recognition mechanism (Gomaa et al., 2014).

Identifying a naked amoeba may be a big problem for a nonspecialist and often also for a specialist. It is therefore nearly impossible to determine the family or genus of an amoeba based on a single observation. The problem is that amoebae do not have a constant body form; they are polymorphic. A single cell can have different shapes, especially when it is stationary or moves in a non-coordinated manner, often changing the direction of locomotion. "One should not expect that every amoeba of a described species will have a form fitting into the illustrated one" (Page, 1976).

Although most amoebae do not have stable morphological characters that could be easily described, there are some light microscopical characteristics, which can be used for identification and describing species, for instance: (1) general pattern and outline of a moving amoeba (see for morphotypes, Smirnov & Brown, 2004); (2) pattern and outline of floating forms; (3) presence and shape of pseudopodia and subpseudopodia; (4) organization of the uroid; (5) shape of an amoeba in cross section; (6) position of the hyaloplasm in the locomotive cell; (7) cyst forming and structure of the cyst; (8) structure and number of the nuclei; (9) locomotive rate in µm/minute; and (10) absence or presence of cilia and their number.

Locomotion, shape, and **pseudopodia** are probably the main characters as the first step of observation of naked amoebae. An amoeba in a freshly made wet mount observed under light microscopy has usually been disturbed during its transfer from the sample vial onto the slide. For some minutes or even hours (depending on the species), the organism does not move and displays a rounded irregular or even a star-shaped form. It bulges, turns, or stays immobile in this stationary form, also called nondirected locomotion. Disturbed amoebae may also float. Usually, they have a more or less star-shaped body, with several pseudopodia radiating from the central body. The shape of these floating forms is characteristic for some taxa (Fig. 2.1 I). After a while, an amoeba in a stationary or floating form starts to move, and during this continuous directional locomotion, the amoeba maintains this shape until it stops or changes direction. During this movement, the shape usually undergoes some minor changes, but this stable locomotive form of an actively, continuously moving amoeba is very characteristic and can be used as a taxonomical instrument (Page, 1988; Smirnov & Brown, 2004). During locomotion, the leading edge of the cell shows an area of transparent, **hyaloplasm**, which looks as clear as water. This anteriorly situated hyaloplasm forms a

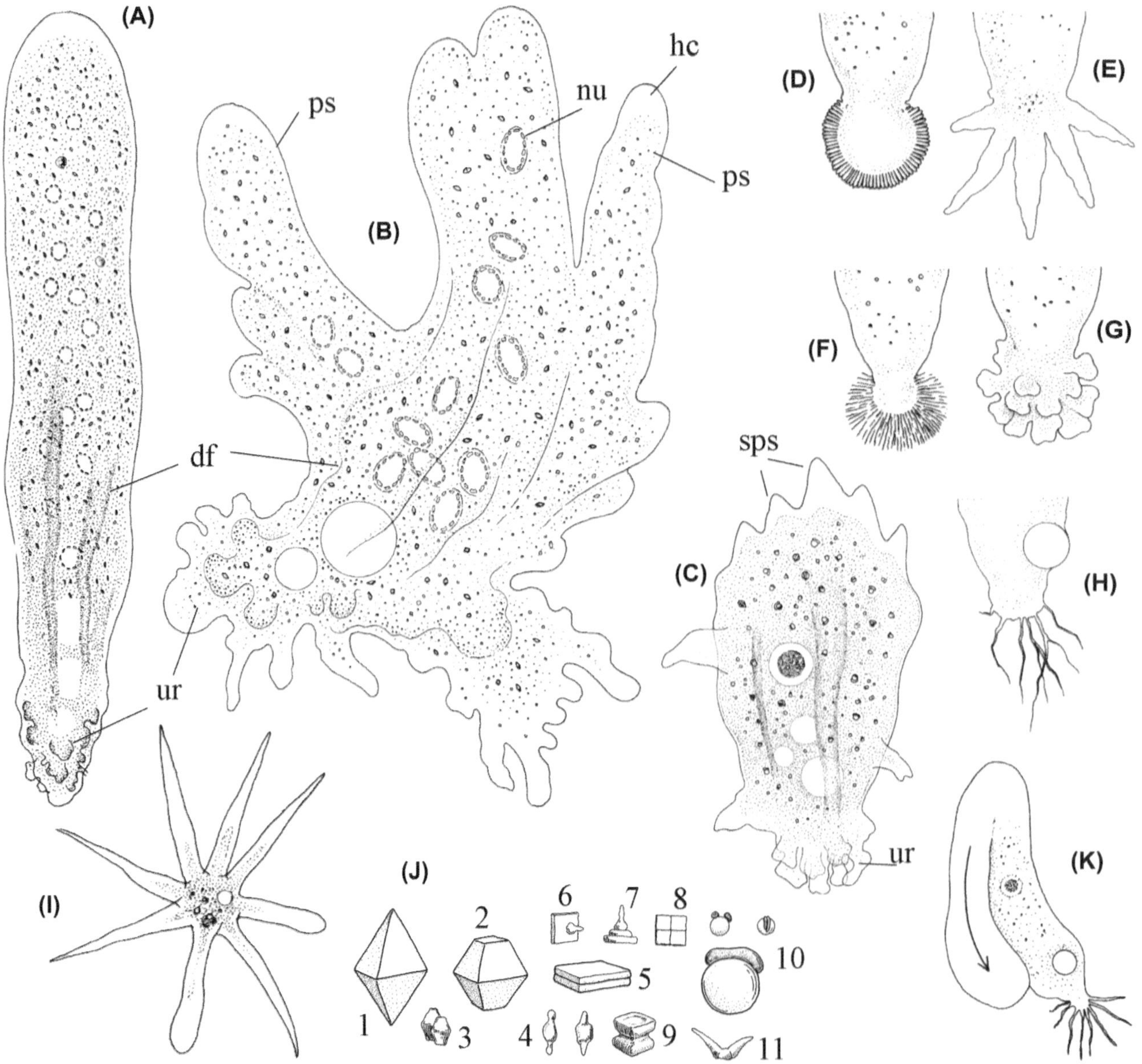

FIGURE 2.1 Line drawings of (A) monopodial form of a *Chaos* amoeba; (B) polypodial form of the same amoeba; (C) *Mayorella* species with typical subpseudopodia; (D) bulbous uroid with papillae; (E) fasciculate uroid, trail remnants of retracted pseudopodia; (F) villous uroid; (G) morulate uroid; (H) uroid with adhesive trailing filaments; (I) floating form; (J) diverse crystals, characteristic for some species; (K) typical eruptive backward flow of hyaloplasm (arrowed) along one side. Abbreviations: *df*, dorsal folds; *hc*, hyaline cap; *nu*, nucleus; *ps*, pseudopodia; *sps*, subpseudopodia; *ur*, uroid. Crystal forms: *1*, bipiramidal shape; *2*, blunt pyramidal shape; *3*, double crystal; *4*, elongated forms as found in *Polychaos fasciculatum*; *5*, strongly flattened crystal with two aberrant forms; *6–7*, as found in *Polychaos dubium*; *8–9*, observed crystals in *Saccamoeba* sp.; *10*, paired crystals, characteristic for some larger *Mayorella* species; 11 croissant-shaped crystal, characteristic for *Saccamoeba wellneri*. *Drawings by F. Siemensma.*

hyaline crescent and is usually called the **hyaline cap** (Fig. 2.1 B). The rest of the amoeba has a granulous aspect due to the presence of many vesicles and is called **granuloplasm**. While the hyaline cap is the anterior part in the locomotive form of the amoeba, the **uroid** is the posterior part. In many species, the uroid has a distinct characteristic shape (Fig. 2.1 D–H). The uroid can be hairy, morulate, pliate, papillated, or bulbous. Some species such as *Rhizamoeba* show trailing filaments, which originate from sticky contact with the substrate (Fig. 2.1 H–K).

A temporary cytoplasmic extension is known as **pseudopodium**, which means literally false foot. Pseudopodia are powered by microfilaments near the cellular membrane. About half of the internal space of an amoeba contains microfilaments. A pseudopodium contains both granuloplasm and hyaloplasm. Form and structure of the pseudopodia are important taxonomical characters. Depending upon their shape, several types of pseudopodia are distinguished:

- **Lobopodia**: more or less cylindrical, finger-shaped pseudopodia; most common type of pseudopodia. Lobose amoebae are by far the largest group.
- **Filopodia**: thin and thread-like pseudopodia. These filopodia can branch or anastomose, depending upon taxa.
- **Reticulopodia**: pseudopodia that anastomose (make contact with another pseudopodium), thus forming a network (reticulum).
- **Granulopodia**: filopodia with a granular structure. These granulopodia can also branch and/or anastomose.
- **Granuloreticulopodia**: filopodia that anastomose and are granulated, known from Foraminifera.
- **Lamellipodia**: broad very thin veil-like pseudopodia.

Most small and medium-sized amoebae do not form distinct pseudopodia, but smaller hyaline projections that lack any locomotive function are called **subpseudopodia** (Fig. 2.1 C). They are completely hyaline and can have different shapes. *Mayorella* species have typical conical subpseudopodia (Fig. 2.1 C). Cylindrical or subcylindrical amoebae that form several distinct pseudopodia are called **polypodial** or **polytactic** (Fig. 2.2 A). These cells can be hand-shaped (palmate), with the pseudopodia as fingers (Fig. 2.3 G). Amoebae without any pseudopodia are usually called **monopodial** (Figs. 2.1 A, 2 C, D). The term monotactic or limax form is applied to amoebae, which are consistently monopodial. Polypodial amoebae can adopt a monopodial form when active, in fast movement. These amoebae are cylindrical or subcylindrical and behave as one large pseudopodium. It may put out a lateral pseudopodium when changing direction (Figs. 2.1 A, 3 F). These forms are called **orthotactic** (Smirnov & Brown, 2004). The distinction between monopodial and polypodial forms can only be applied to amoebae advancing in a constant direction. Some amoebae have **folds** on the dorsal surface of the locomotive form. Some large amoebae may have folds or wings on their lateral surface (Figs. 2.1 A–C, 3 F).

Testate amoebae are apolyphyletic assemblage of at least three major unrelated taxonomic groups, common in freshwater (sediments, plankton) and terrestrial (soil, litter, mosses) habitats. A few forms are known from the supralittoral marine environment. The shape and structural details of the test are important criteria for identification of the testate amoebae taxa such as family, genus, and species. The structure of pseudopodia (lobose or filose) is used to identify the three main groups of testate amoebae: Arcellinida with lobose pseudopods, Euglyphida and Amphitremida with filose pseudopods. The phylogenetic position of these three groups among Amoebozoa, Cercozoa, and Amphitremida, respectively, has been established based on ribosomal DNA sequence data (Bhattacharya et al., 1995; Nikolaev et al., 2005; Gomaa et al., 2013). Recent findings have shown the existence of three other groups within Cercozoa: the Tectofilosida, the Rhizaspididae (Dumack et al., 2016), and the group of *Trachyrhizium* (Shiratori & Ishida, 2016).

Test morphology: Tests can have a broad range of shapes (i.e., circular, semicircular, oval, vase-shaped, etc.). Apertures can be circular or oval (Figs. 2.6–11) and bear indentations of different shapes. The shape, size, and composition of the test are fundamental in species identification.

Dimensions: These organisms cover a relatively broad range of sizes: the smallest (*Paulinella, Pseudodifflugia, Cryptodifflugia, Micropyxidiella*) measure < 10 μm, whereas the largest ones (*Lagenodifflugia, Cyclopyxis*) can reach 500 μm and are sometimes even visible with the naked eye.

Test composition: Based on composition of the test, testate amoebae can be divided into two main groups. The species of one group possess a proteinaceaous organic test. These include members with a flexible membrane (i.e., *Microcorycia*), those with rigid organic material (i.e., *Hyalosphenia*), or those with aerolated surface and with the test constructed with regularly arranged as hollow "building units" (i.e., *Arcella*). The second group includes species with proteinaceous matrix in which mineral elements are embedded; these elements can be self-secreted (i.e., *Euglypha, Lesquereusia*), obtained from the surrounding environment (*Centropyxis, Difflugia*) or can be taken from other testate amoebae (i.e., *Nebela, Apodera*).

MATERIAL PREPARATION AND PRESERVATION

Naked amoebae: These can be collected from nearly any location on earth, in very different ways and from very different habitats. As most amoebae are attached to a substrate, this substrate should be part of the sampled material, e.g., dry moss from trees and roofs, sediment from ditches and rivers, dots of algae, submerged mosses, and pieces of soil.

Naked amoebae should be observed in normal locomotion under normal conditions, e.g., adequate oxygen, minimal evaporation, and freedom of mechanical pressure (Page, 1988). Observation in wet mounts is preferable. These mounts can be kept in moist chambers for many weeks. Most amoebae adhere reasonably well to glass and will move freely on the underside of cover glasses, which is optically an ideal position.

Amoebae should always be allowed to settle down from their disturbed or floating forms or to move out of the debris, which are usually present in material from the wild. Wet mounts can be ready for the first inspection after 2 or

PROTOZOA

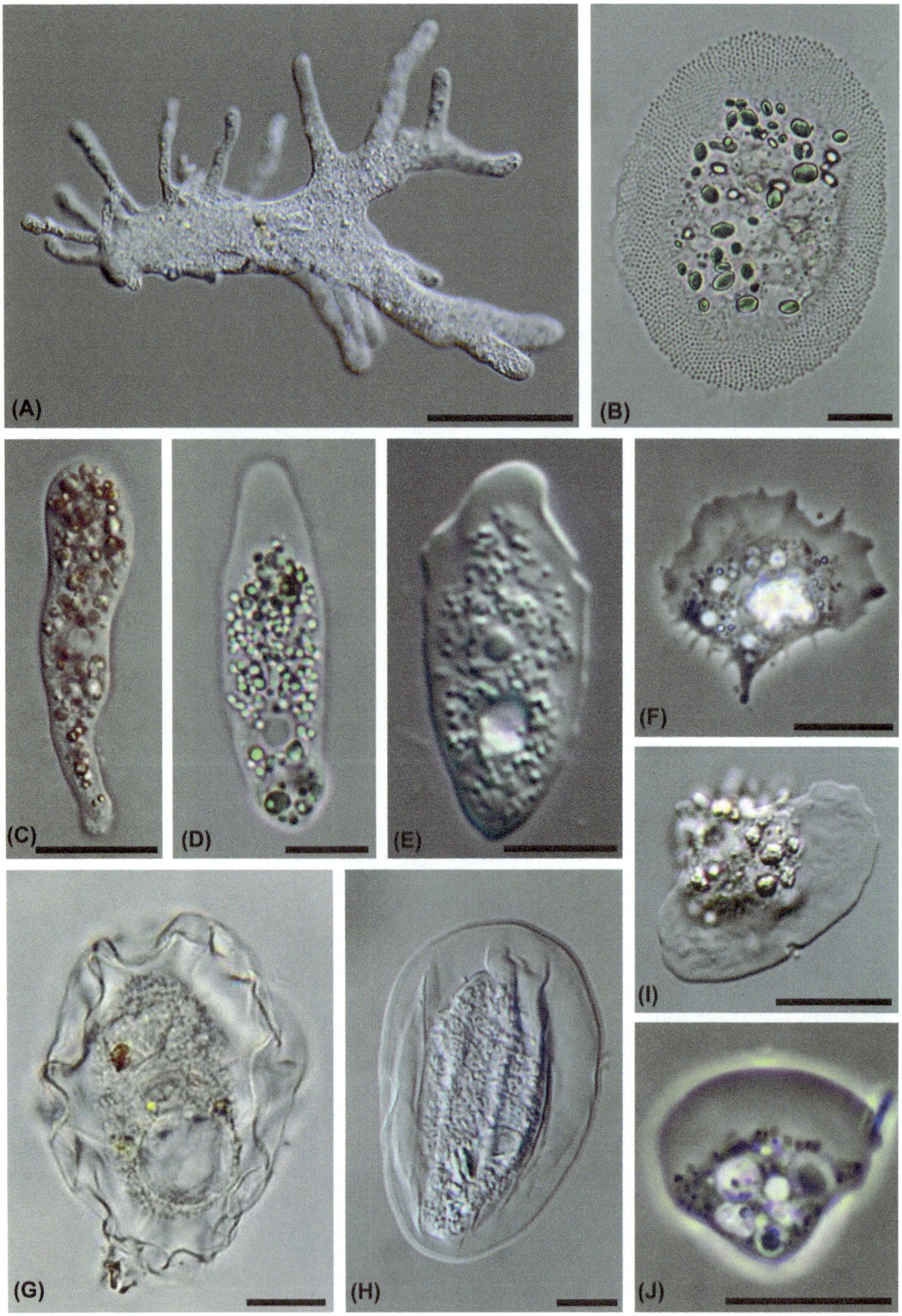

FIGURE 2.2 (A) *Amoeba proteus*; (B) *Cochliopodium bilimbosum*; (C) *Saccamoeba* sp.; (D) *Dermamoeba* sp.; (E) *Paradermamoeba* sp.; (F) *Flamella arnhemensis*; (G) *Thecamoeba verrucosa*; (H) *Thecamoeba striata*; (I) *Vannella simplex*; (J) *Vannella platypodia*. Scale bars: (A) = 100 μm, (I) = 20 μm, (B–H) and (J) = 10 μm. *Images by F. Siemensma.*

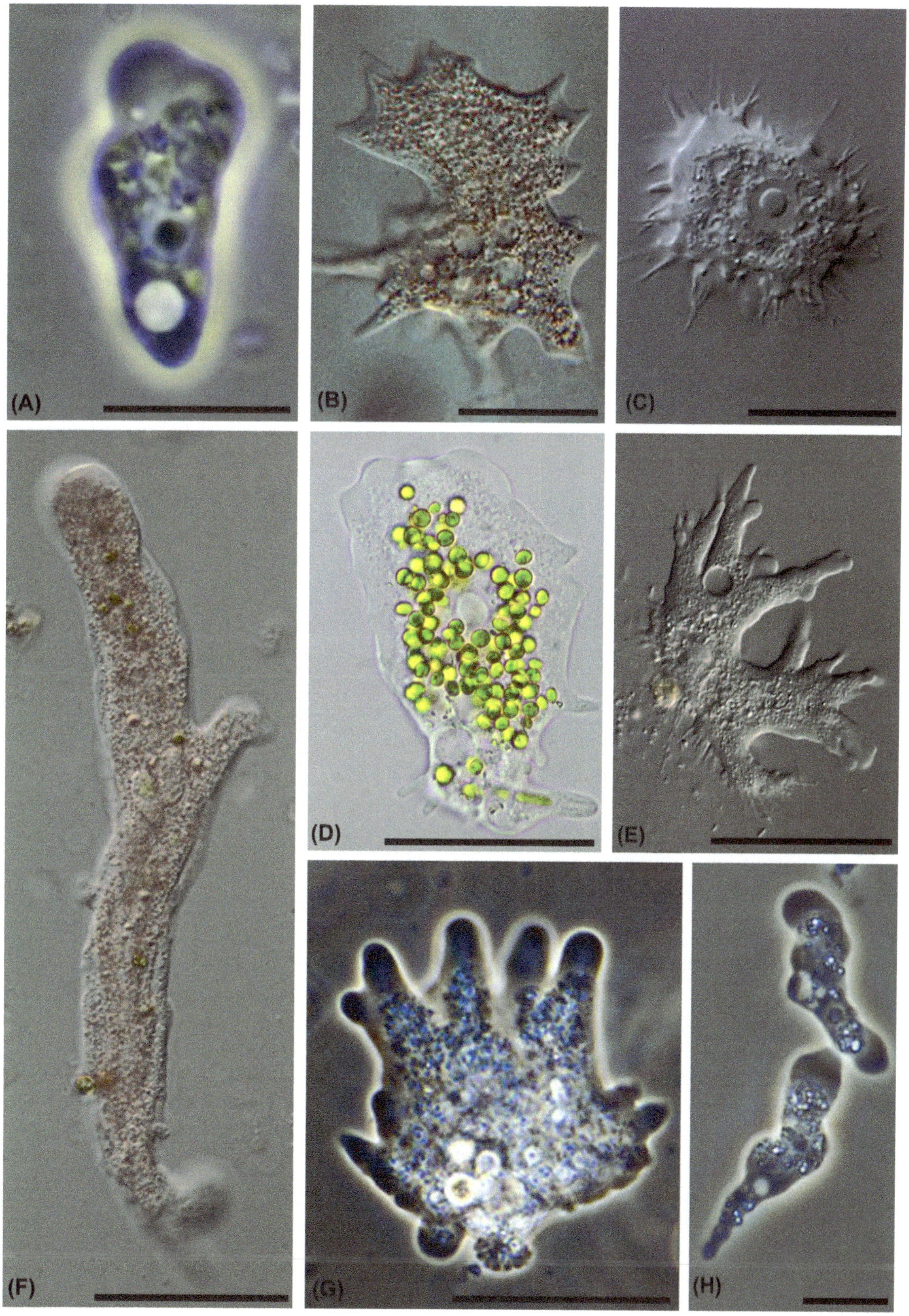

FIGURE 2.3 (A) *Vahlkampfia* sp.; (B) *Mayorella* sp.; (C) *Acanthamoeba* sp.; (D) *Mayorella viridis*; (E) *Rhizamoeba* sp. with adhesive trailing filaments; (F) *Chaos* sp., orthotactic form; (G) *Deuteramoeba algonquinensis*, palmate morphotype; (H) *Hartmannella* sp. Scale bars: (F) = 100 μm, (D, E, and G) = 50 μm, (A–C and H) = 10 μm. *Images by F. Siemensma.*

3 days. In the following days, many amoebae will multiply or form cysts.

The next step is culturing amoebae. Most species can only be identified from cultures, where plenty of cells are present for isolating and for the observation of cysts. For culturing methods and media recipes, refer to Page (1976, 1988) and Page & Siemensma (1991).

Testate amoebae: Various methods of water sampling and a filtration procedure for planktonic testate amoebae have been described in Lahr et al. (2011), Arriera et al. (2015), and Schwind et al. (2016). The filtrate containing testate amoebae is fixed immediately in a formaldehyde solution with final concentration of 2%–4% or alternatively into an ethanol solution of 70% final concentration and stored at room temperature in the dark (Lahr et al., 2011; Schwind et al., 2016). For better conservation of testate amoeba shells, the solution can be buffered with calcium carbonate (Schwind et al., 2016). The shells of testate amoebae can be inspected and quantified by observing filtrates in a Petri dish or on microscopy slides under an optical microscope at 200–1000× magnification. Dead and living testate amoebae can be differentiated after staining with 2% Aniline blue solution (Wanner & Elmer, 2009) or with Rose Bengal (Booth et al., 2010). A Sedgewick-Rafter chamber can be used for quantifying the abundance of testate amoebae in environmental sample. Three subsamples of one sample with a total of 7.5 mL should be analyzed with at least 50 individuals counted per sample (see Schwind et al., 2016). The total abundance should be expressed as individuals per cubic meter. For the preparation of permanent slides, shells of testate amoebae should be collected with a micropipette, washed and placed on a slide with, for example, water-soluble universal mounting mediums (i.e., glycerol), and then covered with a cover glass. Observation with scanning electron microscopy is helpful to identify small species with complex self-secreted scale patterns, such as many Euglyphida or the Arcellinida genus *Quadrulella*. For scanning electron microscopy, shells should be washed several times with distilled water and then washed in a drop of ethanol to remove all salts and mounted on a stub.

KEYS TO AMOEBA

Amoebae: Phyla

1 External shell or test absent, sometimes with a thickened structure bound to the cell membrane 2

1′ With an external test or theca 3

2(1) Amoebae monopodial or polypodial, not markedly eruptive (Figs. 2.2, 3 B–H) .... **Amoebozoa [p. 20]**

2′ Amoebae monopodial, with characteristic, eruptive, cytoplasmic flows; eruptions may be more or less pronounced but are always visible (Fig. 2.3 A) .... Percolozoa, one class: **Heterolobosea [p. 26]**

3(1) Testate amoebae with lobose pseudopods .... **Arcellinida [p. 26]**

3′ Testate amoebae with filose pseudopods .... **Rhizaria [p. 34]**

Amoeba: Amoebozoa: Classes

1 Amoebae with various kinds of lobopodia, often with slender, sometimes filiform subpseudopodia produced from broader hyaline lobopodium; usually uninucleate, sometimes binucleate or multinucleate; with mitochondria 2

1′ Amoebae monopodial, broadly cylindrical or ovoid, sluggish; multinucleate; some species with nonmotile, flagelliform structures; no mitochondria; bacterial endosymbionts; in oxygen-poor habitats .... **Archamoebea [p. 21]**

2(1) Flattened amoebae, never producing tubular, subcylindrical pseudopodia and never altering the locomotive form 3

2′ Tubular amoebae with subcylindrical pseudopodia or capable of altering the locomotive form from a flattened, expanded one to a subcylindrical one, with monoaxial flow of the cytoplasm in every pseudopodium or in the entire cell ..... **Tubulinea (= Lobosea) [p. 21]**

3(2) Amoebae without continuous envelope 4

3(2) Amoebae bounded by a continuous thin, somewhat flexible, envelope having an oval scaliform substructure within a denser matrix; one or many small pores penetrate the envelope, allowing pseudopodia to protrude for very slow, occasional locomotion; marine .... Cutosea

4(3) Nonciliated amoebae, never producing pointed pseudopodia; subpseudopodia, if present, short, never both pointed and branched .... **Discosea [p. 24]**

4′ Ciliated amoebae with temporary pointed pseudopodia or nonciliated amoebae with long, tapering, usually pointed, often branched subpseudopodia .... **Variosea [p. 25]**

Amoebae: Amoebozoa: Archamoebae: Families

Archamoebe are distinguished by the absence of mitochondria, which are converted to nonaerobic organelles. Animal parasites and commensals taxa have been omitted here.

1 Amoebae usually polypodial, with many pseudopodia; single cilium directed forward, with vibrating beat, extending to nucleus; uninucleate, but some species multinucleate (Fig. 2.5 B) .... Mastigamoebidae

1′ Amoebae usually monopodial, spherical to elongated ovoid; cells reaching up to 5 mm long with a large anterior pseudopod and a posterior uroid; one to numerous nuclei; short cilia present in most species; cytoplasm contains several glycogen bodies (Fig. 2.5 A) .... Pelomyxidae

Amoebae: Amoebozoa: Tubulinea (= Lobosea): Orders

1 Tubular or flattened amoebae without subcylindrical pseudopodia 2

1′ Tubular amoebae with subcylindrical pseudopodia .... **Euamoebida [p. 21]**

2(1) Locomotive form altering from a flattened expanded or reticulate one, when slowly moving, to a subcylindrical monopodial one when in rapid movement or under specific conditions; adhesive uroid (Figs. 2.1 K, 3 E, 5 C, E) .... **Leptomyxida [p. 24]**

2′ Flattened monopodial locomotive form, never reticulate or flattened expanded, with or without spiniform subpseudopodia; if spiny subpseudopodia absent, then length/breadth ratio > 6; no adhesive uroid .... **Echinamoebida [p. 24]**

Amoebae: Amoebozoa: Tubulinea (= Lobosea): Euamoebida: Families

Naked amoebae with subcylindrical pseudopodia in locomotion (or the entire cell is monopodial and subcylindrical), without alteration of the locomotive form to a flattened expanded and branched one, without adhesive uroid.

1 Locomotive amoebae not displaying frequent changes in direction or eruptive activity 2

1′ Locomotive amoebae displaying frequent changes in direction, some cells show eruptive activity .... Nolandellidae

2(1) Locomotive amoebae usually > 80 μm in length, with cylindrical pseudopodia; cells orthotactic monopodial and/or polytactic; nucleus ovular (Figs. 2.2 A, 3 F, G) .... **Amoebidae [p. 21]**

2′ Locomotive amoebae usually <80 μm; amoebae monotactic monopodial; nucleus vesicular (Figs. 2.2 C, 3 H) .... **Hartmannellidae [p. 24]**

Amoebae: Amoebozoa: Tubulinea (= Lobosea): Euamoebida: Amoebidae: Genera

1 Amoeba multinucleate 2

1′ Amoeba usually with one nucleus 3

2(1) Amoeba usually polypodial, with longitudinal ridges (Figs. 2.1 A, 3 F) .... *Chaos*

2′ Amoeba usually monopodial, without longitudinal ridges .... *Parachaos*

3(1) Amoeba monopodial, only branching when changing direction 4

3′ Amoeba usually polypodial 5

4(3) Amoeba free-living .... *Trichamoeba*

4′ Amoeba parasitic on freshwater coelenterates (*Hydra*) .... *Hydramoeba*

5(3) One dominant leading pseudopodium with smaller lateral pseudopodia 6

5′ Pseudopodia more or less equal-sized, radiating from a central mass like the fingers of a hand; villous bulb uroid .... *Polychaos*

6(5) Amoeba usually > 200 μm; nucleus granular 7

6′ Amoeba usually < 200 μm; nucleus with few granular pieces (Fig. 2.3 G) .... *Deuteramoeba*

7(6) With longitudinal ridges .... *Amoeba*

7′ Without longitudinal ridges .... *Metachaos*

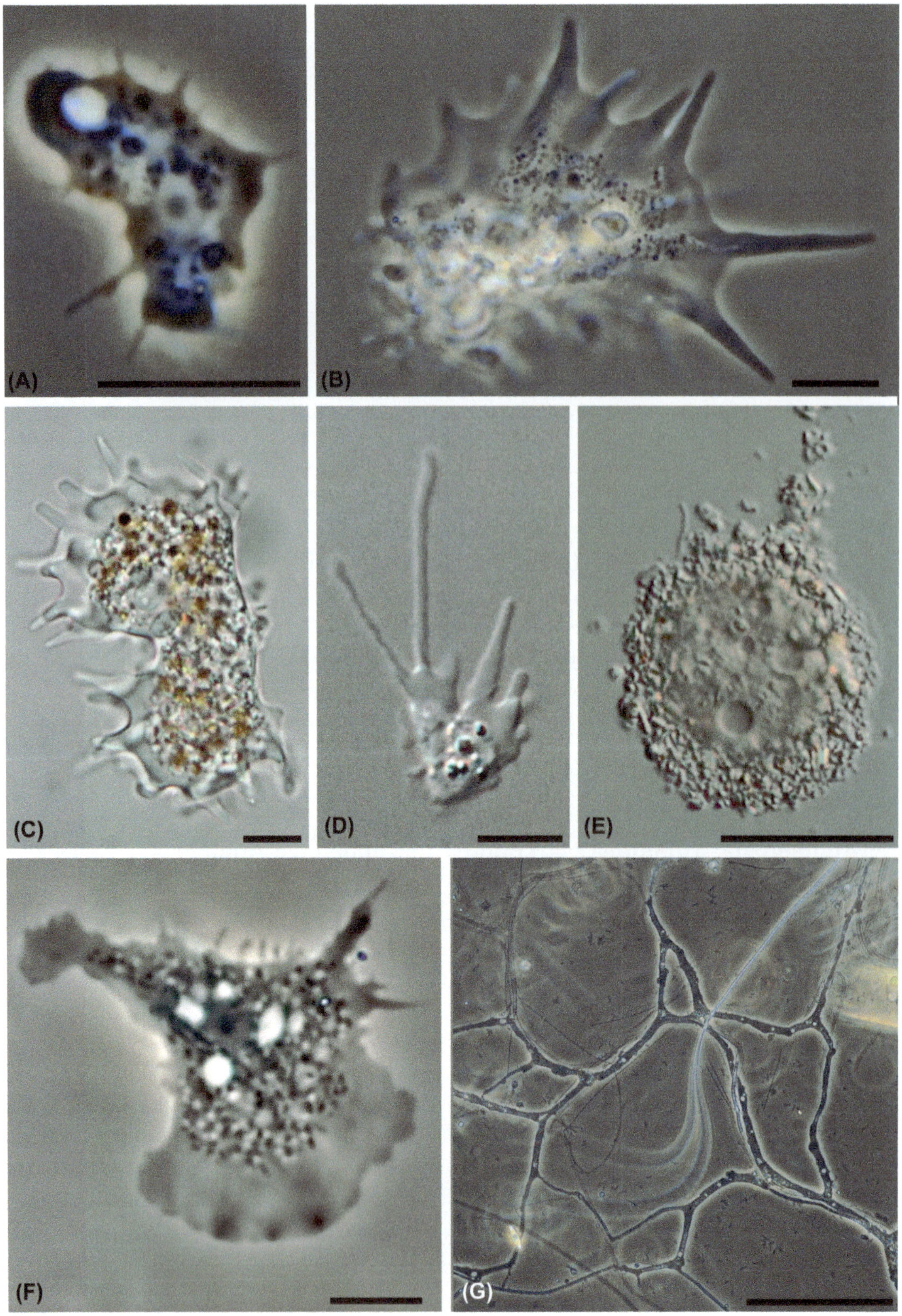

FIGURE 2.4 (A) *Acanthamoeba* sp.; (B) *Korotnevella* sp.; (C) *Dactylosphaerium* sp.; (D) *Vexillifera lemani*; (E) *Gocevia* sp; (F) *Paraflabellula* sp., (G) *Darbyshirella* sp. Scale bars: (A–F) = 10 μm, (G) = 100 μm. *Images by F. Siemensma.*

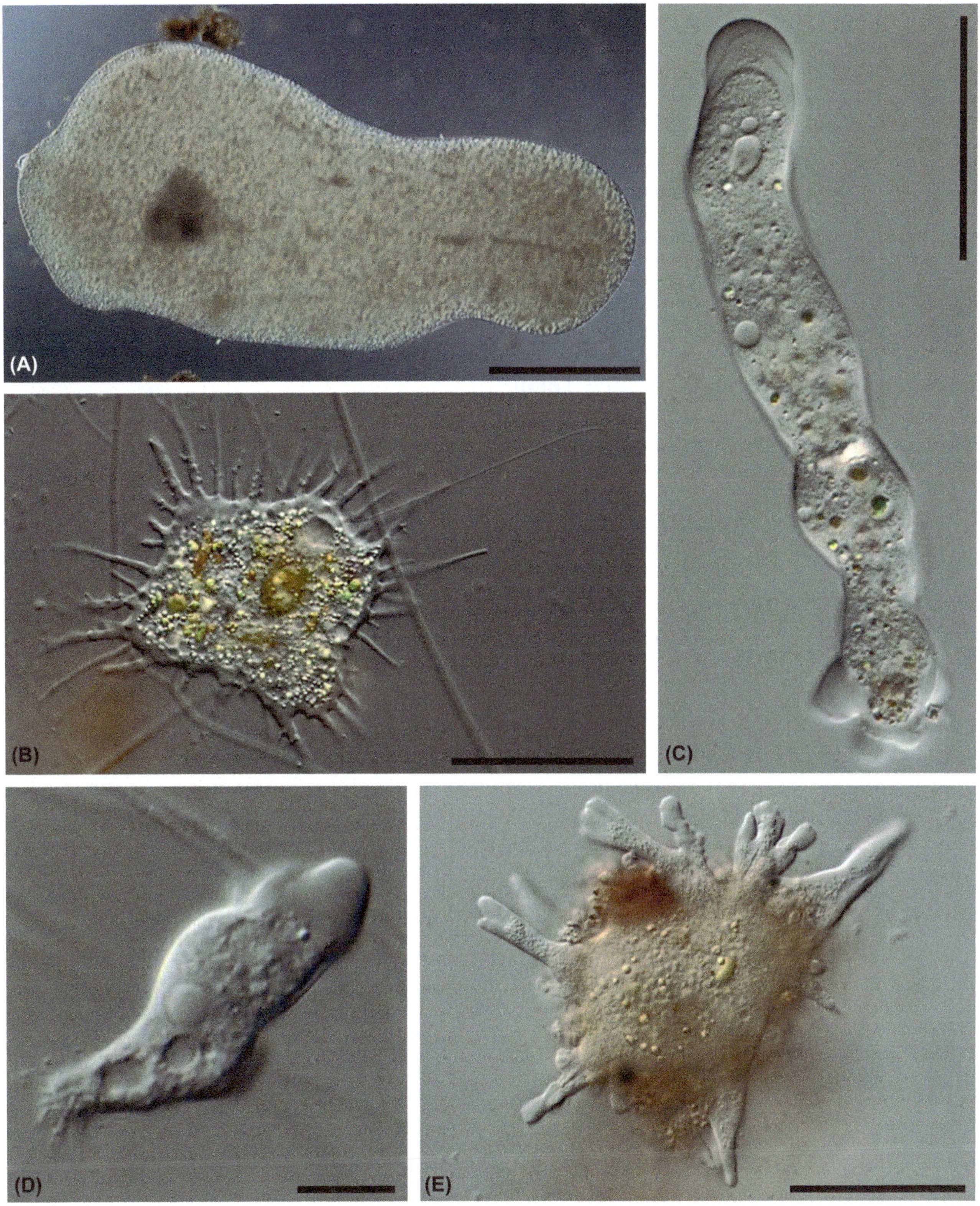

FIGURE 2.5 (A) *Pelomyxa palustris*; (B) *Mastigamoeba* sp.; (C) *Rhizamoeba* sp.; monopodial form; (D) *Willaertia magna*; (E) *Rhizamoeba* sp., stationary form covered by debris. Scale bars: (A) = 500 μm, (B and E) = 50 μm, (C) = 100 μm, and (D) = 10 μm. *Images by F. Siemensma.*

Amoebae: Amoebozoa: Tubulinea (= Lobosea): Euamoebida: Hartmannellidae: Genera

Uninucleate amoebae, monopodial with a cylindrical pseudopodium when active; most floating forms without radiating pseudopodia; nucleus with a prominent central nucleolus.

1	Distinctive hyaline zone in locomotive form	2
1′	Hyaline zone in locomotive form reduced or absent	3
2(1)	Amoebae may aggregate to form fruiting bodies (sorocarps)	*Copromyxa*
2′	Amoebae never aggregate (Fig. 2.3 H)	*Hartmannella*
3(1)	Uroid, often bulbous and villous	4
3′	Uroid, if present, reduced	*Cashia*
4(3)	Cytoplasm often with prominent crystals or inclusions (Fig. 2.2 C)	*Saccamoeba*
4′	Cytoplasm without crystals	*Glaeseria*

Amoebae: Amoebozoa: Tubulinea (= Lobosea): Leptomyxida: Families

Naked with locomotive form altering from a flattened expanded or reticulate one, when slowly moving, to a subcylindrical monopodial one when in rapid movement or under specific conditions; adhesive uroid; uninucleate with tendency to have more and with *Leptomyxa* always multinucleate. Leptomyxid amoebae have a tendency to form multinucleate plasmodia, which can be flabellate, multilobed, or highly branched and reticulate. Amoebae never have anterior subpseudopodia.

1	Amoebae cylindrical during rapid locomotion	2
1′	Amoebae never cylindrical	3
2(1)	Amoebae forming pronounced adhesive uroid or numerous adhesive uroidal filaments in locomotion; cells adopt expanded, fusiform shapes	Leptomyxidae
2′	Amoebae altering locomotive morphology from cylindrical to fabellated and vice versa; cells never adopt expanded, fusiform shapes (Figs. 2.3 E, 5 C, E)	Rhizamoebidae
3(1)	Flattened amoebae with extensive anterior hyaline zones, with or without subpseudopodia; often with posterior trailing uroidal filaments (Fig. 2.4 F)	**Flabellulidae [p. 24]**
3′	Amoebae often much branched but not plasmodial; no mass of trailing filaments	Gephyramoebidae

Amoebae: Amoebozoa: Tubulinea (= Lobosea): Leptomyxida: Flabellulidae: Genera

1	Anterior hyaline with clefts rather than subpseudopodia	*Flabellula*
1′	Anterior hyaline with subpseudopodia short, apically rounded (Fig. 2.4 F)	*Paraflabellula*

Amoebae: Amoebozoa: Tubulinea (= Lobosea): Echinamoebida: Families

1	Amoebae flattened, often spatulate or elongated or very irregular, with or without spiniform subpseudopodia	Echinamoebidae
1′	Amoebae subcylindrical in cross section, vermiform	Vermamoebidae

Amoebae: Amoebozoa: Discosea: Orders

Flattened naked amoebae, never producing tubular, subcylindrical pseudopodia and never altering the locomotive form; cytoplasmic flow polyaxial or without a pronounced axis; no flagellate stage in the life cycle; subpseudopodia, if present, short, never both pointed and branched.

1	Amoebae smooth, without dorsal folds and ridges	2
1′	Amoebae with dorsal folds and/or ridges or with 1–4 pairs of nuclei (Fig. 2.2 G, H)	**Thecamoebida [p. 25]**
2(1)	Amoebae not as below	3
2	Amoebae covered dorsally with a coat independent on the cell membrane or small low mobile globular amoebae with a dorsally irregularly wrinkled and semirigid thick cell coat; without defined aperture; ventral surface naked (Figs. 2.2 B, 4 E)	**Himatismenida [p. 25]**
3(2)	Amoebae not as below	4

3′ Flattened with prominent subpseudopodia, flexible and tapering to a fine tip, and sometimes basally furcated (acanthopodia), or with more distal, broader lobose furcated subpseudopodia, sometimes emerging from a hyaloplasm sheet; without adhesive uroid; flattened sheet without subpseudopodia; some sporocarpic and protosteloid; generally plasma membrane naked without surface scales or styles, although surface coats of varying organization occur in some taxa .... **Centramoebida [p. 25]**

4(3) Flattened amoebae, never producing tubular, subcylindrical pseudopodia and never altering the locomotive form; subpseudopodia, if present, short, never both pointed and branched .... **Glycostylida [p. 25]**

4′ Oblong, lanceolate, or irregularly triangular in locomotion; with smooth cell surface or with few wide ridges, never wrinkled; short, wide triangular pseudopodia and, in some, subpseudopodia of dactylopodial type (Fig. 2.2 D, E) .... **Dermamoebida [p. 25]**

Amoebae: Amoebozoa: Discosea: Thecamoebida: Families

Oblong, flattened cell with dorsal folds and/or ridges; anterior hyaloplasm forms anterolateral crescent and never occupies half or more of the body length; never produce discrete pseudopodia or subpseudopodia; cell coat relatively thin, amorphous, or with extra structures based on amorphous layer.

1 With well-defined dorsal and lateral ridges (Fig. 2.2 G, H) .... Thecamoebidae

1′ During locomotion occasionally with dorsal or lateral ridges that never remain stable .... Stenamoebidae

Amoebae: Amoebozoa: Discosea: Himatismenida: Families

1 Amoebae partially enclosed within a flexible cuticle or wall (tectum) usually covered with microscales or with a dorsal fibrous cell coat without scales. The covering is open along the region of attachment to the substratum 2

1′ Amoebae without such a covering 3

2(1) Locomotive cell without complete hyaloplasmic veil, frontal hyaloplasm with an expanded crescentiform area (Fig. 2.4 E) .... Goceviidae

2′ Locomotive cell with hyaloplasmic veil complete (Fig. 2.2 B) .... Cochliopodiidae

3(1) Locomotive form oval in outline, length ≤ 6 μm; with hyaline pseudopodium and occasionally with fine radiating subpseudopodia .... Parvamoebidae

3′ Locomotive form from rounded to spatulate, length > 10 μm, with short papilliform subpseudopodia .... Pellitidae

Amoebae: Amoebozoa: Discosea: Centramoebida: Families

1 Amoebae with broad pseudopodia and occasionally fingerlike pseudopodia and spiderlike walking movement .... Balamuthiidae

1′ Amoebae with slender, tapering subpseudopodia produced from a broad, hyaline lobopodium (Fig. 2.3 C) .... Acanthamoebidae

Amoebae: Amoebozoa: Discosea: Glycostylida: Families

1 Amoebae with conical or digitiform subpseudopodia 2

1′ Amoebae with long, slender subpseudopodia (usually more than one) frequently giving a spiny appearance to the cell (Fig. 2.4 D) .... Vexilliferidae

2(1) Amoebae have locomotive forms markedly flattened, flabellate, or linguliform with a prominent anterior hyaline zone without subpseudopodia (Fig. 2.2 I, J) .... Vannellidae

2′ Flattened dactylopodial amoebae with blunt, hyaline subpseudopodia, conical or digitiform in outline (Fig. 2.4 B) .... Paramoebidae

Amoebae: Amoebozoa: Discosea: Dermamoebida: Families

1 Oblong, lanceolate, or irregularly triangular in locomotion; with smooth cell surface or with few wide ridges, never wrinkled; short, wide triangular pseudopodia and, in some, subpseudopodia of dactylopodial type (Fig. 2.2 D, E) .... Dermamoebidae

1′ Locomotive digitiform subpseudopodia extending from anterior hyaloplasm; most species with prominent cytoplasmic crystals, truncated, paired, bipyramidal, or clustered with spheroids (Fig. 2.3 B, D) .... Mayorellidae

Amoebae: Amoebozoa: Variosea: Families

Aerobic ciliated amoebae with conical microtubular cytoskeleton and only temporary pointed pseudopodia or nonciliated amoebae with long, tapering, usually pointed, often branched subspseudopodia.

1	Amoebae flattened, sometimes flabellate; with digitiform subpseudopodia from an anterior wide hyaloplasm margin, or with thin pseudopods from the cell body (Fig. 2.2 F)	Flamellidae
1′	Amoebae often very flattened, often spatulate; many filose pseudopodia, several micrometers long, produced from one or more hyaline zones	Filamoebidae

Amoeba: Percolozoa: Heterolobosea: Families

Amoebae commonly cylindrical and monopodial, usually markedly eruptive, with a length usually smaller than 65μm; amoebae rarely flattened, with a few short fine subpseudopodia; usually uninucleate; flagellate stages in many; cyst-forming, a few spore-forming; mitochondrial cristae discoid; mitosis intranuclear, in most with binary division of nucleolus.

1	No fruiting bodies formed	2
1′	Fruiting bodies formed	3
2(1)	Amoebae cylindrical; flagellate stages in most genera (Fig. 2.3 A)	Vahlkampfiidae
2′	Amoebae commonly flattened with a few short, fine subpseudopodia; no flagellate stage known	Gruberellidae
3(1)	Cells of sorocarp differentiated into morphologically distinct spores and stalk cells; flagellate stage present or absent	Acrasidae
3′	Cells throughout sorocarp essentially all alike; no flagellate stages known	Guttulinopsidae

PROTOZOA

Amoeba: Arcellinida: Families and *Incertae sedis*

1	Pseudopods conical, not granular, often anastomosing	2
1′	Pseudopods digitiform, finely granular	3
2(1)	Test hemispherical, with mineral particles embedded in organic matrix; pseudopods conical, sometimes anastomosing; pseudostome not invaginated (Fig. 2.9 C)	Phryganellidae, one genus: *Phryganella*
2′	Test oval, entirely organic, chitineouse or with adhering foreign particles; smooth, transparent, and can be often overlooked due to their small size (Fig. 2.7 E)	Cryptodiffiugiidae, one genus: *Cryptodifflugia*
3(1)	Test rigid	4
3′	Test flexible or semirigid, hemispherical, aperture a slit when folded, with organic chitinoid structure and few mineral debris attached (Fig. 2.11 C)	Microcoryciidae, one genus: *Parmulina*
4(3)	Test rounded, circular, spherical or hemispherical, rounded chitinoid, can be areolar or organic with few mineral debris attached	5
4′	Test oval, fusiform, pyriform, or elongate, with embedded or attached mineral particles, diatom debris, scales or plates recycled from other testate amoebae or self-secreted siliceous particles of various shape (i.e., square, baciliform)	7
5(4)	Test circular or hemispherical, with centered circular aperture, with areolar structure	6
5′	Test circular or rounded, never with areolar structure, organic, often with mineral particles attached to or embedded in the organic cement	10
6(5)	Aperture width < 0.5× test diameter, tests typically composed of square proteinaceous units arranged in a single layer and cemented together, resulting in an areolar surface (Fig. 2.6 C)	Arcellidae, one genus: *Arcella*
6′	Aperture width subequal to test diameter (Fig. 2.9 F)	Arcellinida *incertae sedis*, *Pyxidicula*
7(4)	Aperture circular and invaginated	8
7′	Aperture a slit, with well-developed distinct anterior lip	**Plagiopyxidae [p. 27]**
8(7)	Aperture not central	9
8′	Aperture central	**Trigonopyxidae [p. 27]**
9(8)	Ventral side is flat; tests bilaterally symmetrical, rounded, flattened more at the front than at the rear, ventral part flat, often spines at both sides and at the rear, or circular test or elongated in ventral view, the ventral margin of the aperture is often connected to the dorsal face by some flat pillars or bridges (Fig. 2.7 A)	Centropyxidae, one genus: *Centropyxis*
9′	Ventral side is convex (Fig. 2.11 F)	Arcellinida *incertae sedis*, *Oopyxis*
10(5)	Test laterally compressed, and the aperture is an oval or slit	11
10′	Test laterally not compressed	14
11(10)	Aperture always oval or elliptical	12
11′	Test ovoid, aperture a slit, test composed of rounded or oval plates recycled from other testate amoebae and/or mineral particles embedded in organic cement (Fig. 2.8 B)	Arcellinida *incertae* sedis, *Heleopera*
	[Note: Probably does not belong to standard freshwater communities]	

12(11) Test not deep violet 13

12′ Test deep violet, ovoid, composed of mineral grains or agglutinated particles, aperture oval (Fig. 2.6 E) .... Arcellinida *incertae sedis*; *Awerintzewia cyclostoma* Schouteden, 1906

13(12) Aperture surrounded by siliceous plates giving it a rough outline, oval-pyriform laterally compressed (Fig. 2.6 D) .... Arcellinida *incertae sedis*, *Argynnia*

13′ Aperture always surrounded by ± thin organic rim giving it a smooth outline .... **Hyalospheniidae [p. 33]**

14(10) Test round in cross section, aperture is terminal and circular 15

14′ Test polygonal in cross section, composed of agglutinated mineral particles, shape pyriform, three to five lateral swellings sometimes terminating in a horn on fundus, aperture circular (Fig. 2.11 D) .... Arcellinida *incertae sedis*; *Pentagonia maroccana* Gauthier-Lièvre & Thomas, 1958

15(14) Test symmetrical with symmetrically positioned neck 16

15′ Test with asymmetrical neck, more or less attached to the main body; test mainly composed of self-secreted elongated rod like elements, sometimes mixed with mineral particles (Fig. 2.8 E) .... Arcellinida *incertae sedis*, *Lesquereusia*

16(15) Test divided in two parts by a diaphragm between neck and main body, often visible as a constriction under light microscopy 17

16′ Test not divided by diaphragm 19

17(16) Diaphragm with two openings 18

17′ Internal diaphragm with one opening (Fig. 2.8 D) .... Arcellinida *incertae sedis*, *Lagenodifflugia*

18(17) Diaphragm composed of small mineral grains, two openings are circular (Fig. 2.9 I) .... Arcellinida *incertae sedis*, *Zivkovicia*

18′ Openings formed by organic bridge with few attached mineral particles which connects both broad sides (Fig. 2.9 D) .... Arcellinida *incertae sedis*, *Pontigulasia*

19(16) Not as below 20

19′ Aperture complex, consisting of a central pore and numerous elongated pores radiating around a central pore; the test consists of agglutinated mineral particles (Fig. 2.11 B) .... Arcellinida *incertae sedis*, *Suiadifflugia*

20(19) Aperture lobed or denticulate 21

20′ Aperture round formed by agglutinated mineral particles or diatom frustules, test of different shapes (i.e., oval, pyriform, elongated) agglutinated with mineral particles, diatoms or entire tests of smaller testate amoebae cemented together (Fig. 2.7 G) .... Arcellinida *incertae sedis*, *Difflugia*

21(20) Test is ovoid with lobed aperture surrounded by organic rim or collar 22

21′ Aperture surrounded by organic rim with 3–11 denticulations giving it a lobed appearance; test is fusiform with elongated neck (Fig. 2.9 E) .... Arcellinida *incertae sedis*; *Pseudonebela africana* Gauthier-Lièvre, 1953

22(21) Organic collar incurved, forming a chamber before the aperture with an internal circular opening (Fig. 2.7 C, D) .... Arcellinida *incertae sedis*, *Cucurbitella*

22′ Organic rim not incurved, aperture often presenting several lobes (Fig. 2.8 G) .... Netzeliidae, *Netzelia*

Amoeba: Arcellinida: Plagiopyxiidae: Genera

1 Anterior lip with pores, dark brownish, elliptical test with prolonged ventral lip and shorter dorsal lip with small or larger pores (Fig. 2.6 F) .... *Bullinularia*
[Note: Probably does not belong to standard freshwater communities]

1′ Pores are absent and anterior lip almost covers the aperture (Fig. 2.11 A) .... *Plagiopyxis*
[Note: Probably does not belong to standard freshwater communities]

Amoeba: Arcellinida: Trigonopixidae: Genera

1 Aperture circular, dark yellow or brownish circular or hemispherical species, composed of agglutinated large mineral particles, with central, invaginated, circular, sometimes lobed aperture; length 100–400 μm (Fig. 2.7 F) .... *Cyclopyxis*

1′ Aperture triangular or irregular (Fig. 2.9 H) .... *Trigonopyxis*
[Note: Probably does not belong to standard freshwater communities]

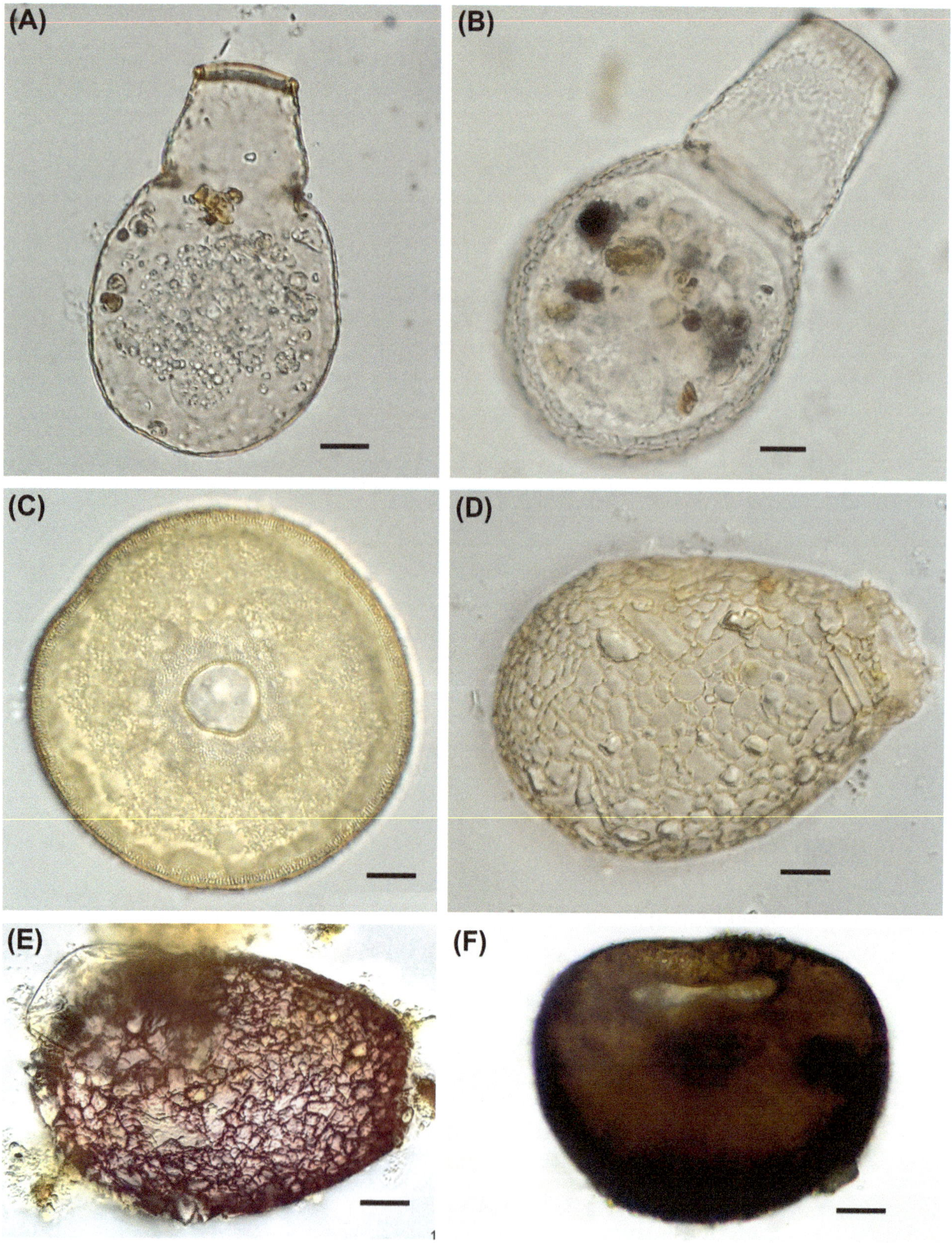

FIGURE 2.6 Light microscope images of (A) *Alocodera cockayni* from Chile; (B) *Apodera vas* from Chile; (C) *Arcella arenaria* from Chile; (D) *Argynnia scotica* from Chile; (E) *Awerintzewia cyclostoma* from Ecuador; and (F) *Bullinularia indica* example from Austria. Scale bares = 10 μm. *Image: (A–D) by L.D. Fernández, (E and F) by F. Siemensma.*

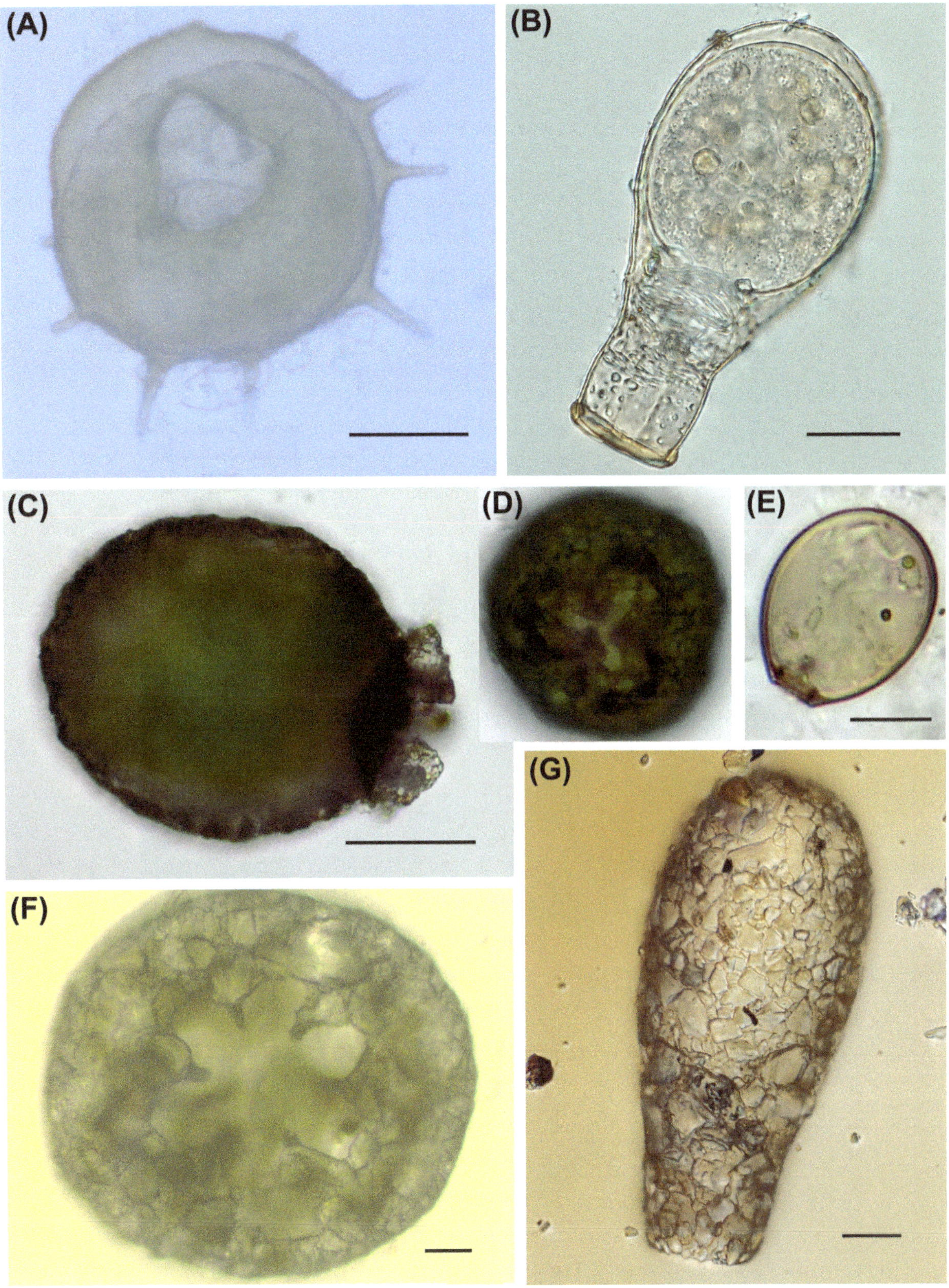

FIGURE 2.7 Light microscope images of (A) *Centropyxis discoides* from Brazil; (B) *Certesella martiali* from Argentina; (C) *Cucurbitella vlasinensis* from Peru and (D) close view of its aperture; (E) *Cryptodifflugia oviformis* example from Austria; (F) *Cyclopyxis impressa* from Brazil; (G) *Difflugia gigantea* from Chile. Scale bars: (A–C and F) = 50 μm, (G) = 20 μm, and (E) = 10 μm. *Images: (A and F) by A. Kosakyan; (B) by E. Lara, (C–E) by F. Siemensma, and (G) L.D. Fernández.*

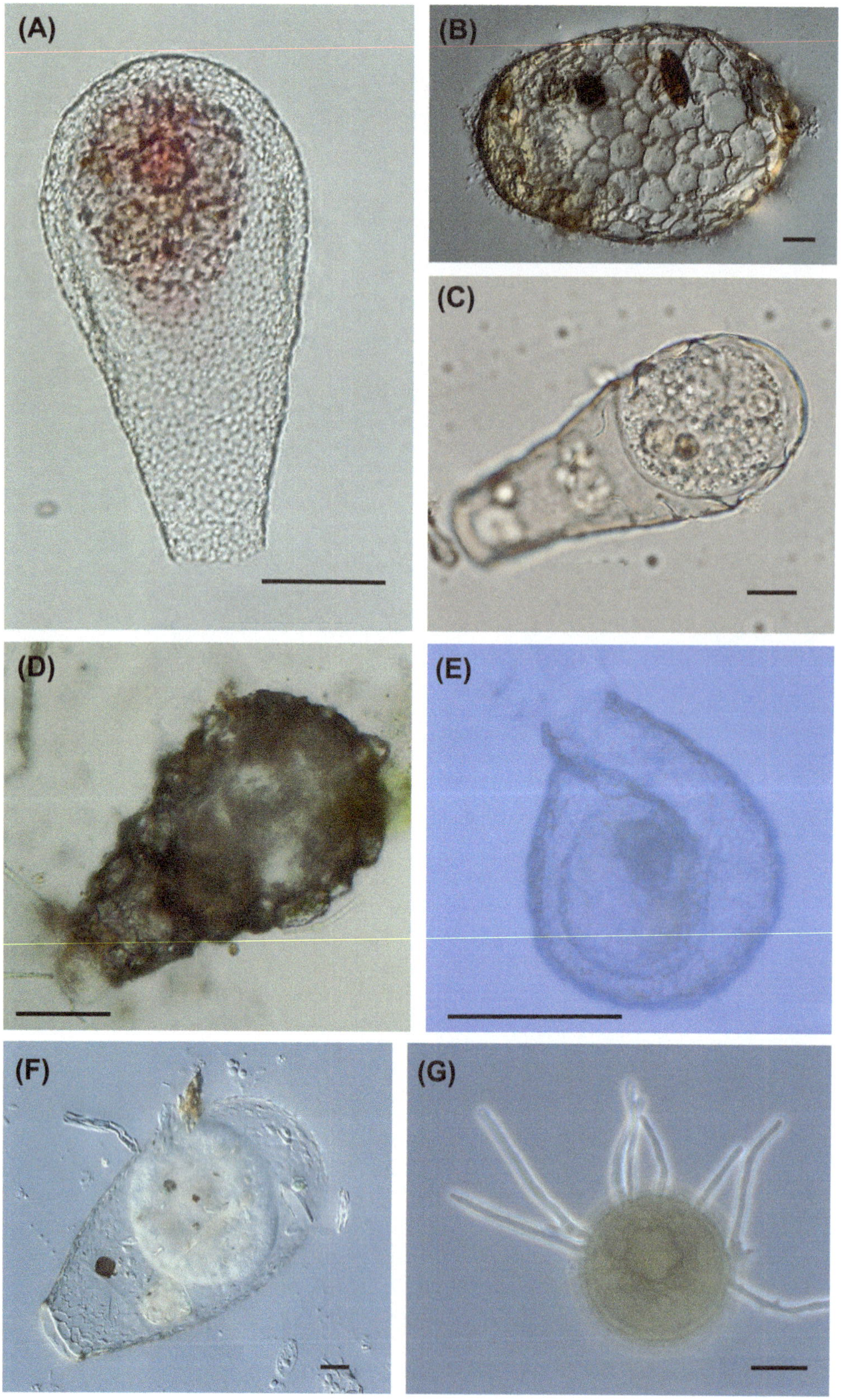

FIGURE 2.8 Light microscope images: (A) *Gibbocarina galeata* digitalized from Penard slide collection 752-2; (B) *Heleopera* sp. from Ecuador; (C) *Hyalosphenia elegans* from Chile; (D) *Lagenodifflugia vas* example from Austria; (E) *Lesquereusia* sp. from Brazil; (F) *Nebela* cf. *collaris* from Chile; and (G) *Netzelia oviformis* from Brazil. Scale bars: (A, D, E, and G) = 50 μm, (B, C, and F) = 10 μm. *Images: (A) by T. Arnet, (B and D) by F. Siemensma, (C and F) by E. Lara, and (E and G) by A. Kosakyan.*

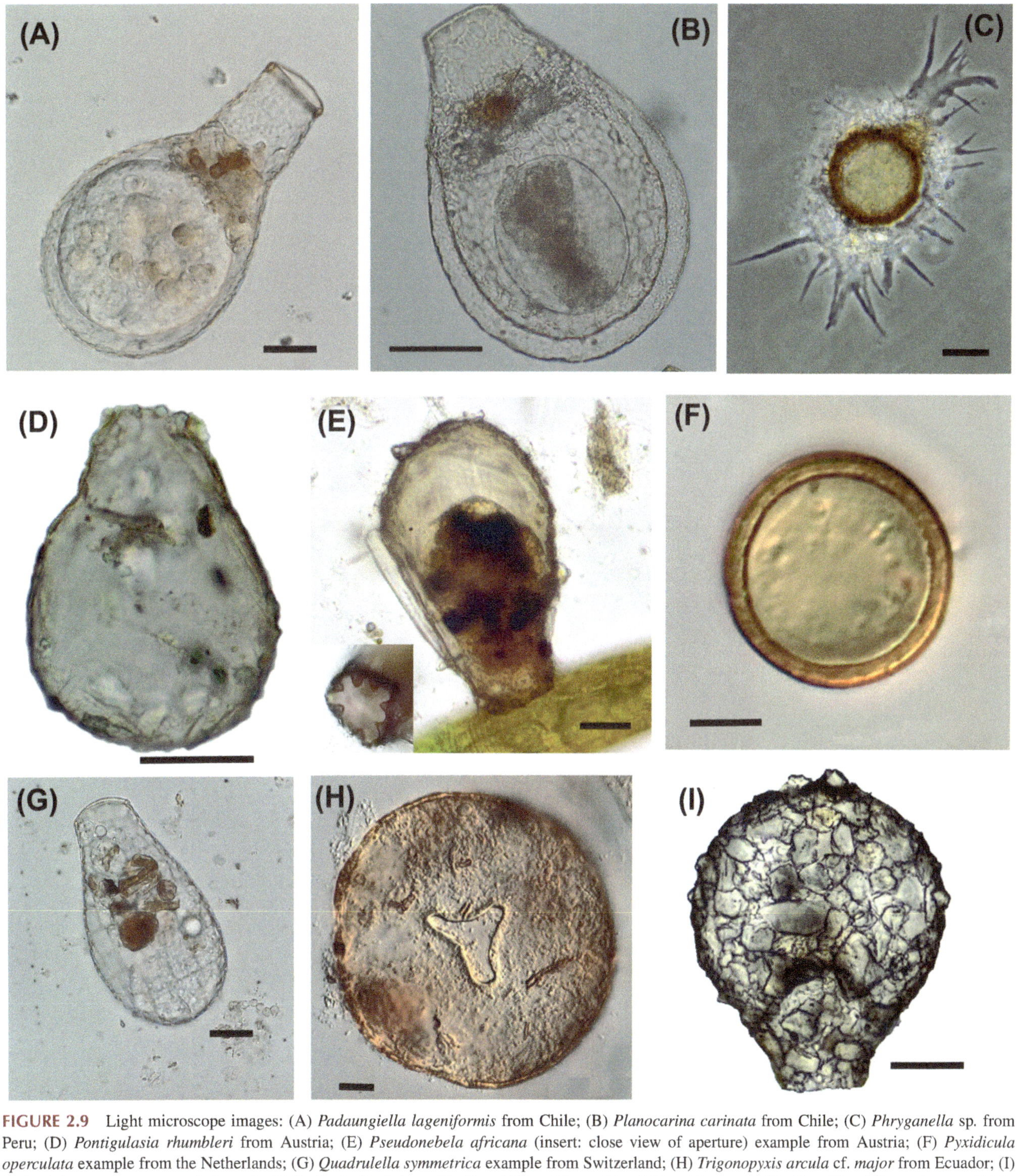

FIGURE 2.9 Light microscope images: (A) *Padaungiella lageniformis* from Chile; (B) *Planocarina carinata* from Chile; (C) *Phryganella* sp. from Peru; (D) *Pontigulasia rhumbleri* from Austria; (E) *Pseudonebela africana* (insert: close view of aperture) example from Austria; (F) *Pyxidicula operculata* example from the Netherlands; (G) *Quadrulella symmetrica* example from Switzerland; (H) *Trigonopyxis arcula* cf. *major* from Ecuador; (I) *Zivkovicia compressa* example from the Netherlands. Scale bars: (A, G, and H) = 10 μm, (B and C) = 20 μm, and (D–F and I) = 50 μm. *Images: (A) by L.D. Fernández, (B and G) by A. Kosakyan, and (C–F, H, and I) by F. Siemensma.*

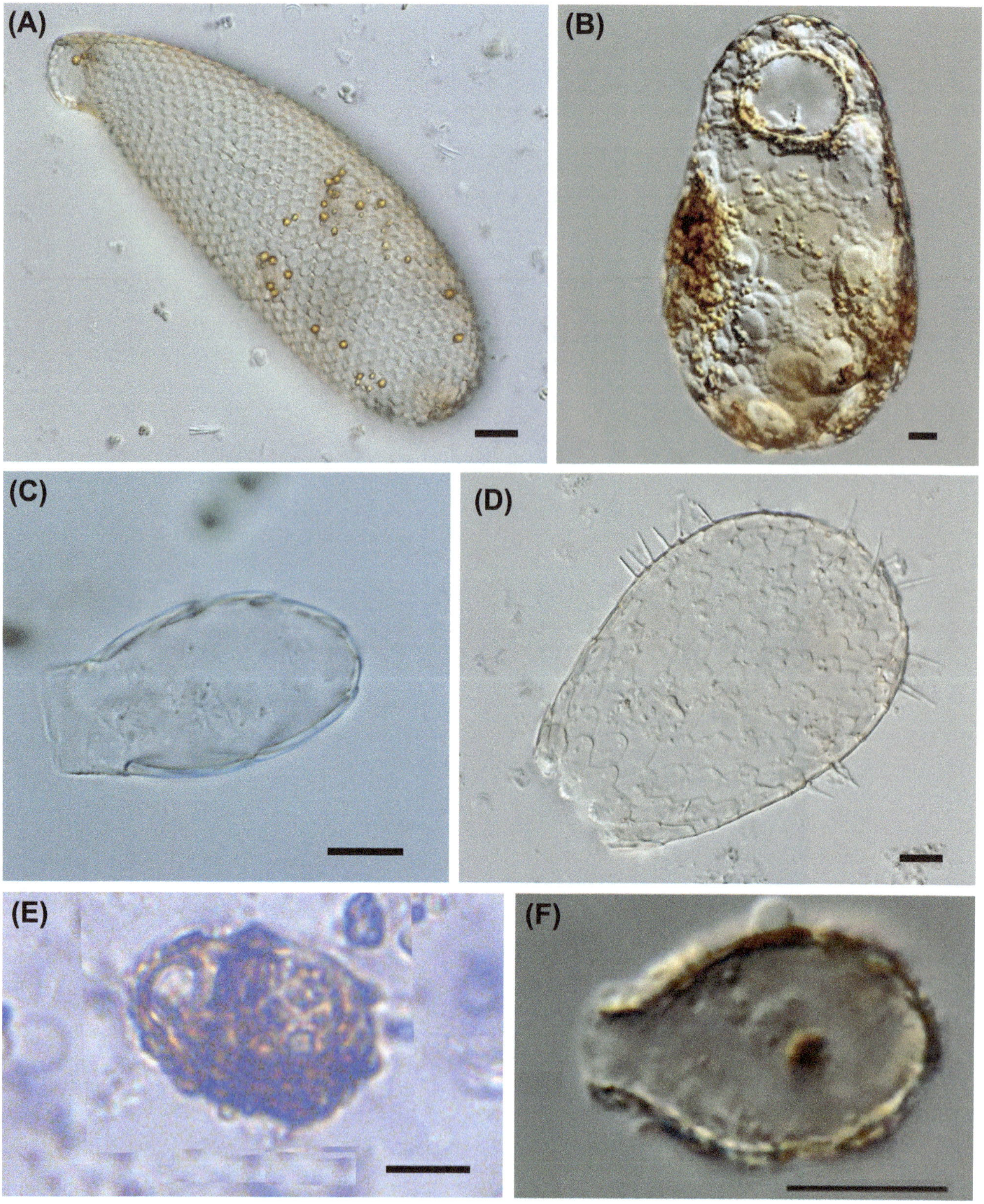

FIGURE 2.10 Light microscope images: (A) *Cyphoderia ampulla* from Chile; (B) *Trinema grandis* from Ecuador; (C) *Sphenoderia valdiviana* from Chile; (D) *Euglypha* sp. from Chile; (E) *Schwabia regularis* from Argentina; (F) *Pseudodifflugia fascicularis* example from Netherlands. Scale bars = 10 μm. *Images: (A) from Kosakyan et al. (2016a), (F) by L.D. Fernández, (B) by F. Siemensma, (C) and (D) by E. Lara, and (E) by L. Burdman.*

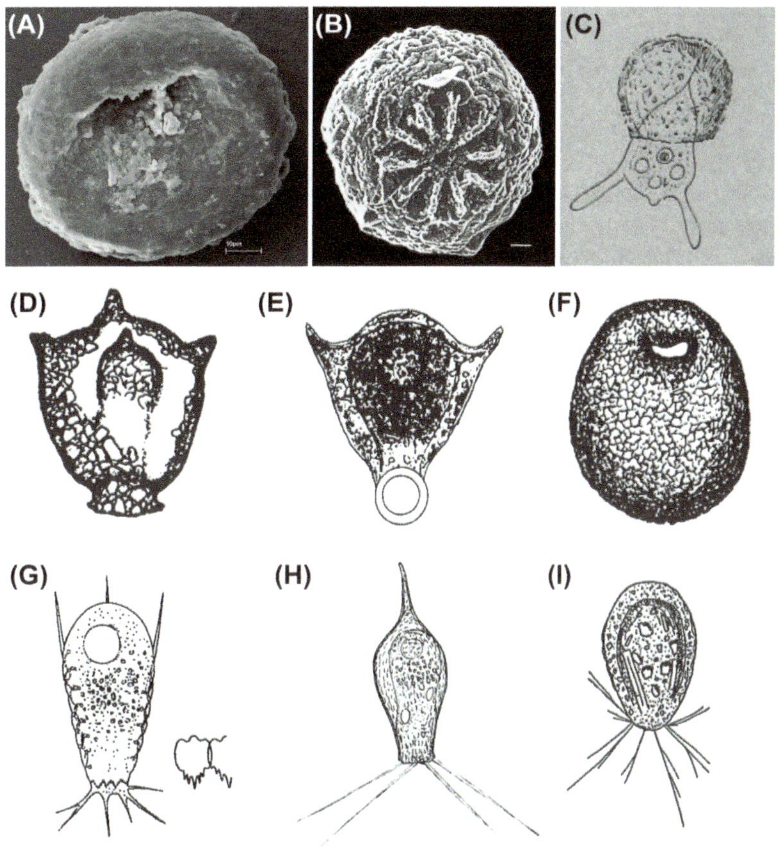

FIGURE 2.11 SEM images or original drawings of taxa for which no LM data was available: (A) SEM image of *Plagiopyxis* sp.; (B) original SEM image of *Suiadifflugia multipora*; (C) original drawing of *Parmulina obtecta*; (D) original drawing of *Pentagonia maroccana*; (E) original drawing of *Campascus cornutus*; (F) original drawing of *Oopyxis cophostoma*; (G) original drawing of *Scutiglypha crenulata*; (H) original drawing of *Pareuglypha reticulata*; (I) original drawing of *Clypeolina marginata*. Scale bars: (A) = 10 μm, (B) = 15 μm. *(A) Image by V. Krashevska; (B) after Green (1975); (C, E, and H) after Pennard (1902); (D) after Lievre & Thomas (1958); (F) after Jung (1942); (G) after Wailes (1912); and (I) from Leidy (1879).*

Amoeba: Arcellinida: Hyalospheniidae: Genera

1	Test not hyaline and has a structure	2
1′	Test hyaline, without structure, can be oval, lageniform, or pyriform; with completely organic test composition (fig. 2.8 C)	*Hyalosphenia*
2(1)	Test composed of particles of exogenous origin (i.e., rounded or elliptic shell plates recycled from other testate amoebae, diatom frustules mineral particles, or finely granular structure resembling as hyaline and organic)	3
2′	Test composed of self-secreted square plates (Fig. 2.9 G)	*Quadrulella*
3(2)	Test ovoid, pyriform, or elongated pyriform, with or without lateral keel; lateral sides tapering toward aperture and the neck is not distinct	4
3′	Test pyriform in broad lateral view with distinct neck	6
4.	Test elongated pyriform with lateral keel	5
4′	Test more ovoid, drop-shaped, sometimes rounded-pyriform in broad lateral view (Fig. 2.8 F)	*Nebela*
5(4)	Lateral keel flat (Fig. 2.9 B)	*Planocarina*
5′	Lateral keel hollow (Fig. 2.8 A)	*Gibbocarina*
6(3)	No constriction at the base of the neck	7
6′	Neck deeply constricted at the junction with the main body of the test (Fig. 2.6 B)	*Apodera*
7(6)	Lateral indentation is absent	8

PROTOZOA

7′ Test with a lateral indentation of both sides in the junction of the main body to the neck (note: the neck is not deeply constricted as in genus *Apodera*; the constriction is only lateral), test often organic, hyaline, but sometimes with a rough granulated structure (visible by SEM) or with siliceous plates (recycled from euglyphid testate amoebae) (Fig. 2.6 A); may be species complex .... *Alocodera cockayni* (Penard, 1911)

8(7) Test composed of plates (scales recycled from euglyphid testate amoebae) embedded in an unstructured organic cement; the neck is very elongate, fusiform (Fig. 2.9 A) .... *Padaungiella*

8′ Test composed of plates (scales recycled from euglyphid testate amoebae) embedded in an unstructured organic cement and often covered with a thick layer of organic cement giving the test a smooth appearance; in broad lateral view, two large pores in depressions are connected by internal tubes situated at the base of the neck (ca. 2/3 from the posterior part of the test); the internal side of the neck with conical protuberances pointing inward perpendicular to the test surface, visible in light microscopy as lines of points (Fig. 2.7 B) ... *Certesella*

Amoeba: Rhizaria: Families and Euglyphida *Incertae Sedis*

1 Test including self-secreted siliceous scales 2

1′ Test reinforced with material recycled from the environment or entirely organic 3

2(1) Scales all discoidal, subequal in size 6

2′ Scales of different sizes and shapes 7

3(1) Test proteinaceous test reinforced with mineral particles or composed of two valves 4

3′ Test including foreign material (agglutinated) 5

4(3) Test curved, pseudostome surrounded by a hyaline collar, proteinaceous test reinforced with mineral particles (Fig. 2.11 I) .... Cyphoderiidae, *Campascus*

4′ Test ovoid, composed of two valves (Fig. 2.11 E) .... Chlamydophryidae, one genus: *Clypeolina*

5(4) Test agglutinated, round, small (less than 30 μm) (Fig. 2.10 F) .... Pseudodiffiugiidae, one genus: *Pseudodifflugia*

5′ Test agglutinated, elongated, large (always > 30 μm), organic collar around the pseudostome (Fig. 2.10 E) .... Euglyphida *incertae sedis*: *Schwabia*

6(2) Test straight, ovoid, tapered toward pseudostome; fundus with a scaly spine (Fig. 2.11 H) .... Euglyphida *incertae sedis*: *Pareuglypha*

6′ Test elongated, ovoid, curved neck (Fig. 2.10 A) .... Cyphoderiidae, one genus: *Cyphoderia*

7(2). Pseudostome in a terminal position 8

7′ Pseudostome in a lateral, eccentric position (Fig. 2.10 B) .... Trinematidae

8(7) Pseudostome often a slit, surrounded by very small scales (Fig. 2.10 C) .... Sphenoderiidae, one genus: *Sphenoderia*

8′ Pseudostome surrounded by denticulate mouth plates .... **Euglyphidae [p. 34]**

Rhizaria: Euglyphidae: Genera

1 Parietal scales scutiform (Fig. 2.11 G) .... *Scutiglypha*

1′ Parietal scales ovoid or rectangular (Fig. 2.10 D) .... *Euglypha*

ACKNOWLEDGMENTS

Financial support from Project 192626868-SFB 990 to V. Krashevska; project "MYXOTROPIC VI" awaerd by the Spanish Government (PGC2018-094660-B-I00) to E. Lara; project FONDECYT N° 11170927 by CONICYT to L.D. Fernández.

REFERENCES

Adl, S.M., A.G.B. Simpson, M.A. Farmer, R.A. Andersen, O.R. Anderson, J.R. Barta, S. S. Bowser, G. Brugerolle, R.A. Fensome, S. Fredericq, T.Y. James, S. Karpov, P. Kugrens, J. Krug, C.E. Lane, L.A. Lewis, J. Lodge, D.H. Lynn, D.G. Mann, R.M. Mccourt, L. Mendoza, O. Moestrup, S.E. Mozley-Standridge, T.A. Nerad, C.A. Shearer, A.V. Smirnov, F.W. Spiegel & M.F.J.R. Taylor. 2005. The new higher level classification of eukaryotes with emphasis on the taxonomy of protists. Journal of Eukaryotic Microbiology 52: 399−451.

Adl, S.M., A.G.B. Simpson, C.E. Lane, J. Lukes, D. Bass, S.S. Bowser, M.W. Brown, F. Burki, M. Dunthorn, V. Hampl, A. Heiss, M. Hoppenrath, E, Lara, L. Le Gall, D.H. Lynn, H. McManus, E.A.D. Mitchell, S.E. Mozley-Stanridge, L.W. Parfrey, J. Pawlowski, S. Reuckert, L. Shadwick, C.L. Schoch, A. Smirnov & F.W. Spiegel. 2012. The revised classification on eukaryotes. Journal of Eukaryotic Microbiology 59: 429−493.

Amacker, N., E.A.D. Mitchell & N. Chèvre. 2018. Development of a new ecotoxicological assay using the testate amoeba *Euglypha rotunda* (Rhizaria; Euglyphida) and assessment of the impact of the herbicide S-metolachlor. Chemosphere 201: 351−360.

Arrieira, R.L., G.M. Alves, L.T.F. Schwind, & F.A. Lansac-Tôha. 2015. Local factors affecting the testate amoeba community (Protozoa: Arcellinida; Euglyphida) in a neotropical floodplain. Journal of Limnology 74: 444−452.

Berney, C. & J. Pawlowski. 2006. A molecular time-scale for eukaryote evolution recalibrated with the continuous microfossil record. Proceedings of the Royal Society B-Biological Sciences 273: 1867−1872.

Bhattacharya, D., T. Helmchen & M. Melkonian. 1995. Molecular evolutionary analyses of nuclear-encoded small subunit ribosomal

RNA identify an independent rhizopod lineage containing the Euglyphina and the Chlorarachniophyta. Journal of Eukaryotic Microbiology 42: 65–69.

Bonnet, L. 1966. Le peuplement thécamoebien de quelques sols du Chili I. Protistologica 2: 113–140.

Booth, R.K., M. Lamentowicz & D.J. Charman. 2010. Preparation and analysis of testate amoebae in peatland palaeoenviron-mental studies. Mires and Peat 7: 1–7.

Chatelain, A.P., R. Meisterfeld, L. Roussel-Delif & E. Lara. 2013. Sphenoderiidae (fam. nov.), a new clade of Euglyphid Testate Amoebae characterized by small, round scales surrounding the aperture. Protist 164:782–792.

Daday, J. 1905. Untersuchungen über die Süsswasser-Mikrofauna Paraguays: Protozoa. Zoologica 18:1–342.

Decloitre, L. 1963. Rhizopodes Thécamoebiens du Vénézuéla. Proceedings of the Zoological Society of London 141: 497–514.

Delaine M., L.D. Fernández, E. Armynot du Chatelet, P. Recourt, J.-L. Potdevin, E.A.D. Mitchell & N. Bernard. 2016. Cinderella's helping pigeons of the microbial world: the potential of testate amoebae for identifying cryptotephra. European Journal of Protistology 55: 152–164.

Dumack, K., M.E. Muller & M. Bonkowski. 2016. Description of *Lecythium terrestris* sp. nov. (Chlamydophryidae, Cercozoa), a soil dwelling protist feeding on fungi and algae. Protist 167: 93–105.

Ehrenberg, 1838. Die infusionthierchen als Vollommene Organismen. Von Leopold Voss. Verlag, Leipzig. 547 pp.

Fernández, L.D., B. Fournier, R. Rivera, E. Lara, E.A.D. Mitchell & C.E. Hernández. 2016. Water-energy balance, past ecological perturbations and evolutionary constraints shape the latitudinal diversity gradient of soil testate amoebae in south-western South America. Global Ecology and Biogeography 25: 1216–1227.

Fernández, L.D., E. Lara & E.A.D. Mitchell. 2015. Checklist, diversity and distribution of testate amoebae in Chile. European Journal of Protistology 51: 409–424.

Friz, C.T. 1992. Taxonomic analyses of seven species of familiy Amoebidae by isozymic characterization of electrophoretic patterns and the descriptions of a new genus and a new species: *Metamaoeba* n. gen. *Amoeba amazonas* n. sp. Archiv für Protistenkunde 142: 29–40.

Finlay, B.J. 2002. Global dispersal of free-living microbial eukaryote species. Science 296: 1061–1063.

Fournier, B., E.E.D. Coffey, W.O. van der Knaap, L.D. Fernández, A. Bobrov & E.A.D. Mitchell. 2016. A legacy of human-induced ecosystem changes: spatial processes drive the taxonomic and functional diversities of testate amoebae in *Sphagnum* peatlands of the Galapagos. Journal of Biogeography 43: 533–543.

Geisen, S., J. Rosengarten, R. Koller, C. Mulder, T. Urich & M. Bonkowski. 2015. Pack hunting by a common soil amoeba on nematodes. Environmental Microbiology 17: 4538–4546.

Gilbert, D., E.A.D. Mitchell, C. Amblard, G. Bourdier & A.J. Francez. 2003. Population dynamics and food preferences of the testate amoeba *Nebela tincta major-bohemica-collaris* complex (Protozoa) in a Sphagnum peatland. Acta Protozoologica 42: 99–104.

Golemansky, V. 1970. Rhizopodes nouveaux du psammon littoral de la Mer Noire. Protistologica 6: 365–371.

Gomaa, F., A. Kosakyan, T.J. Heger, D. Corsaro, E.A.D Mitchell & E. Lara. 2014. One alga to rule them all: unrelated mixotrophic testate amoebae (Amoebozoa, Rhizaria and Stramenopiles) share the same symbiont (Trebouxiophyceae). Protist 165:161–176.

Gomaa F, E.A.D. Mitchell & E. Lara. 2013. Amphitremida (Poche, 1913) is a new major, ubiquitous Labyrinthulomycete Clade. PLoS One 8(1): e53046.

Gomes e Souza, M.B. 2008. Guia das tecamoebas bacia do rio Peruaçu, Minas Gerais: subsídio para conservação e monitoramento da bacia do Rio São Francisco. Editora UFMG, Belo Horizonte, MG. 159 pp.

Haman, D. & B. Kohl. 1994. A thecamoebinid assemblage from Lake Cococha, Tambopata Reserve, Madre de Dios Province, Southeastern Peru. Journal of Foraminiferal Research 24: 226–232.

Heger, T.J., E. Lara & E.A.D. Mitchell. 2011a. Arcellinida testate amoebae (Arcellinida: Amoebozoa): model of organisms for assessing microbial biogeography. Pages 111–129 *in*: D. Fontaneto (ed.), The importance of being small: does size matter in biogeography? Cambridge University Press, Cambridge.

Heger, T.J., J. Pawlowski, V. Golemansky, M. Todorov, E. Lara & E.A.D. Mitchell. 2011b. Comparing potential COI and SSU rDNA barcodes for assessing the diversity and phylogenetic relationships of Cyphoderiid Testate Amoebae (Rhizaria: Euglyphida). Protist 162: 131–141.

Heger, T.J., F. Straub & E.A.D. Mitchell. 2012. Impact of farming practices on soil diatoms and testate amoebae: A pilot study in the DOK-trial at Therwil, Switzerland. European Journal of Soil Biology 49: 31–36.

Hoogenraad, H.R. & A.A. De Groot. 1940. Zoetwaterrhizopoden en -Heliozoën – Fauna van Nederland. Sijthoff, Leiden.

Iudina, T.A. & K.M. Sukhanova. 2000. Cell biology and life cycle of the testate amoeba *Corythion delamarei*. In Russian. Tsitologiia 42: 613–623.

Jung, W. 1942. Illustrierte Thekamöben-Bestimmungstabellen. I. Die Systematik der Nebelinen. Archiv für Protistenkunde 95: 357–390.

Kang, S., A. K. Tice, F. W. Spiegel, J. D. Silberman, T. Panek, I. Cepicka, M. Kostka, A. Kosakyan, D. M. C. Alcantara, A. J. Roger, L. L. Shadwick, A. Smirnov, A. Kudryavtsev, D. J. G. Lahr, and M. W. Brown. 2017. Between a Pod and a Hard Test: The Deep Evolution of Amoebae. Journal of Molecular Biology 34: 2258–2270.

Kosakyan, A., F. Gomaa, E. Lara & D.J.G. Lahr. 2016a. Current and future perspectives on the systematics, taxonomy and nomenclature of testate amoebae. European Journal of Protistology 55: 105–117.

Kosakyan, A., F. Gomaa, E.A.D. Mitchell, E T.J. Heger & E. Lara. 2013. Using DNA-barcoding for sorting out protist species complexes: A case study of the *Nebela tincta-collaris-bohemica* group (Amoebozoa; Arcellinida, Hyalospheniidae). European Journal of Protistology 49: 222–237.

Kosakyan, A., T.J. Heger, B.S. Leander, M. Todorov, E.A.D. Mitchell & E. Lara. 2012. COI barcoding of Nebelid testate amoebae (Amoebozoa: Arcellinida): Extensive Cryptic Diversity and Redefinition of the Hyalospheniidae Schultze. Protist 163: 415–434.

Kosakyan, A., D.J.G. Lahr, M. Mulot, R. Meisterfeld, E.A.D. Mitchell & E. Lara 2016b. Phylogenetic reconstruction based on COI reshuffles the taxonomy of hyalosphenid shelled (testate) amoebae and reveals the convoluted evolution of shell plate shapes. Cladistics 32: 606–623.

Krashevska, V., M. Bonkowski, M. Maraun, & S. Scheu. 2007. Testate amoebae (protista) of an elevational gradient in the tropical mountain rain forest of Ecuador. Pedobiologia (Jena) 51: 319–331.

Krashevska, V., M. Maraun & S. Scheu. 2010. Micro- and macroscale changes in density and diversity of testate amoebae of tropical montane rain forests of Southern Ecuador. Acta Protozoologica 49: 17–28.

Krashevska, V., D. Sandmann, M. Maraun & S. Scheu. 2012. Consequences of exclusion of precipitation on microorganisms and microbial consumers in montane tropical rainforests. Oecologia 170: 1067–1076.

Lahr, D.J.G., T. Bosak, E. Lara & E.A.D. Mitchell. 2015. The Phanerozoic diversification of silica-cycling testate amoebae and its possible links to changes in terrestrial ecosystems. Peer Journal 3: e1234.

Lahr, D.J.G., A. Kosakyan, E. Lara, E.A.D. Mitchell, L. Morais, A.L. Porfiro-Sousa, G.M. Ribeiro, A.K. Tice, T. Pánek, S. Kang & M.W. Brown. 2019. Phylogenomics and morphological reconstruction of Arcellinida Testate Amoebae Highlight Diversity of Microbial Eukaryotes in the Neoproterozoic. Current Biology 29: 991–1001.

Lahr, D.J.G., L.W. Parfrey, E.A.D. Mitchell, L.A. Katz & E. Lara. 2011. The chastity of amoebae: re-evaluating evidence for sex in amoeboid organisms. Proceedings of the Royal Society B-Biological Sciences 278: 2081–2090.

Lamentowicz, Ł., M. Lamentowicz & M. Gabka. 2008. Testate amoebae ecology and a local transfer function from a peatland in western Poland. Wetlands 28: 164–175.

Lamentowicz, M. & M. Obremska. 2010. A rapid response of testate amoebae and vegetation to inundation of a kettle hole mire. Journal of Paleolimnology 43: 499–511.

Lamentowicz, M., M. Obremska & E.A.D. Mitchell. 2008. Autogenic succession, land-use change, and climatic influences on the Holocene development of a kettle-hole mire in Northern Poland. Review of Palaeobotany and Palynology 151: 21–40.

Lansac-Toha, F.A., L.F.M. Velho, L. Costa, N. Simões & G. Alves. 2014. Structure of the testate amoebae community in different habitats in a neotropical floodplain. Brazilian Journal of Biology 74: 181–190.

Lara, E. & F. Gomaa. 2017. Symbiosis between testate amoebae and photosynthetic organisms. Pages 399–419 *in*: M. Grube, L. Muggia & J. Seckenbach (eds.), Algae and Cyanobacteria symbioses. World Scientific Europe.

Lara, E., L. Roussel-Delif, B. Fournier, D.M. Wilkinson & E.A.D. Mitchell. 2016. Soil microorganisms behave like macroscopic organisms: patterns in the global distribution of soil euglyphid testate amoebae. Journal of Biogeography 43: 520–532.

Leclerc, L., 1816. Notes sur Difflugie, nouveau genre de Polype amorph. Mem. Pseudopod morphology: Mémoires du Muséum d'Histoire Naturelle 2: 474–478.

Marin, B., E.C.M Nowack & M. Melkonian. 2005. A plastid in the making: Evidence for a second primary endosymbiosis. Protist 156: 425–432.

Meisterfeld, R., 1991. Vertical distribution of *Difflugia hydrostatica* (Protozoa, Rhizopoda). Vereinigung Verhandlungen International Limnologie 24: 2726–2728.

Meisterfeld, R. 2002. Order Arcellinida Kent 1880. Pages 827–860 *in*: J.J. Lee, G.F. Leedale & P. Bradbury (eds.), The illustrated guide to the protozoa (Second edition), Society of protozoologists, Lawrence, KS.

Meyer, C., M. Diaz-de-Quijano, F. Monna, M. Franchi, M.-L. Toussaint, D. Gilbert & N. Bernard. 2015. Characterization and distribution of deposited trace elements transported over long and intermediate distances in north-eastern France using *Sphagnum* peatlands as a sentinel ecosystem. Atmospheric Environment 101: 286–293.

Mitchell, E.A.D., D.J. Charman & B.G. Warner. 2008. Testate amoebae analysis in ecological and paleoecological studies of wetlands: past, present and future. Biodiversity and Conservation 17: 2115–2137.

Mulot, M, K. Marciszb, L Grandgirarda, E. Lara, A. Kosakyan, B.J.M. Robroek, M. Lamentowiczb, R.J. Payneg & E.A.D. Mitchell. 2017. Genetic determinism *vs*. phenotypic plasticity in Protist morphology. Journal of Eukaryotic Microbiology 64: 729–739.

Nguyen-Viet, H., N. Bernard, E.A.D. Mitchell, P-M. Badot & D. Gilbert. 2008. Effect of lead pollution on testate amoebae communities living in *Sphagnum fallax*: an experimental study. Ecotoxicology and Environmental Safety 69: 130–138.

Nikolaev S.I., E.A.D. Mitchell, N.B. Petrov, C. Berney, J.F. Fahrni & J. Pawlowski. 2005. The testate lobose amoebae (Order Arcellinida Kent, 1880) finally find their home within amoebozoa. Protist 156: 191–202.

Page, F.C. 1976. An Illustrated Key to Freshwater and Soil Amoebae. Freshwater Biological Association, Ambleside. 155 pp.

Page, F. C. 1988. A New Key to Freshwater and Soil Gymnamoebae with instructions for culture – Freshwater Biological Association. 122 pp.

Page, F. C. & F. J. and Siemensma. 1991. Nackte Rhizopoda und Heliozoea – Gustav Fischer Verlag. 170 pp.

Parfrey, L.W., D.J.G. Lahr, A.H. Knoll & L.A. Katz. 2011. Estimating the timing of early eukaryotic diversification with multigene molecular clocks. Proceedings of the National Academy of Sciences of the United States of America 108: 13624–13629.

Pawlowski, J. & F. Burki. 2009. Untangling the phylogeny of amoeboid protists. Journal of Eukaryotic Microbiology 56: 16–25.

Payne, R.J., R.J. Telford, J.J. Blackford, A. Blundell, R.K. Booth, D.J. Charman, Ł. Lamentowicz, M. Lamentowicz, E.A.D. Mitchell, G. Potts, G.T. Swindles, B. Warner & W. Woodland. 2012. Testing peatland testate amoeba transfer functions: Appropriate methods for clustered training-sets. The Holocene 22: 819–825.

Porter, S.M., R. Meisterfeld & A.H. Knoll. 2003. Vase-shaped microfossils from the Neoproterozoic Chuar Group, Grand Canyon: A classification guided by modern testate amoebae. Journal of Paleontology 77: 409–429.

Schmidt, A.R., V. Girard, V. Perrichot & W. Schoenborn. 2010. Testate amoebae from a Cretaceous forest floor microbiocoenosis of France. Journal of Eukaryotic Microbiology 57: 245–248.

Schwind L.T.F., R.L. Arrieira, U.D. Dias, N.R. Simoes, C.C. Bonecker & F.A. Lansac-Toha. 2016. The structure of planktonic communities of testate amoebae (Arcellinida and Euglyphida) in three environments of the Upper Paraná River basin, Brazil. Journal of Limnology 75: 78–89.

Seppey, C.V., B. Fournier, I. Szelecz, D. Singer, E.A.D. Mitchell & E. Lara. 2015. Response of forest soil euglyphid testate amoebae (Rhizaria: Cercozoa) to pig cadavers assessed by high-throughput sequencing. International Journal of Legal Medicine 130: 551–562.

Shiratori, T. & K-I. Ishida. 2016. *Trachyrhizium urniformis* n. g., n. sp., a novel marine filose thecate amoeba related to a Cercozoan environmental clade (Novel Clade 4). Journal of Eukaryotic Microbiology 63: 722–731.

Sigala, I., S. Lozano-Garcia, J. Escobar, L. Perez & E. Gallegos-Neyra. 2016. Testate Amoebae (Amebozoa: Arcellinida) in tropical lakes of central Mexico. Revista De Biologia Tropical 64: 377–397.

Sierra, R., M.V. Matz, G. Aglyamova, L. Pillet, J. Decelle, F. Not, C. Vargas & J. Pawlowski. 2013. Deep relationships of Rhizaria revealed by phylogenomics: A farewell to Haeckel's Radiolaria. Molecular Phylogenetics and Evolution 67(1): 53–59.

Singer, D., A. Kosakyan, A. Pillonel, E.A.D. Mitchell & E. Lara. 2015. Eight species in the *Nebela collaris* complex: *Nebela gimlii* (Arcellinida, Hyalospheniidae), a new species described from a Swiss raised bog. European Journal of Protistology 51: 79–85.

Singer, D., A. Kosakyan, C.V.W Seppey, A. Pillonel, L.D. Fernández, D. Fontaneto, E.A.D. Mitchell & E. Lara 2018. Environmental filtering and phylogenetic clustering correlate with the distribution patterns of cryptic protist species. Ecology 99(4): 904–914.

Smirnov, A.V. & S. Brown 2004. Guide to the study and identification of soil amoebae. Protistology 3: 148–190.

Smith, H.G., A. Bobrov & E. Lara. 2008. Diversity and biogeography of testate amoebae. Biodiversity and Conservation 17: 329–343.

Swindles, G.T., M. Reczuga, M. Lamentowicz, C.L. Raby, T.E. Turner, D.J. Charman, A. Gallego-Sala, E. Valderrama, C. Williams, F. Draper, E.N. Honorio Coronado, K.H. Roucoux, T. Baker & D.J. Mullan. 2014. Ecology of testate amoebae in an Amazonian peatland and development of a transfer function for palaeohydrological reconstruction. Microbial Ecology 68: 284–298.

Szelecz, I., B. Fournier, C. Seppey, J. Amendt & E.A.D. Mitchell. 2014. Can soil testate amoebae be used for estimating the time since death? A field experiment in a deciduous forest. Forensic Science International 236: 90–98.

Tekle, Y.I., J. Grant, O.R. Anderson, T.A. Nerad, J.C. Cole, D.J. Patterson & L.A. Kat. 2008. Phylogenetic placement of diverse amoebae inferred from multigene analyses and assessment of clade stability within 'Amoebozoa' upon removal of varying rate classes of SSU-rDNA. Molecular Phylogenetic and Evolution 47: 339–352.

Vucetich, M.C. 1974. Tecamebianos de turberas de Tierra del Fuego. Neotropica 20: 27–33.

Vucetich, M.C. 1975. Tecamebianos muscícolas y esfagnícolas de Islas Malvinas (Argentina). Neotropica 21: 11–16.

Vucetich, M.C., & E.C. Lopretto. 1995. Rhizopoda: Amoebas Testaceas. Pages 5–54 *in*: Z.A. de Castellanos, S. Coscarón, S. Miquel (eds.), Fauna de agua dulce de la República Argentina 3, Protozoa. Facultad de Ciencias Naturales y Museo, La Plata, Argentina.

Wailes, G.H. 1913. Freshwater Rhizopoda from North and South America. Journal of the Linnean Society of London, Zoology 32: 201–218.

Wanner, M. & M. Elmer. 2009. "Hot spots" on a new soil surface – How do testate amoebae settle down? Acta Protozoologica 48: 281–289.

Wilkinson, D.M. 2001. What is the upper size limit for cosmopolitan distribution in free-living microorganisms? Journal of Biogeography 28: 285–291.

Yeates, G.W. & W. Foissner. 1995. Testate amoebae as predators of nematodes. Biology and Fertility of Soils 20: 1–7.

Ciliophora

Gabriela C. Küppers

División Invertebrados, Museo Argentino de Ciencias Naturales "Bernardino Rivadavia", Ciudad Autónoma de Buenos Aires, Argentina

María Cristina Claps

Instituto de Limnología "Dr. R.A. Ringuelet", La Plata, Buenos Aires, Argentina

Thiago da Silva Paiva

Laboratório de Protistologia, Dept. de Zoologia, Inst. de Biologia, CCS, Universidade Federal do Rio de Janeiro, Rio de Janeiro, Brazil

INTRODUCTION

The phylum Ciliophora comprises a morphologically diverse monophyletic group of protists, which are found in soil, freshwater, and marine environments worldwide. Ciliates occur mostly as free-living species, but many live in association with other organisms as commensals *sensu lato* or parasites (Hausmann & Bradbury, 1996; Lynn, 2008). They play important roles in ecosystems, acting primarily as bacterial filters, but also feeding on other protists and small metazoa (Fenchel, 1987). Freshwater ciliates are regarded as biological indicators of water quality according to the saprobic system (Sládeček, 1973) and are also useful tools in bioassays to detect the presence of heavy metals, such as cadmium (Wanick et al., 2010). Ciliates are abundant in the microscopic communities from activated sludge, and their constant feeding upon bacteria accelerates the process of water clarification in wastewater treatment plants (Al-Shahwani & Horan, 1991).

Most ciliates have widespread geographic distributions, but a moderate degree of endemism is known to exist. Currently, there are more than 8,000 species described in the literature, but Foissner et al. (2008) estimated that only about 4,500 correspond to valid, free-living species. Even though novel species are being discovered every year, recent investigations tell us that about 83%–89% of ciliate diversity remains unknown to science (Foissner et al., 2008). This is frequently attributed to a worldwide lack of interest in taxonomy of ciliates (and of free-living protists in general), mostly because of the difficult methodology to study them and the use of inadequate sampling methods. Within this context, the diversity of ciliates in the Neotropics has been considered the most understudied when compared with other geographic regions. This scenario is slowly changing, and some initiatives are being conducted not only to improve the knowledge of diversity and distribution of ciliates worldwide but also to provide valuable information for ecologists, limnologists, and other people interested in applied research fields.

With that in mind, we prepared taxonomic keys up to genus level for freshwater, some inland slightly brackish waters, and most common freshwater phytotelmic ciliates recorded in the Neotropics and Antarctica after revising a vast list of literature. We based this work on the Protozoa chapter of the previous volume in the same series (Warren et al., 2016) but went one step further, assembling the keys up to genus level. To do so, we followed specific taxonomic works and monographs (Berger, 1999, 2006, 2008, 2011; Foissner et al., 1991, 1992, 1994, 1995, 1999; Foissner & Xu, 2007; Vd'ačný & Foissner, 2012, among others) and adapted information from the keys provided by Curds (1982), Curds et al. (1983) and Taylor & Sanders (2001).

LIMITATIONS

Although we have made some effort to simplify this work for nonspecialists, knowledge of ciliate morphology and general terminology is mandatory for efficient usage of the keys. To become familiarized with the general structural and ultrastructural morphology of ciliates, we strongly recommend the reading of Dragesco & Dragesco-Kernéis (1986), Hausmann & Bradbury (1996), and Lynn (2008). Species level identification of ciliates can be time-consuming, demanding the usage of difficult staining procedures (already necessary for identification of most

PROTOZOA

genera!) and sometimes scanning electron microscopy. Xu et al. (2011) demonstrated that identification of ciliates at genus level provides meaningful resolution for environmental quality assessment; thus, we expect these identification keys will become a useful tool in these kind of studies.

The readers must be aware that when sampling the environment, they will likely find representatives of genera not yet reported to the regions covered by this book. In addition, new species (and genera) of ciliates are constantly being described each year, and as further morphological and phylogenetic discoveries are made, the taxonomic circumscriptions of genera and other categories may change. Therefore, we stress that diligent search of the literature, along with the use of the present keys, is desirable to achieve precise identifications.

TERMINOLOGY AND MORPHOLOGY

This section is intended to be a simple and brief overview of the morphology of ciliates with emphasis on structures that are mentioned in the keys. Terminology is mainly based on Curds (1982) and Lynn (2008), and classification follows Lynn (2008) in most cases.

A great diversity of **body shapes** exists among ciliates. Most of them are ovoid or spherical, but they can also be barrel-shaped (Fig. 2.21 J), reniform (Fig. 2.21 B), pyriform (Fig. 2.18 E′), elongate elliptical (Fig. 2.15 Q), transversely or obliquely truncate (Figs. 2.21 G, 20 I), with an anterior beak or rostrum (Fig. 2.20 L), etc. Ciliates can be round in cross section, dorsoventrally flattened or laterally compressed. The oral opening is situated apically, laterally, ventrally, or posteriorly in some cases. Somatic cilia usually run along the longitudinal body axis but are spiraled in some ciliates. The anterior end of the body is usually held forward while swimming, and right and left refer to the ciliate sides.

Some ciliates live permanently or temporarily inside a **lorica**, which can be membranous, agglutinate particles from the environment, pseudochitinous, etc. This lorica can attach directly to the substrate or by means of a peduncle (Figs. 2.12 L, 13 D, 17 Q).

The cell wall or **pellicle** is underlined by alveoli that in some genera such as *Coleps* and *Euplotes* are filled with inorganic or organic substances giving the ciliate an armored appearance. Among these alveoli, there are subpellicular **extrusomes**. There is a great variety of such extrusive organelles, but the most common are mucocysts, which secrete mucus and form resting or reproductive cysts, and trichocysts that extrude harpon-like fibrous material acting in defensive situations, such as in *Paramecium* and other peniculids. Toxicysts are commonly found in haptorids and discharge enzymatic substances that aid in prey capture and ingestion. Mucocyst-type cortical granules can be pigmented or colorless and arranged in longitudinal bands between somatic kineties. These granules usually occur on the dorsal surface of some flexible hypotrichs or arrange in longitudinal bands between the somatic ciliary rows of heterotricheans.

Most ciliates swim freely, but others, like sessile peritrichs, attach to the substrate by means of a **stalk**, which can be contractile or noncontractile, branched in colonial forms or unbranched. The stalk of peritrichs arises from the aboral region of body or scopula and contracts because the myoneme or spasmoneme runs helically inside it (Fig. 2.17 H).

Tentacles appear in some groups of ciliates but are always present in suctorians (Fig. 2.12 A-L), which adult or trophont forms lack cilia and tentacles function as multiple sites of prey capture and ingestion. These tentacles can be evenly distributed over the body or grouped in fascicles or in very long projections called actinophores. Other ciliates having tentacle-like projections, such as *Actinobolina*, always have cilia, a single oral opening, and tentacles aid in prey capture but do not represent ingestion sites (Fig. 2.23 A, B).

All ciliates have **cilia** at some stage in their lives. These arise from subpellicular basal bodies or kinetosomes that are associated to fibrillar structures constituting the so-called **infraciliature**. Basal bodies and their cilia arrange consecutively in rows along the main axis of the body, forming a **somatic kinety**. The usually dorsoventrally flattened hypotrichs *sensu lato* (hypotrichs and stichotrichs) have compound somatic cilia in the form of **cirri**, arranged in rows or discrete groups mainly on the ventral surface, whereas the dorsal surface bears rows of bristles (Figs. 2.14 K, 15 S). Somatic ciliature is reduced in certain ciliates such as choreotrichs and oligotrichs (Fig. 2.13 G, M). In the case of peritrichs (Fig. 2.17), somatic cilia may be absent, but a posterior ciliated band, the telotrochal band, forms from dormant kinetosomes when the telotroch or phoretic phase detaches from the stalk and swims freely. Suctorians completely lack cilia (Fig. 2.12 A–L), although a ciliated dispersal stage emerges from buds during asexual fission.

In most cases, the **oral opening** lies in anterior body half (apically, subapically, laterally, ventrally), but in others, it may be in the subequatorial or posterior area (Figs. 2.12 U, 18 Q). The **cytostome** is literally the cell "mouth," followed by the **cytopharynx**, which constitutes a nonciliated, tubular passageway into the cytoplasm and food vacuoles forme at its distal end. The cytopharynx can be strengthened by rods of closely packed microtubules (nematodemata) forming a straight basket-like structure or **rhabdos** in litostomateans or a curved and more complex one called **cyrtos** in cyrtophorids and nassophoreans. The oral opening can be at or near the body surface and just encircled by longer somatic cilia. It can also have a preoral and a postoral suture like in some peniculids or a preoral kinety running along a preoral suture like in some cyrtophorids and nassophoreans (Figs. 2.12 P, Q, 18 I). The oral ciliature of nassophoreans is typically arranged as a hypostomial or postoral frange or synhymenium (Fig. 2.12 C′), an extensive ventral band of more complex kineties that courses slightly posterior to the cytostome and may extend onto the dorsal surface. Some litostomateans have an oral bulge or protrusion (Fig. 2.22 M) where the cytostome opens or the oral opening is preceded by one or three "necks" or proboscis (Fig. 2.22 A–E). Spathidid litostomateans have circumoral fragments of kineties or kinetofragments that may derive from the adjacent somatic kineties. In some ciliates, the cytostome is located within a depression called **vestibulum**, lined by somatic cilia. Conversely, the **oral cavity** of spirotricheans and heterotricheans has specialized oral ciliature (Figs. 2.13–16, 20 A–K). In this oral cavity, complex arrangements of more than two kineties give rise to **undulating membranes** or **parorals** of dikinetids (two rows) and **membranelles** or polykineties (more than three rows), which movements couple together and beat as units. Undulating membranes are composed of two rows of closely arranged kinetosomes, and their cilia form a membrane-like structure on the right of the oral area, which can be observed in vivo in many oligohymenophoreans. In spirotricheans, there are two of such membranes, the **endoral** and **paroral**. Conversely, cilia of membranelles appear as triangular or trapezoidal flaps that typically beat on the left of the oral area, forming the **adoral zone of membranelles**. In heterotricheans, various spirotricheans, and peritrichs, the oral ciliature emerges from the oral cavity and encircles the anterior end of body, forming a **peristome** (Figs. 2.13–17, 20 A–K). In heterotricheans and spirotricheans, these membranelles are numerous. Conversely, in oligotricheans, there are usually three membranelles. Peniculids have **peniculi** deep in the buccal cavity, composed of long bands of short cilia that create the feeding currents observed in *Paramecium*. In peritrichs, peristomial ciliature encircles the anterior end of the body, spirals anticlockwise, and plunges deep into an infundibulum toward the cytostome (Fig. 2.17). Armophoreans have a perizonal stripe consisting of five dense kineties running above the zone of membranelles, which serve for the intensification of water currents in the oral area (Fig. 2.19 B). Food vacuoles form at the cytostome–cytopharynx, move in the cytoplasm while digestion occurs, and are exocyted through a permanent pore or **cytoproct**. In some cases, the position of this structure is an important taxonomic feature.

Freshwater ciliates are hyperosmotic with respect to the environment. Consequently, water diffuses constantly through the pellicle toward the cytoplasm where **contractile vacuoles** along with a complex net of canals and tubules eliminate this excess of water through a constant pore. The number and position of contractile vacuoles is often useful to identify some ciliates. Thus, live observation of freshly collected samples during surveys is essential.

A unique characteristic of ciliates is the possession of two kinds of **nuclei**. The **macronucleus** engages in the cell metabolism and controls the organism's phenotype. It divides during reproduction and derives from the micronucleus during conjugation. It can be single or multiple and diverse in shape, such as spherical or ovoid (Figs. 2.12 L, 19 K), vermiform (Fig. 2.17 L), moniliform (Fig. 2.20 K), C- or J-shaped (Figs. 2.13 G, 17 J), etc. The **micronucleus** is involved in genetic changes and sexual processes. It is spherical, smaller than the macronucleus, and can be single or multiple (Figs. 2.12 Q, 13 G, 19 A).

Increase in population size occurs usually by **binary fission**, transversely across the somatic kineties. Although; some ciliates, such as suctorians, produce endogenous or exogenous buds that result in a ciliated, dispersal stage and colpodeans form division cysts. During division, **ontogeny** or **morphogenesis** involves the mitotic division of the micronucleus, whereas macronucleus synthesizes DNA, elongates, and constricts. Cilia and other organelles partly form anew and partly derive by transformation of existing structures. In particular, the formation of the oral ciliature or **stomatogenesis** varies from group to group and is taxonomically important. **Sex** in ciliates takes place by conjugation, when two individuals that fuse temporarily along their oral surfaces exchange gametic nuclei resulting from meiosis and then mitosis. Self-fertilization or **autogamy** also occurs, to reorganize the micronucleus and probably increase the longevity of a clone.

Figure plates illustrate representatives species of most of the genera reported or described for the Neotropics and/or Antarctic regions; although, undetermined species were not included. Figures were assembled in scale according to sizes of each species provided by the authors.

MATERIAL PREPARATION AND PRESERVATION

Until about the middle of the 20th century, descriptions of ciliates relied almost only on live observations. With the popularization of silver impregnation among ciliatologists, a huge amount of new information on the ciliature and morphogenesis became available, and even more morphological refinement were obtained after the application of electron microscopy, which revealed details of the cell surface and the infraciliature (Lynn, 2008). Descriptions became more complex, and alpha-taxonomy of ciliates nowadays requires at least detailed in vivo observations and silver impregnation, with electron microscopy adding important complementary information. Hence, those techniques are indispensable for the reliable identification of many, if not most, ciliate genera and species.

Collecting and Cultivating Freshwater Ciliates

Water samples from shallow environments can be collected manually with variable sized sterilized glass flasks. It is important to obtain also bottom sediments, where many interstitial ciliates inhabit, and organic debris, which are rich in bacteria and thus might contain abundant ciliates. Samples of water from the bottom of rivers or lakes can be obtained with a Van Dorn bottle or similar, and sediments can be further sampled with a corer pipe. Planktonic ciliates can be concentrated using a plankton mesh; however, filtering should be done carefully and slowly to avoid cell losses.

Although the observation of freshly collected samples is important, many ciliate species can be encysted when food is not available or other conditions such as temperature, conductivity, oxygen concentration, or pH are unfavorable. Thus, the establishment of cultures is necessary to promote the excystment of dormant structures and to increase population sizes to have enough material to perform silver impregnation techniques and electron microscopy preparations, facilitating identification.

Most freshwater ciliates are cultivable in raw limnetic cultures, which can be made from enriched aliquots of water plus sediment of homogenized samples. Aliquots are poured from the samples into glass Petri dishes with addition of some crushed rice or wheat grains to promote the growth of bacteria, which serve as initial food source for the ciliates. The cultures can be maintained on a weekly base, by splitting about 50% of the content in the Petri dishes and completing it with water from the samples, removing the old rice fragments with sterilized tweezers and adding new ones. Cultures should also be made from the sediments of temporary ponds, and submerged litter, with addition of distilled and mineral water in separate Petri dishes, to produce different microhabitat conditions and consequently stimulate the development of distinct ciliate communities. The presence of bacterivorous ciliates will stimulate the excystment of carnivores and omnivores, and ecological succession tends to occur within the culture.

Clonal cultures started from single specimens are preferred to study divisional morphogenesis, because they warrant conspecificity of the studied dividers. Such cultures are made by picking a single cell from a raw culture, briefly washing it in a watch glass with mineral water, and then transferring it to a Petri dish with boiled water from the original sample or mineral water enriched with food. The food source can be either bacteria or other protists, and in such cases, it is advisable to check the food vacuoles to see each species preferences. Anaerobic ciliates demand special culture conditions, because exposure to oxygen can inhibit their growth.

To manipulate the ciliates, a capillary micropipette can be made by heating a hollow glass tube or the capillary of a Pasteur pipette with a Bunsen burner, and when the glass softens, stretching it horizontally (Dragesco & Dragesco-Kernéis, 1986). The diameter of the micropipette has to be adjusted under the flame and by breaking the tip of the capillary. This procedure requires some practice. When the micropipette is ready, attach a rubber bulb head to the extremity opposing the capillary.

Identifying the Ciliates

Identification of ciliates begins with observation of freshly collected environmental samples. Live specimens should be examined carefully, because relevant taxonomic features are lost after fixation. Body shape of highly contractile species changes completely, i.e., bell-shaped peritrichs or trumpet-shaped heterotrichs will appear as rounded, unrecognizable "balls." Color must be observed in life. It is often because the cytoplasm is colored, the ciliate has colored cortical granules, or hosts algal endocytobionts. The number and position of contractile vacuoles are also important features to observe in life.

Transfer an aliquot of freshly collected sample to a Petri dish to examine the ciliates under the stereomicroscope. Most ciliates are easily distinguished by their movement; some swim very fast and jump inside the water column, whereas others crawl on the bottom, debris, or water surface, or "walk" on surfaces by means of cirri.

Pick the ciliates carefully with a capillary micropipette, and place a small drop on a slide. First, observe them under the light microscope without a coverslip, to document their behavior and shape without deformation. It can be done at 100–200× magnifications for most species. Overall measurements of body length and width should be taken at this stage because specimens may shrink after fixation, preferably measuring about 10 different sized specimens to sample the intrapopulational variability. Next, add petroleum gel dabs to the corners of a coverslip, and press it down with a mounted needle until ciliates become slightly squeezed. Sometimes, it is necessary to wait for ciliates to slow down. Then, transfer the prepared slide to the microscope to observe the ciliates at higher magnifications. At this stage, the nuclear apparatus, contractile vacuoles, cortical granules, extrusomes, and other cytoplasmic inclusions should be documented. Those are readily seen at 1000× magnification under bright field or differential interference contrast (DIC). The color of cortical granules should be checked first under bright field, to avoid visual artifacts. The nuclear apparatus can usually be observed in living specimens, and it appears as a bright structure. It is easily seen in heavily squeezed specimens, under DIC or phase contrast. Otherwise, supravital staining with 1% methyl green–pyronin can be applied right on the slide, by adding a more or less equal drop to the drop containing the ciliates. This can be done before the placement of the coverslip or through one edge with the aid of a filter paper placed at the other edge of the coverslip. The nuclear apparatus stains blue or violet and the cytoplasm, food vacuoles, and extrusomes appear reddish. The preparation is temporal and lasts a few minutes; thus, it has to be observed immediately (Foissner, 2014).

Several fixatives are available for preservation of ciliates. Acidic Lugol's iodine is usually useful for ecological purposes that imply biomass estimations, since body shapes are conserved and cells are less contracted. However, it is not recommended if one wishes to perform the more sophisticated techniques to evidence the infraciliature, as discussed in the following.

Silver impregnation is mandatory for the correct identification of most ciliates, the techniques based on protargol and silver nitrate being the most frequently used, although silver carbonate is also an option. Protargol and silver carbonate impregnate the bases of cilia, the infraciliature fibers, and nuclear apparatus, whereas silver nitrate reveals the silverline system (argyrome), which is important in the taxonomy of euplotids and colpodeans. The resulting quality of each technique will vary depending on the ciliates, and there are many different protocols. For a detailed explanation on various silver impregnation techniques, we recommend the reading of Foissner (2014). Herein, we provide the description of protocols that we routinely use. Of note, if one wishes to perform quantitative studies, then the "quantitative protargol staining" protocol described by Skibbe (1994) should be preferred. Be aware that different types of silver proteinate/protargol are commercially available. However, the only known to impregnate ciliates properly is light brown-colored, exhibiting the aspect of cappuccino powder. This chemical has become unavailable from suppliers, but fortunately, Pan et al. (2013) published a useful and relatively inexpensive method for synthesizing the effective protargol.

Protargol impregnation, protocol I. This is adapted from Dieckmann (1995) and personal experience. It yields stable permanent preparations of excellent quality. The structures stain in different tonalities of red and purple. The drawbacks are that it is very time-consuming and somewhat difficult for inexperienced researchers.

1. Place 5–10 drops of Bouin's fixative in a watch glass (a square 1 5/8″ watch glass is recommended) with a Pasteur pipette. Pick the ciliates from the cultures using a capillary micropipette, and drop them directly into the fixative. A good fixation should take about 30 min.
2. Concentrate the cells in the middle of the watch glass, and wash them by slowly removing the fixative with a micropipette and replacing it with distilled water, until the liquid inside the watch glass is transparent. There should be no visible traces of the fixative.
3. Add 1 drop of albumen glycerol with a Pasteur pipette. The drop should be placed right above the ciliates, and you should be able to see the albumen glycerol covering them under the stereomicroscope. Mix the fluids gently with the micropipette, avoiding making air bubbles.
4. Pick the ciliates from the watch glass with the micropipette, drop them on clean glass slides, and remove the excess of fluid. Using a mounted eyelash or a needle, gently move the specimens to their desired position. Let the slides dry overnight.
5. After air-dried, transfer the slides to a staining jar with ethanol 100% during 20 min and then successively other jars with ethanol 70% and 30% for 10 min each. Then, transfer the slides to staining jars with distilled water three times for 10 min each.
6. Bleach the ciliates by transferring the slides to a staining jar with 0.2% sodium permanganate during 100–150 s. The bleaching of small ciliates usually takes less time than those with greater size. It is advisable to place them in separate slides. Some experimentation is necessary for the success of this step. After bleaching, quickly wash the slides three times in staining jars with distilled water to remove the excess of sodium permanganate.
7. Immediately transfer the slides to a staining jar with oxalic acid 0.2% for 120–180 s. The amount of time should be proportional to the time spent in the previous step. Remove the slides from the oxalic acid, and quickly wash them in distilled water. Next, wash them with distilled water four times for 5 min each.
8. Place the slides into a staining jar with a preheated (60°C) solution of protargol 0.2%, and impregnate the cells in a water bath at 60°C for 2–3 h.
9. Remove staining jar from the water bath, and take the slides, placing them on a staining rack with the coated side up. With a Pasteur pipette, pour some drops of the protargol solution completely covering the cells.
10. Pour some drops of developer solution on the slides, and let develop for approximately 15 min.
11. Remove the layer of protargol plus developer by tilting the slides on a sheet of absorbing paper, and transfer the slides to a staining jar with distilled water; wash them for approximately 5 min, and repeat this process for another 5 min.
12. Rinse each slide individually in a staining jar with gold chloride 0.5% for 10 s, immediately wash them with distilled water for approximately 3 s, and then rinse in oxalic acid 3% for 30–180 s. Wash the slides again with distilled water for approximately 10 s, and then place them in a staining jar with distilled water.

13. When all the slides were gone through the previous step, transfer them to another staining jar with distilled water for 5 min, and repeat this process four times to thoroughly remove the oxalic acid.
14. Rinse the slides for no longer than 3 s in sodium thiosulfate 2%, and then wash them with distilled water 3 or four times during 5 min each.
15. Dehydrate the cells by transferring the slides to a series of ethanol 30%, 70% for 10 min each, and 100% for 20 min. Air-dry the slides for a few seconds, and mount with Entellan.

Specific Reagents

Bouin's fixative: Prepare in this order, 15 parts saturated aqueous picric acid, 5 parts 37% formaldehyde, 1 part glacial acetic acid.

Albumen glycerol: 15 ml egg white, 15 ml concentrated glycerol.

Developer (storable in a sealed flask for several months): 875 ml distilled water, 11.5 g boric acid, 2.5 g hydroquinone, 17 g sodium sulfite, 125 ml acetone.

Impregnation fixative (storable in a sealed flask for several months): Sodium thiosulfate 5 H_2O.

Remarks: Carefully separate the egg white from the yolk. Use a filter paper to further separate the less viscous part from the egg white, which will be used. The protargol solution should be artificially aged at 60−80°C for 2−4 h, on four to five consecutive days; otherwise, the cells may become overimpregnated. The protargol solution can be used several times, when impregnation finally starts to become faint. The solution can be reactivated by adding 50 mg of protargol powder to 100 ml of the exhausted solution.

Protargol impregnation, protocol II. This is adapted from Wilbert (1975), Song & Wilbert (1995), and personal experience. This protocol is relatively quick and produces good results. It can be used to identify ciliates, as well as to produce stable, permanent material. The structures stain in different brown tonalities. This protocol is less time-consuming and easier than that described anteriorly, but great attention should be paid during the bleaching stage. The entire procedure is performed under the stereomicroscope.

1. Pick ciliates directly from fresh samples or cultures with a micropipette and fix them in an embryo dish containing Bouin's fixative. Fix for at least 10 min. Remove fixative from the embryo dish by aspiring it with a micropipette, maintaining the ciliates in the middle. Then add distilled water drop by drop. Repeat this procedure until the yellow color of the fixative disappears. When adding reagents or during washings, slightly beat with a finger at one edge of the embryo dish to mix and maintain ciliates in the middle.
2. Bleach with sodium hypochlorite by adding a small drop around and over the ciliates with a micropipette with a very narrow capillary. Cells become translucent and inflated. Immediately remove the sodium hypochlorite with a washing micropipette, and replace it with distilled water drop by drop. At least four washing steps are needed to remove the uncolored bleaching solution. This is the critical point of the technique. If ciliates are overbleached, they do not impregnate at all, and if bleaching is too weak, ciliates appear too dark to observe the infraciliature.
3. When the bleaching solution is completely replaced by distilled water, spread slowly the protargol powder directly on the embryo dish containing the ciliates. Just a tip of a spatula is enough to obtain a concentrate solution of protargol. Wait until the protargol dissolves entirely alone, do not mix manually.
4. Seal the embryo dish by adding a drop of water in one corner, and place the glass lid over it. Impregnate for about 45−60 min at 45°C.
5. Reveal the impregnation by adding some drops of 1% hydroquinone dissolved in 5% sodium sulfite. The solution in the embryo dish turns caramel to dark brown. Some impregnated structures can be observed under the stereomicroscope, but it can be accurately checked by picking a drop with ciliates from the embryo dish, and place it in a slide; then add a drop of albumen glycerol and a coverslip. This transient slide can be sealed with nail enamel on each side of the coverslip. Ciliates get clearer after some days, and impregnation lasts for a few weeks.
6. Fix the impregnation by adding a drop of 2% sodium thiosulfate over and around the ciliates in the embryo dish. Immediately wash with distilled water at least 6 or 7 times. The uncolored fixative must be removed entirely since impregnation will disappear after few weeks and the permanent slide will be lost.
7. Pick a drop of impregnated ciliates with a clean micropipette, and transfer it to a grease-free and marked slide. Add a drop of albumen glycerol and mix with a mounted hair in a capillary or in a Pasteur pipette. Remove excess of albumen glycerol between ciliates on the slide until you can see them as tiny spots on the slide when tilting it. Let dry in a hot plate for ca. 10 min.
8. Dehydrate in staining jars with a series of alcohol and xylol. Transfer the slide(s) through ethanol 70%, 95%, two steps of 100%, and finally two steps of xylol, during 5 min each.
9. Mount in fluid synthetic medium (Merckoglas, Eukitt), and allow hardening.

Specific Reagents

Bouin's fixative: Same as for protocol I.

Albumen glycerol: Separate the albumen and filter it through filter paper; use a funnel and an Erlenmeyer. Mix equal parts of egg white and concentrated glycerol in a graduated tube. Add a small crystal of thymol to avoid bacterial growth.

Bleaching solution: Several dilutions of concentrated sodium hypochlorite (55 g/L) should be prepared, according to species requirements.

1 ml sodium hypochlorite: 30 ml distilled water
0.7 ml sodium hypochlorite: 30 ml distilled water
0.5 ml sodium hypochlorite: 30 ml distilled water (most commonly used)

Developer (storable for 1−2 weeks, until it becomes yellow to light brown-colored): Dissolve the hydroquinone in the sodium sulfite solution, in this order: 10 ml distilled water, 0.5 g sodium sulfite, and 0.1 g hydroquinone.

Impregnation fixative (storable in a sealed flask for several months): 10 ml distilled water, 0.2 g sodium thiosulfate 5 H_2O

Remarks. Use a different micropipette for each solution to avoid contamination with reagents as well as one micropipette with a wider capillary for the washings. Commercial sodium hypochlorite has different concentrations of active chloride; this concentration diminishes as the solution gets old. It is required to prepare various concentrations of this solution since ciliates bleach differently depending on species. For this reason, it is also desirable to maintain raw cultures to have great numbers of ciliates. Usually the first time, the impregnation fails, and other concentrations have to be tested. Prepare all solutions and albumen glycerol before starting the staining procedure. The same applies to the dehydrating series of alcohol and xylol in staining jars. Slides should be cleaned and marked with a glass marker to place the drop with impregnated ciliates in this field. Place the drop with ciliates in the opposite surface of the marking; this will allow you to check where the ciliates are during the dehydration series.

Dry silver nitrate impregnation. This method was originally proposed by Klein (1958) and impregnates the silverline system. We present a very simple and quick adaptation that is useful for the ciliates in which the argyrome is of taxonomic relevance.

1. Pick the ciliates from the cultures using a capillary micropipette and briefly wash them in mineral water, using a watch glass.
2. Place very small isolated droplets (with a micropipette), each containing preferably a single specimen, on a clean glass slide and allow air-drying for a few minutes. The ciliates will dehydrate and adhere to the slides. Alternatively, the ciliates may be fixed with osmium vapors before air-drying. It is important to impregnate the ventral side, since most euplotids will dry out with their dorsal side up.
2. Wash carefully the slides with mineral water. This can be done with a Pasteur pipette under the stereomicroscope. Gently tilt the slide on a sheet of absorbing paper. If the cells were fixed with osmium vapors, this process should be repeated at least twice.
3. Place some drops of silver nitrate 2% on the slides, covering the specimens, and let impregnate during approximately 15 min in the dark.
4. Wash the slides again as mentioned earlier.
5. Place the slides at about 10 cm from a UV lamp during 5−10 min. Alternatively, place some drops of developer on the slides, wait for about 1 min, and remove the developer by tilting the slide on absorbing paper.
6. Rinse the slides very briefly (not longer than 2 s!) in an impregnation fixative solution and immediately wash in mineral water.
7. Air-dry the slides and mount with Entellan.

Remarks: The developer and the impregnation fixative solution are the same as in the protargol protocol II.

KEYS TO CILIOPHORA

Ciliophora: Classes

1	With cilia and/or cirri; tentacles may be present but combined with cilia (Actinobolinidae)	2
1′	Adult (trophont) only with tentacles and without cilia (except in free swimming dispersal stages), or only with cilia and usually flattened body; with phylae (microtubular ribbons in the oral apparatus)	**Phyllopharyngea [p. 44]**
2(1)	Somatic ciliature well developed or reduced or absent (Peritrichia); oral ciliature variable	3
2′	Somatic ciliature usually reduced or composed of cirri; oral ciliature conspicuous, with an adoral zone of membranelles (polykineties) often encircling the oral region clockwise before entering the oral cavity, and one or two undulating membranes	**Spirotrichea [p. 47]**
3(2)	Oral ciliature conspicuous, may extend outside the oral cavity, composed of membranelles and undulating membrane(s), cytostome within the oral cavity	4
3′	Oral ciliature usually consists of simple cilia surrounding the oral region, with the cytostome at body surface, preceded by a simple depression or within the buccal cavity	6
4(3)	Oral ciliature composed of numerous membranelles (> 3) and one or two undulating membranes	5

4′ Oral ciliature with 3 left membranelles and right undulating membrane; somatic ciliature may be absent (Peritrichia) .... **Oligohymenophorea [p. 54]**

5(4) Body twisted or laterally compressed and usually armored, sometimes bearing spiny processes somatic ciliature uniform or reduced .... **Armophorea [p. 61]**

5′ Body not as above; somatic ciliature well developed, often highly contractile, sometimes pigmented .... **Heterotrichea [p. 61]**

6(3) Oral region usually apical, subapical, lateral or ventral, with cytostome at or near body surface 7

6′ Cytostome preceded by a simple oral depression or within the buccal cavity 10

7(6) Cytopharynx cyrtos-like 8

7′ Not as above 9

8(7) Ovoid to cylindroid; cytostome apical, subapical or lateral, circumoral ciliature as oral dikinetids .... **Prostomatea [p. 62]**

8′ Body flattened or cylindrical; cytostome ventral; some with a hypostomial frange or synhymenium; others with a simple oral depression .... **Nassophorea [p. 64]**

9(7) Some large, vermiform, contractile and flexible; others with a flattened body and anterior "rostrum", acontractile; somatic ciliature with non- or sparsely ciliated surface region (glabrous zone) .... **Karyorelictea [p. 66]**

9′ Body shape variable; somatic ciliature typically as monokinetids, well developed to sparse; several dorsal somatic kineties differentiated as a dorsal brush (brosse) with specialized dikinetids bearing clavate cilia .... **Litostomatea [p. 66]**

10(6) Body ovoid to elongate ovoid, slightly flattened or laterally compressed, often with spines; somatic ciliature uniform; oral ciliature simple, as extensions of somatic kineties into an oral depression or with inconspicuous polykineties; mainly anaerobic ...... **Plagiopylea [p. 70]**

10′ Body shape variable, many reniform or flattened, some bursiform; oral ciliature within the oral cavity; alveoli well developed as revealed by a typically reticulated argyrome .... **Colpodea [p. 71]**

Ciliophora: Phyllopharyngea: Major Groups

1 With extensible suctorial tentacles, with extrusomes as haptocysts; adults (trophonts) usually sessile; cilia absent (infraciliature present) except in free-swimming dispersal stages; with or without lorica .... **Suctoria [p. 44]**

1′ Cilia present; no suctorial tentacles; conspicuous oral basket or cyrtos lined by microtubular ribbons; body often dorsoventrally flattened, may attach to substrate by ventral thigmotactic cilia; somatic cilia mostly on the ventral side, usually arranged in two somewhat equal fields .... **Cyrtophoria [p. 45]**

Ciliophora: Phyllopharyngea: Suctoria: Families

1 Budding and cytokinesis on surface of trophont (Exogenida) 2

1′ Budding begins in a pouch 3

2(1) Body small, pyriform, or spherical; lorica usually absent; tentacles apical or evenly distributed, rarely arranged in fascicles; stalk usually present, except in planktonic forms .... **Podophryidae [p. 45]**

2′ Usually loricate; lorica opening often with notches or slits through which capitate tentacles protrude; body spherical or saclike, basally not attached to lorica; tentacles arranged in fascicles or rows (Fig. 2.12 A) .... Metacinetidae, *Metacineta*

3(1) Budding and cytokinesis completed in brood pouch (Endogenida) 4

3′ Cytokinesis begins in pouch, completed exogenously (Evaginogenida) 5

4(3) With triangular lorica, sometimes difficult to observe; body trapezium-like or triangular; tentacles in two or three fascicles typically on actinophores .... **Acinetidae [p. 45]**

4′ Without lorica; body ovoid, cylindrical or triangular; tentacles in two (rarely more) fascicles (Fig. 2.12 B) .... Tokophryidae, one genus: *Tokophyra*

5(3) With stalk 6

5′ Without stalk; small, body flattened on the substrate; solitary tentacles arranged in several fascicles (Fig. 2.12 C) .... Heliophryidae, one genus: *Heliophrya*

6(5) Without lorica 7

6′ With lorica; stalk separated from bottom of lorica; body may be laterally flattened (Fig. 2.12 E) Periacinetidae, one genus: *Periacineta*

7(6) Capitate tentacles in fascicles or evenly distributed (Fig. 2.12 D) .... Discophryidae, one genus: *Discophrya*

7′ Capitate tentacles evenly distributed (Fig. 2.12 F) .... Prodiscophryidae, one genus: *Prodiscophrya*

Ciliophora: Phyllopharyngea: Suctoria: Podophryidae: Genera

1 Tentacles without basal widening 2

1′ Tentacles with basal widening (Fig. 2.12 G) .... *Parapodophrya*

2(1) With stalk (Fig. 2.12 I) .... *Podophrya*

2′ Without stalk (Fig. 2.12 H) .... *Sphaerophrya*

Ciliophora: Phyllopharyngea: Suctoria: Acinetidae: Genera

1 With two fascicles of tentacles 2

1′ With three fascicles of tentacles borne upon short actinophores (Fig. 2.12 J) .... *Phyllacineta*

2(1) With stalk (Fig. 2.12 L) .... *Acineta*

2′ With stylotheca (posterior region of lorica elongated forming a stalklike appendage) (Fig. 2.12 K) .... *Acinetides*

Ciliophora: Phyllopharyngea: Cyrtophoria: Families

1 Body dorsoventrally flattened, may attach to substrate by ventral thigmotactic cilia; somatic cilia mostly on ventral surface, usually arranged in two somewhat equal fields (Chlamydodontida) 2

1′ Laterally compressed, dorsal surface rounded; ventral cilia not thigmotactic; attached by an adhesive region or by a conical appendage or podite (Dysteriida) 5

2(1) Without a cross-striated band at the periphery; with left snoutlike projection or beak; right kineties extending to preoral region; oral kineties more than three 3

2′ With a cross-striated band at the periphery of the flattened cell separating the dorsal from the ventral side; with two parallel circumoral kineties and one preoral kinety (Fig. 2.12 M) .... Chlamydodontidae, one genus: *Chlamydodon*

3(2) Longitudinal kineties not extending across complete ventral surface; with central gap 4

3′ Longitudinal kineties extend across complete ventral surface; without central gap, preoral kinety Y-shaped, without a kinety along suture line .... Lynchellidae, one genus: *Chlamydonella*

4(3) Oral opening supported by an oral basket of nematodesmata .... **Chilodonellidae [p. 45]**

4′ Oral opening is a large, elongate, and transverse groove, not supported by nematodesmata; one kinety completely surrounds oral opening (Fig. 2.12 O) .... Gastronautidae, *Gastronauta*

5(1) With extensive left field of kineties in all ventral surface; six nematodesmata armed with "teeth" in the oral opening (Fig. 2.12 N) .... Hartmannulidae, one genus: *Trochilioides*

5′ Somatic ciliature typically reduced, with left ventral somatic kineties as midventral/postoral field, typically separated from an anterior preoral field; with flexible podite for attachment .... **Dysteriidae [p. 45]**

Ciliophora: Phyllopharyngea: Cyrtophoria: Chilodonellidae: Genera

1 Without spines in the dorsal surface 2

1′ With bumplike spines (4–13) in the dorsal surface projecting out the body (Fig. 2.12 S) .... *Odontochlamys*

2(1) With glabrous postoral area (central cilia-free zone) 3

2′ Without glabrous postoral area, preoral kinety sigmoid along suture line (Fig. 2.12 Q) .... *Trithigmostoma*

3(2) Preoral kinety segmented, resulting in many short kineties along suture line (Fig. 2.12 P) .... *Pseudochilodonopsis*

3′ Preoral kinety continuous, located along suture line (Fig. 2.12 R) .... *Chilodonella*

Ciliophora: Phyllopharyngea: Dysteriidae: Genera

1 Apical left field in front of oral aperture; four or more kineties in the right field .... *Dysteria*

1′ Apical left field in the left of oral aperture; always four kineties in the right field (Fig. 2.12 T) .... *Trochilia*

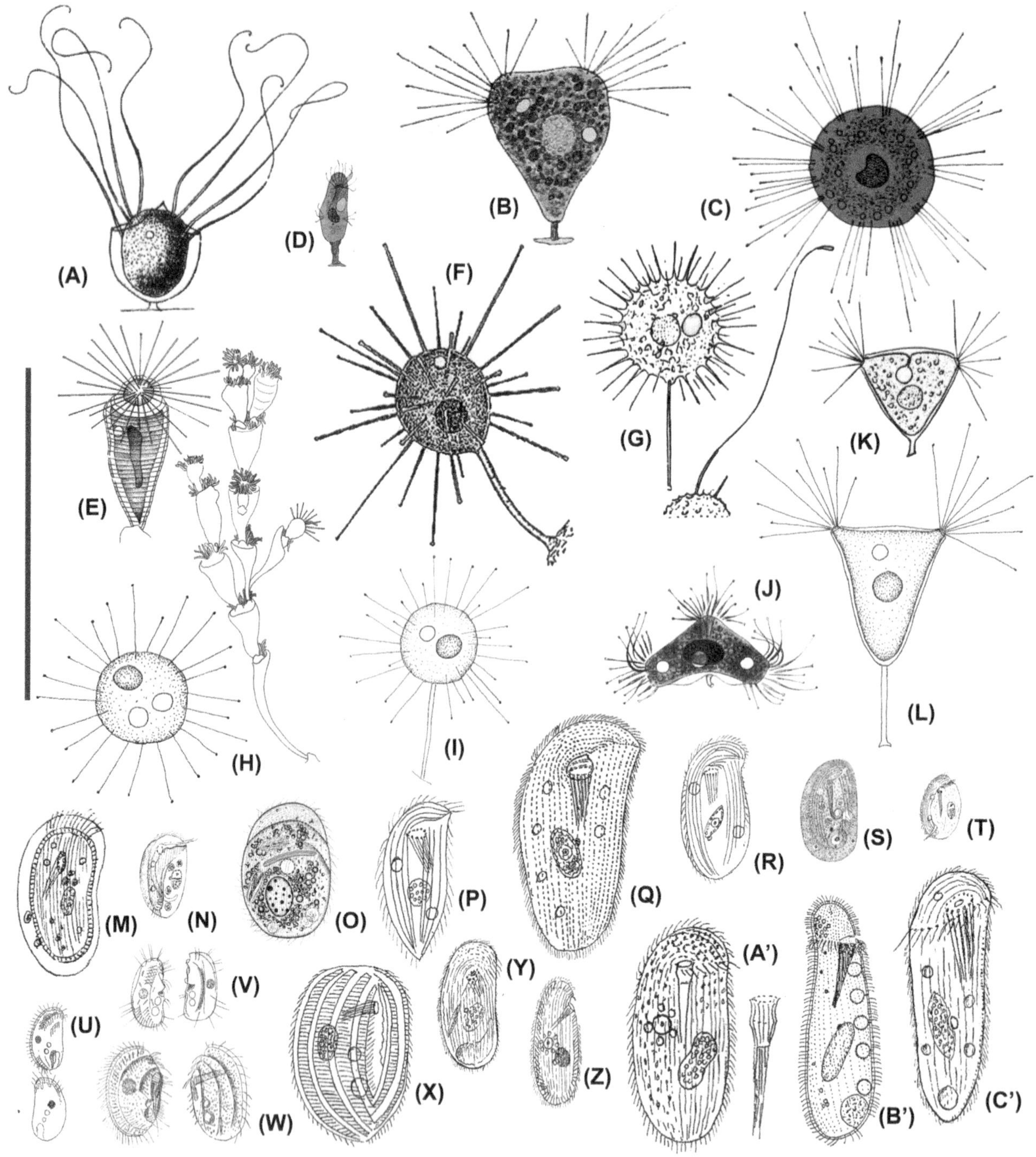

FIGURE 2.12 Representative species of class Phyllopharyngea recorded in the Neotropics: (A) *Metacineta cuspidata* ca. 45 μm; (B) *Tokophrya infusionum* ca. 100 μm; (C) *Heliophrya rotunda* ca. 45 μm; (D) *Discophrya elongata* ca. 50 μm; (E) *Periacineta mexicana* ca. 100 μm; (F) *Prodiscophrya collini* ca. 50 μm; (G) *Parapodophrya soliformis* ca. 60 μm; (H) *Sphaerophrya magna* ca. 70 μm; (I) *Podophrya fixa* ca. 50 μm; (J) *Phyllacineta tripharetrata* ca. 45 μm; (K) *Acinetides triangularis* ca. 60 μm; (L) *Acineta tuberosa* ca. 100 μm; (M) *Chlamydodon mnemosyne* ca. 90 μm; (N) *Trochilioides recta* ca. 45 μm; (O) *Gastronauta multistirata* ca. 60 μm; (P) *Pseudochilodonopsis piscatoris* ca. 75 μm; (Q) *Trithigmostoma cucullulus* ca. 150 μm; (R) *Chilodonella uncinata* ca. 70 μm; (S) *Odontochlamys alpestris* ca. 50 μm; (T) *Trochilia minuta* ca. 25 μm; (U) *Microthorax pusillus* ca. 30 μm; (V) *Drepanomonas revoluta* ca. 35 μm; (W) *Leptopharynx costatus* ca. 35 μm; (X) *Pseudomicrothorax dubius* ca. 100 μm; (Y) *Chilodontopsis depressa* ca. 70 μm; (Z) *Furgasonia rubens* ca. 70 μm; (A′) *Nassula picta* ca. 130 μm; (B′) *Nassulopsis elegans* ca. 140 μm; (C′) *Zosterodasys transversus* ca. 180 μm. Scale bar = 200 μm. *After: (A) modified from Kellikott (1885); (B) and (D) Collin (1911); (C) Hentschel (1916); (E) Mariño Pérez et al. (2010); (F) Root (1914); (G) and (K) Penard (1920); (J) Entz (1907); (M, N, P–R, U–W, Y–A′, and C′) Kahl (1931); (O) Xu et al. (2016); (S) Foissner (1981); (T and X) Penard (1922) after Kahl (1931); (B′) Ehrenberg (1833); (E, O and S) reproduced with permission from copyright holder.*

Ciliophora: Spirotrichea: Major Groups

To follow most of the keys, at least interference contrast for live observation is required, in addition to silver impregnation and ontogenesis investigation in some cases.

1 Body generally ovoid, conical or obconical; mostly planktonic; numerous membranelles in complete or almost complete circle at oral end; usually without cirri (halteriids with cirri-like bristles) 2

1′ Body usually dorsoventrally flattened; with cirri; membranelles forming and adoral zone 3

2(1) Oral cilature clearly divided, forming an open circle of anterior and ventral membranelles .... **Oligotrichida and Halteriida [p. 47]**

2′ Oral ciliature forms a closed circle of external membranelles; some internal membranelles within the oral cavity **Choreotrichida [p. 47]**

3(2) Body rigid or flexible; pellicular alveoli poorly developed; at least one right and one left marginal row of cirri; oral ciliature conspicuous .... **Hypotricha [p. 49]**

3′ Body rigid, usually ribbed; pellicular alveoli well developed; marginal rows of cirri reduced, right row absent; oral ciliature conspicuous or reduced .... **Euplota [p. 52]**

Ciliophora: Spirotrichea: Oligotrichida and Halteriida: Families

1 Somatic ciliature reduced to stiff cilia or bristles, sometimes cirrilike, arranged circumferentially around the equator of body .... **Halteriidae [p. 47]**

1′ Somatic ciliature usually reduced to a horizontal (opened or closed) girdle kinety and a longitudinal ventral kinety with stubby cilia .... **Strombidiidae [p. 47]**

Ciliophora: Spirotrichea: Halteriidae: Genera

1 Somatic kineties very short and reduced to equatorial bristle complexes of vertical and/or horizontal kinetosomes; bristles very long and stiff, sometimes with cirrilike cilia effecting a jumping motion 2

1′ Somatic kineties distinctly long; cilia soft, not arranged in equatorial bristle complexes (Fig. 2.13 H) .... *Meseres*

2(1) Somatic kineties without horizontal row close below each bristle complex (Fig. 2.13 I) .... *Halteria*

2′ Somatic kineties with additional horizontal row close below each bristle complex, forming cirrilike structures (Fig. 2.13 J) *Pelagohalteria*

Ciliophora: Spirotrichea: Strombidiidae: Genera

1 Girdle kinety continuous and horizontally arranged, not extending to posterior pole; ventral kinety at about right angle to girdle kinety, sometimes reduced or absent 2

1′ Girdle kinety conspicuously opened, with left portion extending to posterior pole; ventral kinety arranged longitudinally below right portion of opened girdle kinety (Fig. 2.13 L) .... *Propecingulum*

2(1) Ontogenesis occurs in a permanent tube or neoformation organelle 3

2′ Ontogenesis occurs in a transient tube (Fig. 2.13 K) .... *Strombidium*

3(2) With ventral kinety (Fig. 2.13 N) .... *Limnostrombidium*

3′ Without ventral kinety (Fig. 2.13 M) .... *Pelagostrombidium*

Ciliophora: Spirotrichea: Choreotrichida: Families

1 Loricate 2

1′ Not loricate 3

2(1) Lorica rigid, agglomerate of mineral grains and diatom fragments .... **Codonellidae [p. 49]**

2′ Lorica delicate, tubular, gelatinous or mucoid, with attached debris, often translucent (Fig. 2.13 D) .... Tintinnididae, one genus: *Tintinnidium*

3(1) One or more usually short somatic kineties, somatic cilia short .... **Strobilidiidae [p. 49]**

3′ Somatic kineties uniformly distributed around body and extending along body length; somatic cilia may be long (Fig. 2.13 C) .... Strombidinopsidae, one genus: *Strombidinopsis*

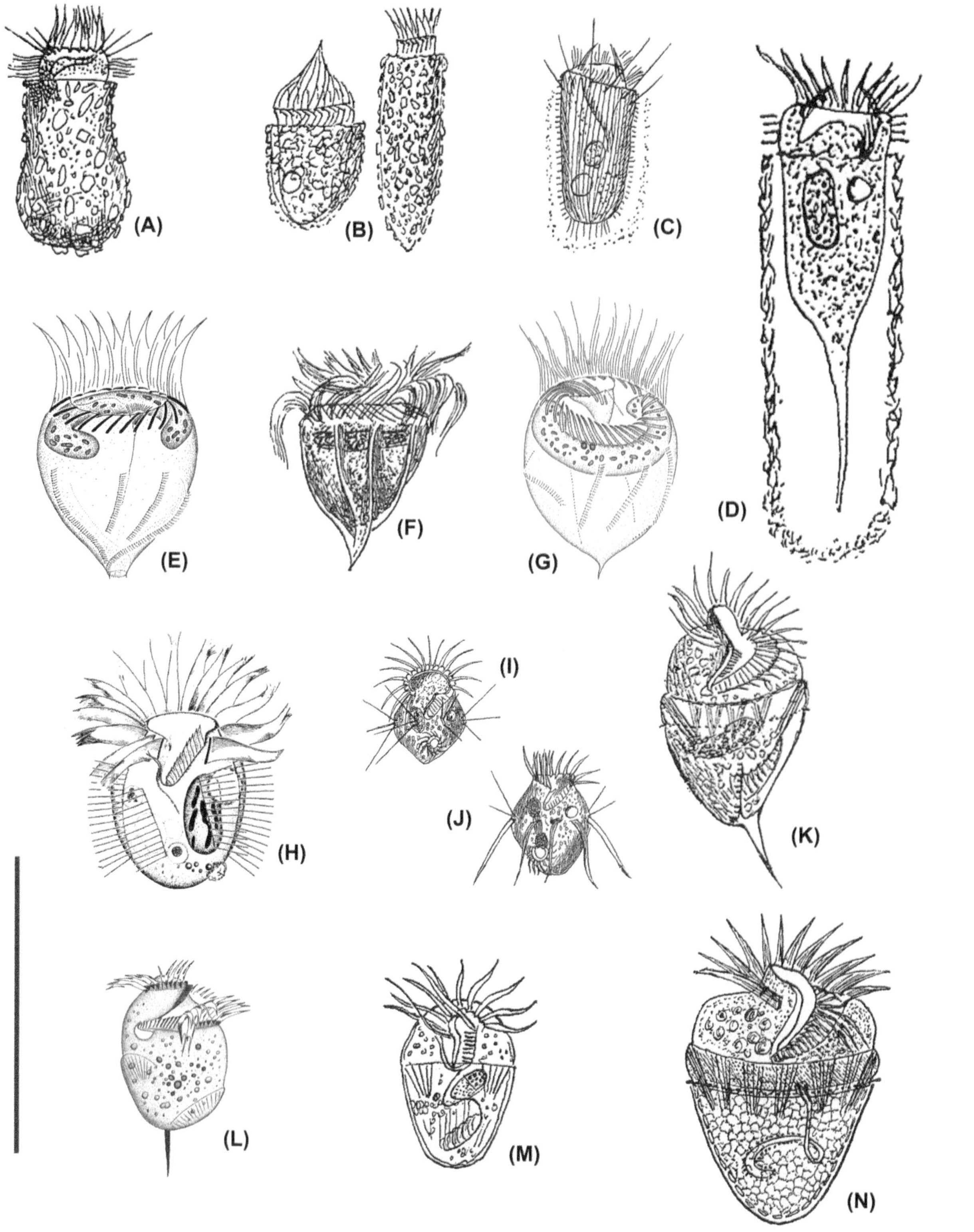

FIGURE 2.13 Representative species of class Spirotrichea recorded in the Neotropics: (A) *Codonella cratera* ca. 70 μm; (B) *Tintinnopsis cylindrata* ca. 50 μm and ca. 70 μm; (C) *Strombidinopsis setigera* ca. 70 μm; (D) *Tintinnidium fluviatile* ca. 150 μm; (E) *Strobilidium caudatum* ca. 60 μm; (F) *Rimostrombidium velox* 50 μm; (G) *Pelagostrobilidium wilberti* ca. 60 μm; (H) *Meseres corlissi* ca. 60 μm; (I) *Halteria grandinella* ca. 30 μm; (J) *Pelagohalteria cirrifera* ca. 40 μm; (K) *Strombidium styliferum* ca. 80 μm; (L) *Propecingulum rufinoi* ca. 45 μm; (M) *Pelagostrombidium mirabile* ca. 65 μm; (N) *Limnostrombidium viride* ca. 80 μm. Scale bar = 80 μm. *After: (A and F) Fauré-Fremiet (1924) after Kahl (1932); (B) Entz (1905) after Kahl (1932); (C and M) Penard (1922) after Kahl (1932); (D, I–K, and N) Kahl (1932); (E) Küppers et al. (2006b); (G) Küppers et al. (2006a); (H) Petz & Foissner (1992); (L) Paiva et al. (2016); (E, G, H, and L) reproduced with permission from the copyright holder.*

Ciliophora: Spirotrichea: Choreotrichida: Codonellidae: Genera

1 Lorica with a hemispherical to broadly conical posterior half and a tubular or helical upper half with spiral turns (difficult to recognize from outside) (Fig. 2.13 A). .... *Codonella*

1′ Lorica elongate, with a broad aperture and a closed aboral end, without an anterior collarlike region; sometimes with an aboral horn (Fig. 2.13 B) .... *Tintinnopsis*

Ciliophora: Spirotrichea: Choreotrichida: Strobilidiidae: Genera

1 Somatic kineties usually more or less longitudinal, not forming a caudal spiral 2

1′ Some somatic kineties form a caudal spiral at posterior truncated end of body (Fig. 2.13 E) .... *Strobilidium*

2(1) All somatic kineties longitudinal (Fig. 2.13 F) .... *Rimostrombidium*

2′ Most somatic kineties longitudinal; with an arched and more or less transversely arranged kinety; shortened adjacent kineties usually abut onto this arched kinety (Fig. 2.13 G) .... *Pelagostrobilidium*

Ciliophora: Spirotrichea: Hypotricha: Families

1 Cirri usually not in oblique spiraled rows around body; transverse cirri may be present 2

1′ Cirri in oblique spiraled rows around body; transverse cirri may be absent; some species with lorica .... **Spirofilidae [p. 50]**

2(1) Ventral cirri arranged in longitudinal files, either linear or zigzag; without fragmented dorsal kineties, with or without dorsomarginal kineties 3

2′ Usually 18 frontoventral cirri, typically arranged in localized groups (frontal, ventral and transverse); dorsally usually with bipolar kineties, fragmented kineties and/or dorsomarginal kineties 10

3(2) Frontoventral cirri arranged in zigzag, forming a midventral complex of paired cirri and sometimes also posterior midventral rows, dorsally with bipolar kineties and with (Uroleptids) or without a dorsomarginal kinety 4

3′ Frontoventral cirri arranged in one or more linear files (not zigzag) 8

4(3) Dorsally with bipolar kineties only 5

4′ Dorsally with bipolar kineties and a dorsomarginal kinety; posterior region usually constricted into a tail (Fig. 2.14 E) .... Uroleptidae, one genus: *Uroleptus*

5(4) Midventral complex composed of cirral pairs only or cirral pairs and midventral rows 6

5′ Midventral complex composed of midventral rows only; many frontal cirri arranged in oblique rows; with transverse cirri; body outline very slender and tripartite, with more or less distinct head, widened central body portion, and tail (Fig. 2.14 F) .... Epiclintidae, one genus: *Epiclintes*

6(5) Usually, many frontal cirri arranged in a bicorona, tricorona, or multicorona; midventral complex composed of cirral pairs only or cirral pairs and midventral rows; more than one left and right marginal rows of cirri 7

6′ Midventral complex composed of cirral pairs only; three enlarged frontal cirri; usually one left and one right marginal, but right marginal rows may be increased .... **Holostichidae [p. 50]**

7(6) Caudal cirri present (Fig. 2.14 U) .... Urostylidae, one genus: *Urostyla*

7′ Caudal cirri absent .... **Pseudourostylidae [p. 50]**

8(3) Frontal cirri not forming an anterior corona 9

8′ Frontal cirri forming a single corona paralleling the anterior arc of the adoral zone of membranelles; transverse cirri may be absent and two shorter oblique frontal rows of smaller cirri may be present (Fig. 2.14 L) .... Keronopsidae, one genus: *Paraholosticha*

9(8) At least with two, typically more than two, ventral cirral files, often not distinctly different from right and left marginal cirral files; presence of transverse and caudal cirri variable .... **Kahliellidae [p. 50]**

9′ With a mixed row of cirri, which originate from different primordia and arrange linearly (ontogenesis needed); transverse and caudal cirri usually present .... **Amphisiellidae [p. 50]**

10(2) Dorsally with bipolar, fragmented and dorsomarginal kineties 11

10′ Dorsally with bipolar kineties, without fragmented and dorsomarginal kineties; middle portion of adoral zone of membranelles straight and extending along left margin of cell; paroral with few, widely spaced cilia, extending far beyond anterior end of endoral (*Gonostomum* pattern) (Fig. 2.15 D) .... Gonostomatidae, one genus: *Gonostomum*

11(10) Somatic ventral ciliature as frontoventral and transverse cirri (FVT), typically 18 in number, large and distinctive, scattered over mid-area of ventral surface, between right and left marginal cirral files, usually with three frontoventral cirri posterior to posterior vertex of the oral region, but several genera with conspicuous files of cirri .... **Oxytrichidae [p. 50]**

11′ Somatic ventral ciliature as frontoventral cirri scattered on anterior near peristomal region, sometimes in posterior, but never on midarea of ventral surface, which is bordered by right and left marginal cirral files (Fig. 2.15 C) .... Trachelostylidae, one genus: *Trachelostyla*

Ciliophora: Spirotrichea: Hypotricha: Spirofilidae: Genera

1 With lorica 2

1′ Without lorica 4

2(1) Sessile, with long peristome in anterior necklike region 3

2′ Planktonic, with anterior snoutlike region; lorica cylindrical, membranous, carried by the ciliate while swimming .... *Spirofilopsis*

3(2) Anterior region corkscrewlike (Fig. 2.14 B) .... *Chaetospira*

3′ Anterior region straight (Fig. 2.14 A) .... *Stichotricha*

4(1) Body usually dorsoventrally flattened, sometimes sigmoid in outline or with a posterior tail-like extension; two or more ventral and two marginal rows of cirri that spiral down the length of body; transverse and caudal cirri present (Fig. 2.14 D) .... *Strongylidium*

4′ Planktonic, body not dorsoventrally flattened, pyriform with posterior end pointed, ovoid- or quadrilateral-shaped; meridional rows connected or not with posterior spiraled rows of cirri; caudal cirri absent (Fig. 2.14 C) .... *Hypotrichidium*

Ciliophora: Spirotrichea: Hypotricha: Holostichidae: Genera

1 One left and one right marginal rows; prominent and numerous transverse cirri forming a roughly longitudinal row; adoral zone of membranelles with gap (Fig. 2.14 G) .... *Holosticha*

1′ Two or more left marginal rows and one right marginal row; few transverse cirri forming a short oblique row; adoral zone of membranelles without gap (Fig. 2.14 J) .... *Diaxonella*

Ciliophora: Spirotrichea: Hypotricha: Pseudourostylidae: Genera

1 With more than one left marginal row and only one right marginal row .... *Trichototaxis*

1′ With more than one left and right marginal rows (Fig. 2.14 K) .... *Pseudourostyla*

Ciliophora: Spirotrichea: Hypotricha: Kahliellidae: Genera

1 With one or two long frontoventral rows; caudal cirri present (Fig. 2.14 I) .... *Cladotricha*

1′ With two or more frontoventral rows; caudal cirri absent (Fig. 2.14 M) .... *Deviata*

Ciliophora: Spirotrichea: Hypotricha: Amphisiellidae: Genera

1 Very small-sized ($\leq 50\ \mu m$) ciliate; with about seven frontal cirri and large and conspicuous dorsal bristles; transverse cirri present (Fig. 2.14 H) (*incertae sedis*) .... *Balladyna*

1′ Size variable, body elongate elliptical; one interrupted frontoventral row with anterior and posterior portions; transverse and post-peristomial cirri absent (Fig. 2.14 O) .... *Nudiamphisiella*

Ciliophora: Spirotrichea: Hypotricha: Oxytrichidae: Genera

1 Without typical 18 FVT groups of cirri 2

1′ With typical 18 FVT groups of cirri 9

2(1) FVT cirri increased (more than 18) or feigning the midventral pattern of urostyloids 3

2′ Frontoventral cirri in rows 4

3(2) FVT cirri increased; three frontal cirri; with dorsal processes or horns (Fig. 2.14 T) .... *Onychodromus*

3′ Several ventral cirral pairs feigning the midventral pattern of urostyloids; two arched rows of frontal cirri; without dorsal processes or horns (Fig. 2.14 S) .... *Neokeronopsis*

4(2) With at least two frontoventral rows 5

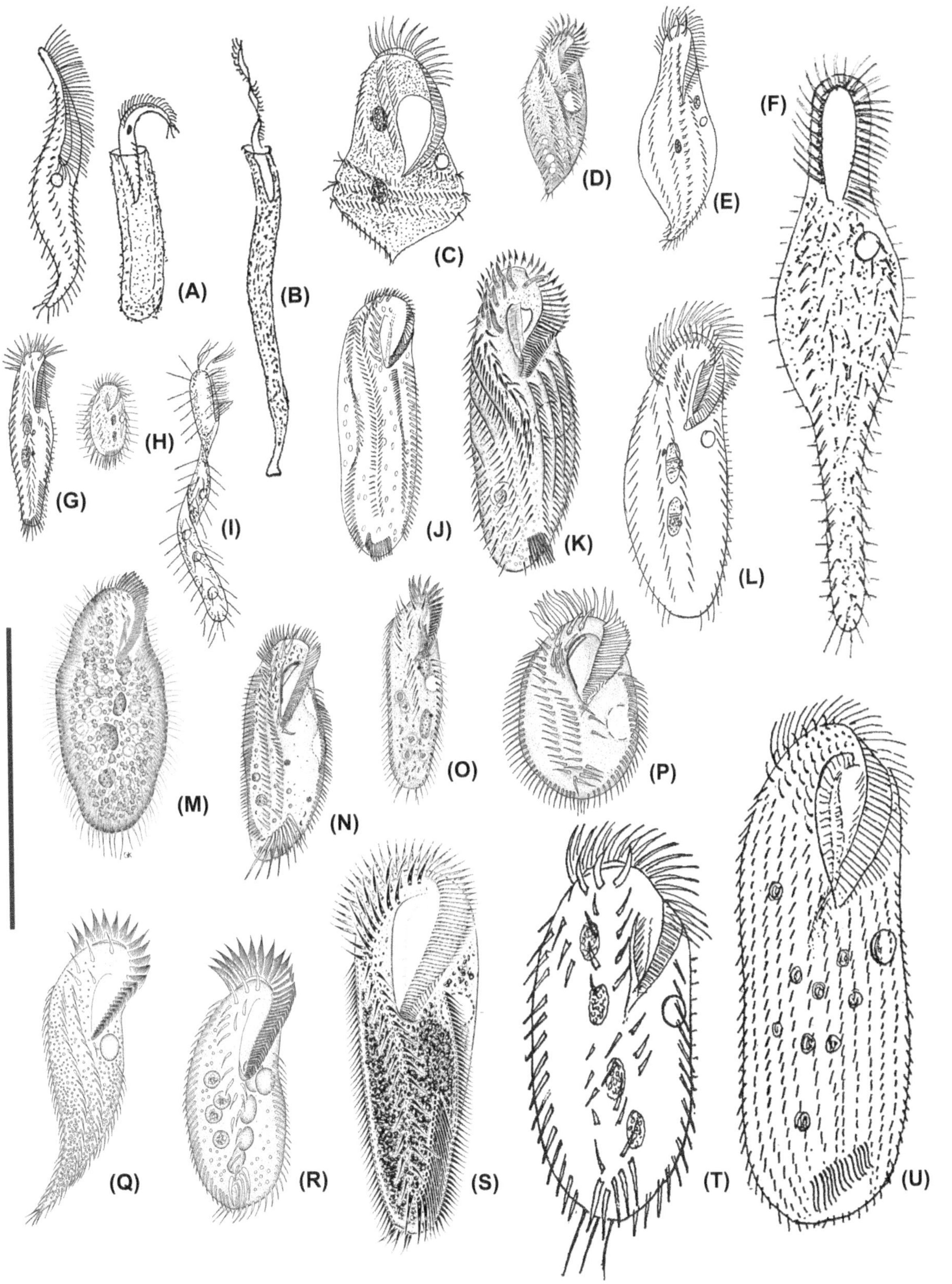

FIGURE 2.14 Representative species of class Spirotrichea recorded in the Neotropics: (A) *Stichotricha secunda* ca. 200 μm; (B) *Chaetospira remex* ca. 280 μm; (C) *Hypotrichidium conicum* ca. 130 μm; (D) *Strongylidium pseudocrassum* ca. 100 μm, © Magnolia; (E) *Uroleptus musculus* ca. 160 μm; (F) *Epiclintes pluvialis* ca. 375 μm; (G) *Holosticha kessleri* ca. 160 μm; (H) *Balladyna parvula* ca. 44 μm; (I) *Cladotricha koltzowii* ca. 200 μm; (J) *Diaxonella pseudorubra* ca. 187 μm; (K) *Pseudourostyla pelotensis* ca. 190 μm, © Magnolia; (L) *Paraholosticha muscicola* ca. 200 μm; (M) *Deviata polycirrata* ca. 180 μm; (N) *Apoamphisiella foissneri* ca. 140 μm, © Magnolia; (O) *Nudiamphisiella interrupta* ca. 120 μm; (P) *Parentocirrus brasiliensis* ca. 100 μm, © Magnolia; (Q) *Pseudouroleptus caudatus* ca. 200 μm; (R) *Gastrostyla steinii* ca. 150 μm, © Magnolia; (S) *Neokeronopsis asiatica* ca. 250 μm; (T) *Onychodromus grandis* ca. 250 μm; (U) *Urostyla grandis* ca. 350 μm. Scale bar = 200 μm. *After: (A, E, and U) Stein (1859) after Kahl (1932); (B) Hudson (1875) after Kahl (1932); (C) Ilowaiski (1921) after Kahl (1932); (D) Paiva & Silva Neto (2007); (F) Smith (1900) after Kahl (1932); (G, I, L, and T) Kahl (1932); (H) Roux (1901) after Kahl (1932); (J) Borror (1972); (K) Paiva & Silva Neto (2006); (M) Küppers & Claps (2010); (N) Paiva & Silva Neto (2004a); (O) Foissner et al. (2002); (P) Paiva & Silva (Neto 2004b); (Q and R) Küppers & Claps (2013); (S) Xu et al. (2016); (D, J, K, and M–S), reproduced with permission from the copyright holder.*

4′ With one more or less continuous frontoventral row (Fig. 2.14 R) ... *Gastrostyla*

5(4) With two frontoventral rows ... 6

5′ With more than three frontoventral rows ... 8

6(5) With two macronuclear nodules ... 7

6′ Usually with four, eight, or more macronuclear nodules (Fig. 2.14 P) ... *Parentocirrus*

7(6) Body outline pisciform with tail-like posterior portion; transverse cirri reduced or absent (Fig. 2.14 Q) ... *Pseudouroleptus*

7′ Body outline different, usually elongate elliptical; transverse cirri present (Fig. 2.14 N) ... *Apoamphisiella*

8(5) With usually three to seven frontoventral rows, one right and one left marginal rows; frontal cirri usually not arranged as an arched corona (Fig. 2.15 A) ... *Paraurostyla*

8′ With more than six ventral rows, multiple (probably) right rows and one left marginal row; anterior arched tricorona of frontal and frontoventral cirri (Fig. 2.15 B) ... *Clapsiella*

9(1) With one marginal row of cirri on the right ... 10

9′ With two or three marginal rows of cirri on the right (Fig. 2.15 E) ... *Pleurotricha*

10(9) Adoral zone of membranelles < 30% of body length ... 11

10′ Adoral zone of membranelles ≥ 30% of body length ... 12

11(10) Frontoventral cirri arranged in a longitudinal row; middle portion of adoral zone of membranelles straight and extending along left margin of cell; paroral with few, widely spaced cilia, extending far beyond anterior end of endoral (*Gonostomum* pattern) (Fig. 2.15 F) ... *Urosoma*

11′ Frontoventral cirri arranged in V; adoral zone of membranelles formed like a question mark; paroral and endoral slightly curved and intersect at about the middle (*Oxytricha* pattern) (Fig. 2.15 G) ... *Urosomoida*

12(10) Undulating membranes not sinusoidal and may intersect optically ... 13

12′ Endoral sinusoidal and fragmented in the middle, does not intersect optically with curved paroral (*Steinia* pattern) (Fig. 2.15 H) *Steinia*

13(12) Undulating membranes conspicuously arched or slightly curved ... 14

13′ Undulating membranes straight or slightly curved, arranged almost in parallel (*Stylonychia* pattern) (Fig. 2.15 R, S) ... *Stylonychia/Tetmemena*

14(13) Undulating membranes conspicuously arched ... 15

14′ Undulating membranes slightly curved and intersect optically at about the middle (*Oxytricha* pattern) ... 16

15(14) Distal portion of paroral semicircularly curved, straight anterior half of endoral extends diagonally, and posterior half is curved, intersecting optically with paroral (*Cyrtohymena* pattern) (Fig. 2.15 I, Q) ... *Cyrtohymena/Rigidohymena*

15′ Arched undulating membranes crossing behind the buccal cirrus, paroral with a hooked distal end (*Notohymena* pattern) (Fig. 2.15 J) ... *Notohymena*

16(14) Caudal cirri absent ... 17

16′ Caudal cirri present ... 18

17(16) Marginal rows clearly separate posteriorly; transverse cirri protrude distinctly beyond posterior end of body; contractile vacuole at about midbody; body very flexible (Fig. 2.15 K) ... *Tachysoma*

17′ Marginal rows confluent posteriorly; body rigid (Fig. 2.15 L) ... *Histriculus*

18(16) Cytoplasm colorless, that is, not distinctly reddish or brownish ... 19

18′ Cytoplasm usually distinctly orange to reddish or brown; only one or two caudal cirri (Fig. 2.15 O) ... *Rubrioxytricha*

19(18) Body flexible ... 20

19′ Body rigid or only slightly flexible ... 21

20(19) Dorsally usually with six rows of bristles consisting in bipolar kineties, fragmented kineties, and dorsomarginals (Fig. 2.15 M) ... *Oxytricha*

20′ Dorsally with four bipolar rows of bristles, dorsomarginal kineties absent ... *Heterooxytricha*

21(19) With two macronuclear nodules (Fig. 2.15 N) ... *Sterkiella*

21′ With three macronuclear nodules (Fig. 2.15 P) ... *Parasterkiella*

Ciliophora: Spirotrichea: Euplota: Families

1 Body parallel-sided; adoral zone of membranelles usually prominent; undulating membranes present ... 2

1′ Small-sized, flattened, and disc-shaped body; reduced adoral zone of membranelles, usually bipartite; undulating membranes absent (Fig. 2.16 A) ... Aspidiscidae, one genus: *Aspidisca*

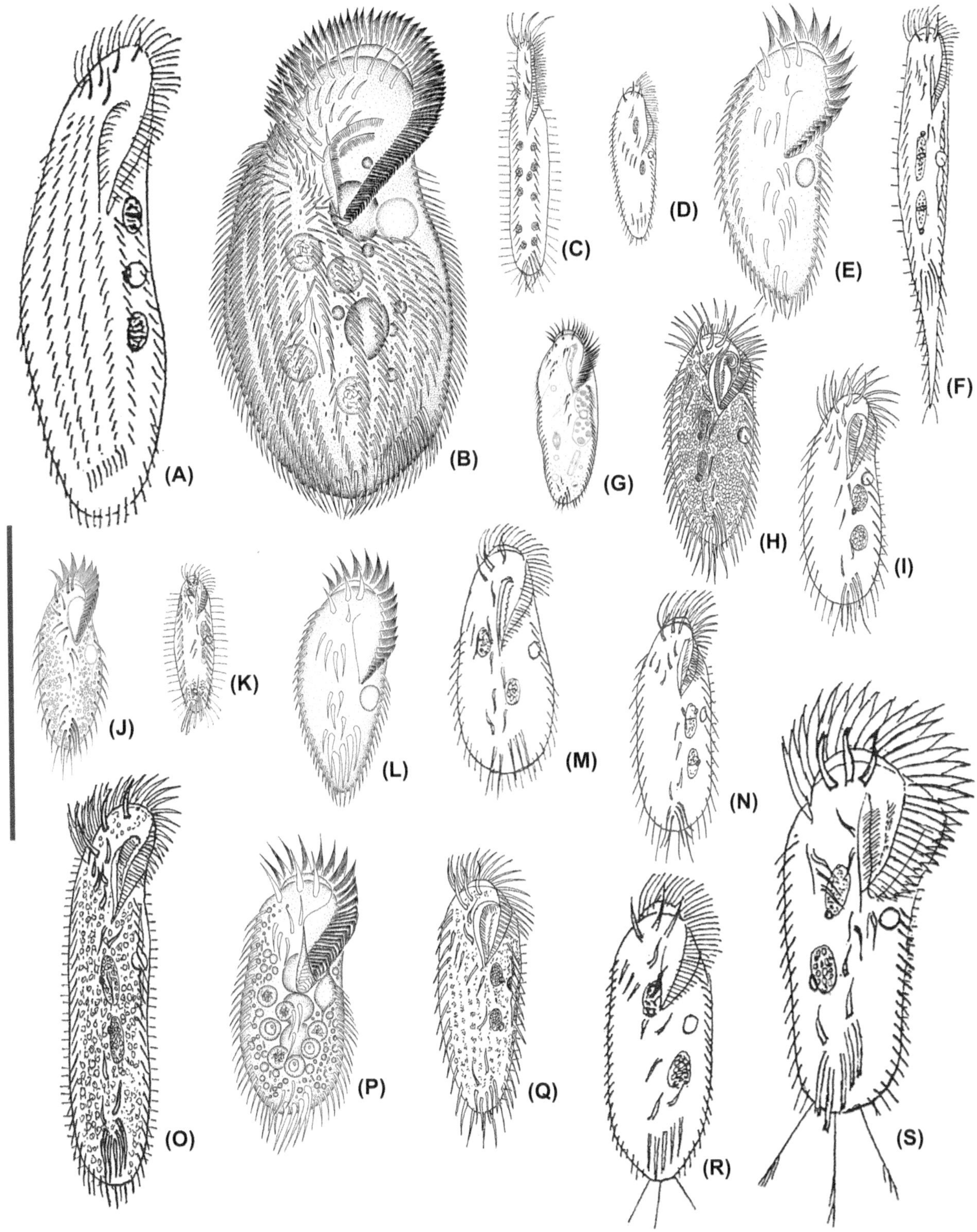

FIGURE 2.15 Representative species of class Spirotrichea recorded in the Neotropics: (A) *Paraurostyla weissei* ca. 300 μm; (B) *Clapsiella magnifica* ca. 300 μm; (C) *Trachelostyla pediculiformis* ca. 180 μm; (D) *Gonostomum affine* ca. 100 μm; (E) *Pleurotricha lanceolata* ca. 200 μm; (F) *Urosoma cienkowskii* ca. 250 μm; (G) *Urosomoida sejongensis* ca. 100 μm, © Magnolia; (H) *Steinia platystoma* 160 μm; (I) *Rigidohymena candens* 150 μm; (J) *Notohymena pampasica* ca. 100 μm; (K) *Tachysoma pellionellum* ca. 80 μm; (L) *Histriculus histrio* ca. 160 μm, © Magnolia; (M) *Oxytricha fallax* ca. 150 μm; (N) *Sterkiella histriomuscorum* ca. 100 μm; (O) *Rubrioxytricha ferruginea* ca. 250 μm; (P) *Parasterkiella thompsoni* ca. 130 μm; (Q) *Cyrtohymena inquieta* ca. 150 μm; (R) *Tetmemena pustulata* ca. 150 μm; (S) *Stylonychia mytilus* ca. 250 μm. Scale bar = 200 μm. *After: (A, D, and M) Stein (1859) after Kahl (1932); (B) Küppers (2014); (C, F, H, I, K, N–O, and Q–S) Kahl (1932); (E) Küppers et al. (2009); (G) Jung et al. (2016); (J) Küppers et al. (2007); (L) Küppers & Claps (2013); (P) Küppers et al. (2011); (B, E, G, J, L, and P) reproduced with permission from the copyright holder.*

2(1) With three hypertrophied right caudal cirri located dorsally; adoral zone continuous or bipartite, with anterior and posterior membranelles .... **Uronychiidae [p. 54]**

2′ Caudal cirri not hypertrophied; adoral zone as a continuous unit; paroral reduced to a patch of cilia at posterior end of buccal cavity, next to a monokinetal endoral .... **Euplotidae [p. 54]**

Ciliophora: Spirotrichea: Euplota: Uronychiidae: Genera

1 Caudal and left marginal cirri hypertrophied; paroral membrane hypertrophied and horseshoe-shaped; adoral zone divided into specialized anterior and posterior sections (Fig. 2.16 E) .... *Uronychia*

1′ Only caudal cirri hypertrophied; adoral zone, paroral, and endoral conspicuous and of usual shape (Fig. 2.16 F) .... *Diophrys*

Ciliophora: Spirotrichea: Euplota: Euplotidae: Genera

1 With less than 10 frontoventral cirri 2

1′ With 10 frontoventral cirri, double or complex argyrome (silver-impregnation needed) (Fig. 2.16 B) .... *Euplotes*

2(1) With 9 frontoventral cirri; double argyrome (Fig. 2.16 D) .... *Euplotoides*

2′ With seven to nine frontoventral cirri; double, triple, or complex argyrome (Fig. 2.16 C) .... *Euplotopsis*

Ciliophora: Oligohymenophorea: Major Groups

1 Body ovoid or elongate, somatic ciliature present 2

1′ Body usually inverted bell-shaped; oral ciliature conspicuous winding in a prominent peristome; somatic ciliature reduced or absent; with aboral scopula that secretes the stalk of sessile species; telotrochal band permanently ciliated on mobile species, temporarily so during dispersal of sessile species; solitary or colonial, some species loricate .... **Peritrichia [p. 55]**

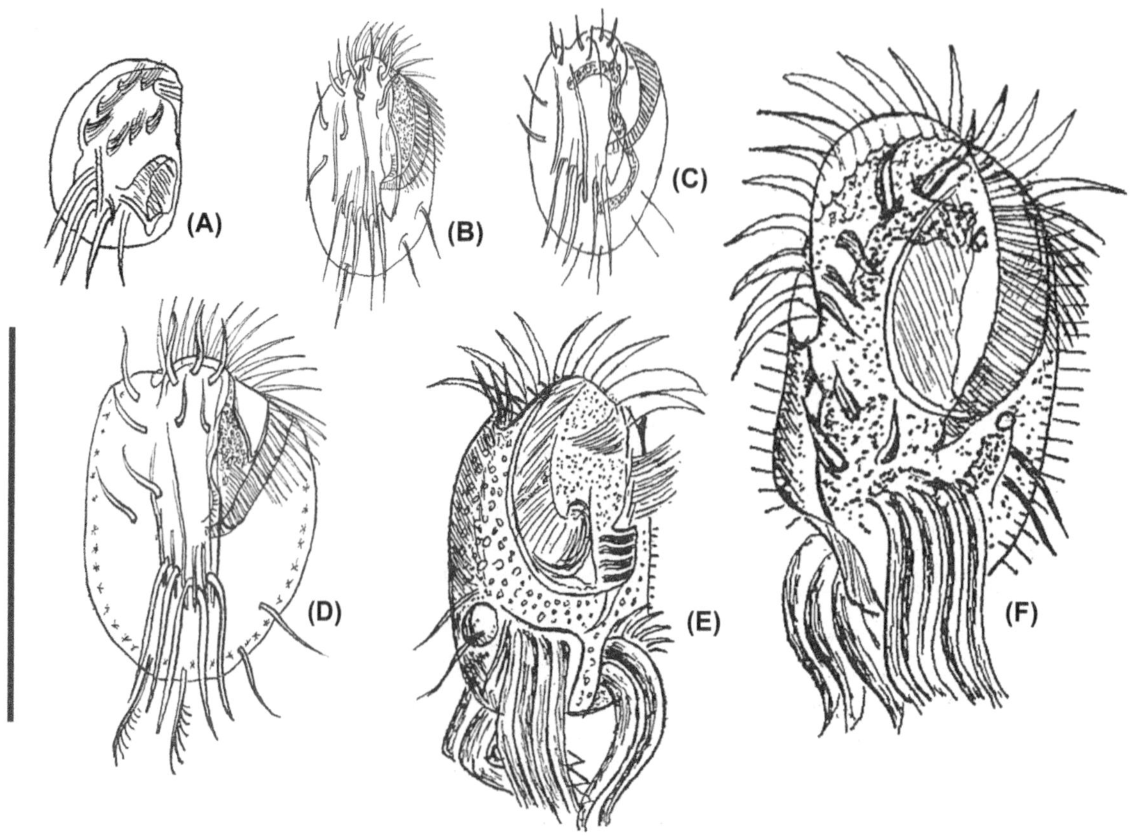

FIGURE 2.16 Representative species of class Spirotrichea recorded in the Neotropics: (A) *Aspidisca turrita* ca. 40 μm; (B) *Euplotes moebiusi* ca. 60 μm; (C) *Euplotopsis affine* ca. 60 μm; (D) *Euplotoides patella* ca. 90 μm; (E) *Uronychia transfuga* ca. 100 μm; (F) *Diophrys scutum* ca. 150 μm. Scale bar = 100 μm. *After Kahl (1932).*

PROTOZOA

2(1) Oral apparatus, typically with a dikinetid paroral and three oral polykinetids 3

2′ Oral polykineties (peniculi) typically elongate and parallel to long axis of oral cavity; somatic ciliature dense with preoral and postoral sutures, with trichocysts .... **Peniculia [p. 57]**

3(2) Somatic ciliature may be sparse, mostly as dikinetids; postoral suture sometimes present; paroral divided into three segments; scutica present .... **Scuticociliatia [p. 59]**

3′ Usually holotrichous, somatic kineties mostly as monokinetids with dikinetids only at anterior tip of body; with preoral suture, but no postoral one; paroral may be unciliated and reduced; scutica absent .... **Hymenostomatia [p. 60]**

Ciliophora: Oligohymenophorea: Peritrichia: Families

1 Trophont motile (without holdfast organelle) or sessile usually with stalk, often attached to other organisms; many gregarious or colonial species 2

1′ Trophont motile; with complex aboral holdfast organelle with hooks and spines (Fig. 2.17 A); ectosymbionts on *Hydra*, fishes, Amphibia .... **Trichodinidae [p. 55]**

2(1) Free-swimming, secondarily stalkless 3

2′ Sessile; attached to substratum directly or by stalk 4

3(2) With permanent ciliated telotrochal band; swims with aboral end forward (Fig. 2.17 D) .... Opisthonectidae, one genus: *Opisthonecta*

3′ Without permanent ciliated telotrochal band; swims with oral end forward using oral ciliature; with one or two rigid caudal bristles or short caudal cilia arising from scopula .... **Astylozoidae [p. 55]**

4(2) Without gelatinous matrix 5

4′ Colonial, with zooids in a gelatinous matrix; green-colored because of algal endosymbionts; peduncle atypical, very thin and translucent (Fig. 2.17 L) .... Ophrydiidae, one genus: *Ophrydium*

5(4) Not loricate 6

5′ With lorica 9

6(5) Stalk contractile; solitary or colonial 7

6′ Stalk not contractile; solitary or colonial 8

7(6) Colonial; continuous myonemes running through the branches of the colony, so that entire colony is contractile; myonemes contract in zigzag in one plane (Fig. 2.17 F) .... Zoothamniidae, one genus: *Zoothamnium*

7′ Colonial or solitary; discontinuous myonemes, so that branches of the colonies contract independently .... **Vorticellidae [p. 57]**

8(6) Peristome with peristomial lip, epistomial disc only slightly projecting .... **Epistylididae [p. 57]**

8′ Peristome without peristomial lip, epistomial disc prominently elevated from the margin of the peristome (Fig. 2.17 N) .... Operculariidae, one genus: *Opercularia*

9(5) Body elongate, always attached to inner posterior surface of lorica .... **Vaginicolidae [p. 57]**

9′ Lorica ovoid, flattened; body rounded, attached laterally to lorica within which the zooid moves freely (Fig. 2.17 R) .... Lagenophryidae, one genus: *Lagenophrys*

Ciliophora: Oligohymenophorea: Peritrichia: Trichodinidae: Genera

1 Incomplete turn of the two adoral ciliary rows in the adoral zone 2

1′ Approximately one complete turn (330°–540°) of the two adoral ciliary rows in the adoral zone, holdfast organelle with denticles with well-developed rays and blades (Fig. 2.17 A) .... *Trichodina*

2(1) The adoral spiral makes a turn of 180°–270°, denticles without centripedal rays that are stunted to form short crooks or platelets .... *Trichodinella*

2′ The adoral spiral makes a turn of 150°–280°, denticles with short rays, centrifugal projections of denticles (blades) attached to the central part almost perpendicularly and denticles inter-locked only by their central conical part .... *Paratrichodina*

Ciliophora: Oligohymenophorea: Peritrichia: Astylozoidae: Genera

1 With long conical pellicular (spines-like) processes (Fig. 2.17 C) .... *Hastatella*

1′ Without these pellicular processes (Fig. 2.17 B) .... *Astylozoon*

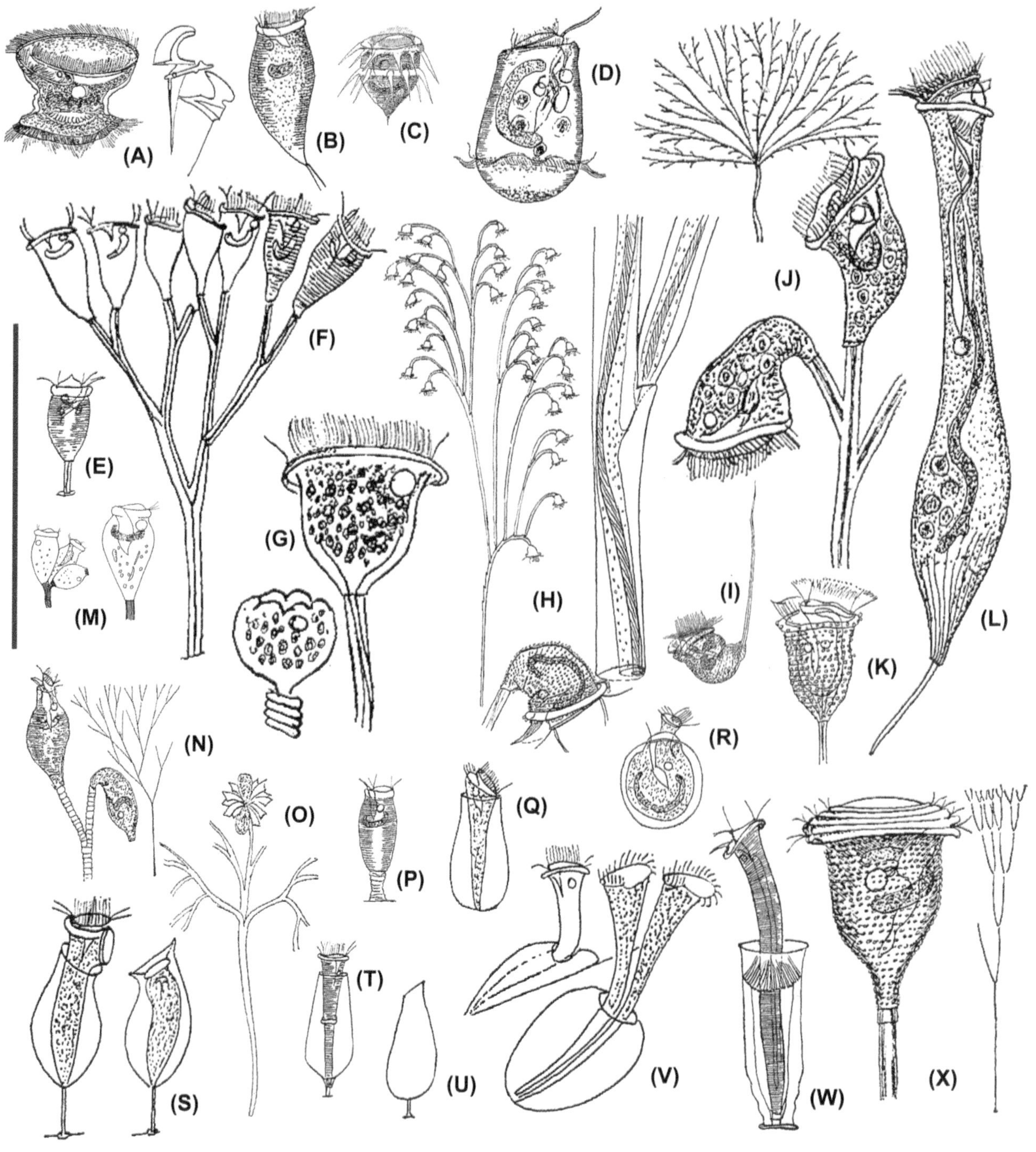

FIGURE 2.17 Representative species of class Oligohymenophorea recorded in the Neotropics: (A) *Trichodina pediculus* ca. 60 μm; (B) *Astylozoon fallax* ca. 100 μm; (C) *Hastatella radians* ca. 40 μm; (D) *Opisthonecta henneguyi* ca. 160 μm; (E) *Intranstylum invaginatum* 36 μm; (F) *Zoothamnium adamsi*, zooid ca. 60 μm; (G) *Vorticella campanula*, zooid ca. 80 μm; (H) *Epicarchesium pectinatum*, zooid ca. 60 μm; (I) *Pelagovorticella mayeri* ca. 45 μm; (J) *Carchesium polypinum*, zooid ca. 100 μm; (K) *Pseudovorticella monilata*, zooid ca. 70 μm); (L) *Ophrydium versatile*, zooid ca. 400 μm; (M) *Epistylis aliciae*, zooid ca. 50 μm; (N) *Opercularia nutans*, zooid ca. 60 μm; (O) *Systylis hoffi*, zooid ca. 200 μm; (P) *Rhabdostyla chironomi* ca. 60 μm; (Q) *Vaginicola ingenita* ca. 60 μm; (R) *Lagenophrys discoidea* ca. 60 μm; (S) *Pyxicola affinis* ca. 80 μm; (T) *Cothurnia annulata* 55 μm; (U) *Cothurniopsis valvata*, lorica 55 μm; (V) *Platycola decumbens*, lorica 90 μm; (W) *Thuricola folliculata* ca. 200 μm; (X) *Campanella umbellaria*, zooid ca. 150 μm. Scale bar = 200 μm. *After: (A–C, E–H, J–L, and N–X) Kahl (1935); (D) Fauré-Fremiet (1906) after Kahl (1935); (I) Fauré-Fremiet (1920) after Kahl (1935); (M) Modenutti & Claps (1986); (M) reproduced with permission from the copyright holder.*

Ciliophora: Oligohymenophorea: Peritrichia: Vorticellidae: Genera

1 Myoneme sinusoidal running in the center of stalk, discontinuous at the bases of branches in colonial forms, usually contracts spirally 2

1′ Myoneme very short restricted to the upper region of each stalk (Fig. 2.17 E) .... *Intranstylum*

2(1) Solitary 3

2′ Colonial 5

3(2) Stalk attached to various substrates 4

3′ Swims freely with stalk directed forward; planktonic (Fig. 2.17 I) .... *Pelagovorticella*

4(3) Pellicular pattern as parallel lines (pellicular striations) encircling body (Fig. 2.17 G) .... *Vorticella*

4′ Pellicular pattern as a brick-work mesh encircling body (Fig. 2.17 K) .... *Pseudovorticella*

5(2) Pellicle with tubercles (Fig. 2.17 H) .... *Epicarchesium*

5′ Pellicle without tubercles (Fig. 2.17 J) .... *Carchesium*

Ciliophora: Oligohymenophorea: Peritrichia: Epistylididae: Genera

1 Colonial; stalk long 2

1′ Solitary; stalk short (Fig. 2.17 P) .... *Rhabdostyla*

2(1) Branches of the stalk terminate in single zooid 3

2′ Branches of the stalk terminate in 40–160 zooids (Fig. 2.17 O) .... *Systylis*

3(2) Adoral membranelles wind 1–1½ times around peristome (Fig. 2.17 M) .... *Epistylis*

3′ Adoral membranelles wind four to six times around peristome (Fig. 2.17 X) .... *Campanella*

Ciliophora: Oligohymenophorea: Peritrichia: Vaginicolidae: Genera

1 Lorica erect and attached to substratum at posterior end 2

1′ Lorica lying on side and attached by side to substratum (Fig. 2.17 V) .... *Platycola*

2(1) Lorica remains open when zooid contracts 3

2′ Lorica closes by means of valves or operculum when zooid contracts 4

3(2) Without stalk (Fig. 2.17 Q) .... *Vaginicola*

3′ With stalk (Fig. 2.17 T) .... *Cothurnia*

4(2) Valve formed from part of lorica 5

4′ Valve or operculum attached to zooid (Fig. 2.17 S) .... *Pyxicola*

5(4) Lorica may be closed by its flexible rim when zooid contracts (Fig. 2.17 U) .... *Cothurniopsis*

5′ Lorica may be closed by an internal valve arrangement (Fig. 2.17 W) .... *Thuricola*

Ciliophora: Oligohymenophorea: Peniculia: Families

1 Body not barrel-shaped, with longitudinal somatic kineties 2

1′ Body barrel-shaped with wider, rounded anterior half, encircled by two distinct transverse ciliary girdles; ciliary tuft at posterior end; oral cavity equatorial, without preoral depression (Fig. 2.18 A) .... Urocentridae, one genus: *Urocentrum*

2(1) Oral cavity huge 3

2′ Oral cavity not huge 4

3(2) Body small to medium-sized, broadly ovoid; oral region extending over most of ventral surface, with conspicuous long polykinety on the left and paroral membrane on the right; with tuft of long caudal cilia (Fig. 2.18 B) .... Lembadionidae, one genus: *Lembadion*

3′ Body very large (ca. 200–700 μm), elongate ovoid, broader at anterior end; V-shaped peristome resembles an adoral zone of membranelles and extends up to equatorial area; without tuft of long caudal cilia (Fig. 2.18 F) .... Neobursaridiidae, one genus: *Neobursaridium*

4(2) Body shape variable, not cone- or heart-shaped 5

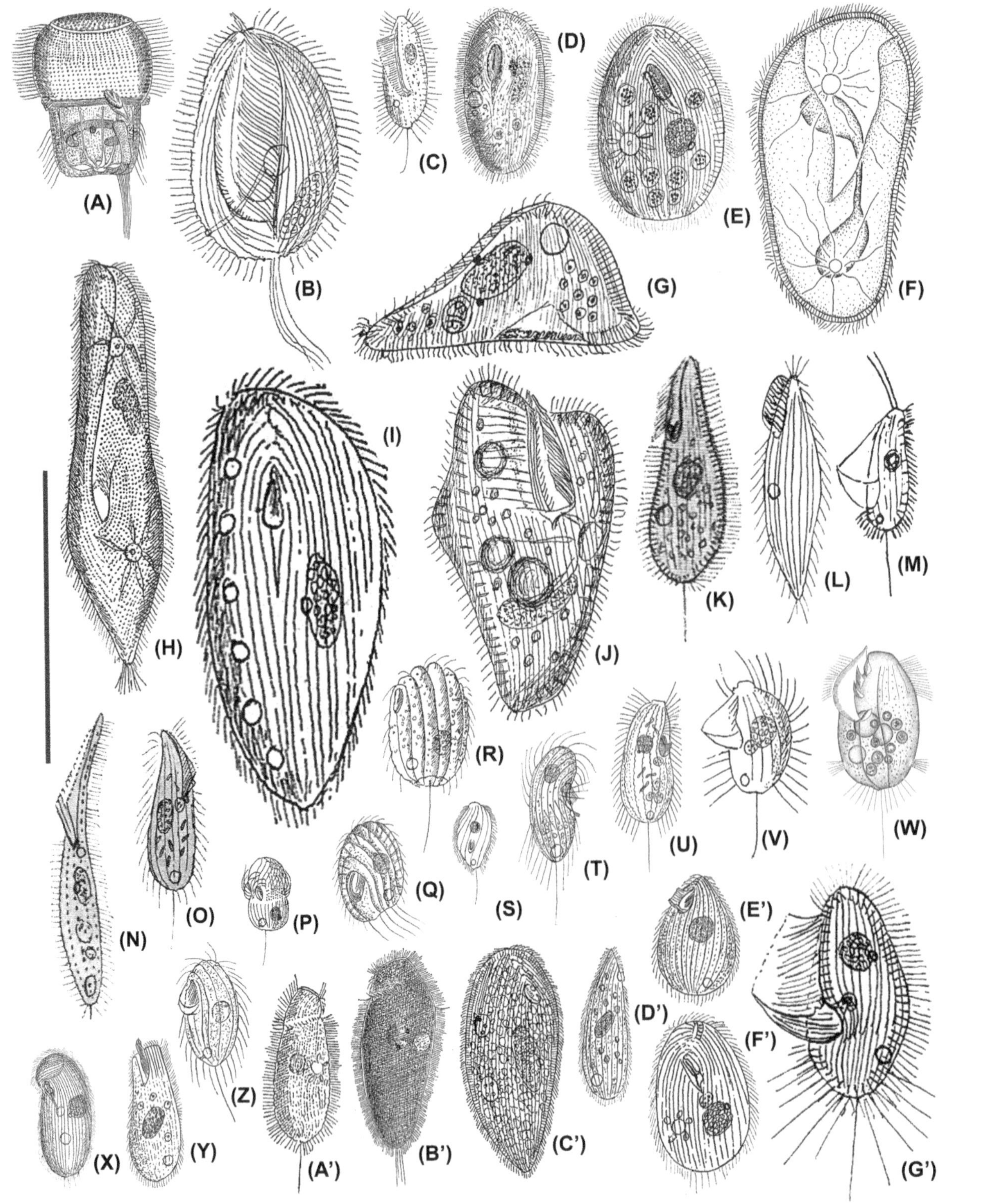

FIGURE 2.18 Representative species of class Oligohymenophorea recorded in the Neotropics: (A) *Urocentrum turbo* ca. 40 μm; (B) *Lembadion lucens* ca. 80 μm; (C) *Ctedoctema acanthocryptum* 35 μm; (D) *Clathrostoma viminale* ca. 50 μm; (E) *Epenardia myriophylli* ca. 120 μm; (F) *Neobursaridium gigas* ca. 200 μm; (G) *Stokesia vernalis* ca. 130 μm; (H) *Paramecium caudatum* ca. 250 μm; (I) *Frontonia vesiculosa* ca. 300 μm; (J) *Disematostoma tetraedricum* ca. 110 μm; (K) *Philasterides armatus* ca. 75 μm; (L) *Kahlilembus attenuatus* 75 μm; (M) *Calyptotricha lanuginosum* 35 μm; (N) *Cohnilembus verminus* 100 μm; (O) *Pseudocohnilembus pusillus* ca. 60 μm; (P) *Urozona buetschlii* 25 μm; (Q) *Cinetochilum margaritaceum* 35 μm; (R) *Platynematum sociale* ca. 40 μm; (S) *Uropedalium pyriformis* ca. 25 μm; (T) *Dexiotrichides centralis* 35 μm; (U) *Uronema nigricans* ca. 40 μm; (V) *Cyclidium glaucoma* 30 μm; (W) *Cristigera hammeri* ca. 40 μm; (X) *Colpidium colpoda* ca. 40 μm; (Y) *Espejoia culex* ca. 50 μm; (Z) *Sathrophilus muscorum* 35 μm; (A′) *Dexiotricha granulosa* ca. 60 μm; (B′) *Loxocephalus luridus* ca. 160 μm; (C′) *Ophryoglena flava* ca. 330 μm; (D′) *Stegochilum acutum* ca. 50 μm; (E′) *Tetrahymena pyriformis* 45 μm; (F′) *Glaucoma scintillans* ca. 60 μm; (G′) *Pleuronema coronatum* ca. 80 μm. Scale bar (A–D and I–G′) = 100 μm, (E–H) = 200 μm, (C′) = 400. *After: (A–V and X–G′) Kahl (1931); (W) Castro et al. (2014).*

4′ Body distinctly conical; oral region on somewhat flattened ventral body surface (Fig. 2.18 G) ... Stokesiidae, one genus: *Stakesia*

5(4) Body ovoid elongate or pyriform; prebucal depression shallow or absent ... 6

5′ Body cigar- or slipper-shaped; conspicuous preoral groove (Fig. 2.18 H) ... Parameciidae, one genus: *Paramecium*

6(5) With postoral suture only; without ring of nematodesmata around the cytopharynx ... **Frontoniidae [p. 59]**

6′ With preoral and postoral suture; nematodesmata form a ring or "basket" around the cytopharynx (Fig. 2.18 D) ... Clathrostomatidae, one genus: *Clathrostoma*

Ciliophora: Oligohymenophorea: Peniculia: Frontoniidae: Genera

1 Body ovoid elongate; postoral suture terminating at posterior end of body (Fig. 2.18 I) ... *Frontonia*

1′ Body pyriform, broadest at anterior portion; postoral suture extending to midbody in dorsal side; oral opening "D"-shaped (Fig. 2.18 J) ... *Disematostoma*

Ciliophora: Oligohymenophorea: Scuticociliatia: Families

1 Conspicuous oral membranes visible from outside the oral cavity ... 2

1′ Oral membranes not prominently visible from outside ... 8

2(1) Somatic ciliature holotrichous, often more sparse posteriorly, one or more caudal cilia; oral region often expansive, with paroral prominent with long cilia, forming a velum as the organism feeds ... 3

2′ Paroral and polykineties extend in parallel toward anterior end of body ... 6

3(2) Without lorica ... 4

3′ Dwells in lorica that is open at both ends; resembles *Pleuronema* or *Cyclidium* (Fig. 2.18 M) Calyptotrichidae, one genus: *Calyptotricha*

4(3) Small to large (30–200 μm long); apical pole ciliated ... 5

4′ Small (15–60 μm long); apical pole unciliated ... **Cyclidiidae [p. 59]**

5(4) Small (ca. 20–30 μm) elongate to oval ciliate; paroral as an open "C," not extending to the left of the cytostome; pellicle notched by short extrusomes, appearing serrated (Fig. 2.18 C) ... Ctedoctematidae, one genus: *Ctedoctema*

5′ Medium to large (80–200 μm) ovoid ciliate; prominent sail-like paroral extends from anterior to well past equator of cell, curling around the subequatorial cytostome; one to many stiff, long caudal cilia; pellicle not serrated (Fig. 2.18 G′) ... Pleuronematidae, one genus: *Pleuronema*

6(2) Body shape oval, pyriform, or fusiform; oral apparatus may extend from apical pole to beyond equatorial area ... 7

6′ Body ovoid to pyriform; oral apparatus in anterior one-fourth of body ... **Philasteridae [p. 60]**

7(6) Small (25–45 μm), slender oval to pyriform body; peristome extends from near the apical pole to half to two-thirds the body length (Fig. 2.18 O) ... Pseudocohnilembidae, one genus: *Pseudocohnilembus*

7′ Small to large (30–150 μm long) elongate ciliate, spindle-shaped or fusiform with anterior end narrowed and tapering; peristomial area extends from the apical pole down toward more or less the middle of body (Fig. 2.18 N) ... Cohnilembidae, one genus: *Cohnilembus*

8(1) Somatic cilia not in girdles, usually in longitudinal rows ... 9

8′ Barrel-shaped body encircled by one girdle of somatic cilia (Fig. 2.18 P) ... Urozonidae, one genus: *Urozona*

9(8) Small to large ciliates; body ovoid elongate or reniform; pellicle not ribbed ... 10

9′ Small to very small ciliates (<60 μm); body ellipsoid, may be flattened; pellicle ribbed; somatic kineties horseshoe-shaped around oral aperture ... **Cinetochilidae [p. 60]**

10(9) Body ovoid to elongate ovoid; anterior end slightly flattened and conspicuously naked ... **Uronematidae [p. 60]**

10′ Body ovoid or reniform ... **Loxocephalidae [p. 60]**

Ciliophora: Oligohymenophorea: Scuticociliatia: Cyclidiidae: Genera

1 Body rounded in cross section; macronucleus at anterior end ... 2

1′ Body flattened, with a ventral groove; macronucleus at midbody (Fig. 2.18 W) ... *Cristigera*

2(1) Oral cilia pattern with membranelles M2 and M3 barely separated and forming a ciliary field composed of almost equidistant, transversely oriented rectangles ... *Protocyclidium*

2′ Membranelles usually nonseparated; paroral membrane "L"-shaped, with the anterior end terminating at or before anterior end of membranelle M1 (Fig. 2.18 V) ... *Cyclidium*

Ciliophora: Oligohymenophorea: Scuticociliatia: Philasteridae: Genera

1 Body long ovoid to pyriform; paroral divided in two, with posterior part bow-shaped (Fig. 2.18 K) .... *Philasterides*

1′ Body fusiform, sometimes more or less S-shaped; paroral continuous, with a flag-shaped appearance in vivo (Fig. 2.18 L) *Kahlilembus*

Ciliophora: Oligohymenophorea: Scuticociliatia: Cinetochilidae: Genera

1 Body flattened 2

1′ Body ellipsoid, circular in cross section; oral opening at right anterior third of body; single caudal cilium (Fig. 2.18 R) . *Platynematum*

2 Body discoid or lenticular, apical pole rounded and terminal end slightly truncate; oral opening displaced to the lower right quadrant of the ventral surface; several caudal cilia (Fig. 2.18 Q) .... *Cinetochilum*

2′ Body ellipsoid; oral region in anterior half of body, with bow-shaped paroral; single caudal cilium (Fig. 2.18 Z) .... *Sathrophilus*

Ciliophora: Oligohymenophorea: Scuticociliatia: Uronematidae: Genera

1 Cytostome in posterior half of body, paroral bow-shaped; body ciliation complete (Fig. 2.18 S) .... *Uropedalium*

1′ Cytostome in middle half of the body, with small paroral and three small internal membranelles; anterior pole unciliated (Fig. 2.18 U) .... *Uronema*

Ciliophora: Oligohymenophorea: Scuticociliatia: Loxocephalidae: Genera

1 Body long-ovoid 2

1′ Body reniform (Fig. 2.18 T) .... *Dexiotrichides*

2(1) Length $\leq$100 µm; apical pole unciliated (Fig. 2.18 A′) .... *Dexiotricha*

2′ Length $>$ 100 µm; apical pole ciliated (Fig. 2.18 B′) .... *Loxocephalus*

Ciliophora: Oligohymenophorea: Hymenostomatia: Families

1 Body ovoid, sometimes tapering posteriorly; somatic kineties may curve 2

1′ Body reniform; right ventral somatic kineties curving left, (abruptly) twisting anterior of the oral region to run parallel to anterior suture .... **Turaniellidae [p. 60]**

2(1) With single or no caudal cilia 3

2′ With tuft of short caudal cilia (Fig. 2.18 D′) .... Spirozonidae, one genus: *Stegochilum*

3(2) Preoral suture parallel to major body axis 4

3′ Preoral suture oblique to major body axis and angled to the left .... **Glaucomidae [p. 60]**

4(3) Oral opening ear-shaped (Fig. 2.18 C′) .... Ophryoglenidae, one genus: *Ophryoglena*

4′ Oral opening pyriform; membranellar system inconspicuous and difficult to see (Fig. 2.18 E′) .... Tetrahymenidae, one genus: *Tetrahymena*

Ciliophora: Oligohymenophorea: Hymenostomatia: Turaniellidae: Genera

1 Preoral suture sigmoid, at anterior pole (Fig. 2.18 X) .... *Colpidium*

1′ Preoral suture straight, slightly displaced toward the ventral side .... *Dexiostoma*

Ciliophora: Oligohymenophorea: Hymenostomatia: Glaucomidae: Genera

1 Preoral suture parallel to major body axis 2

1′ Preoral suture oblique to major body axis and angled to the left (Fig. 2.18 F′) .... *Glaucoma*

2(1) Oral opening oval or pyriform, lying parallel to major body axis; membranelle M3 split into several pieces (Fig. 2.18 Y) .... *Espejoia*

2′ Oral opening 9-shaped; membranelle M3 not split into pieces (Fig. 2.18 E) .... *Epenardia*

Ciliophora: Armophorea: Families

1 Body round or conical, rigid, not prominently twisted to left, posterior spines may be present; somatic cilia as small kineties or cirruslike tufts (Fig. 2.19 D) .... Caenomorphidae, one genus: *Caenomorpha*

1′ Body contorted with anterior part twisted to left, and posterior part sometimes tailed and/or bearing a tuft of longer caudal cilia; somatic cilia uniform .... **Metopidae [p. 61]**

Ciliophora: Armophorea: Metopidae: Genera

1 Body without armor 2

1′ Body with transparent armor, with longitudinally arranged cristate ribs (Fig. 2.19 C) .... *Tropidoatractus*

2(1) Adoral zone of membranelles long and narrow, spiraling S-like from anterior to almost posterior end of body (Fig. 2.19 B) *Brachonella*

2′ Adoral zone of membranelles limited to anterior body region and runs obliquely, not spirally, to equatorial or subequatorial cytostome (Fig. 2.19 A) .... *Metopus*

Ciliophora: Heterotrichea: Families

1 Body not trumpet-shaped 2

1′ Body trumpet-shaped, highly contractile; oral ciliature spirals clockwise around flared anterior end; often pigmented and/or with algal endosymbionts (Fig. 2.20 A) .... Stentoridae, one genus: *Stentor*

2(1); Without lorica 3

2′ Lorica with ampulla and neck; individual with conspicuous pair of "peristomial wings" bearing the oral ciliature, fixed to lorica by a holdfast organelle .... **Folliculinidae [p. 61]**

3(2) Body wormlike, sometimes very large and contractile or with other form 4

3′ Body laterally compressed, pyriform to ellipsoid; noncontractile; often pigmented pink, red, or purple due to cortical granules between somatic kineties; long, narrow peristome on left margin .... **Blepharismidae [p. 61]**

4(3) Oral cavity deep and occupying much of anterior part of body, with prominent buccal ciliature 5

4′ Oral cavity shallow; peristomal area long and narrow, oral ciliature sometimes inconspicuous (Fig. 2.20 E) .... Spirostomidae, one genus: *Spirostomum*

5(4) Paroral membrane inconspicuous; sometimes hosting endosymbiotic algae .... **Climacostomidae [p. 61]**

5′ Paroral membrane prominent .... **Condylostomatidae [p. 62]**

Ciliophora: Heterotrichea: Folliculinidae: Genera

1 Lorica with a short neck; single macronuclear nodule 2

1′ Lorica with double-walled ampulla and long neck; macronucleus beaded (Fig. 2.20 B) .... *Botticula*

2(1) With stalklike holdfast organelle (Fig. 2.20 C) .... *Folliculina*

2′ With broad holdfast organelle (Fig. 2.20 D) .... *Ascobius*

Ciliophora: Heterotrichea: Blepharismidae: Genera

1 Peristome extends about half of body length (Fig. 2.20 F) .... *Blepharisma*

1′ Peristome extends about one-fifth of body length (Fig. 2.20 G) .... *Pseudoblepharisma*

Ciliophora: Heterotrichea: Climacostomidae: Genera

1 Anterior region of body conspicuously pointed (Fig. 2.20 J) .... *Fabrea*

1′ Anterior region of body truncated on the left (Fig. 2.20 I) .... *Climacostomum*

PROTOZOA

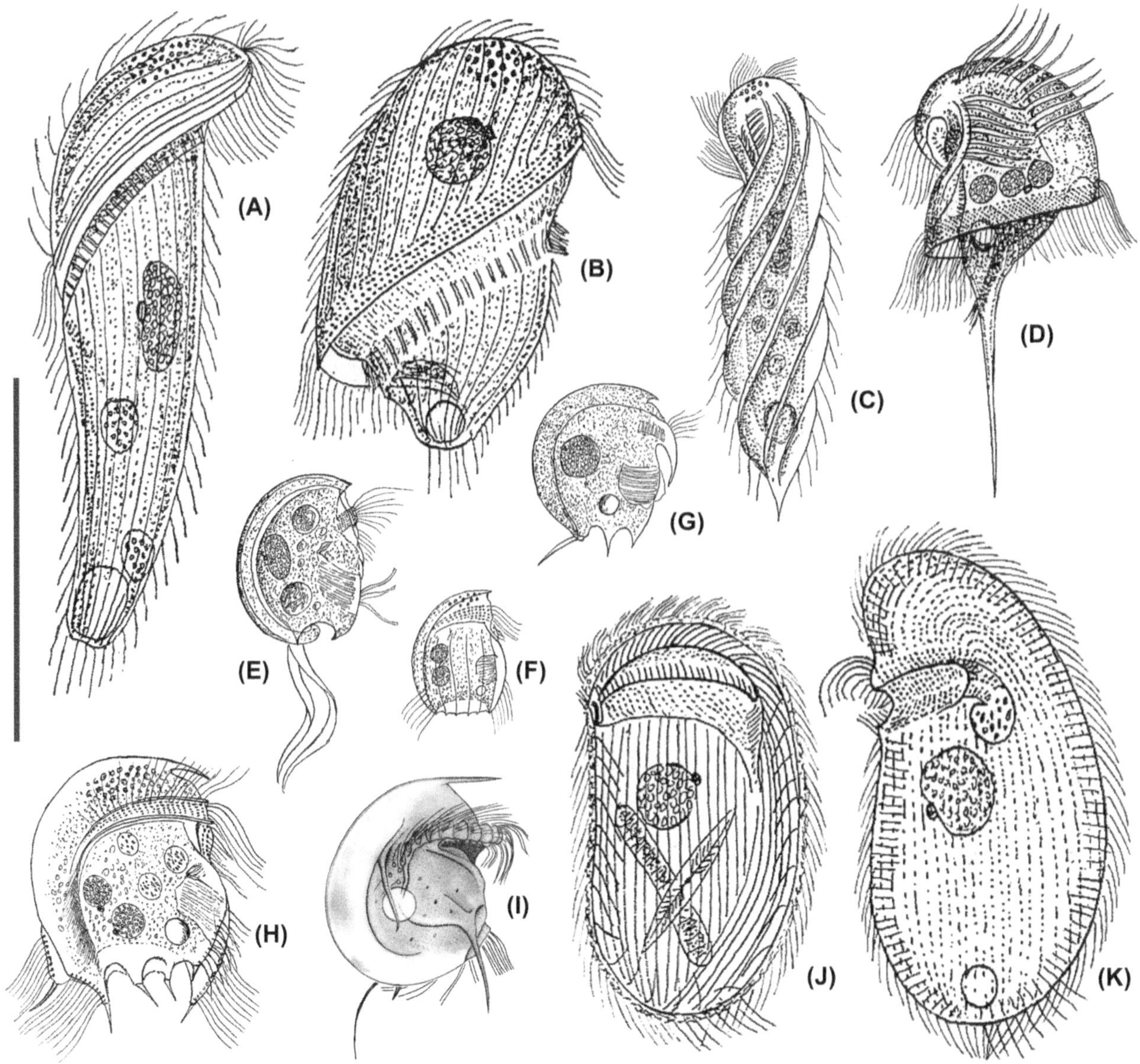

FIGURE 2.19 Representative species of classes Armophorea (A–D) and Plagiopylea (E–K) recorded in the Neotropics: (A) *Metopus es* ca. 150 μm; (B) *Brachonella spiralis* ca. 120 μm; (C) *Tropidoatractus acuminatus* ca. 120 μm; (D) *Caenomorpha medusula* ca. 120 μm; (E) *Mylestoma bipartitum* ca. 40 μm; (F) *Epalxella mirabilis* ca. 35 μm; (G) *Atopodinium fibulatum* ca. 45 μm; (H) *Saprodinium dentatum* ca. 70 μm; (I) *Discomorphella pedroeneasi* ca. 50 μm; (J) *Sonderia vorax* ca. 120 μm; (K) *Plagiopyla nasuta* ca. 140 μm. Scale bar = 100 μm. *After: (A–H) Kahl (1932); (I) Paiva et al. (2017); (J and K) Kahl (1931); (I) reproduced with permission from the copyright holder.*

Ciliophora: Heterotrichea: Condylostomatidae: Genera

1 With frontal cirri; usually in brackish or marine environments (Fig. 2.20 K) .... *Condylostoma*

1′ Without frontal cirri; freshwater (Fig. 2.20 H) .... *Linostomella*

Ciliophora: Prostomatea: Families

1 Somatic cilia uniformly distributed over body 2

1′ Somatic cilia, other than caudals, absent from posterior region 6

2(1) Cortex not armored, without calcium carbonate skeletal plates 3

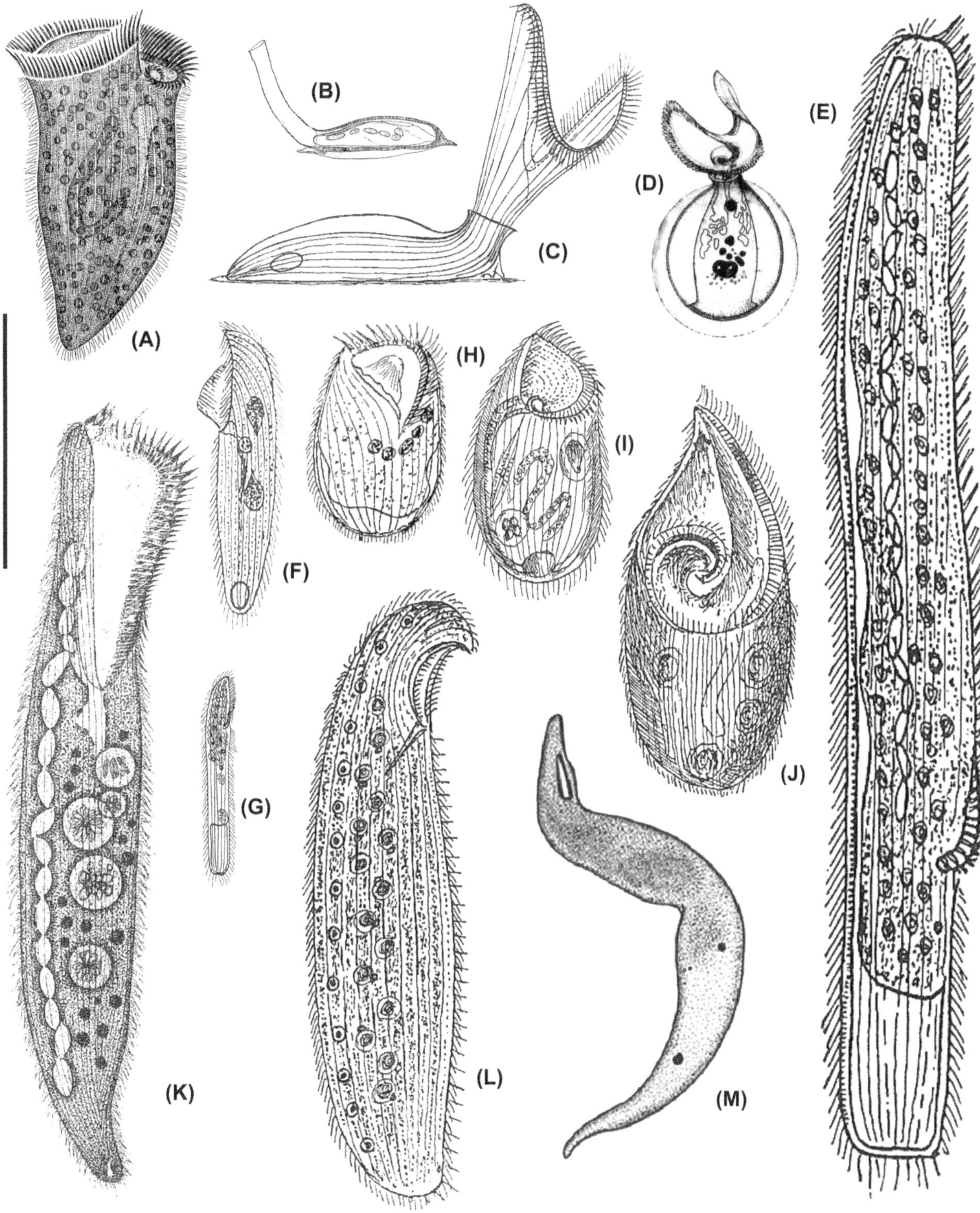

FIGURE 2.20 Representative species of classes Heterotrichea (A–K) and Karyorelictea (L–N) recorded in the Neotropics: (A) *Stentor araucanus* ca. 230 μm; (B) *Botticula ringueleti*, neck of lorica ca. 130 μm; (C) *Folliculina boltoni*, lorica base length ca. 210 μm; (D) *Ascobius lentus*, lorica width ca. 200 μm; (E) *Spirostomum ambiguum* ca. 1.2 mm; (F) *Blepharisma undulans* ca. 220 μm; (G) *Pseudoblepharisma tenue* ca. 150 μm; (H) *Linostomella vorticella* ca. 150 μm; (I) *Climacostomum virens* ca. 200 μm; (J) *Fabrea salina* ca. 300 μm; (K) *Condylostoma spathiosum* ca. 600 μm; (L) *Loxodes magnus* 500 μm; (M) *Corlissina maricaensis* ca. 300 μm. Scale bar = 200 μm. *After: (A) Foissner & Wölfl (1994); (B) Dioni (1972); (C) Thomsen (1922); (D) Mulisch et al. (1998); (E–G and I) Kahl (1932); (H) Penard (1922) after Kahl (1932); (J) Henneguy (1890) after Kahl (1932); (K) Shao et al. (2006); (L) Kahl (1931); (M) Campello-Nunes et al. (2015); (A–D, K, and M) reproduced with permission from the copyright holder.*

2′ Cortex with translucent calcium carbonate skeletal plates (armor); typically barrel-shaped; body frequently spiny, often with prominent anterior and caudal thorns; long caudal cilium (Fig. 2.21 J) .... Colepidae, one genus: *Coleps*

3(2) Oral opening apical, without fingerlike palps or lappets with toxicysts 4

3′ Oral opening apical, slitlike, with a thickened ridge along right edge bearing fingerlike palps or lappets with toxicysts (Fig. 2.21 K) .... Plagiocampidae, one genus: *Plagiocampa*

4(3) Cytopharyngeal basket conspicuous 5

4′ Cytopharyngeal basket delicate .... **Holophryidae [p. 64]**

5(4) Body oval, slightly flattened; somatic ciliature as spiraled rows with striae between them (Fig. 2.21 Q) ..... Placidae, one genus: *Placus*

5′ Body ovoid to cylindrical, sometimes narrowed posteriorly, oval in cross section; somatic ciliature as longitudinal rows, without striae between them (Fig. 2.21 P) .... Prorodontidae, one genus: *Prorodon*

6(1) Without lorica 7

6′ With pseudochitinous lorica and single caudal cilium (Fig. 2.21 L) .... Metacystidae, one genus: *Metacystis*

7(6) Truncate, flattened anterior pole; oral opening as wide as body, with very long oral flaps or palps bearing toxicysts (Fig. 2.21 M) .... Balanionidae, one genus: *Balanion*

7′ Body ovoid, not truncated anteriorly; small oral opening, surrounded by short oral flaps (Fig. 2.21 N) Urotrichidae, one genus: *Urotricha*

PROTOZOA

Ciliophora: Prostomatea: Holophryidae: Genera

1 Anterior end of body truncated obliquely; large circular or oval oral opening leading to a wide shallow depression (Fig. 2.21 I) .... *Bursellopsis*

1′ Anterior end of body not truncated obliquely; circular oral opening as a simple, apical invagination (Fig. 2.21 O) .... *Holophrya*

Ciliophora: Nassophorea: Major Groups

1 Size over 30 μm; body not encased in a rigid pellicle; somatic ciliature usually dense with closely spaced kineties; cyrtos conspicuous; with a synhymenium or hypostomial frange (special ciliary row obliquely or perpendicularly interrupting somatic kineties) 2

1′ Small ciliates ($\leq$ 30 μm); body compressed laterally, always encased in rigid pellicle; somatic ciliature relatively sparse with few, widely separated kineties; cyrtos small; oral ciliature usually as three left polykineties .... **Microthoracida [p. 64]**

2(1) Synhymenium composed of dikinetids or small polykinetids extending from right postoral region to left dorsal body surface, almost encircling the body in some forms .... **Synhymeniida [p. 66]**

2′ Synhymenium beginning in postoral region extending to lateral left onto dorsal surface, sometimes reduced to three to four polykineties restricted to a shallow oral cavity .... **Nassulida [p. 66]**

Ciliophora: Nassophorea: Microthoracida: Families

1 Body shape ellipsoidal to semidiscoidal, oral region near anterior or posterior end of body .... **Microthoracidae [p. 64]**

1′ Body shape oval; oral region in anterior third or in second quarter, on left side of body .... **Leptopharyngidae [p. 64]**

Ciliophora: Nassophorea: Microthoracida: Microthoracidae: Genera

1 Body shape oval to semidiscoidal; oral opening equatorial, as a concavity on left (ventral) side of body (Fig. 2.12 V) .. *Drepanomonas*

1′ Body shape oval; oral opening in posterior region of body (Fig. 2.12 U) .... *Microthorax*

Ciliophora: Nassophorea: Microthoracida: Leptopharyngidae: Genera

1 Left (ventral) edge of body protrudes in buccal area; somatic ciliature reduced, with about four kineties on each side of body (Fig. 2.12 W) .... *Leptopharynx*

1′ Left (ventral) edge of body almost straight; somatic ciliature as more than 10 uninterrupted bipolar kineties, reduced on left surface (Fig. 2.12 X) .... *Pseudomicrothorax*

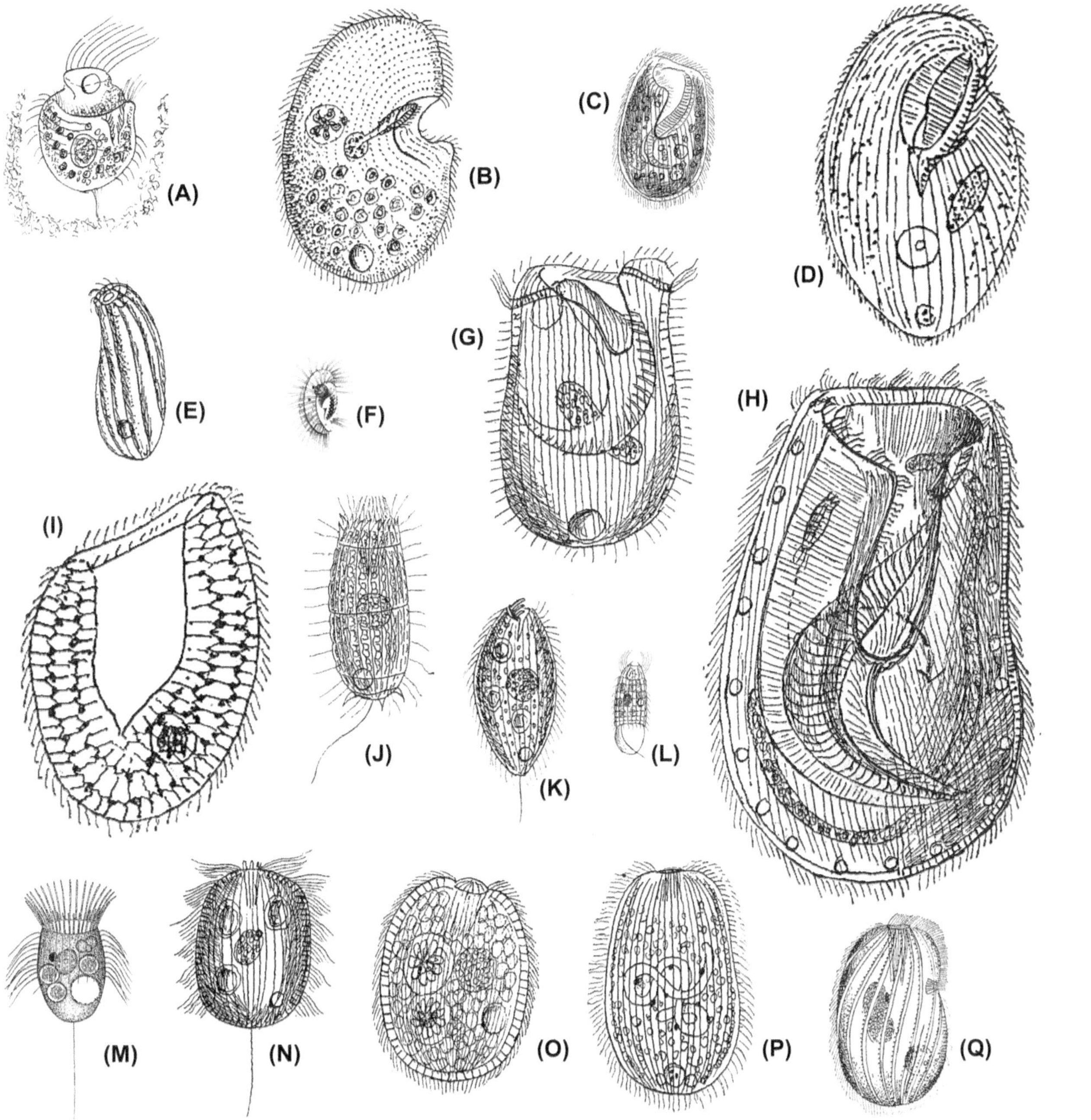

FIGURE 2.21 Representative species of classes Colpodea (A–H) and Prostomatea (I–Q) recorded in the Neotropics: (A) *Mycterothrix erlangeri* ca. 50 μm; (B) *Colpoda cucullus* ca. 80 μm; (C) *Thylakidium truncatum* ca. 70 μm; (D) *Bryometopus sphagni* ca. 100 μm; (E) *Platyophrya vorax* ca. 52 μm; (F) *Kreyella minuta* ca. 25 μm; (G) *Bursaridium pseudobursaria* ca. 135 μm; (H) *Bursaria truncatella* ca. 600 μm; (I) *Bursellopsis spumosa* ca. 400 μm; (J) *Coleps hirtus* ca. 60 μm; (K) *Plagyocampa rouxi* ca. 60 μm; (L) *Metacystis tesselata* ca. 50 μm; (M) *Balanion planctonicum* ca. 20 μm; (N) *Urotricha armata* ca. 50 μm; (O) *Holophrya vesiculosa* ca. 130 μm; (P) *Prorodon ellipticus* ca. 120 μm; (Q) *Placus luciae* ca. 50 μm. Scale bar (A–G and J–Q) = 100 μm; 200 μm; (H and I) = 400 μm. *After: (A–E, and G–L) Kahl (1931); (F) Foissner (1979); (M) Foissner et al. (1990); (N–P) Kahl (1930); (F and M) reproduced with permission from the copyright holder.*

Ciliophora: Nassophorea: Synhymeniida: Families and Genera

1 Body may be flattened, anterior often bent to left; synhymenium extending obliquely on ventral surface, sometimes bent to dorsal surface 2

1′ Body round in cross section, anterior end broadly rounded, never bent toward left; synhymenium extending almost completely around body circumference (Fig. 2.12 B′) .... Nassulopsidae, one genus: *Nassulopsis*

2(1) Body roughly ovoid, not differentiated into a rostrum anteriorly; synhymenium thickly ciliated (Fig. 2.12 C′) .... Orthodonellidae, one genus: *Zosterodasys*

2′ Body slightly dorsoventrally flattened, with broader anterior end, slightly bent to left; sometimes gently tapered posteriorly; synhymenium sparsely ciliated (Fig. 2.12 Y) .... Scaphidiodontidae, one genus: *Chilodontopsis*

Ciliophora: Nassophorea: Nassulida: Families

1 With one to three left oral polykineties and one short, right paroral, not confined to an oral cavity (Fig. 2.12 Z) .... Furgasoniidae, one genus: *Furgasonia*

1′ With many polykineties transversely arranged on the ventral surface (Fig. 2.12 A′) .... Nassulidae, one genus: *Nassula*

Ciliophora: Karyorelictea: Families and Genera

1 Body laterally flattened, noncontractile, with beaklike anterior region (rostrum); ciliated only on right surface of body; oral area in ventral groove behind rostrum bordered by two perioral kineties, plus one intrabuccal kinety (Fig. 2.20 L) ...... Loxodidae, one genus: *Loxodes*

1′ Body elongate with both ends pointed, of large size and highly contractile; somatic ciliature uniform; oral region with more elaborated ciliature, including long adoral membranelles (Fig. 2.20 M) .... Geleiidae, one genus: *Corlissina*

Ciliophora: Litostomatea: Families

1 With a pronounced anterior narrow neck region or proboscis-like elongation 2

1′ Without a pronounced anterior narrow neck region or proboscis-like elongation 7

2(1) Oral opening at apical end of body or is a slit along anterior region of the neck 3

2′ Oral opening at base of neck region .... **Tracheliidae [p. 67]**

3(2) Oral opening at apical end or at the end of a neck 4

3′ Oral opening is a slit along the edge of the neck 6

4(3) With snoutlike oral region, usually at the end of a neck 5

4′ Oral region typically flat .... **Enchelyidae [p. 68]**

5(4) With ciliated snoutlike bump on anterior end consisting of two zones, anterior one unciliated and posterior one with long cilia arising from oblique kineties; neck highly extensible or lacking .... **Lacrymariidae [p. 68]**

5′ With simple, small apical snout (oral bulge); neck not extensible .... **Trachelophyllidae [p. 68]**

6(3) Ciliary rows on right gradually terminating along rightmost perioral kinety, without spica (secant system) .... **Litonotidae [p. 68]**

6′ Ciliary rows on right surface converging on the spica (Fig. 2.22 P) .... Amphileptidae, one genus: *Amphileptus*

7(1) Anterior end not obliquely truncated 8

7′ Anterior end usually obliquely truncated, with distinct oral bulge and slitlike cytostome .... **Spathidiidae [p. 68]**

8(7) Body not extensible 9

8′ Body wormlike, laterally flattened, contractile (Fig. 2.23 Q) .... Homalozoonidae, one genus: *Homalozoon*

9(8) Without tentacles; one or more transverse ciliary bands or girdles sometimes also with long bristles 10

9′ With highly contractile tentacles on body except around oral opening .... **Actinobolinidae [p. 68]**

10(9) With ciliary bands or holotrichous, without long bristles 11

10′ With ciliary bands and long bristles .... **Mesodiniidae [p. 70]**

11(10) With ciliary bands; anterior conical protrusion (proboscis) with extrusomes usually present .... **Didiniidae [p. 70]**

11′ Holotrichous; without anterior conical protrusion with extrusomes .... **Acropisthiidae [p. 70]**

Ciliophora: Litostomatea: Tracheliidae: Genera

1	Elongated body with a long neck (≥ 1/4 of body length)	2
1′	Spherical body with a short neck (< 1/4 of body length) (Fig. 2.22 B)	*Trachelius*
2(1)	Macronucleus moniliform, composed of at least four nodules	3
2′	Macronucleus binodular or in many nodules scattered throughout the cytoplasm (Fig. 2.22 D)	*Dimacrocaryon/Rimaleptus/Dileptus*
3(2)	One perioral kinety right of oral bulge; proboscis conspicuously short; oral bulge opening rounded and located ventrally (Fig. 2.22 C)	*Pseudomonilicaryon/Monilicaryon*
3′	Two perioral kineties right of oral bulge; long spiraled proboscis; oral bulge opening 6-shaped, inverted and located laterally (Fig. 2.22 E)	*Paradileptus*

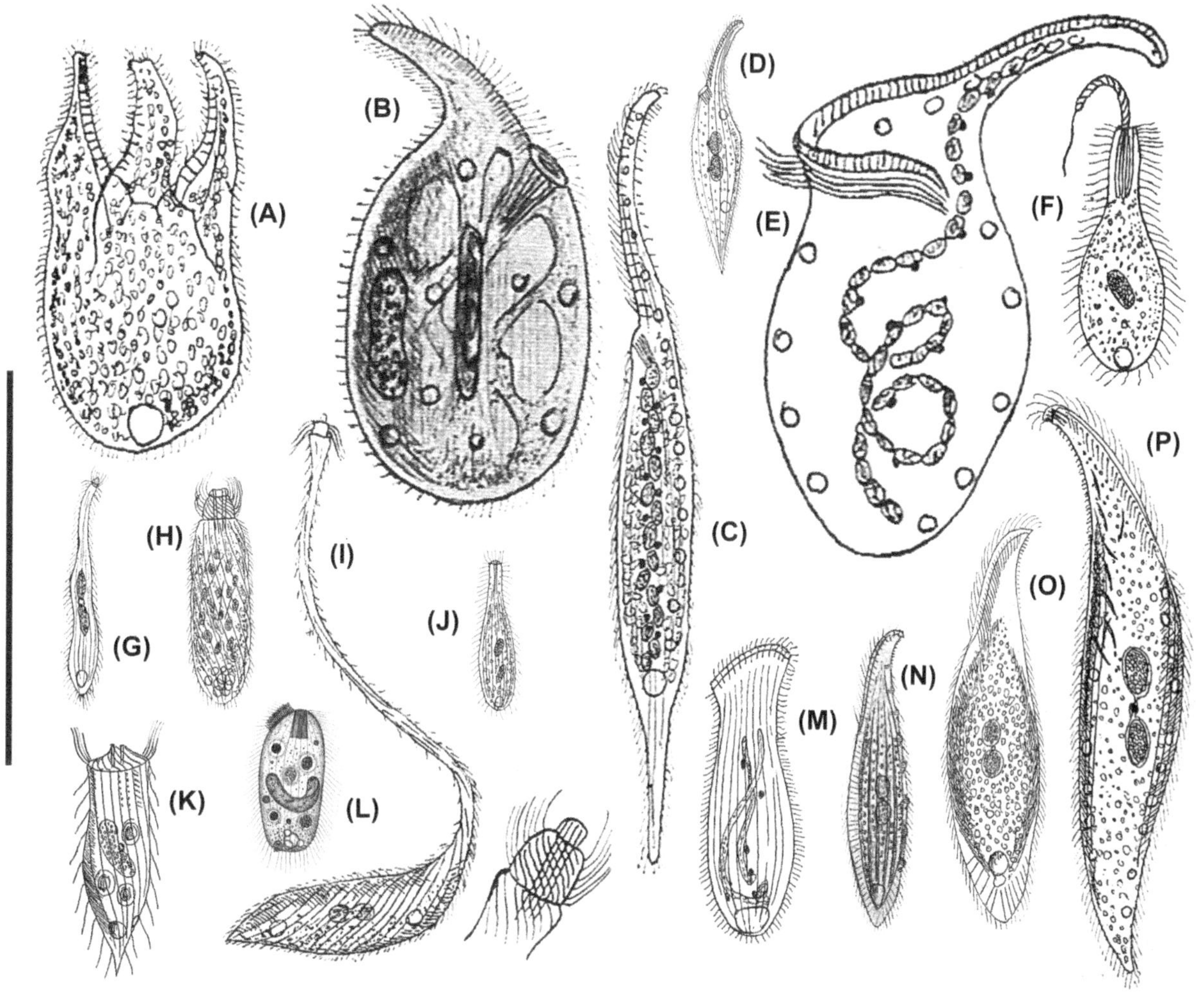

FIGURE 2.22 Representative species of class Litostomatea recorded in the Neotropics: (A) *Teuthophrys trisulca* ca. 200 μm; (B) *Trachelius ovum* ca. 220 μm; (C) *Monilicaryon monilatus* ca. 600 μm; (D) *Rimaleptus bivacuolatus* ca. 130 μm; (E) *Paradileptus elephantinus* ca. 300 μm; (F) *Ileonema dispar* ca. 120 μm; (G) *Epitholiolus chilensis* ca. 100 μm; (H) *Phialina coronata* ca. 100 μm; (I) *Lacrymaria olor* ca. 300 μm; (J) *Trachelophyllum pusillum* ca. 50 μm; (K) *Acropisthium mutabile* ca. 90 μm; (L) *Fuscheria nodosa* ca. 60 μm; (M) *Spathidium spathula* ca. 160 μm; (N) *Loxophyllum helus* ca. 160 μm; (O) *Litonotus lamella* ca. 200 μm; (P) *Amphileptus pleurosigma* 300 μm. Scale bar (A, B, and D–P) = 200 μm; (C) = 300 μm. *After: (A, E, H, and K) Kahl (1930); (B, C, I, and N–P) Kahl (1931); (D) Cunha (1913) after Kahl (1931); (F) Stokes (1884) after Kahl (1930); (J) Roux (1901) after Kahl (1930); (G) Bürger (1906) after Kahl (1930); (L) Foissner (1983); (M) Woodruff & Spencer (1922) after Kahl (1930); (L) reproduced with permission from the copyright holder.*

Ciliophora: Litostomatea: Enchelyidae: Genera

1 Body flasklike to oval, anterior end transversely truncated, without an apical flagellum-like organelle 2
1′ Body flasklike, with long twisted apical flagellum-like organelle and broadly rounded posteriorly (Fig. 2.22 F) .... *Ileonema*
2(1) Oral opening slitlike at slightly concave anterior end (Fig. 2.23 M) .... *Enchelys*
2′ Oral opening surrounded by about 20 fingerlike flaps or lappets (Fig. 2.23 K) .... *Spasmostoma*

Ciliophora: Litostomatea: Lacrymariidae: Genera

1 With highly extensible neck between snoutlike bump and trunk (Fig. 2.22 I) .... *Lacrymaria*
1′ Without neck between snoutlike bump and trunk (Fig. 2.22 H) .... *Phialina*

Ciliophora: Litostomatea: Trachelophyllidae: Genera

1 With epicortical scales 2
1′ Without epicortical scales 3
2(1) With type I scales (a polygonally faceted, dome-shaped superstructure emerges from the flat, ellipsoidal baseplate) (Fig. 2.22 J) .... *Trachelophyllum*
2′ With type III scales (some tiny arcs, which form a dome-shaped superstructure, emerges from a ring-shaped baseplate) (Fig. 2.22 G) .... *Epitholiolus*
3(1) Anterior domed or with very short, but distinct snoutlike region often surrounded by long cilia 4
3′ Apical region ends in an unciliated domelike conical projection (Fig. 2.23 I) .... *Lagynophrya*
4(3) Cylindroidal; oral bulge usually conspicuous, hemispherical, or platelike; dorsal brush composed of three rows of dikinetids (Fig. 2.23 N) .... *Enchelyodon*
4′ Oval or spherical; oral bulge inconspicuous; multirowed dorsal brush (Fig. 2.23 F) .... *Acaryophrya*

Ciliophora: Litostomatea: Litonotidae: Genera

1 Oral bulge extends along ventral side or terminates anteriorly not curved to left 2
1′ Oral bulge and perioral kineties distinctly curved to left anteriorly (Fig. 2.23 L) .... *Acineria*
2(1) Ventral edge lined with clear band of trichocyst warts; left anterior end occupied by a conspicuous field of paired brush cilia (Fig. 2.22 N) .... *Loxophyllum*
2′ Ventral edge never lined with trichocyst warts (Fig. 2.22 O) .... *Litonotus*

Ciliophora: Litostomatea: Spathidiidae: Genera

1 Body spathulate or ovoidal to cylindrical 2
1′ Body with three long necklike extensions (Fig. 2.22 A) .... *Teuthophrys*
2(1) Body usually spathulate; oral bulge elongate elliptical (Fig. 2.22 M) .... *Spathidium*
2′ Body ovoidal to cylindrical; distinct anterior unciliated ridge containing trichocysts that spirals down the body, extending at least one-fourth of body length (Fig. 2.23 O) .... *Perispira*

Ciliophora: Litostomatea: Actinobolinidae: Genera

Requires silver impregnation

1 With a ringlike array in midbody when tentacles are retracted; macronucleus usually elongated and tortuous or consisting of two globular nodules (Fig. 2.23 B) .... *Actinobolina*
1′ Without a ringlike array when tentacles are retracted; single macronucleus, broadly ellipsoidal to slightly reniform (Fig. 2.23 A) .... *Belonophrya*

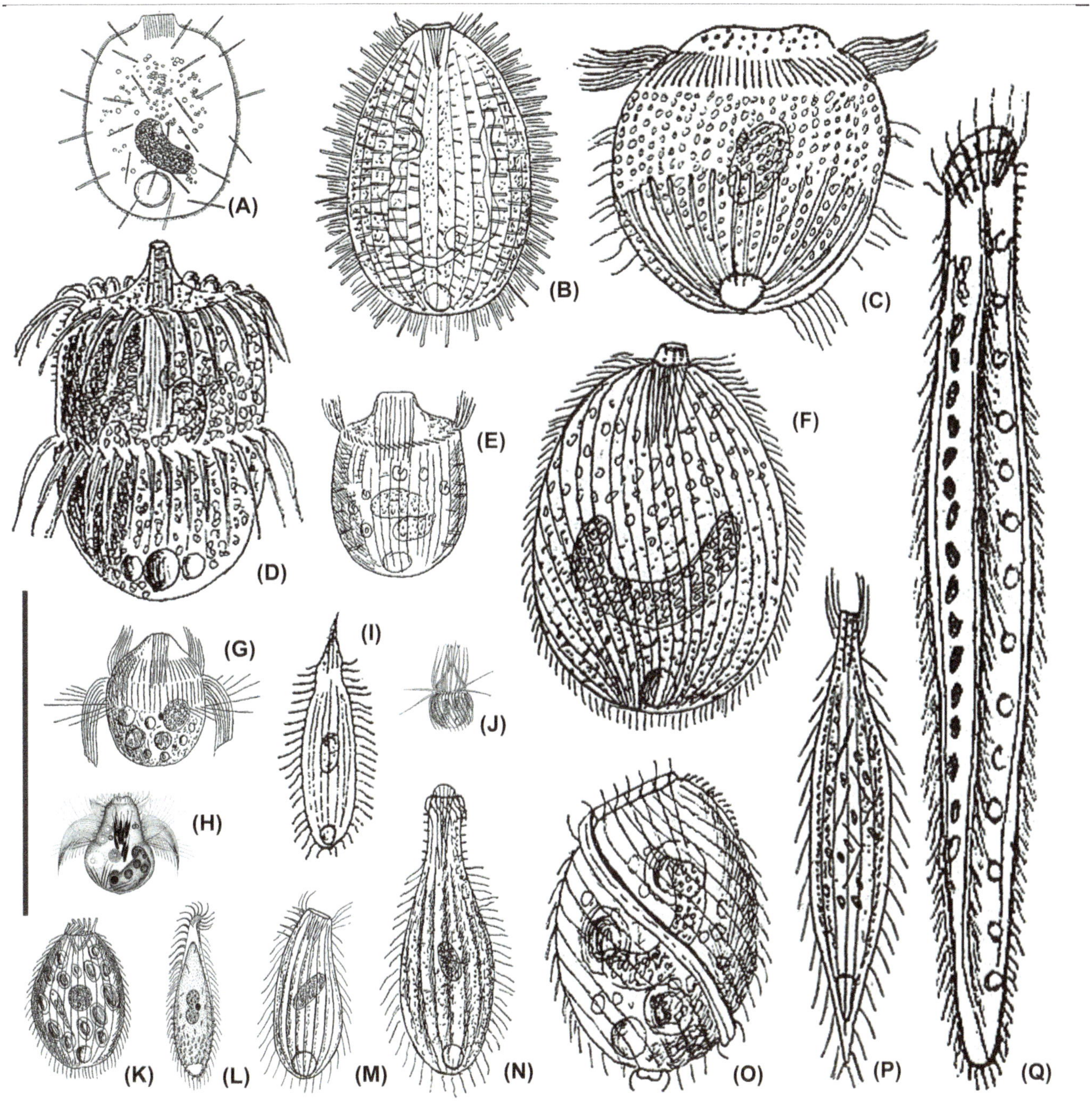

FIGURE 2.23 Representative species of class Litostomatea recorded in the Neotropics: (A) *Belonophrya pelagica* ca. 50 μm; (B) *Actinobolina vorax* 150 μm; (C) *Cyclotrichium viride* ca. 100 μm; (D) *Didinium nasutum* ca. 120 μm; (E) *Monodinium balbiani* ca. 60 μm; (F) *Acaryophrya mamillata* ca. 130 μm; (G) *Askenasia volvox* ca. 40 μm; (H) *Rhabdoaskenasia minima* ca. 30 μm; (I) *Lagynophrya acuminata* 80 μm; (J) *Mesodinium pulex* 25 μm; (K) *Spasmostoma viride* ca. 60 μm; (L) *Acineria uncinata* ca. 60 μm; (M) *Enchelys gasterosteus* ca. 60 μm; (N) *Enchelyodon lasius* ca. 100 μm; (O) *Perispira ovum* ca. 100 μm; (P) *Chaenea teres* ca. 150 μm; (Q) *Homalozoon vermiculare* ca. 500 μm. Scale bar (A–P) = 100 μm, (Q) = 200 μm. *After: (A) André (1914); (B) Wenrich (1929) after Kahl (1930); (C, D, F, G, I–K, and M–Q) Kahl (1930); (E) Fauré-Fremiet (1924) after Kahl (1930); (H) Krainer & Foissner (1990); (L) Foissner & O'Donoghue (1989); (A and H) reproduced with permission from the copyright holder; (L) reproduced from Foissner & O'Donoghue (1989) with permission from CSIRO Publishing.*

Ciliophora: Litostomatea: Mesodiniidae: Genera

1 Three ciliary bands or girdles in anterior body half; anterior girdle composed of moderately large ciliary tufts, middle girdle composed of large and sickle-shaped ciliary tufts, posterior girdle composed of long bristles; without retractable processes on anterior end 2

1′ Two ciliary bands or girdles in anterior body half; anterior girdle composed of small ciliary tufts, middle girdle with large ciliary tufts that extend posteriorly close to the trunk, posterior girdle absent; with retractable processes on anterior end (Fig. 2.23 J) .... *Mesodinium*

2(1) With distinct oral basket or rhabdos (Fig. 2.23 H) .... *Rhabdoaskenasia*

2′ Without distinct oral basket or rhabdos (Fig. 2.23 G) .... *Askenasia*

Ciliophora: Litostomatea: Didiniidae: Genera

1 With anterior conical protrusion (proboscis) with extrusomes 2

1′ Without anterior conical protrusion with extrusomes; with an anterior ciliary band and body ciliature uniform; cells usually broader than long (Fig. 2.23 C) .... *Cyclotrichium*

2(1) With one transverse ciliary band (Fig. 2.23 E) .... *Monodinium*

2′ With two transverse ciliary bands (Fig. 2.23 D) .... *Didinium*

Ciliophora: Litostomatea: Acropisthiidae: Genera

1 Body not contractile, ovoid to cylindrical, sometimes pointed at posterior end 2

1′ Body highly contractile, elongate, with a short, nonextensible snoutlike region with longer cilia, posterior end usually rounded (Fig. 2.23 P) .... *Chaenea*

2(1) Posterior end of body pointed (Fig. 2.22 K) .... *Acropisthium*

2′ Posterior end of body rounded (Fig. 2.22 L) .... *Fuscheria*

Ciliophora: Plagiopylea: Major Groups

1 Body ovoid or ellipsoid, flattened; oral ciliature simple, as extensions of somatic kineties into an oral depression **Plagiopylida [p. 70]**

1′ Body wedge- or helmet-shaped, laterally compressed, with armorlike cuirass and often short posterior spines; quite complex oral cavity with inconspicuous oral polykineties, typically < 10 in number .... **Odontostomatida [p. 70]**

Ciliophora: Plagiopylea: Plagiopylida: Families and Genera

1 Body oval, laterally compressed, with distinct notch on right-hand edge (ventral) in anterior region of body; oral area as a transversely arranged, deep ciliated depression that meets the notch on the right edge (Fig. 2.19 K) .... Plagiopylida, one genus: *Plagiopyla*

1′ Body ovoid to ellipsoid, dorsoventrally flattened; oral area apical or subapical and conical (Fig. 2.19 J) .... Sonderiidae, one genus: *Sonderia*

Ciliophora: Plagiopylea: Odontostomatida: Families

1 Body round or discoidal 2

1′ Body box-shaped, with short posterior spines; short rows of somatic cilia separated into anterior and posterior segments .... **Epalxellidae [p. 70]**

2(1) With two prominent anterior spines and a posterior spine; somatic ciliature sparse, with cirrilike clusters posteriorly and a preoral band of cilia on a conspicuous transverse ridge or frontal awning (Fig. 2.19 I) .... Discomorphellidae, one genus: *Discomorphella*

2′ Somatic ciliature very sparse, with anterior kinetofragments restricted to the ventral (= oral) surface, without frontal awning .... **Mylestomatidae [p. 71]**

Ciliophora: Plagiopylea: Epalxellidae: Genera

1 With a deep, wide terminal invagination surrounded by eight posterior spines with acute needle-like points, usually with anterior spine projecting from dorsal keel (Fig. 2.19 H) .... *Saprodinium*

1′ About six to eight blunt posterior spines surrounding the terminal invagination, sometimes also with an acute anterior spine (Fig. 2.19 F) .... *Epalxella*

Ciliophora: Plagiopylea: Mylestomatidae: Genera

1 With posterior longer cirrilike cilia (Fig. 2.19 E) .... *Mylestoma*

1′ Without posterior longer cirrilike cilia, with four posterior spines and one short spinelike extension of the dorsal keel (Fig. 2.19 G) .... *Atopodinium*

Ciliophora: Colpodea: Families

1 Body shape ovoid, ellipsoidal, bursiform or reniform; oral apparatus subapical, in midbody or in posterior end of body 2

1′ Body outline oval, flattened with delicate keeled armor; oral opening in posterior half of body; ventral surface with two to three semicircular ciliary rows (Fig. 2.21 F) .... Kreyellidae, one genus: *Kreyella*

2(1) Anterior end of body not obliquely truncate and right-slanted to longitudinal body axis 3

2′ Oval oral aperture at obliquely truncate anterior end, right-slanted to longitudinal body axis (Fig. 2.21 E) .... Platyophryidae, one genus: *Platyophrya*

3(2) Body ovoid, reniform or bursiform; oral apparatus in anterior half of body; contractile vacuole(s) at posterior or midbody 4

3′ Body ovoid, circular in cross section; may inhabit a gelatinous lorica; oral apparatus in posterior half of body, opening at base of posterior lobe; contractile vacuole in anterior end; four rows of cilia encircle the body (Fig. 2.21 A) .... Marynidae, one genus: *Mycterothrix*

4(3) Body bursiform or ellipsoid; oral region as a deeply invaginated, funnel-shaped oral depression with oral aperture anterior and extending ventrally forming a cleft; or oral depression oval, lying obliquely across the ventral surface 5

4′ Body ellipsoid to reniform, laterally flattened; oral apparatus usually subapical, with a distinct oral depression (Fig. 2.21 B) .... Colpodidae, one genus: *Colpoda*

5(4) Oral depression not divided internally into two parts by a longitudinal fold 6

5′ Oral depression divided internally into two parts by a longitudinal fold; base of oral funnel turns to left; ventral cleft extends about half of body length (Fig. 2.21 H) .... Bursariidae, one genus: *Bursaria*

6(5) Base of oral funnel turns to right; ventral cleft extends about one-third of body length (Fig. 2.21 G) .... Bursaridiidae, one genus: *Bursaridium*

6′ Oral depression oval elliptical, lying obliquely across the ventral surface, or funnel-shaped, with its base turned to left .... **Bryometopidae [p. 71]**

Ciliophora: Colpodea: Bryometopidae: Genera

1 Body ellipsoid or more or less reniform, may be flattened; oral depression shallow and elliptical; somatic kineties abut V-like forming a postoral suture (Fig. 2.21 D) .... *Bryometopus*

1′ Body bursiform; oral depression deep, funnel-shaped, base of funnel turns to left; without postoral suture (Fig. 2.21 C) .... *Thylakidium*

ACKNOWLEDGMENTS

Financial supports from ANPCyT PICT 2015 1244 PRÉSTAMO BID, PUE 0098 MACN-CONICET, CNPq Universal 485974/2013-4, and FAPESP 2015/17856-0 are acknowledged. We wish to thank the Acta Protozoologica, European Journal of Protistology, The Journal of Eukaryotic Microbiology, Revue suisse de Zoologie, Pan-American Journal of Aquatic Sciences, Naturhistorischen Museum Wien, Denisia, Revista de la Asociación de Ciencias Naturales del Litoral, Institut für Zoologie der Universität Salzburg, Revista Brasileira de Zoociências, Invertebrate Taxonomy, International Review of Hydrobiology, Polar Biology, Revista Mexicana de Biodiversidad, Acta Zoologica Lilloana, Journal of Plankton Research, International Journal of Systematic and Evolutionary Biology, Zootaxa, Prof. Dr. Wilhelm Foissner, Dr. Helmut Berger, and Dr. Krzysztof Wiackowski for granting permissions to reproduce the figures.

REFERENCES

Al-Shahwani, S.M. & N.J. Horan. 1991. The use of protozoa to indicate changes in the performance of activated sludge plants. Water Research 25: 633–638.

André, E. 1914. Recherches sur la faune pélagique du Léman et description de nouveaux genres d'infusoires. Revue suisse de Zoologie 22: 179–193.

Berger, H. 1999. Monograph of the Oxytrichidae (Ciliophora, Hypotrichia). Monographiae Biologicae 78: i-xii, 1–1080.

Berger, H. 2006. Monograph of the Urostyloidea (Ciliophora, Hypotricha). Monographiae Biologicae 85: i-xvi, 1–1304.

Berger, H. 2008. Monograph of the Amphisiellidae and Trachelostylidae (Ciliophora, Hypotricha). Monographiae Biologicae 88: i-xvi, 1–737.

Berger, H. 2011. Monograph of the Gonostomatidae and Kahliellidae (Ciliophora, Hypotricha). Monographiae Biologicae 90: i-xiv, 1–741.

PROTOZOA

Borror, A. 1972. Tidal marsh ciliates (Protozoa): morphology, ecology, systematics. Acta Protozoologica 10: 29–71, plate I.

Campello-Nunes, P.H., N. Fernandes, M. Schlegel & I.D. Silva-Neto. 2015. Description and phylogenetic position of *Corlissina maricaensis* gen. nov., sp. nov. (Karyorelictea, Geleiidae), a novel interstitial ciliate from Brazil, with redefinition of the family Geleiidae. International Journal of Systematic and Evolutionary Microbiology 65: 4297–4308.

Castro, L.A.M., G.C. Küppers, M. Schweikert, M.L. Harada & T. da S. Paiva. 2014. Ciliates from eutrophized water in the northern Brazil and morphology of *Cristigera hammeri* Wilbert, 1986 (Ciliophora, Scuticociliatia). European Journal of Protistology 50: 122–133.

Collin, B. 1911. Etude monographique sur les Acinétiens. I. Recherches experimentales sur l'etendue des variation set les facteurs teratogenes. Archives de Zoologie Expérimentale et Générale 8: 421–497.

Curds, C.R. 1982. British and other freshwater ciliated protozoa. Part 1. Ciliophora: Kinetofragminophora. Cambridge UniversityPress, London. 387 pp.

Curds, C.R., M.A. Gates & D.M. Roberts. 1983. British and other freshwater ciliated protozoa. Part 2. Ciliophora: Oligohymenophora and Polyhymenophora. Cambridge University Press, London. 474 pp.

Dieckmann, J. 1995. An improved protargol impregnation for ciliates yielding reproducible results. European Journal of Protistology 31: 372–382.

Dioni, W. 1972. Un nuevo género de Folliculinidae de agua dulce: *Botticula ringueleti* nov. gen. nov. sp. del río Paraná medio. Acta Zoológica Lilloana 29: 304–313.

Dragesco, J. & A. Dragesco-Kernéis. 1986. Ciliés libres de l'Afrique intertropicale. Introduction á la connaissance et á l'étude des ciliés. Faune Tropicale 26: 1–159.

Ehrenberg, C.G. 1833. Dritter Beitrag zur Erkenntniss grosser Organisation in der Richtung des kleinsten Raumes. Abhandlungen der Königlichen Akademie der Wissenschaften zu Berlin. 203 pp.

Entz, G. 1907. Über einige patagonische Protozoen. Mathematische und Naturwissenschaftliche Berichte aus Ungarn, Leipzig 21: 84–112.

Fenchel, T. 1987. Ecology of Protozoa: The biology of free-living phagotrophic protists. Science Technical Publishers, Madison. 197 pp.

Foissner, W. 1979. Ökologische und systematische Studien über das Neuston alpiner Kleingewässer, mit besonderer Berücksichtigung der Ciliaten. Internationale Revue der gesamten Hydrobiologie und Hydrographie 64: 99–140.

Foissner, W. 1981. Morphologie und Taxonomie einiger neuer und wenig bekannter kinetofragminophorer Ciliaten (Protozoa: Ciliophora) aus alpinen Böden. Zoologische Jahrbücher, Systematik 108: 264–297.

Foissner, W. 1983. Taxonomische Studien über die Ciliaten des Grossglocknergebietes (Hohe Tauern, Österreich) I. Familien Holophryidae, Prorodontidae, Plagiocampidae, Colepidae, Enchelyidae und Lacrymariidae nov. fam. Annalen Des Naturhistorischen Museums in Wien, Serie B 84: 49–85.

Foissner, W. 2014. An update of "basic light and scanning electron microscopic methods for taxonomic studies of ciliated protozoa." International Journal of Systematic and Evolutionary Microbiology 64: 271–292.

Foissner, W., S., Agatha & H., Berger 2002. Soil ciliates (Protozoa, Ciliophora) from Namibia (Southwest Africa), with emphasis on two contrasting environments, the Etosha region and the Namib Desert. Denisia 5: 1–1459.

Foissner, W., H. Berger, H. Blatterer & F. Kohmann. 1995. Taxonomische und ökologische Revision der Ciliaten des Saprobiensystems – Band IV: Gymnostomates, *Loxodes*, Suctoria. Informationsberichte des Bayer, Landesamtes für Wasserwirtschaft 1/95. 540 pp.

Foissner, W., H. Berger & F. Kohmann. 1992. Taxonomische und ökologische Revision der Ciliaten des Saprobiensystems – Band II: Peritrichia, Heterotrichida, Odontostomatida. Informationsberichte des Bayer, Landesamtes für Wasserwirtschaft 5/92. 502 pp.

Foissner, W., H. Berger & F. Kohmann. 1994. Taxonomische und ökologische Revision der Ciliaten des Saprobiensystems – Band III: Hymenostomata, Prostomatida, Nassulida. Informationsberichte des Bayer, Landesamtes für Wasserwirtschaft 1/94. 548 pp.

Foissner, W., H. Berger & J. Schaumburg. 1999. Identification and ecology of limnetic plankton ciliates. Informationsberichte des Bayer, Landesamtes für Wasserwirtschaft, 3/99. 793 pp.

Foissner, W., H. Blatterer, H. Berger & F. Kohmann. 1991. Taxonomische und ökologische Revision der Ciliaten des Saprobiensystems – Band I: Cyrtophorida, Oligotrichida, Hypotrichia, Colpodea. Informationsberichte des Bayer, Landesamtes für Wasserwirtschaft 1/91. 478 pp.

Foissner, W., A. Chao & L.A. Katz. 2008. Diversity and geographic distribution of ciliates (Protista: Ciliophora). Biodiversity and Conservation 17: 345–363.

Foissner, W. & P.J. O'Donoghue. 1989. Morphology and infraciliature of some freshwater ciliates (Protozoa: Ciliophora) from Western and South Australia. Invertebrate Taxonomy 3: 661–696.

Foissner, W., I. Oleksiv & H. Müller. 1990. Morphologie und Infraciliatur einiger Ciliaten (Protozoa: Ciliophora) aus stagnierenden Gewässern. Archiv für Protistenkunde 138: 191–206.

Foissner, W. & S. Wölfl. 1994. Revision of the genus *Stentor* Oken (Protozoa, Ciliophora) and description of *S. araucanus* nov. spec. from South American lakes. Journal of Plankton Research 16: 255–289.

Foissner, W. & K. Xu. 2007. Monograph of the Spathidiida (Ciliophora, Haptoria). Vol I: Protospathidiidae, Arcuospathidiidae, Apertospathulidae. Monographiae biologicae 81: i-xi, 1–485.

Hausmann, K. & P.C. Bradbury (eds.). 1996. Ciliates: Cells as Organisms. Gustav Fischer, Stuttgart. 485 pp.

Hentschel, E. 1916. Biologische Untersuchungen über den tierischen und pflanzlichen Bewuchs im Hamburger Hafen. Mitteilungen des Zoologischen Museums Hamburg 33: 1–176.

Jung J.H., Y.S. Baek, S. Kim & H.G. Choi. 2016. Morphology and molecular phylogeny of a new freshwater ciliate *Urosomoida sejongensis* n. sp. (Ciliophora, Sporadotrichida, Oxytrichidae) from King George Island, Antarctica. Zootaxa 4072: 254–262.

Kahl, A. 1930. Urtiere oder Protozoa I: Wimpertiere oder Ciliata (Infusoria) 1. Allgemeiner Teil und Prostomata. Tierwelt Deutschlands 18: 1–180.

Kahl, A. 1931. Urtiere oder Protozoa I: Wimpertiere oder Ciliata (Infusoria) 2. Holotricha außer den im 1. Teil behandelten Prostomata. Tierwelt Deutschlands 21: 181–398.

Kahl, A. 1932. Urtiere oder Protozoa I: Wimpertiere oder Ciliata (Infusoria) 3. Spirotricha Tierwelt Deutschlands 25: 399–650.

Kahl, A. 1935. Urtiere oder Protozoa I: Wimpertiere oder Ciliata (Infusoria) 4. Peritricha und Chonotricha. Tierwelt Deutschlands 30: 651–886, I–V.

Kellikot, D.S. 1885. Observations on some fresh-water Infusoria: with descriptions of a few species regarded as new. Proceedings of the American Society of Microscopists 7: 38e47.

Klein, B. 1958. The "dry" silver method and its proper use. Journal of Protozoology 5: 99–103.

Krainer K.H. & W. Foissner. 1990. Revision of the genus *Askenasia* Blochmann, 1895, with proposal of two new species, and description of *Rhabdoaskenasia minima* n. g., n. sp. (Ciliophora, Cyclotrichida). Journal of Protozoology 37: 414–442.

Küppers, G.C. 2014. Morphology of a new hypotrichous ciliate, *Clapsiella magnifica* gen. n, sp. n., with a curious dorsal ciliary pattern. European Journal of Protistology 50: 373–381.

Küppers, G.C. & M.C. Claps. 2010. Morphology and notes on morphogenesis during cell division of *Deviata polycirrata* n. sp. and of *Deviata bacilliformis* (Gelei, 1954) Eigner, 1995 (Ciliophora: Kahliellidae) from Argentina. Journal of Eukaryotic Microbiology 57: 273–284.

Küppers, G.C. & M.C. Claps. 2013. Hypotrichous ciliates (Protozoa: Ciliophora) from a temporary pond in Argentina, with redescription of *Apoamphisiella hymenophora* (Stokes, 1886) Berger, 1999. Zootaxa 3626: 55–76.

Küppers, G.C., M.C. Claps & E.C. Lopretto. 2007. Description of *Notohymena pampasica* n. sp. (Ciliophora, Stichotrichia). Acta Protozoologica 46: 221–227.

Küppers, G.C., M.C. Claps & E.C. Lopretto. 2009. Ciliates (Protozoa) from dried sediments of a temporary pond from Argentina. Revista Mexicana de Biodiversidad 80: 581–592.

Küppers, G.C., E.C. Lopretto & M.C. Claps. 2006a. *Pelagostrobilidium wilberti* n. sp. (Oligotrichea, Choreotrichida): morphology and morphogenesis. Journal of Eukaryotic Microbiology 53: 477–484.

Küppers G.C., E.C. Lopretto & M.C. Claps. 2006b. Morphological aspects and seasonal changes of some planktonic ciliates (Protozoa, Ciliophora) from a temporary pond, Buenos Aires province, Argentina. Pan-American Journal of Aquatic Sciences 1: 74–90.

Küppers, G.C., T. da Silva Paiva, B. do Nascimento Borges, M.L. Harada, G. González Garraza & G. Mataloni. 2011. An Antarctic ciliate, *Parasterkiella thompsoni* (Foissner) nov. gen., nov. comb., recorded in Argentinean peat-bogs: morphology, morphogenesis, and molecular phylogeny. European Journal of Protistology 47: 103–123.

Lynn, D.H. 2008. The Ciliated Protozoa. Characterization, Classification, and Guide to the Literature, 3rd ed. Springer, Dordrecht, the Netherlands. 605 pp.

Mariño-Pérez, R., R. Mayén-Estrada & I.V. Dovgal. 2010. *Periacineta mexicana* n. sp. (Ciliophora, Suctoria, Discophryida), epizoic on Mexican backswimmers of the genus *Buenoa* (Insecta, Hemiptera, Notonectidae). Journal of Eukaryotic Microbiology 57: 435–443.

Modenutti, B.E. & M.C. Claps. 1986. Ciliados dulceacuícolas de la Argentina, III: Ciliophora-Peritricha. Revista de la Asociación de Ciencias Naturales del Litoral 17: 71–78.

Mulisch, M., T. Heep, W. Sturm & J. Borcherding. 1998. Redescription of *Ascobius lentus*, a rare freshwater folliculinid (Ciliophora: Heterotrichida) from a pond in Germany. Acta Protozoologica 37: 29–39.

Paiva, T.D.S., G.C. Küppers, D.J.G. Lahr, M. Schweikert & I.D. da Silva-Neto. 2017. *Discomorphella pedroeneasi* sp. nov. (Ciliophora, Odontostomatida): An anaerobic ciliate hosting multiple cytoplasmic and macronuclear endocytobionts. European Journal of Protistology 58: 103–134.

Paiva, T.S., G.C. Küppers & I.D. da Silva Neto. 2016. Morphology and divisional morphogenesis of the brackish water ciliate *Novistrombidium rufinoi* sp. nov. (Ciliophora: Oligotrichia) from Brazil. Revista Brasileira de Zoociências 17: 20–32.

Paiva, T. da S. & I.D. da Silva-Neto. 2004a. Comparative morphometric study of three species of *Apoamphisiella* Foissner, 1997 (Ciliophora: Hypotrichea) from Brazilian locations, including a description of *Apoamphisiella foissneri* sp. n. Zootaxa 505: 1–26.

Paiva, T. da S. & I.D. da Silva-Neto. 2004b. Description of *Parentocirrus brasiliensis* sp. n. (Ciliophora: Spirotrichea), a new ciliate protist present in activated sludge. Zootaxa 504: 1–10.

Paiva, T. da S. & I.D. da Silva-Neto. 2006. *Pseudourostyla pelotensis* sp. nov. (Ciliophora, Stichotrichia, Urostylida): a new psammophilic ciliate from the southern Brasil. Zootaxa 1247: 43–58.

Paiva, T. da S. & I.D. da Silva-Neto. 2007. Morphology and morphogenesis of *Strongylidium pseudocrassum* Wang and Nie, 1935, with redefinition of *Strongylidium* Sterki, 1878 (Protista: Ciliophora: Stichotrichia). Zootaxa 1559: 31–57.

Pan, X., W. A. Bourland & W. Song. 2013. Protargol synthesis: An inhouse protocol. Journal of Eukaryotic Microbiology 60: 609–614.

Penard, E. 1920. Etudes sur les Infusoires tentaculiferes. Mémoires de la Société de Physique et d'Histoire Naturelle de Genève 39: 131–229.

Penard, E. 1922. Études sur les infusoires d'eau douce. Georg & Cie, Genève. 331 pp.

Petz W. & W. Foissner. 1992. Morphology and morphogenesis of *Strobilidium caudatum* (Fromentel), *Meseres corlissi* n. sp., *Halteria grandinella* (Müller), and *Strombidium rehwaldi* n. sp., and a proposed phylogenetic system for oligotrich ciliates (Protozoa, Ciliophora). Journal of Protozoology 39: 159–176.

Root, F.M. 1914. Reproduction and reaction to food in the suctorian *Podophrya collini* n. sp. Archiv für Protistenkunde 35: 114–196.

Shao, C., W. Song, X. Hu, H. Ma, M. Zhu & M. Wang. 2006. Cell division and morphology of the marine ciliate, *Condylostoma spatiosum* Ozaki and Yagiu (Ciliophora, Heterotrichida) based on a Chinese population. European Journal of Protistology 42: 9–19.

Skibbe, O. 1994. An improved quantitative protargol stain for ciliates and other planktonic protists. Archiv für Hydrobiologie 130: 339–347.

Sládeček, V. 1973. System of water quality from the biological point of view. Archiv für Hydrobiologie. Beihefte. Ergebnisse der Limnologie 7: 1–218.

Song, W. & N. Wilbert. 1995. Benthische Ciliaten des Süsswassers. Pages 156–168 *in* R. Röttger (Hrsg.), Praktikum der Protozoologie. Gustav Fischer Verlag, Stuttgart, Jena.

Taylor, W.T. & RW. Sanders. 2001. Chapter 3: Protozoa. Pages 43–96 *in* J. Thorp & A. Covich (eds.), Ecology and classification of North American freshwater invertebrates (Second edition). Academic Press, New York, NY.

Thomsen, R. 1922. Einiges über die Morphologie von *Folliculina boltoni*. Archiv für Protistenkunde 44: 83–98.

Vďačný, P. & W. Foissner. 2012. Monograph of the dileptids (Protista, Ciliophora, Rhynchostomatia). Denisia 31: 1–529.

Wanick, R.C., T.S. Paiva, C.N. Carvalho & I.D. da Silva Neto. 2010. Acute toxicity of cadmium to freshwater ciliate *Paramecium bursaria*. Biociências (Porto Alegre) 16: 104–109.

Warren, A., G.F. Esteban & B.J. Finlay. 2016. Chapter 2: Protozoa. Pages 5–37 *in* J.H. Thorp & D.C. Rogers (eds.), Thorp & Covich's Freshwater Invertebrates. Vol. II. Keys to Nearctic Fauna. Academic Press, Elsevier, Boston, MA.

Wilbert, N. 1975. Eineverbesserte Technik der Protargolimprägnation für Ciliaten. Mikrokosmos 64: 171–179.

Xu, H., Y. Jiang, K.A.S. Al-Rasheid, W. Song & A. Warren. 2011. Spatial variation in taxonomic distinctness of ciliated protozoan communities at genus-level resolution and relationships to marine water quality in Jiaozhou bay, Northern China. Hydrobiologia 665: 67–78.

Xu, Y., C. Shao, X. Fan, A. Warren, K.A.S. Al-Rasheid, W. Song & N. Wilbert. 2016. New contributions to the biodiversity of ciliates (Protozoa, Ciliophora) from Antarctica, including a description of *Gastronauta multistriata* nov. spec. Polar Biology 39: 1439–1453.

Flagellated Protozoa

Cristina Damborenea
División Zoología Invertebrados, La Plata, Argentina

INTRODUCTION

Several flagellates (species that have eukaryotic flagella in the main feeding stage of their life history) are found across eukaryote diversity. Within them, heterotrophic flagellates have one or more flagella per cell acting in both locomotion and food capture. The motion of most flagella occurs by undulatory waves beginning from its base, mostly in one plane. Two flagellar morphologies are recognized: smooth flagella and hairy ones, with rows of flagellar hairs or mastigonemes, which modify their hydrodynamic properties (Fenchel, 1987).

Many have special membranous organelles, the extrusomes, with a variety of functions (e.g., protection, food capture, secretion). These organelles discharge their contents after a stimulus (Brusca et al. 2016).

Several groups, belonging to independent linages, are free living in freshwater habitats: Dinoflagellata, Apicomplexa, Chrysophyta, Cryoptomonada, Diplomonadida, Euglenida, Kinetoplastidea, Choanoflagellata. Unfortunately, keys to the flagellated taxa in the Neotropics could not be developed in detail for this chapter because of the dearth of taxonomists and the generally poor knowledge of this group in the Neotropics.

Chrysophythes are golden-brown mixotrophic or colorless heterotrophic forms that prey on bacteria (Kristiansen & Škaloud, 2017). They are single-celled individuals or colonial forms. Most have two-flagella inserted apically in the cell. One long flagellum, most often forwardly directed, is hairy, with two rows of mastigonemes. It beats with uniplanar sine waves starting from the base. The other flagellum is short and smooth, sometimes only detectable by electron microscopy. It is generally directed laterally and beats in helical waves. A photoreceptor system is present, with a red stigma located anteriorly in a chloroplast. Most are naked cells, but in many, the cells are surrounded by a wall or lorica of different forms (vase-shaped, flat, or globular). The lorica consists of imbricate cellulose or chitin scales. Silica scales produced internally are frequent, and their structure is species specific. The majority are planktonic in freshwater habitats, but some are benthic. Some chrysophytes are very common and cosmopolitan; others are rare with particular disjunct distributions. However, the knowledge of this group of protozoa is very fragmentary in general.

Dinoflagellates are photosynthetic mixotrophic or heterotrophic organisms. They can be found in most aquatic environments including snow, freshwater, marine, or intrazoic habitats. Of about 2,400 living species, 83% are marine, and nearly a half are photosynthetic (Saldarriaga & Taylor, 2017). Dinoflagellates are between 10 and 100 μm in length and of variable shape, but often pyriform. Some have highly complex internal differentiation. They are typically unicellular, very rarely multicellular. They are biflagellated. One ribbonlike transverse flagellum causes turn and forward movement, and a second longitudinal flagellum provides movement in a posterior direction. In most species, the two flagella arise from the side and lie in surface grooves: the transverse in a girdle or cingulum, and the longitudinal in the sulcus. The cingulum divides the cell into an anterior body portion, the episome (or epicone), and a posterior hyposome (hypocone). The cells may be naked (athecate) or possess a wall (thecate, pelliculate). Trichocysts are the most common extrusomes. In the sulcal, close to flagellar roots, an eyespot of different types may be present. In most dinoflagellates, the motile phase (mastigote) is dominant, but some spend their life in nonmotile stage. Dinoflagellates are involved in several important ecological phenomena, such as harmful algal blooms and symbioses (Simpson et al., 2017).

Cryptomonads are protists ubiquitous in freshwater, brackish, and marine habitats. Photosynthetic cryptomonads are planktonic and important primary producers, whereas the phagotrophic genus *Goniomonas* is benthic. They are small protists (approximately 5–50 μm) with a marked asymmetry, and their longitudinal axis may be twisted. The apex is often shifted to the one side of the cell. Two unequal flagella insert subapically (in the vestibulum of the furrow-gullet system) or laterally to the right side of the cell. In most cryptomonads, the longer flagellum directs anteriorly and pulls the cell behind. The cell rotates around the longitudinal axis during forward swimming as a result of their flat form and asymmetry. Refractile ejectosomes (extrusomes) are usually visible.

Cryptomonads exhibit a variety of colors, from brown to olive brown, bright brown, tomato red, brick red, blue green, and sky blue. They usually contain one H-shaped plastid per cell. Several species form palmella stages (an agglomeration of flagellated cells embedded in mucus), which may grow to macroscopically visible size (Hoef-Emden & Archibald, 2017). The diversity of cryptomonads is not yet uncovered, even in well-known regions (Hoef-Emden & Archibald, 2017).

Euglenids are single-celled free-living flagellates with diverse modes of nutrition including phagotrophy, osmotrophy, and photoautotrophy. The organisms usually have two heterodynamic flagella, which originate within an anterior flagellar pocket. One flagellum extends anteriorly or anteriolaterally and is emergent from the pocket. The other flagellum, reduced in length, run posteriorly, is confined to the flagellar pocket, or absent. Freshwater species have contractile vacuoles associated with the reservoir. The storage substance is the paramylon, which granules may be small or extremely large (as in some photoautotrophic species), and tubular extrusomes, mucocysts, or muciferous bodies are common.

Euglenids are distinguishable by their cell surface, which is always supported by a pellicle of parallel protein strips located under the cell membrane. The cell can have diverse modes of motility, including metaboly, substrate-mediated gliding, and swimming (Leander et al., 2017). Under light microscope, periplast plates may be visible on the cell surface. The phagotrophic species engulf bacteria and usually glide along substrates.

Kinetoplastea are colorless flagellates that are divided into Trypanosomatina (all parasite forms with one flagellum) and Bodonina (many marine and freshwater free living with two flagella). All have a kinetoplast, formed by unique large mitochondria with a dark concentration of mitochondrial DNA (Brusca et al., 2016). The phagotrophic bodonids are completely free living, and all feed on bacteria while creeping over substrates. They use the anterior flagellum for locomotion and for propelling food particles toward the cytostome. They are abundant in organic-rich environments, but many species are limited by low oxygen levels.

Choanoflagellates are unicellular or colonial protists found in marine and freshwater environments, in both planktonic and benthic communities. They are heterotrophic phagotrophs (Richter & Nitsche, 2017b). The cells are round and have a single apical flagellum surrounded by a collar of microvilli. The flagellum undulates and creates water currents to transport food particles to the collar. Many choanoflagellates possess an extracellular organic theca; its shape is considered to be diagnostic for species identification. Others possess a distinct siliceous lorica (Richter & Nitsche, 2017b).

Methods to collect flagellates are described in Esteban et al. (2015). For isolating, culturing, and preserving methods, see Warren & Esteban (2019). Cultivation methods are similar to those for ciliate protozoa, as described previously in this chapter.

KEY TO FLAGELLATED PROTOZOA

The key has been modified from Warren & Esteban (2019).

Key to Major Groups

1 Two, one, or four evident flagella, not lying in surface groves 2

1′ Two flagella, one transverse and one longitudinal, both flagella lying in surface groves (Fig. 2.24 D, E) .... Dinoflagellata

2(1) Without a collar of microvilli surrounding the flagellum 3

2′ A collar of microvilli surrounds the single anterior flagellum; solitary or colonial; with or without theca (Fig. 2.25 A, G) Choanoflagellida

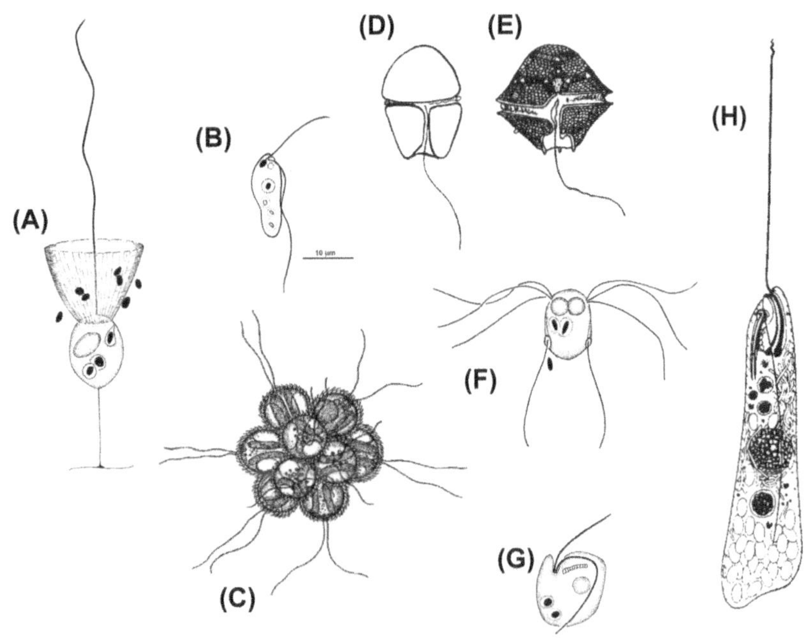

FIGURE 2.24 Variety of flagellated prototists: (A) Choanoflagellate, about 15 μm; (B) bodonid flagellate, about 15 μm; (C) motile colonial chrysophyte, *Synura*, about 30 μm in diameter; (D) and (E) typical dinoflagellata, about 20 μm; (F) anaerobic diplomonadid, about 8 μm; (G) cryptomonadid, about 5 μm; (H) heterotrophi euglenoid, *Peranema*, about 15 μm. *From (A, F, and G) Esteban et al. (2015); (B) Gibson (2017), (C) Kristiansen & Škaloud (2017); (D, E, and H) Warren & Esteban (2019).*

3	Flagella, not arising from an anterior reservoir; without paramylon as reserve substance	4
3′	Two flagella, sometimes only one visible, both arising from an anterior reservoir; contractile vacuole associated with reservoir; paramylon as reserve substance; elongate; body shape often plastic when living (Figs. 2.24 H, 25 A–C, I)	Euglenida
4(3)	Cells without deep furrow-gullet	5
4′	Small cells with a subapical deep gullet where two subequally length flagella arise (Figs. 2.24 G, 25 H)	Cryptomonadida
5(4)	Flagella not inserted anteriorly	6
5′	Two unequal flagella inserted apically; long flagellum often forwardly directed, short flagellum (sometimes not detected) directed laterally, and beats in helical waves (Fig. 2.24 C)	Chrysophyta

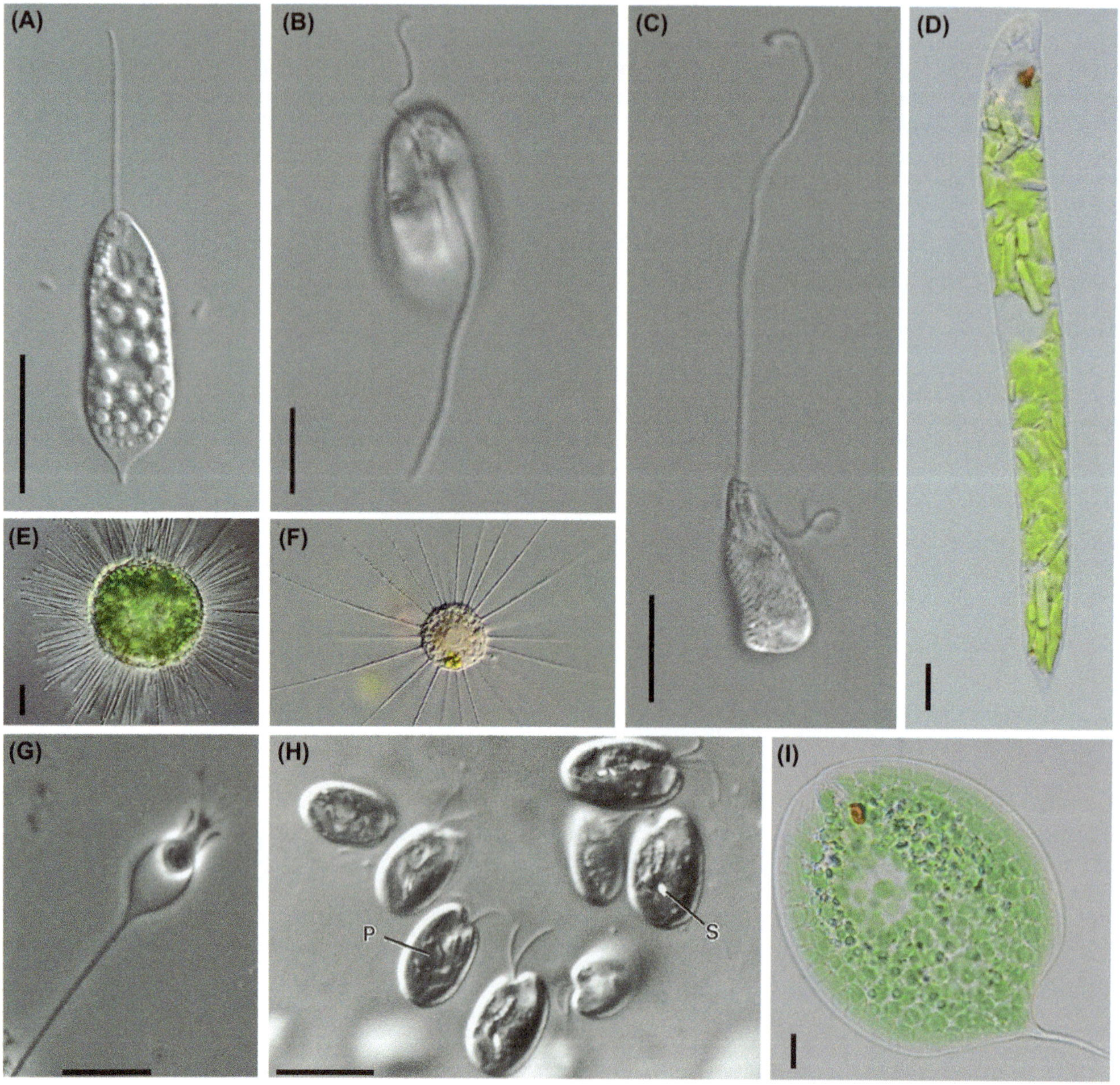

FIGURE 2.25 Light microscope photographs of flagellated protists: (A–C) Phagotrophic euglenoids with different flagella orientations; (D) euglenid, *Euglena*; (E and F) heliozoan protists; (G) stalked choanoflagellate, without theca; (H) cryptomonads, with a pyrenoid (P) and the stigma (S); (I) euglenid, *Phacus*. Scale bars: (A) = 20 μm, (B–D and G–I) = 10 μm, (E) = 30 μm. *From (A–D, and I) Leander et al. (2017); (E and F) Gast (2017); (G) Richter & Nitsche (2017); (H) Clay (2015).*

6(5) Two flagella arising from a flagellar pocket; with a characteristic unique elongate organelle, the kinetoplast, usually associated with the flagella (Fig. 2.24 B) .... Kinetoplastida, Bodonina

6′ Bilateral axial symmetry, with each half with a karyomastigont consisting of a nucleus and four basal bodies; a flagellum emerges from each basal body (Fig. 2.24 F) .... Diplomonadida

REFERENCES

Brusca, R.C., W. Moore & S.M. Shuster. 2016. Invertebrates (Third edition). Sinauer Associates Inc. Sunderland, MA. 1104 pp.

Clay, B.L. 2015. Cryptomonads. Pages 809–850 *in*: J. Wehr, R. Sheath & J.P. Kociolek (eds.), Freswater Algae from North America (End edition). Academic Press. 1066 pp.

Esteban, G.F., B.J. Finlay & A. Warren. 2015. Free-Living Protozoa. Pages 113–132 *in:* J. Thorp & D.C. Rogers (eds.), Ecology and General Biology, Thorp and Covich's Freshwater Invertebrates (Fourth edition). Academic Press, Elsevier, Boston, MA.

Frenchel, T. 1987. Ecology of Protozoa. Springer-Verlag, Berlin. 197 pp.

Gast, R.J. 2017. Centrohelida and other heliozoan-like protists. Pages 955–971 *in:* J.M. Archibald, A.G.B. Simpson & C.H. Slamovits (eds.), Handbook of the Protists. Springer, Cham.

Gibson, W. 2017. Kinetoplastea. Pages 1089–1118 *in:* J.M. Archibald, A.G.B. Simpson & C.H. Slamovits (eds.), Handbook of the Protists. Springer, Cham.

Hoef-Emden, K. & J.M. Archibald. 2017. Cryptophyta (Cryptomonads). Pages 851–891 *in:* J.M. Archibald, A.G.B. Simpson & C.H. Slamovits (eds.), Handbook of the Protists. Springer, Cham.

Kristiansen, J. & P. Škaloud. 2017. Chrysophytha. Pages 331–366 *in:* J.M. Archibald, A.G.B. Simpson & C.H. Slamovits (eds.), Handbook of the Protists. Springer, Cham.

Leander B.S., G. Lax., A. Karnkowska & A.G.B. Simpson. 2017. Euglenida. Pages 1047–1088 *in*: J. Archibald, A. Simpson & C. Slamovits (eds.), Handbook of the Protists. Springer, Cham.

Richter D.J. & R. Nitsche. 2017. Choanoflagellatea. Pages 1479–1496 *in:* J.M. Archibald, A.G.B. Simpson & C.H. Slamovits (eds.), Handbook of the Protists. Springer, Cham. doi:10.1007/978-3-319-28149-0_5.

Saldarriaga, J.F. & J.R.M. Taylor. 2017. Dinoflagellata. Pages 625–678 *in*: J. Archibald, A. Simpson & C. Slamovits (eds.), Handbook of the Protists. Springer, Cham.

Simpson, A.G.B., C.H. Slamovits & J.M. Archibald. 2017. Pages 1–21 *in:* J.M. Archibald, A.G.B. Simpson & C.H. Slamovits (eds.), Handbook of the Protists. Springer, Cham.

Warren, A. & G.F. Esteban. 2019. Protozoa. Pages 7–42 *in:* D.C. Rogers & J.H. Thorp (eds.), Keys to Paleartic Fauna, Thorp and Covich's Freshwater Invertebrates (Fourth edition). Academic Press, Elsevier, Boston, MA.

Chapter 3

Phylum Porifera

Ulisses Pinheiro
Departamento de Zoologia, Universidade Federal de Pernambuco, Recife, Brazil

Ludimila Calheira
Departamento de Zoologia, Universidade Federal de Pernambuco, Recife, Brazil

Chapter Outline

INTRODUCTION

Sponges comprise the earliest groups of Metazoa and are found in diversified aquatic habitats. There are 8975 valid species known worldwide (Van Soest et al., 2018), but only 250 species are found in inland waters. All freshwater sponges are grouped into seven families in the order Spongillida. This low diversity could be related to the paucity of taxonomists and the difficulty to study continental environments (Pinheiro et al., 2015). Despite having only three families, the Neotropical biogeographic region has the richest diversity of freshwater sponges worldwide. Until now, 67 species have been reported from South America and 76 species from the entire Neotropical region (Table 3.1). However, six of nine additional species are shared with the Neartic biogeographic region, and this suggests that only South America should be considered the true home of Neotropical freshwater sponges. Only six species from South America are recorded beyond its boundaries, and these probably need revision because of species complexes. These numbers indicate a high endemism rate of at least 91%. Recently, *Corvomeyenia epilithosa* Volkmer-Ribeiro, Rosa-Barbosa & Machado, 2005 and *Racekiela cavernicola* Volkmer-Ribeiro, Bichuette & De Sousa Machado, 2010 have been included in Brazilian Red List of Threatened Species (Brasil, 2014).

LIMITATIONS

Although the Neotropics has a high diversity of freshwater sponges, the records are concentrated in Brazil and Argentina. The early records came from the Amazon Basin through the material collected by European expeditions in the 19th century. It was not until the second half of the 20th century, that South American researchers started inventorying freshwater sponges. Bonetto and Ezcurra de Drago concentrated efforts on Argentine Mesopotamia (Parana and Uruguay Rivers), while Volkmer-Ribeiro focused her research on inland waters from southern Brazil and the Amazon Basin. Thus, the freshwater sponges distribution is directly related to the location of taxonomists specializing on this phylum. Other regions without researchers have occasional records. For example, Pinheiro (2007) identified northeastern Brazil as the most scarce region for sponge records and surveys. However, after an extensive inventory at Pernambuco State, Nicacio & Pinheiro (2015) found 13

Thorp and Covich's Freshwater Invertebrates. https://doi.org/10.1016/B978-0-12-804225-0.00003-4

FIGURE 3.1 Specimens of *Drulia brownii* in situ, Tupé Lake, Manaus, Amazon Basin, Brazil *Photograph by C.W.C. Branco.*

species from this region. In this sense, inventories are necessary to know the real biodiversity and distribution of freshwater sponges. Even the diversity in the Amazon Basin, with more than 30 species recorded, is probably underestimated if we consider the huge drainage basin.

TERMINOLOGY AND MORPHOLOGY

Neotropical freshwater sponges display a variable morphology that ranges from thinly encrusting, massive, to an arborescent shape. A peculiar phenomenon occurring in the Rio Negro (Brazil) during the dry season in the Amazon Basin is the presence of "arboreal sponges." Some of the trees can tolerate flooding for up to 3 years, and the flood level is marked by resistant stages of the sponges which are attached to the trees (Fig. 3.1) (Manconi & Pronzato, 2016).

Freshwater sponges come in a variety of colors and body consistency. They are black, brown, cream, white, or green. Many sponges develop more pigment in environments with high light levels. The green color, for example, is due to symbiotic green algae, which are common in freshwater sponges when they grow exposed to sunlight (Nicacio & Pinheiro, 2015). The consistency of sponges ranges from soft, compact, or hard, depending upon the amount of **spongin** and **spicules** in the skeleton. An example of extremely hard sponge is *Sterrastrolepis brasiliensis* (Volkmer-Ribeiro & De Rosa-Barbosa, 1978), an endemic sponge from Parana River, that is known locally as "espinho de pedra" (spines of stones) because people get hurt when they step on it.

Sponge spicules consist of **megascleres** and **microscleres**. The former are relatively large and are the skeleton's main structural component. In contrast, microscleres are relatively small and often lack a structural function when present (Manconi & Pronzato, 2002).

A remarkable feature of freshwater sponges is the formation of **gemmules** (Fig. 3.2 A–C). These are resistant bodies with totipotent cells that can be fully coated with special spicules, the **gemmuloscleres** (Fig. 3.2 D). Gemmules can be naked or armed by gemmuloscleres. When armed, the gemmuloscleres are generally radially or tangentially inserted in a gemmular theca. Gemmuloscleres show various morphologies, including smooth to spiny **oxeas** (monaxon diactinal spicule pointed at both tips), **strongyles** (isodiametric diactinal spicule with rounded tips), **birotules** (spicules with diskshaped tips), **pseudobirotules** (spicule without developed rotule, but curved hooks on the tips), **boletiform** (spicule with proximal large rotule supporting a shaft decreasing in thickness toward the distal tip, in shape of rotule or pseudorotule) or **parmuliform** (spicule with a single rotule supporting a short acute conical stem) (Manconi & Pronzato, 2002) (Fig. 3.3). A

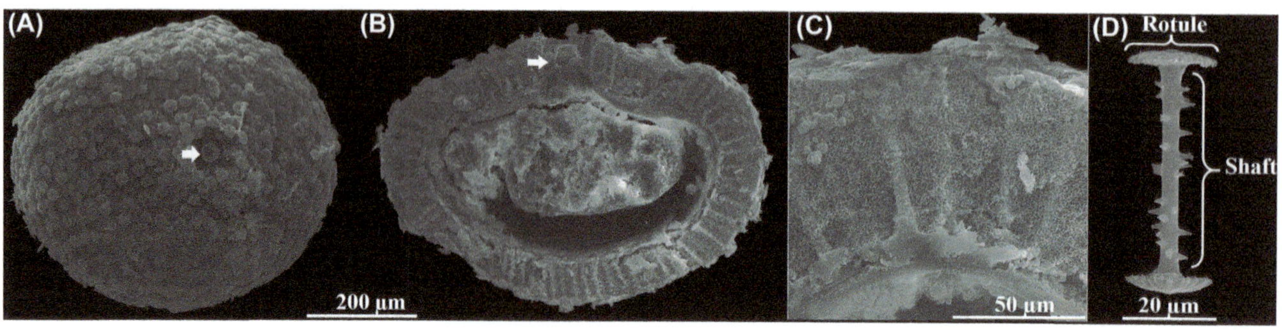

FIGURE 3.2 Gemmule of Spongillidae under Scanning Electron Microscope: (A) entire gemmule; (B) transversal section of a gemmule showing a foramen, gemmuloscleres radially embedded in the theca, and the pneumatic layer; (C) pneumatic layer in detail; and (D) gemmuloscleres birotule. Arrows show the foramen.

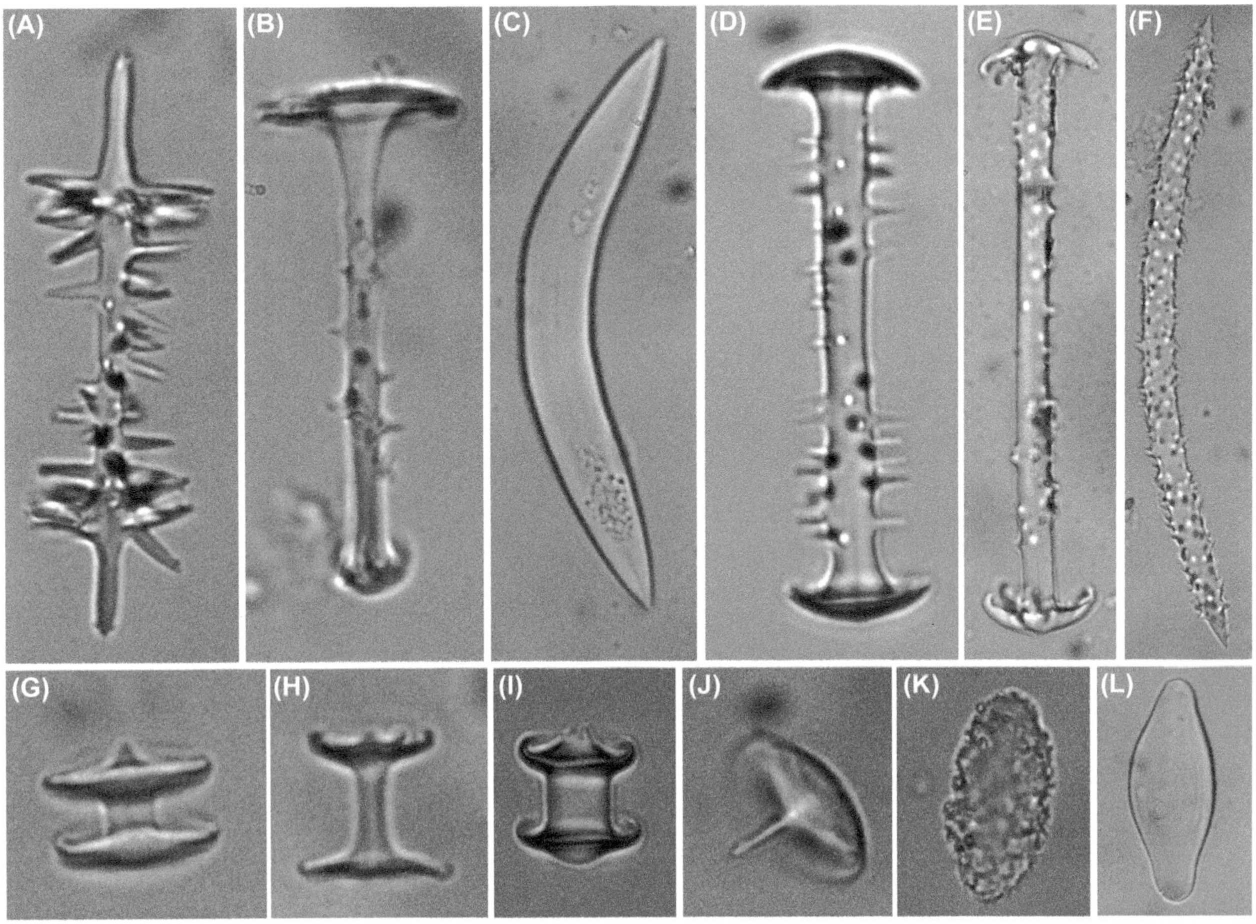

FIGURE 3.3 Gemmuloscleres: (A) sanidaster; (B) tubelliform; (C) oxea boomerang-shaped; (D) birotule; (E) pseudobirotule; (F) acanthoxea; (G–I) birotules; (J) parmuliform; (K) strongyle sterrasters; and (L) amphistrongyle.

foramen could be present in the gemmules, and its morphology varies from being circular, tubular, or rosette-like collar. The number of gemmular layers vary between one to three, and a **pneumatic layer**, if present, provides floatability.

The gemmules are present in three families: Spongillidae, Metaniidae, and Potamolepidae (Manconi & Pronzato, 2002). Species of Spongillidae and Metaniidae have gemmules distributed throughout the sponge with a well-developed pneumatic layer and spined gemmuloscleres. Such features increase dispersion ability for species in these families. For example, *Saturnospongilla carvalhoi* (Volkmer-Ribeiro, 1976) from Amazon Basin, has a ring-shaped pneumatic layer around the gemmule, that led Manconi & Pronzato (2007) to hypothesize an ability of dispersal by anemochory and hydrochory (Fig. 3.4). On the other hand, species of Potamolepidae have limited dispersion ability due to the absence of pneumatic layer and spined gemmuloscleres (Manconi & Pronzato, 2007); their gemmules are always deposited in the sponge-substrate interface.

MATERIAL PREPARATION AND PRESERVATION

Sponge specimens for taxonomic study could be preserved dried or in 80% ethanol. For histological preparation, the sponge should first be fixed in 4% formalin and then preserved in 70% ethanol (Hajdu et al., 2011).

Sponge taxonomy is based on external morphology (consistency, shape, surface, and pattern of arrangement of oscules) and on the types and sizes of spicules (megascleres, microscleres and gemmuloscleres). Sponge identification requires analysis of spicules and gemmules using Light Microscopy (LM) and Scanning Electron Microscopy (SEM), according to the methods described by Hajdu et al. (2011). To prepare the spicule for examination, put an organic portion of a sponge on the slide and drip nitric acid on it while the slide is positioned over a lamp flame. Once the sponge tissue is digested by the acid, the remaining spicules should be mounted on a glass microscope slide. For SEM preparation, put a fragment of the sponge in a centrifuge tube, add nitric acid, digest the tissue over a

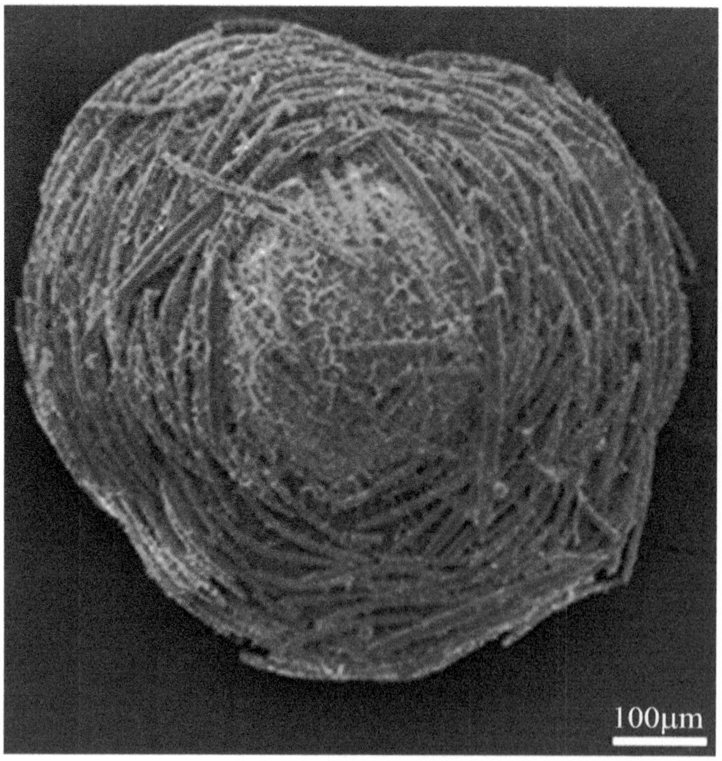

FIGURE 3.4 Gemmules of *Saturnospongilla carvalhoi* (Volkmer-Ribeiro, 1976) under scanning electron microscope.

lamp flame, pour off the acid, and then wash the spicules with distilled water and ethanol. Washing and settling or centrifugation should be repeated at least three times, after which the spicules are mounted on a microscope slide (Reiswig et al., 2010). All categories of spicules should be measured (n = 30), and the minimum—mean—maximum of length/width of shaft and diameter of the rotule when possible (Fig. 3.2 D). Gemmuloscleres and gemmules are usually diagnostic traits for identification.

KEYS TO PORIFERA

There are no synapomorphic characters to distinguish between the Metaniidae and Spongillidae. The Metaniidae was proposed by Volkmer-Ribeiro (1986) in belief of the polyphyletic origin of freshwater sponges. However, molecular data do not support this proposal (Morrow & Cárdenas, 2015). Therefore, we provide a taxonomic key for genera and species of freshwater sponges without splitting them into families. The traditional families are indicated in Table 3.1.

Order Spongillida: Genera

1 Gemmules absent 2

1' Gemmules present 3

2(1) Megascleres acanthotylostrongyles (Fig. 3.5 H) .... *Acanthotylotra alvarengai* Volkmer-Ribeiro, Tavares & Fürstenau-Oliveira, 2009 [Brazil: Pará]

2' Megascleres acanthoxeas (Fig. 3.5 C) .... *Balliviaspongia wirrmanni* Boury-Esnault & Volkmer-Ribeiro, 1991 [Bolivia]

3(1) Gemmuloscleres present 4

3' Gemmuloscleres absent .... ***Pottsiela* [p. 87]**

4(3) Gemmuloscleres with one of the tips rotules, pseudorotules or parmuliform 5

4' Gemmuloscleres exclusively sanidasters, strongyles or oxeas 18

5(4) Gemmuloscleres with rotules or pseudorotules (Fig. 3.3 D, E) 6

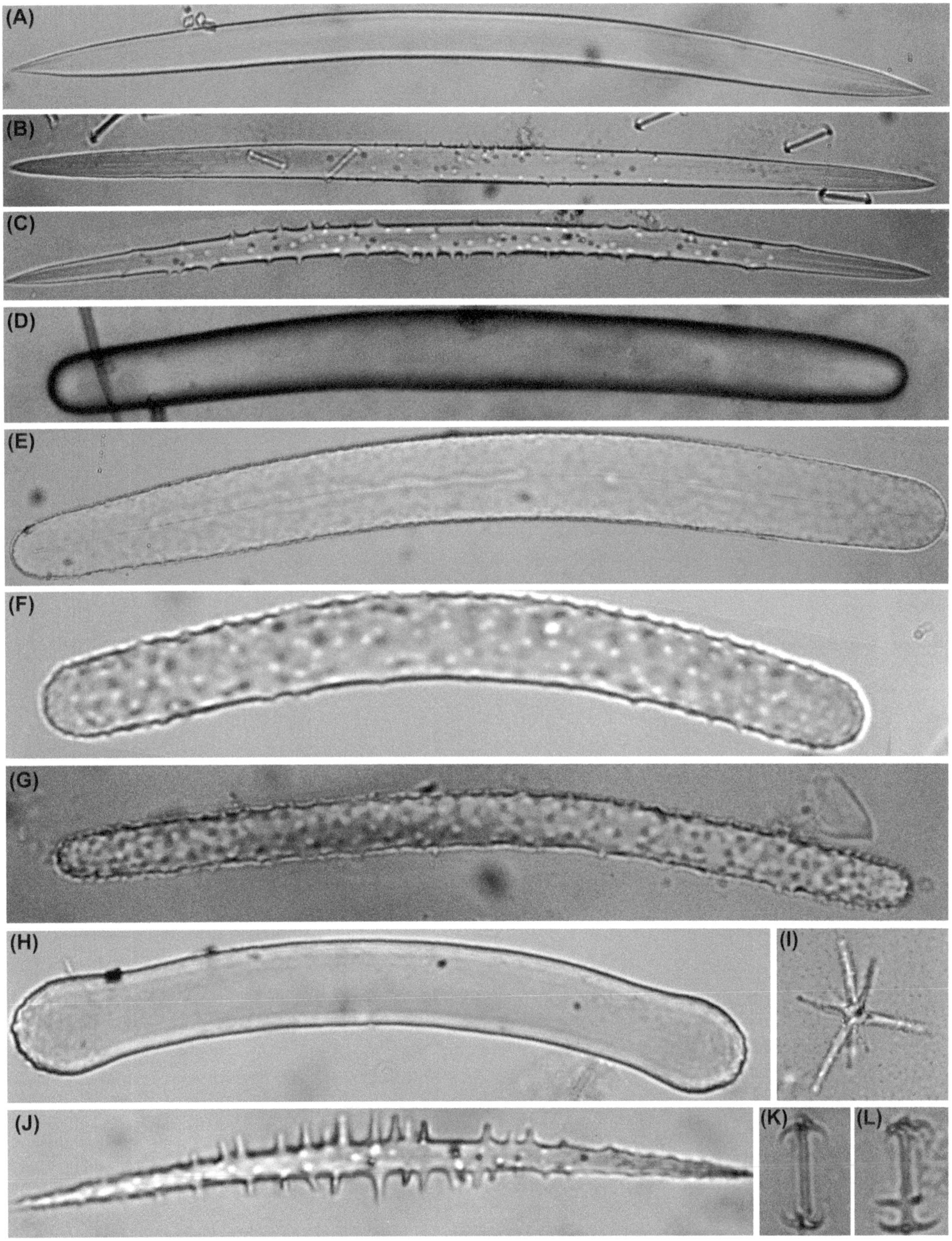

FIGURE 3.5 Megascleres (A–H) and microscleres (I–L): (A) oxea; (B) microspined oxea; (C) acanthoxea; (D) strongyle; (E) microgranulated strongyle; (F) microspined strongyle; (G) acanthostrongyle; (H) acanthotylostrongyle; (I) aster; (J) acanthoxea; and (K–L) pseudobirotules.

5'	Gemmuloscleres parmuliform (Fig. 3.3 J)	***Drulia*** **[p. 87]**
6(5)	Gemmuloscleres birotules with only of the rotules complete	7
6'	Gemmuloscleres birotules with rotules incomplete	12
7(6)	One form of gemmuloscleres	8
7'	Two forms of gemmuloscleres (acanthoxeas and birotules) (Fig. 3.3 F, G)	*Saturnospongilla carvalhoi* Volkmer, 1976
	[Venezuela. Brazil: Amazonas, Goiás, Mato Grosso, Pará, Tocantins]	
8(7)	Gemmuloscleres birotules (Fig. 3.3 D)	9
8'	Gemmuloscleres tubelliforms (Fig. 3.3 B)	***Metania*** **[p. 88]**
9(8)	Microscleres absent	10
9'	Microscleres asters present (Fig. 3.5 I)	11
10(9)	Gemmuloscleres with one category of birotules (Fig. 3.3 H, I)	***Tubella*** **[p. 88]**
10'	Gemmuloscleres with two category of birotules	*Heterorotula fistula* Volkmer & Motta, 1995
	[Brazil: Ceará, Mato Grosso do Sul, Minas Gerias, Rio Grande do Sul]	
11(9)	Gemmuloscleres large birotules (>70 μm)	***Dosilia*** **[p. 89]**
11'	Gemmuloscleres short birotules (17–45 μm)	*Houssayella iguazuensis* Bonetto & Ezcurra de Drago, 1966
	[Brazil: Rio Grande do Sul. Argentina: Misiones]	
12(6)	Gemmuloscleres birotules with only rotules incomplete	13
12'	Gemmuloscleres birotules with rotules incomplete and complete	14
13(12)	Gemmuloscleres only birotules	***Ephydatia*** **[p. 89]**
13'	Gemmuloscleres could be birutules, pseudobirotules, acanthoxeas or acanthostrongyles	***Anheteromeyenia*** **[p. 89]**
14(12)	Microscleres absent	15
14'	Microscleres present	16
15(14)	Gemmuloscleres could be birotules, pseudobirotules or acanthoxeas	***Racekiela*** **[p. 90]**
15'	Gemmuloscleres tubelliforms and pseudobirotules	*Acalle recurvata* (Bowerbank, 1863)
	[Venezuela. Brazil: Amazonas, Goiás, Mato Grosso, Pará, Tocantins. Bolivia]	
16(14)	Microscleres pseudobirotules (Fig. 3.5 K, L)	17
16'	Microscleres only acanthoxeas (Fig. 3.5 J)	***Heteromeyenia*** **[p. 90]**
17(16)	Gemmuloscleres birotules with smooth shaft	***Corvomeyenia*** **[p. 90]**
17'	Gemmuloscleres birotules with spined shaft	***Corvoheteromeyenia*** **(in part) [p. 90]**
18(4)	Microscleres present	19
18'	Microscleres absent	22
19(18)	Microscleres acanthoxeas	20
19'	Microscleres pseudobirotules	21
20(19)	Gemmuloscleres acanthostrongyles	***Spongilla*** **[p. 90]**
20'	Gemmuloscleres strongyles sterrasters (Fig. 3.3 K)	*Sterrastrolepis brasiliensis* Volkmer & Rosa Barbosa, 1978
	[Brazil: Goiás, Paraná]	
21(19)	Gemmuloscleres sanidasters (Fig. 3.3 A)	***Corvoheteromeyenia*** **(in part) [p. 90]**
21'	Gemmuloscleres acanthoxeas and acanthostrongyles	*Corvospongilla seckti* Bonetto & Ezcurra de Drago, 1966
	[Brazil: Goiás, Mato Grosso, Mato Grosso do Sul, Minas Gerias, Pará, Paraná, Rio Grande do Sul, São Paulo, Tocantins. Argentina: Corrientes]	
22(18)	Gemmules with gemmuloscleres tangentially arranged	23
22'	Gemmules with gemmuloscleres radially arranged	***Radiospongilla*** **[p. 91]**
23(22)	Gemmules with pneumatic layer low developed or absent	24
23'	Gemmules with pneumatic layer well developed	***Eunapius*** **[p. 91]**
24(23)	Megascleres could be microgranulated strongyles or acanthostrongyles, oxeas or acanthoxeas	25
24'	Megascleres exclusively smooth strongyles with gemmuloscleres smooth strongyles slightly curved	*Potamophloios guairensis* Volkmer-Ribeiro, Parolin, Fürstenau-Oliveira & De Menezes, 2010
	[Brazil: Paraná]	

TABLE 3.1 List of Neotropical Freshwater Sponge Species

SPONGILLIDA
Spongillidae
***Anheteromeyenia* Schröder, 1927**
A. cheguevarai Manconi & Pronzato, 2005
A. diamantina Calheira & Pinheiro, 2018
A. ornata (Bonetto & Ezcurra de Drago, 1970)
A. vitrea Buso & Volkmer-Ribeiro, 2012
***Corvoheteromeyenia* Ezcurra de Drago, 1979**
C. australis (Bonetto & Ezcurra de Drago, 1966)
C. heterosclera (Ezcurra de Drago, 1974)
C. sanidastosclera Pinheiro, Silva & Calheira, 2015
***Corvospongilla* Annandale, 1911**
C. seckti Bonetto & Ezcurra de Drago, 1966
***Dosilia* Gray, 1867**
D. palmeri (Potts, 1885)
D. pydanieli Volkmer-Ribeiro, 1992
***Ephydatia* Lamouroux, 1816**
E. caatingae Nicacio & Pinheiro, 2015
E. chileana Pisera & Sáez, 2003 †
E. facunda Weltner, 1895
E. fluviatilis (Linnaeus, 1759)
E. robusta (Potts, 1888)
***Eunapius* Gray, 1867**
E. carteri (Bowerbank, 1863)
E. fragilis (Leidy, 1851)
E. igloviformis (Potts, 1884)
***Heteromeyenia* Potts, 1881**
H. barlettai Pinheiro, Calheira & Hajdu, 2015
H. cristalina Batista, Volkmer-Ribeiro & Melão, 2007
H. horsti Kilian & Wintermann-Kilian, 1976[a]
H. insignis Weltner, 1895
***Heterorotula* Penney & Racek, 1968**
H. fistula Volkmer & Motta, 1995
***Houssayella* Bonetto & Ezcurra de Drago, 1966**
H. iguazuensis Bonetto & Ezcurra de Drago, 1966
***Pottsiela* Volkmer-Ribeiro, de Souza Machado, Fürstenau-Olivera & Vieira Soares, 2010**
P. pesae Volkmer-Ribeiro, Souza-Machado, Fürstenau-Oliveira & Vieira-Soares, 2010
P. spoliata (Volkmer-Ribeiro & Maciel, 1983)
***Racekiela* Bass & Volkmer-Ribeiro, 1998**
R. andina Victor Quintero-Hernández & José Barrea, 2017
R. cavernicola Volkmer-Ribeiro, Bichuette & de Sousa Machado, 2010
R. sheilae (Volkmer-Ribeiro, De Rosa-Barbosa & Tavares, 1988)

(*Continued*)

TABLE 3.1 List of Neotropical Freshwater Sponge Species—cont'd

SPONGILLIDA
***Radiospongilla* (Penney & Racek, 1968)** *R. amazonensis* Volkmer & Maciel, 1983
R. crateriformis (Potts, 1882)
R. inesi Nicacio & Pinheiro, 2011
***Saturnospongilla* Volkmer-Ribeiro, 1976** *S. carvalhoi* Volkmer, 1976
***Spongilla* Lamarck, 1816** *S. alba* Carter, 1849
S. cenota Penney & Racek, 1968
***Tubella* Carter, 1881** *T. amazonica* (Weltner, 1895)
T. delicata (Bonetto & Ezcurra de Drago, 1967)
T. gregaria (Bowerbank, 1863)
T. horrida (Weltner, 1893)
T. lanzamirandai (Bonetto & Ezcurra de Drago, 1964)
T. leidii (Bowerbank, 1863)
T. minuta (Potts, 1881)
T. paulula (Bowerbank, 1863)
T. pennsylvanica (Potts, 1882)
T. repens (Hinde, 1888)
T. variabilis Bonetto & Ezcurra de Drago, 1973
Potamolepidae
***Oncosclera* Volkmer-Ribeiro, 1970** *O. atrata* (Bonetto & Ezcurra de Drago, 1970)
O. intermedia Bonetto & Ezcurra de Drago, 1973
O. jewelli (Volkmer, 1963)
O. navicella (Carter, 1881)
O. petricola Bonetto & Ezcurra de Drago, 1967
O. ponsi (Bonetto & Ezcurra de Drago, 1968)
O. rosariae Tavares-Frigo, Volkmer-Ribeiro, Oliveira & Machado, 2015
O. schubarti (Bonetto & Ezcurra de Drago, 1967)
O. spinifera (Bonetto & Ezcurra de Drago, 1973)
O. stolonifera Bonetto & Ezcurra de Drago, 1967
O. tonollii (Bonetto & Ezcurra de Drago, 1968)
***Potamophloios* Brien, 1970** *P. guairensis* Volkmer-Ribeiro, Parolin, Fürstenau-Oliveira & De Menezes, 2010
***Sterrastrolepis* Volkmer-Ribeiro & de Rosa Barbosa, 1978** *S. brasiliensis* Volkmer & Rosa Barbosa, 1978
***Uruguaya* Carter, 1881** *U. corallioides* (Bowerbank, 1863)

TABLE 3.1 List of Neotropical Freshwater Sponge Species—cont'd

SPONGILLIDA
Metaniidae
***Acalle* Gray, 1867** *A. recurvata* (Bowerbank, 1863)
***Corvomeyenia* Weltner, 1913** *C. epilithosa* Volkmer-Ribeiro, Rosa-Barbosa & Machado, 2005
C. thumi (Traxler, 1895)
***Drulia* Gray, 1867** *D. brownii* (Bowerbank, 1863)
D. conifera Bonetto & Ezcurra de Drago, 1973
D. cristata (Weltner, 1895)
D. cristinae Volkmer-Ribeiro, Ezcurra de Drago, Machado & Sabaj, 2017
D. ctenosclera Volkmer & Mothes, 1981
D. uruguayensis Bonetto & Ezcurra de Drago, 1969
***Metania* Gray, 1867** *M. fittkaui* Volkmer-Ribeiro, 1979
M. kiliani Volkmer-Ribeiro & Costa, 1992
M. reticulata (Bowerbank, 1863)
M. spinata (Carter, 1881)
M. subtilis Volkmer, 1979
***Spongillina incertae sedis* Manconi & Pronzato, 2002** *Acanthotylotra alvarengai* Volkmer-Ribeiro, Tavares & Fürstenau-Oliveira, 2009
Balliviaspongia wirrmanni Boury-Esnault & Volkmer-Ribeiro, 1991

[a]*Pinheiro et al. (2015) believe that* Heteromeyenia horsti *is a possible synonym of* H. insignis, *in the sense just* H. insignis *is available in the key, as the two forms cannot be separated.*

25(24) Megascleres microgranulated strongyles .... *Uruguaya corallioides* (Bowerbank, 1863)

[Venezuela. Brazil: Amazonas, Goiás, Mato Grosso de Sul, Rio Grande do Sul, Raraima, São Paulo. Paraguay. Argentina. Uruguay]

25′ Megascleres could be smooth strongyles or acanthostrongyles, oxeas or acanthoxeas .... ***Oncosclera* [p. 91]**

Spongillida: *Pottsiela*: Species

1 Megascleres oxeas large (250–900 μm) varying from smooth to rarely spined .... *Pottsiela spoliata* (Volkmer-Ribeiro & Maciel, 1983)

[Venezuela. Brazil: Amazonas, Mato Grosso, Pará, Rondônia]

1' Megascleres spiny oxeas shorter (<350 μm) .... *Pottsiela pesae* Volkmer-Ribeiro, Souza-Machado, Fürstenau-Oliveira, & Vieira-Soares, 2010

[Venezuela. Brazil: Amazonas, Goiás, Mato Grosso, Pará, Rondônia, Roraima, Tocantins]

Spongillida: *Drulia*: Species

1 Megascleres oxeas 2

1' Megascleres strongyles 3

2(1) Massive sponges .... *Drulia browni* (Bowerbank, 1863)

[Venezuela. Guyana. Brazil: Amazonas, Bahia, Coiás, Mato Grosso, Pará, Paraná, Rio Grande do Sul, Rondônia, Tocantins. Bolivia. Argentina]

2' Encrustant sponges .... *Drulia uruguayensis* Bonetto & Ezcurra de Drago, 1969

[Venezuela. Suriname. Brazil: Amazonas, Goiás, Mato Grosso, Mato Grosso do Sul, Minas Gerais, Pará, Rondônia, Roraima, São Paulo, Tocantins. Argentina]

3(1) Gemmuloscleres parmuliform with stems 4

3' Gemmuloscleres parmuliform without stems .... *Drulia conifera* Bonetto & Ezcurra de Drago, 1973

[Venezuela. Brazil: Amazonas]

4(3) Gemmuloscleres parmuliform with evident stems 5

4' Gemmuloscleres parmuliform with minute stems .... *Drulia cristinae* Volkmer-Ribeiro, Ezcurra de Drago, Machado & Sabaj, 2017

[Brazil: Pará]

5(4) Microscleres with conical spines .... *Drulia cristata* (Weltner, 1895)

[Venezuela. Brazil: Pará, Rondônia, Sergipe, Tocantins]

5' Microscleres with tuberculated spines .... *Drulia ctenosclera* Volkmer & Mothes, 1981

[Brazil: Amazonas, Goiás, Pará, Tocantins]

Spongillida: *Metania*: Species

1 Megascleres oxeas 2

1' Megascleres strongyles .... *Metania reticulata* (Bowerbank, 1863)

[Venezuela. Brazil: Amazonas, Goiás, Mato Grosso, Pará, Rio Grande do Sul, Tocantins]

2(1) More than one category of megascleres 3

2' Only one category of megascleres .... *Metania subtilis* Volkmer, 1979

[Brazil: Amazonas]

3(2) Gemmuloscleres with large shaft (>25 μm) 4

3' Gemmuloscleres with small shaft (<25 μm) .... *Metania fittkaui* Volkmer-Ribeiro, 1979

[Brazil: Amazonas, Rondônia]

4(3) Microscleres birotules present .... *Metania kiliani* Volkmer-Ribeiro & Costa, 1992

[Brazil: Amazonas]

4' Microscleres birotules absent .... *Metania spinata* (Carter, 1881)

[Suriname. Brazil: Amazonas, Goiás, Maranhão, Mato Grosso, Mato Grosso do Sul, Minas Gerais, Pará, Paraná, Roraima, São Paulo]

Spongillida: *Tubella*: Species

1 Megascleres only or predominantly strongyles 2

1' Megascleres only or predominantly oxeas 5

2(1) Megascleres stout (>20 μm) 3

2' Megascleres slender (<20 μm) 4

3(2) Gemmuloscleres with diameter of rotules >14 μm .... *Tubella repens* (Hinde, 1888)

[Panama. Brazil: Goiás, Mato Grosso, Mato Grosso do Sul, Minas Gerais, Pará, Tocantins, Paraná, Rio Grande do Sul, São Paulo, Santa Catarina. Paraguay. Argentina. Uruguay]

3' Gemmuloscleres with diameter of rotules <14 μm .... *Tubella amazonica* (Weltner, 1895)

[Brazil: Amazonas]

4(2) Diameter of rotules 2 times larger than length gemmuloscleres .... *Tubella gregaria* (Bowerbank, 1863)

[Venezuela. Brazil: Amazonas, Goiás, Mato Grosso, Pará, Tocantins. Bolivia]

4' Diameter of rotules less than length gemmuloscleres .... *Tubella lanzamirandai* (Bonetto & Ezcurra de Drago, 1964)

[Brazil: Amazonas, Goiás, Mato Grosso, Minas Gerais, Pará, Rio Grande do Sul, Tocantins. Argentina]

5(1) Megascleres strongly spined 6

5' Megascleres slightly spined or smooth 9

6(5) Gemmuloscleres with two rotules 7

6' Gemmuloscleres with only one rotule (other rotule vestigial) .... *Tubella pennsylvanica* (Potts, 1882)

[Brazil: Amazonas, Mato Grosso, Rondônia. Nearctic]

7(6) Gemmuloscleres with diameter of rotules >10 μm........ 8

7' Gemmuloscleres with diameter of rotules <10 μm........ *Tubella minuta* (Potts, 1881)

[Venezuela. Brazil: Amazonas, Mato Grosso, Rio Grande do Sul. Bolivia. Argentina]

8(7) Megascleres always strongly spined .... *Tubella horrida* (Weltner, 1893)

[Panama]

8' Megascleres can be strongly and slightly spined .... *Tubella variabilis* Bonetto & Ezcurra de Drago, 1973

[Costa Rica. Brazil: Amazonas, Maranhão, Mato Grosso, Mato Grosso do Sul, Minas Gerais, Pará, Paraná, Roraima, Rio Grande do Sul, São Paulo, Tocantins. Argentina]

9(5) Megascleres length <200μm 10

9' Megascleres length >200μm .... *Tubella delicata* (Bonetto & Ezcurra de Drago, 1967)

[Brazil: Amazonas, Bahia, Goiás, Mato Grosso, Rio Grande do Sul, Tocantins. Argentina]

10(9) Gemmuloscleres with rotules unequal diameters .... *Tubella paulula* (Bowerbank, 1863)

[Venezuela. Suriname. Brazil: Alagoas, Amazonas, Bahia, Goiás, Mato Grosso, Minas Gerais, Pará, Rio Grande do Sul, Tocantins. Argentina]

10' Gemmuloscleres with rotules equal diameters .... *Tubella leidii* (Bowerbank, 1863)

[Nearctic: Florida]

Spongillida: *Dosilia*: Species

1 Megascleres only acanthoxeas .... *Dosilia palmeri* (Potts, 1885)

[Mexico. Nearctic]

1' Megascleres only smooth oxeas .... *Dosilia pydanieli* Volkmer-Ribeiro, 1992

[Brazil: Ceará, Goiás, Mato Grosso do Sul, Maranhão, Minas Gerais, Paraná, Rio Grande do Norte, Roraima, São Paulo]

Spongillida: *Ephydatia*: Species

1 Gemmuloscleres with spined shaft 2

1' Gemmuloscleres with smooth shaft .... *Ephydatia fluviatilis* (Linnaeus, 1759)

[Mexico. El Salvador. Virgin Islands. Nearctic]

2(1) Simple and compose spines on the shaft of gemmuloscleres 3

2' Only simples spines on the shaft of gemmuloscleres .... *Ephydatia robusta* (Potts, 1888)

[Mexico. Nearctic]

3(2) Absence of microspines on the shaft of gemmuloscleres 4

3' Presence of microspines on the shaft of gemmuloscleres .... *Ephydatia chileana* Pisera & Sáez, 2003

[Chile]

4(3) Gemmuloscleres with entired rotule .... *Ephydatia facunda* Weltner, 1895

[Cuba. Brazil: Alagoas, Rio Grande do Sul. Peru. Argentina]

4' Gemmuloscleres with toothed rotule .... *Ephydatia caatingae* Nicacio & Pinheiro, 2015

[Brazil: Alagoas, Pernambuco]

Spongillida: *Anheteromeyenia*: Species

1 Only one category of megascleres 2

1' Two categories of megascleres .... *Anheteromeyenia vitrea* Buso & Volkmer-Ribeiro, 2012

[Brazil: Espírito Santo]

2(1) Gemmuloscleres pseudobirotules 3

2' Gemmuloscleres acanthoxeas and acanthostrongyles .... *Anheteromeyenia diamantina* Calheira & Pinheiro, 2018
[Brazil: Bahia]

3(2) Megascleres acanthoxeas gradually pointed .... *Anheteromeyenia cheguevarai* Manconi & Pronzato, 2005
[Cuba]

3' Megascleres acanthoxeas abruptly pointed .... *Anheteromeyenia ornata* (Bonetto & Ezcurra de Drago, 1970)
[Brazil: Amazonas. Argentina: Misiones]

Spongillida: *Racekiela*: Species

1 Gemmuloscleres only birotules 2

1' Gemmuloscleres could be birotules, pseudobirotules or acanthoxeas .... *Racekiela andina* Hernandez & Barreat, 2017
[Venezuela: Mérida]

2(1) Gemmuloscleres with large rotules (20–60 μm) .... *Racekiela cavernicula* Volkmer-Ribeiro, Bichuette & De Sousa Machado, 2010
[Brazil: Bahia]

2' Gemmuloscleres with short rotules (25–35 μm) .... *Racekiela sheilae* (Volkmer-Ribeiro, De Rosa-Barbosa & Tavares, 1988)
[Brazil: Rio Grande do Sul]

Spongillida: *Heteromeyenia*: Species

1 Only one category of gemmuloscleres 2

1' Two categories of gemmuloscleres .... *Heteromeyenia insignis* Weltner, 1895
[Brazil: Rio Grande do Sul]

2(1) Rotule with surface entirely covered with microspines .... *Heteromeyenia cristalina* Batista, Volkmer-Ribeiro & Melão, 2007
[Brazil: Mato Grosso]

2' Rotule with surface entirely smooth .... *Heteromeyenia barlettai* Pinheiro, Calheira & Hajdu, 2015
[Brazil: São Paulo]

Spongillida: *Corvomeyenia*: Species

1 Gemmuloscleres equal rotules .... *Corvomeyenia epilithosa* Volkmer-Ribeiro, Rosa-Barbosa & Machado, 2005
[Brazil: Rio Grande do Sul]

1' Gemmuloscleres unequal rotules .... *Corvomeyenia thumi* (Traxler, 1895)
[Brazil: Bahia, Ceará, Maranhão, Mato Grosso do Sul, Minas Gerais, Pará, Roraima, Rio Grande do Norte, São Paulo]

Spongillida: *Corvoheteromeyenia*: Species

1 Microscleres acanthoxeas absent 2

1' Microscleres acanthoxeas present .... *Corvoheteromeyenia heterosclera* (Ezcurra de Drago, 1974)
[Venezuela. Brazil: Bahia, Ceará, Goiás, Maranhão, Pernambuco]

2(1) Gemmuloscleres birotules .... *Corvoheteromeyenia australis* (Bonetto & Ezcurra de Drago, 1966)
[Curaçao Brazil: Mato Grosso, Rio Grande do Sul. Argentina]

2' Gemmuloscleres sanidasters .... *Corvoheteromeyenia sanidastosclera* Pinheiro, Silva & Calheira, 2015
[Brazil: Bahia]

Spongillida: *Spongilla*: Species

1 Microscleres with simple and compose spines........ *Spongilla alba* Carter, 1849
[Venezuela. El Salvador. Curaçao. Brazil: Alagoas, Amazonas, Rio de Janeiro. Nearctic]

1' Microscleres with only simple spines .... *Spongilla cenota* Penney & Racek, 1968
[Mexico. Costa Rica. Nearctic]

Spongillida: *Radiospongilla*: Species

1 Megascleres only oxeas 2

1' Megascleres strongyles and oxeas .... *Radiospongilla amazonenses* Volkmer & Maciel, 1983

[Brazil: Alagoas, Amazonas, Ceará, Goiás, Maranhão, Mato Grosso, Mato Grosso do Sul, Minas Gerais, Paraná, Rio Grande do Norte, Rio Grande do Sul, Roraima, São Paulo, Tocantins. Argentina]

2(1) Gemmuloscleres pseudobirotules .... *Radiospongilla crateriformis* (Potts, 1882)

[Mexico. Cuba. Barbados. Suriname. Brazil. Nearctic]

2' Gemmuloscleres acanthostrongyles .... *Radiospongilla inesi* Nicacio & Pinheiro, 2011

[Brazil: Pernambuco]

Spongillida: *Eunapius*: Species

1 Megascleres smooth oxeas 2

1' Megascleres acanthoxeas .... *Eunapius igloviformis* (Potts, 1884)

[Nearctic: Florida]

2(1) Short gemmuloscleres (110–140 μm) .... *Eunapius fragilis* (Leidy, 1851)

[Mexico. Brazil: Mato Grosso do Sul; Rio Grande do Sul; São Paulo. Chile. Argentina. Nearctic]

2' Large gemmuloscleres (150–210 μm) .... *Eunapius carteri* (Bowerbank, 1863)

[Panama]

Spongillida: *Oncosclera*: Species

1 Megascleres only oxeas or strongyles 2

1' Megascleres strongyles and oxeas .... *Oncosclera ponsi* (Bonetto & Ezcurra de Drago, 1968)

[Argentina: Entre Ríos]

2 (1) Megascleres only strongyles (Fig. 3.5 D) 3

2' Megascleres only oxeas (Fig. 3.5 A) 5

3(2) Megascleres strongyles microspined/granulated 4

3' Megascleres acanthostrongyles (Fig. 3.5 G) .... *Oncosclera spinifera* (Bonetto & Ezcurra de Drago, 1973)

[Venezuela. Brazil: Amazonas, Goiás, Mato Grosso, Pará, Rondônia, Tocantins. Argentina]

4(3) Megascleres entirely microgranulated (Fig. 3.5 E) .... *Oncosclera atrata* (Bonetto & Ezcurra de Drago, 1970)

[Argentina: Misiones]

4' Megascleres irregularly microspines (Fig. 3.5 F) .... *Oncosclera tonollii* (Bonetto & Ezcurra de Drago, 1968)

[Brazil: Paraná, Tocantins. Argentina: Entre Ríos]

5(2) Megascleres oxeas predominantly smooth 6

5' Megascleres oxeas predominantly spined 9

6(5) Gemmuloscleres predominantly oxeas 7

6' Gemmuloscleres only strongyles 8

7(6) Gemmuloscleres exclusively oxeas boomerang-shaped (Fig. 3.3 C) .... *Oncosclera navicella* (Carter, 1881)

[Venezuela. Brazil: Amazonas, Bahia, Goiás, Mato Grosso, Mato Grosso do Sul, Minas Gerais, Pará, Paraná, Rio Grande do Sul, Rondônia, São Paulo, Tocantins. Argentina. Uruguay]

7' Gemmuloscleres ranging from oxeas to strongyles .... *Oncosclera petricola* Bonetto & Ezcurra de Drago, 1967

[Brazil: Mato Grosso, Rio Grande do Sul. Argentina: Misiones]

8(6) Gemmuloscleres with spines only on the tips (Fig. 3.3 L) .... *Oncosclera jewelli* (Volkmevr, 1963)

[Brazil: Paraná, Rio Grande do Sul]

8' Gemmuloscleres with spines scattered on the surface .... *Oncosclera rosariae* Tavares-Frigo, Volkmer-Ribeiro, Oliveira & Machado, 2015

[Brazil: Rio Grande do Sul]

9(5) Gemmuloscleres ranging from regular to irregular strongyles 10

9' Gemmuloscleres exclusively regular strongyles .... *Oncosclera intermedia* Bonetto & Ezcurra de Drago, 1973
[Venezuela. Brazil: Amazonas, Pará]

10(9) Gemmuloscleres ranging from regular to mucronate strongyles .... *Oncosclera stolonifera* Bonetto & Ezcurra de Drago, 1967
[Argentina]

10' Gemmuloscleres ranging from regular to centrotylote strongyles .... *Oncosclera schubarti* (Bonetto & Ezcurra de Drago, 1967)
[Brazil: Goiás, Rio Grande do Sul, Tocantins. Argentina: Misiones]

REFERENCES

Brasil. 2014. Portaria MMA no 445, de 17 de dezembro de 2014. Reconhece as espécies de peixes e invertebrados aquáticos da fauna brasileira ameaçadas de extinção. Diário Oficial da República Federativa do Brasil, Brasília, DF, 245, 18 dez. Seção I, p. 126–130.

Hajdu, E., S. Peixinho & J.C.C. Fernandez. 2011. Esponjas marinhas da Bahia: guia de campo e laboratório. Série Livros, 45. Museu Nacional, RJ. 276 pp.

Manconi, R. & R. Pronzato. 2002. Chapter 101: Suborder Spongillina subord. nov.: Freshwater sponges. Pages 921–1020 *in:* J.N.A. Hooper and R.W.M. Van Soest (eds), Systema Porifera. A guide to the classification of sponges. Vol. 1, Kluwer Academic/Plenum Publishers, New York.

Manconi, R. & R. Pronzato. 2007. Gemmules as a key structure for the adaptive radiation of freshwater sponges: A morpho-functional and biogeographical study. Pages 61–77 *in*: M.R. Custódio, G. Lôbo-Hajdu, E. Hajdu and M. Muricy (eds), Porifera Research: Biodiversity, Innovation, Sustainability, Serie Livros 28, Museu Nacional, RJ.

Manconi, R. & R. Pronzato. 2016. How to survive and persist in temporary freshwater? Adaptive traits of sponges (Porifera: Spongillida): A review. Hydrobiologia 782: 11–22.

Morrow C. & P Cárdenas. 2015. Proposal for a revised classification of the Demospongiae (Porifera). Frontiers in Zoology 12: 1–27.

Nicacio, G. & U. Pinheiro. 2015. Biodiversity of freshwater sponges (Porifera: Spongillina) from northeast Brazil: new species and notes on systematic. Zootaxa 3981: 220–240.

Pinheiro, U.S. 2007. Contribuições à Taxonomia e Biogeografia das Esponjas de Águas Continentais Brasileiras. Ph.D. Thesis, Universidade Federal do Rio de Janeiro, Rio de Janeiro: Museu Nacional. 220 p.

Pinheiro, U., L. Calheira & E. Hajdu. 2015. A new species of freshwater sponge, *Heteromeyenia barlettai* sp. nov. from an aquarium in São Paulo, Brazil (Spongillida: Spongillidae). Zootaxa 4034: 351–363.

Reiswig, H.M., T.M. Frost & A. Ricciardi. 2010. Porifera. Pages 91–123 *in:* J.H. Thorp and A.P. Covich (eds), Ecology and Classification of North American Freshwater Invertebrates, Academic Press, Elsevier, Boston, MA.

Van Soest, R.W.M., N. Boury-Esnault, J.N.A. Hooper, K. Rützler, N.J. de Voogd, B. Alvarez de Glasby, E. Hajdu, A.B. Pisera, R. Manconi, C. Schoenberg, D. Janussen, K.R. Tabachnick, M. Klautau, B. Picton, M. Kelly & J. Vacelet. 2018. World Porifera Database (WPD). http://www.marinespecies.org/porifera, accessed on 2018-17-05.

Volkmer-Ribeiro, C. 1986. Evolutionary study of the freshwater sponge genus *Metania* Gray, 1867. Metaniidae, new family. Amazoniana 9: 493–509.

Chapter 4

Phylum Cnidaria

María Irene Deserti
Instituto de Investigaciones Marinas y Costeras, Universidad Nacional de Mar del Plata – CONICET, Mar del Plata, Buenos Aires, Argentina

Priscila A. Grohmann
Instituto de Biologia, Departamento de Zoologia, Universidade Federal do Rio de Janeiro, Rio de Janeiro, Brazil

Sérgio N. Stampar
Departamento de Ciências Biológicas, Universidade Estadual Paulista (UNESP), FCL/Campus de Assis, São Paulo, Brazil

Chapter Outline

INTRODUCTION

Cnidaria constitutes a group of about 11,000–13,000 living species and is represented by a large number of very different body forms, from beautiful jellyfish to delicate polyps. They are one of the lower invertebrates of the Eumetazoa branch and considered the sister group of Bilateria. They are also one of the first taxa to exhibit a symmetrical body plan, an intraepithelial nervous system forming a nerve net, a tissue grade of construction and germ cells (Nielsen, 2012), and the most conspicuous apomorphy is the presence of cnidae (only nematocysts in freshwater species). This group shows a large morphological diversity, especially by alternating generations (metagenesis, polyp vs. medusa) in many representatives (Morandini et al., 2016). Cnidarians are almost exclusively marine, with few species living in estuarine, lagoons, and/or freshwater environments. Cnidaria is divided into two subphyla, Medusozoa and Anthozoaria (Collins et al., 2006; Stampar et al., 2014; Zapata et al., 2015). The major difference between these two groups is related to their life cycle, because medusozoan species have an additional stage between polyp and planula—the medusa (Morandini et al., 2016).

All freshwater species are members of Medusozoa (Daly et al., 2007). Worldwide, six genera of the class Hydrozoa are recognized in continental waters: the common and cosmopolitan *Hydra*, the colonial *Cordylophora*, jellyfishes *Craspedacusta* and *Limnocnida* (not represented in Neotropical region), the little polyp *Calpasoma* (Bouillon & Boero, 2000), and the unusual polyp *Polypodium hydriforme* Ussow, 1887, a parasite of fish eggs (Jankowski et al., 2008) (Figs. 4.1–4.4).

In addition, the Myxozoa (Endocnidozoa) (Fig. 4.5), a fairly specialized group of Cnidaria that are microscopic obligate endoparasites of aquatic vertebrates and invertebrates (Foox & Siddall, 2015; Kayal et al., 2018) has traditionally been classified as protists, but early studies with molecular data have shown their affinity with Cnidaria (Siddall et al., 1995; Foox & Siddall, 2015). This group currently includes more than 2,300 species in 64 genera and two subclasses (Morris, 2010; Fiala et al., 2015) and is the largest assemblage of freshwater cnidarians. This group has more than 100 species reported from Brazil, much more diversity than in any other freshwater region of the world (Oliveira et al., 2016). However, endoparasites such as these are outside the scope of this work.

Hydra is the most representative freshwater cnidarian. It occurs on all continents, except Antarctica, inhabiting most permanent meso or eutrophic continental waters (Jankowski et al., 2008). Specimens are usually more abundant in

Thorp and Covich's Freshwater Invertebrates. https://doi.org/10.1016/B978-0-12-804225-0.00004-6

FIGURE 4.1 *Hydra* specimens: (A) brown hydra with one bud and early development of an ovary in the lower half of the column (about 15 mm long); and (B) brown hydra with two buds (about 10 mm long). *Photos by Irene Deserti.*

lentic waters, often attached to submerged and floating macrophytes, or on the underside of leaves and sticks when plants are not abundant. In swift streams and rivers, they are usually found on the underside of stones and sticks. Hydras are strictly carnivorous and very efficient predators. Prey include very small crustaceans, annelids, rotifers, insects, and fish larvae (Schwartz et al., 1983; Elliot et al., 1997).

Cordylophora colonizes brackish waters of estuaries, lagoons, and coastal lakes, usually in salinities from 0 to 30 ppm. It occurs in shallow waters, often in shade, attached to hard substrates and submerged vegetation. The species is invasive and often causes serious clogging problems in piping systems used for the cooling of machines (Folino-Rorem & Indelicato, 2005; Grohmann, 2008).

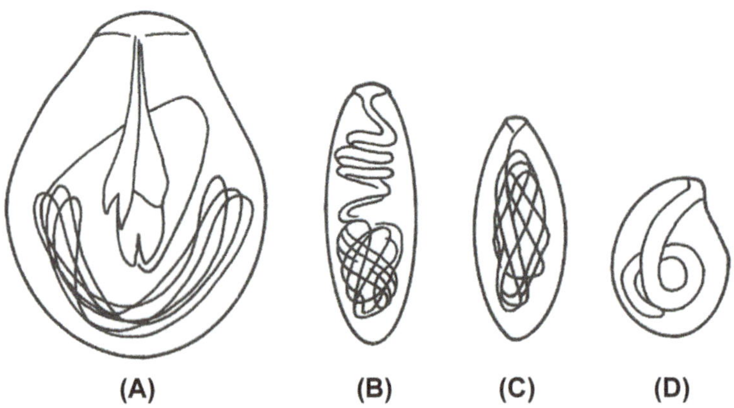

FIGURE 4.2 Types of nematocyst: (A) stenotele; (B) holotrichous isorhiza; (C) atrichous isorhiza; and (D) desmoneme. *Modified from Campbell (1983).*

FIGURE 4.3 *Cordylophora caspia* with gastrozooids and gonophores buds (about 20 mm long). *Photos by Otto M. P. Oliveira.*

Polyps of *Craspedacusta* species are permanently attached to various submerged substrates and grow as single hydranths or, more commonly, as small colonies of two to seven simple hydranths joined at their base. These asexually produce three types of structures: hydrant buds, frustules or planuloids, and medusoid buds, the latter are released as free-swimming medusae (Lytle, 1961). The small polyp can remain undetected for years before a medusa "bloom." These colonies bloom seasonally in subtropical to temperate lentic habitats. They are predators on a variety of zooplankton like cladocerans and copepods.

Calpasoma occurs attached to plant substrates, common in natural and artificial ponds, and feeds on rotifers, nematodes, and other micro invertebrates.

LIMITATIONS

The taxonomy of *Hydra* is characterized by a high number of synonymous names. All *Hydra* species constitute a monophyletic clade (Collins et al., 2006). The species are clustered into four groups: *viridissima* (green hydras), *vulgaris* (common or brown hydras), *oligactis* (stalked hydras), and *braueri* (gracile hydras). These morphological groups are corroborated by DNA analyses (Campbell, 1983; Martínez et al., 2010). However, some morphological features are present only when the animal is healthy and budding. For this reason, cultures are recommended. The externally formed early embryo secretes the embryotheca, and is unique for each group (Campbell, 1983). Unfortunately, it is not common to collect and culture specimens in reproductive condition. Species recognition is not easy in the *vulgaris* group as a result of morphological variability.

Cordylophora caspia Pallas, 1771 may have originated in the Ponto-Caspian region and the genus is currently the focus of numerous invasive species studies. Folino-Rorem et al. (2009) identified several lineages, each with different salinity tolerances, which probably dispersed via ship ballast waters and/or hull biofouling. Some lineages have wide geographic ranges, although different lineages are also frequently sympatric. Molecular evidence suggests multiple cryptic species under the name *C. caspia* (Schable et al., 2008).

Craspedacusta sowerbii Lankester, 1880 is the only species with a medusa stage and is cosmopolitan (Jankowski, 2001). This short-lived pelagic medusa is the most noticeable stage of the species. Its occurrence is sporadic, lasting only a few weeks, usually in late summer and/or autumn. The minute tentacle-less polypoid stage was first described as *Microhydra ryderi* Potts, 1885. Its occurrence is usually based on the presence of the medusae.

Calpasoma dactylopterum Fuhrmann, 1939 is described in some reviews as a special tentaculate form of *Craspedacusta,* and some authors observed the transformation of the *Craspedacusta* polypoid into a *Calpasoma*-like polyp (Holstein, 1995). Some studies support recognition of both as distinct species, although Jankowski (2001) considered at least one form of *Calpasoma* to be congeneric with *Craspedacusta*.

TERMINOLOGY AND MORPHOLOGY

The morphological differences among *Hydra, Cordylophora,* and *Craspedacusta* are obvious. However, the *Craspedacusta* polyp stage and *Calpasoma* are difficult to separate. Nevertheless, they may be different forms of the same species, *Craspedacusta sowerbii*.

Hydra is usually a sedentary **polyp**, with a column length ranging from 1 to 20 mm. A single animal is tubular, with a distal mouth surrounded by one to seven filiform (or moniliform) **tentacles** (Fig. 4.1). Species identification is based on the body form (slenderness of the stalk), presence or absence of symbiotic algae, arrangement and length of tentacles, the young **buds,** and the nature of the **embryotheca** (a secreted chitinous theca which protects soft tissue). The sizes and shapes of undischarged **nematocysts** are another important character. The four types of nematocyst (**stenotele**, desmoneme, atrichous, and **holotrichous isorhiza** (Fig.4.2)) are easily identifiable, since the shapes

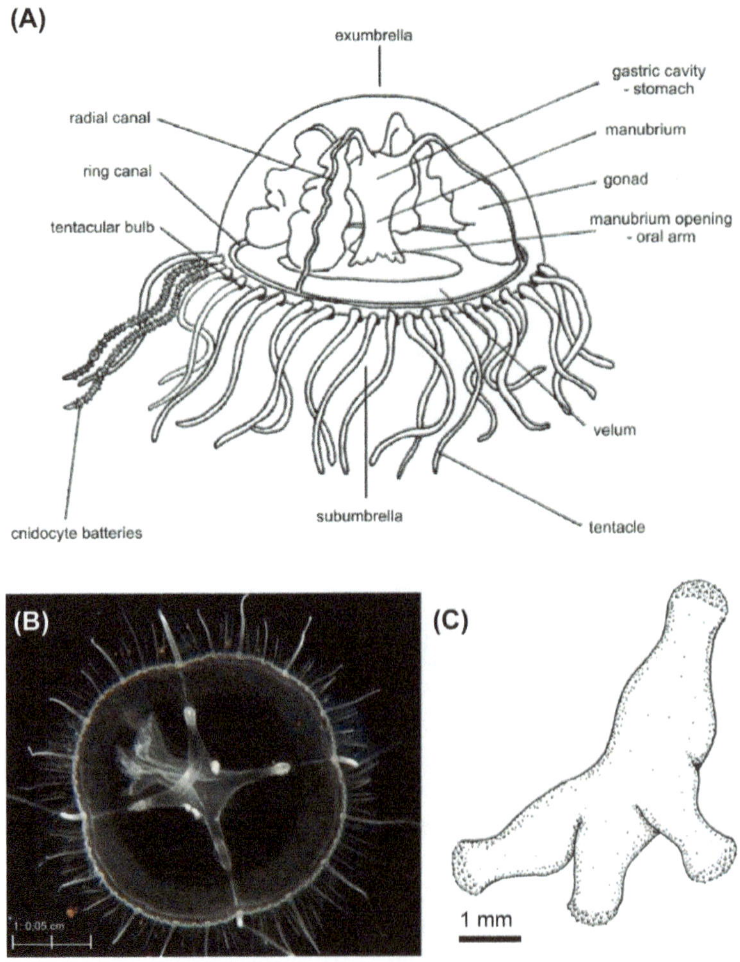

FIGURE 4.4 *Craspedacusta sowerbii*: (A) radial symmetric body plan of the medusa, divided in subumbrella and exumbrella regions from a side projection, with four gonads and up to 80 tentacles, body size is up to 2.5 cm in diameter; (B) aboral view of living specimen, scale bar = 0.05 cm; (C) atentacular hydrants (about 1 mm long). *Modified from: (A) Hroudova et al. (2012); (C) Matthews (1966). (B) Photo by Sergio N. Stampar.*

are very distinct. The holotrichous isorhiza is the less common type, but the most important (Campbell, 1983).

Cordylophora **hydrants** are fusiform, whitish to brownish, and bear 15 to 20 filiform **tentacles** surrounding the mouth (Fig. 4.3). Colonies are polymorphic, with polyps specialized for feeding (**gastrozooids**) or reproduction (**gonophores**). The colony survives during unfavorable conditions via spheres of coenosarc tissue, called **menonts**. *Cordylophora* is a clonal organism that grows asexually, producing additional hydranths to the upright branches (**hydrocauli**) or extending the stolons (hydrorhiza), so producing more hydrocauli and hydranths. This structure is ecologically resilient and regenerates the colonies when favorable conditions are restored.

Craspedacusta is commonly observed in the **medusa** stage, which is the typical jellyfish form. It is very delicate and translucent, varying from 8 to 16 mm. It has a whorl of 40–80 **tentacles** tightly packed around the **bell margin**, a large **manubrium,** and a mouth with four frilly **lips** (Fig. 4.4). The life cycle also includes a nearly transparent microscopic polypoid stage devoid of tentacles.

Calpasoma specimens are small and solitary polyps ranging from 0.2 to 0.3 mm in length, cylindrical in shape, and somewhat similar to the polypoid stage of *Craspedacusta*. The distal part has a lighter color, and tentacles usually appear in two close whorls or are irregularly arranged around the mouth. These specimens reproduce asexually originating propagules from tentacles. These vegetative buds rest only a few hours, and then develop some hyaline tentacles soon after the settlement (Matthews, 1966; Dioni, 1974).

PREPARATION AND PRESERVATION

Cnidarians must be handled carefully, as their tissues are easily damaged. Water movement can also cause the specimen to contract and obscure important characters (mainly in *Hydra*).

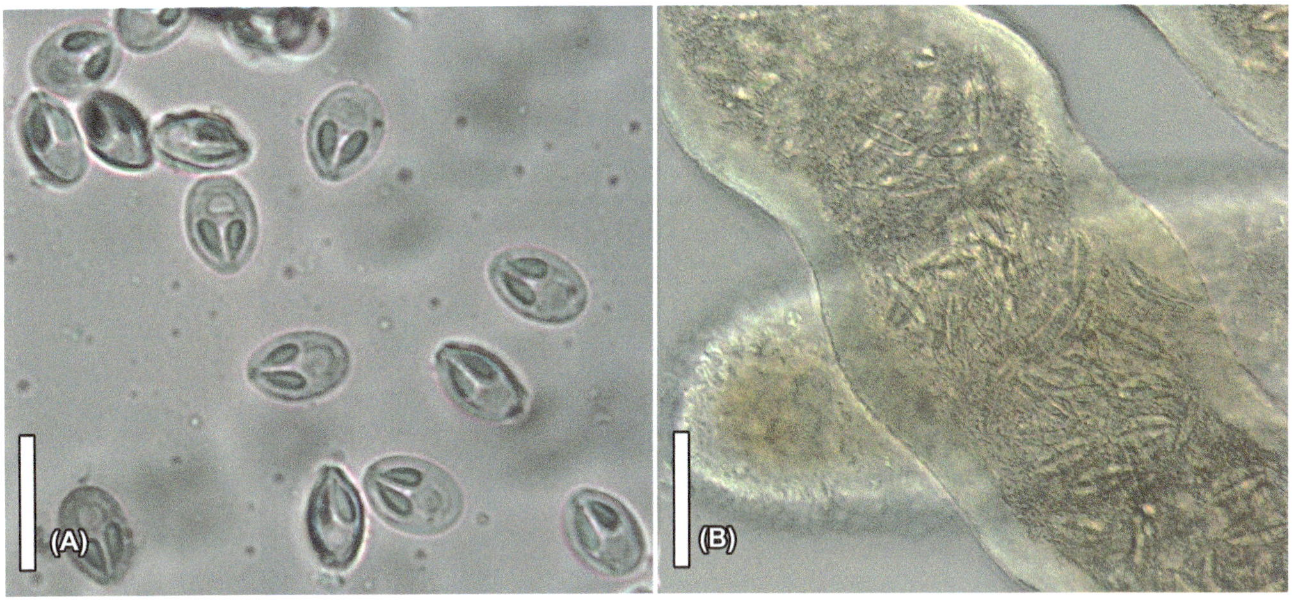

FIGURE 4.5 Myxozoa: (A) *Myxobolus flavus* and (B) *Myxidium amazonense* Scale bar: 50 μm. *Photos by Edson E. Adriano.*

For small organisms flood the specimen quickly in 70%–100% alcohol. Larger animals should be placed individually in a capsule until they relax and then add small amounts of alcohol with a pipette. This process must be done gently, to prevent movement in the surrounding water causing the animal to contract. Once the animal is killed, carefully remove it from the capsule and place it in alcohol. To release the organisms from the substrate where they are attached, use a fine needle. Although the alcohol keeps the morphology and preserves the material for molecular study, the animals lose their coloration.

The final preservation technique involves placing a specimen in an excavated slide, adding a few drops of water, and letting the animal relax. Formalin is added carefully to the edge of the excavation. Then, cover the slide with a cover slip. Remove excess liquid with a small absorbent paper and the edges of the coverslip are sealed. The animal can then be observed under magnifying glass or microscope, but this only allows observation of the body and some structures like ovaries and testes. The animal maintains the color and morphology and is not damaged through manipulation.

The use of a fresh tissue squash is recommended to measure and describe the nematocysts, since fixation destroys the material. In *Hydra* and *Calpasoma* and *Craspedacusta* polyps, use the whole animal; for *Cordylophora* colonies, remove a hydrant that lacks a perisarc (the outer usually chitinous integument), and for *Craspedacusta* medusa cut some tentacles from the medusa. The pressure should be low to prevent discharge of the capsules. A discharged nematocyst usually loses its taxonomic value for measurements but is essential for precise identification.

A drop of water is used to help the disaggregation of material. Use a 100× lens in immersion oil for observation. The use of lactophenol for nematocyst preparations allows the preservation of the material for at least 1 month if the edges of cover slip are sealed. Over time, the limits of the capsules lose sharpness and measurement is almost impossible.

For the two nematocyst types of *Cordylophora,* Wollschlager et al. (2013) used a different treatment for freshwater and brackish populations. For freshwater hydrants, excess water must be removed, and a Giemsa stain added. The excess stain can be removed after 30 min. For brackish populations, rinse the hydrants with deionized water to remove excess salts; the excess water must be removed as well. Then use Giemsa stain to dye the hydrants for at least 4 h. For both populations, rinse the stained hydrants with acid alcohol to help dissociating the tissues and remove the excess stain. Finally, add a drop of Permount to the stained hydrant. In all cases, observations can be made using a 100× lens with immersion oil.

KEYS TO CNIDARIA

Class Hydrozoa: Families

Note: *Cordylophora caspia* is frequently mentioned as its synonym *C. lacustris* Allman, 1844.

1 Medusae or solitary polyps with or without reproductive buds 2

1' Branching colonies with polyps with several tentacles and a stolon attached to a substratum .... Oceaniidae, one species: *Cordylophora caspia* Pallas, 1771
[Mexico. Panama. Brazil. Chile. Argentina. Uruguay.]

2(1) Medusae, or polyps $<$ 1 mm in length .... **Olindiidae [p. 98]**

2' Single polyps $>$1 mm in length .... Hydridae, one genus: ***Hydra* [p. 98]**

Hydrozoa: Olindiidae: Genera

Note: these two forms may be conspecific.

1 Polypoid and medusoid forms; polyps colonial, lacking tentacles .... *Craspedacusta sowerbii* Lankester, 1880
[Mexico. Panama. Venezuela. Brazil. Chile. Argentina]

1' Only polypoid forms; solitary polyps, with tentacles .... *Calpasoma dactylopterum* Furhmann, 1939
[Brazil, Argentina, Uruguay]

Hydrozoa: Hydridae: *Hydra*: Species

1 Polyps not green; stenotele $>$11 μm in length; embryotheca either not piled or not roughly spherical 2

1' Polyps intensively green due to endosymbiotic algae; nematocysts small; *viridissima* group 4

2(1) Holotrichous isorhiza narrowly oval or reniform, width $<$0.5$\times$ length; embryotheca not flattened 3

2' Holotrichous isorhiza broadly fusiform; embryotheca flattened against substratum .... *braueri* group

3(2) Young bud lateral tentacles arising conspicuously early; conspicuous peduncle near foot; embryotheca smooth and thin .... *oligactis* group

3' Young bud tentacles arising in slightly staggered sequence or discontinuous; embryotheca thick with numerous radial spines .... *vulgaris* group

4(1) Holotrichous isorhiza reniform; embryotheca with one narrow chamber proximal to embryo .... *Hydra viridissima* Pallas, 1766
[Cosmopolitan]

4' Holotrichous isorhiza stubby, not reniform; embryotheca with basal chambers in young embryos ..... *Hydra plagiodesmica* Dioni, 1968
[Argentina]

REFERENCES

Bouillon, J. & F. Boero. 2000. The hydrozoa: a new classification in the light of old knowledge. Thalassia Saentina 24: 3-45.

Campbell, R.D. 1983. Identifying *Hydra* species. Pages 19–28 *in:* H.M. Lenhoff (ed.), *Hydra* Research Methods. Plenum Press, New York.

Collins, A.G., P. Schuchert, A.C. Marques, T. Jankowski, M. Medina & B. Schierwater. 2006. Medusozoan phylogeny and character evolution clarified by new large and small subunit rDNA data and an assessment of the utility of phylogenetic mixture models. Systematic Biology 55: 97–115.

Daly, M., M.R. Brugler, P. Cartwright, A.G. Collins, M.N. Dawson, D.G. Fautin, S.C. France, C.S. McFadden, D.M. Opresko, E. Rodriguez & S.L. Romano. 2007. The phylum Cnidaria: A review of phylogenetic patterns and diversity 300 years after Linnaeus. Zootaxa 1668: 127–182.

Dioni, W. 1974. Noticia sobre un raro hidroide de agua dulce de la Argentina y Uruguay, *Calpasoma dactyloptera.* Physis B 33(86): 147–150.

Elliot, J.K., J.M. Elliot & W.C. Leggett. 1997. Predation by *Hydra* on larval fish: Field and laboratory experiments with bluegill (*Lepomis macrochirus*). Limnology and Oceanography 42: 1416–1423.

Fiala, I., P. Bartošová-Sojková & C.M. Whipps. 2015. Classification and Phylogenetics of Myxozoa. Pages 85–110 *in*: B. Okamura, A. Gruhl and J.L. Bartholomew (eds.), Myxozoan Evolution, Ecology and Development. Springer, Heidelberg.

Folino-Rorem, N.C., J.A. Darling & C.A. D'Ausilio. 2009. Genetic analysis reveals multiple cryptic invasive species of the hydrozoan genus *Cordylophora.* Biological Invasions 11: 1869–1882.

Folino-Rorem, N.C. & J. Indelicato. 2005. Controlling biofouling caused by the colonial hydroid *Cordylophora caspia.* Water Research 39: 2731–2737.

Foox, J. & M.E. Siddall, 2015. The Road to Cnidaria: History of Phylogeny of the Myxozoa. Journal of Parasitology 101: 269–274.

Grohmann, P.A. 2008. Bioincrustation caused by a hydroid species in the turbine cooling system at the Funil hydroelectric power plant, Itatiaia, Rio de Janeiro, Brazil. Naturalia 31: 16–21.

Holstein, T. 1995. Cnidaria: Hydrozoa. Pages 67–101 *in:* J. Schwoerbel and P. Zwick (ed.), Süßwasserfauna von Mitteleuropa, Vol. 1–2. Gustav Fischer, Stuttgart.

Hroudova, M., P. Vojta, H. Strnad, Z. Krejcic, J. Ridl, J. Paces, C. Vlcek & V. Paces. 2012. Diversity, phylogeny and expression patterns of Pou and six homeodomain transcription factors in hydrozoan jellyfish *Craspedacusta sowerbyi*. PLoS One 7: e36420.

Jankowski, T. 2001. The freshwater medusae of the world – a taxonomic and systematic literature study with some remarks on other inland water jellyfish. Hydrobiologia 462: 91–113.

Jankowski, T., A.G. Collins & R. Campbell. 2008. Global diversity of inland water cnidarians. Hydrobiologia 595: 35–40.

Kayal, E., B. Bentlage, M.S. Pankey, A.H. Ohdera, M. Medina, D.C. Plachetzki, A.G. Collins & J.F. Ryan. 2018. Phylogenomics provides a robust topology of the major cnidarian lineages and insights on the origins of key organismal traits. BMC Evolutionary Biology 18: 68.

Lytle, C.F. 1961. Patterns of budding in the freshwater hydroid *Craspedacusta*. Pages 317–336 *in*: H.M. Lenhoff and W.F. Loomis (eds.), The Biology of *Hydra* and of some other Coelenterates. University of Miami Press, Coral Gables.

Matthews, D. 1966. A comparative study of *Craspedacusta sowerbyi* and *Calpasoma dactyloptera* life cycles. Pacific Sciences 20: 246–259.

Martínez, D.E., A.R. Iñiguez, K.M. Percell, J.B. Willner, J. Signorovitch & R.D. Campbell. 2010. Phylogeny and biogeography of *Hydra* (Cnidaria: Hydridae) using mitochondrial and nuclear DNA sequences. Molecular Phylogenetics and Evolution 57: 403–410.

Morandini, A.C., A. Schiariti, S.N. Stampar, M.M. Maronna, I. Straehler-Pohl & A.C. Marques. 2016. Succession of generations is still the general paradigm for scyphozoan life cycles. Bulletin of Marine Science 92: 343–351.

Morris, D.J. 2010. Cell formation by myxozoan species is not explained by dogma. Proceedings of the Royal Society B 277: 2565–2570.

Nielsen, C. 2012. Animal Evolution. Oxford, New York.

Oliveira O.M.P., A.E. Migotto, S.N. Stampar, M.V. Kitahara, M.E.A. Santos, J.S. Beneti, E.A. Adriano, L.S. Miranda, A.C. Morandini, T.P. Miranda, M.A. Mendoza-Becerril, A.C. Marques, E.M. Araujo, E.Y. Nishiyama, A.F. Cunha & R.T.S. Cordeiro. 2016. Cnidaria: Catálogo Taxonômico da Fauna do Brasil. Accessed February 2019, http://fauna.jbrj.gov.br/fauna/listaBrasil/FichaPublicaTaxonUC/FichaPublicaTaxonUC.do?id=22.

Schable, N.A., A.M. Kuenzi, C.A. Drake, N.C. Folino-Rorem & J.A. Darlings. 2008. Microsatellite loci for the invasive colonial hydrozoan *Cordylophora caspia*. U.S. Environmental Protection Agency Papers. Paper 151.

Schwartz, S.S., B.J. Hann & P.D.N. Hebert. 1983. The feeding ecology of *Hydra* and possible implications in the structuring of pond zooplankton communities. Biological Bulletin 164: 136–142.

Siddall, M., D.S. Martin, D. Bridge, S.S. Desser & D.K. Cone. 1995. The demise of a phylum of protists: Phylogeny of Myxozoa and other parasitic Cnidaria. Journal of Parasitology 81: 961–967.

Stampar S.N., M.M. Maronna, M.V. Kitahara, J.D. Reimer & A.C. Morandini. 2014. Fast-evolving mitochondrial DNA in Ceriantharia: A reflection of Hexacorallia paraphyly? PLoS One 9: e86612.

Wollschlager, J., N. Folino-Rorem & M.M. Daly. 2013. Nematocysts of the invasive hydroid *Cordylophora caspia* (Cnidaria: Hydrozoa). Biological Bulletin 224: 99–109.

Zapata, F., F.E. Goetz, S.A. Smith, M. Howison, S. Siebert, S.H. Church, S.M. Sanders, C.L. Ames, C.S. McFadden, S.C. France & M. Daly. 2015. Phylogenomic analyses support traditional relationships within Cnidaria. PLoS One 10(10): e0139068.

Chapter 5

Phylum Platyhelminthes

Francisco Brusa
CONICET, División Zoología Invertebrados, Museo de La Plata, La Plata, Argentina

Ana Maria Leal-Zanchet
Instituto de Pesquisas de Planárias and Programa de Pós-Graduação em Biologia, Universidade do Vale do Rio dos Sinos, UNISINOS, São Leopoldo, Rio Grande do Sul, Brazil

Carolina Noreña
Departamento Biodiversidad y Biología Molecular, Museo Nacional de Ciencias Naturales (CSIC), Madrid, Spain

Cristina Damborenea
CONICET, División Zoología Invertebrados, Museo de La Plata, La Plata, Argentina

Chapter Outline

INTRODUCTION

Platyhelminthes comprises a group of obligate parasitic organisms (Monogenea, Digenea, Aspidogastrea, and Cestoda), grouped in Neodermata, and the paraphyletic "Turbellaria." The latter are generally free-living, but symbiotic and parasitic flatworms are not rare. "Turbellaria" includes an amazing variety of forms, but with a similar structural body plan. All have the following characteristics: bilateral symmetry; organs embedded in a solid cellular matrix (the parenchyma); a saclike gut without an anus; a nervous system with an anterior "brain" and lateral nerve chords; and internal fluids regulated by protonephridia. Most are cross-fertilizing hermaphrodites, with morphologically diverse structures like sclerotic parts of the male copulatory organ that forms stylets of different shape and complexity, and with endolecithal or ectolecithal

Thorp and Covich's Freshwater Invertebrates. https://doi.org/10.1016/B978-0-12-804225-0.00005-8

eggs. The location and shape of the pharynx along the main body-axis may also vary.

Most "Turbellaria" are small, with a body size of around 1 mm (microturbellaria), but others, such as Tricladida and Polycladida, are 5–10 cm long (macroturbellaria). They show a worldwide distribution and inhabit most kinds of marine, brackish, freshwater, and terrestrial habitats, including cave environments. Some species of freshwater microturbellarians (e.g., Temnocephalida) have a particular lifestyle, and they live associated with other organisms in commensal relationship.

Based on their simple body structure, turbellarians are placed among the most basal Lophotrochozoa (Littlewood & Bray, 2001). Results obtained from molecular phylogenetic analysis support this hypothesis (Ruiz-Trillo et al., 1999, 2004; Baguñà & Riutort, 2004). "Turbellaria" is a grade that includes a very heterogeneous group of orders. Besides the body plan, all orders share a free-living life style (but some species are symbionts and parasites), similar features of the body wall (simple epithelium, muscle network, absence of cuticle), and regeneration based on stem cell-like neoblasts (Tyler & Hooge, 2004). These characters justified the grouping of two main clades: Catenulida and Rhabditophora (Tyler & Hooge, 2004; Egger et al., 2015).

LIMITATIONS

Although Neotropical freshwater turbellarian studies have only focused on certain geographic areas, diversity is high. The fauna from southern Brazil (São Paulo and Rio Grande do Sul states) and from central Argentina (Santa Fe and Buenos Aires provinces) are best known. There are scattered records for other regions, especially Uruguay and Peru, among others. In the past decades, special interest was paid to triclads (especially in Brazil) and microturbellarians (especially for Argentina). The endemic commensal temnocephalids have been studied in detail, due to their association with other particular species, and their biogeographical relationship with other regions (Martínez-Aquino et al., 2017).

The knowledge of the freshwater turbellarian fauna from Antarctica is limited, even though some researchers mention unidentified free-living platyhelminthes in Antarctic freshwater bodies. A few species of Maricola (Tricladida) (Sluys, 1989) and one species of the microturbellarian Proseriata are known (Ball & Hay, 1977) from the Antarctic region.

Use the keys with caution due to these limitations. In most cases, the identification of higher taxa uses external and general characteristics visible under a stereoscopic microscope, or by "squash" under optic microscopy. For instance, the body shape; coloration; presence of eyes or ciliated pits; type and position of the pharynx; position of genital pores, the gonads, and other genital structures, among other characteristics, are relevant for identification to family. The identification at genus or species levels need more detailed observation and description of special structures of the genital system, and sometimes it is necessary to apply special preparation methods (see Material Preparation and Preservation).

The keys are for those not trained in the freshwater "Turbellaria." They use characters easily seen in whole mounts of living animals. At lower levels, identification requires more detailed observations. For some genera, we developed keys to species level because the characteristics are easy to recognize in live or whole mount preparations (e.g., *Stenostomum*). Once a specimen is identified, it must be checked against the original description or later descriptions in monographs such as Ferguson (1954) and the website created by Schärer (2014) for Macrostomida. Readers should also consult the following texts: Gilbert (1938) and Houben et al. (2014) for *Phaenocora*; Marcus (1946), Kenk (1974, 1989), Sluys (1989, 1990), Leal-Zanchet et al. (2014), Hellmann et al. (2018), and Souza et al. (2018) for Tricladida; Luther (1955), Marcus (1946), and Noreña-Janssen (1995) for Dalyelliidae; Luther (1960) for Catenulida, Macrostomida, Lecithoepitheliata, Prolecithophora, and Proseriata; Luther (1963), Noreña-Janssen & Faubel (1992), and Noreña-Janssen et al. (2016) for Typhloplanida; Noreña et al. (2005) for *Stenostomum* (Catenulida); and Nuttycombe (1956) for Catenulidae; Damborenea & Cannon (2001) for Neotropical *Temnocephala* (Rhabdocoela). More general information is available in Cannon (1986), Tyler et al. (2006–2019), and Noreña et al. (2015). Noreña et al. (2004) and Martínez-Aquino et al. (2014) provide information on Argentine microturbellarian distribution, especially for temnocephalids. Similar information for Brazil can be found in Carbayo & Froehlich (2008) for macroturbellarians, and in Braccini et al. (2016) for microturbellarians.

There is little agreement on the relationships within the "Turbellaria," and the organization of the major groups (e.g., orders) are subject to review and change. The keys mostly follow the organization given by Tyler et al. (2006–2019).

TERMINOLOGY AND MORPHOLOGY

The following terminology will assist in the use of the keys. It follows primarily Cannon (1986), Rieger et al. (1991), and Richter et al. (2010).

Adhesive glands: cell glands found in different parts of the body, which secrete adhesive substances.

Ascus: Glandular pocket near the gonopores.

Auricles: Lateral projections of the head containing sensory organs.

Ciliated pits: Sensory structures (e.g., anterior end of Stenostomidae).

Copulatory bursa: Sac-shaped expansion in the female system, which receives donor sperm. It opens to the exterior through either the gonopore (primary bursa) or a separate pore (secondary bursa).

Copulatory organ: The terminal part of the male reproductive system to deliver the sperm into the body of a partner at copulation. It is often bulbous (copulatory bulb) containing the sperm duct and elements of the prostate glands. There is a penis (penis papilla) if it is capable of protrusion but not armed with hard structures. It may be enclosed in an inner chamber (penis pocket), the walls of which can form a conical prominence that may serve to guide the penis (a penis sheath). The penis is armed when a "cuticular" or "sclerotic" structure is present. A **stylet** is a single sclerotic tube or channel of variable shape; a **cirrus** is an eversible male duct; often spiny.

Excretophore: Large vacuolated cells of the intestine epithelium.

Handles: branches in the basal part of the stylet.

Light refracting bodies: Unpigmented refracting bodies found in species of Stenostomidae. Formed by a variable number of spherical granules; some species have fewer than five and others more than 15. The light refracting bodies are frequently associated with the posterior cerebral lobes, or sometimes with the anterior lobes. Generally, only one pair is found, although in some cases, more pairs are present.

Neoblast: Pluripotent cells responsible for the normal replacement of cells, cell substitution during regeneration, and stem cells for the gonads.

Pharynx: The glandulomuscular structure between mouth and intestine. Used to capture prey and may be of considerable taxonomic value. The main terms are: **simplex** (simple)—a ciliated mouth tube; **plicatus** (folded/plicate)—an annular fold which is variable in length, lacks a delimiting septum, and has a large protrusion capacity; **bulbosus** (bulbous)—a glandulomuscular bulb delimited by a more or less muscular septum, only the end of the pharynx may be protruded; it could be **rosulatus** (rosulate)—bulbous and usually globular and dorsoventrally oriented; or **doliiformis** (doliiform)—bulbous and usually barrel-shaped and horizontally (fronto-caudally) oriented; or **variabilis** (variable)— bulbous, but with variable shape and a weakly differentiated septum (see Noreña et al., 2015).

Proboscis: Anterior protrusible organ, which is a notable feature of the rhabdocoel group Kalyptorhynchia.

Prostatic vesicle: part of the male reproductive system. Structure containing gland cells and its secretion are combined with the sperm during their release.

Prostomium. Body region anterior to mouth; usually containing the brain.

Protonephridium: Organ containing one or more flame cells and connecting ducts, which presumably functions in an osmoregulatory fashion.

Rhabdites: Rod-shaped bodies found in the epidermal cells or below the epidermis. They are characteristic for the platyhelminth taxon Rhabditophora.

Statocyst: An orientation organ of the nervous system; it contains a space enclosing a granule (the statolith) and is surrounded by sensory tissue.

Vitellaria: gland cells that secrete nutrient material, which is included into egg capsules.

Zooid: Individuals formed by paratomy.

MATERIAL PREPARATION AND PRESERVATION

In vivo observations are very important for the study of turbellarians because certain structures are only visible in living specimens. Thus, note size, shape, and color before fixation. The specimens will need to undergo different group-specific treatments for subsequent identification and preservation. For triclads, external characteristics are easily observable under a stereomicroscope. Fix specimens (in Steinman's, 10% formaldehyde or Bouin fixative) for later histological analysis. Fixation of triclads should be done placing the specimens either on an ice cube covered with wet filter paper, or in a small vial, preventing or reducing contraction of the body. Remove the water using the vial and allow the specimen to stretch. Quickly pour fixative over the specimen. For microturbellarians, after measurement and observations on shape and color, place the specimen on a slide with a drop of water and either petroleum jelly or plasticine smeared on the coverslip edges, slowly squash it with the coverslip. Remove the excess water with filter paper. Diagnostic structures of the internal organs are observable in transparent microturbellarians. Some specimens must be fixed and preserved in polyvinyl lactophenol, making them more transparent. Microturbellarians can be fixed in a variety of conventional fixative fluids (e.g., Bouin, 5% formaldehyde) prior to being processed for histological analysis of the internal structures of diagnostic importance.

With few exceptions, the characters used in the following keys are viewable by transparency (squash methodology).

KEYS TO PLATYHELMINTHES

Platyhelminthes: Orders

1	Pharynx simple	2
1′	Pharynx plicatus or bulbous	3
2(1)	Protonephridium unpaired; excretory duct central and excretory pore caudal (Fig. 5.1 A)	**Catenulida [p. 106]**
2′	Protonephridium paired (Fig. 5.1 B)	**Macrostomida [p. 110]**
3(1)	Pharynx plicatus	4
3′	Pharynx bulbous	5
4(3)	Pharynx plicatus directed backwards	6
4′	Pharynx plicatus directed posterioventrally (Fig. 5.1 C)	**Proseriata [p. 110]**
5(3)	Oral pore and genital pore separated	7
5′	Oral pore and male pore joined, frontal; male copulatory organ with stylet; female gonads (ovaries and yolk glands) joined and diffuse (Fig. 5.1 D)	Lecithoepitheliata, one family: **Prorhynchidae [p. 110]**
6(4)	Intestine with anterior and posterior single sacs; with prominent paired ciliated pits (Fig. 5.1 E) [Brazil: São Paulo]	Bothrioplanida, Bothrioplanidae, one species: *Bothrioplana semperi* Braun, 1881
6′	Intestine with an anterior and two posterior branches; without prominent paired ciliated pits; with paired central eyes, sometimes absent (Fig. 5.1 F)	**Tricladida [p. 112]**
7(5)	Pharynx bulbous *variabilis* (with variable shape); oral pore frontal; male pore caudal; without female copulatory organs; female gonads opens into the male atrium (Fig. 5.3 A) [Brazil: São Paulo. Argentina: Buenos Aires]	Prolecithophora, Plagiostomidae, one genus: *Plagiostomum*
7′	Pharynx bulbous *doliiformis* (fronto caudal orientated) (Fig. 5.1 G), or *rosulatus* (dorso ventral orientated) (Fig. 5.1 H); with male and female copulatory organs; male and female pore ventral or caudal; female gonads (ovaries and yolk glands) separated	**Rhabdocoela [p. 114]**

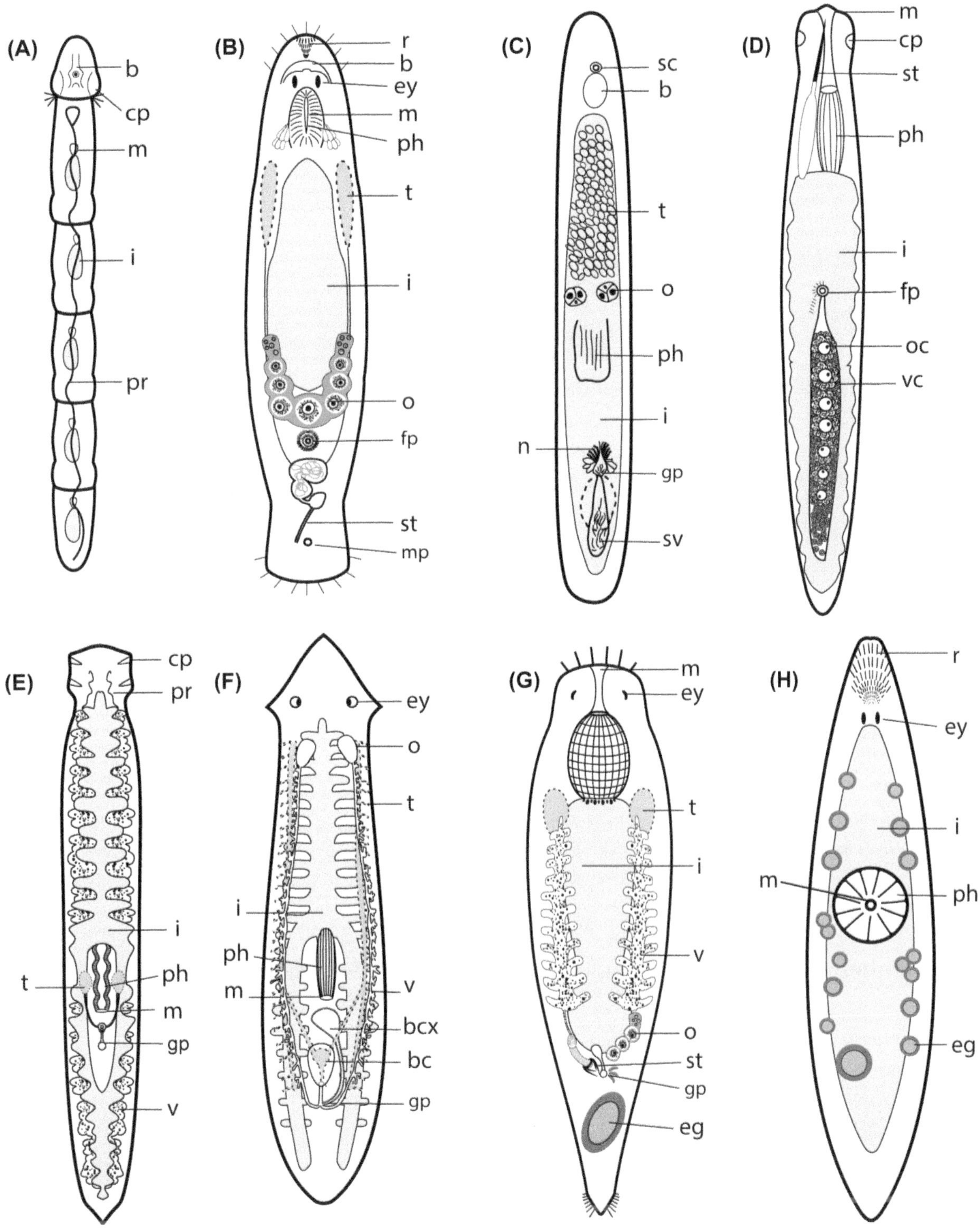

FIGURE 5.1 Schematic representation of Turbellaria orders: (A) Catenulida, Catenulidae, reproducing asexually, approximately 1–2 mm long; (B) Macrostomida, Macrostomidae, 2 mm long; (C) Proseriata, Otoplanidae, 2 mm long; (D) Lecithoepitheliata, Prorhynchidae, 5 mm long; (E) Bothrioplanida, Bothrioplanidae, 5 mm long; (F) Tricladida, Dugesiidae, 10–20 mm long; (G) Rhabdocoela, Dalyelliidae, 2 mm long; and (H) Rhabdocoela, Mesostomidae, 5 mm long. Abbreviations: *b*, brain; *bc*, bulbar cavity; *bcx*, bursa copulatrix; *cp*, ciliated pits; *eg*, egg; *ey*, eye; *fp*, female pore; *gp*, genital pore (hermaphrodite); *i*, intestine; *m*, mouth; *mp*, male pore; *n*, needles; *o*, ovary; *oc*, oocyte; *ph*, pharynx; *pr*, protonephridial duct; *r*, rhabdite tracks; *sc*, statocyst; *st*, stylet; *sv*, seminal vesicle; *t*, testes; *v*, vitellaria; *vc*, vitellocyte. *(A), (B), (D)–(H) from Noreña et al. (2015).*

Platyhelminthes: Catenulida: Families

1 Brain compact; oval; with or without statocyst; ciliated furrow separates the anterior end (*prostomium*) from the posterior region 2

1′ Brain lobed; with anterior and posterior lobes; with or without statocyst; without ciliated furrow 3

2(1) Oral pore near the *prostomium*; often forming chains of numerous and compact zooids .... **Catenulidae [p. 106]**

2′ Oral pore away from the *prostomium*; often forming chains of not more than two or four zooids (Fig. 5.2 A) .... Chordariidae, one genus: *Chordarium*

[Brazil: São Paulo, Paraná]

3(1) With ciliated pits; without statocyst .... **Stenostomidae [p. 106]**

3′ Without ciliated pits; with statocyst (Fig. 5.2 B) .... Retronectidae, one species: *Myoretronectes paranaensis* Noreña-Janssen & Faubel, 1996

[Argentina: Santa Fe]

Platyhelminthes: Catenulida: Catenulidae: Genera

1 Preoral groove cilia are not much longer than other cilia of the *prostomium*; *prostomium* with or without statocyst; with or without epidermal rhabdites (Figs. 5.1 A; 5.2 C–D) .... *Catenula*

[Brazil: São Paulo, Paraná, Rio Grande do Sul. Argentina: Santa Fe, Buenos Aires]

1′ Preoral groove cilia very long; *prostomium* with a medial invagination; gut reaching posterior body region; without rhabdites; zooids chains formed by two zooids, rarely three (Fig. 5.2 E) .... *Dasyhormus*

[Brazil: São Paulo, Paraná, Rio Grande do Sul]

Platyhelminthes: Catenulida: Stenostomidae: Genera

1 With ciliated pits; *prostomium* of variable length; generally with paired light refracting bodies associated to the posterior brain lobes; forming zooids 2

1′ Without ciliated pits; with a long *prostomium*; without light refracting bodies; without zooids (Fig. 5.2 F) .... *Rhynchoscolex*

[Brazil: São Paulo, Paraná, Rio Grande do Sul]

2(1) Ciliated pits not reversible 3

2′ With very specialized reversible paired ciliated pits .... *Xenostenostomum microstomoides* Reisinger, 1976

3(2) Pharynx and intestine not separated by a muscular belt (Fig. 5.2 G) .... ***Stenostomum* [p. 106]**

3′ Pharynx and intestine separated by a muscular belt........ *Myostenostomum*

[Brazil: São Paulo, Rio Grande do Sul]

Platyhelminthes: Catenulida: Catenulidae: *Stenostomum*: Species

1 Specimens with light-refracting bodies 2

1′ Specimens without light-refracting bodies 17

2(1) Light-refracting bodies form by a vesicle with spheres 3

2′ Light-refracting bodies formed by single-granule bodies, with no vesicle .... *Stenostomum simplex* Kepner & Carter, 1931

[Brazil: São Paulo]

3(2) Light-refracting bodies discoid, with more than ten small spheres 4

3′ Light-refracting bodies, bowl-shaped, with one to five spheres 8

4(3) Pharyngeal glands small, rounded 5

4′ Pharyngeal glands rounded and clavate 7

5(4) Pharyngeal glands only in the anterior half of the pharynx 6

5′ Pharyngeal glands uniformly arranged in the whole pharynx .... *Stenostomum tenuicauda* Graff, 1912

[Brazil: São Paulo, Rio Grande do Sul. Peru: Ucayali. Argentina: Buenos Aires]

6(5) With sphincter between pharynx and intestine, oral pore circular .... *Stenostomum grande* Child, 1902

[Brazil: São Paulo, Paraná, Rio Grande do Sul. Peru: Loreto. Argentina: Buenos Aires]

6′ Without a sphincter between pharynx and intestine, oral pore triangular .... *Stenostomum matarazzoi* Marcus, 1949

[Brazil: São Paulo]

7(4) Without a sphincter between pharynx and intestine, opening of pharyngeal glands along the whole pharynx .... *Stenostomum leucops* (Dugès, 1828)

[Brazil: Rio Grande do Sul. Peru: Ucayali. Suriname. Argentina: Santa Fe]

7′ With a sphincter between pharynx and intestine, opening of the pharyngeal glands at the anterior half of the pharynx, internal cerebral lobes associated with the posterior cerebral lobes .... *Stenostomum amphotum* Marcus, 1945

[Brazil: São Paulo, Rio Grande do Sul. Peru: Ucayali. Argentina: Buenos Aires]

8(3) One or more pairs of light-refracting bodies; no caecum dorsal to the pharynx 9

8′ One or more pairs of light-refracting bodies, intestinal caecum dorsal to the pharynx .... *Stenostomum cryptops* Nuttycombe & Waters, 1935

[Brazil: São Paulo]

9(8) One pair of light-refracting bodies 10

9′ More than one pair of light-refracting bodies 14

10(9) Posterior end with no intestine-lacking zone 11

10′ Posterior end with intestine-lacking zone 12

11(10) Pharyngeal glands small rounded, opening uniformly onto the whole pharynx .... *Stenostomum virginianum* Nuttycombe, 1931

[Brazil: São Paulo, Paraná. Argentina: Buenos Aires]

11′ Pharyngeal glands clavate, opening at the transition between mouth and pharynx .... *Stenostomum hemisphericum* Nassonov, 1924

[Suriname. Brazil: Pará, Rio Grande do Sul. Peru: Ucayali]

12(10) Distal end of the protonephridium sinuous and terminal nephridiopore 13

12′ Distal end of the protonephridium straight, and subterminal nephridiopore .... *Stenostomum ciliatum* Kepner & Carter, 1931

[Brazil: São Paulo, Paraná, Rio Grande do Sul. Peru: Ucayali]

13(12) First part of the intestine with high vacuolated cells, separated from the gut by a strong constriction *Stenostomum rosulatum* Marcus, 1945

[Brazil: São Paulo]

13′ Without vacuolated cells in the gut, pharynx glands clavate .... *Stenostomum uronephrium* Nuttycombe, 1931

[Brazil: São Paulo. Peru: Ucayali. Argentina: Santa Fe, Buenos Aires]

14(9) Light-refracting bodies associated only with the anterior brain lobes 15

14′ Light-refracting bodies associated with both anterior and posterior brain lobes 16

15(14) Posterior end without intestine; tail with two appendages; with tentacular pharyngeal organ ("claw-organ") .... *Stenostomum bicaudatum* Kennel, 1889

[Trinidad and Tobago. Brazil: São Paulo, Rio Grande do Sul. Peru: Loreto, Ucayali]

15′ Posterior end without intestine-lacking zone; epidermis with groups of refractive corpuscles .... *Stenostomum glandulosum* Kepner & Carter, 1931

[Suriname. Brazil: São Paulo, Paraná, Rio Grande do Sul]

16(14) Tail simple (one appendage) .... *Stenostomum arevaloi* Gieysztor, 1931

[Brazil: São Paulo, Rio Grande do Sul]

16′ Tail double (two appendages) .... *Stenostomum paraguayense* (Martin, 1908)

[Suriname. Brazil: São Paulo, Rio Grande do Sul. Peru: Loreto, Ucayali. Paraguay. Argentina: Buenos Aires]

17(1) Without pharyngeal glands 18

17′ With pharyngeal glands 19

18(17) With an anterior body constriction 25

18′ Without an anterior body constriction .... *Stenostomum membranosum* Kepner & Carter, 1931

[Brazil: São Paulo, Rio Grande do Sul]

19(17) Posterior end without intestine, tail simple 20

19′ Posterior end with intestine 22

20(19) Tail short; nephridiopore terminal 21

20′ Tail very long forming a dorsal appendage; nephridiopore subterminal ... *Stenostomum pseudoacetabulum* Nuttycombe & Waters, 1935

[Suriname. Brazil: São Paulo, Paraná, Rio Grande do Sul. Peru: Ucayali. Argentina: Buenos Aires]

21(20) Excretophores absent; club-shaped pharyngeal glands; no sphincter between pharynx and intestine .... *Stenostomum tuberculosum* Nuttycombe & Waters, 1938

[Suriname. Brazil: São Paulo, Rio Grande do Sul. Argentina: Buenos Aires]

21′ Excretophores present; rounded pharyngeal glands, with sphincter between pharynx and intestine .... *Stenostomum pegephilum* Nuttycombe & Waters, 1938

[Brazil: São Paulo]

22(19) With excretophores 23

22′ Without excretophores .... *Stenostomum ventronephrium* Nuttycombe, 1932

[Brazil: São Paulo]

23(22) Anterior cerebral lobe with small independent masses 24

23′ Anterior cerebral lobe smooth; intestinal caecum dorsal to the pharynx .... *Stenostomum evelinae* Marcus, 1945

[Suriname. Brazil: São Paulo]

24(23) Anterior end "anatiform" .... *Stenostomum anatirostrum* Marcus, 1945

[Brazil: São Paulo, Rio Grande do Sul]

24′ Anterior end not "anatiform"; pharyngeal glands in the anterior region only .... *Stenostomum corderoi* Marcus, 1945

[Brazil: São Paulo]

25(18) Presence of a constriction between the anterior and posterior brain lobes; radial musculature in the pharynx .... *Stenostomum saliens* Kepner & Carter, 1931

[Suriname. Brazil: São Paulo, Paraná. Argentina: Buenos Aires]

25′ Presence of a constriction between the posterior brain lobes and the mouth; and a small pharynx directed ventrally .... *Stenostomum ashanika* Damborenea, Brusa, Almagro & Noreña, 2011

[Peru: Ucayali]

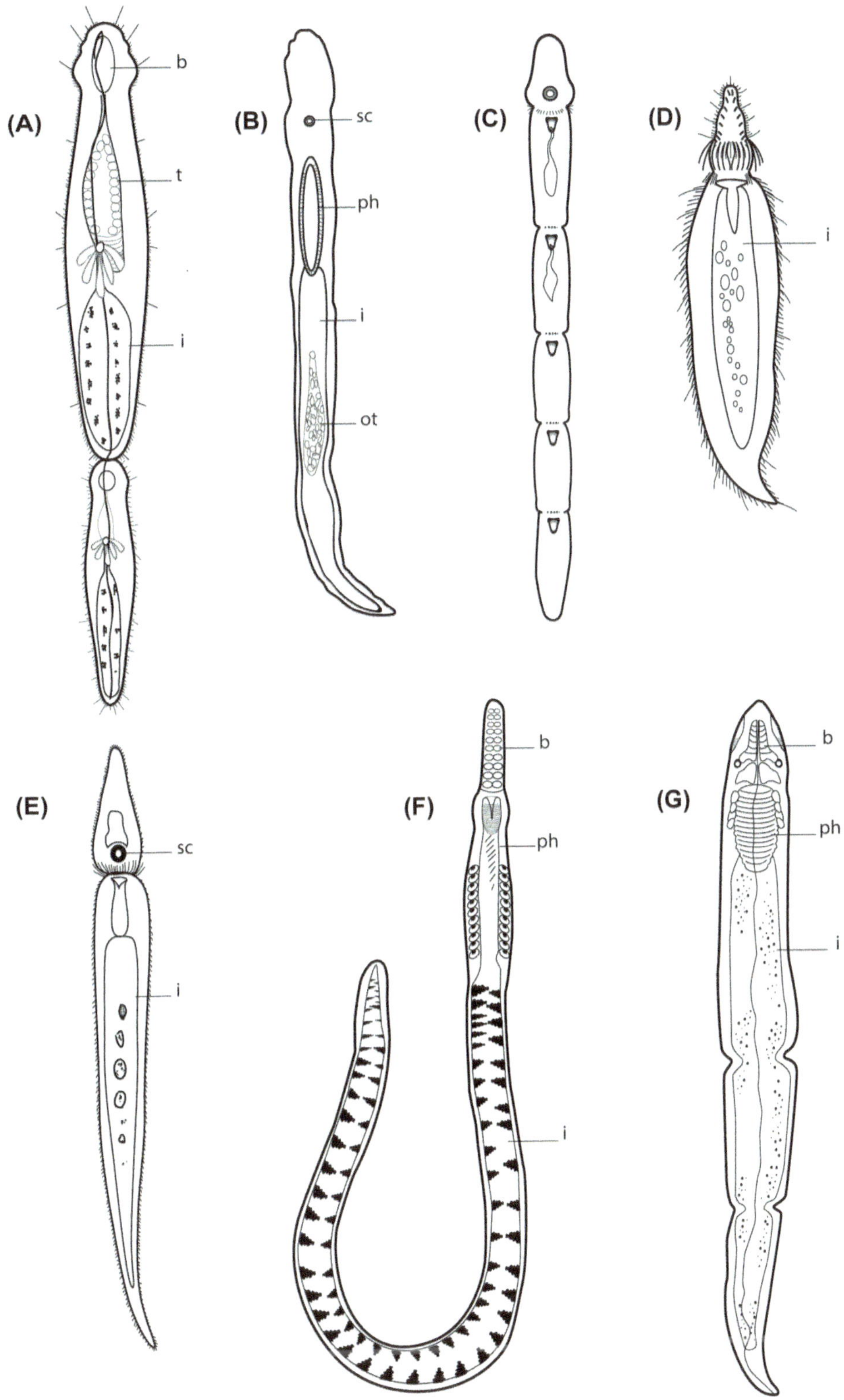

FIGURE 5.2 Schematic representation of some genera of Catenulida from freshwater environments: (A) *Chordarium leucanthum* (Chordariidae), length: 0.5–1.8 mm; (B) *Myoretronectes paranaensis* (Retronectidae), length: up to 2 mm; (C) *Catenula lemnae* (Catenulidae), length: 1–10 mm; (D) *Catenula turgida* (Catenulidae) length: 0.2–0.36 mm; (E) *Dasyhormus lithophorus* (Catenulidae), length: 0.3–0.6 mm; (F) *Rhynchoscolex nanus* (Stenostomidae), length: 0.6–2.0 mm; and (G) *Stenostomum* sp. (Stenostomidae), length: 0.5–4 mm. Abbreviations: *b*, brain; *i*, intestine; *ot*, ovotestis; *ph*, pharynx; *sc*, statocyst; *t*, testes. *(C), (D), (F), (G) from Noreña et al. (2016).*

Platyhelminthes: Macrostomida: Families

1 Body shape oval; with two eyes; without ciliated pits; adhesive posterior end; paired testes and ovaries; stylet tube or funnel like; sexual reproduction (Fig. 5.3 B) .... Macrostomidae, one genus: *Macrostomum*

[Jamaica. Brazil: São Paulo, Rio Grande do Sul. Peru: Lima, Loreto. Argentina: Buenos Aires, Córdoba, Entre Ríos]

1′ Body shape elongated (often forming zooids); anterior and posterior end pointed; without eyes; with ciliated pits; ovaries unpaired; stylet turned; corkscrew like; with sexual and asexual (more common) reproduction (Fig. 5.3 C) .... Microstomidae, one species: *Microstomum lineare* (Müller, 1773)

[Brazil: Rio Grande do Sul. Argentina: Santa Fe]

Platyhelminthes: Proseriata: Families

1 With ciliated pits; with statocyst; without intestinal branches; ovaries before pharynx; common genital pore (Fig. 5.3 D) .... **Otoplanidae [p. 110]**

1′ Without ciliated pits; with statocyst; ovaries before the pharynx; female and male gonopores separated (Fig. 5.3 E) .... **Monocelididae [p. 110]**

Platyhelminthes: Proseriata: Otoplanidae: Genera

1 Pharynx at or near the middle of the body 2

1′ Pharynx in the last third of the body .... *Kata*

[Uruguay: Río Negro; other records are from marine habitats]

2(1) Two rows of testicular follicles in front of the pharynx; two rows of yolk follicles beside the testes, beside and behind the pharynx; cuticular apparatus consisting of needles and a stylet; and unpaired bursa behind the common atrium (Fig. 5.3 D) .... *Itaspiella parana* Noreña, Damborenea & Brusa, 2005

[Argentina: Santa Fe]

2′ Long, digitiform papillae on the body surface; pharynx in the middle of the body; two rows of testicular follicles and paired ovary in front of the pharynx; two rows of yolk follicles behind the pharynx; cuticular apparatus with needles, without stylet; absence of a bursa .... *Philosyrtis rauli* Noreña, Damborenea & Brusa, 2005

[Uruguay: Río Negro]

Platyhelminthes: Proseriata: Monocelididae: Genera

1 With three pores in the genital system (vagina, male pore, female pore) 2

1′ With three genital pores (vagina, male pore, female pore) and an accessory pore of the female system with a sclerotized stylet .... *Minona amnica* Ball & Hay, 1977

[Sub-Antarctic Islands: Macquarie Island]

2(1) Short cirrus in the copulatory organ; seminal and prostatic vesicles not separated by a muscular diaphragm (Fig. 5.3 E) .... *Mesoda gabriellae* Marcus, 1949

[Brazil: São Paulo]

2′ Long cirrus in the copulatory organ; seminal and prostatic vesicles separated by a muscular diaphragm .... *Inaloa scalopura* (Marcus, 1949)

[Brazil: São Paulo]

Platyhelminthes: Lecithoepitheliata: Prorhynchidae: Genera

1 Anterior end flabelliform; with eyes except in some cave forms; male stylet curved; claw-like (Fig. 5.4 A) .... *Geocentrophora applanata* (Kennel, 1888)

[Brazil: São Paulo, Paraná. Peru: Loreto]

1′ Anterior end elongated; sometimes light truncated; not flabelliform; without eyes; male stylet straight (Fig. 5.4 B) .... *Prorhynchus stagnalis* Schultze, 1851

[Brazil: São Paulo, Paraná, Rio Grande do Sul. Peru, Loreto]

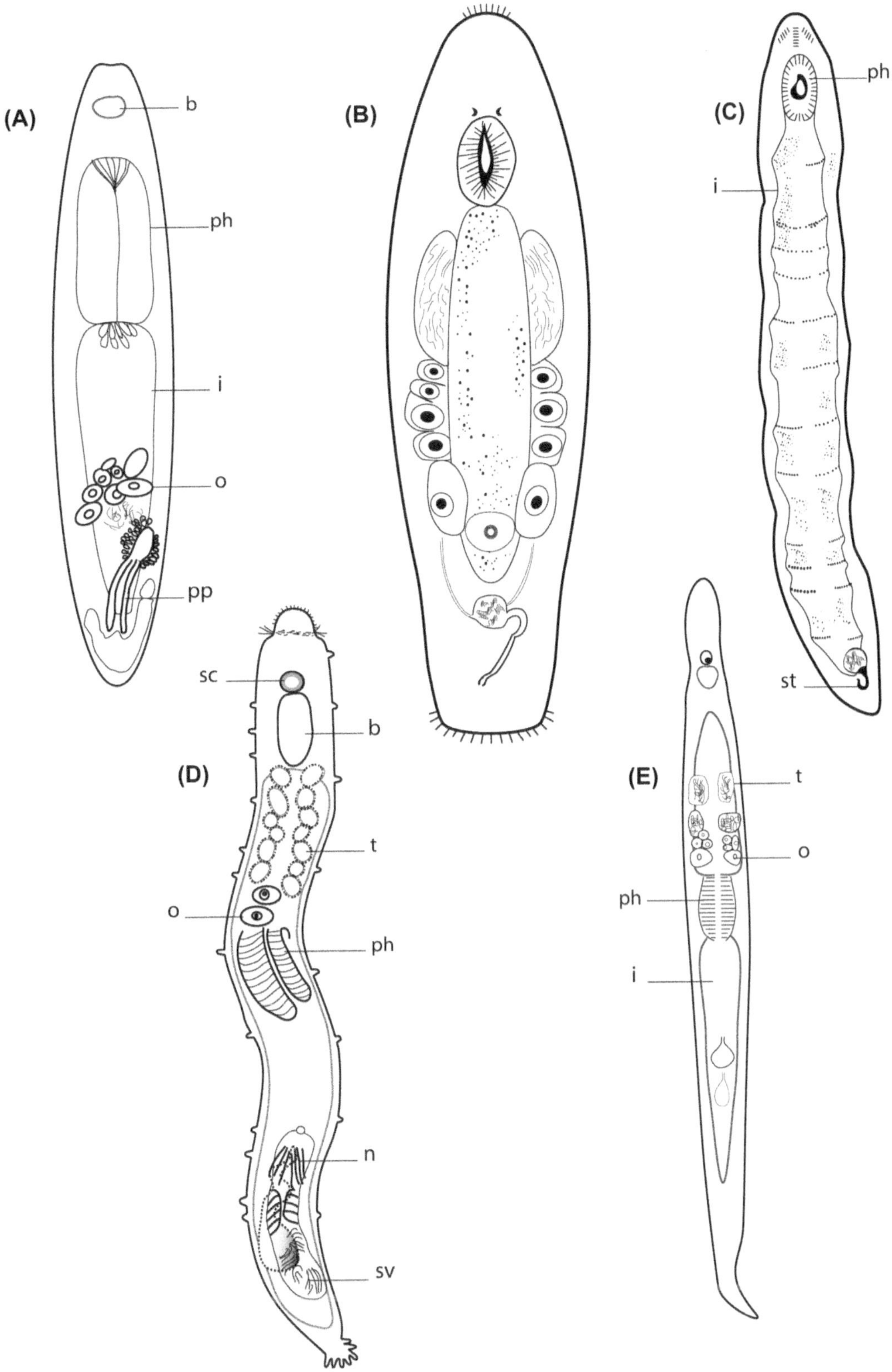

FIGURE 5.3 Schematic representation of some genera of Prolecithophora, Macrostomida, and Proseriata from freshwater environments: (A) *Plagiostomum cilioejaculator* (Prolecithophora, Plagiostomidae), length: about 2.4 mm; (B) *Macrostomum* sp. (Macrostomida, Macrostomidae), length: 1–3 mm; (C) *Microstomum* sp. (Macrostomida, Microstomidae), length: 0.4–4 mm; (D) *Itaspiella parana* (Proseriata, Otoplanidae), length: about 2 mm; and (E) *Mesoda gabriellae* (Proseriata, Monocelididae), length: about 0.8 mm. Abbreviations: *b*, brain; *i*, intestine; *n*, needles; *o*, ovary; *ph*, pharynx; *pp*, penis papilla; *sc*, statocyst; *st*, stylet; *sv*, seminal vesicle; *t*, testes. *(B) and (C) from Noreña et al. (2016).*

Platyhelminthes: Tricladida: Families

1 Head rounded or truncate; without auricles; pigmented or whitish body 2

1′ Head usually triangular; one pair of eyes; white auricles; body usually pigmented; primary bursa copulatrix anterior to copulatory organs (Figs. 5.4 C; 5.8 A, B) .... **Dugesiidae [p. 112]**

2(1) Eyes usually present (one pair); pigmented body; without a clear adhesive organ at the anterior tip; with a primary bursa copulatrix 3

2′ Eyes usually absent; body usually whitish; sometimes with an adhesive organ at the anterior tip; without bursa copulatrix or with a secondary bursa posterior to copulatory organs (Figs. 5.4 D; 5.8 C) .... **Dimarcusidae [p. 112]**

3(2) Primary bursa copulatrix anterior to penis .... Planariidae, one genus: *Phagocata*

[Peru: Loreto]

3′ Primary bursa copulatrix posterior to penis and with a thick circular muscle layer around the bursal canal (Figs. 5.4 E, 8 D) .... **Uteriporidae [p. 112]**

Platyhelminthes: Tricladida: Dugesiidae: Genera

1 Head triangular with pointed or rounded auricles 2

1′ Head rounded, with short and blunt auricles; pharynx unpigmented .... *Romankenkius patagonicus* (Borelli, 1901)

[Chile: Tierra de Fuego; Isla Aracena, Cockburn Channel, Magellan Strait. Argentina: Patagonia]

2(1) Head triangular; pharynx pigmented or not; common gonopore (Figs. 5.4 C; 5.8 A, B) .... *Girardia*

[Mexico. Cuba. Puerto Rico. Jamaica. Guadaloupe. Venezuela. Trinidad and Tobago Colombia. Ecuador. Brazil. Peru. Bolivia. Chile. Paraguay. Argentina. Uruguay]

2′ Head triangular; pharynx pigmented; male and female gonopores .... *Bopsula evelinae* Marcus, 1946

[Brazil: São Paulo]

Platyhelminthes: Tricladida: Dimarcusidae: Genera

1 Eyes absent; body whitish; reproductive organs with one or two openings at the ventral surface; without bursa copulatrix or with a secondary bursa posterior to copulatory organs 2

1′ One pair of eyes and an adhesive organ at the anterior tip; pigmented body; bursa copulatrix absent; female genital duct connecting with the intestine (Figs. 5.4 D; 5.8 C) .... *Rhodax evelinae* Marcus, 1946

[Brazil: São Paulo, Rio Grande do Sul]

2(1) Reproductive organs with one opening at the ventral surface 3

2′ Reproductive organs with two openings at the ventral surface; with a secondary bursa posterior to copulatory organs .... ***Opisthobursa* [p. 112]**

3(2) Head truncate; reproductive organs with one opening at the ventral surface; bursa copulatrix absent; female genital duct connecting with the intestine .... *Hausera hauseri* Leal-Zanchet & Souza, 2014 in Leal-Zanchet et al., 2014

3′ Head rounded, reproductive organs with one opening at the ventral surface; with secondary bursa copulatrix .... *Kawakatsua pumila* Sluys, 2019 in Sluys & Laumer 2019

[Panama: Barro Colorado Island]

Platyhelminthes: Tricladida: Dimarcusidae: *Opisthobursa*: Species

1 Head rounded; hind end pointed; secondary bursa without folds .... *Opisthobursa mexicana* Benazzi, 1972

[Mexico: Tabasco]

1′ Head and hind end truncate; secondary bursa with highly folded walls .... *Opisthobursa josephinae* Benazzi, 1975

[Mexico: Chiapas]

Platyhelminthes: Tricladida: Uteriporidae: Genera

1 With eyes; body width is greater in the median third of the body or in the posterior third of the body 2

1′ Without eyes; margins of the body parallel each other; ovaries on the sides of pharynx; secondary copulatory bursa with two ventral openings posterior to gonopore (Fig. 5.8 D) .... *Sluysia triapertura* Leal-Zanchet & Souza 2018 in Souza et al., 2018

[Brazil: Rio Grande do Norte]

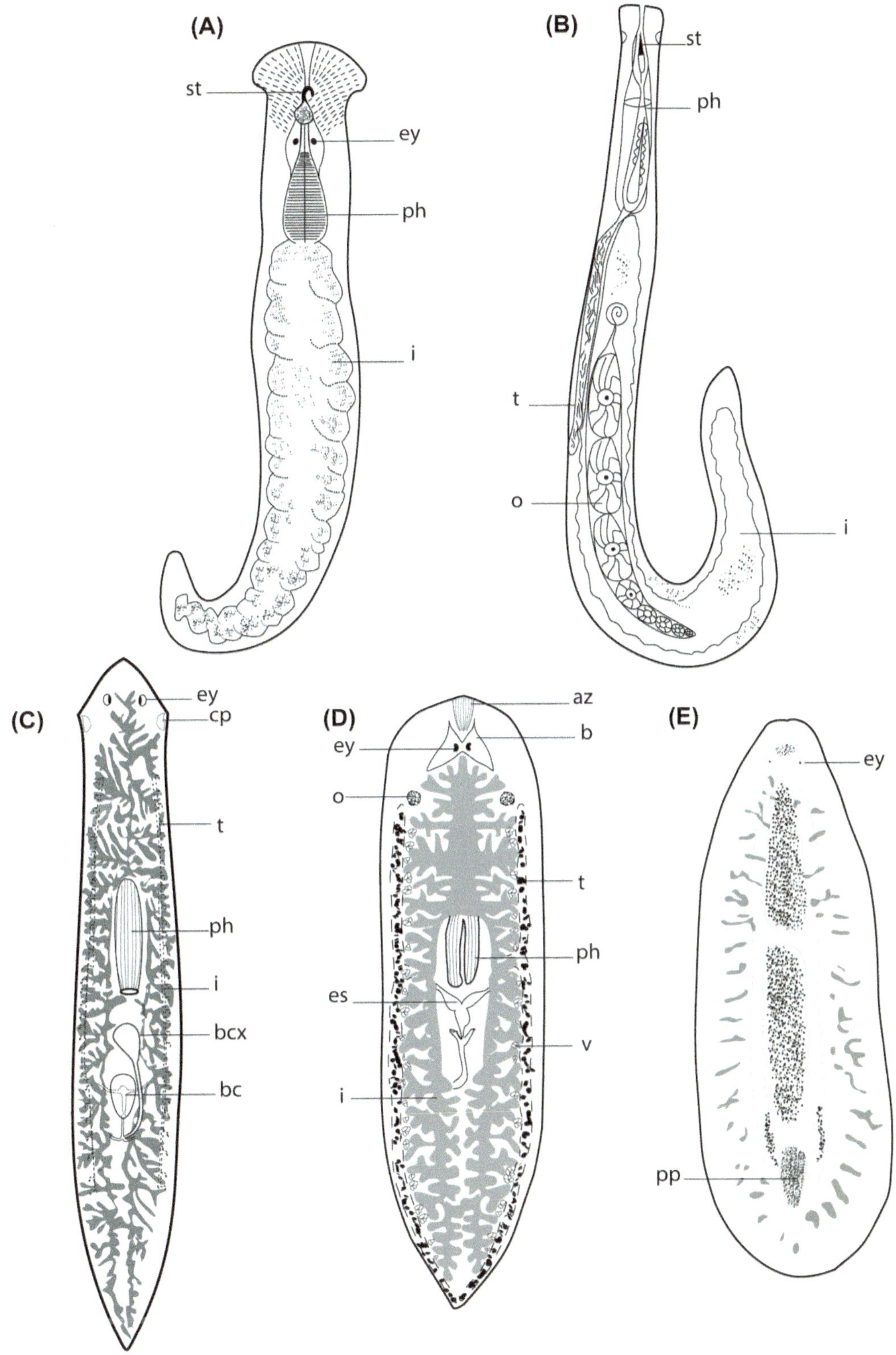

FIGURE 5.4 Schematic representation of some genera of Tricladida and Lecithoepitheliata from freshwater environments: (A) *Geocentrophora* sp. (Lecithoepitheliata, Prorhynchidae) length: 1–4 mm; (B) *Prorhynchus stagnalis* (Lecithoepitheliata, Prorhynchidae) length: 2–7 mm; (C) *Girardia tigrina* (Tricladida, Dugesiidae), length: about 2 cm; (D) *Rhodax evelinae* (Tricladida, Dimarcusidae), length: up to 9 mm; and (E) *Obrimoposthia ohlini* (Tricladida, Uteriporidae), length: up to 11 mm. Abbreviations: *az*, adhesive zone; *b*, brain; *bc*, bulbar cavity; *bcx*, bursa copulatrix; *cp*, ciliated pits; *ey*, eye; *i*, intestine; *o*, ovary; *ph*, pharynx; *pp*, penis papilla; *sp*, spermiducts; *st*, stylet; *sv*, seminal vesicle; *t*, testes; *v*, vitellaria. *(A) and (B) from Noreña et al. (2016); (D) drawn from Marcus (1946); (E) drawn from Sluys (1989).*

2(1) The greatest body width at the level of the median third of the body; bursal canal without a vestibulum; penis papilla provided with chambers delimited by muscular septa (Fig. 5.4 E) .... ***Obrimoposthia* [p. 114]**

2′ The greatest body width at the level of the posterior third of the body; dark pigmented dorsal surface with concentrations of pigment on the head and in a mid-dorsal stripe; bursal canal with a vestibulum ectally to the combined openings of the oviducts .... *Procerodella macrostoma* (Darwin, 1844)
[Chonos Archipelago]

Platyhelminthes: Tricladida: Uteriporidae: *Obrimoposthia*: Species

1 Dorsal and ventral surfaces black with a patchy pattern .... *Obrimoposthia aparala* Sluys & De Vries, 1988
[Crozet Islands, Trinidad Malvinas Island]

1′ Dorsal surface black-brown with two whitish or yellowish, irregularly formed longitudinal bands; ventral surface pale or greyish (Fig. 5.4 E) .... *Obrimoposthia ohlini* (Bergendal, 1899)
[Gough Island, Kerguelen, Marion Island?]

Platyhelminthes: Rhabdocoela: Families

1 Free living species 2

1′ Symbiotic species 6

2(1) Pharynx anterior (at the first body half) 3

2′ Pharynx central; cross-section dorso-ventral flattened; rhabdite roads present; testes dorsal to the vitellaria; female gonad unpaired (Fig. 5.5 A–D) .... **Typhloplanidae [p. 116]**

3(2) Retractable anterior end absent; pharynx type barrel-shaped (doliiformis) frontal orientated; vasa deferentia tubular without dilatations 4

3′ Retractable anterior end present with proboscides; pharynx type rosulatus ventral orientated; vasa deferentia tubular with dilatations (external or false vesicula seminalis) (Fig. 5.5 E) .... **Polycistididae [p. 118]**

4(3) Testes ventral to the lobular vitellaria 5

4′ Testes in front of the smooth vitellaria (Fig. 5.6 A) .... Graffillidae, one species: *Nygulgus evelinae* Marcus, 1954
[Brazil: São Paulo. Argentina: Buenos Aires]

5(4) Rhabdite roads present; copulatory organ pear-shaped, elongated, caudal orientated; ovary unpaired .... **Dalyelliidae [p. 119]**

5′ Rhabdite roads absent; copulatory organ retort or anchor-shaped, frontal orientated; ovary paired (Fig. 5.6 B) .... Provorticidae, one genus: *Baicalellia*
[Brazil: São Paulo, Rio Grande do Sul]

6(1) With tentacles in the anterior region of the body 7

6′ Without tentacles in the anterior region of the body (Fig. 5.7 A) .... Didymorchidae, one genus: *Didymorchis*
[Argentina: Río Negro. Uruguay]

7(6) With two tentacles (Fig. 5.7 B) .... Diceratocephalidae, one species: *Diceratocephala boschmai* (Baer, 1953)
[Introduced in Uruguay]

7′ With five tentacles (Fig. 5.7 C), commensals associated with crustaceans, mollusks, insects, and chelonians .... Temnocephalidae, one genus: *Temnocephala*
[Mexico. Costa Rica. Colombia. Brazil. Peru. Chile. Paraguay. Argentina. Uruguay]

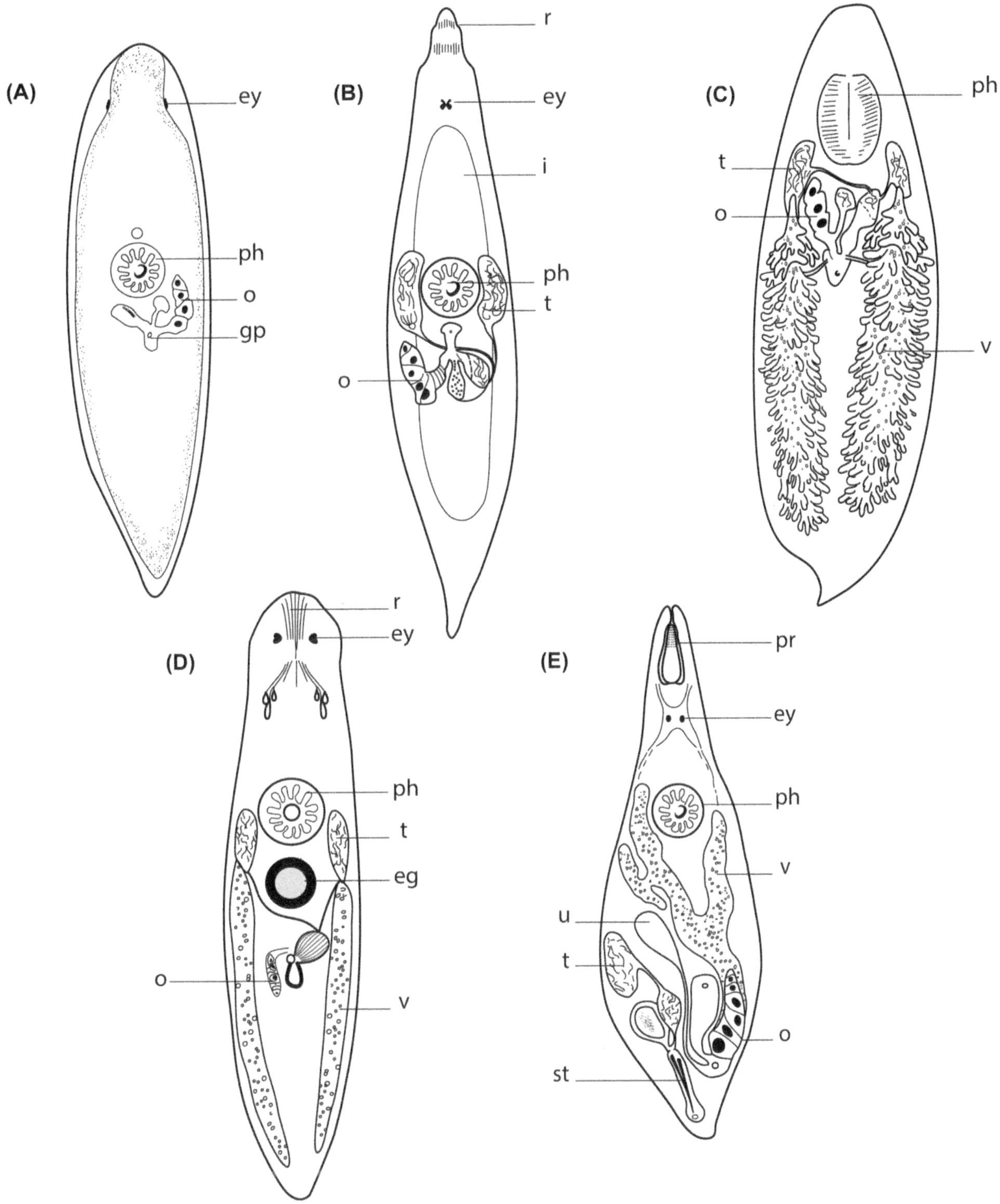

FIGURE 5.5 Schematic representation of some genera of Rhabdocoela Typhloplanidae and Kalyptorhynchia: (A) *Bothromesostoma evelinae* (Typhloplanidae, Mesostominae), length: 3 mm; (B) *Rhynchomesostoma* sp. (Typhloplanidae, Rhynchomesostominae), length: 1–3.5 mm; (C) *Phaenocora* sp. (Typhloplanidae, Phaenocorinae), length: 2–6 mm; (D) *Strongylostoma radiata* (Typhloplanidae, Typhloplaninae), length: 0.75–1.0 mm; and (E) *Gyratrix hermaphroditus* (Kalyptorhynchia, Polycistididae), length: 0.5–12 mm. Abbreviations: *eg*, egg; *ey*, eye; *gp*, genital pore (hermaphrodite); *i*, intestine; *o*, ovary; *p*, proboscis; *ph*, pharynx; *r*, rhabdite tracks; *st*, stylet; *t*, testes; *u*, uteri; *v*, vitellaria. *(A)–(C), and (E) from Noreña et al. (2016).*

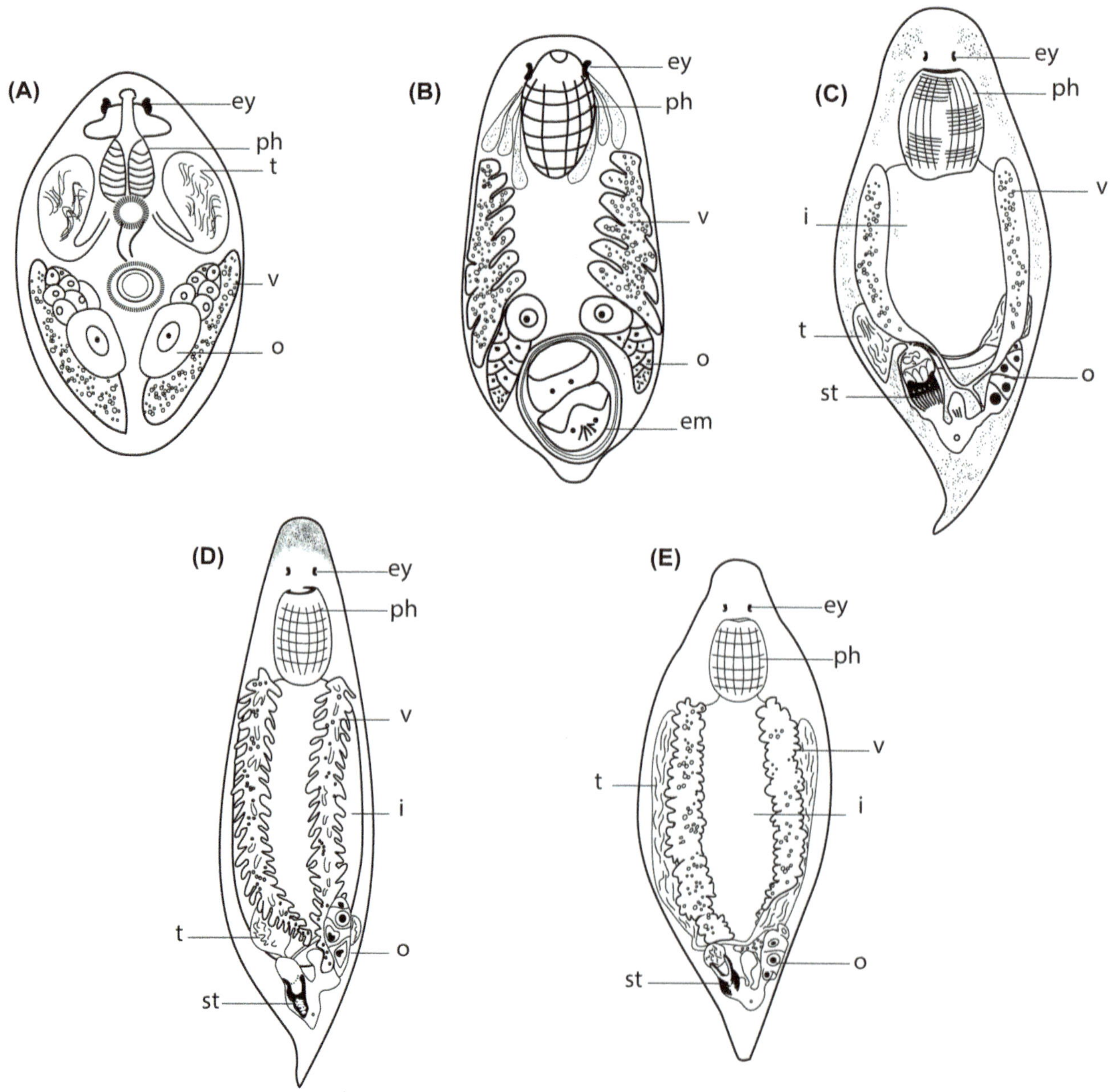

FIGURE 5.6 Schematic representation of some genera of Rhabdocoela: (A) *Nygulgus evelinae* (Graffillidae), length: about 0.5 mm; (B) *Baicalellia evelinae* (Provorticidae), length 0.8–1 mm; (C) *Gieysztoria* sp. (Dalyelliidae), length: 0.7–1.35 mm; (D) *Microdalyellia* sp. (Dalyelliidae), length: 0.5–1.5 mm; and (E) *Dalyellia* sp. (Dalyelliidae), length: 1.5–5 mm. Abbreviations: *em*, embrion; *ey*, eye; *i*, intestine; *o*, ovary; *ph*, pharynx; *st*, stylet; *t*, testes; *v*, vitellaria. *(A) and (B) drawn from Marcus (1954 and 1946, respectively); (C)–(E) from Noreña et al. (2016).*

Platyhelminthes: Rhabdocoela: Typhloplanidae: Genera

1	Pharynx ventral directed	2
1′	Pharynx with other orientation	7
2(1)	Testes dorsal to the vitellaria	3
2′	Testes ventral to the vitellaria	6

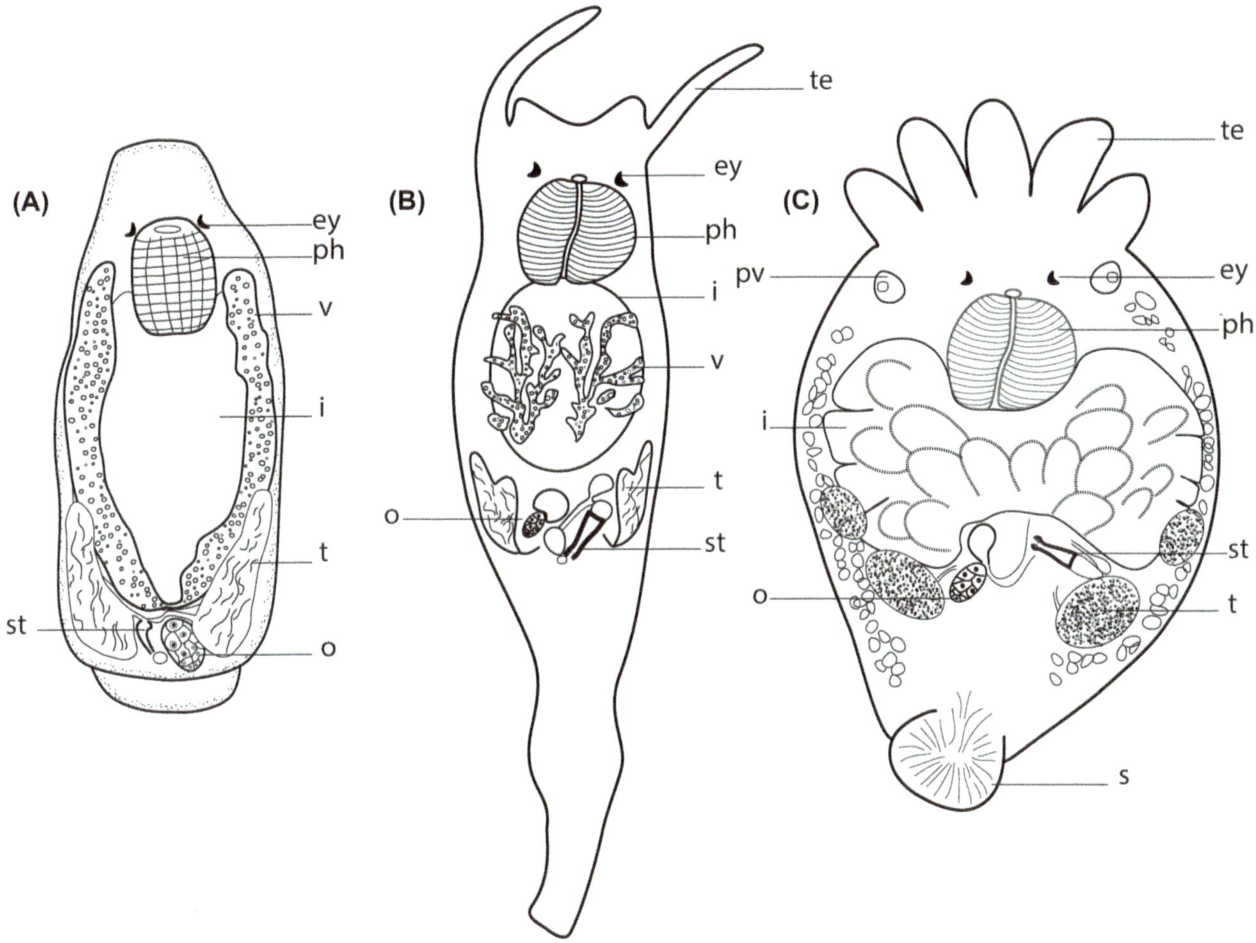

FIGURE 5.7 Schematic representation of some genera of commensal Rhabdocoela: (A) *Didymorchis haswelli* (Didymorchidae), length: 1.40–1.70 mm; (B) *Diceratocephala boschmai* (Diceratocephalidae), length: 2.0–4.5 mm; and (C) *Temnocephala chilensis* (Temnocephalidae), length: 1.2–4.6 mm. Abbreviations: *ey*, eye; *i*, intestine; *o*, ovary; *ph*, pharynx; *pv*, protonephridial vesicle; *s*, sucker; *st*, stylet; *t*, testes; *te*, tentacle; *v*, vitellaria. *(A) Drawn from Mañe-Garzon (1960).*

3(2) Testes sac-like, in front of the pharynx 4

3′ Testes mostly follicular, behind or on both sides of the pharynx, pharynx in the mid-body (second third) with two eyes 5

4(3) With eyes; copulatory organs behind the pharynx .... *Yagua lutheri* du Bois-Reymond Marcus, 1958

[Peru: Lima]

4′ Without eyes (sometimes with pigment patches); without ascus; male and female organ posterior; at the last third of the body .... *Olisthanella parva* Marcus, 1946

[Brazil: São Paulo. Peru: Loreto]

5(3) Dense, dark brownish pigmentation on the dorsal side; with anterior pigment loss eyes fields; eyes masked; with a prepharyngeal glandular organ (Fig. 5.5 A) .... *Bothromesostoma*

[Brazil: São Paulo, Rio Grande do Sul. Peru: Loreto. Argentina: Santa Fe, Río Negro. Uruguay: Canelones]

5′ Different tonalities (transparent; beige; caramel) and pattern (with spots or with thin stripes) of the dorsal pigmentation, eyes visible, without prepharyngeal glandular organ .... *Mesostoma*

[Trinidad and Tobago. Brazil: Minas Gerais, Rio Grande do Sul, São Paulo. Peru: Loreto, Ucayali, Puno. Argentina: Santa Fe, Río Negro. Uruguay: Canelones]

FIGURE 5.8 Photographs of live planarians (Tricladida): (A) *Girardia tigrina* (Dugesiidae), length: about: 15–20 mm; (B) *G. schubarti* (Dugesiidae), length: about 15 mm; (C) *Rhodax* sp. (Dimarcusidae), length: about 4 mm; and (D) *Sluysia triapertura* (Uteriporidae), length: about 6 mm. *Photographs by B. Lares (A), A. Marques (B), J. Braccini (C), and D. Bento (D).*

6(2) Protonephridium pores open in the buccal cavity or near the pharynx; copulatory organ at the middle of the body, close to the pharynx; caudal testes; ejaculatory duct developed as spiny cirrus; testes stretching at the posterior end; with bursa copulatrix; with eyes (Fig. 5.5 D) .... *Strongylostoma*

[Brazil: São Paulo. Peru: Puno]

6′ Protonephridium pores open in the genital atrium; with eyes; male organ with numerous spines in the bursa copulatrix, in the blind sacs or in the male atrium; with two eyes; with a retractable anterior end (telescope-like) (Fig. 5.5 B) .... *Rhynchomesostoma inaliensis* Noreña-Janssen, 1995

[Argentina: Santa Fe]

7(1) Pharynx forward directed; feathery vitellaria; male copulatory organ smooth or with spines (cirrus-like) (Fig. 5.5 C) .... *Phaenocora*

[Brazil: São Paulo, Rio Grande do Sul. Argentina: Río Negro, Córdoba]

7′ Pharynx forward oriented; vitellaria smooth; muscular male copulatory organ with the ejaculatory duct sclerotized .... *Mesophaenocora polyova* Noreña, Brusa, Ponce de León & Damborenea, 2005

[Uruguay: Canelones]

Platyhelminthes: Rhabdocoela: Polycistididae: Genera

1 Copulatory organs in the second third of the body, behind and near the pharynx; paired ovaries; short crooked funnellike stylet; with excretory cup .... *Opisthocystis goettei* (Bresslau, 1906)

[Brazil: Rio Grande do Sul]

1′ Copulatory organs in the last third of the body, behind and clear separate from the pharynx; posteriorly with a long forklike stylet; with eyes (Fig. 5.5 E) .... *Gyratrix hermaphroditus* Graff, 1831

[Puerto Rico. Jamaica. Colombia. Brazil: São Paulo, Paraná, Rio Grande do Sul. Peru: Loreto. Argentina: Santa Fe. Antarctica]

Platyhelminthes: Rhabdocoela: Dalyelliidae: Genera

1	Penis papilla with sclerotized stylet	2
1′	Penis papilla short and small; ductus ejaculatorius no sclerotized, muscular	*Sergia calae* Noreña, Damborenea & Escobedo, 2006
	[Peru: Loreto]	
2(1)	Stylet with two handles and two spiny branches	3
2′	Stylet without handles, with a wide belt, and different kind of spines (Fig. 5.6 C)	*Gieysztoria*
	[Colombia. Brazil: São Paulo, Paraná, Rio Grande do Sul. Peru: Loreto, Puno. Argentina: Buenos Aires, Santa Fe, Río Negro]	
3(2)	Stylet with narrow handles, usually longer than the branches, branches with different kind of spines, with a middle channel; with only one egg; without zoochlorellae (Fig. 5.6 D)	*Microdalyellia*
	[Brazil: São Paulo. Argentina: Río Negro]	
3′	Stylet with wide handles, usually shorter than the branches, branches with robust spines, without a middle channel, several eggs stored in the parenchyma, sometimes with zoochlorellae (Fig. 5.6 E)	*Dalyellia*
	[Argentina: Buenos Aires, Santa Fe]	

ACKNOWLEDGMENTS

This study received support from the PIP 2013-15-0635 CONICET, N11/886 FCNyM (UNLP), and CGL2011-29916 from the Spanish Ministry of Economics. We are very grateful to B. Lares, A. Marques, J. Braccini and D. Bento for the planarians photographs.

REFERENCES

Baguñà, J. & M. Riutort. 2004. Molecular phylogeny of the Platyhelminthes. Canadian Journal of Zoology 82: 168–193.

Ball, I.R. & D.A. Hay. 1977. The taxonomy and ecology of a new monocelid flatworm from Macquarie Island (Platyhelminthes, Turbellaria). Bijdragen tot de Dierkunde 47: 205–214.

Braccini, J.A.L., S.V. Amaral & A.M. Leal-Zanchet. 2016. Microturbellarians (Platyhelminthes and Acoelomorpha) in Brazil: invisible organisms? Brazilian Journal of Biology 76: 476–494.

Cannon, L.R.G. 1986. Turbellaria of the World. A Guide to Families & Genera. Queensland Museum, Brisbane. 186 pp.

Carbayo, F. & E.M. Froehlich. 2008. Estado do conhecimento dos macroturbelários (Platyhelminthes) do Brasil. Biota Neotropica 8: 177–197.

Damborenea, M.C. & L.R.G. Cannon. 2001. On neotropical *Temnocephala* (Platyhelminthes). Journal of Natural History 35: 1103–1118.

Egger, B., F. Lapraz, B. Tomiczek, S. Müller, C. Dessimoz, J. Girstmair, N. Skunca, K.A. Rawlinson, C.B. Cameron, E. Beli, M.A. Todaro, M. Gammoudi, C. Noreña & M.J. Telford. 2015. A Transcriptomic-phylogenomic analysis of the evolutionary relationships of flatworms. Current Biology 25: 1–7.

Ferguson, F.F. 1954. Monograph of the Macrostomine worms of Turbellaria. Transactions of the American Microscopical Society 73: 137–164.

Gilbert, C.M. 1938. A remarkable North American species of the genus *Phaenocora*. Zeitschrift fuer Morphologie und Ökologie der Tiere 33: 53–71.

Hellmann, L., A.M. Leal-Zanchet & R.L. Ferreira. 2018. Two new sympatric troglobitic freshwater flatworms (Platyhelminthes: Dugesiidae) from a hotspot of subterranean biodiversity in the Neotropics. Zootaxa 4438: 561–574.

Houben, A.M., N. Van Steenkiste & T.J. Artois. 2014. Revision of *Phaenocora* Ehrenberg, 1836 (Rhabditophora, Typhloplanidae, Phaenocorinae) with the description of two new species. Zootaxa 3889: 301–354.

Kenk, R. 1974. Chapter 2: History of the study of Turbellaria in North America. Pages 17–22 *in*: N. W. Rise and P. M. Morse (eds.), Biology of the Turbellaria. McGraw-Hill Book Company, New York. NY.

Kenk, R. 1989. Revised list of the North American freshwater planarians (Platyhelminthes: Tricladida: Paludicola). Smithsonian Contributions to Zoology 476: 1–10.

Leal-Zanchet A.M., S.T. Souza & R.L. Ferreira. 2014. A new genus and species for the first recorded cave-dwelling Cavernicola (Platyhelminthes) from South America. ZooKeys 442: 1–15.

Littlewood, D.T.J. & R.A. Bray (eds.) 2001. Interrelationships of the Platyhelminthes. Taylor & Francis, London, New York. 343 pp.

Luther, A. 1955. Die Dalyelliiden (Turbellaria Neorhabdocoela) Eine Monographie. Acta Zoologica Fennica 87: 1–337.

Luther, A. 1960. Die Turbellarien Ostfennoskandiens. I. Acoela; Catenulida; Macrostomida; Lecithoepitheliata; Prolecithophora; und Proseriata. Fauna Fennica 7: 1–155.

Luther, A. 1963. Die Turbellarien Ostfennoskandiens. IV. Neorhabdocoela 2. Typhloplanoida: Typhloplanidae; Solenopharyngidae und Carcharodopharyngidae. Fauna Fennica 16: 1–161.

Mañe-Garzon, F. 1960. *Didymoschis haswelli* n. sp. Un Nuevo Temnocephalida de la cavidad branquial de *Parastacus saffordi* Faxon. Anales del Museo de Historia Natural de Montevideo, 2da serie 7: 1–17.

Marcus, E. 1946. Sôbre Turbellaria brasileiros. Zoologia. Universidade de São Paulo. Boletins da Faculdade de Filosofia, Ciencias e Letras 11: 5–254.

Marcus, E. 1954. Reports of The Lund University Chile Expedition 1948-49. II. Turbellaria. Lunds Universitets Arsskrift 49: 3–115.

Martínez-Aquino, A., F. Brusa & C. Damborenea. 2014. Checklist of freshwater symbiontic temnocephalans (Platyhelminthes: Rhabditophora: Temnocephalida) from the Neotropics. Zoosystematics and Evolution 90: 147–162.

Martínez-Aquino, A., J. Vigliano-Relva, F. Brusa & C. Damborenea. 2017. Historical biogeography of Temnocephalida (Platyhelminthes, Rhabdocoela): testing the Gondwanan hypothesis. Systematics and Biodiversity 15: 327–345.

Noreña, C, C. Damborenea & F. Brusa. 2004. Platyhelminthes de vida libre – Microturbellaria – dulceacuícolas en Argentina. Pages: 225–238 *in*: F.G. Aceñolaza (cordinador), Temas de la Biodiversidad del Litoral fluvial argentino. INSUGEO, Miscelánea, 12, Tucumán.

Noreña, C., C. Damborenea & F. Brusa. 2005. A taxonomic revision of South American species of the genus *Stenostomum* O. Schmidt (Catenulida) based on morphological characters. Zoological Journal of the Linnean Society 144: 37–58.

Noreña, C., C. Damborenea & F. Brusa. 2015. Phylum Platyhelminthes. Pages: 181–203 *in*: J. Thorp & D.C. Rogers (eds.), Ecology and General Biology, Thorp and Covich's Freshwater Invertebrates (4th edition), Academic Press, Elsevier, Boston, MA.

Noreña, C., C. Damborenea & F. Brusa. 2016. Phylum Platyhelminthes. Pages: 91–109 *in*: J. Thorp & D.C. Rogers (eds.), Keys to Neartic Fauna, Thorp and Covich's Freshwater Invertebrates (4th edition), Academic Press, Elsevier, Boston, MA.

Noreña-Janssen, C. 1995. Studies on the taxonomy and ecology of the Turbellaria (Platyhelminthes) in the floodplain of the Paraná River (Argentina). II. Taxonomy and ecology of the Turbellaria. Archiv für Hydrobiologie 2: 211–262.

Noreña-Janssen, C. & A. Faubel. 1992. Revision of the Subfamily Mesostominae, Rhabdocoela, Platyhelminthes. Mitteilungen aus dem Hamburgischen Zoologischen Museum und Institut 89: 7–47.

Noreña-Janssen, C., C. Damborenea & F. Brusa. 2016. Platyhelminthes: Rhabdocoela: Typhloplanidae. Identification Guides to the Plankton and Benthos of Inland Waters 26. Margraf Publishers GmbH, Weikersheim, Germany. 274 pp.

Nuttycombe, J.W. 1956. The Catenulida of the Eastern United States. American Midland Naturalist 55: 419–433.

Richter, S., R. Loese, G. Purschke, A. Schmidt-Rhaesa, G. Scholtz, T. Stach, L. Vogt, A. Wanninger, G. Brenneis, C. Döring, S. Faller, M. Fritsch, P. Grobe, C.M. Heuer, S. Kaul, O.S. Møller, C.H.G. Müller, V. Rieger, B.H. Rothe, M.E.J. Stegner & S. Harzsch. 2010. Invertebrate neurophylogeny: suggested terms and definitions for a neuroanatomical glossary. Frontiers in Zoology 7:1–49.

Rieger, R.M., S. Tyler, J.P.S. Smith III & G. Rieger. 1991. Chapter 2: Turbellaria. Pages 7–140 *in*: F.W. Harrison and B. J. Bogitsh (eds.), Microscopic Anatomy of Invertebrates. Vol. 3: Platyhelminthes and Nemertinea. Wiley-Liss, New York. NY.

Ruiz-Trillo, I., M. Riutort, D.T.J. Littlewood, E.A. Herniou & J. Baguñà. 1999. Acoel flatworms: Earliest exant bilaterian metazoans, not members of Platyhelminthes. Science 283: 1919–1923.

Ruiz-Trillo, I., M. Riutort, H.M. Fourcade, J. Baguñà & J.L. Boore. 2004. Mitochondrial genome data support the basal position of Acoelomorpha and the polyphyly of the Platyhelminthes. Molecular Phylogenetics and Evolution 33: 321–332.

Schärer, L. 2014. Macrostomorpha taxonomy and phylogeny. http://macrostomorpha.info. Consulted 2017-04-17.

Sluys, R. 1989. A Monograph of Marine Triclads. A.A. Balkema, Rotterdam, Brookfield. 463 p.

Sluys, R. 1990. A Monograph of the Dimarcusidae (Platyhelminthes, Seriata, Tricladida). Zologica Scripta 19: 13–29.

Souza, S., M. Riutort, R.L. Ferreira & A. Leal-Zanchet. 2018. An integrative taxonomic approach reveals the first marine triclad (Platyhelminthes) trapped in a cave from a semiarid Neotropical environment. Invertebrate Systematics 32: 627–638.

Tyler, S. & M. Hooge. 2004. Comparative morphology of the body wall in flatworms (Platyhelminthes). Canadian Journal of Zoology 82: 194–210.

Tyler, S., S. Schilling, M. Hooge & L.F. Bush (comp.). 2006–2019. Turbellarian taxonomic database. Version 1.8 http://turbellaria.umaine.edu. Consulted 2020-03-24.

Chapter 6

Phylum Nemertea

Malin Strand
Swedish Species Information Centre, Swedish University of Agricultural Sciences, Uppsala, Sweden

Per Sundberg
Department of Marine Sciences, University of Gothenburg, Gothenburg, Sweden

PHYLUM NEMERTEA

Chapter Outline

INTRODUCTION

Around 1300 species of nemerteans are described (Kajihara et al., 2008), with the vast majority being marine and only 22 species currently known from freshwaters. Of these, only 4 are recorded from the Neotropical region.

Within the two *Prostoma* species reported from South America, *P. eilhardi* (Montgomery, 1894) was identified in Argentina and Brazil (Moretto & Durquet, 1978; Corrêa, 1986), and *P. graecense* (Böhmig, 1892) has been reported from Venezuela (Corrêa, 1951), although this identification is questioned in later literature. *P. graecense* was later reported from Argentina (Tamburi & Cazzaniga, 2006) and Mexico (Lagunas-Calvo et al., 2018). The two species cannot be distinguished from each other based on external features alone, as most members of this genus exhibit a remarkable uniformity in their external appearance. There is no general agreement on the extent of morphological intraspecific variability, which makes it even more difficult. Both appear to be circumglobal, but sporadic. Current DNA data support a cosmopolitan distribution of *P. graecense*. The body length for *P. graecense* reaches up to 5 cm; while for *P. eilhardi*, the literature suggests a maximum length of 0.18 cm.

Koinoporus mapochi Sánchez & Moretto, 1988 is only known from small streams flowing into the Mapocho River (Talagante region) in Chile (Sanchez & Moretto, 1988) and is found under stones in slow-running waters. The species is said to reach a maximum length of 16 mm.

Siolineus turbidus Du Bois-Reymond Marcus, 1948 is only known from four individuals (all females) in the Brazilian river Tapajós (the type locality) found at 28 m depth in the 1940s (Du Bois-Reymond Marcus, 1948). It is the only Neotropical freshwater species representing Heteronemertea.

LIMITATIONS

The number of described nemertean species is most probably lower than the actual number. Genetic evidence shows that sibling and cryptic species are more common than previously recognized, and many regions of the world are yet to be explored to map the nemertean fauna. Three of the documented species are hoplonemerteans (Class Hoplonemertea) and *Siolineus turbidus* is the one species representing heteronemerteans (Class Pilidiophora). The third nemertean class Palaeonemertea has no records documented in the region. The number of known freshwater nemerteans is small; the 22 reported species represent less than 2% of the total number recorded. Few persons have focused on Neotropical freshwater nemerteans, and the lack of data is evident. Only two species have their type locality in this region, and one of these (*Siolineus turbidus*) was described from preserved material, years after the time of collection. The two Neotropical species of *Prostoma* cannot be distinguished by external features.

This chapter should not be used as an overview of the freshwater nemertean fauna of the Neotropics, but rather as

Thorp and Covich's Freshwater Invertebrates. https://doi.org/10.1016/B978-0-12-804225-0.00006-X

a list of existing documentation and a reminder that in this region of the world this group of organisms is principally undiscovered, or at least undocumented.

TERMINOLOGY AND MORPHOLOGY

Only adult nemerteans possess the morphological characters reflected in the identification key. Juveniles are in most cases not identifiable. Some species can be roughly identified by a combined knowledge of shape, size, and color (and, of course, sampling site). To be sure of the species identification one has to go into details. First thing to look for are the **cephalic slits**, which are located on the sides of the head, and are present only in the order Heteronemertea. In Heteronemertea the mouth is always separate from the **proboscis pore** and should always be visible on the ventral side. A caudal cirrus can also be present in heteronemerteans, but not in other groups. If cephalic slits are lacking, there can be **cephalic furrows** (Hoplonemertea). More characters to be observed are presence/absence of **eyes**, number and eye placement pattern if present, color, position of mouth, the presence of a **frontal organ** and the **cerebral ganglia**. In many species it is possible to see through the dorsal body wall and estimate the length of the rhynchocoel. The proboscis of hoplonemerteans is armed with one **stylet** that can be detected under a microscope.

MATERIAL PREPARATION AND PRESERVATION

Nemerteans should always be studied alive when possible, and characters to be observed are presence/absence of eyes, number and placement pattern if present, color, position of mouth, and (in this case), the presence of a frontal organ. For histological sections, it is important to properly anesthetize the animal before fixation to minimize contractions (which will distort anatomical characters). MS 222 or $MgCl_2$ is recommended. A common fixative for nemerteans is Bouin's fluid, and also 3%–4% formalin. Ethanol is required for later DNA extractions, but is less suitable for histology. Another good fixative for RNA and DNA studies is RNAlater, which is easy to carry around (being neither toxic nor inflammable) and especially suited for fieldwork. Long-term storage of specimens preserved for DNA should be kept in freezer.

PHYLUM NEMERTEA

KEYS TO NEMERTEA

Nemertea: Orders

1 Head without cephalic slits .... **Hoplonemertea** **[p. 122]**

1′ Head with cephalic slits (Fig. 6.1) .... Heteronemertea one species: *Siolineus turbidus* Du Bois-Reymond Marcus, 1948 [Brazil]

Nemertea: Hoplonemertea: Genera

1 Semitransparent, whitish, head with four (sometimes six) reddish-brown ocelli arranged in a square .... *Koinoporus mapochi* Sánchez & Moretto, 1988 [Chile]

1′ Uniformly colored, head with up to six eyes distributed along margins .... *Prostoma* [Venezuela? Brazil, Argentina]

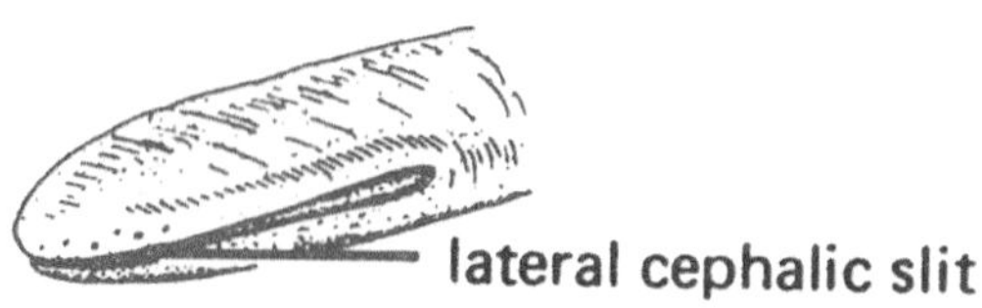

FIGURE 6.1 Cephalic slits in Heteronemertea.

REFERENCES

Corrêa, D.D. 1951. Freshwater nemertines from the Amazon Region and from South Africa. Boletim da Faculdade de Filosofia, Ciências e Letras, Universidade de São Paulo, Zoologia 16: 257–269.

Corrêa, D.D. 1986. Nemertini. Pages 1–20 *in:* R. Schaden (ed.) Manual de identificação de invertebrados límnicos do Brasil 9, Brasilia, CNPq Assessoria Editorial.

Du Bois-Reymond Marcus, E. 1948. An Amazonian heteronemertine. Boletim da Faculdade de Filosofia, Ciências e Letras, Universidade de São Paulo, Zoologia 13: 93–109.

Kajihara, H., A.V. Chernyshev, S-C. Sun, P. Sundberg & F.B. Crandall. 2008. Checklist of Nemertean Genera and Species published between 1995 and 2007. Species Diversity 13: 245–274.

Lagunas-Calvo, O., D.I. Hernández-Mena, G. Rivas, S. Contreras-Mirón, B. Mendoza-Garfias, O.M. Echeverría-Martínez & A. Oceguera-Figueroa. 2018. Rediscovery of the nemertean *Prostoma graecense* from Xochimilco, 75 years after its first and only record. Revista Mexicana de Biodiversidad 89: 39–45.

Moretto, H.J.A. & J. Durquet. 1978. El hoplonemertino *Prostoma eilhardi* (Montgomery, 1894) en el Río de la Plata. Ciencia e Investigación 34: 36–38.

Sanchez, M. & H.J.A. Moretto. 1988. A new genus of freshwater hoplonemertean from Chile. Zoological Journal Linnean Society 92: 193–207.

Tamburi, N.E. & N.J. Cazzaniga. 2006. The freshwater ribbon-worm *Prostoma graecense* (Hoplonemertea: Monostilifera) in South America (Argentina). Zootaxa 1248: 27–34.

Chapter 7

Phylum Gastrotricha

André Rinaldo Senna Garraffoni and Thiago Quintão Araújo

Laboratório de Evolução de Organismos Meiofaunais, Departamento de Biologia Animal, Instituto de Biologia, Universidade Estadual de Campinas, Campinas, São Paulo, Brazil

Chapter Outline

INTRODUCTION

Gastrotricha are microinvertebrates commonly found in marine and freshwater. They are meiofauna that move by ventral ciliary locomotion (Balsamo et al., 2014; Kånneby & Hockberg, 2015). Although they are very small (most less than 0.5 mm), gastrotrichs are recognized for their complex reproductive system and life cycle, with a predominance of hermaphroditism (Weiss, 2001).

The taxon comprises more than 820 species, traditionally divided into the two orders Macrodasyida and Chaetonotida (Balsamo et al., 2015; Kånneby, 2016; Todaro, 2018a, 2018b). The first order comprises 9 families, 33 genera, and more than 365 species, among them 4 freshwater species of the mostly marine macrodasyidans, *Marinellina flagellata* Ruttner-Kolisko, 1955, *Redudasys fornerise* Kisielewski, 1987, *R. neotemperatus* Kånneby & Kirk, 2017 and *R. brasiliensis* Garraffoni, Araújo, Lourenço, Guidi & Balsamo, 2019 (Garraffoni et al., 2010, 2019; Todaro et al., 2012; Araújo et al., 2013; Kånneby & Wicksten, 2014; Kånneby & Kirk, 2017). All feature strap-shaped bodies, pharyngeal pores and, usually, numerous adhesive tubes in the anterior, lateral, and posterior body regions (Todaro & Hummon, 2008; Kånneby & Hockberg, 2015). Within Chaetonotida, 7 families, 31 genera and nearly 480 species are interstitial or epibenthic in marine and brackish waters and two-thirds live in freshwaters (Balsamo et al., 2008). They are fusiform (or tenpin-shaped) and have one pair (rarely two) of adhesive tubes, limited to the posterior end (Kånneby & Hockberg, 2015). The order Chaetonotida is divided into the suborders Multitubulatina (only one genus, *Neodasys*) and Paucitubulatina (25 genera). The Chaetonotidae is the most specious among Paucitubulatina, comprising approximately one-third of the species described for the whole group (Kisielewski, 1991; Balsamo et al., 2009).

Although, the reported number of known species of Neotropical freshwater gastrotrichs is rather inaccurate due

Thorp and Covich's Freshwater Invertebrates. https://doi.org/10.1016/B978-0-12-804225-0.00007-1

to the limited sampling effort, to date approximately 86 species have been described or identified from Argentina, Brazil, Guyana, Paraguay, and Uruguay (Daday, 1905; Cordero, 1918; Grosso, 1973a, 1973b, 1975, 1976, 1982; Grosso & Drahg, 1983, 1984, 1986, 1991; Kisielewski, 1987, 1991; d'Hondt et al., 2005, 2006; Araújo et al., 2013; Garraffoni et al., 2010, 2017b, 2019; Todaro et al., 2012; Kånneby et al., 2013; Garraffoni & Melchior, 2015; Minowa & Garraffoni, 2017). It is important to highlight that this list only includes species from South America, because no study of freshwater gastrotrichs was done in Central America. Thus, the distribution and biodiversity of this taxon are completely unknown in this area in the Neotropical region. Among these 86 species, 42 were originally described in South America and are known only in type locality, 12 were originally described in South America and are known in more than one sampling sites in only one South America country, 4 were originally described in South America and are known in more than one sampling site in two South America countries, 12 were originally described in Europe and are known in one sampling site in South America, and 16 were originally described in Europe and are known in more than one sampling site in South America. The only species originally described in South America and reported in Europe is *Chaetonotus soberanus* Grosso & Drahg, 1983 by Kånneby (2013). This number shows potential bias attributable to an unequal number of sampled sites in each country, because 74 species (85%) were found only in Brazil and Argentina.

LIMITATIONS

Unfortunately, until now there was no attempt to construct a comprehensive key for the Neotropical gastrotrichs. Because of the relatively modest faunistic and taxonomic works done in the Neotropical region (only few sites were sampled from Argentina, Brazil, Guyana, Paraguay, and Uruguay), we feel that this is a great underestimate of the true species richness. It should be emphasized that the majority of the species considered in this report were collected in Brazil, because in the other countries only sporadic samplings were carried out.

Some original descriptions date back more than a century, and thus the researchers involved could not employ the microscopical techniques currently used such as differential interference contrast (DIC) (Garraffoni et al., 2010, 2017b, 2019; Todaro et al., 2012; Garraffoni & Melchior, 2015; Minowa & Garraffoni, 2017). The use of the digital camera to record gastrotrichs around the world started to be used more than 25 years ago (Hummon et al., 2005), but video sequences for the Neotropical fauna only became common a few years ago (Garraffoni et al., 2010, 2017b, 2019; Todaro et al., 2012; Araújo et al., 2013). Furthermore, for taxonomic or systematic purposes, the specimens must be studied alive (Balsamo et al., 2014; Kånneby, 2016).

In addition to the taxonomic knowledge of the group being far from satisfactory, studies on other biological aspects of Neotropical gastrotrichs, such as their ecology, natural history, biogeography, etc., are still very scarce.

TERMINOLOGY AND MORPHOLOGY

In the Neotropics, the general body of a freshwater gastrotrich can be **fusiform (or tenpin-shaped)**, as in chaetonotids, or **vermiform**, as in macrodasyids. The former gastrotrichs' body can be divided into a head, neck, and trunk. The **head** can consist in three (Figs. 7.5 M, 7 O, P) or five **lobes** (Figs. 7.5 L, 8 D) due to the presence of distinct **cephalic plates**: dorsally the **cephalion** (Fig. 7.5 D), laterally the **epi**- and **hypopleurae** (Fig. 7.3 A), ventrally the **hypostomium** (Fig. 7.5 A), and one (Fig. 7.4 C) or two pairs (Fig. 7.4 K) **sensory cephalic** tufts present between these plates. In the semiplanktonic taxa Dasydytidae and Neogosseidae, most of these plates are absent and two or three **ciliary bands** surround the head region. The **mouth** can be **terminal** (Fig. 7.1 B) or **subterminal** (Fig. 7.3 H) and can have a **mouth ring** (Fig. 7.4 B) (segmented or not) surrounding it.

The mouth is connected to a cylindrical muscular **pharynx** (almost uniform width along whole length) or with one (posterior) or two (anterior and posterior) **swellings** (bulbs). One set of **teeth** (Fig. 7.8 O) or two cuticular **pharyngeal rods** (cuticular reinforcement) may occur anteriorly in the pharynx. **Ocellar granules** (Fig. 7.4 P), a pair of dark rounded structures, can be present in the anterior end.

The **neck** can have a varying degree of constriction and, dorsally, a pair of **sensory bristles** is usually present (Fig. 7.4 L) (another pair can be present in the posterior trunk region, Fig. 7.4 I). The trunk can be naked (Fig. 7.2 G) or covered dorsally (Fig. 7.3 C), dorsolaterally, and ventrally (Fig. 7.5 A) with scales (**epidermic ornamentations**). The last includes various types, including trilobate (Figs. 7.3 C, 5 N), pentalobate (Figs. 7.3 J, 5 A), or weakly or strong trilobate. Shapes can be, for example, rounded (Figs. 7.5 N, 6 K), oval (Fig. 7.6 C), or elliptic (Fig. 7.4 J). This structure can bear **spines** (Figs. 7.6 H, 8 F) with one (Fig. 7.6 G) or two (Fig. 7.2 F) **denticles** arising in different positions or also keels (Fig. 7.4 Q).

The **scales** are distributed in columns parallel to the longitudinal body axis (Fig. 7.6 F) or rows perpendicular to the longitudinal body axis (Fig. 7.4 Q). In both cases, the number of scales varies among species. In the ventral side, the **locomotory cilia** (structure that the phylum name was based on: from Greek *gaster*, stomach + *thrix*, hair) are usually distributed in two paired continuous bands through the whole body or in a few tufts. The **interciliary area**

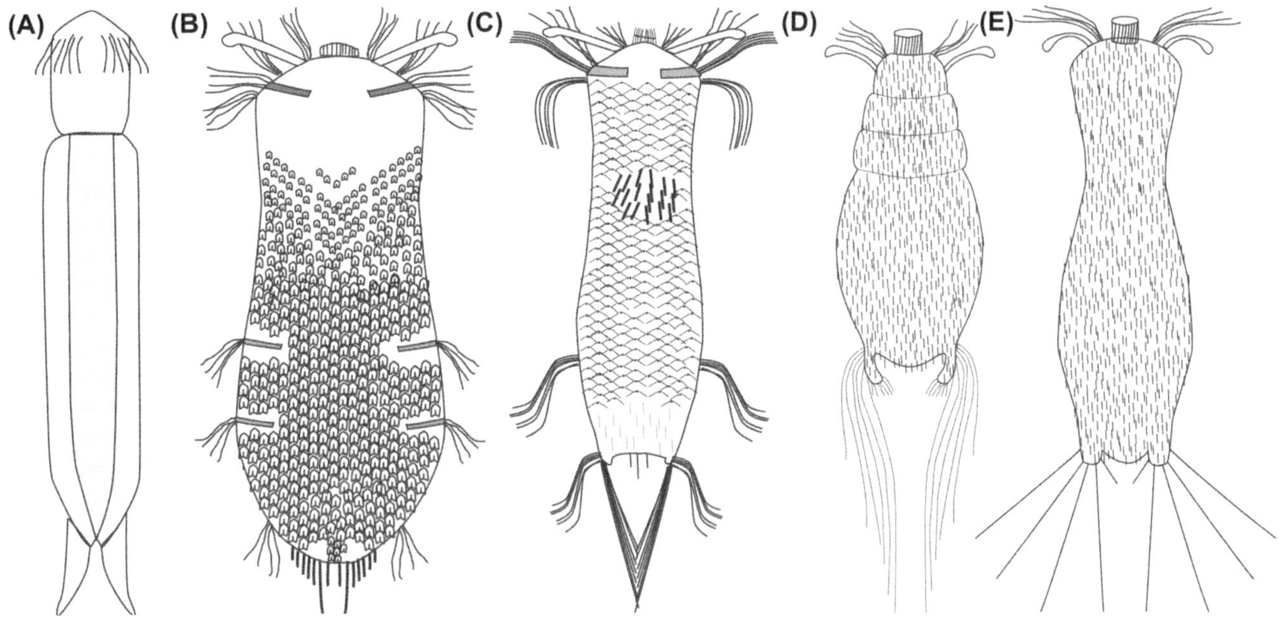

FIGURE 7.1 Schematic drawings of Proichthyidiidae and Neogosseidae species: (A) *Proichthydium coronatum*; (B) *Kijanebalola canina*; (C) *Neogossea acanthocolla*; (D) *Neogossea fasciculata*; and (E) *Neogossea pauciseta*.

(region between the continuous band of cilia) can be naked (Fig. 7.4 H), partially (from posterior region to middle trunk) or fully (from posterior region to close to head) covered with **keeled** or **spined** scales (Fig. 7.5 A) (often smaller than those observed dorsally and dorsolaterally). In the posterior end, one pair of **adhesive tubes** (can vary in curvature, length, and thickness or also be segmented (Figs. 7.5 A–F, 7 F)) arises from a caudal furca (both structures together are normally referred to as the furca) are present, except in the semiplanktonic taxa where they are absent (Fig. 7.2 A–L). A peculiarity observed in most semiplanktonic taxa is the presence of three groups of **long movable lateral spines** (Fig. 7.2 A–L) (cephalic/neck, trunk, and rear) used for locomotion.

Macrodasyid bodies have a smooth cuticle, without scales or spines. The head is rounded bearing several **sensorial cilia**, without **ocellar granules**, and with a terminal or slightly subterminal mouth. Pharynx is cylindrical (almost uniform width along whole length) bearing **pores** at base and opening ventrolaterally. The neck constriction is virtually absent and the whole body has a similar width.

Adhesive apparatus consisting of anterior and posterior tubes. One or two tubes (unequal length) per side arise ventrolaterally in anterior of the animal, whereas the posterior end is bilobate. Ventral ciliation is arranged into a pair of longitudinal bands along the neck and trunk region and forms a single column in the posterior trunk. Alternatively, groups of cilia may be arranged in paired longitudinal series along the anterior part of body and unpaired median ones along the posterior trunk part (Fig. 7.9).

MATERIAL PREPARATION AND PRESERVATION

Qualitative sampling of freshwater gastrotrichs requires very simple equipment, independently of the habitat studied. For interstitial habitats, samples of the upper sediment layer (5–10 cm) can be collected with a manual corer (with 2–5 cm diameter) or with a small shovel. For periphyton and semiplanktonic habitats, water around roots of floating vegetation can be collected with a plastic container or filtered with a net (mesh size 30–40 μm) (Balsamo et al., 2014).

Samples should be processed as soon as possible (30 min–48 h after sediment collection), because good species identification will only be possible when specimens are sorted alive, as main morphological features deteriorate rapidly and general morphology in fixed specimens is difficult to distinguish (Balsamo et al., 2014; Kånneby, 2016). Examination of the collected material can be done until one or two weeks. Live gastrotrichs are usually highly motile and before mounting the slide for observation, the individuals need to be anesthetized (e.g., menthol or 2% $MgCl_2$).

Sorting for Gastrotricha specimens must be done under stereomicroscopes, and detailed observations should occur under high quality compound microscopes equipped with differential interference contrast (DIC) or phase contrast using objectives with 40–100× magnification. The use of a scanning electron microscope (SEM) is one of the most valuable tools for systematic studies. The specimens, after dehydration in a graded ethanol series, are treated with

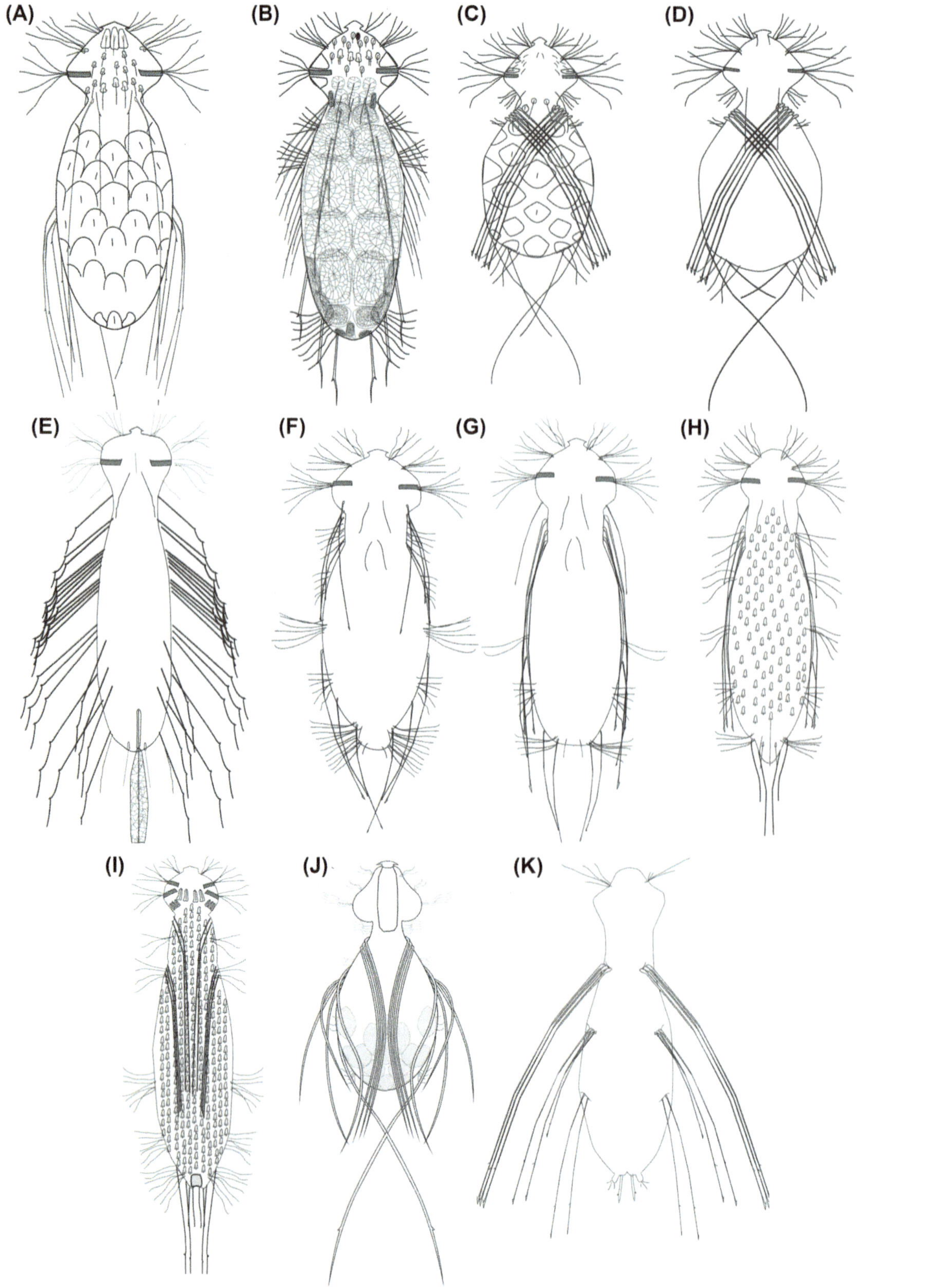

FIGURE 7.2 Schematic drawings of Dasydytidae species: (A) *Setupus aequatorialis*; (B) *Ornamentula paraensis;* (C) *Haltidytes squamosus*; (D) *H. festinans*; (E) *Dasydytes nhumirimensis*; (F) *D. lamellatus*; (G) *D. carvalhoae*; (H) *D. papaveroi*; (I) *D. elongatus*; (J) *Haltidytes pseudosquamosus*, and (K) *Stylochaeta fusiformis.*

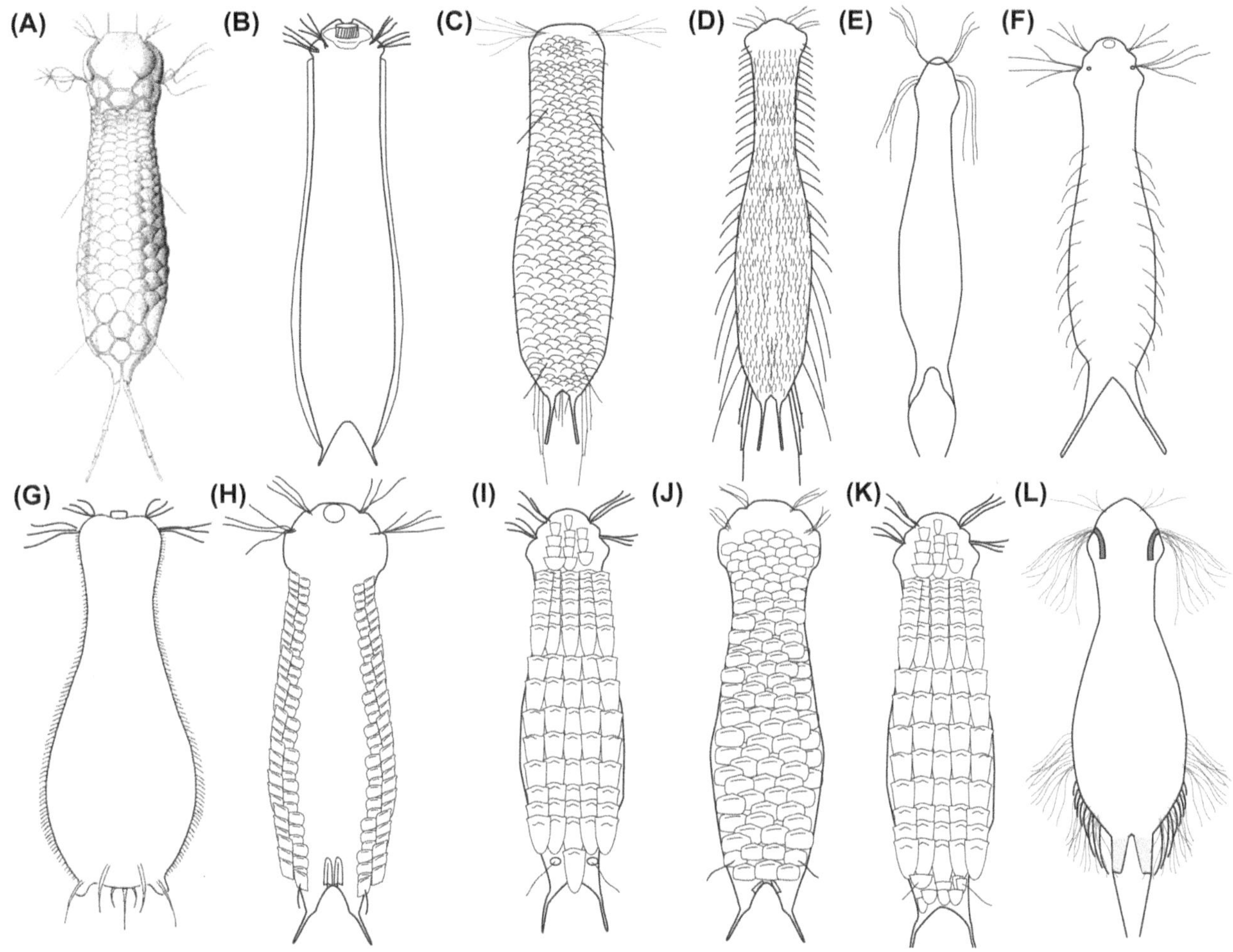

FIGURE 7.3 Schematic drawings of Chaetonotidae species: (A) *Cephalionotus kisielewskii*; (B) *Arenotus strixinoi*; (C) *Lepidochaetus brasilense*; (D) *L. zelinkai*; (E) *Ichthydium forficula*; (F) *I. chaetiferum*; (G) *I. crassum*; (H) *Lepidodermella minus chaetifer*; (I) *L. amazonica*; (J) *L. squamata*; (K) *L. broa*; and (L) *Undula paraensis*.

HMDS (hexamethyldisilazane) or critical point dry, mounted on aluminum stubs, and sputter-coated with gold-palladium (Hochberg & Litvaitis, 2000).

For ecological studies, examination of live specimens is impossible and fixation becomes necessary. However, the use of improper fixation techniques can increase the chance of body damage, preventing good taxonomical identification, or cause contraction and/or disintegration of bodies. Garraffoni et al. (2017a) evaluated the use of three anesthetics and three fixatives for freshwater meiofauna, including gastrotrichs, and they found that the use of $MgCl_2$ or gasified water decreases the possibility of specimen contraction and buffered formalin increases the quality of preservation.

For morphological studies, specimens can be fixed in 2.5% glutaraldehyde or 10% neutralized formaldehyde and kept in microtubes or permanent slides. In the last case, after the fixation, the specimens are put through a series of glycerol, ethanol, and distilled water (Balsamo et al., 2014; Kånneby, 2016). For molecular studies, specimens should be fixed and maintained in 5% ethanol and stored in −20°C. However, it is important to highlight that gastrotrichs are very fragile and they can be destroyed during the observation process. Furthermore, the specimens fixed in the slides will inevitably vanish over time and most of their diagnostic characteristics usually dissolve or are lost immediately after preservation (Balsamo et al. 2014; Garraffoni & Freitas, 2017).

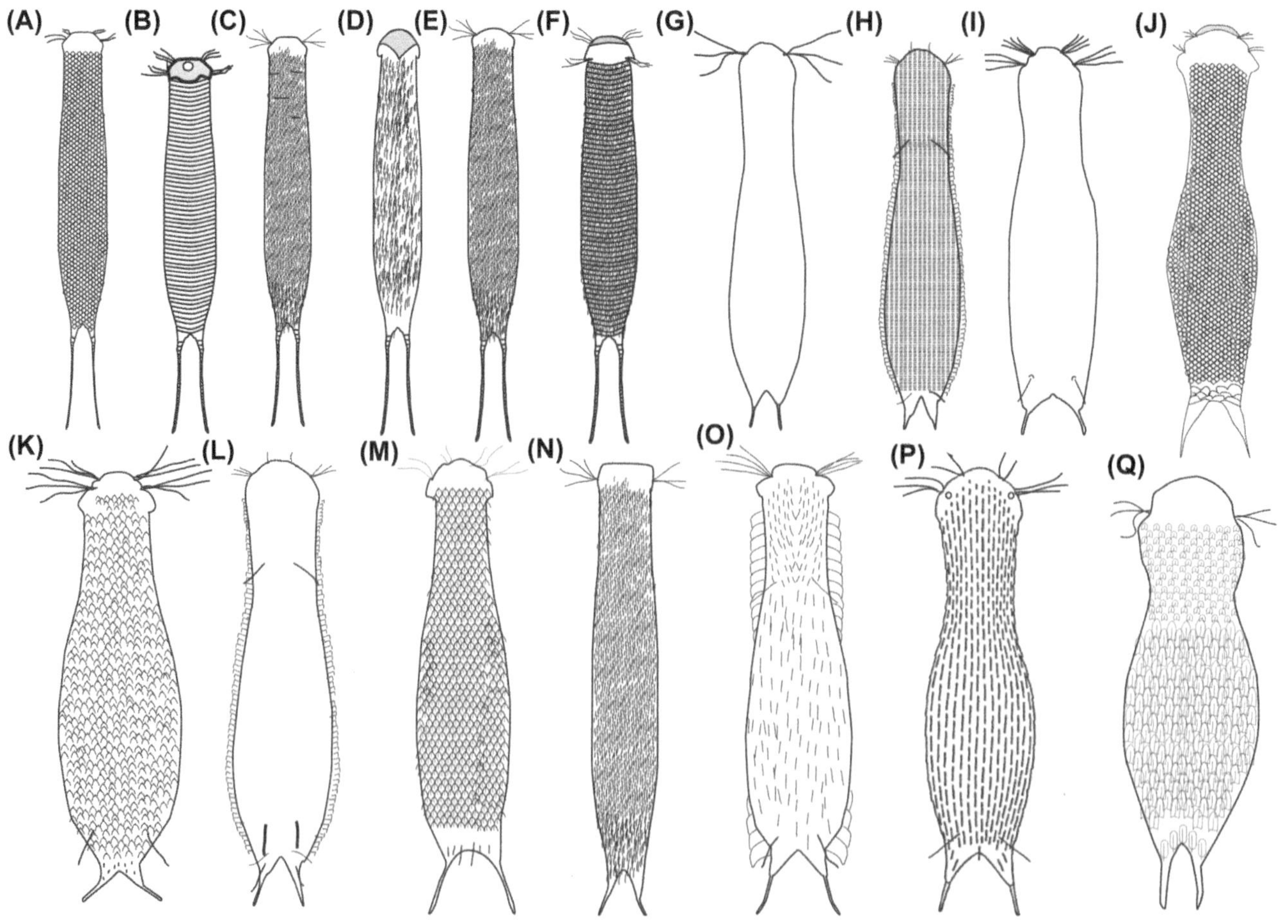

FIGURE 7.4 Schematic drawings of Chaetonotidae species: (A) *Polymerurus rhomboides*; (B) *P. elongatus*; (C) *P. corumbensis*; (D) *P. ringueleti*; (E) *P. nodicaudus*; (F) *P. paraelongatus*; (G) *Aspidiophorus ophiodermus*; (H) *A. oculifer*; (I) *A. slovinensis*; (J) *A. lilloensis*; (K) *A. pleustonicus*; (L) *A.* aff. *tetrachaetus*; (M) *A. bramhsi*; (N) *Heterolepidoderma majus*; (O) *H. jureiense*; (P) *H. ocellatum*; and (Q) *H. mariae*.

KEYS TO GASTROTRICHA

The following keys are most based on the key of Garraffoni & Araújo (2010) for Brazilian gastrotrichs, but several other important works were also checked (Daday, 1905; Murray, 1913; Cordero, 1918; Grosso, 1973a, 1973b, 1975, 1976, 1982; Grosso & Drahg, 1983, 1984, 1986, 1991; Kisielewski, 1987, 1991; Schwank, 1990; d' Hondt et al., 2005, 2006; Kånneby et al., 2009; Garraffoni et al., 2010, 2017b, 2019; Todaro et al., 2012; Araújo et al., 2013; Balsamo et al., 2014; Garraffoni & Melchior, 2015; Kånneby, 2016; Minowa & Garraffoni, 2017). All illustrations and photos used in the present study were based on specimens collected from the Neotropics.

Gastrotricha: Orders

1 Body vermiform; anterior adhesive tubes distributed in two symmetrical groups, each made of two unequal tubes; usually two caudal adhesive tubes; pharyngeal pores usually present; inverted Y-shaped pharyngeal lumen; body cuticle smooth or sculptured (Fig. 7.9) .... Macrodasyida, one family: **Redudasyidae [p. 130]**

1' Body fusiform; anterior adhesives tubes absent; usually two (rarely four) caudal adhesive tubes; pharyngeal pores absent; Y-shaped pharyngeal lumen; smooth or sculptured body cuticle (Figs. 7.1–8) .... **Chaetonotida [p. 133]**

Gastrotricha: Macrodasyida: Redudasyidae: Genera

1 Two anterior adhesive tubes per side, pharyngeal pores present, head weakly marked (Fig. 7.9 A) .... *Redudasys fornerise* Kisielewski, 1987
[Brazil: São Paulo]

1' One anterior adhesive tube per side, pharyngeal pores absent, head strongly marked (Fig. 7.9 B) .... *Redudasys brasiliensis* Garraffoni, Araújo, Lourenço, Guidi & Balsamo, 2019

[Brazil: Minas Gerais]

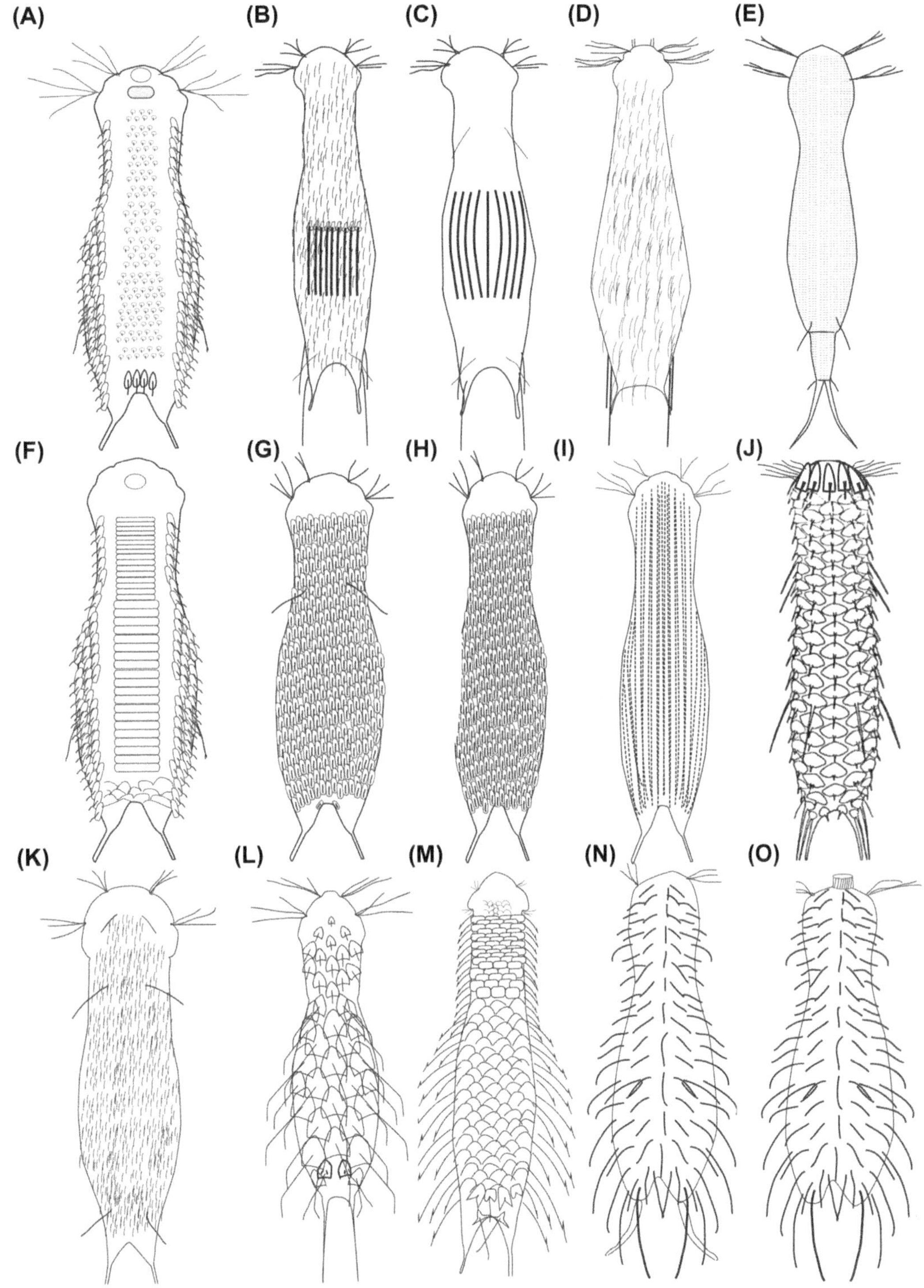

FIGURE 7.5 Schematic drawings of *Chaetonotus* species (Chaetonotidae): (A) *Chaetonotus acanthodes*; (B) *C. bisacer*; (C) *C. succinctus*; (D) *C. corderoi*; (E) *C. silvaticus*; (F) *C. annectens*; (G) *C. intermedius*; (H) *C. breviacanthus*; (I) *C. pseudopolyspinosus*; (J) *C. acanthocephalus*; (K) *C. sanctipauli*; (L) *C. paucisquamatus*; (M) *C. pilaga*; (N) *C. erinaceus*; and (O) *C. paraguayensis*.

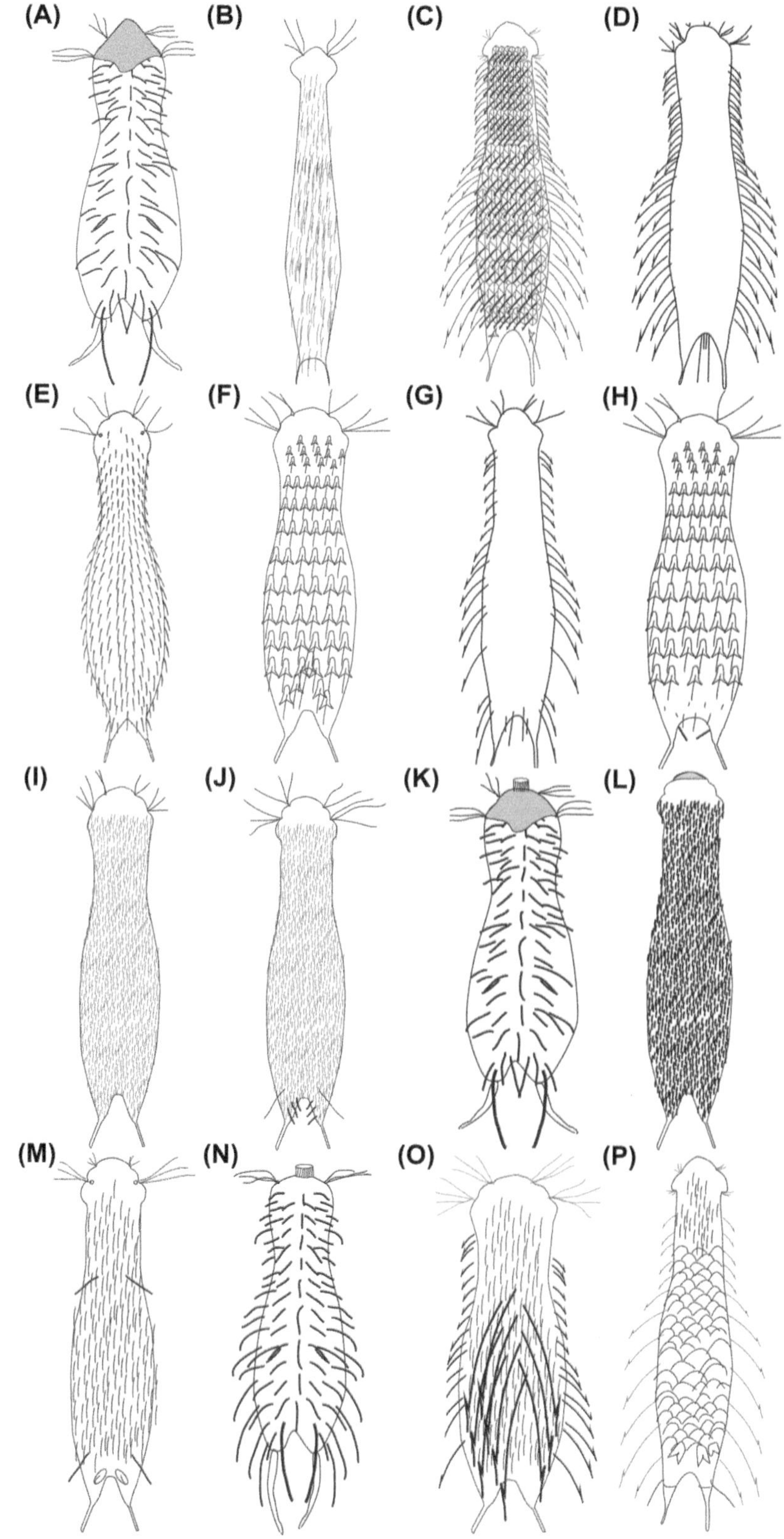

FIGURE 7.6 Schematic drawings of *Chaetonotus* species (Chaetonotidae): (A) *Chaetonotus pusillus*; (B) *C. montevideensis*; (C) *C. tenuisquamatus*; (D) *C. heideri*; (E) *C. oculifer*; (F) *C. persetosus*; (G) *C. heteracanthus*; (H) *C. furcatus*; (I) *C. polyspinosus*; (J) *C. ventrochaetus*; (K) *C. dubius*; (L) *C. odontopharyxns*; (M) *C. lobo*; (N) *C. similis*; (O) *C. balsamoae*; and (P) *C. soberanus*.

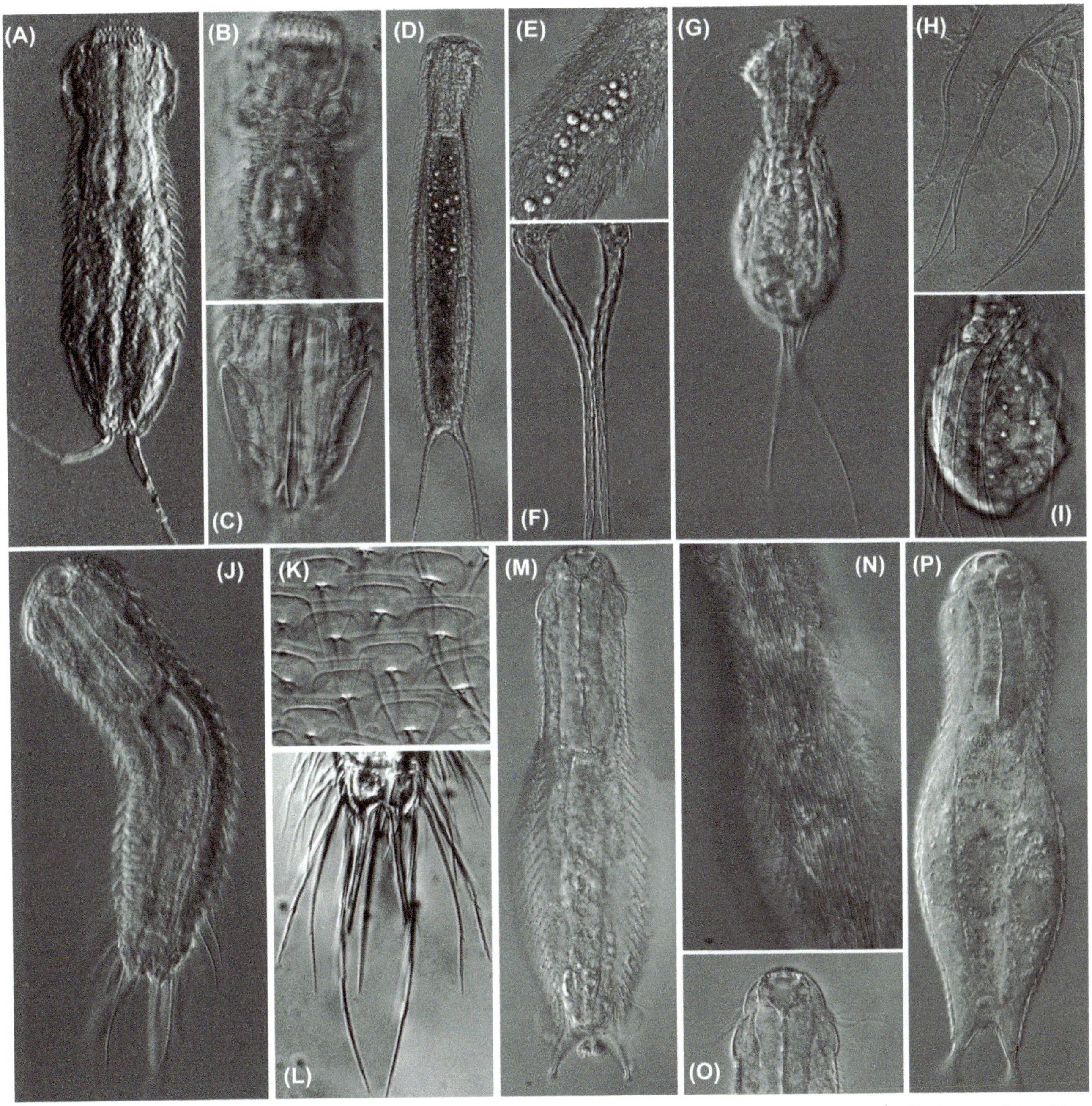

FIGURE 7.7 Microphotographs of live Chaetonotida specimens: (A–C) *Cephalionotus kisielewskii*: (A) ventral habitus, (B) close-up of dorsal head scales complex, (C) close-up of large keeled scales on the ventral posteriormost body; (D–F): *Polymerurus nodicaudus*: (D) dorsal habitus, (E) close-up of dorsal spined body, (F) close-up to segmented scales; (G–I) *Haltidytes pseudosquamosus*: (G) ventral habitus, (H) close-up of "saltatorial" spines, (I) close-up of the insertion of "saltatorial" spines on dorsal trunk; (J–L) *Lepidochaetus zelinkai*: (J) ventral habitus, (K) dorsal spined double-edge scales, (L) close-up of spine complex on the end body; (M–O) *Heterolepidoderma majus*: (M) ventral habitus, (N) small keeled scales on the ventral body, (O) close-up to the three-lobed head; and (P) *Heterolepidoderma mariae*, ventral habitus.

Gastrotricha: Chaetonotida: Families

1 Posterior adhesive tubes absent or, if present, straight 2

1' Posterior adhesive tubes curved and divergent (Fig. 7.1 A) ...... Proichthyidiidae, one species: *Proichthydium coronatum* Cordero, 1918 [Uruguay: Montevideo]

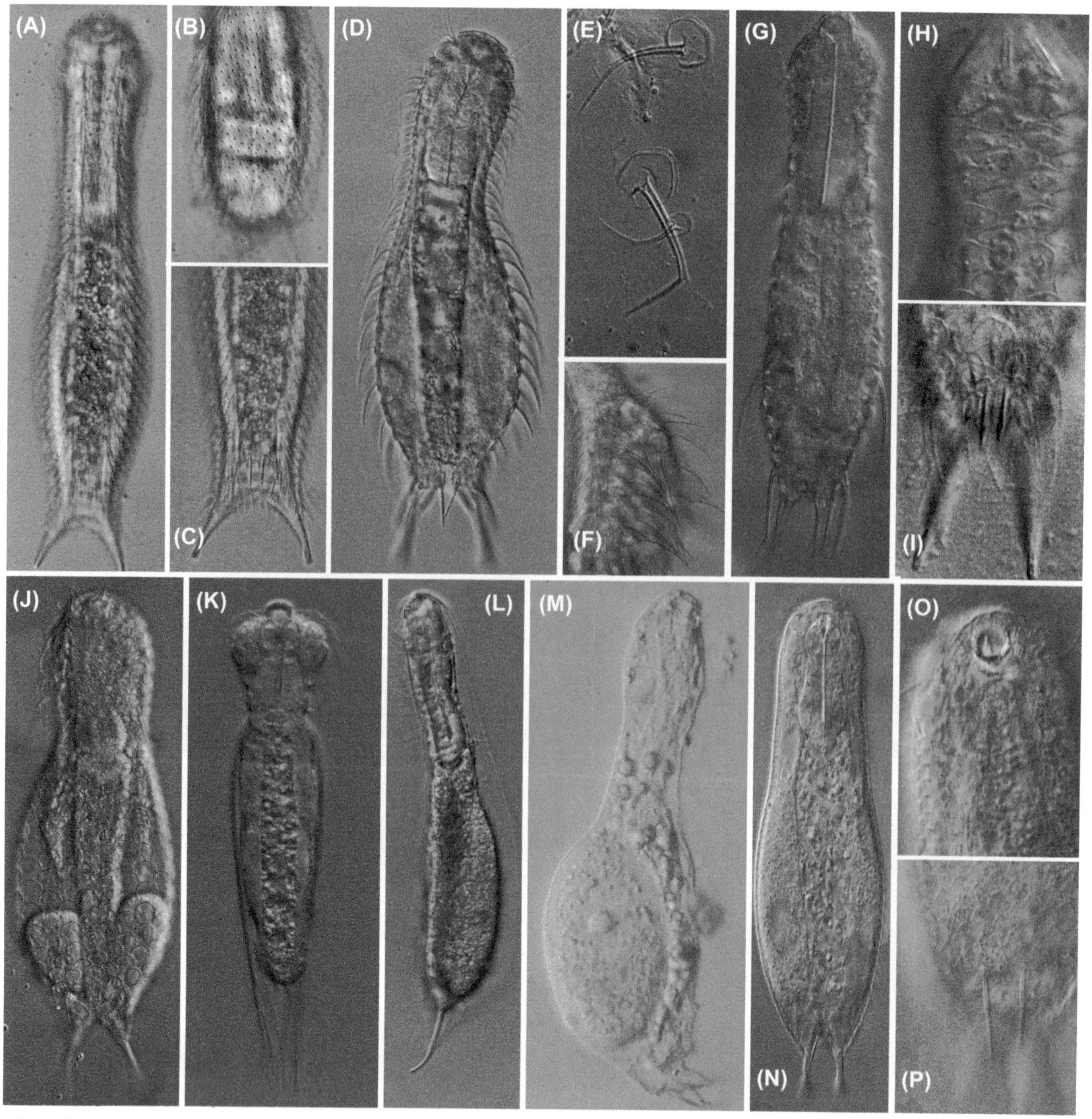

FIGURE 7.8 Microphotographs of live Chaetonotida specimens: (A–C) *Chaetonotus polyspinosus*: (A) ventral habitus, (B) small spined scales on the dorsal body, (C) close-up of the paired keeled scales on the ventral furca base; (D–F) *Chaetonotus* sp.: (D) ventral habitus, (E) close-up of double edged spined scales, (F) close-up of denticulated spines; (G–I) *Chaetonotus acanthocephalus*: (G) ventral habitus, (H) close-up of the dorsal scales and the cephalic spines, (I) close-up of the paired keeled scales on the ventral furca base; (J) *Lepidodermella broa*: ventral habitus; (K) *Stylochaeta* cf. *fusiformis*: ventral habitus; (L) *Arenotus strixinoi*: lateral view; (M) *Ichthydium* sp., lateral view; (N–P) *Aspidiophorus* aff. *tetrachaetus*: (N) ventral habitus, (O) close-up of ventral head (pedunculate scales can be observed on the lateral edge body), and (P) close-up of thick spines in the posterior end of the dorsal side.

2(1) Caudal lobe (furca) reduced or completely absent, always without adhesive tubes; only semiplanktonic or planktonic 3

2' Caudal lobe (furca) usually well-developed, with straight adhesive tubes; mostly epibenthic, periphytic, or interstitial; rarely semiplanktonic (Figs. 7.3–6) .... **Chaetonotidae [p. 135]**

3(2) One pair of clavate cephalic tentacles present; no groups of long spines inserted ventrolaterally (Fig. 7.1 B–E) .... **Neogosseidae [p. 141]**

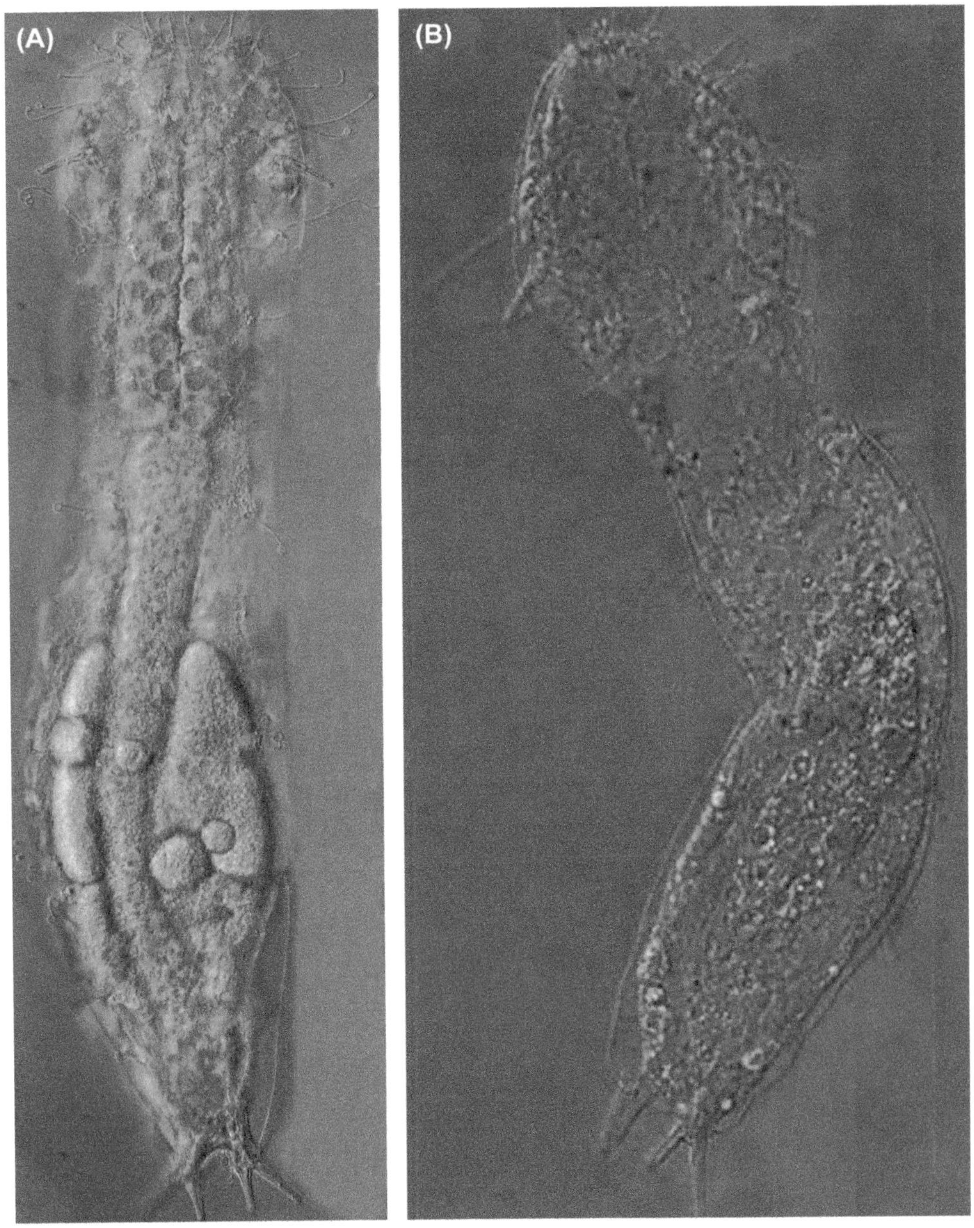

FIGURE 7.9 Microphotographs of live Macrodasyida specimens: (A) ventral habitus of *Redudasys fornerise*; and (B) ventral habitus of *Redudasys brasiliensis*.

3' Clavate cephalic tentacles absent; several lateral groups of long, simple or barbed, motile spines inserted ventrolaterally (Fig. 7.2) .... **Dasydytidae [p. 141]**

Gastrotricha: Chaetonotida: Chaetonotidae: Genera

1 Furca with adhesive tubes, without distal spines; absence of paired undulated transverse band in the cephalic region 2

1' Furca without adhesive tubes, with a pair (occasionally two pairs) of distal spines; presence of paired undulated transverse bands extending from ventral body side dorsolaterally to the cephalic region (Fig. 7.3 L) .... *Undula paraensis* Kisielewski, 1991

[Brazil: Pará]

2(1) Dorsal region naked; scales or spines, if present, only near the base of the furca 3

2' Dorsal region covered by distinct kinds of ornamentation (scales, spines, spined scales, pedunculate scales, or keels) 4

3(2) Body covered with a thick and soft homogeneous cuticular layer, clearly distinct from the epidermis; absence of scales or spines (Figs. 7.3 B, 8 L) .... *Arenotus strixinoi* Kisielewski, 1987

[Brazil: São Paulo]

3' Body covered with a thin cuticular layer, not distinct from the epidermis; cuticle smooth, occasionally with cuticular folds; scales or spines, if present, only near the base of the furca (Figs. 7.3 E–G, 8 M) .... ***Ichthydium* [p. 136]**

4(2) Cephalion, if present, is restricted to head and never fused or juxtaposed to first scale; hypostomium, if present, never covers whole head ventrally; furcal adhesive tubes unsegmented or divided into more than ten parts; if present, one or two paired ventral, small, terminal keeled scales (~ 10 μm length) 5

4' Cephalion extremely wide (Fig. 7.7 B), almost completely fused with anteriormost median scale and juxtaposed to first scale; hypostomium extremely large, covering whole head ventrally; furcal adhesive tubes divided into three parts; two paired, ventral, terminal, large keeled scales (40 and 22 μm in length) (Figs. 7.3 A, 7 A–C) .... *Cephalionotus kisielewskii* Garraffoni, Araújo, Lourenço, Guidi & Balsamo, 2017
[Brazil: Minas Gerais]

5(4) Furca branches not annulate, cephalion compact without any type of expansion 6

5' Furca branches annulate (Fig. 7.7 F), cephalion with posterior and lateral edges free and highly raised (developed) (Figs. 7.4 A–F, 7 D–F) .... ***Polymerurus* [p. 136]**

6(5) Body dorsally covered with nonpedunculate scales 7

6' Body dorsally covered with pedunculate scales (Figs. 7.4 G–N, 8 N–P) .... ***Aspidiophorus* [p. 137]**

7(6) Spines absent in most parts of the body; if present, a few at furca base; keels scales may be present 8

7' Spines and/or spined scales present on whole body, often numerous, with different shapes and sizes (Figs. 7.5, 8 A–I) .... ***Chaetonotus* [p. 137]**

8(7) Posterior end, dorsal and lateral regions without barbed spines 9

8' Posterior end, dorsal and lateral regions with long barbed spines (three to four pairs), increasing in size (posteriormost longer than the adhesive tubes); posterior end dorsal region with few barbed spines (usually five) (Figs. 7.3 C, D, 7 J–L) .... ***Lepidochaetus* [p. 140]**

9(8) Body surface with unkeeled scales; oval, rounded, or slightly squared (Figs. 7.3 H–K, 8 J) .... ***Lepidodermella* [p. 140]**

9' Body surface with keeled scales (with or without short spines); elongate to suboval (Figs. 7.4 N–Q, 7 M–P) ***Heterolepidoderma* [p. 141]**

Gastrotricha: Chaetonotida: Chaetonotidae: *Ichthydium*: Species

1 Spines present ventrolaterally or at posterior end 2

1' Spines absent (Fig. 7.3 E) .... *Ichthydium forficula* Remane, 1927

[Brazil: São Paulo]

2(1) Body posterior end without spines; two ventrolateral columns of spines present; adhesive tubes long and narrow (Fig. 7.3 F) .... *Ichthydium chaetiferum* Kisielewski, 1991

[Brazil: Minas Gerais]

2' Body posterior end with seven spines; ventrolateral columns of spines absence; adhesive tubes short and curved (Fig. 7.3 G) .... *Ichthydium crassum* Daday, 1905

[Paraguay: Central]

Gastrotricha: Chaetonotida: Chaetonotidae: *Polymerurus*: Species

1 Body dorsally covered with nonpedunculate scales; presence of spined scales in the whole body or only near the furcal base 2

1' Body dorsally covered with pedunculated scales; absence of spined scales (Fig. 7.4 A) .... *Polymerurus rhomboides* (Stokes, 1887)

[Brazil: Mato Grosso do Sul; Pará, São Paulo. Argentina: Buenos Aires, Tucumán]

2(1) Body with numerous spined scales 3

2' Body with spined scales only near furcal base (Fig. 7.4 B) .... *Polymerurus elongatus* (Daday, 1905)

[Paraguay: Central]

3(2) Total body length < 500 μm 4

3' Total body length > 700 μm (Fig. 7.4 C) .... *Polymerurus corumbensis* Kisielewski, 1991

[Brazil: Mato Grosso do Sul]

4(3) Furca base without scales 5

4' Furca base with dorsal spineless polygonal scales and lateral external plate .... *Polymerurus squammofurcatus matogrossensis* Kisielewski, 1991
[Brazil: Mato Grosso do Sul]

5(4) Cephalion strongly developed with posterior end roundish; less than 20 longitudinal scales 6

5' Cephalion strongly developed with posterior edge protruding; more than 20 longitudinal scales (Fig. 7.4 D) .... *Polymerurus ringueleti* Grosso, 1975
[Argentina: Buenos Aires, Tucumán]

6(5) Spines near furca much longer than the trunk spines; 13–17 longitudinal rows of scales (Figs. 7.4 E, 7 D–F) .... *Polymerurus nodicaudus* (Voigt, 1901)
[Brazil: São Paulo]

6' Body spines subequal in size; 6–7 columns of scales (Fig. 7.4 F) .... *Polymerurus paraelongatus* Grosso & Drahg, 1986
[Argentina: Tucumán]

Gastrotricha: Chaetonotida: Chaetonotidae: *Aspidiophorus*: Species

1 Furca base dorsally without spines 2

1' Furca base dorsally with spines 6

2(1) Pharyngeal cuticular rods absent 3

2' Pharyngeal cuticular rods present 5

3(2) Pedunculate scales rhomboidal with rounded edges without spiny process 4

3' Pedunculate scales elliptical and keeled with a spiny process (Fig. 7.4 G) .... *Aspidiophorus ophiodermus* Balsamo, 1983
[Brazil: São Paulo]

4(3) Adhesive tubes thin and acute distally; body dorsally covered with pedunculate scales; furca base with nonpedunculate scales (Fig. 7.4 J) .... *Aspidiophorus lilloensis* Grosso & Drahg, 198
[Argentina: Tucumán]

4' Adhesive tubes short, thick, dilated distally; body dorsally with pedunculate scales (Fig. 7.4 H) .... *Aspidiophorus oculifer* Kisielewski, 1981
[Brazil: São Paulo]

5(2) Anterior ciliary area connected with main ventral ciliary bands; body dorsally with only pedunculate scales (Fig. 7.4 I) .... *Aspidiophorus slovinensis* Kisielewski, 1986
[Brazil: São Paulo]

5' Anterior ciliary area separated from main ventral ciliary bands; body dorsally with pedunculate and furca base with nonpedunculate scales (Fig. 7.4 K) .... *Aspidiophorus pleustonicus* Kisielewski, 1991
[Brazil: Mina Gerais, São Paulo]

6(1) Pedunculate scales rhomboidal with rounded edges with the same size on the whole body; two pairs of thick spines (Figs. 7.4 L, 8 N–P) .... *Aspidiophorus aff. tetrachaetus* Kisielewski, 1986
[Brazil: Minas Gerais, São Paulo]

6' Posterior end, dorsally with one transversal row of pedunculate scales much larger than others; posterior end dorsally with three pairs of spines, the medial pair much longer than lateral pairs (Fig. 7.4 M) .... *Aspidiophorus brahmsi* Grosso, 1973
[Argentina: Buenos Aires, Tucumán]

PHYLUM GASTROTRICHA

Gastrotricha: Chaetonotida: Chaetonotidae: *Chaetonotus*: Species

Chaetonotus subgenera are not monophyletic (Kånneby et al., 2012; Kolicka et al., 2016). Distinct forms have been described for some species. Kisielewski (1991) identified four distinct forms of *C. polyspinosus* Kisielewski, 1991 and four distinct forms of *C. similis* Zelinka, 1889; however, we treat them as two species.

1 Middle or posterior trunk bearing a dorsolateral transverse band of long spines of equal size and thickness, simple or bifurcated 2

1' Middle or posterior trunk without dorsolateral transverse band of long spines 5

2(1) Middle trunk bearing a dorsolateral transverse band of long spines; head scales with clear spines 3

2' Posterior trunk bearing a dorsolateral transverse band of long spines; head scales with very short spines (difficult to observe) .... *Chaetonous heterochaetus* Daday, 1905

[Paraguay: Estia Postillon]

3(2) Two pairs of lateral parafurcal spines; interciliary area without transverse scale plates in pharynx region (Fig. 7.5 A) 4

3' No lateral parafurcal spines; interciliary area with transverse scale plates in pharynx region (Fig. 7.5 C) .... *Chaetonotus succinctus* Voigt, 1904

[Brazil: Minas Gerais, Pará]

4(3) Scales posterior to the transverse band of long spines without spines or keel; two anterior most lateral paired spines longer than other cephalic and neck scales (Fig. 7.5 A) .... *Chaetonotus acanthodes* Stokes, 1887

[Brazil: São Paulo]

4' Scales posterior to the transverse band of long spines with spines (one pair of scales with very long spines, subequal to spines in transverse band); all cephalic and neck regions with same size spines (Fig. 7.5 B) .... *Chaetonotus bisacer* Greuter, 1917

[Brazil: Pará, São Paulo. Argentina: Buenos Aires]

5(1) Body without scales; spines arising directly from cuticle 6

5' Body with scales; spines or keels arising from scales 7

6(5) Furca base indistinct from body; adhesive tubes straight (Fig. 7.5 D) .... *Chaetonotus corderoi* Schwank, 1990

[Uruguay: Montevideo]

6' Narrow furca base forming a distinct body segment; adhesive tubes externally bent (Fig. 7.5 E) .. *Chaetonotus silvaticus* (Varga, 1963)

[Brazil: São Paulo]

7(5) Paired hypo- and epipleurae compact without any type of expansion 8

7' Paired hypo- and epipleurae with edges free and highly raised developed .... *Chaetonotus majestuosus* Grosso & Drahg, 1984

[Argentina: Tucumán]

8(7) Dorsal and ventral body sides with keeled or spined scales; interciliary area without transverse plates 9

8' Dorsal scales without spines; ventrolateral scales with simple spines; interciliary area with more than 20 transverse plates in the intestinal region (Fig. 7.5 F) .... *Chaetonotus annectens* Grosso & Drahg, 1991

[Argentina: Tucumán]

9(8) Furcal branch bases without a pair of double-keeled scales surrounded by three elongated trilobate scales 10

9' Furcal branch bases with a pair of round (or squared) double-keeled scales surrounded by three elongated trilobate scales .... *Chaetonotus hystrix* Metschnikoff, 1865

[Brazil: Minas Gerais, São Paulo]

10(9) Body covered with keeled scales 11

10' Body covered with spined scales of different sizes and shapes 13

11(10) Ventral cephalic region with two paired small ciliary areas just below the hypostomion; one pair of pharyngeal rods 12

11' Ventral cephalic region without ciliary areas just below the hypostomion; no pharyngeal rods (Fig. 7.5 G) .... *Chaetonotus intermedius* Kisielewski, 1991

[Brazil: São Paulo]

12(11) Hypostomion with a paired anterior protuberance; furca base without spines (Fig. 7.5 H) .... *Chaetonotus breviacanthus* Kisielewski, 1991

[Brazil: São Paulo]

12' Hypostomion without protuberance; furca base with long spines (Fig. 7.5 I) .... *Chaetonotus pseudopolyspinosus* Kisielewski, 1991

[Brazil: São Paulo]

13(10) Head with cephalion plate 14

13' Head with five cephalic spines (Figs. 7.5 J, 8 G–I) .... *Chaetonotus acanthocephalus* Valkanov, 1937

[Brazil: Mato Gosso do Sul, Minas Gerais, São Paulo]

14(13) Pharyngeal tooth, if present, unpaired; lateral and ventrolateral spines longer than dorsal 15

14' Pharyngeal tooth, if present, paired; ventrolateral spines subequal in size as dorsal ones 16

15(14) Pharyngeal tooth unpaired; 20 columns of spined scales; simple spines (Fig. 7.5 K) .... *Chaetonotus sanctipauli* Kisielewski, 1991

[Brazil: Minas Gerais, São Paulo]

15' Without pharyngeal tooth; nine columns of spined scales; barbed spines (Fig. 7.5 L) .. *Chaetonotus paucisquamatus* Kisielewski, 1991

[Brazil: Mina Gerais, São Paulo]

16(14) Spines increasing gradually in size along half the body 17

16' Spines with similar size on whole body (except for six or nine long spines in the middle dorsum) 28

17(16) One or two lateral caudal spines reaching or projecting beyond the adhesive tubes 18

17' No caudal spines projecting beyond the adhesive tubes 23

18(17) One pair of lateral caudal spines projecting beyond the adhesive tubes 19

18' Two pairs of lateral caudal spines projecting beyond the adhesive tubes 22

19(18) Two pairs of cephalic sensory tufts 20

19' One pair of cephalic sensory tufts 21

20(19) Dorsal scales with only one type: trilobate spined scales .... *Chaetonotus dadayi* Schwank, 1990

[Guyana: Cayenne. Paraguay: Estia Postillon]

20' Dorsal scales of three types: rhomboid without spines, subrectangular with or without spines, semicircular with spines (Fig. 7.5 M) .... *Chaetonotus pilaga* Grosso, 1982

[Argentina: Formosa]

21(19) Circular scales (Fig. 7.5 N) .... *Chaetonotus erinaceus* Daday, 1905

[Guyana: Cayenne. Paraguay: Estia Postillon]

21' Trilobate scales (Fig. 7.5 O) .... *Chaetonotus paraguayensis* Schwank, 1990

[Paraguay: Estia Postillon]

22(18) Adhesives tubes posteriorly rounded; one pair of parafurcal spines (Fig. 7.6 A) .... *Chaetonotus pusillus* Daday, 1905

[Brazil: Mato Grosso do Sul. Paraguay: Estia Postillon]

22' Adhesives tubes posteriorly acute; two pairs of parafurcal spines (Fig. 7.6 B) .... *Chaetonotus montevideensis* Cordero, 1918

[Uruguay: Montevideo]

23(17) Body scales oval with barbed spines 24

23' Body scales trilobate or cordate (pentagonal), with or without barbed spines 25

24(23) Anterior interciliary area without transverse ventral scale bars; mouth to the anterior end of the pharynx without ciliary bands (Fig. 7.6 C) .... *Chaetonotus tenuisquamatus* Grosso, 1982

[Argentina: Formosa]

24' Anterior interciliary area with transverse ventral scale bars; mouth to the anterior end of the pharynx with two pairs of ciliary bands (Fig. 7.6 P) .... *Chaetonotus soberanus* Grosso & Drahg, 1983

[Argentina: Tucumán]

25(23) Body scales trilobate with simple or barbed spines 26

25' Body scales cordate (pentagonal) with denticulate spines (Fig. 7.6 D) .... *Chaetonotus heideri* Brehm, 1917

[Brazil: Minas Gerais, Pará, São Paulo]

26(25) Dorsal scales with denticulate spines; posterior trunk region with spineless trilobate scales; pseudocelli absent 27

26' Dorsal scales with simple spines; whole trunk with spines; pseudocelli (Fig. 7.6 E) .... *Chaetonotus oculifer* Kisielewski, 1981

[Brazil: São Paulo]

27(26) Ventrolateral spines without lamellae .... *Chaetonotus macrochaetus* Zelinka, 1889

[Brazil: São Paulo]

27' Ventrolateral spines with lamellae; ventrolateral spines simple (Fig. 7.6 F) .... *Chaetonotus persetosus* Zelinka, 1889

[Brazil: São Paulo]

28(17) Furcal appendages proximodorsally with spines (parafurcal spines) 29

28' Furcal appendages proximodorsally without spines 32

29(28) Dorsal scales unilobate or trilobate; spines on the whole trunk 30

29' Dorsal scales large, rounded to slightly cordate; trunk pines abruptly end in posteriorly (Fig. 7.6 G) .... *Chaetonotus heteracanthus* Remane, 1927

[Brazil: São Paulo]

30(29) Body with spines sharpening along their length with distal part simple; furca base spines with simple distal part 31

PHYLUM GASTROTRICHA

30' Body with spines equally thick along their length with distal part furcated (not barbed); three dorso medial furcated spines at the furca base (Fig. 7.6 H) .... *Chaetonotus furcatus* Kisielewski, 1991

[Brazil: São Paulo]

31(30) Small trilobate scales with short, straight, simple spines (Fig. 7.8 B); furca base with five pairs of spines (at least two pairs thicker and longer than the others) (Figs. 7.6 I, 8 A–C) .... *Chaetonotus polyspinosus* Greuter, 1917

[Brazil: Pará, São Paulo]

31' Unilobate scales with posterior end concave, and spines strongly bent; furca base with three pairs of spines (Fig. 7.6 J) .... *Chaetonotus ventrochaetus* Kisielewski, 1991

[Brazil: São Paulo]

32(28) Trunk region with some dorsal spines thicker and longer than others 33

32' Trunk region dorsal spines subequal 34

33(32) Posterior end, laterally with six simple long spines (three pairs) (Fig. 7.6 K) .... *Chaetonotus dubius* Daday, 1905

[Brazil: Mato Grosso do Sul]

33' Nine long dorsal double-barbed spines in the middle trunk (Fig. 7.6 O) .... *Chaetonotus balsamoae* Kisielewski, 1997

[Brazil: São Paulo]

34(32) Pharnyx without; teeth up to 18 scales per longitudinal row 35

34' Pharynx anteriorly with teeth; 20–22 scales per longitudinal row (Fig. 7.6 L) .... *Chaetonotus odontopharynx* Grosso & Drahg, 1986

[Argentina: Tucumán]

35(34) Trilobate scales 36

35' Unilobate scales with slightly concave posterior edge; a pair of dorsal, large, keeled, terminal scales (Fig. 7.6 M) .... *Chaetonotus lobo* Kisielewski, 1991

[Brazil: São Paulo]

36(35) Scales with slightly curved, barbed spines; furca base dorsally with a pair of round to squared double-keeled scales; ventral interciliary area with oval keeled scales (Fig. 7.6 N) .... *Chaetonous similis* Zelinka, 1889

[Brazil: São Paulo, Pará]

36' Scales with simple spines; cilia bands united anteriorly; ventral interciliary area smooth in the pharyngeal region .... *Chaetonotus larus* (Müller, 1773)

[Uruguay: Montevideo]

Gastrotricha: Chaetonotida: Chaetonotidae: *Lepidochaetus*: Species

1 One pair of cephalic tufts; anterior and middle body with spineless scales (Fig. 7.3 C) .... *Lepidochaetus brasilense* Kisielewski, 1991

[Brazil: Minas Gerais, Pará, São Paulo]

1' Two pairs of cephalic tufts; anterior and middle body scales with barbed spines (Figs. 7.3 D, 7 J–L) .... *Lepidochaetus zelinkai* (Grünspan, 1908)

[Brazil: Mato Grosso do Sul, Pará, São Paulo, Minas Gerais. Argentina: Buenos Aires]

Gastrotricha: Chaetonotida: Chaetonotidae: *Lepidodermella*: Species

1 Furcal base without lateral spines; no ventrolateral spines 2

1' Furcal base with one pair of lateral spines; trunk with thin ventrolateral spines (Fig. 7.3 H) .... *Lepidodermella minus* chaetifer Kisielewski, 1991

[Brazil: São Paulo]

2(1) Scales with rounded or squared anterior edge 3

2' Scales with convex anterior edge and lateral scales with a pair of corniform protuberances (Fig. 7.3 I) .... *Lepidodermella amazonica* Kisielewski, 1991

[Brazil: Pará]

3(2) Pharyngeal to anterior intestinal region ventrally with longitudinal rows of cuticular plates; anterior end ventrally without light refracting bodies (Fig. 7.3 J) .... *Lepidodermella squamata* (Dujardin, 1841)

[Brazil: Pará, São Paulo. Argentina: Tucumán. Uruguay: Montevideo]

3' Pharyngeal to anterior intestinal region ventrally without longitudinal rows of cuticular plates; anterior end ventrally with a pair of round light-refracting bodies (Figs. 7.3 J, 8 J) .... *Lepidodermella broa* Kisielewski, 1991

[Brazil: São Paulo]

Gastrotricha: Chaetonotida: Chaetonotidae: *Heterolepidoderma*: Species

1 Head trilobate 2

1' Head pentalobate 3

2(1) The two ventral ciliary bands connected with a transverse cephalic cilia band; neck and the rearmost trunk with dorsal scales diagonally arranged (Figs. 7.4 Q, 7 P) .... *Heterolepidoderma mariae* Garraffoni & Melchior, 2015

[Brazil: São Paulo]

2' The two ventral ciliary bands not connected; neck and the rearmost trunk with dorsal scale columns parallel to body axis (Figs. 7.4 N, 7 M–O) .... *Heterolepidoderma majus* Remane, 1927

[Brazil: Mato Grosso do Sul, Pará, São Paulo. Argentina: Tucuman]

3(1) Scales and keels obvious, keels present over entire body; body margin not sinuate 4

3' Scales and keels obscure, keels absent or only in mid-trunk; body margin regularly sinuate .... *Heterolepidoderma gracile* Remane, 1927

[Brazil: Pará. Argentina: Buenos Aires, Tucumán]

4(3) Body without paired ventrolateral column of lamellae 5

4' Body with paired ventrolateral column of lamellae (Fig. 7.4 O) .... *Heterolepidoderma jureiense* Kisielewski, 1991

[Brazil: São Paulo]

5(4) Elliptical keeled scales in 30–32 columns; ventral ciliation as a continuous field of transverse rows covering pharyngeal and anterior intestinal regions ventral surface, then splitting into two separate longitudinal bands .... *Heterolepidoderma famaillensis* Grosso & Drahg, 1991

[Brazil: Minas Gerais. Argentina: Tucumán]

5' Oblong-elliptical keeled scales in 15–20 columns, which taper distally to short simple spine (8–9 μm); ocellar granules present (Fig. 7.4 P) .... *Heterolepidoderma ocellatum* (Metschnikoff, 1865)

[Brazil: São Paulo]

Gastrotricha: Chaetonotida: Neogosseidae: Genera

1 Body end rounded without projections and with an unpaired median group of simple short spines (up to 9 μm) and one paired simple spine (15–20 μm) (Fig. 7.1 B) .... *Kijanebalola canina* Kisielewski, 1991

[Brazil: Pará]

1' Body end truncated with a paired projection(s) bearing several long spines (25–100 μm) (Fig. 7.1 C, D) .... ***Neogossea* [p. 141]**

Gastrotricha: Chaetonotida: Neogosseidae: *Neogossea*: Species

1 Neck without a group of thick barbed spines; dorsal scales nonpedunculated 2

1' Neck with a dorsal group of thick barbed spines (about 15 um); dorsal scales pedunculated (Fig. 7.1 C) .... *Neogossea acanthocolla* Kisielewski, 1991

[Brazil: Mato Grosso do Sul, Pará, Piauí]

2(1) Body bearing two pairs of caudal lobes bearing four groups of caudal spines (Fig. 7.1 D) .... *Neogossea fasciculata* (Daday, 1905)

[Guyana: Cayenne. Paraguay: Estia Postillon]

2' Body bearing one pair of caudal lobes bearing one group of caudal spines (Fig. 7.1 E) .... *Neogossea pauciseta* (Daday, 1905)

[Guyana: Cayenne. Paraguay: Estia Postillon]

Gastrotricha: Chaetonotida: Dasydytidae: Genera

1 No posterior saltatorial spines 2

1' One pair of saltatorial spines (Fig. 7.7 H), saltatorial spines cross in the rest position (Figs. 7.2 C, D, K, 7 G–I) .... ***Haltidytes* [p. 142]**

2(1) Trunk region with dorsal and ventral scales bearing spines or keels; posterior body end without pedunculate lobes 3

2' Trunk region naked; posterior body end with one pair of pedunculate lobes bearing two to three short bristles (styli) (Fig. 7.2 K) .... *Stylochaeta fusiformis* (Spencer, 1890)

[Brazil: Mato Grosso do Sul, Pará]

3(2) Dorsolateral lorica absent; dorsal trunk cuticle with barbed spines 4

3' Dorsolateral region covered with large, rounded ornamented scales forming a rigid lorica; dorsal trunk barbed spines inserted on oval scales (Fig. 7.2 B) .... *Ornamentula paraensis* Kisielewski, 1991

[Brazil: Pará]

4(3) Fusiform body; trilobate head; just posterior to cephalion with dorsal scales bearing three keels; one pair of unequal rear spines (Fig. 7.2 A) .... *Setopus aequatorialis* Kisielewski, 1991

[Brazil: Pará, Belém]

4' Slender body; one, three, or pentalobate head; no keeled dorsal scales on the head just posterior to the cephalion; one or two pair(s) of rear subequal spines (Fig. 7.2 E-I) .... ***Dasydytes*** **[p. 142]**

Gastrotricha: Chaetonotida: Dasydytidae: *Haltidytes*: Species

1 Trunk with well-developed rhombic scales 2

1' Trunk without scales (Fig. 7.2 D) .... *Haltidytes festinans* (Voigt, 1909)

[Brazil: Pará, São Paulo. Argentina: Buenos Aires, Tucumán]

2(1) Dorsal trunk with simple rhombic scales without spines or keels, but ventral and lateral trunk without scales; dorsum with long lateral spines not overlapping (Figs. 7.2 J, 7 G-I) .... *Haltidytes pseudosquamosus* Minowa & Garraffoni, 2017

[Brazil: São Paulo]

2' Dorsal and ventrolateral trunk with keeled rhombic scales; dorsum with long lateral spines overlapping (Fig. 7.2 C) .... *Haltidytes squamosus* Kisielewski, 1991

[Brazil: Pará]

Gastrotricha: Chaetonotida: Dasydytidae: *Dasydytes*: Species

1 Trunk, posteriodorsally, without clavate unpaired structure 2

1' Trunk, posteriodorsally, with clavate, unpaired structure (Fig. 7.2 E) .... *Dasydytes nhumirimensis* Kisielewski, 1991

[Brazil: Mato Grosso do Sul]

2(1) Dorsal body cuticle completely naked; lateral spines double- or triple-barbed 3

2' Dorsal body cuticle covered with scales; lateral spines single-barbed 4

3(2) Posterior trunk with distinct rounded caudal lobe; one pair of caudal spines (Fig. 7.2 F) .... *Dasydytes lamellatus* Kisielewski, 1991

[Brazil: Pará, Piaui]

3' Posterior trunk rounded, without caudal lobe; two pairs of caudal spines with different length (Fig. 7.2 G) .... *Dasydytes carvalhoae* Kisielewski, 1991

[Brazil: Pará]

4(2) Seven ventrolateral paired spines groups up to 45 μm long; paired caudal spines of unequal length (Fig. 7.2 H) .... *Dasydytes papaveroi* Kisielewski, 1991

[Brazil: Pará]

4' Four ventrolateral paired spines groups up to 160 μm long; paired caudal spines of subequal length (Fig. 7.2 I) .... *Dasydytes elongatus* Kisielewski, 1991

[Brazil: Pará]

ACKNOWLEDGMENTS

We express our gratitude to FAPEMIG (ETC 00017-13), Rede ComCerrado (CNPq – 50.6121/2008-09), CNPQ (478825/2013-7), FAPESP (2014/23856-0), and FAEPEX-UNICAMP (2031/15) for financial support. We also thank Dr. Maria Balsamo for meticulous review of the first draft of this text, and Kayla Ann Margaret Wirthwein for editing the English text.

PHYLUM GASTROTRICHA

REFERENCES

Araújo T.Q., F.C. Alcântara & S.A.R. Garraffoni. 2013. New records of Gastrotricha from Minas Gerais, Brazil. Studies on Neotropical Fauna and Environment 48: 68–75.

Balsamo M., P. Grilli, L. Guidi & J. d'Hondt. 2014. Gastrotricha: Biology, ecology and systematics. Families Dasydytidae, Dichaeturidae, Neogosseidae, Proichthydiidae. *in:* H. Dumont (ed.) 24. Backuys and Margraf Publishers GmbH, Weikersheim. 188pp.

Balsamo, M., J-L. d'Hondt, J. Kisielewski & L. Pierboni. 2008. Global diversity of gastrotrichs (Gastrotricha) in fresh waters. Hydrobiologia 595: 85–91.

Balsamo, M., J-L. d'Hondt, J. Kisielewski, M.A. Todaro, P. Tongiorgi, L. Guidi, P. Gilli & Y. de Jong. 2015. Fauna Europaea: Gastrotricha. Biodiversity Data Journal 3: e5800.

Balsamo, M., J-L. d'Hondt, L. Pierboni & P. Grilli. 2009. Taxonomic and nomenclatural notes on freshwater Chaetonotida (Gastrotricha). Zootaxa 2158: 1–19.

Cordero, E.H. 1918. Notes sur les Gastrotriches. Physis 4: 241–244.

Daday von, E. 1905. Untersuchungen über die Süsswasser Mikrofauna Paraguays. Zoologica 44: 1–374.

Garraffoni, A.R., F.C. Alcântara & H.H. Checon. 2017a. Evaluating the anesthetization and fixation efficacy of "soft" and "hard" freshwater benthic meiofauna: what is the best method for specimen preservation? Limnology 18: 209–218.

Garraffoni, A.R.S & T.Q. Araújo. 2010. Chave de identificação de Gastrotricha de águas continentais e marinhas do Brasil. Papéis Avulsos de Zoologia 50: 535–552.

Garraffoni, A.R.S., T.Q. Araújo, A.P. Lourenço & M. Balsamo. 2010. New data on freshwater psammic Gastrotricha from Brazil. Zookeys 60: 1–12.

Garraffoni, A.R., T.Q. Araújo, A.P. Lourenço, L. Guidi & M. Balsamo. 2017b. A new genus and new species of freshwater Chaetonotidae (Gastrotricha: Chaetonotida) from Brazil with phylogenetic position inferred from nuclear and mitochondrial DNA sequences. Systematics and Biodiversity 15: 49–62.

Garraffoni, A.R.S, T.Q. Araújo, A.P. Lourenço, L. Guidi & M. Balsamo. 2019. Integrative taxonomy of a new *Redudasys* species (Gastrotricha: Macrodasyida) sheds light on the invasion of fresh water habitats by macrodasyids. Scientific Reports 9: 2067.

Garraffoni, A.R.S. & Freitas, A.C.L. 2017. Photos belong in the taxonomic Code. Science 355: 805.

Garraffoni, A.R.S. & M.P. Melchior. 2015. New species and new records of freshwater *Heterolepidoderma* (Gastrotricha: Chaetonotidae) from Brazil with an identification key to the genus. Zootaxa 4057: 551–568.

Grosso, L.E. 1973a. Notas sobre gastrotricos Argentinos I. *Chaetonotus (Zonochaeta) gurugutoi* sp. nov. y *Aspidiophrus brahmsi* sp. nov. Physis 32: 133–137.

Grosso, L.E. 1973b. Notas sobre gastrotricos Argentinos II. Neotropica 19: 87–89.

Grosso, L.E. 1975. Especies Argentinas del género *Polymerurus* Remane 1927. Neotropica 21: 46–50.

Grosso, L.E. 1976. *Chaetonotus berissensis* n. sp. (Gastrotricha) relacionada a los órganos sumergidos de plantas flotantes. Neotropica 22: 121–124.

Grosso, L.E. 1982. Nuevas espécies de gastrotricos relacionadas a la vegetación flotante. Neotropica 28: 189-192.

Grosso, L.E. & F. Drahg. 1983. Gastrotricos dulceacuicolas de la provincia de Tucuman I. *Chaetonotus soberanus* sp. nov. y *Aspidiophorus lilloensis* sp. nov. Neotropica 29: 189–193.

Grosso, L.E. & F. Drahg. 1984 Gastrotricos dulceacuicolas de la provincia de Tucuman II. *Chaetonotus majestuosus* n. sp. Acta Zoologica Lilloana 37: 227–230.

Grosso, L.E. & F. Drahg. 1986. Gastrotricos dulceacuicolas de la província de Tucuman III. Acta Zoologica Lilloana 38: 139–149.

Grosso, L.E. & F. Drahg. 1991. Gastrotricos dulceacuicolas de la provincia de Tucuman IV. Acta Zoologica Lilloana 40: 47–51.

Hochberg, R. & M.K. Litvaitis. 2000. Hexamethyldisilazane for Scanning Electron Microscopy of Gastrotricha. Biotechnic and Histochemistry 75: 41–44.

d'Hondt, J. L., R. Pourriot & C. Rougier. 2006. Nouvelles observations sur les Gastrotriches d'eau douce de Guyane francaise. Bulletin Mensuel de la Société Linnéenne de Lyon 75: 239–245.

d'Hondt J.L., R. Pourriot, C. Rougier & D. Guiral. 2005. A propôs de la découverte du gastrotriche d'eau douce *Neogossea fasciculata* (Daday, 1905) en Guyane française. Bulletin mensuel de la Société linnéenne de Lyon 74: 209–216.

Hummon, W. D., M. A. Todaro & W. A. Evans. 2005. Video database for described species of marine Gastrotricha. Meiofauna Marina 14: 23–26.

Kånneby, T. 2013. New species and records of freshwater *Chaetonotus* (Gastrotricha: Chaetonotidae) from Sweden. Zootaxa 3701: 551–588.

Kånneby, T. 2016. Phylum Gastrotricha. Pages: 115–130 *in*: J. Thorp & D.C. Rogers (eds.), Keys to Neartic Fauna, Thorp and Covich's Freshwater Invertebrates (4th edition), Academic Press, Boston, MA.

Kånneby, T. & R. Hochberg. 2015. Phylum Gastrotricha. Pages 211–223 *in*: J.H. Thorp and D.C. Rogers (eds.), Volume I: Ecology and General Biology. Thorp and Covich's Freshwater Invertebrates, Academic Press, Elsevier, Boston.

Kånneby, T. & J.J. Kirk. 2017. A new species of *Redudasys* (Gastrotricha: Macrodasyida: Redudasyidae) from the United States. Proceedings of the Biological society of Washington, 130: 128–139.

Kånneby, T., M.A. Todaro & U. Jondelius. 2009. One new species and records of *Ichthydium* Ehrenberg, 1830 (Gastrotricha: Chaetonotida) from Sweden with a key to the genus. Zootaxa 2278: 26–46.

Kånneby, T., M.A. Todaro & U. Jondelius, 2012. A phylogenetic approach to species delimitation in freshwater Gastrotricha from Sweden. Hydrobiologia 683: 185–202.

Kånneby, T., M.A. Todaro & U. Jondelius. 2013. Phylogeny of Chaetonotidae and other Paucitubulatina (Gastrotricha: Chaetonotida) and the colonization of aquatic ecosystems. Zoologica Scripta 42: 88–105.

Kånneby, T. & M.K Wicksten. 2014. First record of the enigmatic genus *Redudasys* Kisielewski, 1987 (Gastrotricha: Macrodasyida) from the Northern hemisphere. Zoosystema 36: 723–733.

Kisielewski J. 1987. Two new interesting genera of Gastrotricha (Macrodasyida and Chaetonotida) from the Brazilian freshwater psammon. Hydrobiologia 153: 23–30.

Kisielewski, J. 1991. Inland-water Gastrotricha from Brazil. Annales Zoologici 43: 1–168.

Kolicka, M., M. Dabert, J. Dabert, T. Kånneby & J. Kisielewski. 2016. *Bifidochaetus*, a new Arctic genus of freshwater Chaetonotida (Gastrotricha) from Spitsbergen revealed by an integrative taxonomic approach. Invertebrate Systematics 30: 398–419.

Minowa, A.K. & A.R.S. Garraffoni. 2017. A new species of *Haltidytes* Remane, 1936 (Gastrotricha: Chaetonotida: Dasydytidae) from an urban lagoon in Brazil with a phylogenetic reconstruction of the genus based on morphological data. Zoologischer Anzeiger 269:100–109.

Murray, J. 1913. Gastrotricha. Journal of the Quekett Microscopical Club 12: 211–238.

Schwank, P. 1990. Gastrotricha. Pages 1–252 *in*: J. Schwoerbel & P. Zwick (eds.), Süsswasserfauna von Mitteleuropa, Band 3. Gastrotricha und Nemertini. Gustav Fischer Verlag, Stuttgart, Jena, New York.

Todaro, M.A. 2018a. Marine *in*: Gastrotricha World Portal. M.A. Todaro (ed.) [available online at http://www.gastrotricha.unimore.it/marine.htm]. Accessed 08 February 2018.

Todaro, M.A. 2018b. Freshwater *in*: Gastrotricha World Portal. M.A. Todaro (ed.) [available online at http://www.gastrotricha.unimore.it/freshwater.htm]. Accessed 08 February 2018.

Todaro, M.A., M. Dal Zotto, U. Jondelius, R. Hochberg, W.D. Hummon, T. Kånneby & C.E.F. Rocha. 2012. Gastrotricha: A Marine Sister for a Freshwater Puzzle. Public Library of Science One 7: e31740.

Todaro, M.A. & W.D. Hummon. 2008. An overview and a dichotomous key to genera of the phylum Gastrotricha. Meiofauna Marina 16: 3–20.

Weiss, M. 2001. Widespread hermaphroditism in freshwater gastrotrichs. Invertebrate Biology 120: 308–431.

Chapter 8

Phylum Rotifera

Susana B. José de Paggi
Institute of Limnology, Santa Fe, Argentina

Robert Wallace
Department of Biology, Ripon College, Ripon, WI, United States

Diego Fontaneto
Water Research Institute (IRSA), National Research Council (CNR), Verbania, Italy

María Cristina Marinone
University of Buenos Aires, Buenos Aires, Argentina

Chapter Outline

INTRODUCTION

Rotifers, or wheel animals, are a group of tiny (< 2,000 μm) invertebrates that have fascinated people ever since the late 17th century (Wallace, 2002). The monophyletic origin of the three main groups of the phylum (Bdelloidea, Monogononta, and Seisonacea) is widely accepted, as well as their close relationship with phylum Acanthocephala (Sielaff et al., 2016). The latter is a group of endoparasitic worms traditionally considered to be a separate phylum (Wallace et al., 2015). However, presence of the unique intracytoplasmic lamina (Clément, 1993; Wallace et al., 2006), as well as genetic similarities, offers strong support for

Thorp and Covich's Freshwater Invertebrates. https://doi.org/10.1016/B978-0-12-804225-0.00008-3

joining acanthocephalans with rotifers (García-Varela & Nadler, 2006). The current view argues for the following classification. Acanthocephalans and seisonids are sister groups within Pararotatoria (Min & Park, 2009; Fontaneto & Jondelius, 2011; Wey-Fabrizius et al., 2014). Pararotatoria + bdelloids then comprise Hemirotifera (Sørensen & Kristensen, 2015), with Hemirotifera and monogononts comprising the Syndermata (Ahlrichs, 1995, 1997) or Rotifera *sensu lato*. This taxon is then a sister taxon to the Gnathostomulida, creating a group of small, jawed animals, the Gnathifera (Giribet et al., 2000, 2004; Sørensen, 2003). Fontaneto & De Smet (2015) provided a synthesis of previous hypotheses on the phylogenetic relationship within Rotifera. Based on the order of mitochondrial genes, Sielaff et al. (2016) argued for a sister-group relationship of seisonids and acanthocephalans (Pararotatoria hypothesis), together with their inclusion in Hemirotifera with the bdelloids. However, all published sequence-based tree topologies have been rather inconsistent, and the Hemirotifera (= Bdelloidea + Pararotatoria) did not appear in any of the trees. While these studies provide important contributions to better understand the phylogeny of the group, there is a need to build a better classification system.

Given uncertainties in the phylogenetic relationships between the four major groups of rotifers (Acanthocephala, Bdelloidea, Monogononta, and Seisonacea), here we consider them as classes within the phylum, regardless of their relative phylogenetic positions. However, because of the scope of this chapter, the obligatory marine seisonids and the parasitic acanthocephalans are not included.

Rotifers of the Neotropics and Antarctica

The two main groups of rotifers, Bdelloidea and Monogononta, comprise 34 families and 132 genera worldwide (checklists by Segers, 2007, 2013). Of these, 30 families and 94 genera are present in the Neotropics and in Antarctica. The inland waters of the Neotropics host approximately 744 species (623 Monogononta and 121 Bdelloidea); the Antarctica and surrounding islands host 165 species (120 Monogononta and 45 Bdelloidea).

Systematics

Class Bdelloidea

Bdelloids comprise three orders, Adinetida, Philodinida, and Philodinavida (Melone et al., 1998a), with four families, 19 genera, and about 460 species and subspecies (Segers, 2007; Fontaneto & De Smet, 2015). However, due to the fact that cryptic species (sibling species) appear to be very frequent in the group, it is doubtful that this species total reflects actual number of bdelloids species (Fontaneto et al., 2007, 2009, 2011). Bdelloid rotifers can be found in permanent water bodies such as lakes and streams, but several species live in the water film covering terrestrial mosses and soil; tree holes; in traps of the carnivorous plants; and a few are epizoic on insect larvae, crustaceans, molluscs, etc. (e.g., Bateman, 1987; Ricci, 1987; Devetter, 2010). In freshwaters bdelloids, benthic and periphytic are frequent habitats; in addition, they can be found in the hygropsammon and hydropsammon. While occasionally caught in net hauls, bdelloids are rarely planktonic. A distinctive feature of this group is the ability to withstand dry periods through anhydrobiosis, recovering from the dormant stage when water becomes available (Ricci, 1998; Ricci & Fontaneto, 2009; Wilson, 2011).

Class Monogononta

Recent work recorded about 1,600 species, in 3 orders: Ploima, Flosculariaceae, and Collothecacea, comprising 30 families and more than 100 genera (Segers, 2007, 2013). Monogononts are one of the richest groups of freshwater plankton, but they are also found in benthos and littoral communities. They may be solitary or colonial (Wallace, 1987). Most monogonont rotifers are free moving, but some species are sessile (Wallace, 1980); a few species are epizoic and parasitic inhabiting algae, shells of testate amoebae, annelids, and vertebrates like fishes (carp) or Amazonian crocodiles (May, 1989; De Smet, 2006).

LIMITATIONS

Our knowledge of the rotifer fauna of the Neotropical region is still incomplete, and species richness is probably underestimated because of the large bioregion and a dearth of taxonomists working with this group.

There are many unexplored or poorly explored zones in the Neotropics. These include: the Pantanal, part of the floodplains of the three main hydrographic basins, the Magdalena River, and the high-mountain lakes and ponds of the Andean precordillera that stretches from Tierra del Fuego to Ecuador.

Another reason accounting for limitations to our knowledge of species richness of Neotropical rotifers is that many of the studies have used fixed samples, which makes it difficult to identify soft-bodied individuals. Moreover, few publications are accompanied by illustrations that allow the review of characters critical for identification (e.g., trophi). In addition, existence of cryptic species in taxa such as *Brachionus calyciflorus* indicates that we may be underestimating the total number of species of rotifers (Papakostas et al., 2016; Wen et al., 2016).

The records of monogonont rotifers from the Neotropics were previously compiled in extensive reviews by Koste & José de Paggi (1982), and José de Paggi & Koste

(1995). The only updated checklists of monogonont rotifers that have been published are for Argentina (Ferrando & Claps, 2016) and Brazil (Garraffoni & Lourenço, 2012). Fewer investigations have been published on bdelloids than on monogononts. No recent review on bdelloid rotifers from the Neotropics exists, and fewer surveys have been completed in comparison to those from the Nearctic or Palaearctic.

Unfortunately there is no comprehensive key to Rotifer from Neotropical region. One of the first of only a few attempts to assemble a regional key was made by Olivier (1965) for the rotifer fauna of Argentina. Fortunately, former general keys for *Brachionus* and *Keratella* (Ahlstrom, 1940, 1943) included Neotropical species. Regional studies were, therefore, constrained by the lack of updated comprehensive keys for rotifers since that published by Koste (1978). The special series "Guides to the Identification of the Microinvertebrates of the Continental Waters of the World" included some of the Neotropical fauna to the level of genus and species of Monogononta (Segers, 1995a; Nogrady et al., 1995; De Smet, 1996; De Smet & Pourriot, 1997; Nogrady & Segers, 2002). Unfortunately, this large undertaking remains incomplete. The on-line "Rotifer World Catalog" provides species descriptions, references, and useful ecological information at a global scale (Jersabek & Leitner, 2013).

The first observations on Antarctic rotifers are directly related to the exploratory activities carried out by several European countries at the beginning of the 20th century. A recent review (Fontaneto et al., 2015) exists of Antarctic rotifers, including on Subantarctic islands, and another paper described additional new species of bdelloids (Iakovenko et al., 2015).

TERMINOLOGY AND MORPHOLOGY

A critical feature for the identification of rotifers is their trophi. These **jaws** comprise several hard pieces that are often critical for species identification. **Trophi** are typically comprised of seven pieces. An unpaired **fulcrum** and three paired elements: manubrium (manubria), ramus (rami), and uncus (unci) (Fig. 8.1). In bdelloids, the fulcrum is absent. These pieces, together with the muscles that move them, form the **mastax**. The morphology, size, position, and orientation of the pieces are variable, and their arrangement determines eight different types of trophi for Bdelloidea and Monogononta. Each of these morphologies is closely related to a feeding habit in the different families and is thus important in species identification. Malleate, malleoramate, ramate, uncinate, virgate, forcipate, incudate, cardate types of trophi are recognized and a brief description of each type is included in the glossary below. More detailed information can be found Wallace et al. (2015).

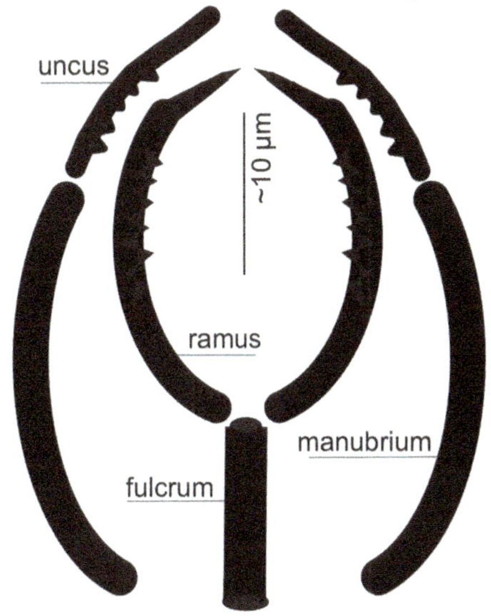

FIGURE 8.1 Schematic view of the basic elements of rotifer trophi. *After Wallace et al. (2016).*

Other important diagnostic features for identification include the lorica and the corona. The **lorica** is a thickened body wall whose thickness is variable. It can be very thin in some species and rather thick in others, and may contain ornamentations like facets, ridges, pustules, and plaques. The lorica may be divided into dorsal and ventral plates; a third plate, the basal one, may also be present in some species. The anterior edge of the dorsal plate often carries spines of variable length.

The **corona**, or rotatory apparatus, is an annular ciliation of the body apical region, although not all rotifers possess a ciliated corona. It helps acquire food and is a means of locomotion. There is great diversity in the typical plan of the corona. De Beauchamp (1907, 1965) and Donner (1956) described several types of coronas in both Bdelloidea and Monogononta families. They are related to different modes of locomotion and feeding behaviors (Wallace et al., 2006; Fontaneto & De Smet, 2015). A good observation of the type of corona is only well achieved from live organisms.

Glossary

- **Alula**: Projection like a wing, at the base of each ramus of the trophi.
- **Antennae**: Dorsal and lateral, single or paired sensory organs (mechanoreceptors).
- **Auricles**: Ciliated appendages on the head region, ear-shaped.
- **Bristles**: Movable appendages of the body, variable in length, lateral or caudal in position (e.g., *Filinia* spp., *Polyarthra* spp.).
- **Fecal pellets**: Defecation reservoirs located in digestive tracts of some species (e.g., *Ascomorpha* spp.).

- **Foot opening**: Aperture through which the foot is projected. Variable outline and position. Especially visible in fixed animals, with the foot retracted (e.g., *Brachionus* spp.).
- **Foot sheath**: Tubular sheath surrounding the foot opening.
- **Head shield**: Membranous cap (semicircular) that covers the corona (e.g., *Squatinella* spp., *Colurella* spp.).
- **Illoricate body**: Rotifer with a very thin body wall.
- **Incus** (trophi): Functional unit composed of fulcrum and rami.
- **Infundibulum**: Funnel-shaped structure present in the Collothecacea corona, rim with long setae.
- **Lamellae**: Thin and transparent cuticle (e.g., located on some pieces of the trophi); this term also applies to a part on the rostrum of bdelloids.
- **Loricate**: Rotifer with lorica (thickened body wall).
- **Malleus** (trophi): Functional unit composed of manubrium and uncus (pl. mallei).
- **Rostrum**: Apical retractable structure, especially in Bdelloidea (used for creeping), sometime elevated on pedicels, and with additional apical lamellae.
- **Spines**: Immobile projections of the lorica, anterior, caudal,or lateral in position (e.g., *Brachionus* spp., *Keratella* spp., *Notholca* spp.).
- **Trophi cardate**: Rami lyre-shaped, similar to the virgate trophi. Specialized for suction action but without piston. Constrained to Lindiidae.
- **Trophi forcipate**: Elongated and dorsoventrally compressed, long manubria. Asymmetrical in *Aspelta* genus. Rami with three chambers like a window opening. It works by gripping. Only in family Dicranophoridae.
- **Trophi incudate**: Fulcrum very short. Rami large, curved inward, pincer-shaped. Manubria and unci degenerate. This trophi type functions by grasping prey with a forceps-like action. Present only in Asplanchnidae.
- **Trophi malleate**: Fulcrum short, rami and manubria well developed; rami with short and thick teeth along the inner margin. The unci with 4–7 large teeth. This type of trophi works by gripping and grinding. Present in Brachionidae, Lepadellidae, Epiphanidae, Euchlanidae, Lecanidae, Mytilinidae, Trichotriidae, and Proalidae.
- **Trophi malleoramate**: Similar to the malleate trophi, but the rami are strongly toothed and unci have many thin teeth. Present in the Order Flosculariaceae.
- **Trophi ramate**: No fulcrum, uncus large as semicircular plates, with many teeth. Constrained to Bdelloidea, almost always homogeneous in structure except for Order Philodinavida (Melone et al., 1998b).
- **Trophi uncinated**: Fulcrum and manubria weak, rami well developed, unci with few teeth, usually with one large and a few small ones. Present only in Collothecidae.
- **Trophi virgate**: Frequently asymmetric. Pieces more or less slender. Long fulcrum and manubria and a powerful domelike hypopharyngeal muscle. Manubria straight or curved inward. Manubria end of variable shape, double or simple crooked, close or open loop, etc. Found only in Gastropodidae, Notommatidae, Scariididae, Synchaetidae, and Trichocercidae.

MATERIAL PREPARATION AND PRESERVATION

Samples for morphological observations (especially of Monogononta) can be fixed in the field with formalin or Lugol's iodine (both at ~5%). Staining with eosin, erythrosine, or Rose Bengal (best) can be useful for enhancing the visualization of preserved individuals, particularly in the case of samples with considerable debris that obscures the individuals. Samples with abundant organic and inorganic detritus are sorted when the individuals are made more clearly visible with a stain. A few drops of glycerol in each sample helps to preserve the material in the case of an accidental drying of the sample. However, for a complete taxonomic study, it is important to deal with unfixed samples containing living and active animals because many species, particularly those of soft-bodied groups, present diagnostic characters that are visible only in living organisms. This is the typical situation for *Notommata*, *Synchaeta,* and *Monommata* among monogononts and for most bdelloids. Then, it is important that live samples in (about 200 ml) plastic or glass bottles with wide mouth and only half-filled be transported to the laboratory, cooled on ice. Some species can survive in a refrigerator for some days, as those of the genus *Synchaeta*.

The observation of individuals may improve through use of narcotics prior to fixation. As pointed out by Wallace et al. (2006), there is no single anesthetic that has proved to be useful for a wide range of species. For a review of the issue, consult Wallace et al. (2006). Carbonated water (soda) is one of the most advisable anesthetics. Boiling water killing (Edmondson, 1959) is useful to avoid the distortion of the body shape caused by use of fixative, particularly in those species in which the use of narcotics is unsuccessful; this is especially true for bdelloids and sessile monogononts (Nogrady et al., 1993). Other chemicals that can be used to try and anesthetize rotifers are: bupivacaine hydrochloride, lidocaine, propranolol, novocaine, menthol, etc (Wallace et al., 2006).

For a good observation of the body shape of preserved rotifers, the investigator should select animals that are not distorted by the contraction caused by fixatives. The selected animal should be separated from the sample and placed on a drop of water-glycerol mixture. In temporary mountings, a mixture of water with 5%–10% of glycerol may be used, because a higher proportion of glycerol contracts the material. A useful technique for obtaining individuals in good conditions for observation is to submit the sample to a treatment of progressively increased concentration of glycerol once a day (5%—10%—25%—50%—75%). The glycerol is eventually embedded within the body tissue.

To work under the microscope with high magnification, individuals should be placed under a coverglass, but avoid crushing the animal. This can be achieved by using small balls of Plasticine or bits of broken coverglass located in the corners underneath the coverglass. Segers (in Wallace et al., 2006) suggested that it is possible to examine an animal under high magnification without a coverglass; this will allow moving it with a fine needle to get a proper view.

Permanent mounts for light microscopy require using a mounting medium, such as glycerine jelly (formula of Kaiser). In countries with warm climates, the preparations must be preserved in air-conditioned rooms. If the preparation is not kept cool, the mounting medium becomes so soft that any tiny flaw in the sealing would allow a leak of the medium. Polyvinyl lactophenol has proved to be a convenient medium of mounting. HydroMatrix mounting medium can be used for loricate rotifers.

Some authors suggest passing the animals directly from the water to the fixing chemical, but soft-bodied individuals undergo deformation. The slides to be used must be very clean, and washing with ethanol is advisable. A serious problem is the presence of dust particles that may become lodged in the mounting medium, rendering a specimen useless for detailed taxonomic work. Thus, we recommend cleaning the slide carefully with a soft, lint-free cloth. Once the individual is placed in the mounting medium, the formation of bubbles should be avoided. This may be achieved by depositing the coverglass obliquely onto the drop, allowing a progressive contact between the glass and the high-density medium. The air bubbles between slides and cover slips can be removed tapping gently on the coverglass with a needle. We recommend using a small, circular cover glass 10 or 12 mm in diameter to ensure a good seal. Information about different techniques can be found in Wallace et al. (2006) and Jersabek et al. (2010).

One of the features to identify rotifers refers to the thickness of their integument: some species have loricate integument (body wall thickened and firm) others illoricate (body wall not thickened or firm). Yet, these are only the extremes of a gradient, and it may not always be possible to understand whether an animal can be considered loricate or illoricate. Wallace et al. (2016) suggested the following procedure: "When working with fresh material, apply gentle pressure from the point of a pencil or probe onto the cover glass while observing the individual. If the body puffs out under pressure and returns to its original shape when the pressure is released, the individual is illoricate. Loricate forms will exhibit much less flexibility of the body wall: i.e., they do not puff out. N.b: If too much pressure is applied to the cover glass, the animal may be crushed, becoming irreversibly damaged, perhaps rendering them unidentifiable. In preserved materials, illoricate forms tend to shrivel up, while the body wall of loricate rotifers will retain its shape even if the inner organs separate from the body wall, collapsing into a central mass of tissue."

Detailed analysis of trophi is indispensable because trophi are an important source of diagnostic characters in many genera (*Trichocerca*, *Cephalodella*, *Notommata*, and *Polyarthra*, among others). A preliminary, but not exhaustive observation is possible without separating the trophi from the animal body. Bleach, sodium hypochlorite, is then used as an effective solvent of soft tissue to extract the trophi. As Wallace et al. (2006) pointed out, there are two main ways to extract and observe the trophi. One of them is to place the animal in a depression slide, where sodium hypochlorite is added, and then to transport the trophi from the depression slide to a flat slide after the soft tissues have disappeared. Alternatively, the animal can be placed on a flat slide in a drop of water. To this a small drop of bleach is added and a cover glass is careful applied to the preparation. The animal is observed as the bleach dissolves its tissues, but observers must be diligent as this can happen quite suddenly. To avoid misinterpretation of the shape of trophi, it is advisable to observe them at different planes of orientation. Observation of the trophi can be improved by means of high magnification ($\times 1000$) using oil-immersion microscope objective lenses. This is particularly useful for smaller trophi. To avoid misidentifications, it is important to document the trophi morphology. Drawing is a necessary tool, but photography is also useful, when many photographic planes are integrated. Obviously, a good picture obtained with the scanning electron microscope (SEM) is very valuable. For example, excellent photos can be seen in papers of several authors (Segers et al., 1994a, 1994b; Segers, 1995a, 1995b; De Smet, 1996, 1998, among others). To prepare trophi for SEM mounting, the procedure is very similar to the one for light microscopy; the only difference is that all excess of bleach should be carefully removed by repeated rinsing before letting the water medium evaporate and obtain dry clean trophi on a slide (Sanoamuang & McKenzie, 1993; De Smet, 1998).

Any of these methods requires practice and patience. For example, the dissolution of the soft tissues can be very rapid, so one must be very watchful. Other difficulties can be encountered in controlling the excess liquid and avoiding bubbles that may prevent adequate visualization of trophi. It is also wise to have several individuals of the species under study for possible material loss.

KEY TO ROTIFERA

The following identification key deals with females and is based on several important works (Donner, 1965; Koste, 1978; Segers, 2004; Ricci & Melone, 2000; Wallace et al., 2006). The number of species recorded by genera is included in brackets; monospecific genera are designated as such. Nomenclature follows Segers (2007, 2013).

The species counts are based on current information as of March 2019 and are undoubtedly underestimates of the total diversity, especially in countries with lower rates of collection.

Rotifera: Classes

1 Trophi ramate generally visible in whole animals without resorting to their isolation (Fig. 8.2); body soft, mostly vermiform; a pair of ovaries, difficult to see; live animals crawl like leeches, attaching alternately with their posterior and anterior ends; preserved bodies contract into cylindrical indeterminate blobs; usually not planktonic in open waters, frequent in vegetated littoral, benthos and mosses .... **Bdelloidea [p. 150]**

1′ Trophi other than ramate (Fig. 8.2); single ovary .... **Monogononta [p. 152]**

Rotifera: Bdelloidea: Orders

Consult Donner (1965) for details and Segers (2007) for nomenclature. Additional species not included in these two works are those from Antarctica described by Iakovenko et al. (2015).

1 Trophi not close to the mouth, deeper, never extruded; esophagus long; proximal and distal minor teeth of uncus plate of the trophi always present; corona with trochi absent or present, but when present never reduced 2

1′ Trophi close to the mouth opening, generally extruded when animal is feeding; esophagus short; proximal minor teeth of uncus plate of the trophi usually reduced or absent; corona reduced to small ciliated field, trochi absent or small .... Philodinavida; one family: **Philodinavidae [p. 150]**

2(1) Corona modified to a ventral ciliated field, no trochi; foot long and extensible, with three toes and two spurs, or short and plump, with discoid distal pseudosegment and spurs absent; common behavior is scraping and browsing the substratum (Fig. 8.3 A) .... Adinetida; one family: **Adinetidae [p. 150]**

2′ Corona frontal, with trochi, mostly elevated on pedicels; foot with two spurs of variable length; zero, two, three, or four toes present (Fig. 8.3 B, C) .... **Philodinida [p. 150]**

Rotifera: Bdelloidea: Philodinavida: Philodinavidae: Genera

1 Foot with two short spurs and four normal toes (Fig. 8.4) .... *Philodinavus*
[One undetermined Antarctic of two species]

1′ Foot with one spur and four large toes (Fig. 8.4) .... *Henoceros*; one species: *Henoceros falcatus* (Milne, 1916)
[One Neotropical of two species]

Rotifera: Bdelloidea: Adinetida: Adinetidae: Genera

1 Foot long and extensible, with three toes and two spurs (Fig. 8.5) .... *Adineta*
[Five Neotropical and ten Antarctic of 22 species]

1′ Foot short with several papillae on the posterior border of the penultimate pseudosegment (Fig. 8.5) .... *Bradyscela*; one species *Bradyscela clauda* (Bryce, 1893)
[One Neotropical of three species]

Rotifera: Bdelloidea: Philodinida: Families

1 Stomach without recognizable lumen, with round pellets in its wall .... **Habrotrochidae [p. 150]**

1′ Stomach with thick wall and visible lumen, usually not filled with round pellets .... **Philodinidae [p. 151]**

Rotifera: Bdelloidea: Habrotrochidae: Genera

1 Upper lip normally developed, trochi visible 2

1′ Upper lip with large lobe, partially covering the trochi (Fig. 8.6) .... *Scepanotrocha*
[Five Neotropical of 11 species]

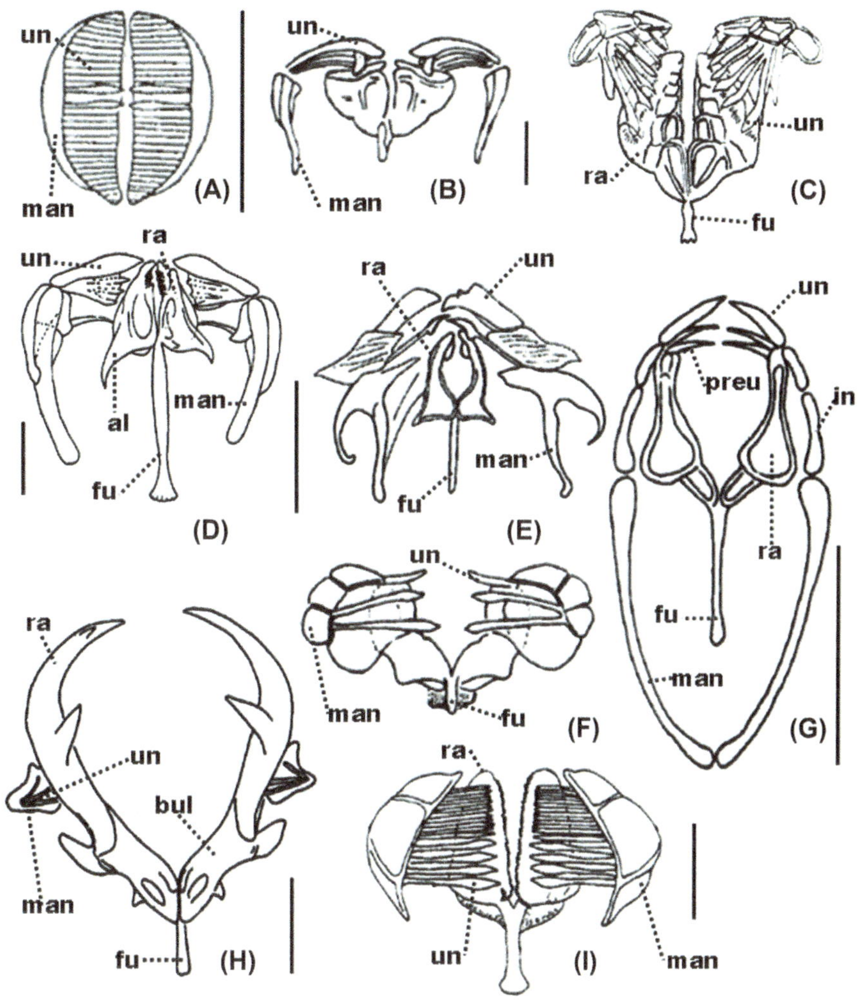

FIGURE 8.2 Trophi types (Bdelloidea and Monogononta): (A) ramate trophi of Bdelloidea; (B and C) malleate trophi of *Epiphanes*, ventrally and ventrally from above; (D) virgate trophi of *Notommata*; (E) cardate trophi of *Lindia*; (F) uncinate trophi of *Collotheca*; (G) forcipate trophi of *Dicranophorus*; (H) incudate trophi of *Asplanchna*; (I) malleoramate trophi of *Ptygura*. Scale bars = 20 μm. Abbreviations: *al*, alulus; *bul*, bulla; *fu*, fulcrum; *in*, intramalleus; *man*, manubrium; *pre*, preuncus teeth; *ra*, ramus; *un*, uncus. *After Koste & Shiel (1986, 1987) and Dartnall (1995).*

2(1) Each trochus with a transversal cuticular ring, incomplete ventrally and dorsally (Fig. 8.6) .... *Otostephanos*
[One Neotropical and one Antarctic of 15 species]

2′ Trochi without cuticular ring (Fig. 8.6) .... *Habrotrocha*
[Thirty-six Neotropical and ten Antarctic of 131 species]

Rotifera: Bdelloidea: Philodinidae: Genera

1 Foot without toes 2

1′ Foot with toes 3

2(1) Foot length > 0.5× trunk; epizoic .... *Anomopus*; one species: *Anomopus chasmagnathi* Mañé-Garzon & Montero, 1973
[One Neotropical of two species]

2′ Foot length < 0.5× trunk; mostly free-living (Fig. 8.7) .... *Mniobia*
[Eleven Neotropical and three Antarctic of 49 species]

3(1) Foot with three toes 4

3′ Foot with four toes 6

4(3) Corona without corniform lateral projections 5

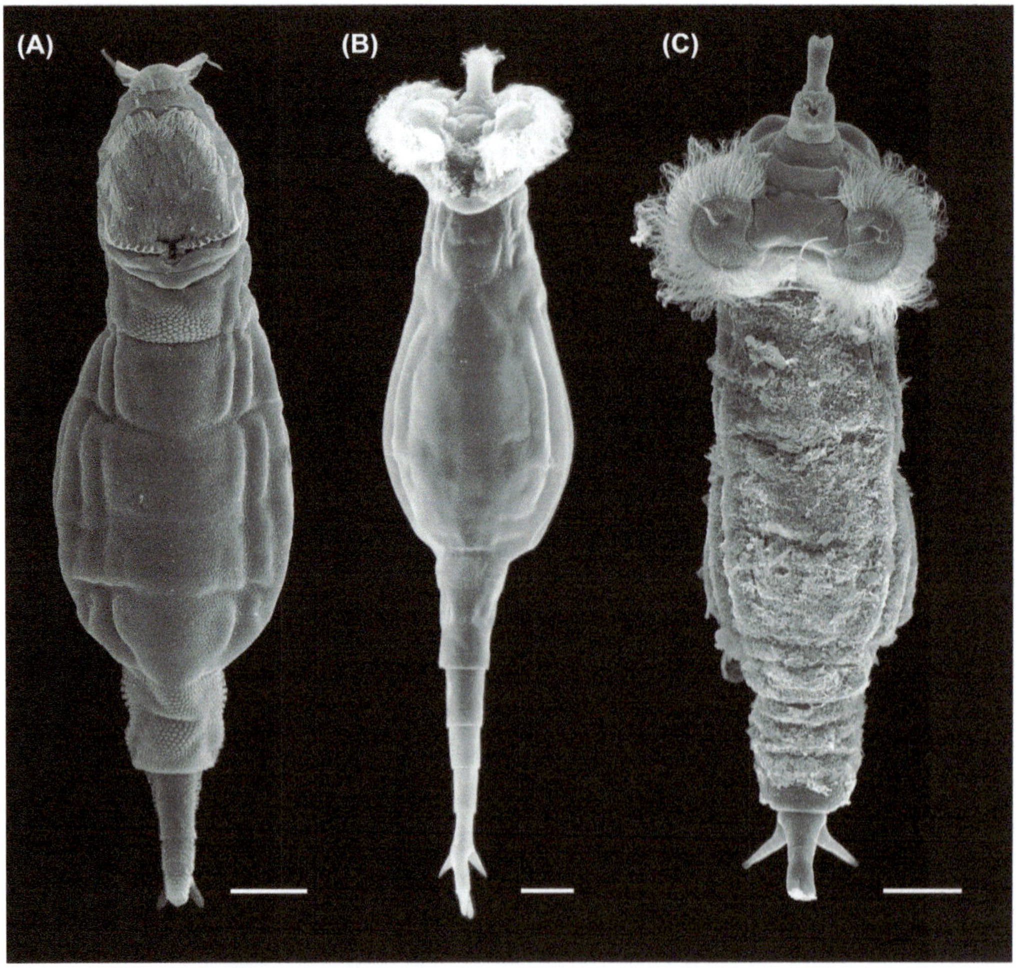

FIGURE 8.3 SEM pictures of Bdelloidea in ventral view: (A) *Adineta tuberculosa*; (B) *Rotaria macrura*; (C) *Dissotrocha aculeate*. Scale bars = 20 μm. *Photo courtesy of Giulio Melone. Reprinted with permission from Fontaneto & De Smet (2015).*

4′ Corona with corniform lateral projections (Fig. 8.7) .... *Ceratotrocha*; one species: *Ceratotrocha cornigera* (Bryce, 1893) [One Neotropical of four species]

5(4) Viviparous, often eye spots on rostrum; rostrum extended also when swimming (Fig. 8.7) .... *Rotaria* [Ten Neotropical and one Antarctic of 27 species]

5′ Oviparous, eye spots absent (Fig. 8.8) .... *Macrotrachela* [26 Neotropical and thirteen Antarctic of 98 species]

6(3) Oviparous, with the exception of *Philodina gregaria* from Antarctica, which is viviparous 7

6′ Viviparous, integument often sculptured (Fig. 8.8) .... *Dissotrocha* [Seven Neotropical of 35 species]

7(6) Integument thick and sculptured; eye spots not evident (Fig. 8.9) .... *Pleuretra* [Three Neotropical of 14 species]

7′ Integument thin and smooth; eye spots, when visible, on brain (Fig. 8.9) .... *Philodina* [Thirteen Neotropical and seven Antarctic of 52 species]

Rotifera: Monogononta: Orders

1 Trophi other than malleoramate 2

1′ Trophi malleoramate (Fig. 8.2 I); corona circular or lobed .... **Flosculariaceae [p. 153]**

2(1) Trophi uncinate; corona without ciliated region, open to a modified wide funnel (infundibulum) with or without setae or tufts of setae around the rim or margin; usually sessile; foot present, toes absent; ambush predators .... **Collothecacea [p. 159]**

2′ Trophi other than uncinate .... **Ploima [p.160]**

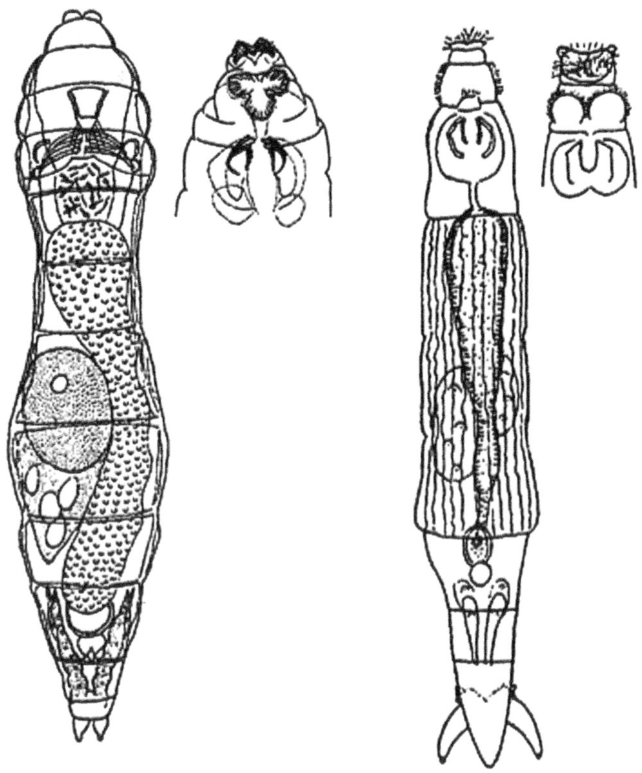

FIGURE 8.4 Philodinavidae: *Philodinavus paradoxus* (left, ~300 μm) and *Henoceros falcatus* (right, 200 μm). *Redrawn with permission from Fontaneto & De Smet (2015).*

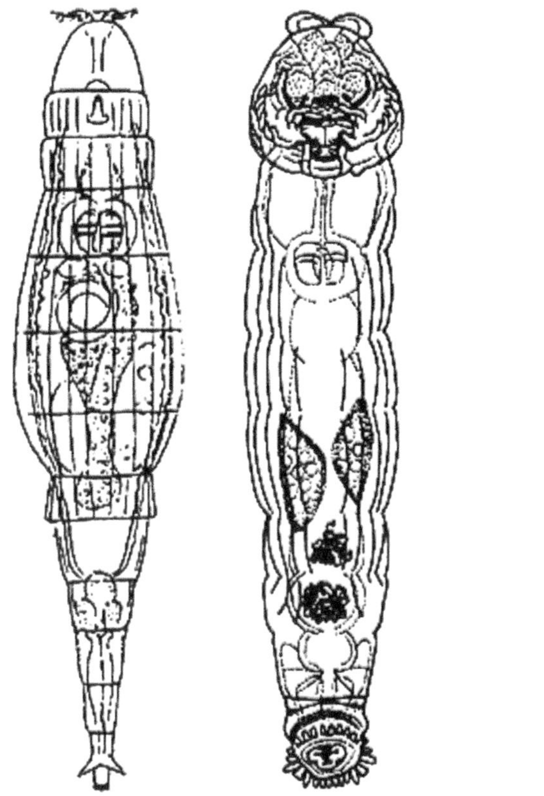

FIGURE 8.5 Adinetidae: *Adineta vaga* (left, 550 μm) and *Bradyscela clauda* (right, 220 μm). *Redrawn with permission from Fontaneto & De Smet (2015).*

Rotifera: Monogononta: Flosculariaceae: Families

Important keys and taxonomic papers on the Flosculariaceae include: Edmondson (1949), Segers (1997), Segers & Wallace (2001), Segers & Shiel (2008), and Segers et al. (2010).

1 Body of different forms; not triangular 2

1′ Body triangular; with six robust armlike appendages with long, pinnate bristles at the tips; foot absent; exclusively planktonic (Fig. 8.10) .... Hexarthridae, one genus: *Hexarthra*
[Eight Neotropical of 21 species]

2(1) Body short, oval or spherical, foot absent or present 3

2′ Body elongate or conical, both with long foot 4

3(2) Body oval with long bristles or long oval or spherical, without bristles; foot absent .... **Trochosphaeridae [p. 153]**

3′ Body oval, without bristles; body mostly compressed dorsoventrally .... **Testudinellidae [p. 154]**

4(2) Body elongate; foot long and large; corona circular (slightly elliptical) to lobate, cordiform, or auriciform, resembling flowers; solitary or colonial; with or without a tube or gelatinous matrix; mostly sessile, but some free swimming .... **Flosculariidae [p. 153]**

4′ Body conical, with long foot, very contractile; corona horseshoe-shaped; antennae visible, even in contracted animals; solitary or colonial .... Conochilidae; one genus: ***Conochilus* [p. 158]**

Rotifera: Monogononta: Flosculariaceae: Trochosphaeridae: Genera

Consult Nogrady & Segers (2002) for details.

1 Setae or bristles absent 2

1′ Body ovate, truncated anteriorly, with two lateral setae (bristles) and one (rarely two) rigid caudal setae; foot absent; excluding setae ~40–325 μm (Fig. 8.11), trophi size very small .... *Filinia*
[Eight Neotropical and one Antarctic of 15 species]

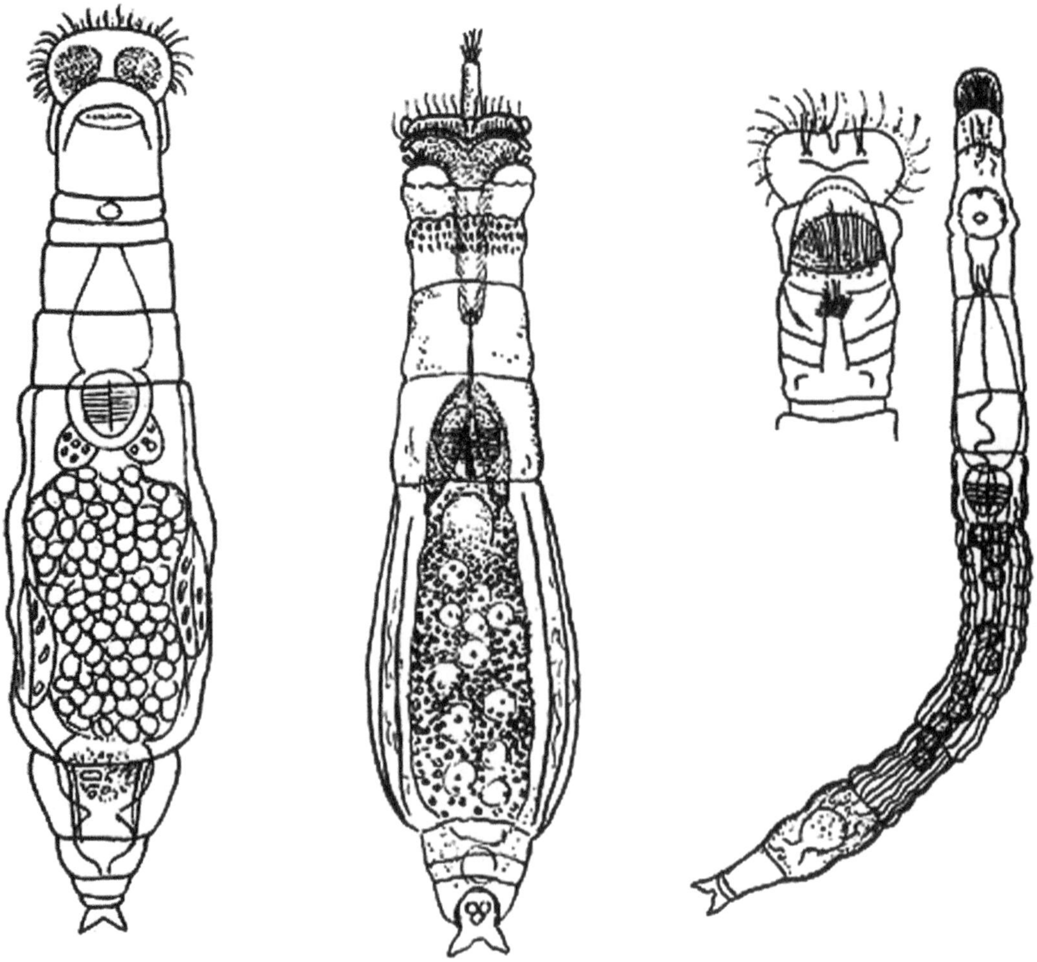

FIGURE 8.6 Habrotrochidae: *Scepanotrocha rubra* (left, ~200 μm), *Otostephanos monteti* (center, 250 μm) and *Habrotrocha serpens* (right, 260 μm). *Redrawn with permission from Fontaneto & De Smet (2015).*

2(1) Body spherical, transparent, corona equatorial or in apical third; foot absent; viviparous; 320–1100 μm (Fig. 8.12) .... *Trochosphaera*; one species: *Trochosphaera aequatorialis* (Semper, 1872) [One Neotropical of 2 species]

2′ Body slightly elongate, oviform, flat; conspicuous lateral antennae at middle of body; corona a slight extension of the body; foot absent; oviparous; 125 × 160 μm (Fig. 8.13) .... *Horaella*; one species: *Horaella thomassoni* Koste, 1973 [One Neotropical of two species]

Rotifera: Monogononta: Flosculariaceae: Testudinellidae: Genera

1 Lorica dorsoventrally flattened, transparent; foot annulated, not visible in contracted animals, foot opening located ventrally near middle of body or in posterior half of body or at its end; ~250 μm (Fig. 8.14) .... *Testudinella* [Seventeen Neotropical of 38 species]

1′ Lorica not greatly flattened; body small, rather possess four nearly equal lobes in cross-section; foot absent; eggs attached to a retractile thread of secretion; ≤ 120 μm (Fig. 8.15) .... *Pompholyx* [Two Neotropical of three species]

Rotifera: Monogononta: Flosculariaceae: Flosculariidae: Genera

Consult Segers (1997) for details on Neotropical Floscularia species and Meksuwan et al. (2011) for additional details on *Lacinularoides*. *Octotrocha speciosa* reported in the Amazon (Koste, 1974a) could pertain to *L. coloniensis*.

1 Dorsal antennae (paired), not (easily) visible in fixed animals 2

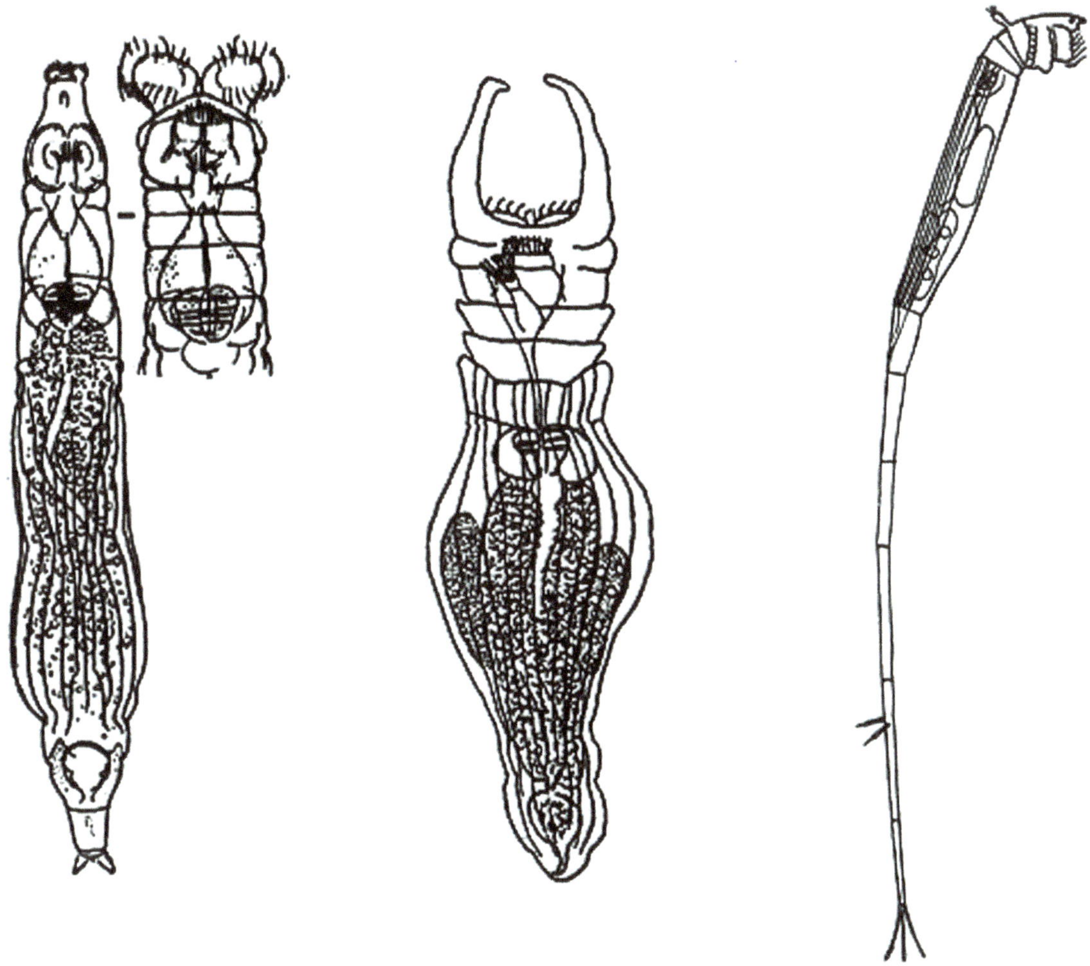

FIGURE 8.7 Philodinidae: *Mniobia orta* (left, 240 μm); *Ceratotrocha cornigera* (center, 330 μm) and *Rotaria neptunia* (right, ca. 1400 μm). *Redrawn with permission from Fontaneto & De Smet (2015).*

1′	Dorsal antenna single, centrally placed, visible when animal is contracted (Fig. 8.16); Segers et al. (2010) highlight the need for a review of this purportedly monospecific, cosmopolitan and eurytopic taxon .... *Beauchampia*; monospecific: *Beauchampia crucigera* (Dutrochet, 1812)
2(1)	Corona with lobes 3
2′	Corona circular, reniform, or cordate 6
3(2)	Corona with more than two lobes 4
3′	Corona bilobate; animals usually solitary (occasionally small colonies, ≤ 12 individuals) in separate straight or slightly curved tubes, each forming a distinct pipe of hardened secretions (Fig. 8.17) .... *Limnias* [Two Neotropical of five species]
4(3)	Corona with more than four lobes 5
4′	Corona with four lobes; sessile, solitary, or forming colonies (Fig. 8.18); ≤ 1500 μm .... *Floscularia* [Eight Neotropical of nine species].
5(4)	Corona with eight lobes, one pair smaller; sessile, solitary in a gelatinous tube; 200–2000 μm (Fig. 8.19) .... *Octotrocha*; monospecific: *Octotrocha speciosa* (Thorpe, 1893)
5′	Corona large with five prominent and two small lobes; sessile, in small to large colonies, occasionally solitary; ≤ 1300 μm (Fig. 8.20) .... *Lacinularoides*; monospecific: *Lacinularoides coloniensis* (Colledge, 1918)
6(2)	Corona circular to reniform 7

PHYLUM ROTIFERA

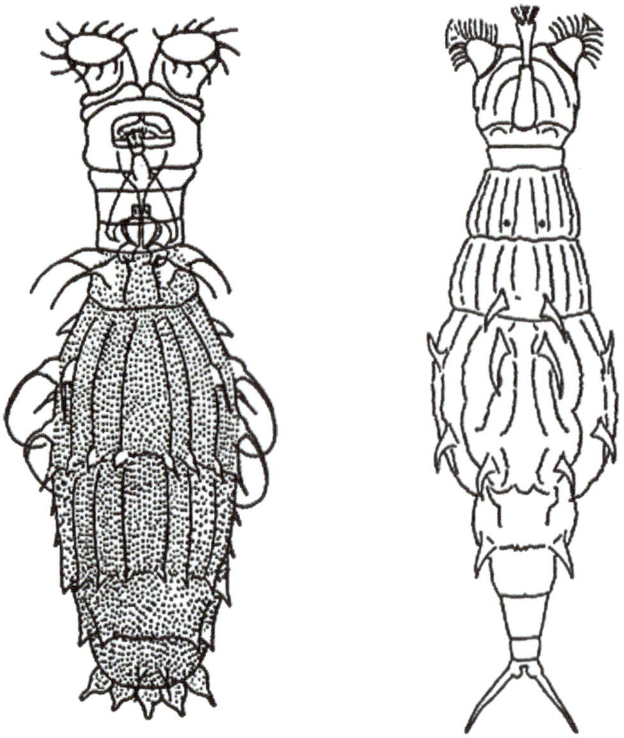

FIGURE 8.8 Philodinidae: *Macrotrachela multispinosa* (left, 300 μm) and *Dissotrocha aculeata* (right, ca. 450 μm). *Redrawn with permission from Fontaneto & De Smet (2015).*

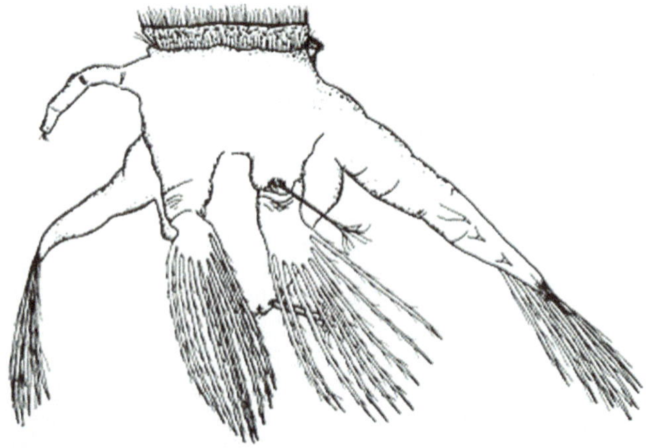

FIGURE 8.10 Hexarthridae: *Hexarthra longicornicula* (~115 μm; Brazil). Known only from Neotropics. After *Turner (1987) reprint by permission from Sringer Nature.*

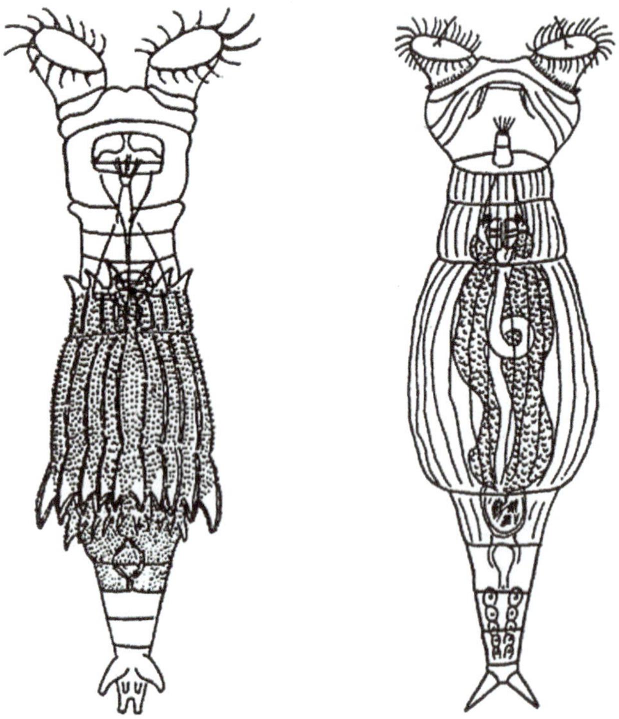

FIGURE 8.9 Philodinida. *Pleuretra brycei* (left, 380 μm) and *Philodina citrina* (right, 450 μm). *Redrawn with permission from Fontaneto & De Smet (2015).*

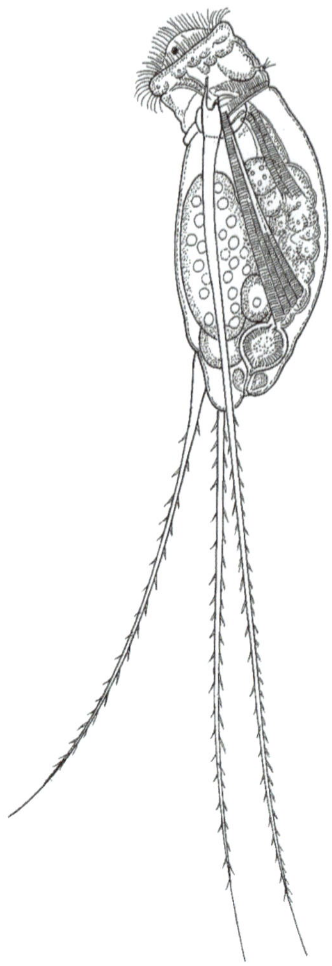

FIGURE 8.11 Trochosphaeridae: Representative form *Filinia* sp. (~520 μm; Brazil). *After Koste et al. (1984).*

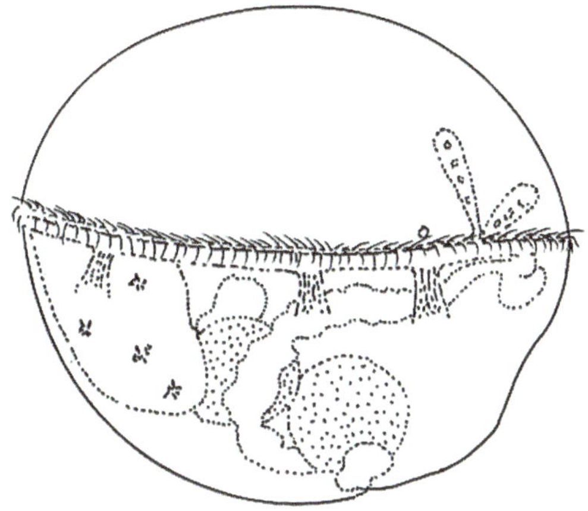

FIGURE 8.12 Trochosphaeridae, *Trochosphaera aequatorialis*, lateral view (~530 μm diameter; Argentina). *After Paggi (1978).*

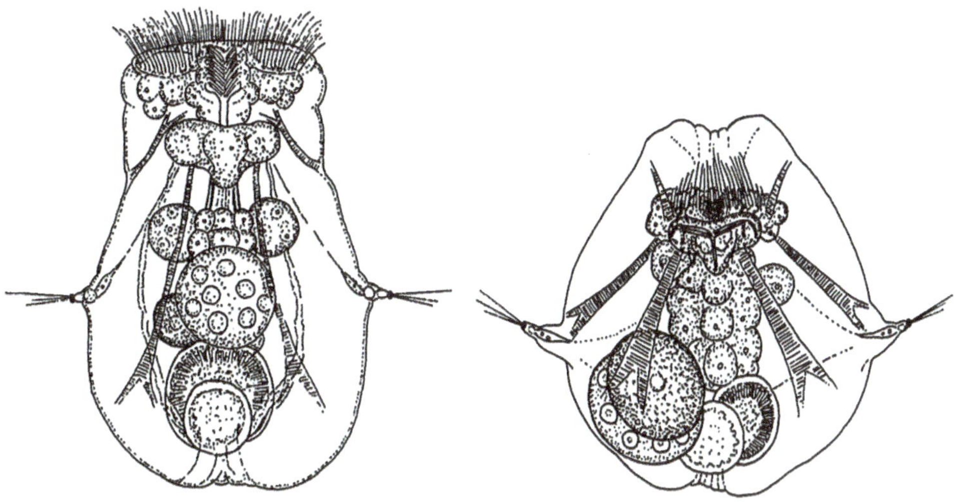

FIGURE 8.13 Trochosphaeridae, *Horaella thomassoni*, extended (left, ~158 μm) and contracted (right) (recorded only from Neotropics, Guyana—Brazil). *After Koste (1973).*

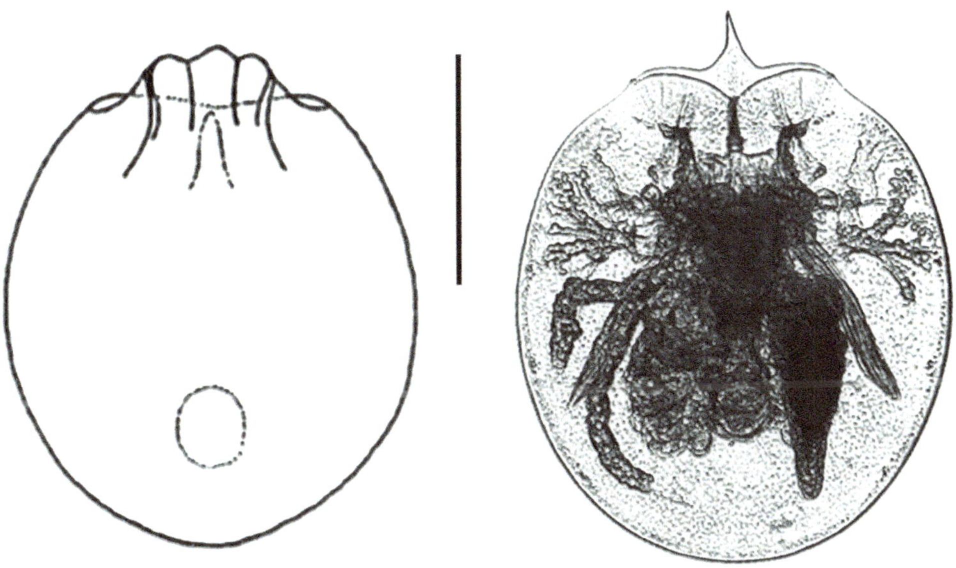

FIGURE 8.14 Testudinellidae, *Testudinella*: *T. ohlei*, ventral view (left; Argentina), scale bar = 50 μm; *T. haueriensis*, ventral view (right, ~ 400 μm; Argentina).

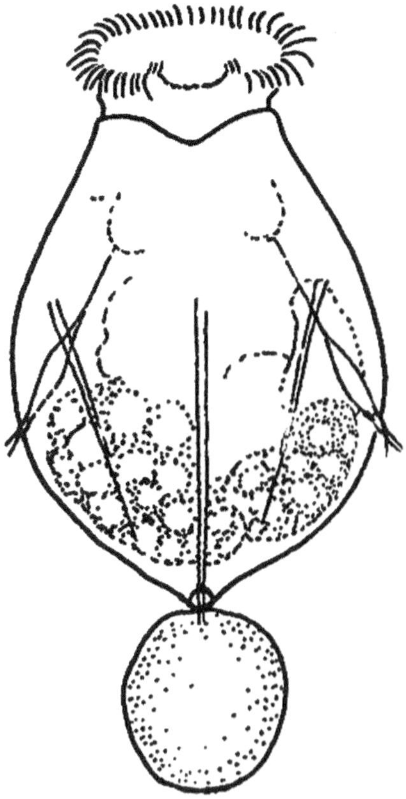

FIGURE 8.15 Testudinellidae, *Pompholyx* sp., ventral view (~ 150 μm; Argentina). *After José de Paggi (1995).*

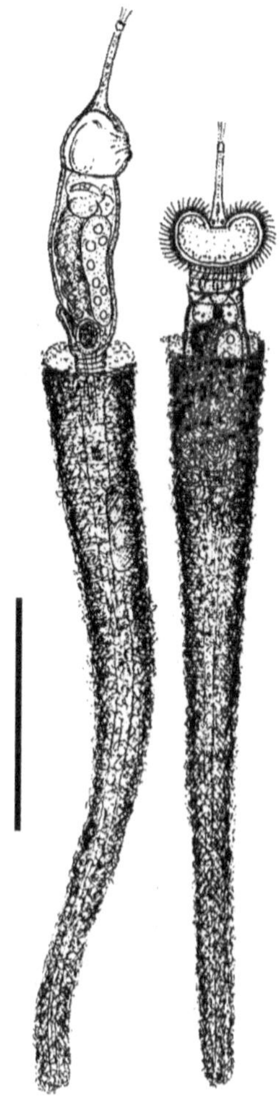

FIGURE 8.16 Flosculariidae, *Beauchampia crucigera*, lateral and ventral view (Jamaica). Scale bar = 100 μm. *After Koste et al. (1995).*

6′ Corona cordate; oviferon absent; sessile or free-living colonies; case gelatinous, well developed; individuals: ≤ 2000 μm, colonies: ≤ 5 mm (Fig. 8.21) .... *Lacinularia*
[Three Neotropical of six species]

7(6) Corona circular to weakly lobed; oviferon absent; attached or free-living, case gelatinous, mostly well developed, 150–1000 μm (Fig. 8.22) .... *Ptygura*
[19 Neotropical, two Antarctic of 31 species]

7′ Corona reniform; oviferon present; sessile or spherical to ellipsoid free-living colonies; case, when present, gelatinous, poorly developed; individuals: 400–2000 μm, colonies: ≤ 4 mm (Fig. 8.23) .... *Sinantherina*
[Four Neotropical of four species]

Rotifera: Monogononta: Flosculariaceae: Conochilidae: *Conochilus:* Subgenera

Diéguez & Balseiro (1998) found an increase in colony size of *Conochilus hippocrepis* (Schrank, 1803) in response to predation by a predaceous calanoid copepod *Parabroteas sarsi*.

1 In colonies, small (upto 35 individuals) or large (upto 100 individuals); dorsal antenna absent; lateral antennae on the apical field, fused completely or quite separated or slightly fused (Fig. 8.24) .... *Conochilus* (Subgenus *Conochilus*)
[Two Neotropical of two species]

1′ Solitary or, one adult individual with one to three young in the same gelatinous hull; dorsal antenna present, small; lateral antennae, situated posterior to the corona, outside of the apical field, not fused or largely or partially fused (Fig. 8.25) .... *Conochilus* (Subgenus *Conochiloides*)
[Three Neotropical of four species]

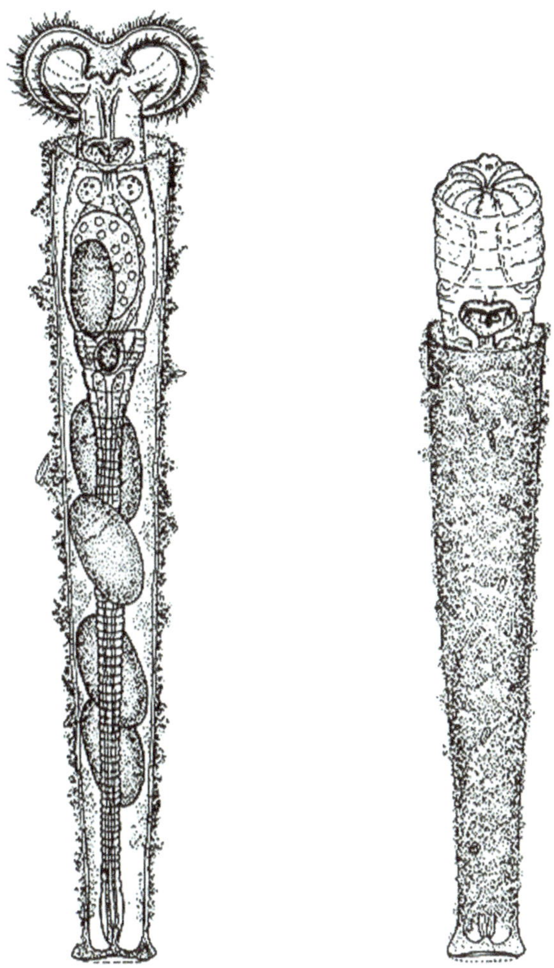

FIGURE 8.17 Flosculariidae, *Limnias ceratophylli*, ventral view, extended (left, 700 μm) and contracted (right, 26 μm) (Jamaica). *After Koste et al. (1991).*

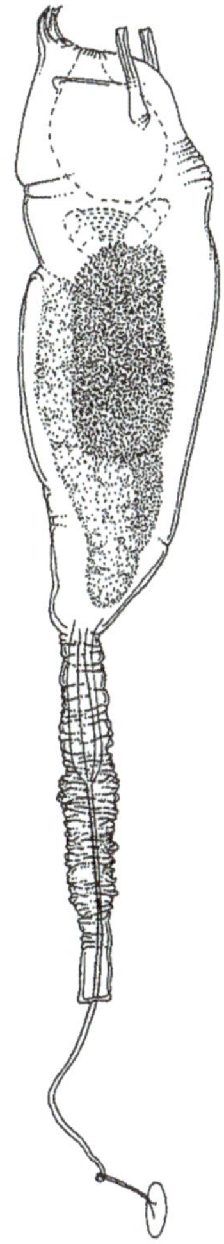

FIGURE 8.18 Flosculariidae, *Floscularia bifida* (~650 μm) lateral view (known only from Neotropics, Brazil). *After Segers (1997).*

Rotifera: Monogononta: Collothecacea: Families

Important keys to the Collothecacea include: Bērziņš (1951) and Meksuwan et al. (2013).

1 Adult corona margin smooth or with long setae or cilia, or lobes; foot long, contractile; sessile, in a gelatinous tube or rarely free-living .... **Collothecidae [p. 159]**

1′ Adult corona margin without setae or cilia; with or without gelatinous tube; foot absent or relatively short (sometimes ventral); sessile or creeping .... **Atrochidae [p. 160]**

Rotifera: Monogononta: Collothecacea: Collothecidae: Genera

Comparatively little is known on the distribution and diversity of sessile rotifers in general, and of Collothecidae in particular, due to the fact that they require live observation for identification and study. This knowledge gap is especially evident regarding sessile rotifers from tropical regions (Meksuwan et al., 2013). Since some members of *Collotheca* also possess elongate lobes, Meksuwan et al. (2013) recommend re-evaluating the status of both genera.

1 Corona extended into elongate lobes edged with whorls of setae; sessile, in a gelatinous tube; foot terminated by a long peduncle, ending in an adhesive disc (Fig. 8.26) .... *Stephanoceros*
[Two Neotropical of two species]

1′ Corona circular, lobed or pointed, setae not arranged in whorls; sessile and nonsessile species; foot terminated by a long peduncle (Fig. 8.27) .... *Collotheca*
[Fourteen Neotropical, two Antarctic of ~50 species]

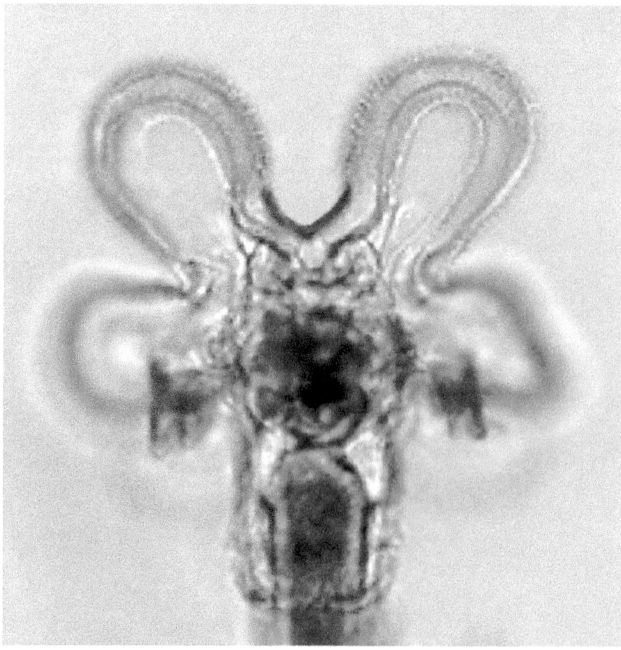

FIGURE 8.19 Flosculariidae, *Octotrocha speciosa*, corona. *After Segers et al. (2010).*

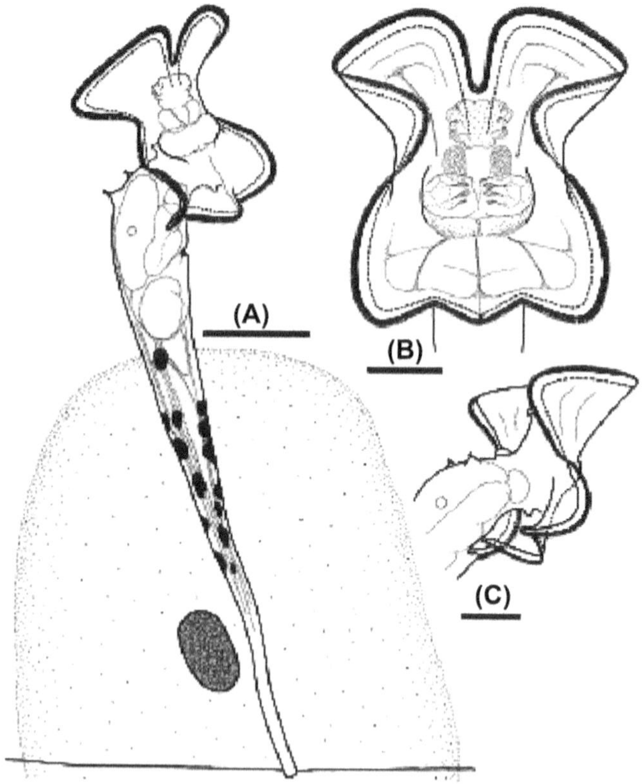

FIGURE 8.20 Flosculariidae, *Lacinularoides coloniensis*: (A) adult; (B) corona, dorsal view; (C) corona ventral view. Scale bars: (A) = 200 μm, (B and C) = 100 μm. *After Meksuwan et al. (2011).*

Rotifera: Monogononta: Collothecacea: Atrochidae: Genera

This family needs revision.

1	Corona with lobe or tentacles, body elongate 2
1′	Corona large and concave, as wide as trunk; body short, cylindrical with a ventral short pedal stalk; usually on aquatic plants (Fig. 8.28) .... *Cupelopagis*; monospecifc: *Cupelopagis vorax* (Leidy, 1857)
2(1)	Corona with a large dorsal lobe bordered by a thin, festooned membrane, in lateral view, crescentiform; body long with a terminal long stalk; observed in association with colonies of *Sinantherina socialis* (Linnaeus) and *Lacinularia flosculosa* (Müller) (Meksuwan et al., 2011); > 1000 μm (Fig. 8.29) .... *Acyclus* [One Neotropical, of ~ two species]
2′	Corona with numerous small peripheral tentacles; body fusiform, foot reduced to a projection at the posterior end; extended individuals > 1000 μm, contracted: 600–700 μm (Fig. 8.30) .... *Atrochus*; monospecific: *Atrochus tentaculatus* (Wierzejski, 1893)

Rotifera: Monogononta: Ploima: Families

This key does not include four families: Birgeidae (*Birgea*), Clariaidae (*Claria*), Cotylegaleatidae (*Cotylegaleata*), all monospecific (De Smet, 2007; Segers, 2008), and Asciaporrectidae that inhabit shells of live testate amoeba. Neither has been reported from the Neotropics or Antarctica.

1	Trophi other than incudate (Fig. 8.2) 2
1′	Trophi incudate; cuticle thin, transparent; body saccate, pyriform or cylindrical; mostly predators .... **Asplanchnidae [p. 167]**
2(1)	Trophi other than cardate (Fig. 8.2) 3
2′	Trophi cardate; body illoricate, fusiform, or vermiform; lateral antennae present, 300–500 μm (Fig. 8.31) .... Lindiidae; one genus: *Lindia* [Four Neotropical, one Antarctic of 13 species]

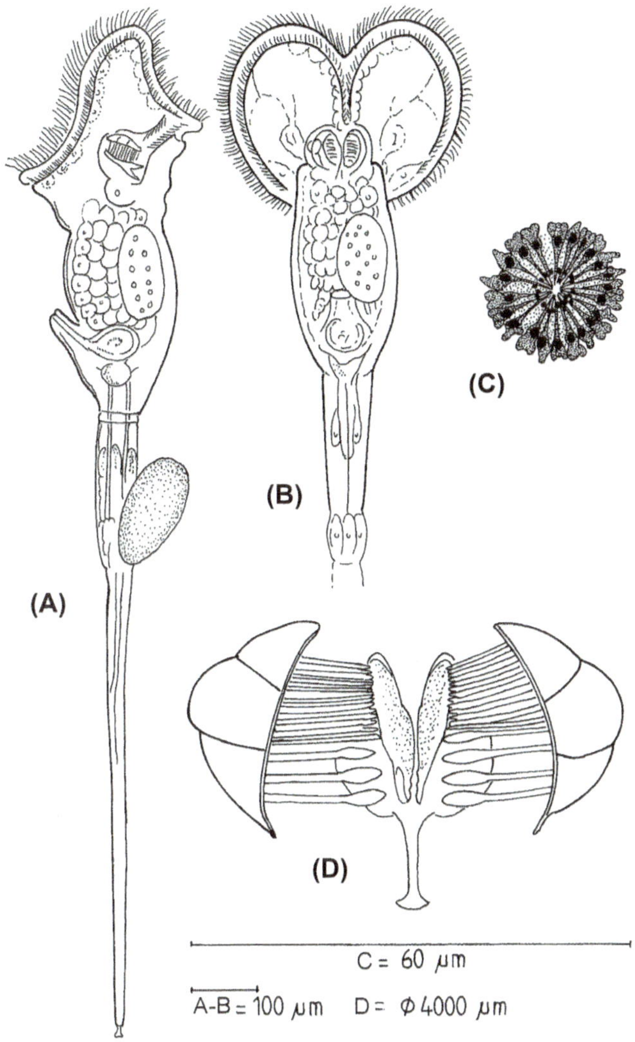

FIGURE 8.21 Flosculariidae, *Lacinularia flosculosa*: (A) whole animal; (B) corona; (C) trophi; (D) colony (Jamaica). *After Koste et al. (1995).*

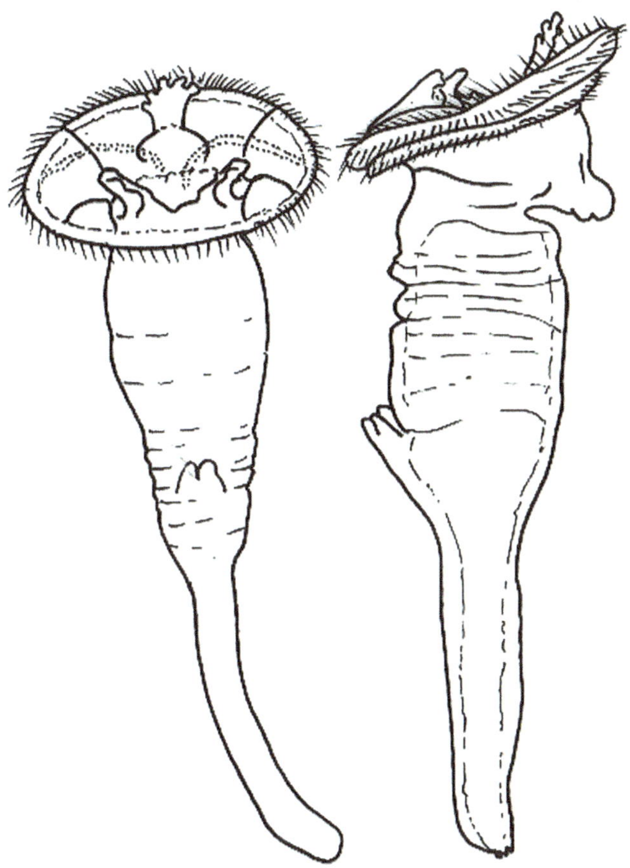

FIGURE 8.22 Flosculariidae, *Ptygura wilsonii*, dorsal and lateral view (~400 µm; Brazil). *After Murray (1913).*

3(2)	Trophi other than forcipate (Fig. 8.2)	4
3′	Trophi forcipate (exceptionally modified as virgate); body illoricate or semiloricate, vermiform to cylindrical; foot present, of variable length; often with two toes	**Dicranophoridae [p. 170]**
4(3)	Trophi malleate	5
4′	Trophi virgate	12
5(4)	Body illoricate, soft, transparent; foot long or short, with one or two toes	6
5′	Body loricate; lorica weak or thick, in some cases with furrows, grooves, sulci, or spines	7
6(5)	Body mostly sacciform or rectangular; mouth set into a funicular buccal field	**Epiphanidae [p. 172]**
6′	Body elongate; mouth set in an oblique, ciliated field on ventral side	**Proalidae [p. 173]**
7(5)	Lorica not extending beyond body to head, foot, and toes	8
7′	Lorica covering body, head, foot, and toes, surface with facets, mostly granulated and with spicules or spines	**Trichotriidae [p. 173]**
8(7)	Corona without membranous cap or semicircular shield	9

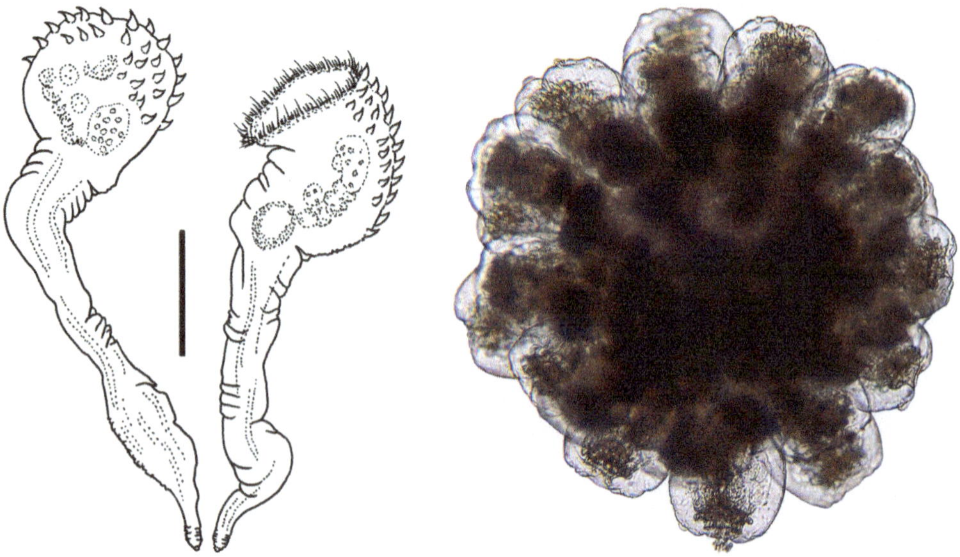

FIGURE 8.23 Flosculariidae, *Sinantherina spinosa*: contracted (Argentina), (left), scale bar = 100 μm; colony of *S. semibullata* (right) (~1400 μm; Argentina). After *José de Paggi (1982).*

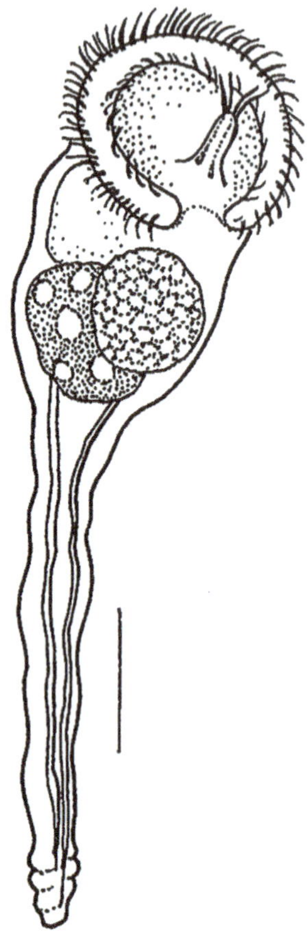

FIGURE 8.24 Conochilidae, *Conochilus* (*Conochilus*) *unicornis* (Argentina); scale bar = 50 μm. *After José de Paggi (1995).*

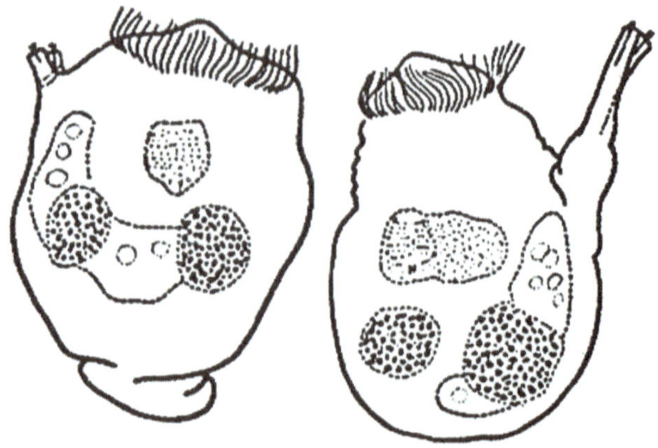

FIGURE 8.25 Conochilidae, *Conochilus (Conochiloides) natans* (left) and *C. coenobasis* (right); contracted specimens (~160 μm; Argentina). *After Paggi & José de Paggi (1973).*

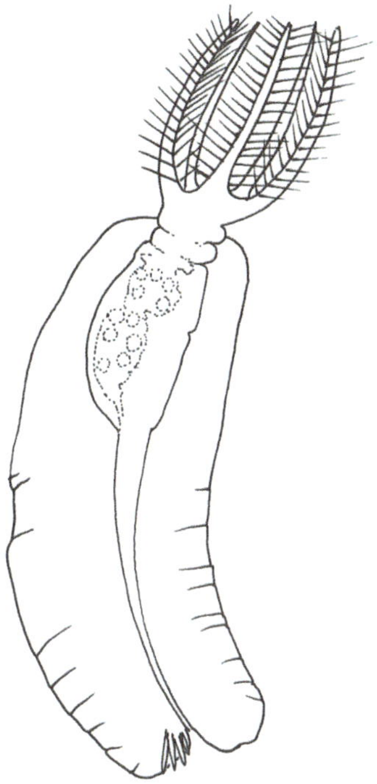

FIGURE 8.26 Collothecidae, *Stephanoceros* (~1000 μm). *After Weber (1898).*

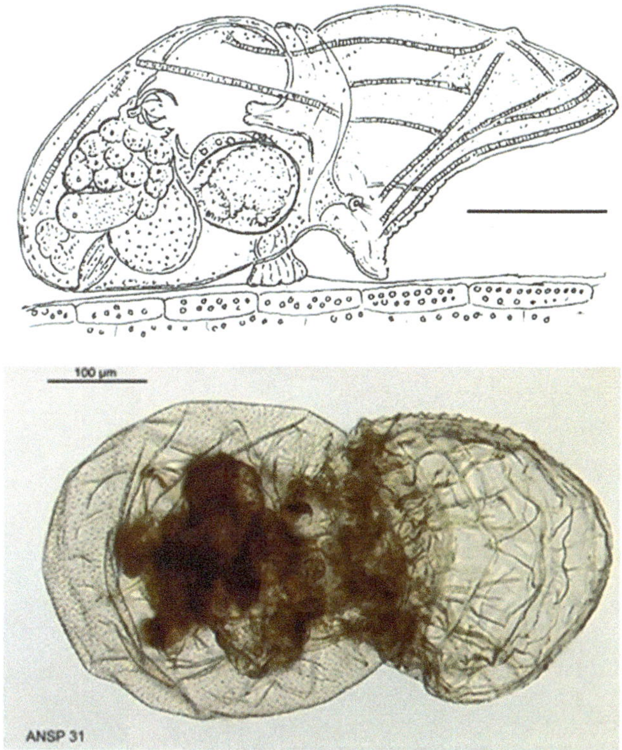

FIGURE 8.28 Atrochidae, *Cupelopagis vorax* (upper; Jamaica), scale bar = 200 μm; *C. vorax* (lower); *photomicrograph after Koste et al. 1995 and after Jersabek (2005)* fide *Jersabek & Leitner (2013).*

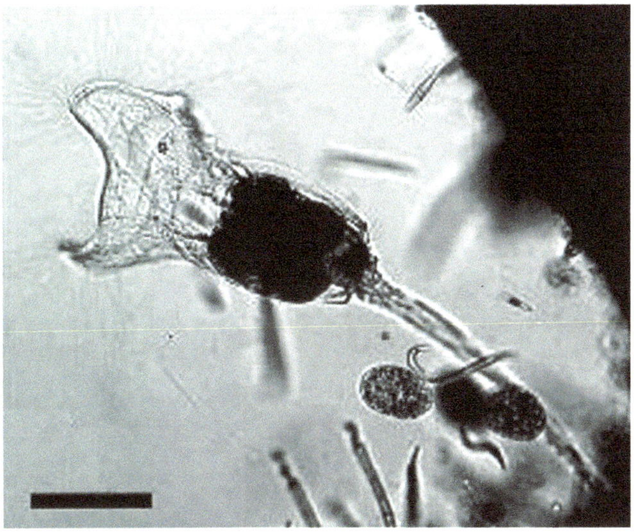

FIGURE 8.27 Collothecidae, *Collotheca ambigua*. Scale bar = 100 μm. *After Meksuwan et al. (2011).*

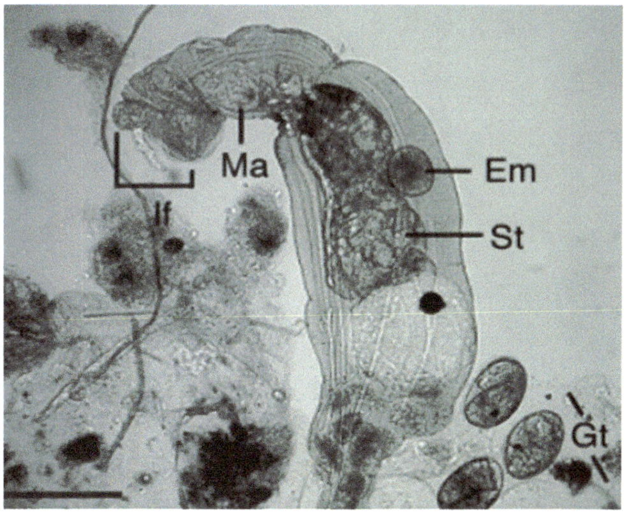

FIGURE 8.29 Atrochidae: *Acyclus inquietus*, lateral view of adult within a colony of *Sinantherina socialis*. Abbreviations: *Em*, embryo; *Gt*, gelatinous tube; *If*, infundibulum; *Ma*, mastax; *St*, stomach; scale bar = 50 μm. *After Hochberg et al. (2010).*

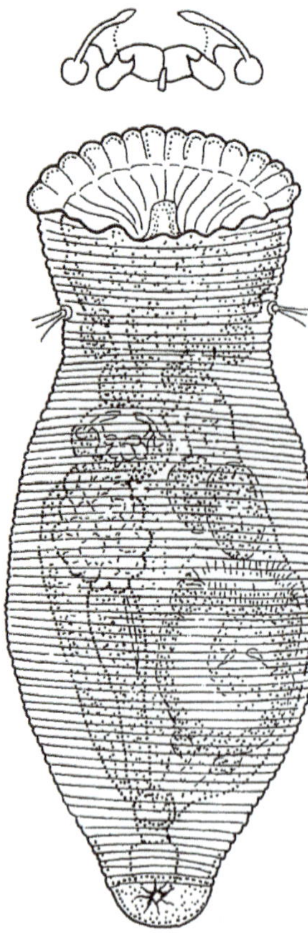

FIGURE 8.30 Atrochidae, *Atrochus tentaculatus*, contracted (~655 μm; Brazil) and trophi. *After Koste et al. (1984).*

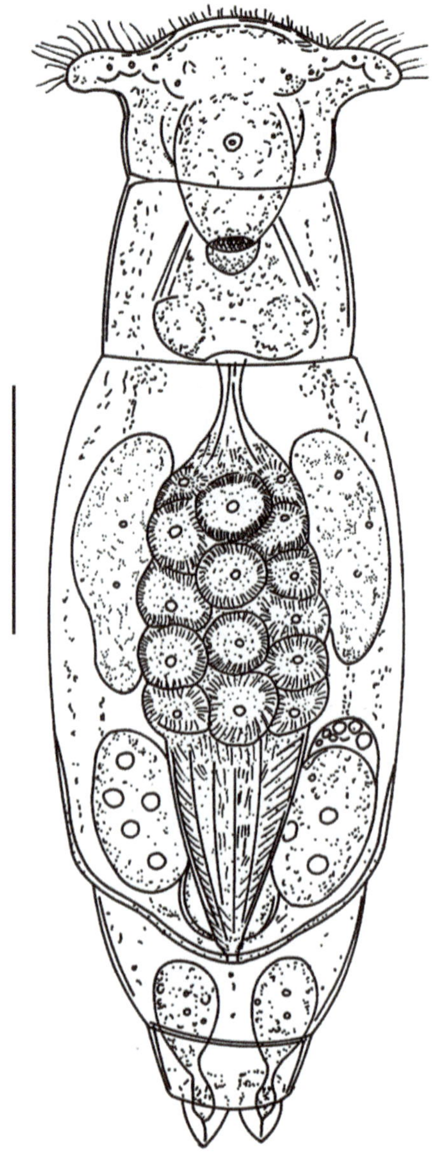

FIGURE 8.31 Lindiidae, *Lindia fulva*, dorsal view (Brazil). Scale bar = 100 μm. *After Koste (2000).*

8′	Corona with a membranous cap or semicircular shield (retractable in *Lepadella* genus); no lateral sulci; ventral sulcus (aperture) or ventral notch in which the foot lies; foot short; toes slender, sharply pointed .... **Lepadellidae [p. 173]**
9(8)	Lorica with dorsal and ventral plates connected laterally by sulci 10
9′	Dorsal and ventral plate fused lorica with a dorsal, medial sulcus (double or single keel); foot with two toes .... **Mytilinidae [p. 179]**
10(9)	Lorica not as below 11
10′	Except in the genus *Beauchampiella*, lorica with dorsal and ventral plates connected laterally by deep sulci; foot and toes projecting between dorsal and ventral plates at the posterior end of the lorica .... **Euchlanidae [p. 179]**
11(10)	Lorica with dorsal and ventral plates connected laterally and caudally by a soft membrane (sulci); foot short, projecting through a foramen (hole) in the ventral plate at the posterior end of the lorica; foot with two relatively long toes, separate or fused (totally or incompletely) in one toe (Figs. 8.32–35) .... Lecanidae; one genus: *Lecane*
	[Ninety-five Neotropical, two Antarctic of ~ 200 species] There are species endemic to the Neotropical region.

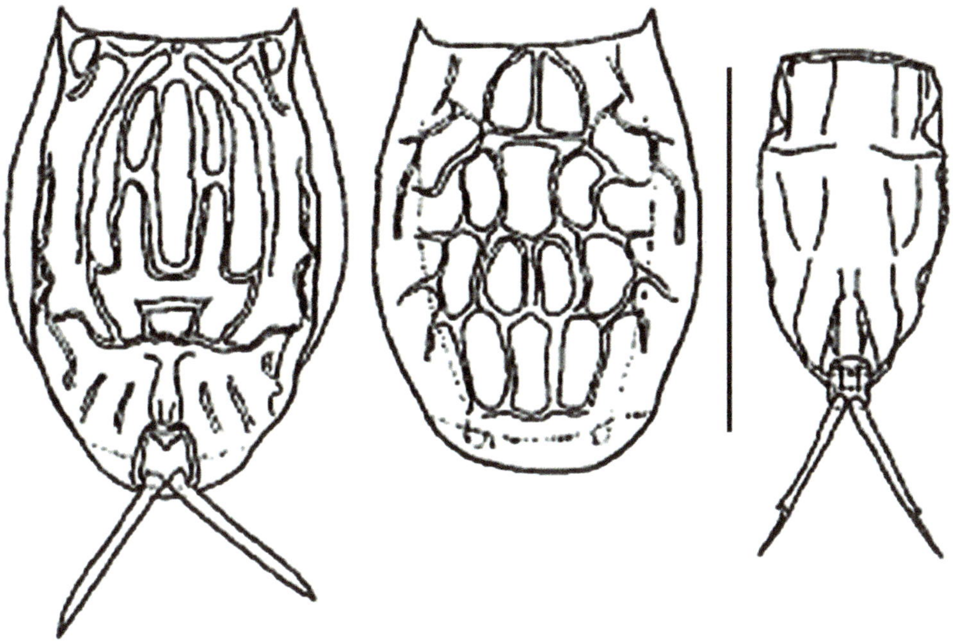

FIGURE 8.32 Lecanidae: *Lecane signifera,* ventral and dorsal views (left); *L. elegans,* ventral view (right) (Argentina); scale bar = 100 μm. *After Martinez & José de Paggi (1988).*

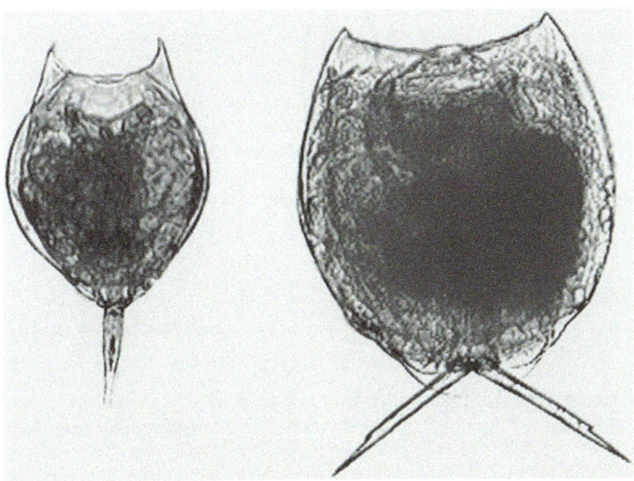

FIGURE 8.33 Lecanidae: *Lecane* sp. (left, ~100 μm); *L. ungulata* (right, ~300 μm; Argentina).

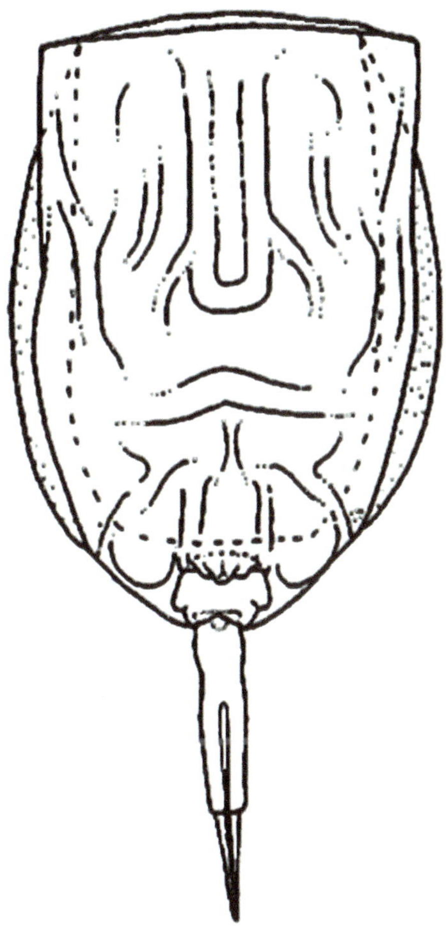

FIGURE 8.34 Lecanidae, *Lecane braziliensis*, ventral view (~110 μm; Brazil); recorded only from Neotropics. *After Segers & Dumont (1995).*

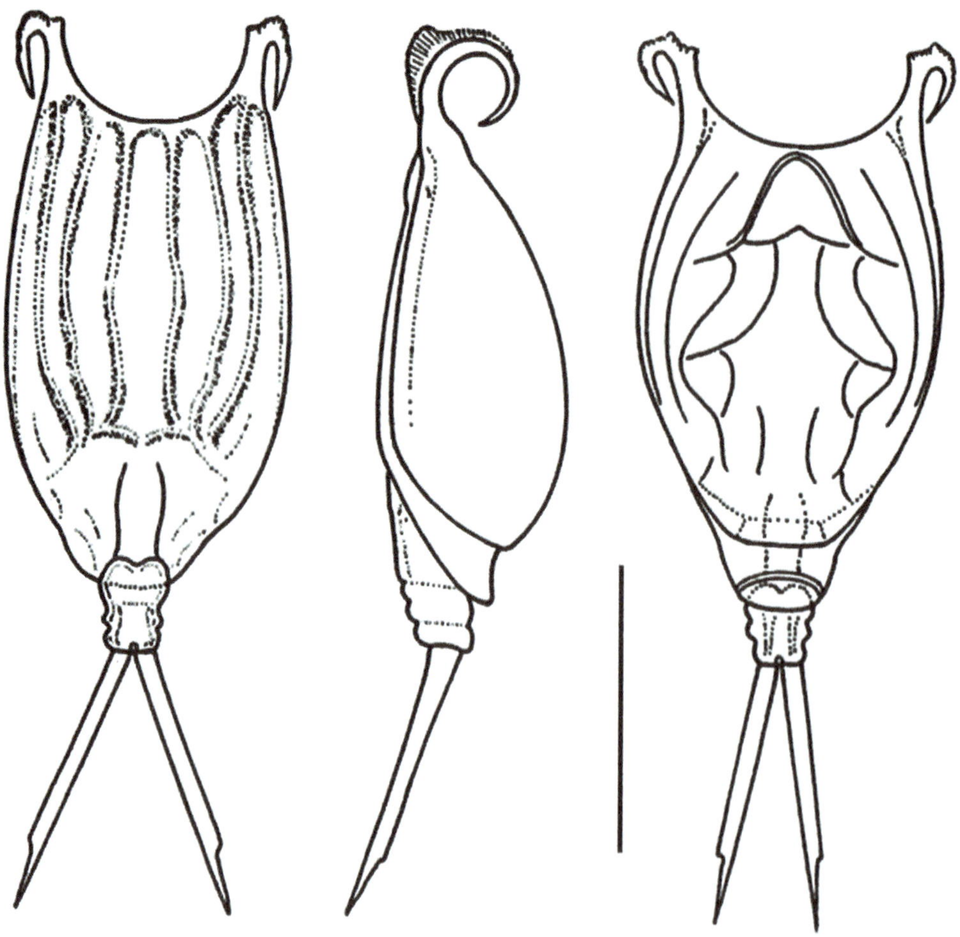

FIGURE 8.35 Lecanidae, *Lecane boettgeri,* ventral, lateral, and dorsal view (Paraguay); scale bar = 100 μm. *After Koste (1986).*

11′	Lorica with dorsal anterior margin with spines; foot present (jointed or ringed) with short toes, or foot absent .... **Brachionidae [p. 182]**
12(4)	Gut without symbiotic zoochlorellae in the stomach wall 13
12′	Gut with symbiotic zoochlorellae in the stomach wall; trophi virgate (Fig. 8.2); body illoricate, spindle-shaped; foot with two short pointed toes; < 400 μm (Fig. 8.36) .... Ituridae; one genus: *Itura* [Four Neotropical of six species]
13(12)	Body not fusiform 14
13′	Body illoricate, fusiform, foot with two pointed toes; knobbed, lateral antennae located near the base of the foot and with long setae, virgate trophi highly modified; in acid waters (pH < 6); ≤ 1000 μm (Fig. 8.37) .... Tetrasiphonidae; monospecific: *Tetrasiphon hydrocora* (Ehrenberg, 1840)
14(13)	Body without bristles or conspicuous lateral auricles 15
14′	Body reniform, loricate with head shield; OR cylindrical or cubical, illoricate, with four groups of appendages (bristles); OR body conical to vasiform with corona with ciliated auricles; foot present or absent .... **Synchaetidae [p. 187]**
15(14)	Characters not as below 16
15′	Lorica stiff, soft; trophi asymmetric; body ovate; foot and toes rudimentary; or cylindrical, often twisted and asymmetrical; foot short, toes setaform, often long .... **Trichocercidae [p. 187]**
16(15)	Characters not as below 17
16′	Body elongate, loricate; foot with three pseudosegments, first short and second and mainly third elongate; two long equal toes; foot and toes longer than rest of body; unci tips pointing outward (Fig. 8.38) .... Scaridiidae; one genus: *Scaridium*

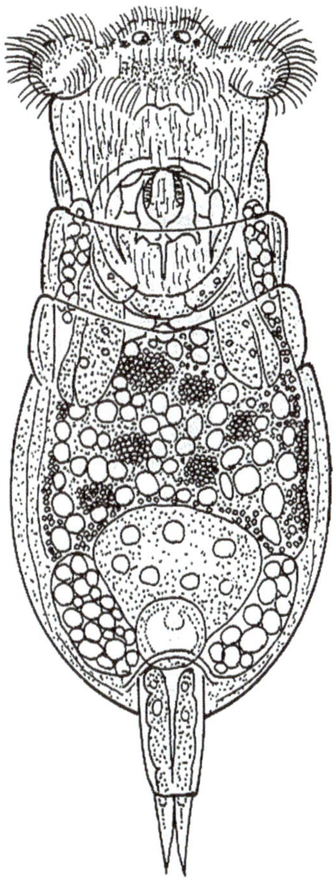

FIGURE 8.36 Ituridae, *Itura chamadis,* ventral view (~330 µm; Paraguay). *After Koste (1986).*

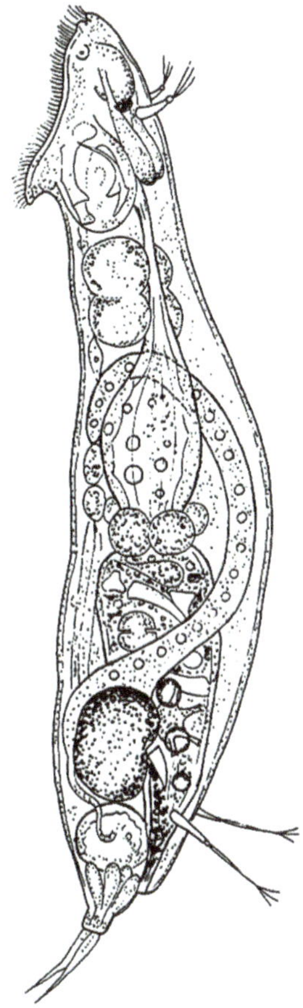

FIGURE 8.37 Tetrasiphonidae, *Tetrasiphon hydrocoral,* lateral view (~450 µm). *After Koste (1978).*

[Four Neotropical, one Antarctic of seven species]

17(16) Body not as below 18

17′ Body conical, illoricate; foot long, with three pseudosegments and single lanceolate toe; 100–200 µm (Fig. 8.39) .... Microcodidae, one genus: *Microcodon*; monospecific: *Microcodon clavus* (Ehrenberg, 1830)

18(17) Body small, ovate to sacciform; lorica smooth, usually colored; foot ventrally located or foot absent .... **Gastropodidae [p. 187]**
Consult Nogrady & Segers (2002) for details.

18′ Body mostly fusiform, elongate to short; foot in the distal end of body, often short, with one or two toes, equal or unequal in length .... **Notommatidae [p. 191]**

Rotifera: Monogononta: Ploima: Asplanchnidae: Genera

Consult Nogrady & Segers (2002) for details.

1 With foot 2

1′ Without foot; intestine and anus absent; 200–2000 µm (Fig. 8.40) .... *Asplanchna*
[Seven Neotropical of nine species]. *Asplanchna herrickii* de Guerne, 1888 does possess an inconspicuous (vestigial) pedal gland and rudimentary toes.

2(1) With intestine and anus (Fig. 8.41) .... *Asplanchnopus*
[Three Neotropical of four species]

2′ Without intestine and anus; 400–700 µm (Fig. 8.41) .... *Harringia*
[One Neotropical of two species]

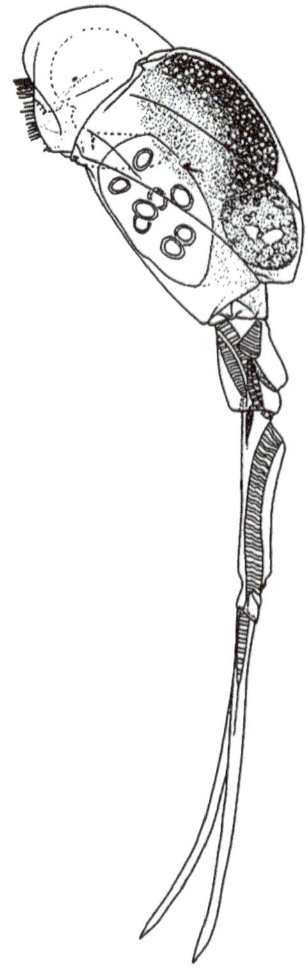

FIGURE 8.38 Scaridiidae, *Scaridium longicaudum*, lateral view (~400 μm). *After Segers (1995b).*

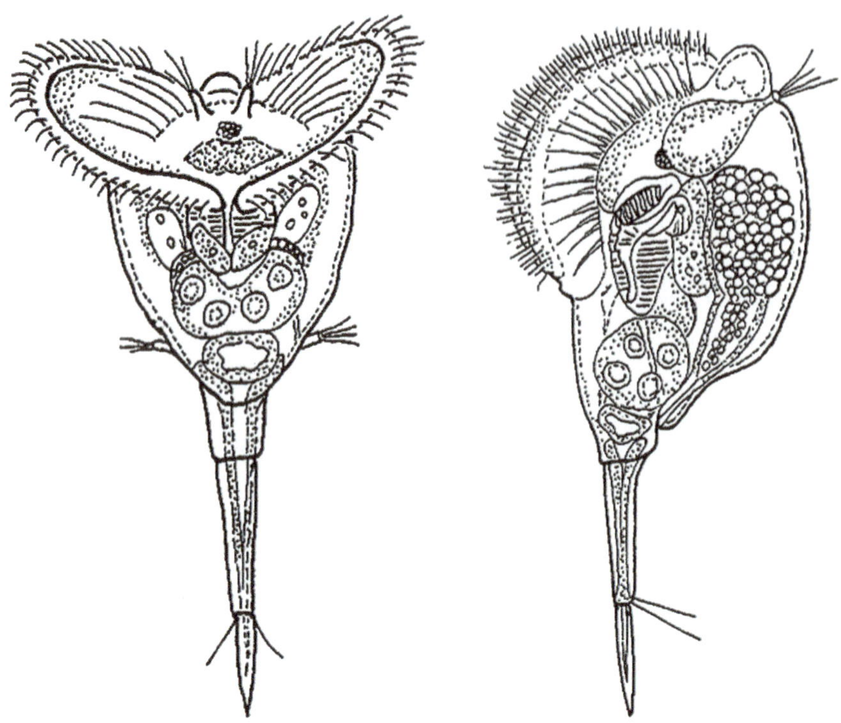

FIGURE 8.39 Microcodidae, *Microcodon clavus*, ventral and lateral views (~200 μm). *After Koste (1978).*

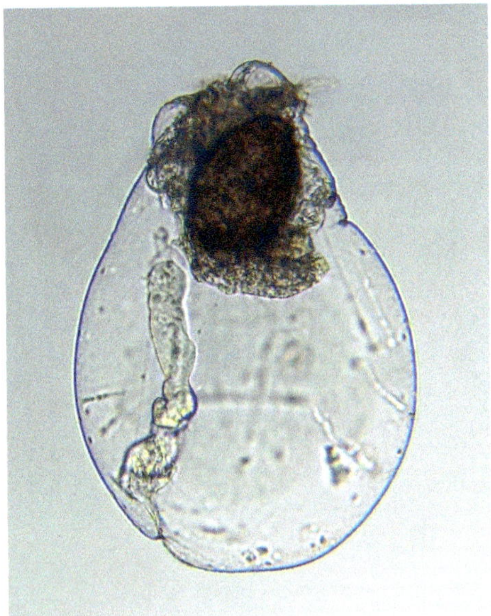

FIGURE 8.40 Asplanchnidae, *Asplanchna brightwelli* (~600 μm; Argentina).

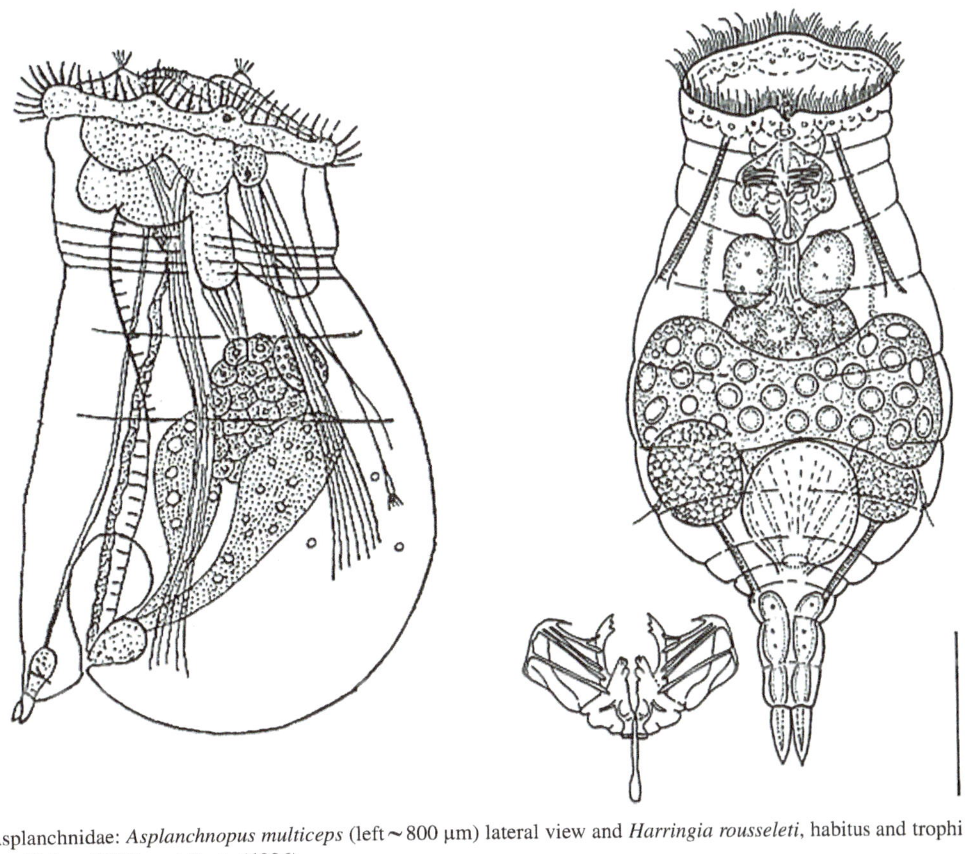

FIGURE 8.41 Asplanchnidae: *Asplanchnopus multiceps* (left ~ 800 μm) lateral view and *Harringia rousseleti*, habitus and trophi (right), scale (right) bar = 100 μm. *After Weber (1898) and Koste (1986).*

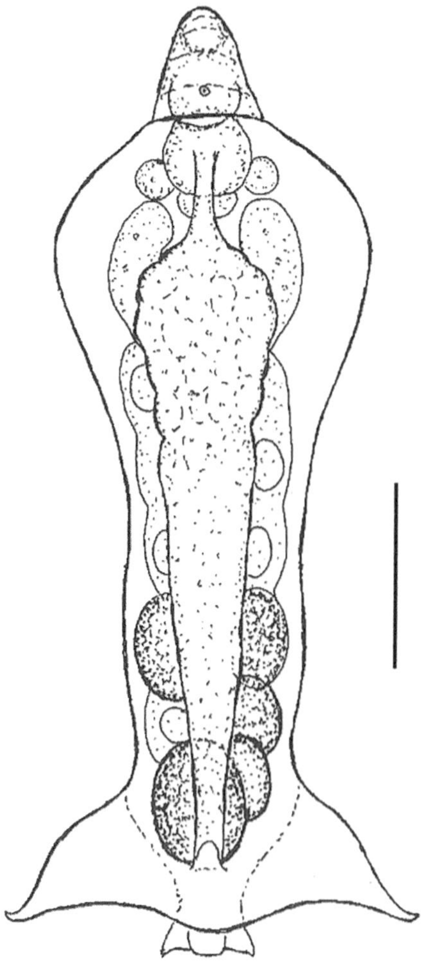

FIGURE 8.42 Dicranophoridae, *Balatro calvus* (Jamaica), dorsal view, scale bar = 50 μm. *After Koste et al. (1995).*

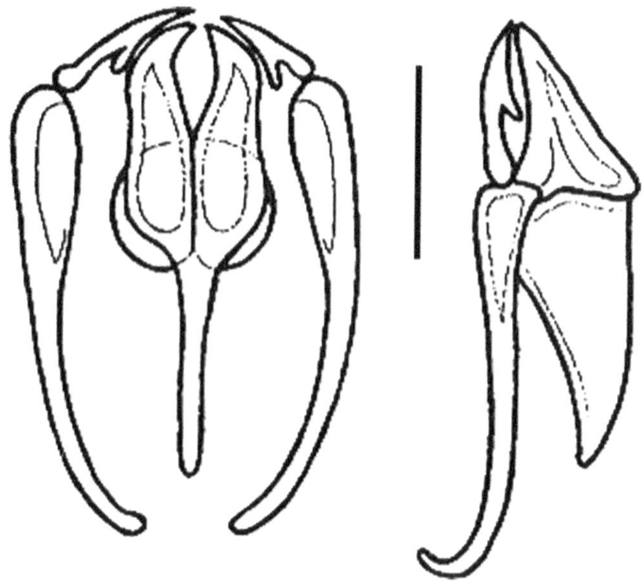

FIGURE 8.43 Dicranophoridae, *Aspelta circinator*, trophi, dorsal and lateral views; scale bar = 10 μm. *After Wulfert (1968).*

Rotifera: Monogononta: Ploima: Dicranophoridae: Genera

The members of this family are mostly soft body, elongate, subcylindrical, or fusiform. They are difficult to identify in preserved samples; therefore, the structure of the trophi is used as the main diagnostic element. Consult De Smet & Pourriot (1997) for details.

1	Corona present	2
1′	Corona absent; head with annulated proboscis; distal end of trunk spatulate; foot posterior end with three lobes, toes absent; endoparasitic in terrestrial Oligochaeta (Fig. 8.42) [One Neotropical of four species]	*Balatro*; one species: *Balatro calvus* Claparède, 1867
2(1)	Trophi symmetrical	3
2′	Trophi asymmetrical (rami and unci), alulae (if present) rounded, triangular, or hamulate; carnivorous; psammic and among algae and mosses (Fig. 8.43) [Three Neotropical of 17 species]	*Aspelta*
3(2)	Corona ventral; incus lyrate	4
3′	Corona oblique; incus lyrate or nonlyrate	5
4(3)	Body covered with mineral particles; outline of rami elliptical, with two teeth on medial margin prior to apex; manubria with lamellae; body: 182−320 μm; trophi: 23−29 μm (Fig. 8.44) [One Antarctic of six species]	*Paradicranophorus*; one species: *Paradicranophorus sordidus* Donner, 1968
4′	Large corona, with two lateral auriform tufts of long cilia; rami lyrate with or without alulae; with short or long teeth on medial margin (Fig. 8.45) [Ten Neotropical of 53 species]	*Dicranophorus*

PHYLUM ROTIFERA

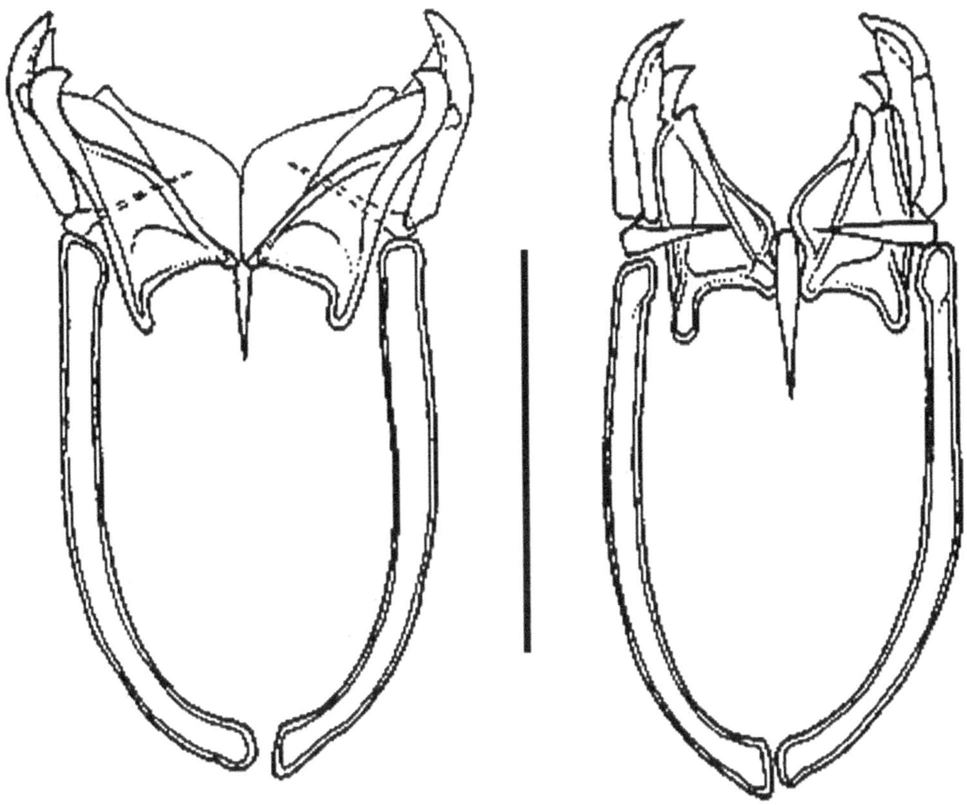

FIGURE 8.44 Dicranophoridae, *Paradicranophorus wockei* trophi: ventral (left) and dorsal views (right) (Bolivia); scale bar = 25 μm. *After Segers et al. (1994b).*

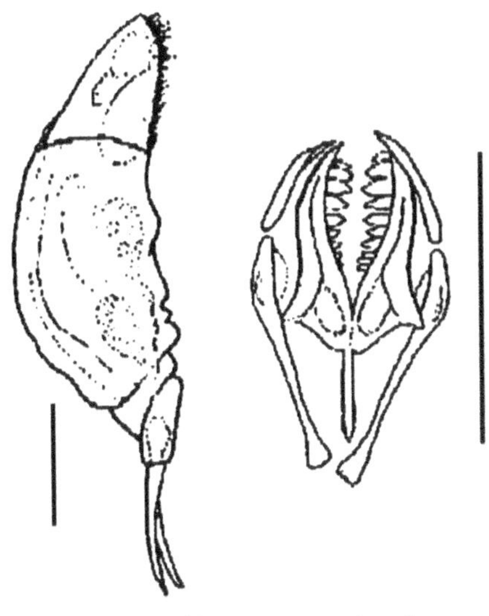

FIGURE 8.45 Dicranophoridae, representative form and trophi of *Dicranophorus* sp. (Argentina); scales bars = 50 μm. *After José de Paggi (2001).*

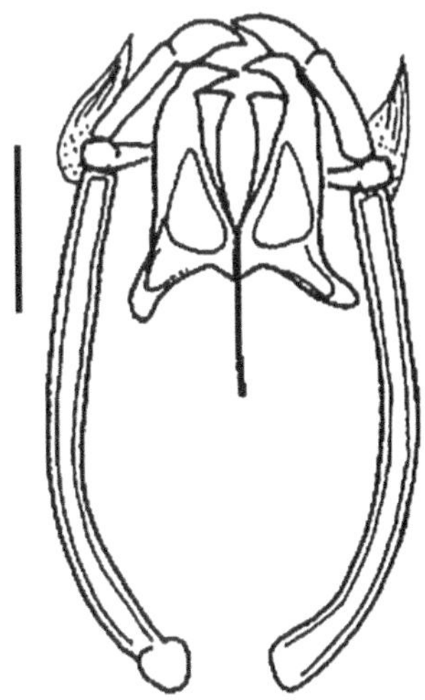

FIGURE 8.46 Dicranophoridae, *Kostea wokei*, trophi, scale bar = 10 μm. *After De Smet (1997).*

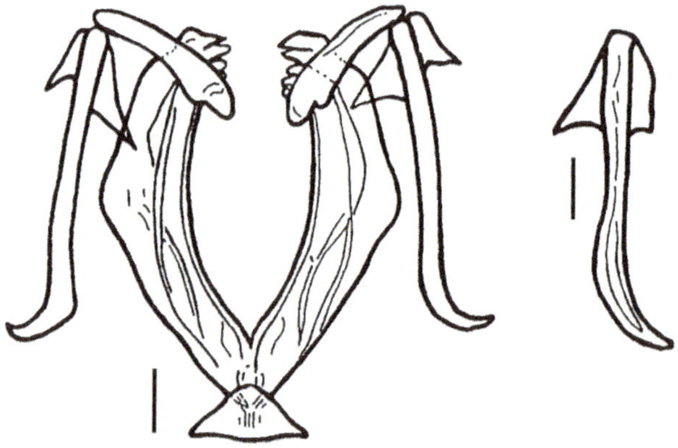

FIGURE 8.47 Dicranophoridae, *Dicranophoroides venezuelensis*, trophi and manubrium, lateral view (Venezuela); scale bars = 10 μm. *After Pourriot & Zoppi de Roa (1981).*

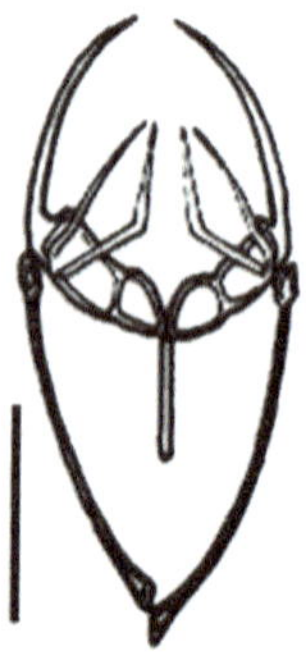

FIGURE 8.48 Dicranophoridae, *Erignatha clastopis*, trophi, ventral view (Bolivia); scale bars = 10 μm. *After Segers et al. (1994b).*

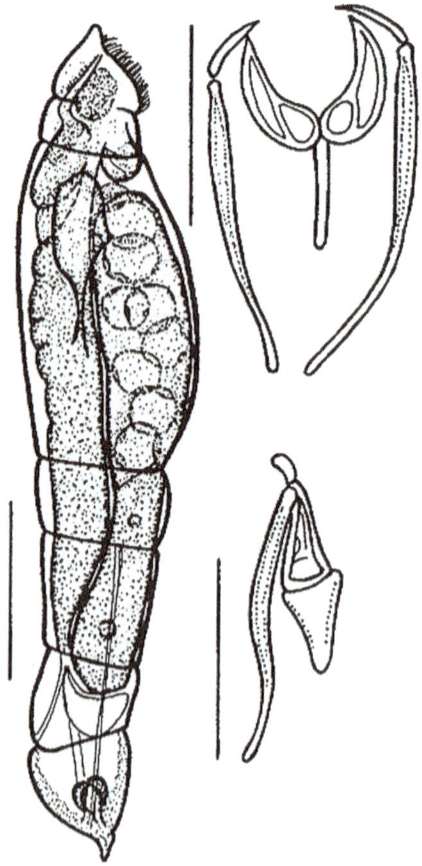

FIGURE 8.49 Dicranophoridae, *Albertia naidis*, lateral view, scale bar = 50 μm; trophi, scale bars = 10 μm. *After Coineau & Kunst (1964).*

5(3) Trunk without lateral protuberances 6

5′ Trunk possessing 3–4 pairs of lateral protuberances (hooks) on trunk; corona oblique, reduced, cilia very short; rami with long posteriolateral alulae intramalleus as a strong spine projecting inwards; ≤ 450 μm (Fig. 8.46), rare .... *Kostea*; monospecific: *Kostea wockei* (Koste, 1961)

6(5) Ramus not as below 7

6′ Rami lyrate, axis of each ramus composed of subbasal chamber only, extended apically, basal chambers situated laterally on external edge of subbasal ones (Fig. 8.47) .... *Dicranophoroides*
[Three Neotropical of four species]

7(6) Ramus not as below 8

7′ Ramus bent at right angles near their midpoint, tapering into acute apex, medial margin without teeth (Fig. 8.48) .... *Erignatha*; one species: *Erignatha clastopis* (Gosse, 1886)
[One Neotropical of six species]

8(7) Ramus medial margin without teeth; manubria without lamellae; foot long, toes papillalike, one or two pseudosegments; endoparasitic in freshwater Oligochaeta (Fig. 8.49) .... *Albertia*; one species: *Albertia naidis* Bousfield, 1886

8′ Ramus medial margin usually with teeth; foot short, conical, toes curved ventrally, ending in pointed tips (Fig. 8.50) .... *Encentrum*
[Three Neotropical, five Antarctic of ~100 species]

Rotifera: Monogononta: Ploima: Epiphanidae: Genera

1 Proboscis lacking 2

1′ Possessing a broad proboscis with two large pigmented eyes; corona as a single band of cilia with a dorsal gap; body fusiform or cylindrical, transparent; foot short with small toes (Fig. 8.51) .... *Rhinoglena*
[Two Antarctic of four species]

2(1) Body sacciform or vermiform, no spur on foot ... 3

2′ Body subquadrate in dorsal view; foot conical, segmented, with a large dorsal spur on the last foot joint (Fig. 8.52) ... *Mikrocodides*
[Two Neotropical of two species]

3(2) Body sacciform or conical ... 4

3′ Body vermiform, cylindrical, with transverse folds; foot rudimentary with retractile toes and pointed posterior end with short cylindrical structures for egg attachment (Fig. 8.53) ... *Proalides*
[Two Neotropical of two species]

4(3) Body sacciform; foot annulated or cylindrical, pseudosegmented, toes short (Fig. 8.54) ... *Epiphanes*
[Five Neotropical, one Antarctic of six species]

4′ Body conical, arched dorsally, in lateral view sigmoid, tapering to a foot with long toes (Fig. 8.54) ... *Cyrtonia*; monospecific: *Cyrtonia tuba* (Ehrenberg, 1834)

Rotifera: Monogononta: Ploima: Proalidae: Genera

A taxonomically unsatisfactory assemblage of diverse taxa (Fontaneto & De Smet, 2015). Consult De Smet (1996) for details.

1 Corona without cirri; head without styli; rostrum lacking or weakly developed ... 2

1′ Corona with cirri; with two long, straight, lateral styli; head with rostrum; two toes curved outwards in dorsal view (Fig. 8.55) *Bryceella*; one species: *Bryceella tenella* (Bryce, 1897)
[One Neotropical of three species]

2(1) Tail with knobbed papilla possessing a long stiff spine; dorsal antennae as knobbed papilla or projection (Fig. 8.56) ... *Proalinopsis*; one species: *Proalinopsis staurus* (Harring & Myers, 1924)
[One Neotropical of six species]

2′ Tail without spine; dorsal antennae never on projection (Fig. 8.57) ... *Proales*
[Five Neotropical of ~35 species]

Rotifera: Monogononta: Ploima: Trichotriidae: Genera

1 Trunk lorica not as below; foot retractable or partly retractable, visible in fixed animals ... 2

1′ Trunk lorica thecaform, longer than wide; minute spines over the lateral antennae, but no marginal spines or spinules; first foot joint dorsally without spines; head and foot entirely retractable, not visible in fixed animals; ~ 300 μm (Fig. 8.58) ... *Wolga*; monospecific: *Wolga spinifera* (Western, 1894)

2(1) Characters not as below ... 3

2′ Trunk lorica ventrally relatively flat; dorsal lorica pustulate with a forked keel; lorica anteriodorsal and anterioventral margins concave; terminal foot segment very large; 200–220 μm (Fig. 8.59) ... *Pulchritia kostei* (José de Paggi, Branco & Kozlowsky-Suzuki, 2000)
[One Neotropical of two species]

3(2) Trunk lorica dorsoventrally compressed, covered with minute spinules, with numerous, large dorsal or caudal spines, except *Macrochaetus aspinus* Segers & Sarma, 1993 (Fig. 8.60); foot with two loricate segments; 90–250 μm ... *Macrochaetus*
[Nine Neotropical of 13 species].

3′ Trunk lorica hexagonal in cross-section, facetted, granulated; foot with three loricate segments; no spines on the trunk; spines on the dorsal side of the first joint; 200–400 μm (Fig. 8.61) ... *Trichotria*
[Four Neotropical of eleven species]

Rotifera: Monogononta: Ploima: Lepadellidae: Genera

1 Lorica dorsoventrally flattened ... 2

1′ Lorica laterally flattened ... 3

2(1) Semicircular retractile dorsal head shield absent; 80–225 μm (Fig. 8.62) ... *Squatinella*
[Two Neotropical of nine species]

2′ Semicircular retractile dorsal semicircular head shield present; foot extends from a deep ventral notch; ~55–200 μm (Figs. 8.63, 64) ... *Lepadella*
[52 Neotropical, seven Antarctic of ~120–130 species]

3(1) Lorica with a longitudinal dorsal sulcus; ventral aperture from which the foot extends; foot + toes > ½ total animal length; terminal foot segment longer than toes; ~100–160 μm (Fig. 8.65) ... *Paracolurella*
[Two Neotropical of two species]

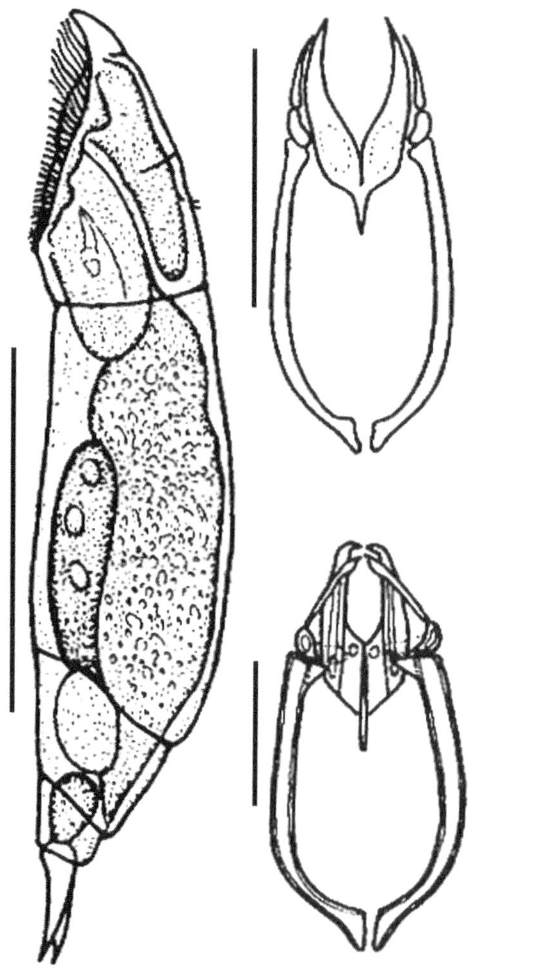

FIGURE 8.50 Dicranophoridae, *Encentrum felis*: lateral view (left), scale bar = 50 μm; trophi, dorsal and ventral views (right), scale bar = 10 μm. *After Harring & Myers (1928) fide De Smet (1997).*
3′

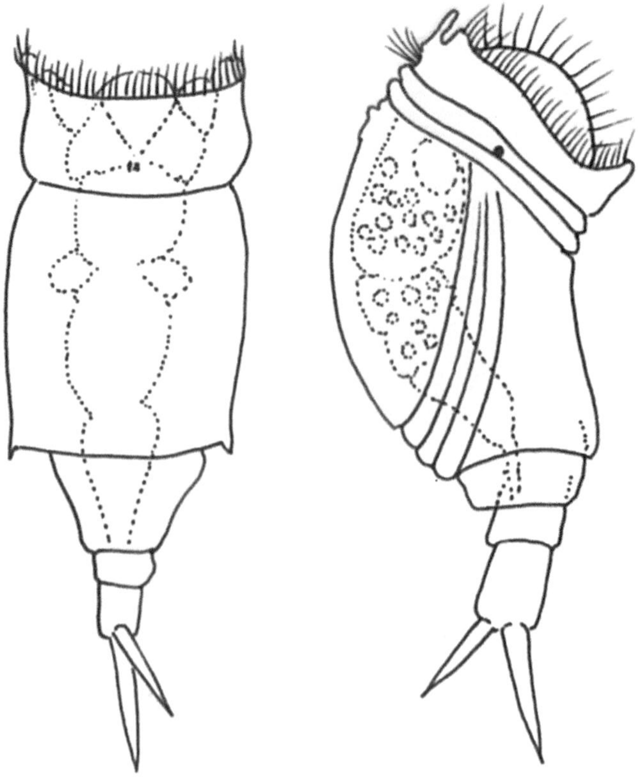

FIGURE 8.52 Epiphanidae, *Mikrocodides chlaena* (~200 μm) dorsal and lateral view. *After Weber (1898).*

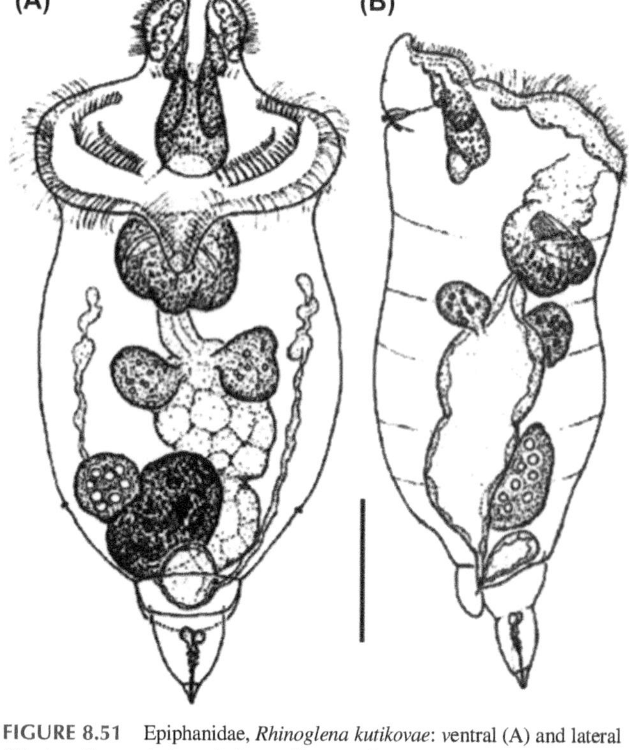

FIGURE 8.51 Epiphanidae, *Rhinoglena kutikovae*: ventral (A) and lateral (B) view (Antarctica), scale bar = 50 μm. *After De Smet & Gibson (2008).*

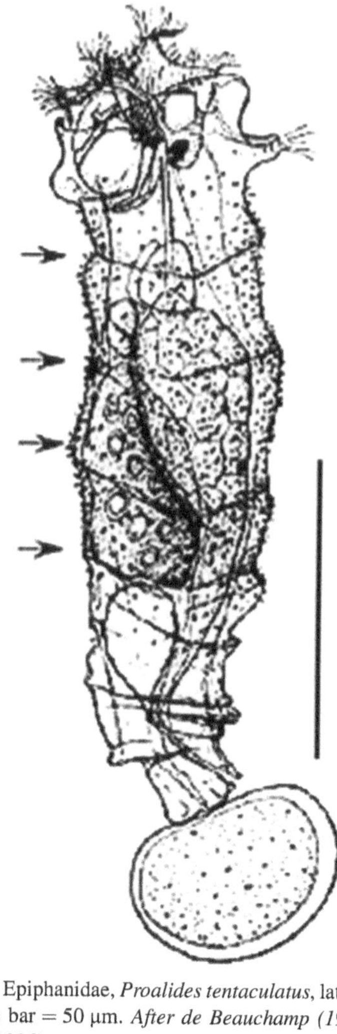

FIGURE 8.53 Epiphanidae, *Proalides tentaculatus*, lateral view: arrows, plications; scale bar = 50 μm. *After de Beauchamp (1907). Modified by Wallace et al. (2016).*

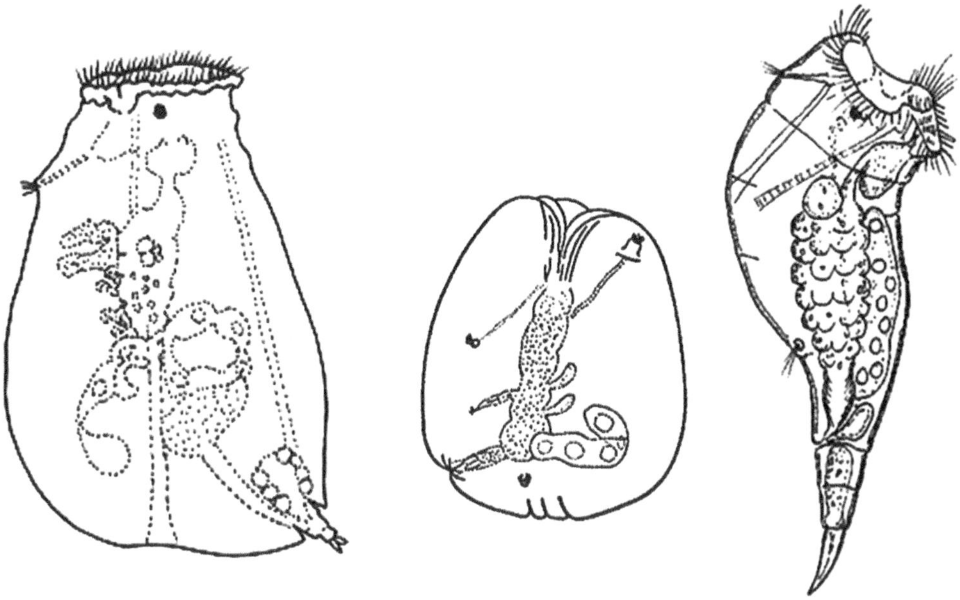

FIGURE 8.54 Epiphanidae: *Epiphanes clavulata*, extended (left, ~500 μm) and contracted (center, ~300 μm; Argentina); *Cyrtonia tuba*, lateral view (right, ~350 μm). *After Hudson & Gosse (1886), Paggi & José de Paggi (1973) and Rousselet (1894) respectively.*

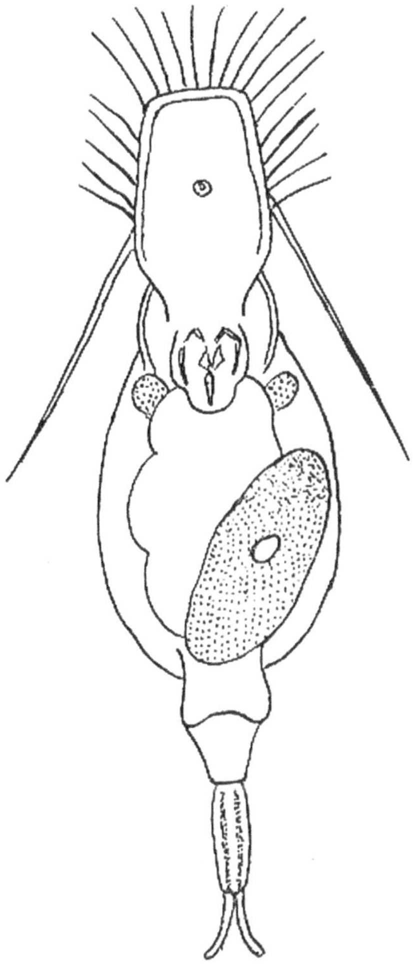

FIGURE 8.55 Proalidae, *Bryceella tenella* dorsal view, (~150 μm, Brazil). *After Murray (1906).*

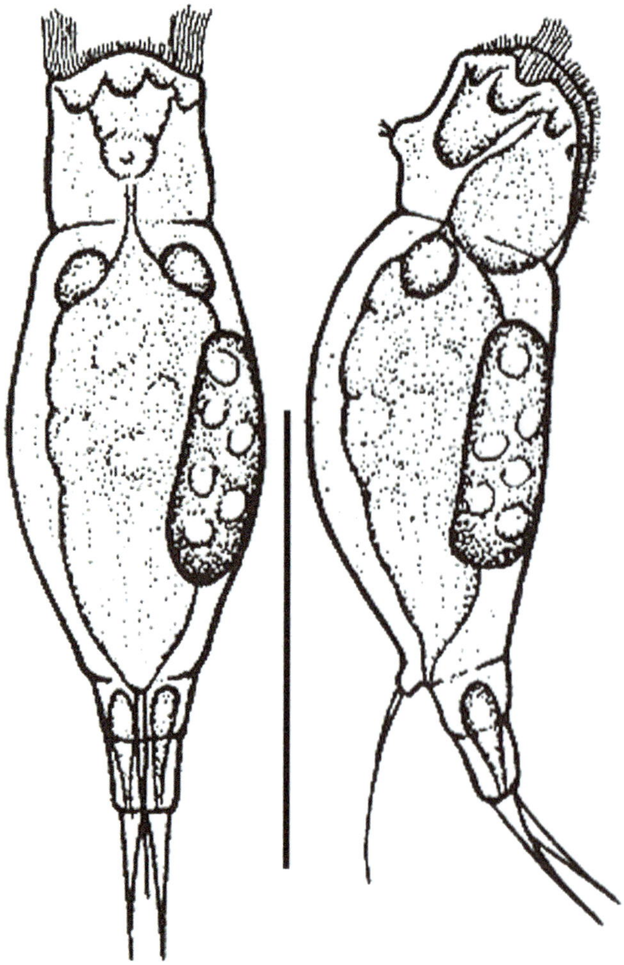

FIGURE 8.56 Proalidae, *Proalinopsis staurus,* dorsal and lateral views; scale bar = 50 μm. *After Harring & Myers (1924).*

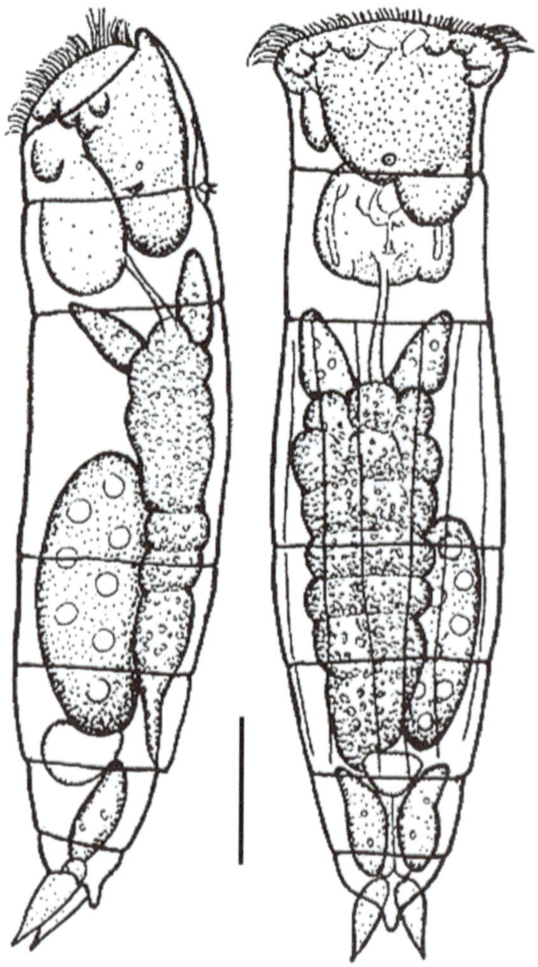

FIGURE 8.57 Proalidae, *Proales fallaciosa*, lateral and dorsal views; scale bar = 50 μm. *After Wulfert (1939).*

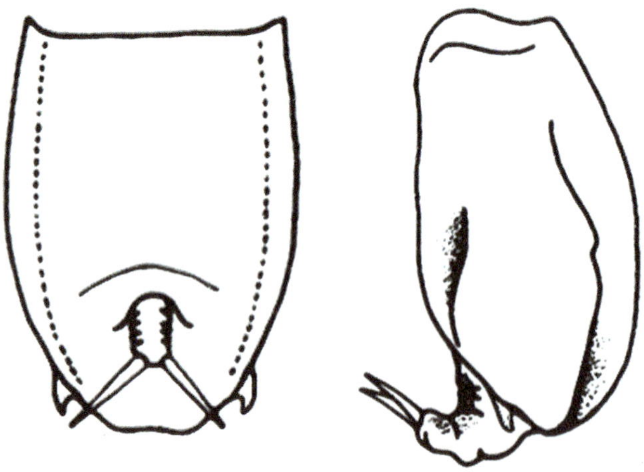

FIGURE 8.58 Trichotriidae, *Wolga spinifera*, ventral and lateral views (~150 μm; Argentina). *After José de Paggi (1982).*

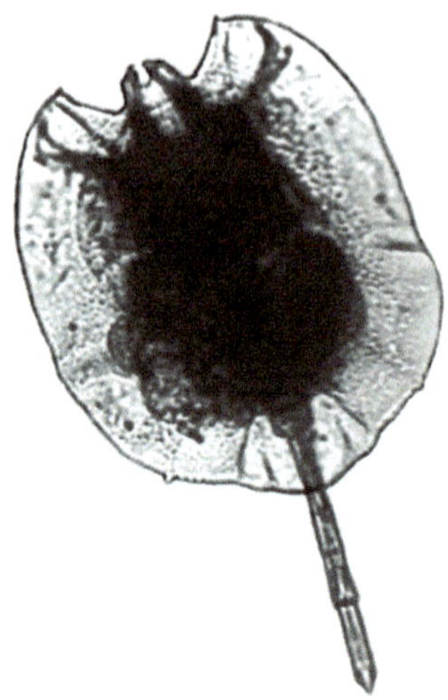

FIGURE 8.59 Trichotriidae, *Pulchritia kostei* (~200 μm; Brazil); photomicrograph courtesy of Christina Castello Branco, Universidad Federal de Rio de Janeiro, Brazil.

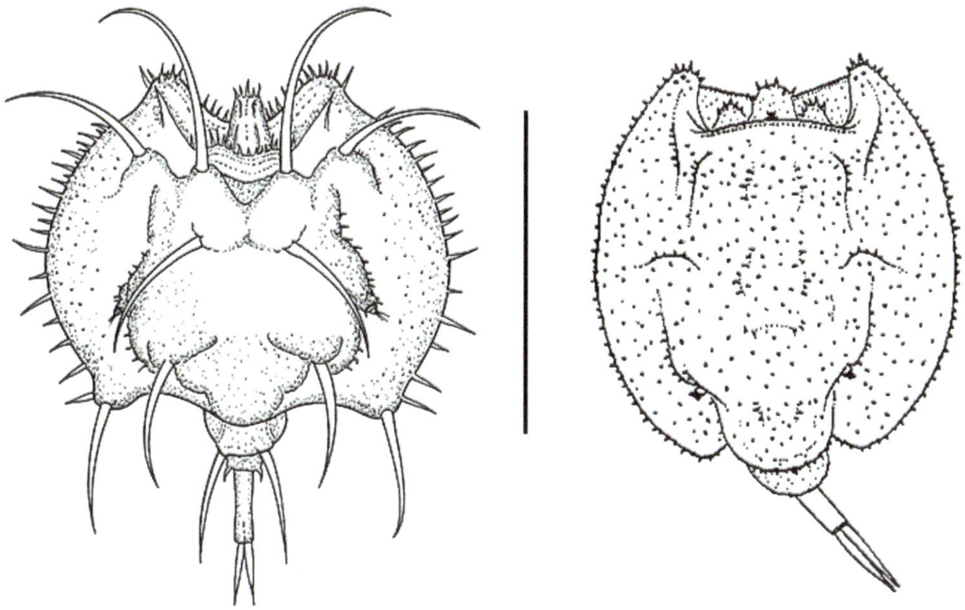

FIGURE 8.60 Trichotriidae: *Macrochaetus multispinosus,* dorsal view (left) (Brazil), scale bar = 50 μm; *M. aspinus,* dorsal view (right) (~72–86 μm; Brazil). *After Koste (2000) and after Segers & Sarma (1993).*

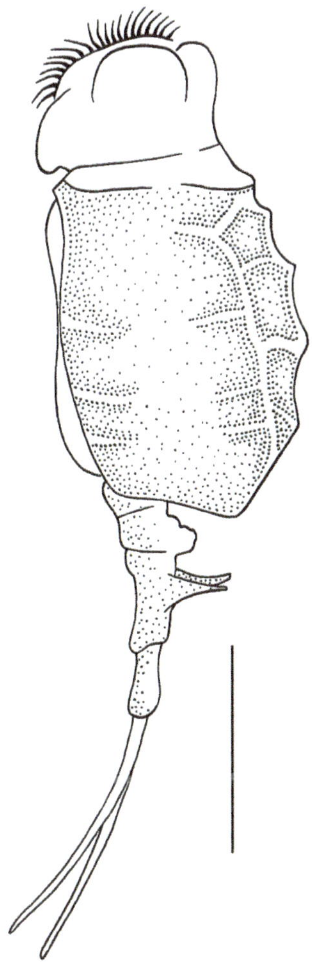

FIGURE 8.61 Trichotriidae, *Trichotria tetractis*, scale bar = 100 μm. *After José de Paggi (1995).*

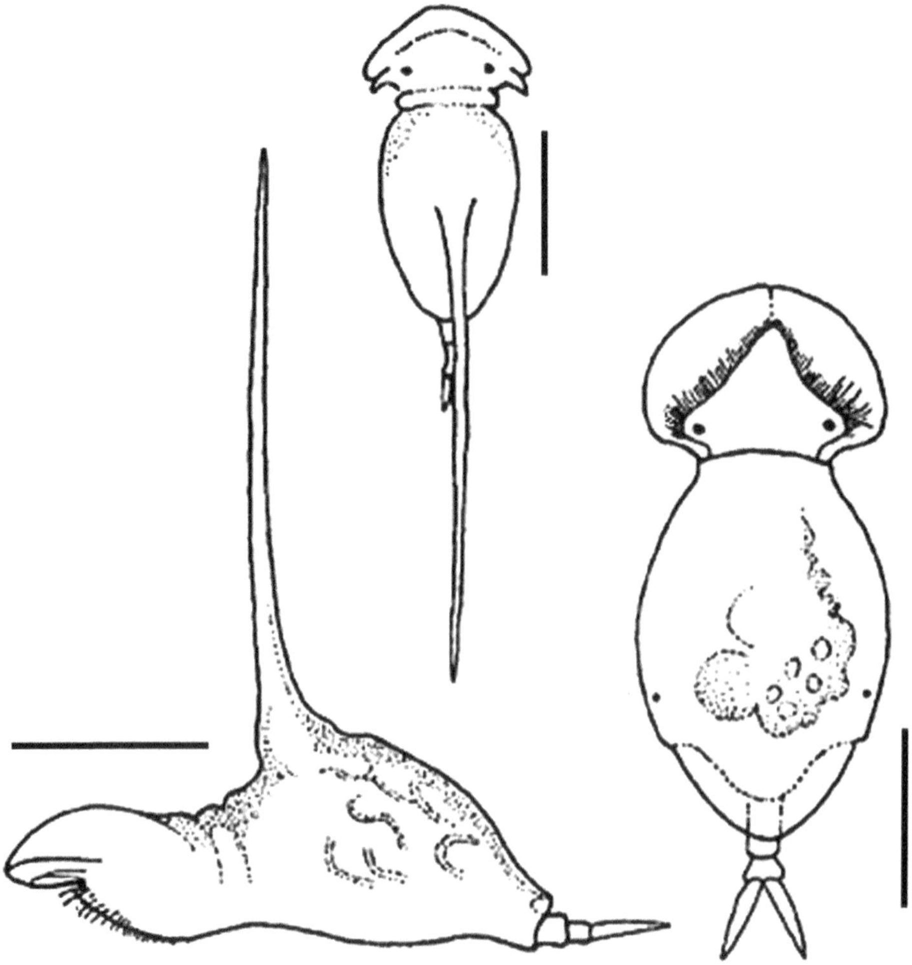

FIGURE 8.62 Lepadellidae: *Squatinella leydigi*, lateral view (left), dorsal view (center); *S. lamellaris*, dorsal view (right) (Argentina), scale bars = 50 μm. *After José de Paggi (1996).*

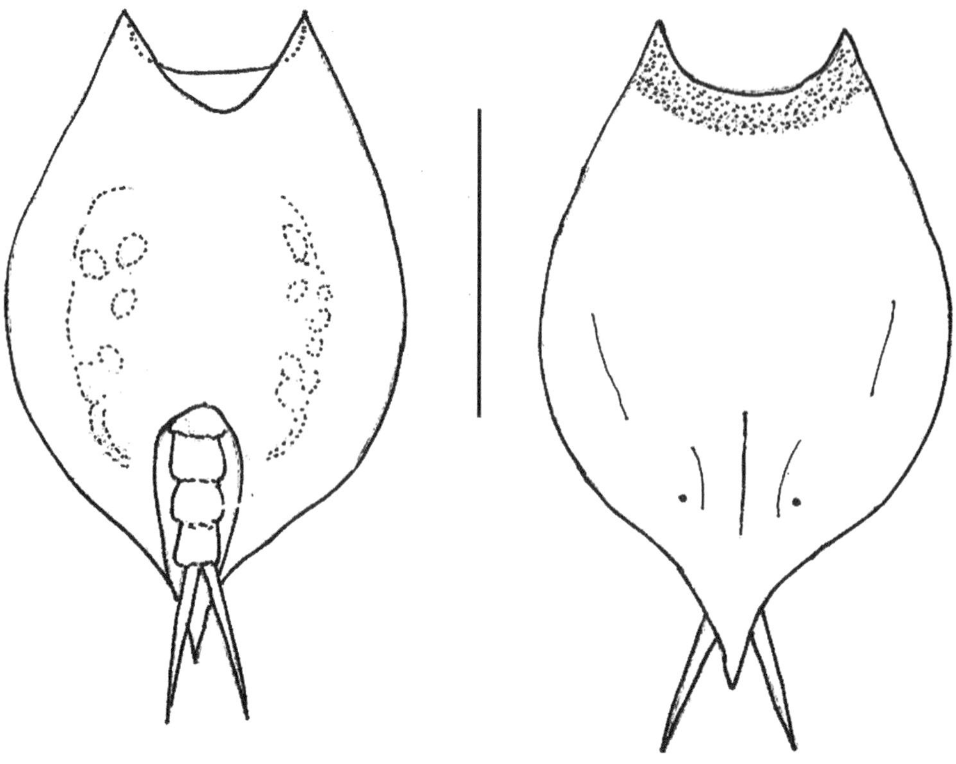

FIGURE 8.63 Lepadellidae: *Lepadella acuminata*, ventral and dorsal views (Argentina). Scale bar = 50 μm. *After José de Paggi (1979).*

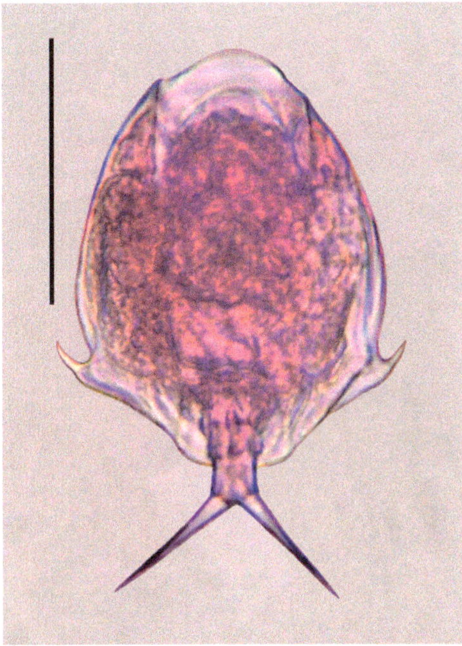

FIGURE 8.64 Lepadellidae, *Lepadella bicornis,* ventral view (Argentina), scale bar = 50 μm.

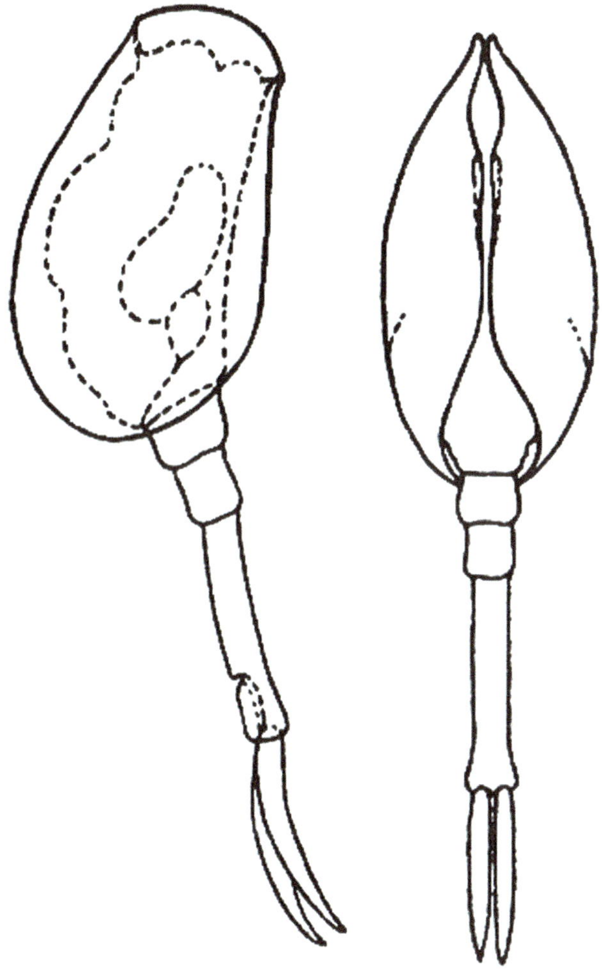

FIGURE 8.65 Lepadellidae, representative form of *Paracolurella* sp., lateral and ventral views (~112–164 μm). *After Koste (1978).*

Body oval or ovoid in lateral view; one piece lorica, without dorsal sulcus; ventral aperture from which the foot extends; foot + toes < ½ total animal length; terminal foot segment not longer than toes; ~50–120 μm (Fig. 8.66) .... *Colurella*
[Fourteen Neotropical, three Antarctic of 23 species]

Rotifera: Monogononta: Ploima: Mytilinidae: Genera

1 Lorica with a dorsal sulcus (in cross-section appears double keeled); sometimes with spines in corner; toes long, thin, mostly ending in pointed tips; 170–380 μm (Fig. 8.67) .... *Mytilina*
[Eleven Neotropical, one Antarctic of ~15 species]

1′ Lorica with a dorsal ridge, ornamented; without dorsal sulcus, one strong keel; toes short; 100–250 μm (Fig. 8.67) .... *Lophocharis*
[Three Neotropical, one Antarctic of 11 species]

Rotifera: Monogononta: Ploima: Euchlanidae: Genera

1 Lorica developed, possessing two plates separated by sulci; toes unequal to body length 2

1′ Lorica thin, body soft, pyriform; foot short with very long toes, equal or longer than the body, superficially resembling *Scaridium*; 420–760 μm (Fig. 8.68) .... *Beauchampiella* monospecific: *Beauchampiella eudactylota* (Gosse, 1886)

2(1) Dorsal plate convex or triangular 3

2′ Dorsal plate concave narrower than ventral plate (*cf. Euchlanis* and *Tripleuchlanis*) toes long ca. 3/4th body length, 230–500 μm (including toes) (Fig. 8.69) .... *Dipleuchlanis*
[Two Neotropical of three species]

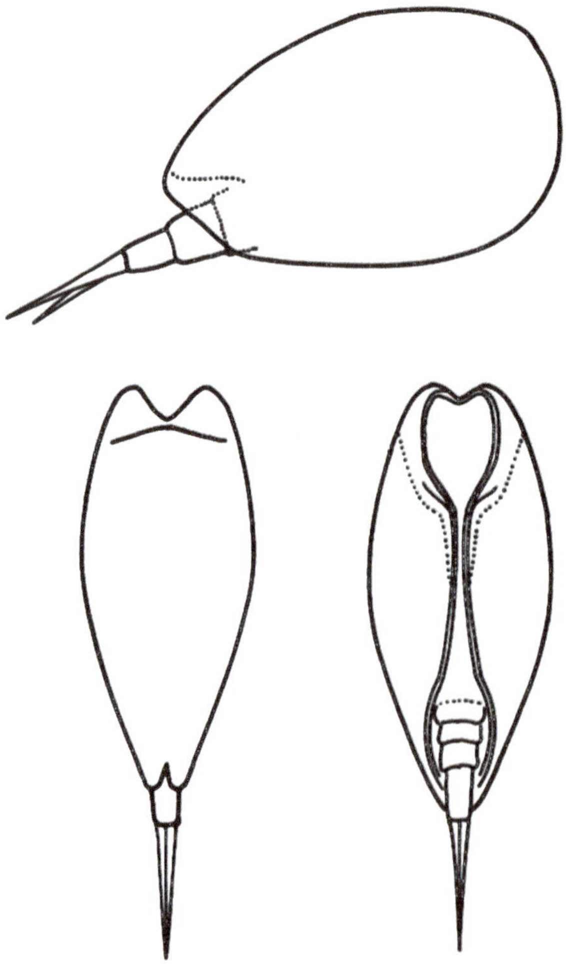

FIGURE 8.66 Lepadellidae, *Colurella hindenburgi*, lateral, dorsal and ventral views (ca. 60 μm; Jamaica). *After Koste et al. (1991).*

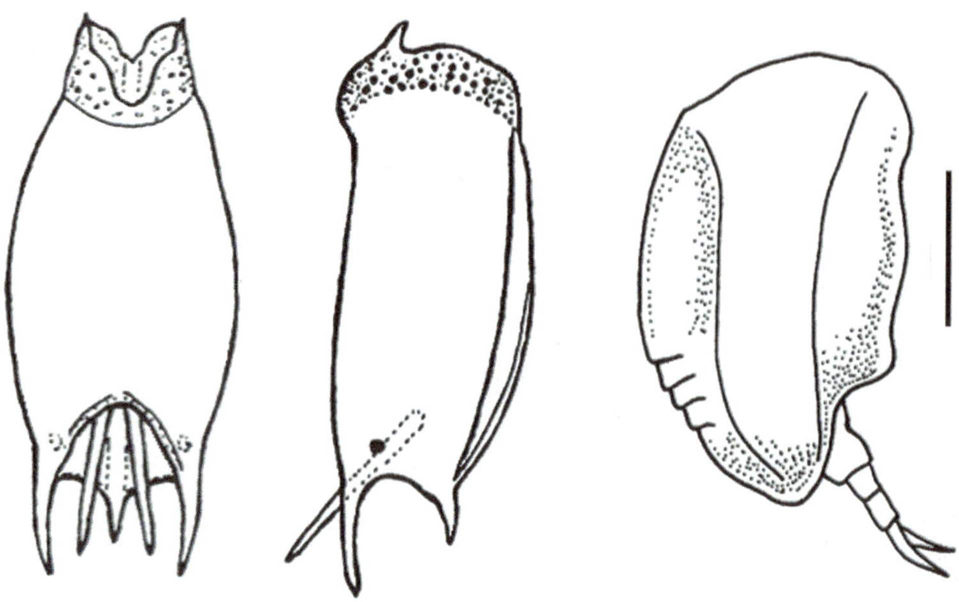

FIGURE 8.67 Mytilinidae. *Mytilina ventralis*, dorsal (left) (380 μm) and lateral (250 μm) views. Representative form of *Lophocharis* sp., lateral view (right; Argentina), scale bar = 100 μm. *After Koste (1974b) and José de Paggi (1995).*

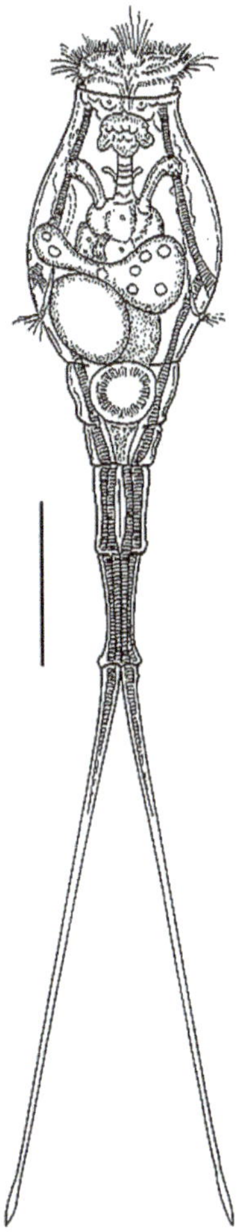

FIGURE 8.68 Euchlanidae, *Beauchampiella eudactylota*, ventral view (Ecuador), scale bar = 100 μm. *After Koste & Böttger (1989).*

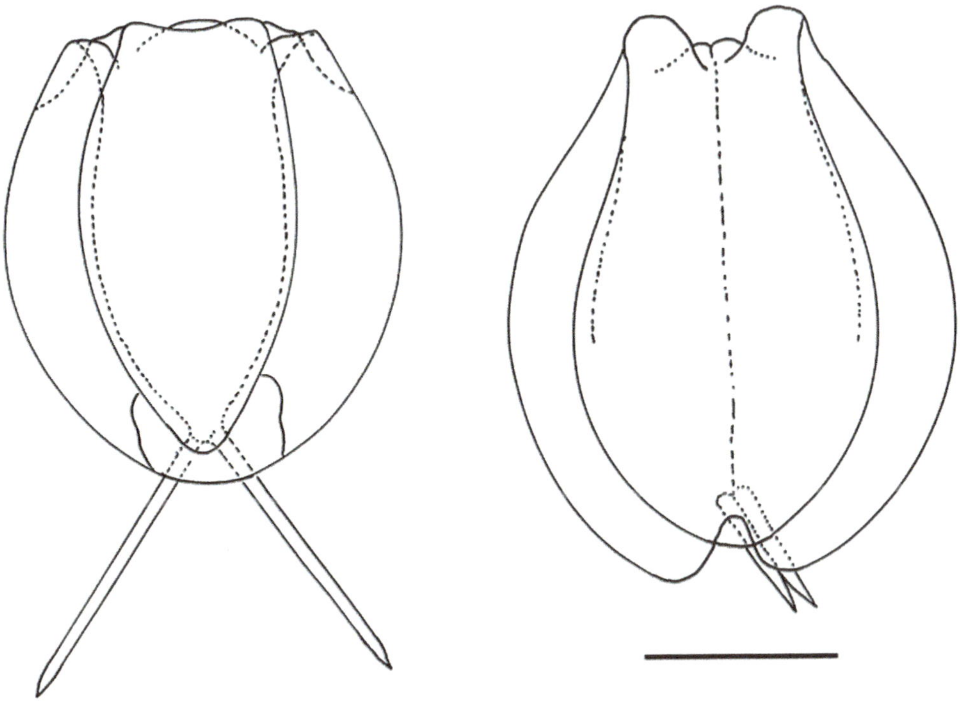

FIGURE 8.69 Euchlanidae: *Dipleuchlanis macrodactyla,* dorsal view on left (Argentina); *Euchlanis*, dorsal view on right; (Argentina). Scale bar = 50 μm.

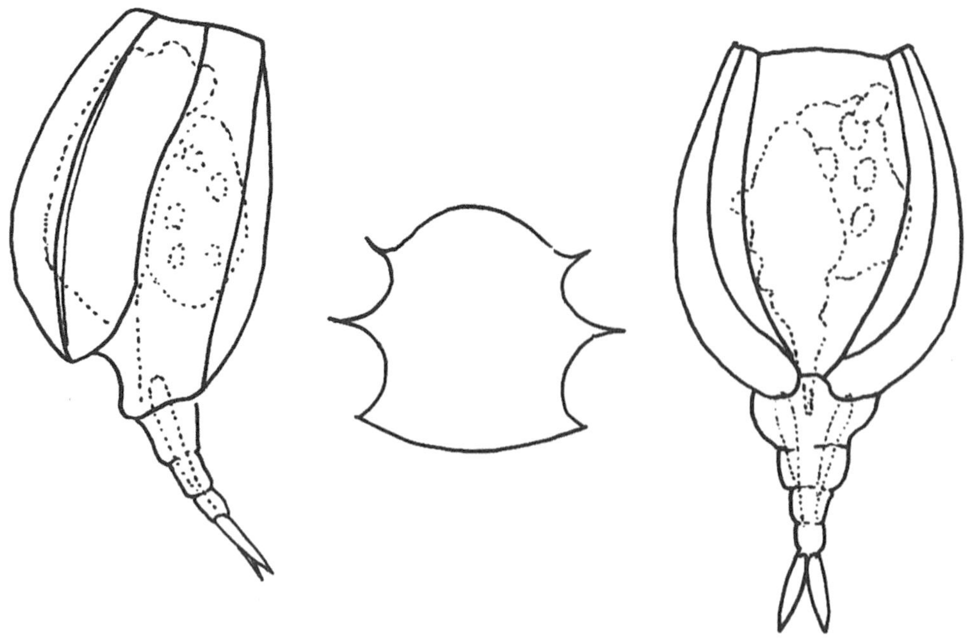

FIGURE 8.70 Euchlanidae, *Tripleuchlanis plicata*, lateral, cross-section of the lorica, and dorsal views (~270 μm). *After Myers (1917).*

3(2) Dorsal plate and ventral plate nearly similar in size, connected by pair of lateral longitudinal membranes with a projection of stiffened cuticle producing an extra fold; 250–270 (including toes) (Fig. 8.70) .... *Tripleuchlanis*; monospecific: *Tripleuchlanis plicata* (Levander, 1894)

3′ Dorsal plate larger than ventral plate, arched or triangular in cross-section by the presence of a median keel or lateral wings; ventral plate flat; foot short, toes with sharp tips; ca. 120–400 μm (Fig. 8.69) .... *Euchlanis*
[Ten Neotropical, one Antarctic of 22 species]

Rotifera: Monogononta: Ploima: Brachionidae: Genera

1 Foot present 2

1′ Foot absent 4

2(1) Foot jointed, partially retractile 3

2′ Foot ringed, tubular, fully retractile, often not visible in fixed animals; ca. 100–400 μm (Figs. 8.71–73) .... *Brachionus*
[Thirty-four Neotropical, six Antarctic of ~75 species]

3(2) Body in dorsal view more or less rectangular, ~200 μm (Fig. 8.74) .... *Plationus*
[Two Neotropical of four species]

3′ Body in dorsal view more or less ovoid, ~165–350 μm (Fig. 8.75) .... *Platyias*
[Three Neotropical of five species]

4(1) Dorsal plate without facets 5

4′ Dorsal plate usually with facets, six anterodorsal spines ~150–450 μm (Figs. 8.76–79) (Lorica smooth, without facets only in *Keratella kostei* (Fig. 8.78); endemic to the Southern zone of the Neotropical region and Antarctica) .... *Keratella*
[Twenty-one Neotropical, five Antarctic of 50 species]

5(4) Lorica dorsal plate anterior margin with spines 6

5′ Lorica dorsal plate anterior margin without spines; with fairly thin lorica; dorsal plate arched, ventral plate nearly flat; ~100 μm (Fig. 8.80) .... *Anuraeopsis*
[Six Neotropical of eleven species]

6(5) Anterior spines asymmetrical (unequal in length), and long caudal spine; ca. 100 μm (excluding spines) (Fig. 8.81) .... *Kellicottia*
[Two Neotropical of two species]

6′ Anterior spines symmetrical; lorica smooth or with longitudinal stripes; ≤200 μm. (Figs. 8.82, 83) .... *Notholca*
[Eight Neotropical, five Antarctic of 44 species]

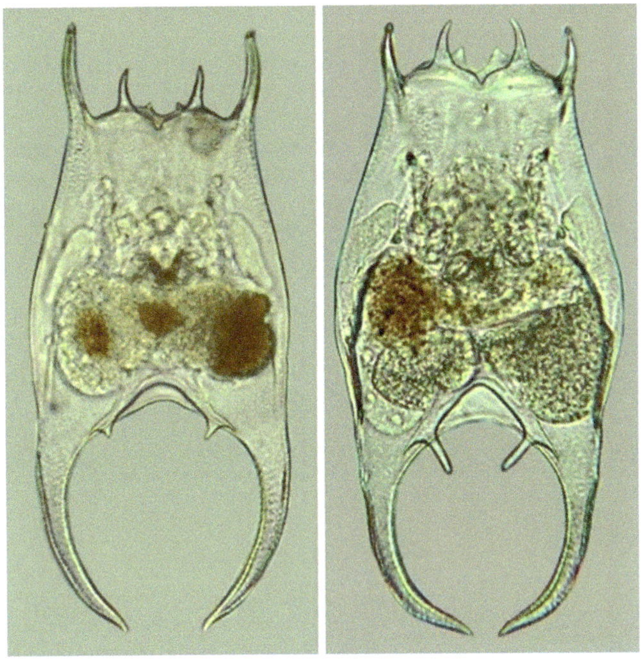

FIGURE 8.71 Brachionidae: *Brachionus angustus*, ventral view (left, ~240 μm); *B.* cf. *laticaudatus*, ventral view (right, ~200 μm; group *B. mirus*, photomicrographs courtesy of Ciro Joko, University of Maringá, Brazil. Comments: Taxonomic status of the *B. mirus*-like forms are not clear and should be revised. Are they morphs of a highly variable species (*B. mirus*) or a group of undefined species?

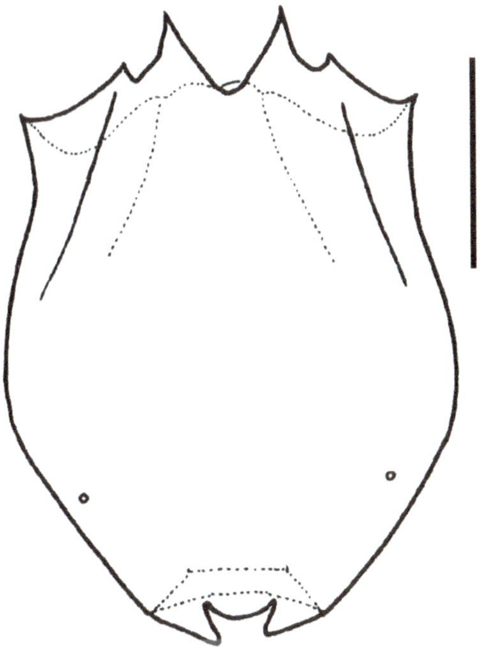

FIGURE 8.72 Brachionidae, *Brachionus kultrum*, dorsal view (Argentina); scale bar = 40 μm. *After Paggi (1981).*

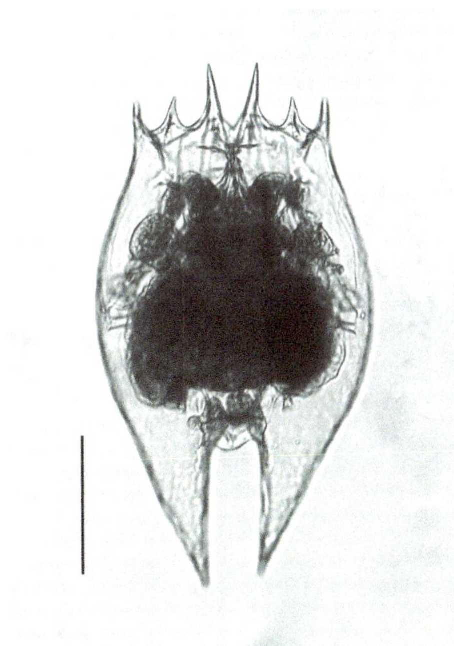

FIGURE 8.73 Brachionidae, *Brachionus variabilis*, dorsal view (Bolivia); scale bar = 100 μm.

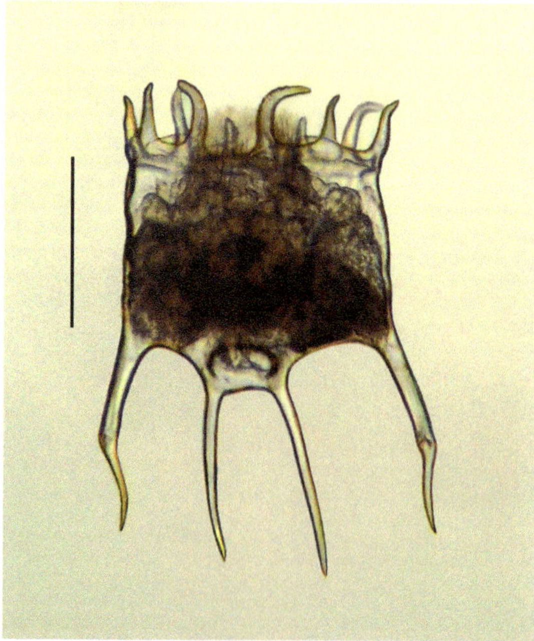

FIGURE 8.74 Brachionidae, *Plationus* sp., dorsal view (Argentina), scale bar = 100 μm.

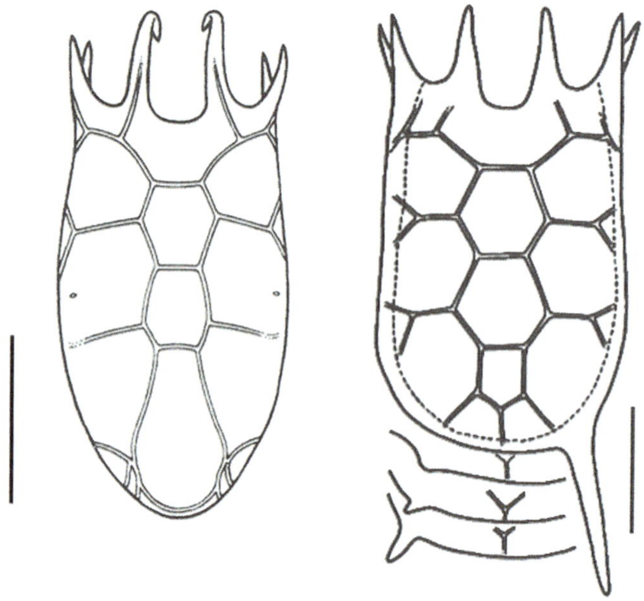

FIGURE 8.76 Brachionidae: *Keratella lenzi*, dorsal view (left), scale bar = 50 μm; *K. procurva*, dorsal view (right; Argentina); scale bar = 30 μm. *After Marinone (1995) and Paggi (1973).*

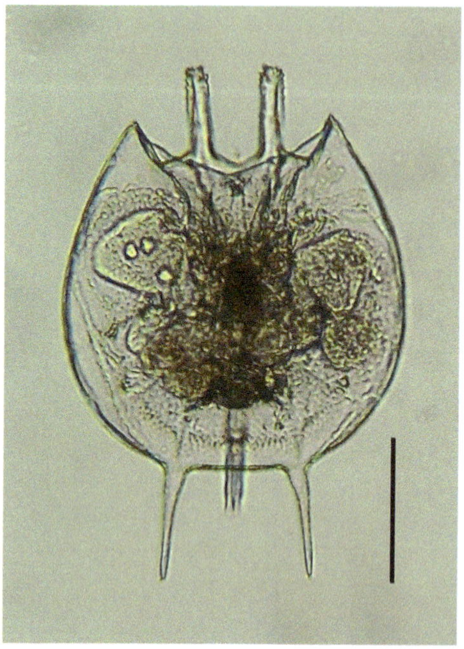

FIGURE 8.75 Brachionidae, *Platyias quadricornis*, dorsal view (Argentina); scale bar = 100 μm.

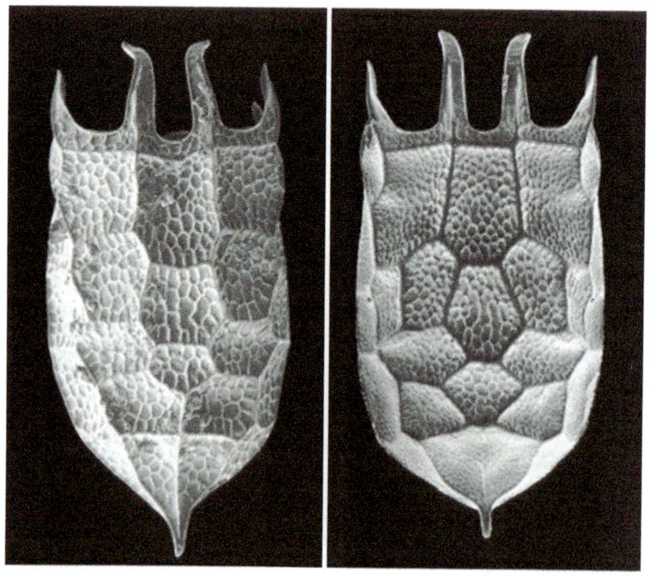

FIGURE 8.77 Brachionidae: *Keratella ona*, dorsal view (left, ~200 μm); *K. yamana*, dorsal view (right, ~150 μm; Argentina). Species endemic to the southern end of the Neotropical region. *After Boltovskoy & Urrejola (1977).*

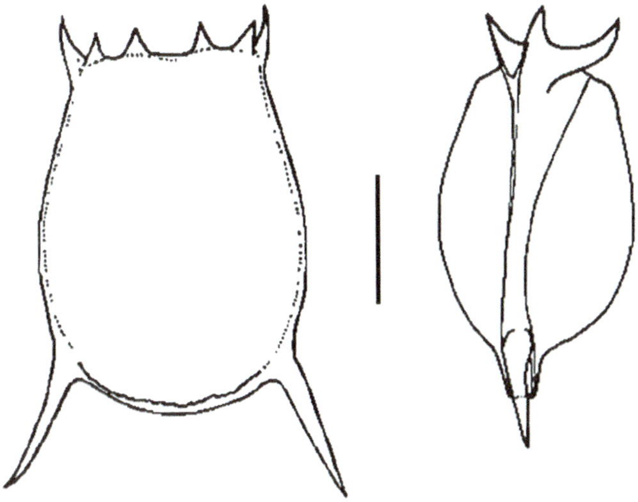

FIGURE 8.78 Brachionidae, *Keratella kostei*, dorsal and lateral views (Argentina), scale bar = 40 μm. *After Paggi (1981).*

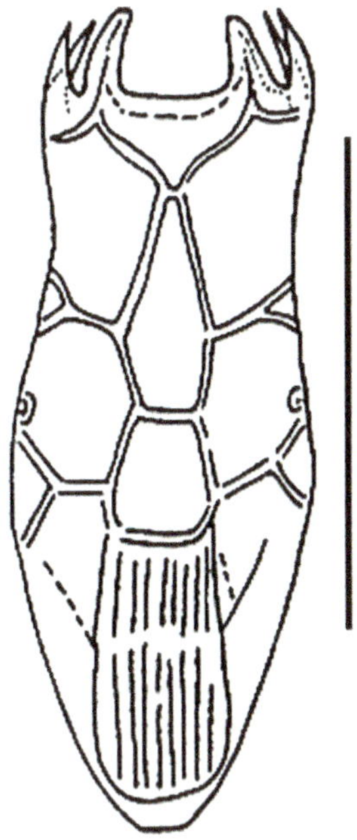

FIGURE 8.79 Brachionidae, *Keratella mexicana*, dorsal view (Mexico), scale bar = 100 μm. *After Kutikova & Silva Briano (1995).*

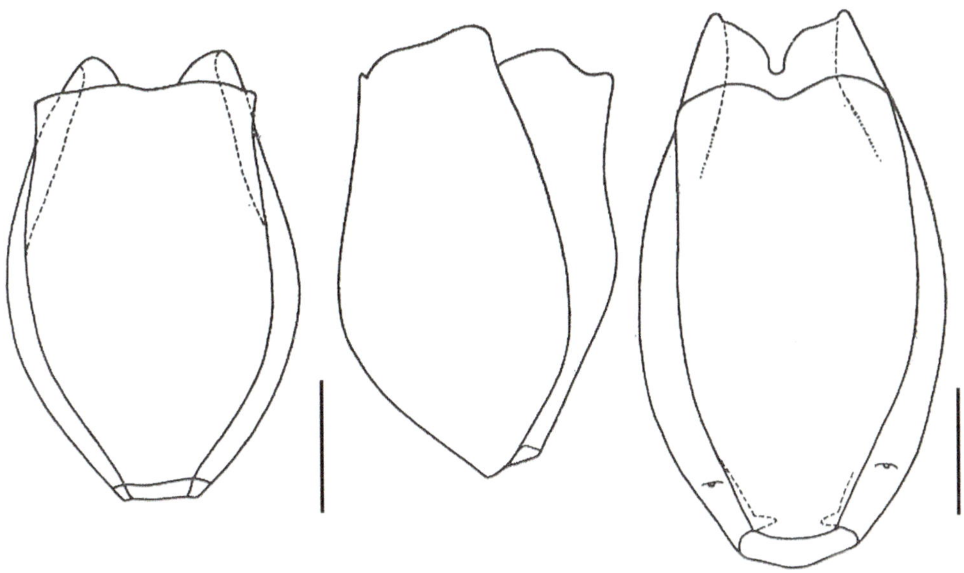

FIGURE 8.80 Brachionidae, *Anuraeopsis fissa*, ventral view (left) and lateral view (center) (Argentina), and *Anuraeopsis navicula* ventral view (right; Argentina), scale bars = 20 μm. *After Paggi (1973).*

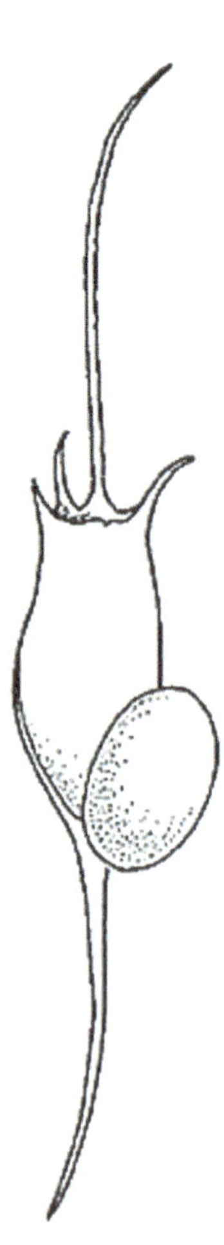

FIGURE 8.81 Brachionidae, *Kellicottia bostoniensis*, ventral and dorsal views (~355–420 μm; Argentina). *After José de Paggi (2002).*

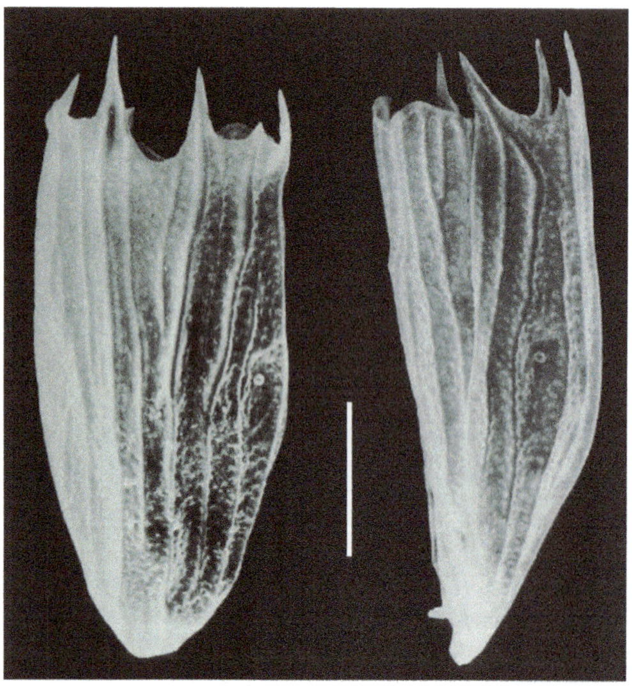

FIGURE 8.82 Brachionidae, *Notholca guidoi*, dorsal and lateral views (Argentina), scale bar = 50 μm. *After Battistoni (1992).*

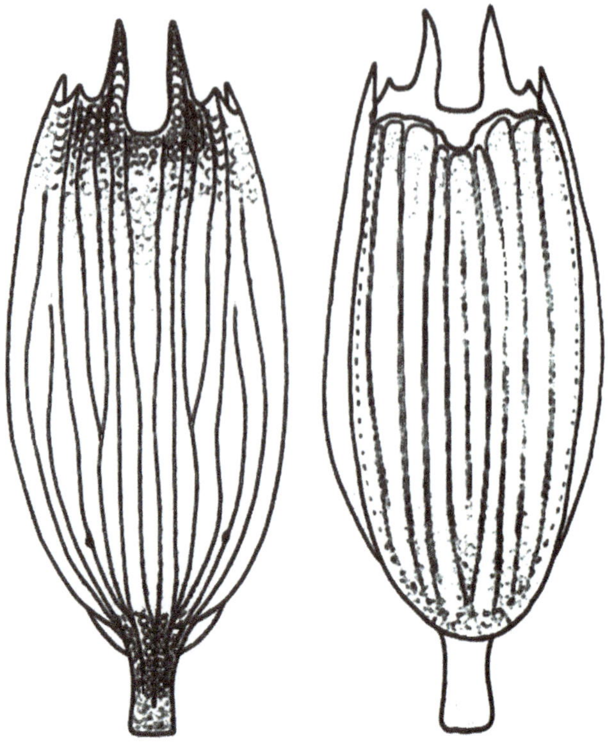

FIGURE 8.83 Brachionidae, *Notholca walterkostei*, dorsal and ventral views (~200 μm; Bolivia). *After Segers et al. (1994b).*

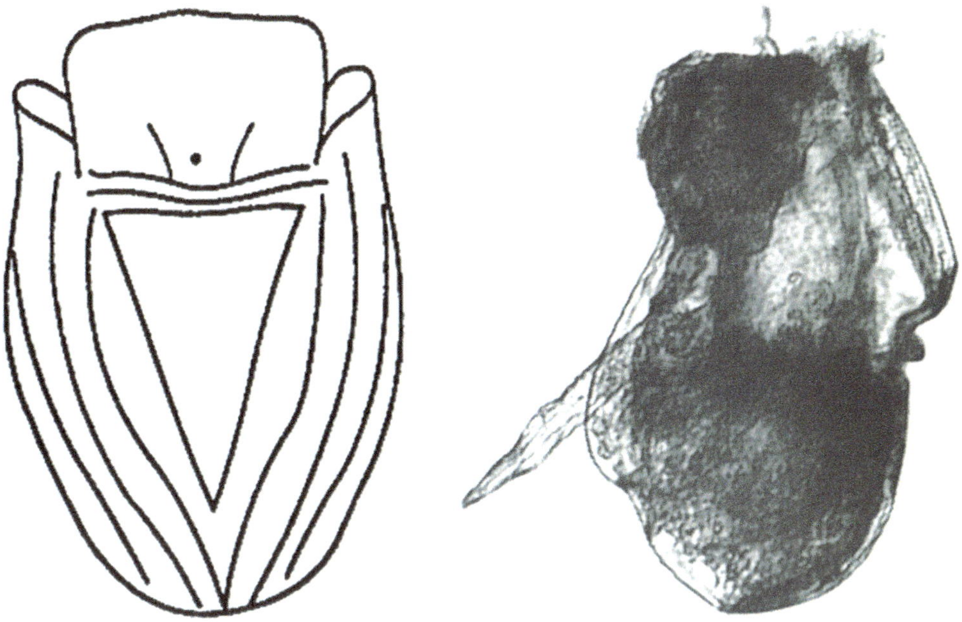

FIGURE 8.84 Synchaetidae: *Ploesoma truncatum*, dorsal view (~200 μm); photomicrograph of *Ploesoma* sp., lateral view (Argentina). *After José de Paggi (1995).*

Rotifera: Monogononta: Ploima: Synchaetidae: Genera

Identification could be difficult in preserved material due to great contraction of the body. *Polyarthra* species are very frequent in the plankton of all freshwater habitats, probably there are more species in Neotropics.

1 Body illoricate 2

1′ Body loricate, ornamented with grooves and ridges, occasionally with overall reticulation; ventrally placed foot, long and flexible, annulated; ca. 130–500 μm (Fig. 8.84) .... *Ploesoma*
[Five Neotropical of eight species]

2(1) Body rectangular, cubical, dorsoventrally compressed, anterior dorsal and ventral corners with four bundles of laminate or remiform appendages (paddles); without foot; 100–225 μm (Fig. 8.85) .... *Polyarthra*
[Seven Neotropical of eleven species]

2′ Body conical, bell or fusiform, with lateral auricles and sensory setae; without remiform appendages; foot retractile, toes minute; ca. 70–600 μm (Figs. 8.86, 87) .... *Synchaeta*
[Eight Neotropical of ~35 species]

Rotifera: Monogononta: Ploima: Trichocercidae: Genera

1 Body short, ovoid, three to four lobes in cross-section (2 lateral, 1 dorsal, 1 ventral [inconspicuous]); foot and toes rudimentary; common in psammon and within *Sphagnum*; ca. 100 μm (Fig. 8.88) .... *Elosa*; one species: *Elosa worrallii* Lord, 1891
[One Neotropical of two species]

1′ Body elongate, loricate, asymmetrical; twisted as a partial helix; foot present, two toes spiniform, of similar or unequal length (right toe reduced); ~100–500 μm (Figs. 8.89, 90) .... *Trichocerca*
[Forty-three Neotropical, four Antarctic of ~70 species]

Rotifera: Monogononta: Ploima: Gastropodidae: Genera

1 Body sacciform to ovate; 1–4 dark defecation reservoirs (like small spheres) in stomach; foot absent; ~75–150 μm (Fig. 8.91) .... *Ascomorpha*
[Six Neotropical of nine species]

1′ Body laterally compressed, dark defecation reservoirs absent; foot present, originating near middle of the ventral side and extending at approximately right angles to the body, ca. 80–400 μm (Fig. 8.92) .... *Gastropus*
[Three Neotropical of three species]

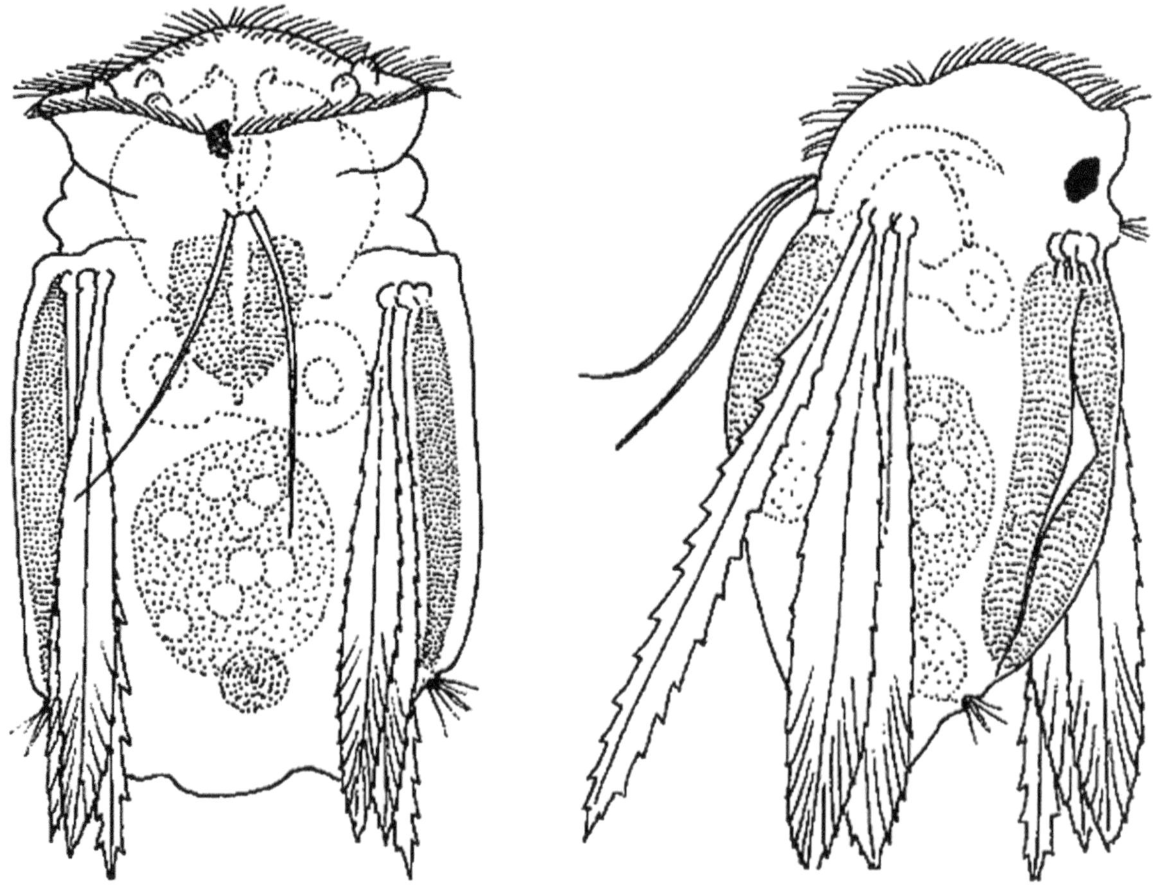

FIGURE 8.85 Synchaetidae, *Polyarthra platensis*, dorsal and lateral views (~90 μm; Argentina). *After José de Paggi & Paggi (2011).*

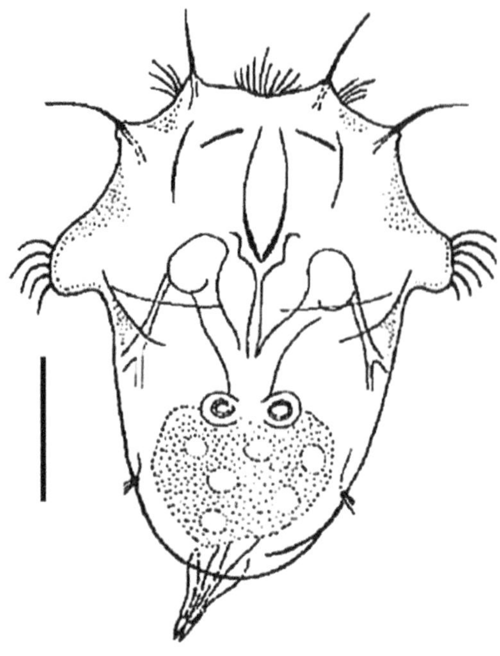

FIGURE 8.86 Synchaetidae, *Synchaeta stylata*, dorsal view (Argentina), scale bar = 50 μm. *After José de Paggi & Paggi (1985).*

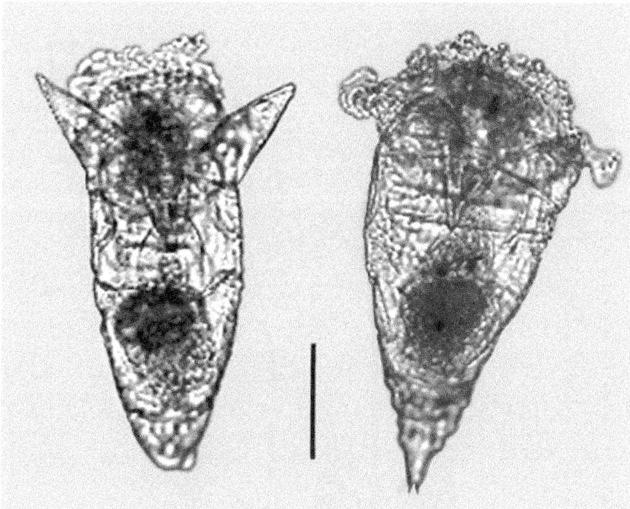

FIGURE 8.87 Synchaetidae, *Synchaeta bicornis*, dorsal view, contracted (left) and extended (right; Argentina); same scale for both sides = 100 μm.

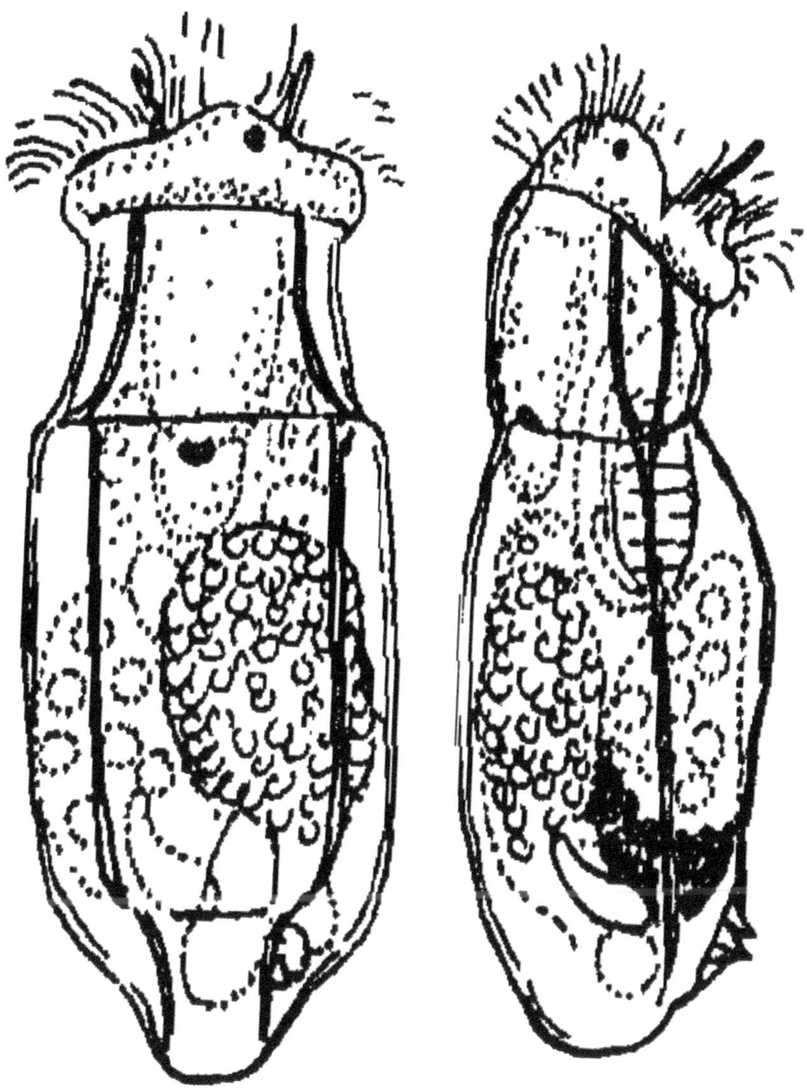

FIGURE 8.88 Trichocercidae, *Elosa worrallii*, dorsal and lateral views (~80–100 μm). *After Wulfert (1960).*

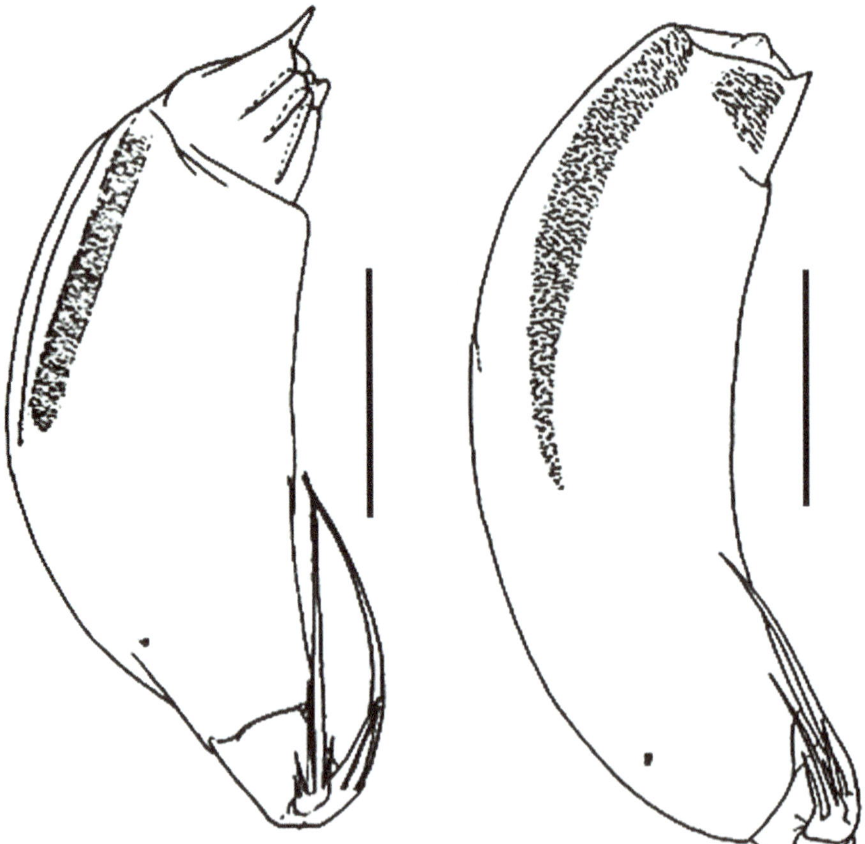

FIGURE 8.89 Trichocercidae: *Trichocerca tigris,* lateral view (left; Bolivia), scale bar = 50 μm; *T. abilioi*, lateral view (right; Brazil), scale bar = 50 μm. *After Segers et al. (1998) and Segers & Sarma (1993).*

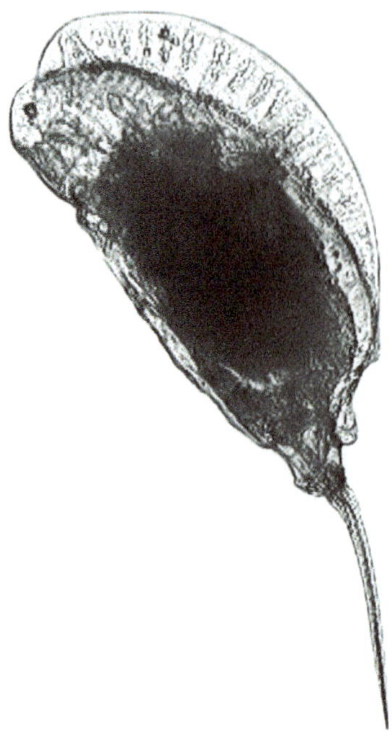

FIGURE 8.90 Trichocercidae, *Trichocerca bicristata*, lateral view (~600 μm; Argentina).

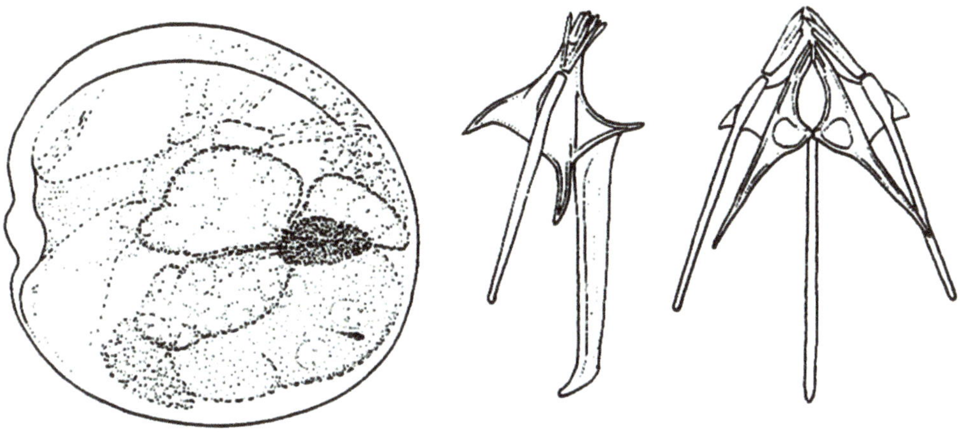

FIGURE 8.91 Gastropodidae: *Ascomorpha tundisii*, contracted body (~80 μm); trophi, lateral and ventral views (~40 μm; Brazil). *After Segers & Dumont (1995).*

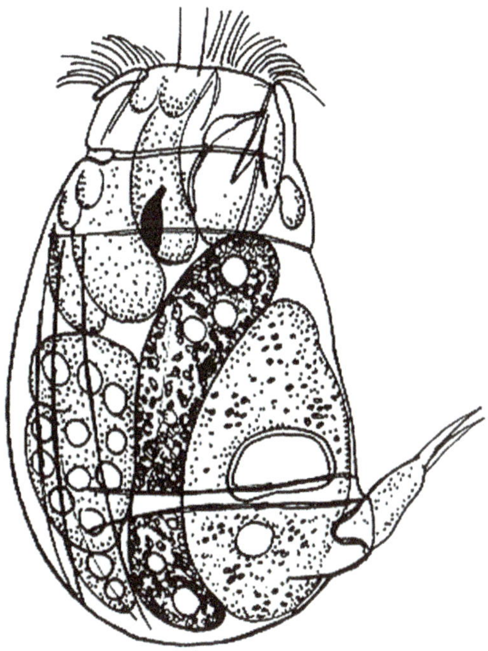

FIGURE 8.92 Gastropodidae, *Gastropus hypotopus*, lateral view (~350 μm). *After Wulfert (1939).*

Rotifera: Monogononta: Ploima: Notommatidae: Genera

Probably the most difficult of the taxonomically unsatisfactory assemblages in the Monogononta (Nogrady et al. 1995); preserved material is of very limited value. The following portion of this key covering the Notommatidae is based on the detailed work of Nogrady et al. (1995).

1	Corona with auricles	2
1′	Corona not as above	3
2(1)	Body dorsally plicate corona oblique; rami with large alulae, foot and toes shorter than body, foot mostly unsegmented; two toes of equal length (Fig. 8.93) [Three Neotropical of four species]	*Taphrocampa*

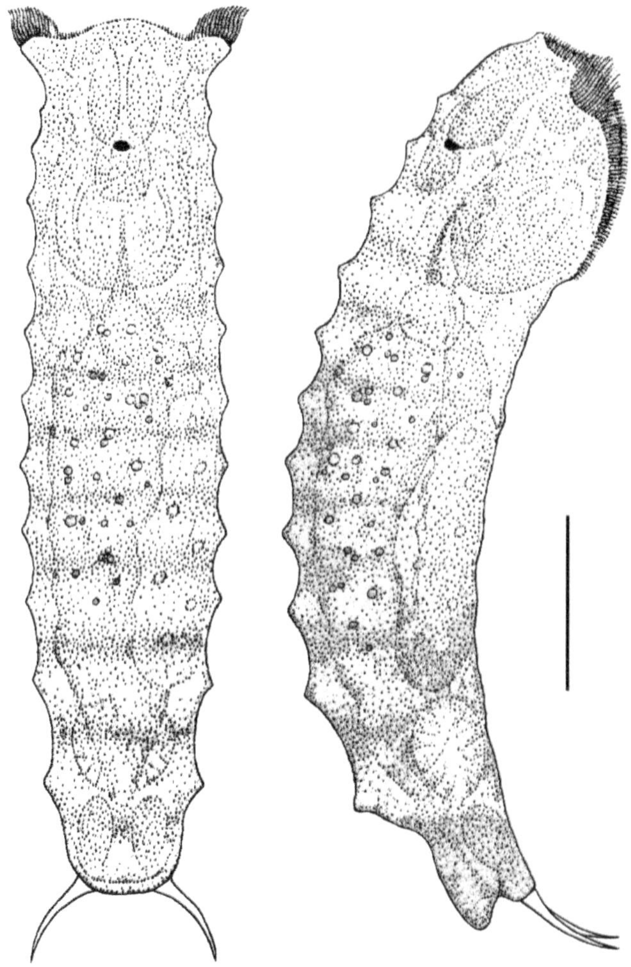

FIGURE 8.93 Notommatidae, *Taphrocampa selenura*, dorsal and lateral views, scale bar = 50 μm. *After Harring & Myers (1924).*

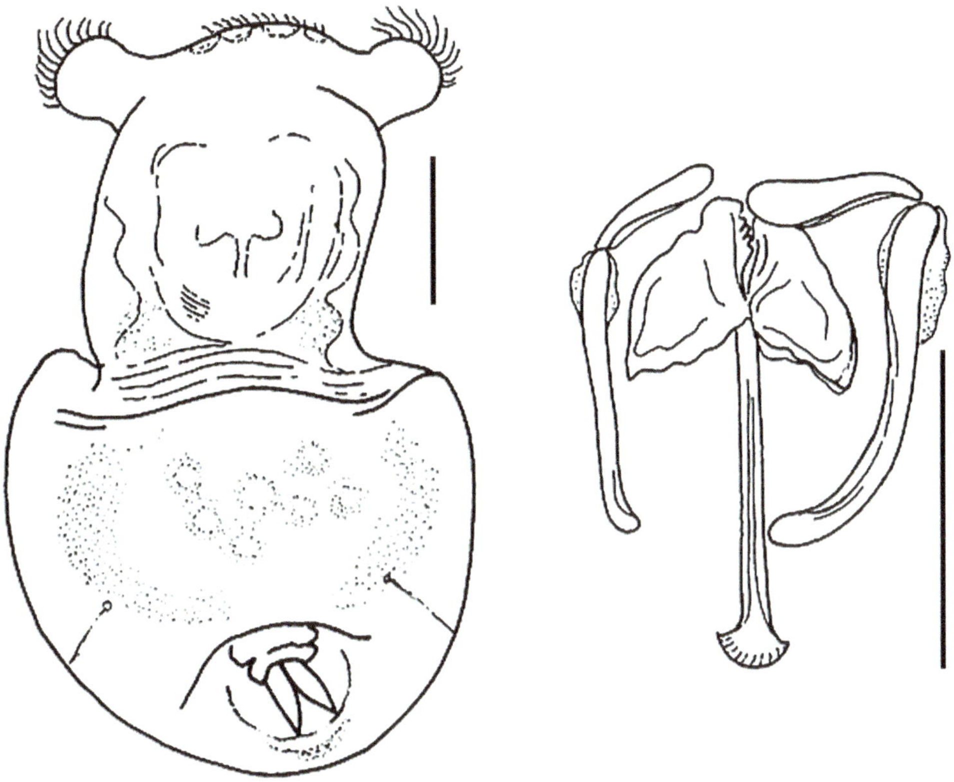

FIGURE 8.94 Notommatidae, *Notommata copeus*, contracted body and trophi (Argentina), scale bar = 50 μm. *After José de Paggi (2001).*

2′ Body (trunk) illoricate, dorsally not plicate as above; corona oblique or ventral; foot and toes shorter than body, foot mostly unsegmented; toes double of equal length (Fig. 8.94) .... *Notommata*
[Twelve Neotropical, one Antarctic of 49 species]

3(1) Corona oblique (tilted ventrally) 4

3′ Corona not oblique 7

4(3) Body illoricate, corona oblique 5

4′ Body loricate (occasionally illoricate), three to five plates separated by furrows; corona with long marginal cilia, but not auricles; two toes of equal length (Fig. 8.95); an account of the trophi types in Fischer & Alrichs (2011). .... *Cephalodella*
[Thirty-seven Neotropical, eight Antarctic of ~156 species]

5(4) Body illoricate, more than one salivary gland 6

5′ Body illoricate, long, slender; single salivary gland; vitellarium oval. (Fig. 8.96) .... *Eothinia*; one species: *Eothinia elongata* (Ehrenberg, 1832)
[One Neotropical of seven species]

6(5) Body illoricate, saccate; double salivary gland; vitellarium ribbon-shaped (Fig. 8.97) .... *Enteroplea*; monospecific: *Enteroplea lacustris* Ehrenberg, 1830

6′ Body illoricate, fusiform in dorsal view, broadest near mid-length, triple salivary gland; (Fig. 8.98) .... *Pourriotia*; one species: *Pourriotia carcharodonta* (De Smet, 2003)
[One Antarctic of two species] position uncertain, the trophi apparently possess elements of both the virgate type, as found in Notommatidae, and the forcipate type, characteristic of Dicranophoridae, consult De Smet (2009).

7(3) Toes not as below 8

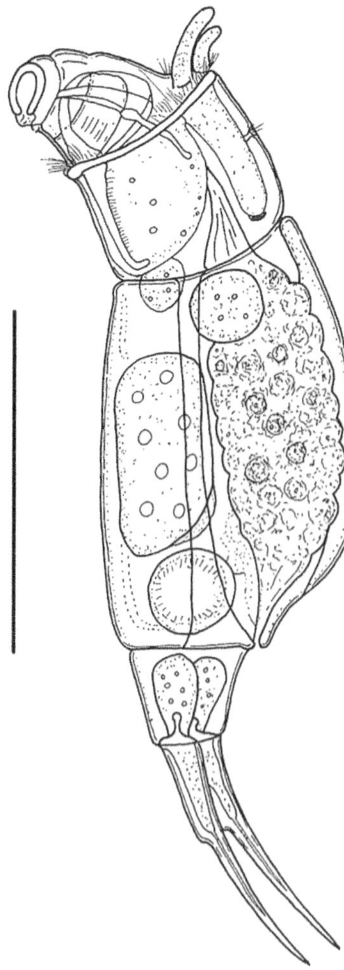

FIGURE 8.95 Notommatidae, *Cephalodella brandorffi*, lateral view (Brazil), scale bar = 50 μm. *After Koste (2000).*

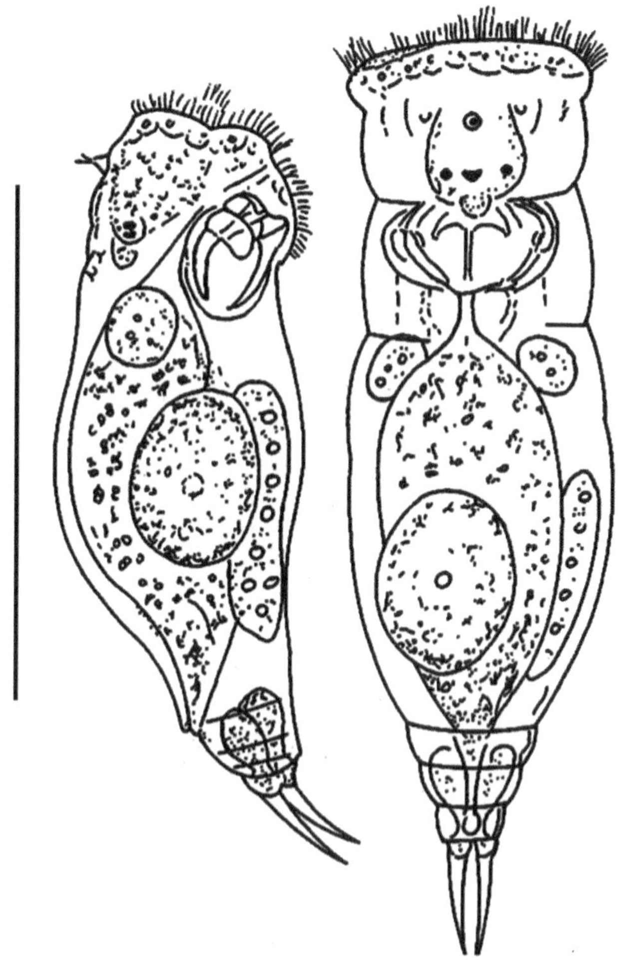

FIGURE 8.96 Notommatidae, representative form of *Eothinia,* lateral and ventral views, scale bar = 100 μm. *After Koste (2000).*

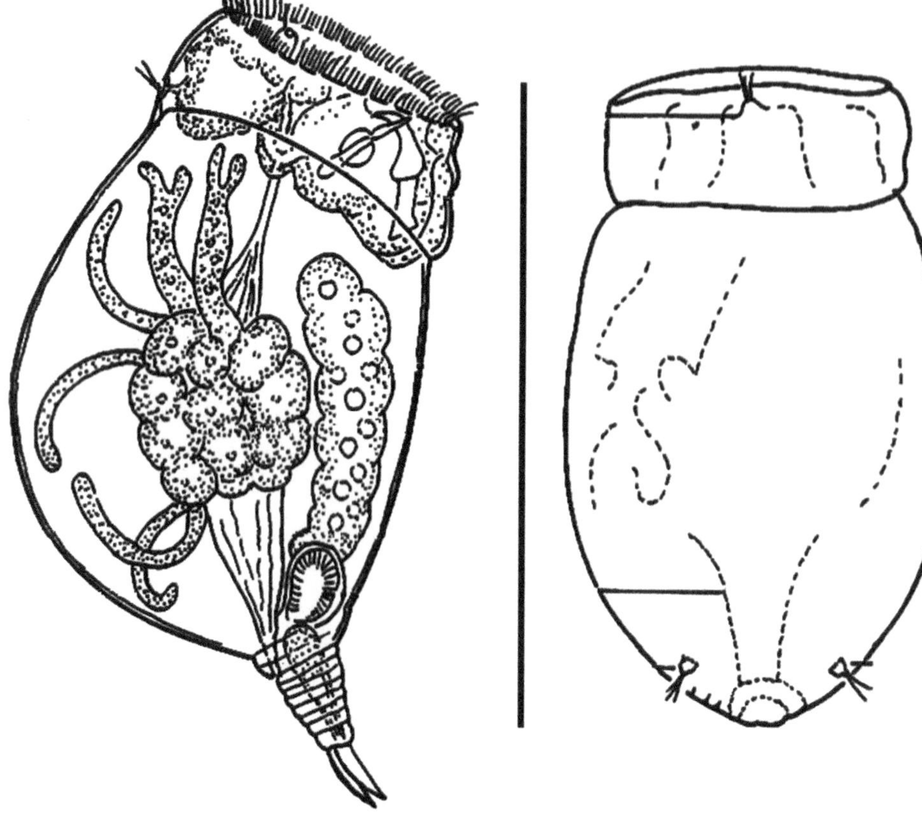

FIGURE 8.97 Notommatidae, *Enteroplea lacustris*, extended animal lateral view and contracted animal (Paraguay), scale bar = 220 μm. *After Koste (1986).*

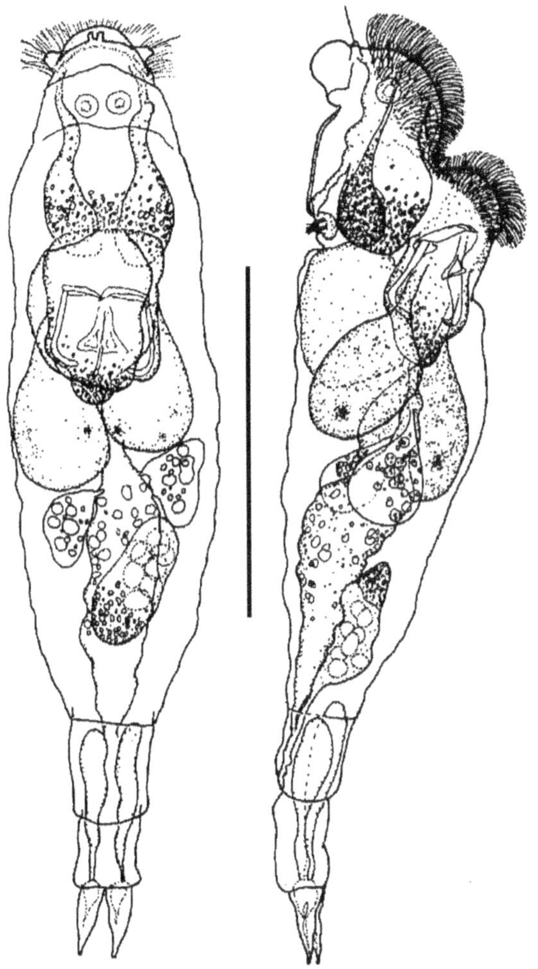

FIGURE 8.98 Notommatidae, *Pourriotia carcharodonta*, dorsal and lateral views (Antarctica), scale bar = 50 μm. *After De Smet (2003).*

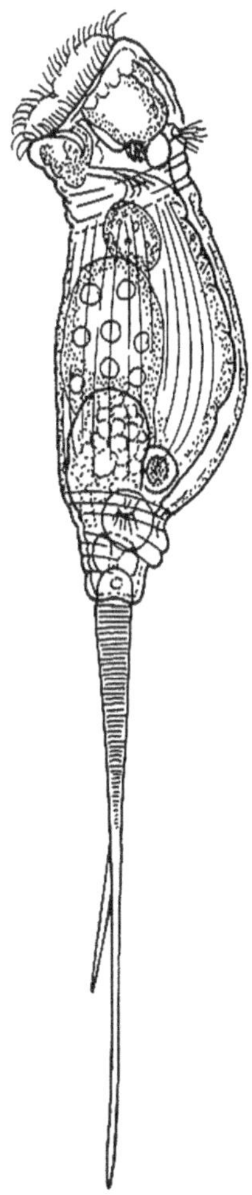

FIGURE 8.99 Notommatidae, *Monommata arndti*, lateral view (~400 μm; Paraguay). *After Koste (1986).*

7′ Two toes, relatively long, mostly longer than body, unequal in length (except *M. aequalis*), corona anterior; body loricate (Fig. 8.99) .... *Monommata*
[Eight Neotropical of 17 species]

8(7) Corona anterior, body illoricate; two toes equal in length 9

8′ Corona anterior, body loricate; two toes wide, mucronate, shorter than body, equal in length (Fig. 8.100) .... *Sphyrias*; monospecific: *Sphyrias lofuana* (Rousselet, 1910)
[Jamaica]

9(8) Salivary glands well formed and symmetrical 10

9′ Salivary glands rudimentary or asymmetrical (Fig. 8.101) .... *Resticula*
[Two Neotropical, one Antarctic of seven species]

10(9) Rami triangular strongly fenestrate, with varying number of teeth, fulcrum wide, rectangular; body stout, saccate; foot with or without segments, two equal toes with long foot glands (Fig. 8.102) .... *Eosphora*
[Five Neotropical of seven species]

10′ Rami triangular, nonfenestrate, fulcrum long, slim, terminally curved; body stout, head and neck well separated; two equal toes (Fig. 8.103) .... *Pleurotrocha*; one species: *Pleurotrocha petromyzon* (Ehrenberg, 1830)
[One Neotropical of seven species]

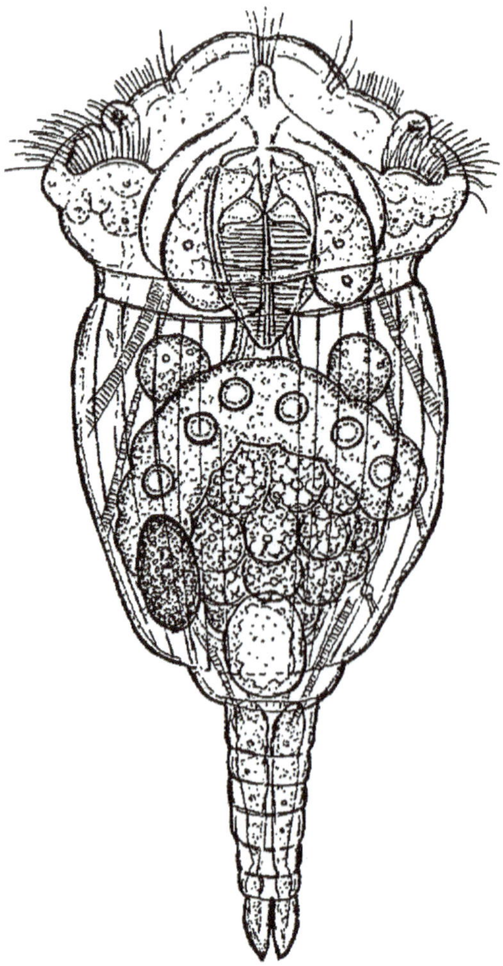

FIGURE 8.100 Notommatidae, *Sphyrias lofuana*, ventral view (~300 μm; Jamaica). *After Koste et al. (1995).*

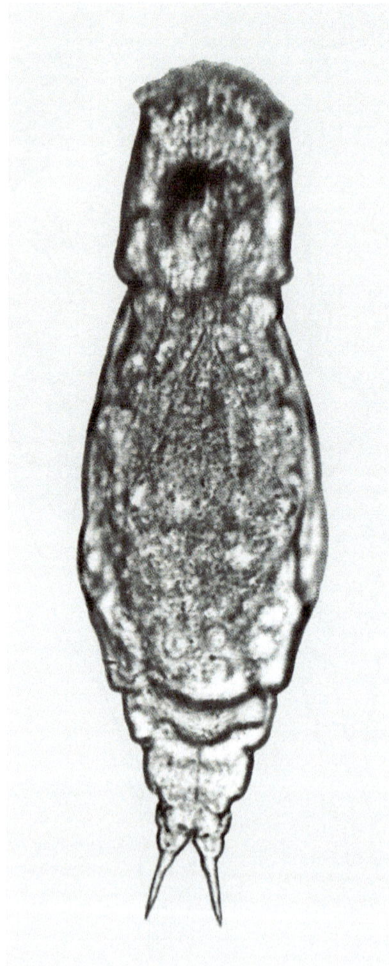

FIGURE 8.101 Notommatidae, *Resticula melandocus*, ventral view (~200 μm; Paraguay). *After Koste (1986).*

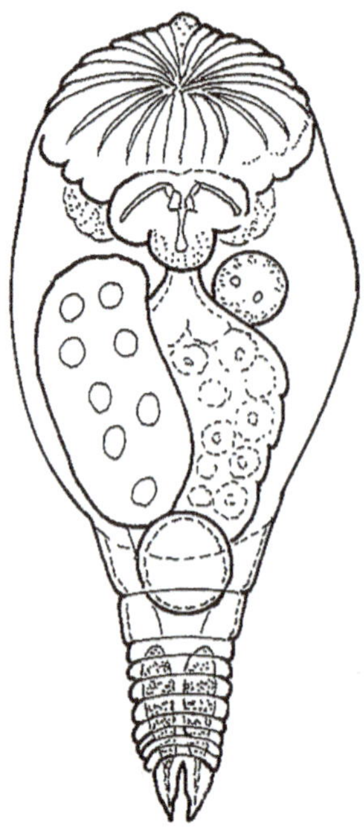

FIGURE 8.102 Notommatidae, *Eosphora anthadis*, contracted female, living animals (~350–400 μm, Brazil). *After Koste et al. (1984).*

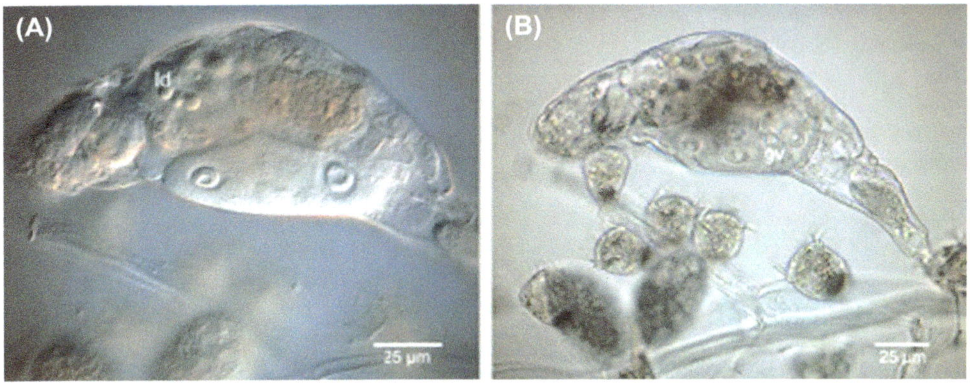

FIGURE 8.103 Notommatidae, *Pleurotrocha petromyzon* in a *Carchesium* colony: (A) adult female; (B) adult female stretching to eat. Abbreviations: *gv*, germovitellarium; *ld*, lipid droplets. *After Wilts et al. (2009).*

REFERENCES

Ahlrichs, W.H. 1995. Ultrastructur und Phylogenik von *Seison nebaliae* (Grube, 1859) und *Seison annulatus* (Grube, 1876). Cuvillier Verlag, Göttingen. 310 pp.

Ahlrichs, W.H. 1997. Epidermal ultrastructure of *Seison nebaliae* and *Seison annulatus*, and a comparison of epidermal structures within Gnathifera. Zoomorphology 117: 41–48.

Ahlstrom, E.H. 1940. A revision of the Rotatorian genera *Brachionus* and *Platyias* with descriptions of one new species and two new varieties. Bulletin of the American Museum of Natural History 77: 148–184.

Ahlstrom, E.H. 1943. A revision of the rotatorian genus *Keratella*, with descriptions of three new species and five new varieties. Bulletin of the American Museum of Natural History 80: 411–457.

Bateman, L.E. 1987. A bdelloid rotifer living as an inquiline in leaves of the pitcher plant, *Sarracenia purpurea*. Hydrobiologia 147: 129–133.

Battistoni, P.A. 1992. Cinco especies del género *Notholca* Gosse, 1886-Rotatoria-de la Argentina, Incluyendo *N. guidoi* sp. n. Iheringia, Zoologia 73: 35–45.

Bērziņš, B. 1951. On the Collothecacean Rotatoria, with special reference to the species found in the Aneboda district, Sweden. Arkiv Zoologi 1: 565–592.

Boltovskoy, A. & R. Urrejola. 1977. Dos nuevas especies del género *Keratella* (Rotatoria) de Tierra del Fuego, Argentina. Limnobios 1(6).

Clément, P. 1993. The phylogeny of rotifers: molecular, ultrastructural and behavioural data. Hydrobiologia 255/256: 527–544.

Coineau, Y. & M. Kunst. 1964. Une nouvelle espèce de rotifère parasite d'oligochète: *Albertia soyeri* n. sp. Vie et Milieu 15: 1007–1015.

Dartnall, H.J.G. 1995. The rotifers of Heard Island: Preliminary survey, with notes on other freshwater groups. Papers and proceedings of the Royal Society of Tasmania. 129: 7–15.

de Beauchamp, P.M. 1907. Morphologie et variations de l'appareil rotateur dans la série des Rotifères. Archives of Zoological Experimental Genetics, Sér. 4, 6: 1–29.

de Beauchamp, P. 1965. Classe des rotifères. Pages 1225–1379 *in*: P.-P. Grassé (ed.), Traité de Zoologie, T. 4, 3: Némathelmintes (Nématodes, Gordiacés), Rotifères, Gastrotriches, Kinorhynques. Masson et Cie, Paris.

De Smet, W.H. (ed.). 1996. Rotifera. Vol. 4: The Proalidae (Monogononta). Guides to the Identification of Microinvertebrates of the Continental Waters of the World 9, SPB Academic Publishing, The Hague. 102 pp.

De Smet, W.H. 1997. Rotifera. Vol. 5: The Dicranophoridae (Monogononta). Guides to the Identification of Microinvertebrates of the Continental Waters of the World 12. SPB Academic Publishing, The Hague. 326 pp.

De Smet, W.H. 1998. Preparation of rotifer trophi for light and scanning electron microscopy. Hydrobiologia 387/388: 117–121.

De Smet, W.H. 2003. *Pourriotia carcharodonta*, a new genus and species of monogonont rotifer from subantarctic Îles Kerguelen (Terres Australes et Antarctiques Françaises). Annales de Limnologie-International Journal of Limnology 39: 273–280.

De Smet, W.H. 2006. Asciaporrectidae, a new family of Rotifera (Monogononta: Ploima) with description of *Asciaporrecta arcellicola* gen. et sp. nov. and *A. difflugicola* gen. et sp. nov. inhabiting shells of testate amoebae (Protozoa). Zootaxa 1339: 31–49.

De Smet, W.H. 2007. Cotylegaleatidae, a new family of Ploima (Rotifera: Monogononta), for *Cotylegaleata perplexa* gen. et sp. nov., from freshwater benthos of Belgium. Zootaxa 1425: 35–43.

De Smet, W.H. 2009. *Pourriotia carcharodonta*, a rotifer parasitic on Vaucheria (Xanthophyceae) causing taxonomic problems. Bulletin de la Société zoologique de France 134: 195–202.

De Smet, W.H. & J. A. Gibson. 2008. *Rhinoglena kutikovae* n. sp. (Rotifera: Monogononta: Epiphanidae) from the Bunger Hills, East Antarctica: a probable relict species that survived Quaternary glaciations on the continent. Polar Biology 31: 595–603.

De Smet, W.H. & R. Pourriot. 1997. Rotifera. Vol. 5: The Dicranophoridae (Monogononta) and the Ituridae (Monogononta). Guides to the Identification of Microinvertebrates of the Continental Waters of the World 12. SPB Academic Publishing, The Hague. 344 pp.

Devetter, M. 2010. A method for efficient extraction of rotifers (Rotifera) from soils. Pedobiologia 53: 115–118.

Diéguez M.C. & E. Balseiro. 1998. Colony size in *Conochilus hippocrepis*: defensive adaptation to predator size. Hydrobiologia 387: 421–425.

Donner, J. 1956. Rädertiere (Rotatorien). W. Ketter & Co. Stuttgart. English Translation and Adaptation by H.G.S. Wright (1966). F. Warne & Co. Ltd., New York. 80 pp.

Donner, J. 1965. Ordnung Bdelloidea (Rotatoria, Rädertiere). Bestimmungsbücher zur Bodenfauna Europas. 6. Akademie-Verlag, Berlin. 297 pp.

Edmondson, W.T. 1949. A formula key to the rotatorian genus *Ptygura*. Transactions of the American Microscopical Society 68: 127–135.

Edmondson, W.T. 1959. Rotifera. Pages 420–494 *in:* W.T. Edmondson (ed.), Freshwater Biology, (Second edition). John Wiley & Sons, Inc., New York.

Ferrando, N.S. & M.C. Claps. 2016. A revised and updated checklist of Monogononta rotifers from Argentina. Check List 12: 1–26.

Fischer, C. & W.H. Ahlrichs. 2011. Revisiting the *Cephalodella* trophi types. Hydrobiologia 662: 205–209.

Fontaneto D. & W.H. De Smet. 2015. Rotifera. Pages 217–300 *in*: A. Schmidt-Rhaesa (ed.), Handbook of Zoology: Gastrotricha, Cycloneuralia and Gnathifera. Vol 3: Gastrotricha and Gnathifera. De Gruyter, Berlin.

Fontaneto, D., E.A. Herniou, T.G. Barraclough & C. Ricci. 2007. On the global distribution of microscopic animals: new worldwide data on bdelloid rotifers. Zoological Studies 46: 336–346.

Fontaneto, D. & U. Jondelius. 2011. Broad taxonomic sampling of mitochondrial cytochrome c oxidase subunit I does not solve the relationships between Rotifera and Acanthocephala. Zoologischer Anzeiger 250: 80–85.

Fontaneto, D., N. Iakovenko, I. Eyres, M. Kaya, M. Wyman & T.G. Barraclough. 2011. Cryptic diversity in the genus *Adineta* Hudson & Gosse, 1886 (Rotifera: Bdelloidea: Adinetidae): a DNA taxonomy approach. Hydrobiologia 662: 27–33.

Fontaneto, D., N. Iakovenko & W.H. De Smet. 2015. Diversity gradients of rotifer species richness in Antarctica. Hydrobiologia (Biology of the Ross Sea) 761: 235–248.

Fontaneto, D., M. Kaya, E.A. Herniou & T.G. Barraclough. 2009. Extreme levels of hidden diversity in microscopic animals (Rotifera) revealed by DNA taxonomy. Molecular Phylogenetics and Evolution 53: 182–189.

Garraffoni A.R.S. & A.P. Lourenço. 2012. Synthesis of Brazilian Rotifera: An updated list of species. Check List 8: 375–407.

García-Varela, M. & S.A. Nadler. 2006. Phylogenetic relationships among Syndermata inferred from nuclear and mitochondrial gene sequences. Molecular Phylogenetics and Evolution 40: 61–72.

Giribet, G., D.L. Distel, M. Polz, W. Sterrer & W.C. Wheeler. 2000. Triploblastic relationships with emphasis on the acoelomates and the position of Gnathostomulida, Cycliophora, Plathelminthes, and Chaetognatha: a combined approach of 18S rDNA sequences and morphology. Systematic Biology 49: 539–562.

Giribet, G., M.V. Sørensen, P. Funch, R.M. Kristensen & W. Sterrer. 2004. Investigations into the phylogenetic position of Micrognathozoa using four molecular loci. Cladistics 20: 1–13.

Harring, H. K. & F.J. Myers. 1924. The Rotifer Fauna of Wisconsin II. A revisión of the Notommatid rotifers, exclusive of the Dicranophoriae. Transactions of the Wisconsin Academy of Sciences, Arts, and Letters 21: 415–549.

Harring, H.K. & F.J. Myers.1928. The rotifer fauna of Wisconsin IV. The Dicranophoridae. Transactions of the Wisconsin Academy of Sciences, Arts, and Letters 23: 667–808.

Hochberg, R., S.O'Brien & A.Puleo. 2010. Behavior, metamorphosis, and muscular organization of the predatory rotifer Acyclus inquietus (Rotifera, Monogononta). Invertebrate Biology 129: 210–219.

Hudson, C.T. & P.H. Gosse. 1886. The Rotifera or Wheel-Animalcules, both british and foreign. London, I: I–VI:1–128 + II: 1–144.

Iakovenko, N.S., J. Smykla, P. Convey, E. Kašparova, I.A. Kozeretska, V. Trokhymets, I. Dykyy, M. Plewka, M. Devetter, Z. Duriš & K. Janko. 2015. Antarctic bdelloid rotifers: diversity, endemism and evolution. Hydrobiologia (Biology of the Ross Sea) 761: 5–43.

Jersabek, C.D., E. Bolortsetseg & H.L. Taylor. 2010. Mongolian rotifers on microscope slides: Instructions to permanent specimen mounts from expedition material. Mongolian Journal of Biological Sciences 8: 51–57.

Jersabek, C.D. & M.F. Leitner. 2013. The Rotifer World Catalog. World Wide Web electronic publication. http://www.rotifera.hausdernatur.at/ (accessed 2018.03.22).

José de Paggi, S. 1979. Contribución al conocimiento de la fauna argentina de Rotíferos. II Algunas especies de los género *Lecane* Nitzch y *Lepadella* Bory de St. Vincents. Neotropica 25: 37–44.

José de Paggi, S. 1982. Contribución al conocimiento de la fauna argentina de Rotíferos. III Nuevos registros específicos en ambientes. Neotropica 28: 117–124.

José de Paggi, S. 1995. Rotíferos. En E. Lopretto y G. Tell (eds.), Ecosistemas de Aguas Continentales. Metodologías para su estudio, Tomo II. Ediciones Sur, Bs. Aires.

José de Paggi, S. 1996. Rotifera (Monogononta) diversity in subtropical waters of Argentina. Annales de Limnologie 32: 209–220.

José de Paggi, S. 2001. Diversity of Rotifera (Monogononta) in wetlands of Rio Pilcomayo national park, Ramsar site (Formosa, Argentina). Hydrobiologia 462: 25–34.

José de Paggi, S. 2002. New data on the distribution of *Kellicottia bostoniensis* (Rousselet, 1908) (Rotifera: Monogononta: Brachionidae): its presence in Argentina. Zoologischer Anzeiger-A Journal of Comparative Zoology 241: 363–368.

José de Paggi, S. & W. Koste. 1995. Additions to the Checklist of Rotifers of the Superorder Monogononta recorded from Neotropis. Internationale Revue der gesamten Hydrobiologie und Hydrographie 80: 133–140.

José de Paggi, S. & J.C. Paggi 1985. Zooplancton de los cuerpos de agua preexistentes en el área del embalse Amutui Quimei (Cuenca del río Futaleufú). Neotropica 31: 119–131.

José de Paggi, S.B. & J.C. Paggi. 2011. A new species of *Polyarthra* Ehrenberg, 1834 belonging to the vulgaris-group (Rotifera: Monogononta: Synchaetidae) from Argentina, with a key to the identification of species in the Neotropical Region. Zootaxa 2828: 51–57.

Koste, W. 1973. *Horaëlla thomassoni* n. sp. ein neues Radertier aus Gewässer der Guiana-Brasilianischen Region der Neotropis. Archiv für Hidrobiologie 72: 375–883.

Koste, W. 1974a. Zur Kenntnis der Rotatorienfauna der "Schwimmenden Weise" einer Ueerlagune in der Varzea Amazonies. Brasilien. Amazoniana 5: 25–60.

Koste, W. 1974b. Rotatorien aus einem Ufersee des unteren Rio Tapajós, dem Lago Paroni (Amazonien). Gëwasser Abwässer 53 (54): 43–68.

Koste, W. 1978. Rotatoria. Die Rädertiere Mitteleuropas. Überordnung Monogononta. Gebrüder Borntraeger, Stuttgart, Vol. I . Textband. 673 pp., Vol. II, Tafelband: II. 234 pp.

Koste, W. 1986. Über die Rotatorienfauna in Gewässern süd Östlich von Concepción, Paraguay, Südamerika. Osnabrücker Naturwissenschaftliche Mitteilungen 12: 129–155.

Koste, W. 2000. Study of the Rotatoria-Fauna of the Littoral of the Rio Branco, South of Boa Vista, Northern Brazil. International Review of Hydrobiology: A Journal Covering all Aspects of Limnology and Marine Biology 85: 433–469.

Koste, W. & K. Böttger. 1989. Rotatorien aus Gewässern Ecuadors. Amazoniana 10: 407–438.

Koste, W., W. Janetzky & E. Vareschi 1991. Über die Rotatorienfauna in Bromelien-Phytotelmata in Jamaika (Aschelminthes: Rotatoria). Osnabrücker Naturwissenschaftliche Mitteilungen 17: 143–170.

Koste, W., W. Janetzky & E. Vareschi. 1995. Zur Kenntnis der limnischen Rotatorienfauna Jamaikas (Rotifera). Teil II. Osnabrücker Naturwissenschaftliche Mitteilungen 20/21: 399–433.

Koste, W. & S. José de Paggi. 1982. Rotifera of the Superorder Monogononta recorded from Neotropis. Gewässer und Abwässer 68: 71–102.

Koste, W, B. Robertson & E. Hardy. 1984. Further taxonomical studies of the Rotifera from Lago Camaleâo, a Central Amazonian varzea lake (Ilha de Marchantaria, Rio Solimôes, Amazonas Brazil). Amazoniana 7: 555–576.

Koste, W. & R.J. Shiel. 1986. Rotifera from Australian inland waters. I. Bdelloidea (Rotifera: Digononta). Australian Journal of Marine and Freshwater Research 37: 765–792.

Koste, W. & R. J. Shiel. 1987. Rotifera from Australian inland waters. II. Epiphanidae and Brachionidae (Rotifera: Monogononta). Invertebrate Systematics 1: 949–1021.

Kutikova, L. A. & M. Silva-Briano. 1995. *Keratella mexicana* sp. nov., a new planktonic rotifer from Aguascalientes, Mexico. Hydrobiologia 310: 119–122.

Marinone, M.C. 1995. A new and phylogenetically suggestive morphotype of *Keratella lenzi* (Rotifer, Monogononta), from Argentina. Hydrobiologia 299: 249–257.

Martínez, C.C. & S. José de Paggi. 1988. Especies de *Lecane* Nitzch (Rotifera Monogononta) en ambientes acuáticos del Chaco oriental y del valle aluvial del río Paraná (Argentina). Revue d'Hydrobiologie Tropicale 21: 279–295.

May, L. 1989. Epizoic and parasitic rotifers. Hydrobiologia 186: 59–67.

Meksuwan, P., P. Pholpunthin & H. Segers. 2011. Diversity of sessile rotifers (Gnesiotrocha, Monogononta, Rotifera) in Thale Noi Lake, Thailand. Zootaxa 2997: 1–18.

Meksuwan, P., P. Pholpunthin & H. Segers. 2013. The Collothecidae (Rotifera, Collothecacea) of Thailand, with the description of a new species and an illustrated key to the Southeast Asian fauna. ZooKeys 315: 1–16.

Melone, G., C. Ricci, H. Segers & R.L. Wallace. 1998a. Phylogenetic relationships of phylum Rotifera with emphasis on the families of Bdelloidea. Hydrobiologia 387/388: 101–107.

Melone, G., C. Ricci & H. Segers. 1998b. The trophi of Bdelloidea (Rotifera): a comparative study across the class. Canadian Journal of Zoology 76: 1755–1765.

Min, G.S. & J.K. Park. 2009. Eurotatorian paraphyly: Revisiting phylogenetic relationships based on the complete mitochondrial genome sequence of *Rotaria rotatoria* (Bdelloidea: Rotifera: Syndermata). BMC Genomics 10: 533.

Murray, J. 1906. The Rotifera of the Scottish Lochs. including descriptions of new species by C.F. Rousselet, F.R.M.S., and D. Bryce, Esq. Transactions of the Royal Society of Edinburgh 45:151–191.

Murray, J. 1913. South American Rotifera. Journal of the Royal Microscopical Society I, 33: 229–246.

Myers, F.J. 1917. Rotatoria of Los Angeles, California, and vicinity, with descriptions of a new species. Proceedings of the United States National Museum 52: 473–478.

Nogrady, T., R. Pourriot & H. Segers. 1995. Rotifera. Vol. 3: The Notommatidae (Monogononta) and the Scaridiidae (Monogononta). Guides to the Identification of Microinvertebrates of the Continental Waters of the World. 8. SPB Academic Publishing, The Hague. 248 pp.

Nogrady, T. & H. Segers (eds.). 2002. Rotifera. Vol. 6: Asplanchnidae, Gastropodidae, Lindiidae, Microcodidae, Synchaetidae, Trochosphaeridae and Filinia (Monogononta). Guides to the Identification of Microinvertebrates of the Continental Waters of the World. 18. Backhuys, Leiden. 264 pp.

Nogrady, T., R.L. Wallace & T.W. Snell. 1993. Rotifera. Vol. 1: Biology, Ecology and Systematics. Guides to the Identification of Microinvertebrates of the Continental Waters of the World. SPB Academic Publishers, The Hague. 142 pp.

Olivier, S.R. 1965. Rotiferos planctónicos de Argentina. Revista Museo La PLata 8 (63): 177–260.

Papakostas, S., E. Michaloudi, K. Proios, M. Brehm, L. Verhage, J. Rota, C. Peña, G. Stamou, V.L. Pritchard, D. Fontaneto & S.A. Declerck. 2016. Integrative taxonomy recognizes evolutionary units despite widespread mitonuclear discordance: evidence from a rotifer cryptic species complex. Systematic Biology 65: 508–524.

Paggi, J. C. 1973. Contribución al conocimiento de los Rotíferos dulceacuícolas planctónicos de la República Argentina. Physis 32(85): 321–330.

Paggi, J. C. 1978. Sobre la presencia de *Trochospaera aequatorialis* Semper y Horaella thomassoni Koste (Rotatoria, Testudinellidae) en lagunas del valle de inundación de Río Paraná. Revista de la Asociación de Ciencias Naturales del Litoral 9: 67–82.

Paggi, J. C. 1981. Observaciones sobre el zooplancton de algunos lagos de la Patagonia extrandina. I Rotíferos. Studies on Neotropical Fauna and Environment 16: 23–33.

Paggi, J.C. & S. José de Paggi. 1973. Sobre algunos rotíferos nuevos para la fauna argentina. Revista de la Asociación de Ciencias Naturales del Litoral 4: 49–60.

Pourriot R. & E. Zoppi de Roa. 1981. Dicranophoridae (Rotiferes Monogontes) Du Venezuela. Bulletin de la Societe Geologique de France 106: 195–199.

Ricci, C. 1987. Ecology of bdelloids: how to be successful. Hydrobiologia 147: 117–127.

Ricci, C. 1998. Anhydrobiotic capabilities of bdelloid rotifers. Hydrobiologia 387: 321–326.

Ricci, C. & D. Fontaneto. 2009. The importance of being a bdelloid: ecological and evolutionary consequences of dormancy. Italian Journal of Zoology 76: 240–249.

Ricci, C. & G. Melone. 2000. Key to the identification of the genera of bdelloid rotifers. Hydrobiologia 418: 73–80.

Rousselet, C.F. 1894. On *Cyrtonia tuba* = *Notommata tuba* Ehrenberg. Journal of the Quekett Microscopical Club 2: 433–435.

Sanoamuang, L.O. & J.C. McKenzie. 1993. A simplified method for preparing rotifer trophi for scanning electron microscopy. Hydrobiologia 250: 91–95.

Segers, H. 1995a. Rotifera. Vol. 2: The Lecanidae (Monogononta). Guides to the Identification of Microinvertebrates of the Continental Waters of the World. 6. SPB Academic Publishing, The Hague. 226 pp.

Segers, H. 1995b. A reappraisal of the Scaridiidae (Rotifera, Monogononta). Zoologica Scripta 24: 91–100.

Segers, H. 1997. Contribution to a revision of *Floscularia* Cuvier, 1798 (Rotifera: Monogononta): notes on some Neotropical taxa. Hydrobiologia 354: 165–175.

Segers, H. 2004. Rotifera: Monogononta. Pages 112–126 *in:* C.M. Yule and H.S. Yong (eds.), Freshwater Invertebrates of the Malaysian

Region. Academy of Sciences Malaysia and Monash University, Kuala Lumpur, Malaysia.

Segers, H. 2007. Annotated checklist of the rotifers (Phylum Rotifera), with notes on nomenclature, taxonomy and distribution. Zootaxa 1564: 1–104.

Segers, H. 2008. Global diversity of rotifers (Rotifera) in freshwater. Hydrobiologia 595: 49–59.

Segers, H. 2013. Rotifera check List (generated Thu Jun 27 16:41:35 +0200 2013 from FADA website). http://fada.biodiversity.be/CheckLists/Rotifera.pdf. 68 pp.

Segers, H. & H.J. Dumont, 1995. 102 + rotifer species (Rotifera: Monogononta) in Broa reservoir (SP., Brazil) on 26 August 1994 with the description of three new species. Hydrobiologia 316: 183–197.

Segers, H., N. L. Ferrufino & L. De Meester. 1998. Diversity and zoogeography of Rotifera (Monogononta) in a flood plain Lake of the Ichilo River, Bolivia, with notes on littl known species. International Review of Hydrobiology 83: 439–448.

Segers, H., D.K. Mbogo & H.J. Dumont. 1994a. New Rotifera from Kenya, with a revision of the Ituridae. Zoological Journal of the Linnean Society 110: 193–206.

Segers, H., L. Meneses & M. Del Castillo. 1994b. Rotifera (Monogononta) from Lake Kothia, a high-altitude lake in the Bolivian Andes. Archiv für Hydrobiologie 132: 227–236.

Segers, H., P. Meksuwan, & L. Sanoamuang. 2010. New records of sessile rotifers (Phylum Rotifera: Flosculariacea, Collothecacea) from Southeast Asia. Belgium Journal of Zoology 140: 235–240.

Segers, H. & S.S. Sarma. 1993. Notes on some new of little known rotifera from Brazil. Revue d'Hydrobiologie tropicale 26: 175–185.

Segers, H. & R.J. Shiel. 2008. Diversity of cryptic Metazoa in Australian freshwaters: a new genus and two new species of sessile rotifer (Rotifera, Monogononta, Gnesiotrocha, Flosculariidae). Zootaxa 1750: 19–31.

Segers, H. & R.L. Wallace. 2001. Phylogeny and classification of the Conochilidae (Rotifera, Monogononta, Flosculariacea). Zoologica Scripta 30: 37–48.

Sielaff, M., H. Schmidt, T.H. Struck, D. Rosenkranz, D.B.M. Welch, T. Hankeln & H. Herlyn. 2016. Phylogeny of Syndermata (syn. Rotifera): Mitochondrial gene order verifies epizoic Seisonidea as sister to endoparasitic Acanthocephala within monophyletic Hemirotifera. Molecular Phylogenetics and Evolution 96: 79–92.

Sørensen, M.V. 2003. Further structures in the jaw apparatus of *Limnognathia maerski* (Micrognathozoa), with notes on the phylogeny of the Gnathifera. Journal of Morphology 255: 131–145.

Sørensen, M.V. & R.M. Kristensen. 2015. Micrognathozoa. Pages 197–216 *in*: A. Schmidt-Rhaesa (ed.), Handbook of Zoology: Gastrotricha, Cycloneuralia and Gnathifera. Vol. 3: Gastrotricha and Gnathifera. De Gruyter, Berlin.

Turner, P.N. 1987. A New Rotifer from a coastal lake in Southeastern Brazil: *Hexarthra logicornicula* n. sp. Hydrobiologia 153: 169–174.

Wallace, R.L. 1980. Ecology of sessile rotifers. Pages 181–193 *in*: H.J. Dumont and J. Green (eds.), Rotatoria, Dr. W. Junk Publishers, The Hague.

Wallace, R.L. 1987. Coloniality in the phylum Rotifera. Hydrobiologia 147: 141–155.

Wallace, R.L. 2002. Rotifers: exquisite metazoans. Integrative and Comparative Biology 42: 660–667.

Wallace, R.L., T.W. Snell, C. Ricci & T. Nogrady. 2006. Rotifera. Vol. 1: Biology, Ecology and Systematics (2nd ed.). Identification Guides to the Microinvertebrates of the Continental Waters of the World. 23, Backhuys, Leiden. 299 pp.

Wallace, R.L., T. Snell & H.A. Smith. 2015. Rotifer: Ecology and General Biology. Pages 225–271 *in:* J.H. Thorp & D.C. Rogers (eds.), Thorp and Covich's Freshwater Invertebrates. Elsevier, Waltham, MA.

Wallace, R.L., T.W. Snell, E.J. Walsh, S.S.S. Sarma & H. Segers. 2016. Phylum Rotifera. Pages: 131–167 *in*: J.H. Thorp & D.C. Rogers (eds.), Keys to Nearctic Fauna, Thorp and Covich's Freshwater Invertebrates. Vol. II. Elsevier, Waltham, MA.

Wen, X., Y. Xi, G. Zhang, Y. Xue & X. Xiang. 2016. Coexistence of cryptic *Brachionus calyciflorus* (Rotifera) species: roles of environmental variables. Journal of Plankton Research 38: 478–489.

Weber E.F. 1898. Faune Rotatorienne du Bassin du Léman. Revue Suisse de Zoologie et Annales du Musée D'Historie Naturelle de Genève 5: 263–785.

Wey-Fabrizius, A.R., H. Herlyn, B. Rieger, D. Rosenkranz, A. Witek, D.B.M. Welch & T. Hankeln. 2014. Transcriptome data reveal syndermatan relationships and suggest the evolution of endoparasitism in Acanthocephala via an epizoic stage. PLoS One 9: e88618.

Wilson, C.G. 2011. Desiccation-tolerance in bdelloid rotifers facilitates spatiotemporal escape from multiple species of parasitic fungi. Biological Journal of the Linnean Society 104: 564–574.

Wilts, E.F., O.R. Bininda-Emonds & W.H. Ahlrichs. 2009. Comparison of the predatory rotifers *Pleurotrocha petromyzon* (Ehrenberg, 1830) and *Pleurotrocha sigmoidea* Skorikov, 1896 (Rotifera: Monogononta: Notommatidae) based on light and electron microscopic observations. Zootaxa 2130: 1–20.

Wulfert, K. 1939. Beiträge zur Kenntnis der Rädertierfauna Deutschlands. Teil IV Die Rädertiere der Saale-Elster-Niederung bei Merseburg in ökologisch-faunistischer Beziehung. Archiv für Hydrobiologie 35: 563–624.

Wulfert, K. 1960. Die Rädertiere saurer Gewässer der Dübener Heide. I. Die Rotatorie des Zadlitzmoors und des Wildenhainer Bruchs. Archiv für Hydrobiologie 56: 261–298.

Wulfert, K. 1968. Rädertiere aus China I. Limnologica (Berlin) 6: 405–416.

Chapter 9

Phylum Nematoda

Marcelo E. Doucet
Instituto de Diversidad y Ecología Animal (CONICET-UNC) and Centro de Zoología Aplicada, Facultad de Ciencias Exactas, Físicas y Naturales, Universidad Nacional de Córdoba, Córdoba, Argentina

Paul De Ley
Department of Nematology, University of California Riverside, Riverside, CA, United States

Paola Lax
Instituto de Diversidad y Ecología Animal (CONICET-UNC) and Centro de Zoología Aplicada, Facultad de Ciencias Exactas, Físicas y Naturales, Universidad Nacional de Córdoba, Córdoba, Argentina

Chapter Outline

Thorp and Covich's Freshwater Invertebrates. https://doi.org/10.1016/B978-0-12-804225-0.00009-5

INTRODUCTION

Numerous species of nematodes live in soil, marine sediments, and freshwater ecosystems worldwide. Hugot et al. (2001) reported a worldwide total of 26,646 recognized species for the entire phylum, including 14,786 species that are either free-living or plant parasites (not parasites of animals). Hodda et al. (2009) reported the latter number slightly lower at 13,838. This represents just a small percentage of the total estimated species richness of extant nematodes, which usually ranges between half a million to 1 million (Hugot et al., 2001).

Some nematode species are thought to be versatile opportunists that can be encountered equally frequently in soil as in freshwater, or as often in freshwater lakes and rivers as in brackish estuaries and coastal wetlands. Others occur only in one of these environments, sometimes with highly specialized adaptations to very particular microhabitats only.

The Neotropical realm harbors overwhelming richness of flora and fauna, forming one of the most important reservoirs of biodiversity on our planet, and yet the references listed at the end of this chapter illustrate the limited number of national or regional nematode inventories published to date from Central or South America.

Nematode ecosystem functions are most often interpreted in terms of their roles in food webs. Depending on their food sources, we can distinguish between the following trophic groups of nematodes: microbivores (always associated with bacteria, also known as saprophages or bacterivores), fungivores (also known as mycophages), predators, omnivores, algivores, phytoparasites, and insect parasites (developing at the expense of a living insect host during part of their life cycle). The existence of these trophic categories indicates the particular importance of these animals as probable regulators of nutrient flows within the ecosystems where they occur.

Large regions of South America flood seasonally, for example, rainforests of the Amazon basin, but also Andean vernal pools, the Pantanal floodplain, as well as extensive savannas such as the Llanos of Bolivia, Venezuela, and Colombia. If we include permanent or (semi)permanent wetlands, shallow lakes, and marshes, Junk (1993) estimated that one-fifth of continental South America consists of wetlands. Our own research shows that "typical" freshwater nematodes can abundantly persist in upper soil layers and in plant litter as anhydrobiotic stages during the dry phase of seasonally flooding habitats.

Conversely, many Neotropical ecosystems are greatly impacted by soil erosion, and this may explain why "typical" soil nematodes are often found in sediment samples from relatively stable rivers and lakes. Moreover, the edges of rivers, lakes, and wetlands represent ecotones with their own peculiarities of variability in time and space, and often also their own plant and animal communities, which may in turn provide specialized niches for nematodes with inherently overlapping terrestrial versus freshwater ranges.

The Antarctic Region and the Andean Puna present their own unique challenges as well, given that water in the marginal soils of their various snow-free valleys or plateaus remains at or near freezing point all year round. Sampling under such conditions is inevitably limited to short timespans, under conditions that make it all but impossible to determine whether a microhabitat should or even can be designated as truly terrestrial versus limnic.

Many assessments of what are typical limnic versus soil habitats for nematodes are based primarily on early literature from almost exclusively temperate zones of the Palearctic and Nearctic Regions. It is likely that many exceptions or extensions of habitat ranges apply to tropical, subtropical, and subantarctic species from these same genera. This is especially true for the Neotropics, because the nematological taxonomic literature of this ecozone consists almost entirely of first reports or descriptions of new species and genera—substantially more so than, for example, literature on nematodes from Africa or Australia.

Given all these considerations, it will be much more informative for freshwater biologists studying Neotropical and Antarctic sites, if we provide a key that allows them to identify all known genera encountered in their samples, without a priori omission of the genera presumed to be true terrestrial forms only. Such a tool will be essential for establishing distribution records, rather than unique points of occurrence, and this in turn will enable us to revisit habitat requirements or plasticity for the genera in question.

LIMITATIONS

This chapter presents an identification key for the main families and all valid genera of soil, epiphyte, and freshwater nematodes recorded to date from the Neotropics and from the Antarctic, including subantarctic islands as well as Pacific or Atlantic islands within 1000 km range of Central or South America. The key has been prepared on the basis of national or regional species compendia, viz., Argentina (Doucet & de Doucet, 1997, 1999), Antarctica (Andrássy, 1998; Velasco-Castrillón et al., 2014), and Costa Rica (Esquivel, 2003). These were supplemented with extensive searches through the taxonomic literature, as well as with genus lists obtained during major taxonomic or ecological surveys of particular sites or ecosystems (Andrássy, 1968; Loof, 1973; Huang & Cares, 2006).

The key is designed for identifying adult nematodes only and excludes groups for which all adult stages occur solely inside an animal host. While not included in this key, it must be noted that juvenile stages of certain animal parasitic, parasitoid, or pathogenic groups can sometimes be encountered in high abundance in a particular sample (most often the infective stages of steinernematids, heterorhabditis, and mermithids), without any identifiable adults cooccurring in the same nematode extract. Mermithid infective stages probably have a range of useful morphological identifying characters, but unfortunately the extended life cycles of many of these nematodes usually preclude presence of such juveniles at times when adults or postparasitic fourth-stage juveniles are found. As a result, the detailed morphology of the infective stage is very rarely included in mermithid species or genus descriptions. More is known about the morphology and morphometrics of steinernematid juveniles, especially so for species described from previous centuries. However, the number of known *Steinernema* species has tripled in the past two decades, and many juvenile characters visible with light microscopy overlap too much among species for reliable identification (it is usually much more practical to identify them using molecular methods).

We have kept technical terms for nematode morphology to a minimum, and wherever possible, we include more than one character per couplet, even if it is only informative for a majority of species or genera, rather than all of them. Given the limitations of light microscopy and the overwhelming diversity of nematodes, as well as their evident propensity for convergent evolution and changes in reproductive or developmental biology, we realize that the use of this key will nevertheless remain challenging. Examples of particular challenges to identification include the following: genera or entire families for which males are extremely hard to find, even though their characters may be essential for identification; occurrence of sexual or individual dimorphism (and even trimorphism; see Ragsdale et al., 2013) in the morphology of lip region and buccal structures; and general lack of information on intraspecific variability. Even when a few species have been subject to a variability study, changes of classification and diagnosis are typically hard to propose without generating greater confusion, particularly so when some populations are known to exceed formal genus or family boundaries.

TERMINOLOGY AND MORPHOLOGY

Example illustrations of diagnostically informative body regions are given in Figs. 9.1–8. More information on nematode anatomy and terminology can be found in Poinar (2015) and in Caveness (1964).

Glossary of Anatomical Terms Used in the Taxonomic Key

- **Amphids**: Paired lateral organs combining chemosensory and glandular functions; they generally open to the exterior on or near the lip region.
- **Amphid opening**: The aperture leading to an internal pouch containing the ciliary nerve endings of the amphid. The opening can be a slit (curved, spiral, loop, arcuate like a question mark, crescentic, elliptic), a circle, or an oval.
- **Amphidelphic**: Having two ovaries, generally one extending anteriad (directed forward) and the other posteriad (directed backward) of the vulva.
- **Basal bulb**: An enlargement of the pharyngeal wall, muscular or glandular, at the posterior of the pharynx.
- **Basal knobs**: Posterior knobs of the stylet shaft, usually there are two subventral and one dorsal knob.
- **Buccal capsule**: *Sensu lato*, the combined vestibulum and buccal cavity. The stoma. *Sensu stricto:* Anterior cuticular widening of the pharyngeal and prepharyngeal tubes, often distinguishable from the lining of the pharynx proper by more developed cuticularizations, differences in transverse lumen shape and/or in arrangement of radial versus interradial features and sectors of the pharyngeal cuticle. The buccal cavity can be barrel-like (Figs. 9.4 D, 5 C), tubular (Fig. 9.4 I), funnel-shaped (Fig. 9.4 G), or like an oil lamp's glass

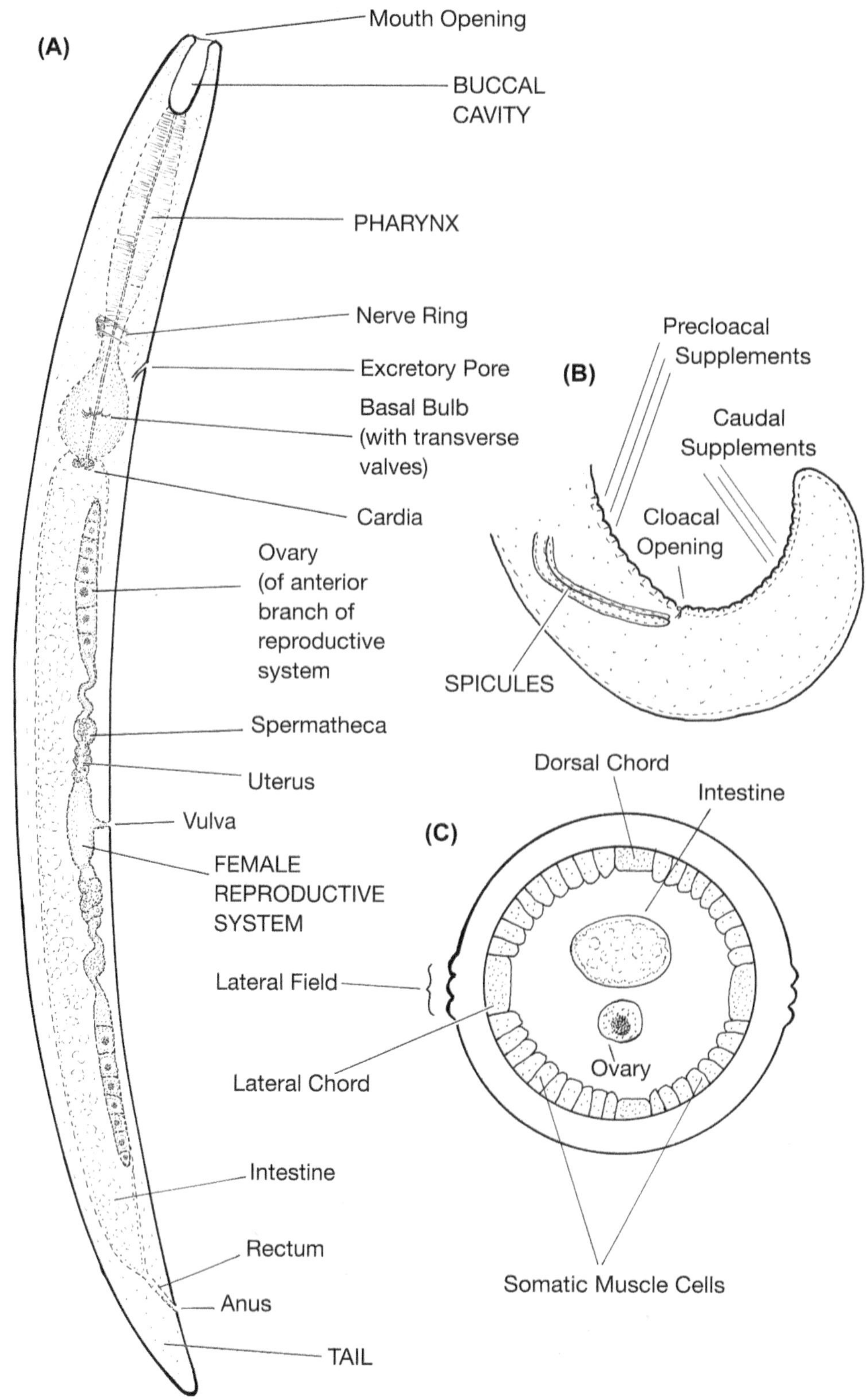

FIGURE 9.1 General anatomy of nematodes: (A) generalized adult female; (B) adult male tail with copulatory spicules; and (C) cross section near the middle of a female's body.

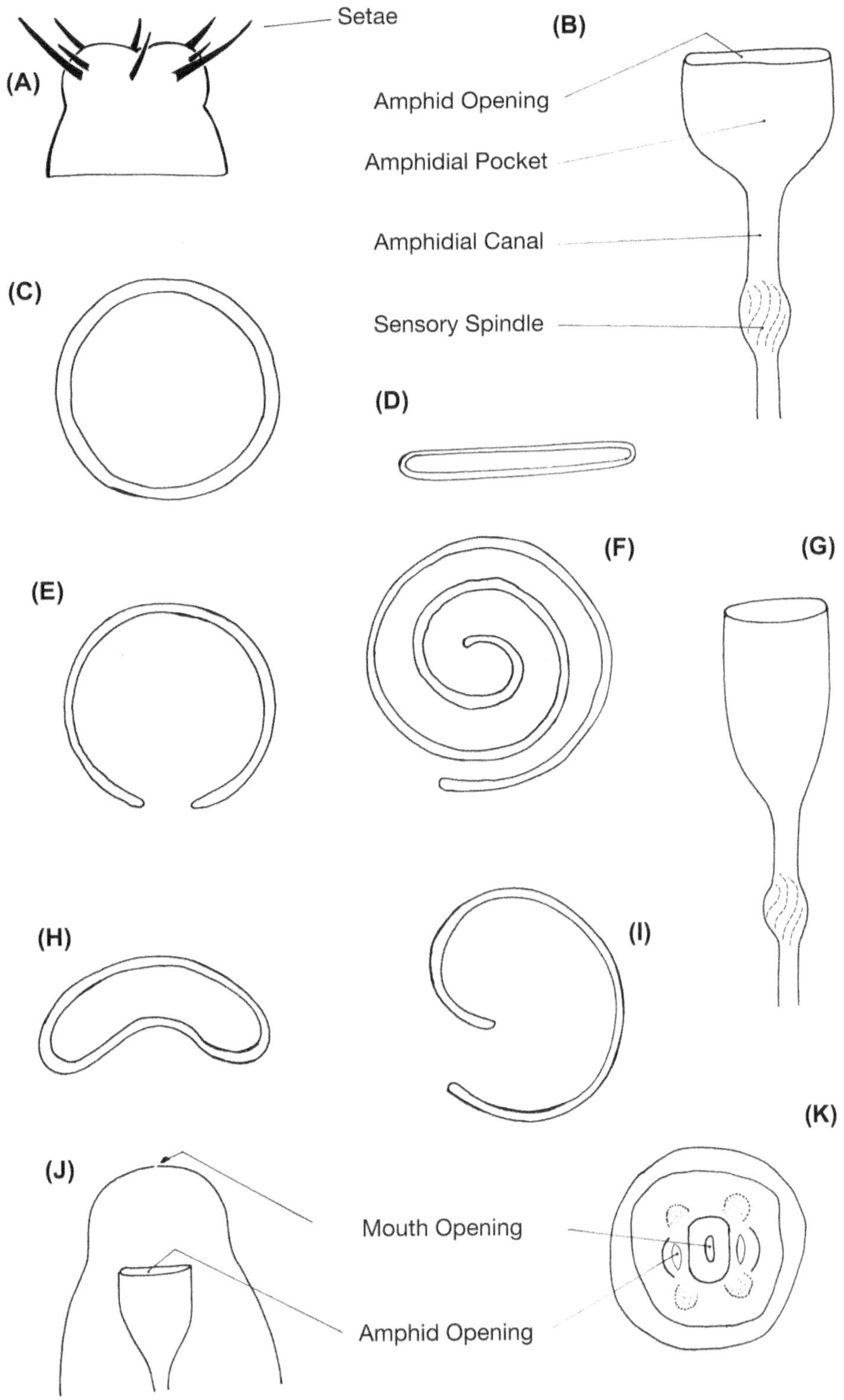

FIGURE 9.2 Examples of anterior sensory organs: (A) setae are elongate examples of sensilla mechanoreceptors; on the lip region, they may occur in up to three circles of 6+ 6+4 (papillae = nonelongate sensilla); (B–L) amphids: (B) major parts of the amphid, a paired chemosensory organ that also has secretory functions, and of which both the opening and pocket can have a variety of shapes, including (B) slit-shaped with stirrup-shaped pocket; (C) closed circular; (D) wide slit; (E) open circular; (F) spiral; (G) narrow slit with goblet-shaped pocket; (H) curved slit or bean-shaped; and (I) open single coil; (J) lateral view of a diagrammatic anterior end showing sideways placement of the amphid opening, as in most free-living species; and (K) face view of a plant parasite (in this case, *Criconemella*), in which the amphids are typically placed closer to the mouth and often face forward rather than sideways.

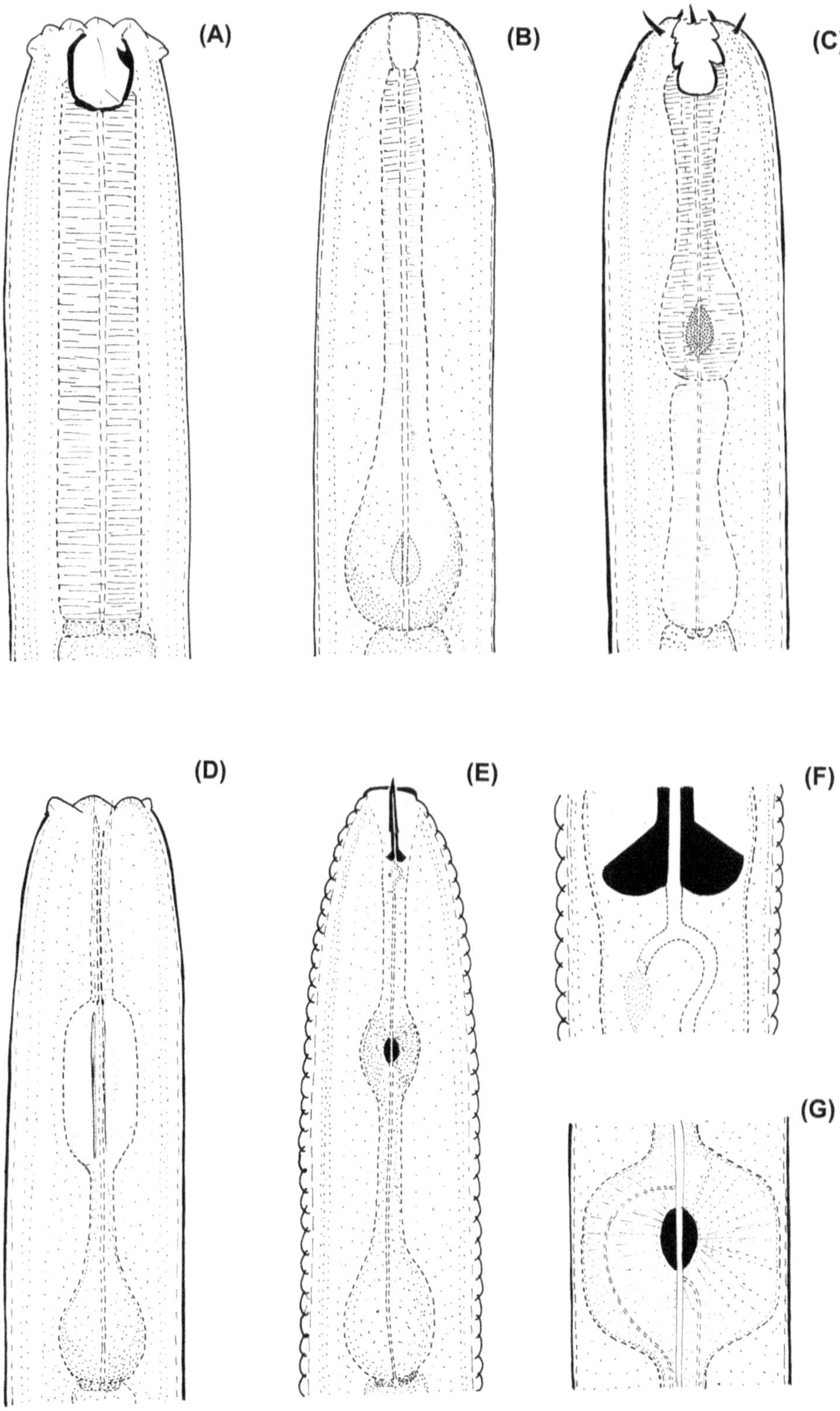

FIGURE 9.3 Diagrammatic examples of different types of pharyngeal shapes: (A) cylindrical (e.g., in Mononchida); (B) two-part pharynx with rounded basal bulb containing longitudinal valves (e.g., in some Chomadorida and in the enoplid genus *Rhabdolaimus*); (C) four-part pharynx with separation of procorpus and median bulb, the latter containing longitudinal valves while the weaker basal bulb lacks valves (e.g., in Rhabditida:

chimney (anteriorly a narrow cylinder that transitions fairly abruptly into a short spherical cavity, which then transitions again into a very short narrow posterior cylinder) as in representatives of some Dorylaimida.

- **Bursa**: Alate (winglike) extensions of the lateral cuticle at the caudal end of the male. Caudal alae.
- **Cardia**: Muscular (sometimes also glandular) structure forming a minute or larger sphincter, the junction of pharynx and intestine.
- **Caudal ala**: Either of the two alate (winglike) sides of the bursa.
- **Cirrus (pl. cirri):** A simple or elaborate cephalic appendage in front of the cephalic setae that lacks innervation.
- **Corpus**: The most anterior segment of the pharynx, typically of cylindrical form and of moderate width. Sometimes, its base can be swollen to form a median bulb—in such instances, the corpus may be divided into procorpus, mesocorpus, and metacorpus.
- **Crescentic**: In the shape of a sickle, or like the new moon, with convex and concave edges.
- **Cyst**: At maturity, the body wall of certain sedentary female plant parasitic nematodes thickens, becomes resistant to decay, and turns brown after the nematode dies, resulting in a protective shell for the eggs that is termed a cyst.
- **Deirids**: A pair of lateral touch receptors situated near the nerve ring; they may present as a swollen nerve ending embedded within the lateral field, or as protruding papillae or short setae.
- **Denticle**: A small tooth or projection. The teeth can be curved, falciform (claw-shaped), triangular, fusiform, or conical.
- **Didelphic**: Having two ovaries.
- **Dorsal gland**: The dorsalmost of three or five pharyngeal glands.
- **Dorsal gland opening**: The aperture of the duct from the dorsal pharyngeal gland into the lumen of the pharynx.
- **Excretory pore**: Ventral opening in the cuticle by which the waste products of the excretory system are emitted to the exterior.
- **Flabellum (pl. flabella)**: Dorsal or ventral extensions of the cuticle with branching shape, like a fan or feather.
- **Fascicle**: A cluster, an organized grouping.
- **Fenestra**: A thin-walled transparent region of cuticle in the vulval zone, which may break down to form the hatching pore.
- **Fimbriated**: Fringed.
- **Gubernaculum**: In males, a grooved cuticularized structure, sometimes paired, which guides the spicule.
- **Guiding ring**: The anterior end of a guiding sheath of the stylet; it is cuticularized or muscular, thus appearing to be denser than surrounding tissue.
- **Hypodermal cords**: Longitudinal hypodermal thickenings lying in the lateral position.
- **Isthmus**: The segment of musculature between the medium bulb and basal bulb of the pharynx.
- **Labial papillae**: Papillae located on the lips.
- **Lateral fields**: A form of cuticular configuration above the lateral cords.
- **Lumen**: The central cavity or channel in a hollow organ or tubular structure. Internal organ space that is bounded by cells or cuticle and filled with fluids, secretions or digesting matter, but not with living cytoplasm or fluid of a body cavity.
- **Monodelphic**: Having one ovary.
- **Mucro**: An abrupt point, tip, or process that terminates the nematode tail.
- **Nerve ring**: A belt, broad, and flat in slender nematodes but narrow in forms with a large pharynx, containing axonal fibers of neurons distributed around it. The nerve cells are generally too diffuse to be termed true ganglia. The nerve ring is anchored to the ventral body wall with a fibrous extension.
- **Odontophore**: Cuticularization of the anteriormost part of the pharynx, it functions as a unit with the odontostyle.
- **Odontostyle**: A stylet terminating with a dorsally oblique aperture. The stylet is derived through evolutionary development from an odontium and originates in the pharynx wall.
- **Papilla**: Sensory organ; minute elevation of the cuticle. Any small nipple-like projection or part.
- **Pharynx**: The partly or wholly muscular tube that leads from the stoma or stylet base to the intestine and is internally lined with cuticle. The esophagus.
- **Probolae**: Small or prominent, sometimes elaborately branching, cuticular appendages encircling the oral aperture; never innervated; situated in some nematode forms in three labial and six cephalic positions.
- **Procorpus**: The anterior subdivision of the corpus, typically of cylindrical form.
- **Prodelphic**: Having a single ovary that extends anterior to the vulva.

Diplogasteridae); (D) narrow tubular buccal cavity leads to a three-part pharynx without division of corpus into procorpus versus median bulb, and with the basal bulb lacking valves (e.g., in Rhabditida: *Myctolaimus*); (E) four-part pharynx with separation of procorpus and median bulb, the latter containing longitudinal valves, whereas the large basal bulb lacks valves (e.g., in Rhabditida: Tylenchoidea); (F) detail of the stylet's basal knobs and nearby dorsal gland opening, as typical in Tylenchoidea; and (G) detail of the very strong median bulb with longitudinal valves just posterior to the dorsal gland opening, typical of some Rhabditida, nearlly all Aphelenchoidoidea, and Aphelenchoidea.

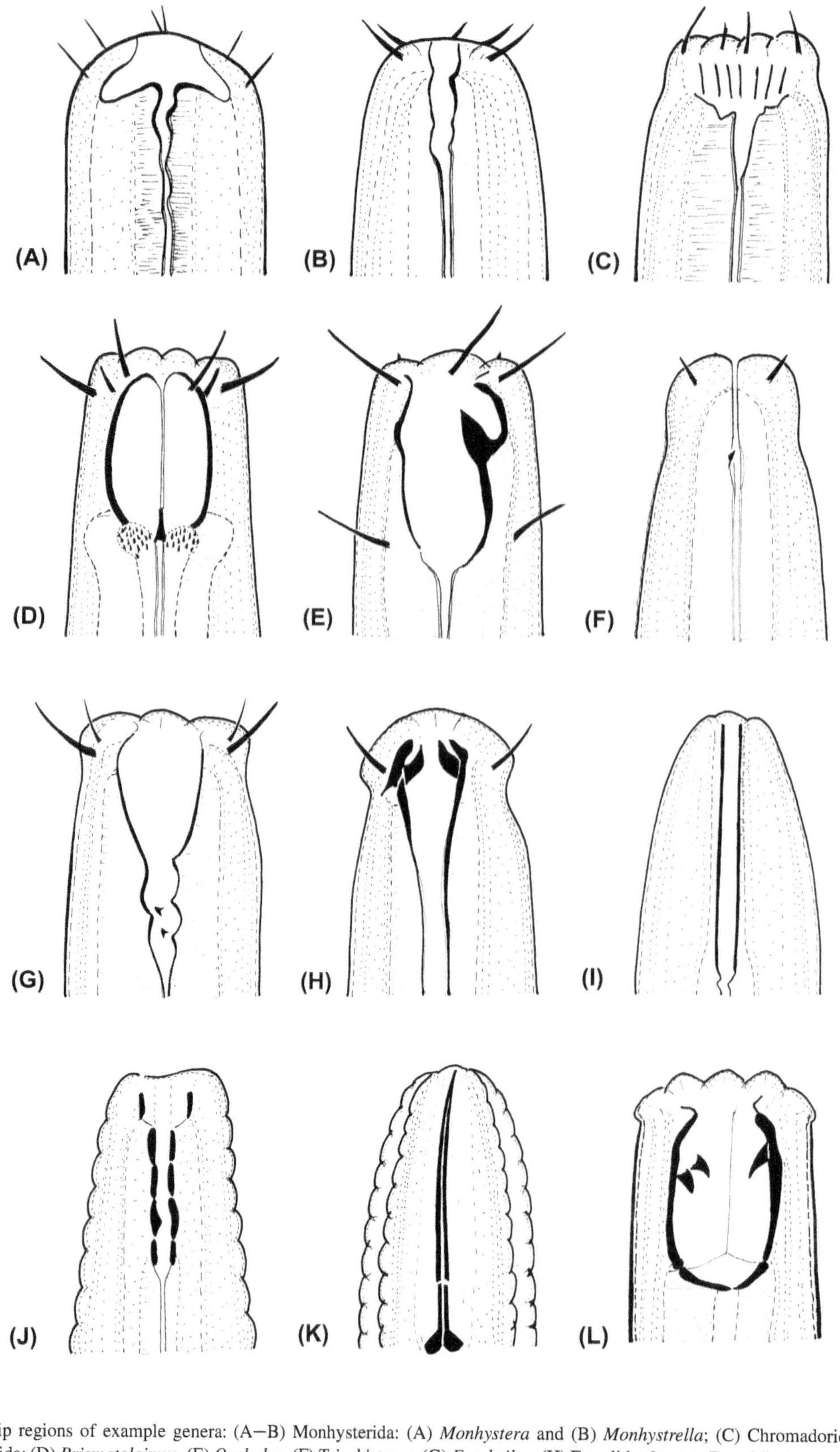

FIGURE 9.4 Lip regions of example genera: (A–B) Monhysterida: (A) *Monhystera* and (B) *Monhystrella*; (C) Chromadorida, *Paracyatholaimus*; (D–G) Triplonchida: (D) *Prismatolaimus*, (E) *Onchulus*, (F) *Trischistoma*, (G) *Eutobrilus*; (H) Enoplida: *Ironus*; (I) and (L) Mononchida: (I) *Cryptonchus* and (L) *Anatonchus*; (J–K) Rhabditida: (J) *Cephalobus* and (K) Hemicycliophora. Not to scale.

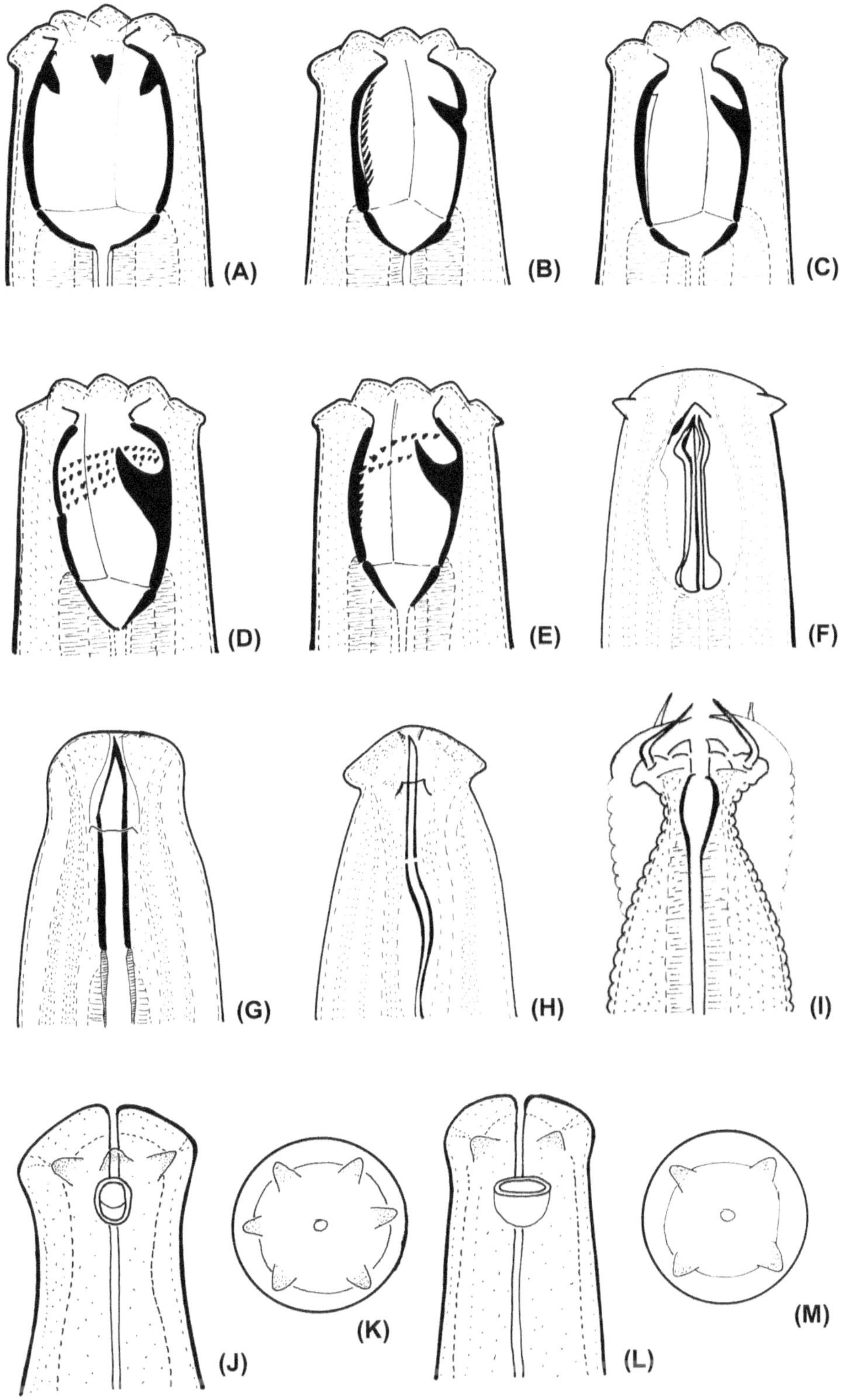

FIGURE 9.5 Lip regions of example genera: (A–E) Mononchida: (A) *Tigronchoides*; (B) *Prionchulus*; (C) *Clarkus*; (D) *Mylonchulus*; and (E) *Oligonchulus*; (F) Triplonchida, *Diphtherophora*; (G–H) Dorylaimida: (G) *Coomansinema* and (H) *Leptonchus*; (I) Plectida, *Tylocephalus*; (J–M) Mermithida: (J) and (K) *Hexamermis*; (L) and (M) *Longimermis*. Not to scale.

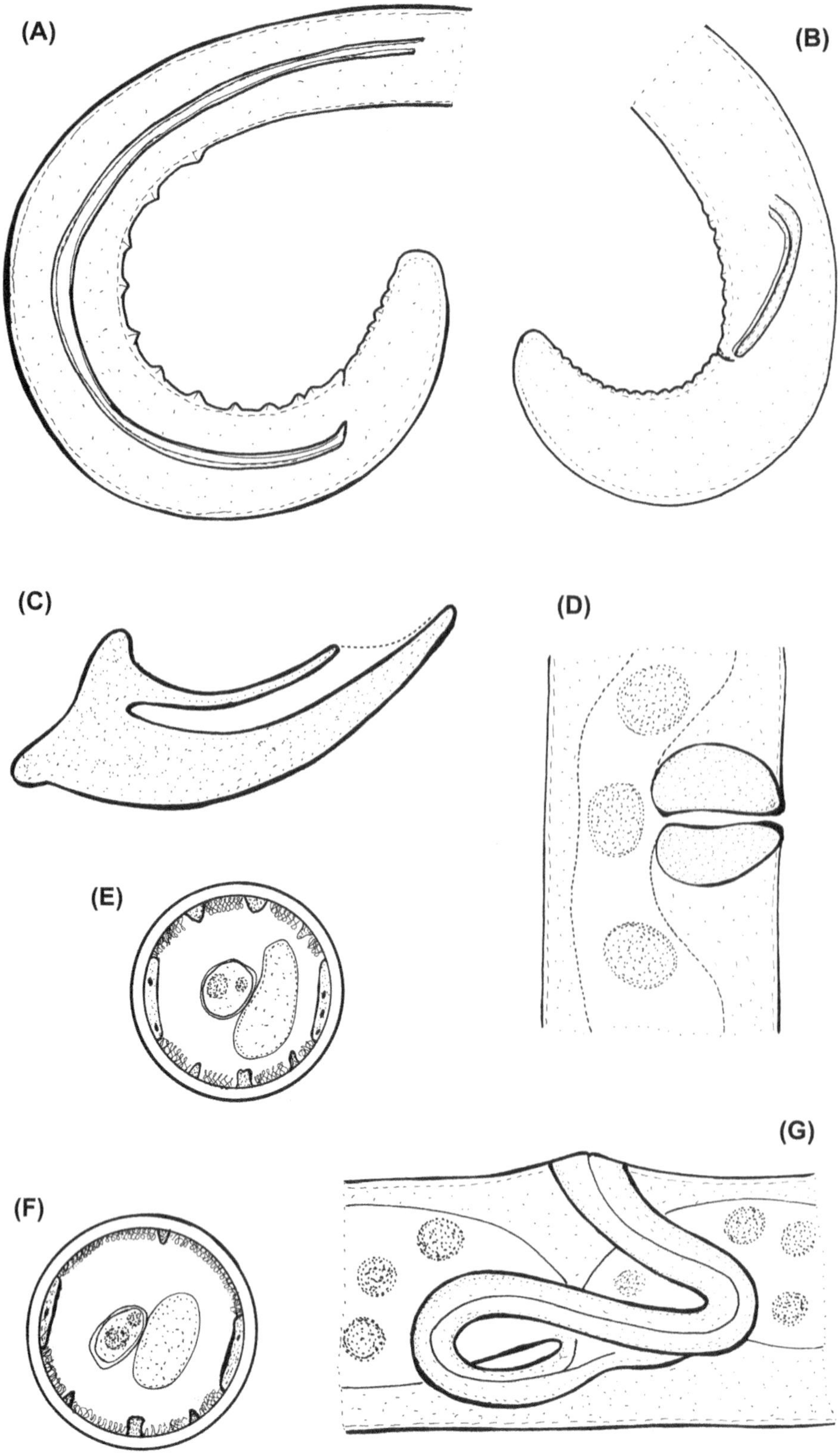

FIGURE 9.6 Reproductive structures and more unusual cross-sectional arrangements (all from Mermithida except C): (A) Male tail with very long spicules extending well anterior of the precloacal supplements, which are in this case spaced apart; (B) male tail with short spicules that do not extend anterior of the precloacal supplements, which in this case are placed in a contiguous series; (C) thorn-shaped spicule typically found in members of order Rhabditida, Aphelenchoidoidea; (D) short transverse vagina with bulbous musculature; (E) cross section of a mermithid with eight hypodermal cords; (F) cross section of a mermithid with six hypodermal cords; and (G) long coiled vagina with cylindrical musculature.

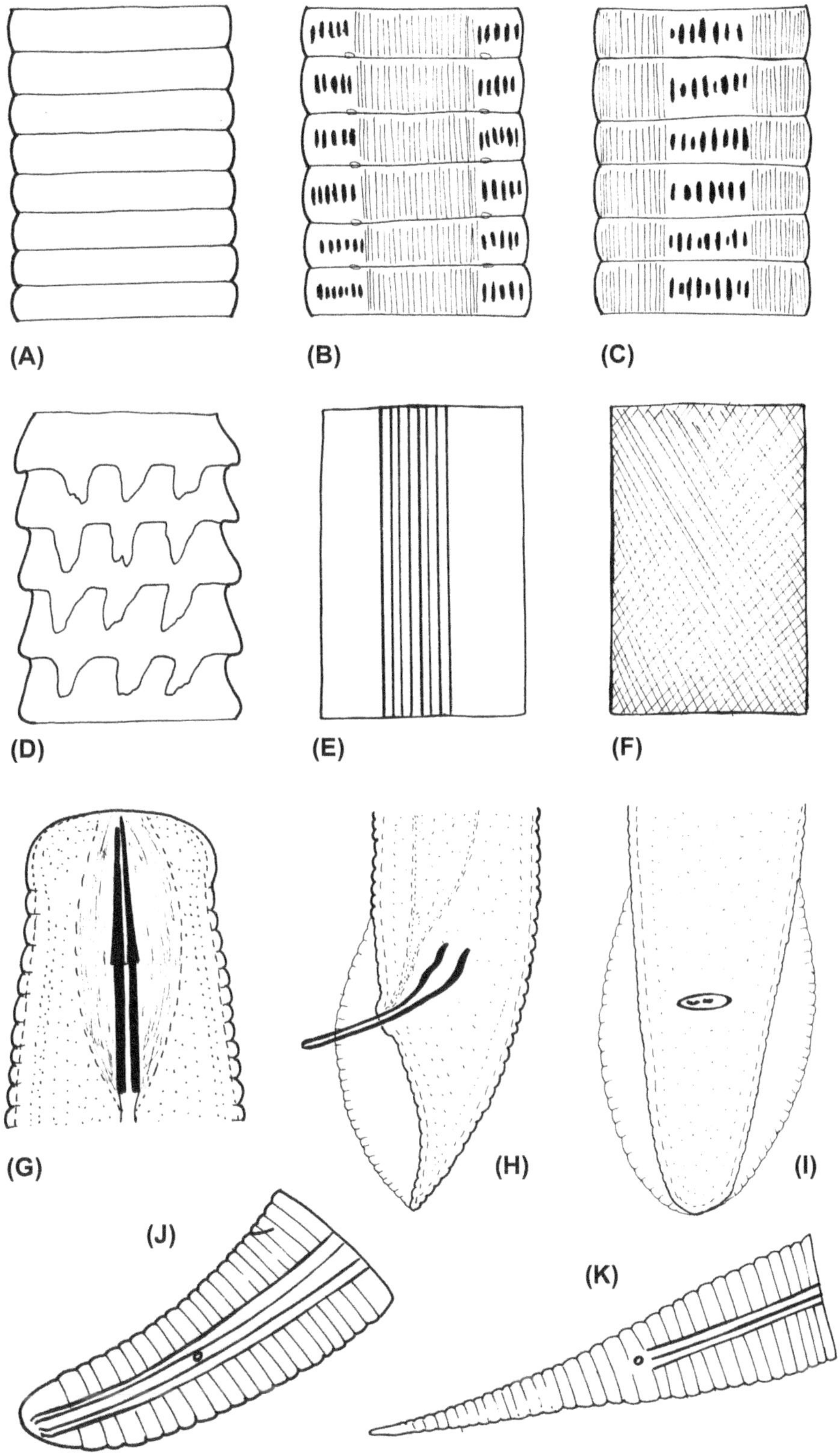

FIGURE 9.7 Cuticular structures and ornamentations: (A) annules lacking punctuations or striae; (B) annules with punctuations on the dorsal and ventral sides, versus striae in the lateral fields; (C) annules with striae on the dorsal and ventral sides, versus punctuations in the lateral fields; (D) posteriorly sloping annules with protruding cuticular scales; (E) smooth cuticle, except for the lateral field that consists, in this example, of eight lines;

- **Renette cell**: Generally, one but, in some forms, two ventral excretory cells.
- **Scutellum (pl. scutella)**: Greatly enlarged phasmids found in some species of the Hoplolaimidae.
- **Sensillum (pl. sensilla)**: A simple sense organ, or one of the structural components of a compound sense organ.
- **Seta (pl. setae)**: Tactile sensory organs of elongate cuticular structure, usually shaped like a fine hair, may or may not articulate with the cuticle proper. A bristle-like protuberance with innervation.
- **Spermatheca**: The enlarged portion of the female reproductive system, which functions as a reservoir in receiving and holding spermatozoa from the male.
- **Spicule**: Male copulatory organ used for the transfer of sperm, usually paired and often of curved shape; an elongate cuticularized piece extrusible through the cloacal opening.
- **Stoma**: The mouth cavity. Segment of the digestive tract between the oral opening and the (often triradiate) beginning of the pharynx including the oral aperture and the stoma walls.
- **Stylet**: A relatively long, rather slender, hollow feeding structure.
- **Supplements**: In males, organs of secretion, attachment, and sensation; ventromedian in position; derived of cuticle; and functioning during copulation.
- **Vestibule**: A distinctly wider anterior part of the buccal cavity in some nematode families; in family Actinolaimidae, it may present as a hollow ring on the surface of the lip region.

MATERIAL PREPARATION AND PRESERVATION

Extraction

To gather data on soil and freshwater nematode communities, it is necessary to first use methods for extracting them from their substrate. Obtaining specimens of large size is relatively simple, as they can already be collected by placing the sample in suspension, leaving the mineral phase to settle for a few seconds, and then pouring the supernatant through a sieve with 250 μm mesh size. Everything retained on the sieve is washed into a Petri dish for observation under a stereo microscope, which easily allows the separated specimens to be recognized.

To extract specimens of smaller size, various methods have been proposed. Some nematodes can be floating in water; in this case, they can be recovered by pouring water samples through a sieve with 40-μm mesh size (Faulkner & Bolander, 1966). Although this method is simple, it has the inconvenience of yielding relatively few specimens. It is therefore appropriate to process samples of a large volume to ensure that a sufficient quantity of nematodes is recovered. Sieving is also useful to obtain clean tap water if no sterile water is locally available in sufficient quantities: Some of the more opportunistic bacterial feeders can occur as contaminants in tap water, so it is prudent to run several liters of tap water through the 40 μm mesh sieve first and wash material from the sieve into a dish to inspect it for nematodes. If it becomes evident that any were transported with the tap water, then it is necessary to always prepare sufficient amounts of filtered tap water before applying any of the following methods.

A sample is typically composed of particles of clay, sand, and other sedimentary material, organic matter, nematodes, and diverse microorganisms such as rotifers, single-celled algae, fungal mycelia, protists, bacteria, and so on. The classical methods used to separate nematodes from the rest of the sample consist of placing the sample in aqueous suspension, leaving it to sediment for a few seconds (allowing the greater part of the mineral fraction to settle) and then pouring the supernatant (containing nematodes along with suspended clay particles and organic matter) through a series of sieves with appropriate mesh sizes, finishing with 40 μm. This allows the nematodes to be concentrated in a greatly reduced volume of water (Ferris et al., 1973).

For samples with elevated proportions of clay and organic matter, it is still very time-consuming to separate out nematodes from such a reduced volume by observation under a stereo microscope. The contents retained by the sieves can instead be placed on a sheet of porous paper (such as paper kitchen towel) inside a Baermann funnel. This is prepared by connecting a rubber tube to the tip of a glass funnel with diameter of 10–20 cm; the end of the rubber tube is kept closed with a Mohr pinchcock clamp. The funnel is filled with water to 1 cm below its edge; the concentrated sample is placed on a sheet of paper towel supported by a coarse steel or plastic mesh that fits within the top of the funnel, in a manner that ensures the paper sheet is in contact with the water. After about 24 h, the nematodes will have abandoned the sample by crawling down through the paper, sinking to the bottom of the closed rubber tube. They can then be collected in a Petri dish by releasing about 10 ml through the Mohr clamp.

(F) cuticle with subsurface pattern of crisscross fibers; (G) smooth lip region of a tylench with postlabial annulations (note that, in this particular example, the stylet lacks knobs); (H) lateral and (I) ventral view of the male tail of a tylench with bursa, the wings reach the tail tip in this case; (J) broadly rounded tail with a lateral field consisting of four lines, the latter extend past the phasmid and reach the tail tip; and (K) conical tail with pointed tip; in this example, the lateral field consists of three lines and stops at the phasmid.

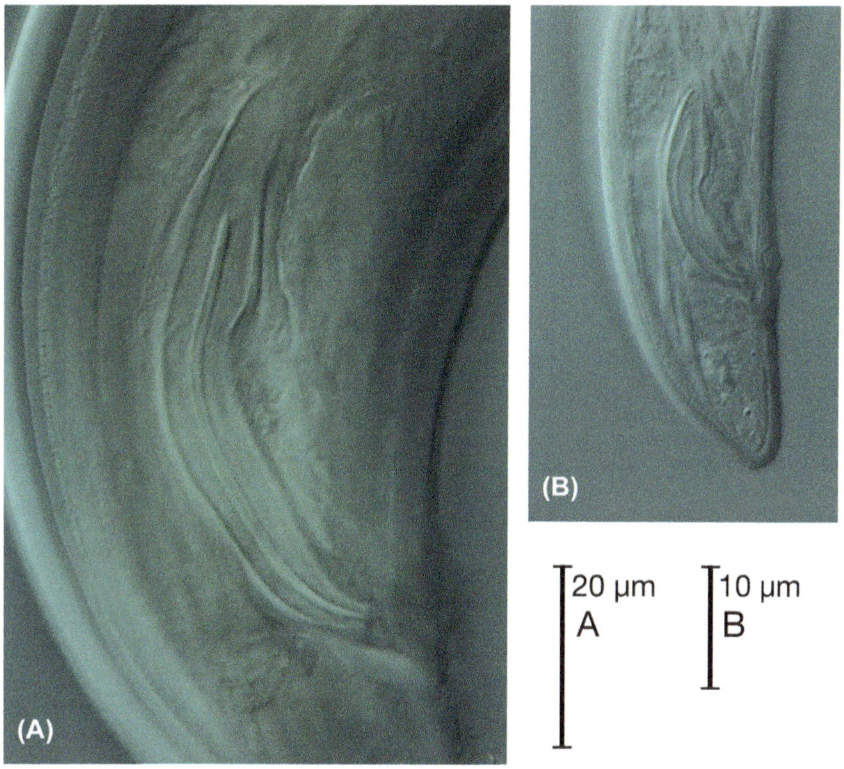

FIGURE 9.8 Two examples of the characteristic shape of the spicule in Dorylaimida, with angular anteroventral silhouette and a (narrow or broader) sclerotized centerline leading to a double tip: (A) *Labronemella*, a genus where spicule length usually exceeds 60 μm; (B) *Discolaimium*, an instance of a much smaller dorylaimid spicule.

Three important limitations apply to the use of Baermann funnels. First, as the method depends on active movement of the organisms, it will not produce meaningful quantitative data and should only be used for qualitative studies. Second, many freshwater nematodes are equipped with adhesive caudal glands or caudal hooks that enable attachment to solid surfaces, and quite a few genera are capable of swimming freely in water. As a result, such nematodes are much less liable to settle out in the rubber tube of the funnel than slower-moving species lacking adhesive caudal structures. Third, leaving a concentrated freshwater sample on a Baermann funnel for 24 h often yields many dead and decomposing individuals, more so than usually occurs with (terrestrial) soil samples. Some of the larger freshwater nematodes like Mermithida and many Dorylaimida are highly susceptible to fungal infections, particularly in confined conditions at the bottom of a funnel with stagnant water. Other groups of freshwater nematodes (e.g., Monhysteridae, Plectidae, and Prismatolaimidae) exhibit bursts of frenetic body movement in well-oxygenated water, which seems to render them more prone to paralysis once hypoxia sets in, compared with much slower-moving soil nematodes such as many Tylenchoidea or Cephaloboidea.

Various other methods have been described for extracting freshwater nematodes, using a series of sieves with increasingly finer mesh sizes (Thorne, 1961; Southey, 1970; Smith, 2001 are among the main sources). Another method allows processing of large samples (500–1000 ml or more) in relatively little time (about 30 min): Dalmasso's settling technique with buckets (1966) followed by flotation–centrifugation, which is commonly used in extraction of soil nematodes as well (Jenkins, 1964). The process works as described in the following:

1) Use a jet of water to flush the sample through a coarse sieve with large mesh size (1–2 mm) and collect everything underneath in a 5–10 l bucket. Coarse material and very large nematodes (typically the post-parasitic and adult stages of Mermithidae) retained on the coarse mesh should be set aside for subsequent observation under the stereo microscope.

Material flushed through the sieve into the bucket will include mineral particles of various sizes, organic matter, nematodes, and various other microorganisms. Within a few minutes, the mineral fraction of sand and loam sized particles will have settled on the bottom, while fine clay, nematodes, and other microorganisms will remain suspended, along with particles of organic matter. Coarse sediments typically settle sufficiently in 60 s, whereas samples with high clay content may require longer, sometimes up to 6 minutes. Next, the supernatant is poured through a 40 μm mesh sieve to separate out all but the tiniest of the foregoing elements. It is advisable to resuspend the sediment at the bottom of the bucket at least once more, to repeat the process and ensure recovery of any nematodes that were previously trapped at the bottom by the first phase of settling material.

2) The contents of the 40 μm sieve are then transferred with a jet of water into appropriately sized centrifuge tubes and submitted to a double centrifugation. The first, in aqueous solution, lasts 5 minutes at 3000 rpm. This cycle causes settling of mineral particles, nematodes, and various other microorganisms of similar size and density, whereas organic matter remains suspended and is subsequently discarded with the supernatant. The second centrifugation step is carried out with a sucrose solution prepared at a density of 1.18–1.20 g/ml. The sediment obtained from the previous step is suspended (in the same centrifuge tube) with the sucrose solution and centrifuged for 1 minute at 1800 rpm. The mineral phase will settle, whereas nematodes remain in suspension in the supernatant, which is then poured through a sieve with 5 μm mesh (to minimize any possibility of losing specimens). The retained nematodes are then washed repeatedly on the sieve, to prevent plasmolysis caused by osmotic effects of the sucrose solution. Finally, the contents of the 5 μm sieve are washed into a Petri dish with a gentle stream of water, leaving the nematodes in a clean suspension that will greatly facilitate all subsequent observations and preparations (Doucet, 1980).

Processing the Nematodes

Nematodes in the water obtained from the last extraction step can be observed under stereo microscope, but identification usually requires higher magnifications and preparation of temporary and permanent slides. The first type consists of specimens transferred into a drop of water or fixative on a glass slide, which is then covered with a cover glass. The second type requires mounting specimens in a more definitive manner in glycerin.

To prepare this type of mount, first kill the nematodes by placing them in a small vial and heating this to 65°C over a small flame or in a preheated water bath. With the aid of a Pasteur pipette, they are then transferred into a small capsule (or watch glass) containing the fixative solution. The most commonly employed is known as F.A.A. (8 ml commercial formalin, 1 ml glacial acetic acid, 20 ml 96% alcohol, 50 ml distilled water). Adding three drops of saturated picric acid will lightly stain the specimens internally, which can help in recognizing organs and structures. The specimens should be left at least 15 days in the fixative.

The following step of the process consists of transferring the nematodes to a watch glass to replace all remaining water and fixative with pure glycerin. Various methods are commonly employed for this purpose, ranging from addition to the dish of an aqueous solution with 1.5% glycerin and 25% ethanol followed by slow evaporation over several weeks, to rapid transfer methods that take only 2 days to complete (Seinhorst, 1962; Netscher & Seinhorst, 1969; Doucet, 1980).

Microscope slides can then be prepared by transferring the nematodes into a drop of glycerin on a glass slide, and placing around the drop four glass fiber fragments with thickness comparable with that of the nematodes. This will prevent that the nematodes get flattened when finally a coverslip is placed on top of the drop and its edges are sealed with nail polish.

KEY TO FRESHWATER AND TERRESTRIAL NEMATODA

Nematoda: Orders

1	Adult female length 0.25–10 mm, almost always with functional rectum and anus; adults have four hypodermal cords	2
1′	Adult female length 8–250 mm, without rectum or anus; adults have six or eight hypodermal cords (Fig. 9.6 E, F)	Mermithida, one family: **Mermithidae [p. 216]**
2(1)	Pharynx cylindrical, largely of uniform width and musculature along its entire length (e.g., Fig. 9.3 A)	3
2′	Pharynx with distinct changes in width and/or presence versus absence of muscle fibers (e.g., Figs. 9.1 A, 3 B–E)	11
3(2)	No setae on lip region (Fig. 9.2 J) or body; cuticle smooth or with many longitudinal ridges	4
3′	At least four setae present on lip region (Fig. 9.2 A), body may have additional setae; cuticle smooth, annulate or punctate (Fig. 9.7 B–D)	5
4(3)	Buccal cavity distinct, tubular or a large capsule, usually contains one or more denticles or teeth (e.g., Figs. 9.4 L, 5 A–E); cuticle without longitudinal ridges	**Mononchida [p. 217]**
4′	Buccal cavity without teeth or denticles, cavity either a tiny pore or narrow tube (e.g., Fig. 9.4 F, G); cuticle with or without prominent longitudinal ridges	**Enoplida (in part) [p. 219]**
5(3)	Amphid openings distinct and located on the lip region	6

5′ Amphids located posterior to lip region by one or more amphidial diameters, or not visible with light microscope ... 7

6(5) Slender to very slender nematodes; body wall without soil or sediment inclusions; body without large setae on conical protrusions ... **Plectida (in part) [p. 220]**

6′ Plump to very plump nematodes; body wall with large to very large annules that often have embedded soil or sediment inclusions; body usually with several pairs of large setae on conical protrusions ... Desmoscolecida, one genus: *Desmoscolex*

7(5) Amphid openings distinct and fully circular (Fig. 9.2 C); tail tip with gland opening ... 8

7′ Amphid openings tiny or larger slits, spirals, loops, question marks, or circles with a distinct break in outline (Fig. 9.2 D–I); tail tip with or without gland opening ... 9

8(7) Cuticle clearly annulate (Fig. 9.7 A); buccal cavity a slender tube (Fig. 9.4 I); female usually with compact and reflexed double ovaries ... Araeolaimida, one genus: *Cylindrolaimus*

8′ Cuticle smooth or finely annulate; buccal cavity is a funnel (Fig. 9.4 A, B); female with single long and outstretched ovary ... Monhysterida, one family: **Monhysteridae [p. 221]**

9(7) Cuticle annulate or smooth; amphid openings are slits, loops, question marks, or circles with a distinct break in their outline (Fig. 9.2, D, E, G–I); buccal cavity lacks rugae ... 10

9′ Cuticle punctate; amphid openings spiral (Fig. 9.2 F); buccal cavity with rugae anteriorly (Fig. 9.4 C) ... Chromadorida (in part), one genus: *Paracyatholaimus*

10(9) Cuticle smooth or very finely annulate; amphids are tiny slits (Fig. 9.2 G); buccal cavity narrow or wide, usually with at least one tiny denticle or a large tooth (Fig. 9.4 D–G); cardia enlarged spherical, or with three rounded gland cells protruding between pharynx and intestine; male supplements not tubular ... **Triplonchida (in part) [p. 221]**

10′ Cuticle clearly annulate; amphids are distinct slits, loops, question marks, or circles with clear break in outline (Fig. 9.2 D, E, H, I) buccal cavity lacks denticles or teeth and is usually a narrow funnel or tube; cardia without protruding gland cells; male supplements tubular ... **Plectida (in part) [p. 220]**

11(2) Pharynx composed of three or four distinct parts differing in width and/or musculature density (e.g., Figs. 9.1 A, 3 C–E) ... 12

11′ Pharynx posteriorly swollen and usually muscular, versus anteriorly slender with or without distinct muscle fibers (e.g., Fig. 9.3 B) ... 14

12(11) Swollen part of pharynx forms a muscular, short, and rounded bulb (e.g., Fig. 9.1 A); anterior part of pharynx usually muscular; tail tip with or without gland opening ... 13

12′ Swollen part of pharynx cylindrical or conical (e.g., Fig. 9.3 C–E); anterior part of pharynx with or without muscle fibers; tail without gland opening ... 17

13(12) Lip region with four setae; amphids distinct; body may have setae; tail almost always with two or more setae but never with phasmids; tail tip with or without gland opening ... **Plectida (in part) [p. 220]**

13′ Lip region rarely with setae; amphids rarely visible; body never has setae; tail lacks setae but almost always has two phasmids, no gland opening at its tip ... **Rhabditida [p. 223]**

14(11) Cuticle finely annulate or punctate; lip region with small setae; amphid circular or spiral; tail is usually ventrally or ventrolaterally curved ... 15

14′ Cuticle smooth; lip region lacks setae; amphids are slits or invisible; tail usually straight. ... **Enoplida (in part) [p. 219]**

15(14) Buccal cavity is a small tube with minute denticles or a funnel with one or two weakly sclerotized teeth; intact tail length usually $< 0.2\times$ body length; tail tip with gland opening ... 16

15′ Buccal cavity is a slender tube with one strongly sclerotized pointed to fusiform protrusible tooth; intact tail length $> 0.2\times$ body length; tail tip without gland opening ... **Triplonchida (in part) [p. 221]**

16(15) Amphids spiral; cuticle finely punctate; buccal cavity moderately sclerotized with one or two curved teeth, its shape is an irregular funnel ... **Chromadorida (in part) [p. 237]**

16′ Amphids small and circular; cuticle finely annulate; buccal cavity is a small tube with very small denticles but no sclerotization ... Desmodorida, one genus: *Prodesmodora*

17(12) Buccal cavity with a straight or curved stylet ... 18

17′ Buccal cavity with three distinct teeth (Fig. 9.4 H) ... **Enoplida (in part) [p. 219]**

18(17) Stylet usually straight, with or without a hollow lumen (Fig. 9.5 G, H); pharyngeal bulb usually long, cylindrical, and strongly muscular; spicules with characteristic shape for dorylaimids (Fig. 9.8 A, B), divided lengthwise by a sclerotized fold, not contained within an ovoid muscle capsule; females often with a more transparent posterior part of the intestine (known as prerectum) ... **Dorylaimida [p. 237]**

18′ Stylet usually curved, its anterior half never with hollow lumen; pharyngeal bulb relatively short, conical, and only weakly muscular; spicules slender, not divided lengthwise, contained within an ovoid capsule of protractor muscle fibers; female intestine lacks a distinctly transparent posterior part ... **Triplonchida (in part) [p. 221]**

Nematoda: Mermithida: Mermithidae: Genera

1	Cuticle with cross-hatch pattern composed of distinct fibers (Fig. 9.7 F)	2
1′	Cuticle without cross-hatch pattern	5
2(1)	Lip region with six papillae (Fig. 9.5 J, K)	3
2′	Lip region with four papillae (Fig. 9.5 L, M)	4
3(2)	All six labial papillae are located at equal distance from mouth; females never have pigmentation near the lip region; mature eggs in utero without brushlike extensions (Fig. 9.5 J, K)	*Hexamermis*
3′	Both lateral papillae are located directly adjacent to mouth opening, the four remaining labial papillae are placed much more peripherally; aging females develop postlabial hypodermal pigmentation that changes from yellowish-brown to bright red in color; mature eggs in utero have large brushlike extensions	*Mermis*
4(2)	Mouth opening closer to ventral side of the lip region than to dorsal side, anterior end clearly asymmetrical in lateral angle view	*Pheromermis*
4′	Mouth opening located centrally on anterior end, which is symmetrical in lateral view (Fig. 9.5 L, M)	*Longimermis*
5(1)	Mouth opening on anterior surface, not posterior to papillae and amphids	6
5′	Mouth opening on ventral side of lip region, clearly posterior to papillae and amphids	*Allomermis*
6(5)	Mouth opening closer to ventral side of the lip region than to dorsal side, anterior end clearly asymmetrical in lateral view	7
6′	Mouth opening located centrally on anterior end, symmetrical in lateral view	12
7(6)	Males with one spicule, no trace of partial fusion of paired spicules in ventral view	8
7′	Males with two separate spicules or with partly fused spicules forming a V- or Y-shaped structure in ventral view (e.g., Fig. 9.6 B)	10
8(7)	Anterior half of spicule almost straight or following body curvature, edges not projecting ventrally (e.g., Fig. 9.6 A); tail usually bluntly rounded	9
8′	Entire spicule curved more than body, anterior edges projecting ventrally and sometimes also dorsally; tail usually conical	*Hydromermis* (in part)
9(8)	Spicule length $< 100\ \mu m$; genital papillae rows extend well anterior of retracted spicule	*Limnomermis* (in part)
9′	Spicule length $250\ \mu m$ to > 1 mm; genital papillae rows not reaching anterior end of retracted spicule (Fig. 9.6 A)	*Gastromermis*
10(7)	Body with six hypodermal cords in cross section (Fig. 9.6 F); spicules separate or distally fused into a V-shaped structure	11
10′	Body with eight hypodermal cords in cross section (Fig. 9.6 E); spicules fused over most of their length into a Y-shaped structure	*Hydromermis* (in part)
11(10)	Vagina sigmoid or outstretched, oriented more or less parallel to body axis, length $> 0.5\times$ maximum body width (e.g., Fig. 9.6 G)	*Strelkovimermis* (in part)
11′	Vagina clearly tapering at one or both ends, transverse, length $< 0.3\times$ maximum body width (e.g., Fig. 9.6 D)	*Bathymermis* (in part)
12(6)	Spicules not helicoid	13
12′	Spicules helically coiled around each other along most of their length	*Amphimermis*
13(12)	Spicules with distinct constriction or articulation at 20%–50% of their length	14
13′	Spicules without constriction or articulation [Spicules of the Nearctic species *Strelkovimermis rubtsovi* Johnson & Kleve, 2002 have a constriction/articulation at 10%–15% of their length. Johnson & Kleve (2002) described the spicules of *S. rubtsovi* as "pinched" at one-third to one-fourth of their length, but their photographs show that this constriction occurs much more anteriorly]	15
14(13)	Spicule length $> 2\times$ body width at cloacal pore, geniculated into two subequal articulating sections with interrupted cuticle at their hinge	*Divisispiculimermis*
14′	Spicules $< 1.5\times$ body width at cloacal pore, with distinct constriction at 20%–30% of their length, sclerotized cuticle of each spicule is not completely geniculated	*Isthmusimermis*
15(13)	Spicules L- or J-shaped, with a straight angle or a tight 90° curve at 50%–80% of their length	16
15′	Spicules more or less gradually curved, without a straight angle or a tight 90° curve at 50%–80% of their length	17
16(15)	Spicules L-shaped, shorter than $200\ \mu m$, with a straight angle at 60%–80% of their length; female body length < 20 mm	*Ditremamermis*
16′	Spicules J-shaped, over $400\ \mu m$ long, with a tight 90° curve at 50%–60% of their length; female body length > 60 mm	*Meximermis*
17(15)	Spicule single or a fused pair forming a single Y-shaped structure	18
17′	Spicules paired, sometimes fused at their distal tips into a V-shaped structure	19

18(17) Spicule single, anterior margins not projecting ventrally; transverse body section with six hypodermal cords ...... *Limnomermis* (in part)

18′ Spicule single or a fused pair forming a single Y-shaped structure, anterior margins projecting ventrally and sometimes dorsally; transverse body section with eight hypodermal cords .... *Hydromermis* (in part)

19(17) Vagina sigmoid or outstretched, oriented more or less parallel to body axis 20

19′ Vagina outline is a barrel or irregular funnel, oriented transversely 21

20(19) Transverse body section with eight hypodermal cords (Fig. 9.6 E) .... *Isomermis*

20′ Transverse body section with six hypodermal cords (Fig. 9.6 F) .... *Strelkovimermis* (in part)

21(19) Precloacal genital papilla series longer than 1.5× precoacal body width 22

21′ Precloacal genital papilla series no longer than 1.25× precloacal body width .... *Agamermis*

22(21) Male tail length > 1.25× body width at cloacal pore 23

22′ Male tail length < 1.0× body width at cloacal pore .... *Bathymermis* (in part)

23(22) Transverse body section with eight hypodermal cords (Fig. 9.6 E) 24

23′ Transverse body section with six hypodermal cords (Fig. 9.6 F) .... *Mesomermis/Neomesomermis* [Nickle (1972) considered *Mesomermis* a *nomen nudum* and proposed the genus *Neomesomermis* to contain species formerly placed in *Mesomermis*. Some authors have followed this proposal (e.g., Vargas et al., 1980), but *Mesomermis* continues to be used frequently as valid genus name in taxonomic literature published since; see e.g., Poinar & Takaoka (1981); Camino et al. (2015)]

24(23) Spicule length < 2× body width at cloacal pore .... *Octomyomermis*

24′ Spicule length > 2× body width at cloacal pore .... *Romanomermis* [The only Neotropical records for *Romanomermis* to date are intentional releases in Mexico of nonnative species *Romanomermis culicivorax* Ross & Smith 1976 and *Romanomermis iyengari* Welch, 1964 during biocontrol experiments (from mass produced inocula reared in laboratory conditions, Pérez-Pacheco et al., 2004; 2009)]

Nematoda: Mononchida: Families

1 Buccal cavity is a slender tube or a moderately wide funnel, weakly to moderately sclerotized, with one to three denticles or one moderately sized tooth; suborder: Bathyodontina 2

1′ Buccal cavity is a wide and strongly sclerotized barrel, usually with one or more large teeth (e.g., Figs. 9.4 L, 5 A–E); suborder: Mononchina 4

2(1) Buccal cavity is a tube, posteriorly with 1–3 minute denticles; gland opening at tail tip small and simple; superfamily: Cryptonchoidea 3

2′ Buccal cavity is a funnel with moderately sized anterior tooth; gland opening at tail tip sclerotized, with a slender or tapering plug .... Superfamily: Mononchuloidea, one family: **Mononchulidae [p. 217]**

3(2) Buccal tube length < 2× lip region width; tail very short, rounded with apical gland opening .... Bathyodontidae, one genus: *Bathyodontus*

3′ Buccal tube length > 4× lip region width (Fig. 9.4 I); intact tail long conoid with gland opening, or filiform without gland opening .... Cryptonchidae, one genus: *Cryptonchus*

4(1) Posterior end of buccal capsule flattened transversely (e.g., Figs. 9.4 L, 5 A); cardia internally with three sclerotized radii that show up as bulbiform "tubercles"; superfamily: Anatonchoidea 5

4′ Posterior end of buccal capsule more or less tapering (e.g., Fig. 9.5 B–E); cardia internally without sclerotized structures; superfamily: Mononchoidea 6

5(4) Buccal capsule with or without dorsal tooth, never with two subventral teeth, although two rows of subventral denticles may occur .... **Iotonchidae [p. 218]**

5′ Buccal capsule with one dorsal tooth and two or four subventral teeth of equal or smaller size, without rows of subventral denticles (e.g., Figs. 9.4 L, 5 A) .... **Anatonchidae [p. 218]**

6(4) Subventral buccal cavity plates with multiple transverse rows of denticles (e.g., Fig. 9.5 D, E) .... **Mylonchulidae [p. 218]**

6′ Subventral buccal cavity plates lack denticles, or denticles not in rows, or denticles only on longitudinal ridges .... **Mononchidae [p. 218]**

Nematoda: Mononchida: Bathyodontina: Mononchuloidea: Mononchulidae: Genera

1 Labial papillae angular, lip region caplike in appearance (outline similar to Fig. 9.5 F–H); tail length < 2× anal body width, tip bluntly rounded with apical or dorsally subapical gland opening .... *Oionchus*

1′ Labial papillae flush with rounded lip region; tail length > 4× anal body width, tip bluntly rounded with gland opening shifted to ventral side .... *Mononchulus*

Nematoda: Mononchida: Mononchina: Anatonchoidea: Iotonchidae: Genera

1	Buccal capsule with dorsal tooth and/or subventral denticles	2
1′	Buccal capsule without teeth or denticles	*Nullonchus*
2(1)	Dorsal tooth located in posterior half of buccal capsule; tail tip with gland opening	3
2′	Dorsal tooth located in anterior half of buccal capsule; tail tip without gland opening	4
3(2)	Dorsal tooth on an outgrowth of sclerotized buccal wall	*Supronchus*
3′	Dorsal tooth directly on sclerotized buccal wall	*Iotonchus*
4(2)	Buccal capsule without longitudinal ridges	5
4′	Buccal capsule with longitudinal sclerotized ridge opposite dorsal tooth	*Jensenonchus*
5(4)	Lip region strongly offset by a deep incisure; tail straight	*Caputonchus*
5′	Lip region weakly offset, without incisure; tail curved	*Mulveyellus*

Nematoda: Mononchida: Mononchina: Anatonchoidea: Anatonchidae: Genera

1	All teeth are angled posteriad (Figs. 9.4 L, 5 A)	2
1′	All teeth are angled anteriad (Fig. 9.5 B–E)	3
2(1)	Teeth placed posterior to anterior edge of sclerotized buccal capsule, at approximately one-third to half its length (Fig. 9.4 L)	*Anatonchus*
2′	Teeth placed at anterior edge of sclerotized buccal capsule, or very close (Fig. 9.5 A)	*Tigronchoides*
3(1)	Dorsal tooth and both subventral teeth located far posterior in buccal capsule	4
3′	Dorsal tooth located in anterior half of buccal capsule	5
4(3)	Females with two ovaries and very long tail; male tail <2× anal body width; tail tip without gland opening	*Doronchus*
4′	Females with one or two ovaries and tail equal to male tail; tail tip with or without gland opening	*Miconchus*
5(3)	Buccal capsule with two pairs of subventral teeth	6
5′	Buccal capsule with one pair of subventral teeth located far posterior	*Promiconchus*
6(5)	Two pairs of subventral teeth, respectively, located just anterior and just posterior to middle of buccal capsule	*Paracrassibucca*
6′	Two pairs of subventral teeth, respectively, located opposite dorsal tooth and far posterior in buccal capsule	*Crassibucca*

Nematoda: Mononchida: Mononchina: Mononchoidea: Mylonchulidae: Genera

1	Subventral buccal sectors anteriorly with five or more transverse denticle rows forming two rasps	2
1′	Subventral buccal sectors bearing two transverse rows with just a few denticles each (Fig. 9.5 E)	*Oligonchulus*
2(1)	Buccal capsule posteriorly without denticles (Fig. 9.5 D)	*Mylonchulus*
2′	Buccal capsule posteriorly with two additional clusters of denticles, these are unordered	*Brachonchulus*

Nematoda: Mononchida: Mononchina: Mononchoidea: Mononchidae: Genera

1	Buccal capsule with one dorsal tooth and two or more subventral teeth	2
1′	Buccal capsule with one dorsal tooth but no subventral teeth (there may be rows of subventral denticles)	5
2(1)	Dorsal tooth in buccal capsule posterior half; only two subventral teeth	3
2′	Dorsal tooth in of buccal capsule anterior half; eight or more subventral teeth are placed in two or four irregular rows	4
3(2)	Subventral teeth all placed in buccal capsule posterior half	*Comiconchus*
3′	Subventral teeth are all placed in buccal capsule anterior half	*Cobbonchus*
4(2)	Subventral teeth in two irregularly spaced longitudinal rows of four–six each	*Actus*
4′	Subventral teeth in four irregularly spaced longitudinal rows of four–six each	*Sporonchulus*
5(1)	Buccal capsule without transverse ridges; tail tip usually without gland opening	6
5′	Buccal capsule with a pair of short transverse ridges roughly even with dorsal tooth; tail tip with gland opening	*Mononchus*
6(5)	Buccal capsule without longitudinal ridges opposite dorsal tooth, or ridges present but not denticulated	7

6′ Buccal capsule with a pair of well-developed longitudinal denticulated ridges in ventral corner, opposite dorsal tooth (Fig. 9.5 B) .... *Prionchulus*

7(6) Subventral plates without longitudinal ridges opposite dorsal tooth, or longitudinal ridges present but not demarcated anteriorly by a denticle .... *Coomansus*

7′ Subventral plates with longitudinal ridges opposite dorsal tooth, the anterior ends of these ridges form a denticle (Fig. 9.5 C) .... *Clarkus*

Nematoda: Enoplida: Families

1 Buccal cavity tiny, very short, or an indistinct narrow tube; tail tip with or without gland opening 2

1′ Buccal cavity prominent, if tubular then very distinct; tail tip usually lacks gland opening 4

2(1) Body slender to very slender, 0.4–3.5 mm long; buccal cavity is a tiny pore or short and narrow with minute posterior denticles; pharynx long and lacking a rounded basal bulb; females monodelphic with one distinct ovary; tail tip with or without gland opening and mucro 3

2′ Body not especially slender, length < 0.7 mm; buccal cavity very short or an indistinct narrow and long tube; pharynx short with rounded basal bulb; females didelphic, both ovaries are often tiny; tail tip with gland opening or mucro .... **Rhabdolaimidae [p. 219]**

3(2) Body very slender; lip region lacks setae but can have distinct pores; buccal cavity is a tiny pore; tail tip without gland opening or mucro .... **Alaimidae [p. 219]**

3′ Body slender; lip region with short setae; buccal cavity short and narrow with tiny posterior denticles (Fig. 9.4 F); tail tip with gland opening or mucro .... **Trischistomatidae [p. 220]**

4(1) Cuticle smooth; lateral fields never present; lip region offset; buccal cavity with three sclerotized teeth 5

4′ Cuticle with many longitudinal ridges or striae, many longitudinal and transverse rows of punctations, and/or many lateral lines in two lateral fields; lip region not offset but may have thin oral disk; buccal cavity is a long tube without teeth 6

5(4) Buccal cavity cylindrical, posteriorly with three triangular and large teeth; tail tip with gland opening .... Thalassogeneridae, one genus: *Thalassogenus*

5′ Buccal cavity with three anterior claw-shaped teeth (Fig. 9.4 H) and narrow posterior tube shorter than pharynx; tail tip lacks gland opening .... Ironidae, one genus: *Ironus*

6(4) Body 3.5–5 mm long; cuticle may have transverse and longitudinal rows of small punctations; lateral fields wide but faint or distinct and composed of over 20 incisures each; tail short, rounded .... Isolaimidae, one genus: *Isolaimium*

6′ Body shorter than 1.5 mm; cuticle with longitudinal ridges but no punctations or lateral fields; tail long and slender with tapering or star-shaped tip .... **Aulolaimidae [p. 220]**

Nematoda: Enoplida: Rhabdolaimidae: Genera

1 Buccal cavity is a short tube with one anterior dorsal denticle and two posterior subventral denticles .... *Udonchus*

1′ Buccal cavity is an indistinct narrow tube with three anterior denticles that may appear as three tiny granules .... *Rhabdolaimus*

Nematoda: Enoplida: Alaimidae: Genera

1 Amphid openings are not visible, or are tiny pores located far posterior to lip region 2

1′ Amphid openings are slits, crescents, elliptic, or circular, their location varies from edge of lip region to far posterior of lip region 3

2(1) Labial sensilla papilliform; female with posterior ovary only .... *Alaimus*

2′ Labial sensilla large and cup-shaped; female with two ovaries .... *Caviputa* (females)

3(1) Cuticle without longitudinal ridges 4

3′ Cuticle with many longitudinal ridges .... *Cristamphidelus*

4(3) Labial sensilla large and cup-shaped; females usually with one ovary only; vagina with thin cuticular walls 5

4′ Labial sensilla papilliform or tiny pits; females usually have two ovaries; vagina with thick cuticular walls .... *Amphidelus*

5(4) Amphids posterior to labial sensilla by a distance of at least one lip region width 6

5′ Amphids located immediately posterior to labial sensilla .... *Caviputa* (males)

6(5) Amphids oval with sclerotized margins; females with anterior ovary only .... *Etamphidelus*

6′ Amphids crescentic; females usually with posterior ovary only (rarely with two ovaries or with anterior ovary only) ..... *Paramphidelus*

Nematoda: Enoplida: Trischistomatidae: Genera

1 Tail weakly ventrally curved, its posterior half is much thinner than anterior half .... *Tripylina*

1′ Posterior body and tail curved ventrolaterally, compression by coverslip often causes both to appear as if dorsally curved; tail tapers evenly (Fig. 9.4 F) .... *Trischistoma*

Nematoda: Enoplida: Aulolaimidae: Genera

1 Buccal tube much shorter than the pharynx; tail tip brush or star-shaped .... *Pseudoaulolaimus*

1′ Buccal tube much longer than the pharynx; tail tip pointed .... *Aulolaimus*

Nematoda: Plectida: Families

1 Pharynx without a well-defined basal bulb, or with fusiform basal bulb that lacks valves 2

1′ Pharynx with short and well-defined basal bulb containing sclerotized valves 3

2(1) Mouth opening placed on a minute projection; pharynx moderately slender with two fusiform expansions; excretory pore posterior to cardia; renette cell length > 3× body width, located far posterior to cardia .... Ohridiidae, one genus: *Domorganus*

2′ Mouth opening not projecting; pharynx robust to very long and slender, without swellings or with one weak terminal expansion; excretory pore located in buccal cavity; renette cell about two body widths long, adjacent to cardia .... **Aphanolaimidae [p. 220]**

3(1) Basal bulb ovoid to spherical with distinct transverse valves 4

3′ Basal bulb fusiform, spherical, or ovoid with longitudinal valves 6

4(3) Lips usually lack sclerotized margins; cuticle annulate; males with one preanal row of tubular and/or recessed genital supplements as well as paired caudal setae; tail tip almost always with gland opening 5

4′ Lips with sclerotized margins that give the lip region a crown-shaped appearance; cuticle punctate; males with two rows of subventral genital papillae or setae (never tubular or recessed); tail tip never with gland opening .... **Metateratocephalidae [p. 220]**

5(4) Buccal cavity anteriorly with short sclerotized section and a distinctly longer unsclerotized posterior section; females with two ovaries; tail tip almost always with gland opening. 6

5′ Buccal cavity is a slender cylindrical tube followed by a closed unsclerotized section of similar length; females with one anterior ovary; tail tip hardly ever with gland opening .... Chronogasteridae, one genus: *Chronogaster*

6(3, 5) Basal bulb valves are cup-shaped, folded, or quarter circles, often covered in parallel striae or fine denticles; tail short to long and slender .... **Plectidae [p. 221]**

6′ Basal bulb valves flat and smooth; tail always long and slender .... **Leptolaimidae [p. 221]**

Nematoda: Plectida: Aphanolaimidae: Genera

1 Buccal cavity is small to minute, without strongly sclerotized arches; pharynx very long and slender; amphid opening width 0.3× lip region width 2

1′ Buccal cavity well-developed, anteriorly with strongly sclerotized arches and posteriorly robust cylindrical; pharynx robust and short; amphid opening width <0.2× lip region width .... *Anonchus*

2(1) Buccal cavity without vestibule; males with tubular supplements only 3

2′ Buccal cavity with a small anterior vestibule or antechamber; male usually with a long series of cup-shaped supplements preceding tubular supplements .... *Aphanonchus*

3(2) Buccal cavity short and narrow everywhere, with or without a short sclerotized cylindrical part; amphid with or without central elevation; males with 2–13 tubular supplements .... *Aphanolaimus*

3′ Buccal cavity always with short sclerotized cylindrical part whose wide is 0.3–0.5× lip region width; amphid never with central elevation; males with 8–19 tubular supplements .... *Paraphanolaimus*

Nematoda: Plectida: Metateratocephalidae: Genera

1 Lip region distinctly offset .... *Metateratocephalus*

1′ Lip region not offset .... *Euteratocephalus*

Nematoda: Plectida: Plectidae: Genera

The anterior features of several genera in this family require differential interference contrast or scanning electron microscopy (SEM) for correct interpretation of their true shape and configuration.

1	Amphid shape resembles an open circle or question mark; sclerotized section of buccal cavity has straight walls and is roughly as long as wide; lateral cuticle lacks pores or has unsclerotized glandular pores	2
1′	Amphids are slits; sclerotized stoma section very short with semicircular walls; lateral cuticle with two irregular rows of sclerotized gland openings	*Anaplectus*
2(1)	Lip region bilaterally symmetrical with greatly modified lips that may have fimbriated projections or edges; cephalic setae modified into elongate horns or flaps; postlabial region expanded dorsally and ventrally	3
2′	Lip region radially symmetrical with all lips of equal shape: flattened, rounded or drop-shaped, in a few species with sclerotized tips or edges; cephalic sensilla setiform; postlabial region not expanded	*Plectus*
3(2)	Lip region with fimbriated edges or projections; cephalic setae modified into transparent flaps with incised edges; postlabial expansions roughly hemispherical, joining the body cuticle at a nearly straight angle	4
3′	Lip region lacks fimbriated features; cephalic setae modified into long horns; postlabial expansions taper or join the body cuticle at an oblique angle (Fig. 9.5 I)	*Tylocephalus*
4(3)	Lip region with fimbriated projections; postlabial expansions smooth	5
4′	Lip region with fimbriated edges; postlabial expansions annulate	*Ereptonema*
5(4)	One pair of fimbriated projections; cephalic flaps each have four or five tips; each lateral sensilla is placed on a long process that is moderately sclerotized	*Wilsonema*
5′	Two pairs of fimbriated projections; cephalic flaps each have three tips; each lateral sensilla is placed on a short process that does not appear sclerotized	*Neotylocephalus*

Nematoda: Plectida: Leptolaimidae: Genera

1	Amphids located clearly posterior to lip region; lateral field indistinct, without lines	*Leptoplectonema*
1′	Amphids located on lip region or first body annule; lateral field with two lines	*Paraplectonema*

Nematoda: Monhysterida: Monhysteridae: Genera

1	Vulva not close to anus; female tail usually straight	2
1′	Vulva very close to anus; female tail curved	*Geomonhystera*
2(1)	Body 0.5–1.2 mm long; buccal cavity width 0.5–0.8× lip region width (Fig. 9.4 A)	3
2′	Body <0.55 mm long; buccal cavity width 0.2–0.3× lip region width (Fig. 9.4 B)	*Monhystrella*
3(2)	Buccal cavity without denticles	4
3′	Buccal cavity with three denticles	*Tridentulus*
4(3)	Distance from anterior end to amphid > lip region width; body cavity with numerous rod-shaped crystals	*Eumonhystera*
4′	Distance from anterior end to amphid < lip region width; body cavity without rod-shaped crystals (Fig. 9.4 A)	*Monhystera*

Nematoda: Triplonchida: Families

1	Cuticle smooth or finely annulate; buccal cavity never with large teeth; amphids are small slits, width <0.25× corresponding body width; tail tip with or without gland opening.	2
1′	Cuticle finely to distinctly annulate; buccal cavity with or without large teeth; amphids are slits, circles, or loops, width 0.14–0.7× corresponding body width; tail tip usually without gland opening	5
2(1)	Lip region without setae, cap-shaped (Fig. 9.5 F) in lateral view; anterior end contains a stylet shaped like a needle or chisel; tail very short and blunt; tail tip without gland opening	3
2′	Lip region usually with setae, not cap-shaped (Fig. 9.4 G); anterior end without stylet; stoma usually with two posterior denticles; tail usually long and often curved; tail tip with gland opening	4
3(2)	Anterior end contains a slender, curved stylet that is over 20 μm long	**Trichodoridae [p. 222]**

3′ Anterior end contains a straight, complex sclerotized structure with bulbous multilayered tip or tooth (Fig. 9.5 F) ... Diphterophoridae, one genus: *Diphtherophora*

4(2) Buccal cavity closed or nearly so, without sclerotized walls; labial setae present but often short; tail curved ventrally ... **Tripylidae [p. 222]**

4′ Buccal cavity usually open, tapering at one or both ends, typically with sclerotized walls; labial setae usually present and long; tail straight or curved ventrally ... **Tobrilidae [p. 222]**

5(1) Buccal cavity length > lip region width, with sclerotized lining and/or teeth ... 6

5′ Buccal cavity minute, much shorter than lip region width, without any sclerotized structures ... Bastianiidae, one genus: *Bastiania*

6(5) Pharynx cylindrical along its entire length; amphid openings are slits; buccal cavity usually wide with sclerotized walls, with or without large triangular teeth ... 7

6′ Pharynx with a short basal bulb; amphid openings circular; buccal cavity a slender tube with one sclerotized pointed to fusiform protrusible tooth ... Odontolaimidae, one genus: *Odontolaimus*

7(6) Buccal cavity with one or more large teeth (Fig. 9.4 E) ... **Onchulidae [p. 223]**

7′ Buccal cavity without large teeth but often with many denticles posteriorly (Fig. 9.4 D) ... Prismatolaimidae, one genus: *Prismatolaimus*

Nematoda: Triplonchida: Trichodoridae: Genera

1 Females monodelphic, prodelphic ... 2

1′ Females didelphic, amphidelphic ... 4

2(1) Basal bulb overlaps with anterior intestine; vulval region without lateral body pores ... 3

2′ Basal bulb not overlapping intestine; vulval region has lateral body pores ... *Monotrichodorus*

3(2) Vulva positioned at 60%–66% of body length; caudal pores absent ... *Ecuadorus*

3′ Vulva positioned at over 80% of body length; caudal pores present ... *Allotrichodorus*

4(1) Basal bulb usually overlaps with anterior intestine; vulval region does not have a pair of lateral body pores ... 5

4′ Basal bulb usually does not overlap with anterior intestine; vulval region usually with a pair of lateral body pores ... *Trichodorus*

5(4) Vulva positioned at 60%–64% of body length; caudal pores absent ... *Nanidorus*

5′ Vulva positioned at less than 60% of body length; caudal pores present ... *Paratrichodorus*

Nematoda: Triplonchida: Tripylidae: Genera

1 Cephalic setae placed slightly posterior to labial setae; cardia with three rounded but separate gland cells ... *Tripyla*

1′ Cephalic and labial setae are placed at same level; cardia large and rounded, composed of three stacked pairs of cells ... *Tripylella*

Nematoda: Triplonchida: Tobrilidae: Genera

1 Lip region with setae; adult body length > 0.9 mm ... 2

1′ Lip region with papillae only; adult body length < 0.7 mm ... *Tobrilia*

2(1) Posterior end of buccal cavity has one widened area with one or two denticles (Fig. 9.4 G) ... 3

2′ Posterior end of buccal cavity with two wider pockets that are clearly separated from one another, each pocket contains one denticle ... *Neotobrilus*

3(2) No denticles in the anterior part of buccal cavity, widened area at posterior end of buccal cavity has two denticles (Fig. 9.4 G) ... 4

3′ Anterior part of buccal cavity has one denticle, widened area at posterior end of buccal cavity has one denticle ... *Epitobrilus*

4(3) Anterior part of buccal cavity tapering gradually toward the posterior widening with two denticles (Fig. 9.4 G); longest setae on lip region measure 40%–60% of lip region width; male supplements often enlarged, protruding, and covered with minute spines ... *Eutobrilus*

4′ Anterior part of buccal cavity bottoming out transversely at transition to the posterior widening with two denticles; longest setae on lip region measure up to 40% of lip region width; male supplements always small, recessed, and lacking a spinose surface ... *Tobrilus*

Nematoda: Triplonchida: Onchulidae: Genera

1 The largest tooth or teeth are triangular, conical, or needle-shaped ... 2

1′ The buccal cavity contains three claw-shaped teeth ... *Stenonchulus*

2(1) Buccal cavity with one large dorsal tooth plus many rows of hooks or denticles ... 3

2′ Buccal cavity with one large dorsal tooth but without other teeth, hooks, or denticles ... *Onchulus*

3(2) Anterior walls of buccal cavity with numerous obliquely transverse rows of denticles, the dorsal tooth is triangular in shape, no other teeth or lancets are present ... *Limnonchulus*

3′ Walls of buccal cavity with 12 longitudinal rows of hooks or denticles, the dorsal tooth is elongate conical and accompanied by multiple pairs of slender lancets ... *Kinonchulus*

Nematoda: Rhabditida: Suborders

This key includes the suborders of Rhabditida as well as three families of uncertain classification at suborder level.

1 Buccal cavity is a pore, tube, or funnel, with or without denticles or teeth, never with a two- or three-part stylet; basal bulb muscular or glandular, not overlapping intestine; rectum and anus well developed ... 2

1′ Buccal cavity contains a stylet shaped like a dart, with two (cone + shaft) or three parts (cone + shaft + knobs); basal bulb glandular, often overlaps intestine; rectum and anus often poorly developed, sometimes wholly absent ... **Tylenchina (in part) [p. 224]**

2(1) Lip region rarely crown-shaped, usually lacking sclerotized lip margins; pharynx almost always with a distinct isthmus preceding a muscular or glandular basal bulb, the latter respectively has or lacks transverse valves; tail usually with a pair of phasmids ... 3

2′ Lip region crown-shaped with sclerotized lip margins; cylindrical part of pharynx lacks separation of corpus and isthmus, basal bulb muscular with transverse valves; tail lacks phasmids ... Teratocephalidae, one genus: *Teratocephalus*

3(2) Deirids absent or present; if present, then they either emerge laterally as papillae or short setae near level of nerve ring, or they are embedded in the lateral chords as transverse nerve endings, never modified into longitudinal rods; median bulb present or absent; females have one or two ovaries; spicules fused or paired, equal in size and shape, or completely absent ... 4

3′ Deirids consist of a pair of longitudinal rods measuring 8−12 μm, embedded in lateral chords near level of nerve ring; median bulb absent; females with one anterior ovary only; spicules paired, of unequal size and shape ... Brevibuccidae, one genus: *Brevibucca*

4(3) Buccal cavity symmetrical in lateral view, usually lacks denticles, never has teeth ... 5

4′ Buccal cavity weakly or strongly asymmetrical in lateral view, at least in its posterior part, usually with one or more denticles, sometimes also with distinct teeth ... 9

5(4) Buccal cavity length $< 0.2\times$ pharynx length; basal bulb muscular, usually with transverse valves; median bulb present or absent ... 6

5′ Buccal cavity length $\geq 0.2\times$ pharynx length; basal bulb glandular without transverse valves; median bulb absent ... **Rhabditina (in part) [p. 234]**

6(5) Lip region almost never has setae or curved brushes; median bulb present or absent; male has three pairs of precloacal genital sensilla or fewer; tail tip rarely sclerotized and/or a hook, very rarely dorsally curved, never with a double claw ... 7

6′ Lip region with six or six + four stout setae; median bulb absent; male with two−seven pairs of precloacal genital sensilla; tail tip often is a sclerotized mucro terminating as a dorsally curved hook or double claw ... **Chambersiellidae [p. 236]**

7(6) Median bulb present; females didelphic; males with or without bursa; if present, then the bursal wings always have rays ... 8

7′ Median bulb present or absent; females monodelphic; males with or without bursa; if present, then bursal wings usually lack rays ... **Tylenchina (in part) [p. 224]**

8(7) Buccal cavity evenly tubular along most of its length; median bulb single; males often have a bursa, its wings always have rays ... **Rhabditina (in part) [p. 234]**

8′ Buccal cavity consists of a short anterior tube followed by a longer posterior funnel, the latter has muscular walls along its entire length; median bulb double; males never have a bursa ... **Tylenchina (in part) [p. 224]**

9(4) Cuticle usually without detached layer; vagina without loose cuticular extension along ventral body surface; spicules present, paired or partly fused ... 10

9′ Cuticle with detached outer layer that loosely folds back on itself around the anterior body and/or near the anus around the tail; vagina has loose cuticular extension along ventral body surface; spicules entirely absent ... Myolaimina, Myolaimidae, one genus: *Myolaimus*

10(9) Cuticle rarely with distinct annulation; buccal cavity evenly tubular along its length; median bulb usually present; females usually didelphic; male bursa present or absent; if absent, then genital sensilla are setae .... **Rhabditina (in part) [p. 234]**

10′ Cuticle often clearly annulate; buccal cavity consists anteriorly of a short tube; its remainder forms an equally long or clearly longer narrow canal or funnel; this posterior canal or funnel has muscular walls along its entire length; median bulb usually absent; females always monodelphic; males never with bursa, their genital sensilla are indistinct or shaped as papillae .... **Tylenchina (in part) [p. 224]**

Nematoda: Rhabditida: Tylenchina: Families

This key includes the families of Tylenchina as well as one genus of uncertain classification at family level.

1 Buccal cavity is a pore, tube, or funnel with or without denticles or teeth, never with a two- or three-part stylet; basal bulb muscular or glandular, not overlapping intestine; rectum and anus well developed; males never have a bursa and always have several pairs of papilliform genital sensilla; spicules anteriorly usually with a subapical fenestra located on their dorsal side 2

1′ Buccal cavity contains a dart-shaped stylet with two (cone + shaft) or three parts (cone + shaft + knobs: Fig. 9.4 K); basal bulb glandular, often overlaps intestine; rectum and anus often poorly developed, sometimes wholly absent; males with or without bursa, genital sensilla only very rarely visible; spicules anteriorly often with an apical fenestra 8

2(1) Cuticle usually very finely annulate; lateral fields usually absent; median bulb absent or present; females with two ovaries or with anterior ovary only; gonad(s) usually reflexed in or near area with largest oocytes; spermatheca usually absent or in line, rarely offset; spicule width often $> 0.25\times$ their length 3

2′ Cuticle usually with distinct annules that are 1 μm wide or more; lateral fields present; median bulb usually absent; females with anterior gonad only, reflexed at oviduct with an offset spermatheca projecting anteriorly, entire ovary directed posteriad; spicule width $\leq 0.25\times$ their length 4

3(2) Lip region moderately sclerotized, bilaterally symmetrical with two or four rounded "lips"; buccal cavity consists of a short wider anterior section and a clearly longer posterior funnel lined entirely by muscle cells; pharynx corpus with double swelling; females have two ovaries .... Alloionematidae, one genus: *Rhabditophanes*

3′ Lip region rarely sclerotized, its symmetry is tri- or hexaradiate; buccal cavity consists of a short and moderately wide tube or chamber anteriorly followed by a short posterior funnel, the latter part is lined by muscle but usually no longer than the anterior part; pharynx with one median bulb or none; females with anterior ovary only .... **Panagrolaimidae [p. 226]**

4(2) Lip region with or without anterior projections; buccal cavity at least as long as lip region width; pharynx with muscular basal bulb and distinct transverse valves 5

4′ Lip region lacks anterior projections; buccal cavity very small or just a pore; pharynx slender, basal bulb more or less reduced, lacks transverse valves .... **Osstellidae [p. 226]**

5(4) Buccal cavity is a narrow tube or a funnel with four or more segments posteriorly; lip region width ≥ 5 μm, without oral disc; tail short, long, or very long 6

5′ Buccal cavity is slender tubular; lip width 3 μm with a minute oral disc; tail very long and slender .... Alirhabditidae, one genus: *Alirhabditis*

6(5) Lip region width usually > 5 μm, without short labial setae (there may, however, be setiform projections without innervation); tail short, long, or very long 7

6′ Lip region width 5 μm, smoothly rounded with six short labial setae; tail very long and slender .... *Macrolaimellus*

7(6) Lip region with two thin ventral flaps, two small fimbriated projections, or three groups of small filaments; buccal cavity posteriorly without denticle, anteriorly with wide section that is as long as one lip region width .... **Tricirronematidae [p. 226]**

7′ Lip region without cuticular projections, or with various cuticular projections arranged in groups of six and/or groups of three; buccal cavity anteriorly with a wider section that is usually shorter than lip region width, its posterior section usually has one dorsal denticle in the second posteriormost segment (Fig. 9.4 J) .... **Cephalobidae [p. 227]**

8(1) Cuticular annulations usually invisible or minute; stylet often lacks basal knobs; dorsal gland opening located in a strongly developed median bulb, far posterior from stylet; females with anterior ovary only; males either have two to five pairs of genital papillae or a bursa with three pairs of rays 9

8′ Cuticle often clearly annulate; stylet usually with basal knobs; dorsal gland opening located in procorpus, less than two corresponding body widths from stylet; median bulb absent, weak, or strongly developed; females with two ovaries or with anterior ovary only; males never have genital papillae, if they have a bursa, then it lacks rays 10

9(8) Pharyngeal isthmus longer than a few μm; basal bulb abuts intestine or overlaps with it; spicules slender without a distinct ventral limb; males have four to five pairs of genital papillae or a well-developed bursa with three pairs of rays .... **Aphelenchidae [p. 228]**

9′ Pharyngeal isthmus usually just a few μm long; this can cause the median bulb to appear as if it is in basal position, because the true basal bulb completely overlaps with the intestine; spicules often broad and shaped like a rose thorn, with protruding rostrum and/or limb extending ventrally from their anterior end (Fig. 9.6 C); males have two–five pairs of genital papillae, a small bursa is only rarely present and never has rays .... **Aphelenchoididae [p. 229]**

10(8) Annulation often very strong, annules often sloping posteriad in females; cuticle with or without loose extra layer; female stylet often elongate and strongly sclerotized with cone usually much longer than shaft, knobs often with anterior surface concave; distance between female stylet and median bulb is often less than twice the stylet length; basal bulb usually smaller than median bulb; males often incapable of feeding, with reduced stylet and pharynx; male tail usually lacks bursa 11

10′ Annulation fine to strong, annules not sloping; cuticle never with loose extra layer; stylet usually not very long or very strongly sclerotized; its cone shorter than shaft or about equal, rarely longer, knobs usually with anterior surface flat or convex; distance between female stylet and median bulb usually exceeds twice the stylet length; basal bulb often comparable in volume to median bulb, or larger; males usually capable of feeding, only rarely with reduced stylet and pharynx; male tail often with bursa 12

11(10) Annulation on most of female body > 2 μm wide; stylet cone length $\geq 4\times$ shaft length; distance between female stylet and median bulb $< 2\times$ stylet length .... **Criconematidae [p. 229]**

11′ Annules < 2 μm wide in females; stylet cone length $> 1\times$ shaft length; distance between female stylet and median bulb 0.16–6× stylet length .... **Tylenchulidae [p. 230]**

12(10) Slender, often very small nematodes; lip region rarely sclerotized; tail usually elongate conical to filiform; phasmids usually absent 13

12′ Robust nematodes; lip region more or less sclerotized; tail never filiform; phasmids usually present 15

13(12) Lip region usually flattened; median bulb moderately developed, basal bulb often cylindrical and several body widths long, with or without intestinal overlap; female tail usually conical with pointed tip, varying in length from less than 2 anal body widths to over 12 body widths long, but only rarely filiform; male tail with bursa 14

13′ Lip region flattened, rounded, or conical; median bulb weakly to moderately developed, basal bulb often pyriform and less than two body widths long, not overlapping intestine; female tail usually over 10 anal body widths long, elongate conical to filiform; male tail with or without bursa .... **Tylenchidae [p. 230]**

14(13) Males with shorter and much thinner pharynx than females; median bulb lacks valves; life cycle alternates between fungivorous free-living adults and insect parasitic adults. .... **Neotylenchidae [p. 232]**

14′ Both sexes with equally long and wide pharynx; median bulb absent, weak without valves, or moderately developed with small longitudinal valves; life cycle does not include insect parasitic adults .... **Anguinidae [p. 232]**

15(12) Females do not become strongly swollen; males without helical torsion of lateral field, their tail as long as one or more cloacal body widths and mostly with bursa 16

15′ Females are sedentary parasites that become strongly swollen before and after their last moult; males often with a very short blunt tail that lacks a bursa and has the cloacal opening placed subterminally 18

16(15) Lip region usually strongly sclerotized; stylet length > 10 μm, usually robust, knobs often higher than wide; basal bulb always overlaps intestine; females usually with one or two ovaries, vulva varying in position between half and four-fifths of body length; female tail with or without mucro, hemispherical to conical or subcylindrical, its length 0.5–8× anal body width 17

16′ Lip region rarely strongly sclerotized; stylet length > 10 μm, usually slender, knobs often wider than high; basal bulb rarely overlaps intestine; females usually with two ovaries and vulva at or near midbody; female tail usually conical or subcylindrical, broadly rounded without mucro, its length 2–5x anal body width .... **Dolichodoridae (= Belonolaimidae) [p. 232]**

17(16) Body straight or coiled into a loose spiral; height of lip region 0.5–1× lip region width; stylet length $\geq 3\times$ lip region width; length of basal bulb overlap with intestine $\leq 2\times$ corresponding body widths; females almost always with two ovaries; female tail usually quite short and broadly rounded; male tail conical or rounded with bursa usually reaching tip .... **Hoplolaimidae [p. 233]**

17′ Body straight or curved; lip region height usually $<0.5\times$ its width; stylet length $\leq 2\times$ lip region width; length of basal bulb overlap with intestine usually $> 2\times$ corresponding body width; females with anterior ovary only or with two ovaries; tail in both sexes usually conical to subcylindrical and longer than two anal body widths; male tail with bursa reaching tip or terminating slightly anterior to tip .... **Pratylenchidae (in part) [p. 233]**

18(15) Newly moulted females are already swollen; males have or lack a bursa 19

18′ Newly moulted females vermiform, becoming irregularly fusiform or strongly swollen as they mature; males have a bursa .... **Pratylenchidae (in part) [p. 233]**

19(18) Mature females mostly external to root tissue; male stylet well developed or very slender; male tail with or without bursa 20

19′ Mature females entirely embedded in root tissue (typically a root gall); males have fully functional stylet and pharynx; male tail lacks bursa and has subterminal cloacal opening .... Meloidogynidae, one genus: *Meloidogyne*

20(19) Mature females have one or more concave curves or folds in their body outline, usually on their ventral side; males have a very slender stylet and pharynx; male tail is curved and conical, length $\geq 2\times$ anal body width, bursa does not envelop tail tip .... Rotylenchulidae, one genus: *Rotylenchulus*

20′ Mature females are convexly rounded on all sides; males have well-developed and fully functional stylet and pharynx; male tail dome-shaped, length $< 1\times$ cloacal body width, bursa absent .... **Heteroderidae [p. 234]**

Nematoda: Rhabditida: Tylenchina: Panagrolaimidae: Genera

Live material of *Halicephalobus* and *Tricephalobus* is easily cultured, but culturing requires proper biohazard containment procedures and protective equipment: *Halicephalobus gingivalis* (also known by its synonym *Micronema deletrix*) is occasionally implicated in severe or fatal cases of granuloma and meningoencephalitis in humans, cattle, and horses (see, e.g., Anwar et al., 2015).

1 Median bulb present, distinct from procorpus 2

1′ Corpus without distinctly swollen median bulb 3

2(1) Adult size usually > 350 μm; ovary tip without flexure .... *Tricephalobus*

2′ Adult size often < 350 μm; ovary tip reflexed, cap cell directed back toward vagina instead of toward tail .... *Halicephalobus*

3(1) Lip region more or less crown-shaped, due to presence of six cuticular flaps and sometimes also 12 fine cuticular filaments; female gonad with offset spermatheca protruding anteriad from the point of gonad flexure, the latter occurs at junction of oviduct and uterus 4

3′ Lip region more or less rounded; female gonad lacks a spermatheca, or the spermatheca is present but tubular and in line with the outstretched oviduct and uterus; in both cases, the gonad flexure occurs entirely within the ovary 5

4(3) Lip region crown-shaped with six U-shaped sclerotizations; muscle-lined posterior part of buccal cavity clearly longer than the sclerotized wider anterior part; body cuticle moderately annulate; tail long conical with pointed tip .... *Panagrobelium*

4′ Lip region crown-shaped, with or without U-shaped sclerotizations; muscle-lined posterior part of buccal cavity is as long as, or shorter than, the sclerotized and wider anterior part; annulation very fine, body cuticle usually appears smooth; tail short or long conical with pointed tip .... *Panagrobelus*

5(3) Anterior part of buccal cavity relatively narrow, the sclerotized section of its lining is at least as long as the unsclerotized section preceding it; spicules never with bifid tip 6

5′ Anterior part of buccal cavity relatively wide, sclerotized section of its lining much shorter than the unsclerotized part preceding it; spicules usually with bifid tip .... *Panagrellus*

6(5) Slender to very slender nematodes, body length/width ratio usually > 30; tail elongate conical, usually > 6× anal body width; typically found in freshwater or fermenting liquids, or in saturated substrates 7

6′ Body length/width ratio rarely > 30; tail conical, usually < 6× anal body width; found in an extremely broad range of substrates including, for example, decaying wood, beetle frass and galleries, bird nests, broken plant stems, leaf litter, soil, brackish water, etc. 8

7(6) Spicules elongate, mostly curved ventrally but with slender anterior knob directed dorsally; specialists of fermenting liquids such as vinegar .... *Turbatrix*

7′ Spicules relatively broad, ventrally curved with anterior knob plump and rounded; specialists of freshwater environments .... *Propanagrolaimus*

8(6) Sclerotized part of buccal cavity with indented or thickening cuticle halfway along its length, posterior part of buccal cavity tapering, surrounded by pharyngeal muscle; spicules broad, usually with one or more sclerotized center lines .... *Panagrolaimus*

8′ Sclerotized part of buccal cavity without clear indentation or thickening of its cuticle, posterior part of buccal cavity either flat and stopping at junction with pharynx, or tapering and surrounded by pharyngeal muscle; spicules often slender without sclerotized center lines .... *Plectonchus*

Nematoda: Rhabditida: Tylenchina: Osstellidae: Genera

1 Buccal cavity consisting of more than just a pore; distinctions between pharyngeal corpus, isthmus and basal bulb well defined, basal bulb with valve rudiments .... *Deficephalobus*

1′ Buccal cavity is just a pore in the apical cuticle; pharynx with corpus and basal bulb only slightly wider than isthmus, lacking distinct boundaries, basal bulb without valves .... *Drilocephalobus*

Nematoda: Rhabditida: Tylenchina: Tricirronematidae: Genera

1 Lip region without ventral flaps, mouth opening not ventrally angled 2

1′ Lip region with two thin ventral flaps, mouth opening slightly ventrally angled .... *Trualaimus*

2(1) Lip region with one middorsal and one midventral fimbriated projection .... *Bicirronema*

2′ Lip region with one midventral and two subdorsal groups of small and sometimes irregularly branching cuticular filaments .... *Tricirronema*

Nematoda: Rhabditida: Tylenchina: Cephalobidae: Genera

The anterior features of several genera in this family require SEM for correct interpretation of shape and configuration. Moreover, intraspecific variability can exceed genus boundaries, and some species simply defy cohesive genus diagnosis. For example, two minute species placed in *Nothacrobeles* have unbranched labial probolae without tines, one of these also lacks basal ridges (see Tandingan De Ley et al., 1999).

1 Pharynx compact to very slender, isthmus rarely folded or coiled; postvulval sac does not extend past ovary tip 2

1′ Pharynx very slender, isthmus usually folded or coiled; postvulval sac usually extends past ovary tip ...... *Heterocephalobellus* (in part)

2(1) Lip region without cuticular projections 3

2′ Lip region with cuticular projections arranged in a group of six and/or a group of three 5

3(2) Lip region with six small and rounded lips, or with fused pairs of lips that appear as three very low rounded ridges (Fig. 9.4 J) 4

3′ Lip region with four low rounded lips, lateral lips reduced ..... *Pseudacrobeles (Bunobus)*

4(3) Female tail short-conical to elongate with pointed tail tip, never with mucro; lateral fields stop at phasmids in females only ..... *Pseudacrobeles (Pseudacrobeles)* (in part)

4′ Female tail usually cylindrical with rounded tip, the latter often carries an acute or blunt mucro; lateral fields extend past phasmids in both sexes ..... *Cephalobus* (in part)

5(2) Lip region without three cuticular protrusions placed around the mouth 6

5′ Lip region with one dorsal and two subventral cuticular protrusions placed around the mouth, each of these may be short or long, simple or branched, smooth or serrated 9

6(5) Lip region crown-shaped and strongly rounded, lips often with U-shaped sclerotizations connecting each pair of neighboring lips; female tail tip with a hook, which usually has one straight tip and one recurved tip 7

6′ Lip region sometimes crown-shaped but never strongly rounded, U-shaped sclerotizations present or absent; female tail tip with or without hook 8

7(6) Lip region moderately offset; lips form small converging flaps of equal size and carry one setiform projection each, with or without U-shaped sclerotizations ..... *Teratolobus*

7′ Lip region strongly offset; lips are larger converging flaps without setiform projections, lateral lips only half as wide as the subventral and subdorsal lips, U-shaped sclerotizations connect each pair of neighboring lips ..... *Panagrolobus*

8(6) Lip region no wider than first body annule, radially symmetrical with six equal lips projecting anteriad as thin serrate flaps, their combined edges contain a total of 18 U-shaped sclerotizations; female tail tip usually with a hook that has one straight tip and one recurved tip ..... *Cribronema*

8′ Lip region moderately offset, slightly wider than first body annule, bilaterally symmetrical with reduced lateral lips, the subdorsal plus subventral lips form four large transparent flaps without setiform projections or U-shaped sclerotizations; tail tip with simple pointed mucro ..... *Acrolobus*

9(5) Three labial probolae: one dorsal and two subventral projections adjacent to mouth opening; the lips themselves often carry six additional projections that are placed more peripherally 10

9′ No cuticular structures project anteriorly immediately next to the mouth opening; lips fused into three peripheral pairs of large, transparent cuticular flaps that project forward and have serrated edges ..... *Metacrolobus*

10(9) Lip region has exactly six cuticular projections; these are pointed or finely rounded 11

10′ Lip region has more or fewer than six cuticular projections 13

11(10) The three labial probolae are low ridges that carry two projections each; lateral fields continue past phasmids in both sexes 12

11′ The six lips are conical acute, whereas the three labial probolae are low rounded; lateral fields stop at phasmids in females only; female tail conical to elongate with pointed tail tip ..... *Pseudacrobeles (Pseudacrobeles)* (in part)

12(11) Female tail conical to cylindrical with rounded tip, the latter often carries an acute or blunt mucro ..... *Cephalobus* (in part)

12′ Female tail long or moderately long with conical pointed tip or rounded club-shaped tip ..... *Eucephalobus* (in part)

13(10) Lips arranged in three pairs, these are separated from each other by three pairs of U-shaped sclerotizations 14

13′ Lip region without pairs of U-shaped sclerotizations 15

14(13) Labial probolae low-rounded or short and broad with double tips; no more than six U-shaped sclerotizations; cuticle often punctated, never with longitudinal striae; female tail with elongate sigmoid ventral outline, curved dorsal outline and pointed tip ..... *Zeldia*

14′ Labial probolae thin and Y-shaped; three pairs of U-shaped sclerotizations often alternate with three additional unpaired sclerotizations; cuticle never punctated but may have longitudinal striae; female tail conical or cylindrical, its tip has a spike-, harpoon- or brush-shaped mucro .... *Stegelletina*

15(13) All three labial probolae are low-rounded or conical-pointed 16

15′ Usually, all three labial probolae are double-tipped, branched, and/or ornamented with tines (very rarely, this is only true for the dorsal probola) 21

16(15) Vulva placed at 45%−85% of body length; spicules < 50 μm 17

16′ Vulva placed at over 90% of body length; spicules > 50 μm long .... *Metacrobeles*

17(16) Female tail short-conical to elongate with pointed tip, never with mucro 18

17′ Female tail conical or cylindrical with broadly or finely rounded tip; the latter may carry an acute or blunt mucro 19

18(17) Each lip carries a thin cuticular filament that is as long as the lip region is wide, lip edges serrated elsewhere; lateral fields extend past phasmids in both sexes .... *Scottnema*

18′ Each lip carries a minute cuticular filament, lip edges not serrated; lateral fields stop at phasmids in females only .... *Pseudacrobeles* (*Pseudacrobeles*) (in part)

19(17) Anterior end of intestine without a deep fold or overlapping pouch 20

19′ Anterior end of intestine forms a distinct pouch that is one to two times as long as basal bulb and overlaps with it or is folded posteriad from it .... *Placodira*

20(19) Labial probolae low rounded to conical pointed; corpus weakly swollen to fusiform; lateral field with three to five lines .. *Acrobeloides*

20′ Labial probolae low rounded; corpus cylindrical or weakly swollen; lateral field always with three lines .... *Cephalobus* (in part)

21(15) Female tail conical to cylindrical with rounded tip; the latter may carry an acute or blunt mucro 22

21′ Female tail conical with pointed tip, never with distinct mucro 25

22(21) Cuticle without deep longitudinal incisures; finer longitudinal striae are sometimes present, but these are never continuous across successive annules 23

22′ Cuticle deeply divided by longitudinal incisures that are mostly continuous across successive annules .... *Stegelleta*

23(22) Labial probolae at least 4 μm long; tail rarely with mucro 24

23′ Labial probolae just a few μm long; tail often carries an acute or blunt mucro .... *Cephalobus* (in part)

24(23) Lips fused in pairs and/or only weakly separated, without long filaments .... *Chiloplacus*

24′ Lips spherical and deeply separated, grouped in pairs that alternate with three pairs of cuticular filaments, which are almost as long as the labial probolae .... *Chiloplacoides*

25(21) Base of each labial probola with a flattened ridge or broad ledge projecting aborally just below its bifurcation (this structure may or may not be continuous with the outside edges of the primary branches) 26

25′ Base of each labial probola without a ridge or ledge projecting aborally just below its bifurcation (in a few cases, a rounded knob projects aborally from base) 27

26(25) Lip edges with short serrations; primary branches of labial probolae very short to quite long, often with rows of tines and/or forked tips; corpus never widely distended or with strongly sclerotized lining .... *Nothacrobeles*

26′ Lip edges with three−five long and transparent tines; primary branches of labial probolae quite long and pointed, never with tines or forked tips; corpus in dead animals often widely distended, giving the false appearance of strongly sclerotized lining .... *Paracrobeles*

27(25) Labial probolae filamentous to robust, with short or long primary branches that may carry scattered tines but never have contiguous rows; lips grouped in pairs or separate, short triangular with strongly or weakly serrated edges; there may be 30 U-shaped sclerotizations; amphid openings small, slits, or ovals .... *Cervidellus*

27′ Labial probolae robust with long primary branches, their inside and outside edges each with a row of at least five contiguous tines; lips separate, elongate triangular with nine or more tines each, never with U-shaped sclerotizations; amphid openings small, circular. .... *Acrobeles*

Nematoda: Rhabditida: Tylenchina: Aphelenchidae: Genera

1 Basal bulb overlapping intestine as a long glandular lobe; female tail usually without mucro; male tail has a bursa with three pairs of genital rays 2

1′ Basal bulb not overlapping intestine; female tail short cylindrical with rounded or flattened tip, which often has an indentation or irregularly rounded mucro; male tail with paired genital papillae .... *Paraphelenchus*

2(1) Female tail elongate conical with pointed and weakly curved tip .... *Pseudaphelenchus*

2′ Female tail short cylindrical with rounded or flattened tip .... *Aphelenchus*

Nematoda: Rhabditida: Tylenchina: Aphelenchoididae: Genera

Davies et al. (2015) proposed that all Neotropical and Nearctic species of *Schistonchus* be transferred to a separate genus, which they named *Ficophagus*.

1 Stylet very slender, > 15 μm long, knobs entirely absent 2

1′ Stylet length usually <15 μm, slender or robust, knobs distinct, rarely minute or absent 4

2(1) Lip region continuous to moderately offset, without perioral ring; stylet 15–25 μm long; annulation invisible or very fine; female tail elongate conical with straight and pointed or filiform tip, no mucro 3

2′ Lip region clearly offset, with distinct perioral ring; stylet length > 25 μm; annulation distinct and over 1 μm wide along most of body; female tail short or long, conical with weakly curved tip or mucro .... *Anomyctus*

3(2) Vulva with or without thickened cuticle; female tail usually shorter than three anal body widths; male tail without spike-shaped mucro .... *Ektaphelenchus*

3′ Vulva without thickened cuticle; female tail usually longer than three anal body widths; male tail with a spike-shaped mucro .. *Seinura*

4(1) Stylet with distinct knobs 5

4′ Stylet knobs minute or absent 11

5(4) Excretory pore small; pharyngeal isthmus just a few μm long; female tail often curved, with or without variously shaped mucro(s) ... 6

5′ Excretory pore enlarged, located near junction of procorpus and median bulb; pharyngeal isthmus 15 μm long; female tail straight conical with a single broad and stubby mucro that has frayed edges .... *Punchaulus*

6(5) Lip region confluent to strongly offset; stylet knobs usually smaller than valves in median bulb; female tail conical pointed, with or without mucro 7

6′ Lip region strongly offset; stylet knobs as large as valves in median bulb; female tail weakly curved, conical with rounded tip .... *Robustodorus*

7(6) Lip region confluent to strongly offset; stylet knobs rounded; male tail tip without flattened sides; female tail often with mucro 8

7′ Lip region strongly offset; stylet knobs often somewhat elongated longitudinally and flattened transversely; male tail tip with flattened sides forming a small bursa; female tail curved with pointed or fine rounded tip, usually without mucro .... *Bursaphelenchus*

8(7) Excretory pore varying in position from near nerve ring to level of median bulb; female tail with split or frayed mucro(s) 9

8′ Excretory pore varying in position from near isthmus to just posterior to lip region; female tail weakly to strongly curved, conical pointed, with or without a short ventrally curved mucro at tip; male tail varying from blunt and much shorter than female tail, to similarly shaped and proportioned .... *Schistonchus*

9(8) Lip region angular or rounded, not with distinct labial disc 10

9′ Lip region high, moderately to strongly offset, with visible labial disc; female tail conical with frayed mucro .... *Tylaphelenchus*

10(9) Anterior vulval lip with or without flap; mucro not split into tubercles, it may be star-shaped, but the tips of the star are not frayed 11

10′ Anterior vulval lip usually forming a flap that covers the posterior lip; mucro split into three or four thin tubercles with frayed edges .... *Laimaphelenchus*

11(4 and 10) Females from 250 μm to 1 mm in length; stylet slender; excretory pore anterior to nerve ring; tail often mucronate; both sexes live in soil, sediment, or plant tissue .... *Aphelenchoides*

11′ Females < 300 μm; stylet robust; excretory pore placed posterior to median bulb at a distance of two corresponding body widths; female tail short, conical pointed; males only occur as parasitic adults .... Free-living generation of *Praecocilenchus*

Nematoda: Rhabditida: Tylenchina: Criconematidae: Genera

Multiple criconematid genera can often be found in the same soil sample. In that case, great caution is advised for use of juvenile characters in identification, for example, by using molecular diagnostics to match adults with juveniles, or by relying only on molting fourth-stage juveniles in which the adult cuticle has already formed even if the juvenile cuticle has not yet been shed.

Four genus names are commonly used in the literature for *Criconemella* [=*Criconemoides* = *Macroposthonia* = *Mesocriconema*]. Despite many attempts at clarification, resolution of this ambiguity has proven all but impossible. The original proposal of *Macroposthonia* was, for example, based on a male only, and male criconematids have far fewer informative morphological characters than females.

1 Cuticle with loose extra layer (Fig. 9.4 K) 2

1′ Cuticle without loose extra layer 3

2(1) Postvulval region bluntly rounded or conical-pointed, its length usually < 3× anal body width; spicules not semicircular or with a slender J shape .... *Hemicriconemoides*

2′ Postvulval region conical-pointed, its length often > 4× anal body width; males with thin and long spicules curved into a half-circle or a slender J or U shape (Fig. 9.4 K) .... *Hemicycliophora*

3(1) Lip region with rounded projections on anterior surface, usually without widely separated anterior disk or first annule; the lip region is only rarely more strongly sclerotized than subsequent body annules 4

3′ Lip region usually without rounded projections on anterior surface, with anterior disk or first annule that is separated from subsequent annules by a gap that is often as wide as two body annules; lip region often sclerotized much more strongly than adjacent body annules .... *Discocriconemella*

4(3) Annules without longer posterior projections; lip region often has four small rounded lobes (submedian lobes) or six bulges of the first annule (pseudolips) directed anteriorly; juveniles have annules without cuticular scales 5

4′ Annules posteriorly decorated with various cuticular projections and/or lip region has four or six protruding rounded lips, but it lacks four or six small rounded lobes that are modifications of the first annule (submedian lobes or pseudolips); juveniles have annules with cuticular scales 7

5(4) Submedian lobes with equal sized gap from lateral versus dorsal or ventral angles; first annule protruding or recessed into the second annule; third annule has greater diameter than the second 6

5′ Submedian lobes grouped into a dorsal and a ventral pair, with distinct gap between them in lateral view only; first annule narrow and recessed into the much thicker and wide second annule, followed by a third annule of smaller transverse diameter than the second .... *Nothocriconemoides*

6(5) Stylet length ≥ 90 μm .... *Neobakernema*

6′ Stylet length < 85 μm .... *Criconemella* [= *Criconemoides* = *Macroposthonia* = *Mesocriconema*]

7(4) Cuticle with scales in rows 8

7′ Cuticle without scales, or scales are not arranged in rows 9

8(7) Scales alternating on successive annules .... *Croserinema*

8′ Scales are aligned on successive annules .... *Ogma*

9(7) Annulation on most of body 3–20 μm wide; vulva usually separated from tail tip by more than six annules; if anus is visible, then it is usually separated from tail tip by over three annules; female tail conical-pointed to bluntly rounded .... *Criconema*

9′ Annulation at middle of body 8–18 μm wide; vulva separated from tail tip by less than six annules; if anus is visible, then it is usually separated from tail tip by less than three annules; female tail irregularly shaped to bluntly rounded .... *Lobocriconema*

Nematoda: Rhabditida: Tylenchina: Tylenchulidae: Genera

1 Females are sedentary semiendoparasites, males are vermiform; excretory pore placed very far posterior in females and sometimes also in males; males with stylet, their pharynx is reduced or has median bulb larger than basal bulb 2

1′ Females are not sedentary parasites, or they do not have excretory pore placed very far posteriorly; males without stylet and with reduced pharynx, usually with median bulb not larger in volume than basal bulb 3

2(1) Male with narrowly conical lip region, very thin stylet with minute knobs and very slender pharynx with rudimentary bulbs; excretory pore placed at over 65% of body length in females, at 50%–60% of body length in males; female tail blunt; male tail long-conical with fine rounded tip; females do not become encapsulated .... *Tylenchulus*

2′ Male with broadly conical lip region, its stylet has distinct knobs and its pharynx appears functional with median bulb larger than basal bulb; excretory pore placed at 33%–60% of body length in females, not posteriorly displaced in males; tails conical-pointed in both sexes; females usually become encapsulated by root material .... *Trophotylenchulus*

3(1) Females < 0.5 mm long, body length/width ratio < 30; bursa absent or not reaching posterior half of male tail; tails finely rounded .. 4

3′ Females > 0.5 mm long, slender, length/width ratio > 30; bursa narrow but long, extends from precloacal region to tail tip; tails pointed .... *Tylenchocriconema*

4(3) Females invade below root epidermis and become cylindrically swollen as they mature; female stylet 90–102 μm long; vulval body width often ≥ distance from vulva to anus .... *Cacopaurus*

4′ Females do not invade root tissues and usually remain vermiform as they mature; female stylet 12–120 μm long; vulval body width < distance from vulva to anus .... *Paratylenchus*

Nematoda: Rhabditida: Tylenchina: Tylenchidae: Genera

Morphological distinction between *Filenchus* and *Malenchus* may require SEM, especially of the lip region. Moreover, *Filenchus* is the most speciose genus in Tylenchidae by far, with 95 valid species worldwide according to Geraert (2008), most of which have not yet been studied with SEM.

1 Females with two ovaries; phasmids may be present 2

1′ Females with anterior ovary only; phasmids very rarely present 3

2(1) Female vulva not flanked by subventral flaps; male without projecting cloacal lips; phasmids may be present in both sexes; tail very long, often with drop-shaped or rounded tip ..... *Psilenchus*

2′ Female vulva flanked by small subventral flaps; male with projecting cloacal lips; phasmids present in female but not reported in male; tail very long, often with recurved tail tip, which is finely rounded ..... *Antarctenchus*

3(1) Cuticle with longitudinal ridges 4

3′ Cuticle without longitudinal ridges (except for the lateral field) 7

4(3) Lip region rarely offset, without four long setae; bursal wings usually form thin borders that extend along a shorter or longer part of the tail 5

4′ Lip region offset with four long setae; cuticle with longitudinal ridges; bursa with curved wings that project ventrally but flank the cloacal region only ..... *Eutylenchus*

5(4) Annules divide the longitudinal ridges into blocks; lateral field with three or usually four lines 6

5′ Longitudinal ridges not divided into blocks by the annules; lateral field with only two lines forming a single lateral ridge ..... *Ridgellus*

6(5) Lip region confluent or moderately offset; stylet cone as long as shaft; vulva with small flanking subventral wings ..... *Coslenchus*

6′ Lip region confluent; stylet cone length usually 0.5× shaft length; vulva without small flanking subventral wings ..... *Neothada*

7(3) Stylet cone length = shaft length 8

7′ Stylet cone length 0.5× shaft length 11

8(7) Lateral field usually with two or six lines; tail mostly straight and often with filiform tip. 9

8′ Lateral field with four barely visible lines; tail usually curved with curvature increasing toward its tip, which is pointed or finely rounded but never filiform ..... *Tylenchus*

9(8) Stylet knobs not particularly large; male without cloacal tube 10

9′ Stylet knobs large relative to stylet length and thickness; male with lips of cloacal opening raised into a short tube ..... *Aglenchus*

10(9) Annulation invisible with light microscope; stylet 9–15 μm long, with small spherical or diverging knobs; distance from stylet to median bulb 2–5× stylet length; lateral field with two barely visible lines ..... *Polenchus*

10′ Annulation distinct; stylet 14–25 μm long, knobs distinct and wider than high; distance from stylet to median bulb 1–1.5× stylet length; usually six lateral lines, rarely four ..... *Cephalenchus*

11(7) Lip region with or without dorsoventral flattening/lateral widening; intact tail almost always < 0.3× body length (tail can be longer in two Neotropical *Filenchus* species, however both have a lip region without flattening or widening as well as a relatively wide and sclerotized excretory pore) 12

11′ Lip region dorsoventrally clearly flattened and laterally slightly widened compared with postlabial body width; intact tail 0.3–0.5× body length (excretory pore usually visible but never wide or sclerotized) ..... *Lelenchus*

12(11) Body extremely slender, length/width ratio 50–200 13

12′ Body length/width ratio < 50 15

13(12) Stylet with knobs; intact tail ≤ 0.2× total body length 14

12′ Stylet without knobs; intact tail 0.2–0.35× total body length ..... *Chilenchus*

14(13) Body clearly narrowing immediately posterior to vulva or cloaca; male has bursa composed of two small rounded flaps flanking the cloaca only ..... *Ecphyadophora*

14′ Body not noticeably narrowing immediately posterior to vulva or cloaca; male has bursa composed of two small squarish flaps flanking the cloaca only ..... *Tenunemellus*

15(12) Lip region anteriorly flattened, with a small oral disc or a laterally curved crescentic plate that contains the mouth opening 16

15′ Lip region anteriorly flattened or rounded, never with a disc- or crescent-shaped plate around the mouth opening 17

16(15) Lip region with small oral disc; tail length 5–6× anal body width ..... *Discotylenchus*

16′ Lip region with a crescent-shaped plate that curves laterally around the lip region and gives it an offset appearance; filiform tail 10–15× anal body width ..... *Cucullitylenchus*

17(15) Lip region with tiny but usually discernible oblique, transverse, or lambda-shaped amphid opening in lateral view 18

17′ Lip region with amphid opening longitudinal or not discernible in lateral view 19

18(17) Body not curved along its entire length (including tail) ..... *Basiria*

18′ Entire body curved, with curvature increasing gradually toward tail tip ..... *Boleodorus*

19(17) Annulation < 2 μm wide, rarely deeply incised into cuticle; lip region not usually wider or flatter in dorsal/ventral view; basal bulb pyriform or cylindrical ..... *Filenchus*

19′ Annulation distinct, 0.8–3 μm wide, often deeply incised into cuticle; anterior surface of lip region appears more flattened, and lip region is wider in dorsal or ventral view compared with lateral view; basal bulb always pyriform .... *Malenchus*

Nematoda: Rhabditida: Tylenchina: Neotylenchidae: Genera

1 Basal bulb overlaps intestine over distance < body width; female tail blunt .... *Deladenus*

1′ Basal bulb overlaps intestine over distance > 2× body width; female tail pointed .... *Hexatylus*

Nematoda: Rhabditida: Tylenchina: Anguinidae: Genera

1 Median bulb absent or weak without valves; basal bulb overlaps intestine over a distance ≥ corresponding body width 2

1′ Median bulb has valves; basal bulb without overlap or overlaps intestine over a distance < body width 5

2(1) Basal bulb overlaps intestine over a distance subequal to one body width 3

2′ Basal bulb overlaps intestine over a distance > 2× body width 4

3(2) Corpus wider than isthmus but lacking a median bulb; vulva flanked by annulate subventral flaps; males unknown .... *Pterotylenchus*

3′ Median bulb wider than procorpus and isthmus; vulva without sublateral flaps; males abundant .... *Ficotylus*

4(2) Clear distinction between procorpus, median bulb, and isthmus; median bulb with valves; basal bulb overlaps intestine over a distance of two to four body widths .... *Pseudhalenchus*

4′ No distinction between procorpus, median bulb, or isthmus; basal bulb overlaps intestine over a distance well in excess of four body widths .... *Orrina*

5(1) Females swell up as they mature (without losing their vermiform shape); median bulb without valves 6

5′ Females do not swell up; median bulb with or without valves 7

6(5) Females become coiled full circle or more as they mature; ovary filled with multiple rows of oocytes .... *Anguina*

6′ Females usually do not become strongly coiled as they mature; ovary with a single row of oocytes .... *Subanguina*

7(5) Median bulb distinct, with longitudinal valves .... *Ditylenchus*

7′ Median bulb weak, without longitudinal valves .... *Nothotylenchus*

Nematoda: Rhabditida: Tylenchina: Dolichodoridae: Genera

Morphological distinction between *Paratrophurus* and *Tylenchorhynchus* is not possible for a significant minority of species in both genera. *Tylenchorhynchus* is the most speciose genus in Dolichodoridae with 133 valid species worldwide according to Geraert (2011). In addition to *Paratrophurus* and *Tylenchorhynchus*, half a dozen other extremely similar and sometimes species-rich genera are known from other parts of the world but have not been reported from Neotropical or Antarctic locations to date.

1 Lip region composed of four distinct sectors, lateral lips reduced; stylet longer than five lip region widths 2

1′ Lip region composed of six distinct sectors or not noticeably subdivided, all lips roughly equally developed; stylet usually shorter than three lip region widths 5

2(1) Stylet cone length > 2× shaft; pharyngeal bulb with intestinal overlap, distance from its junction with intestine to median bulb equaling roughly one body width 3

2′ Stylet cone length 1–1.8× shaft; pharyngeal bulb without intestinal overlap, distance from its junction with intestine to median bulb ≥ 2× body width 4

3(2) Lateral field marked by a single line; stylet knobs wider than high .... *Belonolaimus*

3′ Lateral field has four lines, with annulation crossing the outer two; stylet knobs usually spherical .... *Ibipora*

4(2) Lateral field with three lines; female tail pointed, its outline resembles a funnel .... *Dolichodorus*

4′ Lateral field marked by four lines; female tail usually bluntly rounded .... *Neodolichodorus*

5(1) Lateral field with two to five lines 6

5′ Lateral field with six lines .... *Amplimerlinius*

6(5) Female tail bluntly rounded; male tail usually of comparable length; bursa reaches male tail tip 7

6′ Female tail elongate conical with pointed or finely rounded tip; male pointed and much shorter; bursa with deeply excavated posterior edges, does not reach tail tip .... *Meiodorus*

7(6) Lateral field with two-four lines 8

6′ Lateral field with five lines .... *Quinisulcius*

8(7) Females with two ovaries 9

8′ Females with anterior ovary only .... *Trophurus*

9(8) Cuticle without longitudinal lines (except for the lateral field) 10

9′ Cuticle with longitudinal lines .... *Neodolichorhynchus*

10(9) Lip region continuous or offset, usually with conical outline and very fine annules; female tail length usually $< 2.5\times$ anal body width, with tail tip as wide as anal body width and thickness of tail tip cuticle greater than half of anal body width .... *Paratrophurus*

10′ Lip region continuous or offset, with conical, rounded, or flattened outline, with fine to distinct annules; female tail length is usually $> 2.5\times$ anal body width, its tip usually less wide than anal body width, and thickness of tail tip cuticle usually less than half of anal body width .... *Tylenchorhynchus*

Nematoda: Rhabditida: Tylenchina: Hoplolaimidae: Genera

1 Phasmids may be located well anterior to anus, present in females and often also in males 2

1′ Phasmids absent in both sexes .... *Aphasmatylenchus*

2(1) Phasmids pore-shaped and usually located on tail; female tail often with mucro 3

2′ Phasmids enlarged into circular scutella, located on tail or (much) more anteriorly; female tail never with mucro 4

3(2) Distance from dorsal gland opening to stylet knobs $\leq 0.25\times$ stylet length; basal bulb overlaps intestine more on its (sub)dorsal side .... *Rotylenchus*

3′ Distance from dorsal gland opening to stylet knobs $0.25-0.67\times$ stylet length; basal bulb overlaps intestine more on its (sub)ventral side .... *Helicotylenchus*

4(2) Scutella placed well anterior to anus; there may be substantial longitudinal distance between the right-side versus left-side scutellum 5

4′ Scutella placed on or near tail, with minimal separation between their right-side and left-side locations .... *Scutellonema*

5(4) Stylet knobs higher than wide, with one or more anterior projections .... *Hoplolaimus*

5′ Stylet knobs more or less rounded, without anterior projections .... *Aorolaimus*

Nematoda: Rhabditida: Tylenchina: Pratylenchidae: Genera

1 Females are markedly swollen and sedentary throughout the adult stage, or they are vermiform at first but become swollen and sedentary as they mature 2

1′ Females are vermiform and motile throughout the adult stage 3

2(1) Females are swollen and sedentary throughout the adult stage; female body often irregularly folded back on itself in one or more places, posteriorly tapering evenly to a conical pointed tail; males have a bursa that does not envelop tail tip .... *Achlysiella*

2′ Newly moulted females vermiform, becoming fusiform and finally strongly swollen while retaining a noticeably constricted posterior end with blunt tail tip; males have a bursa that envelops the tail tip .... *Nacobbus*

3(1) Males with spherical or irregular lip region with reduced stylet and pharynx; bulb overlap mostly (sub)dorsal 4

3′ Both sexes with flattened lip region and functional stylet and pharynx; bulb overlap mostly (sub)ventral 5

4(3) Females with one ovary; vulva at over 80% of body length; tail length $< 3\times$ anal body width .... *Hoplotylus*

4′ Females with two ovaries (posterior one may be much smaller than the other); vulva at 55%–70% of body length; tail length $> 3\times$ anal body width .... *Radopholus*

5(3) Females with one or two ovaries; tail length $< 3\times$ anal body width; never with pointed mucro 6

5′ Females always with two ovaries; tail length $> 3\times$ anal body width; often with pointed mucro .... *Hirschmanniella*

6(5) Females with two ovaries 7

6′ Females with anterior ovary only .... *Pratylenchus*

7(6) Deirids indistinct; basal bulb overlap well-developed, mostly (sub)ventral; posterior ovary reduced in size compared with the anterior one .... *Apratylenchoides*

7′ Deirids distinct in lateral fields near nerve ring; basal bulb overlap very short, mostly (sub)dorsal; both ovaries of roughly equal size .... *Pratylenchoides*

Nematoda: Rhabditida: Tylenchina: Heteroderidae: Genera

1	Mature females with or without vulval cone, their entire body turning into a protective cyst full of eggs; spicules > 30 μm long	2
1′	Mature female body pyriform with subterminal anus and median vulva, no protruding tail or vulval cone, does not become a cyst; spicules < 30 μm	*Meloidodera*
2(1)	Female body lemon- or lime-shaped to pyriform, always with a short vulval cone; spicules with bifid or single tip	3
2′	Female body ovoid or spherical, without vulval cone; spicule tip never bifid	4
3(2)	Female body lemon-shaped to pyriform, cysts with vulval cone located at opposite end of embedded lip region; spicules with bluntly bifid tip or single tip	*Heterodera*
3′	Female body lime-shaped to perfectly spherical, cysts with vulval cone variously located at opposite end of embedded lip region, or shifted to a more sideways position; spicules with pointed bifid tip	*Cactodera*
4(2)	Female body spherical; male excretory pore can be located posterior to overlapping basal bulb	5
4′	Female body ovoid with subterminal apex, not truly spherical; males unknown	*Atalodera*
5(4)	Cyst cuticle finely wrinkled, not punctated; vulva surrounded by a circular fenestra, but not the anus	*Globodera*
5′	Cyst cuticle finely wrinkled and punctated; vulva and anus are each surrounded by a circular fenestra	*Punctodera*

Nematoda: Rhabditida: Rhabditina: Families

Alloionema is a facultative pathogen of snails and slugs and type genus of Alloionematidae in Tylenchina, not Rhabditina. Its free-living stages are morphologically so convergent with Rhabditina as to require placing them in the Rhabditina part of this key. However, DNA sequence data and morphology of pathogenic stages show that the genus is in fact closest to *Rhabditophanes* and clearly belongs in Tylenchina for classification purposes.

Morphology of *Aenigmenchus* is very similar to the genus *Shahnema* and its junior synonym *Baldwinema,* both known only from Asia and placed in Tylenchina: Panagrolaimoidea by Atighi et al. (2015) on the basis of DNA sequence analysis. No sequence data are available for *Aenigmenchus*, and its combination of morphological characters is ambiguous enough that it too may belong in Tylenchina instead of Rhabditina.

Most genera of Neodiplogasteridae and some Diplogasteridae include species with pronounced dimorphism—or even trimorphism—of the buccal cavity. These species often have a "stenostoma" form with buccal features fitting the diagnosis of Diplogasteridae and having few or no plicae, whereas the "eurystoma"/"megastoma" forms typically have many plicae, as well as the claw-shaped tooth or teeth considered typical of Neodiplogasteridae (e.g., Ragsdale et al., 2013). Attempting identification of just a single adult specimen to species or even to genus is not advisable. Conversely, a varied series of adults with two or more distinct buccal morphologies may in fact all represent a single species.

1	Lip region usually without setae or setiform projections, always with left–right bilateral symmetry; cuticle usually smooth or finely annulate only, nearly always identical on right versus left side of body; male tail symmetrical with both wings of bursa developed, or with matching genital sensilla on left versus right side	2
1′	Lip region with multiple setiform projections and dorsoventral (not left–right) bilateral symmetry; cuticle with complex ornamentations that are often differently structured on right versus left side of body; male tail usually asymmetrical with only the left wing of bursa developed, the tail's right side has genital setae	**Bunonematidae [p. 235]**
2(1)	Dorsal sector of buccal cavity identical to subventral sectors; basal bulb always with transverse valves	3
2′	Dorsal sector of posterior buccal cavity often swollen compared with subventral sectors; basal bulb usually without transverse valves	6
3(2)	Females with two ovaries; vulva placed at 50%–70% of body length; spicules almost always separate	4
3′	Females with anterior ovary only; vulva placed at > 75% body length; spicules separate or partly fused	**Mesorhabditidae [p. 235]**
4(3)	Lip region not narrow and elongated; males with bursa and nine to 10 pairs of genital rays	5
4′	Lip region narrow and elongated, lips fused into three rounded pairs; males without bursa, cloacal region has eight pairs of genital papillae	*Stomachorhabditis*
5(4)	Posterior end of buccal cavity constricted and carries denticles	**Rhabditidae [p. 235]**
5′	Posterior end of buccal cavity not constricted, it lacks denticles	**Diploscapteridae [p. 236]**
6(2)	Dorsal sector of stoma posteriorly with only one denticle placed on a swelling; basal bulb without transverse valves, rarely with valve rudiments; males with or without bursa	7
6′	Dorsal sector of stoma posteriorly with three or four denticles placed in a longitudinal series on a swelling; basal bulb with transverse valves; males without bursa (see comment above)	Free-living adults of *Alloionema*
7(6)	Corpus strongly muscular and cylindrical, without median bulb	8
7′	Corpus weakly to strongly muscular, weakly to clearly divided into a cylindrical procorpus versus fusiform to cylindrical median bulb	9
8(7)	Buccal cavity is a long and narrow tube without teeth; female tail conical pointed, not filiform	Cylindrocorporidae, one genus: *Myctolaimus*

8′	Buccal cavity with a large tooth, anteriorly wide and posteriorly a long tube; female tail very long and filiform	Odontopharyngidae, one genus: *Zullinius*
9(7)	Buccal cavity with or without teeth; basal bulb without any valve rudiments	10
9′	Buccal cavity with tiny denticles but never with teeth; basal bulb contains small valve rudiments	Pseudodiplogasteroididae, one genus: *Pseudodiplogasteroides*
10(9)	Body slender or not, tail often filiform; buccal cavity variously shaped but always with distinct lumen; females with one or two ovaries; males with six or more pairs of genital sensilla, or a bursa with rays	11
10′	Small and slender nematodes with very long filiform tail; buccal cavity without distinct lumen, appears to contain a minute stylet; females with anterior ovary only; males with two pairs of genital papillae, no bursa (see comment above)	*Aenigmenchus*
11(10)	Body length < 1.2 mm; spicules with or without distinct offset part, its diameter $< 0.17\times$ spicule length; tails filiform or conical with pointed tip but no mucro	12
11′	Body 1.5–2.8 mm long; males with enlarged, clearly offset ovoid "head" of spicules, its diameter is often $> 0.17\times$ spicule length; tails in both sexes conical with rounded terminus bearing a short-pointed mucro	**Neodiplogasteridae (in part) [p. 236]**
12(11)	Mouth opening can usually open or close due to presence of numerous flexible folds (plicae), which show as parallel longitudinal striae in the anteriormost buccal cavity; buccal cavity is a funnel or barrel, it has a small to large dorsal tooth and often also smaller subventral teeth or denticles	13
12′	Anteriormost buccal cavity lacks numerous parallel longitudinal striae; buccal cavity tubular without teeth, or with denticle(s) only	**Diplogasteroididae [p. 236]**
13(12)	Dorsal tooth triangular (see comment above)	**Diplogasteridae [p. 236]**
13′	Dorsal tooth curved (see comment above)	**Neodiplogasteridae (in part) [p. 236]**

Nematoda: Rhabditida: Rhabditina: Bunonematidae: Genera

1	Right side of body with thickened cuticle containing polygonal web of punctations, as well as elevated ridges, warts, and/or large sclerotized plates	2
1′	Right side of body with thickened cuticle containing a grid of transverse, longitudinal, and/or triangular rods and sclerotizations	*Pterygorhabditis*
2(1)	Anteriormost edge of thickened right-side cuticle forms a distinctly raised and more strongly sclerotized ledge just posterior to the lip region	*Bunonema*
2′	Anteriormost edge of thickened right-side cuticle is not raised or more strongly sclerotized just posterior to the lip region	*Rhodolaimus*

Nematoda: Rhabditida: Rhabditina: Mesorhabditidae: Genera

1	Lip region with well-separated, rounded or angular lips that carry papillae only; internal lining of the procorpus has no series of refringent dashes	2
1′	Lip region confluent with weakly separated lips carrying short setae only; internal lining of the procorpus bordered by a series of narrowly spaced refringent dashes; bursa varying from very poorly developed to prominent and enveloping male tail	*Mesorhabditis*
2(1)	Lips clearly separated, the gaps between them are not sclerotized, or only weakly	3
2′	Lips well separated with distinctly sclerotized gaps between them	*Teratorhabditis*
3(2)	Lip region strongly offset with six rounded lips that are equal in shape	*Cruznema*
3′	Lip region continuous with six flattened lips that are alternately shaped like shorter and longer wedges in face view	*Distolabrellus*

Nematoda: Rhabditida: Rhabditina: Rhabditidae: Genera

1	Bursa reaches tail tip	2
1′	Bursa does not reach tail tip	*Rhabditis*
2(1)	Female rectum $< 2\times$ body width	3
2′	Female rectum $> 2\times$ body width	*Oscheius*
3(2)	Bursa with two spaced genital rays placed at level anterior to cloacal opening; female tail conical pointed, length $< 3\times$ anal body width	*Pelodera*
3′	Bursa with three grouped genital rays placed at level anterior to cloacal opening; female tail conical pointed, length $> 3\times$ anal body width	*Pellioditis*

Nematoda: Rhabditida: Rhabditina: Diploscapteridae: Genera

1	Lips not modified into strongly sclerotized hooks; tubular part of buccal cavity anteriorly with denticles, length > 7× width	*Odontorhabditis*
1′	Lips modified into two strongly sclerotized hooks; tubular part of buccal cavity without anterior denticles, length < 7× width	*Diploscapter*

Nematoda: Rhabditida: Rhabditina: Neodiplogasteridae: Genera

In *Neodiplogaster*, the eurystoma form can be all but indistinguishable from *Mononchoides* (see Kanzaki, 2016).

1	Body length <1.2 mm; spicules with or without distinct offset part, diameter < 0.17× spicule length; tails filiform or conical with pointed tip but no mucro	2
1′	Body 1.5–2.8 mm long; males with enlarged, offset ovoid anterior "head" of spicules, diameter > 0.17× spicule length; tails in both sexes conical with rounded terminus bearing a short-pointed mucro	*Parasitodiplogaster*
2(1)	Buccal cavity lacks a narrow posterior part, or its wide anterior part is clearly longer	3
2′	Narrow posterior part of buccal cavity well developed, usually longer than the wide anterior part	4
3(2)	Narrower posterior part of buccal cavity is essentially absent	*Pristionchus*
3′	Narrower posterior part of buccal cavity is small and lined with two or more small cuticular knobs	*Koerneria*
4(2)	Posterior part of buccal cavity is almost as wide as the anterior part; dorsal tooth large, with an opposing subventral tooth of similar size and shape as well as one or more serrated ridges	*Mononchoides*
4′	Posterior part of buccal cavity is a slender tube with projecting terminal knob on its dorsal side only; dorsal tooth is small and opposed by an equally small and claw-shaped subventral tooth; there are no serrated ridges	*Neodiplogaster*

Nematoda: Rhabditida: Rhabditina: Diplogasteroididae: Genera

1	Buccal cavity posteriorly with angular dorsal sector but no denticle	2
1′	Buccal cavity posteriorly with distinct dorsal denticle	3
2(1)	Lip edges finely sclerotized; males with amphids usually distinct, ovoid	*Demaniella*
2′	Lip edges not sclerotized; amphids minute slits in both sexes	*Rhabdontolaimus*
3(1)	Buccal tube present, with continuous lining for > 0.67× length of buccal cavity	*Rhabditidoides*
3′	Buccal cavity subdivided transversely into four or five short sections; there is no single and distinct cuticular tube	*Diplogastrellus*

Nematoda: Rhabditida: Rhabditina: Diplogasteridae: Genera

1	Lip region with six long setae; buccal cavity is a barrel with large tooth (strongly resembles the unrelated genera shown in Figs. 9.4 E, 5 C)	2
1′	Lip region without six long setae; buccal cavity is only moderately wide and has a small- to medium-sized tooth	3
2(1)	Females with two ovaries	*Butlerius*
2′	Females with anterior ovary only	*Monobutlerius*
3(1)	Posterior narrow part of buccal cavity < 0.5× its length; dorsal tooth medium-sized; junction of vagina and uteri with offset dorsal pouch	*Acrostichus*
3′	Posterior narrow part of buccal cavity 0.5× its length; dorsal tooth is of very modest size; junction of vagina with uteri lacks a dorsal pouch	*Diplogasteritus*

Nematoda: Rhabditida: Chambersiellidae: Genera

1	Lip region with six + four stout setae; tail tip in both sexes with shorter or longer sclerotized hook, its tip is bifid, curved dorsally and shaped like a double claw	2
1′	Lip region with six stout setae and four cephalic papillae; tail with short mucro that can be straight or curved, neither sclerotized nor bifid at its tip	*Macrolaimus*
2(1)	In addition to six + four stout setae, the lip region has six projecting brushlike cirri; mouth opening surrounded by six projecting, curved and strongly sclerotized flaps that resemble jawlike structures; buccal cavity posteriorly with a narrow pharyngeal section roughly as long as corresponding body width	3

2′ Lip region lacks projecting brushlike cirri; mouth opening surrounded by six low and moderately sclerotized flaps; buccal cavity posteriorly with or without discernible pharyngeal section .... *Diastolaimus*

3(2) Females with anterior ovary only .... *Chambersiella*

3′ Females with two ovaries .... *Geraldius*

Nematoda: Chromadorida: Genera

1 Pharynx with rounded basal bulb and faint or distinct longitudinal valves; buccal cavity longer than wide, irregularly tapering with one or two curved teeth .... *Achromadora*

1′ Pharynx cylindrical to weakly swollen; buccal cavity short and wide, dorsally with a prominent conical tooth and subventrally with two small denticles .... *Paracyatholaimus*

Nematoda: Dorylaimida: Families

1 Stylet is a mural tooth or needle without anterior opening; buccal lumen is usually visible passing next to stylet; cardia usually surrounded by three rounded gland cells; vagina usually lacks sclerotizations .... **Nygolaimidae [p. 238]**

1′ Stylet is a hollow tooth or needle with very small to very large opening near its tip; buccal lumen is usually visible passing through the stylet itself; cardia never surrounded by three rounded gland cells; vagina with or without sclerotizations 2

2(1) Lip region contains strong circular sclerotizations, often accompanied by paired teeth positioned near tip of stylet at rest 3

2′ Lip region without circular sclerotizations, at most with sclerotized granules or a sclerotized labial framework 4

3(2) Cuticle never with longitudinal ridges, often with irregular rows of sclerotized gland openings; vagina without sclerotizations; female tail short and bluntly rounded; males have genital supplements roughly equally spaced in a single row .... **Carcharolaimidae [p. 239]**

3′ Cuticle with or without longitudinal ridges, with numerous body pores but no sclerotized gland openings; vagina with sclerotizations; female tail usually elongate with filiform or finely rounded tip; males often have genital supplements grouped in two or three fascicles .... **Actinolaimidae [p. 239]**

4(2) Body may either be very long or very slender, but not both; stylet length $< 65\ \mu m$; vagina with or without sclerotizations 5

4′ Body usually very long and very slender; stylet elongate needle-shaped, length $> 70\ \mu m$; vagina never with sclerotizations .... **Longidoridae [p. 239]**

5(4) Pharynx and intestine usually touching or just a few μm apart (very rarely, they may be separated more significantly, but then the tail is not elongate or filiform)........ 6

5′ Pharynx and intestine separation $\geq 10\ \mu m$, due to presence of an elongate cardia; tail elongated with finely rounded or filiform tip .... **Belondiridae (in part) [p. 240]**

6(5) Lip region not cap-shaped; basal bulb longer than one half of pharynx length 7

6′ Lip region often cap-shaped (Fig. 9.5 H); basal bulb represents only half of the total pharynx length at most, it is usually shorter than one-third of the pharynx 12

7(6) Stylet usually straight, shorter or longer than lip region width; male supplements are either grouped tightly together in one or more fascicles, or they are spaced more or less evenly along a single file of fewer than 40 total 8

7′ Stylet slightly curved, length $> 1.5\times$ lip region width; basal bulb forms $> 60\%$ of pharynx length; males with over 40 precloacal supplements, mostly placed close together in a double file with zigzag pattern .... **Tylencholaimidae (in part) [p. 240]**

8(7) Pharyngeal bulb without sheath 9

8′ Pharyngeal bulb completely covered by a sheath composed of one layer of long and flat muscle cells, which often form a spiral pattern .... **Belondiridae (in part) [p. 240]**

9(8) Stylet robust or needle-shaped with opening $< 0.17\times$ its length, or lip region width $< 0.67\times$ its length, but not both 10

9′ Stylet needle-shaped with opening $< 0.17\times$ its length, lip region width $< 0.67\times$ its length .... **Nordiidae [p. 241]**

10(9) Lip region usually offset; cuticle without longitudinal ridges; stylet robust or slender with its opening occupying more or less than half its length, lip region width subequal to its length; guiding ring single or double; female tail short or elongate 11

10′ Lip region usually confluent; cuticle with or without ridges; stylet robust with opening $< 0.5\times$ its length and lip region width varying 0.5–$0.67\times$ its length (e.g., Fig. 9.5 G); guiding ring double; female tail usually elongate .... **Dorylaimidae [p. 241]**

11(10) Body length 0.4–3 mm; stylet smaller or more robust, its opening usually $< 0.5\times$ length; guiding ring single or double; female tail rounded, conical, or elongate .... **Qudsianematidae [p. 242]**

11′ Body length 1.5–11 mm; stylet robust with opening $>0.5\times$ length; guiding ring single; female tail short conical or hemispherical .... **Aporcelaimidae [p. 243]**

12(6) Stylet neither sigmoid, nor constricted 13

12′ Stylet with sigmoid outline and usually slightly constricted posterior to the opening .... **Mydonomidae [p. 243]**

13(12) Lip region usually cap-shaped and clearly offset (Fig. 9.5 H); cuticle often develops irregular folds and blisters upon fixation; buccal cavity lacks striae; slender part of pharynx is not swollen near nerve ring; bulb length 0.17–0.5× pharynx length 14

13′ Lip region rounded, not offset; cuticle not prone to forming folds and blisters upon fixation; buccal cavity anteriorly with numerous longitudinal fine striae; slender part of pharynx is swollen near nerve ring yet narrowest between nerve ring and basal bulb; bulb length 0.2–0.25× pharynx length .... **Aulolaimoididae [p. 244]**

14(13) Stylet needle- or bristle-shaped, often without distinct lumen; odontophore often slightly curved; basal bulb ovoid, pyriform, or cylindrical, bulb length ≤ 0.3× pharynx length, often with fusiform or valvelike thickening of its cuticular lining just anterior to the cardia .... **Leptonchidae [p. 244]**

14′ Stylet needle-shaped, can be very short and thin but usually still has a recognizable lumen; odontophore often straight; basal bulb cylindrical, bulb length usually 0.3–0.5× pharynx length, without modified lining near the cardia .... **Tylencholaimidae (in part) [p. 240]**

Nematoda: Dorylaimida: Nygolaimidae: Genera

1 Lip region confluent to offset, its anterior surface is not sclerotized 2

1′ Lip region clearly offset, its anterior surface recessed and sclerotized, forming a ribbed basketlike structure .... *Mylodiscus*

2(1) Mural tooth variously shaped, length often subequal to lip region width; pharyngeal bulb not subdivided; females with one or two ovaries, vagina with or without sclerotizations 3

2′ Mural tooth resembles a narrow Greek letter lambda, length < lip region width; pharyngeal bulb may be subdivided into two or three parts; females with two ovaries, vagina with sclerotizations .... *Nygolaimellus*

3(2) Pharyngeal bulb lacks or has a thin sheath; if present, then it is not muscular and spiral 4

3′ Pharyngeal bulb with a sheath of spiral muscles, at least around its posterior end; females with posterior ovary only, vagina lacks sclerotizations .... *Nygellus*

4(3) Lip region offset by a deep constriction; cardia without three rounded gland cells 5

4′ Lip region confluent or offset by constriction; cardia often with three rounded gland cells 7

5(4) Buccal cavity without denticles; vagina with or without sclerotizations 6

5′ Buccal cavity laterally and dorsally with numerous tiny denticles opposite mural tooth; vagina lacks sclerotizations .... *Nygolaimium*

6(5) Mural tooth obliquely inclined at rest; vagina lacks sclerotizations .... *Scapidens*

6′ Mural tooth longitudinal at rest; vagina with or without sclerotizations .... *Aporcelaimoides*

7(4) Mural tooth bristle-shaped 8

7′ Mural tooth needle-, lambda-, or delta-shaped 9

8(7) Body curvature not S-shaped in fixed specimens; lip region confluent or very weakly offset, vulva longitudinal .... *Paravulvus* (in part)

8′ Body curvature S-shaped in fixed specimens; lip region offset, vulva transverse .... *Solididens*

9(7) Tail usually straight, length < 3× body width, its tip broadly rounded 10

9′ Tail curved, length > 5× body width, tapering to a finely rounded tip .... *Afronygus*

10(9) Vulva transverse, vulval region never with enlarged ventral pores 11

10′ Vulva longitudinal, usually near one or more enlarged ventral pores .... *Paravulvus* (in part)

11(10) Lip region offset, not ventrally inclined 12

11′ Lip region almost always confluent, often slightly ventrally inclined 13

12(11) Mural tooth needle-shaped; males common, gubernaculum present .... *Paranygolaimus*

12′ Mural tooth delta-shaped; males very rare, gubernaculum absent .... *Nygolaimus*

13(11) Mural tooth usually needle-shaped, as long as or longer than lip region width 14

13′ Mural tooth needle-shaped or narrow delta-shaped, shorter than lip region width .... *Laevides*

14(13) Mural tooth needle-shaped or elongate delta-shaped; tail hemispherical or club-shaped (= greatest tail width > anal body width) .... *Clavicaudoides*

14′ Mural tooth usually needle-shaped; its length equals or exceeds lip region width; tail hemispherical .... *Aquatides*

Nematoda: Dorylaimida: Carcharolaimidae: Genera

1 Sclerotized labial basket either subdivided into paired plates or finely lined with numerous parallel ridges; sublabial sclerotized rods connecting guiding ring to the labial basket are absent or shorter than height of labial basket .... *Carcharolaimus*

1′ Sclerotized labial basket subdivided into paired plates; sublabial sclerotized rods connecting guiding ring to labial basket are clearly longer than height of labial basket .... *Caribenema*

Nematoda: Dorylaimida: Actinolaimidae: Genera

1 Cuticle with 30–40 longitudinal ridges; stylet slender to very slender, its opening shorter than one-fifth of stylet length 2

1′ Cuticle with or without longitudinal ridges; stylet robust to slender, its opening longer than one fifth of stylet length 3

2(1) Lip region offset by a moderate to deep depression, > 10 μm wide and often larger in males than in females .... *Practinocephalus*

2′ Lip region confluent, ≤ 10 μm wide, equal in both sexes .... *Brasilaimus* (in part)

3(1) Buccal structures posterior to vestibular ring are covered with numerous denticles; males have genital supplements roughly equally spaced in a single row 4

3′ Buccal structures between vestibular ring and guiding ring lack denticles; males usually have genital supplements grouped in two or three fascicles 6

4(3) Males with short rounded tail; females with conical to elongated tail and finely rounded tip 5

4′ Both sexes with elongated tail and finely rounded tail tip .... *Trachactinolaimus*

5(4) Buccal cuticle between vestibular ring and denticles smooth or only with very faint longitudinal striae .... *Paractinolaimus*

5′ Buccal cuticle between vestibular ring and denticles with numerous parallel sclerotized longitudinal striae .... *Westindicus*

6(3) The strongly sclerotized toothlike ridges immediately posterior to denticulated buccal cuticle have one pointed tip each 7

6′ The strongly sclerotized toothlike ridges immediately posterior to denticulated buccal cuticle have two pointed tips each .... *Neoactinolaimus*

7(6) Cuticle with 14–45 longitudinal ridges; anteriormost part of the slender section of the pharynx is not muscular; males have genital supplements grouped in two or three fascicles 8

7′ Cuticle without longitudinal ridges; entire pharynx is muscular; males have genital supplements roughly equally spaced in a single row .... *Egtitus*

8(7) Nonmuscular anterior section of pharynx < 0.3× pharynx length; vulva longitudinal .... *Parastomachoglossa*

8′ Nonmuscular anterior section of pharynx > 0.3× pharynx length; vulva transverse .... *Brasilaimus* (in part)

Nematoda: Dorylaimida: Longidoridae: Genera

1 Guiding ring positioned near posterior end of stylet at rest; odontophore distinct, swollen at base 2

1′ Guiding ring positioned near tip of stylet at rest; odontophore linear, not swollen at base 5

2(1) Odontophore base without moderately developed flanges; dorsal gland nucleus ≥ 10 μm posterior to dorsal gland opening; amphids are pores or slits equaling 0.2–0.9× lip region width 3

2′ Odontophore base with strongly developed flanges; dorsal gland nucleus at or near dorsal gland opening; amphids are slits equaling about 0.5× lip region width .... *Xiphinema*

3(2) Lip region offset by constriction; amphids are slits almost as wide as lip region 4

3′ Lip region confluent or offset by a depression; amphids are pores or slits <0.6× wide as lip region, sides of amphidial pockets rounded .... *Xiphidorus*

4(3) Amphidial pockets have two rounded posteriad extensions flanking amphidial canal .... *Australodorus*

4′ Amphidial pockets have straight tapering sides converging on the amphidial canal .... *Paraxiphidorus*

5(1) Amphids are pores .... *Longidorus*

5′ Amphids are slits .... *Paralongidorus*

Nematoda: Dorylaimida: Belondiridae: Genera

We consider *Swangeria longicauda* Ahmad & Naz, 2010 (described from Nicaragua) to belong in *Qudsiella*.

1 Pharynx and intestine separation $\geq$ 10 µm, due to presence of an elongate cardia; tail elongated with finely rounded or filiform tip ... 2

1′ Pharynx and intestine usually touching or just a few µm apart (very rarely they may be separated more significantly but then the tail is not elongate or filiform) ... 3

2(1) Pharyngeal bulb without sheath ... *Paraqudsiella*

2′ Pharyngeal bulb covered by a sheath of (usually spiral) muscles ... *Qudsiella*

3(1) Stylet very short and fusiform with both of its sides slightly bulging outward; slender part of pharynx mostly muscular ... 4

3′ Stylet very short with sides that are straight, curved on the ventral side only, or both slightly curved parallel to one another; slender part of pharynx usually nonmuscular ... 5

4(3) Lip region confluent; vagina without sclerotizations ... *Axonchium*

4′ Lip region offset by a constriction; vagina with sclerotizations ... *Metaxonchium*

5(3) Lip region without sclerotized granules adjacent to mouth opening; odontostyle lining thin and straight ... 6

5′ Lip region with sclerotized granules adjacent to mouth opening; odontostyle slightly widening posteriorly ... *Dorylaimellus*

6(5) Females with two ovaries ... 7

6′ Females with posterior ovary only ... 11

7(6) Female tail elongate conical to filiform ... 8

7′ Female tail short and rounded, its greatest width exceeds anal body width ... 9

8(7) Muscular sheath cells spiral sinistrally around the basal bulb; the latter is shorter than one-third of pharynx length ... *Paraoxydirus*

8′ Muscular sheath cells spiral dextrally around the basal bulb, the latter is longer than one-third of pharynx length ... *Oxydirus* (in part)

9(7) Junction of slender part of pharynx with basal bulb is not constricted ... 10

9′ Junction of slender part of pharynx with basal bulb is deeply constricted ... *Anchobelondira*

10(9) Lip region lacks six small swellings of the lips adjacent to the mouth ... *Amphibelondira*

10′ Lip region with six small swellings of the lips protruding adjacent to the mouth (these swellings are known as liplets) ... *Belondirella*

11(6) Intact female tail is $>4\times$ body width ... 12

11′ Female tail $<3\times$ body width ... *Belondira*

12(11) Lip region confluent or weakly offset and cap-shaped; stylet length 3–5x width; female tail elongate conical to filiform ... *Oxydirus* (in part)

12′ Lip region confluent, its sides internally sclerotized; stylet length 5–6$\times$ width; female tail length 4–6$\times$ anal body width with swollen rounded terminus ... *Oxybelondira*

Nematoda: Dorylaimida: Tylencholaimidae: Genera

1 Stylet straight, length $<0.5–2.5\times$ lip region width; basal bulb forms half of pharynx length or much less; male supplements are few, scattered, and placed in single file ... 2

1′ Stylet slightly curved, length $>1.5\times$ lip region width; basal bulb length $>0.6\times$ pharynx length; males with over 40 precloacal supplements, most of which are placed close together in a double file with zigzag pattern ... *Vanderlindia*

2(1) Stylet needle-shaped, length 0.5–2.5$\times$ lip region width; females with one or two ovaries ... 3

2′ Stylet fusiform, length 0.75$\times$ lip region width; females with two ovaries ... *Promumtazium*

3(2) Stylet length 0.7–1.25$\times$ lip region width; odontophore often with sclerotized basal granules or knobs ... 4

3′ Stylet length 1.5–2.2$\times$ lip region width; odontophore without distinct basal granules or knobs ... *Chitwoodius*

4(3) Buccal cavity not sclerotized, indistinct or widening between mouth and guiding ring, forming an inverted cone or funnel ... 5

4′ Buccal cavity finely sclerotized, widening and then narrowing again between mouth and guiding ring ... *Capilonchus*

5(4) Lip region without oral disc; odontophore without knobs or with small basal knobs; tail short and rounded ... 6

5′ Lip region with oral disc; stylet length subequal to lip region width; odontophore with large sclerotized basal knobs; tail elongate with finely rounded tip ... *Discomyctus*

6(5) Amphid slit-shaped; females with one or two ovaries ... *Tylencholaimus*

6′ Amphid pore-shaped; females with posterior ovary only ... *Tantunema*

Nematoda: Dorylaimida: Nordiidae: Genera

1	Lip region confluent or offset; amphid located at posterior edge of lip region	2
1′	Lip region offset by a constriction; amphid located on lateral lip	*Kochinema*
2(1)	Lip region anteriorly flattened, often with angular papillae and usually offset; females with two ovaries, vagina usually with sclerotized pieces	3
2′	Lip region conical-rounded, confluent or weakly offset; females with posterior ovary only, vagina lacks sclerotized pieces	*Actinolaimoides*
3(2)	Lip region with anterior concavity that may contain rounded swellings protruding immediately adjacent to the mouth (these swellings are known as liplets)	4
3′	Lip region not concave anteriorly, without liplets	5
4(3)	Lip region concavity without liplets; male with five pairs of genital papillae on curved conical tail	*Chrysonemoides*
4′	Lip region concavity contains rounded liplets; males unknown	*Baqriella*
5(3)	Buccal cavity without sclerotizations; stylet straight unless it is longer than four lip region widths	6
5′	Buccal cavity with distinctly sclerotized granules or walls; stylet often weakly dorsally curved, never longer than four lip region widths	*Pungentus*
6(5)	Postlabial region not sclerotized; buccal cavity not sclerotized; tail bluntly rounded or conical	7
6′	Postlabial region with sclerotized subcuticle; buccal cavity postorally with sclerotized walls; tail filiform or nearly so	*Sicaguttur*
7(6)	Odontophore base not swollen; tail rounded or conical	8
7′	Odontophore weakly to strongly swollen; tail rounded	*Enchodelus*
8(7)	Lips with angular papillae or short setae; female tail curved conical with pointed or finely rounded tip	9
8′	Lips with rounded or angular papillae; female tail rounded or conical	10
9(8)	Lips with distinctly angular papillae; stylet length 1−3× lip region width; vagina usually with sclerotizations	*Oriverutus*
9′	Lips with six plus four triangular setae; stylet length 1.7-3× lip region width; vagina without sclerotizations; female tail curved conical with pointed or finely rounded tip	*Echinodorus*
10(8)	Body evenly curved along most of its length, anteriorly only moderately tapering; lip region width > 0.3× cardiac body width	11
10′	Body forms a series of tight coils posteriorly, anteriorly very strongly tapering; lip region width 0.17× cardiac body width	*Acunemella*
11(10)	Body < 1 mm; vagina usually with sclerotizations, its cuticle is not especially thicker near vulva than elsewhere	*Longidorella*
11′	Body > 1 mm; vagina without sclerotizations, its cuticle can be much thicker near vulva than near its junction with the uteri	*Heterodorus*

Nematoda: Dorylaimida: Dorylaimidae: Genera

1	Females with two ovaries	2
1′	Females with posterior ovary only	*Opisthodorylaimus*
2(1)	Guiding ring single; both sexes with filiform tail or both sexes with short tail	3
2′	Guiding ring almost always double; males with blunt or short pointed tail; female tail conical to filiform	5
3(2)	Vulva usually longitudinal; vagina usually without sclerotizations; both sexes with filiform tail	4
3′	Vulva longitudinal or transverse; vagina with sclerotizations; female tail short with digitate tip, male with blunt tail (Fig. 9.5 G)	*Coomansinema*
4(3)	Vulva at 34%−41% body length; vagina without sclerotizations; spicules without sclerotized central line or fold; only one or two precloacal supplements	*Amphidorylaimus*
4′	Vulva at 35%−56% body length; vagina with sclerotizations; spicules with sclerotized central line or fold; 13−31 precloacal supplements	*Prodorylaimus*
5(2)	Cuticle without annulate internal layer; vulva longitudinal or transverse; males with contiguous or spaced supplements	6
5′	Cuticle with distinctly annulate internal layer; vulva longitudinal; male with 19−37 contiguous supplements	*Idiodorylaimus*
6(5)	Buccal cavity without sclerotized granules; vulva longitudinal or transverse	7
6′	Buccal cavity flanked by large sclerotized granules; vulva transverse; male unknown	*Fuscheila*
7(6)	Cuticle without longitudinal ridges; males with spaced or contiguous supplements	8

7′ Cuticle with 28–60 longitudinal ridges; males with 22–62 contiguous supplements; vulva longitudinal or transverse .... *Dorylaimus*

8(7) Guiding ring always double; prerectum length equals four anal body widths or more; male tail blunt 9

8′ Guiding ring single or weakly double; prerectum length shorter than four anal body widths; male tail pointed .... *Afrodorylaimus*

9(8) Vulva longitudinal or transverse; male with 20–47 contiguous supplements; female tail conical to filiform .... *Laimydorus*

9′ Vulva usually longitudinal; male with 10–18 contiguous supplements; female tail filiform .... *Paradorylaimus*

Nematoda: Dorylaimida: Qudsianematidae: Genera

1 Postlabial region and buccal cavity without sclerotized parts 2

1′ Postlabial region with sclerotized subcuticle; buccal cavity with partly sclerotized walls .... *Sclerolabia*

2(1) Females either very small or with filiform tail, but not both; lips rounded or angular; odontophore not swollen at base, or if swollen, then adjacent pharynx cuticle is equally thick; vulval sclerotizations present or absent 3

2′ Females very small with filiform tail; lips with pointed angular papillae; odontophore swollen at base; vulval sclerotizations present; males unknown .... *Minidorylaimus*

3(2) Odontostyle either short or strongly sclerotized but not both; odontophore and its adjacent pharyngeal cuticle not thickened; males known for most species, often less abundant than females 4

3′ Odontostyle short and very strongly sclerotized; odontophore and the adjacent pharyngeal cuticle distinctly thickened at their junction; males unknown .... *Pachydorylaimus*

4(3) Both sexes with conical or rounded tail, sometimes female tail with a short caudal projection that is absent in male 5

4′ Females with conical to filiform tail; guiding ring single; vulval sclerotizations present; male tail bluntly rounded .... *Mesodorylaimus* (in part)

5(4) Lip region with smaller or more prominent swellings of the lips protruding immediately adjacent to the mouth (these swellings are known as liplets); guiding ring double 6

5′ Lip region usually without liplets; guiding ring single 9

6(5) Body length 0.3–4 mm; stylet length 1.5–2× lip region width, the opening covers 0.25–0.3× length; tail in both sexes usually rounded 7

6′ Body length 1.5–3 mm; stylet 20 μm long or subequal to lip region width, its opening about 0.2× length; tail in both sexes curved conical .... *Pararhyssocolpus*

7(6) Liplets prominent; vulva longitudinal; tail tip identical in both sexes 8

7′ Liplets low; vulva transverse; female tail has a short projection that is absent in the male .... *Talanema*

8(7) Lip region weakly to strongly offset, its anterior surface not concave; stylet distinctly wider than thickness of postlabial cuticle; tail rounded in both sexes .... *Labronema*

8′ Lip region weakly to strongly offset, with distinctly concave anterior surface that contains prominent liplets; stylet relatively slender, only as wide as thickness of postlabial cuticle; tail rounded in both sexes .... *Labronemella*

9(5) Body anteriorly very tapering; lip region width < 0.25× cardiac body width 10

9′ Body anteriorly less narrow; lip region width usually > 0.3× cardiac body width 12

10(9) Stylet straight and slender with narrow opening 11

10′ Stylet weakly sigmoid, opening spans 0.5× length but is only revealed when stylet protrudes (at rest, the opening is mostly covered by flexible refringent lining of the dorsal buccal cavity, creating the impression of a narrow opening and spout-shaped stylet) .... *Paraxonchium*

11(10) Stylet at rest positioned slightly off center in lip region, 14–16 μm long, 1.5× lip region width, narrow opening spans 0.2× length; odontophore straight; basal bulb length 0.5× pharynx length; tail in both sexes conical and ventrally curved .... *Parapalus*

11′ Stylet centered, 13–14 μm long, 1.2–1.4× lip region width, opening spans 0.3× length; odontophore curved; basal bulb length 0.3× pharynx length; female tail short conical and ventrally curved, male unknown .... *Cerviconema*

12(9) Body wall with numerous finely sclerotized ventral pores and two to three dorsal ones; vagina without sclerotizations 13

12′ Body pores not sclerotized; vagina with or without sclerotizations 14

13(12) Stylet opening > 0.5× length .... *Moshajia*

13′ Stylet opening 0.25–0.3× length .... *Lordellonema*

14(12) Stylet opening 0.25–0.5× length; vagina lacks sclerotizations 15

14′ Stylet opening spans 0.25–0.3× length; vagina with sclerotizations 17

15(14) Stylet slender, opening spans 0.25–0.5× length; lateral fields each with a row of ovoid gland cells 16

15′ Stylet relatively squat, opening spans 0.5× length; lateral fields without gland cells .... *Sicorinema*

16(15) Lip region not transversely expanded, usually with rounded liplets flanking mouth opening; tail conical with convex dorsal outline and finely rounded tip .... *Discolaimium*

16′ Lip region transversely expanded and dish- or suction cup shaped, with anterior surface forming a parenthesis-shaped outline in lateral view; tail bluntly rounded .... *Discolaimus*

17(14) Lip region confluent or weakly offset by a depression 18

17′ Lip region offset by a constriction 23

18(17) Female tail straight or mostly straight 19

18′ Female tail conical and ventrally curved 22

19(18) Tail length in both sexes < 1.5× anal body width, tip broadly rounded 20

19′ Female tail length ≥ 2× anal body width, usually with finely rounded tip 21

20(19) Cardia connected with three gland cells .... *Torumanawa*

20′ Cardia without three gland cells .... *Takamangai*

21(19) Amphids located at posterior edge of lip region; females with two ovaries; female tail length 2–10× anal body width, anteriorly wide but narrows fairly abruptly to a short or long slender part with fine tip; male tail bluntly rounded .... *Mesodorylaimus* (in part)

21′ Amphids located on lateral lips; females with posterior ovary only; female tail length 2–3× anal body width, conical with convex dorsal outline and slightly projecting finely rounded tip; males unknown .... *Ecumenicus*

22(18) Vulva usually transverse; female tail length < 3× anal body width .... *Eudorylaimus*

22′ Vulva often longitudinal or pore-shaped; female tail length 2.5–8× anal body width .... *Epidorylaimus*

23(17) Body length 0.3–3 mm; vulva usually transverse; tail usually > anal body width; tail tip usually finely rounded to pointed 24

23′ Body length 2.5–3.3 mm; vulva longitudinal; tail length subequal to anal body width; tail tip narrow rounded .... *Amblydorylaimus*

24(23) Body length 0.3–0.8 mm; tail length 1–3× anal body width; tail tip usually finely rounded to pointed .... *Microdorylaimus*

24′ Body length 2.5–3.3 mm; vulva longitudinal; tail length 1–2× anal body width; tail tip narrow rounded .... *Allodorylaimus*

Nematoda: Dorylaimida: Aporcelaimidae: Genera

1 Lip region offset by a deep constriction, lips usually clearly separate from each other 2

1′ Lip region confluent and rounded with all lips merged .... *Makatinus*

2(1) Lips always clearly separate from each other 3

2′ Lip region offset by a deep constriction, rounded with all lips merged .... *Akrotonus*

3(2) Vagina usually with sclerotizations 4

3′ Vagina without sclerotizations .... *Aporcella*

4(3) Dorsal side of tail usually concave, occasionally notched; tip broadly or narrowly rounded, not directed obliquely dorsally (except when dorsal side is notched) 5

4′ Dorsal side of tail weakly to strongly sigmoid; tip pointed or finely rounded, directed obliquely dorsally .... *Aporcelinus*

5(4) Body length usually < 3 mm, never reaching 4 mm; vulval body width rarely exceeds 100 μm 6

5′ Body length 3–11 mm; vulval body width > 100 μm 7

6(5) Stylet opening spans 0.67× stylet length; males fairly common; female tail length 1–2× anal body width, narrowly rounded, conical with convex dorsal outline and usually slightly projecting tip .... *Metaporcelaimus*

6′ Stylet opening spans 0.5× stylet length; males very rare; female tail as long as 0.3–1× anal body width, broadly rounded, conical, or hemispherical, sometimes notched .... *Aporcelaimellus*

7(5) Stylet more or less trowel-shaped, its opening spans its length: only the posterior edges of its cuticle join circumference of buccal lining; female tail hemispherical to dome-shaped, its length < 0.8× anal body width .... *Sectonema*

7′ Stylet opening spans 0.67–0.75× length; female tail conical rounded, its length 0.7–1.2× anal body width .... *Aporcelaimus*

Nematoda: Dorylaimida: Mydonomidae: Genera

1 Pharyngeal bulb with muscular sheath .... *Mydonomus*

1′ Pharyngeal bulb without muscular sheath .... *Dorylaimoides*

Nematoda: Dorylaimida: Aulolaimoididae: Genera

1 Female tail elongate to filiform .... *Aulolaimoides*
1′ Female tail short and rounded .... *Adenolaimus*

Nematoda: Dorylaimida: Leptonchidae: Genera

1 Stylet thin like a bristle, with no discernible lumen or opening 2
1′ Stylet with visible lumen and opening (Fig. 9.5 H) 4
2(1) Odontophore swollen at base 3
2′ Odontophore linear, without knobs or swelling .... *Coronatyleptus*
3(2) Odontophore swellings not sclerotized .... *Basirotyleptus*
3′ Odontophore swellings sclerotized .... *Sclerostylus*
4(1) Stylet often with a convex expansion along its dorsal side; odontophore with offset rounded knobs at its base 5
4′ Stylet simple; odontophore without offset knobs 7
5(4) Lip region without oral disc; stylet with a convex expansion along its dorsal side 6
5′ Mouth opening placed on a small oral disc; stylet without expansion .... *Doryllium*
6(5) Lip region confluent with sclerotized sides, its overal shape plain conical or anteriorly with a ring-shaped or multilobed projection .... *Margollus*
6′ Lip region usually offset and cap-shaped, never with sclerotized sides .... *Tylencholaimellus*
7(4) Stylet needle-shaped, thinner than postlabial body cuticle 8
7′ Stylet at least as wide as postlabial body cuticle, often slightly curved or fusiform .... *Tyleptus*
8(7) Buccal cavity not sclerotized, indistinct or widening between mouth and guiding ring, forming an inverted cone or funnel 9
8′ Buccal cavity finely sclerotized, widening and then narrowing again between mouth and guiding ring, shaped like an oil lamp's glass chimney .... *Proleptonchus*
9(8) Lip region cap-shaped (Fig. 9.5 H) with or without oral disc, confluent or offset; amphid slit-shaped and wide $<0.5\times$ body width; stylet length $\leq 1.5\times$ lip region width; odontophore with or without swellings at base 10
9′ Lip region hemispherical, offset by a constriction; amphid slit-shaped and almost as wide as constriction; stylet length $2-3\times$ lip region width; odontophore with well-developed swollen flanges at its base .... *Zalophidera*
10(9) Lip region without oral disc (Fig. 9.5 H); basal bulb length $\sim 2\times$ corresponding body width 11
10′ Lip region with distinct oral disc; basal bulb length $\sim 1-2\times$ corresponding body width 12
11(10) Vulva transverse, without sclerotizations (Fig. 9.5 H) .... *Leptonchus*
11′ Vulva longitudinal, with sclerotizations .... *Funaria*
12(10) Stylet length < 20 μm; odontophore without swellings at base 13
12′ Stylet length 40–100 μm; odontophore with well-developed swollen flanges at its base .... *Xiphinemella*
13(12) Guiding ring prominent; junction of slender part of pharynx with basal bulb is constricted, the latter length $\sim 2\times$ corresponding body width and lacks a fusiform modification of the lining .... *Meylis*
13′ Guiding ring faint; junction of slender part of pharynx with basal bulb is not constricted, the latter length $\sim 1\times$ corresponding body width and has a fusiform modification of the lining .... *Caveonchus*

REFERENCES

Andrássy, I. 1968. Fauna Paraguayensis 2. Nematoden aus den Galeriewäldern des Acaray-Flusses. Opuscula Zoologica Budapest 8: 167–315.

Andrássy, I. 1998. Nematodes in the Sixth Continent. Journal of Nematode Morphology and Systematics 1: 107–186.

Anwar, M.A., H.N. Gokozan, M.K. Ball, J. Otero & B.S. McGwire. 2015. Fatal human eosinophilic meningo-encephalitis caused by CNS co-infection with *Halicephalobus gingivalis* and West Nile virus. Parasitology International 64: 417–420.

Atighi, M.R., E. Pourjam, W. Bert, M. Pedram, R. Ghaemi & Y. Panahandeh. 2015. An aberrant nematode, *Baldwinema* n. gen. (Nematoda: Panagrolaimomorpha: Panagrolaimidae), from Iran with description of three new species. Nematology 17: 877–896.

Camino, N.B., S.E. González & G.R. Reboredo. 2015. Biodiversity of Mermitids (Nematoda, Mermithidae) parasitoids of aquatic Diptera (Diptera, Simuliidae, Culicidae and Chironomidae) in Argentina. Asociación Peruana de Helmintología e Invertebrados Afines (APHIA). Neotropical Helminthology 9: 65–71.

Caveness, F.E. 1964. A Glossary of Nematological Terms. Pacific Printers, Ibadan, Nigeria. 83 pp.

Davies, K.A., W. Ye, N. Kanzaki, F. Bartholomaeus, Y. Zeng & R.M. Giblin-Davis. 2015. A review of the taxonomy, phylogeny, distribution and co-evolution of *Schistonchus* Cobb, 1927 with proposal of *Ficophagus* n. gen. and *Martininema* n. gen. (Nematoda: Aphelenchoididae). Nematology 17: 761–829.

Dalmasso, A. 1966. Méthode simple d'éxtraction des nematodes du sol. Revue d'Écologie et Biologie du Sol 3: 473–478.

Doucet, M.E. 1980. Técnicas básicas en nematología del suelo. IDIA 387–388: 34–43.

Doucet, M.E. & M.M.A. de Doucet. 1997. Nematodes and agriculture in continental Argentina. An overview. Fundamental and Applied Nematology 20: 521–539.

Doucet, M.E. & M.M.A. de Doucet. 1999. Supplement to the list of soil and freshwater nematodes recorded in continental Argentina. Nematology 1: 217–220.

Esquivel, A. 2003. Nematode fauna of Costa Rican protected areas. Nematropica 33: 131–145.

Faulkner, L.R. & W.J. Bolander. 1966. Occurrence of large nematode populations in irrigation canals of South Central Washington. Nematologica 12: 591–600.

Ferris, V., J.M. Ferris & J.T. Tjepkema. 1973. Genera of freshwater nematodes (Nematoda) of Eastern North America. Biota of Freshwater Ecosystems Identification Manual - The Environmental Protection Agency. 40 pp.

Geraert, E. 2008. The Tylenchidae of the World – Identification of the Family Tylenchidae (Nematoda). Academia Press. Ghent, Belgium. 540 pp.

Geraert, E. 2011. The Dolichodoridae of the World – Identification of the Family Dolichodoridae (Nematoda). Academia Press. Ghent, Belgium. 520 pp.

Hodda, M., L. Peters & W. Traunspurger. 2009. Nematode Diversity in Terrestrial, Freshwater Aquatic and Marine Systems. Pages 45–93. *in*: M. Wilson & T. Kakouli-Duarte (eds.), Nematodes as Environmental Indicators. CABI Publishing. Wallingford, UK.

Huang, S.P. & J.E. Cares. 2006. Nematode communities in soils under different land use systems in Brazilian Amazon and Savannah vegetation. Pages 163–183. *in:* F.M.S. Moreira, J.O. Siqueira & L. Brussaard (eds.), Soil Biodiversity in Amazonian and Other Brazilian Ecosystems. CABI Publishing, Wallingford, UK.

Hugot, J.P., P. Baujard & S. Morand. 2001. Biodiversity in helminths and nematodes as a field of study: An overview. Nematology 3: 199–208.

Jenkins, W.R. 1964. Rapid centrifugal-flotation technique for separating nematodes from soil. Plant Disease Reporter 48: 692.

Johnson, A.A. & M.G. Kleve. 2002. *Strelkovimermis rubtsovi* n. sp. and *Strelkovimermis ozawindibi* n. sp. (Nematoda: Mermithidae) parasitizing chironomid (Insecta: Diptera) adults eclosing from northern Minnesota Glacial Lakes. Journal of Parasitology 88: 942–946.

Junk, W.J. 1993. Wetlands of tropical South America. Pages 679–739 *in*: D.F. Whigham, D. Dykyjová & S. Hejný (eds.), Wetlands of the World: Inventory, Ecology and Management Volume I: Africa, Australia, Canada and Greenland, Mediterranean, Mexico, Papua New Guinea, South Asia, Tropical South America, United States. Kluwer Academic Publishers. Dordrecht, the Netherlands.

Kanzaki, N. 2016. Stomatal Dimorphism of *Neodiplogaster acaloleptae* (Diplogastromorpha: Diplogastridae). PLoS One 11: e0155715.

Loof, P.A.A. 1973. Freshwater nematodes from Suriname collected by J. van der Land. Zoologische Verhandelingen 129: 1–46.

Netscher, C. & J.W. Seinhorst. 1969. Propionic acid better than acetic acid for killing nematodes. Nematologica 15: 286.

Nickle, W.R. 1972. A contribution to our knowledge of Mermithidae (Nematoda). Journal of Nematology 4: 113–146.

Pérez-Pacheco, R., C. Rodriguez-Hernandez, J. Lara-Reyna, R. Montes-Belmont, G. Ramirez-Valverde, & L. Martinez-Martinez. 2004. Parasitism of *Romanomermis iyengari* in larvae of three species of mosquito in laboratory, and of *Anopheles pseudopunctipennis* in the field. Agrociencia 38: 413–421.

Pérez-Pacheco, R., A. Santamarina-Mijares, A. Vásquez-López, S.H. Martínez-Tomás & J. Suárez-Espinosa. 2009. Efectividad y supervivencia de *Romanomermis culicivorax* en criaderos naturales de larvas de mosquitos. Agrociencia 43: 861–868.

Poinar Jr, G.O. 2015. Nemata. Pages 273–302 *in:* J.H. Thorp & D.C. Rogers (eds.), Ecology and General Biology, Volume I: Ecology and General Biology. Thorp and Covich's Freshwater Invertebrates, Academic Press, Elsevier, London.

Poinar, G.O. & H. Takaoka. 1981. Three new mermithids (Nematoda) from Guatemalan blackflies. Systematic Parasitology 3: 13.

Ragsdale, E.J., N. Kanzaki, W. Röseler, M. Herrmann & R.J. Sommer. 2013. Three new species of *Pristionchus* (Nematoda: Diplogastridae) show morphological divergence through evolutionary intermediates of a novel feeding-structure polymorphism. Zoological Journal of the Linnean Society 168: 671–698.

Seinhorst, J.W. 1962. On the killing, fixation and transferring to glycerin of nematodes. Nematologica 8: 29–32.

Smith, D.G. 2001. Pennak's Freshwater Invertebrates of the United States, 4th ed. John Wiley & Sons, New York. 648 pp.

Southey, J.F. 1970. Laboratory Methods for Work With Plant and Soil Nematodes. London, UK, Her Majesty's Stationery Office. 148 pp.

Tandingan De Ley, I., P. De Ley, J.G. Baldwin, M. Mundo-Ocampo & S.A. Nadler. 1999. Three new species of *Nothacrobeles* (Nemata: Cephalobidae) from the Mojave Desert, California. Journal of Nematology 31: 482–497.

Thorne, G. 1961. Principles of nematology. McGraw-Hill, New York. 553 pp.

Vargas, M., I.A. Rubtsov & F. Fallas-Barrantes. 1980. Bionomics of black flies (Diptera: Simuliidae) in Costa Rica. V. Description of *Neomesomermis travisi* sp. n. (Nematoda: Mermithidae). Revista de Biología Tropical 28: 73-89.

Velasco-Castrillón, A., J.A.E. Gibson & M.I. Stevens. 2014. A review of current Antarctic limno-terrestrial microfauna. Polar Biology 37: 1517-1531.

Chapter 10

Phylum Nematomorpha

Fernanda Zanca
Centro de Estudios Parasitológicos y de Vectores (CEPAVE, CONICET-UNLP), La Plata, Argentina

Cristina De Villalobos
Museo de Ciencias Naturales, La Plata, Argentina

Andreas Schmidt-Rhaesa
Zoological Museum, University Hamburg, Hamburg, Germany

Matthew G. Bolek
Department of Integrative Biology, Oklahoma State University, Stillwater, OK, United States

Ben Hanelt
Department of Biology, University of New Mexico, Alburquerque, NM, United States

PHYLUM NEMATOMORPHA

Chapter Outline

INTRODUCTION

About 350 species of Nematomorpha have been described; the majority (20 genera) are freshwater (Gordiida), and 1 genus, *Nectonema*, includes 5 marine species. Gordiida, commonly known as horsehair worms, live in several types of freshwater habitats and have an interesting life cycle including a parasitic phase, which includes insects and other invertebrates as host, and a free-living aquatic phase as adults and larvae. After emerging from their host, dioecious species form Gordian knots, mate, and females deposit egg strings. Additionally, a parthenogenetic species exists (Hanelt et al., 2012). Larvae infect paratenic (transport) hosts to reach the definitive host. Recently, reviews of the biology and ecology of Nematomorpha were published by Bolek et al. (2015) and Schmidt-Rhaesa et al. (2016).

At present a total of 77 species are known from the Neotropical region. They belong to 10 genera. A fossil species, *Palaeochordodes protus* Poinar, 1999, has been described from a piece of Dominican amber (15−45 million years old) in association with its host, a cockroach (Poinar, 1999).

LIMITATIONS

The traditional methods used for determination and description of gordiid species based on earlier light microscopy (LM) investigations are often very difficult. These

Thorp and Covich's Freshwater Invertebrates. https://doi.org/10.1016/B978-0-12-804225-0.00010-1

methods are still important, but they should be supplemented with scanning electron microscopy (SEM) techniques to develop a more complete description of the general morphology and cuticle features of gordiids. Such is the case of the two Neotropical species, *Parachordodes capitosulcatus* (Montgomery, 1898) and *P. ciferrii* Sciacchitano, 1932 which were described using only histological sections and where megareoles, characteristics of the genus *Parachordodes*, were not recorded. Similarly, the descriptions of *Chordodesauranthiacus* Linstow, 1906 and *C. pilosus* Möbius, 1855 are very brief and supplemented by insufficient figures. Because of this, these species are not included in the *Chordodes* key, and more detailed structural investigations using LM and SEM are required.

For morphological analyses, the potential intraspecific variation of characters is a substantial problem. These variations are observable especially when large numbers of specimens are investigated, such as *Chordodes festae* Camerano, 1897 and *C. peraccae* (Camerano, 1894) (de Villalobos & Zanca, 2001). In many cases, character states are only known from a single or few male or female specimens, for example, in some species of *Beatogordius* (de Villalobos et al., 2003). Therefore, there is nearly a complete lack of knowledge concerning the intraspecific variability for most species of gordiids.

Currently the possibility of using molecular techniques (Hanelt et al., 2015) should allow more clarity in resolving frequent taxonomic instabilities. A clear distinction between the genera *Neochordodes* and *Pseudochordodes*, which are not considered to belong to a monophyletic clade (Schmidt-Rhaesa, 2002a) is necessary, because some species have areoles in a megareolar pattern only slightly different from regular areoles (Schmidt-Rhaesa et al., 2016).

Considering that free-living adult gordiids are found seasonally and sometime are difficult to collect, the finding of cysts or larvae in paratenic hosts can confirm and extend the geographical distribution and discovery of new species (Harkins et al., 2016).

The monophyly of many taxa are questioned. The reader will find that the keys are separated by genders as it is easier to distinguish between sexes, and there are many genera in which species have great sexual dimorphism at the posterior end and in the cuticle, and in some species only one gender is known.

TERMINOLOGY AND MORPHOLOGY

Measurements such as length and diameter as well as coloration of the body are not reliable characters to identify species within the Gordiida. The diagnostic characters for genera and species identification are the anterior and posterior ends and cuticular structures. The anterior end is tapering or spherical. Some species show a white tip, followed by a dark collar. From the collar, dark ventral and dorsal lines can extend over the whole body length. The posterior end differs among sexes. The majority of females have a round posterior end (Fig. 10.1 A); only the females of the genus *Paragordius* have a trilobed posterior end (Fig. 10.1 B). In males of several genera (*Acutogordius, Beatogordius, Gordius, Gordionus, Parachordodes*, and *Paragordius*) the posterior end is divided into two lobes (Fig. 10.2 A–E), but males of *Chordodes, Neochordodes, Noteochordodes, Palaeochordodes*, and *Pseudochordodes* have a rounded terminal end (Fig. 10.2 F–I).

The cloacal opening is ventral in all males and slightly subterminal in females. The cuticular structures such as bristles, spines, a **postcloacal crescent** (**semicircular fold**) (Fig. 10.2) or **adhesive warts** are important for identification, especially in the posterior end of males.

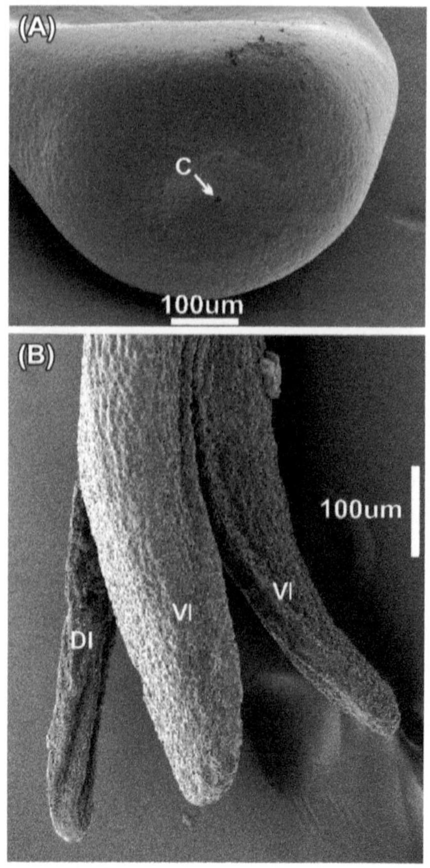

FIGURE 10.1 Posterior end of females: (A) *Gordius robustus*; and (B) *Paragordius* sp. Abbreviations: *C*, terminal cloacal opening; *DI*: dorsal lobe; *VI*, two ventral lobes. All images SEM.

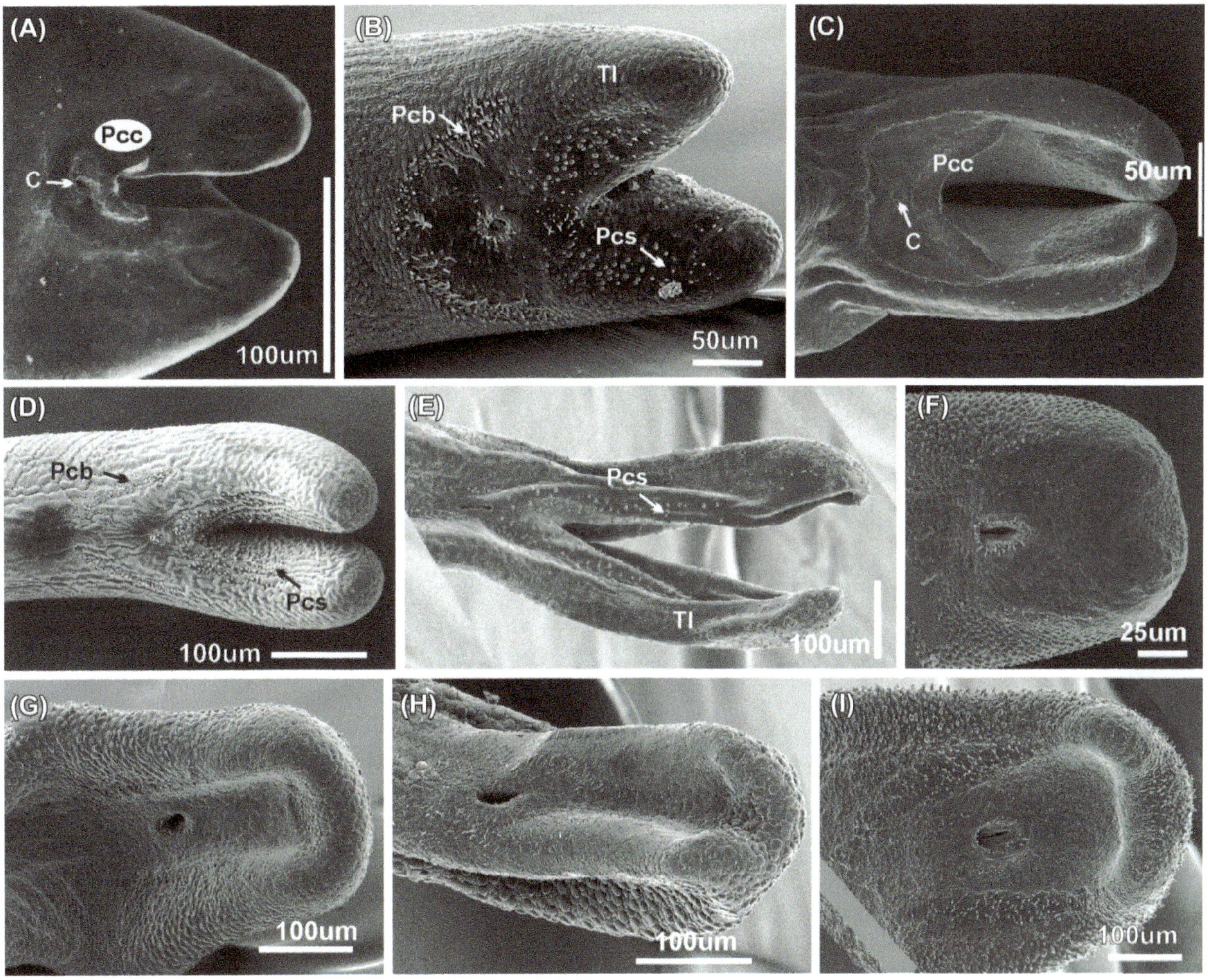

FIGURE 10.2 Ventral side of posterior end in males: (A) *Acutogordius obesus*; (B) *Beatogordius* sp.; (C) *Gordius austrinus*; (D) *Gordionus* sp.; (E) *Paragordius andreasii*; (F) *Chordodes nobilii*; (G) *Neochordodes* sp.; (H) *Pseudochordodes bedriagae*; and (I) *Noteochordodes talensis*. Abbeviations: *c*, cloacal opening; *Pcb*, precloacal bristles; *Pcc*, postcloacal crest; *Pcs*, postcloacal spines; *Tl*, tail lobes. All images SEM.

The cuticle of the body can be smooth (Fig. 10.3 C) or structured into **cuticular areoles** (Fig. 10.3 A–B, D–I). Areoles could be more or less elevated cuticular structures of variable number and shape and exist in one or more types. The basic differences between species are the shape, distribution pattern, and the abundance or even absence of areoles (Schmidt-Rhaesa et al., 2008). In species of some genera, such as *Pseudochordodes* and *Parachordodes*, two types of areoles are present, **micrareoles** and **megareoles** (Fig. 10.3 F). The cuticle of the genus *Chordodes* displays a high diversity of areolar types, but is characterized by a particular type of areole called the **crowned areole**, which has a circle of numerous filaments on top (Figs. 10.3 H–I, 4) (de Villalobos & Zanca, 2005). In the genus *Chordodes*, in several cases, two types of crowned areoles occur on the cuticle of the same specimen. In these cases, crowned areoles on the lateral side of the body have short apical projections, but paired rows of crowned areoles with very long projections occur along the ventral and in some species also along the dorsal midline (Fig. 10.4 A,B). This characteristic is most distinctive in females, but in addition, males of *Chordodes alticuspis* de Villalobos, Zanca & Yanez, 2009 from the Neotropical region (de Villalobos et al., 2009) and *Chordodes queenslandi* Schmidt-Rhaesa, 2002 from Australia (Schmidt-Rhaesa, 2002b) also contain such areoles.

It is possible to differentiate three types of gordiid larvae and cysts for six Neotropical gordiid genera using

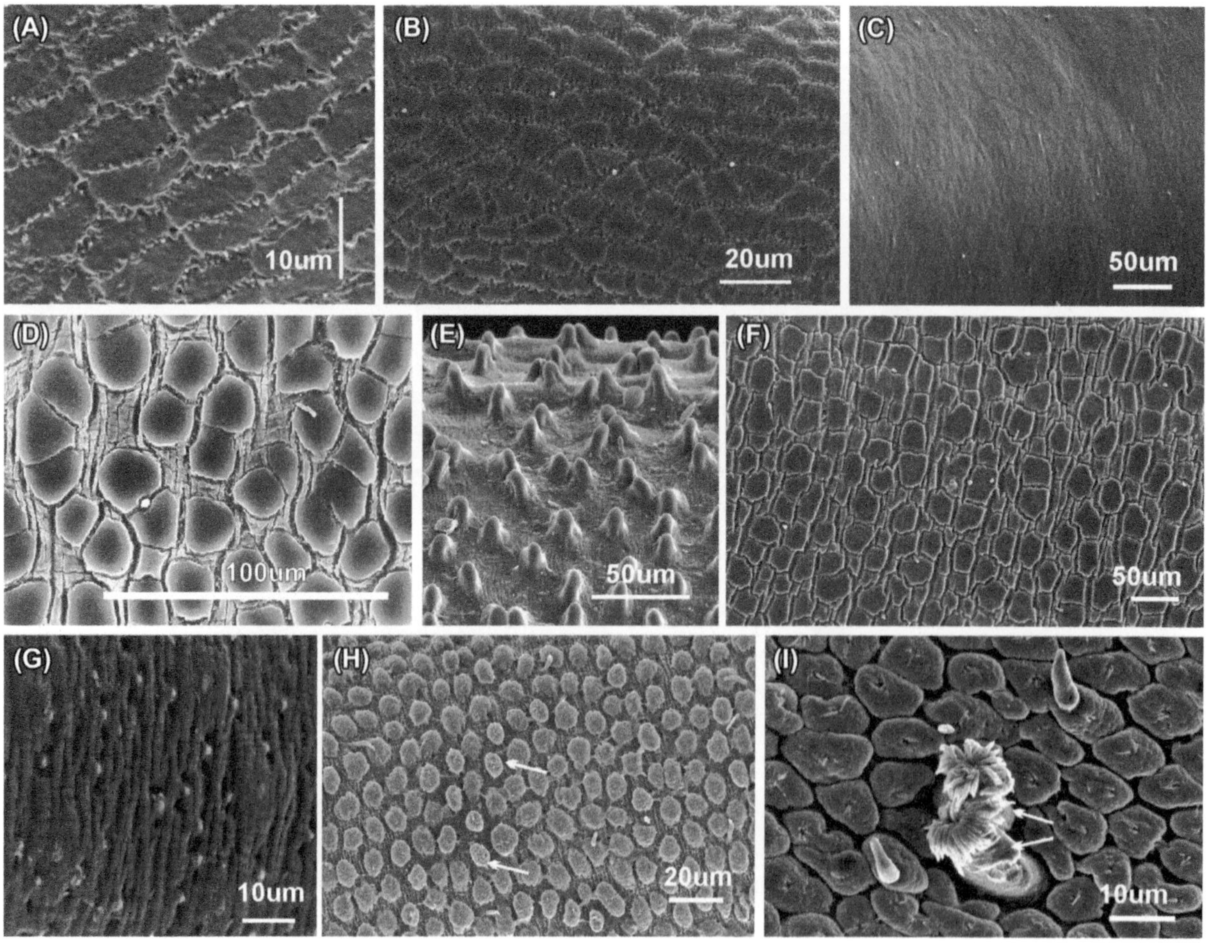

FIGURE 10.3 Areoles on the cuticle surface: (A) *Beatogordius funis*; (B) *Gordionus ondulatus*; (C) *Gordius robustus*; (D) *Neochordodes* sp.; (E) *Noteochordodes talensis*; (F) *Pseudochordodes bedriagae*; (G) *Paragordius esavianus*; (H) *Chordodes nobilii*; and (I) *C. bouvieri*. Note the crowned areoles (arrows) in *Chordodes* species. All images SEM.

morphological characters (Poinar et al., 2004; Szmygiel et al., 2014; Chiu et al., 2016). The larvae are morphologically distinct from the adults and are 50–100 μm in length. They have two body regions, a preseptum and a postseptum (Fig. 10.5). The **preseptum** contains rings of hooks and three stylets to penetrate the intestinal wall of the paratenic host. A distinct glandular structure, the **pseudointestine**, is located in the **postseptum** and varies morphologically among genera and species. In the paratenic host, the larvae are folded inside cysts (Fig. 10.5 B–D, F–G, I–J). Within the cyst, larvae of *Acutogordius, Gordius,* and *Paragordius* species always fold twice. However, the postseptum of *Acutogordius* and *Gordius* species is two to three times as long as the preseptum and, consequently, the posterior end of the postseptum always reaches the posterior end of the preseptum (Fig. 10.5 B–C). In contrast, the postseptum of *Paragordius* species is only 1.3 times as long as the preseptum, and the posterior end of the postseptum never reaches the posterior end of the preseptum. Additionally, larvae of *Paragordius* contain distinctly longer hooks on the outer ring of the larval preseptum, and these are clearly visible in the cyst (Fig. 10.5 E–G). Finally, larvae of *Chordodes, Neochordodes*, and *Parachordodes* have a postseptum to preseptum ratio almost equal in length. As a result, larvae of species in these genera only fold once within the cyst (Fig. 10.5 H–J).

MATERIAL PREPARATION AND PRESERVATION

The best way to preserve nematomorph specimens is in 70%–90% ethanol. For DNA studies material should be placed into 100% ethanol and into the freezer. Before

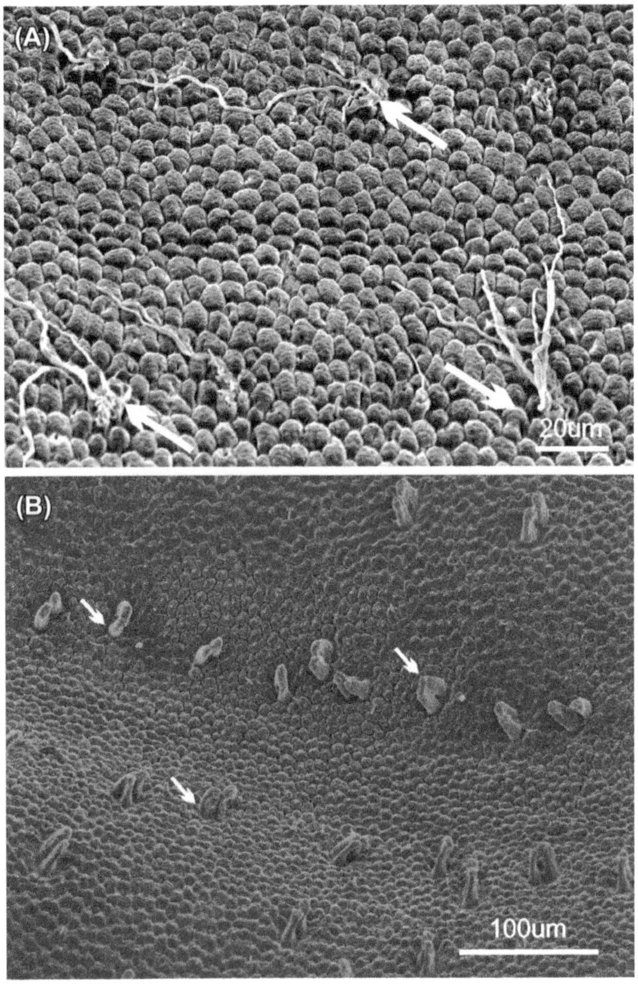

FIGURE 10.4 Ventral groove of females: (A) *Chordodes nobilii* with paired rows of crowned areoles with very long projections along the ventral midline (large arrow); and (B) *C. bouvieri* with one type of crowned areoles with very long projections (small arrow). All images SEM.

processing for morphological studies, specimens should be washed and cleaned ultrasonically for a few minutes in water mixed with a few drops of Clinique rinse-off eye solvent (Clinique Laboratories, New York, USA) followed by rinsing with distilled water (Salas et al., 2011). Then the material should be prepared for light microscopy (LM) and Scanning Electron Microscopy (SEM).

For LM, part of the midbody (about 5–10 mm) is cut with a scalpel longitudinally and the internal tissue is removed from the cuticle. The cuticle is then placed in lactophenol for 3–6 h or in a 3M solution of NaOH for 15 min. Following three rounds of rinsing with distilled water, only the outer cuticular layer remains, which can be observed on a slide in glycerol. For SEM, the posterior and anterior ends and parts of the midbody region (about 5–10 mm in length) are dehydrated in an increasing ethanol series, critical point dried, and coated with gold or gold-palladium in a sputter coater.

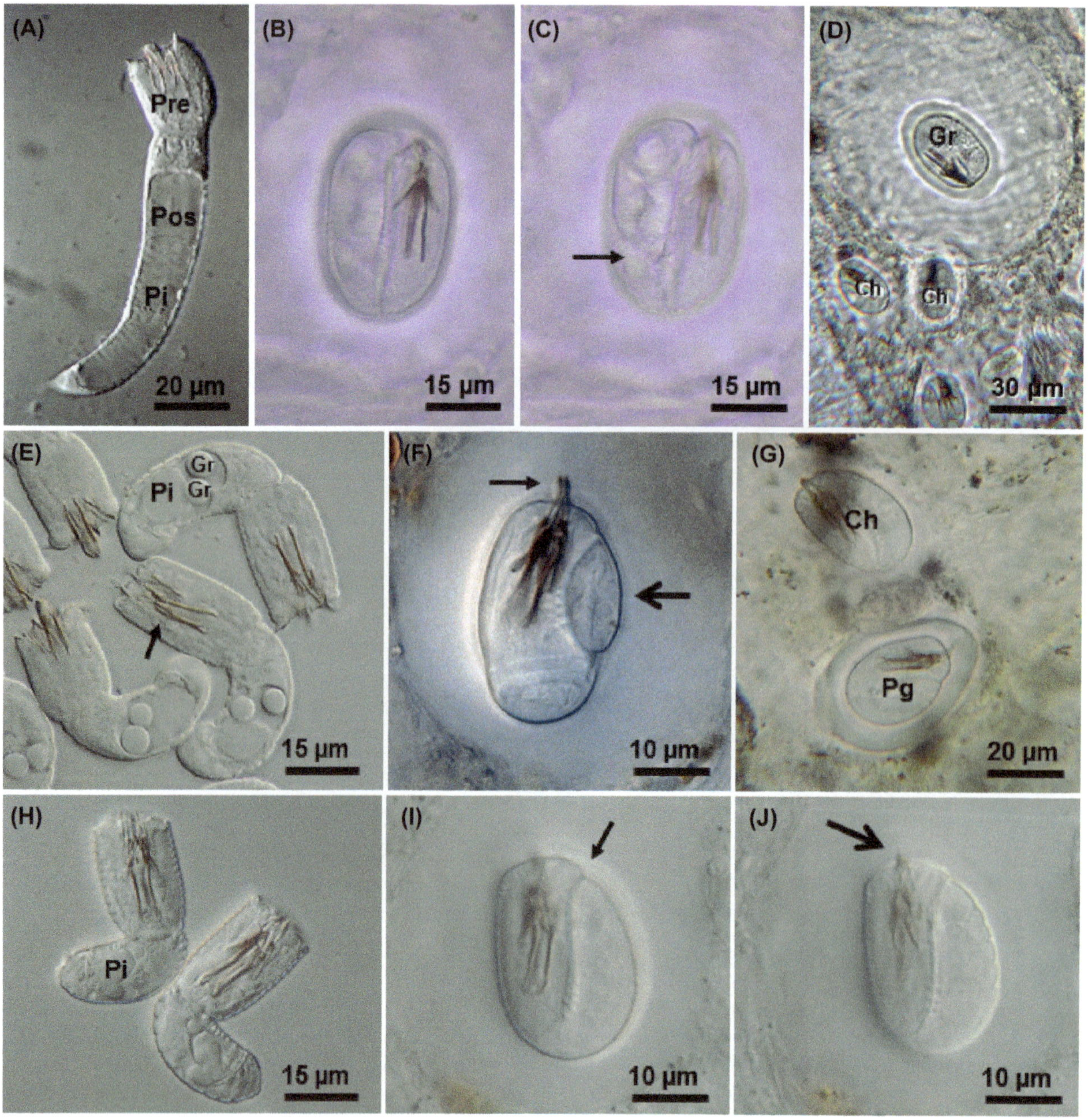

FIGURE 10.5 Larval and cyst types of Neotropical gordiids: (A) lateral view of *Gordius* type larva, note the preseptum (Pre) and the relatively long postseptum (Pos) containing an elongate pseudointestine (Pi); (B) and (C), right and left lateral views of a *Gordius* type cyst, note that there are no visible spines on the anterior region of the double folded larva and the posterior end of the postseptum reaches the posterior end of the preseptum (arrow); (D) *Gordius* type (Gr) and *Chordodes* type (Ch) cysts in the tissue of a freshwater snail collected in Mexico, note the clear halolike structure around the cysts and difference in size between *Gordius* and *Chordodes* type cysts; (E) lateral view of a *Paragordius* type larva, note the complex pseudointestine (Pi) containing two anterior granules (Gr) and clearly visible stylets (arrow) in the preseptum; (F) dorsal view of a *Paragordius* type cyst, note the relatively long spines on anterior end of the larva (small arrow) which is folded twice, with the postseptum wrapped around the preseptum (large arrow); (G) lateral view of a *Chordodes* type cyst (Ch) and *Paragordius* type cyst (Pg) in the tissue of a freshwater snail collected in Mexico; (H) lateral view of *Chordodes* type larvae; note the distinct v-shaped pseudointestine (Pi); and (I–J) lateral views of a *Chordodes* type cyst, note the larva only folds once, with the posterior end of the postseptum reaching the anterior end of the preseptum (small arrow) and relatively small spines (large arrow) on the anterior region of the folded larva.

KEYS TO GORDIIDA

Gender

1 Cloacal opening ventral (Fig. 10.2) .... male **[p. 253]**

1′ Cloacal opening terminal (Fig. 10.1) .... female **[p. 253]**

Gordiida Males: Genera

1 Tail lobes absent, posterior end rounded (Fig. 10.2 F–I) 2

1′ Paired tail lobes present (Fig. 10.2 A–E) 4

2(1) More than one type of areoles present 3

2′ Only one type of areoles present (Fig. 10.3 D) .... ***Neochordodes*** **[p. 253]**

3(2) Crowned areoles present (Fig. 10.3 H–I) .... ***Chordodes*** **[p. 254]**

3′ Crowned areoles absent (specimens should be tried in both keys) .... ***Pseudochordodes*** **[p. 256]** or ***Noteochordodes*** **[p. 256]**

4(1) Tail lobes rounded 5

4′ Tail lobes pointed (Fig. 10.2 A) .... ***Acutogordius*** **[p. 256]**

5(4) Postcloacal crescent absent 6

5′ Postcloacal crescent present (Fig. 10.2 C) .... ***Gordius*** **[p. 257]**

6(5) Tail lobe longer than wide 7

6′ Tail lobe length < 2x width 8

7(6) Cuticle with areoles arranged pairwise in lines parallel to the body axis (Fig. 10.3 A) .... ***Beatogordius*** **[p. 257]**

7′ Cuticle with irregular cords (Fig. 10.3 G) .... ***Paragordius*** **[p. 258]**

8(6) Cuticle with one type of areoles (Fig. 10.3 B) .... ***Gordionus*** **[p. 258]**

8′ Cuticle with two types of areoles .... ***Parachordodes*** **[p. 259]**

Gordiida Females: Genera

1 Posterior end rounded (Fig. 10.1 A) 2

1′ Three tail lobes present (Fig. 10.1 B) .... ***Paragordius*** **[p. 258]**

2(1) Cuticle with areoles 3

2′ Cuticle smooth (Fig. 10.3 C) .... *Gordius robustus* Leidy, 1951
[Mexico: Chiapas, Nuevo León, Tamanlipas. Trinidad and Tobago: Mt. Samana. Brazil. Chile. Paraguay. Argentina: Buenos Aires, Chubut, Corrientes, Misiones, Río Negro, Salta, San Luis. Uruguay]

3(2) Crowned areoles absent 4

3′ Crowned areoles present (Figs. 10.3 H–I, 4) .... ***Chordodes*** **[p. 255]**

4(3) More than one type of areoles present 5

4′ Only one type of areoles present (Fig. 10.3 B, D) (specimens should be tried in both keys) ***Gordionus*** **[p. 258]** or ***Neochordodes*** **[p. 253]**

5(4) Interareolar furrows narrow 6

5′ Interareolar furrows wide (Fig. 10.3 E) .... ***Noteochordodes*** **[p. 256]**

6(5) Cuticle with two areolar type, megareoles present 7

6′ Cuticle with one areolar type, megareoles absent (Fig. 10.3 B) .... ***Gordionus*** **[p. 257]**

7(6) Megareoles large........ ***Parachordodes*** **[p. 259]**

7′ Megareoles small (Fig. 10.3 F) .... ***Pseudochordodes*** **[p. 256]**

Gordiida: *Neochordodes*: Species

This key applies to both genders.

1 Areoles distributed equally 2

1′ Areoles clustered, with broad spaces between clusters 11

2(1) Areoles arranged transversally 3

2′ Areoles arranged longitudinally ..... *Neochordodes semiluna* de Villalobos & Camino, 1999
[Argentina: Salta]

3(2) Areoles rounded or oval 4

3′ Areoles polygonal or quadrangular 8

4(3) Areoles rounded 5

4′ Areoles oval 7

5(4) Areolar surface structured 6

5′ Areolar surface smooth ..... *Neochordodes colombianus* Faust & Ramos, 1960
[Colombia: Salamina]

6(5) Apical surface with bristles ..... *Neochordodes chordodides* Schmidt-Rhaesa & Menzel, 2005
[Belize: Cayo. Costa Rica: Santa Rosa]

6′ Apical surface wrinkled ..... *Neochordodes australis* Miralles & de Villalobos, 1996
[Argentina: Buenos Aires]

7(4) Areoles smooth ..... *Neochordodes serranensis* Miralles & de Villalobos, 1996
[Argentina: Buenos Aires]

7′ Areoles with thin, joined cuticular projections ..... *Neochordodes puntatus* Miralles & de Villalobos, 1996
[Argentina: San Luis]

8(3) Areoles polygonal 9

8′ Areoles quadrangular 10

9(8) Surface smooth ..... *Neochordodes uniareolatus* Carvalho, 1946
[Venezuela: Maracay. Argentina: Buenos Aires, Córdoba, San Juan, San Luis]

9′ Areoles with small spots and granules ..... *Neochordodes occidentalis* (Montgomery, 1898)
[Mexico: Veracruz. Guatemala: Quezaltenango (as *Chordodes griffinii*)]

10(8) Areolar surface verrucous ..... *Neochordodes bonaerensis* Miralles & de Villalobos, 1996
[Argentina: Buenos Aires]

10′ Areoles surface smooth ..... *Neochordodes torrenticola* Miralles & de Villalobos, 1996
[Argentina: Buenos Aires]

11(1) Areolar clusters transverse ..... *Neochordodes serranensis* Miralles & de Villalobos, 1996
[Argentina: Buenos Aires]

11′ Areolar clusters in circular or semicircular arrays ..... *Neochordodes nietoi* (Cavallero, 1936)
[Mexico: México D.F.]

Gordiida: Male *Chordodes*: Species

1 Areoles widely spaced; differences between these species not clear 2

1′ Areoles more or less close together 3

2(1) Crowned areoles in pairs or in clusters of three ..... *Chordodes lenti* Carvalho, 1944 or *Chordodes moraisi* (Carvalho, 1942)
[Brazil: Minas Gerais]

2′ Crowned areole clusters composed of a pair of crowned areoles, surrounded by 8–10 circumcluster areoles ..... *Chordodes peraccae* (Camerano, 1894)
[Mexico (as *Chordodes griffinii*). Brazil: Mato Grosso do Sul. Bolivia: Aguairenda. Argentina: Tucumán]

3(1) Simple areoles with irregular or structured surface 4

3′ Simples areoles with smooth surface ..... *Chordodes balzani* Camerano, 1896
[Venezuela: Río Cuchivero, Caicara. Trinidad and Tobago: Central Range Mountains. Brazil: São Pedro do Sul. Bolivia: Basso Beni. Argentina: Misiones]

4(3) Simple areoles with fine bristles 5

4′ Simple areoles without bristles 6

5(4) Crowned areoles surrounded by circumcluster areoles ..... *Chordodes festae* Camerano, 1897
[Venezuela: Merida. Ecuador: Cuenca]

5′ Crowned areoles not surrounded by circumcluster areoles (Fig. 10.3 I) ..... *Chordodes bouvieri* (Villot, 1885)
[Venezuela: Chama, Escurial, Merida. Colombia: Gimenez. Ecuador: Gualáquiza. Argentina: Tucumán]

6(4) Simple areoles strongly structured into "blackberry" areoles or with verrucous surface 7

6′ Simple areoles roundish with irregular surface .... *Chordodes matensis* de Villalobos & Miralles, 1997
[Argentina: Misiones]

7(6) Crowned areoles present on lateral body side 8

7′ Crowned areoles absent on lateral body sides .... *Chordodes brasiliensis* Janda, 1893
[Brazil: Espirito Santo, Minas Gerais, Rio de Janeiro. Argentina: Buenos Aires, Córdoba, Corrientes, Entre Ríos, Jujuy, Misiones, Salta, San Luis. Uruguay: Colonia, Mercedes, Paysandú, Tacuarembó]

8(7) Crowned areoles not dimorphic 9

8′ Crowned areoles dimorphic .... *Chordodes alticuspis* de Villalobos, Zanca & Yanez, 2009
[Ecuador: Orellana]

9(8) Crowned areoles in groups of five to seven 10

9′ Crowned areoles in small groups of two to four .... *Chordodes ecuatoriensis* de Villalobos, Zanca & Yanez, 2009
[Ecuador: Orellana]

10(9) Crowned areoles height > 2x simples areoles .... *Chordodes cornuta* De Villalobos & Camino, 1999
[Argentina: Salta]

10′ Crowned areoles height < 2x simple areoles (Fig. 10.3 H) .... *Chordodes nobilii* Camerano, 1901
[Argentina: Buenos Aires, Córdoba, San Luis]

Gordiida: Female *Chordodes*: Species

1 Ventral groove without long filaments (Fig. 10.4 B) 2

1′ Ventral groove with long filaments (Fig. 10.4 A) 5

2(1) Simple areoles irregular or polygonal 3

2′ Simple areoles rounded with smooth surface .... *Chordodes moraisi* (Carvalho, 1942)
[Brazil: Minas Gerais]

3(2) Simples areoles elevated 4

3′ Simple areoles flat (Fig. 10.3 I) .... *Chordodes bouvieri* (Villot, 1885)
[Venezuela: Chama, Escurial, Merida. Colombia: Gimenez. Ecuador: Gualáquiza. Argentina: Tucumán]

4(3) Simple areoles with verrucous surface .... *Chordodes polytuberculatus* Schimidt-Rhaesa & Menzel, 2005
[Costa Rica: Río Tempisquito]

4′ Simple areoles with fine bristles .... *Chordodes festae* Camerano, 1897
[Venezuela: Merida. Ecuador: Cuenca]

5(1) Crowned areoles isolated or in pairs 6

5′ Crowned areole clusters with pair of crowned areoles surrounded by circumcluster: ventral or ventral and dorsal areoles furrow 8

6(5) Crowned areoles isolated 7

6′ Crowned areoles in pairs .... *Chordodes carmelitanus* Carvalho & Feio, 1950
[Brazil: Minas Gerais]

7(6) Simple areoles smooth with fine bristles .... *Chordodes corderoi* (Carvalho, 1946)
[Venezuela: Naiguatá]

7′ Simple areoles surface structured .... *Chordodes lotus* Miralles & de Villalobos, 1997
[Argentina: Misiones]

8(5) Ventral furrow crowned areole clusters with a pair of crowned areoles surrounded by circumcluster 9

8′ Dorsal and ventral furrow crowned areole clusters with a pair of crowned areoles surrounded by circumcluster areoles 10

9(8) Simples areoles shape irregular, surface verrucous .... *Chordodes ruginosus* de Villalobos, Zanca & Yanez, 2009
[Ecuador: Orellana, Taracoa]

9′ Simples areoles surface smooth .... *Chordodes peraccae* (Camerano, 1894)
[Mexico (as *Chordodes griffinii*). Brazil: Urucum. Bolivia: Aguairenda. Argentina: Tucumán]

10(8) Lateral crowned areoles in cluster surrounded by circumcluster areoles 11

10′ Lateral crowned areoles in cluster without circumcluster areoles 12

11(10) Crowned areole clusters with 10–12 areoles .... *Chordodes staviarskii* Carvalho & Feio, 1950
[Brazil: Río de Janeiro]

11′ Crowned areole clusters with19–23 areoles .... *Chordodes balzani* Camerano, 1896
[Venezuela: Río Cuchivero, Caicara. Trinidad and Tobago: Central Range Mountains. Brazil: São Pedro do Sul. Bolivia: Basso Beni. Argentina: Misiones]

12(10) Crowned areoles lows forming groups of 3–10 areoles (Fig. 10.3 H) .... *Chordodes nobilii* Camerano, 1901
[Argentina: Buenos Aires, Córdoba, San Luis]

12′ Crowned areoles high and isolated .... *Chordodes brasiliensis* Janda, 1893
[Brazil: Espirito Santo, Minas Gerais, Rio de Janeiro. Argentina: Buenos Aires, Córdoba, Corrientes, Entre Ríos, Jujuy, Misiones, Salta, San Luis. Uruguay: Colonia, Mercedes, Paysandú, Tacuarembó]

Gordiida: *Pseudochordodes*: Species

This key covers both genders. Male specimens should also be tried in the *Noteochordodes* key below.

1 Areoles clustered 2

1′ Areoles isolated .... *Pseudochordodes guatemalensis* Miralles, de Villalobos & Rodriguez, 1997
[Guatemala]

2(1) Clusters with more than four areoles most abundant 3

2′ Most clusters contain 2-4 areoles 4

3(2) Areoles surrounding clusters flat and covered with cuticular lines .... *Pseudochordodes meridionalis* Carvalho & Feio, 1950
[Mexico: Tamanlipas. Chile: Linares. Argentina: Salta]

3′ Areoles surrounding clusters rounded with rough surface .... *Pseudochordodes bulbareolatus* Schmidt-Rhaesa & Menzel, 2005
[Mexico: Vera Cruz. Guatemala]

4(2) Areoles surrounding clusters large with smooth surface 5

4′ Areoles surrounding clusters small, irregular, with rugged surface .... *Pseudochordodes manteri* Carvalho, 1942
[Mexico: Veracruz]

5(4) Polygonal areoles (Fig. 10.3 F) .... *Pseudochordodes bedriagae* (Camerano, 1896)
[Argentina: Buenos Aires, Córdoba, La Rioja]

5′ Rounded areoles with minute pores .... *Pseudochordodes dugesi* (Camerano, 1898)
[Mexico: Veracruz. Argentina: Catamarca, Salta]

Gordiida: *Noteochordodes*: Species

Both genders are identified in this key. Male specimens should also be tried in the *Pseudochordodes* key above.

1 Areoles arranged perpendicular to body longitudinal axis 2

1′ Areoles of the midbody scattered .... *Noteochordodes achosmosus* de Villalobos, Zanca & Salas, 2008
[Argentina: Catamarca]

2(1) Areoles isolated with a tubercle in a lateral concavity 3

2′ Areoles isolated with a tubercle resting on one lateral sides 4

3(2) Areoles conical, surface smooth (Fig. 10.3 E) .... *Noteochordodes talensis* (Camerano, 1897)
[Brazil: South Mato Grosso. Bolivia: Tarija. Chile: Salto Quilpué. Argentina: Catamarca, Jujuy, Neuquén, Salta, Tucumán]

3′ Areoles elongated, narrow, surface flaking .... *Noteochordodes desantisi* Miralles & de Villalobos, 2000
[Argentina: Catamarca]

4(2) Areoles smooth or with a central depression and sinuate borders .... *Noteochordodes cymatium* de Villalobos, Zanca & Salas, 2008
[Argentina: Catamarca, Salta]

4′ Areoles smooth or with a central depression and smooth border .... *Noteochordodes saltae* Miralles & de Villalobos, 2000
[Argentina: Catamarca, Salta]

Gordiida: *Acutogordius*: Species

Only males can be identified.

1 Cuticular fold unpaired 2

1′ Cuticular fold paired, extending onto both tail lobes (Fig. 10.2 A) .... *Acutogordius obesus* (Camerano, 1895)
[Ecuador: Gualaquiza]

2(1) Cuticle with irregular roundish depressions .... *Acutogordius acuminatus* Miralles & De Villalobos, 1998
[Brazil: Xiruca]

2′ Cuticle with rounded areole .... *Acutogordius americanus* Miralles & De Villalobos, 1998
[Costa Rica: Tunivialba]

Gordiida: Male *Gordius*: Species

1 Cloacal opening on flat surface 2

1′ Cloacal opening in depression (Fig. 10.2 C) .... *Gordius austrinus* de Villalobos, Zanca & Ibarra-Vidal, 2005
[Chile: Valparaiso]

2(1) Cloacal opening lacking anterior semicircular bristle row 3

2′ Cloacal opening with anterior semicircular bristle row, posteriorly with semicircular crescent with lateral bristle rows .... *Gordius paranensis* Camerano, 1892
[Brazil: Paraná. Chile: Casablanca, Puyehue, Salto. Paraguay: Asunción.]

3(2) Midbody cuticle smooth 4

3′ Midbody cuticle with elliptic depression .... *Gordius deltaensis* Miralles & De Villalobos, 1996
[Argentina: Buenos Aires]

4(3) With scarce tubercles (Fig. 10.3 C) .... *Gordius robustus* Leidy, 1851
[Mexico. Costa Rica. Trinidad and Tobago: Trinidad. Chile: VIII Región, Valparaiso. Argentina: Chubut, Corrientes, Misiones, Salta, San Luis]

4′ With circular structures .... *Gordius guatemalensis* Linstow, 1902
[Guatemala: Chiacam]

Gordiida: Female *Gordius*: Species

1 Cuticle with areoles 2

1′ Cuticle smooth with abundant spiniform tubercles .... *Gordius spinosus* Miralles & de Villalobos, 1996
[Argentina: Río Negro]

2(1) Areoles rectangular, porous, with short tubercles .... *Gordius labidus* Miralles & de Villalobos, 1996
[Argentina: San Luis]

2′ Areoles polygonal with scattered tubercles .... *Gordius attoni* Redlich, 1980
[Argentina: Salta]

Gordiida: Male *Beatogordius*: Species

1 Tale lobes thick and parallel 2

1′ Tale lobes thin .... *Beatogordius latastei* (Camerano, 1895)
[Chile: Santiago de Chile. Argentina: Córdoba]

2(1) Cloacal opening anterioventrally lacking adhesive warts 3

2′ Cloacal opening with anterioventral adhesive warts .... *Beatogordius abaiconus* Carvalho, 1946
[Brazil: Minas Gerais]

3(2) Areoles smooth 4

3′ Areoles rugged .... *Beatogordius variabilis* Miralles, 1981
[Argentina: Buenos Aires, Córdoba]

4(3) Areoles rectangular or quadrangular 5

4′ Areoles oval; areoles of same pair separated by narrow furrow with abundant tubercles .... *Beatogordius alfredi* (Camerano, 1894)
[Brazil: Minas Gerais. Argentina: Buenos Aires, Córdoba, Salta, Tucumán]

5(4) Areoles rectangular 6

5′ Areoles quadrangular with tubercles forming a continuous longitudinal stripe (Fig. 10.3 A) .... *Beatogordius funis* de Villalobos, Schmidt-Rhaesa & Zanca, 2003
[Argentina: Salta]

6(5) Areoles of same pair appear united by small cuticular bridges .... *Beatogordius inesae* (Cavalieri, 1961)
[Argentina: Santiago del Estero]

6′ Areoles of same pair with a narrow furrow .... *Beatogordius palustre* de Villalobos, Schmidt-Rhaesa & Zanca, 2003
[Argentina: Salta]

Gordiida: Male *Paragordius*: Species

1 Depressions if present, not arranged in pairs or forming perpendicular lines 2

1′ Depressions arranged in pairs or forming perpendicular lines to longitudinal body axis .... *Paragordius andreasii* Zanca & de Villalobos, 2006
[Argentina: Buenos Aires, San Luis]

2(1) Irregular cuticular elevations, if present, not disposed transversally 3

2′ Irregular cuticular elevations disposed transversally to longitudinal body axis (Fig. 10.3 G) .... *Paragordius esavianus* Carvalho, 1942
[Brazi: Estado Espirito Santo. Argentina: Misiones, Salta]

3(2) Cuticle surface covered with bristles (appearing hirsute), in some sectors remnants of areolar borders .... *Paragordius minusculus* Carvalho, 1944
[Brazil: Minas Gerais]

3′ Irregular areoles with small, central tubercle or pore present .... *Paragordius varius* (Leidy, 1851)
[Cuba. Mexico: Baja California Sur, Mexico D.F., Monterrey, Oaxaca, Tumaulipas, Vera Cruz. Guatemala. Costa Rica. Venezuela. Colombia. Brazil: Minas Gerais. Perú. Bolivia. Argentina: Buenos Aires, Catamarca, Córdoba, Jujuy, Mendoza, Misiones, Río Negro, Salta, San Luis, Tucumán. Uruguay]

Gordiida: Female *Paragordius*: Species

1′ Posterior lobes length equal 2

1 One of the posterior lobes is shorter than others .... *Paragordius diversolobatus* Heinze, 1935
[Costa Rica]

2(1) Areoles present 3

2′ Areoles absent (Fig. 10.3 G) .... *Paragordius esavianus* Carvalho, 1942
[Brazil: Espirito Santo. Argentina: Misiones, Salta]

3(2) Areoles irregular, close together, with central small tubercle or pore .... *Paragordius varius* (Leidy, 1851)
[Cuba. Mexico: Baja California Sur, Mexico D.F., Monterrey, Oaxaca, Tumaulipas, Vera Cruz. Guatemala. Costa Rica. Venezuela. Colombia. Brazil: Minas Gerais. Perú. Bolivia. Argentina: Buenos Aires, Catamarca, Córdoba, Jujuy, Mendoza, Misiones, Río Negro, Salta, San Luis, Tucumán. Uruguay]

3′ Areoles square or rectangular, scattered; interareolar space with scarce tubercles .... *Paragordius flavescens* Linstow 1906
[Costa Rica. Brazil: Sao Paulo. Argentina: Tucumán]

Gordiida: G*ordionus* Species

Both genders can be identified here.

1 Areolar surface smooth 2

1′ Areolar surface verrucous .... *Gordionus platycephalus* (Montgomery, 1898)
[Guatemala]

2(1) Interareolar furrow wide, with interareolar structures 3

2′ Interareolar furrow narrow, with granules .... *Gordionus prismaticus* (Villot, 1874)
[Colombia: Bogotá. Argentina: Buenos Aires, Córdoba]

3(2) Oval or polygonal areoles 4

3′ Rectangular areoles with undulating margins (Fig. 10.3 B) .... *Gordionus ondulatus* Miralles, 1984
[Argentina: Buenos Aires]

4(3) Oval areoles; interareolar furrow with rounded structures with small tubercle on top .... *Gordionus enigmaticus* de Villalobos, Zanca & Ibarra-Vidal, 2005
[Chile: X Region, Corral]

4′ Polygonal areoles; interareolar furrow with pores .... *Gordionus porosus* de Villalobos, Zanca & Ibarra-Vidal, 2005
[Argentina: Buenos Aires]

Gordiida: *Parachordodes*: Species

Both genders can be identified here.

1 Areoles small and large, rounded, the larger with a tubercle, smooth surface .... *Parachordodes ciferrii* Sciacchitano, 1932 [Dominican Republic: Haina]

1′ Areoles small, elevated, situated close together with rough surface .... *Parachordodes capitosulcatus* (Montgomery, 1898) [Cuba]

ACKNOWLEDGMENT

We thank the National Science Foundation for funding our Nematomorph work (award numbers DEB-0949951 to MGB and DEB-0950066 to BH and AS-R).

REFERENCES

Bolek, M., A. Schmidt-Rhaesa, C. De Villalobos & B. Hanelt. 2015. Chapter 15: Phylum Nematomorpha. Pages 303–326 *in*: J.H Thorp and D.C. Rogers, (eds.), Ecology and General Biology: Thorp and Covich's Freshwater Invertebrates, Volume I, Academic Press.

Chiu, M.C., C.G. Huang, W.J. Wu & S.F. Shiao. 2016. Annual survey of horsehair worm cysts in northern Taiwan, with notes on a single seasonal infection peak in chironomid larvae (Diptera: Chironomidae). Journal of Parasitology 102: 319–326.

de Villalobos, C., A. Schmidt-Rhaesa & F. Zanca. 2003. Revision of the genus *Beatogordius* (Gordiida, Nematomorpha). II. South American species with inclusion of two new species. Memorias do Instituto Oswaldo Cruz 98: 115–128.

de Villalobos, C. & F. Zanca. 2001. Scanning electron microscopy and intraspecific variation of *Chordodes peraccae* and *Chordodes festae* (Gordiacea, Nematomorpha). Systematic Parasitology 50: 117–125.

de Villalobos, C. & F. Zanca. 2005. Ultrastructural redescription of *Chordodes moraisi* (Carvalho, 1942) and *C. staviarskii* Carvalho & Feio, 1950 and reinterpretation of *C. gestri* Camerano, 1904 and *Pseudochordodes griffinii* (Camerano, 1898) (Gordiida, Nematomorpha). Journal of Natural History 39: 597–606.

de Villalobos, C., F. Zanca & A. Barragan Yanez. 2009. Three new species of *Chordodes* a new data on hairworms (Nematomorpha) from Ecuador. Zootaxa 2205: 37–52.

Hanelt, B., M.G. Bolek & A. Schmidt-Rhaesa. 2012. Going Solo: discovery of the first parthenogenetic gordiid (Nematomorpha: Gordiida). PLoS One 7: e34472.

Hanelt, B., A. Schmidt-Rhaesa & M.G. Bolek. 2015. Cryptic species of hairworm parasites revealed by molecular data and crowdsourcing of specimen collections. Molecular Phylogenetics and Evolution 82: 211–218.

Harkins, C., R. Shannon, M. Papes, A. Schmidt-Rhaesa, B. Hanelt & M.G. Bolek. 2016. Using Gordiid cysts to discover the hidden diversity, potential distribution, and new species of Gordiids (Phylum Nematomorpha). Zootaxa 4088: 515–530.

Poinar, G.O. 1999. *Palaeochordodes protus* n.g., n.sp. (Nematomorpha, Chordodidae), parasites of a fossil cockroach, with a critical examination of other fossil hairworms and helminths of extant cockroaches (Insecta: Blattaria). Invertebrates Biology 118: 109–115.

Poinar, G.O., J. Rykken & J. La Bonte. 2004. *Parachordodes tegonotus* n. sp. (Gordioidea: Nematomorpha), a hairworm parasite of ground beetles (Carabidae: Coleoptera), with a summary of gordiid parasites of carabids. Systematic Parasitology 58: 139–148.

Salas, L., C. de Villalobos & F. Zanca. 2011. Sexual size dimorphism, sex ratio and the relationship between seasonality and water quality in four species of Gordiida, Nematomorpha) from Catamarca, Argentina. Journal of Helminthology 85: 319–324.

Schmidt-Rhaesa, A. 2002a. Are the genera of Nematomorpha monophyletic taxa? Zoologica Scripta 31: 185–200.

Schmidt-Rhaesa, A. 2002b. Australian species of *Chordodes* (Nematomorpha) with a description of two new species, remarks on the genus and its life history. Journal of Natural History 36: 1569–1588.

Schmidt-Rhaesa, A., C. de Villalobos & F. Zanca. 2008. Summary of *Chordodes* species (Nematomorpha, Gordiida), with a discussion of their diagnostic characters. Verhandlungen des Naturwisenschaftlichen Vereins inHamburg 44: 37–114.

Schmidt-Rhaesa, A., C. de Villalobos, F. Zanca, B. Hanelt & M.G. Bolek. 2016. Chapter 10: Phylum Nematomorpha. Pages 181–188 *in*: J.H. Thorp and D.C. Rogers, (eds.), Keys to Neartic Fauna: Thorp and Covich's Freshwater Invertebrates, Volume II, Academic Press.

Szmygiel, C., A. Schmidt-Rhaesa, B. Hanelt & M.G. Bolek. 2014. Comparative descriptions of non-adult stages of four genera of Gordiids (Phylum: Nematomorpha). Zootaxa, 376: 101–118.

Chapter 11

Phylum Mollusca

María Gabriela Cuezzo, Diego Eduardo Gutiérrez Gregoric, Jean-Pierre Pointier, Antonio Alejandro Vázquez, Cristián Ituarte, Maria Cristina Dreher Mansur, Janine Oliveira Arruda, Gary M. Barker, Sonia Barbosa dos Santos, Ximena Maria Constanza Ovando, Luiz Eduardo Macedo de Lacerda, Monica Ammon Fernandez, Silvana Carvalho Thiengo, Aline Carvalho de Mattos, Elizangela Feitosa da Silva, Maria I. Berning, Gonzalo A. Collado, Igor Christo Miyahira, Thiago Nunes Antoniazzi, Daniel Mansur Pimpão and Cristina Damborenea

Chapter Outline

Thorp and Covich's Freshwater Invertebrates. https://doi.org/10.1016/B978-0-12-804225-0.00011-3

Introduction to the Phylum

Cristina Damborenea
CONICET. División Zoología Invertebrados, Museo de La Plata, La Plata, Argentina

INTRODUCTION

Mollusca is the second largest metazoan phylum and is one of the most morphologically diverse. The body comprises three distinguishable regions: a **head** that may bear different sensory structures, a muscular **foot**, and a visceral mass which is covered by a thick epidermal layer, called the **mantle** or **pallium**. The mantle usually secretes a hard calcareous skeleton, the **shell**, and forms the **mantle or pallial cavity**, which houses the gills, the openings of the digestive, nephridial, and genital systems. The water circulates through the cavity. Another unique feature is the presence of a **radula** in the fore gut (except in Bivalvia), which is a toothed rasping strap using in feeding.

Two Mollusca classes—Gastropoda and Bivalvia—have colonized freshwater environments of every continent except Antarctica. Gastropods are the best studied mollusks, and the only ones that undergo torsion (a rotation of 90–180 degrees of the visceral mass in relation to the foot). Freshwater gastropods have variable diets and substrate preferences, and despite the uniform structural plan and habits, they possess striking ecological specializations. Many gastropod species belong to the "prosobranchs" (breathe through gills, carry an operculum, and are usually gonochoric and occasionally parthenogenetic or androgenetic). On the other hand, the "pulmonates" snails respire over the mantle surface, generally are hermaphrodites, and have developed diverse reproductive strategies (e.g., long-term sperm storage, multiple insemination, sperm sharing) in the continental environment (Dillon, 2000; Pyron & Brown, 2015).

The bivalve shell has two valves connected dorsally by an elastic ligament, which enclose a spacious mantle cavity; all freshwater species are filter feeders. Freshwater bivalves have diverse reproduction modes, some being gonochoristic and others hermaphroditic. Many species are adapted to hold developing larvae, brooding or delaying juveniles (Dillon, 2000; Pyron & Brown, 2015).

Most descriptions are based on shell morphology and on few specimens, without considering intraspecific variability, or modifications due to variable environmental conditions. For this reason, it is a good practice to contrast identifications with museum collections, or with specimens identified by specialists.

Material preparation and preservation techniques are presented below in each group.

KEYS TO MOLLUSCA

Mollusca: Classes

1	Shell single, coiled (snails) or uncoiled (freshwater "limpets"), radula present	**Gastropoda [p. 266]**
1′	Shell with two valves connected by a ligament hinge (mussels and clams), without radula	**Bivalvia [p. 390]**

REFERENCES

Dillon, R.T. 2000. The Ecology of Freshwater Mollucs. Cambridge University Press. Cambridge. 509 pp.

Pyron, M. & K.M. Brown. 2015. Introduction to Mollusca and Class Gastropoda. Pages 383–421 *in:* J.H. Thorp & D.C. Rogers (eds.), Volume I Ecology and General Biology, Thorp and Covich's Freshwater Invertebrates, 4th edition. Elsevier.

Class Gastropoda

María Gabriela Cuezzo
Instituto de Biodiversidad Neotropical (CONICET-UNT, Facultad de Ciencias Naturales), Tucumán, Argentina

INTRODUCTION

Gastropoda is the largest molluskan class. There are more species of freshwater mollusks than all the mammal species of the world. The global freshwater gastropod fauna is estimated at approximately 4000 valid described species, but the probable number is in the order of 8000 species (Strong et al., 2008; Rosenberg, 2014). At the same time, freshwater mollusks are also the most endangered fauna on the planet (Lydeard & Cummings, 2019).

Among the higher groups of gastropods generally accepted, the Caenogastropoda, Neritimorpha, and Heterobrachia includes freshwater taxa. Only the Viviparoidea, Glacidorboidea, and nearly all Hygrophila comprise superfamilial (or above) groupings with members represented exclusively in freshwater (Strong et al., 2008).

The Neotropical freshwater gastropod fauna is dominated by two main components: the Caenogastropoda (with Ampullaridae and Cochliopidae as the richest families), and pulmonate heterobranchs (with Planorbidae as the most speciose group). Strong et al. (2008) estimated the valid freshwater gastropod species to be 440–533 for the Neotropical region, but the Neotropical caenogastropods alone (this chapter) may equal this number. Large areas remain undersampled such as rivers and streams of the Amazon Basin, and as a consequence the actual diversity may be much greater than currently recognized.

Freshwater gastropods occur in every Neotropical ecoregion, and in every aquatic habitat. There are high variabilities in habitat specialization; some groups are only found in particular habitats while others are more tolerant, even living for long periods out of water or tolerating highly saline environments or elevated ranges of temperatures.

Caenogastropods show a significant radiation in the Neotropical region, especially the Cochliopidae, Ampullaridae, and Pachychilidae. The heterobranch component is less diverse and not comparable to the high diversity found in the Palearctic and Nearctic regions. Neotropical freshwater snails show varied reproductive and life history strategies. Most Caenogastropods are exclusively dioecious and reproduction is sexual except for parthenogenetic species among Thiaridae. However, many species are brooders (all Viviparidae, some Cerithioidea, Rissooidea, and some Heterobranchia) and retain their young in brood pouches that represent modifications of the oviduct, mantle cavity, or cephalic hemocoel (Strong et al., 2008). Heterobranch freshwater gastropods are hermaphrodites and oviparous. Neritinimorpha lives and reproduces in rivers and estuaries. Egg masses are stuck on rocks and release planktotrophic larvae downstreams to the sea, where salinities are required for their development. Later juveniles migrate upstream to their freshwater habitats (Pointier & Noya, 2015).

Several species of freshwater gastropods are of medical and veterinary importance because they are intermediary hosts for trematodes and nematodes infecting humans and cattle.

LIMITATIONS

Taxonomic knowledge of all Neotropical freshwater gastropods is similarly limited. The low numbers of taxonomists and monographic publications are a major problem, and there are still large unexplored geographical areas. There are no comprehensive catalogs for most of the Neotropical freshwater taxa. Simultaneously, little attention has been given to phylogenetic relationships, especially of the Neotropical freshwater pulmonates. Shell features are currently the basis for Neotropical species identification, but intraspecific variation has been rarely studied. This situation has led to the proliferation of species names and misidentifications. In addition, most of the morphological studies are limited to shell morphology, and anatomical information is missing in most species. Despite how much work has been done, the shell has few characters useful to discriminate species due to convergence in shape and size. Moreover, shell shape varies depending on environmental factors. The jaw and radula have also limited importance for species identification.

Future studies should integrate and complement morphological and molecular studies combined with a new appreciation of ecological and geographical patterns to a better understanding of the Neotropical fauna.

TERMINOLOGY AND MORPHOLOGY

Caenogastropods includes some of the largest (17 cm in shell height in Ampullariidae) and smallest gastropods (<1 cm in shell height, Truncatelloidea).

Neotropical Neritimorpha (Neritidae) have characteristic globose shell shapes usually called neritiform with a short spire, a wide columellar area, and a calcified operculum that is ornamented with an internal apophysis (Cuezzo, 2009). Heterobranch gastropods are also a diverse group and include a wide array of shell forms, although in general smaller and thinner than most of caenogastropods. Shell shape in both, caenogastropods and heterobranchs, can be planispiral, helicoid, turriform, ovate-conic, elongate-conic, cylindrical, globose, trochiform, conic (ancyliform), etc. Much of the gastropod identification relies on their shell structure, but some of their characters have a high degree of overlap, so that most are not unique to a particular group.

General gastropod shell structures are illustrated in Fig. 11.1. The **shell** may be **dextral** (or right-coiled) or **sinistral** (or left-coiled). Whorls can be flattened, slightly convex, or convex. The shell opening located in the last whorl is termed the **aperture**. The operculate groups (Caenogastropoda, Neritimorpha, and Truncatelloidea) possess an **operculum** overlapping the aperture, while Hygrophyla snails (heterobranchs) lack this structure. The operculum could be calcareous as in Neritidae, or corneous as in Trunchateidoidea and Ampullaridae. Opercula can be circular, ovate, or ellipsoidal; they can have a central nucleus and concentric whorls (multispiral operculum) or have a nucleous eccentric (paucispiral operculum). Shells can also be smooth with shallow lines, or sculptured especially among the Cerithioidea, which can have conspicuous structures such as tubercles, knots, or keels.

Freshwater snails consist of two major groups, one that possesses **gills** (or a single gill) and an **operculum**, and another group of taxa (pulmonate snails) that have a "**lung**," can breathe atmospheric air, and lack an operculum. Most operculate freshwater gastropods are typically dioecious, with the males having an enlarged right tentacle as a copulatory organ, or they possess a penis, a specialized verge, or are aphallic. They can have different reproductive strategies, ranging from oviparity, viviparity, or ovoviviparity. Females can be oviparous laying calcareous eggs outside the water or they can lay clusters of gelatinous eggs in the water. Some groups are parthenogenetic and can be viviparous. Pulmonates, in contrast, are all monoecious, and self-fertilization is possible. They are oviparous and can lay clusters of gelatinous eggs on rocks, on the shell of other snails, or over any surface.

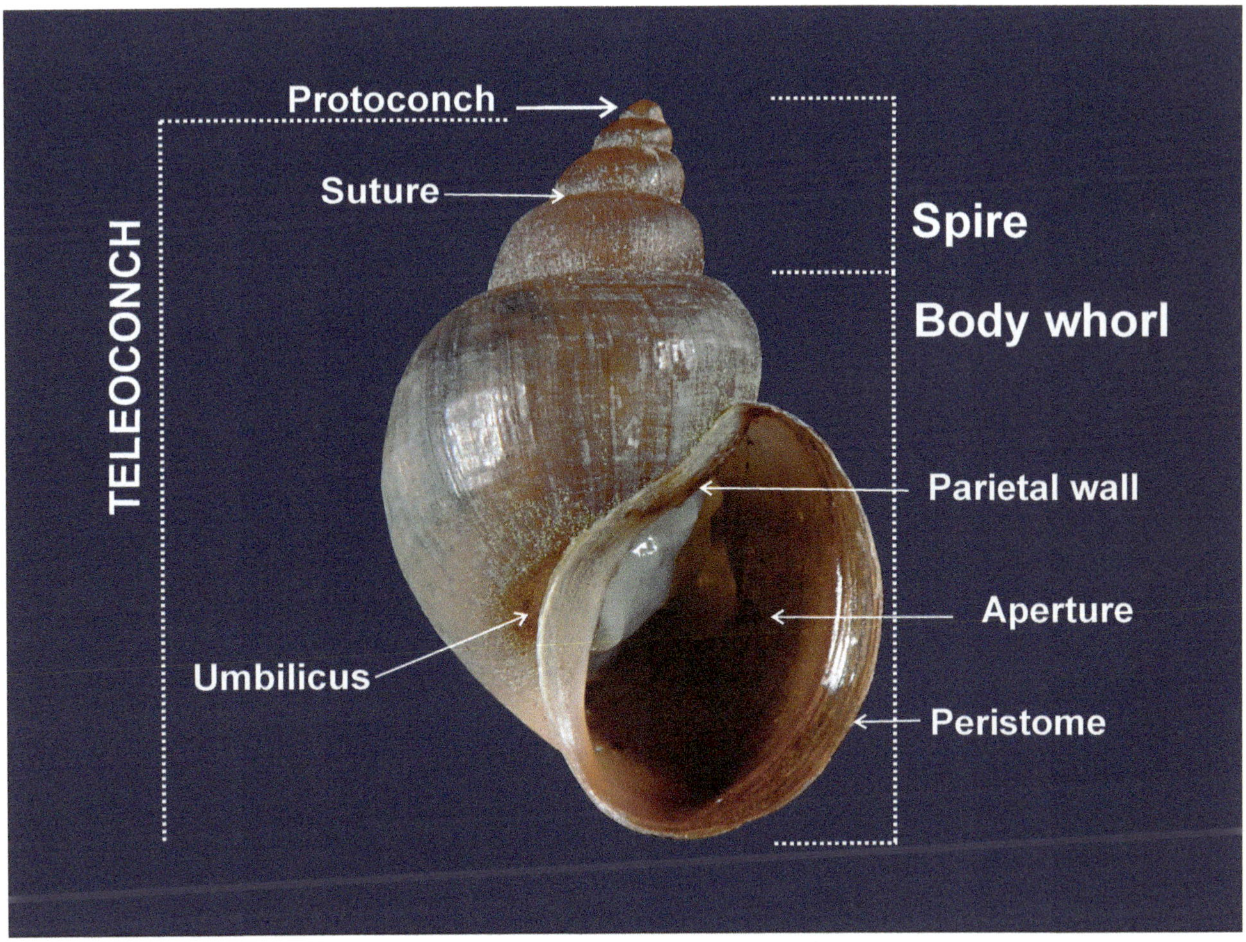

FIGURE 11.1 External morphology of gastropod shells.

PHYLUM MOLLUSCA

MATERIAL PREPARATION AND PRESERVATION

Preparation of all freshwater snails for anatomical purposes requires a relaxation process. There are different ways of carrying out this process, but most common is to use menthol crystals or Nembutal (0.25% pentobarbital salt). Relaxation process can last 10–48 h, depending on the method, specimen's size, and number of specimens per container. Fixation is also important, especially for histological and molecular studies. Different fixative solutions are used depending on the aim of the research and the taxonomic group. Ethanol or isopropyl are good fixatives because they do not damage the shell, but others such as Railliet-Henry's fixative is better for soft tissues. When planning to use electron microscopy a good fixative is Karnovsky solution (solution of paraformaldehyde plus glutaraldehyde) at 4°C in phosphate buffer and postfixation in 1% osmium tetroxide. Although different fixatives can be used depending on the different species, it is recommended not to use formalin as it dissolves the shells. Preservation can be done in 75% ethanol.

KEYS TO GASTROPODA

Gastropoda: Subclasses

1	Shell and operculum present; cephalic tentacles typically long and narrow, with eyes on basolateral swellings	2
1′	Operculum absent, except in Glacidorbidae and Valvatoidea (adult shell less than 10 mm, Figs.11.9, 10), shell present or absent; cephalic tentacles various, with eyes basomedial or absent	**Heterobranchia [p. 271]**
2(1)	Shell globose, neritiform, with very short spire, body whorl wide, operculum calcareous, lower margin with internal projections (apophysis) (Fig. 11.52)	Neritimorpha, one superfamily: **Neritidea [p. 332]**
2′	Shell form variable (turriform, planispiral, ovate, elongate-conic, cylindrical, trochiform), spire variable, operculum generally corneous, without apophysis, with central or eccentric nucleous (Figs. 11.62–67, 75)	**Caenogastropoda [p. 334]**

REFERENCES

Cuezzo, M.G. 2009. Mollusca Gastropoda. Pages 595–629 *in*: E. Domínguez & H.R. Fernández (eds.), Macroinvertebrados bentónicos sudamericanos. Sistemática y biología, Fundación Miguel Lillo, Tucumán, Argentina.

Lydeard, C. & K. Cummings (eds.). 2019. Freshwater mollusks of the World: A distribution atlas. John Hopkins University Press, Baltimore 242pp.

Pointier, J.P. & O. Noya. 2015. Family Neritidae. Pages 25–31 *in*: J.P. Pointier (ed.), Freshwater Molluscs of Venezuela. Conch Books, Germany.

Rosenberg, G. 2014. A New critical estimate of named species-level diversity of the recent Mollusca. American Malacological Bulletin 32: 308–322. https://doi.org/10.4003/006.032.0204.

Strong, E.E., O. Gargominy, W.F. Ponder & P. Bouchet. 2008. Global diversity of gastropods (Gastropoda; Mollusca) in freshwater. Hydrobiologia 595: 149–166.

Subclass Heterobranchia

Maria Gabriela Cuezzo
Instituto de Biodiversidad Neotropical (CONICET-UNT, Facultad de Ciencias Naturales), Tucumán, Argentina

INTRODUCTION

Heterobranchia, meaning "different-gilled snails," forms one of the most successful and diverse groups within Gastropoda, but has poorly resolved phylogenetic relationships especially in basal taxa. This diverse taxon comprises the Euthyneura (with the Opisthobranchia and Pulmonata), and the "Lower Heterobranchia" (with several "basal" members such as Valvatoidea, Architectonicoidea, Mathildoidea, Omalogyroidea, and Cimoidea). Although the monophyly of the heterobranch clade is supported by numerous autapomorphies, the monophyly of several currently recognized subgroupings is equivocal.

In Heterobranchia shell morphology is very diverse. Most Pulmonata are distinguished by their pulmonary cavity and pneumostome (the orifice that connects the lung with the exterior), and all supposedly have a procerebrum (Van Mol, 1967; Haszprunar, 1988; Haszprunar & Huber, 1990). Nevertheless, neither a pneumostome nor a procerebrum is present in Glacidorbidae, a family included in the pulmonates by Ponder (1986).

The main freshwater Pulmonate group in Heterobranchia is Hygrophila, a clade that comprises most of the limnic families such as Planorbidae, Lymnaeidae, Physidae, and Chilinidae, all present in the Neotropical region. Hygrophila gastropods are simultaneous hermaphrodites, but they outcross, and some have the capacity of self-fertilization. Eggs are laid in gelatinous egg cases and usually attached to plants, rocks, or even other snail shells. Males have an introvertible penis and fertilization is internal. They are mainly herbivores or detritivores. Species of Hygrophila live in a variety of environments, both lotic and lentic, from rivers, streams, ponds, lakes, etc., even in temporary aquatic habitats. Most are found in large populations and because they are easy to collect and identify, they could be indicator organisms (Strong et al., 2008).

The cosmopolitan Succineidae (Succineoidea, Eupulmonata) are found in marshes, swamps, and lake margins but can also be found on moist soil or emergent vegetation (Barker, 2001a). Chilinidae and Glacidorbidae have restricted geographical ranges, while Planorbidae, Lymnaeidae, and Physidae are widely distributed.

The Acochlidioidea conform a group of species mostly living interstitially in marine sands, while few of them have conquered freshwater systems and brackish environments. *Tantulum elegans* (Rankin, 1979) is one of the few species inhabiting freshwater and was found in a mountain spring marsh isolated from the sea in the Caribbean island of Saint Vincent. Most other freshwater species of Acochlidioidea are distributed over different tropical Indo-Pacific islands. These species have sizes between 25 and 30 mm, considerably bigger than the Caribbean species, and live in rivers and streams close to the sea. The reproductive system of acochlidians is monaulic but, within the group, there is a wide variety of special reproductive features. These include modification or loss of the copulatory organs, and sperm transfer by hypodermal impregnation or spermatophores. Some of the marine species have separate sexes, contrary to the general hermaphroditic condition of the opisthobranchs. *Tantulum elegans* has copulatory organs while penial stylets are absent, reasons to believe that, in this species, the mode of sperm transfer is by copulation (Neusser & Schrödl, 2007). This species shows a separate vas deferens to accommodate autosperm and an otherwise undivided pallial gonoduct for allosperm and eggs.

Overall Neotropical species richness is estimated as more than 100 species [50 species Chilinidae, 8 Lymnaeidae, > 26 Planorbinae, 18 Ancylinae, ≥ 20 Physidae, 3 Glacidorbidae, 1 Tantulidae]. Final numbers are estimates only because of the large number of synonyms in all genera, lack of proper studies, and many understudied areas especially in Bolivia, Paraguay, Ecuador, Colombia, and Central America. Inventories and taxonomic revisions involving morphological and molecular tools are needed.

LIMITATIONS

Most Heterobranchia taxa are easily identified to family level. Planorbidae has received most attention, due to most species having medical importance. However, taxonomic revisions on the Neotropical taxa are scarce or generally lacking in most families.

One of the main limitations is that species descriptions have been frequently based almost exclusively on shell characters, and on few specimens. Shell variability has rarely been studied, and anatomical information is usually scarce or lacking. Consequently, in some families the same species has been described under different names in different countries. The identification of close species with anatomical affinities has been extremely difficult, when only shell shape is analyzed. The reproductive system provides the best morphological characters, but is limited when used alone. Species or species groups show similar or identical reproductive and shell morphologies and in these cases molecular identification seems to be unavoidable. More recently, molecular markers have proved to be an important tool, and a good complement of anatomical characters. However, approaches using single genetic sequences or genes, restricted taxon sampling, or restricted areas have been of limited impact.

Another limitation is that vast areas remained poorly surveyed, and inventories are often lacking. Public malacological collections documenting local fauna do not always exist. Moreover, alcohol preserved specimens are either old or completely lacking.

Some taxonomic keys have been published for different taxa distributed in the Neotropics (Taylor, 2003; Cuezzo, 2009; Pointier & Nova, 2015a), but no key is comprehensive enough to cover all the Neotropics and the variations in size and morphology that are sometimes found within a species.

TERMINOLOGY AND MORPHOLOGY

Heterobranch gastropods have a single **shell** with variable morphologies. Hygrophila lack an operculum. If we consider the shell as a tube, it is possible to understand the different shell forms. This tube is generally coiled around a central axis, called the **columella**. When the shell is not coiled around an axis, the shape is a cone with a single or few whorls, and is called **patelliform** (Fig. 11.2 A). The different portions of the shell are the **spire**, formed by all the whorls, except for the body whorl, the widest whorl with the aperture (Fig. 11.1). Another distinctive shape is the **planospiral** shell, coiled around an axis, but each whorl overlapping the previous whorl. The result is a shell in which the spire has almost the same height as the body whorl or is depressed. When each whorl is shifted over the other during shell coiling, the resultant shell has a projected spire over the body whorl, as in the Lymnaeidae or Physidae. The base of the columella is the **umbilicus**, most obvious in perforate shells when it opens to the exterior, and less distinguishable when it is closed in imperforate shells. The **aperture** is generally to the right, except in Physidae and some Planorbidae. Limits among the different whorls are marked by the suture. The apical portion of the spire is the **protoconch**, while the rest of the shell is the **teleoconch**. The aperture margin is the **peristome**, which can be expanded or sharp. **Teeth** or **columellar thickenings** are found in Chilinidae and some Planorbidae. In Succineidae a trend to shell reduction and slug form has occurred and the shell in some genera is reduced to the extent that it is internalized beneath the mantle shield (Barker, 2001b).

Shell sculpture in Hygrophila is in general simple; it can be smooth or with growth lines or spiral/axial grooves. Shell coloration is generally uniform, and not very striking, except in the Chilinidae where conspicuous color bands zig-zag in some species, contrasting with the color background.

Traditional linear measurements in spiral and patelliform shells are total shell length and width, spire and body whorl length, aperture length and width (Fig. 11.3).

MATERIAL PREPARATION AND PRESERVATION

Dry shells should be preserved separated from the soft body, whenever is possible, because in general shells are fragile or thin and the alcohol or other preserving solutions damage them. The best storage is in glass or plastic vials plugged with cotton.

For DNA studies, specimens should be placed directly into absolute or 96% ethanol and preserved in a freezer. Formaldehyde is not recommended, not only for molecular studies, but also for general fixation and preservation, because it causes severe damage and even completely destroys the shell over time.

For anatomical and histological studies, specimens must be relaxed before fixation. Menthol crystals are the anesthetic most commonly used, but others, such as chloretone, Nembutal, or chloral hydrate are good for relaxation. A simple procedure is to transfer specimens into vessels of water with floating menthol crystals on the surface, and placed in the refrigerator, undisturbed for 12–24 h. The tentacle of the specimens should be touched periodically until the animal does not show movements or retracts its body. Succineidae can take up to 4 days. Then, transfer the specimens to the preferred fixative solution. Another procedure is to first submerge the snails in hot water (70°C) in a vessel or a Petri dish for 30 s to a minute (variable time according to different groups), and then transfer them into cold water previous to adding the fixative solution. The body is then removed from the shell by gentle rotating the foot in different directions, then submerging only the soft parts into the fixative. A good fixative solution is modified Railliet-Henry (93% distilled water, 2% glacial acetic acid, 5% formaldehyde, and 6 g sodium chloride per liter) modified for freshwater mollusks and extensively used with Planorbids.

Groups with special methodologies include planorbids and ancylids, as described below.

[1]Planorbids should be placed in a 0.1% or 0.05% solution of sodium pentobarbital (Hypnol or Nembutal) in tap water for 4–6 h, depending on the size of the specimen. The specimens are then killed by plunging into hot water at 70°C for about 40 s and then transferred to cool water. Each specimen (under water) is pulled by the foot with forceps, so as to disconnect the columellar muscle from the shell. If it is not possible to relax the specimens before preservation, the shell is plunged to the aperture into hot water (70°C) for 15 s, and gradually immersed in the water until complete 40 s, then placed in cold water. The whole animal is drawn out of the shell and fixed in modified Railliet-Henry's fluid. The shell should be kept in water for a week until the blood

1 M. Ammon Fernandez, E. Feitosa da Silva and S. Carvalho Thiengo contributed with Material Preparation and Preservation of planorbids.

PHYLUM MOLLUSCA

FIGURE 11.2 Heterobranchids: (A) *Omalonyx convexus* (Succineidae); (B) *Pseudosuccinea columella* (Lymnaeidae) note triangular, flat tentacles; (C) *Stenophysa minor* (Physidae), note filiform ocular tentacles and mantle margins partially overlapping the shell; (D) *Chilina lilloi* (Chilinidae), note lamellar ocular tentacles; (E) *Biomphallaria* sp. (left) and *Drepanotrema* sp.(right) (Planorbidae), note filiform ocular tentacles and discoidal shells. Scale bars: (A) and (C) = 5 mm; (B) and (D) = 2 mm; and (E) 1 mm.

and tissue residue are diluted, then it should be washed and dried at room temperature (Fernandez et al., 2008). Species that have many whorls or lamellae should not be removed from the shell to avoid body ruptures and shell damage. After anesthetizing and heating at 70°C, the sample must be separated: (1) some specimens should be left in a vial covered with water until the bodies dissolve and only shells remain; (2) some should be preserved in Railliet-Henry's fixative, which will destroy the calcium of the shell and the animal remains intact for dissection.

[2]Ancylids should be immersed in iodine solution (lugol's iodine) for 30 min to 1 h. The muscles will acquire orange or reddish color with a metallic sheen. This

2 S. Barbosa dos Santos, X.M.C. Ovando and L. E. Macedo de Lacerda contributed with Material Preparation and Preservation of ancylids.

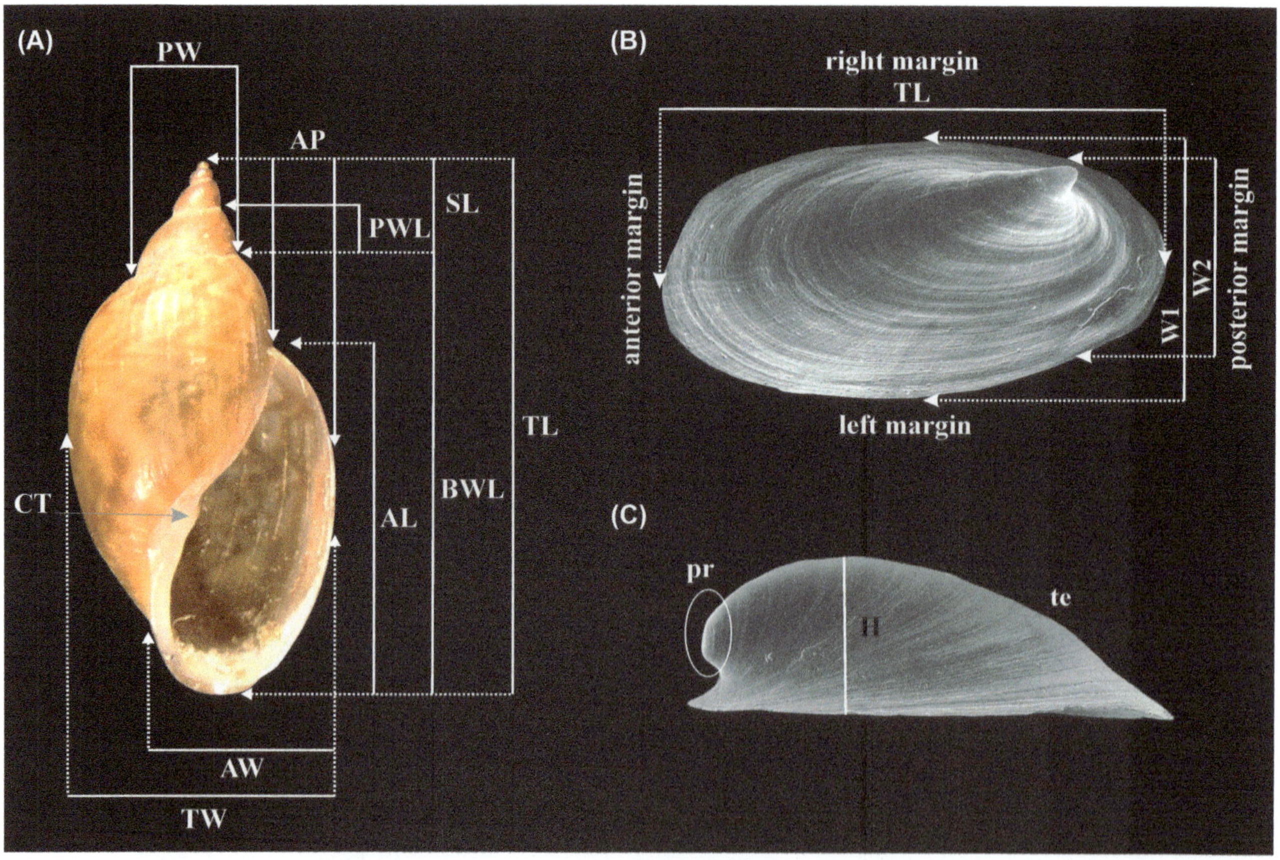

FIGURE 11.3 Traditional linear shell measurements in heterobranchids: (A) *Chilina* sp., shell in ventral view; (B) patelloid shell in dorsal view (Ancylinae); and (C) lateral view of a patelliform shell (Ancylinae). Abbreviations: *AL*, aperture length; *AP*, aperture projection; *AW*, aperture width; *BWL*, body whorl length; *CT*, columellar teeth; *H*, total height; *pr*, protoconch; *te*, teleoconch; *PW*, penultimate whorl width; *PWL*, penultimate whorl length; *SL*, spire length; *TL*, total length; *TW*, total width; *W*, shell width. *Photo credits: (A) D. Gutiérrez Gregoric; (B) and (C) X. Ovando.*

allows one to differentiate between connective tissue and muscles. The ejaculatory organ is complex. After isolating it from body wall, dehydrate in alcohol and without coloring, mount in pure glycerin on a microscope slide, observing it under different light intensities. Another method is after isolating the ejaculatory complex, using fine needles, carefully remove the prepuce, being careful not to break the penial sheath. Then, dehydrate it in an ethanol series with two final washes in absolute ethanol; color with acetic Carmine for 12–20 h. Differentiate the tissues with acetic acid for 5–10 s. Then, dehydrate the specimen on a slide in a drop of phenol, quickly removing the excess with a filter paper under the stereomicroscope. Clarify the specimen with a drop of creosote, mount with entelan or pure glycerin, and examine under the microscope. Serial histological cuts of the ejaculatory complex can be used.

Ancylid shells should be cleaned and dried for analysis under binocular stereoscopic microscope with a strong light. Each shell must be placed with the opening down and the rear facing the observer. The left and the right sides of shell correspond to the left and right of the observer. Apical microsculpture must be studied by scanning electron microscopy (SEM), but, it is possible to observe details in well cleaned shells under a light microscope using different light intensities. To observe the protoconch microsculpture, choose uneroded shells. Put the shells in a small Petri dish with weak sodium hypochlorite solution (5%) for 2–12 h, until the periostracum detaches; wash the shells 4 to 5 times in distilled water, dry on filter paper in a Petri dish, protecting them from dust. The process can be hastened by placing the shells in an oven or on a hot plate at 50°C. It is possible to see bubbles leaving the shell under a stereoscopic microscope. The periostracum can be easily removed with a brush, but is important to not force its detachment.

For morphological studies of the radula, this structure must be dissected out from the buccal mass, cleaned with an ultrasonic cleaner, and studied with SEM (Holznagel, 1998). Also, the radula can be cleaned by putting it in a small Petri dish with dilute sodium hypochlorite solution (5%) for 10–15 min (the time depends on buccal bulb size). This process must be observed under microscope and finished when the radula is free from any muscular tissue. The radula should then be washed with distilled water twice to eliminate all hypochlorite solution.

Both radula and shells can be examined using SEM. The radula is extended on an SEM-stub covered by a double-side carbon tape. To extend the radula, put a drop of 70% alcohol on the stub and transfer the radula from the Petri dish to the stub with needles. The orientation of a mounted radula can be checked under a light microscope with a 25–40× objectives to determine the side with denticles. Let the alcohol evaporate.

KEY TO HETEROBRANCHIA

Gastropoda: Heterobranchia: Superfamilies and Families

1 Shell form variable, reduced or not; freshwater or terrestrial; ocular tentacles with eyes basal or distal; Pulmonata 2

1′ Shell reduced, internal spicular skeleton posterior to tentacles in cerebral ganglia region; freshwater, semilentic, interstitial; total length ~2 mm, width ~0.29 mm; two pairs tentacles, anterior pair larger, slightly flattened proximally, moving mostly in horizontal plane, the posterior pair smaller, moving mostly in vertical plane; mouth anterior, between head and foot; eyes absent; colorless to whitish; foot with muscle fibers and internal hemocoelic spaces, demarcated by moderate cephalopedal and visceropedal grooves from anterior hump; anus ventral; "Opisthobranchia" .... Acochlidioidea, one species: *Tantulum elegans* (Rankin, 1979)
[Saint Vincent and Grenadines: Golden Grove]

2(1) Shell form variable but never reduced, housing entire animal; ocular tentacles with eyes basal (Fig. 11.2 B–E) 3

2′ Shell thin, ovate, may be reduced, not capable of housing entire animal; ocular tentacles with eye distal (Fig. 11.2 A), freshwater and terrestrial .... Succineoidea, one family: **Succineidae [p. 274]**

3(2) Adult shell < 8 mm, with operculum 4

3′ Adult shell > 3 mm, without operculum; Hygrophila 5

4(3) Shell 5–8 mm in diameter, depressed, subglobose smooth to transversely wrinkle, 3½ whorls and 3 to 3.5 mm of height, spire not prominent, and apex obtuse; aperture and operculum circular; shell base is indistinctly subangular around the umbilicus; gill plumose, pallial tentacle present .... Valvatoidea, Valvatidae, one species: *Valvata humeralis* (Say, 1829)
[Central America: Mexico]

4′ Shell < 3 mm, flat or near flat spire to planispiral, operculum circular or oval, multispiral to paucispiral, with central to eccentric nucleus, unknown anatomy .... Glacidorboidea, one family: **Glacidorbidae [p. 276]**
[South America: southern Argentina and Chile]

5(3) Tentacles filiform or lamellar; shell coiled, sinistral, or dextral (Fig. 11.2 C) 6

5′ Tentacles broad; flat, triangular; shell coiled, always dextral (Fig.11.2 B) .... Lymnaeoidea, one family: **Lymnaeidae [p. 282]**

6(5) Tentacles short, shape variable; shell mostly dextral; thickness and form variable, apertural teeth may be present 7

6 Tentacles long, filiform; shell sinistral, generally thin, without apertural teeth .... Lymnaeoidea, one family: **Physidae [p. 294]**

7(6) Tentacles lamellar; shell dextral, elongate with dilated aperture, thick, with columellar thickenings or apertural teeth; mantle margins simple, not overlapping shell; foot wide and short, rounded posteriorly (Fig. 11.2 D) .... Chilinoidea; one family: **Chilinidae [p. 296]**

7′ Tentacles filiform; shell sinistral or dextral, elongate, patelliform, planospiral or fusiform, usually thin, without columellar thickenings or apertural teeth, foot generally thin (Fig.11.2 E) .... Planorbioidea, one family: **Planorbidae [p. 302]**

Succineoidea, Succineidae

Janine Oliveira Arruda
Museu de Ciências Naturais de Porto Alegre, Porto Alegre, Rio Grande do Sul, Brazil

Gary M. Barker
Landcare Research, Lincoln, Canterbury, New Zealand

INTRODUCTION

Succineidae, nearly a cosmopolitan family, has been traditionally classified in the order Stylommatophora (Patterson, 1971; Solem, 1978), however, their affinities within this group has long being contentious due to some peculiar morphological characters (e.g., Burch & Patterson, 1969). In a review of the diversity and adaptive morphology of terrestrial gastropods, Barker (2001b) considered Succineoidea (Succineidae + Athoracophoridae) a well-supported clade, basal to Stylommatophora, with several morphological synapomorphies. Recent molecular phylogenetic analyses did not fully resolve the phylogenetic relationships of the Succineoidea, but showed a more-or-less basal position within the "non-achatinoid," one of the two principal clades within the Stylommatophora (Wade et al., 2001).

Most Succineidae taxonomic studies rely on shell morphology. However, due to the limited diversity in shell form, lack of ornamentation or apertural lamellae or other characters useful for identification, many shell-based taxonomic decisions are considered unreliable (Miles, 1958; Patterson, 1971). Recently, besides the shell, morphological characters of radula, jaw, reproductive system, and patterns of pigmentation are being used for generic and specific characterization (Patterson, 1971; Tillier, 1981; Arruda & Thomé, 2008a, 2008b).

Many *Succinea* species are known only from the shell, which is unreliable for specific identification. In the Neotropics, four species are considered to be obligate associates of freshwater systems. Most other Neotropical *Succinea* species for which biology is known are not closely associated with freshwaters but occur in mesic to seasonally moist arid terrestrial habitats. However, the biology of many species in *Succinea* remains poorly known. Species such as *S. meridionalis* d'Orbigny, 1846 are often associated with aquatic environments but also occur in a wide range of terrestrial habitats. Due to the poor knowledge of Neotropical *Succinea*, it is impossible to present an identification key for species within this genus.

Oxyloma is widely distributed (Eurasia, Africa, the Americas). Also many species are only known from shell characters, and there is no phylogenetic analysis to determine monophyly. In the Neotropical region, only one species, *O. salleana* (Pfeiffer, 1849), is known to be associated with freshwater systems. The biology of many species remains poorly known and may include additional inhabitants of freshwater systems.

The systematics of *Omalonyx* is now well established after recent studies (Arruda et al., 2006; Arruda, 2011; Arruda & Thomé, 2008a, 2008b, 2011), although molecular phylogenetic studies are lacking to confirm specific-level limits and establish infragenus relationships.

Succineids possess a thin and ovate shell, with up to four whorls and usually short spire; the aperture is large, ovate, and with a thin and simple peristome. The jaw possesses a large, quadrangular accessory plate on its cutting edge (elasmognathous); the radula teeth possess ectocones, usually splitting to multiple cusps on the marginal teeth, the endocone is united with the mesocone on the lateral teeth, but is distinct on the marginal teeth (Pilsbry, 1948). The lung is short. The kidney is orientated transversally in respect to the body's longitudinal axis, extending from pericardium to the rectum, and lying at right angle to the heart axis (Baker, 1995). The tentacular and buccal muscles are free or nearly so from their original in the columella; the right ocular retractor passes between the male and female branches of the genitalia, or passes entirely to the left of the genitalia. The genital atrium is usually shallow, behind and below the right eye-stalk. The pallial gonoducts are united as a spermoviduct. The seminal vesicle is usually bilobed but occasionally indistinctly multilobed. The prostate gland is shorter than the free oviduct (Pilsbry, 1948).

Succinea, *Oxyloma*, and *Omalonyx* possess an external shell, with its edge covered by the mantle border, except in *Omalonyx brasiliensis* (Simroth, 1896), where the shell is completely covered by the mantle.

The shell of *Succinea* and *Oxyloma* species are very similar, which generally means species are unreliably assigned to genus based on shell morphology. Their shells are ovate with 2.5 to 4 whorls (Fig. 11.4), into which the animal can take shelter. In *Omalonyx*, the unguiform shell (Fig. 11.5) does not enable the animal housing into it.

Concerning the reproductive system, there are some specific characteristics that allow generic distinctions. In *Oxyloma* there is a phallus appendix (Fig. 11.6), that is present neither in *Succinea* nor *Omalonyx*. *Omalonyx* possess a unique epiphallic sphincter (Fig. 11.7).

PHYLUM MOLLUSCA

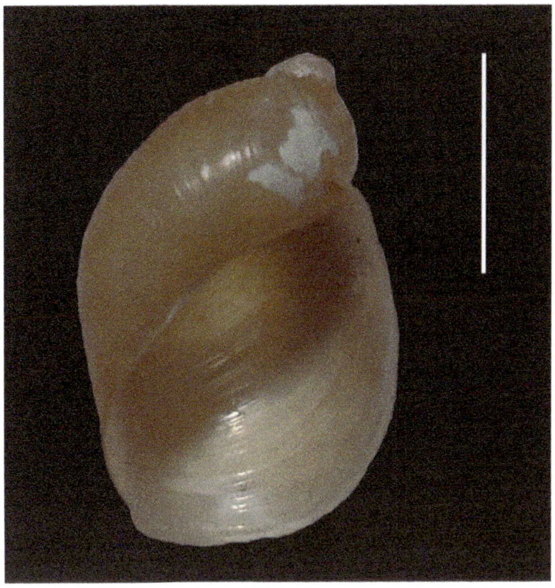

FIGURE 11.4 Shell of *Succinea* sp. (Succineidae) in frontal view. Scale bar = 5 mm.

FIGURE 11.5 Shell of *Omalonyx matheronii* (Succineidae) in dorsal (left) and ventral (right) view. Scale bar = 5 mm.

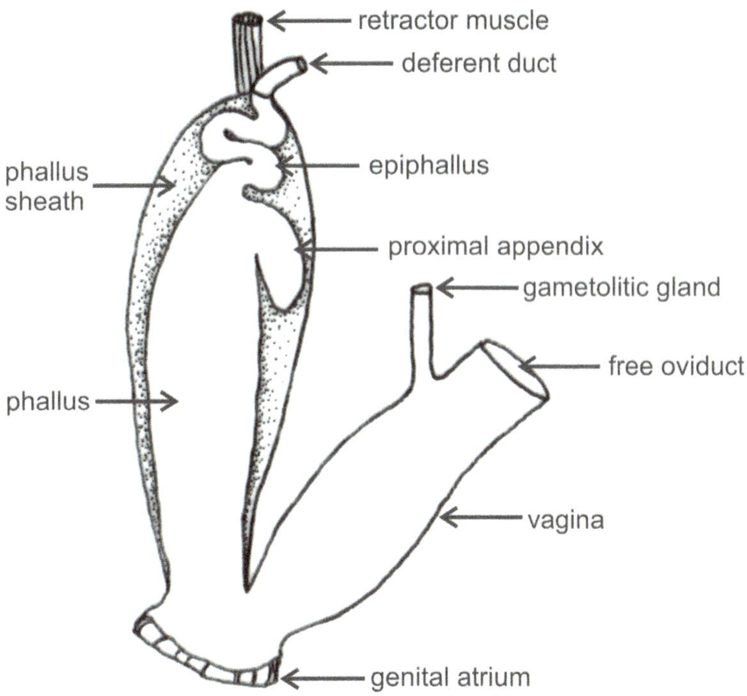

FIGURE 11.6 Phallus complex *Oxyloma* sp. (Succineidae) based on Franzen (1966) and Patterson (1973).

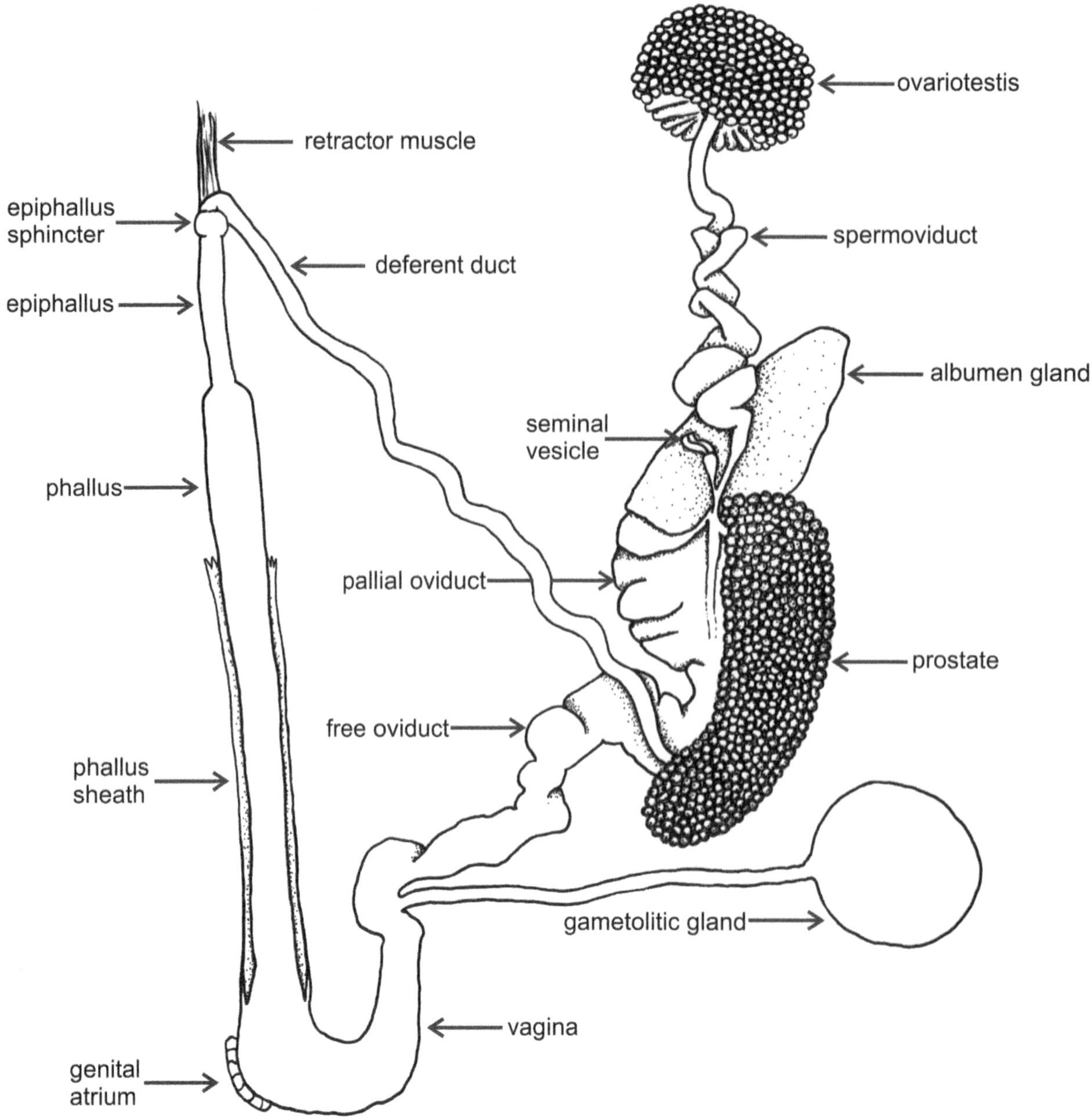

FIGURE 11.7 Reproductive system of an *Omalonyx* sp. (Succineidae) depicted with a torn phallus sheath and extended phallus.

KEY TO SUCCINEIDAE

Gastropoda: Heterobranchia: Succineoidea: Succineidae: Genera

Four *Succinea* species are exclusively associated with freshwater systems: *S. guatemalensis* Morelet, 1849, *S. virgata* Martens, 1865, *S. andecola* Crawford, 1939, and *S. gabrieli* Tomlin, 1932. These species are poorly known.

1 Shell, of several whorls, high spired, capable of containing entire animal; phallus short, with or without appendix; epiphalic sphincter absent 2

1′ Shell reduced, not capable of containing entire animal; shell border or entire shell enclosed in mantle, comprising a near obsolete spire (~ 0.5 whorl) and a rapidly expanding, body whorl with oval aperture; phallus exceptionally long, with epiphallic sphincter (Fig. 11.7) .... ***Omalonyx* [p. 275]**

2(1) Phallus with proximal appendix; shell 2.5 whorls, length 19–20 mm, width 10–11 mm, aperture ~0.75× shell length .... *Oxyloma salleana* (Pfeiffer, 1849)
[USA, Mexico, Guatemala. Nearctic]

2′ Phallus without an appendix; shell of 2.5–3 whorls, length 6–26 mm, spire short, atop of a rapidly expanding body whorl; oval aperture ~ 0.65–0.85 shell length (Fig. 11.4) .... *Succinea*
[Bermuda, Costa Rica, Nicaragua, Mexico, Guatemala, Peru]

Gastropoda: Heterobranchia: Succineoidea: Succineidae: *Omalonyx*: Species

1 Phallus internal surface proximal half with papillae 2

1′ Phallus internal surface proximal half with folds .... *Omalonyx convexus* (Heynemann, 1868)
[Brazil, Bolivia, Argentina, Uruguay]

2(1) Epiphallus surface smooth 3

2′ Epiphallus with duct running over surface .... *Omalonyx unguis* (d'Orbigny, 1837)
[Brazil, Bolivia, Paraguay, Argentina, Uruguay]

3(2) Phallus wider than epiphallus; phallus internal surface with papillae 4

3′ Epiphallus and phallus subequal; phallus internal surface proximal extremity with wrinkled longitudinal folds .... *Omalonyx geayi* Tillier, 1980
[Suriname, French Guiana, Brazil, Ecuador, Bolivia]

4(3) Free oviduct length 2 pallial oviduct; phallus proximal region wider than remainder 5

4′ Free oviduct length 3× pallial oviduct, inner wall with robust longitudinal folds; phallus internal surface with discoid or polygonal papillae .... *Omalonyx pattersonae* Tillier, 1981
[Guadeloupe, Suriname, French Guiana, Brazil, Colombia, Ecuador, Peru, Bolivia]

5(4) Ovariotestis hemispherical; phallus internal surface with elliptical papillae .... *Omalonyx matheronii* (Potiez & Michaud, 1835)
[Lesser Antilles, Trinidad, Dominica, Guadeloupe, Venezuela, Suriname, French Guiana, Brazil, Peru, Ecuador, Bolivia]

5′ Ovariotestis spherical; phallus internal surface with rhombus papillae; vagina wall very thick, internal surface with robust rhombus papillae .... *Omalonyx brasiliensis* (Simroth, 1896)
[Brazil]

Glacidorboidea, Glacidorbidae

Diego Eduardo Gutiérrez Gregoric
División Zoología Invertebrados, Museo de La Plata, Facultad de Ciencias Naturales y Museo, Universidad Nacional de La Plata, CONICET, La Plata, Buenos Aires, Argentina

INTRODUCTION

Glacidorbidae is restricted to the Gondwanan realm (Tasmania; southeastern and southwestern Australia; southern Argentina and Chile (Ponder & Avern, 2000; Rumi et al., 2015). Ponder (1986) studied the anatomy of the SE Australian *Glacidorbis hedleyi* (Iredale, 1943) and classified the group as operculate basommatophoran pulmonates. He proposed Glacidorbidae and Glacidorbacea to include *Glacidorbis* and its relatives; all of which are freshwater taxa. There are 23 described species (American and Australian). Recent molecular studies link the family with the basal "Pulmonata" (Dinapoli & Klussmann-Kolb, 2010; Holznagel et al., 2010; Dayrat et al., 2011; Golding, 2012), as also suggested by anatomy (Ponder, 1986) and sperm ultrastructure (Healy, 1995).

Until 2015, only one species was known from South America, *Gondwanorbis magallanicus* (Meier-Brook & Smith, 1976) from southern Chile. Rumi et al. (2006, 2008) and Núñez et al. (2010) included *G. magallanicus* in their revision of Argentinean freshwater mollusks based on records in the province of Tierra del Fuego (Southern Patagonia, Gondwanan realm). Rumi et al. (2015) conducted a review of this family in South America with the description of two new species for *Gondwanorbis* and a description of a new genus and species, (*Patagonorbis nahuelhuapensis* Rumi & Gutiérrez Gregoric, 2015). These authors based their research on recently collected material, analyzing shell, operculum, and radula.

The short age of records may be due to the small size (to 3 mm) of samples or possibly erroneous determination in some instances, as glacidorbids can be confused with juvenile shells of Planorbidae *Antillorbis*, *Drepanotrema*, and *Biomphalaria*. More intensive field work in Patagonian lakes, swamps, and small streams would allow confirmation of the actual distribution, diversity, and abundance of this family in South America.

The family presents with a small to minute shell (up to 3 mm in South American species), dextrally coiled, orthostrophic to hyperstrophic, with a flat or near flat spire to planispiral, with a wide, shallow umbilicus. The operculum is circular or oval, multispiral to paucispiral, with central to eccentric nucleus. A radula with large central teeth bearing several sharp cusps on a pointed mesoconehas an arched and broader base lacking additional cusps. The lateral teeth are vestigial, narrow, and small; the marginal teeth are absent (Ponder & Avern, 2000). The anatomy characteristics in South American species are not known yet.

PHYLUM MOLLUSCA

KEY TO GLACIDORBOIDEA

Gastropoda: Heterobranchia: Glacidorboidea: Glacidorbidae: Genera

1 Shell with keel or peripheral angulation, planispiral; operculum circular and paucispiral; central radular teeth basal width ~ 2× mesocone; radula with or without pair of vestigial lateral teeth .... **Gondwanorbis** **[p. 276]**

1′ Shell without keel, slightly hyperstrophic; operculum oval, exterior edges with pustulae; radula with semivestigial lateral teeth (Fig. 11.8) .... *Patagonorbis nahuelhuapensis* Rumi & Gutiérrez Gregoric, 2015 in Rumi et al., 2015
[Argentina: Neuquén, Río Negro]

Gastropoda: Heterobranchia: Glacidorboidea: Glacidorbidae: *Gondwanorbis*: Species

1 Shell with keels 2

1′ Shell planispiral with weak peripheral angulation, without keels; protoconch with apical and secondary sculpture with spiral lines (Fig. 11.9) .... *Gondwanorbis fueguensis* Rumi & Gutiérrez Gregoric, 2015 in Rumi et al., 2015
[Argentina: Tierra del Fuego]

2(1) Shell with three keels, two strong (peripheral and ventral), and one weak (dorsal); protoconch apical and secondary sculpture with fine spiral lines; radula: central tooth with five or six cusps of similar size, strong, long and pointed, positioned laterally to mesocone; vestigial lateral teeth simple 40% height of central teeth (Fig. 11.10) .... *Gondwanorbis tricarinatus* Rumi & Gutiérrez Gregoric, 2015 in Rumi et al., 2015
[Chile: Southern]

2′ Shell smooth, almost glossy with weak peripheral keel; radula: central tooth with short mesocone with four or five cusps on each side; vestigial lateral teeth reduced, 6–7 μm wide and 15–16 μm long, rhombic and lacking denticles (Fig. 11.11) .... *Gondwanorbis magallanicus* (Meier-Brook & Smith, 1976)
[Chile: Southern]

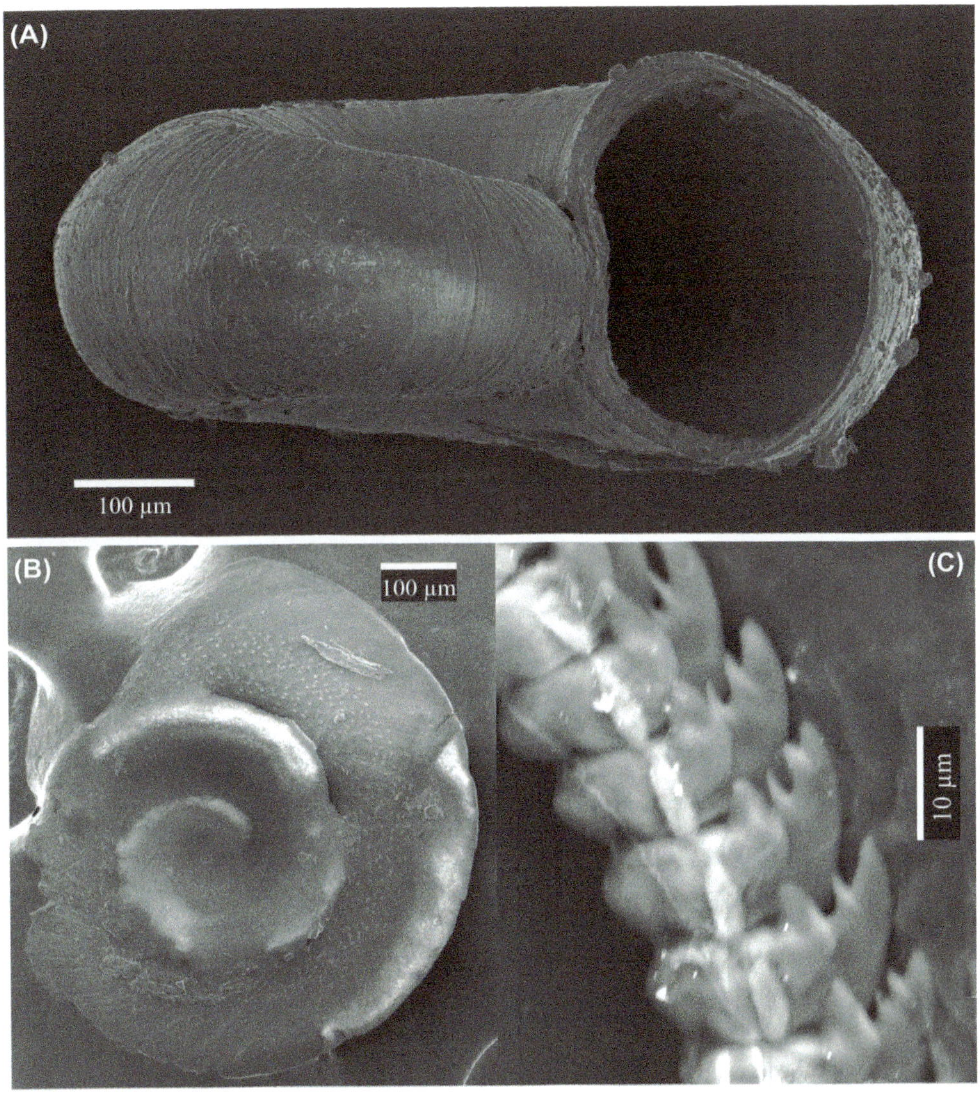

FIGURE 11.8 *Patagonorbis nahuelhuapensis* (Glacidorbidae): (A) Lateral view of shell slightly hyperstrophic; (B) oval operculum; (C) radula with vestigial lateral teeth. *Images from La Plata Museum, MLP-Ma-13656.*

FIGURE 11.9 Shell of *Gondwanorbis fueguensis* (Glacidorbidae): (A) Lateral view of shell with weak peripherical angulation; (B) ventral view; (C) partial oblique dorsal view; (D) dorsal view; (E) and (F) dorsal view of protoconch with apical and secondary sculpture with spiral lines; (G) and (H), punctures on inner surface of shell. *Images authorized for Taylor & Francis from Rumi et al. (2015) (www.tandfonline.com).*

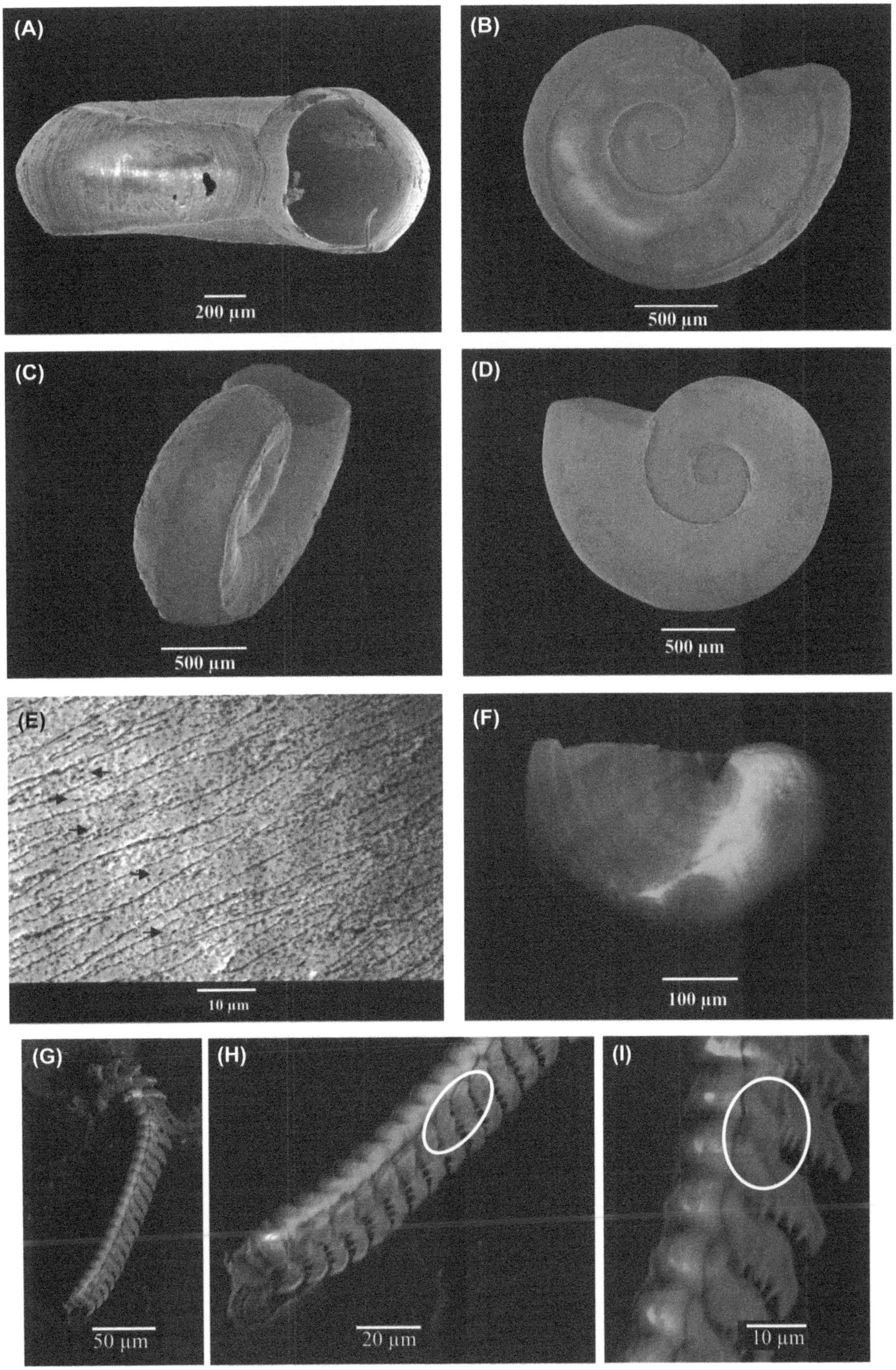

FIGURE 11.10 *Gondwanorbis tricarinatus* (Glacidorbidae), shell (A–F) and radula (G–I): (A) Lateral view; (B) ventral view; (C) oblique ventral view with two strong keels (periphery and ventral); (D) dorsal view with one weak keel (dorsal); (E) punctures on inner surface of shell, *arrows* show location of punctures; (F) dorsal view of an embryonic shell from inside adult specimen with fine spiral lines sculpture; (G–I) Radula, circle indicates the vestigial lateral tooth. *Images authorized for Taylor & Francis from Rumi et al. (2015) (www.tandfonline.com).*

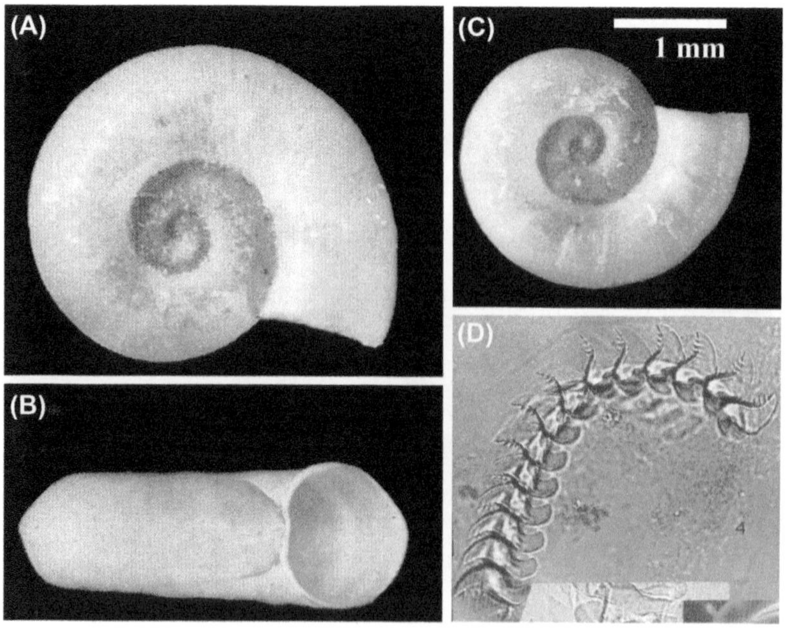

FIGURE 11.11 *Gondwanorbis magallanicus* (Glacidorbidae), shell (A–C) and radula (D): (A) dorsal view; (B) lateral view with weak peripherical keel; (C) ventral view; (D) radula. *Images authorized for Schweizerbart publishing from Meier-Brook & Smith (1976) (www.schweizerbart.de).*

Lymnaeoidea, Lymnaeidae

Jean-Pierre Pointier
PSL Research University, CRIOBE, USR 3278 EPHE-CNRS-UPVD, Perpignan, France

Antonio A. Vázquez
Laboratory of Malacology, Institute of Tropical Medicine Pedro Kourí, Havana, Cuba

INTRODUCTION

In freshwater pulmonates shell characters are for the most part absent and shell shape alone has been used to describe new taxa in the 19th and 20th centuries and even as recently as the 21st century. Using only shell shape as a diagnostic character is of limited use for separating closely related species because of its wide phenotypic variation (Jarne et al., 2011). Consequently, a plethora of names for freshwater pulmonates is found in the literature of which many are likely the same species, i.e., synonyms. The case of the Lymnaeidae family is caricatural with about 1,200 described species and several dozens genera and subgenera (Hubendick, 1951; Burch, 1982), whereas recent studies suggest that the family contains approximately 100 species (Strong et al., 2008) and a limited number of genera (Correa et al., 2010; Vinarski, 2013). Characteristics of the radula have also been used in the lymnaeid systematics; but as for shell shape, a wide variation has been observed within and between genera and species and has also proved of limited taxonomic interest (Hubendick, 1951; Paraense, 1976c). The anatomy of the reproductive apparatus (e.g., shape and size of prostate, penial complex) is more informative but still not fully diagnostic (Jarne et al., 2011). That some species of lymnaeids (e.g., *Galba* spp.) mainly reproduce through self-fertilization sets a practical limit to investigations on reproductive isolation. To further complicate things, several lymnaeid species are excellent travellers and have now invaded new areas and habitats throughout the world presumably following human activities over the last few decades (Madsen & Frandsen, 1989; Pointier et al., 2005, 2011). Consequently, a survey of their respective type localities described more than a century ago could be totally misleading (Jabbour-Zahab et al., 1997; Correa et al., 2011).

In the last decades, several genetic studies using allozymes and DNA markers have been carried out and brought a new insight to the systematics of lymnaeids. However, each of these studies used only one genetic sequence or gene (e.g., Remigio, 2002) or considered limited geographic zones (e.g., Puslednik et al., 2009). A more complete study was carried out by performing a phylogenetic analysis using a four gene approach (Correa et al., 2010). This approach included sequences of 50 taxa from the five continents (i.e., approximately half of the diversity of the family's estimated global diversity) and covered most Neotropical species. This phylogenetic analysis demonstrated the existence of three deeply rooted clades of Lymnaeidae representing the main geographic origin of species (America, Eurasia, and the Indo-Pacific region). Moreover, two highly supported subclades can be recognized within the American and the Indo-Pacific clades, respectively. In addition, these results indicated that genera in Lymnaeidae do not reflect phylogenetic relationships, to the notable exception of *Radix* (Correa et al., 2010). These authors have then proposed to use genera for each clade and subclade according to the principle of priority of the International Code of Zoological Nomenclature (ICZN). *Pseudosuccinea columella* (Say, 1817) is however a species weakly supported in the American clade and consequently the genus *Pseudosuccinea* has been retained in the nomenclature until more information becomes available (Correa et al., 2010).

Presently, in the Neotropical region eight taxa may be considered as valid species according to recent studies: *Pseudosuccinea columella* (Say, 1817), *Pectinidens diaphanus* (King & Broderip, 1832), *Lymnaea rupestris* Paraense, 1982, *Galba cousini* (Jousseaume, 1887), *Galba cubensis* (L. Pfeiffer, 1839), *Galba truncatula* (O.F. Müller, 1774), *Galba viator* (d'Orbigny, 1835), and *Galba schirazensis* (Küster, 1862). Scientific knowledge about these taxa is highly variable. Some species have been the subject of numerous morphological and molecular studies (e.g., *G. truncatula*, see Jabbour-Zahab et al., 1997; Samadi et al., 2000; Meunier et al., 2001) while others have been largely neglected (e.g., *L. rupestris*, see Paraense, 1982).

KEY TO LYMNAEIDAE

Gastropoda: Heterobranchia: Lymnaeoidea: Lymnaeidae: Genera

Neotropical lymnaeids may be divided into two main groups: (i) a group of four taxa including species in which morphology can be used for a tentative identification—*Pseudosuccinea columella, Pectinidens diaphana, Lymnaea rupestris, Galba cousini*—and (ii) a group of four closely related species where molecular markers must be used for a reliable identification—*Galba truncatula, G. viator, G. cubensis, G. schirazensis.* All Neotropical species have been included in this key; but in this context, users must consider the tentative identification key presented below, which is only based on morphological characters, with reserves. Distribution in the Neotropical region is included in the key.

1 Periostracum without spiral sculptures 2

1′ Periostracum ornamented with spiral and axial sculptures; spire short, last whorl usually larger than the spire; ureter with two distinct flexures (Fig. 11.12) .... *Pseudosuccinea columella* (Say, 1817)
[Worldwide]

2(1) Shell with shallow sutures 3

2′ Shell with deep sutures, shouldered and channeled whorls; adults usually less than 6 mm; ureter straight; prostate voluminous without fissure (Fig. 11.13) .... *Lymnaea rupestris* Paraense, 1982
[Brazil: Santa Catarina State]

3(2) Spermiduct variable but never sinuous .... ***Galba* [p. 282]**

3′ Spermiduct sinuous; ureter straight; penis sheath as long as, but never larger, than preputium; prostate with a longitudinal fissure, adult shells up to 15 mm (Fig. 11.14) .... *Pectinidens diaphanus* (King & Broderip, 1832)
[Chile, Argentina, Malvinas (Falkland) Islands]

Gastropoda: Heterobranchia: Lymnaeoidea: Lymnaeidae: *Galba*: Species

1 Ureter without two distinct flexures 2

1′ Ureter with two distinct flexures; vagina bulbous and swelling; spermiduct flat with a granular appearance; shell up to 17 mm (Fig. 11.15) .... *Galba cousini* (Jousseaume, 1887)
[Venezuela, Colombia, Ecuador]

2(1) Prostate not pear shaped 3

2′ Prostate pear shaped; penis sheath shorter than preputium; adult shells length 6—12 mm; rounded whorls (Fig. 11.16) .... *Galba truncatula* (O.F. Müller, 1774)
[Venezuela, Colombia, Peru, Bolivia, Brazil, Chile, Argentina]

3(2) Prostate small ovate, more or less rectangular; penis sheath shorter than preputium; adult shell length 4—9 mm; whorls rounded (Fig. 11.17) .... *Galba schirazensis* (Küster, 1862)
[Dominican Republic, Venezuela, Colombia, Ecuador, Peru]

3′ Prostate small, mainly ovate; penis sheath shorter than preputium; adult shell length 6—12 mm, whorls rounded (Figs. 11.18, 19); these species cannot be morphologically differentiated and only molecular markers may allow for some resolution (Alda et al., 2018), but they may be synonyms .... *Galba cubensis* (L. Pfeiffer, 1839) / *Galba viator* (d'Orbigny, 1835)
[*G. cubensis*: Cuba, Dominican Republic, Puerto Rico, Guadeloupe, Martinique, Venezuela, Colombia, Ecuador, Peru, Brazil, Argentina, Uruguay]
[*G. viator*: Brazil, Chile, Argentina]

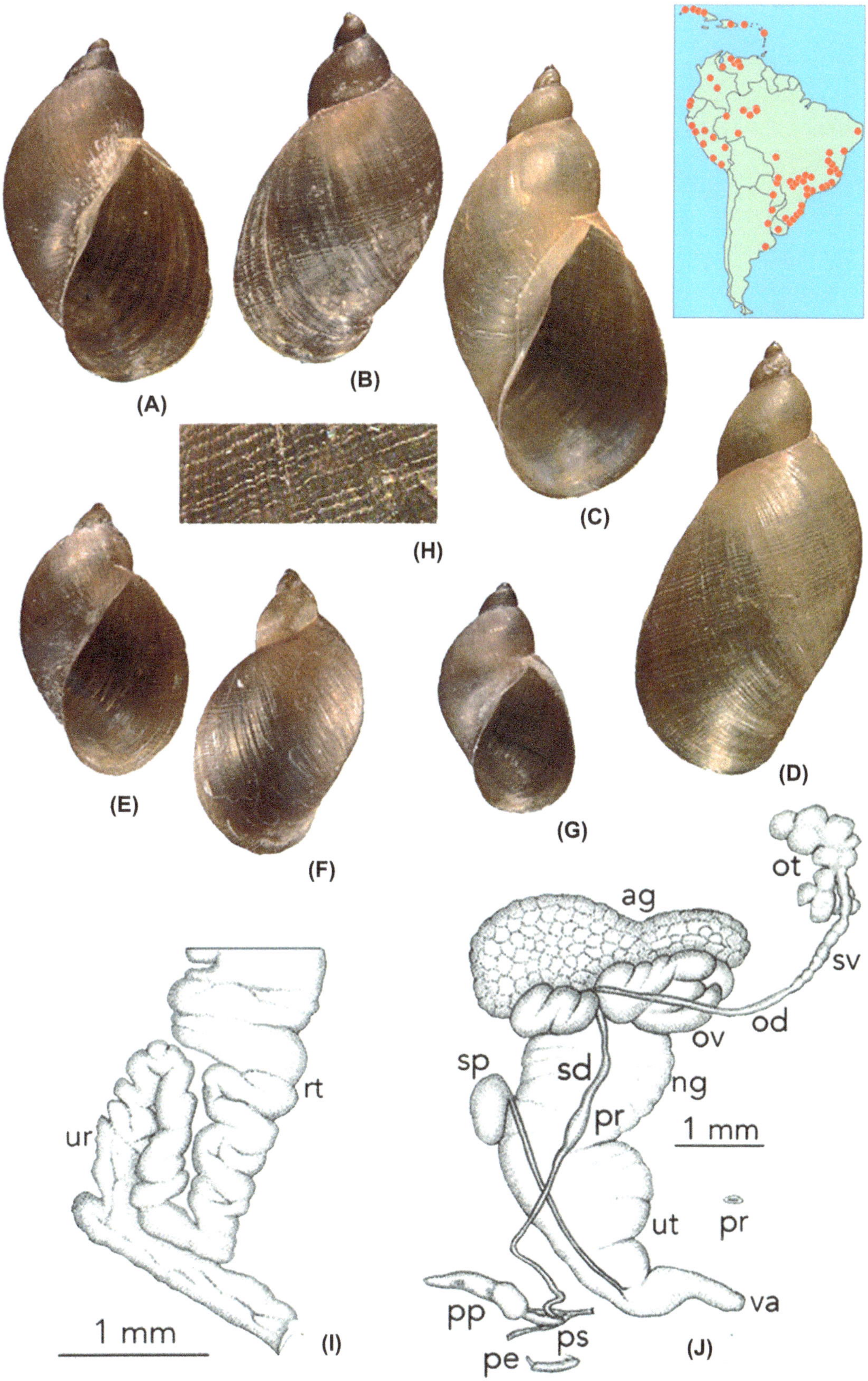

FIGURE 11.12 *Pseudosuccinea columella* (Lymnaeidae): (A) and (B) from Willow, Texas, USA, H (height) = 8.4 mm; (C) and (D) from Caracas, Venezuela, H = 16.2 mm; (E) and (F) from La Elvira, Ecuador, H = 7 mm; (G) from Los Dos Cerritos, Venezuela, H = 5 mm; (H) detail of periostracum showing characteristical striae; (I) kidney from La Linda, Venezuela; (J) anatomy of the reproductive system, from El Ayer, Colombia. Abbreviations: *ag*, albumen gland; *ng*, nidamental gland; *od*, ovispermiduct; *ot*, ovotestis; *ov*, oviduct; *pe*, penis; *pp*, preputium; *pr*, prostate; *ps*, penis sheath; *rt*, renal tube; *sd*, spermiduct; *sp*, spermatheca; *sv*, seminal vesicle; *ur*, ureter *ut*, uterus; *va*, vagina. The map shows the range of *P. columella*; each *red circle* indicates approximately an area where the species has been reported in the literature.

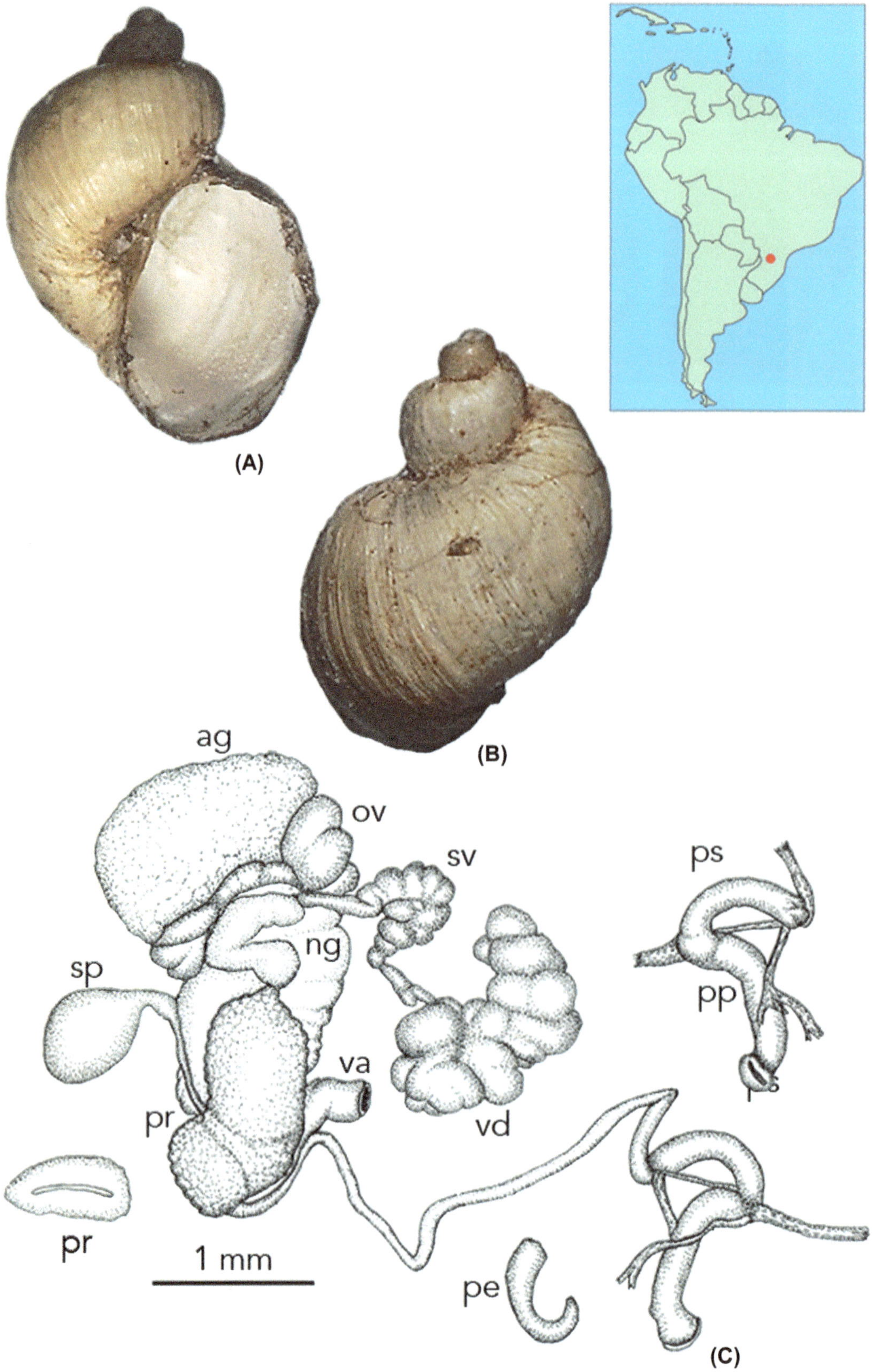

FIGURE 11.13 *Lymnaea rupestris* (Lymnaeidae): (A) and (B) lectotype, Nova Teutônia, Santa Catarina, Brazil, CMIOC, photo courtesy S. Thiengo, H = 4.2 mm; (C) anatomy of the reproductive system. Abbreviations: *ag*, albumen gland; *ng*, nidamental gland; *ov*, oviduct; *ot*, ovotestis; *pe*, penis; *pp*, preputium; *pr*, prostate; *ps*, penis sheath; *sd*, spermiduct; *sp*, spermatheca; *sv*, seminal vesicle; *va*, vagina; *vd*, vas deferens. The map shows the range of *L. rupestris*; the *red circle* indicates the type locality which is the only site where the species has been reported in the Santa Catarina State, Brazil. (C) *Redrawn from Paraense (1982).*

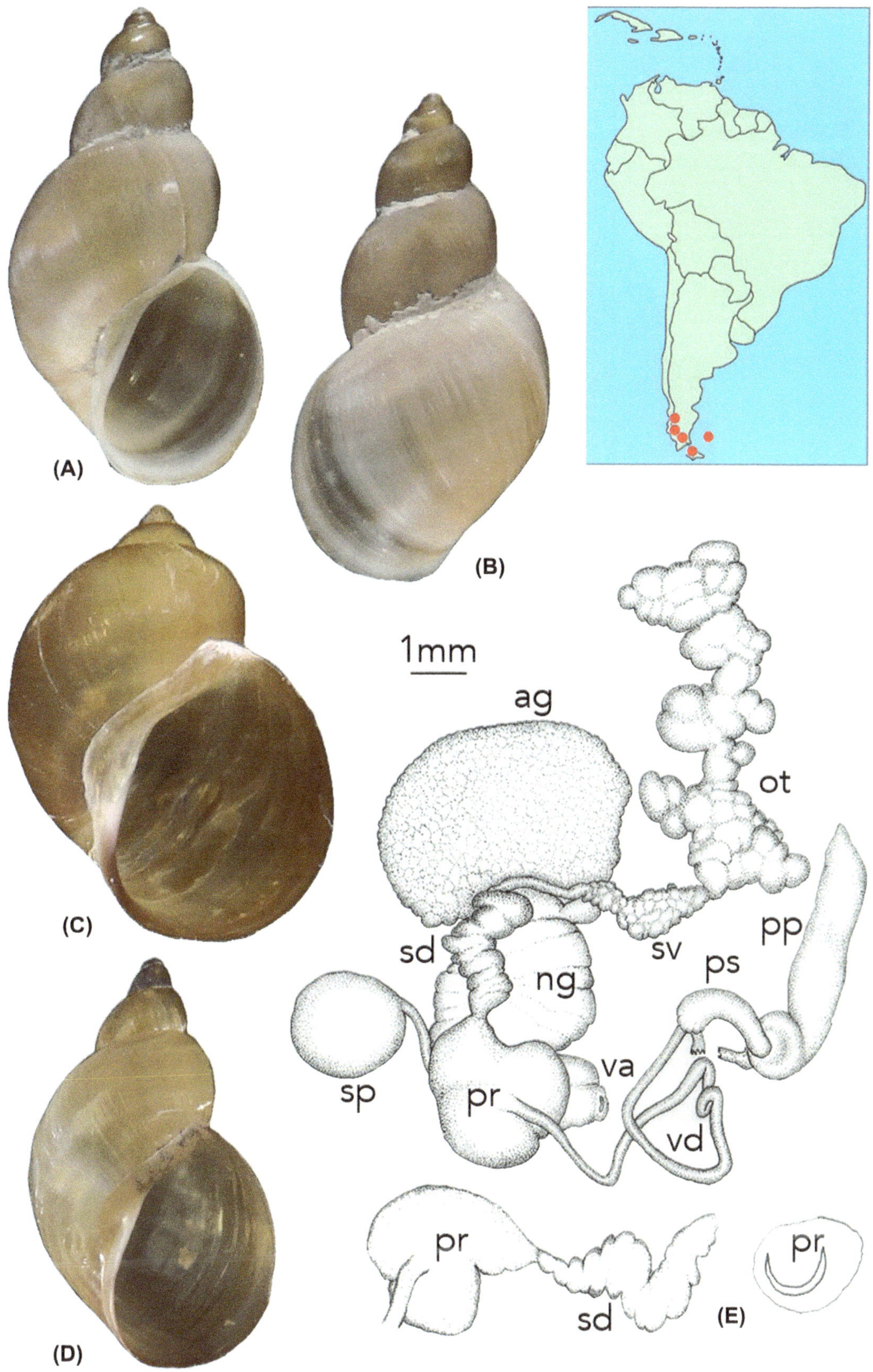

FIGURE 11.14 *Pectinidens diaphanus* (Lymnaeidae): (A) and (B) from Arroyo Marisol, Argentina, H (height) = 13.1 mm; (C) from Laguna Azul, Argentina, H = 12.3 mm; (D) from Arroyo el Surdo, Argentina, H = 11.1 mm; (E) anatomy of the reproductive system, from Ewan Norte, Argentina. Abbreviations: *ng*, nidamental gland; *ot*, ovotestis; *pp*, preputium; *pr*, prostate; *ps*, penis sheath; *sd*, spermiduct; *sp*, spermatheca; *sv*, seminal vesicle; *va*, vagina; *vd*, vas deferens. The map shows the range of *P. diaphanus*; each *red* circle indicates approximately an area where the species has been reported.

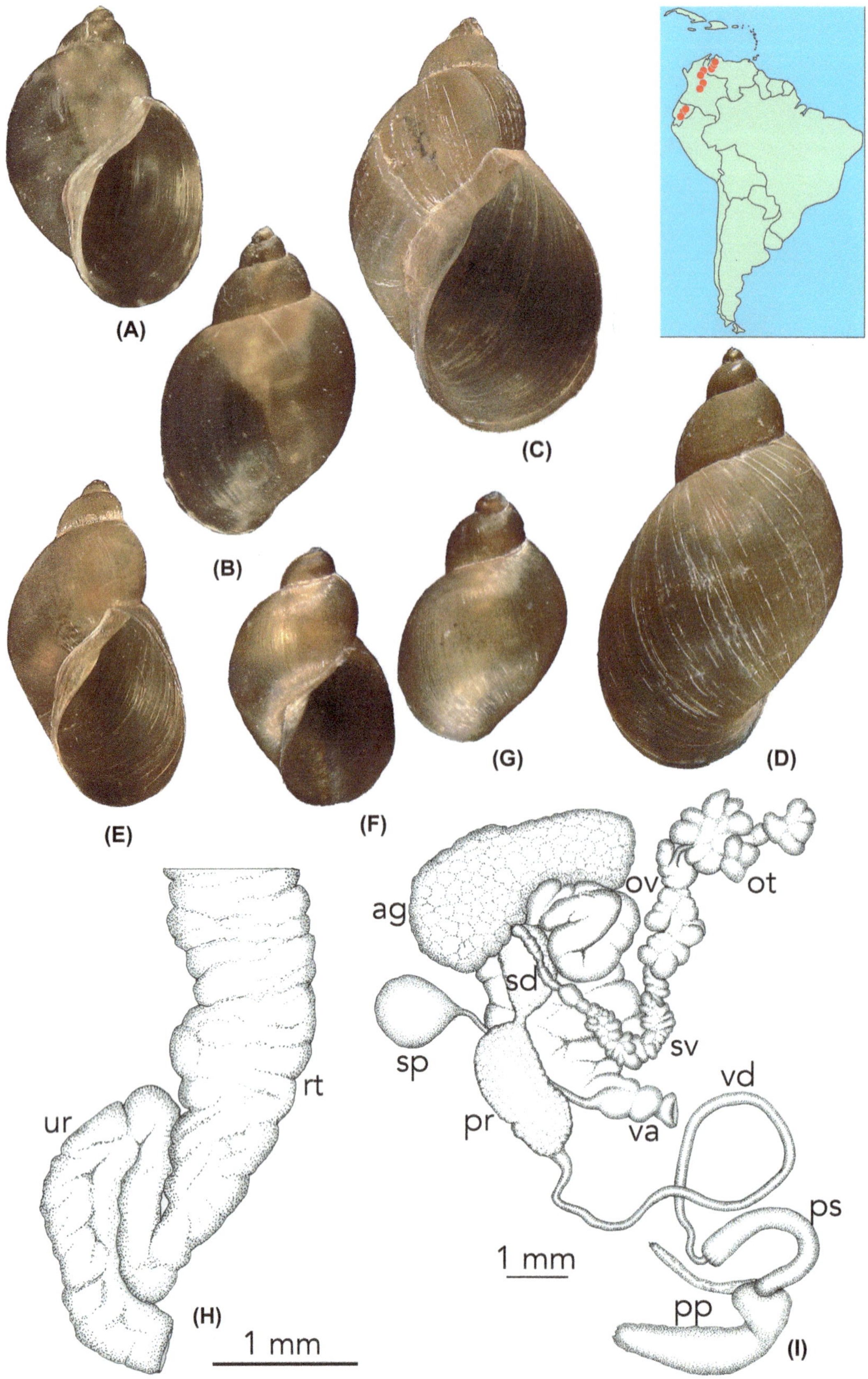

FIGURE 11.15 *Galba cousin* (Lymnaeidae): (A) and (B) from La Esperanza, Ecuador, H (height) = 8.2 mm; (C) and (D) from Mal Abrigo, Colombia, H = 13.4 mm; (E) form El Tisure, Venezuela, H = 8.7 mm; (F) and (G) Micarache, Venezuela, H = 4.8 mm; (H) kidney, from Mucubaji, Venezuela; (I) anatomy of the reproductive system, from El Topon, Colombia. Abbreviations: *ag*, albumen gland; *ov*, oviduct; *ot*, ovotestis; *pp*, preputium; *pr*, prostate; *ps*, penis sheath; *rt*, renal tube; *sd*, spermiduct; *sp*, spermatheca; *sv*, seminal vesicle; *va*, vagina; *vd*, vas deferens; *ur*, ureter. The map shows the range of *G. cousini*; each *red circle* indicates approximately an area where the species has been reported in the literature; only references based on a molecular identification of the species have been taken into account.

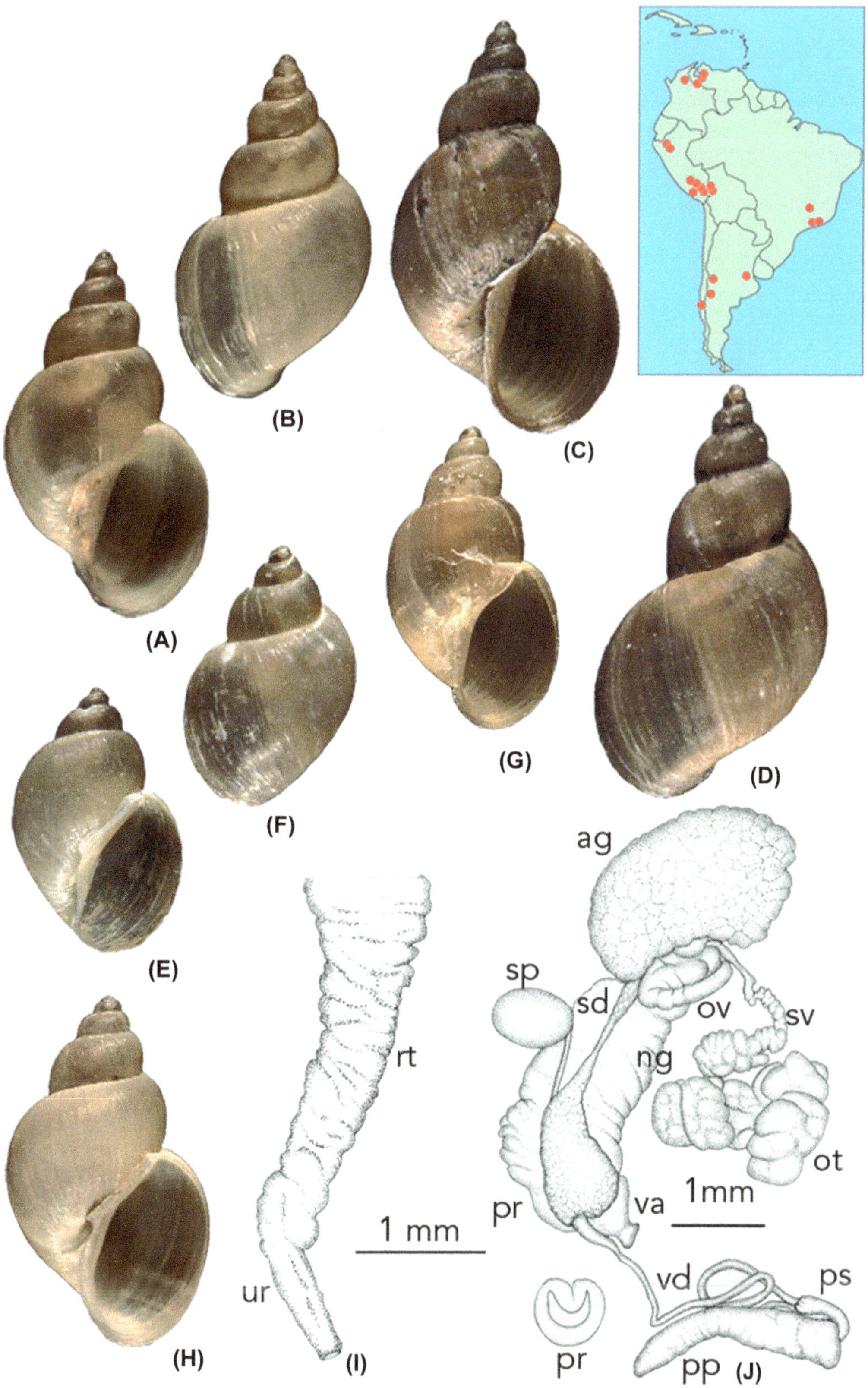

FIGURE 11.16 *Galba truncatula* (Lymnaeidae): (A) and (B) from Copancara, Bolivia, H = 8.9 mm; (C) and (D) from La Joya, Peru, H = 10.1 mm; (E–F) from El Sapo, Venezuela, H = 5.5 mm; (G) from Paso el Condor, Venezuela, H = 6.0 mm; (H) from Valdivia, Chile, H = 7.4 mm; (I) kidney, from Batallas, Bolivia; (J) anatomy of the reproductive system, from Batallas, Bolivia. Abbreviations: *ag*, albumen gland; *ng*, nidamental gland; *ov*, oviduct; *ot*, ovotestis; *pp*, preputium; *pr*, prostate; *ps*, penis sheath; *rt*, renal tube; *sd*, spermiduct; *sp*, spermatheca; *sv*, seminal vesicle; *va*, vagina; *vd*, vas deferens; *ur*, ureter. The map shows the range of *G. truncatula*; each *red circle* indicates approximately an area where the species has been reported in the literature; only references based on a molecular identification of the species have been taken into account.

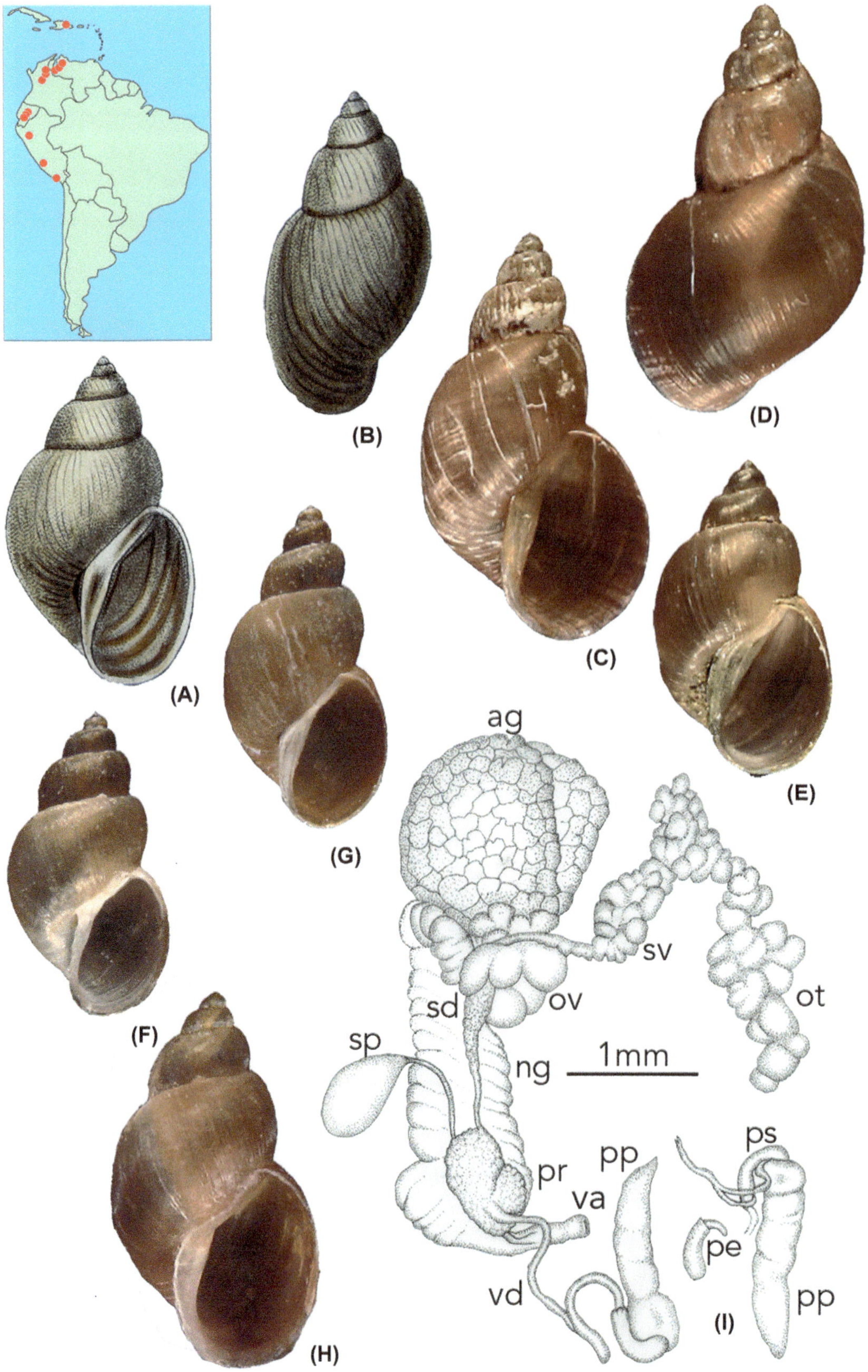

FIGURE 11.17 *Galba schirazensis* (Lymnaeidae): (A) and (B) original drawings of *Limnaeus schirazensis* Küster, 1862; (C) and (D) from Alborada, Colombia, H = 9.0 mm; (E) from Alborada, Colombia, H = 5.2 mm; (F) Bedico, Louisana, U.S.A., H = 5.4 mm; (G) Manto de la Novia, Ecuador, H = 5.7 mm; (H) from La Trampa, Venezuela, H = 7.4 mm; (I) anatomy of the reproductive system, from La Trampa, Venezuela. Abbreviations: *ag*, albumen gland; *ng*, nidamental gland; *ov*, oviduct; *ot*, ovotestis; *pe*, penis; *pp*, preputium; *pr*, prostate; *ps*, penis sheath; *sd*, spermiduct; *sp*, spermatheca; *sv*, seminal vesicle; *va*, vagina; *vd*, vas deferens. The map shows the range of *G. schirazensis*; each *red circle* indicates approximately an area where the species has been reported in the literature; only references based on a molecular identification of the species have been taken into account.

PHYLUM MOLLUSCA

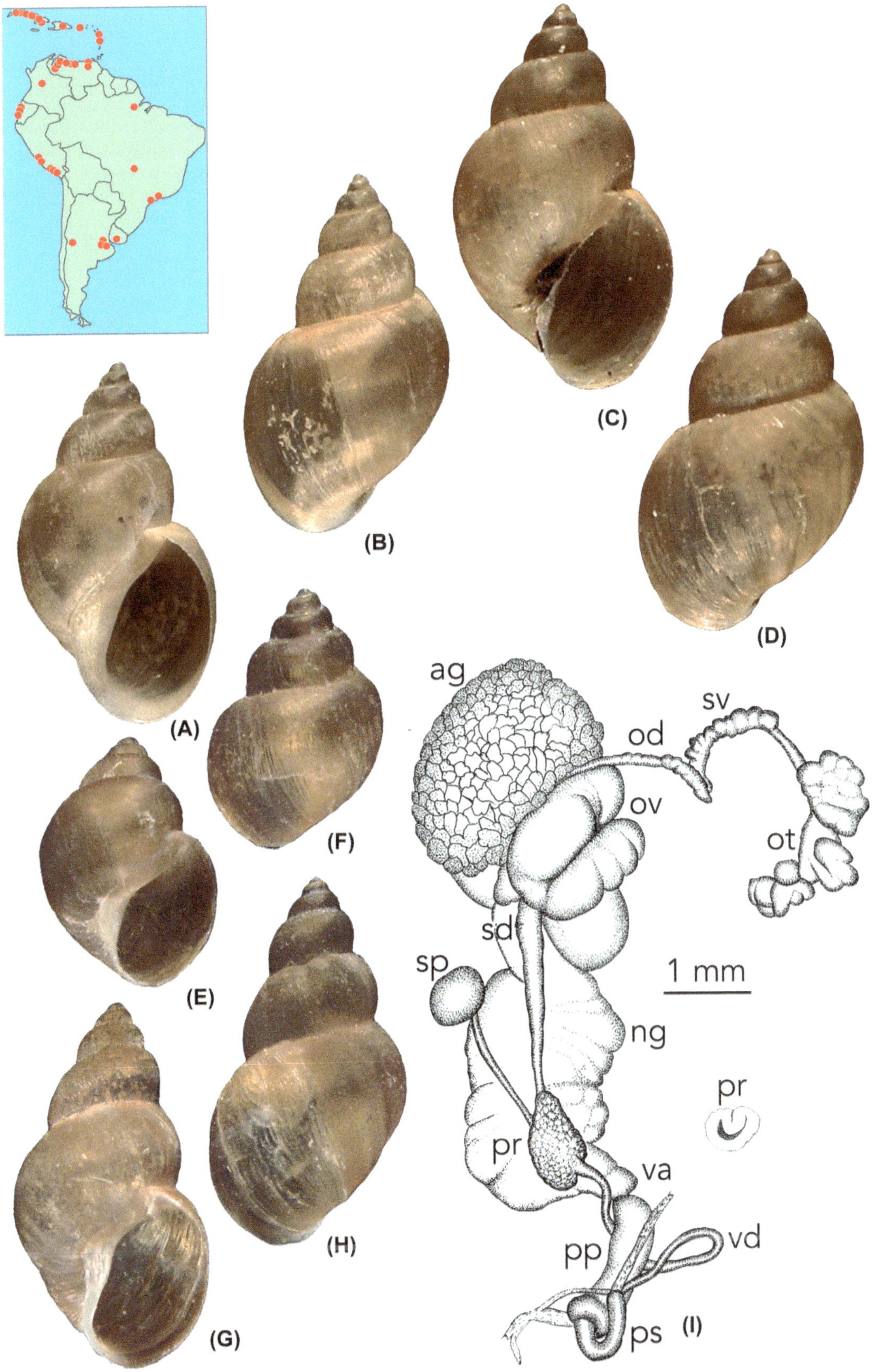

FIGURE 11.18 *Galba cubensis* (Lymnaeidae): (A) and (B) from Havana, Cuba, H (height) = 9 mm; (C) and (D) from La mesa de Esnujaque, Venezuela, H = 10.1 mm; (E–F) from Urao, Venezuela, H = 5 mm; (G–H) from Zavallo, Argentina, H = 7.8 mm; (I) anatomy of the reproductive system, Havana, Cuba. Abbreviations: *ag*, albumen gland; *ng*, nidamental gland; *od*, ovispermiduct; *ov*, oviduct; *ot*, ovotestis; *pp*, preputium; *pr*, prostate; *ps*, penis sheath; *sd*, spermiduct; *sp*, spermatheca; *sv*, seminal vesicle; *va*, vagina; *vd*, vas deferens. The map shows the range of *G. cubensis*; each *red circle* indicates approximately an area where the species has been reported in the literature; only references based on a molecular identification of the species have been taken into account.

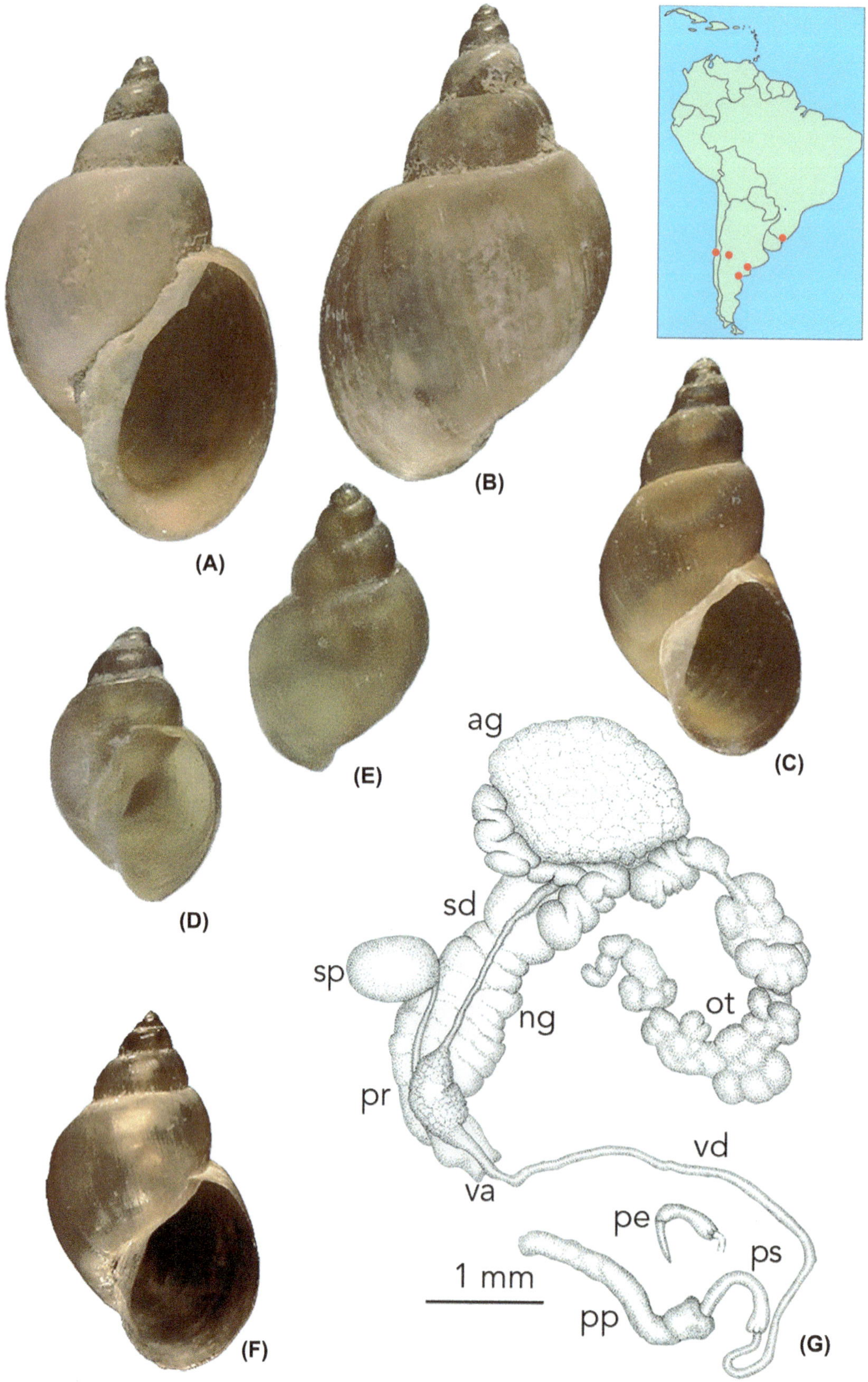

FIGURE 11.19 *Galba viator* (Lymnaeidae): (A) and (B) from Río Atuel, Argentina, H (height) = 11.9 mm; (C) Manzano Grande, Argentina, H = 8.1 mm; (D) and (E) from Las Minas, Argentina, H = 4.2 mm; (F) from Frias, Rio Negro, Argentina, H = 6.5 mm; (G) anatomy of the reproductive system, from Bahia Blanca, Argentina. Abbreviations: *ag*, albumen gland; *ng*, nidamental gland; *ot*, ovotestis; *pe*, penis; *pp*, preputium; *pr*, prostate; *ps*, penis sheath; *sd*, spermiduct; *sp*, spermatheca; *sv*, seminal vesicle; *va*, vagina; *vd*, vas deferens. The map shows the range of *G. viator*; each *red circle* indicates approximately an area where the species has been reported in the literature; only references based on a molecular identification of the species have been taken into account.

Lymneoidea, Physidae

Maria Gabriela Cuezzo
Instituto de Biodiversidad Neotropical (CONICET-UNT, Facultad de Ciencias Naturales), Tucumán, Argentina

INTRODUCTION

Physidae is cosmopolitan as it has been introduced in many areas around the world, but is mainly diverse in the Nearctic region (Burch, 1982). According to Taylor (2003), 80 species of Physidae have been reported worldwide based on morphological characters. However, these characters are variable and very often overlap among the species. Recent molecular studies have shortened the total number of species between 15 and 20 (Wethington & Lydeard, 2007).

Physid classification has been extremely controversial and is currently not resolved. Species misidentifications appear to be a common problem. Nineteenth-century workers (Walker, 1918) generally recognized two genera, the monotypic *Aplexa* and the speciose *Physa* (Wethington & Lydeard, 2007). Thiele (1931–1935) and Zilch (1959–1960) also recognized only the two older genera, *Aplexa* and *Physa*, but with two subgenera under the former (*Aplexa s.s.* and *Stenophysa*) and four subgenera under the latter (*Physa s.s.*, *Alampetista*, *Costatella*, and *Petrophysa*). Starobogatov (1967) created two new subfamilies, Aplexinae with three genera and Physinae with four. Posteriorly, Te (1978) provided a phenetically based classification scheme for the family using mainly the penial morphology, and other associated characters to construct his groupings, and proposed to classify Physidae in two subfamilies, Aplexinae with *Aplexa* and *Stenophysa*, and Physinae with genera *Physa* and *Physella*. In 2003, Taylor made a revision of Physidae and reclassified them by using characters of the penial complex, assuming that their changes were progressive. These progressive characters were the principal basis to arrange the family into 23 genera, four grades, and four clades within the family. The two established subfamilies were divided into seven new tribes including 11 new genera. Taylor (2003) supported the idea that the area of greatest diversity for Physidae is Central and South America instead of North America, as suggested by Te (1978).

Most recently, Wethington & Lydeard (2007) in their molecular phylogeny based on mitochondrial sequences supported the monophyly of Physidae, recovering six major clades corresponding to differences in penial morphology. They concluded that the absence of a preputial gland is common to *Stenophysa* and *Aplexa* and the difference between them lies mainly in the penial sheath; in *Aplexa* it is glandular while in *Stenophysa* it is muscular (Te, 1978; Taylor, 2003; Wethington & Lydeard, 2007).

Physidae can be distinguished from other pulmonates by the following characters: a high-spired sinistral shell, radula with teeth in V-shaped rows, simple jaw with no lateral processes, and lack of both hemoglobin and pseudobranchia (Wethington & Lydeard, 2007). Typical morphology also includes the long and filiform cephalic tentacles, muscular foot with a long triangular pointed end, and mantle edges that can be extended partially overlapping the shell. Typically, the shell surface is glossy and thin and does not provide important characters for generic or species identification. It is very variable in size at species level. Physids are hermaphroditic snails and diaulic with both reproductive pores located at the left side of the body. The feminine pore is found close to the pneumostome and the masculine pore behind the ocular tentacle.

Physids are frequently found in a variety of aquatic ecosystems, not only in lentic habitats such as streams and rivers. They are a natural intermediate host of *Trichobilharzia* (Trematoda: Schistosomatidae), a potential causative agent of avian cercarial dermatitis in Brazil (Pinto et al., 2014). Their rate of crawling is much more rapid than most gastropods, and this along with their early age at maturity and high fecundity explains why they are so widespread (Brown, 1991).

The invasive species, *Physa acuta* Draparnaud, 1805 has received different names in many newly occupied countries (*Physa chilensis* Clessin, 1886; *Physa cubensis* Pfeiffer, 1839; *P. guadaloupensis havanensis "Pfeiffer"* Clench, *P. bermudezi* Aguayo, 1935) (Paraense & Pointier, 2003; Taylor, 2003; Paraense, 2005; Albrecht et al., 2009). Life-cycle characteristics like high proliferation rates, high passive dispersal capacities, and high tolerance to polluted water highly favored its invasive capacity. This exotic species is very variable in shell shape and this morphological variability caused confusion with the taxonomic identification of the native fauna. *Physa acuta* was an early invader in Europe, i.e., in the 19th century, and is established in many regions of Asia, Australia, and Africa (Albretch et al., 2009).

Physidae has received great attention, and has been the focus of several studies, mainly in the Nearctic and European regions. However, along the Neotropical region, with the exception of some countries such as Costa Rica (Taylor, 2003) and Venezuela (Pointier & Noya, 2015b), this family still needs to be surveyed and investigated through morphology and molecular techniques.

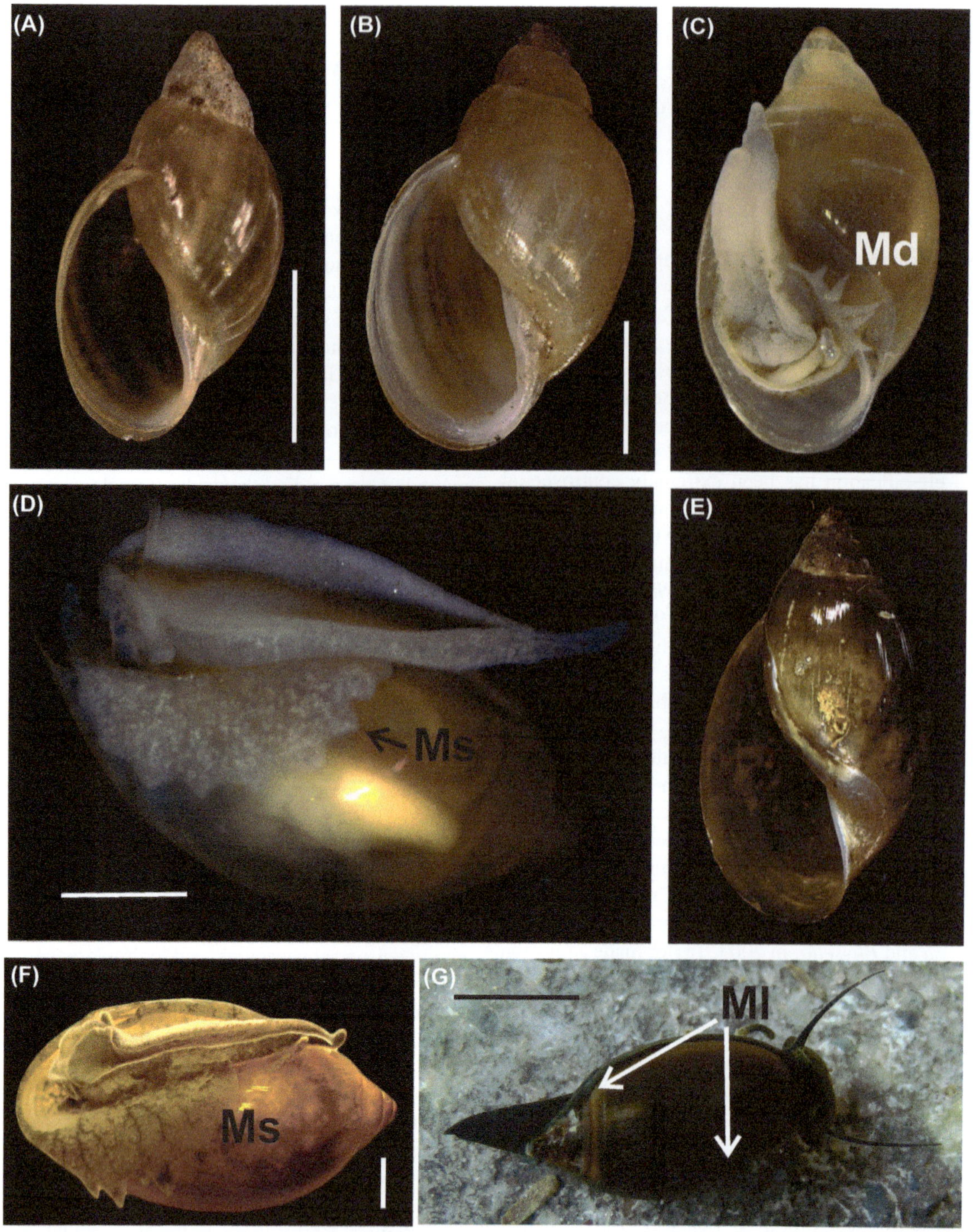

FIGURE 11.20 Shell and mantle in Physidae: (A) and (B), shell variability in *Physa acuta* from the same locality of occurrence, Argentina, Tucumán; (C) *Physa acuta* showing the digitiform mantle edge; (D) *Stenophysa panamensis*, Darien National Park, Panamá; (E) shell shape of *Stenophysa minor*, note the shallow sutures and triangular spire, Argentina, Tucumán; (F) *Stenophysa peruviana*, Los Amigos, Madre de Dios, Peru; (G) live specimen of *Stenophysa minor*. Abbreviations: *Md*, digitiform mantle margin; *Ml*, mantle margin with rounded lobes; *Ms*, serrate mantle margin. Scale bars: (A), (B) and (G) = 5 mm; (D) = 2 mm; (F) = 1 mm. (F) Photograph of Cristian Ituarte.

PHYLUM MOLLUSCA

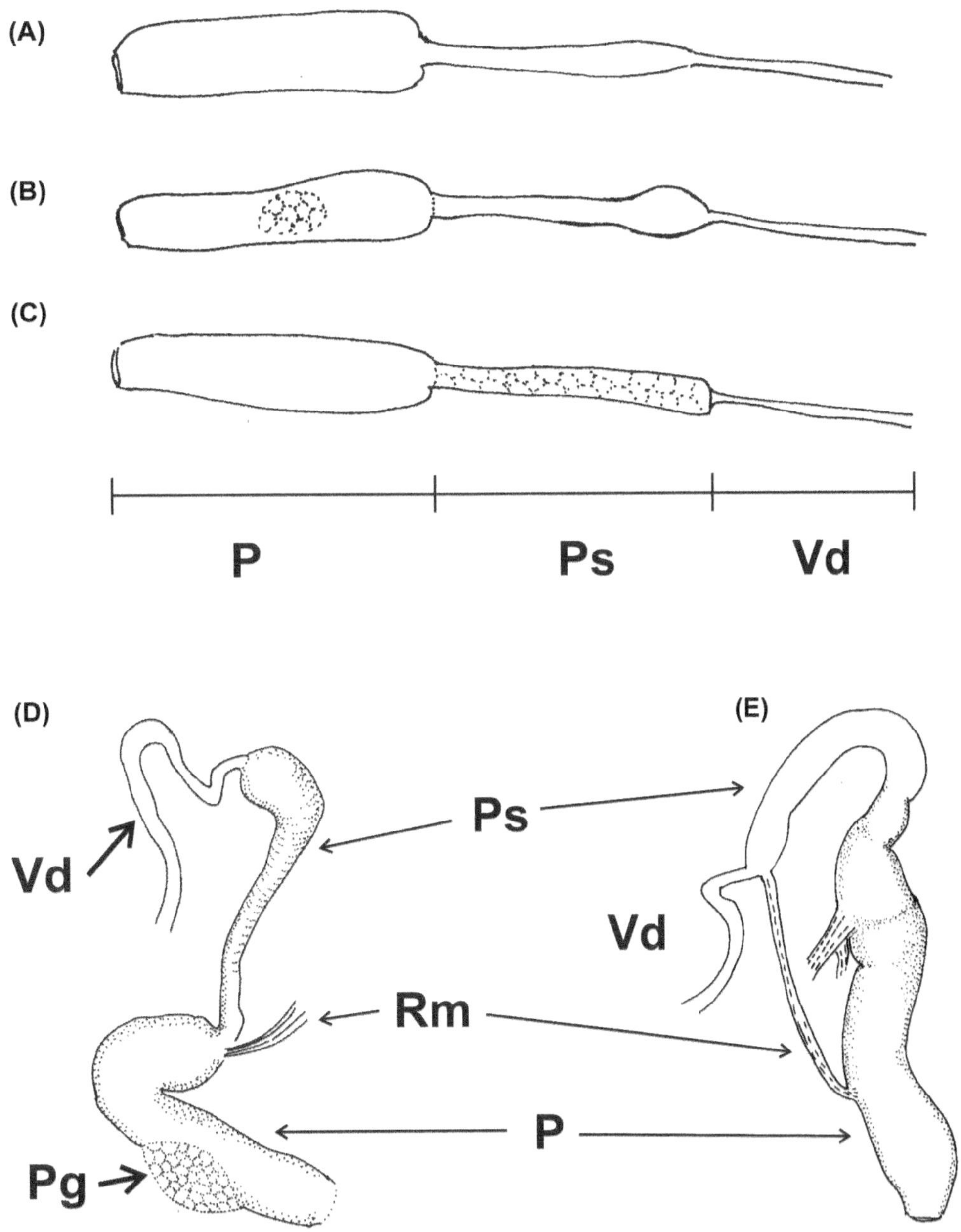

FIGURE 11.21 Common penial complex morphologies in Neotropical Physidae: (A) penis without preputial gland, muscular penial sheath; (B) penis with preputial gland, muscular penial sheath; (C) penis without preputial gland, glandular penial sheath; (D) *Physa acuta*, penial complex; (E) *Stenophysa minor* penial complex. Abbreviations: *P*, penis; *Pg*, preputial gland; *Ps*, penis sheath; *Rm*, retractor muscle; *Vd*, vas deferens. (A–C) after Wethington & Lydeard (2007).

KEY TO PHYSIDAE

Gastropoda: Heterobranchia: Physidae: Genera

Due to the high intraspecific variability in morphological characters existing in physids, accurate species identification is not possible until molecular genetic studies are conducted to better resolve taxonomy. For this reason, only genera are included

1	Shell with conic, short spire, glossy or shiny surface, shallow sutures; mantle margin with variable shapes, overlapping or not the shell; penis morphology without preputial gland, with muscular or glandular penial sheath 2
1	Shell with acute, pointed, tall spire; dull to silky surface; shallow to deep sutures (Fig. 11.20 A, B); mantle margin with triangular to thin digitations, not overlapping the shell, and divided into two regions, left-posterior and columellar parts (Fig. 11.20 C); penial morphology with preputial gland along with one part muscular penial sheath (Fig. 11.21 B, D) .... *Physa* (*Physa acuta* group) [Caribbean Islands, Central and South America]
2(1)	Shell narrowly subovate to conic, glossy; mantle not overlapping the shell, with margins without lobes or digitations; penial morphology with glandular penial sheath (Fig. 11.21 C) .... *Aplexa* [Central America, northern South America]
2′	Shell ovoid to fusiform, shiny (Fig. 11.20 D, E); mantle margin reflects broadly on both sides over the shell, in general with margins with rounded lobes but can be widely serrate (Fig. 11.20 D, F, G); penial morphology with muscular penial sheath (Fig. 11.21 A, E) .... *Stenophysa* [Caribbean Islands, South America]

PHYLUM MOLLUSCA

Chilinoidea, Chilinidae

Diego Eduardo Gutiérrez Gregoric
División Zoología Invertebrados, Museo de La Plata, Facultad de Ciencias Naturales y Museo, Universidad Nacional de La Plata. CONICET, La Plata, Buenos Aires, Argentina

INTRODUCTION

The Chilinidae contains one genus, *Chilina*, and includes almost 50 species. This family is endemic in southern South America, extending through Argentina, Chile, Uruguay, Brazil, Paraguay, and the Malvinas (Falkland) Islands (Pilsbry, 1911; Castellanos & Gaillard, 1981; Brown & Pullan, 1987) (Fig. 11.22). Some records also mention the family in Peru, but that has not been authenticated.

The Chilinidae has primitive features such as the presence of one chiastoneuric nervous system, horizontal lamellar tentacles, noncontractile pneumostome, and an incomplete division of both male and female ducts (Haeckel, 1911; Harry, 1964). By these features and molecular contributions, Chilinidae is a basal group within Hygrophila beside the Latidae family of New Zealand (Klussmann-Kolb et al., 2008; Jörger et al., 2010).

FIGURE 11.22 Distribution of Chilinidae.

Unlike other gastropod families, Chilinidae reach larger sizes in colder regions than in hotter regions. Sizes up to 50 mm are recorded in South Argentina and South Chile, while species near the tropics do not exceed 20 mm (Castellanos & Miquel, 1991). *Chilina* species can be found in quite different types of habitats such as lakes, lagoons, dams, waterfalls, streams, rivers, canals, and estuaries generally with clean oxygenated water and variable temperature ranges (Gutiérrez Gregoric, 2008; Cuezzo, 2009).

Historically, the species descriptions were based almost exclusively on shell characters. Due to huge polymorphism, anatomical and molecular studies are needed in most species. Recently, a review of the species from Argentina, with descriptions of the shell, radula, reproductive and nervous systems, and genetics analyses was begun (Ituarte, 1997; Gutiérrez Gregoric & Rumi 2008; Gutiérrez Gregoric, 2008, 2010; Ovando & Gutiérrez Gregoric 2012; Gutiérrez Gregoric et al., 2014; Gutiérrez Gregoric & de Lucía, 2016). These works describe four new species and redescribe other seven. In Argentina 24 chilinid species are present. However, the species present in the southern Argentina and Chile need to be studied.

The family and genus has a globular to bulimoide shell with an expanded last whorl. The last whorl is dilated, and sometimes it is the only part of the shell that is preserved. The aperture occupies two-third of the totality of the last whorl. The spire is erect or immersed, although in rapid environments, the spire can be eroded. In the aperture, up to two oblique, toothlike folds of different appearance (strong or weak) that characterize the shells of this family are shown. The periostracum is opaque with a greenish-yellow color and typical dark brown color bands in zigzag that contrast with the periostracum background color (Fig. 11.23 A–F).

The anatomy is characterized by a wide and short foot. The nervous system shows a vestigial chiastoneury (Fig. 11.23 I). The pulmonary roof is pigmented, and the kidney occupies almost its entire length. The internal wall of the kidney has numerous transverse trabeculae of irregular contour. The rectum is on the right side of the mantle cavity, and the anus is near the pneumostome. Male and female ducts are incompletely divided; the common duct opens to the hermaphrodite duct, with irregular contours on both sides. The female portion is characterized by the calcareous granules that it has in the vaginal lumen and, the presence of secondary bursa copulatrix or accessory seminal receptacle. The male portion has a penis with cuticularized teethlike structures and, the penis sheath with different structures as ornamentation (folds, pustules) (Fig. 11.23 G). The radula is characterized is this family by the rows arranged in "V" with one central tooth and numerous lateral teeth (Fig. 11.23 H) (Castellanos & Gaillard, 1981; Ovando & Gutiérrez Gregoric, 2012; Gutiérrez Gregoric et al., 2014).

ACKNOWLEDGMENTS

I want to thank to the curators of the malacological collections of Museo de La Plata, Argentina (MLP), Museo Argentino de Ciencias Naturales, Argentina (MACN), Muséum National d'Histoire Naturelle, France (MNHN), Academy of Natural Sciences of Philadelphia, USA (ANSP), and Smithsonian National Museum of Natural History, USA (USNM).

KEY TO *CHILINA* SPECIES

Gastropoda: Heterobranchia: Chilinoidea: Chilinidae: *Chilina*: Species

Due to the scarce information about internal morphology of many species, the dichotomous key is based mainly on shell characters, especially for the species from Argentina, Brazil, Uruguay, and Paraguay. The *Chilina* spp. from southern Argentina and Chile need to be revised.

1	Shell thick or thin; without last whorl angular or subangular; one or two columellar teeth (Fig. 11.23 A–D)	2
1′	Shell thick; with last whorl angular or subangular; with almost parallel edges; one or two columellar teeth (Fig. 11.23 E–F)	18
2(1)	Shell fluminoide or globular, thick; generally with two columellar teeth; variable color; low or immersed spire; ≤ 20 mm in total length (Fig. 11.23 A–C)	3
2′	Shell bulimoide; thin, generally one columellar tooth; spire developed; shells can exceed 20 mm in total length (Fig. 11.23 D)	13
3(2)	Shell globular; inconspicuous teeth; dark periostracum; with weak reddish bands, well-defined growth lines crossed by spiral grooves; forming a reticulate shell ornamentation	4
3′	Shell fluminoide; marked teeth; periostracum light olive color; strong reddish bands; without spiral grooves	6
4(3)	Shell thin; spire short	5
4′	Shell thick; purplish; with an immersed spire in adult specimens; length of aperture equal to length of last whorl; total width 80% aperture length; interior of shell markedly iridescent; one columellar tooth (Fig. 11.24 A) *Chilina iguazuensis* Gutiérrez Gregoric & Rumi, 2008 [Brazil and Argentina: Iguazú River]	
5(4)	Length of the aperture shorter than the length of the last whorl; total width 87% of the length of the aperture; brown and uniform color; two columellar teeth (Fig. 11.24 B) *Chilina megastoma* Hylton Scott, 1958 [Brazil and Argentina: Iguazú River and tributaries]	
5′	Shell small, aperture projection 50% of total width; last whorl large, 97% of the total length; with strong white callus; light brown periostracum with strong, thin longitudinal reddish bands; one columellar tooth (Fig. 11.24 C) *Chilina santiagoi* Gutiérrez Gregoric & de Lucía, 2016 [Brazil and Argentina: Uruguay River and tributaries]	
6(3)	Two columellar teeth	7
6	One columellar tooth (Fig. 11.24 D) *Chilina perrieri* Mabille, 1833 [Argentina: Santa Cruz]	
7(6)	Last whorl without keel	8
7′	Last whorl with keel	12
8(7)	Shell with two strong teeth	9
8′	Shell with two weak teeth, the lower one more developed	11
9(8)	Shell with strong red bands; periostracum olive color	10
9′	Shell with light reddish periostracum with some dark brown spots; spire eroded; width 74% of last whorl length; strong white callus (Fig. 11.24 E) *Chilina luciae* Gutiérrez Gregoric & de Lucía, 2016 [Argentina: Uruguay River and tributaries]	
10(9)	Penis complex curved; penis sheath can be almost the same length or else twice the prepuce; left subintestinal parietal connective is longer than the right visceral one (Fig. 11.24 F) *Chilina fluminea* (Maton, 1809) [Brazil: Rio Grande do Sul. Argentina: Buenos Aires, Corrientes, Entre Ríos. Uruguay]	
10′	Penis complex cylindrical and slender; penis sheath is longer than the prepuce; right visceral connective is longer than the one left subintestinal parietal connective (Fig. 11.24 G) *Chilina tucumanensis* (Castellanos & Miquel, 1980) [Argentina: Córdoba, Tucumán]	
11(8)	Periostracum light brown with weak dark reddish zigzag bands; spire immersed; aperture 90% of last whorl; penis sheath inner sculpture with triangular regular pustules (Figs. 11.24 H, 25 A) *Chilina nicolasi* Gutiérrez Gregoric & de Lucía, 2016 [Brazil and Argentina: Uruguay River]	

PHYLUM MOLLUSCA

11′ Periostracum light brown without reddish zigzag bands; spire short; aperture 85% of last whorl; penis sheath inner sculpture displaying star-shaped pustules; pustules arranged on left and right margins; penis sheath middle region smooth, without sculpture (Figs. 11.24 I, 25 B) .... *Chilina portillensis* Hidalgo, 1880
[Chile. Argentina: Jujuy, Tucumán]

12(7) Keel well developed in all whorls; aperture oval, and about 80% ~ last whorl length; four to five bands in zigzag reddish brown varying in intensity (Fig. 11.24 J) .... *Chilina rushii* Pilsbry, 1896
[Brazil: Rio Grande do Sul. Argentina: Buenos Aires, Corrientes, Entre Ríos, Misiones. Uruguay]

12′ Keel developed or undeveloped; aperture oval, ~ 75% last whorl length; periostracum dark brown with faint dark bands (Fig. 11.24 K) .... *Chilina gallardoi* Castellanos & Gaillard, 1981
[Brazil: Rio Grande do Sul. Argentina: Corrientes, Entre Ríos, Misiones. Uruguay]

13(2) Elongated shell with subparallel edges, spire developed; one columellar tooth (Fig. 11.23 C) 14

13′ Elongated shell with convex aperture edges, spire low developed; one or two columellar teeth 15

14(13) Shell slender; apex acute; numerous reddish bands in zigzag (Fig. 11.24 L) .... *Chilina parchappii* (d'Orbigny, 1835)
[Argentina: Buenos Aires, Mendoza]

14′ Shell not slender; apex obtuse; few reddish bands in zigzag (Fig. 11.24 M) .... *Chilina patagonica* (Sowerby, 1874)
[Chile. Argentina: Santa Cruz, Tierra del Fuego; Malvinas (Falkland) Island]

15(13) One columellar tooth 16

15′ Two columellar teeth 17

16(15) Shell with zigzag bands; suture of the last whorl oblique; aperture length ~ 75% last whorl length; penultimate whorl developed (15% of the length of the last whorl) (Fig. 11.24 N) .... *Chilina fulgurata* (Pilsbry, 1911)
[Argentina: Chubut, Neuquén, Río Negro, Santa Cruz]

16′ Shell without or few zigzag bands; suture of the last whorl perpendicular to the longitudinal axis; aperture length ~ 80% last whorl length; penultimate whorl developed (7% last whorl length) (Fig. 11.24 O) .... *Chilina aurantia* Marshall, 1924
[Chile. Argentina: Neuquén, Río Negro]

17(15) Two columellar teeth, lower tooth more prominent and developed than upper, innermost on columella; longitudinal reddish bands (Fig. 11.26 A) .... *Chilina lilloi* Ovando & Gutiérrez Gregoric, 2012
[Argentina: Tucumán]

17′ Two columellar teeth of equal development; reddish bands in zigzag (Fig. 11.26 B) .... *Chilina strebeli* Pilsbry, 1911
[Argentina: Chubut, Santa Cruz]

18(1) Shell with a pronounced shoulder; spire of low development; aperture > 67% total length 19

18′ Shell with a less pronounced shoulder; spire developed; aperture < 67% total length 23

19(18) Shell (>12 mm); last whorl angular; one tooth columellar 20

19′ Shell (<12 mm); last whorl dilated; spire conical 22

20(19) Shell subangular < 20 mm 21

20′ Shell with shoulder marked on the last whorl, angular; > 20 mm (Fig. 11.26 C) .... *Chilina gibbosa* (Sowerby, 1841)
[Chile. Argentina: Chubut, Neuquén, Río Negro]

21(19) Radula with central tooth bicuspid and asymmetric with strong serrated edges (Fig. 11.26 D) .... *Chilina mendozana* Strobel, 1874
[Argentina: Mendoza, San Juan]

21′ Radula with central tooth tricuspid without serrates edges (Fig. 11.26 E) .... *Chilina guaraniana* Castellanos & Miquel, 1980
[Paraguay. Argentina: Corrientes]

22(19) One columellar tooth (Fig. 11.26 F) .... *Chilina sanjuanina* Gutiérrez Gregoric, Ciocco & Rumi, 2014
[Argentina: San Juan]

22′ Two columellar teeth (Fig. 11.26 G) .... *Chilina cuyana* Gutiérrez Gregoric, Ciocco & Rumi, 2014
[Argentina: Mendoza]

23(18) Shell total width 50% total length; aperture length 80% of the last whorl (Fig. 11.26 H) .... *Chilina dombeiana* (Bruguiere, 1789)
[Argentina: Chubut, Río Negro, Neuquén. Chile]

23′ Shell total width 36% total length; aperture length 50% last whorl (Fig. 11.26 I) .... *Chilina neuquenensis* Marshall, 1933
[Argentina: Neuquén]

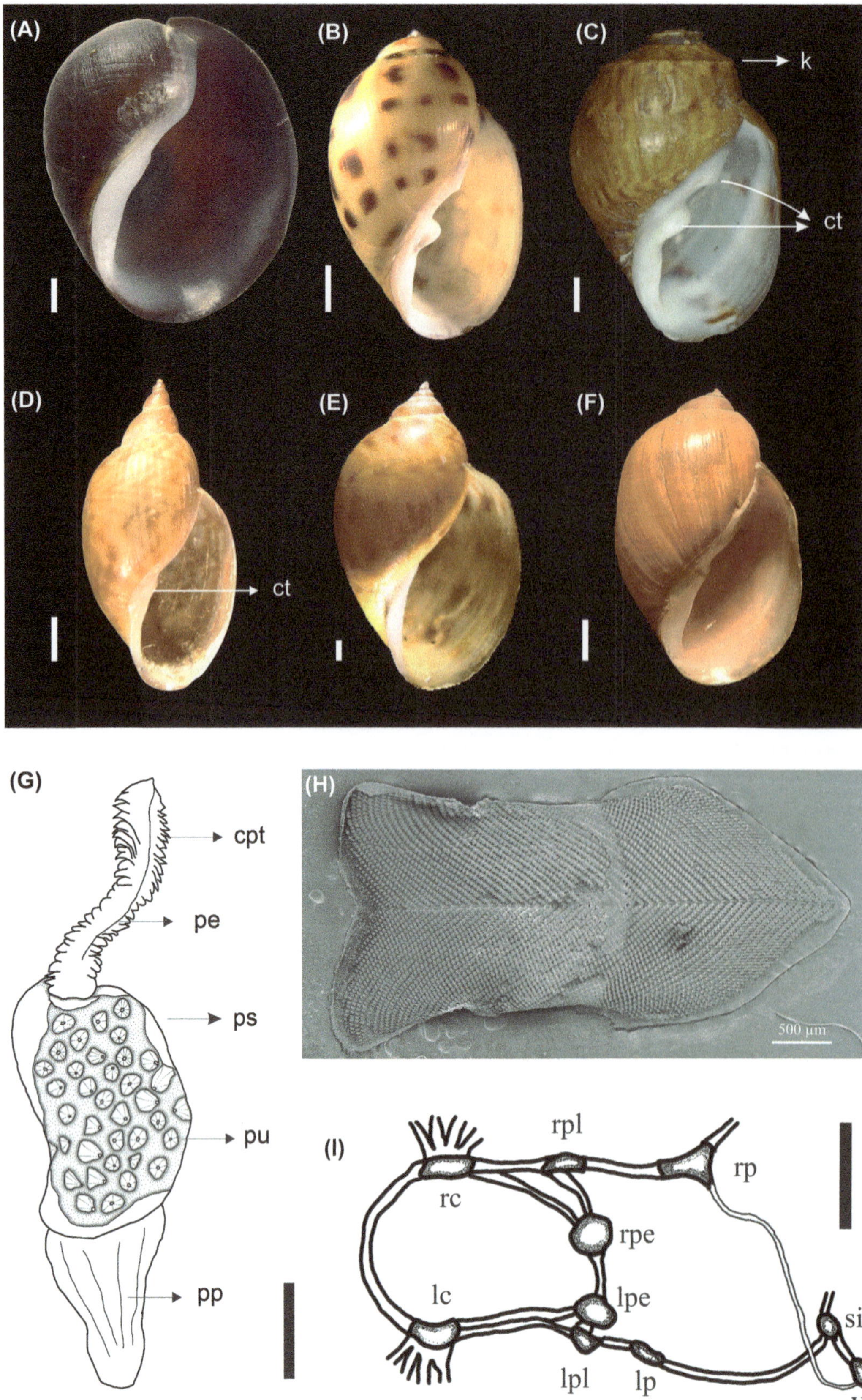

FIGURE 11.23 Chilinidae. (A–F) Shell features: (A) Globular; (B) fluminoide; (C) fluminoide with keel; (D) bulimoide with subparallel edges; (E) angular; (F) subangular; abbreviations: *ct*, columellar teeth; *k* keel; scale bar = 2 mm. (G) Penis sheath inner wall; abbreviations: *cpt*, cuticular penis tooth; *pe*, penis; *pp*, preputium; *ps*, penis sheath; *pu*, pustules; scale bar = 1 mm. (H) Radula of Chilinidae in a general view. (I) Diagram of nervous system; abbreviations: *lc*, left cerebral; *lpe*, left pedal; *lp*, left parietal; *lpl*, left pleural; *rc*, right cerebral; *rp*, right parietal; *rpe*, right pedal; *rpl*, right pleural; *si*, subintestinal; *v*, visceral; scale bar = 1 mm.

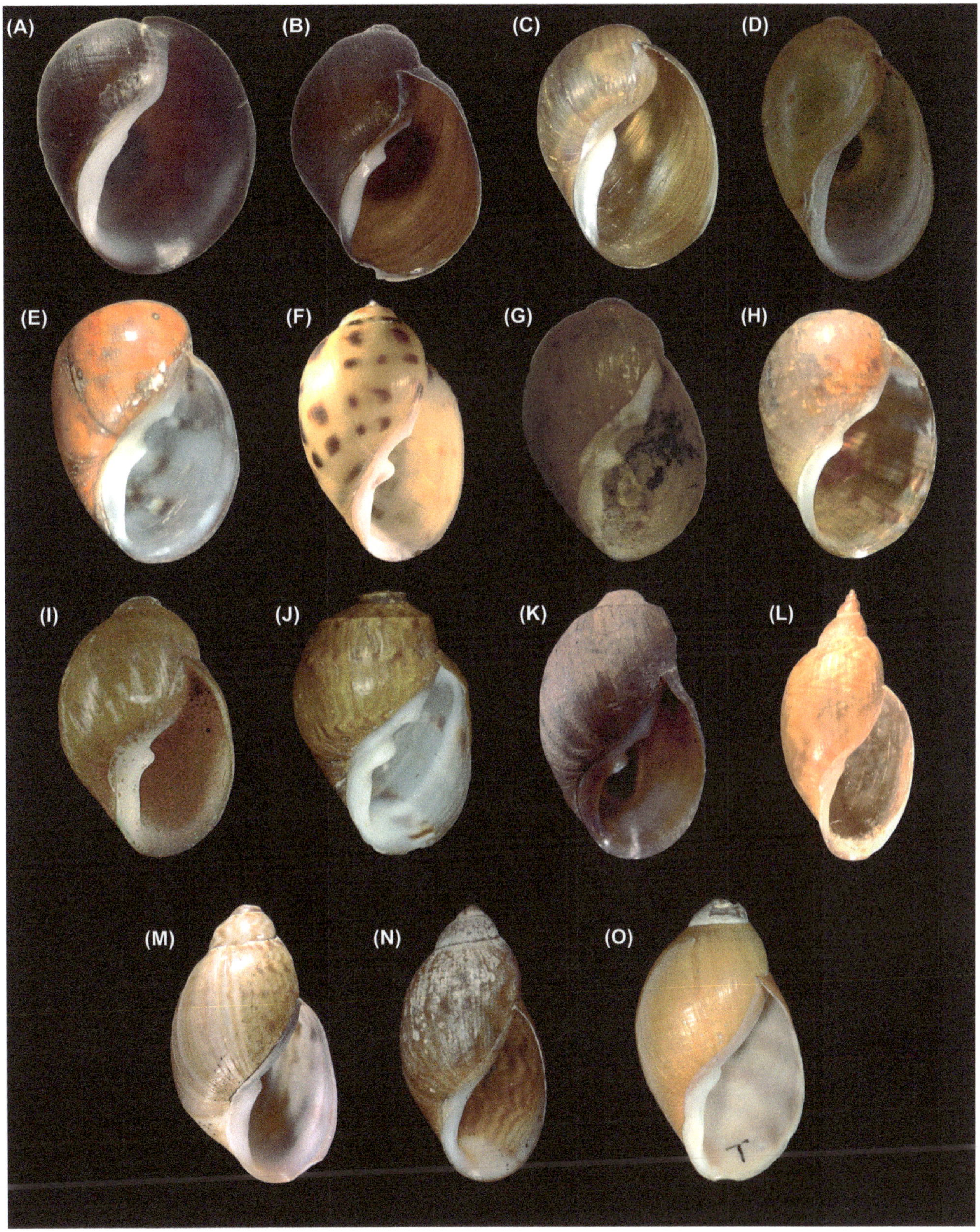

FIGURE 11.24 Shell of Chilinidae species: (A) *Chilina iguazuensis* (total length −TL = 18 mm, MLP 12,526); (B) *C. megastoma* (TL = 22 mm, Author's material); (C) *C. santiagoi* (last whorl length −LWL = 8.42 mm, MLP 14135); (D) *C. perrieri* (TL = 14 mm, MNHN 23107); (E) *C. luciae* (LWL = 10.62 mm, MLP 14141); (F) *C. fluminea* (TL = 13.3 mm, MLP 12578); (G) *C. tucumanensis* (TL = 14.4 mm, MACN 9927); (H) *C. nicolasi* (LWL = 13.50 mm, MLP 13412); (I) *C. portillensis* (TL = 12 mm, MNHN 23108); (J) *C. rushii* (TL = 17.3 mm, ANSP 69677); (K) *C. gallardoi* (TL = 14.8 mm, MLP w/n); (L) *C. parchappii* (TL = 14 mm, MLP 15449); (M) *C. patagonica* (TL = 19.7 mm, Author's material); (N) *C. fulgurata* (TL = 27.1 mm, ANSP 88661); (O) *C. auranita* (TL = 20.6 mm, USNM 360163). *ANSP*, Academy of Natural Sciences of Drexel University, Philadelphia; *MACN*, Museo Argentino de Ciencias Naturales "Bernardino Rivadavia"; *MLP*, Museo de La Plata, Argentina; *MNHN*, Muséum National d'Histoire Naturelle, Paris; *USNM*, National Museum of Natural History, Smithsonian Institution, Washington D.C.

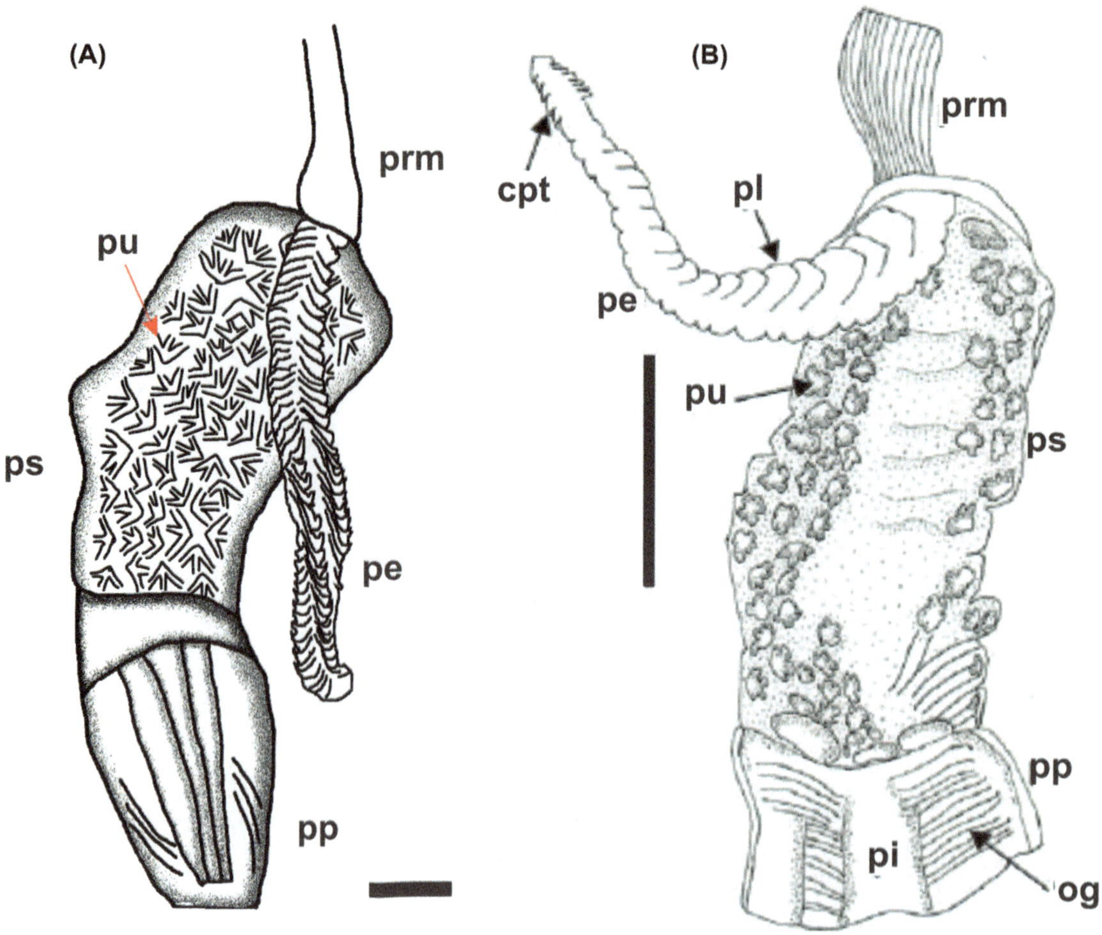

FIGURE 11.25 Chilinid penis sheath inner wall: (A) *Chilina nicolasi*; (B) *C. portillensis*. Abbreviations: *cpt*, cuticular penis tooth; *l*, lamellas; *og*, oblique grooves; *pe*, penis; *pi*, pilasters; *pl*, penis lamella; *pp*, preputium; *prm*, penis retractor muscle; *ps*, penis sheath; *pu*, pustules; *vd*, vas deferens. Scale bars = 1 mm. *Modified from: (A) Gutiérrez Gregoric & de Lucía (2016); (B) Ovando & Gutiérrez Gregoric (2012).*

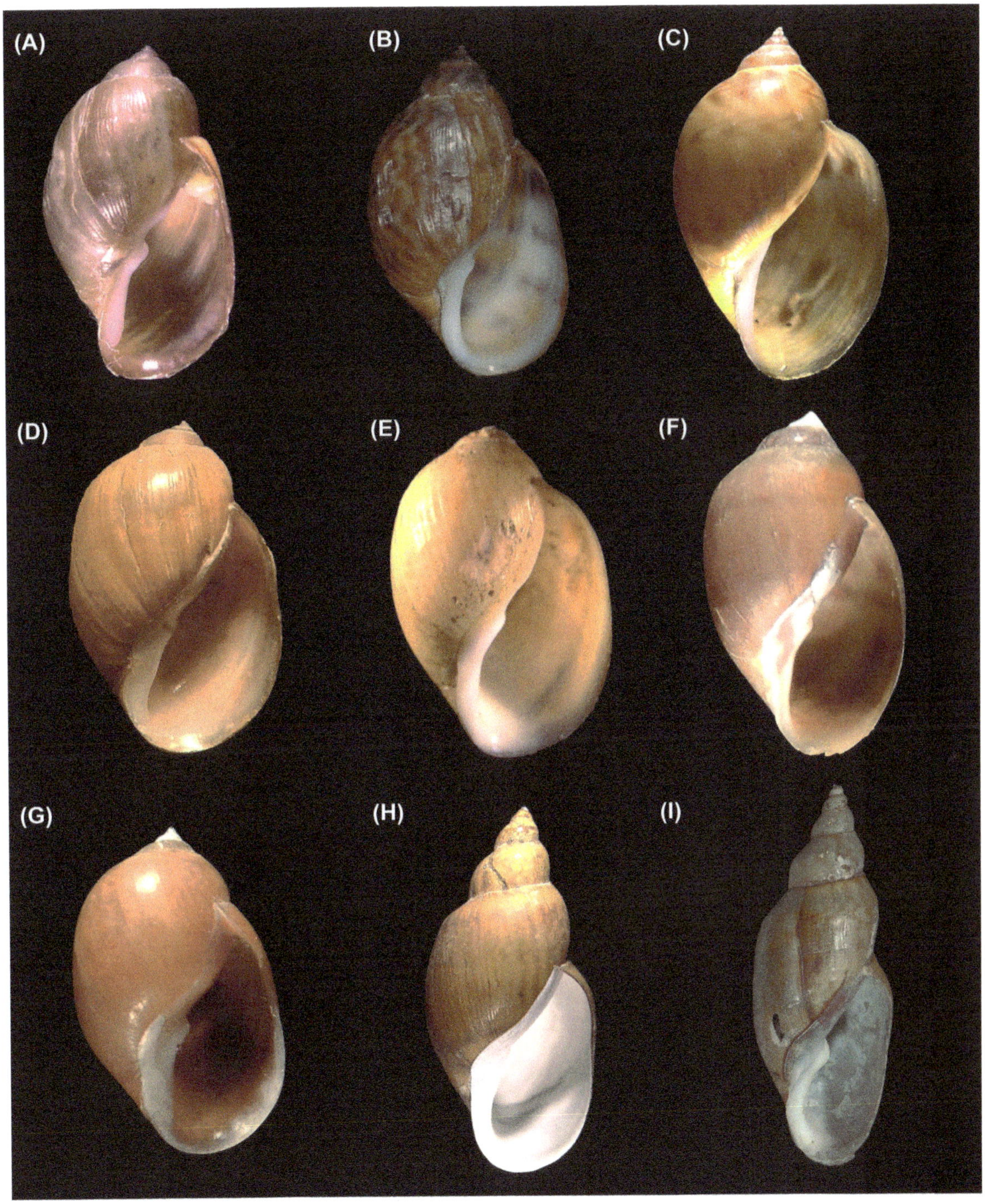

FIGURE 11.26 Shell of Chilinidae species: (A) *Chilina lilloi* (total length –TL = 12 mm, MLP 13343); (B) *C. strebeli* (TL = 24.1 mm, ANSP 88734); (C) *C. gibbosa* (TL = 30 mm, MLP 12743); (D) *C. mendozana* (TL = 15.1 mm, MLP 7391); (E) *C. guaraniana* (TL = 17.3 mm, MACN 568); (F) *C. sanjuanina* (TL = 10.8 mm, MLP 13545); (G) *C. cuyana* (TL = 11.1 mm, MLP 13660); (H) *C. dombeiana* (TL = 37 mm, Author's material); (I) *C. neuquenensis* (TL = 38 mm, USNM 414168). *ANSP*, Academy of Natural Sciences of Drexel University, Philadelphia; *MACN*, Museo Argentino de Ciencias Naturales "Bernardino Rivadavia"; *MLP*, Museo de La Plata, Argentina; *USNM*, National Museum of Natural History, Smithsonian Institution, Washington D.C.

PLANORBIOIDEA, PLANORBIDAE

Gastropoda: Heterobranchia: Planorbioidea: Planorbidae: Subfamilies

1	Shell patelliform	**Ancylinae [p. 305]**
1′	Shell spiral	**Planorbinae [p. 315]**

Planorbioidea, Planorbidae, Ancylinae

Sonia Barbosa dos Santos, Ximena Maria Constanza Ovando and Luiz Eduardo Macedo de Lacerda

Laboratório de Malacologia Límnica e Terrestre, Departamento de Zoologia, Universidade do Estado do Rio de Janeiro (UERJ), Rio de Janeiro, Brazil

INTRODUCTION

Freshwater air-breathing limpets are found in lentic environments attached to fallen decayed leaves and tree branches, and flowing water attached to stones. They are small, around 5 mm length, but can reach 15 mm; the visceral mass is domed, with a folded triangular pseudobranch with one or two lobes at the left side on the inferior margin of palial cavity. There are currently more than 18 nominal species of Ancylinae reported from Neotropical region (Lanzer, 1996; Santos, 2003).

Since the recognition of "Ancylidia" of Rafinesque (1815), a group created to house all patelliform freshwater snails (the European *Ancylus* as type genus), the systematic of the group has changed over time. Studies in detail on shell morphology, microsculpture, and soft parts were conducted to increase the knowledge of the group. The artificial nature of the family "Ancylidae" was long established, and several taxonomic arrangements were proposed, most of them based on shell microsculpture, protoconch, radula, and morphological characters of reproductive and muscular system.

The external shell morphology, shell microsculpture, apex shape, and the details of its sculpture are one of the most important characters used in the diagnosis of genera and species. The terms anterior margin, posterior margin, right margin, left margin, teleoconch, protoconch (or apex) are used to describe the shell (Fig. 11.27). Several features

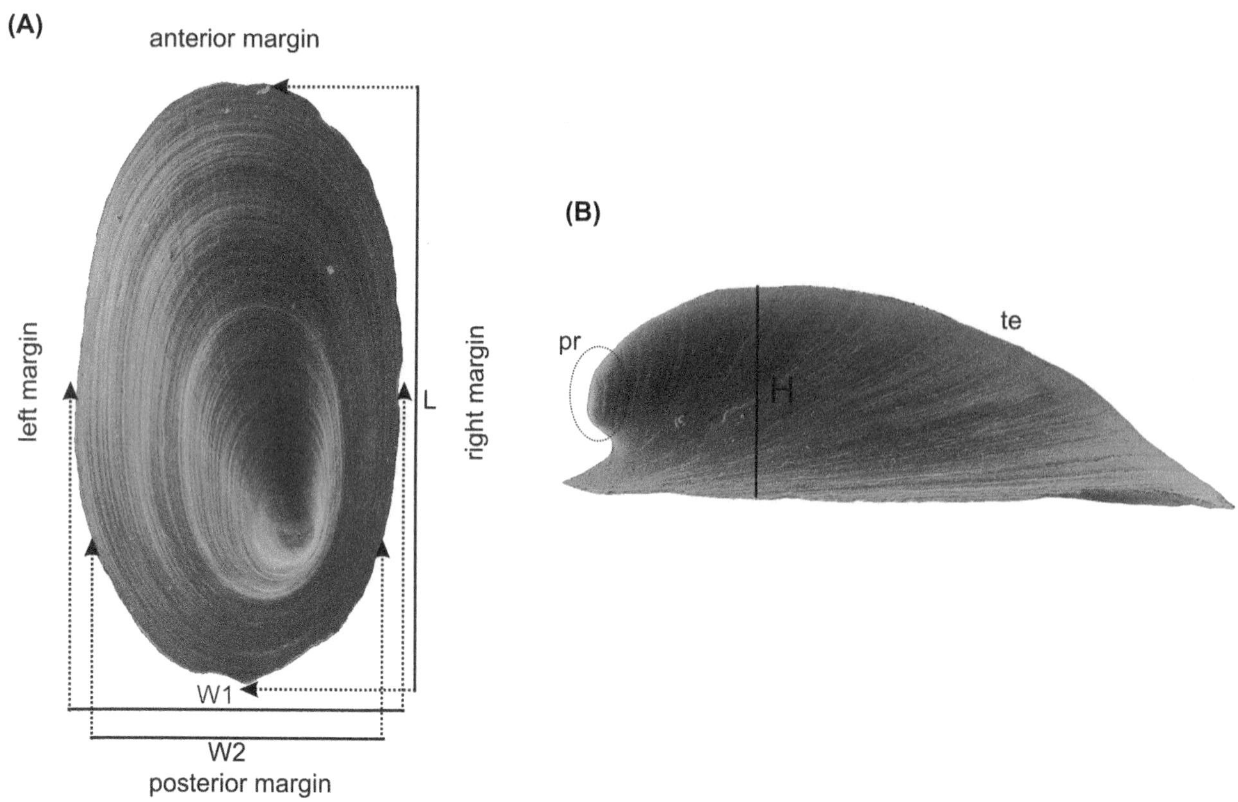

FIGURE 11.27 Ancylid shell morphology: (A) Dorsal view; (B) right lateral view. Abbreviations: *L*, length; *pr*, protoconch; *te*, teleoconch; *W1*, anterior width; *W2*, posterior width.

PHYLUM MOLLUSCA

of the apex are important: its position on the shell (central and posterior, central and anterior, decentralized and posterior); and its tilt (without tilt, backwards to the right, or aboard the shell) (Fig. 11.28). Also, the protoconch can be without sculpture (*Hebetancylus*, *Laevapex*, and *Sineancylus*); with radial lines (*Ferrissia*); with pits and smooth area (*Gundlachia*); only with pits (*Anisancylus*); with pits arranged as lines (*Burnupia*); with pits and radial lines (*Anisancylus* and *Uncancylus*) (Fig. 11.29). In the teleoconch, the presence (or absence) of radial lines as well as their disposition and proximity between them, the concentric lines, and the presence or absence of periostracal hairs can be observed. The shell aperture can be elliptical, rounded or elongated.

Some species develop a basal septum that partially closes the shell aperture; these are known as "gundlachioids shells" (Fig. 11.30 A). The formation of the septum is linked to adverse environmental conditions (drying), in order to preserve body water, helping survival. After this period, the shell continues growing around the remaining aperture in the anterior portion of shell (Fig. 11.30 B).

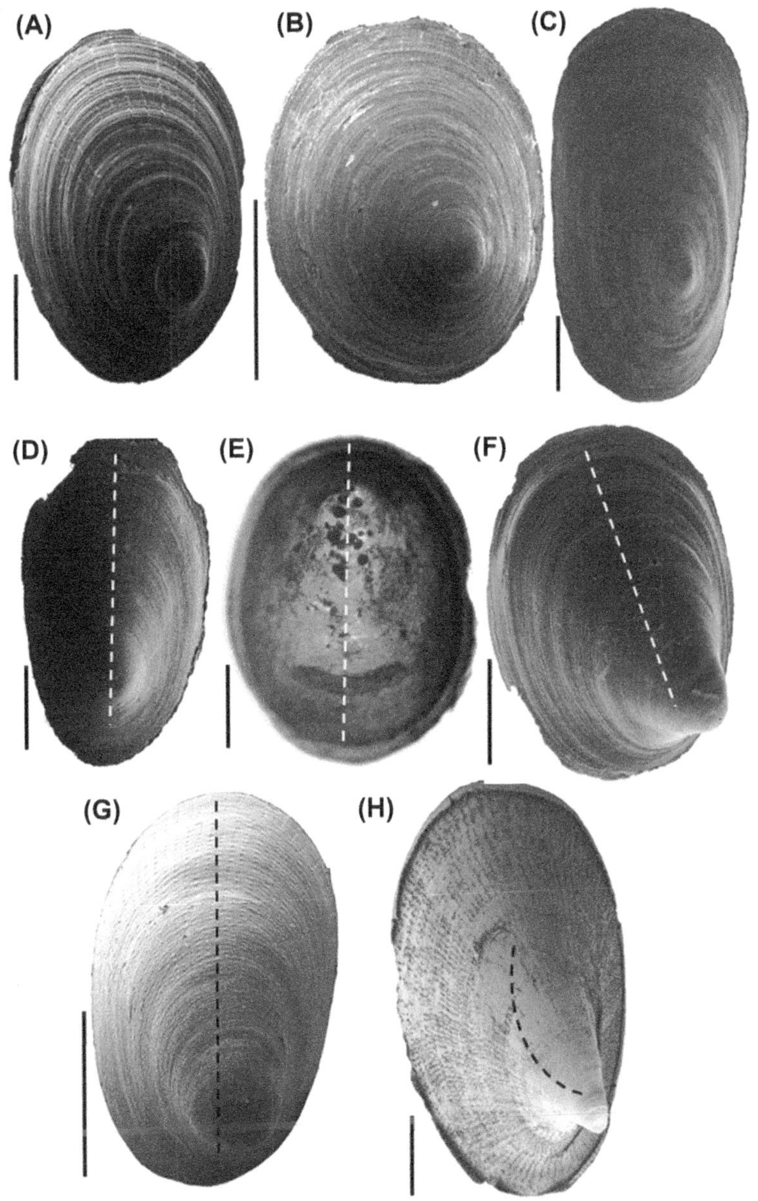

FIGURE 11.28 Comparative morphology of ancylid shells. (A–C) Shape of the aperture: (A) Elliptical; (B) rounded; and (C) elongated. (D–F) Position of the apex: (D) Central and posterior; (E) central and anterior; and (F) decentralized and posterior; the dotted lines indicate the longitudinal axis of the body and the position of the apex in relation to it. (G–H) Tilt of the apex: (G) Without tilt; and (H) backwards to the right, or aboard the shell; the dotted lines indicate the direction of the deflected apex. Presence of periostracal hairs can be observed in the teleoconch. Scale bars = 1 mm.

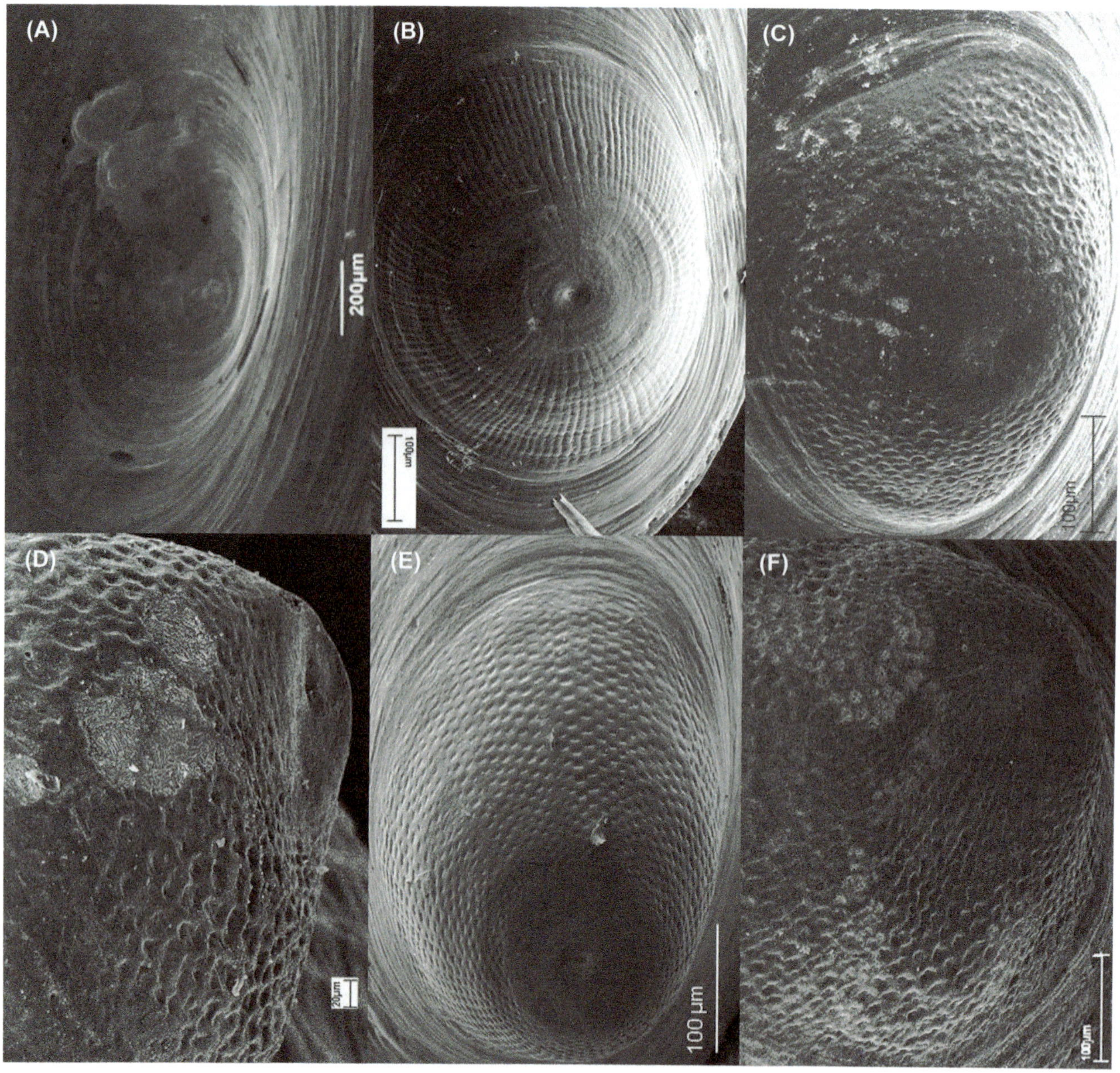

FIGURE 11.29 Detail of ancylid microsculpture of protoconch by SEM: (A) Without sculpture (*Hebetancylus*); (B) with radial lines (*Ferrissia*); (C) with pits and smooth area (*Gundlachia*); (D) only pits (*Anisancylus*); (E) pits arranged as lines (*Burnupia*); and (F) pits and radial lines (*Anisancylus*).

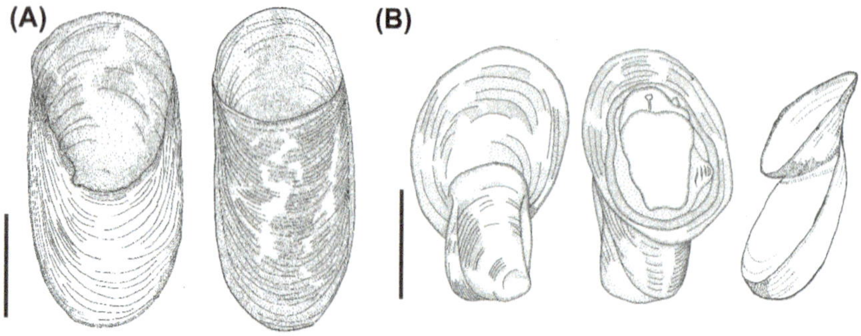

FIGURE 11.30 Gundlachiods shells with septum in different stages of development: (A) aperture partially closed shell (under water stress); and (B) different views of gundlachioids shells after a period of hydric stresses. Scale bars = 1 mm. *(A) Modified from Ovando et al. (2011).*

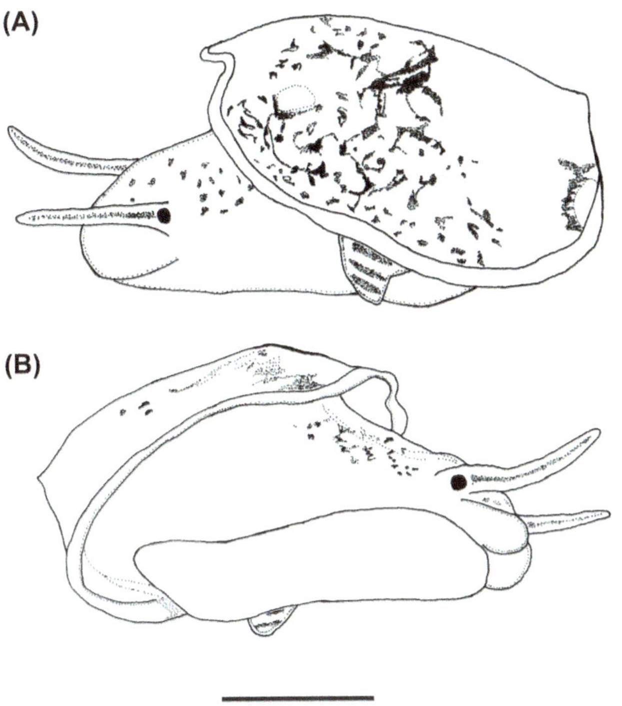

FIGURE 11.31 External morphology of ancylids: (A) Dorsal view showing mantle pigmentation; and (B) right lateral view. Scale bar = 1 mm.

The animal body shows three regions: head, foot, and visceral mass (Fig. 11.31). The head is bilaterally symmetric, with long tentacles (*Gundlachia*, *Anisancylus*, *Hebetancylus*, *Laevapex*, *Uncancylus*) or short (*Burnupia*, *Ferrissia*), subcylindrical, slightly triangular at the base, forming a well-defined lapel that extends abreast of the male genital pore, situated near the base of the left tentacle. The tentacles usually have a median axis of dark pigmentation. Two short and rounded oral palps protrude in front of the mouth, when the animal is moving. The oral opening is roughly T or U-shaped, limited by united platelets that form the yellowish jaw. The foot is rounded or slightly oval, wider on the anterior portion, with the rear end rounded, rarely tapered. The mantle covers the visceral mass; dorsal mantle surface can show different patterns of pigmentation (intensity and disposition of the pigmented spots), commonly concentrated on the right side. However, some species lack pigmentation, as *Ferrissia* sp., while others are heavily pigmented (*Anisancylus obliquus* and *Laevapex* sp.). The margin of the mantle is thin and usually not pigmented. The yellow and rounded ovotestis is at the posterior apical portion of visceral mass, the digestive gland on the right side and, the pericardium, the flexed excretory system and nidamental gland on the left side of the body. On the dorsal mantle surface, three muscles scars of the dorsoventral adductor muscles that attach the animal to the shell are present. The morphology of the muscle scars also helps in identification of genera and species.

Different measurements and ratios from shells are used to analyze the intraspecific variation (Fig. 11.27): length (L: greater distance from anterior and posterior margin); anterior width (W1: greater distance between left and right margins); posterior width (W2: distance between the right and left margins at the level of the posterior part of the apex); height (H: perpendicular distance from the highest point of the shell to the opening); width in relation to the length (W1/L); posterior width in relation to the length (W2/L); ratio of height in relation to length (H/L); ratio of height in relation to width (H/W1); posterior width in relation to anterior width (W1/W2).

KEY TO ANCYLINAE

The taxonomy of Neotropical region ancylids is currently being revised because some nominal species are known only by original descriptions based on the shell morphology. Some species such as *Laevapex vazi* Santos, 1989 is not included in this identification key as its shell sculpture is different from a typical *Laevapex*, although ejaculatory complex with penis and ultrapenis and absence of flagellum are similar characters. This species is being taxonomically revised.

Gastropoda: Heterobranchia: Planorbioidea: Planorbidae: Ancylinae: Genera

1	Shell rounded in dorsal view; apex obtuse, slightly projected, smooth	2
1′	Shell elliptical in dorsal view	3
2(1)	Shell apex on posterior third of right side (Fig. 11.32 A–D) [Brazil: Bahia; Panama]	*Laevapex*
2′	Shell apex on anterior side [Argentina: Misiones]	*Sineancylus rosanae* (Gutiérrez Gregoric, 2012)
3(1)	Shell apex generally on posterior third of right side, deflected to right or posteriorly	4
3′	Shell apex generally close to midline, slightly deflected to right side	7

4(3) Protoconch with pits in radial arrangement or pits slightly ordered 5

4′ Protoconch with irregular arranged pits 6

5(4) Apex acute prominent, deflected posteriorly .... ***Anisancylus* [p. 306]**

5′ Apex rounded; muscular impression elongated to the right, > 0.5x mantle length (Fig. 11.33 A–C) .... *Burnupia ingae* Lanzer, 1991
[Brazil: Alagoas, Espírito Santo, Goiás, Mato Grosso, Pará, Santa Catarina, Rio de Janeiro, Rio Grande do Sul]

6(4) Apex acute, hamulate .... ***Uncancylus* [p. 306]**

6′ Apex rounded on posterior side, not in midline .... ***Gundlachia* [p. 306]**

7(3) Protoconch only with radial parallel lines .... ***Ferrissia* [p. 307]**

7′ Protoconch smooth, without microsculpture (lines or pits), obtuse .... ***Hebetancylus* [p. 307]**

Gastropoda: Heterobranchia: Planorbioidea: Planorbidae: Ancylinae: *Anisancylus*: Species

1 Protoconch sculpture with irregular pits, without radial lines; mantle roof pigmentation absent to very dark; right muscular scar with a medial constriction (Fig. 11.34 A, B) .... *Anisancylus dutrae* (Santos, 1994)
[Brazil: Bahia, Pernambuco]

1′ Protoconch sculpture with irregular, obvious pits and radial lines; mantle roof pigmentation generally dark; right muscular scar vertically elongate, parallel to mantle edge (Fig. 11.34 C, D) .... *Anisancylus obliquus* (Broderip & Sowerby, 1832)
[Brazil: Rio Grande do Sul. Chile: Aconcagua, Coquimbo, Santiago, Concepción, Valdivia, Lhanquihue. Argentina: Córdoba, Neuquén, Río Negro, Salta. Uruguay: Artigas, Lavalleja, Payssandú]

Gastropoda: Heterobranchia: Planorbioidea: Planorbidae: Ancylinae: *Uncancylus*: Species

1 Shell high, not laterally compressed; apex not surpassing shell lateral margin 2

1′ Shell slightly high, laterally compressed; apex surpassing shell lateral margin .... *Uncancylus crequi* (Bavay, 1904)
[Peru: Titicaca Lake. Bolivia. Chile]

2(1) Teleoconch without radial lines and weak growth lines .... *Uncancylus foncki* (Philippi, 1866)
[Chile: Llanquihue Lake]

2′ Teleoconch with obvious radial lines; anterior right muscle scar elliptical located transversally in relation to body axis; mantle pigmentation diffuse (Fig. 11.35 A, B) .... *Uncancylus concentricus* (d'Orbigny, 1835)
[Costa Rica: Cartago, Coris. Venezuela: Guanare, Los Teques. Colombia: Guaca. Ecuador: Guayaquil. Brazil: Bahia, Mato Grosso, Mato Grosso do Sul, Minas Gerais, Rio Grande do Sul, Santa Catarina, São Paulo. Peru: Lima, Trujillo. Paraguay: Asunción, Caaguazú, Villarica. Argentina: Buenos Aires, Córdoba, Entre Rios, Salta, Tucumán. Uruguay: Canelones, Florida, Maldonado, Montevideo, Tacuarembó]

Gastropoda: Heterobranchia: Planorbioidea: Planorbidae: Ancylinae: *Gundlachia*: Species

1 Aperture rounded (Fig. 11.28 B) 2

1′ Aperture elliptical (Fig. 11.28 A) 4

2(1) Teleoconch with sculpture 3

2′ Teleoconch without sculpture 5

3(2) Teleoconch with radial lines; mantle pigmentation present 4

3′ Teleoconch with spiral lines; mantle pigmentation absent (Fig. 11.36 A–C) .... *Gundlachia bakeri* (Pilbry, 1913)
[Brazil: Pará]

4(1) Shell apex with wide depression; right anterior muscular scar inner region with an elongated adhesive area (Fig. 11.36 G–I) .... *Gundlachia radiata* (Guilding, 1828)
[Central America, Caribbean Islands. Colombia. Brazil: Alagoas, Amazonas, Pará, Paraíba, Pernambuco, Rio de Janeiro. Argentina: Jujuy, Tucumán]

4′ Shell apex with narrow depression; right anterior muscular scar broadly ovate (Fig. 11.36 D–F) .... *Gundlachia ticaga* (Marcus & Marcus, 1962)
[Brazil: Goiás, Mato Grosso, São Paulo, Rio de Janeiro. Argentina: Jujuy, Salta, Tucumán]

5(2) Right anterior muscle scar elliptical (Fig. 11.36 K, L) 6

5′ Right anterior muscle elongated similar to others muscle scars (Fig. 11.36 J) .... *Gundlachia lutzi* (Walker, 1925)
[Brazil: Minas Gerais]

6(5) Posterior muscular scar elongated and oblique to body axis; right anterior muscle scar narrowing in inner region; mantle pigmentation concentrated on left side (Fig. 11.36 K) .... *Gundlachia leucaspis* (Ancey, 1901)
[Brazil: Mato Grosso]

6′ Posterior muscle scar elongated and transverse to body axis; mantle pigmentation absent (Fig. 11.36 L) .... *Gundlachia saulcyana* (Bourguignat, 1853)
[Venezuela. Brazil: Bahia, Pernambuco, Rio de Janeiro]

Gastropoda: Heterobranchia: Planorbioidea: Planorbidae: Ancylinae: *Ferrissia*: Species

1 Protoconch with close radial lines (Fig. 11.37) 2

1′ Protoconch with separated radial lines .... *Ferrissia gentilis* Lanzer, 1991
[Brazil: Rio Grande do Sul, Santa Catarina]

2(1) Right anterior muscle scar elongated crescent; mantle pigmentation absent (Fig. 11.37 A–C) .... *Ferrissia irrorata* (Guilding, 1828)
[Saint Vincent. Grenada. Colombia: Antioquia. Argentina: Tucumán]

2′ Right anterior muscle scar elongated; small dark pigmentation spot between anterior muscle scars (Fig. 11.37 D–F) .... *Ferrissia fragilis* (Tryon, 1863)
[Brazil: Minas Gerais, Rio de Janeiro. Nearctic]

Gastropoda: Heterobranchia: Planorbioidea: Planorbidae: Ancylinae: *Hebetancylus*: Species

1 Aperture elongate with posterior margin narrowest; apex blunt, smooth; muscle scars rounded (Fig. 11.38 A, C) .... *Hebetancylus moricandi* (d'Orbigny, 1837)
[Venezuela. Brazil: Amazonas, Bahia, Ceará, Espírito Santo, Goiás, Mato Grosso do Sul, Minas Gerais, Pará, Paraíba, Pernambuco, Rio Grande do Sul, Rondônia. Peru. Bolivia. Paraguay. Argentina. Uruguay]

1′ Aperture elliptical; apex depressed; radial lines weak, shallow, anterior left and posterior muscle scars rounded, anterior right muscle with small inner projection (Fig. 11.38 D–F) .... *Hebetancylus excentricus* (Morelet, 1851)
[Mexico to Costa Rica. Nearctic]

ACKNOWLEDGMENTS

To Conselho Nacional de Desenvolvimento Científico e Tecnológico (CNPq) and Coordenação de Acompanhamento do Pessoal de Ensino Superior (CAPES) for finnancial support.

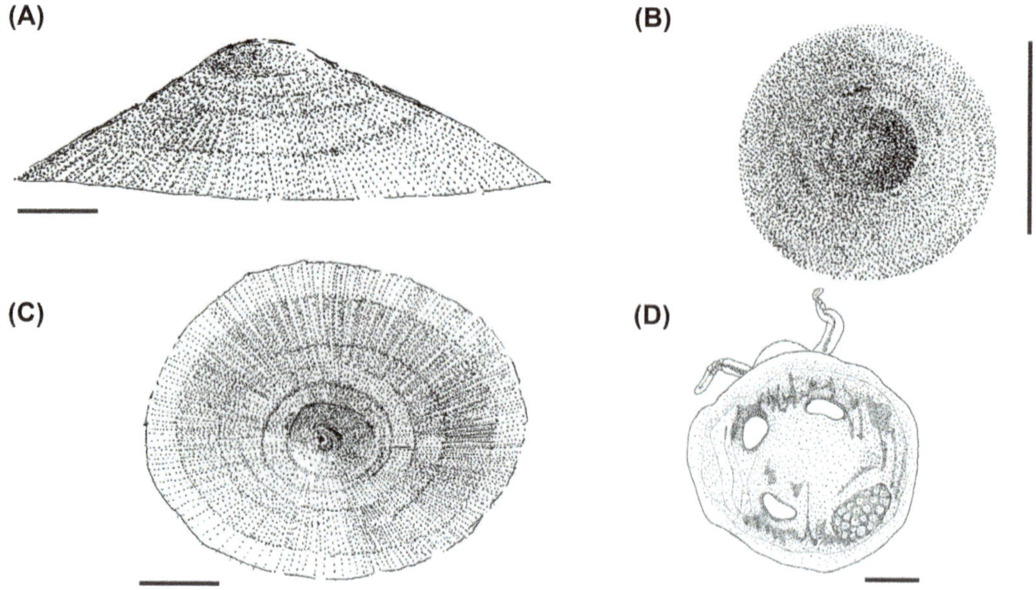

FIGURE 11.32 Species of *Laevapex* (Ancylinae): (A) Shell morphology in lateral view; (B) detail of protoconch sculpture; (C) shell morphology in dorsal view; and (D) soft parts in dorsal view showing morphology of muscles scars and tentacles. Scale bars = 1 mm.

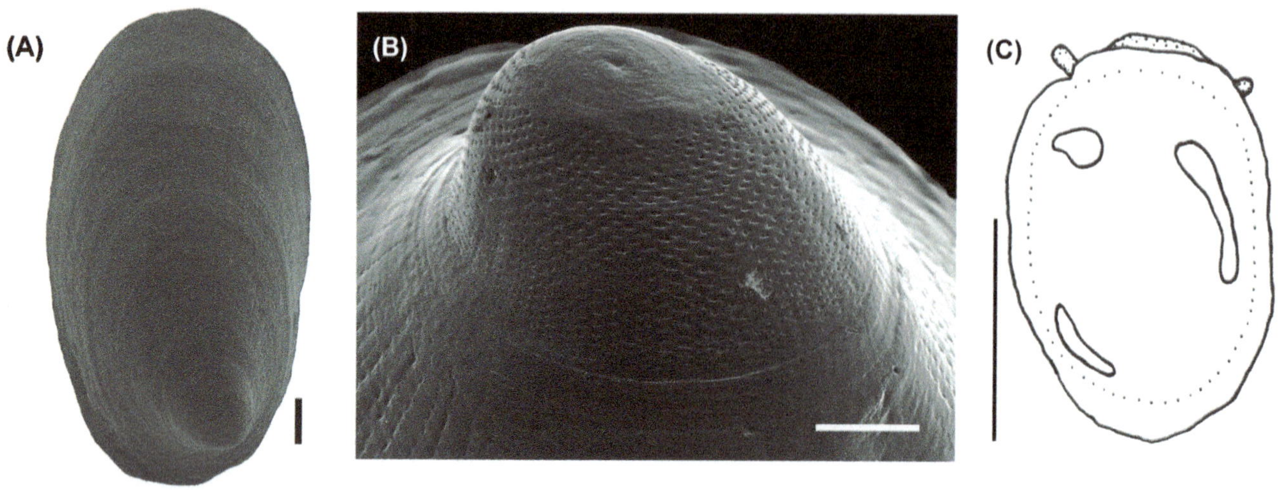

FIGURE 11.33 *Burnupia ingae* (Ancylinae): (A) Shell general morphology under SEM; (B) detail of protoconch sculpture; (C) soft parts in dorsal view showing the pigmentation pattern and morphology of muscle scars. Scale bar (A) and (C) = 1 mm, (B) = 100 μm. *Modified: (A) and (C) from Lacerda & Santos (2011); and (B) from Ovando et al. (2016).*

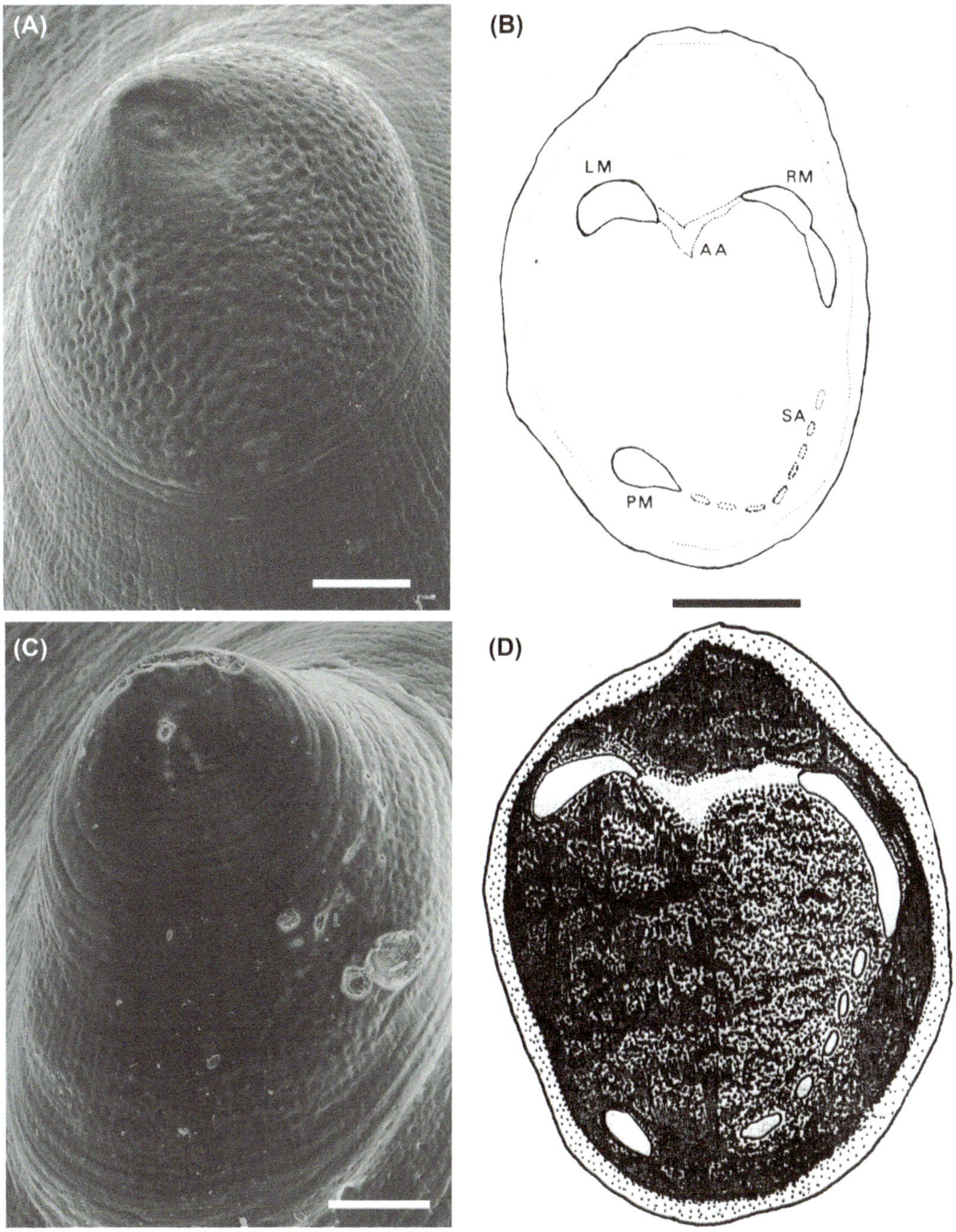

FIGURE 11.34 Species of *Anisancylus* (Ancylinae). *A. dutrae*: (A) Detail of protoconch sculpture by SEM of a paratype (B) soft parts in dorsal view showing the pigmentation pattern and morphology of muscle scars. *A. obliquus*: (C) Detail of protoconch sculpture by SEM; (D) soft parts in dorsal view showing the pigmentation pattern and morphology of muscle scars. Abbreviations: *AA*, adhesive area; *LM*, left muscle; *PM*, posterior muscle; *RM*, right muscle; *SA*, small adhesive area. Scale bars (A) and (C) = 100 μm; (B) and (D) = 1 mm. *Modified (A) and (C) from Ovando et al. (2014); (B) from Santos (1994); and (D) from Santos (2000).*

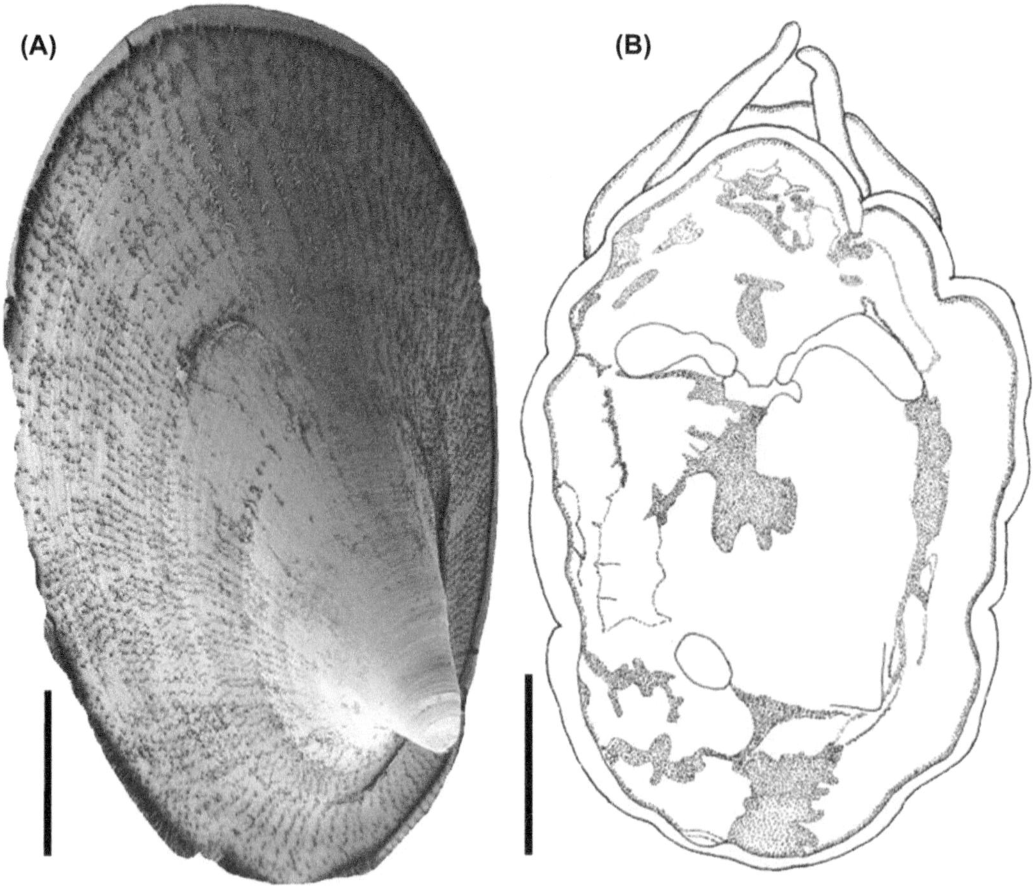

FIGURE 11.35 *Uncancylus concentricus*: (A) Shell morphology under SEM; (B) soft part in dorsal view showing the pigmentation pattern and morphology of muscle scars. Scale bars = 1 mm. *Modified from Ovando et al. (2014).*

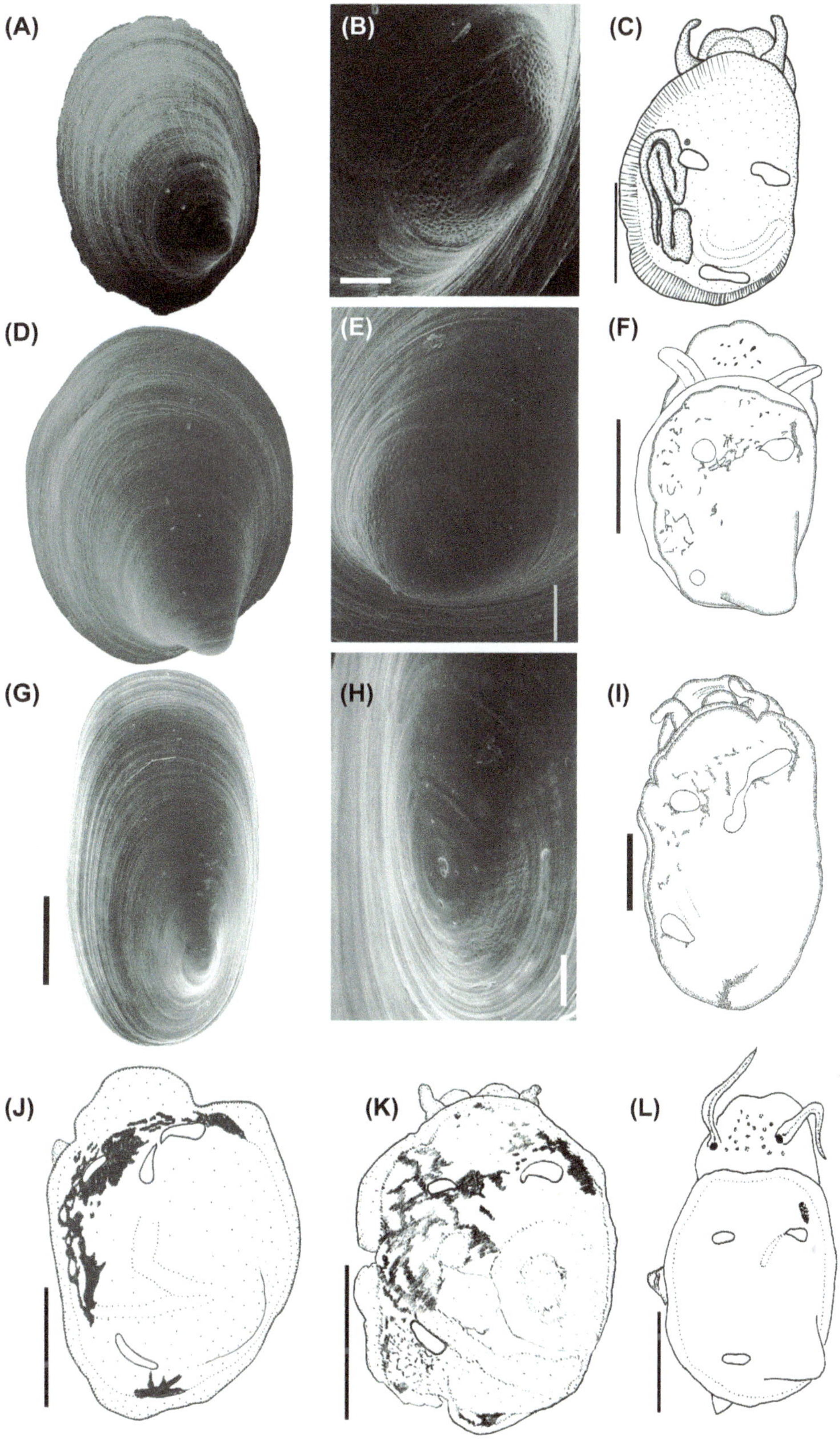

FIGURE 11.36 Species of *Gundlachia* (Ancylinae): (A–C) *G. bakeri*; (D–F) *G. ticaga*; (G–I) *G. radiata*; (J) *G. lutzi*; (K) *G. leucaspis*; (L) *G. saulcyana.* Scale bars (A), (C), (D) and (F) = 1 mm, (B), (E) and (H) = 100 µm. *(D-F) Modified from Ovando et al. (2014).*

PHYLUM MOLLUSCA

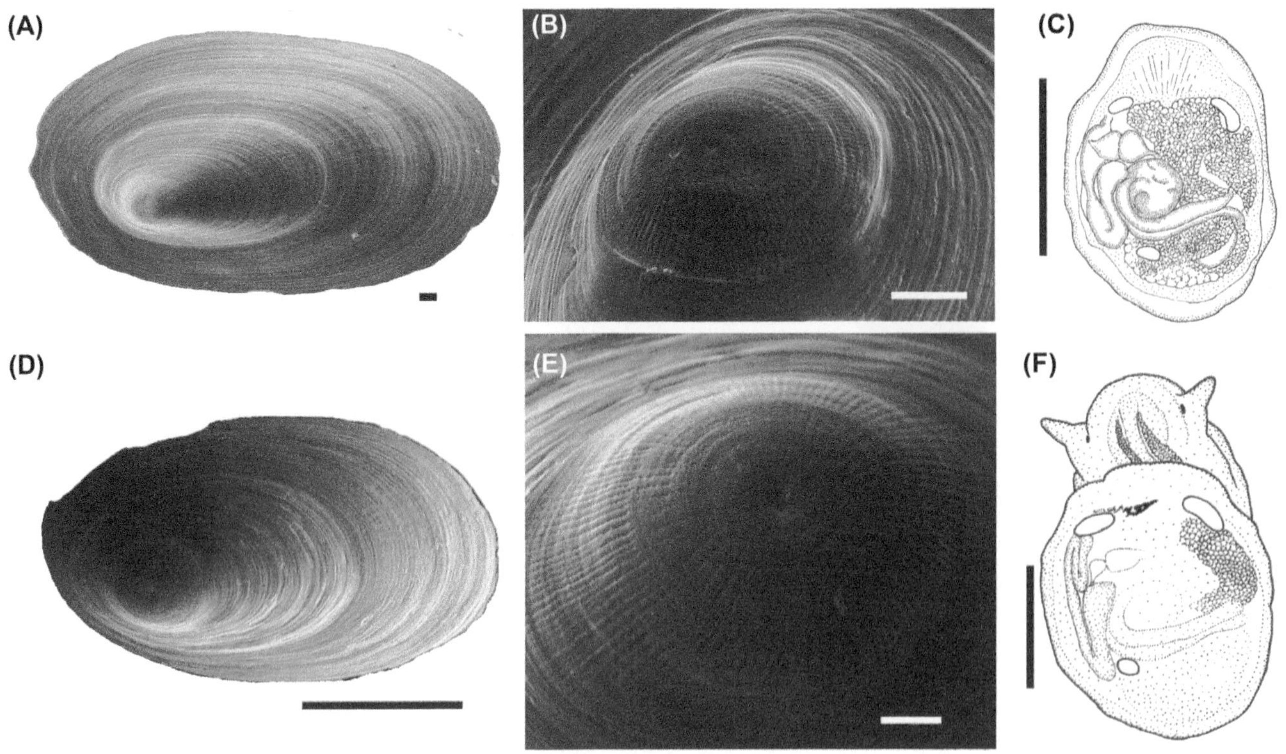

FIGURE 11.37 Species of *Ferrissia* (Ancylinae): (A–C) *F. irrorata*; (D–F) *F. fragilis*. Scale bars (A), (B), and (E) = 100 μm; (C), (D) and (F) = 1 mm. *Modified (A–C) from Ovando et al. (2014); and (D-F) from Lacerda et al. (2015).*

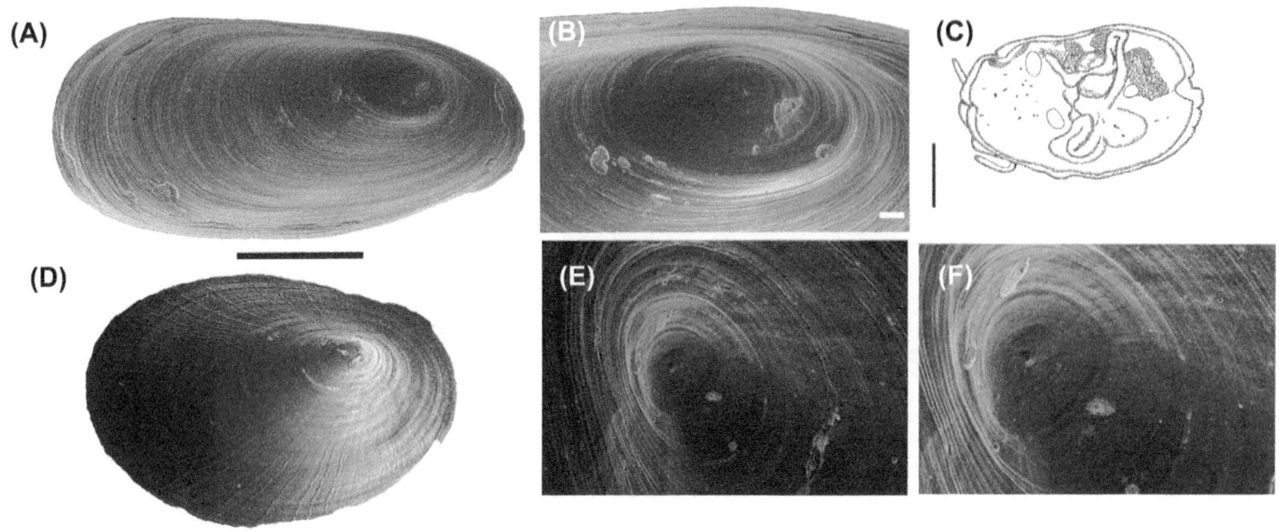

FIGURE 11.38 Species of *Hebetancylus* (Ancylinae). (A–C) *Hebetancylus moricandi*: (A) Shell morphology under SEM; (B) detail of protoconch sculpture; and (C) soft parts in dorsal view showing the pigmentation pattern and morphology of muscle scars. (D–F) *Hebetancylus excentricus*: (D) shell morphology under SEM, showing radial lines in the teleoconch; (E) detail of the concentric growth lines of the teleoconch; (F) soft parts in dorsal view showing the pigmentation pattern and morphology of muscle scars. Scale bars (A) = 1000 μm; (B), (E), and (F) = 100 μm; (C) and (D) = 1 mm. *(A-C) Modified from Ovando et al. (2014).*

Planorbioidea, Planorbidae, Planorbinae

Monica Ammon Fernandez, Elizangela Feitosa da Silva and Silvana Carvalho Thiengo

Laboratório de Malacologia / Laboratório de Referência Nacional para Esquistossomose - Malacologia, Instituto Oswaldo Cruz, Fundação Oswaldo Cruz, Rio de Janeiro, Brazil

INTRODUCTION

Planorbids taxonomy still needs further sound morphological and molecular studies with samples preferably collected from the type-localities. Few species have shell as the only diagnostic characteristics when they reach maturity, e.g., *Biomphalaria schrammi* (Crosse, 1864) and *Biomphalaria helophila* (d'Orbigny, 1835) whose adult specimens bear internal lamellae. However, as the shell may vary due to environmental conditions its use for taxonomic purpose is limited. It is also important to point out that the taxonomic status of some old available names is still undefined by lack of knowledge of the anatomy of topotypic specimens, as reported by Paraense (2003a) to *Plesiophysa granulata* (Sowerby, 1873).

The position of the salivary gland in relation to the nervous ring distinguishes some genera of Planorbinae (Fig. 11.39), but the reproductive system (Fig. 11.40) bears the most important characters.

Regarding the shell, the most important characters for identification are: (a) shell shape; (b) shell aperture, whether it is angular or rounded, sinistral or dextral; (c) the sides right and left, whether it has carina or it is rounded, and (d) the number of whorls. The protoconch, mantle pigmentation, the presence of a set of lamellae (Fig. 11.41), grooves, and growth lines are also important diagnostic features.

The anatomical identifications should be carefully based on the diagnostic characteristics, such as renal ridge, the shape and number of the prostatic diverticula, vagina smooth or expanded in small pouch, the shape and length of the prepuce and its relation to the penis sheath, flagella size, among others. For example, *Biomphalaria occidentalis* Paraense, 1981 is nearly indistinguishable from *Biomphalaria tenagophila* (d'Orbigny, 1835), by shell and most genital organ characteristics, except for the absence of a vaginal pouch and the shape of the prepuce (Paraense, 1981).

The use of molecular techniques is also an important tool. Regarding Brazil, the work of Vidigal et al. (1996) was a pioneer, followed among others, by Caldeira et al. (2009) who presented species-specific profiles of the Brazilian species of *Biomphalaria*. Recently studies have considered Cytochrome Oxidase (COI) and 16S good as molecular markers to elucidate the relationships among members of the *B. tenagophila* complex or differentiate species of *Biomphalaria* from a specific geographic region of Brazil (Tuan et al., 2012; Palasio et al., 2017). However, these results still need to be extended based on analysis of populations from different regions. Paraense (2007), undoubtedly the specialist on Neotropical planorbids, commenting on the limits of the use of shell and morphology to species discrimination, pointed out that the molecular and morphological studies must have the same purpose, and if properly applied, reach the same objective: the specific identification.

Regarding the Cuban species, for example, Yong et al. (2001) compared conchological, anatomical, and genetic characteristics from several populations of the *Biomphalaria havanensis* (Pfeiffer, 1839) complex from Cuba and other localities in the Caribbean region and separated two species: *B. havanensis* from type locality and *Biomphalaria* sp. On the other hand, Vidigal et al. (2001), using morphological and molecular analysis, showed that among the studied Cuban *Biomphalaria* populations (five nominal species in total) the occurrence of only two species

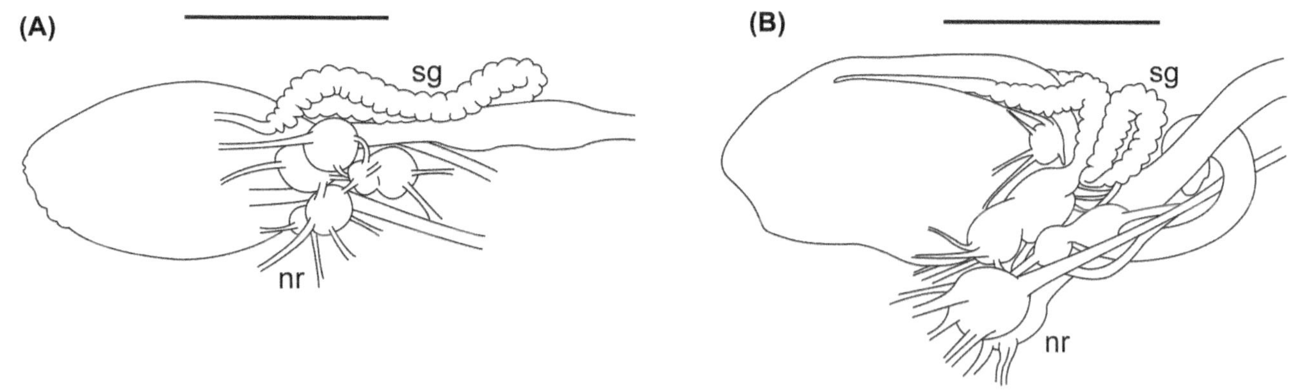

FIGURE 11.39 Planorbinae salivary glands position: (A) Salivary gland out of the nervous ring; (B) salivary gland inside of the nervous ring. Abbreviations; *nr*, nervous ring; *sg*, salivary gland. Scale bar (A) = 1 mm and (B) 0.2 mm. *Modified from Paraense (1975).*

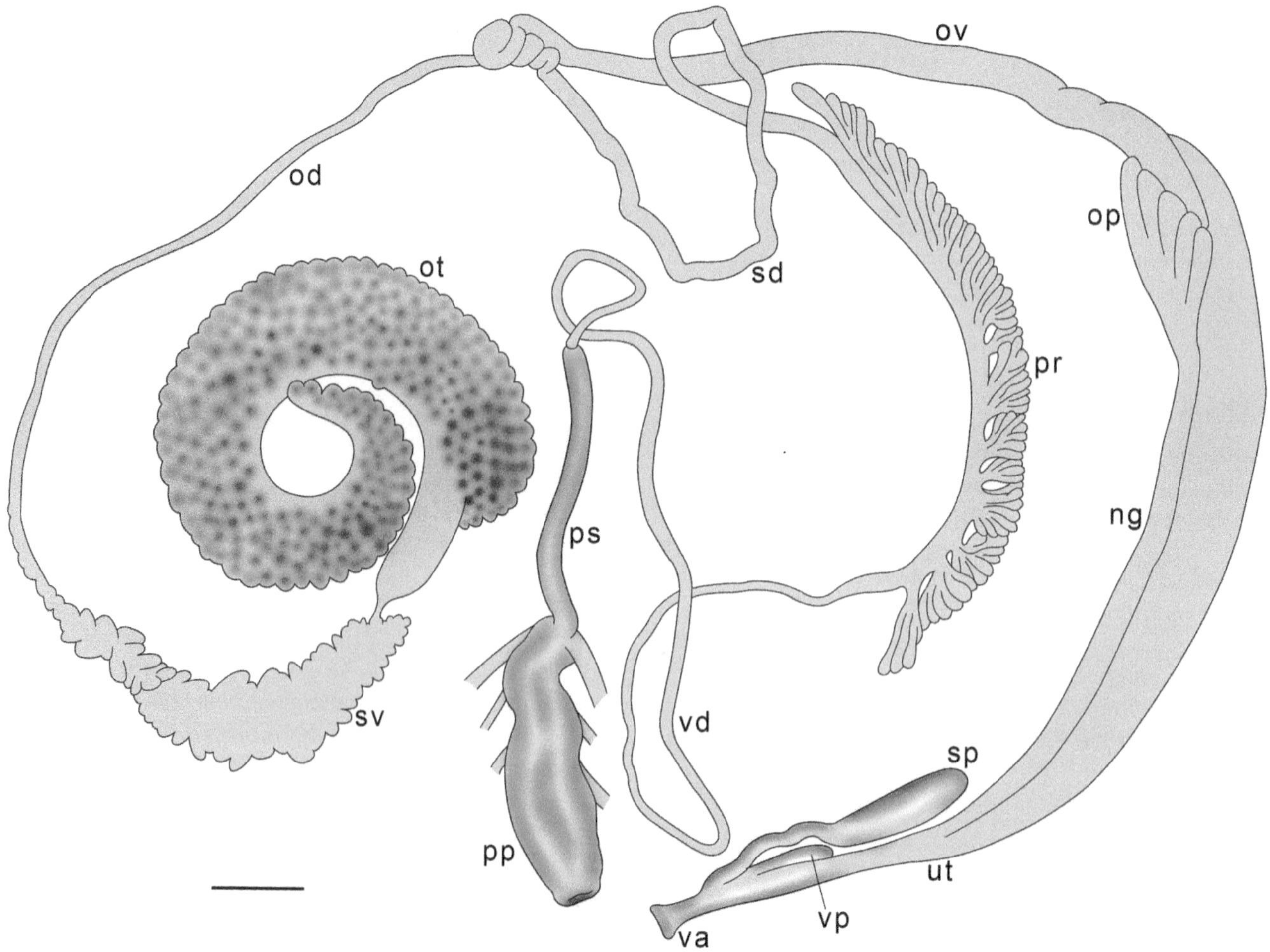

FIGURE 11.40 Reproductive system of *Biomphalaria* spp. Abbreviations: *ng*, nidamental gland; *od*, ovispermiduct; *op*, oviduct pouch; *ot*, ovotestis; *ov*, oviduct; *pp*, prepuce; *pr*, prostate; *ps*, penis sheath; *sd*, spermiduct; *sp*, spermatheca; *sv*, seminal vesicle; *ut*, uterus; *va*, vagina; *vd*, vas deferens; *vp*, vaginal pouch. Bar scale = 1 mm. *Modified from Paraense (1975).*

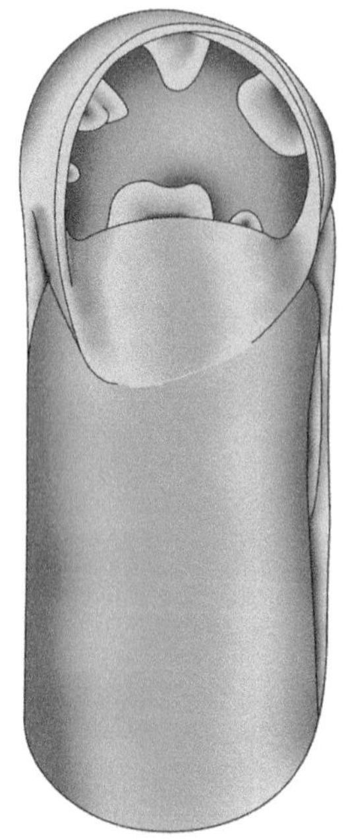

FIGURE 11.41 Planorbid shell shows the presence of a set of lamellae, two parietal and four palatal. Scale bar = 1 mm. *Modified from Paraense & Deslandes (1956).*

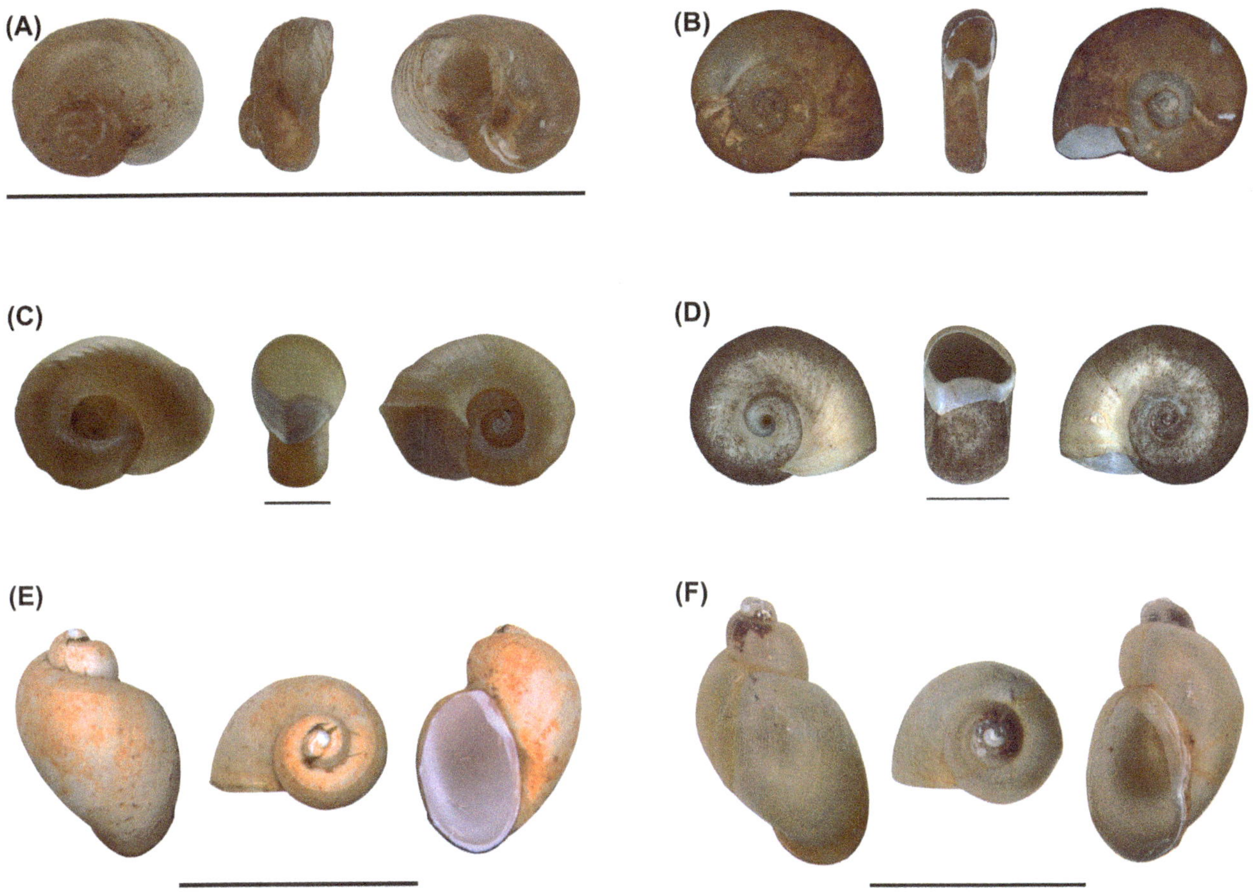

FIGURE 11.42 (A) Planorbid shells of *Acrorbis petricola*; (B) *Antillorbis nordestensis*; (C) *Helisoma duryi*; (D) *Helisoma trivolvis*; (E) *Plesiophysa guadeloupensis*; (F) *Plesiophysa dolichomastix*. Scale bar = 5 mm.

was confirmed: *B. havanensis* and *Biomphalaria obstructa* (Morelet, 1849).

The preparation of planorbid specimens for study requires special care. The preparation and preservation techniques are detailed in the general section of Heterobranchia.

KEY TO PLANORBINAE

For morphological data of species see Fernandez et al. (2012), Paraense (1966, 1975, 1976a, 1976b, 1981, 1984, 1988, 2002, 2003a), and Paraense & Deslandes (1956, 1962).

Gastropoda: Heterobranchia: Planorbioidea: Planorbidae: Planorbinae: Genera

Gyraulus hindsianus (Dunker, 1848) from Ecuador (Paraense, 2004) and *Gyraulus percarinatus* Paraense, 2003 from Panama (Paraense, 2003b) are in need of complementary morphological observations, so the genus is not included in the key.

1 Shell not discoidal 2

1′ Shell discoidal (growth is in one plane) 3

2(1) Shell helicoid (length ≤ 1.6 mm, width 2.6 mm), spire elevated (Fig. 11.42 A) .... *Acrorbis*; one species: *Acrorbis petricola* Odhner, 1937 [Brazil, Argentina]

2′ Shell elongate-ovate (length ≤ 9 mm, width 5 mm), sinistral .... ***Plesiophysa*** **[p. 317]**

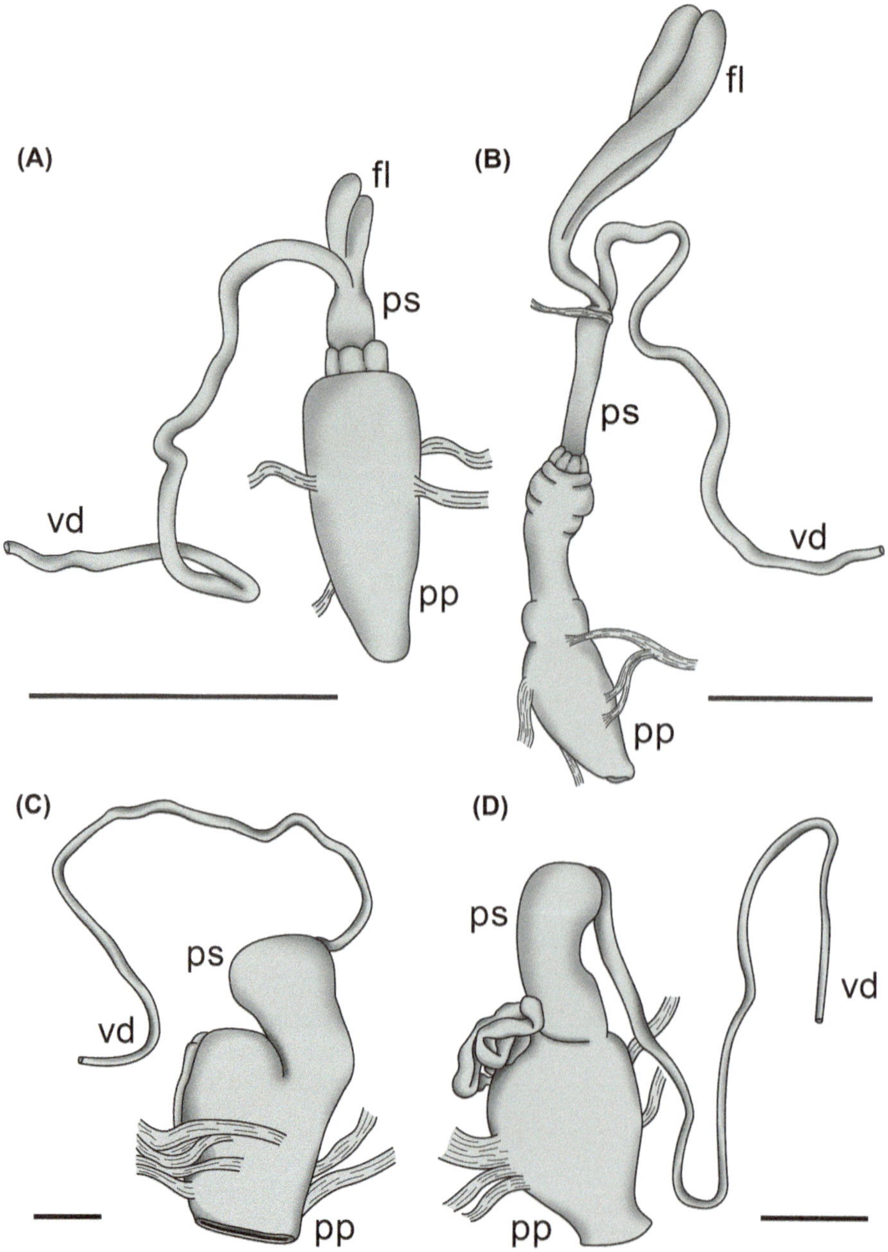

FIGURE 11.43 Planorbid reproductive system (male): (A) *Plesiophysa guadeloupensis*; (B). *Plesiophysa dolichomastix*; (C) *Helisoma duryi*; and (D) *Helisoma trivolvis*. Abbreviations: *fl*, flagella; *pp*, prepuce; *ps*, penis sheath; *vd*, vas deferens. Scale bars = 1 mm. *Modified from Paraense (1976a, 1976b, 2002, 2003a).*

3(1)	Salivary gland outside nervous ring	4
3′	Salivary gland inside nervous ring	5
4(3)	Shell very thick, aperture widened relative to whorls; shell with carinae on both sides; large prepuce with an external duct (duct of preputial organ) connecting with penial sheath	***Helisoma* [p. 317]**
4′	Shell slightly thick, aperture proportionally increasing relative to whorls; penial complex with prepuce, penis sheath and penis, well defined, without preputial organ duct [Neotropics. Nearctic]	***Biomphalaria* [p. 318]**

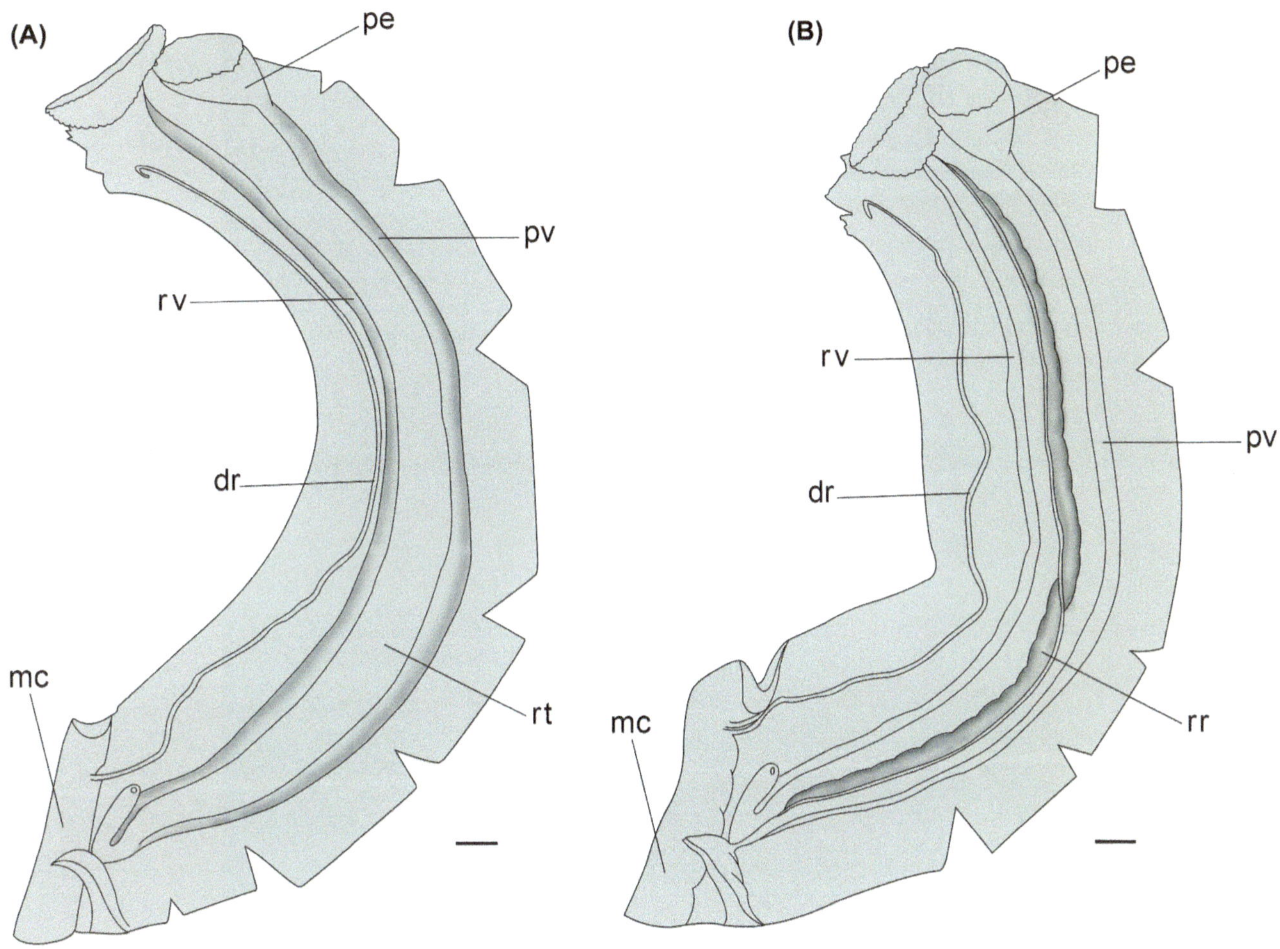

FIGURE 11.44 Planorbid renal region of (A) *Biomphalaria* spp. and (B) *Biomphalaria glabrata*. Abbreviations: *dr*, dorsolateral ridge; *mc*, mantle collar; *pe*, pericardium; *pv*, pulmonary vein; *rr*, renal ridge; *rt*, renal tube; *rv*, renal vein. Scale bar = 1 mm. *Modified from Paraense (1975).*

5(2) Whorls increasing rapidly (larger outer whorl); aperture rounded and expanded (Fig. 11.42 B) .... *Antillorbis*
[Dominican Republic, Puerto Rico, Guatemala, Brazil, Argentina, Uruguay]

5′ Whorls increase proportionally, aperture rounded or elongated .... ***Drepanotrema* [p. 325]**

Gastropoda: Heterobranchia: Planorbioidea: Planorbidae: Planorbinae: *Plesiophysa*: Species

1 Flagella length < penial complex (penis sheath plus prepuce) (Fig. 11.42 E, 43 A) .... *Plesiophysa guadeloupensis* (Fischer in Mazé, 1883)
[Guadeloupe, Dominican Republic, Brazil]

1′ Flagella length ≥ penial complex (Figs. 11.42 F, 43 B) .... *Plesiophysa dolichomastix* Paraense, 2002
[Brazil]

Gastropoda: Heterobranchia: Planorbioidea: Planorbidae: Planorbinae: *Helisoma*: Species

1 Penis sheath pyriform with a dorsal globular and a basal portion narrower continuous with prepuce (Figs. 11.42 C, 43 C) .... *Helisoma duryi* (Wetherby, 1879)
[Central America. Nearctic]

1′ Penis sheath is nearly cylindrical, with a somewhat wider proximal portion; prepuce approximately fusiform, proximal portion bulging gradually narrowing distally (Figs. 11.42 D, 43 D) .... *Helisoma trivolvis* (Say, 1817)
[Cuba, Mexico, Guatemala, Belize, Costa Rica, El Salvador, Nicaragua, Ecuador, Brazil, Peru. Nearctic]

Gastropoda: Heterobranchia: Planorbioidea: Planorbidae: Planorbinae: *Biomphalaria* Species and Subspecies of Brazil

This key is only for material collected in Brazil. The taxonomy of species from other regions requires morpho-molecular revisions and the identification must be verified with the descriptions. In the case of species outside of Brazil, identification should only go to the genus level. Table 11.1 gives the list of species and the country where they have been recorded

1 Renal tube ventrally smooth or unpigmented (Fig. 11.44 A) 2

1′ Renal tube ventrally with a renal ridge (pigmented longitudinal ridge) (Fig. 11.44 B) or a pigmented line in young specimens (diameter < 10 mm); vaginal pouch present (Figs. 11.44 B, 45 A) .... *Biomphalaria glabrata* (Say, 1818) [Haiti, Puerto Rico, Antigua and Barbuda, Guadeloupe, Dominican Republic, Martinique, Saint Lucia, Venezuela, Guyana, French Guiana, Suriname. Brazil. Nearctic]

TABLE 11.1 List of *Biomphalaria* species and subspecies from Neotropical region, with type, locality, and occurrence.

Species	Figure	Distribution
Biomphalaria amazonica Paraense, 1966	11.45 K, 47 C	TL: Brazil: Igarapé da Cachoeirinha, Amazonas. O: Bolivia. Brazil. Colombia.
Biomphalaria andecola (d'Orbigny, 1935)	11.49 A	TL: Bolivia: Isla del Sol, Lake Titicaca O: Bolivia. Chile. Peru.
Biomphalaria chilensis (Anton, 1839)		TL: Chile: Maypo. O: Chile.
Biomphalaria cousini Paraense, 1966	11.45 L, 47 D	TL: Ecuador: Santo Domingo de Los Colorados, Pichincha. O: Brazil. Ecuador
Biomphalaria edisoni Estrada, Velásquez, Caldeira, Bejarano, Rojas & Carvalho, 2006		TL: Colombia: Cali. O: Colombia.
Biomphalaria equatoria (Cousin, 1887)		TL: Ecuador: Guayaquil. O: Ecuador.
Biomphalaria glabrata (Say, 1818)	11.44 B, 45 A	TL: Guadeloupe. O: Antigua and Barbuda. Brazil. Dominican Republic Guadeloupe. Guyana. French Guiana. Haiti. Martinique. Puerto Rico. Saint Lucia. Suriname. Venezuela.
Biomphalaria havanensis (Pfeiffer, 1839)	11.49 B	TL: Cuba: Havana (?) O: Cuba. Dominican Republic. Haiti. Puerto Rico.
Biomphalaria helophila (d'Orbigny, 1835)	11.49 C	TL: Peru: Callao. O: Barbados. Belize. Costa Rica. Cuba. Dominican Republic. El Salvador. Guadeloupe. Guatemala. Haiti. Jamaica. Nicaragua. Peru. Puerto Rico.
Biomphalaria intermedia Paraense & Deslandes 1962	11.45 F, 48 B	TL: Brazil: Valparaiso, São Paulo. O: Argentina. Brazil. Paraguay.
Biomphalaria kuhniana (Clessin, 1883)	11.45 H, 48 D	TL: Suriname, Paramaribo. O: Brazil. Colombia. Dominican Republic. Guadeloupe. Martinique. Panama. Suriname. Venezuela.
Biomphalaria nicaraguana (Morelet, 1851)		TL: Nicaragua: Lake Nicaragua. O: Nicaragua.
Biomphalaria obstructa (Morelet, 1849)	11.49 D	TL: Mexico: Isla del Carmen, Campeche. O: Belize. Cuba. El Salvador. Guatemala. Haiti. Mexico. Nicaragua. Puerto Rico. United States of America.

TABLE 11.1 List of *Biomphalaria* species and subspecies from Neotropical region, with type, locality, and occurrence.—cont'd

Species	Figure	Distribution
Biomphalaria occidentalis Paraense, 1981	11.45 D, 46 C	TL: Brazil: Campo Grande, Mato Grosso do Sul. O: Argentina. Brazil. Paraguay.
Biomphalaria oligoza Paraense, 1975	11.45 I, 47 B	TL: Brazil: Morretes, Paraná. O: Argentina. Brazil. Colombia.
Biomphalaria orbignyi Paraense, 1975	11.49 E	TL: Argentina: Paraná, Entre Ríos. O: Argentina.
Biomphalaria peregrina (d'Orbigny, 1835)	11.45 J, 48 C	TL: Argentina: Negro river, Patagonia. O: Argentina. Bolivia. Brazil. Chile. Colombia. Ecuador. Paraguay. Peru. Puerto Rico. Uruguay. Venezuela.
Biomphalaria prona (Martens, 1873)	11.49 F	TL: Venezuela: Lake of Valencia O: Venezuela.
Biomphalaria schrammi (Crosse, 1864)	11.45 E, 47 A	TL: Guadeloupe: Pointe-à-Pitre. O: Brazil. Guadeloupe. Guyana. Venezuela.
Biomphalaria sericea (Dunker, 1848)	11.49 G	TL: (?) O: Ecuador.
Biomphalaria straminea (Dunker, 1848)	11.45 G, 48 A	TL: Venezuela: Lagunillas and Caracas. O: Argentina. Brazil. Colombia. Costa Rica. Dominican Republic. El Salvador. Guadeloupe. Guyana. Haiti. Martinique. Paraguay. Suriname. Trinidad and Tobago. Uruguay. Venezuela.
Biomphalaria subprona (Martens, 1899)	11.49 H	TL: Mexico: Amatitlán, Tabasco. O: Guatemala. Mexico.
Biomphalaria riisei (Dunker in Clessin, 1883)		TL: Puerto Rico. O: Antillean Islands. Puerto Rico.
Biomphalaria temascalensis Rangel-Ruiz, 1987		TL: Mexico: Temascal, Oaxaca. O: Mexico.
Biomphalaria tenagophila tenagophila (d'Orbigny, 1835)	11.45 C, 46 B	TL: Argentina: Corrientes. O: Argentina. Brazil. Paraguay. Peru. Uruguay.
Biomphalaria tenagophila guaibensis Paraense, 1984	11.45 B, 46 A	TL: Brazil: Guaíba, Rio Grande do Sul. O: Brazil. Uruguay.
Biomphalaria trigyra (Philippi, 1869)	11.49 I	TL: Peru: Pimentel. O: Ecuador. Peru.

O, Occurrences listed in alphabetic order; *TL*, Type locality. Species in bold are those included in the keys.

2(1) Shell with carinae on both sides 3

2′ Shell with no carinae on both sides 5

3(2) Penial sheath and prepuce of similar lengths 4

3′ Prepuce length 2× penis sheath in most populations; vaginal pouch on ventral wall (Figs. 11.45 B, 46 A) .... *Biomphalaria tenagophila guaibensis* Paraense, 1984 [Brazil, Uruguay]

4(3) Prepuce width ≥ penial sheath, width increasing to the distal region; vaginal pouch present (Figs. 11.45 C, 46 B) .... *Biomphalaria tenagophila tenagophila* (d'Orbigny, 1835) [Brazil, Peru, Paraguay, Argentina, Uruguay]

4′ Prepuce keeps about the same width; vaginal pouch absent (Figs. 11.45 D, 46 C) .... *Biomphalaria occidentalis* Paraense, 1981 [Brazil, Paraguay, Argentina]

FIGURE 11.45 Shell of *Biomphalaria* (Planorbinae): (A) *Biomphalaria glabrata*; (B) *B. tenagophila guaibensis*; (C) *B. tenagophila tenagophila*; (D) *B. occidentalis*; (E) *B. schrammi*; (F) *B. intermedia*, (G) *B. straminea*, (H) *B. kuhniana*, (I) *B. oligoza*, (J) *B. peregrina*, (K) *B. amazonica*; and (L) *B. cousini*. Scale bar = 5 mm.

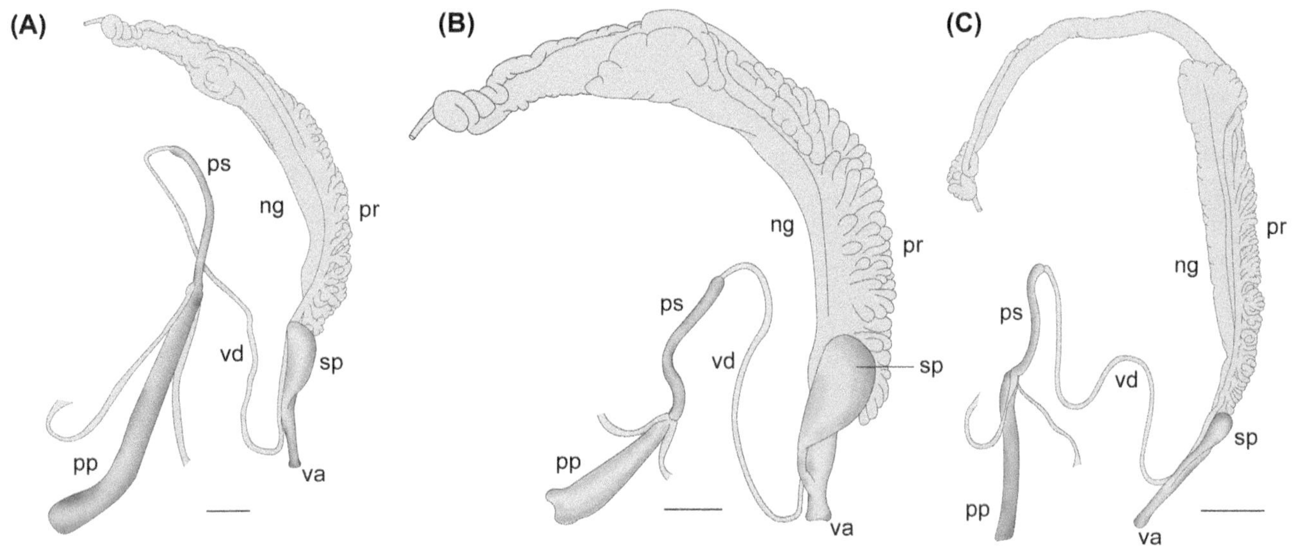

FIGURE 11.46 *Biomphalaria* (Planorbinae) reproductive system (male and female): (A) *Biomphalaria tenagophila guaibensis*, (B) *B. tenagophila tenagophila*; and (C) *B. occidentalis*. Abbreviations: *ng*, nidamental gland; *pp*, prepuce; *pr*, prostate; *ps*, penis sheath; *sp*, spermatheca; *va*, vagina; *vd*, vas deferens. Scale bar = 1 mm. *Modified from Paraense (1981, 1984).*

5(2) Ratio of penial sheath and prepuce < three 6

5′ Ratio of penial sheath and prepuce > three; the last whorl, internally, shows the presence of lamellae (Figs. 11.45 E, 47 A) .... *Biomphalaria schrammi* (Crosse, 1864)
[Guadeloupe, Venezuela, Guyana, Brazil]

6(5) Vagina with vaginal pouch or lateral corrugations, sometimes poorly developed 7

6′ Vagina with vaginal pouch and lateral corrugations, sometimes poorly developed; region of the vagina with morphological variations in the degree of corrugation and pouch (Figs. 11.45 F, 48 B) .... *Biomphalaria intermedia* Paraense & Deslandes, 1962
[Brazil, Paraguay, Argentina]

7(6) Vagina with a series of lateral corrugations above the spermatheca 8

7′ Vagina with vaginal pouch, sometimes poorly developed 9

8(7) Vagina robust lateral corrugation; penis middle cross-section with three muscle layers (Figs. 11.45 G, 48 A) .... *Biomphalaria straminea* (Dunker, 1848)
[Haiti, Dominican Republic, El Salvador, Guadeloupe, Martinique, Costa Rica, Venezuela, Trinidad and Tobago, Colombia, Guyana, Suriname, Brazil, Paraguay, Argentina, Uruguay]

8′ Vaginal corrugation weak, in some specimens with a gradation of increasing wrinkles; penis middle cross-section with two muscle layers; prostate with four to nine diverticula, shorter and less branched (Figs. 11.45 H, 48 D) .... *Biomphalaria kuhniana* (Clessin, 1883)
[Dominican Republic, Guadeloupe, Martinique, Panama, Colombia, Venezuela, Suriname, Brazil]

9(7) Vaginal pouch well-developed and prostate diverticula little interspaced 10

9′ Vaginal pouch poorly developed, difficult to recognize; few prostate diverticula (0–7) and much interspaced (Figs. 11.45 I, 47 B) .... *Biomphalaria oligoza* Paraense, 1975
[Colombia, Brazil, Argentina]

10(9) Penis sheath length < prepuce; prostate with 5–12 diverticula arranged in a row 11

10′ Penis sheath and prepuce show similar lengths; prostate with ~ 8 to 22 diverticula, the prostate almost always overlying spermatheca top (Figs. 11.45 J, 48 C) .... *Biomphalaria peregrina* (d'Orbigny, 1835)
[Puerto Rico, Venezuela, Colombia, Ecuador, Brazil, Peru, Bolivia, Chile, Paraguay, Argentina, Uruguay]

11(10) Vas deferens widest portion with width equal to penis sheath; vagina is short, cylindrical (Figs. 11.45 K, 47 C) .... *Biomphalaria amazonica* Paraense, 1966
[Colombia, Brazil, Bolivia]

11′ Vas deferens widest point with width < penis sheath; vagina shows a swollen cephalic portion, separated from the caudal portion by a more or less marked constriction (Figs. 11.45 L, 47 D) .... *Biomphalaria cousini* Paraense, 1966
[Ecuador, Brazil]

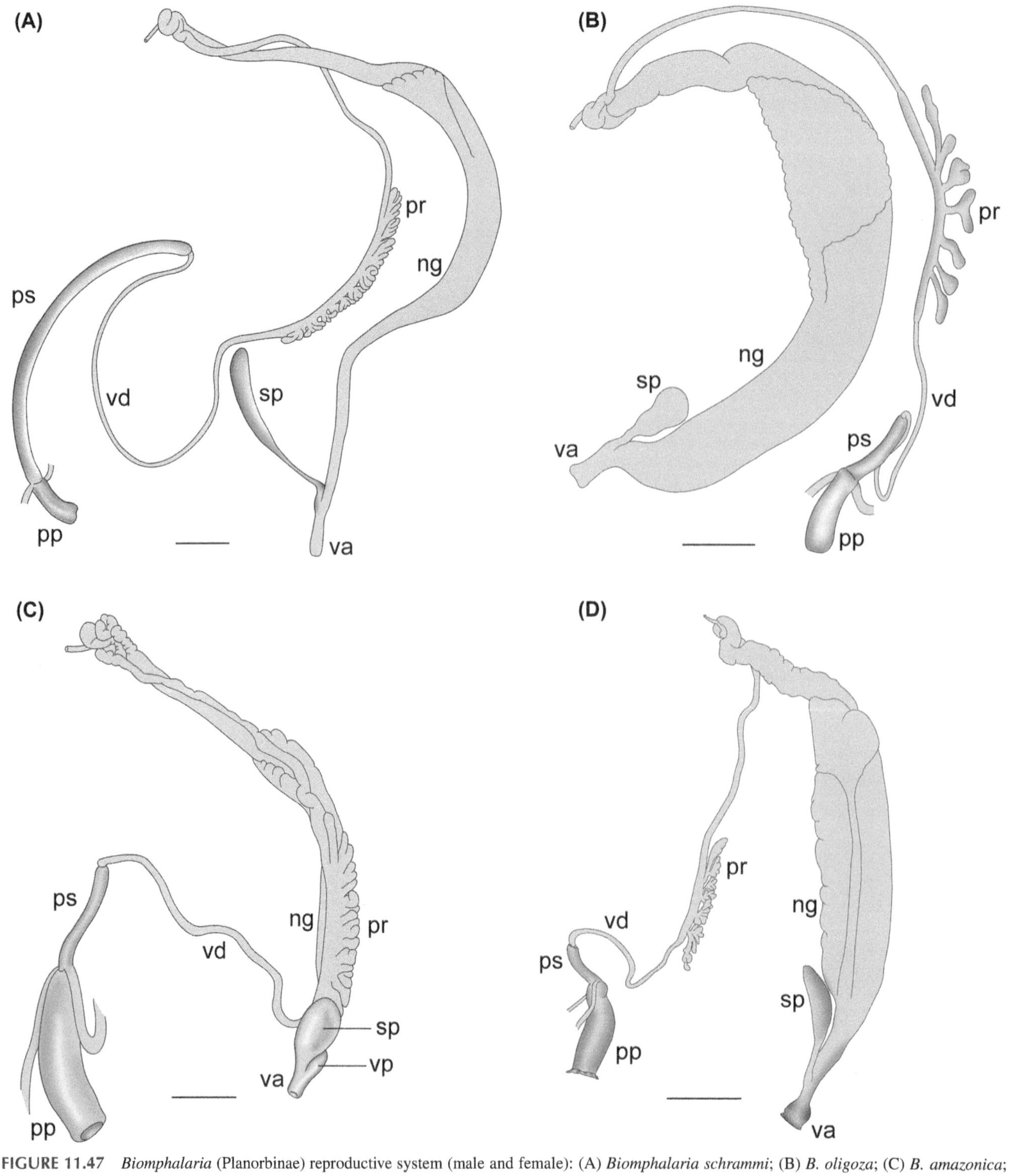

FIGURE 11.47 *Biomphalaria* (Planorbinae) reproductive system (male and female): (A) *Biomphalaria schrammi*; (B) *B. oligoza*; (C) *B. amazonica*; and (D) *B. cousini*. Abbreviations: *ng*, nidamental gland; *pp*, prepuce; *pr*, prostate; *ps*, penis sheath; *sp*, spermatheca; *va*, vagina; *vd*, vas deferens; *vp*, vaginal pouch. Scale bar = 1 mm. *Modified from Paraense (1966, 1975).*

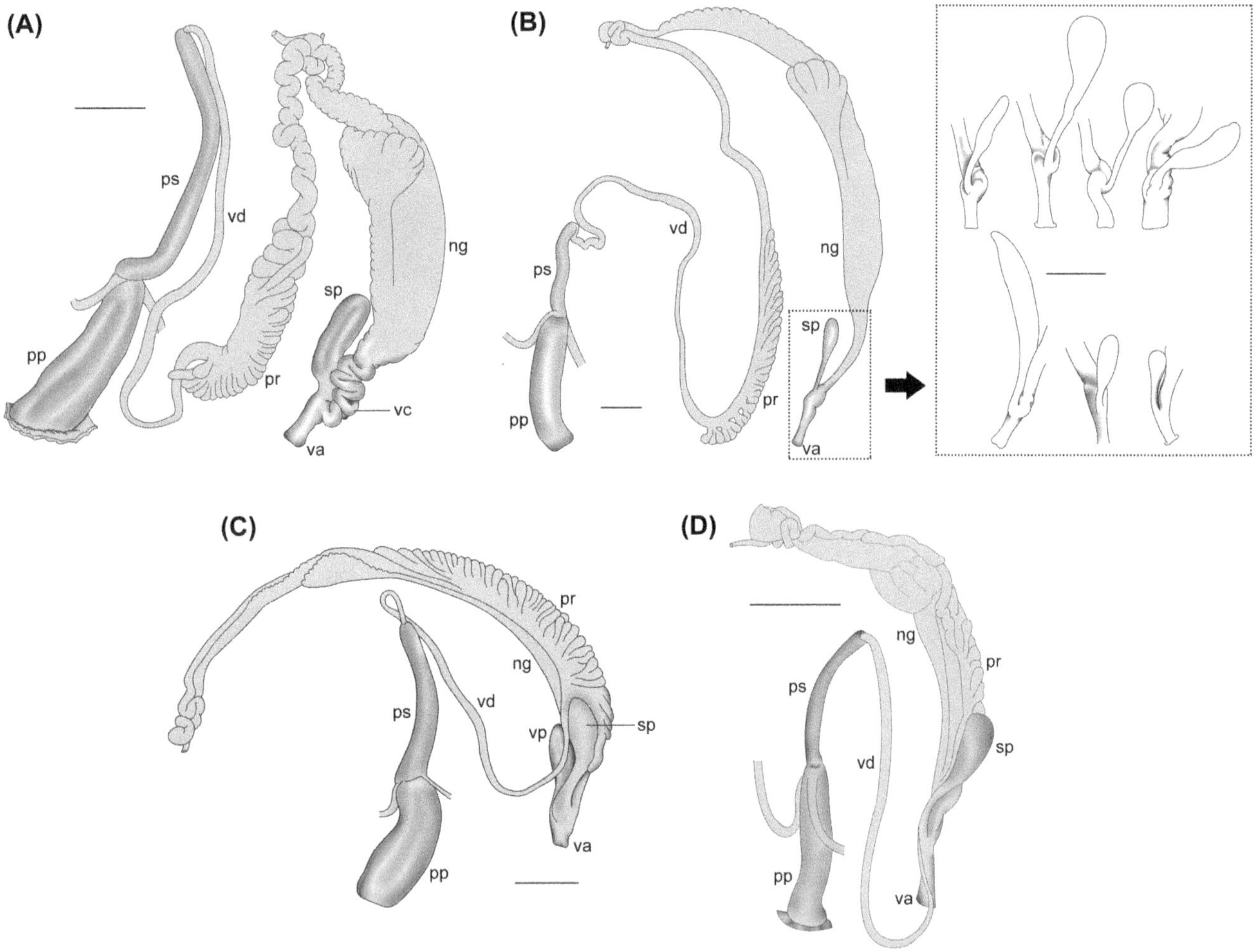

FIGURE 11.48 *Biomphalaria* (Planorbinae) reproductive system (male and female) of (A) *Biomphalaria straminea*; (B) *B. intermedia* and variation of diagnostic morphological characteristics; (C) *B. peregrina*; and (D) *B. kuhniana*. Abbreviations: *ng*, nidamental gland; *pp*, prepuce; *pr*, prostate; *ps*, penis sheath; *sp*, spermatheca; *va*, vagina; *vc*, vaginal corrugations; *vd*, vas deferens; *vp*, vaginal pouch. Scale bar = 1 mm. *Modified from Paraense (1975, 1988) and Paraense & Deslandes (1962).*

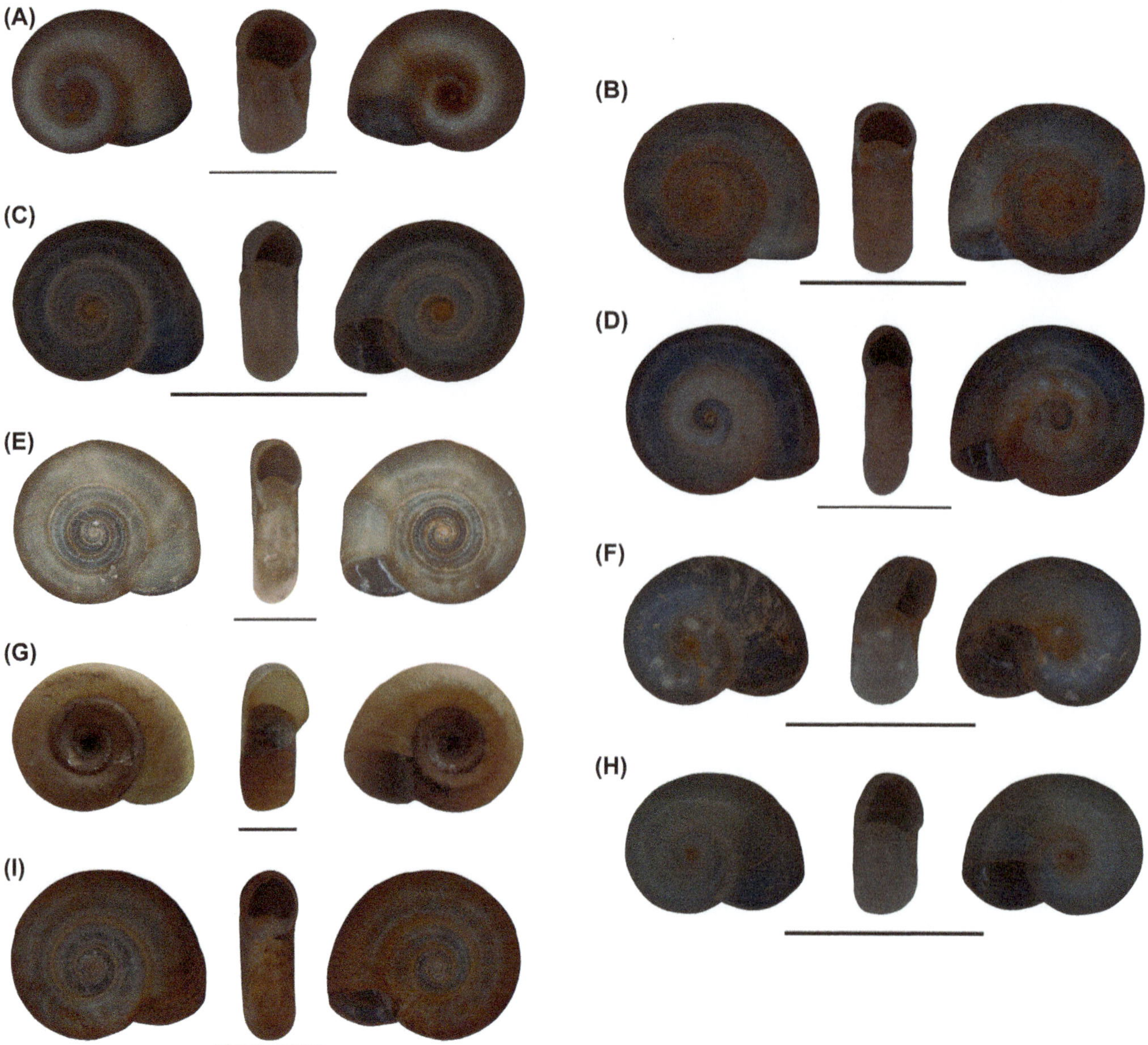

FIGURE 11.49 Shells of species of *Biomphalaria* (Planorbinae): (A) *Biomphalaria andecola*; (B) *B. havanensis*; (C) *B. helophila*; (D) *B. obstructa*; (E) *B. orbignyi*; (F) *B. prona*, (G) *B. sericea*, (H) *B. subprona*; and (I) *B. trigyra*. Scale bar = 5 mm.

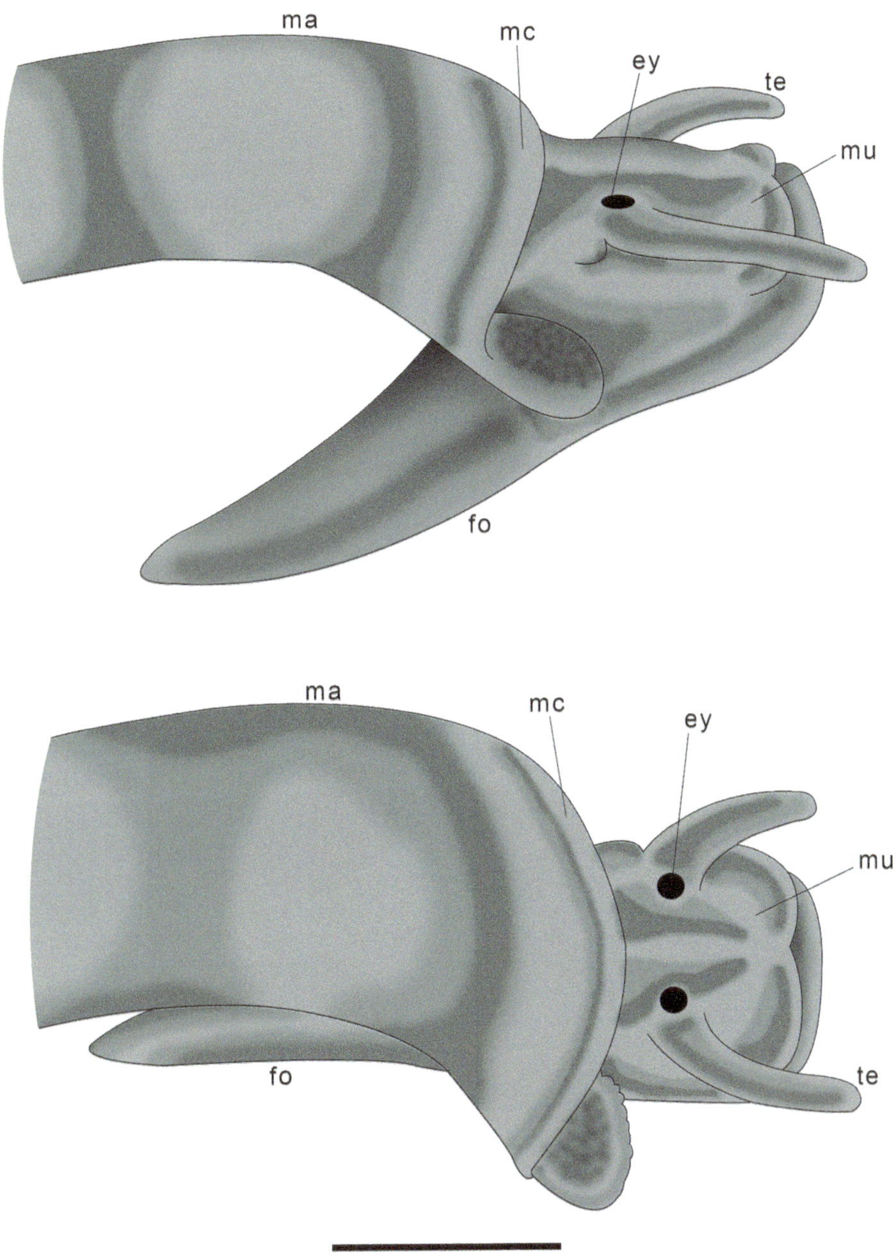

FIGURE 11.50 Cephalopodal region of species of *Drepanotrema* (Planorbinae). Abbreviations: *ey*, eyes; *fo*, foot; *ma*, mantle; *mc*, mantle collar; *mu*, muffle; *te*, tentacles. Scale bar = 1 mm. *Modified from Paraense (1975).*

Gastropoda: Heterobranchia: Planorbioidea: Planorbidae: Planorbinae: *Drepanotrema*: Species from Brazil

This key is only for material collected in Brazil. In the case of species outside of Brazil, identification should only go to the genus level. Only two species are not found in Brazil: *Drepanotrema limayanum* (Lesson, 1830) (Fig. 11.51 E) from Peru, and *D. surinamense* (Clessin, 1884) (Fig. 11.51 D) from Costa Rica, Ecuador, Guadeloupe, Guyana, Panama and Suriname; both species are in need of revision.

1 Cephalopodal region with pigment strips (Fig. 11.50), disposed as follows: a dorsal pair on the neck and head which pass tangentially to the medial sides of the eyes; a lateral pair which duplicates the dorsal one from the eyes to the cephalic end; a pair which borders the muffles; and a pair along the margins of the foot; tentacle dark pigmentation is concentrated in an axial line 2

1′ Cephalopodal region without a pigmentation pattern as described above 6

2(1) Shell > 6 mm length, depressed; the whorls increase regularly 3

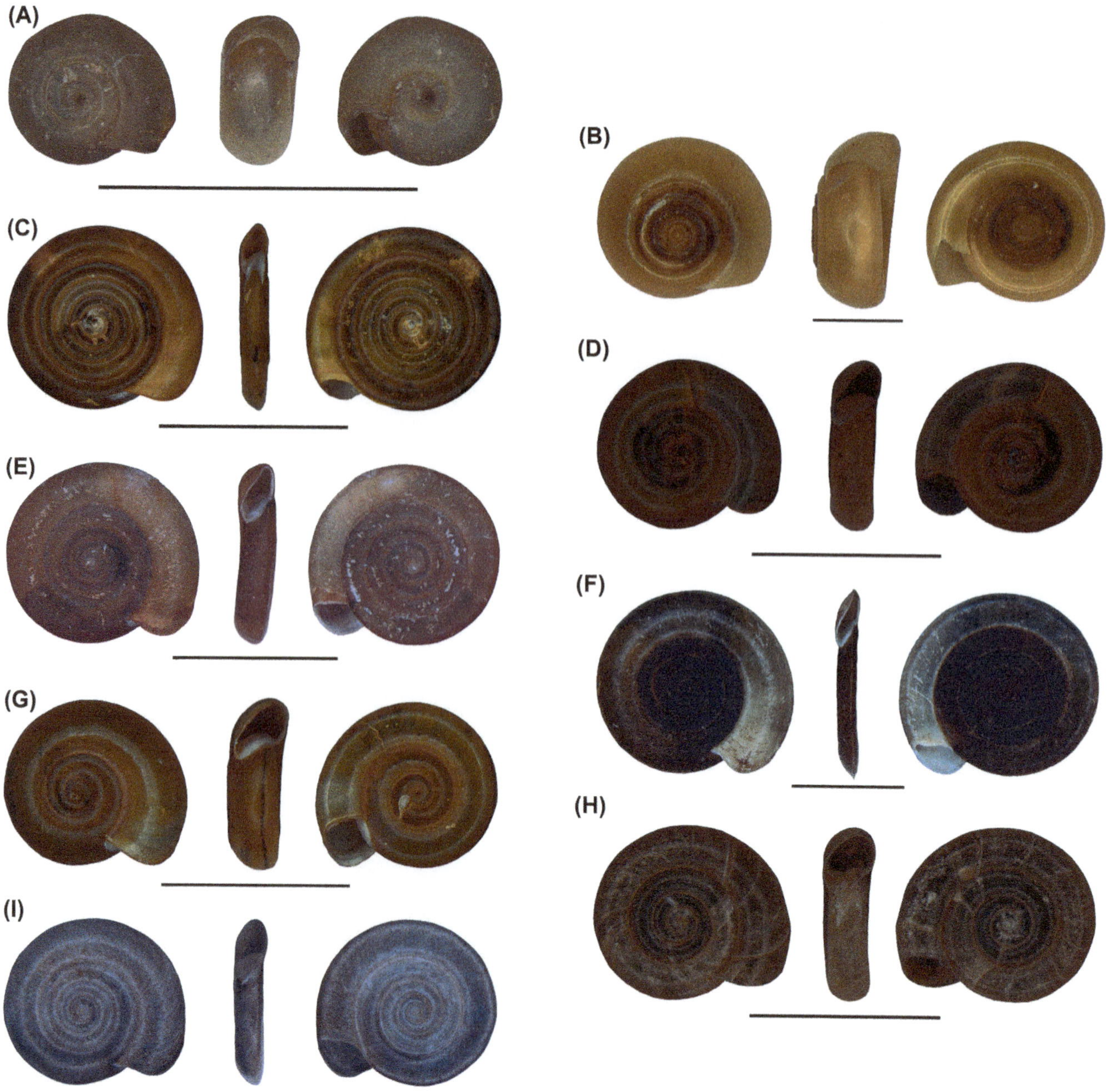

FIGURE 11.51 Shells of species of *Drepanotrema* (Planorbinae): (A) *D. anatinum*; (B) *D. pileatum*; (C) *D. cimex*; (D) *D. surinamense*; (E) *D. limayanum*; (F) *D. depressissimum*; (G) *D. lucidum*; (H) *D. heloicum*; and (I) *D. kermatoides*. Scale bar = 5 mm.

2′ Shell < 6 mm length, subglobose; body whorl exceedingly predominates over the remaining whorls (Fig. 11.51 A) .... *Drepanotrema anatinum* (d'Orbigny, 1835)
[Mexico, Antilles, Cuba, Haiti, Dominican Republic, Puerto Rico, Jamaica, Saint Lucia, Guatemala, Belize, Nicaragua, Costa Rica, Venezuela, Trinidad and Tobago, Guyana, Suriname, Ecuador, Brazil, Argentina, Uruguay]

3(2) Flagella length < penis sheath 4

3′ Flagella > 1.5× penis sheath 5

4(3) Shell flattened, shallowly concave on both sides; prostate diverticula (varies from 17 to 26) short (Fig. 11.51 C) .... *Drepanotrema cimex* (Moricand, 1837)
[Mexico, Antilles, Cuba, Haiti, Puerto Rico, Jamaica, Venezuela, Brazil, Peru, Argentina, Uruguay]

4′ Shell right side convex, left side deeply concave; prostate diverticula (about 100) long (Fig. 11.51 B) .... *Drepanotrema pileatum* Paraense, 1971

[Guadeloupe, Costa Rica. Panama, Guyana, Suriname, Ecuador, Brazil]

5(3) Shell carinate; shell aperture oblique and very narrow; prostate diverticula (29–52) are long (Fig. 11.51 F) .... *Drepanotrema depressissimum* (Moricand, 1839)

[Antigua and Barbuda, Barbados, Costa Rica, Guadeloupe, Nicaragua, Panama, Saint Lucia, Venezuela, Brazil, Peru, Paraguay, Argentina, Uruguay]

5′ Shell not carinate; shell aperture crescentiform; prostate diverticula (13–29) short (Fig. 11.51 G) .... *Drepanotrema lucidum* (Pfeiffer, 1839)

[Mexico, Haiti, Jamaica, Dominican Republic, Puerto Rico, Guatemala, Belize, Guadeloupe, Nicaragua, Antigua and Barbuda, Saint Lucia, Barbados, Venezuela, Saint Vincent, Trinidad and Tobago, Ecuador, Brazil, Paraguay, Argentina, Uruguay]

6(1) Aperture rounded (Fig. 11.51 H); flagellae (two) length 0.6–4× as penis sheath .... *Drepanotrema heloicum* (d'Orbigny, 1835)

[Brazil, Argentina, Uruguay]

6′ Aperture oblique, very narrow (Fig. 11.51 I); flagellae (two or one) very short, sometimes rudimentary .... *Drenanotrema kermatoides* (d'Orbigny, 1835)

[Puerto Rico, Ecuador, Brazil, Peru, Paraguay, Argentina, Uruguay]

ACKNOWLEDGMENTS

To Mr Eduardo da Silva Cinilha for the illustrations and Dr. Wladimir Lobato Paraense (in memoriam) for the teachings in malacology.

REFERENCES

Albrecht C., O. Kroll, E. Moreno Terrazas & T. Wilke. 2009. Invasion of ancient Lake Titicaca by the globally invasive *Physa acuta* (Gastropoda: Pulmonata: Hygrophila). Biological Invasions 11:1821–1826.

Alda, P., M. Lounnas, A.A. Vázquez, R. Ayaqui, M. Calvopiña, M. Celi-Erazo, R.T. Dillon Jr, P. Jarne, E.S. Loker, F.C. Muñiz Pareja, J. Muzzio-Aroca, A.O. Nárvaez, O. Noya, L. Martini Robles, R. Rodríguez-Hidalgo, N. Uribe, P. David, J-P. Pointier & S. Hurtrez-Boussès. 2018. A new multiplex PCR assay to distinguish among three cryptic *Galba* species, intermediate hosts of *Fasciola hepatica*. Veterinary Parasitology 15: 101–105.

Arruda, J.O. 2011. Revisão taxonômica, sistemática e análise cladística de Omalonyx d'Orbigny, 1837 (Mollusca, Gastropoda, Succineidae). Ph. D. Thesis, Programa de Pós Graduação em Zoologia, Faculdade de Biociências, Pontifícia Universidade Católica do Rio Grande do Sul, Porto Alegre, Brazil. 114 pp.

Arruda, J.O., S.R. Gomes, R. Ramírez & J.W. Thomé. 2006. Morfoanatomia de duas espécies do gênero *Omalonyx* (Mollusca, Pulmonata, Succineidae) com novo registro para Minas Gerais, Brasil. Biociências 14: 61–70.

Arruda, J.O. & J.W. Thomé. 2008a. Revalidation of *Omalonyx convexus* (Heynemann, 1868) and emendation of the type locality of *Omalonyx unguis* (Orbigny, 1837). Archiv für Molluskenkunde 137: 159–166.

Arruda, J.O. & J.W. Thomé. 2008b. Synonymization of *Neohyalimax* Simroth, 1896, and *Omalonyx* d'Orbigny, 1837, with a redescription of *Omalonyx brasiliensis* (Simroth, 1896) (Gastropoda, Succineidae). The Nautilus 122: 94–98.

Arruda, J.O. & J.W. Thomé. 2011. Biological aspects of *Omalonyx convexus* (Mollusca, Gastropoda, Succineidae) from the Rio Grande do Sul State, Brazil. Biotemas 24: 95–101.

Baker, H.B. 1955. Heterurethrous and aulacopod. The Nautilus 68:109-112.

Barker, G.M. 2001a. The biology of terrestrial Molluscs. CABI Publishing, Wallingford.

Barker, G.M. 2001b. Chapter 1: Gastropods on land: phylogeny, diversity and adaptive morphology. Pages 1–146 *in*: G.M. Barker (ed.), The biology of terrestrial mollusks, CABI Publishing, Wallingford.

Brown, D.S. & N.B. Pullan. 1987. Notes on shell, radula and habitat of *Chilina* (Basommatophora) from the Fakland Island. Journal of Molluscan Studies 53: 105–108.

Brown, K.M. 1991. Mollusca Gastropoda. Pages 285–314, *in*: K.M. Brown & C. Lydeard (eds.), Ecology and Classification of North American Freshwater Invertebrates. Academic Press. 911 pp.

Burch, J.B. 1982. North American freshwater snails. Walkerana 1: 217–365.

Burch, J.B. & C.M. Patterson. 1969. The systematic position of the Athoracophoridae (Gastropoda: Euthyneura). Malacologia 9: 259–260.

Caldeira, R.L., L.K. Jannotti-Passos & O.S. Carvalho. 2009. Molecular epidemiology of Brazilian *Biomphalaria*: A review of the identification of species and the detection of infected snails. Acta Tropica 111: 1–6.

Castellanos, Z.A. de & M.C. Gaillard. 1981. Fauna de agua dulce de la República Argentina, 15(4). Mollusca Gasterópoda: Chilinidae. PROFADU (CONICET), Buenos Aires. 32 pp.

Castellanos, Z.A. de & S. Miquel. 1991. Fauna de agua dulce de la República Argentina, 15(9). Distribución de los Pulmonata Basommatophora. PROFADU (CONICET), Buenos Aires. 11 pp.

Correa, A.C., J.S. Escobar, P. Durand, F. Renaud, P. David, P. Jarne, J.P. Pointier & S. Hurtrez-Boussès. 2010. Bridging gaps in the molecular phylogeny of the Lymnaeidae (Gastropoda: Pulmonata), vectors of Fascioliasis. BMC Evolutionary Biology 10: 381.

Correa, A.C., J.S. Escobar, O. Noya, L.E. Velásquez, C. González-Ramírez, S. Hurtrez-Boussès & J.P. Pointier. 2011. Morphological and molecular characterization of Neotropic Lymnaeidae (Gastropoda: Lymnaeoidea), vectors of fasciolosis. Infection Genetics and Evolution 11: 1978–1988.

Cuezzo, M.G. 2009. Mollusca Gastropoda. Pages 595–654 *in*: E. Domínguez & H. Fernández, (eds), Macroinvertebrados bentónicos sudamericanos, sistemática y biología. Fundación Miguel Lillo, Tucumán, Argentina.

PHYLUM MOLLUSCA

Dayrat, B., M. Conrad, S. Balayan, T.R. White, C. Albrecht, R. Golding, S.R. Gomes, M.G. Harasewych & A.M. de Frias Martins. 2011. Phylogenetic relationships and evolution of pulmonate gastropods (Mollusca): New insights from increased taxon sampling. Molecular Phylogenetics and Evolution 59: 425–437.

Dinapoli, A. & A. Klussmann-Kolb. 2010. The long way to diversity – Phylogeny and evolution of the Heterobranchia (Mollusca: Gastropoda). Molecular Phylogenetics and Evolution 55: 60–76.

Fernandez, M.A., S.B. Santos, I.C. Miyahira, I.C.B. Gonçalves, R.F. Ximenes & S.C. Thiengo. 2012. Gastrópodes límnicos invasores: morfologia comparada. Pages 125–136, *in*: M.C.D. Mansur, C.P. Santos, D. Pereira, I.C.P. Paz, M.L.L. Zurita, M.T.R. Rodriguez, M.V. Nehrke & P.E.A. Bergonci (Orgs.), Moluscos límnicos invasores no Brasil: biologia, prevenção e controle. Redes Editora, Porto Alegre, Brasil.

Fernandez, M.A., S.C. Thiengo & R.S. Amaral. 2008. Técnicas malacológicas. Pages 34–70 *in*: R.S. Amaral, S.C. Thiengo & O.S. Pieri (orgs.), Vigilância e Controle de Moluscos de Importância Médica: Diretrizes Técnicas. MS, Secretaria de Vigilância em Saúde, Editora do Ministério da Saúde, Brasília.

Franzen, D.S. 1966. Anatomy of the succineid gastropod *Oxyloma salleana* (Pfeiffer). The Nautilus 80: 59–69.

Golding, R. 2012. Molecular phylogenetic analysis of mudflat snails (Gastropoda: Euthyneura: Amphiboloidea) supports an Australasian centre of origin. Molecular Phylogenetics and Evolution 63: 72–81.

Gutiérrez Gregoric, D.E. 2008. Estudios morfoanatómicos y tendencias poblacionales en especies de la familia Chilinidae Dall, 1870 (Mollusca: Gastropoda) en la Cuenca del Plata. Ph. D. Thesis Facultad de Ciencias Naturales y Museo, Universidad Nacional de La Plata, Argentina. 174 pp.

Gutiérrez Gregoric, D.E. 2010. Redescription of two endemic species of Chilinidae (Gastropoda: Basommatophora) from Del Plata Basin (South America). Journal of Conchology 40: 321–332.

Gutiérrez Gregoric, D.E., N.F. Ciocco, & A. Rumi. 2014. Two new species of *Chilina* Gray from Cuyo Malacological Province, Argentina. (Gastropoda: Hygrophila: Chilinidae). Molluscan Research 34: 84–97.

Gutiérrez Gregoric, D.E., & M. de Lucía. 2016. Freshwater gastropods diversity hotspots: three new species from the Uruguay River (South America). PeerJ 4: e2138.

Gutiérrez Gregoric, D.E, & A. Rumi. 2008. *Chilina iguazuensis* (Gastropoda: Chilinidae), new species from Iguazú National Park, Argentina. Malacologia 50: 321–330.

Haeckel, W. 1911. Beiträge zur Anatomine der Gattung *Chilina*. Zoologische Jahrbücher 13: 89–136.

Harry, W.H. 1964. The anatomy of *Chilina fluctuosa* Gray reexamined, with prolegomena on the phylogeny of the higher limnic Basommatophora (Gastropoda: Pulmonata). Malacologia 1: 355–385.

Haszprunar, G. 1988. On the origin and evolution of major gastropod groups, with special reference to the Streptoneura. Journal of Molluscan Studies 54: 367–441.

Haszprunar, G. & G. Huber. 1990. On the central nervous system of Smeagolidae and Rhodopidae, two families of questionably allied with the Gymnomorpha. Journal of Zoology, London 220: 185–199.

Healy, J.M. 1995. Molluscan sperm ultrastructure: correlation with taxonomic units within Gastropoda, Cephalopoda and Bivalvia. Pages 99–113 *in:* J.D. Taylor (ed.) Origin and evolutionary radiation of the Mollusca. Oxford University Press, Oxford.

Holznagel, W. E. 1998. A nondestructive method for cleaning gastropod radulae from frozen, alcohol-fixed, or dried material. American Malacological Bulletin 14: 181–183.

Holznagel, W.E., D.J. Colgan & C. Lydeard. 2010. Pulmonate phylogeny based on 28S rRNA gene sequences: A framework for discussing habitat transitions and character transformation. Molecular Phylogenetics and Evolution 57: 1017–1025.

Hubendick, B. 1951. Recent Lymnaeidae, their variation, morphology, taxonomy, nomenclature and distribution. Kungl Svenska Vetenskapsacademiens Handlingar 3: 1–223.

Ituarte, C.F. 1997. *Chilina megastoma* Hylton Scott, 1958 (Pulmonata: Basommatophora): A study of topotypic specimens. American Malacological Bulletin 14: 9–15.

Jabbour-Zahab R., J.P. Pointier, J. Jourdane, P. Jarne, J. Oviedo, M.D. Bargues, S. Mas-Coma, R. Anglés, G. Perera, C. Balzan, K. Khallayoune & F. Renaud. 1997. Phylogeography and genetic divergence of some lymnaeid snails, intermediate hosts of human and animal fascioliasis with special reference to lymnaeids from the Bolivian Altiplano. Acta Tropica 64: 191–203.

Jarne P., J.P. Pointier & P. David. 2011. Biosystematics of *Biomphalaria* spp. with an emphasis on *Biomphalaria glabrata*. Pages 1–32 *in*: R. Toledo & B. Fried (eds.) *Biomphalaria* snails and larval Trematodes. Springer.

Jörger, K.M., I. Stöger, Y. Kano, H. Fukuda, T. Knebelsberger & M. Schrödl. 2010. On the origin of Acochlidia and other enigmatic euthyneuran gastropods, with implications for the systematics of Heterobranchia. BMC Evolutionary Biology 10: 323–343.

Klussmann-Kolb, A., A. Dinapolil, K. Kuhn, B. Streit & C. Albrecht. 2008. From sea to land and beyond. New insights into the evolution of euthyneuran Gastropoda (Mollusca). BMC Evolutionary Biology 8: 57–73.

Lacerda, E.L.M. de & S.B. Santos. 2011. *Burnupia ingae* Lanzer, 1991 (Gastropoda: Ancylidae): current distribution in Brazil. Check List 7: 862–864.

Lacerda, L.E.M., C.S. Richau, C.R.L. Amaral, D.A. da Silva, E.F. Carvalho & S.B. Santos. 2015. *Ferrissia fragilis* (Tryon, 1863): a freshwater snail cryptic invader in Brazil revealed by morphological and molecular data. Aquatic Invasions 10: 157–168.

Lanzer, R.M. 1996. Ancylidae (Gastropoda: Basommatophora) na América do Sul: sistemática e distribuição. Revista Brasileira de Zoologia 13: 175–210.

Madsen H. & F. Frandsen. 1989. The spread of freshwater snails including those of medical and veterinary importance. Acta Tropica 46: 139–146.

Meier-Brook, K. & B.J. Smith. 1976. *Glacidorbis* Iredale, 1943, a genus of freshwater prosobranchs with a Tasmanian-Southeast Australian-South Andean distribution. Archive für Molluskenkunde 106: 191–198.

Meunier, C, C. Tirard C, S. Hurtrez-Boussès, P. Durand, M.D. Bargues, S. Mas-Coma, J.P. Pointier, J. Jourdane & F. Renaud. 2001. Lack of molluscan host diversity and the transmission of an emerging parasitic disease in Bolivia. Molecular Ecology 10: 1333–1340.

Miles, C.D. 1958. The family Succineidae (Gastropoda: Pulmonata) in Kansas. The University of Kansas Science Bulletin 38: 1499–1543.

Neusser, T.P. & M. Schrödl. 2007. *Tantulum elegans* reloaded: A computer-based 3D-visualization of the anatomy of a Caribbean freshwater acochlidian gastropod. Invertebrate Biology 126: 18–39.

Núñez, V., D.E. Gutiérrez Gregoric & A. Rumi. 2010. Freshwater gastropod provinces from Argentina. Malacologia 53: 47–60.

Ovando, X.M.C. & D.E. Gutiérrez Gregoric. 2012. Systematic revision of *Chilina* Gray (Gastropoda: Pulmonata) from northwestern Argentina and description of a new species. Malacologia 55: 117–134.

Ovando, X.M.C., L.E.M. Lacerda, R.R. Dutra & S.B. Santos. 2016. *Burnupia ingae* Lanzer, 1991 (Gastropoda: Hygrophila): Contributions to the Knowledge on its morphology and first records in Rio De Janeiro State, Brazil. Malacologia, 59:333–339.

Ovando, X.M.C., L.E.M. Lacerda & S.B. Santos. 2011. Mollusca, Gastropoda, Heterobranchia, Ancylidae, *Gundlachia radiata* (Guilding, 1828): First record of occurrence for the northwestern region of Argentina. Check List 7: 263–266.

Ovando X.M.C., L.E.M. Lacerda & S.B. Santos. 2014. Taxonomy, morphology and distribution of Ancylinae (Gastropoda: Pulmonata: Planorbidae) in Argentina. Journal of Conchology 41: 707–729.

Palasio, R.G.S., M.C.A Guimarães, F.P. Ohlweiler & R. Tuan, 2017. Molecular and morphological identification of *Biomphalaria* species from the stare of São Paulo. ZooKeys, 668: 11–32.

Paraense, W.L. 1966. "*Biomphalaria amazonica*" and "*B. cousini*", two new species of Neotropical planorbid molluscs. Revista Brasileira de Biologia 26: 115–126.

Paraense, W.L. 1975. Estado atual da sistemática dos planorbídeos brasileiros. Arquivos do Museu Nacional 55: 105–128.

Paraense, W.L. 1976a. A natural population of *Helisoma duryi* in Brazil. Malacologia 15: 369–376.

Paraense, W.L. 1976b. *Helisoma trivolvis* and some of its synonyms in the Neotropical región (Mollusca: Planorbidae). Revista Brasileira de Biologia 36: 187–204.

Paraense, W.L. 1976c. *Lymnaea viatrix*: a study of topotypic specimens (Mollusca: Lymnaeidae). Revista Brasileira de Biologia 36: 419–428.

Paraense, W.L. 1981. *Biomphalaria occidentalis* sp.n. from South America (Mollusca, Basommatophora, Pulmonata). Memórias do Instituto Oswaldo Cruz 76: 199–211.

Paraense, W.L. 1982. *Lymnaea rupestris* sp. n. from southern Brazil (Pulmonata: Lymnaeidae). Memórias do Instituto Oswaldo Cruz 77: 437–443.

Paraense, W.L. 1984. *Biomphalaria tenagophila guaibensis* ssp. n. from Southern Brazil and Uruguay (Pulmonata: Planorbidae). I. Morphology. Memórias do Instituto Oswaldo Cruz 79: 465–469.

Paraense, W.L. 1988. *Biomphalaria kuhniana* (Clessin, 1883), planorbid mollusc from South America. Memórias do Instituto Oswaldo Cruz 83: 1–12.

Paraense, W.L. 2002. *Plesiophysa dolichomastix* sp. n. (Gastropoda: Planorbidae). Memórias do Instituto Oswaldo Cruz 97: 505–508.

Paraense, W.L. 2003a. *Plesiophysa guadeloupensis* ("Fischer" Mazé, 1883). Memórias do Instituto Oswaldo Cruz 98: 519–521.

Paraense, W.L. 2003b. A bird's eye survey of Central American Planorbid molluscs. Memorias do Instituto Oswaldo Cruz 98: 51–67.

Paraense, W.L. 2004. Planorbidae, Lymnaeidae and Physidae of Ecuador (Mollusca: Basommathophora). Memórias do Instituto Oswaldo Cruz 99(4): 357–362.

Paraense, W.L. 2005. Planorbidae, Lymnaeidae and Physidae of Argentina (Mollusca: Basommatophora). Memorias do Instituto Oswaldo Cruz 100: 491–493.

Paraense, W.L. 2007. Taxonomia molecular e taxonomia morfológica. Pages 203–206, *in*: S.B. Santos, A.D. Pimenta, S.C. Thiengo, M.A. Fernandez & R.S. Absalão (Orgs.), Tópicos em Malacologia – Ecos do XVIII EBRAM, Sociedade Brasileira de Malacologia, Rio de Janeiro.

Paraense, W.L. & N. Deslandes. 1956. Observations on "*Australorbis janeirensis*" (Clessin, 1884). Revista Brasileira de Biologia 16: 81–102.

Paraense, W.L. & N. Deslandes. 1962. "*Australorbis intermedius*" sp. n. from Brazil (Pulmonata, Planorbidae). Revista Brasileira de Biologia 22: 343–350.

Paraense, W.L. & J.P. Pointier. 2003. *Physa acuta* Draparnaud, 1805: a study of topotypic specimens. Memorias do Instituto Oswaldo Cruz 98: 513–517.

Patterson, C.M. 1971. Taxonomic studies of the land snail family Succineidae. Malacological Review 4: 131–202.

Patterson, C.M. 1973. Generic and specific characters in the land snail family Succineidae. Malacological Review 6: 54–56.

Pilsbry, H.A. 1911. Non-marine Mollusca of Patagonia. Princeton University Expedition to Patagonia 3: 513–633.

Pilsbry, H.A. 1948. Land Mollusca of North America. Vol. II, part 2. Monographs of the Academy of Natural Science of Philadelphia 3: 771–847.

Pinto, H.A., S.V. Brant & A. L. Melo. 2014. *Physa marmorata* (Mollusca: Physidae) as a natural intermediate host of *Trichobilharzia* (Trematoda: Schistosomatidae), a potential causative agent of avian cercarial dermatitis in Brazil. Acta Tropica 138: 38–43.

Pointier J.P., P. David & P. Jarne. 2005. Biological invasions: the case of planorbid snails. Journal of Helminthology 79: 249–256.

Pointier J.P., P. David & P. Jarne. 2011. The biological control of the snail hosts of schistosomes: the role of competitor snails and biological invasions. Pages 215–238 *in*: R. Toledo & B. Fried (eds.), *Biomphalaria* snails and larval Trematodes. Springer.

Pointier J.P., C. González, O. Noya & B. Alarcón de Noya. 2015. Family Lymnaeidae. Pages 101–124. *in*: J.P. Pointier (ed.) Freshwater molluscs of Venezuela and their medical and veterinary importance. Conchbooks, D-55296, Harxheim, Germany.

Pointier, J.P. & O. Noya. 2015a. Family Neritidae. Pages 25–31 *in*: J.P. Pointier (ed.), Freshwater Molluscs of Venezuela. Conchbooks, D-55296, Harxheim, Germany.

Pointier, J.P. & O. Noya. 2015b. The family Physidae. Pages 124–129 *in*: J.P. Pointier (ed.), Freshwater Molluscs of Venezuela and their medical and veterinary importance. Conchbooks, D-55296, Harxheim, Germany.

Ponder, W.F. 1986. Glacidorbidae (Glacidorbaceae: Basommatophora), a new family and superfamily of operculate freshwater gastropods. Journal of the Linnean Society of London 87: 53–83.

Ponder, W.F. & G.J. Avern. 2000. The Glacidorbidae (Mollusca: Gastropoda: Heterobranchia) of Australia. Records of the Australian Museum 52: 307–535.

Puslednik, L., W.F. Ponder, M. Dowton & A.R. Davis. 2009. Examining the phylogeny of the Australian Lymnaeidae (Heterobranchia: Pulmonata: Gastropoda) using mitochondrial, nuclear and morphological markers. Molecular Phylogenetics and Evolution 52: 643–659.

Remigio, E.A. 2002. Molecular phylogenetic relationships in the aquatic snail genus *Lymnaea*, the intermediate host of the causative agent of fascioliasis: insights from broader taxon sampling. Parasitology Research 88: 687–696.

Rumi, A., D.E. Gutiérrez Gregoric, N. Landoni, J. Cárdenas Mancilla, S. Gordillo, J. González & D. Alvarez. 2015. Glacidorbidae in South America: Revision and description of a new genus and new species from Argentinean Patagonia and Chile. Molluscan Research 35: 143–153.

Rumi, A., D.E. Gutiérrez Gregoric, V. Núñez, I.I. Cesar, M.A. Roche, M.P. Tassara, S.M. Martín & M.F. López Armengol. 2006. Freshwater gastropoda from Argentina: species richness, distribution patterns, and an evaluation of endangered species. Malacologia 49: 189–208.

Rumi, A., D.E. Gutiérrez Gregoric, V. Núñez & G.A. Darrigran. 2008. Malacología Latinoamericana. Moluscos de agua dulce de Argentina. Revista de Biología Tropical 56: 77–111.

Samadi, S., A. Roumégoux, M.D. Bargues, S. Mas-Coma & J.P. Pointier. 2000. Morphological studies of Lymnaeid snails from the human fascioliasis endemic zone of Bolivia. Journal of Molluscan Studies 66: 31–44.

Santos, S.B. 1994. *Gundlachia dutrae*: n. sp. from northwest Brazil (Mollusca: Basommatophora: Ancylidae). Memória Instituto Oswaldo Cruz 89: 153–160.

Santos, S.B. 2000. Morfologia do sistema muscular e do sistema reprodutor de *Anisancylus obliquus* (Broderip & Sowerby) do Chile, como contribuição a sistemática de Ancylidae (Gastropoda, Pulmonata, Basommatophora). Revista Brasileira de Zoologia 17: 995–1000.

Santos, S.B. 2003. Estado atual do conhecimento dos Ancilídeos na América do Sul (Mollusca, Gastropoda, Pulmonata, Basommatophora). Revista de Biologia Tropical 50: 1–33.

Solem, A. 1978. Classification of the land Mollusca. Pages 49–97 *in*: V. Fretter & J. Peake (eds.), Pulmonates, Vol. 2A, Systematics, Evolution and Ecology. Academic Press, London.

Starobogatov, Y.I. 1967. On the systematization of freshwater Pulmonate Molluscs. Trudy Zoologicheskogo Instituta Leningrad 42: 280–304.

Strong, E.E., O. Gargominy, W.F. Ponder & P. Bouchet. 2008. Global diversity of gastropods (Gastropoda; Mollusca) in freshwater. Hydrobiologia 595: 149–166.

Taylor, D.W. 2003. Introduction to Physidae (Gastropoda: Hygrophila) biogeography, classification, morphology. Revista de Biologia Tropical 51 (Suppl. 1): 1–287.

Te, G. A. 1978. The systematics of the family Physidae (Basommatophora: Pulmonata). Ph. D. Dissertation, University of Michigan, Ann Arbor. 325 p.

Thiele, J. 1931–1935. Handbuch der Systematischen Weichtierkund. Part 1. Gustav Fischer, Jena. Pages 1–376.

Tillier, S. 1981. South American and Juan Fernández succineid slugs (Pulmonata). Journal of Molluscan Studies 47: 125–146.

Tuan, R., F.P Ohlweiler, M.C.D.A Guimarães, R.G.S. Palasio & R. Dalla Zanna. 2012. Pattern of genetic divergence of mitochondrial DNA sequences in Biomphalaria tenagophila complex species based on barcode and morphological analysis. *In:* B. Rokni (ed.) INTECH Open Access Publisher.

Van Mol, J.J. 1967. Evolution phylogénétique du ganglion cérébroïde des Gastéropodes Pulmonées. Mémoires de l'Académie Royale de Belgique 37: 1–168.

Vidigal, T.H.D.A., E.D. Neto, A.J.C. Simpson & O. S. Carvalho. 1996. A low stringency polymerase chain reaction approach to the identification of *B. glabrata* and *B. tenagophila*, intermediate snail hosts of *S. mansoni* in Brazil. Memórias do Instituto Oswaldo Cruz 91: 739–744.

Vidigal, T.H.D.A., R.L. Caldeira, A.J.C. Simpson & O.S. Carvalho. 2001. Identification of *Biomphalaria havanensis* and *Biomphalaria obstructa* populations from Cuba using polymerase chain reaction and restriction fragment length polymorphism of the ribosomal RNA intergenic spacer. Memórias do Instituto Oswaldo Cruz 96: 661–665.

Vinarski, M.V. 2013. One, two or several? How many lymnaeid genera are there? Ruthenica 23: 41–58.

Wade, C.M., P.B. Mordan & B. Clarke. 2001. A phylogeny of the land snails (Gastropoda: Pulmonata). Proceedings of the Royal Society of London B 268: 413–422.

Walker, B. 1918. A synopsis of the classification of the freshwater Mollusca of North America, North of Mexico. Miscellaneous Publications of the Museum of Zoology. University of Michigan 6: 1–213.

Wethington, A.R. & C. Lydeard. 2007. A molecular phylogeny of Physidae (Gastropoda: Basommatophora) based on mitochondrial DNA sequences. Journal of Molluscan Studies 73(3): 1–17.

Yong, M., A. Gutierrez. G. Perera, P. Durand & J.P. Pointier. 2001. The *Biomphalaria havanensis* complex (Gastropoda: Planorbidae) in Cuba: a morphological end genetic study. Journal Molluscan Studies 67: 103–111.

Zilch, A. 1959–1960. Euthyneura, Part 2. Pages 1–12 *in*: O.H. Schindewolf (ed.), Handbuch der Paläozoologie, Band 6, Gastropoda. Borntraeger, Berlin.

Subclass Neritimorpha

Jean-Pierre Pointier
PSL Research University, CRIOBE, USR 3278 EPHE-CNRS-UPVD, Perpignan, France

Antonio A. Vázquez
Laboratory of Malacology, Institute of Tropical Medicine Pedro Kourí, Havana, Cuba

INTRODUCTION

Neritimorpha is a group of gastropod mollusks including several families of land snails (e.g., Helicinidae), deep water marine limpets (e.g., Phenacolepadidae), and freshwater and marine snails and limpets (e.g., Neritidae). According to Uribe et al. (2016), as other larger clades within the Gastropoda, Neritimorpha has independently evolved snail-, slug- and limpetlike forms comprising approximately 2000 living species. There are certain typical features that characterize the Neritimorpha such as the highly convoluted larval shell and the usually absorbed internal structure during the late ontogeny allowing for more limpetlike shells (Kano et al., 2002). Most Neritimorpha are grazers that feed on algae, but detritivores and carnivorous species also occur. There are no important threats to freshwater Neritimorpha with most species having healthy populations. However, many species are a direct target of the aquarium pet trade due to their colorful, polished shells (Perera & Walls, 1996), a practice that may lead to undesired introductions.

Neritids (Neritidae) are mainly marine and brackish, but several species occur in Neotropical and Indo-Pacific freshwaters. Within the subfamily Neritininae (brackish and freshwater species), the type genus *Neritina* is characterized by very polished globose shells (neritiform) with a very short spire and a wide columellar area. Shells lack umbilicus and have a calcified operculum with internal apophyses for muscular attachment that perfectly closes the aperture. Adults live in rivers or estuaries and lay their ~2 mm egg capsules on almost any available surface (plants, rocks, or even each other). Depending on the species, each capsule may contain a variable number of eggs reaching up to 45 in *Neritina virginea* (Linnaeus, 1758) (Mattews-Cascon & Martins, 1999). Eggs release planktotrophic larvae downstream to the sea, where salinities are required for their development. Many neritines are amphidromous, which means that after metamorphosis and recruitment in river mouths, juveniles migrate upstream to their freshwater habitats (Crandall, 1999). Genetic studies showed that freshwater neritids may have a pelagic larval duration of at least 8 weeks promoting dispersal at very long distances (Crandall et al., 2010). Neritids are microphagous and graze upon periphyton.

LIMITATIONS

Taxonomy in this group is complicated and morphological descriptions sometimes do not allow for a clear species differentiation. Several genera of freshwater neritids (e.g., *Neritina, Vitta, Vittina*) based on conchological characters have been considered (Eichhorst, 2016). However, molecular analyses using CO1 and 16S rNDA markers on a large spectrum of species demonstrates the presence of a single genus within the freshwater *Neritina* (Quintero-Galvis & Castro, 2013). Consequently, pending published molecular results including Neotropical and other worldwide neritines, *Neritina* will be the single genus recognized for the species occurring in the Neotropical area. Descriptions of closely related species such as *N. virginea* and *N. clenchi* do not allow for a clear distinction and the validity of the latter is doubtful, suggesting a synonymy with *N. virginea* (Blanco et al., 2014). Similar concerns have been raised at the molecular level between *N. usnea* and *N. piratica* with no differences in the COI region suggesting both should be considered a single taxon (Quintero-Galvis & Castro, 2013).

Before 2005, Neritilidae was included as a subfamily within the Neritidae, but it is now given full family status (Bouchet et al., 2017). The Neritilidae includes several small tropical species (less than 7 mm). According to Symonds & Tracey (2014), it is difficult to find reliable characters that allow clear separation from the Neritidae. The shell is small, uniformly white to cream or orange-brown without patterns. The operculum is calcified with an internal apophysis like in Neritidae. Some authors consider certain characters as unique for the Neritilidae, i.e., larval shell having spiral microstriae on its last whorl; globular embryonic shell sometimes wholly obscured by a subsequent, thin larval shell layer of vertical prisms; the axis of the protoconch being significantly tilted relative to the teleoconch axis; and all planktotrophic *Neritilia* having a much smaller protoconch than other recent neritimorph groups. However, Symonds & Tracey (2014) argue these features to be unique and state that diagnosis should be based upon a combination of several shell characters for each genus.

Neritilids are small mollusks usually inhabiting the upper part of rivers where they commonly co-occur with *Neritina* spp. In some rivers, they may reach high densities under stones. Several genera are accepted in this family, but only one genus is recognized in the Neotropics (*Neritilia*).

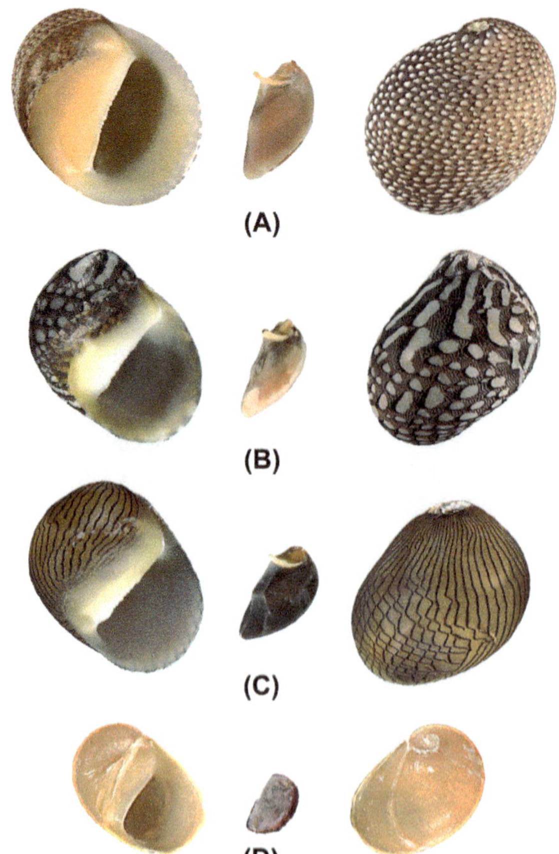

FIGURE 11.52 Neotropical Neritidae and Neritiliidae, showing front, operculum and back: (A) *Neritina punctulata*, Latouche River, Martinique, H (height) = 22.2 mm; (B) *Neritina virginea*, Latouche River, Martinique, H = 20.7 mm; (C) *Neritina usnea*, Valle de la Cruz, Venezuela, H = 19.9 mm; (D) *Neritilia succinea*, Vauclin River, Martinique, H = 5.9 mm.

Except for *N. succinea*, which can be found in several islands in the middle part of the Lesser Antilles (Pointier, 2008; Espinosa et al., 2017), the three remaining species are reported only from Cuba and found in subaquatic sea caves and river mouths. The new Cuban species are only supported by morphological characters that according to the original descriptions (Espinosa et al., 2017; Espinosa & Ortea, 2018) cannot be accurately separated (most features are repeated with only slight variations in size serving as diagnostic). For instance, descriptions of the new species are based on very few examined individuals (n = 4 in *N. abeli*; n = 2 in *N. serrana*; n = 1 in *N. mariaella*) which, taking into consideration that habitat-related polymorphism is commonly present in mollusks, molecular studies are very much needed to validate the species.

KEY TO NERITIMORPHA

1 Adult shell length > 7 mm, color pattern present, sometimes highly variable, found in fresh and brackish habitats (Fig. 11.52 A–C) .. .. Neritidae, one genus: *Neritina*
[Mexico, Cuba, Dominican Republic, Puerto Rico, Lesser Antilles, Colombia, Venezuela, French Guiana, Brazil. Nearctic]

1′ Adult individuals shell high ≤ 7 mm, single color without pattern, found in fresh, brackish and marine habitats (Fig. 11.52 D) .... Neritiliidae, one genus: *Neritilia*
[Cuba, Lesser Antilles]

REFERENCES

Blanco, J.F., S. Tamayo & F.N. Scatena. 2014. Phenotypic variability of the shell in Neritinidae (Gastropoda: Neritimorpha) in Puerto Rican rivers. Revista de Biología Tropical 62: 53–68.

Bouchet, P., J.P. Rocroi, B. Hausdorf, A. Kaim, Y. Kano, A. Nützel, P. Parkhaev, M. Schrödl & E.E. Strong. 2017. Revised classification, nomenclator and typification of gastropods and monoplacophoran families. Malacologia 61: 1–526.

Crandall, E.D. 1999. Early life history aspects of amphidromous neritids snails in Moorea, French Polynesia. Berkeley Scientific 3: 98–103.

Crandall, E.D., J.R. Taffel & P.H. Barber. 2010. High gene flow due to pelagic larval dispersal among South Pacific archipelagos in two amphidromous gastropods (Neritomorphs, Neritidae). Heredity 104: 563–572.

Eichhorst, T.E. 2016. Neritidae of the world. Vol 2, Conchbooks. 672 pp.

Espinosa, J. & J. Ortea. 2018. Una nueva especie del género *Neritilia* von Martens, 1879 (Gastropoda: Neritiliidae) de la Península de Guanahacabibes, Cuba, con una lista de taxones recientes en el área protegida. Avicennia 22: 13–18.

Espinosa, J., J. Ortea & Y.L. Diez-García. 2017. El género *Neritilia* von Martens, 1879 (Mollusca: Gastropoda: Neritilidae) en Cuba, con la descripción de dos nuevas especies. Avicennia 20: 49–52.

Kano, Y., S. Chiba & T. Kase. 2002. Major adaptive radiation in neritopsine gastropods estimated from 28S rRNA sequences and fossil records. Proceedings of the Royal Society, London, Series B 269: 2457–2465.

Mattews-Cascon, H. & I.X. Martins. 1999. Notes on the reproduction of *Neritina virginea* (Linnaeus, 1758) in northeast Brazil (Mollusca: Gastropoda: Neritidae). Arquivos de Ciências do Mar 32: 129–132.

Perera, G. & J.G. Walls. Applesnails in the aquarium. T.F.H. Publications. 121 pp.

Pointier, J.P. 2008. Guide to the freshwater molluscs of the Lesser Antilles. Conchbooks. 128 pp.

Quintero-Galvis, J. & L.R. Castro. 2013. Molecular phylogeny of the Neritidae (Gastropoda: Neritimorpha) based on the mitochondrial genes cytochrome oxidase 1 (CO1) and 16S rDNA. Acta Biológica Colombiana 18: 307–318.

Symonds, M.F. & S. Tracey. 2014. *Neritilia* (Gastropoda, Neritopsina, Neritiliidae): pushing back the timeline. Cainozoic Research 14: 3–7.

Uribe, J.E., D. Colgan, L.R. Castro, Y. Kano & R. Zardoya. 2016. Phylogenetic relationships among superfamilies of Neritimorpha (Mollusca: Gastropoda). Molecular Phylogenetics and Evolution 104: 21–31.

Subclass Caenogastropoda

María Gabriela Cuezzo
Instituto de Biodiversidad Neotropical (CONICET-UNT, Facultad de Ciencias Naturales), Tucumán, Argentina

KEY TO CAENOGASTROPODA

Gastropoda: Caenogastropoda: Superfamilies

1 Operculum concentric, corneous 2

1′ Operculum paucispiral, or multispiral, calcareous or corneous, shell with medium to high spire, ovate, elongate-conic, cylindrical to turriform, > 100 mm height 3

2(1) Shell globose with short spire or planospiral in form, 15–170 mm height .... Ampullaroidea, one family: **Ampullariidae [p. 341]**

2′ Shell oval with conic, tall spire, 25–50 mm height (Fig. 11.53) .... Viviparoidea, one species: *Sinotaia quadrata* (Benson, 1842) [Argentina: Buenos Aires, Córdoba. Exotic species]

3(1) Shell planispiral to trochoid, conic, or elongate-conic (Fig. 11.68), smooth surface or slightly sculptured, in general shell < than 1 cm, males if present with a simple penis or with lobes or accessory appendices .... **Truncatelloidea [p. 351]**

3′ Shell ovoid, oblong-turriform, conic in spire, generally elevated, strong, surface strongly sculptured (Fig. 11.75), shell < 100 mm, males if present without a verge .... **Cerithioidea [p. 358]**

FIGURE 11.53 *Sinotaia quadrata* (Viviparoidea): (A) Adult crawling showing the cylindrical snout and tentacles, with eyes located externally at the base, note the short rounded margin of the foot, scale bar = 10 mm; (B) juvenile specimen showing periostracal hairs and triangular lamellae in spiral rows, these structures later will be absent in adults; (C) adult solid shell showing the outer margin of aperture blackish (arrow), and operculum sculptured with concentric growth lines with nucleus subcentral closer to inner margin.

Ampullarioidea: Ampullariidae

Silvana Carvalho Thiengo, Monica Ammon Fernandez and Aline Carvalho de Mattos

Laboratório de Malacologia, Instituto Oswaldo Cruz/Fiocruz, Referência Nacional para Esquistossomose-Malacologia, Rio de Janeiro, Brazil

INTRODUCTION

The ampullariids predominantly occur in inland waters in tropical and subtropical regions of Africa, Asia, and the Americas. Currently, there are eight extant genera recognized, of which *Afropomus*, *Saulea*, and *Lanistes* are African, *Pila* occurs in both Africa and Asia, and *Asolene* d'Orbigny, 1838, *Felipponea* Dall, 1919, *Marisa* Gray, 1824, and *Pomacea* Perry, 1810 are American (Hayes et al., 2015).

The shells of ampullariids vary greatly in size, with adults ranging from 15 to 170 mm in height as *Pomacea maculata* Perry, 1810, the largest freshwater gastropod (Cowie et al., 2006). They are generally globose and dextrally coiling, although the shell of *Marisa* is planispiral and that of *Lanistes* appears sinistral (actually dextral but hyperstrophic) (Thiengo et al., 2011). The whorls present dark brown spiral bands in many species and an umbilicus is usually present. The aperture is closed by an operculum. Regarding the thickness, the shell may vary from very thick (*Asolene* and *Felipponea*) to very thin, as in *Pomacea papyracea* (Spix in Wagner, 1827), which is almost devoid of calcium carbonate. The animal has a large lanceolate foot and the proboscis bears two small tentacles (labial palps). The two cephalic tentacles are long, tapering, and the eyes are located at their bases, on the eyestalks. There are two nuchal lobes, the left one forming a respiratory siphon, which may be long, short or even absent, according to adaptations to an amphibious life (Fig. 11.54). This extensible siphon, long in *Pomacea* and shorter in *Lanistes* and *Felipponea*, is used to direct air to the lung opening when the animal is submerged (Hayes et al., 2012). A remarkable characteristic is the division of the pallial cavity into two parts, the left one functions as a lung and the right one has an elongated monopectinate gill, which makes aerial and aquatic breathing possible. A clearly visible bipectinate osphradium is located next to the mantle edge on the left side. According to Hayes et al. (2015) *Asolene, Felipponea, Marisa, Lanistes*, and presumably *Afropomus* and *Saulea* are remarkably amphibious, enabling them to survive long periods buried in the mud whereas *Pomacea* and *Pila* are less amphibious. The radula is taenioglossate (2:1:1:1:2) and the jaw is well developed (Fig. 11.55).

Ampullariids are dioecious, oviparous and females lay clusters of calcareous eggs above water (in *Pomacea* and *Pila*) or gelatinous eggs in water (*Asolene, Lanistes*, and *Marisa*). They are voracious generalist macroherbivores, for which some species of *Pomacea* and *Marisa* were used to control planorbid vectors of schistosomiasis (Pointier et al., 1988) as well as to control aquatic macrophytes (Simberloff & Stiling, 1996). However, ampullariids may also be agricultural pests (rice, taro, and watercress), primarily *Pomacea*, and in some Asian countries at least two species, *Pomacea maculata* and *P. canaliculata* (Lamarck, 1822), have been introduced for food or through the aquarium trade and have become pests (Hayes et al., 2012).

Ampullariids generally occur in large populations and play an important role in food webs. In addition, ampullariids may transmit the nematode *Angiostrongylus cantonensis* (Chen, 1935), which causes eosinophilic meningitis in humans. This zoonosis, endemic to the southeast of Asia and the Pacific Islands, has been dispersed within the Neotropics during the last two decades. In Brazil, different species of terrestrial gastropods as well as specimens of *Pomacea lineata* (Spix in Wagner, 1827) have been found naturally infected with *A. cantonensis* (Thiengo et al., 2010). Other Neotropic ampullariids reported as vectors are *Pomacea* sp. in Ecuador (Pincay et al., 2009) and *P. canaliculata* in China and Hawaii, where the species were introduced primarily for food (Lv et al., 2009; Kim et al., 2014).

Approximately 90 nominally valid species are currently listed to the Neotropics according to Cowie & Thiengo (2003). Recent studies using morphological, molecular and biological data (Hayes, 2009; Hayes et al., 2012; Thiengo et al., 2011) have divided Ampullariidae into five clades: *Flagellata, Effusa, Pomacea, Limnopomus*, and *Marisa*. *Flagellata* clade occupies the most basal position in the phylogenetic tree presented by Hayes (2009), similarly to *P. papyracea*, although it is considered out of the mentioned clades here due to its peculiar corneous shell. *Marisa* and *Pomacea* clades are more derived and present contiguous distribution in the Neotropics (Fig. 11.56).

LIMITATIONS

The taxonomy is based primarily on the shell characteristics, and remains confused. Considerable intraspecific variation has led to the descriptions of a large number of nominal species, misidentification, and a great number of

PHYLUM MOLLUSCA

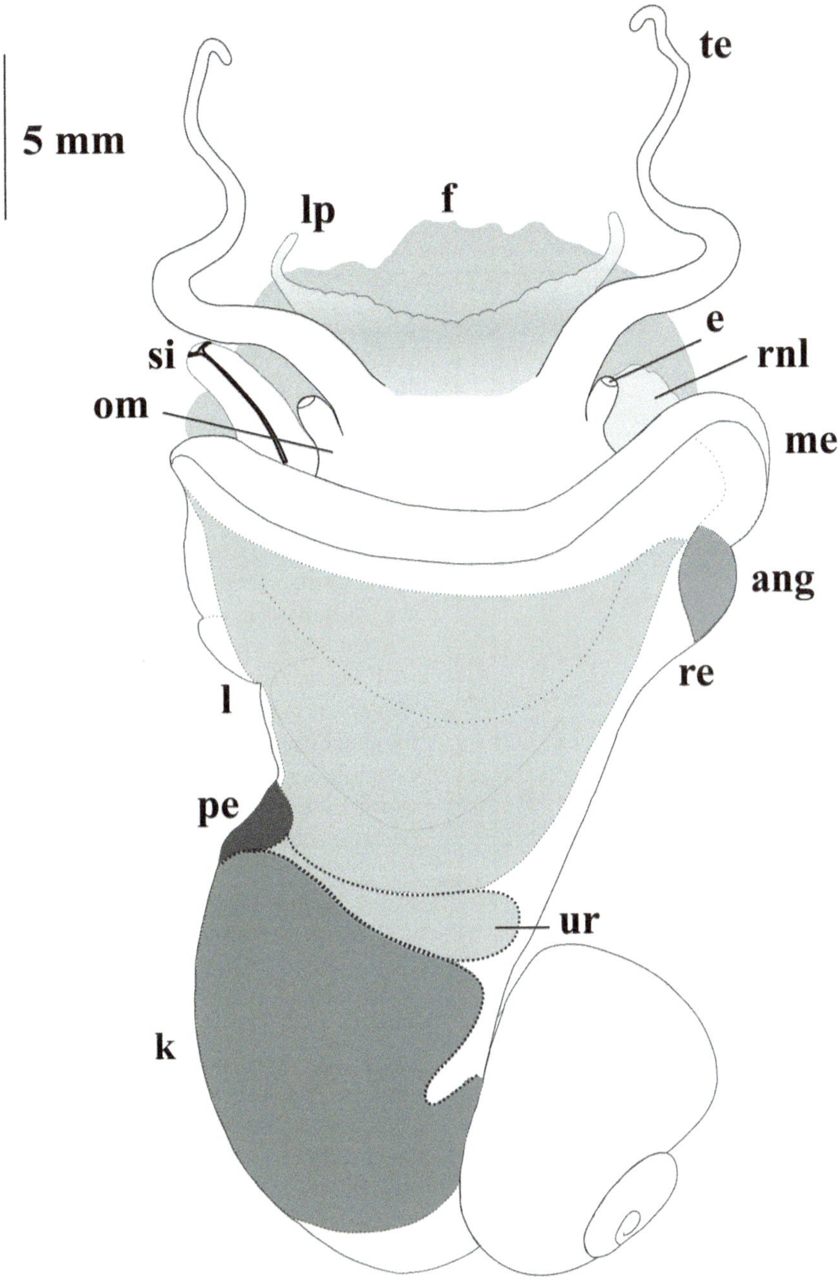

FIGURE 11.54 Specimen of *Pomacea* (Ampullariidae) removed from the shell. Abbreviations: *ang*, anal gland; *e*, eye; *f*, foot; *l*, lung; *lp*, labial palp; *me*, mantle edge; *om*, ommatophore; *pe*, pericardium; *re*, rectum; *rnl*, right neck lobe; *si*, siphon; *te*, tentacle; *ur*, ureter.

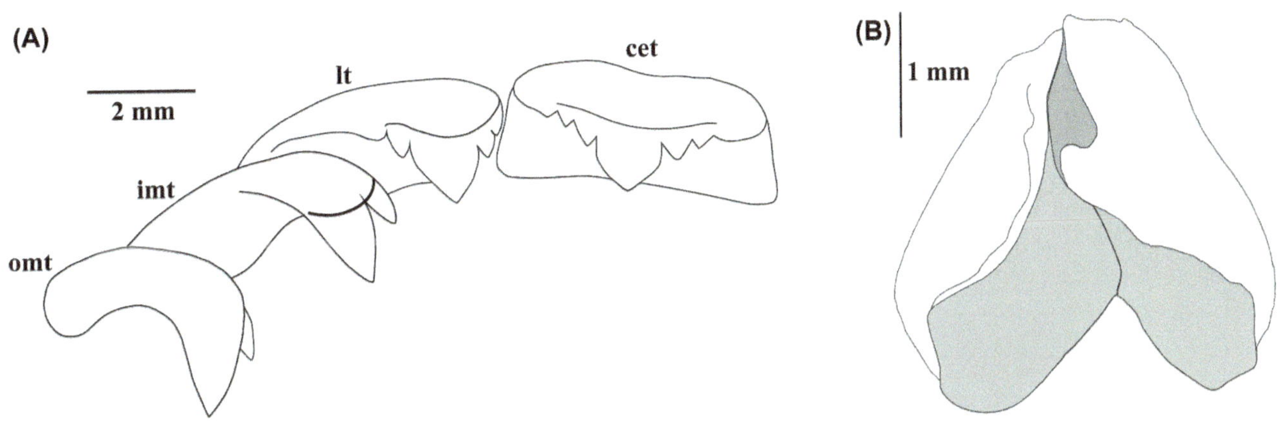

FIGURE 11.55 Radula and jaws of ampullariids: (A) Radula taenioglossate (2:1:1:1:2); (B) jaw. Abbreviations: *cet*, central tooth; *imt*, inner marginal tooth; *lt*, lateral tooth; *omt*, outer marginal tooth.

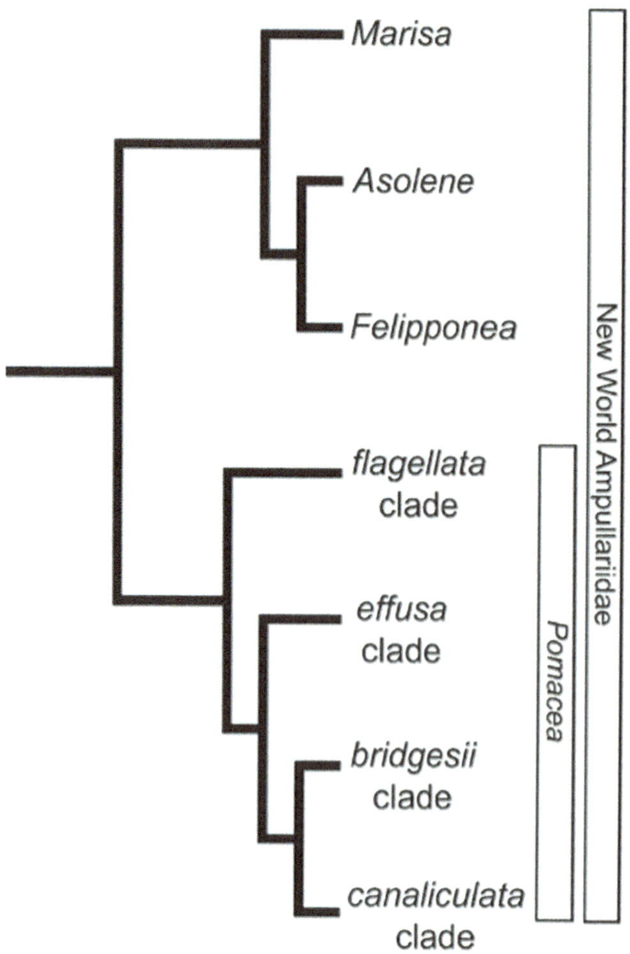

FIGURE 11.56 Phylogenetic tree of New World Ampullariidae showing genetic relationships, according to Hayes (2009).

inconsistently synonymized species. The shell has taxonomic value at genus level, but has limited value for species, as it may vary in shape, thickness, color, and size according to habitat and age. In addition, even distantly related species may present very similar shells. Hayes et al. (2012) reported that the main differences between *P. canaliculata* and *P. maculata* shells are not in adult specimens but in newly hatched juveniles. Similarly, the usefulness of the radula and jaw is limited to genus level, although there may be subtle differences among distantly related species within species and genera (Berthold, 1991; Martin & Negrete, 2007; Hayes et al. 2012). Molecular data associated with sound morphological studies have been carried out in recent years and the results are promising for clarification of the taxonomy of this group (Hayes et al., 2012). The lack of species discrimination data makes identification keys to species impossible.

TERMINOLOGY AND MORPHOLOGY

The morphology is treated in previous works such as Souza-Lopes (1955, 1956), Hylton-Scott (1958), Michelson (1961), Andrews (1964, 1965), Thiengo (1987, 1989), Berthold (1991), Thiengo et al. (1993; 2011), Catalan et al. (2002), Hayes et al. (2012), and Gamarra-Luques et al. (2013). Most studies deal with *Pomacea*, the genus with the largest number of species and the largest geographic distribution in the Neotropics.

The main features used to distinguish the genera are the **operculum** (corneous in the Neotropical species), the **siphon** (long, short, or absent) and the **sperm canal** (closed in the Neotropical species). The operculum may not seal the aperture completely in some *Pomacea* species. *Asolene* is brevisiphonate whereas *Marisa* and *Pomacea* are longisiphonate (Michelson, 1961). The **kidney** is composed of two interconnected chambers, an anterior (=ureter) located in the mantle roof, and a larger visceral posterior chamber. In some genus as *Asolene, Marisa*, and *Pomacea* the anterior chamber of the kidney is transversal to the longitudinal axis of the animal (Fig. 11.57). Diagnostic characters of species are mainly in the reproductive system such as: **penial sheath** anatomy, shape and size of **prostate** and **penis** (Thiengo, 1989; Hayes et al., 2012). The testis (Fig. 11.58) lies in the three whorls of the spire, short and numerous efferent ducts fuse into two main branches that converge into the spermiduct, which is very narrow and runs to the base of the spire until opening into the seminal vesicle. The prostate follows the seminal vesicle and continues along the left side of the rectum until it reaches the anal papilla where it bends slightly left, opening into the seminal groove below the penis pouch (Fig. 11.59).

In *Pomacea* spp. the distal end of the penis pouch forms a U-shaped channel which leads the penis to the penial sheath during copulation. The penial sheath is generally well developed and rises near the mantle edge. It has a longitudinal channel and three glands may be present on its inner surface (apical, medial, and basal). The position and shape of these glands are useful for species discrimination as well as the shape of the penial sheath. An outer basal gland is also present in the penial sheath of all species, and it may vary in shape and size among the species. The arborescent **ovary** (Fig. 11.60) occupies a similar position to the testis, and it is composed of multibranched tubules lying superficially on

PHYLUM MOLLUSCA

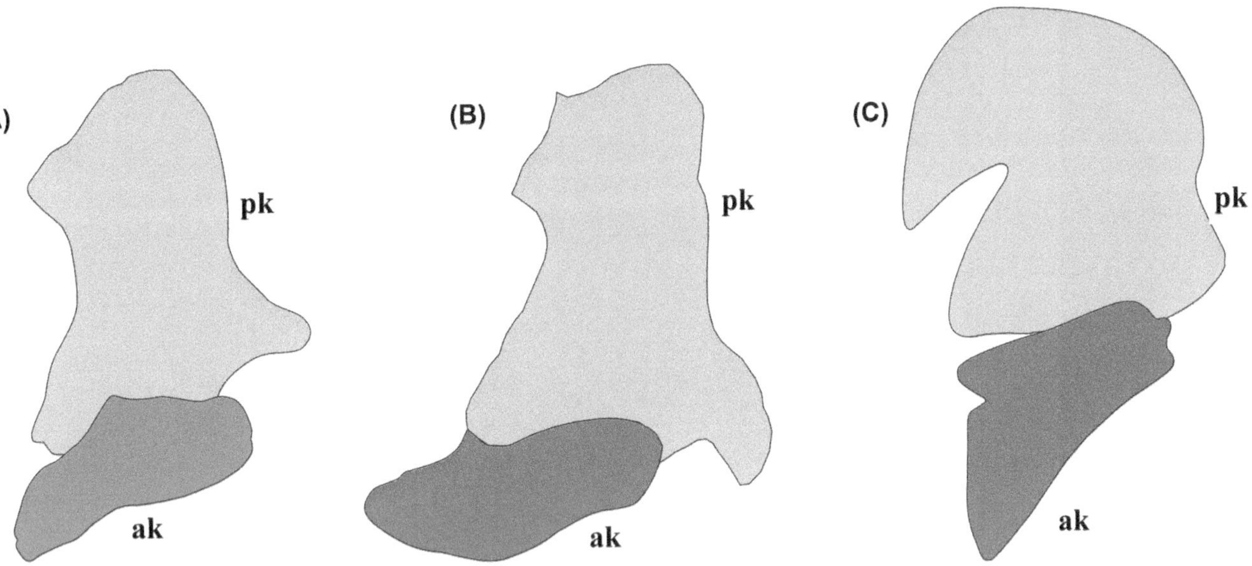

FIGURE 11.57 Scheme of the ampullariid kidney: (A) *Asolene*; (B) *Marisa*; (C) *Pomacea*. Abbreviations: *ak*, anterior kidney; *pk*, posterior kidney.

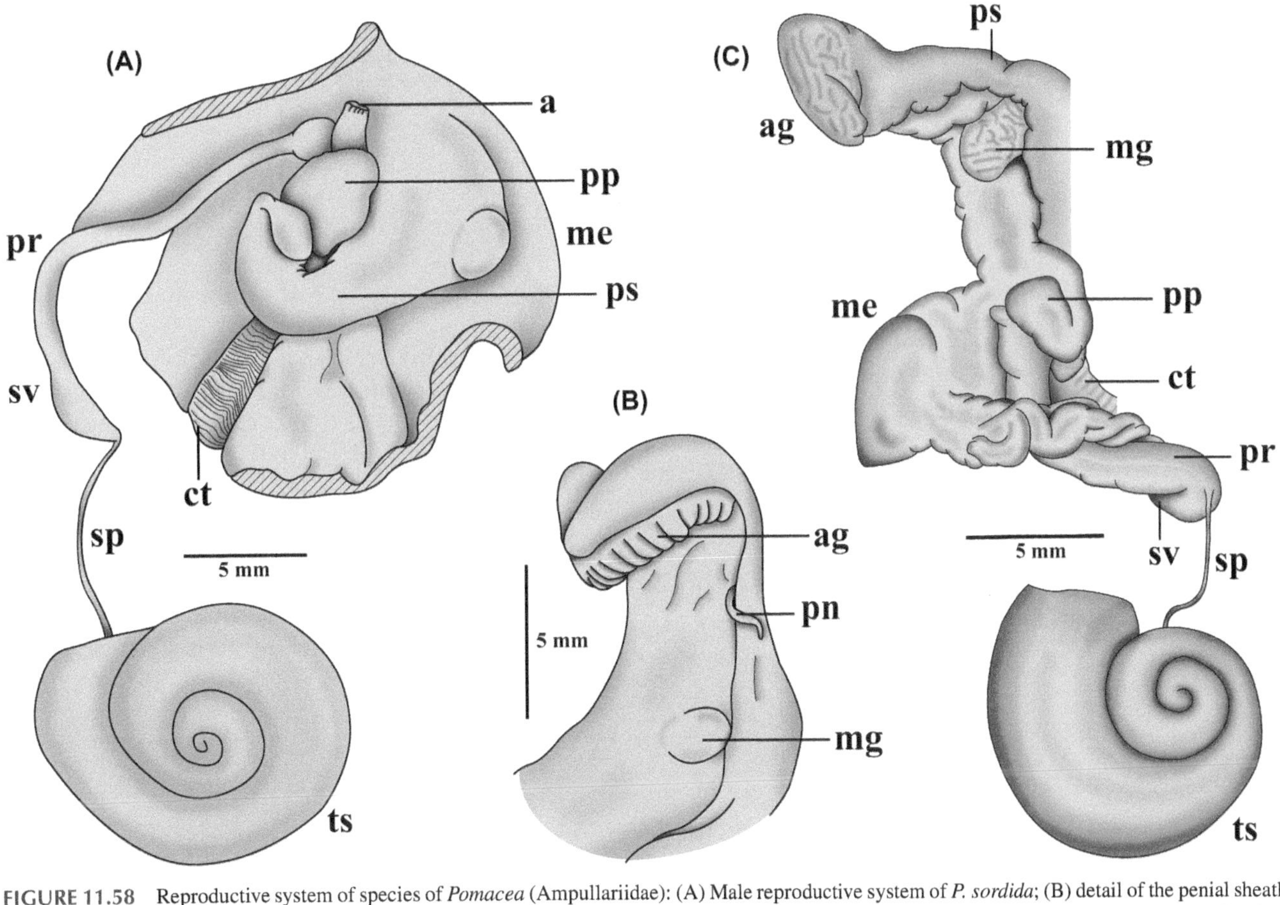

FIGURE 11.58 Reproductive system of species of *Pomacea* (Ampullariidae): (A) Male reproductive system of *P. sordida*; (B) detail of the penial sheath of *P. sordida* in copulation position; (C) male reproductive system of *P. figulina*. Abbreviations: *a*, anus; *ag*, apical gland; *ct*, ctenidium; *me*, mantle edge; *mg*, medial gland; *pn*, penis; *pp*, penis pouch; *pr*, prostate; *ps*, penial sheath; *sp*, spermiduct; *sv*, seminal vesicle; *ts*, testis.

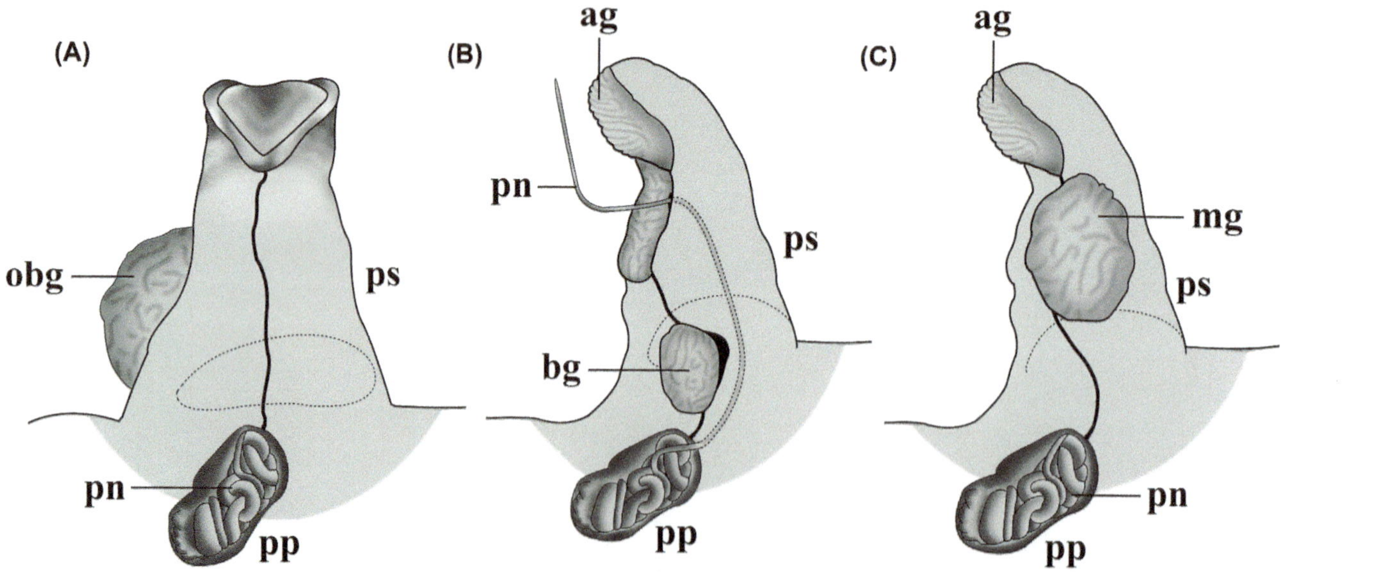

FIGURE 11.59 Scheme of penial pouch and penial sheath of species of *Pomacea* (Ampullariidae) showing glands of penial sheath in different positions. Abbreviations: *ag*, apical gland; *bg*, basal gland; *mg*, medial gland; *pn*, penis; *pp*, penis pouch; *ps*, penial sheath; *obg*, outer basal gland.

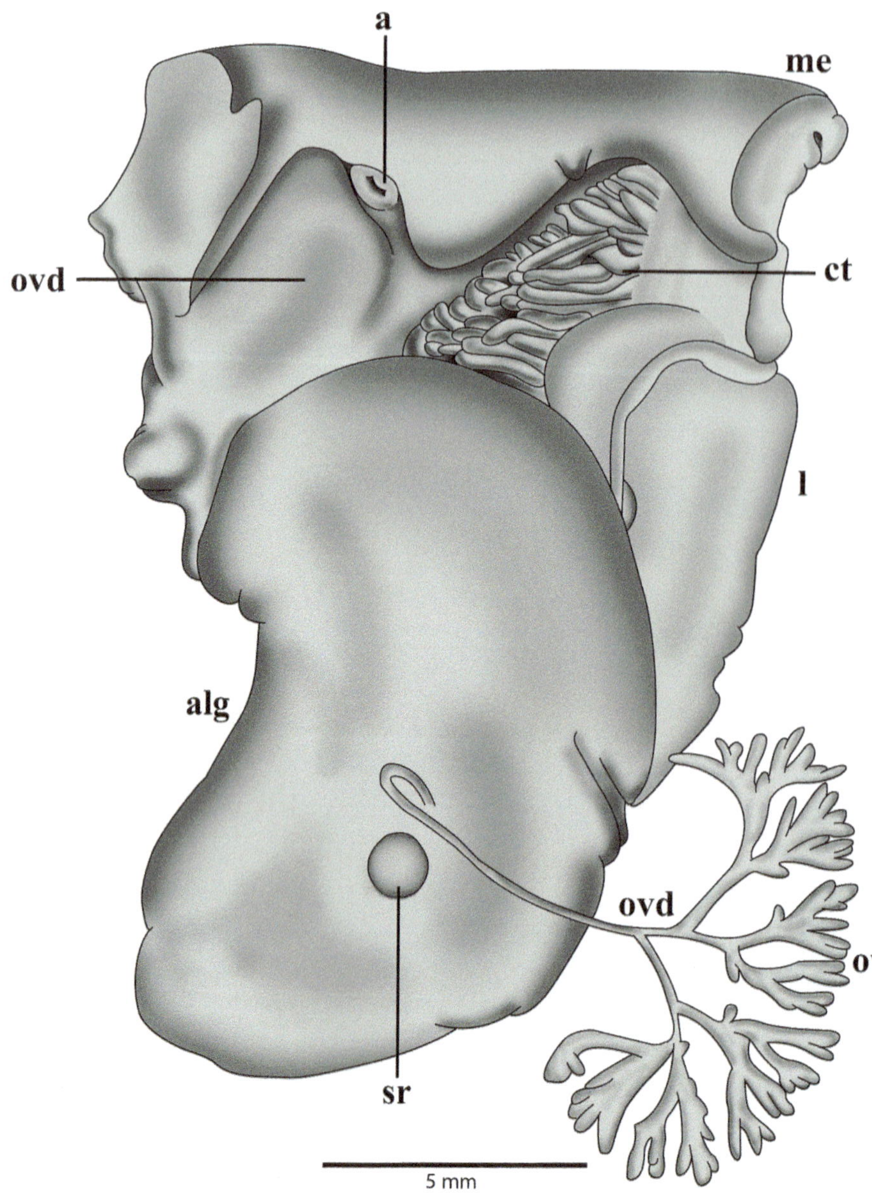

FIGURE 11.60 Female reproductive system of *Pomacea figulina* (Ampullariidae). Abbreviatures: *a*, anus; *alg*, albumen gland; *ct*, ctenidium; *l*, lung; *me*, mantle edge; *ov*, ovary; *ovd*, oviduct; *sr*, seminal receptacle.

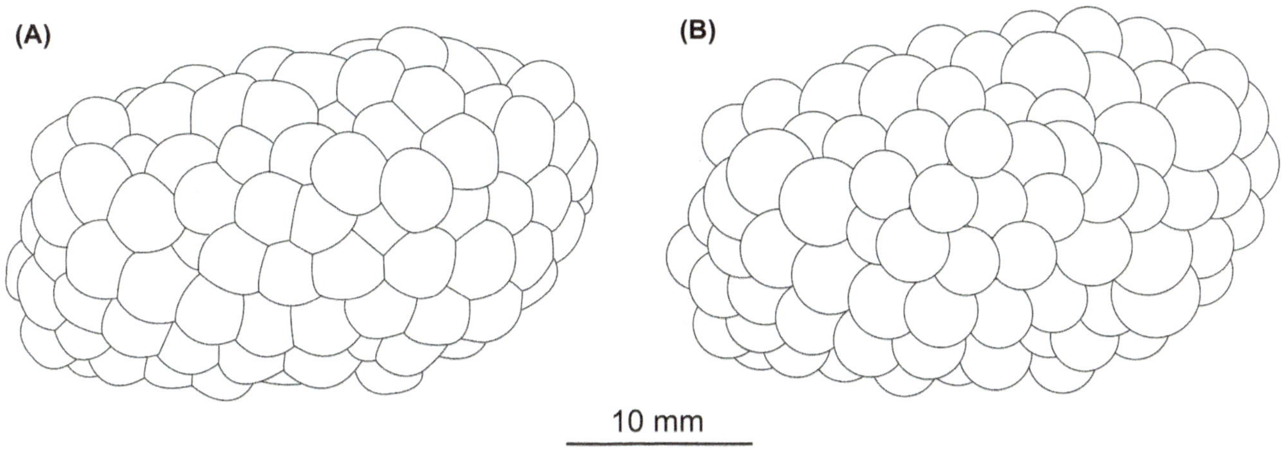

FIGURE 11.61 Ampullarid egg mass: (A) Polygonal eggs; (B) spherical eggs.

the digestive gland. The **visceral oviduct** follows the ovary and runs along the columellar axis to the spire base until it merges into the albumen gland. **Pallial oviduct** located in the floor of the mantle cavity comprises muscular seminal receptacle, large copulatory bursa, diverticulae of the albumen gland, and spiral lamellae of capsulae gland, all embedded within the mass of parenchymal tissue (Thiengo, 1987; Hayes et al., 2012). Females of *Marisa* and *Pomacea* usually present vestigial male copulatory apparatus at the mantle edge beside the gonopore. In *Pomacea* a characteristic in females that helps species discrimination is the number of the ovary tubules that vary among species (Thiengo, 1989; Thiengo et al., 2011).

Oviposition and egg cluster present characters useful to distinguish genera, such as oviposition location (if under or above water surface), whereas the shape (spherical or polygonal), the color (pink, whitish, green, and orange) and the size of eggs (Fig. 11.61) are characteristics used to distinguish species (Hayes et al., 2012).

MATERIAL PREPARATION AND PRESERVATION

Specimens prior dissection must be relaxed in 2% hypnol, and after 10–14 h removed from their shells with the aid of forceps, in order to tear off the columellar muscle. The soft bodies are then preserved in Railliet-Henry's fixative (2% acetic acid, 5% formol, 0.6% NaCl) or 70% ethanol, for morphological studies. A piece of foot tissue is sampled and stored in 95% ethanol for molecular studies. If it is not possible to relax the specimens before preservation, the heat-shock technique (Fukuda et al., 2008) is recommended to kill specimens and make easier to remove them from the shell.

ACKNOWLEDGMENTS

To Raquel de Souza Leal and Eduardo Cinilha for their help with the illustrations.

KEY TO AMPULLARIIDAE

Gastropoda: Caenogastropoda: Ampullarioidea: Ampullariidae: Genera

The morphology of the shell allows discrimination between genera, but has limited value to distinguish species.

1 Eggs deposited in gelatinous masses in the water, generally on macrophyte 2

1′ Calcareous egg mass deposited above water; shell globose, ranging from thin to medium thickness, generally with dark brown spiral bands and angulated whorls separated by impressed sutures; rarely very thick; long siphon (Figs. 11.62, 63) .... *Pomacea*
[Widely distributed in the Neotropics. Nearctic]

FIGURE 11.62 Shell of *Pomacea* spp. (Ampullariidae): (A) *P. glauca*; (B) *P. maculata*; (C) *P. papyracea*; (D) *P. urceus*.

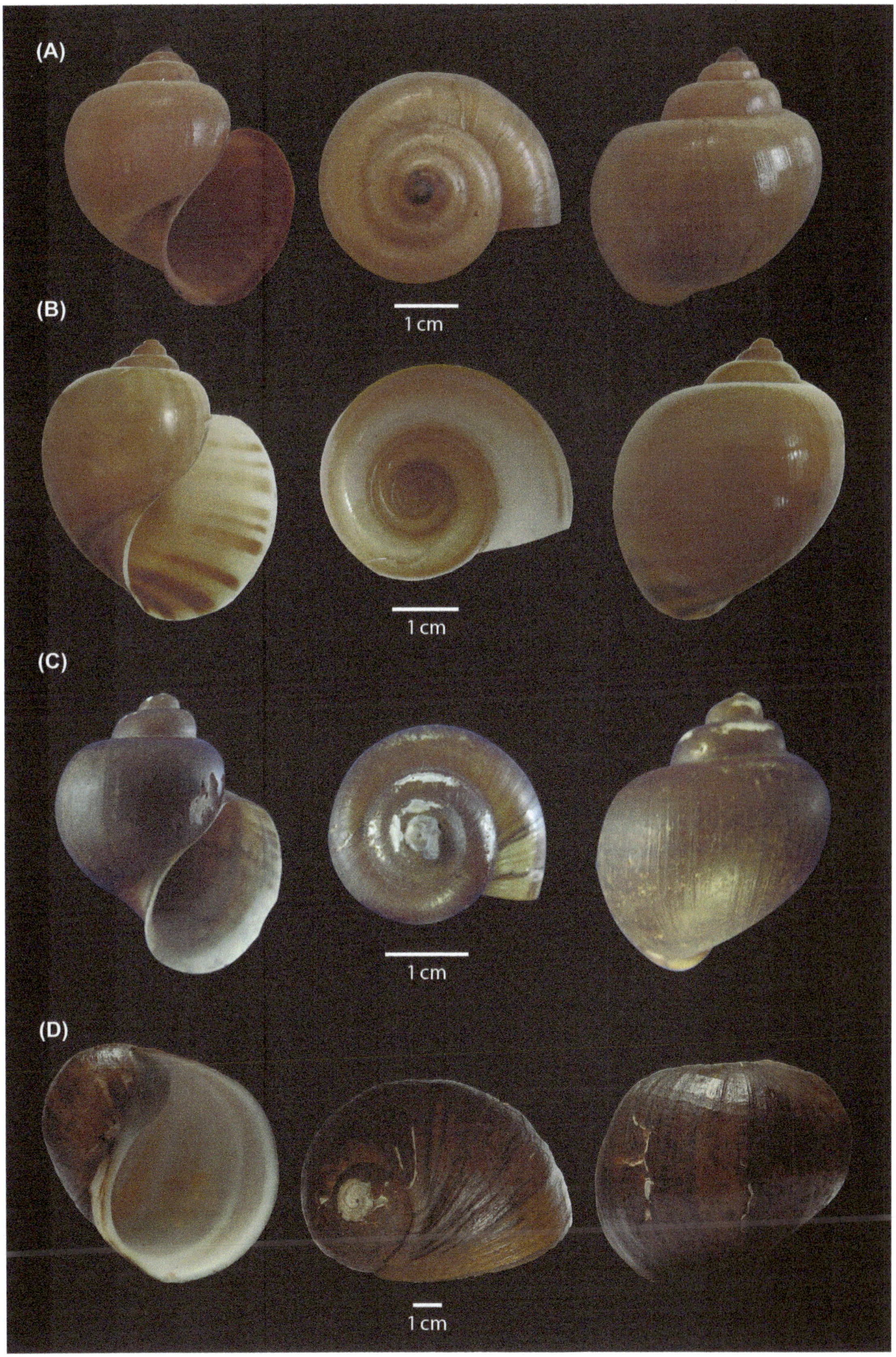

FIGURE 11.63 Shell of *Pomacea* spp. (Ampullariidae): (A) *P. sordida*; (B) *P. lineata*; (C) *P. diffusa*; (D) *P. megastoma*.

FIGURE 11.64 Shell of *Marisa* spp. (Ampullariidae): (A) *M. cornuarietis*; (B) *M. planogyra*.

2(1) Shell globose or ovate; short siphon 3

2′ Shell planispiral; generally with dark spiral bands (Fig. 11.64) .... *Marisa*
[Mexico, Costa Rica, Panamá, Trinidad and Tobago, Venezuela, Colombia, Guyana, French Guiana, Suriname, Brazil, Bolivia, Paraguay, Argentina. Nearctic]

3(2) Shell globose, thick, yellow to brown, with dark brown spiral bands variable in number and thickness (Fig. 11.65) .... *Asolene*
[Venezuela, Colombia, Ecuador, Brazil, Guyana, Suriname, French Guiana, Peru, Bolivia, Paraguay, Argentina, Uruguay; seven species]

3′ Shell ovate, small to medium size, very thick, porcelainous, decollated, with flattened whorls; attached to rocks in lotic systems (Fig. 11.66) .... *Felipponea*
[Brazil, Argentina, Uruguay; three species]

FIGURE 11.65 Shell of *Asolene* spp. (Ampullariidae): (A) *A. meta*; (B) shell of *A. spixii.*

FIGURE 11.66 Shell of *Felipponea neritiniformis* (Ampullariidae).

REFERENCES

Andrews, E.B. 1964. The functional anatomy and histology of the reproductive system of some pilid gastropod molluscs. Proceedings of the Malacological Society of London 36: 121–140.

Andrews, E.B. 1965. The functional anatomy of the mantle cavity, kidney and blood system of some pilid gastropods (Prosobranchia). Proceedings of the Zoological Society of London 146: 70–94.

Berthold, T. 1991. Vergleichende Anatomie, Phylogenie und historische Biogeographie der Ampullariidae (Mollusca, Gastropoda). Abhandlungen des Naturwissenschaftlichen Vereins in Hamburg 29: 1–256.

Catalán, M.N., S.N. Fernandez & B.C. Winik. 2002. Oviductal structure and provision of egg envelops in the apple snail *Pomacea canaliculata* (Gastropoda, Prosobranchia, Ampullariidae). Biocell 26: 91–100.

Cowie, R.H., K.A. Hayes & S.C. Thiengo. 2006. What are apple snails? Confused Taxonomy and Some Preliminary Resolution. Pages 3–24 *in*: C.J. Ravindra & S.S. Leocadio. Global Advances in Ecology and Management of Golden Apple Snails, Nueva Ecija, Philippine Rice Research Institute.

Cowie, R.H. & S.C. Thiengo. 2003. The apple snails of the Americas (Mollusca: Gastropoda: Ampullariidae: *Asolene, Felipponea, Marisa, Pomacea, Pomella*): a nomenclatural and type catalog. Malacologia 45: 41–100.

Fukuda, H., T. Haga & Y. Tatara. 2008. Niku-nuki: a useful method for anatomical and DNA studies on shell-bearing. Zoosymposia 1: 15–38.

Gamarra-Luques, C., M. Giraud-Billoud & A. Castro-Vazquez. 2013. Reproductive organogenesis in the apple snail *Pomacea canaliculata* (Lamarck, 1822), with reference to the effects of xenobiotics. Journal of Molluscan Studies 79: 147–162.

Hayes, K.A. 2009. Evolution, Molecular Systematics and Invasion Biology of Ampullariidae. Thesis: Doctor of Philosophy in Zoology (Ecology, Evolution and Conservation Biology), University of Hawaii, USA, 196p.

Hayes, K.A., R.L. Burks, A. Castro-Vazquez, P.C. Darby, H. Heras, P.R. Martín, J-W Qiu, S.C. Thiengo, I.A. Vega, T. Wada, Y. Yusa, S. Burela, M.P. Cadierno, J.A. Cueto, F.A. Dellagnola, M.S. Dreon, M.V. Frassa, M. Giraud-Billoud, M.S. Godoy, S. Ituarte, E. Koch, K. Matsukura, M.Y. Pasquevich, C. Rodriguez, L. Saveanu, M.E. Seuffert, E.E. Strong, J. Sun, N.E. Tamburi, M.J. Tiecher, R.L. Turner, P.L. Valentine-Darby & R.H. Cowie. 2015. Insights from an integrated view of the biology of apple snails (Caenogastropoda: Ampullariidae). Malacologia 58: 245–302.

Hayes, K.A., R.H. Cowie, S.C. Thiengo & E. Strong. 2012. Comparing apples with apples: clarifying the identitiesof two highly invasive Neotropical Ampullariidae (Caenogastropoda). Zoological Journal of the Linnean Society 166: 723–753.

Hylton-Scott, M. 1958. Estudio morfológico y taxonómico de los ampulláridos de la República Argentina. Revista del Museo Argentino de Ciencias Naturales Bernardino Rivadavia 3: 233–333.

Kim, J.R., K.A. Hayes, N.W. Yeung & R.H. Cowie. 2014. Diverse Gastropod Hosts of *Angiostrongylus cantonensis*, the Rat Lungworm, Globally and with a Focus on the Hawaiian Islands. PLoS One 9(5): e94969.

Lv, S., Y. Zhang, H.X. Liu, L. Hu, K. Yang, P. Steinmann, Z. Chen, L.Y. Wang, J. Utzinger & X.N. Zhou. 2009. Invasive snails and an emerging infectious disease: results from the first national survey on *Angiostrongylus cantonensis* in China. PLoS Neglected Tropical Diseases 3(2): e368.

Martin, S.M. & L.H.L. Negrete. 2007. Radular ultrasctructure of South American Ampullariidae (Gastropoda: Prosobranchia). Brazilian Journal of Biology 67: 721–726.

Michelson, E.H. 1961. On the generic limits in the family Pilidae (Prosobranchia: Mollusca). Breviora 133: 1–10.

Pincay T., L. García, E. Narváez, O. Decker, L. Martini & J.M. Moreira. 2009. Angiostrongyliasis due to *Parastrongylus (Angiostrongylus) cantonensis* in Ecuador. First report in South America. Tropical Medicine & International Health 14 (Supplement 2): S37.

Pointier, J.-P., A. Théron & D. Imbertestablet. 1988. Decline of a sylvatic focus of *Schistosoma mansoni* in Guadeloupe (French West Indies) following competitive displacement of the snail host *Biomphalaria glabrata* by *Ampullaria glauca*. Oecologia 75: 38–43.

Simberloff, D. & P. Stiling. 1996. Risks of species introduced for biological control. Biological Conservation 78: 185–192.

Souza-Lopes, H. 1955. Sôbre duas espécies do gênero *Pomacea* Perry com um estudo da genitália em ambos os sexos (Mollusca). Revista Brasileira de Biologia 15: 203–210.

Souza-Lopes, H. 1956. Sôbre *Pomacea lineata* (Spix, 1827) (Mollusca, Mesogastropoda Architaenioglossa). Revista Brasileira de Biologia 16: 375–380.

Thiengo, S.C. 1987. Observations on the morphology of *Pomacea lineata* (Spix, 1827) (Mollusca, Ampullariidae). Memórias do Instituto Oswaldo Cruz 82: 563–570.

Thiengo, S.C. 1989. On *Pomacea sordida* (Swainson, 1823) (Prosobranchia, Ampullariidae). Memórias do Instituto Oswaldo Cruz 84: 351–355.

Thiengo, S.C., C.E. Borda & J. L. Barros-Araújo. 1993. On *Pomacea canaliculata* (Lamarck, 1822) (Mollusca; Pilidae: Ampullariidae). Memórias do Instituto Oswaldo Cruz 88: 67–71.

Thiengo, S.C., K.A. Hayes, A.C. Mattos & M.A. Fernandez. 2011. A Família Ampullariidae no Brasil: Aspectos Morfológicos, Biológicos e Taxônomicos. *In* M.L.A. Fernandez, S.B. Santos, A.D. Pimenta & S.C. Thiengo (eds.) Tópicos em Malacologia (Ecos do XIX Encontro Brasileiro de Malacologia). Technical Books, Rio de Janeiro.

Thiengo, S.C., A. Maldonado, E.M. Mota, E.J. Torres, R.L. Caldeira, O.S. Carvalho, A.P. Oliveira, R.O. Simões, M.A. Fernandez & R.M. Lanfredi. 2010. The giant African snail *Achatina fulica* as natural intermediate host of *Angiostrongylus cantonensis* in Pernambuco, northeast Brazil. Acta Tropica 115: 194–199.

Truncatelloidea

Gonzalo A. Collado
Facultad de Ciencias, Departamento de Ciencias Básicas, Universidad del Bío-Bío, Chillán, Chile

INTRODUCTION

Truncatelloidea are tiny caenogastropods, with most species less than 1 cm in length. These animals occur in marine, freshwater, and terrestrial ecosystems worldwide, in many cases dominating the secondary productivity. Marine and freshwater species are characterized by having a gill and an operculum. In the Neotropical region, truncatelloideans can be found in a variety of aquatic habitats including streams, rivers, lakes, ponds, springs, canals, reservoirs, saltpans, estuaries, and caves. In this realm, the superfamily is represented by 34 native genera containing more than 260 species currently assigned to Cochliopidae, Tateidae, and Pomatiopsidae (Hershler & Thompson, 1992; Criscione & Ponder, 2013). Of these three families, Cochliopidae is the most representative group, with 31 genera: *Andesipyrgus* Sketi, Hershler & Velkovrh, 1993, *Aphaostracon* Thompson, 1968, *Aroapyrgus* Baker, 1931, *Cochliopa* Stimpson, 1865, *Cochliopina* Morrison, 1946, *Durangonella* Morrison, 1945, *Dyris* Conrad, 1871, *Emmericiella* Pilsbry, 1909, *Feliconcha* Wesselingh, 2006, *Glabertryonia* Wesselingh, 2006, *Heleobia* Stimpson, 1865, *Heleobops* Thompson, 1968, *Lithococcus* Pilsbry, 1911, *Littoridina* Souleyet, 1852, *Littoridinops* Pilsbry, 1952, *Longosoma* Wesselingh, 2006, *Mesobia* Thompson & Hersler, 1991, *Mexicenotica* Grego, Angyal and Liévano-Beltrán, 2019, *Nanivitrea* Thiele, 1928, *Onobops* Thompson, 1968, *Pterides* Pilsbry, 1909, *Pyrgophorus* Ancey, 1888, *Rachipteron* Thompson, 1964, *Sioliella* Haas, 1949, *Subcochliopa* Morrison, 1946, *Tepalcatia* Thompson & Hershler, 2002, *Texadina* Abbott & Ladd, 1951, *Toxosoma* Conrad, 1874, *Tropidebora* Pilsbry, 1944, *Tryonia* Stimpson, 1865 and *Zetekina* Morrison, 1947. The shell shapes vary from low to highly conic-elongated. They are found over the entire Neotropical region and represent the more ecologically diverse group of the superfamily. The Neotropical Pomatiopsidae, represented by two genera, *Idiopyrgus* Pilsbry, 1911 and *Spiripockia* Simone, 2012, in general have a turreted shell with a distinctive aperture. They are restricted to the Amazonas River Basin. Only one endemic genus of Tateidae, *Potamolithus* Pilsbry, 1896, is represented in the Amazonas River Basin, La Plata River Basin, and Patagonia, with about 50 species that inhabit epigean and subterranean ecosystems.

LIMITATIONS

The classification and identification of the Neotropical freshwater truncatelloideans to species, genus, and family levels is difficult. Besides the high diversity, many are morphologically similar, and the taxonomic challenges are heightened by their simple and constant shells, small size, few available characters to delineate species, and convergent and parallel evolution (Liu et al., 2003; Criscione & Ponder, 2013; Collado et al., 2016). It is important to mention that the low number of specialists, the limited number of monographic works published so far, and still large unexplored geographical areas combined with the limited taxonomic knowledge accumulated in some taxa of the group also creates difficulties. Because species identification is mainly based on reproductive traits, known only in very few species, only the families and genera are used in the keys.

Species descriptions have historically based on shell morphology. However, the size, protoconch, aperture shape, operculum, male and female reproductive systems, development type, radula structure, and other morphological and anatomical features have also been sometimes included. Of these, penis, radula, and reproductive systems have been intensively used in all taxonomic levels. Recently, species separation has been supported with new technological advances especially in scanning electron microscopy.

TERMINOLOGY AND MORPHOLOGY

Terminology provided by Hershler & Ponder (1998) for "hydrobioid" snails is used here. Hershler & Thompson (1992) also provided a comprehensive morphological study of Cochliopidae taxa. A brief summary is provided here. The general shape of the **shell** varies widely within the truncatelloideans. Fig. 11.67 illustrates the external shell features. The shell shape can be planispiral, helicoid, valvatiform, trochiform, ovate-conic, elongate-conic, or turriform (Fig. 11.68 A–G). However, it is important to note that intermediate forms may be present. The whorls of an adult form the typical shell or **teleoconch**. The earliest shell is the **protoconch**, also called embryonic or larval shell, which can be found at the shell apex. In some individuals, it may be eroded. The **operculum** (Fig. 11.69 A–D), a corneous plate that serves as a gateway into the shell, may be circular, ovate, or ellipsoidal. It is usually thin, flat, transparent, semitranslucent, yellow-beige, brown, light brown, brown, or orange-brown. The structure has a nuclear region from which it grows a new material that is added in a spiral or circular mode. In the inner face, an attachment scar indicates where the operculum adheres to the muscle of the foot.

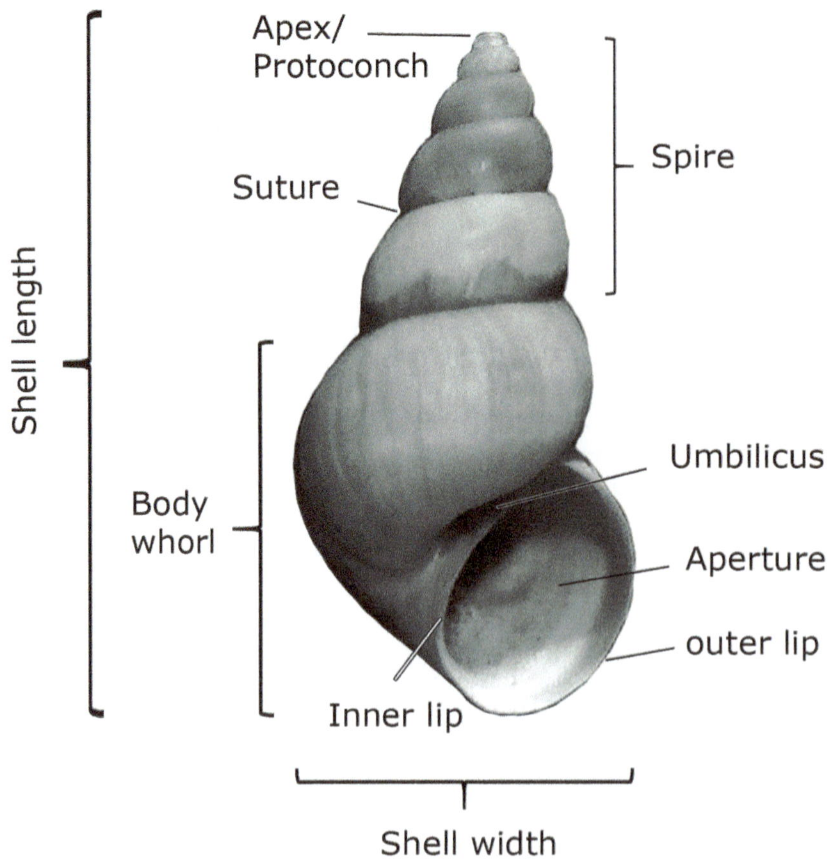

FIGURE 11.67 External view of a specimen of the genus *Heleobia* (Cochliopidae) to show shell characters.

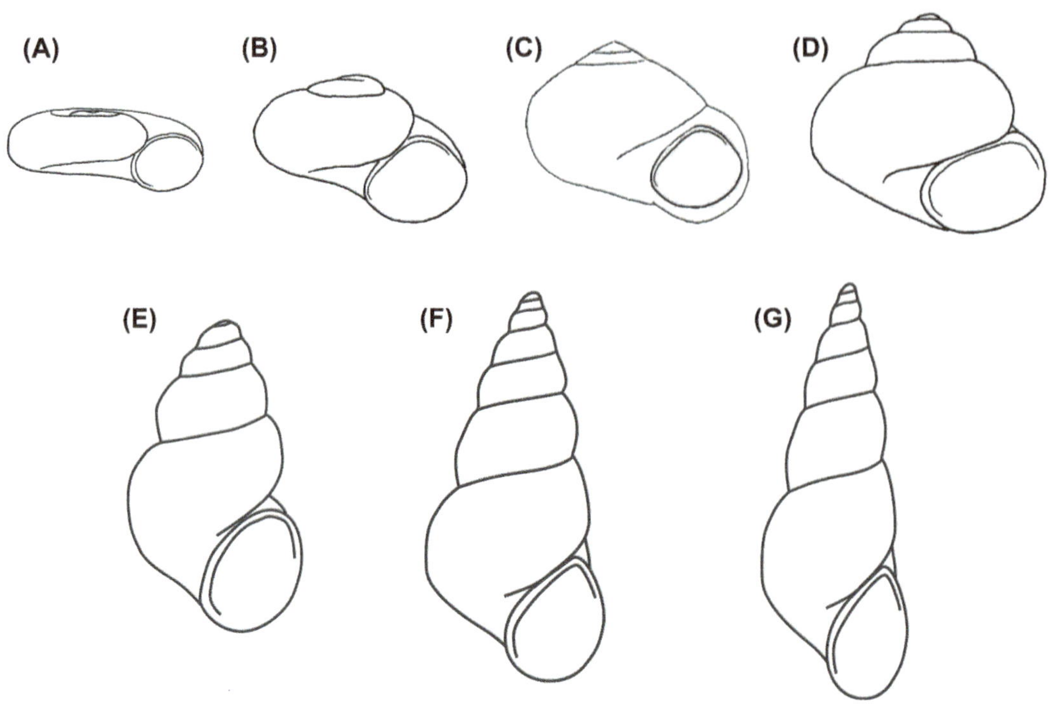

FIGURE 11.68 Shell shapes in Truncatelloidea: (A) Planispiral; (B) helicoid; (C) valvatiform; (D) trochoid; (E) ovate-conic; (F) elongate-conic; (G) turreted. *Modified from Hershler & Ponder (1998).*

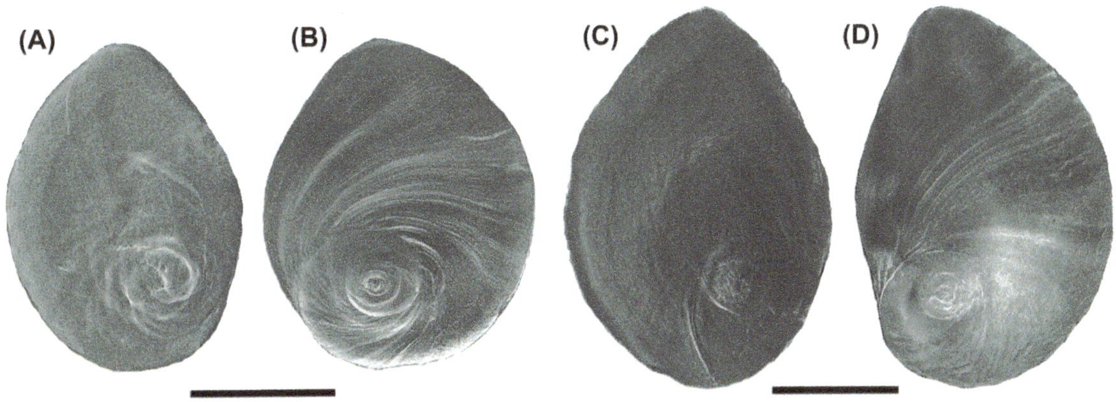

FIGURE 11.69 Operculum of truncatelloidean taxa: (A) *Potamolithus* sp., internal view; (B) *Potamolithus* sp., external view; (C) *Heleobia* cf. *cumingii*, internal view; (D) *Heleobia* sp., external view. Scale bars = 400 μm.

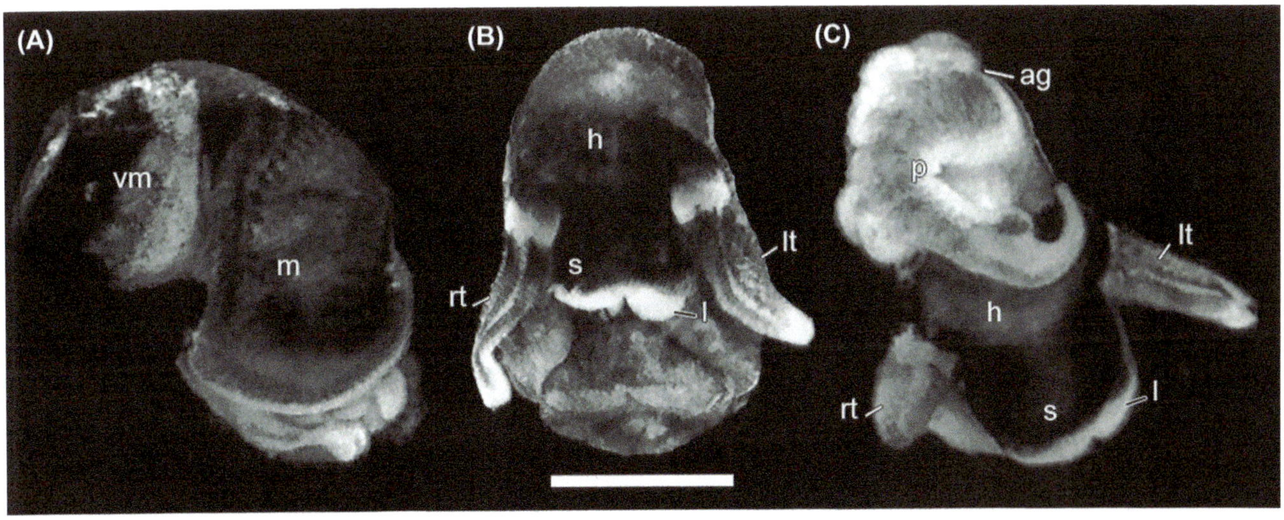

FIGURE 11.70 External anatomical characters of a typical truncatelloidean represented by *Heleobia*: (A) Part of soft body showing the mantle and proximal portion of visceral mass of an unsexed individual; (B) female specimen after mantle dissection; (C) male specimen after mantle dissection. Abbreviation: *ag*, apocrine gland; *h*, head; *l*, lip; *lt*, left tentacle; *m*, mantle; *p*, penis; *rt*, right tentacle; *s*, snout; *vm*, visceral mass. Scale bar = 1 mm.

Fig. 11.70 A–C illustrate some external body features. A **penis**, located in the neck of the snail, a short distance of the right cephalic tentacle, can be viewed in males. Several characters of the penis are used in the keys. This organ can be simple, without accessory structures, or more complex, including conspicuous accessory lobes, aprocrine glands, or papillae. In the distal section, a terminal papilla is usually present. In a few species, males are rare while in others the females develop a nuchal node.

The **reproductive system** of truncatelloideans varies among genera and species depending on the reproductive strategy. Fig. 11.71 A illustrates a simplified model of the female reproductive system of a typical oviparous species while Fig. 11.71 B shows that of a male. Development can be either direct or indirect. Most of species are oviparous.

The **radula** of a truncatelloidean is of the taenioglossate type, with seven teeth in each row, including one lateral tooth and two marginal teeth, inner and outer, on each side of the central tooth (Fig. 11.72). The central tooth, also called median or rachidian tooth, frequently has several cusps, the largest in the middle. The type, number, size, and organization of the cusps of the teeth differ among taxa, mainly at supraspecific level.

MATERIAL PREPARATION AND PRESERVATION

Truncatelloidean snails can be treated according to the methods described in Geiger et al. (2007) for small microgastropods. However, a brief protocol is provided here according to the experience of the author.

Truncatelloidean snails inhabit aquatic ecosystems containing sediment, salts, and others suspended materials that may adhere strongly to the shell making difficult to

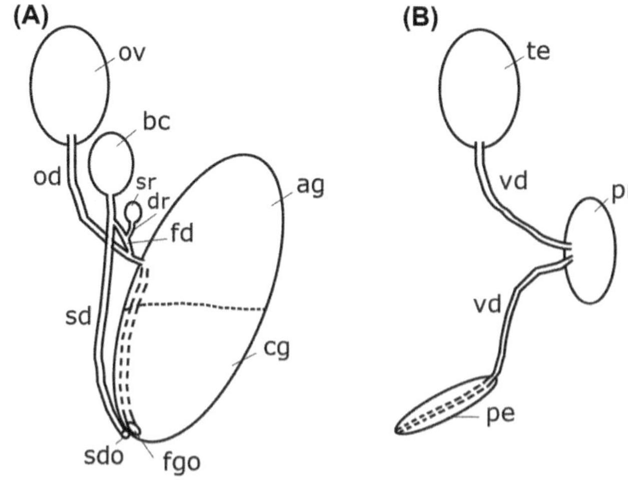

FIGURE 11.71 Schematic representation of a truncatelloidean female (A) and male (B) reproductive system. Abbreviations: *ag*, albumen gland; *bc*, bursa copulatrix; *cg*, capsule gland; *dr*, duct from seminal receptacle; *fd*, fertilization duct; *fgo*, female genital opening; *od*, oviduct; *ov*, ovary; *pe*, penis; *pr*, prostate; *sd*, spermathecal duct; *sdo*, spermathecal duct opening; *sr*, seminal receptacle; *te*, testis; *vd*, vas deferens.

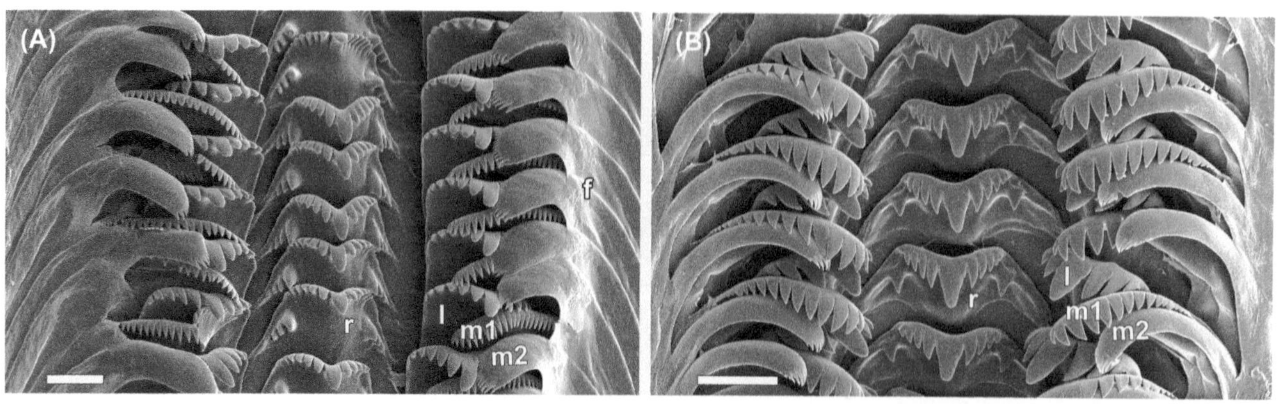

FIGURE 11.72 Truncatelloidean radula showing the central or rachidian tooth (r), one lateral tooth (l) and two marginal teeth, the inner (m1) and the outer (m2): (A) *Potamolithus* sp. (Tateidae); (B) *Heleobia* sp. (Cochliopidae). The flange of the external marginal teeth is more developed in Tateidae. Scale bars: (A) = 10 μm, (B) = 20 μm.

obtain good images of animals. Due to its small size, they need to be cleaned using a stereomicroscope. For this, the snails should be transferred to a small Petri dish containing water or 70% ethanol. To avoid breaking the shell, the tip of the forceps should be wrapped with parafilm. The material adhered can be removed from the shell with a paint brush while gently holding the snail with the tweezers. The material strongly adhered can be removed during cleaning with the help of a bent needle. For cleaning, snails can also be immersed in a diluted sodium hypochlorite solution for five to ten minutes to remove the periostracum, although chlorine damages tissue. Prior to the transference of hypochlorite solution to alcohol (and vice versa), the snails should be rinsed with distilled water. Time can be reduced by removing the periostracum with the paint brush during transient washings in water because the hypochlorite may also damage the bristles of the brush. Because the shell generally has an elongated conical shape, it is difficult to focus at the same time on the apex, the body whorl, and the aperture. This can be solved by placing a small hard object under the apex of the shell to tilt it upward. Metal objects are not recommended because they shine much with the incident light. Good quality images can be obtained by photographing snails immersed in water or alcohol using the stereo microscope. Photographs of dry specimens can also be obtained using this instrument or a scanning electron microscope (SEM).

For anatomical studies it is advisable to put the snails in solution of menthol crystals for relaxation for about 12 h, rinse the animals in distilled water, and then preserve in 70% ethanol (Bowler et al., 1996). Snails can also be relaxed in an anesthetizing isomolar solution of 10% magnesium chloride for about 30 min. Dissection should be conducted under magnification in a Petri dish containing water or ethanol. To isolate the soft body, the snails should get between the sheets of the tweezers and then gently press to break the shell. The snail should be returned to the liquid to remove the remains of the shell. The operculum can be isolated from the foot making a straight cut with a scalpel while holding the head-foot of the animals with the tweezers. In a similar way, the penis can also be isolated from the head after removing the mantle. The operculum can be cleaned immersing it in the diluted hypochlorite solution for about three to five minutes. It is important to rotate continuously this structure with the bent needle while it is immersed in this solution for cleaning both surfaces at the same time. A prolonged period of time in the hypochlorite solution can damage the operculum, so it has to be transferred to distilled water with a dropper immediately after cleaning. To study the radula, the radular sac needs to be dissected from the soft body. This is achieved performing a cut with a scalpel in the frontal area of the head to separate the cephalic tissue and remove the sac containing the structure. The radula can be obtained immersing the radular sac for about 12–24 h in 10% sodium hydroxide to remove the organic material. This can also be made holding the sac in a diluted sodium hypochlorite solution for three to 10 minutes. After being rinsed with distilled water, the radula can be preserved in 70% ethanol. The opercula and radulae can be transferred from one solution to another or mounting means with a dropper. Observation can be performed in a light microscope or SEM. For this, the specimen is suspended in a drop of liquid (usually water) and oriented on the mounting medium. For light microscopy, the specimen is mounted and oriented directly on the slide. For SEM, the specimen can be mounted on double-sided sticky tape, or double-sided carbon tabs. In both cases, the water will evaporate in a few minutes.

For DNA analysis where gross morphology is not of immediate interest, specimens sampled can be placed directly in 95% ethanol, absolute ethanol, or be frozen. Formalin is not recommended. To obtain tissue, the soft body should be isolated from the shell as already mentioned. All the soft body of the snail can be used for DNA extraction; however, even in small specimens, the gill and mantle may provide sufficient amounts of genomic DNA to perform standard amplifications of different molecular markers.

KEY TO TRUNCATELLOIDEA

Gastropoda: Caenogastropoda: Truncatelloidea: Families

1	Shell planispiral to trochoid, globose, conic, or elongate conic	2
1′	Shell weakly elongated or turreted, peristome expanded or having a lip with a rounded sinus	**Pomatiopsidae [p. 351]**
2(1)	Radula with outer marginal teeth large and thin, bearing a very narrow flange	**Cochliopidae [p. 351]**
2′	Radula with outer marginal teeth bearing a wide flange	**Tateidae [p. 353]**

Gastropoda: Caenogastropoda: Truncatelloidea: Pomatiopsidae: Genera

1	Shell fragile, translucent, with pustules; peristome highly expanded; subterranean [Brazil]	*Spiripockia punctata* Simone, 2012
1′	Shell solid, turreted, spire long [Brazil]	***Idiopyrgus* [p. 351]**

Gastropoda: Caenogastropoda: Truncatelloidea: Pomatiopsidae: *Idiopyrgus*: Species

1	Whorls strongly convex [Brazil]	*Idiopyrgus souleyetianus* Pilsbry, 1911
1′	Whorls very lightly convex [Brazil]	*Idiopyrgus rudolphi* (Haas, 1938)

Gastropoda: Caenogastropoda: Truncatelloidea: Cochliopidae: Genera

1	Shell planispiral to elongate-conic	2
1′	Shell long and narrow, with many convex whorls (7 to 10), aperture very expanded; freshwater [Northern Mexico]	*Pterides*
2(1)	Shell with short spire	3
2′	Shell with high spire	7

3(2) Shell helicoid 4

3′ Shell near planispiral to trochoid 5

4(3) Penis small, basally with single nonglandular lobe on inner curvature; Pacific coastal streams, brackish, freshwater .... *Cochliopa*
[Costa Rica, Panama]

4′ Penis, large, slender, distally pointed, with dark internal pigment; Pacific coastal streams .... *Subcochliopa*
[Costa Rica, Panama]

5(3) Shell valvatiform to slightly trochoid 6

5′ Shell trochoid; Pacific coastal streams; freshwater .... *Lithococcus*
[Costa Rica, Colombia, Ecuador]

6(5) Penis simple, without prominent folds, freshwater .... *Cochliopina*
[Neotropics. Nearctic]

6′ Penis slender, inner curvature with folds proximally and low, nonglandular lobe distally, freshwater .... *Nanivitrea*
[Cuba, Jamaica]

7(2) Shell ovate-conic to pupiform 8

7′ Shell most commonly elongate-conic 11

8(7) Aperture length < 0.5× shell length 9

8′ Aperture length ≈ 0.5× shell, with siphonal notch; freshwater .... *Sioliella effusa* Haas, 1949
[Brazil: Pará]

9(8) Penis tip simple 10

9′ Penis tip bilobed, dorsal surface at base of right terminal lobe with a dark spot; fresh and brackish water .... *Rachipteron philopelum* Thompson, 1964
[Costa Rica: Puntarenas]

10(9) Penis distal portion with a tapering, smooth filament; freshwater .... *Aroapyrgus*
[Mexico, Guatemala, El Salvador, Nicaragua, Costa Rica, Panama, Colombia, Brazil]

10′ Penis distal tip strongly tapered, without filament; freshwater .... *Andesipyrgus sketi* Hershler & Velkovrh, 1993
[Colombia: Santander. Ecuador: Napo Basin]

11(7) Aperture not expanded 12

11′ Aperture expanded; fresh and brackish water .... *Emmericiella*
[Mexico: Tamaulipas]

12(11) Teleoconch smooth, with spiral lines or striae 13
[Neotropics. Nearctic]

12′ Teleoconch with two or three prominent sharp keels on the first two teleoconch whorls; freshwater ...... *Dyris amazonicus* (Haas, 1949)
[Brazil]

13(12) Penis simple 14

13′ Penis complex, with distal appendices; fresh and brackish water .... *Pyrgophorus*
[Neotropics. Nearctic]

14(13) Penis distal tip bulbous 15

14′ Penis distal tip tapered 17

15(14) Prostate gland small, with short pallial section 16
[Mexico, Guatemala. Nearctic]

15′ Prostate gland elongate, with large pallial section .... *Zetekina*
[Central America]

16(15) Females oviparous; freshwater .... *Aphaostracon*
[USA: Florida]

16′ Females ovoviviparous; fresh and brackish water .... *Tryonia*
[Mexico, Guatemala. Nearctic]

17(14) Surface of the penis with accessory structures 18

17′ Surface of the penis with small folds in the first two-thirds of the organ (without papillae nor apocrine glands) ...... *Onobops/Tepalcatia*
[*Onobops*: Atlantic Coast from Maryland, USA to Veracruz, Mexico]
[*Tepalcatia*: Rio Balsas Basin, central Mexico]

18(17) Penis surface with apocrine gland 19

18′ Penis surface with papillae 21

19(18) Penis apocrine gland nonstalked 20

19′ Penis apocrine gland stalked .... *Heleobia/Texadina*
[*Heleobia*: Neotropics. Europe, Asia, and Africa]
[*Texadina*: Gulf of Mexico, Jamaica. Nearctic]

20(19) Distributed in Caribbean Island; brackish water .... *Heleobops*
[Caribbean Islands. Nearctic]

20′ Distributed in Central America; freshwater .... *Mesobia pristina* Thompson & Hershler, 1991
[Honduras: Lago de Yajoa]

21(18) Penis papillae arranged in series along outer curvature and distal half of inner curvature; fresh and brackish water .... *Littoridinops*
[Caribbean Island. North America]

21′ Penis papillae single, paired, in clusters or on low lobular swellings; fresh and brackish water .... *Littoridina*
[Neotropics. Nearctic]

Gastropoda: Caenogastropoda: Truncatelloidea: Tateidae: Genera

1 Shell thin, variable in shape; operculum orange-brown, with a white smear on attachment area .... *Potamopyrgus antipodarum* (Gray, 1843)
[Chile: Northern and central region. Invasive]

1′ Shell relatively solid, operculum without white smear .... *Potamolithus*
[Brazil, Paraguay, Chile, Argentina, Uruguay]

REFERENCES

Bowler, P.A., T. Johnson & W.J. Mautz. 1996. Flash-freezing using cold fluids: field and laboratory methods for preventing retraction of snails during fixation. Journal of Molluscan Studies 62: 124–126.

Collado, G.A., M.A. Valladares & M.A. Méndez. 2016. Unraveling cryptic species of freshwater snails (Caenogastropoda, Truncatelloidea) in the Loa River basin, Atacama Desert. Systematics and Biodiversity 14: 417–429.

Criscione, F. & W.F. Ponder. 2013. A phylogenetic analysis of rissooidean and cingulopsoidean families (Gastropoda: Caenogastropoda). Molecular Phylogenetics and Evolution 66: 1075–1082.

Geiger, D.L., B.A. Marshall, W.F. Ponder, T. Sasaki & A. Warén. 2007. Techniques for collecting, handling, preparing, storing and examining small molluscan specimens. Molluscan Research 27: 1–50.

Hershler, R. & W.F. Ponder. 1998. A review of morphological characters of hydrobioid snails. Smithsonian Contributions to Zoology 600: 1–55.

Hershler, R. & F.G. Thompson. 1992. A review of the freshwater gastropod subfamily Cochliopinae (Prosobranchia, Hydrobiidae). Malacological Review (Supplement) 5: 1–140.

Liu H-P, R. Hershler & K. Clift. 2003. Mitochondrial DNA sequences reveal extensive cryptic diversity within a western American springsnail. Molecular Ecology 12: 2771–2782.

Cerithioidea

Maria I. Berning
Bavarian State Collection of Zoology (ZSM), Section Mollusca, Munich, Germany

INTRODUCTION

The Cerithioidea is one of the largest gastropod groups and currently includes 20 recent families, approximately 168 extant genera, and roughly 890 extant valid species (WoRMS, 2019). Cerithioidea has representatives occurring in marine, estuarine, and freshwater habitats around the world (Ponder & Warrén, 1988). Among limnic taxa, several families have successfully radiated in ancient lakes as well as lotic systems in North, Central, and South America; the Mediterranean Basin; Southeast Asia and Madagascar (Strong et al., 2011). It is also a very diverse taxon regarding reproductive morphology and brooding modes, which vary considerably in their complexity and general organization (Köhler & Glaubrecht, 2005). In Neotropical freshwaters, two cerithioidean families are present: Pachychilidae and Thiaridae.

The freshwater Pachychilidae, with a worldwide distribution, is still one of the most poorly known cerithioid groups (Strong et al., 2011). The original name "Pachychili" and its family status have changed several times during the last two centuries. Most Neotropical members were first placed into the genus *Melania* Lamarck, 1799 and then later moved to other cerithioidean families like Thiaridae, Paludomidae, Melanopsidae, and Pleuroceridae. Based on morphological and molecular data, they are now members of the monophyletic Pachychilidae (Strong et al., 2011).

Pachychilidae includes 11 genera with about 250 recognized species distributed in the tropical regions of America, Africa, Asia, and Madagascar (Köhler & Glaubrecht, 2007). In the Neotropics, the taxon comprises only two genera: *Doryssa* H. Adams & A. Adams, 1854 and *Pachychilus* I. Lea & H.C. Lea, 1850. Pachychilidae has a disrupted distribution, with *Pachychilus* spread from San Luis Potosí (Mexico) to Nicaragua and Cuba, and *Doryssa* ranging from the Andean border between Colombia and Venezuela, through the Guianas to southern Brazil. There are about 50 valid described species of *Doryssa* and approximately 102 *Pachychilus*. These species inhabit both lotic and lentic ecosystems and are still widely used by native communities as lime and supplementary protein sources (e.g., Nations, 1979; Healy et al., 1990; Royero, 1994–1996; McKillop, 2010).

Thiaridae is a worldwide-distributed family, primarily circumtropical, but with many species extremely widespread due to their "traveler" abilities (Glaubrecht, 1996). These species can either host parasites affecting humans and domestic animals, or can successfully control other gastropods that are intermediate parasite hosts (Faria-Vaz et al., 1986; Ferrer-Lopez et al., 1991; Pointier et al., 1994; Gutierrez et al., 1997; Simone, 2001). They are also economically important because they can block water pipes and damage equipment (Appleton et al., 2009).

As early as 1898, Moore raised the idea of thiarid polyphyly based on morphology. Recent molecular studies confirm that Thiaridae *sensu lato* is a polyphyletic group

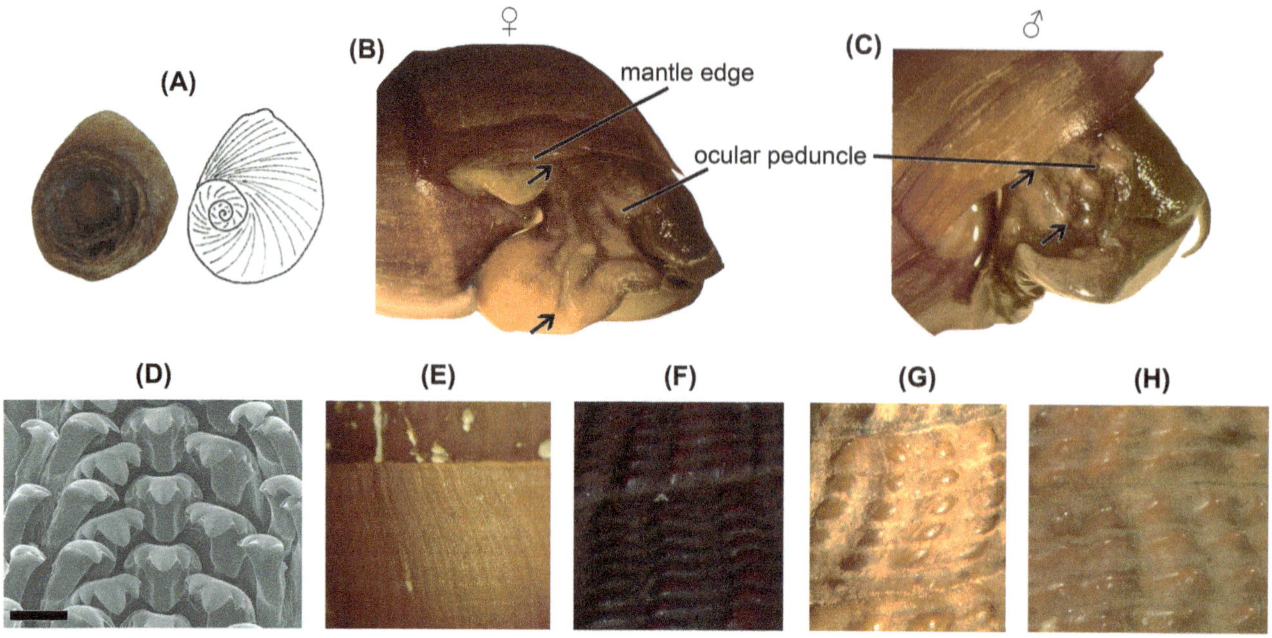

FIGURE 11.73 Diagnostic features of the Pachychilidae: (A) multispiral operculum; (B) female, *arrows* showing the extension of egg-laying sinus; (C) male, *arrows* showing the extension of the genital groove; (D) radula detail, scale bar = 100 μm; (E) crosswise sculpture; (F) threadlike axial elements; (G) tubercles; (H) crenulation.

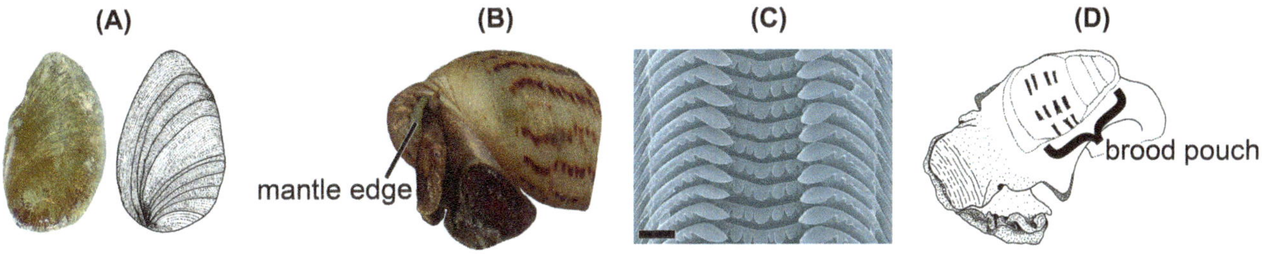

FIGURE 11.74 Diagnostic features of the Thiaridae: (A) paucispiral operculum; (B) mantle edge papillate; (C) radula detail, scale bar = 50 μm; (D) brood pouch with embryonic shell. Picture (B) courtesy of Bill Frank, www.jaxshells.org.

comprising at least three independent evolutionary lineages. Thiaridae *sensu stricto* includes only animals with a brood pouch in the neck region, as well as a midgut with a deep spiral caecum and a large accessory pad (Gomez, 2009). There are approximately 110 thiarid species worldwide (Strong et al., 2011), ranging from rocky, fast-moving streams, to broad, slow moving rivers and quiet lakes (Houbrick, 1988). All native Neotropical thiarids belong to the genus *Aylacostoma* Spix, 1827 with some South American species being classified under the subgenera *Basistoma* Lea, 1852, *Longiverena* Pilsbry & Olsson, 1935 and *Verena* Moricand, 1841. *Aylacostoma* comprises about 35 nominal species (Vogler et al., 2014) that occur in Central and South America as well as in the Antilles. The introduced *Melanoides tuberculata* (Müller, 1774) and *Tarebia granifera* (Lamarck, 1816) are now widespread in the Neotropics, occurring in almost every freshwater ecosystem. Both species are usually found sharing habitats with local or endemic species, with population densities well above those of its native relatives (Vázquez Perera & Perera Valderrama, 2010).

LIMITATIONS

Freshwater species of the cerithioidean have particularly highly polymorphic shells, contributing to such a proliferation of names during the 18th and 19th centuries that the estimated ratio of available names to valid species is as high as 10:1 (Strong et al., 2011). There are no comprehensive catalogs or taxonomic lists including all species of Neotropical thiarids or the South American pachychilids. Most of the existing research is incomplete and old, with mainly cryptic specific descriptions and only a few generic monographs. The rest of the literature just focused on biological and ecological observations as a result of broad regional studies or different types of inventories (e.g., Parodiz, 1982; Pérez & López, 2003; Naranjo Garcia, 2003a, 2003b). The only studies on the biogeography of *Hemisinus* and *Pachychilus* are from Simpson (1894) and Goodrich & van der Schalie (1937). Publications of Lydeard et al. (2002) and Köhler et al. (2004) are until now the only ones using morphological and molecular evidence to depict the relationships between some of the Central and South American species belonging to Pachychilidae and Thiaridae.

Shell features are the base for Neotropical cerithioidean species descriptions. Because shell shape varies depending on environmental factors, the only way to unravel the true number of species or cryptic species complexes is the use of integrative taxonomic methods.

As far as possible, the keys include valid names for introduced species, trying to avoid synonymy, in which case, the older name is preferred. Species features are based on the original descriptions, and the figures include several type specimens. Type locality of each species appears in brackets under each name in the figure legends. Specimens depicted may come from localities different than the type locality. Because Pachychilidae and Thiaridae species are relatively poor dispersers, shells with the same or very similar appearance from distant type localities were often originally treated as different species. Shells not included in the key could be morphological variations of already described species or new species.

The Pachychilidae needs a modern revision, and there are no previous keys to species. Their taxonomy is poorly resolved mainly because of the inconsistent usage of available names and the paucity of integrative studies. Gomez (2009) found that shell and radula characteristics are unsuitable for distinguishing most *Pachychilus* species, and the radulae of *Pachychilus* and *Doryssa* are not separable. The works of Fischer & Crosse (1870–1902) and Martens (1890–1901) represent the first comprehensive investigations of the Mesoamerican pachychilids. Fischer and Crosse (1892) split Mexican and Guatemalan species into five different "sections" (i.e., *Pachychilus s.s*, *Cercimelania*, *Glyptomelania*, *Oxymelania*, and *species incertae*), assigning type species to each, which has led to numerous subsequent misinterpretations (Gomez-Berning et al., 2012). More recently, Simone's (2006) catalog of land and freshwater mollusks of Brazil and Thompson's (2011) checklist of land and freshwater snails from Mexico and Central America tried to compile all available names and type material. Both works need to be approached very carefully.

Simone (2006), for example, wrongly assigned Pachychilidae species to the Thiaridae and vice versa, as well as *Doryssa* to the Pleuroceridae. Thompson (2011) maintained Fischer and Crosse's type species for each "section" (or subgenus), included some Nearctic taxa, and assigned several thiarids and pleurocerid species to Pachychilidae. The last thorough revision of *Pachychilus* (Gomez-Berning et al., 2012) provided all valid and available names, including their type location and type specimens, as the starting point for future integrative taxonomical studies. Fischer and Crosse (1892) described in detail the foregut anatomy as well as the reproductive and nervous system of *Pachychilus* in general. More recently, Simone (2001) thoroughly described one Mexican specimen.

Doryssa is a group that has fallen into oblivion despite its abundance in South American lotic environments. Species of *Doryssa* appear in Reeve's monograph of the genus *Melania* (Reeve, 1851–1870), and since then mainly in species lists of biological assessments of aquatic ecosystems (Alonso & Berrenstein, 2006; de Jesus et al., 2008) or in anthropological studies (Royero, 1994–1996). Tucker-Abbot (1955) and Simone (2001) are the only anatomical descriptions of *Doryssa* species.

The Thiaridae is both one of the richest gastropod components of freshwater faunas and one of the least known. There are numerous species descriptions, of which the taxonomic statuses remain unresolved, mainly because they only take into account shell features (Köhler & Glaubrecht, 2003). Also, phylogenetic relationships cannot be determined due to the lack of sufficient, adequately preserved body samples and molecular studies (Gomez, 2009).

There is no comprehensive revision of *Aylacostoma* or taxonomic keys (Vogler et al., 2014). Gomez et al. (2011) provided the only exhaustive revision of a Caribbean thiarid species. There are some checklists for Argentinian *Aylacostoma* and their distribution (Gutiérrez Gregoric et al., 2006; Rumi et al., 2006, 2008; Cuezzo, 2009). Also, some descriptions of Argentinian and Brazilian species are available (Hylton Scott, 1953, 1954; Simone, 2001; Vogler et al., 2014). Increasing numbers of introduced species like *Melanoides* and *Tarebia*, and hydroelectric power projects (Townsend & Newell, 2006; Vogler et al., 2014) represent a serious threat to Neotropical thiarids.

TERMINOLOGY AND MORPHOLOGY

The plesiomorphic traits characterizing cerithioideans are aphallate males, open gonoducts, and reproduction via spermatophores. They are also unique in the range of midgut morphologies; in reproductive (ovipary vs. vivipary or ovoviviparity) and life history strategies (gonochorism vs. parthenogenesis); and in the complexity of the female pallial reproductive tract, which bears specialized sperm-storage pouches of uncertain homology to those of other caenogastropods (Strong et al., 2011). Cerithioideans usually have short radulae characterized by central teeth with serrated edges, middle teeth that migrate to the side, and marginal teeth with dentate edges (Thiele, 1929–1935). In general, they have a thick and well-developed hypobranchial gland (as can be observed in *Pachychilus*), and epiathroid dialyneurous nervous systems with zygoneury occurring only in some estuarine and freshwater families (which may indicate some selective advantage in this environment) (Houbrick, 1988; Glaubrecht, 1996).

All Neotropical Caenogastropoda have a spiral **shell** opening to the right (Fig. 11.75 A). For description purposes, a **whorl** is each complete rotation of the shell, and the overlapping of whorls form the **spire** (Fig. 11.75 A). Whorls can be flattened (Fig. 11.75 H, J), slightly convex (Fig. 11.75 E, F), or convex (Figs. 11.75 A, 81 G). The indentation line between whorls is the **suture** (Fig. 11.75 A). Whorl number is counted with the aperture of the shell looking to the front. The last whorl of the shell includes the **aperture** (Fig. 11.75 B). The aperture is divided into three areas: the **external margin or lip**, the area next to the penultimate whorl of the shell or **callus**, and the **columella**, which is the tip of the central axis of the shell (Fig. 11.75 B). Shells can be complete (Fig. 11.75 A), have an eroded apex (Fig. 11.75 D, E) or be decollated (Fig. 11.75 C, H). Caenogastropoda shells can also be smooth (Figs. 11.73 E, 75 A, E) or sculptured. The shell sculpture can include different grades of spiral elements (Fig. 11.75 D) or axial elements (Fig. 11.75 C) or both (Fig. 11.75 C). The degree of spiral and axial elements combination determines the presence of threads (Fig.11.73 F), tubercles (Fig. 11.73 G), knots or keels (Figs. 11.73 H, 87 C), and plaits (Figs. 11.75 G, 77 E). Shell shape is a very variable feature. It can be ovate-conic (Fig. 11.75 E), oblong-conic (Fig. 11.75 F), conic (Fig. 11.75 G), elongate-conic or turriform (Fig. 11.75 H), cylindrical (Fig. 11.75 I), fusiform (Fig. 11.75 J), or subcylindrical (Fig. 11.75 K). Shells can have color spots either scattered (Fig. 11.76 E) or organized in lines or bands (Figs. 11.76 C, 85 A, B).

Pachychilidae are dioecious, egg-laying organisms. The animals are characterized by the presence of an operculum with a central nucleus and concentric whorls (multispiral operculum, Fig. 11.73 A); smooth mantle edge (Fig. 11.73 B); short ocular peduncles (Fig. 11.73 B, C); very long radula, and numerous otoliths (Morrison, 1954). The kidney morphology of pachychilids is very different from other freshwater taxa. Like marine forms, the organ has no internal subdivisions, and in most members of the family, the nephropore is concealed beneath a unique flap

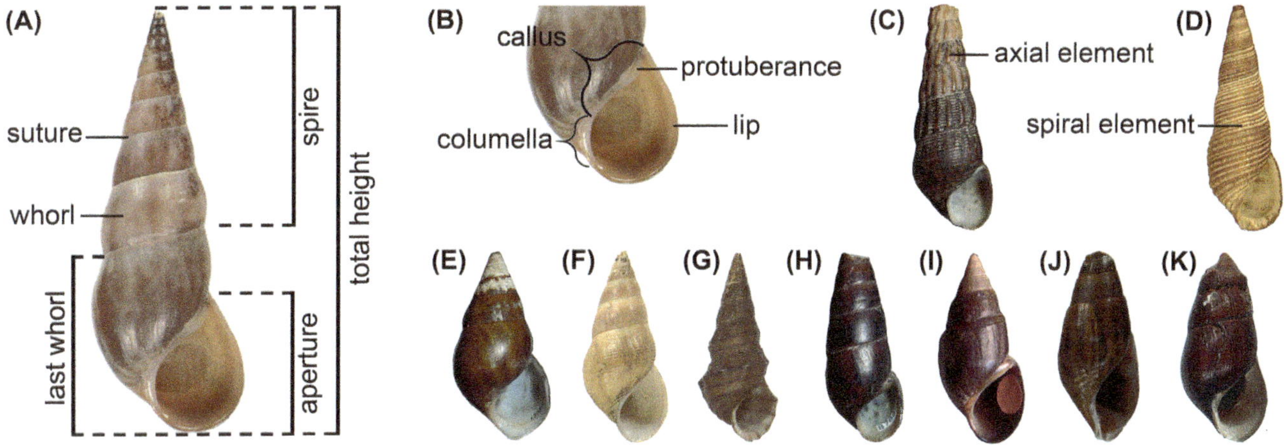

FIGURE 11.75 Cerithioidea shell features: (A) shell total height, spire, suture, last whorl, aperture; (B) aperture detail showing callus, callus protuberance, columella, lip. Shell sculpture: (C) axial element and spire decollated; (D) spiral elements: spiral lines (elevated) separated by spiral grooves (depressed). Shell shapes: (E) ovate-conic; (F) oblong-conic; (G) conic (or pyramidal); (H) elongate-conic (or turriform); (I) cylindrical; (J) fusiform (or spindle-shape); (K) subcylindrical.

extending from the gonoduct. In contrast, all other freshwater taxa (Melanopsidae, Paludomidae, Pleuroceridae, Semisulcospiridae, and Thiaridae) have an internally subdivided kidney, with a bladder communicating externally via the nephropore (Strong et al., 2011).

Neotropical Pachychilidae are nonparthenogenetic species with sexual dimorphism, the later evidenced by the shape and length of a genital groove in the right side of the head-foot (Fig. 11.73 B, C). All *Pachychilus* species share a radula with a square middle tooth with seven denticles at the edge; lateral tooth with five to six cusps, and strong marginal teeth with two to three small cusps (Fig. 11.73 D). *Doryssa* animals have eyes not visible externally; ctenidial vein very broad in its posterior half; gill filaments long and sharp; stomach without complex folds, and pallial gonoducts with massive laminas, without chambers (Simone, 2001).

Thiaridae are defined by a paucispiral operculum with a small subterminal nucleus (Fig. 11.74 A); mantle edge with papillae (which can be quite long, Fig. 11.74 B); thin and weak hypobranchial gland; large osphradium the length of which may be a little less than the ctenidial length; outer marginal radular tooth with large lateral flange (which is an unusual feature of some cerithioidean radulae, Fig. 11.74 C); salivary glands above and anterior to nerve ring; absence of zygoneury; and a head-foot brood pouch (Fig. 11.74 D) innervated by extensions of the pedal ganglia. The group shows a reproductive strategies spectrum ranging from true parthenogenesis to bisexual reproduction, with parthenogenesis as an autapomorphy of only some thiarid genera. Organisms in this group also give birth to small-shelled snails (Houbrick, 1988; Gomez, 2009). Regarding *Aylacostoma*, there are only anatomical descriptions of the Jamaican and Cuban species (Gomez, 2009; Gomez et al., 2011), and of five animals from Brazil and Argentina (Simone, 2001; Vogler et al., 2014).

Aphallate: Without copulatory organ.

Autapomorphy: Derived trait unique to a given taxon.

Ctenidial vein: Supplies and drains blood from the gill.

Ctenidium: Respiratory organ or gill.

Dialyneury: Arrangement of the nervous system where each pleural ganglion connects with the opposite half of the visceral commissure via the pallial nerve.

Dioecious: Male and female individual organisms.

Epiathroid: State in which the pleural and the pedal ganglia lie close to the cerebral ganglia, forming a neural cluster.

Gonochorism: State of having just one of at least two distinct sexes in any one individual organism.

Gonoduct: Duct leading from the gonad to the exterior.

Hypobranchial gland: Thickening of the tissue located in the roof of the animal's mantle cavity, between the ctenidium and the rectum.

Mantle edge: Dorsal body wall that covers the visceral mass.

Midgut or **stomach**: Part of the digestive tract specialized for storing and digesting food Items. The typical cerithioid midgut includes a well-formed crystalline style and style sac, a large gastric shield, a broad sorting area, large central pad, one or two digestive gland ducts (sometimes including a reduced or vestigial spiral caecum), and an extension of the minor typhlosole.

Nephropore or **nephridiopore**: The External opening of the kidney or the bladder.

Operculum: Corneous or calcareous anatomical structure, like a lid, attached to the upper surface of the foot. It serves to close the aperture of the shell when the soft parts of the animal are retracted.

Osphradium: The olfactory organ in the roof of the pallial cavity.

Otoliths or **statoliths**: Minute calcareous particles found in the statocysts (balance organ).

Pallial: Relating to or produced by the mantle of a mollusk.

Parthenogeny: Reproduction in which an unfertilized egg develops into a new individual.

Paucispiral: Spiral with few turns.

Periostracum: Thin organic coating which is the outermost layer of the shell of many mollusks.

Plesiomorphy: Ancestral character.

Radula: Toothed, chitinous ribbon used for scraping or cutting food.

Spermatophore: Mass or a capsule containing spermatozoa.

Zygoneury: Arrangement of the nervous system, usually on the right side of the body, in which the mantle-nerve of the pleural ganglion runs directly to the supra- or subintestinal ganglion of its side.

MATERIAL PREPARATION AND PRESERVATION

Preparation of freshwater operculate snails (like Pachychilidae and Thiaridae), for anatomical purposes, requires a relaxation process. In the field, fill a plastic container with the water where the mollusks are living and rinse each specimen to eliminate sediment and debris. Put the cleaned samples into another container filled with more clean water and transport the containers in coolers to the laboratory for relaxation and fixation (Duncan, 2008; Cuezzo, 2009). Relaxation takes 10–48 h depending on the method, the snail size, and the specimens amount pro container. The containers should be all time in a quiet place, free from loud noises or vibration. For anatomical purposes, submerge the specimens few minutes in 60–70°C water until they do not react to touch and then carefully pour some 96% ethanol into the water to start the fixation. Continue adding ethanol little by little, avoiding a retraction of the snails into the shells. For mollusks relaxation, one can also use menthol crystals or Nembutal (0.25% pentobarbital salt). Spread the crushed menthol crystals on the water surface of the containers or pour the Nembutal into the water and wait until the animals are out of their shells (Cuezzo, 2009). Preservation should be in at least 75% ethanol.

Fix immediately in the field specimens intended for molecular analyses. Separate the snails from the sediment as described above. Make a picture of the front and back of the shells with scale. Proceed to open a hole in the upper part of the shell spire using a sharp knife, stone, or hammer. Preserve snails with holed spires in 96%–99% ethanol. Label the samples thoroughly.

Use soft bodies relaxed and preserved in 75% ethanol for anatomical and histological analyses, radulae extraction, and embryonic shell observations. Also, extract the radulae from the molecular analyses specimens to have more integrative data. Use K-proteinase as described by Holznagel (1998) for radula extraction; mount it on aluminum stubs with adhesive pads, and coating for scanning electron microscopy (SEM) examination. For observations of soft structures on SEM, use the Hexamethyldisilazane (HMDS) method (Nation, 1983; Barré et al., 2006; Gomez, 2009). For histological studies of females containing embryonic shells in the brood pouch, decalcify the bodies in successive solutions of 7% nitric acid (HNO_3, 3 days), sodium sulfate (Na_2SO_4, 1 day), and distilled water (1 day) before inclusion (Romeis, 1989). For molecular analyses, isolate Genomic DNA from muscle tissue (preferably from the internal side of the foot) and proceed to extraction using a CTAB extraction protocol (Winnepenninckx et al., 1993) or an available extraction kit.

KEY TO CERITHIODEA

Gastropoda: Caenogastropoda: Cerithioidea: Families

1 Shell spire usually decollated (Fig. 11.75 C); operculum rounded, with a central nucleus and concentric whorls (multispiral, Fig. 11.73 A); mantle edge smooth (Fig. 11.73 B); tentacles long and bulky, with short ocular peduncles on the external base (Fig. 11.73 C); dioecious, oviparous; females with long egg-laying sinus on the right side of the foot (Fig. 11.73 B), males also with genital groove but half the length (Fig. 11.73 C); very long radula with square radular teeth (Fig. 11.73 D) .... **Pachychilidae [p. 358]**

1′ Shell usually turriform (Fig. 11.75 H); operculum corneous, thin, and pliable, with a small subterminal nucleus and few whorls (paucispiral, Fig. 11.74 A); mantle edge papillated (Fig. 11.74 B); short tentacles, broad at the base and narrow at the tips; animals are primarily parthenogenetic; females with a brood pouch posterior to head (Fig. 11.74 D); brood pouch pore present on right side of head-foot, close to the tip of the pallial oviduct; radula with rectangular teeth (Fig. 11.74 C) .... **Thiaridae [p. 368]**

Gastropoda: Caenogastropoda: Cerithioidea: Pachychilidae: Genera

1 South American species .... ***Doryssa* [p. 359]**

1′ Mesoamerican and Cuban species .... ***Pachychilus* [p. 364]**

Gastropoda: Caenogastropoda: Cerithioidea: Pachychilidae: *Doryssa*: Species

1 Shell smooth with or without crosswise (Fig. 11.73 E) or spiral growth lines ornamentation (Fig. 11.75 D) 2

1′ Shell conspicuously sculptured with grooves (Fig. 11.75 D), tubercles (Fig. 11.73 G), crenulated (Fig. 11.73 H), or keels (Fig. 11.77 E) 11

2(1) Shell color light or pale 3

2′ Shell color dark 7

3(2) Shell completely smooth or with visible spiral lines in the base of the last whorl 4

3′ Shell with shallow impressed or faint spiral lines throughout 6

4(3) Shell with dark color spots organized in lines or bands 5

4′ Shell color light chestnut to dark brown, lighter specimens marked by small, irregularly placed black spots; shell elongate-conical (Fig. 11.75 H), solid, decollated, usually five to six whorls, about 38 mm height; spiral whorls flattened, ornamented with faint axial ribs and spiral growth lines; periostracum rather heavy, aperture oblique, narrow, bluish-white within (Fig. 11.76 A) .... *Doryssa iheringi* Baker, 1913

[Brazil: St. Antonio da Cachoeira, Jari River]

5(4) Very smooth shell, oblong, acutely conical (Fig. 11.75 G), solid, with seven to eight slightly convex whorls, about 30 mm height, suture shallow, last whorl gibbously rounded; shell color yellowish-olive with interrupted axial dark brown bands; aperture ovate, reddish-chestnut within, columellar lip thickened particularly at its upper extremity (Fig. 11.76 B) .... *Doryssa laevissima* (Sowerby 1824)

[Venezuela: La Guaira River]

5′ Shell conic (Fig. 11.75 G), rather thin, apex barely decollated, spire whorls flat with last whorl carinated; shell color olive green with dark brown spots and stripes, forming interrupted spiral bands; aperture ovate, wide, columella twisted (Fig. 11.76 C) .... *Doryssa kappleri* Vernhout, 1914

[Suriname: Lucie River]

6(3) Shell oblong-conical (Fig. 11.75 F), seven slightly convex whorls, pale beneath the sutures, ridged spiral striations throughout the shell; color light yellow-olive; aperture ovate, effused at the base (Fig. 11.76 D) .... *Doryssa lumbricus* (Reeve, 1860)

[Brazil: Pará]

6′ Shell turriform (Fig. 11.75 H), somewhat thin, four to six flattened and imbricated whorls, apex truncate, about 32 mm height; shell ornamented with fine spiral grooves; color honey-yellow with small, irregularly scattered or abundant black spots arranged in axial rows or nearly covering the surface in some shells (Fig. 11.76 E) .... *Doryssa jaryensis* Baker, 1913

[Brazil: St. Antonio da Cachoeira, Jari River]

7(2) Shell without dark color spots, aperture plain white or bluish-white 8

7′ Shell or aperture showing reddish or black spots 9

8(7) Shell stoutly ovate (Fig. 11.75 E), strong decollated (usually only the remaining last two whorls), rather thick, whorls stepped and slightly convex, smooth, superficially spirally grooved, color brown-black; aperture ovate, very acute above, columella short and callus thick and white (Fig. 11.76 F) .... *Doryssa batesii* (Reeve, 1860)

[Brazil: Amazonas mouth]

8′ Shell ovate-conic (Fig. 11.75 E), solid, three to four almost flat whorls, decollated, last whorl angulated at the periphery, 25–28 mm height, suture crenulated, a little impressed above; color dark purplish-black; aperture acuminate-ovate, bluish-white; columella curved, slightly twisted, light brown in color (Fig. 11.76 G) .... *Doryssa geijskesi* Pain, 1956

[Suriname: Makambo Island, Coppename River]

9(7) Shell elongate-conic (Fig. 11.75 H) with flattened or slightly convex whorls, last whorl not angulated 10

9′ Shell conic (Fig. 11.75 G), spire turriform, eroded, four to five whorls, about 20 mm height, last whorl periphery sharply angulate; shell covered with dense periostracum marked throughout with fine, close incised spiral striae and flat spiral cords in the base of the last whorl; the color is varying from light yellow-olive in young shells to nearly black in the adults; everywhere marked with irregularly placed and shaped reddish-brown spots (Fig. 11.76 H) .... *Doryssa cachoeirae* (Baker, 1913)

[Brazil: St. Antonio da Cachoeira, Jari River]

FIGURE 11.76 Species of *Doryssa* (Pachychilidae): (A), *D. iheringi* (ANSP 109354); (B) *D. laevissima* (ANSP 63652 and USNM 170006, respectively); (C) *D. kappleri* (USNM 783189); (D) *D. lumbricus* (NHMUK 20050169-1); (E) *D. jaryensis* (ANSP 109359); (F) *D. batesii* (ANSP 26771); (G) *D. geijskesi* (MCZ 247046); (H) *D. cachoeirae* (ANSP 109356). Scale bar = 1 cm. Museum acronyms: *ANSP*, Academy of Natural Sciences of Drexel University, Philadelphia; *MCZ*, Museum of Comparative Zoology, Harvard University; *NHMUK*, Natural History Museum, London; *USNM*, National Museum of Natural History, Smithsonian Institution, Washington D.C.

10(9) Shell cylindrical (Fig. 11.75 I), solid, decollated, with eight convex imbricated whorls, last whorl slightly convex; sculptured with faint axial carved lines; shell color olive, revealing dark spots under black periostracum; aperture broad ovate, external lip pale bluish, aperture with conspicuous dark red spots which are transversely arranged for the most part, with some longitudinal rows converging (Fig. 11.77 A) .... *Doryssa decollata* (Lamarck, 1843)

[Suriname]

10′ Shell turriform (Fig. 11.75 H), smooth, thick, apex eroded, eight whorls slightly convex, about 46 mm height, last whorl inflated; color olive-yellow with brown-black periostracum, dark red spots irregularly distributed; aperture ovate with thick, bluish lips; inferior edge oblique and projected; superior angle callused and narrowly channeled, columella perpendicular to the aperture base (Fig. 11.77 B) .... *Doryssa petechialis* (Brot, 1860)

[Venezuela: Orinoco River]

11(1) Adult shell > 50 mm total height, turreted, solid and heavy 12

11′ Adult shell 20–50 mm total height 15

12(11) Shell sculptured with spiral grooves crossed by strong, prominent axial folds 13

12′ Shell sculpture with elevated spiral plaits or folds, without conspicuous axial elements 14

13(12) Shell with spire very elevated but truncated, eight to nine whorls; upper whorls nearly flat, last two whorls more convex; the suture becoming deeper in last whorl, which is produced in a prominent keel defining a horizontal shoulder below the suture; color dull olive-citrine indistinctly maculated with black (Fig. 11.77 C) .... *Doryssa rex - regina* complex Baker, 1913

[Brazil: Jari River]

13′ Spire eroded, with four to five flattened whorls remaining; color dark brown; shell sculptured with fine and numerous spiral rings crossed by not very pronounced, tightly arranged axial ribs; aperture, ovate, wide, angle above a bit protruded, columella and callus thickened (Fig. 11.77 D) .... *Doryssa maranhaoensis* Haltenorth & Jaeckel, 1940

[Brazil: Maranhão]

14(12) Shell with stepped whorls, apex eroded, about seven whorls remaining, color olive-chestnut; first whorls of the spire rounded, then longitudinally plicate, base of the last whorl spirally ridged; aperture ovate slightly produced at the base (Fig. 11.77 E) .... *Doryssa plicata* (Reeve, 1859)

[Andean mountains between Colombia and Venezuela]

14′ Shell turriform, spire decollated with six overlapping whorls, upper whorls flattened last whorl globose, about 54 mm height; sculptured with elevated spiral folds; color yellow-olive with black spots arranged in spiral lines; aperture ovate, acuminate above and below (Fig. 11.77 F) .... *Doryssa rixosa* Ihering, 1902

[Brazil: Paranapanema, Paraná]

15(11) Shell partially or totally sculptured with spiral striae (Fig. 11.75 F), grooves (Fig. 11.75 D), ridges (Fig. 11.77 H) or ribs (Fig. 11.73 F) 16

15′ Shell sculptured with tubercles (Fig. 11.73 G) or keeled (Fig. 11.75 G) 20

16(15) Shell dark 17

16′ Shell color light brown with a few dark, irregular, longitudinal streaks and covered by innumerable light chestnut-brown dots, very distinct within the aperture; shape ovate-conical (Fig. 11.75 E), rather thin, apex eroded, four to six whorls rapidly increasing in diameter, suture well impressed, the last whorl globosely inflated and with a shoulder in the middle—only visible in adult specimens, about 16–20 mm total height; surface covered with raised, rounded, spiral striae, intersected on the first whorls by axial ribs; aperture rather small, oval, outer lip somewhat curved forwards, turning obliquely to the left at the base, columella slightly curved (Fig. 11.77 G) .... *Doryssa millepunctata* (Tryon, 1865)

[Amazonas River]

PHYLUM MOLLUSCA

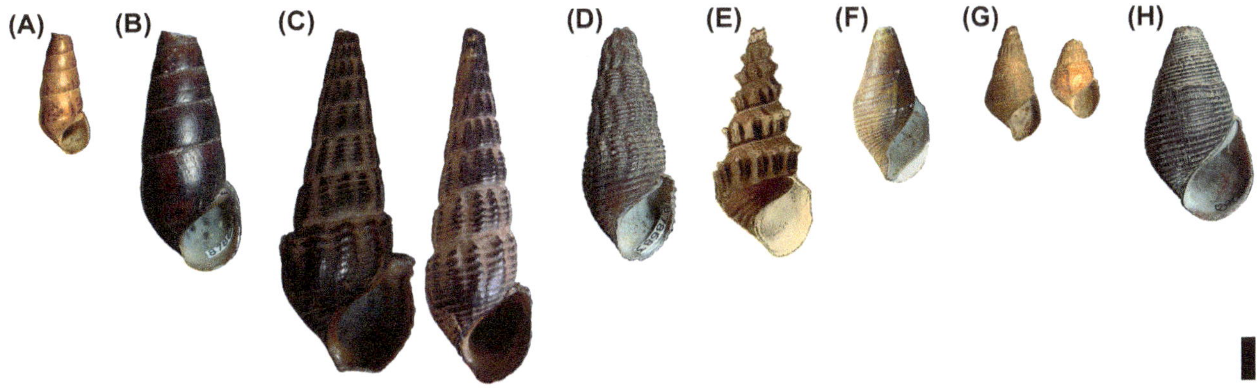

FIGURE 11.77 Species of *Doryssa* (Pachychilidae): (A) *D. decollata* (USNM 878678); (B) *D. petechialis* (USNM 878677); (C) *D. rex-regina* complex (ANSP 109349 and ANSP 109348 respectively); (D) *D. maranhaoensis* (USNM 878683); (E) *D. plicata* after Reeve 1859; (F) *D. rixosa* (ANSP 166612); (G) *D. millepunctata* (ANSP 26757 and ANSP 26775, respectively); (H) *D. consolidata* (ANSP 166611). Scale bar = 1 cm. Museum acronyms: *ANSP*, Academy of Natural Sciences of Drexel University, Philadelphia; *USNM*, National Museum of Natural History, Smithsonian Institution, Washington D.C.

17(16) Shell short but slender, with oval, narrow aperture, spire decollated 18

17′ Shell short, globose, grooved, broad aperture 19

18(17) Shell ovate-conic (Fig. 11.75 E), spire turriform, thin but solid, with four flattened whorls, suture deep, ~34 mm total height; color black, aperture showing interrupted lines of reddish-brown spots; shell adorned with densely growing, elevated spiral stripes (Fig. 11.77 H) .... *Doryssa consolidata* (Bruguiere, 1790)

[Brazil: Maranhão and Pará] Figure 11.77

18′ Shell fusiform (Fig. 11.75 J), thin but solid, with four slightly convex whorls, suture deep, about 28 mm height; color dark olive to black, shiny; sculpture consisting of flat, narrow grooves in the base of the last whorl and fine axial growth lines; aperture oblique, columella very simple (Fig. 11.78 A) .... *Doryssa heathi* (Baker, 1913)

[Brazil: Jari River]

19(17) Shell ovoid (Fig. 11.75 E), strongly inflated, thick, short decollated spire, three to four slopingly convex whorls and suture simple; ~40 mm height; color yellow-greenish with dark, shiny periostracum, aperture white with irregularly arranged reddish-brown spots; shell grooved throughout with obsolete axial ridges only visible in the first whorls; aperture broadly sinuated above, lip and columella thick and yellowish (Fig. 11.78 B) .... *Doryssa ventricosa* (I. Lea, 1856)—*bullata* (I. Lea, 1856)—*scarabus* (Reeve, 1860) complex

[Brazil: Amazonas River, Pará]

19′ Shell solid, apex barely eroded; spire with four convex whorls, deep suture; 14–17 mm total height; color yellow-green; sculptured with spiral ribs and obsolete axial threads forming small tubercles in the last whorl; aperture ovate (Fig. 11.78 C) .... *Doryssa schuppi* (Ihering, 1902)

[Brazil: Rio Grande do Sul]

20(15) Shell keeled with spiral crenellations (Fig. 11.73 H) 21

20′ Shell partially or totally sculptured with axial lines of tubercles and spiral grooves not raised (Fig. 11.73 G) 25

21(20) Shell short, globose, aperture broad and ovate 22

21′ Shell elongate-conic (Fig. 11.75 E), sculptured with keels (Fig. 11.73 H) 23

22(21) Shell rather light, short spire with four to five whorls, suture deep, first whorls flattened and the last whorl globose, conspicuously sculptured till the middle, ~40 mm height; periostracum yellow-olive to dark brown; shell with well-developed spiral keels and axial ribs, their intersection producing tubercles markedly elongate; columella thick, shading yellowish (Fig. 11.78 D) .... *Doryssa globosa* Baker, 1913

[Brazil: Pará, Amazonas River, Igarapé de Paituná near Monte Alegre]

22′ Shell solid and heavy, apex eroded, ovate-conic (Fig. 11.75 E), four whorls, ~45 mm height, spire whorls flattened, last whorl convex; color dark olive brown to nearly black, shiny; shell showing spiral keels throughout the whole shell and axial ribs more marked than the spiral keels; the intersection of the ribs and keels producing slightly produced tubercles; aperture subangular above and below (Fig. 11.78 E) .... *Doryssa starski* Baker, 1913

[Brazil: Pará, Rivers Irirí-Xingú]

23(21) Spire whorls convex, aperture small and rounded 24

23′ Shell conic (Fig. 11.75 G), apex eroded, five flattened whorls, ~26 mm height; covered with black periostracum, shell underneath yellow-greenish, last whorl with axial blackish flames; whole shell with strong spiral ribs and weak axial folds forming small knots with the ribs; keeled upwards, flattened over the keel; aperture ovate, brown with three narrow black bands; columella white or reddish-brown, twisted (Fig. 11.78 F) .... *Doryssa cingulata* (Jonas, 1844)

[Brazil]

FIGURE 11.78 Species of *Doryssa* (Pachychilidae): (A) *D. heathi* (ANSP 109351); (B) *D. ventricosa-bullata-scarabus* complex (USNM 119591); (C) *D. schuppi* (MZSP 7361); (D) *D. globosa* (ANSP 109352); (E) *D. starski* (ANSP 109350); (F) *D. cingulata* (ANSP 26796); (G) *D. chloris* (USNM 569014); (H) *D. pernambucensis* (NHMUK 1984204-1). Scale bar: (A), (B), (D–H) = 1 cm; for (C) = 0.25 cm. Picture (C) courtesy of Luiz Ricardo L. Simone. Museum acronyms: *ANSP*, Academy of Natural Sciences of Drexel University, Philadelphia; *MZSP*, Museu de Zoologia da Universidade de São Paulo, Brazil; *NHMUK*, Natural History Museum, London; *USNM*, National Museum of Natural History, Smithsonian Institution, Washington D.C.

24(23) Shell conic (Fig. 11.75 G), five convex whorls, suture well marked; color yellow-green to light brown, light spiral bands above the suture and the base of the last whorl; spiral keels crossed by axial bulges, the later forming knobs at the suture; the columella forming a rounded peak at its base (Fig. 11.78 G) .... *Doryssa chloris* (Troschel, 1848)
[Guyana: Essequibo]

24′ Shell oblong-conic (Fig. 11.75 F), rather solid, six whorls convex, apex barely eroded, suture deep impressed, about 50 mm height; color olive brown; shell densely sculptured with narrow axial ribs passing over concentric folds, interstices granulated; columella arched, aperture with scattered, irregular black spots (Fig. 11.78 H) .... *Doryssa pernambucensis* (Reeve, 1861)
[Brazil: Pernambuco] Figure 11.78

25(20) Shell sculptured with axial lines of tubercles (Fig. 11.73 F) never reaching the last whorl 26

25′ Shell sculptured with axial lines of tubercles throughout, last whorl lower middle grooved (Fig. 11.75 C) 29

26(25) Shell elongate, spire turriform, whorls slightly convex, only first whorls with axial lines of tubercles, aperture ovate, longer than wide 27

26′ Shell ovate-conic (Fig. 11.75 E), last whorl without tubercles, aperture very broad 28

27(26) Shell thick, spire decollated, six to eight whorls, about 35 mm height; color dark brown, with small chestnut colored spots marking the interior of the aperture; spiral creases throughout the shell, in the upper whorls crossed by axial, rather distant stout ribs forming a mesh; aperture transversely ovate obliquely notched at the base, columella arched and thick (Fig. 11.79 A) .... *Doryssa transversa* (I. Lea & H.C. Lea, 1850)
[Guyana]

27′ Shell rather solid, apex barely eroded, eight stepped whorls; color tawny olive; surface spirally grooved throughout with concentric plaits, distinct towards the apex; aperture lip slightly sinuate (Fig. 11.79 B) .... *Doryssa nicotiana* (Reeve, 1860)
[Brazil]

28(26) Shell decollated, solid, 6–10 stepped plano-convex whorls; color black or blackish-olive, sparingly black-dotted toward the apex; shell corded throughout with spiral ridges intercepted by prominent axial ribs never reaching the last whorl; aperture lunar-ovate, sinuated above, notched below (Fig. 11.79 C) .... *Doryssa atra* (Bruguiere, 1792)
[French Guiana: Cayenne]

28′ Shell solid, five to nine convex, somehow imbricate whorls, decollated, height ~40 mm, color dark olive with irregular, spiral rows of black dots; sculptured with shallow spiral grooves and wavy, tight axial folds forming narrow ribs that look like tubercles; very short columella (Fig. 11.79 D) .... *Doryssa capillaris* (Brot, 1872)
[South America]

29(25) Shell ovate-conic (Fig. 11.75 E) or conic (Fig. 11.75 G), spire turriform 30

29′ Shell stoutly conical (Fig. 11.75 G), very low spire barely eroded, five convex whorls, thick and solid, between 21–27 mm height; color olive-black with dark axial lines following the growth lines; sculptured with big, deep, spiral grooves and ruffled axial ribs given the whole surface a tuberculous appearance; aperture broad, squarely ovate, sinuated at each end (Fig. 11.79 E) .... *Doryssa brevior* (Troschel, 1848)
[Guyana: Essequibo]

30(29) Shell slender, spire whorls flattened or convex, the last whorl never globose nor shouldered 31

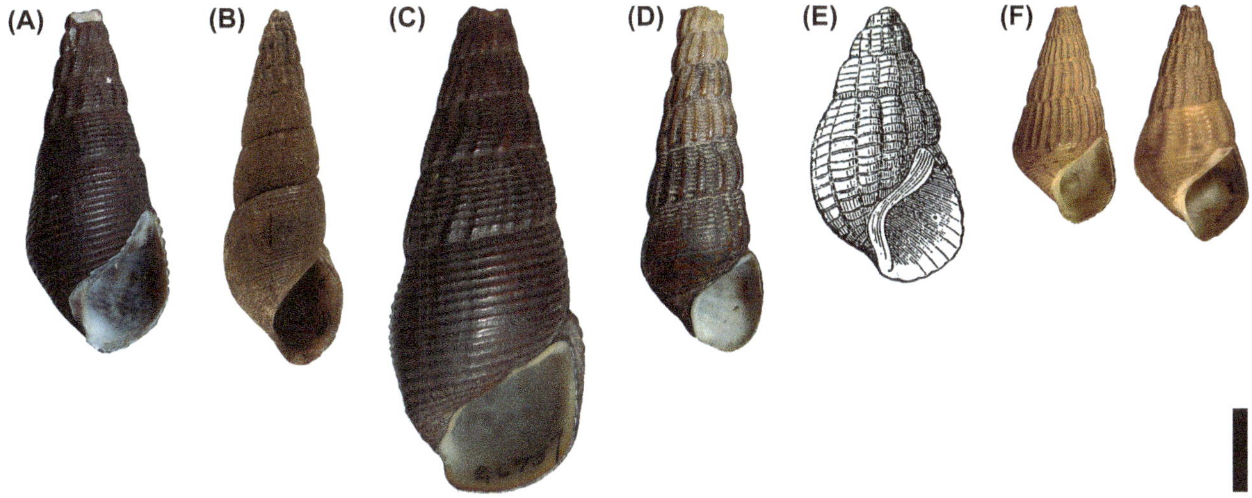

FIGURE 11.79 Species of *Doryssa* (Pachychilidae): (A) *D. transversa* (USNM 119619); (B) *D. nicotiana* (CFS 15758); (C) *D. atra* (ANSP 26757); (D) *D. capillaris* (USNM 119688); (E) *D. brevior*, after Chenu (1859); (F) *D. ipupiara* (USNM 656149: Pto. Ordaz—Venezuela and USNM 119591: Guiana, respectively). Scale bar = 1 cm. Museum acronyms: *ANSP*, Academy of Natural Sciences of Drexel University, Philadelphia; *CFS*, Biologiezentrum, Linz, Austria; *USNM*, National Museum of Natural History, Smithsonian Institution, Washington D.C.

30′ Shell conic (Fig. 11.75 G), apex eroded, ≤30 mm total height, seven slightly convex whorls, suture shallow, last whorl globose and shouldered in the middle; color yellow to dark brown; sculpture consisting of fine spiral striae with very conspicuous, strong axial threads arranged at regular intervals, reaching the middle of the last whorl; the base of the last whorl grooved; aperture broad, elliptical, angled above, forming shallow channel below, peristome with irregularly arranged dark brown spots (Fig. 11.79 F) .... *Doryssa ipupiara* Simone, 2001
[Brazil: Roraima] Figure 11.79

31(30) Spire whorls convex, aperture rounded with superior angle exhibiting a thick white bulge 32

31′ Spire whorls flattened or slightly convex, sculpture giving the shell a granulated appearance 33

32(31) Shell with eight to nine whorls, 33–38 mm height, sutures impressed; color brown; spirally ridged with intercrossing plicated folds; the spiral-corded ridges are close-set, obtuse, and flattened in the interstices between the axial folds; aperture ovate, channeled at the base, and showing reddish-brown spots (Fig. 11.80 A) .... *Doryssa macapa* (Reeve, 1860)
[Brazil: Macapa]

32′ Shell solid, apex decollated, six slowly increasing whorls, ∼21 mm height, dark greenish-brown, with olive spots visible at the base; sculpture consisting of spiral wrinkles and strongly developed ribs, vanishing in the base of the last whorl; columella twisted, callus thick and bluish (Fig. 11.80 B) .... *Doryssa gracilis* Vernhout, 1914
[Suriname: Corantijn]

FIGURE 11.80 Species of *Doryssa* (Pachychilidae): (A) *D. macapa* (USNM 515487); (B) *D. gracilis* (USNM 119737); (C) *D. aquatilis—charpentieri—branca* complex (USNM 100365 and ANSP 26760 respectively); (D) *D. inconspicua* (ANSP 109362). Scale bar = 1 cm. Museum acronyms: *ANSP*, Academy of Natural Sciences of Drexel University, Philadelphia; *USNM*, National Museum of Natural History, Smithsonian Institution, Washington D.C.

PHYLUM MOLLUSCA

33(31) Shell solid, eight to ten whorls, suture well marked, about 33 mm height, yellow to dark brown; spirally strongly corded with regular ridges throughout and axially plicated; aperture ovate, effused above and below, columella excavated (Fig. 11.80 C) .... *Doryssa aquatilis* (Reeve, 1859)—*charpentieri* (Reeve, 1859)—*branca* (Reeve, 1860) complex
[Brazil: Rio Branca]

33′ Shell solid, decollated, five to six whorls, 28 mm total height, suture well-defined, color dark olive, numerous axial grooves and narrow spiral folds arranged crosswise (Fig. 11.80 D) .... *Doryssa inconspicua* (Brot, 1872)
[Brazil] Figure 11.80

Gastropoda: Caenogastropoda: Cerithioidea: Pachychilidae: *Pachychilus*: Species

1 Shell smooth or slightly crosswise marked (Fig. 11.73 E) 2

1′ Shell conspicuously sculptured (Figs. 11.73 F–H) 19

2(1) Adult shell 50–100 mm total height 3

2′ Adult shell 10–45 mm total height 7

3(2) Shell conic (Fig. 11.75 G) 4

3′ Shell oblong (Fig. 11.75 F) or subcylindrical (Fig. 11.75 K) 5

4(3) Adult shell reaching 10 cm, shell rather slender and acutely conical (Fig. 11.75 G), outline of the spire straight, 10–11 spire whorls, slightly eroded apex; color olive green or olive brown; aperture ovate narrowed above (Fig. 11.81 A) .... *Pachychilus potamarchus* Pilsbry, 1893
[Mexico: Tabasco]

4′ Shell ovate-conic (Fig. 11.75 E), solid, apex eroded, six to seven slightly convex spire whorls, adult shell 52–54 mm height, last whorl big and globose; surface covered by dusky olive-yellowish strong periostracum, shell beneath white adorned with more or less distinct irregular and interrupted longitudinal black streaks; aperture decorated with brown or purplish tracts (Fig. 11.81 B) .... *Pachychilus dalli* (Pilsbry, 1896)
[Mexico: Huilotopec (Oaxaca)]

5(3) Shell oblong-turriform with conic spire (Fig. 11.75 F), strong 6

5′ Subcylindrical shell (Fig. 11.75 K), apex strong eroded, thick, four to five hardly convex spire whorls, ~52 mm shell height, suture edges poorly distinguishable; color intense dark olive; aperture ovate, lip reflected and thickened, columella thick and white (Fig. 11.81 C) .... *Pachychilus chrysalis* (Brot, 1872)
[Mexico: Isthmus of Tehuantepec]

6(5) Shell bellied, strong, conic spire, apex barely truncated, up to 12 whorls with a flat outline (usually eight to nine spire whorls), ~60 mm height, suture narrow with thick edges; dark periostracum peeling off, shell white-blue and violet-red mottled; aperture sharp-oval, with a protuberance in the superior angle merging with the callus, violet with milky margins (Fig. 11.81 D) .... *Pachychilus indiorum* (Morelet, 1849)
[Mexico: Chiapas, Palenque]

6′ Shell with nine to ten flattened whorls, ~54 mm height; color olive red to dark brown; aperture oval, sharp above, base round, lip heavy thickened, fused with callus and columella (Fig. 11.81 E) .... *Pachychilus hellerii* (Brot, 1872)
[Mexico: Tehuantepec (Oaxaca)]

7(2) Shell oval (Fig. 11.75 E) or conic (Fig. 11.75 G) 8

7′ Shell oblong (Fig. 11.75 F) or globose 17

8(7) Shell color very light 9

8′ Shell dark 12

9(8) Shell strong but not thick 10

9′ Shell strong and very thick 11

10(9) Shell oval-conic (Fig. 11.75 E), spire turriform often complete, ten spire whorls with a flat outline, last whorl bellied, suture pale with a rust-color bend in the edge, ~32 mm; color ash-violet with dark little flames, apex whorls pale violet-red; aperture rounded, callus fused with the columella (Fig. 11.81 F) .... *Pachychilus cinereus* (Morelet, 1849)
[Guatemala: Vera Paz, River in Coban City]

10′ Shell conic-turriform (Fig. 11.75 H), spire sharp elevated with up to 11 convex, regular whorls, ~35 mm height, threadlike suture; color light yellow, very bright; aperture small, round, outer lip expanded (Fig. 11.81 G) .... *Pachychilus cumingii* I. Lea & H.C. Lea, 1850
[Honduras: Large rivers of Copan]

11(9) Shell oblong-conic (Fig. 11.75 F), apex frequently eroded, four to six slightly convex whorls, ~42 mm height, suture simple, last whorl oval; color olive brown, aperture big, inner with brown flame-shape spots (Fig. 11.81 H) .... *Pachychilus larvatus* (Brot, 1877)
[Mexico: Tehuantepec (Oaxaca)]

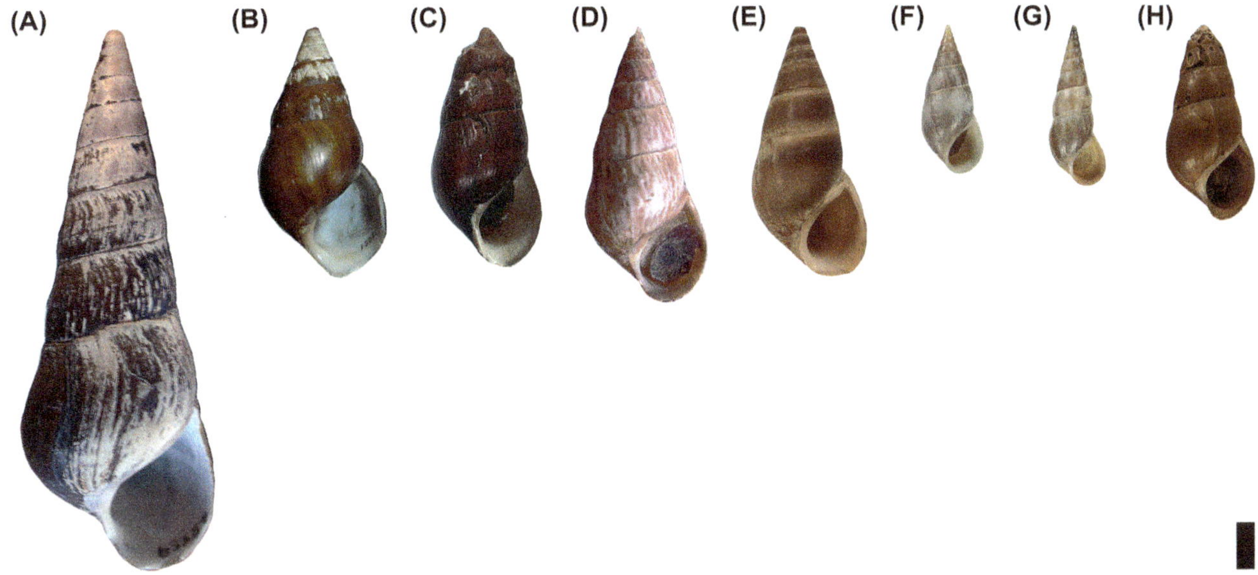

FIGURE 11.81 Species of *Pachychilus* (Pachychilidae): (A) *P. potamarchus* (ANSP 63387a); (B) *P. dalli* (USNM 133214); (C) *P. chrysalis* (MHNG 134); (D) *P. indiorum* (ANSP 63566); (E) *P. hellerii* (MHNG 154); (F) *P. cinereus* (NHMUK 1893.2.4.1799); (G) *P. cumingii* (NHMUK 20070009/1); (H) *P. larvatus* (MHNG 191). Scale bar = 1 cm. All pictures after Gomez-Berning et al. (2012). Museum acronyms: *ANSP*, Academy of Natural Sciences of Drexel University, Philadelphia; *MHNG*, Muséum d'Histoire Naturelle de la Ville de Genève; *NHMUK*, Natural History Museum, London; *USNM*, National Museum of Natural History, Smithsonian Institution, Washington D.C.

11′ Fusiform shell (Fig. 11.75 J), apex barely eroded, 12 subconvex whorls, ~37 mm height, suture conspicuous, last whorl shouldered in the middle; color yellow-green, not bright; aperture ovate, callus and columella thickened (Fig. 11.82 A) .... *Pachychilus rusticulus* (von dem Busch, 1858)
[Central America]

12(8) Shell spire short 13

12′ Shell spire elevated (spire turriform) 14

13(12) Shell oval (Fig. 11.75 E), apex eroded, six to seven flattened whorls, 28–39 mm height, the last whorl bellied; color brown-black; aperture with straight columella, lip very thick (Fig. 11.82 B) .... *Pachychilus corvinus* (Morelet, 1849)
[Guatemala: Streams in the province of Vera-Paz]

13′ Shell oval-conic (Fig. 11.75 E), very short and dark, apex more or less eroded, suture well marked, four to five flattened whorls, ~16 mm height, last whorl a bit longer than the rest of the spire, angular in the middle; aperture oval, brown-violet, peristome strongly developed (Fig. 11.82 C) .... *Pachychilus sargi* (Crosse & Fischer, 1875)
[Guatemala]

14(12) Material from Cuba 15

14′ Material from continental Mesoamerica; shell conic (Fig. 11.75 G), seven convex whorls, about 33 mm height, suture impressed; aperture widely ovate, purple, callus and columella strongly thickened (Fig. 11.82 D) .... *Pachychilus vallesensis* (Hinkley, 1907)
[Mexico: San Luis Potosi, Valles River]

15(14) Shell conic, spire elevated with flattened whorls, the last whorl carinated or subcarinated 16

15′ Shell turriform (Fig. 11.75 H), strong, with six to eight convex whorls, sutures impressed, ~35 mm height; adult shells often truncated; shell dark violet with a white subsutural band; sculptured with very fine, wavy, spiral striae; aperture rounded, columella arched diffused into the callus, white protuberance below the insertion of the lip with the parietal wall, inner violet (Fig. 11.82 E) .... *Pachychilus violaceus* Preston, 1911
[Cuba]

16(15) Shell elongated, turriform (Fig. 11.75 H), apex could be eroded, with seven to nine flattened whorls, 17–25 mm height, last whorl subcarinated; color dark olive green or chestnut; surface smooth ornamented with numerous thin spiral grooves rapidly disappearing in the last whorls; aperture ovate, lip sinuated (Fig. 11.82 F) .... *Pachychilus conicus* (Orbigny, 1845)
[Cuba: San Juan de los Remedios]

16′ Shell pyramidal, spire with nine flattened whorls, 23–25 mm height, apex entire, last whorl carinated in the middle; color chestnut to dark brown; crosswise sculpture very conspicuous (Fig. 11.73 E); aperture ovate, angled above, channeled below (Fig. 11.82 G) .... *Pachychilus fuentesi* Aguayo, 1936
[Cuba: Baracoa, El Yunque]

PHYLUM MOLLUSCA

17(7) Shell oblong-conic (Fig. 11.75 F) 18

17′ Shell globose, spire very short mainly eroded in adults, two to three whorls remaining, ~19 mm height, last whorl very globose; adult shell smooth and young shells having fine striated, densely packed spiral growth lines; aperture large, black within, outer lip thin; columellar margin black with white lining on the inner edge and heavily callused (Fig. 11.82 H) .... *Pachychilus pila* (Pilsbry & Hinkley, 1910) [Mexico: San Luis Potosi, Tamasopo River]

FIGURE 11.82 Species of *Pachychilus* (Pachychilidae): (A) *P. rusticulus* (GSUB 14882); (B) *P. corvinus* (MNHN 24444); (C) *P. sargi* (MNHN 24457); (D) *P. vallesensis* (ANSP 96592); (E) *P. violaceus* (NHMUK 1854.10.4.196); (F) *P. conicus* (USNM 467264 and USNM 742092 respectively); (G) *P. fuentesi* (USNM Sabaneta, Cuba); (H) *P. pila* (ANSP 99559). Scale bar = 1 cm. All pictures after Gomez-Berning et al. (2012), except for pictures (E–G). Museum acronyms: *ANSP*, Academy of Natural Sciences of Drexel University, Philadelphia; *GSUB*, Geowissenschaftliche Sammlung der Universität Bremen; *MNHN*, Muséum National d'Histoire Naturelle, Paris; *NHMUK*, Natural History Museum, London; *USNM*, National Museum of Natural History, Smithsonian Institution, Washington D.C.

18(17) Spire elevated, about nine slightly convex whorls, ~36 mm height, suture briefly impressed, last whorl shouldered in the middle; color brown, shabbily spotted; aperture big, oval, narrow, angular above, external lip sharp, columella whitish thickened above, inner dark (Fig. 11.83 A) .... *Pachychilus planensis* (I. Lea, 1858) [Honduras: Plan and Omoa, Valley of Ulua River, Atlantic slope]

18′ Shell with apex often eroded, spire with eight flattened whorls, ~45 mm height, last whorl inflated; color olive or dark yellow, sometimes with dark bands; aperture oval, inner blueish with external lip violet or chestnut-colored, callus ending as protuberance above (Fig. 11.83 B) .... *Pachychilus panuculus* (Morelet, 1851) [Guatemala: Petén Department]

19(1) Shell size 20–36 mm 20

19′ Shell 50–90 mm 26

20(19) Shell cylindrical (Fig. 11.75 I), ovate (Fig. 11.75 E), or pyramidal (Fig. 11.75 G), partially sculptured 21

20′ Shell ovate or conic (Fig. 11.75 E), completely sculptured 24

21(20) Shell with deep suture, dark in color (black or brown) 22

21′ Shell short, ovate, adults with deeply eroded conic spire, three to four whorls, ~23 mm height, last whorl convex being smooth in adults; color dark brown; young specimens are having spire sculptured with spiral carina and ridges; aperture bluish-white, outer lip thin but strongly arched forming a circle (Fig. 11.83 C) .... *Pachychilus tristis* (Pilsbry & Hinkley, 1910) [Mexico: San Luis Potosi, Tamasopo River]

22(21) Shell sculptured with spiral cords (Fig. 11.75 D) 23

22′ Shell pyramidal (Fig. 11.75 G), strong but not thick, apex eroded, ten convex whorls, ~26 mm height, suture very conspicuous and undulate; shell sculptured with three nodular lines in the upper part of each whorl, superior whorls smooth and last whorl with three to four moderately elevated spiral lines; color brown or dark brown; aperture ovate (Fig. 11.83 D) .... *Pachychilus saussurei* (Brot, 1860) [Mexico: Hidalgo, Forest of the Rio Grande]

23(22) Shell ovate-conic (Fig. 11.75 E), solid, spire acuminate, five to six convex whorls, ~30 mm height; covered with a very smooth, dense, glossy, brownish-black periostracum; early whorls smooth, intermediate and last whorls showing a few spiral cords below the suture and a weak angle at the shoulder; aperture rounded with projected external lip (Fig. 11.83 E) . *Pachychilus atratus* (Pilsbry & Hinkley, 1910) [Mexico: San Luis Potosi, Tamasopo River]

23′ Shell solid, spire stepped, five to seven slightly convex whorls, ~32 mm height, last whorl heavily shouldered underneath the suture; color black, first whorls sculptured with evenly spaced sharp spiral cords, last two whorls with less sculpture; aperture ovate, callus very thick (Fig. 11.83 F) .... *Pachychilus monachus* (Pilsbry & Hinkley, 1910) [Mexico: San Luis Potosi, Coy River]

24(20) Shell conic (Fig. 11.75 G) 25

24′ Shell ovate-conic (Fig. 11.75 E), slender, ~33–35 mm height, apex eroded, five to six convex whorls sculptured with three to four spiral, elevated bands each, the last whorl bellied; color black; aperture rounded, callus and columella thick and fused (Fig. 11.75 G) .... *Pachychilus rasconensis* Thiele, 1928
[Mexico: San Luis Potosi]

25(24) Shell long, conic (Fig. 11.75 G), apex eroded, spire with eight to nine flattened whorls, 34–36 mm height; color dark olive with small blackish spots sparsely scattered; sculptured with strong spiral cords with sometimes smaller interstitial spirals, fine axial striation (Fig. 11.83 H) .... *Pachychilus olssoni* (Pilsbry, 1950)
[Guatemala: Petén, Rio de la Pasion]

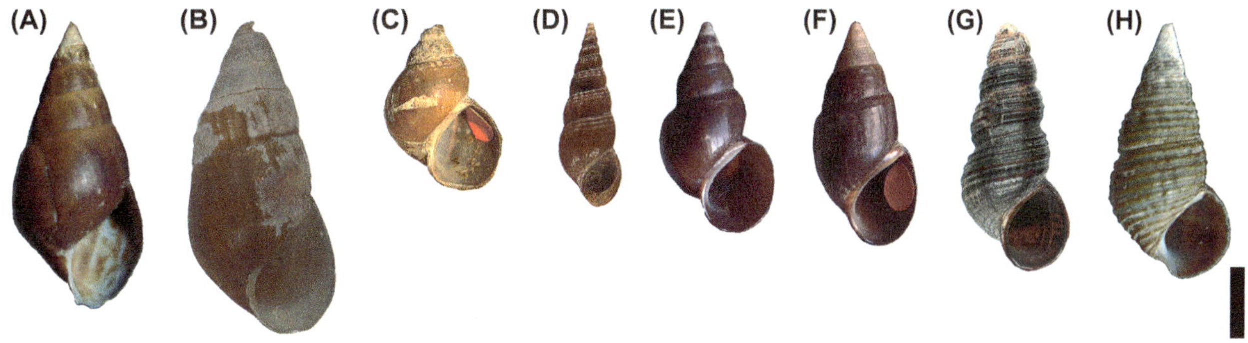

FIGURE 11.83 Species of *Pachychilus* (Pachychilidae): (A) *P. planensis* (USNM 119725); (B) *P. panuculus* (MNHN 24451); (C) *P. tristis* (ANSP 99566); (D) *P. saussurei* (MHNG 125); (E) *P. atratus* (ANSP 99570); (F) *P. monachus* (ANSP 99581); (G) *P. rasconensis* (ZMB 109.653); (H) *P. olssoni* (ANSP 186101). Scale bar = 1 cm. All pictures after Gomez-Berning et al. (2012). Museum acronyms: *ANSP*, Academy of Natural Sciences of Drexel University, Philadelphia; *MHNG*, Muséum d'Histoire Naturelle de la Ville de Genève; *MNHN*, Muséum National d'Histoire Naturelle, Paris; *USNM*, National Museum of Natural History, Smithsonian Institution, Washington D.C.; *ZMB*, Museum für Naturkunde, Berlin.

25′ Shell conic (Fig. 11.75 G), thick, spire subelevated, 11 flattened whorls, shell about 33 mm height, the color of iron rust, last whorl bellied and keeled; shell grooved throughout; aperture large, round, inner suffused with red (Fig. 11.84 A) .... *Pachychilus rubidus* I. Lea, 1856
[Mexico: San Luis Potosi]

26(19) Shell partly sculptured 27

26′ Shell throughout conspicuously sculptured 29

27(26) Shell turriform (Fig. 11.75 H) with very acute spire 28

27′ Shell slender, thick, spire usually stark decollated, with five flattened whorls remaining, ~50 mm height, suture well defined; color brown, covered with dark periostracum; last whorl ovoid, sculptured with concentric spiral cords in the base; aperture ovate, lip protruded below, short columella (Fig. 11.84 B) .... *Pachychilus subexaratus* Crosse & Fischer, 1891
[Guatemala: tributaries of Lakes Izabal and Amatitlán]

28(27) Shell spire with 10–11 convex whorls, ~60 mm height, apex barely eroded; color olive brown, last whorl sculptured with *ca.* five elevated spiral threads and strong, extruded axial bands; aperture oval, external lip thickened merging with the columella (Fig. 11.84 C) .... *Pachychilus largillierti* (Philippi, 1843)
[Central America, probably Nicaragua]

28′ Shell spire with 11 flattened whorls, apex barely eroded, ~86 mm height; shell yellow-green with or without dark brown periostracum, first whorls colored with axial reddish lines, last whorl sculptured with many small spiral ribs; aperture ovate, white-blue (Fig. 11.84 D) .... *Pachychilus pyramidalis* (Morelet, 1849)
[Mexico: Tabasco]

29(26) Shell sculptured with spiral ribs 30

29′ Shell very thick, apex eroded, about seven whorls, ~62 mm height, suture almost obsolete; color black, sculptured with protuberances and grooves, middle whorls with a single row of protuberances that gives each whorl an angular outline; the tubercles appear to degenerate into elevated ribs toward the first and last whorls, giving them a grooved appearance; last whorl big, base densely grooved; aperture greatly acute above (Fig. 11.84 E) .... *Pachychilus polygonatus* (I. Lea & H.C. Lea, 1850)
[Honduras: Copan]

30(29) Shell spire with nine slightly convex whorls, ~63 mm height; sculptured with threadlike granulate spiral ribs and spiral reddish-brown color lines; middle rib on each whorl elevated than the rest, giving the whorl an angular outline; shell color yellow-green; aperture ovate with external lip flexed below, interior white-blue (Fig. 11.84 F) .... *Pachychilus glaphyrus* (Morelet, 1849)
[Guatemala: Usumacinta River]

30′ Shell with seven to eight convex whorls, about 55 mm height, apex eroded; shell color yellow, reddish-brown or blackish; shell sculptured throughout with spiral wavy threads; aperture oval-acute, bluish (Fig. 11.84 G) .... *Pachychilus lacustris* (Morelet, 1849)
[Guatemala: Izabal Lake]

FIGURE 11.84 Species of *Pachychilus* (Pachychilidae): (A) *P. rubidus* (USNM 119713); (B) *P. subexaratus* (MNHN 24464); (C) *P. largillierti* (ZMB 112.944); (D) *P. pyramidalis* (MNHN 24452); (E) *P. polygonatus* (USNM 119596); (F) *P. glaphyrus* (MNHN 24454); (G) *P. lacustris* (NHMUK 1893.2.4.1776). Scale bar for (A) = 0.5 cm; for (B–G) = 2 cm. All pictures after Gomez-Berning et al. (2012). Museum acronyms: *MNHN*, Muséum National d'Histoire Naturelle, Paris; *NHMUK*, Natural History Museum, London; *USNM*, National Museum of Natural History, Smithsonian Institution, Washington D.C.; *ZMB*, Museum für Naturkunde, Berlin.

Gastropoda: Caenogastropoda: Cerithioidea: Thiaridae: Genera

In the Neotropics, this family is represented by 35 native *Aylacostoma* species and two introduced species: *Melanoides tuberculata* and *Tarebia granifera.* The three genera only differ in their internal anatomy, which inhibits creating a key to the genera based on external characteristics. The key below includes both native and introduced species.

Gastropoda: Caenogastropoda: Cerithioidea: Thiaridae: Species

1 Shell completely smooth or sculptured with either fine growth lines or spiral elements or both (Fig. 11.73 E) 2

1′ Shell totally or partially sculptured with conspicuous spiral striations (Fig. 11.76 D), curved axial ribs (Fig. 11.75 C), prominent spiral ridges (Fig. 11.77 H), keels (Fig. 11.77 E), or crenulated (Fig. 11.73 H) 17

2(1) Shell slightly globose, color yellow-olive to light brown with reddish to dark dots, stains or bands often arranged in spiral lines, the last whorl bigger than the spire, never exceeding 20 mm total height 3

2′ Shell conic with turriform spire (Fig. 11.75 H) or turriform and slender, reaching more than 40 mm total height 9

3(2) Material from northern South America 4

3′ Material from Cuba; shell globose, thin but solid spire with up to four convex whorls, apex not eroded, 18 mm total height; suture narrow without subsutural depression; color light brown, with dark brown spiral bands in the entire shell; aperture round, slightly channeled below (Fig. 11.85 A) .... *Aylacostoma brevis* (d'Orbigny, 1842) [Cuba]

4(3) Material from Colombia or Ecuador 5

4′ Material from Brazil; shell thick, strong, five convex whorls sculptured with faint spiral growth lines; covered with reddish, spiral, dotted lines; columella and callus very thick and fused, aperture ovate scarcely channeled, with chestnut colored bands (Fig. 11.85 B) .... *Aylacostoma aspersum* (Reeve, 1860) [Brazil]

5(4) Material from Ecuador 6

5′ Material from Colombia; shell thin, spire moderately elevated with obtuse apex, suture well impressed, five slightly convex whorls, ~13 mm height; color bright olive with darker bands and spiral rows of dark red dots, red bands within the aperture; aperture big, angulated above and below, columella curved, slightly notched at the base (Fig. 11.85 C) .... *Aylacostoma binneyi* (Tryon, 1866) [Colombia]

6(5) Material from North Ecuador including Quito 7

6′ Material from east and south Ecuador 8

7(6) Shell globose (as in Fig. 11.78 B), spire acute and short, ≤ seven whorls, usually strong decollated with two to four remaining whorls, 13–21 mm height; adult specimens colored black; juveniles light brown with red bands and dots; aperture dark red or with disrupted color bands, callus conspicuous, white with some violet .... *Aylacostoma nigrum* (Miller, 1878)
[Ecuador: Esmeraldas, Rivers Blanco by San Jose and Cachabi]

7′ Shell elongate-conic (Fig. 11.75 H), thick, spire with five subconvex whorls, suture simple, ~17 mm height; color olive-yellow with series of reddish dots or spiral bands; aperture ovate, lip sharp and simple, columella bent to the left (Fig. 11.85 D, E) .... *Aylacostoma osculati* (Villa, 1855)—*simplex* (Tryon, 1866) complex
[Ecuador: Quito]

FIGURE 11.85 Species of *Aylacostoma* (Thiaridae): (A) *A. brevis* (NHMUK 1854.10.4.194 and USNM 53636 respectively); (B) *A. aspersum* (NHMUK 20070086); (C) *A. binneyi* (ANSP 26786); (D) *A. osculati* (NHMUK 1878.1.28.584, ANSP 26785, and ANSP 122968 respectively); (E) *A. simplex* (ANSP 26782). Scale bar = 1 cm. Museum acronyms: *ANSP*, Academy of Natural Sciences of Drexel University, Philadelphia; *NHMUK*, Natural History Museum, London; *USNM*, National Museum of Natural History, Smithsonian Institution, Washington D.C.

8(6) Shell subinflated, decollate, four to five whorls, 14–17 mm height; color dark olive green, periostracum light brown, with small black spots in spiral series; aperture large, forming a small channel at the base of the columella (Fig. 11.86 A) .... *Aylacostoma guayaquilensis* (Petit, 1853)
[Ecuador: Guayaquil]

8′ Shell elongate-conic (Fig. 11.75 H), thick, apex barely eroded, up to seven whorls, 15–17 mm height, conspicuous growth lines; shell color yellow with spiral series of reddish dots, periostracum light brown; aperture ovate, columella thick and twisted, inner showing three to four reddish-brown bands (Fig. 11.86 B) .... *Aylacostoma fuscopunctatum* (von dem Busch, 1859)
[Ecuador: Napo River]

9(2) Material from the Greater Antilles 10

9′ Material from the mainland 11

10(9) Adult shell conic (Fig. 11.75 G), thin but solid, apex usually not eroded; spire with four to six flattened whorls, suture narrow and subsutural depression shallow, both forming colored band, up to 43 mm height; visible faint axial and spiral growth lines; color from yellow-brown to black, with reddish, spiral, interrupted color bands; aperture ovate, angled above, expanded below, inside white (Fig. 11.86 C) .... *Aylacostoma lineolatum* (W. Wood, 1828)
[Jamaica]

10′ Shell conical-broad (Fig. 11.75 G), decollate, four to eight slightly convex whorls, last whorl broadening at the middle, suture narrow and simple, faint but visible axial growth lines; the color is ranging from greenish-yellow to dark brown, uniform or with reddish continuous or interrupted bands all over the shell; aperture angled above, fused with the columella below, outer lip simple (Fig. 11.86 D) .... *Aylacostoma cubanianum* (d'Orbigny, 1842)
[Cuba]

11(9) Shell ovate-conic (Fig. 11.75 E), with big last whorl or whorls with convex outline 12

11′ Shell slender, spire outline flattened 16

12(11) Shell delicate with convex whorls, last whorl proportionate to the spire 13

12′ Last whorl globose, large in comparison to the spire 14

13(12) Shell oblong-conic (Fig. 11.75 F), solid, spire can have up to seven whorls (≤44 mm height) but is usually strong decollated, with three to four whorls remaining (13–14 mm height); suture deep with two to three deep carved lines underneath; color dark brown to black; aperture ovate, narrowing at the base, with black-brown bands, columella almost straight (Fig. 11.86 E) ...... *Aylacostoma muzensis* (Brot, 1878)
[Colombia: Muzo]

13′ Shell decollated with four to five whorls remaining, 25–30 mm height, faint spiral and axial growth lines; color yellow-olive with reddish-brown spots spirally arranged but forming an irregular pattern; aperture ovate, white, columella yellowish, twisted (Fig. 11.86 F) .... *Aylacostoma panamensis* (Pilsbry, 1939)
[Panama]

PHYLUM MOLLUSCA

14(12) Last whorl large and globose 15

14′ Shell with seven whorls, apex not eroded, 29–36 mm height, suture deep with a light color band underneath, last whorl slightly convex; color yellow-olive to brown with reddish-brown spots arranged in axial rows; aperture ovate, columella fused with the callus, thick and white, forming a small protuberance (Fig. 11.86 G) .... *Aylacostoma acutissimum* (von dem Busch, 1858) [Venezuela]

15(14) Shell shape as in Fig. 11.78 B, relatively strong, spire decollated, seven flat whorls with deep suture, ~29 mm height; color yellowish-olive, with numerous threadlike spiral dark spots; aperture elongate, fusiform, strongly sinuate above .... *Aylacostoma planogyrum* (Brot, 1878) [Peru: Lima]

15′ Shell greenish-brown with spiral rows of dark brown square spots, a light color band without spots under the suture; ~30 mm height; aperture narrowly ovate or lanceolate, outer margin thin, little arcuate, basal margin narrowly rounded, columellar margin nearly straight, a little thickened, of a violaceous hue, notched below, younger specimens showing light brown bands in aperture (Fig. 11.86 H) .... *Aylacostoma maculatum* (Martens, 1899) [Costa Rica: Confluence of the Shei and Tararia Rivers]

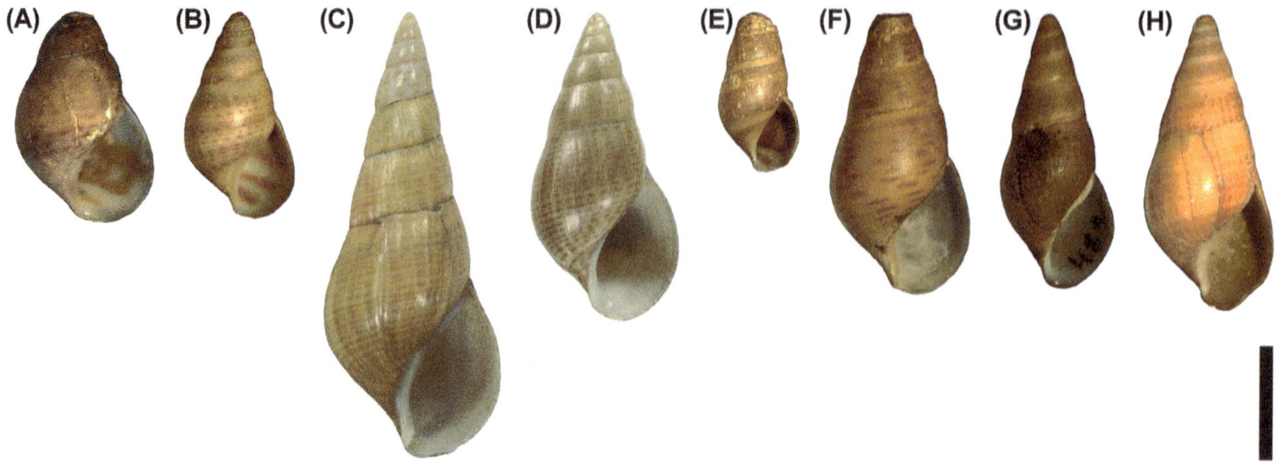

FIGURE 11.86 Species of *Aylacostoma* (Thiaridae): (A) *A. guayaquilensis* (ANSP 188811); (B) *A. fuscopunctatum* (ANSP 322927); (C) *A. lineolatum* (NHMUK 1984206/1); (D), *A. cubanianum* (MCZ 94734); (E) *A. muzensis* (ANSP 166104); (F) *A. panamensis* (ANSP 178116); (G) *A. acutissimum* (ANSP 120738); (H) *A. maculatum* (ANSP 26817). Scale bar = 1 cm. Picture (C) after Gomez et al. (2011). Museum acronyms: *ANSP*, Academy of Natural Sciences of Drexel University, Philadelphia; *MCZ*, Museum of Comparative Zoology, Harvard University; *NHMUK*, Natural History Museum, London.

16(11) Shell elongate turriform (Fig. 11.75 H), eight to ten flattened stepped whorls, sometimes sculptured with superficial spiral grooves, ~34 mm height, color dark olive; suture deep, with subsutural ridge; aperture lunar-ovate, truncated at the base, columella twisted (Fig. 11.87 A) .... *Aylacostoma strigilatum* (Dunker, 1843) [Venezuela: Orinoco River]

16′ Shell strong, spire high, apex sometimes completely eroded, three to five flattened sloping whorls ≤42 mm height, suture simple with subsutural furrow; the last whorl sculptured by broad spiral cords; color dark brown interleaved with yellowish bands and decorated with reddish-dark brown, minute, spiral spots regularly spaced, periostracum persistent; aperture ovate, angled above, channeled below, in young specimens showing dark, linear spiral spots (Fig. 11.87 B) .... *Aylacostoma brunneum* Vogler & Peso, 2014 [Paraguay: High Paraná River, Itapúa, Río Beach]

17(1) Shell sculptured with keels, tubercles (Fig. 11.73 G), or crenulated (Fig. 11.73 H) 18

17′ Shell sculptured with spiral ribs through the shell (similar to Fig. 11.77 H) or only in the last whorl or the base of the last whorl 21

18(17) Shell sculptured with tubercles (Fig. 11.73 G) or crenulated (Fig. 11.73 H) 19

18′ Shell keeled, globosely turbinated, solid, five convex whorls, 35–40 mm total height; sculpture very conspicuous, made up of strong, wavy, and crenate spiral ribs crossed by axial folds; the last whorl broadly angled at the upper part, the ridge on the angle building more or less protruding nodules; color light brown to blackish-chestnut; aperture squarely oblong, large; columella thick and twisted (Fig. 11.87 C) .... *Aylacostoma crenocarinum* (Moricand, 1841) [Brazil: Bahia, Rio Pedra Branca]

19(18) Shell sculpture with tubercles (Fig. 11.73 G) 20

PHYLUM MOLLUSCA

19′ Shell crenulated, turriform, 20–27 mm total height, spire whorls flattened and stepped; sculptured with prominent spiral ridges and nodes, nodes overlapping suture; color varies from yellowish to light brown, never mottled with reddish spots (Fig. 11.87 D) .... *Tarebia granifera* (Lamarck, 1822)
[Widespread, invasive. Oriental]

20(19) Shell ovate or fusiform (Fig. 11.75 E), up to seven whorls but usually decollated, 24–35 mm height; sculptured with axial ribs and spiral grooves the intersection forming tubercles; color greenish with series of fine dark lines; aperture very narrow, elongate, exceeding more than the half of the shell height, with two dark bands (Fig. 11.87 E) .... *Aylacostoma goyazinus* (Ihering, 1909)
[Brazil: Goias]

20′ Shell slender and elongate, with 7–11 rounded whorls, 10–40 mm total height; sculptured with axial, weakly curved ribs and much fine spiral striations; color yellowish to dark brown usually marked with red or rust-colored spots, flames, or bands (most apparent on juvenile specimens); suture simple, not crenulated (Fig. 11.87 F) .... *Melanoides tuberculata* (Müller, 1774)
[Widespread, invasive. Oriental]

21(17) Shell sculptured throughout with deep or shallow spiral ribs (like the sculpture of the shell depicted in Fig. 11.77 H) or ridges (Fig. 11.75 D) 22

21′ Last whorl or base of the last whorl displaying spiral ridges 38

22(21) Shell ≤25 mm height, globose 23

22′ Shell elongated, turriform (Fig. 11.75 H) 24

23(22) Shell globosely turbinated, thin, three to four convex whorls, last whorl ventrose; close spirally ridged sculpture; color brownish-olive, obliquely streaked with black; aperture broadly ovate, columella thick, broad, callus twisted (Fig. 11.87 G) .... *Aylacostoma pulcher* (Reeve, 1860)
[Brazil: Pernambuco]

23′ Shell ovate turbinated, solid, three to four whorls, sutures channeled; color burnt brown, unspotted, covered with a rich swarthy periostracum (Fig. 11.87 H) .... *Aylacostoma obesum* (Reeve, 1860)
[Brazil]

24(22) Shell elongated and slender 25

24′ Shell oval-conic (Fig. 11.75 E) or conic (Fig. 11.75 G) 31

25(24) Adult shell total height < 30 mm 26

25′ Adult shell total height ≥ 50 mm 29

26(25) Spire whorls stepped 27

26′ Spire with convex whorls, last whorl inflated 28

27(26) Shell oblong-conic (Fig. 11.75 F), apex eroded, six to seven whorls, ~27 mm height, finely striated; color green-yellow or dark brown, striae often ornamented with filiform rust colored spots; aperture acute, ovate, small, columella almost straight, lip acute (Fig. 11.87 I) .... *Aylacostoma ruginosum* (Morelet, 1849)
[Guatemala: Izabal Lake]

FIGURE 11.87 Species of *Aylacostoma*, *Tarebia* and *Melanoides* (Thiaridae): (A) *A. strigilatum* (USNM 522002); (B) *A. brunneum* (holotype MLP 13757); (C) *A. crenocarinum* (ANSP 122766); (D) *T. granifera* (CMA-UCO 1518); (E) *A. goyazinus* (ANSP 120641); (F) *M. tuberculata* (ZMB 113162); (G) *A. pulcher*; (H) *A. obesum* after Reeve, 1860 (I) *A. ruginosum* (USNM 128783). Scale bar = 1 cm. Picture (B), courtesy of Roberto Vogler and (D), courtesy of Esteban Peláez. Museum acronyms: *ANSP*, Academy of Natural Sciences of Drexel University, Philadelphia; *CMA-UCO*, Universidad Catolica de Oriente, Colombia; *MLP*, Museo de La Plata, Argentina; *USNM*, National Museum of Natural History, Smithsonian Institution, Washington D.C.; *ZMB*, Museum für Naturkunde, Berlin.

27′ Shell thin, six flattened whorls, apex eroded, suture deep; color light brown; sculptured with deep spiral grooves building elevated ribs, two first ribs underneath the suture more conspicuous than the rest, giving the shell a stepped appearance; aperture small, rounded, columella indistinct (Fig. 11.88 A) .... *Aylacostoma circumsulcatum* (von dem Busch, 1858) [Brazil]

28(26) Shell turreted, rather solid, eight whorls sloping round the upper part, then convex; color olive with thick periostracum, encircled with two narrow, distant chestnut bands barely apparent externally, but plainly shown in the aperture; spirally corded throughout with ridges, the interstices between excavated; aperture rather small, lip sinuated at the upper part, columella rigidly curved (Fig. 11.88 B) .... *Aylacostoma bicinctum* (Reeve, 1860) [Brazil]

28′ Shell narrow-turriform, solid, greatly elevated, 42 mm height, six to seven convex whorls, truncated, color light to dark brown; ornamented with spiral regular furrows and growth lines crosswise; aperture elliptical, base wide truncated and sinuated (Fig. 11.88 C) .... *Aylacostoma wesseli* (Brot, 1864) [Panama]

29(25) Aperture with lip simple, without notches or deflections 30

29′ Aperture with lip conspicuously sinuate, notched at upper part; shell turreted with spire very much elongate, nine to ten flattened whorls, apex eroded, and sutures slightly channeled; olive-yellow to dark horn color; aperture small, somewhat squarely ovate, columella incurved, slightly truncated at the base (Fig. 11.88 D) .... *Aylacostoma edwardsii* (I. Lea, 1852) [Brazil: Tocantins River]

30(29) Shell fusiform (Fig. 11.75 J), spire elevated but truncated, solid, five convex whorls, ~ 33 mm height, last whorl greatly elongate; color light olive, sculptured with sharp spiral furrows crossed by growth striae; aperture fusiform-oblong, base channeled, columella a bit twisted (Fig. 11.88 E) .... *Aylacostoma gealei* (Brot, 1872) [Colombia]

30′ Shell solid, six to seven flattened whorls, apex conserved, suture deep but barely distinguishable from the shell sculpture; color brown, thickly grooved all over the whorls, looking like a succession of coils of rope piled on each other, ribs between the grooves flattened; aperture ovate, columella indistinct, lip wavy due to the spiral striation (Fig. 11.88 F) .... *Aylacostoma obruta* (I. Lea, 1850) [Brazil]

31(24) Last whorl big, globose and spire narrow 32

31′ Last whorl slightly convex, not much different than the spire 35

32(31) Very deep spiral sculpture 33

32′ Spiral sculpture superficial 34

33(32) Shell solid, decollate, three to five flattened whorls remaining, 33–36 mm height, periostracum dark and shiny; shell sculptured with regularly arranged, elevated spiral threads and subsutural band; aperture ovate, angled above channeled below, lip sinuous, inner bluish with parallel dark red bands and spiral interrupted dot lines (Fig. 11.88 G) .... *Aylacostoma guaraniticum* (Hylton-Scott, 1953) [Argentina: High Paraná, Misiones, Ombú]

33′ Shell turbinate, eight convex whorls, apex barely eroded; shell color white with scattered reddish-brown dots, periostracum olive yellow; aperture ovate, columella thickened and twisted (Fig. 11.88 H) .... *Aylacostoma tenellum* (Reeve, 1860) [Brazil: Pernambuco]

34(32) Shell turbinate, seven to eight flattened whorls, 25–35 mm height; color yellowish-olive, with axial series of dashed reddish-brown dots, shiny; aperture wide, ovate, columella thick, excavated and twisted (Fig. 11.88 I) .... *Aylacostoma punctatum* (Reeve, 1860) [Brazil: Pernambuco]

34′ Shell thin but strong, six to eight convex whorls, apex eroded, suture deep and narrow; color dark brown, regularly grooved throughout; grooves deepening after suture forming a band lighter in color; aperture rounded, slightly sinuous above, channeled below, columella twisted, thick (Fig. 11.88 J) .... *Aylacostoma kochii* (Bernardi, 1856) [Brazil]

35(31) Specimens from Brazil 36

35′ Specimens from Venezuela; spire with seven to eight slightly convex whorls, shell white with olive green periostracum, spirally grooved, grooves somewhat distant; aperture small, effused at the base (Fig. 11.88 K) .... *Aylacostoma venezuelensis* (Reeve, 1859) [Puerto Cabello, Venezuela]

FIGURE 11.88 Species of *Aylacostoma* (Thiaridae): (A) *A. circumsulcatum* (ANSP 120644); (B) *A. bicinctum* (NHMUK 20050148); (C) *A. wesseli* (USNM 882004); (D) *A. edwardsii* (ANSP 122956); (E) *A. gealei* (ANSP 26795); (F) *A. obruta* (USNM 119595); (G) *A. guaraniticum* (paratype MLP 11213); (H) *A. tenellum* (NHMUK 1984216/1); (I) *A. punctatum* (NHMUK 1984209/2); (J) *A. kochii* (ANSP 26788); (K) *A. venezuelensis* (NHMUK 1984215). Scale bar = 1 cm. Picture (G) courtesy of Roberto Vogler. Museum acronyms: *ANSP*, Academy of Natural Sciences of Drexel University, Philadelphia; *MLP*, Museo de La Plata, Argentina; *NHMUK*, Natural History Museum, London; *USNM*, National Museum of Natural History, Smithsonian Institution, Washington D.C.

36(35) Shell slim and elegant 37

36′ Shell turriform, about five whorls, apex eroded, ≤25 mm height; color light brown with some sparse dark red spiral spots; sculptured with broad, uniform in size spiral ribbons separated by a narrow and shallow furrow; aperture elliptical with shallow channel below (Fig. 11.89 A) .... *Aylacostoma exoplicata* Simone, 2001
[Brazil, Pará, Caraipé River]

37(36) Shell conic (Fig. 11.75 G), apex acute, eight to nine whorls flat or slightly convex in outline, 35 mm total height; color olive green with interrupted, reddish lines; sculptured with more or less deep spiral grooves, flattened ribs between the grooves give the aperture lip a crenulated appearance, subsutural band of much lighter color; aperture oval, channeled below, showing regularly distributed brown spiral lines (Fig. 11.89 B) .... *Aylacostoma brasiliensis* (Moricand, 1837)
[Brazil: Bahia, Villa de Barra]

37′ Shell strong, apex barely eroded, seven slightly convex whorls, suture simple, last whorl big, ~21 mm height; yellowish, ornamented with fine, interrupted dark spiral lines; aperture ovate channeled below, columella twisted (Fig. 11.89 C) to justify to right .... *Aylacostoma schneideri* (Brot, 1878)
[Brazil: Maranon]

38(21) Shell fusiform (Fig. 11.75 J) 39

38′ Shell conic-turriform (Fig. 11.75 H) 41

39(38) Shell narrow with an elongate, narrow aperture 40

39′ Shell turriform, solid, seven flattened whorls, decollated, ~30 mm height, spire whorls growing partially over the previous whorl give the shell a stepped shape, last whorl very big; last two whorls sculptured with little spiral ribs; color olive green with serial dark red points, carved line above suture; aperture ovate, angled above (Fig. 11.89 D) .... *Aylacostoma scalaris* (Wagner, 1827)
[Brazil: Rio de Janeiro]

40(39) Shell fusiform oblong (Fig. 11.75 F), three to four stepped flattened whorls, last whorl contracted and spirally grooved at the base, 35–40 mm total height; color olive brown, longitudinally striped with black; aperture narrowly oblong-ovate, with reddish-brown spiral bands, columella thin and twisted (Fig. 11.89 E) .... *Aylacostoma tenuilabris* (Reeve, 1860)
[Brazil]

40′ Shell very elongate, four to five whorls, decollate, suture incised and of much lighter color, ≤30 mm total height; color yellowish with short dark axial lines, periostracum dark olive; spiral ornamentation almost limited to the last whorl base; aperture very long and narrow, reddish-brown; columella white, thin and truncated (Fig. 11.89 F) .... *Aylacostoma francanum* (Ihering, 1909)
[Brazil: São Paulo]

FIGURE 11.89 Species of *Aylacostoma* (Thiaridae): (A) *A. exoplicata* (ANSP 109368); (B) *A. brasiliensis* (NHMUK 1984214/1 and NHMUK 1984216/2-3 respectively); (C) *A. schneideri* after Brot, 1878 (D) *A. scalaris* (NHMUK 1903.12.5.1-6 and USNM 103727 respectively); (E) *A. tenuilabris* (NHMUK 1984222/1 and NHMUK 1903.12.5.1-6 respectively); (F) *A. francanum* (USNM 172947). Scale bar = 1 cm. Museum acronyms: *ANSP* = Academy of Natural Sciences of Drexel University, Philadelphia; *NHMUK*, Natural History Museum, London; *USNM*, National Museum of Natural History, Smithsonian Institution, Washington D.C.

41(38) Shell < 30 mm total height 42

41′ Shell > 30 mm total height 46

42(41) Spire whorls with sparse spiral grooves 43

42′ Subsutural groove forming an elevated band between the suture and the groove 44

43(42) Shell subglobosely ovate, thin, apex eroded, three to five flattened whorls, ≤ 30 mm height, last whorl shouldered, suture deep; color dark yellow with axial brownish-black stripes; aperture ovate-elongate; columella straight, thickened (Fig. 11.90 A) .... *Aylacostoma flammeum* (Baker, 1913)
[Brazil]

43′ Shell slender, whorls flat, stepped arranged, strong decollate, three to four remaining whorls, about 21 mm height; spiral sulci persistent and deep; color olive yellow with blackish, irregular flames; aperture broad and short, columella thick and a little twisted (Fig. 11.90 B) .... *Aylacostoma elongatum* (Baker, 1913)
[Brazil: Rio Jamauchim]

44(42) Shell thick and strong 45

44′ Shell fusiform (Fig. 11.75 J), spire decollated, four to five flattened, stepped whorls, ~30 mm height, suture impressed, subsutural furrow forming a broad, slightly elevated band; color yellow-olive green, reddish spots forming spiral interrupted lines throughout the shell; sculptured with spiral, flattened ribs in the base of the last whorl; aperture ovate, longer than wide, channeled below; columella white and twisted, inner showing brown spiral bands (Fig. 11.90 C) .... *Aylacostoma araguaryanum* (Ihering, 1909)
[Brazil: Minas Gerais, Araguari River]

45(44) Shell 23–28 mm total height, decollated, three to four remaining flattened whorls, last whorl inflated; periostracum yellow, irregular black spots throughout the shell; lower middle of the last whorl with spiral grooves separating parallel ribbons; aperture ovate, lip sinuous (Fig. 11.90 D) .... *Aylacostoma stigmaticum* (Hylton-Scott, 1954)
[Paraguay: High Parana, Isla Ibicuy]

45′ Shell turriform (Fig. 11.75 H), apex eroded, five flattened whorls, ≤22 mm height; color dark brown with some sparse, spiral, reddish spots; sculptured with spiral, uniform in size, broad cords separated by shallow and narrow furrow; aperture ovate, angled above and channeled below (Fig. 11.90 E) .... *Aylacostoma ci* Simone, 2001
[Brazil: Rondonia, Jamari River]

46(41) Shell thick, last whorl globose, big, aperture ovate-rounded 47

46′ Shell subglobosely ovate, rather solid, strong decollated, three whorls remaining, spire whorls slightly convex, last whorl grooved at the base; color mustard yellow, conspicuously adorned with chestnut axial stripes; aperture wide, oblong-ovate, lip sinuated at the upper part, columella twisted (Fig. 11.90 F) .... *Aylacostoma zebra* (Reeve, 1860)
[Brazil: Amazonas River]

47(46) Shell turriform, solid, eight to ten flattened stepped whorls, > 35 mm, the last whorl sometimes concavely angled around upper part; color olive yellow to dark brown sprinkled with linear red dots; spirally grooved toward the aperture; aperture narrowly ovate, columella very thick and twisted forming a channel with the base of the lip, callus and lip thickened, white or light brown colored, sometimes showing

spiral reddish-brown bands or lines (Fig. 11.90 G) .... *Aylacostoma behnii* (Reeve, 1860)
[Brazil: Pernambuco]

47′ Shell conic-fusiform, four to six slightly convex whorls, apex eroded, 30–37 mm total height, suture simple, subsutural ribbons wide and pale in color; color lemon yellow with dark brown periostracum; juveniles showing dark red dotted spiral lines; shell sculptured with conspicuous spiral threads and deep grooves in the last whorls, being moderate or faint in the first whorls; aperture ovate-oblique channeled below (Fig. 11.90 H) .... *Aylacostoma chloroticum* (Hylton-Scott, 1954)
[Paraguay: High Parana, Isla Ibicuy]

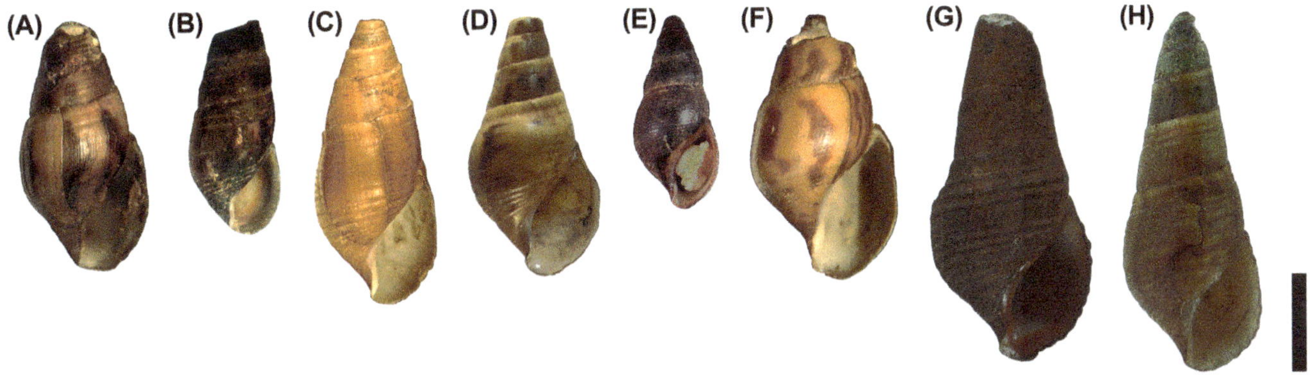

FIGURE 11.90 Species of *Aylacostoma* (Thiaridae): (A) *A. flammeum* (ANSP 109366); (B) *A. elongatum* (ANSP 109367); (C) *A. araguaryanum* (USNM 171449); (D) *A. stigmaticum* (syntype MLP 10964); (E) *A. ci* (holotype MZSP 28691); (F) *A. zebra* (ANSP 122940); (G) *A. behnii* (NHMUK 1903.12.5.1-6); (H) *A. chloroticum* (holotype MLP 10958,). Scale bar = 1 cm. Pictures (D) and (H) courtesy of Roberto Vogler; picture (E) courtesy of Luiz Ricardo L. Simone. Museum acronyms: *ANSP*, Academy of Natural Sciences of Drexel University, Philadelphia; *MLP*, Museo de La Plata, Argentina; *MZSP*, Museu de Zoologia da Universidade de São Paulo, Brazil; *NHMUK*, Natural History Museum, London; *USNM*, National Museum of Natural History, Smithsonian Institution, Washington D.C.

REFERENCES

Alonso, L.E. & H.J. Berrenstein (eds.). 2006. A Rapid Biological Assessment of the Aquatic Ecosystems of the Coppename River Basin, Suriname. RAP Bulletin of Biological Assessment 39. Conservation International, Washington, DC. 118 pp.

Appleton, C.C., A.T. Forbes & N.T. Demetriades. 2009. The occurrence, bionomics and potential impacts of the invasive freshwater snail *Tarebia granifera* (Lamarck, 1822) (Gastropoda: Thiaridae) in South Africa. Zoologische Mededelingen 83: 525–536.

Barré, C., D. ONeil & V.M. Bricelj. 2006. Preparation of large bivalve specimens for scanning electron microscopy using Hexamethyldisilazane (HMDS). Journal of Shellfish Research 25: 639–641.

Chenu, J.C. 1859. Manuel de Conchyliologie et de Paleontologie Conchyliologique, Band 1. Victor Masson, Paris. 508 pp.

Cuezzo, G. 2009. Chapter 19: Mollusca Gastropoda. Pages 595–629 *in*: E. Domínguez & H.R. Fernández (eds.), Macroinvertebrados bentónicos sudamericanos. Sistemática y biología. Fundación Miguel Lillo, Tucumán, Argentina. 656 pp.

de Jesus, A.J.S., T.C.P. das Neves Costa & M. Camargo. 2008. Registros de moluscos gastropoda no médio Rio Xingu – Pará. Uakari 3: 96–103.

Duncan, N. 2008. Survey Protocol for Aquatic Mollusk Species: Preliminary Inventory and Presence/Absence Sampling, Version 3.1. Portland, OR. Interagency Special Status/Sensitive Species Program. U.S. Department of Interior, Bureau of Land Management, Oregon/Washington and U.S. Department of Agriculture, Forest Service, Region 6. 52 pp.

Faria-Vaz, J., H.M. Santana-Teles, M.A. Correa & S.P. da Silva-Leite. 1986. Ocorrencia no Brasil de *Thiara (Melanoides) tuberculata* (O.F. Muller, 1774) (Gastropoda, Prosobranchia), primeiro hospedeiro intermediário de *Clonorchis sinensis* (Cobbold, 1875) (Trematoda, Plathyhelmintes). Revista de Saúde Pública, Sao Paulo 20: 318–322.

Ferrer-Lopez, J.R., G. Perera de Puga & M. Yong-Cong. 1991. Estudio de la morfometría de los juveniles de cuatro poblaciones de *Tarebia granifera* en condiciones de laboratorio. Revista Cubana de Medicina Tropical 43: 26–30.

Fischer, P.H. & J.H.C. Crosse. 1870–1902. Études sour les Mollusque terrestres et fluviatiles du Mexique et du Guatemala. Mission scientifique au Mexique et dans l'Amérique Centrale. Imprimerie Nationale, Paris. 731 pp.

Glaubrecht, M. 1996. Evolutionsökölogie und Systematik am Beispiel von Süß- und Brackwasserschnecken (Mollusca: Caenogastropoda: Cerithioidea): Ontogenese-Strategien, paläontologische Befunde und Historische Zoogeographie. Backhuys Publishers, Leiden, 499 pp.

Goodrich, C. & H. van der Schalie. 1937. Mollusca of Petén and north Alta Vera Paz, Guatemala. Miscellaneous Publications, University of Michigan, Museum of Zoology 34. 47 pp.

Gomez, M.I. 2009. Systematics, phylogeny, and biogeography of Mesoamerican and Caribbean freshwater gastropods (Cerithioidea: Thiaridae and Pachychilidae). Dr. rer. Nat. Thesis, Mathematisch-

PHYLUM MOLLUSCA

Naturwissenschaftlichen Fakultät I, Humboldt University of Berlin, Berlin, Germany. 220 pp.

Gomez, M.I., E.E. Strong & M. Glaubrecht. 2011. Redescription and Anatomy of the Viviparous Freshwater Gastropod *Hemisinus lineolatus* (W. Wood, 1828) from Jamaica (Cerithioidea, Thiaridae). Malacologia 53: 229–250.

Gomez-Berning, M., F. Köhler & M. Glaubrecht. 2012. Catalogue of the nominal taxa of Mesoamerican Pachychilidae (Mollusca: Caenogastropoda). Zootaxa 3381: 1–44.

Gutierrez, A., G. Perera, M. Yong & J.A. Fernandez. 1997. Relationships of the prosobranch snails *Pomacea paludosa, Tarebia granifera* and *Melanoides tuberculata* with the abiotic environment and freshwater snail diversity in the central region of Cuba. Malacological Review 30: 39–44.

Gutiérrez Gregoric, D.E., V. Núñez, A. Rumi & M.A. Roche. 2006. Freshwater gastropods from del Plata Basin, Argentina. Checklist and new locality records. Comunicaciones de la Sociedad Malacologica del Uruguay 9: 51–60.

Healy P.F, E. Kitty & L.E. Wright. 1990. Ancient and Modern Maya Exploitation of the Jute Snail (*Pachychilus*). Latin American Antiquity 1: 170–183.

Holznagel, W.E. 1998. Research Note: A nondestructive method for cleaning gastropod radulae from frozen, alcohol-fixed, or dried material. American Malacological Bulletin 14: 181–183.

Houbrick, R. 1988. Cerithioidean Phylogeny. Pages 1–330 *in*: W.F., Ponder, D.J. Eernisse & J.H. Waterhouse (eds.), Prosobranch Phylogeny, Malacological Review, Ann Arbor, Michigan.

Hylton-Scott, M.I. 1953. El género "*Hemisinus*" (Melaniidae) en la costa Fluvial Argentina. Physis 20: 438–443.

Hylton-Scott, M.I. 1954. Dos Nuevos Melanidos del Alto Paraná (Mol. Prosobr.). Neotropica 1: 45–48.

Köhler, F. & M. Glaubrecht. 2003. Morphology, reproductive biology and molecular genetics of ovoviviparous freshwater gastropods (Cerithioidea, Pachychilidae) from the Philippines, with description of a new genus *Jagora*. Zoologica Scripta 32: 35–39.

Köhler, F. & M. Glaubrecht. 2005. Fallen into oblivion - the systematic affinities of the enigmatic *Sulcospira* Troschel, 1858 (Cerithioidea: Pachychilidae), a genus of viviparous freshwater gastropods from Java. The Nautilus 119: 15-26, 90–91.

Köhler, F. & M. Glaubrecht. 2007. Out of Asia and into India: on the molecular phylogeny and biogeography of the endemic freshwater gastropod *Paracrostoma* Cossmann, 1900 (Caenogastropoda: Pachychilidae). Biological Journal of the Linnean Society 91: 627–651.

Köhler, F., T. von Rintelen, A. Meyer & M. Glaubrecht. 2004. Multiple origin of viviparity in Southeast Asian gastropods (Cerithioidea: Pachychilidae) and its evolutionary implications. Evolution 58: 2215–2226.

Lydeard, C., W.E. Holznagel, M. Glaubrecht & W.F. Ponder. 2002. Molecular phylogeny of a circum-global, diverse gastropod superfamily (Cerithioidea: Mollusca: Caenogastropoda): pushing the deepest phylogenetic limits of mitochondrial LSU rDNA sequences. Molecular Phylogenetics and Evolution 22: 399–406.

Martens, E.Y. 1890–1901. Land and Freshwater Mollusca. Biologia Centrali-Americana. R.H. Porter, London. 706 pp.

McKillop, H. 2010. Ancient Maya canoe navigation and its implications for Classic to Postclassic Maya economy and sea trade: A view from the south coast of Belize. Journal of Caribbean Archaeology. Special Publication 3: 93–105.

Moore, J.E.S. 1898. On the hypothesis that Lake Tanganyika represents an old Jurassic sea. Quarterly Journal of Microscopical Science, new series 41: 303–321.

Morrison, J.P.E. 1954. The relationships of old and new world melanians. Proceedings of the United States National Museum 103: 357–394.

Naranjo Garcia, E. 2003a. Historia de la malacología en México con énfasis en la malacología continental. Revista de Biología Tropical 51: 463–471.

Naranjo Garcia, E. 2003b. Moluscos continentales de México: Dulceacuícolas. Revista de Biología Tropical 51: 495–505.

Nation, J.L. 1983. A new method using Hexamethyldisilazane for preparation of soft insect tissues for scanning electron microscopy. Stain Technology 58: 347–351.

Nations, J.D. 1979. Snail Shells and Maize Preparation: A Lacandon Maya Analogy. American Antiquity 44: 568–571.

Parodiz, J.J. 1982. Distribution and origin of the continental South America malacofauna. Malacologia 22: 421–425.

Pérez, A.M. & A. López. 2003. Listado de la malacofauna continental (Mollusca: Gastropoda) del Pacifico de Nicaragua. Revista de Biología Tropical 51: 405–451.

Pointier, J-P., R.N. Incani, C. Balzan, P. Chrosciechowski & S. Prypchan. 1994. Invasion of the rivers of the littoral central region of Venezuela by *Thiara granifera* and *Melanoides tuberculata* (Mollusca: Prosobranchia: Thiaridae) and the absence of *Biomphalaria glabrata*, snail host of *Schistosoma mansoni*. The Nautilus 10: 124–128.

Ponder, W.F. & A. Warrén. 1988. Classification of the Caenogastropoda and Heterostropha - a list of the family-group names and higher taxa. Pages 288–317 *in*: W.F. Ponder (ed.), Prosobranch Phylogeny "Symposium on Prosobranch Phylogeny", Malacological Review, Ann Arbor, Michigan. 346 pp.

Reeve, L.A. 1851–1870. Conchiologia Iconica: or, Illustrations of the shells of molluscous animals. Monograph of the genus *Melania* (Vol. 12). Lovell Reeve, London. 232 pp.

Romeis, B. 1989. Mikroskopische Technik. Böck, P. (ed.), Urban und Schwarzberg, Munich.

Royero, R. 1994-1996. Algunos aspectos de la etnoictiología y la historia natural de los Yanomami del alto río Siapa y río Mavaca, Estado Amazonas, Venezuela. Antropologica 84: 73–96.

Rumi, A., D.E. Gutiérrez Gregoric, V. Núñez, I.I. César, M.A. Roche, M.P. Tassara, S.M. Martín & M.F. López Armengol. 2006. Freshwater Gastropoda from Argentina: Species richness, distribution patterns, and an evaluation of endangered species. Malacologia 49: 189–208.

Rumi, A., D.E. Gutiérrez Gregoric, V. Núñez & G.A. Darrigran. 2008. Malacología latinoamericana. Moluscos de agua dulce de Argentina. Revista de Biología Tropical 56: 77–111.

Simone, L.R. 2001. Phylogenetic analyses of Cerithioidea (Mollusca, Caenogastropoda) based on comparative morphology. Arquivos de Zoologia Museu de Zoologia da Universidade de São Paulo 36: 147–263.

Simone, L.R.L. 2006. Land and freshwater molluscs of Brazil. EGB, Fapesp, São Paulo. 390 pp.

Simpson, C.T. 1894. Distribution of the land and freshwater mollusks of the West-Indian region and their evidence with regard to past changes of land and sea. Proceedings of the United States National Museum 17: 423–450.

Strong, E.E., D.J. Colgan, J.M. Healy, C. Lydeard, W.F. Ponder & M. Glaubrecht. 2011. Phylogeny of the gastropod superfamily Cerithioidea using morphology and molecules. Zoological Journal of the Linnean Society 162: 43–89.

Thiele, J. 1929–1935. Handbuch der Systematischen Weichtierkunde. Gustav Fischer, Jena. 376 pp.

Thompson, F.G. 2011. An annotated checklist and bibliography of the land and freshwater snails of Mexico and Central America. Bulletin of the Florida Museum of Natural History 50: 1–299.

Townsend, S. & D. Newell. 2006. Technical Progress Report IABIN Invasive species thematic network content building project implement, Update and maintain an I3N IAS Database in Jamaica. Kingston: Institute of Jamaica, Natural History Division. Available: http://www.jamaicachm.org.jm/Document/Interim%20Technical%20%20Financial%20Report.pdf. Acceded 2019-04-23.

Tucker-Abbot, R. 1955. Anatomy of the Venezuelan gastropod, *Doryssa kappleri*. The Nautilus 69: 44–46.

Vázquez Perera, A.A & S. Perera Valderrama. 2010. Endemic Freshwater molluscs of Cuba and their conservation status. Tropical Conservation Science 3: 190–199.

Vogler, R.E., A.A. Beltramino, J.G. Peso & A. Rumi. 2014. Threatened gastropods under the evolutionary genetic species concept: redescription and new species of the genus *Aylacostoma* (Gastropoda: Thiaridae) from High Paraná River (Argentina–Paraguay). Zoological Journal of the Linnean Society 172: 501–520.

Winnepenninckx, B., T. Backeljau & R. de Wachter. 1993. Extraction of high molecular weight DNA from molluscs. Trends in Genetics 9: 407.

WoRMS Editorial Board. 2019. World Register of Marine Species. World Register of Marine Species. Available: http://www.marinespecies.org at VLIZ. Accessed 2019-05-31. doi:10.14284/170.

Class Bivalvia

Cristián Ituarte
Division of Invertebrate Zoology, Museo Argentino de Ciencias Naturales, Buenos Aires, Argentina

Maria Cristina Dreher Mansur
Grupo de Pesquisa do CNPq, Biodiversidade de Moluscos Continentais, Museu de Ciências Naturais, Porto Alegre, RS, Brazil

INTRODUCTION

Four freshwater bivalve orders occur in the Neotropical region: Unionida, Venerida, Mytilida, and Myida.

The Unionida are exclusively freshwater mollusks without Recent marine ... (upper case) marine relatives. They are known as naiads, mussels, "cucharas" (in Spanish), "ostras-de-rio" or "mariscos-de-rio" (in Brazilian Portuguese), or as "itã" in northern South American Amerind language. Naiads vary from 2 to 10 cm but may be larger. They live commonly on, or partially buried in, sandy or muddy sandy bottoms of ponds and rivers. The majority are active filter-feeders, for which they prefer habitats rich in organic matter. Only two species in South America live attached to rocks or embedded in small spaces between rocks or hollows of rocky or lateritic substrates. Unionids have strategies that allow survival in extreme conditions such as waterfalls, drought, and flood pulses. Larvae of most species are external fish parasites, allowing them to disperse upstream. The parasitic larvae, known as glochidium or lasidium, are modified veligers (the larva typical for bivalves) (Mansur, 1999; Wächtler et al., 2001; Cummings & Graff, 2015). In South America there are 64 species of Hyriidae and 47 Mycetopodidae, covering 64% of the species richness of freshwater bivalves (Pereira et al., 2014). In addition, three species of Mycetopodidae were reported by Haas (1969) from Central America (Trinidad and Tobago, Guatemala, and Nicaragua), and a representative number of nominal species of Nearctic Unionida listed from south of the Tropic of Cancer by Graf & Cummings (2019). There are no reviews for these species or precise information about their occurrences, so they are not included in the keys.

The Venerida, with the families Corbiculidae, Sphaeriidae, and Dreissenidae, includes a great number of freshwater species. Sphaeriidae, with 4 genera (*Pisidium*, *Musculium*, *Sphaerium*, *Eupera*, and *Byssanodonta*) and about 40 species, is characteristic of small ponds and watercourses with soft muddy bottoms. *Pisidium* is the most speciose genus in this family. Dreissenidae is represented by 4 freshwater species with the recently new described genus *Rheodreissena* (Geda et al., 2018; Mansur et al., 2019).

The Mytilida are represented by only one nonindigenous species: *Limnoperna fortunei* (Dunker, 1857), reported in the early 90s from the Río de la Plata estuary. From there it rapidly spread into the Río de la Plata Basin and regions of the western Atlantic river drainages including São Francisco River in Brazil.

The Amazonian *Anticorbula fluviatilis* (H. Adams, 1860) is a peculiar component of the freshwater bivalve fauna, characterized by a confusing set of diagnostic features for which it has been tentatively placed within the order Myida. Also included in the order Myida is the eurioic *Erodona mactroides* Daudin in Bosc, 1801, a species of clear marine lineage, which is widely distributed in the fluvial zone of the estuary of the Río de la Plata and southern Brazil Atlantic drainages.

LIMITATIONS

The overall knowledge of the Neotropical freshwater bivalve fauna is rather limited. Many regions are still unexplored or are poorly studied. The better surveyed and studied South American areas are in Argentina, Brazil, and Chile. Vast areas of northern South America and the Caribbean need surveys, and many species reported from these regions are only known from imprecise original descriptions. Data on Central America and the Caribbean are scant, and comprehensive studies are needed. Graf & Cumming (2019) provide a complete list of the species. Some species of *Mytilopsis* are not included here due to their definite saline water habitat, as well as other *Polymesoda* species.

Overall distribution knowledge is strongly limited. Many species are only known from their type localities. Also, the occurrence of species in a water course is usually punctual; a single record of a species in a river does not imply its presence throughout the entire basin. Further collecting and taxonomic research is needed before having an acceptable sketch of the geographic diversity and distribution of the bivalve fauna in the Neotropics. Sphaeriidae are in need of a deep revision (see comments below). The genus *Neocorbicula* (also referred to as *Cyanocyclas*), that together with *Polymesoda* are the native Neotropic corbiculids, for which a plethora of nominal species were described in the 1920s, also needs a modern revision in order to clarify its actual diversity.

The bivalve taxa include representatives of heterogeneous orders that are difficult to arrange in a nonartificial taxonomic key based on phylogenetically significant characters. The following keys are based on features useful

PHYLUM MOLLUSCA

for identification by a generalist collector, not necessarily acquainted with bivalve taxonomy.

The high taxonomic ranks of the Unionida are relatively clear with rare exceptions; however, species identification within *Diplodon*, *Castalia* (Hyriidae), and *Anodontites* (Mycetopodidae) is particularly problematic. Most species diagnostic characters are from shell morphology, which may be variable, particularly along a geographical gradient or from one basin to another, and sometimes even along the course of a river (Mansur, 1974; Mansur & Veitenheimer, 1979). Furthermore, shell shape may show a tendency to accentuate the roundness and thickness of the margins and hinge teeth over ontogenetic development. Environmental dynamics may also cause shell alterations, including the erosion of the umbos and beaks, masking or erasing early shell sculpture, which is important in Hyriidae taxonomy.

Comprehensive taxonomic and comparative studies are still pending for most Unionida. The main gaps relate to detailed shell morphology and its variations during growth, larval morphology, adult comparative anatomy, reproduction and life cycle details, host specificity, habitat preferences, molecular biology, and distribution. As an example, there are no well-documented diagnostic characters to distinguish adult shells of *Diplodon* versus *Rhipidodonta* (Miyahira et al., 2017). However, Miyahira et al. (2019) described some differences in morphology and shell for some species regarding those genera. Nonetheless, further studies are still needed in order to allow comparisons among species, especially within *Rhipidodonta*. It is known that the shells of the latter are generally rounded in outline and their larvae have no hooks (nonparasitic). Nevertheless, there are species of Rhipidodontini whose larvae have never been described, and that there are species of *Diplodon* with rounded shells. Usually, larvae are not preserved in collections and are difficult to find in nature. Little is known about the life cycle of many *Rhipidodonta* species (Mansur & da Silva, 1999), and nothing is known about when juveniles are released from encapsulated eggs. Genetic relationships at genus and species level are still incipient. Santos-Neto et al. (2016) based on information on COI nucleotide sequences verified that very similar shell morphologies of the hyriid *Castalia ambigua* Lamarck, 1819 are quite distant genetically, probably representing cryptic species; conversely, the same study found that four sets of specimens of *C. ambigua* with conspicuous shell differences were genetically very close. Molecular information will certainly provide new clues for species delimitation and identification.

TERMINOLOGY AND MORPHOLOGY

A general understanding of some morphological and anatomical details is needed to identify freshwater bivalves, particularly for representatives of the Hyriidae, Mycetopodidae, Corbiculidae, and Sphaeriidae. In the same way, information on detailed morphology of the larval shell is relevant and significant. The following information is complementary to that offered by Cummings and Graf (2015).

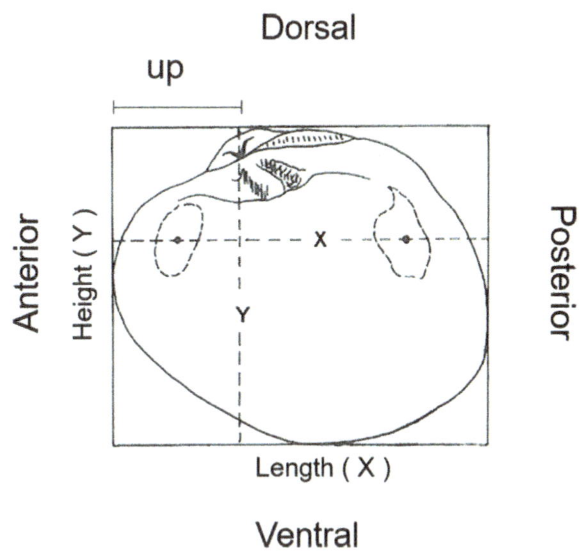

FIGURE 11.91 Right valve of a generalized Unionida. Abbreviations: *up*, umbo position determined by the distance between beak and the anterior end, perpendicular to Y axis; *X*, line bisecting both adductor muscle scars at their center, dividing the shell in dorsal and ventral portions; *Y*, line bisecting umbo perpendicular to the anteroposterior axis (X). *Adapted from Mansur & Pereira (2006).*

It is necessary to orient the **shell** in space (Fig. 11.91). The two valves of a shell are united dorsally by an elastic structure, the **ligament** (Figs. 11.92, 97), which is always posterior. The anterior end is generally shorter and lower than posterior. Placing the shell with the ligament upwards and the anterior end of the shell directed away from the observer, the right valve corresponds to its right side and vice versa.

The two valves composing a shell may be identical in shape and size (equivalve shell) or unequal (unequivalve shell). The shell outline and morphometry are the main features in identifying Sphaeriidae, but also very useful in the other groups. Several measurements such as shell length (L), shell height (H), and shell width (W) (Figs. 11.91, 92) are obtained either with a caliper or under stereoscopic microscope, and relationships among them calculated and expressed as L/H, H/L ratios or indexes, as the **convexity index** Ci = W/H.100. Also important is the **ligament-length index** (Li) (length of ligament or ligament pit expressed as the percentage of shell length), and the position of the umbos expressed as the percentage of the shell length left forward a perpendicular line that intercepts the central point of beaks and the anterioposterior axis (Fig. 11.92 B).

To describe the highly diverse Unionida shell shapes, the shell outline is assimilated to geometric figures (Fig. 11.93,

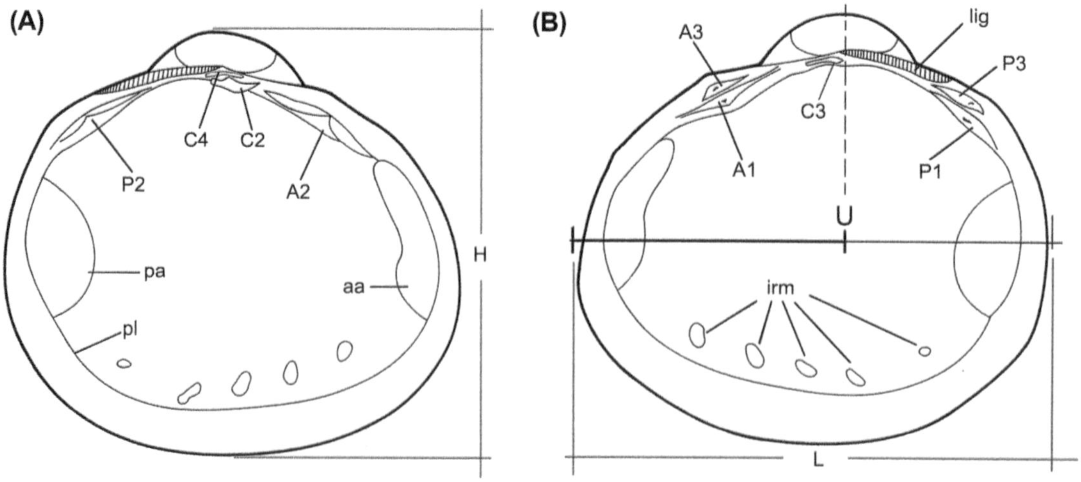

FIGURE 11.92 Inner morphology of left (A) and right (B) valves of a generalized Sphaeriidae. Abbreviations: *aa*, anterior adductor muscle scar; *A1*, *A3*, inner and outer right anterior lateral teeth; *A2*, left anterior lateral tooth; *C2*, *C4*, inner and outer left cardinal teeth; *C3*, right cardinal tooth; *H*, height; *irm*, inner radial muscles; *L*, maximum length; *lig*, ligament; *pa*, posterior adductor scar; *pl*, pallial line; *P1*, *P3*, inner and outer right posterior lateral teeth; *P2*, left posterior lateral tooth; *U*, umbo distance from anterior end.

adapted from Mansur & Pereira, 2006). It may have flattened anterior and/or posterior expansions of the dorsal margins known as **wings**. In some species, the shell is capitate, i.e., the anterior region is differentiated from the rest of the shell having the shape of either a handle or a head (Fig. 11.93). The outer shell surface is covered by the **periostracum**, an organic layer which may be smooth, or forming simple periostracal folds, lamellae of variable shape and size. The shell color, usually corresponding to that of the periostracum, may be useful as a complementary diagnostic character; also the glossiness of the periostracum is commonly used for taxonomic identification in the Unionida. In Mycetopodidae (Unionida) the periostracum may present radial rows of fine plates [*Monocondylaea costulata* (Moricand, 1858)], Fig.11.125 A), and peculiar reticulate mesoscopic ornamentation which shape should help in species diagnosis, mainly in Monocondylaeinae.

The **outer shell surface** may be ornamented with different structures that may be relevant as diagnostic characters. In species of Hyriidae (Unionida) the ornamentations may be in the form of pustules (blisterlike prominences), nodules (small projections like knots or slumps), and undulations on the shell surface. Other features of the outer shell surface are useful in taxonomy. The **carina**, a ridge on the outer surface at posterior half of shell, extends from the umbo toward the shell angle, formed by posterior and ventral shell margins; it may be extremely sharp as in most species of the tribe Castaliini (Hyriidae) (Figs. 11.94 A, 119 E, I–L, 120 A–G) and in *Paxyodon syrmatophorus* (Meuschen in Gronovius, 1781) (Fig. 11.119 A). The commarginal **undulations,** sculpture of the shell formed by ripples parallel to the margin of shell on the outer surface, are conspicuous in *Corbicula* spp. (Fig 11.95 A) and *Castaliella* (Fig. 11.119 F), but should not to be confused with commarginal growth arrest ridges or periostracal folds. The **growth lines** are fine lines of the periostracum related to minute resting periods of growth and the **commarginal growth arrest ridges** (Fig. 11.94 A) are deeper marks originated in periodic growth arrests (long growth stops leave deeper marks, for example, those related to reproductive periods during which the animal mainly directs its energy to gamete production). The shape and separation of commarginal **undulations,** ridges, **periostracal folds** (involve only the macro or mesoscopic folds of periostracum and not the underlying shell structures, Fig. 11.94 A), the radial sculpture of juvenile shells; the length, number, and orientation of **wrinkles**, and the degree of development of the carina (or posterior ridge) are relevant for species identification in Unionida.

The more or less pointed, usually, conical portion of each valve which corresponds to the initial point of the valve growth, is named **beak**. The **umbo** is the portion of the valve around beak. The degree of prominence of the umbos is a variable character; it may be inflated, outstanding of the shell surface or low, not prominent. According to their position, the umbos may be placed centrally (mesogyrous) or displaced in a variable degree to the anterior (prosogyrous) or posterior (opistogyrus) end of the valve (Fig. 11.95). Anterior to the umbos there may be a more or less demarcated lanceolate area, the **lunule**; posterior to the umbos a depressed elongate or heart shaped area, the **escutcheon**, may be present (Figs. 11.96 E, 97 E). The escutcheon is also known as **shield**, a much flattened area in so me species of Castaliini, and particularly sculptured in *Castalia undosa* Martens, 1885 (Fig. 11.119 L). In cases of a not flattened but posteriorly projected escutcheon, the area situated between the carina and the dorsal/posterior margins of each valve is named **dorsal**

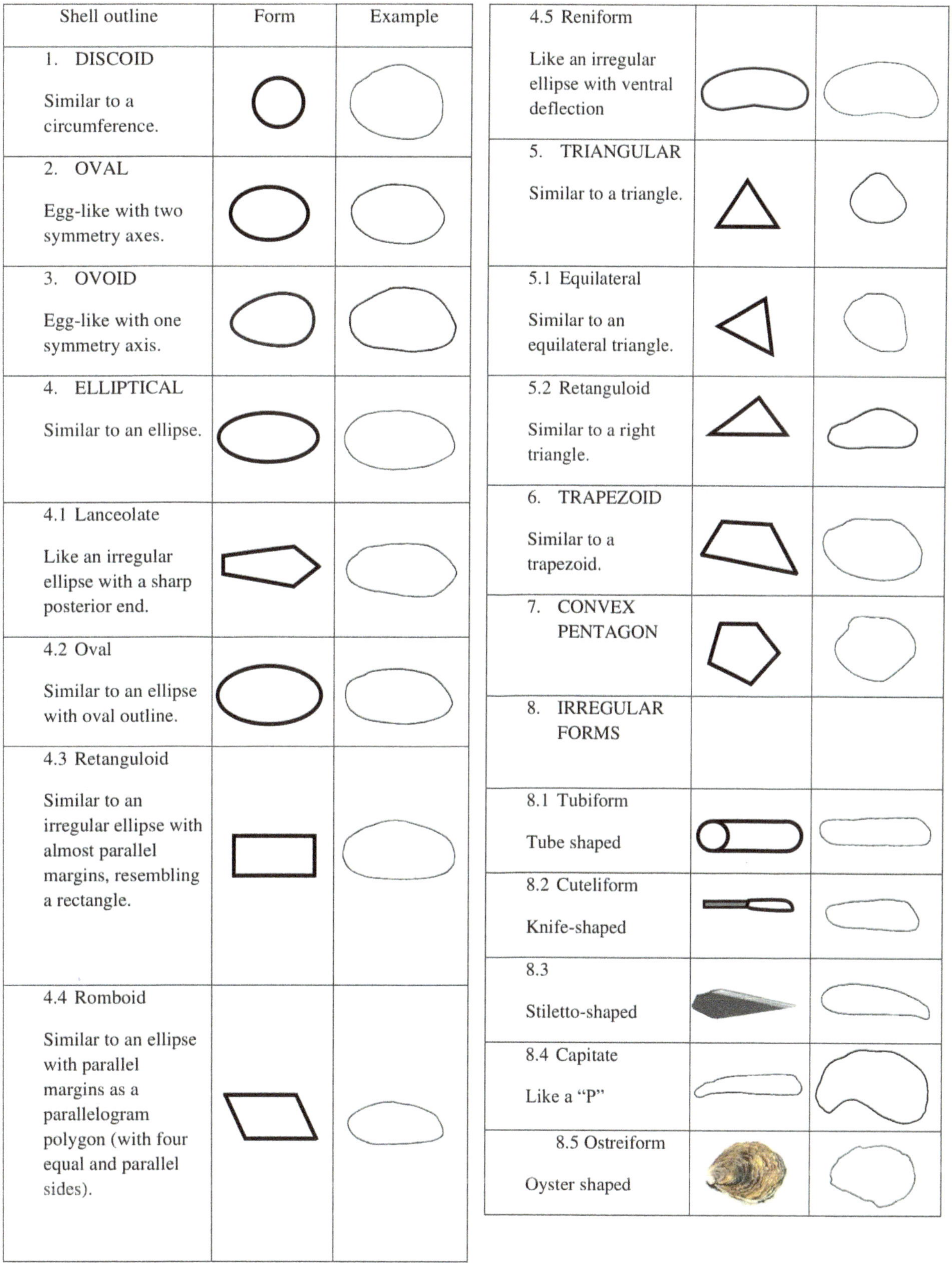

Shell outline	Form	Example
1. DISCOID Similar to a circumference.		
2. OVAL Egg-like with two symmetry axes.		
3. OVOID Egg-like with one symmetry axis.		
4. ELLIPTICAL Similar to an ellipse.		
4.1 Lanceolate Like an irregular ellipse with a sharp posterior end.		
4.2 Oval Similar to an ellipse with oval outline.		
4.3 Retanguloid Similar to an irregular ellipse with almost parallel margins, resembling a rectangle.		
4.4 Romboid Similar to an ellipse with parallel margins as a parallelogram polygon (with four equal and parallel sides).		
4.5 Reniform Like an irregular ellipse with ventral deflection		
5. TRIANGULAR Similar to a triangle.		
5.1 Equilateral Similar to an equilateral triangle.		
5.2 Retanguloid Similar to a right triangle.		
6. TRAPEZOID Similar to a trapezoid.		
7. CONVEX PENTAGON		
8. IRREGULAR FORMS		
8.1 Tubiform Tube shaped		
8.2 Cuteliform Knife-shaped		
8.3 Stiletto-shaped		
8.4 Capitate Like a “P”		
8.5 Ostreiform Oyster shaped		

FIGURE 11.93 Unionida: nomenclature of the shell shape based on shell outlines. *Adapted from Mansur & Pereira (2006).*

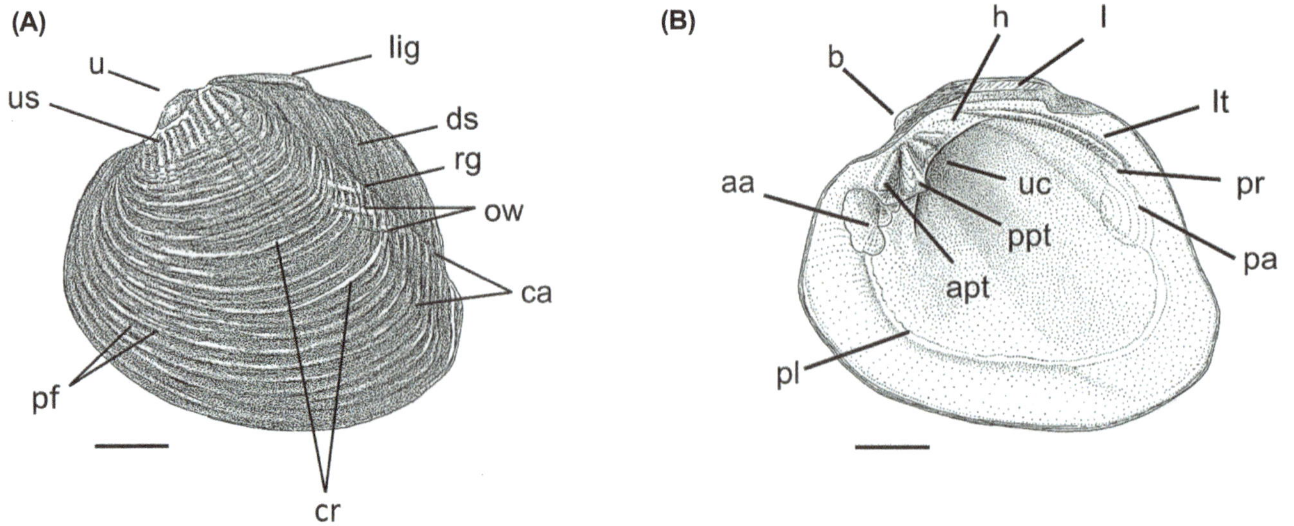

FIGURE 11.94 Drawings of an outer view of a left valve (A) and inner view of a right valve (B) of the hyriid *Castalia martensi*. Abbreviations: *aa*, anterior adductor muscle scar; *apt*, anterior pseudocardinal tooth; *b*, beak; *ca*, carina; *h*, hinge plate; *gl*, growth lines; *ds*, dorsal slope; *l*, ligament; *lt*, lateral tooth; *ow*, oblique wrinkles; *pa*, posterior adductor muscle scar; *pl*, pallial line; *ppt*, posterior pseudocardinal tooth; *pr*, posterior retractor muscle scar; *rg*, radial groove; *u*, umbo; *uc*, umbonal cavity; *us*, umbonal sculpture. Scale bar: 10 mm. Drawings M.C.D. Mansur, *adapted from Mansur et al. (1988)*.

FIGURE 11.95 Shell shape and position of beaks and umbos: (A) central in Corbiculidae; (B) displaced anteriorly in *Eupera* (Sphaeriidae); (C) displaced posteriorly in *Pisidium* (Sphaeriidae); (D) subterminal in *Limnoperna* (Mytilidae); (E) terminal in *Mytilopsis* (Mytilidae). Scale bars: (A) = 5 mm; (B) and (C) = 1 mm; (D) and (E) = 10 mm.

slope (= posterior slope). The umbonal sculpture, usually formed by radial or convergent bars on the surface of the umbos, sometimes extending toward a part or the shell surface, may be diagnostic in species of Hyriidae, as well as the shape and deepness of the umbonal cavity (i.e., the depression on the inner and dorsal part of each valve under the umbo in direction to the beak).

The two valves of a shell are interlocked by the **hinge** apparatus. The hinge plate bears a variable number of teeth; in edentulous hinges, teeth are absent. Teeth placed just below the umbos are the cardinal teeth, and those located anterior and posterior to the umbos are the lateral teeth (Fig. 11.96 A–F). Lateral teeth may be serrated (Fig. 11.96 B) or smooth (Fig. 11.96 F). In Unionida Hyriidae, the

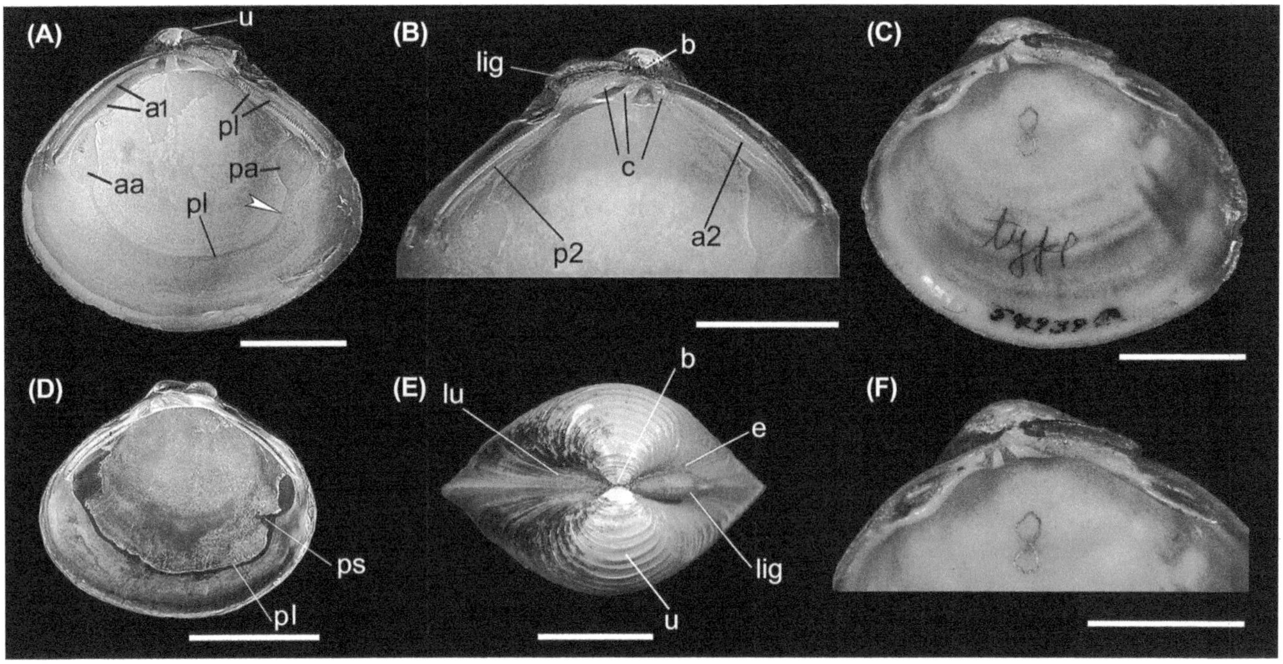

FIGURE 11.96 Shell morphology of Corbiculidae: (A) and (B) inner views of a right and left valves of *Corbicula*; (C) inner view of a right valve of *Polymesoda*; (D) inner view of a right valve of *Neocorbicula*; (E) dorsal view of shell of *Corbicula*; (F) detail of hinge of a right valve of *Polymesoda*. Abbreviations: *aa*, anterior adductor scar; *a1*, anterior right lateral teeth; *a2*, anterior left lateral tooth; *b*, beak; *c*, cardinal teeth; *e*, escutcheon; *lig*, ligament; *lu*, lunule; *pa*, posterior adductor scar; *pl*, pallial line; *ps*, pallial sinus; *p2*, posterior lateral tooth (left); *u*, umbo; *white arrow* in (A) indicates the point of attachment of siphon retractors not forming a sinus in the genus *Corbicula*. Scale bars: (A–F) = 10 mm.

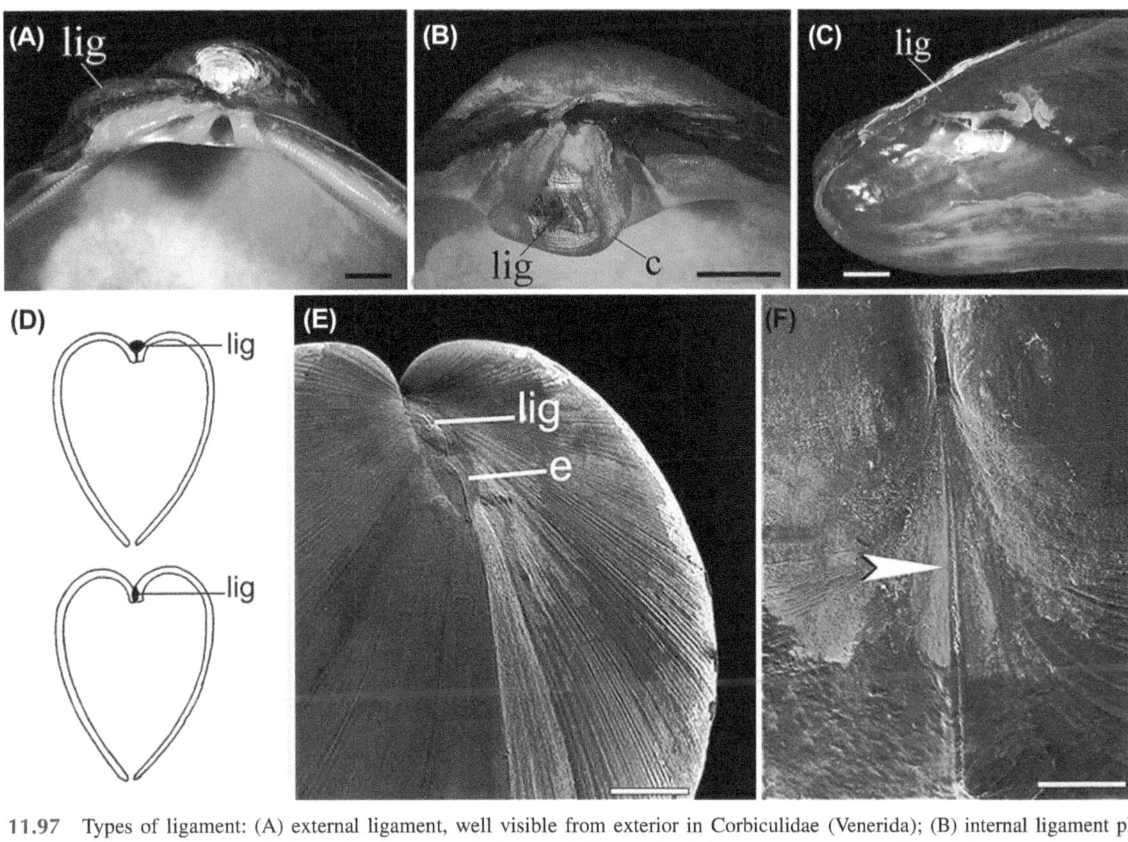

FIGURE 11.97 Types of ligament: (A) external ligament, well visible from exterior in Corbiculidae (Venerida); (B) internal ligament placed in a chondrophore in Erodonidae (Myida); (C) internal ligament in Mytilidae (Mytilida); (D–F) ligament in the Sphaeriidae; (D) schematic representation of outer (upper figure) ligament in an extroverted ligament pit and internal ligament (bottom figure) in an introverted ligament pit; (E) intermediate position of ligament, internal but slightly visible from exterior; (F) internal ligament (the *arrow* indicates the position of the ligament). Abbreviations: *c*, chondrophore; *e*, escutcheon; *lig*, ligament. Scale bars: (A) and (B) = 2 mm, C = 1 mm, (E) and (F) = 200 μm.

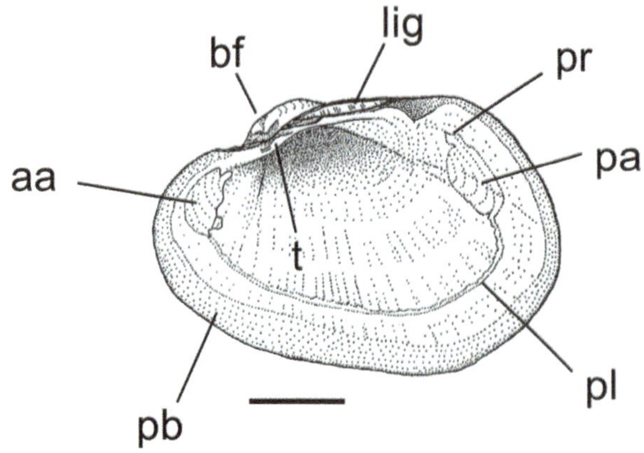

FIGURE 11.98 Drawing of an inner view of a right valve of a mycetopodid *Monocondylaea minuana*. Abbreviations: *aa*, anterior adductor muscle scar; *bf*, beak furrow; *lig*, ligament; *pa*, posterior adductor muscle scar; *pb*, prismatic border; *pl*, pallial line; *pr*, posterior retractor muscle scar; *t*, tooth. Scale bar = 10 mm. Drawing M.C.D. Mansur, adapted *from Mansur et al. (1988).*

anterior teeth are named pseudocardinal, they are usually short and compact, varying in number and shape from rounded to lamellate or split (Fig. 11.94 B); posterior to the pseudocardinal teeth are lateral teeth in the form of the lamellae, two teeth in the right valve and one tooth in the left. In the Mycetopodidae only pseudocardinal teeth are present in the genera *Monocondylaea, Haasica, Fossula*, and *Tamsiella* (Fig. 11.98). In the remaining Mycetopodidae (*Anodontites, Leila, Acostaea*, and *Bartlettia*) the hinge plate is edentulous, i.e., without teeth.

The **ligament** is an organic structure attached to a more or less extended socket, the ligament pit of the hinge plate of left and right valves; in freshwater bivalves it is placed behind the umbos. The ligament may be external (Fig. 11.96 B, E), more or less visible from exterior (sometimes covered by minute and thin calcareous plates) (Fig. 11.97 E) or internal (Fig. 11.97 D, F). In shells with edentulous hinge, as for example in Mytilidae, the ligament is strong and long (Fig. 11.97 C). In Unionida, the ligament is usually external (Fig. 11.94 A, B). Lastly, the ligament may be internal, stout, being enclosed in a chondrophore, as in the genus *Erodona* (Myida) (Figs. 11.97 B, 108 B, C).

In the Unionida, the **inner shell surface** is usually formed by a layer of nacre, a pearly white or iridescent substance. It is also named "mother of pearl", and often colored in many species. In some species of Mycetopodidae the inner border of the shell is nonnacreous, matte, forming the **prismatic border** (Figs. 11.98, 126 H, 127 E, I).

Considering the soft parts, two lateral expansions of the tegument (epidermis) at sides of the body, called the **mantle**, cover the inner surface of each valve and are attached to it by the orbicular mantle muscle and the inner radial mantle muscles. The position of the **pallial line** (the scar of the insertion of the orbicular mantle muscle) (Figs. 11.92, 94, 98), relative to the shell margin (close to it or variably separated from the shell margin), may be a relevant feature in the identification of *Pisidium* species; the number and position of the scars of inner radial mantle muscles are also relevant (Fig. 11.92).

At the posterior end, the right and left inner mantle folds fuse to form the **inhalant** and **exhalant apertures** (the sites at which the water enters and leaves the mantle cavity, respectively). The mantle at these points may extend to form the inhalant and exhalant siphons, which can bear sensitive papillae (Fig. 11.99 A). The number and position of the siphon retractor muscles may be important for the identification of *Pisidium* species (Fig. 11.99 B, C). In species with well-developed siphons, the scar of the siphon retractor muscles at the posterior end of the inner shell surface form the **pallial sinus** (Fig. 11.96 D), the presence or absence of the pallial sinus is diagnostic for corbiculids (Venerida) at genus level. In Unionida, two true siphons are present only in the genus *Leila* (Mycetopodidae).

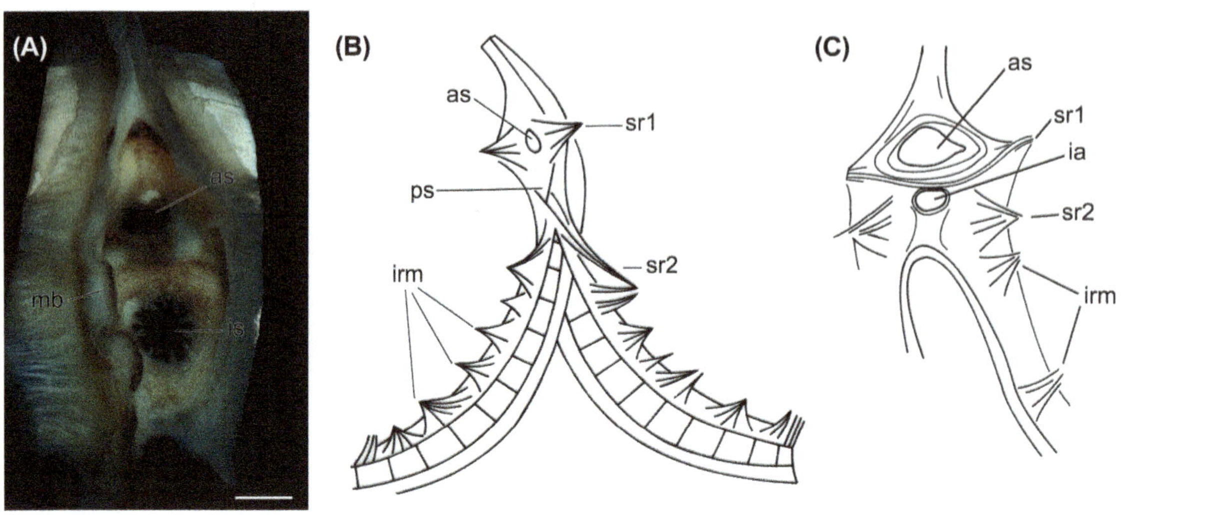

FIGURE 11.99 Siphons and siphonal apertures: (A) Corbiculidae; (B) and (C) Sphaeriidae. Abbreviations: *as*, anal siphon; *ia*, inhalant aperture; *irm*, inner radial mantle muscles; *ps*, presiphonal suture; *sr1*, siphon retractor 1; *sr2*, siphon retractor 2. Scale bar = 1 mm.

PHYLUM MOLLUSCA

Cummings & Graf (2015) describe in detail the morphology of siphonal apertures in other Unionida.

The **foot** is a muscular organ located in the anteroventral region of the visceral mass. It stretches out from the shell between the valves for locomotion and burial. Regarding the general shape, the foot is a structure laterally flattened, sometimes of cylindrical section as in some Sphaeriidae (Venerida). In Hyriidae and Mycetopodidae the foot may be tonguelike or axelike, and rarely cylindrical with a distal intumescence as in the genus *Mycetopoda* (Fig. 11.100). In adults of *Bartlettia stefanensis* (Moricand, 1856) it is much reduced in size, and absent in *Acostaea.* In adult phase of Mytilida and Dreissenidae, the foot is greatly reduced in size.

In Sphaeriidae of the genus *Pisidium* the type of **nephridium** (the renal organ), in combination with the type of ligament (see below), number of demibranchs, and mantle apertures, defines the suprageneric taxonomic ordering. If the dorsal loop of the nephridium is visible in dorsal view, the nephridium is said to be of "open type," if not, is of "closed type," in the latter case the lateral loop may or may not be visible (Fig. 11.101).

The number and degree of development of demibranchs, which form the **gills** (Fig. 11.102), and the presence or absence and type of incubatory sacs, brooding sacs or "marsupium," for brooding of embryos are diagnostic features at genus level for Sphaeriidae and Corbiculidae (Fig. 11.103).

Sphaeriidae and Corbiculidae, except in some introduced species of Asian origin like *Corbicula fluminalis* (Müller, 1774) brood their embryos within inner demibranchs (see particular details on the brooding process in Cummings & Graf, 2015). Native corbiculids of the genera *Neocorbicula* and *Polymesoda* brood their embryos to an advanced stage of development, being released as minute juveniles (of up to 4–5 mm length) (Figs. 11.103, 104 A–C). The number of brooded embryos is relatively low, at most 150 offspring per maternal individual.

Among the Sphaeriidae, *Pisidium* species are synchronous brooders, developing in a large brood sac with individual chambers for embryos (Fig. 11.104 E) (Cooley & Ó Foighil, 2000). Also synchronous brooders are *Eupera* and *Byssanodonta*, which brood their embryos loose within the branchial space without forming brood sacs (Fig. 11.104 D). *Musculium* and *Sphaerium* are sequential brooders; their embryos develop in several thin-walled sacs each one containing embryos of different generations (Fig. 11.104 F) (for details of the process, see Cummings & Graf, 2015; Park & O'Foighil, 2000).

Hyriidae and Mycetopodidae (Unionida) brood modified veliger larvae within the inner demibranchs which

FIGURE 11.100 Photographs showing the different shapes of the foot in Hyriidae and Mycetopodidae: (A) tonguelike, *Diplodon suavidicus*; (B) axelike, *Leila blainvilliana*; (C) cylindrical with a distal intumescence, *Mycetopoda siliquosa*. Scale bars: 10 mm. *Adapted from Pereira et al. (2012).*

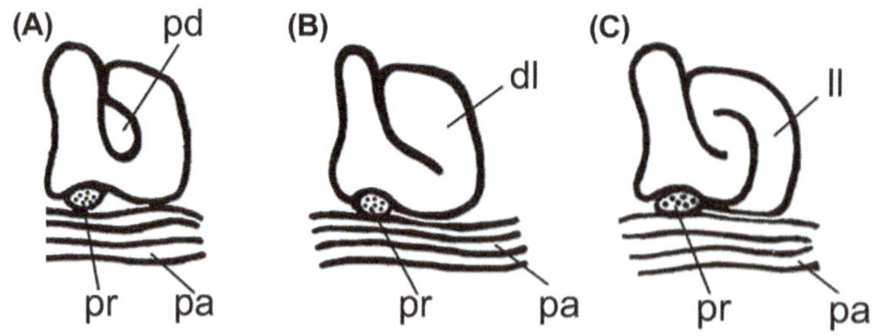

FIGURE 11.101 Types of nephridia in *Pisidium* (dorsal view): (A) closed type with lateral loop nonvisible; (B) closed type with lateral lobe visible; (C) open type, with pericardial duct visible in dorsal view. Abbreviations: *pa*, posterior adductor; *dl*, dorsal lobe; *ll*, lateral lobe; *pd*, pericardial duct; *pr*, pedal retractor.

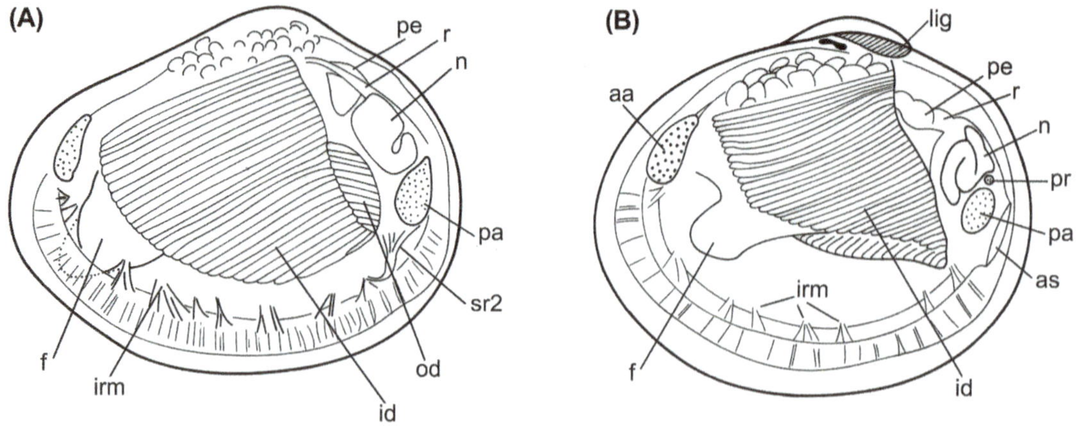

FIGURE 11.102 Anatomical sketches of *Pisidium* species with two (A) and one (B) demibranchs. Note the differences in the type of nephridia and arrangement of inner mantle radial muscles. Abbreviations: *aa*, anterior adductor muscle; *as*, anal siphon; *f*, foot, *id*, inner demibranch; *irm*, inner radial mantle muscles; *lig*, ligament; *n*, nephridium; *od*, outer demibranch; *pa*, posterior adductor muscle; *pe*, pericardium; *pr*, pedal retractor muscle; *r*, rectum; *sr2*, siphon retractor muscle 2.

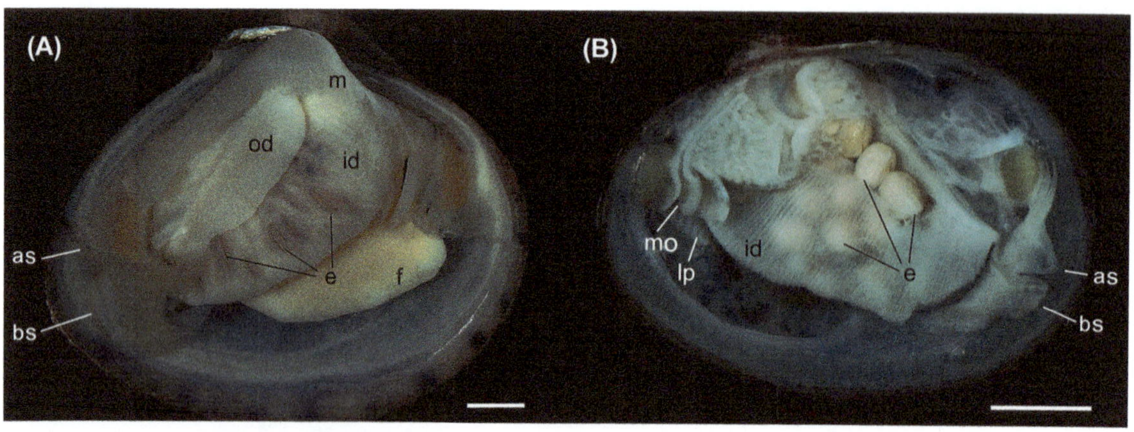

FIGURE 11.103 Anatomy: (A) *Neocorbicula* sp. showing the inner demibranch serving as marsupium (the right valve and mantle lobe were removed); (B) *Eupera simoni* sagittal section showing the visceral mass and embryos brooded in the branchial space of inner demibranch. Abbreviations: *as*, anal siphon; *bs*, branchial siphon; *e*, embryos; *f*, foot; *id*, inner demibranch; *lp*, labial palp; *m*, mantle; *mo*, mouth; *od*, outer demibranch. Scale bars: (A) = 2 mm, (B) = 1 mm.

after being liberated lead, except in the genus *Rhipidodonta*, a parasitic life (see details in Cummings & Graf, 2015). The larva of the Hyriidae is the **glochidium**, a bivalved larva provided with a relatively large flagellum (absent in *Castalia*); two types of glochidia are recognized: the glochidium is provided with hooks, a parasitic larva with a strong hook at the ventral margin of each larval valve (Fig. 11.105 A), and the glochidium without hooks of the *Rhipidodonta* (Hyriidae), a nonparasitic larva completing its life cycle protected within the

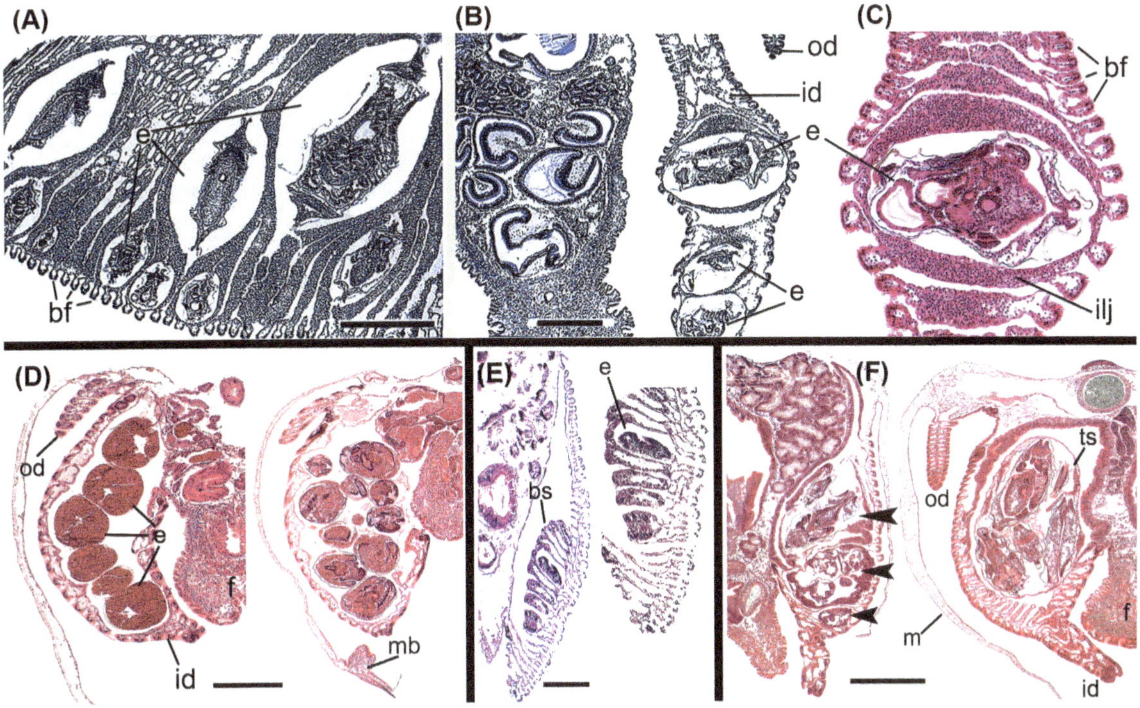

FIGURE 11.104 Different types of brooding in Corbiculidae and Sphaeriidae: (A–C) histological sections of *Neocorbicula* sp. in which embryos are brooded within the water tubes of both inner demibranchs, (A) sagittal section and (B) transverse section of inner demibranch, (C) detail of (B); (D) *Eupera* sp. transverse sections showing embryos in an early (left) and advanced (right) stage of development, brooded within the suprabranchial chamber and branchial space of the inner demibranch without forming brood sacs (synchronous brooder); (E) *Pisidium* sp., transverse section showing a single brooding sac (a detail at right) with compartments for each embryo, brooding is synchronous; (F) multiple brooding sacs in *Musculium* sp. (asynchronous brooding), the image at left shows three brood sacs (arrows) with embryos in different developmental stages; the image at right shows a third order sac with embryos in an advanced stage of development. Abbreviations: *bf*, branchial filaments; *bs*, brood sac; *e*, embryos; *id*, inner demibranch; *ilj*, interlamellar junction; *m*, mantle; *mb*, free mantle border; *od*, outer demibranch. Scale bars: (A–F) = 100 μm.

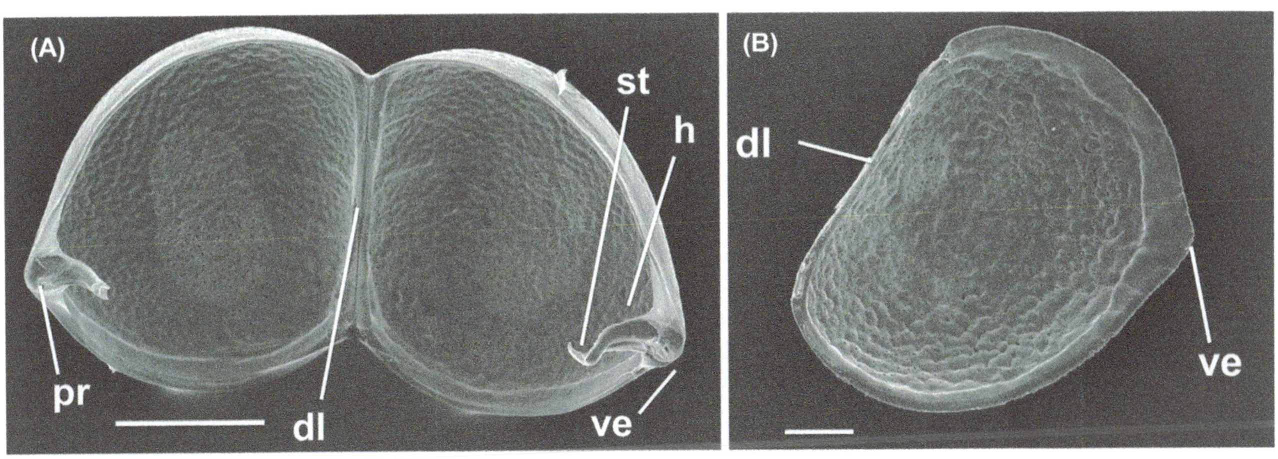

FIGURE 11.105 Scanning electron microscope photographs of inner views of glochidium larvae (Hyriidae): (A) glochidium with hooks of *Diplodon*; (B) glochidium left valve without hook of *Rhipidodonta*. Abbreviations: *dl*, dorsal line; *h*, hook; *pr*, protuberance; *st*, stylet (lost in the left hook); *ve*, ventral end. Scale bars (A) and (B) = 50 μm. Photographs and edition by D.M. Pimpão and M.C.D. Mansur.

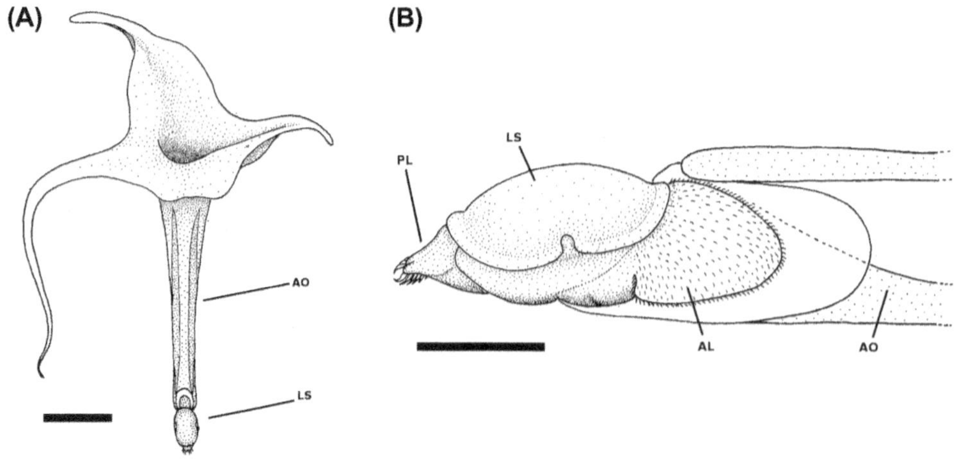

FIGURE 11.106 Schematic drawing of lasidium larvae of Mycetopodidae: (A) dorsal view; (B) detail of the shell in lateral view. Abreviations: *AL*, anterior lobe; *AO*, adhesive organ; *LS*, larval shell; *PL*, posterior lobe. Scale bars: (A) = 150 μm, (B) = 50 μm. Drawings M.C.D. Mansur *adapted from Veitenheimer-Mendes & Mansur (1978).*

egg capsule until the release of embryos as minute juveniles (Fig. 11.105 B). The larva of the Mycetopodidae is the **lasidium** (Figs. 11.106, 107), a larva with a central body with two ventral lobes, a single dorsal shell, and a posterior bilobate tail with terminal hooks; frontally there is a lobe with a transparent distensible adhesive organ (Silva-Souza et al., 2011). Among the Mycetopodidae, details of the lasidium morphology are only known for very few species; the anterior adhesive organ may be tapered (*Monocondylaea*), or in the form of a lily flower (*Anodontites, Mycetopoda*, and *Acostaea*).

MATERIAL PREPARATION AND PRESERVATION

The better fixation procedure for general morphological studies is to fix specimens in 10% formalin for 24–48 h and then, after a two-hour rinse in tap water, preserve whole specimens or tissues in ethanol 70%. It is always advisable to fix and preserve some specimens of each sample in ethanol 96 or 100% for eventual molecular studies. In both cases, sectioning the posterior adductor muscle prior fixation will improve the process facilitating the penetration of the fixative.

The Sphaeriidae should be relaxed prior to fixation; this can be achieved by adding some menthol crystals to a small vial containing living specimens for several hours. Other effective and fast method is to add warm water at about 50–70°C to the vial containing specimens. Specimens for scanning electron microscopy study of shells should be treated with a diluted commercial sodium hypochlorite solution to eliminate remaining soft parts after dissection and the periostracum adhered to valves. After rinsing in distilled water and drying at room temperature (or in a stove at no more than 30°C), specimens are ready to be mounted and coated for scanning.

The Unionida considering their usual large size needs some special treatment. For morphological studies the relaxation of the specimens before the fixation and preservation is recommended. Good results are obtained by leaving the animals in a flask with water of their habitat during one or two days with menthol crystals (Graf & Cumming, 2019). Prior to fixation, it is recommended to insert a small object (for example, tooth pick) between the valves to avoid the closing of valves; this procedure facilitates the contact of the soft parts with the fixative. For improving the preservation of soft parts, it is suggested to separate the flesh from the shell, keeping the shell dry, apart from the soft parts, fixed and preserved in liquid. This prevents corrosion and loss of the shell color derived from a long period of exposure to alcohol. For a good preservation of glochidia, the best method is to separate them from branchial tissues (gills) before anesthetizing the maternal animal, and then following the treatment recommended for SEM and for light microscopy by Mansur & da Silva (1999).

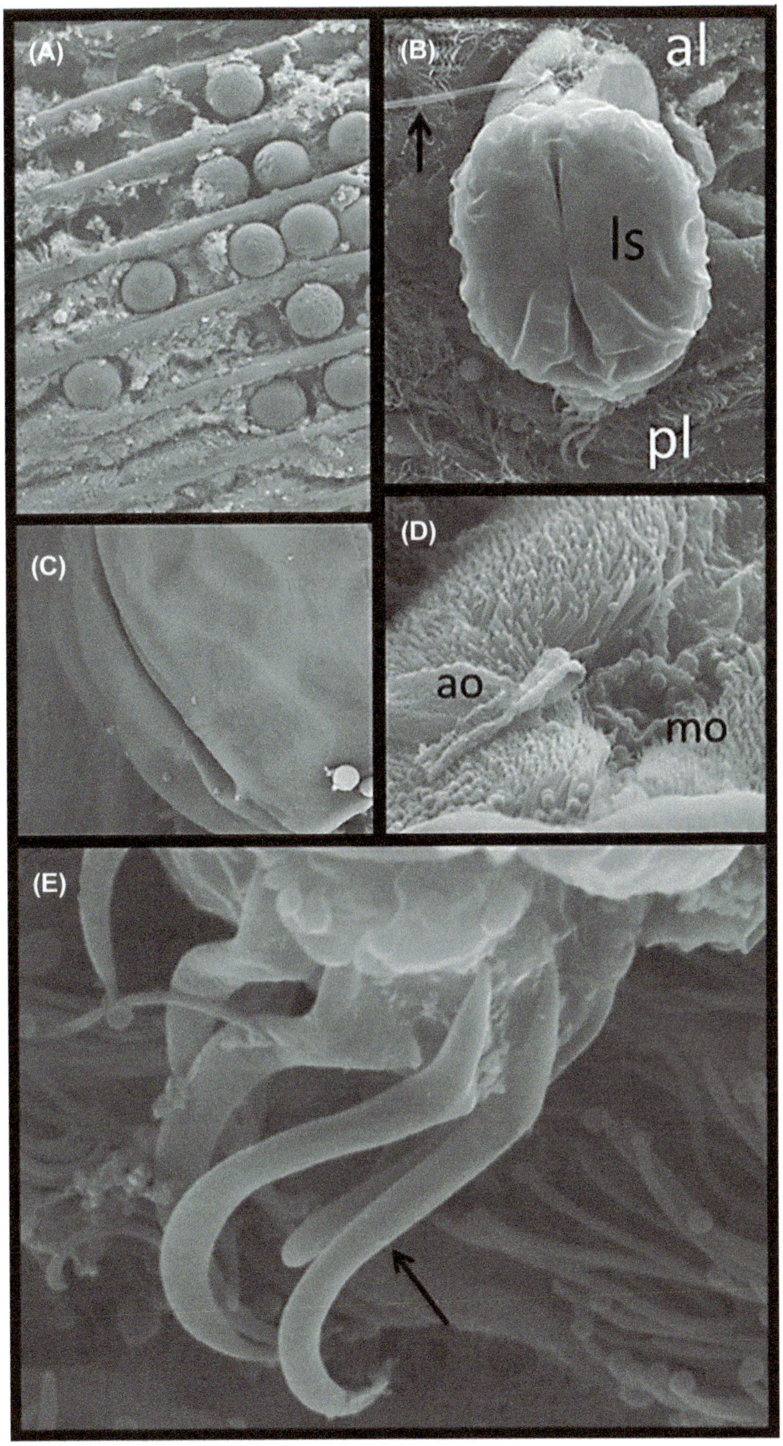

FIGURE 11.107 Scanning electron microscope photograph of lasidium larvae of Mycetopodidae: (A) eggs incubation inside the gills of *Anodontites trapesialis*; (B) dorsal view of the lasidium with the *arrow* showing the adhesive organ; (C) detail of the dorsal face of the shell; (D) dorsal view of the anterior lobe; (E) detail of the terminal hooks at the posterior lobe with *arrow* showing a hook. Abbreviations: *al*, anterior lobe; *ao*, adhesive organ; *ls*, larval shell; *mo*, larval mouth; *pl*, posterior lobe. Magnification: (A) = 560X, (B) = 1500X, (C) = 5000X, (D) and (E) = 12,000X. *From Silva-Souza et al. (2011) with authorization.*

KEY TO BIVALVIA

Bivalvia: Orders

1 Shell equivalve 2

1′ Shell inequivalve, right valve larger, overlapping left valve at ventral margin or irregular .... **Myida [p. 391]**

2(1) Umbos at the anterior third in relation to the shell length, in the middle or behind 3

2′ Umbos subterminal, shell mytiloid, hinge plate simple, without teeth (in freshwater representatives), ligament long, in a platform behind umbones, periostracum smooth, shiny (Fig. 11.95 D) .... Mytilida, one family Mytilidae, one species *Limnoperna fortunei* (Dunker, 1857) [Introduced. Atlantic drainages in Brazil, Bolivia, Paraguay, Argentina, Uruguay]

3(2) Hinge plate with cardinal and lateral teeth anterior and posterior to the umbos, if not, with a conspicuous septum and apophysis below terminal umbos .... **Venerida [p. 392]**

3′ Hinge plate without teeth or one to three pseudocardinal teeth and lateral teeth, if present, only posterior .... **Unionida [p. 409]**

Key to Bivalvia, Myida

Cristián Ituarte
Division of Invertebrate Zoology, Museo Argentino de Ciencias Naturales, Buenos Aires, Argentina

Bivalvia: Myida: Families

Within Corbulidae, *Anticorbula* is an enigmatic monotypic genus, not related to any freshwater bivalve family and has been treated in Lyonsiidae (Anomalodesmata, Pholadomyoida) (Simone, 1999). *Anticorbula fluviatilis* inhabit slow streams; adhered to rocks and limbs by its strong tapered byssus.

1 Shell inequivalve, right valve larger than left valve; anterior end rounded, posterior end elongated in a rostrum; inner shell surface porcelainous; hinge with a strong internal ligament enclosed in a chondrophore in left valve, right valve with two cardinal teeth at sides the chondrophore; living attached by a byssus to gravel or pebbles in soft sandy and muddy bottoms (Fig. 11.108 A–C) .... Erodonidae, one species: *Erodona mactroides* Daudin in Bosc, 1801 [Brazil: Paraná State coastal systems to Lagoa dos Patos and Guaîba Lake. Argentina and Uruguay: Rio de la Plata estuary, up to Ensenada]

1′ Shell inequivalve irregular, ellipsoid or subquadrate, with posterior end truncate; posterodorsal margin parallel to ventral margin; umbos low, wide; hinge edentulous; byssus strong, tapered (Fig. 11.108 D, E) Corbulidae, one species: *Anticorbula fluviatilis* (H. Adams, 1860) [Brazil: Pará, Trombetta, and Xingu rivers, Tapajos River. Peru: Marañón River, Amazonas River Basin]

FIGURE 11.108 Representatives of Myida. (A–C) *Erodona mactroides*: (A) Outer view of a left valve; (B) inner view of hinge in a right valve; (C) inner view of hinge in a left valve. (D) and (E) *Anticorbula fluviatilis*: (D) Live specimen viewed from right side attached to a hard substrate; (E) live specimen showing the well-developed tapered byssus. Scale bars: (A) = 10 mm, (B–E) = 5 mm. (D) and (E) photographs by Laurie C. Anderson.

PHYLUM MOLLUSCA

Key to Bivalvia, Venerida

Cristián Ituarte
Division of Invertebrate Zoology, Museo Argentino de Ciencias Naturales, Buenos Aires, Argentina

Bivalvia: Venerida: Families

1 Hinge with cardinal and lateral teeth 2

1′ Hinge edentulous, with a septum and apophysis for the attachment of anterior byssus retractor muscle, shell elongate, mytiloid, umbos terminal, epibysate .... **Dreissenidae [p. 392]**

2(1) Hinge with three cardinal teeth in each valve, lateral teeth usually serrate; shell solid, surface with commarginal low ribs or fine striae; yellowish to brown, sometimes greenish with purple radial stripes; adults > 1.5 cm .... **Corbiculidae [p. 395]**

2′ Hinge with one cardinal tooth in the right valve and one or two in left valve, lateral teeth always smooth; shell thin, fragile, whitish or creamy, shell surface smooth or faintly striate; adults <1 cm .... **Sphaeriidae [p. 397]**

Bivalvia: Venerida: Dreissenidae: Genera

1 Adult shell usually > 20 mm, apophysis well differentiated from the apical septum, somewhat underneath, and only partially fused to the dorsal margin of the septum; inhalant siphon with a distal crown of densely arranged tentacles, and inner ring of few tentacles; outer demibranchs reduced in size, representing more than 50% the surface of the inner one; larval development: nonbrooding, planktonic; living in brackish coastal waters, including river estuaries, mangroves, lagoons and canals .... ***Mytilopsis* [p. 392]**

1′ Adult shell < 15 mm; apophysis strongly associated to the apical septum; inhalant siphon with distal edge smooth and inner ring of 20–40 tentacles; outer demibranchs reduced in size, representing 24%–45% the surface of the inner one; larval development: biphasic brooding (branchial and pallial); living in clear, fast-flowing freshwaters in large inland river systems .... ***Rheodreissena* [p. 393]**

Bivalvia: Venerida: Dreissenidae: *Mytilopsis*: Species

1 Shell mytiliform, ventrally concave, umbos curved downwards; apophysis usually lanceolate or hamulate and more closely associated with hinge plate than septum (Fig. 11.109 E–H) .... *Mytilopsis sallei* (Récluz, 1849)
[Northern South America.Colombia, Venezuela, Brazil: Recife (introduced, referred to as *Mytilopsis cf. sallei*)]

1′ Adult shell < 30 mm, shell outline low, oblong, subrectangular or oblong-elongate (date shaped); ventral and posterodorsal margin subrects; umbos not curved; apophysis small, rounded, closely associated with septum (Fig. 11.109 A–D) .... *Mytilopsis leucophaeata* (Conrad, 1831)
[Caribe. Introduced in Brazil: Rio de Janeiro, Rodrigo de Freitas Lagoon]

FIGURE 11.109 *Mytilposis* species (Dreissenidae). (A–D) *M. leucophaeta*: (A) and (B) Outer view of left valves showing different shell outlines; (C) ventral view; (D) inner view of a left valve). (E–H) *M. sallei*: (E) Outer view of a left valve; (F) inner view of a right valve; (G) detail of anterior end of a right valve showing septum and apophysis; (H) a dissected specimen showing anatomy. Scale bars: (A–F), and (H) = 5 mm; (G) = 2.5 mm. *(A–D)* © *Amgueddfa Cymru—National Museum of Wales, with permission; (E–H)* © *Tan Koh Siang, with permission.*

Bivalvia: Venerida: Dreissenidae: *Rheodreissena*: Species

1 Shell outline with greatest height at anterior half of shell length; dorsal and lateral areas of shell surface with conspicuous mesoscopic sculpture of wide undulations, transverse to growth lines 2

1′ Shell outline with greatest height at median or posterior to the half of shell length from the umbo, shell surface without mesoscopic sculpture 3

2(1) Adult shell 2.1–6.3 mm length (average 4.3 mm), mytiliform, dorsoventrally flattened, bicarinate in extremely flattened specimens; greatest height < 0.5x shell length; posterodorsal margin subrect; byssal notch shallow; periostracum smooth or weakly lamellate at posterior end of shell, showing a contrasting pattern of light and dark areas in regular bands (Fig. 11.110 A–F); area of outer demibranch

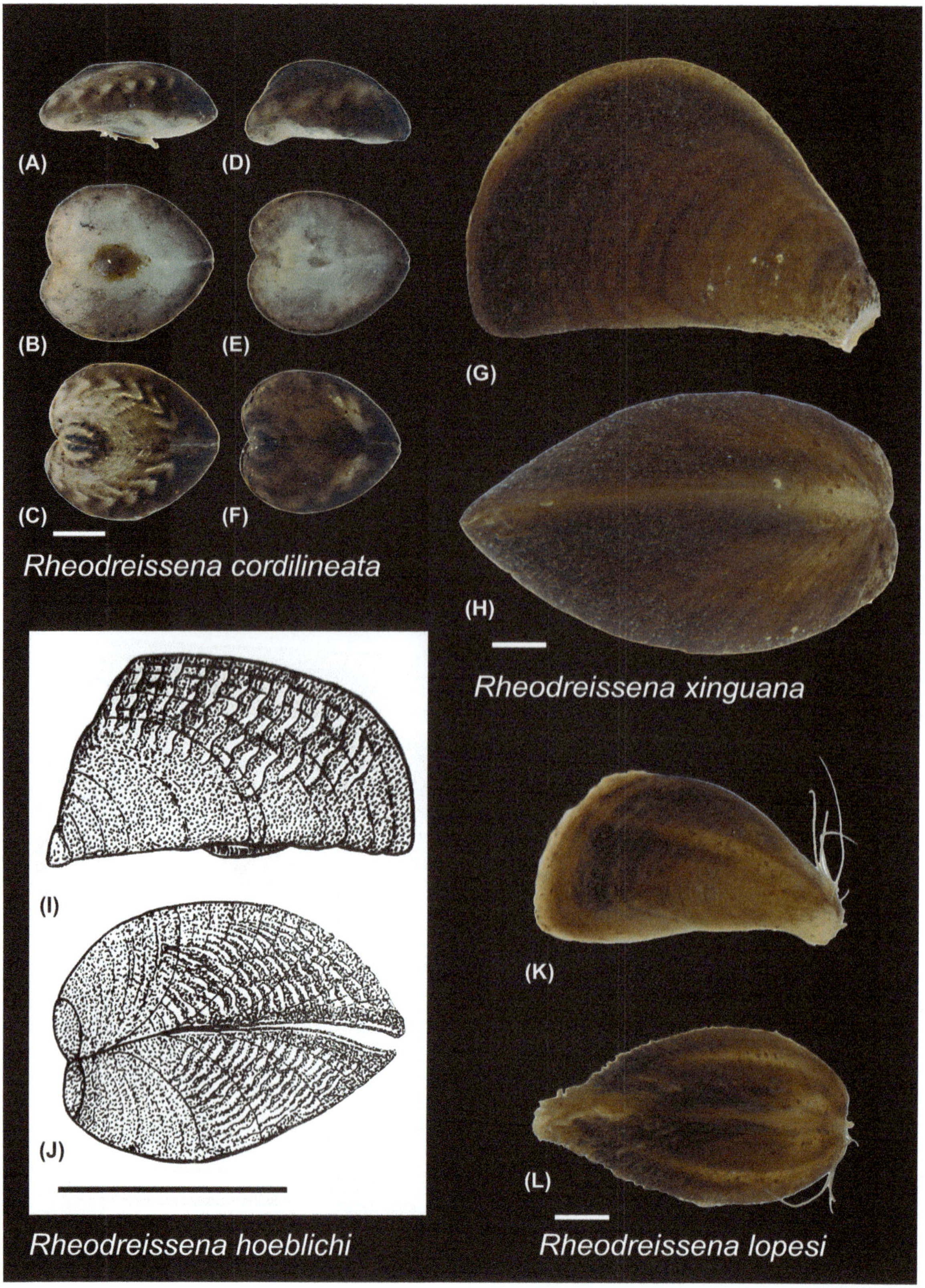

FIGURE 11.110 Species of *Rheodreissena* (Dreissenidae). (A–F): *R. cordilineata*, (A) and (D) left lateral views, (B) and (E) ventral views, (C) and (F) dorsal views. (G) and (H): *R. xinguana*, (G) right lateral view and (H) dorsal view. (I) and (J): *R. hoeblichi* (I) left lateral view and (J) dorsal view. (K) and (L): *R. lopesi*: (K) right lateral view and (L) dorsal view. (A–H), (K) and (L): photographs by Mark Sabaj, Academy of Natural Sciences Philadelphia; Xingu Project (NSF DEB-1,257,813). Scale bars: (A–F), (G), (H), (K) and (L) = 1 mm; (I) and (J) = 10 mm.

about 24.0% of the inner one .... *Rheodreissena cordilineata* Mansur, Pereira, Pimpão & Sabaj, 2019
[Brazil: Para, Trombetas, Xingu River Basin; Rondônia, Madeira River Basin]

2′ Adult shell ≤ 15 mm shell outline rather triangular, sharply carinate; posterodorsal margin subrect; periostracum stout; byssal notch particularly marked in left valve (Fig. 11.110 I, J) .... *Rheodreissena hoeblichi* (Schütt, 1991)
[Venezuela: Caroni River (tributary of lower Orinoco)]

3(1) Adult shell ≤ 13.3 mm length (average 9.6 mm), laterally flattened, flabelliform; posteriodorsal margin high, rounded; byssal notch present, outer shell surface with regular very fine commarginal undulations, periostracum nearly smooth, with few very fine radial ribs mostly ramified in plumiform pattern; outer demibranch surface representing 44.6% of inner one (Fig. 11. 110 G, H) .... *Rheodreissena xinguana* Mansur, Pereira, Pimpão & Sabaj, 2019
[Brazil: Para, Tapajos and Xingu River basins]

3′ Adult shell < 8.8 mm length (average 6.5 mm); shell elongated, mytiliform, ventral margin flat, sometimes concave or sinusoidal; byssal notch absent or very slightly marked, sometimes only present in one valve; periostracum lamellose except for a smooth dorsal shield and small pseudolunular area; outer demibranch surface representing 27.7% of inner (Fig. 11.110 K, L) .... *Rheodreissena lopesi* (Alvarenga & Ricci, 1989)
[Brazil: Para, Tocantins, Araguaia, Xingu Basin]

Bivalvia: Venerida: Corbiculidae: Genera

1 Pallial sinus well-marked, deep (Fig. 11.96 C, D); inhalant and exhalant siphons long; lateral teeth serrated or not 2

1′ Pallial sinus absent, represented by a slightly marked, shallow and wide curve (Fig. 11.96 A); both inhalant and exhalant siphons short, with papillae (Fig. 11.3 A); lateral teeth serrated .... ***Corbicula* [p. 395]**

2(1) Lateral teeth serrated (Fig. 11.96 D), shell thin or moderately thick, shell surface finely striated; papillae only in inhalant siphon .... ***Neocorbicula* [p. 395]**

2′ Lateral teeth smooth (Figs. 11.96 C, F), shell solid; shell surface with well-marked, coarse, commarginal ribs ..... ***Polymesoda* [p. 397]**

Bivalvia: Venerida: Corbiculidae: *Corbicula*: Species

Three species of *Corbicula* are recognized in the Neotropical region. A fourth "morph" from Brazil (Guaíba Lake and Lagoa dos Patos) and Argentina (Iguazú Falls), reported as *Corbicula* sp. (Fig. 11.111 D) is characterized by a solid shell, with a more ovate outline, with low, small umbos, and shell surface with fine, low ribs (see Lee et al., 2005; Mansur et al., 2011)

1 Shell equilateral, without rostrum 2

1′ Shell robust, inequilateral, outline trigonal, high, with posterior rostrum, umbos inflated, outstanding from shell surface; commarginal ribs coarse, strongly marked (Fig. 11.111 M); adult size ≤ 5 cm (Fig. 11.111 A) .... *Corbicula fluminea* (Müller, 1774)
[Invasive. Venezuela, Colombia, Ecuador, Brazil, Peru, Argentina Uruguay. Oriental]

2(1) Shell somewhat fragile, umbos central, low, not inflated, beaks minute; commarginal ribs low, somewhat fine (Fig. 11.111 N); inner surface bluish; adult size ≤ 2.5 cm (Fig. 11.111 B) .... *Corbicula largillierti* (Philippi, 1844)
[Invasive. Brazil: in south, central and NE drainages. Argentina: Río de la Plata drainage, Buenos Aires, Córdoba, Catamarca. Oriental]

2′ Shell robust, high, nearly equilateral, umbos rounded, inflated; commarginal ribs medium; adult size ≤ 3 cm (Fig. 11.111 C) .... *Corbicula fluminalis* (Müller, 1774)
[Invasive. Brazil: Rio Grande do Sul, at Jacuí River, Guaíba Lake and Lagoa dos Patos. Uruguay: Merin Lagoon]

Bivalvia: Venerida: Corbiculidae: *Neocorbicula*: Species

Cyanocyclas was, in the recent years, frequently used as synonym of *Neocorbicula* according to Parodiz (1996); however, *Neocorbicula* is the valid name. Parodiz & Hennings (1965) in an attempt to clarify the taxonomy of the native corbiculids from the Paraná and Uruguay River basins reduced 30 nominal species to only 2 valid names: *Neocorbicula limosa* (Maton, 1809) and *N. paranensis* (d'Orbigny, 1835). Clavijo (2015), based on a molecular phylogenetic analysis recovered *N. exquisite* (Marshall, 1924) (Fig. 11.21 I), *N. fortis* (Marshall, 1924) (Fig. 11.21 J), and *N. guahybensis* (Marshall, 1927) (Fig. 11.21 K) as valid species, together with *N. limosa* and *N. paranensis*. However, these nomenclatorial acts are still pending of validation by an official publication.

Other nominal species: *Corbicula amazonica* (Prime 1870) (Amazonas River), *Neocorbicula bavayi* (Ancey, 1880) (Venezuela and French Guiana), *N. cuneata* (Jonas, 1844) (Venezuela, Orinoco River), *N. rotunda* (Prime, 1860) (Suriname, Suriname River), and *Cyanocyclas surinamica* (Clessin, 1879) (South America, from the stomach of fish), need revision

1 Shell ovate, not solid; umbos not inflated, beaks pointing 2

1′ Shell trigonal, high, very solid; umbos inflated, outstanding from shell surface (Fig. 11.111 H) .... *Neocorbicula paranensis* (d'Orbigny, 1835)
[Argentina: Santa Fe, at Paraná River; Formosa, at Paraguay River]

2(1) Shell surface finely striated, sometimes with radial color stripes (Fig. 11.111 E, O) .... *Neocorbicula limosa* (Maton, 1809)
[Brazil: Lakes and ponds from SE Atlantic drainage. Argentina: Buenos Aires and southeastern provinces in ponds and small streams that flow into the Uruguay, Paraná, and Río de la Plata rivers. Uruguay: Colonia, Florida, Lavalleja, Río Negro, Rivera, Tacuarembó, and Yi]

2′ Shell surface with regularly spaced commarginal low ribs (Fig. 11.111 L) .... *Neocorbicula brasiliana* (Deshayes, 1854)
[Brazil: Para, Piauí Parnaiba River delta]

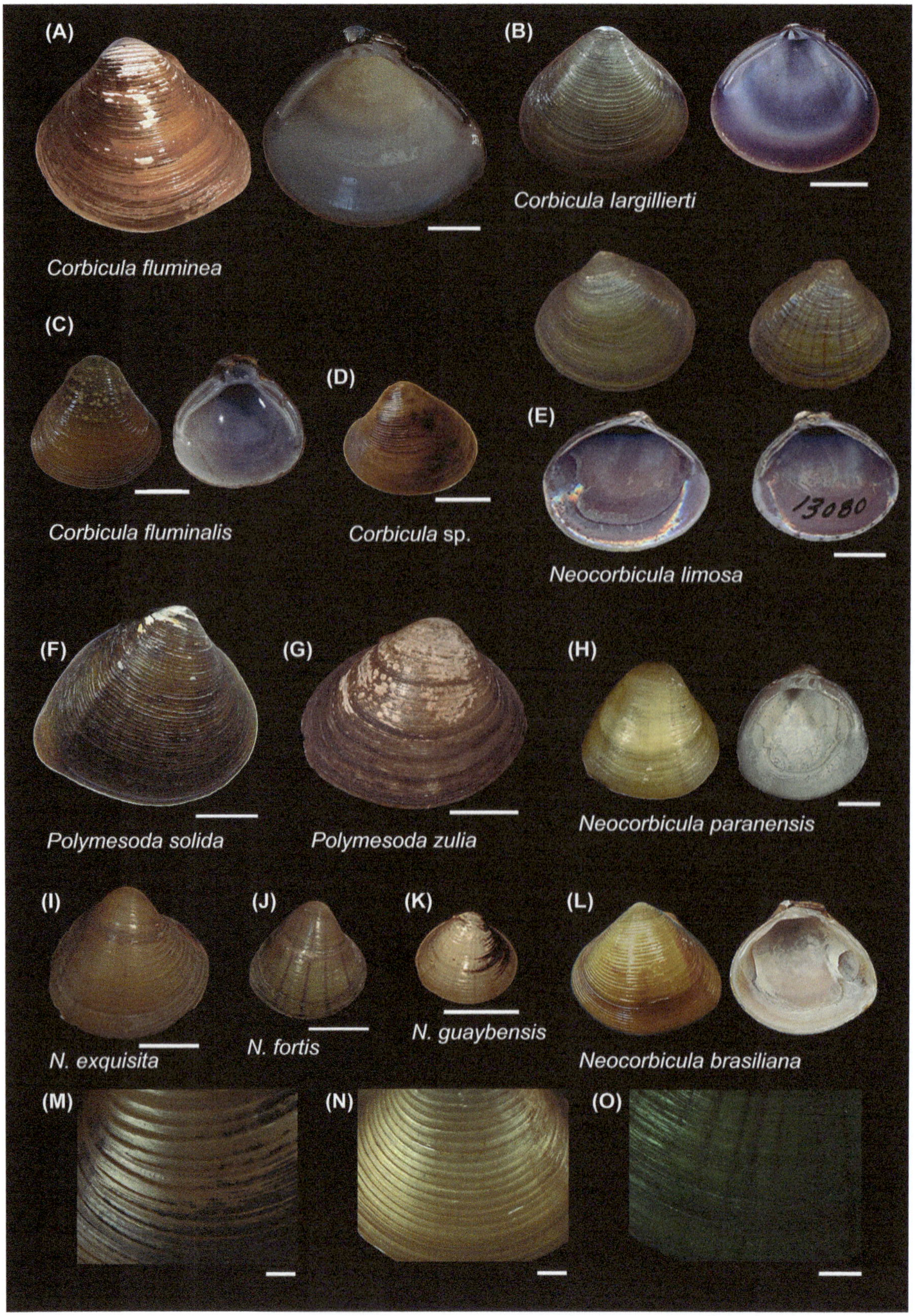

FIGURE 11.111 Species of Corbiculidae: (A) *Corbicula fluminea*, outer (left) and inner (right) views of a right valve; (B) *C. largillierti*, outer (left) and inner (right) views of a right valve; (C) *C. fluminalis*; (D) *Corbicula* sp.; (E) *Neocorbicula limosa*, two outer views (above) showing different color pattern, and two inner views (below) of left and right valves; (F) *Polymesoda solida*, outer view of right valve; (G) *P. zulia*, right valve; (H) *Neocorbicula paranensis*; (I–K) Morphotypes of three nominal species of *Neocorbicula*: (I) *N. exquisita*, (J) *N. fortis* and (K) *N. guahybensis*; (L) *N. brasiliana*, left and right valves. (M–O) details of shell surface: (M) *C. fluminea*, (N) *C. largillierti*; (O) *N. limosa*. Scale bars: (A–G), (I–K) = 10 mm, (H) = 5 mm, (M–O) = 0.5 mm. (F) Photograph by M.V. de la Hoz; (G) photograph by Paul Callomon ANSP); (I–K) photograph by C. Clavijo.

Bivalvia: Venerida: Corbiculidae: *Polymesoda*: Species

Two *Polymesoda* species have been reported from freshwater. Other species have been reported exclusively from brackish environments (e.g., *Polymesoda mexicana* (= *P. aequilatera*) reported from Venezuela, Guyana, Suriname, French Guiana, and Brazil (Pereira et al., 2014)

1 Shell nearly equilateral, solid, shell outline trigonal high, posterior end widely arcuate, sloping toward a slightly produced rostrum; outer surface with coarse commarginal ribs (Fig. 11.111 F) .... *Polymesoda solida* (Philippi, 1846)
[Venezuela: Maracaibo Lake. Colombia: ponds and streams close to the Magdalena River]

1′ Shell slightly inequilateral, umbos displaced anteriorly, shell outline ovate, low, commarginal ribs moderately coarse (Fig. 11.111 G) . ... *Polymesoda zulia* H.B. Baker, 1930
[Venezuela: Maracaibo Lake, an inland saline habitat]

Bivalvia: Venerida: Sphaeriidae: Genera

1 Umbo central or variably displaced to anterior end, both inhalant and exhalant siphons present 2

1′ Umbo in the posterior half of shell variably displaced backward (Fig. 11.112 A, B); one cardinal tooth in the right valve, two in the left (Fig. 11.112 C); shell smooth or faintly striated with very low commarginal ribs; adult shell small, adult shell < 10 mm length; only inhalant aperture or both inhalant aperture and exhalant siphon present .... ***Pisidium* [p. 398]**

2(1) Two cardinal teeth in left valve and one in the right valve (Fig. 11.112 F), periostracum not lamellate; byssus gland absent in adults 3

2′ One cardinal tooth in each valve or valves without teeth; periostracum variably lamellate, shell surface yellowish with purple spots, sometimes absent, on the inner shell surface usually visible from exterior; byssus gland present and functional in adults 4

3(2) Adult shell outline trapezoidal, high, umbos inflated, nepionic shell usually marked forming a cup (Fig. 11.112 D), inhalant and exhalant siphons united at the bases and along nearly entire longitude, only separated at the tips .... ***Musculium* [p. 402]**

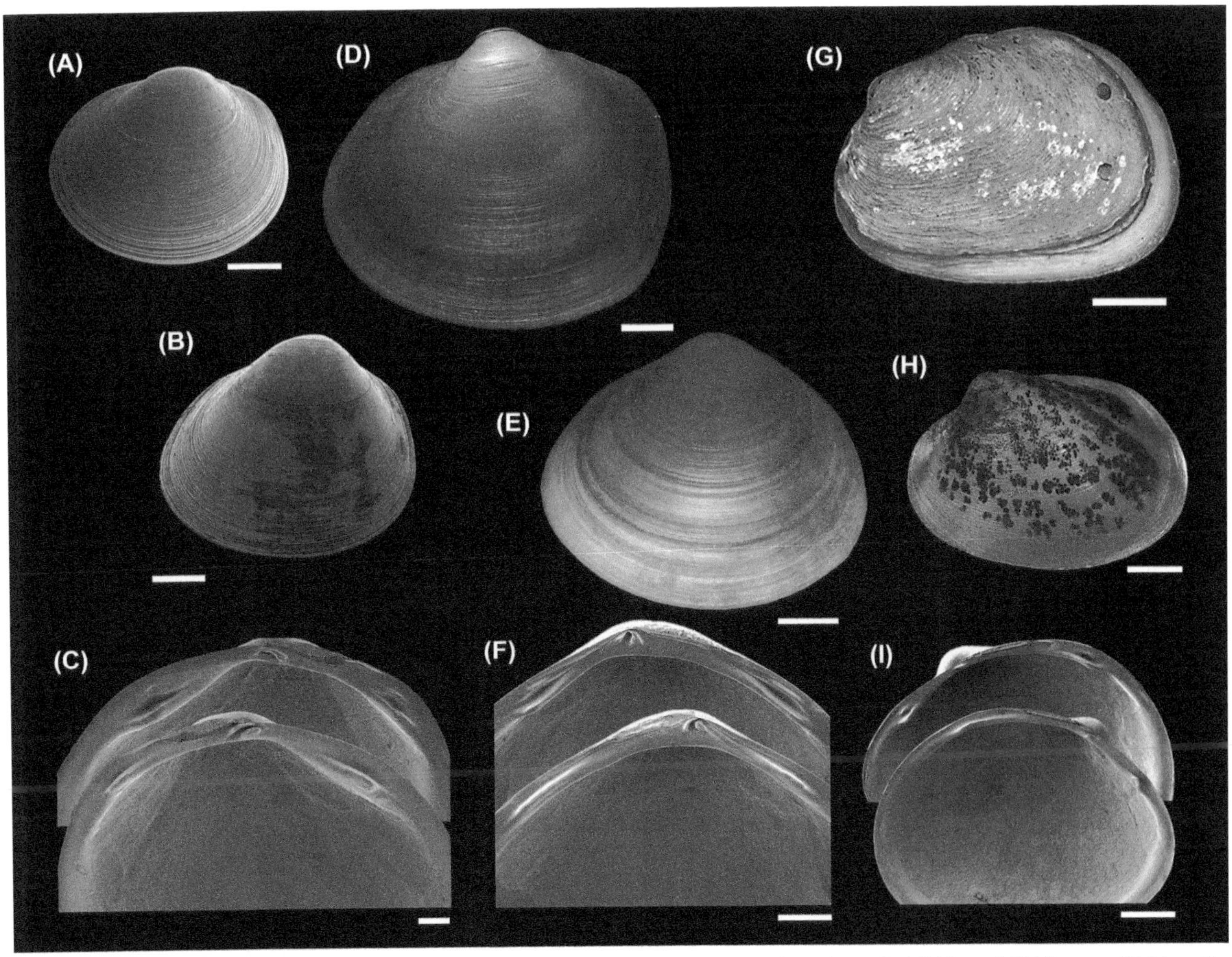

FIGURE 11.112 Sphaeriidae genera, outer shell views and hinge details: (A) *Pisidium* sp.; (B) *Pisidium vile*; (C) hinge of *Pisidium* sp.; (D) *Musculium argentinum*; (E) *Sphaerium cambaraense*; (F) hinge of *Sphaerium* sp.; (G) *Byssanodonta paranensis*; (H) *Eupera simoni*; (I) hinge of *Eupera primei*. Scale bars: (A) and (B) = 0.5 mm; (C), (F), (H) and (I) = 1 mm; (D), (E) and (G) = 2 mm.

PHYLUM MOLLUSCA

3′ Adult shell outline rounded or slightly quadrangular, umbos low, not outstanding from shell surface (Fig. 11.112 E); inhalant and exhalant siphons well separated along its entire length .... ***Sphaerium* [p. 404]**

4(2) One cardinal tooth in each valve, the left always reduced in size (Fig. 11.112 I); inner and outer shell surface usually with purple spots (Fig. 11.112 H); adult shell < 5 mm .... ***Eupera* [p. 405]**

4′ Cardinal teeth present as rudimentary reliefs in juveniles, absent in adults, inner surface of adult shell usually without color spots; adult shell ≤ 8.2 mm (Figs. 11.112 G, 118 A) .... *Byssanodonta paranensis* d' Orbigny, 1835 [Argentina: Corrientes, upper Paraná River at Paso de la Patria and Itá Corá]

Bivalvia: Venerida: Sphaeriidae: *Pisidium*: Species

The highly variable and scant diagnostic characters in *Pisidium* (particularly when available data are restricted to shell features) make difficult to elaborate identification keys to the species diversity from wide geographic regions (Table 11.2). Diagnostic differences among species are often subtle, making impossible discerning among species based only on a restricted dichotomous combination of shell characters.

The relatively well-known diversity of Neotropical *Pisidium* species concentrates in southern South America, mainly Argentina and Chile. Within this vast geographic realm, three areas may be defined and their *Pisidium* diversity arranged in identification keys for each of them. The keys that follow are to be considered as a first approach to the identification of a species, a task that should be complemented with details of full description of species given in the current literature on *Pisidium* taxonomy (Ituarte, 2007 and references therein).

TABLE 11.2 Neotropical *Pisidium* species currently recognized as valid and their type localities.

Species	Figure	Type Locality
Pisidium bejumae H. B. Baker, 1930*		Venezuela: Carabobo, near Bejuma
Pisidium chicha Ituarte, 2005	11.114 A	Argentina: Jujuy, from unnamed stream flooding to Zapla River
Pisidium chilense (d'Orbigny, 1846)	11.113 F	Chile: Concepción
Pisidium chiquitanum Ituarte, 2001	11.114 E	Bolivia: Santa Cruz de la Sierra, La Siberia (west of Comarapa)
Pisidium davisi (Bartsch, 1908)*		Ecuador: Chanchan River
Pisidium dorbignyi (Clessin, 1879)	11.115 D	Uruguay: Maldonado
Pisidium forense Meier-Brook, 1967*	11.115 H	Brazil: Minas Gerais, Juiz de Fora
Pisidium globulus Clessin, 1888*	11.115 G	Brazil: Rio Grande do Sul,Taquara
Pisidium huillichum Ituarte, 1999	11.113 G	Chile: Puerto Cárdenas, Yelcho lake
Pisidium inacayali Ituarte, 1996	11.113 B	Argentina: Chubut, Trevelin
Pisidium iquito Ituarte, 2004*	11.115 C	Peru: Loreto, Pacaya-Samiria Reserve
Pisidium lebruni Mabille, 1884	11.113 C	Chile: Punta Arenas
Pisidium llanquihuense Ituarte, 1999	11.113 A	Chile: Llanquihue Lake at Puerto Octay
Pisidium magellanicum (Dall, 1908)	11.113 H	Magellan Strait (described from a Shell drained into the sea)
Pisidium meierbrooki Kuiper & Hinz, 1984	11.114 D	Bolivia: La Paz, Batalla
Pisidium ocloya Ituarte, 2005	11.114 B	Argentina: Jujuy, swampy areas close to Burrumayo River
Pisidium omaguaca Ituarte, 2005	11.114 C	Argentina: Jujuy, Terma de Reyes
Pisidium patagonicum Pilsbry, 1911	11.113 D	Argentina: Santa Cruz, springs on the Chico River
Pisidium pipoense Ituarte, 2000*	11.115 E	Argentina: Misiones, San Ignacio, Santo Pipó Rivulet
Pisidium plenilunium (Melville &Standen, 1907)	11.113 E	Malvinas (Falkland) islands, Cape Pembroke
Pisidium punctiferum (Guppy, 1867)*		Trinidad and Tobago: Port of Spain
Pisidium sterkianum Pilsbry, 1897	11.115 F	Uruguay: Montevideo, creek in the Prado (Miguelete Rivulet)
Pisidium taraguyense Ituarte, 2000	11.115 A	Argentina: Corrientes, San Martín, La Cruz
Pisidium vegae Pilsbry, 1926		Mexico: Querétaro
Pisidium vile Pilsbry, 1897	11.115 B	Uruguay: Montevideo, creek in the Prado (Miguelete Rivulet)
Pisidium wolfii Clessin, 1879		Ecuador: Chillo valley, Rio Pedro

Asterisk (*) denotes species known only from type locality; in bold those species included in the keys.

Key to *Pisidium* species from Chilean and Argentinean Patagonia

1 Adult shell > 2.5 mm 2

1′ Adult shell < 2.5 mm (Fig. 11.113 G) .... *Pisidium huillichum* Ituarte, 1999
[Chile: Puerto Cárdenas at Lago Yelcho]

2(1) Adult shell nearly as high as long 3

2′ Adult shell longer than higher 4

3(2) Shell outline subtrapezoidal 7

3′ Shell orbicular, umbos not surpassing dorsal margin in inner view (Fig. 11.113 E) ... *Pisidium plenilunium* (Melville & Standen, 1907)
[Malvinas (Falkland) Islands: Cape Pembroke; Roy Cove, Herbert Stream]

4(2) Adult shell < 5.5 mm length 5

4′ Adult shell > 5.5 mm length 6

5(4) Shell compressed (Ci = 51 ± 4), umbos subcentral (Fig. 11.113 F) .... *Pisidium chilense* (d'Orbigny, 1846)
[Chile: Concepción; Lautaro City on Río Cautín, Villa Rica Lake; Valdivia, Chaiten]

5′ Shell slightly inflated (Ci 61 ± 4), umbos displaced backward (Fig. 11.113 B) .... *Pisidium inacayali* Ituarte, 1996
[Argentina: Chubut, streams between Trevelin and Futaleufú dam]

6(4) Shell outline subquadrate, dorsal margin subrect, forming angles when it joins anterior and posterior margins, posterior end truncate (Fig. 11.113 A) .... *Pisidium llanquihuense* Ituarte, 1999
[Chile: Puerto Octay at Llanquihue Lake]

6′ Shell outline rounded-ovate, dorsal margin arcuate, not forming angles with anterior and posterior margins, posterior end evenly curved (Fig. 11.113 C) .... *Pisidium lebruni* Mabille, 1884
[Chile: Punta Arenas]

7(3) Shell solid, length ≤ 5.1 mm, H/L index 92 ± 1.6, moderately inflated; hinge plate short, deeply arched, strong; umbos subcentral (Fig. 11.113 D) .... *Pisidium patagonicum* Pilsbry, 1911
[Argentina: Santa Cruz, Río Chico, Arroyo Ecke]

7′ Shell not solid, length ≤ 5.7 mm, H/L index 88 ± 2, slightly compressed (Fig. 11.113 H) .... *Pisidium magellanicum* (Dall, 1908)
[Chile: Punta Arenas; Cabeza de Mar. Argentina: Santa Cruz, small streams at the base of Andes, Río Chico; Neuquén, San Martín de los Andes at Laguna Verde]

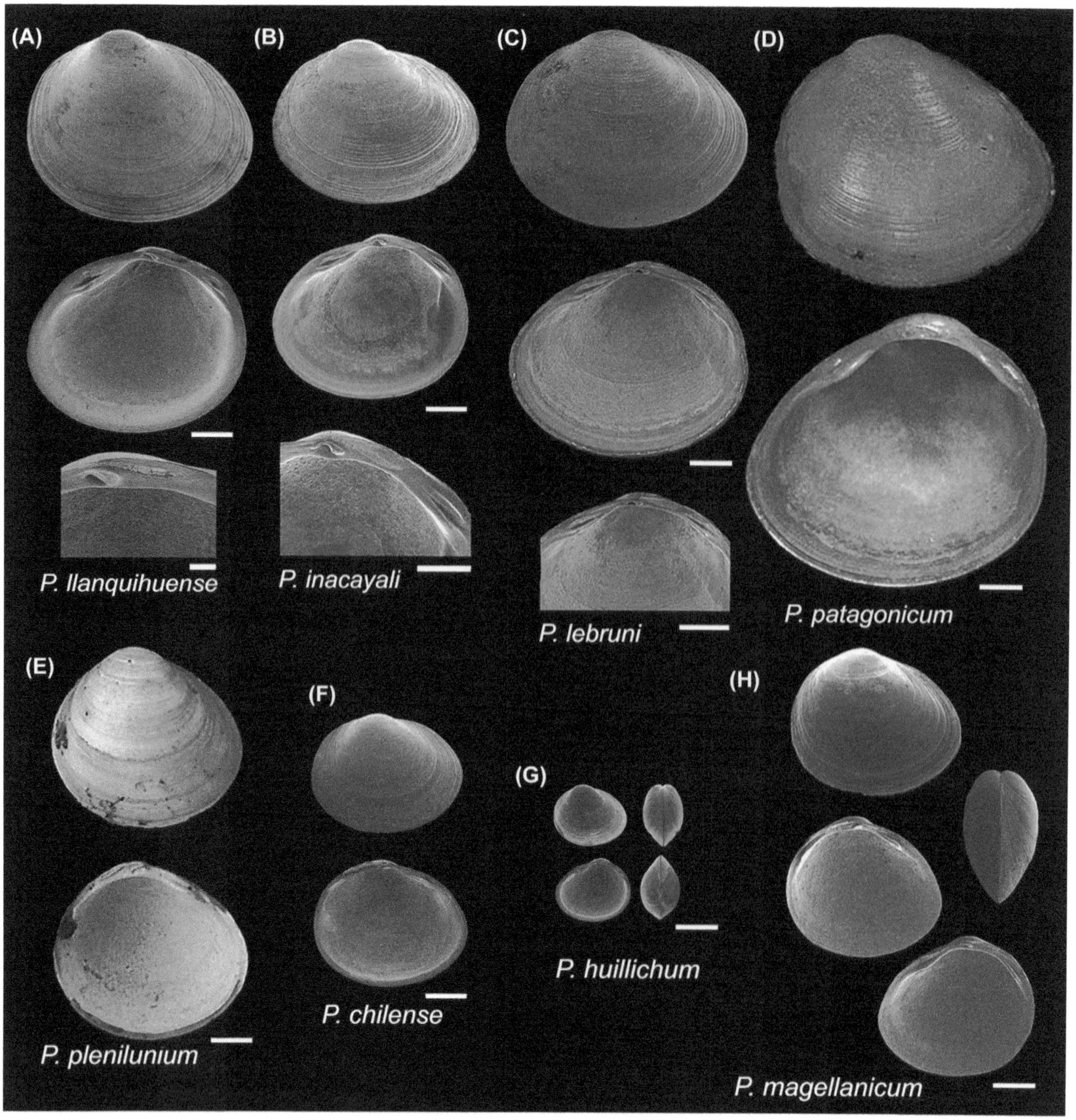

FIGURE 11.113 Species of *Pisidium* (Sphaeriidae) from Chilean and Argentinean Patagonia: (A) *Pisidium llanquihuense*; (B) *P. inacayali*; (C) *P. lebruni*; (D) *P. patagonicum*; (E) *P. plenilunium*; (F) *P. chilense*; (G) *P. huillichum*; (H) *P. magellanicum*. Scale bars for inner and outer shell views: (A–H) = 1 mm; for detail of hinges: (A) = 200 μm; (B) = 500 μm; (C) = 1 mm.

Key to *Pisidium* species from Central South America (NW Argentina, Central Bolivia, and Northern Chile)

1 Shell outline rounded-ovate, variably high 2

1′ Shell outline high, sharply trapezoidal, shell fragile, translucent; small adult size (< 3.5 mm length), two siphonal apertures present, ligament included, not visible from exterior (Fig. 11.114 B) .... *Pisidium ocloya* Ituarte, 2005
[Argentina: Jujuy, Salta, Catamarca, Tucumán (between 880 and 2155 m altitude)]

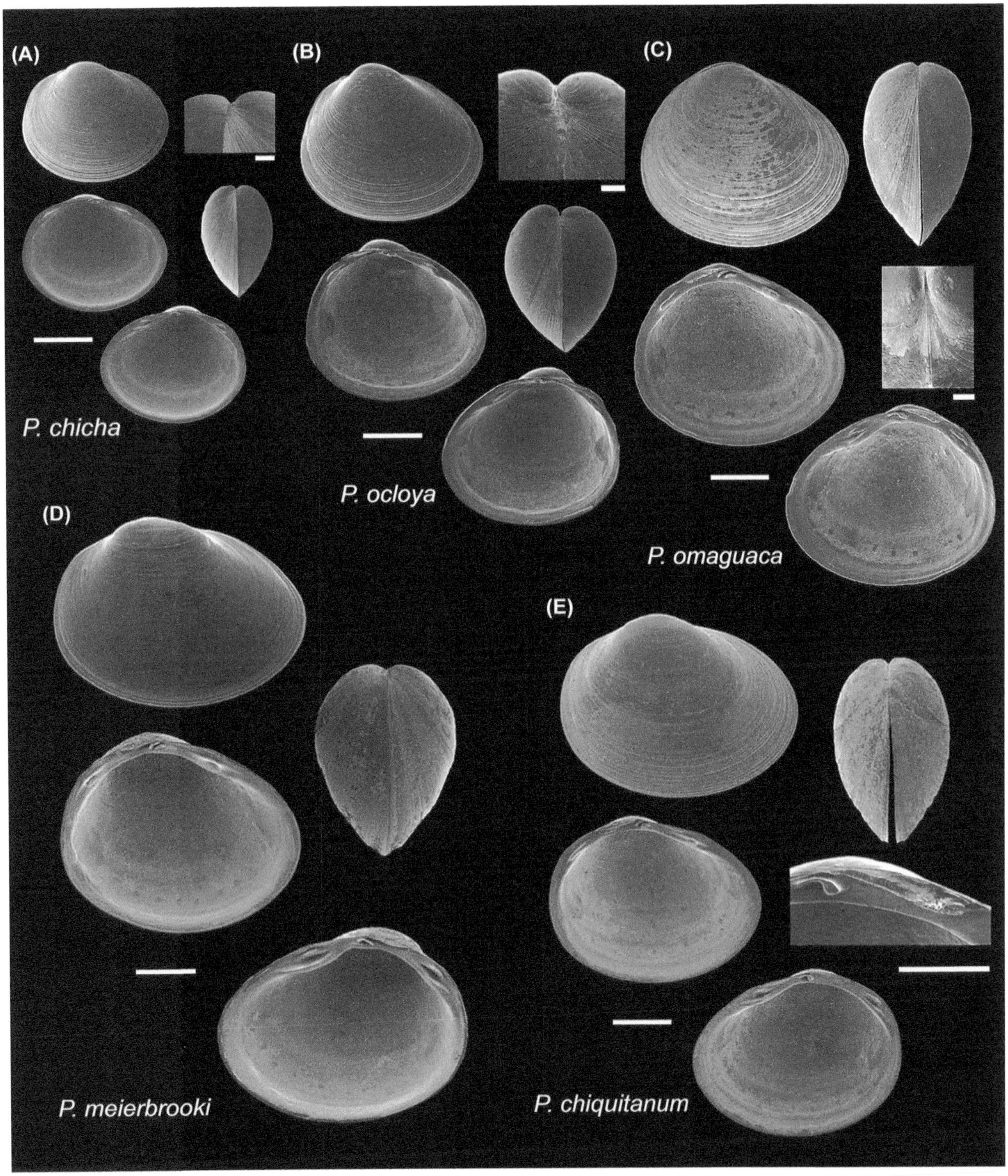

FIGURE 11.114 Species of *Pisidium* (Sphaeriidae) from central South America: (A) *Pisidium chichi*; (B) *P. ocloya*; (C) *P. omaguaca*; (D) *P. meierbrooki*; (E) *P. chiquitanum*. Scale bars for inner and outer shell views: (A) and (B) = 0.5 mm; (C–E) = 1 mm; for detail of ligament area and hinge: (A) = 100 μm; (B) and (C) = 200 μm; (E) = 500 μm.

2(1) Adult shell > 3 mm length, shell outline ovate, high or transversely elongated; hinge plate moderately strong; ligament pit included, ligament may or may not be visible from exterior 3

2′ Adult shell < 3 mm length, not inflated (Ci = 68 ± 5); shell outline markedly oval; umbos low, rounded, only slightly displaced backwards; hinge plate thin; ligament slightly visible from exterior; only anal siphon present (Fig. 11.114 A) .... *Pisidium chicha* Ituarte, 2005
[Argentina: Jujuy, Catamarca, Tucumán (between 940 and 2,740 m altitude)]

3(2) Shell outline ovate transversely elongated 4

3′ Shell outline ovate high, tending to trigonal, posterior end very short, truncated; umbos very low, only slightly visible above dorsal margin; ligament never visible from exterior; anal siphon and branchial aperture present (Fig. 11.114 C) .... *Pisidium omaguaca* Ituarte, 2005 [Argentina: Jujuy, Salta (between 1400 and 4100 m altitude)]

4(3) Shell solid, moderately inflated; hinge plate robust, umbos wide, prominent, variably displaced backward (located between 54% and 63% of shell length), both branchial and anal apertures present (Fig. 11.114 D) .... *Pisidium meierbrooki* Kuiper & Hinz, 1984 [Peru: between Titicaca Lake and Cuzco (from 3500 to 4450 m altitude). Bolivia: Alto-Andean regions in the Altiplano Basin. Chile: Alto-Andean regions of Northern Chile]

4′ Shell thin, umbos very low, slightly displaced (located about 62% of shell length) (Fig. 11.114 E) .... *Pisidium chiquitanum* Ituarte, 2001 [Bolivia: La Siberia (west of Comarapa). Argentina: Jujuy, Salta, Tucumán, Mendoza (between 1360 and 2400 m altitude)]

Key to *Pisidium* Species from South American Atlantic Drainage Systems

1 Adult size > 5 mm length 2

1′ Adult shell size < 5 mm length 3

2(1) Shell outline ovate, moderately protruded anteriorly; ligament external, protruded, well-visible from the exterior; umbos well-marked from shell surface, displaced posteriorly (located at 59 ± 6% of shell length) (Fig. 11.115 F) .... *Pisidium sterkianum* (Pilsbry, 1897) [Perú: Iquitos. Bolivia. Paraguay. Argentina: Buenos Aires, Río de la Plata; Corrientes, Entre Ríos, Santa Fe. Uruguay: Montevideo, Salto, Lavalleja, Canelones]

2′ Adult shell length 7 mm; shell outline subquadrangular or trapezoidal, high (H/L index 82 ± 2); umbos inflated, pointed (Fig. 11.115 C) .... *Pisidium iquito* Ituarte, 2004 [Peru: Loreto, Pacaya-Samiria Reserve]

3(1) Adult shell length ~4 mm, shell outline subquadrangular 4

3′ Adult shell < 4 mm length, shell outline ovate or subtrigonal 5

4(3) Adult shell ≤ 4.1 mm, shell outline subquadrangular, nearly equilateral, moderately high, ventral margin evenly and slightly curved; umbos subcentral (Fig. 11.115 E) .... *Pisidium pipoense* (Ituarte, 2000) [Argentina: Misiones, Arroyo Santo Pipó]

4′ Adult shell ≤ 4.7 mm length, shell outline subquadrangular high, inequilateral; ventral margin markedly curved, anterior end angled (Fig. 11.115 H) .... *Pisidium forense* Meier-Brook, 1967 [Brazil: Minas Gerais, Juiz de Fora]

5(3) Shell outline ovate variably projected anteriorly 6

5′ Shell outline subtrigonal, high (H/L index 89 ± 2.3), inflated (Ci = 82 ± 4.6), dorsal margin very short; hinge plate solid (Fig. 11.115 B) .... *Pisidium vile* Pilsbry, 1897 [Brazil: Rio Grande do Sul. Bolivia: Santa Cruz de la Sierra, Estación Parapetí. Argentina: Buenos Aires, streams related to the Río de la Plata Basin. Uruguay: Montevideo]

6(5) Shell outline ovate short 7

6′ Shell outline ovate anteriorly protruded, shell thin, fragile, strongly inflated (Ci = 84 ± 3.6); hinge plate extremely thin (cardinal teeth hanging from dorsal margin), dorsal margin very short, posterior end truncated (Fig. 11.115 A) ...... *Pisidium taraguyense* Ituarte, 2000 [Argentina: Corrientes, La Cruz, Uruguay: northwestern coast, in small streams flooding into Uruguay River]

7(6) Shell outline with no marked angles; anterior, posterior, and ventral margins evenly curved; shell not high, moderately inflated; umbos wide, low, blunt (Fig. 11.115 G) .... *Pisidium globulus* Clessin, 1888 [Brazil: Rio Grande do Sul, Taquara]

7′ Shell outline with angles at the joining of dorsal margin with anterior and posterior margins; umbos narrow, inflated (Fig. 11.115 D) .... *Pisidium dorbignyi* (Clessin, 1879) [Argentina: Martín García Island. Uruguay: Artigas, Florida, Maldonado]

Bivalvia: Venerida: Sphaeriidae: *Musculium*: Species

1 Shell outline sharply trapezoidal, posterior end bluntly truncate; umbos inflated, well-outstanding from the shell surface, beaks with a distinct small cup, corresponding to the nepionic shell; shell surface finely striated (Fig. 11.116 A, B) .... *Musculium argentinum* (d'Orbigny, 1835) [Argentina: Mendoza, Buenos Aires, Tierra del Fuego. Uruguay: around Montevideo. Chile: between 38° and 41°S]

1′ Shell outline rounded-ovate, if subquadrate, with no marked angles; anterior, ventral, and posterior margins evenly arcuate, umbos not inflated, relatively low, with well-marked nepionic shell (Fig. 11.116 C, D) .... *Musculium patagonicum* (Pilsbry, 1911) [Chile: only reported from a lake at 51°42′S. Argentina: Chubut, Neuquén and Santa Cruz]

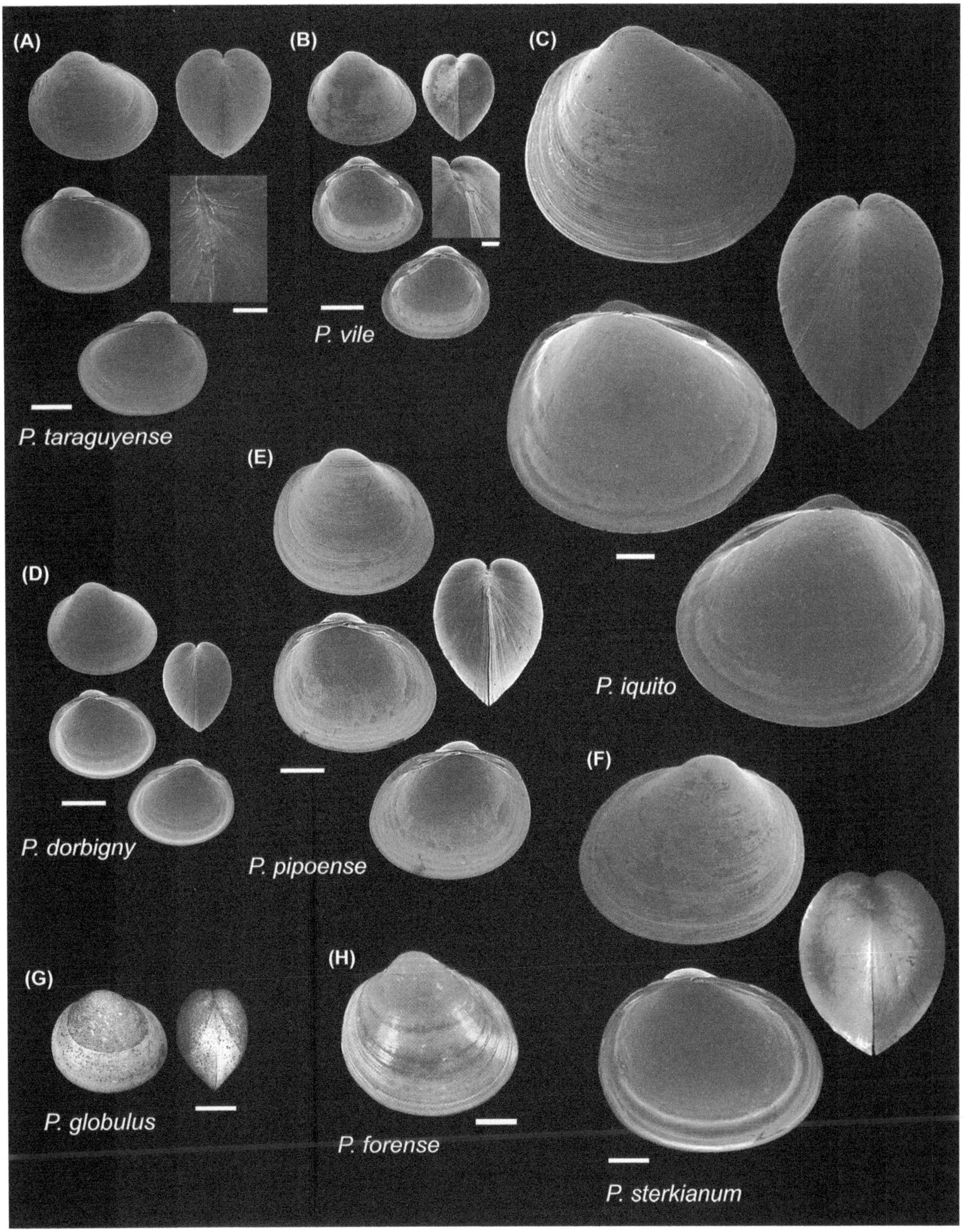

FIGURE 11.115 Species of *Pisidium* (Sphaeriidae) from South American eastern drainages: (A) *Pisidium taraguyense*; (B) *P. vile*; (C) *P. iquito*; (D) *P. dorbigny*; (E) *P. pipoense*; (F) *P. sterkianum*; (G) *P. globulus*; (H) *P. forense*. Scale bars for inner and outer shell views: (A–H) = 1 mm; for detail of ligament area: (A) = 100 μm, (B) = 200 μm.

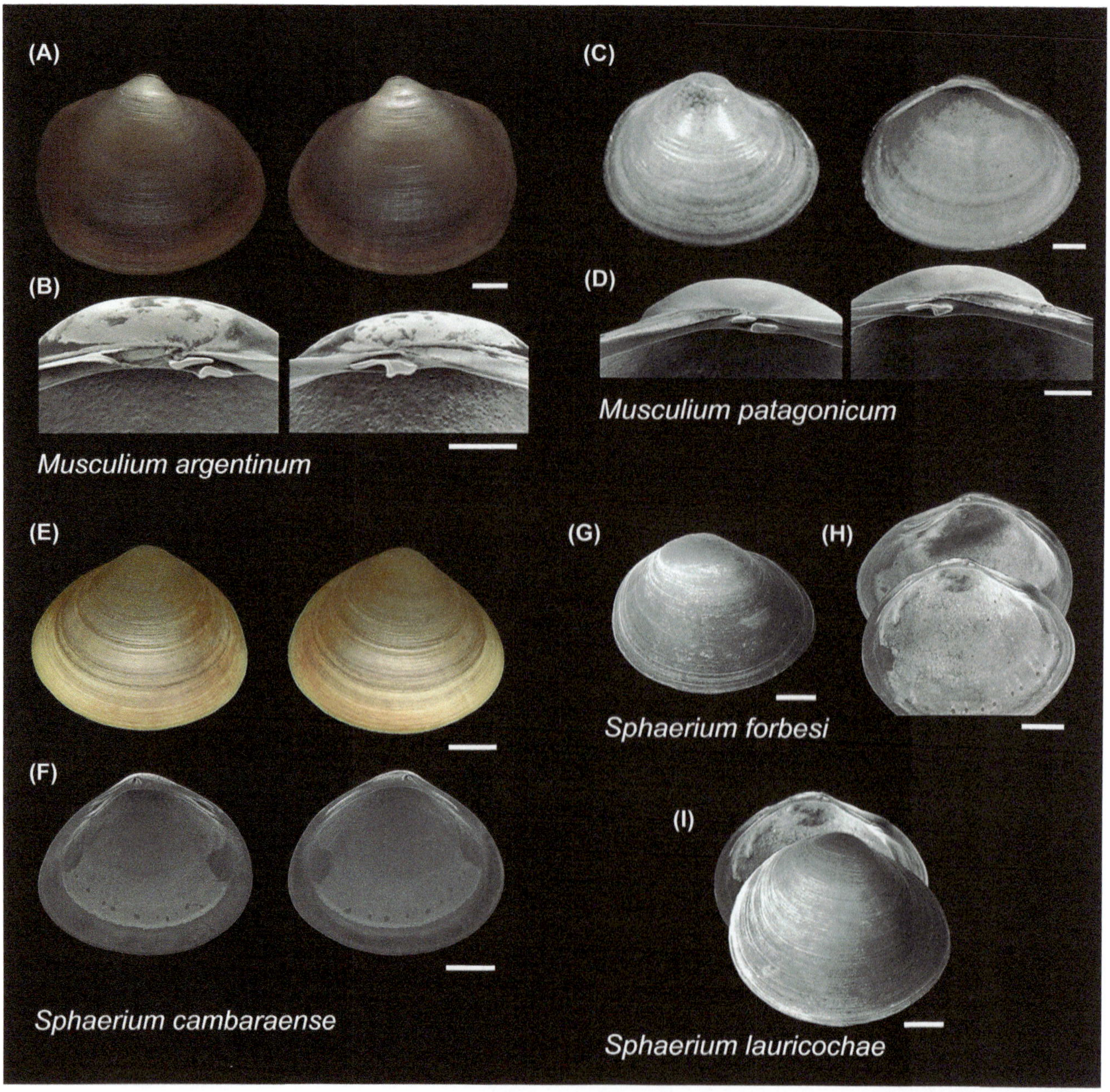

FIGURE 11.116 Sphaeriidae species. (A) and (B) *Musculium argentinum*: (A) External view of right and left valves; (B) detail of left and right cardinal teeth. (C) and (D) *Musculium patagonicum*: (C) External view of right and left valves; (D) detail of left and right cardinal teeth. (E) and (F) *Sphaerium cambaraense*: (E) left and right valves; (F) inner views of right and left valves. (G) and (H) *Sphaerium* forbesi: (G) Outer view of left valve; (H) inner view of left (front) and right (behind) valves. (I) *Sphaerium lauricochae*: outer view of a right valve (front), inner view of the hinge of a left (behind) valve. Scale bars: (A), (C), (E), (F–I) = 2 mm; (B) and (D) = 0.5 mm.

Bivalvia: Venerida: Sphaeriidae: *Sphaerium*: Species

There are a some old, poorly known, species of *Sphaerium* described from Central America and Mexico (considering those described from localities below the Tropic of Cancer) whose taxonomic status is uncertain, needing a comprehensive revision (Table 11.3). At present, the scant knowledge on the shell morphology and anatomy, as well as on their intra and interspecific variability, prevents from systematizing them in an identification key. For details on the taxonomic history of these species as well as for some proposed synonymies, see Graff & Cummings (2019).

TABLE 11.3 Nominal species of *Sphaerium* from Central America and Southern North America and their type localities

Species	Type locality
Sphaerium aequatoriale Clessin, 1879	Ecuador: Quito, Chillogallo
Sphaerium costaricanum Martens, 1900	Costa Rica: La Palma; Alajuela
Sphaerium davisi (Bartsch, 1908) (possibly a *Pisidium* species)	Ecuador: Chanchan River valley
Sphaerium jalapensis Pilsbry, 1899	Mexico: Veracruz, from a pond at Jalapa
Sphaerium luridum Martens, 1900	Mexico: Jalisco
Sphaerium martensi Pilsbry, 1899	Mexico: Michoacan, Tzintzuntzan, Patzcuaro lake
Sphaerium mexicanum (Dall, 1905)	Mexico: San Luis Potosí
Sphaerium novoleonis Pilsbry, 1899	Mexico: Nuevo León, Monterrey
Sphaerium subtransversum Prime, 1860	Mexico: Tabasco
Sphaerium titicacense Pilsbry, 1924	Bolivia: Lake Titicaca at Yunguyo
Sphaerium triangulare (Say, 1829)	Mexico: Around Mexico City (lake Texcoco?)

Key to *Sphaerium* Species from Southern South America

1 Shell outline rounded or ovate, umbos low, wide, slightly outstanding from shell surface 2

1′ Shell outline subtrapezoidal tending to trigonal, anterior margin sharply curved; umbos nearly central, low, not inflated, hinge plate solid (Fig. 11.116 E, F) .... *Sphaerium cambaraense* Dreher-Mansur, Ituarte & Meier-Brook, 2008 [Brazil: Lajeado da Margarida, at Camisas River, Rio Grande do Sul]

2(1) Shell ovate, umbos slightly displaced backward, anterior and posterior ends equally high (Fig. 11.116 G, H) .... *Sphaerium forbesi* (Philippi, 1869) [Colombia. Bolivia: several high altitude localities. Chile: Cotacotani]

2′ Shell outline rounded, high, posterior margin insensibly connected with dorsal margin, posterior end higher than anterior end, umbos very low, pointing forward (Fig. 11.116 I) .... *Sphaerium lauricochae* (Philippi, 1869) [Perú: La Torna, Cachimayo, Cashca-Maihei, Acomayo, Titicaca Lake, Lauricocha lagoon. Bolivia: Suches, Apolobamba at Lago Nubi, Belén. Chile: Antofagasta, Ojos de Ascotan. Argentina: Jujuy, Tinate, Cochinoca]

Bivalvia: Venerida: Sphaeriidae: *Eupera*: Species

At present, 25 *Eupera* species are currently recognized as valid (Table 11.4), however, for most, the original descriptions are poor, lacking of details or ambiguous, and not always well illustrated (sometimes, available illustrations show mismatching shell morphologies for the same species (e.g., Figs. 11.117 J, 118 F). Moreover, 18 species are only known from the localities given in the original description (many of them without precision) and have never been reported again since then. In particular, from Central America and the Caribbean, a plethora of species have been described and most of them were not reviewed or even reported again after the nongood old, devoid of illustrations, original descriptions. Drawings available for all putative species from these regions are offered in Fig. 11.117, only as a pictorial guide to the identification of materials coming from Central America samplings. A deep review of the many nominal taxa from northern South America, Central America, Caribbean Islands, and southern North America is strongly needed and still pending. The best online source for references and relevant information on original descriptions is the Mussel Project (MUSSELp) by Graff & Cummings (2019).

Regarding South American species, the situation is similar; there are several nonwell studied species, e.g., *Eupera bahiensis* (Spix, 1827, described from Bahía, Brazil, *Eupera gravis* H.B. Baker, 1930 from Venezuela, *Eupera modioliforme* (Anton, 1837) (= ? *Eupera moquiniana* Bourguignat, 1854) reported from Brazil and Venezuela, and *Eupera tumida* (Clessin, 1879) described from Bahía, Brazil.

However, from the early 1900s, more proper descriptions of new species were provided, giving a better knowledge of species morphology and variability, allowing their arrangement in an identification key.

PHYLUM MOLLUSCA

TABLE 11.4 Neotropical species of *Eupera* currently considered as valid and their type localities.

Species	Figure	Type locality
Eupera bahamensis (Clench, 1938)*		Bahamas
Eupera bahiensis (Spix, 1827)	11.117 A	Brazil: Bahia, Peruaguaçú River
Eupera barbadensis (Prime, 1861)*	11.117 B	Barbados
Eupera cubensis (Prime, 1865)	11.117 C	Cuba: Santa Catalina de Guantánamo
Eupera doellojuradoi Klappenbach, 1962	11.118 B	Uruguay: Colonia at Puerto Platero
Eupera elliptica Ituarte &Mansur, 1993*	11.118 G	Argentina: Salto Dos Hermanas, Iguazú Falls
Eupera gravis H.B. Baker, 1930*	11.117 D	Venezuela: near Palma Sola
Eupera guaraniana Ituarte, 1994*	11.118 H	Argentina: Entre Ríos, Uruguay River at Colón
Eupera haitiensis Klappenbach, 1970*	11.117 E	Haiti: Département du Sud, near Los Cayes
Eupera iguazuensis Ituarte, 1989*	11.118 I	Argentina: Salto Dos Hermanas, Iguazú Falls
Eupera insignis Pilsbry, 1925*		Mexico: Veracruz, La Laja
Eupera klappenbachi Mansur & Veitenheimer, 1975	11.118 E	Brazil: Rio Grande do Sul, Porto Alegre, Guaiba River
Eupera meridionalis (Prime, 1861)*	11.117 F	Panama
Eupera modioliforme (Anton, 1837)		Sudamerica
Eupera parvula (Prime, 1865)*	11.117 G	Puerto Rico: Hamacao
Euperapittieri Martens, 1900*	11.117 H	Costa Rica
Eupera platensis Doello Jurado, 1921	11.118 C	Argentina: Buenos Aires, Río Santiago
Eupera portoricensis (Prime, 1861)*	11.117 I	Puerto Rico
Eupera primei Klappenbach, 1967*	11.118 F	Peru: Pequino, near Ucayali River
Eupera simoni (Jousseaume, 1889)	11.118 D	Venezuela: Caracas, Laguna de Espino
Eupera tumida (Clessin, 1879)*	11.117 J	Brazil; Bahia
Eupera veatleyii (Adams, 1849)*	11.117 L	Jamaica
Eupera ventricosa (Clessin, 1882)*	11.117 K	Haiti
Eupera viridantis (Prime, 1865)*	11.117 M	Guadeloupe
Eupera weinlandi (Clessin, 1882)*	11.117 N	Haiti

Asterisk (*) denotes species known only from location reported in the original description; in bold those species included in the key.

Key to South American *Eupera* Species

1 Shell rounded-ovate, elliptic or subtrapezoidal, posterior end equal or moderately higher than anterior end 2

1′ Shell sharply trapezoidal, posterior end markedly higher than anterior, truncate; dorsal margin oblique; adult size $\leq$ 7.2 mm (Fig. 11.118 H) .... *Eupera guaraniana* Ituarte, 1994
[Argentina: Entre Ríos, Uruguay River at Colón]

2(1) Shell outline rounded ovate or elliptic, umbos displaced anteriorly 3

2′ Shell outline subtrigonal, umbos central or subcentral; posterior end equal or lower than anterior, obliquely truncate (Fig. 11.118 C) .. .. *Eupera platensis* Doello Jurado, 1921
[Argentina: Buenos Aires close to Río de la Plata and inland waters. Uruguay: Río de la Plata Basin]

3(2) Shell outline rounded-ovate, posterior end higher than anterior end 4

3′ Shell outline markedly elliptic, umbos subterminal, located at about 15% of maximum length; adult size 4.6 mm (Fig. 11.118 G) .... *Eupera elliptica* Ituarte & Mansur, 1993
[Argentina: Misiones, Salto Dos Hermanas, Iguazú Falls]

4(3) Shell high, dorsal margin sharply sloping toward anterior end 5

PHYLUM MOLLUSCA

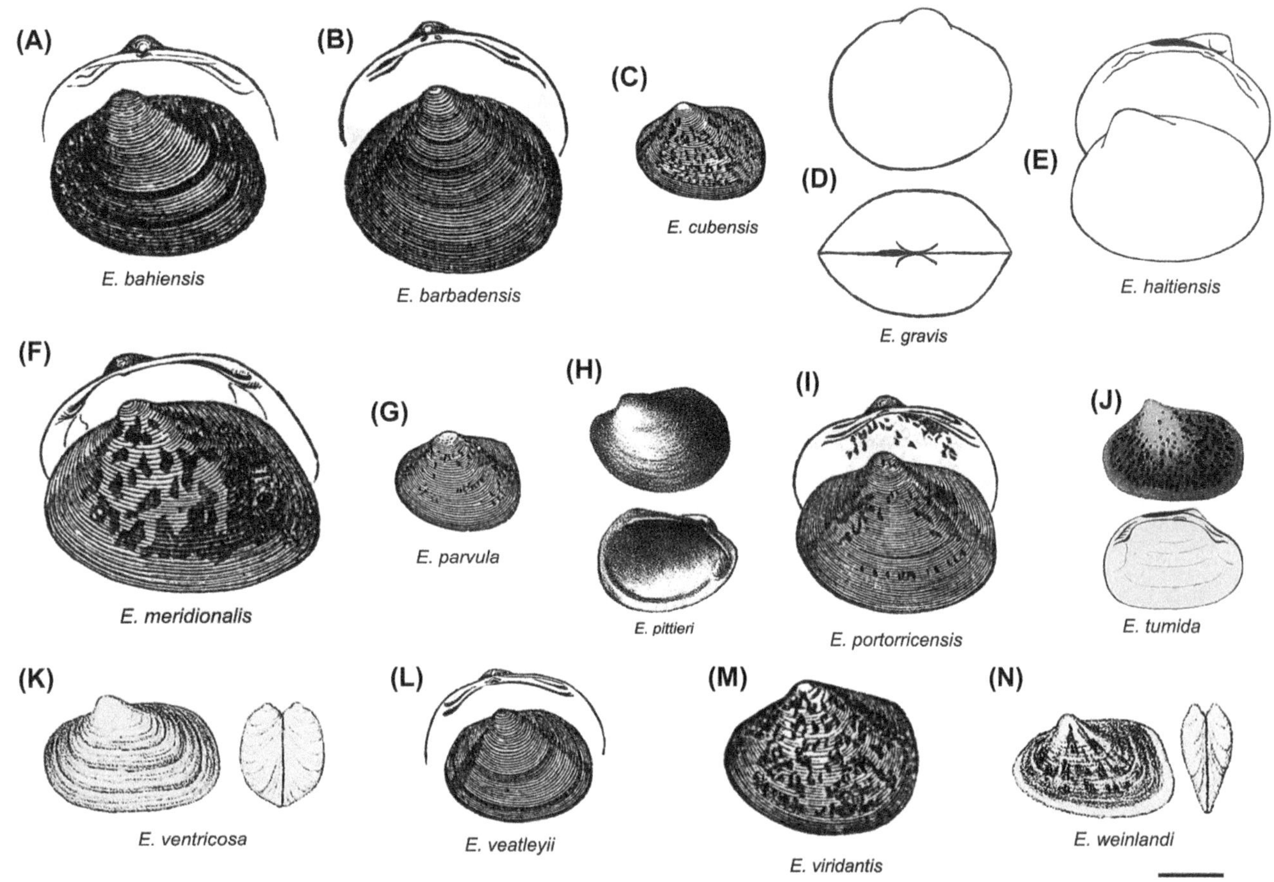

FIGURE 11.117 Species of *Eupera* (Sphaeriidae) from South and Central America: (A–C), (F), (G), (I), (L) and (M): monograph by Prime (1865); (H), (J), (K), (N): from original descriptions; (E), (F): redrawn from original descriptions. Scale bars: (A–D), (F–N) = 2 mm; (E) = 1 mm.

4′ Shell low, dorsal margin somewhat parallel with ventral margin 6

5(4) Periostracum lamellate, adult shell > 5 mm 7

5′ Periostracum smooth, adult shell < 4 mm (Fig. 11.118 I) .... *Eupera iguazuensis* Ituarte, 1989
[Argentina: Misiones, Salto Dos Hermanas, Iguazú Falls]

6(4) Umbos relatively high, well outstanding from dorsal margin, pointed; adult shell < 6 mm (Fig. 11.118 D) .... *Eupera simoni* (Jousseaume, 1889)
[Venezuela: Caracas; Orinoco Basin. Brazil: Roroaima, Rio Uraricoera; Amazonas Basin. Peru: Madre de Dios River Basin]

6′ Umbos low, slightly surpassing dorsal margin, rounded; adult size ≤ 6.75 mm (Fig. 11.118 B) .... *Eupera doellojuradoi* Klappenbach, 1962
[Uruguay: Colonia, Puerto Platero; Salto, Salto Chico]

7(5) Dorsal margin forming a marked angle at the joining posterior margin, posterior margin short, subrect (Fig. 11.118 E) .... *Eupera klappenbachi* Mansur & Veitenheimer, 1975
[Brazil: Rio Grande do Sul, Guaiba River Basin, Santa Catarina. Argentina: streams related to lower Paraná River Basin]

7′ Dorsal margin evenly arcuate, posterior margin curved, not forming angle with posterior end, posterior end evenly curved (Fig. 11.118 F) .... *Eupera primei* Klappenbach, 1967
[Peru: Pequino, near Ucayali River]

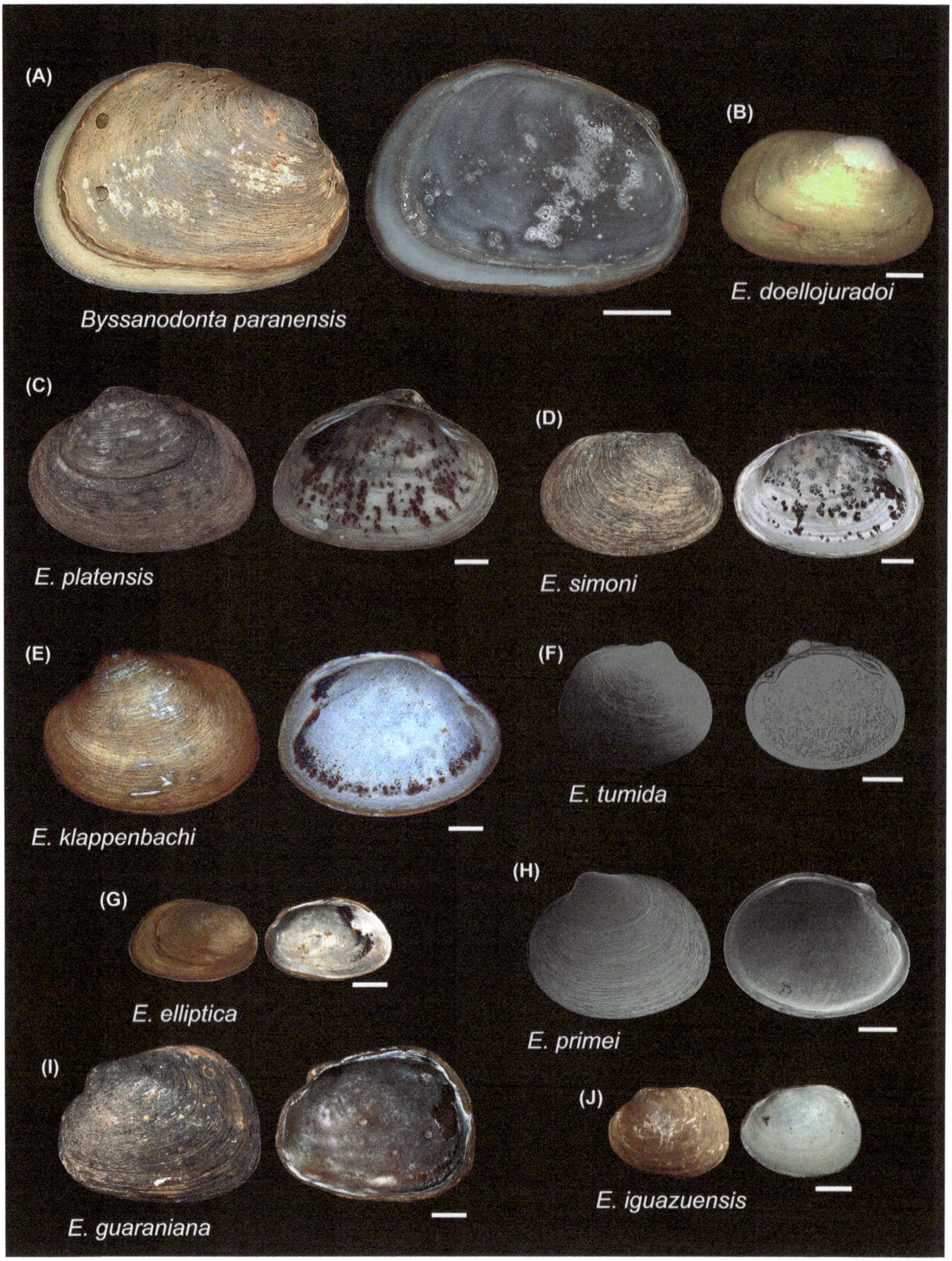

FIGURE 11.118 South American species of *Byssanodonta* and *Eupera* (Sphaeriidae): (A) *Byssanodonta paranensis*, outer and inner views of right and left valves; (B) *Eupera doellojuradoi*, outer view of a right valve; (C) *E. platensis*, outer and inner views of left valves; (D) *E. simoni*, outer and inner views of right valves; (E) *E. klappenbachi*, outer and inner views of left valves; (F) *E. tumida*, outer and inner view of right valves; (G) *E. elliptica*, outer and inner views of right valves; (H) *E. primei*, outer and inner views of left valves; (I) *E. guaraniana*, outer and inner views of left and right valves; (J) *E. iguazuensis*, outer and inner views of left and right valves. Scale bars: (A) = 20 mm, (B–I) = 10 mm.

Key to Unionida

Maria Cristina Dreher Mansur
Grupo de Pesquisa do CNPq, Biodiversidade de Moluscos Continentais, Museu de Ciências Naturais, Porto Alegre, Brazil

Igor Christo Miyahira
Universidade Federal do Estado do Rio de Janeiro, Rio de Janeiro, Brazil

Janine Oliveira Arruda
Grupo de Pesquisa do CNPq, Biodiversidade de Moluscos Continentais, Museu de Ciências Naturais, Porto Alegre, Brazil

Thiago Nunes Antoniazzi
Universidade Federal do Rio Grande do Sul, Porto Alegre, Brazil

Daniel Mansur Pimpão
Instituto Brasileiro do Meio Ambiente e dos Recursos Naturais Renováveis, Brasilia, Brazil

The species displayed in the keys are partially based on the ists of Pereira et al. (2014) and on the catalog of Haas (1969). Within the Hyriidae, we grouped species into tribes following Parodiz & Bonetto (1963), Bonetto (1967), and updated the Rhipidodontini according Graf & Cummings (2007) and Santos-Neto et al. (2016). Santos-Neto et al. (2016), based on COI nucleotide sequences, support the validity of *Callonaia* within Castaliini. Species of *Castalina* (cataloged by Haas, 1969) were placed in *Castalia*, due to the variability and inconsistency of interspecific and intergeneric characters. Species distribution was compiled from Pereira et al. (2014) and updated using information from Musselp (Graf & Cummings, 2019).

The following acronyms are used on this key: ANSP (Academy of Natural Sciences of Philadelphia, Philadelphia); CIAMT-Biv (Universidade Federal do Mato Grosso, Cuiabá); CMNH (Carnegie Museum of Natural History, Pittsburgh); CZPB (Coleção de Zoologia Paulo Bührnheim da Universidade Federal do Amazonas, Manaus); INPA (Instituto Nacional de Pesquisa da Amazonia, Manaus); MCN (Museu de Ciências Naturais de Porto Alegre, Porto Alegre); MCP (Museu de Ciência e Tecnologia da Pontifícia Universidade Católica do Rio Grande do Sul, Porto Alegre); MNHG (Muséum d'Histoire Naturelle Genève, Geneva); MNHM (Museo Nacional de Historia Natural, Montevideo); MNHN (Muséum National d'Histoire Naturelle, Paris); MLP (Museo de La Plata, La Plata); MP (Mussel Project website); MZL coll Charpentier (Musée Cantonal de Zoologie, Lausanne); MZSP (Museu de Zoologia da Universidade de São Paulo, São Paulo); NBCN (Naturalis Biodiversity Center Leiden, Leiden); NHNUK (British Natural History Museum, London); SMF (Senckenberg Museum, Frankfurt); SMNS (Stuttgart State Museum of Natural History, Stuttgart); UERJ (Universidade do Estado do Rio de Janeiro, Rio de Janeiro); USNM (National Museum of Natural History - Smithsonian Institution, Washington); UZH (Zoological Museum at the University of Zurich, Zurich); ZMB (Museum für Naturkunde Berlin, Berlin); and ZSM (Zoologische Staatssammlung München, Munich).

Bivalvia: Unionida: Families

1 Shell with pseudocardinal teeth anterior to beaks and lateral teeth posterior to beaks; usually with umbonal sculpture, larval type glochidium (Figs. 11.94, 119–123) .... **Hyriidae [p. 409]**

1′ Shell without lateral teeth; pseudocardinal teeth may be present or absent, when present, up to three, near beaks; without umbonal sculpture; larval type lasidium (Figs. 11.98, 124–128) .... **Mycetopodidae [p. 419]**

Bivalvia: Unionida: Hyriidae: Tribes

1 Shell without wings 2

1′ Shell dorsally winged (Fig. 11.119 A–D) .... **Prisodontini [p. 409]**

2(1) Shell triangular to subtriangular, greatly inflated, carina well developed (Figs. 11.119 E–L, 120 A–G) .... **Castaliini [p. 410]**

2′ Shell discoid, oval, ovoid, rhomboid, or elliptic; carina very low and rounded (Figs. 11.120 H–L, 121–123) .... **Rhipidodontini [p. 411]**

Bivalvia: Unionida: Hyriidae: Prisodontini: Genera

1 Umbo without sculpture 2

1′ Umbo strongly sculptured by large bars, pustules, and/or nodules (Fig. 11.119 C, D) .... ***Triplodon* [p. 410]**

2(1) Shell high, large and heavy; outline triangular to subquadrangular, umbos almost at the middle of the length; posterior margin perpendicular to dorsal border (Fig. 11.119 A) .... *Paxyodon syrmatophorus* (Meuschen in Gronovius, 1781)
[Guyana and Brazil: Amazonas and Tocantins basins]

PHYLUM MOLLUSCA

2′ Shell elongated and moderately inflated; outline triangular-rectangulоid to rhomboid; posterior margin oblique to dorsal border and concave under posterior wing (Fig. 11.119 B) .... *Prisodon obliquus* Schumacher, 1817
[Ecuador, Brazil, and Peru: Amazonas and Tocantins basins]

Bivalvia: Unionida: Hyriidae: Prisodontini: *Triplodon*: Species

1 Umbo strongly sculptured by large convergent oriented bars and pustules, anterior and posterior wings present, the anterior smaller (Fig.11.119 C) .... *Triplodon corrugatus* (Lamarck, 1819)
[Venezuela, Guyana, Brazil, and Peru: Orinoco and Amazonas basins]

1′ Most part of the shell surface sculptured by convergent rows of small nodules; only posterior wing is present (Fig. 11.119 D) .... *Triplodon chodo* Mansur & Pimpão, 2008
[Brazil: Madeira River and other tributaries of the lower Amazonas Basin]

Bivalvia: Unionida: Hyriidae: Castaliini: Genera

1 Shell moderately truncated or with evident carina; surface in general strongly sculptured by radial, parallel bars or commarginal undulations 2

1′ Shell triangular, very high, heavy and obliquely truncated; surface smooth, umbo very high, inflated and without sculpture in adults; carina extremely pronounced, and behind it is a flattened large escutcheon (Fig. 11.119 E) .... *Callonaia duprei* (Récluz, 1842)
[Brazil: Tocantins, Araguaia and Xingu rivers]

2(1) Shell's surface matte, smooth, or sculptured by parallel or radial bars (Figs. 11.119 G–L, 120 A–G) .... ***Castalia*** **[p. 410]**

2′ Shell's surface glossy, with conspicuous and regular commarginal undulations; periostracum reddish-brown; umbonal sculpture with around five short and feeble parallel bars, frequently eroded (Fig. 11.119 F) .... *Castaliella sulcata* (Krauss, 1849)
[Suriname, French Guiana: Maroni River]

Bivalvia: Unionida: Hyriidae: Castaliini: *Castalia*: Species

Castalia species with wide distribution in different hydrographic basins in South America, like *C. ambigua* Lamarck, 1819, *C. inflata* d'Orbigny, 1835, *C. multisulcata* Hupé, 1857 and *C. cordata* Swainson, 1840, are the most difficult to identify due to the wide gradient of morphological variations. Many basins connect springs through in times of flood, which could hinder the speciation process over time. See comments in the introductory part of Chapter 11.

1 Shell relatively fragile, moderately inflated; umbo and carina low and rounded; umbonal sculpture expanded and finely radiated 2

1′ Shell strong, inflated; umbo high, carina sharp to rounded; umbo sculpture by feeble, strong or coarse radial or parallel bars 3

2(1) Shell triangular elongated, inequilateral, ventral margin nearly straight, posterior end near the base; cardinal teeth almost lamellar and lateral teeth nearly straight (Fig. 11.119 G) .... *Castalia orinocensis* Morrison, 1943
[Venezuela and Colombia: Magdalena and Orinoco rivers]

2′ Shell triangular, nearly equilateral; ventral margin rounded, posterior end near the middle of height, posterior pseudocardinal tooth strongly split, and lateral teeth gently curved (Fig. 11. 119 H) .... *Castalia ecarinata* Mousson, 1869
[Colombia: Magdalena Basin]

3(1) Umbonal sculpture with vestigious, feeble, relatively short bars or absent in adults 4

3′ Umbonal sculpture strong, bars elongated; surface matte, dark brown to black 10

4(3) Umbonal sculpture with vestigious or relatively short bars 5

4′ Umbonal sculpture absent in adults; shell large, heavy, with the anterior part very expanded in height and length, and a little upwards; posterior region short strongly carinate and dorsal slope a little extended; beak inflated and projected (Fig. 11.119 I) .... *Castalia orbignyi* (Deville & Hupé, 1850)
[Brazil: Tocantins and lower Amazonas basins]

5(4) Umbonal sculpture with short, few, very thin and low parallel bars 6

5′ Umbonal sculpture with relatively short parallel or radial bars 7

6(5) Shell's surface relatively smooth, pale yellow; carina elevated and sharp, hinge plate large and teeth strong (Fig. 11.119 J) .... *Castalia quadrata* Sowerby, 1867
[According to the species' author the type locality is unknown; the type label mentions "Para" (Brazil); but Simpson (1914) considered that the species is from Guyana]

6′ Shell's surface irregular, commarginally wrinkled, and a little shiny, brown to yellow; carina relatively low and rounded (Fig. 11.119 K) .... *Castalia schombergiana* Sowerby, 1869
[Guyana, Brazil: coastal rivers of the north of South America and Amazonas Basin]

7(5) Umbonal sculpture with short parallel or convergent bars 8

7′ Umbonal sculpture with short radial bars 9

8(7) Carina and flattened escutcheon sculptured by a series of strong, oblique, and curved parallel bars (Fig. 11.119 L) .... *Castalia undosa* Martens, 1885
[Brazil: Upper Paraná Basin]

8′ Carina and dorsal slope sculptured by thin irregular oblique wrinkles (Fig. 11.120 A, arrows) .... *Castalia nehringi* (Ihering, 1891)
[Brazil: Upper Paraná Basin]

9(7) Umbonal sculpture with few, wide, and rather coarse radial bars, separated by narrow grooves; carina rounded with few oblique wrinkles; dorsal slope separated from carina by a deep radial groove (Figs. 11.94, 120 B, arrow) .... *Castalia martensi* (Ihering, 1891)
[Uruguay and south Brazil: lower Uruguay River, South Atlantic coastal rivers basins]

9′ Umbonal sculpture with few wide low bars, separated by prolongated wide grooves; carina rounded (Fig. 11.120 C) .... *Castalia psammoica* (d'Orbingy, 1835)
[Brazil, Argentina: Uruguay and Rio de la Plata basins]

10(3) Shell inflated, anterior part short, umbo inflated, conspicuous; umbonal sculpture with relatively strong and elongated bars 11

10′ Shell inflated; anterior region slightly projected but narrow, tapering above and below; umbo low, carina not sharp; sculpture tending to be radial, bars in front forming distally a zig-zag arrangement (Fig. 11.120 D) .... *Castalia multisulcata* Hupé, 1857
[Brazil: Amazonas and Paraguay basins. Note: Hupé (1857) mentions Brazil, but some syntypes label mentions "rio Mondégo," which is a branch of the Paraguay River near Corumbá at Mato do Grosso do Sul, Brazil, also near the border with Bolivia and Paraguay]

11(10) Shell's carina not sharp; sculpture just to the half of the shell height or more, tending to be radial, bars distally forked and flattened on the anterior region 12

11′ Shell's carina very sharp; anterior region short, escutcheon flattened, sculpture expanding through the disk, straight parallel and protruding bars, surface brown to black, matte (Fig. 11.120 E) .... *Castalia inflata* d'Orbigny, 1835
[Brazil, Bolivia, Paraguay, Argentina, Uruguay: Amazonas, Tocantins, Plata basins]

12(11) Shell medium sized, escutcheon flattened, umbo inflated but low (Fig. 11.120 F) .... *Castalia ambigua* (Lamarck, 1819)
[South America, from the Guyana to Plata Basin]

12′ Shell very large, escutcheon with a little expanded dorsal slope, umbo inflated and high, periostracum brown matte, showing ventrally strong commarginal undulations (Fig. 11.120 G) .... *Castalia cordata* Swainson, 1840
[Guyana, Brazil: Central Amazonas Basin and coastal rivers of north of South America]

Bivalvia: Unionida: Hyriidae: Rhipidodontini: Genera

There are two Rhipidodontini genera, which differ mainly on the glochidium larva. The larvae of some species of this tribe are not known, representing an unresolved taxonomic problem. Thus, provisionally Rhipidontini with unknown glochidia were not included in the key, but placed in the genus *Diplodon*, in a separate list with respective figures (Table 11.5; Fig. 11.123 E–K)

1 Glochidium without hooks (Figs. 11.120 H–L, 121 A–F) .... ***Rhipidodonta* [p. 411]**

1′ Glochidium with hooks (Figs. 11.121 G–L, 122, 123) .... ***Diplodon* [p. 414]**

Bivalvia: Unionida: Hyriidae: Rhipidodontini: *Rhipidodonta*: Species

1 Shell's outline discoid 2

1′ Shell's outline from ovoid to oval and from elliptic-oval to elliptic-rhomboid 4

2(1) Shell strong; umbo not in central position; pseudocardinal teeth double in each valve; posterior pseudocardinal tooth robust and split 3

2′ Shell not heavy; umbo inflated, almost central; pseudocardinal teeth elongated and lamellate, only one at left valve and double at right valve, lateral teeth lamellate; periostracum greenish-brown (Fig. 11.120 H) .... *Rhipidodonta variabilis* (Maton, 1809)
[Argentina, Uruguay: Rio de la Plata Basin]

3(2) Shell large; carina rounded and separated from dorsal slope by a deep radial groove; umbonal sculpture radial and expanded over the umbonal region; wrinkles can be present on the posterior region of the shell; periostracum light brown to brown (Fig. 11.120 I) .... *Rhipidodonta paranensis* (Lea, 1834)
[Brazil, Paraguay, Argentina, and Uruguay: Plata Basin]

3′ Shell's outline discoid (rarely ovoid), posterior margin tending to be oblique truncated; carina rounded and low without posterior groove behind; umbonal sculpture short, often with continuing rows of low wrinkles over the central part of the valves; periostracum brown to lustrous black (Fig. 11.120 J) .... *Rhipidodonta decepta* (Simpson, 1914)
[Brazil: Lower Jacuí Basin]

FIGURE 11.119 External and partial internal view of mussel species of Hyriidae: (A) *Paxyodon syrmatophorus* (INPA 1187); (B) *Prisodon obliquus* (SMNS ZI0076332); (C) *Triplodon corrugatus* (MCN 38); (D) *Triplodon chodo* (CZPB 12); (E) *Callonaia duprei* (INPA 1656); (F) *Castaliella sulcata* (SMNS ZI0076336); (G) *Castalia orinocensis* (MCN 6235); (H) *C. ecarinata* (MP/SMF/UZH 350035 syntype); (I) *C. dorbignyi* (MCN 41947); (J) *C. quadrata* (NHMUK 18491.5-5); (K) *C. schombergiana* (MCN 8441); (L) *C. undosa* (MCN 7689) with detail of the escutcheon's sculpture on the right. Scale bars = 10 mm. (A–G) (I), (K) and (L) Photographs and edition by the authors of Unionida; (H) MP/SMF/UZH with permission; (J) copyright NHMUK.

4(1) Shell's outline elliptic-oval, elliptic-rhomboid, and oval 5

4′ Shell's outline ovoid; anterior part tapering above and below, but a little projected upwards; periostracum dark brown (Fig. 11.120 K) .... *Rhipidodonta peraeformis* (Lea, 1860) [Argentina, Uruguay: Uruguay Basin]

5(4) Shell's outline oval 6

PHYLUM MOLLUSCA

FIGURE 11.120 External and partial internal view of mussel species of Hyriidae: (A) *Castalia nehringi* (MCN 7688); (B) *C. martensi* (MCN 41948); (C) *C. psammoica* (NHMUK 1854.12.4.849); (D) *C. multisulcata* (MP/MNHN Paris Syntype 0285); (E) *C. inflata* (INPA 316); (F) *C. ambigua* (MP/MNHN Paris IM-2000-27235 holotype); (G) *C. cordata* (NHMUK 1965197); (H) *Rhipidodonta variabilis* (SMF 11427); (I) *R. paranensis* (ZMB - Paetel collection); (J) *R. decepta* (MCN 41329); (K) *R. peraeformis* (USNM 85774); (L) *R. burroughiana* (MNHN Montevideo - Taguató 686). Arrow in (A) shows oblique wrinkles on carina and dorsal slope. Arrow in (B) shows a deep radial groove that separate dorsal slope from carina. Scale bar = 10 mm. (A), (B), (E), (H–J and (L) Photographs and edition by the authors of Unionida; (C) and (G) copyright NHMUK; (D) and (F) MP/MNHN Paris with permission; (K) USNM with permission.

5′	Shell's outline elliptic-oval to elliptic-rhomboid	8
6(5)	Shell < 5 cm	7
6′	Shell > 5 cm; ventral margin convex, more expanded after the middle of length; posterior margin pointed with the distal extremity near the middle of the shell's height; sculpture radial and restricted to umbonal region (Fig. 11.120 L) .... *Rhipidodonta burroughiana* (Lea, 1834) [Brazil, Argentina, and Uruguay: Plata Basin]	

TABLE 11.5 Mollusca: Bivalvia: Unionida: Hyriidae: Rhipidodontini with unknown glochidia, provisionally included as *Diplodon*.

Species		Distribution
Diplodon fontainianus (d'Orbigny, 1835)	11.123 E	Brazil: East Atlantic coastal rivers basins
Diplodon. funebralis (Lea, 1860)	11.123 F	Brazil, Uruguay: Lower Uruguay Basin
Diplodon greeffeanus (Ihering, 1893)	11.123 G	Brazil: Upper Paraná Basin
Diplodon rotundus Spix in Wagner, 1827	11.123 H	Brazil: Southern Brazil
Diplodon solisianus (d'Orbigny, 1835)	11.123 I	Argentina, Uruguay: Lower Plata Basin
Dipldon uruguayensis (Lea, 1860)	11.123 J	Brazil, Uruguay: Uruguay Basin
Diplodon voltzi Vernhout, 1914	11.123 K	French Guiana, Suriname, Venezuela: coastal rivers of these countries

7(6) Length ~3 cm; ventral margin expanded, posterior margin with a sharp end; short radial bars in umbonal sculpture; periostracum light brown to brown; ligament robust (Fig. 11.121 A) .... *Rhipidodonta iheringi* (Simpson, 1914)
[Brazil: Lower Jacui Basin]

7′ Pseudocardinal teeth short, posterior curved and stronger, posterior margin of the shell with a blunt point at half of the height; umbonal sculpture with very short bars (Fig. 11.121 B), shell outline usually oval, rarely elliptic .... *Rhipidodonta hildae* (Ortmann, 1921)
[Brazil: Lower Jacui Basin]

8(5) Shell's outline elliptic-oval 9

8′ Shell's outline elliptic-rhomboid 10

9(8) Shell heavy and relatively large, posterior pseudocardinal tooth robust, high and a little split behind, posterior margin of the shell gently truncated (Fig. 11.121 C) .... *Rhipidodonta charruana* (d'Orbigny, 1835)
[Brazil, Argentina, Uruguay: Uruguay, Jacuí, and Rio de La Plata basins]

9′ Shell relative heavy, but small; pseudocardinal teeth short, thin, and lamellar; umbonal sculpture composed by few and short bars (Fig. 11.121 D) .... *Rhipidodonta koseritzi* (Clessin, 1888)
[Brazil: Guaíba Lake Basin]

10(8) Shallow adductor muscles scars, sculpture usually covering all or great part of the shell, composed by robust bars without grains; posterior cardinal tooth short and curved (Fig. 11.121 E) .... *Rhipidodonta hylaea* (d'Orbigny, 1835)
[Peru, Ecuador, Brazil, Bolivia, Uruguay, and Argentina: Amazonas and Plata basins]

10′ Deep and large adductor muscle scars, pallial line straight or almost straight, pseudocardinal teeth robust, one to three radial grooves running from umbonal cavity to ventral and/or posterior margin, shell thick, umbonal sculpture composed by very short and thin bars (Fig. 11.121 F) .... *Rhipidodonta garbei* (Ihering, 1910)
[Brazil: Paraíba do Sul and Doce River basins]

Bivalvia: Unionida: Hyriidae: Rhipidodontini: *Diplodon*: Species

Some species needs taxonomic revision. According to Miyahira et al. (2019), *D. dunkerianus* (Lea 1856), and *D. besckeanus* (Dunker, 1848) are synonymous of *D. ellipticus* Spix in Wagner, 1827 (Fig. 11.122 C). *Diplodon expansus* (Krüster, 1856) had its type redescribed (Miyahira et al., 2013) and recently added to the synonymy of *D. multistriatus* (Lea, 1831) (Miyahira et al., 2019). Miyahira et al. (2020) after morphological and taxonomical revision of *D. granosus* (Bruguière, 1782) redescribed the species and restricted its occurrence area.

1 Glochidium hooks with distal stylets (*Diplodon* s.s) 2

1′ Glochidium hooks without distal stylets; shell outline ovoid; teeth robust, right pseudocardinal tooth bifid; periostracum light brown (Fig. 11.121 G) .... *Diplodon* (*Australis*) *solidulus* (Philippi, 1869)
[Chile: southern Chile]

2(1) Shell outlines from ovoid to elliptic-lanceolate, from elliptic-oval to elliptic-rhomboid 3

2′ Shell's outline discoid to oval, more inflated in front than behind, umbo low, dorsal margin oblique and curved, posterior margin rounded; hinge curved, pseudocardinal teeth heavy, compressed and bifid; periostracum brown (Fig. 11.121 H) .... *Diplodon rhombeus* Spix in Wagner (1827)
[Brazil: São Francisco and Amazonas basins]

3(2) Shell's outline oval 4

PHYLUM MOLLUSCA

3′ Shell outlines from ovoid to elliptic-lanceolate, from elliptic-oval to elliptic-rectanguloid and elliptic-rhomboid 6

4(3) Shell large (length usually > 7 cm) 5

4′ Shell short (length ~4 cm), strong, inflated, periostracum black not shining, pseudocardinal teeth small, lamellar, single in left and double in right valve; carina a little raised and rounded, posterior slope narrow; posterior extremity at the middle of the height (Fig. 11.121 I) .... *Diplodon piceus* (Lea, 1860)
[Brazil, Argentina, Uruguay: Uruguay Basin]

5(4) Shell posterior margin lightly truncated with the posterior extremity near the middle of the height; pseudocardinal teeth in general, bifid, lamellar, and long in each valve, the posterior tooth larger, coarser, lower, and poorly split (Fig. 11.121 J) .... *Diplodon delodontus* (Lamarck, 1819)
[Brazil, Paraguay, Argentina, and Uruguay: Lower Plata Basin]

5′ Shell thick, posterior margin obliquely truncated with the posterior extremity near the base; posterior pseudocardinal tooth strong and radially split (Fig. 11.121 K) .... *Diplodon wymanii* (Lea, 1860)
[Brazil, Argentina, Uruguay: Uruguay Basin]

6(3) Shell's outline ovoid 7

6′ Shell's outlines from elliptic-oval to elliptic-rectanguloid and elliptic-rhomboid to elliptic-elongated 8

7(6) Shell ovoid, length > 5 cm, pseudocardinal teeth double in right valve and single in left, very long and near parallel to hinge line; shell very inflated; umbo large but low; umbonal sculpture unknown, posterior slope ornamented by oblique undulated narrow wrinkles; periostracum matte, greenish-black posteriorly and brown at the anterior part (Fig. 11.121 L) .... *Diplodon caipira* (Ihering, 1893)
[Brazil: Upper Paraná Basin]

7′ Shell ovoid, ~ 3 cm length, with anterior region tapering above and below, truncated behind; periostracum yellow to olive brown, matte; umbo strongly sculptured by long radial bars, convergent at the center and forked behind; pseudocardinal teeth small, compressed, and double in each valve (Figs. 11.100 A, 122 A) .... *Diplodon suavidicus* (Lea, 1856)
[Venezuela, Brazil: Orinoco, and Amazonas basins; Upper Paraguay River]

8(6) Shell's outline elliptic lanceolate 9

8′ Shell outlines from elliptic-oval to elliptic-rectanguloid, from elliptic-rhomboid to elliptic-elongated 10

9(8) Shell inflated, relatively strong, umbo large but low; umbonal sculpture with short bars at the center and elongated before and after the central bars; anterior bars slightly curved pointing down or backwards; carina large and rounded; posterior end situated at half height or near the base; shiny periostracum, from black to greenish-black; pseudocardinal teeth double and compressed in each valve, the posterior in general split dorsally; impressions of the anterior adductor muscle deep and relatively large (Fig. 11.122 B) .... *Diplodon aethiops* (Lea, 1860)
[Brazil, Uruguay: South Atlantic coastal rivers basins]

9′ Shell similar to the previous species but more fragile, posterior end situated at half height; periostracum lustrous, from black to brown; umbonal sculpture composed of two short bars at center and longer bars at anterior and posterior ends; pseudocardinal teeth lamellate and very thin; anterior adductor impressions relative short and not deep (Fig. 11.122 C) .... *Diplodon ellipticus* Spix in Wagner, 1827
[Brazil: São Francisco and East Atlantic coastal rivers basins]

10(8) Shell's outline elliptic-oval 11

10′ Shell outlines from elliptic-rectanguloid to elliptic-rhomboid 17

11(10) Umbonal sculpture composed by bars, with or without granules 12

11′ Sculpture composed only by granules, usually covering almost the whole shell; periostracum brown to dark brown (Fig. 11.122) .... *Diplodon granosus* (Bruguière, 1792)
[Venezuela, Guyana, French Guiana, Suriname, Brazil: Orinoco and Amazonas basins, as also coastal rivers between both catchments]

12(11) Umbonal sculpture composed by bars with granules 13

12′ Umbonal sculpture composed by bars without granules 15

13(12) Narrow bars on umbonal sculpture 14

13′ Large bars on umbonal sculpture, extending over the half of the shell surface; dorsal and posterior margins rounded; posterior extremity near the base line (Fig. 11.122 E) .... *Diplodon guaranianus* (d'Orbigny, 1835)
[Ecuador, Bolivia, Paraguay, Argentina: Upper Amazonas and Paraguay basins]

14(13) Bars of umbonal sculpture that cross each other, forming a net pattern, extending from umbo to, sometimes, half of the shell; and often with granules on the sculpture; the bars are usually thin, rarely larger, but thickened where the bars cross each other (Fig. 11.122 F) .... *Diplodon multistriatus* (Lea, 1831)
[Brazil: coastal rivers of Bahia, Espirito Santo, Rio de Janeiro, and São Paulo states; Upper Paraná Basin]

14′ Bars of umbonal sculpture juxtaposed and short, anterior bars around five extended by granular bars, two converging at the center and eight posterior; shell robust and compressed; umbos small slightly protruding the dorsal line; pseudocardinal teeth lamellar and short; periostracum green to dark brown (Fig. 11.122 G) .... *Diplodon paulista* (Ihering, 1893)
[Brazil: Upper Paraná Basin]

15(12) Numerous bars on the umbonal sculpture (10 or more) 16

15′ Few (six to eight) short bars on the umbonal sculpture; anterior bars between two to four, one or two convergent at the center and three posterior; shell outline very variable, from elliptic-oval to elliptic-rhomboid; very small and low umbo; pseudocardinal teeth thin and short (Fig. 11.122 H) .... *Diplodon chilensis* (Gray, 1828)
[Chile, Argentina: Chilean, and Patagonian basins. There is a dubious record for Lima, Peru]

16(15) Umbonal sculpture composed by radial bars of the same size, including the central ones; shell relatively inflated and pseudocardinal teeth bifid, lamellate, and relative elongated, umbo projected above the hinge line (Fig. 11.122 I) ... *Diplodon rhuacoicus* (d'Orbigny, 1835)
[Brazil, Argentina, Uruguay: Paraná, Uruguay, and South Atlantic coastal rivers basins of Southern Brazil and Uruguay]

16′ Umbonal sculpture composed by radial bars of different sizes, the central are convergent and shorter; shell with a large convexity behind the middle of the length; at the internal face, under the mentioned convexity, there is an oblique large depression, as if the surface were pressed by a human thumb; anterior region of the valves thicker than posterior region (Fig. 11.122 J) .... *Diplodon berthae* Ortmann, 1921
[Brazil: South Atlantic coastal rivers basins]

17(10) Shell's outline elliptic-rectanguloid 18

17′ Shell's outline elliptic-rhomboid 20

18(17) Ventral margin of the shell more or less convex 19

18′ Ventral margin slightly concave at the middle of the length; dorsal and ventral margins more convex behind the middle of the length; periostracum black or dark brown, little bright at the center and serous at the borders; umbo low; hinge and ligament very narrow; subumbonal cavity very low; pseudocardinal tooth on right valve bifid and oblique, the posterior larger with a rounded surface bearing some granules, single pseudocardinal on left valve; lateral teeth relative thick (Fig. 11.122 K) .... *Diplodon martensi* (Ihering, 1893)
[Brazil: South and East Atlantic coastal rivers basins from Rio de Janeiro to Rio Grande do Sul states; but never recorded again in RJ after the original description]

19(18) Shell fragile, slightly inflated; periostracum copper brown and matte; pseusocardinal and lateral teeth lamellar (Fig. 11.122 L) .... *Diplodon imitator* Ortmann, 1921
[Brazil: South Atlantic coastal basins at Rio Grande do Sul]

19′ Shell very similar to the anterior species, except that the dorsal and ventral margins are more convex, and the species is relatively more oval and compressed; pseudocardinal teeth are double in both valves and the posterior radially split on right valve (Fig. 11.123 A) .... *Diplodon vicarius* Ortmann, 1921
[Brazil: Only know from the type locality at Ribeira do Iguape River (São Paulo)]

20(17) Shell's outline elliptic-rhomboid, elongated 21

20′ Shell's outline elliptic-rhomboid, short; shell inflated, umbo large but low; surface with a raised and rounded carina ending at the posterior point situated above the base; ventral margin convex more expanded at the middle of the length, sculpture radial and strong, with convergent central bars; pseudocardinal teeth compressed, relative short, double in both valves (Fig. 11.123 B) .... *Diplodon parodizi* Bonetto, 1962
[Brazil, Argentina, Paraguay, Uruguay: Lower Plata Basin]

21(20) Shell strong, inflated; umbo full and elevated above the hinge line, sculpture of short radial bars; posterior carina high ending in a point near the base line, periostracum lustrous, bottle green to black (Fig. 11.123 C) .... *Diplodon parallelopipedon* (Lea, 1834)
[Brazil, Bolivia, Paraguay, Argentina, Uruguay: Madeira River (Upper Amazonas) and Plata Basin]

21′ Shell fragile, compressed; umbo low; expanded umbonal sculpture over the most part of the surface, composed of very thin bars and close to each other forming several V; anterior part of the shell smaller and tapering above; posterior margin truncated (Fig. 11.123 D, with a young shell right above to show the sculpture) .... *Diplodon obsolescens* F. Baker, 1914
[French Guiana, Guyana, Venezuela, Brazil: coastal rivers of north South America and Amazonas Basin]

FIGURE 11.121 External and partial internal view of mussel species of Hyriidae, Tribe Rhipidodontini: (A) *Rhipidodonta iheringi* (MCN 6609); (B) *R. hildae* (SMF 4643); (C) *R. charruana* (MNHN Montevideo 1210); (D) *R. koseritzi* (SMF 11182); (E) *R. hylaea* (MLP 6220); (F) *R. garbei* (SMF 11262); (G) *Diplodon (Australis) solidulus* (MLP 6051); (H) *D. rhombeus* (MCP 8343); (I) *D. piceus* (MCN 30828); (J) *D. delodontus* (MHNG INVE 50565); (K) *D. wymanii* (USNM 84378); (L) *D. caipira* (SMF 3875). Scale bars = 10 mm. (A–I) and (L) Photographs and edition by the authors of Unionida; (J) MHMG INVE with permission; (K) USNM with permission.

FIGURE 11.122 External and partial internal view of mussel species of Hyriidae, Tribe Rhipidodontini: (A) *Diplodon suavidicus* (INPA 195); (B) *D. aethiops* (MCN 31130); (C) *D. ellipticus* (UERJ 10555); (D) *D. granosus* (UERJ 10480); (E) *D. guaranianus* (NHMUK 1854124841 holotype); (F) *D. multistriatus* (MLP 6158); (G) *D. paulista* (SMF 3873); (H) *D. chilensis* (MCN 9826); (I) *D. rhuacoicus* (NHMUK 1854124838 holotype); (J) *D. berthae* (SMF 4642); (K) *D. martensi* (SMF 3929); (L) *D. imitator* (CMNH 619.248). Scale bars = 10 mm. (A-D, F-H, J, K) Photographs and edition by the authors of Unionida; (E) and (I) The Trustees of the NHMUK; (L) CMNH with permission.

FIGURE 11.123 External and partial internal view of mussel species of Hyriidae, Tribe Rhipidodontini: (A) *D. vicarious* (CMNH 619.251); (B) *D. parodizi* (MLP 6564); (C) *D. parallelopipedon* (ZMB - Paetel collection); (D) *D. obsolescens* (MCN 30624), on the right above, detail of a young shell; (E) *D. fontainianus* (ZMB 45976); (F) *D. funebralis* (MNHN Montevideo – no number); (G) *D. greeffeanus* (SMF 3878); (H) *D. rotundus* (ZSM 20041236); (I) *D. solisianus* (ZSM 12288); (J) *D. uruguayensis* (USNM 85166); (K) *Diplodon voltzi* (NBCN 289451). Scale bars = 10 mm. (B–I) Photographs and edition by the authors of Unionida; (A) CMNH with permission; (J) USNM with permission; (K) copyright NBCN.

Bivalvia: Unionida: Mycetopodidae: Genera

1 Shell not like an oyster (anterior and posterior adductor muscle scars present) 2

1′ Shell like an oyster, external surface lamellar, one valve fixed on hard substrate; in adults only posterior adductor muscle scar present (Fig. 11.124 A) .... *Acostaea rivolii* (Deshayes, 1827) [Colombia: Magdalena Basin]

2(1) Anterior part capitate, narrower and bending down in a different angle of the remaining part of the shell 3

2′ Anterior part not capitate 4

3(2) Outline variable and irregular, shell outline resembling a "P"; anterior part immersed and adapted to the space of small gaps of the substrate, or between stones or even wedged in laterite; external shell's surface irregular greenish foliated (Fig.11.124 B) .... *Bartlettia stefanensis* (Moricand, 1856)
[Peru, Brazil, Paraguay: Upper Amazonas and Paraguay basins]

3′ Shell elongated with the anterior part shortened and separated from the rest of the shell by a radial groove that starts at the umbo and descends ventrally; behind the umbo the shell keeps dorsally straight, carinate, and sometimes ventrally concave; external shell's surface regular not foliated, brownish (Fig. 11.124 C) .... *Mycetopodella falcata* (Higgins, 1868)
[Colombia, Ecuador, Brazil, Peru, Bolivia: Upper Amazonas Basin]

4(2) Shell's outline variable, from oval to trapezoid or like a tube to knife-shaped, impression of the posterior retractor muscle united to the posterior adductor muscle scar 5

4′ Shell's outline like a stiletto, wider in front, impression of the posterior retractor muscle separated from the posterior adductor muscle scar (arrow and asterisk respectively in Fig. 11.124 D) .... *Lamproscapha ensiformis* (Spix in Wagner, 1827)
[South America: Orinoco to Plata basins, except Chile, Colombia, French Guiana and Suriname]

5(4) Absence of pallial sinus 6

5′ Presence of pallial sinus (Fig.11.124 G, H) .... ***Leila* [p. 420]**

6(5) Shell with pseudocardinal teeth or denticles on hinge plate 7

6′ Shell without teeth on hinge plate 11

7(6) Hinge with rudimentary teeth represented by small lamella/ae parallel to the hinge line, or one or more low denticles at one valve and, low grooves partially nacreous or formed by brown conchioline on the opposite face 8

7′ Hinge with strong teeth and deep nacreous grooves 9

8(7) Hinge plate with two or more flattened teeth and the respective grooves at the opposite valve, distributed oblique or perpendicularly to the hinge line (arrow in Fig. 11.124 E) .... *Fossula fossiculifera* (d'Orbigny, 1835)
[Brazil, Bolivia, Uruguay: São Francisco, Paraná, Paraguay, and Uruguay basins]

8′ Hinge plate narrow or vestigious, one or two small pseudocardinal denticles at one valve and short compressed lamella/ae on opposed valve, bordered behind and under the beak by a brown elongated oblique shallow groove (arrow) on each valve (Fig. 11.124 I–K) .... ***Tamsiella* [p. 420]**

9(7) Two or three radial teeth 10

9′ One nacreous rounded or compressed tooth in each valve, rarely split (Fig. 11.125 A–H) .... ***Monocondylaea* [p. 422]**

10(9) Two teeth on each valve (Fig. 11.124 F) .... *Haasica balzani* (Ihering, 1893)
[Brazil: Paraguay River]

10′ Three teeth on each valve, the anterior one stronger (Fig. 11.124 I–K) .... ***Diplodontites* [p. 422]**

11(6) Shell's outline elongated like a tube, length > 2x height, foot very long with a distal intumescence (Fig.11.126 A–C) .... ***Mycetopoda* [p. 424]**

11′ Shell outline variable but length < 2× height, foot linguiform or securiform, without distal intumescence (Figs. 11.100, 126 D–J, 127, 128) .... ***Anodontites* [p. 424]**

Bivalvia: Unionida: Mycetopodidae: *Leila*: Species

1 Shell's outline discoid, inflated, subumbonal cavity very deep, situated at or very close to the center of the dorsal line, which is a little sinuous (Fig. 11.124 G) .... *Leila esula* (d'Orbigny, 1835)
[Guyana, Venezuela, Brazil, Colombia, Peru, Bolivia, Paraguay: Amazonas and Paraguay basins]

1′ Shell's outline ovoid, subumbonal cavity low, anterior to the middle of the dorsal line, which is nearly straight (Figs. 11.100 B, 124 H) .... *Leila blainvilliana* (Lea, 1834)
[Brazil, Paraguay, Uruguay, Argentina: Paraná, Paraguay, Uruguay, and Jacuí basins]

Bivalvia: Unionida: Mycetopodidae: *Tamsiella:* Species

1 Shell's outline elliptic to elliptic-rhomboid 2

1′ Shell oval, rounded, heavy, compressed, large and rounded posteriorly, tapering in front; umbo low, lamella on left valve compressed but high (arrow in Fig. 11.124 I), periostracum thick, matte, greenish-black .... *Tamsiella amazonica* Bonetto, 1972
[Brazil, Bolivia, Peru: Upper Amazonas Basin]

2(1) Shell relatively small but heavy, inflated; umbo relatively large, posterior region higher, oblique truncated; tooth (arrow in Fig. 11.124 J) and lamella evident, periostracum thick, matte, lamellose, pale greenish to brown .... *Tamsiella tamsiana* (Dunker, 1858)
[Venezuela, Colombia: Orinoco Basin]

PHYLUM MOLLUSCA

2′ Shell, elongated; anterior region short, tapering and low; umbo narrow and low; teeth rudimentary, sometimes hard to detect, represented by a denticle on right valve and a short lamella (arrow in Fig. 11.124 K) parallel to the hinge line on left valve; periostracum reticulate, pale yellow to greenish brown .... *Tamsiella schroeteriana* (Lea, 1852)
[Brazil, Peru: Amazonas, Tocantins, Paraguay basins]

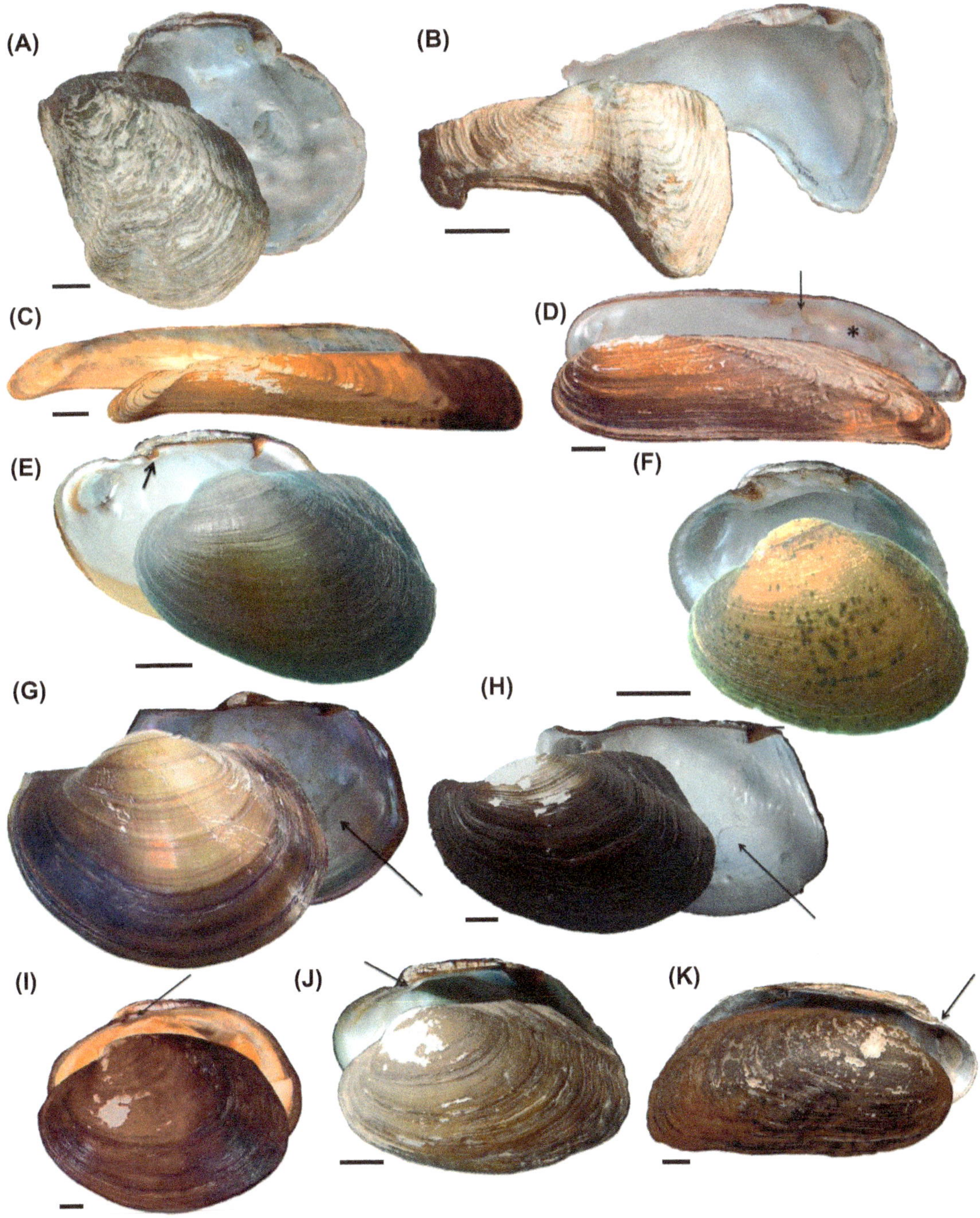

FIGURE 11.124 External and partial internal view of mussel species of Mycetopodidae s.s.: (A) *Acostaea rivolii* (MCN 7904); (B) *Bartlettia stefanesis* (MCN 30751); (C) *Mycetopodella falcata* (INPA 1606); (D) *Lamproscapha ensiformis* (MCN 9906); (E) *Fossula fossiculifera* (CIAMT-Biv 262); (F) *Haasica balzani* (CIAMT-Biv 269); (G) *Leila esula* (INPA 905); (H) *Leila blainvilliana* (MCN 2337); (I) *Tamsiella amazonica* (MZUSP 36610); (J) *Tamsiella tamsiana* (INPA 1461); (K) *Tamsiella schoeteriana* (ZSM). Arrow in (D): posterior retractor muscle impression, in (G) and (H): the pallial sinus, in (I) and (J): compressed lamella; asterisk in (D): posterior retractor muscle impression. Scale bars = 10 mm. (A–J) Photographs and edition by the authors of Unionida.

PHYLUM MOLLUSCA

Bivalvia: Unionida: Mycetopodidae: *Monocondylaea:* Species

Monocondylaea costulata (Moricand, 1858) needs taxonomic revision; *Iheringella isocardioides* (Lea, 1856) and *I. semisulcata* (Adams, 1870) are probable synonyms of this species

1 Shell's outline trapezoid or like a convex pentagon 2

1′ Shell's outline discoid, oval, ovoid, or subtriangular 3

2(1) Shell outline trapezoid; ventral margin convex at the middle, with a concavity between this convexity and carina (asterisk in Fig. 11.125 A), where a radial and shallow depressions ends; posterior end at or near the base; cardinal tooth robust in general split in great extension; periostracum lamellose, olive to pale brown, matte, with ornamentation forming dark rows of scales like radial ribs (arrow in Fig. 11.125 A) ..... *Monocondylaea costulata* (Moricand, 1858)
[Brazil, Peru, Bolivia: Amazonas Basin]

2′ Shell's outline like a convex pentagon; ventral margin convex at the middle, with a shallow radial depression behind and in front of this convexity; posterior end at the middle of the shell height; cardinal tooth not split and periostracum without rows of scales (Fig. 11.125 B) ..... *Monocondylaea guarayana* d'Orbigny, 1835
[Bolivia: Amazonas Basin; Paraguay: Upper Paraguay River]

3(1) Shell's outline discoid 4

3′ Shell's outline oval, ovoid to subtriangular 5

4(3) Shell fragile, relatively small and compressed, nearly equilateral; dorsal slope squarely subtrucated, periostracum pale green to light brown, nacre bluish-white (Fig. 11.125 C) ..... *Monocondylaea franciscana* (Moricand, 1836)
[Brazil: São Francisco River]

4′ Shell relatively strong, large, and compressed, inequilateral; dorsal slope rounded, periostracum dark olive, nacre in general greenish-white with large green spots (Fig.11.125 D) ..... *Monocondylaea corrientesensis* d'Orbigny, 1835
[Brazil, Paraguay, Argentina, Uruguay: Plata Basin and South Atlantic coastal rivers of Brazil and Uruguay]

5(3) Shell's outline oval to ovoid 6

5′ Shell's outline subtriangular to subtriangular rectanguloid 7

6(5) Shell's outline oval, rounded, strong, solid, inflated; umbo inflated and high, beak prosogyrous, lunula large and short; hinge thick, tooth split radially in nodules, base convex rounded (Fig. 11.125 E) ..... *Monocondylaea jaspidea* (Hupé in Castelnau, 1857)
[Guyana, Venezuela, Brazil: Amazonas Basin]

6′ Shell's outline ovoid, elongated, shell relatively small and compressed, umbo low, beak centrogyrous without furrow, lunula not evident, hinge narrow with one tooth in each valve, base nearly straight (Fig. 11.125 F) ..... *Monocondylaea parchappii* d'Orbigny, 1835
[Brazil, Argentina: Paraná and Paraguay basins]

7(5) Shell's outline subtriangular, very inflated; posterior margin high and vertically truncated forming a right angle with the dorsal line; umbo high, beak prosogyrous; lunula short and large; nacre purple (Fig. 11.125 G) ..... *Monocondylaea paraguayana* d'Orbigny, 1835
[Brazil, Paraguay, Argentina, Uruguay: Plata Basin and South Atlantic coastal rivers of Brazil and Uruguay]

7′ Shell's outline rectanguloid (sometimes subtriangular), inflated; posterior margin obliquely truncated; umbo high and inflated; beak prosogyrous divided by a short furrow, lunula evident; nacre white to greenish (Figs. 11.98, 125 H) (the shell form varies along the species distribution from west to east; the subtriangular-rectanguloid outline from Plata Basin to elliptic-oval at Jacuí and Patos-Mirim drainages; the beaks' central furrow is present in all young exemplars or bigger shells with preserved umbo) ..... *Monocondylaea minuana* d'Orbigny, 1835
[Brazil, Argentina, Uruguay: Plata and Patos-Mirim basins]

Bivalvia: Unionida: Mycetopodidae: *Diplodontites:* Species

1 Shell's outline elliptic similar to an isosceles triangle 2

1′ Shell's outline elliptic-ovoid, short; base straight at the middle; frontal part tapering but pointing up; teeth relatively short and small; periostracum green (Fig. 11.125 I) ..... *Diplodontites olssoni* Pilsbry, 1933
[Colombia: Sino River]

(1) Shell strong; dorsal and ventral margins regularly arched, little flattened dorsally in front; umbo inflated but barely overlying the dorsal margin; periostracum thick, matte, pale yellow to olive green, with around 10 radial green rays (Fig. 11.125 J) ..... *Diplodontites cookei* Marshall, 1922
[Colombia, Ecuador, Peru: Upper Amazonas River]

2′ Shell heavy, ventrally flattened; dorsally arched and higher behind the umbo; umbo inflated and high; surface with green rays on the dorsal slope (Fig. 11.125 K) ..... *Diplodontites pilsbryana* Olsson & Wurtz, 1951
[Colombia: Magdalena Basin]

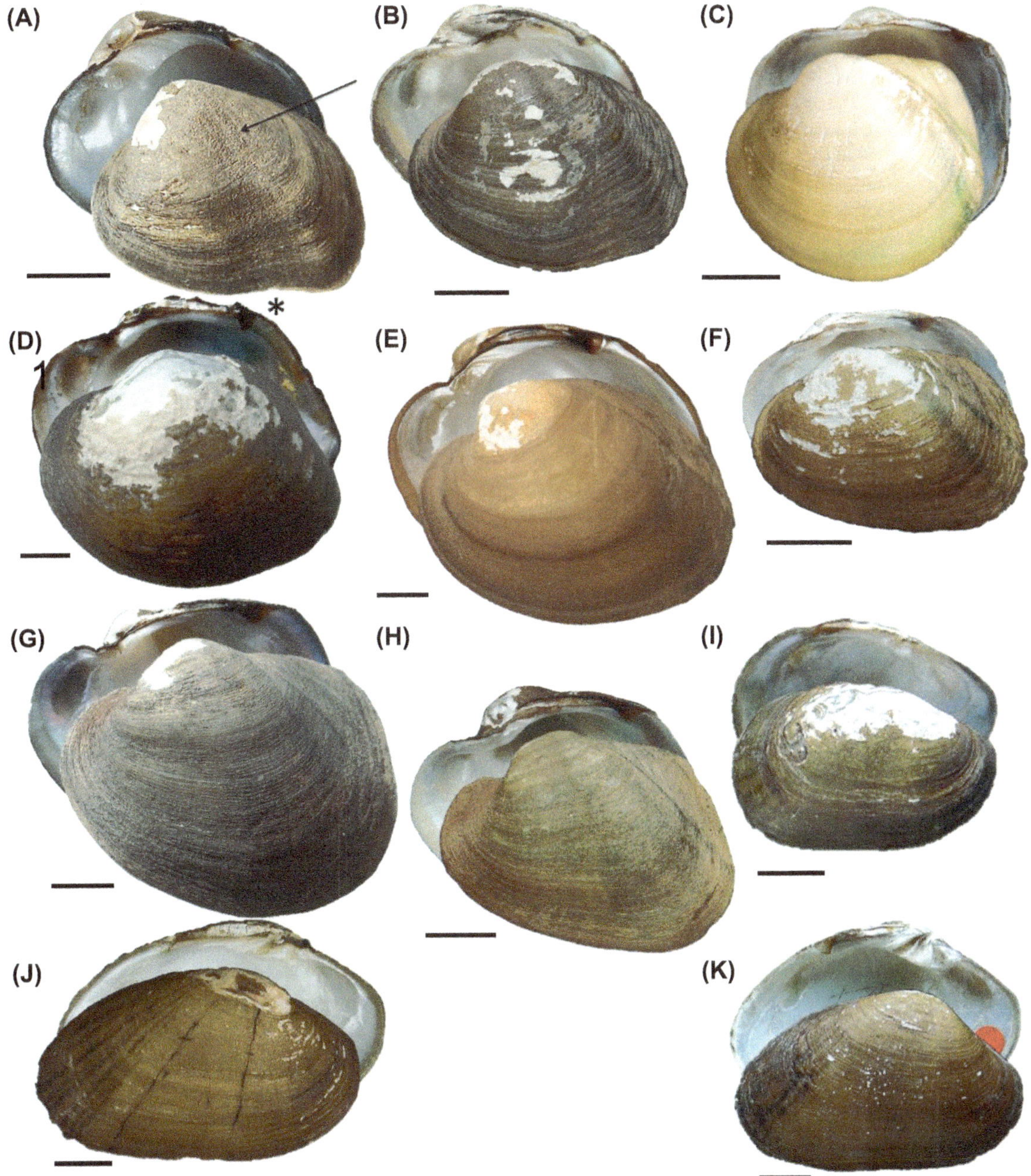

FIGURE 11.125 External and partial internal view of mussel species of Mycetopodidae s.s. (A) *Monocondylaea costulata* (MCN 41949); (B) *M. guarayana* (NHMUK 1854.9.4.17); (C) *M. franciscana* (MCN 31350); (D) *M. corrientesensis* (MCN 30762); (E) *M. jaspidea* (MZUSP 15348); (F) *M. parchappii* (MCN 7812); (G) *M. paraguayana* (MCN 31121); (H) *M. minuana* (MCN 30820); (I) *Diplodontites olssoni* (MP/ANSP 160187); (J) *D. cookei* (MP/USNM 341473); (K) *D. pilsbryana* (MP/ANSP 187180 holotype). Arrow in (A): periostracum ornamentation forming dark rows of scales like radial ribs; asterisk in (A): ventral concavity before carina. Scale bars = 10 mm. (A), (C–H) Photographs and edition by the authors of Unionida; (B) copyright NHMUK; (I, K) MP/ANSP with permission; (J) MP/USNM with permission.

Bivalvia: Unionida: Mycetopodidae: *Mycetopoda:* Species

1 Shell scarcely inflated, inequilateral, with the beaks at circa 25%−30% of the shell length, with or without a moderate anterior ventral gap (arrows in Fig. 126 B, C) 2

1′ Shell inflated, near or exactly equilateral, with the beaks at circa 43% of the shell length (arrow in Fig. 11.126 A); anterior half cylindrical with a rounded frontal intervalvular gap; foot elongated and projects anteriorly (Fig. 11.126 A) .... *Mycetopoda soleniformis* d'Orbigny, 1835
[Venezuela, Colombia, Brazil, Peru, Bolivia, Paraguay, Argentina, Uruguay: Plata and Amazonas basins]

2(1) Shell's ventral border slightly convex, posterior margin rounded, foot able to be completely withdrawn inside shell (Fig. 11. 126 B) .... *Mycetopoda legumem* (Martens, 1888)
[Brazil, Paraguay, Argentina, Uruguay: Southern Plata basin and South Atlantic coastal rivers of Brazil and Uruguay]

2′ Shell's ventral border oblique and divergent in relation to the dorsal line, posterior margin oblique, truncated, foot very long and oblique, not able to be withdrawn inside shell (Figs. 11.100 C, 126 C) .... *Mycetopoda siliquosa* (Spix in Wagner, 1827)
[Costa Rica to South America, east of Andes to the Plata Basin]

Bivalvia: Unionida: Mycetopodidae: *Anodontites:* Species

Several *Anodontites* species need taxonomic revision. Haas (1969) considered the species described for Central America, *A. lenticularis* (Lea, 1868), *A. montesianus* (Sowerby, 1879), and *A. depexus* (Martens, 1900) as synonyms or nearly related to *Anodontites inaequivalvis* (Lea, 1868); and also Haas (1969) associated *A. montezuma* (Lea, 1841) from Central America and *A. aroanus* (Baker, 1930), from Venezuela, to *A. trigonus* (Spix in Wagner, 1827) synonym of *A. elongatus* (Swainson, 1823) both from South America. The shell of *Anodontites cylindraceus* (Lea, 1836) is very similar to *A. guanarensis* Marshall, 1927, from Venezuela, also related to *A. elongatus* (Swainson, 1823) (Fig. 11.128 E). A morphological comparative study is needed between these species.

Anodontites carinatus (Dunker, 1858) is similar to *Tamsiella schroeteriana* (Fig. 11.124 K). *Anodontites puelchanus* (d'Orbigny, 1835) is sometimes related to *A. elongatus* or to *A. patagonicus* (Lamarck, 1819). The following nominal species: *A. amazonensis* (Lea, 1860), *A. georginae* (Griffith & Pidgeon, 1833), *A. guanarensis* Marshall, 1927, *A. irisans* Marshall, 1926, *A. leotaudi* (Guppy, 1864), *A. mortonianus* (Lea, 1834), *A. trautwinianus* (Lea, 1852), and *A. trigonus* (Spix in Wagner, 1827) were considered as synonyms of *Anodontites elongatus* that should be revised. *Anodontites tehuantepecensis* (Crosse & Fischer, 1893) is probably a synonym of *A. montezuma* (Lea, 1841)

1 Shell's outline from discoid, oval, ovoid, elliptic-lanceolate, elliptic-oval to reniform 2

1′ Shell's outline trapezoid 23

2(1) Shell's outline discoid to oval 3

2′ Shell's outline from ovoid to reniform 7

3(2) Outline oval 4

3′ Outline discoid, shell inflated, umbo inflated and projecting above the hinge line; hinge line near straight, periostracum smooth, the color varies from greenish-dark brown to pale straw color, prismatic border large, carina rounded and not pronounced, with green rays. (Fig. 11.126 D) .... *Anodontites trapezeus* (Spix in Wagner, 1827)
[Venezuela, Brazil, Bolivia, Paraguay, Argentina, Uruguay: Amazonas, Plata and São Francisco basins. Note: The form is variable, being ovoid with a strong shell in the south to discoid and thinner shell in the north of South America]

4(3) Shell oval, length ~ 0 mm, with narrow prismatic border 5

4′ Shell oval, length ~ 60 mm, strong and with large prismatic border 6

5(4) Shell relatively fragile, little convex, inequilateral and inequivalve with the left valve a little more inflated than right, beak inflated but low, periostracum yellow and green, matte (Fig. 11. 126 E) .... *Anodontites montezuma* (Lea, 1841)
[From Mexico to Venezuela]

5′ Shell moderately solid and inflated, inequilateral, slightly inequivalve; umbo inflated but low; ligament imbedded in a sort of groove, which extends to the anterior point of the shell, posterior ridge very low, rounded; periostracum lamellose posteriorly, at shell center it is bright with many radial lines composed of rows of small ribs in arcs from umbo to base; color mostly yellow, varying from greenish-tawny to straw color (Fig. 11.126 F) .... *Anodontites inaequivalvis* (Lea, 1868)
[Guatemala, Nicaragua]

6(4) Shell's umbos scarcely inflated, color pale yellow and periostracum lamellose posteriorly, at shell center it is bright with many radial lines composed of rows of small ribs in arcs from umbo to base (Fig. 11.126 G) .... *Anodontites pittieri* Marshall, 1922
[Venezuela: known only from the type locality "Guaremales" (?)]

6′ Periostracum with radiating, dichotomous or simple lines of narrow "V"s of green rays and, two long rays on the dorsal slope; nacre pale pink to white, frequently with black zig-zag lines near pallial line and or at muscles scars (Fig. 11.126 H) .... *Anodontites ferrarisi* (d'Orbignyi, 1835)
[Brazil, Argentina, Uruguay: Lower Uruguay and Paraná basins]

7(2) Shell's outline ovoid 8

7′ Shell's outline from elliptic to elliptic reniform 9

8(7) Shell outline ovoid with the posterior region vertical, sometimes obliquely truncated or rounded; shell small and fragile; beaks low; color yellow to greenish or tawny; external surface completely or in great part sculptured by radial juxtaposed rows of arches (detail in Fig. 11.126 I), nacre variable from blue to silver, often with dark lines forming "V"s (Fig. 11. 126 I) .... *Anodontites crispatus* Bruguière, 1792
[French Guiana, Surinam, Guyana, Venezuela, Ecuador, Colombia, Brazil, Peru, Bolívia: Amazonas Basin and other basins of north South America]

8′ Shell short, scarcely inflated, fragile; beak moderately high, periostracum with radial plication and often with series of brown or green rays mainly on the dorsal slope; nacre bluish or greenish tinted (Fig. 11.126 J) .... *Anodontites obtusus* (Spix in Wagner, 1827)
[Brazil: São Francisco and Amazonas basins]

9(7) Shell's outline elliptic (anterior and posterior ends at the middle of the shell height) 10

9′ Shell's outline from elliptic-lanceolate to elliptic-oval 11

10(9) Shell's outline elliptic, varying from elongated to elliptic-rhomboid, subinflated, anterior end rounded and relatively projected, beak small, rather sharp and a little projected over the dorsal line, valves thin, with shallow muscles scars, with dark green periostracum, smooth, shiny and marked with broken rays, nacre blue-gray with dark, zig-zag markings (Fig. 11.127 A) .... *Anodontites lucidus* (d'Orbigny, 1835)
[Brazil, Uruguay, Argentina: Rio de la Plata, Uruguay, and Jacuí-Patos-Mirin basins]

10′ Shell's outline elliptic, with the anterior end narrow and tapering; dorsal outline strongly and evenly curved, beak not projected, umbonal region and hinge large, valves thick with deep anterior muscle scars; periostracum pale at the center and dark brown on the borders, with radial rows of fine mesosculpture, nacre silver to greenish (Fig. 11.127 B) .... *Anodontites napoensis* (Lea, 1868)
[Ecuador: Napo River]

11(9) Shell's outline elliptic-lanceolate 12

11′ Elliptic oval to elliptic-reniform 13

12(11) Shell long varying from elliptic to rhomboid, subsolid, convex, inequilateral, dorsal outline slightly curved, base nearly straight, anterior end narrowed and rounded, posterior end obliquely subtrucate above, rounded below, beak rather low, periostracum with faint, looped wrinkles arranged in radiating rows, scarcely wrinkled behind the middle of the shell, tawny green with delicate, faint, darker rays, somewhat shiny, nacre bluish, violet-tinted and richly iridescent (Fig. 11.127 C) .... *Anodontites cylindraceus* (Lea, 1836)
[Mexico]

12′ Shell elongated; anterior region short, tapering and low; umbo low and narrow, posterior end near the middle of the shell height, carina low but evidenced by a conspicuous ray, periostracum thick, matte, greenish-dark brown (Fig. 11.127 D) .... *Anodontites carinatus* (Dunker, 1858)
[Colombia: Magdalena Basin]

13(11) Shell's outline elliptic-oval 14

13′ Shell's outline elliptic-rectanguloid to elliptic-reniform 18

14(13) Shell's outline elliptic-oval (> 7 cm) with shallow anterior muscle scars 15

14′ Shell's outline elliptic-oval ($\sim$6 cm), with deep anterior muscle scars often lined by many concentric dark lines; anterior region relatively short but high and the posterior region oblique truncated with the posterior end under the middle of the height, dorsal line regularly arched, umbonal region and hinge plate large and thick, periostracum relatively smooth, varying from chestnut to dark olive and frequently with fine radiating green rays (Fig. 11.127 E) .... *Anodontites iheringi* (Clessin, 1882)
[Brazil: Jacuí Basin at Rio Grande do Sul]

15(14) Shell elliptic-oval with relatively narrow prismatic border 16

15′ Shell elliptic-oval with relatively large prismatic border 17

16(15) External surface smooth and bright with many long radial rows of arcs; internal surface with minute muscular impressions at subumbonal cavity and posterior to the anterior adductor muscle impressions, more evident in young exemplars (arrow in Fig. 11.127 F) (form changes from ovoid to elliptic-rhomboid and the color from greenish-dark brown to pale straw color) (Fig.11.127 F) .... *Anodontites tortilis* (Lea, 1852)
[From Costa Rica to Guyana]

16′ External surface matte, with lamellar periostracum in front and behind the shell, color olive green, in youngs pale green, in front and behind umbo is a reticular area organized in radial rows (arrow in Fig. 11.127 G), more ventrally are relatively short radial rows of arches; carina low but conspicuous, eventually a secondary rudimentary carina; internal surface bright, strongly iridescent, mainly from blue-green to pink, prismatic border dark olive colored (Fig. 11.127 G) .... *Anodontites schomburgianus* (Sowerby, 1870)
[Colombia, Venezuela, Guyana, Brazil, Peru: Magdalena, Orinoco, Amazonas, and Tocantins basins]

17(15) Shell solid, bright, inflated, sometimes with a yellow or reddish brown periostracum; internal surface with large prismatic border (arrow in Fig. 11.127 H), mainly at the shell base; without minute muscles impressions in the umbonal cavity or more frontally in youngs; external surface usually in general rayless, with strongly marked commarginal arrest ridges (Fig. 11.127 H) .... *Anodontites patagonicus* (Lamarck, 1819)
[Brazil, Paraguay, Argentina, Uruguay: Plata, Uruguay, and Patos-Mirim basins]

17′ Shell with large prismatic border (arrow in Fig. 11.127 I), rather thin, subinflated; beak lower; anterior end narrower, shorter and lower; nacre purplish (Fig. 11.127 I) .... *Anodontites puelchanus* (d'Orbigny, 1835)
[Argentina: Negro Basin at Patagonia]

18(13) Shell's outline elliptic-rectanguloid 19

18′ Shell's outline elliptic-rhomboid to elliptic-reniform 20

19(18) Shell's outline elliptic-rectanguloid, short, tending to be rhomboid or reniform in older shells; shell strong and heavy; anterior end narrowed, rounded and projecting upward; periostracum lamellate, matte, olive brown to black; nacre grayish-green to blue, with brown spots, strongly iridescent (Fig. 11.128 A) .... *Anodontites tenebricosus* (Lea, 1834)
[Venezuela, Brazil, Bolivia, Peru, Paraguay, Uruguay, Argentina: Amazonas and Plata basins]

19′ Shell's outline elliptic-rectanguloid, extremely elongated tending to be rhomboid-reniform in older shells; shell light and thin, beak low; periostracum lightly lamellous, greenish-brown (Fig. 11.128 B) .... *Anodontites soleniformis* (d'Orbingy, 1835)
[Brazil, Peru, Bolivia, Paraguay, Argentina: Amazonas and Plata basins]

20(18) Shell's outline elliptic-rhomboid 21

20′ Shell's outline elliptic-reniform, shell strong, inflated; valves thick, gaping; deep anterior adductor scars; periostracum of fine mesosculpture formed by fine radiating threads (Fig. 11.128 C) .... *Anodontites colombiensis* (Marshall, 1922)
[Colombia: Magdalena Basin]

21(20) Shell thin and light, periostracum chestnut brown, nacre purple to red, strongly iridescent (Fig. 11.128 D) .... *Anodontites leotaudi* (Guppy, 1864)
[Trinidad and Tobago]

21′ Shell's outline form very variable from rhomboid to elliptic-lanceolate or to elliptic-oval; shell in general very thick and heavy, inflated to subinflated; ventral margin varying from straight to curved with a major convexity at the middle, beak inflated and large but relatively low; external surface bright and glossy, color varying from black, brown, to emerald green; internal surface blue, varying to salmon pink, strongly iridescent; narrow prismatic border (Fig. 11.128 E) .... *Anodontites elongatus* (Swainson, 1823)
[Central and South America]

22(1) Shell's outline trapezoid; shell inflated and large in adults (length $<$ 200 mm); hinge line straight or nearly; anterior base gapping, prismatic border narrow (Fig. 11.128 F) .... *Anodontites trapesialis* (Lamarck, 1819)
[From Mexico to Argentina, except Guyana, Suriname, French Guiana, and Chile in South America. Note: The shell outline changes from trapezoid in the south to oval in the north, and the color from greenish-dark brown to pale yellow]

22′ Very similar to the former species, but the hinge line raised and rounded especially behind (Fig. 11.128 G) .... *Anodontites moricandi* (Lea, 1860)
[Brazil: São Francisco Basin]

FIGURE 11.126 External and partial internal view of mussel species of Mycetopodidae s.s. (A) *Mycetopoda soleniformis* (SMNS ZI0076323); (B) *M. legumen* (MCP 8625); (C) *M. siliquosa* (MCN 31613); (D) *Anodontites trapezeus* (MZL 1); (E) *A. montezuma* (MP/SMF 5164 paratype); (F) *A. inaequivalvis* (MCN 8896); (G) *A. pittieri* (INPA 1462); (H) *A. ferrarisi* (MCN 7958); (I) *A. crispatus* (INPA 185); (J) *A. obtusus* (MCN 31365). Arrow in (A), (B) and (C): shows the position of the umbones in relation to the shell length; Arrow in (H): prismatic border; Arrow in (I): detail of the periostracum sculpture. Scale bars = 10 mm. (A–D) and (F–J) Photographs and edition by the authors of Unionida; (E) MP/SMF with permission.

FIGURE 11.127 External and partial internal view of mussel species of Mycetopodidae s.s. (A) *Anodontites lucidus* (MCN 7937); (B) *A. cylindracea* (MP/USNM 86712); (C) *A. guanarensis* (MP/USNM 365225); (D) *A. carinatus* (MP/ZMB 108725); (E) *A. iheringi* (MCN 6458); (F) *A. napoensis* (MP/USNM 25429); (G) *A. tortilis* (MCN 6568 and 6471), right valve of a young and left valve of an adult specimen; (H) *A. schomburgianus* (MCN 30577); (I) *A. patagonicus* (MCN 7876); (J) *A. puelchanus* (MLP 4155). Arrow in (F): minute muscular impressions at subumbonal cavity; arrow in (G): reticular area organized in radial rows. Scale bars = 10 mm. (A), (E), (G–I) Photographs and edition by the authors of Unionida; (B), (D) and (F) MP/USNM with permission; (D) ZMB with permission; (J) MLP with permission.

FIGURE 11.128 External and partial internal view of mussel species of Mycetopodidae s.s. (A) *Anodontites tenebricosus* (MCN 9870); (B) *A. soleniformis* (INPA 490); (C) *A. colombiensis* (MP/USNM 341472); (D) *A. leotaudi* (MP/USNM 359920); (E) *A. elongatus* (ZSM); (F) *A. trapesialis* (MCN 8307); (G) *A. moricandi* (UERJ 10243). Scale bars = 10 mm. (A), (B) and (F–G) Photographs and edition by the authors of Unionida; (C) and (D) MP/USNM with permission; (E) ZSM with permission.

ACKNOWLEDGMENTS

Our thanks to the curators and technicians for supporting the museum surveys and/or providing us with photographs: Gary Rosenberg and Mark Sabaj (ANSP); Timothy Pearce (CMNH); Nair Otaviano Aguiar and Fabio Siqueira Pitaluga de Godoi (CZPB); Márcio Luiz de Oliveira and Célio Magalhães (INPA); Janine Oliveira Arruda (MCN); Carlos Lucena and Lúcia Maria Zani Richinitti (MCP); Cristhian Clavijo and Javier González (MHNM); Jean Mariaux and Yves Finet (MNHG); Virgine Heroes and Philippe Maestrati (MNHN); Gustavo Darrigran and Mônica Tassara (MLP); Anne Freitag (MZL); Daniel Graf and Kevin Cummings (MP); Luiz Ricardo Lopes Simone (MZSP); Jonathan Ablett (NHNUK); Julia Sigwart and Ronald Janssen (SMF); Ira Richling (SMNS); Sonia Barbosa dos Santos and Maria Regiana Salgado (UERJ); Claudia Callil (UFMT); Robert Hershler and John Pfeiffer (USNM); Christine Zorn (ZMB); and Michael Schröld and Enrico Schwabe (ZSM).

REFERENCES

Bonetto, A.A. 1967. La superfamilia Unionacea en la cuenca amazónica. Atas do Simpósio sobre a Biota Amazônica 3: 3–62.

Clavijo, C. 2015. Diversidad de Corbiculidae en Uruguay. MSc Thesis, Universidad de la República, Facultad de Ciencias. Montevideo, Uruguay.

Cooley, L.R. & D. Ó Foighil. 2000. Phylogenetic analysis of the Sphaeriidae (Mollusca: Bivalvia) based on partial mitochondrial 16s rDNA gene sequences. Invertebrate Biology 119: 299–308.

Cummings, K.S. & D.L. Graff. 2015. Class Bivalvia. Ecology and General Biology. Pages 423–506 *in* J.H. Thorp & D.C. Rogers (eds.), Thorp and Covich's Freshwater Invertebrates. Elsevier, Waltham, MA.

Graf, D.L. & K.S. Cummings. 2007. Review of the systematics and global diversity of freshwater mussel species (Bivalvia: Unionoida). Journal of Molluscan Studies 73: 291–314.

PHYLUM MOLLUSCA

Graf, D.L. & K.S. Cummings. 2019. The Freshwater Mussels (Unionoida) of the World (and other less consequential bivalves), updated 7 March 2019. MUSSEL Project Web Site, http://mussel-project.uwsp.edu/fmuotwaolcb/. Accessed 3rd April, 2019.

Geda, S.R., N.K. Lujan, M. Perkins, E. Abernethye, M.H. Sabaj & M. Gangloff. 2018. Multilocus phylogeny of the zebra mussel family Dreissenidae (Mollusca: Bivalvia) reveals a fourth Neotropical genus sister to all other genera. Molecular Phylogenetics and Evolution 127: 1020–1033.

Haas, F. 1969. Superfamilia Unionacea. Pp. i-x, 1–663 *in*: R. Martens, W. Hennig & H. Wermuth (eds.), Das Tierrich, Lieferung, Vol 88. Walter de Gruyter and Co., Berlin.

Hupé, M.H. 1857. Mollusques. Pages: 1–96 *in:* F. Castelnau (ed.), Animaux nouveaux ou rares récuellis pendant l'expedition dans les parties centrales de l'Amérique du Sud, de Rio de Janeiro à Lima au Peru. Paris 7(3).

Ituarte, C. 2007. Las especies de *Pisidium* C. Pfeiffer, 1821 de Argentina, Bolivia, Chile, Perú y Uruguay (Bivalvia Sphaeriidae). Revista Museo Argentino de Ciencias Naturales 9: 169–203.

Lee, T., S. Siripattrawan, C. Ituarte & D.O' Foighil. 2005. Invasion of the clonal clams: *Corbicula* lineages in the New World. American Malacological Bulletin 20: 113–122.

Mansur, M.C.D. 1974. *Monocondylaea minuana* Orb., 1835, variabilidade de concha e morfologia do sistema digestive (Bivalvia, Mycetopodidae). Iheringia, Série Zoologia 45: 3–25.

Mansur, M.C.D. 1999. Gloquídio de *Diplodon martensi* (Ihering) (Molluscs, Bivalvia, Hyriidae) a seu ciclo parasitário. Revista Brasileira de Zoologia 16: 185–194.

Mansur, M.C.D. & M.G.O. da Silva.1999. Description of glochidia of five species of freshwater mussels (Hyriidae: Unionoidea) from South America. Malacologia 41: 475–483.

Mansur, M.C.D. & D. Pereira. 2006. Bivalves límnicos da bacia do rio dos Sinos, Rio Grande do Sul, Brasil (Bivalvia, Unionoida, Veneroida e Mytiloida). Revista brasileira de Zoologia 23: 1123–1147.

Mansur, M.C.D., D. Pereira, P.E. Aydos Bergonci, D. Mansur Pimpão, J.R. de Souza Barradas & M.H. Sabaj. 2019. Morphological assessment of *Rheodreissena* (Bivalvia: Veneroida: Dreissenidae) with an updated diagnosis of the genus, descriptions of two new species, redescription of *R. lopesi*, and the first account of larval brooding in New World dreissenids. Proceedings of the Academy of Natural Sciences of Philadelphia 166: 1–45.

Mansur, M.C., C.P. Santos & M.V. Nehrke. 2011. Corbiculidae na América do Sul, espécies nativas e invasoras, dispersão e a situação das pesquisas no Brasil (Mollusca: Bivalvia). Pp. 348–362 *in*: M.A. Fernandez, S.B. Santos, A.D. Pimenta & S.C. Thiengo (eds), Topicos em Malacologia – Ecos do XIX EBRAM, Sociedade Brasileira de Malacologia, Rio de Janeiro.

Mansur, M.C.D., C. Schulz & L.M.M.P. Garces. 1987. Moluscos bivalves de água doce: Identificação dos gêneros do sul e leste do Brasil. Acta Biológica Leopoldência 9: 181–202.

Mansur, M.C.D. & I.L. Veitenheimer. 1979. Redescrição de *Mycetopoda legumen* (Martens, 1888) Revista Brasileira de Zoologia 39(3): 695–702.

Mansur, M.C.D., I.L. Veitenheimer-Mendes & J.E.M. Almeida Caon. 1988. Mollusca, Bivalvia de um trecho do curso inferior do Rio Jacuí Rio Grande do Sul, Brasil. Iheingia, Série Zoologia 67: 87–108.

Miyahira, I.C., M.C.D. Mansur, D.M. Pimpão, S.R.M. Couceiro & S.B. Santos. 2020. Morphology and distribution of the freshwater mussel *Diplodon granosus*, a rare and poorly understood species. Acta Amazonica 50(1): 44–53.

Miyahira, I.C., M.C.D. Mansur & S.B. Santos. 2013. Revision of the type specimens of *Diplodon ellipticus* and *Diplodon expansus* (Bivalvia, Unionoida, Hyriidae). Spixiana 36: 173–182.

Miyahira, I.C., M.C.D. Mansur & S.B. Santos. 2019. Redescription of *Diplodon ellipticus* Spix in Wagner, 1827, *Diplodon multistriatus* (Lea, 1831), and *Rhipidodonta garbei* (Ihering, 1910) (Bivalvia: Hyriidae) from coastal rivers of eastern and northeastern of Brazil. Archiv für Molluskenkunde 148: 9–34.

Miyahira, I.C., S.B. Santos & M.C.D. Mansur 2017. Freshwater mussels from South America: state of the art of Unionida, especially Rhipidodontini. Biota Neotropica 17: 1–13.

Park, J-K. & D. O'Foighil. 2000. Sphaeriid and Corbiculid clams represent separate heterodont bivalve radiations into freshwater environments. Molecular Phylogenetics and Evolution 14: 75–88.

Parodiz, J.J. 1996. The taxa of fossil mollusca introduced by Hermann von Ihering. Annals of the Carnegie Museum 65: 183–296.

Parodiz, J.J. & A.A. Bonetto. 1963. Taxonomy and Zoogeography relationships of the South American naiads (Pelecypoda: Unionacea and Mutelacea). Malacologia 1: 179–213.

Parodiz, J.J. & L. Hennings. 1965. The *Neocorbicula* (Mollusca, Pelecypoda) of the Paraná-Uruguay basin, South America. Annals of the Carnegie Museum 38: 69–96.

Pereira, D., M.C. Dreher Mansur, L.D.S. Duarte, A. Schramm de Oliveira, D. Mansur Pimpão, C. Tasso Callil, C. Ituarte, E. Parada, S. Peredo, G. Darrigran, F. Scarabino, C. Clavijo, G. Lara, I.C. Miyahira, M.T. Raya Rodriguez & C. Lasso. 2014. Bivalve distribution in hydrographic regions in South America: historical overview and conservation. Hydrobiologia 735: 15–44.

Pereira, D., M.C.D. Mansur & D.M. Pimpão. 2012. Identificação e diferenciação dos bivalves límnicos invasores dos demais bivalves nativos do Brasil. Cap 5. Pp. 75–94 *in:* M.C.D. Mansur, C.P. Santos, D. Pereira, I.C.P. Paz, M.L.L. Zurita, M.T.R. Rodriguez, M.V. Nehrke & P.E.A. Bergonci (eds), Moluscos Límnicos Invasores no Brasil: biologia, prevenção, controle. Porto Alegre, Redes Editora.

Prime, T. 1865. Monograph of American Corbiculadae. Smithsonian Miscellaneous Collections 7(145): 1–80.

Santos-Neto, G. da C., C.R. Beasley, H. Schneider, D.M. Pimpão, W.R. Hoeh, L.R.L. Simone & C.H. Tagliaro. 2016. Genetic relationships among freshwater mussel species from fifteen Amazonian rivers and inferences on the evolution of the Hyriidae (Mollusca: Bivalvia: Unionida). Molecular Phylogenetics and Evolution 100: 148–159.

Silva-Souza, A.T., P. Guardia-Felipi &N.R. Arrebola. 2011. Embryonic development of *Anodontites trapesialis* (Lamarck, 1819) (Bivalvia: Mycetopodidae). Brazilian Journal of Biology 71: 139–144.

Simone, L.R. 1999. Anatomy and systematics of *Anticorbula fluviatilis* (H. Adams, 1860) (Bivalvia: Lyonsiidae) from the Amazon Basin, Brazil and Peru. Nautilus 113: 48–55.

Simpson, C.T. 1914. A descriptive catalogue of the naiades or pearly freshwater mussels. Michigan, Bryant Walker, 1540p.

Veitenheimer-Mendes, I.L. & M.C.D. Mansur. 1978. *Mycetopoda legumen* (Martens, 1888): lasidio e desenvolvimento parasitário (Bivalvia, Mycetopodidae). Revista Brasileira de Biologia 38: 531–536.

Wächtler, K., M.C.D. Mansur & T. Ritcher. 2001. Larval type and early postlarval biology in Naiads (Unionoida). Pages 93–125 *in*: G. Bauer & K. Wächtler (eds.), Ecology and Evolution of the Freshwater Mussels Unionoida, Springer-Verlag, Berlin.

Chapter 12

Phylum Annelida

Mercedes R. Marchese, Roberto G. Alves, Alejandro Oceguera-Figueroa, Christopher J. Glasby, João Gil, Daniel Martin, Tarmo Tim, Stuart R. Gelder and Cristina Damborenea

Chapter Outline

Thorp and Covich's Freshwater Invertebrates. https://doi.org/10.1016/B978-0-12-804225-0.00012-5

Introduction to the Phylum

Cristina Damborenea

CONICET, División Zoología Invertebrados, Museo de La Plata, FCNyM, UNLP, Argentina

INTRODUCTION

Annelids comprise more than 17,000 species, which can be found in marine, limnic, and terrestrial habitats (Zhang, 2011). Traditionally, Annelida has been split into Clitellata (earthworms and leeches), which are common in inland freshwater systems, and Polychaeta (bristle worms), a large group of primarily marine worms with few inland water species. Two other groups inhabit freshwater in the Neotropics: Branchiobdellida and Aphanoneura. Branchiobdellidans are an independent clitellate taxon of equal rank to the oligochaetes and leeches. Aphanoneurans are oligochaete-like aeolosomatids that lack clitellum and for which was created a second aclitellate taxon, the Aphanoneura (Timm, 1981). However, the relationships within Annelida are still poorly understood, and a robust reconstruction of annelid evolutionary history is needed (Struck et al., 2011). Recent studies have recovered two main annelid clades: Errantia and Sedentaria (which includes Clitellata) along with several minor marine basal clades (Struck et al., 2011). Consequently, the taxonomic status of Clitellata is currently uncertain, and polychaetes are paraphyletic.

The Neotropical Region supports a high diversity of freshwater annelids, with 233 species in 82 genera, and only the Palaearctic Region has more species reported (Glasby & Timm, 2008; Martin et al., 2008).

LIMITATIONS

Annelids from the Neotropical Region are under sampled, and knowledge about this group is limited. Large geographical areas lack records, posing a great task for future generations of biologists.

External characters allow the separation of main groups of Annelida. In addition, their recorded habitats are useful in separating taxonomic groups (Thorp & Lovell, 2016).

TERMINOLOGY AND MORPHOLOGY

Annelids show a huge diversity of body plans, and it is difficult to describe a consistent anatomy (Fig. 12.1). Most annelids possess multiple repetitive **segments** along the anterior–posterior body axis. The annelid body is divided into a **prostomium**, a segmented trunk, and a **pygidium**. The head of the annelids may bear appendages, as **palps** or **antennae**, but these are lacking in a number of taxa. The first segment, termed **peristomium**, has the mouth, and several members have sclerotized mandibular structures. The segments of many annelids can bear ventral and dorsal groups of chitinous **chaetae,** which may be organized in **parapodia,** the paired lateral appendages of polychaetes. The pygidium contains the anus, which is usually either dorsally or terminally located and often bears pairs of cirri. Both sexual and asexual reproductive strategies have developed in annelids. For sexual reproduction, different types of free spawning, brooding, and encapsulation of embryos in cocoons can be distinguished. A glandular epidermal girdle, partly behind the female pores, forms the **clitellum** that secretes a cocoon where eggs are laid. Asexual paratomy is frequent in some groups, forming chains of several individuals. For details of the morphology and biology of freshwater annelids, see Verdonschot (2015).

MATERIAL PREPARATION AND PRESERVATION

Long-term preservation of annelids requires fixation in 5%–10% buffered formalin solution. Better results are obtained relaxing the specimens in a 7% solution of $MgSO_4$ for some minutes before fixation. After a few days, the specimens must be transferred to 70%–80% ethyl alcohol. For molecular studies, the specimens must be preserved in 95% or 100% ethyl alcohol or frozen at −80°C.

Detailed techniques for preservation and preparation of specimens are given below in each annelid group.

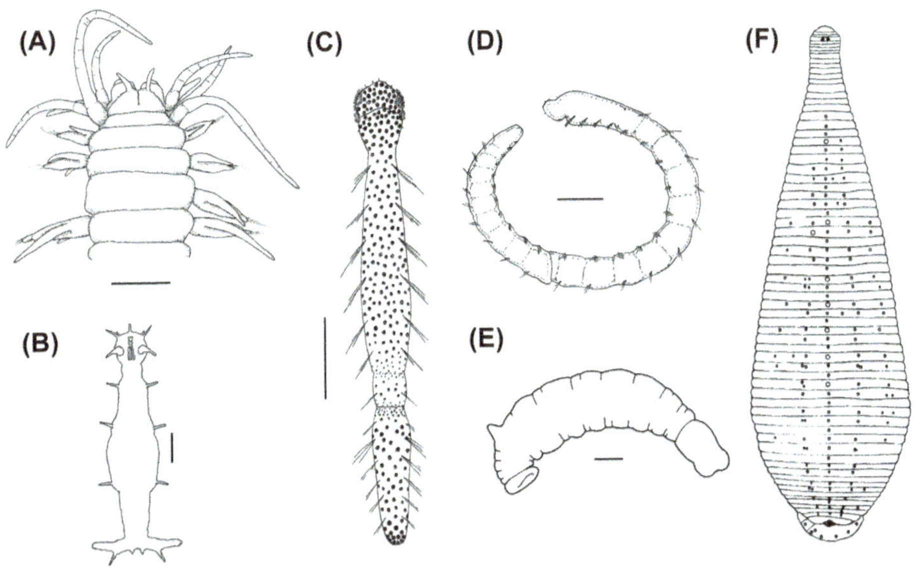

FIGURE 12.1 Freshwater annelids. (A) *Namanereis minuta* (Polychaeta); (B) *Stratiodrilus vilae* (Polychaeta); (C) *Aeolosoma evelinae* (Aeolosomatidae); (D) *Allonais paraguayensis* (Oligochaeta); (E) *Sathodrilus prostates* (Branchiobdellidae); (F) *Helobdella brasiliensis* (Hirudinea); Scale bar for (A)–(E) = 0.2 mm. *Modified (A) from Glasby (1999), (B) from Amato (2001), (C) and (D) from Marcus 1944; (E) from Holt (1973); (F) from Ringuelet (1985).*

KEYS TO ANNELIDA

This higher classification is under review, and a wide consensus is lacking. Therefore, we follow here the classification presented in the previous volumes (Verdonschot, 2015; Thorp & Lovell, 2016, 2019).

Annelida: Classes

1	Prostomium bearing cilia, papillae, or tentacles; chaetae relatively abundant; parapodia well developed, reduced, or absent (Fig. 12.1 A–C) 2
1′	Prostomium simple; parapodia absent; a glandular region that partially or completed encircled the animal in a swollen region called clitellum, sometimes only visible during reproduction (Fig. 12.1 D–E) .... **Clitellata [p. 433]**
2(1)	Minute worms, 1–2 mm or chains of animals up to 10 mm; prostomium with cilia only; head appendages and segmental parapodia absent; worms move in a gliding motion using cilia; colored epidermal glands all over the body surface (Fig. 12.1 C) .... Aphanoneura, one family: Aeolosomatidae, one genus: ***Aeolosoma* [p. 476]**
2′	Minute to large worms; prostomium with palps or tentacles, head with appendages, segmental parapodia bearing numerous chaetae, sometimes absent (Fig. 12.1 A, B) .... **Polychaeta [p. 483]**

Annelida: Clitellata: Subclasses

1	With one (posterior) or two (anterior and posterior) suckers 2
1′	Without suckers (Fig. 12.1 D) .... **Oligochaeta [p. 441]**
2(1)	One posterior sucker; head distinct followed by 11 body segments; ectosymbiontic on crayfishes, isopods and crabs (Fig. 12.1 E) .... Branchiobdellidea, one family: **Branchiobdellidae [p. 461]**
2′	Two suckers; anterior sucker of four segments, mouth on sucker ventral surface; 32 postoral segments, subdivided superficially into 2–16 annuli; posterior sucker of seven segments (Fig. 12.1 E) .... **Hirudinida [p. 465]**

REFERENCES

Amato, J.F.R. 2001. A new species of *Stratiodrilus* (Polychaeta, Histriobdellidae) from freshwater crayfishes of southern Brazil. Iheringia Série Zoologia Porto Alegre 90: 37–44.

Glasby, C.J. 1999. The Namanereidinae (Polychaeta: Nereididae). Part 1, taxonomy and phylogeny. Records of the Australian Museum, Supplement 25: 1–129.

Glasby, C.J. & T. Timm. 2008. Global diversity of polychaetes (Polychaeta; Annelida) in freshwater. Hydrobiologia 595: 107–115.

Holt, P.C. 1973. A Summary of the branchiobdellid (Annelida: Clitellata) fauna of Mesoamerica. Smithsonian Contribution to Zoology 142: 1–40.

Marcus, E. 1944. Sôbre Oligochaeta limnicos do Brasil. Boletim da Faculdade de Filosofia, Ciências e Letras, Universidade de São Paulo, Zoologia 43: 5–135.

Martin, P., E. Martínez-Ansemil, A. Pinder, T. Timm & M.J. Wetzel. 2008. Global diversity of oligochaetous clitellates ("Oligochaeta"; Clitellata) in freshwater. Hydrobiologia 595: 117–127.

Ringuelet, R.A. 1985. Annulata, Hirudinea. Pages 10-321 *in*: Z.A. de Castellanos (dir.) Fauna de Agua dulce de la República Argentina, Vol 17, Fundacion para la Educación, la Ciencia y la Cultura, Buenos Aires.

Struck, T.H., C. Paul, N. Hill, S. Hartmann, C. Hösel, M. Kube, B. Lieb, A. Meyer, R. Tiedemann, G. Purschke & C. Bleidorn. 2011. Phylogenomic analyses unravel annelid evolution. Nature 471: 95–98.

Thorp, J.H. & L.L. Lovell. 2016. Introduction to Annelida. Pages 223–225 *in*: J.H. Thorp & D.C. Rogers (eds.), Keys to Nearctic Fauna, Thorp and Covich's Freshwater Invertebrates, Vol. II, 4th Edition. Academic Press, New York.

Thorp, J.H. & L.L. Lovell. 2019. Introduction to Annelida. Pages 360–364 *in*: D.C. Rogers & J.H. Thorp (eds.), Keys to Palaearctic Fauna, Thorp and Covich's Freshwater Invertebrates, Vol. IV, 4th Edition. Academic Press, New York.

Timm, T. 1981. On the origin and evolution of aquatic Oligochaeta. Proceedings of the Estonian Academy of Sciences, Biology 30: 174–181.

Verdonschot, P.F.M. 2015. Introduction to Annelida and the class Polychaeta. Chapter 20. Pages 509–528 *in*: J.H. Thorp & D.C. Rogers (eds.), Ecology and General Biology, Thorp and Covich's Freshwater Invertebrates, Vol. I, 4th Edition. Academic Press, New York.

Zhang, Z-Q. 2011. Animal biodiversity: an outline of higher-level classification and survey of taxonomic richness. Zootaxa 3147: 1–237.

Class Clitellata: Subclass Oligochaeta

Mercedes R. Marchese
Instituto Nacional de Limnología (CONICET-UNL), Santa Fe, Argentina

Roberto G. Alves
Departamento Zoologia, Universidade Federal de Juiz de Fora, Juiz de Fora, MG, Brazil

INTRODUCTION

The aquatic representatives of the oligochaetous Clitellata appear to be a paraphyletic assemblage, which have been classified on the basis of the segmental position of their genital organs (Erséus, 2005).

Oligochaetes are common and usually very abundant in freshwater environments and are associated with different substrates, including different sediment types and sizes, emergent or submersed aquatic vegetation, artificial substrata, periphyton, and leaf packs. Some worms can exploit both aerobic and anaerobic sediments rich in sulfur. Although they reach their highest densities and richness in fine sediments, where the majority inhabit the first 2–3 cm, many tubificids will burrow as deep as to 10 cm depth. Some species inhabit phytothelmata. Many naidid species are associated with freshwater sponges (Kahl & Konopacka, 1981; Melão & Rocha, 1996; Corbi et al., 2005; Gorni & Alves, 2008). They are usually ignored, or overlooked in samples of small headwater streams, but some species (especially *Pristina, Nais,* and *Chaetogaster* species) occur in sandy, leaf litter patches, in epilithon and detritus retained among the trichopteran cases. Several species can swim and are good colonizers of submerged natural and artificial substrata.

A total of 8,400 species of oligochaetes have been described worldwide, and nearly 5,000 are considered valid species. Of the 1,700 aquatic oligochaetes, about 1,100 are freshwater, 100 are restricted to groundwater habitats, and 600 are marine (Erséus, 2005; Wetzel et al., 2006, Martin et al., 2008). Martínez-Ansemil et al. (2012) reported that more than 300 species live in ground waters, although most are limited to the karst region of southern Europe (Martin et al., 2008; Martínez-Ansemil & Sambugar, 2010).

The number of species from South America is imprecise: 70 freshwater species are reported from Brazil (Righi, 1984), and 110 species are recorded from Argentina (Marchese & Paggi, 2003). Christoffersen (2010) reported 171 from South America, and Martin et al. (2008) reported 178 species from the Neotropical Region and an endemism rate of 51%.

LIMITATIONS

The knowledge of oligochaetes from the Neotropical Region is very limited, mainly because large areas have not been sampled. Argentina is the country with the highest number of species recorded, followed by Brazil. The studies from Brazil are mainly from Paraná, Sao Paulo, and Minas Gerias states, whereas in Uruguay, they are from environments around Montevideo city. In Argentina, a broader geographic area has been sampled, including the austral Patagonia, Antarctic, and Subantarctic islands, which were studied at beginning the 20th century. Scientists later focused on the Northwest and on Buenos Aires province and in the Paraná River basin. In the other countries of the Neotropical Region, there are scattered records, mainly in Paraguay, Bolivia, Chile, Colombia, Peru, Venezuela, Guyana, Suriname, Ecuador, Costa Rica, Guatemala, Nicaragua, Cuba, West Indies, and South of Mexico.

Howmiller (1974), Gavrilov (1977, 1981), Di Persia (1980), Harman (1982), Righi (1984), Brinkhurst & Marchese (1989, 1992), Gaviria (1993), Righi & Hamoui (2002), Marchese (2009), and Christoffersen (2007, 2010) provided an inventory and compiled the literature for Neotropical oligochaetes. In addition, Righi (1984), Brinkhurst & Marchese (1989, 1992), and Marchese (2009) provided identification keys for the aquatic oligochaetes. More general information on oligochaetes is available in Reynolds & Wetzel (2018).

Many Neotropical aquatic oligochaetes are cosmopolitan or at least widely distributed. However, some patterns seem evident: Phreodrilidae is representative of Andean Patagonic environments, whereas Naididae, Opistocystidae, Narapidae, and Alluroididae are more representative of tropical areas. On the other hand, genera such as *Dero, Allonais, Slavina, Pristina, Opistocysta, Branchiura*, and *Bothrioneurum* are common and generally well represented in the Neotropics. The monospecific Narapidae is only reported from South America. Phreodrilidae of Gondwanic origin has been reported from the Andean Patagonic region and Antarctic islands although there is also a record from Lake Titicaca. The Naididae is widely distributed globally and shows generic and specific endemicity in Neotropical Region. The family Haplotaxidae also has a worldwide distribution, and *Haplotaxis aedeochaeta* Brinkhurst & Marchese, 1987 is endemic to South America.

We have excluded the Enchytraeidae because enchytraeids are mainly terrestrial, and only few groups can be classified as aquatic forms (Christoffersen, 2009). However, Schmelz & Alves (2017) described and illustrated genera and species of enchytraeids from first-order streams in Brazil.

TERMINOLOGY AND MORPHOLOGY

By convention, the oligochaete segments are identified and labeled in Roman numerals, whereas Arabic numerals are used to denote septa between segments. For example, septum 3/4 is the septum that separates segments III and IV.

The first segment, or **peristomium**, surrounding the mouth is always devoid of **chaetae** (Fig. 12.2). The **prostomium** is attached to the peristomium and may or may not have a **proboscis**. Some species have eyes. The first numbered region is labeled I. Chaetae are arranged in four **bundles** on each segment: two dorsolateral (termed dorsal) and two ventrolateral (termed ventral) bundles (Fig. 12.2).

The principal morphological features used to identify oligochaetes are the **chaetae** or setae (Fig. 12.3). A single specimen may exhibit several kinds of chaetae in particular forms and positions. The dorsal chaetae usually begin in segment II or one or more segments behind II, whereas the ventral chaetae always begin in II (Fig. 12.4).

Hair chaetae or **setaform** (Fig. 12.3) (**capilliforms**) are elongate, thin structures present on the dorsal bundles (except for the Capilloventridae, Parvidrilidae, and one species of Phreodrilidae where they are also ventral) of some naidids (including tubificines), opistocystids, and phreodrilids. They may bear fine lateral hairs that are often only visible at high magnification (400× or 1000×). The term hispid will be used here for this hair chaetae type giving a plumose appearance. Some hair chaetae have serrations on one side of the chaetae like a saw blade. The term serrate will be used for this chaetae form. The hair chaetae in *Slavina* (mainly *Slavina evelinae* Marcus, 1942) are thick and present bifid tips visible at 400×.

Bifid chaetae (Figs. 12.3–4) are found in both dorsal and ventral bundles in most worms. They are sigmoid, with the distal (upper) ends bifurcating to a varying degree. The form and relative lengths of these teeth are commonly referred to in the keys. Sometimes the upper distal tooth becomes so reduced as to be almost invisible, and this may be enhanced by wear. Dorsal chaetae in the naidids (Naidinae and Pristininae) often differ from ventrals and are referred to as **aciculate**. These are generally much smaller than the bifid ventral chaetae and usually occur with hair chaetae. High magnification is normally needed to see the form of the tips of needle chaetae. **Pectinate** or **palmate chaetae** (Fig. 12.3) are usually found in the presence of any hair chaetae in dorsal bundles in the Naididae. They consist of bifids with a set of fine intermediate teeth, which may merge to form a web between the main teeth. When the intermediate and outer teeth become equally large, the chaetae are termed **palmate**. Ventral chaetae in some tubificines may show traces of intermediate teeth in a few anterior bundles, but little or no emphasis is placed on this feature.

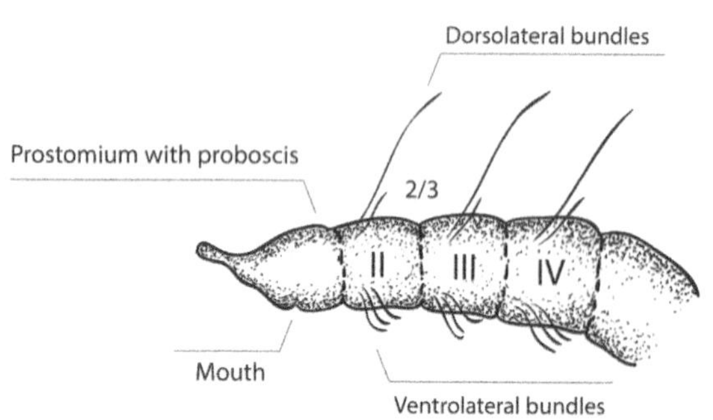

FIGURE 12.2 Anterior end of a generalized oligochaete illustrating prostomium with proboscis, types and common location of somatic chaetae, segment labeled in Roman numerals and septa in Arabic numerals.

Simple pointed (or unicuspids) chaetae (Fig. 12.3) are less frequent in the freshwater worms but are found in lumbriculids, haplotaxids, alluroidids, enchytraeids, some anterior ventral bundles of tubificines like *Spirosperma* species, and dorsal bundles of some opistocystids and naidids in association or not with hair chaetae.

Unusual chaetae make some species of tubificines instantly recognizable. There are clavate chaetae in middorsal bundles in *Aulodrilus pigueti* Kowalewski, 1914 (Fig. 12.3) and bifids with lateral flanges in *Aulodrilus limnobius* Bretscher, 1899.

Some species have no chaeta (e.g., *Narapa bonettoi* Righi & Varela, 1983, *Achaeta* sp.), and dorsal bundles are absent in *Chaetogaster* or may be missing from one or more segments in some species.

Genital chaetae (Fig. 12.3) are usually associated with the genital pores of mature specimens. Those accompanying or adjacent to the spermathecal pores (spermathecal chaetae) are usually solitary and large and have blade or gutter-shaped ectal ends made of the greatly modified teeth of their bifid ancestral form. Spermathecal chaetae are commonly found in segment X (ninth chaetal bundle) in place of the normal chaetae of the ventral bundles in tubificines, or anteriorly (segment V) in naidines. These spermathecal chaetae look as though they act to facilitate the transfer of sperm into the spermathecae of the concopulant (Cuadrado & Martínez-Ansemil, 2001). Spermathecal chaetae are found further forward in naidids (e.g., *Pristina*) where they may be multiple and simply larger than normal chaetae. Penial chaetae are often numerous in each ventral bundle, commonly on XI in tubificines, and the distal ends are blunt, often swollen with the teeth more or less fused together. The inner ends may be elongate, straight, and arranged fanwise such that the outer tips are brought together so that they probably act as claspers during the copulation. Penial chaetae in the naidines and pristinines are located further forward (on VI), as are their genital pores.

PHYLUM ANNELIDA

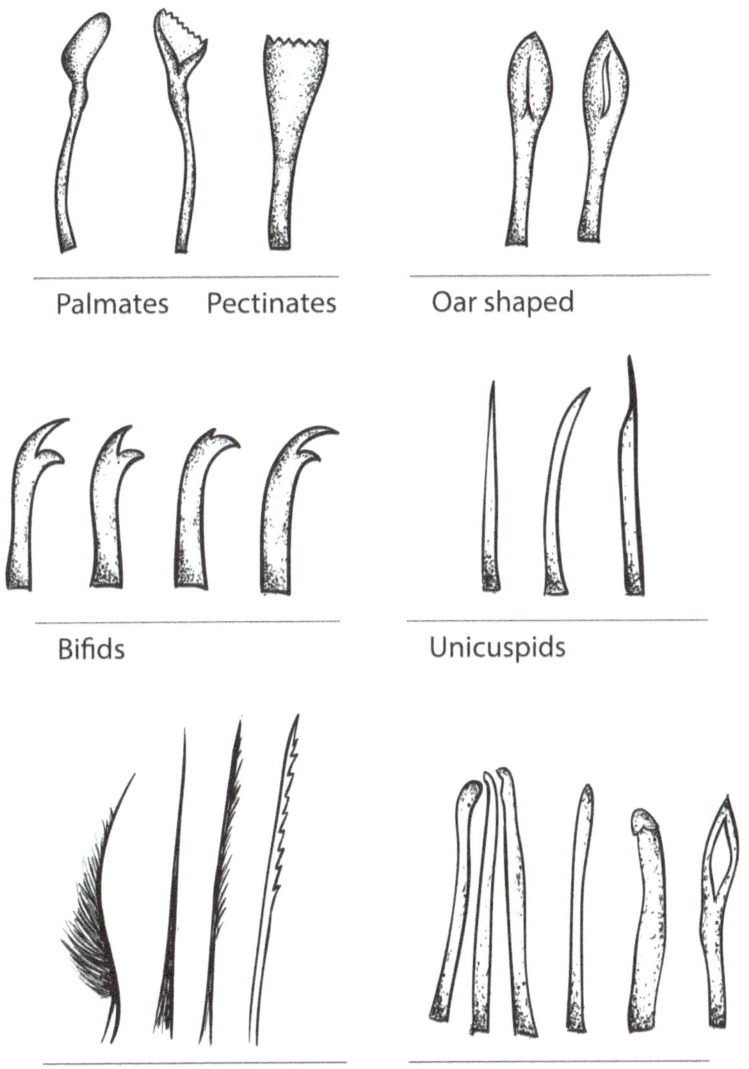

FIGURE 12.3 Types of somatic and genital chaetae.

Note that genital chaetae (and other reproductive organs) are only present in mature specimens. Most bifid chaetae (except in Parvidrilidae) have a more or less median swelling referred to as the **nodulus** (Fig. 12.5). This is absent from the hair chaetae and from simple pointed chaetae in many Enchytraeidae.

Some oligochaetes have **gills**; for example, *Dero* (Naididae) is characterized by the presence of a branchial fossa on the terminal segment with gills in different number and forms depending on the species (Fig. 12.6). *Branchiura sowerbyi* Beddard, 1892 (Rhyacodrilinae) has middorsal and midventral gills on each of the posterior segments (Fig. 12.27), whereas *Phreodrilus branchiatus* Beddard, 1891 (Phreodrilidae) has lateral gills on posterior segments (Fig. 12.18).

Opistocystidae are characterized by the presence of a small dorsomedian projection and two elongate lateral projections of the terminal segments (Fig. 12.7).

Some genera have papillae with or without accumulated foreign particles adhering on the body wall (*Slavina, Stephensoniana, Spirosperma, Peloscolex*). Species of *Slavina* (mainly *S. evelinae*) are characterized by an opaque body wall with prominent cuticularized papillae, with secretions that trap bacteria, fungus, and organic matter, making difficult the observation of dorsal needle chaeta (Fig. 12.8). In *Stephensoniana,* the papillae are present as rings of small granule.

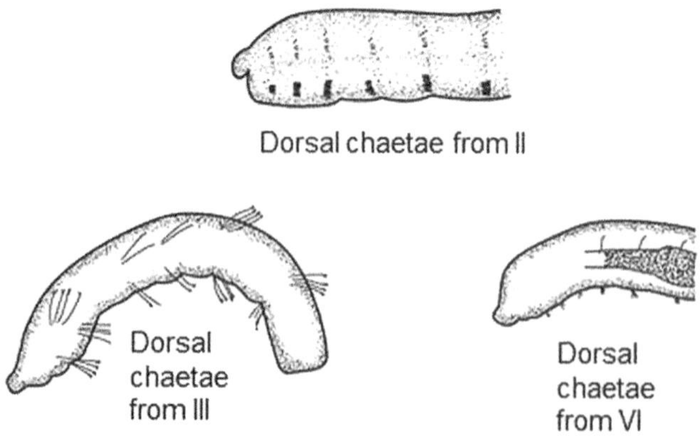

FIGURE 12.4 Generalized scheme noting the beginning of the dorsal chaetae. *From Brinkhurst & Marchese (1989) with permission.*

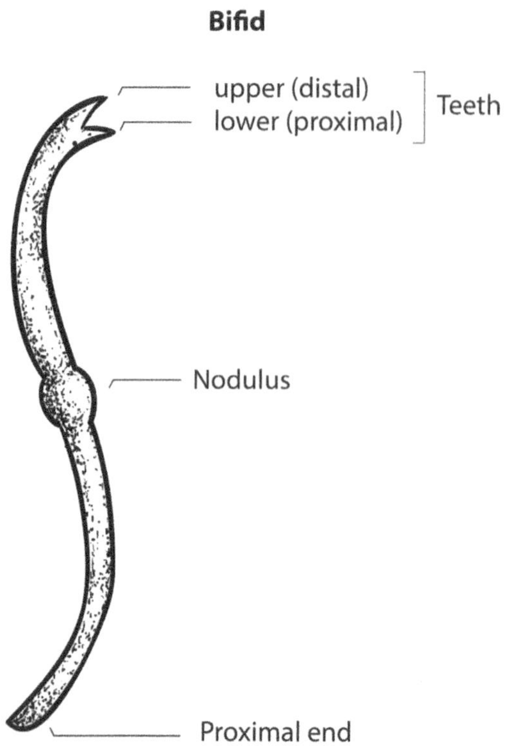

FIGURE 12.5 Typical bifid chaetae with nodulus. *Reprinted from Brinkhurst & Marchese (1989) with permission.*

Reproductive Organs

A mature worm can usually be recognized by the enlargement of the genital area (e.g., in segments V–VIII in naidines or X–XII in tubificines) (Fig. 12.9). The clitellum is a thickened and more opaque body wall with a different texture resulting from development of a glandular layer of cells around two or more segments in the region of the genitalia that differs between families. This cell layer secretes the cocoon into which the sperm and eggs are deposited and the embryo develops (Fig. 12.10).

There are one or more pairs of spermathecae located in or near the gonadal segments, commonly in front of the testicular segments. The number and position of the spermathecae may vary within the Lumbriculidae, but there is usually one pair in the testicular segment (Naididae) or separated from the other genitalia in V (Enchytraeidae). In a few species, the spermathecae may be absent. Sperm may be as an amorphous mass in the spermathecae or organized into bundles of different complexity called spermatozeugmata (Fig. 12.11). In *Bothrioneurum* and *Paranadrilus*, the sperm are placed in spermatophores that attach externally to the partner's body surface (Fig. 12.12).

The eggs and sperm are released from the gonads (one pair of each in the opistocystids, naidids, and phreodrilids, often more in the lumbriculids and haplotaxids) into the coelom. The female funnels open directly to the exterior. Each male funnel usually leads into a ciliated vas deferens that terminates in a muscular atrium. This may end in a penial structure of some description, sometimes at the end of an ejaculator duct. Parts of these structures may bear prostate glands. Sperm is stored and fed by the male ducts and is transferred to the spermathecae of a mate (or may be packaged and attached to the body wall in *Bothrioneurumn* or even placed into a secondary pouch in *Paranadrilus*).

There may or not be a true **penis**, which may be unsheathed in cuticle. The penes of tubificines are formed from an infolding of the body wall and are therefore covered by a layer of cuticle. When the cuticle is thick enough to be seen in cleared whole mounts as a distinct structure, it is termed a cuticular penis sheath (Fig. 12.13). Additional inversions of the body wall produce protrusible pseudopenes in which the male pore empties into a sac formed by a single inversion of the body wall. Muscles around this sac and attached to the body wall allow this sac to be protruded.

The differences in the reproductive systems of the major families are illustrated in Fig. 12.14.

MATERIAL PREPARATION AND PRESERVATION

Identification of oligochaetes is most effective alive when all tissues are more or less transparent and the internal organs can be easily observed. Any kind of fixation or preservation turns their body opaque. When many species are present, this method is not appropriate, and it is necessary to fix the samples (Timm, 2009). The slow addition of 10–30 ml of ethanol (15%) can be used for holding the live field samples of worms. This procedure allows the relaxation of specimens and subsequent reduction of contraction of oligochaetes. This method of relaxation, for a period of 15–30 min prior to fixation, improves your ability to orient the worm body

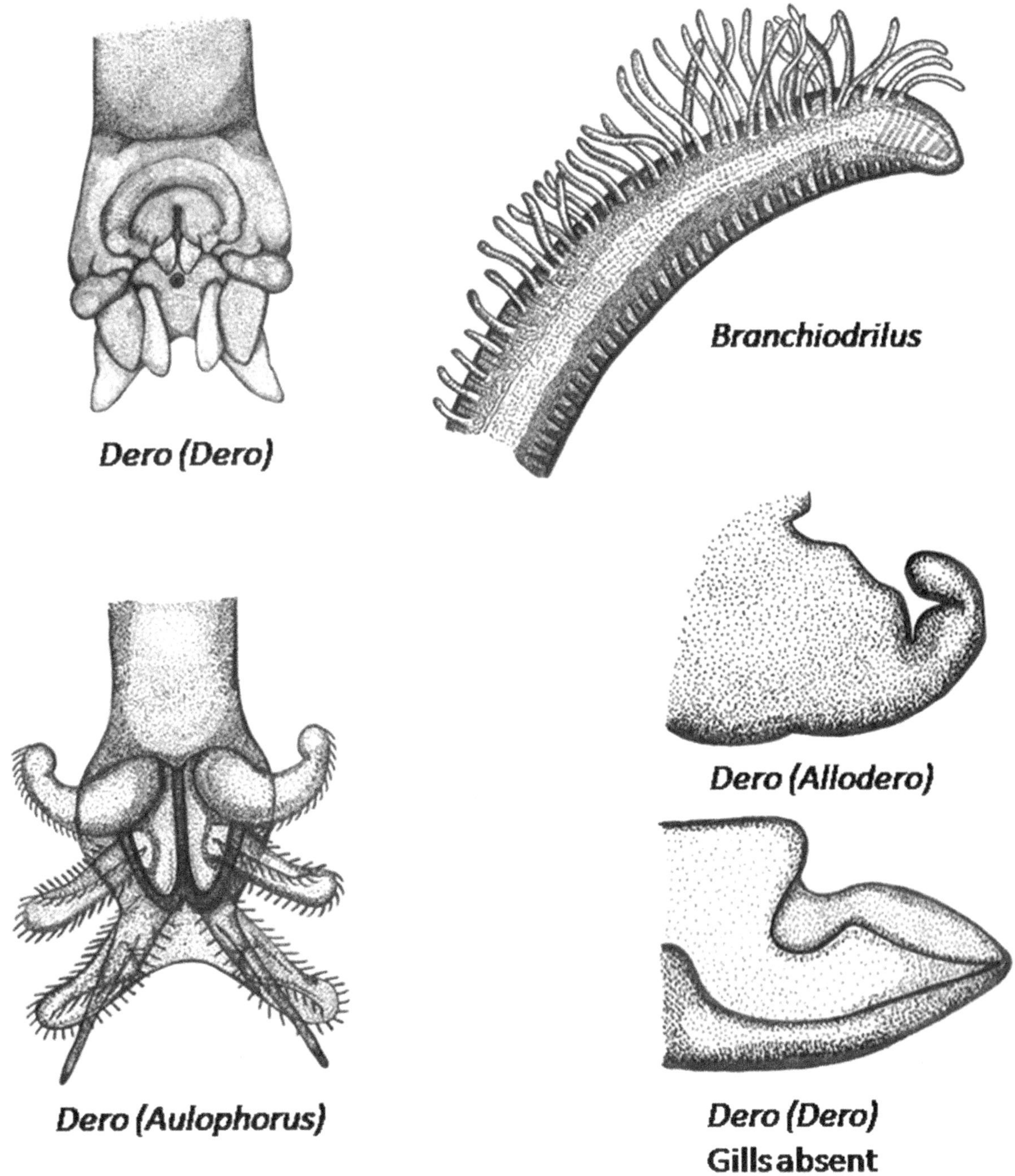

FIGURE 12.6 Morphological characteristics of gills in Naidinae: Branchial fossa with gills and without palps in *Dero* (*Dero*); branchial fossa with both gills and elongate palps in *Dero* (*Aulophorus*); branchial fossa rudimentary or absent in *Dero* (*Allodero*); dorsolateral gills in *Branchiodrilus*.

correctly for temporary or permanent slide mounts (Wetzel et al., 2006). Freshwater oligochaetes are best fixed and stored after collection in 5%–10% buffered formalin. When the specimens are removed from the sediment and debris, they can be rinsed with tap water and then placed into >80% ethyl alcohol for long-term storage. Ethyl alcohol should not be used because it is a preservative and not a fixative. Its use will result in the destruction or deterioration of the specimen. Some researchers use Bouin as fixative solution because it is a good fixative when the histological staining and dissection of tissue is planned. If genetic or molecular analysis is required, the worms must be fixed and preserved in alcohol (90%–96%) but never in formalin. After fixation, worms intended for molecular analyses should be kept as soon as possible at low temperatures (−20 °C) (Timm & Martin, 2015).

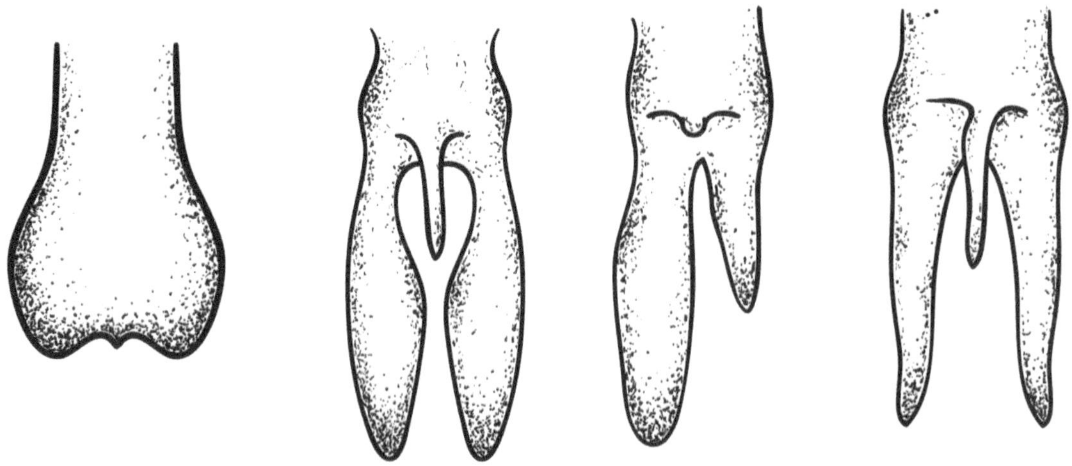

FIGURE 12.7 Caudal appendages characteristics of Opistocystidae (100×). *Reprinted from Brinkhurst & Marchese (1989) with permission.*

Stain such as erythrosin, rose bengal, and phloxine B can be added to facilitate the sorting of the specimens from the sediment. Sorting fixed samples under a stereoscopic microscope helps avoid the loss of little specimens. Slide preparation requires dehydration through a series of alcohol washes at 80%, 90%, and 100% followed by a clearing of specimens in a xylene/Canada Balsam series with proportions of 70/30, 50/50, and 30/70 until 100% Canada Balsam is achieved. Other mounting media also may be used such as Permount, CMC, Euparal, and polyvinyl lactophenol. Such slides should usually be sealed with a ring of material (Glyceel, nail varnish, etc.) around the edges of the coverslips to prevent the media from drying up. Specimens for permanent mounts should be stained before the dehydratation process. Common carmine stains such as borax carmine and Mayer's paracarmine increase the visibility of reproductive organs.

Preserved worms can be studied as temporary mounts in Amman's lactophenol, which is normally used for clearing (few hours or days, depending on the specimens) and mounting medium for temporary observation and identification of fixed oligochaetes. However, Amman's lactophenol is effective in revealing chitinous structures such as chaetae and penial sheaths, but it will destroy all soft tissue (Timm, 2009). This substance should be handled with extreme care because it contains phenol, a known carcinogen. It consists of 20% phenol, 20% acid lactic, 20% water, and 40% glycerine and should be stored in dark bottles. However, it is hygroscopic and does not have a good shelf life (Brinkhurst & Gelder, 1991). The other option is glycerine medium that softens the worms; they can then be flattened by exerting slight pressure on the cover glass to better expose the chaetae and some other details (Timm, 2009). A microscope with good illumination is very important for species identifications, but a compound microscope with phase contrast and differential interference contrast (DIC) optic is better for the identification of aquatic oligochaetes because it allows a good differentiation of internal structures. The observation of chaetal morphology of many species (e.g., *Pristina, Nais, Dero*, etc.) requires oil immersion lenses (100×) for accurate species level determinations. To separate genital region of each worm in two (sagitally) or even when dissecting out the male ducts, it may be necessary to identify taxa to the generic level, when details of reproductive organs need to be established. Serial sections are also employed in taxonomic work, but not in routine identification.

In samples with abundant oligochaetes, subsampling in quantitative studies is recommended. There are numerous techniques described in the literature, but the more common technique used for aquatic oligochaetes is to empty the entire vial of specimens into a Petri dish and swirl the worms until they cover the entire area of the dish. Once they uniformly cover the dish, they divide it into the appropriate number of quadrats (two, four, or eight) and randomly remove specimens from the necessary number of quadrats.

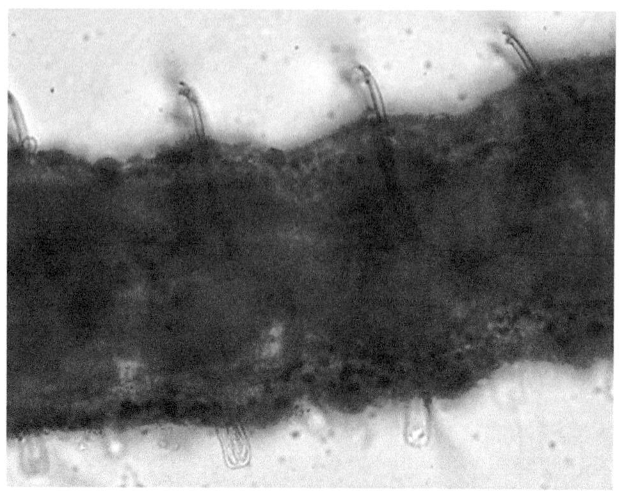

FIGURE 12.8 Body wall with prominent cuticularized papillae in *Slavina* sp. (400×).

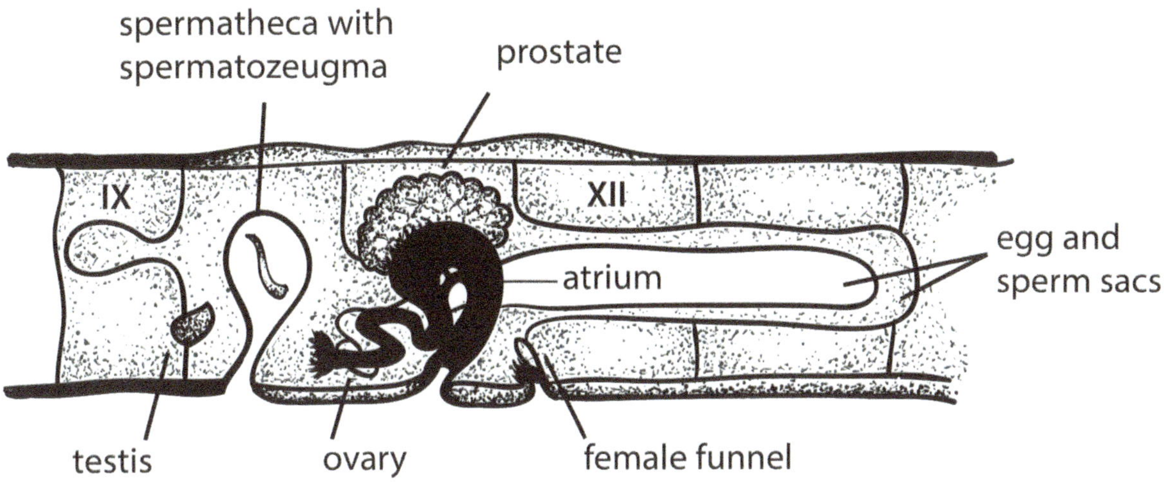

FIGURE 12.9 Scheme of genitalia location in the majority of Tubificinae. *Redraw from Brinkhurst & Marchese (1989).*

FIGURE 12.10 Representative forms of eggs cocoon of tubificids. *Photo: L. Montalto (INALI-CONICET-UNL, Argentina).*

KEYS TO OLIGOCHAETA

Clitellata: Oligochaeta: Families

1 Body length variable, distinctly segmented; chaetae present or absent; hair chaetae if present restricted to dorsal bundles with other types of chaetae 2

1′ Body length 2.0–4.5 mm, with indistinct external segmentation; chaetae absent; internal septa well developed; no pharyngeal glands in IV–VI; asexual reproduction by fission common; sexually mature specimens with clitellum in VI–VIII, testes in V, atria and male pores in VI, single ovary and paired spermathecae in VII; sandy river bottoms (Fig. 12.15) .... Narapidae, one species: *Narapa bonettoi* Righi & Varela, 1983
[Brazil: Paraná River, Paraguay River, Gouveia Stream. Paraguay: Paraguay River. Argentina: Paraná River System. Uruguay: Uruguay River]

PHYLUM ANNELIDA

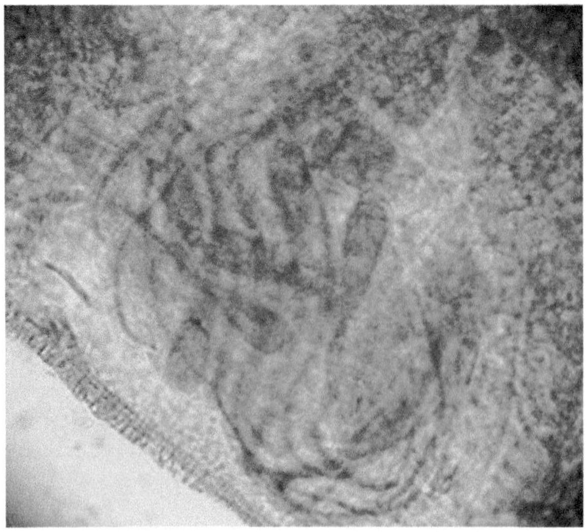

FIGURE 12.11 Spermatheca with spermatozeugma of *Tubifex blanchardi* (400×). *Photo: M. Marchese (INALI-CONICET-UNL, Argentina).*

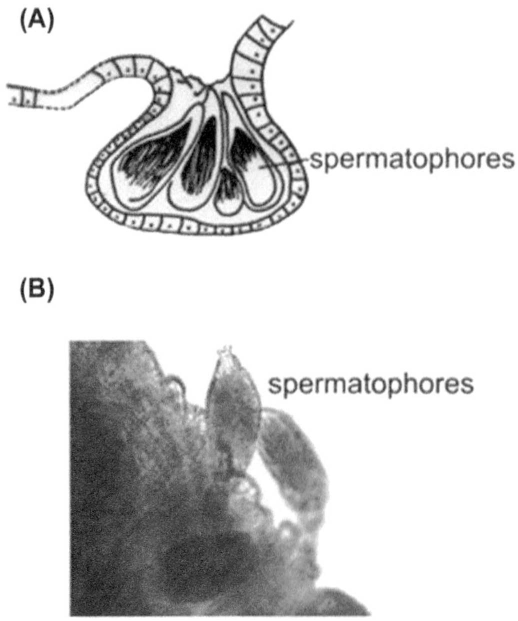

FIGURE 12.12 Spermatophores external of the body wall: (A) *Paranadrilus descolei* and (B) *Bothrioneurum* sp. (400×).

2(1) Worms not very long and thin, mostly small enough to mount whole on a slide; chaetae absent, or two or more per bundle; prostomium and segment I normal 3

2′ Body elongate, length at least 4–5 cm (preserved), width (slide mounts) 0.16 mm anteriorly, 0.11 posteriorly, with long segments; prostomium and peristomium elongate, projecting in front of a large mouth and sometimes divided by a transverse groove; ventral chaetae large, arcuate, solitary; dorsal chaetae small, straight rods that may be missing in a number of segments; these worms are predators and have a well-developed pharynx (Fig. 12.16) .... Haplotaxidae, one species: *Haplotaxis aedeochaeta* Brinkhurst & Marchese, 1987 [Brazil: Paraná River, Paraguay River. Paraguay: Paraguay River. Argentina: Paraná River System]

3(2) Chaetae different than below, two or more per bundle, sigmoid with nodulus 4

3′ Chaetae simple pointed, straight or sigmoid, commonly not nodulate, one or more per bundle, or totally absent, dorsal chaetae mostly from II; dorsal bundles shifted laterally; dorsal and ventral chaetae alike; no genital chaetae; prostomium short, no proboscis; compact pharyngeal glands in IV–VI; spermathecae in V, clitellum in XII–1/2 XIII, testes in XI, ovaries in XII (Fig. 12.14) ...... Enchytraeidae [Mainly terrestrial, not identified further]

PHYLUM ANNELIDA

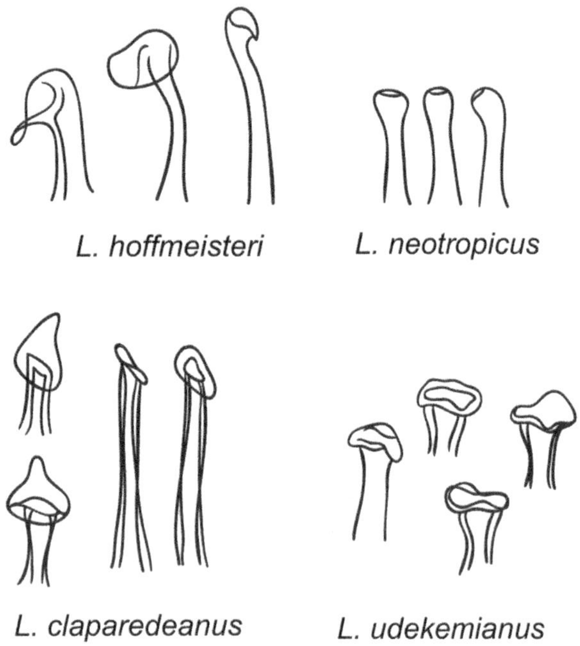

FIGURE 12.13 Types of cuticule penial sheaths of *Limnodrilus* spp. *From Brinkhurst & Marchese (1989) with permission.*

4(3) Chaetae two per bundle, simple pointed or bifid with the upper tooth minute 5

4′ Chaetae more than two per bundle, all simple pointed or bifid or ventral chaetae bifid and dorsal with hairs plus modified chaetae 6

5(4) Prostomium with or without proboscis; dorsal and ventral chaetae begin in segment II with two simple pointed or bifid chaetae with upper tooth reduced; hair and genital chaetae absent; gills or caudal appendages absent (Fig. 12.17) .... Lumbriculidae one species: *Lumbriculus variegatus* (Müller, 1774) [Introduced. Mexico: Guaracha Reservoir. Brazil: Minas Gerais. Argentina: Patagonia. Native to Holarctic Region]

5′ Prostomium without proboscis; simple pointed chaetae; one or two sterile segments between testes and ovaries in *Brinkhurstia*; genital chaeta modified in XIII; testes in X, ovaries in XIII; male pore in XIII and female pore in XIV; spermathecal pore single median dorsal in IX .... **Alluroididae [p. 443]**

6(4) Tail end without processes unless accompanied by gills; prostomium with proboscis present or absent 7

6′ Tail end with three ciliated appendages (Fig. 12.7), one median and two lateral; prostomium with a proboscis; dorsal chaetae begin in segment II with two to four hair chaetae accompanied by simple pointed aciculate or hair chaetae; ventral chaetae bifid; genital chaetae absent; testes behind XIV, ovaries behind XV .... **Opistocystidae [p. 447]**

7(6) Dorsal chaetae begin in segment III; ventral chaetae paired, one simple pointed and one bifid chaetae or a pair of one chaetal type; testes in XI and spermathecae in XIII (Fig. 12.18) .... **Phreodrilidae [p. 448]**

7′ Dorsal chaetae begin in segment II or more posteriorly; chaetae rarely simple pointed; dorsal bundles may be bifid, bifid plus hair chaetae or pectinate to palmate chaetae with hair chaetae, usually numerous; ventral chaetae usually bifid and more than two per bundle .... **Naididae [p. 448]**

Clitellata: Oligochaeta: Alluroididae: Genera

1 Penial chaetae present 2

1′ Penial chaetae absent; male pores in XIII, testes in XI, spermathecal pore middorsal in 7/8 .... *Righiella jamiesoni* Omodeo & Coates, 2001 [Guyana]

2(1) Male pores in XIII, testes in X, spermathecal pore single, dorsal in IX .... ***Brinkhurstia* [p. 446]**

2′ Male pores in XIV, testes in XI, spermathecal pore lateral in 6/7 .... *Kathrynella guyanae* Omodeo, 1996 [Guyana]

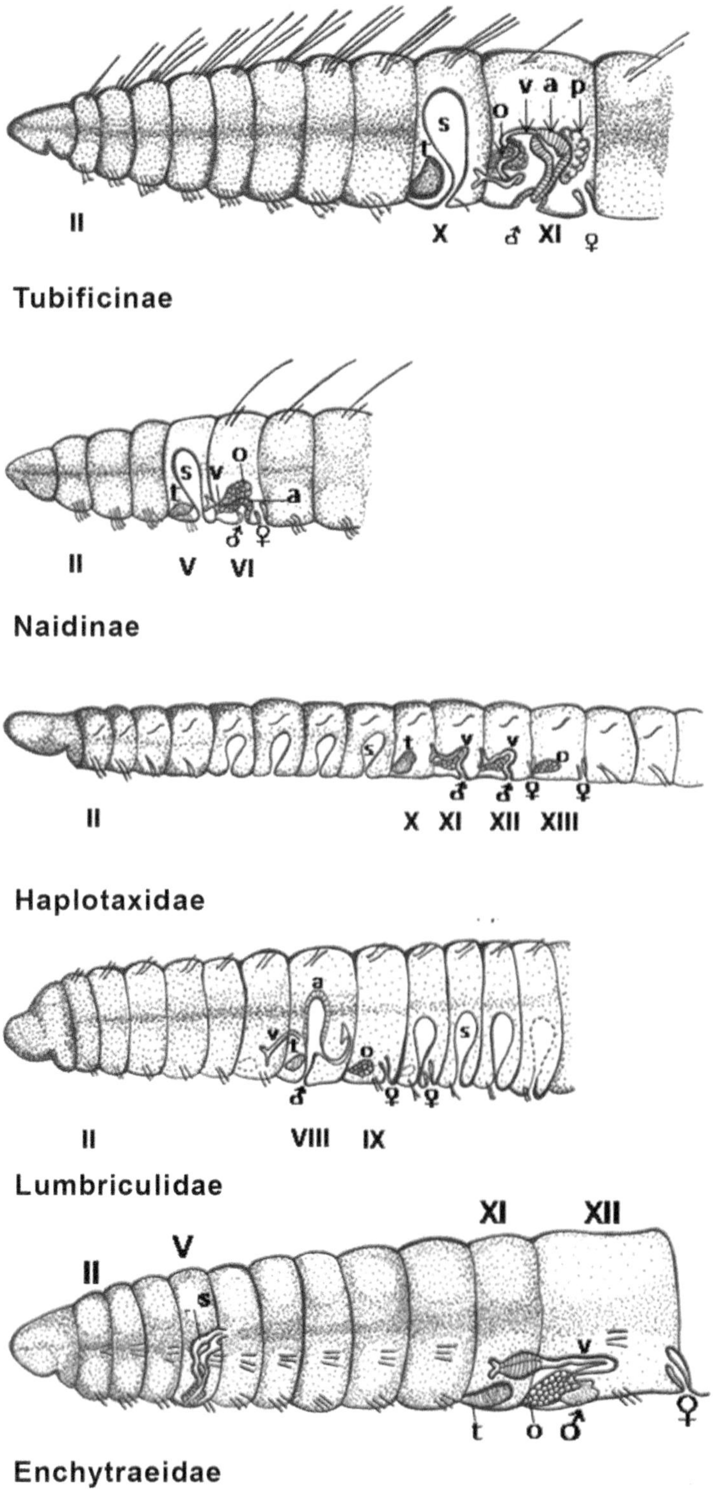

FIGURE 12.14 Location of reproductive organs in different families of Oligochaeta. Abbreviations: *a*, atrium; *o*, ovary; *p*, prostate; *s*, spermathecae; *t*, testis; *v*, vas deferens.

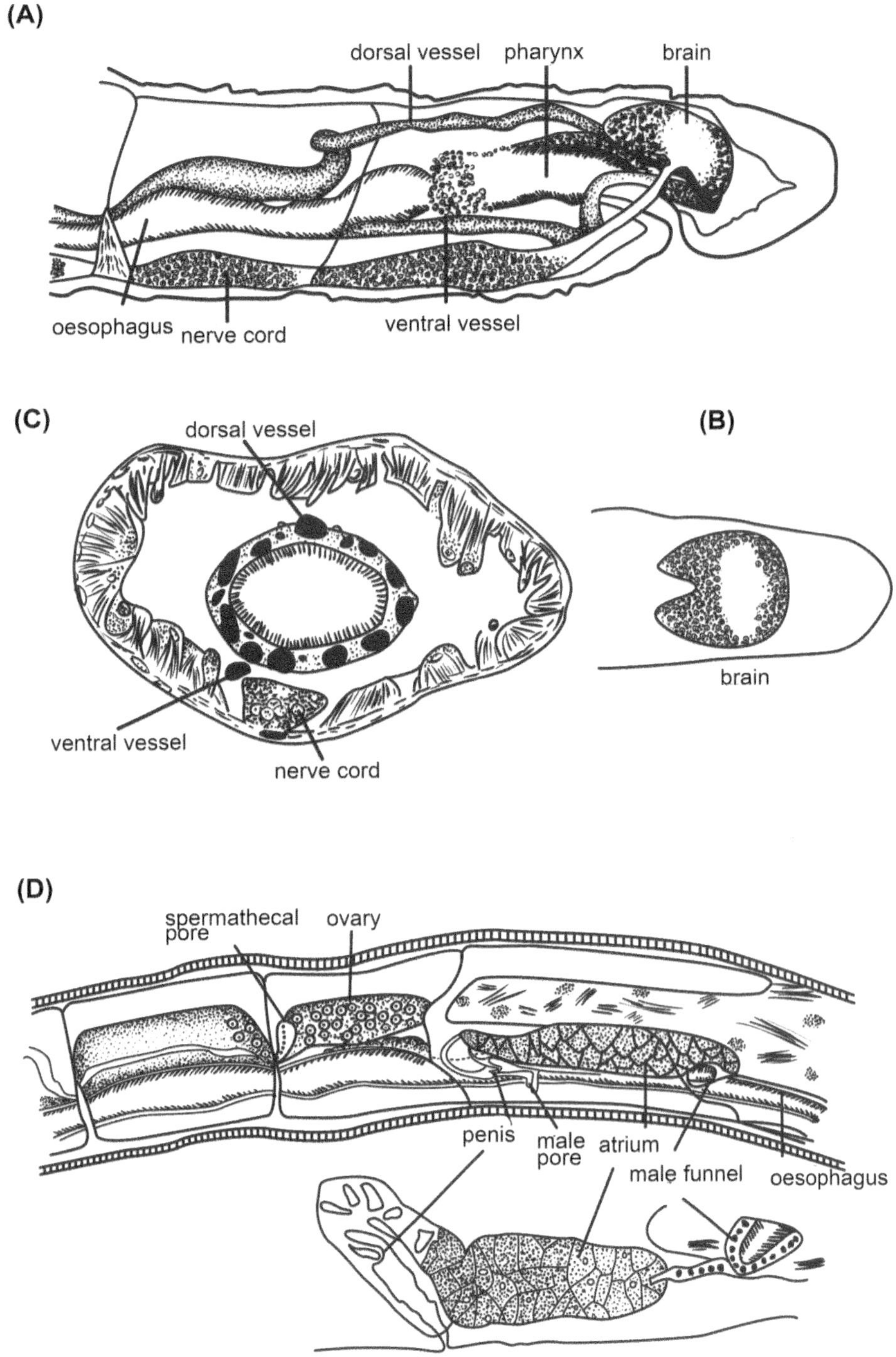

FIGURE 12.15 *Narapa bonettoi*: (A) lateral view of prostomium and first segments; (B) dorsal view of prostomium and brain; (C) transverse section; and (D) reproductive organs. *From Brinkhurst & Marchese (1989) with permission.*

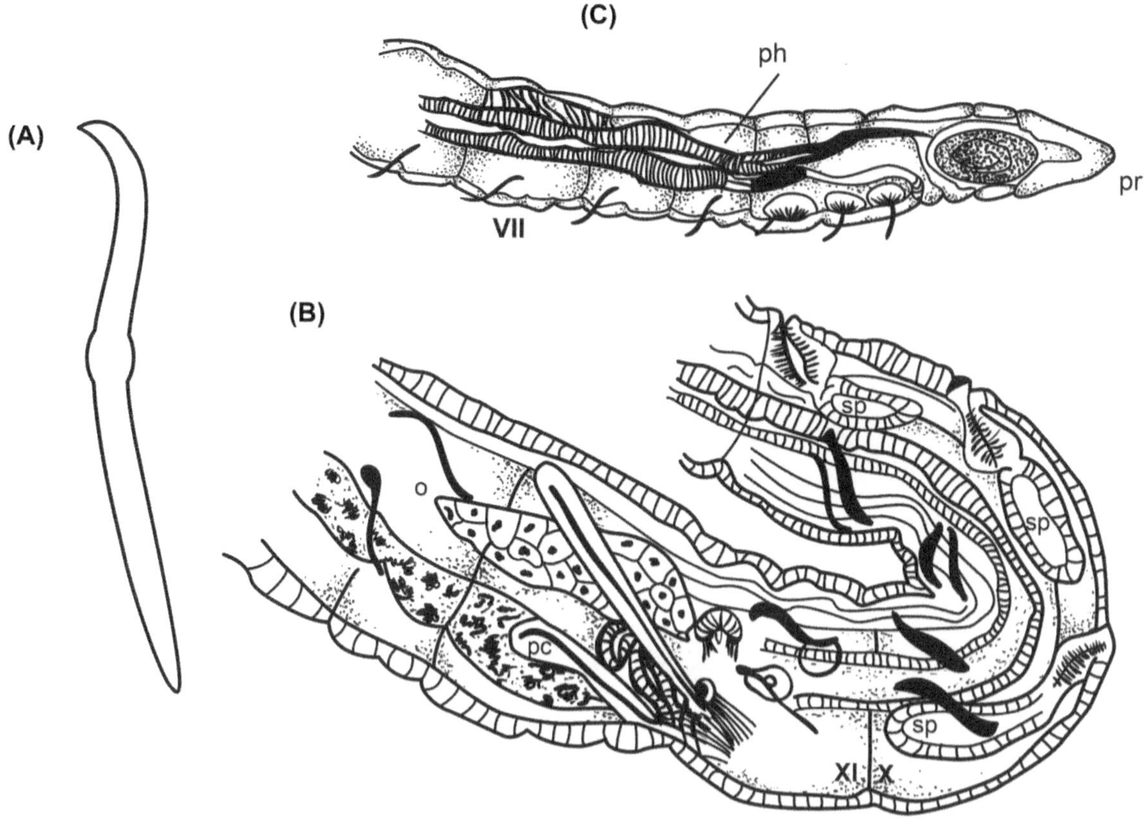

FIGURE 12.16 *Haplotaxis aedeochaeta*: (A) anterior ventral chaetae; (B) reproductive organs; and (C) prostomium and anterior segments. Abbreviations: *o*, ovary; *pc*, penial chaeta; *ph*, pharynx; *pr*, prostomium; *sp*, spermathecae. *Redraw from Brinkhurst & Marchese (1989).*

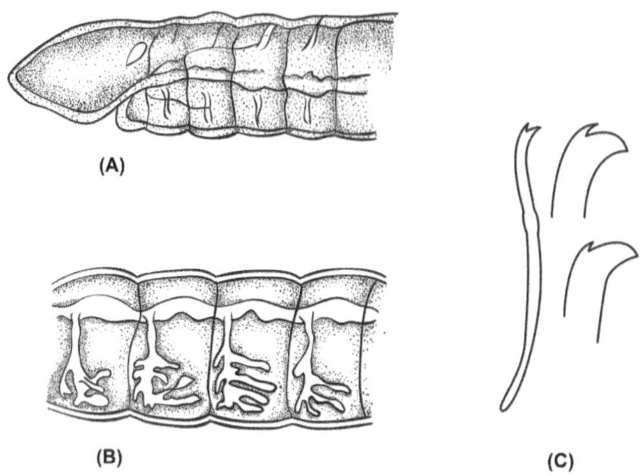

FIGURE 12.17 *Lumbriculus variegatus*: (A) anterior segments showing the conical prostomium and the brain; (B) posterior segments showing the lateral blood vessels; and (C) chaetae. *Redraw from Brinkhurst & Marchese (1989).*

Clitellata: Oligochaeta: Alluroididae: *Brinkhurstia*: Species

1 Penial chaetae straight (Fig. 12.19) .... *Brinkhurstia americana* (Brinkhurst, 1964)
[Panama: Coiba Island. Guyana. Brazil. Paraguay. Argentina]

1′ Penial chaetae taeniform .... *Brinkhurstia donaldi* Omodeo & Coates, 2001
[Guyana]

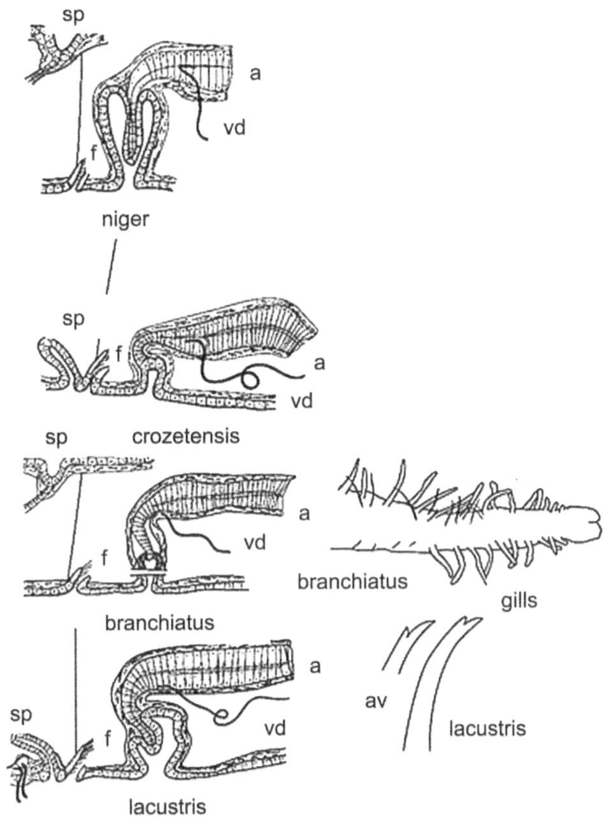

FIGURE 12.18 Phreodrilidae. *Antarctodrilus niger, Nesodrilus crozetensis, Phreodrilus branchiatus,* and *Insulodrilus lacustris.* Anterior ventral chaetae of *I. lacustris.* Lateral gills on posterior segments of *Phreodrilus branchiatus.* Abbreviations: *a,* atrium; *f,* funnel; *sp,* spermathecae; *vd,* vas deferens. *Reprinted from Brinkhurst & Marchese (1989) with permission.*

Clitellata: Oligochaeta: Opistocystidae: Genera

1 Dorsal bundles with hispid hair chaetae 2

1′ Dorsal bundles with smooth hair chaetae and one to two needles setaform; ventral chaetae, four to five with the upper tooth thinner but of equal length than the lower; testes in XIV or XV, ovaries in XV or XVI, spermathecae in XVI or XVII .... *Trieminentia corderoi* (Harman, 1970)
[Costa Rica. Paraguay. Argentina. Uruguay]

2(1) Dorsal bundles with one to four hispid hairs and one to four simple pointed aciculate chaetae without nodulus; ventral chaetae, three to five with the upper tooth longer and thinner than the lower, posterior bundles with teeth almost equal length; testes in XI, ovaries in XII, atria pear shape in XII with diffuse prostata, penes present; spermathecae in XIII .... *Crustipellis tribranchiata* (Harman, 1969)
[Guyana]

2′ Dorsal bundles with hispid or serrate hair chaetae; testes in XXI, ovaries in XXII, spermathecae with ventral pores in XXIII .... ***Opistocysta*** **[p. 447]**
[Colombia. Guyana. Brazil. Paraguay. Argentina. Uruguay]

Clitellata: Oligochaeta: Opistocystidae: *Opistocysta*: Species

1 Dorsal bundles with hair chaeta with thick serrations, ventral chaetae three to five with teeth equally long, lower tooth thicker than the upper .... *Opistocysta serrata* Harman, 1970
[Brazil. Paraguay. Argentina]

1′ Dorsal bundles with two to three simple pointed aciculate chaetae with nodulus and two to three thin hispid hair chaetae, ventral chaetae three to five with the upper tooth longer and thinner than lower .... *Opistocysta funiculus* Cordero, 1948
[Colombia. Guyana. Brazil. Argentina. Uruguay]

Clitellata: Oligochaeta: Phreodrilidae: Genera

1 Gills absent 2

1′ With lateral gills posteriorly; ventral chaetae, one bifid and one simple pointed per bundle; spermathecal chaetae absent; pseudopenes short, eversible (Fig. 12.18) .... *Phreodrilus branchiatus* Beddard, 1891
[Chile]

2(1) Spermathecal chaetae absent 3

2′ A pair of modified spermathecal chaetae; ventral chaetae both with nodulus, one slightly bifid, one simple pointed anteriorly and may be bifid posteriorly (Fig. 12.18) .... *Insulodrilus lacustris* (Benham, 1903)
[Argentina]

3(2) Ventral chaetae one thick bifid with small upper tooth and one thin simple pointed per bundle; penes large within sacs (Fig. 12.18) .... *Antarctodrilus niger* Beddard, 1894
[Argentina. Antarctic Islands: South Georgia]

3′ Ventral chaetae bifids of equal shape and length; pseudopenes small and eversible (Fig. 12.18) .... *Nesodrilus crozetensis* (Michaelsen, 1905)
[Argentina. Antarctic Islands: South Georgia]

Clitellata: Oligochaeta: Naididae: Genera

Three species of *Aulodrilus* are included in this key.

1 Dorsal chaeta absent or present, bundles may begin somewhere between II and VI or even posteriorly; hair chaetae when present accompanied by chaetae, which usually differ markedly from the ventrals (= aciculates) and may be bifid, simple pointed, spatulate or pectinate; ventral bundles with numerous bifid chaetae, those of II or II–V often quite different from the rest; asexual reproduction by budding common; clitellum in V–VIII in mature specimens; mature forms with spermathecae in IV, or V, or VII, male pores one segment behind them; eyes may be present; prostomium may bear a proboscis; paired ciliated gills may be present around the anus (Fig. 12.14); Naidinae and Pristininae 2

1′ Dorsal chaetae begin in II; hair chaetae when present, usually accompanied by pectinate chaetae, sometimes by bifids; in either instance, the dorsal chaetae resemble the ventrals in general form and the ventrals may even have rudimentary pectination; ventral bundles with numerous bifid chaetae, very occasionally some simple pointed chaetae anteriorly; reproduction usually sexual, with spermathecal pores usually on X, in some species associated with modified large chaetae and sometimes displaced forward by one or two segments; male pores on XI in most instances; modified penial chaetae present or absent; penis sheaths present or absent; eyes absent; no proboscis; gills absent or present on the posterior segments as single middorsal and midventral projections (*Branchiura*) (Fig. 12.14); Tubificinae and Rhyacodrilinae 16

2(1) Dorsal chaetae present 3

2′ Dorsal chaetae absent (Fig. 12.20) .... *Chaetogaster*
[Cosmopolitan]

3(1) Hair chaetae present in dorsal bundles 4

3′ Hair chaetae absent in dorsal bundles 13

4(3) Prostomium with proboscis 5

4′ Prostomium without proboscis 6

5(4) Dorsal chaetae begin in II or III .... *Pristina* (in part)
[Cosmopolitan]

5′ Dorsal chaetae begin in VI (Fig. 12.21) .... *Stylaria*
[Mexico. Brazil. Argentina]

6(4) Dorsal chaetae begin in II or III 7

6′ Dorsal chaetae begin in posterior segments 9

7(6) Dorsal chaetae begin in II or III; body wall naked 8

7′ Dorsal chaetae begin in II; body wall with glandular secretions and encrusted with foreign matter (Fig. 12.22) .... *Stephensoniana trivandrana* (Aiyer, 1926)
[Mexico. Haiti. Panama. Colombia. Suriname. Brazil. Paraguay. Argentina]

8(7) Dorsal chaetae begin in II or III, hair chaetae longer and very thin in median segments, may be missing from many bundles and one stout simple pointed nodulate needle .... *Bratislavia dadayi* (Michaelsen, 1905)
[Suriname. Brazil. Argentina]

8′ Dorsal chaetae begin in II, hair chaetae not very thin and present in all dorsal bundles .... *Pristina* (in part)
[Cosmopolitan]

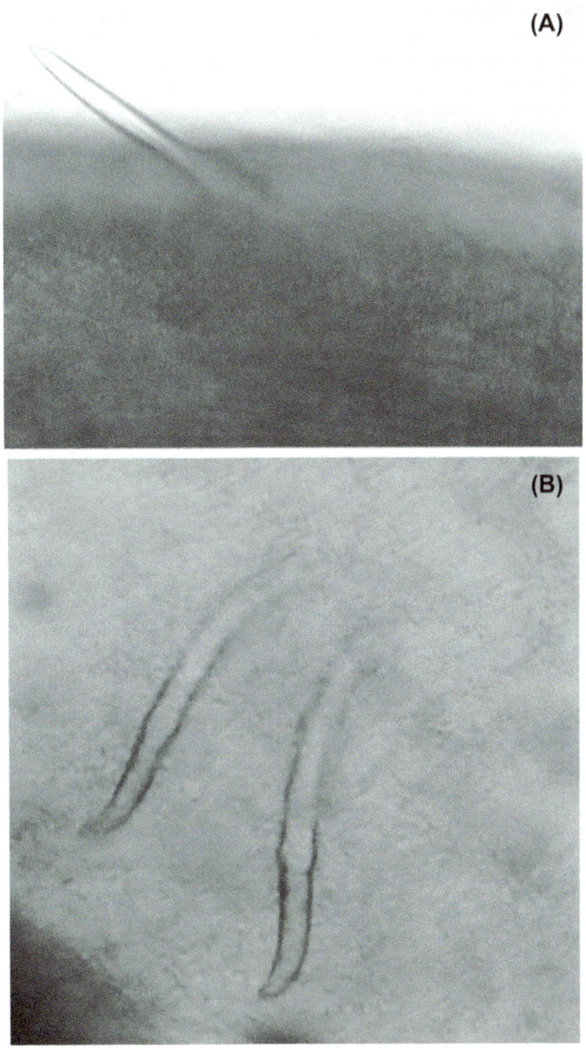

FIGURE 12.19 *Brinkhurstia americana* (1000×): (A) penial chaetae and (B) somatic chaetae.

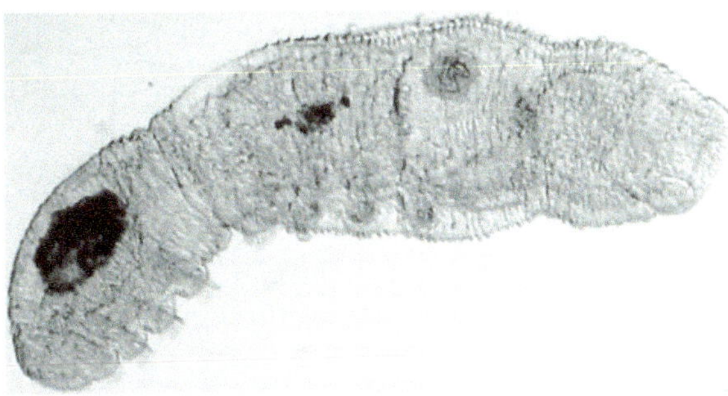

FIGURE 12.20 *Chaetogaster* (100x). *Photo: L. Falci Theza Rodrigues (UFJF, Brazil).*

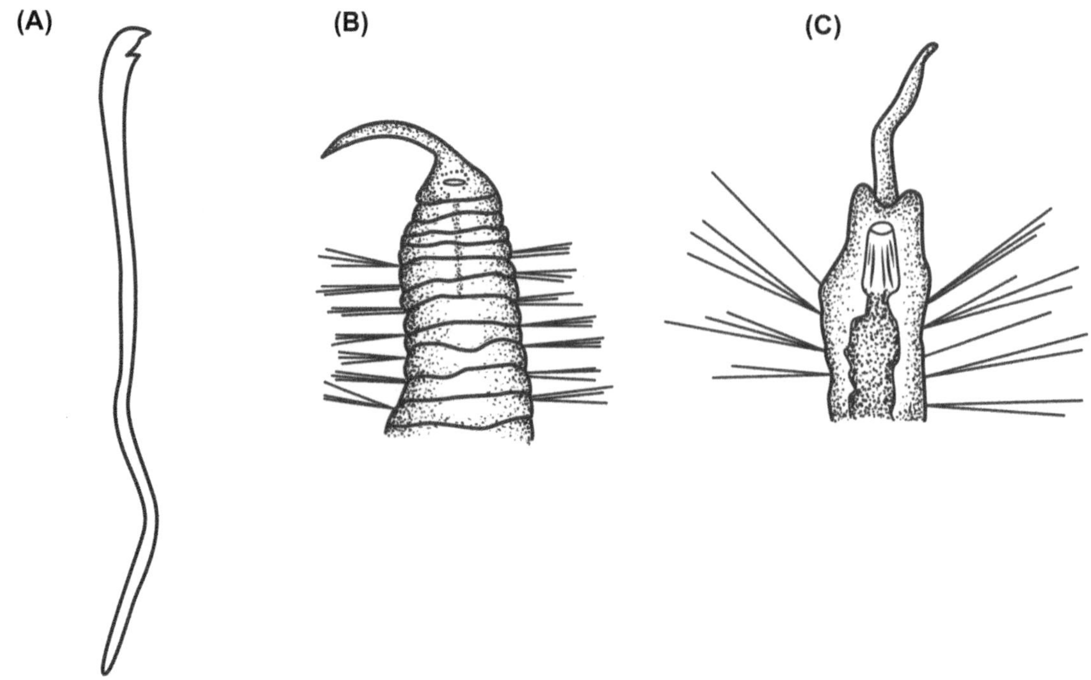

FIGURE 12.21 Prostomium with proboscis of *Stylaria*: (A) ventral chaeta; (B) *Stylaria fossularis*; and (C) *Stylaria lacustris.*

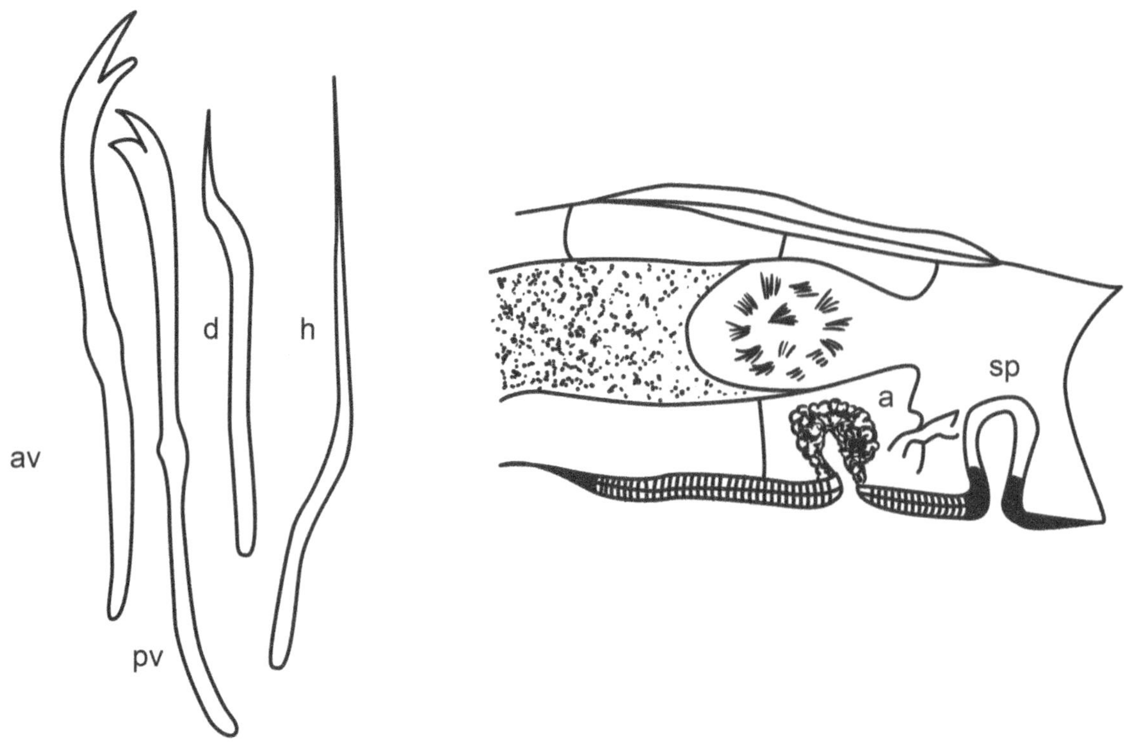

FIGURE 12.22 *Stephensoniana trivandrana*: dorsal and ventral chaetae and reproductive organs. Abbreviations: *a*, atrium; *av*, anterior ventral chaeta; *d*, dorsal chaeta, *h*, hair chaeta; *pv*, posterior ventral chaeta; *sp*, spermathecae. *From Brinkhurst & Marchese (1989) with permission.*

9(6)	Dorsal chaetae begin in V or VIII; branchial fossa and gills absent in terminal segments	10
9′	Dorsal chaetae begin in IV or VI; branchial fossa with gills present or rudimentary in terminal segments (Fig. 12.23) [Cosmopolitan]	*Dero*
10(9)	Body wall naked	11
10′	Sensory papillae present, organic matter adhered to body wall present or absent	12
11(10)	Dorsal aciculate chaetae of distinctive shape, with lateral teeth of unequal length, upper tooth longer or shorter or pectinate or simple pointed; ventral chaetae change in form between V and VI, the differences between anterior and posterior chaetae are not great [San Salvador. Cuba. Barbados. Martinique. Jamaica. Puerto Rico. Aruba. Bonaire. Curaçao. Venezuela. Ecuador. Suriname. Peru. Brazil. Paraguay. Argentina]	*Allonais*
11′	Dorsal aciculate chaetae simple pointed or bifid, ventral chaetae change in form between V and VI, the differences being considerable in most species, involving total length and width, relative sizes of the teeth and nodulus position [Cosmopolitan]	*Nais*

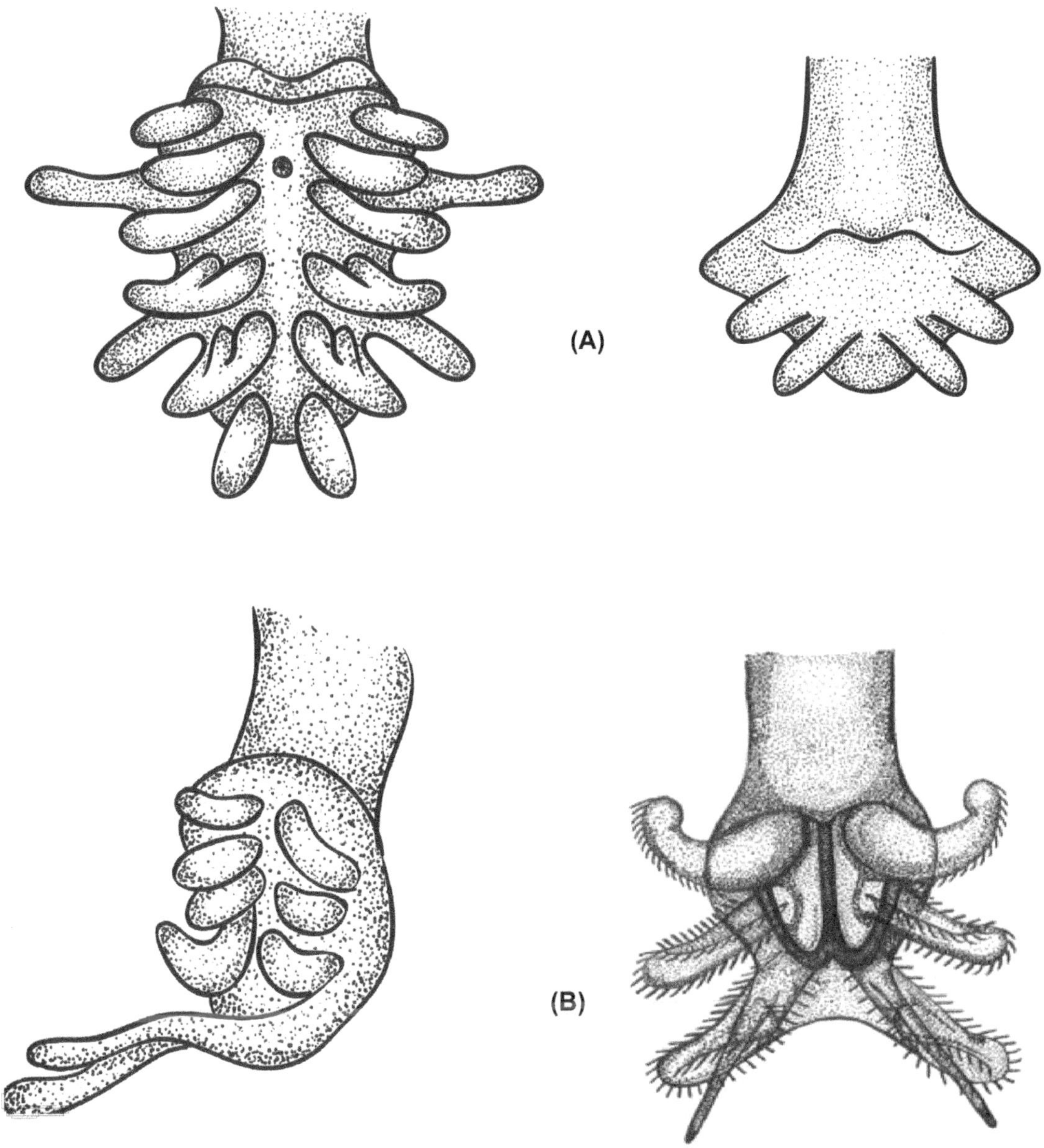

FIGURE 12.23 Branchial fossa of *Dero*: (A) *Dero (Dero)* and (B) *Dero (Aulophorus)*.

12(10) Dorsal aciculate chaetae thin distally, with slightly swollen or minutely bifid tips; ventral chaetae either alike, or those of II differ from the rest, or II–V shorter and thinner than the rest; elongate hairs on V or VI in two species; body wall covered with papillae, with secreted matter that incorporates bacteria and other foreign particles making chaetae hard to see .... *Slavina*
[Venezuela. Suriname. Peru. Brazil. Uruguay. Argentina]

12′ Dorsal chaetae begin in XVIII or XX; body wall naked .... *Haemonais waldvogeli* Bretscher, 1900
[Suriname. Martinique. Guadeloupe. Brazil. Argentina]

13(3) Dorsal chaetae begin in posterior segments 14

13′ Dorsal chaetae begin in II .... *Homochaeta lactea* (Černosvitov, 1937)
[Argentina]

14(13) Dorsal chaetae begin in V or VI 15

14′ Dorsal chaetae begin in III (Fig. 12.24) .... *Amphichaeta leydigi* Tauber, 1879
[Brazil. Argentina]

15(14) Dorsal chaetae begin in V .... *Paranais*
[Brazil. Peru. Bolivia. Argentina]

15′ Dorsal chaetae begin in VI (Fig. 12.25) .... *Ophidonais serpentina* (Müller, 1773)
[Argentina. Uruguay]

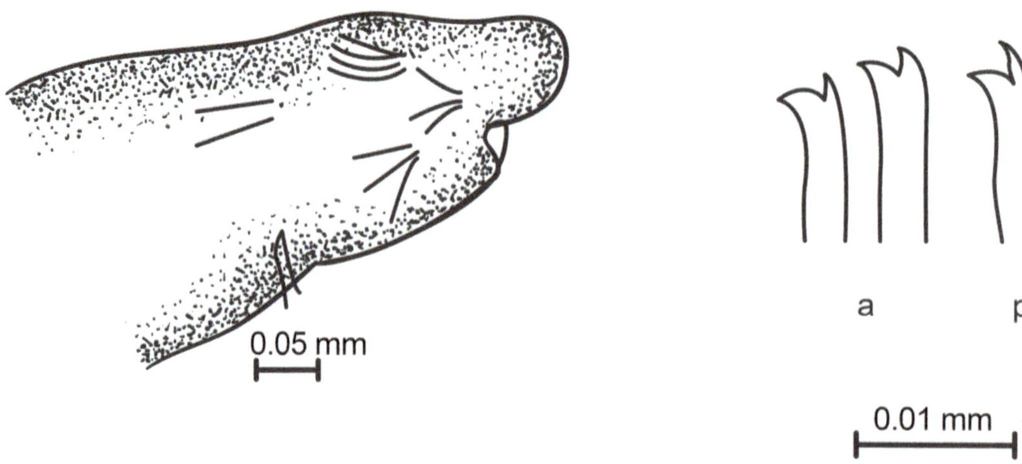

FIGURE 12.24 *Amphichaeta leydigi*: anterior (a) and posterior (p) chaetae. *From Brinkhurst & Marchese (1989) with permission.*

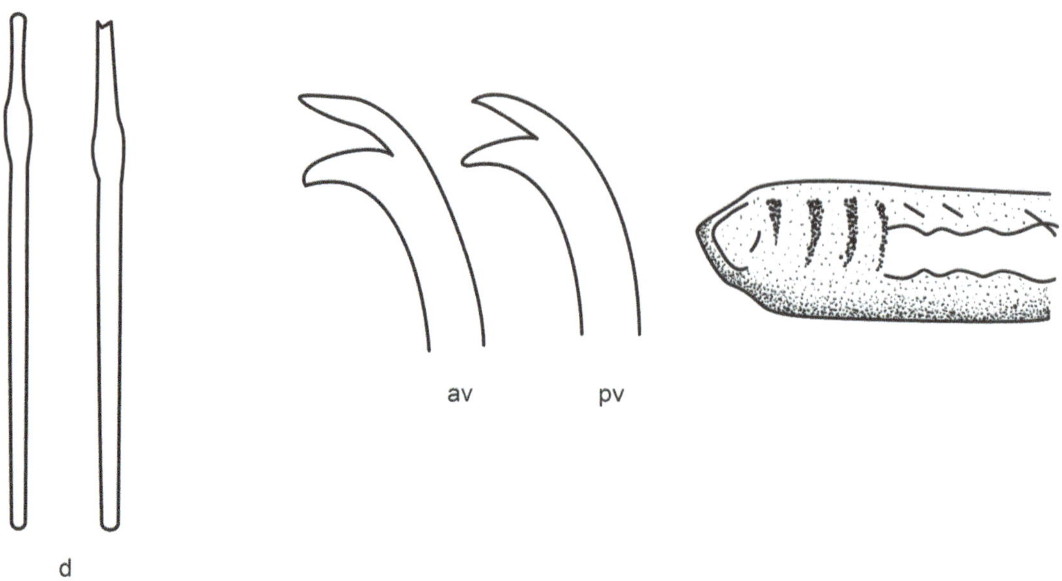

FIGURE 12.25 *Ophidonais serpentina*: dorsal and ventral chaetae. Abbreviations: *av*, anterior ventral; *d*, dorsal; *pv*, posterior ventral. *From Brinkhurst & Marchese (1989) with permission.*

16(1)	Dorsal hair chaetae present	17
16′	Dorsal hair chaetae absent	24
17(16)	Body wall not covered with papillae	18
17′	Body wall covered with small papillae (Fig. 12.26) [Venezuela]	*Spirosperma velutina* (Grube, 1879)

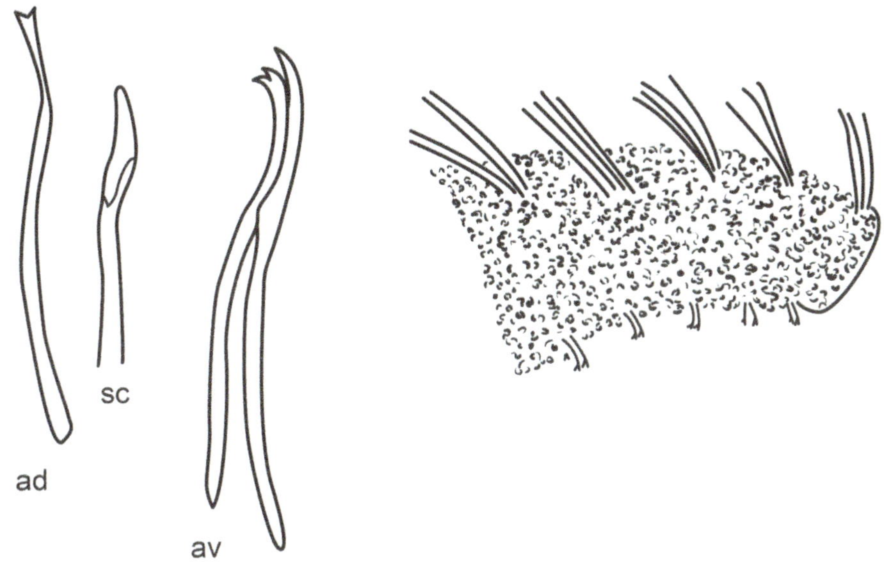

FIGURE 12.26 *Sprirosperma velutinus. ad*, anterior dorsal chaetae; Abbreviations: *av*, anterior ventral chaetae; *sc*: spermathecal chaetae.

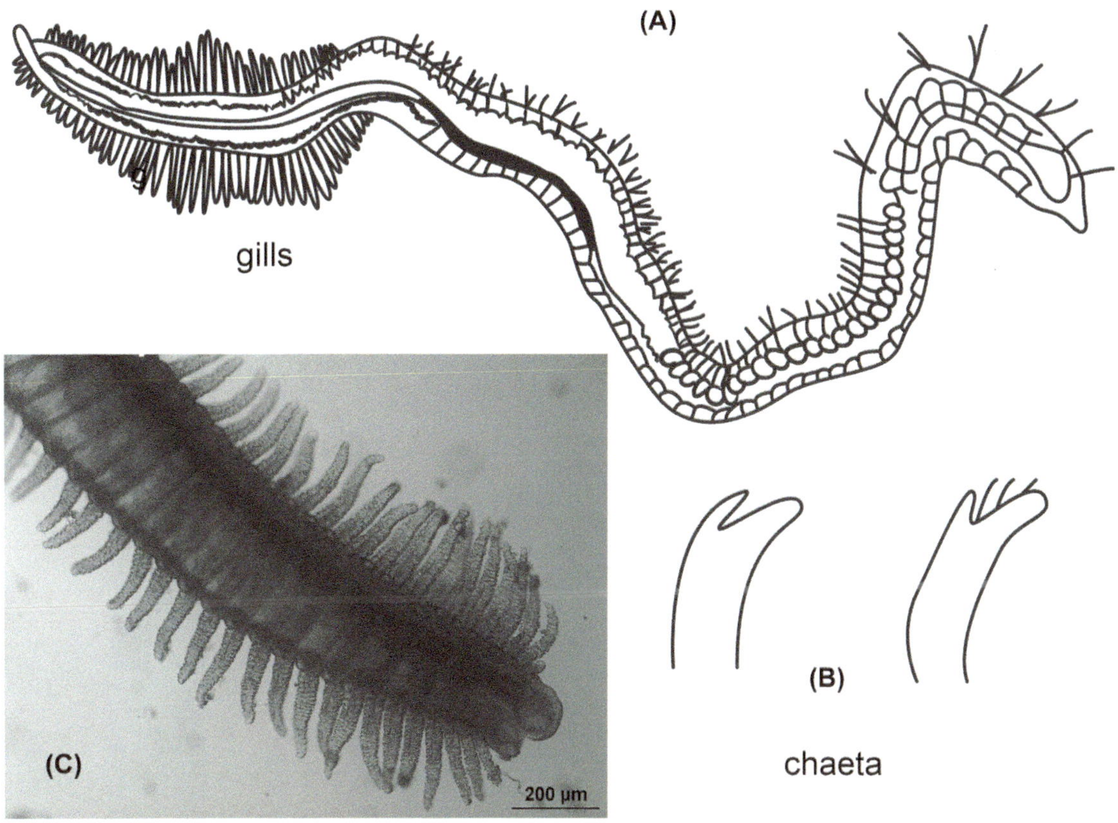

FIGURE 12.27 *Branchiura sowerbyi*: (A) illustration of the anterior and posterior segments; (B) pectinate dorsal and ventral chaetae; and (C) gills in posterior segments.

18(17)	Hair chaetae, accompanied by pectinate chaetae	19
18′	Hair chaetae accompanied by bifid chaetae, or chaetae with the thin short upper teeth duplicated, or oar-shaped chaetae behind VI or posteriorly	20
19(18)	No modified genital chaetae; thin cuticular penis sheaths present in mature specimens in XI	21
19′	Chaetae genital modified; no penis sheaths	22
20(18)	Usually large worms with middorsal and midventral gill filaments on each posterior segments (Fig. 12.27)	*Branchiura sowerbyi* Beddard, 1892 [Mexico. Colombia. Brazil. Argentina. Uruguay]
20′	Small worms without gills; dorsal hair and bifid chaetae with upper tooth shorter and thinner than the lower, from VII dorsal bifid become oar-shaped (Fig. 12.28)	*Aulodrilus pigueti* Kowalewski, 1914 [Colombia. Guyana. Brazil. Peru. Paraguay. Argentina]

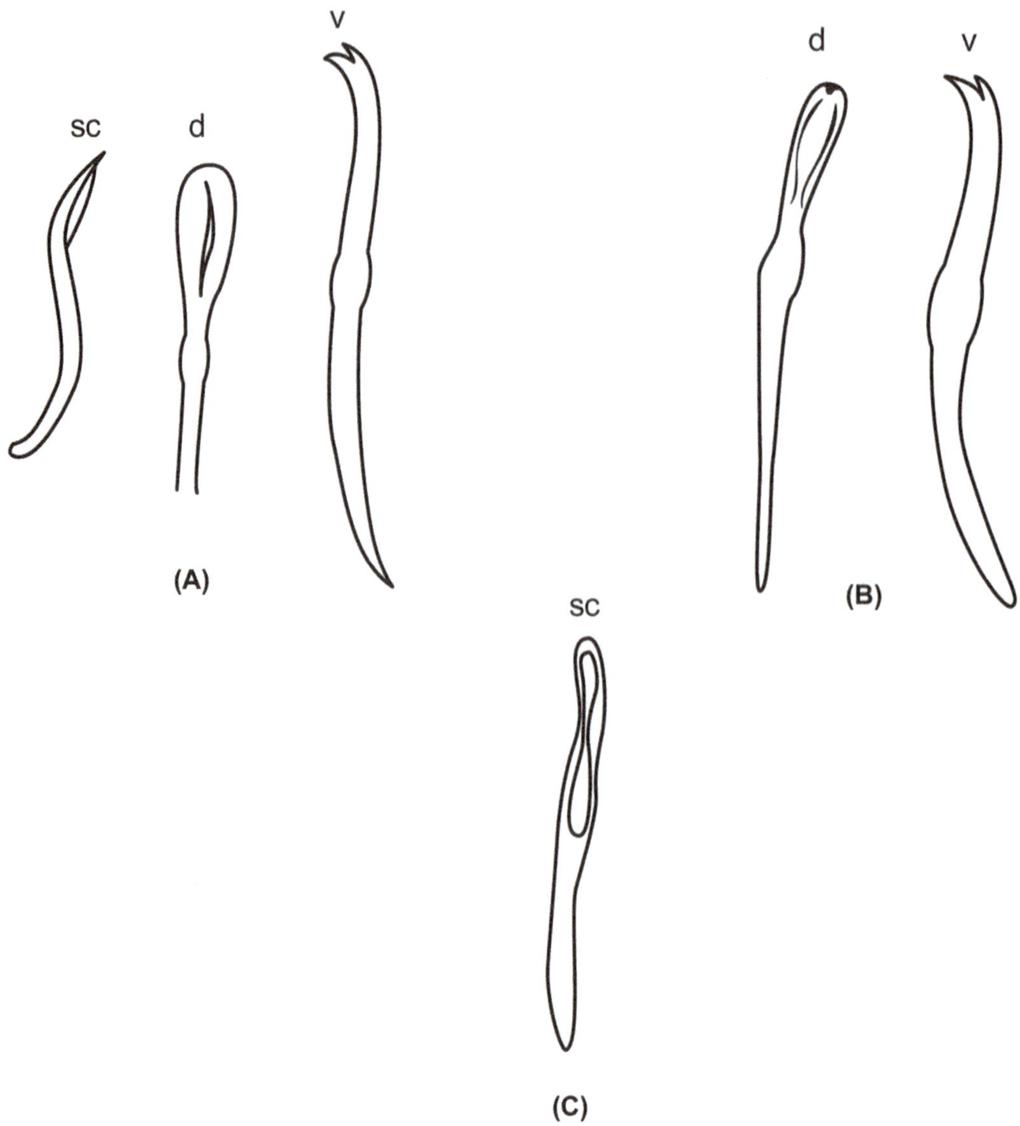

FIGURE 12.28 *Aulodrilus* sp: (A) *Aulodrilus pigueti*; (B) *Aulodrilus limnobius*; and (C) *Aulodrilus adetus*. Abbreviations: *d*, dorsal chaetae; *v*, ventral chaetae; *sc*, spermathecal chaetae.

21(19) Truncated cone-shaped thin cuticular penis sheaths in XI of mature specimens .... *Ilyodrilus frantzi* Brinkhurst, 1965
[Argentina: Río de la Plata]

21′ Tube shaped or conical cuticular penis sheaths in XI of mature specimens .... *Tubifex* (in part)
[Cosmopolitan]

22(19) Penial chaetae modified in XI; no modified spermathecal chaetae 23

22′ Spermathecal chaetae broader and longer than the normal ventral chaetae replace ventral chaetae on X in mature specimens (or in one instance on VII–VIII or between VI and XII); no cuticular penis sheaths .... *Potamothrix*
[Peru. Uruguay]

23(22) Penial chaetae, three to five per bundle more or less straight, in a fanlike bundle with three short, knobbed distal ends close together .... *Rhyacodrilus*
[Paraguay. Argentina]

23′ Penial chaetae, three per bundle, sharply simple pointed .... *Ainudrilus dartnalli* Erséus & Grimm, 2002
[Antarctic Islands: South Georgia]

24(16) Modified penial chaetae present or absent; no cuticular penis sheaths in XI 25

24 No modified genital chaetae; cylindrical cuticular penis sheaths present in XI (Fig. 12.13) .... *Limnodrilus*
[Cosmopolitan]

25(24) Prostomial pit absent. 26

25′ Prostomium with a median dorsal sense organ (pit) and stalked spermatophores attached the body wall in mated specimens (Fig. 12.29) .... *Bothrioneurum*
[Mexico. Venezuela. Brazil. Peru. Paraguay. Argentina]

26(25) Modified genital chaetae present 27

26′ Modified genital chaetae absent 30

27(26) Modified penial chaetae present and spermathecal chaetae absent 28

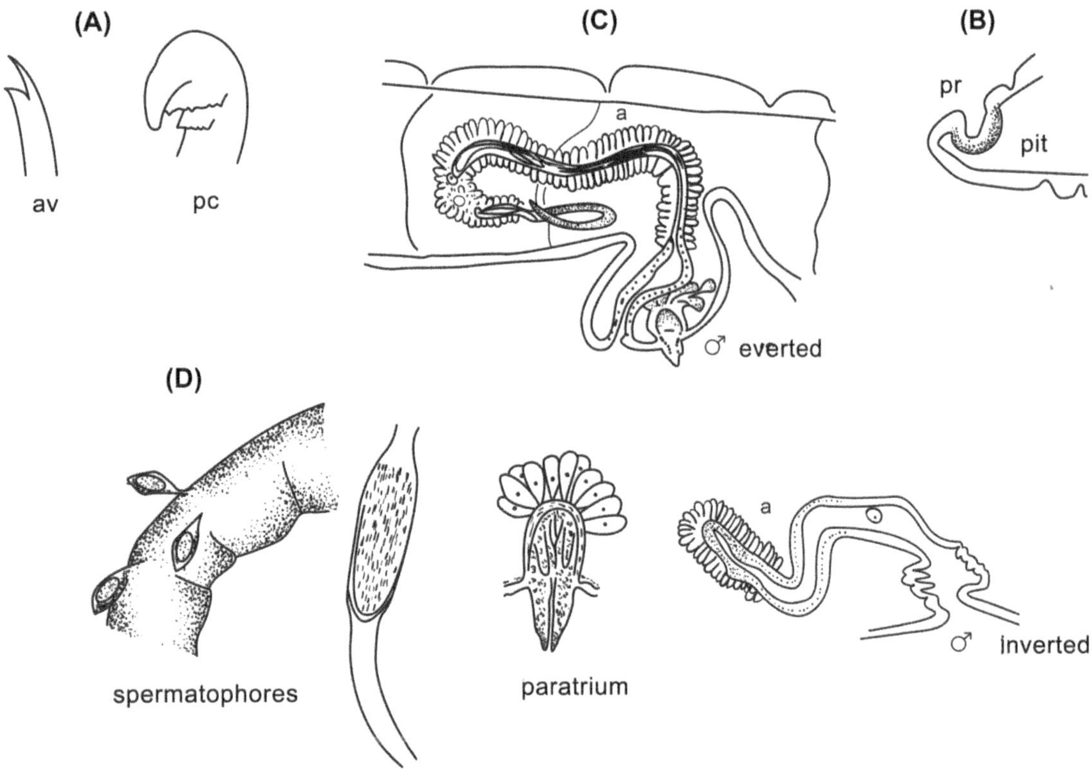

FIGURE 12.29 *Bothrioneurum* sp: (A) anterior ventral (av) and penial chaetae (pc) of *Bothrioneurum vejdovskyanum*; (B) prostomium with middorsal sensory organ of *Bothrioneurum* spp.; (C) reproductive organs; and (D) spermatophores in 100× and 400×. Abbreviations: *a*, atrium; *pr*, prostomium.

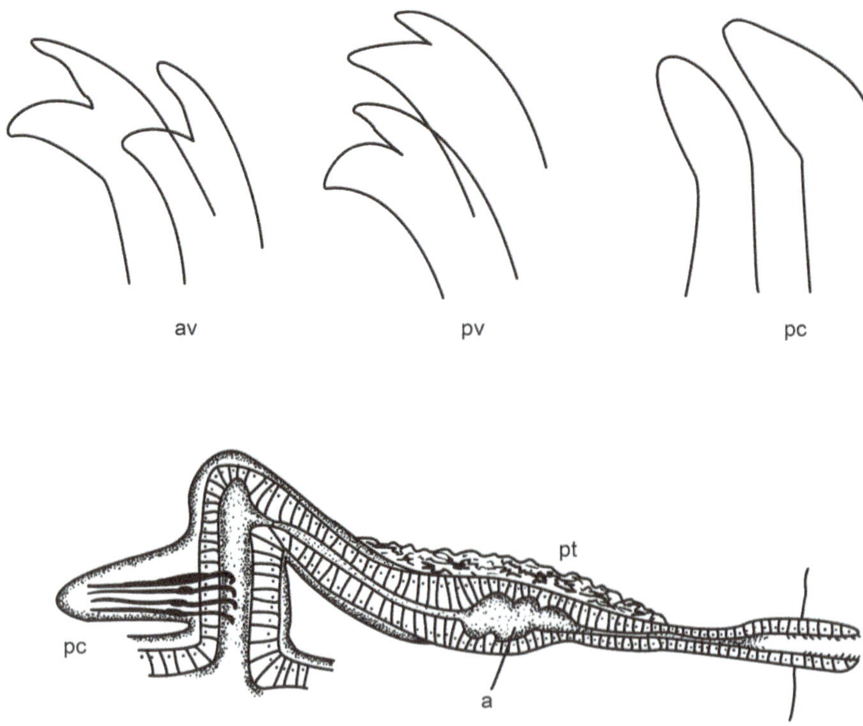

FIGURE 12.30 *Monopylephorus moleti*. Abbreviations: *a*, atrium; *av*, anterior ventral chaetae; *pc*, penial chaetae; *pt*, prostate; *pv*, posterior ventral chaetae. *Redrawing from Brinkhurst & Marchese (1989).*

27′	Modified penial chaetae absent and spermathecal chaetae present 29
28(27)	Spermathecae single in VII; median male pore on VIII leads to separate penial chaetal sacs containing up to four straight simple pointed thick penial chaetae (Fig. 12.30) .... *Monopylephorus moleti* Brinkhurst & Marchese, 1987 [Argentina: Paraná River System]
28′	Spermathecae paired in X; male pores separate on XI; penial chaetae three to four per bundle in XI, hooked, simple pointed tips, thick .... *Epirodrilus antipodum* Černosvitov, 1939 [Peru and Bolivia: Lake Titicaca basin]
29(27)	Spermathecal chaetae straight and hollow-tipped in ventral bundles of X in mature specimens .... *Isochaetides lacustris* (Černosvitov, 1939) [Peru and Bolivia: Lake Titicaca basin]
29′	Spermathecal chaetae spoon-shaped in ventral bundles of IX in mature specimens .... *Aulodrilus adetus* du Bois-Reymond Mrcus, 1947 [Brazil: Sao Paulo]
30(26)	Dorsal and ventral chaetae bifid with teeth about equal or the upper tooth much longer than the lower anteriorly; spermatheca absent; vasa deferentia very long and thin, atria irregular tubes (Fig. 12.31) .... *Paranadrilus descolei* Gavrilov, 1955 [Brazil: Paraná River, Paraguay River. Peru: Amazon drainage. Paraguay: Paraguay River. Argentina: Paraná River System]
30′	Dorsal and ventral chaetae bifid with upper tooth shorter than the lower; dorsal chaetae of mid and posterior bundles with lateral wings, give the chaetae a narrow oar-shaped appearance .... *Aulodrilus limnobius* Bretscher, 1899 [Brazil. Argentina.]

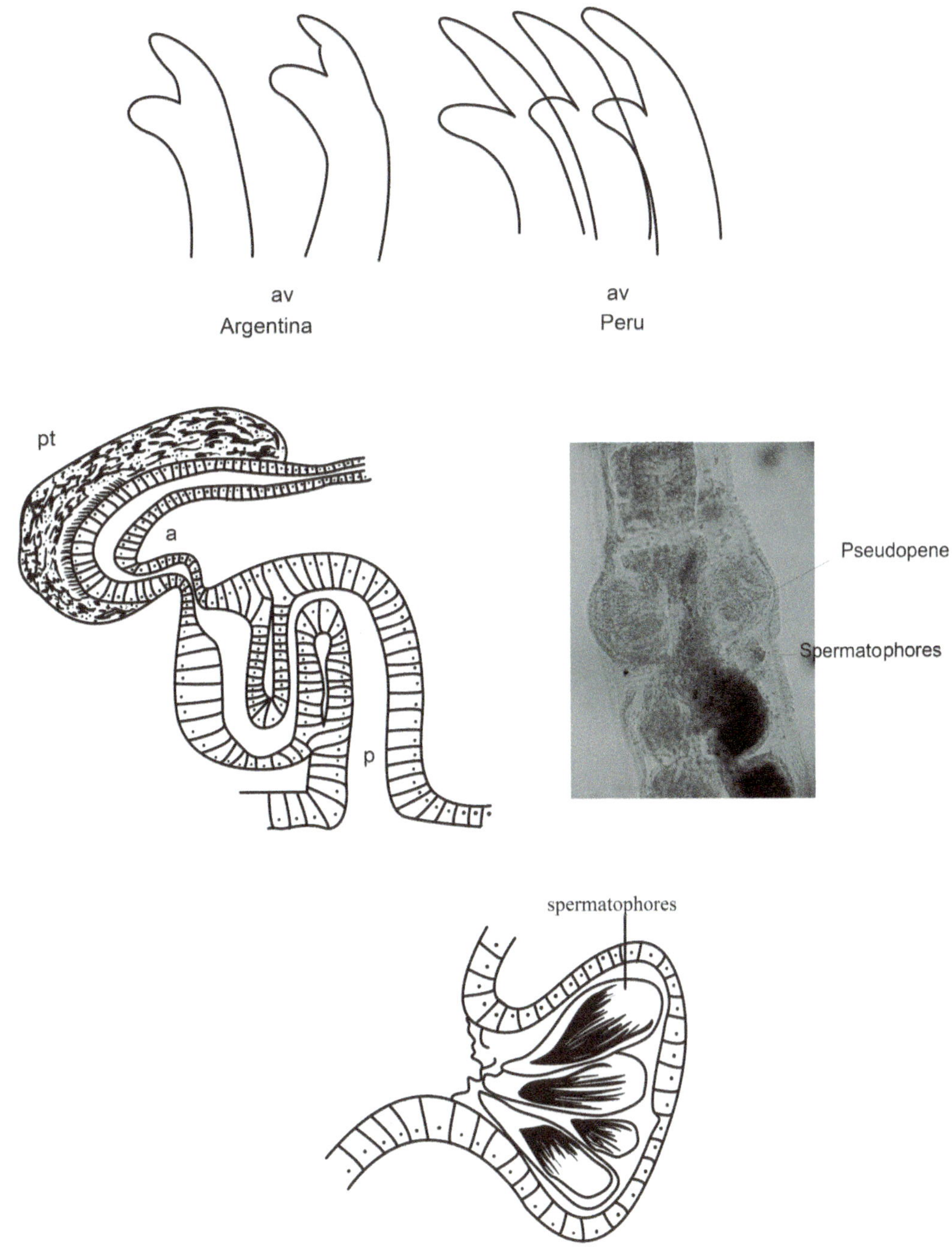

FIGURE 12.31 *Paranadrilus descolei*: Anterior ventral chaetae and reproductive organs. Abbreviations: *a*, atrium; *p*, penis; *pt*, prostomium. *Photo: M.Marchese (INALI-CONICET-UNL).*

REFERENCES

Brinkhurst, R.O. & S.R. Gelder.1991. Annelida: Oligochaeta and Branchiobdellida. Pages 401–435 *in*: J.H. Thorp & A.P. Covich (eds.), Ecology and classification of North American freshwater invertebrates. Academic Press, San Diego, USA.

Brinkhurst, R.O. & M.R. Marchese. 1989. Guide to the freshwater aquatic microdrile oligochaetes of South and Central America. Colección Climax, n. 6. Asociación de Ciencias Naturales de Litoral, Santo Tomé, Argentina. 179 pp.

Brinkhurst, R.O. & M.R. Marchese. 1992. Guía para la Identificación de Oligoquetos acuáticos Continentales de Sud y Centroamérica. Colección Climax, n. 6 Segunda edición. Asociación de Ciencias Naturales de Litoral, Santo Tomé, Argentina 207 pp.

Cuadrado, S. & E. Martínez-Ansemil. 2001. External structures used during attachment and sperm transfer in tubificids (Annelida,

Oligochaeta). *in*: P. Rodriguez & P.F.M. Verdonschot (eds), Aquatic Oligochaete Biology VIII. Hydrobiologia 463: 107–113.

Corbi, J.J., S. Trivinho-Strixino & R.G. Alves. 2005. Records of oligochaetes in freshwater sponges, on bryozoarians and on colonial hydrozoans from Brazil. Brazilian Journal Biology 65: 187–188.

Christoffersen, M.L. 2007. A catalogue of aquatic microdrile oligochaetes (Annelida: Clitellata) from South America. Acta Hydrobiologica Sinica 31: 59–86.

Christoffersen, M.L. 2009. Species diversity and distributions of microdrile earthworms (Annelida, Clitellata, Enchytraeidae) from South America. Zootaxa 2065: 51–68.

Christoffersen, M.L. 2010. Continental biodiversity of South American oligochaetes: the importance of inventories. Acta Zoológica Mexicana 2: 35–46.

Di Persia, D.H. 1980. The aquatic Oligochaeta of Argentina. Pages 70–113 *in*: R.O. Brinkhurst & D.G. Cook (eds.), Aquatic oligochaete biology. Plenum Press, New York.

Erséus, C. 2005. Phylogeny of oligochaetous Clitellata. Hydrobiologia 535/536: 357–372.

Gaviria, E.A. 1993. Claves para las especies colombianas de las familias Naididae y Tubificidae (Oligochaeta, Annelida). Caldasia 17: 237–247.

Gavrilov, K. 1977. Oligochaeta. Pages 99–121 *in*: S.H. Hurlbert (ed.), Biota acuática de Sudamerica Austral. San Diego State University, San Diego.

Gavrilov, K. 1981. Oligochaeta. Pages 170–190 *in*: S.H. Hurlbert, G. Rodriguez & N.D. dos Santos (eds.), Aquatic biota of tropical South America. Part 2. Anarthropoda. San Diego University, San Diego.

Gorni, G.R. & R.G. Alves. 2008. Naididae species (Annelida: Oligochaeta) associated with the sponge *Metania spinata* (Carter, 1881) (Porifera: Metaniidae) from a Southeastern Brazilian reservoir. Acta Limnologica Brasiliensia 20: 261–263.

Harman, W.J. 1982. Oligochaeta. Pages 162–165 *in*: S.H. Hurlbert & A. Villalobos-Figueroa (eds.), Aquatic Biota of Mexico, Central America and the West Indies. San Diego State University, San Diego, USA.

Howmiller, R.P. 1974. Some Naididae and Tubificidae from Central America. Hydrobiologia 44: 1–12.

Kahl, K. & A. Konopacka. 1981. Oligochaeta inhabiting the colonies of the sponge *Spongilla lacustris* (L.) in the River Gac. Acta Hydrobiologica 23: 243–249.

Marchese, M.R. 2009. Annelida Oligochaeta. Pages 551–565 *in*: E. Dominguez & H.R. Fernandez (eds.), Macroinvertebrados bentónicos sudamericanos. Fundación Miguel Lillo, Tucumán, Argentina.

Marchese, M.R. & A. Paggi. 2003. Diversidad de Oligochaeta (Annelida) y Chironomidae (Diptera) del litoral fluvial Argentino. INSUGEO, Miscelánea 12: 217–223.

Martin, P., E. Martínez-Ansemil, A. Pinder, T. Timm & M.J. Wetzel. 2008. Global diversity of oligochaetous clitellates ("Oligochaeta"; Clitellata) in freshwater. Hydrobiologia 595: 117–127.

Martínez-Ansemil, E. & B. Sambugar. 2010. Annelida, an often neglected component of groundwater ecosystems. Page 77 *in*: A. Moškric & P. Trontelj (eds.), Abstract Book, 20th international conference on subterranean biology. Postojna, Slovenia. 20th International Conference on Subterranean Biology.

Martínez-Ansemil, E., M. Creuzé des Châtelliers, P. Martin & B. Sambugar. 2012. The Parvidrilidae – a diversified groundwater family: description of six new species from southern Europe, and clues for its phylogenetic position within Clitellata (Annelida). Zoological Journal of the Linnean Society 166: 530–558.

Melão, M.G.G. & O. Rocha. 1996. Macrofauna associada a *Metania spinata* (Cárter, 1881), Porifera, Metaniidae. Acta Limnologica Brasiliensia 8: 59–64.

Reynolds, J.W. & M. Wetzel. 2018. Nomenclatura Oligochaetologica – A catalogue of names, descriptions and type specimens. Editio Secunda. URL: http://www.inhs.illinois.edu/people/mjwetzel/nomenoligo . Accessed 3 May 2019.

Righi, G. 1984. Oligochaeta *in*: R. Schaden (ed.), Manual de identificação de invertebrados límnicos do Brasil. Vol. 17. Conselho Nacional de Desenvolvimento Científico e Tecnológico, Brasilia. 48pp.

Righi, G. & V. Hamoui. 2002. Oligochaeta, Naididae of the West Indies and Adjacent Regions. Papéis Avulsos de Zoologia Sao Paulo 42: 119–167.

Schmelz, R.M. & R.G. Alves. 2017. On a collection of enchytraeids (Oligochaeta) from first order streams in São Paulo State, Brazil. Opuscula Zoologica Budapest, 48 (Supplementum 2): 21–35.

Timm, T. 2009. A guide for the freshwater Oligochaeta and Polychaeta of Northern and Central Europe. Lauterbornia 66: 1-235.

Timm, T. & P.J. Martin. 2015. Clitellata: Oligochaeta. Pages 529–549 *in*: J.H. Thorp & D.C. Rogers (eds.), Ecology and General Biology, Thorp and Covich's Freshwater Invertebrates, Vol. I, 4th Edition. Academic Press, New York.

Wetzel, M.J., S. Fend, K.A. Coates, R.D. Kathman & S.R. Gelder. 2006. Taxonomy, Systematics and Ecology of the aquatic Oligochaeta and Branchiobdellida (Annelida, Clitellata) of North America, with emphasis of the Fauna Occurring in Florida. A Workbook. Florida Department of Environmental Protection (FDEP), 269 pp.

Class Clitellata: Subclass Branchiobdellidea

Stuart R. Gelder

Department of Science and Math, University of Maine at Presque Isle, Presque Isle, ME, United States

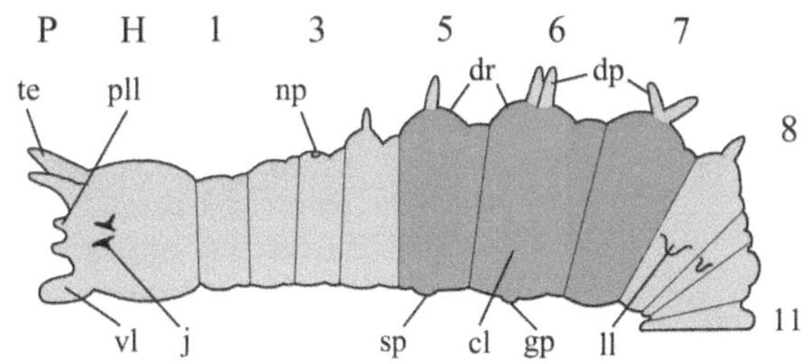

FIGURE 12.32 Lateral view of generalized branchiobdellidan to show most of the external features. Abbreviations: *1–11*, body segment numbers; *cl*, clitellum; *dp*, dorsal projections; *dr*, dorsal ridges; *gp*, genital pore; *H*, head; *j*, jaw; *ll*, lateral segmental lobes; *np*, anterior nephridial pore; *P*, peristomium; *pll*, peristomial lateral lobe; *sp*, spermatheca pore; *te*, dorsal peristomial lip tentacle; *vl*, ventral peristomial lip. *Copyright © 2015 Dr Stuart R. Gelder. Published by Elsevier Inc. All rights reserved.*

INTRODUCTION

Branchiobdellidans (crayfish worms) are small (adults: 0.8–12.0 mm long) annelids that form an ectosymbiotic association on the exposed surface or in branchial chambers of freshwater crustacean hosts, principally astacoidean crayfishes, but also shrimps, crabs, and isopods (Gelder & Williams, 2015). Branchiobdellidans have been reported from central Mexico to Costa Rica, and this forms the southern portion of their North American distribution (Holt, 1973; Holt & Opell, 1993; Gelder, 2011). Neotropical branchiobdellidans consist of five genera with those in southern Mexico being found on crayfish, crabs, and isopods, whereas in Nicaragua and Costa Rica, they are reported only on crabs. Branchiobdellidans are considered to form an order, Branchiobdellida, containing a family, Branchiobdellidae, with the five Neotropical genera residing in the subfamily Cambarincolinae; however, to be consistent with the other clitellate taxa in this chapter, branchiobdellidans have been raised in rank to subclass, Branchiobdellidea Sawyer, 1986.

Information on the distribution of branchiobdellidans in the region is based on published reports, whereas some unpublished morphological observations are used to make the key less ambiguous. However, it should be remembered that future investigations will extend known species boundaries, and a number of new species are waiting to be described. Future research will involve collecting crustacean hosts, and these may be restricted in some areas. Therefore, it is a collectors' responsibility to be acquainted with and follow local regulations.

LIMITATIONS

With few exceptions, identification of crayfish worms requires microscopical examination of either preserved or live mounted specimens. As many species are morphologically similar, determination of several morphological characters is necessary for identification, including body length and shape, jaw structure, anterior nephridial pore number, male and female reproductive structures, and other morphological features. Species identification depends on recognition of various reproductive organ structures located in body segments 5 and 6 in adults; identification of juveniles is inferential at best. There are no legally restricted branchiobdellidans in the Neotropical Region.

TERMINOLOGY AND MORPHOLOGY

Terms found in branchiobdellidan morphological descriptions differ from those used for other annelids and can lead to confusion. Fig. 12.32 illustrates the location of external body features and jaws used in species identification. Branchiobdellidan segment numbering differs from that of both oligochaetes and leeches. Branchiobdellidans have a distinct **head**, consisting of a **peristomium** and 3 segments, followed by 11 numbered body segments. These numbers are given as Arabic numerals to avoid confusion with the Roman numeral–based segment numbering systems used in other annelids. Body length and shape are important features as are ornamentation on the peristomium (lobes and tentacles) and segments (dorsal ridges formed by supernumerary muscles and dorsal projections) (Fig. 12.32).

Internal identifying characters include the **paired jaws** and the size, shape, and number of **teeth** (Fig. 12.33). These features are best observed from a dorsal or ventral view

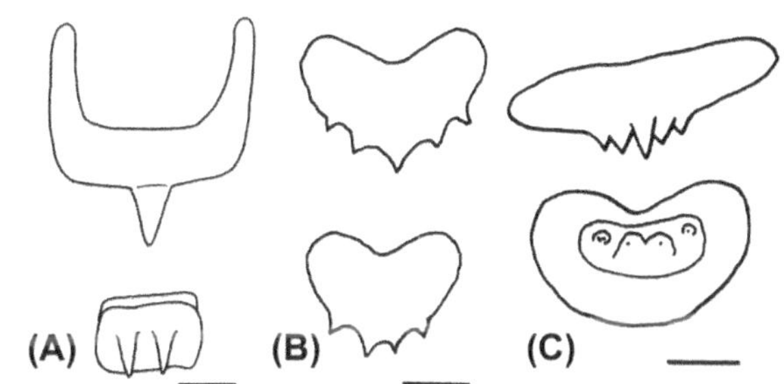

FIGURE 12.33 Jaws, dorsal over ventral, teeth pointing posteriorly: (A) *Cambarincola hoffmani*, dental formula 1/2; (B) *Oedipodrilus cuetzalanae*, dental formula 5/4; (C) *Sathodrilus veracruzicus*, dental formula 5/4. Scale bar (A) and (B) = 20 μm, (C) = 30 μm. *Redrawn: (B) from Holt (1984); (C) from Holt (1973).* *Copyright © 2015 Dr Stuart R. Gelder. Published by Elsevier Inc. All rights reserved.*

when each jaw is slightly displaced. The dental formula consists of the numbers of teeth on dorsal and ventral jaws, shown as 1/2 and 5/4 in Fig. 12.33. However, dentation varies slightly in some species.

Figure 12.34 shows all reproductive characters used in the key except for the penis. The relative position of a structure can be referred to as "ectal"—nearest the exterior (= proximal)—or "ental"—farthest from the exterior (= distal). For example, vasa deferentia enter glandular atrium ectally, as shown in Fig. 12.34. The glandular atrium has a free ental end; however, it may be bifurcated to form two deferent lobes. Granule-containing cells line the **glandular atrium**, and where these also line the **prostate gland**, the gland is called "undifferentiated." In contrast, when the prostate gland is lined with vacuolar cells, it is referred to as "differentiated". Some species have a prostatic ental bulb, but its size can vary greatly depending on preservation method. A **muscular atrium** and its ectal **bursa** connect the glandular atrium with the ventral epidermis. A **penis** is located in the bursa, but in some species, it extends entally into the muscular atrium. The penis is characterized by its method of extension: either "protrusible" or "eversible" with the typical appearance of each shown in Fig. 12.35 A, B, respectively. In a few species, penial length approaches total body length, e.g., *Oedipodrilus cuetzalanae* Holt, 1984.

The **spermatheca** or sperm receptacle (Fig. 12.34) is the female organ in segment 5 and consists of a pore (sp), duct (sd), bulb (sb) and, when present, an ental process (ep). In some species, the spermatheca is highly variable in size

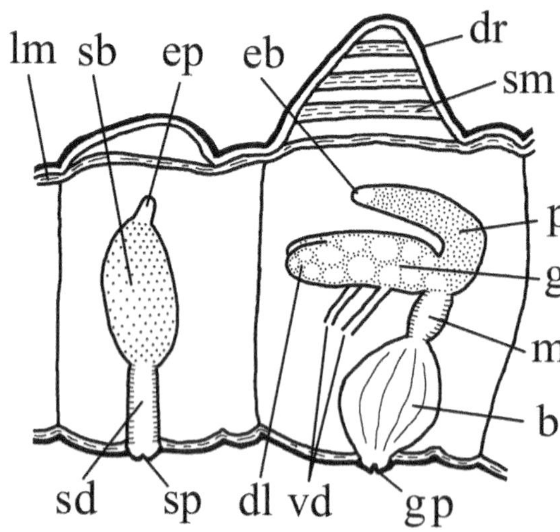

FIGURE 12.34 Lateral view of body segments 5 and 6 showing spermatheca (or female spermatozoa receptacle) and male genitalia. Abbreviations: *b*, bursa; *dl*, deferent lobes; *dr*, dorsal ridge; *eb*, ental bulb; *ep*, ental process; *ga*, glandular atrium; *gp*, genital pore; *lm*, longitudinal body muscles; *ma*, muscular atrium; *p*, prostate gland (shown as "differentiated"); *sb*, spermatheca bulb; *sd*, spermatheca duct; *sm*, supernumerary muscle; *sp*, spermatheca pore; *vd*, vasa deferentia. *Copyright © 2015 Dr Stuart R. Gelder. Published by Elsevier Inc. All rights reserved.*

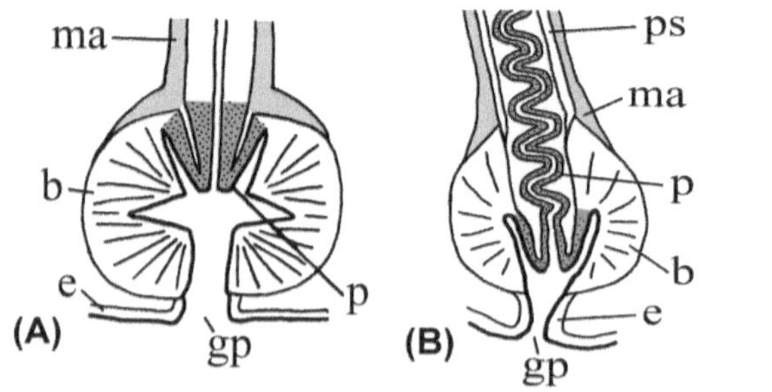

FIGURE 12.35 Schematic longitudinal section through bursa showing a generalized retracted protrusible (A) and eversible (B) penes. Abbreviations: *b*, bursa; *e*, epidermis; *gp*, genital pore; *ma*, muscular atrium (light gray); *p*, penis (dark gray); *ps*, penial sheath. *Copyright © 2015 Dr Stuart R. Gelder. Published by Elsevier Inc. All rights reserved.*

and shape due to the volume of sperm it contains and preservation conditions.

MATERIAL PREPARATION AND PRESERVATION

Although branchiobdellidan identification is frequently accomplished with preserved material, several distinguishing characters are most easily determined using wet-mount preparations of live worms. To prepare a live branchiobdellidan for wet mounting, place the worm in a drop of water on a microscope slide and gently add a cover glass. The amount of water on the slide, and thus the mobility of the worm, can be controlled by withdrawing water with a paper tissue or adding drops from a pipette. Too little water, however, will squash and damage the worm. Examine the specimen under a compound light microscope. A specimen must be observed from both dorsoventral and lateral aspects to see all important characters. Changing the aspect simply requires pushing the cover glass gently so the specimen rolls through a quarter turn. Although a compound microscope with DIC illumination is best for seeing the structures, ordinary bright and dark field illumination will enable all features to be adequately observed. Examination of live branchiobdellidans can greatly reduce the time needed for species identification and leaves the specimen available for subsequent processing (e.g., establishing a culture, for museum curation, histology, electron microscopy, molecular sequencing).

The best general preservative for studying branchiobdellidan morphology is 70% ethanol. It is also ideal for long-term preservation, not carcinogenic, and allows specimens to be mounted for either identification or molecular sequencing. Ten percent formalin is a less desirable, as it is carcinogenic, needs prolonged washing, and can promote formation of insoluble crystals on the surface of the worm that can obscure characters. Occasionally, live branchiobdellidans need to be immobilized or relaxed, and this is most easily achieved by immersing them briefly, 0.5–4.0 min, in carbonated water, e.g., soda water.

To produce permanent museum standard preparations, dehydrate unstained specimens in ethanol solutions, clear in clove oil or oil of wintergreen (methyl salicylate) to remove preservation opacity, and mount them in Canada balsam on microscope slides. The clearing agents also add contrast to a specimen's internal organs. As all permanent mounting methods only allow each worm to be seen from one aspect, multiple mounted specimens are needed to see all the identifying characters. Genitalia in large whole branchiobdellidans (6–10 mm long) may be too thick to be seen on slide preparations. In such cases, the organs should be removed while they are immersed in water, 70% ethanol, or glycerine by microdissection. Unpreserved specimens should then be irrigated with preservative and slide mounted, as described earlier. Instruments for this procedure can be made from razor blade fragments and insect (very small) pins glued to the end of tooth picks.

Alternative preparation methods for examining branchiobdellidans include staining and using clearing mixtures that make semipermanent microscopical mounts (see microtechnique textbooks or online). Specimens in these mountants last from a few days to several months, as some of the component chemicals damage soft tissues making the branchiobdellidans questionable or useless for future reference.

KEYS TO BRANCHIOBDELLIDEA

Clitellata: Branchiobdellidea: Branchiobdellidae: Genera

1	Penis protrusible (Fig. 12.35 A)	2
1′	Penis eversible (Fig. 12.35 B)	4
2(1)	Prostate gland arises from glandular atrium not adjacent to muscular atrium	3
2′	Prostate gland arises from glandular atrium adjacent to muscular atrium (Fig. 12.34)	***Cambarincola* [p. 461]**
3(2)	Dorsal ridge absent across segment 8	***Forbesodrilus* [p. 462]**
3′	Dorsal ridge present across segment 8 [Mexico: Veracruz]	*Pterodrilus mexicanus* Ellis, 1919
4(1)	Bursa large, one transverse constriction	***Sathodrilus* [p. 462]**
4′	Bursa large, two transverse constrictions [Mexico: Puebla]	*Oedipodrilus cuetzalanae* Holt, 1984

Clitellata: Branchiobdellidea: Branchiobdellidae: *Cambarincola*: Species

1	Prostate gland cells undifferentiated	2
1′	Prostate gland cells differentiated	9
2(1)	Ental bulb absent	3
2′	Ental bulb present	5
3(2)	Glandular atrium length equal or longer than prostate gland	4
3′	Glandular atrium length shorter than prostate gland [Mexico: Veracruz]	*Cambarincola carcinophilus* Holt, 1973
4(3)	Glandular atrium length equal to prostate gland [Mexico: San Luis Potosí]	*Cambarincola speocirolanae* Holt, 1984
4′	Glandular atrium length equal longer than prostate gland [Mexico: Nuevo Leon]	*Cambarincola ellisi* Holt, 1973
5(2)	Dental formula 5/4	6
5′	Dental formula 1/2	8
6(5)	Glandular atrium length about equal prostate gland	7
6′	Glandular atrium length longer than prostate gland [Mexico: Campeche, Nuevo Leon, Puebla, San Luis Potosí, Veracruz]	*Cambarincola susanae* Holt, 1973
7(6)	Ventral jaw semicircular [Mexico: Puebla, Veracruz]	*Cambarincola jamapaensis* Holt, 1973
7′	Ventral jaw triangular [Mexico: San Luis Potosí, Veracruz]	*Cambarincola olmecus* Holt, 1973
8(5)	Glandular atrium shorter than muscular atrium [Mexico: Puebla]	*Cambarincola micradenus* Holt, 1973

8′ Glandular atrium about equal to muscular atrium .... *Cambarincola hoffmani* Holt, 1973
[Mexico: Puebla]

9(1) Dental formula 5/4 .... *Cambarincola toltecus* Holt, 1973
[Mexico: Veracruz]

9′ Dental formula 6/6 .... *Cambarincola smalleyi* Holt, 1964
[Costa Rica]

Clitellata: Branchiobdellidea: Branchiobdellidae: *Forbesodrilus*: Species

1 Dental formula 5/4 .... *Forbesodrilus nanognathus* (Holt, 1973)
[Mexico: Veracruz. Nicaragua]

1′ Dental formula 7/7 .... *Forbesodrilus acudentatus* (Holt, 1973)
[Mexico: Tamaulipas]

Clitellata: Branchiobdellidea: Branchiobdellidae: *Sathodrilus*: Species

1 Prostate gland present 2

1′ Prostate gland absent .... *Sathodrilus veracruzicus* (Holt, 1968)
[Mexico: Puebla, Veracruz]

2(1) Prostate gland length equal to glandular atrium .... *Sathodrilus prostates* Holt, 1973
[Mexico: Puebla, Veracruz]

2′ Prostate gland length 1/8 of glandular atrium .... *Sathodrilus villalobosi* Holt, 1968
[Mexico: Hidalgo, Puebla]

REFERENCES

Gelder, S.R. 2011. Reassignment of a Central American species of the Branchiobdellida (Annelida: Clitellata) to *Forbesodrilus* n.g. Acta zoologica bulgarica 63: 119−123.

Gelder, S.R. & B.W. Williams. 2015. Clitellata: Branchiobdellida. Pages 551−563 *in*: J.H. Thorp & D.C. Rogers (eds.), Ecology and General Biology, Thorp and Covich's Freshwater Invertebrates, Vol. I, 4th Edition. Academic Press, New York.

Holt, P.C. 1973. A summary of the branchiobdellid (Annelida: Clitellata) fauna of Mesoamerica. Smithsonian Contributions to Zoology 142: 1−40.

Holt, P.C. 1984. On some branchiobdellids (Annelida: Clitellata) from Mexico with the descriptions of new species of the genera *Cambarincola* and *Oedipodrilus*. Proceedings of the Biological Society of Washington 97:34−42.

Holt, P.C. & B.D. Opell. 1993. A checklist of and illustrated key to the genera and species of the Central and North American Cambarincolidae (Clitellata: Branchiobdellida). Proceedings of the Biological Society of Washington 106: 251−295.

Class Clitellata: Subclass Hirudinida

Alejandro Oceguera-Figueroa
Laboratorio de Helmintología, Instituto de Biología, Universidad Nacional Autónoma de México, Ciudad Universitaria, Coyoacán, Ciudad de México, México

INTRODUCTION

Leeches, together with earthworms (Oligochaeta), crayfish worms (Branchiobdellida), and acanthobdellids (Acanthobdellida), form a well-supported monophyletic group. Leeches are easily recognized by their segmented and annulated body as well as by the presence of muscular suckers at the ends of the body. Leeches have 34 somites, of which only 27 are distinguishable externally and are designated anteriorly through posteriorly as I–XXVII. Somites are superficially divided by furrows; midbody somites (complete somite) display the full number of annuli per somite, ranging from 2 to 16. The number of annuli in complete somites is important in leech taxonomy. Suckers function as locomotion organs as well as to attach to hosts and prey. Leeches display different ecological roles in their ecosystems, functioning as ectoparasites (blood-feeders on vertebrates) and predators mainly of invertebrates and function as vectors of blood parasites of fish, reptiles, and amphibians (Moser et al., 2016). Leeches are often abundant in lakes, ponds, swamps, irrigations ditches, or "acequias" as well as in rivers and streams. Some species inhabit moist soil or rotting logs.

There are around 110 species of leeches known in the Neotropics, including some enigmatic terrestrial leeches such as *Americobdella valdiviana*, *Cylicobdella* spp., and *Diestecostoma* spp., as well as an understudied group of marine taxa (which are not included in this section). Current knowledge of Neotropical leeches is based largely on the work of a small number of taxonomists; probably the most important is the Argentinian Raúl Ringuelet (1914–82). However, some other taxonomists have contributed, including John Percy Moore (1869–1965) and Eduardo Caballero y Caballero (1904–74). The keys presented here include updated taxonomy based on recent studies, focusing on phylogenetic analyses based on morphologic and molecular data.

LIMITATIONS

The taxonomy of Neotropical freshwater leeches is understudied. To the best of our knowledge, only Argentina, Brazil, and Mexico have had professional leech taxonomists, rendering the other countries largely unstudied. Some of the descriptions of the local leech fauna were done by European taxonomists that studied material collected during the scientific expeditions of the 19th century, and these studies in some cases represent the only record of the species. The extensive areas without any leech records, together with the recognition of several poorly described and understood groups, strongly suggest that the number of species is still unknown. Even though generic assignment based on external morphology is relatively easy, species identification requires the study of internal anatomy. With the recent development of phylogenetic methods and the acceptance that only monophyletic groups should be recognized and named, some of the taxonomic categories historically accepted by leech taxonomists have to be modified to some extent. The order Rhynchobdellida, defined by the presence of a protrusible proboscis, appears to be a paraphyletic assemblage based on DNA sequences. Even though we recognize that some advances in nomenclatural aspects have been conducted recently (see Tessler et al., 2018), for practical and historical reasons, we use the name Rhynchobdellida in this chapter. Furthermore, current classification of the families of the order Arhynchobdellida also requires major rearrangement. Sawyer (1986) recognized seven families within Arhynchobdellida mainly defined by a combination of morphological and ecological characteristics as well as by their feeding preferences (blood-feeders vs. macrophagous). Very few of the families conceived by Sawyer (1986) and early authors were recovered as monophyletic in modern analyses (Apakupakul et al., 1999; Borda et al., 2008; Phillips & Siddall, 2009; Phillips et al., 2010); for this reason, classification under the order level is problematic in Arhynchobdellida and is not used in the keys for this order.

TERMINOLOGY AND MORPHOLOGY

The following are terms found in leech taxonomy needed to use the keys. **Agnathous**: Lack of jaws in the pharynx, as in members of *Erpobdella, Orchibdella, Patagoniobdella,* and *Semiscolex.* **Annuli** (singular annulus): External transverse rings separated by furrows. Midbody somites display the full or complete annulation pattern consisting basically in three annuli (a1, a2, and a3) as in members of *Helobdella* and *Placobdella.* Each annuli can be subdivided into secondary annuli (a1 into b1 and b2; a2 into b3 and b4, and a3 into b5 and b6); in most members of Arhynchobdellida, the complete somite is five-annulate (b1, b2, a2, b5, and b6). **Anterior sucker** (also cephalic sucker or oral sucker): Attachment or suction organ located at the anterior end of the leech body. In most cases, it is smaller than the posterior sucker, located at the posterior end; it contains the mouth or mouth pore at the ventral surface. **Bacteriomes**: Specialized organs that harbor endosymbiotic bacteria; saclike structures or spheroidal organs attached to the esophagus in *Placobdella* and *Haementeria* species, respectively. **Blood-feeding** (haematophagy): Feeding practice based on vertebrate blood; species of *Haementeria* and *Placobdella* are exclusively blood-feeders, whereas species of *Oxyptychus* and *Limnobdella* may alternate blood with other food sources, such as frog eggs or invertebrates. **Cephalic bulb:** Anterior end enlarged, forming a distinct disc, as in *Helobdella chilensis* R. Blanchard 1900. **Cocoon**: A protective case that contains fertilized eggs. It is secreted by the clitellum. Cocoons are either deposited in submerged hard objects (i.e., *Erpobdella*), moist areas (i.e., *Limnobdella* and *Oxyptychus*) or brooded by adult leeches (i.e., members of Glossiphoniidae). **Compact salivary glands**: Cluster of salivary cells as in *Haementeria*, *Placobdella*, and some species of *Helobdella*. **Complete somite**: Midbody somite, normally display the complete annulation pattern. **Crop with lateral caeca:** Crop provided with lateral digitiform or branched projections. In most cases, each somite with one pair of caeca. **Diffuse salivary glands:** Salivary cells dispersed or diffused in the parenchyma. **Digitiform lateral branchiae**: Fingerlike extension (gills) of the body wall, present in *Bogabdella ringueleti* (Mañe-Garzón, 1973) and *Unoculubranchiobdella expansa* Solano Lobo Peralta, Rodrigues Matos & Maues Serra-Freire, 1998. **Ejaculatory bulbs:** Enlargement of the paired ejaculatory ducts as in *Erpobdella* spp. **Eyespots**: Photoreceptors located on the dorsal surface of the anterior end of the body. The number and arrangement of eyespots are important in higher classification of leeches; eyespots in four pairs as in *Theromyzon* species; eyespots in one pair as in *Helobdella* and *Haementeria* species; eyespots in two pairs as in *Placobdella* species; eyespots in five pairs as in *Oxyptychus, Limnobdella,* and *Hirudinaria*. **Gonopores**: External openings of the male and female reproductive systems; both gonopores are located on the ventral surface in the area of the clitellum; male gonopore is anterior to the female; in most cases, the male gonopore is larger. **Liquidosomatophagy**: Feeding practice based on soft tissues and internal fluids of aquatic invertebrates, in particular snails. **Macrophagy**: Feeding practice with emphasis on invertebrates and relatively large particles of food. **Marsupium**: Ventral pouch in which cocoons are protected, in *Maiabdella batracophila* Ringuelet, 1980. **Nuchal scute**: Chitinous structure located on the dorsal region about somite XII. **Oviducts**: Paired ducts that connect the ovisacs with the female gonopore. **Ovisac**: Celomic sacs that produce and store eggs. **Papillae**: Sensory organs typically present on the dorsal surface. In most cases, forming longitudinal rows metamerically arranged. **Parental care**: Behavioral strategy that includes investment of the parents. In leeches, consists on the attachment of eggs and young leeches to the ventral surface of adult leeches. **Pharynx**: Anterior muscular section of the digestive system, between the mouth and the esophagus. In *Haementeria* species, it is recurved. **Posterior sucker**: Attachment or suction organ located at the posterior end of the leech body. **Postcaeca**: The most posterior pair of caeca in the crop; in most cases, it has a posterior distribution. **Proboscis**: Feeding organ characteristic of the members of Rhynchobdellida; a tubular protrusible pharynx with blunt tip and triradiate lumen. **Pulsative vesicles**: Respiratory organs of some piscicolids; lateral extensions of the coelomic system. **Testisacs**: Celomic sacs that produce and store spermatozoa; in most leeches, one pair of testisacs per reproductive somite are present. **Tubercles**: Large papillae typically present on the dorsal surface; in most cases, they form longitudinal rows metamerically arranged. **Vagina**: Enlargement of the female gonopore; in most hirudiniforms, this is U-shaped.

MATERIAL PREPARATION AND PRESERVATION

Some leeches display vivid colors in life, with dorsal and ventral pigmented stripes and metameric dots and spots. Pigmentation usually vanishes after fixation, and quite often, museum specimens are colorless. For this reason, it is important to keep records (preferably both written and photographic) of the pigmentation before fixation. Proper fixation can be difficult and time-consuming, especially in the field. If leeches are placed in the fixative too quickly, they will contract making morphological traits difficult, if not impossible, to observe. To properly fix leeches, it is important to narcotize and relax them. Several methods have been proposed, all require keeping the leeches in containers with water. The main method consists of adding drops of 95%–100% ethanol to the receptacle with the leeches and water. This process is done slowly, with gradual increases in concentration until the leech's movements stop and it gives no reaction to touch. This process can take up to 30 min, depending on the specimen's size. Then, remove the mucus with paper towels. Lower concentrations of ethanol might be used, but the amount of ethanol and time will increase. Instead of ethanol, carbonated water or 6% nembutal can be used. Once relaxed, leeches should be straightened and placed in containers between paper towels and covered with 95%–100% ethanol for 24 h or more, depending on the size of the specimens. Alternatively, one can use 5%–10% buffered formalin. Large leeches (>10 cm) should be injected with formalin to preserve internal organs. For long-term storage, specimens should be kept in 75% ethanol with low light and controlled temperature. For molecular analyses, some specimens or at least their posterior sucker should be placed directly in 96% ethanol and kept at least at 4°C or lower if possible. Analyses of transcriptomic data (RNA sequences) require the use of specialized fixatives, such as RNAlater or similar buffers using whole specimens or particular tissues; however, this fixative is not recommended for morphological studies. For permanent slide preparations, particular for small leeches, specimens should be flattened between two glass slides immediately after narcotization. Staining should be done with a mixture of Mayer's paracarmine and Ehrlich's hematoxylin and mounted on slides with Canada balsam. Alternatively, staining with Semichon's acetocarmine and clearing with methyl salicylate also works well. For histological preparations, the use of 4% paraformaldehyde, 2.5% glutaraldehyde, or instead, Fleming's or Bouin's fixatives is recommended.

KEYS TO HIRUDINIDA

Clitellata: Hirudinida: Orders

1 Mouth large, occupying entire anterior sucker cavity (in *Tyrannobdella rex*, the mouth is a small pore in center of anterior sucker); body elongate, not flattened, with lateral margins parallel (Fig. 12.36 A); no protrusible proboscis; jaws with teeth present or absent; five pairs of eyespots arranged in an arc on somites II–VI (Fig. 12.37 A) or three pairs with one pair labial and two pairs buccal (Fig. 12.37 B) .... **Arhynchobdellida [p. 465]**

1′ Mouth a small pore on anterior sucker ventral surface; body dorsoventrally flattened with anterior end narrower than posterior (some exceptions in *Helobdella*) (Fig. 12.36 B–D); muscular pharyngeal protrusible proboscis; no jaws or teeth; eyespots in pairs paramedially, not arranged in an arc on somites II–VI (Fig. 12.37 C–F) .... **Rhynchobdellida [p. 468]**

Clitellata: Hirudinida: Arhynchobdellida: Genera

1 Jaws present, one or three; parasitic [Arhynchobdellida with jaws for which no suprageneric name is available] 2

1′ Jaws absent (agnathous); macrophagous [Arhynchobdellida without jaws for which no suprageneric name is available] 6

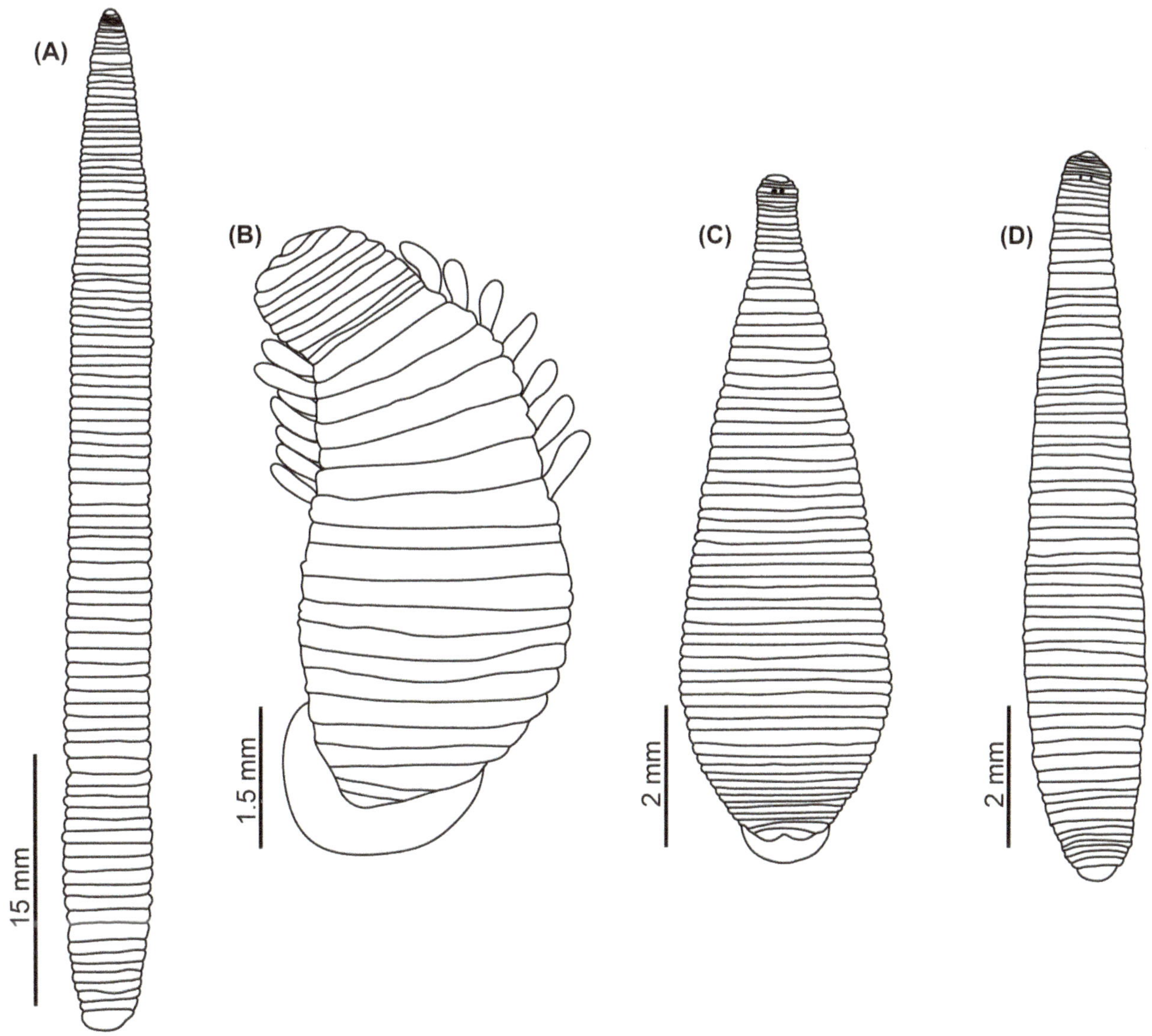

FIGURE 12.36 Dorsal view of the general body shape of leeches: (A) *Semiscolex* spp. (Arhynchobdellida); (B) *Bogobdella ringueleti* (Ozobranchidae); (C) dorsoventrally flattened Glossiphoniidae; and (D) semicircular in cross section, Glossiphoniidae.

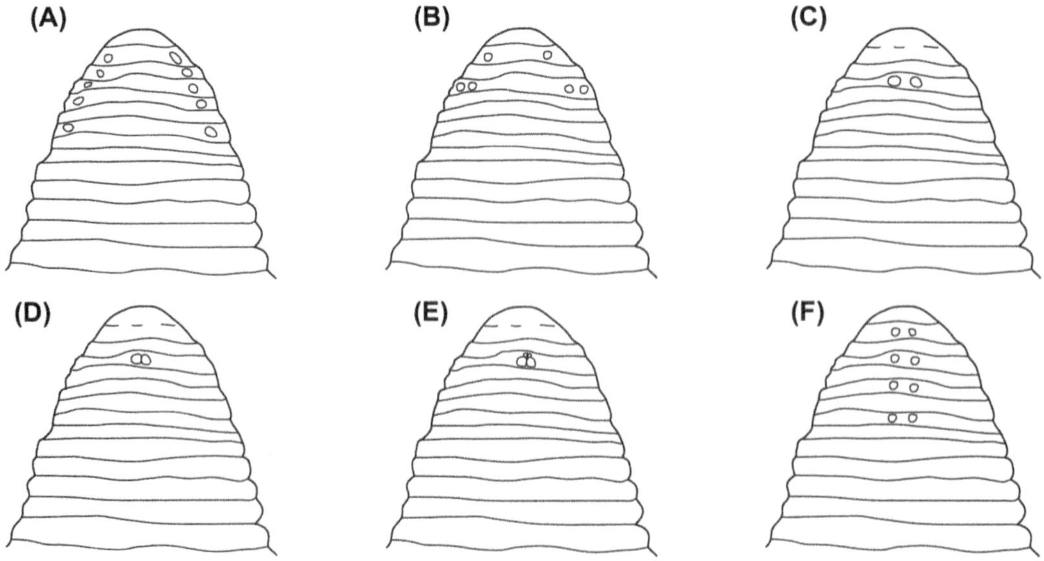

FIGURE 12.37 Dorsal view of the anterior end of (A) *Semiscolex lamothei*; (B) *Erpobdella triannulata*; (C) *Helobdella triserialis*; (D) *Haementeria* spp; (E) *Placobdella ringueleti*; and (F) *Theromyzon* spp.

2(1) Posterior sucker large, wider than posterior body; mouth velar, forming a small pore; one or three jaws; few large denticular teeth; median reproductive organs micromorphic; typically parasitic in mammal nasal passages 3

2′ Posterior sucker narrower than posterior body; mouth large, occupying entire anterior sucker cavity; three jaws, each jaw with a single row (monostichodont) of small denticles; median reproductive organs mesomorphic 4

3(2) Gonopores separated by ½ + 4 + ½ annuli; single jaw with eight large teeth; parasite in human nasal passages [monospecific] .... *Tyrannobdella rex* Phillips, Arauco-Brown, Oceguera-Figueroa, Gómez, Beltrán, Lai & Siddall, 2010 [Peru: East of the Andes]

3′ Gonopores separated by five annuli; three jaws [monospecific] .... *Pintobdella chiapasensis* Caballero, 1957 [Mexico: Chiapas]

4(2) Complete somite (midbody somite) three-annulate 5

4′ Complete somite (midbody somite) three-annulate [monospecific] .... *Mesobdella gemmata* (E. Blanchard, 1849) [Chile: Southern region]

5′ Dorsum gray or olive green; crop with two pairs of gastric caeca per somite; male reproductive system without ejaculatory bulbs (Fig. 12.38 A) .... ***Oxyptychus* [p. 467]**

5(4) Dorsum reddish or orange; crop with one pair of gastric caeca per somite; male reproductive system with ejaculatory bulbs .... *Hirudinaria manillensis* Lesson, 1842 [Asiatic. Introduced: Antigua and Barbuda, Haiti, Puerto Rico]

6(1) Eyespots in five pairs, arranged in a parabolic arc on somites II–VI (Fig. 12.37 A); pharynx short 7

6′ Eyespots in three pairs, one pair labial and two pairs buccal (Fig. 12.37 B); testisacs small and very numerous; pharynx ~0.3× body length; vagina absent 9

7(6) Vagina present, arcuate; ovisacs simple, spheroidal; oviducts short uniting into vagina (Fig. 12.38 C) 8

7′ Vagina absent; oviducts long, tubular and reflected; ovisacs on somite XV or XV/XVI behind penis region; testisacs in two pairs per somite .... ***Orchibdella* [p. 467]**

8(7) Testisacs in two pairs per somite .... ***Patagoniobdella* [p. 468]**

8′ Tesitisacs in one pair per somite (Fig. 12.38 C) .... ***Semiscolex* [p. 468]**

9(6) Gastropores (accessory copulatory pores) in one pair, one anterior to the male gonopore, one posterior to the female gonopore; gonopores separated by 4 + ½ annuli; pharyngeal stylets in three pairs; postcaeca in one pair in the posterior part of the caeca .... *Barbronia weberi* (Blanchard, 1897) [Orienta, Palaearctic. Introduced: Brazil, Costa Rica, Mexico]

9′ Gonopores separated by three annuli; pharyngeal stylets absent; crop tubular, without postcaeca .... ***Erpobdella* [p. 468]**

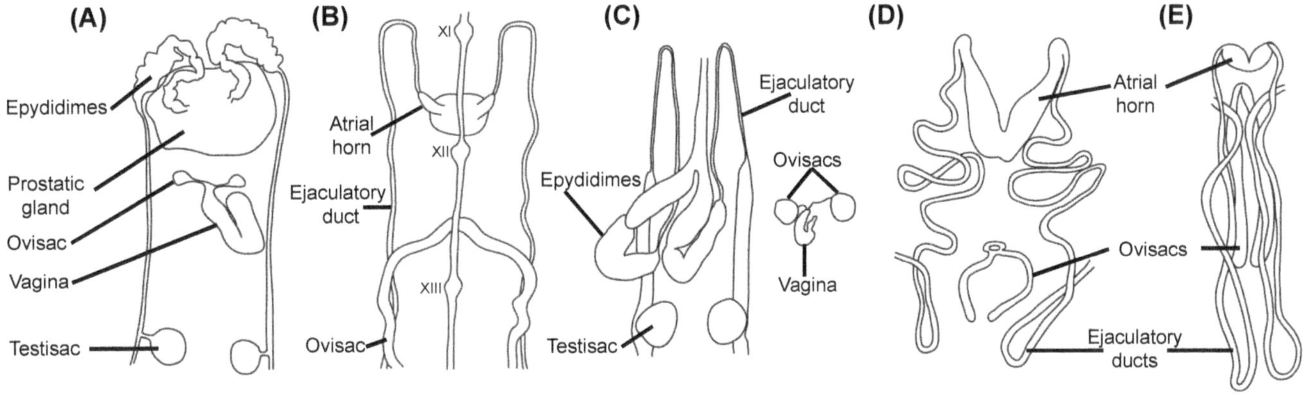

FIGURE 12.38 Dorsal view of the reproductive systems of (A) *Oxyptychus bora*; (B) *Erpobdella triannulata*; (C) *Semiscolex lamothei* (figure of the female reproductive system to the right of the male reproductive system); (D) *Haementeria acuecueyetzin*; and (E) *Helobdella* spp.

Clitellata: Hirudinida: Arhynchobdellida: *Oxyptychus*: Species

1 Gonopores separated by four or five annuli 2

1′ Gonopores separated by less than four annuli 4

2(1) Gonopores separated by four annuli 3

2′ Gonopores separated by five annuli ..... *Oxyptychus bora* Oceguera-Figueroa, Barrio, Aldea-Guevara & Siddall, 2012 [Peru: Iquitos]

3(2) Male gonopore at XIIb1/b2; female gonopore at XII/XIII ..... *Oxyptychus festai* (Dequal, 1916) [Ecuador: Guayas, Los Ríos]

3′ Male gonopore at XI/XII; female gonopore at XIIb5/b6 ..... *Oxyptychus brasiliensis* (Pinto, 1920) [Brazil: Goiás, Minas Gerais, Rio de Janeiro, São Paulo. Argentina: Buenos Aires]

4(1) Gonopores not on contiguous annuli, separated by more than one annuli 5

4′ Gonopores on contiguous annuli, separated by ½ + ½; male gonopore on XIIb2; female gonopore on XIIa2 ..... *Oxyptychus inexpectatus* Ringuelet, 1945 [Argentina: Buenos Aires, Córdoba, San Luis. Uruguay: Maldonado, Rivera]

5(4) Gonopores separated by 3 + ½ annuli; male gonopore at XI/XII; female gonopore at XIIb5. 6

5′ Gonopores separated by ½ + 2 + ½ or by 1 + ½ annuli 7

6(5) Dorsum with 107–108 annuli, somite VII four-annulate; somites VIII–XXIV five-annulate ..... *Oxyptychus antellarum* (Moore, 1901) [Panama. Puerto Rico]

6′ Dorsum with 103 annuli, somite VII three-annulate; somites VIII and XXIV four-annulate; somites IX to XXIII five-annulate ..... *Oxyptychus strenuus* Ringuelet, 1948 [Paraguay: Guaira. Argentina: Formosa]

7(5) Gonopores separated by ½ + 2 + ½ annuli ..... *Oxyptychus ornatus* (Weyenbergh, 1883) [Brazil: Rio Grande do Sul. Uruguay: Maldonado, Montevideo, San Jose. Argentina: Buenos Aires, Cordoba, Formosa, La Rioja, Salta]

7′ Gonopores separated by 1 + ½ annuli ..... *Oxyptychus striatus* Grube, 1851 [Venezuela: Caracas, Río Negro. Uruguay: Montevideo. Argentina: Buenos Aires, Córdoba, Entre Ríos, Misiones, San Luis]

Clitellata: Hirudinida: Arhynchobdellida: *Orchibdella*: Species

1 Female gonopore between somites XII/XIII 2

1′ Female gonopore on somite XIIIb1; male gonopore variable on somite XIb6, XI/XII, or XIIb1; anus on XXV/XXVI; four annuli between anus and posterior sucker ..... *Orchibdella pampeana* Ringuelet, 1945 [Argentina: Buenos Aires, Entre Ríos, Jujuy, Santa Fe. Uruguay: Canelones, Maldonado]

2(1) Ejaculatory bulbs robust; ovisacs with a terminal recurrent section on somite XV ..... *Orchibdella diagita* Ringuelet, 1978 [Argentina: Catamarca]

2′ Ejaculatory bulbs weakly developed; ovisacs simple without folds ..... *Orchibdella peruviensis* Ringuelet, 1976 [Peru: Cusco]

Clitellata: Hirudinida: Arhynchobdellida: *Patagoniobdella*: Species

1 Gonopores separated by more than two annuli 2

1′ Gonopores separated by ½ + 1 + ½ annuli .... *Patagoniobdella fraterna* Ringuelet, 1976
[Chile: Valdivia. Argentina: Chubut, Neuquén]

2(1) Gonopores separated by ½ + 6 or 6 + ½ annuli .... *Patagoniobdella ademonia* Ringuelet, 1976
[Argentina: Río Negro]

2′ Gonopores separated variably, by ½ + 3, 3 + ½, or ½ + 2; male gonopore on XII b1; female gonopore on XII a2/b5, XII b5, or XII b5/b6 .... *Patagoniobdella variabilis* (Blanchard, 1900)
[Chile: Concepción, Magallanes, Valdivia. Argentina: Chubut, Neuquén, Río Negro, Santa Cruz, Tierra del Fuego]

Clitellata: Hirudinida: Arhynchobdellida: *Semiscolex*: Species

1 Gonopores situated on annuli 2

1′ Gonopores in furrows, between annuli 4

2(1) Gonopores separated by less than ½ + 6 + ½ 3

2′ Gonopores separated by ½ + 6 + ½ .... *Semiscolex intermedius* Ringuelet, 1942
[Argentina: Buenos Aires, Capital Federal, Corrientes. Uruguay: Colonia]

3(2) Gonopores separated by ½ + 5 + ½ .... *Semiscolex lamothei* Oceguera-Figueroa, 2006
[Mexico: Veracruz]

3′ Gonopores separated by ½ + 4 + ½ .... *Semiscolex zonatus* Oka, 1930
[Brazil: Parana, São Paulo]

4(1) Gonopores separated by seven or less annuli 5

4′ Gonopores separated by eight annuli .... *Semiscolex notatus* Cordero, 1937
[Brazil: Parahyba]

5(4) Gonopores separated by seven annuli .... *Semiscolex juvenilis* Kinbergh, 1867
[Brazil: Rio Grande do Sul, São Paulo. Argentina: Buenos Aires. Uruguay: Maldonado, Montevideo, Treinta y Tres]

5′ Gonopores separated by six annuli .... *Semiscolex similis* (Weyenberh, 1879)
[Brazil: Rio de Janeiro. Paraguay: Careguatay. Bolivia: Cochabamba. Argentina: widespread. Uruguay: widespread]

Clitellata: Hirudinida: Arhynchobdellida: *Erpobdella*: Species

1 Ejaculatory ducts not forming a preatrial loop extending to XI before entering the atrial horns .... *Erpobdella adani* Tessler, Siddall & Oceguera-Figueroa, 2018
[Mexico: Chiapas]

1′ Ejaculatory ducts forming a preatrial loop extending to XI before entering the atrial horns (Fig. 12.38 B) .... *Erpobdella triannulata* Moore, 1908
[Mexico: Southern region. Guatemala. Honduras. Costa Rica. Panama]

Clitellata: Hirudinida: Rhynchobdellida: Families

1 Body margins without lateral branchiae 2

1′ Body margins with simple, undivided, digitiform branchiae (Fig. 12.36 B) .... **Ozobranchidae [p. 468]**

2(1) Body cylindrical in cross section, long and narrow; anterior sucker robust and distinct from body; eggs in hard, chitinous cocoons; no brooding cocoons or parental care; blood-feeders almost exclusively on fish and some aquatic invertebrates .... **Piscicolidae [p. 469]**

2′ Body flattened dorsoventrally, or subcylindrical (Fig. 12.36 C, D); anterior sucker weak, narrower than body; eggs in membranous cocoons, young brooded on parent ventral surface; free living and blood-feeders on vertebrates, mostly mammals and amphibians .... **Glossphoniidae [p. 469]**

Clitellata: Hirudinida: Rhynchobdellida: Ozobranchidae: Genera

Ozobranchidae are ectoparasite of the turtle genus *Podocnemys*.

1 Lateral branchiae in seven pairs .... *Bogabdella ringueleti* (Mañe-Garzón, 1973)
[Colombia: Meta]

1′ Laterial branchiae in 10 pairs [monospecific] .... *Unoculubranchiobdella expansa* Solano Lobo Peralta, Rodrigues Matos & Maues Serra-Freire, 1998. [Brazil: Manaus. Peru: Iquitos]

Clitellata: Hirudinida: Rhynchobdellida: Piscicolidae: Genera

1 Pulsative vesicles present (at least nine pairs); eyespots in two pairs in anterior sucker; compete somite seven-annulate; ectoparasite of the fish *Rhamdia guatemalensis* and *Petenia splendida* ....*Cystobranchus moorei* Sawyer, 1986 [Mexico: Yucatan. Guatemala: Peten]

1′ Pulsative vesicles absent; eyespots in one or two pairs in the anterior sucker .... ***Myzobdella* [p. 469]**

Clitellata: Hirudinida: Rhynchobdellida: Piscicolidae: *Myzobdella*: Species

1 Eyespots in one pair in anterior sucker 2

1′ Eyespots in two pairs in anterior sucker lateral branches in five pairs; ectoparasite of the fish *Rhamdia quelen* .... *Myzobdella uruguayensis* Mañe Garzón & Montero, 1977 [Argentina; Buenos Aires. Uruguay: Maldonado]

2(1) Complete somite 18-annulate; ectoparasite of the fish *Rhamdia guatemalensis, Megalops atlanticus, Coptodon rendalli,* and *Mugil curema* .... *Myzobdella lugubris* Leidy, 1851 [Mexico: Yucatan. Puerto Rico: Santa Teresa Lagoon]

2′ Complete somite 12-annulate; ectoparasite of the fish *Hoplias malabaricus* .... *Myzobdella platensis* (Cordero, 1933) [Argentina: Rio de la Plata basin]

Clitellata: Hirudinida: Rhynchobdellida: Glossiphoniidae: Genera

The Neotropics have the highest diversity for the group and may be the center of origin (Siddall & Borda, 2003), and a major revision is needed. Nearly half of the species belong to *Helobdella*. The number of species will probably increase in the future. Riguelet (1949) recorded *Glossiphonia mesembrina* in Argentina based on a single specimen with three paired eyespots, an obviously atypical trait but similar to *Helobdella* in all the other characters; the status of this material needs to be reinvestigated.

Ringuelet (1985) listed three *Batracobdella* species in the Neotropics: *Batracobdella dubia* Ringuelet, 1958; *Batracobdella gemmata* Blanchard, 1900; and *Batracobdella magnidiscus* (Moore, 1938). Both first two species may be species of *Haementeria* given the digestive system characteristics and *B. magnidiscus* from Yucatan and Guatemala may be young *Pintobdella*. *Batracobdella xenoica* Ringuelet, 1972, later transferred to the monotypic *Dacnobdella,* exhibits a mixture of characters of *Helobdella* (nuchal gland at somite VIII) and *Haementeria* (compact salivary glands and recurved esophagus) and is parasitic on the fish *Oncorhynchus mykiss*. With current information, no conclusions about its taxonomic status can be drawn.

Neotropical records of the Palaearctic *Helobdella stagnalis* Linnaeus, 1758, probably correspond to *Helobdella adiastola* Ringuelet 1972 from which it is unsatisfactorily distinguished or alternatively, to an undescribed species.

At least three species of *Placobdella* have been reported in Brazil: *Placobdella striata* Oka, 1932, *Placobdella bistriata* (Pinto, 1920), and *Placobdella maculata* Weber, 1915; however, they do not display the classical "placobdellid" eyespot arrangement. Most likely, they belong to *Haementeria*; however, a major revision is needed. *Placobdella parasitica*, originally from the Nearctic, has been reported in Panama, from the Canal area.

1 Ventral surface with three openings; male and female gonopores plus a third opening posterior to the genital area 2

1′ Ventral surface with two openings, male and female gonopores 3

2(1) Ventral surface with a marsupium (opening for brooding eggs) on somites XVII/XVIII; ectoparasite of frogs *Telmatobius* [monospecific] .... *Maiabdella batracophila* Ringuelet, 1980 [Argentina: Jujuy]

2′ Ventral surface with a third opening on XII/XIII; female and posteriormost open into a female bursa [monospecific] .... *Tribothrynobdella andicola* Ringuelet, 1978 [Peru: Apurimac, Junin]

3(1) Eyespots in one to three pairs, most commonly two (Fig. 12.37 C–E) 4

3′ Eyespots in four pairs on paramedial lines of somites II–V (Fig. 12.37 F), body soft and translucent .... ***Theromyzon* [p. 470]**

4(3) Eyespots in one pair well separated or two pairs coalesced forming a single spot (Fig. 12.36 C, E) 5

4′ Eyespots in one pair, coalesced forming a single spot (Fig. 12.37 D); compact salivary glands in two pairs; bacteriomes spherical in two pairs (Fig. 12.39 A); ovisacs simple, forming a ring around the ventral nerve cord (Fig. 12.38 D) .... ***Haementeria* [p. 470]**

5(4) Eyespots in two pairs, inconspicuous confluent pair on II and larger, conspicuous pair on III, also confluent (placobdellid arrangement) (Fig. 12.37 E); compact salivary glands; bacteriomes on pair, in form of sac connecting to the esophagus (Fig. 12.39 B); blood-feeding .... *Placobdella ringueleti* López-Jimenez & Oceguera-Figueroa, 2008 [Mexico: Chiapas, Oaxaca. Panama: Chiriqui. Costa Rica: Puntarenas].

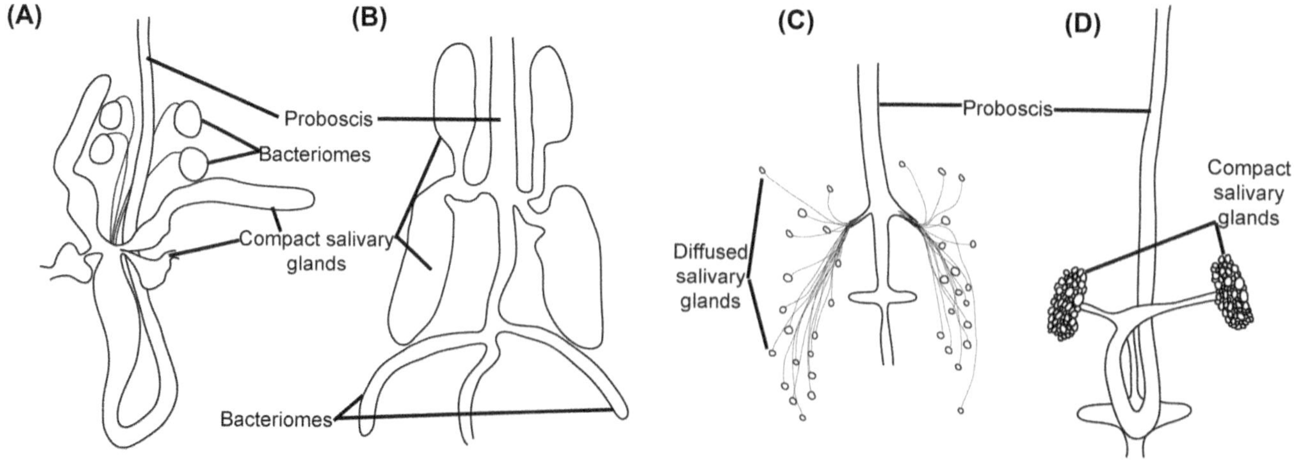

FIGURE 12.39 Dorsal view of the salivary glands of (A) *Haementeria acuecueyetzin*; (B) *Placobdella ringueleti*; (C) *Helobdella virginiae*; and (D) *Helobdella buzz* and *Helobdella longicolis.*

5′ Eyespots in one pair (Fig. 12.37 C); typically a single pair of diffused salivary glands (Fig. 12.39 C); however, some species with one pair of compact glands (Fig. 12.39 D); no bacteriomes; ovisacs tubular, simple; male atrial horns robust (Fig. 12.38 E); digestive system highly variable (Fig. 12.40 A–H) liquidosomatophagous .... ***Helobdella* [p. 472]**

Clitellata: Hirudinida: Rhynchobdellida: Glossiphoniidae: *Theromyzon*: Species

1 Gonopores separated by three annuli; male gonopore XI/XII, female gonopore on XII/XIII .... *Theromyzon propinquum* Ringuelet, 1947
[Peru. Chile. Argentina]

1′ Gonopores separated by four annuli, male gonopore on XI a2/a3, female gonopore on XII/XIII .... *Theromyzon pallens* Philippi, 1867
[Chile: Puerto Montt]

Clitellata: Hirudinida: Rhynchobdellida: Glossiphoniidae: *Haementeria*: Species

1 Complete somite two annulate (formerly under the genus *Oligobdella*) 2

1′ Complete somite three- or five-annulate on dorsal surface, five-annulate on ventral surface 4

2(1) Complete somites with anterior annuli larger than posterior annuli (a1 + a2) > a3 3

2′ Complete somites with anterior and posterior annuli about the same size, (a1 + a2) = a3; ectoparasite of the turtle *Phrynops geoffroanus* .... *Haementeria brasilensis* Cordero, 1937
[Brazil: Pernambuco]

3(2) Complete somites with anterior annuli partially subdivided ventrally; ectoparasite of the reptile *Paleosuchus trigonatus* .... *Haementeria columbensis* Ringuelet, 1972
[Colombia: Amazonas]

3′ Complete somites with anterior annuli partially subdivided on both surfaces; ectoparasite of amphibians and reptiles .... *Haementeria cheloniae* Ringuelet, 1981
[Argentina: Formosa, Santa Fe. Uruguay: Colonia, Montevideo]

4(1) Venter with papillae or tubercles 5

4′ Venter smooth, without papillae or tubercles 8

5(4) Ventral surface without paramarginal lines of black tubercles on a2 6

5′ Ventral surface with paramarginal lines of black tubercles on a2; ventral surface with annuli with 10–12 irregularly distributed papillae; ectoparasite of an "alligator" .... *Haementeria maculata* (Weber, 1915)
[Brazil: Porto Alegre]

6(5) Dorsum without a middorsal longitudinal row of tubercles on each annulus 7

6′ Dorsum with a middorsal longitudinal row of tubercles on each annulus; dorsum with a pair of paramedial rows of large tubercles only on a2 .... *Haementeria tuberculifera* (Grube, 1871)
[Cuba: Guantanamo. Colombia: Meta. Venezuela: Apure, Guárico. Suriname. Brazil: Pernambuco. Argentina: Cordoba]

7(6) Ventral surface with conspicuous longitudinal rows of tubercles in four or six pairs; ectoparasite of the caiman *Caiman crocodilus* .... *Haementeria paraguayensis* (Weber, 1915)

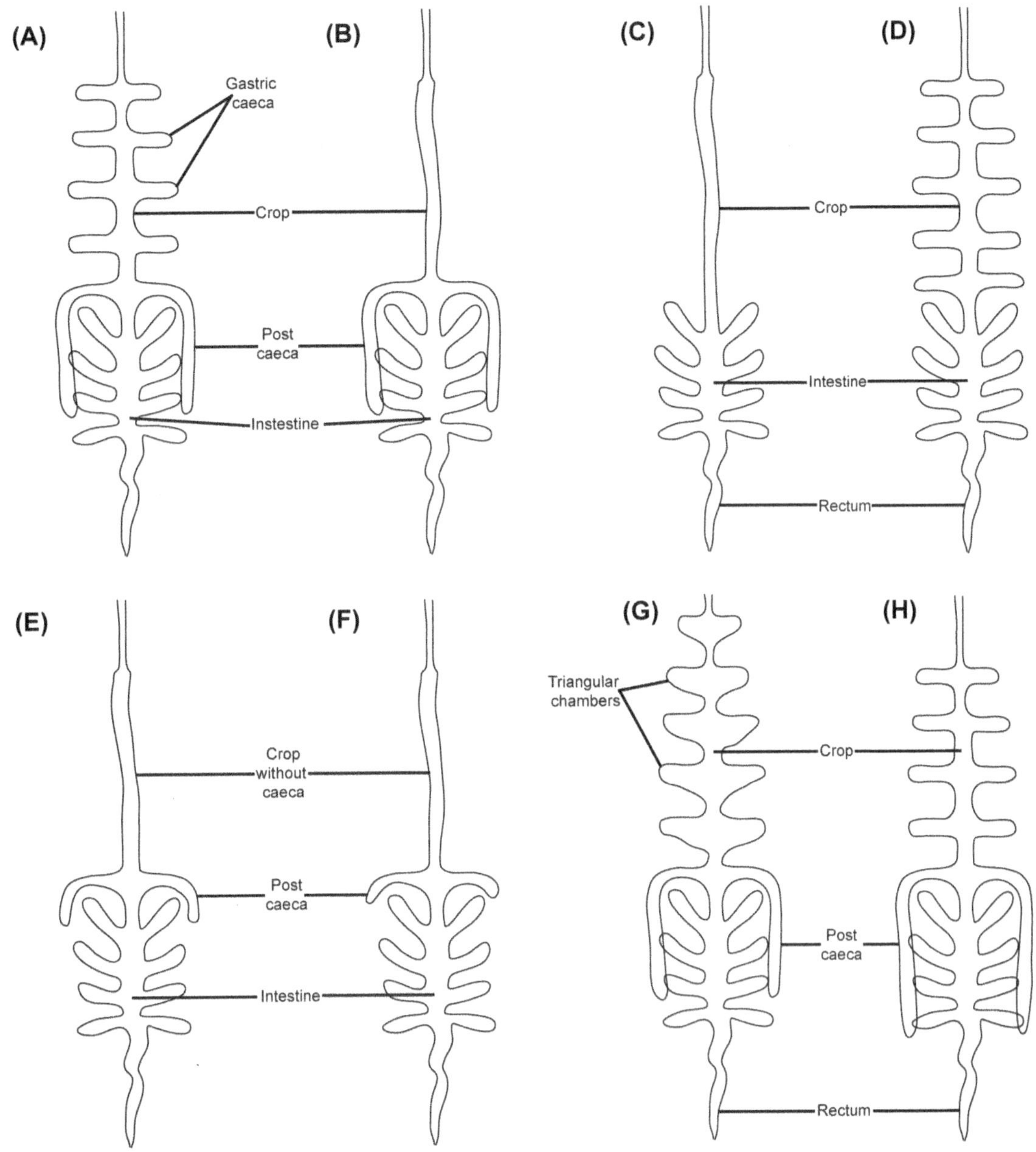

FIGURE 12.40 Dorsal view of the crop and intestine of *Helobdella* species (A) *Helobdella triserialis* and *Helobdella paranensis*; (B) *Helobdella nahuelhuapensis*; (C) *Helobdella elongate*; (D) *Helobdella ampllulariae*; (E) *Helobdella similis*; (F) *Helobdella nununununojensis*; (G) *Helobdella schlenzae*; and (H) *Helobdella socimulcensis*.

[Venezuela. Brazil: Parahyba, Pernambuco, Rio de Janeiro, São Paulo. Paraguay: Region Central. Argentina: Buenos Aires, Corrientes, Santa Fe]

7′ Ventral surface without large tubercles but fully covered with abundant papillae; ectoparasite of *Trichechus manatus* and other mammals .... *Haementeria acuecueyetzin* Oceguera-Figueroa, 2008
[Mexico: Chiapas, Oaxaca, Veracruz]

8(4) Dorsal annuli subdivided 9

8′ Dorsal annuli not subdivided (a1, a2, a3), heavily papillated .... *Haementeria ghilianii* de Filippi, 1849
[French Guiana: Sinnamary. Brazil: Pará, Rio de Janeiro]

9(8) Sensillae baciliform, conspicuous on dorsal surface 10

9′ Sensillae circular; small leeches, less than 32 mm; occasional parasites of fish (Siluriformes) .... *Haementeria eichhorniae* Ringuelet, 1978
[Argentina: Buenos Aires. Uruguay: Rocha]

10(9) Dorsum with a middorsal longitudinal row of tubercles on a2; ectoparasite of frogs and mammals .... *Haementeria molesta* (Cordero, 1934)
[Brazil: São Paulo. Uruguay: Canelones, Colonia, Florida, Montevideo]

10′ Dorsum without middorsal longitudinal row of tubercles; two pairs of lateral rows of tubercles more conspicuous on a2 and a3 .... *Haementeria depressa* (E. Blanchard, 1849)
[Chile: Santiago. Paraguay: Central, Guaira. Argentina: widespread. Uruguay: Artigas, Canelones, Colonia, Paysandú]

Clitellata: Hirudinida: Rhynchobdellida: Glossiphoniidae: *Helobdella*: Species

This group needs a thorough revision. The key includes 37 well-characterized species, and their identification is not problematic. Another 12 species are not included due to poor and equivocal descriptions and thus are treated as nomina dubia (i.e., *Helobdella anoculis* Weber, 1915, from Brazil: Rio Grande do Soul and Paraguay: Guaira; *Helobdella budgei* (Grube, 1871) from Brazil: Florianopolis, Santa Catarina; *Helobdella columbiensis* Weber, 1913 from Colombia: Cundinamarca; *Helobdella fuhrmanni* Weber, 1913 from Colombia: Cundinamarca; *Helobdella hemispherica* Weber, 1913 from Colombia: Cundinamarca; *Helobdella huaroni* Weber, 1916 from Peru: Junín; *Helobdella luteopunctata* Weber, 1915 from Chile: Región de los Lagos; *Helobdella paraguayensis* Weber, 1915 from Paraguay: Central; *Helobdella peruvensis* Weber, 1916 from Peru: Lima; *Helobdella* sp. Ringuelet, 1953 from Bolivia: Cochabamba; *Helobdella villarsi* Weber, 1916 from Peru: Junin; *Helobdella votuporanguensis* Castro, 1971 from Brazil: São Paulo).

Neotropical records of *H. stagnalis* Linnaeus, 1758 (now restricted to Europe), probably correspond to *H. adiastola* or alternatively, to an undescribed species. Ringuelet (1985) recognized six subspecies of *Helobdella triserialis* (E. Blanchard, 1849) for which no satisfactory diagnosis is available, some with wide distribution in the Neotropical Region. DNA analyses of some Mexican populations demonstrated that they should be considered as an independent species. The taxonomic status of *Helobdella punctatolineata* Moore, 1939 from Puerto Rico is either an independent species or a variety of *H. triserialis* and needs to be investigated (see Sawyer, 1986).

1 Somite VIII dorsally with a nuchal (chitinous) scute or gland (museum specimens might lose the scute leaving only the gland) 2

1′ Somite VIII dorsally without nuchal scute nor gland at VIII 10

2(1) Annuli subdivided, typically complete somite six annulate [(b1 + b2) + (b3 + b4) + (b5 + b6)] 3

2′ Annuli not subdivided, complete somite three annulate [a1 + a2 + a3], nuchal scute present .... *Helobdella adiastola* Ringuelet 1972
[Widespread. Colombia. Ecuador. Brazil. Peru. Paraguay. Chile. Argentina. Uruguay]

3(2) Somite VIII dorsally with a conspicuous nuchal scute 4

3′ Somite VIII dorsally with a nuchal gland 6

4(3) Metameric transverse coloration absent 5

4′ Metameric transverse darker coloration present on [b1 + b2] .... *Helobdella duplicata* Moore, 1911
[Brazil. Peru. Chile. Bolivia. Argentina. Uruguay. Ringuelet (1985) recognized two subspecies: *Helobdella duplicata duplicata* Moore, 1911 for specimens without dorsal papillae and *Helobdella duplicata tuberculata* Ringuelet, 1958 for specimens with dorsal papillae]

5(4) Anterior third violet blending to yellow posteriorly when alive; compact salivary glands at the base of the proboscis as well as diffuse salivary tissue in the parenchyma .... *Helobdella bolivianita* Siddall, 2001
[Bolivia: Santa Cruz]

5′ Dorsum dark gray median to light brown/beige with pale beige broad middorsal longitudinal zone and one pair of dorsal paramedial black stripes; with salivary cells diffused in the parenchyma .... *Helobdella pichipanan* Siddall & Borda, 2004
[Chile: Región de la Araucanía]

6(3) Crop with lateral caeca (Fig. 12.40 A) 7

6′ Crop without lateral caeca, only postcaeca (Fig. 12.40 B) 9

7(6) Pharynx recurved, "U"-shaped (Fig. 12.39 D) 8

7′ Pharynx straight .... *Helobdella diploides* Ringuelet, 1948
[Paraguay: Guaira, Villarica]

8(7) Salivary glands compact (Fig. 12.39 D); bromeliad phytotelmata .... *Helobdella buzz* Iwama, Nogueira & Gonçalves, 2017
[Brazil: São Paulo]

8′ Salivary glands diffuse .... *Helobdella scutifera* Blanchard, 1900
[Chile: Llanquihue, Magallanes, Valdivia. Argentina: Chubut, Formosa, Neuquén, Río Negro, Santa Cruz, Tierra del Fuego]

9(6) Dorsum and venter with three pairs of fine supramarginal longitudinal lines .... *Helobdella ringueleti* Siddall, 2001
[Bolivia: La Paz]

9′ Dorsum with one pair of supramarginal longitudinal lines, venter with one pair of inframarginal longitudinal lines .... *Helobdella nahuelhuapensis* Gullo, 2006
[Argentina: Río Negro]

10(1) Annuli subdivided, typically complete somite six-annulate [(b1 + b2) + (b3 + b4) + (b5 + b6)] 11

10′ Annuli not subdivided, complete somite three-annulate [a1 + a2 + a3] 13

11(10) Salivary glands diffuse, connecting to proboscis base (Fig. 12.39 C) 12

11′ Salivary glands compact in somite XIII connecting to proboscis base (Fig. 12.39 D); color yellowish, without stripes .... *Helobdella longicolis* Weber, 1915
[Paraguay: Central. Argentina: Jujuy]

12(11) Gastric crop with triangular chambers (Fig. 12.40 G) .... *Helobdella schlenzae* Iwama & Arruda, 2016
[Brazil: São Paulo]

12′ Gastric crop with digitiform caeca (Fig. 12.40 A) .... *Helobdella paranensis* Oka, 1930
[Brazil: Parana]

13(10) Gastric crop simple, tubular, without caeca, postcaeca may or may not be present (Fig. 12.40 B, C) 14

13′ Gastric crop with chambers or digitiform caeca 20

14(13) Postcaeca absent (Fig. 12.40 C) 15

14′ Postcaeca present (Fig. 12.40 B) 17

15(14) Testisacs in six pairs 16

15′ Testisacs in seven pairs .... *Helobdella michaelseni* Blanchard, 1900
[Brazil. Peru. Chile. Paraguay. Argentina. Uruguay]

16(15) Color gray-violet or green-violet .... *Helobdella obscura* Ringuelet, 1942
[Brazil: Río Grande do Sul, São Paulo. Peru: Junin. Bolivia: Cochabamba. Argentina: Buenos Aires, Río Negro]

16′ Dorsal surface with one or two pairs of transverse lines per annulus .... *Helobdella fantasmae* Gullo, 2009
[Argentina: Río Negro]

17(14) Postcaeca well developed, reaching XXII (Fig. 12.40 B) 18

17′ Postcaeca very short, restricted to XIX or XX (Fig. 12.40 F) 19

18(17) Dorsum and venter whitish, without any pigmentation .... *Helobdella elongata* (Castle, 1900)
[Mexico: Tabasco, Veracruz. Cuba. Costa Rica. Panama. Puerto Rico]

18′ Dorsum and venter irregularly mottled .... *Helobdella wodzickiorum* Siddall & Borda, 2004
[Chile: Santiago]

19(17) Postcaeca very short, only on XIX (Fig. 12.40 F) .... *Helobdella nununununojensis* Siddall, 2001
[Bolivia: La Paz]

19′ Postcaeca short, reaching XX (Fig. 12.40 E) .... *Helobdella similis* Ringuelet, 1942
[Brazil: São Paulo. Chile: Tarapacá, Valdivia. Argentina: Chubut, Cordoba, Neuquén, Río Negro, Tucumán]

20(13) Body thin, circular, or semicircular in cross section (Fig. 12.36 D); gastric crop with six digitiform caeca 21

20′ Body lanceolate, dorsoventrally flattened (Fig. 12.36 C) 22

21(20) Most posterior pair of caeca not forming postcaeca (Fig. 12.40 D); parasite of apple snails (Ampullaridae) .... *Helobdella ampullariae* Ringuelet, 1945
[Argentina: Buenos Aires, Entre Rios]

21′ Most posterior pair of caeca forming postcaeca (Fig. 12.40 A) .... *Helobdella sorojchi* Siddall, 2001
[Bolivia: La Paz]

22(20) Dorsal surface without tubercles or papillae 23

22′ Dorsal surface with tubercles or papillae 30

23(22) Postcaeca only on XIX 24

23′ Postcaeca well developed, reaching XX or further (Fig. 12.40 H) 25

24(23) Salivary glands compact .... *Helobdella cryptica* (Ringuelet, 1978)
[Chile: Tarapacá. Argentina: Jujuy]

24′ Salivary glands diffuse .... *Helobdella hyalina* Ringuelet 1942
[Peru: Junin. Argentina: Buenos Aires, Chaco, Corrientes, Jujuy. Uruguay: Canelones]

25(23) Dorsal surface with multiple longitudinal lines of pigment 26

25′ Dorsum without longitudinal stripes 27

26(25) Dorsum green−gray with around 14 pairs of longitudinal lines of pigment, esophagus short and straight .... *Helobdella cordobensis* Ringuelet, 1942
[Chile: Cautin. Argentina: Córdoba]

26′ Dorsum yellowish, esophagus recurved .... *Helobdella titicacensis* Ringuelet, 1959
[Peru: Cusco, Junin, Pasco, Puno]

27(25) Dorsum without spots 28

27′ Dorsum yellowish with black spots irregularly arranged .... *Helobdella araucana* Ringuelet, 1978
[Argentina: Neuquén, Río Negro]

28(27) Midbody margin annuli without lateral lobes 29

28′ Midbody margin annuli with lateral lobes covered with minute papillae, rendering serrated appearance .... *Helobdella festai* (Dequal, 1916)
[Ecuador: Cañar]

29(28) Anterior end forming a cephalic bulb .... *Helobdella chilensis* R. Blanchard 1900.
[Chile: Valparaiso, Llanquihue. Argentina: Santa Cruz, Tierra del Fuego]

29′ Anterior end not forming a cephalic bulb; compact salivary glands .... *Helobdella malvinensis* (Ringuelet, 1978)
[Malvinas (Falkland) Islands]

30(22) Salivary glands diffuse at proboscis base 31

30′ Salivary glands compact at proboscis base .... *Helobdella chaquensis* Ringuelet, 1978
[Argentina: Formosa]

31(30) Dorsum with tubercles on every annuli of midbody somites, forming a continuous medial stripe 32

31′ Dorsum with tubercles on a2 of midbody somites, forming a discontinuous medial stripe 33

32(31) Dorsally with single medial tubercle stripe .... *Helobdella simplex* (Moore, 1911)
[Peru: Puno. Argentina: widespread. Uruguay: Montevideo]

32′ Dorsally with one medial and two lateral tubercle stripes; four lateral stripes in total, with tubercles only on a2 and a3 .... *Helobdella brasilensis* (Weber, 1915)
[Brazil: Rio Grande do Sul. Argentina: Buenos Aires, Córdoba, Entre Ríos. Uruguay: Canelones, Colonia, Flores, Maldonado, Montevideo, Soriano]

33(31) Dorsum with multiple longitudinal stripes 34

33′ Dorsum and venter with whitish spots irregularly arranged .... *Helobdella virginiae* Oceguera-Figueroa, 2007
[Mexico: Veracruz]

34(33) Tubercles tinged with black or gray, pointed, occupying less than annulus length 35

34′ Tubercles light colored, rounded occupying almost all the length of the annulus .... *Helobdella striata* Ringuelet, 1953
[Brazil: São Paulo. Argentina: Buenos Aires, Córdoba, Jujuy, Salta, Tucumán. Uruguay: Colonia]

35(34) Common salivary ducts length $\sim 1\times$ midbody somite (Fig. 12.39 C) 36

35′ Common salivary ducts length $\sim 3\times$ midbody somite .... *Helobdella chaviensis* Iwama & Arruda, 2016
[Brazil: São Paulo]

36(35) Dorsal longitudinal midline stripe uninterrupted, from eyespots to anus. Postcaeca well developed, reaching XXIII (Fig. 12.40 H) .... *Helobdella socimulcensis* Caballero, 1931

36′ Dorsal longitudinal midline stripe interrupted at the cephalic region in some specimens. Postcaeca reaching XXII (Fig. 12.40 A) .... *Helobdella triserialis* (E. Blanchard, 1849)
[Widespread. Ringuelet (1985) recognized six subspecies for which no satisfactory diagnosis is available]

REFERENCES

Apakupakul, K., M.E. Siddall & E.M. Burreson. 1999. Higher level relationships of leeches (Annelida: Clitellata: Euhirudinea) based on morphology and gene sequences. Molecular Phylogenetics and Evolution 12: 350–359.

Borda, E., A. Oceguera-Figueroa & M.E. Siddall. 2008. On the classification, evolution and biogeography of terrestrial haemadipsoid leeches (Hirudinea: Arhynchobdellida: Hirudiniformes). Molecular Phylogenetics and Evolution 46: 142–154.

Moser, E.M., F.R. Govedich, A. Oceguera-Figueroa, D.J. Richardson & A. J. Phillips. 2016. Subclass Hirudinida. Pages 244–259 *in:* J.H. Thorp & D.C. Rogers (eds.), Keys to Neartic Fauna, Thorp and Covich's Freshwater Invertebrates, Vol. II, 4th Edition. Academic Press, New York.

Phillips, A.J., R. Arauco-Brown, A. Oceguera-Figueroa, M. Beltrán, G.P. Gómez & M.E. Siddall. 2010. *Tyrannobdella rex* n. gen. n. sp. and the evolutionary origins of mucosal leech infestations. PLos One 5: e10057. Doi:10.1371/journal.pone.0010057.

Phillips, A.J. & M.E. Siddall. 2009. Poly-paraphyly of Hirudinidae: many lineages of medicinal leeches. BMC Evolutionary Biology 9: 246. Doi.org/10.1186/1471-2148-9-246.

Riguelet, R.A. 1949. Notas sobre Hirudíneos neotropicales. VI. Presencia del género Glossiphonia en Argentina y otras adiciones al conocimiento de la Hirudofauna de los paises del Plata. Notas Museo de La Plata 14, Zoología 122: 141–159.

Ringuelet, R.A. 1985. Annulata. Hirudinea. En Castellanos, Z. (ed.). Fauna de agua dulce de la Republica Argentina 27: 1-321.

Sawyer, R.T. 1986. Leech biology and behaviour. Claredon Press, Oxford. 793 pp.

Siddall, M.E. & E. Borda. 2003. Phylogeny and revision of the leech genus *Helobdella* (Glossiphoniidae) based on mitochondrial gene sequences and morphological data and a special consideration of the *triserialis* complex. Zoologica Scripta 32: 23–33.

Tessler, M., D. de Carle., M.L. Voiklis., O.A. Gresham., J.S. Neumann., S. Cios & M. E. Siddall. 2018. Worms that suck: phylogenetic analysis of Hirudinea solidifies the position of Acanthobdellida and necessitates the dissolution of Rhynchobdellida. Molecular Phylogenetics and Evolution 127: 129–134.

Class Aphanoneura

Tarmo Timm
Estonian University of Life Sciences, Centre for Limnology, Tartumaa, Estonia

INTRODUCTION

Aphanoneura is a small clade of annelids adapted to freshwaters where they inhabit mainly on the bottom surface or on aquatic vegetation. The Family Aeolosomatidae has long been treated as primitive oligochaetes, due to their convergent similarities with small oligochaetes of the subfamilies Naidinae and Pristininae in their external appearance, vegetative reproduction by budding, and mode of life. The Aeolosomatidae were separated from Oligochaeta and all Clitellata by Brinkhurst & Jamieson (1971). The name Aphanoneura, derived from their nerve chain hidden in the body wall, was coined by Vejdovský (1884) and resurrected for the order Aeolosomatida (including Aeolosomatidae) by Timm (1981).

LIMITATIONS

A thorough description of several *Aeolosoma* species from southern Brazil was published by Marcus (1944). Single species descriptions and faunistic records from the Neotropical Region are in several other papers. Bunke (1967) revised the synonymy of the species previously described by Marcus (1944). Another survey of the group, including synonymy and world literature, was compiled by Van der Land (1971).

Sixteen valid species of *Aeolosoma* are recorded from Neotropics along with some poorly described species. Two of the latter are known as members of *Hystricosoma*, a genus characterized by the arrangement of hair chaetae in each dorsal bundle in two transverse rows or in an ellipse. However, this character was not mentioned in the meagre descriptions of two species recorded from the Neotropical Region by Michaelsen (1900, 1933), *Aeolosoma pictum* Schmarda, 1861 from Colombia, and *Hystricosoma insularum* Michaelsen, 1933 from Bonaire and Aruba Islands. Both may be *Aeolosoma* but remain species *incertae sedis*.

Aeolosoma macrogaster Schmarda, 1861 from Nicaragua, San Juan del Norte, and *A. cubana* Botea, 1983 described on fixed material from a cave in Cuba are *species dubia*. *A. macrogaster* and *Aeolosoma ternarium* Schmarda, 1861, Neotropical species later affiliated with a separate aeolosomatid genus *Pleurophleps* (now invalid) (Michaelsen, 1900; Van der Land, 1971), are poorly described and may represent naidid oligochaetes. *Nais ternaria* Schmarda, 1861, from Central America, Cuba, Jamaica, and Nearctic, is unfortunately confused in some subsequent literature with the Oriental aphanoneuran *A. ternarium* Schmarda, 1861.

TERMINOLOGY AND MORPHOLOGY

Aeolosoma are small (about one to several mm) with a rather delicate, transparent body (the genus name can be translated from Greek as "airy body"). **Chaetae** are located in four bundles per segment (excluding the mouth segment), as in the oligochaetes. Parapodia, which are characteristic of most polychaetes, are absent in aphanoneurans. The chaetae are similar in dorsal and ventral bundles and usually of two lengths. The longer chaetae are always **hairlike**, whereas the shorter ones can be also **sigmoid** and with a modified apex in some species. The **prostomium** lacks eyes and is large and flat. Its ciliated ventral surface serves for gliding translocation of the worm. Paired **ciliated pits** are present laterally on the posterior portion of prostomium and may be homologous with the nuchal organs of polychaetes. Colored **epidermal glands** are scattered over the whole body in most species.

The main mode of reproduction is paratomy (budding), resulting in a **chain** of new **zooids** arising posteriorly. The segment number of the first (mother) zooid is limited, seldom reaching up to 18. Sexual maturation rarely occurs. However, testes can develop in several segments, with spermatozoa ripening freely in body cavity (the dissepiments being rudimentary). The spermatozoa are emitted at copulation by metanephridia without special male ducts. Several paired, unicellular spermathecae occur in the forebody. Only one functioning ovarium and a single midventral female pore are present. The female pore is surrounded by a glandular pad (copulatory gland). A clitellum is absent, and individual eggs are laid in a chitinlike protective shell.

MATERIAL PRESERVATION AND PREPARATION

Due to their very delicate body, most aphanoneurans will shrink or burst and become opaque during fixation, leaving only the chaetae easily visible under a microscope. Preserved material alone is hardly suitable for a correct species identification. They should be studied alive, if possible. The rapid vegetative reproduction of several species in cultures enables a study of many individuals of the same clone. Details of chaetae, in contrast, are distinguishable only with immersion microscopy, which is hardly possible on live individuals due to their mobility.

KEY TO APHANONEURA

Aphanoneura: Aeolosomatida: Aeolosomatidae: *Aeolosoma*: Species

Modified from Van der Land (1971).

1 Chaetal bundles with long, flexible hair chaetae and short, stiff sigmoid chaetae 2

1′ Chaetal bundles with hair chaetae of equal or different length 6

2(1) Sigmoid chaetae smooth or with only one row of very small denticles on distal portion 3

2′ Sigmoid chaetae with two rows of 5–10 distinct denticles on concave side of distal end, 0–4 per bundle, often lacking in segment II; hair chaetae 0–4 per bundle, subequal; epidermal glands scarce, colorless; intestine dilated in segments IV–VI; first zooid 0.6–1 mm, 7–8 segments; chains up to four zooids, 1–2.5 mm (Fig. 12.41 A–D) ..... *Aeolosoma travancorense* Aiyer, 1926 [Suriname. Brazil. Argentina. Uruguay. Palaearctic]

3(2) Sigmoid chaetae in dorsal and ventral bundles 4

3′ Sigmoid chaetae only in ventral bundles, mostly in posterior segments, 0–2 per bundle, smooth or with a number of small denticles on concave side of distal end; hair chaetae 0–7 per bundle, unequal, long by 0–2 and short 0–5 per bundle; epidermal glands orange-red; intestine dilated in segments III–VI; first zooid 0.5–1 mm, 6–8 segments; chains up to six zooids, up to 3 mm (Fig. 12.41 E–H) ..... *Aeolosoma evelinae* Marcus, 1944 [Suriname. Brazil. Argentina. Uruguay. Palaearctic]

4(3) Sigmoid chaetae mostly with denticles (sometimes invisible), not sharp; epidermal glands colored 5

4′ Sigmoid chaetae smooth, distal end sharp and only slightly bent, 0–2 per bundle; hair chaetae 0–5 per bundle, unequal, long 0–2(3) and short 0–2 per bundle; epidermal glands colorless (a few can be light orange or yellow); intestine dilated in segments III–VI; first zooid (5)6 segments; chain of two zooids 1–2 mm ..... *Aeolosoma beddardi* Michaelsen, 1900 [Suriname. Brazil. Argentina. Uruguay. Holarctic and Oriental]

5(4) Sigmoid chaetae with 2–3 very small, sometimes invisible denticles on concave side of distal part, 0–2 per bundle, most often absent in segment II; hair chaetae 0–3(4) per bundle, unequal; epidermal glands lemon; intestine dilated in segments III–V; first zooid 6–7(8) segments; chains up to four zooids, 1–2 mm (Fig. 12.41 I, J) ..... *Aeolosoma sawayai* Marcus, 1944 [Brazil. Argentina. Uruguay. Palaearctic]

5′ Sigmoid chaetae strongly bent, with 6–7 denticles on concave side of distal end, 0–2 per bundle, lacking in segment II and sometimes in segment III; hair chaetae 0–5 per bundle, sometimes lacking in posterior bundles; epidermal glands red; intestine dilated in segments III–VI; first zooid 7–8 segments; chains of two zooids up to 1 mm (Fig. 12.41 K–M) ..... *Aeolosoma corderoi* Marcus, 1944 [Brazil]

6(1) Freshwater; hair chaetae of unequal length 7

6′ Marine supralittoral; uniform hair chaetae 6–7(9) per bundle in forebody, less posteriorly; epidermal glands orange with variable yellowish or reddish tinge; lateral ciliated pits not observed; first zooid 8–9 segments, about 1 mm; chains 1.6–1.85 mm ..... *Aeolosoma maritimum dubiosum* Westheide & Schmidt, 1974 [Ecuador: Islands]

7(6) Relatively small, body width < 300 μm, chaetal length < 350 μm; dissepiments lacking; no evident external segmentation 8

7′ Relatively very large, body width > 350 μm; dissepiments present, external segmentation evident; hair chaetae (4)5–7(12) per bundle, up to 600 μm; epidermal glands green or greenish blue; intestine dilated in segments IV–XI(XIII); first zooid 13–18 segments; chains up to five zooids, up to 10 mm (Fig. 12.41 N, O) ..... *Aeolosoma marcusi* Van der Land, 1971 [Brazil. Argentina. Uruguay]

8(7) First zooid 7–11 segments 9

8′ First zooid 14–16 segments; relatively large, body width 150–200 μm; hair chaetae (3)5–7(8) per bundle, long by 0–2(3) and short by 2–5 per bundle; epidermal glands orange to dark red, partially colorless; intestine dilated in segments IV–IX(XI); chains of 2–3 zooids, 2–4 mm (Fig. 12.41 P) ..... *Aeolosoma gertae* Marcus, 1944 [Brazil. Palaearctic]

9(8) Hair chaetae mostly longer than the body width, the shortest not resembling sigmoid chaetae 10

9′ Hair chaetae mostly considerably shorter than body width, by 1–5(6) per bundle, subequal or unequal, long by 1–3 and short, 1–4 per bundle, the latter somewhat resembling sigmoid chaetae; epidermal glands orange-red; intestine dilated in segments III–VI; first zooid 7–8 segments; chains up to five zooids, to 2.5 mm; in sexually mature individuals copulatory gland in intersegmental furrow V/VI or VI/VII, and up to five pairs of spermathecae in segments II–VI (Fig. 12.42 A–D) ..... *Aeolosoma quaternarium* Ehrenberg, 1831 [Colombia(?). Argentina. Uruguay. Palaearctic]

10(9) Longest hair chaetae > 160 μm; body width ~ 150–300 μm 11

10′ Longest hair chaetae < 160 μm; body width 40–175 μm 12

11(10) Body width 200–300 μm; hair chaetae 2–8 per bundle, long, 1–4 per bundle, 190–250 μm, short 0–5 per bundle, 80–100 μm; posterior end bilobed; epidermal glands green with various tinge; intestine dilated in segments IV–VI(VII); first zooid 7–8(9) segments, about 2 mm; chains up to eighth zooids and 8 mm (Fig. 12.42 E) ..... *Aeolosoma viride* Stephenson, 1911 [Brazil. Palaearctic]

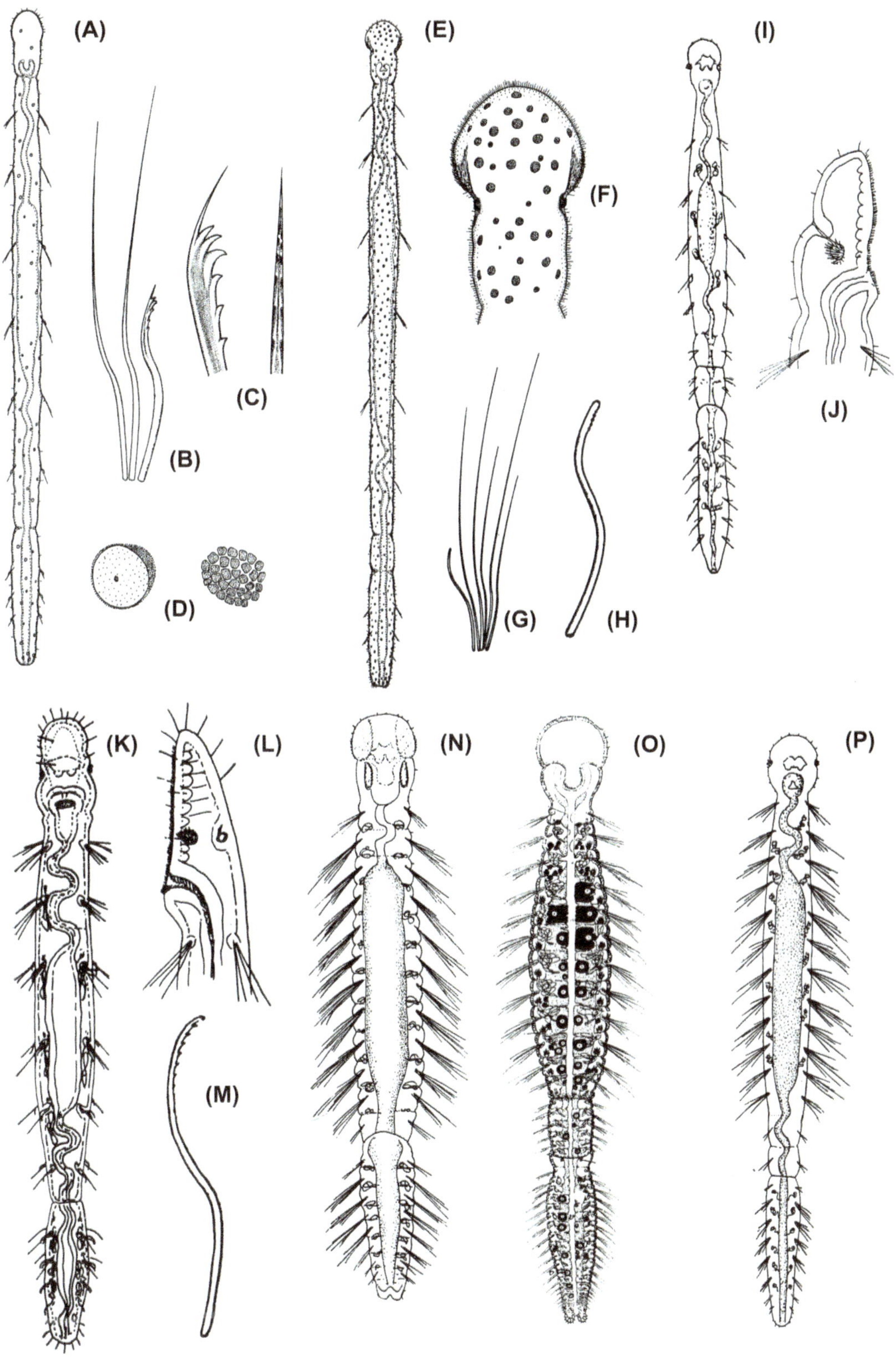

FIGURE 12.41 Species of *Aeolosoma. Aeolosoma travancorensis*: (A) dorsal view; (B) bundle of chaetae; (C) distal end of short chaeta; (D) epidermal glands. *Aeolosoma evelinae*: (E) dorsal view; (F) anterior end in dorsal view; (G) bundle of chaetae; (H) short chaeta. *Aeolosoma sawayai*: (I) dorsal view; (J) anterior end in right view. *Aeolosoma corderoi*: (K) dorsal view; (L) anterior end in left view; (M) short chaeta. *Aeolosoma marcusi*: (N) dorsal view of immature individual; (O) ventral view of sexually mature individual. *Aeolosoma gertae*: (P) dorsal view. *(A–H) After Bunke (1967); (I–P) after Marcus (1944).*

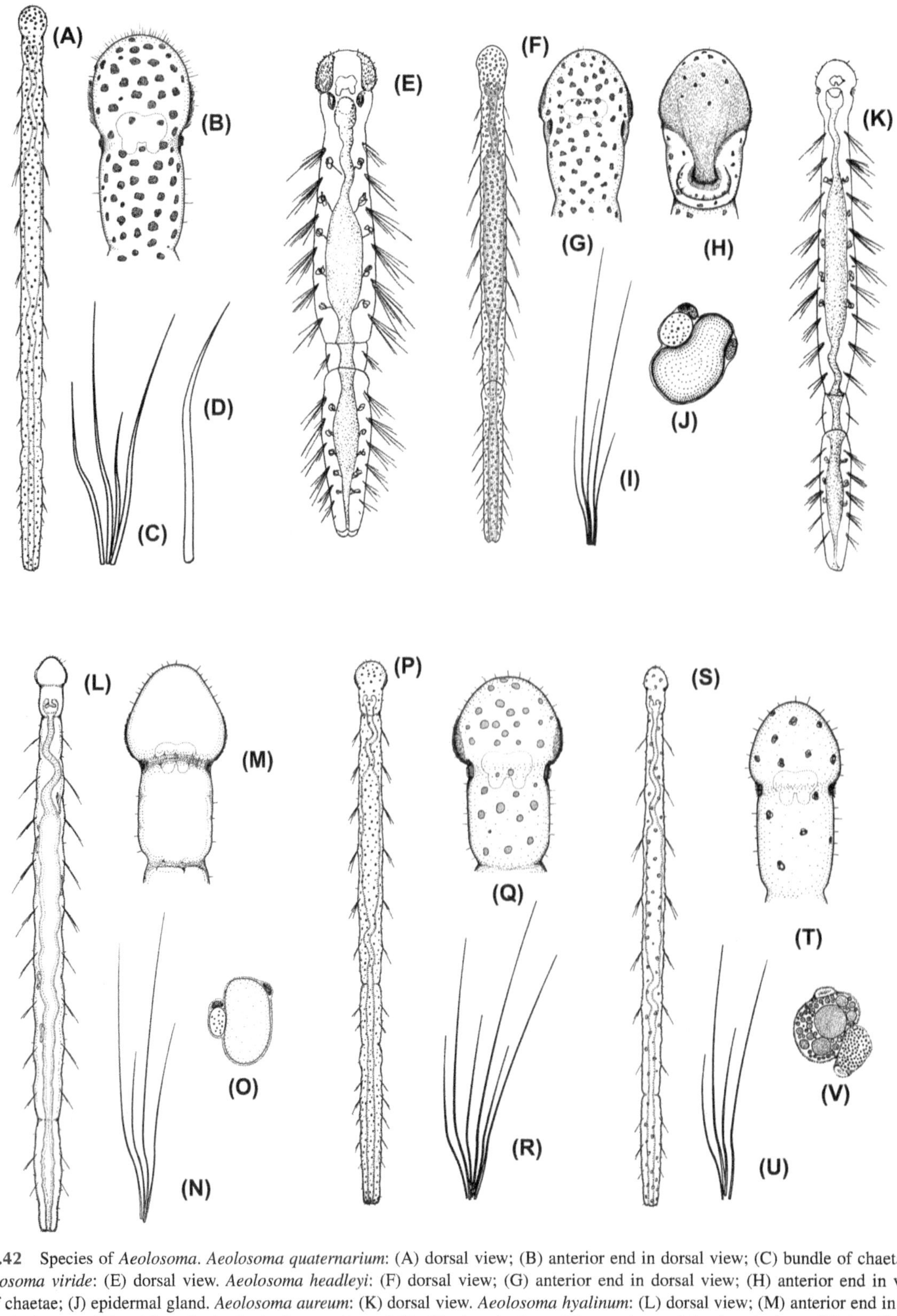

FIGURE 12.42 Species of *Aeolosoma*. *Aeolosoma quaternarium*: (A) dorsal view; (B) anterior end in dorsal view; (C) bundle of chaetae; (D) short chaeta. *Aeolosoma viride*: (E) dorsal view. *Aeolosoma headleyi*: (F) dorsal view; (G) anterior end in dorsal view; (H) anterior end in ventral view; (I) bundle of chaetae; (J) epidermal gland. *Aeolosoma aureum*: (K) dorsal view. *Aeolosoma hyalinum*: (L) dorsal view; (M) anterior end in dorsal view; (N) bundle of chaetae; (O) epidermal gland. *Aeolosoma hemprichi*: (P) dorsal view; (Q) anterior end in dorsal view; (R) bundle of chaetae. *Aeolosoma variegatum*: (S) dorsal view; (T) anterior end in dorsal view; (U) bundle of chaetae; (V) epidermal gland. *(A–D), (F–J), and (L–V) after Bunke (1967); (E) and (K) after Marcus (1944).*

11′ Body width 150–200 μm; hair chaetae up to 6(9) per bundle, long 0–4 per bundle, 150–400 μm, short 2–5 per bundle; ventral chaetae usually shorter than dorsal; indistinct intersegmental furrows usually present; posterior end often bilobed; epidermal glands green with various tinge, partially colorless; intestine dilated in segments IV to VIII–X; first zooid 9–11 segments, up to 2 mm; chains up to four zooids and 4 mm; in sexually mature individuals glandular pad with gonopore can lie in segments V, VI, VII, or VIII (Fig. 12.42 F–J) .... *Aeolosoma headleyi* Beddard, 1888
[Costa Rica. Suriname. Brazil. Paraguay. Uruguay. Palaearctic]

12(10) Epidermal glands colored 13

12′ Epidermal glands colorless; hair chaetae 2–5(6) per bundle, long 0–2, 100–130(140) μm, short 1–4(5) per bundle; intestinal dilatation with gradual limits in segments IV to VI–VIII; first zooid (8)9–11 segments, 0.5–1 mm; chains up to four zooids and 2 mm (Fig. 12.42 L–O) .... *Aeolosoma hyalinum* Bunke, 1967
[Suriname. Palaearctic]

13(12) Epidermal glands green or yellow 14

13′ Epidermal glands orange to dark red; hair chaetae 2–6(8) per bundle, unequal but in smaller individuals subequal, long 0–3(4) per bundle, 70–120 μm, short 1–3(4) per bundle; intestine dilating from segment VI; first zooid (6)7(8) segments, 0.3–1 mm; chains up to six zooids and 2 mm; in sexually mature individuals glandular pad with female pore in intersegmental furrow IV/V or V/VI and three pairs of spermathecae in segments II–IV (Fig. 12.42 P–R) .... *Aeolosoma hemprichi* Ehrenberg, 1828
[Suriname. Brazil. Argentina. Uruguay. Cosmopolitan]

14(13) First zooid 6–9 segments 15

14′ First zooid 10 or 11 segments; hair chaetae (2)4–6(8) per bundle, up to 150 μm; epidermal glands brilliant golden-green; intestine dilated in segments III–IX; chains up to four zooids and 3 mm (Fig. 12.42 K) .... *Aeolosoma aureum* Marcus, 1944
[Brazil. Uruguay. Palaearctic]

15(14) Epidermal glands golden yellow; intestinal dilatation beginning from segments III–V; hair chaetae 2–5 per bundle, dorsally up to 140 μm, ventrally less; first zooid (7)8(9) segments; chains of three zooids 1.5–2 mm .... *Aeolosoma flavum* Štolc, 1903
[Argentina. Palaearctic]

15′ Epidermal glands greenish yellow, granular, partially colorless; hair chaetae by 2–6 per bundle, long 0–2 per bundle, up to 135 μm, short 1–4 per bundle; intestine dilating from segments IV–VI; first zooid 7–9 segments, 0.8–1.2 mm; chains up to four zooids and 2 mm (Fig. 12.42 S–V) .... *Aeolosoma variegatum* Vejdovský, 1885
[Cuba(?).Venezuela(?). Suriname. Afrotropical, Nearctic, and Palaearctic]

REFERENCES

Brinkhurst, R.O. & B.G.M. Jamieson. 1971. Aquatic Oligochaeta of the World. Oliver & Boyd, Edinburgh. 860 pp.

Bunke, D. 1967. Zur Morphologie und Systematik der Aeolosomatidae Beddard 1895 und Potamodrilidae nov. fam. (Oligochaeta). Zoologische Jahrbücher, Abteilung für Systematik, Ökologie und Geographie der Tiere 94: 187–368.

Marcus, E. 1944. Sôbre Oligochaeta limnicos do Brasil. Boletim da Faculdade de Filosofia, Ciências e Letras, Universidade de São Paulo, Zoologia 43: 5–135.

Michaelsen, W. 1900. Oligochaeta. Das Tierreich, 10. R. Friedländer und Sohn, Berlin. 575 pp.

Michaelsen, W. 1933. Süß- und Brackwasser-Oligochäten von Bonaire, Curaçao und Aruba. Zoologische Jahrbücher, Abteilung für Systematik, Ökologie und Geographie der Tiere 64: 327–350.

Timm, T. 1981. On the origin and evolution of aquatic Oligochaeta. Proceedings of the Estonian Academy of Sciences, Biology 30: 174–181.

Van der Land, J. 1971. Family Aeolosomatidae. Chapter 13: Family Aeolosomatidae. Pages 665–706 *in*: R.O. Brinkhurst & B.G.M. Jamieson (eds.), Aquatic Oligochaeta of the World, Oliver & Boyd, Edinburgh.

Vejdovský, F. 1884. System und Morphologie der Oligochaeten. Franz Řivnač, Prag. 170 pp.

Class Polychaeta

Christopher J. Glasby
Museum & Art Gallery of the Northern Territory, Darwin, NT, Australia

João Gil
CCMAR, Universidade do Algarve, Faro, Portugal

Daniel Martin
Centre d'Estudis Avançats de Blanes, CEAB-CSIC, Catalunya, Spain

INTRODUCTION

Polychaeta are mainly marine and brackish water annelids. Recent studies using independent molecular data sets have shown that the class is a catchall for many unrelated annelid groups, and as a consequence, the name is likely to be formally abandoned in the near future. The number of polychaete taxa living in the Neotropical fresh waters is very few compared with their marine counterparts, comprising only 22 known species in six genera. The 22 species belong to Nereididae (11 species) and Histriobdellidae (8 species), followed by Capitellidae, Fabriciidae, and Sigalionidae each with one species. This number of species would nearly triple if we included estuarine species known to tolerate exposure to freshwater (Glasby et al., 2009).

The 22 species may be found in habitats ranging from rivers and streams to various types of groundwater aquifers, including springs and sinkholes in karstic environments (Glasby et al., 2009, 2014). Unlike other realms where large lakes and endorheic water bodies (including saline lakes) provide other habitats for polychaete fauna, there are no records of polychaetes in inland lakes of the Neotropics including lakes Titicaca, Nicaragua, and Mar Chiquita. Furthermore, although there are reports of nonnative polychaete introductions into estuaries of the Neotropics (e.g., *Neanthes succinea* (Leuckart, 1847), *Ficopomatus enigmaticus* (Fauvel, 1923), and *Ficopomatus uschakovi* (Pillai, 1960) (Schwindt & Iribarne, 1998; Arteaga-Flórez et al., 2014)), none appear to have migrated to freshwater bodies or have been secondarily introduced into inland lakes as has happened in other realms. In this regard, it is worth noting that the record of *F. enigmaticus* in Mar Chiquita (Schwindt & Iribarne, 1998) refers to a coastal lagoon in the south of Buenos Aires province, rather than to the endorheic lake of the same name in Córdoba province.

With the exception of *Stratiodrilus* species, most freshwater polychaete species in the Neotropics are benthic, living in unconsolidated sediments. Members of *Stratiodrilus* are tiny commensals of freshwater decapod crustaceans of the families Aeglidae, Parastacidae, and Trichodactylidae (Amato et al., 2004). The North American fabriciid, *Manayunkia speciosa* (Leidy, 1859), was recently reported to have been introduced to the Uruguay River, perhaps by ships via the Río de la Plata (Armendáriz et al., 2011). In its native range, it is an intermediate host for myxozoan parasites that cause the fatal disease ceratomyxosis in salmonid fishes, but its faunal associations and ecosystem effects in the Neotropics are unknown.

Apart from the introduced *M. speciosa*, most species are endemic to the Neotropics; however, none of the genera are endemic. The exceptional species are *Namalycastis senegalensis* (Saint-Joseph, 1901) and *Namanereis hummelincki* (Augener, 1933), which have amphi-Atlantic distributions. The most diverse and characteristic genera of freshwater polychaetes in the Neotropics are the decapod crustacean commensals *Stratiodrilus*, with eight species, and the free-living groundwater members of *Namanereis*, with six species. Indeed, the former only occurs in the southern South America (Argentina, Brazil, Chile, and Uruguay), Australia (two species), and Madagascar (one species), suggesting a Gondwanan origin along with their crustacean hosts. While *Namanereis* is more widely distributed, those Neotropical species all belong to the bifid-jaw subgroup, which have an Afro-Caribbean distribution suggesting a more recent Tethyan origin for this group (Glasby et al., 2014).

LIMITATIONS

The Neotropical freshwater polychaetes are well known, and for the most part, they are readily identified to species level, as the result of recent revisionary studies on the two most speciose families, Nereididae and Histriobdellidae. Keys to these two groups are based primarily on characters of the chaetae and posterior locomotor appendages, respectively. Unfortunately, the habitat limitations of some groups, particularly Nereididae, are less clear. Many of those species are supralittoral and/or euryhaline, so they can potentially occur in freshwater. In this treatment, we have included in the key only the nereidid species specifically reported from freshwater.

The amphi-Atlantic nereidid species *Namalycastis senegalensis* and *Namanereis hummelincki* need to be reviewed, as many recent systematic studies with widespread polychaete species indicate that multiple cryptic species could be involved. Also, the genus *Manayunkia* (Fabriciidae) needs to be revised, particularly to confirm the identity of the record of *M. speciosa* from the Uruguay River.

TERMINOLOGY AND MORPHOLOGY

Polychaetes are characterized by having a body formed by multiple small, repeated units (segments) arranged longitudinally along the body. Often, the only differentiated regions are the presegmental **prostomium**, which may include sensory and feeding appendages, and the postsegmental **pygidium**, which includes the anus and often, sensory and attachment organs. However, some forms may have the body divided into, at least, two regions: an anteriormost thorax and a posteriormost abdomen. Most polychaetes have characteristic bundles of **chaetae** on each segment (one or two on each side). These chaetae are often carried by the **parapodia**, which are variously well developed and assist with locomotion. Other paired internal and external structures are feeding-related (palps, tentacular crown; Fig. 12.43), sensory (palps, cirri), protective (elytrae), respiratory (branchiae), excretory (nephridia), or reproductive (gonopores).

The morphology of polychaetes is extraordinarily diverse, as it corresponds to a group of organisms that are adapted to virtually all aquatic environments. This section will highlight the main terms and morphological structures characterizing the Neotropical freshwater polychaetes.

Nereidids are typical errant polychaetes, with multi-segmented, long bodies and a well-developed cephalic region, which bears two bilobed palps, two antennae, tentacular cirri (three or four pairs; Fig. 12.45 A, C), and a muscular eversible pharynx equipped with a pair of lateral jaws, multiple accessory denticles (called paragnaths), and/or papillae; however, denticles and papillae are entirely absent in the subfamily Namanereidinae, which are the only representatives of the family in Neotropical freshwater. Also unlike typical nereidids, namanereidids have parapodia lacking the dorsal (**notopodial**) ramus, and the ventral (**neuropodial**) ramus comprises a single lobe

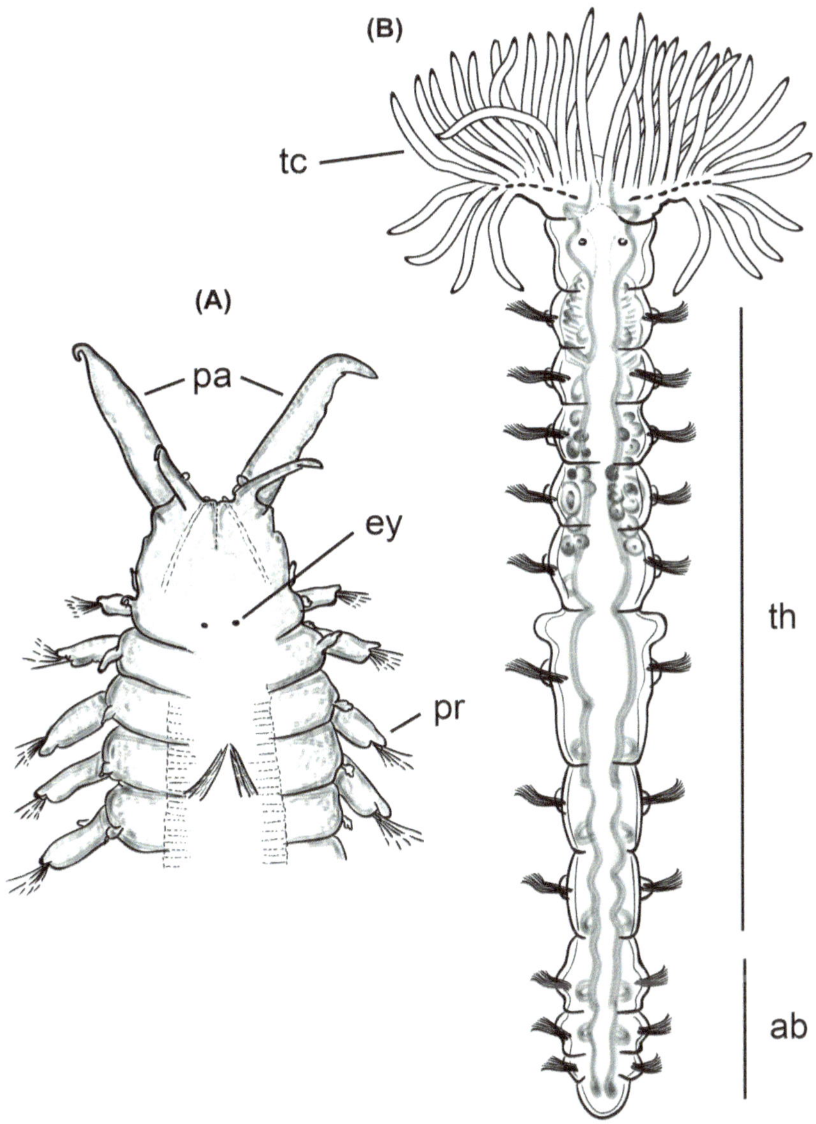

FIGURE 12.43 (A) *Pisione garciavaldecasasi*; and (B) *Manayunkia speciosa*. Abbreviations: *ab*, abdomen; *ey*, eyes; *pa*, palps; *pr*, parapodia; *tc*, tentacular crown; *th*, thorax. *Redrawn (A) from San Martin et al. (1998), and (B) from Leidy (1859).*

PHYLUM ANNELIDA

bearing the chaetae. **Dorsal cirri** may be short, cylindrical, and tapered throughout (*Namanereis*), or bear basal cirrophores anteriorly and become flattened posteriorly (*Namalycastis*). Chaetae are compound and comprise one or more types with the terminal blades either short (falcigers; Fig. 12.45 E–N), long (spinigers), or intermediate in length (pseudospingers, or simply "elongated falcigers" in the more recent literature). The postsegmental pygidium may have many incisions (*Namalycastis*) or be tripartite (*Namanereis*), and it bears one pair of pygidial, or anal, cirri (Fig. 12.45 B, D).

Sigalionids are errant polychaetes more typical of marine habitats. Indeed, the single Neotropical species, *Pisione garciavaldecasasi* San Martín, López & Camacho, 1998, was formerly placed in the family Pisionidae, now incorporated in the family Sigalionidae. Like other sigalionids, *Pisione* have multisegmented, long bodies and a well-developed cephalic region, which bears long, unarticulated ventrolateral palps (Fig. 12.43 A). Unlike typical sigalionids, *Pisione* species lack a facial tubercle, have uniramous parapodia (notopodial lobes and chaetae are absent), their dorsal cirri are tapered, and none are modified as elytrae, and the capillary chaetae are smooth.

Histriobdellids are atypical errant polychaetes, reflecting their commensal lifestyle. Their body has few segments, the head lacks eyes, and parapodia are absent or poorly developed and lack chaetae. The head end is fairly uniform across the group bearing three prostomial antennae, paired palps, a peristomium with a pair of cirri, and a muscular ventral pharynx equipped with paired mandibles and four to five pairs of maxillae. The posterior end is divided into two large **posterior locomotory appendages** ("feet"). These appendages bear lobes, cirri, and tubercles of various shapes and sizes that are taxonomically highly significant (Fig. 12.44 A–H).

Capitellids are vermiform "sedentary" polychaetes with numerous segments, a distinct thorax and abdomen, and a head lacking antennae, tentacles, or palps. The proboscis is saclike and nonmuscular. The first chaetiger may be biramous or uniramous, but thereafter they are biramous; parapodial lobes are absent or low throughout, and dorsal and ventral cirri are absent; abdominal neuropodial lobes are low ridges. Thoracic chaetae are typically capillary, and abdominal ones are hooks bearing a distal hood. Posterior parapodia may bear retractile branchiae. The pygidium is a simple ring or cone, or rarely, lamellar. Pygidial appendages are absent or, if present, comprise one pair of cirri, or a single medial cirrus.

Fabriciids are a group of small-sized, tube-dwelling (sedentary) polychaetes formerly considered as a subfamily of the feather-duster worms, Sabellidae. Like the latter, they have reduced prostomia and an anterior tentacular crown, which functions as both a feeding and respiratory organ (Fig. 12.43 B). But fabriciids have a smaller body and separate branchial lobes comprising the tentacular crown, whereas sabellids have branchial lobes that are fused middorsally. Fabriciids have a ring-shaped peristomium, with or without an anterior collar. Their bodies are segmented and weakly divided into a thorax and abdomen, and their parapodia are usually biramous. Dorsal and ventral cirri, as well as branchiae, epidermal papillae, and pygidial cirri, are absent.

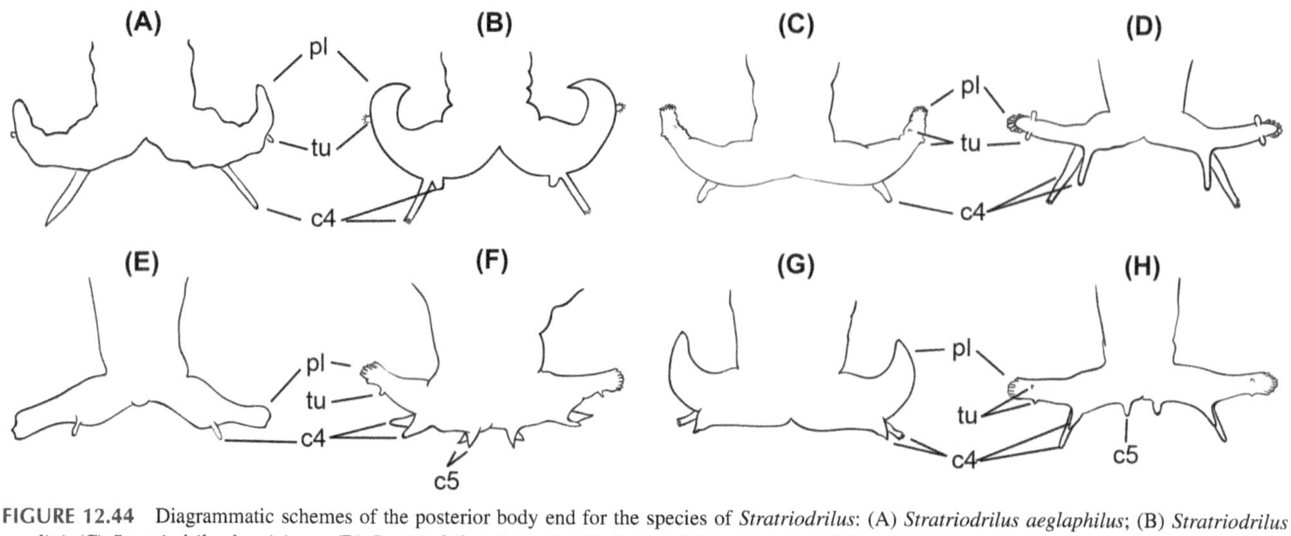

FIGURE 12.44 Diagrammatic schemes of the posterior body end for the species of *Stratriodrilus*: (A) *Stratriodrilus aeglaphilus*; (B) *Stratriodrilus arreliai*; (C) *Stratriodrilus brevicirrus*; (D) *Stratriodrilus circensis*; (E) *Stratriodrilus platensis*; (F) *Stratriodrilus pugnaxi*; (G) *Stratriodrilus robustus*; (H) *Stratriodrilus vilae*. Abbreviations: *c4*, posterior cirri 4; *c5*, posterior cirri 5; *pl*, posterior limbs; *tu*, tubercle. *Redrawn from: (A) and (F) Vila & Bahamonde (1985); (B) (D) and (G) Steiner & Amaral (1999); (C) Amato et al. (2004); (E) Harrison (1928); (H) Amato (2001). Nomenclature according to Amaral & Morgado (1997). Note that Vila & Bahamonde (1985), and Amaral & Morgado (1997) pictured* S. platensis *with the C4 with an extra pair of conical ventral lobes, which does not concur with the type description and the original drawing.*

MATERIAL PREPARATION AND PRESERVATION

Polychaetes are most easily identified after fixation, although information on coloration, ciliation, etc., which may be useful at the species level, may be lost or obscured. Optimal specimen condition for morphological observations is achieved through first relaxing the animals using a few grains of $MgCl_2$ or $MgSO_4$ dissolved in freshwater for a few minutes to several hours (depending on the group). After relaxation, freshwater polychaetes should be fixed in a 4% formaldehyde solution made by diluting stock formaldehyde (i.e., 40%) 10 times with tap water. After one to a few days, specimens should be washed gently in freshwater and transferred to 70%–80% ethanol for identification and long-term storage. Alternatively, fixation in osmium tetroxide after relaxing is advisable for observations under scanning electron microscopy.

KEYS TO POLYCHAETA

Polychaeta: Families

1	Head bearing paired palps and/or antennae (Fig. 12.43 A); free-living and commensal forms; Errantia 2
1′	Head lacking appendages or bearing a tentacular crown (Fig. 12.43 B); free-living forms only; Sedentaria 4
2(1)	Free-living, benthic forms with well-developed parapodia, undivided posterior end bearing paired cirri 3
2′	Commensal on Crustacea, parapodia absent or poorly developed, chaetae absent; posterior end divided into two large locomotory appendages (Fig. 12.44) .... Histriobdellidae, one genus: ***Stratiodrilus* [p. 483]**
3(2)	Palps short, biarticulated; eyes, if present, two pairs, prominent (Fig. 12.45 A, C) .... **Nereididae [p. 484]**
3′	Palps long, unarticulated; eyes, if present, subdermal (Fig. 12.43 A); in gravel and stone beds .... Sigalionidae, one species: *Pisione garciavaldecasasi* San Martín, López & Camacho, 1998 [Panama: Coiba Island]
4(1)	Tentacular crown present; body short (Fig. 12.43 B); tubiculous .... Fabriciidae, one species *Manayunkia speciosa* (Leidy, 1859) [Argentina and Uruguay: silty-clay sediments of the lower Uruguay River]
4′	Tentacular crown absent (head appendages lacking); body earthworm like; free-living; in blue mud of marshes .... Capitellidae, one species: *Notomastus ouanaryensis* (Gravier, 1901) [French Guiana: Ouanary River]

Polychaeta: Histriobdellidae: *Stratiodrilus*: Species

Occurring on decapod crustaceans. Key modified after Amato et al. (2004)

1	Segmental cirri on chaetiger 4 (C4) without associated conical ventral lobe (Fig. 12.44 A, C, E) 2
1′	Segmental cirri on chaetiger 4 (C4) with associated conical ventral lobe (Fig. 12.44 B, D, F–H) 4
2(1)	Posterior locomotory appendages with tubercles 3
2′	Posterior locomotory appendages without tubercles (Fig. 12.44 E); jaw apparatus reaching the limit between head and first trunk segment .... *Stratiodrilus platensis* Cordero, 1927 [Brazil: São Paulo State on *Trichodactylus* sp. Argentina. Uruguay: Canelones, Solís Chico and Miguelete streams on *Aegla laevis* (Latreille)]
3(2)	Posterior locomotory appendages with two large tubercles, segmental cirri (C1 to C4) long (Fig. 12.44 A) .... *Stratiodrilus aeglaphilus* Vila & Bahamonde, 1985 [Chile: Maipo River, Bellavista, Petorca to Chiloé, on species of Aeglidae and Parastacidae]
3′	Posterior locomotory appendages with two inconspicuous tubercles, short lateral segmental cirri (C1–C3), short C4 with a dilation at base (Fig. 12.44 C) .... *Stratiodrilus brevicirrus* Amato, Daudt & Amato, 2004 [Brazil: Rio Grande do Sul, Taquara, Fazenda Fialho, Arroio do Mineiro on *Aegla platensis* Schmitt]
4(1)	Anal lobes (= C5) present (Fig. 12.44 F, H) 5
4′	Anal lobes absent (Fig. 12.44 B, D, G) 6
5(4)	With two pairs of anal lobes (Fig. 12.44 F) .... *Stratiodrilus pugnaxi* Vila & Bahamonde, 1985 [Chile: Reúmen, Chaimávida, Petorca to Chiloé, on several species of Parastacidae and Aeglidae]
5′	With a single pair of anal lobes (Fig. 12.44 H) .... *Stratiodrilus vilae* Amato, 2001 [Brazil: Rio Grande do Sul, Taquara, Mineiro Creek on species of *Parastacus*]

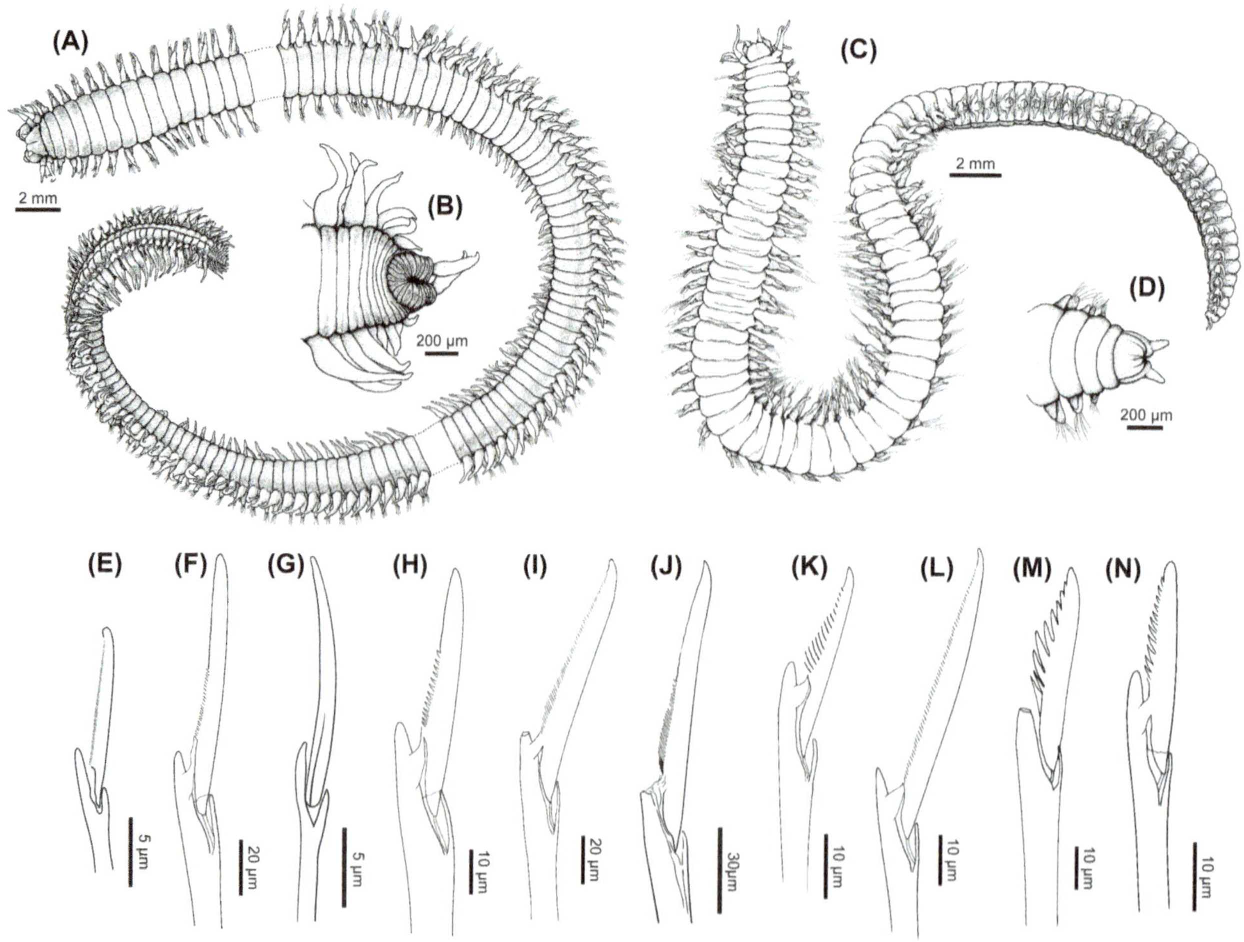

FIGURE 12.45 *Namalycastis abiuma*: (A) typical body plan, dorsal view; (B) pygidium. *Namanereis cavernicola*: (C) typical body plan, dorsal view; (D) pygidium. Supraacicular falcigers from chaetiger 10: (E) *Namalycastis caetensis*; (F) *Namalycastis macroplatis*; (G) *Namalycastis siolii*; (H) *Namalycastis senegalensis*; (I) *Namanereis cavernicola*; (J) *Namanereis christopheri*; (K) *Namanereis hummelincki*; (L) *Namanereis minuta*; (M) *Namanereis serratis*; (N) *Namanereis stocki*. *Redrawn from: (A–D), (F), (H), (I) and (K–N) Glasby (1999); (E) Alves & Santos (2016); (G) Corrêa (1948); (J) Conde-Vela (2017).*

6(4) Posterior locomotory appendages with tubercles (Fig. 12.44 B, D) 7

6′ Posterior locomotory appendages without tubercles, associated lobe close to C4 pyriform (Fig. 12.44 G); body robust .... *Stratiodrilus robustus* Steiner & Amaral, 1999
[Brazil: São Paulo on *Trichodactylus* sp.]

7(6) Posterior locomotory appendages with one tubercle, papilliform lobe associated with C4 (Fig. 12.44 B) and jaw apparatus very long with posterior end exceeding first trunk segment .... *Stratiodrilus arreliai* Amaral & Morgado, 1997
[Brazil: São Paulo State, São Pedro, Peroba Cave, Jaraguá Peak, on *Aegla* sp.]

7′ Posterior locomotory appendages with two conspicuous tubercles, one conical ventral lobe associated with C4 (Fig. 12.44 D) and jaw apparatus not reaching first trunk segment .... *Stratiodrilus circensis* Steiner & Amaral, 1999
[Brazil: Paraná, Piraquara, Piraquara River. Chile: Lake Riñihue, Tólten River, Petorca to Chiloé. Argentina: Río Negro, Nahuel Huapi Lake; Tucumán, Chico River. On several species of Aeglidae and Parastacidae]

Polychaeta: Nereididae: Genera

Modified after Glasby (1999), Glasby et al. (2014) and Conde-Vela (2017).

1 Anterior dorsal cirri cylindrical with cirrophores, posterior ones foliate and lacking cirrophores; pygidium multi-incised (Fig. 12.45 A, B) .... ***Namalycastis* [p. 485]**

1′ Dorsal cirri cylindrical throughout, lacking cirrophores; pygidium tripartite (Fig. 12.45 C, D) .... ***Namanereis* [p. 485]**

Polychaeta: Nereididae: *Namalycastis*: Species

1 Compound falcigerous chaetae present in anterior parapodia, but absent in posterior parapodia 2

1′ Compound falcigerous chaetae absent throughout .... *Namalycastis geayi* (Gravier, 1901)
[French Guiana: Ouanary River, muddy banks and sandy sediments of a spring]

2(1) Acicular neuropodial ligule subconical; compound falcigers with blades smooth or serrated 3

2′ Acicular neuropodial ligule bilobed; compound falcigers with blades serrated 4

3(2) Compound falcigers with blades smooth (Fig. 12.45 G) .... *Namalycastis siolii* (Corrêa, 1948)
[Brazil: Amazon Region, Tapajóz River]

3′ Compound falcigers with blades finely serrated from base to near tip (Fig. 12.45 E) .... *Namalycastis caetensis* Alves & Santos, 2016
[Brazil: Pará, Caeté River]

4(2) Anteriormost dorsal cirri length 2.0–5.0× parapodial lobe; dorsalmost compound falcigers in subacicular fascicle of anterior parapodia with 5–12 teeth (Fig. 12.45 H) .... *Namalycastis senegalensis* (Saint-Joseph, 1901)
[Suriname: Parimaribo. French Guiana: Ouanary River in mud of marshes, Ilet-la-Mère, Mahury, Rivière de Montsinéry. Brazil: Ilha de Marajó]

4′ Anteriormost dorsal cirri length 1.0–1.9× parapodial lobe; dorsalmost compound falcigers in subacicular fascicle of anterior parapodia with 16–30 teeth (Fig. 12.45 F) .... *Namalycastis macroplatis* Glasby, 1999
[French Guiana. Suriname: Surinam River on river bank and Pedro Miguel Locks. Brazil: Ilha de Marajó]

Polychaeta: Nereididae: *Namanereis*: Species

1 Four pairs of tentacular cirri 2

1′ Three pairs of tentacular cirri 3

2(1) Antennae cirriform; supraacicular compound falcigers with 50–60 teeth (Fig. 12.45 L); springs and wells .... *Namanereis minuta* Glasby, 1999
[Hispaniola Island. Haiti, La Vallée, Grand'Anse, and Grande Ravine du Sud]

2′ Antennae subconical; supraacicular compound falcigers with 9–14 teeth (Fig. 12.45 N); sometimes in groundwater aquifers .... *Namanereis stocki* Glasby, 1999
[Cuba. Jamaica: St. Ann's Great River, St. Ann's Bay. Hispaniola Island. Haiti]

3(1) Subacicular neurochaetae include elongated falcigers and spinigers 4

3′ Subacicular neurochaetae comprise falcigers only (Fig. 12.45 M); in wells .... *Namanereis serratis* Glasby, 1999
[Hispaniola Island. Haiti, Étang Saumâtre]

4(3) Supraacicular compound falcigers with 35–80 minute teeth (Fig. 12.45 I, J) 5

4′ Supraacicular compound falcigers with 15–20 teeth (Fig. 12.45 K); in supply gutter of water basin (in limestone region) .... *Namanereis hummelincki* (Augener, 1933)
[Cuba. Jamaica. Netherlands Antilles: Bonaire, Curaçao, and Aruba (Fontein). Hispaniola Island. Haiti. Barbados. Montserrat. Venezuela: Blanquilla]

5(4) Anterior margin of prostomium incised; antennae longer than prostomium; in a freshwater pond within a limestone rock cave .... *Namanereis cavernicola* Solís-Weiss & Espinasa, 1991
[Mexico: Guerrero, Taxco de Alarcón, Isote Cavern, 1650 m asl]

5′ Anterior margin of prostomium entire; antennae shorter than prostomium; in a spring pool .... *Namanereis christopheri* Conde-Vela, 2017
[Saint Vincent and the Grenadines: Saint Vincent, Golden Grove, near Chateaubelair Bay]

REFERENCES

Alves, P.R. & C.S.G. Santos. 2016. Description of a new species of *Namalycastis* (Annelida: Nereididae: Namanereidinae) from the Brazilian coast with a phylogeny of the genus. Zootaxa 4144: 499–514.

Amato, J.F.R. 2001. A new species of *Stratiodrilus* (Polychaeta, Histriobdellidae) from freshwater crayfishes of southern Brazil. Iheringia Série Zoologia, Porto Alegre 90: 37–44.

Amato, J.F.R., L.C.C. Daudt & S.B Amato. 2004. New species of *Stratiodrilus* (Polychaeta, Histriobdellidae) from freshwater anomuran crustaceans of southern Brazil. Biociências 12: 121–127.

Amaral, A.C. & E.H. Morgado. 1997. *Stratiodrilus* (Annelida; Polychaeta; Histriobdellidae) associated with a freshwater decapod, with the description of a new species. Proceedings of the Biological Society of Washington 110: 471–475.

Armendáriz, L.C., A. Paola & A.R. Capítulo. 2011. *Manayunkia speciosa* Leidy (Polychaeta: Sabellidae): introduction of this nonindigenous species in the Neotropical Region (Uruguay River, South America). Biological Invasions 13: 281–284.

Arteaga-Flórez, C., V. Fernández-Rodríguez & M.H. Londoño-Mesa. 2014. First record of the polychaete *Ficopomatus uschakovi* (Pillai,

PHYLUM ANNELIDA

1960) (Annelida, Serpulidae) in the Colombian Caribbean, South America. ZooKeys 371: 1–11.

Conde-Vela, V.M. 2017. The troglomorphic adaptations of Namanereidinae (Annelida, Nereididae) revisted, including a redescription of *Namanereis cavernicola* (Solís-Weiss & Espinasa, 1991), and a new Caribbean species of *Namanereis* Chamberlin, 1919. Subterranean Biology 23: 19–46.

Corrêa, D.D. 1948. A polychaete from the Amazon-region. Boletim da Faculdade de Filosofia, Ciências e Letras, Universidade de São Paulo. Zoologia 13: 245–257.

Glasby, C.J. 1999. The Namanereidinae (Polychaeta: Nereididae). Part 1. Taxonomy and phylogeny. Records of the Australian Museum, Supplement 25: 1–129.

Glasby, C.J., D. Fiege & K. Van Damme. 2014. Stygobiont polychaetes: notes on the morphology and the origins of groundwater *Namanereis* (Annelida: Nereididae: Namanereidinae), with a description of two new species. Zoological Journal of the Linnean Society 171: 22–37.

Glasby, C.J., T. Timm, A.I. Muir & J. Gil. 2009. Catalogue of non-marine Polychaeta (Annelida) of the World. Zootaxa 2070: 1–52.

Harrison, L. 1928. On the genus *Stratiodrilus* (Archiannelida: Histriobdellidae), with a description of a new species from Madagascar. Records of the Austalian Museum 16: 116–122.

Leidy, J. 1859. *Manayunkia speciosa*. Journal of the Academy of Natural Sciences of Philadelphia 10: 90.

San Martín, G., E. López & A.I. Camacho. 1998. First record of a freshwater Pisionidae (Polychaeta): description of a new species from Panama with a key to the species of *Pisione*. Journal of Natural History 32: 1115–1127.

Schwindt, E. & O.O. Iribarne. 1998. Reef of *Ficopomatus enigmaticus* (Polychaeta: Serpulidae) in the Mar Chiquita Coastal Lagoon, Argentina. Butlletí de la Societat d'Història Natural de les Balears 41: 35–40.

Steiner, T.M. & A.C.Z. Amaral. 1999. The family Histriobdellidae (Annelida, Polychaeta) including descriptions of two new species from Brazil and a new genus. Contributions to Zoology 68: 95–108.

Vila, P. & N. Bahamonde. 1985. Two new species of *Stratiodrilus, S. aeglaphilus* and *S. pugnaxi* (Annelida: Histriobdellidae) from Chile. Proceedings of the Biological Society of Washington 98: 347–350.

Chapter 13

Phylum Ectoprocta (Bryozoa)

Timothy S. Wood
Department of Biological Sciences, Wright State University, Dayton, OH, United States

Maria Cristina Orellana Liebbe
Facultad de Ciencias, Universidad Católica de la Santísima Concepción, Concepción, Chile

Chapter Outline

INTRODUCTION

The Ectoprocta, or Bryozoa, is primarily a marine group with over, 8,000 recognized species and a rich fossil record with many thousands more. The 100 or so bryozoan species known from fresh water would seem to be overshadowed by such a robust family tree. In fact, however, freshwater bryozoans are an important part of the benthic community in lakes, ponds, and rivers worldwide. They filter suspended particles from the water, and their fecal pellets provide nourishment for a wide variety of scavengers. Bryozoans are also significant biofouling organisms that often interfere with the function of irrigation, water treatment, and industrial cooling systems (Wood, 2005b).

The phylum Entoprocta is sometimes grouped with the phylum Ectoprocta (Bryozoa) because of certain similarities; but because these are superficial, this group is discussed in the separate Chapter 14.

LIMITATIONS

In this chapter we recognize 35 species of freshwater bryozoans in the Neotropical region. Many are well known and easily identified, while others are distinguished only by minute and sometimes tentative features. Most of the difficult taxonomic issues occur in the family Plumatellidae, which comprises over 60% of the Neotropical species. Over the past several decades, much progress has been made in the taxonomy of the entire group, but many questions remain unresolved.

Freshwater bryozoans are unequally classified among two major groups: (1) the exclusively freshwater class Phylactolaemata; and (2) the mostly marine order Ctenostomata within the class Gymnolaemata. Until recently the number of known phylactolaemate species in the Neotropical region was 24 with another 6 species of ctenostomes (Massard & Geimer, 2008). In the preparation of this chapter, we eliminated 4 species as errors in literature and added 11 that could be fully documented. These numbers will surely be revised upwards in the years to come. Our current estimate of bryozoans in the Neotropical region is 30 phylatolaemate species and 5 ctenostomes, the largest total number of freshwater bryozoan species in any biogeographic region.

The soft tissues of bryozoans offer few reliable features for species identification. Hard parts, when they are available, tend to be much more consistent and verifiable. In fact, this general principle holds true for most invertebrate animals. It is surely not by coincidence that animal groups lacking hard anatomical features have relatively few recognized species. The likely reason is that many species simply cannot be distinguished by morphology alone, and the tools of molecular genetics may not be sufficiently

Thorp and Covich's Freshwater Invertebrates. https://doi.org/10.1016/B978-0-12-804225-0.00013-7

refined to detect cryptic species. Among freshwater bryozoans, the ctenostomes, with no hard anatomical parts, are often very difficult to identify with certainty. It is not surprising that the species number in this group is relatively low. The traditional taxonomist is forced to rely on such variable features as tentacle counts and the layout of the gut.

By contrast, phylactolaemate bryozoans have taxonomically useful hard parts in the encapsulated products of asexual reproduction called statoblasts. These provide a wealth of characters, such as relative shapes and dimensions, as well as tiny tubercles and netlike lines that are believed to be species specific. In most cases, the statoblast alone is sufficient to identify a species. In fact, when statoblasts are absent, many phylactolaemates cannot be positively identified at the species level.

One difficulty in preparing an identification key for Neotropical phylactolaemates is that some species are known only by their statoblasts. The colonies have never been described, and in most cases the sessile statoblasts (sessoblasts) are unknown as well. It was therefore essential to build the key around the features of free statoblasts (floatoblasts) and to use colony morphology only at a secondary level.

However, even statoblasts have their limitations. The search for ever more elusive taxonomic features has inevitably led to important features being detected only through scanning electron microscopy (SEM). This becomes a challenge to those investigators without access to such an instrument. In the following key, we have tried to avoid using characters that rely on SEM examination, but in a few cases it simply was not possible.

The reported species distributions naturally depend on the accuracy of species identification. Our understanding of taxonomically significant features has evolved over many years. Even now the value of floatoblast nodules, for example, is not universally accepted. Partly for this reason, the species distributions stated in the key below should be regarded as tentative.

As explained below, it must be remembered that different preservation techniques will alter certain key morphological characters.

TERMINOLOGY AND MORPHOLOGY

Bryozoan colonies have two basic forms: globular and tubular (Fig. 13.1). Both are composed of numerous zooids, all physically and physiologically connected (Figs. 13.1, 3). The feeding organ is a cluster of ciliated tentacles called a **lophophore,** which captures particles suspended in the water. To accommodate many tentacles the dorsal side of the lophophore is deflected inward to make a horseshoe shape (Fig. 13.2 A). In species with few tentacles the lophophore is circular in outline (Fig. 13.2 B).

The outermost part of a zooid is a nonliving layer called an **ectocyst**, composed mostly of chitin. The ectocyst ranges from transparent to opaque, sometimes within the same species, depending on age, growth rate, and other factors. The inner part of the zooid is the **polypide**, which includes the lophophore and the entire digestive tract. The polypide can be shifted toward the zooid tip to extend the lophophore. When the zooid is alarmed, the entire polypide is quickly retracted and the lophophore becomes fully protected (Fig. 13.3).

Statoblasts are essentially tiny packages of yolky material and germinal cells enclosed within a **bilayered case**. The case is the "hard part" mentioned earlier, and it can be essential for species identification. There are several types of statoblasts: floatoblasts, sessoblasts, and piptoblasts. **Floatoblasts** are freely released and have a peripheral ring of gas-filled chambers for buoyancy (Figs. 13.4 A, 5). **Sessoblasts** are firmly cemented to the

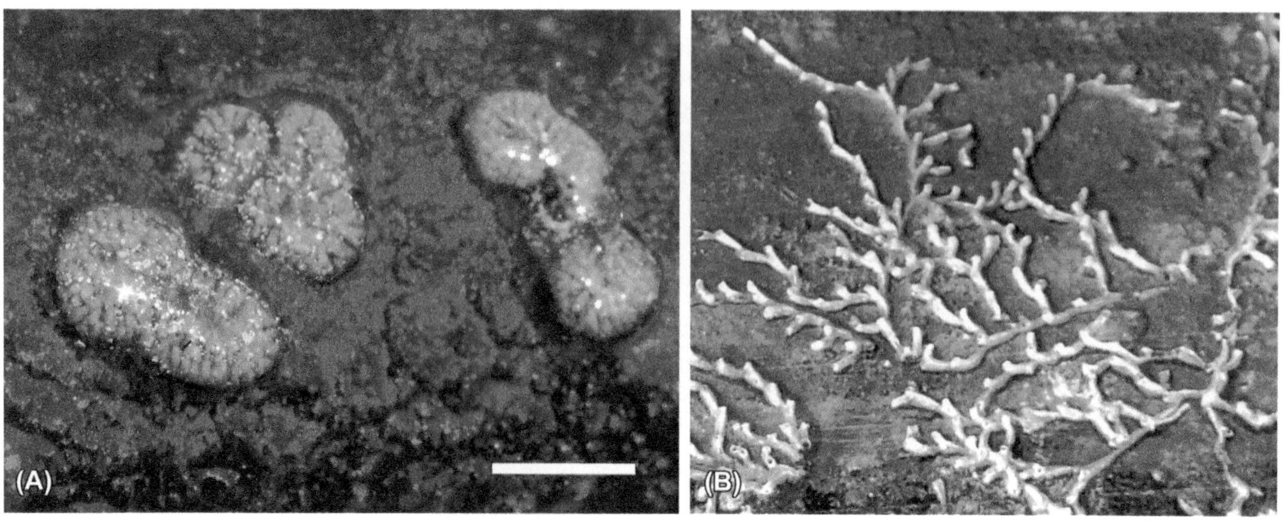

FIGURE 13.1 Two basic morphologies of bryozoan colonies: (A) globular colonies of *Asajirella gelatinosa*; and (B) tubular colonies of *Plumatella javanica*. Scale bar for (A) and (B) = 1 cm.

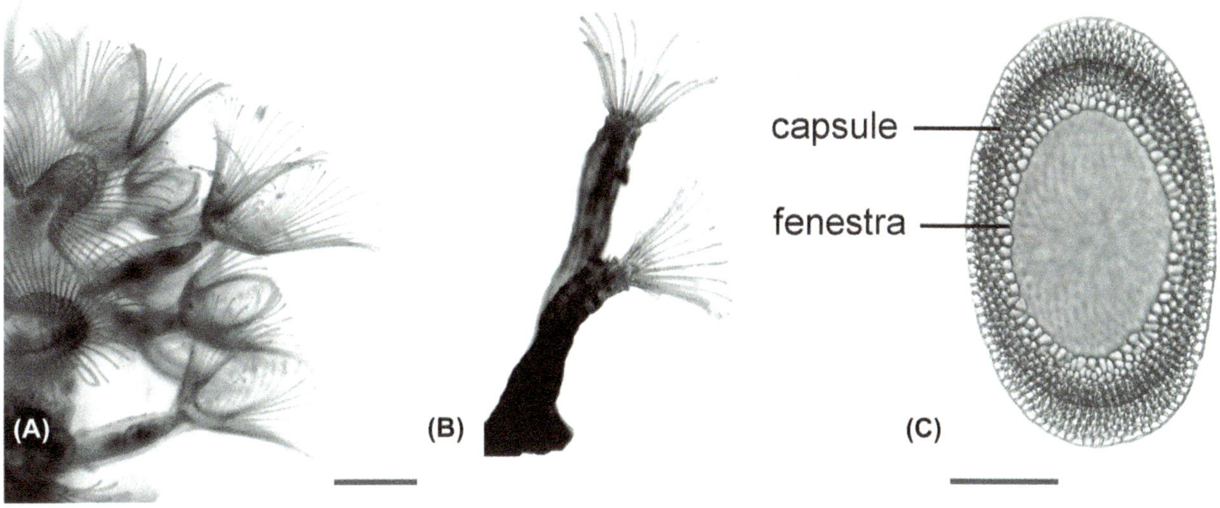

FIGURE 13.2 Aspects of bryozoan colonies: (A) horseshoe-shaped lophophores typical of most phylactolaemate species; (B) circular lophophores of fredericellid species; and (C) uninflated floatoblast of *Plumatella casmiana* distinguishing the capsule from the fenestra. Scale bar for (A) and (B) = 0.5 mm, for (C) = 0.01 mm.

substrate (Figs. 13.4 B, 10). Zooids in the family Fredericellidae produce beanlike **piptoblasts** that generally remain with the substrate but are not cemented in place (Fig. 13.9).

All statoblasts are formed like two halves of a clam shell, which split apart at the time of germination. Each half, or valve, is normally composed of two layers: an inner capsule and an outer **periblast** (Fig. 13.4). The capsule has no taxonomic value, but the periblast bears a wide variety of morphological features. In floatoblasts, the periblast includes the ring of gas-filled chambers. In the center of this ring is an area called the **fenestra**. The relative dimensions of the periblast and its fenestra can be useful taxonomic features (Figs. 13.12–15). By convention, the valve with the smallest fenestra is considered dorsal, while the opposite side is ventral (Fig. 13.5).

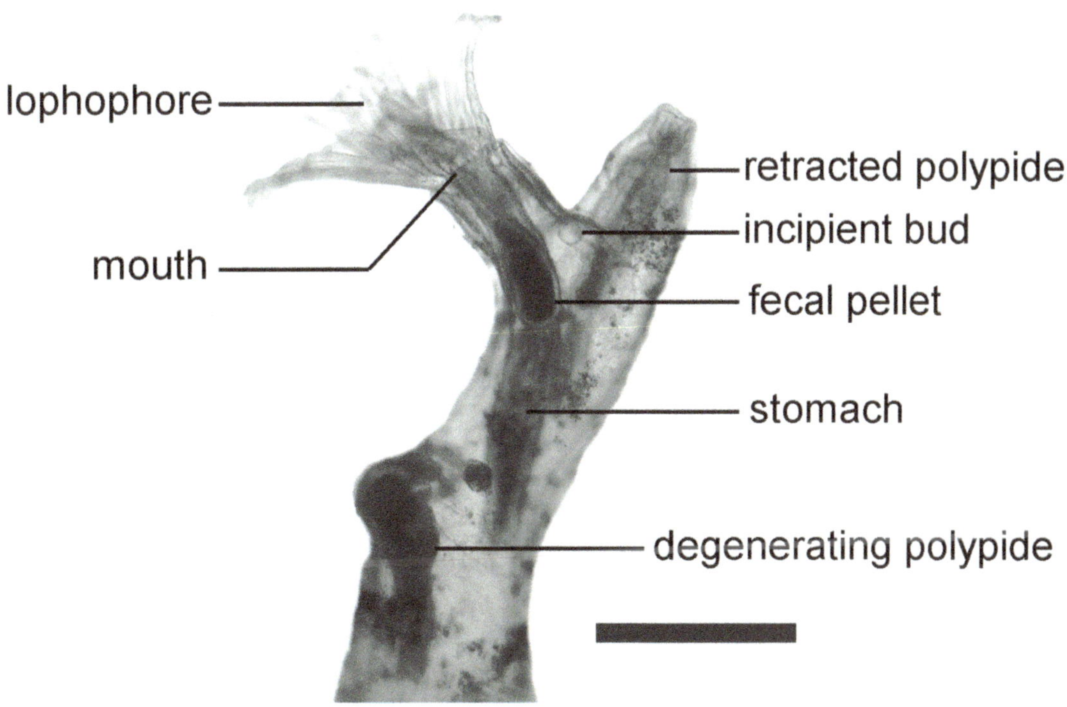

FIGURE 13.3 Zooids of a phylactolaemate bryozoan. Scale bar = 0.5 mm.

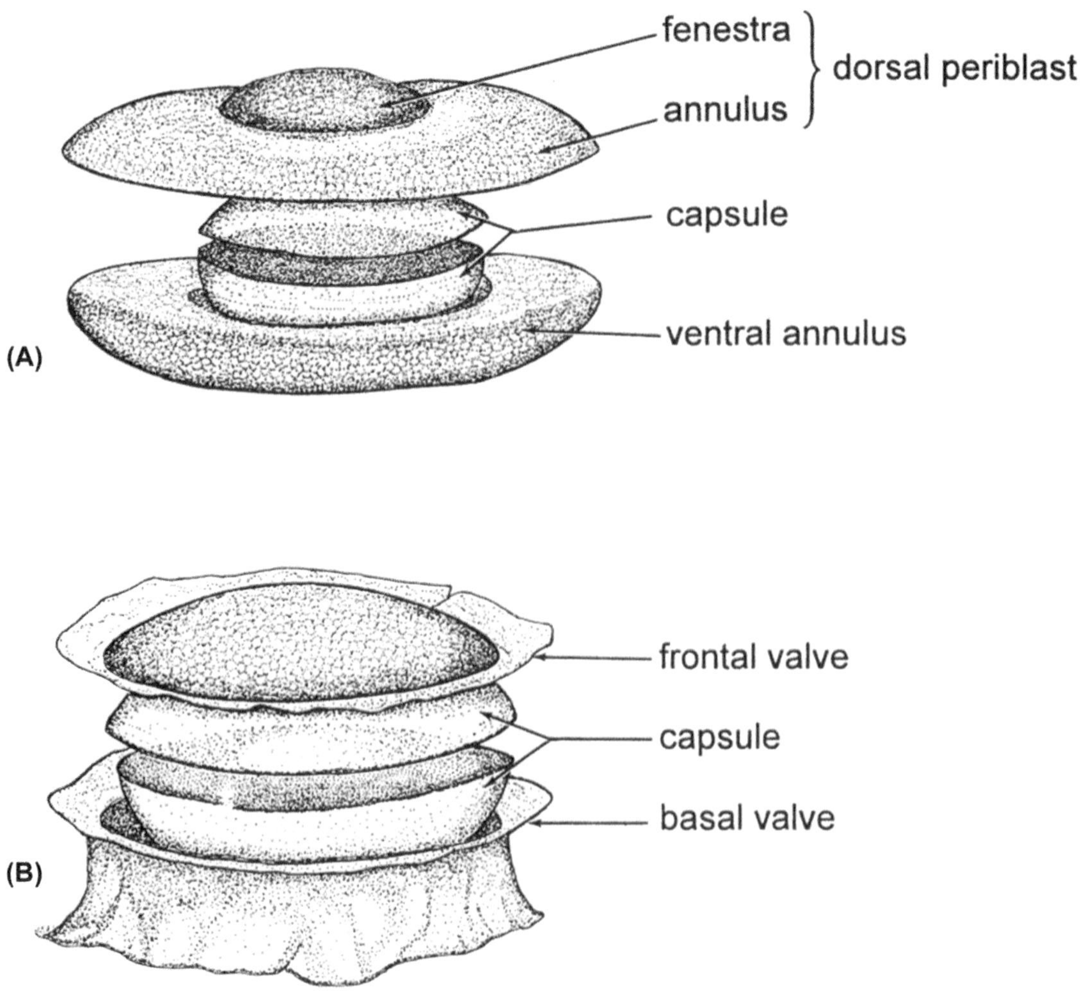

FIGURE 13.4 Sclerotized components of generalized phylactolaemate statoblasts in exploded view: (A) floatoblast; and (B) sessoblast.

Several terms are useful in describing statoblasts. The line where the valves join together is called a **suture** (Fig. 13.6). **Tubercles** are pimplelike protuberances that may occur on various regions of the statoblast surface (Fig. 13.7 A, B). The **reticulum** is a netlike pattern of raised lines (Figs. 13.5, 8 E). **Pitting** is the appearance of small depressions seen in certain fredericellid piptoblasts (Fig. 13.9 A, B). Examining a floatoblast in side view (suture toward the viewer) can help determine whether or not the valves are laterally symmetrical (Fig. 13.6).

Statoblasts are never formed by freshwater ctenostome bryozoans. Instead, the dormant phase of the life cycle is represented by modified structures called **hibernaculae**. In Neotropical species, the hibernaculae have limited taxonomic value.

MATERIAL PREPARATION AND PRESERVATION

Standard procedure is to fix and preserve bryozoans in either 5% formalin or 70% ethyl alcohol. For molecular genetics, 100% ethyl alcohol is the preferred preservative. For histological work, the best fixative is buffered glutaraldehyde, with the buffer concentration at 0.01M, not 0.1M as for many other invertebrates.

Many investigators like to preserve specimens with the lophophores extended; and for this to happen, the material must be narcotized. Menthol (crystals or molded wafers) is very effective and convenient, although it is not recommended for fine histological work. Also, menthol has a low melting point and tends to melt under hot field conditions (Wood, 2005a). The solid menthol is floated on the water surface in a small, closed container with the specimen. Exposure time varies with temperature and species but seldom exceeds 1 hour. The other alternative is to administer a chloral hydrate solution, drop by drop, and mix it over several hours. This is much less damaging to sensitive tissues than menthol, but chloral hydrate is difficult to obtain in countries where it is listed as a controlled substance.

For close examination of statoblasts, the valves must often be separated and cleaned in a strong solution of potassium (or sodium) hydroxide. If you are in a hurry,

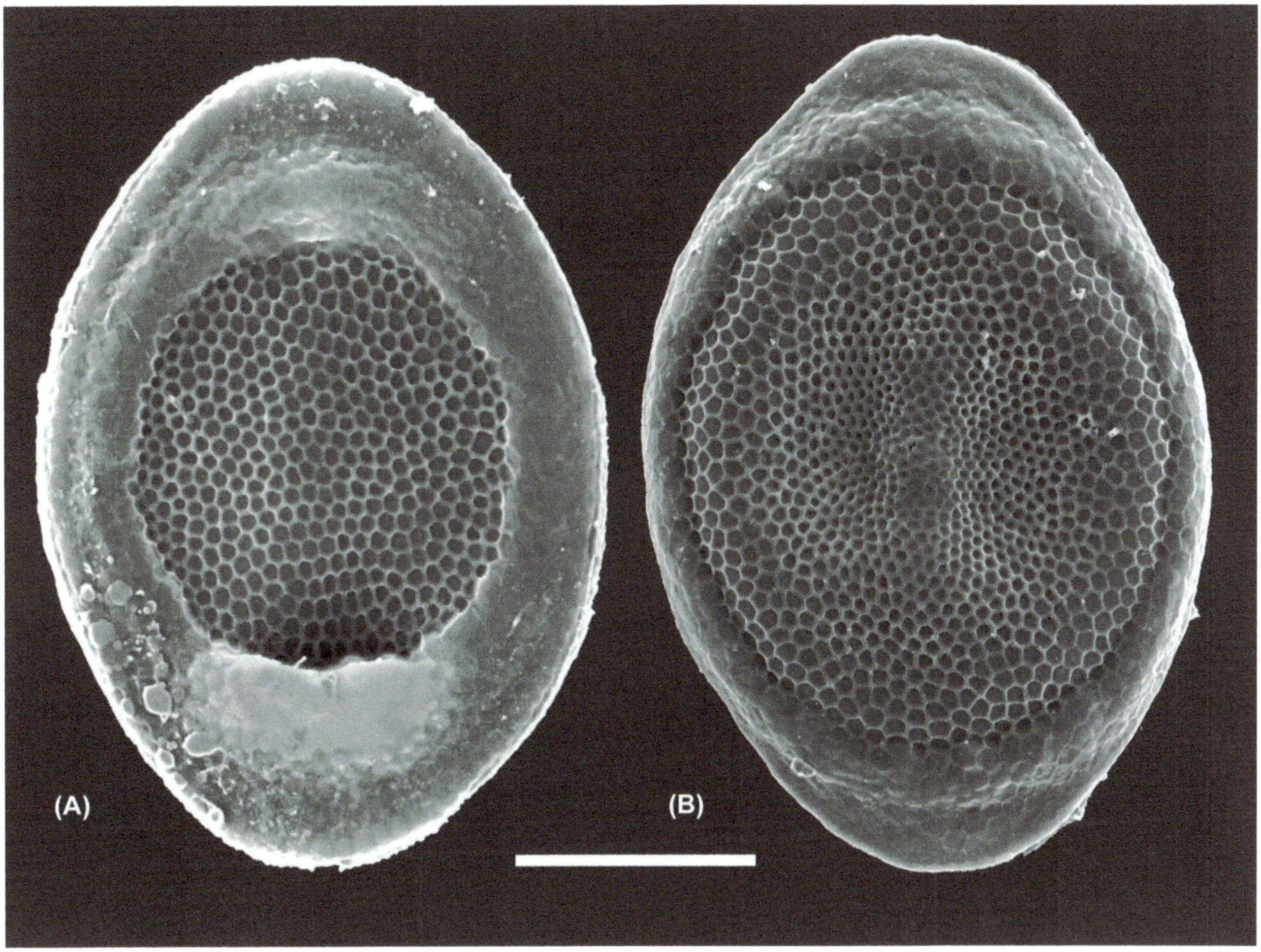

FIGURE 13.5 Floatoblast valves of *Plumatella vaihiriae*: (A) dorsal valve; and (B) ventral valve. Scale bar = 100 μm.

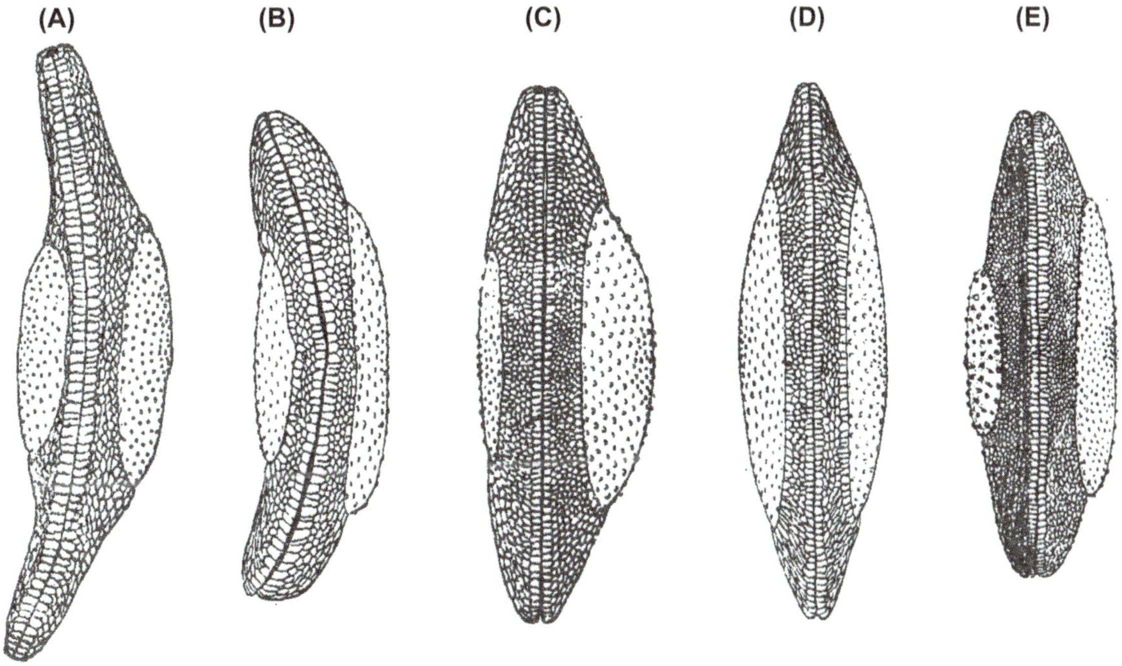

FIGURE 13.6 Floatoblast lateral view showing various suture shapes and positions (the dorsal valve faces left in all images): (A) *Stolella agilis tica* (curved suture); (B) *Plumatella carvalhoi* (curved); (C) *S. agilis* (asymmetrical); (D) *S. evelinae* (symmetrical); and (E) *S. agilis iheringi* (symmetrical). *(A) Modified from Du Bois-Reymond Marcus (1953), (B–E) Modified from Marcus (1942).*

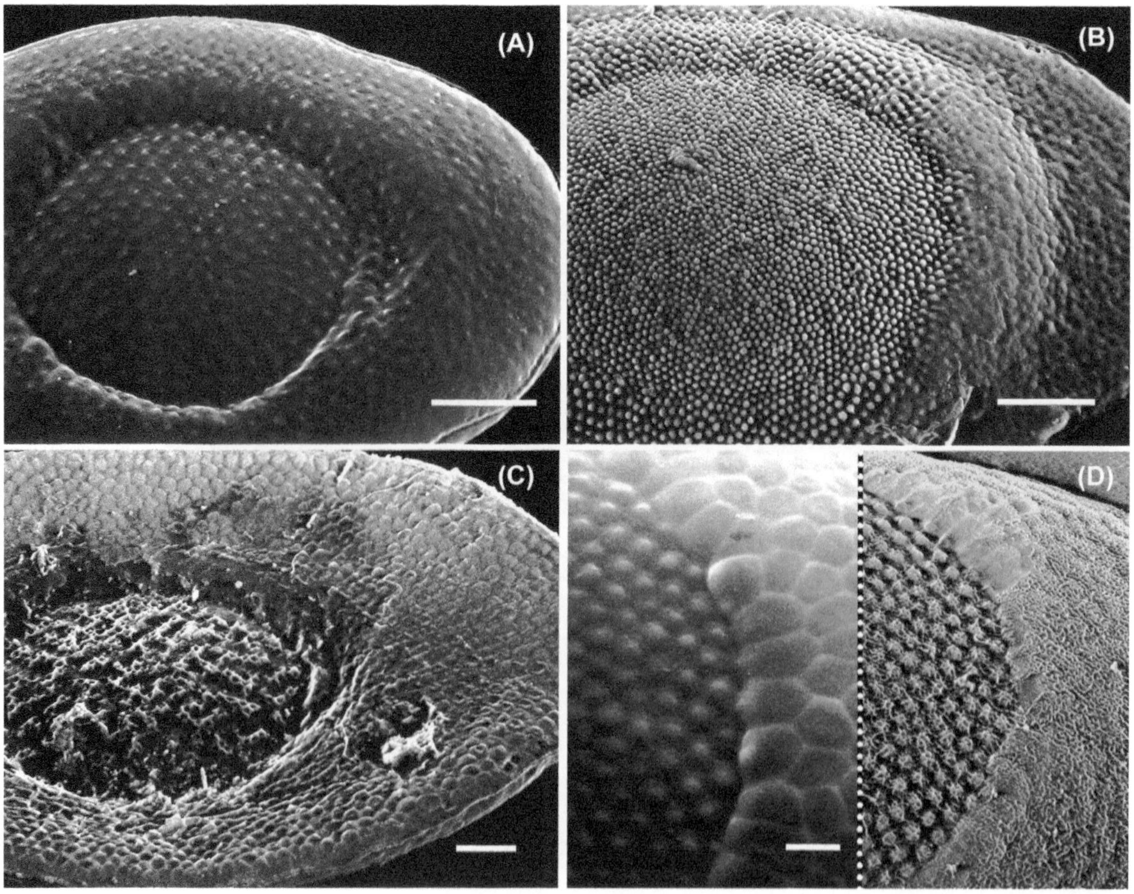

FIGURE 13.7 Floatoblast surface features revealed by scanning electron microscopy: (A) *Plumatella fungosa* showing tubercles on annulus as well as on fenestra, scale bar = 50 μm; (B) *P. patagonica* showing fenestra tubercles spilling over onto the annulus, scale bar = 25 μm; (C) *P. siolii* showing irregular craters and ridges on the dorsal fenestra, scale bar = 50 μm; and (D) two images comparing typical features of *P. emarginata* on the left and unusual wrinkled features of *P. mukaii* on the right, scale bar for both images = 10 μm.

heat the statoblasts in a few drops of the solution in a spoon over a flame for no more than a minute. Then transfer the statoblasts to fresh water and tease apart the valves. The capsule valves may separate from the periblast or not, depending on the species.

When examining statoblast valves with light microscopy, keep in mind that you can detect structures only on the fenestra, not on the annulus. Even then, you may not detect faint reticulation. Tubercles and reticulation on the fenestra can be easily confused because under the light microscope they appear very similar. However, tubercles often act as tiny lenses; and as you focus up and down, you can reduce them to little points of light. With most reticulation you can focus up and down and the pattern will just go in and out of focus without forming points of light. Nevertheless, there are times when you really cannot tell the difference between tubercles and reticulation without scanning electron microscopy.

It is a common mistake to confuse the statoblast capsule with the fenestra. This can lead to errors in species identification. In fresh material, the amber-colored capsule is generally hidden by gas-filled cells of the annulus. But when the gas is lost in preserved specimens, the capsule is fully visible and the outlines of the fenestra are less distinct (Fig. 13.2 C).

A key feature in some species is the lateral symmetry of the statoblast. To achieve a good lateral view under the microscope it is helpful to add a very small amount of cotton wool to the water and allow the cotton fibers to hold the statoblast in any desired position. It usually takes a little practice to get this right.

Statoblasts can be prepared for scanning electron microscopy by simply drying and sputtering on aluminum pins. Normally the ventral valve will collapse, which does not affect species identification. However, if you are finicky about such things you can always use critical point drying or HMDS (1,1,1,3,3,3,-hexamethyldisilazane) to retain the original shape.

Identifying bryozoan species often requires taking measurements of the statoblast using an optimal micrometer, and this brings up a sensitive subject. The literature is filled with incorrect measurements because statoblast valves being measured had been either dried or flattened under a coverslip when the data were taken. Simply separating the valves also can alter their shape, usually by broadening one or both valves. Even under normal

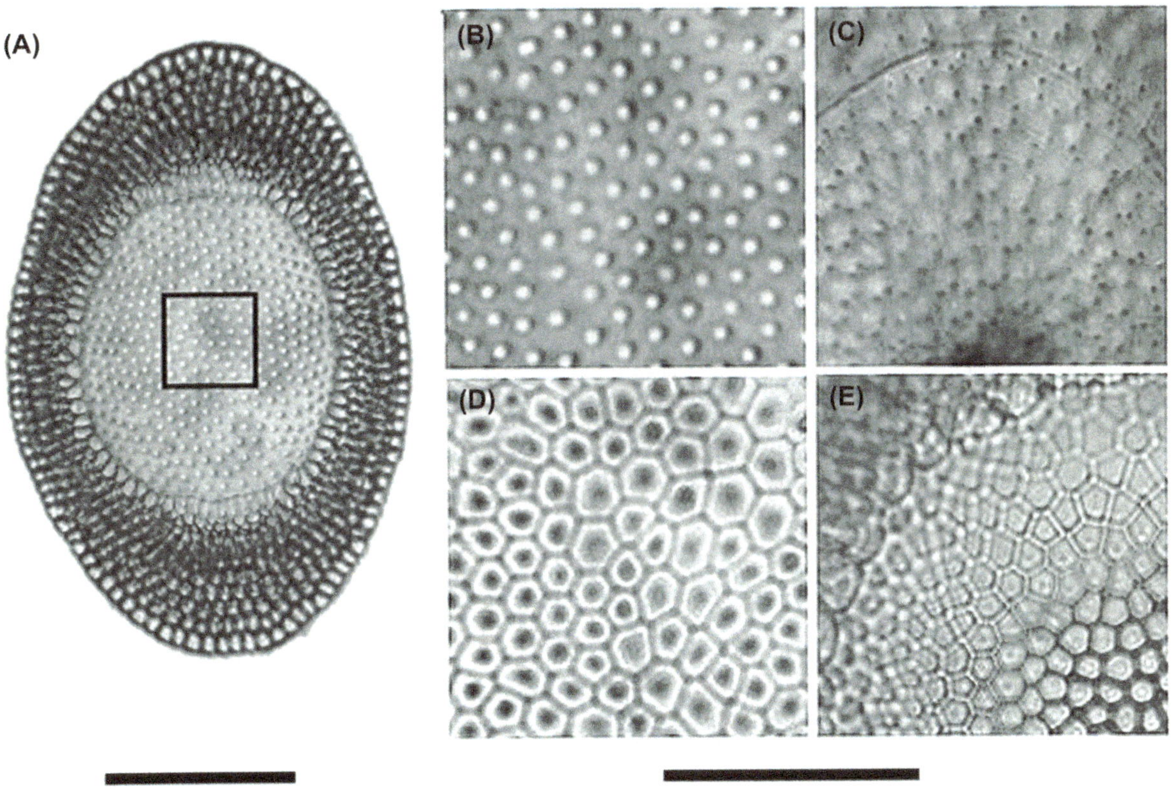

FIGURE 13.8 Features of the floatoblast dorsal fenestra visible with light microscopy: (A) floatoblast of *Plumatella bushnelli* showing general area of interest, scale bar = 100 μm; (B) *P. bushnelli* showing small, well-spaced tubercles; (C) *P. javanica* showing crowded tubercles and porelike holes created where adjacent tubercles meet; (D) *P. repens* showing tubercles surrounded by reticulated lines; and (E) *P. vaihiriae* showing reticulation without tubercles, scale bar for (B–E) = 10 μm.

conditions statoblast size can vary considerably within a species and often within a single colony. There are several things to be learned from this:

- whenever possible take measurements from fresh material in water;
- measure numerous statoblasts and run the statistics;
- never rely on dimensions from separated valves or from statoblasts prepared for SEM;
- approach published statoblast dimensions with some caution.

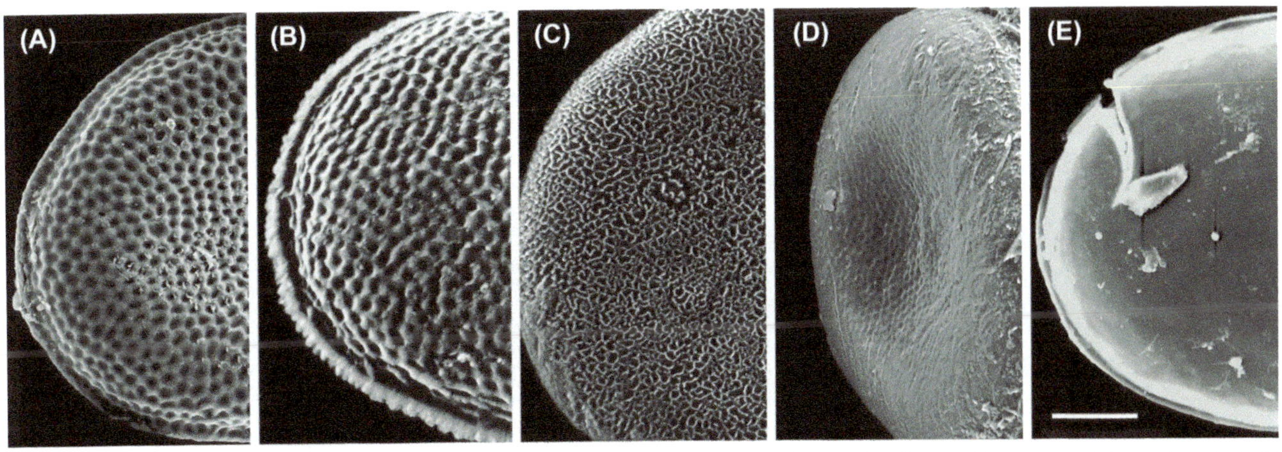

FIGURE 13.9 Four fredericellid species: (A) *Fredericella indica*; (B) *F. crenulata*; (C) *F. browni*; (D) *F. tenax* (shown at an oblique orientation); (E) *F. sultana*. Scale bar for all images = 50 μm. *Modified in part from Hirose & Mawatari (2011).*

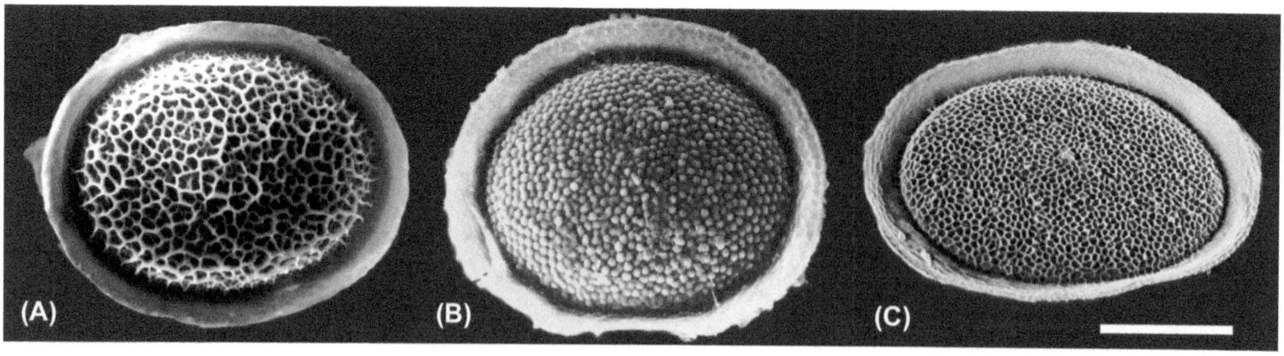

FIGURE 13.10 Sessoblast surface features: (A) irregular reticulation of *Plumatella reticulata*; (B) tubercles common in many species, in this instance *P. repens*; and (C) regular reticulation of *P. vaihiriae*, which sometimes thickens to form pores. Scale bar = 100 μm.

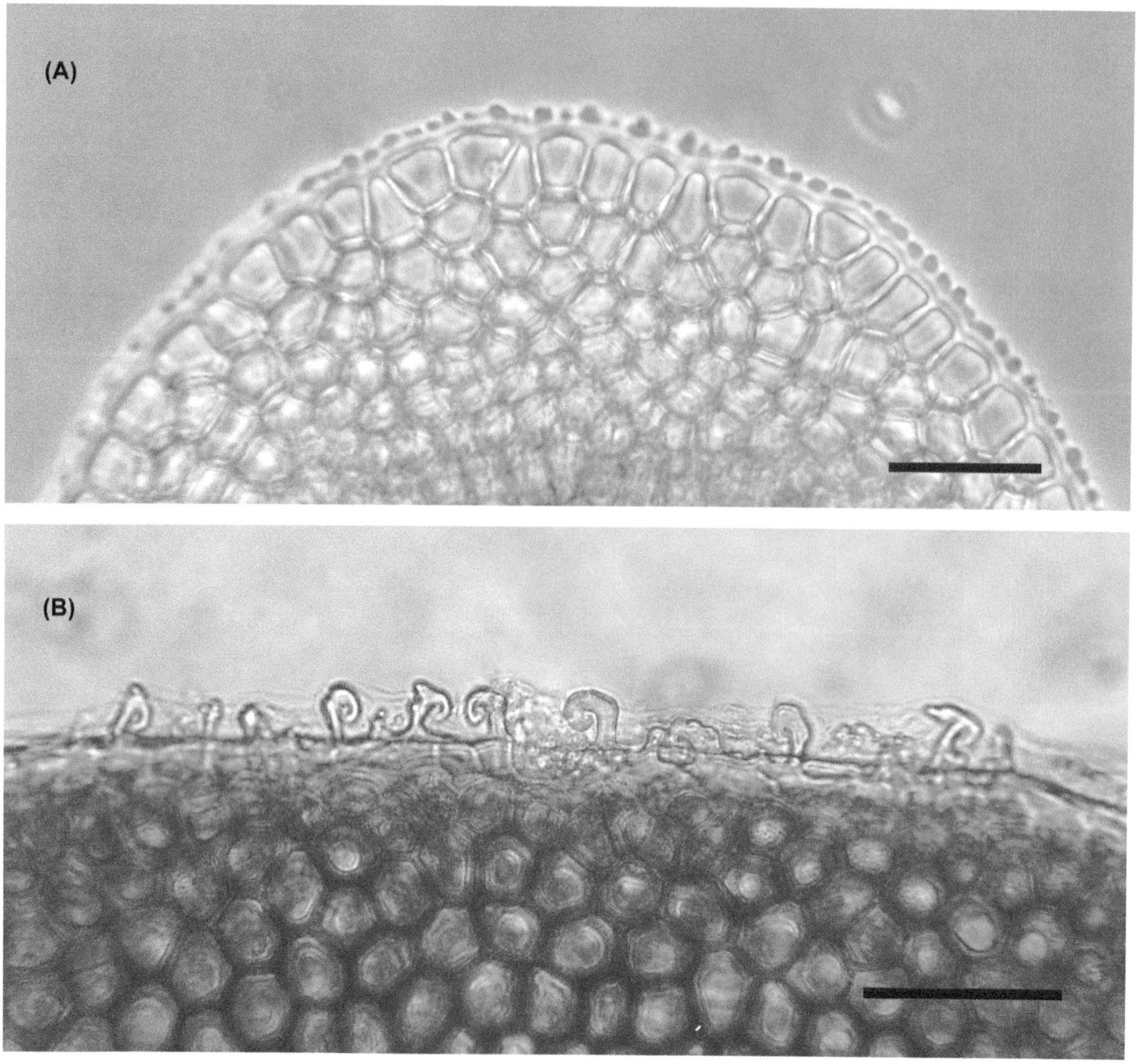

FIGURE 13.11 Marginal statoblast features: (A) small protuberances from the suture in *Plumatella jariensis*, scale bar = 25 μm; and (B) hooks from the periblast of *Asajirella gelatinosa*, scale bar = 50 μm.

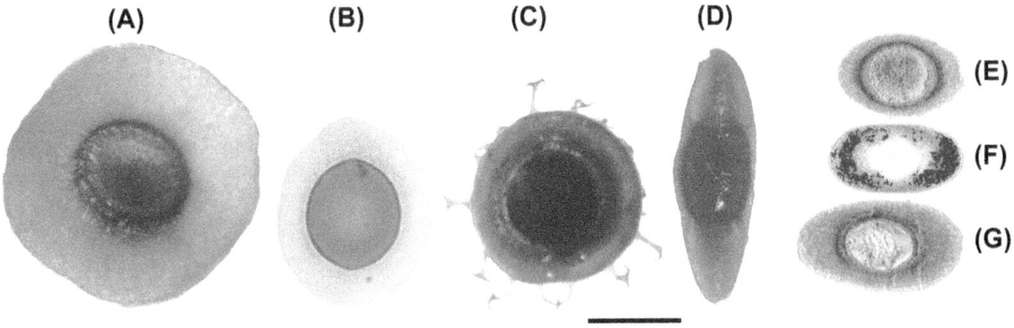

FIGURE 13.12 Large floatoblasts: (A), *Asajirella gelatinosa*; (B) *Lophopus jheringi* (immature); (C) *Pectinatella magnifica*; (D) *Tapajosella elongata*; (E) *Stolella agilis tica*; (F) *Plumatella patagonica*; and (G) *P. siolii*. Scale bar for all images = 500 μm.

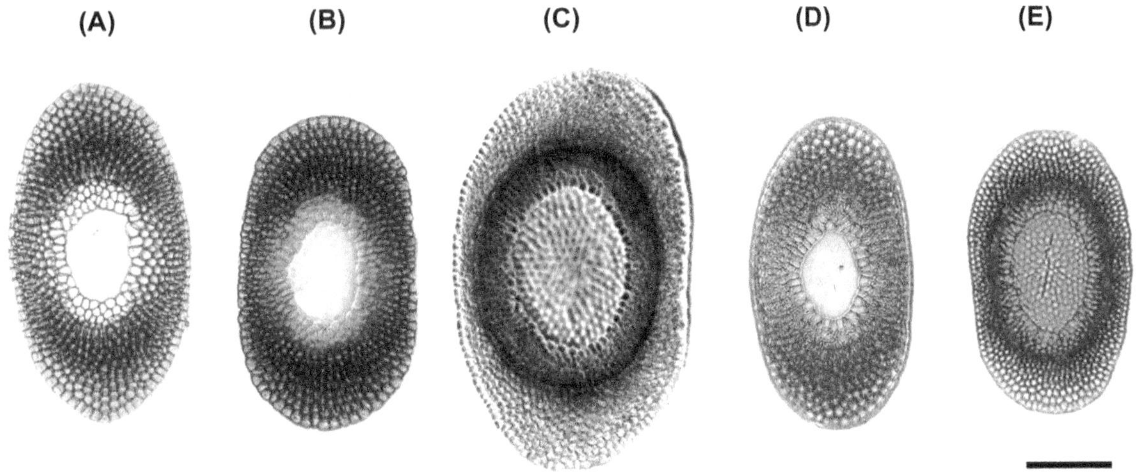

FIGURE 13.13 Floatoblasts with small dorsal fenestrae: (A) *Plumatella emarginata*; (B) *P. reticulata*; (C) *Stolella agilis ihrengi*; (D) *P. mukaii*; and (E) *P. osburni*. Scale bar for all images = 100 μm.

KEYS TO ECTOPROCTA (BRYOZOA)

Ectoprocta: Classes

1 Extended lophophore U-shaped in outline (Fig. 13.2 A) with the exception of Fredericellidae (Fig. 13.2 B); lophophore with more than 20 tentacles; orifice is circular when lophophore is withdrawn; internal septa infrequent; statoblasts usually present .... **Phylactolaemata [p 495]**

1′ Extended lophophore always circular in outline with fewer than 20 tentacles; orifice appears quadrangular when lophophore is withdrawn; statoblasts never present; individual zooids separated by internal septa .... Gymnolaemata, one order **Ctenostomata [p 500]**

Ectoprocta: Phylactolaemata: Families

1 Floatoblast length <3x width 2

1′ Floatoblast dimensions ~ 1 mm long and 0.3 mm wide (Fig. 13.12 D) .... Tapajosellidae, one species: *Tapajosella elongata* Wood & Okamura, 2017 [Amazon Basin]

2(1) Colony composed of a linear, branching series of zooids (Fig. 13.1 B); branches sometimes adhering to each other; tentacles fewer than 65; statoblasts <800 μm in any dimension 3

2′ Colony globular and either lobed or entire in outline Fig. 13.1 A); tentacles more than 65; statoblasts at least 800 μm in longest dimension 4

3(2) Lophophore circular in outline, tentacles fewer than 25 (Fig. 13.2 B); statoblasts (piptoblasts) reniform and retained within the colony, sometimes cemented to the substrate but never released freely into the water, never buoyant .... Fredericellidae, one genus: ***Fredericella* [p. 498]**

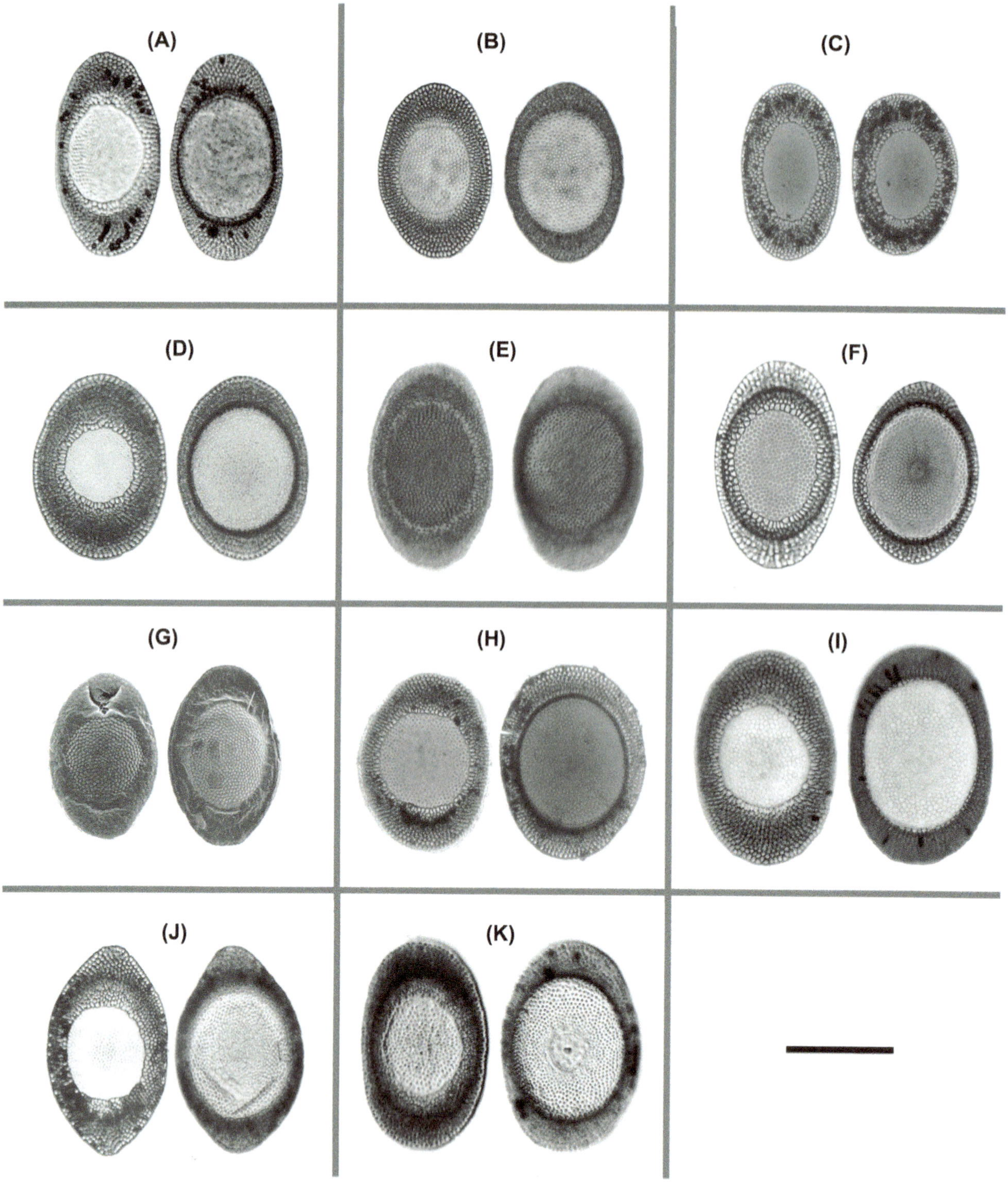

FIGURE 13.14 Miscellaneous floatoblasts (dorsal valves always on the left): (A) *P. annulata* of Rogick (1942); (B) *Plumatella bushnelli;* (C) *P. casmiana*; (D) *P. fungosa*; (E) *P. jariensis*; (F) *P. javanica*; (G) *P. pirassununga*; (H) *P. repens*; (I) *P. rugosa*; (J) *P. vaihiriae*; and (K) *Stolella agilis.* Scale bar for all images = 200 μm.

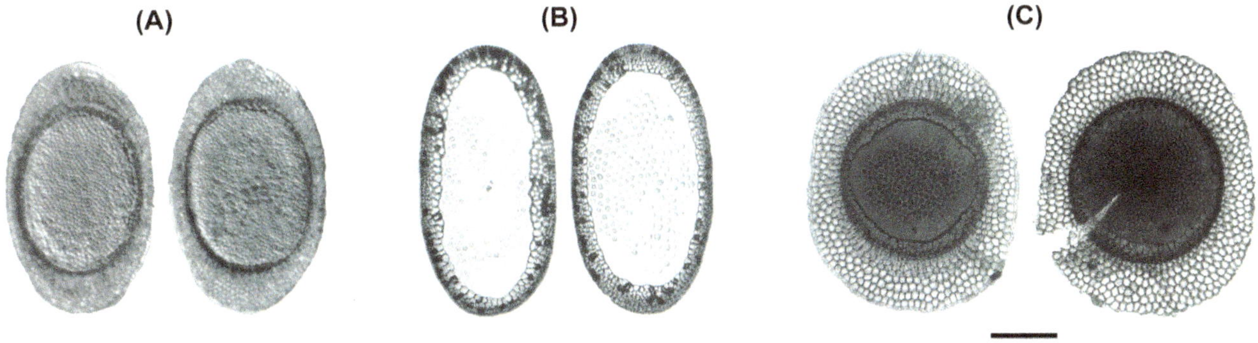

FIGURE 13.15 Floatoblasts with similarly sized fenestrae on dorsal and ventral valves (dorsal valve always on the left): (A) *Stolella evelinae*; (B) *Plumatella casmiana* leptoblast; and (C) *P. marcusi*. Scale bar for all images = 100 μm.

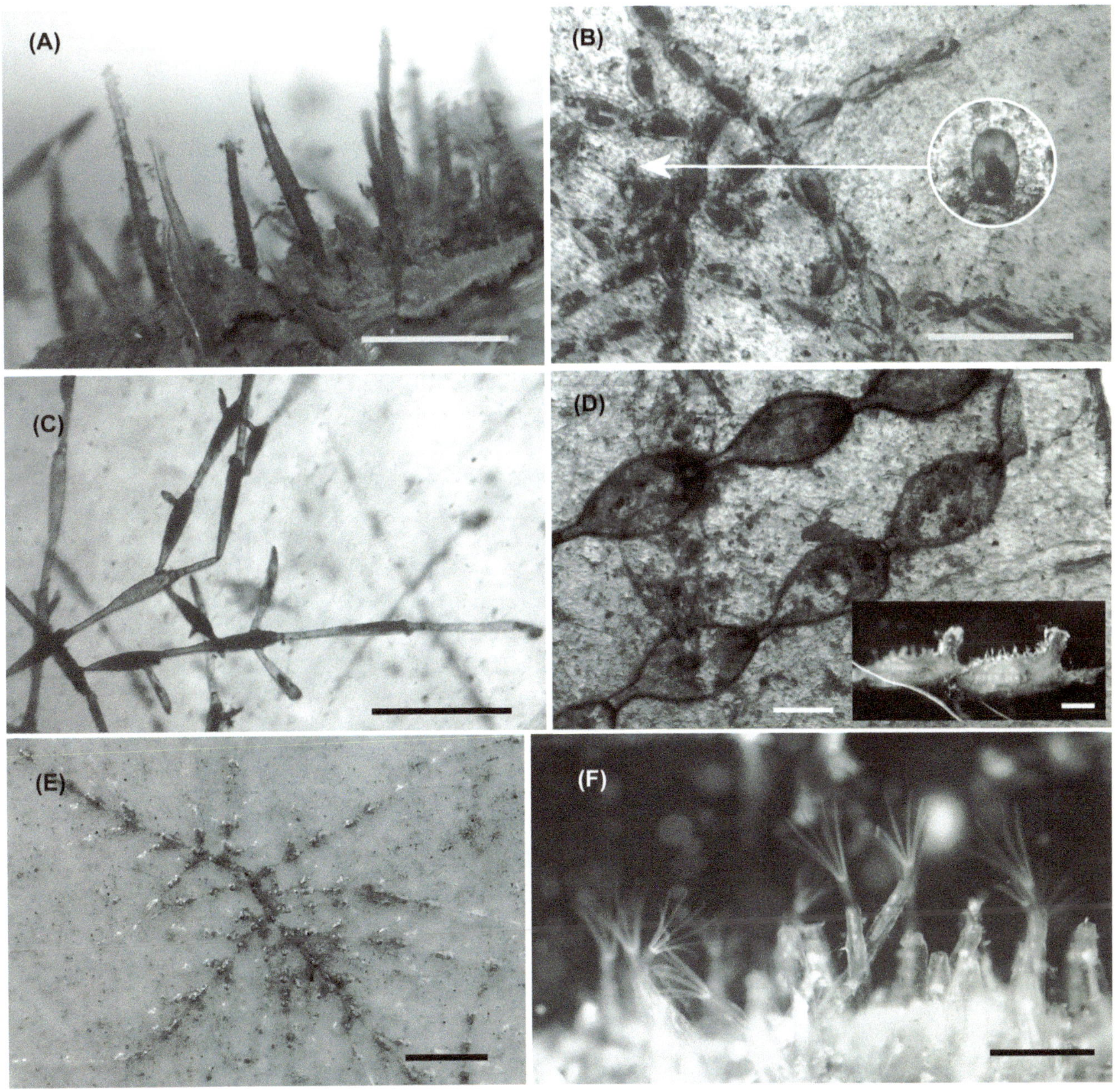

FIGURE 13.16 Ctenostome bryozoan colonies: (A) *Pottsiella erecta*, scale bar = 1 mm; (B) *Timwoodiellina natans*, scale bar = 0.5 mm, inset showing a developing nautizooid bud; (C) *Paludicella articulata*, scale bar = 0.5 mm; (D) *Hislopia corderoi*, scale bar = 0.25 mm, inset showing two zooids in lateral view, scale bar = 0.2 mm; (E) *Victorella pavida* prostrate colony, scale bar = 5 mm; and (F) *V. pavida* erect zooids, each with only 8 tentacles, scale bar = 1 mm.

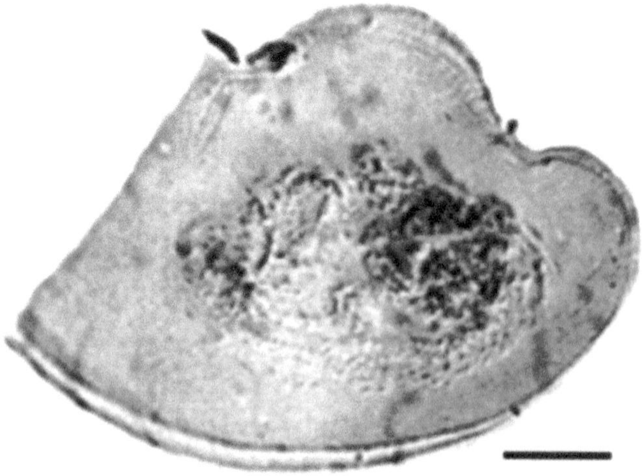

FIGURE 13.17 Cyphonautes larva of *Hislopia corderoi* from the Amazon Basin, scale bar = 50 μm. *Photo by Gerd-Oltmann Brandorff.*

3′ Lophophore U-shaped in outline (Fig. 13.2 A), tentacles more than 22; statoblasts including a freely released, usually buoyant form and a sessile form (sessoblast) cemented to the substratum .... Plumatellidae, two genera: ***Stolella*** and ***Plumatella*** **[p. 498]**

4(2) Colony gelatinous and slimy, ranging from a flat sheet to large, globular mass; mouth region with red pigmentation; a prominent pair of white spots at end of each arm of lophophore; statoblasts round with hooked spines radiating from outer margin of annulus (Fig. 13.12 C) .... Pectinatellidae, one species: *Pectinatella magnifica* (Leidy, 1851) [Guatemala (known only from a single statoblast). Nearctic, invasive in Palearctic and Oriental]

4′ Colony, zooids, and statoblasts not as above; mouth region may be without red pigmentation; lophophore lacking pair of white spots; floatoblasts buoyant only after being dried .... **Lophopodidae [p. 500]**

Ectoprocta: Phylactolaemata: Fredericellidae: *Fredericella*: Species

1 Piptoblast appears dull when dry and viewed with reflected light 2

1′ Piptoblast appears shiny when dry and viewed with reflected light 4

2(1) Piptoblast elongate, length:width ratio at least 1.5 3

2′ Piptoblast broad, nearly circular; valves smooth but covered by a minutely wrinkled membrane soluble in warm alkali solution (Fig. 13.9 C) .... *Fredericella browni* (Rogick & Brown, 1945) [Mexico. Panama. Afrotropical and Nearctic]

3(2) Piptoblast margins minutely toothed (Fig. 13.9 B) .... *Fredericella crenulata* Marcus, 1953 [Brazil: São Paulo]

3′ Piptoblast margins smooth, not minutely toothed (Fig. 13.9 A) .... *Fredericella indica* Annandale, 1909 [Worldwide but not verified from South America]

4(1) Piptoblast broad, nearly circular, frontal valve faintly pitted when viewed with transmitted light (Fig. 13.9 D) (Note: To see this faint texture feature it is best to separate and clean the valves in hot KOH or NaOH, then place them in clean fresh water and examine the frontal valve with substage illumination); statoblast held firmly to the substratum .... *Fredericella tenax* Wood & Okamura, 2017 [Brazil: Amazon Basin]

4′ Piptoblast elongate, length:width at least 1.5, surface completely smooth (Fig. 13.9 E) .... *Fredericella sultana* (Blumenbach, 1779) [Mexico. Brazil. Chile. Paraguay. Argentina. Australian, Oriental and Palearctic]

Ectoprocta: Phylactolaemata: Plumatellidae: *Plumatella* and *Stolella*: Species

1 Floatoblast length >500 μm 2

1′ Floatoblast length <500 μm 3

2(1) Floatoblast fenestrae bearing small tubercles which spill over onto the annulus (Figs. 13.7 B, 12 F) .... *Plumatella patagonica* Wiebach, 1974 [Chile: near Concepción; Patagonia]

2′ Floatoblast fenestrae bearing a rough and irregular pattern (Figs. 13.7 C, 12 G) .... *Plumatella siolii* Wiebach, 1970 [Amazon Basin]

3(1) Dorsal fenestra width <0.5 floatoblast width (Fig. 13.13) 4

3′ Dorsal fenestra width >0.5 floatoblast width (Fig. 13.14) 8

4(3) Floatoblast laterally asymmetrical: ventral valve convex, dorsal valve flat or nearly so (Fig. 13.6 A–C) 5

4′ Floatoblast laterally symmetrical: dorsal and ventral valves about equally convex (Fig. 13.6 D, E) 6

5(4) Floatoblast with finely wrinkled ornamentation over entire surface (Figs. 13.7 D-right, 13 D). (Note: The wrinkled surface feature is best seen with scanning electron microscopy. However, it can sometimes be detected with a compound microscope using reflected, not transmitted, light) .... *Plumatella mukaii* Wood, 2001
[Chile. Nearctic and Palearctic]

5′ Floatoblast without wrinkled ornamentation (Figs. 13.7 D-left, 13 A) .... *Plumatella emarginata* Allman, 1844
[Brazil. Argentina. Widely distributed]

6(4) Floatoblast roughly rectangular, the long sides almost parallel (Fig. 13.13 B, C); colony wall tan to darkly pigmented 7

6′ Floatoblast elliptical (Figs. 13.6 E, 13 C); colony wall colorless, and transparent .... *Stolella agilis ihiringi* Marcus, 1942
[Brazil: Sao Paulo]

7(6) Sessoblasts with reticulated pattern on frontal valve, the reticulated lines appearing dark with reflected light but pale with scanning electron microscopy (Figs. 13.10 A, 13 B) .... *Plumatella reticulata* Wood, 1988
[Panama (Wood, "Pers. Observation"). Nearctic and Palearctic]

7′ Sessoblasts with tubercles on frontal valve as well as on the floatoblast fenestra (Fig. 13.13 E) ....
.... *Plumatella osburni* Rogick & Brown, 1942
[Panama. Amazon Basin]

8(3) Floatoblast dorsal and ventral fenestrae similar in size and shape (Fig. 13.15) 9

8′ Floatoblast dorsal fenestra distinctly smaller than ventral fenestra (Fig. 13.14) 11

9(8) Floatoblast includes an inner capsule, which makes the fenestra appear brown 10

9′ Floatoblast capsule absent; valves thin, fragile, transparent. (Note: This is a unique floatoblast known as a "leptoblast" which germinates immediately after release from the colony) (Fig. 13.15 B) .... *Plumatella casmiana* Oka, 1907
[Worldwide]

10(9) Floatoblast circular or nearly so, length <1.1x width (Fig. 13.15 C), colony unknown .... *Plumatella marcusi* Wiebach, 1970
[Amazon Basin]

10′ Floatoblast oblong, length >1.1x width (Figs. 13.6 D, 15 A); body wall thin, colorless, transparent ....
.... *Stolella evelinae* Marcus, 1941
[Brazil: São Paulo State, Pirassununga]

11(8) Floatoblast suture is distinctly curved in lateral view (Fig. 13.6 A, B) 12

11′ Floatoblast suture is straight in lateral view (Fig. 13.6 C–E) 13

12(11) Floatoblast length <400 μm (Fig. 13.6 B) .... *Plumatella carvalhoi* Marcus, 1942
[Brazil: São Paulo (only from an aquarium)]

12′ Floatoblast length >400 μm (Figs. 13.6 A, 12 E) .... *Stolella agilis tica* Marcus, 1953
[Vicinity of Titicaca Lake]

13(11) Floatoblast dorsal fenestra length is <1.5x fenestra width 14

13′ Floatoblast dorsal fenestra length is >1.5x fenestra width (Fig. 13.14 C); body wall seldom transparent, ectocyst with a carina, tentacles <30 .... *Plumatella casmiana* Oka, 1907
[Worldwide]

14(13) Floatoblast dorsal fenestra bearing a reticulated pattern without tubercles 15

14′ Floatoblast dorsal fenestra bearing tubercles 16

15(14) Floatoblast dorsal annulus bulging around inner capsule to form a distinct shoulder; entire floatoblast surface covered with raised, reticulate pattern, most easily seen in the fenestra area (Figs. 13.5, 8 E, 14 J). (Note: Several species have a "capsule bulge" when the floatoblast is dry but only *Plumatella vaihiriae* has such a bulge when wet; the feature is best seen on the floatoblast dorsal side. Sessoblast frontal valve densely reticulated or pitted and without tubercles, Fig. 13.10 C, 14J) .... *Plumatella vaihiriae* Hastings, 1929
[Argentina. Nearctic, Oriental, and Palearctic]

15′ Floatoblast dorsal annulus not as above, when viewed with SEM floatoblast annulus cells are concave, giving the annulus a roughened appearance; sessoblast frontal valve tuberculated; floatoblast is shown in Fig. 13.14 I; colony wall often reddish in color ....
.... *Plumatella rugosa* Wood et al., 1998

[Panama. Afrotropical, Australia, and Nearctic]

16(14) Floatoblast dorsal valve and fenestra dimensions not corresponding to Fig. 13.14 D; floatoblast annulus lacking tubercles as seen with SEM 17

16′ Floatoblast dorsal valve and fenestra dimensions corresponding to Fig. 13.14 D; SEM shows tubercles on the floatoblast annulus (Fig. 13.7 A); colonies may form dense cushions of crowded, contiguous zooids (Fig. 13.14 D) .... *Plumatella fungosa* (Pallas, 1768)
[Chile (Orellana Liebbe, personal observation). Nearctic, Palearctic, and Australia]

17(16) Colony wall dark, well sclerotized, and stiff 18

17′ Colony wall soft, transparent to lightly encrusted, but not heavily sclerotized 19

18(17) Floatoblast bearing minute projections from the suture (Figs. 13.11 A, 14 E) .... *Plumatella jariensis* Wood & Okamura, 2017
[Amazon Basin]

18′ Floatoblast suture without minute projections; dorsal tubercles, if present, only weakly expressed (Fig. 13.14 A) .... *Plumatella annulata* Hozawa & Toriumi, 1940 *sensu* Rogick & Brown 1942
[Guatemala]

19(17) Colonies entirely adherent to the substratum, statoblasts <400 μm long 20

19′ Portions of the colony growing erect and free from the substratum, free statoblasts about 400 μm long (Figs. 13.6 C, 14 K) .... *Stolella agilis* Marcus, 1942
[Brazil]

20(19) Floatoblast dorsal fenestra with tubercles only and no reticulation 21

20′ Floatoblast dorsal fenestra with light reticulation and with a single tubercle in each cell (Figs. 13.8 D, 14 H); sessoblast with crowded tubercles (Fig. 13.10 B) .... *Plumatella repens* (Linnaeus, 1758)
[Neotropics (?). Nearctic, Palearctic, Oriental, and Australian]

21(20) Floatoblast tubercles well-spaced on dorsal fenestra 22

21′ Floatoblast tubercles low and densely crowded on dorsal and ventral fenestra, careful focusing will reveal small porelike structures between tubercles (Figs. 13.8 C, 14 F) .... *Plumatella javanica* Kraepelin, 1906
[Amazon Basin]

22(21) Fenestra tubercles appearing as sharply defined beads (Figs. 13.8 A, B, 14 B); scanning electron microscopy shows the annulus covered with rashlike nodules .... *Plumatella bushnelli* Wood, 2001
[Mexico: South (Wood, "Pers. Observation"). Afrotropical, Nearctic, Oceana, Palaearctic

22′ Fenestra tubercles less sharply defined (Fig. 13.14 G); scanning electron microscopy shows no rashlike nodules on the annulus .... *Plumatella pirassununga* Wood & Okamura, 2017
[Brazil: São Paulo, Pirassununga]

Ectoprocta: Phylactolaemata: Lophopodidae: Genera

1 Statoblast margins with minute hooks (Figs. 13.11 B, 12 A) .... *Asajirella*, one species: *Asajirella gelatinosa* Oka, 1891
[Panama. Oriental]

1′ Statoblast margins smooth (Fig. 13.12 B) .... *Lophopus*, one species: *Lophopus jheringi* Meissner, 1893
[Brazil: Rio Grande do Sul]

Ectoprocta: Gymnolaemata: Ctenostomata: Families

1 Tentacles numbering more than eight 2

1′ Tentacles numbering exactly eight (Fig. 13.16 E, F); young colonies straight lines of serial zooids grow in a pinnate pattern with pairs of lateral branches extending in opposite directions, all with uniform diameter (Fig. 13.16 E); under more crowded conditions zooids become erect but retain uniform diameter; entire colony fully attached to the substratum except for occasional long, narrow, adventitious branches extending from the midpoint of erect zooids (Fig. 13.16 F); hibernaculae are irregularly shaped, slightly convex bodies bearing short, bifurcating processes around the periphery, formed as lateral buds; brackish and freshwater .... Victorellidae, one species: *Victorella pavida* Saville Kent, 1870
[Brazil: São Paulo. Uruguay: Montevideo. Australian, Nearctic, and Palaearctic]

2(1) Colony adhering tightly to the substratum except in cases of extreme overcrowding 3

2′ Colony branches often growing free of the substratum (Fig. 13.16 C); large portions of a vigorous colony often free of the substratum; tentacles numbering 14–23 (Fig. 13.16 C); hibernaculae formed as terminal elements of a branch or else internally .... Paludicellidae, one species: *Paludicella articulata* (Ehrenberg, 1831) [Paraguay. Argentina. Nearctic Oriental, and Palearctic]

3(2) Zoids more or less flattened, except for the small portion through which the lophophore is extended 4

3′ Each zooid typically includes an erect element up to 2.5 mm long tapering towards the base and the tip (Fig. 13.16 A), also a slender, stolonlike part connected to another zooid; however, the erect portion may be significantly reduced in small colonies; tentacles numbering 19–20; hibernaculae occur as thick-walled dilated portions of the narrow stolons........ Pottsiellidae, one species: *Pottsiella erecta* (Potts, 1884) [Central America. Argentina. Nearctic]

4(3) Certain zooids bearing bulbous dorsolateral outgrowths from the lateral margin, which elongate to become fully functional zooids, eventually detaching from the colony (Fig. 13.16 B); tentacles numbering 12–14; hibernaculae formed as lateral or distal buds, each bearing a single pair of short projections on one side .... Natanidae, one species: *Timwoodiellina natans* (Wood, 2006) [Brazil: São Paulo. Oriental]

4′ Zooids without bulbous outgrowths (Fig. 13.16 D); hibernaculae formed directly from old zooids; fertilized eggs released into the water developing into planktivorous cyphonautes larvae (Fig. 13.17); tentacles numbering 13,14–16; older zooids with up to 26 spines on the frontal surface and around the orifice; cyphonautes larvae with an indentation in the margin of one side (Fig. 13.17) .... Hislopidae, one species: *Hislopia corderoi* (Mané-Garzón, 1959) [Brazil. Argentina. Uruguay]

REFERENCES

Du Bois-Reymond Marcus. 1953. Bryozoa from Lake Titicaca. Boletim da Faculdade de Filosofia, Ciencias e Letras, Universidade de São Paulo, Zoologia 18: 149–163.

Hirose, M. & S.F, Mawatari, 2011. Freshwater Bryozoa of Lake Biwa, Japan. Species Diversity 16: 1–37.

Marcus, E. 1942. Sôbre Bryozoa do Brasil 2. Boletim da Faculdade de Filosofia, Ciencias e Letras, Universidade de São Paulo, Zoologia. 6: 57–96.

Massard, J.A. & G. Geimer. 2008. Global diversity of bryozoans (Bryozoa or Ectoprocta) in freshwater: an update. Bulletin de la Société des Naturalistes luxembourgeois 109: 139–148.

Rogick, M. & C. Brown. 1942. Studies on fresh–water Bryozoa XII. A collection from various sources. Annals of the New York Academy of Sciences 43: 123–144.

Wood, T. 2005a. Study methods for freshwater bryozoans. Denisia 16: 103–110.

Wood, T. 2005b. The pipeline menace of freshwater bryozoans. Denisia 16: 203–208.

Chapter 14

Phylum Entoprocta

Timothy S. Wood
Department of Biological Sciences, Wright State University, Dayton, OH, United States

PHYLUM ENTOPROCTA

Chapter Outline

INTRODUCTION

Entoprocta is a small phylum of about 150 described species (Nielsen, 2012) known almost entirely from marine habitats, with several species occurring in estuarine areas. Only two species have been described from fresh water: *Loxosomatoides sirindhornae* Wood, 2005 in Thailand and *Urnatella gracilis* Leidy, 1847, first reported from southeastern United States and now distributed worldwide (Oda, 1982). At least one additional freshwater species is known from Thailand but not yet described.

In general, entoprocts are small, sessile, filter feeding animals. The freshwater species grow as colonies of interconnected zooids. They are most commonly found in flowing or lightly turbulent water. A dense population of *Urnatella* can cover all available hard substrates in a felty layer (Emschermann, 1987).

For many years entoprocts were included in the phylum Bryozoa along with a larger group called Ectoprocta (see Chapter 13). Similarities between the groups include sessile colonies (although many marine species are solitary), a U-shaped gut, and the use of ciliated tentacles to gather food. Potentially more significant are the apparent similarities in larval development (see Nielsen, 2012). However, despite much debate, the relationship has never been conclusive, and the weight of new evidence now places the entoprocts in a wholly independent phylum (Waeschenbach et al., 2012).

The discussion below focuses on *Urnatella gracilis*, the best known and most widely distributed freshwater entoproct. In the Neotropical region, the species has been reported from Argentina and Uruguay (Bonetto & Cordiviola de Yuan, 1963).

TERMINOLOGY AND MORPHOLOGY

Urnatella gracilis grows as a stalk of articulating segments capped by a bulbous head bearing ciliated tentacles. The stalk is often branched, with each segmented branch terminating in its own head, also called an atrium or calyx (Fig. 14.1). The stalk is almost entirely muscular except for a branching protonephridial system. The head includes all the organs and tissues required for the normal life functions of feeding, digestion, excretion, sexual reproduction, sensory perception, and nervous coordination. Tentacles bear several rows of cilia of varying sizes. The feeding current flows in a direction opposite from ectoproct bryozoans, drawn from near the base of the head and discharged away from the mouth. Particles trapped by cilia are thrown into a food groove along which they travel down the tentacle to the mouth opening.

Reproduction appears to be mostly asexual. Under certain optimal conditions, the branching colonies of *Urnatella* develop special buds of 1–3 zooids that detach from the parent and drift away. These free buds may also creep slowly along the substrate by alternating contractions of longitudinal and transverse muscles (Emschermann, 1987). By this means the local population can grow very rapidly.

Sexual reproduction in *Urnatella* also has been described, but it appears to be a rare occurrence (Emschermann, 1965). Both ova and sperm have been observed in the same colony. It is believed that embryos are brooded one at a time and released as free-swimming larvae.

When environmental conditions become unfavorable, the head drops off and each of the stalk segments becomes a dormant hibernaculum with powers of regeneration.

Thorp and Covich's Freshwater Invertebrates. https://doi.org/10.1016/B978-0-12-804225-0.00014-9

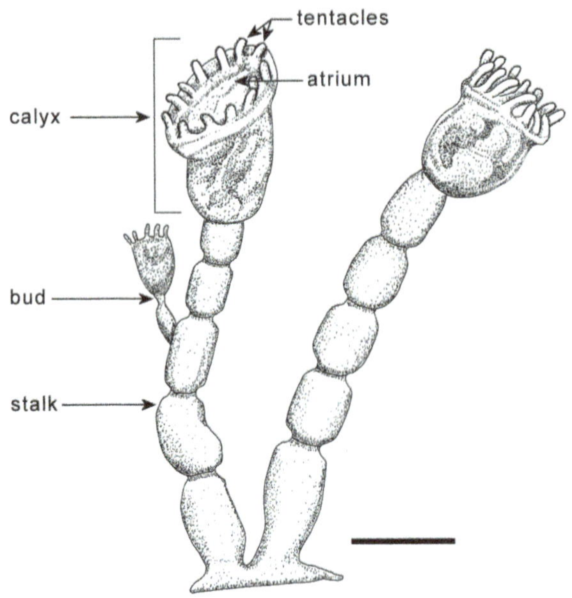

FIGURE 14.1 External morphology of an entoproct colony, *Urnatella gracilis*. Scale bar = 0.5 mm.

Overwintering occurs in this form. Resistance of the hibernaculum to desiccation and other environmental extremes has not been tested.

MATERIAL PREPARATION AND PRESERVATION

In the laboratory dormant stalks of *Urnatella* colonies can remain viable in water for at least 3 years at 0–4°C. After germination at 15–20°C, they should be kept in circulating water and fed twice weekly from a culture of *Chlamydomonas* (Emschermann, 1987).

Entoproct colonies may be fixed and preserved in either 5% formalin or 70% ethyl alcohol. Attempts to narcotize with menthol are seldom very successful. For histological work, the best fixative is 2% glutaraldehyde solution in 0.01M sodium-cacodylate buffer at a pH of 7.4 for 1 h at around 23°C (Schwaha et al., 2010). Molecular genetics requires preservation in 100% ethyl alcohol.

REFERENCES

Bonetto, A. & E. Cordiviola de Yuan. 1963. Notas sobre Briozoos (Endoprocta y Ectoprocta) del río Paraná, I. *Urnatella gracilis* Leidy y *Hislopia lacustris* Carter en el Paraná medio. Physis 24: 81–85.

Emschermann, P. 1965. Über die sexuelle Fortpflanzung und die Larve von *Urnatella gracilis* Leidy (Kamptozoa). Zeitschrift für Morphologie und Ökologie der Tiere 55: 100–114.

Emschermann, P. 1987. Creeping propagation stolons – an effective propagation system of the freshwater entoproct *Urnatella gracilis* Leidy (Barentsiidae). Archiv fur Hydrobiologie 108: 439–448.

Nielsen, C. 2012. Animal evolution: interrelationships of the living phyla. Oxford University Press, Oxford. 464 pp..

Oda, S. 1982. *Urnatella gracilis*, a freshwater kamptozoan occurring in Japan. Annotationes Zoologicae Japonenses 55: 151–166.

Schwaha, T., T.S. Wood & A. Wanninger. 2010. Trapped in freshwater: the internal anatomy of the entoproct *Loxosomatoides sirindhornae*. Frontiers in Zoology 7: 7.

Waeschenbach, A., P.D. Taylor & D.T.J. Littlewood. 2012. A molecular phylogeny of bryozoans. Molecular Phylogenetics and Evolution 62: 718–735.

Chapter 15

Phylum Tardigrada

Diane R. Nelson
Department of Biological Sciences, East Tennessee State University, Johnson City, TN, United States

Roberto Guidetti and Lorena Rebecchi
Department of Life Sciences, University of Modena and Reggio Emilia, Modena, Italy

Łukasz Kaczmarek
Department of Animal Taxonomy and Ecology, Adam Mickiewicz University in Poznań, Poznań, Poland

Sandra McInnes
British Antarctic Survey, Natural Environment Research Council, Cambridge, United Kingdom

Chapter Outline

PHYLUM TARDIGRADA

INTRODUCTION

Tardigrades are xerophilous, hygrophilous, and hydrophilous micrometazoans categorized into two main classes (Eutardigrada and Heterotardigrada) with over 1,300 species described from terrestrial, freshwater, and marine habitats. Commonly called "water bears," they must be surrounded by a film of water to be active. Cryptobiosis allows tardigrades to survive periods of desiccation in terrestrial habitats, but few freshwater and marine species are known to have this adaptation. Encystment occurs in some freshwater, moss dwelling, and soil tardigrades, possibly in response to stressful environmental conditions (e.g., oxygen depletion, pH alteration, and temperature variation). Some typically terrestrial species may inhabit freshwater environments, but most are probably "accidentals" washed into water during precipitation events (rainfall, flooding), although several are limnoterrestrial and colonize both terrestrial and freshwater habitats. The fewest number of taxa are obligate limnic species; however, the freshwater habitat has been investigated less frequently so it is difficult to estimate the total number of species. In the literature, the number of Neotropic and Antarctic species in freshwater habitats is over 50, with about 15 species found exclusively or primarily in aquatic environments, although we expect the actual number to be much higher. These species are primarily eutardigrades that belong to 2 orders and 8 families (see Degma et al., 2019, an online checklist that updates the systematic list of the genera, families, and superfamilies found in Nelson et al. (2015), as a result of additional systematic revisions). The total number of species and records reported in the Neotropic and Antarctic regions is much less than in other zoogeographic realms primarily due to fewer researchers and difficulty in access, resulting in very limited studies conducted in these regions, except for Argentina, Chile, Costa Rica, and the Antarctic Peninsula (Table 15.1). We expect the global

Thorp and Covich's Freshwater Invertebrates. https://doi.org/10.1016/B978-0-12-804225-0.00015-0

TABLE 15.1 Tardigrade taxa reported from freshwater habitats and substrates in the Neotropical region, Antarctica, and sub-Antarctic islands

	Taxa	Locations	Habitats	Substrates
	Neotropical region			
Eutardigrada				
Apochela, Milnesiidae	*Milnesium* cf. *tardigradum*	Argentina, Brazil, Uruguay	Terrestrial, found in aquatic habitats	11
Parachela				
Macrobiotoidea, Macrobiotidae	*Macrobiotus coronatus* (de Barros, 1942)	Uruguay	Terrestrial, probably misidentification	11
	Macrobiotus echinogenitus Richters, 1904	Colombia, Uruguay	Terrestrial, probably misidentification	11
	Macrobiotus hufelandi Schultze, 1834	Argentina, Brazil, Colombia, Uruguay	Terrestrial, probably misidentification	11
	Macrobiotus furcatus Ehrenberg, 1859	Brazil, Uruguay	Terrestrial, probably misidentification	11
	Macrobiotus polyopus Marcus, 1928	Uruguay	Terrestrial, probably misidentification	11
	Macrobiotus primitivae de Barros, 1942	Brazil	Terrestrial, probably misidentification	11
	Mesobiotus harmsworthi (Murray, 1907)	Brazil, Colombia, Uruguay, Venezuela	Terrestrial, probably misidentification	4, 11
	Mesobiotus orcadensis (Murray, 1907)	Argentina, Uruguay	Terrestrial, probably misidentification	5
	Mesobiotus stellaris (du Bois-Reymond Marcus, 1944)	Uruguay	Terrestrial, probably misidentification	11
	Minibiotus acontists (de Barros, 1942)	Uruguay	Terrestrial, probably misidentification	11
	Minibiotus intermedius (Plate, 1888)	Argentina, Brazil, Colombia, Uruguay, Venezuela	Terrestrial, probably misidentification	8
	Minibiotus marcusi (de Barros, 1942)	Uruguay	Terrestrial, probably misidentification	11
	Paramacrobiotus areolatus (Murray, 1907)	Uruguay	Terrestrial, probably misidentification	11
	Paramacrobiotus richtersi (Murray, 1911)	Argentina, Brazil, Uruguay, Venezuela	Terrestrial, probably misidentification	5
Murrayidae	*Dactylobiotus ambiguus* (Murray, 1907)	Argentina, Brazil, Venezuela	True freshwater	2
	Dactylobiotus caldarellai Pilato & Binda, 1994	Chile	True freshwater	11
	Dactylobiotus dispar (Murray, 1907)	Argentina	True freshwater	3
	Dactylobiotus lombardoi Binda & Pilato, 1999	Chile	True freshwater	11
	Dactylobiotus macronyx (Dujardin, 1851)	Paraguay	True freshwater	11
	Dactylobiotus parthenogeneticus Bertolani, 1982	Argentina, Bolivia	True freshwater	8
	Murrayon pullari (Murray, 1907)	Argentina, Brazil	True freshwater	7, 11

TABLE 15.1 Tardigrade taxa reported from freshwater habitats and substrates in the Neotropical region, Antarctica, and sub-Antarctic islands—cont'd

	Taxa	Locations	Habitats	Substrates
Isohypsibioidea, Isohypsibiidae	*Doryphoribius doryphorus* (Binda & Pilato, 1969)	Uruguay	Freshwater found in wet habitats	11
	Doryphoribius evelinae (Marcus, 1928)	Argentina, Brazil	Freshwater found in wet habitats	11, 12
	Doryphoribius zappalai Pilato, 1971	Argentina, Uruguay	True freshwater	8
	Isohypsibius deflexus (Mihelcic, 1960)	Argentina	True freshwater	3
	Isohypsibius granulifer Thulin, 1928	Brazil	True freshwater	11
	Isohypsibius myrops (du Bois-Reymond Marcus, 1944)	Brazil	True freshwater	8
	Isohypsibius papillifer (Murray, 1905)	Brazil	True freshwater	11
	Isohypsibius asper (Murray, 1906)	Argentina, Uruguay	True freshwater	8
	Pseudobiotus longiunguis (Iharos, 1968)	Chile	True freshwater	1
	Pseudobiotus megalonyx (Thulin, 1928)	Argentina, Brazil, Peru, Suriname, Uruguay	True freshwater	1, 3
	Thulinius stephaniae (Pilato, 1974)	Argentina	True freshwater	11
Hypsibioidea, Hypsibiidae	*Adropion scoticum* (Murray, 1905)	Brazil, Colombia, Uruguay	Terrestrial, found in aquatic habitats	8
	Hypsibius dujardini (Doyère, 1840)	Argentina, Chile, Uruguay	Freshwater found in wet habitats	8
	Mixibius saracenus (Pilato, 1973)	Argentina	True freshwater	6
	Pilatobius recamieri (Richters, 1911)	Argentina	Terrestrial and freshwater	1, 8
	Antarctica and sub-Antarctic islands			
Heterotardigrada, Echiniscidae	*Echiniscus jenningsi* Dastych, 1984	SG, SI	Terrestrial, found in aquatic margins	1, 9
Eutardigrada				
Apochela, Milnesiidae	*Milnesium antarcticum* Tumanov, 2006	[PAL]; SI	Terrestrial, found in aquatic margins	1, 9
Parachela				
Macrobiotoidea, Macrobiotidae	*Mesobiotus furciger* (Murray, 1906)	END, GRA, MAUD, PAL; SI, SG	Terrestrial, found in aquatic margins	1, 9
Murrayidae	*Dactylobiotus* cf. *ambiguus*	GRA, [PAL]; [HD], KGI, [MAC], [PE], SHD, SHL, SG, SI,	True freshwater (also in terrestrial sites)	1, 9, 10
	Murrayon hastatus/pullari	MAC	Freshwater	1, 9, 10
Hypsibioidea, Hypsibiidae	*Acutuncus antarcticus* (Richters, 1904)	END, GRA, MAUD, [PAL], RON, SCOT, WILK; HD, KGI, SG, SHD, SI	Freshwater and terrestrial	1, 9, 10

(*Continued*)

PHYLUM TARDIGRADA

TABLE 15.1 Tardigrade taxa reported from freshwater habitats and substrates in the Neotropical region, Antarctica, and sub-Antarctic islands—cont'd

	Taxa	Locations	Habitats	Substrates
	Adropion greveni (Dastych, 1984)	[?PAL], SI	Terrestrial, found in aquatic habitats	1, 9
	Diphascon dastychi Pilato & Binda, 1999	GRA, SCOT	Freshwater	1
	Diphascon cf. *pingue* (Marcus, 1936) *[alpinus]*	MAUD, [PAL], [SCOT], WILK; SG, SI	Terrestrial and freshwater	1, 9, 10
	Diphascon ongulense (Morikawa, 1962)	END, MAUD	Terrestrial and freshwater	1, 9
	Diphascon victoriae Pilato & Binda, 1999	SCOT	Freshwater	1
	Hypsibius cf. *dujardini*	[PAL]; MAC, PE	Terrestrial, found in aquatic margins	1, 9
Isohypsibioidea, Isohypsibiidae	*Isohypsibius asper* (Murray, 1906)[a]	GRA, PAL; KE, KGI, SG, SI	True freshwater (also in terrestrial sites)	1, 9, 10
	Isohypsibius laevis McInnes, 1995	SI	True freshwater	1, 9, 10
	Isohypsibius sp.	END (2 species), WILK; PE	True freshwater	1, 9
	Isohypsibius cf. *papillifer* (Murray, 1905)	KGI, SG, SI	True freshwater	1, 9, 10
	Ramajendas renaudi (Ramazzotti, 1972)	KE, KGI	Terrestrial, found in aquatic margins	1

Locality abbreviations: **Antarctica**: *END*, Enderby Sector; *GRA*, Graham Land Peninsula; *MAC*, Maquarie Island; *MAUD*, Dronning Maud Sector; *PAL*, Palmer Land Peninsula; *RON*, Ronne Sector; *SCOT*, Scott Sector; *WILK*, Wilkes Sector; **sub-Antarctic Islands**: *HD*, Heard Island; *KE*, Iles Kerguelen; *KGI*, King George Island (S. Shetlands); *PE*, Prince Edward Islands; *SG*, South Georgia; *SHD*, Deception Island (S. Shetlands); *SHL*, Livingston Island (S. Shetlands); *SI*, Signy Island (S. Orkneys); [initials], Questionable identification or not aquatic habitat. **Substrates**: 1, aquatic sediment; 2, peat bog + inside of bromeliads; 3, plankton sample; 4, peat bog mud; 5, peat bog moss; 6, peat moss near glacier; 7, peat bog *Sphagnum*; 8, moss; 9, cyanobacterial algal mat; 10, Antarctic submerged moss; 11, unspecified substrate or aquatic and terrestrial samples collected but species not linked with habitat; 12, *Selaginella* sp.
[a]See Dastych (2016).
Modified from Kaczmarek et al. (2014, 2015) for Neotropical region, Convey & McInnes (2005), Czechowski et al. (2012), Velasco-Castrillón et al. (2015), and personal notes (SJM) for Antarctica/sub-Antarctic Islands.

zoogeographic distribution of limnoterrestrial tardigrades, as previously reported in the literature, to undergo further analysis and revision as species groups and cryptic species are investigated more thoroughly (McInnes, 1994; McInnes & Pugh, 2007; Meyer, 2013; Kaczmarek et al., 2014, 2015; Guidetti et al., 2015).

Hydrophilous tardigrades are benthic organisms that inhabit permanent freshwater habitats, crawling on aquatic plants or in the interstitial spaces of sediments in lakes, rivers, streams, groundwater, and temporary or permanent ponds. They also live in subterranean caves, the hyporheic zone, activated sludge of sewage treatment plants, and in cryoconite holes in glaciers. These tardigrades serve as predators, prey, or consumers in food webs.

The continent of Antarctica and the sub-Antarctic islands are ice-dominated, therefore all terrestrial habitats, and in particular freshwater habitats, are fragmented (Chown & Convey, 2007). Most of the Antarctic lakes and pools are found in the maritime Antarctic (Peninsular region) and coastal Continental Antarctica. Glacier surface melt water pools, ice-dammed pools abutting nunataks, and cryoconite holes form temporary freshwater habitats in and on glacial environments. Shallow lakes, lake margins, and pools are ice-free for less than 4 months of the year. During the summer, perennial cyanobacterial mats (predominantly *Phormidium* spp.) and filamentous green algal assemblages proliferate in the shallow lakes, which are depauperate in fauna, with tardigrades forming an important element of the biomass (Convey & McInnes, 2005).

In Antarctic regions, there are two endemic genera: *Ramajendas* and the monospecific *Acutuncus*. *Ramajendas* species (currently attributed to the family Isohypsibiidae) were found in freshwater sediment, mosses, and lichens of Antarctica and the sub-Antarctic Islands. In the family Hypsibiidae, *Acutuncus antarcticus* (Richters, 1904), considered a pan-Antarctic species, is the most abundant and common tardigrade in Antarctica, living in freshwater ecosystems (lakes, ponds, streams, and cryoconite holes) and terrestrial soils, grass, algae, mosses, and lichens in non-glacial areas. Its distribution covers continental Antarctica, maritime Antarctic, sub-Antarctic islands, and new studies indicate that the genus may contain cryptic species.

Based on a few long-term studies on tardigrades in Antarctica, tardigrade species show distinct depth-related associations within lakes, and the minimal predation pressure and perennial abundance of cyanobacterial mats (food source) can allow very high population numbers (McInnes & Pugh, 1999). As a component of the meiofauna, tardigrades may be indicators of pollution in freshwater, but they have been used in only a few toxicity or environmental monitoring studies in Europe (Nelson et al., 2015). Antarctica is considered a global pollution indicator (Huiskes et al., 2006).

The Neotropical region offers very diverse freshwater habitats for tardigrades, from rivers, streams, and lakes in tropical lowlands (in Central America and northern South America) up to high mountain lakes, ponds, and glaciers in the Andean Mountains on the west coast of South America. The central regions of South America provide freshwater habitats in temporary, dry, or very hot climatic zones. On the southern end of the South American continent, freshwater tardigrades can be found in very cold ponds, rivers, lakes, and in cryoconite holes on the glaciers. In the Neotropical region, some unusual freshwater habitats also available for tardigrades include tree holes and bromeliads.

We know little about the population dynamics of freshwater (epigean, hyporheic, or groundwater) tardigrades in the Neotropics, however tardigrades in the psammon community of some lakes and rivers may show a distinct seasonality in population densities. The only confirmed endemic genus is the monospecific *Schusterius*, family Macrobiotidae. *Schusterius tridigitus* (Schuster, 1983) was found once in terrestrial cryptogams in Tierra del Fuego (Argentina). There is a single record of *Acutuncus antarcticus* for South America (Argentina), but the identification is dubious and needs confirmation.

LIMITATIONS

Tardigrade identification is based primarily on morphological variations in the claws, buccal-pharyngeal apparatus, cuticle, and egg shell. These characters are included in the key and shown in Figs. 15.1–13. Our key is limited to genera since many species are difficult to identify even by tardigrade taxonomists. Cryptic species are present and many species groups need taxonomic revision. Although adult specimens are recommended for use in identification, juvenile eutardigrades are usually similar to adults though much smaller. Sexual dimorphism in characters is known

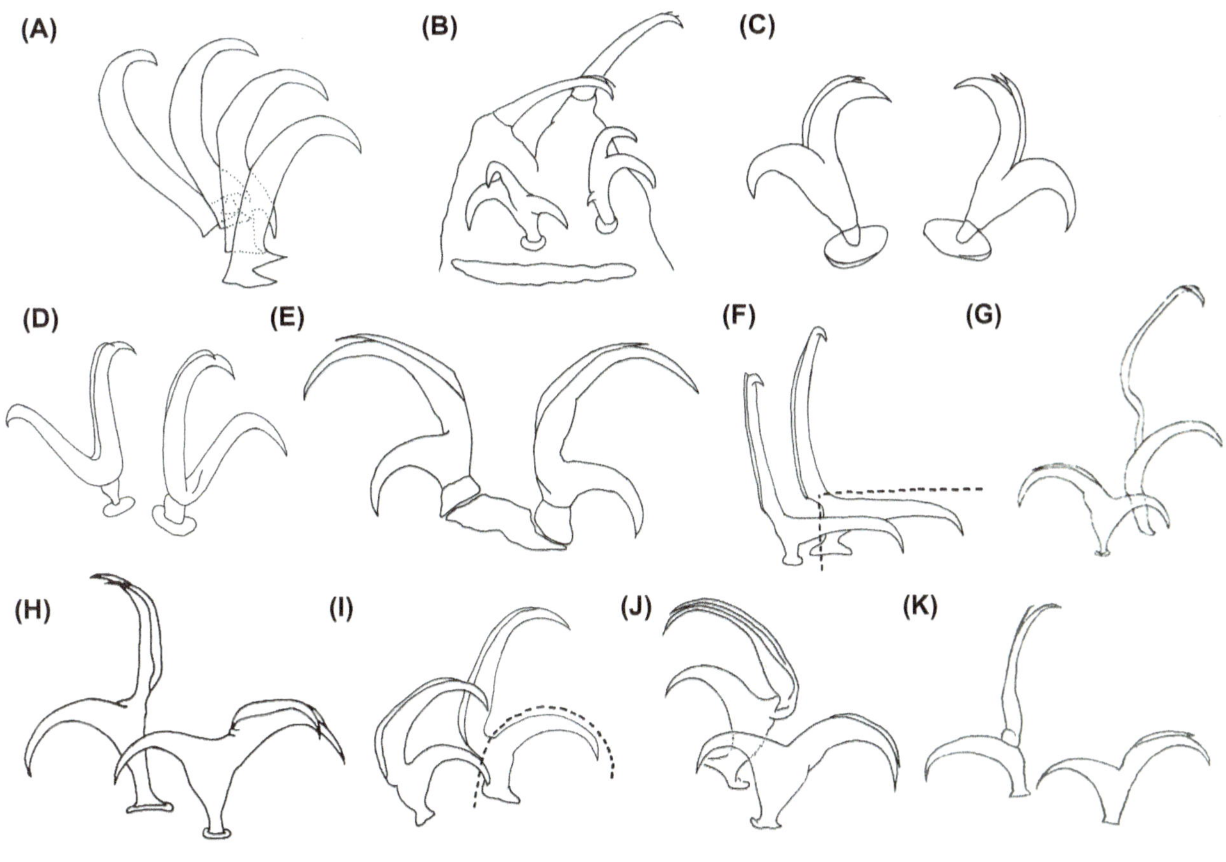

FIGURE 15.1 Diagrammatic representation of claw morphologies in tardigrades. (A) Heterotardigrada: *Echiniscus*; (B–M) Eutardigrada: (B) *Milnesium*; (C) *Macrobiotus*; (D) *Murrayon*; (E) *Dactylobiotus*; (F) *Isohypsibius*; (G) *Ramajendas*; (H) *Thulinius*; (I) *Hypsibius*; (J) *Acutuncus*; (K) *Mixibius*. *(A–F), (H) and (I), Redrawn from Bertolani (1982); (G), Redrawn from Pilato & Binda (1990); (J), Redrawn from Dastych (1991); (K), Redrawn from Pilato & Binda (1996).*

PHYLUM TARDIGRADA

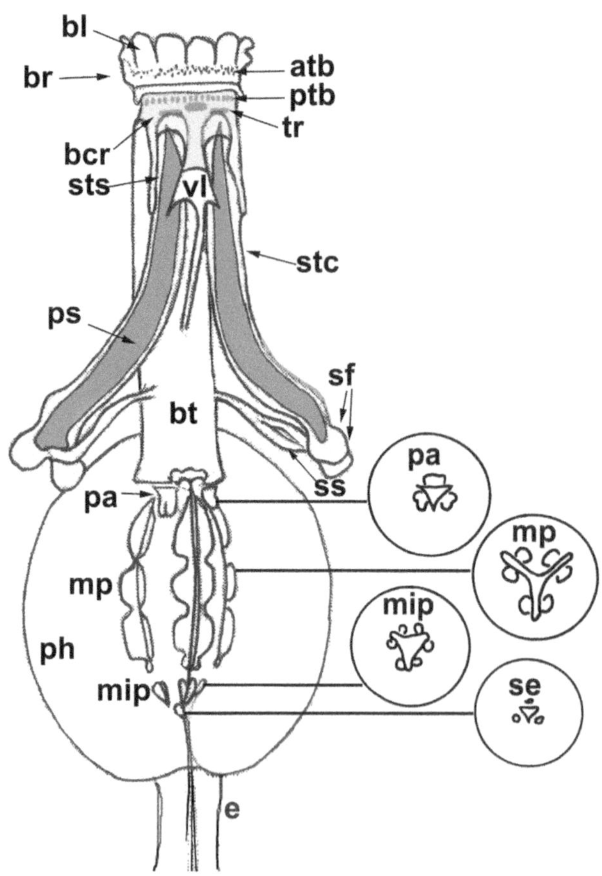

FIGURE 15.2 Diagrammatic representation of a tardigrade buccal-pharyngeal apparatus, with cross-sectional view of the pharynx containing the sclerified cuticular structures. Abbreviations: *atb*, anterior band of teeth; *bcr*, buccal crown; *bl*, peribuccal lamellae; *br*, buccal ring; *bt*, buccal tube; *e*, esophagus; *mip*, microplacoid; *mp*, macroplacoid; *pa*, pharyngeal apophyses; *ph*, pharynx; *ptb*, posterior band of teeth; *ps*, piercing stylets; *se*, septulum; *sf*, stylet furca; *ss*, stylet supports; *stc*, stylet coat; *sts*, stylet sheaths; *tr*, transverse ridges; *vl*, ventral lamina. *Redrawn from Guidetti et al. (2012).*

for Heterotardigrada as well as apochelan eutardigrades, but only for a few species of parachelan eutardigrades. Revisions of genera, families and superfamilies in recent years are based on a combination of molecular and morphological evidence (Sands et al., 2008; Guidetti et al., 2009; Jørgensen et al., 2010; Marley et al., 2011; Vicente et al., 2013; Bertolani et al., 2014).

TERMINOLOGY AND MORPHOLOGY

In echiniscid heterotardigrades (occasionally found in freshwater), each leg terminates in four unbranched claws in adults (Fig. 15.1 A) but only two claws per leg in the first instar. The limnoterrestrial eutardigrades usually possess two double claws on each leg, arranged as external and internal claws on the first three pairs of legs and as anterior and posterior claws on the hind legs. The claws on each leg may be similar or different in shape and size (Figs. 15.1 B–K, 3). In true limnic species of eutardigrades, the claws are usually long to very long. Each double claw consists of a longer main (primary) branch with two accessory points, and a secondary branch without accessory points (Fig. 15.4) (Guidetti & Bertolani, 2005; Pilato & Binda, 2010).

In the eutardigrade order Apochela (Milnesiidae), the two claw branches are separate from each other. The primary branch is long and thin and has accessory points, while the secondary branch is short, stout, and usually bears two or three hooks or spurs; the two claws of each leg are symmetrical (Fig. 15.1 B). In the eutardigrade order Parachela, the primary and secondary branches are connected and usually arise from a common basal tract; the secondary branch has no hooks. In many cases, the base of the claw is surrounded by a cuticular lunule (Figs. 15.1 C, D, 3 C, 4).

Within the Parachela, the families Macrobiotidae and Murrayidae have symmetrical claws with a lateral-medial sequence of claw branches as follows: secondary branch of the external claw and its primary branch, primary branch of the internal claw and its secondary branch ("2-1-1-2 claws") (Fig. 15.1 C–E). The basal tract may consist of three parts: a proximal stalk connecting the basal tract to the lunule, a distal portion (which may be delimited by a transverse septum), and a common tract after which the two claw branches diverge (Fig. 15.4) (Guidetti & Bertolani, 2005; Pilato & Binda, 2010). Macrobiotidae have Y-shaped double claws (Figs. 15.1 C, 3 C), with the claw branches fused over a basal tract of variable length; Murrayidae have double claws with the two branches diverging immediately after the basal tract (V-shaped, Fig. 15.1 D; L-shaped, Figs. 15.1 E, 3 D). In the limnic *Dactylobiotus* (Figs. 15.1 E, 3 D) (Murrayidae), a cuticular link between the bases of the claw pair is present.

All other families are characterized by asymmetrical claws ("2-1-2-1" claw sequence: secondary branch of the external claw and its primary branch, secondary branch of the internal claw and its primary branch) (Fig. 15.1 F–K). In Isohypsibiidae, the secondary branch is inserted perpendicularly (or at an obtuse angle) to the common basal tract of the claw; the external and internal claws are often similar in size and shape (*Isohypsibius* type claw, Figs. 15.1 F, 3 B). In *Mixibius*, the external claw is of the *Isohypsibius* type; the internal claw is a modified *Isohypsibius* type in which the secondary branch and the basal portion of the internal claw form an angle slightly larger than 90 degrees and therefore, in some positions, this claw may seem to be the *Hypsibius* type, but it systematic position is debatable as to whether this genus is in Isohypsibiidae or Hypsibiidae (Fig. 15.1 K). In Hypsibiidae (e.g., *Acutuncus, Adropion, Diphascon, Hypsibius*), the secondary branch of the external double claw forms a common arc with the basal tract; the external and internal claws are different in shape and size (*Hypsibius* type claw, Figs. 15.1 I, 3 A) (Guidetti & Bertolani, 2005; Pilato & Binda, 2010). In *Acutuncus* (Figs. 15.1 J, 13 A, C), the outer claws of the first three pairs of legs and posterior claws of the hind legs are the

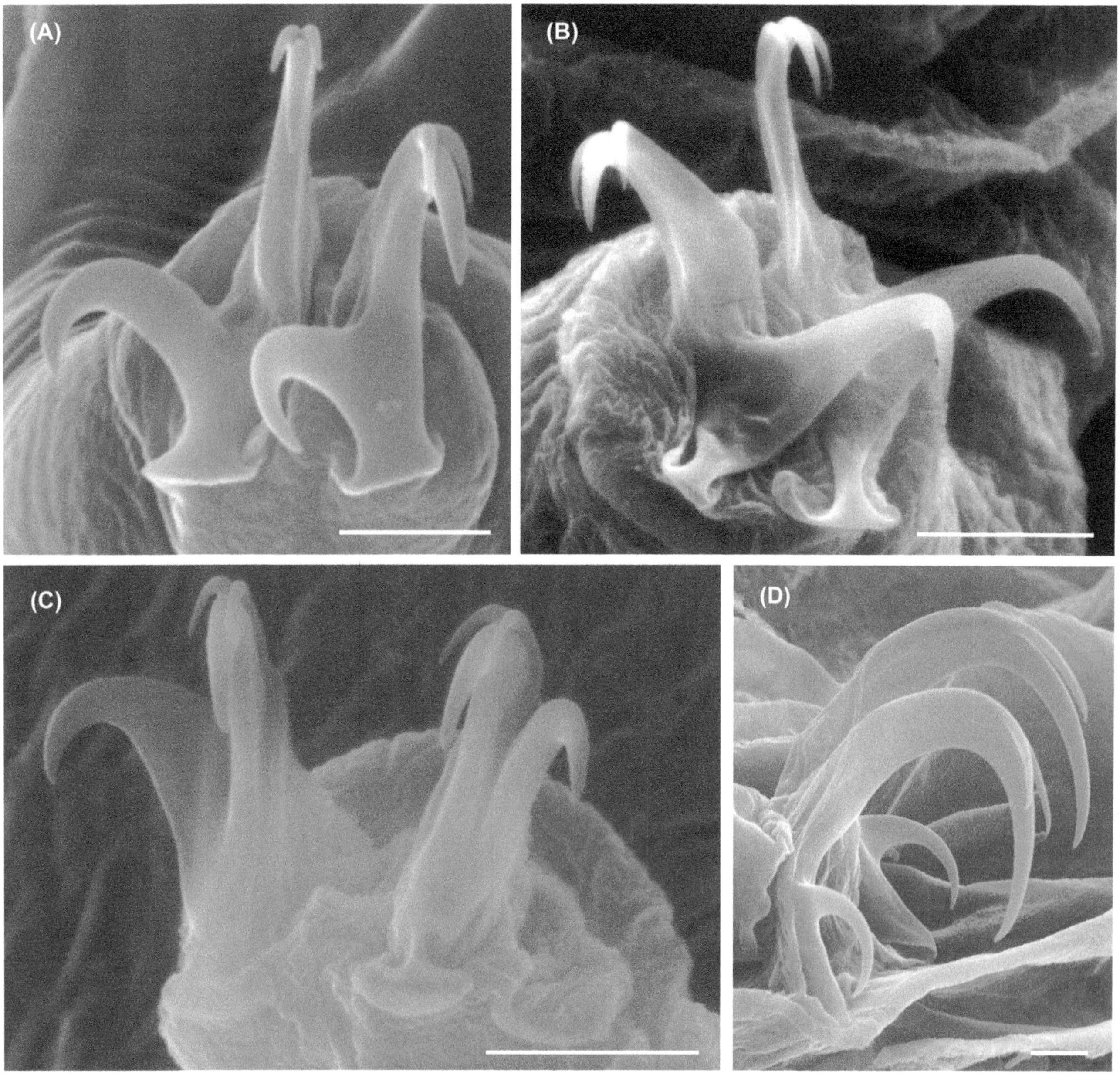

FIGURE 15.3 Scanning electron micrographs of eutardigrade claws: (A) *Diphascon* sp.; (B) *Thulinius* sp.; (C) *Macrobiotus* sp.; (D) *Dactylobiotus* sp. Scale bars: (A) = 2.5 μm, (B–D) = 5 μm.

Isohypsibius type; however, the inner claws of the first three pairs of legs and anterior claws of the hind legs are the *Hypsibius* type. In *Ramajendas* (Figs. 15.1 G, 13 D, F), the claws of each leg are very different from each other; the external claw is the *Hypsibius* type with an extremely long, slender primary branch, while the internal claws are the *Isohypsibius* type.

The buccal-pharyngeal apparatus is taxonomically important, especially in the eutardigrades (Figs. 15.2, 5–8, 13 B, E). The presence and the number of peribuccal structures around the mouth (Fig. 15.5) are significant at the generic level (Guidetti & Bertolani, 2005; Pilato & Binda, 2010; Guidetti et al., 2012, 2013).

Most eutardigrades have a completely rigid buccal tube, which is supported by a ventral lamina (strengthening bar) (Figs. 15.2, 6 B, H, I, 7 A, 8 A, B) in macrobiotids, murrayids, and several isohypsibiids (*Apodibius, Doryphoribius, Haplomacrobiotus, Haplohexapodibius, Hexapodibius, Parhexapodibius*). The murrayids *Dactylobiotus* and *Murrayon* have an evident hook in the ventral lamina in lateral view (Fig. 15.8 A) (Guidetti & Bertolani, 2005; Pilato & Binda, 2010; Guidetti et al., 2012, 2013).

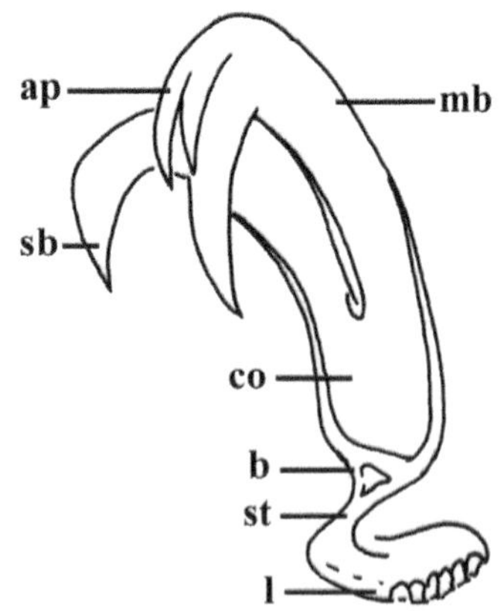

FIGURE 15.4 Diagrammatic representation of tardigrade 2claw structure. Abbreviations: *ap*, accessory points; *b*, basal portion; *co*, common tract; *l*, lunula; *mb*, main (primary) branch; *sb*, secondary branch; *st*, stalk. *Redrawn from Nelson & McInnes (2002).*

The buccal crown is composed of cuticular crests and laminae on which the stylet protractor muscles are attached (Figs. 15.2, 7, 8) (Guidetti et al., 2012, 2013). In lateral view, the buccal crown shows apophyses that are taxonomically significant and vary in size and shape (e.g., crests as in *Isohypsibius*, *Ramajendas,* and *Thulinius*, hooks as in *Hypsibius*, and others; Fig. 15.8).

In some genera in different eutardigrade families (Milnesiidae, Hypsibiidae, Isohypsibiidae, Ramazzottiidae, Macrobiotidae) and in some echiniscid heterotardigrades, a rigid buccal tube between the buccal ring and the stylet supports is followed by a flexible, annulated pharyngeal tube from just below the stylet supports to the pharynx (Figs. 15.6 J–L, 7 C).

Lining the lumen of the pharynx are three cuticular bars in heterotardigrades (Fig. 15.6 A), or three double rows of cuticular thickenings called placoids (macroplacoids, microplacoids) (Figs. 15.2, 6 B–L), which are taxonomically significant in the Parachela (Eutardigrada). In Apochela (e.g., *Milnesium*) and a few Parachela (e.g., *Astatumen, Itaquascon*), the cuticular thickenings within the pharynx are reduced or absent (Guidetti & Bertolani, 2005; Pilato & Binda, 2010; Guidetti et al., 2012, 2013).

The two extant classes, Heterotardigrada and Eutardigrada are distinguished by the presence (in Heterotardigrada) or absence (in Eutardigrada) of a pair of cuticular cephalic cirri (called cirri A) (Figs. 15.9 A, 10 A).

Eggs are often essential for the identification of freshwater and terrestrial species (Fig. 15.11). Ornamented eggs, usually deposited freely and singly or in small groups, are present in *Acutuncus*, Macrobiotidae (e.g., *Macrobiotus*, *Paramacrobiotus*, *Mesobiotus*), and Murrayidae (*Dactylobiotus*), which contain freshwater species. Other eutardigrades and limnoterrestrial heterotardigrades in freshwater environments lay smooth eggs within the exuvium (Figs. 15.11, 12).

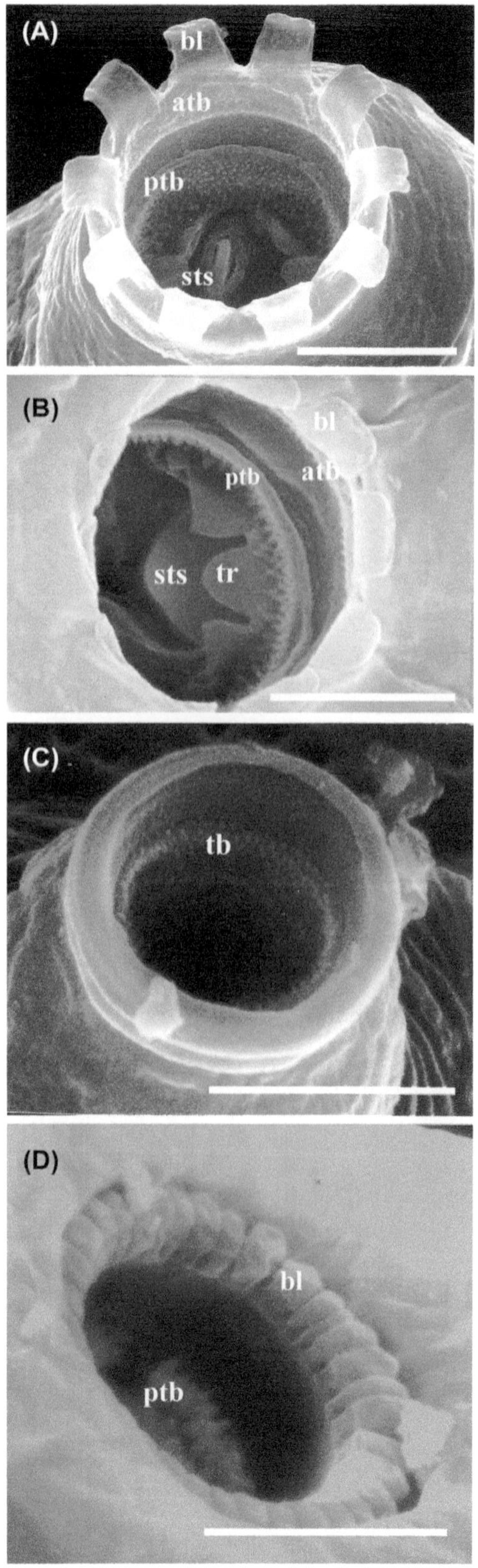

FIGURE 15.5 Scanning electron micrographs of eutardigrade mouth openings: (A) *Dactylobiotusselenicus*; (B) *Macrobiotus* sp.; (C) *Isohypsibiuslunulatus*; (D) *Pseudobiotus* sp. Abbreviations: *atb*, anterior band of teeth; *bl*, peribuccal lamellae; *ptb*, posterior band of teeth; *sts*, stylet sheaths; *tb*, teeth band; *tr*, transverse ridges. Scale bars = 5 μm.

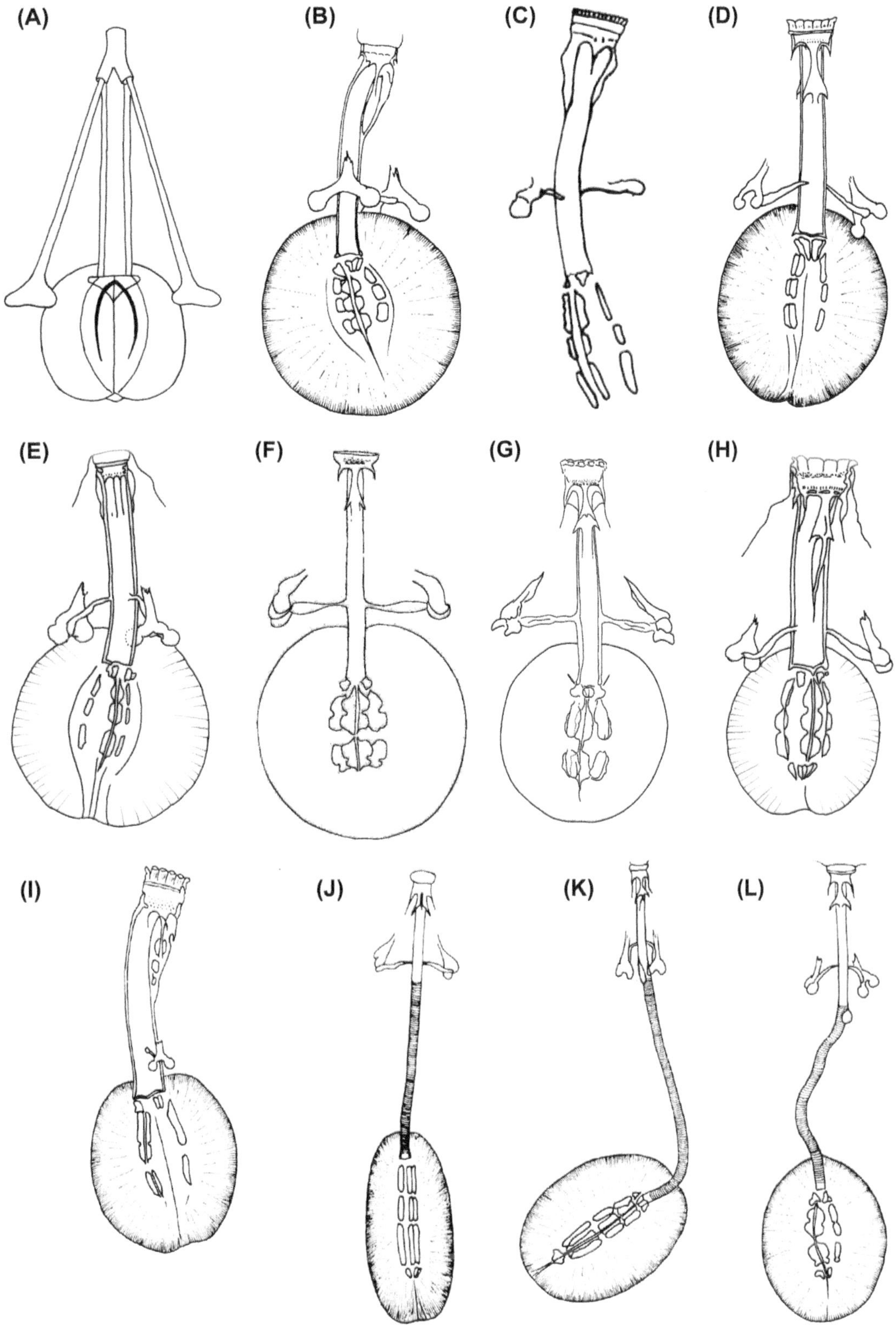

FIGURE 15.6 Diagrammatic representation of buccal-pharyngeal apparatuses: (A) Heterotardigrada: *Echiniscus*; (B-L) Eutardigrada: (B) *Doryphorybius*; (C) *Pseudobiotus*; (D) *Thulinius*; (E) *Isohypsibius*; (F) *Ramajendas*; (G) *Acutuncus*; (H) *Mesobiotus*; (I) *Dactylobiotus*; (J) *Adropion*; (K)*Diphascon*; (L) *Pilatobius*. *(A) Redrawn from Kristensen (1987); (B–E), (H–L) Redrawn from Bertolani (1982); (F) Redrawn from Pilato & Binda (1990); (G) Redrawn from Dastych (1991).*

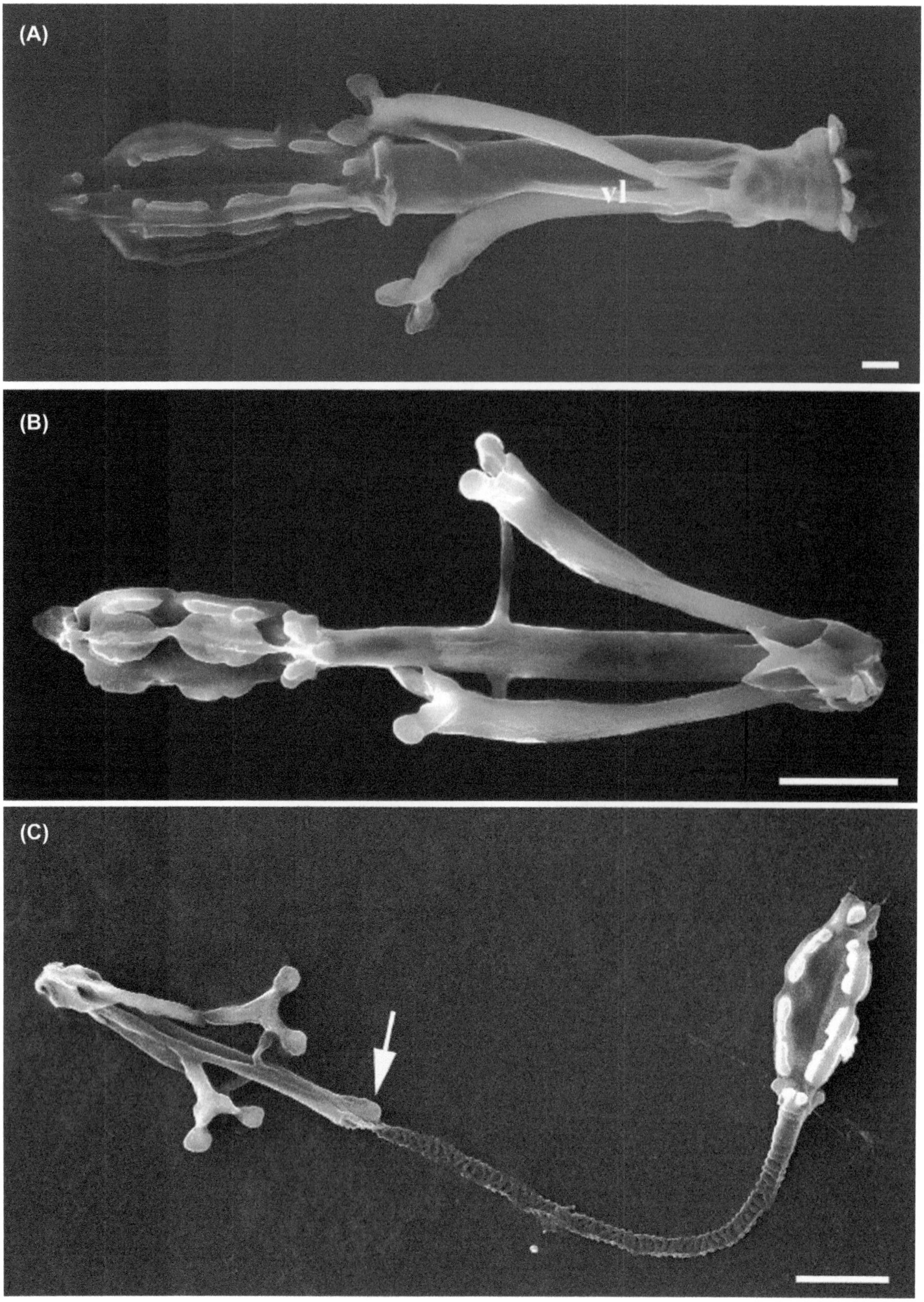

FIGURE 15.7 Scanning electron micrographs of eutardigrade buccal-pharyngeal apparatuses: (A) *Parmacrobiotusrichtersi* with a ventral lamina (*vl*) on the buccal tube; (B) *Hypsibius dujardini* and (C) *Diphascon* (*Diphascon*) sp. without a ventral lamina on the buccal tube. Arrow, drop-like thickening. Scale bars = 5 μm.

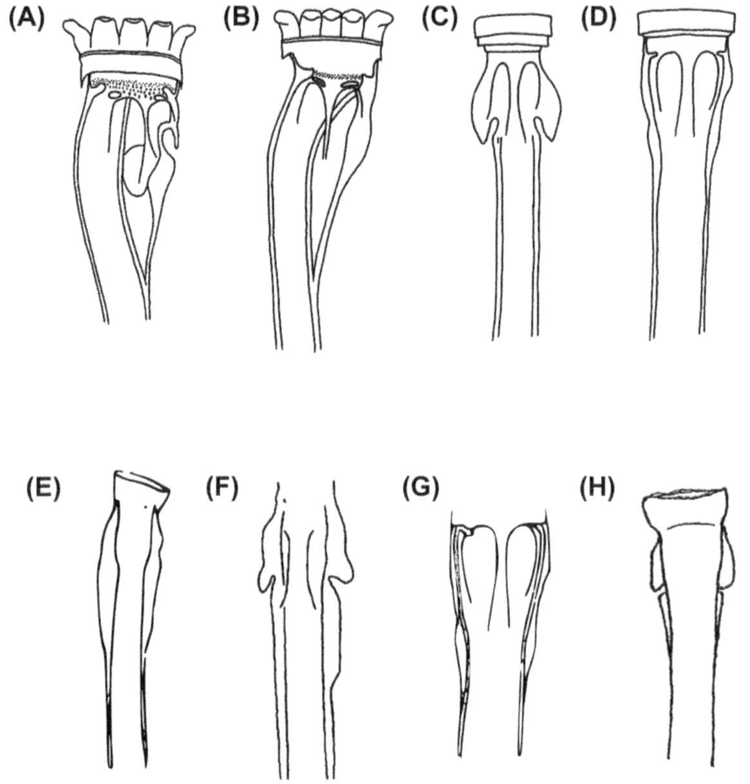

FIGURE 15.8 Diagrammatic representation of apophyses for the insertion of the stylet muscles on the wall of the buccal tube: (A) *Dactylobiotus*; (B) *Macrobiotus*; (C) *Hypsibius* and *Diphascon*; (D) *Isohypsibius*; (E) *Ramajendas*; (F) *Acutuncus*; (G) *Thulinius;* (H) *Mixibius*. *(A–D) and (G) Redrawn from Bertolani (1982); (E) Redrawn from Pilato & Binda (1990); (F) Redrawn from Dastych (1991); (H) Redrawn from Pilato & Binda (2010).*

MATERIAL PREPARATION AND PRESERVATION

The collecting techniques for tardigrade were described extensively in Nelson et al. (2015). However, all scientific Antarctic research first requires an Environmental Impact Assessment, which is conducted by the National Operator for the region. These assessments are guided by the SCAR Codes of Conduct (SCAR, 2017), which for lake sampling states: "Lakes and streams: Choose sampling equipment which is the least destructive, when the aquatic or coastal environment is to be sampled. Sample carefully and avoid cumulative impact. Dredges, trawls and box corers should be used as little as possible, avoiding excessive and unnecessary sampling." So, samples can be taken by scuba divers, who can hover carefully above the algal mat/moss/sediment and siphon small samples.

If possible, mount individual specimens collected in the field on microscope slides immediately and observe with phase contrast and/or differential interference contrast (DIC) microscopy. Hoyer's mounting medium (distilled water, 50 cm^3; gum arabic, 30 g; chloral hydrate, 150 g; glycerol, 20 cm^3) produces cleared, distended specimens, but seal the coverslip (e.g., with epoxy paint or clear fingernail polish) to ensure permanence. The amount of chloral hydrate varies from 50 to 220 g in different formulas. Reduce the amount of chloral hydrate to decrease the extent of clearing of specimens, and add potassium iodide (KI, 1 g) to stain specimens slightly to increase the resolution of morphological structures. Hoyer's medium is now available commercially. Variants of the gum arabic/chloral hydrate media that have been used include Heinze's, Faure's, and Faure-Berlese's media. Since chloral hydrate is a controlled substance, other recommended media include glycerin, polyvinyl alcohol (PVA), and polyvinyl lactophenol (PVLP).

If not mounted on slides after extraction, preserve tardigrades in small amounts of water, with Carnoy's fluid, boiling water, or boiling alcohol (85% or higher) for tardigrade fixation, but use alcohol to (70% or higher) preserve specimens to be mounted later. Fixation shrinks the tardigrades making them more difficult to identify, and specimens lose color in alcohol and some characters (e.g., fine cuticular structures) are less visible. For molecular analysis (DNA extraction) preserve tardigrades in a very small amount of distilled water and freeze at −20°C, or preserve in 100% (absolute) alcohol.

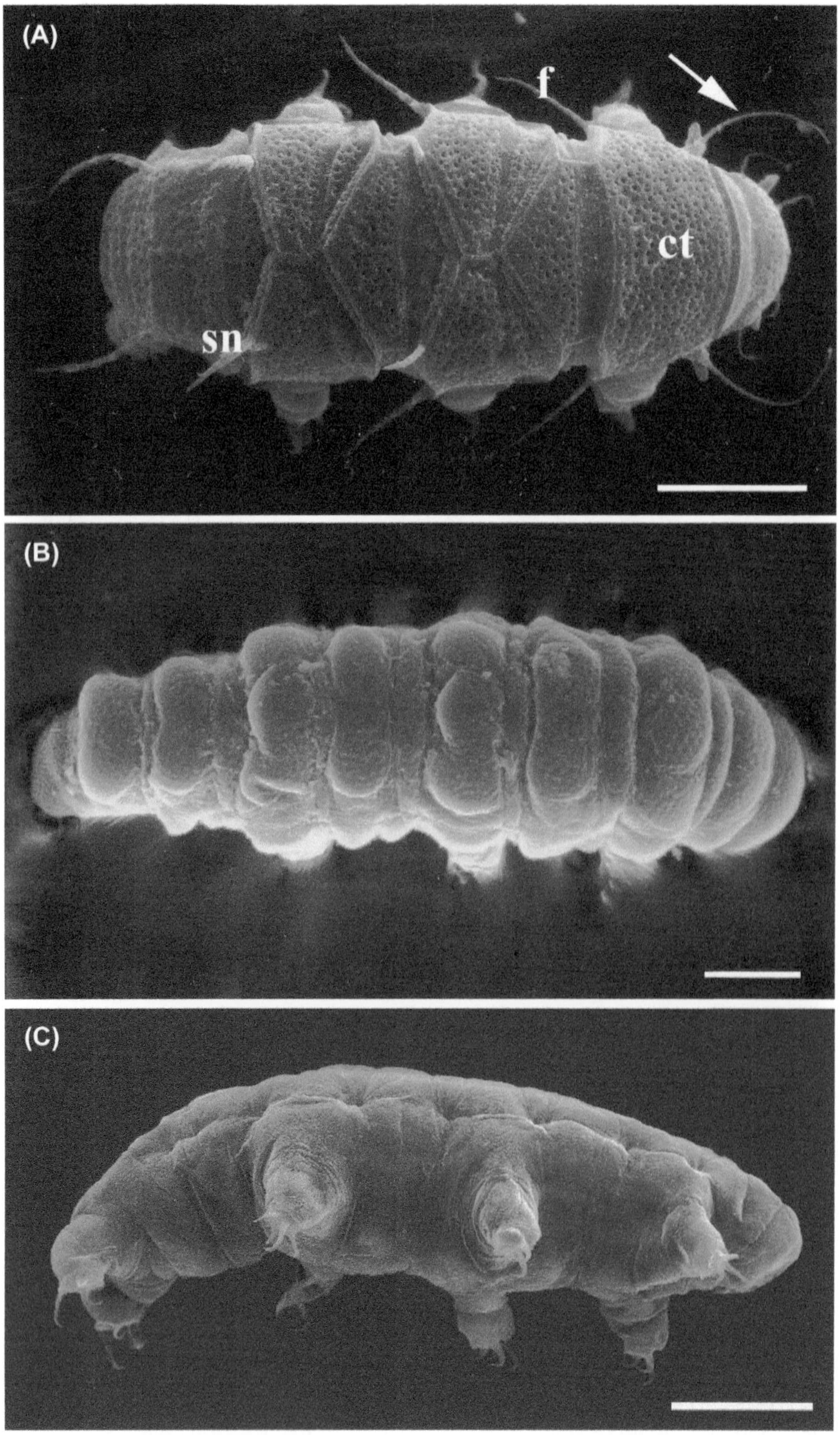

FIGURE 15.9 Scanning electron micrographs of tardigrades: (A) dorsal view of the heterotardigrade *Echiniscusduboisi* (arrow, cirrus A); (B) dorsal view of the eutardigrade *Isohypsibiuslunulatus*; (C) lateral view of the eutardigrade *Acutuncusantarcticus*. Abbreviations: *ct*, cuticular plate; *f*, filament; *sn*, spine. Scale bars = 25 μm.

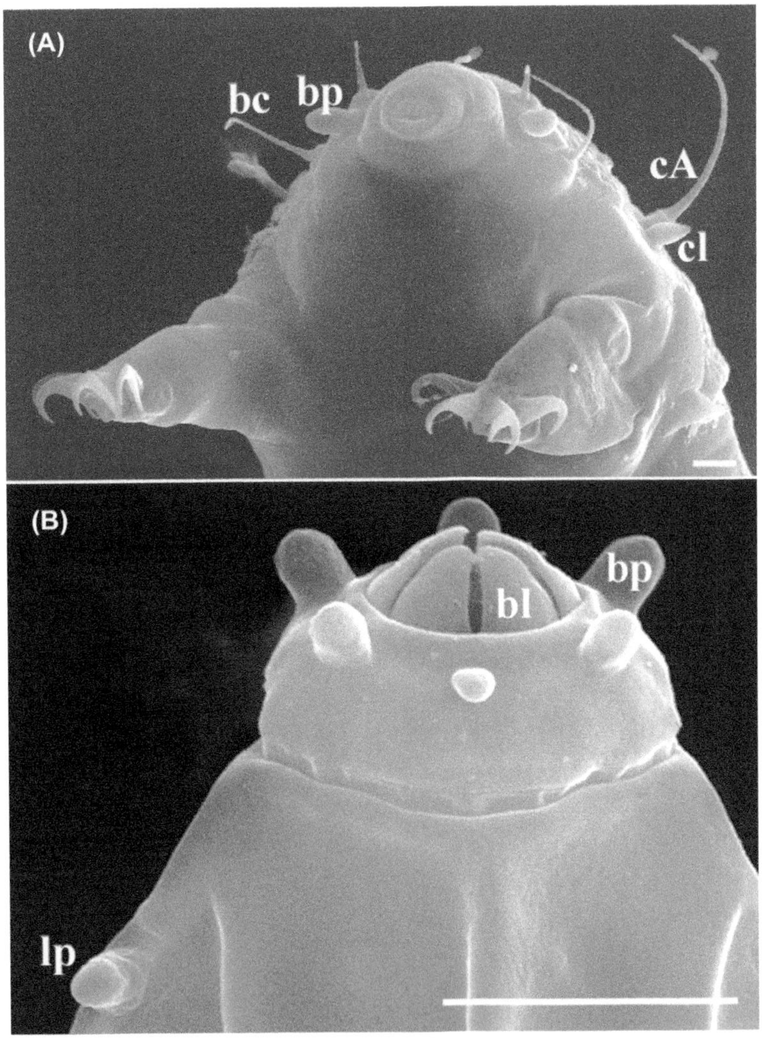

FIGURE 15.10 Scanning electron micrographs of tardigrade heads: (A) the echiniscid heterotardigrade *Antechiniscus parvisentus*; (B) the milnesiideutardigrade *Milnesium* sp. Abbreviations: *bc*, buccal cirri; *bl*, peribuccal lamellae; *bp*, buccal papillae; *cA*, cirrus A; *cl*, clava; *lp*, lateral papilla. Scale bars = 10 μm.

KEYS TO TARDIGRADA

In the following keys, we use "obligate" for the genera with species that typically live in freshwater habitats, "limnoterrestrial" for the genera with species that may be found in both freshwater and terrestrial habitats, and "accidental" for the genera with species that typically do not live in freshwater habitats but may be rarely found or are accidently washed or blown into the aquatic habitat. The species of each genus found in the Neotropic and Antarctic regions are listed in Table 15.1, together with their distributions.

Tardigrada: Classes

1 Lateral cirri A present (Figs. 15.9 A, 10 A); accidental .... Heterotardigrada, one genus: *Echiniscus*

1′ Lateral cirri A absent .... **Eutardigrada [p. 517]**

Tardigrada: Eutardigrada: Orders

1 Head without cephalic papillae; usually two double claws per leg (Fig. 15.1 C–K) .... **Parachela [p. 520]**

1′ Head with cephalic papillae, including two lateral papillae (Fig. 15.10 B); claws with the secondary branch not connected to the primary branch (Fig. 15.1 B); smooth eggs laid within exuvium (Fig. 15.11 E); limnoterrestrial .... Apochela, one genus: *Milnesium*

FIGURE 15.11 Tardigrade eggs (A–D) scanning electron micrographs (SEM) and (F) light micrograph (DIC): (A) *Paramacrobiotus* sp.; (B) *Acutuncus antarcticus*; (C) *Macrobiotus* sp.; (D) magnification of (C); (E) *Milnesium* sp. smooth egg extracted from exuvium; (F) ornamented egg of *Dactylobiotus* sp. Scale bars (A–D), (F) = 25 μm, (E) = 8 μm.

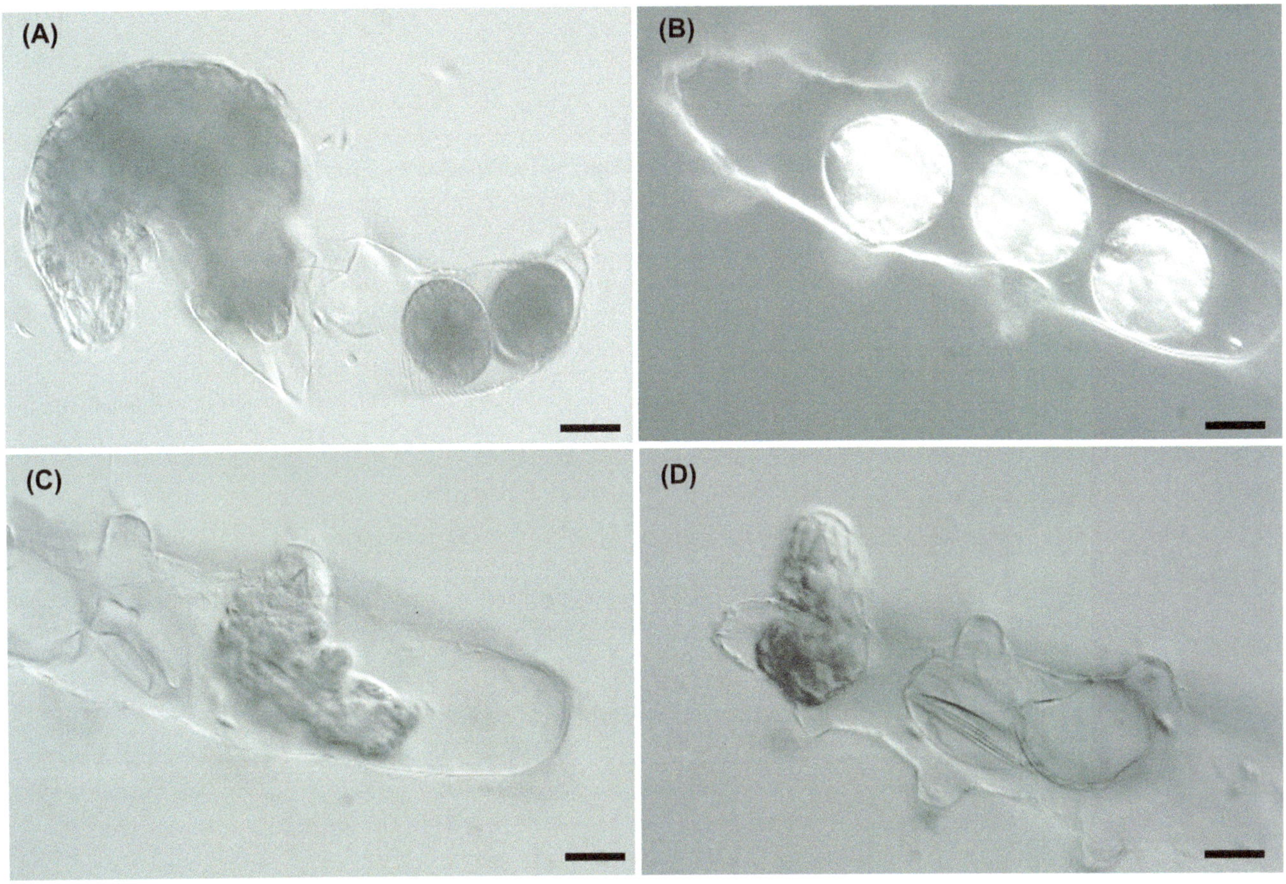

FIGURE 15.12 *In vivo* micrographs of *Diphascon* sp. (DIC). (A) animal releasing the eggs in the exuvium; (B) exuvium with three embryonated eggs; (C) newborn (juvenile) trying to find the exit from the exuvium; (D) newborn (juvenile) crawling out of exuvium. Scale bars = 50 μm.

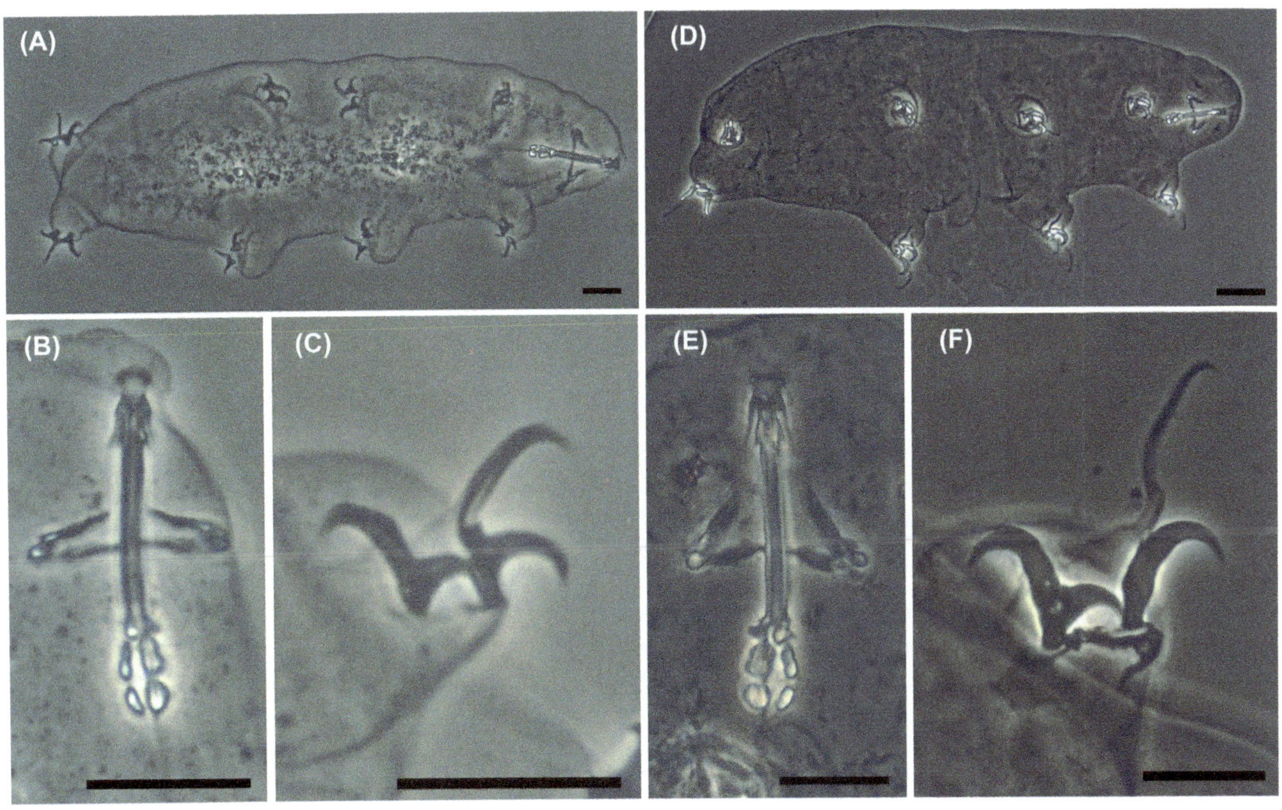

FIGURE 15.13 Micrographs of *Acutuncus antarcticus* (A–C) and *Ramajendas* sp. (D-F). (A, D) Habitus; (B, E) buccal-pharyngeal apparatus; (C, F) claws of hind leg. Scale bars (A–C), (E), (F) = 20 μm, (D) = 50 μm.

Tardigrada: Eutardigrada: Parachela: Superfamilies

1 Two double claws on each leg "asymmetrical," with sequence 2-1-2-1 (Figs. 15.1 F–K, 3 A, B); smooth eggs laid within the exuvium (Fig. 15.12) or ornamented eggs laid freely (Fig. 15.11 B) 2

1′ Two double claws on each leg similar in size and shape and "symmetrical," with sequence 2-1-1-2 (Figs. 15.1 C–E, 3 C–D); buccal tube with ventral lamina (Figs. 15.2, 7 A, 8 A, B); ornamented eggs laid freely (Fig. 15.11 A, C, D, F) .... **Macrobiotoidea [p. 520]**

2(1) Internal and external claws of similar size and shape (except in *Ramajendas*; Fig. 15.1 G); claws with primary and secondary branch continuous with the basal portion; secondary branch of each claw forming a right angle (or wider) with claw base (Fig. 15.1 F, H); eggs laid within the exuvium .... Isohypsibioidea, one family: **Isohypsibiidae [p. 520]**

2′ Internal and external claws of clearly dissimilar size and shape (Figs. 15.1 I, J, 3 A); most with external claws of *Hypsibius* type, with secondary branch forming a common arc with the basal portion; most with smooth eggs laid within the exuvium .... Hypsibioidea, one family: **Hypsibiidae [p. 521]**

Tardigrada: Eutardigrada: Parachela: Macrobiotoidea: Families

1 L- or V-shaped claws, i.e., with the two claw branches diverging immediately after the basal portion (Fig. 15.1 D, E); evident hook on the ventral lamina (Fig. 15.8 A) .... **Murrayidae [p. 520]**

1′ Y-shaped claws, i.e., with claw branches fused over a tract of variable length (common tract) (Figs. 15.1 C, 3 C, 4); no evident hook on the ventral lamina (Fig. 15.8 B) .... **Macrobiotidae [p. 520]**

Tardigrada: Eutardigrada: Parachela: Macrobiotoidea: Murrayidae: Genera

1 Very well-developed L-shaped claws, with cuticular thickenings connecting the base of the claws on each leg (Figs. 15.1 E, 3 D); without lunules; obligate .... *Dactylobiotus*

1′ V-shaped claws, without cuticular thickenings connecting the base of the claws on each leg, with lunules (Fig. 15.1 D); limnoterrestrial .... *Murrayon*

Tardigrada: Eutardigrada: Parachela: Macrobiotoidea: Macrobiotidae: Genera

1 Three macroplacoids 2

1′ Two macroplacoids; microplacoid, if present, close to the second macroplacoid; cuticle with or without pores; eggs with processes of various types; limnoterrestrial .... *Macrobiotus*

2(1) Rod-shaped macroplacoids; buccal lamellae present; cuticle without pores 3

2′ Round macroplacoids; microplacoid present; buccal lamellae absent; papulae present .... *Minibiotus*

3(2) Microplacoid always present and clearly close to the third macroplacoid (less than its length, Fig. 15.6 H); claw common tract characterized by an internal septum defining a distal part; limnoterrestrial .... *Mesobiotus*

3′ Microplacoid, if present, distant from the third macroplacoid more than its length (Fig. 15.7 A); eggs always with large processes that appear reticulated with light microscope (Fig. 15.11 A); limnoterrestrial .... *Paramacrobiotus*

Tardigrada: Eutardigrada: Parachela: Isohypsibiidae: Genera

1 Secondary branch of each claw forming right (or wider) angle with base; *Isohypsibius* type claws (Fig. 15.1 F, H) 2

1′ External claw of the *Hypsibius* type with an extremely long and slender primary branch; internal claw of the *Isohypsibius* type (Figs. 15.1 G, 13 D, F); dorsal and ventral apophysis for the insertion of the stylet muscles shaped as a very long "ridge" and symmetrical (Figs. 15.6 F, 8 E); limnoterrestrial .... *Ramajendas*

2(1) Ventral lamina absent 3

2′ Ventral lamina present (Fig. 15.6 B); limnoterrestrial .... *Doryphoribius*

3(2) Peribuccal lamellae around the mouth opening present (caution, may be difficult to see) (Fig. 15.5 A, B, D); very long claws (Fig. 15.1 H) 4

3′ Peribuccal lamellae around the mouth opening absent (Figs. 15.5 C, 6 E, 8 D); limnoterrestrial .... *Isohypsibius*

4(3) About 30 peribuccal lamellae present (Fig. 15.5 D); placoid row in a straight line (Fig. 15.6 C); obligate .... *Pseudobiotus*

4′ Twelve peribuccal lamellae present, often fused (caution, may be difficult to see); placoid row in the shape of a Grecian urn (Fig. 15.6 D); one or two cuticular bars under (but not connected to) the claw bases on legs I–III; limnoterrestrial .... *Thulinius*

Tardigrada: Eutardigrada: Parachela: Hypsibioidea: Hypsibiidae: Genera

The systematic position of *Mixibius* is under debate as to whether this genus is in Isohypsibiidae or Hypsibiidae.

1 External claws of the *Isohypsibius* type 2

1′ External claws of the *Hypsibius* type 3

2(1) External claws of the *Isohypsibius* type, internal claws of modified *Isohypsibius* type with the secondary branch and the basal portion forming an angle slightly larger than 90 degrees; apophyses for the insertion of the muscles of the stylets on buccal tube hook-shaped and asymmetrical (Fig. 15.8 H); limnoterrestrial .... *Mixibius*

2′ External claws of the *Isohypsibius* type, internal claws of the *Hypsibius* type (Figs. 15.1 J, 13 A, C); apophyses for the insertion of the stylets muscles in form of "acute hooks" and symmetrical (Fig. 15.8 F); eggs provided with processes (more visible with light microscope than Scanning Electron Microscope) (Fig. 15.11 B) and laid freely; limnoterrestrial .... *Acutuncus*

3(1) Buccal tube with a posterior part of flexible and spiral composition (pharyngeal tube) (Figs. 15.6 J–L, 7 C) 4

3′ Buccal tube rigid, without posterior part of flexible and spiral composition; limnoterrestrial .... *Hypsibius*

4(3) Droplike thickening between the rigid and flexible (pharyngeal) portions of the buccal tube present (Fig. 15.7 C); limnoterrestrial 5

4′ Droplike thickening between the rigid and flexible (pharyngeal) portions of the buccal tube absent (Fig. 15.6 J); limnoterrestrial .... *Adropion*

5(4) Two macroplacoids similar in length, organized in rows that look like parentheses; septulum present (Fig. 15.6 L); limnoterrestrial .... *Pilatobius*

5′ Two macroplacoids without septulum or three macroplacoids with or without septulum (Fig. 15.6 K); limnoterrestrial .... *Diphascon*

REFERENCES

Bertolani, R. 1982. Tardigradi. Guide per il Riconoscimento delle Specie Animali delle Acque Interne Italiane. Consiglio Nazionale Delle Ricerche, Verona, Italy. 104 pp..

Bertolani, R., T. Altiero & D.R. Nelson. 2009. Tardigrada (Water Bears). Pages 443–455 *in*: G.E. Likens (ed.), Encyclopedia of Inland Waters, Volume 2. Elsevier, Oxford, UK.

Bertolani, R., R. Guidetti, T. Marchioro, T. Altiero, L. Rebecchi & M. Cesari. 2014. Phylogeny of Eutardigrada: New molecular data and their morphological support lead to the identification of new evolutionary lineages. Molecular Phylogenetics and Evolution 76: 110–126.

Bertolani, R. & L. Rebecchi. 1999. Tardigrada. Pages 703–717 *in:* E. Knobil & J.D. Neill (eds.). Encyclopedia of Reproduction 4: Academic Press, San Diego, CA, USA.

Chown, S.L. & P. Convey. 2007. Spatial and temporal variability across life's hierarchies in the terrestrial Antarctic. Philosophical Transactions of the Royal Society B 362: 2307–2331.

Convey, P. & S.J. McInnes. 2005. Exceptional, tardigrade dominated ecosystems in Ellsworth Land, Antarctica. Ecology 86: 519–527.

Czechowski, P.L., C.J. Sands, B.J. Adams, C.A. D'Haese, J.A.E. Gibson, S.J. McInnes & M.I. Stevens. 2012. Antarctic Tardigrada: a first step in understanding molecular operational taxonomic units (MOTUs) and biogeography of cryptic meiofauna. Invertebrate Systematics 26: 526–538.

Dastych, H. 1991. Redescription of *Hypsibius antarcticus* (Richters, 1904), with some notes on *Hypsibius arcticus* (Murray, 1907) (Tardigrada). Mitteilungen aus dem Hamburgischen Zoologischen Museum und Institute 88: 141–159.

Dastych, H. 2016. Redescription and taxonomic status of the Antarctic water-bear *Isohypsibius tetradactyloides* (Richters, 1907), as concluded from the rediscovered type material (Tardigrada). Acta Biologica Benrodis 18: 21–43.

Degma, P., R. Bertolani & R. Guidetti. 2019. Actual checklist of Tardigrada species. (34th ed.). https://doi.org/10.25431/11380_1178608.

Guidetti, R., T. Altiero, T. Marchioro, L. Sarzi Amadè, A.M. Avdonina & L. Rebecchi. 2012. Form and function of the feeding apparatus in Eutardigrada (Tardigrada). Zoomorphology 131: 127–148.

Guidetti, R. & R. Bertolani. 2005. Tardigrade taxonomy: an updated check list of the taxa and a list of characters used in their identification. Zootaxa 845: 1–46.

Guidetti, R., R. Bertolani & L. Rebecchi. 2013. Comparative analysis of the tardigrade feeding apparatus: adaptive convergence and evolutionary pattern of the piercing stylet system. Journal of Limnology 72(s1): 24–35.

Guidetti, R., K.I. Jönsson & R.M. Kristensen. 2015. Tardigrades of Sweden: an updated check-list. Zootaxa 3981: 491–507.

Guidetti, R., R.O. Schill, R. Bertolani, T. Dandekar & M. Wolf. 2009. New molecular data for tardigrade phylogeny, with the erection of *Paramacrobiotus* gen. n. Journal of Zoological Systematics and Evolutionary Research 47: 315–321.

Huiskes, A.H.L., P. Convey & D. Bergstrom. 2006. Trends in Antarctic terrestrial and limnetic ecosystems: Antarctica as a global indicator. Pages 1–13. *in*: D. Bergstrom, P. Convey & A.H.L. Huiskes, (eds.), Trends in Antarctic Terrestrial and Limnetic Ecosystems: Antarctica as a Global Indicator. Springer-Verlag, Heidelberg, Germany.

Jørgensen, A., S. Faurby, J.G. Hansen, N. Møbjerg & R.M. Kristensen. 2010. Molecular phylogeny of Arthrotardigrada (Tardigrada). Molecular Phylogenetics and Evolution 54: 1006–1015.

Kaczmarek, Ł., Ł. Michalczyk & S.J. McInnes. 2014. Annotated zoogeography of non-marine Tardigrada. Part I: Central America. Zootaxa 3763: 1–62.

Kaczmarek, Ł., Ł. Michalczyk & S.J. McInnes. 2015. Annotated zoogeography of non-marine Tardigrada. Part II: South America. Zootaxa 3923: 1–107.

Kristensen, R.M. 1987. Generic revision of the Echiniscidae (Heterotardigrada) with a discussion of the origin of the family. Pages 261–335. *in*: R. Bertolani (ed.), Biology of Tardigrades. Selected Symposia and Monographs, Unione Zoologica Italiana 1. Mucchi, Modena, Italy.

Marley, N.J., S.J. McInnes & C.J. Sands. 2011. Phylum Tardigrada: a re-evaluation of the Parachela. Zootaxa 2819: 51–64.

McInnes, S.J. 1994. Zoogeographic distribution of terrestrial/freshwater tardigrades from current literature. Journal of Natural History 28: 257–352.

McInnes, S.J. & P. Pugh. 1999. Zonation in Antarctic lake-dwelling benthic meiofauna with emphasis on the Tardigrada. Zoologischer Anzeiger 238: 283–288.

McInnes, S.J. & P. Pugh. 2007. An attempt to revisit the global biogeography of limnoterrestrial Tardigrada. Journal of Limnology 66 (Suppl. 1): 90–96.

Meyer, H.A. 2013. Terrestrial and freshwater Tardigrada of the Americas. Zootaxa 3747: 1-71.

Nelson, D.R., R. Guidetti & L. Rebecchi. 2015. Chapter 17. Tardigrada. Pages 347-380 *in*: J.H. Thorp & D.C. Rogers (eds.), Thorp and Covich's Freshwater Invertebrates, 4th Edition, Vol. I, Ecology and General Biology. Academic Press, Elsevier, Boston, MA, USA.

Nelson, D.R. & S.J. McInnes. 2002. Tardigrada. Pages 177–215. *in:* S.D. Rundle & J.M. Schmid-Araya (eds.), Freshwater Meiofauna: Biology and Ecology. Backhuys Publications, Leiden, The Netherlands.

Pilato, G. & M.G. Binda. 1990. Tardigradi dell'Antartide. I. *Ramajendas*, nuovo genere di Eutardigrado. Nuova posizione sistematica di *Hypsibius renaudi* Ramazzotti, 1972 e descrizione di *Ramajendas frigidus* n. sp. Animalia 17: 61–71.

Pilato, G. & M.G. Binda. 1996. *Mixibius fueginus*, nuova specie di eutardigrado della Terra del Fuoco. Bollettino delle sedute dell'Accademia Gioenia di Scienze Naturali, Catania 29 (351): 27-32.

Pilato, G. & M.G. Binda. 2010. Definition of families, subfamilies, genera and subgenera of the Eutardigrada, and keys to their identification. Zootaxa 2404: 1–54.

Sands, C.J., S.J. McInnes, N.J. Marley, W.P. Goodall-Copestake, P. Convey & K. Linse. 2008. Phylum Tardigrada: an "individual" approach. Cladistics 24: 1–11.

SCAR (Scientific Committee on Antarctic research). 2017. SCAR's Environmental code of conduct for terrestrial scientific field research in Antarctica (revision 12 January, 2009). https://www.scar.org/library/policy/codes-of-conduct/3407-code-of-conduct-terrestrial-scientific-field-research-in-antarctica/file/. Accessed 07-04-2020.

Velasco-Castrillón, A., S.J. McInnes, M.B. Schultz, O.M. Arróniz-Cresp, C.A. D'Haese, J.A.E. Gibson, B.J. Page, T.J. Page, A.D. Austin, S.J.B. Cooper & M.I. Stevens. 2015. Mitochondrial DNA analyses reveal widespread tardigrade diversity in Antarctica. Invertebrate Systematics 29: 578–590.

Vicente, F., P. Fontoura, M. Cesari, L. Rebecchi, R. Guidetti, A. Serrano & R. Bertolani. 2013. Integrative taxonomy allows the identification of synonymous species and the erection of a new genus of Echiniscidae (Tardigrada, Heterotardigrada). Zootaxa 3613: 557–572.

PHYLUM TARDIGRADA

Chapter 16

Phylum Arthropoda: Introduction and Arachnida

D. Christopher Rogers, Natalia A. Fredes, Pablo A. Martínez, Nelson Ferretti, Gabriel Pompozzi, Almir Rogério Pepato and Pedro Henrique da Silva Conceição

Chapter Outline

Introduction to Arthropoda

D. Christopher Rogers

Kansas Biological Survey, and The Biodiversity Institute, University of Kansas, Lawrence, KS, United States

INTRODUCTION

Arthropoda constitutes the largest and most morphologically diverse group of animals. Specific relationships between the taxa of this clade are only partially understood, as are the relationships between the Arthropoda and other possibly related phyla. One hypothesis joins Arthropoda

Thorp and Covich's Freshwater Invertebrates. https://doi.org/10.1016/B978-0-12-804225-0.00016-2

and Annelida as the "Articulata" (reviewed in Scholtz, 2002), uniting the two groups based upon articulation. More recently, the Arthropoda and the Nemata have been united in a larger group as the "Ecdysozoa," being all animals that molt (Aguinaldo et al., 1997).

Aquatic Arthropods include the Chelicerata (mites and spiders), Myriapoda (aquatic millipedes), and the Crustacea (crustaceans, apterygotans, and insects). The Chelicerata are discussed below, while Crustacea are described in Chapters 18–23.

The Diplopoda are the most diverse group within subphylum Myriapoda, and aquatic forms are exceedingly rare and localized (Geoffry, 2015). Most forms encountered in freshwater have fallen into, or live in very moist terrestrial habitats associated with freshwater systems (Antić et al., 2017).

TERMINOLOGY AND MORPHOLOGY

Arthropods all bear segmented bodies with articled limbs. The Chelicerata body is divided into a cephalothorax, which contains the head, limbs, and thoracic organs, and an abdomen. The articulation may be obscured due to body articles fusing, as in mites. Chelicerates lack antennae, although the oral palpi may be elongated and serve a similar sensory function.

The Myriapoda have elongated bodies with serially homologous body articles, each bearing two pairs of limbs. In reality, each article only has one limb pair, but the body "segments" actually consist of two articles fused into one. Only one pair of antennae is present.

The Crustacea have the body divided into three divisions: head, thorax and abdomen. The thorax typically may have three to seven pairs of ambulatory limbs; however, in the Neotropical Branchiopoda there may be up to 11 pairs. The abdomen may have limbs as well, called pleopods or "swimmerettes," which can be used to propel the animal. Crustacea *sensu stricto* generally has two pairs of antennae, and the Hexapoda have one pair.

KEYS TO ARTHROPODA

Arthropoda: Subphyla

1 Less than 20 paired limbs; never more than 10 pairs for walking 2

1′ More than 10 pairs of identical walking limbs; millipedes .... Myriopoda: **Diplopoda [p. 524]**

2(1) Animal body variously separated into two divisions, cephalothorax and abdomen in spiders and gnathosoma and idiosoma in mites; antennae absent; four pairs of legs (except larval mites which have three); never with a bivalved carapace that can close entirely around the animal .... Chelicerata, one class: **Arachnida [p. 525]**

2′ Animal body divided into three or more divisions, if fused, then body completely enclosed in a bivalved carapace; one or two pairs of antennae present .... **Crustacea** (Chapter 17) **[p. 565]**

Diplopoda: Species

Millipedes frequently fall into aquatic habitats, but only two species are considered aquatic in the Neotropical region, both belonging to the order Polydesmida, the largest of all millipede orders. Only the juveniles and subadult stages of *Myrmecodesmus adisi* (Hoffman, 1985) are hygrophilus; the adults are unable to withstand inundation.

1 Collum at most slightly projecting over head base dorsally; each body article with anterior leg pair longer than posterior leg pair; stigmata on slender tubercles; metatergal spines digitiform to spiniform .... Chelodesmidae, one species: *Pandirodesmus disparipes* (Silvestri, 1932) [Guyana]

1′ Collum flabellate, completely concealing head from above; legs in each pair subequal; stigmata sessile .... Pyrgodesmidae, one species: *Myrmecodesmus adisi* (Hoffman, 1985) [Amazon Basin]

REFERENCES

Aguinaldo, A.M., J.M. Turbeville, L.S. Linfoot, M.C. Rivera, J.R. Garey, R.A. Raffe & J.A. Lake. 1997. Evidence for a clade of nematodes, arthropods and other molting animals. Nature 387: 489–493.

Antić, D.Ž., T. Rađa & S.E. Makarov. 2017. Dalmatosomatidae, a new monotypic family, and *Dalmatosoma agaricum* gen. et sp. nov. (Diplopoda: Chordeumatida: Craspedosomatidea) from Croatia, Balkan Peninsula. Zootaxa 4403: 289–306.

Geoffry, J.-J. 2015. Subphylum Myriopoda, Class Diplopoda. Pages 661–669. *in*: J.H. Thorp & D.C. Rogers (eds.) Ecology and General Biology, in Thorp and Covich's Freshwater Invertebrates, Vol. I, Fourth ed., Academic Press, Boston.

Scholtz, G. 2002. The Articulata Hypothesis – or what is a segment? Organisms, Diversity & Evolution 2: 197–215.

Arachnida (Introduction)

ARACHNIDA: SUBCLASSES

Arachnids are a large group of mainly terrestrial organisms classified in 15 extant orders (see Proctor et al., 2015). Representatives of four of these orders, Araneae (true spiders) and three orders of the subclass Acari (mites)—namely Mesostigmata, Sarcoptiformes, and Trombidiformes—are the only chelicerates habitually found in and around freshwater habitats. Among these, members of only a few groups of Sarcoptiformes and Trombidiformes are truly aquatic. Phylogenetic relationships of arachnid orders are not clearly understood, and the classification is undergoing substantial revision based on interpretation of morphological, behavioral, and genetic data. Here we follow the classification and morphological terminology outlined by Proctor et al. (2015) in Volume 1 of this series and Krantz & Walter (2009).

1 Body less conspicuously divided into gnathosoma (capitulum) and idiosoma, spinnerets absent, mites .... **Acari [p. 525]**

1′ Body conspicuously divided into cephalothorax and abdomen, spinnerets present, spiders .... **Araneae [p. 546]**

ARACHNIDA: ACARI: ORDERS AND SUBORDERS[1]

1 Chelicerae typically chelate and dentate; pedipalps simple; prodorsum usually with a pair of well-defined bothridia bearing distinctively modified bothridial setae; opisthosoma usually bearing paired lateral glands .... Order Sarcoptiformes, suborder **Oribatida [p. 535]**

1′ Chelicerae with fixed digit regressed and movable digit a hook or stylifrom structure; pedipalps simple or with distal articles variously modified; prodorsum lacking well-defined bothridia and with sensillae setiform; opisthosoma lacking paired lateral glands.................... .... Order Trombidiformes, Suborder **Prostigmata** [Only Halacaroidea is treated here]

REFERENCES

Krantz, G.W. & Walter, D.E. 2009. A Manual of Acarology, Third ed. Texas Tech University Press, Lubbock, TX.

Proctor, H.C., I.M. Smith, D.R. Cook & B.P. Smith. 2015. Subphylum Chelicerata, Class Arachnida. Pages 599–660 *in*: J.H. Thorp & D.C. Rogers (eds.) Vol. I: Ecology and General Biology, in Thorp and Covich's Freshwater Invertebrates, Fourth ed., Academic Press, Boston, MA.

1 Note from the editors: We were unable to locate any authors willing to treat Neotropical mites as a whole while following the guidelines for our volumes. However, we are eternally grateful to Natalia A. Fredes, Pablo A. Martínez, Almir Rogério Pepato, and Pedro Henrique da Silva Conceição for their production of keys to the Orbatida and Halacariodea.

Acari: Oribatida

Natalia A. Fredes and Pablo A. Martínez

Departamento de Biología, Universidad Nacional de Mar del Plata, Mar del Plata, Buenos Aires, Argentina

INTRODUCTION

Oribatid mites are one of the most diversified microarthropod groups, with a diversity estimated at more than 110,000 species (Walter & Proctor, 2013), of which only 9% are known. They are present in a variety of habitats including frozen and tropical environments, caves, deserts, suspended soils, and freshwater and marine habitats (Walter & Proctor, 2013).

Only 1% are truly aquatic (completing their life cycle in water) (Schatz & Behan-Pelletier, 2008). Most genera belong to the cohorts Nothrina (40%) and Brachypylina (60%) (Behan-Pelletier & Eamer, 2007). The actual knowledge of aquatic oribatids in Neotropical and Antarctic regions is considered poor; Schatz & Behan-Pelletier (2008) reported members of Trhypochthoniellidae, Tegeocranellidae, Hydrozetidae, Limnozetidae, and Zetomimidae, while Pugh & Dartnall (1994) found in freshwater Antarctic environments a few members of Camisiidae, Ceratozetidae, Malaconothridae, Metrioppiidae, Oppiidae, Podacaridae, and Galumnidae.

Oversaturated soils such as those on the edge of rivers and lakes, peatlands, and temporary overflow soils also contain an aquatic and semiaquatic oribatid community. From these environments, members of Malaconothridae, Haplozetidae, Crotoniidae, Nothridae, Eremobelbidae, and Euzetidae are reported. Although not truly aquatic, they can survive in high-content water environments and, in some cases, are tolerant of submergence. *Eremobelba foliata* (Hammer, 1958) (Eremobelbidae) and *Rostrozetes ovulum* (Berlese, 1908) (Haplozetidae), studied by Beck (1969), synchronized their life cycle with the Amazonia forest hydroperiods, entering in diapause during the temporary floods and producing eggs after this period. Another taxon, *Trimalaconothrus cajamarcensis* (Hammer, 1961) (Malaconothridae), can tolerate submergence and, probably to avoid predation, retains the eggs throughout embryogenesis until the larval stage. *Euzetes globulus* (Nicolet, 1855) (Euzetidae) is another species associated with oversaturated edges of streams (Fredes & Martínez, 2011) with a reproductive strategy for surviving oversaturated environments. Michael (1882) and Fredes (personal observation) observed dead females containing larvae. This reproductive tactic plus a high synchronization with environment may be a strategy to enhance offspring survival in oversaturated environments.

Oribatid mites associated with emergent plants have been reported by Covarrubias & Mellado (1998) and Villagrán-Mella et al. (2005) from aquatic environments of Chile. Among them, *Trhypochthoniellus longisetus* (Berlese, 1904) [= *T. excavatus* (Willmann 1919)] (Trhypochthoniidae), *Malaconothrus translamellatus* (Hammer, 1958), *M. angulatus* (Hammer, 1958), *Fossonothrus latus* (Hammer, 1962) (Malaconothridae), *Suctobelbella ornatissima* (Hammer, 1958) (Suctobelbidae), and *Williamszetes elsosneadensis* (Hammer, 1958) (Tegoribatidae) were present on floating plants. The mite *Orthogalumna terebrantis* (Wallwork, 1965) (Galumnidae) feeds and reproduces on the water hyacinth *Eichhornia crassipes* (Mart.) (Solms, 1883) (Cordo & DeLoach, 1975, 1976) and its immature stages live in burrows in the leaves. Aquifers and groundwaters, one of the greatest freshwater reservoirs, represent the least explored lotic environment. Taxonomically these environments contain a poor fauna of crustaceans, annelids, and some oribatid mites. Tione et al. (2011), analyzing the community of invertebrates in a loessic phreatic aquifer in Argentina, found members of early-derivate Oribatida, which were identified as *Aphelacarus acarinus* Berlese, 1910 (Palaeosomata, Aphelacaridae) and *Haplochthonius clavatus* (Hammer, 1958) (Enarthronota, Haplochthoniidae) (Martínez, unpublished).

LIMITATIONS

Throughout the present bibliographic revision of freshwater oribatid mites from Neotropical and Antarctic regions, we found a variety of taxonomical resolutions, ranging from the sole mention of "Oribatida" to subspecies level records. In this work we only consider the records that appear at the family level and at levels below that.

TERMINOLOGY AND MORPHOLOGY

The following is a brief description of their morphological characters; for more detailed information, see Norton & Behan-Pelletier (2009). The integument is a sclerotized cuticle, with thinner areas, which represent respiratory surfaces or gland secretory areas. Among the glands, a complex of dermal glands called the **octotaxic system** are paired and usually located in the proximity of notogastral setae insertions (Fig. 16.1 A). This system, when present, consists of **porose areas** (Figs. 16.1 D, 4 I, Aa, A1, A2, A3) or, if they are invaginated, of saccules (Fig. 16.1 A Sa, S1, S2, S3, C). The level of sclerotization varies from a hard and dark cuticle, such as Galumnidae and Euzetidae, to a weak and light cuticle, such as Enarthronota. In Palaeosomata the integument generally is not sclerotized, but some taxa may have small dorsal plates, or a wrinkling cuticular region, called the **aesthenic zone** (Fig. 16.4 A, C, black arrows). Sclerotized oribatid mites may be covered

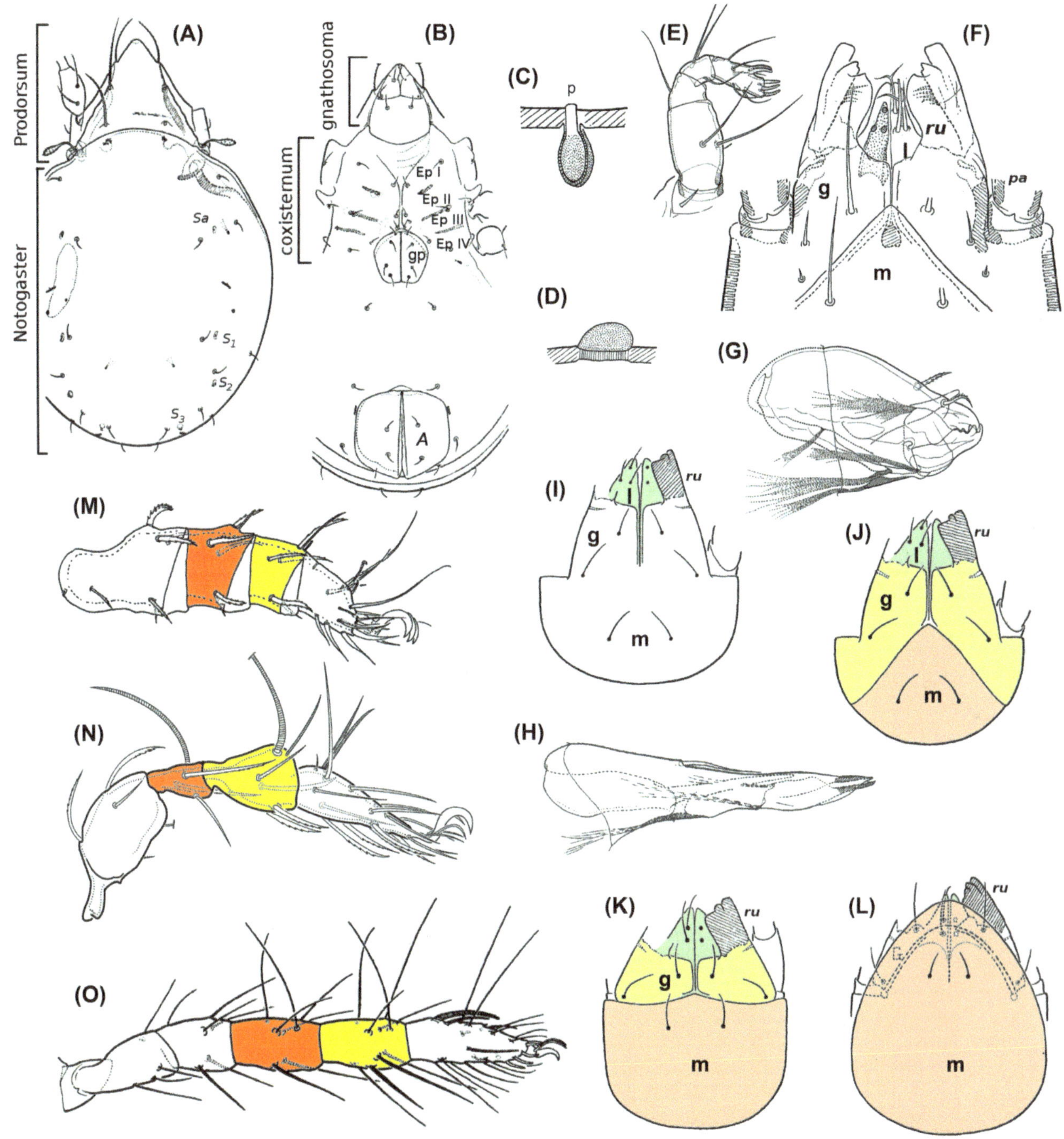

FIGURE 16.1 Morphology of oribatid mites: (A) Dorsal and (B) ventral division of the body; (C) sacculus and (D) porose area in sagittal view; (E) brachypylina palp; (F) infracapitulum; (G) typical chaelate-dentate chaelicera; (H) suctorial chaelicera. Types of infracapituli: (I) anarthric; (J) stenarthric; (K) diarthric; and (L) diarthric modified (*green*, lateral lip; *yellow*, genae; *orange*, mentum; *hatched*, rutellum). Legs of oribatid mites (*yellow*, tibiae; *orange*, genua): (M) Malaconothroidea-type leg; (N) Brachypylina-type leg, (O) Palaeosomata-type leg. Abbreviations: *Ep I–IV*, epimeres; *g*, gena; *gp*, genital plates; *l*, lateral lip; *m*, mentum; *p*, pori; *pa*, palpal insertion; *ru*, rutellum; *S*, saccules). *(A), (B), and (E) from Grandjean (1953); (B) and (C) from Norton et al. (1997); (F) from Granjean (1957); (G) from Grandjean (1947); (H) from Grandjean (1951); (I–L) modified from Grandjean (1957); (M) modified from Weigmann (1997); (N) modified from Fredes & Martínez (2016); and (O) modified from Grandjean (1954).*

with a secretory layer primarily composed of wax, called **cerotegument**. This layer can be thin or thick and structured (in granules, columns, etc.) or amorphous. In some *Hydrozetes* the cerotegument layer is composed of 60% of airspace for gas exchange.

The body of sclerotized oribatids is generally covered by a series of plates, often divided by **scissures** or furrows, and frequently a rooflike plate called the **tectum** protects the articulation. The main body division lies along the sejugal furrow; in dorsal view it divides the anterior region, called the **prodorsum**, from the posterior **notogaster** (Fig. 16.1 A). The prodorsum is an entire plate that generally bears five pairs of setae: **rostral** (ro), **lamellar** (le), **interlamellar** (in), **exobothridial** (ex) and **bothridial** or **sensilla** (ss). In Palaeosomata and some Nanhermanniidae, there is an extra pair of exobothridial setae, totaling six prodorsal setal pairs. Among these, sensilla are the most characteristic setae, emerging from a broad, cup-shaped insertion (the **bothridum**) and may exhibit a variety of shapes, such as setiform, spatulate, ciliate, and clavate, among others (Fig. 16.3 E–H). The prodorsum anterior region, the **rostrum**, frequently projects and curves ventrally to cover the mouthparts (**stegasime condition**). In early derivate oribatids, such as *Aphelacarus*, the prodorsum ends abruptly leaving the mouthparts exposed (**astegasime condition**). On the prodorsum, and particularly in Brachypylina, cuticular folds or laminar expansions may extend generally from the bothridia to the rostrum anterior border. The former, called **costulae** (Fig. 16.2 F), are common in members of the Oppiidae. The second ones, called **lamellae** (Fig. 16.2 B–D), are very common in Ceratozetidae, Limnozetidae, and Euzetidae; with the shape varying from a narrow projection, such as in *Hydrozetes* and *Euzetes*, to a wide and fully developed lamella that can cover all the prodorsum such as in members of Tegoribatidae (Fig. 16.3 C). Transverse ridges, called the transcostulae and translamellae may distally connect the costulae and the lamellae, respectively. In the majority of Galumnidae, the prodorsum has delicate ridges called lamellar and sublamellar lines (Figs. 16.2 A, 4 Q, 7 I). Finally, the majority of Suctobelbidae has a singular prodorsum ornamented with lateral flat regions called **tectopedial fields** and an unpaired knobbed tubercle (Fig. 16.2 E).

The **notogaster** is a plate that covers the body dorsally and laterally, behind the sejugal furrow. Frequently it is an entire plate, but in the early derivate infraorder Enarthronota is often subdivided into two (Fig. 16.5 D) or more plates (Fig. 16.7 K) yielding transverse articulation denominated **scissures**. In the notogaster are a variable number of setae, frequently arranged in more or less discernible rows. These rows are coded conventionally as **c**, **d**, **e**, **f**, **h**, and **p** from sejugal furrow to the posterior end of notogaster (Figs. 16.4 D, 5 D). In members of cohort Brachypylina, this plate may have lateral, cuticular expansions called pteromorphs, variable in shape and degree of articulation with the notogaster (Figs. 16.3 C, 4 O, 5 M, 6 F). The notogaster presents a series of slitted mechanoreceptor organs called lyrifissures or cupules; and in many Brachypylina, an octotaxic system in the form of porose areas (Figs. 16.4 I, K, 7 I) or saccules (Figs. 16.1 A, 6 J) is present. Other important structures are the **opisthonotal glands** (= opisthosomal glands) located in the midlateral region of the notogaster. Frequently, this structure is inconspicuous, but in some Parhypochthoniidae (Fig. 16.5 A) and Crotoniidae, is clearly visible. Some Limnozetoidea and Cymbaeremaeoidea may have a light area on the notogaster anterior region, called a **lenticulus** (Figs. 16.4 F, 6 H). This structure is a photoreceptor organ, generally consisting of a cuticular cornea, receptors, pigment cells, and fat body cells (Alberti & Coons, 1999). In derivative Brachypylina, notogaster could have a posterior tectum that protects circumgastric scissure (Fig. 16.3 D).

In ventral view, the body of sclerotized taxa consists of an anterior region where the mouthparts (gnathosoma) are held inside a cavity, the **camerostoma** (Fig. 16.1 B). The gnathosoma includes a pair of **chelicerae** and **palpi** (Fig. 16.1 E, G) and an unpaired structure called **subcapitulum** (Fig. 16.1 F). Chelicerae are modified appendages composed of two segments, generally ending in a dentate chela (Fig. 16.1 G). In some taxa, the chelicerae are slender and elongate, and the chelae loose their dens (Fig. 16.1 H). This adaptation, very common in Suctobelbidae, also may be coupled with modifications in the rostrum shape and other gnathosoma parts. The subcapitulum usually has a proximal base, the mentum, paired anterior lobes (genae) and, a pair of lateral lips (Fig. 16.1 I). Based on the nature of the labiogenal articulation, three basic subcapitular forms are recognized: **anarthric**, where labiogenal articulation is absent (typical in many Enarthronota) (Fig. 16.1 I); **stenarthric**, where there is a V-shaped labiogenal articulation that runs obliquely from a central point in the midline to each posterolateral corner (typical in Eniochthoniidae and Gehypochthoniidae) (Fig. 16.1 J) and **diarthric**, where the labiogenal articulation runs transversely at a level just posterior to the palpi (typical of most Brachypylina) (Fig. 16.1 K). Diarthric subcapitula may have a mental tectum that projects forward and covers the labiogenal articulation (Fig. 16.1 L) (e.g., Galumnidae and Euzetidae). Distally on each gena there is a modified seta, the **rutellum** (Fig. 16.1 F). The rutella may be setiform (in some Enarthronota), but in most species they are enlarged with teeth and cutting blades.

Behind the camerostome is the coxisternal region or simply **coxisternum**. This plate, which forms the floor of the podosoma and supports the legs, is composed of four **epimera**, conventionally called I–IV (Fig. 16.1 B). The coxisternum typically has a setation of 3-1-3-3 setae pairs

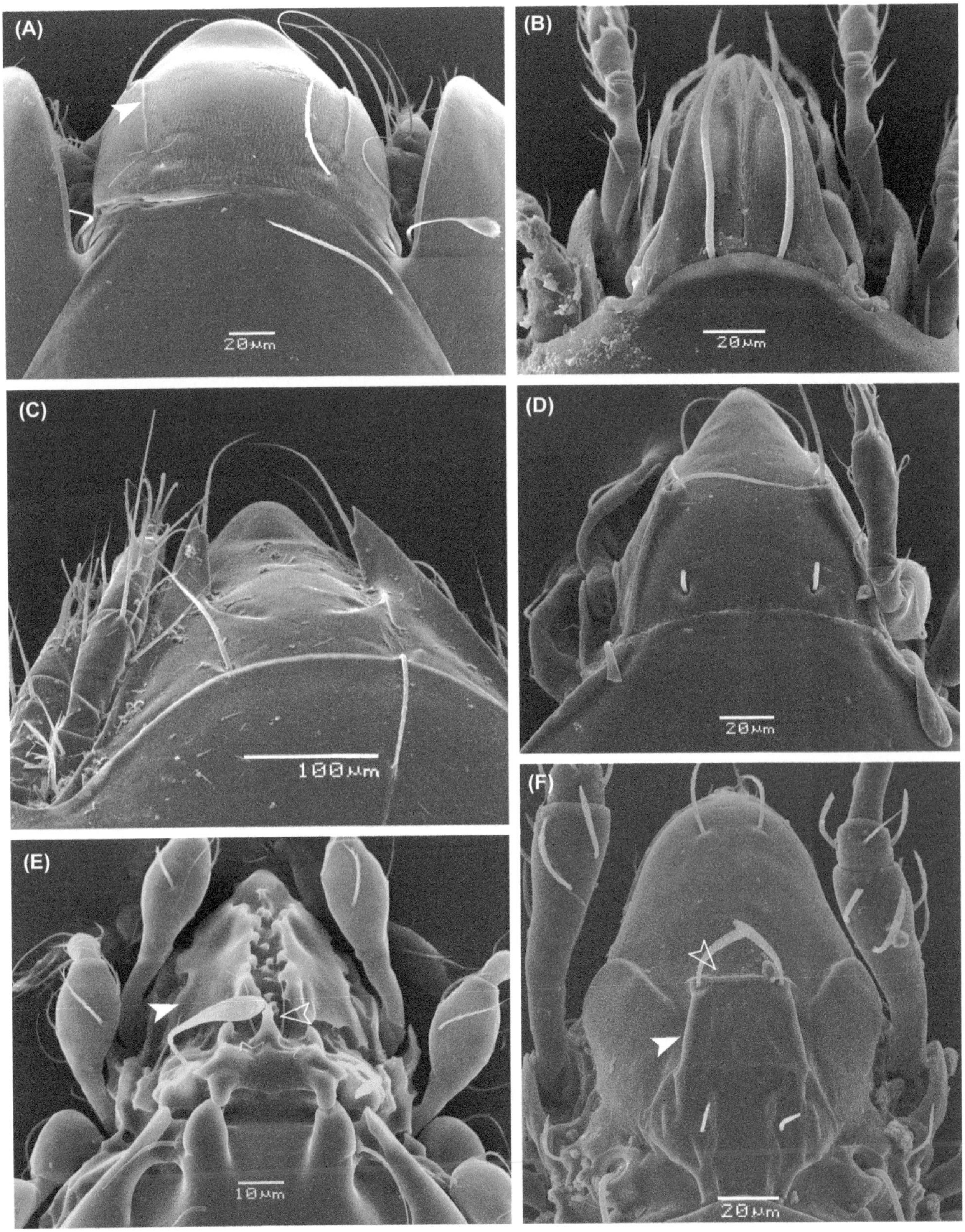

FIGURE 16.2 Prodorsal structures of oribatid mites: (A) Lamellar lines (*Galumna*, Galumnidae) (*white arrow*, line L). (B–D) Types of lamellae: (B) *Tectoribates* (Tegoribatidae); (C) *Euzetes* (Euzetidae); and (D) *Hemileius* (Oribatuloidea); (E) Tectal fields (*white arrow*) and knoblike tubercle (*outline arrow*) (*Suctobelbella*, Suctobelbidae); (F) Costulae (*white arrow*) and transcostulae (*outline arrow*) (*Karenella*, Oppiidae).

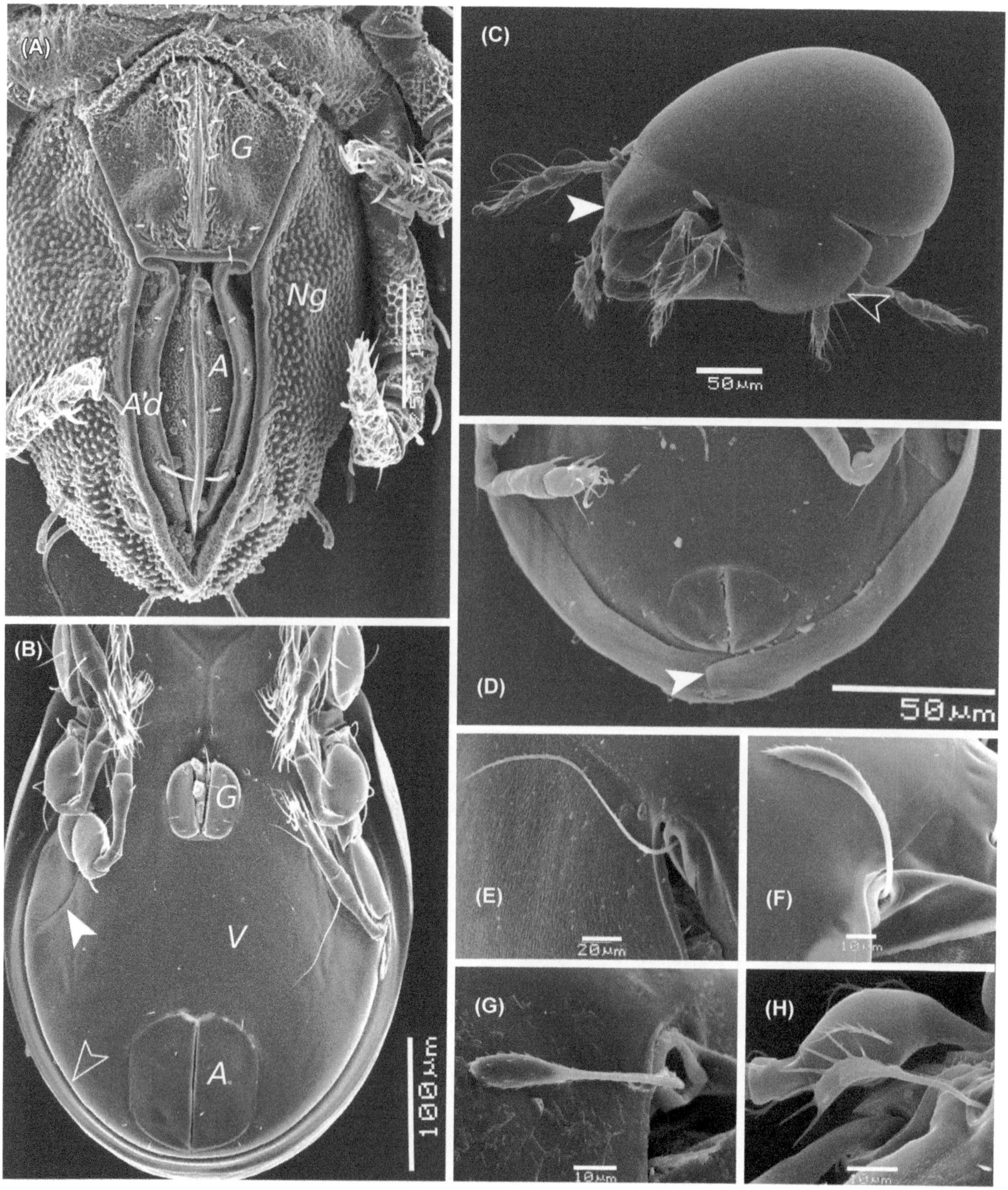

FIGURE 16.3 Morphology of oribatid mites: (A) Macropylina condition (*Nothrus*, Nothridae); (B) Brachypylina condition (*Scheloribates*, Oribatuloidea) (*white arrow*, circumpedal carina; *outline arrow*, circumgastric scissure); (C) *Physobates* (Tegoribatidae) (*white arrow*, lamellae completely fused; *outline arrow*, pteromorph); (D) posterior notogastral tectum (*white arrow*) (*Paralamellobates*, Ceratozetoidea). (E–H) Sensilla types: (E) setiform (*Galumna*, Galumnidae); (F) fusiform (*Scheloribates*, Oribatuloidea); (G) clavate (*Mancoribates*, Oribatuloidea); (H) pectinate (*Oxyoppia*, Oppiidae). Abbreviations: *A*, anal plates; *Ad*, adanal plates; *G*, genital plates; *Ng*, notogaster; *V*, ventral plate.

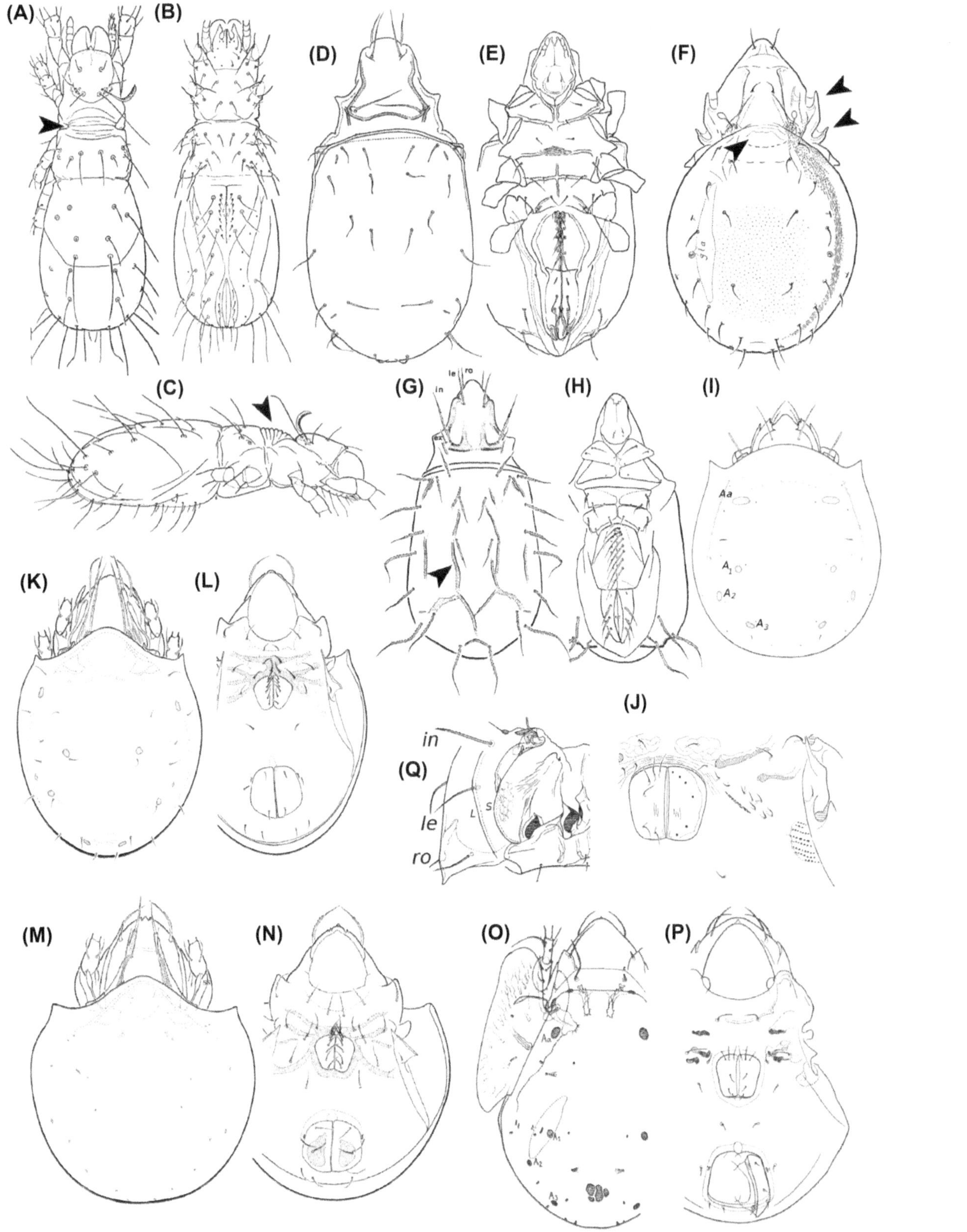

FIGURE 16.4 Oribatid mites: (A–C) *Aphelacarus acarinus* (Palaeacaridae) in dorsal, ventral, and lateral view (*black arrows*, aesthenic zone); (D) and (E) *Malaconothrus japonicus* (Malaconothridae) in dorsal and ventral view; (F) *Hydrozetes incisus* (Hydrozetidae) in dorsal view (*left black arrow*, lenticulus; *right black arrows*, pedotecta); (G) and (H) *Trimalaconothrus yunnanensis* (Malaconothridae) in dorsal (*black arrow*, longitudinal ridge) and ventral view; (I) *Euzetes globulus* (Euzetidae) in dorsal view; (J) detail of epimeric region of *Euzetes*; (K) and (L) *Zetomimus francisi* (Zetomimidae) in dorsal and ventral view; (M) and (N) *Heterozetes aquaticus* (Zetomimidae) in dorsal and ventral view, (O–Q) *Galumna alata* (Galumnidae) in dorsal, ventral, and lateral view. Abbreviations: *A*, porose areae; *in*, interlamellar seta; *L*, lamellar line; *le*, lamellar seta; *ro*, rostral seta, *S*, sublamellar line. *(A–C) from Grandjean (1954); (D) and (E) from Aoki (1966); (F) from Grandjean (1948); (G) and (H) from Yamamoto & Aoki (1998); (I) from Pérez-Iñigo (1993); (J) from Grandjean (1970); (K–N) from Behan-Pelletier & Eamer (2003a); (O–Q) from Grandjean (1936).*

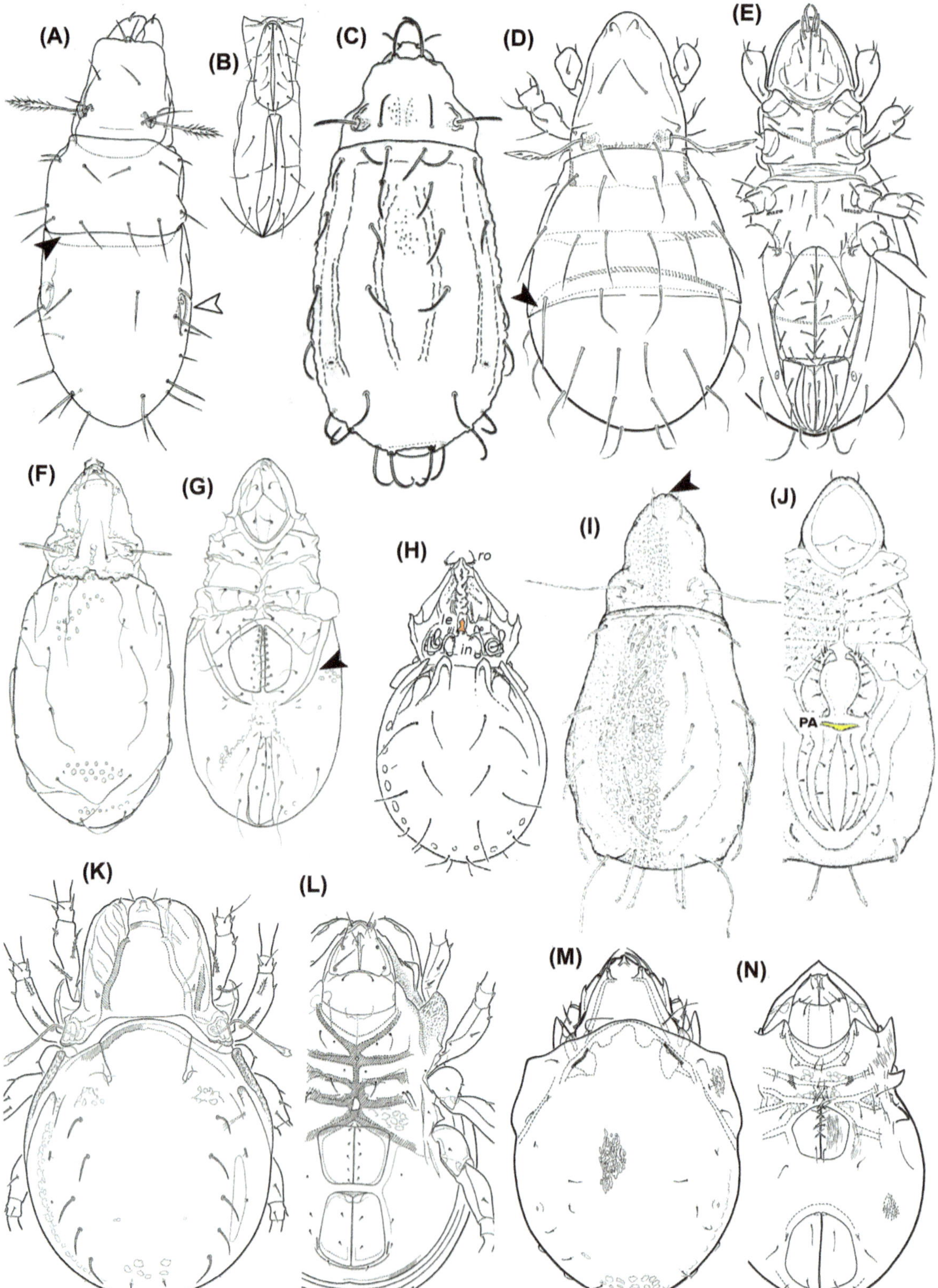

FIGURE 16.5 Oribatid mites: (A) and (B) *Gehypochthonius japonicus* (Parhypochthoniidae) in dorsal view and detail of anogenital region, respectively (*black arrow*, transversal scissure; *outline arrow*, opisthonotal gland); (C) *Platynothrus robustior* (Camisiidae) in dorsal view; (D) and (E) *Eniochthonius mahunkai* (Eniochthoniidae) in dorsal and ventral view (*black arrow*, transversal scissure); (F) and (G) *Nanhermannia bifurcata* (Nanhermanniidae) in dorsal and ventral view (*black arrow*, crescentiform scissure); (H) *Suctobelbella ornatissima* (Suctobelbidae) in dorsal view (*in*, interlamelar seta; *le*, lamellar seta, *ro*, rostral seta; *in orange*, knoblike tubercule); (I) and (J) *Nothrus silvestris* (Nothridae) in dorsal (*black arrow*, rostral incision) and ventral view (*PA*, preanal plate); (K) and (L) *Tegeocranellus laevis* (Tegeocranellidae) in dorsal and ventral view; (M) and (N) *Limnozetes borealis* (Limnozetidae) in dorsal and ventral view. *(A) and (B) from Aoki (1975); (C) from Mahunka (1980); (D) and (E) from Norton & Behan-Pelletier (2007); (F) and (G) from Fujikawa (1990); (H) from Hammer (1958); (I) and (J) from Olszanowski (1996); (K) and (L) from Grandjean (1962); (M) and (N) from Behan-Pelletier (1989).*

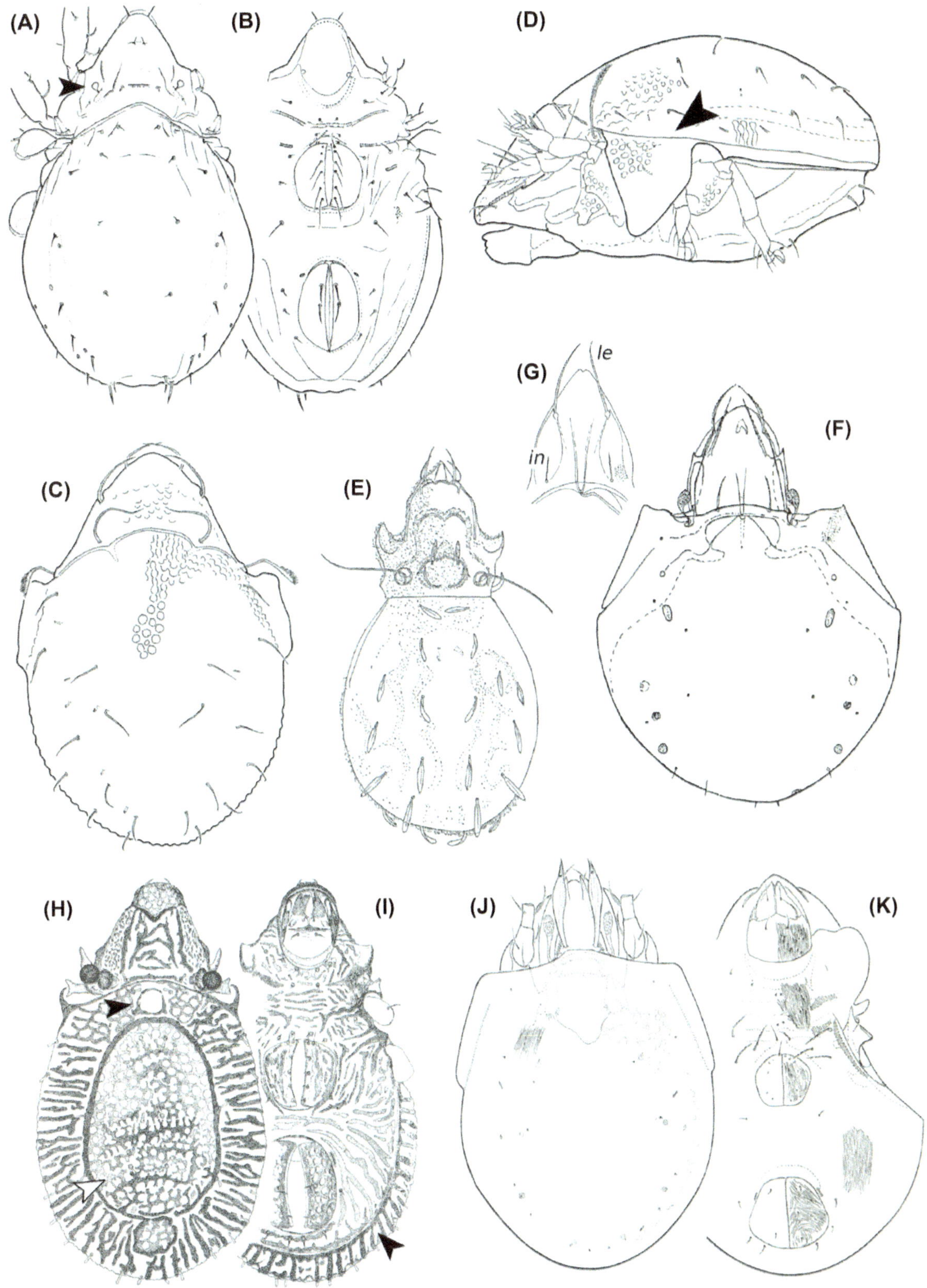

FIGURE 16.6 Oribatid mites: (A) and (B) *Podacarus auberti* (Ameronothridae) in dorsal and ventral view (*black arrow*, sensillus); (C) and (D) *Rostrozetes ovulum* (Haplozetidae) in dorsal and lateral view (*black arrow*, pteromorph articulation); (E) *Eremobelba foliata* (Eremobelbidae) in dorsal view; (F) and (G) *Williamszetes elsosneadensis* (Tegoribatidae) dorsal view and detail of prodorsum (*le*, lamellar seta; *in*, interlamellar seta); (H) and (I) *Scapheremaeus corniger* in dorsal (*black arrow*, lenticulus; *white arrow*, centrodorsal plate) and ventral view (*black arrow*, circumnotogastral plate); (J) and (K) *Pelopsis baloghi* (Punctoribatidae) in dorsal and ventral view. *(A) and (B) from Grandjean (1955); (C) and (D) from Beck (1965); (E–G) from Hammer (1958); (H) and (I) from Pérez-Iñigo & Subías (1974); (L) and (M) from Behan-Pelletier & Eamer (2003b).*

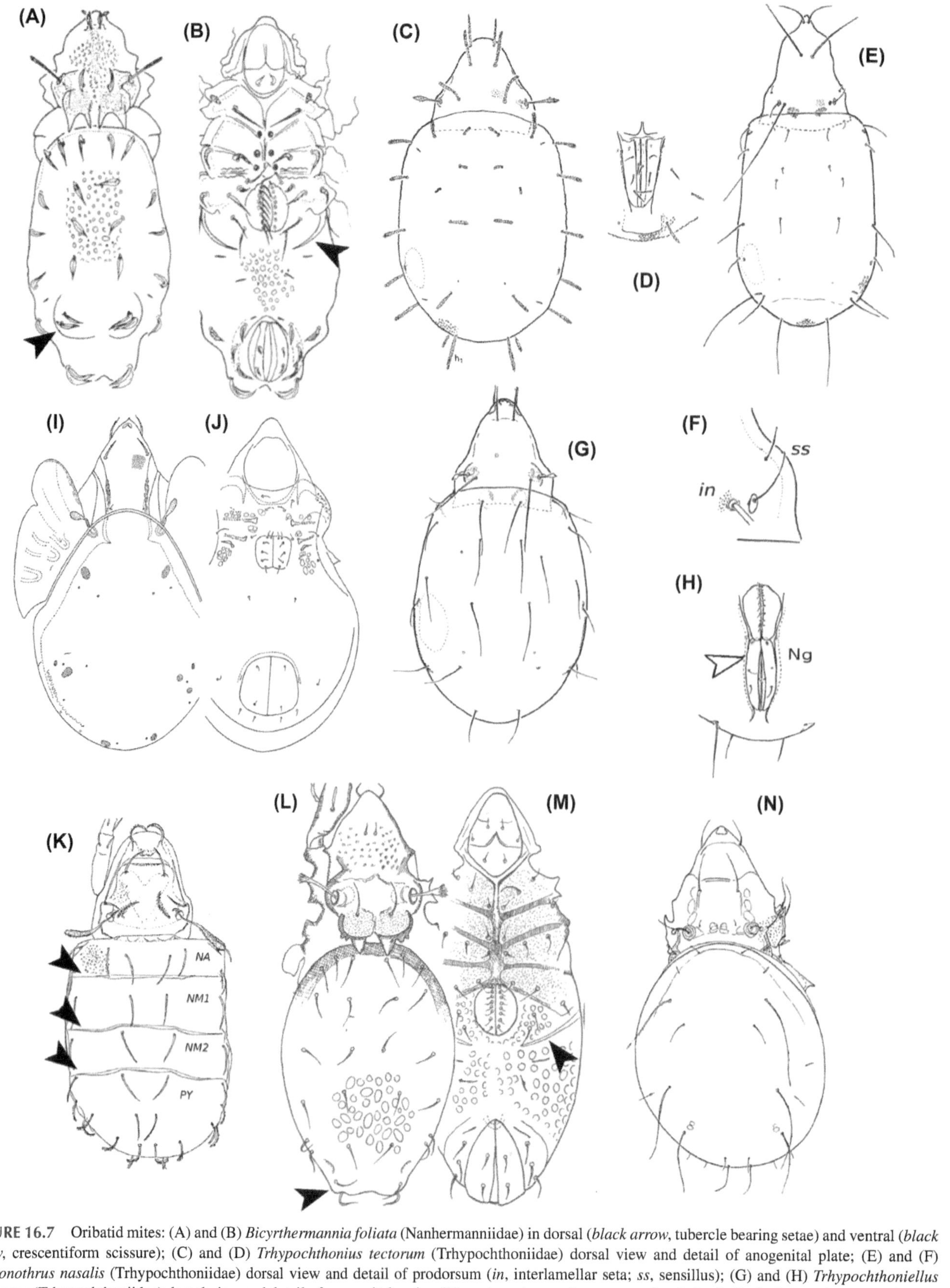

FIGURE 16.7 Oribatid mites: (A) and (B) *Bicyrthermannia foliata* (Nanhermanniidae) in dorsal (*black arrow*, tubercle bearing setae) and ventral (*black arrow*, crescentiform scissure); (C) and (D) *Trhypochthonius tectorum* (Trhypochthoniidae) dorsal view and detail of anogenital plate; (E) and (F) *Mucronothrus nasalis* (Trhypochthoniidae) dorsal view and detail of prodorsum (*in*, interlamellar seta; *ss*, sensillus); (G) and (H) *Trhypochthoniellus longisetus* (Trhypochthoniidae) dorsal view and detail of anogenital region (*Ng*, notogaster; *outline arrow*, adanal plate); (I–J) *Orthogalumna terebrantis* (Galumnidae) dorsal and ventral view; (K) *Haplochthonius clavatus* in dorsal view (*black arrows*, scissures; *NA*, *NM1*, *NM2*, *PY*: notogastral plates); (L) and (M) *Cyrthermannia baloghorum* (Nanhermanniidae) in dorsal (*black arrows*, tubercle bearing setae) and ventral view (*black arrows*, crescentiform scissure); (N) *Globoppia intermedia* (Oppiidae) dorsal view. *(A) and (B) from Mahunka (1985); (C–H) from Weigmann (1997); (I–J) from Wallwork (1965); (K) from Hammer (1958); (L) and (M) from Pérez-Iñigo & Baggio (1988); (N) from Hammer (1962).*

from epimere I to IV. In some members of Cymberemaeidae and Ceratozetoidea, the coxisternum are flanked by a spine—or carinate tectum, the **discidium**, that frequently ends in a sharp projection, the **custodium**. Protecting from the back the inserts of the legs I and II there are cuticular expansions called **pedotecta** (Fig. 16.4 F).

Finally, behind the coxisternum, the posterior region is composed of several ventral plates that vary in degree of fusion. According to the arrangement of these ventral plates, two conditions occur: **macropylina** and **brachypylina**. In macropylina, the venter is formed by the notogaster that folds ventrally and connects with **aggenital** and **adanal** plates (Fig. 16.3 A). In this condition, genital and anal apertures are generally closer to each other. In brachypylina condition, there is an expansion and fusion of both aggenital and adanal plates delimiting a ventral plate where genital and anal plates are embedded (Fig. 16.3 B). Generally, in the brachypylina condition the dorsal notogaster and ventral plates are separated by a saggital articulation called **circumgastric scissure** (Fig. 16.3 B). One exception is Nanhermanniidae which has a variation of the brachypylina condition: the venter is formed by the expansion and fusion of notogaster and aggenital and adanal plates (Figs. 16.5 G, 7 B, M). Other common structures frequently found in the brachypylina ventral region are paired circumpedal carina that curve anteromedially from the ventral plate edge, running behind leg IV and forward across epimera IV and III (Fig. 16.3 B).

Adult Palaeosomata legs have six free articles, while the remaining oribatid mites typically have five articles: **trochanter**, **femur** (divided in Palaeosomata), **genu**, **tibia**, and **tarsus** (Fig. 16.1 N). In the early derivate oribatids, Palaeosomata and Enarthronota, all articles are quite similar in shape (Fig. 16.1 M, O), in Brachypylina there is a level of specialization, mainly in genua and tibiae (Fig. 16.1 N). Ventral **carinae** are frequent in femora and genua. In some taxa, joints between segments have a protective tectum denominated **retrotecta**. The legs end in segments, called the **pretarsus**. Frequently the pretarsus ends in one (monodactylous) or three **claws** (tridactylous); bidactyly is very rare.

KEYS TO ORIBATIDA

Arachnida: Acari: Oribatida: Families

1 Legs with five free articles, femora undivided; prodorsum with or without rostral tectum; bothridial setae diverse, smooth or barbed; body usually tan or brown, with noticeable sclerotization, rarely white or brightly colored; epimere II usually with one pair of setae 2

1′ Legs with six free articles, femora divided (Fig. 16.1 O); prodorsum astegasime, with no rostral tectum; bothridium and base of bothridial setae simple, never barbed, ciliate, or pectinate; body white; coxisternum II with three pairs of setae (Fig. 16.4 A–C) .... Aphelacaridae, one species: *Aphelacarus acarinus* (Berlese, 1910) [Semicosmopolitan]

2(1) All legs with genua similar to tibiae in size and shape, genua not geniculate (Fig. 16.1 M); venter of macropyline type (Fig. 16.3 A) with paired aggenital and adanal plates often distinguishable; coxisternal region often transversely divided by sejugal articulation and usually not fused to aggenital region 3

2′ All legs with genua noticeably shorter than tibiae, differently shaped, geniculate (Fig. 16.1 N); venter of the brachypylina type (Fig. 16.3 B) with an unified rigid plate in which only the subcapitulum and the paired genital and anal plates are distinct and movable. 10

3(2) Notogaster subdivided transversely by a single complete scissure (Fig. 16.5 A, D) 4

3′ Notogaster subdivided by three scissures (Fig. 16.7 K) or not subdivided transversely 5

4(3) Prodorsum without rostral tectum; setal rows *d* and *e* separated by a transverse scissure; pronotaspis with six pairs of setae, without sulcus; opisthonotal gland present (Fig. 16.5 A); legs bi- or tridactylous, with well-developed lateral claws, with or without reduced, hamulate empodial claw (Fig. 16.5 A, B) .... Gehypochthoniidae, one genus: *Gehypochthonius* [Semicosmopolitan]

4′ Prodorsum with rostral tectum, with a transverse scissure posterior to setal row *e*; pronotaspis with eight pairs of setae and a transverse, medially incomplete sulcus at level of setal row *d*; opisthonotal gland absent; legs monodactylous (Fig. 16.5 D, E) .... Eniochthoniidae, one genus: *Eniochthonius* [Cosmopolitan]

5(3) Notogaster not subdivided transversely 6

5′ Notogaster with three transverse scissures, delimiting four plates (NA, NM1, NM2, PY); NA with four pairs of setae (row *c* only); genital plates twice the size of anal plates; aggenital setae absent, anal plates with four pairs of setae; preanal plate and seta absent (Fig. 16.7 K) .... Haplochthoniidae, one species: *Haplochthonius clavatus* (Hammer, 1958) [Argentina. India. Spain]

6(5) Body not cylindrical, often globose or dorsally flattened; prodorsum without posteriorly directed tubercles; notogaster distinct from ventral region throughout its length; ventral plate presents, adanal plates distinct, genital and anal plates adjacent or only narrowly separated 7

6′ Body cylindrical, often elongated; prodorsum posteriorly with paired tooth or carinate tubercles projecting over dorsosejugal groove; notogaster completely fused to ventral region, without ventral plate; ventral aspect such as Brachypylina type, with a pair of crescentiform scissures between genital and anal plates (Figs. 16.5 F, G, 7 A, B) .... **Nanhermanniidae [p. 537]**
[Cosmopolitan]

7(6) Rostrum without medial incision; epimere II usually with one pair of setae; genital plate with 4–24 pairs of setae, but all near medial edge; aggenital setae present or absent; preanal plate present or absent; integument various, rarely foveolate 8

7′ Rostrum with short medial incision; epimere II with three or more pairs of setae; genital plate with nine pairs of setae, one posterior seta far lateral of others, away from medial edge of plate; aggenital setae absent; preanal plate distinct; integument distinctly foveolate (Fig. 16.5 I, J) .... Nothridae, one genus: *Nothrus*
[Cosmopolitan]

8(7) Genital plate without paraxial carina; anal plates much narrower than adanal plate, sometimes inconspicuous; aggenital setae absent 9

8′ Genital plate with a distinct paraxial carina bearing all setae; anal plate wider or only slightly narrower than adanal plate, clearly not greatly reduced; aggenital setae present (Fig. 16.5 C) .... Camisiidae, one species: *Platynothrus robustior* Berlese, 1916
[Argentina. Uruguay]

9(8) Sensilla and bothridium absent; gnathosoma with subcapitulum anarthric (Fig. 16.1 I); cuticle of notogaster not imbricate; cerotegument present in form of large, thin plates that appear waxy, without excrescences, strongly birefringent in polarized light (Fig. 16.4 D, E, G, H) .... **Malaconothridae [p. 537]**

9′ Sensilla present, usually with well-developed bothridium (vestigial in some *Trhypochthoniellus*); gnathosoma with subcapitulum diarthric (Fig. 16.1 K), cuticle of notogaster at least partly imbricate; cerotegument, if present, of different form, not birefringent (Fig. 16.7 C–H) .... **Trhypochthoniidae [p. 537]**

10(2) Notogaster without octotaxic system; pteromorphs absent; discidium and circumpedal carina usually absent 11

10′ Notogaster with octotaxic system (Fig. 16.1 A); pteromorphs present or absent; discidium and circumpedal carina usually present (Fig. 16.3 B) 18

11(10) Prodorsum without tectopedial fields or knoblike costulae; setae *ro* curved, not geniculate; gnathosoma not above; chelicerae chelate-dentate (Fig. 16.1 G) 12

11′ Prodorsum with tectopedial fields and short costulae or knobbed tubercle (Fig. 16.2 E); setae *ro* geniculate; gnathosoma modified, subcapitulum anarthric (Fig. 16.1 I), rutella forming tube; chelicerae attenuate-edentate (Figs. 16.1 H, 5 H) .... Suctobelbidae, one species: *Suctobelbella ornatissima* (Hammer, 1958)
[Neotropical]

12(11) Adanal neotrichy absent; tibiae and tarsi I–IV without retrotecta, femora I–IV and trochanters III–IV with porose areas 13

12′ Adanal neotrichy present (more than three setal pairs); tibiae and tarsi I–IV with retrotecta, femora I–IV and trochanters III–IV with saccules (Fig. 16.6 E) .... Eremobelbidae, one species: *Eremobelba foliata* (Hammer, 1958)
[Neotropical]

13(12) Integument well sclerotized; sensilla reduced or normally developed; coxisternal setation 3-1-3-3 14

13′ Integument weakly sclerotized and easily deformed; sensilla very small or absent (Fig. 16.6 A); coxisternal setation 3-1-2-2, 3-1-2-3, or 3-1-3-2 (Fig. 16.6 A, B) .... Ameronothridae, one species: *Podacarus auberti* Grandjean, 1955
[Antarctic]

14(13) Notogaster with lenticulus (Figs. 16.4 F, 6 H) 15

14′ Notogaster without lenticulus 16

15(14) Prodorsum normally without setae *in*; costulae generally present; notogaster normally with an oval centrodorsal plate encompassed by a circumnotogastral plate; rough sculpture usually presents; epimera III–IV not distinctly delineated; border of epimere III not visible (Fig. 16.6 H, I) .... Cymbaeremaeidae, one genus: *Scapheremaeus*
[Cosmopolitan, except Antarctic]

15′ Prodorsum with setae *in*; short lamellae and tutorial ridges present; notogaster without centrodorsal plate; epimeral border III visible (Fig. 16.4 F) .... Hydrozetidae, one genus: *Hydrozetes*
[Cosmopolitan, except Antarctic]

16(14) Integument of genital area same color as anal region; well-developed lamellae present or absent 17

16′ Integument of genital area usually darker than surrounding area; well-developed lamellae present (Fig. 16.5 K, L) .... Tegeocranellidae, one genus: *Tegeocranellus*
[Cosmopolitan, except Antarctic]

17(16) Prodorsum with lamellae and often with translamella .... Peloppiidae, one species: *Macquarioppia striata* (Wallwork, 1963)
[Antarctic]

17′ Prodorsum without lamellae; costulae present or absent (Fig. 16.7 N) .... **Oppiidae [p. 538]**

18(10) Pteromorphs, movable or not, never auriculate; lamellae generally present 19

18′ Pteromorphs movable, auriculate; lamellae and translamella absent; lamellar lines usually present (Figs. 16.2 A, 4 Q, 7 I); dark brown specimens (Figs. 16.4 O–Q, 7 I, J) .... **Galumnidae [p. 538]**

19(18) Sensilla well developed; bothridium uncovered 20

19′ Sensilla weakly developed; bothridium covered by a dorsal expansion; notogaster with immovable pteromorphs (Fig. 16.5 M, N) .... Limnozetidae, one genus: *Limnozetes* [Semicosmopolitan]

20(19) Lamellae, if present, touching medially (Fig. 16.2 B) or separate (Fig. 16.2 C, D) 21

20′ Lamellae completely fused medially (Figs. 16.3 C, 6 F, G) .... Tegoribatidae, one species: *Williamszetes elsosneadensis* (Hammer, 1958) [Neotropical]

21(20) Pedotectum I well developed as large lamina; notogaster with immovable pteromorphs 22

21′ Pedotectum I most weakly developed as small lamina; pteromorphs movable, with hinge (Fig. 16.6 C, D) .... Haplozetidae: one species: *Rostrozetes ovulum* (Berlese, 1908) [Pantropical, subtropical]

22(21) Pretarsi I and IV with same number of claws 23

22′ Pretarsi I and IV usually with different number of claws (Fig. 16.4 J, K) .... **Zetomimidae [p. 538]**

23(22) Notogaster without posterior tectum 24

23′ Notogaster with posterior tectum (Figs. 16.3 D, 6 J, K) .... Punctoribatidae, one genus: *Pelopsis* [Nearctic, Neotropical]

24(23) Infracapitulum diarthric modified (mentum with tectum, Fig. 16.1 L); epimeric neotrichy presents; dark brown specimens (Fig. 16.4 I, J) .... Euzetidae, one species: *Euzetes globulus* (Nicolet, 1855) [Neotropical, Palaearctic]

24′ Infracapitulum diarthric not modified (without tectum, Fig. 16.1 K); epimeric neotrichy absent; rostrum wide with three apices .... Ceratozetidae, one genus: *Magellozetes* [Antarctic, South America]

Arachnida: Acari: Oribatida: Nanhermanniidae: Genera

1 Notogaster with blunt tubercles bearing part of notogastral setae (Fig. 16.7 A, L) 2

1′ Notogaster without blunt tubercles; notogastral setae long, setiform, or slightly dilated (Fig. 16.5 F, G) .... *Nanhermannia* [Cosmopolitan]

2(1) Posterior protuberance of prodorsum on each side with two spiniform processes (Fig. 16.7 A, B) .... *Bicyrthermannia* [Neotropical, Oriental]

2′ Posterior protuberance of prodorsum with a quadridentate crest on each side (Fig. 16.7 L, M) .... *Cyrthermannia* [Pantropical]

Arachnida: Acari: Oribatida: Malaconothridae: Genera

1 Legs monodactylous; leg tarsi short, broad; notogaster without longitudinal ridges (Fig. 16.4 D, E) .... *Malaconothrus* [Cosmopolitan]

1′ Legs tridactylous; leg tarsi elongated, gradually narrowed distally; notogaster with two longitudinal ridges (Fig. 16.4 G, H) .... *Trimalaconothrus* [Cosmopolitan]

Arachnida: Acari: Oribatida: Trhypochthoniidae: Genera

1 Rostral setae well separated, mutual distance usually equal to or greater than that of lamellar setae (Fig. 16.7 C, G); rostrum not excised laterally to form an anterior mucro; sensillus and bothridium usually well formed (rarely regressed); epimere III with three pairs of setae; legs tridactylous 2

1′ Rostral setae closely adjacent, inserted on anterior mucro formed by laterally excised rostrum (Fig. 16.7 E); sensillus small, setiform; bothridium simple, not cuplike, and located on small, disciform elevation; epimere III with two pairs of setae; legs monodactylous (Fig. 16.7 E, F) .... *Mucronothrus* [Neotropical, Australian, and Holarctic regions]

2(1) Subcapitulum stenarthric (Fig. 16.1 J); anal plates well formed, only slightly narrower than adanal plates; three pairs of adanal setae; epimere IV with two or three pairs of setae (Fig. 16.7 C, D) .... *Trhypochthonius* [Cosmopolitan]

2′ Subcapitulum diarthric (Fig. 16.1 K); anal plates unusually narrow, straplike (Fig. 16.7 H); two pairs of adanal setae; epimere IV with two pairs of setae (Fig. 16.7 G, H) .... *Trhypochthoniellus*
[Cosmopolitan]

Arachnida: Acari: Oribatida: Oppiidae: Genera

1 Sensillus globular to fusiform, smooth, or with tiny bristles; notogastral heterotrichy present (3-4 pairs of setae considerably longer than the rest) (Fig. 16.7 N) .... *Globoppia intermedia* (Hammer, 1962)
[Pantropical, Southern Hemisphere]

1′ Sensillus fusiform, unilaterally ciliate; notogastral heterotrichy absent .... *Austroppia crozetensis* (Richters, 1908)
[Antarctic, South America]

Arachnida: Acari: Oribatida: Galumnidae: Genera

1 Lamellar setae located between lines L and S (Fig. 16.4 O–Q) .... *Galumna alata* (Hermann, 1804)
[Semicosmopolitan]

1′ Lamellar setae located between lines L; line L straight, therefore lines L and S diverging (Fig. 16.7 I, J) .... *Orthogalumna terebrantis* (Wallwork, 1965)
[Argentina, Uruguay, Nearctic]

Arachnida: Acari: Oribatida: Zetomimidae: Genera

1 Notogaster without porose areas; notogastral setae generally reduced to their alveoli (Fig. 16.4 M, N) .... *Heterozetes*
[Holarctic, Neotropical, Oriental]

1′ Notogaster with porose areas; notogastral setae medium length (Fig. 16.4 K, L) .... *Zetomimus*
[Neotropical, Holarctic]

REFERENCES

Alberti, G. & L.B. Coons. 1999. Acari: mites. Pages 515–1265. *in*: F.W. Harrison & R.F. Foelix (eds.), Microscopic anatomy of invertebrates, Vol. 8, Wiley & Sons Publication, New York.

Aoki, J.-I. 1966. A new species of soil Mite, *Malaconothrus japonicus*, from Central Japan (Cryptostigmata; Malaconothridae). Annotationes Zoologicae Japonenses 39: 169–172.

Aoki, J.-I. 1975. Two species of the primitive oribatid genus *Gehypochthonius* from Japan. Annotationes Zoologicae Japonenses 48: 55-59.

Beck, L. 1965. Über Variabilität und Wertigkeit morphologischer Merkmale-bei adulten Oribatiden (Arachnida, Acari) am Beispiel der Gattung *Rostrozetes* Sellnick, 1925. Abhandlungen der Senckenbergischen Naturforschenden Gesellschaft 508: 1–64.

Beck, L. 1969. Zum jahreszeitlichen Massenwechsel zweier Oribatidenarten (Acari) im neotropischen Überschwemmungswald. Verhandlungen der Deutschen Zoologischen Gesellschaft, Innsbruck, Vol. 1968: 535–540.

Behan-Pelletier, V.M. 1989. *Limnozetes* (Acari: Oribatida: Limnozetidae) of Northeastern North America. The Canadian Entomologist 121: 453–506.

Behan-Pelletier, V. & B. Eamer. 2003a. Zetomimidae (Acari: Oribatida) of North America. Pages 21–56. *in*: Smith, I.M. (ed.) An Acarological Tribute to David R. Cook (from Yankee Springs to Wheeny Creek). Indira Publishing House, West Bloomfield, Michigan, USA.

Behan-Pelletier, V.M. & B. Eamer. 2003b. Redefinition of *Pelopsis* (Acari: Oribatida: Mycobatidae), with description of *Pelopsis baloghi* sp. n. from Costa Rica. Acta Zoologica Academiae Scientiarum Hungaricae 49: 5–15.

Behan-Pelletier, V.M. & B. Eamer. 2007. Aquatic Oribatida: Adaptations, constraints, distribution and ecology. Pages 71–82. *in*: J.B. Morales-Malacara, V. Behan-Pelletier, E. Ueckermann, T.M. Pérez, E.G. Estrada-Venegas & M. Badii (eds.), Acarology XI: Proceedings of the International Congress, Instituto de Biología and Facultad de Ciencias, Universidad Nacional Autónoma de México; Sociedad Latinoamericana de Acarología, México.

Cordo, H.A. & C.J. DeLoach. 1975. Ovipositional specificity and feeding habits of the waterhyacinth mite, *Orthogalumna terebrantis* in Argentina. Enviromental Entomology 4: 561–565.

Cordo, H.A. & C.J. DeLoach. 1976. Biology of the waterhyacinth mite in Argentina. Weed Science 24: 245–249.

Covarrubias, R. & I. Mellado. 1998. Oribátidos de Chile (Acarina: Oribatida): II. Especies asociadas a plantas acuáticas. Acta entomológica chilena 22: 37–44.

Fredes, N.A. & P.A. Martínez. 2011. First record of *Euzetes globulus* (Nicolet, 1855) from Neotropical region (Acari: Oribatida). Genus 22: 175–179.

Fredes, N.A. & P.A. Martínez. 2016. Revision of *Physobates* (Acari, Oribatida, Tegoribatidae) and redescription of *Physobates spinipes* Hammer, 1962 with comments on tegoribatid genera. Zootaxa 4098: 191–200.

Fujikawa, T. 1990. Oribatid mites from *Picea glehni* forest at Mo-Ashoro, Hokkaido (3) Two new species of the family Nanhermanniidae. Edaphologia 43: 5–15.

Grandjean, F. 1936. Les Oribates de Jean Frédéric Hermann et de son pére. Annales de la Société entomologique de France 105: 27–110.

Grandjean, F. 1947. L'origine pileuse des mors et la chaetotaxie de la mandibule chez les Acariens actinochitineux. Comptes rendus des séances de l'Academie des Sciences 224: 1251–1254.

Grandjean, F. 1948. Sur les *Hydrozetes* (Acariens) de l'Europe occidentale. Bulletin du Muséum national d'Histoire naturelle (2e sér.) 20: 328–335.

Grandjean, F. 1951. Observation sur les Oribates (22e série). Bulletin du Museum, 2e série 23:91–98.

Grandjean, F. 1953. Sur les genres "*Hemileius*" Berl. et "*Siculobata*" ng. (Acariens, Oribates). Mémoires du Muséum National dHistoire Naturelle, Nouvelle série, Série A, Zoologie 6: 117–137.

Grandjean, F. 1954. Etude sur les Palaeacaroïdes (Acariens, Oribates). Mémoires du Muséum National dHistoire Naturelle, Nouvelle série, Série A, Zoologie 7: 179–274.

Grandjean, F. 1955. Sur un acarien des lies Kerguelen. *Podacarus auberti* (Oribate). Mémoires du Muséum National dHistoire Naturelle, Nouvelle série, Série A, Zoologie 8: 109–150.

Grandjean, F. 1957. L'infracapitulum et la manducation chez les Oribates et d'autres Acariens. Annales des Sciences Naturelles, Zoologie 11: 233–281.

Grandjean, F. 1962. Le genre *Tegeocranellus* Berl. 1913 (Oribates). Acarologia 4: 78–100.

Grandjean, F. 1970. Nouvelles observations sur les Oribates (7e série). Acarologia 12: 432–460.

Hammer, M. 1958. Investigations on the Oribatid Fauna of the Andes Mountains I. The Argentine and Bolivia. Biologiske Skrifter Kongelige Danske Videnskabernes Selskab 10: 1–262.

Hammer, M. 1962. Investigations on the oribatid fauna of the Andes Mountains. III. Chile. Biologiske Skrifter Kongelige Danske Videnskabernes Selskab 13: 1–96.

Mahunka, S. 1980. Data to the knowledge of mites preserved in the Berlese collection (Acari: Tarsonemina, Oribatida). II. Folia Entomologica Hungarica 41: 105–121.

Mahunka, S. 1985. Neue und interessante Milben aus dem Genfer Museum LVII. Oribatida Americana 9: Antilles I (Acari). Revue suisse de Zoologie 92: 119–144.

Michael, A.D. 1882. I – Further notes on British Oribatidae. Journal of the Royal Microscopical Society 2:1–18. Source: http://biodiversitylibrary.org/page/31213836.

Norton, R.A., G. Alberti, G. Weigmann & S. Woas. 1997. Porose integumental organs of oribatid mites (Acari, Oribatida). 1. Overview of types and distribution. Zoologica (Stuttgart) 146:1–33.

Norton, R.A. & V.M. Behan-Pelletier. 2007. *Eniochthonius mahunkai* sp. n. (Acari: Oribatida: Eniochthoniidae), from North American peatlands, with a redescription of *Eniochthonius* and a key to North American species. Acta Zoologica Academiae Scientiarum Hungaricae 53: 295–333.

Norton, R.A. & V.M. Behan-Pelletier. 2009. Oribatida. Chapter 15. Pages 430–564. *in*: G.W. Krantz & D.E. Walter (eds.), A Manual of Acarology. Texas Tech University Press, Lubbock.

Olszanowski, Z. 1996. A monograph of the Nothridae and Camisiidae of Poland (Acari: Oribatida: Crotonioidea). Genus. International Journal of Invertebrate Taxonomy. Supplement: 1–201.

Pérez-Iñigo, C. 1993. Acari, Oribatei, Poronota. In: Fauna Iberica. CSIC, Madrid. 320 pp.

Pérez-Iñigo, C. & D. Baggio. 1988. Oribates édaphiques du Brésil (IV) oribates de l'Etat de São Paulo (première partie). Acarologia 29: 189–204.

Pérez-Iñigo, C. & L.S. Subías. 1974. Redescription de *Scapheremaeus corniger* (Berlese, 1908) (Acari, Oribatei). Acarologia 16: 739–745.

Pugh, P. & H. Dartnall. 1994. The Acari of fresh-and brackish water habitats in the Antarctic and sub-Antarctic regions. Polar Biology 14: 401–404.

Schatz, H. & V.M. Behan-Pelletier. 2008. Global diversity of oribatids (Oribatida: Acari: Arachnida). Hydrobiologia 595: 323–328.

Tione, M.L., J.C. Bedano & M. Blarasin. 2011. Comunidades de invertebrados en aguas subterráneas y su relación con variables ambientales. Ecología Austral 21: 87–100.

Villagrán-Mella, R., M.E. Casanueva & L.E. Parra. 2005. Mites in the parenchyma of *Juncus procerus* in marshey wetlands in the Bio Bio region, Chile. Gayana 69: 22–26.

Wallwork, J.A. 1965. A leaf-boring galumnoid mite (Acari: Cryptostigmata) from Uruguay. Acarologia 7: 758–764.

Walter, D.E. & H.C. Proctor. 2013. Mites: ecology, evolution and behaviour. Life at a microscale. Springer, New York. 494 pp.

Weigmann, G. 1997. New and old species of Malaconothroidea from Europe. Spixiana 20: 199–218.

Yamamoto, Y. & Aoki, J.-I. 1998. Two new oribatid mites of the genus *Trimalaconothrus* from Yunnan Province in China (Acari. Oribatida: Malaconothridae). Edaphologia 61: 15–21.

Acari: Halacaroidea

Almir Rogério Pepato and Pedro Henrique da Silva Conceição
Departamento de Zoologia, Universidade Federal de Minas Gerais, Belo Horizonte, Brazil

INTRODUCTION

The family Halacaridae (Prostigmata, Halacaroidea) is unique among Acari in being exclusively aquatic and mainly marine. Perhaps, it is the family with the broadest altitudinal range among animals, since it inhabits from the abyssal depths to alpine snow melting streams and high altitude lakes. The family Pezidae, the other family of Halacaroidea, also includes two freshwater species from Australia (Harvey, 1990).

Among the 1118 species and 63 genera reported in the last through revision (Bartsch, 2007, 2009), only 56 species classified into 18 genera were obtained from fresh- and oligohaline waters. Despite being much more diversified in marine environments, Halacaridae may be locally abundant in freshwater and some species seem to be widespread (found in several continents).

Halacaroidea mites were traditionally regarded as monophyletic and sister group to the superfamily Bdelloidea (=Bdellidae + Cunaxidae) (Lindquist et al., 2009). Recent phylogenetic studies including sequence data recovered them as sister group of Parasitengonina (which includes chiggers and fresh water mites) (Pepato & Klimov, 2015; Dabert et al., 2016). At first glance, it may appear as a major conflict with morphological data, upon which traditional classification relies, but long ago Witte (1991) proposed this very same association based solely on morphology, pointing to six putative synapomorphies uniting Halacaroidea to Parasitengona.

History and Current Knowledge in the Neotropics

Our knowledge on freshwater Halacaridae from Neotropics reflects the absence of a systematic effort in sampling across the region. The records are scattered in the literature, and most reports resulted from samples taken targeting other groups. For example, first report of Halacaridae mites, came from collections targeting zooplankton in an Andean lake in Peru at an elevation of 5400 m. Walter (1919) described the single nymph obtained as *Halacarus processifer*, which was later reclassified as *Lobohalacarus* by Viets (1939).

Lobohalacarus hummelincki Viets, 1940 and *Limnohalacarus cultellatus* Viets, 1940 were reported from Venezuela. The later species proved to be widespread, with ranging distribution from northern USA to southern Brazil (Viets, 1940; Bartsch, 1984, 2011; Pepato & Costa, 2015; Ojeda et al., 2016).

Two surprising occurrences came from the isolated Malvinas (Falkland) Islands. *Peregrinacarus falklandensis* Bartsch, 2001 was obtained from a freshwater creek 200 m from the seashore (Bartsch, 2001), while *Soldanellonyx monardi* Walter, 1919 was found in lentic waters (Pugh & Dartnall, 1994). Pešić et al. (2010) found both species along with *Lobohalacarus weberi* (Romijn & Viets, 1924) from the Chilean Patagonia.

The last species to be reported from the Neotropics was *Limnohalacarus mamillatus* Fain & Lambrechts, 1987. This was originally described from an artificial tank present outside of the Neotropics but with material from there (Fain & Lambrecht, 1987). Later this mite species was collected in southern Brazil (Pepato & Costa, 2015).

Finally, during the preparation of this chapter, the authors found slide mounted individuals belonging to the genera *Porohalacarus* and *Soldanellonyx* from Brazil in the taxonomical collection at the University of Minas Gerais (CCT-UFMG). They are included in the keys and pictures below.

TERMINOLOGY AND MORPHOLOGY

Similar to other mites, Halacaridae has a pseudotagma, the **gnathosoma**, composed by two principal parts: the **cheliceral frame**, bearing the chelicerae, and the **infracapitulum** (also termed **subcapitulum**), bearing the **pedipalps** detached from the rest of body, the **idiosoma** (Alberti et al., 2011). Palps may be inserted laterally or dorsally. They are composed by four segments, variously fused or reduced, usually numbered from proximal to distal. Chelicerae are devoid of setae, and lack a fixed digitus (they are not chelate).

Besides its ecological uniqueness, Halacaridae diverges from other Prostigmata by: highly reductive idiosomal chaetotaxy (up to seven pairs of dorsal setae); **glandularia** restricted to the dorsum (up to five glands); and first and second legs directed anteriorly and third and fourth directed posteriorly. They also share a typical set of four dorsal (anterior, ocular, and posterior plates) and ventral plates (anterior and posterior epimeral plates, largely derived from coxae and genitoanal plates) (Fig. 16.8 A, B) that may have their homologous areas recognized even when fused, reduced, or divided. The presence of medial and lateral eyes is detectable by its dark pigments (Fig. 16.8 C) or the presence of lens (corneas).

Legs have six segments, comprising trochanter, basi- and telofemur, genu, tibia, and tarsus. The pretarsus usually has a medial piece that bears a variously developed claw-like empodium, and lateral claws (Fig. 16.8 B–D). The leg

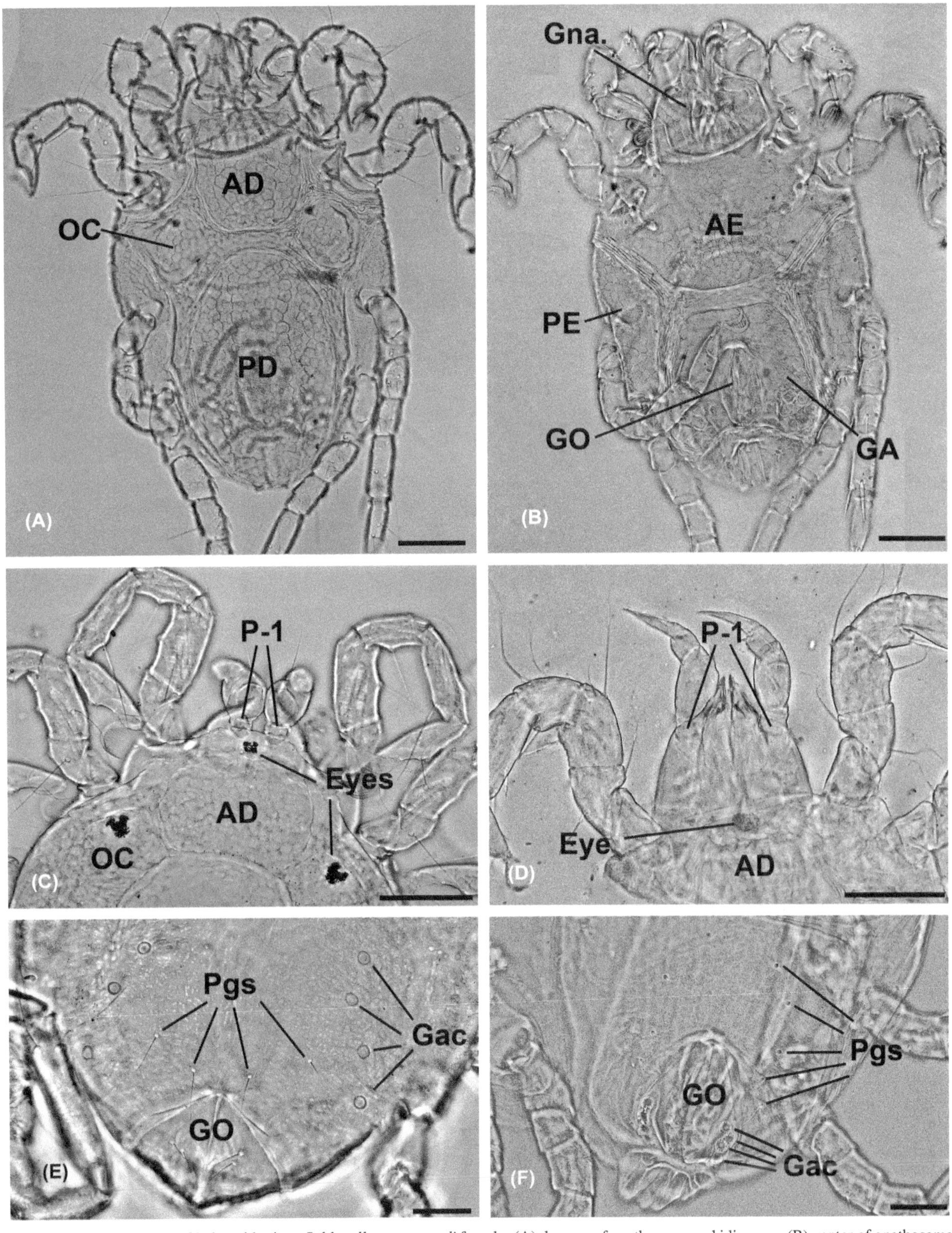

FIGURE 16.8 Morphology of halacarid mites. *Soldanellonyx monardi* female: (A) dorsum of gnathosoma and idiosoma; (B) venter of gnathosoma and idiosoma. *Limnohalacarus mamillatus* female: (C) dorsum of gnathosoma and anterior idiosoma, note the dorsal attachment of palps; (E) genital area. *Porohalacarus alpinus* female: (D) dorsum of gnathosoma and anterior idiosoma, note the lateral attachment of palps; (F) genital area. Abbreviations: *AD*, anterior dorsal plate; *AE*, anterior epimeral plate; *GA*, genitoanal plate; *Gac*, genital acetabula; *GO*, genital opening; *OC*, ocular plate; *P-1*, proximal segment of palp; *PD*, posterior dorsal plate; *PE*, posterior epimeral plate; *Pgs*, perigenital setae. Scale bars: (A–C) and (E) = 50 μm; (D) and (F) = 25 μm.

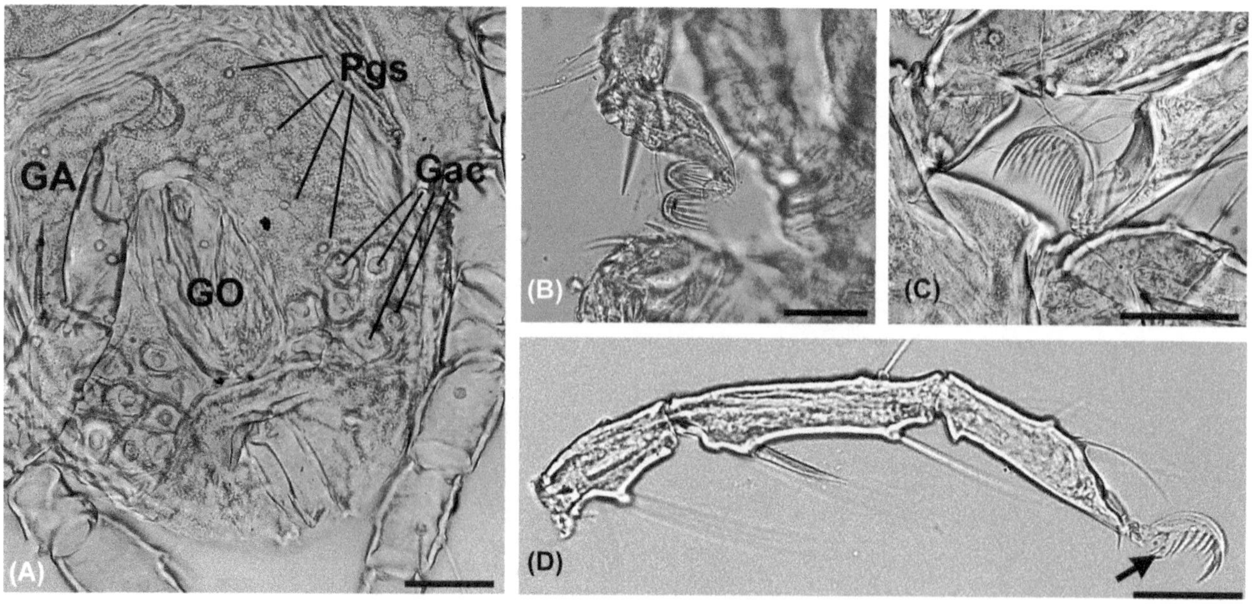

FIGURE 16.9 Morphology of halacarid mites. *Soldanellonyx monardi* female: (A) genital area; (B) claws of tarsi I. *Limnohalacarus mamillatus* female: (C) Claws of tarsi I; (D) leg III, arrow points to the lamellar ventral process. Abbreviations: *GA*, genitoanal plate; *Gac*, genital acetabula; *GO*, genital opening; *Pgs*, perigenital setae. Scale bars = 25 μm.

chaetotaxy (i.e., number, position, and morphology of setae) is an important source of characters for Halacaridae identification, especially in specific level. The setae vary in length, thicknesses (filiform to blunt spines), and shape (smooth, pectinate, or plumose). Solenidia, chemosensory structures usually similar to thick, blunt setae with immovable bases and banded patterns of minute "pores," are restricted in halacarid to tarsi I and II. Tarsi I also bear a famulus, another chemosensory structure, which varies from a pore to a setalike structure similar to a solenidion.

Sexual dimorphism in Halacaridae mites is mainly confined to the genital area (Chatterjee & Guru, 2012). Males are rare among freshwater taxa and distinguishable by a more hirsute genital area and the presence of an internal genital skeleton, or **spermatophorotype**. Females have a protrusible **ovipositor**, bearing variously shaped spines.

Except for observations of eggs attached to the hind legs of Pezidae and *Limnohalacarus*, there are no reports on brood care in Halacaroidea. Life cycle comprises prelarvae, larvae, and up to three nymphal stages (usually only two, protonymph and deutonymph) and adults. Nymphal stages are distinguished by the progressive gain of setae, genital acetabula, and sclerotization. Larval legs and leg IV in protonymphs have fused femora.

Epimeral pores and genital acetabula are serial homologs consisting of a porose cap and underlying transport epithelium (Alberti & Coons, 1999). Indeed, their role in osmoregulation was determined experimentally for the first time in *Porohalacarus alpinus* (Thor, 1910), a freshwater halacarid (Bartsch, 1973). Epimeral pores are usually larval organs, since larvae lack structures on the genital region. They may be retained until adulthood in many Halacaridae genera, being very large in freshwater *Copidognathus* (Benfatti et al., 1989).

KEYS TO HALACARIDAE

Arachnida: Acari: Halacaridae: Genera

There are three nominal species belonging to *Lobohalacarus* recorded for South America: *L. processifer* (Walter, 1919), *L. hummelincki* Viets, 1940, and *L. weberi* (Romijn & Viets, 1924). According to Bartsch (2008), the combination of characters in *L. processifer* and *L. hummelincki* lie within the range of variation of *L. weberi*, so it is impossible to key them, and raises a question of their species validity.

1	Palps inserted laterally on gnathosoma in a way that trochanters are discernible ventrally (Fig. 16.8 D); genital acetabula on margins (Fig. 16.8 F) or beneath genital sclerites in females	2
1′	Palps inserted dorsally on gnathosoma in a way that trochanters are not discernible ventrally (Fig. 16.8 C); genital acetabula on genital plate in females (Figs. 16.8E, 9 A)	4
2(1)	Genital acetabula external, posterior on genital sclerites (Fig. 16.9 A)	3

2′ Genital acetabula internal, beneath genital sclerites .... *Peregrinacarus falklandensis* Bartsch, 2001
[Chile: Patagonia. Malvinas (Falkland) Islands]

3(2) Seven pairs of dorsal setae, first leg with tarsus lateral claw fossa membrane enlarged relative to dorsomedial membrane, and bearing one ventromedial spiniform seta and two ventral setae .... *Lobohalacarus* Viets, 1939
[Venezuela: Northeastern. Peru: Andean region. Chile: Patagonia]

3′ Six pairs of dorsal setae, first leg with tarsus lateral claw fossa membrane not enlarged relative to dorsomedial membrane, and bearing one ventral seta and pair of parambulacral setae .... *Porohalacarus* Thor, 1922
[Brazil: Rio Grande do Sul, Imbé (CCT-UFMG: 29°57′30.63" S, 50°09′14.50"W)]

4(1) First leg with tarsal claws strongly pectinated, tines in an umbrellalike arrangement (Fig. 16.9 B); genital acetabula on posterolateral regions of genital plate (Fig. 16.9 A); no eggs attached to leg IV in females .... *Soldanellonyx monardi* Walter, 1919
[Brazil: São Paulo, Campos do Jordão CCT-UFMG: 22°46′50.93"S, 45°36′41.78" W. Malvinas (Falkland) Islands]

4′ First leg with tarsal claws bearing slender tines (Fig. 16.9 C) and or lamellar ventral process (Fig. 16.9 D); genital acetabula on lateral margins of genital plates (Fig. 16.8 E); eventually, eggs may be found attached to leg IV in females .... *Limnohalacarus* Walter 1917

Arachnida: Acari: Halacaridae: *Limnohalacarus*: Species

1 Lateral claws on tarsus I pectinate along their entire length, tines long (Fig. 16.9 C); eye pigments discernible (Fig. 16.8 C) ventral plates fused; tibia I with nine setae .... *Limnohalacarus mamillatus* Fain & Lambrechts, 1987
[Brazil: South]

1′ Lateral claws on tarsus I tines short and delicate, restricted to their distal half; claws of tarsus I with well-developed basal lamellar process similarly to posterior legs; ventral plates separated by strips of striated cuticle; tibia I numbering seven setae .... *Limnohalacarus cultellatus* Viets, 1940
[Caribbean region, Mexico, Brazil: West and South]

REFERENCES

Alberti, G. & L.B. Coons. 1999. Acari—Mites. Pages 515–1265. *in*: F.W Harrison (ed.), Microscopic Anatomy of Invertebrates, Vol. 8c, New York, Wiley-Liss.

Alberti, G., M. Heethoff, R.A. Norton, S. Schmelzle, A. Seniczak & S. Seniczak. 2011. Fine structure of the gnathosoma of *Archegozetes longisetus* Aoki (Acari: Oribatida, Trhypochthoniidae). Journal of Morphology 272: 1025–1079.

Bartsch, I. 1973. *Porohalacarus alpinus* (Thor) (Halacaridae, Acari), ein morphologischer Vergleich mit marinen Halacariden nebst Bemerkungen zur Biologie dieser Art. Entomologisk Tidskrift 74: 116–123.

Bartsch, I. 1984. Halacaridae (Acari) von den Westindischen Inseln. Bijdragen tot de Dierkunde 54: 185–196.

Bartsch, I. 2001. A new freshwater halacarid mite, genus *Peregrinacarus* (Halacaridae, Acari) from the Falklands. Hydrobiologia 452: 139–144.

Bartsch, I. 2007. Global diversity of halacarid mites (Halacaridae: Acari: Arachnida) in freshwater. Hydrobiologia 595: 317–322.

Bartsch, I. 2008. Freshwater halacarid mites (Halacaridae: Prostigmata: Acari) from Tunisia, three new records and notes on geographical distribution of these species. Entomologische Mitteilungen aus dem zoologischen Museum Hamburg 15: 15–27.

Bartsch, I. 2009. Checklist of marine and freshwater halacarid mite genera and species (Halacaridae: Acari) with notes on synonyms, habitats, distribution and descriptions of the taxa. Zootaxa 1998: 1–170.

Bartsch, I. 2011. North American freshwater Halacaridae (Acari): literature survey and new records. International Journal of Acarology 37: 490–510.

Chatterjee, T. & B.C. Guru. 2012. Sexual dimorphism in halacarid mites (Acari, Halacaridae): an updated overview. Acta Biologica 19 (728): 97–112.

Dabert, M., H. Proctor & J. Dabert. 2016. Higher-level molecular phylogeny of the water mites (Acariformes: Prostigmata: Parasitengonina: Hydrachnidiae) Hydrobiologia 595: 317–322.

Fain, A. & L. Lambrechts. 1987. Observations on the acarofauna of fish aquariums. II. A new oribatid and two new halacarid mites. Bulletin et Annales de la Société Royal Entomologique de Belgique 123: 103–118.

Harvey, M.S. 1990. Pezidae, a new freshwater mite family from Australia (Acarina: Halacaroidea). Invertebrate Taxonomy 3: 771–781.

Lindquist, E.E., G.W, Krantz & D.E. Walter. 2009. Classification. Pages 97–103. *in*: G.W. Krantz & D.E. Walter (eds.), A Manual of Acarology, Lubbock, Texas Tech University Press.

Ojeda, M., G. Rivas & F. Álvarez. 2016. First record of the genus *Limnohalacarus* (Acari: Halacaridae) from Mexico. Revista Mexicana de Biodiversidad 87: 1131–1137.

Pepato, A.R. & S.G.S. Costa. 2015. New records of the genus *Limnohalacarus* (Halacaridae, Trombidiformes) from southern Brazil. Check List 11: 1500.

Pepato, A.R. & P.B. Klimov. 2015. Origin and higher-level diversification of acariform mites – evidence from nuclear ribosomal genes, extensive taxon sampling, and secondary structure alignment. BMC Evolutionary Biology 15: 178.

Pešić, V., H. Smit & T. Datry. 2010. New records of water mites (Acari: Hydrachnidia, Halacaroidea) from Patagonia (Chile). Systematic & Applied Acarology 15: 151–160.

Pugh, P.J.A. & H.J.G. Dartnall. 1994. The Acari of fresh- and brackish water habitats in the Antarctic and sub-Antarctic regions. Polar Biology 14: 401–404.

Viets, K. 1939. Über die Milbengruppe der Porohalacaridae (Acari). Zugleich ein Beitrag zur Kenntnis der Milbenfauna der Bremer Umgebung und des Plöner Sees. Abhandlungen. Naturwissenschaftlicher Verein zu Bremen 31: 502–514.

Viets, K. 1940. Zwei neue Porohalacaridae (Acari) aus Südamerika. Zoologischer Anzeiger 130: 191–201.

Walter, C. 1919. Hydracarinen aus den peruanischen Anden und aus Brasilien. Revue de Suisse Zoologie 27: 19–59.

Witte, H. 1991. The phylogenetic relationships within the Parasitengonae. Pages 171–182. *in*: F. Dusbábek & V. Bukva (eds.), Modern Acarology, vol. 2, The Hague, SPB Academic Publishing, Prague.

Arachnida: Araneae

Nelson Ferretti

Instituto de Ciencias Biológicas y Biomédicas del Sur (INBIOSUR-CONICET-UNS), Bahía Blanca, Buenos Aires, Argentina

Gabriel Pompozzi

Sección Entomología, Facultad de Ciencias, Universidad de la República, Montevideo, Uruguay

INTRODUCTION

Spiders (Order Araneae) are among the most familiar arthropods to people as well as being very abundant in most environments. Most spiders live in strictly defined environments, and the limitations are usually set by physical conditions, such as temperature, humidity, wind, and light intensity, as well as by biological factors, such as vegetation structure, prey availability, and predators (Foelix, 2011). Freshwater spiders live on specific habitats near but not in the water where they can walk on its surface and dive after tadpoles or small fishes. Neotropical spiders associated with freshwater are quite diverse, and are found in four families: Ctenidae, Lycosidae, Pisauridae, and Trechaleidae, all within Lycosoidea. The families Ctenidae, Lycosidae, and Pisauridae have a few freshwater representatives including *Ancylometes* (Ctenidae) and nursery web spiders (Pisauridae). The most diverse and widespread family associated with Neotropical freshwater environments is Trechaleidae, with 15 genera (World Spider Catalog, 2019) found almost exclusively in the Neotropics. Spiders in this family live in vegetation near the margins of rocky streams and small rivers (Carico, 1993). Aquatic spiders possess flexible tarsi that may be helpful in locomotion on the surface of the water during foraging. They also can crawl underwater to capture prey.

LIMITATIONS

The taxonomy of the Neotropical families (mainly Lycosidae and Trechaleidae) that includes some freshwater associated spider species is currently being revised, and many upcoming changes will not be reflected in this edition. One clear example of this is a recent, unpublished revised systematics of Trechaleidae, which involves many new species, synonymies, and even a new family status. The identification key for Trechaleidae mostly ends at genera, and keys for species are only provided when the taxonomic knowledge of a genus is sufficient. For example, the genus *Diapontia* (Lycosidae), contains four species, but only two are known from only one sex (World Spider Catalog, 2019), and thus the key ends at the genera. Many of the family, genus, and species definitions are provisional and will likely be further clarified with new analyses and taxonomic reviews. Identification of spider species requires microscopic examination of preserved specimens. The initial step in separating families of Araneae by the eyes arrangement is quite easy and should pose few problems since the relative position of the eyes is very important for the systematic classification of spiders. As many species are morphologically similar, determination of several genitalic characters is necessary for identification, including the structures and shape of the palpal tibia and bulb in males and the shape of the epigyne in adult females.

TERMINOLOGY AND MORPHOLOGY

A spider's body consists of two main parts: an anterior portion, the **prosoma** or cephalothorax, and a posterior part, the **opisthosoma** or abdomen. These are connected by a narrow stalk, the **pedicel**. The prosoma serves as the place of attachment for six pairs of appendages: one pair of **chelicerae** and one pair of leglike **pedipalps** or palps are situated in front of four pairs of walking legs. The anterior part of the prosoma bears the eyes. Most spiders have eight eyes, whose specific arrangement patterns varies between families. Usually the eyes are placed in two rows (sometimes three), and accordingly they are referred to as anterior lateral, anterior median, posterior lateral, and posterior median eyes (Foelix, 2011). The articles of the male palp are similar to those of the female, except that the male palps lack a pretarsus and a claw, and has a copulatory bulb attached to the tarsus. A retrolateral process, known as the **retrolateral tibial apophysis** is on the palpal tibia of adult male spiders in a great number of families and is thought to stabilize the expanded genital bulb internally during mating. On some families, the retrolateral tibial apophysis can be divided into two prominent ectal (basal) and ental (distal) divisions. The tarsus of the male is modified to accommodate the copulatory bulb, and is called the **cymbium**. The dorsal surface of the cymbium carries many chemosensory setae. The copulatory bulb has one or more sclerotised pieces called sclerites, and an internal blind duct, the spermophore. The copulatory bulb is not connected to the testis; hence, the male must charge his bulb by absorbing a drop of sperm through the embolus. In entelegynes spiders, the movement of the bulb is entirely hydraulic and rotates as it inflates. The more conserved sclerites define the main divisions or sections of the copulatory bulb as follows: (a) a basal division includes the **subtegulum**, and within it, the blind end of the spermophore, called the **fundus**; (b) a median division includes the **tegulum**, and may have a conductor and median apophysis attached to it; the spermophore runs through the tegulum (in Pisauridae, a distal tegular projection is present); and (c) the embolic division includes the intromittent structure with the sperm outlet, called the **embolus** (Ramírez, 2014).

The female ovaries connect to the epigastric fold via the oviduct. The uterus externus ends in the gonopore, usually hidden inside the epigastric furrow. Female spiders store the sperm in receptacles called spermathecae; hence, fertilization does not occur during mating. Females with entelgyne genitalia have unidirectional sperm flow; the sperm is inserted through a copulatory duct and drawn from the spermathecae through a fertilization duct. The copulatory openings are exposed and associated with the externally sclerotized epigyne. This structure is generally composed of three plates: the median or middle field and two lateral lobes or plates (Ramírez, 2014). The borders between those plates are marked by the epigynal folds, usually ending in or near the copulatory openings. The receptacles for sperm storage are generically referred to as spermathecae, and at least one spermatheca (entelegynes) on each side connects to both the copulatory and the fertilization ducts. These spiders usually have two pairs of spherical receptacles, the primary and secondary spermathecae (Ramírez, 2014). The copulatory opening receives the embolus of the male copulatory bulb. The copulatory duct is usually correlated in shape and length with the embolus, and the fertilization duct connects the primary spermathecal to the uterus externus.

MATERIAL PREPARATION AND PRESERVATION

Spiders can be preserved in 80% ethyl alcohol for long periods. However, specimen preparation for molecular purposes should be preserved in 99% ethyl alcohol and preferably frozen at −80°C. Due to the small size of some taxonomic characters, mainly the genitalia, identification and observation require a stereoscopic microscope with 10−80× magnification. The procedure used for a careful examination of the female genitalia is to make two cuts around the genital area. Fold back the epigynal plate and seminal receptacles before or during submersion in clove oil. Then anchor the specimen on fibrous paper tissue for examination. After examination, bend the genitalia back. Observe the internal female genitalia after digestion in potassium hydroxide (KOH). Make a wide opening through a superficial incision in the spider abdomen to allow solution to contact internal tissues. The abdomen must then be immersed in a 10%−20% solution of KOH at 100°C in a double boiler, until appropriate digestion occurs, and then observed in alcohol. Finally, the remaining pieces of cuticle should be mounted in an excavated glass slide with alcohol or lactic acid and a glass cover slip, and then observed under a compound microscope (Ramírez, 2014). The male copulatory bulb can be clarified using clove oil. This will, for example, expose the path of the spermophore. Alternatively, the path can be revealed by cycling between 10% KOH and distilled water for a few to 30 min until the palps are fully expanded (Ramírez, 2014)

ACKNOWLEDGMENT

Special thanks to Luis Piacentini for his help and advice about the Neotropical Lycosidae genera associated with freshwater environments.

KEYS TO ARANEAE

Arachnida: Araneae: Families

1	Eyes arranged in three rows (4-2-2), the first row having small anterior median and lateral eyes; two large and forward-facing posterior median eyes, and behind these a pair of large posterior lateral eyes (Fig. 16.10 B, C)	2
1′	Eyes arranged in three rows (2-4-2) with small anterior lateral eyes positioned far back in a middle row with the posterior median eyes, and the posterior lateral eyes behind these (Fig. 16.10 A)	Ctenidae, one genus: ***Ancylometes* [p. 546]**
2(1)	Posterior eyes forming a trapeze in shape (Fig. 16.10 C); male palp with retrolateral tibial apophysis	3
2′	Posterior eyes forming a square in shape (Fig. 16.10 B); male palp without retrolateral tibial apophysis	**Lycosidae [p. 548]**
3(2)	Tarsi of legs nonflexible with a membranous distal area that conforms an apical pseudosegment; females produce a nursery web surrounding the egg sac, and the egg sac is carried underneath the sternum	**Pisauridae [p. 552]**
3′	Tarsi of legs usually flexible but without a distinct apical pseudosegment; females carry the egg sac with the spinnerets	**Trechaleidae [p. 557]**

Arachnida: Araneae: Ctenidae: *Ancylometes*: Species

Identification keys are constructed separately for males and females since the main morphological distinguishing characters are genitalic. Males of *Ancylometes japura* (Höffer & Brescovit, 2000), *A. pantanal* (Höffer & Brescovit, 2000) and *A. riparius* (Höffer & Brescovit, 2000) are unknown.

1	Males	2
1′	Females	8
2(1)	Prosoma of males with white marginal bands	3

PHYLUM ARTHROPODA: INTRODUCTION AND ARACHNIDA

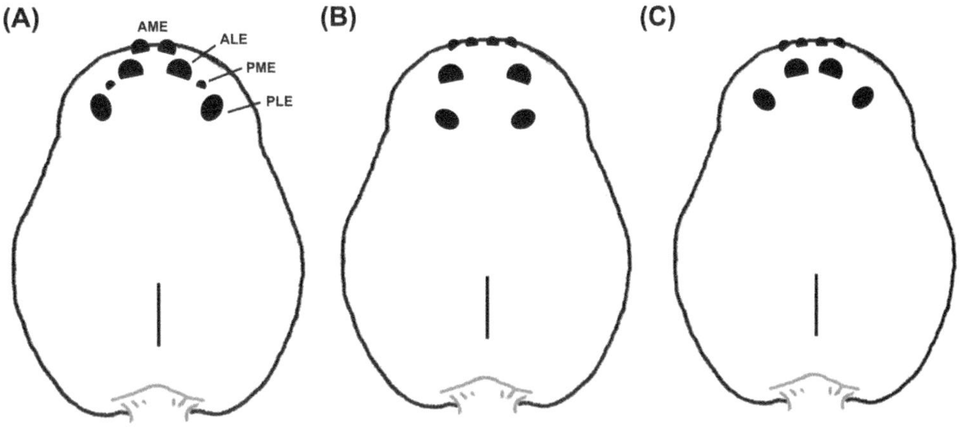

FIGURE 16.10 Eyes arrangement of Neotropical freshwater spider families: (A) Ctenidae; (B) Lycosidae; (C) Pisauridae—Trechaleidae. Abbreviations: *ALE*, anterior lateral eyes; *AME*, anterior median eyes; *PLE*, posterior lateral eyes; *PME*, posterior median eyes.

2′ Prosoma of males without white marginal bands .... *Ancylometes terrenus* (Höffer & Brescovit, 2000) [Brazil: Amazonas, Mato Grosso]

3(2) Small spiders of about <14 mm of total body length 4

3′ Large spiders of about >20 mm of total body length 5

4(3) Male palp with large ovoid membranous lobe covering base of embolus; short median apophysis ending in a hook; retrolateral tibial apophysis very large and sharp .... *Ancylometes jau* (Höffer & Brescovit, 2000) [Brazil: Central Amazonia]

4′ Male palp with small transparent rectangular membranous lobe; short median apophysis oblique, hammerlike; retrolateral tibial apophysis small and domed .... *Ancylometes amazonicus* (Simon, 1898) [Peru: Huanuco. Brazil: Acre, Mato Grosso]

5(3) Median apophysis of male palp without an internal bulge 6

5′ Male palp with a median apophysis having a characteristic internal bulge .... *Ancylometes bogotensis* (Keyserling, 1877) [Honduras to Brazil and Bolivia]

6(5) Median apophysis of male palp very large and long in relation to conductor 7

6′ Median apophysis of male palp short, straight, and broad (Fig. 16.11 A) .... *Ancylometes concolor* (Perty, 1833) [Brazil. Bolivia. Paraguay. Argentina]

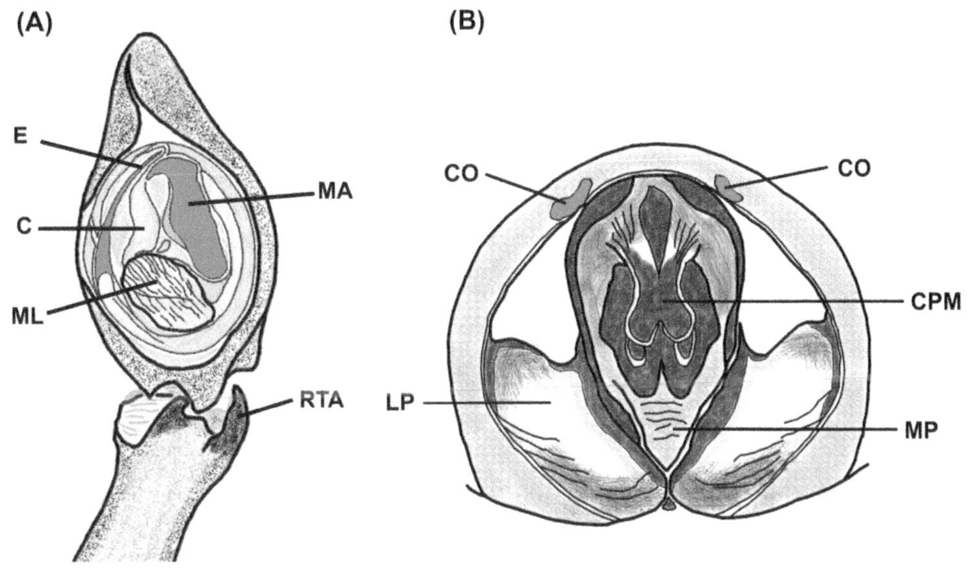

FIGURE 16.11 *Ancylometes* (Ctenidae): (A) Male palpal bulb, ventral view; (B) female epigyne, ventral view. Abbreviations: *C*, conductor; *CO*, copulatory opening; *CPM*, central projection of medial plate; *E*, embolus; *LP*, lateral plate; *MA*, median apophysis; *ML*, membranous lobe; *MP*, median plate; *RTA*, retrolateral tibial apophysis. *Redrawn and modified from Höfer & Brescovit (2000).*

7(6) Male palp with median apophysis straight and with broad apex; palp with membranous lobe large and straight ... *Ancylometes hewitsoni* (Pickard-Cambridge, 1897)
[Brazil: Amazonia. Bolivia]

7′ Male palp with median apophysis curved; palp with membranous lobe small and triangular ... *Ancylometes rufus* (Walckenaer, 1837)
[Amazon Basin and Atlantic coast of Brazil]

8(1) Female epigyne with copulatory ducts that open on lateral sides ... 9

8′ Female epigyne with copulatory ducts that open on anterior-lateral edge (Fig. 16.11 B) ... 10

9(8) Epigynal median plate with central projection narrow and triangular ... *Ancylometes jau* (Höffer & Brescovit, 2000)
[Brazil: Central Amazonia]

9′ Epigynal median plate with a spoonlike and broad projection ... *Ancylometes pantanal* (Höffer & Brescovit, 2000)
[Brazil: Pantanal]

10(8) Epigynal median plate with protuberance ... 11

10′ Epigynal median plate without protuberance ... *Ancylometes rufus* (Walckenaer, 1837)
[Amazon Basin and Atlantic coast of Brazil]

11(10) Total body length >20 mm; epigynal median plate wider in its anterior part ... 12

11′ Total body length <14 mm; epigynal median plate wider in its posterior half ... *Ancylometes amazonicus* (Simon, 1898)
[Brazil: Acre, Mato Grosso. Peru: Huanuco]

12(11) Epigynal median plate with a short and small central projection ... 13

12′ Epigynal median plate with a large and long central projection ... 14

13(12) Epigynal median plate oval-pointed (cordate) ... *Ancylometes terrenus* (Höffer & Brescovit, 2000)
[Brazil: Amazonas, Mato Grosso]

13′ Epigynal median plate narrow trapezoidal ... *Ancylometes hewitsoni* (Pickard-Cambridge, 1897)
[Brazil: Amazonia. Bolivia]

14(12) Epigyne with median plate triangular almost piriform ... 15

14′ Epigyne with median plate trapezoidal... *Ancylometes concolor* (Perty, 1833)
[Brazil. Bolivia. Paraguay. Argentina]

15(14) Epigyne with median plate narrow, triangular with narrow projection ... 16

15′ Epigyne with median plate like typical bike-saddle form and bearing large broad projection ... *Ancylometes japura* (Höffer & Brescovit, 2000)
[Brazil: Amazonas]

16(15) Female epigyne nearly quadratic ... *Ancylometes riparius* (Höffer & Brescovit, 2000)
[Brazil: Central Amazonia]

16′ Female epigyne nearly oval ... *Ancylometes bogotensis* (Keyserling, 1877)
[Honduras to Brazil and Bolivia]

Arachnida: Araneae: Lycosidae: Genera

1 Terminal apophysis present (Fig. 16.13 A, B) ... 2

1′ Terminal apophysis lost (Fig. 16.12) ... ***Aglaoctenus* [p. 551]**

2(1) Male palpal bulb with terminal apophysis minute and scalelike; females with epigynal hood ... 3

2′ Male palpal bulb with terminal apophysis long and arcuate (Fig. 16.13 A); females lack epigynal hood (Fig. 16.13 C) ... *Arctosa*, one species: *Arctosa sapiranga* (da Silva & Lise, 2009)
[Brazil: Rio Grande do Sul, Sapiranga]

3(2) Males with membranous median apophysis and palpal tibia with a basal retrolateral apophysis covered with strong setae (Fig. 16.14 A, B); females with the anterior part of the median septum not narrowed (Fig. 16.14 C) ... ***Agalenocosa* [p. 551]**

3′ Males with highly sclerotized median apophysis and palpal tibia without retrolateral apophysis (Fig. 16.15 A); females with the anterior part of the median septum narrowed (Fig. 16.15 B) ... *Diapontia*
[Argentina. Brazil. Peru. Uruguay]

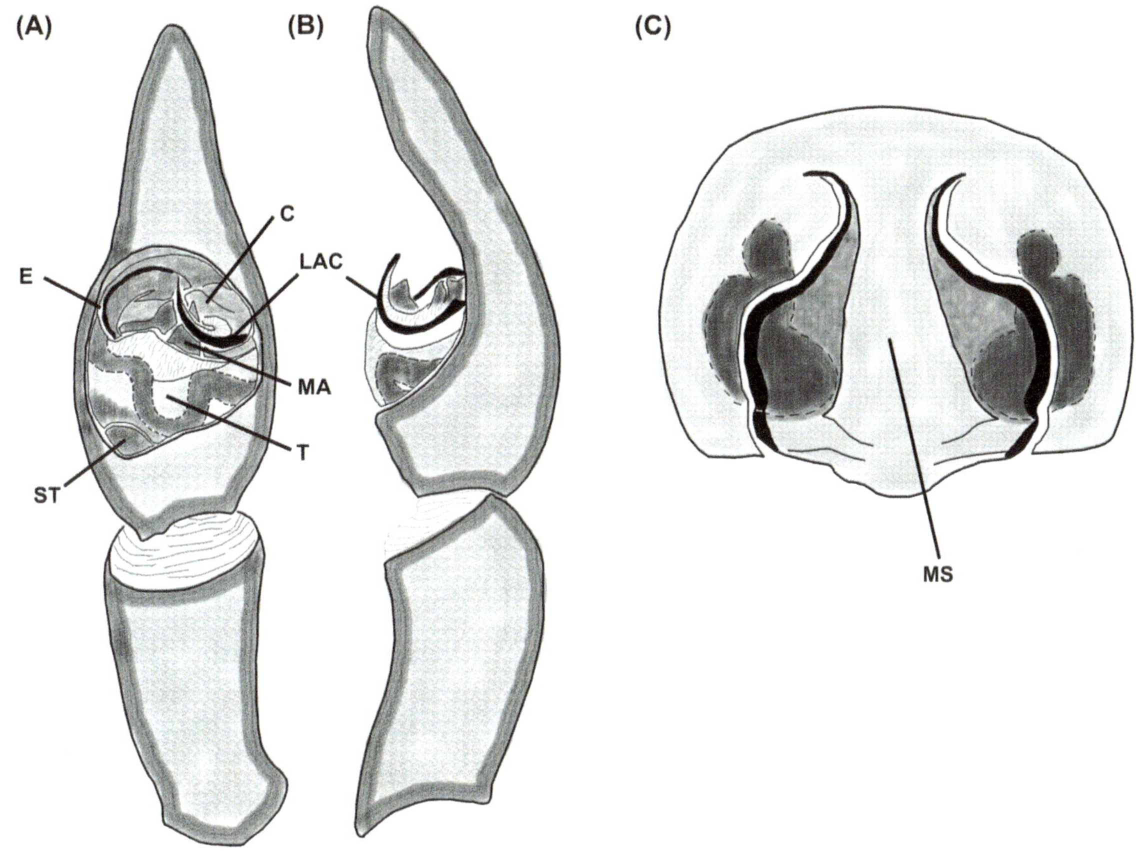

FIGURE 16.12 *Aglaoctenus* (Lycosidae): (A) Male palpal bulb, ventral view; (B) male palpal bulb, lateral view; (C) female epigyne, ventral view. Abbreviations: *C*, conductor; *E*, embolus; *LAC*, lateral apophysis of conductor; *MA*, median apophysis; *MS*, median septum; *ST*, subtegulum; *T*, tegulum. *Redrawn and modified from Santos et al. (2003).*

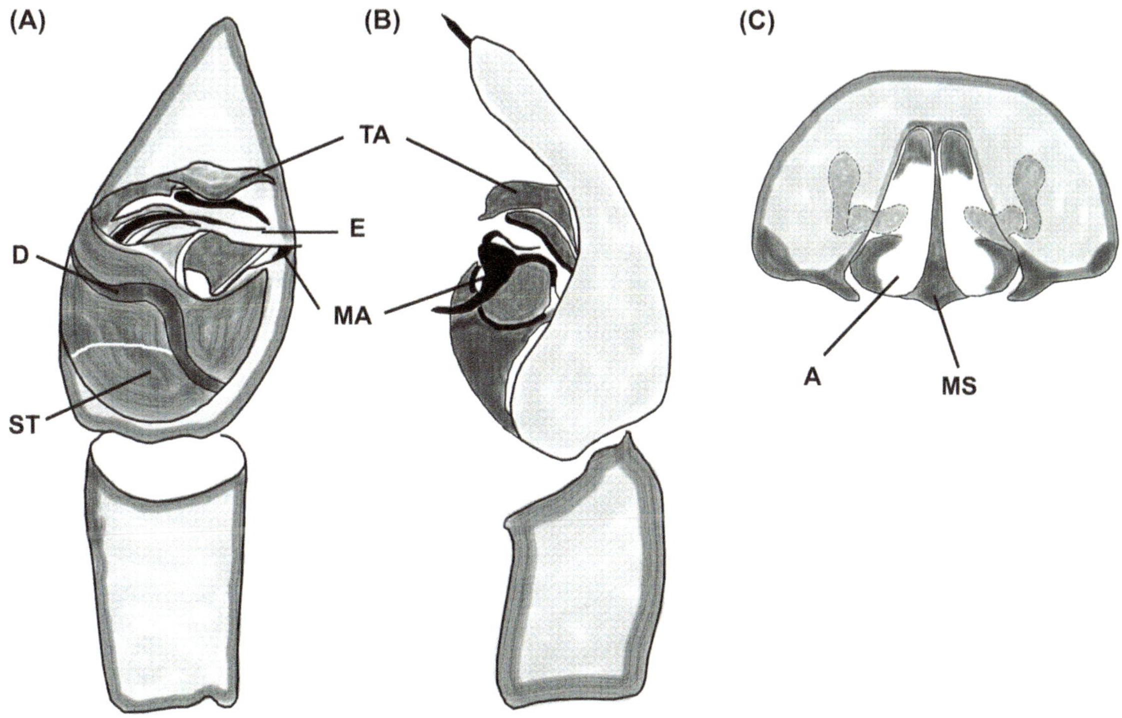

FIGURE 16.13 *Arctosa sapiranga* (Lycosidae): (A) Male palpal bulb, ventral view; (B) male palpal bulb, lateral view; (C) female epigyne, ventral view. Abbreviations: *A*, atrium; *D*, duct; *E*, embolus; *MA*, median apophysis, *MS*, median septum; *ST*, subtegulum; *TA*, terminal apophysis. *Redrawn and modified from da Silva & Lise (2009).*

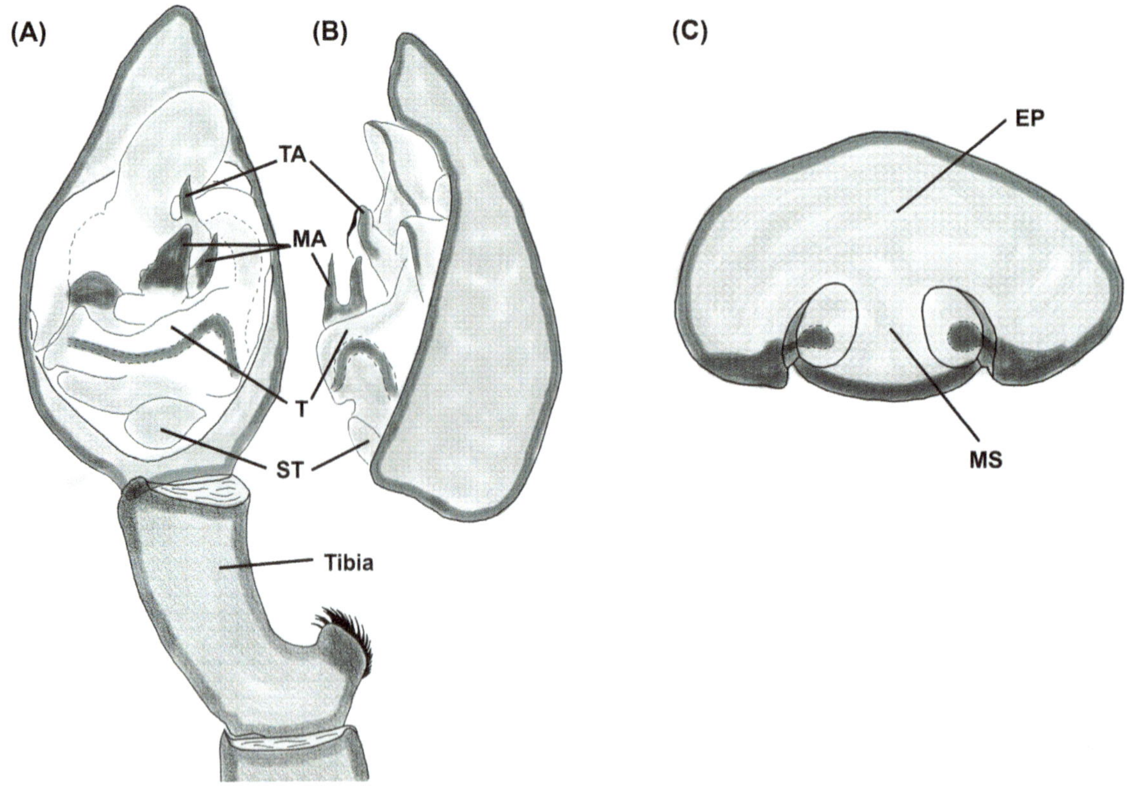

FIGURE 16.14 *Agalenocosa* (Lycosidae): (A) Male palpal bulb, ventral view; (B) male palpal bulb, lateral view; (C) female epigyne, ventral view. Abbreviations: *EP*, epigynal plate; *MA*, median apophysis; *MS*, median septum; *ST*, subtegulum; *T*, tegulum; *TA*, terminal apophysis. *Redrawn and modified from Piacentini (2014).*

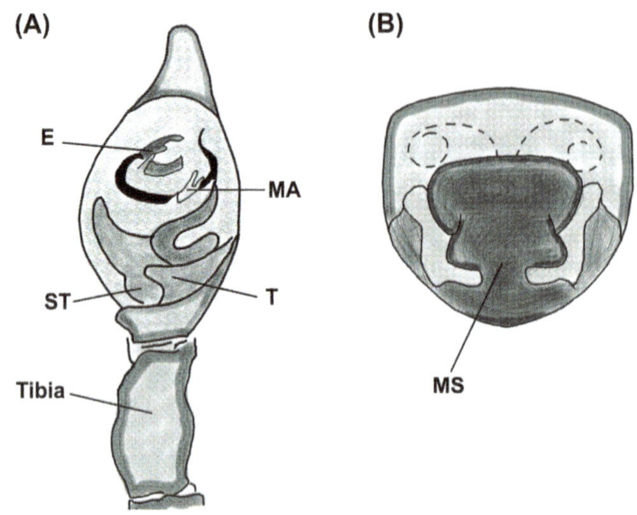

FIGURE 16.15 *Diapontia* (Lycosidae): (A) Male palpal bulb, ventral view; (B) female epigyne, ventral view. Abbreviations: *E*, embolus; *MA*, median apophysis; *MS*, median septum; *ST*, subtegulum; *T*, tegulum. *Redrawn and modified from Mello-Leitão (1944).*

Arachnida: Araneae: Lycosidae: *Aglaoctenus*: Species

Identification keys are constructed separately for males and females since the main morphological distinguished characters are genitalic. Males of *A. puyen* (Piacentini, 2011) and *A. yacitata* (Piacentini, 2011) are unknown.

1 Males 2

1′ Females 4

2(1) Cymbium with anterior portion not elongated 3

2′ Cymbium with anterior portion elongated ..... *Aglaoctenus oblongus* (Koch, 1847) [Brazil. Argentina. Uruguay]

3(2) Median apophysis wide and base of cymbium conspicuous ..... *Aglaoctenus castaneus* (Mello-Leitão, 1942) [Ecuador. Brazil. Peru. Argentina]

3′ Median apophysis narrow and base of the cymbium inconspicuous ..... *Aglaoctenus lagotis* (Holmberg, 1876) [Colombia to Argentina]

4(1) Carapace with narrow marginal brown bands 5

4′ Carapace with wide marginal light brown bands ..... *Aglaoctenus puyen* (Piacentini, 2011) [Argentina: Río Negro, Neuquén]

5(4) Median septum of epigyne with inverted T-shape 6

5′ Median septum of epigyne subquadrangular ..... *Aglaoctenus oblongus* (Koch, 1847) [Brazil. Argentina. Uruguay]

6(5) Carapace with light yellow/brown submarginal bands 7

6′ Carapace with orange submarginal bands ..... *Aglaoctenus yacitata* (Piacentini, 2011) [Argentina: Misiones]

7(6) Median septum T-shaped, shorter longitudinally and wider transversally ..... *Aglaoctenus castaneus* (Mello-Leitão 1942) [Ecuador. Brazil. Peru. Argentina]

7′ Median septum T-shaped, longer longitudinally and narrower transversely ..... *Aglaoctenus lagotis* (Holmberg 1876) [Colombia to Argentina]

Arachnida: Araneae: Lycosidae: *Agalenocosa*: Species

Identification keys are constructed separately for males and females since the main morphological distinguished characters are genitalic. Males of *A. gamas* (Piacentini, 2014) and *A. gentilis* (Mello-Leitão, 1944) are unknown.

1 Males 2

1′ Females 7

2(1) Body coloration light brown with a light brown median band and a conspicuous Y-shaped dark mark that extends from the fovea to the posterior lateral eyes, submarginal bands pale 3

2′ Body coloration dark ..... *Agalenocosa luteonigra* (Mello-Leitão, 1945) [Argentina. Uruguay]

3(2) Male palp with tibial apophysis 4

3′ Male palp without tibial apophysis ..... *Agalenocosa grismadoi* (Piacentini, 2014) [Paraguay. Argentina]

4(3) Body size 4–6.5mm 5

4′ Body size 7–10mm 6

5(4) Tibial apophysis short ..... *Agalenocosa punctata* (Mello-Leitão, 1944) [Argentina: Buenos Aires, Entre Ríos]

5′ Tibial apophysis large and straight ..... *Agalenocosa velox* (Keyserling, 1891) [Brazil. Argentina]

6(4) Terminal apophysis broad ..... *Agalenocosa tricuspidata* (Tullgren, 1905) [Argentina: Jujuy, Catamarca, Salta, Tucumán]

6′ Terminal apophysis narrow ..... *Agalenocosa pirity* (Piacentini, 2014) [Argentina: Buenos Aires, Entre Ríos, Santa Fe]

7(1) Body coloration light brown with a light brown median band and a conspicuous Y-shaped dark mark that extends from the fovea to the posterior lateral eyes, submarginal bands pale 8

7′ Body coloration dark .... *Agalenocosa luteonigra* (Mello-Leitão, 1945) [Argentina. Uruguay]

8(7) Straight carapace in lateral view 9

8′ Domed carapace in lateral view .... *Agalenocosa grismadoi* (Piacentini, 2014) [Paraguay. Argentina]

9(8) Epigynal plate not covering the anterior ventral side of the abdomen 10

9′ Epigynal plate large covering almost the anterior ventral side of the abdomen .... *Agalenocosa velox* (Keyserling, 1891) [Brazil. Argentina]

10(9) Body size <6 mm; tibia I ventrally with 2-0-0 or 2-2-0 spines 11

10′ Body size 6.4–10mm; tibia I ventrally with 0-2-0 spines .... *Agalenocosa pirity* (Piacentini, 2014) [Argentina: Buenos Aires, Entre Ríos, Santa Fé]

11(10) Head of spermatheca rounded 12

11′ Head of spermatheca enlarged .... *Agalenocosa tricuspidata* (Tullgren, 1905) [Argentina: Catamarca, Jujuy, Salta, Tucumán]

12(11) Stalk of spermathecae almost of the same length than the head 13

12′ Stalk of spermathecae nearly parallel and longer than the head .... *Agalenocosa gamas* (Piacentini, 2014) [Argentina: Santa Fé]

13(12) Stalk of spermathecae with a protrusion .... *Agalenocosa gentilis* (Mello-Leitão, 1944) [Argentina: Northern Buenos Aires]

13′ Stalk of spermathecae without a protrusion .... *Agalenocosa punctata* (Mello-Leitão, 1944) [Argentina: Buenos Aires, Entre Ríos]

Arachnida: Araneae: Pisauridae: Genera

1 Males with distal tegular projection and median apophysis slightly developed, and very short embolus (Fig. 16.16 A, B); epigyne of females with lateral lobes medially projected over the middle field well developed (Fig. 16.16 C) .... ***Architis* [p. 552]**

1′ Males with a well-developed distal tegular projection and median projection, and very long embolus (Fig. 16.17 A, B); epigyne of females with slightly developed lateral lobes and middle field (Fig. 16.17 C) .... ***Thaumasia* [p. 555]**

Arachnida: Araneae: Pisauridae: *Architis*: Species

Identification keys are constructed separately for males and females since the main morphological distinguished characters are genitalic. Males of *A. altamira* (Santos, 2007), *A. dianasilvae* (Santos, 2007), *A. ikuruwa* (Carico, 1981), *A. turvo* (Santos, 2007), and *A. fritzmuelleri* (Santos, 2007) are unknown.

The classification is modified and updated from the identification key presented by Santos (2007).

1 Males 2

1′ Females 12

2(1) Palpal tibia with dorsal and ventral branches of retrolateral tibial apophysis fused at the base 3

2′ Dorsal and ventral branches of retrolateral tibial apophysis not fused at the base 6

3(2) Coxae I venter with sparse blunt spines 4

3′ Coxae I without such spines, embolus robust .... *Architis maturaca* (Santos, 2007) [Brazil: Maturacá]

4(3) Retrolateral tibial apophysis ventral branch not fused to ventral tibial apophysis 5

4′ Retrolateral tibial apophysis ventral branch, fused to ventral tibial apophysis and folded dorsally .... *Architis tenuis* (Simon, 1898) [From Panama to Brazil]

5(4) Ventral retrolateral tibial apophysis almost as long as dorsal branch .... *Architis comaina* (Santos, 2007) [Peru: Amazonas]

PHYLUM ARTHROPODA: INTRODUCTION AND ARACHNIDA

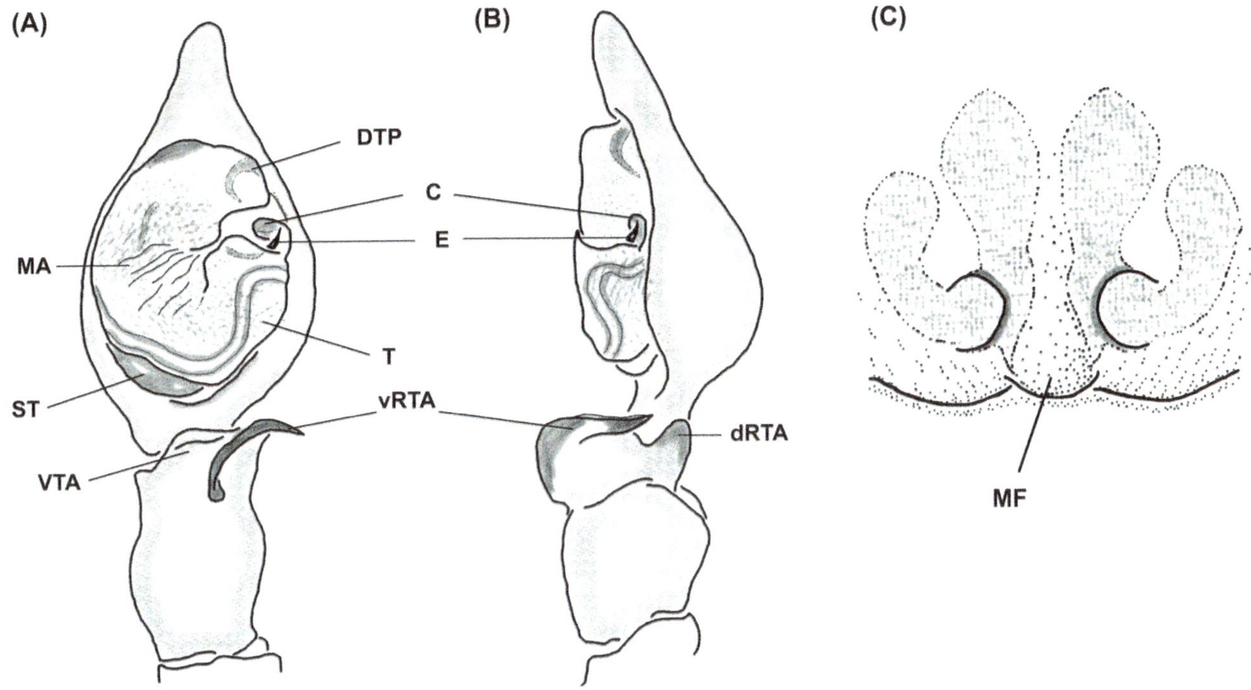

FIGURE 16.16 *Architis* (Pisauridae): (A) Male palpal bulb, ventral view; (B) male palpal bulb, lateral view; (C) female epigyne, ventral view. Abbreviations: *C*, conductor; *dRTA*, dorsal branch of retrolateral tibial apophysis; *DTP*, distal tegular projection; *E*, embolus; *MA*, median apophysis; *MF*, median field; *ST*, subtegulum; *T*, tegulum; *vRTA*, ventral branch of retrolateral tibial apophysis; *VTA*, ventral tibial apophysis. *Redrawn and modified from Santos (2007).*

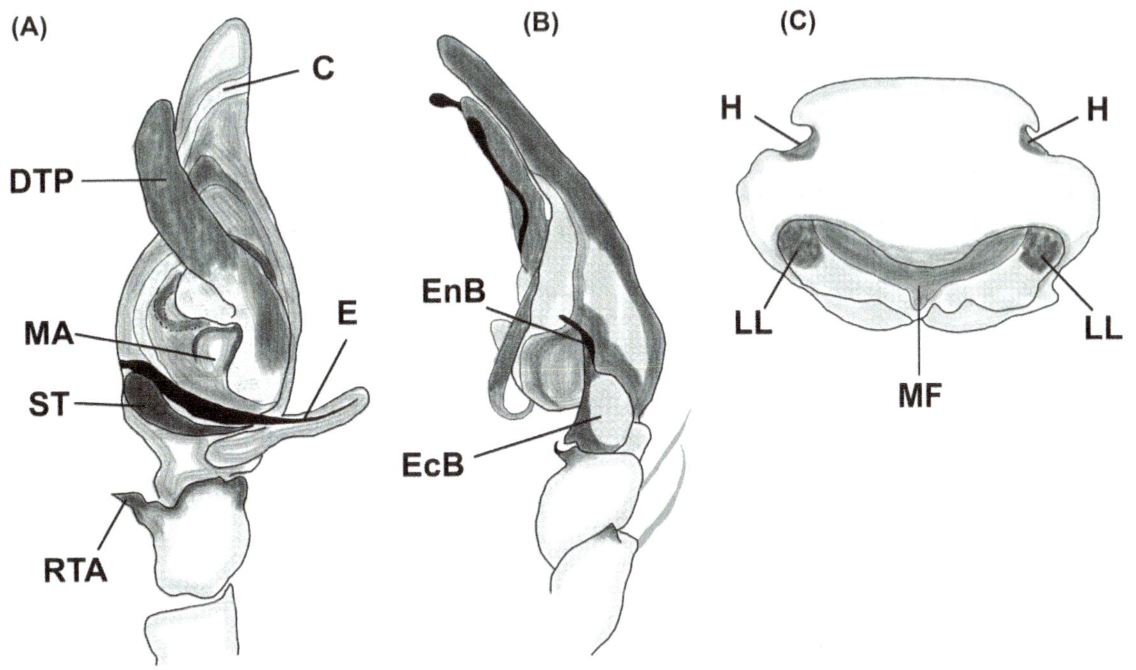

FIGURE 16.17 *Thaumasia* (Pisauridae): (A) Male palpal bulb, ventral; (B) male palpal bulb, lateral; (C) female epigyne, ventral. Abbreviations: *C*, conductor; *DTP*, distal tegular projection; *E*, embolus; *EcB*, ectal branch of retrolateral tibial apophysis; *EnB*, ental branch of retrolateral tibial apophysis; *H*, hood; *LL*, lateral lobes; *MA*, median apophysis; *MF*, middle field; *RTA*, retrolateral tibial apophysis; *ST*, subtegulum.

5′ Ventral retrolateral tibial apophysis length <0.5x dorsal branch .... *Architis cymatilis* (Carico, 1981)
[Colombia to Brazil]

6(2) Ventral tibial apophysis articulated with pedipalpal tibia apex, subtegulum hidden under tegulum 7

6′ Ventral tibial apophysis fused with pedipalpal tibia apex, subtegulum visible in ventral view 9

7(6) Retrolateral tibial apophysis dorsal branch almost as large or larger than ventral branch 8

7′ Retrolateral tibial apophysis dorsal branch smaller than ventral branch small .... *Architis tenuipes* (Simon, 1898)
[Brazil: Acre, Amazonas, Mato Grosso, Pará]

8(7) Ventral tibial apophysis hidden under anterior projection of palpal tibia in ventral view .... *Architis erwini* (Santos, 2007)
[Ecuador: Francisco de Orellana]

8′ Ventral tibial apophysis entirely visible in ventral view .... *Architis helveola* (Simon, 1898)
[Colombia. Ecuador. Brazil]

9(6) Distal tegular projection without such apex; ventral tibial apophysis with a retrolateral projection 10

9′ Distal tegular projection with a narrow apex, pointing anteriorly .... *Architis colombo* (Santos, 2007)
[Brazil: Parana]

10(9) Conductor with a prolateral lobe 11

10′ Conductor without prolateral lobe .... *Architis robusta* (Carico, 1981)
[Brazil: Bahía, Pará]

11(10) Ventral tibial apophysis deeply notched prolaterally .... *Architis capricorna* (Carico, 1981)
[Brazil: Sao Pablo]

11′ Ventral tibial apophysis without prolateral notch .... *Architis brasiliensis* (Mello-Leitão, 1940)
[Brazil: Central and Southern]

12(1) Copulatory duct visible in dorsal view, free or encapsulated 13

12′ Copulatory duct inconspicuous .... *Architis tenuis* (Simon, 1898)
[Panama to Brazil]

13(12) Copulatory duct encapsulated inside a sclerotised case or in epigynal wall folds 14

13′ Copulatory duct not encapsulated 21

14(13) Copulatory duct inside subcylindrical, sclerotised cases 15

14′ Copulatory duct inside differently shaped case or epigynal wall folds 16

15(14) Epigyne with wide, lateral copulatory openings .... *Architis cymatilis* (Carico, 1981)
[Colombia to Brazil]

15′ Copulatory openings covered by a wide transverse keel .... *Architis dianasilvae* (Santos, 2007)
[Peru: Pasco]

16(14) Epigyne without posteriolateral atria 17

16′ Epigyne with a pair of large, posteriolateral atria and median field triangular .... *Architis maturaca* (Santos, 2007)
[Brazil: Maturacá]

17(16) Copulatory duct inside epigynal wall folds in at least half their trajectory 18

17′ Copulatory duct inside globular to ovoid sclerotised cases 20

18(17) Copulatory openings covered with laterally directed, sclerotised hoods; copulatory ducts inside epigynal wall folds that are not sinuous 19

18′ Copulatory openings large, exposed; copulatory duct inside sinuous epigynal wall folds .... *Architis comaina* (Santos, 2007)
[Peru: Amazonas]

19(18) Epigynal wall folds covering only the posterior half of copulatory duct .... *Architis ikuruwa* (Carico, 1981)
[Peru. Bolivia]

19′ Epigynal wall folds completely covering copulatory duct .... *Architis altamira* (Santos, 2007)
[Brazil: Pará]

20(17) Epigyne with median field as long as wide and with a pair of posterior lateral pockets . ... *Architis turvo* (Santos, 2007)
[Brazil: Rio Grande do Sul]

20′ Epigyne with median field longer than wide, without posterior lateral pockets .... *Architis tenuipes* (Simon, 1898)
[Brazil: Acra, Amazonas, Mato Grosso, Pará]

21(13) Epigynal lateral lobes not projected medially; epigyne median field not sagittate 22

21′ Epigynal lateral lobes projecting medially, forming rounded keels over median field; median field sagittate .... *Architis helveola* (Simon, 1898)
[Colombia. Ecuador. Brazil]

22(21) Fertilization duct short, not wrapping copulatory duct 23

22′ Fertilization duct long, wrapping long and sinuous copulatory duct .... *Architis erwini* (Santos, 2007)
[Ecuador: Francisco de Orellana]

23(22) Copulatory duct almost as wide as head of spermathecae and with gently sinuous trajectory 24

23′ Copulatory duct visibly narrower than head of spermathecae and with spiral trajectory 25

24(23) Median field restricted to anterior half of epigyne, copulatory openings covered by posteriorly directed sclerotized hoods .... *Architis colombo* (Santos, 2007)
[Brazil: Parana]

24′ Median field large and rhomboid; sclerotised hoods over copulatory openings absent .... *Architis fritzmuelleri* (Santos, 2007)
[Brazil: Santa Catarina]

25(23) Spermathecae stalk long, forming loops 26

25′ Spermathecae stalk short, gently curved .... *Architis robusta* (Carico, 1981)
[Brazil: Bahía, Pará)

26(25) Copulatory duct flattened, not spiraled between spermathecae .... *Architis brasiliensis* (Mello-Leitão, 1940)
[Brazil: Central and Southern]

26′ Copulatory duct cylindrical spiraled between spermathecae .... *Architis capricorna* (Carico, 1981)
[Brazil: Sao Pablo]

Arachnida: Araneae: Pisauridae: *Thaumasia*: Species

Identification keys are constructed separately for males and females since the main morphological distinguished characters are genitalic. Males of *T. caracarai* (da Silva & Carico, 2012) are unknown. Females of *T. heterogyna* (Chamberlin & Ivie, 1936), *T. scoparia* (Simon, 1888), *T. diaisi* (da Silva & Carico, 2012), *T. lisei* (da Silva & Carico, 2012), and *T. peruana* (da Silva & Carico, 2012) are unknown.

1 Males 2

1′ Females 17

2(1) Median apophysis small and short 3

2′ Median apophysis large and elongated 5

3(2) Median apophysis apex rounded 4

3′ Median apophysis apex acute and with triangular tip .... *Thaumasia heterogyna* (Chamberlin & Ivie, 1936)
[Panama to Brazil]

4(3) Retrolateral tibial apophysis ental branch short and ventrally curved at tip .... *Thaumasia velox* (Simon, 1898)
[Panama to Argentina]

4′ Retrolateral tibial apophysis ental branch long and curved on distal half .... *Thaumasia scoparia* (Simon, 1888)
[Venezuela: Orinoco]

5(2) Median apophysis divided or bifurcated 6

5′ Median apophysis entire 12

6(5) Median apophysis bifurcated 7

6′ Median apophysis divided 10

7(6) Median apophysis bifurcated just in the apex 8

7′ Median apophysis entirely bifurcated .... *Thaumasia argenteonotata* (Simon, 1898)
[Mexico to Brazil]

8(7) Median apophysis apex acute 9

8′ Median apophysis apex rounded .... *Thaumasia oriximina* (da Silva & Carico, 2012)
[Brazil: Distrito Federal, Pará]

9(8) Retrolateral tibial apophysis with ventral branch ectal branch of retrolateral tibial apophysis long and slender tip .... *Thaumasia annulipes* (Pickard-Cambridge, 1903)
[Suriname. Brazil. Peru]

9′ Retrolateral tibial apophysis without ventral branch anterior half of the ental branch of retrolateral tibial apophysis lamellar .... *Thaumasia xingu* (da Silva & Carico, 2012)
[Colombia to Brazil]

10(6) Median apophysis without such curvature 11

10′ Median apophysis with a median curvature .... *Thaumasia onca* (da Silva & Carico, 2012)
[Colombia to Brazil]

11(10) Median apophysis with a long and slender apex .... *Thaumasia diaisi* (da Silva & Carico 2012)
[Ecuador. Brazil]

11′ Median apophysis without slender apex .... *Thaumasia acreana* (da Silva & Carico, 2012)
[Brazil: Acre, Amazonas]

12(5) Median apophysis with rounded apex 13

12′ Median apophysis with slender or acute apex 16

13(12) Median apophysis without grooved apex 14

13′ Median apophysis with grooved apex .... *Thaumasia senilis* (Perty, 1833)
[Costa Rica to Paraguay]

14(13) Median apophysis apex without curvature 15

14′ Median apophysis apex prolaterally curved .... *Thaumasia lisei* (da Silva & Carico, 2012)
[Brazil: Amazonas]

15(14) Median apophysis long; retrolateral tibial apophysis ectal branch long and triangular .... *Thaumasia abrahami* (Mello-Leitão, 1948)
[Honduras to Brazil]

15′ Median apophysis shorter; retrolateral tibial apophysis ectal branch shorter and domed .... *Thaumasia caxiuana* (da Silva & Carico, 2012)
[Brazil: Pará]

16(12) Body size 15mm .... *Thaumasia peruana* (da Silva & Carico, 2012)
[Peru: Huanuco, San Martin]

16′ Body size 7mm .... *Thaumasia hirsutochela* (da Silva & Carico, 2012)
[Costa Rica to Brazil]

17(1) Epigyne middle field without a septum 18

17′ Epigyne middle field with a median septum .... *Thaumasia argenteonotata* (Simon, 1898)
[Mexico to Brazil]

18(17) Epigyne middle field with large central projection 19

18′ Epigyne middle field with small projection 20

19(18) Copulatory ducts straight; secondary copulatory duct enlarged .... *Thaumasia senilis* (Perty, 1833)
[Costa Rica to Paraguay]

19′ Copulatory ducts with undulated trajectory .... *Thaumasia oriximina* (da Silva & Carico, 2012)
[Brazil: Distrito Federal, Pará]

20(18) Epigyne middle field anterior projection rounded 21

20′ Epigyne middle field anterior projection triangular 22

21(20) Epigyne middle field cordate .... *Thaumasia xingu* (da Silva & Carico, 2012)
[Colombia to Brazil]

21′ Epigyne middle field not cordate .... *Thaumasia annulipes* (Pickard-Cambridge, 1903)
[Suriname. Brazil. Peru]

22(20) Epigyne anterior middle field noticeably projected 23

22′ Epigyne anterior middle field slightly projected 25

23(22) Epigyne middle field without such hoodlike shape 24

23′ Epigyne middle field hoodlike and conspicuous .... *Thaumasia abrahami* (Mello-Leitão, 1948)
[Honduras to Brazil]

24(23) Epigyne middle field with deep lateral excavation .... *Thaumasia onca* (da Silva & Carico, 2012)
[Colombia to Brazil]

24′ Epigyne middle field normal lateral excavation .... *Thaumasia hirsutochela* (da Silva & Carico, 2012)
[Costa Rica to Brazil]

25(22) Epigyne middle field excavated 26

25′ Epigyne middle field not excavated 27

26(25) Spermathecae with slender head and wide copulatory ducts .... *Thaumasia acreana* (da Silva & Carico, 2012)
[Brazil: Acre, Amazonas]

26′ Spermathecae with stout head and thinner copulatory ducts .... *Thaumasia caracarai* (da Silva & Carico, 2012)
[Mexico to Brazil]

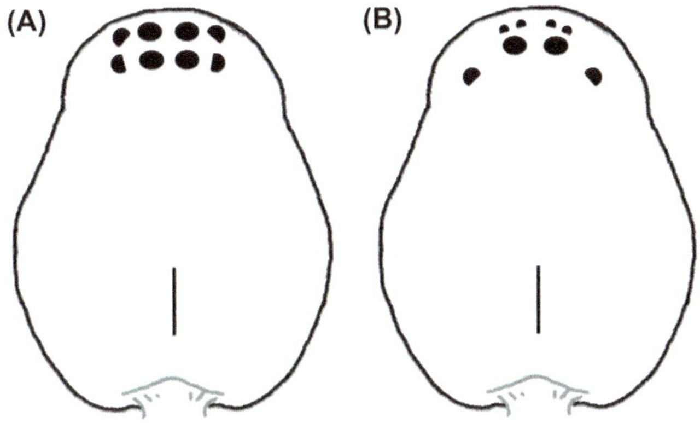

FIGURE 16.18 Eyes arrangement of spider subfamilies within Trechaleidae: (A) Rhoicininae; (B) Trechaleinae.

27(25) Epigyne middle field laterally, slightly excavated reaching about 1/3 of epigyne length, short and folded copulatory ducts .... *Thaumasia caxiuana* (da Silva & Carico, 2012)
[Brazil: Pará]

27 Epigyne middle field laterally deeply excavated reaching almost 1/2 of the epigyne length, short and unfolded copulatory ducts .... *Thaumasia velox* (Simon, 1898)
[From Panama to Argentina]

Arachnida: Araneae: Trechaleidae: Subfamilies

1 Width of anterior eye row distinctly larger than the length of the posterior median eyes (Fig. 16.18 A) .... **Rhoicininae [p. 557]**

1′ Width of anterior eye row equal to or only slightly larger than the length of the posterior median eyes (Fig. 16.18 B) .... **Trechaleinae [p. 557]**

Arachnida: Araneae: Trechaleidae: Rhoicininae: Genera

1 Posterior eye row straight or slightly recurved with posterior lateral eyes next to the posterior median eyes 2

1′ Posterior eye row strongly recurved with posterior lateral eyes widely separated from the posterior median eyes .... *Heidrunea*
[Brazil: Amazonas]

2(1) Males with median apophysis short and coiled; female epigyne middle field with a pair of circular membranous areas .... *Barrisca*
[Colombia. Venezuela. Peru]

2′ Males with median apophysis larger and slightly curved; female epigyne middle field without such circular membranous areas .... *Rhoicinius*
[Guayana. Ecuador. Brazil. Peru]

Arachnida: Araneae: Trechaleidae: Trechaleinae: Genera

The classification is modified and updated from the provisional identification key presented by Carico & da Silva (2010).

1 Tarsi flexible and usually bent 2

1′ Tarsi not flexible and straight 6

2(1) Metatarsi flexible and usually bent 3

2′ Metatarsi not flexible and always straight 4

3(2) Tibia I with six or more ventral pairs of macrosetae .... *Syntrechalea*
[Venezuela. Colombia. Suriname. Brazil. Peru. Bolivia. Paraguay]

3′ Tibia I with four or less ventral pairs of macrosetae .... *Hesydrus*
[Venezuela. Colombia. Ecuador. Peru. Bolivia]

4(2) Males with ventral division of median apophysis rounded in outline; females with middle field of epigyne divided into two lateral parts or wider posteriorly if not divided 5

4′ Males with ventral division of median apophysis angular in outline (Fig. 16.19 A); females with middle field of epigyne not divided posteriorly and usually wider anteriorly (Fig. 16.19 B) .... *Trechalea*
[Colombia. Brazil. Peru. Argentina]

5(4) Males with ventral division of median apophysis small; females with middle field of epigyne divided into two lateral parts .... *Trechaleoides*
[Brazil. Paraguay. Argentina. Uruguay]

5′ Males with ventral division of median apophysis large; females with middle field of epigyne not divided into two lateral parts, wider at posterior margin .... *Paratrechalea*
[Brazil. Argentina. Uruguay]

6(1) Both sexes with median band absent or, if present, with a uneven edge and not bold 7

6′ Both sexes with a broad bold dark band extending the length of body and limited laterally by narrow white lines .... *Dossenus*
[Colombia. Peru]

7(6) Leg pattern not as above; male palp with guide tip directed retrolaterally 8

7′ Legs with several distinct parallel longitudinal dark lines; male palp with guide tip directed distally .... *Dyrines*
[Venezuela. Guyana. Brazil. Peru]

8(7) Retrolateral tibial apophysis of males composed of one division or if two; the ectal division located proximally along rim distant from ental division 9

8′ Retrolateral tibial apophysis of males composed of two divisions located at or near the distal rim of tibia .... *Enna*
[Venezuela. Guyana. Brazil. Peru. Bolivia]

9(8) Male palp with tegulum without projection 10

9′ Male palp with tegulum with projection directed distally .... *Amapalea*
[Brazil: Amapá. Pará]

10(9) Male palp with tegulum with transverse groove; middle field of female epigyne short. T-shaped and not situated between a distinct pair of lateral lobes .... *Caricelea*
[Peru: Cusco]

10′ Male palp with tegulum without groove; middle field of female epigyne large; straight and situated between a distinct pair of lateral lobes .... *Paradossenus*
[From Colombia to Argentina]

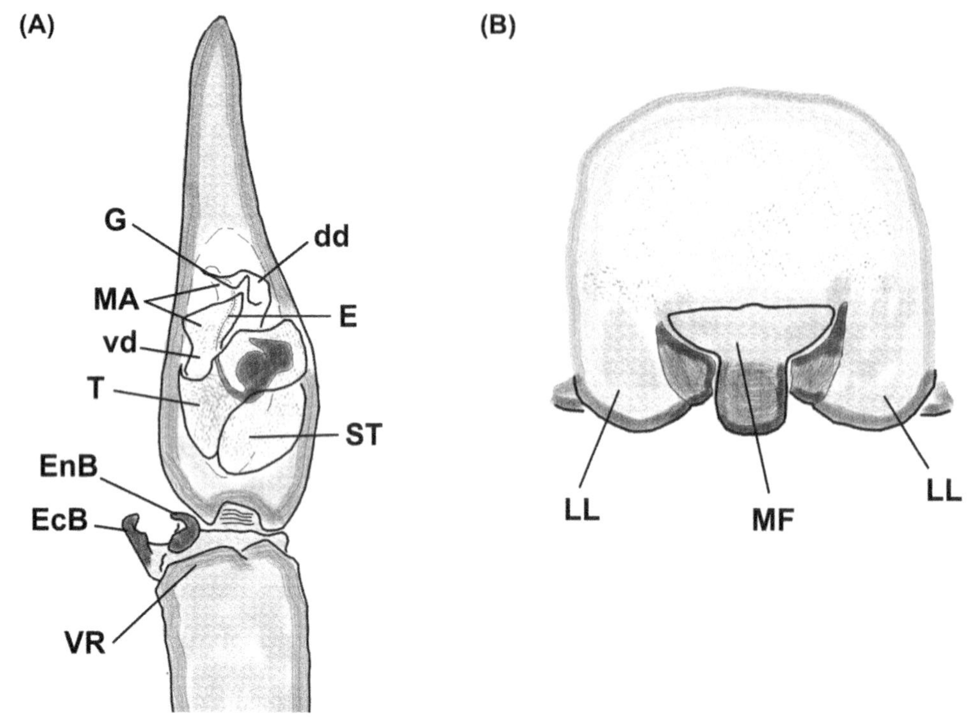

FIGURE 16.19 Trechaleidae: (A) Male palpal bulb, ventral view; (B) female epigyne, ventral view. Abbreviations: *dd*, dorsal division; *E*, embolus; *EcB*, ectal branch of retrolateral tibial apophysis; *EnB*, ental branch of retrolateral tibial apophysis; *G*, guide; *LL*, lateral lobes; *MA*, median apphysis; *MF*, middle field; *ST*, subtegulum; *T*, tegulum; *vd*, ventral division; *VR*, ventrodistal rim. *Redrawn and modified from Carico (1993).*

REFERENCES

Carico, J.E. 1993. Revision of the genus *Trechalea* Thorell (Araneae. Trechaleidae) with a review of the taxonomyof the Trechaleidae and Pisauridae of the western hemisphere. The Journal of Arachnology 21: 226–257.

Carico, J.E. & E.C. da Silva. 2010. Taxonomic review of the Neotropical spider genus *Paradossenus* (Araneae: Lycosoidea: Trechaleidae: Trechaleinae) with a new erection of the subfamily Trechaleinae and a key to included genera. The Journal of Arachnology 38: 212–236.

da Silva, E.C. & A.A. Lise. 2009. A new species of the world spider genus *Arctosa* (Araneae. Lycosidae) from Southern Brazil. Revista Peruana de Biología 15: 27–32.

Foelix, R. 2011. Biology of Spiders. Third ed. Oxford University Press. Oxford. U.K. 417 pp.

Höfer, H. & A.D. Brescovit. 2000. A revision of the Neotropical spider genus *Ancylometes* Bertkau (Araneae: Pisauridae). Insect Systematics & Evolution 31: 323–360.

Mello-Leitão, C.F. de. 1944. Arañas de la provincia de Buenos Aires. Revista del Museo de La Plata (N.S. Zool.) 3: 311–393.

Piacentini, L.N. 2014. A taxonomic review of the wolf spider genus *Agalenocosa* Mello-Leitão (Araneae. Lycosidae). Zootaxa 3790: 1–35.

Ramírez, M.J. 2014. The morphology and phylogeny of dionychan spiders (Araneae: Araneomorphae). Bulletin of the American Museum of Natural History 390: 1–374.

Santos, A.J. 2007. A revision of the Neotropical nursery-web spider genus *Architis* (Araneae: Pisauridae). Zootaxa 1578: 1–40.

Santos, A.J., E.S.S. Álvares & A.D. Brescovit. 2003. On the third valid species of the genus *Aglaoctenus* tullgren (Araneae. Lycosidae). Revista Ibérica de Aracnología 8: 89–92.

World Spider Catalog. 2019. World Spider Catalog. Version 20.5. Natural History Museum Bern. online at http://wsc.nmbe.ch. Accessed 2 September 2019. doi: 10.24436/2.

Chapter 17

Arthropoda: Introduction to Crustacea and the Class Hexapoda

James H. Thorp, Neusa Hamada and D. Christopher Rogers

Chapter Outline

Introduction to Crustacea

D. Christopher Rogers

Kansas Biological Survey, and The Biodiversity Institute, University of Kansas, Lawrence, KS, United States

INTRODUCTION

Within the Arthropoda recent molecular, morphological, and paleontological studies have suggested many classification hypotheses that upend traditional views. The preponderance of evidence demonstrates that the Hexapoda (Apterygota + Insecta) represent a clade nested within the Crustacea (Reiger et al., 2005). This Crustacea + Hexapoda clade has been referred to as "Tetraconata" or "Pancrustacea" (Reiger et al., 2005; Oakley et al., 2013), and this nomenclature was followed in the Thorp and Covich (fourth edition) Volume I of the present series. However, in Volume II, we treated the Crustacea + Hexapoda clade under the name "Crustacea." We think that creating a new *nomen* for a previously named clade just because some other clade is found to nest within it will only create confusion. Instead we prefer to apply the principles of the International Code of Zoological Nomenclature (which do not officially apply to higher level taxonomy) and treat the Hexapoda as a clade within the Crustacea, as would be done for lower taxonomic levels.

Relationships within the Crustacea are clearer than ever before, and for higher categories we follow Schwentner et al. (2018). However, we revert to a more traditional organization (Ahyong et al., 2011) for the lower categories encompassed within this clade.

TERMINOLOGY AND MORPHOLOGY

Hexapods and the traditional "Crustacea" are fundamentally separated by the position of the genitalia. In the "crustaceans" the genitalia are distoventrally positioned on the thorax. Malacostracan males typically have the first pleopod pair (abdominal limbs) modified to transfer sperm from the thoracic gonopods to the female. The Hexapoda have the genitalia terminally positioned at the end of the abdomen, often withdrawn internally.

Thorp and Covich's Freshwater Invertebrates. https://doi.org/10.1016/B978-0-12-804225-0.00017-4

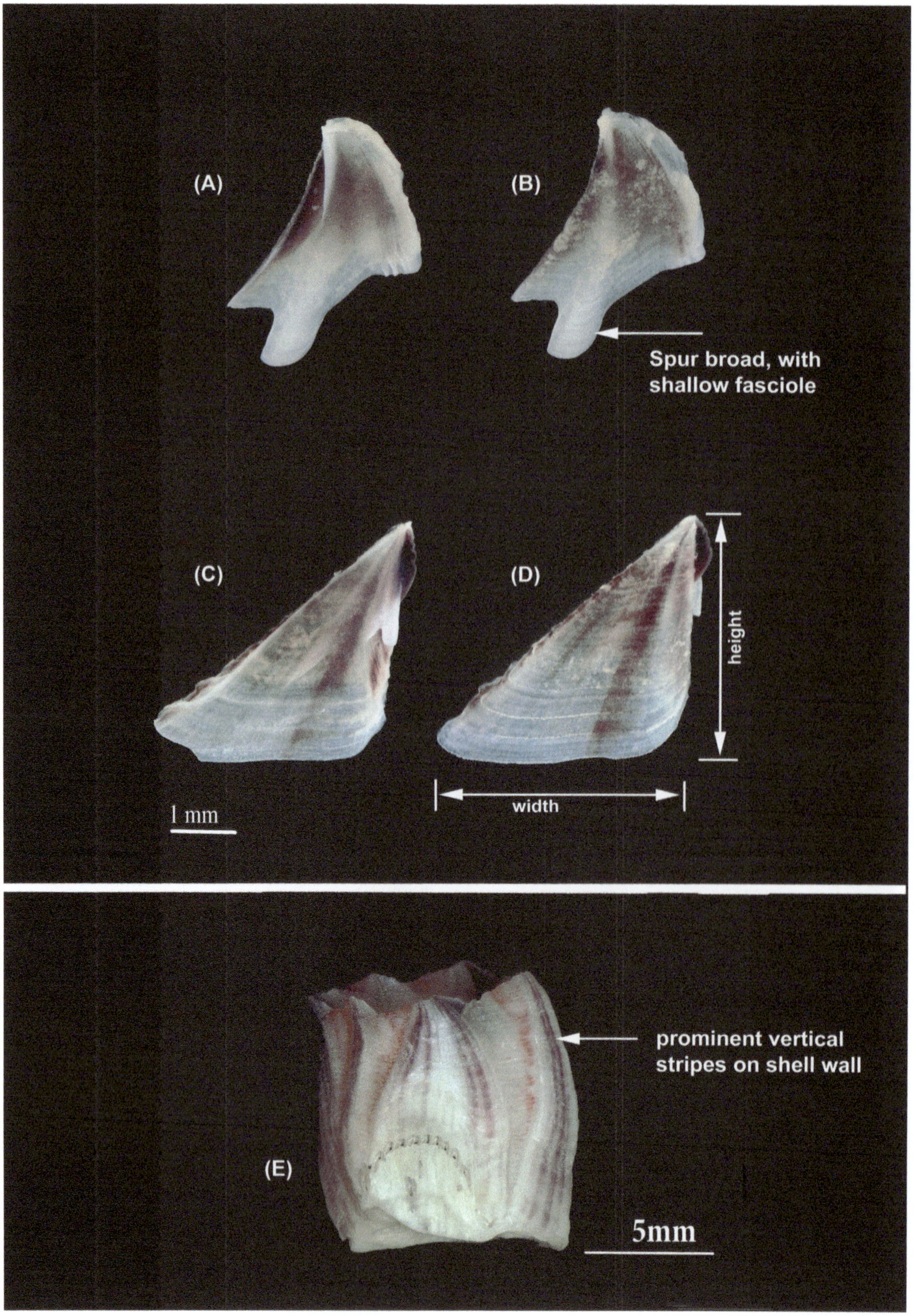

FIGURE 17.1 Example of an inland water barnacle (Cirripedia, *Amphibalanus* sp.): (A) tergum interior; (B) tergum exterior; (C) scutum interior; (D) scutum exterior; (E) whole shell exterior, side view. *From Van Syoc (2016).*

FIGURE 17.2 (A) Fairy shrimp *Artemia* sp.; (B) fairy shrimp *Thamocephalus* sp.; (C) cladoceran *Daphnia* sp.; (D) tadpole shrimp *Lepidurus bilobatus*. *Photographs from Thorp & Rogers (2011). Photographer Matthew A. Hill.*

FIGURE 17.3 Photograph of seed shrimp (Ostracoda) from Thorp & Rogers (2011). *Photographer Robert Moeller.*

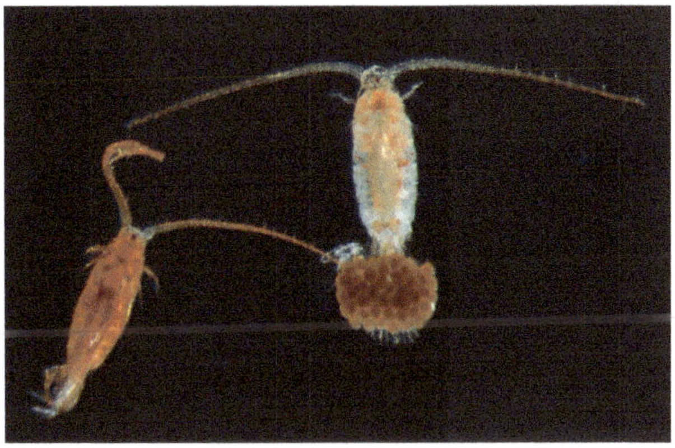

FIGURE 17.4 Male (left) and female (right) calanoid copepods from Thorp & Rogers (2011). *Photographer Mathew Moeller.*

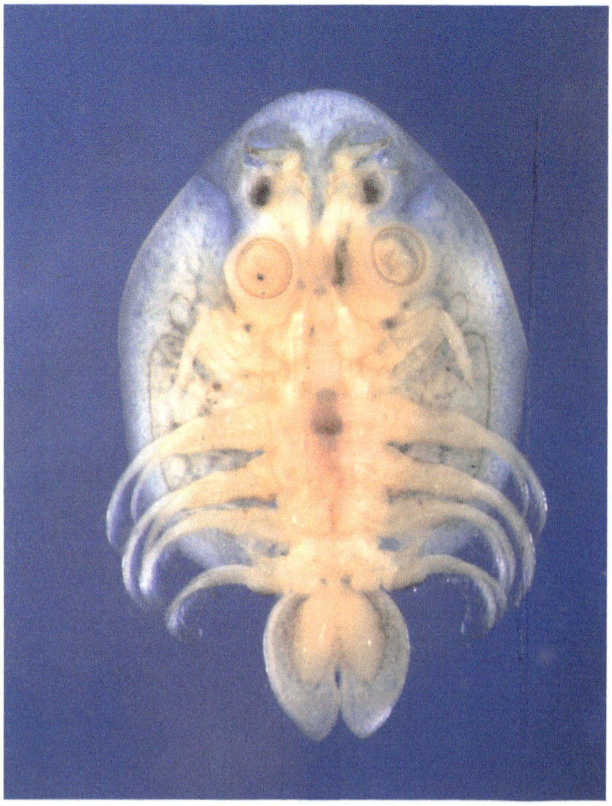

FIGURE 17.5 Fish lice *Argulus* (Branchiura). *Photographer John Pfieffer.*

FIGURE 17.6 Some members of the large class Malacostraca: (A) *Gammarus* sp. (Amphipoda); (B) aquatic sow bug (Isopoda); (C) the freshwater crab *Valdivia serrata* (Trichodactylidae); (D) mysid or opossum shrimps (Mysida). *Photographs by Mathew Hill (A, D), John Pfieffer (B) and Célio Magalhães (C).*

KEYS TO CRUSTACEA

Crustacea classes

1 Adults with genitalia positioned at posterior end of thorax; if limbs present then more than three pairs; two pairs of antennae; free living or parasitic; wings always absent 2

1′ Adults with genitalia positioned at abdominal terminus; larvae with limbs present or absent, if present, not more than three pairs (pseudolegs may be present in some larval dipterans and lepidopterans); one pair of antennae; wings present or absent .... **Hexapoda [p. 571]**

2(1) Limbs and eyes or at least eye stalks present; free living or parasitic, never entirely sessile 3

2′ Limbs present or absent; eyes absent; animal entirely sessile, immovably fused to substrate or host; barnacles (Fig. 17.1; see also Chapter 18) .... **Cirripedia [p. 582]**

3(2) Thoracopods segmented, never lamellar; carapace bivalved or not 4

3′ Thoracopods lamellar, not segmented (Fig. 17.2 A–B); carapace bivalved or not (Fig. 17.2 C–D; see also Chapter 19) .... **Branchiopoda [p. 590]**

4(3) Carapace not bivalved 5

4′ Carapace bivalved, enclosing entire animal (Fig. 17.3; see also Chapter 20) .... **Ostracoda [p. 636]**

5(4) Adult with naupliar eye present or absent, if present, then eyes compound, or at least eyestalks present 6

5′ Naupliar eye present, compound eyes and eyestalks absent (Fig. 17.4; see also Chapter 21) .... **Copepoda [p. 671]**

6(5) Four pairs of thoracopods present; mouth situated apically on a proboscis; telson always absent, but abdomen may be forked (Fig. 17.5; see also Chapter 22) .... Branchiura, one family: **Argulidae [p. 804]**

6′ Five or more thoracopod pairs; mouthparts never projected on a proboscis; telson or pleotelson present, uropods present (many taxa, including amphipods, isopods, decapods, and mysids; Fig. 17.6 A–D; see also Chapter 23) .... **Malacostraca [p. 812]**

REFERENCES

Ahyong, S.T., J.K. Lowry, M. Alonso, R.N. Bamber, G.A. Boxshall, P. Castro, S. Gerken, G.S. Karaman, J.W. Goy, D.S. Jones, K. Meland, D.C. Rogers & J. Svavarsson. 2011. Subphylum Crustacea Brünnich, 1772. Pages 1–237 *in*: Z-Q Zhang (ed.), Animal biodiversity: An outline of higher-level classification and survey of taxonomic richness. Zootaxa 3148: 1-237.

Oakley, T.H., J.M. Wolfe, A.R. Lindgren & A.K. Zaharoff. 2013. Phylotranscriptomics to bring the understudied into the fold: monophyletic Ostracoda, fossil placement, and pancrustacean phylogeny. Molecular Biology & Evolution 30: 215–233.

Reiger, J.C., J.W. Schultz & R.E. Kambic. 2005. Pancrustacean phylogeny: hexapods are terrestrial crustaceans and maxillopods are not monophyletic. Proceedings of the Royal Society B, 272: 395–401.

Schwentner, M., S. Richter, D.C. Rogers & G. Giribet. 2018. Tetraconatan phylogenomics with special focus on Malacostraca and Branchiopoda. Proceedings of the Royal Society B: 285: 20181524.

Thorp, J.H. & D.C. Rogers. 2011. Field Guide to Freshwater Invertebrates of North America. Academic Press, Elsevier. 274 p.

Van Syoc, R.J. 2016. Subclass Cirripedia. Pages 516–524 *in*: J.H. Thorp & D.C. Rogers (eds.), Volume II: Keys to Nearctic Fauna. Thorp and Covich's Freshwater Invertebrates, Academic Press, Elsevier, London.

Class Hexapoda

James H. Thorp
Kansas Biological Survey and Department of Ecology and Evolutionary Biology, University of Kansas, Lawrence, KS, United States

Neusa Hamada
Coordenação de Biodiversidade, Instituto Nacional de Pesquisas da Amazônia-INPA, Manaus, AM, Brazil

D. Christopher Rogers
Kansas Biological Survey, and The Biodiversity Institute, University of Kansas, Lawrence, KS, United States

INTRODUCTION[1]

This chapter includes taxa of the crustacean class Hexapoda that are adapted in some life stage for living on or under the water surface: the subclasses Entognatha (or Collembola, the "springtails"; Fig. 17.7) and Insecta. Although springtails occur predominately in moist terrestrial habitats, a few taxa occupy surfaces of freshwater lentic and lotic waters. Of the 14 insect orders with aquatic or semiaquatic members, five consist almost exclusively of species having aquatic larvae (a term also including nymphs and naiads): Ephemeroptera (mayflies), Megaloptera (alderflies, dobsonflies, fishflies, and hellgrammites), Odonata (damselflies and dragonflies), Plecoptera (stoneflies), and Trichoptera (caddisflies). The remaining 10 insect orders range from those with numerous to very few aquatic species, some of which are barely aquatic. Five orders (Coleoptera, Diptera, Hemiptera, Lepidoptera, and Neuroptera) are primarily terrestrial, but three (Coleoptera, Diptera, and Hemiptera) often represent a substantial proportion of aquatic insect diversity. Four other orders have so few aquatic, or perhaps semiaquatic, species that they are often not listed among the truly aquatic orders, though they are occasionally reported from tropical freshwaters. These four taxa are the orders Blattodea (roaches), Hymenoptera (mostly parasitoid wasps attacking aquatic eggs and larvae of other orders), Mecoptera (scorpionflies), and Orthoptera (grasshoppers).

The 14 insect orders discussed here vary in the extent of aquatic life stages. Six (Blattodea, Ephemeroptera, Hemiptera, Odonata, Orthoptera, and Plecoptera) have a hemimetabolous life cycle with three developmental stages: egg, larva, and adult (usually terrestrial). Eight orders (Coleoptera, Diptera, Hymenoptera, Lepidoptera, Mecoptera, Megaloptera, Neuroptera, and Trichoptera) have a holometabolous life cycle with four developmental stages: egg, larva, pupa, and adult.

Only about 4% of the known species of springtails in the tropics and southern hemisphere in general are associated with freshwater (Deharveng et al., 2008). Of those with some aquatic system link, all are associated with the surface film of aquatic systems. More detailed information on their phylogeny and Neotropical diversity can be found in Cipola et al. (2018), while general information on their biology and ecology of these small (0.12–1 mm) invertebrates are described in Thorp & O'Neill (2015) and on websites devoted to springtails, such as one maintained by Peter Bellinger, Kenneth Christiansen, and Janssens (www.collembola.org).

Classification of insects and springtails is limited here to the ordinal level because a companion volume "Keys to Neotropical Hexapoda" (Hamada et al., 2018) focused exclusively on Hexapoda, with one chapter on Collembola (Cipola et al., 2018) and 41 chapters on Insecta. Almost all Hexapoda covered in this volume are tropical or subtropical, but a few freshwater species occur on islands near Antarctica (see Chapter 1). These include mostly species with short generation times (e.g., midges) but also a few others, such as diving beetles.

LIMITATIONS

The ordinal level key below encompasses only the aquatic phase of the life cycle, including both adult and larvae where appropriate. More detailed keys to Hexapoda throughout the Neotropics are available in Volume III of this invertebrate book series (Hamada et al., 2018). While a number of texts describing the insects of a given country or region have been published, the only recent series that covers most of the Neotropics is "Aquatic Biodiversity in Latin America" (Pensoft Publishers). As of July 2018, however, only six volumes on insects had been released: four on Diptera and one each on Ephemeroptera and Plecoptera. Unfortunately, taxonomic knowledge has tended to focus on certain groups and countries, thereby potentially misleading students new to insect identification who are working in another region of the Neotropics. Moreover, many Neotropical habitats have not been adequately surveyed for new species.

TERMINOLOGY AND MORPHOLOGY

Entognatha

Detailed information on the morphology and taxonomy of Neotropical Collembola are discussed in Cipola et al. (2018), from which the following discussion is derived in part. These springtails are minute, wingless hexapods that

1 This chapter benefitted substantially from chapters on Hexapoda in the general coverage of Nearctic taxa in Vol. II (DeWalt & Resh, 2016) and from detailed chapters on Neotropical Hexapoda in Vol. III (Hamada et al., 2018).

can be identified by the presence of three structures (Fig. 17.7 A–D): (1) the **furcula** (or furca), a springing mechanism on abdominal segment IV; (2) the **collophore**, a terminal, tubelike and somewhat wedge-shaped structure on the ventral side of the first abdominal segment; and (3) the **tenaculum** (retinaculum), a ventral appendage of abdominal segment III which retains the furcula when not in use. The furcula is present in most species and in all taxa considered at least semiaquatic. It consists of a basal, unpaired piece (**manubrium**) and two arms. Each arm consists of a pair of basal **dentes** (singular dens) and a pair of distal **mucrones** (singular mucro). The furcula is typically folded under the abdomen by a midventral, clasplike structure (**tenaculum**). When the tenaculum releases the furcula, the latter springs backward and propels the animal several centimeters through the air (hence, the common name "springtail"). The furculae of aquatic species are dorsoventrally flattened with an enlarged mucro. This creates a broad contact area with the water surface, thereby allowing a more efficient strike on the water's surface and lessening any tendency to break through the surface tension. The furculae primarily functions in the springtail's escape from predators rather than serving in conventional locomotion.

Most aquatic individuals are less than 3 mm long. They have fewer abdominal segments (6) than possessed by insects (which number as high as 11 in primitive insects).

Insecta

Insects have three main body regions: the head, thorax, and abdomen, as illustrated for stoneflies (order Plecoptera) by DeWalt & Resh (2016) in Fig. 17.8 A, B.

The head contains a single pair each of antennae and compound eyes and three or fewer pairs of simple eyes (**ocelli**), as viewed from a dorsal perspective (Fig. 17.8 A). Mouthparts include an upper lip or **labrum**, paired **mandibles** (typically heavily sclerotized and toothlike), and maxillae. Both mandibles and maxillae are multiarticulate and may feature a pair of elongated **palps**. Structures for feeding and sometimes defense are located ventrally on the head (Fig. 17.8 B).

The 3-segmented thorax is the tagma between the head and abdomen. From front to back, it consists of the **prothorax**, **mesothorax**, and **metathorax**. Sclerites on each thoracic segment are termed the **notum** (dorsal), **sternum** (ventral), multiple **pleura** (lateral). Each of these three thoracic segments contains one pair of legs divided from the body outward (mesial to distal) into five parts: a **coxa** (plural coxae), **trochanter**, **femur** (femora), **tibia** (tibiae), and **tarsus** (tarsi). One to five "articles" occur on the femur and tibia in some insect orders. A tarsus may contain one to two claws. Wingpads (developing wings) are commonly found dorsally on the meso- and metathorax in hemimetabolous insect larvae (i.e., those insects characterized by incomplete metamorphosis with an egg, nymph, and adult but no pupal stage). Gills may be present on the thorax and abdomen and rarely on the head.

The abdomen has a similar arrangement of sclerites as found on the thorax, but pleura may be absent in some major groups. Up to 10 visible segments are present, and these are numbered starting at "one" nearest the thorax. Abdominal "appendages" occur in many groups including terminal **cerci** (singular = cercus).

MATERIAL PREPARATION AND PRESERVATION

Chapter 3 (Rogers & Thorp, 2015) in Volume I of this book series provides detailed advice on collecting, preserving, and culturing invertebrates in general. Below we summarize some aspects of preparing samples of Hexapoda in particular. For more detailed information, see related sections in chapters on various hexapod taxa in Hamada et al. (2018).

The first step in preserving specimens is to identify their ultimate intended use. Store aquatic organisms in ethanol (EtOH) for most uses and temporarily in formalin unless DNA is going to be extracted, in which case only EtOH should be used. Initial preservation can be in 95% EtOH but eventually they should be shifted to 70%–80% EtOH for reference collections or use in teaching or various life history studies because permanent storage in 95% EtOH results in brittle specimens. Kahle's fluid or even 5%–10% buffered formalin produces a supple specimen with better preserved internal structures, but these fluids are not recommended for long-term storage and prevent disposal problems.

The type of storage vial and stopper you use will depend on availability, but we recommend the use of 3 or 4 dram vial with either glass patent lip vials with either neoprene or rubber stoppers (preferably the latter to avoid the neoprene swelling and/or bleeding color) to prevent desiccations. To avoid gradual evaporation of the EtOH, do not use screw cap vials or shell vials with cork stoppers or plastic snap caps.

Some insects, such as odonates and some beetles and true bugs, are more appropriately pinned as dry specimens (especially adults), though they can be stored in EtOH. In the latter case, adult dragonflies and damselflies, for example, can be placed on paper triangles and soaked in acetone to preserve colors and extract the fats that would otherwise make the insect appear dull and greasy looking. Afterward this extraction process, the specimen should be dried with the wings folded above the thorax, stored in clear, glycine envelopes with labeled acid free cards, and placed in pest proof containers.

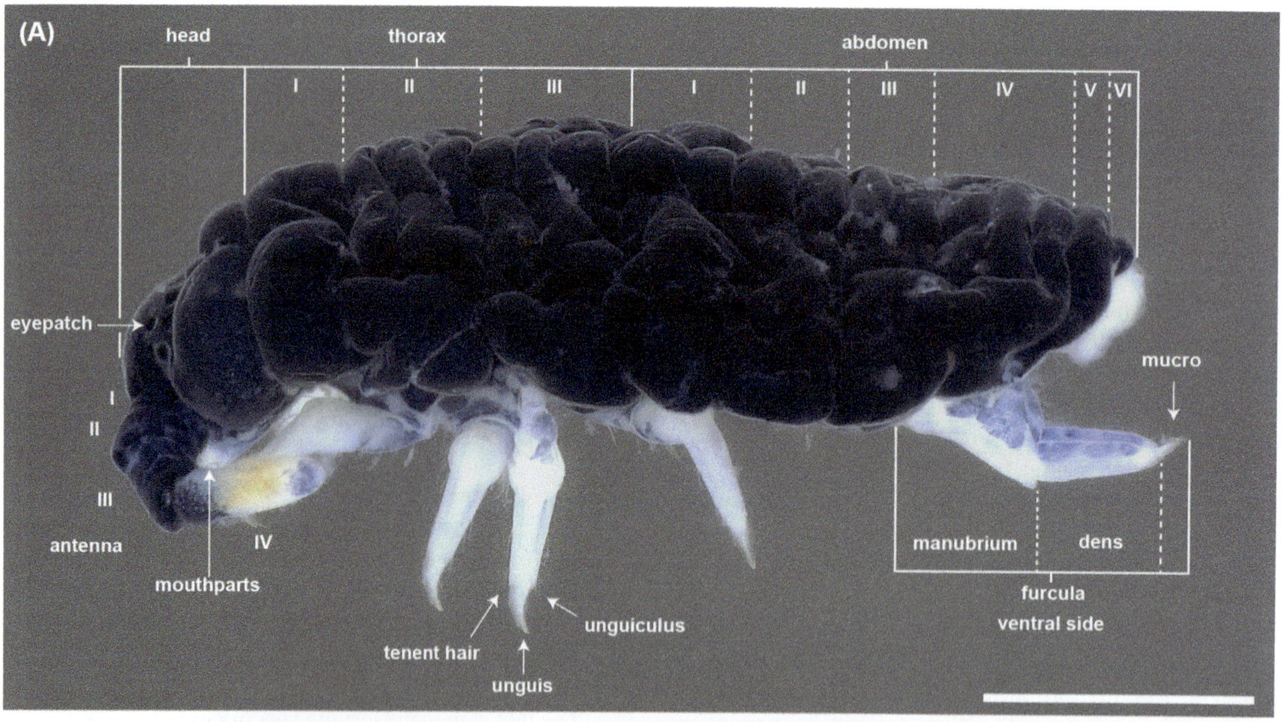

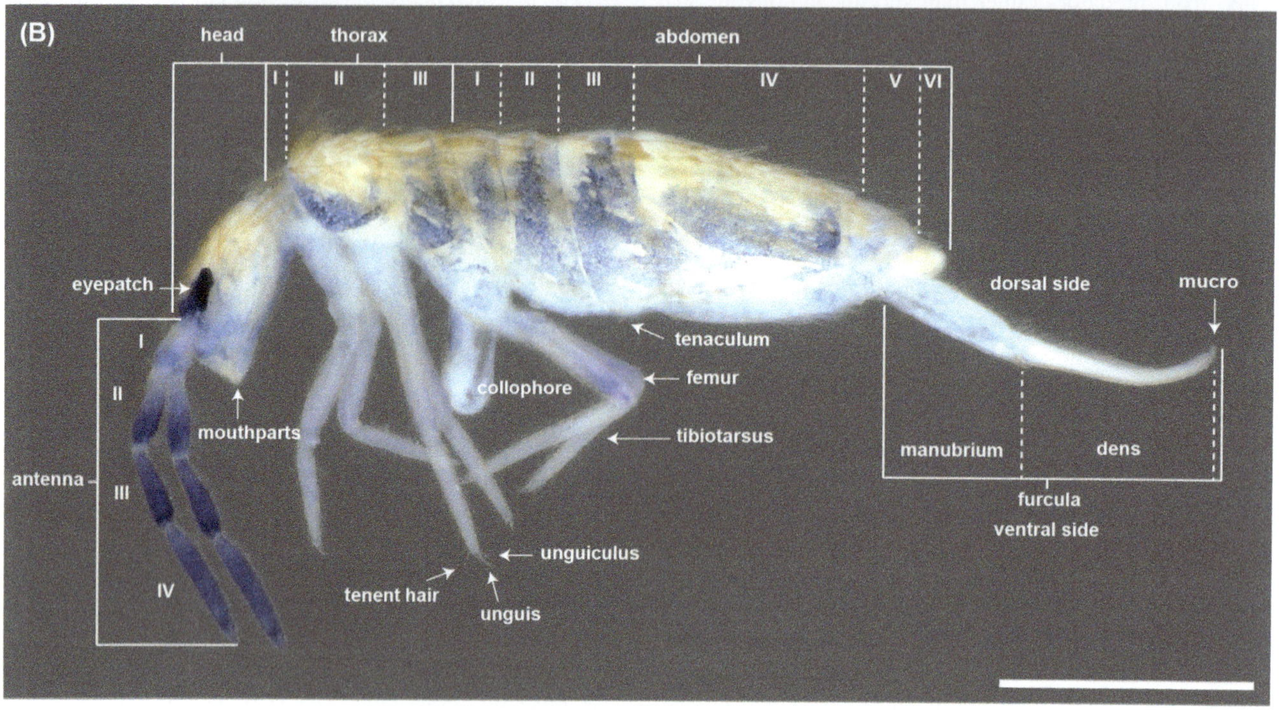

FIGURE 17.7 Cont'd.

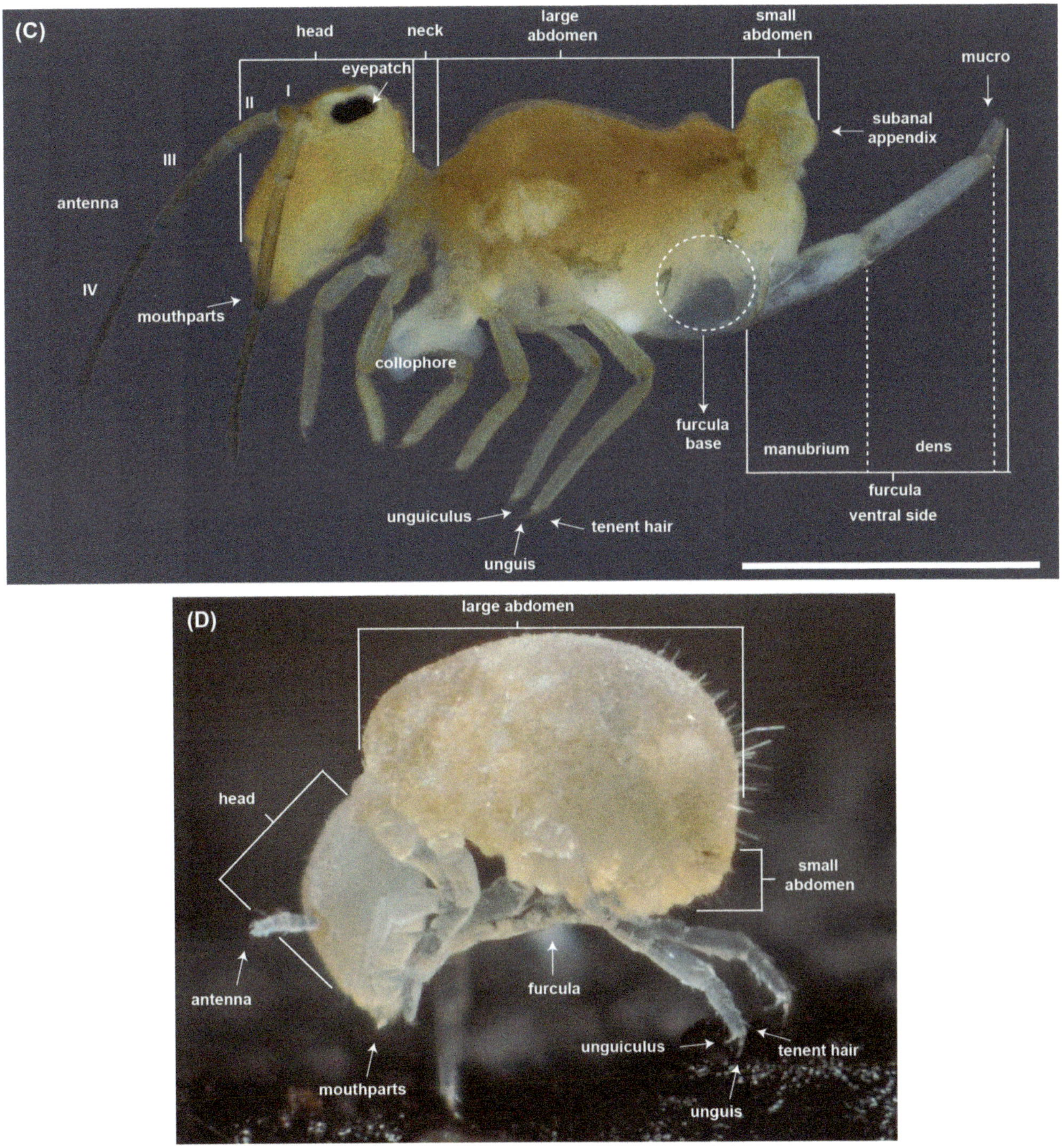

FIGURE 17.7 Lateral view of the external anatomy of representatives of four primary orders of Entognatha (Collembola) in Neotropical freshwaters: (A) Poduromorpha *(Neotropiella carli)*; (B) Entomobryomorpha (*Seira dowlingi*); (C) Symphypleona (*Arlesminthurus sp.*); (D) Neelipleona (*Neelus murinus*). Scale bars: (A) = 1 mm, (B–C) = 0.5 mm. *Photographs (A–C) from Cipola et al. (2018); photograph (D) courtesy of Andy Murray (https://www.chaosofdelight.org).*

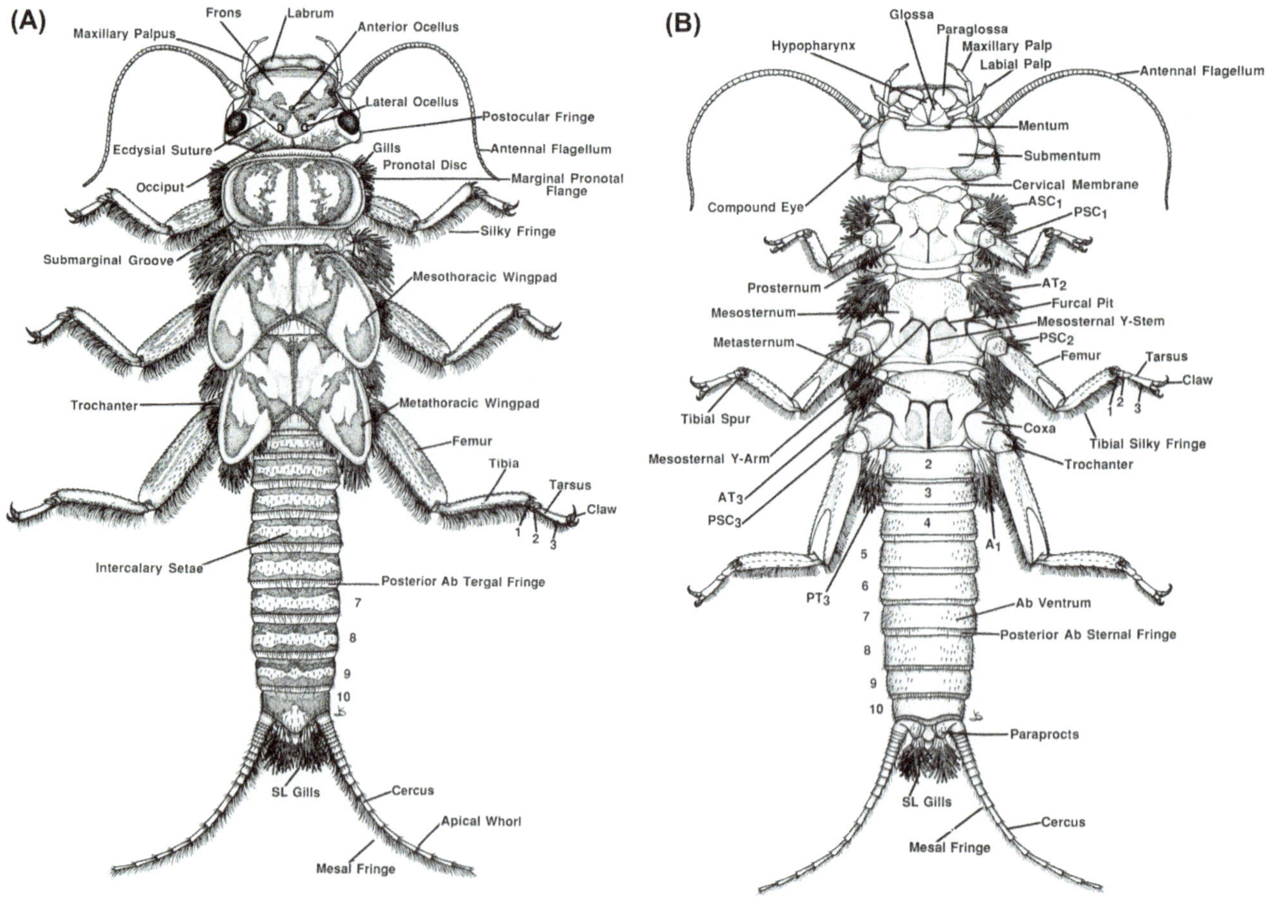

FIGURE 17.8 Dorsal (A) and ventral (B) views of a stonefly nymph (order Plecoptera). *(A) and (B) courtesy of DeWalt & Resh (2016).*

FIGURE 17.9 Representative fly larvae of aquatic Diptera from four of many families: (A) cranefly (Tipulidae); (B) mosquito (Culicidae); (C) midge (Chironomidae); (D) blackfly (Simuliidae). Scale bars = 1 mm. *(A–D) courtesy of Fusari et al. (2018).*

FIGURE 17.10 Semiaquatic wasps (order Hymenoptera): (A) photograph of a preserved adult wasp, Eulophidae; (B) female wasp *Tanychela* (Ichneumonidae) underwater on a rock. *Photograph courtesy of Robert A. Deans.*

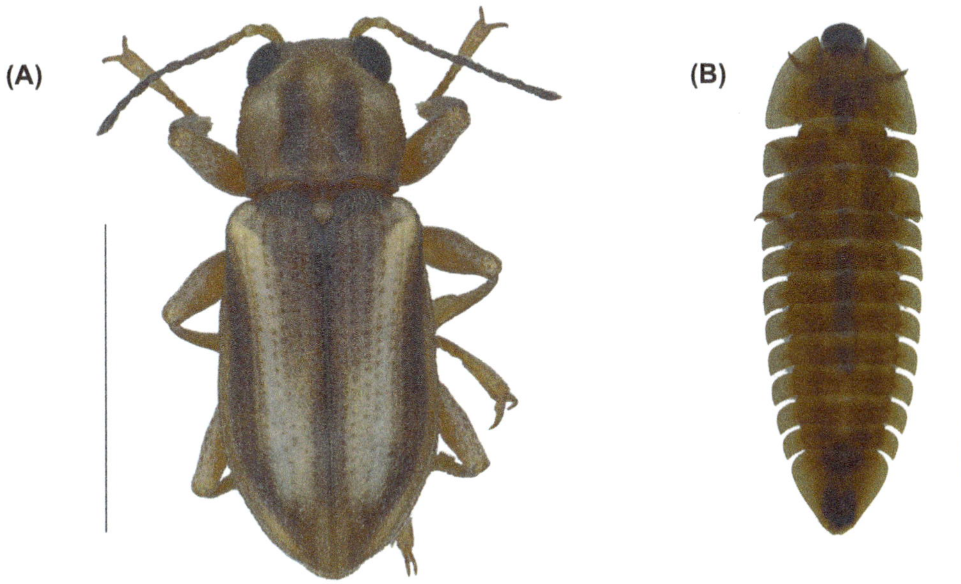

FIGURE 17.11 Coleoptera beetles: adult (A) and larval (B) beetles, Elmidae. Scale bars = 1 mm. *(A) and (B) courtesy of Benetti et al. (2018).*

KEYS TO HEXAPODA

The following keys to Hexapoda cover only species whose lives involve aquatic or semiaquatic larvae and/or adults, including those that parasitize aquatic stages of other taxa. Pupal stages, where present, are not keyed here. Keys to Entognatha orders are based primarily on Cipola et al. (2018). Keys to Neotropical insect orders are derived nearly *verbatim* from Thorp et al. (2018) and based in part on DeWalt & Resh (2016).

Hexapoda: Subclasses

1 Mouthparts internal; wings absent; abdomen forked distally; abdominal collophore present (Fig. 17.7) .... **Entognatha (Collembola) [p. 572]**

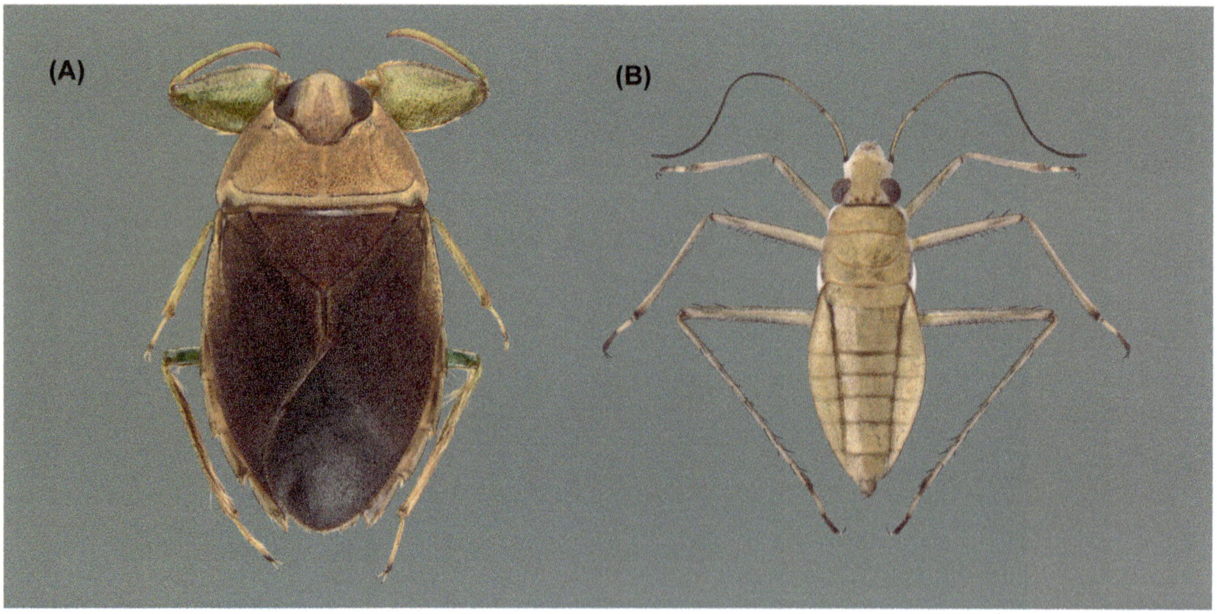

FIGURE 17.12 Hemiptera bugs: (A) *Ambrysus* (Naucoridae), dorsal view; (B) *Mesovelia* (Mesoveliidae), dorsal view. *(A) and (B) courtesy of Figueiredo Moreira et al. (2018).*

1′	Mouthparts external (Fig. 17.8 B); wings present or not; collophore absent	**Insecta [p. 572]**

Hexapoda: Entognatha (Collembola): Orders

Key to Neotropical Entognatha based on Cipola et al. (2018).

1	Body elongated; thoracic and abdominal segmentation distinct (Fig. 17.7 A, B)	2
1′	Body globular; thoracic and abdominal segments fused (Fig. 17.7 C, D)	3
2(1)	First thoracic segment distinct dorsally, with dorsal setae or papillae (Fig. 17.7 A)	Poduromorpha
2′	First thoracic reduced, without setae or papillae (Fig. 17.7 B)	Entomobryomorpha
3(1)	Eyes usually present; antennae longer than head; small abdomen distinct; size and color variable (Fig. 17.7 C)	Symphypleona
3′	Eyes absent; antennae shorter than head; small abdomen indistinct; very minute individuals; usually lacking pigmentation (Fig. 17.7 D)	Neelipleona

Hexapoda: Insecta: Orders

1	Thorax without jointed legs	2
1′	Thorax with three pairs of jointed legs	3
2(1)	Mummy-like stage, with developing adult structures enveloped in a case; holometabolous insects (Coleoptera, Diptera, Hymenoptera, Lepidoptera, Mecoptera, Megaloptera, Neuroptera, and Trichoptera); not keyed below	pupae
2′	Not completely enclosed in a case; mobile larvae, mostly with prolegs, pseudopods, or creeping welts on one or more segments; larvae (Fig. 17.9)	Diptera
3(1)	With large functional wings; forewings may be rigid and chitinized, at least at the base	4
3′	Wingless, or with developing wings (wingpads) or brachypterous	8
4(3)	Not with the combination of characteristics described below	5
4′	Abdomen narrowly constricted at junction with thorax; basal half of forewings parallel to leading edge; usually <3 mm; enter water only to deposit eggs on or in aquatic insect hosts; adults (Fig. 17.10 A–B)	Hymenoptera

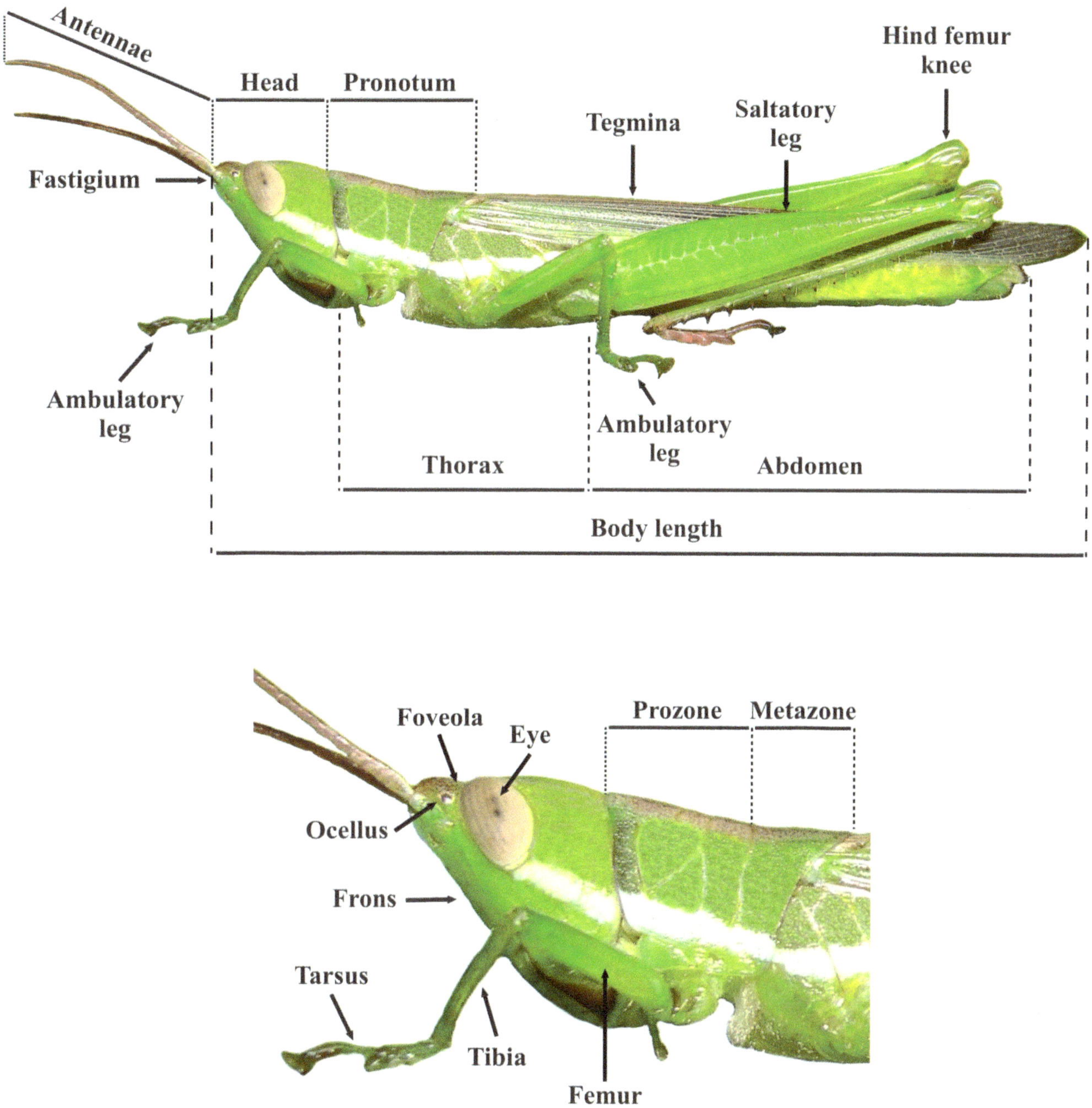

FIGURE 17.13 Representation of the external morphological structures of a semiaquatic grasshopper (Orthoptera, Acrididae). *Figure courtesy of Nunes-Gutjahr & de Souza Braga (2018).*

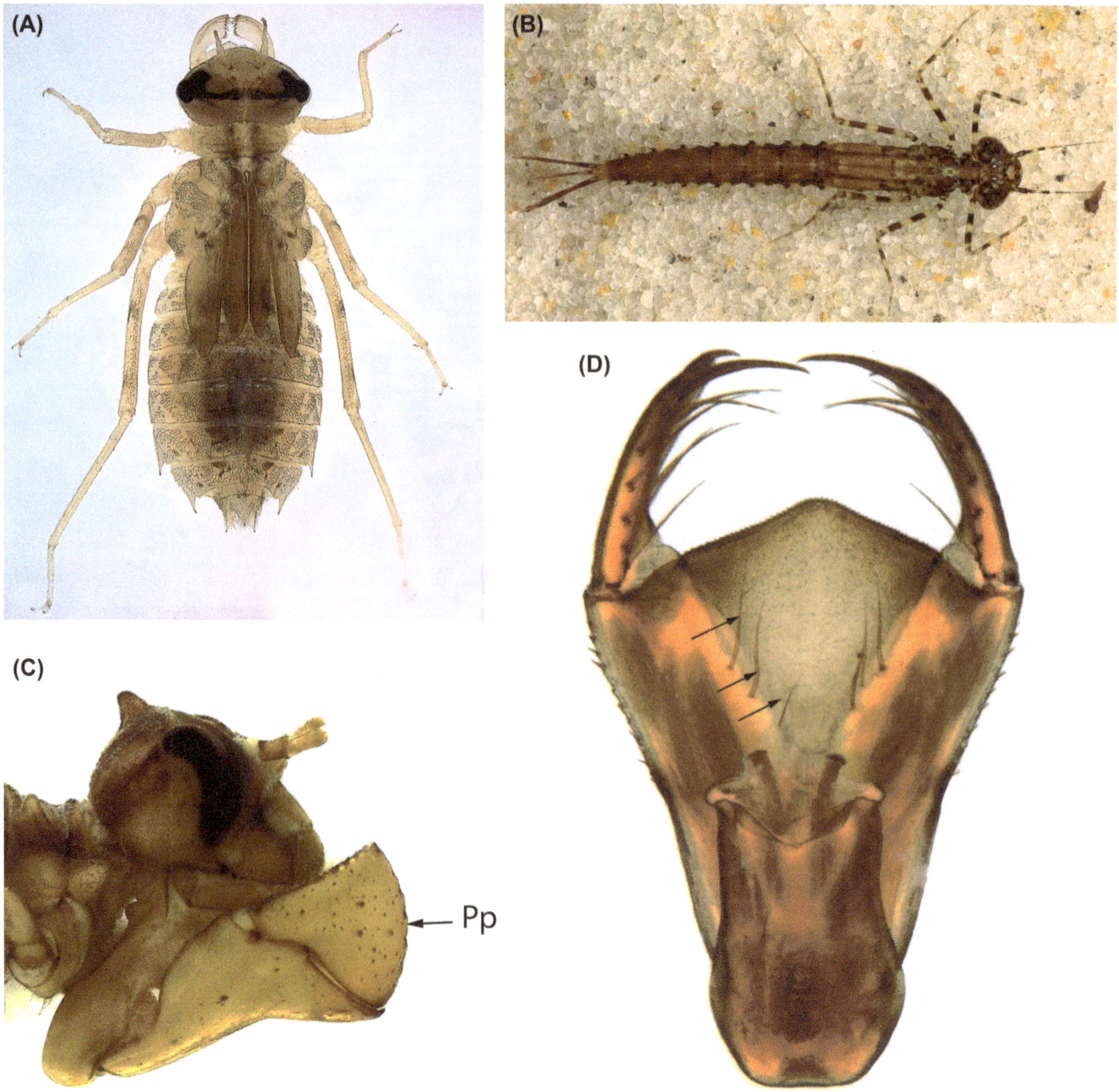

FIGURE 17.14 Larval Odonata: (A) dorsal view of the dragonfly *Perithemis*; (B) dorsal view of the damselfly *Perilestes attenuates*; (C) larval head (lateral view) of the dragonfly *Rialla villosa* (Libelluloidea), Pp: premental palp; (D) labium (dorsal view) of the damselfly *Idioneura ancilla*, arrows indicate premental setae. *(A) and (C) courtesy of Neiss et al. (2018); (B) photograph courtesy of Neiss & Hamada (2014); and (D) courtesy of Lozano et al. (2018).*

FIGURE 17.15 Dorsal (A) and lateral (B) views of the lacewing insect *Climacia* (Neuroptera, Sisyridae). Scale bars = 0.5 mm. *Courtesy of Martins & Ardila-Camacho (2018).*

FIGURE 17.16 Lateral (A) and dorsal (B) views of larvae of the aquatic moth *Petrophila* (Lepidoptera, Crambidae). *Courtesy of Graça & Solis (2018).*

FIGURE 17.17 Lateral view of the caddisfly larva of *Diplectrona* (Trichoptera, Hydrosychidae). *Courtesy of Pes et al. (2018).*

FIGURE 17.18 Dorsal view of the alderfly larva *Ilyobius* (Megaloptera, Sialidae). *Courtesy of Neusa Hamada.*

FIGURE 17.19 Lateral view of a scorpionfly larva of *Nannochorista* (Mecoptera, Nannochoristidae) larva a scorpionfly larva. Scale bar = 500 μm. *Courtesy of Elena Lukashevich and Dmitry Shcherbakov, Paleontological Institute, Moscow and from Machado (2018).*

FIGURE 17.20 Nymph of the mayfly *Cloeon smaeleni* Lesyage, 1924 (Ephemeroptera, Baetidae). *Courtesy of Salles et al. (2018).*

FIGURE 17.21 Dorsal view of the stonefly nymph *Paragripopteryx* (Plecoptera, Gripopterygidae). Scale bar 1 mm. *Courtesy of Avelino-Capistrano et al. (2018).*

5(4)	Mesothoracic wings rigid and chitinized (Fig. 17.11 A) or thickened in basal half (Fig. 17.12 A)	6
5′	Meso- and metathoracic wings completely membranous with numerous distinct veins, forewings leathery	7
6(5)	Chewing mouthparts; or if appearing as a tube, then antennae are situated halfway down the tube; mesothoracic wings as elytra (rigid and chitinized) or elytra may be reduced; adults (Fig. 17.11 A)	Coleoptera (in part)
6′	Sucking mouthparts formed into a broad or narrow tube; mesothoracic wings hardened in basal half; adults (Fig. 17.12 A)	Hemiptera (in part)
7(5)	Hind legs with enlarged femora for jumping (Fig. 17.13)	Orthoptera (in part)
7′	Hind legs subequal in size to the other legs and ambulatorial as the other pairs	Blattodea (in part)
8(3)	With rudimentary wings, wing pads, or brachypterous; body shape not vermiform	9
8′	Wingless; body shape vermiform	14
9(8)	Without extensible labium; abdomen not terminating in lamellae or pointed structures	10
9′	Labium extensible, geniculate; abdomen terminating in lamellae or pointed structures (Fig. 17.14)	Odonata
10(9)	Mouthparts not modified for sucking, not formed into a tube or pair of stylets	11
10′	Mouthparts modified for sucking, formed into a tube or pair of stylets; adults and nymphs	Hemiptera (in part)
11(9)	Abdomen without terminal appendages, or if terminal appendages present, only two, length <0.3x abdomen; abdominal and thoracic segments without gills	12
11′	Abdomen terminating in two or three elongate, multiarticulate appendages, length ≥0.5x abdomen; abdominal (and/or rarely thoracic) segments with gills, or gills terminal on abdomen	13
12(11)	Hind legs with enlarged femora for jumping; (Fig. 17.13)	Orthoptera (in part)
12′	Hind legs subequal in size to the other legs, ambulatorial as the other pairs	Blattodea (in part)
13(11)	Abdomen with plumose, forked, or lamellar gills; with two cerci and often a median caudal filament occasionally reduced to stub; tarsus with one claw; larvae (Fig. 17.20)	Ephemeroptera
13′	Abdomen with or without gills; if present, gills on abdominal segments 1−2 or 3, and/or at cerci bases; two cerci, no median filament; tarsi with two claws; larvae (Figs. 17.8, 21)	Plecoptera
14(9)	Mouthparts with developed mandibles to chew	15
14′	Mouthparts as a pair of long stylets to feed on Porifera; larvae (Fig. 17.15)	Neuroptera
15(14)	Ventral prolegs absent on abdominal segments 3−6	16
15′	Ventral prolegs on abdominal segments 3−6, each with a ring of fine hooks; larvae (Fig. 17.16)	Lepidoptera
16(15)	Antennae elongate, with three or more antennomeres	17
16′	Antennae extremely small, inconspicuous, with one antennomere; larvae (Fig. 17.17)	Trichoptera
17(16)	Antennae with three or > five antennomeres	18
17′	Antennae with four antennomeres; larvae (Fig. 17.18)	Megaloptera
18(17)	Abdomen posterior end with a pair of hooks inserted dorsally, curved ventrally with the apices pointing anteriorly; larvae (Fig. 17.19)	Mecoptera
18′	Abdomen posterior end not as described above larvae	Coleoptera (in part)

REFERENCES

Avelino-Capistrano, F., P. Pessacq & L.S. Barbosa. 2018. Order Plecoptera. Pages 119−141 *in*: N. Hamada, J.H. Thorp & D.C. Rogers (eds.), Volume III: Keys to Neotropical Hexapoda. Thorp and Covich's Freshwater Invertebrates, Elsevier, London.

Benetti, C.J., M.C. Michat & M. Archangelsky. 2018. Order Coleoptera: introduction. Pages 497−517 *in*: N. Hamada, J.H. Thorp & D.C. Rogers (eds.), Volume III: Keys to Neotropical Hexapoda. Thorp and Covich's Freshwater Invertebrates, Elsevier, London.

Cipola, N.G., D.D. da Silva & B.C. Bellini. 2018. Class Collembola. Pages 11−55 *in*: N. Hamada, J.H. Thorp & D.C. Rogers (eds.), Volume III: Keys to Neotropical Hexapoda. Thorp and Covich's Freshwater Invertebrates, Elsevier, London.

Deharveng, L., C.A. D'Haese & A. Bedos. 2008. Global diversity of springtails (Collembola; Hexapoda) in freshwater. Hydrobiologia 595: 329−338.

DeWalt, R.E. & V.H. Resh. 2016. Class Hexapoda. Pages 413−437 *in*: J.H. Thorp & D.C. Rogers (eds.), Volume II: Keys to Nearctic Fauna. Thorp and Covich's Freshwater Invertebrates, Academic Press, Elsevier, London.

Figueiredo Moreira, F.F., H.D. Duarte Rodrigues, R.W. Sites, I. da Rocha Silva Cordeiro & O. Martins Magalhaes. 2018. Orden Hemiptera. Pages 175−216 *in*: N. Hamada, J.H. Thorp & D.C. Rogers (eds.), Volume III: Keys to Neotropical Hexapoda. Thorp and Covich's Freshwater Invertebrates, Academic Press, Elsevier, London.

Fusari, L.M, G.P.S Dantas & L.C. Pinho. 2018. Orden Diptera. Pages 607−623 *in*: N. Hamada, J.H. Thorp & D.C. Rogers (eds.), Volume

III: Keys to Neotropical Hexapoda. Thorp and Covich's Freshwater Invertebrates, Academic Press, Elsevier, London.

Graça, M.B. & M.A. Solis. 2018. Order Lepidoptera. Pages 325–338 *in*: N. Hamada, J.H. Thorp & D.C. Rogers (eds.), Volume III: Keys to Neotropical Hexapoda. Thorp and Covich's Freshwater Invertebrates, Academic Press, Elsevier, London.

Hamada, N, J.H. Thorp & D.C. Rogers (eds.). 2018. Volume III: Keys to Neotropical Hexapoda. Thorp and Covich's Freshwater Invertebrates, Academic Press, Elsevier, London.

Lozano, F., J. Muzón, D.A. Anjos dos Santos & P. Pessacq. 2018. Superfamily Coenagrionoidea. Pages 475–494 *in*: N. Hamada, J.H. Thorp & D.C. Rogers (eds.), Volume III: Keys to Neotropical Hexapoda. Thorp and Covich's Freshwater Invertebrates, Academic Press, Elsevier, London.

Machado, R.J.P. 2018. Order Mecoptera. Pages 349–351 *in*: N. Hamada, J.H. Thorp & D.C. Rogers (eds.), Volume III: Keys to Neotropical Hexapoda. Thorp and Covich's Freshwater Invertebrates, Academic Press, Elsevier, London.

Martins, C.C. & A. Ardila-Camacho. 2018. Order Neuroptera. Pages 229–236 *in*: N. Hamada, J.H. Thorp & D.C. Rogers (eds.), Volume III: Keys to Neotropical Hexapoda. Thorp and Covich's Freshwater Invertebrates, Academic Press, Elsevier, London.

Neiss, U.G., G. Fleck, P. Pessacq & K. Tennessen. 2018. Odonata: Superfamily Libelluloidea. Pages 399–447 *in*: N. Hamada, J.H. Thorp & D.C. Rogers (eds.), Volume III: Keys to Neotropical Hexapoda. Thorp and Covich's Freshwater Invertebrates, Academic Press, Elsevier, London.

Neiss, U.G. & N. Hamada. 2014. Ordem Odonata, Chapter 14, pages 217–284 *in*: N. Hamada, J.L. Nessimian & R.B. Querino (eds.), Insetos Aquáticos na Amazônia brasileira: taxonomia, biologia e ecologia. Editora do Instituto Nacional de Pesquisas da Amazônia, Manaus.

Nunes-Gutjahr, A.L. & C.E. de Souza Braga. 2018. Order Orthoptera. Pages 143–167 *in*: N. Hamada, J.H. Thorp & D.C. Rogers (eds.), Volume III: Keys to Neotropical Hexapoda. Thorp and Covich's Freshwater Invertebrates, Academic Press, Elsevier, London.

Pes, A.M., R.W. Holzenthal, L.M. Camargoes, J.V. Sganga, A.P.M. Santos & P. Barcelos-Silva. 2018. Order Trichoptera. Pages 237–324 *in*: N. Hamada, J.H. Thorp & D.C. Rogers (eds.), Volume III: Keys to Neotropical Hexapoda. Thorp and Covich's Freshwater Invertebrates, Academic Press, Elsevier, London.

Rogers, D.C. & J.H. Thorp. 2015. Collecting, preserving, and culturing invertebrates. Pages 57–62 *in*: J.H. Thorp & D.C. Rogers (eds.), Volume I: Ecology and General Biology. Thorp and Covich's Freshwater Invertebrates, Academic Press, Elsevier, London.

Salles, F.F., E. Domínguez, C. Molineri, C. Nieto, R. Boldrini & L.G. Dias. 2018. Order Ephemeroptera. Pages 61–117 *in*: N. Hamada, J.H. Thorp & D.C. Rogers (eds.), Volume III: Keys to Neotropical Hexapoda. Thorp and Covich's Freshwater Invertebrates, Academic Press, Elsevier, London.

Thorp, J.H. & B.J. O'Neill. 2015. Hexapoda — Introduction to insects and Collembola. Pages 849–871 *in*: J.H. Thorp & D.C. Rogers (eds.), Volume I: Ecology and General Biology. Thorp and Covich's Freshwater Invertebrates, Academic Press, Elsevier, London.

Thorp, J.H. & D.C. Rogers. 2011. Field Guide to Freshwater Invertebrates of North America. Academic Press, Elsevier, London.

Thorp, J.H., D.C. Rogers & N. Hamada. 2018. Introduction. Pages 3–9 *in*: N. Hamada, J.H. Thorp & D.C. Rogers (eds.), Volume III: Keys to Neotropical Hexapoda. Thorp and Covich's Freshwater Invertebrates, Elsevier, London.

Chapter 18

Class Cirripedia

Fábio Bettini Pitombo

Departamento de Biologia Marinha, Universidade Federal Fluminense, Niterói, Rio de Janeiro, Brazil

Chapter Outline

INTRODUCTION

Among approximately 1,000 described species of Cirripedia, only a few occur in low salinity environments and continental waters. Most live as sessile suspension feeders, attached to various substrates, from the intertidal zone to the abyssal plains or on floating objects. Although the majority are marine, a few tolerate brackish or freshwaters.

In the Neotropical region, seven amphibalanines are known to occur in continental waters, mostly on marginal areas with saline influence. Most live in estuaries, but some also occur in high saline coastal lagoons (Arenas, 1971; Henry, 1973; Henry & McLaughlin, 1975; Poirrier & Partridge, 1979; Dineen & Hines, 1992; Young, 1994).

LIMITATIONS

This group is poorly known in inland habitats, with little dedicated collecting and research. Little is known about the morphological variation. As older specimens of the free-living taxa become eroded, characters may be obscured or lost.

TERMINOLOGY AND MORPHOLOGY

There are few specific structures required for inland aquatic cirripede identification.

The **shell** is formed by a set of plates: **rostrum**, **carina,** and **carinolaterals**, interconnected by superimposed margins, radii, and alae (Fig. 18.1). The color, texture, and structure of the shell can vary, and it can be solid or formed by two laminae (inner and outer laminae) interconnected by a longitudinal septa (Fig. 18.1 D, E). The lower end of the septa normally bears lateral denticles that form the connection of the parietal plates with the calcareous basis (Fig. 18.1 C, E). The shell may be formed by one or more rows of longitudinal tubes, normally crossed by transverse septa (Fig. 18.1 D). When more than one row of tubes are present, they are formed by a bifurcation of longitudinal septa near the outer wall and/or by a secondary inner lamina (Fig. 18.1 F). The calcareous basis, when present, may be solid or tubular, formed by two laminae connected by radiating septa.

Two pairs of **opercular plates**, **scuta** and **terga** (Fig. 18.2), close the shell aperture (Fig. 18.1 A, B). Both plates are united to the shell by a membrane that connects their basal margins to the lower end of the sheath that surrounds the shell aperture. The outer face of the scutum has growth lines/ridges that run parallel to the basal margin (Fig. 18.2 A, E) and can be crossed by deep longitudinal striae (Fig. 18.2 E). Both scutum and tergum have specific names on each margin according to what is adjacent to it, such as occludent margin facing the aperture present between the two scuta, as well as basal, scutal, tergal, and carinal margins (Fig. 18.2). Both plates present an articular ridge and furrow that allow movement among them. The scutum articular ridge has a projection on its lower margin present in all *Amphibalanus* (Fig. 18.2 B) and *Fistulobalanus*, but is absent on *Elminius kingii* Gray, 1831 (Fig. 18.2 F). The

Thorp and Covich's Freshwater Invertebrates. https://doi.org/10.1016/B978-0-12-804225-0.00018-6

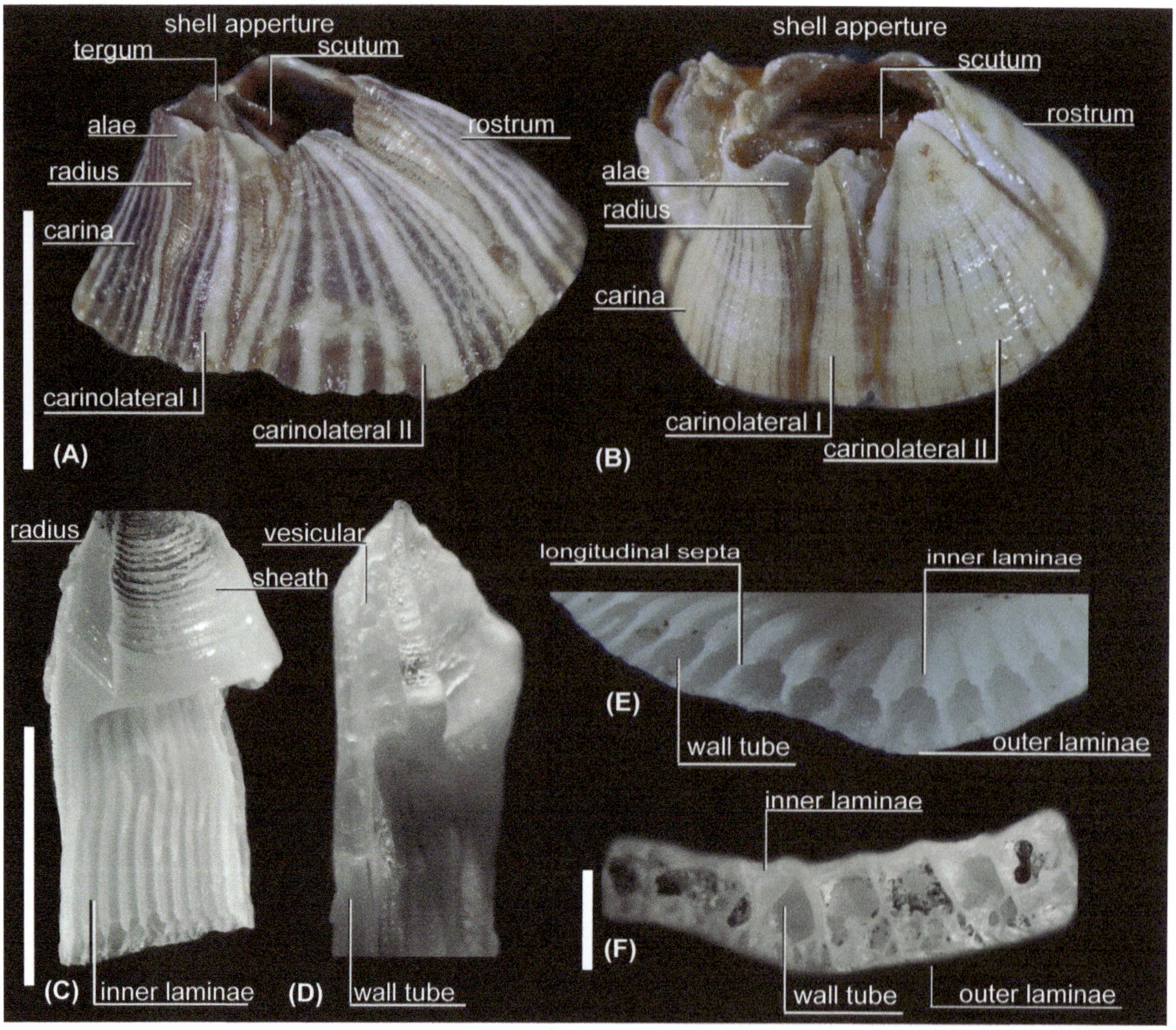

FIGURE 18.1 Shell nomenclature: (A) Lateral view of *Amphibalanus amphitrite*; (B) lateral view of *Fistulobalanus citerosum*; (C) inner view of carinolateral plate of *F. dentivarians*; (D) lateral view of a broken carinolateral plate of *A. subalbidus*, showing the vesicular construction of the sheath; (E) inferior view of *A. eburneus* wall plate, with one row of tubes; and (F) inferior view of *F. dentivarians* wall plate, with more than one row of tubes. Scale bars (A–D) = 1 cm; (E) and (F) = 0.4 cm.

scutum presents on its inner face the insertion scar of the adductor muscle that connects both scuta. This insertion can be elevated by the adductor ridge (Fig. 18.2 B) or can be plain (Fig. 18.2 F). The tergum has a projection on its basal margin called a spur (Fig. 18.2 C, D, G, H); the projection is formed by downward bending of the basal margin, and varies on width, position, and length and is an important feature on species delimitation. The tergum basal margin is divided on carinal and scutal side (Fig. 18.2 G), and the carinal side can be straight or concave. Near the basicarinal angle on inner face are longitudinal crests used for the attachment of the tergal depressor muscle (Fig. 18.2 D).

MATERIAL PREPARATION AND PRESERVATION

Specimens should be preserved in 70% or stronger alcohol for morphological study. It is necessary to detach the somatic part, from the shell, and each part must be treated differently. The opercular and parietal plates must be dissected separately and cleaned with a mild solution of sodium hypochlorite; and after rinsing to remove any residue, store the dry plates. Keep the body in alcohol, while the cirri and mouthparts can be dissected and slide mounted for microscopic examination when necessary.

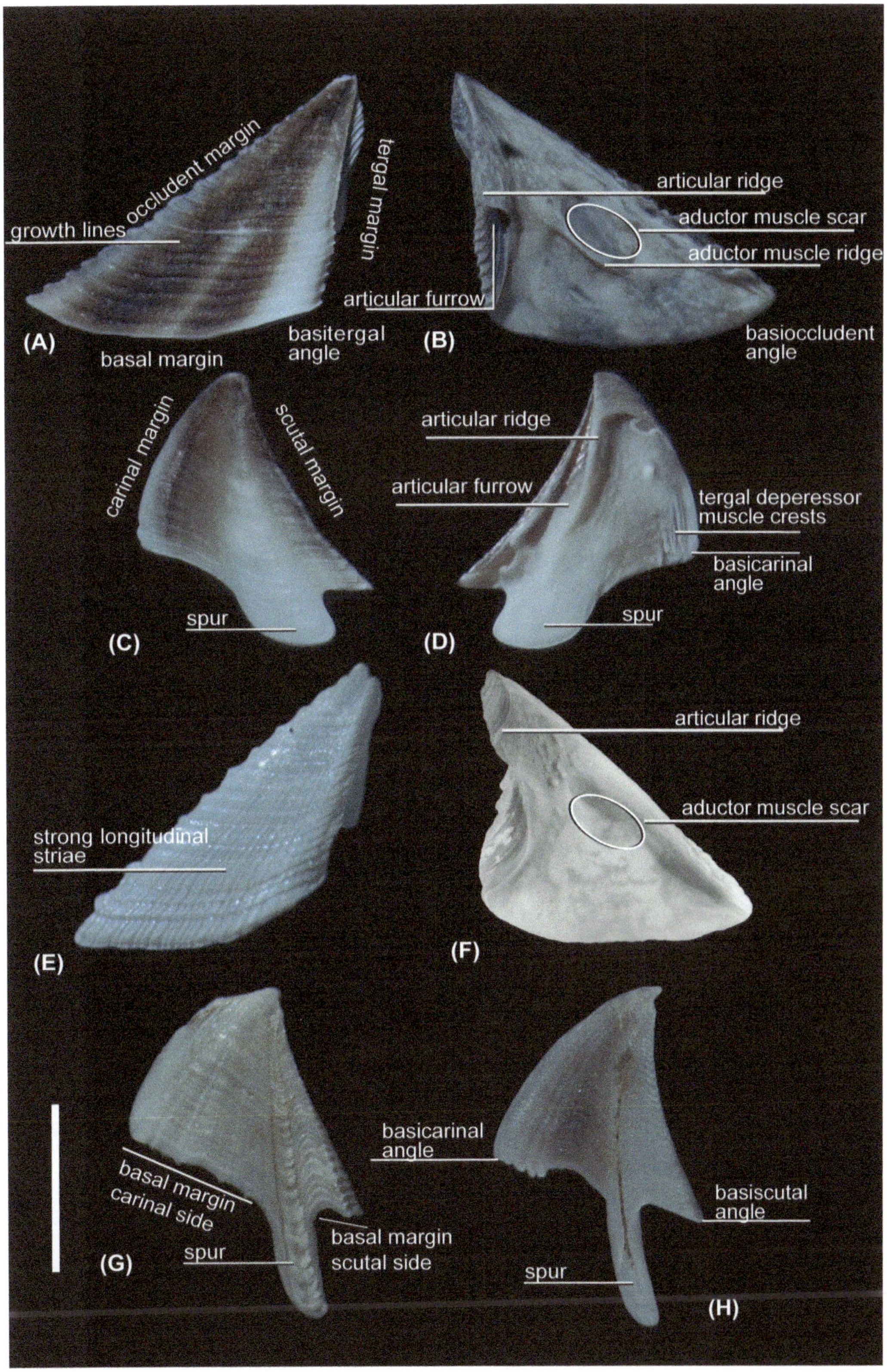

FIGURE 18.2 Opercular plates nomenclature: (A) and (B) Scuta of *Amphibalanus amphitrite*; (C) and (D) terga of *A. amphitrite*; (E) scutum of *A. eburneus*; (F) scutum of *Elminius kingii*; (G) tergum of *Fistulobalanus citerosum*; and (H) tergum of *F. dentivarians*. Scale bar = 0.5 cm.

KEYS TO CIRRIPEDIA

Cirripedia: Superorders

1 Sessile, attached to hard substrates; not parasitic .... Thoracica, **Balanomorpha [p. 582]**

1′ Endoparasites of the fiddler crab *Uca mordax*, with only the female egg sac extending from the ventral surface of the host .... Rhizocephala, one species: *Ptychascus barnwelli* Andersen, Bohn, Høeg & Jensen, 1990

[Puerto Rico]

Cirripedia: Thoracica: Balanomorpha: Families

1 Shell smooth, white, formed of four solid plates (Fig. 18.3 I); basis membranous; scutum without adductor ridge (Fig. 18.2 F); labrum deeply cleft; tergum with spur removed from basiscutal angle .... Austrobalanidae, one monotypic genus: *Elminius kingii* Gray, 1831

[Chile, Argentina]

1′ Shell porous formed by six plates (Fig. 18.1 A, B) (only *Tetrabalanus* with four) each formed by two laminae longitudinally connected by a series of longitudinal septa (Fig. 18.1 C–F); each septa provides a firm attachment with its calcareous basis by a series of transverse basal denticles on basal end (Fig. 18.1 C, E) .... **Balanidae [p. 582]**

Cirripedia: Thoracica: Balanomorpha: Balanidae: Genera

1 Shell wall of six plates, with one or more row of tubes; sheath solid or vesicular 2

1′ Shell wall of four plates with one row of tubes (Fig. 18.3 G); sheath solid .... *Tetrabalanus polygenus* Cornwall, 1941

[Costa Rica, Panama, Ecuador]

2(1) Shell wall with more than one row of tubes formed by a secondary inner laminae or a bifurcation of the longitudinal septa; sheet vesicular (Fig. 18.1 F) .... ***Fistulobalanus* [p. 582]**

2′ Shell wall with one row of tubes; sheet with or without vesicles (Fig. 18.1 E) .... ***Amphibalanus* [p. 582]**

Cirripedia: Thoracica: Balanomorpha: Balanidae: *Fistulobalanus*: Species

1 Shell wall white, sometimes with longitudinal red stripes; additional row of tubes formed by bifurcation of longitudinal septa near outer lamina with or without secondary inner laminae; radii oblique not narrow 2

1′ Shell wall varying from white with thin longitudinal red stripes to nearly covered by red/purple longitudinal bands; shell wall with secondary row tubes formed by a secondary inner lamina and the bifurcation of longitudinal septa; radii narrow 3

2(1) Shell wall with secondary row of tubes formed mostly by longitudinal septa bifurcations; tergal spur narrow, separated from basicarinal angle by 1–2x its width (Fig. 18.3 F) .... *Fistulobalanus pallidus* (Darwin, 1854)

[Panama, Venezuela, Trinidad, Suriname, French Guiana]

2′ Shell wall with secondary row tubes formed by secondary inner laminae and the bifurcation of longitudinal septa; tergal spur widely separated from basicarinal angle by<its width (Figs. 18.2 G, 3 H) .... *Fistulobalanus citerosum* (Henry, 1973)

[Brazil: Para to Rio Grande do Sul]

3(1) Terga basal margins on carinal side not deeply excavated (Figs. 18.2 H, 3 E), inner surface not projecting below basal margin on carinal side .... *Fistulobalanus dentivarians* (Henry, 1973)

[Mexico, Panama, Ecuador]

3′ Terga basal margins deeply excavated on carinal side, with inner surface not projecting below basal margin on carinal side .... *Fistulobalanus suturaltus* (Henry, 1973)

[Costa Rica]

Cirripedia: Thoracica: Balanomorpha: Balanidae: *Amphibalanus*: Species

1 Shell white 2

1′ Shell with white and red longitudinal stripes; radii nearly even and red pigmented near adjacent plate (may be faded in sun bleached, dead shells) (Figs. 18.2 A-D, 3 A) .... *Amphibalanus amphitrite* (Darwin, 1854)

[Mexico, Jamaica, Panama, Curaçao, Bonaire, Trinidad, Peru, Brazil, Argentina. Nearctic]

FIGURE 18.3 Right view of barnacle shells: (A) *Amphibalanus amphitrite*; (B) *A. improvisus*; (C) *A. eburneus*; (D) *A. subalbidus*; (E) *Fistulobalanus dentivarians*; (F) *F. pallidus*; (G) *Tetrabalanus polygenus*; (H) *F. citerosum*; and (I) *Elminius kingii*. Scale bar = 1 cm.

2(1) Scutum growth lines smooth or crossed by fine striae 3

2′ Shell homogeneously white, radii and alae oblique; scutum outer face, with growth ridges marked by deep longitudinal striae (Fig. 18.2 E); tergum spur wide, carinal side basal margin concave (Fig. 18.3 C) .... *Amphibalanus eburneus* (Gold, 1841)

[Mexico, Caribbean Islands, Panama, Venezuela, Colombia, Brazil. Nearctic]

3(2) Shell white, with hyaline longitudinal lines, radii narrow, alae level (Fig. 18.3 B); tergal spur narrow close to basiscutal angle .... *Amphibalanus improvisus* (Darwin, 1854)

[Mexico: Jalisco, Nayarit, Veracruz. Puerto Rico. Honduras: Caribbean coast. Costa Rica: Caribbean coast. Trinidad, Ecuador, Peru, Brazil, Argentina. Nearctic]

3′ Shell homogeneous white (Fig. 18.3 D); tergum with wide spur; sheath vesicular (Fig.18.1 D) *Amphibalanus subalbidus* (Henry, 1973)

[Mexico: Colorado Delta. Brazil: Paraiba. Nearctic]

REFERENCES

Arenas, J. 1971. Distribution de *Elminius kingii* (Cirr.) en el estuario del Rio Valdivia. Beitrage zur Neotropischen Fauna 6: 199–206.

Dineen, J.F. & A.H. Hines. 1992. Interactive effects of salinity and adult extract upon settlement of the estuarine barnacle *Balanus improvisus* (Darwin, 1854). Journal of Experimental Marine Biology and Ecology 156: 239–252.

Henry, D.P. 1973. Descriptions of four new species of the *Balanus amphitrite*-complex (Cirripedia, Thoracica). Bullettin Marine Science 23: 964–1001.

Henry, D.P. & P.A. McLaughlin. 1975. The barnacles of the *Balanus amphitrite* complex (Cirripedia, Thoracica). Zoologische Verhandelingen 141: 1–254, 22 pl.

Poirrier, M.A. & M.R. Partridge. 1979. The barnacle, *Balanus subalbidus*, as a salinity bioindicator in the oligohaline estuarine zone. Estuaries 2: 204–206.

Young, P.S. 1994. Superfamily Balanoidea Leach (Cirripedia, Balanomorpha) from the Brazilian coast. Boletim do Museu Nacional, Serie Zoologia 356: 1–36.

Chapter 19

Class Branchiopoda

D. Christopher Rogers
Kansas Biological Survey, and The Biodiversity Institute, University of Kansas, Lawrence, KS, United States

Rosa Graciela Cohen
Universidad de Buenos Aires, Facultad de Ciencias Exactas y Naturales, Departamento de Biodiversidad y Biología Experimental, Buenos Aires, Argentina

Brenda J. Hann
Department of Biological Sciences, University of Manitoba, Winnipeg, MB, Canada

Chapter Outline

Thorp and Covich's Freshwater Invertebrates. https://doi.org/10.1016/B978-0-12-804225-0.00019-8

INTRODUCTION

The Neotropical bioregion has approximately 300 described species of branchiopod crustaceans, although this number is certainly an underestimate. They are known from fresh and saline temporary and permanent aquatic habitats, from small seasonal wetlands and bofedals, to lakes, ponds, and large rivers. Most are planktonic or semibenthic and are often among the most abundant invertebrates in those habitats. Branchiopods are often used as indicators of aquatic ecosystem health, which ultimately reflects on land use practices and management (Brendonck et al., 2008; Rogers, 2009). Several species are economically important (Martínez-Pantoja et al., 2002).

LIMITATIONS

The Neotropical Branchiopoda are still in an age of discovery, with new species described every year. Much of the Neotropical Region is still unexplored, and we expect these keys to be out of date soon. The anostracan keys presented here are based primarily on characters of the males, but females can be identified in some taxa.

Neotropical notostracans are represented by two genera: *Lepidurus* and *Triops*, both in need of closer examination and revision. One species is reported for both genera in this region; however, neither is reliably identifiable beyond genus and both are in need of revision. Both male and female laevicaudatan clam shrimp and the Cyclestherida are easily identified. The spinicaudatan clam shrimp are more difficult: all the Neotropical fauna are in need of revision, and the *Eulimnadia* species require specimens with mature eggs for identification.

One constraint in anostracan identification is that the most reliable taxonomic characteristics, such as the male second antenna, can change in size and shape, according to the degree of development during the postembryonic life and subsequent moults. Similarly, adult anostracan size and time to sexual maturity are not constant for any species and depend on local environmental conditions. Thus, different populations may have adults of very different sizes, with a corresponding variable degree of development across structures. A third complication is endogamy in each population due to low genetic flux among neighboring ponds (monopolization hypothesis, De Meester et al., 2002); this results in local population variation in particular features.

Cladocera keys are primarily based upon adult females, which are readily identified to genus. Species-level identifications typically require use of a compound microscope, and many genera need to be revised or have incompletely known species. Neotropical cladocerans in the Daphniidae (e.g., *Daphnia, Simocephalus, Ceriodaphnia*) are particularly variable morphologically due to seasonal polymorphisms (cyclomorphosis). Through rapid successive parthenogenetic generations, phenotypic changes (decrease in size, head becoming enlarged and rounded, lengthening of tail spines, etc.) protect individuals from predation. Many *Daphnia* species are known to hybridize, further complicating identification beyond genus or species group. Several families (e.g., Ilyocryptidae, Chydoridae) are prevalent in wetlands and other littoral habitats, which can be difficult to sample. Hence, these taxa also need revision.

TERMINOLOGY AND MORPHOLOGY

Neotropical branchiopods range in size from less than a millimeter to 10 cm.

Anostracan second antennae are nonlocomotory, uniramous, sexually dimorphic appendages (Figs. 19.1 A–C, 2 H). The **male second antennae** morphology is the most useful taxonomical tool; however, they are usually reduced in females. In males, they are robust (or enlarged), biarticulate, variously ornamented appendages, which are part of the sexual recognition system and are used to amplex the female prior to mating. The second antennal proximal antennomere may have diverse ornamentation, such as pulvilli (mediobasal spiny patches), apophyses, medial spiny crests, rows of spines, and so on. Similarly, the distal antennomere morphology, particularly of the distal end, may be discriminant among species. The male second antennae may each have a prehensile or cheliform antennal appendage, which depends from the proximal antennomere distomedially. In *Streptocephalus*, the antennal appendages are cheliform, divided into two rami, referred to as a "thumb" and "finger" (Fig. 19.1 B). The "thumb" is typically longer than the finger and may have a posterior spur. The "finger" may have one or more "teeth" on the anterior margin. Furthermore, anostracan males may have a frontal appendage, which projects anteriorly from the frons. This frontal appendage may be variously branched and ornamented (Thamnocephalidae) (Fig. 19.1 C) or simple (Streptocephalidae) (Figs. 19.1 A, B, 12) and is also part of the mate recognition system. In Thamnocephalidae, the frontal appendage comprises a right and left "arm." The branches of the "arm" are defined by their position relative to the orientation of the animal, with the frontal appendage projecting anteriorly. The branches may be mediodorsal (MD), dorsal (D), apical (A), or ventral (V) (Figs. 19.1 C, 8–10). Thus the first dorsal branch is referred to as 1D, and the second is referred to as 2D (Belk & Pereira, 1982; Pereira, 1983). However, a new terminology based on the successive appearance of the first, second, and third branches during the postembryonic development in *Thamnocephalus* was recently proposed (Obregón-Barboza et al., 2015).

Diplostracan **second antennae** are typically branched, with an anterior endopod and a posterior exopod, and are used for propulsion (Fig. 19.2 A–G). In notostracans (Fig. 19.1 F, G), both pairs of antennae are reduced, and sensory functions are accomplished by elongated endites of the first and second thoracopods. The head is proportioned to the body in all branchiopods but is massive in the laevicaudatans (Fig. 19.2 D). A rostrum is absent in anostracans and notostracans.

Adult branchiopods have a pair of **compound eyes** and, sometimes, an additional single median eye (**nauplius eye**) persistent from the larval stages. Compound eyes are pedunculate in anostracans and sessile in the remaining branchiopods. In Notostraca, the compound eyes are close together; in Diplostraca, they are close together or coalescent. Some cave-dwelling cladocerans lack eyes entirely.

Branchiopods also have a nuchal organ that lies on the dorsal surface of the head, although in the Limnadiidae (Spinicaudata), it may be borne on a short stalk. Nuchal organs are probably involved in ion transport.

It is difficult to distinguish between notostracan sexes, as they are very similar; females or hermaphrodites are only distinguishable from males by the modifications on thoracopod 11 for carrying eggs. Anostracans show a marked sexual dimorphism mainly in the second antenna (male claspers) and sometimes also in the thoracopods. The anostracan gonopod morphology is genus specific. Laevicaudatan males have an abbreviated rostrum with the apex truncated, whereas the females have a longer more pointed rostrum. Furthermore, males have the first thoracopod pair modified to amplex the female, as is also the case in Cyclestherida (Fig. 19.2 E). Spinicaudata (Fig. 19.2 A, B) shows subtle sexual dimorphisms, but males always have the first two thoracopod pairs modified as claspers. Anostraca, Notostraca, Laevicaudata, and Spinicaudata are mainly oviparous (excluding some ovoviviparous populations of *Artemia*), whereas Cyclestherida and Cladocera females have a dorsal brood chamber where juveniles develop (direct development) except for the Haplopoda. Some cladocerans (Anomopoda) and the Cyclestherida produce an ephippium, which is an envelopment that contains the eggs; the remaining branchiopods release the eggs without a case. In many taxa, egg morphology is very useful in taxonomy, at genus and specific level.

MATERIAL PREPARATION AND PRESERVATION

All branchiopods can be preserved in ethyl alcohol long term; for purposes of identification, all can be manipulated in alcohol. Dissection is seldom necessary, but it is a good idea to have several animals available for identification. This is especially true for the cladocerans because these

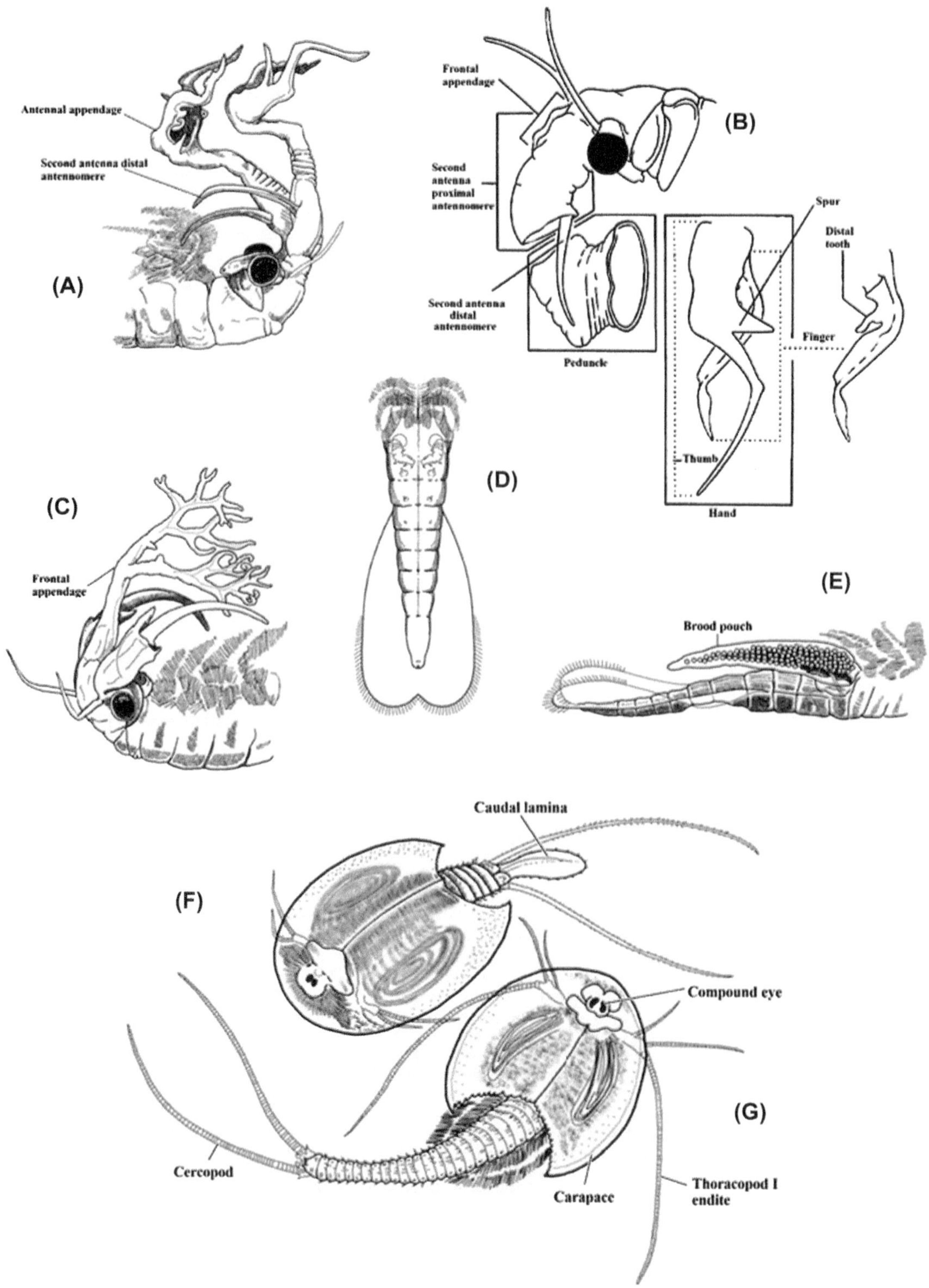

FIGURE 19.1 (A and B): *Streptocephalus* sp.; (A) male, anterior end, right lateral view of head; (B) generalized morphology of the second antennae. (C–E): *Thamnocephalus platyurus*; (C) male, anterior end, left lateral view of head; (D) male genital region and abdomen, ventral view; (E) female genital region and abdomen, left lateral view. (F) *Lepidurus* sp. (G) *Triops* sp.

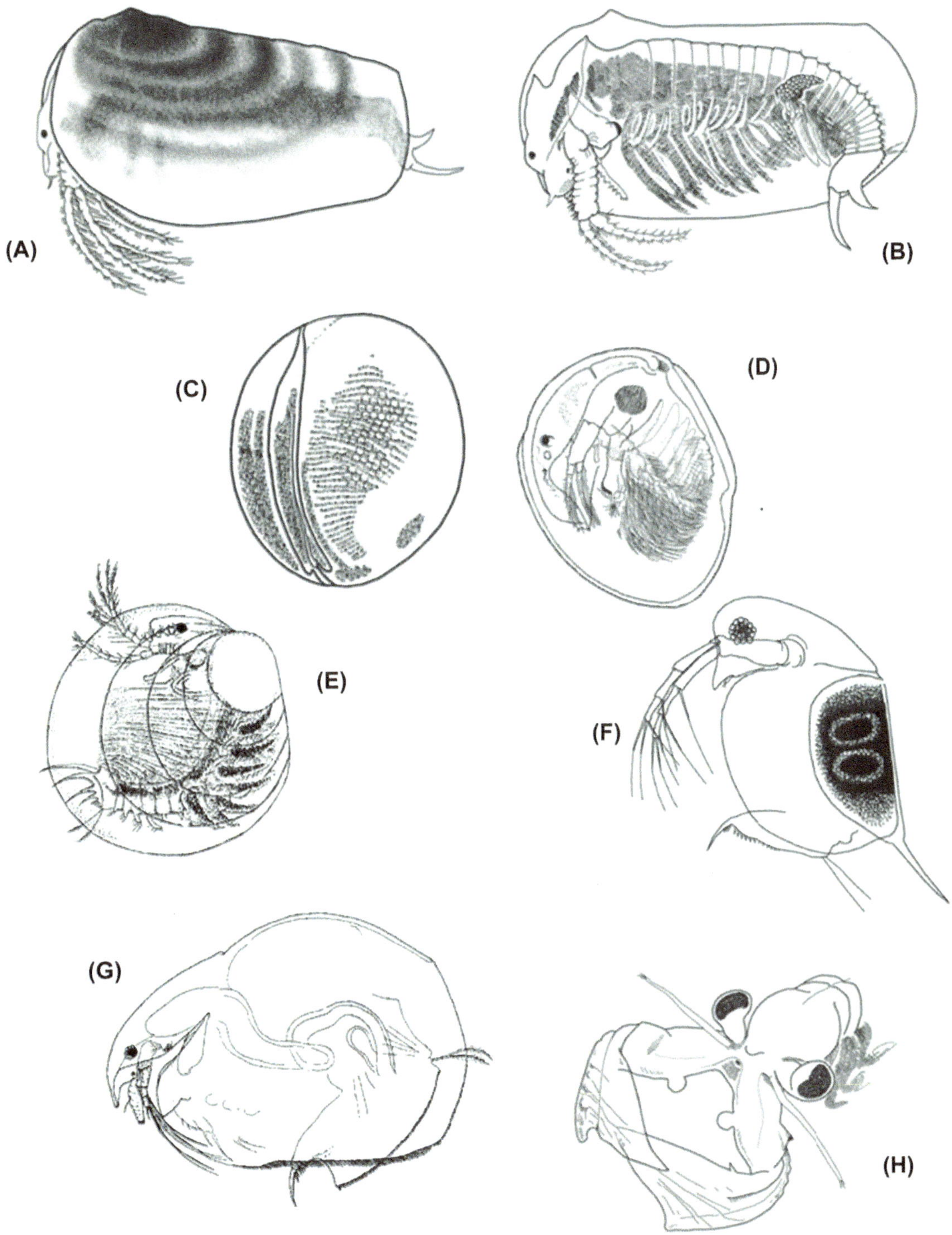

FIGURE 19.2 (A and B) *Leptestheria* sp.; (A) habitus, left lateral view; (B) same view with left half of carapace removed. (C and D) *Lynceus* sp.; (C) habitus, left anterolateral view; (D) left lateral view, with left half of carapace removed; (E) *Cyclestheria* sp.; (F) *Daphnia* sp.; (G) *Eurycercus* sp.; (H) *Artemia franciscana*, male, head, anterior-oblique view.

small, delicate animals are easily crushed. Cladocerans smaller than about 1 mm should be slide mounted for high magnification viewing; larger species can often be identified without being mounted, using a dissecting microscope with a magnification of $\leq$50 diameters. The best cladoceran mounting medium is one that is more or less permanent and soluble in both water and alcohol solutions, so specimens can be added directly to the medium without tedious dehydration. If you have access to chloral hydrate (currently a controlled substance), a good medium is Hoyer's. Divide the Hoyer's into two batches. Let one sit in a warm place until it is as thick as honey, and keep the second batch in a closed jar so it remains thin. You can always add more water to thin either batch. Keep at least some of the thick and thin Hoyer's in screw top eye dropper bottles.

To make a slide, use the eye dropper to put a small streak of dilute Hoyer's (about a quarter of a drop) toward one end of the slide. Arrange the specimens in this streak. It is wise to place four or so of what you think are the same kind of animal on a slide to allow for comparison and viewing of difficult characters. Delicate specimens can be protected from compression by putting a few pieces of broken cover slips around the specimens. Let the Hoyer's streak dry on a slide warmer or in a warm place (below 100°C). When the streak is dry, add a drop or two of the thick Hoyer's and gently lower a cover glass over the Hoyer's. Put the slide back into a warm place to dry again. Label the slide as to the date and location of the collection. It is important to use thickened Hoyer's in the last step; otherwise, the Hoyer's will shrink as it dries and produces large bubbles in the final preparation. However, if you use thickened Hoyer's, small bubbles produced will disappear as the medium dries. These slides are semipermanent. If the climate is humid, the Hoyer's will thin so that cover glasses will slip off the slide. If the climate is arid, the Hoyer's will desiccate enough so that the cover glasses pop off the slide. These problems may be avoided by storing the slides horizontally in a climate that has moderate humidity and by ringing the slide with clear nail polish.

An alternate mounting method, especially useful for bosminids, chydorids, and macrothricids, is polyvinyl lactophenol stained with lignin pink. Temporary slides can be made using glycerol or glycerine jelly. Best of all for museum specimens is possibly Canada Balsam, although this requires the specimens be dehydrated through an alcohol series.

KEYS TO BRANCHIOPODA

Branchiopoda: Orders

1 Compound eyes sessile, close together or fused; carapace present, sometimes reduced or absent (Figs. 19.1 F, G, 2 A–G) 2

1′ Compound eyes on stalks separated; carapace absent (Figs. 19.1 A–C, 2 H, 7) .... **Anostraca [p. 590]**

2(1) Carapace broad, never folded or bivalved; eyes projected dorsally through carapace (Fig. 19.1 F, G) .... Notostraca, one family: **Triopsidae [p. 605]**

2′ Carapace bivalved, folded, or reduced; eyes not projecting dorsally through carapace (Fig. 19.2 A–G) .... **Diplostraca [p. 606]**

Branchiopoda: Anostraca: Families

1 Male second antennae with distal antennomere not broadly triangular, or if triangular, not tapering to apex; brood pouch without pair of spines (Figs. 19. 1 A–C, E, 4–10, 12) 2

1′ Male second antennae with distal antennomere broadly triangular, apex acute (Figs. 19.2 H, 3 A, B); brood pouch with two pairs of spines: one lateral and one ventral, the last pair sometimes reduced (Fig. 19.3 E); halophiles .... Artemiidae, one genus: ***Artemia* [p. 591]**

2(1) Gonopods not visible in dorsal view, rigid basal portions parallel, directed ventrally, with the bases closely united; male antennal and/or frontal appendages present or not; female second antennae lamellar, 2–3× as broad as thick, short, or elongate; vas deferens always with a dorsal loop (usually visible through the integument) 3

2′ Gonopods visible in dorsal view, directed ventrolaterally, widely separated at the base; male frontal and antennal appendages always absent; female second antennae subconical with apex spiniform, or conical curved counterclockwise and with a medial projection; vas deferens always without a dorsal loop (Figs. 19.4–7 A, B) .... **Branchinectidae [p. 591]**

3(2) Male second antennal proximal antennomere with or without appendages, if present, then appendages cylindrical or lamellar, but never cheliform; second antennae and appendages not laterally compressed (Figs. 19.7 C–E, 8–10) .... **Thamnocephalidae [p. 597]**

3′ Male second antennal proximal antennomere terminating in a large, cheliform, multiramal appendage directed ventrally (Figs. 19.1 A, B, 12); second antennae and appendages laterally compressed .... Streptocephalidae, one genus: ***Streptocephalus* [p. 603]**

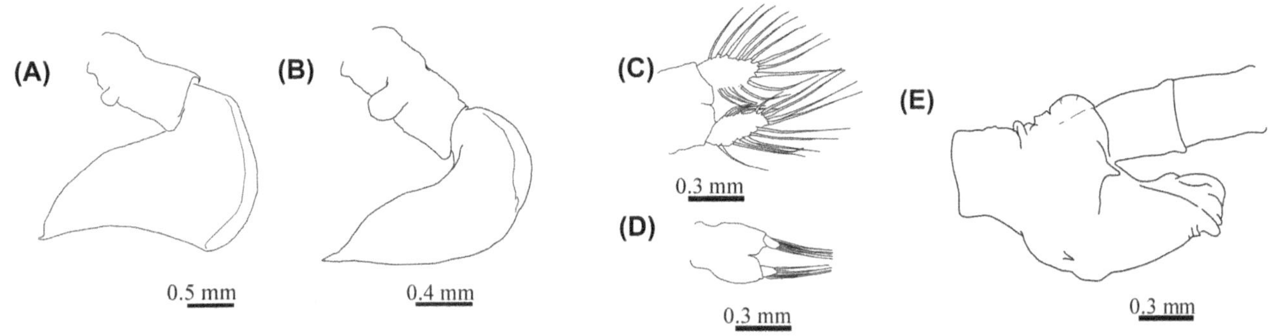

FIGURE 19.3 *Artemia franciscana* (A and C): (A) male antenna 2; (C) cercopods; *Artemia persimilis* (B, D, E): (B) male antenna 2; (D) cercopods; (E) brood pouch, lateral view.

Branchiopoda: Anostraca: Artemiidae: *Artemia*: Species

1 Male second antennal proximal antennomere with a large basomedial frontal knob; second antennal distal antennomere falciform; cercopods very short or rudimentary, with one or a few plumose setae each; female with a pair of elongate dorsal roughened projections following the lateral contour of the second genital segment (Fig. 19.3 B, D, E) .... *Artemia persimilis* Piccinelli & Prosdocimi, 1968
[Peru, Chile, Argentina. Introduced in Palaearctic]

1′ Male second antennal proximal antennomere with a small basomedial rounded frontal knob; second antennal distal antennomere securiform; cercopods developed, with a mean number of 11 plumose setae on each; female without dorsal projections on the second genital segment (Figs. 19.2 H, 3 A,C) .... *Artemia franciscana* Kellogg, 1906
[Neotropical. Nearctic. Invasive in Afrotropical, Australia, Oriental, and Palaearctic regions]

Branchiopoda: Anostraca: Branchinectidae: Genera

1 Male abdominal segment II lacking a pair of ventrolateral sclerotized plates; gonopod rigid basal portion bearing a single basomedial spine or projection (Figs. 19.4–6) .... ***Branchinecta* [p. 591]**

1′ Male abdominal segment II bearing a pair of ventrolateral sclerotized plates densely covered in spinules; gonopod rigid basal portion bearing a sclerotized medial bifurcated projection, the proximal portion linguiform, the distal, spiniform; high-altitude species (Fig. 19.7 A, B) .... ***Archaebranchinecta* [p. 595]**
[Peru, Bolivia, Argentina]

Branchiopoda: Anostraca: Branchinectidae: *Branchinecta*: Species

1 Male second antenna basal antennomere with (Figs. 19.4 E, G, I, 5 A, D, E, F) or without (Fig. 19.6 C, D) a basomedial tubercle or apophysis, but always with additional ornamentations such as spines, protuberances, papillae, mounds, spinose crests, and/or processes 2

1′ Male second antenna basal antennomere smooth, bearing a posteriomedial, smooth tubercle; distal antennomere curved medially in proximal third, apex medially directed; apex anteriolateral edge at bend with a small lateral lobe perpendicular to apex (Fig. 19.4 A, B); apex posterior margin with a rasp .... *Branchinecta rocaensis* Cohen, 1982
[Argentina]

2(1) Male second antenna basal antennomere with 3–10 anterior and medial spines (Fig. 19.4 C, F, H) 3

2′ Male second antenna basal antennomere with many anterior and medial spines and/or with many spines or other structures on a medial longitudinal projection (Figs. 19.4 G, 5, 6) 5

3(2) Male second antenna basal antennomere with 3–10 distomedial spines in a single row; distal antennomere apex rounded, directed medially, or truncated and straight (Fig. 19.4 C, D, F); some female thoracic segments with lateral or dorsolateral projections; brood pouch not reaching telson 4

3′ Male second antenna basal antennomere with 7–8 medial spines arranged in two or three parallel rows in the middle of the article; distal antennomere apex subquadrangular, directed medially (Fig. 19.4 H); with posteriodistal margin bearing a rasp; female thorax without lateral ornamentations; brood pouch fusiform, often extending beyond telson .... *Branchinecta leonensis* César, 1987
[Argentina]

4(3) Male second antenna basal antennomere with medial row with 3–7 spines, the more distal ones may be grouped on a small mound; distal antennomere apex rounded, bent medially, with a small blunt lump in the distal part of the posterior margin (Fig. 19.4 C, D) .... *Branchinecta iheringi* Lilljeborg, 1889
[Brazil, Argentina, Uruguay]

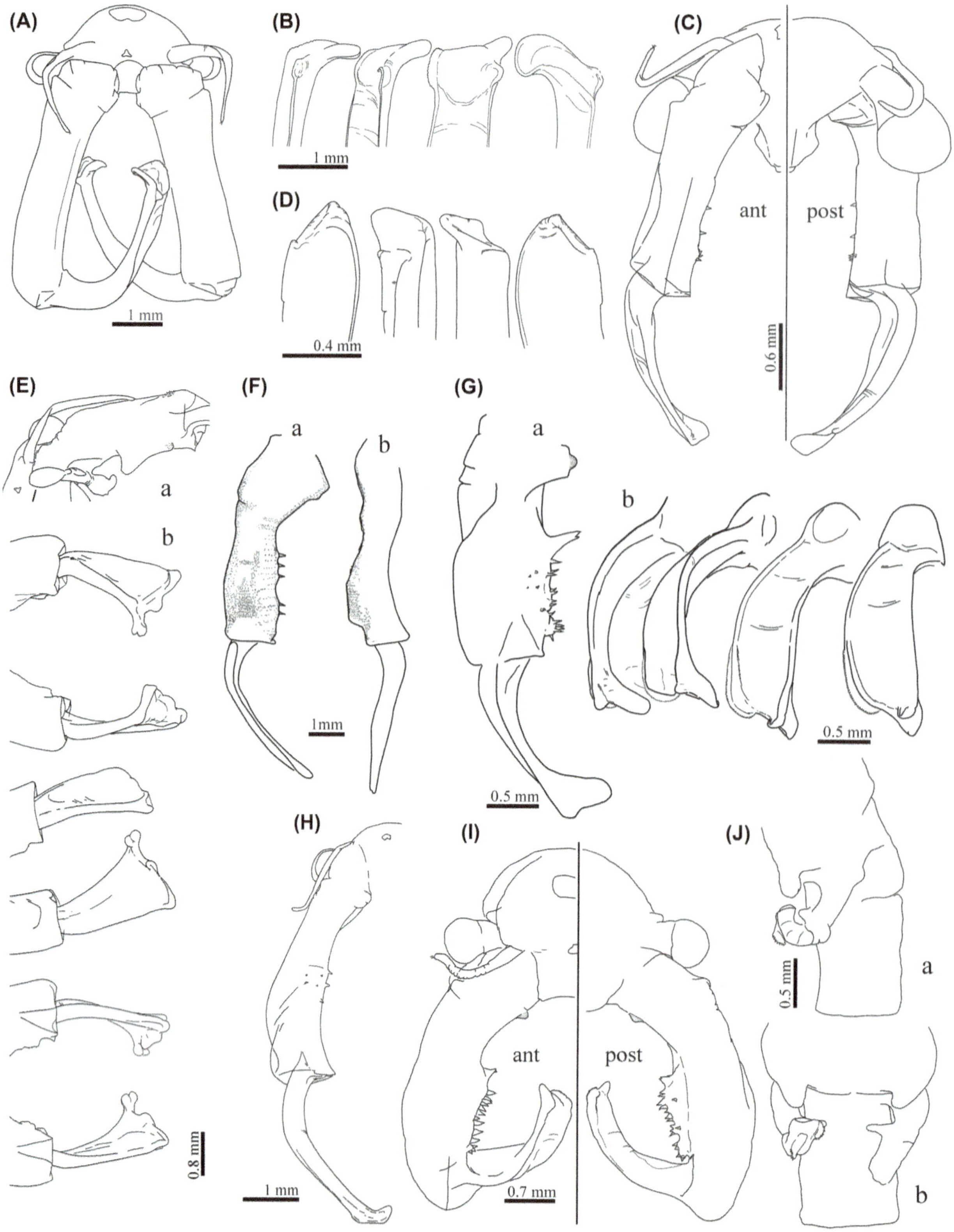

FIGURE 19.4 *Branchinecta*: (A) *Branchinecta rocaensis* male head, anterior view; (B) *B. rocaensis* tip of antenna 2 distal antennomere rotated clockwise from anterior to posterior edge; (C) *Branchinecta iheringi* male head, anterior view (ant) and posterior view (post); (D) *B. iheringi* tip of antenna 2 distal antennomere rotated clockwise from concave to convex surface; (E) *Branchinecta fueguina* male antenna 2, a, basal antennomere, lateral—anterior view; b, pair of distal antennomeres rotated anticlockwise from anterior to posterior view; (F) *Branchinecta ferrolimneta* male antenna 2, a. anterior view; b: lateral view; (G) *Branchinecta granulosa*, a, male antenna 2 anterior view; b, distal antennomere rotated clockwise; (H) *Branchinecta leonensis* male antenna 2 anterior view; (I) *Branchinecta gaini* male head, ant and post; (J) *Branchinecta gaini* male genital segments: a, lateral view, b, ventral view. *Modified: (B) from Cohen (1982); (C) from Cohen (1995); (D) from Cohen (1993); (E) from Cohen (2008); (F) from Rogers & Ferreira (2007); (G) from Cohen (1992).*

4′ Male second antenna basal antennomere anterior, medial and posterior surfaces densely covered with fine denticles, which are larger and denser on a low, anterodistal, longitudinal ridge, and 6–10 large spines in distal half of medial surface; distal antennomere laterally flattened, arcing medially at basal third, with apex truncate and not directed medially (Fig. 19.4 F) .... *Branchinecta ferrolimneta* Rogers & Ferreira, 2007
[Brazil]

5(2) Male second antenna basal antennomere with a medial, longitudinal crest, bulge, or peg shaped projection bearing blunt or acute spines, scattered setae, or small flat projections; crest or bulge sometimes preceded by a proximal row of spines (Figs. 19.4 G, I, 5, A–F, 6 F) 6

5′ Male second antenna basal antennomere with many spines of varying size in varying arrangements but not borne on a medial, longitudinal crest, bulge, projection or, if a medial bulge with spines is present, the antennomere lacks a proximomedial tubercle (Figs. 19.5 E–G, 6 A–E) 15

6(5) Male second antenna proximal antennomere with a basal anteriomedial welt and posteriomedial ornamented apophysis (Fig. 19.4 E) 7

6′ Male second antenna proximal antennomere with a basomedial wide spine, projection or variously ornamented tubercle, without posteriomedial apophysis (Figs. 19.4 A, C, F–I, 5, A, D, E, F, 6 A–C, E, F) 9

7(6) Male second antenna basal antennomere with a medial squamate peglike projection at the middle of the article; distal antennomere curved, not triangular, apex not trilobed 8

7′ Male second antenna basal antennomere with a medial crest more prominent proximally, bearing scattered sensorial setae, without medial squamate projection; distal antennomere strongly rotated, articulating almost perpendicular to basal antennomere, triangular in shape, widening distally, and slightly curved medially; apex trilobed, medium lobe not projecting medially; posterior lobe cleft (Fig. 19.4 E) .... *Branchinecta fueguina* Cohen, 2008
[Argentina]

8(7) Male second antenna basal antennomere with a medial squamate projection, extending along ~ 50% of the article medial surface, distally with two to five parallel transverse indentions .... *Branchinecta mexicana* Maeda-Martínez, Obregón-Barboza & Dumont, 1993
[Mexico]

8′ Male second antenna basal antennomere with a medial squamate projection, extending along ~ 20% of the article medial surface, and lacks indentions .... *Branchinecta packardi* Pearse, 1912
[Mexico. Nearctic]

9(6) Male second antenna distal antennomere apex directed medially and trilobed, with anterior, median, and posterior lobes (Figs. 19.4 G, 5 A, D) 10

9′ Male second antenna distal antennomere apex directed medially, not trilobed (Figs. 19.4 I, 5 E, F, 6 F) 12

10(9) Male second antenna basal antennomere medial crest more prominent proximally but not produced as a strong spiny process (Fig. 19.5 A, D) 11

10′ Male second antenna basal antennomere medial crest produced proximally as a strong medioposteriorly directed spiny process; crest with strong, acute, conical spines slightly decreasing in size distally; crest distal end with a small mound bearing a group of few spines (Fig. 19.4 G) .... *Branchinecta granulosa* Daday, 1902
[Chile, Argentina. Antarctic and Sub-Antarctic islands records need verification]

11(10) Male second antenna basal antennomere with a basomedial tubercle low and rounded; prominent medial rounded protuberance basal to a smooth crest; basal tubercle, protuberance, and crest provided with similar small rounded sensory mounds with setae; distal antennomere apex with medium lobe rounded; male genital segments not prominent in lateral view (Fig. 19.5 A, B) .... *Branchinecta prima* Cohen, 1983
[Argentina]

11′ Male second antenna basal antennomere with a proximomedial large and ornamented spine or projection; medial surface crest lacking a proximal rounded protuberance; crest with scattered small flat projections or small blunt spines; distal antennomere apex with medial lobe obliquely truncate; male genital segments prominent in lateral view (Fig. 19.5 C, D) .... *Branchinecta valchetana* Cohen, 1981
[Chile: Tierra del Fuego. Argentina]

12(9) Male second antenna basal antennomere without a row of spines proximal to a medial crest or bulge (Figs. 19.4 I, 5 F, 6 F) 13

12′ Male second antenna basal antennomere with a medial, longitudinal row of small, short spines, proximal to a medial bulge (sometimes poorly defined) bearing larger acute spines, increasing in size distally; distal antennomere apex rounded and bent medially (Fig. 19.5 E) .... *Branchinecta somuncurensis* Cohen, 1983
[Argentina]

13(12) Male second antenna basal antennomere with proximomedial tubercle bearing spinulae; medial surface crest with acute, subequal spines, although most proximal spine may be larger, with one to three points; crest distally with a small mound bearing a cluster of few spines; cercopods in both sexes longer than telson (Figs. 19.4 I, 5 F) 14

13′ Male second antenna basal antennomere with anteriomedial tubercle bearing small rounded sensorial mounds, each with a small central seta; medial surface crest covering antennomere middle third, and bearing small, blunt, rounded spines; distal antennomere apex bent medially slightly; cercopods in both sexes very short, subequal in length to telson; high-altitude species (Fig. 19.6 F) .... *Branchinecta brushi* Hegna & Lazo-Wasem, 2010
[Chile, Argentina]

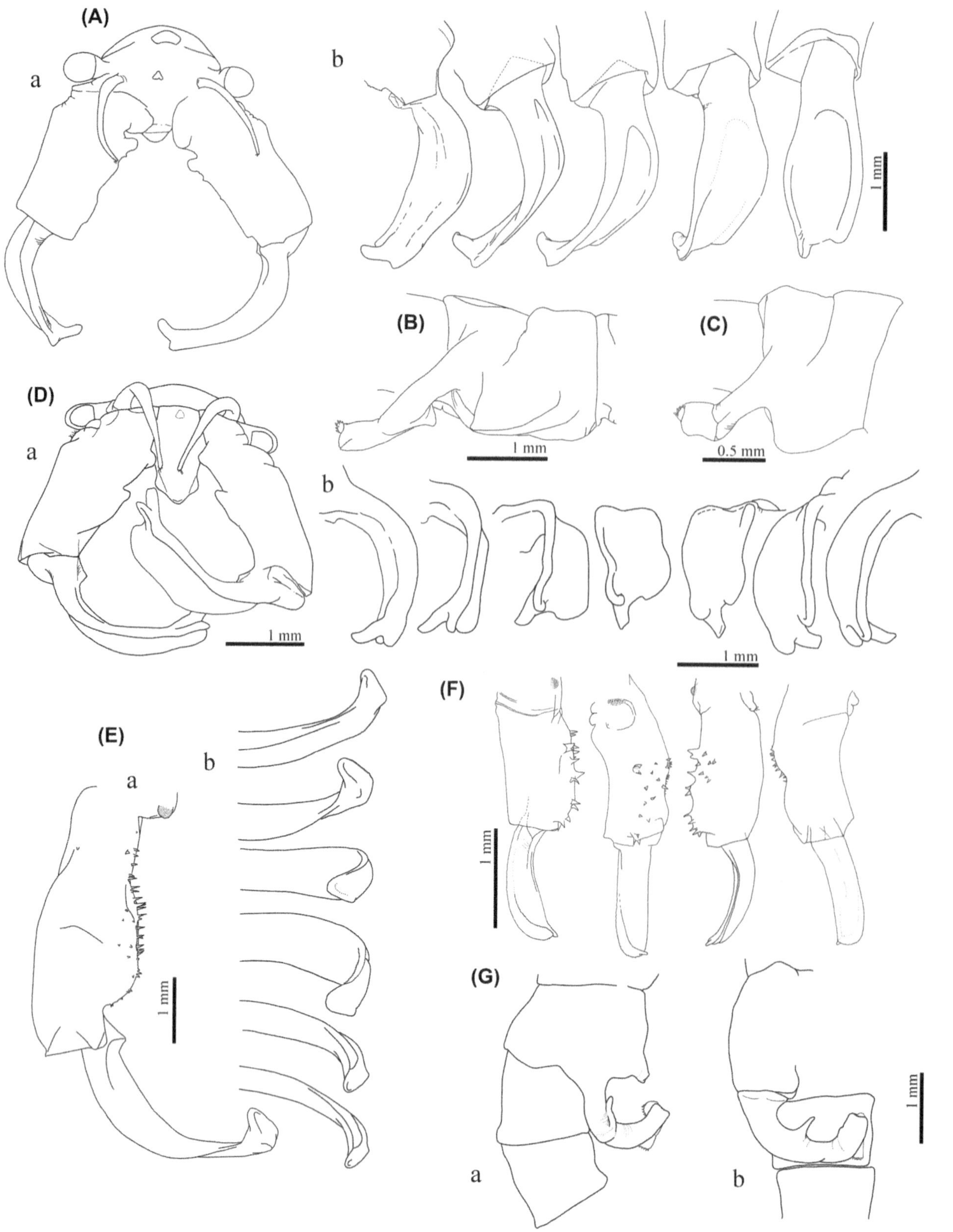

FIGURE 19.5 *Branchinecta*. (A) *Branchinecta prima*: a, male head anterior view; b, male antenna 2 distal antennomere rotated anticlockwise; (B) *B. prima* male genital segments, lateral view; (C) *Branchinecta valchetana* male genital segments, lateral view; (D) *B. valchetana*: a, male head anterior view; b, male antenna 2 distal antennomere rotated anticlockwise from anterior to posterior edge; (E) *Branchinecta somuncurensis*: a, male antenna 2 anteromedial view; b, male antenna 2 distal antennomere rotated from posterior to anterior edge; (F) *Branchinecta vuriloche*: a, male antenna 2 distal antennomere rotated anticlockwise from anterior to posterior view; (G) *B. vuriloche* male genital segments: a, lateral view; b, ventral view. *Modified: (B and E) from Cohen (1983); (C and D) from Cohen (1981); (F and G) from Cohen (1985).*

14(13) Male second antenna distal antennomere apex rounded, anterior lobe protruding distally, posterodistal margin bearing a rasp that sometimes protrudes distally giving the apex bilobed appearance; gonopod proximal rigid projection not hamulate, apically rounded; genital segments prominent in lateral view, bearing a pair of ventral lobes, apically rounded (Fig. 19.4 I–J)... *Branchinecta gaini* Daday, 1910 [Antarctica: Petermann Island, South Shetland Islands, South Orkney Islands. Continental records need verification]

14′ Male second antenna distal antennomere apex rounded, with an anteriomedial, minute digitiform projection; gonopod proximal rigid projection slightly curved, hamulate; genital segments prominent in lateral view, bearing a pair of ventral conical lobes (Fig. 19.5 F, G) .... *Branchinecta vuriloche* Cohen, 1985 [Chile, Argentina]

15(5) Male second antenna basal antennomere bearing a proximomedial tubercle, which is smooth or ornamented with sensorial structures (setae, small projections, tiny spines) (Figs. 19.5 E, F, 6 A, B, E) 16

15′ Male second antenna basal antennomere without a proximomedial tubercle (Fig. 19.6 C, D) 20

16(15) Male second antenna distal antennomere apex bent medially and ending in a single widened rounded lobe, or projecting in an anterior triangular lobe (Figs. 19.5 E, 6 B, E) 17

16′ Male second antenna distal antennomere slightly curved with almost parallel margins, apex not bent medially, apex rounded or truncate (Figs. 19.5 F, 6 A) 19

17(16) Male second antenna basal antennomere without a wide papilla proximal to the spiny armature; female abdomen not markedly elongated (Figs. 19.5 E, 6 E) 18

17′ Male second antenna basal antennomere proximal third bearing a medial, wide papilla with two or three points; medial and anterior surfaces with acute spines, becoming more numerous distally; distal article apex a single rounded lobe, bent medially; male genital segments protrude slightly in lateral view; female abdomen longer than head plus thorax; brood pouch very large and thick, extending to telson (Fig. 19.6 B) .... *Branchinecta papillosa* Birabén, 1946 [Argentina]

18(17) Male second antenna basal antennomere anteromedial surface with a proximal longitudinal row of small, short spines, extending distally to a patch of spines, bearing larger acute spines and scattered small spines; spines increase in size distally; distal antennomere apex rounded and bent medially (Fig. 19.5 E) .... *Branchinecta somuncurensis* Cohen, 1983 [Argentina]

18′ Male second antenna basal antennomere anteriomedial surface with a longitudinal row of 10–20 blunt, subequal spines not extending to antennomere distal end; distal antennomere apex a triangular lobe, bent medially (Fig..19.6 E) .... *Branchinecta palustris* Birabén, 1946 [Argentina, Bolivia?, Chile?]

19(16) Male second antenna basal antennomere with anterior surface bearing a proximal elongate protuberance that extends to half the article, provided with small rounded sensorial mounds and short, straight spines; basal antennomere distomedial surface with oblique rows of stout spines reaching the distal end; basal antennomere with a prominent anteriolateral sensorial area; distal antennomere often with a basal anteriolateral corniform projection; distal antennomere sinuate, lacking distal rasp; female second antenna with a prominent medial sensorial conical mound with hyaline setae or tiny spines (Fig. 19.6 A) .... *Branchinecta achalensis* César, 1985 [Argentina, Uruguay]

19′ Male second antenna basal antennomere anteriomedial surface with scattered, robust spines or rarely inserted on a poorly defined longitudinal ridge; distal antennomere not sinuate, with distal rasp, apex with a small digitiform projection; female second antenna without markedly prominent medial sensorial area (Fig. 19.5 F) .... *Branchinecta vuriloche* Cohen, 1985 [Chile, Argentina]

20(15) Male second antenna basal antennomere anteriomedial surface with a smooth proximal bulge (may bear short spines) and a distal spiny bulge; basal antennomere posteromedial surface with scattered spines increasing in length and density distally; distal antennomere subtriangular in cross section, with medial surface concave, margins lined with denticles; distal antennomere often with a basoanterior linguiform projection; high-altitude species (Fig. 19.6 C) .... *Branchinecta papillata* Rogers, De los Ríos & Zúñiga, 2008 [Chile, Argentina]

20′ Male second antenna basal antennomere anteriomedial surface with several rows of pointy spines, each tipped with a seta, spines increasing in length distally; distal third without spines; distal antennomere slightly curving medially, not concave, without marginal denticles, lacking a basal projection, apex rounded (Fig. 19.6 D) .... *Branchinecta tarensis* Birabén, 1946 [Argentina]

CLASS BRANCHIOPODA

Branchiopoda: Anostraca: Branchinectidae: Archaebranchinecta: Species

1 Male second antenna proximal antennomere with long digitiform processes projecting medially, nearly touching each other; between gonopod bases with a ventral pair of close, prominent bulges; female second genital segment with a ventrolateral spine; brood pouch with pair of small ventrolateral outgrowths (Fig. 7A, B)........ *Archaebranchinecta pollicifera* (Harding, 1940) [Peru]

1′ Male second antenna basal antennomere with short digitiform processes usually hidden behind the basal antennomere; between gonopod bases with a pair of obscure low bulges; female first and second genital segments with a lateral spine and a ventrolateral spine, respectively; brood pouch lacking ventrolateral outgrowths........ *Archaebranchinecta aimara* Cohen, Marinone & Adamowicz, 2019 [Bolivia]

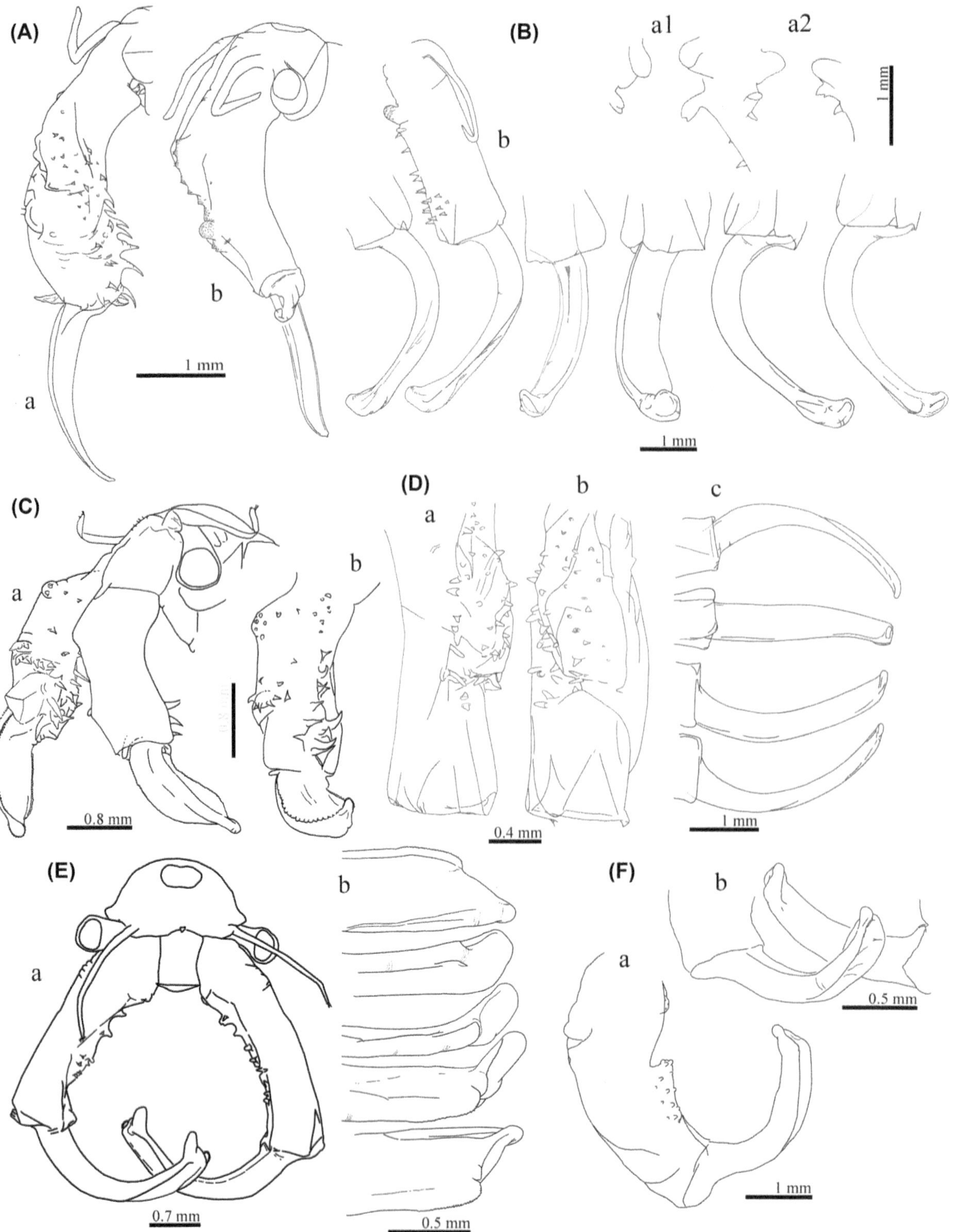

FIGURE 19.6 (A) *Branchinecta achalensis* male antenna 2: a, lateromedial view; b, lateral view; (B) *Branchinecta papillosa* male antenna 2: a1, a2, pairs of basal–medial papillae in the basal antennomere; b, A2 rotated clockwise from anterior to posterior edge; (C) *Branchinecta papillata*: a, male head. lateral view; b, male antenna 2, medial view; (D) *Branchinecta tarensis* male antenna 2: a, basal antennomere, medial view; b, basal antennomere anterior view; c, distal antennomere rotated clockwise (up to down) from anterior to posterior edge; (E) *B. palustris*: a, anterior view; b, male antenna 2 distal antennomere rotated clockwise from convex to concave surface; (F) *Branchinecta brushi* male antenna 2: a, lateral view; b, distal antennomeres. *Modified: (A) from Cohen (1987); (C) from Cohen (2012); (E) from Cohen (1981).*

Branchiopoda: Anostraca: Thamnocephalidae: Genera

1	Cercopods free, subcylindrical	2
1′	Cercopods broadly transverse and fused with telson and the flattened sides of a variable number of abdominal segments, into a broad, flat paddle (Fig. 19.1 D, E)	***Thamnocephalus* [p. 598]**
2(1)	Cercopods shorter than 50% of abdomen length; eye peduncle not abnormally short	3
2′	Cercopods more than 50% as long as abdomen; eye peduncle shorter than compound eye diameter (Fig. 19.7 C) [Argentina]	*Aeginecta longicauda* Rogers & Pessacq, 2018
3(2)	Male without an "antenna-like" appendage (Fig. 19.7 D, E)	4
3′	Male with an "antenna-like" appendage between eyestalk and antennae	***Dendrocephalus* [p. 599]**
4(3)	Male frontal appendage with a short basal trunk bifurcated into two primary long branches, each with subbranches; gonopods with medial surface of rigid proximal portion without a dense patch of denticles; female second antenna lamellar and oval, not apically acute; freshwater (Fig. 19.7 E) [Brazil, Paraguay]	*Spiralifrons mira* (Gurney, 1931)

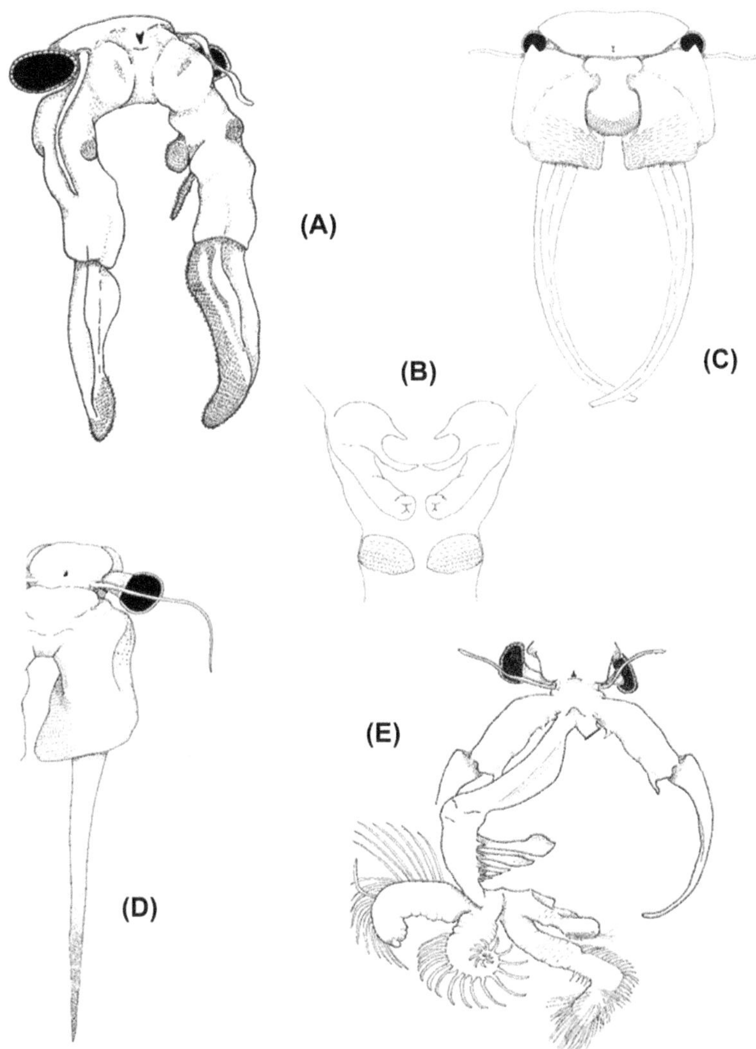

FIGURE 19.7 Representative Anostraca genera: (A) *Archaebranchinecta pollicifera*, male, head, right anterior oblique view; (B) *A. pollicifera*, retracted gonopods, ventral view; (C) *Aeginecta longicauda*, male, head, anterior view; (D) *Phallocryptus wrighti*, male, left half of head, anterior view; (E) *Spiralifrons mira*, male, head, anterior view, left half of frontal appendage removed.

CLASS BRANCHIOPODA

4′ Male lacking frontal appendage; gonopods with medial surface of rigid proximal portion with a dense cornified patch of denticles; female second antenna lamellar and triangular, slightly produced medially, apically acute or angled; halophilic (Fig. 19.7 D) .... *Phallocryptus wrighti* (Smirnov, 1948) [Argentina]

Branchiopoda: Anostraca: Thamnocephalidae: *Thamnocephalus*: Species

Terminology for the frontal appendage follows Obregón-Barboza et al. (2015) (Fig. 19.8), terminology following Belk & Pereira (1982) in brackets.

1 Male second antenna basal antennomere with a conspicuous distolateral conical spine; frontal appendage large, multiramal, with first, second, and third branches (= 1MD, 2D, and 3D–7D branches) (Fig. 19.1 C) 2

1′ Male second antenna basal antennomere without a distolateral conical spine; frontal appendage reduced, with two short unbranched arms; halophilic .... *Thamnocephalus* (*Simplicephalus*) *salinarum* Cohen, 2002 [Argentina]

2(1) Frontal appendage second branch with dorsal and ventral subbranches (= branches 3D–7D and 1V–5V); all or some dorsal sub-branches arborescent (Fig. 19.8 A) 3

2′ Frontal appendage second branch without ventral subbranches; dorsal subbranches (= branches 3D–7D) unramified, but sometimes forked apically (Fig. 19.8 A) 4

3(2) Frontal appendage first branch (= branch 2D) with three unramified subbranches, the most proximal one coiled, distally flattened and folded; third branch (= branch 1MD) broadly triangular, flattened and folded, with a small apical spine (Fig. 19.8 E) .... *Thamnocephalus* (*Thamnocephalus*) *chacosaltensis* Cohen, 2016 [Argentina]

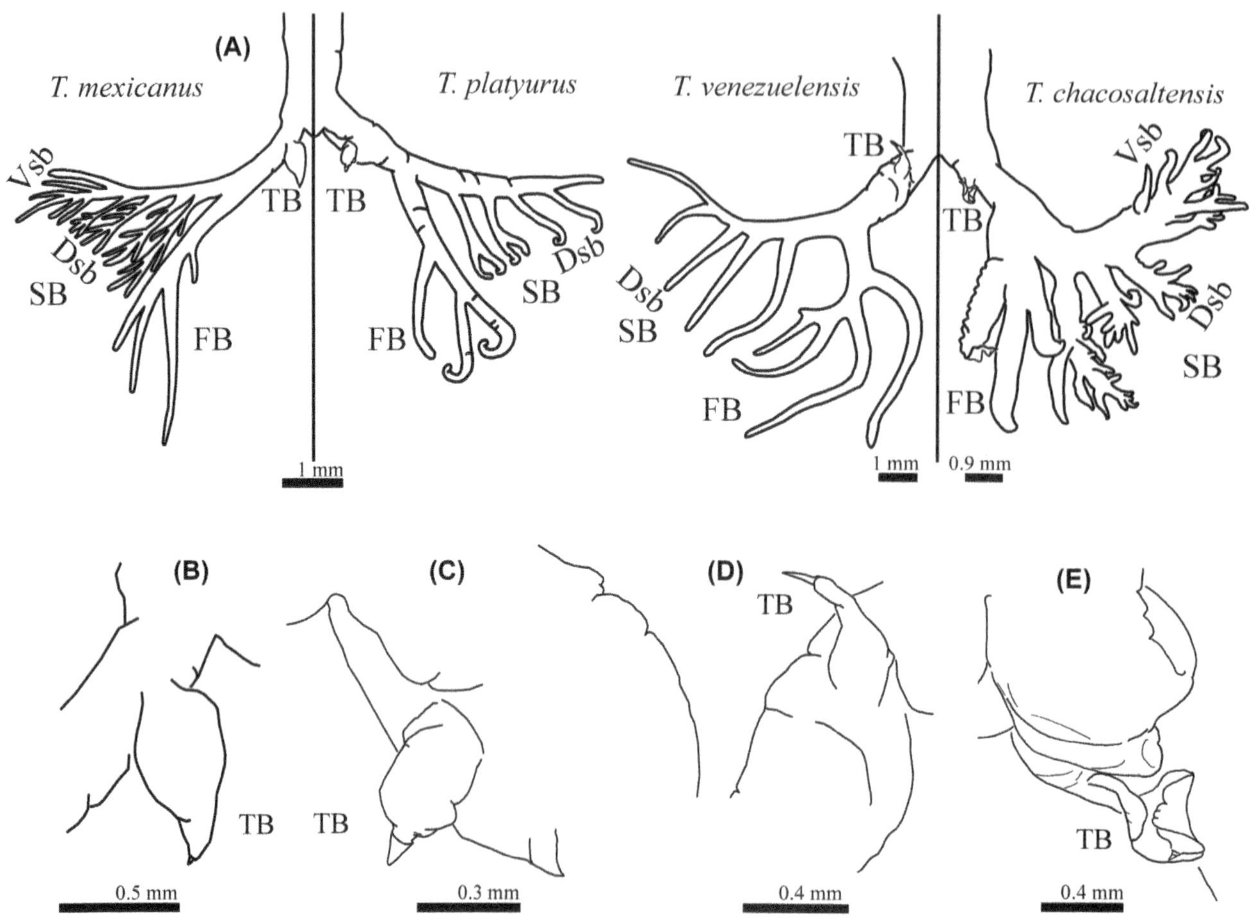

FIGURE 19.8 *Thamnocephalus* sp.: (A): nomenclatural system of the frontal appendage (according to Obregón-Barboza et al., 2015) for *Thamnocephalus mexicanus*, *Thamnocephalus platyurus*, *Thamnocephalus venezolanus*, and *Thamnocephalus chacosaltensis*; (B–E) third branch: (B) *T. mexicanus*; (C) *T. platyurus*; (D) *T. venezuelensis*; (E) *T. chacosaltensis*. Abbreviations: *Dsb*, dorsal subbranches; *FB*, first branch; *SB*, second branch; *TB*, third branch; *Vsb*, ventral subbranches. *(A) Modified from Belk & Pereira (1982), Obregón-Barboza et al. (2015) and Cohen (2016).*

3′ Frontal appendage first branch (= branch 2D) with four to five unramified subbranches, never distally flattened nor folded; third branch (= 1 MD) flat, foliform, with a small apical spine (Fig. 19.8 B) .... *Thamnocephalus (Thamnocephalus) mexicanus* Linder, 1941 [Mexico. Nearctic]

4(2) Frontal appendage second branch (= branches 3D-7D), with three to four unramified subbranches not bifurcated distally; third branch (= branch 1MD) conical, with a medial coniform projection, and apex laterally directed and bearing a spine (Fig. 19.8 D) .... *Thamnocephalus (Thamnocephalus) venezuelensis* Belk & Pereira, 1982 [Aruba, Venezuela]

4′ Frontal appendage second branch (= branches 3D-7D), with three to seven unramified subbranches, the first two apically bifurcated; third branch (= branch 1MD) cylindrical with a distal spine directed medially (Figs. 19.1 C, 8 C) .... *Thamnocephalus (Thamnocephalus) platyurus* Packard, 1877 [Mexico. Nearctic]

Branchiopoda: Anostraca: Thamnocephalidae: *Dendrocephalus*: Species

Terminology follows Pereira (1983) (Fig. 19.9).

1 Frontal appendage branch 1V, anterior subbranch with apex acute (e.g., Fig. 19.9 B) 2

1′ Frontal appendage branch 1V, anterior subbranch with apex expanded (e.g., Fig.19.9 A, C) 8

2(1) Frontal appendage branch 2D, subbranch III without a multiramal palmate projection (e.g., Fig. 19.10 A, C) 3

2′ Frontal appendage branch 2D, subbranch III posterior margin distal third with a multiramal palmate projection (e.g., Figs. 19.9 A, 10 B); branch 2A apex rounded on one side, and with prominent spine on opposite side (Fig. 19.9 F); male thoracopod 1 endopodite with proximolateral lobe not margined with spines; male thoracopod 2 endopodite without proximolateral lobe .... *Dendrocephalus conosuris* Pereira & Ruiz, 1995 [Argentina]

3(2) Male thoracopod 1 endopodite lateral margin not projecting or with a reduced proximolateral outgrowth, lateral margin with small spines without distolateral region extended into a lobe (e.g., Fig. 19.11 A) 4

3′ Male thoracopod 1 endopodite with a proximolateral outgrowth and a distolateral row of stout spines or distolateral region extended into lobe margined with spines (e.g., Fig. 19.11 D–G) 5

4(3) Frontal appendage branch 2V ending in two cell pads (e.g., Fig. 19.10, E); branch 2D subbranch III most robust of the three subbranches .... *Dendrocephalus sarmentosus* Pereira & Belk, 1987 [Ecuador: Galapagos Islands]

4′ Frontal appendage branch 2V ending in four or five cell pads (e.g., Fig. 19.10, D), branch 2D subbranch I longest of the three subbranches, subbranch III wider and slightly shorter than subbranch I .... *Dendrocephalus xikrini* Rabet & Bozelli 2018 [Brazil]

5(3) Male thoracopod 1 endopodite proximolateral outgrowth elongate, smooth, and spiniform (e.g., Fig. 19.11 D, E) 6

5′ Male thoracopod 1 endopodite proximolateral outgrowth a prominent either margined with blunt, sometimes scattered, spines, or margined by stout spines (e.g., Fig. 19.11 G) 7

6(5) Male thoracopods 1–3 with endopodite proximolateral outgrowth elongate and spiniform; frontal appendage branch 2A podiform with a basal cell pad .... *Dendrocephalus spartaenovae* Margalef, 1961 [Aruba Venezuela]

6′ Male thoracopod 1 with endopodite proximolateral outgrowth elongate and spiniform, thoracopods 2 and 3 without proximolateral outgrowth; frontal appendage branch 2A podiform without basal cell pads .... *Dendrocephalus riograndensis* Rogers & Volcan, 2016 [Brazil]

7(5) Male thoracopod 1 endopodite proximolateral outgrowth tapered, subacute, margined with short, blunt spines, surface sometimes ornamented with short rounded spines; male thoracopod 2 endopodite margined with blunt or acute spines, its surface sometimes ornamented with short rounded spines (Fig. 19.11 G); thoracopods 1-3 with endopodite distolateral corner lobiform, margined with spines; frontal appendage branch 2A apex hamulate, forming right angle with the branch main axis (Fig. 19.9 D)... .... *Dendrocephalus argentinus* Pereira & Belk, 1987 [Bolivia, Paraguay, Argentina]

7′ Male thoracopod 1 endopodite proximolateral outgrowth with a few smooth, stout spines, distolateral border with spines; frontal appendage branch 2A apex podiform .... *Dendrocephalus goiasensis* Rabet & Thiéry, 1996 [Brazil]

8(1) Frontal appendage branch 2D, subbranch III posterior margin distal third with a multiramal palmate projection (e.g., Figs. 19.9, A, 10 B) 9

8′ Frontal appendage branch 2D, subbranch III posterior margin lacking a multiramal palmate projection (e.g., Fig. 19.10 A, C) 11

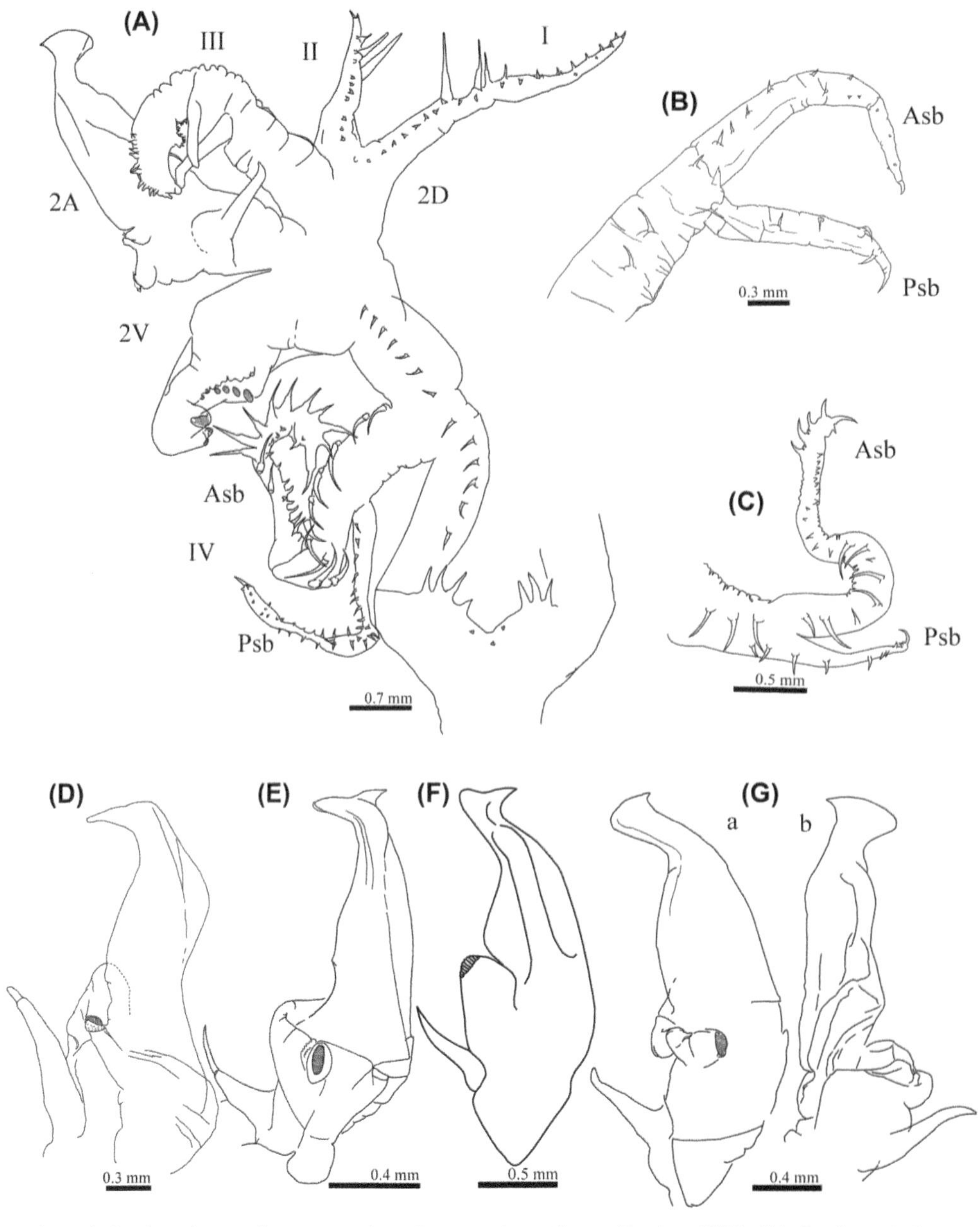

FIGURE 19.9 *Dendrocephalus* frontal appendage nomenclatural system (according to Pereira, 1983). (A) *Dendrocephalus cervicornis*; (B) *Dendrocephalus argentinus* branch IV, anterior subbranch ending acutely; (C) *Dendrocephalus brasiliensis* branch IV, anterior subbranch ending expanded; (D) *D. argentinus* branch 2A ending hamulate; (E) *Dendrocephalus affinis* branch 2A ending podiform; (F) *Dendrocephalus conosuris* branch 2A ending rounded at one side and showing prominent opposite distoventral spine; (G) *D. cervicornis* branch 2A ending rounded flattened: a, b different views. Abbreviations: *1V*, most proximal ventral branch; *2V, 2A, 2D*, most distal ventral, anterior, and dorsal branches, respectively; *Asb* and *Psb*, anterior and posterior subbranches of 1V; *I, II, III*, anterior, medium and posterior subbranches of 2D. *Modified: (C) from de Paiva Barros-Alves et al. (2016); (E) from Cohen et al. (2014), (F) from Pereira & Ruiz (1995).*

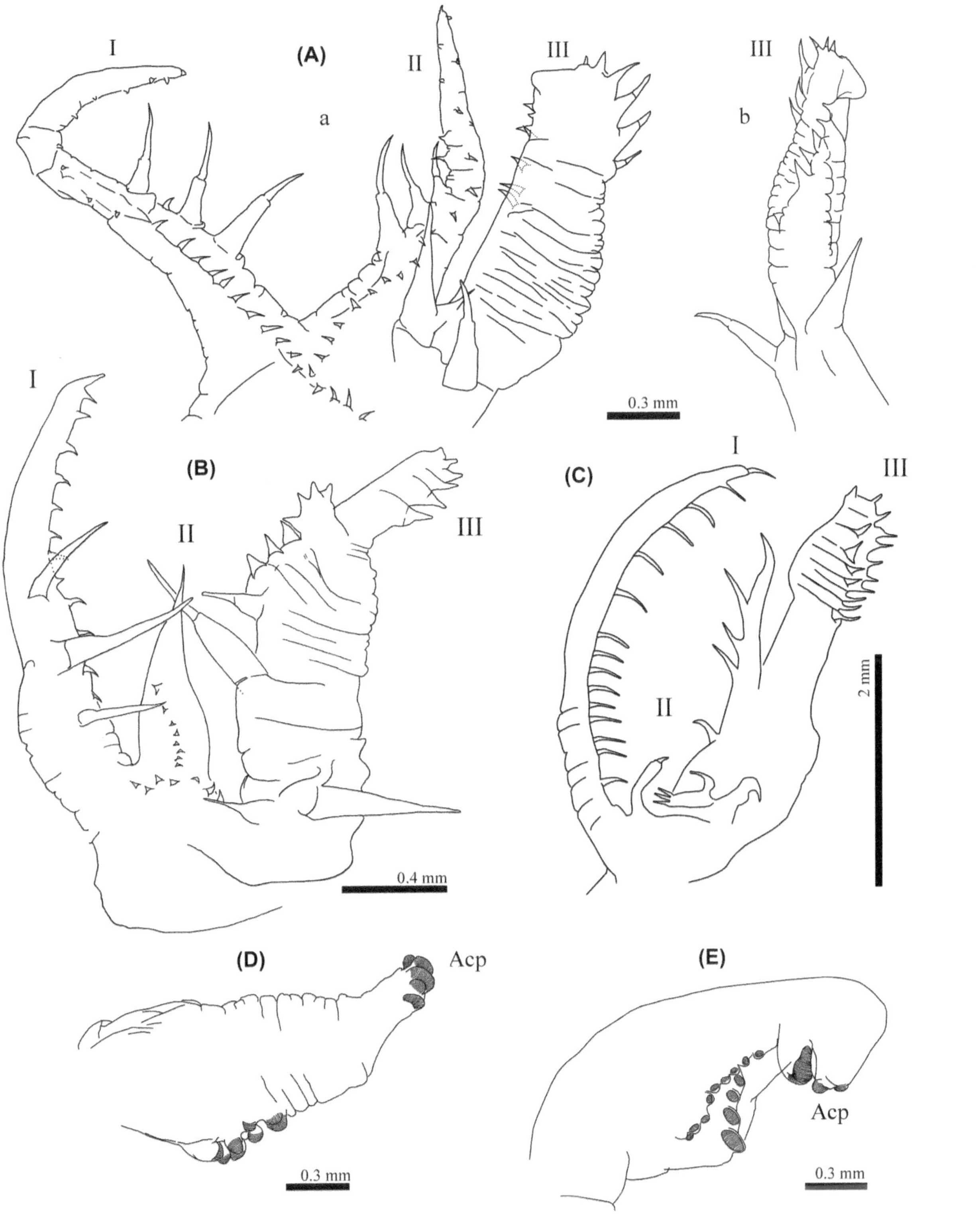

FIGURE 19.10 *Dendrocephalus* frontal appendage branch 2D. (A) *Dendrocephalus argentinus*: a, subbranch III without multiramal palmate projection on posterior margin; b, another view of subbranch III; (B) *Dendrocephalus affinis* subbranch III with multiramal palmate projection on posterior margin; (C) *Dendrocephalus carajaensis* subbranch II reduced, digitiform. *Dendrocephalus* frontal appendage branch 2V; (D) *D. argentinus* with more than two to three apical cell pads; (E) *Dendrocephalus cervicornis* with two to three apical cell pads. Abbreviation: *Acp*, apical cell pads. *Modified: (B) from Cohen et al. (2014); (C) from Rogers et al. (2008).*

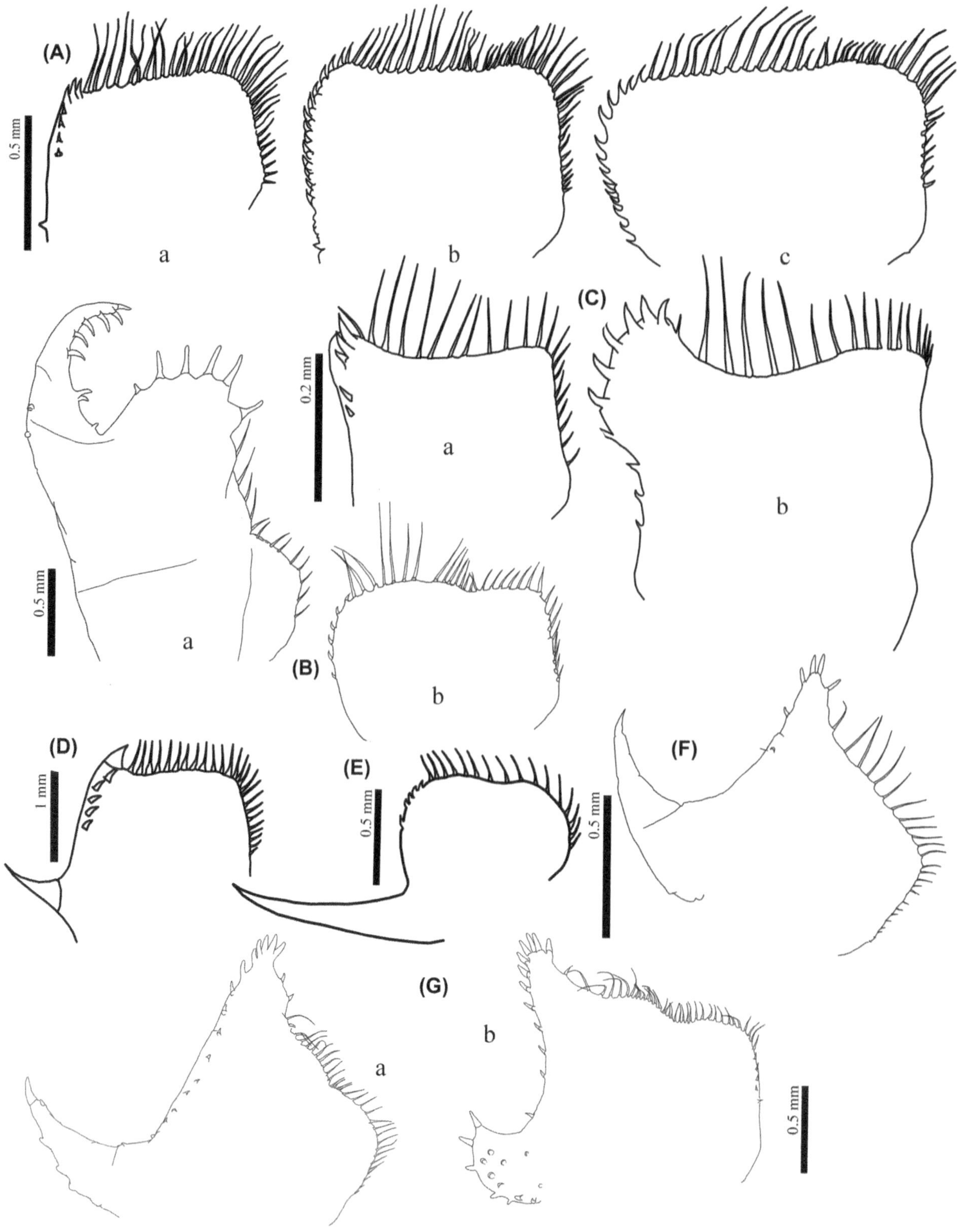

FIGURE 19.11 *Dendrocephalus* male endopodites. (A) a, b, c: *Dendrocephalus brasiliensis* thoracopods 1-3, without proximolateral outgrowth, with distolateral row of spines; (B) *Dendrocephalus cervicornis* thoracopods 1 and 2: a, endopodite 1 very extended longitudinally, with proximolateral outgrowth beset with strong spines and distolateral corner extended into a spine margined lobe; b, endopodite cordate, with only spine row on lateral border; (C) *Dendrocephalus carajaensis* a, b: thoracopods 1 and 2, without proximolateral outgrowth, with distolateral corner extended as a lobe and bearing spines; (D) *Dendrocephalus venezolanus* endopodite 1, with proximolateral spiniform projection without spines; (E) *Dendrocephalus cornutus* endopodite 1 with proximolateral spiniform projection without spines; (F) *Dendrocephalus affinis* endopodite 1, with proximolateral spiniform projection smooth or with a few spines, and prominent distolateral corner extended into a spine-margined lobe; (G) *Dendrocephalus argentinus* thoracopods 1 and 2: a, endopodite 1 with proximolateral outgrowth edged with short spines rounded at apex; b, endopodite 2 with proximolateral projection margined with strong spines apically rounded or acute, and frequently with short rounded spines on outgrowth surface. All figures oriented with their lateral edge toward the left. *Modified: (A) from Rabet & Thiéry (1996); (C) from Rogers et al. (2012); (D) from Pereira (1984); (E) from Pereira & Belk (1987).*

9(8) Frontal appendage branch 2V with several apical cell pads (e.g., Fig. 19.10, D); male thoracopod 1 not projecting distoventrally, proximolateral outgrowth tapered, curved, acute, and inerm or with few spines, distolateral corner prominent, lobiform, margined with spines; endopodite of thoracopod 2 of similar size and outgrowths as thoracopod 1 (e.g., Fig. 19.11, F) 10

9′ Frontal appendage branch 2V usually with only two apical cell pads; branch 2A rounded apically, flattened (Fig. 19.9 A, G); male thoracopod 1 endopodite elongate, distoventrally projecting, length 2× thoracopod 2, with a curved, tapering, distolateral outgrowth, margined with stout spines, distomedial corner prominent, margined with stout spines; thoracopod 1 endopodite lateral margin without or with few short blunt spines; thoracopod 2 endopodite, cordate, without projections, lateral margin with a spines row (Fig. 19.11 B) .... *Dendrocephalus cervicornis* (Weltner, 1890) (*sensu* Pereira, 1983)
[Argentina]

10(9) Frontal appendage branch 2A podiform (Fig. 19.9 E) .... *Dendrocephalus affinis* Pereira, 1984
[Venezuela, Colombia]

10′ Frontal appendage branch 2A sharply hamulate, apex bent 45 degrees or greater (e.g., Fig. 19.9 D) .... *Dendrocephalus geayi* Daday, 1908 (*sensu* Pereira, 1983)
[Venezuela]

11(8) Male thoracopod 1 endopodite with a proximolateral outgrowth (Fig. 19.11 D–G) 12

11′ Male thoracopod 1 endopodite without a proximolateral outgrowth (Fig. 19.11 A, C) 16

12(11) Male thoracopod 1 endopodite with proximolateral outgrowth margined with stout spines (e.g., Fig. 19.11G) 13

12′ Male thoracopod 1 endopodite with a proximolateral outgrowth smooth (e.g., Fig. 19.11 D, F) 15

13(12) Frontal appendage 2A apex podiform (e.g., Fig. 19.9 E) 14

13′ Frontal appendage 2A apex hamulate (e.g., Fig. 19.9, D) .... *Dendrocephalus orientalis* Rabet & Thiéry, 1996
[Brazil]

14(13) Frontal appendage branch 2D with long spines at base of subbranches I and II .... *Dendrocephalus aranai* Rabet & Lacau 2018
[Brazil]

14′ Frontal appendage branch 2D without long spines at base of subbranches I and II .... *Dendrocephalus thieryi* Rabet, 2006
[Brazil]

15(12) Male thoracopod 1 endopodite with distolateral corner bearing a single large spine subequal in length to distal margin setae, distolateral margin with spines (Fig. 19.11 D); frontal appendage branch 2A apex hamulate (e.g., Fig. 19.9, D); branch 2D subbranch II shorter than subbranches I and III (e.g., Fig. 19.10, B); subbranch III with three basal spines, plus one very large spine two-thirds the length of the subbranch .... *Dendrocephalus venezolanus* Pereira, 1984
[Venezuela, Colombia]

15′ Male thoracopod 1 endopodite with distolateral corner bearing spines shorter than the distal margin setae (Fig. 19.11 E); frontal appendage branch 2A apex podiform (e.g., Fig. 19.9 E); branch 2D subbranches I and II subequal; subbranch III with two long, basal spines, lacking additional very large spine .... *Dendrocephalus cornutus* Pereira & Belk, 1987
[Costa Rica]

16(11) Male thoracopods 1, 2, and 3 endopodites with distolateral corner thickened, produced, and bearing spines (Fig. 19.11 C); frontal appendage branch 1V, posterior subbranch digitiform, lacking spines; branch 2A apex hamulate (e.g., Fig. 19.9 D); branch 2D subbranch II reduced, digitiform, apex bearing a small spine; subbranch III forked (Fig. 19.10 C) .... *Dendrocephalus carajaensis* Rogers, Corrêa Gomes & Vieira, 2012.
[Brazil]

16′ Male thoracopods 1, 2, and 3 endopodites with a distolateral row of spines (Fig. 19.11 A); frontal appendage branch 1V, posterior subbranch shorter than anterior and not digitiform, bearing spines, and with an apical spine (Fig. 19.9 C); branch 2A apex podiform(e.g., Fig. 19.9 E); branch 2D, subbranch II not reduced, cylindrical, apex rounded and bearing two spines .... *Dendrocephalus brasiliensis* Pesta, 1921 (*sensu* Rabet & Thiéry, 1996)
[Brazil]

Branchiopoda: Anostraca: Streptocephalidae: *Streptocephalus*: Species

See Fig. 19.1 A and B for terminology.

1 Male second antennal appendage with peduncle subequal to or shorter than second antennal proximal antennomere 2

1′ Male second antennal appendage with peduncle 1.5 times longer than second antennal proximal antennomere 4

2(1) Peduncle subequal in length to second antennal proximal antennomere; "thumb" with spur present (Fig. 19.12 B, D, F, H) 3

2′ Peduncle shorter than second antennal proximal antennomere; "thumb" without a spur (Fig. 19.12 A) .... *Streptocephalus kargesi* Spicer, 1985
[Mexico: Veracruz]

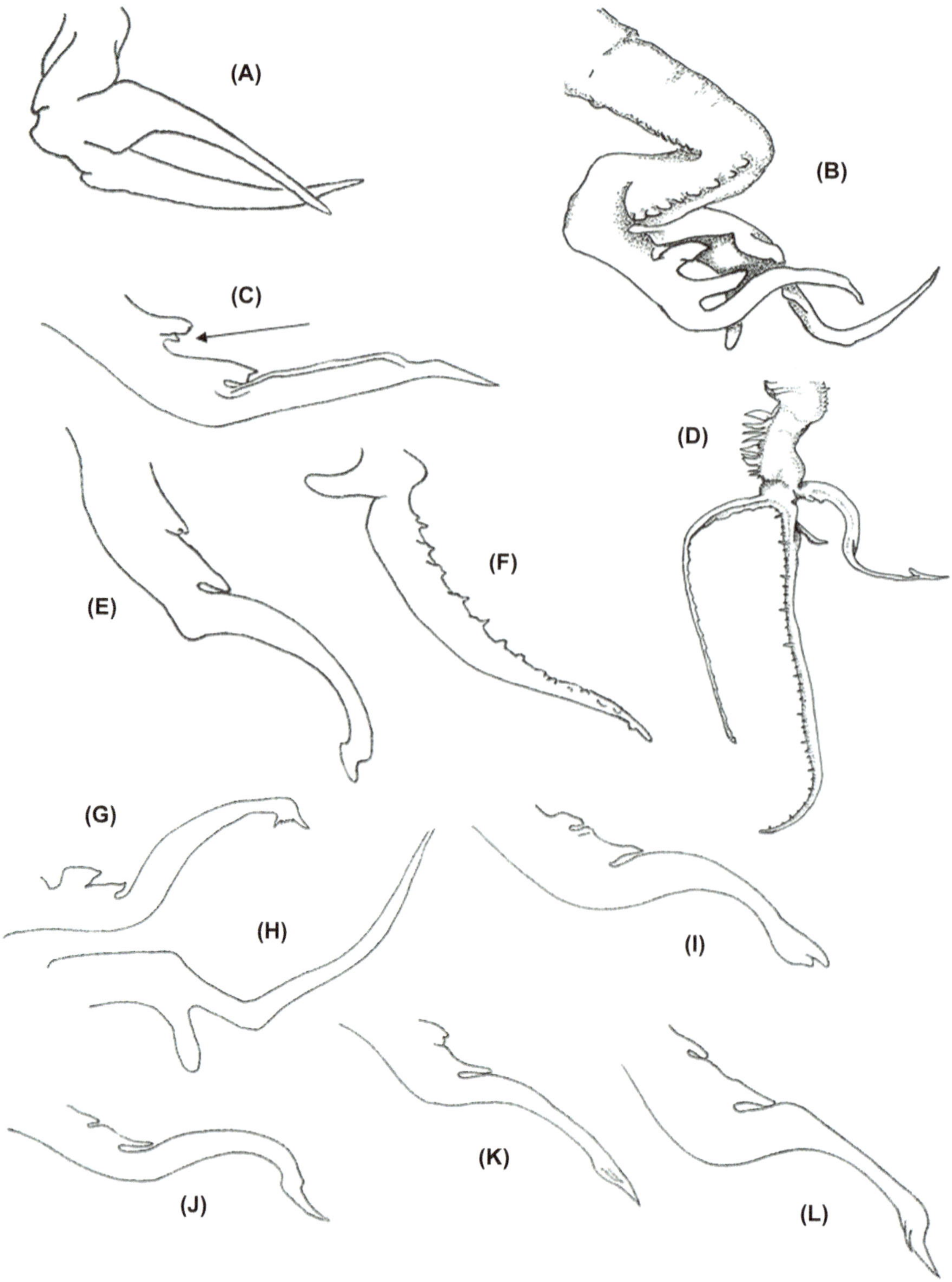

FIGURE 19.12 *Streptocephalus* species: (A) *Streptocephalus kargesi*, antennal appendage, right lateral view; (B) *Streptocephalus sealii*, antennal appendage, right lateral view; (C) *Streptocephalus similis*, left "finger," medial view, arrow points to third projection, lateral to proximal "tooth"; (D) *Streptocephalus moorei*, antennal appendage, right anterior view; (E) *Streptocephalus antillensis*, left "finger," medial view; (F) *S. antillensis*, left "thumb," medial view; (G) *Streptocephaluslinderi*, left "finger," medial view; (H) *S. linderi*, left "thumb," medial view; (I) *Streptocephalus texanus*, left "finger," medial view; (J) *Streptocephalus dorothae*, left "finger," medial view; (K) *Streptocephalus mackini*, left "finger," medial view; (L) *Streptocephalus henridumontis*, left "finger," medial view.

3(2) "Finger" with two "teeth"; genitalia with linguiform outgrowths; frontal appendage bilobed (Fig. 19.12 B) .... *Streptocephalus sealii* Ryder, 1879
[Mexico: Veracruz. Nearctic]

3' "Finger" with three "teeth"; genitalia without linguiform outgrowths; frontal appendage truncate; mature specimens with cercopods bearing stout marginal spines on the distal half (Fig. 19.12 C) .... *Streptocephalus similis* Baird, 1852
[Mexico: Nuevo León, San Luis, Potosí, Tamaulipas. Haiti. Dominican Republic. Puerto Rico. Jamaica. Belize. Antigua and Barbuda. Nearctic]

4(1) "Thumb" with anterior margin spinose (Fig. 19.12 D, F) 5

4' "Thumb" with anterior margin inerm (Fig. 19.12 B, H) 6

5(4) "Thumb" bearing a basal dorsolateral curved process; spur short, acuminate; peduncle with lateral vermiform processes (Fig. 19.12 D) .... *Streptocephalus moorei* Belk, 1973
[Mexico: Chihuahua. Nearctic]

5' "Thumb" without a basal process; "thumb" with two digitiform projections; spur truncate (Fig. 19.12 F) .... *Streptocephalus antillensis* Mattox, 1950
[Puerto Rico]

6(4) Spur truncated apically 7

6′ Spur angulate apically 10

7(6) Apex of "finger" with a lateral lobe or lamella (Fig. 19.12 I) 8

7′ Apex of "finger" with a posterior row of subtending spines, the largest spine being furthest from the apex; "finger" strongly recurved (Fig. 19.12 G); eight conical projections on medial surface of peduncle; "thumb" long and acuminate .... *Streptocephalus linderi* Moore, 1966
[Mexico: Coahuila, Nuevo León, Tamaulipas. Nearctic]

8(7) Apex of "finger" with a lateral lamella, lying along the apex (Fig. 19.12 I) 9

8′ Apex of "finger" with a podiform lateral lobe; peduncle without conical projections; proximal "tooth" on "finger" rounded .... *Streptocephalus guzmani* Maeda-Martínez, Belk, Obregón-Barboza, & Dumont, 1995
[Mexico: Coahuila]

9(8) Proximal "tooth" on "finger" larger than distal "tooth" .... *Streptocephalus potosinensis* Maeda-Martínez, Belk, Obregón-Barboza & Dumont, 1995
[Mexico: San Luis Potosí]

9′ Proximal "tooth" on "finger" smaller than distal "tooth" (Fig. 19.12 I) .... *Streptocephalus texanus* Packard, 1871
[Mexico: Baja California Sur, Chihuahua, Coahuila, Durango, Nuevo León, Oaxaca, Potosí, San Luis, Tamaulipas. Guadeloupe: Désirade. Antigua and Barbuda: Island of Barbuda. Nearctic]

10(6) Apex of "finger" with a subtending spur (Fig. 19.12 K, L) 11

10′ Apex of "finger" without a subtending spur (Fig. 19.12 J) .... *Streptocephalus dorothae* Mackin, 1942
[Mexico: Baja California Sur. Nearctic]

11(10) Proximal "tooth" truncate; distal "tooth" with basal width 1.5× height; "finger" apex with lateral subtending spur; "finger" anterior margin evenly curved (Fig. 19.12 K) .... *Streptocephalus mackini* Moore, 1966
[Mexico: Aguascalientes, Chihuahua, Coahuila, Distrito Federal, Durango, Estado de México, Guanajuato, Hidalgo, Jalisco, Morelos, Nuevo León, Puebla, Querétaro, San Luis Potosí, Sinaloa, Sonora, Tlaxcala, and Zacatecas. Nearctic]

11′ Proximal "tooth" subtriangular; distal "tooth" with base width subequal to height; "finger" apex with posterolateral subtending spur; "finger" anterior margin flattened (Fig. 19.12 L) .... *Streptocephalus henridumontis* Maeda-Martínez, Obregón-Barboza, Prieto-Salazar & García-Velazco, 2005
[Mexico: Baja California Norte, Sinaloa, Sonora. Nearctic]

Branchiopoda: Notostraca: Triopsidae: Genera

The Neotropical Triopsidae need to be revised.

1 Telson without a caudal lamina projecting between cercopods (Fig. 19.1 G) .... *Triops*
[Widespread]

1′ Telson with a caudal lamina projecting between cercopods (Fig. 19.1 F) .... *Lepidurus patagonicus* Berg, 1900
[Argentina]

Branchiopoda: Diplostraca: Suborders

The Cyclestherida needs revision.

1 Carapace folded, no true hinge present, growth lines present or absent laterally flattened or carapace reduced or absent (Fig. 19.1 A, B, E–G) 2

1′ Carapace with a true, interlocking dorsal hinge; carapace subglobular, smooth, without growth lines, containing entire animal including head (Fig. 19.1 C, D) .... Laevicaudata, one family: **Lynceidae [p. 606]**

2(1) Carapace encompassing head, growth lines present, although may be faint (Fig. 19.1 A, B) 3

2′ Carapace, if present, not encompassing head, growth lines usually absent (Fig. 19.1 F, G). .... **Cladocera [p. 607]**

3(2) First antenna smooth, without lobes along anterior surface (Fig. 19.1 E) .... Cyclestherida, *Cyclestheria* [Widespread. Circumtropical]

3′ First antenna with a longitudinal series of lobes on anterior surface (Fig. 19.1 A, B) .... **Spinicaudata [p. 626]**

Branchiopoda: Diplostraca: Laevicaudata: Lynceidae: Genera

Modified from Martin & Belk (1988) and Rogers & Olesen (2016). *Paralimnetis* is in great need of revision: *Paralimnetis rapax* Gurney 1931 reported from Paraguay, *Paralimnetis mapimi* Maeda-Martínez, 1987 from northern Mexico, and there are two undetermined species reported from Colombia in 1995 by Roessler (Rogers & Olesen, 2016).

1 Male thoracopod II unmodified, or, if modified, never with hooked process on protopodite .... ***Lynceus* [p. 606]**

1′ Male thoracopod II with protopod modified on right or left side as large sclerotized hamulate projection (Fig. 19.13 A); thoracopod II, endite 6 distal apex with a minute hook on one or both sides .... *Paralimnetis* [See comments above]

Branchiopoda: Diplostraca: Laevicaudata: Lynceidae: *Lynceus*: Species

This key is to males only, which have thoracopod I obviously modified as a clasper. The Neotropical *Lynceus* is probably underrepresented in the literature (Rogers & Olesen, 2016). We fully expect that there are undescribed species to be discovered. Furthermore, both *Lynceus rotundirostris* (Daday, 1902) and *Lynceus tropicus* Daday, 1927 are incompletely known and may be more widespread.

1 Carapace lacking an umbone 2

1′ Carapace with an umbone .... *Lynceus rotundirostris* (Daday, 1902) [Argentina]

2(1) Male rostrum with medial keel line bifurcated (Fig. 19.13 B) 3

2′ Male rostral keel not bifurcated (Fig. 19.13 C) 5

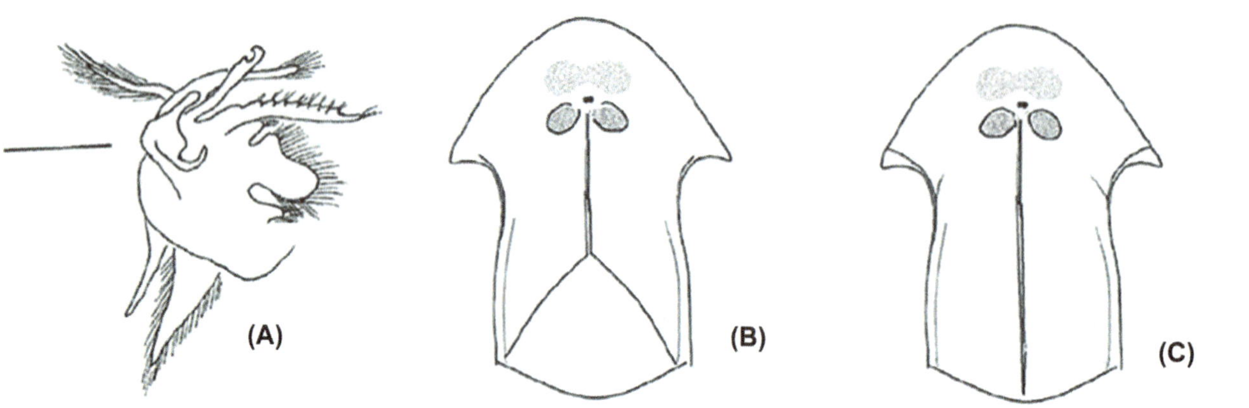

FIGURE 19.13 Laevicaudata: (A) *Paralimnetis* sp. male thoracopod II, protopod indicated; (B) *Lynceus mallinensis*, male, head, anterior view; (C) *Lynceus* sp., head, anterior view.

3(2) Male rostral keel bifurcated midway between eyes and rostral apex (Fig. 19.13 B) 4

3′ Male rostral keel bifurcated near eyes .... *Lynceus tropicus* Daday, 1927 [Colombia]

4(3) Rostral keel bifurcation point ~50% of keel length from eyes .... *Lynceus mallinensis* Pessacq, Epele & Rogers, 2011 [Argentina]

4′ Rostral keel bifurcation point ~30% of keel length from eyes .... *Lynceus huentelauquensis* Sigvardt, Rogers, De Los Ríos, Palero, & Olesen, 2019. [Chile]

5(2) Male claspers unequal; left clasper with a pronounced dorsal protrusion on endopod ("finger") .... *Lynceus aequatorialis* Daday, 1927 [Colombia]

5′ Male claspers equal, clapser endopod lacking a dorsal protrusion .... *Lynceus brevifrons* (Packard, 1877) [Mexico. Nearctic]

Branchiopoda: Diplostraca: Cladocera: Infraorders

Based upon females.

1 Trunk and four to six pairs of thoracopods enclosed in a carapace that opens ventrally 2

1′ Trunk and four pairs of cylindrical thoracopods not enclosed by a carapace .... Onychopoda, one genus: *Polyphemus* [Brazil]

2(1) Six pairs of foliaceous limbs; antennae with >10 natatory setae on both rami (except *Holopedium*, which has only one ramus) .... **Ctenopoda [p. 607]**

2′ Four to six differentiated limb pairs; antennae with <10 natatory setae on both rami .... **Anomopoda [p. 608]**

Branchiopoda: Diplostraca: Cladocera: Ctenopoda: Families

1 Second antenna (the swimming antenna) uniramous .... Holopedidae, one species: *Holopedium amazonicum* Stingelin, 1904 [Brazil: Amazon River basin]

1′ Second antenna with two rami .... **Sididae [p. 607]**

Branchiopoda: Diplostraca: Cladocera: Ctenopoda: Sididae: Genera

1 Second antenna dorsal ramus with two antennomeres, large process; ventral ramus with three antennomeres; carapace valves with long marginal setae and forming a ventral inflexion directed medially; postabdominal claw with two basal spines 2

1′ Second antennae rami each with three antennomeres (proximal antennomere of lower ramus may be very small), no process; carapace valves lacking long marginal setae and not forming a ventral inflexion directed medially .... *Sida americana* Korovchinksy, 1979 [Mexico, Colombia, Brazil, Peru, Argentina]

2(1) Postabdomen with 20 marginal denticles; no clusters of submarginal spines (fascicles) 3

2′ Postabdomen without dorsal marginal denticles; clusters or patches of submarginal spines (fascicles) present .... ***Diaphanosoma* [p. 607]**

3(2) Female first antennae long, almost reaching or exceeding length of second antenna basal antennomere 4

3′ Female first antennae short, less than half length of basal antennomere of second antenna .... *Latonopsis australis* Sars, 1888 [Cuba, Colombia, Brazil, Peru, Chile]

4(3) Postabdominal claw with three long, thin basal spines .... ***Sarsilatona* [p. 608]**

4′ Postabdominal claw with two long, thin basal spines, and a short basal spine with length <0.5x claw base width .... ***Pseudosida* [p. 608]**

Branchiopoda: Diplostraca: Cladocera: Ctenopoda: Sididae: *Diaphanosoma*: Species

1 Carapace ventral margins forming a narrow inflexion 2

1′ Carapace ventral margins forming a wide, often lamellar inflexion 4

2(1) Carapace posteroventral margin with variable sized denticles 3

2′ Carapace posteroventral margin with uniform denticles; posterior valve margin with 1-3 dorsal spines; posteroventral valve margin with very small denticles ..... *Diaphanosoma fluviatile* Hansen, 1899
[Haiti, Nicaragua, Venezuela, Brazil, Argentina]

3(2) Postabdominal terminal claw basodistal spine much longer than remaining spines; valve posteroventral margin intermediate denticles small and numerous ..... *Diaphanosoma chilense* Daday, 1902
[Chile, Argentina]

3′ Postabdominal terminal claw basodistal spine not much longer than others; valve posteroventral margin intermediate denticles relatively large and scanty ..... *Diaphanosoma birgei* Kořinek, 1981
[Cuba, Guatemala, Venezuela, Brazil, Argentina]

4(1) Distal antennomere of dorsal antennal ramus with seven setae 5

4′ Distal antennomere of dorsal antennal ramus with eight setae 7

5(4) Valves posteroventral margin with denticles, two submarginal spines on each valve 6

5′ Posteroventral margin of valves without denticles, one submarginal spine on each valve ..... *Diaphanosoma bergamini* Paggi & Falavigna da Rocha, 1999.
[Mexico, Venezuela, Brazil]

6(5) Carapace posteroventral margins with denticle rows different sizes; valve ventral inflexion with marginal setae and spines ..... *Diaphanosoma brevireme* Sars, 1901
[Cuba. Guatemala. Nicaragua. Venezuela. Brazil. Paraguay. Argentina]

6′ Carapace posteroventral margins with denticles small, numerous, in even rows; valve ventral inflexion with marginal setae only ..... *Diaphanosoma polyspina* Korovchinsky, 1982
[Venezuela, Colombia, Brazil]

7(4) Head rectangular, not protruding dorsally; posteroventral valve margins with small denticles interspersed between large ones ..... *Diaphanosoma spinulosum* Herbst, 1975
[Neotropics]

7′ Head conical, protruding dorsally; posteroventral valve margins with denticles decreasing in size dorsally ..... *Diaphanosoma dentatum* Herbst, 1968
[Venezuela, Colombia]

Branchiopoda: Diplostraca: Cladocera: Ctenopoda: Sididae: *Sarsilatona*: Species

1 Postabdomen dorsal protuberances large; ventral valve margins with small denticles ..... *Sarsilatona serricauda* (Sars, 1901)
[Neotropics]

1′ Postabdomen dorsal protuberances small and rounded; ventral valve margins with thin setulae ..... *Sarsilatona behningi* Korovchinsky, 1985
[Brazil]

Branchiopoda: Diplostraca: Cladocera: Ctenopoda: Sididae: *Pseudosida*: Species

1 Female first antennae with sensory papillae laterally or on basal outgrowth apex 2

1′ Female first antennae large basolateral outgrowth apex with sensory papillae ..... *Pseudosida ramosa* Daday, 1904
[Cuba, Guatemala, Venezuela, Colombia, Brazil, Paraguay]

2(1) Postabdomen with one marginal row of anal teeth ..... *Pseudosida bidentata* Herrick, 1884
[Cuba, Colombia]

2′ Postabdomen with two marginal rows of anal teeth ..... *Pseudosida variabilis* Daday, 1904
[Paraguay]

Branchiopoda: Diplostraca: Cladocera: Anomopoda: Families

1 First antennae not fused with the rostrum 2

1′ First antennae fused with the rostrum to form two long and pointed tusks ..... **Bosminidae** **[p. 609]**

2(1) First antennae of one antennomere each, which may be several times longer than wide to so short it is reduced to a mound with a cluster of setae 3

2′ First antennae with two antennomeres (a nearly square basal antennomere) ..... Ilyocryptidae, one genus: ***Ilyocryptus*** **[p. 609]**

3(2) Second antenna rami with three antennomeres 4

3′ Second antenna ventral ramus with three antennomeres, dorsal ramus with 4 antennomeres (look for a small basal antennomere) ..... **Daphniidae [p. 610]**

4(3) First antenna originating on head ventral margin, and more or less covered by rostrum 5

4′ First antenna originating just ventral to the anterior tip of the head, rostrum lacking ..... **Macrothricidae [p. 614]**

5(4) Postabdomen strongly flattened, broad, with a single row of 80–150 marginal denticles, resulting in a serrate appearance; anus distal ..... Eurycercidae, one genus: ***Eurycercus*** **[p. 616]**

5′ Postabdomen thick, not so broad, and usually with two rows of marginal denticles, mostly fewer than 20 on each side; anus proximal ..... **Chydoridae [p. 616]**

Branchiopoda: Diplostraca: Cladocera: Anomopoda: Bosminidae: Genera

1 Lateral head pores present; postabdominal claw basal accessory claw absent; first antennae separately attached to the head, straight, or curved toward the body ..... ***Bosmina*** **[p. 609]**

1′ Lateral head pores absent; postabdominal claw basal accessory claw present; first antennae fused together in basal half, separated distally and curved laterally away from head and body ..... ***Bosminopsis*** **[p. 609]**

Branchiopoda: Diplostraca: Cladocera: Anomopoda: Bosminidae: *Bosmina*: Species

There is extensive phenotypic variability within species. *Bosmina brehmi* and *Bosmina chilensis* require further investigation.

1 Lateral head pore nearer mandible proximal tip than to second antenna base; mucro (carapace posteroventral spine) with a row of microspinules 2

1′ Lateral head pore at head shield margin, above second antennae base; mucro lacking microspinules ..... *Bosmina* (*Bosmina*) *longirostris* (O.F. Müller, 1776) species complex [Cuba, Colombia, Brazil, Peru, Chile]

2(1) Rostrum moderately long; sensory seta somewhat displaced from first antenna base; first antenna length $>$ 2× mucro 3

2′ Rostrum short; first antenna with basal sensory seta; first antenna length $\leq$ 1.5× mucro ..... *Bosmina* (*Liederobosmina*) *tubicen* (Brehm, 1953) [Cuba, Mexico, Colombia, Brazil]

3(2) Carapace dorsal margin forming continuous arc with first antennae ..... *Bosmina* (*Liederobosmina*) *huaronensis* Delachaux, 1918 [Mexico, Brazil, Peru. Bolivia. Argentina]

3′ Carapace dorsal margin forming indented arc with first antennae ..... *Bosmina* (*Liederobosmina*) *hagmanni* (Stingelin, 1904) [Cuba, Colombia, Brazil, Chile]

Branchiopoda: Diplostraca: Cladocera: Anomopoda: Bosminidae: *Bosminopsis*: Species

1 Carapace without long tail spine 2

1′ Carapace with long tail spine ..... *Bosminopsis negrensis* Brandorff, 1976 [Venezuela, Brazil]

2(1) Rostrum present, head distinctly convex with a dome ..... *Bosminopsis deitersi* Richard, 1895 [Colombia, Brazil, Peru]

2′ Rostrum absent, head smoothly rounded, well-developed eye ..... *Bosminopsis brandorffi* Rey & Vasquez, 1989 [Venezuela, Brazil]

Branchiopoda: Diplostraca: Cladocera: Anomopoda: Ilyocryptidae: *Ilyocryptus*: Species

Kotov & Štifter (2006) provide keys for all species; however, morphological variation is not yet understood; minute differences between species may be due to phenotypic plasticity.

1 Postabdominal claw basoventral setulae short, slightly longer than claw base width 2

1′ Postabdominal claw basoventral setulae long, much longer than claw base width 5

2(1) Adult body triangular-ovoid in outline, posterior margin setae long, geniculate 3

2′ Adult body ovoid-elongate in outline, valve with five anterior most setae long, posterior margin with 21–28 defensive spines .... *Ilyocryptus elegans* Paggi, 1992
[Argentina]

3(2) Valves without lateral horn 4

3′ Valves with lateral horn .... *Ilyocryptus paranaensis* Paggi, 1989
[Mexico, Venezuela, Colombia, Brazil, Paraguay, Argentina]

4(3) Valve posterior margin setae each with a robust spiniform setule (rarely two) .... *Ilyocryptus spinifer* Herrick, 1882
[Widespread]

4′ Valve posterior margin setae without spiniform setulae .... *Ilyocryptus plumosus* Kotov & Štifter, 2004
[Nicaragua, Venezuela, Colombia]

5(1) Body length > 600 μm; limb 1 with large seta near ejector hooks, or single ejector hook. 6

5′ Body length < 550 μm; no seta near ejector hooks .... *Ilyocryptus sarsi* Stingelin, 1913
[Colombia, Brazil]

6(5) Anus medial, or somewhat closer to postabdomen base than distal extremity 7

6′ Anus subdistal, preanal margin considerably longer than postanal margin .... *Ilyocryptus brevidentatus* Ekman, 1905
[Chile, Argentina, Falkland (Malvinas) Islands, South Georgia Island]

7(6) Preanal teeth partly or fully doubled, lacking adjacent spinules or setulae 8

7′ Preanal teeth single, adjacent spinules, or setulae present 9

8(7) Second antenna dorsal ramus, with second antennomere spine length $< 0.5\times$ antennomere width; postabdomen with preanal portion longer than postanal .... *Ilyocryptus cuneatus brasiliensis* Kotov & Elmoor-Loureiro, 2008
[Brazil]

8′ Second antenna dorsal ramus, with second antennomere spine length $0.5\times$ antennomere width; postabdomen with preanal portion shorter than postanal .... *Ilyocryptus silvaeducensis paraensis* Kotov & Elmoor-Loureiro, 2008
[Brazil]

9(7) Head shield with lateral projection and mandibular articulation wide; preanal teeth straight .... *Ilyocryptus denticulatus* Delachaux, 1918
[Chile, Argentina]

9′ Head shield without lateral projection, mandibular articulation narrow; preanal teeth curved slightly posteriorly; lateral swimming setae with long setules on proximal and distal articles .... *Ilyocryptus sordidus* (Liévin, 1848)
[Brazil, Peru, Chile. Palaearctic]

Branchiopoda: Diplostraca: Cladocera: Anomopoda: Daphniidae: Genera

1 First antenna length <0.5 head width; postabdomen teeth always single 2

1′ First antenna length subequal to head width; postabdomen posterior angle (near claw) usually bearing a tooth with two points 5

2(1) Carapace ventral margin rounded and not pigmented 3

2′ Carapace ventral margin straight and usually black .... ***Scapholeberis* [p. 611]**

3(2) Adult tail spine, if present, length $< 3\times$ breadth 4

3′ Adult with a tail spine, length $\geq 3\times$ breadth, acute .... ***Daphnia* [p. 611]**

4(3) Second antenna, four antennomere ramus, with second antennomere bearing an apical spine, length $\sim 1/4$ s antennomere .... ***Simocephalus* [p. 612]**

4′ Second antenna, four antennomere ramus without an apical spine .... ***Ceriodaphnia* [p. 613]**

5(1) Second antenna, four antennomere ramus with four terminal setae; ocellus absent .... ***Moina* [p. 613]**

5′ Second antenna, four antennomere ramus with three terminal setae and a spine shorter than its antennomere; ocellus present .... *Moinodaphnia macleayi* (King, 1853)
[Cuba, Nicaragua, Colombia, Brazil, Peru]

Branchiopoda: Diplostraca: Cladocera: Anomopoda: Daphniidae: *Scapholeberis*: Species

1 Carapace distal margin with conspicuous hyaline membrane; submarginal membrane denticulated 2

1′ Carapace distal margin with denticulated membrane 3

2(1) Carapace and head reticulated but smooth; rostrum triangularly produced anteriorly; depression between eyes and sides of rostrum; second groove along head behind first antenna insertion ...... *Scapholeberis rammneri* Dumont & Pensaert, 1983
[Chile]

2′ Carapace and head with numerous rows of spines; rostrum trilobite with narrow, rounded middle lobe ...... *Scapholeberis spinifera* Nicolet, 1849
[Brazil, Peru, Chile]

3(1) Valve posterior rim denticulated membrane with or without an underlying hyaline membrane 4

3′ Valve posterior rim with denticulated membrane conspicuous, without underlying hyaline membrane; mucro length 0.1–0.2 valve rim. ...... *Scapholeberis armata* Herrick, 1882
[Mexico, Venezuela, Colombia, Brazil, Paraguay]

4(3) Valve posterior rim with denticulated membrane markedly thick, with conspicuous underlying hyaline membrane ...... *Scapholeberis duranguensis* Quiroz-Vazquez & Elias-Gutierrez, 2009
[Mexico]

4′ Valve posterior rim with denticulated membrane inconspicuous and with equally inconspicuous underlying hyaline membrane, mucro length 0.1–0.2 valve rim ...... *Scapholeberis freyi* Dumont & Pensaert, 1983
[Brazil]

Branchiopoda: Diplostraca: Cladocera: Anomopoda: Daphniidae: *Daphnia*: Species

1 Tail spine absent 2

1′ Tail spine present 4

2(1) Body rounded, head width $>$ 0.5 carapace width 3

2′ Body rounded, head width $<$ 0.5 carapace width ...... *Daphnia chilensis* (Hann, 1986)
[Chile: Andes]

3(2) Carapace ventral margin anterior 2/3 bearing a long setal fringe ...... *Daphnia studeri* (Rühe, 1914)
[Sub-Antarctic Islands]

3′ Carapace ventral margin anterior midregion bearing a long setal fringe ...... *Daphnia marcahuasensis* (Valdivia & Burger, 1989)
[Peru]

4(1) Postabdominal claw teeth in three groups (pectens); middle pectin middle teeth length $\geq$ 1.5 times as proximal pecten teeth 5

4′ Postabdominal claw with fine teeth decreasing regularly in size from base to apex, middle pecten teeth length $<$ 1.5 times as proximal pecten 11

5(4) Rostrum very short; first antennae aesthetascs exceeding apex 6

5′ Rostrum longer; first antennae aesthetascs not exceeding apex 7

6(5) Ocellus absent; caudal spine length $<$ 0.2 body length ...... *Daphnia menucoensis* Paggi, 1996
[Argentina]

6′ Ocellus present; no caudal spine ...... *Daphnia paggii* Kotov, Sinev & Berrios, 2010
[Chile: Andes]

7(5) Fornices well developed, pointed, secondary fornices, dorsal ridge penetrates head shield anteriorly 8

7′ Head shield broad in dorsal view, fornices rounded, no secondary fornix ...... *Daphnia dadayana* Paggi, 1999
[Chile, Argentina]

8(7) Tail spine length $>$ 0.5 carapace length; carapace posteroventral and dorsal margins with stout spinulae 9

8′ Tail spine length $<$ 0.5 carapace length; carapace posteroventral and dorsal margin with long spinulae ...... *Daphnia inca* Kořinek & Villalobos, 2003
[Peru, Chile]

9(8)	Head without large crest	10
9′	Head with large laminate dorsal crest [Argentina]	*Daphnia ornithocephala* Birabén, 1953
10(9)	Tail spine length 0.5−1.0 carapace; no gap between postabdominal claw and row of anal teeth [Cosmopolitan]	*Daphnia similis* Claus, 1876
10′	Tail spine length 0.5× carapace; small gap between postabdominal claw and row of anal teeth [Argentina]	*Daphnia spinulata* Birabén, 1917
11(4)	Body oval, head posterior width slightly ≤ to greatest carapace width	12
11′	Body round, head posterior width < greatest carapace width [Ecuador, Chile, Argentina]	*Daphnia ambigua* Scourfield, 1947
12(11)	Second abdominal process < 0.3 length of first, often shorter than third process	13
12′	Second abdominal process intermediate in size between first and third, usually > 0.3 length of first	14
13(12)	Carapace ventral margin posterior half with spinulae [Mexico, Venezuela, Colombia, Brazil, Argentina]	*Daphnia laevis* Birge, 1879
13′	Carapace ventral margin posterior 3/4 with spinulae [Venezuela, Brazil]	*Daphnia gessneri* Herbst, 1967
14(13)	Carapace internal ventral margin near midline with long setal fringe	15
14′	Carapace internal ventral margin near midline without long setal fringe	16
15(14)	Rostrum apex projecting, an acute or small snout, aesthetascs may extend beyond apex, first antennae tubular [Widespread]	*Daphnia obtusa*-group
15′	Rostrum short, blunt, rounded, aesthetascs extending beyond apex, first antennae placed subapically, small [Peru, Chile, Argentina]	*Daphnia peruviana* Harding, 1955
16(14)	Tail spine variable, oriented posterodorsally, never posteroventrally	17
16′	Tail spine small, oriented posteriorly in body midline, often posteroventrally [Colombia, Peru, Argentina]	*Daphnia parvula* Fordyce, 1901
17(16)	Postabdominal processes II and III slightly setose apically or glabrous	18
17′	Postabdominal processes II and III hirsute [Includes many hybrids; widespread]	*Daphnia pulex*-group
18(17)	Carapace melanized, most clearly dorsally [Argentina. Holarctic]	*Daphnia middendorrfiana* Fischer, 1851
18′	Carapace not melanized [Includes *pulex x pulicaria* hybrids, see Adamowicz et al., 2004; widespread]	*Daphnia pulicaria* Forbes, 1893

Branchiopoda: Diplostraca: Cladocera: Anomopoda: Daphniidae: *Simocephalus*: Species

1	Postabdominal claw with basal pecten length either 2× or 0.5 the more distal teeth length	2
1′	Postabdominal claw with tooth size decreasing evenly from the base to apex	5
2(1)	Postabdominal claw basal pecten teeth length ~2× the more distal teeth; head rounded or pointed, but without apical corner spines	3
2′	Postabdominal claw basal pecten teeth length ~0.5× the more distal teeth; head with a right angled anterior corner, often with spines near the angle [Cuba, Colombia, Brazil, Peru, Chile]	*Simocephalus serrulatus* (Koch, 1841)
3(2)	Adult carapace posterior margin with a low rounded mound	4
3′	Adult carapace with a pointed triangular protrusion [Colombia, Brazil]	*Simocephalus daphnoides* Herrick, 1883
4(3)	Head pointed, rostrum short [Venezuela, Colombia, Brazil]	*Simocephalus acutirostratus* (King, 1853)

4′ Head rounded, rostrum very long .... *Simocephalus latirostris* Stingelin, 1906
[Colombia, Brazil, Paraguay]

5(1) Carapace posterior end with a protruding, rounded mound .... *Simocephalus mirabilis* Orlova-Bienkowskaja, 1998
[Neotropics. Nearctic]

5′ Carapace posterior end evenly curved .... *Simocephalus mixtus* Sars, 1903
[Neotropics. Nearctic]

Branchiopoda: Diplostraca: Cladocera: Anomopoda: Daphniidae: *Ceriodaphnia*: Species

Key modified from Dodson & Frey (2001).

1 Cervical fenestra absent 2

1′ Cervical fenestra present 5

2(1) Carapace posteroventral margin with submarginal microspinulae subequal in size, without occasional longer microspinulae (looking like exclamation points) 3

2′ Carapace posteroventral margin with submarginal microspinulae subequal in size, with occasional longer microspinulae 4

3(2) Carapace either not reticulated, or if reticulated, then reticulation line nodes without projecting setae or spines; head never with an anterior horn (observation at different seasons in the year are necessary to confirm absence); fornices may have rather stout, straight, or hooked acute spines .... *Ceriodaphnia quadrangula* (O.F. Müller, 1785)
[Cuba, Costa Rica, Venezuela, Colombia, Peru]

3′ Carapace often reticulated, with reticulation line nodes bearing microsetae or spines (length usually $< 0.5\times$ reticulation cell width); head sometimes with an anterior, acute horn; fornices often with sharp spines .... *Ceriodaphnia cornuta* Sars, 1885
[Cuba, Colombia, Brazil, Peru]

4(2) Carapace ventral margin with seven to nine long marginal setae (difficult to see, but always present); postabdominal claw with teeth decreasing in size evenly; proximal anal denticles alternating with long, fine, setaform accessory spines; postabdomen dorsal margin proximal to anal denticles smooth (not serrate); carapace posteroventrally with $\sim$40 submarginal microspinulae between each larger microspinulae; fornices smoothly rounded or sometimes with small, laterally pointed spines .... *Ceriodaphnia pulchella* Sars, 1862
[Colombia, Brazil, Argentina]

4′ Carapace ventral margin with 15 long marginal setae near the margin center (subequal in length to postabdominal claw); postabdominal claw first and second pectens with teeth length $1.5\times$ teeth in third and distal pectin .... *Ceriodaphnia laticaudata* P.E. Müller, 1867.
[Mexico, Brazil]

5(1) Postabdominal claw, middle pecten varies from 22 fine spinulae, slightly longer and heavier than those of the third pecten to 8–14 heavy, prominent, ovate teeth with length $> 2\times$ as those of third pecten; first antenna long, thick with anterior sensory setae arising from distinct peduncle near midpoint; anal denticles nearly straight .... *Ceriodaphnia dubia* Richard, 1894
[Invasive. Widespread]

5′ Postabdominal claw, middle pecten with two to eight prominent, straight teeth (similar to large toothed morph of *C. dubia*); first antenna with anterior sense setae nearly terminal; anal denticles curved .... *Ceriodaphnia reticulata* (Jurine, 1820)
[Colombia, Brazil]

Branchiopoda: Diplostraca: Cladocera: Anomopoda: Daphniidae: *Moina*: Species

1 Body laterally flattened, carapace with slight dorsal keel, ocellus present 2

1′ Carapace rotund without a dorsal keel; ocellus usually absent 3

2(1) First antennae long, cylindrical, on slight protuberance; second antenna four-antennomere ramus, with terminal antennomere bearing four distal setae .... *Moinodaphnia macleayi* (King, 1853)
[Cuba, Nicaragua, Colombia, Brazil, Peru]

2′ First antennae short, usually folded back against head; second antenna four-antennomere ramus, with terminal antennomere bearing three distal setae .... *Moina reticulata* (Daday, 1905)
[Cuba, Colombia, Brazil]

3(1) Postabdomen dorsal margin tooth row with distal tooth bident 4

3′ Postabdomen dorsal margin tooth row with distal tooth simple 5

4(3) First thoracic leg last and penultimate articles with marginal plumose setules; ephippium with one or two eggs 6

4′ First thoracic leg last and penultimate articles with marginal spines; ephippium with two eggs .... *Moina macrocopa* (Straus, 1820)
[Invasive. Palearctic]

5(3) Body length 1.2–1.6 mm; no supraocular depression; ephippium with two eggs .... *Moina hutchinsoni* Brehm, 1937
[Mexico]

5′ Body length ~ 1 mm; supraocular depression present; ephippium with one egg .... *Moina eugeniae* Oliver, 1954
[Colombia, Argentina]

6(4) Body length < 1 mm; rostrum pronounced; supraocular depression present, carapace with distinct reticulation .... *Moina rostrata* McNair, 1980

6′ Body length 0.5–0.7 mm; first antenna basal sensory setae very long .... *Moina minuta* Hansen, 1899
[Colombia, Brazil]

7(4) Head and carapace with long setae 8

7′ Head and carapace glabrous 9

8(3) Body length ≤ 1.2 mm long; one egg per ephippium .... *Moina affinis* Birge, 1893
[Cuba, Mexico]

8′ Some adults > 1.2 mm; two eggs per ephippium .... *Moina wierzejskii* Richard, 1895
[Mexico, Argentina]

9(5) Ocellus absent .... *Moina micrura* Kurz, 1875
[Cuba, Colombia, Brazil, Peru, Chile, Argentina]

9′ Ocellus present .... *Moina dumonti* Kotov, Elias-Gutierrez & Granados-Ramirez, 2005
[Cuba, Mexico]

Branchiopoda: Diplostraca: Cladocera: Anomopoda: Macrothricidae: Genera

1 Second antenna ramus with four antennomeres with four setae (0-0-1-3) 2

1′ Second antenna ramus with four antennomeres with two, three, or five setae 4

2(1) Intestine not looped (convoluted) 3

2′ Intestine looped (convoluted); postabdomen bilobed, preanal lobe with serrations .... *Streblocerus*
[Venezuela]

3(2) Postabdomen dorsal margin bilaterally compressed; postanal margin distally with widely splayed spines; anal opening proximal side with single large spine .... *Grimaldina brazzai* Richard, 1892
[Widespread]

3′ Postabdomen dorsal margin without a long spine; antennal seta formula (0-0-1-3/1-1-3) .... ***Macrothrix* [p. 614]**

4(1) Second antenna ramus with four antennomeres bearing two setae 5

4′ Second antenna ramus with four antennomeres bearing 5 (0-1-1-3) setae .... *Guernella raphaelis* Richard, 1892
[Mexico, Colombia, Brazil, Argentina]

5(4) Carapace globular, without dorsal keel; carapace dorsal margin smooth, no spine; postabdomen preanal margin serrated .... *Onchobunops tuberculatus* (Fryer & Paggi, 1972)
[Mexico, Nicaragua, Colombia, Argentina]

5′ Carapace with moderate dorsal keel; dorsal carapace margin with a spine behind cervical depression; postabdomen preanal margin serrated .... *Drepanothrix dentata* (Eurén, 1861)
[Bolivia]

Branchiopoda: Diplostraca: Cladocera: Anomopoda: Macrothricidae: *Macrothrix*: Species

1 Carapace valves dorsally with a tooth and a keel; body subovoid elongate without lateral bulges 2

1′ Carapace valves dorsally without a tooth 3

2(1) Carapace valves dorsally with a prominent, strongly sclerotized dentiform projection on finely serrated dorsal keel .... *Macrothrix mexicanus* Ciros Pérez, Silva-Briano & Elías Gutiérrez, 1996.
[Mexico]

2′ Carapace valves dorsally with a small dentiform projection on smooth dorsal keel .... *Macrothrix camjatae* Harding, 1955
[Peru, Bolivia]

3(1) Trunk limb II with no lobe near scrapers four and five 4

3′ Trunk limb II with lobe bearing stout conical seta, densely ciliated distally between scrapers four and five .... *Macrothrix smirnovi* Ciros Pérez & Elías Gutiérrez, 1997.
[Mexico]

4(3) Limbs I, II, and III setae without robust pectens 5

4′ Limbs I, II, and III setae with robust pectens .... *Macrothrix marthae* Elías Gutiérrez & Smirnov, 2000.
[Mexico]

5(4) First antenna with two or more robust spines along its length 6

5′ First antenna with spinulae but no spines along its length 9

6(5) First antenna with five or more robust spines along its length 7

6′ First antenna with two robust spines along its length .... *Macrothrix flabelligera* Smirnov, 1992
[Brazil]

7(6) First antenna with five to seven robust spines along its length 8

7′ First antenna with eight to nine robust spines along its length .... *Macrothrix brandorffi* Kotov & Hollwedel, 2004
[Brazil]

8(7) Postabdominal claw lateral surface with four or five denticles .... *Macrothrix paulensis* Sars, 1900
[Mexico. Venezuela. Colombia. Brazil]

8′ Postabdominal claw lateral surface with 15 or 16 denticles .... *Macrothrix sioli* (Smirnov, 1982)
[Brazil]

9(5) Natatory setae distal article much shorter than proximal article 10

9′ Natatory setae distal article length 0.3× proximal article .... *Macrothrix atahualpa* Brehm, 1936
[Bolivia, Chile]

10(9) Valves not squamose, with lines or other ornamentation 11

10′ Valves covered by scales .... *Macrothrix spinosa* King, 1853
[Neotropics]

11(10) First antenna baciliform, not dilated distally 12

11′ First antenna dilated distally; carapace dorsal margin with fine serrations 14

12(11) Carapace posterodorsal angle without sharp spine 13

12′ Carapace posterodorsal angle with sharp spine .... *Macrothrix superaculeata* (Smirnov, 1982)
[Brazil]

13(12) Carapace without striations 15

13′ Carapace finely striated .... *Macrothrix elegans* Sars, 1901
[Neotropics]

14(11) Postabdominal claw denticles widely spaced, occupying middle third of claw length .... *Macrothrix magna* Daday, 1902
[Chile and Argentina: Patagonia]

14′ Postabdominal claw denticles closely spaced, occupying distal half of claw length .... *Macrothrix agsensis* Dumont, Silva-Briano & Babu, 2002
[Brazil]

15(13) Strong dorsal keel, no dorsal marginal serrations 16

15′ Weak or no dorsal keel 17

16(15) Carapace dorsal margin breached by a "step" at head posterior margin; carapace dorsal margin with serrations .... *Macrothrix sierrafriatensis* Silva-Briano, Nguyen & Dumont, 1999
[Mexico]

16′ Carapace dorsal margin and head posterior margin outline contiguous .... *Macrothrix oviformis* Ekman, 1900
[Colombia. Southern South America. Falkland (Malvinas) Islands, South Georgia Islands, S. Orkney Island]

17(15) Antenna I with six to eight transverse spinule rows with distinct reticulation .... *Macrothrix boergeni* Studer, 1878
[Kerguelen Islands]

17′ Antenna I with five to seven transverse spinule rows, no reticulation .... *Macrothrix ruehei* Kotov, 2007
[Crozet Island, Marion Island, New Amsterdam Island]

Branchiopoda: Diplostraca: Cladocera: Anomopoda: Eurycercidae: *Eurycercus*: Species

1 Carapace broadly rounded, no dorsal keel except in ephippial females; head shield rounded behind; head pore projecting like a bubble; intestine with a single loop 2

1′ Carapace with prominent, sharp median dorsal keel; median pore projecting from the head and directed posteriorly with a prominent indentation immediately behind; head shield pointed posteriorly .... *Eurycercus* (*Eurycercus*) *microdontus* Frey, 1978
[Nearctic. Mexico?]

2(1) Labral keel length = first antenna 3

2′ Labral keel length < first antenna .... *Eurycercus* (*Bullatifrons*) *norandinus* Aranguren, Monroy & Gaviria, 2010
[Colombia, Brazil]

3(2) Lateral head pores elongated longitudinally .... *Eurycercus* (*Bullatifrons*) *meridionalis* Bekker, Kotov & Elmoor-Loureiro, 2010
[Brazil]

3′ Lateral head pores not elongated .... *Eurycercus* (*Bullatifrons*) *longirostris* Hann, 1982
[Mexico]

Branchiopoda: Diplostraca: Cladocera: Anomopoda: Chydoridae: Genera

Much progress has been achieved in the past decade in resolving the polyphyletic genus *Alona*, with the delineation of new genera and species groups. More changes are expected.

1 Mandible proximal tip articulates where head shield and carapace touch; midline with two or three major head pores, which usually are connected by a double sclerotized ridge, resembling a channel; carapace free posterior margin not greatly less than maximum height of animal; postabdominal claw with one (sometimes absent) basal spine; postabdomen nearly always with lateral fascicles; subfamily Aloninae 2

1′ Mandible proximal tip articulates in a special pocket on head shield, some distance in a dorsal direction from head shield and carapace point of contact; two major head pores on midline, completely separated from one another, with no connecting sclerotized ridge; minor pores most commonly near midline between major pores; carapace free posterior margin usually considerably less than maximum height of animal; postabdominal claw typically with two basal spines; postabdomen lacking articulated lateral fascicles; subfamily Chydorinae 19

2(1) Compound eye absent; ocellus present or absent 3

2′ Compound eye and ocellus both present 5

3(2) Ocellus present 4

3′ Ocellus absent .... *Spinalona anophtalma* Ciros-Pérez & Elías-Gutiérrez, 1997
[Mexico]

4(2) Carapaces from previous instars firmly nested dorsally; head shield transversely truncate posteriorly; single median head pore .... ***Monospilus* [p. 619]**

4′ Carapaces not nested; two separated median pores without sclerotized connecting ridge; minor pores located far laterally; very short antennae .... *Bryospilus repens* Frey, 1980
[Venezuela]

5(2) Compound eye larger than ocellus; limb III exopodites with six or seven setae and limb IV with six setae 6

5′ Compound eye smaller than ocellus; limb III and IV exopodites greatly reduced, with four setae .... *Geoffreya fryeri* Kotov, Sinev & Berrios, 2010
[Chile: Andes]

6(5) Body strongly compressed laterally, at times almost wafer thin; postabdominal claw concave margin with a secondary spine midway, usually with a pecten of fine setulae decreasing in length proximally between it and the basal spine 7

6′ Body not so strongly compressed; postabdominal claw having only a basal spine of variable length (except in *Euryalona*); sometimes none at all 9

7(6) Carapace weakly keeled, and head without a keel 8

7′ Carapace strongly keeled, head also usually strongly keeled; postabdomen narrow, elongate, tapered distally, with many (generally about 15 or more) marginal denticles .... ***Camptocercus*** **[p. 619]**

8(6) Postabdomen shorter and broader, parallel sided, without any distinct marginal denticles .... *Acroperus tupinamba* Sinev & Elmoor-Loureiro, 2010 [Brazil]

8′ Postabdomen rather slender, elongate, somewhat tapered distally; body high; carapace widely open behind; postabdomen dorsal margin extends distal to claw base and basal claw spine .... ***Kurzia*** **[p.619]**

9(5) Rostrum narrowly rounded; carapace may be weakly reticulated, but not head 10

9′ Rostrum very broad in dorsal view, semicircular, wider than body; carapace and head strongly reticulated; carapace ventral margin with a dense setal fringe, generally with two posteroventral, large, sharp, triangular spines; postabdomen tapered distally, marginal denticles small, lateral fascicles minute .... *Graptoleberis testudinaria occidentalis* Sars, 1901 [Colombia, Brazil, Peru, Bolivia]

10(9) Postabdomen elongate, rather narrow, marginal denticles robust 11

10′ Postabdomen much less elongate (except in *Leydigia*), marginal denticles usually robust, but in some species greatly reduced 12

11(10) Postabdomen dorsal margin straight or slightly convex; marginal denticles very short proximally, increasing in length distally to three very long curved denticles; four median head pores (middle one divided in two), usually completely separated from one another, or with middle pores joined by a sclerotized ridge; postabdominal claw basal spine long, slender, attached some distance from claw base; head and carapace generally have a yellowish, hyaline appearance .... ***Oxyurella*** **[p. 619]**

11′ Postabdomen dorsal margin distinctly concave; marginal denticles increase in length distally, gradually; one median head pore, sometimes completely absent; postabdominal claw proximal half concave margin with a setal pecten, increasing in length distally; trunk limb I with a stout spine on distomedial lobe, with coarse, rounded tubercles on concave edge .... ***Euryalona*** **[p. 620]**

12(10) Postabdomen large, broad, flattened, almost semicircular 13

12′ Postabdomen smaller, with shorter setae in lateral fascicles; head pores variable 15

13(12) Postabdomen armed laterally with long, spiniform setae in groups, distal groups projecting beyond postabdomen margin; postabdomen margin with short, fine spinulae; rostrum short 14

13′ Postabdomen armed laterally with short setae in groups, not projecting beyond postabdomen margin; postabdomen margin with stout denticles; rostrum elongated .... ***Leydigiopsis*** **[p. 620]**

14(11) Labrum anterior margin setulated; limb I ventral margin with setae; three median head pores close together, and with minor pores close-in laterally; carapace free posterior margin very long; ocellus as large as compound eye or larger .... ***Leydigia*** **[p. 620]**

14′ Labrum anterior margin smooth; limb I ventral margin with robust denticles .... *Parvalona parva* (Daday, 1905) [Colombia, Brazil, Paraguay]

15(12) Carapace usually not sculptured at all, or if so, then rather weakly; postabdomen usually with well-developed marginal denticles and lateral fascicles; two or three median head pores, most commonly united by a sclerotized ridge, although sometimes completely separated 16

15′ Carapace strongly sculptured, with longitudinal striations posteriorly and about five vertical striations anteriorly, roughly paralleling anterior margin; postabdomen dorsal edge minutely serrate; about 14 fascicles laterally; no median head pores, but instead there are two comma-shaped thickenings, which may contain the minor pores .... ***Notoalona*** **[p. 621]**

16(15) Carapace valve posteroventral corners without denticles, or bearing one to three denticles, separated by the width of the denticle base or less 17

16′ Carapace valves with posteroventral corners bearing three to five denticles, distance between them > 2× denticle base width .... ***Karualona*** in part **[p. 621]**

17(16) Rostrum very blunt, truncated or with a subapical depression in lateral view; three primary head pores 18

17′ Rostrum acute in lateral view; two or three primary head pores 28

18(17) Lateral head pores tiny, dots; antenna I with terminal aesthetasc, not more than 1.5× longer than terminal setae .... ***Leberis*** in part **[p. 621]**

18′ Two lateral head pores, floraform; antenna I with terminal aesthetasc length 2× terminal setae *Nicsmirnovius fitzpatricki* (Chen, 1970) [Brazil]

19(1) Head shield not elongated posteriorly; if two head pores are present, postpore (PP) distance between posterior head pore and head shield posterior margin usually considerably less than interpore (IP) distance 20

19′ Head shield elongated posteriorly, extending well beyond heart to dorsum middle; PP distance usually greater than IP distance, sometimes by several fold 22

20(19) Zero or one median head pore 21

20′ Two median head pores, well separated and not connected by a sclerotized ridge 23

21(20) One median head pore only in first instar; none in later instars; postabdomen with long proximal and distal marginal denticles, and with shorter ones in between; labral keel large, with one to four anterior teeth .... ***Ephemeroporus*** **[p. 621]**

21′ One median head pore in all instars ... *Dadaya macrops* (Daday, 1898) [Mexico, Nicaragua, Colombia, Brazil]

22(19) Postabdomen of more typical chydorid structure, with pre- and postanal angles well-developed series of postanal marginal denticles ... ***Alonella* [p. 622]**

22′ Postabdomen consisting of a much expanded postanal portion, having a series of short spines along dorsal margin and many small clusters of spinules laterally ... ***Dunhevedia* [p. 622]**

23(20) Body elongate, somewhat flattened; carapace posteroventral angle with one or more spines at the posteroventral angle ... 24

23′ Body globular, nearly round; carapace posteroventral angle seldom with spines ... 25

24(23) Postabdominal length/width ratio (distance from the base of the postabdominal claw to the base of the natatory setae divided by the postabdomen width at anus distal edge) 3.7 to 5.2; postabdomen dorsal margin with two distal most spines (near claw) length ~2× the more proximal spines ... ***Picripleuroxus* [p. 622]**

24′ Postabdominal length/width ratio 2.3 to 3.3; postabdomen dorsal margin with two distal most teeth (near the claw) no longer than more proximal teeth ... ***Pleuroxus* [p. 622]**

25(23) Carapace ventral margin without a hook; carapace marginal setae arise submarginally posteriorly, forming a distinct duplicature ... 26

25′ Carapace ventral margin anteriorly with a hamulate projection containing a groove, in which is located a stout seta with spinulae from trunk limb I; postabdomen distal angle with a cluster of long, slender setae ... *Anchistropus ominosus* Smirnov, 1985 [Brazil]

26(25) Postabdomen short, with fewer than 15 marginal denticles, and laterally with crescentic clusters of short spinulae, which do not resemble fascicles; carapace is clear to light yellow ... 27

26′ Postabdomen elongate, dorsal and ventral sides parallel; postanal portion with ~15 slender marginal denticles; lateral surface with fascicles of setae; carapace heavily sclerotized, golden in color ... *Pseudochydorus globosus* (Baird, 1843) [Cuba, Brazil]

27(26) Carapace lacks a posteroventral spine; carapace ventral setae submarginal, in an arc inside carapace margin; ephippium with one or two eggs ... ***Chydorus* [p. 623]**

27′ Carapace posteroventral angle with a spine; carapace ventral setae marginal; first antennae with two submarginal aesthetascs (blunt sensory setae); ephippium with two eggs; carapace often dimpled ... *Paralona pigra* (Sars, 1862) [Venezuela, Colombia, Peru, Bolivia, Chile]

28(17) With other combinations of characters ... 29

28′ Large postabdomen with merged marginal teeth, long terminal claw; limb I anteroventral stiff setulae reduced to one or two per group; exopodite III with seven setae ... ***Alona* in part [p. 623]**

29(28) Six limb pairs; filter plate on limb V of three setae; limb III with seven exopodite setae; postabdomen marginal denticles merged; labral keel with ventral setule groups; *Hexalona* branch ... 30

29′ Five limb pairs, other features with many reductions; *Coronatella* branch ... 32

30(29) Labrum anterior margin without notch ... 31

30′ Labrum anterior margin with notch, posterior margin with 2 setulae clusters ... *Prenda arvensis* Sousa, Elmoor-Loureiro & Santos, 2016 [Brazil]

31(30) Main head pores two to three, lateral head pores minute ... ***Alona* in part [p. 623]**

31′ Main head pores two to three, lateral head pores slit shaped or rounded, with sack ... ***Flavalona* [p. 624]** [Brazil]

32(29) Body with dorsal keel ... 33

32′ Body without dorsal keel ... 34

33(32) Postabdomen basal spine length 0.3× claw length; lateral fascicles with weakly developed setae ... ***Leberis* in part [p. 621]**

33′ Postabdomen with basal spine length <0.2–0.25× postabdominal claw ... ***Celsinotum* [p. 624]**

34(32) Valves at posteroventral angle with zero to two small denticles ... 35

34′ Valves at posteroventral angle with one to five large denticles; carapace ventral margin with distinctive rounded bulge at anterior third; postabdominal claw basal spine length > 2/3 claw; postabdomen with weakly developed setae in lateral fascicles ... *Magnospina dentifera* (Sars, 1901) [Brazil]

35(34) Postabdomen lateral fascicles with robust setae ... 36

35′ Postabdomen lateral fascicles with weakly developed setae; trunk limb I with long anterior seta on endite 1 .... *Bergamina lineolata* Elmoor-Loureiro, Santos-Wisniewski & Rocha, 2013
[Brazil]

36(35) Two connected head pores 37

36′ Three connected head pores or connection interrupted or absent 39

37(36) Valves without denticles 38

37′ Valves with short denticles; trunk limb I mediodistal lobe with well-developed seta 1 .... ***Karualona*** in part **[p. 621]**

38(37) Trunk limb III with well-developed seta on distomedial lobe .... ***Alpinalona*** **[p. 624]**

38′ Trunk limb III with no seta .... ***Anthalona*** **[p. 624]**

39(13) Limb III exopodite with seven setae .... ***Ovalona*** **[p. 625]**

39′ Limb III exopodite with six setae .... ***Coronatella*** **[p. 626]**

Branchiopoda: Diplostraca: Cladocera: Anomopoda: Chydoridae: *Monospilus*: Species

Key modified from Sousa et al. (2017).

1 Postabdominal claw with distalmost basal spine robust; proximalmost basal spine short and thin 2

1′ Postabdominal claw with both basal spines robust .... *Monospilus dispar* Sars, 1862
[Neotropics. Nearctic]

2(1) Postabdominal claw with two pectens, proximal of few robust denticles, distal of spines. .... *Monospilus brachyspinus* Sousa, Elmoor-Loureiro & Panarelli, 2017
[Brazil]

2′ Postabdominal claw with single, serrate pecten of robust denticles .... *Monospilus macroerosus* Sousa, Elmoor-Loureiro & Panarelli, 2017
[Brazil]

Branchiopoda: Diplostraca: Cladocera: Anomopoda: Chydoridae: *Camptocercus*: Species

1 Head and valves with well-developed dorsal keel .... *Camptocercus dadayi* Stingelin, 1913
[Neotropics]

1′ No dorsal keel .... *Camptocercus aloniceps* Ekman, 1900
[Peru, Bolivia, Chile]

Branchiopoda: Diplostraca: Cladocera: Anomopoda: Chydoridae: *Kurzia*: Species

Species reviewed in Hudec (2000).

1 Postabdominal distal corner with lobe not protruding beyond terminal claw base, with three thick denticles at apex, up to nine marginal clusters of spiculiform denticles .... *Kurzia media* Birge, 1879
[Mexico, Colombia]

1′ Postabdominal distal corner distinct, sharp, never forming a lobe, with one apical cluster of spiculiform denticles .... *Kurzia polyspina* Hudec, 2000
[Neotropics]

Branchiopoda: Diplostraca: Cladocera: Anomopoda: Chydoridae: *Oxyurella*: Species

1 Labrum without setulae; postabdomen long, slender; may retain moulted carapace .... *Oxyurella longicaudis* (Birge, 1910)
[Cuba, Colombia, Brazil, Paraguay, Argentina]

1′ Labrum with setulae; postabdomen, short, wider .... *Oxyurella ciliata* Bergamin, 1939
[Venezuela, Colombia, Brazil]

Branchiopoda: Diplostraca: Cladocera: Anomopoda: Chydoridae: *Euryalona*: Species

Euryalona brasiliensis Brehm & Thomsen, 1936 is atypical of the genus and may belong in *Kurzia* (see Rajapaksa & Fernando, 1987).

1 Trunk limb I distomedial lobe, with longest seta heavily sclerotized, arcuate, tip flexible and with six to nine denticles on concave surface 2

1′ Trunk limb I distomedial lobe, with longest seta with no denticles on concave surface .... *Euryalona brasiliensis* Brehm & Thomsen, 1936 [Brazil]

2(1) Postabdomen elongate, tapering distally, length/width > 4; distal lobe with three to five long denticles .... *Euryalona orientalis* (Daday, 1898) [Neotropics]

2′ Postabdomen wider and shorter, length/width ~ 3; distal lobe with six to eight denticles .... *Euryalona fasciculata* Daday, 1905 [Paraguay, Argentina]

Branchiopoda: Diplostraca: Cladocera: Anomopoda: Chydoridae: *Leydigiopsis*: Species

Key based on Van Damme & Sinev (2013).

1 Postabdomen broad, with maximum height near postanal margin middle 2

1′ Postabdomen narrow, widening distally, with maximum height close to postabdomen end. .... *Leydigiopsis megalops* Sars, 1901 [Neotropics]

2(1) Rostrum reaching beyond the first antenna aesthetascs 3

2′ Rostrum not reaching beyond the first antenna aesthetascs .... *Leydigiopsis brevirostris* Brehm, 1838 [Brazil]

3(2) Rostrum weakly curved .... *Leydigiopsis ornata* Daday, 1905 [Cuba, Colombia, Brazil, Paraguay]

3′ Rostrum strongly curved medially .... *Leydigiopsis curvirostris* Sars, 1901 [Neotropics]

Branchiopoda: Diplostraca: Cladocera: Anomopoda: Chydoridae: *Leydigia*: Species

Kotov (2009) revised *Leydigia*.

1 Postabdominal claw basal spine length = to claw base; carapace without fine striation 2

1′ Postabdominal claw basal spine rudimentary or absent; fine striation on valves 3

2(1) Ocellus triangular-rounded or subquadrangular; limb distal articles with swimming setae insertions basally chitinous .... *Leydigia* (*Leydigia*) *louisi louisi* Jenkin, 1934 [Cuba, Chile]

2′ Ocellus irregular in shape; limb distal articles with swimming setae chitinous insertions far from article bases .... *Leydigia* (*Leydigia*) *louisi mexicana* Kotov, Elias-Gutierrez & Nieto, 2003 [Mexico]

3(1) Labral keel lateral fascicles with setulae fine, same length as marginal fascicles; PP distance three to five times IP distance 4

3′ Labral keel lateral fascicles with setulae robust, longer than marginal fascicles; PP distance seven to eight times IP distance .... *Leydigia* (*Neoleydigia*) *ipojucae* Brehm, 1938 [Brazil]

4(3) Postabdomen with maximum height at midpoint, with distal portion sloping toward postabdominal claw; PP distance three to four times IP distance .... *Leydigia* (*Neoleydigia*) *lourdesae* Kotov & Fuentes-Reines, 2014 [Colombia]

4′ Postabdomen with maximum height not at midpoint, ventral margin straight; PP distance four to five times IP distance .... *Leydigia* (*Neoleydigia*) cf. *striata* Birabén, 1939 [Mexico, Cuba, Colombia, Brazil]

Branchiopoda: Diplostraca: Cladocera: Anomopoda: Chydoridae: *Notoalona*: Species

Genus reassignment in Rajapaksa & Fernando (1987).

1	Head shield lateral margin with a distinct notch	2
1′	Head shield lateral margin without a notch [Venezuela, Colombia, Brazil]	*Notoalona sculpta* (Sars, 1901)
2(1)	Postabdomen curved medially; preanal margin $\leq$ anal length; labrum anterior margin with one to five serrations [Mexico, Cuba, Colombia, Brazil]	*Notoalona globulosa* (Daday, 1898)
2′	Postabdomen straight; preanal margin $>$ anal length; labrum anterior margin with 1-3 serrations [Mexico. Nearctic]	*Notoalona freyi* Rajapaksa & Fernando, 1987

Branchiopoda: Diplostraca: Cladocera: Anomopoda: Chydoridae: *Karualona*: Species

1	Carapace valve posteroventral angle with one to four setulae between robust denticles [Mexico, Colombia, Brazil]	*Karualona penuelasi* Dumont & Silva-Briano, 2000
1′	Carapace valve posteroventral angle with five to eight setulae between robust denticles [Colombia, Brazil, Chile]	*Karualona muelleri* (Richard, 1897)

Branchiopoda: Diplostraca: Cladocera: Anomopoda: Chydoridae: *Leberis*: Species

1	Labrum anterior margin convex	2
1′	Labrum anterior margin sinuate; valve ventral margin with 84–95 setae [Neotropics]	*Leberis davidi* (Richard, 1895)
2(1)	Valve ventral margin setae 51–55 [Mexico: Durango]	*Leberis chihuahuaensis* Sinev & Silva-Briano, 2012
2′	Valve ventral margin setae 67–75 [Colombia]	*Leberis colombiensis* Kotov & Fuentes-Reinés, 2015

Branchiopoda: Diplostraca: Cladocera: Anomopoda: Chydoridae: *Ephemeroporus*: Species

1	Carapace without dorsal keel	2
1′	Carapace with high dorsal keel; labrum smooth with no crenulations or teeth [Brazil]	*Ephemeroporus quasimodo* Elmoor-Loureiro, 2014
2(1)	Carapace posteroventral angle without teeth	3
2′	Carapace posteroventral angle with one tooth	5
3(2)	Labrum with distinct teeth	4
3′	Labrum smooth without teeth or crenulations [Peru, Chile]	*Ephemeroporus poppei* (Richard, 1897)
4(3)	Labrum with three to four teeth; valve posteroventral angle with two to three spiniform setae [Colombia, Brazil]	*Ephemeroporus tridentatus* (Bergamin, 1939)
4′	Labrum with one tooth [Venezuela, Brazil, Peru, Bolivia]	*Ephemeroporus acanthodes* Frey, 1982
5(2)	Labrum with one tooth [Neotropics]	*Ephemeroporus hybridus* (Daday, 1905)
5′	Labrum with four distinct teeth [Neotropics]	*Ephemeroporus barroisi* (Richard, 1894)

Branchiopoda: Diplostraca: Cladocera: Anomopoda: Chydoridae: *Alonella*: Species

1 Rostrum and first antennae subequal ... 2

1′ Rostrum much longer than first antennae ... 4

2(1) Carapace posteroventral angle with one to three indentations ... 3

2′ Carapace posteroventral angle with two to six small spines; postabdomen with six to eight small sharp postanal spines ... *Alonella pulchella* Herrick, 1884
[Mexico]

3(2) Carapace polygons with internal striae; valve length: height 1.5 ... *Alonella excisa* (Fischer, 1854)
[Colombia, Chile]

3′ Carapace polygons with internal striae; valve length: height 1.8 ... *Alonella clathratula* Sars, 1896
[Brazil, Chile]

4(1) Postabdomen much longer than wide, many postanal spines ... 5

4′ Postabdomen short ... *Alonella dadayi* Birge, 1910
[Cuba, Colombia, Brazil, Peru]

5(4) Rostrum very long, slender, recurved ... *Alonella leptorhyncha* (Smirnov, 1996)
[Brazil]

5′ Rostrum not recurved; postabdomen with 9–11 postanal spines ... *Alonella hamulata* (Birge, 1879)
[Brazil]

Branchiopoda: Diplostraca: Cladocera: Anomopoda: Chydoridae: *Dunhevedia*: Species

1 Labral plate with many teeth ... 2

1′ Labral plate with one tooth ... *Dunhevedia odontoplax* Sars, 1901
[Cuba, Colombia, Brazil, Peru, Chile]

2(1) Labral plate with three to four teeth ... *Dunhevedia colombiensis* Stingelin, 1914
[Colombia, Brazil]

2′ Labral plate with 8–10 teeth; posteroventral angle of valves with two to three denticles ... *Dunhevedia americana* Rajapaksa & Fernando, 1987
[Cuba, Haiti]

Crustacea: Branchiopoda: Cladocera: Anomopoda: Chydoridae: *Picripleuroxus*: Species

1 Postabdomen dorsal margin with more than 20 postanal denticles ... *Picripleuroxus similis* (Vavra, 1900)
[Colombia]

1′ Postabdomen dorsal margin with fewer than 15 postanal denticles ... *Picripleuroxus quasidenticulatus* Smirnov, 1996
[Brazil]

Branchiopoda: Diplostraca: Cladocera: Anomopoda: Chydoridae: *Pleuroxus*: Species

1 Body with no medial keel, labral keel protruding beyond rostrum apex; gnathobase III filter plate with eight to nine setae ... 2

1′ Body with medial keel, rostrum protruding beyond labral keel apex; gnathobase III filter plate with nine setae ... *Pleuroxus scopuliferus* (Ekman, 1900)
[Chile, Argentina, Sub-Antarctic Islands]

2(1) First antenna with well-developed peg at base; valve ventral margin setae all marginal ... 3

2′ First antenna with rudimentary peg at the base; valve ventral margin setae all submarginal ... *Pleuroxus hardingi* Smirnov, Kotov & Coronel, 2006
[Peru, Bolivia, Chile]

3(2) Valves brown, not transparent, with prominent reticulation; valve posterior margin setulae marginal; gnathobase III filter plate with 8 setae 4

3′ Valves light, with obscure reticulation; valve posterior margin setulae submargina .... *Pleuroxus varidentatus* Frey, 1993 [Mexico, Chile, Argentina]

4(3) Valve posteroventral angle with no teeth, carapace ridged 5

4′ Valve posteroventral angle with one to two teeth; postabdomen postanal teeth strong, single; gnathobase III filter plate with 8 setae .... *Pleuroxus paraplesius* Frey, 1993 [Argentina]

5(4) Gnathobase III filter plate with nine setae; body size 0.69–1.16 mm .... *Pleuroxus fryeri* Kotov, Sinev & Berrios, 2010 [Chile: Andes]

5′ Gnathobase III filter plate with eight setae; body size to 0.6 mm .... *Pleuroxus caca* Harding, 1955 [Argentina]

Branchiopoda: Diplostraca: Cladocera: Anomopoda: Chydoridae: *Chydorus*: Species

1 Valves with no pecteniform meshes 2

1′ Valves with numerous small meshes .... *Chydorus parvireticulatus* Frey, 1987 [Venezuela]

2(1) Body shape globular 3

2′ Body shape elongated; valve posteroventral angle with a denticle; labrum large .... *Chydorus nitidulus* (Sars, 1901) [Cuba, Nicaragua, Venezuela, Colombia, Brazil]

3(2) Valve posteroventral angle without denticles 4

3′ Valve posteroventral angle with rounded, blunt denticles, posteroventral angle bent outward forming a flange .... *Chydorus dentifer* Daday, 1905 [Colombia, Brazil]

4(3) Labral plate without clearly elongated apex 5

4′ Labral plate oval, with elongated apex; differentiated only by small differences the males .... *Chydorus sphaericus* (O.F. Müller, 1776) and *Chydorus patagonicus* Ekman, 1900 [Brazil, Chile]

5(4) Valves without setulae 6

5′ Valves with setulae and wavy polygons .... *Chydorus pubescens* Sars, 1901 [Cuba, Nicaragua, Colombia, Brazil]

6(5) Body globular; valve posteroventral angle not bent laterally .... *Chydorus eurynotus* Sars, 1901 [Neotropics]

6′ Body globular; valve posteroventral margins bent laterally forming a flange .... *Chydorus ventricosus* Daday, 1898 [Colombia, Brazil]

Branchiopoda: Diplostraca: Cladocera: Anomopoda: Chydoridae: *Alona*: Species

1 Large postabdomen with merged marginal teeth, long terminal claw; limb I anteroventral stiff setulae reduced to one or two per group; exopodite III with seven setae 2

1′ With other combinations of characters 3

2(1) Body widening distally, valves posterodorsal angle high; 0.68–0.79 mm body length .... *Alona boliviana* Sinev & Coronel, 2006 [Bolivia]

2′ Body oval, valves postdorsal angle low; 0.52–0.61 mm body length .... *Alona yara* Sinev & Elmoor-Loureiro, 2010 [Brazil]

3(1) Labral keel posterior margin with setules 4

3′ Labral keel anterior margin with setules; *Alona intermedia* group 5

4(3) Main head pores two; postabdomen elongated, distal portion rounded; *Alona affinis* group .... *Alona ossiani* Sinev, 1998 [Neotropics]

4′ Main head pores three; postabdomen short, distal portion rounded or truncated; *Alona guttata* group .... *Alona guttata* Sars, 1862 [Neotropics]

5(3) Labral keel glabrous; postabdomen lateral fascicles length ≤ marginal denticles .... *Alona elisae* Sousa, Elmoor-Loureiro & Santos, 2016 [Brazil]

5′ Labral keel anterior margin with two to four short spinulae; postabdomen lateral fascicles longer than marginal denticles .... *Alona isabellae* Sousa, Elmoor-Loureiro & Santos, 2016 [Brazil]

Branchiopoda: Diplostraca: Cladocera: Anomopoda: Chydoridae: *Flavalona*: Species

This key is based on Sinev & Dumont (2016).

1 Lateral head pores long, length > 0.5 IP distance; postabdomen distal angle acute 2

1′ Lateral head pores short, length < 0.5 IP distance; postabdomen distal angle rounded 3

2(1) Three major head pores; lateral head pore pockets deep, rounded, depth ~1.5–2× lateral pore length .... *Flavalona margipluma* Sousa, Santos, Güntzel, Diniz, de Melo Jr. & Elmoor-Loureiro, 2015. [Brazil]

2′ Two major head pores .... *Flavalona setigera* (Brehm, 1931) [Neotropics]

3(1) Adult female body length ≤ 0.7 mm; exopodite III seta 6 length ~2/3× seta 3; labrum anterior margin without denticles .... *Flavalona hudeci* (Sinev, 1999) [Neotropics]

3′ Adult female body length ≤ 0.52 mm; exopodite III seta 6 length almost = to seta 3 .... *Flavalona iheringula* (Kotov & Sinev, 2004) [Neotropics]

Branchiopoda: Diplostraca: Cladocera: Anomopoda: Chydoridae: *Celsinotum*: Species

1 Postabdomen postanal portion length 1.2–1.3× anal portion; postabdomen distalmost denticles single .... *Celsinotum laticaudatum* Smirnov & Santos-Silva, 1995 [Brazil]

1′ Postabdomen postanal portion length 2–2.3× anal portion; postabdomen distalmost denticles in groups of 2–5 .... *Celsinotum candango* (Richard, 1897) [Brazil]

Branchiopoda: Diplostraca: Cladocera: Anomopoda: Chydoridae: *Alpinalona*: Species

1 Carapace posterodorsal angle with prominence formed by union of valves; head shield length/width ~ 1.4 .... *Alpinalona cajasi* Alonso & Sinev, 2017 [Ecuador: Andes]

1′ Carapace posterodorsal angle broadly rounded; head shield length/width ~ 1.6 .... *Alpinalona manueli* (Sinev & Zawisza, 2013) [Mexico]

Branchiopoda: Diplostraca: Cladocera: Anomopoda: Chydoridae: *Anthalona*: Species

Key modified from Van Damme et al. (2011).

1 Second antenna with all terminal setae thin and long 2

1′ Second antenna exopod with three strongly thickened and acute terminal setae .... *Anthalona acuta* Van Damme, Sinev & Dumont, 2011 [Venezuela, Brazil]

2(1) Lateral fascicles with distal spine reaching 1/3 to ½ its length beyond marginal teeth; labral keel without proximal denticle or small denticle 3

2′ Lateral fascicles with distal spine not reaching beyond marginal teeth; labral keel with proximal denticle (sometimes blunt) 4

3(2) Carapace margin with conspicuous, long, widely spaced setae in posterior portion .... *Anthalona brandorffi* (Sinev & Hollwedel, 2002) [Brazil]

3′ Carapace margin with short setae in posterior portion, not widely spaced; labral keel smooth .... *Anthalona neotropica* Sousa, Elmoor-Loureiro & Debastiani-Junior, 2015 [Brazil]

4(2) Postabdominal claw with basal spinulae reaching over half width of basal spine; limb I longest distomedial lobe seta with basal spine length ≥ apical part .... *Anthalona verrucosa verrucosa* (Sars, 1901) [Neotropics]

4′ Postabdominal claw with basal spinulae reaching over half width of basal spine; limb I longest distomedial lobe seta with basal spine length < apical part .... *Anthalona verrucosa pectinata* (Elías-Gutiérrez & Suárez-Morales, 1999) [Mexico]

Branchiopoda: Diplostraca: Cladocera: Anomopoda: Chydoridae: *Ovalona*: Species

This key is modified from Sinev (2015).

1 Postabdominal claw basal spine length > claw base width 2

1′ Postabdominal claw basal spine length < 1/3 claw base width; postabdomen curved, with convex ventral and concave dorsal margin .... *Ovalona bromelicola* (Smirnov, 1998) [Nicaragua]

2(1) Middle abdominal segment glabrous 3

2′ Middle abdominal segment densely hirsute .... *Ovalona nigra* (Smirnov, 1996) [Bolivia, Chile]

3(2) Postabdomen elongated, narrow, with parallel margins (height-length ratio > 2.5) 4

3′ Postabdomen shorter and wide, height−length ratio < 2.4 5

4(3) Postanal marginal denticles decrease in size basally; valves posteroventral angle without denticles; head shield posterior margin without notches or with a single notch in the middle .... *Ovalona glabra* (Sars, 1901) [Neotropics]

4′ Postanal marginal denticles subequal in size; valves posteroventral angle with one to two denticles .... *Ovalona aguascalientensis* (Sinev & Silva-Briano, 2012) [Mexico]

5(4) Valves anterior group setae distalmost seta length equal to previous setae; postabdomen with maximum width at the middle of postanal margin 6

5′ Valves anterior group setae distalmost seta longer than previous setae; postabdomen evenly narrowing distally, with obtuse distal angle .... *Ovalona anamariae* (Sinev & Silva-Briano, 2012) [Mexico]

6(5) All main head pores fully connected 7

6′ Connection between some pores interrupted or absent 8

7(6) Postanal marginal denticles subequal in size 9

7′ Postanal marginal denticles short, decrease in size proximally; most denticles length < postabdominal claw base width .... *Ovalona pascua* Van Damme, 2016 [Chile: Easter Island]

8(6) Posterior margin of labrum with clusters of setules .... *Ovalona setulosa* (Megard, 1967) [Mexico]

8′ Posterior margin of labrum without clusters of setules .... *Ovalona setulosa arangureni* Sinev & Fuentes-Reines, 2016 [Colombia]

9(7) Postanal marginal denticles short, subequal to claw base width 10

CLASS BRANCHIOPODA

9′ Postanal marginal denticles long, length > postabdominal claw base width .... *Ovalona altiplana* (Kotov, Sinev & Berrios, 2010) [Chile: Andes]

10(9) Postabdomen postanal margin with clusters of three to five short denticles; adult female length 0.5–0.65 mm; major head pores located far from head shield posterior margin, PP distance about 0.6–0.7 IP distance .... *Ovalona weinecki* (Studer, 1878) [Sub-Antarctic Islands]

10′ Postabdomen postanal margin with robust composite denticles; adult female length 0.33–0.4 mm; major head pores located close to head shield posterior margin, PP less than 0.2 IP .... *Ovalona kaingang* (Sousa, Elmoor-Loureiro & Santos, 2015) [Brazil]

Branchiopoda: Diplostraca: Cladocera: Anomopoda: Chydoridae: *Coronatella*: Species

Sousa et al. (2015) revised the genus, and their key is modified here.

1 Distomedial lobe setae with thick proximal denticles; carapace setal row reaches end of carapace 2

1′ Distomedial lobe setae with short and thin proximal denticles; carapace setal row not reaching end of carapace .... *Coronatella paulinae* Sousa, Elmoor-Loureiro & Santos, 2015 [Brazil]

2(1) Carapace posteroventral corner with denticles 3

2′ Carapace posteroventral corner without denticles 4

3(2) Carapace posteroventral corner with one denticle; labral keel with one denticle .... *Coronatella monacantha* (Sars, 1901) [Neotropics]

3′ Carapace posteroventral corner with two denticles; labral keel without denticles .... *Coronatella undata* Sousa, Elmoor-Loureiro & Santos, 2015 [Brazil]

4(3) Carapace posterior margin medial spinules arranged in groups 5

4′ Carapace posterior margin medial spinules not arranged in groups .... *Coronatella poppei* (Richard, 1897) [Chile, Argentina]

5(4) Postabdomen terminal claw basally with short spinules 6

5′ Postabdomen terminal claw basally with long and thin setules .... *Coronatella* cf. *circumfimbriata* [Chile: Andes]

6(5) Postabdomen postanal margin armed with six to eight clusters of denticles .... *Coronatella circumfimbriata* (Megard, 1967) [Mexico, Cuba]

6′ Postabdomen postanal margin armed with five to six clusters of denticles .... *Coronatella serratalhadensis* Sousa, Elmoor-Loureiro & Santos, 2015 [Brazil]

Branchiopoda: Diplostraca: Spinicaudata: Families

The taxonomy of the genus *Cyzicus* is confused, the genus needs revision.

1 Fornicies extending to rostrum (Fig. 19.14 A, B) 2

1′ Fornicies not extending to rostrum (Fig. 19.14 C) .... **Limnadiidae [p. 626]**

2(1) Rostrum lacking an apical spine (Fig. 19.14 A) .... Cyzicidae, one genus; *Cyzicus* [Widespread]

2′ Rostrum with a single apical spine (may be broken off, look for scar) (Fig. 19.14 B) .... Leptestheriidae, one genus: ***Leptestheria* [p. 628]**

Branchiopoda: Diplostraca: Spinicaudata: Limnadiidae: Genera

The genus *Metalimnadia* needs revision.

1 Frontal organ pedunculate (Fig. 19.14 C) .... ***Eulimnadia* [p. 628]**

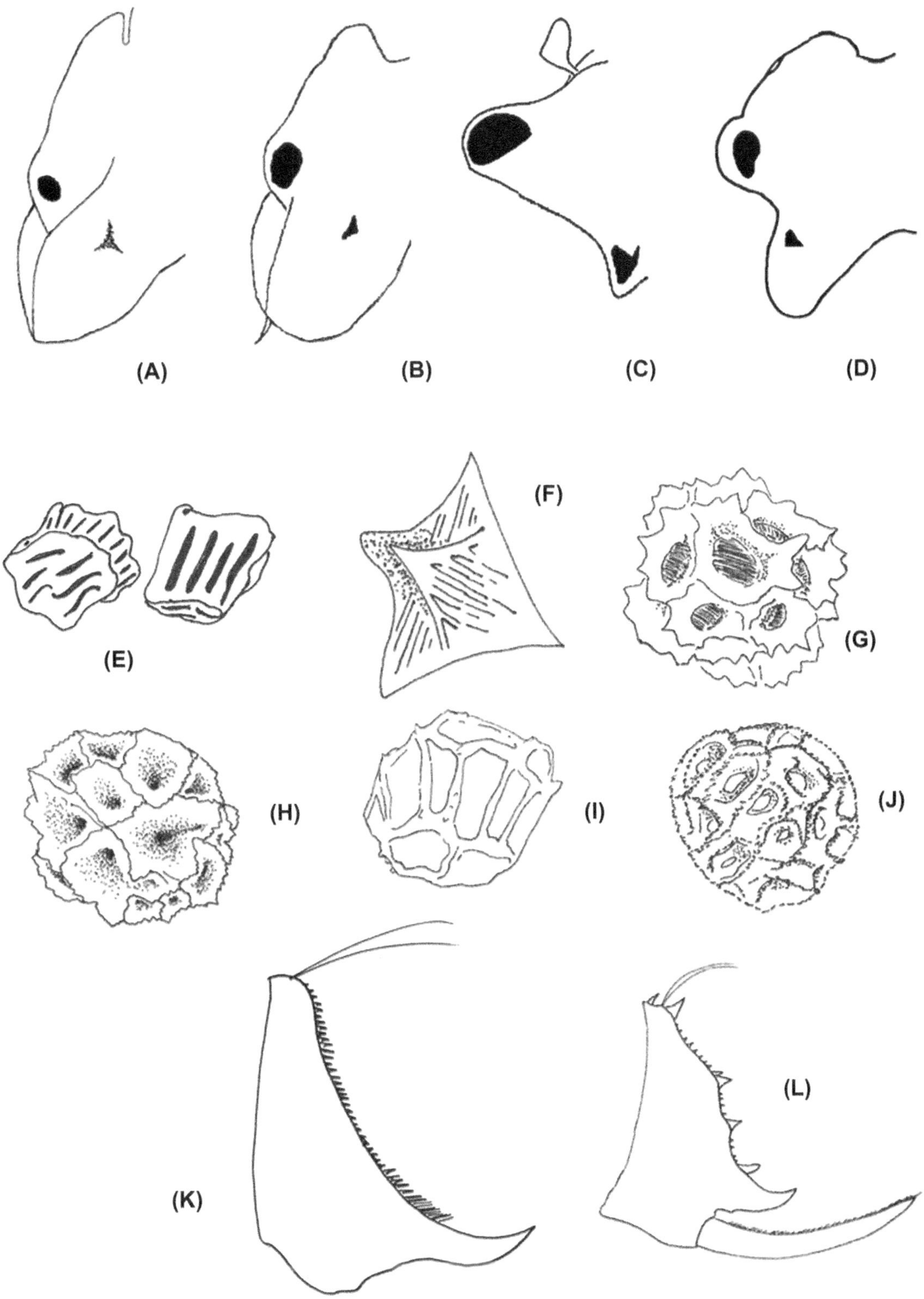

FIGURE 19.14 Spinicaudata: (A) *Cyzicus* sp. head, left lateral view; (B) *Leptestheria* sp. head, left lateral view; (C) *Eulimnadia* sp. head, left lateral view; (D) *Metalimnadia* sp. head, left lateral view; (E) *Eulimnadia texana* sensu lato eggs; (F) *Eulimnadia orinoquiensis* egg; (G) *Eulimnadia magdalensis* egg; (H and I) *Eulimnadia pampa* eggs (variation); (J) *Eulimnadia ovilunata* egg; (K) *Leptestheria venezuelica* telson, left lateral view; (L) *Leptestheria titicacae* telson, left lateral view.

1′ Frontal organ sessile (Fig. 19.14D) .... *Metalimnadia*
[Venezuela, Colombia, Suriname, Brazil]

Branchiopoda: Diplostraca: Spinicaudata: Limnadiidae: *Eulimnadia*: Species

The *Eulimnadia* species are only identifiable using egg morphology (Belk, 1989; Rabet, 2010), and we follow Rabet (2010), Rogers et al. (2012), and Marinone et al. (2016) for the taxonomy. *Eulimnadia texana* sensu lato includes *Eulimnadia belki, Eulimnadia colombiensis, Eulimnadia cylindrova*, and *Eulimnadia ovisimilis*.

1 Egg subspherical (Fig. 19.14 G–J) 2

1′ Egg not subspherical (Fig. 19.14 E, F) 6

2(1) Polygonal depressions without a domed central projection (Fig. 19.14 G–I) 3

2′ Polygonal depressions with a domed central projection (Fig. 19.14 J) .... *Eulimnadia ovilunata* Martin & Belk, 1989
[Argentina]

3(2) Ridge intersections not projecting (Fig. 19.14 G) 4

3′ Ridge intersections projecting as spines, flanges, or rounded points (Fig. 19.14 H, I) .... *Eulimnadia pampa* Marinone, Urcola & Rabet, 2016
[Argentina]

4(3) Polygonal depressions not narrow slits 5

4′ Polygonal depressions long narrow slits, with ends widened, broader than width of slit; slit ends rounded .... *Eulimnadia follisimilis* Pereira & García, 2001
[Venezuela. Nearctic]

5(4) Ridges defining polygon walls subequal in width to the polygons .... *Eulimnadia brasiliensis* Sars, 1902
[Venezuela, Brazil]

5′ Ridges defining polygon walls narrow, much narrower than polygons (Fig. 19.14 G) .... *Eulimnadia magdalensis* Roessler, 1990
[Venezuela, Colombia, Brazil]

6(1) Egg not cylindrical, or at most subcylindrical with one end inflated 7

6′ Egg cylindrical (Fig. 19.14 E) .... *Eulimnadia texana* Packard, 1871 sensu lato
[Mexico, Venezuela, Colombia, Brazil, Ecuador, Bolivia, Argentina, Nearctic]

7(6) Egg pyriform (Fig. 19.14 F) .... *Eulimnadia orinoquiensis* (Roessler, 1991)
[Colombia]

7′ Egg subcylindrical with one end inflated .... *Eulimnadia geayi* Daday, 1926
[Mexico, Venezuela, Colombia]

Branchiopoda: Diplostraca: Spinicaudata: Leptestheriidae: *Leptestheria*: Species

1 Carapace dorsal margin straight 2

1′ Carapace dorsal margin with growth lines projecting as small carinae .... *Leptestheria cristata* García & Pereira, 2003
[Venezuela]

2(1) Male telsonal spines evenly arranged, increasing in size from proximal to distal end (Fig. 19.14 K) 3

2′ Male telsonal spine rows with large and small spines mixed (Fig. 19.14 L) 4

3(2) Male telsonal spine rows with distal most spines larger than proceeding spines, but similar in form .... *Leptestheria compleximanus* (Packard, 1877)
[Mexico. Nearctic]

3′ Male telsonal spine rows with distalmost spines setaform (Fig. 19.14 K) .... *Leptestheria venezuelica* Daday, 1923
[Venezuela]

4(2) Occipital condyle elongate, acute (Fig. 19.14 L) .... *Leptestheria titicacae* Harding, 1940
[Perú, Chile, Bolivia, Argentina]

4′ Occipital condyle triangular .... *Leptestheria brevispina* García & Pereira, 2003
[Venezuela]

REFERENCES

Adamowicz, S., P.D.N. Hebert & M.C. Marinone. 2004. Species diversity and endemism in the *Daphnia* of Argentina: a genetic investigation. Zoological Journal of the Linnean Society 140: 171–205.

Belk, D. 1989. Identification of species in the conchostracan genus *Eulimnadia* by egg shell morphology. Journal of Crustacean Biology 9: 115-125.

Belk, D. & G. Pereira. 1982. *Thamnocephalus venezuelensis*, new species (Anostraca: Thamnocephalidae), first report of *Thamnocephalus* in South America. Journal of Crustacean Biology 2: 223–226.

Brendonck, L., D.C. Rogers, J. Olesen, S. Weeks & R. Hoeh. 2008. Global diversity of large branchiopods (Crustacea: Branchiopoda) in freshwater. Hydrobiologia 595: 167–176.

Cohen, R.G. 1981. Notas sobre anostracos neotropicales (Crustacea). I. *Branchinecta valchetana* sp. nov. y redescripcion de *Branchinecta palustris* Birabén, 1946. Physis (B) 40: 1–13.

Cohen, R.G. 1982. Notas sobre anostracos neotropicales (Crustacea). II. *Branchinecta rocaensis* sp. nov. Physis (B) 40: 1–13.

Cohen, R.G. 1983. Notas sobre anostracos neotropicales (Crustacea). III. *Branchinecta somuncurensis* y *Branchinecta prima* spp. nov. Physis (B) 41: 69–80.

Cohen, R.G. 1985. Notas sobre anostracos neotropicales (Crustacea). IV. *Branchinecta vuriloche* sp. nov. Physis (B) 43: 1–6.

Cohen, R.G. 1987. Notas sobre anostracos neotropicales (Crustacea). V. Nueva observaciones morphologicas sobre *Branchinecta achalensis* César, 1986. Physis (B) 45: 13–20.

Cohen, R.G. 1992. Redescription of *Branchinecta granulosa* Daday, 1902 from Argentina (Crustacea: Branchiopoda). Hydrobiologia 228: 195–202.

Cohen, R.G. 1993. Nuevos aportes a la morfologia y distribucion de *Branchinecta iheringi* Lilljeborg, 1889 (Crustacea: Anostraca). Physis 48: 114–115.

Cohen, R.G. 1995. Intraspecific variability in *Branchinecta iheringi* Lilljeborg, 1889 (Crustacea: Anostraca). Studies on Neotropical Fauna and Environment 30: 61–64.

Cohen, R.G. 2008. A new species of *Branchinecta* (Crustacea: Branchiopoda: Anostraca) from the Argentinian Tierra del Fuego. Zootaxa 1884: 60–68.

Cohen, R.G. 2012. Morphological diversity displayed by high altitude *Branchinecta papillata* (Anostraca) and additional morphological comparisons with *Branchinecta achalensis*. Journal of Crustacean Biology 32: 1–13.

Cohen, R.G. 2016. *Thamnocephalus chacosaltensis*, a new species of Anostraca (Crustacea: Branchiopoda) from Salta province (Argentina). Zootaxa 4088: 76–90.

Cohen, R.G., D. Serna-Macías & P. Eslava-Eljaiek. 2014. Redescription of *Dendrocephalus affinis* (Anostraca: Thamnocephalidae): first record of the species from Colombia and additional morphological observations useful in taxonomy. Journal of Crustacean Biology 34: 82–89.

De Meester, L., A. Gómez, B. Okamura & K. Schwenk. 2002. The Monopolization Hypothesis and the dispersal–gene flow paradox in aquatic organisms. Acta Oecologica 23: 121–135.

Dodson, S.I. & D.G. Frey. 2001. Cladocera and other Branchiopoda. Pages 849–913 *in:* J.H Thorp & A.P. Covich (eds.), Ecology and Systematics of North American Freshwater Invertebrates, 2nd edn., Academic Press, New York.

Hudec, I. 2000. Subgeneric differentiation within *Kurzia* (Crustacea: Anomopoda: Chydoridae) and a new species from Central America. Hydrobiologia 421: 165–178.

Kotov, A.A. 2009. A revision of *Leydigia* Kurz, 1875 (Anomopoda, Cladocera, Branchiopoda), and subgeneric differentiation within the genus. Zootaxa 2082: 1–68.

Kotov, A.A. & P. Štifter. 2006. Cladocera: Family Ilyocryptidae (Branchiopoda: Cladocera: Anomopoda. Guides to the Identification of the Microinvertebrates of the Continental Waters of the World. Backhuys Publishers, Leiden.

Marinone, M.C., J.I. Urcola & N. Rabet. 2016. Review of the *Eulimnadia* (Branchiopoda: Spinicaudata: Limnadiidae) from Argentina with a description of a new species. Zootaxa 4158: 419–432.

Martin, J.W. & D. Belk. 1988. A review of the clam shrimp family Lynceidae Stebbing, 1902 (Branchiopoda: Conchostraca), in the Americas. Journal of Crustacean Biology 8: 451–482.

Martínez-Pantoja, M.A., J. Alcocer & A. Maeda-Martínez. 2002. On the Spinicaudata (Branchiopoda) from Lake Cuitzeo, Michoacán, México: first report of a clam shrimp fishery. Hydrobiologia 486: 207–213.

Obregón-Barboza, H., G. Murugan, H. García-Velazco & A.M. Maeda-Martínez. 2015. A review of the *Branchinecta* (Branchiopoda: Anostraca) from the Baja California Peninsula: first record of the giant fairy shrimp *B. gigas* Lynch, 1937 from Mexico. Journal of Crustacean Biology 35: 433–440.

de Paiva Barros-Alves, S., D.F. Rodrigues Alves, E.A. Bolla Jr., N. Rabet & G.L. Hirose. 2016. Morphological review of the freshwater fairy shrimp *Dendrocephalus brasiliensis* Pesta, 1921 (Anostraca: Thamnocephalidae). Nauplius 24: e2016008.

Pereira, G. 1983. Taxonomic importance of the frontal appendage in the genus *Dendrocephalus* (Anostraca: Thamnocephalidae). Journal of Crustacean Biology 3: 293–305.

Pereira, G. 1984. Two new species of *Dendrocephalus* (Anostraca: Thamnocephalidae) from Venezuela. Journal of Crustacean Biology 4: 147–153.

Pereira, G. & D. Belk. 1987. Three new species of *Dendrocephalus* (Anostraca: Thamnocephalidae) from Central and South America. Journal of Crustacean Biology 7: 572–580.

Pereira, G. & L. Ruiz B. 1995. A New Species of *Dendrocephalus* (Anostraca, Thamnocephalidae) from Argentina. Crustaceana 68: 567–574.

Rabet, N. 2010. Revision of the egg morphology of *Eulimnadia* (Crustacea, Branchiopoda, Spinicaudata). Zoosystema 32: 373–391.

Rabet, N. & A. Thiéry. 1996. The neotropical genus *Dendrocephalus* (Crustacea: Anostraca: Thamnocephalidae) in Brazil (South America), with a description of two new species. Journal of Natural History 30: 479–503.

Rajapaksa, R. & C.H. Fernando. 1987. Redescription and assignment of *Alona globulosa* Daday, 1898 to a new genus *Notoalona* and a description of *Notoalona freyi* sp. nov. Hydrobiologia 144: 131–153.

Rogers, D.C. 2009. Branchiopoda (Anostraca, Notostraca, Laevicaudata, Spinicaudata, Cyclestherida). Pages 242–249 *in*: G.F. Likens (ed.), Encyclopedia of Inland Waters 2, Academic Press, New York.

Rogers, D.C., P. de los Ríos & O. Zúñiga. 2008. Fairy shrimp (Branchiopoda: Anostraca) of Chile. Journal of Crustacean Biology 28: 543–550.

Rogers, D.C. & A. Ferreira. 2007. A new species of *Branchinecta* (Crustacea: Anostraca) from Brasil. Zootaxa 1445: 27–34.

Rogers, D.C. & J. Olesen. 2016. Laevicaudata catalogus (Crustacea: Branchiopoda): an overview of diversity and terminology. Arthropod Systematics & Phylogeny 74: 221–240.

Rogers, D.C., N. Rabet & S.C. Weeks. 2012. Revision of the extant genera of Limnadiidae (Branchiopoda: Spinicaudata). Journal of Crustacean Biology 32: 827–842.

Sinev, A.Y. 2015. Revision of *pulchella*-group of *Alona* s. lato leads to its translocation to the genus *Ovalona* Van Damme et Dumont, 2008 (Branchiopoda: Anomopoda: Chydoridae). Zootaxa 4044: 451–492.

Sinev, A.Y. & H.J. Dumont. 2016. Revision of the *costata*-group of *Alona* s. lato (Cladocera: Anomopoda: Chydoridae) confirms its generic status. European Journal of Taxonomy 223: 1–38.

Sousa, F.D.R., L.M.A. Elmoor-Loureiro, J.R. Debastiani-Júnior, R. Mugnai & A. Senna. 2015. New records of *Anthalona acuta* Van Damme, Sinev & Dumont 2011 and *Anthalona brandorffi* (Sinev & Hollwedel, 2002) in Brazil, with description of a new species of the simplex-branch (Crustacea: Cladocera: Chydoridae). Zootaxa 4044: 224–240.

Sousa, F.D.R., L.M.A. Elmoor-Loureiro & E.A. Panarelli. 2017. The amazing diversity of the genus *Monospilus* Sars, 1862 (Crustacea: Branchiopoda: Aloninae) in South America. Zootaxa 4242: 467–492.

Van Damme, K. & A.Y. Sinev. 2013. Tropical Amphi-Pacific disjunctions in the Cladocera (Crustacea: Branchiopoda). Journal of Limnology 70: 209–244.

Van Damme, K., A.Y. Sinev & H.J. Dumont. 2011. Separation of *Anthalona* gen.n. from *Alona* Baird, 1843 (Branchiopoda: Cladocera: Anomopoda): morphology and evolution of scraping stenothermic alonines. Zootaxa 2875: 1–64.

Chapter 20

Class Ostracoda

Janet Higuti

State University of Maringá, Centre of Research on Limnology, Ichthyology and Aquaculture, Graduate Programme in Ecology of Inland Water Ecosystems, Maringá, Paraná, Brazil

Koen Martens

Royal Belgian Institute of Natural Sciences (RBINS), Brussels, Belgium, and Department of Biology, University of Ghent, Ghent, Belgium

Chapter Outline

INTRODUCTION

Ostracoda are small, bivalved Crustacea that form an important part of the biodiversity of inland waters, but which are often neglected in ecological surveys. This aversion to include ostracods in general ecological and biodiversity research mostly reflects the fact that identification often requires complete dissection, experience with these organisms, and a good microscope. Another reason, however, is the lack of good identification keys for the different zoogeographical regions. We hope that the present key to the nonmarine ostracods of the Neotropical region will remedy this problem for South and Central America, Antarctica, and adjacent islands.

There are 331 species in 77 genera of recent nonmarine ostracods known from the Neotropical region (Meisch et al., 2019), but many new taxa remain to be described, while the status of several old doubtful species needs to be reassessed. For example, various nominal species in large genera, such as *Chlamydotheca* and *Strandesia s.l.* might eventually be considered synonymous with others. On the other hand, there is increasing evidence for the existence of large numbers of cryptic species, which may only be detected with molecular methods. The most spectacular example is that of Bode et al. (2010) where close to 40 cryptic species were identified in the Holarctic species *Eucypris virens* (Jurine, 1820) in Europe alone. Some examples have also been found in the Neotropics, such as in darwinulids (Schön et al., 2012) and in *Strandesia* (Higuti et al., 2013; Schön et al., 2018). The present key allows the identification of 63 freshwater and terrestrial genera. A further 14 genera which were included in the index by Kempf (1980, 1997) are not considered here, as they are either genera that only have marine or brackish water

Thorp and Covich's Freshwater Invertebrates. https://doi.org/10.1016/B978-0-12-804225-0.00020-4

species in the Neotropics, or belong to the family Entocytheridae which are epibiontic on other Crustacea (not free living).

The monograph by Sars (1901), the series of papers on Brazil by Klie (e.g., 1940), the comprehensive papers on Colombian ostracods by Roessler between 1982 and 1990 and the groundbreaking work on the West Indies by Broodbakker with 10 papers between 1982 and 1984 (e.g., Broodbakker, 1984) under the guidance of D.L. Danielopol (Broodbakker & Danielopol, 1982) are all milestones in the history of taxonomy and ecology of Neotropical nonmarine ostracods.

I.D. Pinto published extensively illustrated monographs between 1961 and 1993 (e.g., Pinto & Kotzian, 1961; Würdig & Pinto, 1990). Two more recent series of papers have contributed greatly to the taxonomy and ecology of previously unknown biomes of South American ostracods. R.L. Pinto and colleagues described a variety of previously unknown terrestrial ostracods from São Paulo State between 2003 and 2013 (e.g., Pinto et al., 2008). Higuti and colleagues showed that the floating aquatic macrophytes, typical of South American floodplains, contain highly diverse pleuston communities in their submerged root systems, in which ostracods abound (e.g., Higuti et al., 2007; Higuti & Martens, 2014; Campos et al., 2017; Conceição et al., 2018). It is also of note that Higuti et al. (2013) and Schön et al. (2018) introduced molecular phylogenetic work on South American nonmarine ostracods.

There are also several recent papers on the Argentinian fauna by different authors (Cusminsky et al., 2005; Fontana & Ballent, 2005; Laprida, 2006; Díaz & Lopretto, 2011; D'Ambrosio et al., 2015).

The checklist of Ramírez (1967) was useful for a long time but was updated by Martens & Behen (1994). Martens & Savatenalinton (2011) and Meisch et al. (2019) tabulated the nonmarine ostracods of the world and their occurrence in the different zoogeographical regions, so that the ostracod fauna of the Neotropical region can be extracted from these lists.

LIMITATIONS

The best-known parts of the Neotropics are the West Indian Islands, several floodplains of North and Central Brazil, and lakes from southern Brazil. Recently, several (semi-) terrestrial taxa were discovered, while also bromeliad phytothelmata received new attention. However, large parts of southern and western South America remain scantly explored. Dozens of new species have been collected from the ancient Lake Titicaca and adjacent high-altitude lakes but await formal description. Genuine Antarctic lakes appear to be devoid of nonmarine ostracods (De Deckker, Personal communication), but extensive surveys are needed to confirm this. Very little is known about the ostracods from Antarctic islands.

Only West Indian Islands have been explored for subterranean fauna, and this remains a challenging field for future studies on the main continent and other islands. Temporary pools from arid areas also deserve more attention, as they can have highly endemic faunas of sometimes gigantic forms (>5 mm) because of the absence of fish predation. Estuaries can sometimes support brackish water species of otherwise fully marine genera (e.g., *Xestoleberis*, *Semicytherura*—see below) as well as some Paracypridinae. These species are not included here, as they are outside the scope of this work. On the other hand, the few (semi-) terrestrial species that can be found far inland are mostly closely related to genuine inland water ostracods and as some of them can on occasion be found in fully aquatic environments (e.g., some darwinulids), they are included in the present chapter.

There are several genera that require taxonomic revision. For example, it is highly unlikely that all species presently assigned to *Candona* and *Pseudocandona* actually belong there, as these (often old) generic assignments ignore recent revisions and divisions of these genera in small, monophyletic clades. In addition, genera such as *Chlamydotheca* and *Strandesia* comprise several older uncertain species of which identity and/or position is uncertain. The present key is not the place to conduct such a revision, and only few taxonomic remarks are entered before the keys (see below).

TERMINOLOGY AND MORPHOLOGY

As the present chapter has been in production for over three years, one new non-marine ostracod genus has been described since, which is not in the keys, nor in the Table 20.1, nl. *Neostrandesia striata* by Ferreira et al. (2019). This genus and species shows a striking case of convergent evolution with *Bradleytriebella lineata* (Victor & Fernando, 1981) and will be identified using the present key as the latter taxon. Also, a revision of Brazilian *Strandesia* was published by Ferreira et al. (2020).

Valve and appendage morphology, anatomy, and terminology have been extensively described by Smith et al. (2015) as an introduction to all ostracod chapters in the present series of identification guides, and we refer to this text for details on the terminology used in the keys below. Here, we present a summary of the most used terms. The most important structures are shown in Fig. 20.1 A–F.

Ostracods have a bivalved **carapace (Cp)**, reduced body segmentation, and reduced number of appendages. The valves are impregnated with calcium and dorsally attached along the **hinge**, which can either consist solely out of ligaments (e.g., Cypridoidea) or can include teeth and matching sockets (e.g., in Cytheroidea). In or close to the center of both valves are the **central adductor muscles**, which close the valves. The attachments leave **adductor muscle scars (AMS)** (Fig. 20.1 A, D) that are visible on the inside, and often also in transparent light on the outside of the valves. These muscle scar patterns are important to distinguish higher taxonomic levels of ostracods.

Valves consist of outer and **inner lamellae (il)** (Fig. 20.1 A, B). Around the **valve margins (vm)**, these lamellae are fused over a distance and thus create the **fused zone (fz)**, through which **radial pore canals (rpc**, visible in transparent light) can run. The inner lamella is in part calcified and the edge of this calcified inner lamella is called the **inner margin (im)**. Valve margins can be simple and unornamented, showing **marginal setae** and in some groups **marginal tubercles** or **spines**. Other taxa can have a complex of lists and ridges that allow the animal to close the valves more tightly. A distinction has to be made between selvages and inner lists. **Selvages (sl)** are old valve margins that have been displaced inwardly during evolution. Mostly, they still have traces of old radial pore canals and setae. **Inner lists (il)** do not have remnants of such structures.

There are seven pairs of true appendages, four cephalic and three thoracic (Fig. 20.1 C–F).

- The **First Antennae (A1)** are at the front of the body and are mostly curved in dorsal direction. They are uniramous (all post antenullar appendages are originally biramous, although in some appendages this biramous aspect is largely lost) and mostly consist of seven segments. If the animal is a swimmer, the last four segments will have long natatory setae (e.g., cyprids). Nonswimmers will have spines or spiniform setae there (darwinulids, cytherids).
- The **Second Antennae (A2)** are in second position and are turned toward the ventral side. There are two (fused) basipodal segments, large three-segmented endopods, and small, reduced exopods. Swimming species also have long natatory setae on the first endopodal segment; in nonswimming species these setae are either strongly reduced or fully absent. The last two endopodal segments have distal claws and setae in a pattern that can be sexually dimorphic. Antennae are mainly used for locomotion (swimming, crawling), for clinging to surfaces and by males during copulation.
- The **Mandibulae (Md)** consist of a sclerotized coxa with strong teeth and a palp. The coxa is used for chewing and crushing food segments. The palp is the endopod with a small branchial plate (exopod) used for respiration. In darwinulids, the palps have long and strong setae that form a basket which is presumably for filter feeding. Otherwise the Md-palps are used for the manipulation of food segments.
- A pair of **rake-like organs**, used to crush food, is situated between the third and the fourth pair of cephalic appendages. Their homology remains unknown.
- The **Maxillulae (Mx or Mx1)** consist of a plate to which three endites and a two-segmented palp as well as a large respiratory plate are attached. The endites are used for chewing food segments and the palp for food manipulation.

In between the cephalic and thoracic appendages, there is a pair of brushlike organs in male cytherids. No other ostracods have these organs there and their homology remains unknown.

- The **first thoracopod (T1)**; in other literature sometime referred to as Maxilla—Mx2, differs considerably among the three superfamilies. In darwinulids and cypridoids they are reduced and sexually dimorphic. In these groups, the female appendage consists of a plate with distal setae, a one or two-segmented palp, and a small respiratory plate. Here, the appendage is used mostly for food manipulation. In male cypridoids, the palps are transformed in strong, two-segmented, and mostly asymmetrical prehensile palps that open the female valves during copulation. In darwinulid males, which were only found once (Smith et al., 2006), these palps are less developed but still more so than in the female. In cytherids, this appendage is a walking leg.
- The **second thoracopod (T2)** is a walking leg in all groups, although its main function could rather be clinging to substrate than actual walking.
- The **third thoracopod (T3)** is a walking leg in darwinulids and cytherids and an upturned cleaning leg in cypridoids. In Cyprididae, the last two segments are fused to form a pincer-shaped organ, used to clean the long (natatory) setae of the A1 and the A2. In the other three groups of the Cypridoidea, these segments remain unfused.

Nearly all ostracods have a pair of **caudal rami (CR)**, which can consist of a strong ramus and two distal setae and claws or can be reduced to a pair of flagellate setae. In some literature, the caudal rami are called furcae, uropods, or uropodal rami. When fully developed, as in most Cyprididae (not in Cypridopsinae), they are used for locomotion or for clinging to surfaces. The CR are not considered true appendages that are associated to a particular segment but are thought to be secondary structures. CR are supported by an attachment, which distally of the CR, can have a sclerotized loop, called **Triebel's loop (TL)**.

Female copulatory organs are generally externally unornamented shallow protrusions with a genital operculum (except for two African giant genera, *Afrocypris* and *Liocypris*). The internal structure of female copulatory organs is not well studied, and they are generally not used for identification of species and genera, although in some candonids protruding lobes can occur.

Male copulatory appendages, called **hemipenes (Hp)**, are mostly large to very large. In cytherids, the hemipenes can be up to one-third of the body mass, because all muscles needed to pump the giant sperm cells into the female are incorporated in the actual hemipenes (ostracod sperms can be 10 times longer than the body length). In cytherids, copulatory processes and associated structures are also externally positioned. In cypridoids, the muscles are outside of the actual hemipenes in a special sperm pump called the Zenker organ. Copulatory processes and associated structures are internal, enveloped by hemipenal sheets, and are only external during erection.

Males and female can be recognized by the presence of the prehensile palps and the hemipenes in males, but also by the sometimes large and conspicuous **testical tubes** (four pairs) which can be visible in the posterior part of the body in males.

MATERIAL PREPARATION AND PRESERVATION

Each type of habitat where ostracods occur requires its own methods for qualitative and quantitative sampling, such as hand nets, box-corers and grabs, dredges, pipettes, etc. For an overview of various collecting techniques for living nonmarine ostracods, see Smith et al. (2015) and Martens & Horne (2016).

A 4% formaldehyde solution was the most common preservative for aquatic animals, and is still used for some, including crustacean zooplankton. However, formaldehyde can be toxic, allergenic, and carcinogenic and its use is therefore strongly discouraged. Preserving ostracods in formaldehyde also makes them unsuitable for classic molecular research. When using formaldehyde solutions to preserve ostracods, it is also vital that the formalin is neutralized by adding sodium bicarbonate crystals, so that the pH of the sample is between 9 and 10. If ostracods are kept in acid formalin for a number of days, the valves will rapidly decalcify, and identification then becomes extremely hazardous and illustration of valves impossible. The failure to neutralize the formalin when preserving nonmarine ostracods is one of the reasons that this group remained ill known (more than half of existing ostracod collections in museums are virtually useless because if it) and studied by so few biologists.

Ostracods are best killed and fixed in ethanol, but to attain a proper fixation, a concentration of at least 70%–80% much reach the animal. If 95% ethanol is added to a wet sample, often concentrations of less than 50% are attained, and this is insufficient to properly fix the tissues. If samples are fixed immediately after collecting, then most ostracods will have their valves tightly closed, which hampers dissection considerably (see below). If samples are left closed and unfixed for some time (depending on climate and temperature this could be anything from a few hours to a day), then ostracods will die because of lack of oxygen with valves spread open. Ethanol can then be added to the sample. This will facilitate subsequent dissection (see below).

However, if you wish to perform molecular studies on your ostracods, specimens should be stored in pure ethanol (i.e., not denatured by camphor or other additives that make the alcohol undrinkable) at 4°C. DNA will often be well preserved only for individuals preserved in ~100% ETOH. Therefore, fixing whole samples in pure ethanol might not yield good material for DNA-based research. Whether for molecular or morphological work, it is always wise to change the alcohol in the sample after a few days or weeks.

Proper identification and description of ostracods nearly invariably demand a complete dissection of the soft parts. This is not easy and will require practice. The dissection is performed with two needles, either steel insect pins (the smallest type: 000) fitted in convenient handles (for example, large rechargeable pencil holders, of which the opening is stuffed with a flexible rubber) or pieces of tungsten wire. The latter gives finer points to the needles, but these are so vulnerable that to touch the valves slightly may be enough to destroy the needle. Tungsten needles are better for dissection of cladocerans and copepods; steel needles are better for ostracods, except for very small taxa less than 500 μm long.

The actual dissection requires two major steps: (1) opening the valves and separating valves and soft body; and (2) dissecting the soft parts.

The first step is performed in 96% ethanol in an embryo dish:

- insert the needles between the valves and cut the central muscle scars;
- separate valves and soft parts;
- dry valves in the air and store in micropaleontological (cavity) slides;
- place the soft body in a drop of glycerine or another dissection medium (see below).

When the carapace is globular and tightly closed, step 1 (opening the valves) might be a challenge. One can choose to crush one valve; but as valves mostly have asymmetrical anatomies, one loses information. A more elegant method is to fix the specimen with the ventral side up on a small piece of double-sided sticky tape. The outside of the animal will dry, and one can then better attempt to open the valves. Because the valves are tightly closed, the soft parts will not dry immediately. This method easily has a success rate of 75% or more. If few specimens are available, it is advisable to note down some external features, prior to dissection (e.g., width-length ratio, overlap of valves, shape of ovarium, colors).

Valves can be stored in ethanol in separate tubes, but it is more convenient to store them dry in so-called microcavity or micropalaeontological slides. Often, a water-based glue (called tragacanthin) is used to fix valves in these cavities; but unless these slides need to be transported or sent by mail, it is advisable not to use this glue for two reasons: (1) if the glue is pure, it is prone to become infected by fungi; or

(2) if traces of formalin are used to prevent this, then the valves will be decalcified in the long run, even when dry (air moisture will be sufficient). If formalin in tragacanthin is neutralized with sodium (bi)carbonate, then the valves risk being covered with a white powder.

For the second part of the dissection, frustration will be the price paid for experience, but it is good to take the following suggestions into account:

- Divide the body in an anterior and a posterior part. The anterior part will contain A1, A2, Md, Mx1 and in Cypridoidea sometimes also the T1; the posterior part will contain the thoracic legs, the caudal rami, the hemipenes, and the Zenkers organs.
- Leave the large A1 and A2 attached to the anterior body part until the very last, as they will facilitate orientation of the pieces.
- T1 in most taxa, and cypridopsine caudal rami, are hard to find and several specimens will have to be dissected before you will be able to locate and examine them.
- When dealing with large specimens (3 mm or more), divide the appendages over at least two slides, making sure that large and smaller parts are segregated over the two slides.

Other accounts on ostracod dissection are in Danielopol (1982) and Namiotko et al. (2011).

Glycerine Method

The dissected appendages need then be arranged to take a maximum of space in the glycerine drop, so that no limb will come to lie on top of another one and are then covered with a cover slip. The preparation is sealed with glycol. Recently, however, glycol was taken out of commercial trade in several countries. Normal nail-varnish will also seal the dissection, although these seals are known eventually to crack. Other sealing mediums are Canada balsam, Eukit, and Murrayite. For long-term storage, it is necessary that the preparation is sealed, as any contact with air will cause the preparation to dry out completely.

Glycerine-jelly (Gj)

Add several drops of methylene blue to the glycerine-jelly. Before starting to mount the limbs, the Gj has to be heated in a water bath in order to make it fluid. In this case, the limbs will not be mounted between a slide and a cover slip, but between two cover slips, which are subsequently mounted in a metal holder. The following description is based on notes of Danielopol (personal communication).

- Fix a cover slip (24 × 24 mm) with plasticine on a normal microscope slide. Prepare a second cover slip (18 × 18 mm), putting small points of plasticine at the four corners.
- Put one drop of hot Gj in the center of the large, fixed cover slip. Spread it in order to get a thin, rectangular film of Gj.
- Insert the different limbs in the Gj film, arranged in two horizontal rows, containing the sequence of left and right limbs. Let the preparation dry.
- After 3–6 h, add a new film of medium hot Gj, which has to cover the entire area covered by the first drop. Put the small cover slip on top and press gently at the corners. The plasticine will help prevent any deformation of the limbs. Let the preparation dry.
- Seal the slide (see above for the sealing mediums).
- Remove the preparation, consisting of the two coverslips sealed to each other from the microscope slide and slip it in the metal holder. Fix carton labels on both sides of the preparation.

This method is rather more time consuming, but has a number of advantages: (1) the use of the glycerine-jelly allows the limbs to be arranged in proper sequence and in the desired orientation (with normal glycerine-preparations, this is not possible); and (2) the use of two coverslips (which can also be used with normal glycerine) allows the limbs to be observed with the microscope from both sides of the preparation.

Other mediums

Glycerine-based mediums are known to slowly dissolve the limbs after a long time (50–100 years) in a number of cases. For ostracods which were fixed in ethanol, the limbs can be mounted in other mediums like euparal and polyvinyl-lactophenol. These mediums are not (yet?) known to dissolve limbs, but for a while, at least the latter medium was reported to be carcinogenic. As these mediums polymerize when exposed to the air, the preparations do not require additional sealing but dissecting needs to be done relatively fast.

KEYS TO OSTRACODA

All freshwater Ostracoda belong to the Podocopida. Myodocopida and Platycopida are both exclusively marine. Table 20.1 presents a full list of genera that have been cited in the literature on Neotropical, nonmarine ostracods. However, the keys only treat real (inland) freshwater or terrestrial ostracods. So, the commensal or ectoparasitic Entocytheridae, brackish water taxa, and some genera with doubtful identification (e.g., *Bradleystrandesia, Chrissia*) are not treated in the keys, but are listed in Table 20.1 for completeness.

Ostracoda: Podocopida: Superfamilies

1 Valves not elongated or subquadrate in lateral view, valve surface smooth or ornamented; AMS not arranged in a rosette pattern (Fig. 20.1 A, D) 2

1′ Valves elongated or subquadrate in lateral view, surface smooth; AMS arranged in a rosette-pattern (Fig. 20.2 B, C, E, F); appendages T2 and T3 walking legs (Fig. 20.1 E) .... Darwinuloidea, one family: **Darwinulidae [p. 636]**

2(1) Last three appendage pairs (T1−T3, not CR) subequal walking legs (Fig. 20.1 F) exception of T3 in male *Cytheridella*—see below); four AMS arranged in a vertical row (Fig. 20.1 A) .... **Cytheroidea [p. 636]**

2′ Last three appendage pairs (T1−T3, not CR) with different shapes and function (Fig. 20.1 C); AMS with different pattern, mostly resembling a "pawprint" (Fig. 20.1 D) .... **Cypridoidea [p. 637]**

Ostracoda: Podocopida: Darwinuloidea: Darwinulidae: Genera

Darwinulids are small- to medium-sized ostracods (0.4−0.8 mm), with carapaces elongate, subsquarish, or rounded in lateral view. The hinge is adont or with a medial groove, anterior and posterior cardinal teeth on the right valve, cardinal ridge and sockets on the left valve. Valve margins are devoid of selvages or (continuous) inner or outer lists. The fused zones are very narrow with few straight and very short marginal pore canals. *Vestalenula* has a ventro-caudal keel on the right valve (a remnant of an outer list). Other genera like *Penthesilenula, Microdarwinula,* and *Alicenula* have internal ventral or caudal teeth on the left valve (remnants of inner lists). *Darwinula s.s.* is devoid of any such features. All species (except those in *Microdarwinula*) have an externally visible brooding cavity.

A frontal eye is present. The A2 is devoid of natatory setae. Md and Mx have large respiratory plates. T1 is a maxilliped, with three-segmented palp, in the female. T2 and T3 are walking legs. The CR, if present, is reduced to single seta. A postabdomen is present or absent.

Although in general not very common, the species can be found in a wide range of habitats, including lakes, rivers, springs, groundwater, etc., but also in more marginal biotopes such as (semi-)terrestrial habitats. Drought-resistant stages (present in Cytheroidea and Cypridoidea) have not yet been described for Darwinulidae. There are only about 30 living species thus far described.

1 Cp elongate or subquadrate in lateral view, brood pouch in dorsal view externally visible; AMS positioned anterior to the center (Fig. 20.2 B, C, E, F) 2

1′ Cp rounded in lateral view, brood pouch in dorsal view not externally visible; AMS centrally placed (Fig. 20.2 A) ..... *Microdarwinula*
[Cosmopolitan. Neotropical: Brazil]

2(1) Right valve without posterioventral keel (Fig. 20.2 F), left valve with or without internal posterioventral teeth (Fig. 20.2 B, E); A2 exopodite with two setae and a spine (Fig. 20.2 G) 3

2′ Right valve with posterioventral keel (Figs. 20.2 C, 17 A), left valve with anterioventral internal tooth (Fig. 20.2 B); A2 exopodite with one seta and a spine (Fig. 20.2 D) .... *Vestalenula*
[Cosmopolitan. Neotropical: Brazil]

3(2) Cp elongated (Fig. 20.1 E), left valve without anterior inner tooth; Md-palp last segment with <five claws (three or four, Fig. 20.2 I, J), penultimate segment with seta *y* shorter than seta *z* (Fig. 20.2 I) or absent (Fig. 20.2 J) 4

3′ Cp in lateral view subquadrate; left valves with both anterior and posterior internal tooth (Figs. 20.2 E, 17 B); Md-palp last segment with five claws, penultimate segment with seta *y* as long as seta *z* (Fig. 20.2 H) .... *Penthesilenula*
[Cosmopolitan. Neotropical: Brazil, Chile, Peru]

4(3) Right valve in dorsal view overlapping left valve, length >0.65 mm; Md-palp: penultimate segment with seta *z* long, seta *y* shorter, setae *a* and *b* present (Fig. 20.2 I) .... *Darwinula stevensoni*(Brady & Robertson, 1870)
[Cosmopolitan. Neotropical: West Indies, Nicaragua, Brazil]

4′ Left valve overlapping right valve, length < 0.65 mm; Md-palp: penultimate segment with seta *z* short, seta *y* absent, seta *a* absent (Fig. 20.2 J) .... *Alicenula*
[Cosmopolitan. Neotropical: Brazil]

Ostracoda: Podocopida: Cytheroidea: Families

Cytheroids are mostly small; the length of the carapace is 1 mm or less. Valves can be either smooth or ornamented, with reticulation, nodes, tubercles, and spines all being possible. Marginal valve structures mostly consist of fused zone and well-developed calcified inner lamellae; inner lists and selvages can be either present or absent. Hinge is mostly well developed, rarely adont. Valve overlap is variable. AMS invariably consists of four isolated scars in a vertical row.

A2 lacks natatory setae, the exopodite of A2 is a long, hollow seta connected to a gland at the base of the A2. Respiratory plate on Md is small. T1−T3 are generally all walking limbs (except for the parasitic Entocytheridae). Male reproductive organs (hemipenes) have the sperm pump incorporated in the main body (i.e., there is no separate Zenker organ as in Cypridoidea—see below); the copulatory complex is fully external, i.e., not enveloped by a peniferal sheath. UR absent or strongly reduced.

Most species occur in larger epigean waterbodies, but there are also some interstitial taxa, while at least some limnocytherids can produce drought-resistant stages and can occur in temporary pools. One genus (*Elpidium*) only occurs phytothelmata (mostly bromeliads), while another genus (*Intrepidocythere*) is the only known fully terrestrial limnocytherid.

1 Cp well calcified, with or without external ornamentation; T1—T3 mostly with long terminal claws (Fig. 20.3 B) (always free living) 2

1′ Cp weakly calcified, externally without ornamentation; T1—T3 with terminal claws short, hamulate (Fig. 20.3 A) (only commensal or ectoparasitic on Amphipoda, Malacostraca) .... Entocytheridae
[Cuba. Mostly Nearctic and western Palaearctic]

2(1) T1 and T2 basal segment posterior seta stout and annulated (Fig. 20.3 C) (mostly in slightly saline/brackish water) .... Cytherideidae, one genus: *Cyprideis*
[Cosmopolitan]

2′ T1 and T2 basal segment posterior setae slender or absent, never annulated (Fig. 20.3 B) .... **Limnocytheridae [p. 637]**

Ostracoda: Podocopida: Cytheroidea: Limnocytheridae: Genera

Cohuo-Durán et al. (2014) report the genus *Limnocytherina* from Central America. However, it is uncertain that these species indeed belong to this genus. Thus, *Limnocytherina* is for the time being not included.

1 Female carapace with externally visible (especially in dorsal and ventral views) posterior brood pouch (Fig. 20.3 D, E); males without such swollen posterior and often smaller than females; Timiriaseviinae 2

1′ Female carapace without posterior brood pouch; males mostly distinctly longer and larger than females (to accommodate very large hemipenes—see female in Fig. 20.3 F, male in Fig. 20.3 G); Limnocytherinae 4

2(1) External valve surface smooth, T3 a normal walking limb (Fig. 20.4 B) 3

2′ External valve surface set with pits and colored patches, T3 transformed in a prehensile limb, especially in females (Figs. 20.4 A, 17 C, D); very common in lakes and ponds .... *Cytheridella*
[Neotropical]

3(2) Carapace mostly rounded in dorsal view, with width more than half the length (Figs. 20.3 E, 4 E); male A2 terminal segment with one serrated claw (Fig. 20.4 D); right valve internal view in Fig. 20.4 C; in phytotelmata .... *Elpidium*
[Neotropical. Sometimes found in exported bromeliads in other parts of the world]

3′ Carapace narrower in dorsal view, width less than half the length (Fig. 20.4 E); male A2 terminal segment with three serrated claws (Fig. 20.4 F); fully terrestrial, in leaf litter .... *Intrepidocythere*
[Brazil]

4(1) Hemipenes with very large UR (Fig. 20.4 G, H) 5

4′ UR on hemipenes smaller, mostly reduced to a few setae (Fig. 20.4 I) .... *Limnocythere*
[Cosmopolitan. Neotropical: Andean lakes, southern Brazil]

5(4) Valves with large lateral spines or with numerous small spines (Fig. 20.4 J, K); UR on hemipenes large and membranous (Fig. 20.4 G) .... *Neolimnocythere*
[Peru]

5′ Valves without spines (Fig. 20.4 L); UR on hemipenes large and chitinized (Fig. 20.4 H) .... *Paracythereis*
[Peru]

Ostracoda: Podocopida: Cypridoidea: Families

Cypridoidea contains both very small (*ca.* 0.4 mm—*Danielocandona*) and very large (7-8 mm—*Megalocypris*) species. Valves can externally be smooth or be set with a variety of ornamentations; valve shape extremely variable (narrow and elongated, round and broad) and can be symmetrical or widely asymmetrical (e.g., with dorsal humps, or with position of valves skewed in frontal view). Valves can marginally be very simple, or have complex structures involving a set of selvages, inner lists, etc. (e.g., *Chlamydotheca*). Basic AMS shows a "pawprint" pattern, with a variable number of scars (but mostly six) in a variable pattern.

The A2 exopodite is mostly reduced to a small plate, bearing two to three small setae; many species with robust natatory setae (5 + 1 in adults) on A2. T1—T3 with different morphologies and functions. T1 with strong sexual dimorphism, forming a small plate with a weak one to two segmented palp, the latter developed into string (mostly asymmetrical) clasper organs (also one to two segmented) in males. T2 a walking limb. T3 a cleaning limb. UR well developed in some groups (strong ramus, two claws, two setae) and reduced in some groups. Males with a well-developed sperm pump (Zenker's organ) outside of the actual hemipenes; the latter with copulatory organs internally and enveloped by peniferum in nonerect state. Females never with an externally visible and inflated brood pouch.

Three-quarters of nonmarine ostracods belong to this superfamily. Cypridoideans occur in a large variety of aquatic and (semi-)terrestrial habitats. The majority of Cyprididae can occur in temporary habitats, many candonids are interstitial while notodromatids are generally hyponeustic (living upside down, attached to the water surface).

1 Cp of variable shape, smooth or ornamented; eye cups fused or absent; Mx1 3rd masticatory process with two to three stout claws and various setae (Fig. 20.5 C) 2

1′ Cp mostly subglobular, set with pustules and ventral ridges, eye-cups dorsally not fused (Fig. 20.5 A—slightly exaggerated); Mx1 3rd masticator process with four to six stout claws and various setae (Fig. 20.5 B) .... Notodromadidae
[Cosmopolitan Note: *Argentodromas* (Argentina) and *Newnhamia* (Argentina, Uruguay]

2(1) Cp of variable shape, smooth or ornamented, without mediodorsal sulcus; T1 palp in females (mostly) unarticulate (Fig. 20.5 G) 3

2′ Cp subquadrate, with mediodorsal sulcus (Fig. 20.5 D, E); dorsal margin straight over *c* half its length, valve surface mostly heavily ornamented (reticulated, set with pits, spines); female T1 palp (mostly) consisting of two segments (Fig. 20.5 F) .... Ilyocyprididae, one genus: *Ilyocypris*
[Cosmopolitan]

3(2) T3 without a terminal pincer, i.e., terminal and penultimate segments clearly separate (Fig. 20.5 H) .... **Candonidae [p. 638]**
[Cosmopolitan]

3′ T3 with a terminal pincer, i.e., terminal and penultimate segments partly fused (Fig. 20.5 I, J) (exceptions: *Neocypridopsis, Callistocypris*—see below) .... **Cyprididae [p. 639]**
[Cosmopolitan]

Ostracoda: Podocopida: Cypridoidea: Candonidae: Genera

We follow the classical division of cyclocypridinid genera based on the valve features, where species with marginal tubercles are united in *Physocypria*, those without the marginal tubercles in *Cypria*.

Callistocypridinae are at present still classified in the Cyprididae. However, the T3 has no terminal pincer (Fig. 20.8 L), so following the above keys the genus will also be sorted in the Candonidae. The exact position of this subfamily and genus needs to be investigated.

1 Cp short, dorsal margin in lateral view rounded (Fig. 20.6 A); A2 with (long) natatory setae (Fig. 20.6 B); male prehensile palps with two segments (Fig. 20.6 C); Cyclocypridinae 2

1′ Cp with variable shapes, but mostly subquadrate or elongate, rarely (never?) with dorsal margin rounded (Fig. 20.6 D); A2 without swimming setae (Fig. 20.6 E); male prehensile palps with one segment (Fig. 20.6 F); Candoninae 3

2(1) Right valve with marginal tubercles (Fig. 20.6 G) .... *Physocypria*
[Cosmopolitan. Neotropical: Brazil]

2′ Right valve without marginal tubercles (Fig. 20.6 H) .... *Cypria*
[Cosmopolitan. Neotropical: Colombia]

3(1) CR with large ramus, two claws, and one or two setae (Fig. 20.7 A, B) 4

3′ CR with less claws or setae and with ramus possibly reduced (Fig. 20.7 C, D) 8

4(3) Length 0.3–0.4 mm; CR short; ramus and claws robust; terrestrial (Fig. 20.7 A) 5

4′ Length >0.5 mm; CR ramus and claws elongate; aquatic (Fig. 20.7 B) 6

5(4) T3 terminal segment with one claw and two setae (Figs. 20.7 E, 8 L); T2 without seta d1 (Fig. 20.7 F); CR proximal seta distally positioned (Fig. 20.7 G) .... *Caaporacandona*
[Brazil]

5′ T3 terminal segment with one claw only (Fig. 20.7 H); T2 seta d1 present (Fig. 20.7 I); CR proximal seta centrally positioned (Fig. 20.7 A) .... *Terrestricypris*
[Brazil]

6(4) Cp not covered in stiff setae (Fig. 20.7 L); T3 basal segment with two setae (Fig. 20.7 M) 7

6′ Cp usually covered in long and stiff setae (Fig. 20.7 J); T3 basal segment with three setae (Fig. 20.7 K) .... *Pseudocandona s.l.*
[West Indies, Venezuela, Brazil]

7(6) Cp in subovate in dorsal view, W/L ratio > 0.4 .... *Candona s.s.*
[West Indies, Brazil]

7′ Cp laterally compressed in dorsal view, W/L < 0.4 .... *Fabaeformiscandona*
[Holarctic. Neotropical: invasive?]

8(3) CR with large ramus, two claws, and one distal seta, proximal (ventral) seta missing (Fig. 20.8 A) 9

8' CR with small ramus and only one claw (Fig. 20.7 C, D) 11

9(8) Md-palp with terminal segment length less than twice the basal width (Fig. 20.8 C, setae and claws not drawn) 10

9' Md-palp with terminal segment length at least three times the basal width) (Fig. 20.8 B, setae and claws not drawn) .... *Candonopsis*
[Cosmopolitan]

10(9) Both valves with broad anterior calcified inner lamellae, inner margin sinuous (Figs. 20.8 D, 17 E); hemipenes with protruding lobes "a" only (Fig. 20.8 E) .... *Candobrasilopsis*
[Brazil]

10′ Both valves with narrow calcified inner lamella, inner margin evenly rounded (Fig. 20.8 F); hemipenes with protruding lobes "a" and "b" (Fig. 20.8 G); surface water .... *Latinopsis*
[Brazil]

11(8) CR with elongated ramus and a single end claw, both distinct and separated from each other (Fig. 20.7 C) 12

11′ CR flagellate (but sclerified), with ramus and end claw fused and without additional setae (Fig. 20.7 D); length = *ca.* 0.4 mm (Fig. 20.8 H); subterranean .... *Danielocandona*
[Venezuela]

12(11) All appendages quite elongate and slender; CR with elongated ramus, a single end claw and one to two small setae (Fig. 20.7 C); right valve dorsally sometimes projecting beyond left valve (Fig. 20.8 I, J); length = 0.6–0.8 mm; subterranean 13

12′ All appendages stout, with short and compact claws and segments; CR with elongated, narrow ramus and one distinct end claw, without additional setae (Fig. 20.8 K); left valve overlapping right valve on all sides; length = 0.4–0.5 mm; terrestrial; Callistocypridinae .... *Callistocypris*
[Brazil]

13(12) A1 with six segments (segments 3 and 4 fused); A2 in male with four endopodal segments (Fig. 20.8 I, J) .... *Caribecandona*
[West Indies, probably endemic to Haiti]

13′ A1 with seven segments; A2 in male with three endopodal segments (segments 2 and 3 fused) .... *Cubacandona*
[West Indies, probably endemic to Cuba]

Ostracoda: Podocopida: Cypridoidea: Cyprididae: Subfamilies

There are more than 20 subfamilies in the Cyprididae, but only 10 have thus far been reported from the Neotropical region.

1 CR with ramus, two terminal claws, and with two to three setae (Fig. 20.9 B); length <0.5–5 mm 2

1′ CR in females reduced to a flagellum, usually with an additional basal seta (Fig. 20.9 A), absent in males (incorporated in hemipenis); length mostly < 1 mm .... **Cypridopsinae [p. 640]**

2(1) CR attachment without apical TL (Fig. 20.9 D) 3

2′ CR attachment with apical TL (Fig. 20.9 C) .... **Cypricercinae [p. 640]**

3(2) T2 endopod first segment with two lateral setae (Fig. 20.9 E); Cp in lateral view 3–5 mm, either rounded (Fig. 20.9 G) or elongate (Fig. 20.9 H) 4

3′ T2 endopod first segment with one lateral seta (Fig. 20.9 F); Cp generally smaller and of different shape 5

4(3) Both valves with anterior selvage inwardly displaced; either relatively high and rounded, with anterioventral beak (Fig. 20.9 G) or elongated without beak and possibly with posterioventral spine (Figs. 20.9 H, 17 F) .... Cypridinae; one genus: *Chlamydotheca*
[Guatemala, Venezuela, Columbia, Ecuador, Brazil, Peru, Bolivia, Chile, Paraguay, Argentina, Uruguay]

4′ Both valves without inwardly displaced selvages or inner lists; Cp elongated and laterally flattened (Fig. 20.9 I, J) .... **Eucypridinae (partim) [p. 641]**

5(3) CR attachment without basal sclerotized reinforcements (Fig. 20.9 M) 6

5′ CR attachment at the basis with sclerotized reinforcement, either robust triangular structure (indicated in Fig. 20.9 L), or reduced to a trabecula (Fig. 20.9 K, arrow) .... **Herpetocypridinae [p. 641]**

6(5) Left valve without long anterior and posterior spines 7

6′ Carapace elongate, left valve with very large anterior and posterior spines (Fig. 20.9 N, O) .... Rudjakoviellinae, one species: *Rudjakoviella prolongata* (Triebel, 1962)
[Venezuela]

7(6) T1 without "c"-seta, only "d" seta present 8

7′ T1 with "c" and "d" setae (Fig. 20.10 A) .... **Eucypridinae partim [p. 641]**

8(7) At least one valve without marginal tubercles 9

8′ Valves subrectangular, laterally compressed, anterior and posterior margins set with large, spinal tubercles (Fig. 20.10 B–D) .... Pelocypridinae, one genus: *Pelocypris*
[Brazil]

9(8) Valves globular, laterally flattened or elongate, without marginal septae 10

9’ Carapace globular (Fig. 20.10 E), valves at least anteriorly set with a row of marginal septae (Figs. 20.10 F, 17 G, H) .... Cyprettinae; one genus: *Cypretta*
[Circumtropical to cosmopolitan.]

10(9) Cp elongate, selvage largely inwardly displaced in one valve only or neither valve; T2 with penultimate segment divided (Fig. 20.10 J) 11

10′ Cp globular, frontal selvage largely inwardly displaced in both valves (Fig. 20.10 G, H), T2 with penultimate segment undivided (Fig. 20.9 F) .... Cypridinae s.s.; one genus: *Cypris*
[Cosmopolitan. Neotropical: West Indies]

11(10) Cp length *ca.* 3x height, selvage in right valve submarginal, in left valve largely inwardly displaced, both along anterior and posterior margins (Fig. 20.10 K–M); in need of revision. .... Dolerocypridinae; one genus: *Dolerocypris*
[Holarctic. Neotropical: Guatemala, Chile, Paraguay, Argentina]

11′ Cp length <2× height; posterior selvage in left valve absent or not so strikingly inwardly displaced (Fig. 20.16 B, E, H) .... **Cyprinotinae [p. 641]**

Ostracoda: Podocopida: Cypridoidea: Cyprididae: Cypridopsinae: Genera

Cuminsky et al. (2005) reported *Kapcypridopsis megapodus* from Argentina. However, the occurrence of this South African genus in South America is highly unlikely.

1 A2 of female with claw G2 large, stout, and serrated (Fig. 20.11 A); Zonocypridini 2

1′ Claw G2 of female A2 of normal shape and size (Fig. 20.11 B) 3

2(1) Cp rounded and with pronounced external valve ornamentation (spines, ridged) but never overly hirsute (Fig. 20.11 C, D) .... *Zonocypris*
[Not yet recorded in Neotropics with certainty]

2′ Cp smooth, but highly hirsute (Figs. 20.11 E, 17 R) .... *Cabelodopsis hispida* (Sars, 1901)
[Brazil]

3(1) Terminal segment of Mx1-palp elongated (Fig. 20.11 F); Cypridopsini 4

3′ Terminal segment of Mx1-palp spatulate (Fig. 20.11 G, H, I); Potamocypridini .... *Potamocypris*
[Cosmopolitan]

4(3) T3 with an apical pincer, i.e., segments 3 and 4 largely fused (Fig. 20.12 D, E) 5

4′ T3 with fourth segment separate from third and bearing three setae (Fig. 20.12 A–C) .... *Neocypridopsis*
[Neotropical]

5(4) Cp elongate, in dorsal view width ≤0.5× length; right valve overlapping left valve (Fig. 20.12 I, J) 6

5′ Cp rounded, in dorsal view width >0.5× length; left valve overlapping right valve (Fig. 20.12 F, G) .... *Cypridopsis*
[Cosmopolitan]

6(5) CR with stem short and triangular (Fig. 20.12 H–J) .... *Sarscypridopsis aculeata* (Costa, 1847)
[Cosmopolitan. Neotropical: Chile, Uruguay]

6′ CR with stem elongated and with parallel sides (Fig. 20.12 K) .... *Plesiocypridopsis*
[Not yet recorded in Neotropics with certainty]

Ostracoda: Podocopida: Cypridoidea: Cyprididae: Cypricercinae: Genera

1 TL triangular, in middle or distal part of CR attachment or in dorsal branch (Fig. 20.13 B–D); Cp rounded or elongate; external valves smooth or set with ornamentations 2

1′ TL an oval loop in dorsal branch of CR attachment (Fig. 20.13 A), Cp rounded in dorsal view (Fig. 20.17 O) .... ***Bradleytriebella* [p. 640]**

2(1) TL ventral branch short and stubby (Fig. 20.13 B, C); Cp in dorsal view flattened, width <0.3x length; Cp without caudal or dorsal protuberances 3

2′ TL ventral branch robust (Fig. 20.13 D); Cp width ≥0.3x length, both caudal and dorsal protuberances possible 4

3(2) TL divided into two to three compartments (Fig. 20.13 B); valves elongate and laterally flattened (Figs. 20.13 E, 17 I) .... *Diaphanocypris*
[West Indies, Ecuador, Brazil]

3′ TL a single cell (Fig. 20.13 C) .... *Nealecypris clavigera* (G.W. Müller, 1898)
[Caribbean]

4(2) Cp in lateral view elliptical, subovate, or elongate (Fig. 20.13 F, G); one species (*S. bicuspis*, Fig. 20.17 J) with dorsal large, pointed helmet (Fig. 20.13 H, I); never with anterior spine, two species with small posterior spine (S. psittacea and S. colombiensis)' (Fig. 20.17 K–M) *Strandesia*
[Cosmopolitan]

4′ Cp in lateral view elongate, often with posterior spines (Figs. 20.13 J, K, 17 P) .... *Cypricercus*
[Neotropical]

Ostracoda: Podocopida: Cypridoidea: Cyprididae: Cypricercinae: *Bradleytriebella* Species

1 CP externally with fine, longitudinal ridges; no marginal spines or dorsal helmet ... *Bradleytriebella lineata* (Victor & Fernando, 1981)
[Brazil]

1′ Cp externally smooth, with marginal three spines (two short anterior on LV, one long posterior on right valve) and a dorsal helmet on right valve .... *Bradleytriebella trispinosa* (Pinto & Purper, 1965)
[Brazil]

Ostracoda: Podocopida: Cypridoidea: Cyprididae: Herpetocypridinae: Genera

1 Anterior margins of both valves set with marginal septae (Fig. 20.14 A, B) 2

1′ Anterior margins of both valves without marginal septae 3

2(1) Cp in lateral view rounded, anterior marginal septae small, anterior inner calcified lamella wide, posterior calcified inner lamella absent in both valves (Fig. 20.14 A); CR symmetrical .... *Isocypris beauchampi* (Paris, 1920)
[Cosmopolitan. Neotropical: Brazil, Chile]

2′ Cp in lateral view elongate, anterior marginal septae large, posterior calcified inner lamella narrow but present in both valves (Fig. 20.14 B); CR asymmetrical, at least one with large spines on ramus and claws (Fig. 20.14 C) .... *Stenocypris*
[Circumtropical. Neotropical: West Indies, Brazil]

3(1) Anterior overlap small, inverse or absent 4

3′ Anterior overlap of right valve by left valve very large (Fig. 20.17 Q) .... *Paranacypris*
[Brazil]

4(3) CR with two claws and two setae, proximal seta is a seta (Fig. 20.14 D) .... *Herpetocypris*
[Cosmopolitan]

4′ CR with three claws and one seta, proximal seta is a claw (Fig. 20.14 E) .... *Ilyodromus*
[Australasia. Afrotropical. Palaearctic. Neotropical: Paraguay, Chile]

Ostracoda: Podocopida: Cypridoidea: Cyprididae: Eucypridinae: Genera

1 Species max 2 mm, valves less compressed; T2 first endopodal segment with one apical seta (Fig. 20.9 F) 2

1′ Cp large (2.5–4 mm), elongated, and laterally compressed (Fig. 20.9 I, J); T2 first endopodal segment with two apical setae (Fig. 20.9 E) .... *Amphicypris/Cypriconcha*
[Holarctic. Neotropical: Paraguay, Argentina. South Georgia]

2(1) Both valves without selvages (Fig. 20.15 A, B) 3

2′ Right valve with inwardly displaced anterior selvage (Fig. 20.15 C) 4

3(2) Valves short and high, anteriorly often set with *Porenwarzen*, left valve without internal tooth .... *Eucypris*
[Cosmopolitan. Not yet recorded in Neotropics]

3′ Valves elongated, without *Porenwarzen*, left valve ventrally with an internal tooth .... *Tonnacypris*
[Holarctic, with invasions in the southern hemisphere]

4(2) Left valve with pronounced inner list (Fig. 20.15 D) .... *Trajancypris*
[Holarctic, with invasions in the southern hemisphere]

4′ Left valve without pronounced inner list (Fig. 20.15 E) .... *Argentocypris*
[Argentina]

Ostracoda: Podocopida: Cypridoidea: Cyprididae: Cyprinotinae: Genera

The ubiquitous genus *Heterocypris* is in need of revision and probably composed of several genera.

1 At least one valve with marginal tubercles, one valve always overlapping another anteriorly and posteriorly, anteriorly in dorsal view without pointed rostrum (Fig. 20.16 D) 2

1′ Both valves without marginal tubercles, no valve overlapping the other anteriorly or posteriorly; in dorsal view anteriorly with a pointed rostrum (Fig. 20.16 A–C) .... *Riocypris*
[Uruguay]

2(1) Left valve overlapping right valve, right valve with marginal tubercles 3

2′ Right valve overlapping LV, left valve with marginal tubercles (Fig. 20.16 D–F) .... *Hemicypris*
[Tropical and subtropical. Neotropical: West Indies, Brazil, Paraguay]

3(2) Right valve without large dorsal expansion (Fig. 20.16 G–I) .... *Heterocypris*
[Cosmopolitan]

3′ Cp in left lateral view right valve with large dorsal expansion, overarching left valve (arrowed in Fig. 20.11 J) .... *Cyprinotus*
[Tropical and subtropical]

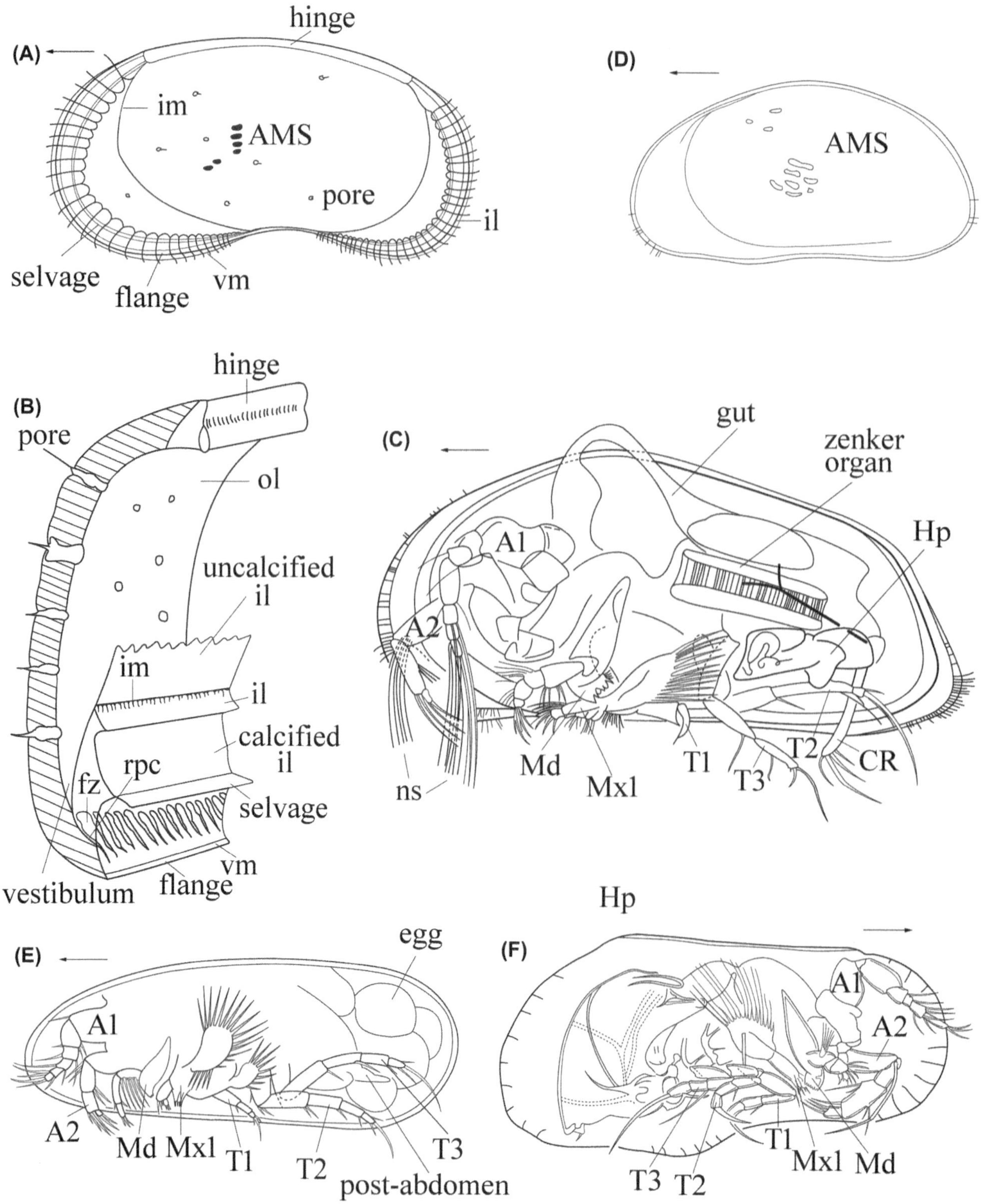

FIGURE 20.1 Podocopid superfamilies, general features: (A) Cytheroidea, hypothetical ostracod, right valve internal view showing various anatomical details, including AMS consisting of four scars in a vertical row; (B) Cytheroidea, same valve as above, but D-V cut and slightly rotated, showing anatomical details; (C) Cypridoidea, male, Cp left lateral view, left valve removed and only one of each pair of appendages shown, T1 a palp, T2 a walking limb, T3 a cleaning limb; (D) Cypridoidea, right valve internal view showing pawprintlike AMS; (E) Darwinuloidea, female, Cp left lateral view, left valve removed and only one of each pair of appendages shown, T1 a palp, T2–T3 walking limbs; (F) Cytheroidea, male, Cp right lateral view, right valve removed and only one of each pair of appendages shown, T1–T3 walking limbs. Abbreviations: *A1*, antennule; *A2*, antenna; *AMS*, adductor muscle scars; *CR*, caudal ramus; *fz*, fused zone; *Hp*, hemipenis; *il*, inner lamella; *im*, inner margin; *Md*, mandibulae; *Mx1*, maxillula; *ns*, natatory setae; *ol*, outer lamella; *rpc*, radial pore canal; *T1*, first thoracopod; *T2*, second thoracopod; *T3*, third thoracopod; *vm*, valve margin.

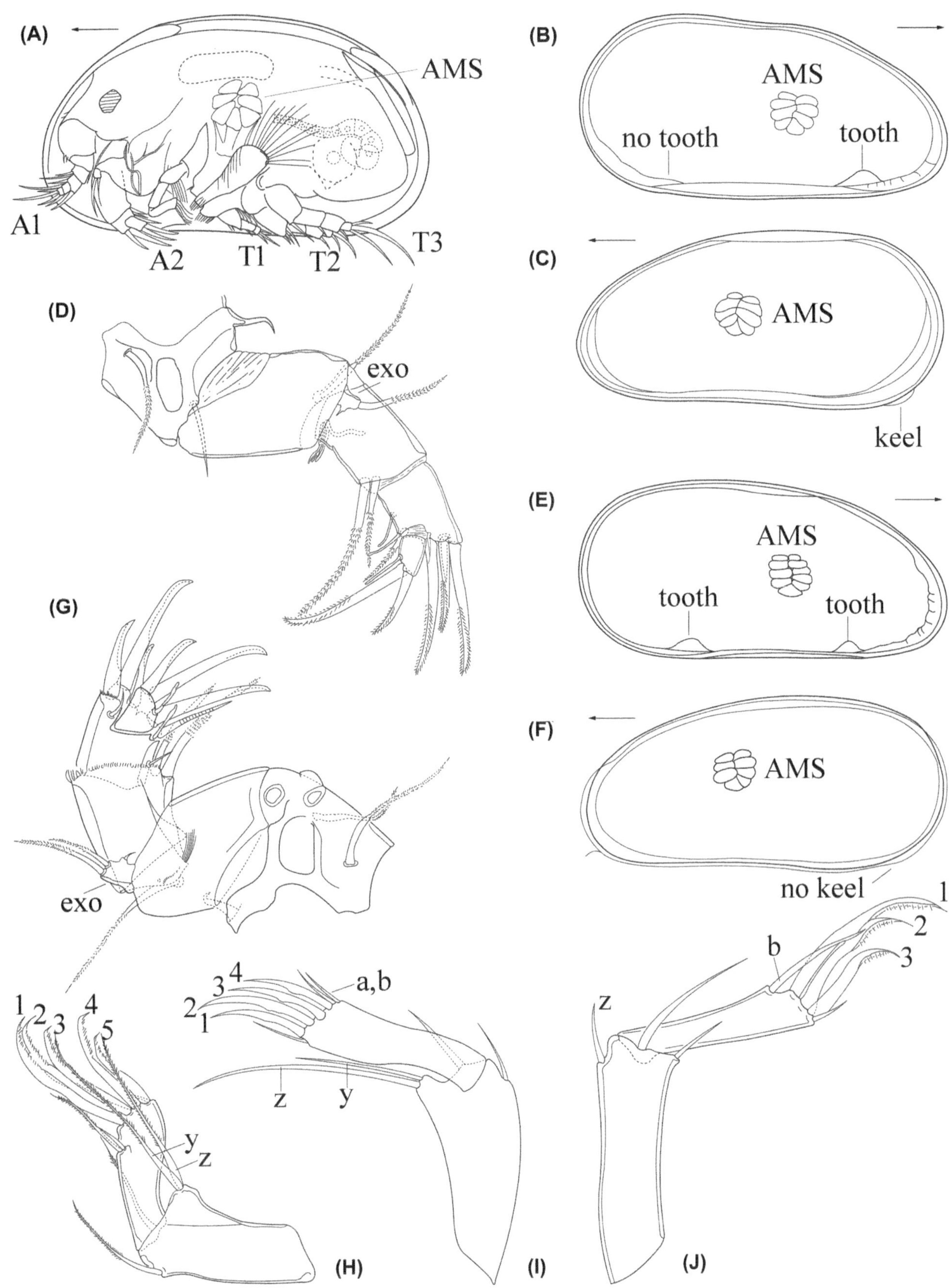

FIGURE 20.2 Darwinuloidea, Darwinulidae: (A) *Microdarwinula* sp., female, Cp left lateral view with left valve removed and only one of each pairs of appendages shown; (B) *Vestalenula* sp., left valve internal view showing rosettelike AMS and absence of internal posteroventral tooth; (C) *Vestalenula* sp., right valve internal view showing presence of posteroventral, external keel, and rosettelike AMS; (D) *Vestalenula* sp., A2 showing exopodite with one seta and one spine; (E) *Penthesilenula* sp., left valve internal view showing presence of internal posteroventral tooth and rosettelike AMS; (F) *Penthesilenula* sp., right valve internal view showing rosettelike AMS and absence of external posteroventral keel; (G) *Penthesilenula* sp., A2 showing exopodite with two setae and one spine; (H) *Penthesilenula*, Md-palp with five apical claws, and setae *y* and *z* both long; (I) *Darwinula*, Md-palp with four apical claws, and seta *y* much shorter than seta *z*; setae *a* and *b* both present; (J) *Alicenula*, Md-palp with three apical claws, seta *z* small and seta *y* absent; seta *a* absent. Abbreviations: *A1*, antennule; *A2*, antenna; *AMS*, adductor muscle scars; *exo*, exopodite; *T1*, first thoracopod; *T2*, second thoracopod; *T3*, third thoracopod.

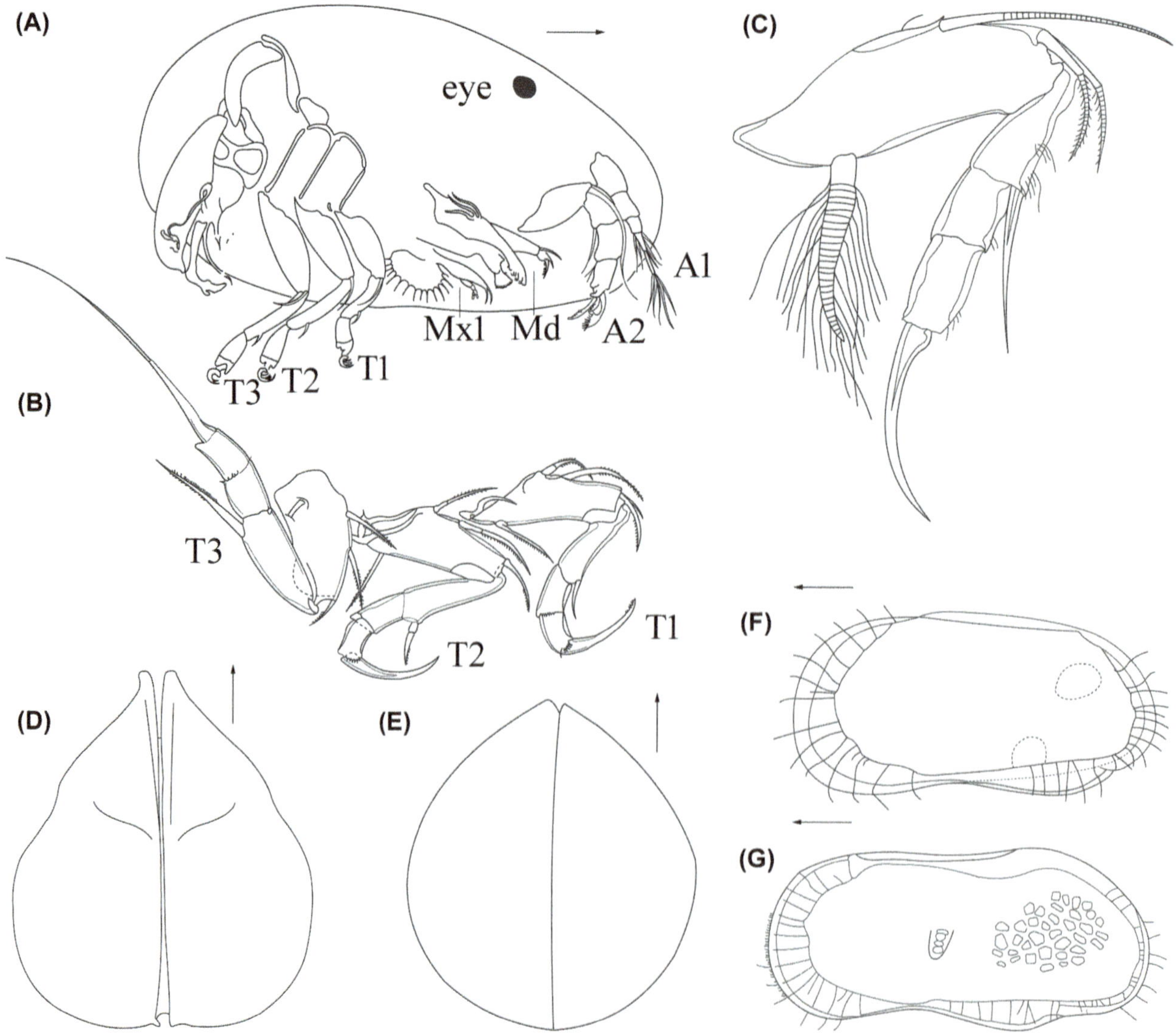

FIGURE 20.3 Cytheroidea: (A) Entocytheridae, female, Cp right lateral view with right valve removed and only one of each pairs of appendages shown, T1–T3 with terminal hooklike claws; (B) Limnocytheridae, three thoracic appendages all walking legs, terminal claws not hooklike; (C) Cytherideidae, *Cyprideis*, male, T2 with large and annulated ventral seta on first segment; (D) *Cytheridella*, female, Cp dorsal view showing dorsolateral sulcus, and expanded posterior brooding pouch; (E) *Metacypris*, female, Cp dorsal view showing absence of dorsolateral sulcus and different shape of posterior brooding pouch; (F) *Limnocythere*, female, right lateral view; (G) *Limnocythere*, male, right lateral view. Abbreviations: *A1*, antennule; *A2*, antenna; *Md*, mandibulae; *Mx1*, maxillula; *T1*, first thoracopod; *T2*, second thoracopod; *T3*, third thoracopod.

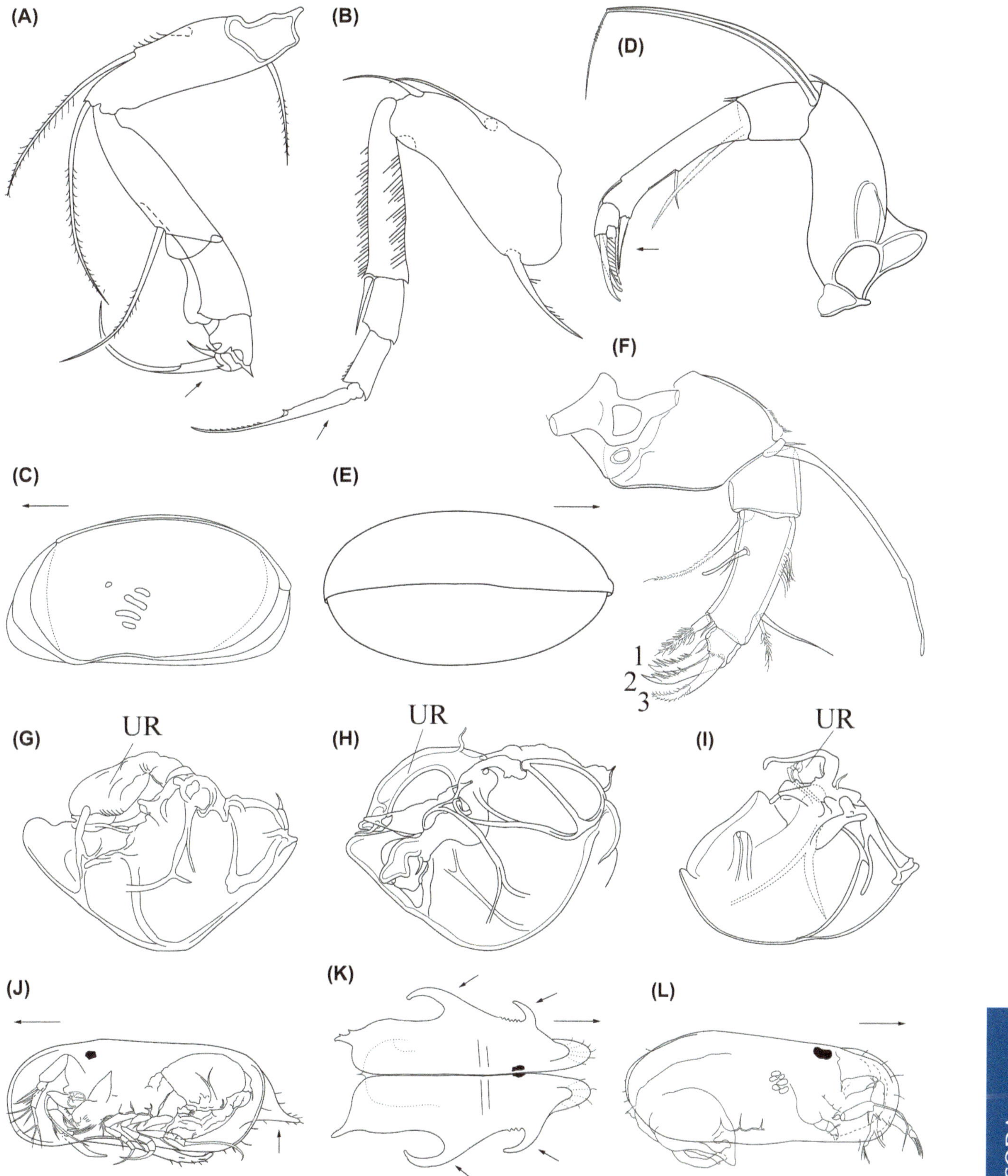

FIGURE 20.4 Cytheroidea: (A) *Cytheridella*, T3 transformed in a cleaning leg; (B) *Elpidium*, T3 a normal walking leg; (C) *Elpidium*, right valve internal view; (D) *Elpidium*, male, A2 with one serrated end claw; (E) *Intrepidocythere*, Cp dorsal view; (F) *Intrepidocythere*, male, A2 with three serrated end claws; (G) *Neolimnocythere*, hemipenis showing very large and membranous upper ramus; (H) *Paracythereis*, hemipenis showing very large and chitinized upper ramus; (I) *Limnocythere*, hemipenis showing "normal," hooklike upper ramus; (J) *Neolimnocythere*, Cp left lateral view with left valve removed and only one of each pairs of appendages shown, note large posterior spine; (K) *Neolimnocythere*, Cp dorsal view showing large lateral spines; (L) *Paracythereis*, Cp right lateral view, note absence of large posterior spine. Abbreviation: *UR*, upper ramus of the cytheroid hemipenis.

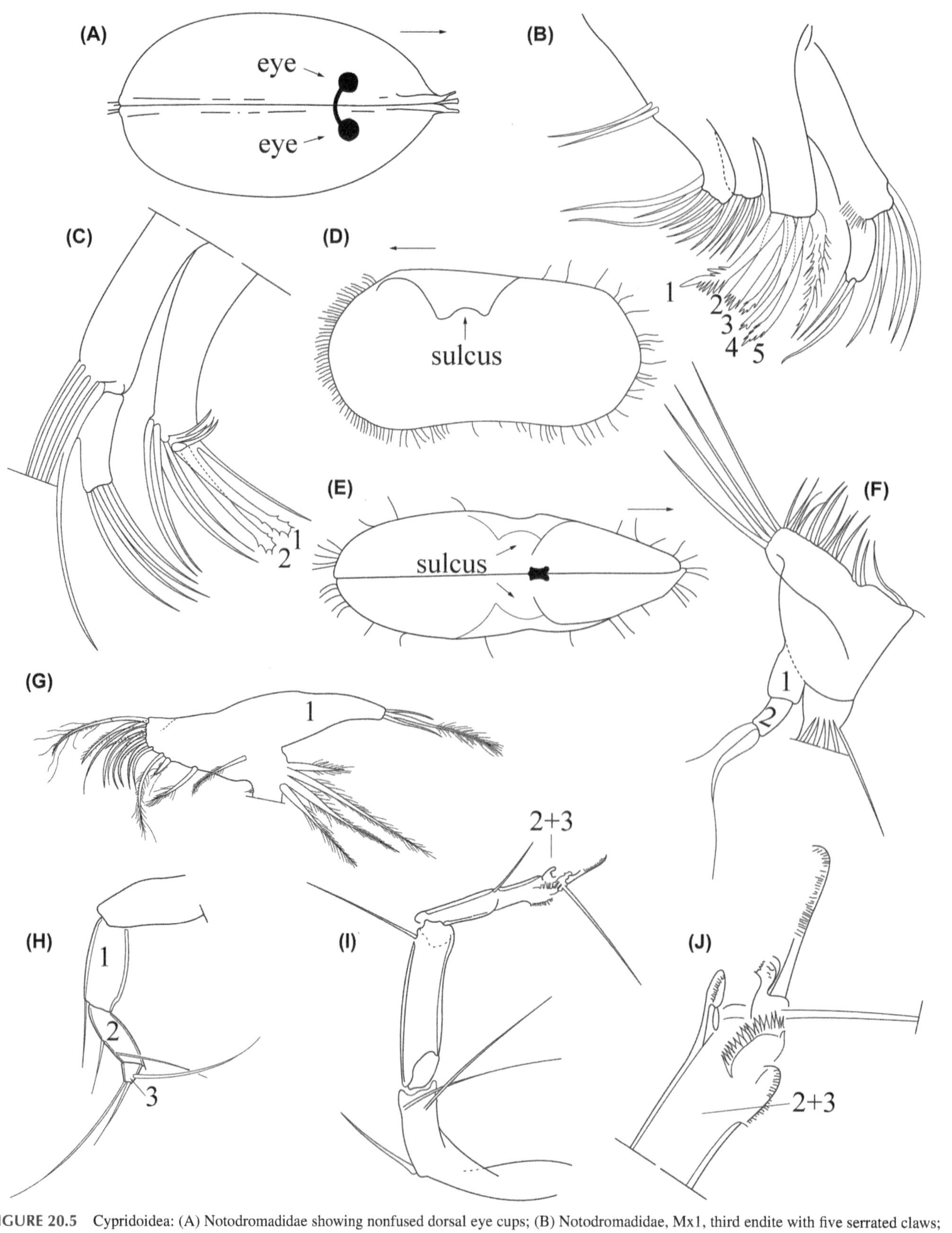

FIGURE 20.5 Cypridoidea: (A) Notodromadidae showing nonfused dorsal eye cups; (B) Notodromadidae, Mx1, third endite with five serrated claws; (C) Cyprididae, Mx1, third endite with two serrated claws; (D) Ilyocyprididae, Cp right lateral view showing laterodorsal sulci; (E) Ilyocyprididae, Cp dorsal view showing laterodorsal sulci; (F) Ilyocyprididae, female, T1 with two-segmented palp; (G) Cyprididae, female, T1 with one-segmented palp; (H) Candonidae, T3 with segment 3 clearly separated from segment 2; (I) Cyprididae, T3 with segments 2 and 3 fused into a pincer; (J) Cyprididae, T3, detail of distal pincer.

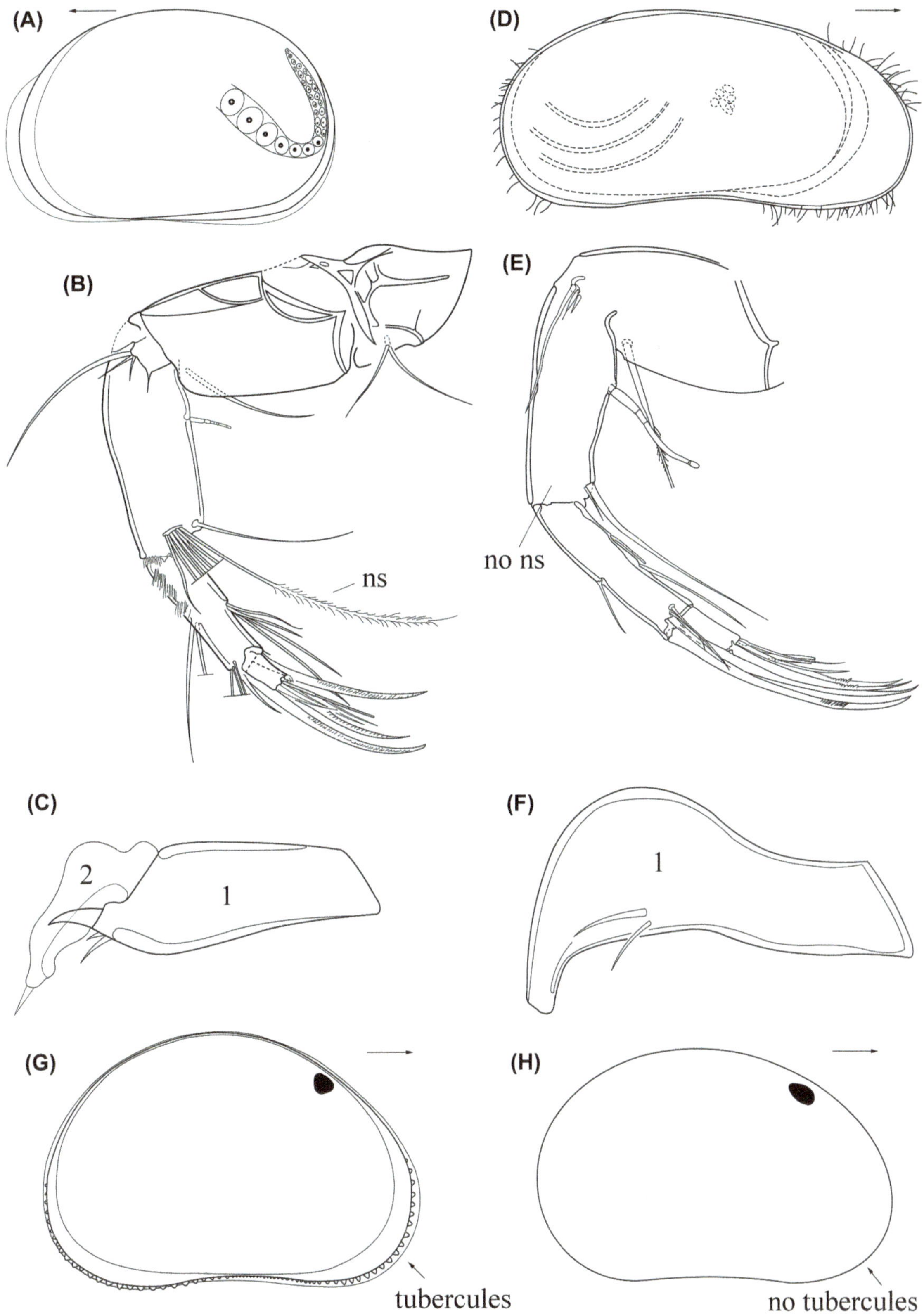

FIGURE 20.6 Candonidae: (A) Cyclocypridinae, Cp in lateral view rounded and highly arched; (B) Cyclocypridinae, A2 with natatory setae; (C) Cyclocypridinae, male, T1 consisting of two segments; (D) Candonidae, Cp in lateral view elongated; (E) Candonidae, A2 without natatory setae; (F) Candonidae, male, T1 consisting of one segment only; (G) *Physocypria*, right valve with marginal tubercles; (H) *Cypria*, right valve without marginal tubercles. Abbreviation: *ns*, natatory setae.

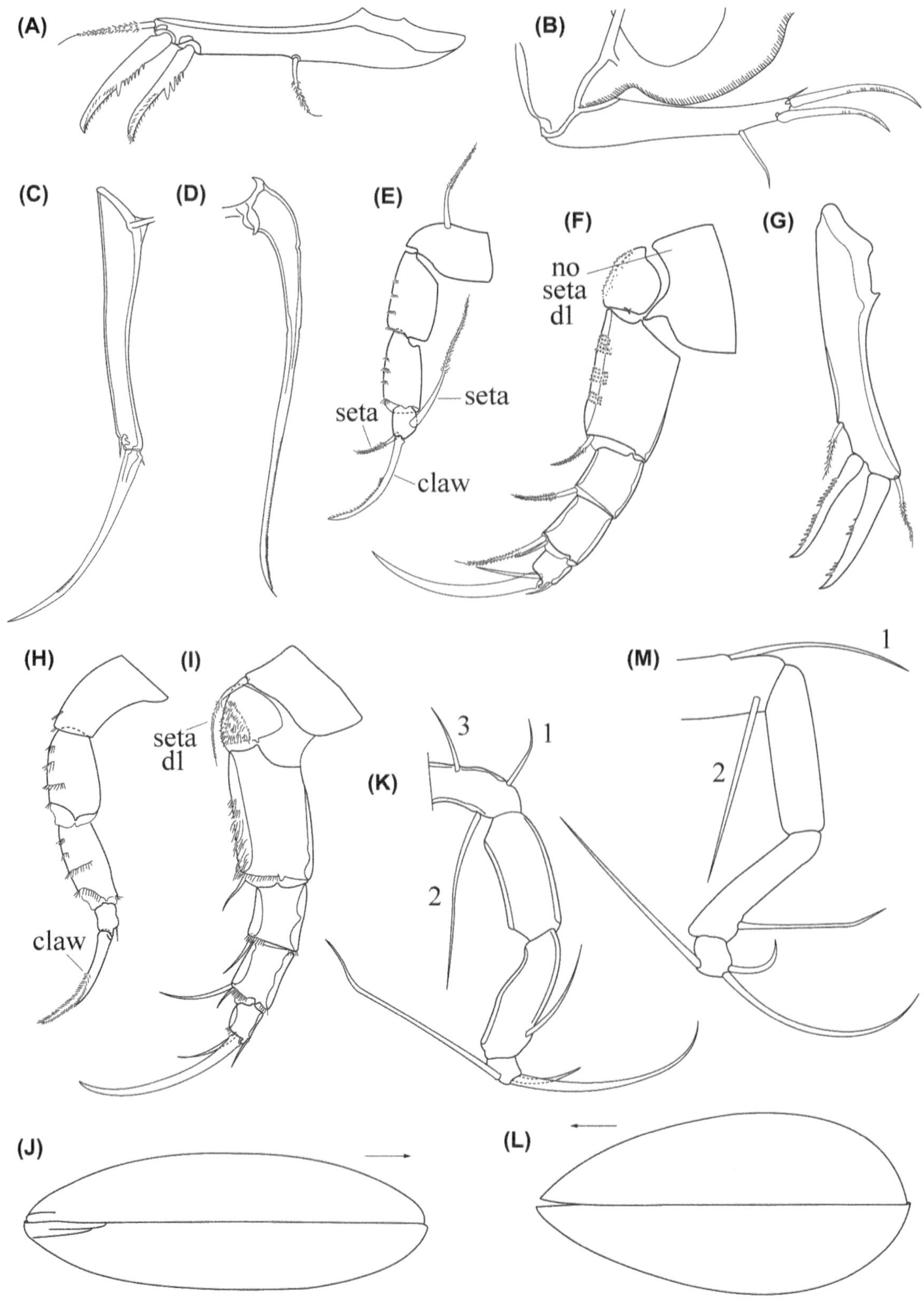

FIGURE 20.7 Candonidae: (A) *Terrestricypris*, caudal ramus; (B) Candonidae, caudal ramus unreduced (two claws, two setae); (C) Candonidae, caudal ramus reduced (one ramus, one claw); (D) Candonidae, caudal ramus reduced (one ramus); (E) *Caaporacandona*, T3 showing one claw and two setae on last segment; (F) *Caaporacandona*, T2 without seta d1; (G) *Caaporacandona*, caudal ramus; (H) *Terrestricypris*, T3 showing a single claw on last segment; (I) *Terrestricypris*, T2 with seta d1; (J) *Pseudocandona*, Cp dorsal view; (K) *Pseudocandona*, T3 showing three setae on first segment; (L) *Candona*, Cp dorsal view; (M) *Candona*, T3 showing two setae on first segment.

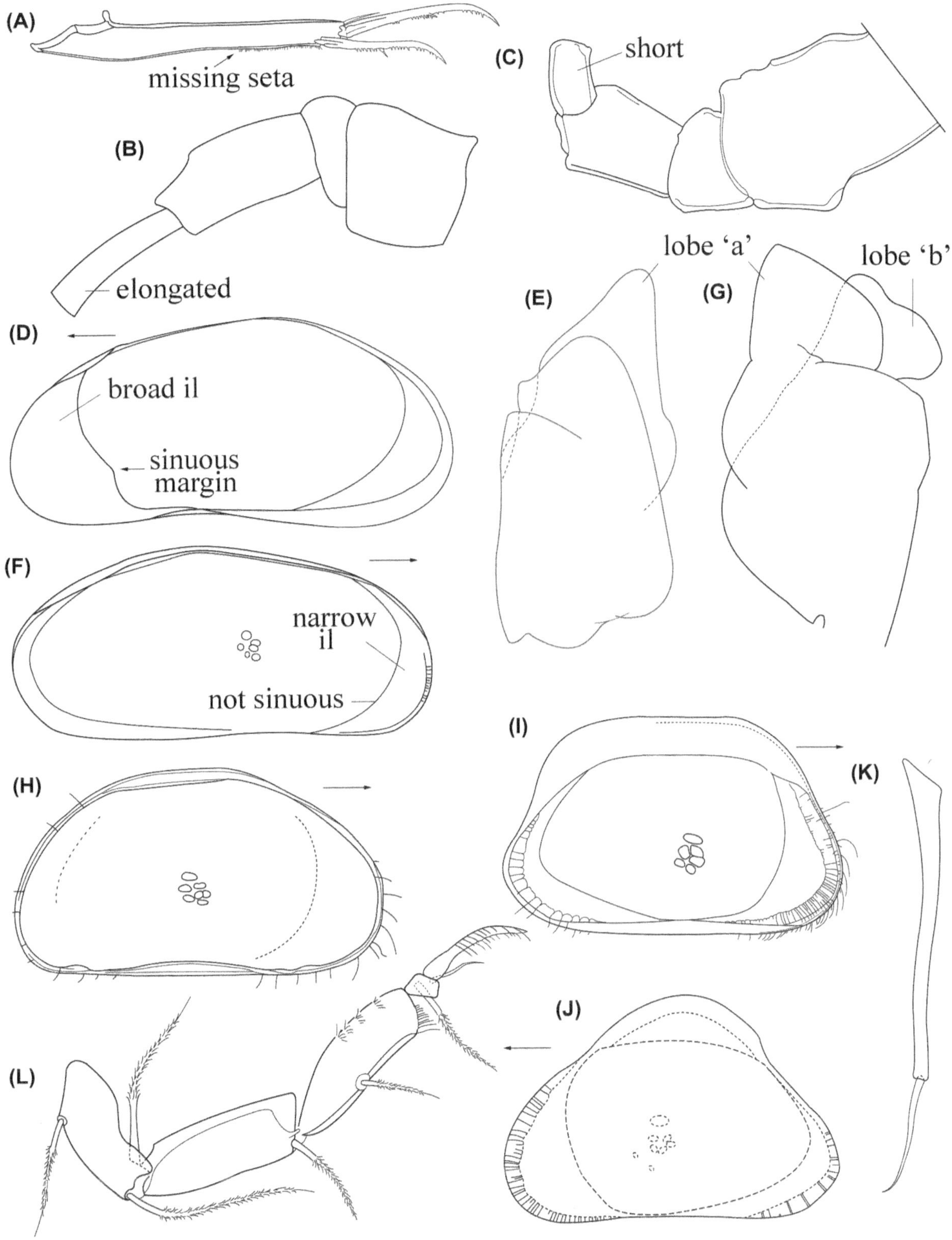

FIGURE 20.8 Candonidae: (A) *Candonopsis*, caudal ramus with proximal seta missing; (B) *Candonopsis*, Md-palp, without setae and claws, showing elongated terminal segment; (C) *Candobrasilopsis*, Md-palp, without setae and claws, showing short terminal segment; (D) *Candobrasilopsis*, right valve internal view showing broad inner lamella and sinuous inner margin; (E) *Candobrasilopsis*, hemipenis outline; (F) *Latinopsis*, right valve outer view showing narrow inner lamella and evenly rounded anterior inner margin; (G) *Latinopsis*, hemipenis outline; (H) *Danielocandona*, Cp right lateral view; (I) *Caribecandona*, Cp right lateral view; (J) *Caribecandona*, Cp left lateral view; (K) *Caaporacandona*, caudal ramus; (L) *Caaporacandona*, T3. Abbreviation: *il*, inner lamella.

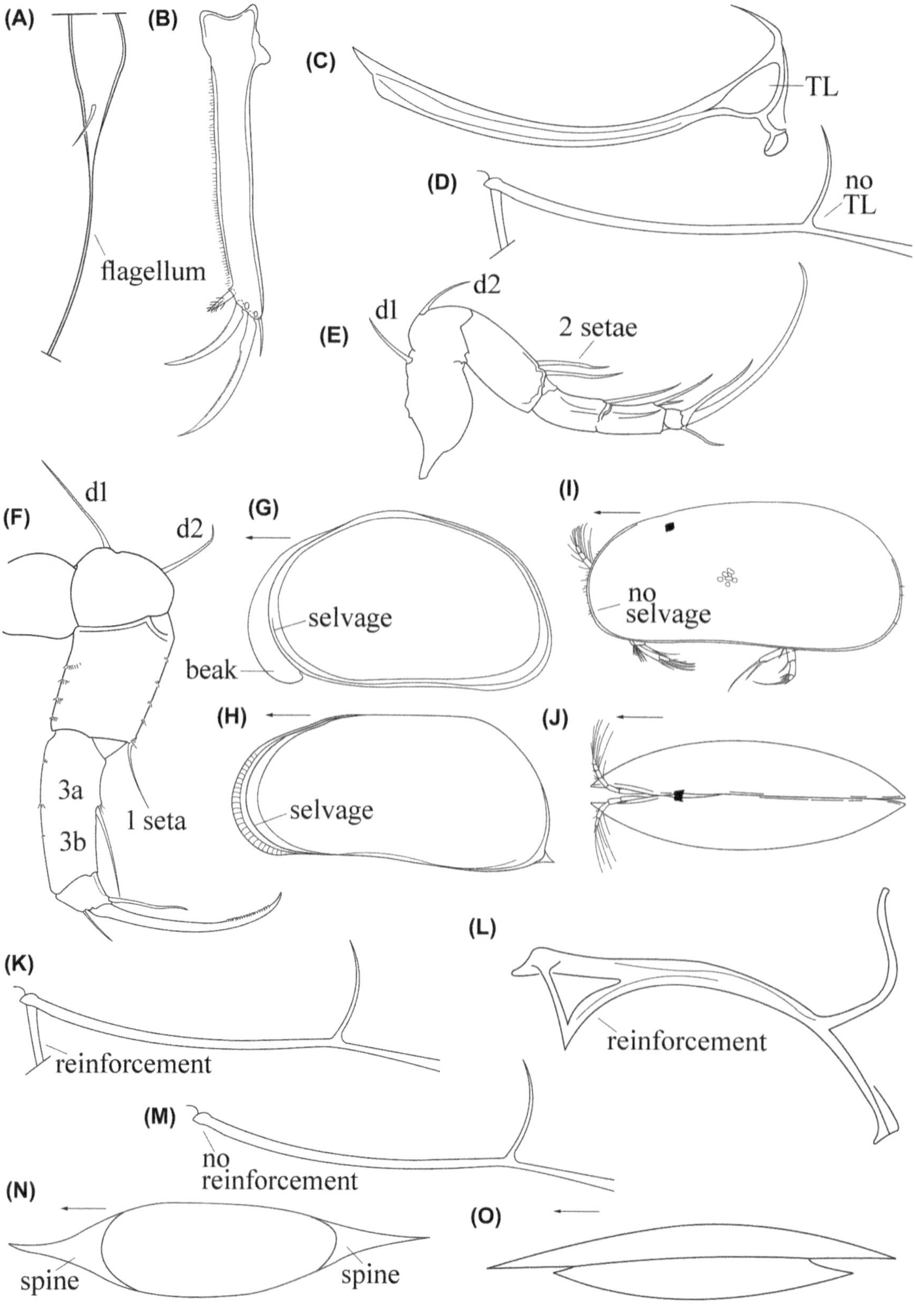

FIGURE 20.9 Cyprididae: (A) Cypridopsinae, flagellumlike caudal ramus; (B) Herpetocypridinae (*Paranacypris*), fully developed caudal ramus; (C) Cypricercinae, attachment of caudal ramus with distal Triebels' loop; (D) Cyprididae, attachment of caudal ramus without distal Triebels' loop; (E). *Chlamydotheca*, T2 with two distal setae on first endopodal segment; (F) Cyprididae, T2 with one distal seta on first edopodal segment; (G) *Chlamydotheca incisa* (Claus, 1892), right valve internal view showing presence of anteroventral beaklike structure and absence of posteroventral spine; (H) *Chlamydotheca unispinosa* (Baird, 1862), right valve internal view showing absence of anteroventral beak and presence of posteroventral spine; (I) *Amphicypris*, Cp left lateral view; (J) *Amphicypris*, Cp dorsal view; (K) Herpetocypridinae, attachment of caudal ramus with proximal trabecule as reinforcement; (L) Herpetocypridinae, attachment of caudal ramus with proximal triangular reinforcement; (M) Cyprididae, attachment of caudal ramus without proximal reinforcements; (N) *Rudjakoviella*, left valve external view showing large anterior and posterior spines; (O) *Rudjakoviella*, Cp dorsal view showing large anterior and posterior spines on LV. Abbreviation: *TL*, Triebels' loop.

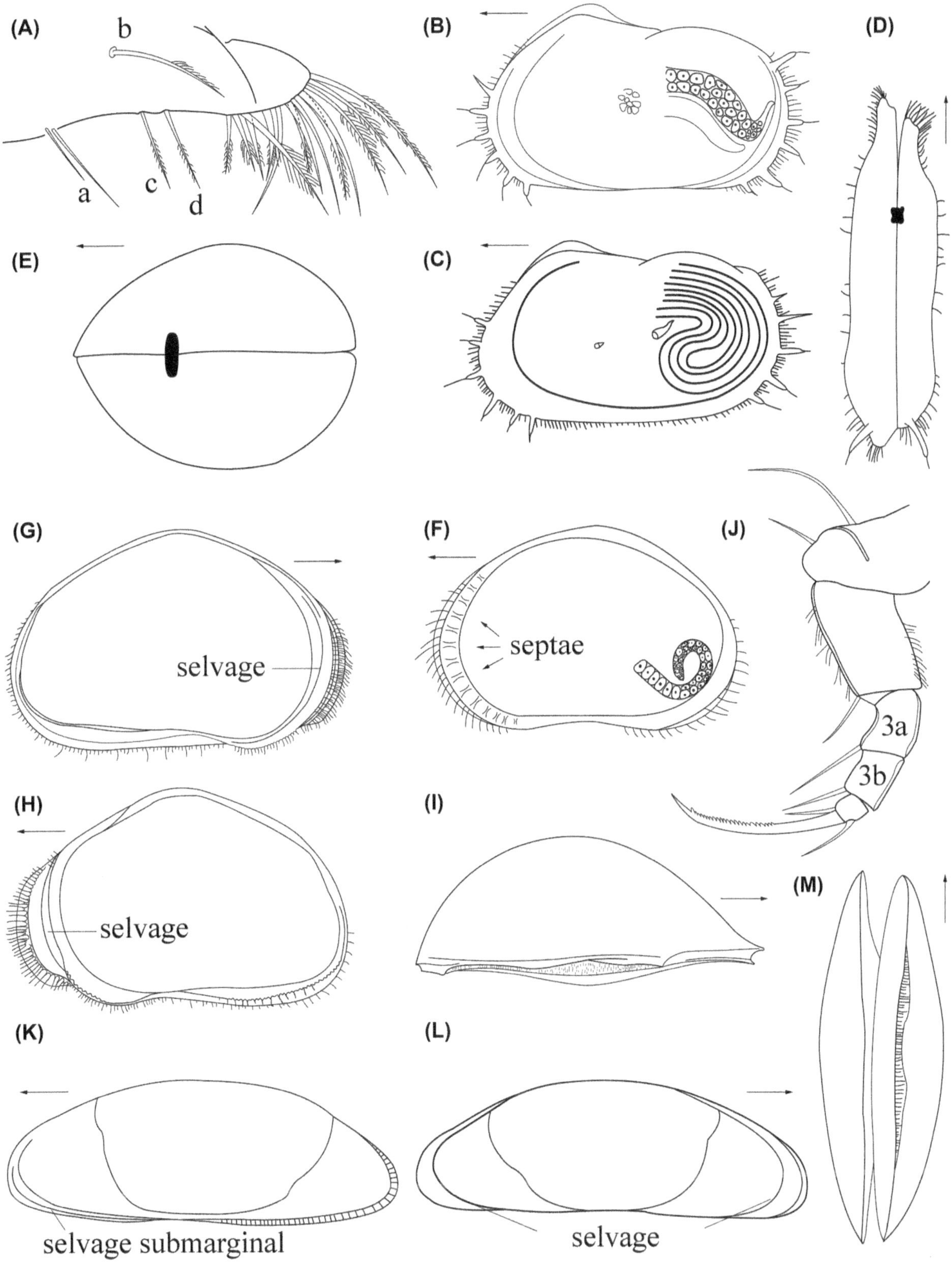

FIGURE 20.10 Cyprididae: (A) Eucypridinae, T1 showing presence of "d" and "c" setae; (B) *Pelocypris*, female, Cp left lateral view; (C) *Pelocypris*, male, Cp left lateral view; (D) *Pelocypris*, female, Cp dorsal view; (E) *Cypretta*, Cp dorsal view; (F) *Cypretta*, Cp left lateral view showing anterior marginal septae, present in both valves; (G) *Cypris*, left valve internal view showing submarginal selvage; (H) *Cypris*, right valve internal view showing largely inwardly displaced selvage; (I) *Cypris*, left valve dorsal view; (J) *Eucypris*, T2 showing divided penultimate segment; (K) *Dolerocypris*, left valve external view with submarginal selvages; (L) *Dolerocypris*, right valve external view with largely displaced selvages; (M) *Dolerocypris*, Cp ventral view (slightly opened).

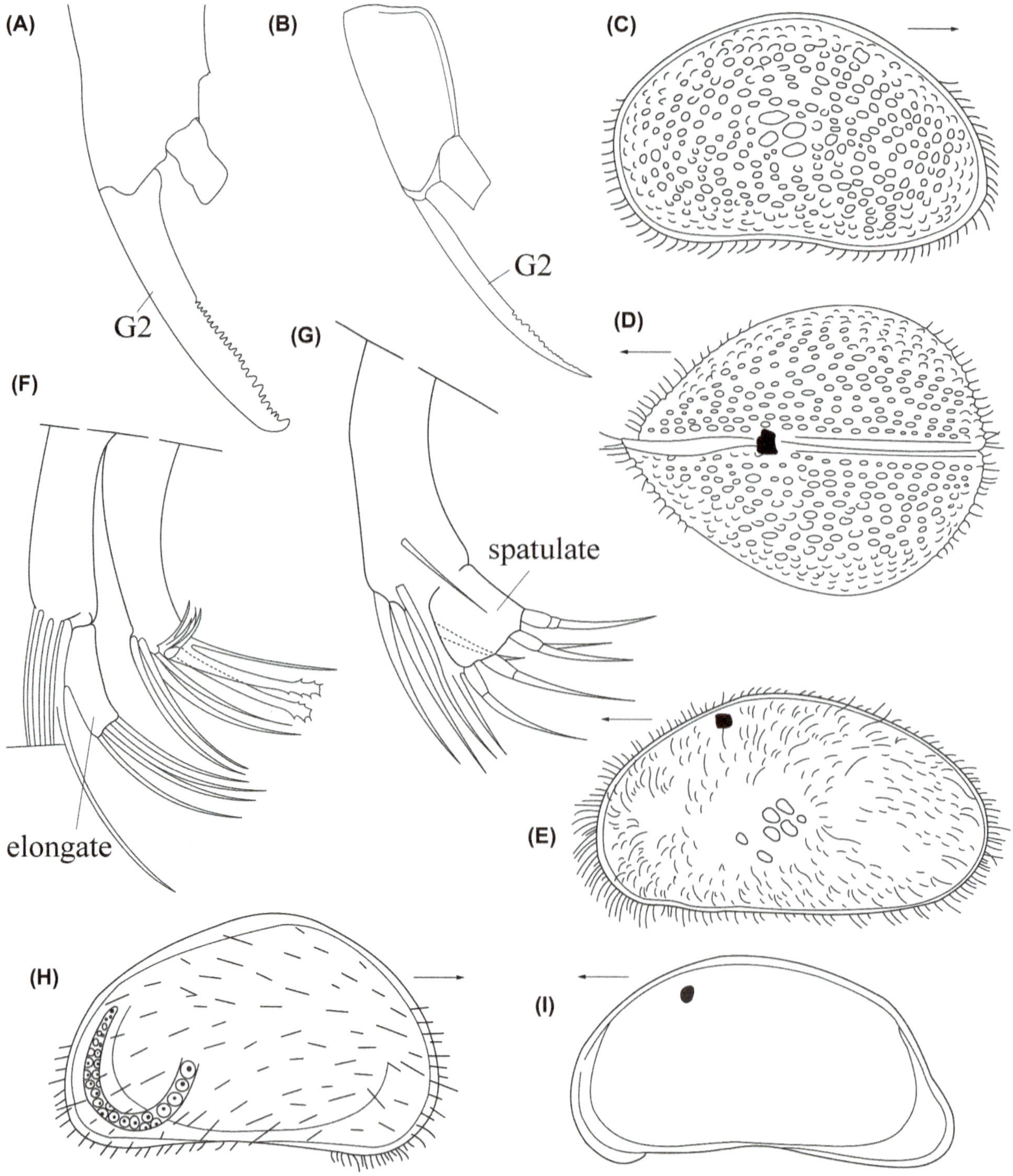

FIGURE 20.11 Cyprididae, Cypridopsinae: (A) Zonocypridini, overdeveloped claw G2 on A2 in females; (B) Cypridopsini, normal claw G2 on A2 in females; (C) *Zonocypris cordata* (Sars, 1924), Cp right lateral view showing striking external valve ornamentation; (D) *Zonocypris cordata* (Sars, 1924), Cp dorsal view showing wide Cp and striking external valve ornamentation; (E) *Cabelodopsis hispida* (Sars, 1901), Cp left lateral view showing hirsute Cp; (F) Cypridopsini, Mx1-palp showing elongated distal segment; (G) Potamocypridini, Mx1-palp showing spatulate distal segment; (H) *Potamocypris*, right valve external view; (I) *Potamocypris*, left valve external view.

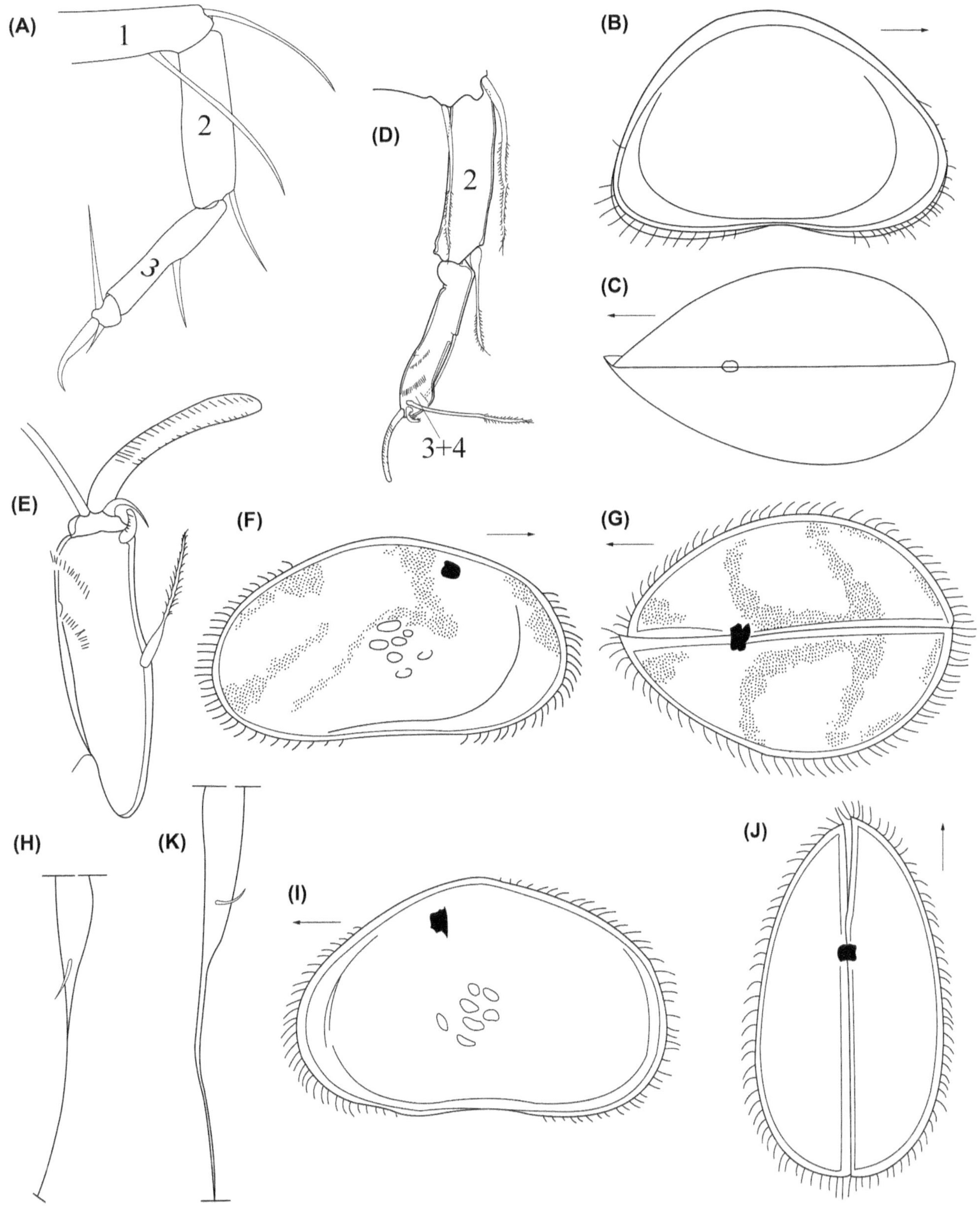

FIGURE 20.12 Cyprididae, Cypridopsinae: (A) *Neocypridopsis*, T3 with separate distal (fourth) segment; (B) *Neocypridopsis*, Cp right lateral view; (C) *Neocypridopsis*, Cp dorsal view; (D) Cypridopsinae, T3 with segments 3 and 4 fused; (E) Cypridopsinae, T3 with segments 3 and 4 partly fused; (F) *Cypridopsis s.s.*, Cp right lateral view; (G) *Cypridopsis s.s.*, Cp dorsal view; (H) *Sarscypridopsis*, caudal ramus showing short and triangular base; (I) *Sarscypridopsis*, Cp left lateral view; (J) *Sarscypridopsis*, Cp dorsal view; (K) *Plesiocypridopsis*, caudal ramus showing elongated base with largely parallel sides.

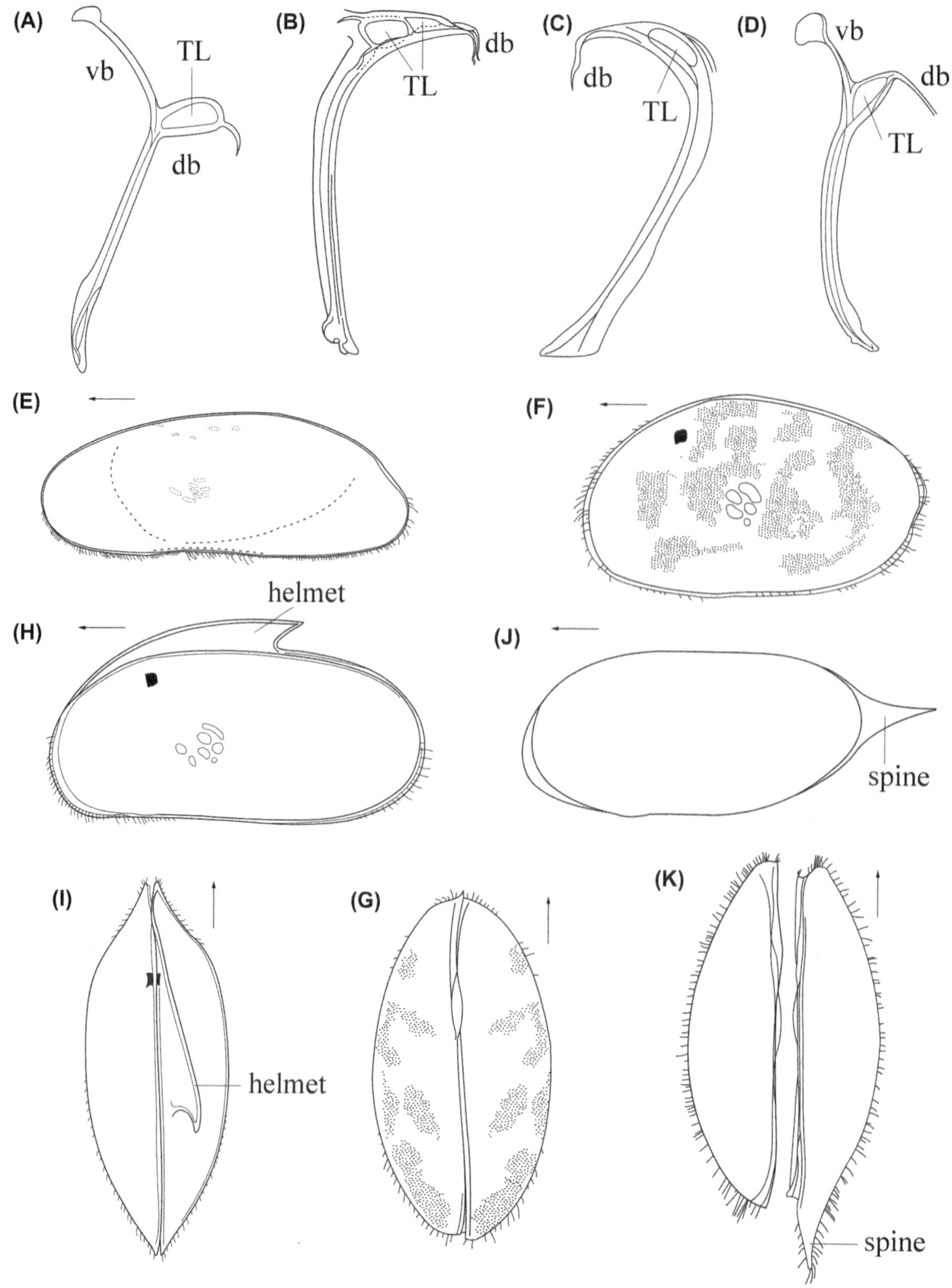

FIGURE 20.13 Cyprididae, Cypricercinae: (A) Cypricercinae, attachment of caudal ramus showing Triebels' loop inside the dorsal branch; (B) *Diaphanocypris*, attachment of caudal ramus showing centrally positioned Triebels' loop with multiple cells; (C) *Nealecypris*, attachment of caudal ramus showing centrally positioned Triebels' loop with single cell and ventral branch absent; (D) *Strandesia*, attachment of caudal ramus showing centrally positioned Triebels' loop with single cell and long ventral branch; (E) *Diaphanocypris*, right valve internal view; (F) *Strandesia obtusata* (Sars, 1901), Cp left lateral view; (G) *Strandesia obtusata* (Sars, 1901), Cp dorsal view; (H) *Strandesia bicuspis* (Claus, 1892), Cp left lateral view showing dorsal "helmet"; (I) *Strandesia bicuspis* (Claus, 1892), Cp dorsal view showing dorsal "helmet"; (J) *Cypricercus*, Cp left lateral view showing large posterior spine on right valve; (K) *Cypricercus*, Cp dorsal view showing large posterior spine on right valve. Abbreviations: *db*, dorsal branch; *TL*, Triebels' loop; *vb*, ventral branch.

CLASS OSTRACODA

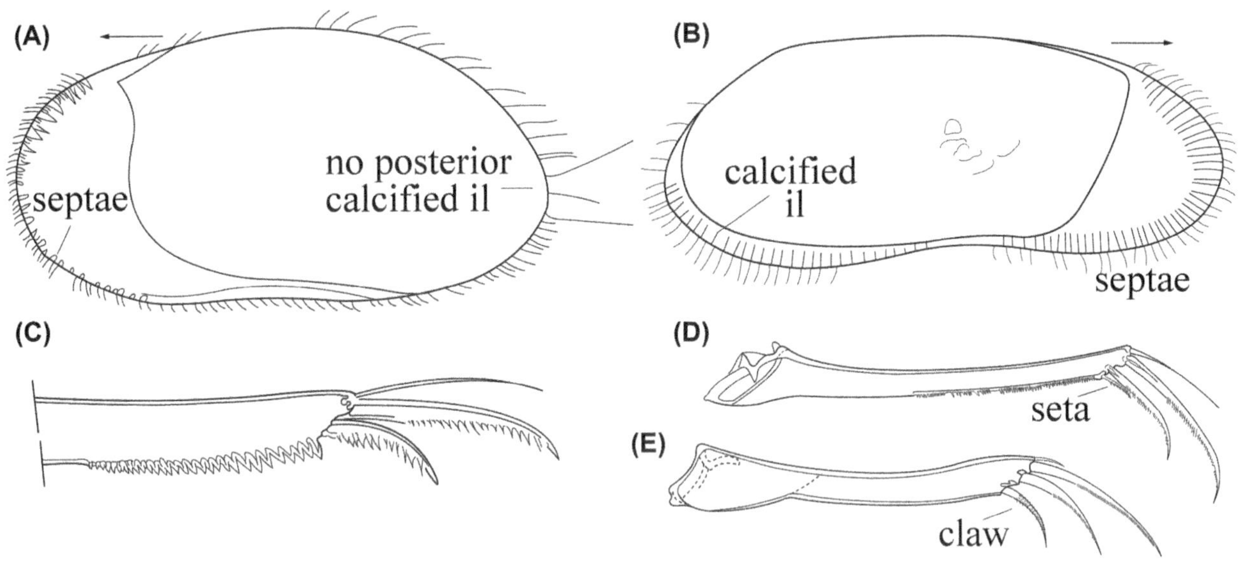

FIGURE 20.14 Cyprididae, Herpetocypridinae: (A) *Isocypris*, right valve internal view showing anterior marginal septae and absence of posterior calcified inner lamella; (B) *Stenocypris*, left valve internal view showing presence of posterior calcified inner lamella and presence of marginal septae; (C) *Stenocypris*, distal part of right caudal ramus showing large ventral spines (absent on left caudal ramus); (D) *Herpetocypris*, caudal ramus showing proximal seta; (E) *Ilyodromus*, caudal ramus showing proximal seta transferred into a third claw. Abbreviation: *il*, inner lamella.

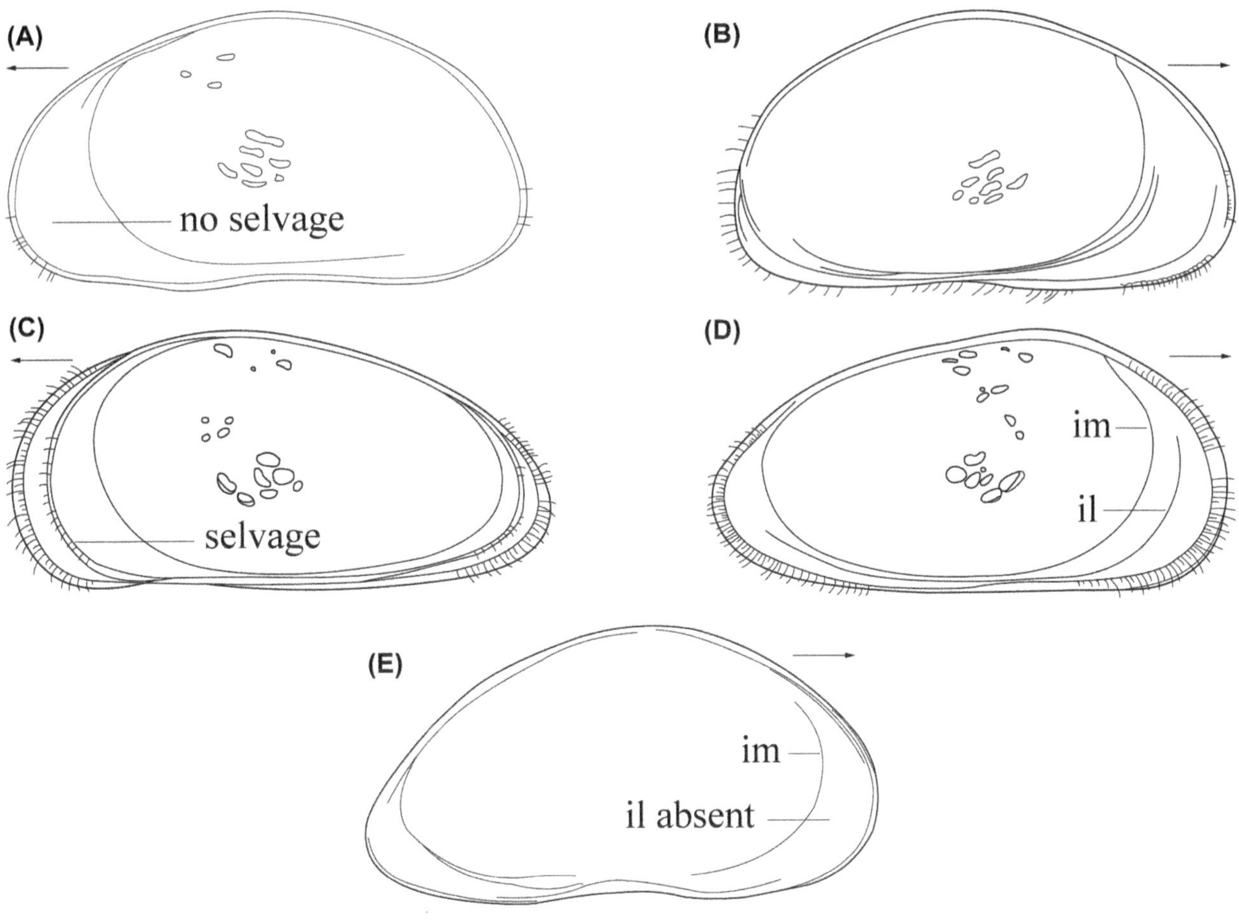

FIGURE 20.15 Cyprididae, Eucypridinae: (A) *Eucypris*, right valve internal view showing absence of anterior selvage; (B) *Eucypris*, left valve internal view showing absence of anterior selvage and presence of short anterior inner lists; (C) *Trajancypris*, right valve internal view showing presence of anterior selvage; (D) *Trajancypris*, left valve internal view showing presence of anterior inner list; (E) *Argentocypris*, left valve internal view showing absence of anterior inner list. Abbreviations: il, inner lamella; im, inner margin.

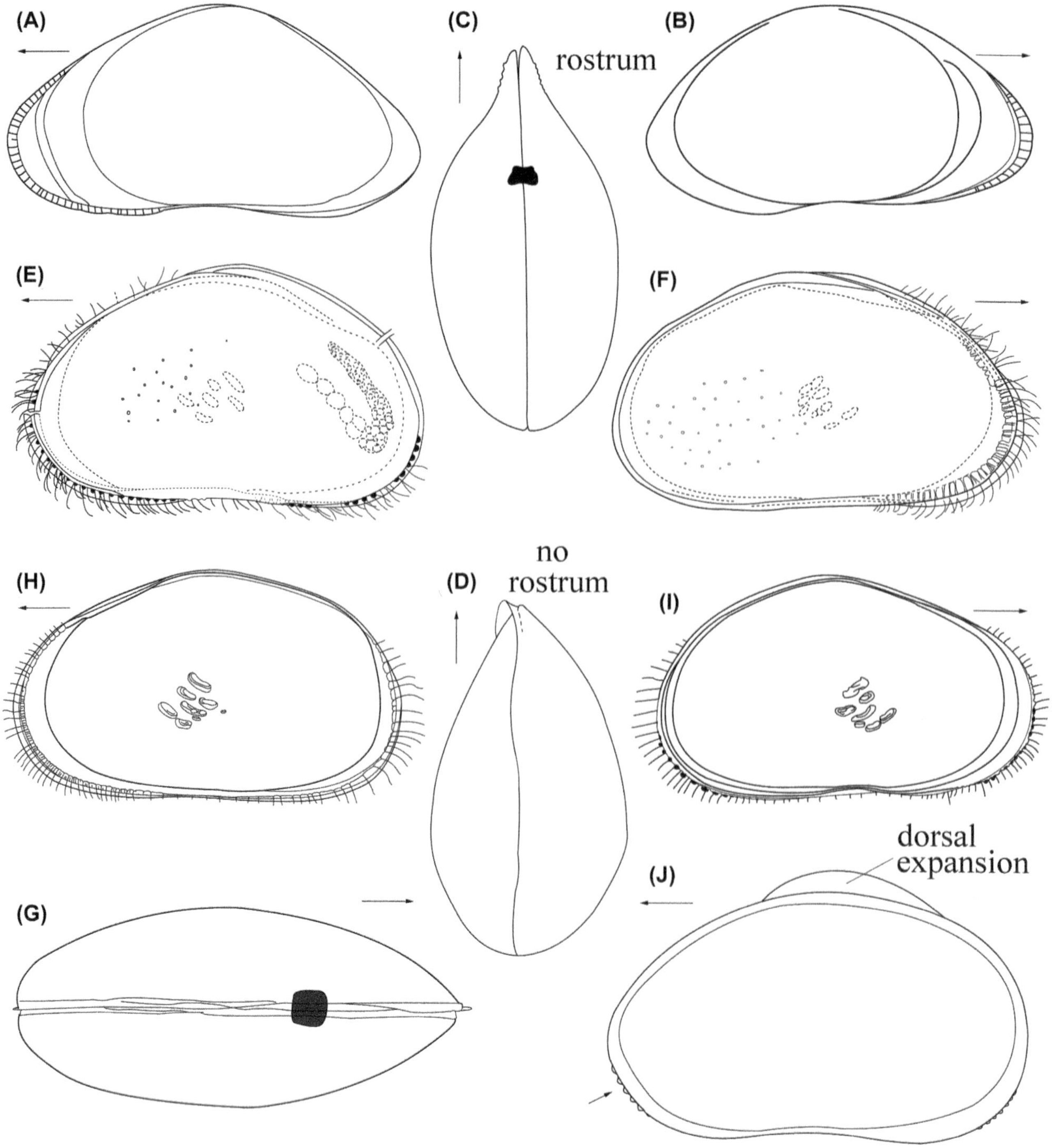

FIGURE 20.16 Cyprididae, Cyprinotinae: (A) *Riocypris*, right valve internal view showing absence of marginal tubercles; (B) *Riocypris*, left valve internal view showing absence of marginal tubercles; (C) *Riocypris*, Cp dorsal view showing absence of anterior overlap and presence of anterior rostrum; (D) *Hemicypris*, Cp dorsal view showing clear anterior right valve > left valve overlap and absence of rostrum; (E) *Hemicypris*, left valve external view showing presence of marginal tubercles; (F) *Hemicypris*, right valve external view showing absence of marginal tubercles; (G) *Heterocypris*, Cp dorsal view showing anterior left valve > right valve overlap and absence of rostrum; (H) *Heterocypris*, left valve external view showing absence of marginal tubercles; (I) *Heterocypris*, right valve external view showing presence of marginal tubercles; (J) *Cyprinotus*, Cp left lateral view showing dorsal expansion (possibly larger) and marginal tubercles on right valve (arrow).

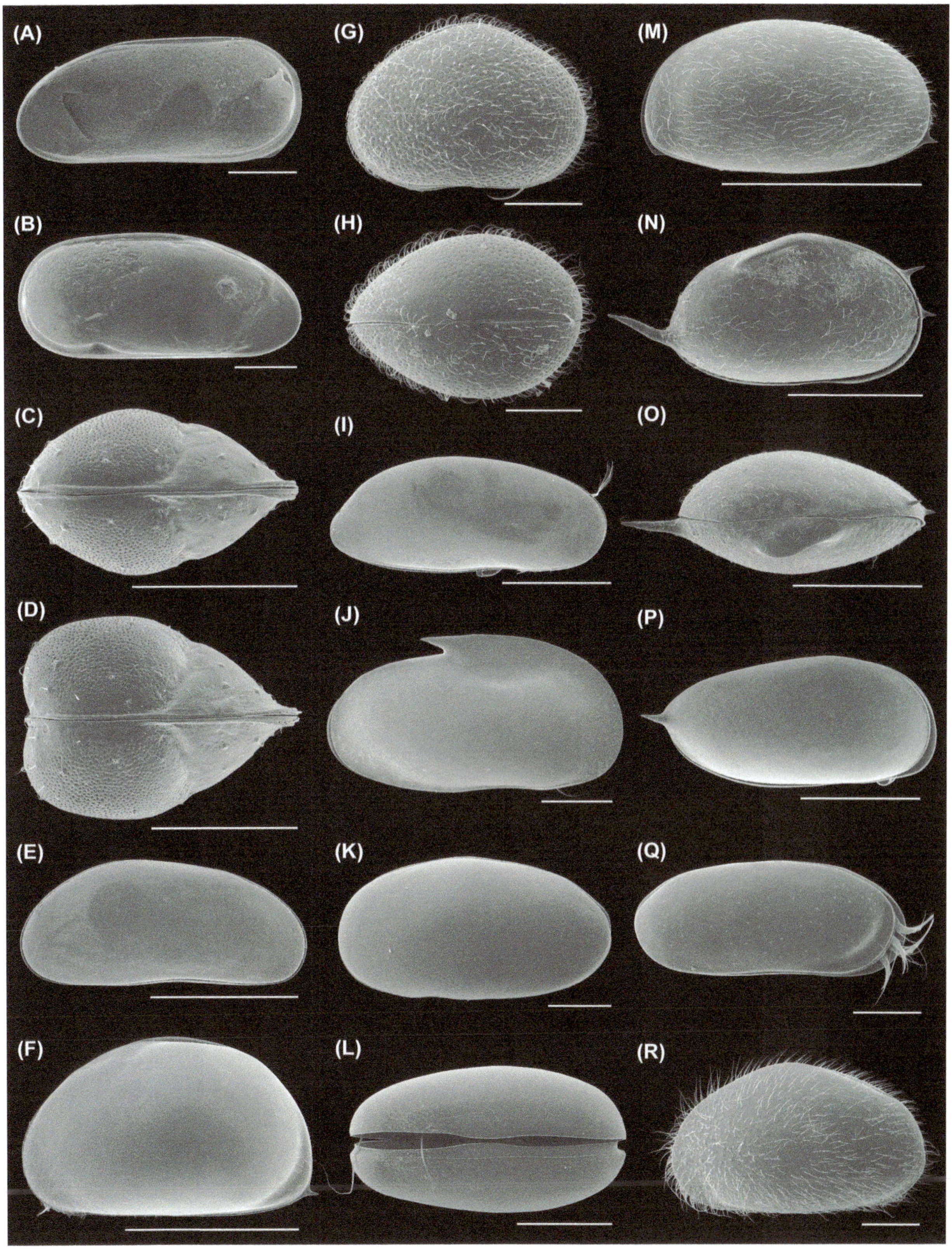

FIGURE 20.17 Scanning electron microscopy images of valves and carapaces of selected Neotropical ostracod species: (A) *Vestalenula pagliolii* (Pinto & Kotzian, 1961); (B) *Penthesilenula brasiliensis* (Pinto & Kotzian, 1961); (C and D) *Cytheridella ilosvayi* Daday, 1905, male (C) and female (D); (E) *Candobrasilopsis elongata* Higuti & Martens, 2014; (F) *Chlamydotheca iheringi* (Sars, 1901); (G and H) *Cypretta vivacis* Würdig & Pinto, 1993; (I) *Diaphanocypris meridana* (Furtos, 1936); (J) *Strandesia bicuspis* (Claus, 1892); (K) *Strandesia lansactohai* Higuti & Martens, 2013; (L) *Strandesia mutica* (Sars, 1901); (M) *Strandesia psittacea* (Sars, 1901); (N and O) *Bradleytriebella trispinosa* (Pinto & Purper, 1965); (P) *Cypricercus centrurus* (Klie, 1940); (Q) *Paranacypris samambaiensis* Higuti, Meisch & Martens, 2009; (R) *Cabelodopsis hispida* (Sars, 1901). Scale bars: (A) and (B) = 100 μm; (G), (H), (K), (Q), and (R) = 200 μm; (C–E), (I), (J), (L), and (N–P) = 500 μm; (F) = 2000 μm; (M) = 1000 μm. (A), (B), (E–L), (Q), (R): Upper Paraná River floodplain, Brazil; and (C), (D), (M–P): Amazon River floodplain, Brazil.

TABLE 20.1 A Taxonomic Checklist of Neotropical Freshwater and Terrestrial Ostracod Genera

Class Ostracoda Latreille, 1806
  Order Podocopida Sars, 1866
    Superfamily Darwinuloidea Brady & Norman, 1889
      Family Darwinulidae Brady & Norman, 1889
        Genus ***Alicenula*** Rossetti & Martens, 1998
        Genus ***Darwinula*** Brady & Robertson, 1885
        Genus ***Microdarwinula*** Danielopol, 1968
        Genus ***Penthesilenula*** Rossetti & Martens, 1998
        Genus ***Vestalenula*** Rossetti & Martens, 1998
    Superfamily Cytheroidea Baird, 1850
      Family Limnocytheridae Klie, 1938
        Subfamily Limnocytherinae Klie, 1938
          Genus ***Limnocythere*** Brady, 1968 (including ***Limnocytherina*** Negadaev-Nikonov, 1967)
          Genus ***Neolimnocythere*** Delachaux, 1928
          Genus ***Paracythereis*** Delachaux, 1928
        Subfamily Timiriaseviinae Mandelstam, 1960
          Genus ***Cytheridella*** Daday, 1905
          Genus ***Elpidium*** O.F. Müller, 1880
          Genus ***Intrepidocythere*** Pinto, Rocha & Martens, 2008
      Family Cytherideidae Sars, 1925
        Subfamily Cytherideinae Sars, 1925
          Genus ***Cyprideis*** Jones, 1857
      Family Xestoleberididae Sars, 1928 (Not treated in this key)
        Genus ***Xestoleberis*** Sars, 1866
      Family Entocytheridae Hoff, 1942 (Not treated in this key)
        Subfamily Entocytherinae Hoff, 1942
          Genus ***Ankylocythere*** Hart, 1962
          Genus ***Entocythere*** Marshall, 1903
          Genus ***Uncinocythere*** Hart, 1962
      Family Cytheruridae G.W. Müller, 1894 (Not treated in this key)
        Subfamily Cytherurinae G.W. Müller, 1894
          Genus ***Cytherura*** Sars, 1866
          Genus ***Semicytherura*** Wagner, 1957
      Family Cytheridae Baird, 1850 (Not treated in this key)
        Subfamily Cytherinae Baird, 1850
          Genus ***Pericythere*** Hartmann, 1957
          Genus ***Perissocytheridea*** Stephenson, 1938 [Syn.: *Ilyocythere* Klie, 1939]
    Superfamily Cypridoidea Baird, 1845
      Family Ilyocyprididae Kaufmann, 1900
        Genus ***Ilyocypris*** Brady & Norman, 1889
      Family Notodromadidae Kaufmann, 1900
        Genus ***Argentodromas*** Diaz & Martens, 2018
        Genus ***Newnhamia*** King, 1855
      Family Candonidae Kaufmann, 1900
        Subfamily Candoninae Kaufmann, 1900
          Tribe Candonini Kaufmann, 1900
            Genus ***Candona*** Baird, 1845
            Genus ***Fabaeformiscandona*** Kristc, 1972
            Genus ***Pseudocandona*** Kaufmann, 1900
            Genus ***Caribecandona*** Broodbakker, 1983
            Genus ***Cubacandona*** Broodbakker, 1983
          Tribe Candonopsini Karanovic, 2004
            Genus ***Candobrasilopsis*** Higuti & Martens, 2012
            Genus ***Candonopsis*** Vavra, 1891
            Genus ***Latinopsis*** Karanovic & Datry, 2009
          Tribe Namibcypridini Martens, 1992
            Genus ***Danielocandona*** Broodbakker, 1983
          Tribe Terrestricypridini Schornikov, 1969
            Genus ***Caaporacandona*** Pinto, Rocha & Martens, 2005
            Genus ***Terrestricypris*** Schornikov, 1980
        Subfamily Paracypridinae Sars, 1923 (Not treated in this key)
          Genus ***Dolerocypria*** Tressler, 1937
          Genus ***Paracypria*** Sars, 1910

CLASS OSTRACODA

TABLE 20.1 A Taxonomic Checklist of Neotropical Freshwater and Terrestrial Ostracod Genera—cont'd

- Genus ***Pontoparta*** Vavra, 1901
- Genus ***Thalassocypria*** Hartmann, 1957
- Subfamily Cyclocypridinae Kaufmann, 1900
  - Genus ***Cypria*** Zenker, 1854 [Syn.: *Keysercypria* (partim)]
  - Genus ***Physocypria*** Vavra, 1898 [Syn.: *Keysercypria* (partim)]
- Family Cyprididae Baird, 1845
  - Subfamily Callistocypridinae Schornikov, 1980
    - Genus ***Callistocypris*** Schornikov, 1980
  - Subfamily Cyprinotinae Bronstein, 1947
    - Genus ***Cyprinotus*** Brady, 1886
    - Genus ***Hemicypris*** Sars, 1903
    - Genus ***Heterocypris*** Claus, 1893
    - Genus ***Riocypris*** Klie, 1935
  - Subfamily Eucypridinae Bronstein, 1947
    - Tribe Eucypridini Bronstein, 1947
      - Genus ***Amphicypris*** Sars, 1901
      - Genus ***Argentocypris*** Diaz & Martens, 2014
      - Genus ***Cypriconcha*** Sars, 1926
      - Genus ***Eucypris*** Vavra, 1891
      - Genus ***Tonnacypris*** Diebel & Pietrzeniuk, 1975
      - Genus ***Trajancypris*** Martens, 1989
  - Subfamily Cypricercinae Mckenzie, 1971
    - Tribe Cypricercini Mckenzie, 1971
      - Genus ***Bradleystrandesia*** Broodbakker, 1983 (Not treated in this key)
      - Genus ***Bradleytriebella*** Savatenalinton & Martens, 2009
      - Genus ***Cypricercus*** Sars, 1895
      - Genus ***Strandesia*** Stuhlmann, 1888 [Syn.: *Acanthocypris* Claus, 1892; Syn.: *Neocypris* Sars, 1901]
    - Tribe Nealecypridini Savatenalinton & Martens, 2009
      - Genus ***Diaphanocypris*** Würdig & Pinto, 1990
      - Genus ***Nealecypris*** Savatenalinton & Martens, 2009
  - Subfamily Rudjakoviellinae Triebel, 1973 In Malz
    - Genus ***Rudjakoviella*** Triebel, 1973 In Malz [Syn.: *Xenocypris* Triebel, 1962]
  - Subfamily Pelocypridinae Triebel, 1962
    - Genus ***Pelocypris*** Klie, 1939
  - Subfamily Herpetocypridinae Kaufmann, 1900
    - Tribe Herpetocypridini Kaufmann, 1900
      - Genus ***Herpetocypris*** Brady & Norman, 1889
      - Genus ***Ilyodromus*** G.W. Müller, 1908
    - Tribe Psychrodromini Martens, 2001
      - Genus ***Paranacypris*** Higuti, Meisch & Martens, 2009
    - Tribe Stenocypridini Ferguson, 1964
      - Genus ***Chrissia*** Hartmann, 1957 (Not treated in this key)
      - Genus ***Stenocypris*** Sars, 1889
    - Tribe Isocypridini Hartmann & Puri, 1974
      - Genus ***Isocypris*** G.W. Müller, 1908
  - Subfamily Dolerocypridinae Triebel, 1961
    - Genus ***Dolerocypris*** Kaufmann, 1900
  - Subfamily Cypridinae Baird, 1845
    - Genus ***Chlamydotheca*** Saussure, 1858 [Syn.: *Pachycypris* Claus, 1892]
    - Genus ***Cypris*** O.F. Müller, 1776
  - Subfamily Cyprettinae Hartmann, 1963
    - Genus ***Cypretta*** Vavra, 1895
  - Subfamily Cypridopsinae Kaufmann, 1900
    - Tribe Cypridopsini Kaufmann, 1900
      - Genus ***Cypridopsis*** Brady, 18,687
      - Genus ***Neocypridopsis*** Klie, 1940 [Syn.: *Notiocypridopsis* De Deckker, 1981]
      - Genus ***Plesiocypridopsis*** Rome, 1965
      - Genus ***Sarscypridopsis*** Mckenzie, 1977
    - Tribe Potamocypridini Ghetti & Mckenzie, 1981
      - Genus ***Potamocypris*** Brady, 1870
    - Tribe Zonocypridini Higuti & Martens, 2012
      - Genus ***Cabelodopsis*** Higuti & Martens, 2012
      - Genus ***Zonocypris*** G.W. Müller, 1898

ACKNOWLEDGMENTS

We thank Eduardo Grou (Maringá, Brazil) for inking all the drawings electronically and Julien Cillis (Brussels, Belgium) for technical help with the scanning electron micrographs. We owe many thanks to Cristina Damborenea (La Plata, Argentina) for guiding us through the editorial process with this chapter, and for showing us much more patience than we deserved! Large parts of the taxonomic revisions that are reported on herein were made possible by grants allocated to either of the present authors by GTI, Belspo, CAPES, CNPq and UEM. The State University of Maringá (UEM, Maringá) and the Royal Belgian Institute of natural Sciences (RBINS, Brussels) have a bilateral Memorandum of Understanding regarding collaborative Scientific Research.

REFERENCES

Bode, S.N.S., S. Adolfsson, D.K. Lamatsch, M.J.M. Martins, O. Schmit, J. Vandekerkhove, F. Mezquita, T. Namiotko, G. Rossetti, I. Schön, R.K. Butlin & K. Martens. 2010. Exceptional cryptic diversity and multiple origins of parthenogenesis in a freshwater ostracod. Molecular Phylogenetics and Evolution 54: 542–552.

Broodbakker, N.W. 1984. The distribution and zoogeography of freshwater Ostracoda (Crustacea) in the West Indies, with emphasis on species inhabiting wells. Bijdragen tot de Dierkunde 54: 25–50.

Broodbakker, N.W. & D.L. Danielopol . 1982. The chaetotaxy of Cypridacea (Crustacea, Ostracoda) limbs: proposals for a descriptive model. Bijdragen tot de Dierkunde 52: 103–120.

Campos, R., E.O. Conceição, M.B.O. Pinto, A.P.S. Bertoncin, J. Higuti & K. Martens. 2017. Evaluation of quantitative sampling methods in pleuston: an example from ostracod communities. Limnologica 63: 36–41.

Cohuo-Durán, S., L. Pérez & I. Karanovic. 2014. On *Limnocytherina axalapasco,* a new freshwater ostracod (Podocopida: Limnocytheridae) from Mexican crater lakes. Revista de Biologia Tropical 62: 15–32.

Conceição, E.O., J. Higuti, R. Campos & K. Martens. 2018. Effects of flood pulses on persistence and variability of pleuston communities in a tropical floodplain lake. Hydrobiologia 807: 175–188.

Cusminsky, G.C., P.A. Pérez, A. Schwalb & R. Whatley. 2005. Recent lacustrine ostracods from Patagonia, Argentina. Revista Española de Micropaleontología 37: 431–450.

DÁmbrosio, D.S., A.R. Díaz, A. García & M.C. Claps. 2015. First description of the soft part anatomy of *Ilyocypris ramirezi* Cusminsky & Whatley (Crustacea, Ostracoda) from Argentina, South America. Zootaxa 3957: 59–68.

Danielopol, D.L. 1982. The preparation of ostracods limbs for optical microscopy (dissection, staining and mounting procedures). Pages 41–42 *in*: R. Maddocks (ed.): Abstracts of the 8th International Symposium on Ostracoda, University of Houston, Houston.

Díaz, A.R. & E.C. Lopretto. 2011. A new species of *Keysercypria* Karanovic (Crustacea: Ostracoda) from Argentina. Zootaxa 3063: 64–68.

Ferreira, V.G., J. Higuti & K. Martens 2019. A striking case of convergent evolution in two species of Cypricercinae (Crustacea, Ostracoda), with the description of a new genus and species from Brazil. Zoologischer Anzeiger 283:1-11. (doi: 10.1016/j.jcz.2019.06.007)

Ferreira, V.G., J. Higuti & K. Martens 2020. Taxonomic revision of *Strandesia* s.s. (Crustacea, Ostracoda) from four Brazilian floodplains, with the description of three new species. Zootaxa 4760:1-74. (DOI:10.11646/zootaxa.4760.1.1)

Fontana, S.L. & S. Ballent. 2005. A new giant cypridid ostracod (Crustacea) from southern Buenos Aires province, Argentina. Hydrobiologia 533: 187–197.

Higuti, J. & K. Martens. 2014. Five new species of Candoninae (Crustacea, Ostracoda) from the alluvial valley of the Upper Paraná River (Brazil, South America). European Journal of Taxonomy 106: 1–36.

Higuti, J., I. Schön, L. Audenaert & K. Martens. 2013. On the *Strandesia obtusata/elliptica* – lineage (Ostracoda, Cyprididae) in the alluvial valley of the Upper Paraná River (Brazil), with the description of three new species. Crustaceana 86: 182–211.

Higuti, J., L.F.M. Velho, F.A. Lansac-Tôha & K. Martens. 2007. Pleuston communities are buffered from regional flood pulses: the example of ostracods in the Paraná River floodplain, Brazil. Freshwater Biology 52: 1930–1943.

Kempf E.K. 1980. Index and Bibliography of Nonmarine Ostracoda. 1. Index A. Sonderveröffentlichungen des Geologischen Instituts der Universität zu Köln 35: 1–188.

Kempf E.K. 1997. Index and Bibliography of Nonmarine Ostracoda. 6. Index A, Supplement 1. Sonderveröffentlichungen des Geologischen Instituts der Universität zu Köln 109: 1–142.

Klie, W. 1940. Süsswasserostracoden aus Nordostbrasilien IV. *Eucypris* oder *Heterocypris*? Zoologischer Anzeiger 129: 113–121.

Laprida, C. 2006. Ostrácodos recientes de la llanura pampeana, Buenos Aires, Argentina: ecología e implicancias Paleolimnológicas. Ameghiniana 43: 181–204.

Martens, K. & F. Behen. 1994. A Checklist of the recent non-marine ostracods (Crustacea, Ostracoda) from the inland waters of South America and adjacent islands. Travaux scientifiques du Musée d'Histoire Naturelle de Luxembourg 22: 1–81.

Martens, K. & D.J. Horne. 2016. Collecting and processing living non-marine ostracods. Journal Crustacean Biology 36: 849–854.

Martens, K. & S. Savatenalinton. 2011. A subjective checklist of the Recent, free-living, non-marine Ostracoda (Crustacea). Zootaxa 2855: 1–79.

Meisch, C., R.J. Smith & K. Martens. 2019 A subjective global checklist of the recent non-marine Ostracoda (Crustacea). European Journal of Taxonomy 492: 1–135.

Namiotko T., D.L. Danielopol & A. Baltanás. 2011. Soft body morphology, dissection and slide-preparation of Ostracoda: a primer. Joannea Geologie und Paläontologie 11: 327–343.

Pinto, I.D. & C.B. Kotzian. 1961. Novos Ostracodes de Familia Darwinulidae e a variação das impressões musculares. Boletim do Instituto de Ciências Naturais de Porto Alegre 11: 5–64.

Pinto, R.L., C.E.F. Rocha & K. Martens. 2008. On the first terrestrial ostracod of the Superfamily Cytheroidea (Crustacea, Ostracoda): description of *Intrepidocythere ibipora* n. gen. n. sp. from forest leaf litter in São Paulo State, Brazil. Zootaxa 1828: 29–42.

Ramírez, F.C. 1967. Ostrácodos de lagunas de la Provincia de Buenos Aires. Revista Museo de La Plata (Nueva Serie), Zoologia (73)10: 5–54.

Roessler, E.W. 1982. Estudios taxonómicos, ontogenéticos, ecológicos y etológicos sobre los ostrácodos de agua dulce en Colombia. I. Estudio morfológico de una nueva especie Colombiana del género *Heterocypris* Claus, 1892 (Ostracoda, Podocopa, Cyprididae). Caldasia 13: 429–452.

Sars, G.O. 1901. Contributions to the knowledge of the freshwater Entomostraca of South America. Part II. Copepoda-Ostracoda. Archiv for Mathematik og Naturvidenskab B 14: 1–52, 8 plts.

Schön, I., J. Higuti, T. Patel & K. Martens. 2018. High connectivity through long distance aquatic dispersal of (cryptic) non-marine ostracod species (Ostracoda, Crustacea) in four major Brazilian floodplains. Molecular Phylogenetics and Evolution 121: 86–97.

Schön I., R.L. Pinto, S. Halse, A.J. Smith, K. Martens & C.W.Birky. 2012. Cryptic species in putative ancient asexual darwinulids (Crustacea, Ostracoda). Plos One Biology 7: e39844.

Smith, A.J., D.J. Horne, K. Martens & I. Schön. 2015. Class Ostracoda. Pages 757–780 *in*: J. Thorp & D.C. Rogers (eds.), Ecology and general biology: Thorp & Covich's Freshwater Invertebrates, Elsevier.

Smith, R.J., T. Kamiya & D.J Horne. 2006. Living males of the 'ancient asexual' Darwinulidae (Ostracoda: Crustacea). Proceedings of the Royal Society B 273: 1569–1578.

Würdig, N.L. & I.D. Pinto. 1990. *Diaphanocypris* a new Ostracoda genus occurring in South and Central America. Pesquisas 17: 31–38.

Chapter 21

Class Copepoda

Eduardo Suárez-Morales, Martha A. Gutiérrez-Aguirre, Samuel Gómez, Gilmar Perbiche-Neves, Daniel Previattelli, Edinaldo Nelson dos Santos-Silva, Carlos E.F. da Rocha, Nancy F. Mercado-Salas, Taisa M. Marques, Yanis Cruz-Quintana and Ana María Santana-Piñeros

Chapter Outline

Thorp and Covich's Freshwater Invertebrates. https://doi.org/10.1016/B978-0-12-804225-0.00021-6

Introduction to Copepoda

Eduardo Suárez-Morales
El Colegio de la Frontera Sur (ECOSUR), Chetumal, Quintana Roo, México

Martha A. Gutiérrez-Aguirre
Universidad de Quintana Roo, Cozumel, Quintana Roo, México

INTRODUCTION

The crustacean class Copepoda contains over 14,000 known species. Of these, about 3,000 are freshwater forms (Boxshall & Defaye, 2008). As a group, copepods are the most abundant metazoans of the world, mainly because of their success in the marine zooplankton; they have also an amazing diversity of forms (Boxshall & Halsey, 2004). Continental freshwater environments have been successfully colonized by copepods, and these crustaceans can be found in most aquatic and semiaquatic habitats ranging from moist soils, leaf packs, damp moss, groundwater, anchialine waters, wetlands, and phytotelmata, to lakes, rivers, ephemeral ponds, hypersaline systems, and estuaries (Boxshall & Halsey, 2004). General aspects of the freshwater copepod biology and ecology, including internal anatomy, reproduction, feeding, behavior, population regulation, and distribution, can be found in Dussart & Defaye (2001), Reid & Williamson (2010), and Suárez-Morales (2015).

Currently, the Copepoda, recently recognized as a monophyletic group, is divided into 10 orders based on molecular data (Khodami et al., 2017). Copepod orders are contained in two groups, the Progymnoplea and the Neocopepoda. The latter includes the Calanoida, Cyclopoida, and Harpacticoida, with representatives in freshwater habitats; parasitic forms of the family Ergasilidae (Poecilostomatoida, a large lineage of the Cyclopoida), are frequently found infecting freshwater fish. Overall, relatively few families of copepods have successfully colonized freshwater habitats (Boxshall & Jaume, 2000). The keys presented in this chapter include both free-living and parasitic families, genera, and species of these four orders that are known from the Neotropical region. Of the ~2,300 species of Calanoida, about 550 are freshwater forms; the calanoid family Diaptomidae, containing 59 genera, is the most diverse, particularly in the Neotropics. The order Cyclopoida includes the family Cyclopidae, the most successful and widespread group among the free-living freshwater copepods; it contains more than 60 genera and more than 800 species. The order Harpacticoida is the least known of the three main groups in freshwater environments, but it is highly diverse with approximately 3,600 species, most of them marine, but some 1,120 can be found inland, mostly in fresh or brackish waters (Boxshall & Defaye, 2008). Only a few species of freshwater cyclopoid copepods, particularly of the genus *Diacyclops*, are known in the Antarctic (Karanovic et al., 2014).

Among the Calanoida, some primarily marine forms have freshwater derivatives; in the Neotropical Region, Centropagidae is highly diverse, and the genus *Boeckella* is represented by 41 species in Asia and the Southern Hemisphere. The coastal marine Pseudodiaptomidae includes the highly diverse genus *Pseudodiaptomus* with few records in full freshwater habitats; these are deemed as recent habitat invasions (Boxshall & Jaume, 2000; Suárez-Morales, 2003).

Among the harpacticoids, the Canthocamptidae is clearly the dominant group in freshwater benthic habitats, with a high degree of endemism in the Americas (Reid, 1994). It contains some 35 freshwater genera, including the highly diverse and widespread genera *Attheyella* and *Elaphoidella*. The Parastenocarididae contains five genera chiefly inhabiting groundwaters and interstitial habitats; containing more than 200 species, the genus *Parastenocaris* is the most diversified in this harpacticoid family. Members of primarily marine harpacticoid families such as Ameiridae, Harpacticidae, Huntemanniidae, Laophontidae, and Tachidiidae also occur in continental waters (Reid & Williamson, 2010).

The analysis of the regional distributional patterns confirms that the Palaearctic Region, with more than 1,200 freshwater species recorded, harbors the highest diversity of freshwater Copepoda (including both free-living and associated forms), and it is followed by the Neotropical (561 species) region. The distributional patterns of regional endemism of the Copepoda differ from the species richness–related ones. Based on estimations presented by Boxshall & Defaye (2008), the Neotropical Region has the highest proportion of endemic genera (54 endemic/104 genera = 51%). Of the major groups of Copepoda, the Diaptomidae contains the highest number of endemic genera in the Neotropical Region (14/18 = 77%).

LIMITATIONS

It was expected that the diversity of these copepods in tropical latitudes would be clearly greater than in temperate areas, but results of a latitudinal analysis of diversity distributional patterns did not support this assumption (Reid, 1994). The transitional areas near the boundaries of different biogeographical zones tend to be more speciose, as species from both regions can coexist. Some of these patterns of diversity could be a reflection of asymmetrical taxonomic efforts in the different regions. For instance, general limnobiological surveys have a tendency to overlook the Harpacticoida because of their complex taxonomy, which is not easily handled by general biologists or technicians. The number of genera and species of representative taxa included in previous efforts in reference to Nearctic freshwater copepods (Reid & Williamson, 2010; Reid, 2016) appears to reflect these patterns in terms of diversity. These previous contributions contain 14 genera and 84 species of Diaptomidae versus 23/105 in the present keys for the Neotropical diaptomid fauna, 15/85 of Cyclopinae (vs. 23/118), and 7/28 of Eucyclopinae (vs. 5/79).

TERMINOLOGY AND MORPHOLOGY

Free-living freshwater copepods have various shapes derived from a generally cylindrical pattern, ranging from fusiform to globose or pear-shaped. Copepods have a chitinous exoskeleton, conspicuous multiarticulated antennules, and a single, simple, anterior eye. One of the defining apomorphies of the Copepoda is the structure of their swimming legs, each pair of which is connected at the base by a "coupler" or "intercoxal sclerite," which apparently increases the energetic efficiency of the legs (Reid & Williamson, 2010). The body has articulated appendages on the cephalic region and the thorax, plus two setose caudal rami on the posterior end of the urosome.

Currently, the morphologic terminology proposed by Huys and Boxshall (1991) is the most widely accepted among copepodologists, and it is used herein to characterize this group. A single major articulation usually divides the body functionally into an anterior **prosome** and a posterior **urosome**, and there are different body plans representing each of the three main groups of freshwater copepods (Figs. 21.1–3). In the podoplean pattern (Cyclopoida, Harpacticoida), the major body articulation is located between the fifth and sixth thoracic somites (or body segments), bearing the fourth and fifth legs, respectively. In the gymnoplean pattern (Calanoida), the major articulation between the prosome and urosome lies between the fifth pedigerous (leg-bearing) somite and the limbless genital somite. The head, including five cephalic somites, is fused with the first and sometimes second thoracic segment to form the **cephalosome**. The thorax consists of seven postcephalic segments, i.e., the second to sixth thoracic somites, starting with the fused somite bearing the maxillipeds and ending with the genital somite (seventh thoracic somite) in both sexes. The posterior part of the body has four somites, the last one being the anal somite. These somites bear no limbs and are usually narrower than the prosome, except in the harpacticoids, in which the urosomal and thoracic somites are equally wide. The observable number of somites in the urosome is variable as a result of distinct fusion patterns, but it is usually between three and five. The anal somite has a pair of terminal caudal rami (Reid & Williamson, 2010). The terminology used to identify some of the cephalic appendages has been variable; hence, in this chapter we used the following terms as morphologic synonyms: antennules = first antenna = A1; antenna = second antenna = A2. The head has five pairs of segmented biramous and uniramous appendages: the **first antennae** (antennules) **(A1)**, **second antennae** (antennae) **(A2)**, **mandibles (Md)**, **first maxillae** (maxillules) **(Mxl)**, and **second maxillae** (maxillae) **(Mx)**; the structure and armature of these appendages has some variation among the Calanoida, Cyclopoida, and Harpacticoida (Figs. 21.1–3). The **first antennae** are uniramous, with a variable number of articles, from 6 to more than 20. The male first antennae have a reduced segmentation, with a distinct fusion pattern, and they are geniculate and modified for grasping the female during copulation. In male harpacticoids and cyclopoids, both first antennae are geniculate (Figs. 21.2, 3); in male calanoids, usually only the right first antenna is geniculate (Fig. 21.1). The **antenna** of calanoids and harpacticoids is a biramous appendage, but in most freshwater cyclopoids, the exopodal ramus is reduced to a single seta or it can be absent in some groups. Some minor differences have been reported between male and female antennae of cyclopoids, particularly in reference to their ornamentation, which is a relevant character in some genera.

The mouth opening is lined by an anterior **labrum** and a posterior **labium**; the former structure is usually ornamented with setae or with teethlike processes and bears the pores of the labral glands (Dussart & Defaye, 2001). The biramous **mandibles (Md)** are situated next to the oral aperture. In calanoids, the mandible consists of a biramous palp with a five-articulated exopod and a 2-articulated endopod. In cyclopoids, the palp is reduced, with an endopod represented by a protuberance bearing two long setae. The inner part of the basal mandibular articles is elongate and has a strongly chitinized distal cutting blade armed with a series of teeth that are variable in shape and size, depending on the feeding habits of the copepod. The **first maxilla** is a biramous appendage adapted to grasp and break up food particles. It is more complex and primitive in the Calanoida, with heavily armed coxal and basal processes, but is relatively reduced in Cyclopoida and Harpacticoida (Dussart & Defaye, 2001). The next cephalic appendages are the **second maxillae** and the **maxillipeds (Mxp)**; both are uniramous as a result of the loss of the exopodal ramus. These appendages have a similar basic structure and function, related to the capture and handling

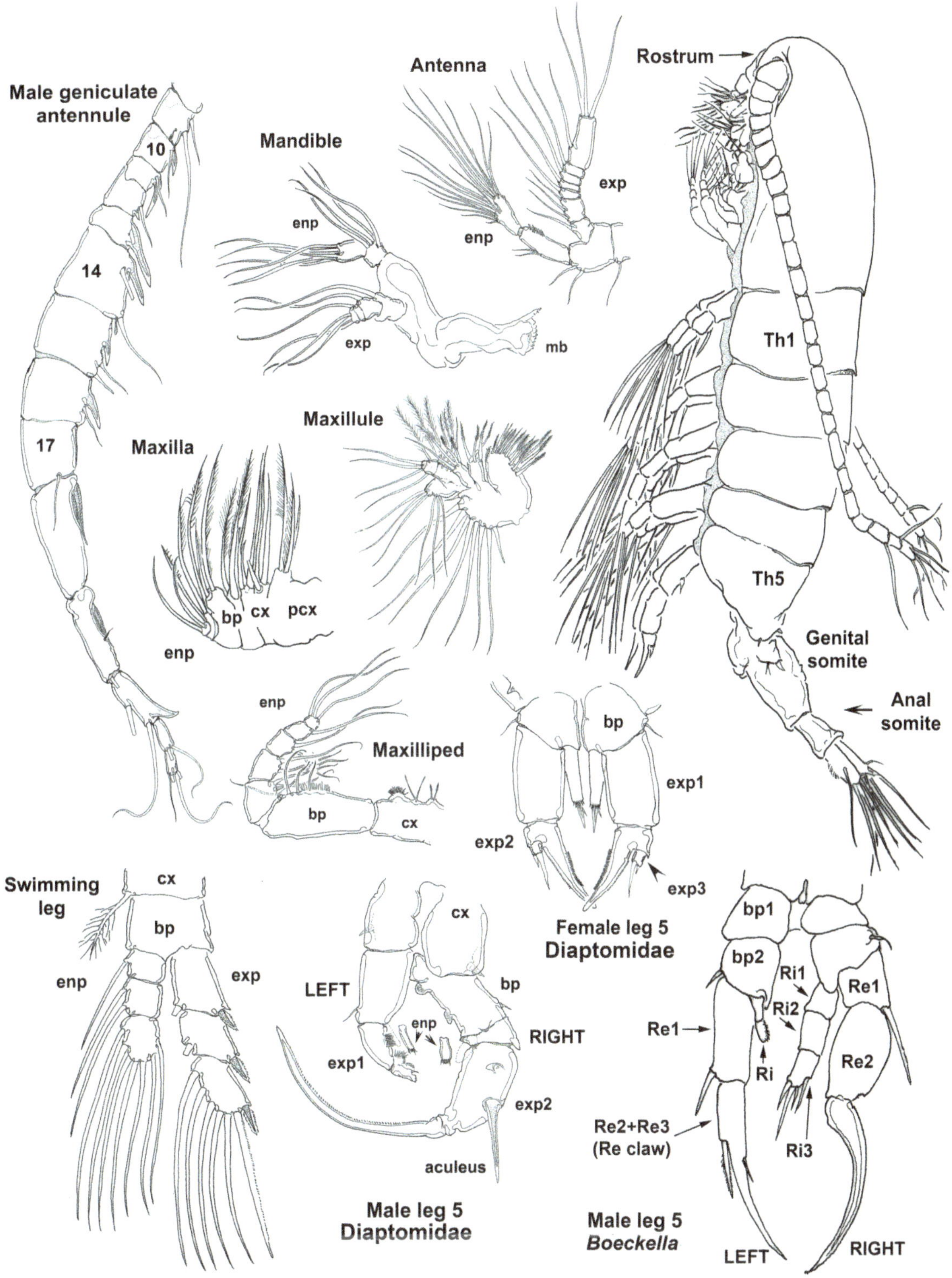

FIGURE 21.1 Representative structures of calanoid copepods. See text for terminology and abbreviations.

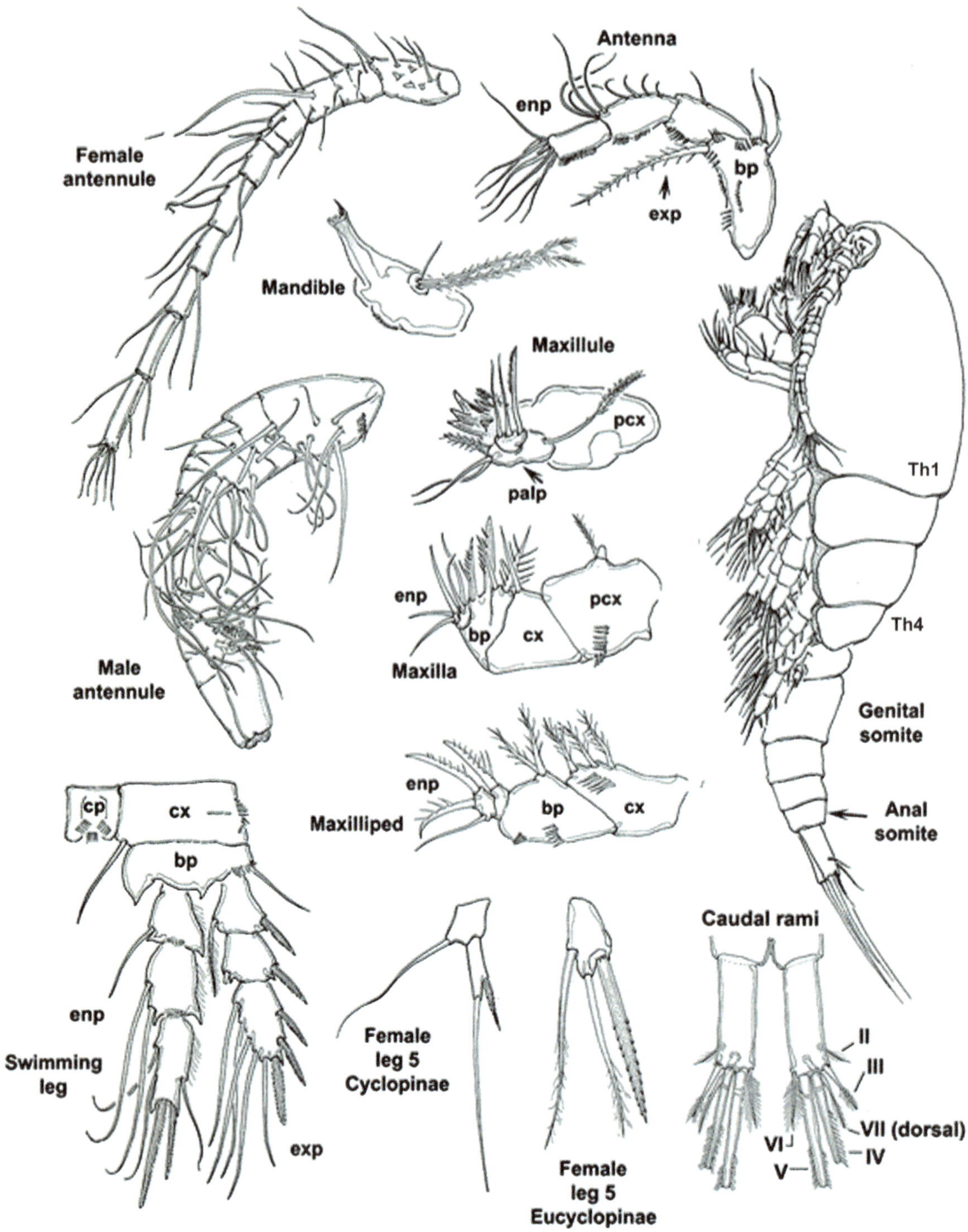

FIGURE 21.2 Representative structures of cyclopoid copepods. See text for terminology and abbreviations.

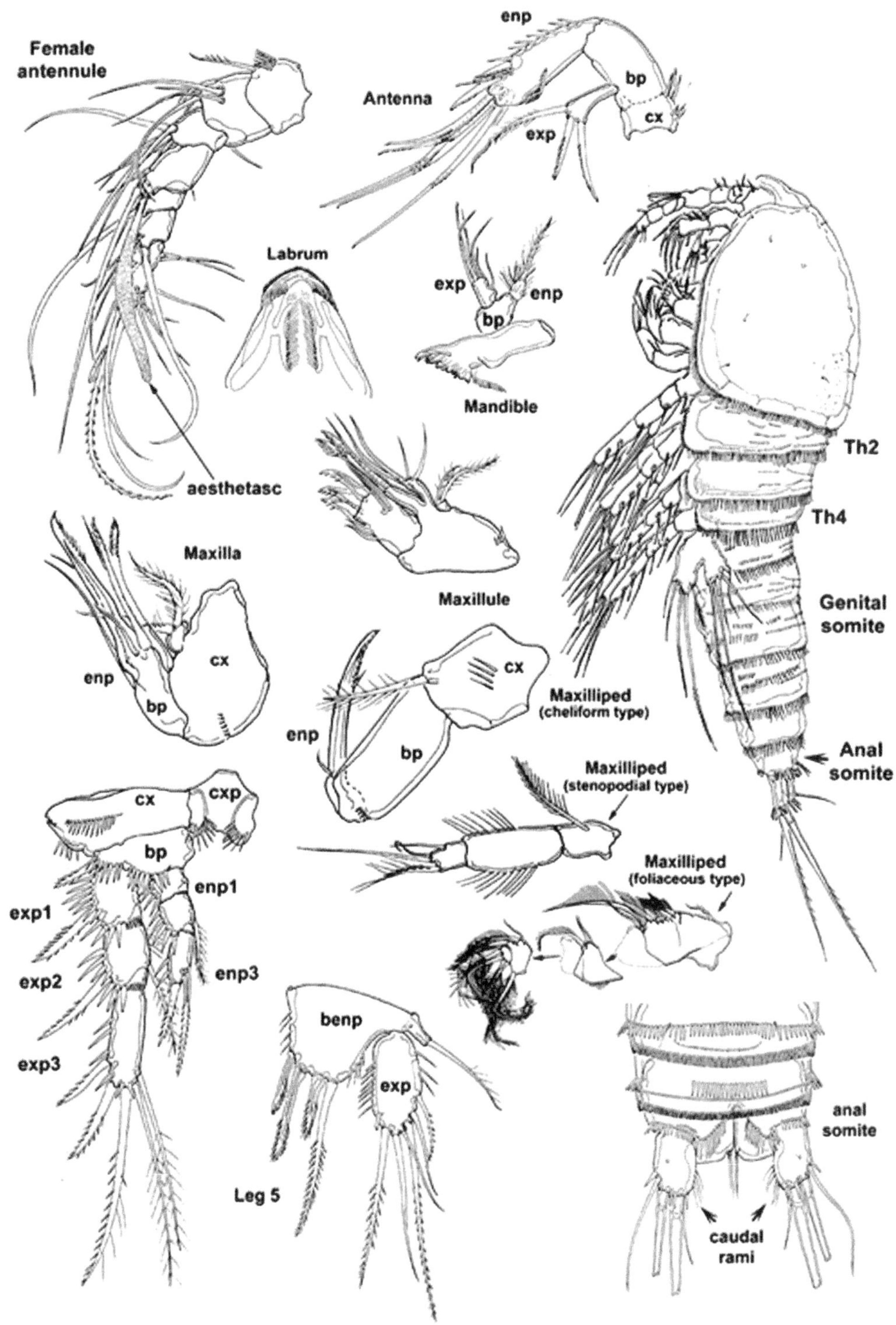

FIGURE 21.3 Representative structures of harpactocoid copepods. See text for terminology and abbreviations.

of food items. In the Cyclopidae, the maxillary coxa bears spine ornamentations in different patterns that can be taxonomically useful in some genera. The openings of the excretory maxillary glands are present at its base. Maxillipeds are slender than second maxillae, which are usually more heavily armed in calanoids, cyclopoids, and harpacticoids (Figs. 21.1–3).

Five or six postmaxillipedal pairs of well-developed **thoracic appendages** are generally present. The first thoracic somite (**Th**), fused to the cephalosome, bears the maxillipeds, whereas the body somites 2–5 bear four pairs of biramous swimming legs (legs 1–4), each pair basally conjoined by a coupler (**cp**) or coxal sclerite (**cxp**) that can be simple and smooth or have different processes or ornamentation patterns with taxonomic value, especially among the Cyclopidae (Figs. 21.2–3). The **swimming legs** consist of two broad basal segments, the **coxa (cx)** and the **basis** or **basipod (bp)**, from which the inner (**endopod** = **enp**) and outer (**exopod** = **exp**) rami arise, each rami with one to three articles (Figs. 21.1–3). Leg articles are armed with setae and spinae in distinct patterns. The number of articles in the swimming legs and their armature are variable in the different groups and are taxonomically valuable characters.

The **fifth legs** are, in general, one of the most important appendages for the identification of free-living freshwater copepods. In calanoids, the fifth legs are usually symmetrical in the female, and highly asymmetrical and prehensile in the male (Fig. 21.1). In cyclopoids and harpacticoids, the fifth pair of legs is reduced and symmetrical; their setation pattern and armature in both sexes has taxonomic value (Figs. 21.2, 3). Cyclopoids and harpacticoids have vestigial **sixth legs**, which are larger in males. The **urosome** of copepods has no appendages; instead, it terminates in two distal caudal rami armed with a basic pattern of seven caudal setae (Fig. 21.2) showing distinct reductions (Huys & Boxshall, 1991; Dussart & Defaye, 2001; Reid & Williamson, 2010). The proportions and armature of the caudal rami have taxonomic relevance in some groups.

MATERIAL PREPARATION AND PRESERVATION

Copepods can be fixed in a 10%–40% solution of formalin (53.7% formaldehyde) or 70% ethanol. Molecular studies, however, require fixation in 96%–100% ethanol to avoid DNA damage. Formalin is toxic, and specimens become brittle over time. This fixative solution should be buffered to reduce acidity. Meiofaunal samples should be fixed within 5 min after collection with a 3%–5% formalin buffered solution (Huys & Boxshall, 1991). The pH of the fixative should be raised to about 8.0–8.2. After fixation (hours to a few days), the samples should be transferred to the definitive preservation medium, either 70% ethanol, 2% formalin, or a solution of propylene phenoxetol (0.5 ml), propylene glycol (4.5 ml), and distilled water (95 ml). In all instances, it is useful to add a few drops of glycerine to the vial; this will keep the body flexible and protects the specimen from desiccation.

Several techniques have been developed for the separation of copepods in samples. Staining samples using rose bengal or neutral red, before or during fixation with formalin, aids the visual separation of specimens from sediment samples. However, staining must be avoided if specimens are to be taxonomically examined by Nomarski differential interference contrast (DIC) microscopy.

Dissection of taxonomically important appendages is required for positive identification of copepods. Small insect pins or electrolytically sharpened tungsten needles can be mounted on the end of wooden holders or glass tubing for use in dissection. The specimen is taken from the sample with a forceps or with a looped needle and placed on a glass microscope slide with a drop of glycerol (10%–20% glycerine + 80%–90% ethanol). The ethanol will evaporate in a few hours leaving the specimen in pure glycerine; gentle warming will speed evaporation. An excavated slide can be used for dissection and then transfer the appendages obtained to a regular slide so that a coverslip can be placed over the final preparation for examination under a compound microscope. Compression of the body may conceal some structures. To avoid this problem, the coverslide may be padded with pieces of a broken coverslide around the appendage or body. Small quantities of modeling clay can also be placed under the corners of the coverslide. This will also allow some lateral movement to roll the specimen into different positions for examination of ventral, lateral, and dorsal surfaces.

In some cases, it is useful to clear the specimens to perform more accurate observations of microcharacters. The specimen can be placed in lactic acid and examined in this medium, but it must be removed after a short period because it softens the tissues and eventually destroys the specimen. Integumental structures (sensilla, pores) are important characters in different groups of copepods, particularly in cyclopoids. The internal tissues must first be removed by heating the specimen in a 10% solution of KOH for 30 min and up to 3–4 h, as needed. The integument is washed in water and 70% ethanol and placed in a solution of Chlorazol Black (1%) and examined in a drop of glycerine (Dussart & Defaye, 2001). When the examination of microcharacters outside the observation range of light microscopy is needed, specimens can be prepared for observation under a scanning electron microscope (SEM).

This process includes dehydration in progressive concentrations of ethanol, critical point drying, coating with gold or gold–palladium, and plating for examination.

Semipermanent mounts can be made by sealing a glycerin coverslip with clear fingernail acrylic polish or a polymerous varnish. One of several standard mounting media can be used for more permanent mounts. A small amount of neutral red dye in glycerin or a mounting medium with a dye will color the dissected parts so that they are easier to locate after dissection. The dissection can be performed directly in the viscous medium, the dissected appendages appropriately arranged in order, and the coverslip put in place. Parts of larger species can be arranged along a small streak of medium. When using a mounting medium other than glycerin, it is useful to let the mounting medium dry somewhat after dissection before adding more medium and adding a coverslip. This helps the dissected parts remain in place. If the medium is allowed to dry for a week and the edges sealed with a long-lasting sealant such as spar varnish, the mount will last indefinitely (Reid & Williamson, 2010).

More details on techniques for copepod preparation, dissection, staining, mounting, and preservation can be found in Dussart & Defaye (2001), Reid & Williamson (2010), and Reid (2016). When morphological traits that can distinguish species or lineages within nominal species are not observable, molecular techniques are a valuable tool, particularly by estimations of genetic divergences of the DNA sequences of the mitochondrial DNA from the COI genes (Lee & Frost, 2002).

KEY TO COPEPODA

Copepoda: Orders

1 Body with podoplean tagmosis, first antennae short, with 6–17 articles; both male first antennae geniculate 2

1′ Body with gymnoplean tagmosis; first antennae long, with 20–26 articles reaching beyond half of total body length, male with one first antenna geniculate (Fig. 21.1) .... **Calanoida [p. 673]**

2(1) Podoplean tagmosis moderately or strongly defined, cephalothorax clearly widened, robust; first antennae short, with 6–17 articles in female, second antennae prehensile or simple, male first antennae geniculate; fifth leg with one to three articles, without baseoendopod 3

2′ Podoplean tagmosis poorly defined, body cylindrical or wormlike, often with long caudal setae; first antennae short, with less than 9 articles in the female and up to 14 in males; fifth leg broad, with baseoendopod (Fig. 21.3) .... **Harpacticoida [p. 713]**

3(2) Podoplean tagmosis well defined; first antennae with 11–17 articles in the female; second antennae not prehensile, male first antennae geniculate; fifth leg with 1–3 articles, without baseoendopod (Fig. 21.2) .... **Cyclopoida [p. 745]**

3′ Podoplean tagmosis moderately defined; first antennae short, usually with six articles; second antennae elongated, distinctively prehensile; fifth leg reduced to a lobe armed with one or two setae .... Poecilostomatoida, one family: **Ergasilidae [p. 782]**

Order Calanoida

Gilmar Perbiche-Neves
Departamento de Hidrobiologia, Universidade Federal de São Carlos, São Carlos, São Paulo, Brazil

Eduardo Suárez-Morales
El Colegio de la Frontera Sur, Chetumal, Quintana Roo, México

Daniel Previattelli
Departamento de Zoologia, Instituto de Biociências, Universidade de São Paulo, São Paulo, Brazil

Edinaldo Nelson dos Santos-Silva
Instituto Nacional de Pesquisas da Amazônia, Manaus, Amazonas, Brazil

Carlos E.F. da Rocha
Departamento de Zoologia, Instituto de Biociências, Universidade de São Paulo, São Paulo, Brazil

INTRODUCTION

The origin of the Calanoida as a group is old, probably from the Silurian period (Selden et al., 2010). The colonization of continental waters by the original marine forms occurred in different geographic regions and at different times (Boxshall & Jaume, 2000). The number of species of Diaptomidae shows a positive correlation with spatial area (Perbiche-Neves et al., 2014b). The Palaearctic Region shows the highest diversity (144 species), followed by the Oriental Region (92 species); the Neotropical Region harbors the third richest diaptomid fauna in the world, with 82 species (Boxshall & Defaye, 2008). After these estimations, the number of species in the Neotropics has grown, almost matching that of the Oriental Region.

The Diaptomidae is the most diverse among the freshwater calanoid copepods. Members of this group occur almost in all kinds of freshwater environments. Diaptomids have been studied in the Neotropical Region since the end of the 18th century by European expeditions, mainly sampling localities where access by land or navigation was possible. Some of the described species were never found again (e.g., *Argyrodiaptomus neglectus* Wright, 1938, *Argyrodiaptomus granulosus* Brehm, 1933, *Odontodiaptomus michaelseni* Mrázek, 1901, *Notodiaptomus dubius* Dussart, 1985, and *Notodiaptomus pseudodubius* Defaye & Dussart, 1988). Other species were not recorded again for a long time, but were collected recently (Perbiche-Neves et al., 2011, 2013).

The Diaptomidae contains four subfamilies: Diaptominae, the most diversified and widely distributed throughout the Americas (Suárez-Morales et al., 2005; Perbiche-Neves et al., 2014a); Microdiaptominae with only one representative in the Neotropical region, *Microdiaptomus cokeri* Osorio-Tafall, 1942, and another in Europe, both exclusively troglobitic forms (Elías-Gutiérrez & Suárez-Morales, 1998); Paradiaptominae is not as diverse as the Diaptomina, it contains a few genera only and is mostly restricted to the African continent (Rayner, 1999; Suárez-Morales & Rayner, 2004; Suárez-Morales et al., 2015); and Speodiaptominae contains only one genus distributed in Eurasia.

The biogeographic processes of the Neotropical diaptomids reveal a high level of endemism visible in spatial units, such as provinces, ecoregions, hydrographic basins, etc., which are also related to the regional geological history (Suárez-Morales, 2003; Suárez-Morales et al., 2005; Previattelli et al., 2013; Perbiche-Neves et al., 2014b). Spatial barriers, climate, and habitat changes (e.g., man-made reservoirs) are advanced as relevant factors acting on diaptomid diversity, probably even more than water quality and eutrophication, which clearly affects cyclopoids from artificial reservoirs (Suárez-Morales et al., 2005; Perbiche-Neves et al., 2014a, 2016). Further taxonomic studies are likely to reveal more undescribed species, particularly from localities that are difficult to access and sample. Molecular studies have also revealed cryptic species in at least two Neotropical genera (Montiel-Martínez et al., 2008; Gutiérrez-Aguirre et al., 2014).

The family Centropagidae, essentially marine, has some genera with representatives in inland waters. In the Americas, there are four genera known to dwell in freshwater: *Boeckella*, *Neoboeckella*, *Limnocalanus* (=*Parabroteas*), and *Osphranticum*. Of these, the most diverse is *Boeckella*, with more than 14 species in the continent, most of them distributed in the Andean and Patagonian regions; only a few species are known from the Neotropical region (Menu-Marque et al., 2000; Adamowicz et al., 2010). *Neoboeckella* is also an Andean group that is represented by two species only, *Neoboeckella kinseli* and *Neoboeckella loeffleri*, both from high-altitude waters (Bayly, 1992a).

The Pseudodiaptomidae consists mainly of coastal and estuarine waters forms. Only some species of *Pseudodiaptomus*, with more than 85 species known, have invaded freshwater environments. In the Neotropical region only *Pseudodiaptomus marshi* Wright, 1936, *Pseudodiaptomus gracilis* (Dahl, 1894), and *Pseudodiaptomus richardi* (Dahl, 1894) have been reported from full freshwater habitats (Walter, 1989; Suárez-Morales et al., 1996; Santos-Silva, 2008).

LIMITATIONS

These keys were prepared considering that the knowledge on the Neotropical calanoids has grown as a result of improved methods for morphologic analysis and supported by new technology facilities. Previous studies by several authors were valuable as background of the keys because they include detailed revisions of some groups, unpublished laboratory guides, and local or regional inventories of species in the Neotropics (e.g., Wright, 1927; Kiefer, 1936b; Brehm, 1957; Brandorff, 1976; Dussart, 1984b, 1985; Matsumura-Tundisi, 1986; Suárez-Morales et al., 1996; Previattelli, 2006; Elías-Gutiérrez et al., 2008; Brandorff, 2012; Santos-Silva et al.,1989, 2013; Perbiche-Neves et al., 2015; Gutiérrez-Aguirre & Cervantes-Martínez, 2016a).

A key was prepared to Calanoida families, followed by a key to genera and separate keys to the species of the most specious genera (>three species). Those genera with a few species only were set as endings of the corresponding options of the genus keys. The largest diaptomid genus in these regions is *Notodiaptomus*; with 36 species that we separated into two groups: one including the *nordestinus* complex (Santos-Silva et al., 2013) with 13 species, and the other with the remaining 23 species of *Notodiaptomus*.

The appendages and structures used in the key are presented in Figs. 21.4 and 5, and it was designed to ease the identification process for nonspecialists and students. Overall, this contribution is expected to motivate future studies and further improvements or modifications. For "*Diaptomus*" *sensu lato*, which currently contains 11 species, the same strategy was applied, adding separately each species in the genus key for further analysis.

MORPHOLOGY AND TERMINOLOGY

Diaptomidae follows the general calanoid morphology (Fig. 21.1). The group shows a conspicuous sexual dimorphism. Males have a highly modified right first antenna (antennule) (A1R) and asymmetrical fifth leg (P5). Females bear symmetrical first antenna and P5, but there are species with asymmetrical last pediger (Ped) and genital double-somite (GS). The body consists of an anterior tagma, the **prosome**, composed of **cephalothorax** (Fig. 21.4, light blue), bearing **first antennae (A1)** and **second antennae (A2)**, followed by subsequent mouthparts, and five **pedigers**, each with a pair of **legs**; a total or partial fusion may occur between pediger 4 and 5. The posterior tagma is represented by a slender **urosome (Ur)**, which includes the GS, the **postgenital urosomites**, and distal **caudal rami (CR)** (Reid, 2016).

The morphological structures used to distinguish the species of Diaptomidae are pointed out in Figs. 21.4 and 5. Both traditional (e.g., fifth leg, right antennule) and recently explored characters (e.g., detailed setal formula of antennules, rows of spinules) were chosen to guide the identification of each species, trying to emphasize the most distinctive characters in all cases. This is why several figures are used to illustrate each species, and some of the options presented in the key include many characters. This is expected to aid the readers to efficiently compare and eventually identify their specimens. Most of the illustrations presented in this document were redrawn or taken from original descriptions, many of them older than 50 years.

ACKNOWLEDGMENT

This guide of identification was produced by putting together parts of other studies. Each author acknowledges financial sources and persons: GPN is grateful to FAPESP (proc. process number 2008/02015-7; 2009/00014-6; 2011/18358-3) and Marcos Nogueira, Danilo Naliato, Silvia Casanova, and Jorge Laço Portinho for help in fieldwork during samplings at La Plata River Basin; GPN and ESM is grateful to the editors of this book. We dedicate this study to the many copepodologists that contributed at different times and circumstances to the knowledge of the morphology, taxonomy, and systematics of these crustaceans in the Neotropical region.

KEY TO CALANOIDA

Calanoida: Families

1 Female genital double-somite without rows and tufts of hairlike elements, caudal ramus not particularly elongate 2

1′ Female genital double-somite modified, with rows and tufts of long hairlike elements or spinules arranged in different patterns; long, narrow caudal ramus, 3.5–7× longer than wide; estuarine or coastal forms, few freshwater species .... Pseudodiaptomidae, one genus: ***Pseudodiaptomus*** **[p. 676]**

2(1) Caudal ramus subequally long, relatively short; endopod of legs 2–4 3-articulated; fifth legs of females and males with endopods reduced to one or two articles .... **Diaptomidae [p. 676]**

2′ Caudal setae unequally long, fourth setae from outer margin longer and thicker than the others; endopods of male and female legs 1–5 always 3-articulated .... **Centropagidae [p. 699]**

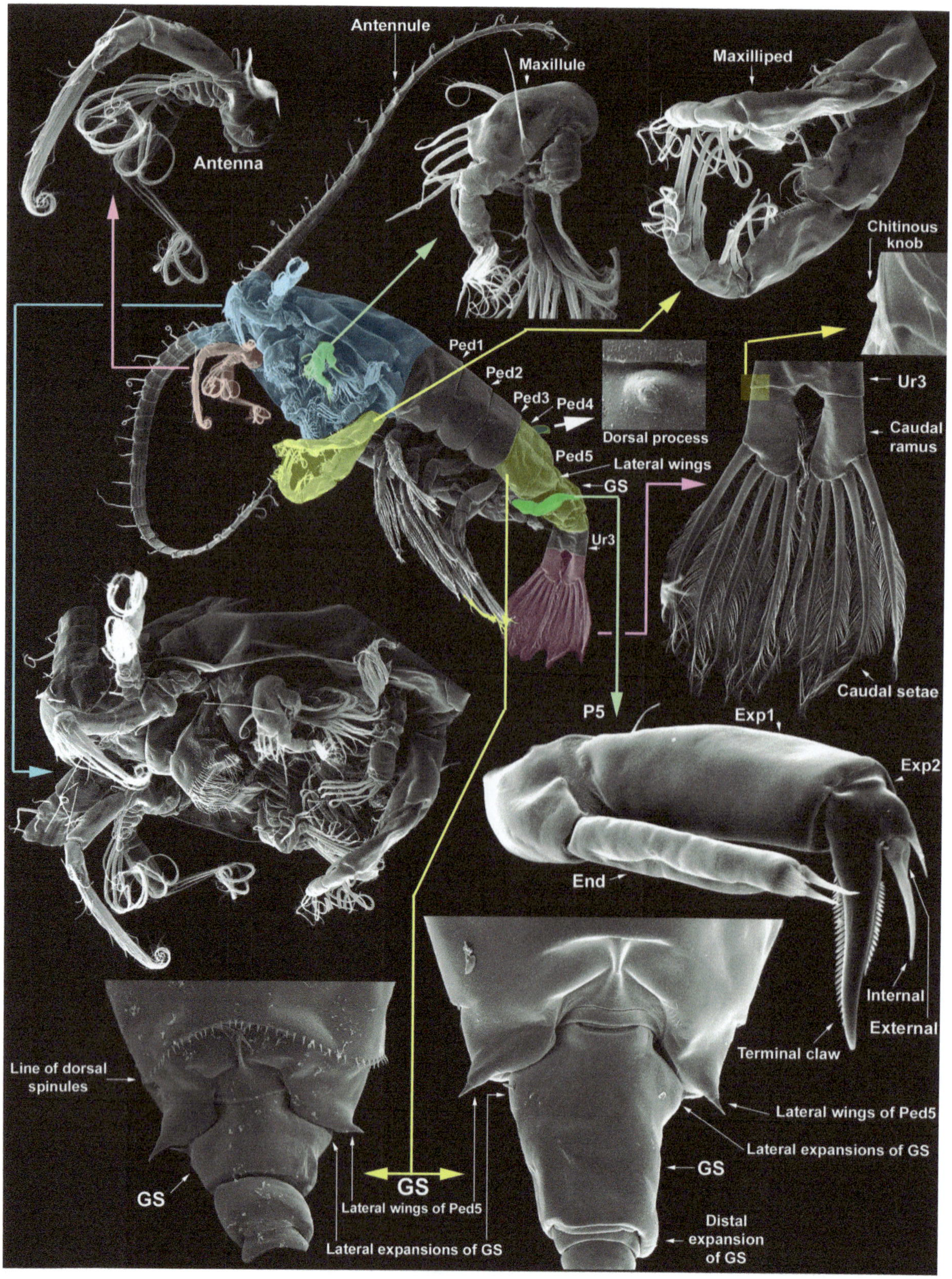

FIGURE 21.4 Body parts and appendages of female Diaptomidae. Abbreviations: *P5*, fifth leg; *End (endopod)*, basipod inner ramus of the swimming legs; *Exp*, basipod outer ramus of the swimming legs; *GS*, genital double-somite; *Ped*, pediger; *Ur*, urosome.

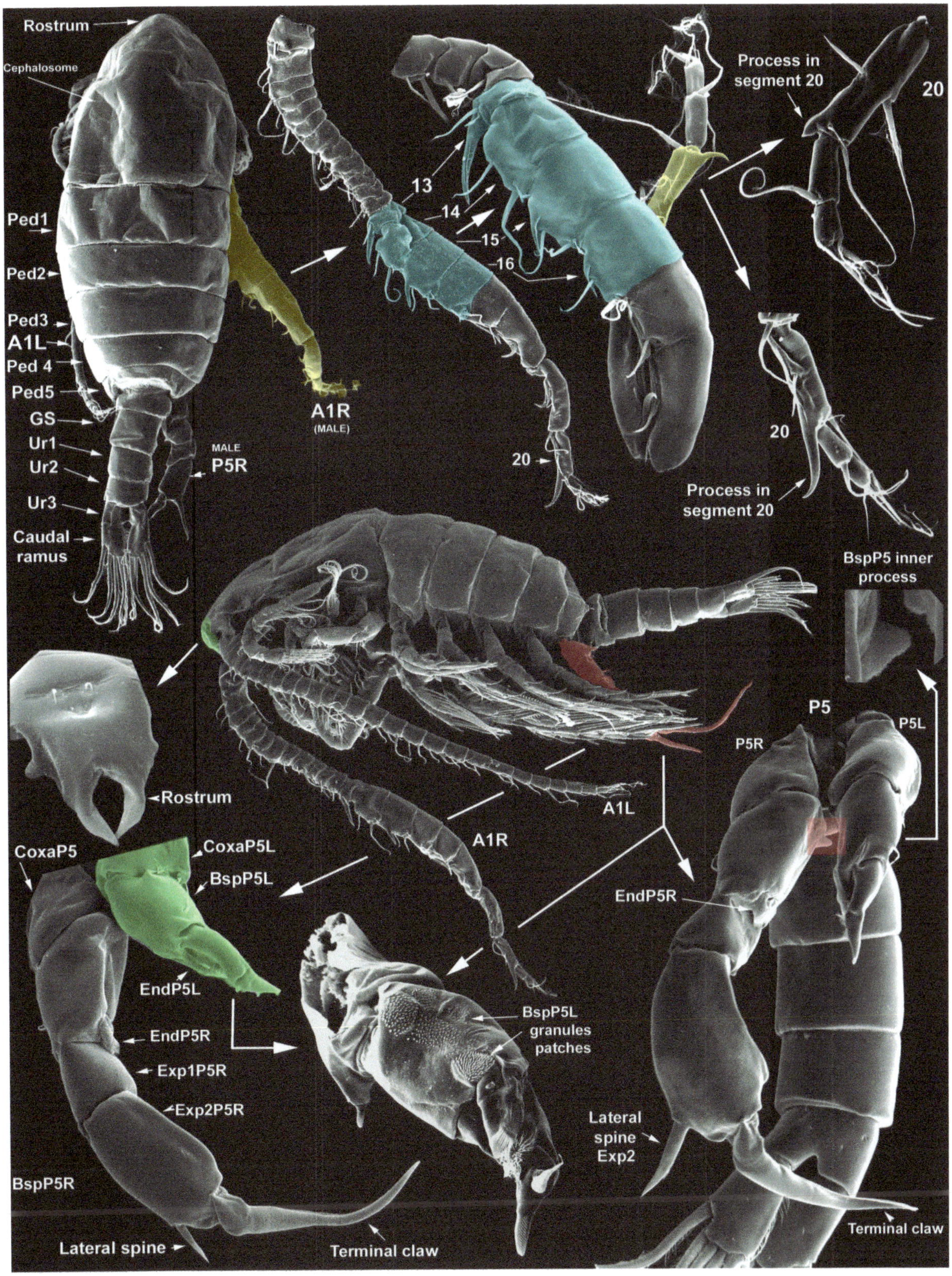

FIGURE 21.5 Body parts and appendages of male Diaptomidae. Abbreviations: *A1L*, left first antenna; *A1R*, right first antenna; *Bsp*, swimming leg basipod; *End*, (endopod) basipod inner ramus of the swimming legs; *Exp*, basipod outer ramus of the swimming legs; *GS*, genital double-somite; *P5L*, left fifth leg; *P5R*, right fifth leg; *Bsp*, basepodite; *Ped*, pediger; *Ur*, urosome.

Calanoida: *Pseudodiaptomus*: Species

This key to species is based on the illustrated key of the American species provided by Walter (1989).

1 Female with paired egg sacs; male fifth leg with left endopodal ramus 2

1′ Female with one egg sac; male fifth leg lacking left endopodal ramus .... *Pseudodiaptomus gracilis* (Dahl, 1894) [Brazil]

2(1) Female first urosomite lacking long hairs, second urosomite with right anterior surface furnished with long surface spinae: right caudal ramus longer that left .... *Pseudodiaptomus marshi* Wright, 1936 [Mexico, Central America, Colombia, Brazil]

2′ Female first urosomite with tuft of long hairs on right posterior corner, right and left caudal rami equally long .... *Pseudodiaptomus richardi* (Dahl, 1894) [Brazil]

Calanoida: Diaptomidae: Genera

A key to the Neotropical diaptomid genera is presented here, but additional keys are available from Dussart & Defaye (1995, 2001) and Suárez-Morales et al. (1996). Other keys for specific groups of species or South American hydrographic basins are provided by Santos-Silva et al. (2013) and Perbiche-Neves et al. (2015).

Notodiaptomus is the largest genus in the Neotropical Region and needs revision. It includes the *nordestinus* complex deemed as the true *Notodiaptomus* (Wright, 1935; Santos-Silva et al., 2013).

1 Endopods of P2-P4 3-articulated; P1 endopod with more than one article; P4 exopod 3-articulated 2

1′ Endopods of P1-P4 1-articulated; P1-P4 exopod 2-articulated (Fig. 21.6 A); cave dwelling .... *Microdiaptomus*, monotypic: *Microdiaptomus cokeri* (Osorio-Tafall, 1942) [Mexico]

2(1) Females with rounded or triangular dorsal keel-like process on pediger 4 3

Note: *Notodiaptomus coniferoides* Wright, 1927, whose females have a dorsal process on pediger 4, was excluded from this option because its congener *Notodiaptomus simillimus* lacks such a process, and both are grouped and distinguished from all other species by a semicircular process on right P5 exopod 2. These species are included in the key (see options 40 and 42).

2′ Females without dorsal process on pediger 4 13

3(2) Spinous process on article 14 of right first antenna absent; right P5 basepodite internal margin not smooth in males; female left lateral wing not larger than right; female sensilla of genital article not positioned 90 degrees to body axis 4

3′ Spinous process on article 14 of right first antenna; male right P5 basepodite with smooth internal margin except for proximal projection reaching left P5 basepodite; lateral wings of pediger 5 asymmetrical, left-side conical and larger than right; expansions of female genital article slightly asymmetrical, with left sensilla positioned posteriorly to right, and both sensilla at 90 degrees in relation to body axis (Fig. 21.6 B) .... *"Diaptomus" ohlei* Brandorff, 1978 [Brazil]

4(3) Internal margin of male right P5 basepodite not smooth; female genital article not both in semicircular and elliptical shape 5

4′ Internal margin of males, right P5 basepodite with smooth margin; lateral spine of right P5 exopod 2 not curved, measuring up to 0.3× as long as terminal claw; in females, the right lateral wing of pediger 5 slightly longer and narrower than left side; genital double-somite asymmetrical, with semicircular left expansion, right-side elliptical, right sensillum of genital double-somite not inserted on semicircular projection (Fig. 21.6 C) .... *Notodiaptomus transitans* Kiefer, 1929 [Brazil, Paraguay, Argentina]

5(4) Male right P5 basepodite without longitudinal fissure ornamented with small sclerotized granulations 6

5′ Male right P5 basepodite with longitudinal fissure ornamented with small sclerotized granulations, typical of the "*nordestinus* complex" of *Notodiaptomus*; lateral spine of exopod 2 inserted close to terminal claw, short, length $< 3\times$ of terminal claw, curved medially; female sensilla of left genital double-somite expansion inserted posteriorly to top of expansion, right-side sensilla inserted on semicircular ornamentation (projection), posteriorly positioned (Fig. 21.6 D) .... *Notodiaptomus carteri* Lowndes, 1934 [Brazil]

6(5) Females without a prominent spine at the anterior surface of P5 exopod 1; right P5 exopod 2 of males not triangular 7

6′ Females with prominent spine on anterior surface of P5 exopod 1; lateral spine of right P5 exopod 2 inserted distally, length $\sim 3\times$ terminal claw; P5 right exopod 2 triangular, proximal width of article twice as widest part (Fig. 21.6 E) .... *Scolodiaptomus*, monotypic: *Scolodiaptomus corderoi* (Wright, 1936) [Brazil]

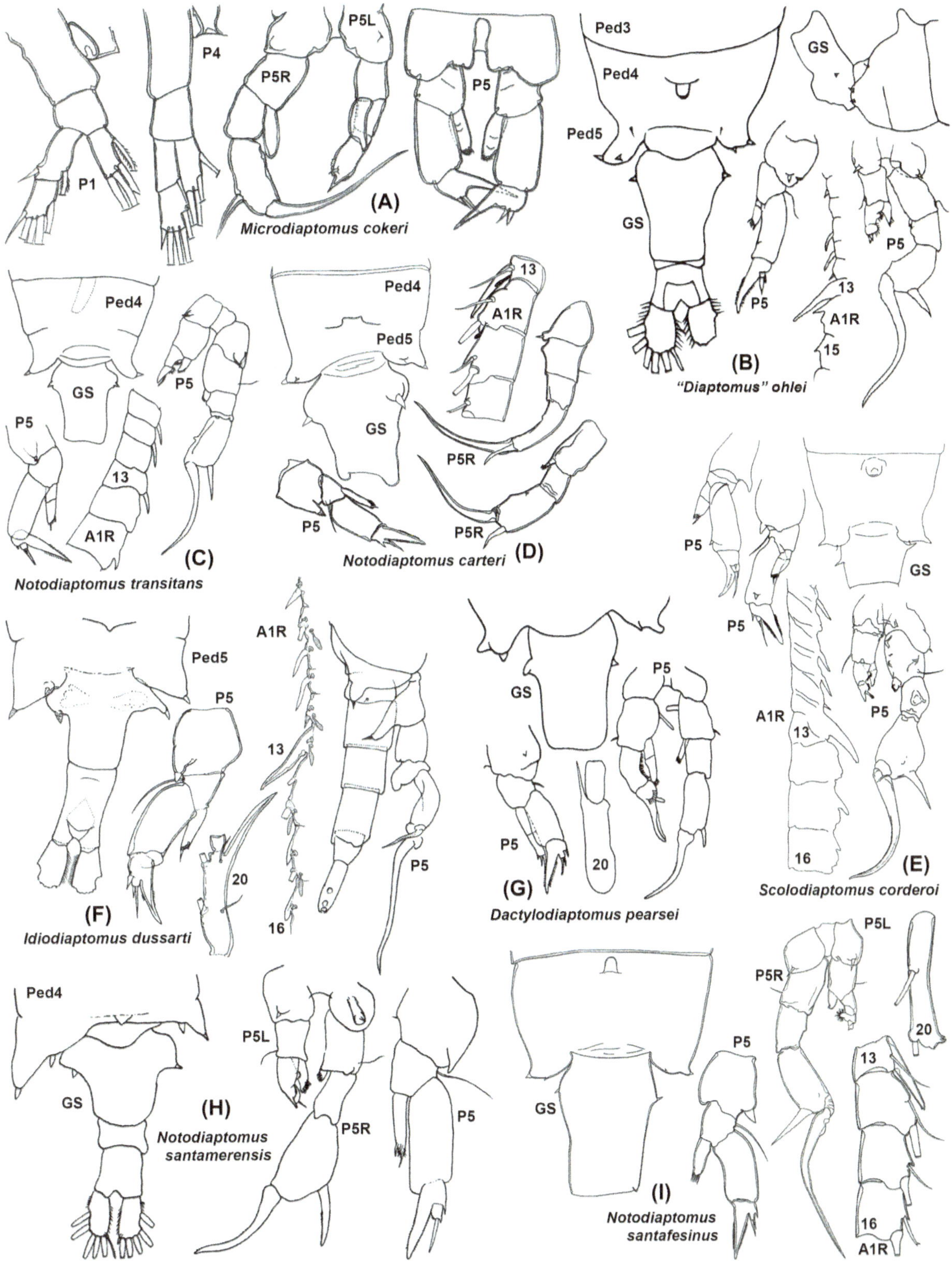

FIGURE 21.6 Diaptomidae species: (A) *Microdiaptomus cokeri*; (B) "*Diaptomus*" *ohlei*; (C) *Notodiaptomus transitans;* (D) *Notodiaptomus carteri*; (E) *Scolodiaptomus corderoi*; (F) *Idiodiaptomus dussarti*; (G) *Dactylodiaptomus pearsei*; (H) *Notodiaptomus santamerensis*; (I) *Notodiaptomus santafesinus*. Abbreviations: *A1R*, right first antenna; *GS*, genital double somite; *P5*, fifth leg; *P5L*, left fifth leg; *P5R*, right fifth leg; *Ped*, pediger. *Redrawn from: (A) Osorio-Tafall (1942); (B) Brandorff (1978); (C) Kiefer (1939); (E) Reid (1987b); (F) Paggi (2011); (G) and (H) Wright (1927); and (I) Cicchino et al. (2001).*

7(6) Article of male right and left P5 in different proportions of length and width; lateral spine of right P5 exopod 2 not inserted distally on article 8

7′ In males, all articles of right and left P5 always longer than wide, with lateral spine of right P5 exopod 2 inserted distally on articles, close to insertion of terminal claw; female lateral wings of pediger 5 well developed, reaching at least midlength of genital double-somite; genital double-somite asymmetrical, with at least one rounded expansion (Fig. 21.6 F) .... *Idiodiaptomus* [Brazil, Argentina]

8(7) Males with left P5 exopod 1 plus terminal claw not remarkably long, reaching beyond the coxal length; females without posterior curved sensilla 9

8′ Length of male left P5 exopod 1 plus terminal claw > 2× the length of the coxa, forming a remarkably long structure compared with other forms; well-developed inner process on right P5 coxa; presence of a thin distal process on article 20 of male right first antenna; females with two processes on each side of lateral wings at pediger 5, genital double-somite with a posterior curved sensillum on left margin (Fig. 21.6 G) .... *Dactylodiaptomus*, monotyoic: *Dactylodiaptomus pearsei* (Wright, 1927) [Brazil]

9(8) Males lacking a long process on right P5 coxa; female left lateral wing of pediger 5 not remarkably larger than right; female genital double-somite not asymmetrical as described below 10

9′ Males with long, thin process furnished with small apical sensillum on right P5 coxa; right P5 endopod short; lateral spine of right P5 exopod 2 as wide as article, terminal claw short, 1.5× long as lateral spine; females with long, triangular lateral left wing; two large sensilla on lateral wings; female genital double-somite asymmetrical, proximally expanded, with caudal curved sensilla on left side, right side with sensillum at 90 degrees (Fig. 21.6 H) .... *Notodiaptomus santamerensis* Wright, 1927 [Brazil]

10(9) Lateral spine of male right P5 exopod 2 not inserted very close to insertion point of terminal claw; lateral sensilla of female genital double-somite not inserted in the same position, left sensillum not curved posteriorly 11

10′ Article 20 of male right first antenna with small bifid distal process; inner margin of rigth P5 basepodite and left P5 basepodite smooth; well-developed sensilla on both sides of P5 coxa; right P5 exopod 2 twice as long as wide; lateral spine of right P5 exopod 2 inserted close to terminal claw basis, 3× long as terminal claw; female genital double-somite symmetrical, lateral sensilla inserted in same position, left curved posteriorly (Fig. 21.6 I) .... *Notodiaptomus santafesinus* Ringuelet & Ferrato, 1967 [Paraguay, Argentina]

11(10) Female with dorsal keel-like process on last thoracic somite; characters not combined as below 12

11′ Female pediger 5 lateral wings asymmetrical, left side larger than right, rounded, right side acute; female genital double-somite strongly asymmetrical, right side with semicircular expansion; male with spinous process on right first antenna articles 14–16, membrane and small expansion at distal margin; male right P5 basepodite wider than it is long, inner margin smooth; right P5 endopod very short; right P5 exopod 1 with posteriorly curved expansion; right P5 exopod 2 2× longer than wide, considering the width of the proximal margin; lateral spine of right P5 exopod 2 short, not longer than the proximal width of article, lateral spine inserted close the distal margin of article (Fig. 21.9 J) .... *Notodiaptomus gibber* Poppe, 1889 [Brazil. It is likely that this species represents an undescribed genus]

12(11) Male right first antenna with spiniform processes on articles 10, 11, 13–15 (sometimes on 16); right P5 exopod with claw less than twice as long as article; female P5 endopod with two relatively long terminal spinae (Fig. 21.7 A) .... ***Mastigodiaptomus* [p. 687]**

12′ Male right first antenna with spiniform processes on articles 8, 10, 11, and 13 (sometimes 14); right P5 exopod with long claw, twice as long as bearing article; female P5 endopod without long terminal spinae, usually ornamented with apical row of hairlike elements (Fig. 21.7 B) .... *Arctodiaptomus dorsalis* (Marsh, 1907) [Southern United States, Mexico]

13(2) Process of penultimate article of male right first antenna present, usually longer or of the same length as the succeeding article........ 14

13′ Process of the penultimate article of the male right first antenna absent or relatively short, not longer than the succeeding article 22

14(13) Absence of small sclerotized granulations on inner margin of male coxa of right and/or left P5; female genital double-somite not slightly asymmetrical 15

14′ Small sclerotized granulations on inner margin of male coxa of right and/or left P5; female lateral wings bilobed (Fig. 21.7 C) .... ***Argyrodiaptomus* [p. 689]**

15(14) Lateral wings in males and females not as long, asymmetrical; male modified setae of right first antenna article 11 not longer than on article 13; female P5 exopod 2 with lateral spine 16

15′ Lateral wings of males reaching beyond midlength of genital double-somite in most species; modified setae on right first antenna article 11 (XIII) longer than modified setae in article 13 (XV); process on article 20 of right first antenna reaching distal margin of or beyond distal margin of succeeding article; well-developed process on right P5 coxa, reaching midlength of basis; inner margin of P5 basis with semicircular lamella; large expansions on genital double-somite, right side reaching midlength of somite; lateral spine of female P5 exopod 2 absent (Fig. 21.7 D) .... ***Rhacodiaptomus* [p. 692]**

16(15) Male right P5 exopod with lateral spine not inserted medially on article, not as long as article, not concave along its length; right P5 endopod not wider than it is long; female urosome 2 without triangular process, right genital double-somite sensillum not larger than left one 17

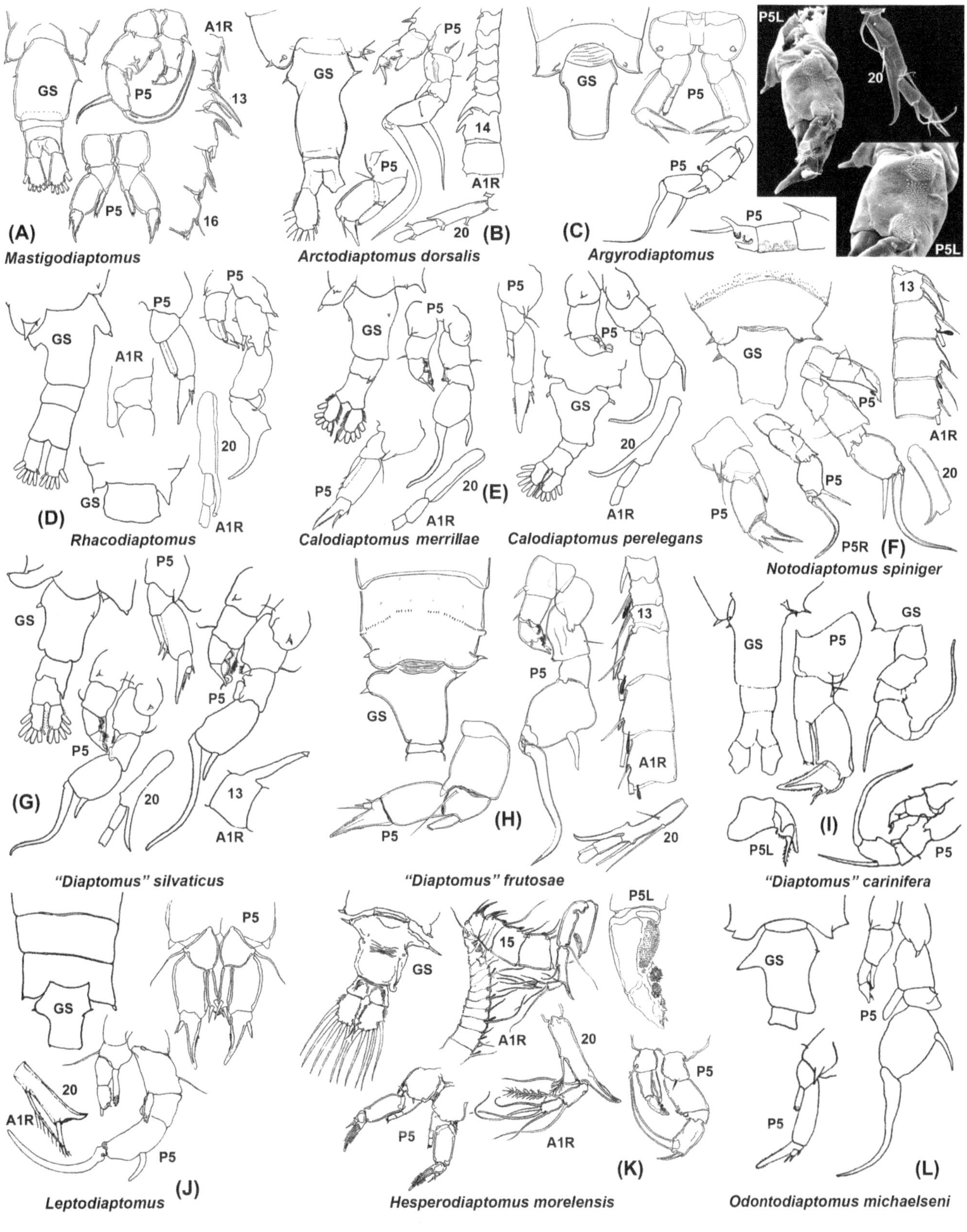

FIGURE 21.7 Diaptomidae species: (A) *Mastigodiaptomus* sp.; (B) *Arctodiaptomus dorsalis*; (C) *Argyrodiaptomus* sp.; (D) *Rhacodiaptomus*; (E) *Calodiaptomus merrillae* and *Calodiaptomus perelegans*; (F) *Notodiaptomus spiniger*; (G) "*Diaptomus*" *silvaticus*; (H) "*Diaptomus*" *frutosae*; (I) "*Diaptomus*" *carinifera*; (J) "*Leptodiaptomus*" sp.; (K) *Hesperodiaptomus morelensis*; (L) *Odontodiaptomus michaelseni*. Abbreviations: *A1R*, right first antenna; *GS*, genital double somite; *P5*, fifth leg. *Redrawn from: (B) Gutiérrez-Aguirre & Cervantes-Martínez (2013); (D), (E) and (G) Wright (1927); (I) Lowndes (1934); (K) Granados-Ramírez & Suárez-Morales (2003); and (L) Mrázek (1901).*

16′ Male right P5 exopod 2 with lateral spine inserted medially on article, being at least as long as article, with concave curvature along its length; right P5 endopod wider than it is long; female urosome 2 with small dorsal triangular process at distal margin, or two close sensilla on pediger 5; right sensillum larger than left one in lateral expansions of genital double-somite (Fig. 21.7 E) .... *Calodiaptomus* [Brazil]

17(16) Dorsal surface of male and female pedigers 4 and 5 without rows of spinules; absence of process on medial margin of male rigth P5 basepodite; female genital double-somite with different configuration on both sides, left side not hemispherical 18

17′ Dorsal surface of male and female pedigers 4 and 5 with rows of spinules; distal margin of male right first antenna article 20 with simple distally tapering process, article 15 with spinous process on anterior margin; mammiliform process present on medial margin of male right P5 basepodite; female lateral wings of pediger 5 slightly asymmetrical, not elongate; genital double-somite with hemispherical swelling on left side only, left sensilla directed posteriorly; P5 endopod long, reaching beyond midway of P5 exopod 1 inner margin (Fig. 21.7 F) ....*Notodiaptomus spiniger* Brian, 1925 [Brazil, Uruguay]

18(17) Male lacking middle/distal rectangular expansion on right P5 exopod 1; female Ur2 without expansion on right distal margin 19

18′ Male right first antenna article 13 with long spinous process reaching article 15; right P5 exopod 1 with inner medial/distal rectangular expansion; female pediger 5 with asymmetrical lateral wings, left side thinner and longer than right side, which is conical, with inner sensillum inserted on small expansion; genital double-somite slightly asymmetrical; urosome 2 with expansion on right distal margin reaching at least midlength of urosome 3 (Fig. 21.7 G) .... *"Diaptomus" silvaticus* Wright, 1927 [Trinidad, Brazil]

19(18) Male right P5 exopod 2 in many shapes except subtriangular, longer than wide or subquadrate; female genital double-somite left side lacking semicircular expansion with apical sensillum 20

19′ Male right P5 exopod subtriangular, 1.1−1.2× wider than long; female pediger 5 with transverse row of spinules marking position of posterior margin; genital double-somite asymmetrical, left swelling larger than right, with hemispherical shape; sensillum on left swelling directed slightly posteriorly, right sensillum anteriorly directed (Fig. 21.7 H). ....*"Diaptomus" frutosae* Perbiche-Neves & Boxshall, 2013 [Paraguay, Argentina]

20(19) Male right P5 exopod 2 lateral spine not remarkably long, not reaching beyond the length of this article; female P5 endopod not too long, barely reaching the distal margin of right P5 exopod 1 21

20′ Male P5R basis with inner expansion on proximal margin, adjacent to medial knob; surface of right P5 exopod 2 ornamented, lateral spine long, at least 1.5× as long as bearing article; female with asymmetrical lateral wings, left size larger than right; genital double-somite asymmetrical, with rounded expansion on left margin; left sensillum more distally inserted than that on right side which is conical, acute; P5 endopod long, reaching distal margin of right P5 exopod 1 (Fig. 21.7 I) .... *"Diaptomus" carinifera* Lowndes,1934 [Paraguay]

21(20) Female P5 endopod being 3−4× smaller than inner margin of first exopodal article; right first antenna with strong spiniform processes on articles 10, 11, and 13 (rarely on 8); second article of left male P5 shorter than first, not sharply divided into two distal pilose processes (Fig. 21.7 J) .... ***Leptodiaptomus* [p. 694]**

21′ Female P5 endopod almost as long as inner margin of first exopodal article; right first antenna with strong spiniform processes on articles 10, 11, and 13; second article of left male leg 5 longer than the first, clearly divided into two pilose processes (Figs. 21.7 K, 14 G) .... *Hesperodiaptomus morelensis* Granados-Ramírez & Suárez-Morales, 2003 [Mexico]

22(13) Absence of small knob in all body surface 23

22′ Males and females with chitinous knobs on body surface; P5 and dorsal surface of pedigers and urosome; male right P5 endopodwell developed, longer than right P5 exopod 1; female genital double-somite asymmetrical, left side with acute projection and apical sensillum, right side with distal lateral expansion (Fig. 21.7 L) .... ***Odontodiaptomus* [p. 694]**

23(22) Absence of knobs on right P5 basepodite and body 24

23′ Chitinous knobs present only on the right P5 basepodite and absent on body (Fig. 21.8 A) .... *Tumeodiaptomus* [Chile]

24(23) P5 exopod 1 with lateral spine 25

24′ Male right P5 basepodite ornamented along inner margin; rigth P5 endopodwell developed, longer than right P5 exopod 1; P1 exopod 1 without lateral spine (0-0-1); female lateral wings symmetrical; genital double-somite slightly asymmetrical, left sensillum inserted posteriorly, right one horizontally (Fig. 21.8 B) .... *Aspinus acicularis* Brandorff, 1973 [Brazil]

25(24) Pediger 5 with small dorsal spinae or smooth, other characters not as described below 26

25′ Large dorsal spinules on distal margin of pediger 5, visible in 40×, especially on the lateral surface, dorsal surface furnished with small spinules or smooth; female pediger 5 lateral wings well developed, posteriorly projected, reaching beyond midlength of genital double-somite; internal setae of P5 exopod 2 long, almost reaching distal end of terminal claw (Fig. 21.8 C) .... *Dasydiaptomus coronatus* (G.O. Sars, 1901) [Brazil]

26(25) Absence of a cleft with small granulations along male right P5 basepodite, lack of sclerotized process on right P5 exopod 1; female pediger 5 lateral wings and genital double-somite lateral expansions clearly asymmetrical 27

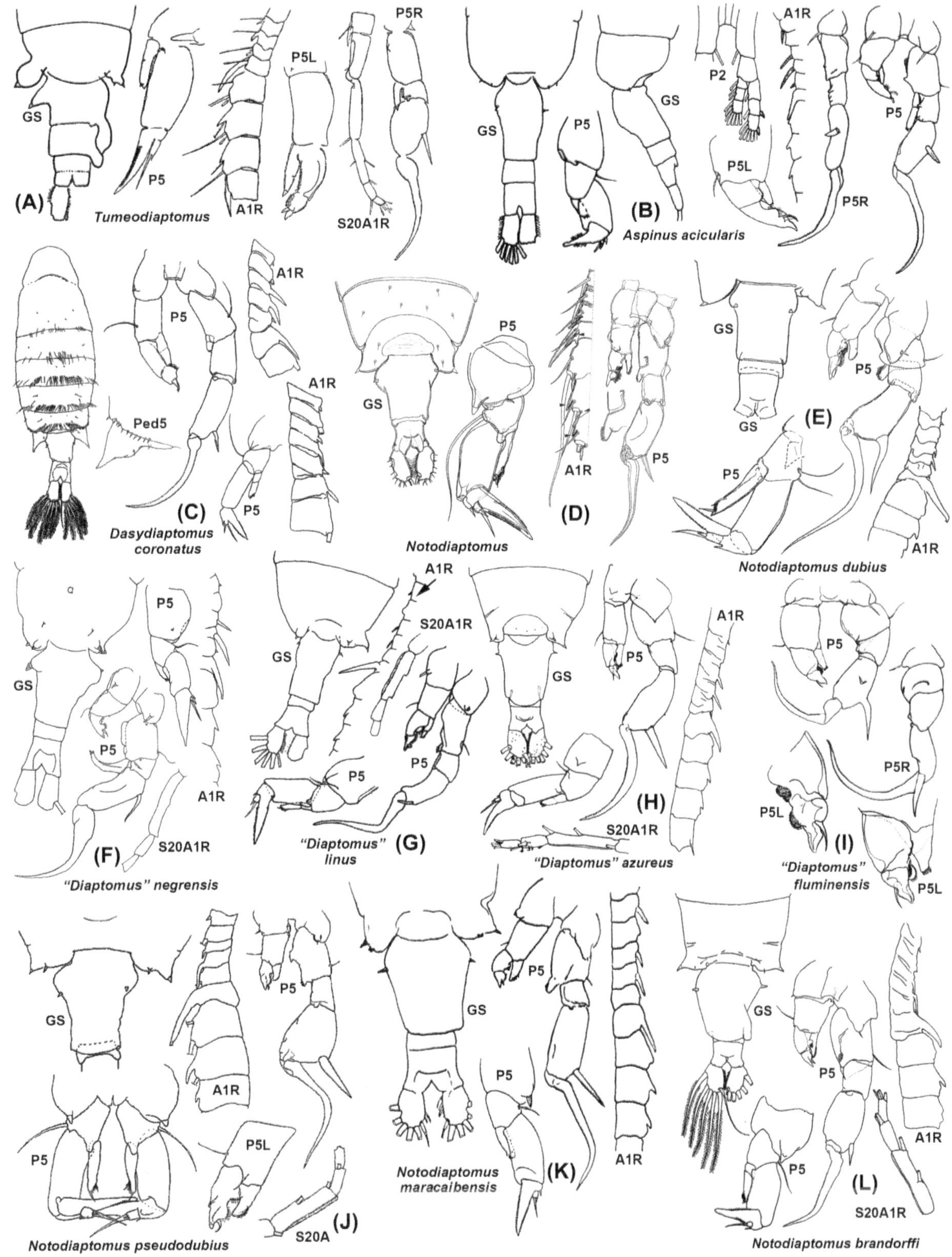

FIGURE 21.8 Diaptomidae species: (A) *Tumeodiaptomus*; (B) *Aspinus acicularis*; (C) *Dasydiaptomus coronatus*; (D) *Notodiaptomus nordestinus* complexs; (E) *Notodiaptomus dubius*; (F) "*Diaptomus*" *negrensis*; (G) "*Diaptomus*" *linus*; (H) "*Diaptomus*" *azureus*; (I) "*Diaptomus*" *fluminensis*; (J) *Notodiaptomus pseudodubius*; (K) *Notodiaptomus maracaibensis*; (L) *Notodiaptomus brandorffi*. Abbreviations: *A1R*, right first antenna; *GS*, genital double-somite; *P5L*, left fifth leg; *P5R*, right fifth leg; *Ped*, pediger; *S*, article. *Redrawn from: (A) Brehm (1935); (B) and (G) Brandorff (1973); (C) Sars (1901); (E) Dussart (1985); (F) Andrade & Brandorff (1975); (H) and (I) Reid (1985a); (J) Defaye & Dussart (1988); (K) Kiefer (1954); (L) Reid (1987b).*

26′ Cleft with small granulations along male right P5 basepodite plus sclerotized process in distal margin of posterior surface of right P5 exopod 1; female pediger 5 lateral wings and genital double-somite lateral expansions relatively symmetrical (Fig. 21.8 D) .... ***Notodiaptomus sensu stricto (nordestinus*** **complex) [p. 696]**

27(26) In males, without well-developed projection in the distal portion of right P5 coxa; other male and female characters not as described below 28

27′ Spinous process of male right first antenna article 13 reaching distal margin of article; well-developed projection at distal part of right P5 coxa; inner margin of right P5 basepodite smooth; lateral spine of right P5 exopod 2 inserted medially, proximally curved, length 2.5× the terminal claw; in females, lateral wings of pediger 5 symmetrical; genital double-somite with symmetrical expansions and sensilla posteriorly positioned (Fig. 21.8 E) .... *Notodiaptomus dubius* Dussart, 1985

28(27) Male right P5 endopod short, not triangular or trapezoidal; female lateral wings and lateral expansions of genital double-somite not as described below 29

28′ Male right P5 endopod long, trapezoidal, tapering distally into acute tip, as long as bearing article; lateral spine of right P5 exopod 2 inserted proximally on article, reaching almost its midlength; female lateral wings of pediger 5 small, asymmetrical, left not conical shape as right, inner sensilla at tip of expansion, right sensilla larger than left; genital double-somite asymmetrical, left sensilla curved posteriorly and larger than right side, which is smaller, curved anteriorly (Fig. 21.8 F) .... *"Diaptomus" negrensis* Brandorff, 1976 [Brazil]

29(28) Male right first antenna with spinous process in article 13 not reaching distal end of article 14, lateral spine of male right P5 exopod 2 not short; left sensillum of genital double-somite lateral expansion inserted on projection with varied shapes except circular or without expansion; both sensilla of genital double-somite directed in other directions except posteriorly 30

29′ Male right first antenna with spinous process in article 13 reaching distal end of article 14, with spinous process on articles 15 and 16; internal margin of right P5 basepodite smooth; lateral spine of right P5 exopod 2 inserted close to terminal claw, shorter than segmental width; female lateral wings asymmetrical, left shorter than right, inserted on semicircular projection, differing from right side; genital double-somite slightly asymmetrical, with both sensilla directed posteriorly (Fig. 21.8 G) .... *"Diaptomus" linus* Brandorff, 1973 [Brazil]

30(29) Inner margin of male right P5 basepodite not smooth, with different combination of characters as described below; specimens without blue pigmentation when alive 31

30′ Inner margin of male right P5 basepodite smooth, rigth P5 endopod not conical; female lateral wings of pediger 5 asymmetrical, right larger than left; genital double-somite slightly asymmetrical; specimens showing blue pigmentation (Fig. 21.8 H) .... *"Diaptomus" azureus* Reid, 1985 [Brazil]

31(30) Male right P5 coxa lacking a long, thin spine on expansion; right P5 basepodite smaller than 1.6 and larger than 2× as wide as long; lateral spine of right P5 exopod 2 inserted on other parts of the external margin of right P5 exopod 2, not in the middle 32

31′ Male right P5 coxa with long, thin spine on large expansion; right P5 basepodite approximately 1.8× as wide as long; presence of a well-developed sclerotized process on surface of right P5 exopod 2; lateral spine of right P5 exopod 2 inserted slightly beyond midlength of outer margin of right P5 exopod 2, as long or longer than segmental width (Fig. 21.8 I) (female unknown) .... *"Diaptomus" fluminensis* Reid, 1985 [Brazil]

32(31) Female genital double-somite sensilla not curved posteriorly, combination of characters of male different from those described below 33

32′ Male with small spinous process in articles 14 and 15 of right first antenna; knob on inner margin of right P5 basepodite; lateral spine of right P5 exopod 2 measuring 2.5× the length of terminal claw; female lateral wings slightly asymmetrical; genital double-somite asymmetrical, left expansion larger than right, both sensilla curved posteriorly; lateral setae of P5 basepodite long, longer than endopod length; exopod 2 plus terminal claw long, equaling the P5 exopod 1 length (Fig. 21.8 J) .... *Notodiaptomus pseudodubius* Dussart & Defaye, 1988 [French Guiana]

33(32) Male right first antenna article 15 with spinous process; male right P5 exopod 2 not 2.5× longer than wide; lateral spine of P5 exopod 2 not inserted subdistally, with different proportions and characters than those described below 34

33′ Male right first antenna article 15 lacking spinous process; inner margin of right P5 basepodite smooth; right P5 endopod short; right P5 exopod 2 almost 2.5× longer than article width; lateral spine of P5 exopod 2 inserted subdistally, 0.25× as long as terminal claw; female lateral wings of pediger 5 asymmetrical, left side with both sensilla inserted closer to each other than on right side; genital double-somite 1.5× longer than wide, slightly asymmetrical, both sensilla arranged horizontally to body axis (Fig. 21.8 K) .... *Notodiaptomus maracaibensis* Kiefer, 1954 [Venezuela]

34(33) Characters not combined as below 35

34′ Male with granulose ornamentation along inner margin of right P5 basepodite, which is 1.5× longer than wide; right P5 endopod short, inserted on lateral expansion of right P5 coxa; P5 exopod 2 about 1.7× longer than wide, lateral spine short, about half the article width and inserted subdistally at 2/3 of outer margin of right P5 exopod 2; female lateral wings of pediger 5 slightly asymmetrical; genital double-somite asymmetrical, with larger expansion on right side, this sensillum slightly curved anteriorly, torsion of distal dorsal surface of genital double-somite is observable; P5 endopod not reaching the terminal claw, external seta of P5 exopod 2 not reaching distal end of terminal claw (Fig. 21.8 L) .... *Notodiaptomus brandorffi* Reid, 1987 [Brazil]

35(34) Characters not combined as below 36

35′ Inner margin of male right P5 basepodite smooth; right P5 endopod long, reaching distal margin of right P5 exopod 1, diamond-shaped; female lateral wings asymmetrical, left side bifid with sensilla at tips; genital double-somite asymmetrical, with semicircular expansion on the left side, sensillum inserted dorsally on small projection; left sensillum inserted on small conical expansion arising from another curved projection; P5 endopod long, reaching beyond the insertion of terminal claw; internal and external setae of P5 exopod 2 short (Fig. 21.9 A) .... *Colombodiaptomus brandorffi* Gaviria, 1989 [Colombia]

36(35) Characters not combined as below 37

36′ Male right P5 basepodite with semicircular knob on inner margin, adjacent to coxa; right P5 basepodite twice as long as wide; lateral spine of right P5 exopod 2 inserted close to insertion of terminal claw, as wide as article; female pediger 5 lateral wings slightly asymmetrical, left side slenderer than right; genital double-somite asymmetrical, left sensillum inserted more distally than right one (Fig. 21.9 B) .... *"Diaptomus" dilatatus* Dussart, 1984 [Venezuela]

37(36) Characters not combined as below 38

37′ Small process on distal margin of male right first antenna article 20; external margin of right P5 exopod 1 twice as long as wide; lateral spine of right P5 exopod 2 inserted beyond midlength of external margin, measuring half the width of bearing article; female with conical coxal sensillum inserted; P5 endopod long, reaching beyond middle of P5 exopod 1 (Fig. 21.9 C) .... *"Diaptomus" leoninicollinus* Marsh, 1913 [Panama]

38(37) Characters not combined as below 39

38′ Male right first antenna articles 14, 15 and 16 lacking spinous processes; right P5 basepodite 1.6−1.8× as long as wide, with inner proximal knob, lateral seta of article long, at least as long as bearing article; right P5 endopod long, almost reaching distal margin of right P5 exopod 1; right P5 exopod 2 2.5× longer than wide; lateral spine of right P5 exopod 2 inserted beyond midlength of outer margin of article, 2.5× the length of terminal claw; left P5 basepodite more than twice as long as wide; left P5 endopod long, as wide as proximal width of left P5 basepodite; female pediger 5 lateral wings asymmetrical, conical; genital double-somite twice as long as wide, slightly asymmetrical, small sensilla; P5 with large projection on coxa, 2.5× as long as wide; lateral seta of P5 basepodite long, almost reaching distal end of P5 exopod 1; P5 endopod reaching middle of P5 exopod 15 (Fig. 21.9 D) .... *Notodiaptomus susanae* (Paggi, 1976) [Argentina, Uruguay]

39(38) Characters not combined as below 40

39′ Female with slightly asymmetrical lateral wings, expansions well developed; genital double-somite slightly asymmetrical, with both sensilla inserted posteriorly, genital double-somite about 1.5× longer than wide; P5 endopod long, reaching distal end of P5 exopod 1; external seta of P5 exopod 2 long, almost reaching distal end of terminal claw; male right P5 coxa with long expansion reaching right P5 basepodite distal margin; inner margin of right P5 basepodite ornamented; right P5 endopod twice as long as wide, reaching distal end of right P5 exopod 1; right P5 exopod 1 almost twice as wide as long; right P5 exopod 3 about 2.5× longer than wide at its narrowest part; lateral spine of right P5 exopod 2 inserted beyond midlength of outer margin, curved proximally, at least 2.5× as long as terminal claw (Fig. 21.8 E) .... *Notodiaptomus dubius* Dussart, 1985 [Brazil]

40(39) Characters not combined as below 41

40′ Female genital double-somite with rounded expansions and lacking dorsal projection, right side of genital double-somite expanded posteriorly over succeeding urosomite; P5 exopod 2 lacking lateral spine; male right P5 exopod 1 broader than long; inner margin of right P5 exopod 2 with protruding projection and inner knob on posterior surface; left P5 exopod 2 without spiniform projection, inner setose pad of left P5 exopod 2 not prominently rounded (Fig. 21.9 E) .... *Notodiaptomus coniferoides* Wright, 1927 [Brazil, Argentina, Uruguay]

41(40) Characters not combined as below 42

41′ Female genital double-somite asymmetrical, left side represented by conical projection and adjacent rounded expansion, right side with conical projection only; P5 with short lateral spine; male right P5 exopod 1 longer than broad; right P5 exopod 2 lacking knob; left P5 exopod 2with spiniform projection and prominently rounded inner setose pad (Fig. 21.9 F) .... *Notodiaptomus simillimus* Cicchino, Santos-Silva & Robertson, 2001 [Venezuela, Brazil]

42(41) Distal left exopod of male P5 acute or bulbous, without hairy setae at its base, with one or two ornamented terminal setae; left P5 exopod 2 not forming chela-like structure; process on antepenultimate article of male right first antenna present 43

42′ Male left P5 exopod 2 rectangular, inner margin pilose; ramus ends in chela-like structure, with two distinct processes, one short, denticulate, the other digitiform; female first antennae with one seta on article 11, 15−17; process on antepenultimate article of male right first antenna absent (Fig. 21.9 G) .... *Skistodiaptomus* [United States]

43(42) Characters not combined as below 44

43′ Female lateral wings slightly asymmetrical, left longer and wider than right; genital double-somite symmetrical, less than 1.5× longer than wide; lateral seta of right P5 basepodite reaching midlength of P5 endopod; P5 endopod reaching beyond midlength of P5 exopod 1; inner seta of P5 exopod 2 reaching beyond midlength of terminal claw; male right first antenna with spinous process on articles 14, 15, and 16,

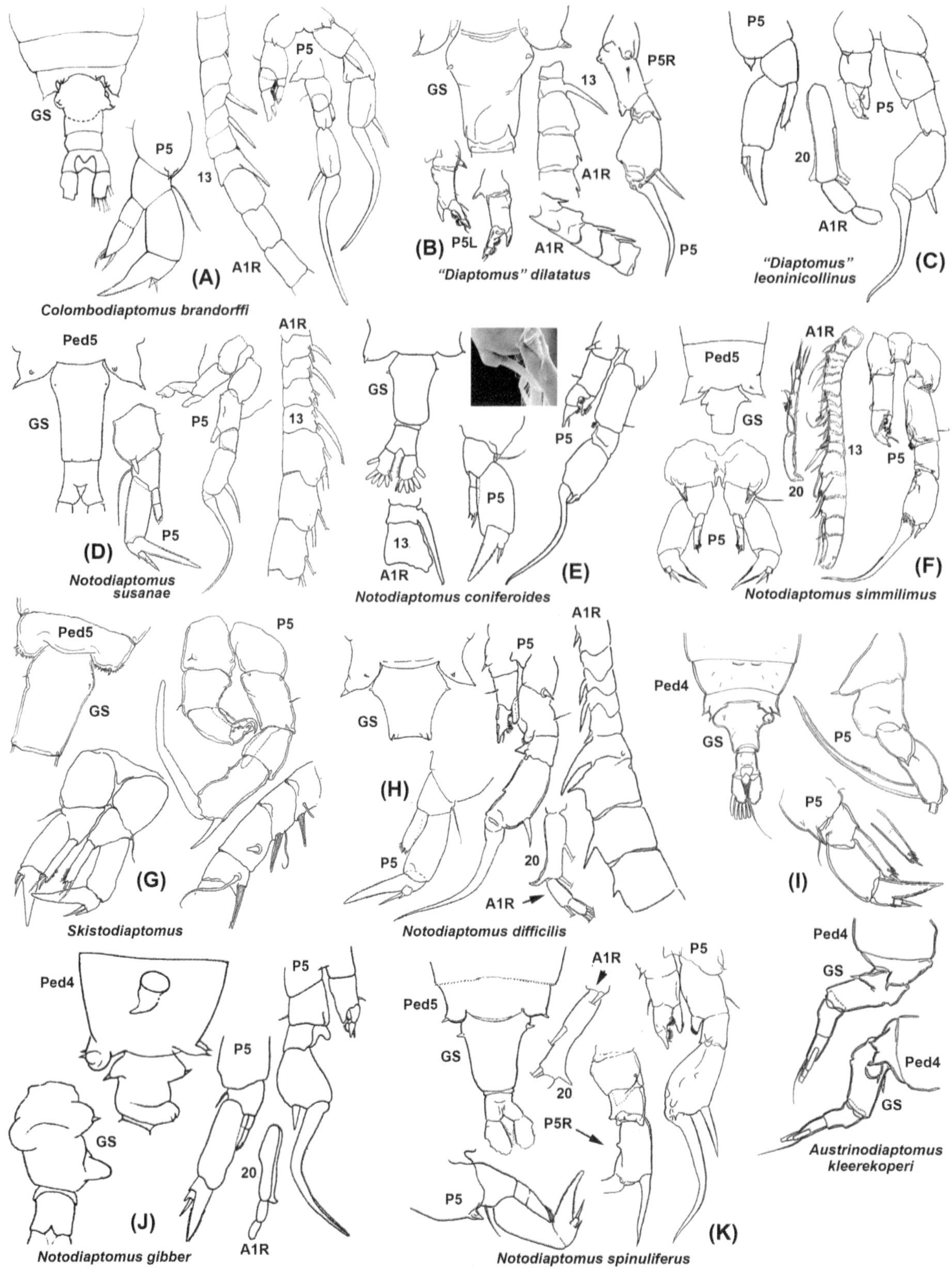

FIGURE 21.9 Diaptomidae species: (A) *Colombodiaptomus brandorffi*; (B) "*Diaptomus*" *dilatatus*; (C) "*Diaptomus*" *leoninicollinus*; (D) *Notodiaptomus suzanae*; (E) *Notodiaptomus coniferoides*; (F) *Notodiaptomus simmilimus*; (G) *Skistodiaptomus*; (H) *Notodiaptomus difficilis*.; (I) *Austrinodiaptomus kleerekoperi*. (J) *Notodiaptomus gibber*; (K) *Notodiaptomus spinuliferus*. Abbreviations: *A1R*, right first antenna; *GS*, genital double-somite; *P5*, fifth leg; *Ped*, pediger.

longest on article 15; small spinous process on right first antenna article 20, 2.5× longer than wide; right P5 basepodite 1.5× longer than wide; right P5 endopod 1.5–1.7× longer than wide, reaching midlength of right P5 exopod 1; inner acute expansion on right P5 exopod 1; right P5 exopod 2 about 2.5× longer than wide; lateral spine of right P5 exopod 2 short, inserted at 2/3 of article, as long as the width of the basis of right P5 exopod 2; left P5 basepodite 1.5× longer than wide (Fig. 21.9 H) .... *Notodiaptomus difficilis* Dussart & Frutos, 1986
[Argentina]

44(43) Characters not combined as below 45

44′ Male right P5 endopod 1.5× longer than wide; external margin of right P5 exopod 2 2.5× longer than article width; lateral spine of right P5 exopod 2 inserted on proximal 1/3 of article, shorter than the width of article and at least 5× as long as terminal claw; female lateral wings of pediger 5 asymmetrical, double on each side, conical; genital double-somite asymmetrical, both sides with rounded expansions, right side larger; P5 endopod reaching distal end of P5 exopod 1; outer seta of P5 exopod 2 reaching midlength of P5 exopod 2 inner seta (Fig. 21.9 I) .... *Austrinodiaptomus kleerekoperi* Reid, 1997
[Brazil]

45(44) Characters not combined as below 46

45′ Male and female with rows of spinules along distal margin of pediger 5; male right first antenna with short spinous process on article 20; inner margins of right P5 basepodite and left P5 basepodite smooth; right P5 endopod short, 1.5× longer than wide; right P5 exopod 2 about 1.7× longer than wide; lateral spine of right P5 exopod 2 inserted distally on article, slightly longer than segmental width; female lateral wings slightly asymmetrical; genital double-somite expansions asymmetrical, right side larger than left, rounded, sensillum at tip curved, anteriorly directed; inner setae of P5 exopod 2 reaching middle of terminal claw (Fig. 21.9 K) .... *Notodiaptomus spinuliferus* Dussart & Matsumura-Tundisi (1986)
[Brazil, Argentina]

46(45) Characters not combined as below 47

46′ Male right first antenna with spinous processes on articles 13, 15, and 16; inner margin of P5 basepodite smooth; lateral spine of right P5 exopod 2 inserted in the middle part of the article, about 2.5–3× as long as terminal claw; female lateral wings of pediger 5 symmetrical, with sensilla curved posteriorly; genital double-somite slightly asymmetrical, rounded, with sensilla curved posteriorly, left one inserted more distally than right one (Fig. 21.10 A) .... *Notodiaptomus paraensis* Dussart & Robertson, 1984
[Brazil]

47(46) Characters not combined as below 48

47′ Male right first antenna with well-developed spinous process on articles 15 and 16; lateral spine of right P5 exopod 2 inserted medially portion on article, measuring between 2.5–3× the length of terminal claw; left lateral expansion of male genital double-somite curved posteriorly; females with three spinules on distal margin of pediger 5 on middle portion on dorsal view; lateral wings of pediger 5 asymmetrical, left side longer than right; genital double-somite asymmetrical, with conical expansion curved posteriorly on left side, with a sensilla at tip; right side without expansion (Fig. 21.10 B) .... *Notodiaptomus kieferi* Brandorff, 1973
[Brazil]

48(47) Characters not combined as below 49

48′ Male with inner conical process on coxa of right P5; basis as long as wide, with surface ornamented; lateral spine of right P5 exopod 2 inserted medially on article, about as long as segmental width; female P5 endopod articulated, inner setae of P5 exopod 2 reach midlength of terminal claw (Fig. 21.10 C) .... *Notodiaptomus echinatus* Lowndes (1934)
[Paraguay, Argentina]

49(48) Characters not combined as below 50

49′ Male urosome 3 with triangular dorsal process with apical chitinous knobs; right first antenna with well-developed spine process on articles 10, 11, and 13, process on article 11 longer than on article 13; right P5 endopod well developed, triangular; lateral spine of right P5 exopod 2 inserted proximally, curved posteriorly measuring more than the length of article; left P5 coxa wider than long; female lateral wings of pediger 5 asymmetrical, left longer than right, sensillum curved anteriorly; genital double-somite asymmetrical, with sensilla of both sides inserted on semicircular projections, left side longer than right; lateral seta of P5 coxa long, almost reaching the end of terminal claw (Fig. 21.10 D) .... *"Diaptomus" curvatus* Perbiche-Neves, Boxshall & Paggi, 2013
[Argentina]

50(49) Characters not combined as below 51

50′ Male right first antenna with small process at distal end of article 20, 2× longer than wide; inner margin of right P5 basepodite smooth; right P5 endopod well developed, 2× longer than wide; right P5 exopod 2 2× longer than wide, lateral spine of article inserted close to terminal claw, measuring 0.2× as long as terminal claw; female genital double-somite slightly asymmetrical; P5 with P5 exopod 1 almost wider than it is long (Fig. 21.10 E) .... *"Diaptomus" gatunensis* Marsh, 1911
[Panama]

51(50) Characters not combined as below 52

51′ Males lateral spine of right P5 exopod 2 shorter than the length of right P5 exopod 1, spine inserted distally; female with single row of dorsal spinules on pediger 4; lateral wings of pediger 5 symmetrical; genital double-somite symmetrical except for left sensilla positioned posteriorly and right sensilla anteriorly (Fig. 21.10 F) .... *Notodiaptomus anceps* Brehm, 1958
[Argentina]

52(51) Male and female without teeth on outermost caudal setae; male lateral spine of right P5 not short and curved; female genital double-somite not asymmetrical as described below 53

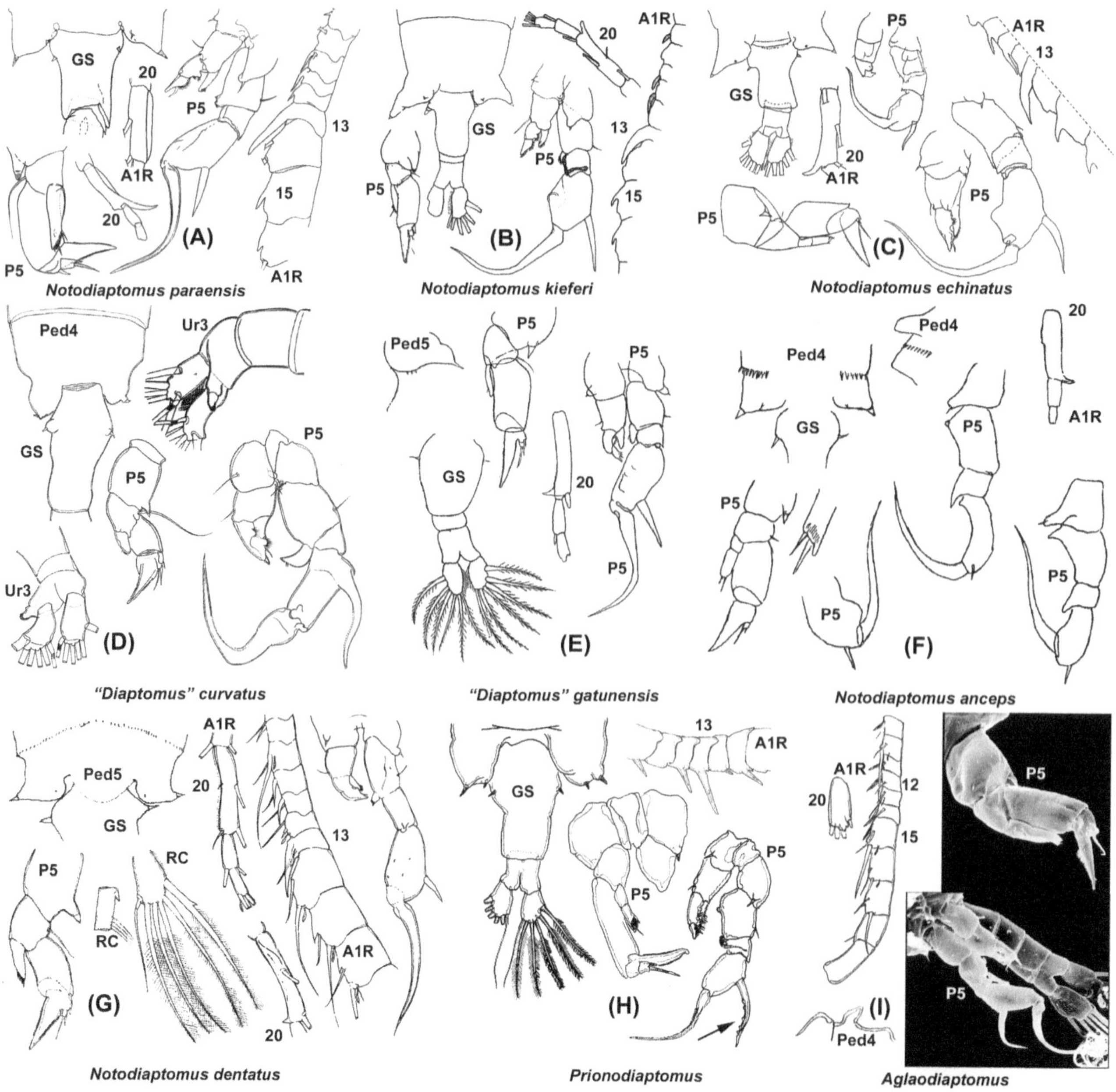

FIGURE 21.10 Diaptomidae species: (A) *Notodiaptomus paraensis*; (B) *Notodiaptomus kieferi*; (C) *Notodiaptomus echinatus*; (D) "*Diaptomus*" *curvatus*; (E) "*Diaptomus*" *gatunensis*; (F) *Notodiaptomus anceps*; (G) *Notodiaptomus dentatus*; (H) *Prionodiaptomus* sp.; (I) *Aglaodiaptomus* sp. Abbreviations: *A1R*, right first antenna; *GS*, genital double-somite; *Ped*, pediger; *RC*, caudal ramus; *Ur*, urosome.

52′ Male with a tooth on the outermost caudal seta; lateral spine of right P5 exopod 2 short, curved, ~ 5× smaller than terminal claw, inserted on distal 1/3 of article; female with dorsal row of spinules on pediger 4; lateral wings of pediger 5 asymmetrical, sensillum on left side posteriorly directed, inserted on a projection, right sensillum transverse to body axis; genital double-somite asymmetrical, with left expansion rounded and longer than right, sensilla equally long as wide, right side less developed than left, triangular, with sensilla 2× longer than wide (Fig. 21.10 G) .... *Notodiaptomus dentatus* Paggi, 2001
[Argentina]

53(52) Male right first antenna articles 14–16 without spinous processes; female P5 with aculeus inserted on proximal third of article furnished with several distinctive denticles along inner margin; basis of male right P5 with two papilliform processes on the inner margin; right P5 endopod present (Fig. 21.10 H) .... *Prionodiaptomus*
[Mexico, Guatemala, Venezuela, Colombia, Peru]

53′ Male right first antenna with strong spiniform processes on articles 10, 11, and 13; antepenultimate article with spiniform lamella or short process; female P5 not particularly long or thin, right P5 endopod vestigial or absent (Fig. 21.10 I) .... *Aglaodiaptomus clavipes* (Schacht, 1897) [Mexico]

Calanoida: Diaptomidae: *Mastigodiaptomus*: Species

1 Dorsal surface of urosomites smooth; male right P5 with aculeus and terminal claw not as below; female characters not as below 2

1′ Dorsal surface of urosomites ornamented with rows of spinules; male right P5 with aculeus and terminal claw with whiplike, looped tips; female urosome wide, strong; genital double-somite with deep transverse wrinkles on posterodorsal and posteroventral surface and with sinuous right posterior margin (Fig. 21.11 G) .... *Mastigodiaptomus suarezmoralesi* Gutiérrez-Aguirre & Cervantes-Martínez, 2013 [Mexico]

2(1) Spine on antenular article 16 absent or reduced; P5 basepodite 2 with processes different to those described below 3

2′ Spine on male right first antenna article 16 strongly developed; P5 basepodite 2 with proximal subrectangular projection plus two bulbous processes on inner margin female genital double-somite elongated, 2.2× longer than wide, with weakly developed lateral spinae (Fig. 21.11 A) .... *Mastigodiaptomus reidae* Suárez-Morales & Elías-Gutiérrez, 2003 [Mexico: Yucatan. Some morphological variability of this species was reported by Brandorff (2012)]

3(2) Large copepods (more than 2.2 mm, excluding caudal ramus) 4

3′ Small or medium-sized copepods (less than 2.0 mm) 5

4 Right P5 exopod 1 with a transverse cuticular ridge; right first antenna with fanglike process; sensillum on P5 basepodite 1 long, strong, length/width ratio > 2 .... *Mastigodiaptomus purpureus* (Marsh, 1907) [Cuba, Haiti]

4′ Right P5 exopod 1 with tongue-shaped process; antepenultimate article of right antennule with rounded process, distally knob-shaped (Fig. 21.11 B) .... *Mastigodiaptomus maya* Suárez-Morales & Elías-Gutiérrez, 2003 [Mexico: Yucatan]

5(3) Male right P5 exopod 2 with long lateral spine, its length equals $^3/_4$ of article or is as long as article 6

5′ Male right P5 exopod 2 with lateral spine relatively short, less than half of article length; female P5 with unsegmented endopod 8

6(5) Same structure than below, without process, female P5 endopod 2-articulated 7

6′ Male right P5 basepodite 2 with butterfly-shaped sclerotized process on the inner margin; female P5 endopod 2-articulated 10

7(6) Male left P5 exopod lacking distal denticles; rigth P5 endopod short, not reaching distal margin of exopod; female first antennae reaching beyond distal end of caudal setae (Fig. 21.11 C), P5 endopod not reaching beyond midlength of exopod 1 .... *Mastigodiaptomus nesus* Bowman, 1986 [Mexico, Central America]

7′ Left P5 exopod with distal denticles; right P5 endopod long, reaching distal margin of first exopodal article, female first antenna not exceeding distal end of caudal setae (Fig. 21.11 H), P5 endopod reaching beyond midlength of exopod 1 .... *Mastigodiaptomus texensis* Wilson, 1953 [Mexico, Southern United States]

8(5) Right P5 exopod 2 with inner oblique ridge or smooth 9

8′ Right P5 exopod 2 with semicircular transverse sclerotized lamella on proximal outer surface and inner longitudinal "Y"-shaped ridge (Fig. 21.11 D); lateral spine on distal exopodal article inserted on 2/3 of article; female urosome 2 with straight posterior margin (Fig. 21.11 D), P5 endopod 1-articulated .... *Mastigodiaptomus montezumae* (Brehm, 1955) [Mexico]

9(8) Right P5 exopod 2 with inner oblique ridge (Fig. 21.11 I, arrow); lateral spine on P5 exopod 2 subterminal; female urosome 2 with posterodistal protrusion, P5 endopod 1-articulated .... *Mastigodiaptomus amatitlanensis* (M.S. Wilson, 1941) [Guatemala]

9′ Right P5 exopod 2 smooth, lacking lamellae or ridges on surface; .lateral spine on P5 exopod 2 subterminal, right caudal ramus with ventral wedgelike process (Fig. 21.11 J, arrow), P5 endopod 2-articulated, genital double-somite with process on right posterior margin (Fig. 21.11 J, arrow) .... *Mastigodiaptomus cuneatus* Gutiérrez-Aguirre & Cervantes-Martínez, 2016 [Mexico]

10(6) Process on article 20 of male right first antenna with smooth hyaline membrane, spine on right first antenna article 10 not reaching beyond distal margin of article, short rostral processes; female rostral spinae short, first antennae not reaching distal margin of caudal ramus, P5 endopod with relatively long apical setae, about half the length of endopod 2 (Fig. 21.11 E) .... *Mastigodiaptomus albuquerquensis* (Herrick, 1985) [Mexico, Central America, Southern United States]

10′ Male right antenna article 20 process with serrate hyaline membrane, spine on right first antenna article 10 reaching beyond distal margin of article, long rostral processes; female rostral spinae long, first antennae reaching beyond distal margin of caudal ramus, P5 endopod with short apical setae, less than 1/3 the length of endopod 2 (Fig. 21.11 F) .... *Mastigodiaptomus patzcuarensis* Gutiérrez-Aguirre, Cervantes-Martínez & Elías-Gutiérrez, 2014 [Mexico]

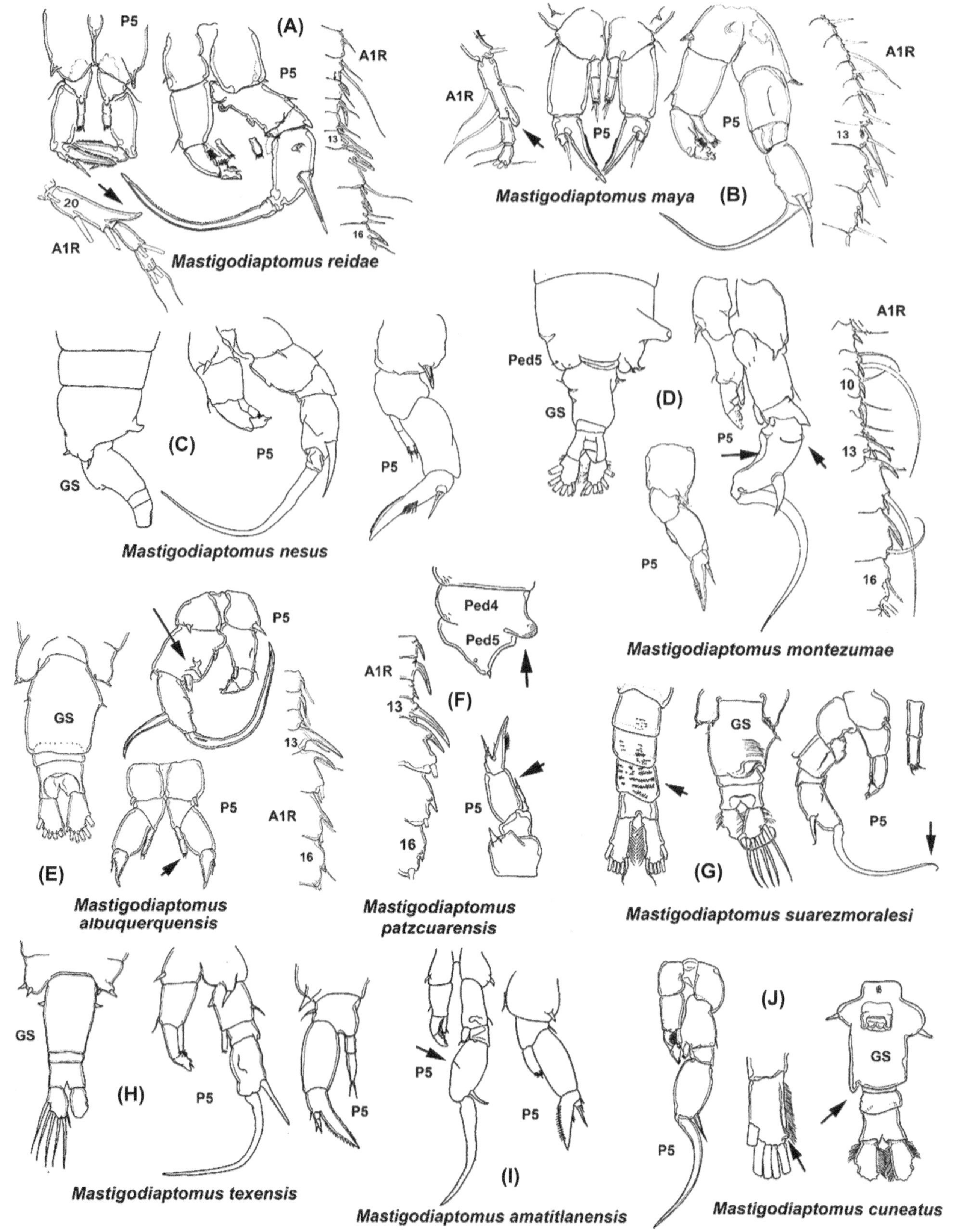

FIGURE 21.11 *Mastigodiaptomus* species: (A) *Mastigodiaptomus reidae*; (B) *Mastigodiaptomus maya;* (C) *Mastigodiaptomus nesus*; (D) *Mastigodiaptomus montezumae*; (E) *Mastigodiaptomus albuquerquensis*; (F) *Mastigodiaptomus patzcuarensis*; (G). *Mastigodiaptomus suarezmoralesi*; (H) *Mastigodiaptomus texensis*; (I) *Mastigodiaptomus amatitlanensis*; (J) *Mastigodiaptomus cuneatus*. Abbreviations: *A1R*, right first antenna; *GS*, genital double-somite; *P5*, fifth leg; *Ped*, pediger. *Modified from: (C) Bowman (1986), (D) Santos-Silva et al. (1996), (E) and (F) Gutiérrez-Aguirre et al. (2014), (G) Gutiérrez-Aguirre & Cervantes-Martínez (2013), (H) and (I) from Wilson (1941), (J) from Gutiérrez-Aguirre & Cervantes-Martínez (2016a).*

Calanoida: Diaptomidae: *Argyrodiaptomus*: Species

A key to the species of the genus is provided by Previattelli (2006). *Argyrodiaptomus neglectus* was not added in the key, because it was not found any more since its description.

Calanoida: Diaptomidae: *Argyrodiaptomus*: Species (Females)

1 P5 endopod no too short, passing the middle of P5 exopod 1 2

1′ P5 endopod short, not reaching the middle of P5 exopod 1 (Fig. 21.12 F) .... *Argyrodiaptomus granulosus* (Brehm, 1933) [Uruguay]

2(1) Female with bilobed lateral wings 3

2′ Lateral wings limited to one main expansion and continuing to the posterior margin of pediger 5 6

3(2) Spinules on dorsal outer margin of endopod 1 of second antenna; absence of spinules between first and second coxal endites of maxilliped; absence of setules on surface of P1 4

3′ Patches of spinules on dorsal outer margin of endopod 1 of second antenna; one group of spinules between first and second coxal endites of maxilliped; setules on anterior surface of the P1 distributed in two patches, longer and more numerous than those on basis 8

4(3) One or more rows of spinules on endopod endopod 1 of second antenna; lines of spinules in the anterior surface of P1 exopod 3 absent; lateral seta of P5 exopod 3 short, not reaching proximal 1/3 of medial seta 5

4′ Lines of spinules on endopod 1 of second antenna absent; anterior surface of third article of P1 exopod 3 with a line of spinules; lateral setae of P5 exopod 3, not reaching midway of medial seta 9

5(4) Absence of spinae on posterior surface of coxa P1 and P2; smooth lateral and posterior margins of the basis P2; P2 coxa without spinae on posterior surface (Fig. 21.12 D) .... *Argyrodiaptomus falcifer* (Daday, 1905) [Argentina]

5′ Spinae on posterior surface of coxa P1; spinae on posterior surface of P2; setules on lateral and posterior surfaces of P2 basepodite; spinae on posterior surface of the coxa (Fig. 21.12 C) .... *Argyrodiaptomus denticulatus* (Pesta, 1927) [Argentina]

6(2) P5 exopod 2 plus terminal claw short than the same length of P5 exopod 1 7

6′ P5 exopod 2 plus terminal claw in the same length of P5 exopod 1; urosome 2 smaller than succeeding somites (Fig. 21.12 I) .... *Argyrodiaptomus nhumirim* Reid, 1997 [Brazil]

7(6) P5 exopod 2 plus terminal claw slightly short than the length of P5 exopod 1; usosome 2 clearly smaller than the succeeding somites (Fig. 21.12 J) .... *Argyrodiaptomus paggi* Previattelli & Santos-Silva, 2007 [Brazil]

7′ P5 exopod 2 plus terminal claw clearly short than the length of P5 exopod 1 (Fig. 21.11 B). .... *Argyrodiaptomus bergi* (Richard, 1897) [Brazil, Argentina, Uruguay]

8(3) Absence of ornamentation on exopod 1; exopod of maxilliped with six distal setae and patches of spinules in the anterior face (Fig. 21.12 K) .... *Argyrodiaptomus robertsonae* Dussart, 1985 [Brazil]

8′ Presence of ornamentation on exopod 1; with spinules between first and second coxal endites of maxilliped; antennules passing the genital double-somite, but not the urosome 2; P5 endopod reaching the end of P5 exopod 1 (Fig. 21.12 A) .... *Argyrodiaptomus azevedoi* (Wright, 1935) [Brazil]

9(4) Left side of lateral wings larger than right (double, at least); antennules symmetrical reaching beyond halfway of genital double-somite (Fig. 21.12 E) .... *Argyrodiaptomus furcatus* (G.O. Sars, 1901) [Brazil]

9′ Lateral wings not as above; antennules length not reaching beyond halfway of genital double-somite; Right P5 endopod 1 with two setae in the inner margin; rigth P5 endopod 1 with a pore between the line of spinules and setae; second article bilobed, with a groove between lobes; internal lobe with eight distal setae (Fig. 21.12 G) .... *Argyrodiaptomus macrochaetus* Brehm, 1937 [Brazil]

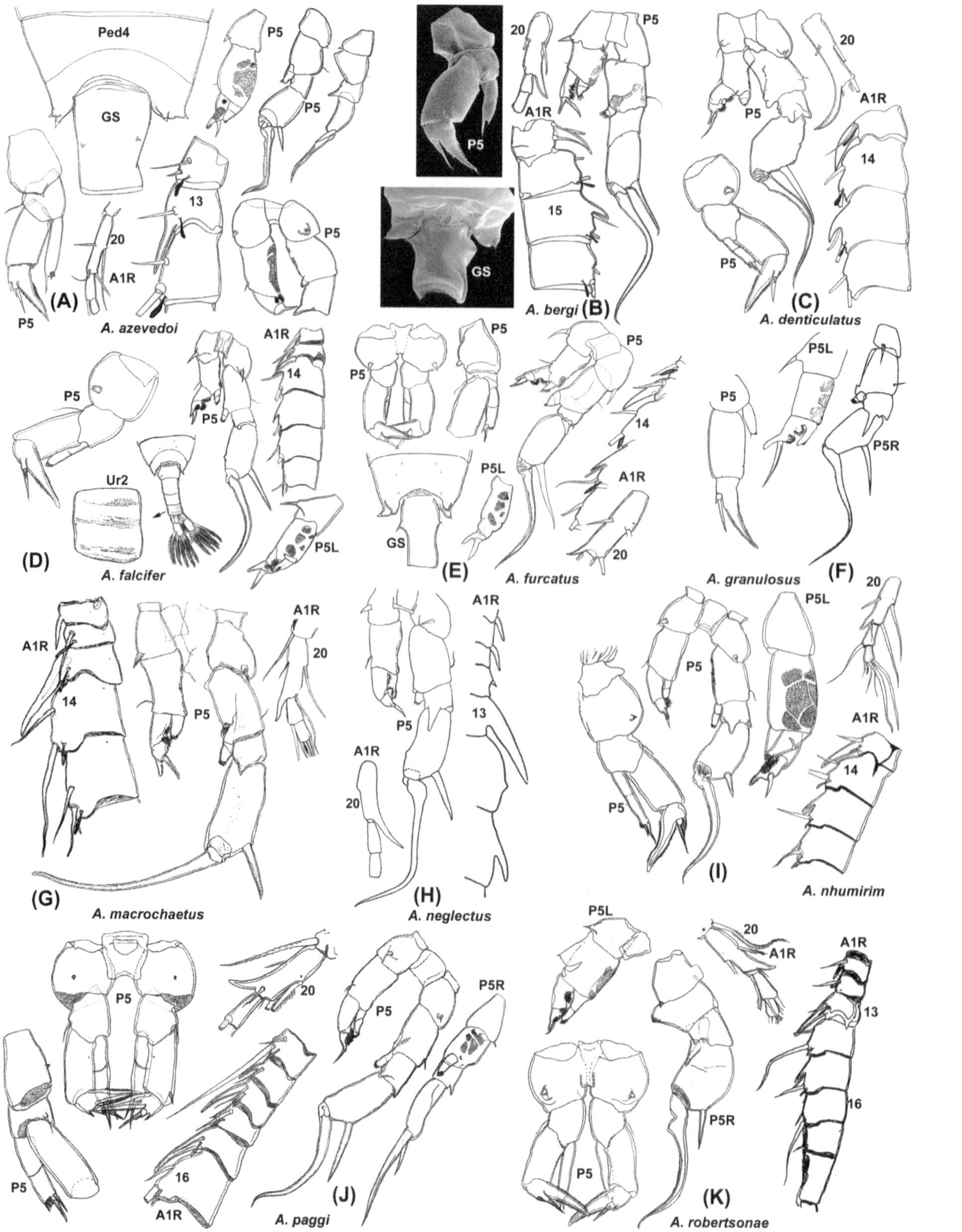

FIGURE 21.12 *Argyrodiaptomus* species: (A) *Argyrodiaptomus azevedoi*; (B) *Argyrodiaptomus bergi*; (C) *Argyrodiaptomus denticulatus*; (D) *Argyrodiaptomus falcifer*; (E) *Argyrodiaptomus furcatus*; (F) *Argyrodiaptomus granulosus*; (G) *Argyrodiaptomus macrochaetus*; (H) *Argyrodiaptomus neglectus*; (I) *Argyrodiaptomus nhumirim*; (J) *Argyrodiaptomus paggi*; (K) *Argyrodiaptomus robertsonae*. Abbreviations: *A1R*, right first antenna; *GS*, genital double somite; *P5*, fifth leg; *P5L*, left fifth leg; *P5R*, right fifth leg; *Ped*, pediger; *Ur*, urosome.

Calanoida: Diaptomidae: *Argyrodiaptomus*: Species (Males)

1 Presence of many patch of spinules in the inner margin of P5 basepodite of both legs 2

1′ Absence or few spinules in right P5 basepodite or botth leg, or presence in left basepodite P5 4

2(1) Lateral spine length of right P5 exopod 2 measuring more than three parts of the terminal claw length 3

2′ Lateral spine length of right P5 exopod 2 approximately 4× smaller than terminal claw length; length of spinous seta on left P5 exopod 2 reaching beyond the digitiform process, but not twice as long as process (Fig. 21.12 I) .... *Argyrodiaptomus nhumirim* Reid, 1997 [Brazil]

3(2) Lateral spine length of right P5 exopod 2 measuring 3× less than the terminal claw length, absence of spinous process on external margin of right first antenna article 15 (XVII); antennules not reaching beyond distal end of prosome (Fig. 21.12 A) .... *Argyrodiaptomus azevedoi* (Wright, 1935) [Brazil]

3′ Lateral spine of right P5 exopod 2 approximately a half-length of terminal claw (Fig. 21.12 J) .... *Argyrodiaptomus paggi* Previattelli & Santos-Silva, 2007 [Brazil]

4(1) Lateral spine of right P5 exopod 2 inserted in the most distal part middle part of this article 5

4′ Lateral spine of right P5 exopod 2 inserted in the middle part of this article, approximately 2.3 smaller than terminal claw length; terminal setae of left P5 exopod 2 long, with similar length of the article left P5 exopod 1 (Fig. 21.12 F) .... *Argyrodiaptomus granulosus* Brehm, 1933 [Uruguay]

5(4) With an expansion or protuberance in the inner margin of right P5 basepodite 6

5′ Without an expansion or protuberance in the inner margin of right P5 basepodite 8

6(5) Apex of expansion of right P5 basepodite in the inner margin smooth 7

6′ Apex of expansion of right P5 basepodite in the inner margin with patch of spinules; spinous process on right first antenna article 15 (XVII); antennules reaching beyond posterior margin of prosome, but not beyond distal portion of urosome 2 (Fig. 21.12 K) .... *Argyrodiaptomus robertsonae* Dussart, 1985 [Brazil]

7(6) Vestigial seta on right first antenna articles 2 (III), 3 (V), and 5 (VII) of right first antenna; absence of spine on right first antenna article 14 (XVI); presence of spinae on posterior surface of P1 and P2 coxa; expansion on right P5 basepodite reaching basis of left P5 basepodite forming digitiform protuberance; lateral spine of right P5 exopod 2 2× smaller than terminal claw length (Fig. 21.12 C) .... *Argyrodiaptomus denticulatus* (Pesta, 1927) [Argentina]

7′ Expansion on right P5 basepodite small and semicircular, lateral spine of right P5 exopod 2 3× smaller than terminal claw length (Fig. 21.12 G) .... *Argyrodiaptomus macrochaetus* Brehm, 1937 [Brazil]

8(5) Lateral spine length of P5 exopod 2 less than 3× smaller than terminal claw length 9

8′ Lateral spine of P5 exopod 2 3× smaller than terminal claw length; terminal setae of left P5 exopod 2 short, measuring a half of the length of left P5 exopod 1; absence of spine in the article 14 of right first antenna (Fig. 21.12 H) .... *Argyrodiaptomus neglectus* (Wright, 1938) [Brazil]

9(8) Spinous process in article 15 of right first antenna not large; P5 exopod 2 in the same length of right P5 coxa plus P5 exopod 2 10

9′ Large spinous process in article 15 of rigth first antenna; P5 exopod 2 smaller in length than coxa plus P5 exopod 2 (Fig. 21.12 B) .... *Argyrodiaptomus bergi* (Richard, 1897) [Brazil, Argentina, Uruguay]

10(9) Vestigial seta on right first antenna articles 2 (III), 3 (V), 5 (VII), and 6 (VIII); spinous process in article 14 (XVI) of right first antenna; absence of spinae on posterior surface of coxa of P1 and P2; expansion on the right P5 basepodite not reaching basis of left P5 basepodite; lateral spine of right P5 exopod 2 2.1× smaller than terminal claw length (Fig. 21.12 D) .... *Argyrodiaptomus falcifer* (Daday, 1905) [Argentina]

10′ Right P5 endopod 1 without proximal inner digitiform protuberance; expansion on left P5 basepodite positioned medially on article; absence of spinous process in article 14 (XVI) of right first antenna; lateral spine of right P5 exopod 2 2× smaller than terminal claw length (Fig. 21.12 E) .... *Argyrodiaptomus furcatus* (G.O. Sars, 1901) [Brazil]

Calanoida: Diaptomidae: *Rhacodiaptomus*: Species (Females)

Female of the species Rhacodiaptomus calatus Brandorff, 1973 was not added in this key because a redescription is needed. See more details in Cicchino et al. (2001).

1 Genital double-somite with large lateral wings 2

1′ Genital double-somite with discrete expansions, right side slightly longer than left (less than double its size); lateral wings of pediger 5 slightly asymmetrical (Fig. 21.13 G) .... *Rhacodiaptomus ringueleti* Cicchino & Dussart, 1991 [Venezuela]

2(1) Genital double-somite with large lateral wings, right side larger than left (almost or more the double its length) 3

2′ Genital double-somite with large lateral wings, right side larger than left (less than double its length), left genital double-somite expansion conical, reaching midlength of genital double-somite; right expansion with large semicircular elongated proximal expansion, reaching beyond midlength of genital double-somite, plus another small distal semicircular expansion (Fig. 21.13 E) .... *Rhacodiaptomus insolitus* (Wright, 1927) [Brazil]

3(2) Left genital double-somite expansion rounded or conical 4

3′ Left genital double-somite expansion cylindrical, reaching beyond midlength of genital double-somite, with a sensilla posteriorly positioned; right genital double-somite expansion semicircular, not reaching midlength of genital double-somite, laterally positioned (Fig. 21.13 D) .... *Rhacodiaptomus flexipes* (Wright, 1927) [Brazil]

4(3) Left expansion of genital double-somite conical 5

4′ Left genital double-somite expansion rounded, not reaching midlength of genital double-somite; right expansion of genital double-somite conical, reaching midlength of article (Fig. 21.13 F) .... *Rhacodiaptomus retroflexus* Brandorff, 1973 [Brazil]

5(4) Left expansion of genital double-somite conical, not reaching the middle of genital double-somite; right expansion of genital double somite also conical, reaching beyond midlength of genital double-somite (Fig. 21.13 C) .... *Rhacodiaptomus calamensis* (Wright, 1927) [Brazil]

5′ Left genital double-somite expansion conical, passing the middle of genital double-somite; right expansion of genital double-somite flat, with a leaf shape and distal portion thin (similar a cone), reaching the end of urosome (Fig. 21.13 A) .... *Rhacodiaptomus besti* Santos-Silva & Robertson, 1993 [Brazil]

Calanoida: Diaptomidae: *Rhacodiaptomus*: Species (Males)

1 Process at distal margin of right P5 basepodite not bifid and thin, lateral spine of right P5 exopod 2 not thin and long 2

1′ Process at distal margin of right P5 basepodite bifid, similar, and thin; lateral spine of right P5 exopod 2 thin and long (Fig. 21.13 A) .... *Rhacodiaptomus besti* Santos-Silva & Robertson, 1993 [Brazil]

2(1) Process at distal margin of right P5 basepodite not bifid, thin, without a large projection; lateral spine of right P5 exopod 2 not thin, long, small and curved 3

2′ Process at distal margin of right P5 basepodite asymmetrical, with a large projection and a small resembling the bifid type; lateral spine of right P5 exopod 2R small and curved (Fig. 21.13 C) .... *Rhacodiaptomus calamensis* (Wright, 1927) [Brazil]

3(2) Right P5 coxa with large process (reaching at least the midlength of basepodite) 4

3′ Right P5 coxa with a medium-sized process not reaching midlength of basepodite, long, forming rectangular lobe with slightly bilobed tips, positioned on inner distal surface of article (Fig. 21.13 G) .... *Rhacodiaptomus ringueleti* Cicchino & Dussart, 1991 [Venezuela]

4(3) Right P5 coxa with large or long process not expanding in two tips 5

4′ Right P5 coxa with a large process (reaching midlength of basepodite) expanding in two tips resembling a fish tail, at distal inner surface of article (Fig. 21.13 E) .... *Rhacodiaptomus insolitus* (Wright, 1927) [Brazil]

5(4) Right P5 coxa with a long process, expanding in the extremity directed to the distal portion 6

5′ Right P5 coxa with a large process (reach the middle of right P5 basepodite length); "armlike" expanding on basepodite and with a constriction in the extremity, positioned posterolaterally for the external region of this article (Fig. 21.13 D) .... *Rhacodiaptomus flexipes* (Wright, 1927) [Brazil]

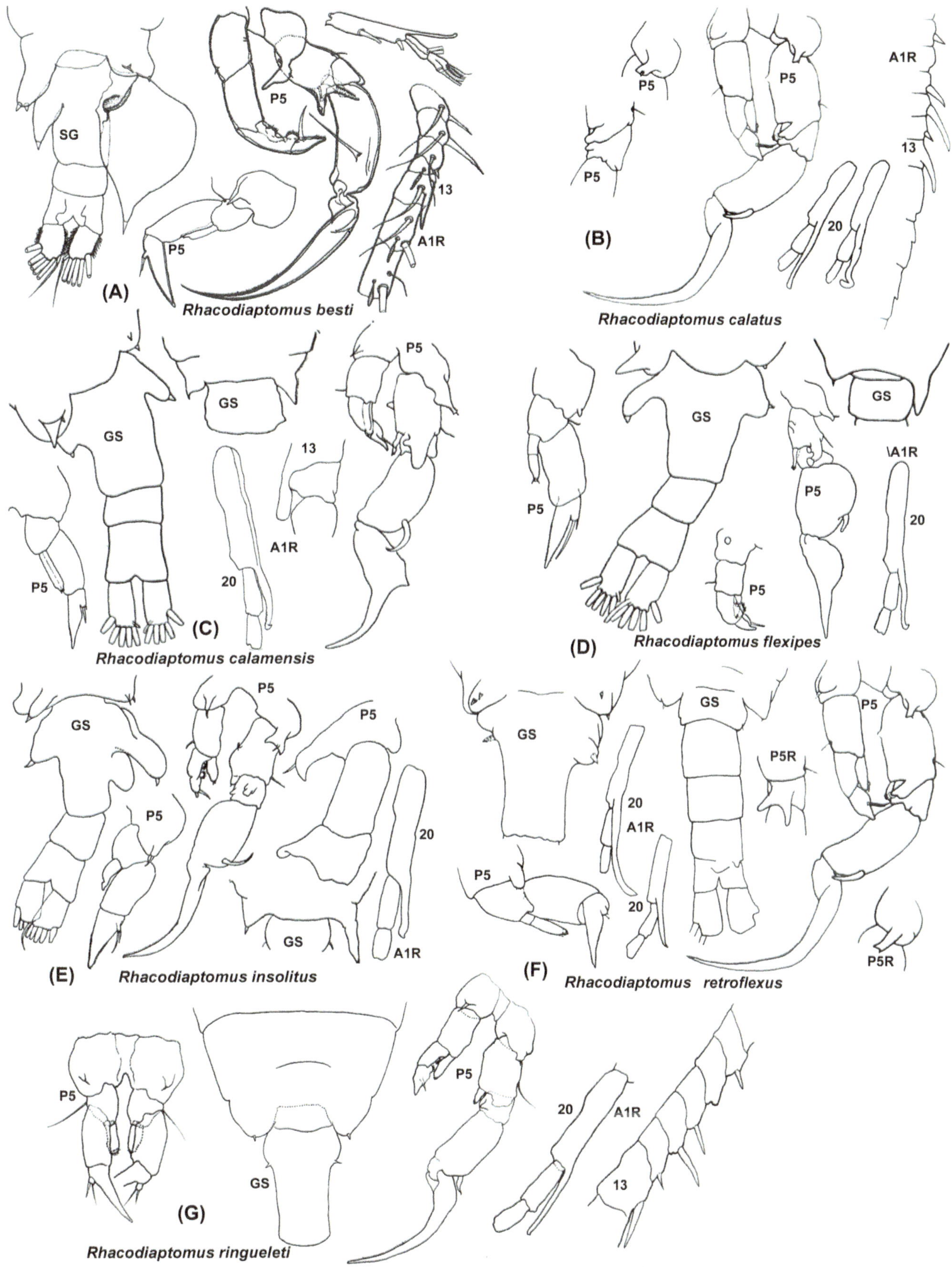

FIGURE 21.13 *Rhacodiaptomus* species: (A) *Rhacodiaptomus besti*; (B) *Rhacodiaptomus calatus*; (C) *Rhacodiaptomus calamensis*; (D) *Rhacodiaptomus flexipes*; (E) *Rhacodiaptomus insolitus*; (F) *Rhacodiaptomus retroflexus*; (G) *Rhacodiaptomus ringueleti*. Abbreviations: *A1R,* right first antenna; *GS,* genital double somite; *P5,* fifth leg.

6(5) Posterior surface of basis P5R large and resembling a pincer of different length, almost equal width and length (Fig. 21.13 A) .... *Rhacodiaptomus retroflexus* Brandorff, 1973
[Brazil]

6′ Elongated process on posterior surface of right P5 basepodite of middle size, process longer than wide (Fig. 21.13 B) .... *Rhacodiaptomus calatus* Brandorff, 1973
[Brazil]

Calanoida: Diaptomidae: *Leptodiaptomus*: Species

A key to the Nearctic species is presented by Reid (2016).

1 Male right first antenna spine on article 13 weakly developed, straight, usually not reaching distal margin of article 14; article 20 of right first antenna with straight or curved distally acute process, which can range from half the length of next article to reach beyond its length; female characters not as described above 2

1′ Male right first antenna with spine on article 13 strongly developed, sinusoidal, reaching half-length of article 16; article 20 of right first antenna with bulbous distal process on inner margin reaching halfway of next article; female genital double-somite asymmetrical, right side with lateral projection with terminal spine; apical outer seta of P5 endopod long, 1.8× longer than inner seta (Fig. 21.14 D) .... *Leptodiaptomus dodsoni* Elías-Gutiérrez, Suárez-Morales & Romano, 1999
[Mexico]

2(1) Male and female characters not as described below 3

2′ Process on article 20 of male right first antenna not reaching halfway of article 21; posterodorsal process on the right side of genital double-somite weakly developed, represented by low protuberance not reaching halfway of succeeding anal somite (Fig. 21.14 E) .... *Leptodiaptomus garciai* (Osorio-Tafall, 1942)
[Mexico]

3(2) Male right first antenna process on article 20 inserted subapically; female characters different from described below 4

3 Process of article 20 of male right first antenna reaching at the middle of article; female genital double-somite with right postero dorsal process well-developed, subtriangular, reaching halfway of succeeding anal somite (Fig. 21.14 A) .... *Leptodiaptomus novamexicanus* (Herrick, 1895)
[Southern USA, Mexico]

4(3) Male right first antenna with spine on article 8; P5 exopod 1 without inner hyaline projection; female genital double-somite without dorsal posterodistal process on right side 5

4′ Male right first antenna lacking spine on article 8; P5 exopod 1 with inner hyaline projection; female characters different from above 6

5(4) Male right first antenna spine on article 8 longer than article and twice as long as spine on article 12; right first antenna with spinae on articles 8, 10, 11, and 13; female genital double-somite pilose along posterodorsal margin, with lateral projections: right one strongly protuberant, left one barely noticeable; genital double-somite 1.1× as long as wide (Fig. 21.14 C) .... *Leptodiaptomus connexus* Light, 1938
[Mexico, USA]

5′ Male right first antenna spine on article 8 shorter than article; right first antenna with spinae on articles 8, 10, 11, 12, and 15, the latter being noticeably prominent; right first antenna article 20 with curved process, longer than succeeding article; urosome 1 with spine on right margin, left side with protruding lobe; urosome 4 with expansion on right side, visible in dorsal position; female genital double-somite different from above; female caudal ramus with setules along outer and inner margins. P5 endopod 1.4× longer than inner margin of exopod 1, with apical setae 0.8–1× the length of the endopod (Fig. 21.14 F) .... *Leptodiaptomus mexicanus* (Marsh, 1929)
[Mexico]

6(4) Coxa of male left P5 with strong spine; process on distal half of right P5 exopod 1 subquadrate; distal claw of right P5 exopod distally acute, lacking inner lamella; female genital operculum proximally straight, lateral projections of genital double-somite posteriorly directed; apical setae of P5 endopod relatively short, unequally long, one being twice as long as the other .... *Leptodiaptomus siciloides* (Lilljeborg in Guerne & Richard, 1889)
[United States, Nearctic]

6′ Coxa of male left P5 without spine; process on distal half of right P5 exopod 1 anvil-shaped; distal claw of terminal right P5 exopod distally rounded, with inner lamella; female genital operculum proximally rounded, lateral projections of genital double-somite directed diagonally; apical setae of female P5 endopod relatively long, with similar length and width (Fig. 21.14 B) .... *Leptodiaptomus cuauhtemoci* (Osorio-Tafall, 1941)
[Mexico]

Calanoida: Diaptomidae: *Odontodiaptomus*: Species (Females)

1 Right genital double-somite side with small distal expansion, not reaching the succeeding urosomite 2

1′ Right margin of genital double-somite with large distal expansion posteriorly directed, almost reaching midlength of succeeding urosomite .... *Odontodiaptomus thomseni* (Brehm, 1933)
[Uruguay]

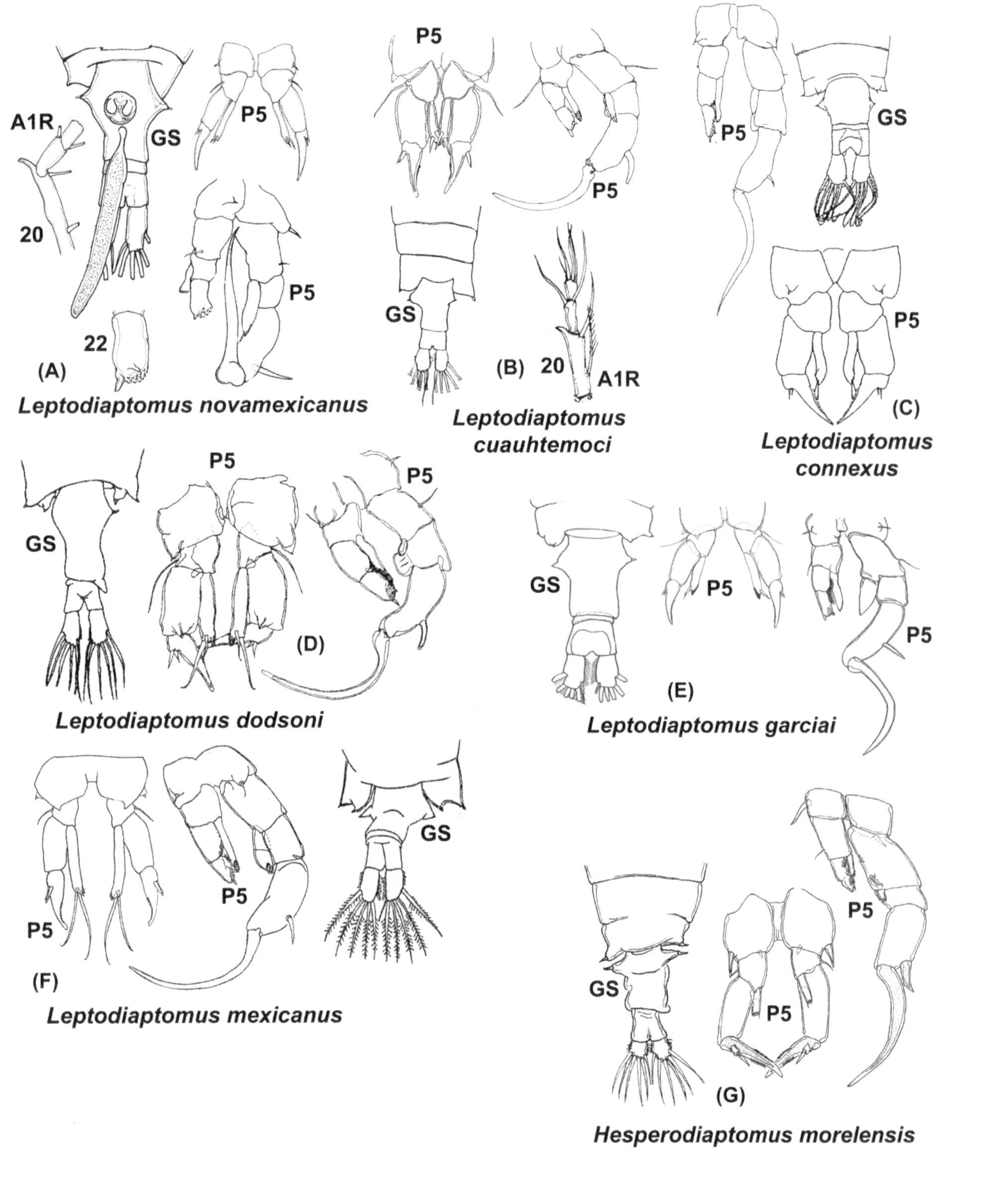

FIGURE 21.14 *Leptodiaptomus* sp. and *Hesperodiaptomus morelensis*: (A) *Leptodiaptomus novamexicanus*; (B) *Leptodiaptomus cuauhtemoci*; (C) *Leptodiaptomus connexus*; (D) *Leptodiaptomus dodsoni*; (E) *Leptodiaptomus garciai*; (F) *Leptodiaptomus mexicanus*; (G) *Hesperodiaptomus morelensis*. Abbreviations: *A1R*, right first antenna; *GS*, genital double-somite; *P5*, fifth leg. *Modified from: (B) Suárez-Morales et al. (2000), (D) Silva-Briano & Suárez-Morales (2010); (E) Montiel-Martínez et al. (2008); (F) Grimaldo-Ortega et al. (1998).*

2(1) Lateral seta of P5 exopod 2 long, reaching beyond midlength of terminal claw; distal projection of right side of genital double-somite with a small semicircular process .... *Odontodiaptomus paulistanus* (Wright, 1936)
[Brazil]

2′ Lateral seta of EP5 exopod 2 short, not reaching midlength of terminal claw (Fig. 21.7 L) .... *Odontodiaptomus michaelseni* (Mrázek, 1901)
[Argentina]

Calanoida: Diaptomidae: *Odontodiaptomus*: Species (Males)

1 Right P5 exopod 1 without large rounded process; lateral spine of right P5 exopod 2 not inserted proximally on article 2

1′ Right P5 exopod 1 with large rounded process as long as bearing article; lateral spine of right P5 exopod 2 inserted proximally on article .... *Odontodiaptomus thomseni* (Brehm, 1933)
[Uruguay]

2(1) Lateral spine of right P5 exopod 2 shorter than the distance between its insertion point and terminal claw Fig. 21.7 L) .... *Odontodiaptomus michaelseni* (Mrázek, 1901)
[Argentina]

2′ Lateral spine of right P5 exopod 2 as long as the distance between its insertion point and terminal claw .... *Odontodiaptomus paulistanus* (Wright, 1936)
[Brazil]

Calanoida: Diaptomidae: *Notodiaptomus*: *Notodiaptomus nordesticus* complex: Species (Females)

A key and revision to this genus is presented by Santos-Silva et al. (2013). *Notodiaptomus inflatus* (Kiefer, 1933) and *N. dahli* (Wright, 1936) were not included in this key because no specimens were available for examination and available descriptions are incomplete.

1 Pedigers with transverse rows and/or groups of spinules at least on one somite 2

1′ Pedigers without rows or groups of spinules on pedigerous somites 7

2(1) Pediger 5 without lateral constrictions 3

2′ Pediger 5 with lateral constrictions 5

3(2) Distal portion of pediger 5 without spinules 4

3′ Distal portion of pediger 5 with very small spinules (dorsally and ventrally) (Fig. 21.15 F) .... *Notodiaptomus henseni* (Dahl, 1894)
[Brazil]

4(3) Pedigers 3 and 4 with rows of spinules on distal margin (Fig. 21.15 I) .... *Notodiaptomus isabelae* (Wright, 1936)
[Brazil]

4′ Pedigers 3 and 4 with distal margin smooth, showing single row of spinules along suture of pedigers4 and 5 (Fig. 21.15 G) .... *Notodiaptomus iheringi* (Wright, 1935)
[Brazil]

5(2) Dorsal surface of pediger 5 with transverse row of spinules between lateral constrictions of last somite 6

5′ Dorsal surface of pediger 5 without rows of spinules between lateral constrictions of last somite (Fig. 21.15 J).......... *Notodiaptomus jatobensis* (Wright, 1936)
[Brazil]

6(5) Pedigers 3, 4, and 5 with multiple rows of spinules on distal margin (Fig. 21.15 B) .... *Notodiaptomus anisitsi* (Daday, 1905)
[Brazil, Paraguay, Argentina]

6′ Pedigers 3, 4, and 5 without multiple rows of spinules on distal margin (Fig. 21.15 H) .... *Notodiaptomus incompositus* (Brian, 1926)
[Brazil, Argentina, Uruguay]

7(1) Right first antenna article 11 of with single seta 8

7′ Right first antenna article 11 with two setae 10

8(7) Incomplete suture between pedigers 4 and 5 9

8′ Complete intersomital suture between pedigers 4 and 5 (Fig. 21.15 E) .... *Notodiaptomus deitersi* (Poppe, 1891)
[Brazil]

9(8) Genital double-somite with rugose integument on lateral/ventral distal surfaces (Fig. 21.15 D) .... *Notodiaptomus conifer* (Sars, 1901)
[Brazil, Paraguay, Argentina]

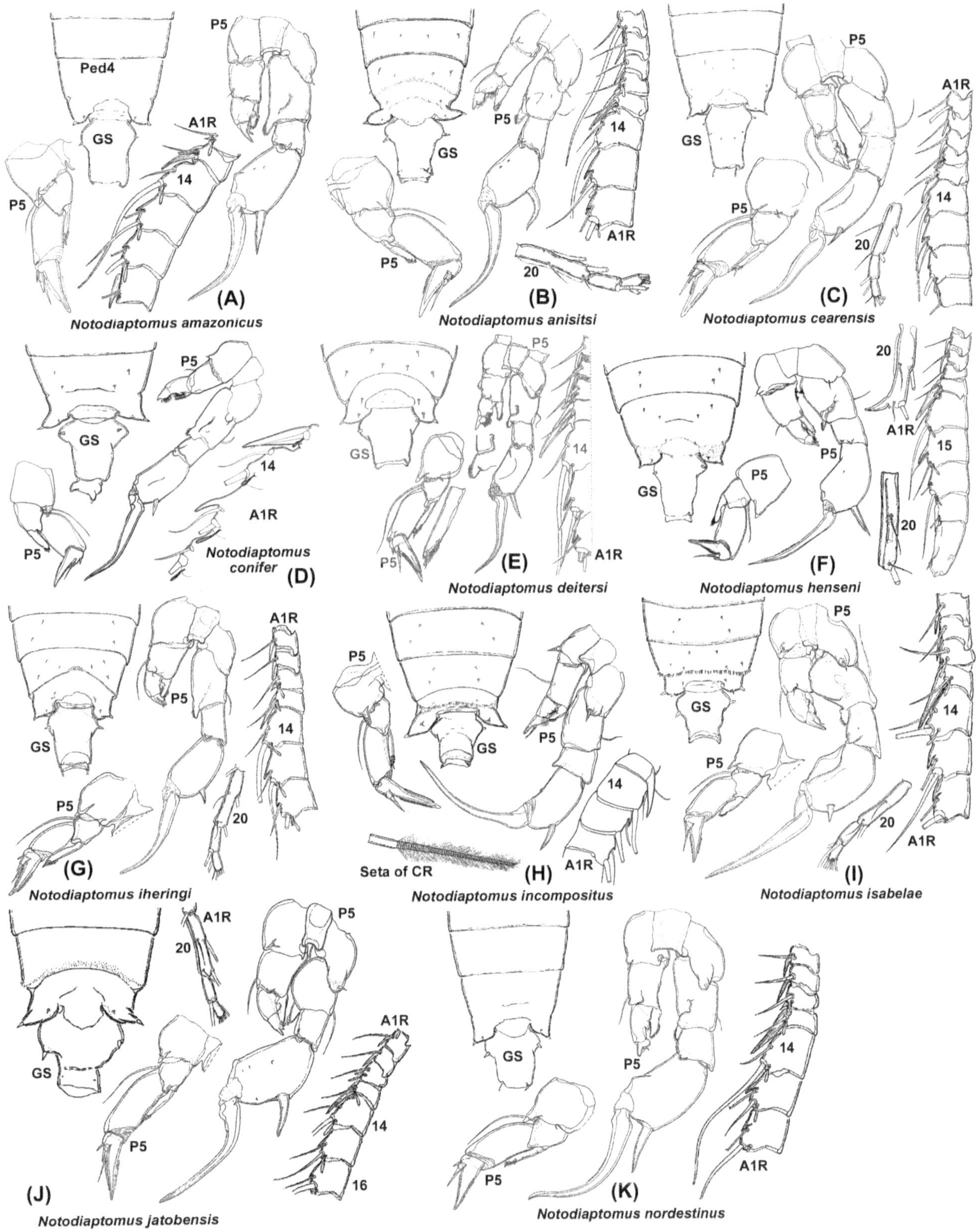

FIGURE 21.15 *Notodiaptomus* species: (A) *Notodiaptomus amazonicus*; (B) *Notodiaptomus anisitsi*; (C) *Notodiaptomus cearenses*; (D) *Notodiaptomus conifer*; (E) *Notodiaptomus deitersi*; (F) *Notodiaptomus henseni*; (G) *Notodiaptomus iheringi*; (H) *Notodiaptomus incompositus*; (I) *Notodiaptomusisabelae*; (J) *Notodiaptomus jatobensis*; (K) *Notodiaptomus nordestinus*. Abbreviations: *A1R*, right first antenna; *CR*, caudal ramus; *GS*, genital double-somite; *P5*, fifth leg; *Ped*, pediger; *Ur*, urosome.

9′ Genital double-somite with rugose integument only on medial/lateral right surface (Fig. 21.15 K) .... *Notodiaptomus nordestinus* (Wright, 1935) [Brazil]

10(7) P5 basepodite outer setae reaching distal margin of exopod 1; P5 endopod 1-articulated, but with discontinuity on the cuticle and suture in part of inner posterior surface, indicating plane of fusion of articles (Fig. 21.15 C) .... *Notodiaptomus cearensis* (Wright, 1936) [Brazil]

10′ P5 basepodite external setae reaching midlength of exopod 1; P5 endopod 1-articulated, without discontinuity in the cuticle or suture (Fig. 21.15 A) .... *Notodiaptomus amazonicus* (Wright, 1935) [Venezuela, Brazil]

Calanoida: Diaptomidae: *Notodiaptomus*: *Notodiaptomus nordestinus* complex: Species (Males)

1 Caudal seta with smooth rachis 2

1′ Caudal seta with reticulate rachis (Fig. 21.15 H) .... *Notodiaptomus incompositus* (Brian, 1926) [Brazil, Paraguay, Argentina]

2(1) At least one pediger with rows or groups of spinules 3

2′ Pedigers without rows or groups of spinules 7

3(2) Distal portion of last pediger without spinules 4

3′ Distal portion of last pediger with spinules (dorsally and ventrally) (Fig. 21.15 F) .... *Notodiaptomus henseni* (Dahl, 1894) [Brazil]

4(3) Pedigers 3 and 4 with distal margin smooth 5

4′ Pedigers 3 and 4 with multiple lines of spinules on surface of distal margin (Fig. 21.15 I) .... *Notodiaptomus isabelae* (Wright, 1936) [Brazil]

5(4) Suture between pedigers 4 and 5 with rows of spinules all along 6

5′ Suture between pedigers 4 and 5 with rows of spinules restricted to lateral/dorsal surface (Fig. 21.15 J) .... *Notodiaptomus jatobensis* (Wright, 1936)

6(5) Pedigers 4 and 5 with multiple lines of spinules along suture (Fig. 21.15 B) .... *Notodiaptomus anisitsi* (Daday, 1905) [Brazil, Paraguay, Argentina]

6′ Pedigers 4 and 5 with single line of spinules along the suture of fusion (Fig. 21.15 G) .... *Notodiaptomus iheringi* (Wright, 1935) [Brazil]

7(2) Right first antenna article 11 with two setae 8

7′ Right first antenna article 11 with one seta 9

8(7) Modified seta of right first antenna article 13 forming spiniform process with bifid apex (Fig. 21.15 C) .... *Notodiaptomus cearensis* (Wright, 1936) [Brazil]

8′ Modified seta of right first antenna article 13 forming spiniform process with rounded apex (Fig. 21.15 A) .... *Notodiaptomus amazonicus* (Wright, 1935) [Venezuela, Brazil]

9(7) Spiniform process on right first antenna article 15 not reaching distal margin of article, also in individuals with distal process on article right first antenna 20 10

9′ Spiniform process on right first antenna article 15 very long, reaching beyond the distal margin of article (Fig. 21.15 D) .... *Notodiaptomus conifer* (Sars, 1901) [Brazil, Paraguay, Argentina]

10(9) P5 pair slender; right P5 basepodite with semicircular protuberance covered by small tubercles on inner proximal surface (Fig. 21.15 E) .... *Notodiaptomus deitersi* (Poppe, 1891) [Brazil]

10′ P5 pair robust; right P5 basepodite without semicircular protuberance, with very small group of tubercles on proximal inner surface (Fig. 21.15 K) .... *Notodiaptomus nordestinus* (Wright, 1935) [Brazil]

Calanoida: Centropagidae

Eduardo Suárez-Morales
El Colegio de la Frontera Sur, Chetumal, Quintana Roo, México

Calanoida: Centropagidae: Genera

The keys to the freshwater Centropagidae are based on the works and criteria by Bayly (1992a, 1992b), but the contributions by Menu-Marque & Zúñiga (1994), Locascio de Mitrovich & Menu-Marque (1997), Menu-Marque & Balseiro (2000), and Samanez & López (2014) were useful to complete the distributional data included herein.

1 Second antenna with 5-articulated exopod, articles 2–4 short 3

1′ Second antenna with seven-articulated exopod, article 1 and 3–6 short 2

2(1) Caudal ramus short, about 2× longer than wide; female urosome with four somites .... ***Osphranticum labronectum*** Forbes, 1882 **[p. 699]**

2′ Caudal ramus narrow, long, about 6–7× longer than wide; female urosome with three somites ..... *Limnocalanus sarsi* (Mrázek, 1901) [Chile, Argentina]

3(1) Male P5R with 1–3-articulated endopod; female urosome with three somites, P5 with 3-articulated endopod .... ***Boeckella*** **[p. 699]**

3′ Male P5R endopod absent; female urosome with two somites, P5 with 2-articulated endopod .... ***Neoboeckella*** **[p. 701]**

Calanoida: Centropagidae: *Osphranticum labronectum*: Subspecies

1 Dorsal caudal seta 0.4–0.7× long as caudal ramus; body length of females 1.7–2.5 mm, males 1.4–2.3 mm .... *Osphranticum labronectum labronectum* Forbes, 1882
[Southern United States]

1′ Dorsal caudal seta 1.6–1.7× long as caudal ramus; body length of females 1.4 mm, males 1.2 mm .... *Osphranticum labronectum mexicanum* Gutiérrez-Aguirre & Suárez-Morales, 1999
[Mexico, Guatemala]

Calanoida: Centropagidae: *Boeckella*: Species

Identification based primarily on males, key follows Bayly (1992a, 1992b), abbreviations as in Fig. 21.1. The key is based almost exclusively on the structure of the male fifth leg; additional data can be found in Bayly (1992a, 1992b).

1 Right basepodite 1 with no projection at inner distal corner 2

1′ Right basepodite 1 with prominent projection at inner distal corner (Fig. 21.16 A, arrow) .... *Boeckella bergi* Richard, 1897
[Brazil, Argentina, Uruguay]

2(1) Inner margins of right and left basepodite 2 at proximal attachment less than (or at most equal to) proximal width of either right or left basepodite 2 3

2(1) Inner margins of right and left basepodite 2 at proximal insertion considerably greater than proximal width of either right or left basepodite 2 (Fig. 21.16 B, double arrow) .... *Boeckella gracilis* (Daday, 1902)
[Colombia, Ecuador, Peru, Bolivia, Argentina]

3(2) Right exopod 1 without projections on inner edge; left exopod 1 without laminar expansion on posterior face of article 4

3′ Right exopod 1 with 2-pronged, forklike process on inner edge (Fig. 21.16 C, arrow); left exopod 1 with longitudinal laminar expansion (Fig. 21.16 C, broken line) .... *Boeckella vallentini* (Scott, 1914)
[Argentina, Malvinas (Falkland) Islands, Kerguelen Islands]

4(3) Right exopod 2 with outer distal spine not extending more than halfway along right exopod claw 5

4′ Right exopod 2 with outer subdistal spine extending more than halfway along right exopod claw (Fig. 21.16 D, arrow) .... *Boeckella occidentalis* Marsh, 1906
[Peru, Bolivia, Chile]

5(4) Right endopod reaching distal margin of right exopod 2 or beyond 6

5′ Right endopod not reaching distal margin of right exopod 2 14

6(5) Left exopod claw with inner spine (Figs. 21.16 F, G, arrows) 7

6′ Left exopod claw lacking inner spine 9

7(6) Left endopod claw lacking an outer spine or with only one outer spine 8

7′ Inner distal margin of basepodite 2 with two acute spiniform processes; left exopod claw with two outer spinae (Fig. 21.16 E, arrows), right endopod 2-articulated, inner margin of proximal article inflated (Fig. 21.16 E) .... *Boeckella palustris* (Harding, 1955)
[Peru, Bolivia, Argentina]

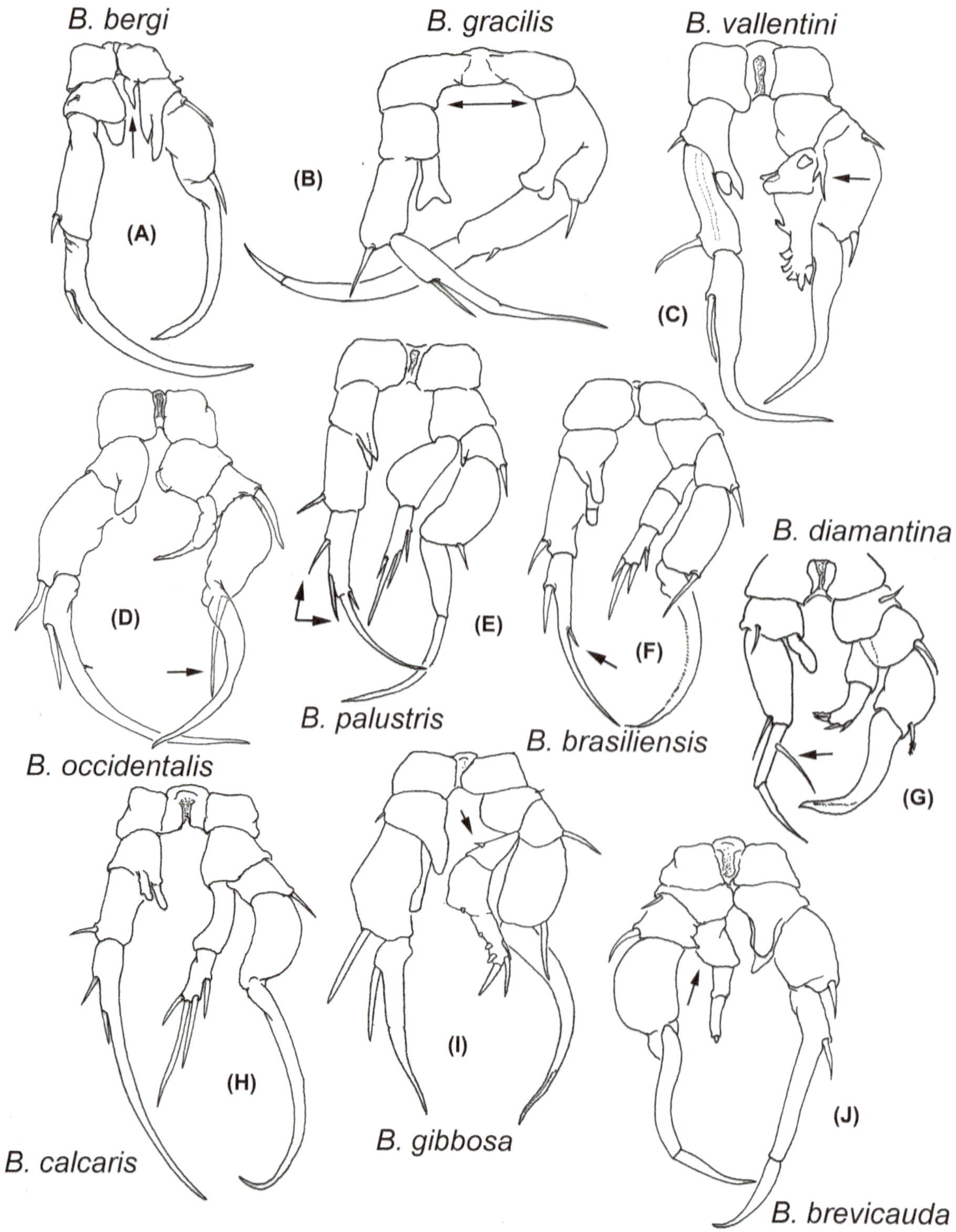

FIGURE 21.16 Male fifth legs of species of *Boeckella*, anterior view: (A) *Boeckella bergi*; (B) *Boeckella gracilis*; (C) *Boeckella vallentini*; (D) *Boeckella occidentalis*; (E) *Boeckella palustris*; (F) *Boeckella brasiliensis*; (G) *Boeckella diamantina*; (H) *Boeckella calcaris*; (I) *Boeckella gibbosa*; (J) *Boeckella brevicaudata*. *Figures redrawn: (A–F), (H), (J) from Bayly (1992b), (G) from Menu-Marque & Zúñiga (1994), and (I) from Locascio de Mitrovich & Menu-Marque (1997).*

8(7) Left exopod claw with one short inner spine (Fig. 21.16 F, arrow); right endopod 3-articulated, distal article with four short spiniform setae (Fig. 21.16 F) .... *Boeckella brasiliensis* (Lubbock, 1855) [Brazil, Chile, Argentina]

8′ Left exopod claw without outer spine, with long inner spine (Fig. 21.16 G, arrow); left endopod reduced, 1-articulated, unarmed; right endopod 2-articulated; distal article with 4 short spinae (Fig. 21.16 G) .... *Boeckella diamantina* Menu-Marque & Zúñiga, 1994 [Argentina]

9(6) Right exopod claw not particularly long, left endopod 1–3 articulated; right endopod 2- or 3-articulated, with terminal article armed with three or more subequally long setae or spinae 10

9′ Right exopod claw long, slender, about 2.5× long as exopod 2; right endopod 2-articulated, proximal article elongate, distal article with four elements, middle two longest (Fig. 21.16 H) .... *Boeckella calcaris* (Harding, 1955) [Peru, Bolivia, Argentina]

10(9) Left endopod 2 or 3-articulated, proximal article endopod 1 no longer and wider than endopod 2 and endopod 3. 11

10 Right endopod 3-articulated, proximal article clearly longer and wider than succeeding endopod 2 and 3 (Fig. 21.16 J, arrow) .... *Boeckella brevicaudata* (Brady, 1875) [Argentina]

11(10) Left endopod distinctly 3-articulated, proximal endopod article unarmed, distal article with three or four spinae or setae 12

11′ Right endopod 3-articulated, proximal article with inner spine (Fig. 21.16 I, arrow) articles 2–3 partially fused, article 2 with expanded inner margin; distal article armed with six spinae .... *Boeckella gibbosa* (Brehm, 1935) [Chile, Argentina]

12(11) Left endopod with two or three articles, right exopod claw not thickened proximally, curved, or bent in angle 13

12′ Left endopod 1-articulated, right endopod 3-articulated, inner margin of second article moderately expanded; exopod claw strongly thickened proximally and bent medially almost through right angle (Fig. 21.17 A) .... *Boeckella silvestrii* Daday, 1901 [Argentina]

13(12) Left endopod 3-articulated, distal article smooth (Fig. 21.17 B, arrow); right endopod with three distal setae; left basepodite 2 with inner thumblike process .... *Boeckella poppei* Mrázek, 1901 [Chile, Argentina]

13′ Left endopod 2-articulated, distal article slender, armed with four apical setae (Fig. 21.17 C, arrow); right endopod 2-articulated, with four distal setae .... *Boeckella antiqua* Menu-Marque & Balseiro, 2000 [Argentina]

14(5) Right endopod extending more than halfway along right exopod 2 15

14′ Right endopod not reaching distal edge of right exopod 1 or barely extending past line of segmentation between right exopod 1 and exopod 2 16

15(14) Right endopod 3-articulated (Fig. 21.17 D, arrow); left endopod 2-articulated, unarmed .... *Boeckella meteoris* Kiefer, 1928 [Chile, Bolivia, Argentina.]

15′ Right endopod 1-articulated, thumblike (Fig. 21.17 E, broken line); right terminal claw proximally thickened, medially bent .... *Boeckella poopoensis* Marsh, 1906 [Peru, Bolivia, Chile, Argentina]

16(14) Left exopod 1 without semicircular expansion on inner edge 17

16′ Inner margin of left exopod 1 with rounded expansion (Fig. 21.17 F, G, arrow) .... *Boeckella michaelseni* (Mrázek, 1901) [Chile, Argentina]

17(16) Intersegmental division between left exopod claw and left exopod 1 almost directly opposite to division between right exopod claw and right exopod 2 (Fig. 21.17 H, arrows); right endopod 2-articulated (Fig. 21.17 H); left endopod 1-articulated; outer spine of exopod 1 short, not reaching half-length of exopod 2 .... *Boeckella gracilipes* Daday, 1901 [Peru, Argentina]

17′ Intersegmental division between left exopod claw and left exopod 1 directly opposite to a point distinctly distal to division between right exopod claw and right exopod 2 (Fig. 21.17 I, tilted double arrow); right endopod 2-articulated (Fig. 21.17 I); left endopod 1-articulated; outer spine of exopod 1 long, reaching beyond half-length of exopod 2 .... *Boeckella titicacae* Harding, 1955 [Peru, Bolivia]

Calanoida: Centropagidae: *Neoboeckella*: Species

The key is based almost exclusively on the structure of the male fifth leg. Additional data can be found in Bayly (1992a, 1992b).

1 Inner margin of male P5 basepodite 2 and exopod 1 without pronged expansions; exopod 1 with three large projections on inner margin (Fig. 21.17 K, arrows); female/male size ratio 1.32 .... *Neoboeckella kinzeli* (Löffler, 1955) [Peru]

1′ Inner margin of male P5R basepodite 2 and exopod 1 with three pronged forklike expansions (Fig. 21.5 J, arrows); inner margin of right basepodite 2 with large hooked spur (gray area, Fig. 21.17 J) .... *Neoboeckella loffleri* Bayly, 1992 [Peru, Bolivia]

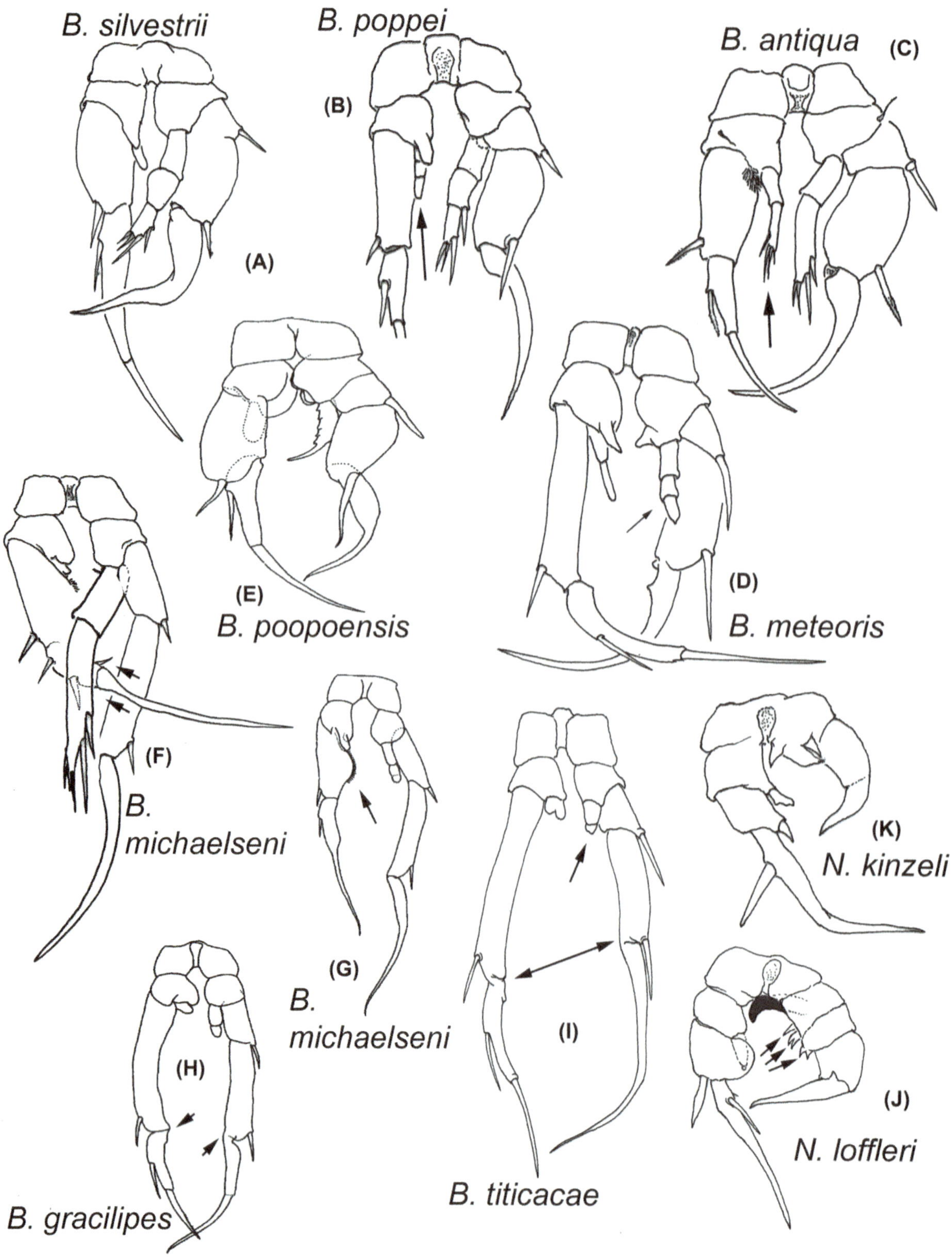

FIGURE 21.17 Male fifth legs of species of *Boeckella* and *Neoboeckella*, anterior view: (A) *Boeckella silvestrii*; (B) *Boeckella poppei*; (C) *Boeckella antiqua*; (D) *Boeckella meteoris*; (E) *Boeckella poopoensis*; (F) *Boeckella michaelseni*; (G) *Boeckella michaelseni*; (H) *Boeckella gracilipes*; (I) *Boeckella Titicacae*; (J) *Neoboeckella loffleri*; (K) *Neoboeckella kinzeli. Figures redrawn: (A), (B), (D–H), (J) from Bayly (1992b), (C) from Menu-Marque & Balseiro (2000).*

Order Harpacticoida

Samuel Gómez

Instituto de Ciencias del Mar y Limnología, UNAM. Unidad Académica Mazatlán. Mazatlán, Sinaloa, México

INTRODUCTION

Most harpacticoids can be found in benthic marine and brackish ecosystems, but some forms live in the water column and can be found in plankton samples. Also, some harpacticoids can be found associated with a variety of invertebrates and vertebrates (Huys, 2016). It is generally accepted that harpacticoids originated in marine waters (Huys & Boxshall, 1991). For example, Boxshall & Jaume (2000) recognized at least 10 independent colonization events of freshwater habitats for the Parastenocarididae, Chappuisidae, Phyllognathopodidae, Ectinosomatidae, Harpacticidae, Miraciidae, Ameiridae, Canthocamptidae, Cletodidae, and Laophontidae. Given their cosmopolitan distribution, the Canthocamptidae and the Parastenocarididae (except for the later that is absent from New Zealand) are assumed to have colonized freshwater habitats in Pangea (Boxshall & Jaume, 2000). Following the same reasoning, Boxshall & Jaume (2000) suggested that the main lineage of the Ameiridae originated in Laurasia and underwent several colonization events of freshwater habitats during the postGondwana/Laurasia split. On the other hand, given the lack of diversification, the remaining families, the Cletodidae and Laophontidae, might represent modern incursions into fresh water habitats and could represent potential colonizers (Boxshall & Jaume, 2000).

About 1384 species of harpacticoids can be found inland, mostly in freshwater habitats (Defaye & Dussart, 2011), and some 640 species are truly stygobionts (Galassi et al., 2009). The harpacticoid inland fauna consists mainly of members of the Canthocamptidae, Parastenocarididae, and Phyllognathopodidae, although some other families can be found also in freshwater habitats (e.g., Ameiridae, Miraciidae, Ectinosomatidae, Harpacticidae, Chappuisidae, Nannopodidae, Laophontidae, Cletodidae, and Darcythompsoniidae) (Boxshall & Jaume, 2000; Boxshall & Defaye, 2008; Defaye & Dussart, 2011).

Epigean forms can be found not only in rivers, solution holes, marshes, lakes, streams, crater and hypersaline lakes, springs, ditches, and artificial containers such as tanks and discarded tires, but also in semiterrestrial situations such as damp moss and leaf litter, as well as in very unusual habitats such as water-filled tree hollows and phytotelmata (e.g., bromeliad tanks, pitchers in pitcher plants, and axil waters) (Kitching, 2000; Boxshall & Jaume, 2000).

Stygobiotic copepods originated from marine, freshwater, or semiterrestrial ancestors (Galassi et al., 2009) and have colonized many different subterranean habitats. All the hypogean representatives of the Maxillipoda belong to the class Copepoda, with hypogean forms of seven orders and about 1000 species from around the globe (Galassi et al., 2009; Romero, 2009). With about 500 hypogean species (Romero, 2009), harpacticoids have successfully colonized wells, caves, cenotes, and the hyporheic zone (the region beneath and alongside a stream bed, where shallow groundwater and surface water mix) (Boxshall & Defaye, 2008; Suárez-Morales, 2015; Reid, 2016). Harpacticoids dominate the benthos, and no records of this order from plankton in groundwaters are available, being gravel and sand the preferred substrata for epibenthic and interstitial forms, which also seem to prefer places with low water flow (Galassi et al., 2009). Along with other orders of Copepoda, harpacticoids may play an important role in the decomposition of organic matter, provision of food for macroinvertebrates and meiofauna, and promotion of nutrient flow between surface and subterranean environments (Galassi et al., 2009).

LIMITATIONS

Several keys have been published so far, and it is suggested here to use more than one key for the identification of harpacticoids. Useful keys are those proposed by Suárez-Morales et al. (1996) for the copepods from the Yucatan Peninsula, Wells (2007) for the harpacticoids from the world, Bruno et al. (2005) for freshwater copepods (Cyclopoida, Calanoida, and Harpacticoida) from the Everglades, Reid (1982) for the males and females of the genus *Forficatocaris*, Gómez & Gee (2009) for the species of the genus *Cletocamptus*, Gaviria & Defaye (2012, 2015) for the Colombian species of Canthocamptidae and of the genus *Elaphoidella*, respectively, and Reid (2016) for the freshwater copepods from the Nearctic region. Also it is highly advisable to consult the original descriptions when trying to identify species.

The most evident limitation in the identification of harpacticoid copepods is, obviously, their size (most commonly < 500 μm), and the concomitant need to dissect and mount their mouth parts and swimming legs for proper inspection. The dissection of mouth parts and swimming legs might not be necessary for the identification of those taxa with important reductions in the number of setae and spinae, and articulation of mouth parts and swimming legs (e.g., Argestidae, Ancorabolidae, Darcythompsoniidae, Paramesochridae, Cletodidae), because the different appendages do not overlap and can be observed clearly in ventral position. Unfortunately, most freshwater families of harpacticoids have not undergone this kind of reductions, and mouth parts and swimming legs cannot be observed clearly in whole specimens because they do overlap. The dissection of harpacticoids can be frustrating, and one certainly needs some skills that can be achieved only with practice and patience. Some guidelines for the study of copepods can be found in Huys & Boxshall (1991).

It is sometimes difficult to define the salinity limits to define an organism as freshwater, brackish, or marine, especially because the salinity data are often not available and because it is not uncommon that salinity in nature can be observed as a continuum from fresh to full marine (Boxshall & Defaye, 2008) especially in estuaries and lagoons. On the other hand, oligo- (salinity from 0.5 to 5 ppm), meso- (salinity from 5 to 18 ppm), and polyhaline (salinity 18–30 ppm) inland ecosystems can harbor a mixture of marine and freshwater fauna (Kiefer, 1934, 1936a) and are therefore included in this chapter.

TERMINOLOGY AND MORPHOLOGY

Harpacticoida is primitively an epibenthic marine order with over 4300 species distributed in about 681 genera and 57 families and can be recognized and identified by (1) its **podoplean tagmosis**, in which the major body articulation is situated between the fourth and fifth pedigerous segments (Figs. 21.18, 27, 31), contrary to the gymnoplean body plan found only in the orders Platycopioida and Calanoida, in which the major articulation is behind the fifth pedigerous segment; (2) by the fusion, in most groups, of the first pedigerous segment to the cephalosome to form the **cephalothorax** (Figs. 21.18 B, 27–30) (exceptions of this are members of the families Canuellidae, Phyllognathopodidae (see Figs. 21.31–34), Chappuisiidae, Latiremidae, and Cerviniidae in which the first pedigerous segment is free from the cephalosome); (3) by the presence of a **double genital segment** produced by the fusion of the second (genital) and third urosomal segments in the female of most groups (Figs. 21.18, 27, 31); (4) by a maximum of 9 and 14 articles in the female and male first antennae, respectively (Fig. 21.19); (5) by the presence of **haplocer** (Figs. 21.19 B, C), **subchirocer** (Fig. 21.19 D), or **chirocer** (Fig. 21.19 E) first antennae in males, defined by the degree of modification, being the chirocer type (with one article

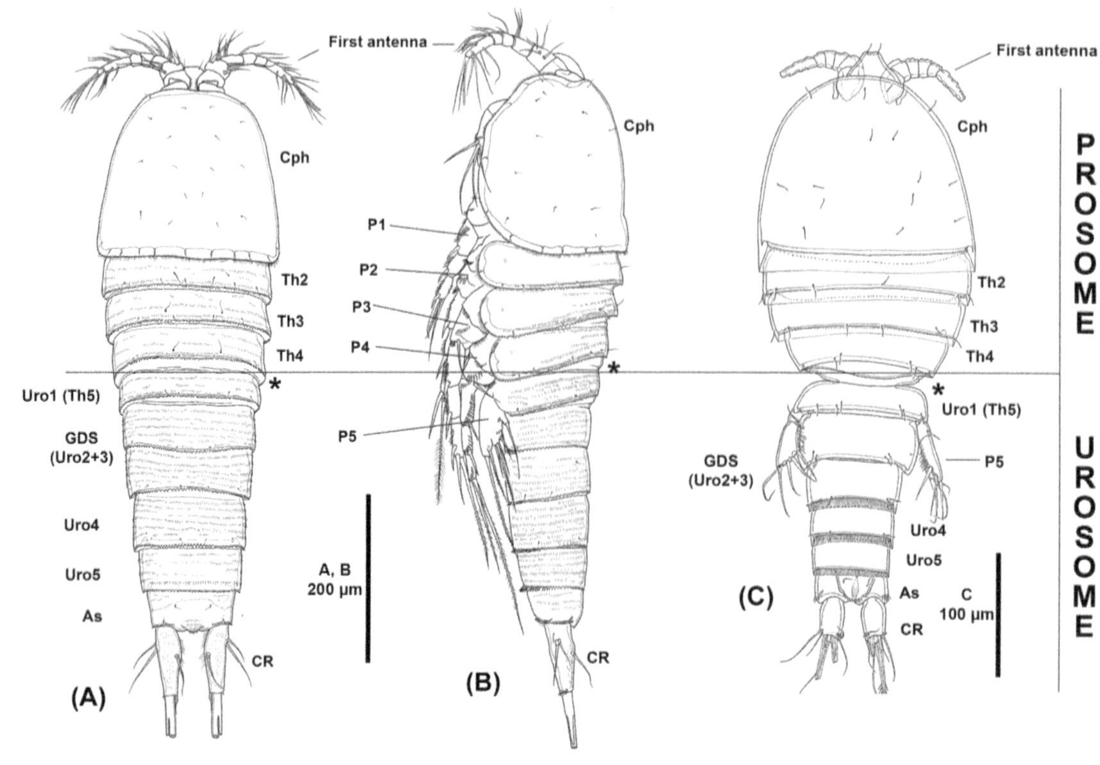

FIGURE 21.18 General external morphology of harpacticoids showing podoplean tagmosis: (A) and (B) *Cletocamptus albuquerquensis* (Canthocamptidae), dorsal and lateral; (C) *Delavalia* sp. (Miraciidae). Abbreviations: *As,* anal segment; *Cph,* cephalothorax; *CR,* caudal rami; *Th2–Th4,* second–fourth thoracopod (note that Th1 is fused to the cephalothorax); *Uro1–5,* first–fifth urosomal segments (note that the first urosomal segment is the fifth thoracopod); *GDS,* genital double-segment (formed by the fusion, in the female, of the genital—second urosomal—and the third urosomal segments); *P1–P5,* legs 1–5. *(A) and (B) Courtesy of Ray Gerber (Department of Biology, Saint Joseph's College of Maine).*

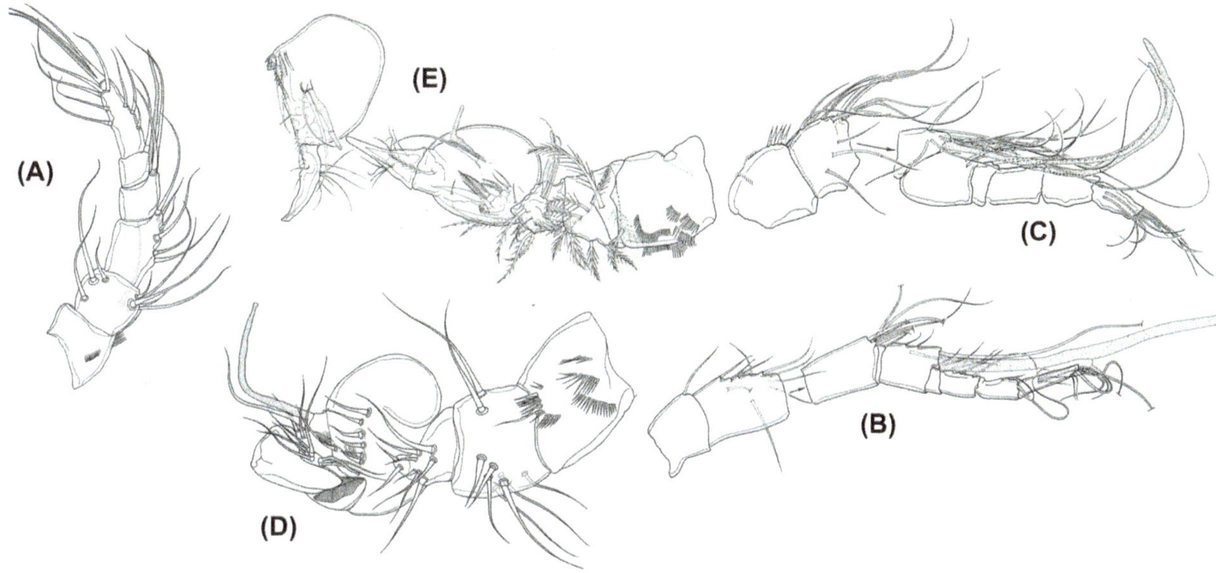

FIGURE 21.19 General structure of the first antenna of harpacticoids showing the different types of first antennae in the males: (A) *Cletocamptus albuquerquensis* (Canthocamptidae), female; (B) *Tisbe* sp. (Tisbidae), male, haplocer; (C) *Paradactylopodia* sp. (Dactylopusiidae), male, haplocer; (D) *Cletocamptus albuquerquensis* (Canthocamptidae), male, subchirocer; (E) *Microarthridion* sp. (Tachidiidae), male, chirocer, showing lateral view of penultimate and last articles. *(D) Courtesy of Ray Gerber (Department of Biology, Saint Joseph's College of Maine).*

distal to geniculation) the most modified, followed by the subchirocer (with two articles distal to geniculation), and the haplocer type (only weakly modified, with a variable number of articles distal to geniculation); (6) by the presence of a second antennae **endopod** of two (Fig. 21.20 A, F, G) or three (Fig. 21.20 C, E) articles, and an exopod with a maximum of eight articles as in Fig. 21.20 C, E (usually not more than four articles as in Fig. 21.20 A, B, D, F, G, and sometimes absent), and with coxa and basis; the basis and the first endopodal article may be completely or partially fused into an allobasis (Fig. 21.20 B, D); (7) presence of **biramous mandibles** (Fig. 21.21); basis with four inner setae at the most; endopod with 1 (Fig. 21.21 A, B, E, F, G) or (sometimes) two articles (Fig. 21.21 C, D), very reduced (Fig. 21.21 H) or absent; exopod with four articles at the most or absent; when strongly reduced or when each ramus cannot be recognized, it is commonly called **palp** (Fig. 21.21 H); (8) **first maxillae** (Fig. 21.22), **second maxillae** (Fig. 21.23), and **maxillipeds** (Fig. 21.24), with praecoxa, coxa, and basis; (9) **biramous first maxillae** (Fig. 21.22) with coxal endite and epipodite fused to coxa and represented by five setae at the most; basis with two basal endites (often fused to basis), exopod with one article and endopod with one or two articles (as in the family Canuellidae, Fig. 21.22 F); (10) second **maxillae** (Fig. 21.23) uniramous (exopod absent), syncoxa (praecoxa and coxa fused) each with two endites at the most; most harpacticoids with allobasis (basis and first endopodal article fused) bearing an endite often transformed into a strong, curved claw (Fig. 21.23 C, E); (11) **uniramous maxillipeds** (Fig. 21.24) with syncoxa (praecoxa and coxa fused); endopod with one article or reduced to a vestige with a strong claw, or with two articles in primitive groups such as Cerviniidae, Tisbidae, Chappuisiidae, and Paramesochridae; can be of three types: **subchelate** (in most harpacticoids, Fig. 21.24 C, D), **stenopodial** (as in Ectinosomatidae and Zosimeidae, Fig. 21.24 E, F) or **phyllopodial** (as in Canuellidae, Longipediidae, Neobradyidae, and Phyllognathopodidae, Fig. 21.24 A, B); (12) **swimming legs** 1–4 (Figs. 21.25, 27–34) basically biramous (although in some groups the endopod may be absent), with three articles at the most; with three protopods (a small praecoxa, a well-developed coxa and a basis with an outer seta or a spine, and an inner spine in the basis of the first leg); the members of each leg pair united by an intercoxal sclerite (Fig. 21.25 C); in some groups, the first leg is noticeably modified and can be of great taxonomic importance (see, for example, Fig. 21.25 A, C, D); (13) **leg 5** (Fig. 21.26) is biramous in few primitive species, whereas in most species, the basis and the endopod are fused into a baseoendopod where an abaxial part bearing the outer seta and the endopodal lobe can be discerned (Fig. 21.26 A); sometimes the endopodal lobe may be reduced completely, or the left and right baseoendopods can be fused to form a continuous plate (Fig. 21.26 D); the exopod and the baseoendopod may be completely fused (as in Fig. 21.26 B); coxa absent, but with a minute intercoxal sclerite only in some species; the exopod possesses one article in females,

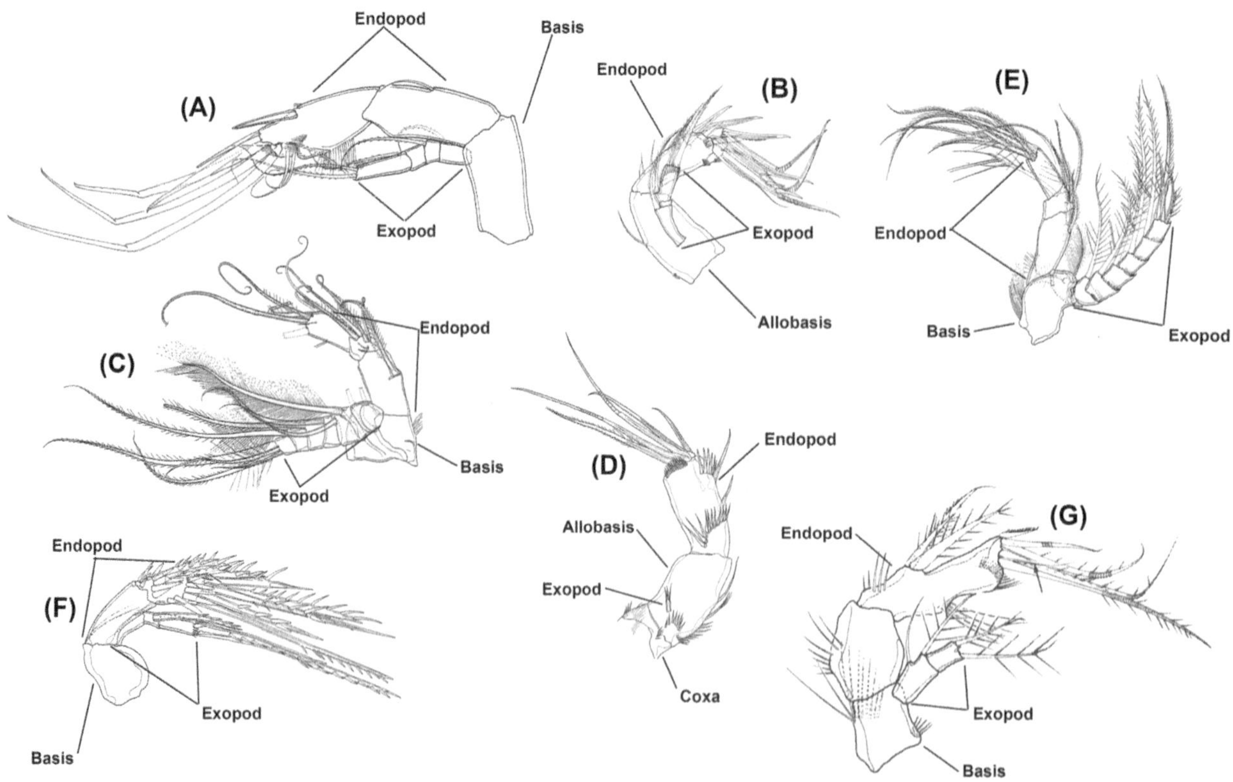

FIGURE 21.20 General structure of the second antenna of harpacticoids showing the basis and the allobasis (fusion of the basis and first endopodal segment): (A) *Tisbe* sp. (Tisbidae); (B) *Amphiascopsis* sp. (Miraciidae); (C) *Scottolana* sp. (Canuellidae); (D) *Cletocamptus albuquerquensis* (Canthocamptidae); (E) *Longipedia* sp. (Longipediidae); (F) *Ectinosoma* sp. (Ectinosomatidae); (G) *Microarthridion* sp. (Tachidiidae). *(D) Courtesy of Ray Gerber (Department of Biology, Saint Joseph's College of Maine).*

but in males of some species of Cerviniidae and in the parastenhelid genus *Parastenhelia*, it can possess two or three articles (for a general review of external morphology of harpacticoid copepods, see Huys & Boxshall, 1991, Huys et al., 1996, Suárez-Morales, 2015, and Reid, 2016).

The swimming legs are typically armed with setae and spinae (collectively called armature) and are also ornamented with setulae and spinulae, surface pores, and tube pores (collectively called ornamentation). Also, there are different types of setae and spinae depending on their shape and ornamentation (see, for example, Bodin, 1968 and Watling, 1989). The number, shape, and position of the setae and spinae of the swimming legs is taxonomically important, and two different kinds of tables are typically used in specialized literature to describe this. Sewell (1949) proposed to denote the setae of swimming legs with Arabic numerals and spinae with Roman numerals. For example, the exopod and endopod of leg 2 in Fig. 21.25 B possess three articles, and each of them possesses a number of outer, inner, and apical setae and spinae (the latter only for the case of the third article of each ramus). The first and second exopodal articles possess an outer spine and an inner seta, and the third article possesses three outer spinae, two apical setae, and two inner setae; on the other hand, the first and second endpodal articles possess one and two inner setae, respectively, and the third endopodal article possesses one outer spine, two apical setae, and two inner setae. Thus, the armature of the exopod and endopod should be expressed as:

Ramus	First (Proximal) Article	Second (Middle) Article	Third (Distal) Article
Exopod	I-1	I-1	III,2,2
Endopod	0-1	0-2	I,2,2

Note that (1) the armature complements (setae and spinae) are read from the outside to the inside, (2) the outer and inner armature of each article are denoted by Roman numerals (for spinae) and Arabic numerals (for setae) separated by a hyphen, (3) the relative position of the armature complements of the last article is read from the outside to the inside, and the outer, apical, and inner complements are

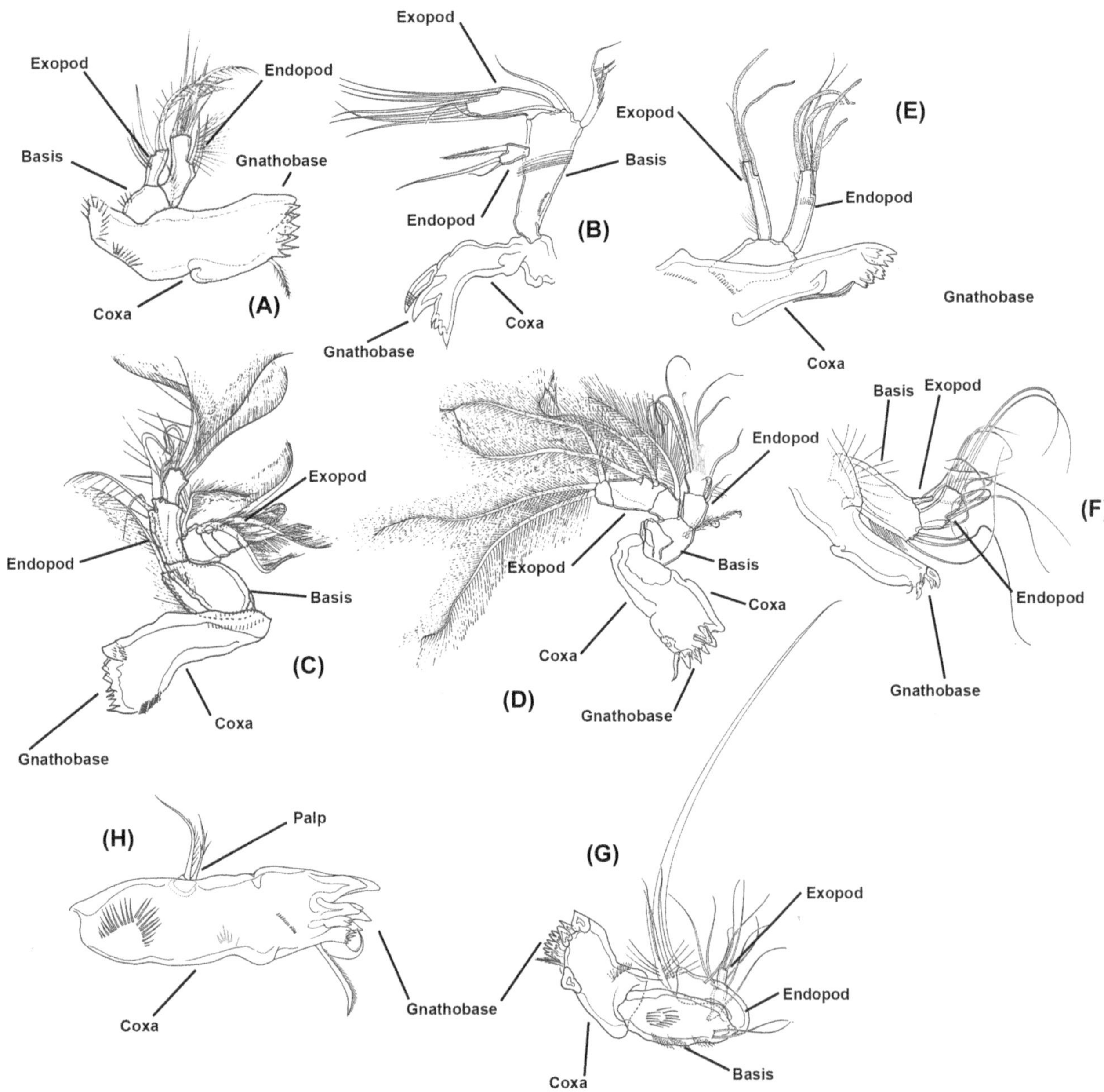

FIGURE 21.21 General structure of the mandible of harpacticoids: (A) *Microarthridion* sp. (Tachidiidae); (B) *Ectinosoma* sp. (Ectinosomatidae); (C) *Longipedia* sp. (Longipediidae); (D) *Scottolana* sp. (Canuellidae); (E) *Tisbe* sp. (Tisbidae); (F) *Amphiascopsis* sp. (Miraciidae); (G) *Delavalia* sp. (Miraciidae); (H) *Cletocamptus albuquerquensis* (Canthocamptidae). *(H) Courtesy of Ray Gerber (Department of Biology, Saint Joseph's College of Maine).*

separated by a colon. An inner seta can be found on the coxa of some primitive harpacticoids, the basis of leg 1 typically possesses an inner and outer seta or spine, and the basis of legs 2–4 possesses typically an outer seta; all this can be tabulated likewise. Another simpler way to describe the position of the armature complements without making any distinction between setae and spinae is that proposed by Lang (1934). Using the same example, the armature of the exopod and endopod should be expressed as:

Ramus	First (Proximal) Article	Second (Middle) Article	Third (Distal) Article
Exopod	1	1	2,2,3
Endopod	1	2	2,2,1

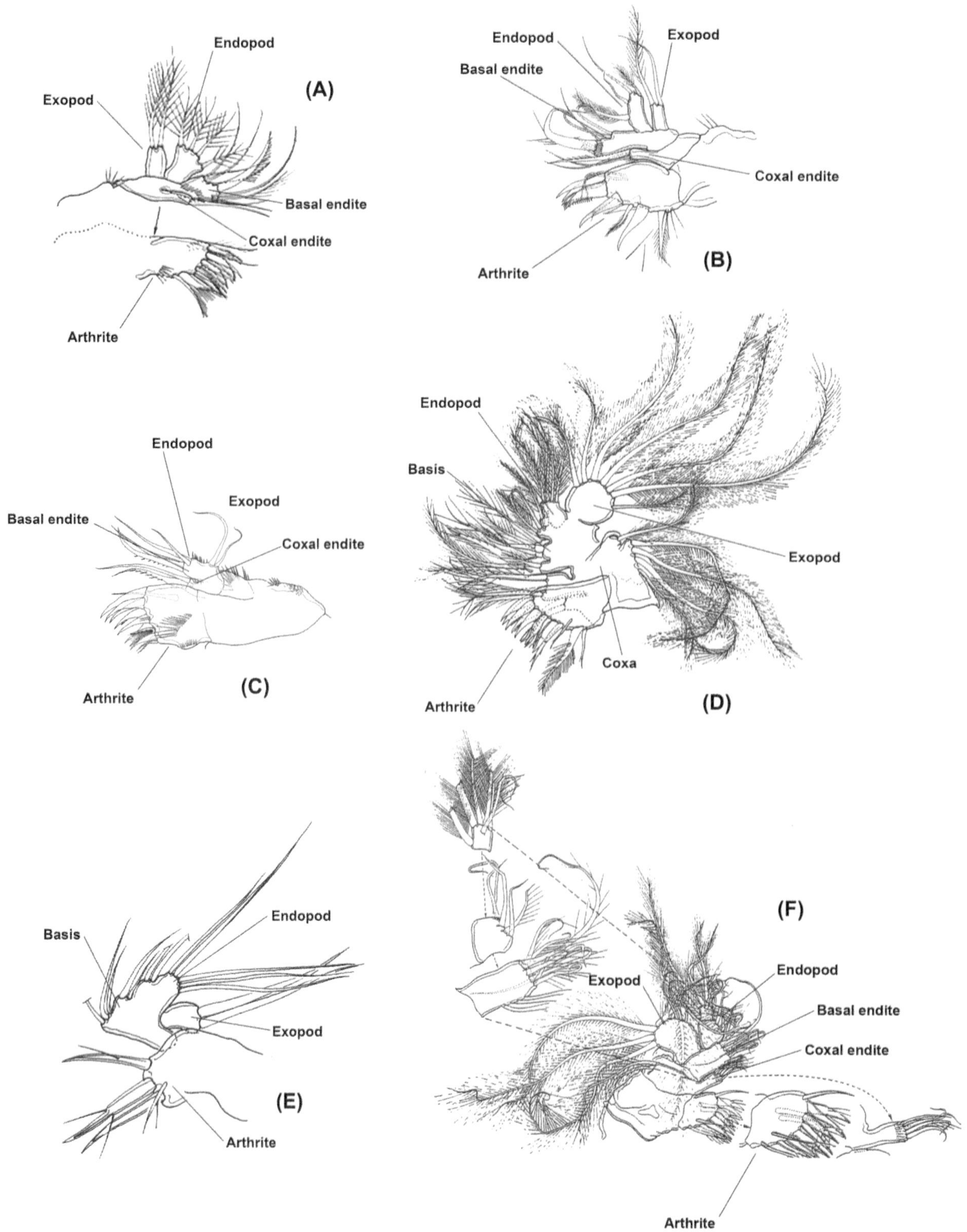

FIGURE 21.22 General structure of the first maxilla of harpacticoids: (A) *Robertsonia* sp. (Miraciidae); (B) *Amphiascopsis* sp. (Miraciidae); (C) *Cletocamptus albuquerquensis* (Canthocamptidae); (D) *Longipedia* sp. (Longipediidae); (E) *Halectinosoma* sp. (Ectinosomatidae); (F) *Scottolana* sp. (Canuellidae). *(C) Courtesy of Ray Gerber (Department of Biology, Saint Joseph's College of Maine).*

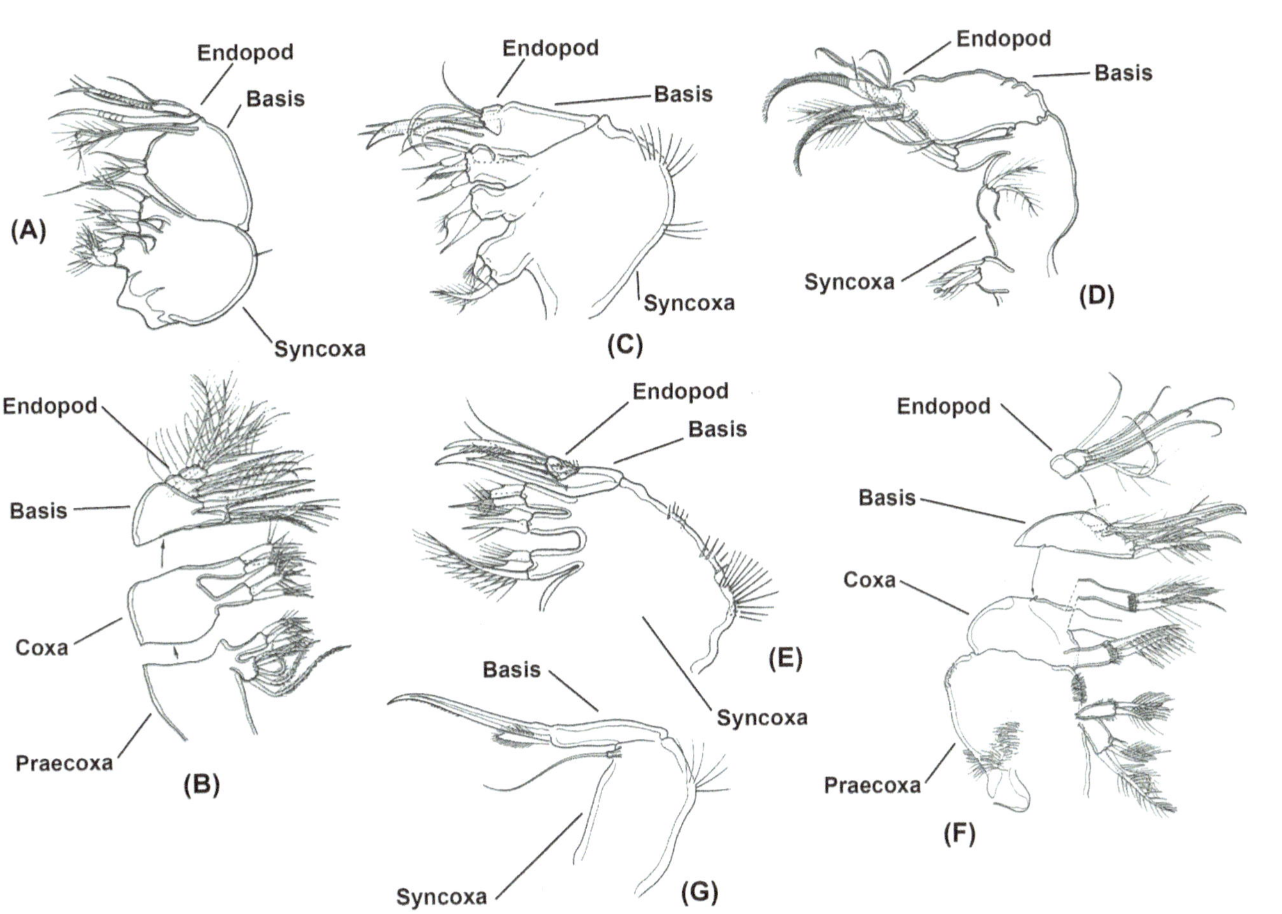

FIGURE 21.23 General structure of the second maxilla of harpacticoids: (A) *Halectinosoma* sp. (Ectinosomatidae); (B) *Longipedia* sp. (Longipediidae); (C) *Amphiascopsis* sp. (Miraciidae); (D) *Pseudectinosoma* sp. (Ectinosomatidae); (E) *Robertsonia* sp. (Miraciidae); (F) *Scottolana* sp. (Canuellidae); (G) *Tisbe* sp. (Tisbidae).

Note that (1) the armature complements (setae and spinae) are read from the inside to the outside, (2) the outer spinae of the first and second exopodal articles are not considered, (3) the relative position of the armature complements of the last article is read from the inside to the outside, and the outer, apical, and inner complements are separated by a colon. Sewells'(1949) nomenclature is deemed more appropriate for some groups with unusual armature complements such as the family Superornatiremidae whose genera possess supernumerary elements on the proximal and middle exopodal articles and the middle and distal endopodal articles (see Huys, 1996).

The caudal rami are located at the posterior end of the anal segment and are composed of one article only, and their shape, ornamentation, and armature is taxonomically important. Each caudal ramus bears a maximum of seven setae that, in the specialized literature, are numbered using Roman numerals as follows: seta I (anterolateral, usually ventral to seta II), seta II (anterolateral, usually dorsal to seta I), seta III (posterolateral), seta IV (outer distal), seta V (medial distal), seta VI (inner distal), and seta VII (dorsal, typically on a biarticulated socle) (see also Huys & Boxshall, 1991).

Sexual dimorphism is typically expressed in body segmentation (the second and third urosomal segments are typically fused in the females of most harpacticoids, but separated in the males), first antennae (normal in the female, but haplocer, subchirocer, or chiorocer in the male), in the basis of leg 1 (with a short blunt inner outgrowth or a variable number of lamellae in the males of some genera), in the number of articles and general shape of the endopods of leg 2 or 3, in leg 5, and leg 6. Sexual dimorphism can also be expressed in the shape of the caudal rami as in some species of the genus *Cletodes*, and in some members of the family Tetragonicipitidae, in the strength and thickness of some setae and spinae of swimming legs as in the genus *Cletocamptus*, in the general shape of the leg 3 exopod as in the family Parastenocarididae, or in leg setation. The modification of some appendages due to the expression of sexual dimorphism is sometimes difficult to describe, and

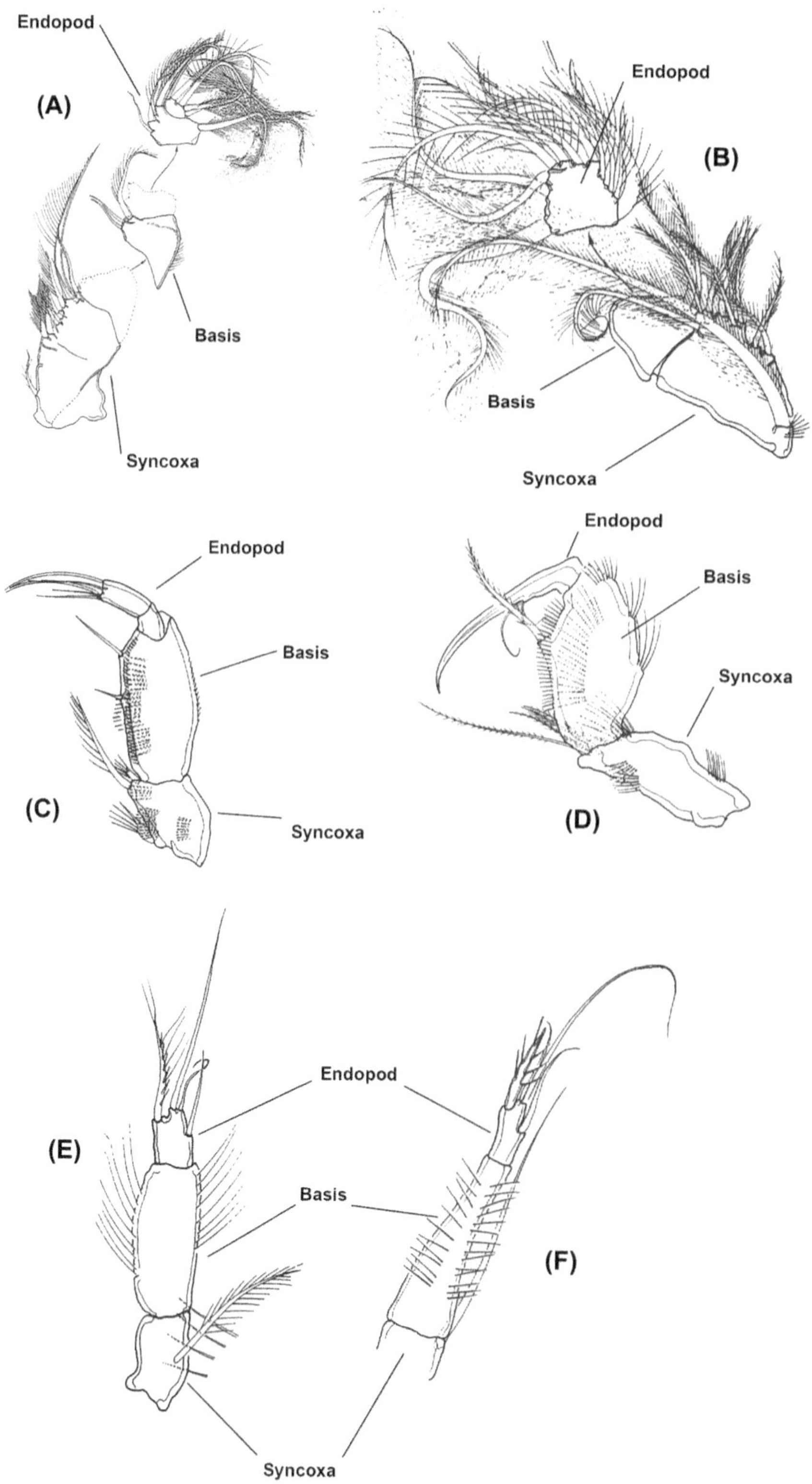

FIGURE 21.24 General structure of three types of maxilliped of harpacticoids: (A) and (B) phyllopodial; (C) and (D) subchelate; (E) and (F) stenopodial. (A) *Scottolana* sp. (Canuellidae); (B) *Longipedia* sp. (Longipediidae); (C) *Robertsonia* sp. (Diosaccinae); (D) *Amphiascopsis* sp. (Miraciidae); (E) *Pseudectinosoma* sp. (Ectinosomatidae); (F) *Halectinosoma* sp. (Ectinosomatidae).

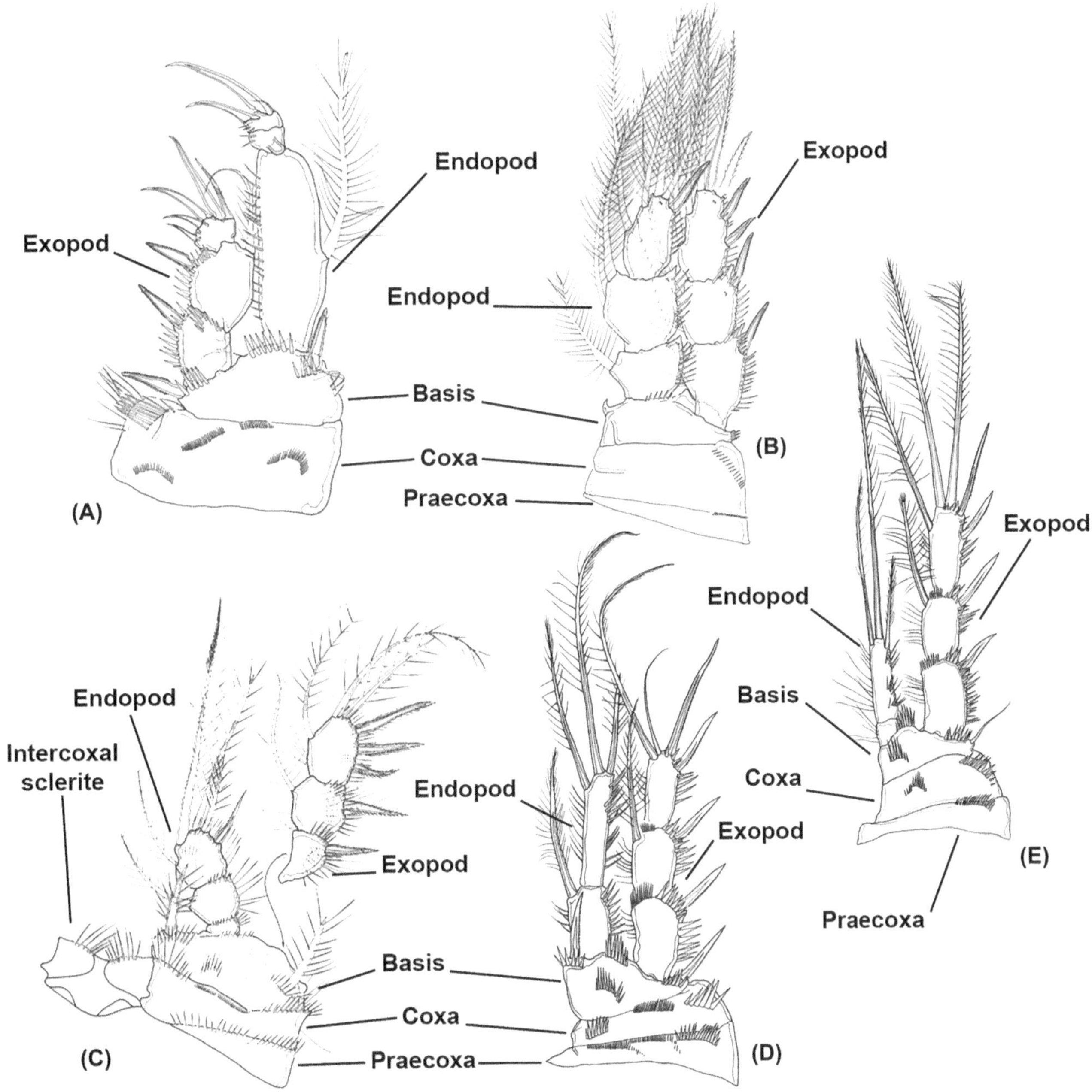

FIGURE 21.25 General structure of swimming legs of harpacticoids: (A) and (B) *Paradactylopodia* sp. (Dactylopusiidae), legs 1 and 2, respectively; (C) *Microarthridion* sp. (Tachidiidae), leg 1; (D) and (E) *Cletocamptus albuquerquensis* (Canthocamptidae), legs 1 and 2, respectively. *(D) and (E) Courtesy of Ray Gerber (Department of Biology, Saint Joseph's College of Maine).*

researchers use some specialized terms for very specific genera only. For example, the male leg 3 exopod is strongly modified in the males of the family Parastenocarididae and possesses only one article with an apical blunt apophysis and an outer curved short spine called thumb.

ACKNOWLEDGMENT

To Dr. Alexandes Kieneke from the Deutsches Zentrum für Marine Biodiversitätsforschung (DZMB; Senckenberg am Meer Wilhelmshaven) for his help with the confocal microphotographs. This is publication number 43 that uses data from the Senckenberg am Meer Confocal Laserscanning Microscope Facility.

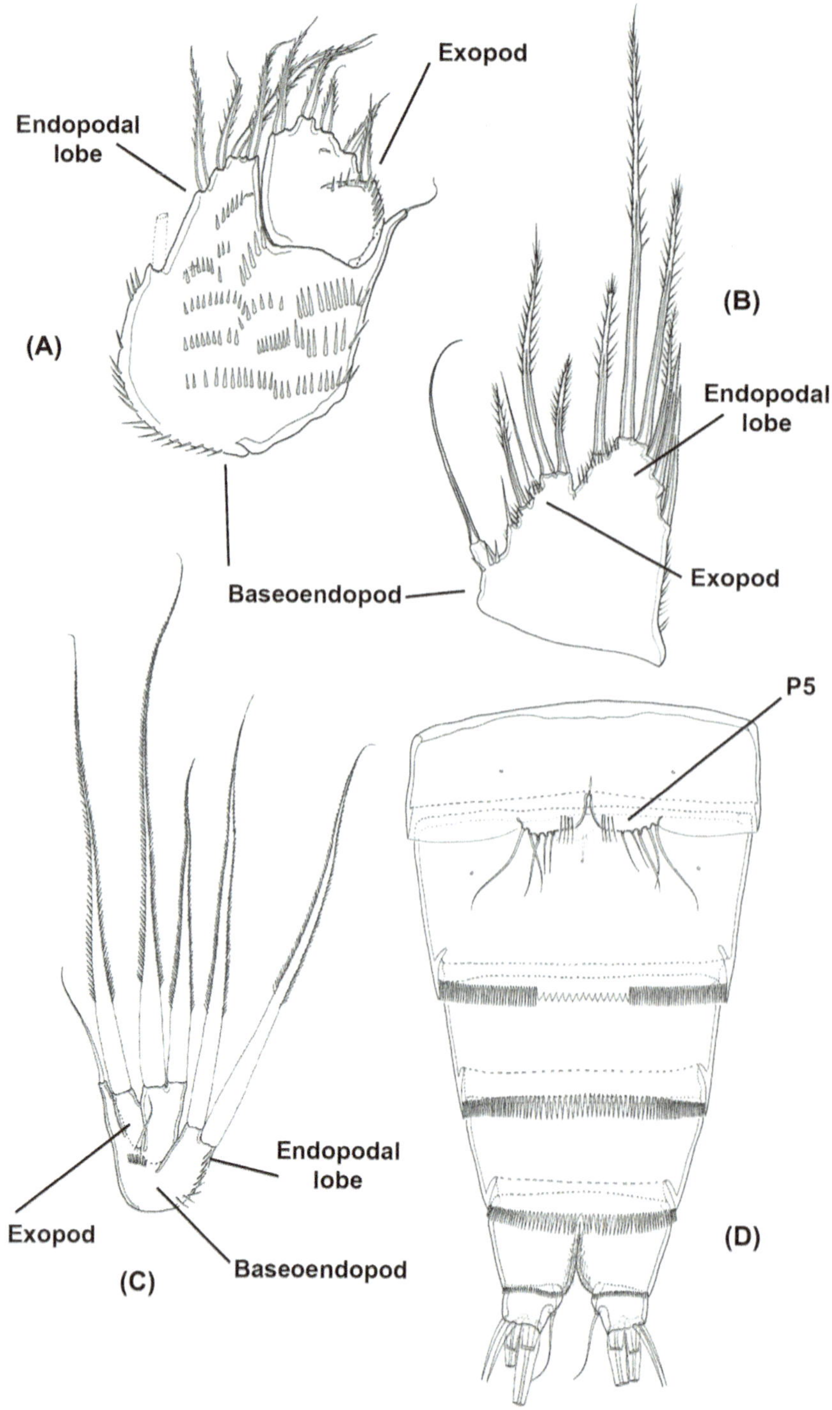

FIGURE 21.26 General structure of leg 5 of harpacticoids: (A) *Neozausodes sextus* (Harpacticidae); (B) *Cletocamptus albuquerquensis* (Canthocamptidae); (C) *Halectinosoma* sp. (Ectinosomatidae); (D) *Ectinosoma* sp. (Ectinosomatidae). *(B) courtesy of Ray Gerber (Department of Biology, Saint Joseph's College of Maine).*

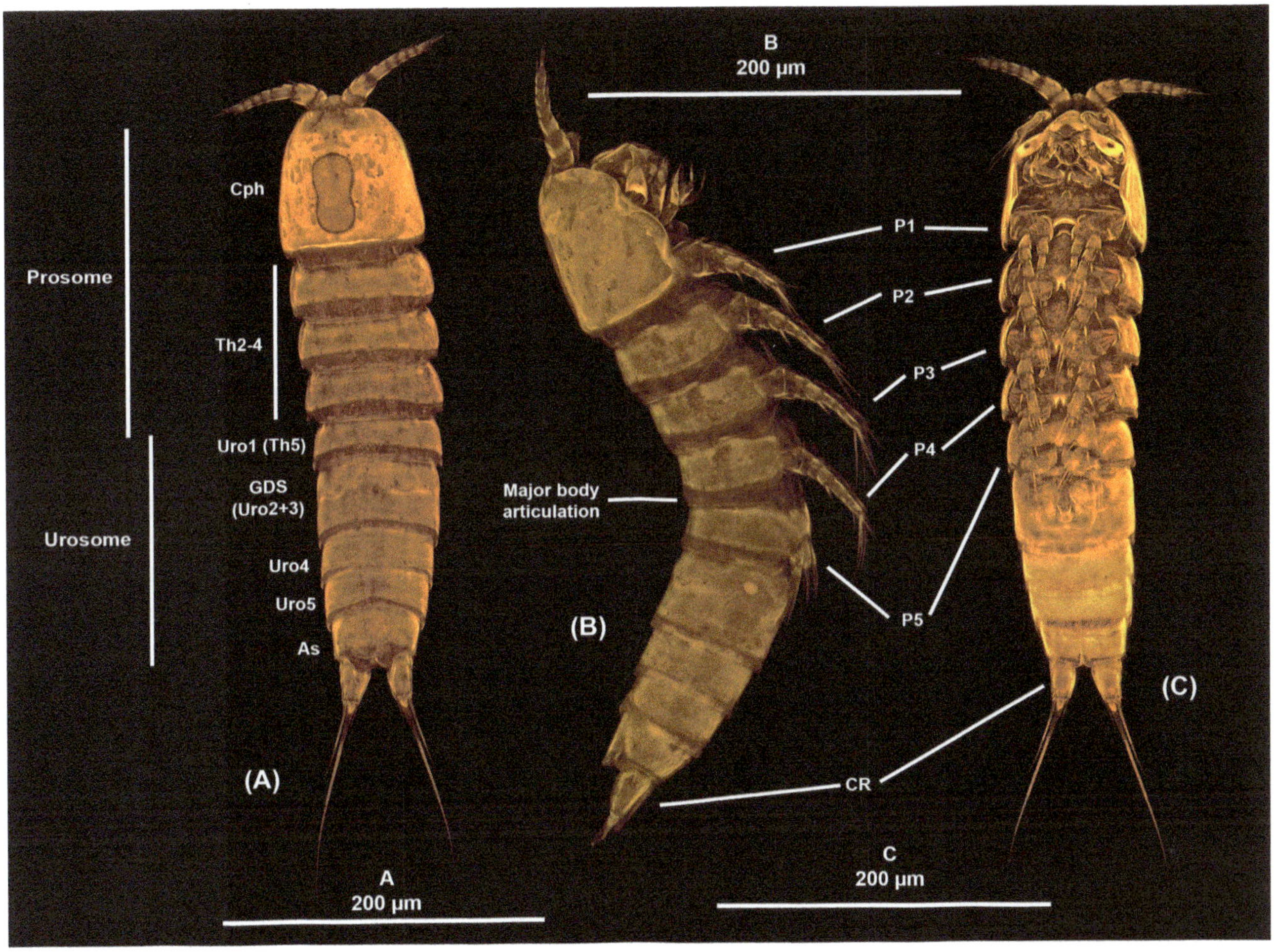

FIGURE 21.27 General aspect of *Epactophanes* sp. (Canthocamptidae, female: (A) dorsal; (B) lateral; (C) ventral. Abbreviations: *Cph,* cephalothorax; *CR,* caudal rami; *GDS,* genital double-segment; *P1–P5,* swimming legs 1–5; *Th1–4,* first–fourth thoracopod; *Uro1–5,* urosomal segments 1–5. Material collected by Nancy Mercado Salas, confocal laser scanning microphotographs courtesy of Terue C. Kihara and Gerardo Guerrero.

KEYS TO HARPACTICOIDA

Several sections of the following keys have been taken directly from several papers, with interpolations and updates where appropriate. Authoritative sources of information used in this keys include Boxshall (2008), Huys (2009), Defaye & Dussart (2011), and Walter & Boxshall (2019). Relevant sources of information include Boxshall & Defaye (2008), Bruno et al. (2000, 2002a, 2002b, 2003a, 2003b, 2005), Corgosinho & Martínez Arbizu (2005), Corgosinho et al. (2007a, 2007b, 2008, 2010a, 2010b, 2012a, 2012b), Defaye & Dussart (2011), Fuentes-Reinés & Suárez-Morales (2014), Fuentes-Reinés & Zoppi de Roa (2013), Gaviria & Aranguren (2007), Gómez & Morales-Serna (2014), Gómez et al. (2017), Hribar & Reid (2008), Pinto-da-Rocha (1994, 1995), Reeves & Reid (2002), Reid (1982, 1984, 1987a, 1988, 1990, 1992, 1993b, 1998, 2016), Reid & Hribar (2006), Suárez-Morales (2015), Suárez-Morales & Reid (1998), and Suárez-Morales et al. (1996, 2006, 2010).

Harpacticoida: Families

1 Segment bearing leg 1 completely fused to cephalothorax (Figs. 21.18, 27–30); maxilliped subchelate (Fig. 21.24 C, D) or absent, never flat and foliform 2

1′ Segment bearing leg 1 not fused to cephalothorax (total of 10 body segments in female) (Figs. 21.31–34); maxilliped flat and foliform, with several setae .... **Phyllognathopodidae [p. 719]**

FIGURE 21.28 General aspect of *Epactophanes* sp. (Canthocamptidae), male: (A) dorsal; (B) lateral; (C) ventral. Material collected by Nancy Mercado Salas, confocal laser scanning microphotographs courtesy of Terue C. Kihara and Gerardo Guerrero.

2(1) Outer margin of praecoxa and coxa of leg 1 without incised crest; other characters not combined 3

2′ Outer margin of praecoxa and coxa of leg 1 with deeply incised crest; first antennae with well-developed outer spinous process in the first article, second article with weakly developed process; mandibular palp uniramous, with two articles; proximal maxillary endite with specialized seta; second antennae without exopod, without abexopodal seta; swimming legs 1–4 exopods with three articles; leg 1 first endopodal article without inner seta; legs 2–4 endopods with one article; legs 2, 3, and 4 endopods with four, five, and four setae, respectively .... Cristacoxidae, one species: *Cubanocleta noodti* Petkovski, 1977
[Cuba: Provincia Oriente, Río Baconao]

3(2) Characters not combined as below 4

3′ Body cylindrical and elongate; maxilliped absent; second antennae with exopod represented by two setae; leg 1 endopod not prehensile, with two articles, first article with anteriorly directed seta; leg 5 very reduced, with one article; legs 1–4 third exopodal article with 4, 4, 5, 5 setae/spinae; leg 1 first endopodal article with inner seta, legs 2–4 first endopodal article without inner seta; legs 1–4 second endopodal article with 3, 4, 4, 4 setae, respectively .... Darcythompsoniidae, one species: *Leptocaris brevicornis* (Douwe, 1904)
[Cuba. El Salvador: Laguna de Ilopango. Brazil]

4(3) Characters not combined as below 5

4′ Body as in Fig. 21.36 A; leg 1 exopod with two articles (Fig. 21.36 C); leg 1 endopod with two articles, prehensil, much longer than exopod, second endopodal article with one strong claw and a reduced seta (Fig. 21.36 C); legs 2–4 exopods with three articles; leg 4 third exopodal article with five spinae/setae (with three outer spinae); legs 2–4 endopods with two articles; leg 2 second endopodal article with two inner setae; leg 4 second endopodal article with three setae; leg 5 exopod and baseoendopod separated in female (Fig. 21.36 D), each with three setae (Fig. 21.36 D); second antennae exopod with one article bearing four setae (Fig. 21. 36 B) .... Laophontidae, one species: *Onychocamptus mohammed* (Blanchard & Richard, 1891)
[Mexico: Yucatan. Cuba: Cueva De Jagüey. Caribbean. Brazil]

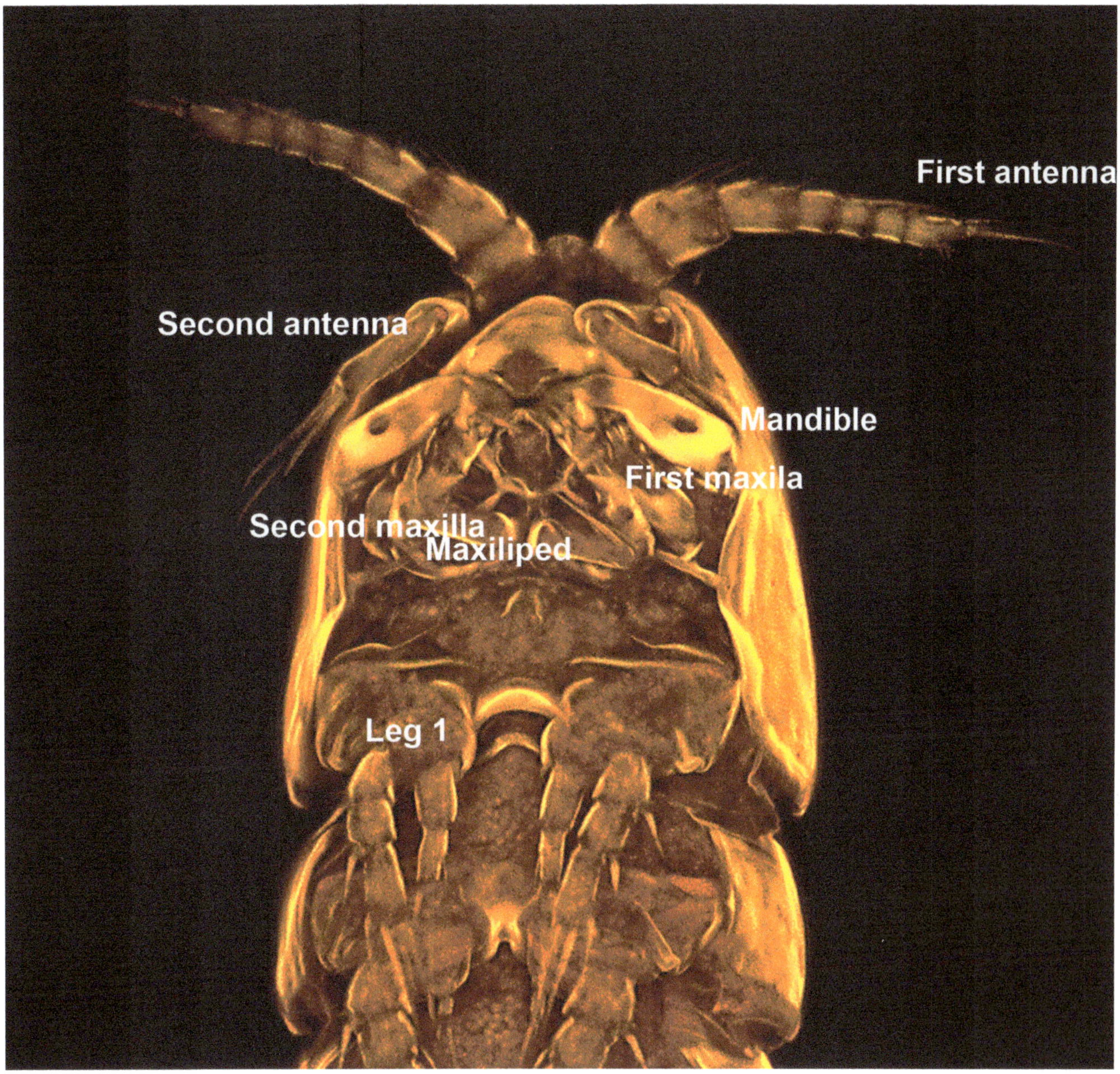

FIGURE 21.29 Cephalothorax of *Epactophanes* sp. (Canthocamptidae), ventral view of a female. Material collected by Nancy Mercado Salas, confocal laser scanning microphotographs courtesy of Terue C. Kihara and Gerardo Guerrero.

5(4) Characters not combined as below 6

5′ First antennae of female robust, short, compact, with five articles (similar to Fig. 21.37 B); second antennae with allobasis, exopod with one article bearing four setae (Fig. 21.37 C); female legs 1–4 exopods with three articles; legs 2–4 second exopodal articles with one inner seta; legs 2–4 third exopodal article with P2-P4 exopod 3 with 6, 7, 7 setae/spinae; legs 1–3 endopods with two articles; leg 4 endopod with one article; legs 2–3 second endopodal articles with three setae; leg 4 endopod with two setae (Fig. 21.37 D–G); female leg 5 with platelike baseoendopod and small separate exopod, endopodal lobe with four, exopod with five setae (Fig. 21.37 H); male leg 5 exopod with five setae, endopodal lobe with three setae (Fig. 21.37 I), highly variable .... Nannopodidae, one species: *Nannopus paulstris* Brady, 1880 [Brazil: Paraná, São Paulo]

6(5) Characters not combined as below 7

6′ Vermiform, without clear demarcation between pro- and urosome; rostrum minute; second antennae with allobasis; second antennae exopod with one articlebearing one seta; leg 1 exopod with three articles, second article without outer spine or inner seta; leg 1 endopod with two articles; legs 2–4 endopods with one article at most .... **Parastenocarididae [p. 721]**

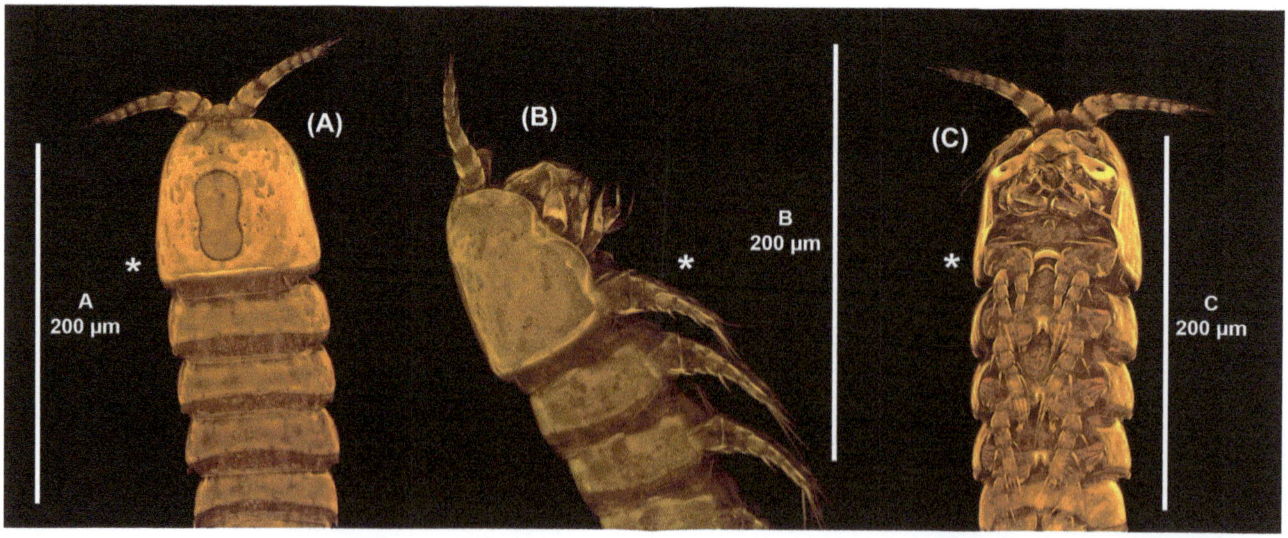

FIGURE 21.30 *Epactophanes* sp. (Canthocamptidae), female, showing fused leg 1 bearing somite: (A) dorsal; (B) lateral; (C) ventral. Asterisk indicates leg 1 fused to the cephalosome. Material collected by Nancy Mercado Salas, confocal laser scanning microphotographs courtesy of Terue C. Kihara and Gerardo Guerrero.

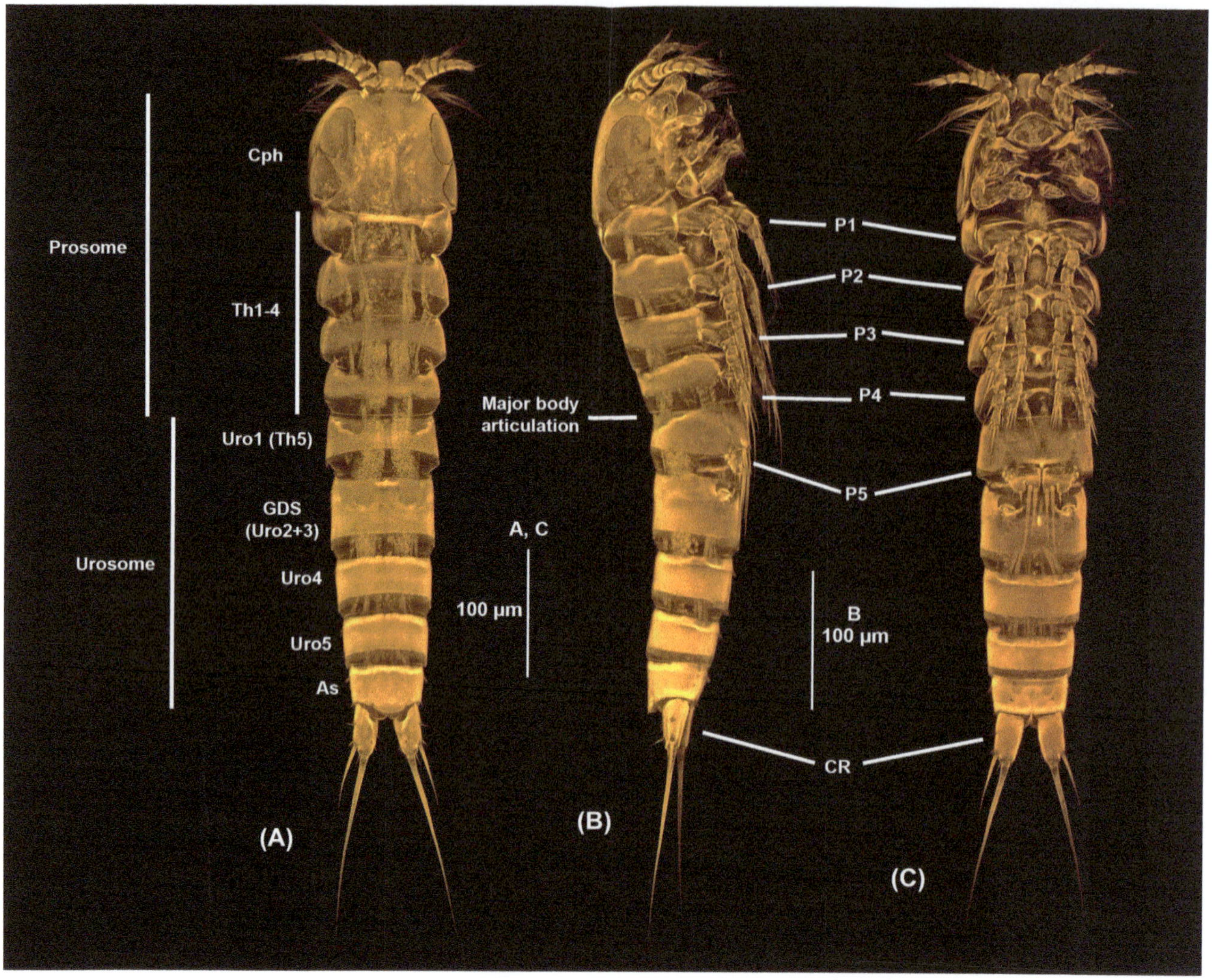

FIGURE 21.31 General aspect of *Phyllognathopus* sp. (Phyllognathopodidae), female: (A) dorsal; (B) lateral; (C) ventral. Abbreviations: *Cph*, cephalothorax; *CR*, caudal rami; *GDS*, genital double-segment; *P1–P5*, swimming legs 1–5; *As*, anal segment; *Th1–4*, first–fourth thoracopod; *Uro1–5*, urosomal segments 1–5. Material collected by Nancy Mercado Salas, confocal laser scanning microphotographs courtesy of Terue C. Kihara.

FIGURE 21.32 General aspect of *Phyllognathopus* sp. (Phyllognathopodidae), male: (A) dorsal; (B) lateral; (C) ventral. Material collected by Nancy Mercado Salas, confocal laser scanning microphotographs courtesy of Terue C. Kihara.

7(6) Characters not combined as below 8

7′ Rostrum large, reaching at least distal margin of the first article of the first antennae; maxillipedal endopod elongated, with one to three setae; male leg 2 endopod with two articles .... **Miraciidae [p. 730]**

8(7) Rostrum small; leg 1 exopod with three articles, third exopodal article with four or five setae/spinae; leg 1 endopod with three articles, first and second endopodal articles with or without inner seta, third endopodal article with three setae; leg 5 exopod and baseoendopod not

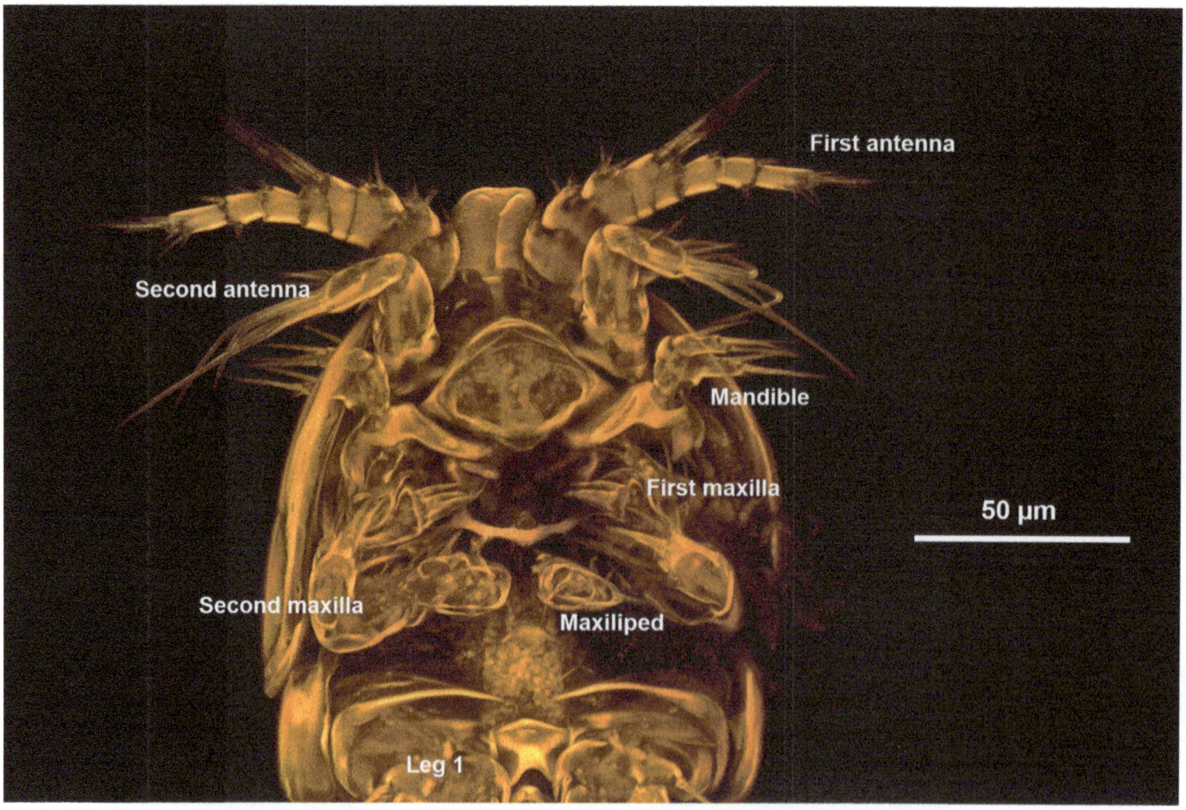

FIGURE 21.33 Cephalothorax of *Phyllognathopus* sp. (Phyllognathopodidae) ventral view of a female. Material collected by Nancy Mercado Salas, confocal laser scanning microphotographs courtesy of Terue C. Kihara.

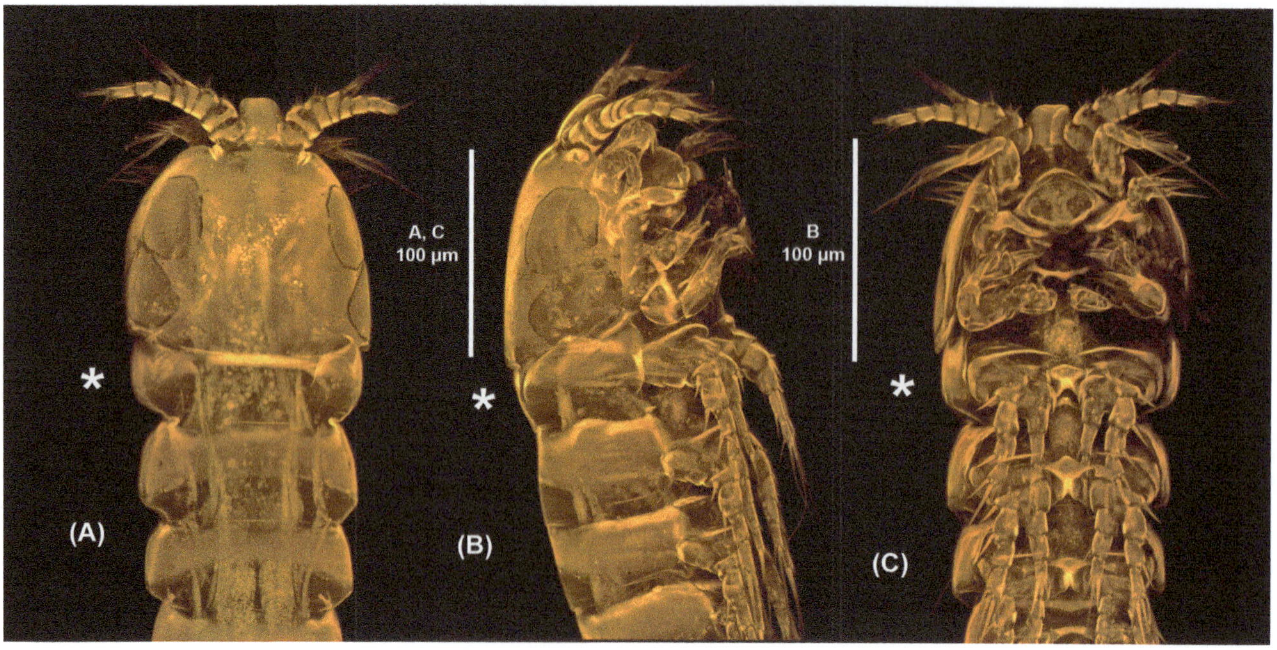

FIGURE 21.34 *Phyllognathopus* sp. (Phyllognathopodidae), female, showing separated leg 1 bearing somite: (A) dorsal; (B) lateral; (C) ventral. Asterisk indicates leg 1 separated from the cephalosome. Material collected by Nancy Mercado Salas, figures courtesy of Terue C. Kihara.

fused, rarely without exopod; female leg 5 exopod typically with six or three setae, baseoendopod with four or five setae, rarely unarmed; male leg 5 exopod typically with five to seven setae, endopodal lobe with one to four setae, rarely unarmed; caudal rami length < 6× width .... **Ameiridae [p. 731]**

8′ Rostrum fused to cephalothorax or distinct, sometimes sexually dimorphic; prosomal and/or urosomal segments with posterior margin plain, crenulated or with slender spinulae; second antennae exopod with one or two articles, typically with four setae in all, sometimes represented by single seta; mandibular palp uniramous, with one or two articles, sometimes represented by single seta; legs 1–4 exopod with three articles; leg 1 endopod with two or three articles; legs 2–3 endopod with two articles; leg 4 endopod with one or two articles; legs 2–3 with first endopodal article typically small, wider than long, without armature; male leg 3 endopod with three or two articles with inner apophysis; anal operculum rounded or with medial acute projection conferring the anal operculum a triangular shape, with or without teeth, setulae or spinulae; female and male leg 5 exopod and baseoendopod fused or separated; leg 5 baseoendopods fused in male .... **Canthocamptidae [p. 733]**

Harpacticoida: Phyllognathopodidae: Genera

The family has a complex systematics due to morphological and ecological variability. For an in depth revision, see Karanovic & Ranga Reddy (2004), Defaye & Dussart (2011) and Galassi et al. (2011). It is highly recommended to check the available descriptions of the genera and species within this family during the identification process.

1 Leg 4 exopod with three articles; leg 4 endopod with two articles; anal operculum spinulose; male leg 5 exopod with seven setae .... *Allophyllognathopus*, one species: *Allophyllognathopus brasiliensis* Kiefer, 1967 [Brazil: Amazonas, Rio Negro, Caatinga]

1′ Leg 4 exopod with three articles; leg 4 endopod with two articles; anal operculum naked (similar to Figs. 21.31 A, 32 A); male leg 5 exopod with five or six setae (similar to Fig. 21.34 B); female leg 5 exopod with four setae (similar to Fig. 21.35 A) .... ***Phyllognathopus* [p. 719]**

Harpacticoida: Phyllognathopodidae: *Phyllognathopus*: Species

1 Male leg 5 exopod with six setae .... *Phyllogntahopus viguieri* (Maupas, 1892) [Cuba, El Salvador: Laguna De Ilopango, Jamaica: Trelawny Parish, Puerto Rico, Colombia: Río Ariari, Puerto Caldas, Brazil: Paraná, Argentina]

1′ Male leg 5 exopod with five setae .... *Phyllognathopus chappuisi* (Delachaux, 1924) [Suriname]

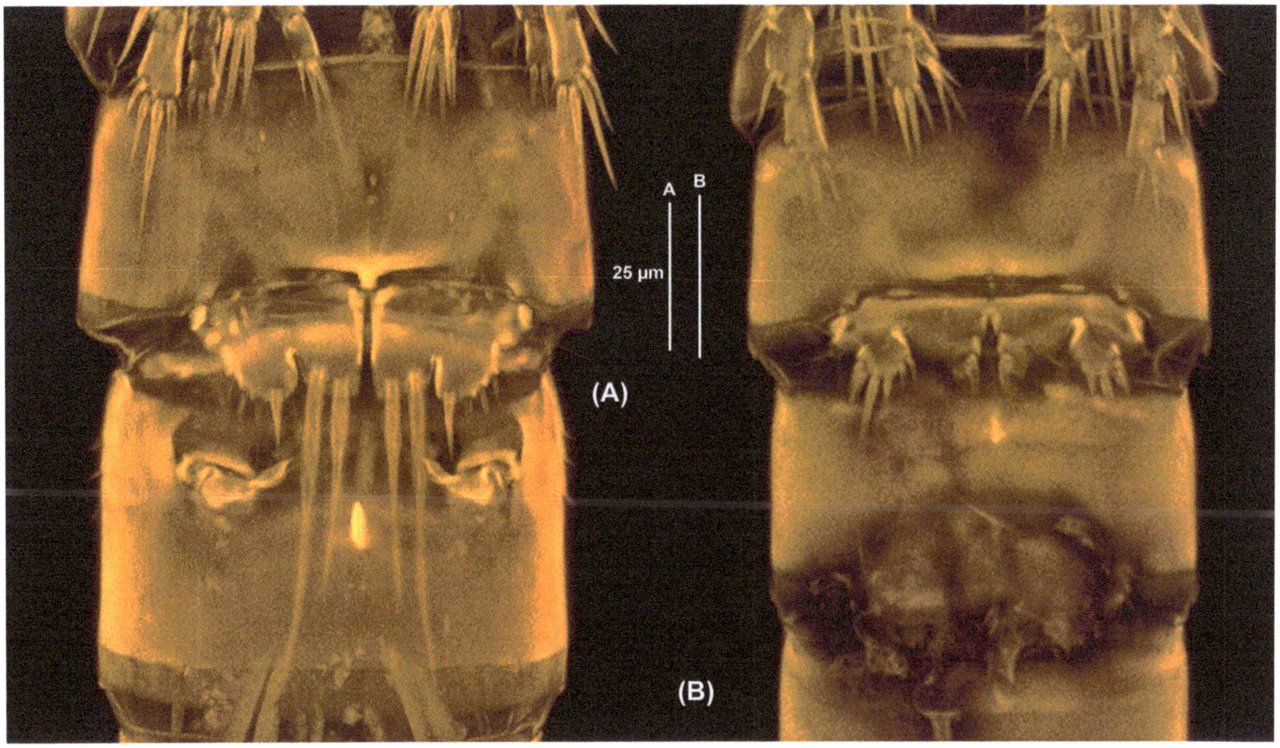

FIGURE 21.35 *Phyllognathopus* sp. (Phyllognathopodidae) fifth thoracopod showing leg 5: (A) female; (B) male. Material collected by Nancy Mercado Salas, confocal laser scanning microphotographs courtesy of Terue C. Kihara.

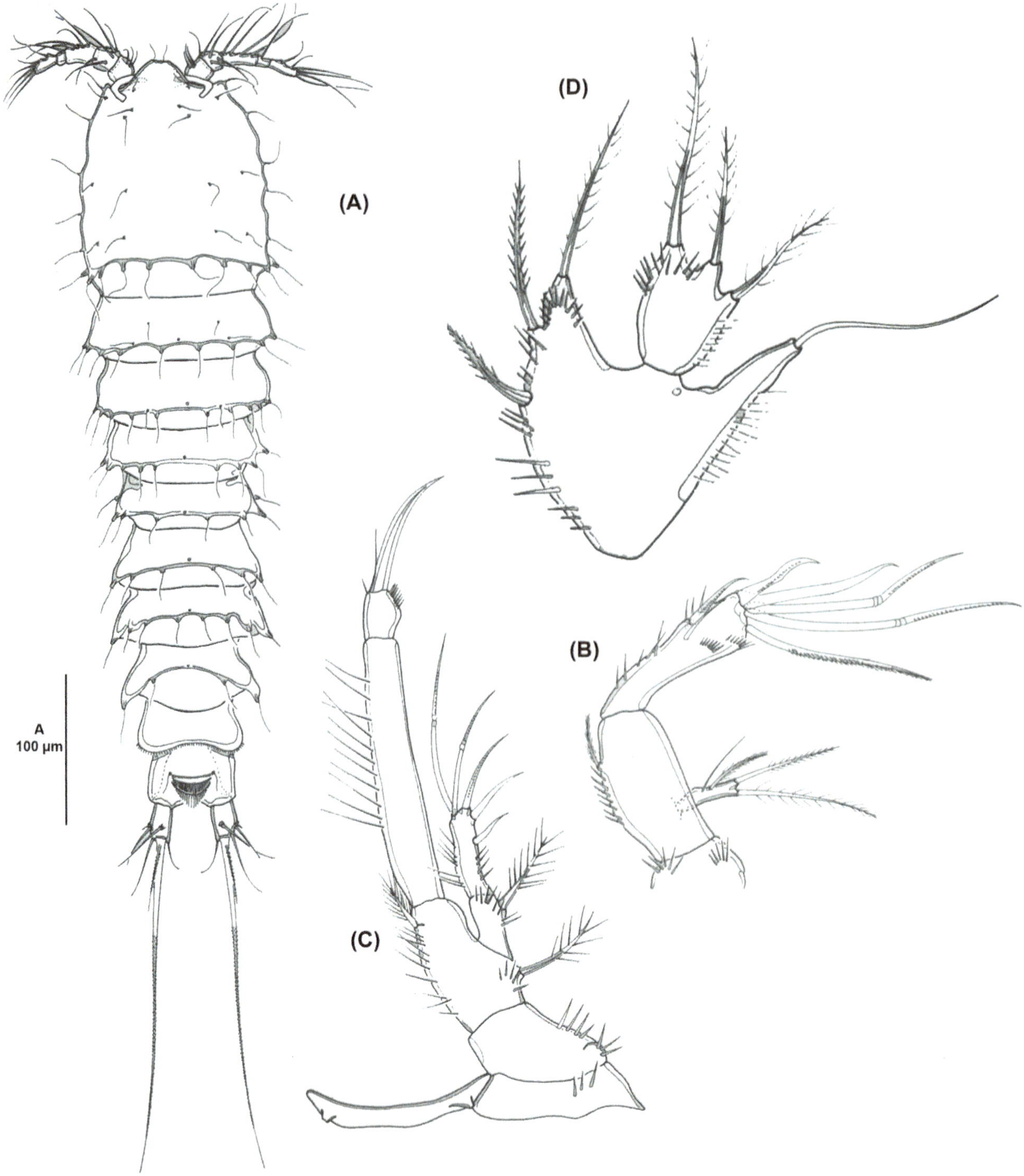

FIGURE 21.36 *Onychocamptus* sp. (Laophontidae): (A) female habitus, dorsal; (B) second antenna; (C) female leg 1; (D) female leg 5.

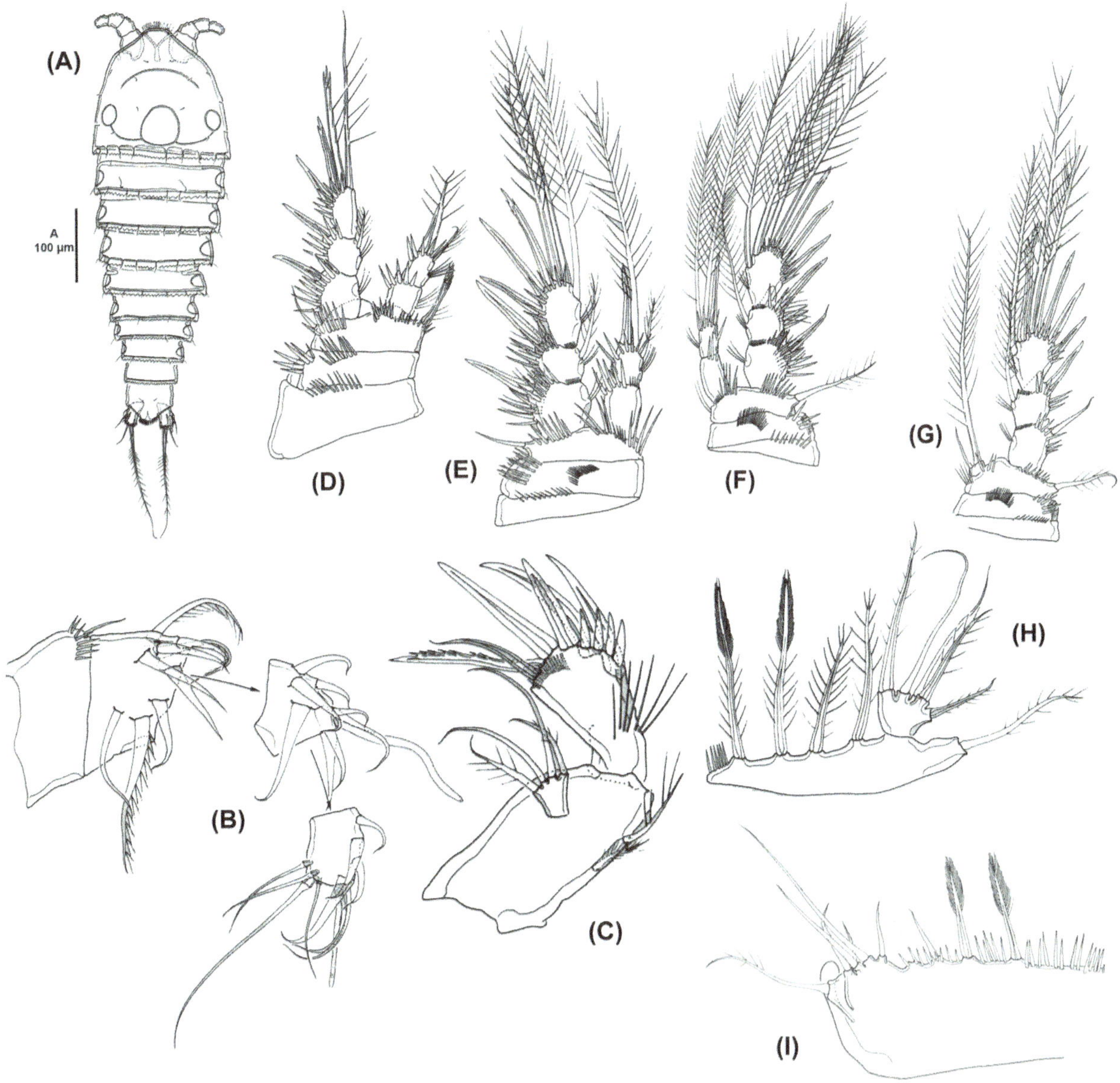

FIGURE 21.37 *Nannopus palustris* (Nannopodidae): (A) habitus, dorsal; (B) female first antenna; (C) second antenna; (D) female leg 1; (E) female leg 2; (F) female leg 3; (G) female leg 4; (H) female leg 5; (I) male leg 5.

Harpacticoida: Parastenocarididae: Subfamilies

1 Male first antennae with articles 5–8 forming a functional unit for clasping the female; leg 1 basis with or without inner seta/spine; female leg 3 endopod with a stout terminal seta fused to or separated from supporting article, both elements combined shorter, as long as, or longer than leg 3 first exopodal article .... **Parastenocaridinae [p. 722]**

1′ Male first antennae with articles 5–7 forming a functional unit for clasping the female, article 8 points medially; leg 1 basis without inner armature; leg 3 endopod without apical seta (except for *Murunducaris loyolai* with 1 minute seta), always short, never longer than leg 3 first exopodal article, with rounded apex .... **Fontinalicaridinae [p. 727]**

Harpacticoida: Parastenocarididae: Parastenocaridinae: Genera

1 Leg 4 third exopodal article with outer spine located subdistally .... ***Remaneicaris* [p. 722]**

1′ Leg 4 third exopodal article with outer spine located distally .... ***Parastenocaris* [p. 726]**

Harpacticoida: Parastenocarididae: Parastenocaridinae: *Remaneicaris*: Species

1 Anal segment with spinulae dorsally, and/or anal operculum with spinulae in both sexes 2

1′ Anal segment without spinulae dorsally in both sexes, or male anal segment without, but female anal segment with dorsal spinular row; anal operculum without spinulae in both sexes 16

2(1) Characters not combined as below 3

2′ Anal segment without spinulae dorsally, anal operculum with spinular row close to posterior margin; caudal rami with small spinulae proximally; male leg 3 exopod elongate, outer margin with longitudinal, continuous row of subequal spinulae, inner margin concave, apophysis short, rounded, outer subdistal seta strongly modified into a sclerotized, recurved, long element (thumb) reaching beyond apophysis, endopod a long hyaline element, with 1 apical seta; female leg 4 endopod a strong spike reaching distal fourth of first exopodal article; male leg 4 endopod transformed into a cylindrical segment with short apical seta barely reaching middle of first exopodal article .... *Remaneicaris tageae* (Noodt, 1972)
[Brazil: São Paulo, Rio Cubatão near Santos]

3(2) Characters not combined as below 4

3′ Anal segment with transverse row of small spinulae on proximal half dorsally; dorsal surface of anal operculum with row of small spinulae; caudal rami with some small spinulae medially; female and male leg 5 without inner unguiform projection, with four setae in all; male leg 3 exopod elongate, outer margin with proximal row of long spinulae, and with subdistal row of comparatively smaller spinulae, inner margin concave, apophysis short, rounded, slightly bilobed, thumb elongate, short, not reaching beyond apophysis, endopod a single spinule; basis with an inner foliform structure close to endopod; male leg 4 endopod a strong spike reaching tip of first exopodal article .... *Remaneicaris analuizae* Corgosinho & Martínez Arbizu, 2005.
[Brazil: Minas Gerais, Florestal, Sitio Do Corgosinho, Riberão Do Ouro]

4(3) Characters not combined as below 5

4′ Anal segment without spinulae dorsally; anal operculum with dorsal continuous row of small spinulae; inner margin of female caudal rami strongly convex proximally; female and male caudal rami:anal segment length ratio, about 1.5; female and male caudal rami length:width ratio, about 6.1; male leg 3 exopod with slightly concave inner margin, outer margin with longitudinal continuous row of subequal spinulae, apophysis short, rounded, thumb slender, very long, nearly about half the length of exopod, endopod oval, with a small apical seta; female leg 4 endopod a strong spike reaching beyond first exopodal article; male leg 4 endopod transformed into cylindrical article with long apical seta reaching tip of first exopodal article .... *Remaneicaris divae* (Noodt, 1972)
[Brazil: São Paulo, Rio Cubatão near Santos]

5(4) Characters not combined as below........ 6

5′ Penultimate segment with proximal short row of spinulae dorsally; anal segment with very long and strong spinulae flanked on each side by much smaller spinulae; caudal rami cylindrical or foliform in lateral view or female leg 4 endopod a strong spike barely reaching middle of first exopodal article; male leg 4 endopod transformed into lanciform element not reaching middle of first exopodal article; male leg 3 exopod elongate, inner margin concave, or deeply concave medially, apophysis slightly elongate, hyaline, thumb sigmoid, short, not reaching beyond apophysis; male leg 3 endopod an oval article reaching slightly distal margin of basis 14

6(5) Anal segment with proximal row of small spinulae dorsally, with continuous or medially interrupted row of subequal spinulae close to anal operculum, the later without spinulae 7

6′ Anal segment with proximal or medial row of spinulae, without spinulae close to or on anal operculum 9

7(6) Antepenultimate and penultimate urosomal segments without spinulae dorsally; dorsal surface of anal segment with transverse row of spinulae proximally, with medially interrupted transverse row of comparatively stronger spinulae close to anal operculum; male leg 4 endopod lanciform, hirsute, with distal spine; male leg 3 exopod not as below, leg 3 endopod a short article with subdistal or distal armature or spinular ornamentation 8

7′ Antepenultimate and penultimate urosomal segments with short spinule row dorsally; dorsal surface of anal segment with transverse row of four pairs of spinulae proximally, with continuous transverse row of comparatively stronger spinulae close to anal operculum; male leg 4 endopod oval, blunt, and broad, with blade-shape distal spine; male leg 3 exopod elongated, with concave inner margin, with one longitudinal row of spinulae barely reaching middle of article, outermost two subdistal spinulae modified into strong spinulae (proximalmost

strongest), thumb strong and curved, leg 3 endopod a short article, unarmed .... *Remaneicaris juliae* Corgosinho, Martínez Arbizu & dos Santos-Silva, 2007
[Brazil: Minas Gerais, Florestal, Sitio Do Corgosinho, Riberão Do Ouro]

8(7) Female and male leg 5 with four setae (including basal seta); first and second distalmost setae as long or slightly longer than inner unguiform projection in female, much shorter in male; caudal rami about 5× as long as wide in the female, 6× in the male; male leg 3 exopod with elongated, conical apophysis, thumb highly modified, with two long, rather slender processes, leg 3 endopod with three distal spinulae .... *Remaneicaris tridactyla* Corgosinho, Martínez Arbizu & dos Santos-Silva, 2007.
[Brazil: Minas Gerais, Florestal, Sitio Do Corgosinho, Riberão Do Ouro]

8′ Female and male leg 5 with four setae (including basal seta); first and second distalmost setae longer than inner unguiform projection in female and male; caudal rami about 4× as long as wide in the female, 3.5× in the male; male leg 3 exopod with slightly protruded apophysis, thumb highly modified, with two short, nearly oval processes, leg 3 endopod with one distal seta .... *Remaneicaris paratridactyla* Corgosinho, Martínez Arbizu & dos Santos-Silva, 2007
[Brazil: Minas Gerais, Florestal, Sitio Do Corgosinho, Riberão Do Ouro]

9(6) Anal operculum rounded 10

9′ Anal operculum slightly concave 11

10(9) Caudal rami as long as anal segment, cylindrical in dorsal and lateral view, apical setae well developed; male leg 2 endopod a short, bilobed article, outer lobe with apical seta, basis of leg 3 with inner, blunt protuberance proximally; leg 3 exopod with inner margin concave, with small protuberance proximally, outer margin with two longitudinal rows of spinulae, apophysis rounded, with 2 minute pointed protuberances, thumb slightly sigmoid, tapering distally, with bifid tip, not reaching beyond apophysis; leg 4 endopod a strong spike, hirsute proximally, with small inner spinulae in distal half, barely reaching middle of first exopodal article; female unknown .... *Remaneicaris euniceae* Corgosinho & Martínez Arbizu, 2005
[Brazil: Minas Gerais, Florestal, Riberão Do Ouro]

10′ Caudal rami much shorter than anal segment, with dorsal keel, cylindrical in dorsal view, foliform in lateral view, apical setae dwarfed; male unknown .... *Remaneicaris hexacantha* (Kiefer, 1936)
[Brazil: Paraiba, Rio Serido]

11(9) Female and/or male leg 5 with inner unguiform process; male leg 3 endopod an elongated article 12

11′ Anal somite with dorsal, short spinular row medially (four spinulae); female and male leg 5 without inner unguiform process; female leg 4 endopod barely reaching level of insertion of outer spine of first exopodal article; male leg 4 endopod only slightly modified into strong spike with some inner and apical spinulae; male leg 3 exopod elongated, inner margin concave, outer margin with continuous row of spinulae, apophysis elongated, hyaline, thumb sigmoid, slender, reaching far beyond apophysis; male leg 3 endopod a short, foliform article 13

12(11) Anal segment with medially interrupted dorsal row of spinulae (two sets of five spinulae each); male leg 5 with long, inner, unguiform process; male leg 4 endopod modified into strong, foliform article, hirsute, barely reaching proximal fourth of first exopodal article; male leg 3 exopod elongated, inner margin concave, outer margin with row of spinulae on proximal half, with long setulae along distal half, apophysis elongated, thumb sigmoid, with inner hyaline membrane, not reaching tip of apophysis; male leg 3 endopod an elongated article with apical seta and few spinulae; female unknown .... *Remaneicaris itacambirucui* Corgosinho, Martínez Arbizu & do Santos-Silva, 2010
[Brazil: Minas Gerais, Grão Mogol City, Jequitinhonha River, Itacambiriçu River]

12′ Anal segment with continuous row of spinulae dorsally, medial spinulae very small and flanked by comparatively stronger spinulae; female leg 5 with long, inner, unguiform process; unguiform process of the male leg 5 very small; female leg 4 endopod reaching distal margin of first exopodal article; male leg 4 endopod modified into strong, foliform article, hirsute, reaching middle of first exopodal article; male leg 3 exopod elongated, inner margin concave, outer margin with proximal and subdistal row of spinulae, apophysis elongated, thumb sigmoid, slender, reaching slightly beyond apophysis; male leg 3 endopod an elongated article, with setulae apically .... *Remaneicaris persephone* (Noodt, 1965)
[Argentina: Córdoba, Río Cuarto]

13(11) Female caudal rami cylindrical in lateral view .... *Remaneicaris paraensis paraensis* (Noodt, 1963)
[Brazil: Mouth of Pará River, near Icoaraci]

13′ Female caudal rami clavate in lateral view .... *Remaneicaris paraensis bulbifera* (Noodt, 1963)
[Brazil: Mouth of Pará River, near Icoaraci]

14(5) Characters not combined as below 15

14′ Male leg 3 exopod with inner margin deeply concave medially; female leg 5 with long inner unguiform projection; male leg 5 with small unguiform projection .... *Remaneicaris remanei* (Noodt, 1963)
[Paraguay: Lake Yparacai near San Bernardino]

15(14) Male caudal rami elongate in dorsal view, concave proximally, cylindrical in lateral view .... *Remaneicaris paraguayensis paraguayeneis* (Noodt, 1963)
[Paraguay: Lake Yparacai near San Bernardino]

15′ Male caudal rami oval in dorsal view, foliform in lateral view .... *Remaneicaris paraguayensis phylloides* (Noodt, 1963)
[Paraguay: Lake Yparacai near San Bernardino]

16(1) Characters not combined as below 17

16′ Outer margin of male leg 3 exopod with one or two longitudinal rows of spinulae (one proximal, one subdistal), distalmost two outer exopodal spinulae modified into robust spinulae, or distalmost outer spinule modified into robust spinule with or without hyaline membrane, apophysis slightly elongated, thumb sigmoid or straight, at least as long as apophysis 30

17(16) Characters not combined as below 18

17′ Male leg 3 exopod elongated, outer margin with one proximal and one subdistal longitudinal row of spinulae, apophysis and thumb not visible, wrapped into a membrane, strongly bent inwards at an angle of 90 degrees, endopod short, foliform, with distal hyaline membrane; male leg 4 endopod a short blunt article with hyaline membrane apically; male and female leg 5 unguiform; anal operculum convex .... *Remaneicaris membranacea* (Noodt, 1965)
[Argentina: San Luis, near San Luis city; Córdoba, near Río Cuarto]

18(17) Characters not combined as below 19

18′ Outer margin of male leg 3 exopod slightly convex, with one proximal and one subdistal longitudinal row of spinulae, apophysis characteristic, a long hook, thumb arising from strong base, with distal hyaline part, endopod a long bipinnate seta reaching far beyond distal margin of basis; male leg 4 endopod a strong spiniform element with subdistal constriction, with outer setulae before constriction, bipinnate after constriction, reaching slightly beyond first exopodal article; male leg 5 with inner unguiform projection; female unknown .... *Remaneicaris oncophora* (Noodt, 1965)
[Argentina: Córdoba, Río Cuarto]

19(18) Characters not combined as below 20

19′ Inner margin of male leg 3 exopod convex medially, with one row of outer spinulae, apophysis and thumb hyaline apically; male leg 3 endopod a short article not reaching distal margin of basis and armed with one apical seta, or a long article reaching far beyond distal margin of basis and armed with one distal seta, or a spiniform element reaching far beyond distal margin of basis; male leg 4 endopod a small, thick and rounded article, or a strong bipinnate spine; anal operculum straight or convex; male anal segment without, female anal segment with or without dorsal spinulae; caudal rami with or without spinular ornamentation dorsally 24

20(19) Characters not combined as below 21

20′ Male leg 3 exopod with slightly concave inner margin, outer margin with or without continuous longitudinal row of subequal spinulae, apophysis rounded or slightly elongated inwards, thumb very strong, about half as thick as the width of apophysis, or noticeably thinner, reaching far beyond or as long as apophysis; male leg 3 endopod a long article with one seta fused apically, or an oval article with apical spinulae, or club-shaped without armature or ornamentation; male leg 4 endopod a single, irregular, hirsute article with two inner protuberances and with a distal hyaline seta, or a strong, wide, hirsute article reaching beyond middle of first exopodal article, or a strong spike with long proximal and small subdistal spinulae along inner margin; male leg 5 with or without inner unguiform projection 27

21(20) Characters not combined as below 22

21′ Male leg 3 exopod elongated, nearly straight, with few spinulae proximally and subdistally, apophysis short, hyaline, thumb long and slender, reaching far beyond apophysis; male leg 3 endopod a spinelike element; male leg 4 endopod small, with spinulae basally, armed with 1 apical spinule bent outwards; female leg 4 endopod barely reaching middle of first exopodal article; female leg 5 with inner, small, unguiform projection, with three innermost setae situated at the same level, conferring leg 5 a subquadrate shape; caudal rami tapering distally .... *Remaneicaris meyerabichi* (Noodt, 1962)
[El Salvador: Rio Zacatiapa]

22(21) Male leg 3 exopod nearly straight, outer margin with one longitudinal row of spinulae, apophysis short, hyaline, thumb short, reaching slightly beyond apophysis; male leg 3 endopod a single article fused to a seta apically; female leg 3 endopod a strong spike barely reaching middle or slightly beyond middle of first exopodal article; female leg 4 endopod reaching slightly beyond middle or tip of first exopodal article; female and male leg 5 with inner unguiform projection, with three or four setae in all, including outer basal seta; anal operculum straight or slightly concave 23

22′ Male leg 3 exopod with concave inner margin, outer margin with one longitudinal row of spinulae proximally and subdistally, apophysis moderately elongated, thumb sigmoid, reaching far beyond apophysis, with or without endopod, if present, an elongated article with apical ornamentation; male leg 4 endopod reaching slightly beyond or reaching proximal fourth of first exopodal article; female leg 3 endopod a strong spike slightly shorter than first exopodal article; female leg 4 endopod reaching tip of first exopodal article; female and male leg 5 with inner unguiform projection, with four setae in all, including outer basal seta; anal operculum straight 29

23(22) Female leg 3 endopod a strong spike barely reaching middle of first exopodal article; female leg 4 endopod reaching middle of first exopodal article; female and male leg 5 with three setae in all, including outer basal seta; anal segment slightly concave .... *Remaneicaris icoaraci* (Noodt, 1963)
[Brazil: Mouth of Pará River, near Icoaraci]

23′ Female leg 3 endopod a strong spike reaching slightly beyond middle of first exopodal article; female leg 4 endopod reaching slightly beyond first exopodal article; female and male leg 5 with four setae in all, including outer basal seta; anal segment straight .... *Remaneicaris pluto* (Noodt, 1965)
[Argentina: Córdoba, Río Cuarto]

24(19) Inner margin of male leg 3 exopod extremely convex, conferring the inner margin a W-shape, apophysis and thumb with hyaline structure distally, outer margin of exopod with two or three small spinulae, endopod a small article not reaching beyond distal margin of basis and armed with small apical seta; male leg 4 endopod modified into a small, thick and rounded article; anal operculum straight; caudal rami with or without dorsal spinular ornamentation 25

24′ Inner margin of male leg 3 exopod slightly convex, apophysis and thumb with hyaline structure distally, outer margin of exopod with four to seven spinulae, endopod styliform, or long article reaching far beyond distal margin of basis and armed with one apical, slender seta; male leg 4 endopod modified into strong bipinnate spine; male anal segment without, female anal segment with or without transverse spinular row dorsally; anal operculum slightly convex; caudal rami with or without dorsal spinular ornamentation 26

25(24) Male leg 4 endopod not curved outwards; female leg 4 endopod reaching distal margin of first exopodal article; caudal rami with some spinulae dorsally; shape of the female caudal rami variable from cylindrical to oval in lateral view .... *Remaneicaris ciliata* (Noodt, 1965)
[Argentina: Córdoba Province, La Bolsa, Río Sierra]

25′ Male leg 4 endopod curved outward; female leg 4 endopod not reaching distal margin of first exopodal article; caudal rami without spinular ornamentation dorsally .... *Remaneicaris cordobaensis* (Noodt, 1965)
[Argentina: Córdoba, La Bolsa, Río Sierra]

26(24) Outer margin of male leg 3 exopod with seven spinulae, endopod styliform, reaching far beyond distal margin of basis; male leg 4 endopod almost as long as first exopodal article; female and male anal segment and anal operculum without transverse spinular ornamentation dorsally; shape of the female caudal rami variable from cylindrical to oval in lateral view .... *Remaneicaris sanctiludovici* (Noodt, 1965)
[Argentina: San Luis Province, near San Luis]

26′ Outer margin of male leg 3 exopod with four spinulae, endopod a long article with one distal seta reaching far beyond distal margin of basis; male leg 4 endopod barely reaching middle of first exopodal article; female anal segment with, male anal segment without transverse row of spinulae close to anal operculum .... *Remaneicaris sierrae* (Noodt, 1965)
[Argentina: Córdoba, La Bolsa, Río Sierra]

27(20) Male leg 3 exopod with continuous longitudinal row of subequal spinulae along outer margin, apophysis rounded or slightly elongated inwards, thumb slender, as long as, or longer than apophysis; male leg 3 endopod an oval article with apical spinulae, or club-shaped without armature or ornamentation; male leg 4 endopod a strong, wide, hirsute article reaching beyond middle of first exopodal article, or a strong spike with long proximal and small subdistal spinulae along inner margin; male leg 5 with inner unguiform projection .. 28

27′ Male leg 3 exopod without spinulae along outer margin, apophysis rounded, thumb very strong, nearly as thick as the width of apophysis, reaching slightly beyond the later; male leg 3 endopod a long article with one seta fused apically; male leg 4 endopod a single, irregular, hirsute article with two inner protuberances and with a distal hyaline seta, reaching middle of first exopodal article; female leg 3 endopod spiniform, reaching middle of first exopodal article; female leg 4 endopod a strong spine reaching slightly beyond middle of first exopodal article; male leg 5 without inner unguiform projection, with three innermost setae situated at the same level conferring leg 5 a subquadrate shape; female leg 5 with small, inner, unguiform projection, with two outermost setae at the same level, innermost seta arising subdistally .... *Remaneicaris* ignotus (Dussart, 1983)
[Brazil, Argentina: Corrientes, Laguna Sirena]

28(27) Male leg 2 endopod barely reaching middle of first exopodal article; male leg 3 exopod with apophysis slightly elongated inwards, thumb as long as apophysis, with inner hyaline membrane; male leg 3 endopod club-shaped without armature or ornamentation; male leg 4 endopod a strong spike with long proximal and small subdistal spinulae along inner margin; male leg 5 with inner unguiform projection, all setae situated at different levels conferring leg 5 a triangular shape; female unknown .... *Remaneicaris ivoneae* Corgosinho, Martínez Arbizu & do Santos-Silva, 2010
[Brazil: Paraná, Moretes City, Nhundiaquara River]

28′ Male leg 2 endopod reaching distal fourth of first exopodal article; male leg 3 exopod with apophysis rounded, thumb slender, reaching far beyond the apophysis, without hyaline membrane; male leg 3 endopod an oval article with apical spinulae; male leg 4 endopod a strong, wide, hirsute article reaching beyond middle of first exopodal article; male leg 5 with inner unguiform projection, all setae situated at the same level conferring leg 5 a subquadrate shape; female unknown .... *Remaneicaris insolitus* Corgosinho, Martínez Arbizu & do Santos-Silva, 2010
[Brazil: São Paulo, São Pedro, Paraná/Prata]

29(22) Male leg 3 without endopod; male leg 4 endopod reaching slightly beyond first exopodal article .... *Remaneicaris rhizophora* (Noodt, 1965)
[Argentina: Córdoba, La Bolsa, Río Sierra]

29′ Male leg 3 endopod an elongated article with apical ornamentation; male leg 4 endopod reaching proximal fourth of first exopodal article .... *Remaneicaris hecate* (Noodt, 1965)
[Argentina: Córdoba, Río Cuarto]

30(16) Characters not combined as below 31

30′ Male leg 3 exopod with thumb more than twice as long as apophysis, sigmoid, leg 3 endopod as long as three fourths the length of the entire exopod 34

31(30) Male leg 3 endopod a long spiniform element reaching far beyond distal margin of basis; leg 3 exopod with one proximal and one subdistal longitudinal row of spinulae, distalmost outer spinule of male leg 3 exopod modified into strong spinule with hyaline inner membrane or modified into simple strong, hyaline or normal spinule, thumb straight, twice as long or as long as apophysis; caudal rami with or without spinulae dorsally; female leg 4 endopod slightly beyond middle or reaching tip of first exopodal article; male leg 4 endopod a strong spike reaching tip or about middle or proximal third of first exopodal article or very small barely reaching base of first exopodal article .. 32

31′ Male leg 3 endopod a minute segment not reaching distal margin of basis, with one apical tiny seta; leg 3 exopod with one proximal, short row of spinulae, subdistally with one spinule only, distalmost outer spinule of male P3EXP modified into strong simple spinule, thumb sigmoid, slightly beyond apophysis; caudal rami without spinulae dorsally; male P4ENP a strong spike reaching middle of EXP1 .... *Remaneicaris clandestina* (Noodt, 1963)
[Brazil: Pará, Amazon River near Santarém]

32(31) Caudal rami without spinulae dorsally; female leg 4 endopod slightly beyond middle of first exopodal article; male P3EXP with distalmost outer spinule modified into strong spinule with hyaline inner membrane, thumb straight, twice as long as apophysis; male leg 4 endopod a strong spike reaching tip of first exopodal article 33

32′ Caudal rami with spinulae dorsally; female leg 4 endopod reaching tip of first exopodal article; male P3EXP with distalmost outer spinule modified into simple strong spinule, thumb straight, as long as apophysis; male leg 4 endopod a strong spike reaching proximal third of first exopodal article .... *Remaneicaris drepanephora* (Kiefer, 1967)
[Brazil: Pará, Rio Tapajos]

33(32) Male leg 3 exopod with distalmost outer spinule modified into strong spinule with hyaline inner membrane, thumb straight, nearly as half the length of the entire exopod; male leg 4 endopod a strong spike reaching tip of first exopodal article .... *Remaneicaris argentina* (Noodt, 1965)
[Argentina: Córdoba, La Bolsa, Río Sierra]

33′ Male leg 3 exopod with distalmost outer spinule of modified into strong hyaline spinule, thumb straight, noticeably shorter than half the length of the entire exopod; male leg 4 endopod a strong spike barely reaching basal part of first exopodal article .... *Remaneicaris palaciosi* (Noodt, 1962)
[El Salvador: Barranco Tovar, Citala, Río Lempa, Río Acehuafa]

34(30) Male leg 3 exopod with one proximal and one medial row of spinulae along outer margin; female leg 4 endopod reaching middle of first exopodal article; female leg 5 with inner unguiform projection .... *Remaneicaris jujuyensis* (Noodt, 1965)
[Argentina: Jujuy, Quebrada Honda]

34′ Male leg 3 exopod with one subdistal row of spinulae along outer margin; female leg 4 endopod reaching tip of first exopodal article; female leg 5 without inner unguiform projection .... *Remaneicaris hurdi* (Jakobi & Silva, 1962)
[Brazil: Santa Catarina]

Harpacticoida: Parastenocarididae: Parastenocaridinae: *Parastenocaris:* Species

The genus *Parastenocaris* has been used as a repository for species that belong to different evolutionary lineages (Corgosinho et al., 2010b), and its systematics is still confusing. Several species need to be redescribed in detail, and their systematical position revisited. Some authors (e.g., Reid, 1995; Galassi & de Laurentiis, 2004) have suggested that the use of the genus name *Parastenocaris* should be restricted to the species around the *Parastenocaris brevipe* group.

1 Caudal rami and other characters not combined as below 2

1′ Dorsal surface of anal segment and anal operculum without spinular ornamentation; anal operculum convex; leg 2 endopod a long article reaching middle of first exopodal article, with one apical seta; leg 3 endopod very small, about 4× as long as wide, with few apical spinulae, unarmed; leg 4 endopod a strong bipinnate spine slightly longer than first exopodal article; leg 5 with three setae in all, including outer basal seta; female caudal rami pear-shaped in dorsal view; with three setae located laterally halfway along outer margin of rami; male unknown .... *Parastenocaris bolbodes* Kiefer, 1968
[Brazil: Pará, Central Amazonia, lower Rio Tapajos]

2(1) Male leg 3 exopod not as below; basis of male leg 1 without inner spine 3

2′ Caudal rami with three lateral setae halfway outer margin of rami; basis of male leg 1 with inner, strong spine; male leg 3 exopood characteristic, apophysis elongated, with long, proximal stalk, distally golf-club-shaped, hyaline, thumb long, reaching beyond apophysis, straight, hyaline; female leg 3 endopod a very small article; male leg 4 endopod modified into strong spine; female and male leg 5 with inner unguiform projection, with three setae in all, outer basal seta included; all setae situated at different levels conferring leg 5 a triangular shape in female, very elongated in male .... *Parastenocaris santaremensis* Noodt, 1963
[Brazil: Pará, Amazon River near Santarém]

3(2) Female leg 3 without endopod; other characters not combined as below 4

3′ Anal somite twice as long as caudal rami, the later narrow at its base, much wider medially conferring each ramus a conical shape; male leg 4 endopod a very small article .... *Parastenocaris tapajosensis* Noodt, 1963
[Brazil: Pará, Amazon River near Santarém]

4(3) Male leg 3 exopod straight or with inner margin concave, with inner pointed protuberance proximally, apophysis elongated, hyaline, thumb sigmoid, bent outwards distally; female leg 3 with or without endopod; male leg 4 endopod a long article with hyaline apical part; male leg 3 with or without endopod 5

4′ Outer spinae of female leg 3 exopod somewhat dwarfed, rather thick; female leg 4 endopod globose, with pointed apical protuberance; female leg 5 characteristic, reduced to a trapezoidal plate with three outer setae, outermost basal seta included; with minute inner pointed projection; male unknown .... *Parastenocaris guyanensis* Defaye & Dussart, 1988
[French Guiana: Swamp near Saint-Jean]

5(4) Leg 1 exopod reaching middle of second endopodal article or leg 1 exopod and endopod of about the same length; male leg 4 endopod of zig-zag shape; male leg 3 endopod a small spiniform element; anal operculum convex 6

5′ Leg 1 exopod and endopod of about the same length; male leg 4 endopod straight, with hyaline apical part; male leg 3 without endopod; anal operculum rather straight .... *Parastenocaris kubitzkii* Noodt, 1972 [Colombia: Río Ariari, Puerto Caldas]

6(5) Leg 1 exopod reaching middle of second endopodal article; caudal rami divergent .... *Parastenocaris roettgeri* Noodt, 1972 [Colombia: Río Ariari, Puerto Caldas]

6′ Leg 1 exopod and endopod of about the same length; caudal rami parallel .... *Parastenocaris columbiensis* Noodt, 1972 [Colombia: Río Ariari, Puerto Caldas]

Harpacticoida: Parastenocarididae: Fontinalicaridinae: Genera

1 Characters not combined as below 2

1′ Female leg 3 without endopod; male leg 3 exopod robust, strongly incurved, with a very strong thumb, the later longer than apophysis; male leg 4 endopod reduced to a blunt article ornamented with three tiny apical spinulae or digitiform with distal outer corner produced into setiform structure, or completely absent; male and female leg 5 trapezoidal, all setae inserted distally .... ***Siolocaris* [p. 727]**

2(1) Characters not combined as below 3

2′ Male leg 3 exopod with two articles, first exopodal article short, rectangular, second exopodal article an articulated apophysis, thumb hypertrophic, nearly as long as first and second exopodal articles combined, inserted on the distal edge of first exopodal article, second exopodal article (apophysis) strongly sclerotized, articulated with first article on its inner margin and curved against thumb; leg 5 with three setae; seta II of caudal rami missing .... *Iticocaris*, monotypic: *Iticocaris itica* (Noodt, 1962) [El Salvador: Barranco Tovar, Zacatiapa, Río Jiboa (between San Salvador and Zacatecoluca); Río Acehuafa; Río Las Cañas; Izalco (Volcán Izalco) Atecozol. Brazil]

3(2) Characters not combined as below 4

3′ Basis of male leg 1 with inner spine; male leg 2 endopod sexually dimorphic, stronger in male; male leg 3 exopod with inner proximal protuberance bearing one papilla apically, apophysis with an apical claw; male leg 4 endopod lamelliform, with inner margin serrated; male leg 4 with first exopodal article ornamentedwith proximal row of tiny spinulae along inner margin; male leg 5 a quadrate plate, fused at base with intercoxal sclerite, with hypertrophied distal spine and strong proximal spine, intercoxal sclerite strongly developed, with medial process; anterolateral setae I, II, and III, situated laterally, proximal to dorsal seta VII .... ***Murunducaris* [p. 728]**

4(3) Characters not combined as below 5

4′ Sexual dimorphism expressed in legs 1–5; second antennae with exopod armed with one or two apical setae; anal operculum flanked by strong spinae on each side (absent in *Forficatocaris amazonensis* Noodt, 1963 only); female leg 3 endopod reduced or absent .... ***Forficatocaris* [p. 728]**

5(4) Cephalothorax and urosomal segments 2–5 with dorsal integumental windows in male, on cephalothorax and urosomal segments 2–4 in female; caudal seta II reduced, often not visible; male leg 2 endopod sexually dimorphic; basis of male leg 3 with strong, inner spinule; male leg 3 exopod strong, ending in a long forceps formed by apophysis and thumb, both with hyaline margin, endopod a slender seta; coxa of male leg 4 with two long, anterior spinulae; male leg 4 with first exopodal article reduced, with inner invagination proximally; male leg 4 endopod sexually dimorphic .... ***Brasilibathynellocaris* [p. 729]**

5′ Female and male leg 2 endopod with one article; female leg 3 without endopod; female leg 4 endopod with one article; female leg 5 subquadrate; male leg 1 with first endopodal article armed with one to three curved inner spinae; male leg 3 endopod a small blunt process or a small spine or absent; male leg 4 endopod with one or two articles; male leg 5 long, subquadrate, with or without laterally directed unguiculate extension .... ***Potamocaris* [p. 730]**

Harpacticoida: Parastenocarididae: Fontinalicaridinae: *Siolocaris*: Species

1 Male leg 3 exopod with one article, with three outer spinulae, apophysis short, rounded, robust, thumb strong, much longer than apophysis, with pointed tip and roundish distal hyaline membrane; leg 3 endopod a tiny seta; leg 4 without endopod; leg 5 trapezoidal, with three setae in all, with spinulae along distal half of inner margin; female unknown .... *Siolicaris sioli* (Noodt, 1963) [Brazil: Pará, Amazon River, near Santarém]

1′ Male leg 3 exopod very complex, with two articles, first exopodal article robust, with two blunt processes, second exopodal article an apophysis, strong, drawn out into an outward curved claw, thumb strong, longer than apophysis, with distal hyaline membrane and a proximal spur; leg 3 endopod a tiny seta; female unknown .... *Siolicaris jakobi* (Noodt, 1963) [Brazil: Pará, Amazon River, near Santarém]

Harpacticoida: Parastenocarididae: Fontinalicaridinae: *Murunducaris*: Species

1 Male leg 2 endopod different from below; other characters not combined as below 2

1′ Male leg 3 without endopod, apophysis of exopod a curved distal spine not incorporated to segment, thumb slender, hyaline, bent laterally at midlength, twice as long as apophysis; male leg 2 endopod a foliform oval article with one slender subapical inner seta and seven small spinulae .... *Murunducaris juneae* Reid, 1994
[Brazil: campo úmido Da Onça, Fazenda Agua Limpa, Distrito Federal]

2(1) Male leg 3 endopod a spine, exopodal thumb a strong spine or a short lateral spine, apophysis shorter or longer than thumb; female leg 5 with three innermost setae situated at very different levels along outer margin of leg; male leg 5 quadrate, fused to intercoxal sclerite, with outer seta and three distal elements, distalmost spine hypertrophied, appearing as a cuneiform or rectangular process, proximalmost element a strong spine hidden behind distal hypertrophied spine, intercoxal sclerite strongly developed into a spiniform process or into a keyhole-like process ornamented with a medial spinule 3

2′ Male leg 3 without endopod, exopod with two sets of outer spinulae, thumb a strong spine, apophysis shorter than thumb, ending in long curved spine, about 3x as long as apophysis; female leg 5 with three innermost setae situated nearly at the same level; male leg 5 quadrate, fused to intercoxal sclerite, with an outer seta and three distal elements, distalmost spine hypertrophied, appearing as a bulb with a distal crown, proximalmost seta developed as a strong spine, intercoxal sclerite strongly developed .... *Murunducaris noodti* Corgosinho, Martínez Arbizu & Reid, 2008
[Peru: Arequipa]

3(2) Male leg 3 exopod straight, with inner protrusion proximally, concave medially, thumb a strong spine, apophysis slightly shorter than thumb, ending in curved spine; male leg 4 endopodite lamelliform, curved outwards, with inner hump on distal third; outer margin with serrated lamella, with subdistal row of setulae and two subdistal spinulae; distalmost spine of male leg 5 hypertrophied, a cuneiform process, intercoxal sclerite strongly developed into a keyhole-like process ornamented with a medial spinule .... *Murunducaris loyolai* Corgosinho, Martínez Arbizu & Reid, 2008
[Brazil: Florianopolis]

3′ Male leg 3 exopod curved inwards, proximal third with inner hump; outer margin with row of spinulae on proximal third, thumb resented by a short lateral spine, apophysis longer than thumb, with 1 apical spine; basis of male leg 4 with inner modified spinule, endopod lamelliform, straight, outer margin serrated; distalmost spine of male leg 5 hypertrophied, rectangular, intercoxal sclerite developed into a spiniform process .... *Murunducaris dactyloides* (Kiefer, 1967)
[Brazil: Pará, Tapajós River, near Santarém]

Harpacticoida: Parastenocarididae: Fontinalicaridinae: *Forficatocaris*: Species and Subspecies

1 Second antennae exopod with two apical setae 2

1′ Second antennae exopod with one apical seta 6

2(1) Female and male anal operculum flanked by one or two strong spinae on each side, outermost spine on each side smaller 3

2′ Female and male anal operculum without such spinae .... *Forficatocaris amazonensis* Noodt, 1963
[Brazil]

3(2) Female caudal rami with inner apical seta smoothly tapering; outer margin of male leg 5 without well-developed lobes 4

3′ Female caudal rami with inner apical seta strongly bulbous proximally; outer margin of male leg 5 developed as three large lobes .... *Forficatocaris guarani* Noodt, 1963
[Brazil, Paraguay: Lake Ipacaria near San Bernardino]

4(3) Female caudal rami with longitudinal row of five to seven small spinulae and three strong spinae dorsally, inner margin unornamented; female anal operculum flanked by one or two strong spinae on each side 5

4′ Female caudal rami with two small spinulae dorsally, inner margin hirsute .... *Forficatocaris tetracantha* Noodt, 1963
[Peru: Río Huacamayo]

5(4) Anal operculum flanked on each side by two spinae, inner spinae much longer than outer ones; female leg 5 without inner ornamentation .... *Forficatocaris forficata forficata* Noodt, 1963
[Brazil: Pará, Amazon River, near Santarém]

5′ Anal operculum flanked on each side by one spine; inner margin of female leg 5 slightly hirsute .... *Forficatocaris forficata crenensis* Dussart, 1984
[Venezuela: El Carmen, Hato Becerra near Santarém]

6(1) Female leg 3 endopod with one article, reaching at least middle of first exopodal article, with or without apical seta 7

6′ Female leg 3 endopod reduced to small knob, unarmed 10

7(6) Female leg 5 with outer subdistal spine reaching beyond article 8

7′ Female leg 5 with outer subdistal spine barely reaching distal margin of article 9

8(7) Female leg 5 with outer subdistal element bearing a spinule; lateral seta with five articles; male leg 5 with two setae, without outer spine .... *Forficatocaris noodti* Jakobi, 1969
[Brazil: Três Rios, Rio Paraiba do Sul]

8′ Female leg 5 with outer subdistal element and lateral seta not as above, normal; male leg 5 with three outer setae .... *Forficatocaris fittkaui* Jakobi, 1972
[Brazil: Paraná, Rio Nhundiaquara]

9(7) Female leg 5 with outer subdistal spine lanciform, round in cross section; male leg 3 with first exopodal article lacking membranous structure; apical setae of male leg 4 with third exopodal article not articulated, simple; male leg 5 with three outer setae; male leg 5 with outer margin not lobed ventrally; male caudal rami at least twice as long as broad .... *Forficatocaris lilianae* Noodt, 1972
[Brazil: Rio Mineiro, Villa Atlantica/Mongagua, between Santos and Itanhaém]

9′ Female leg 5 with outer subdistal spine very small, triangular, flat, not separated from segment; male leg 3 with first exopodal article bearing a tapering hyaline membrane longer than and inserted medially to second exopodal article; male leg 4 with third exopodal article bearing nonarticulated, simple; male leg 5 with three outer setae; outer margin of male leg 5 not lobed ventrally; male caudal rami at least twice as long as broad .... *Forficatocaris schadeni* Reid, 1991
[Brazil: campo úmido, Fazenda Agua Limpa, Distrito Federal; Paraná]

10(6) Female caudal rami without ornamentation on inner proximal third 11

10′ Female caudal rami with inner proximal third hirsute; male leg 5 with outer, strong spine; male leg 3 with second exopodal article slender, about 6× as long as broad, slightly longer than first exopodal article .... *Forficatocaris claudii* Noodt, 1972
[Brazil: Rio Mineiro, Villa Atlantica/Mongagua, between Santos and Itanhaém]

11(10) Female caudal rami gradually tapering, with two strong inner spinae dorsolaterally; male leg 5 with three outer setae; male caudal rami short, about 1.6× as long as broad .... *Forficatocaris jakobii* Noodt, 1972
[Brazil: São Paulo, Rio Cubatão]

11′ Female caudal rami bulbous, without inner spinae; male leg 5 with strong, outer spine; male leg 3 with second exopodal article stout, about 4× as long as broad, slightly shorter than first exopodal article .... *Forficatocaris evelinae* Noodt, 1972
[Brazil: São Paulo, Rio Cubatão]

Harpacticoida: Parastenocarididae: Fontinalicaridinae: *Brasilibathynellocaris*: Species groups

1 Leg 2 endopod as long as first exopodal article; male leg 4 with first exopodal article lacking anterior row of strong spinulae .... ***Brasilibathynellae* group [p. 729]**

1′ Leg 2 endopod more ornamented in male, shorter than first exopodal article in both sexes; male leg 4 with first exopodal article bearing an anterior row of strong spinulae, endopod with hyaline region .... ***Salvadorensis* group [p. 729]**

Harpacticoida: Parastenocarididae: Fontinalicaridinae: *Brasilibathynellocaris*: *Brasilibathynellae* Group: Species

1 Last urosomal segment without well-developed serrated hyaline frill; male leg 3 with first exopodal article lacking inner hyaline cushion; male leg 5 with row of spinulae on distal inner margin .... *Brasilibathynellocaris brasilibathynellae* (Jakobi & Loyola e Silva, 1962)
[Brazil: Santa Catarina State, Florianopolis, Santa Catarina Island, Lagõa Village]

1′ Last urosomal segment with well-developed serrated hyaline frill; male leg 3 with first exopodal article bearing an inner hyaline cushion; male leg 5 without ornamentation on distal inner margin .... *Brasilibathynellocaris paranaensis* (Jakobi, 1972)
[Brazil: Paraná, City of Morretes, Nhundiaquara River]

Harpacticoida: Parastenocarididae: Fontinalicaridinae: *Brasilibathynellocaris*: *Salvadorensis* Group: Species and Subspecies

1 Male leg 4 endopod not transformed into a grasping organ, without proximal ornamentation; male leg 2 with second exopodal article ornamented with hairlike spinulae on inner distal corner; basis of male leg 4 with row of spinulae between outer seta and first exopodal article 2

1′ Male leg 4 endopod longer than broad, modified into a grasping organ, with serrated proximal membrane; male leg 2 with second exopodal article bearing strong spinulae on inner distal corner; basis of male leg 4 without ornamentation on outer margin 4

2(1) Male leg 4 endopod as long as broad in its proximal half, with hyaline membrane on distal margin only 3

2′ Male leg 4 endopod longer than broad, with hyaline membrane along entire surface .... *Brasilibathynellocaris cuscatlanensis* (Noodt, 1962)
[El Salvador: Río Lempa at Citala and Cuscatlan Bridge, border with Honduras]

3(2) Female and male caudal rami about 3.7× as long as wide; all setae well developed; seta IV long; seta V normal, long .... *Brasilibathynellocaris salvadorensis salvadorensis* (Noodt, 1962)
[El Salvador: Río Zacatiapa and Río Guasala, between San Salvador and Zacatecoluca; Laguna De Ilopango, Balneario Apulo; Río Acehuafa; Río Las Cañas]

3′ Female caudal rami ~ 2.6× as long as wide; all setae dwarfed, seta IV very short, strong, plumose, seta V lanciform, nearly as long as half the length of caudal ramus; male unknown .... *Brasilibathynellocaris salvadorensis lanceolata* (Noodt, 1962)
[El Salvador: Laguna De Ilopango, Balneario Apulo, near San Salvador]

4(1) Caudal rami nearly cylindrical, tapering distally in dorsal view; ~ 2.8× as long as wide, all setae well-developed, normal .... *Brasilibathynellocaris panamericana panamericana* (Noodt, 1962)
[El Salvador: Río Guasala, between San Salvador and Zacatecoluca, Río Grande, Acajutla. Honduras: Río Tela. Peru (?)]

4′ Caudal rami rectangular in dorsal view, about 1.8× as long as wide; trapezoidal in lateral view, all setae very reduced, plumose .... *Brasilibathynellocaris paramericana truncata* (Noodt, 1962)
[El Salvador: rocky sea bay, near La Fochota, South of La Union. Honduras: Río Tela]

Harpacticoida: Parastenocarididae: Fontinalicaridinae: *Potamocaris*: Species

Three species of *Potamocaris* (*Potamocaris bidentata* (Dussart, 1981), *Potamocaris tridentata* (Dussart, 1981), and *Potamocaris estevesi* (Reid, 1991)) are known only from the males, whereas two species (*Potamocaris bidens* (Noodt, 1955) and *Potamocaris cuiabaensis* Reid, 1991) are known only from the females, and only *Potamocaris bifida* Dussart, 1979 is known from both the male and female. Therefore, the reader is suggested to check the available descriptions for the six species. Note that *Potamocaris dentata* is synonym of *P. bifida* (Reid, 1991b).

1 Caudal rami shorter than anal somite, and 2.5× as long as broad at most 2

1′ Caudal rami as long or nearly as long as anal somite, and more than 2.5× as long as broad 3

2(1) Male caudal rami less than twice as long as broad; male leg 1 with first endopodal article bearing three inner spinae; female unknown .... *Potamocaris tridentata* (Dussart, 1981)
[Argentina: upper Paraná River]

2′ Female caudal rami about 2.5× as long as broad; male unknown .... *Potamocaris cuiabaensis* Reid, 1991
[Brazil: Cuiabá, Mato Grosso, Rio Coxipó; Paraná]

3(1) Anal operculum crescentic, protruding or not protruding beyond posterior margin of anal segment; male leg 1 with first endopodal article bearing one or two inner spinae; male leg 5 with two, female leg 5 with four setae (but note that the female of *P. bidentata* below is unknown); male leg 5 with or without laterally directed spiniform process; caudal rami from 3 to 3.5× as long as wide 4

3′ Anal operculum rounded; female leg 5 with two setae; caudal rami about 4× as long as broad, and slightly longer than anal segment; male unknown .... *Potamocaris bidens* (Noodt, 1955)
[Brazil: Paraiba River, near Caeté, Minas Gerais]

4(3) Male leg 1 with first endopodal article armed with one or two inner spinae; male leg 5 with laterally directed spiniform process; male leg 4 with one or two articles 5

4′ Male leg 1 with first endopodal article noticeably longer than second endopodal article, with two inner spinae; male leg 2 endopod a small ovate article with two setae; male leg 4 with endopod of one article, with subdistal inner, blunt projection; male leg 5 without laterally directed spiniform process; female unknown .... *Potamocaris bidentata* Dussart, 1981
[Argentina: upper Paraná River]

5(4) Anal operculum protruding beyond posterior margin of anal segment; caudal rami about 3× as long as broad; male leg 1 with first endopodal article bearing one inner spine; male leg 3 endopod a small process; thumb of male leg 3 exopod slightly curved, without toothlike inner process; male leg 4 with endopod of two articles, of which second article a long, lanciform spike with irregular inner margin .... *Potamocaris bifida* Dussart, 1979
[Brazil: upper Paraná. Paraguay: Paraguay River. Argentina: Santa Fe]

5′ Anal operculum not protruding beyond posterior margin of anal segment; caudal rami about 3× as long as broad; male leg 1 with first endopodal article armed with two inner spinae; male leg 3 endopod reduced to a knoblike expansion with bilobed tip, incompletely separated from basis; thumb of male leg 3 exopod with short toothlike inner process; male leg 4 endopod a single irregularly sclerotized, lanciform article; female unknown .... *Potamocaris estevesi* Reid, 1991
[Brazil: Rio de Janeiro, Lagoa De Cima; Minas Gerais, Belo Oriente, Rio Doce; São Paulo, Rio Ribeira de Iguape Basin, mouth of Rio Jacupiranga, Lagoa Jipuvura, mouth of River Una da Aldeia, and Rio Mogi Guassu]

Harpacticoida: Miraciidae: Genera

1 Characters not combined as below 2

1′ Second antennae with exopod, leg 1 endopod and leg 4 endopod of three articles; leg 1 with third exopodal article bearing four setae/spinae; leg 2–4 with third exopodal articles armed with 3, 3, 2 outer spinae .... *Paramphiascella,* one species: *Paramphiascella aquaedulcis* Dussart, 1984
[Venezuela: Río Portuguesa, Caño Falcon, Guarico]

2(1) Second antennae exopod, legt 1 endopod and leg 4 endopod with two articles; leg 1 exopod with third article armed with four setae/spinae; legs 2–4 exopod with third articles bearing 1, 1, 1 outer spinae .... *Schizoperopsis* (*Psammoschizoperopsis*), monotypic: *Schizoperopsis* (*P.*) *gauldi* (Chappuis & Rouch, 1961) [Brazil: Rio Doce, North of Recife]

2′ Second antennae exopod with two articles; leg 1 endopod and leg 4 endopod with three articles; leg 1 exopod with third article armed with four setae/spinae; legs 2–4 exopod with third articles bearing 2, 2, 2 outer spinae .... ***Schizopera* [p. 731]**

Harpacticoida: Miraciidae: *Schizopera*: Species

1 Characters not combined as below 2

1′ Legs 2–4 exopod with third article bearing four setae/spinae each; legs 2–4 exopod with second article without inner armature; leg 3 endopod with third article armed with three, of leg 4 with two elements; legs 2–4 endopod with first article without inner armature .... *Schizopera* (*Schizopera*) *noodti* Rouch, 1962 [Argentina: Punta Canteras, Buenos Aires]

2(1) Legs 2–4 exopod with third articles bearing 4, 5, 4 setae/spinae each 3

2′ Legs 2–4 exopod with third articles bearing 4, 4, 4 setae/spinae each 4

3(2) Leg 1 endopod slightly longer than exopod; female leg 5 endopodal lobe with three setae, exopod with five setae; male leg 5 endopodal lobe with two, exopod with five setae .... *Schizopera haitiana* Kiefer, 1934 [Haiti, Venezuela: Lago Valencia]

3′ Leg 1 endopod shorter than exopod; female leg 5 endopodal lobe with five, exopod with six setae; male unknown .... *Schizopera triacantha* Kiefer, 1934 [Haiti]

4(2) Length ratio of leg 1 first endopodal article: second and third endopodal articles combined ~ 2.2; female leg 5 exopod with setae 2 and 3, being seta 1 the outermost element and seta 6 the innermost element, much shorter than basal outer seta; female leg 5 with outermost seta of endopodal lobe longer than half the length of the distalmost seta 5

4′ Length ratio of leg 1 first endopodal article: second and third articles combined ~ 1.7; female leg 5 exopod with setae 2 and 3, being seta 1 the outermost element and seta 6 the innermost element, as long as basal outer seta; female leg 5 with outermost seta endopodal lobe about half the length of the distalmost seta .... *Schizopera vicina* Herbst, 1960 [Peru: Río Majes]

5(4) Female leg 5 exopod with innermost seta 1 about as long as half the length of the innermost seta 2, being seta 1 the innermost element and seta 6 the outermost element; female leg 5 with outermost seta of the endopodal lobe and innermost exopodal seta subequal .... *Schizopera tobae cubana* Petkovski, 1973 [Mexico: cenote Zacate and cenote Viejo, Quintana Roo. Cuba: Cueva De Jagüey, Río Baconao, Provincia Oriente]

5′ Female leg 5 exopod with innermost seta 1 much shorter than half the length of the innermost seta 2, being seta 1 the innermost element and seta 6 the outermost element; outermost seta of the female leg 5 endopodal lobe longer than innermost exopodal seta .... *Schizopera tobae tobae* Chappuis, 1931 [Cuba: Cueva del Tunel, La Salud, Quivicán, Mayabeque Province]

Harpacticoida: Ameiridae: Genera

1 Female leg 4 endopod with two or three articles 2

1′ Female leg 4 endopod with one article; female leg 1 with second exopodal article bearing outer spine; second antennae exopod with one article, with two or three setae; Legs 1–4 with outer basal seta .... ***Stygonitocrella* [p. 732]**

2 Female legs 2–4 endopods with three articles; legs 2–4 with first endopodal articles visibly shorter than second; other characters not combined as below 3

2′ Leg 1 endopod with first and second article unarmed; female legs 2–3 endopods with two articles; leg 4 endopod with one or two articles; legs 2–4 with first exopodal article lacking inner seta; leg 2 with third exopodal article bearing two outer spinae; female and male leg 5 with baseoendopods fused medially .... ***Nitocrella* [p. 732]**

3(2) Leg 3 with third exopodal article bearing two outer spinae .... ***Parapseudoleptomesochra* [p. 732]**

3′ Leg 3 with third exopodal article bearing three outer spinae .... ***Nitokra* [p. 732]**

Harpacticoida: Ameiridae: *Stygonitocrella*: Species

1 Female legs 2–3 endopods with two articles, leg 4 endopod with 1 article; second antennae exopod with three setae; female and male leg 5 without endopodal lobe; male leg 5 exopod with three, female leg 5 exopod with four setae .... *Stygonitocrella (Stygonitocrella) montana* (Noodt, 1965)
[Argentina: Mendoza, Uspallata Valley]

1′ Female legs 2–4 endopods with one article; second antennae exopod with two setae; female leg 5 a single plate; male leg 5 exopod with five, endopodal lobe with two setae .... *Stygonitocrella (Stygonitocrella) orghidani* (Petkovski, 1973)
[Cuba: Cueva de Emilio, Pinar del Río Province]

Harpacticoida: Ameiridae: *Nitocrella*: Species

1 Anal operculum with row of strong spinulae along posterior margin; male leg 5 exopod with four or five setae; male and female leg 5 endopodal lobe with or without setae 2

1 Anal operculum with minute spinulae close to posterior margin; male leg 5 exopod with five, endopodal lobe with two setae; female unknown .... *Nitocrella negreai* Petkovski, 1973
[Cuba: Cueva del Agua, Baracoa, Guantánamo Province]

2(1) Female leg 5 exopod with four, endopodal lobe with three setae; male leg 5 exopod with four or five, endopodal lobe with two setae 3

2′ Female and male leg 5 exopod with four setae; female and male leg 5 endopodal lobe without setae; female leg 4 with third exopodal article bearing six setae .... *Nitocrella motasi* Petkovski, 1976
[Cuba: Arroyo Seboruquito, Mayari, Oriente Province]

3(2) Male leg 5 exopod with five setae; female leg 5 endopodal lobe more or less well developed; leg 1 exopod reaching slightly beyond second endopodal article; female leg 4 with third exopodal article bearing five setae; legs 2–4 with first and second exopodal articles subequal .... *Nitocrella cubanorum* Petkovski, 1976
[Cuba: Cueva del Segundo Cauce and Gran Caverna de Santo Tomas, Pinar del Río Province]

3′ Male leg 5 exopod with four setae; female leg 5 endopodal lobe only slightly developed; leg 1 exopod not reaching third endopodal article; female leg 4 with third exopodal article bearing six setae; legs 2–4 with first exopodal articles clearly shorter than second .... *Nitocrella caraioni* (Petkovski, 1976)
[Cuba: Río Sabanilla and Río Miel, Baracoa, Oriente Province]

Harpacticoida: Ameiridae: *Parapseudoleptomesochra*: Species

1 Female leg 5 exopod with six setae, endopodal lobe with four setae 2

1′ Female and male leg 5 exopod with three setae; female leg 5 endopodal lobe with four, of male with 1 seta .... *Parapseudoleptomesochra subterranea subterranea* (Chappuis, 1928)
[Mexico: Yucatán]

2(1) Female leg 2 with third exopodal article bearing five setae; leg 1 with second exopodal article and first endopodal article lacking inner armature; male unknown .... *Parapseudoleptomesochra botosaneauni* (Petkovski, 1973)
[Mexico (?): Yucatán. Cuba: Cueva La Patana, Guantánamo Province and Cueva del Tunel, La Salud, Quivicán, Mayabeque Province]

2′ Female leg 2 with third exopodal article bearing six setae; leg 1 with second exopodal article and first endopodal article bearing inner seta .... *Parapseudoleptomesochra incerta* (Chappuis, 1933)
[Netherlands Antilles: Bak Ariba, Curaçao]

Harpacticoida: Ameiridae: *Nitokra*: Species

1 Leg 1 with first endopodal article as long or longer than entire exopod 2

1′ Leg 1 with first endopodal article shorter than entire exopod 3

2(1) Anal operculum without spinulae; female leg 3 endopod shorter than first and second exopodal articles combined .... *Nitokra pusilla* Sars G. O., 1911
[Mexico: Cueva San Bulha, Motul, Yucatán]

2′ Anal operculum with spinulae; female leg 3 endopod longer than first and second exopodal articles combined .... *Nitokra typica typica* Boeck, 1865
[Mexico: Progreso near Piste, Yucatán]

3(1) Leg 2 with third endopodal article bearing three setae; caudal rami about twice as long as broad 4

3′ Leg 2 with third endopodal article bearing four setae; caudal rami wider than long .... *Nitokra spinipes spinipes* Boeck, 1865
[Mexico: Yucatan, Progreso. Guatemala. Neartic]

4(3) Leg 1 with first endopodal article slightly longer than first and second exopodal articles combined 5

4′ Leg 1 with first endopodal article shorter than first and second exopodal articles combined .... *Nitokra lacustris lacustris* (Schmankevich, 1875)
[Mexico: Yucatan, Progreso. Cuba. Belize. Aruba: Pos Di Noord. Bonaire: Fontein and Pos Di Pepe. Neartic]

5(4) Caudal rami slightly longer than broad; inner margin without medial spinulae .... *Nitokra lacustris sinoi* Marcus & Por, 1961
[Mexico: Chetumal, Quintana Roo; Celestun, Yucatan. Cuba: Baconao River, Provincia Oriente. Colombia: Laguna de Navio Quebrado, Camarones, La Guajira; Ciénaga Grande de Santa Marta. Neartic]

5′ Caudal rami subquadrate; inner margin with medial spinulae .... *Nitokra lacustris colombianus* Reid, 1988
[Colombia: Bahia Solano, Chocó]

Harpacticoida: Canthocamptidae: Subfamilies

The family Canthocamptidae is in urgent need of revision, and many of its genera should be rediagnosed. Borutzky (1952) divided the family into four subfamilies: Halocampthocamptinae, Canthocamptinae, Morariinae, and Epactophaninae, and Por (1986) created the subfamily Hemimesochrinae. This view was adopted here for identification purposes.

Following Bruno & Cottarelli (1999), the genus *Epactophanes* is still controversial, and its definition is circumstantial. Some authors (e.g., Lang, 1948; Dussart, 1967; Shen et al., 1979) support the cosmopolitanism and high variability of *Epactophanes richardi* Mràzek, 1893. The later view was adopted here.

1 Anal operculum rounded, with strong spinulae; legs 2–4 exopods with subquadrate articles conferring the legs a more or less compact shape; second antennae exopod with 1 article, with two setae; mandibular palp reduced to a seta; male leg 3 endopod with 2 articles, second endopodal article with inner apophysis not reaching tip of third exopodal article; leg 1 with second expodal articles with unarmed inner margin, of legs 2–4 with second exopodal articles bearing inner seta; female legs 1–3 endopods with two articles, leg 4 endopod with 1 article; female leg 5 exopod and baseoendopod separated, endopodal lobe with five, exopod with four setae; male leg 5 with three setae (two exopodal + one basal seta), baseoendopods fused, without armature .... Epactophaninae, one species: *Epactophanes richardi* Mrázek, 1893
[Jamaica: Trelawny Parish. Honduras: El Cusco National Park. Colombia: Monserrate, Santa Fe, Bogota. Guyana. Brazil: Paraná]

1′ Anal operculum rounded or with medial part an acute, long expansion, with variable number of teeth or with setulae; legs 2–4 exopods with rectangular segments conferring the legs an elongated shape; second antennae exopod with one or two articles, with four setae at most; mandibular palp with one or two articles, or represented by a seta; male leg 3 endopod with three articles, rarely with two articles, if with three articles, then second endopodal article with inner, long apophysis, if with two articles, then first endopodal article with inner long apophysis; legs 1–4 with second exopodal article typically bearing inner seta; leg 1 endopod with two or three articles; female legs 2–4 endopods with two articles, leg 4 endopod rarely with one article; female leg 5 exopod and baseoendopod fused or separated; male leg 5 exopod and baseoendopod fused or separated, exopod with five setae at most, endopodal lobe with three setae at most .... **Canthocamptinae [p. 733]**

Harpacticoida: Canthocamptidae: Canthocamptinae: Genera

1 Female legs 1–3 endopods with three articles, leg 4 endopod with two articles; female leg 5 exopod with five, endopodal lobe with five or six setae; male leg 5 exopod with five or six, endopodal lobe with two setae; male leg 2 endopod with two articles 2

1′ Female legs 1–3 endopods with two articles, leg 4 endopod with one or two articles; male leg 2 endopod with two articles; other characters not combined 3

2(1) Armature formula of female leg 2 endopod 1,1,121; male leg 4 with second endopodal article lacking outer apophysis .... ***Bryocamptus* [p. 734]**

2′ Armature formula of female leg 2 endopod 1,1,221; male leg 4 with second endopodal article bearing outer apophysis .... ***Canthocamptus* [p. 734]**

3(1) Characters not combined as below 4

3′ Secon antennae exopod with one article, with three setae; female maxilliped subchelate, syncoxa with stout spinulose distal seta, endopod with tiny seta and slender claw; male maxilliped dimorphic, endopod without seta, claw very large, heavily sclerotized, recurved, and blunt; female leg 5 baseoendopods partly fused medially, mediodistal margin of endopodal lobe deeply crenate and expanded; male legs 2–4 endopods and caudal rami sexually dimorphic; anal operculum produced posteriorly, posterior margin rounded and crenate .... *Pindamoraria*, monotypic: *Pindamoraria boraceiae* Reid & Rocha, 2003
[Brazil: Trilha de Clareira; Poço Verde Waterfall on Rio Claro, Ribeirão do Campo, São Paulo]

4(3) Characters not combined as below 5

4′ Female leg 5 exopod with three setae, endopodal lobe absent; female legs 2–4 endopods subquadrate, about as long as wide; legs 2–4 with first endopodal articles bearing inner seta; legs 2–4 with second endopodal articles bearing 4, 5, 3 setae, respectively; caudal rami slightly longer than anal segment, dorsoventrally depressed .... *Spelaeocamptus*, one specie: *Spelaeocamptus neotropicus* Noodt, 1963 [Peru: Río Guacamayo]

5(4) Characters not combined as below 6

5′ Second antennae exopod with one article, with four setae; anal operculum produced, posterior margin without ornamentation; posterior margin of pro- and urosomal segments plain or serrate; legs 2–4 with first and second endopodal articles subequal; first endopodal article slightly longer than wide, with one inner seta; legs 2–4 with second endopodal articles bearing 3, 4, 4 setae; male legs 2–4 endopods with two articles; leg 2 with first endopodal article bearing one inner seta and two outer strong spinulae; leg 4 with second endopodal article bearing an apical element cork-screw-like and hyaline; male leg 3 endopod with two articles; first endopodal article with outer strong spinule and inner pectinate element; second endopodal article with proximal inner and distal outer elements fused to segment, with distal inner seta in-between .... ***Moraria* [p. 735]**

6(5) Characters not combined as below 7

6′ Second antennae exopod with one or two articles, with four setae; anal operculum rounded, posterior margin plain or with small spinulae; posterior margin of pro- and urosomal segments plain; leg 1 exopod and endopod with two articles; second exopodal article with five setae/spinae; legs 2–4 exopods with three articles; female legs 2–4 endopods with two articles; legs 2–4 with first endopodal article bearing an inner seta, slightly wider than long, smaller than second endopodal article; male leg 3 endopod with three articles, second endopodal article with inner apophysis, third endopodal article with inner long spine and reduced outer element; female leg 5 exopod with three setae, endopodal lobe with five subequal setae .... ***Maraenobiotus* [p. 735]**

7(6) Characters not combined as below 8

7′ Rostrum not fused to cephalothorax, with distal row of spinulae ventrally; female rostrum broad; male rostrum sexually dimorphic, slender; posterior margin of prosomal segments typically with slender spinulae, sometimes serrate; anal operculum rounded; typically, with spinulae, sometimes naked; second antennae exopods typically with one article, with up to three setae, rarely with five setae, sometimes represented by minute segment with one seta only; mandibular palp reduced, with one article with up to three setae, sometimes with a small seta inserted on coxa; female and male leg 5 exopod and baseoendopod fused; both legs fused medially in the male; legs 1–4 exopods with three articles; female legs 1–3 endopods with two articles; male legs 1–2 endopods with two articles; male leg 3 endopod dimorphic, typically with three, rarely with two articles, second endopodal article with inner stout, short apophysis; leg 4 endopod with one or two articles; when two articles legs 2–4 with first endopodal article small, wider than long, without armature; caudal rami with seven setae; seta I very reduced .... ***Cletocamptus* [p. 735]**

8(7) Posterior margin of segments plain or serrate; anal operculum rounded, broadly triangular or lanciform, posterior margin serrate, with teeth, setulae, or smooth; setae of caudal rami normal, seta VI sometimes missing; female leg 5 exopod with four or five setae, endopodal lobe with six, rarely with four or five setae; leg 1 with first endopodal article shorter, as long as, or longer than exopod; male leg 5 exopod with four or five, rarely with three setae, endopodal lobe without or with two, rarely with one seta .... ***Attheyella* [p. 736]**

8′ Posterior margin of segments plain or serrate; anal operculum rounded or strongly crescentic, posterior margin serrate, with teeth, spinulae, setulae, or smooth; setae of caudal rami normal, sometimes setae II and III spiniform, or all setae reduced, seta IV sometimes reduced or missing; female leg 5 exopod with three or four, rarely with five setae, endopodal lobe with four, rarely with two setae; leg 1 with first endopodal article at least as long as exopod; male leg 5 exopod with three or four, rarely with two setae, endopodal lobe without, rarely with one seta .... ***Elaphoidella* [p. 740]**

Harpacticoida: Canthocamptidae: Canthocamptinae: *Bryocamptus*: Species

1 Posterior margin of anal operculum with about 10 single spinulae; caudal rami subquadrate, about as long as broad; second antennae exopod with one article, with four setae; leg 1 exopod as long as first and second endopodal articles combined; male unknown .... *Bryocamptus (Bryocamptus) campaneri* (Reid, 1993) [Brazil: Distrito Federal, Fazenda Agua Limpa]

1′ Posterior margin of anal operculum with bifurcated spinulae; caudal rami rectangular, slightly longer than broad; second antennae exopod with two articles, first article without, second article with four setae; leg 1 exopod longer than first and second endopodal articles combined; male leg 5 exopod with six, endopodal lobe with two setae .... *Bryocamptus (Bryocamptus) minutus minuts* (Claus, 1863) [Mexico: Río Champoton, Yucatan]

Harpacticoida: Canthocamptidae: Canthocamptinae: *Canthocamptus*: Species

1 Caudal rami with length:width ratio of about 1.7; leg 1 with second exopodal article lacking inner seta; leg 1 with first endopodal article reaching middle of third exopodal article; female leg 5 exopod with five, endopodal lobe with six setae; male leg 5 exopod without inner seta close to baseoendopod .... *Canthocamptus (Canthocamptus) microstaphylinus microstaphylinus* Wolf, 1905 [Venezuela: Río Portuguesa, Caño Falcon, Guarico]

1′ Caudal rami with length:width ratio of about 2.1; leg 1 with second exopodal article bearing inner seta; leg 1 with first endopodal article as long as exopod; female leg 5 exopod and endopodal lobe with five setae each; male leg 5 exopod with inner seta close to baseoendopod .... *Canthocamptus* (*Canthocamptus*) *staphylinus* (Jurine, 1820)
[Honduras: La Ceiba. Panama: Lago Gatun near Gold Hill]

Harpacticoida: Canthocamptidae: Canthocamptinae: *Moraria*: Species

1 First antennae with eight articles; posterior margin of pro- and urosomal segments plain; leg 1 endopod with two articles, second endopodal article with one inner strong and pinnate spine, one inner subdistal slender and small seta, distally with one very long element and one outer spine; female leg 5 exopod with one inner spinule at most 2

1′ First antennae with seven articles; posterior margin of pro- and urosomal segments serrate; leg 1 endopod with two articles; second endopodal article with one inner strong spinule, one inner subdistal slender and small seta, distally with one very long element and one outer spine; female leg 5 exopod with longitudinal row of spinulae along inner margin; male leg 5 with endopodal lobe bearing two spinae of which innermost twice as long as outermost, with pointed process between them .... *Moraria* (*Moraria*) *cristata* Chappuis, 1929
[Mexico: Progreso and Río Champoton, Yucatan]

2(1) Female leg 5 exopod without inner spinulae, endopodal lobe with innermost seta clearly shorter than adjoining element; female leg 4 endopod with three elements in all; male unknown .... *Moraria* (*Moraria*) *cusuca* Fiers & Jocque, 2013
[Honduras: El Cusco National Park, West of San Pedro Sula]

2′ Female leg 5 exopod with only one inner spinule at the base of the innermost seta, endopodal lobe with two innermost setae subequal; female leg 4 endopod with four elements in all; male leg 5 with endopodal lobe bearing two subequal spinae, without pointed process between them .... *Moraria* (*Moraria*) *catracha* Fiers & Jocque, 2013
[Honduras: El Cusco National Park, West of San Pedro Sula]

Harpacticoida: Canthocamptidae: Canthocamptinae: *Maraenobiotus*: Species

1 Legs 2–4 with third exopodal articles bearing 5, 6, 5 setae/spinae, other characters not combined as below 2

1′ Legs 2–4 with third exopodal articles bearing 4, 6, 5 setae/spinae; male caudal setae IV and V originate side-by-side, seta VI ventral to setae IV and V, partially obscured in dorsal view .... *Maraenobiotus fontinaloides* Löffler, 1960
[Peru]

2(1) Female caudal rami with origin of caudal seta V overlapping dorsally origin of caudal seta IV; male caudal seta IV, V, and VI originate side-by-side; female anal operculum without, male anal operculum with spinulae along posterior margin .... *Maraenobiotus naticochensis* Delachaux, 1918
[Peru: Lago Naticocha]

2′ Female caudal rami with caudal setae IV, V, and VI originating side-by-side; female anal operculum with small spinulae along posterior margin; male unknown .... *Maraenobiotus fontinalis* Harding, 1955
[Peru: Lagunillas, Lago Titicaca]

Harpacticoida: Canthocamptidae: Canthocamptinae: *Cletocamptus*: Species

1 Second antennae exopod a long article with more than one seta; mandibular palp with one article, minute, with one or two setae, with or without seta on coxa; other characters below not combined 2

1′ Posterior margin of cephalothorax and prosomal segments serrate; second antennae exopod with 1 minute article bearing one seta; mandibular palp with one article, with three setae, without seta on coxa; leg 4 endopod with one article; male leg 3 endopod with two articles, with constriction indicating original division between second and third articles, inner apophysis of second endopodal article reaching beyond tip of second exopodal article, nearly as long as supporting segment, male leg 2 with second endopodal article bearing inner seta slenderer than in female; female and male leg 5 exopod and endopodal lobe barely discernible, separated only by small notch .... *Cletocamptus dominicanus* Kiefer, 1934
[Haiti: Lago Enriquillo]

2(1) Leg 2 with second endopodal article bearing four setae 3

2′ Leg 2 with second endopodal article bearing three setae 5

3(2) Second antennae exopod with three setae 4

3′ Second antennae exopod with 2two setae .... *Cletocamptus kummleri* (Delachaux, 1918)
[Peru: Lago Huaron and Lago Titicaca, Argentina: Laguna de Junin]

4(3) Mandibular palp minute, with one seta plus one seta on coxa; leg 4 with third exopodal article bearing five setae/spinae; male leg 2 with second endopodal article bearing an inner seta transformed into strong, stout, and short spine; male leg 3 with second endopodal article bearing apophysis reaching tip of first exopodal article .... *Cletocamptus pilosus* Gómez & Gee, 2009
[Argentina: Mendoza, Embalse Nihuil]

4′ Mandibular palp with two setae, without seta on coxa; leg 4 with third exopodal article bearing four setae/spinae; male leg 2 with second endopodal article bearing a normal inner seta; male leg 3 with second endopodal article bearing apophysis reaching middle of first exopodal article .... *Cletocamptus gomezi* Suárez-Morales, Barrera-Moreno & Ciros-Pérez, 2013
[Mexico: Lago Alchichica, Puebla]

5(2) Mandibular palp minute, with two setae plus one seta on coxa; female leg 3 with second endopodal article bearing three setae; male leg 3 endopod with three articles; female and male leg 5 exopod and endopodal lobe clearly discernible, separated by deep notch 6

5′ Mandibular palp with two setae, without seta on coxa; female leg 3 with second endopodal article bearing five setae; male leg 3 endopod with two articles, inner apophysis small, originating subdistally; female and male leg 5 exopod and endopodal lobe barely discernible, separated only by small notch .... *Cletocamptus albuquerquensis* (Herrick, 1894)
[Netherlands Antilles: Curaçao, Hispaniola, Aruba, Argentina: Córdoba, Lucio V. Mansilla and Totoralejos; La Pampa, Santa Rosa, Laguna Luro]

6(5) Female leg 6 with two setae; caudal rami up to 2× as long as wide 7

6′ Female leg 6 with three setae; caudal rami about 3× long as wide .... *Cletocamptus tertius* Gómez & Gee, 2009
[Argentina: La Pampa, Santa Rosa, Laguna Don Tomas]

7(6) Female leg 5 exopod reaching about middle of endopodal lobe, well below insertion site of outermost endopodal seta; female leg 5 endopodal lobe with fourth innermost seta twice as long as third innermost seta; caudal rami about as long as wide; female genital double-segment wider than long; fourth and fifth urosomal segments with several rows of small spinulae ventrally .... *Cletocamptus cecsurirensis* Gómez, Scheihing & Labarca, 2007
[Chile: Salar de Surire, Andean Plateau]

7′ Female leg 5 exopod reaching far beyond middle of endopodal lobe, nearly reaching insertion site of outermost endopodal seta; female leg 5 endopodal lobe with fourth innermost seta 3× as long as third innermost seta; caudal rami nearly twice as long as wide; female genital double-segment nearly as long as wide; fourth and fifth urosomal segments with one or two rows of long spinulae; male unknown .... *Cletocamptus assimilis* Gómez & Gee, 2009
[Argentina: Formosa, Monte Lindo]

Harpacticoida: Canthocamptidae: Canthocamptinae: *Attheyella*: Subgenera

1 Segments with posterior margins serrate, rarely smooth or with slender spinulae; anal operculum rounded, with several small teeth, setulae, or smooth, or broadly triangular or lanciform; leg 1 endopod with three articles, longer than exopod; leg 1 with second exopodal article bearing inner seta; legs 2–4 with third exopodal articles bearing three outer spinae; leg 4 endopod with two articles; female leg 5 exopod and endopodal lobe with four or five, and four to six setae, respectively; male leg 5 exopod with three, four, or five, endopodal lobe without, or with one or two setae; male leg 3 with second exopodal article with outer spine normal, leg 4 with third exopodal article withouter spine normal or modified 2

1′ Segments with posterior margins smooth; anal operculum rounded with three to 10 strong teeth along posterior margin; leg 1 endopod with two articles, shorter or as long as exopod; leg 1 with second exopodal article lacking inner seta; legs 2–4 with third exopodal articles bearing two outer spinae; leg 4 endopod with one or two articles; female leg 5 exopod and endopodal lobe with five and six setae, respectively; male leg 5 exopod with four, endopodal lobe without seta; male leg 3 with second exopodal article bearing normal outer spine, male leg 4 with third exopodal article bearing normal outer spine .... ***Attheyella (Canthosella)*** **[p. 737]**

2(1) Characters not combined as below 3

2′ Anal operculum rounded, with posterior margin frilled; caudal rami with caudal seta V curved basally, falciform, arising from bifid lappetlike projection, caudal seta VI arising from squarish dorsal lappet; second antennae exopod with 1 article, with two or three setae; female leg 5 exopod and endopodal lobe with four setae each; male leg 5 exopod with three setae, endopodal lobe without armature .... *Attheyella* (*Attheyella*), one species: *Attheyella* (*A.*) *jureiae* Por & Hadel, 1986
[Brazil: Amazonas, Paraná; São Francisco; Represa Do Broa and Serra da Juréia, São Paulo. Argentina: Santa Fe; Buenos Aires]

3(2) Characters not combined as below 4

3′ Anal operculum rounded, with several small teeth along posterior margin; caudal rami with caudal setae normal; leg 1 with first endopodal article longer than first and second exopodal articles combined; leg 1 with second endopodal article lacking inner seta; female leg 4 with second endopodal article bearing six, of male with five setae; female leg 5 exopod and endopodal lobe with five setae each; male leg 5 exopod with five, endopodal lobe with two setae .... *Attheyella* (*Ryloviella*) *pilosa* Chappuis, 1929
[Mexico: cave and cenote Luchil, Yucatan]

4(3) Segments with posterior margins serrate or smooth dorsally; anal operculum rounded, with teeth, setulae, or smooth; leg 1 with second exopodal article bearing inner seta; leg 2 with third exopodal article bearing three outer spinae; second endopodal article with three, four, or five setae; leg 3 with second endopodal article bearing three inner setae; female leg 5 exopod with five, endopodal lobe with six setae; male leg 3 endopod with three articles; male leg 4 with first endopodal article lacking inner seta; second endopodal article with two or three setae; male leg 5 exopod with five, endopodal lobe with two setae; male leg 3 with second exopodal article bearing outer hooklike spinae; leg 4 with third exopodal article bearing outer apical element modified into a strong spine with one or two acute outgrowths, or into a strong deeply bifurcated spine .... ***Attheyella (Chappuisiella)*** **[p. 737]**

4′ Segments with posterior margins serrate, with slender spinulae or smooth; anal operculum broadly triangular, rarely lanciform or rounded; leg 1 with second exopodal article with or without inner seta; leg 2 with third exopodal article bearing two or three outer spinae; second endopodal article with three, four, or five setae; leg 3 with second endopodal article bearing two or three inner setae; female leg 5 exopod with four, endopodal lobe with six setae; male leg 3 with endopod of two or three articles; male leg 4 with first endopodal article bearing inner seta, second endopodal article with three, four, or five setae; male leg 5 exopod with four, rarely with three setae, endopodal lobe with two, rarely with one seta; male leg 3 with second exopodal article bearing normal outer spine, leg 4 with third exopodal article bearing normal outer apical element .... ***Attheyella (Delachauxiella)*** **[p. 738]**

Harpacticoida: Canthocamptidae: Canthocamptinae: *Attheyella* (*Canthosella*): Species

1 Caudal rami about 2× as long as wide; leg 4 endopod with 2 articles 2

1′ Caudal rami at most 1.6× as long as wide; leg 4 endopod with one or two articles 3

2(1) Anal operculum rounded, with five strong teeth; leg 1 with first endopodal article as long as exopod; leg 4 endopod with three setae/spinae; male unknown .... *Attheyella* (*Canthosella*) *pilagaensis* Janetzky, Martínez Arbizu & Reid, 1996
[Argentina: Río Pilagá]

2′ Anal operculum rounded, with four strong teeth; leg 1 with first endopodal article shorter than exopod; leg 4 endopod with two setae/spinae; female unknown .... *Attheyella* (*Canthosella*) *striblingi* (Reid, 1990)
[Costa Rica: Monteverde Cloud Forest Reserve]

3(1) Female leg 4 endopod with two articles; first endopodal article without, second with one, two, or five setae; male leg 4 endopod with 1 or 2 articles, if two articles, then first article without seta, second article with two or three setae, if one article, then with one seta 4

3′ Female and male leg 4 endopod with one article, bearing two setae .... *Attheyella* (*Canthosella*) *aliena* Noodt, 1965
[Brazil: Fazenda Inglesa, Petrópolis, Rio de Janeiro]

4(3) Anal operculum with nine or 10 small teeth 5

4′ Anal operculum with up to six strong teeth 6

5(4) Anal operculum with 10 small teeth; leg 2 with second endopodal article bearing three setae; female leg 3 with second endopodal article bearing four setae; female leg 4 endopod with 2 articles; second endopodal article with one seta; male leg 4 endopod with one article, with one seta .... *Attheyella* (*Canthosella*) *antillica* (Petkovski, 1973)
[Cuba: Cueva del Abono, Bahía Honda, Pinar del Río Province]

5′ Anal operculum with 9 small teeth; leg 2 with second endopodal article bearing four setae; female leg 3 with second endopodal article bearing five setae; female leg 4 endopod with two articles; second endopodal article with two setae; male leg 4 endopod with two articles, second article with two setae .... *Attheyella* (*Canthosella*) *chocoensis* Gaviria & Defaye, 2012
[Colombia: Nuquí, Jurubida, Utría, Chocó]

6(4) Anal operculum with three strong teeth; leg 1 with first endopodal article shorter than exopod 7

6′ Anal operculum with four, five, or six strong teeth; leg 1 with first endopodal article shorter or as long as exopod 8

7(6) Male leg 2 with second endopodal article bearing three setae and two tiny humps; male leg 3 endopod with 2 articles; second endopodal article with inner apophysis and two apical setae .... *Attheyella* (*Canthosella*) *mervini* Janetzky, Martínez Arbizu & Reid, 1996
[Jamaica: Trelawny Parish, wet limestone forest near Windsor]

7′ Male leg 2 with second endopodal article bearing three setae, without humps; female unknown; male leg 3 endopod with three articles, second article with inner apophysis, third article with two apical setae .... *Attheyella* (*Canthosella*) *siolii* (Kiefer, 1967)
[Brazil: Rio Negro, Caatingas, Amazonas]

8(6) Anal operculum with five strong teeth; leg 1 endopod as long as exopod; leg 3 with third exopodal article bearing one inner seta, with five setae/spinae in all; leg 4 with second endopodal article bearing five setae; male unknown .... *Attheyella* (*Canthosella*) *kalima* (Delachaux, 1924)
[Brazil, Argentina: Paraná medio]

8′ Anal operculum with four to six strong teeth; leg 1 endopod shorter than exopod; leg 3 with third exopodal article bearing two inner setae, with six setae/spinae in all; leg 4 with second endopodal article bearing two setae .... *Attheyella* (*Canthosella*) *vera* Por & Hadel, 1986
[Brazil: Serra Da Juréia, São Paulo]

Harpacticoida: Canthocamptidae: Canthocamptinae: *Attheyella* (*Chappuisiella*): Species

1 Posterior margin of abdominal segments serrated dorsally; posterior margin of anal operculum with small spinulae or with setulae 2

1′ Posterior margin of abdominal segments plain dorsally; anal operculum with strong spinulae along posterior margin .... *Attheyella* (*Chappuisiella*) *huaronensis* (Delachaux, 1918)
[Peru: Lago Titicaca, Lago Huaron and La Raya, Ayaviri. Argentina: Río de la Plata]

2(1) Posterior margin of anal operculum with setulae 3

2′ Posterior margin of anal operculum with small spinulae. Note: the following taxa do not have both sexes described, caution must be used when assigning species 6

3(2) Caudal rami without sharp outgrowth 4

3′ Dorsal seta VII of caudal rami arising from sharp outgrowth of ramus, clearly visible in lateral view, at about midlength of ramus; caudal rami about 1.7× as long as wide .... *Attheyella* (*Chappuisiella*) *godeti* (Delachaux, 1918) [Brazil: Porto Alegre. Peru. Argentina: Río Quia]

4(3) Leg 1 with first and second endopodal articles combined as long as or slightly longer than exopod; caudal seta IV barely 1.5× as long as ramus; caudal seta VI very reduced, noticeably shorter than ramus; other characters not combined as below 5

4′ Leg 1 with first and second endopodal articles combined visibly shorter than exopod; male leg 3 endopod barely reaching tip of second exopodal article, apophysis on second endopodal article not reaching tip of exopod; male leg 4 with second endopodal with innermost and medial setae very long and subequal; male leg 4 with third exopodal article small, ~1.4× as long as wide; caudal seta IV about twice as long as ramus; caudal seta VI as long or slightly longer than ramus; female unknown .... *Attheyella* (*Chappuisiella*) *subdola* (Brian, 1927) [Brazil: Porto Alegre, Rio Grande Do Sul. Argentina: Buenos Aires]

5(4) Female genital double-segment with spinulae; ventral spinulae of female fifth urosomal segment of moderate length; female anal segment without spinulae ventrally close to inner corner of caudal rami; male leg 3 endopod reaching about middle of third exopodal article, apophysis of second endopodal article nearly as long or noticeably longer than exopod; male leg 4 with second endopodal article with innermost seta as long as outer element, both visibly shorter than medial seta; male leg 4 with third exopodal article ~3× as long as wide .... *Attheyella* (*Chappuisiella*) *fuhrmanni fuhrmanni* (Thiébaud, 1912) [Costa Rica, Guatemala, Venezuela, Colombia: Atlantico, Magdalena, Bogota, Cundinamarca, Antioquia, Brazil: Recife, Rio Grande do Sul, Rio de Janeiro, Paraná, São Paulo, Argentina: Buenos Aires, Chaco, Formosa, Mendoza, Santa Fe]

5′ Female genital double-segment without spinulae; ventral spinulae of female fifth urosomal segment exceedingly long, subequal (except for three or four medial spinulae), reaching about middle of anal segment; female anal segment with two long spinulae ventrally close to inner corner of each caudal ramus, inner spinule longer, nearly reaching posterior margin of caudal ramus; male unknown .... *Attheyella* (*Chappuisiella*) *fuhrmanni spinosa* Dussart, 1983 [Argentina: Corrientes]

6(2) Males 7

6′ Female: leg 3 with second endopodal article bearing four setae/spinae; male unknown .... *Attheyella* (*Chappuisiella*) *orinocoensis* Dussart, 1984 [Venezuela: Rio Portuguesa, Caño Falcon, Guarico; Caño Near Orinoco, Barrancas]

7(6) Male: leg 4 with second endopodal article bearing two subequal setae; leg 3 with first endopodal article without inner seta; female unknown .... *Attheyella* (*Chappuisiella*) *guyanensis* (Delachaux, 1924) [Guyana. Suriname]

7′ Male: leg 4 with second endopodal article bearing three setae, of which medial element longer, inner and outer setae subequal; leg 3 with first endopodal article with inner seta; female: leg 3 with second endopodal article bearing six setae/spinae .... *Attheyella* (*Chappuisiella*) *pichilafquensis* Löffler, 1961 [Guatemala. Ecuador: Lago Papallacta, Antisana, Napo Region, Carchi Region. Colombia: Laguna Los Tunjos 2 and See Pozo Azul, Cundinamarca]

Harpacticoida: Canthocamptidae: Canthocamptinae: *Attheyella* (*Delachauxiella*): Species

1 Anal operculum rounded, smooth 2

1′ Anal operculum broadly triangular or lanciform 3

2(1) Female leg 1 with first endopodal article bearing two inner setae/spinae, as long as exopod; female leg 2 with third exopodal article and second endopodal article with five and three setae/spinae, respectively; female leg 4 with first endopodal article bearing an inner seta, second article with five setae/spinae; male leg 3 endopod with 2 articles; first endopodal article with inner apophysis reaching beyond tip of exopod; second endopodal article with three apical setae only .... *Attheyella* (*Delachauxiella*) *aculeata* (Thiébaud, 1914) [Argentina: Mendoza, mountain range]

2′ Female leg 1 with first endopodal article bearing one inner element, barely reaching middle of second exopodal article; female leg 2 with third exopodal article and second endopodal article with six and four setae/spinae, respectively; female leg 4 with first endopodal article without inner seta, second article with five setae/spinae (apical outer seta minute); male unknown .... *Attheyella* (*Delachauxiella*) *nuda* Löffler, 1961 [Argentina: Tucumán, Las Azucenas]

3(1) Female anal operculum lanciform, narrow exceedingly longer than caudal rami 4

3′ Female anal operculum broadly triangular, reaching posterior margin of caudal rami at most 5

4(3) Anal operculum with a lateral deep constriction at about 1/3 of its length on both sides, about 3× as long as caudal rami .... *Attheyella* (*Delachauxiella*) *lanceolata* (Delachaux, 1918) [Peru: Lago Huaron]

4′ Anal operculum without lateral constriction, ~2× as long as caudal rami at most .... *Attheyella* (*Delachauxiella*) *schindleri* Kiefer, 1957
[Peru: Lago Titicaca]

5(3) Anal operculum different; other characters not combined as below 6

5′ Margin of anal operculum with coarse, sclerotized teeth; leg 1 with second exopodal article without inner seta; female leg 2 with second endopodal article with four, of male with three setae; male leg 4 with first endopodal article without inner seta, second article with three setae; female leg 5 exopod with four, endopodal lobe with six setae; male leg 5 exopod with three, endopodal lobe with one seta .... *Attheyella* (*Delachauxiella*) *yemanjae* Reid, 1993
[Brazil: Distrito Federal, Fazenda Agua Limpa]

6(5) Female leg 5 exopod with four, endopodal lobe with six setae 7

6′ Female leg 5 exopod with four, endopodal lobe with five setae; male unknown ... *Attheyella* (*Delachauxiella*) *dadayi* (Chappuis, 1924)
[Paraguay]

7(6) Caudal rami length 1.7–2× breadth 8

7′ Caudal rami length ≤ 1.4× breadth 10

8(7) Female leg 5 exopod with third seta from outer to inner margin shorter than exopod 9

8′ Female leg 5 exopod with third seta from outer to inner margin longer than exopod ...... *Attheyella* (*Delachauxiella*) *freyi* Löffler, 1963
[Ecuador: Antisana, between Napo and Pichincha Provinces; El Angel, Carchi Province, Colombia: Laguna de San Rafael, Puracé, Cauca]

9(8) Outer margin of caudal rami regular; anal operculum with one lateral pointed projection on each side; margin smooth; male unknown .... *Attheyella* (*Delachauxiella*) *maxima* (Delachaux, 1918)
[Peru: Lago Titicaca, Lago Lagunilla, Lago Huaron]

9′ Outer margin of caudal rami irregular; anal operculum without lateral pointed projections; margin with fine setulae .... *Attheyella* (*Delachauxiella*) *insignis* (Delachaux, 1918)
[Peru: Lago Huaron]

10(7) Female leg 5 exopod with second seta from inner to outer margin shorter or as long as exopod 11

10′ Female leg 5 exopod with second seta from inner to outer margin longer than exopod 14

11(10) Female leg 5 with endopodal lobe well developed; female leg 5 exopod length:width ratio ~ 1.7–2; female leg 5 exopod with first and second setae from inner to outer margin arising more or less at about the same level, third seta arising at a lower level conferring the exopod an oval appearance; other characters not combined 12

11′ Female leg 5 with endopodal lobe only weakly developed; female leg 5 exopod length:width ratio about 1.3; female leg 5 exopod with first, second, and third setae from inner to outer margin arising more or less at the same level conferring the exopod a subquadrate appearance; female genital double-segment with four spinulae laterally; anal segment with short spinular row close to inner corner of each caudal ramus ventrally .... *Attheyella* (*Delachauxiella*) *horvathi* (Chappuis. 1924)
[Paraguay]

12(11) Leg 1 with first endopodal article shorter or as long as exopod; leg 2 with second endopodal article bearing five setae/spinae; outermost seta of female leg 5 endopodal lobe shorter than exopod; other characters not combined as below 13

12′ Leg 1 with first endopodal article clearly longer than exopod; leg 2 with second endopodal article bearing four setae/spinae; outermost seta of female leg 5 endopodal lobe longer than exopod; female genital double-segment with one row of subequal spinulae laterally; anal somite with row of about 8–9 spinulae close to inner corner of each caudal ramus ventrally; anal operculum reaching middle of caudal rami .... *Attheyella* (*Delachauxiella*) *broiensis* Reid, 1993
[Brazil: Broa (Lobo) Reservoir, São Paulo]

13(12) Female genital double-segment with two spinular rows laterally, of which proximalmost spinulae shorter and stronger; female and male anal segment with two spinulae close to inner corner of each caudal ramus ventrally; female and male anal operculum reaching posterior margin of caudal rami or slightly shorter .... *Attheyella* (*Delachauxiella*) *wieseri* Löffler, 1961
[Argentina: Tucumán, Las Azucenas]

13′ Female genital double-segment with one spinular row laterally; female anal segment without, male anal segment with three spinulae close to each caudal ramus ventrally; female anal operculum with pointed tip, reaching slightly beyond caudal rami; male anal operculum with blunt tip reaching distal margin of caudal rami .... *Attheyella* (*Delachauxiella*) *clavigera* Harding, 1955
[Peru: Lago Lagunilla, Bahia Taman, Lago Titicaca]

14(10) Female anal operculum reaching middle of caudal rami at most 15

14′ Female anal operculum reaching slightly beyond posterior margin of caudal rami .... *Attheyella* (*Delachauxiella*) *ferox* (Delachaux, 1918)
[Peru: Lago Huaron]

15(14) Female anal operculum reaching middle of caudal rami; leg 1 with first endopodal article longer than first and second exopodal articles combined; caudal rami with strong inner spinulae; anal segment without spinulae ventrally close to each caudal ramus; male anal operculum reaching posterior margin of caudal rami, with or without proximal deep constriction, distal tip pointed or rounded 16

15′ Female anal operculum barely reaching proximal third of caudal rami; leg 1 with first endopodal article nearly as long as, or as long as exopod; caudal rami without inner spinulae; anal segment with crescentic row of spinulae ventrally close to each caudal ramus; male unknown .... *Attheyella* (*Delachauxiella*) *lanata* (Mrázek, 1901)
[Argentina: Neuquén, Lago Currhué Grande, Uruguay: Montevideo]

16(15) Line of demarcation between the two components of the female genital double-segment (second and third urosomal segments) with dorsolateral spinulae; second seta of leg 5 exopod from inner to outer margin considerably more than half the length of the innermost seta, with setulae .... *Attheyella* (*Delachauxiella*) *inconstans inconstans* Harding, 1955
[Peru: La Raya; Lagunillas, Puno; San Antonio de Esquilache, Puno; Unduavi Pass; Lago Titicaca]

16′ Line of demarcation between the two components of the female genital double-segment (second and third urosomal segments) without dorsolateral spinulae; second seta of leg 5 exopod from inner to outer margin scarcely half the length of the innermost seta, bare .... *Attheyella* (*Delachauxiella*) *inconstans egena* Harding, 1955
[Peru: La Raya; Lagunillas, Puno; San Antonio de Esquilache, Puno; Unduavi Pass; Lago Titicaca]

Harpacticoida: Canthocamptidae: Canthocamptinae: *Elaphoidella*: Species Groups

Lang (1948) subdivided the genus *Elaphoidella* into 10 groups. Petkovski & Brancelj (1988) created group XI for *Elaphoidella serbica* Petkovski & Brancelj, 1988. Gaviria & Defaye (2015) noted that Lang's (1948) groups I, II, VII, VIII, and X are represented in the Neotropics, presented a key for the Colombian species of *Elaphoidella*, and suggested that *Elaphoidella pintoae* Reid & Jose, 1987 should be placed provisionally into a subgroup of group II. Gaviria (1993) suggested that *Elaphoidella colombiana* Gaviria, 1993 should be placed in a new group, and Gaviria & Defaye (2015) formally defined group XII for that species. Also, Gaviria & Defaye (2015) attributed provisionally *Elaphoidella negroensis* Kiefer, 1967 and *Elaphoidella prohumboldti* Petkovski, 1908, to group VII until the males of these species are described. At present, groups I, II, VII, VIII, X, and XII are represented in the neotropics. It is suggested to pay special attention to the characteristics of each group and to compare the biological material under identification with the available descriptions. Given the complex combinations of character states of the different groups of *Elaphoidella*, and the possibility of finding new species of potentially new groups, it is highly advisable to check the diagnoses of the groups of *Elaphoidella* in the above references. The diagnoses of the groups present in Neotropical freshwater habitats are given in the following:

Group I. Leg 1 endopod with 3 article; female: leg 5 endopodal lobe poorly developed, with four setae, outermost seta shorter, exopod with four or five setae; caudal rami with or without dorsal carina, if present not pointed; leg 3 with third exopodal article bearing two inner setae; leg 2 with second endopodal article bearing five setae; leg 3 with second endopodal article bearing six setae; leg 4 with second endopodal article bearing four setae; male leg 2 with first endopodal article bearing one inner seta, second article with four setae; male: leg 4 with second endopodal article bearing three setae.

Group II. Leg 1 endopod with two or three articles; female: leg endopodal lobe poorly developed, with four setae, exopod with four to six setae; caudal rami with dorsal carina, sometimes pointed; leg 3 with third exopodal article bearing two inner setae; leg 4 with third exopodal article with two inner setae; leg 2 with first endopodal article with or without inner seta; leg 2 with second endopodal article bearing five setae; leg 3 with first endopodal article with or without inner seta, second article with five or six setae; leg 4 with first endopodal article with or without inner seta; male leg 2 with first endopodal article with one inner seta, second article with four setae; male: leg 4 with second endopodal article bearing three or four setae; leg 4 with third exopodal article with proximal outer spine modified.

Group VII. Leg 1 endopod with three articles; female: leg 5 endopodal lobe reaching at least middle of exopod, with four very long setae, exopod with four setae; caudal rami with well-developed carina, pointed; anal operculum with strong teeth; leg 3 with third exopodal article bearing two inner setae; leg 4 with third exopodal article bearing two inner setae; leg 2 with first endopodal article without inner seta, second article with four setae; leg 3 with first endopodal article without inner seta, second article with three or four setae; leg 4 with first endopodal article without inner seta, second article with three or four setae; male: leg 2 with first endopodal article with or without inner seta, second article with three setae; leg 4 with second endopodal article with three setae; leg 4 with third exopodal article with outer distal spine and outer apical seta modified.

Group VIII. Leg 1 endopod with three articles; female: leg 5 endopodal lobe reaching at least middle of exopod, with four well-developed setae; leg 5 exopod with four setae; caudal rami with well-developed carina, pointed; anal operculum smooth or with blunt teeth; leg 3 with third exopodal article bearing two inner setae; leg 4 with third exopodal article bearing two inner setae; leg 2 with first endopodal article without inner seta, second article with three or four setae; leg 3 with first endopodal article with or without inner seta, second article with four or five setae; leg 4 with first endopodal article without inner seta, second article with three setae; male: leg 2 with first endopodal article with or without inner seta, second article with two or three setae; leg 4 with second endopodal article bearing two or three setae; leg 4 with third exopodal article with outer apical seta modified.

Group X. Leg 1 endopod with two articles; female: leg 5 endopodal lobe reaching at least middle of exopod, with four setae of which outermost and innermost shorter, exopod with three or four well-developed setae; caudal rami with well-developed carina, pointed; leg 3 with third exopodal article bearing two inner setae; leg 4 with third exopodal article bearing one or two inner setae; leg 2 with first endopodal article lacking inner seta, second article with three or four setae; leg 3 with first endopodal article lacking inner seta, second article with three to five setae; leg 4 with first endopodal article lacking inner seta, second article with one to three setae; male: leg 2 with first endopodal article lacking inner seta; leg 4 with first endopodal article lacking inner seta, second article with one to three setae; leg 4 with third exopodal article bearing normal spinae/setae; leg 4 with third exopodal article as long as, or shorter than preceding articles.

Group XII. Leg 1 endopod with three articles; female: leg 5 endopodal lobe reaching at least middle of exopod, with four well-developed setae, exopod with five well-developed setae; caudal rami with dorsal carina, not pointed; anal operculum with about 19 spinulae; leg 3 with third exopodal article bearing two inner setae; leg 4 with third exopodal article bearing two inner setae; leg 2 with first endopodal article bearing inner seta, second article with five setae; leg 3 with first endopodal article bearing inner seta, second article with six setae; leg 4 with first endopodal article bearing inner seta, second article with four setae; male: leg 2 with first endopodal article bearing inner seta; leg 2 with second endopodal article bearing three setae; leg 4 with first endopodal article lacking inner seta, second article with two setae; leg 4 with third exopodal article with distal outer spine and outer apical seta modified.

1 Female leg 5 baseoendopod poorly developed 2

1′ Female leg 5 baseoendopod reaching at least middle of exopod 3

2(1) Male leg 4 with third exopodal article bearing spinae/setae as in female, without sexual dimorphism .... **Group I [p. 741]**

2′ Male leg 4 with third exopodal article with proximal outer spine modified, other setae/spinae not modified .... **Group II [p. 741]**

3(1) Leg 1 endopod with three articles 4

3′ Leg 1 endopod with two articles .... **Group X [p. 744]**

4(3) Female leg 5 exopod with four setae 5

4′ Female leg 5 exopod with five setae .... Group XII one species: *Elaphoidella colombiana* Gaviria, 1993 [Colombia: Laguna Los Tunjos 3, Páramo de Chisaca, Cundinamarca. Eastern Andean mountain range]

5(4) Anal operculum with strong teeth; male leg 4 with third exopodal article with distal outer spine and outer apical seta modified, sexually dimorphic .... **Group VII [p. 742]**

5′ Anal operculum smooth or with blunt teeth; male leg 4 with third exopodal article with outer apical seta modified, sexually dimorphic, other elements without sexual dimorphism .... **Group VIII [p. 743]**

Harpacticoida: Canthocamptidae: Canthocamptinae: *Elaphoidella*: Group I: Species

1 Posterior margin of cephalothorax, pro-, and/or urosomal segments serrated; posterior margin of anal operculum with spinulae; setae of female and male caudal rami setiform; other characters not combined as below 2

1′ Posterior margin of cephalothorax, pro-, and/or urosomal segments plain; posterior margin of anal operculum with or without spinulae; female caudal rami with, male caudal rami without small inner tooth; lateral setae of female and male caudal rami spiniform; female legs 2−4 with second endopodal articles bearing 5, 5, 4 setae, respectively; female leg 5 endopodal lobe with three innermost setae subequal, outermost setae minute; leg 1 with first endopodal article shorter than first and second exopodal articles combined; male leg 2 with second endopodal article bearing four setae; male leg 3 endopod with two articles, first article with inner apophysis; male leg 4 endopod with three setae; outermost seta of male leg 5 exopod minute, innermost element shorter than exopod .... *Elaphoidella cabezasi* Petkovski, 1982 [Cuba: Rio Cuyaguateje, Cabezas, Pinar del Río Province]

2(1) Leg 1 with first endopodal article as long as first and second exopodal articles combined; other characters not combined as below 3

2′ Leg 1 endopod shorter than first and second exopodal articles combined, female caudal rami with, male caudal rami without inner hyaline process on inner distal corner; female legs 2−4 with second endopodal articles bearing 4, 5, 4 setae, respectively; female leg 5 endopodal lobe with three innermost setae subequal, outermost seta well developed, as long as exopod; male leg 2 with second endopodal article bearing three setae, leg 4 with second endopodal article bearing two setae .... *Elaphoidella quemadoi* Petkovski, 1982 [Cuba: Cueva del Segundo Cauce, Quemado Pineda, Pinar del Río Province]

3(2) Female and male caudal rami without inner tooth; female leg 3 with second endopodal article bearing six setae; female leg 4 with second endopodal article bearing four setae; female leg 5 exopod with four elements, without strong inner spinule; female leg 5 endopodal lobe with three innermost setae subequal, innermost element nearly as long as exopod; male legs 2 and 4 with second endopodal articles bearing four and three setae, respectively; male leg 3 endopod with three articles .... *Elaphoidella jojoi* Petkovski, 1982 [Cuba: Río Jojoi, Segua Baracoa, Oriente Province]

3′ Female caudal rami with inner tooth; female leg 3 with second endopodal article bearing five setae; female leg 4 with second endopodal article bearing three setae; female leg 5 exopod with four elements and a strong inner spinule; female leg 5 endopodal lobe with three innermost setae subequal, outermost seta minute; male unknown .... *Elaphoidella sabanillae* Petkovski, 1982 [Cuba: Rio Sabanilla, Río Miel, Rio Baconao, Oriente Province]

Harpacticoida: Canthocamptidae: Canthocamptinae: *Elaphoidella*: Group II: Species

1 Anal operculum strongly crescentic, posterior margin with setulae or with hyaline fringed border 2

1′ Anal operculum rounded, posterior margin naked, with setulae or with spinulae 3

2(1) Posterior margin of anal operculum with setulae; armature formula of female leg 2 exopod and endopod 0,1,122 and 1,221, respectively; armature formula of female leg 4 endopod 0,220; female leg 5 exopod with five, endopodal lobe with four setae; male leg 2 with first endopodal article bearing inner seta, second article with four setae; male leg 5 exopod with three setae, endopodal lobe unarmed .... *Elaphoidella grandidieri* (Guerne & Richard, 1893) [Brazil: São Francisco, Amazonas. Colombia: Laguna de Navío Quebrado, Santuario de Fauna y Flora "Los Flamencos", Camarones, La Guajira; Ciénaga Grande de Santa Marta, Universidad de Antioquía, Medellín. Cuba: Cueva de Bellamar, Matanzas Province. Mexico: Cuautla, Morelos State. French Antilles: Guadeloupe]

2′ Anal operculum with hyaline fringed border; armature formula of female leg 2 exopod and endopod 0,1,123 and 0,220, respectively; armature formula of female leg 4 endopod 1,221; female leg 5 exopod and endopodal lobe with four setae each; male unknown .... *Elaphoidella pintoae* Reid & José, 1987 [Brazil: Gruta de Clarona, Goiás; Lapa de Clarona, Amazonas]

3(1) Female leg 5 exopod and endopodal lobe with five and four setae, respectively 4

3′ Female leg 5 exopod and endopodal lobe with number of setae different 6

4(3) Characters not combined as below 5

4′ Female first antennae with nine articles; female leg 2 with second endopodal article bearing three setae; female leg 3 with first endopodal article lacking inner seta, second article with six setae; female caudal rami with caudal seta IV displaced ventrally to seta V; male leg 2 with first endopodal article bearing one inner seta, second article with three setae; male leg 4 with second endopodal article bearing three setae; male leg 5 exopod with four setae, endopodal lobe without armature .... *Elaphoidella parvifurcata* Petkovski, 1980
[Cuba: Gran Piedra, Río Baconao, Oriente Province]

5(4) Female first antennae with eight articles; female leg 1 with first and second endopodal articles with or without inner seta, third article with three setae; female leg 4 with first endopodal article with or without inner seta, second article with three, four, or five setae; female leg 5 exopod with four, endopodal lobe with 5 setae (this subspecies is highly variable and it is suggested to check the descriptions available, see also Wells (2007: 93, point 256) and Hamond (1987: 1081)) .... *Elaphoidella bidens bidens* (Schmeil, 1894)
[Argentina: Guaycurú, San Hilario, Brazil: Paraná and San Francisco, Colombia: Ciénaga Grande de Santa Marta, Laguna Cerro del Padre Amaya, Palmitos, Cuba, El Salvador, Guatemala, Paraguay, Venezuela: Caño Falcon, Río Portuguesa, Guarico]

5′ Female first antennae with nine articles; female leg 1 with first and second endopodal articles bearing one inner seta, third article with three setae; female leg 4 with first endopodal article without inner seta, second article with four setae; female leg 5 exopod with 5, endopodal lobe with 4 setae; female leg 5 exopod with 5, endopodal lobe with 4 setae; male unknown .... *Elaphoidella bidens subterranea* Nogueira M. H., 1959
[Brazil: Gruta de Itapiriçu, Río Branco]

6(3) Cephalothorax, pro- and/or urosomal segments with posterior margin plain; anal operculum rounded, posterior margin with setulae; female leg 3 with third exopodal article bearing five setae; female leg 3 with first endopodal article bearing inner seta, second article with five setae; female leg 5 exopod with three, endopodal lobe with four setae; male unknown .... *Elaphoidella laciniata* (Douwe, 1911)
[Brazil: Itapura, São Paulo]

6′ Cephalothorax, pro- and/or urosomal segments with posterior margin serrated; anal operculum rounded, posterior margin naked; female leg 3 with third exopodal article with six setae; female leg 3 with first endopodal article without inner seta, second article with six setae; female leg 5 exopod and endopodal lobe with four setae each; male leg 2 with first endopodal article lacking inner seta, second article with four elements; male leg 4 with first endopodal article lacking inner seta, second article with three elements. Male leg 5 exopod with four setae, endopodal lobe without armature .... *Elaphoidella neotropica* Petkovski, 1973
[Cuba: Cueva de Pio Domingo, Pinar del Río Province]

Harpacticoida: Canthocamptidae: Canthocamptinae: *Elaphoidella*: Group VII: Species

1 Cephalothorax, pro- and/or urosomal segments with posterior margin plain 2

1′ Cephalothorax, pro- and urosomites with posterior margin serrated 3

2(1) Male leg 2 with second endopodal articles bearing two setae; female unknown .... *Elaphoidella humboldti* Löffler, 1963
[Ecuador: Antisana]

2′ Male leg 2 with second endopodal article bearing three setae; female legs 2–4 with second endopodal articles bearing 4, 5, 3 setae, respectively; female leg 5 exopod and endopodal lobe with four setae each .... *Elaphoidella schubarti* (Chappuis, 1936)
[Brazil: Paraná and Amazonas. Colombia]

3(1) Leg 1 endopod with two articles 4

3′ Leg 1 endopod with three articles 6

4(3) Caudal rami with caudal seta IV not crossing seta V; other characters not combined as below 5

4′ Caudal rami with caudal seta IV crossing seta V; female legs 2–4 with second endopodal articles bearing 2, 3, 2 setae, respectively; female leg 5 exopod and endopodal lobe with four setae each; male leg 4 with third exopodal article bearing one inner seta .... *Elaphoidella einslei* Petkovski, 1973
[Cuba: Cueva Oscura, Pinar del Río Province]

5(4) Male leg 2 with first endopodal articleunarmed, second article with three setae; male leg 4 with third exopodal article bearing one inner seta; male leg 4 with first endopodal article unarmed, second article with one setea; female legs 2–4 with second endopodal articles bearing 1, 2, 1 setae, respectively; female leg 5 exopod with four, endopodal lobe with two setae .... *Elaphoidella crenobia* Petkovski, 1973
[Cuba: Cueva de Pio Domingo, Pinar del Río Province]

5′ Male leg 2 with first endopodal article unarmed, second article with one seta and three apical spinulae; male leg 4 with third exopodal article bearing two inner setae; male leg 4 with first endopodal article unarmed, second article with two setae; female unknown .... *Elaphoidella subcrenobia* Petkovski, 1980
[Cuba: Cueva del Segundo Cauce and Gran Caverna de Santo Tomás, Pinar del Río Province]

6(3) Female leg 5 exopod with three, endopodal lobe with four setae; female leg 1 with first endopodal article bearing inner seta; other characters not combined as below 7

6′ Female leg 5 exopod and endopodal lobe with four setae each; female leg 1 with first endopodal article lacking inner seta; female legs 2–4 with second endopodal articles with 3, 5, 3 setae, respectively; male unknown .... *Elaphoidella negroensis* Kiefer, 1967
[Brazil: Caatingas, Rio Negro, Amazonas]

7(6) Characters not combined as below ... 8

7′ Female leg 2 with second endopodal article with three setae; female leg 3 with third exopodal article with six, and second endopodal article with four setae; female leg 4 with third exopodal article bearing six, second endopodal articles with three setae, respectively; male unknown ... *Elaphoidella prohumboldti* Petkovski, 1980
[Cuba: Cueva de Atabex, Oriente Province; Cueva El Rabón, Cueva Banega, Santiago de Cuba Province]

8(7) Female legs 2–4 with second endopodal articles bearing 2, 1, 1 setae, respectively. Caudal rami with caudal seta IV displaced ventrally to seta V; male leg 3 with third exopodal article bearing six setae; male leg 4 with second endopodal article bearing two setae; male leg 5 exopod with two setae, endopodal lobe unarmed ... *Elaphoidella turgisetosa* Petkovski, 1980
[Cuba: Cueva Seboruco, Barrio Seboruco, Oriente Province]

8′ Female legs 2–4 with second endopodal articles bearing three setae each; caudal rami with caudal seta IV situated laterally to seta V; male leg 3 with third exopodal article with five setae; male leg 4 with second endopodal article bearing three setae; male leg 5 exopod with three setae, endopodal lobe unarmed ... *Elaphoidella botosaneauni* Petkovski, 1973
[Cuba: Potrero del Molino, Cueva Los Golondrinos, Las Cuatrocientas Rosas, Cueva de Los Panaderos, Cueva La Campana, Cueva del Guano, Cueva de la Virgen, Cueva del Indio, Cueva de Pio Domingo, and Cueva del Abono]

Harpacticoida: Canthocamptidae: Canthocamptinae: *Elaphoidella*: Group VIII: Species

1 Cephalothorax, prosomal and urosomal segments with posterior margin serrated ... 2

1′ Cephalothorax, prosomal and urosomal segments with posterior margin plain ... 6

2(1) Female legs 2 and 4 with second endopodal article bearing three setae each ... 3

2′ These characters not combined ... 5

3(2) Anal operculum with posterior margin with setulae or spinulae ... 4

3′ Anal operculum with posterior margin smooth; leg 1 with second exopodal article lacking inner seta; male unknown ... *Elaphoidella suarezi* Reid, 1987
[Brazil: Paraná and São Francisco, Bahia. Colombia: Valle Buenaventura, Valle del Cauca; Valle Magdalena, Tumaco]

4(3) Cephalothorax and prosomal segments with posterior margin smooth, of urosomal segments finely serrated; anal operculum with posterior margin setulose ... *Elaphoidella sewelli minuta* Chappuis, 1932
[Colombia: Ciénaga Grande de Santa Marta]

4′ Cephalothorax and prosomal and urosomal segments with posterior margin strongly serrated; anal operculum with posterior margin bearing about 18 spinulae ... *Elaphoidella sewelli americana* (Chappuis, 1933)
[Netherlands Antilles: Pos Ichi, Bonaire and Curaçao. Peru]

5(2) Anal operculum with posterior margin bearing about seven strong teeth/spinulae; female legs 2–4 with second endopodal articles bearing two setae each ... *Elaphoidella bispina* Dussart, 1984
[Venezuela: El Carmen near Calabozo, Hato Becerra]

5′ Anal operculum with posterior margin bearing about 14 small teeth/spinulae; female legs 2–4 with second endopodal articles bearing 4, 4, 3 setae, respectively ... *Elaphoidella paraplesia* Kiefer, 1967
[Brazil: Caatingas, Rio Negro, Amazonas]

6(1) Anal operculum rounded, posterior margin either with long teeth/spinulae in both sexes, or smooth in female; male leg 2 with second endopodal article pointed apically, with two or three setae ... 7

6′ Anal operculum rounded, posterior margin naked in both sexes; caudal rami with caudal setae II and III spinelike; armature formula of male leg 3 exopod 0,0,222; male leg 5 exopod with two setae, endopodal lobe unarmed ... *Elaphoidella surinamensis* (Delachaux, 1924)
[French Guiana, Suriname]

7(6) Caudal rami with caudal seta VI as long or slightly longer than ramus; male anal operculum with posterior bearing strong crescentic teeth; male leg 2 with second endopodal article bearing two setae, pointed apically; male leg 3 with first endopodal article bearing inner seta, second article with inner apophysis, third article with one stout distal seta and acute terminal projection; male leg 4 endopod with one article, with three setae; female unknown ... *Elaphoidella radkei* Reid, 1987
[Colombia: Valle Buenaventura, Valle del Cauca; Valle Magdalena, Tumaco]

7′ Caudal rami with caudal seta VI reduced; female anal operculum with posterior margin smooth, with eight long teeth/spinulae in male; male leg 2 with second endopodal article bearing three setae, pointed apically; male leg 3 with first endopodal article lacking inner seta, second article with inner apophysis, third article with two apical setae, without acute terminal projection; male leg 4 endopod with two articles, second article with three setae; female leg 2 with second endopodal article bearing four setae; female leg 3 with first endopodal article bearing inner seta, second article with five elements ... *Elaphoidella malayica* (Chappuis, 1928)
[French Antilles: Martinique near Le Morne-Rouge]

Harpacticoida: Canthocamptidae: Canthocamptinae: *Elaphoidella*: Group X: Species

1 Female leg 5 exopod and endopodal lobe separated by broad gap 2

1′ Female leg 5 exopod and endopodal lobe not separated by broad gap 7

2(1) Female anal operculum with posterior margin naked, or with setulae or spinulae/teeth, of male with setulae or spinulae/teeth; caudal rami with caudal seta IV slightly longer or shorter than ramus, minute or missing; female legs 1–4 with second endopodal articles bearing a complement armature different from 4, 3, 4, 3 setae 3

2′ Anal operculum with posterior margin smooth, with hyaline membrane; female legs 1–4 with second endopodal articles bearing 4, 3, 4, 3 setae, respectively; caudal rami with caudal seta IV slightly longer than ramus; male unknown .... *Elaphoidella parajakobi* Reid & José, 1987
[Brazil: Buraco do Dedé, Goiás; Paraná and Amazonas]

3(2) Female and male leg 4 endopod with two articles 4

3′ Female and male leg 4 endopod with one article 5

4(3) Female leg 5 exopod and endopodal lobe with four setae each .... *Elaphoidella armata* (Delachaux, 1918)
[Brazil: Iguazu falls, Paraná-Argentina-Paraguay, Peru: Lago Naticocha]

4′ Female leg 5 exopod with three, endopodal lobe with four setae .... *Elaphoidella paramuna* Gaviria & Defaye, 2015
[Colombia: Laguna Buitrago, Chingaza region, Cundinamarca]

5(3) Female and male leg 4 endopod with three setae 6

5′ Female and male leg 4 endopod with one seta .... *Elaphoidella neoarmata* Petkovski, 1973
[Cuba: Cueva La Patana, Guantánamo Province; Cueva Grande de Caguanes, Sancti Spíritus, Las Villas Province]

6(5) Female leg 3 with second endopodal article bearing four setae; male leg 5 exopod with three setae, endopodal lobe unarmed .... *Elaphoidella synjakobii* Petkovski, 1980
[Cuba: Río Sabanilla, Baracoa, Guantánamo Province; Rio Jojo, Guantánamo Province; Cueva Banega, Santiago de Cuba Province]

6′ Female leg 3 with second endopodal article bearing three setae; male leg 5 exopod with three, endopodal lobe with one seta .... *Elaphoidella jakobii* Nogueira M. H., 1959
[Brazil]

7(1) Female caudal rami at most 1.5×as long as wide; other characters not combined as below 8

7′ Female caudal rami elongate, about 2.6× as long as wide, lateral setae II and III, and dorsal seta VII reduced, seta VI visibly shorter than ramus, seta IV missing; male caudal rami dimorphic, about 1.6× as long as wide, lateral setae II and III well developed, dorsal seta VII reduced, seta VI as long as, or slightly longer than ramus, seta IV missing; female first antennae with seven articles; female anal operculum with posterior margin smooth, of male with spinulae; innermost and outermost seta of female and male leg 5 exopod subequal, medial seta longest .... *Elaphoidella karllangi* Petkovski, 1973
[Cuba: Cueva de las Majáes, Pinar del Río Province; Cueva Seboruco, Holguín Province]

8(7) Male (and probably female) caudal rami about as long as wide; male leg 5 exopod with three setae (one inner, and two apical elements); female unknown .... *Elaphoidella brevifurcata* Chappuis, 1936
[Brazil, French Guiana]

8′ Female (and probably male) caudal rami about 1.5× as long as wide; male leg 5 exopod with four setae (two inner, and two apical emenets) .... *Elaphoidella pectinata* (Delachaux, 1924)
[French Guiana, Suriname, Costa Rica]

The taxonomy and systematics of harpacticoids are in constant flux. New taxa are described each year, some are redescribed, and some others are reallocated in new or preexistent taxa. The taxonomy and systematics of harpacticoids, and of freshwater and inland forms in particular, have undergone important changes since the preparation of this material. These are summarized in the following.

Khodami et al. (2017) excluded the families Longipediidae and Canuellidae from the order Harpacticoida and placed them into a new order, the Canuelloida.

Gómez et al. (2017) redescribed *Cletocamptus albuquerquensis* and *Cletocamptus dominicanus.*

Corgosinho et al. (2017) described the genus *Eirinicaris* Corgosinho, Schizas, Previattelli, da Rocha & Santos-Silva, 2017 of the family Parastenocarididae. They found this new genus in samples taken from the Tocantins River basin (Brazil). This new genus belongs to the subfamily Parastenocaridinae and can be identified by the unique sexually dimorphic anal segment and caudal rami. This new genus possesses only one species, *Eirinicaris antonioi* Corgosinho, Schizas, Previattelli, da Rocha & Santos-Silva, 2017.

Gaviria et al. (2017) proposed the genus *Colombocaris* Gaviria, Defaye & Corgosinho, 2017 for *Colombocaris isabellae* Gaviria, Defaye & Corgosinho, 2017 from submerged mosses of a high Andean lake in Colombia. This genus belongs to the subfamily Fontinalicaridinae. They also proposed the genus *Noodtcaris* Gaviria, Defaye & Corgosinho, 2017 for *Parastenocaris columbiensis*, *Parastenocaris kubitzkii*, and *Parastenocaris roettgeri* from the Colombian Llanos Orientales and for *Parastenocaris tapajosensis* from the Brazilian Amazonas, and gave a redescription of the latter. This genus belongs also to the subfamily Fontinalicaridinae and is defined by the ornamentation of the caudal rami and morphology of legs 2–5. They gave a key to the species of *Colombocaris* and *Noodtcaris.*

Corgosinho et al. (2019) presented an updated checklist of freshwater Copepoda from Ecuador and the Galapagos Archipelago. They commented on *Attheyella* (*Chappuisiella*) *pichilafquensis*, *Attheyella* (*Delachauxiella*) *freyi*, *Elaphoidella humboldti*, and an unidentified species of *Bryocamptus.* They also commented on some species of *Cletocamptus* that were not considered as freshwater harpacticoids in this chapter.

Order Cyclopoida

Eduardo Suárez-Morales
El Colegio de la Frontera Sur, Chetumal, Quintana Roo, México

Martha A. Gutiérrez-Aguirre
Universidad de Quintana Roo, Cozumel, Quintana Roo, México

INTRODUCTION

The keys for Neotropical Cyclopoida and distributional data on the species followed the valuable contributions of several authors; these works include worldwide and regional keys and listings by Dussart (1984a), Dussart & Frutos (1985, 1986), Reid (1985b, 1987b, 1990, 2016), Suárez-Morales et al. (1996), Dussart & Defaye (2006), Rocha & Botelho (1998), Silva (2008), and Elías-Gutiérrez et al. (2008).

The terminology used for the anatomical structure of the Cyclopoida body (females and males adults, free living copepods) are based on those used by Huys & Boxshall (1991) and Ueda & Reid (2003). The criteria proposed by Fiers (2006) in reference to the taxonomic arrangement of *Allocyclops*, and related cyclopine genera were followed in this document.

Illustrations of selected characters are provided for most of the species included in this key, and in some cases, more than one figure is presented for each species in order to ease the identification process.

KEYS TO CYCLOPOIDA

Cyclopoida: Families

1 Mandibular palp composed of more than one article 2

1′ Mandibular palp absent or reduced to a rounded protuberance, armed with 2–3 setae .... **Cyclopidae [p. 745]**

2(1) Second antenna with two or three articles, typical of marine coastal waters, some reach brackish waters .... Oithonidae (not keyed further)

2′ Second antenna with four articles mainly marine and interstitial forms, some reach brackish waters .... Cyclopinidae (not keyed further)

Cyclopoida: Cyclopidae: Subfamilies

1 Fifth leg terminal article with one to three setae and/or spinae; thoracic legs 2- or 3-articled 2

1′ Fifth leg terminal article with four or five setae and/or spinae; thoracic legs 3-articled .... **Halicyclopinae [p. 745]**

2(1) Fifth leg terminal or only article with one or two setae and/or spinae; thoracic legs 2- or 3-articled .... **Cyclopinae [p. 747]**

2′ Fifth leg terminal article with three setae and/or spinae; thoracic legs always 3-articled .... **Eucyclopinae [p. 767]**

Cyclopoida: Cyclopidae: Halicyclopinae: Genera

Mostly marine, coastal and brackish water forms, few species in freshwater.

1 With mouthparts, particularly the second antennae raptorial (Fig. 21.38 A) .... ***Prehendocyclops* [p. 745]**

1′ Second antennae and mouthparts not raptorial .... ***Halicyclops* [p. 747]**

Cyclopoida: Cyclopidae: Halicyclopinae: *Prehendocyclops*: Species

1 Dorsal caudal seta length ≥2.5× caudal ramus; body relatively robust; maxilliped terminal article well developed, cylindrical, with one to three setae 2

1′ Dorsal caudal seta length subequal to caudal ramus; body relatively slender; maxilliped terminal article reduced to a small knob with a distal seta (Fig. 21.38 B) .... *Prehendocyclops abbreviatus* Rocha, 2000
[Mexico: Yucatán]

2(1) Maxilliped terminal article armed with three setae, one medial, two apical (Fig. 21.38 C); dorsal caudal seta length 2.5× caudal ramus .... *Prehendocyclops monchenkoi* Rocha, 2000
[Mexico: Yucatán]

2′ Maxilliped terminal article armed with one or two apical setae (Fig. 21.38 D); dorsal caudal seta length ∼ 5× caudal ramus .... *Prehendocyclops boxshalli* Rocha, 2000
[Mexico: Yucatán]

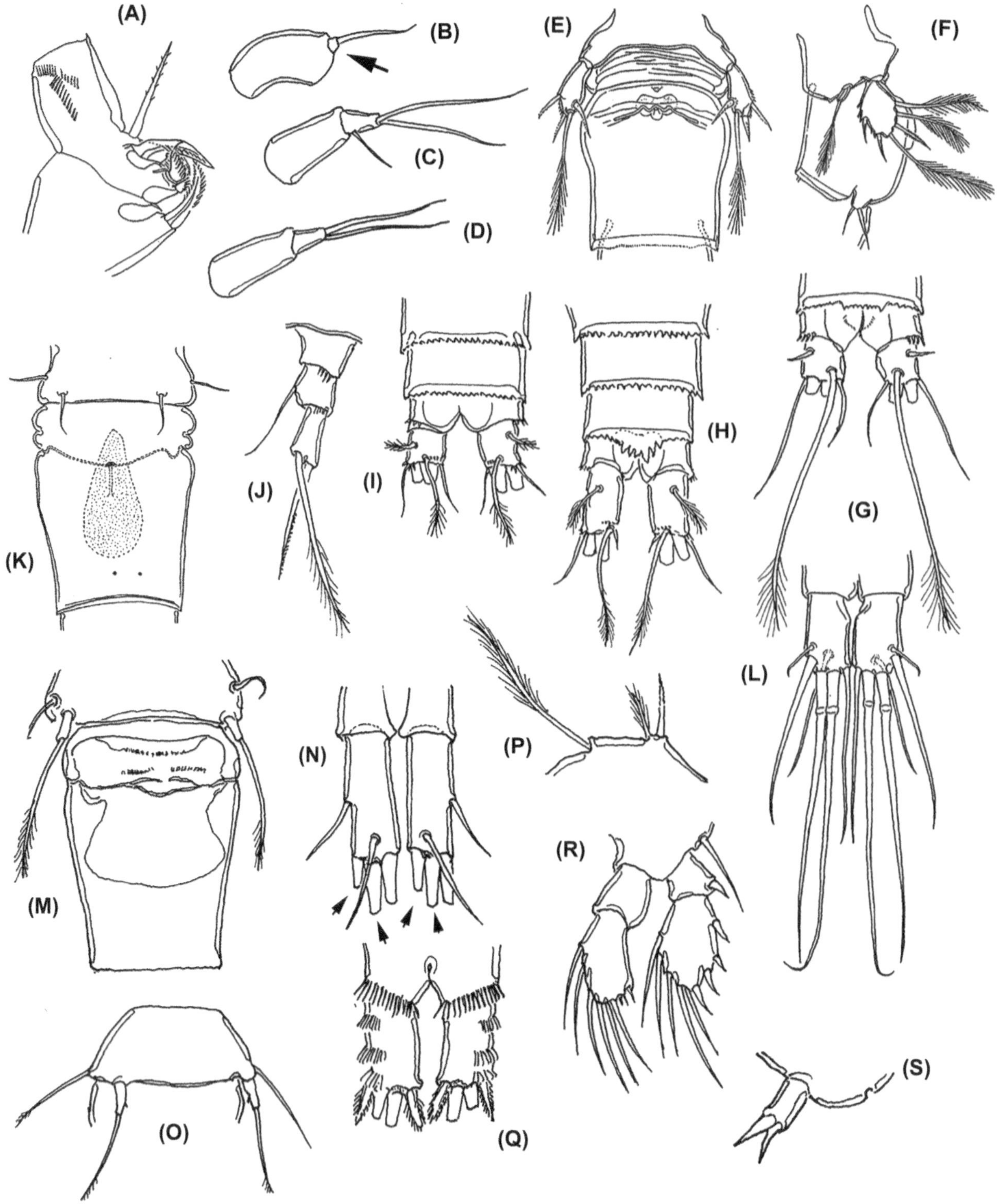

FIGURE 21.38 Morphology of Cyclopoida, Halicyclopinae: (A) *Prehendocyclops monchenkoi* second antennae showing raptorial endopod (distal part of segment omitted); (B) *Prehendocyclops abbreviatus* maxilliped showing reduced distal segment (arrow); (C) *P. monchenkoi* maxilliped; (D) *Prehendocyclops boxshalli* maxilliped; (E) *Halicyclops cenoticola* (female) fifth pedigerous and genital double-somite ventral view; (F) *H. cenoticola* (male) fifth pedigerous and genital double-somite lateral view; (G) *H. cenoticola* anal somite and caudal rami dorsal view; (H) *Halicyclops lindbergi* urosome and caudal rami dorsal view; (I) *Halicyclops dussarti* urosome and caudal rami dorsal view; (J) *Orthocyclops modestus* fifth leg; (K) *Neutrocyclops brevifurca* fifth pedigerous and genital double-somite ventral view; (L) *N. brevifurca* anal somite and caudal rami ventral view; (M) *Idiocyclops stocki* fifth pedigerous and genital double-somite ventral view; (N) *I. stocki* caudal rami dorsal view showing terminal setae (arrows); (O) *Ponticyclops boscoi* fifth pedigerous somite and fifth legs; (P) *Virbiocyclops silvaticus* fifth leg; (Q) *Menzeliella staheli* anal somite and caudal ramus dorsal view; (R) *M. staheli* leg 4; (S) *M. staheli* fifth leg.

Cyclopoida: Cyclopidae: Halicyclopinae: *Halicyclops*: Species

Includes freshwater species only. Based on Rocha et al. (1998a, 1998b).

1	Combination of characters not as below	2
1′	Dorsal caudal seta (VII) as long as lateral most terminal seta (III); male sixth leg with two short spiniform elements; male fifth leg exopod with six elements (spinae or setae) (Fig. 21.38 E–G) ... *Halicyclops cenoticola* Rocha, Iliffe, Reid & Suárez-Morales, 1998 [Mexico: Yucatán]	
2(1)	Dorsal caudal seta (VII) length ~ 2× caudal ramus; sixth leg with one short seta	3
2′	Dorsal caudal seta (VII) length ~ 0.5× lateral most terminal seta (III); sixth leg with three elements ... *Halicyclops caneki* Fiers, 1995 [Mexico: Yucatán]	
3(2)	Preanal somite posterior margin expanded dorsally into coarse, irregularly serrate pseudoperculum reaching halfway along anal somite (Fig. 21.38 H); caudal ramus subrectangular ... *Halicyclops lindbergi* Rocha, 1995 [Brazil]	
3′	Preanal somite posterior margin not expanded dorsally, with serrate hyaline frill straight, not forming a pseudoperculum; caudal ramus subquadrate (Fig. 21.38 I) ... *Halicyclops dussarti* Rocha, 1995 [Brazil]	

Cyclopoida: Cyclopidae: Cyclopinae: Genera

For details on *Allocyclops*, *Bryocyclops*, *Fimbricyclops*, *Haplocyclops*, *Muscocyclops*, and *Teratocyclops*, see Rocha & Bjornberg (1987, 1988), Reid (1993a), Iepure & Defaye (2003) and Fiers (2006).

1	Fifth leg with less than three free articles	2
1′	Fifth leg with three free articles (Fig. 21.38 J) ... *Orthocyclops*, one species: *Orthocyclops modestus* (Herrick, 1883) [Mexico]	
2(1)	Fifth leg with one free article or completely fused to body	3
2′	Fifth leg 2-articulated, proximal article with a seta, second (terminal) with two elements	19
3(2)	Fifth leg represented by a single articulate free or fused article armed with one or two terminal setal elements	4
3′	Fifth leg completely fused to fifth pedigerous somite, represented by three setae	15
4(3)	Fifth leg armed with single seta	5
4′	Fifth leg with more than one setal element (spine + seta)	6
5(4)	Antennule with 11 articles, four terminal caudal setae, fourth leg 2-articulated (Fig. 21.38 K–L) ... *Neutrocyclops*, monotypic: *Neutrocyclops brevifurca* (Lowndes, 1934) [Mexico, Venezuela, Colombia, Brazil]	
5′	Antennule with 10 articles, only three terminal caudal setae (Fig. 21.38 M, N arrow) ... *Idiocyclops*, monotypic: *Idiocyclops stocki* Herbst, 1987 [Suriname]	
6(4)	Fifth leg article with an apical seta and a minute spine on medial position, or only one apical seta	7
6′	Fifth leg article with a seta and a subequal spine; legs 1–4 with 2-articled rami	8
7(6)	Antennule with 17 articles; legs 1–4 with 3-articulated rami, fifth leg free article cylindrical, with a long apical seta and one tiny spine at distal third (Fig. 21.38 O) ... *Ponticyclops*, monotypic: *Ponticyclops boscoi* Reid, 1987 [Brazil]	
7′	Antennule with 10–12 articles; legs 1–4 with 2-articulated rami	***Microcyclops* [p. 748]**
8(6)	Legs 1–4 exopodites spine formula: 3, 4, 4, 3	9
8′	Legs 1–4 exopodites spine formula not as above	13
9(8)	Fifth leg article with two setal elements: one lateral setae and one medial spine, both inserted terminally	10
9′	Fifth leg article with two setal elements: one seta and one spine widely separated	***Apocyclops* [p. 752]**
10(11)	Leg 4 with 2-articulated endopod	11
10′	Leg 4 with 1-articulated endopod	***Hesperocyclops* [p. 752]**
11(10)	Seminal receptacle not strongly expanded; antennule with 11 articles	12
11′	Seminal receptacle strongly expanded posteriorly; antennule with 11, 12, or 13 articles	***Metacyclops* [p. 752]**

12(11) Caudal ramus lateral caudal seta (II) inserted posterodorsally on distal third; leg 4 endopodite with six setal elements; fifth leg free article with one spine and one seta, both short, subequally (Fig. 21.38 Q–S) .... *Menzeliella*, one species: *Menzeliella staheli* Chappuis, 1917 [Suriname]

12′ Caudal ramus lateral caudal seta (II) inserted on half of ramus; leg 4 endopodite with four setal elements; fifth leg free article with one short spine and one long (length ~ 6× spine) slender seta (Fig. 21.39 A, B) .... *Teratocyclops*, monotypic: *Teratocyclops cubensis* Plesa, 1981 [Cuba]

13(8) Legs 1–4 exopodites spine formula: 2, 3, 3, 2 14

13′ Legs 1–4 exopodites spine formula: 3, 3, 3, 3; fifth leg free article width > length, with relatively long medial spiniform spinulate seta, and one lateral plumose seta (Fig. 21.39 J, K) .... *Fimbricyclops*, monotypic: *Fimbricyclops jimhensoni* Reid, 1992 [Puerto Rico]

14(13) Fifth leg article quadrate, length = breadth, with two distal, short subequal setae .... ***Muscocyclops* [p. 756]**

14′ Fifth leg on rectangular expansion armed with two setal elements: one short slender seta, shorter than medial robust, serrate spiniform element (Fig. 21.38 P) .... *Virbiocyclops*, monotypic: *Virbiocyclops silvaticus* (Rocha & Björnberg, 1988) [Brazil]

15(3) Coxa of leg 4 without medial coxal seta 16

15′ Coxa of leg 4 with medial coxal seta 17

16(15) Leg 4 exopodite 1 or 2-articulated; anal operculum strongly developed, usually ornamented with spinulae; legs 1–4 exopodites spine formula: 3, 3, 3, 3 with some variation on leg 4; caudal ramus lateral caudal seta (II) inserted on distal half .... ***Bryocyclops* [p. 758]**

16′ Leg 4 exopodite 1-articulated; anal operculum not produced or weakly developed; legs 1–4 exopodites spine formula: 2, 2, 2, 2 or 2, 3, 3, 2; caudal ramus lateral caudal seta (II) inserted on proximal half (Fig. 21.39 C, I) .... *Haplocyclops*, one species: *Haplocyclops torresi* Rocha, Torres & Maia-Barbosa, 1998 [Brazil]

17(15) Antennule with 11 articles; caudal rami not particularly elongate; legs 1–4 exopodite spine formula: 3, 3, 3, 2 or 2, 3, 3, 2 18

17′ Antennule with 10 articles, caudal rami length ~ 0.4× urosome; legs 1–4 exopodites spine formula: 3, 4, 4, 3; leg 4 second endopod with one apical seta and one apical spine .... ***Yansacyclops* [p. 758]**

18(17) Legs 1–4 exopodites spine formula: 2, 3, 3, 2; leg 4 second endopod with one apical seta and one spine (Fig. 21.39 D–G, arrow) .... *Olmeccyclops*, one species: *Olmeccyclops veracruzanus* (Suárez-Morales, Mendoza & Mercado-Salas, 2010) [Mexico: Southeast]

18′ Legs 1–4 exopodites spine formula: 3, 3, 3, 2; leg 4 second endopod with two apical spinae (Fig. 21.39 H) .... *Allocyclops*, one species: *Allocyclops botosaneanui* Plesa, 1981 [Cuba]

19(2) Fifth leg terminal article with two long, similar setae; antennule always with 17 articles 20

19′ Fifth leg terminal article with one seta and a shorter spiniform element; antennule with 17 or fewer articles 21

20(19) Fifth leg terminal article with medial seta inserted at the middle; female antennule distal article length 3–5× than width; antennule, two distalmost articles with hyaline membrane........ ***Mesocyclops* [p. 758]**

20′ Fifth leg terminal article, with medial seta inserted distally or subdistally; female antennule distal article length 2× width; antennule, two distalmost articles without hyaline membrane .... ***Thermocyclops* [p. 762]**

21(19) Antennule with 11–17 articles; fifth leg terminal article with two setal elements, one apical slender seta and one apical or subapical spiniform seta 22

21′ Antennule with 17 articles; fifth leg proximal article ≥ 1.5× wider than long; fifth leg terminal article with long apical seta and a medial or subapical minute spiniform process, shorter than article, never reaching beyond article distal margin; caudal ramus medial surface pilose .... ***Megacyclops* [p. 764]**

22(21) Antennule with 11–13 articles; fifth leg terminal article with apical spiniform seta, length > article, and >0.2× adjacent apical seta .... ***Diacyclops* [p. 764]**

22′ Antennule with 11–17 articles; fifth leg terminal article with subapical spiniform seta, length < article, and <0.2× adjacent apical seta, but extending beyond article distal margin .... ***Acanthocyclops* [p. 765]**

Cyclopoida: Cyclopidae: Cyclopinae: *Microcyclops*: Species

Key is based on Dussart (1984a), Dussart & Frutos (1986), Rocha (1998) and Gutiérrez-Aguirre & Cervantes-Martínez (2016b).

1 Fifth leg free article smooth, lacking medial spine (Fig. 21.40 Q) 2

1′ Fifth leg free article with medial spine 5

2(1) Caudal ramus lateral most terminal seta (III) insertion with row of spinulae (Fig. 21.40 P, arrow) 3

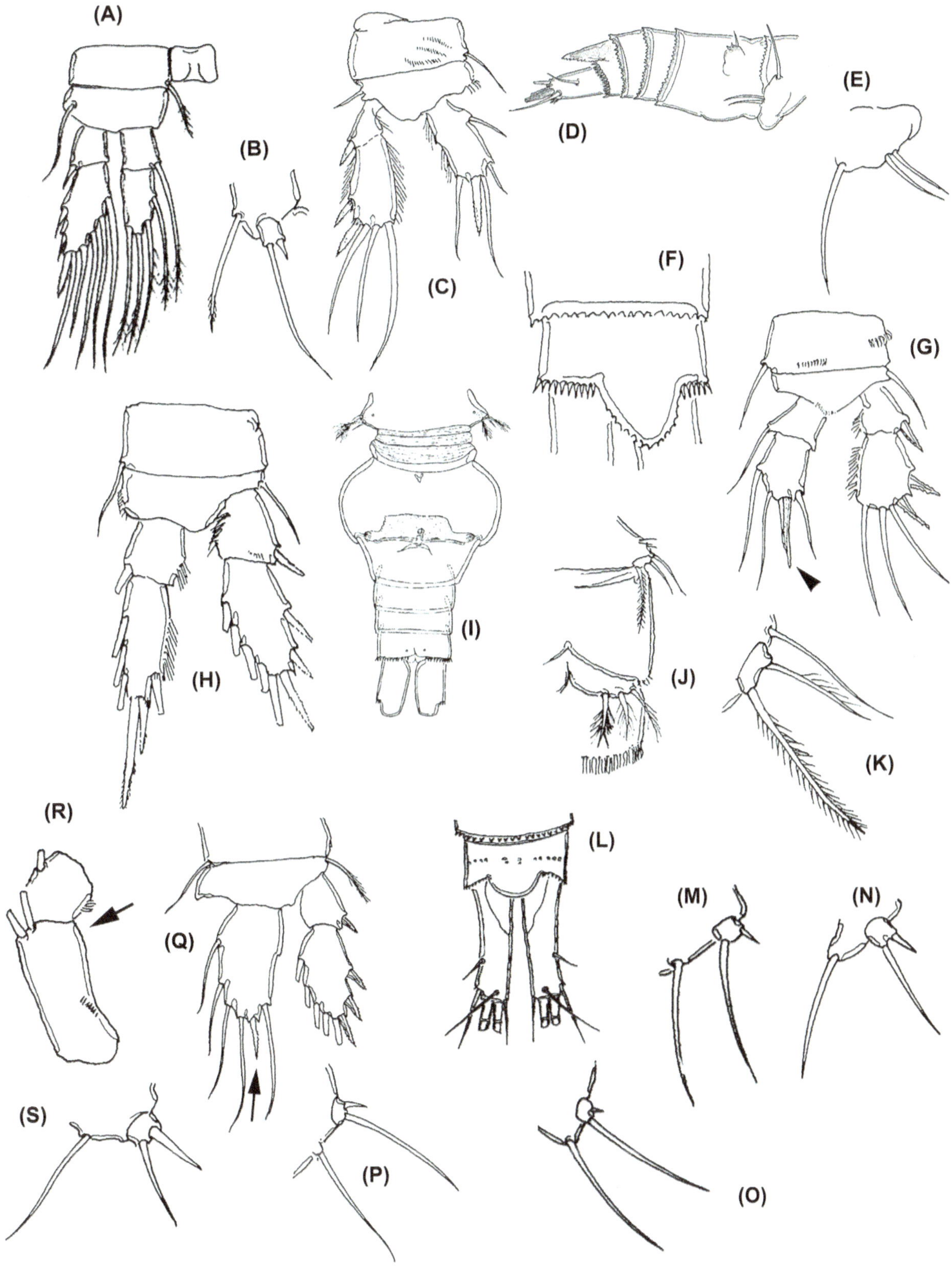

FIGURE 21.39 Morphology of Cyclopoida, Cyclopinae: (A) *Teratocyclops cubensis* leg 4; (B) *T. cubensis* fifth leg; (C) *Haplocyclops torresi* leg 4; (D) *Olmeccyclops veracruzanus* urosome lateral view; (E) *O. veracruzanus* fifth leg; (F) *O. veracruzanus* anal operculum dorsal view; (G) *O. veracruzanus* leg 4; (H) *Allocyclops botosaneanui* leg 4; (I) *H. torresi* urosome ventral view; (J) *Fimbricyclops jimhensoni* fifth pedigerous somite and genital double-somite; (K) *F. jimhensoni* fifth leg; (L) *Hesperocyclops inauditus* anal somite and caudal rami dorsal view; (M) *H. inauditus* fifth leg; (N) *Hesperocyclops improvisus* fifth leg; (O) *Hesperocyclops pescei* fifth leg; (P) *Hesperocyclops herbsti* fifth leg; (Q) *H. herbsti* leg 4; (R) *Hesperocyclops venezuelanus* second antenna basipodite lacking exopodal seta (arrow); (S) *H. venezuelanus* fifth leg.

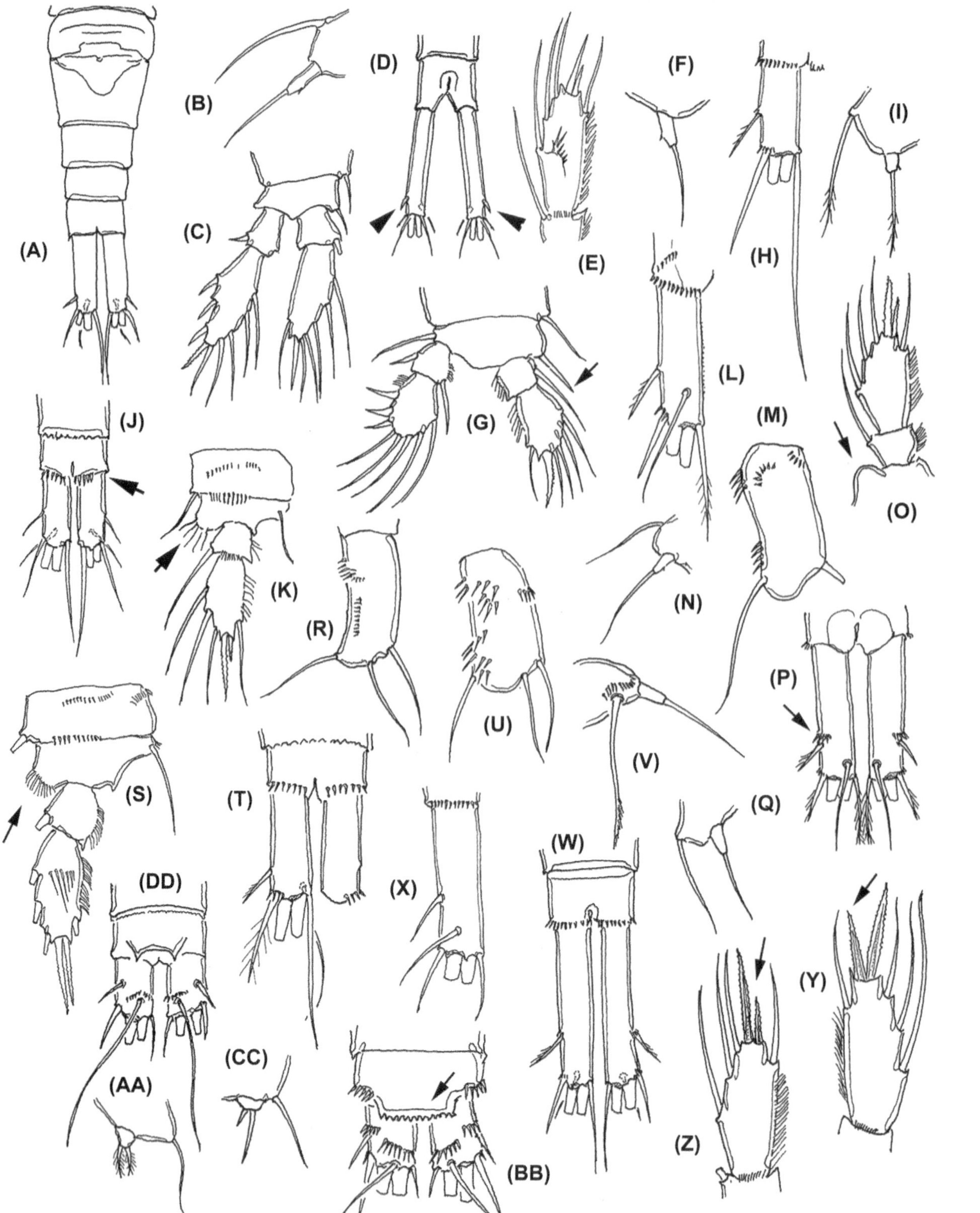

FIGURE 21.40 Morphology of Cyclopoida, Cyclopidae, Cyclopinae: (A) *Microcyclops finitimus* urosome ventral view; (B) *M. finitimus* fifth leg; (C) *Microcyclops furcatus* leg 3; (D) *M. furcatus* anal somite and caudal rami ventral view showing lateral caudal setae (arrows); (E) *Microcyclops elongatus* leg 4 second endopodal segment; (F) *M. elongatus* fifth leg; (G) *M. elongatus* leg 1 showing basipodite spine (arrow); (H) *Microcyclops mediasetosus* caudal ramus; (I) *M. mediasetosus* fifth leg; (J) *Microcyclops inarmatus* anal somite and caudal rami ventral view showing incomplete spinulation (arrow); (K) *M. inarmatus* leg 4 endopod showing ornamentation of basipodite (arrow); (L) *Microcyclops ceibaensis* caudal ramus dorsal view; (M) *M. ceibaensis* second antenna basipodite caudal view; (N) *M. ceibaensis* fifth leg; (O) *Microcyclops medius* leg 4 endopod showing smooth medial basipodite margin (arrow); (P) *M. medius* caudal rami dorsal view showing spinules at insertion of lateral caudal seta (arrow); (Q) *M. medius* fifth leg; (R) *Microcyclops dubitabilis* second antenna basipodite; (S) *M. dubitabilis* leg 4 endopod showing hairy medial basipodite margin (arrow); (T) *M. dubitabilis* caudal rami ventral view; (U) *Microcyclops echinatus* second antenna basipodite; (V) *M. echinatus* fifth leg; (W) *M. echinatus* caudal rami ventral view; (X) *Microcyclops pumilis* caudal ramus dorsal view; (Y) *Microcyclops anceps anceps* leg 4 second endopodal segment showing lateral spine (arrow); (Z) *M. anceps* var. *minor* leg 4 second endopodal segment showing lateral spine (arrow); (AA) *Muscocyclops operculatus* fifth leg; (BB) *Muscocyclops therasiae* anal somite and caudal rami, dorsal view showing operculum (arrow); (CC) *M. therasiae* fifth leg; (DD) *Muscocyclops bidentatus* anal somite and caudal rami, dorsal view showing operculum.

2′ Caudal ramus lateral most terminal seta (III) insertion without spinulae 4

3(2) Caudal ramus length/width ratio = 4.3 (Fig. 21.40 P); leg 4 basipodite medial margin smooth (Fig. 21.40 O, arrow) ..... *Microcyclops medius* Dussart & Frutos, 1986
[Argentina]

3′ Caudal ramus length/width ratio = 3.4 (Fig. 21.40 T); leg 4 basipodite medial margin with short, slender filiform setae (Fig. 21.40 S, arrow); second antenna basipod ornamentation as in Fig. 21.40 R ..... *Microcyclops dubitabilis* Kiefer, 1934
[United States, Mexico, Haiti, Brazil, Uruguay]

4(2) Caudal ramus length/width ratio = 4.0–5.0; lateral caudal seta (II) inserted subdistally at 80% of caudal ramus length (Fig. 21.40 D, arrows); legs 2–4 endopod second article without spinulae (Fig. 21.40 C) ..... *Microcyclops furcatus* (Daday, 1905)
[Widespread in South America]

4′ Caudal ramus length/width ratio = 1.7–2.9, lateral caudal seta (II) inserted medially, at 55% of caudal ramus length (Fig. 21.40 X) ..... *Microcyclops pumilis* Pennak & Ward, 1985
[United States]

5(1) Fifth leg medial spine small, articulated, inserted medially, not reaching article distal margin (Fig. 21.40 B, F) 6

5′ Fifth leg medial spine robust, unarticulated, inserted terminally or subterminally, reaching beyond article distal margin 11

6(5) Caudal ramus medial most terminal (VI)/lateral most terminal (III) caudal setae = 1.6–2.0; caudal ramus length/width ratio = 2.7–6.0 (Fig. 21.40 A); lateral caudal seta (II) inserted at 60%–75% of the caudal ramus length; fifth leg free article elongate, clearly longer than wide 7

6′ Caudal ramus medial most terminal (VI)/lateral most terminal (III) caudal setae = 3.0; caudal ramus length/width ratio = 2.3; lateral caudal seta (II) inserted medially, at 57% of the caudal ramus length (Fig. 21.40 H); fifth leg free article short, subquadrate (Fig. 21.40 I) ..... *Microcyclops mediasetosus* Dussart & Frutos, 1985
[Argentina]

7(6) Leg 1 basipodite with medial spine; leg 4 basipodite medial margin hirsute or with robust spiniform setae, or a combination of both conditions; leg 1 intercoxal sclerite smooth or armed; leg 4 intercoxal sclerite smooth or ornamented 8

7′ Leg 1 basipodite medial margin with filiform setae, medial spine absent; leg 4 basipodite medial margin hirsute; leg 1 intercoxal sclerite smooth, leg 4 intercoxal sclerite ornamented with spinulae (Fig. 21.40 A, B) ..... *Microcyclops finitimus* Dussart, 1984
[Venezuela, Colombia, French Guiana, Brazil]

8(7) Leg 1 basipodite medially hirsute, medial spine reaching distal half of leg 1 second endopod; caudal ramus length/width ratio = 2.5–6.0, with or without spinae at insertion of both lateral caudal (II) and lateral most terminal (III) setae 9

8′ Leg 1 basipodite medially smooth, medial spine reaching proximal half of leg 1 second endopod (Fig. 21.40 G, arrow); surface of leg 4 second endopod with spinules (Fig. 21.40 E); caudal ramus length/width ratio = 5–6, with row of spinae at insertion of lateral caudal seta (II) that extends dorsally; no spinae at insertion of lateral most terminal seta (III) ..... *Microcyclops elongatus* (Lowndes, 1934)
[Brazil, Paraguay]

9(8) Anal somite with spinae along both ventral and dorsal posterior margins; caudal setae II and terminal setae III basally with spinulae; legs 1–4 intercoxal sclerites with rows of spinulae; second antenna basipodite caudal surface with more than one group of spinulae; caudal ramus length/width ratio =3.2–5.5 10

9′ Anal somite with spinae on ventral posterior margin only (Fig. 21.40 J, arrow); caudal setae II and terminal setae III without basal spinulae; legs 1–4 intercoxal sclerites smooth; leg 4 basipodite medial margin with long filiform setae (Fig. 21.40 K, arrow); second antenna basipodite caudal surface with one group of spinulae; caudal ramus length/width ratio = 2.1–2.9 ..... *Microcyclops inarmatus* Gutiérrez-Aguirre & Cervantes-Martínez, 2016
[Mexico: Tabasco, Haiti]

10(9) Fifth leg basal seta insertion with two spinulae rows (Fig. 21.40 V); caudal ramus length/width ratio = 5.2–5.5 (Fig. 21.40 W); second antenna basipodite caudal surface with three spinule clusters (Fig. 21.40 U) ..... *Microcyclops echinatus* Fiers, Ghenne & Suárez-Morales, 2000
[Mexico: Yucatan]

10′ Fifth leg basal seta insertion smooth (Fig. 21.40 N); caudal ramus length/width ratio = 3.2–4.0 (Fig. 21.40 L); second antenna basipodite caudal surface with two arcuate spinule rows and two lateral rows of long spinules (Fig. 21.40 M) ..... *Microcyclops ceibaensis* (Marsh, 1919)
[Mexico, Cuba, Honduras, Costa Rica, Venezuela, Brazil]

11(5) Caudal rami length/width ratio = 3.5–4.0; leg 1 basipodite medial margin ornamented 12

11′ Caudal rami length/width ratio = 2.4; leg 1 basipodite medial margin smooth; leg 4 basipodite medial margin with setulae; dorsal caudal seta (VII) length 1.5× caudal rami ..... *Microcyclops anceps pauxensis* Herbst, 1962
[Brazil]

12(11) Body length 0.8–0.9 mm; fifth leg basal seta insertion usually lacking ornamentation; caudal ramus lateral most terminal seta (III) insertion with spinulae; leg 4 second endopodal article lateral spine reaching beyond 50% medial spine (Fig. 21.40 Y, arrow) ..... *Microcyclops anceps anceps* (Richard, 1897)
[Mexico, Guatemala, Venezuela, Colombia, Guyana, Uruguay]

12′ Body length < 0.75 mm; fifth leg basal seta insertion with spinule row; caudal ramus lateral most terminal seta (III) insertion smooth; leg 4 second endopodal article lateral spine not reaching 50% of medial spine length (Fig. 21.40 Z, arrow) ..... *Microcyclops anceps* var. *minor* Dussart, 1984
[Venezuela]

Cyclopoida: Cyclopidae: Cyclopinae: *Apocyclops*: Species

Species keys for North America and the Caribbean region are in Reid et al. (2002).

1 Second antenna (including coxa) 5-articulated; legs 2–4 basipodites medial margin produced into spiniform processes; legs 3 and 4 endopod apical spine length > endopod; body with cuticular ornamentation (sculpturing, reticulation) 2

1′ Second antenna (including coxa) 4-articulated; legs 2–4 basipodites medial margin rounded; legs 3 and 4 endopod apical spine length < endopod, body unornamented 3

2(1) Caudal ramus length/width ratio = 9; dorsal caudal seta (VII) length 0.6× ramus (Fig. 21.41 A); leg 1 basipodite posterior margin unornamented .... *Apocyclops dimorphus* Kiefer, 1931
[Mexico, Haiti, Nearctic: Southern United States]

2′ Caudal ramus length/width ratio = 4.5; dorsal caudal seta (VII) length 0.3× ramus (Fig. 21.41 B); leg 1 basipodite posterior margin with spinule rows between rami insertions (Fig. 21.41 C, arrow) .... *Apocyclops dengizicus* (Lepeshkin, 1900)
[Probably in Mexico]

3(1) Fifth leg free article smooth 4

3′ Fifth leg free article with spinulae (Fig. 21.41 D) .... *Apocyclops spartinus* Ruber, 1968
[Note: Probably represent a misidentification in South America]

4(3) Dorsal caudal seta (VII) length = caudal ramus (Fig. 21.41 E) 5

4′ Dorsal caudal seta (VII) length ~ 1.8× caudal ramus (Fig. 21.41 F, arrows) .... *Apocyclops distans* (Kiefer, 1956)
[Venezuela, Caribbean Islands]

5(4) Leg 4 basipodite anteriomedial surface lacking circular patch of spinulae: fifth leg apical seta length > 2.5× spine (Fig. 21.41 G, H) .... *Apocyclops panamensis* (Marsh, 1913)
[Central America, Mexico, Caribbean Islands, Nearctic: United States]

5′ Leg 4 basipodite anteriomedial surface with circular patch of spinulae (Fig. 21.41 I, arrow); fifth leg apical seta length <2.2× spine (Fig. 21.41 J) .... *Apocyclops procerus* Herbst, 1955
[Nicaragua, Brazil, Peru]

Cyclopoida: Cyclopidae: Cyclopinae: *Hesperocyclops*: Species

Based on Galassi & Pesce (1992).

1 Legs 1–4 second exopodal articles spine formula: 3, 4, 3, 3 2

1′ Legs 1–4 second exopodal articles spine formula: 3, 4, 4, 3; fifth leg free article spine slightly longer than article (Fig. 21.39 N) .... *Hesperocyclops improvisus* Herbst, 1984
[Antigua, Guadeloupe]

2(1) Second antenna basipodite lacking exopodal seta (Fig. 21.39 R, arrow); fifth leg free article spine length ~0.5× adjacent seta (Fig. 21.39 S) .... *Hesperocyclops venezuelanus* Galassi & Pesce, 1992
[Venezuela]

2′ Second antenna basipodite with exopodal seta; fifth leg free article spine length < 0.3× adjacent seta 3

3(2) Leg 4 endopod lateral spine length 0.5× article, medial setae slightly longer than spine 4

3′ Leg 4 endopod lateral spine length < 0.5× article, medial setae clearly longer than spine (Fig. 21.39 Q, arrow, P) .... *Hesperocyclops herbsti* Rocha & Bjornberg, 1987
[Brazil]

4(3) Anal operculum well developed, semicircular (Fig. 21.39 L); fifth leg free article spine longer than article (Fig. 21.39 M) .... *Hesperocyclops inauditus* Dussart & Frutos, 1986
[Argentina]

4′ Anal operculum low, rounded; fifth leg free spine minute, shorter than article (Fig. 21.40 O) .... *Hesperocyclops pescei* Petkovski, 1988
[Colombia]

Cyclopoida: Cyclopidae: Cyclopinae: *Metacyclops*: Species

Based on Reid (1987b), Herbst (1988) and Rocha (1994).

1 Antennule with 13 or 17 articles; leg 4 intercoxal sclerite with rounded processes bearing spinae (Fig. 21.42 A) and additional spinule rows 2

1′ Antennule with 11 or 12 articles; leg 4 intercoxal sclerite with or without processes, ornamented or smooth 3

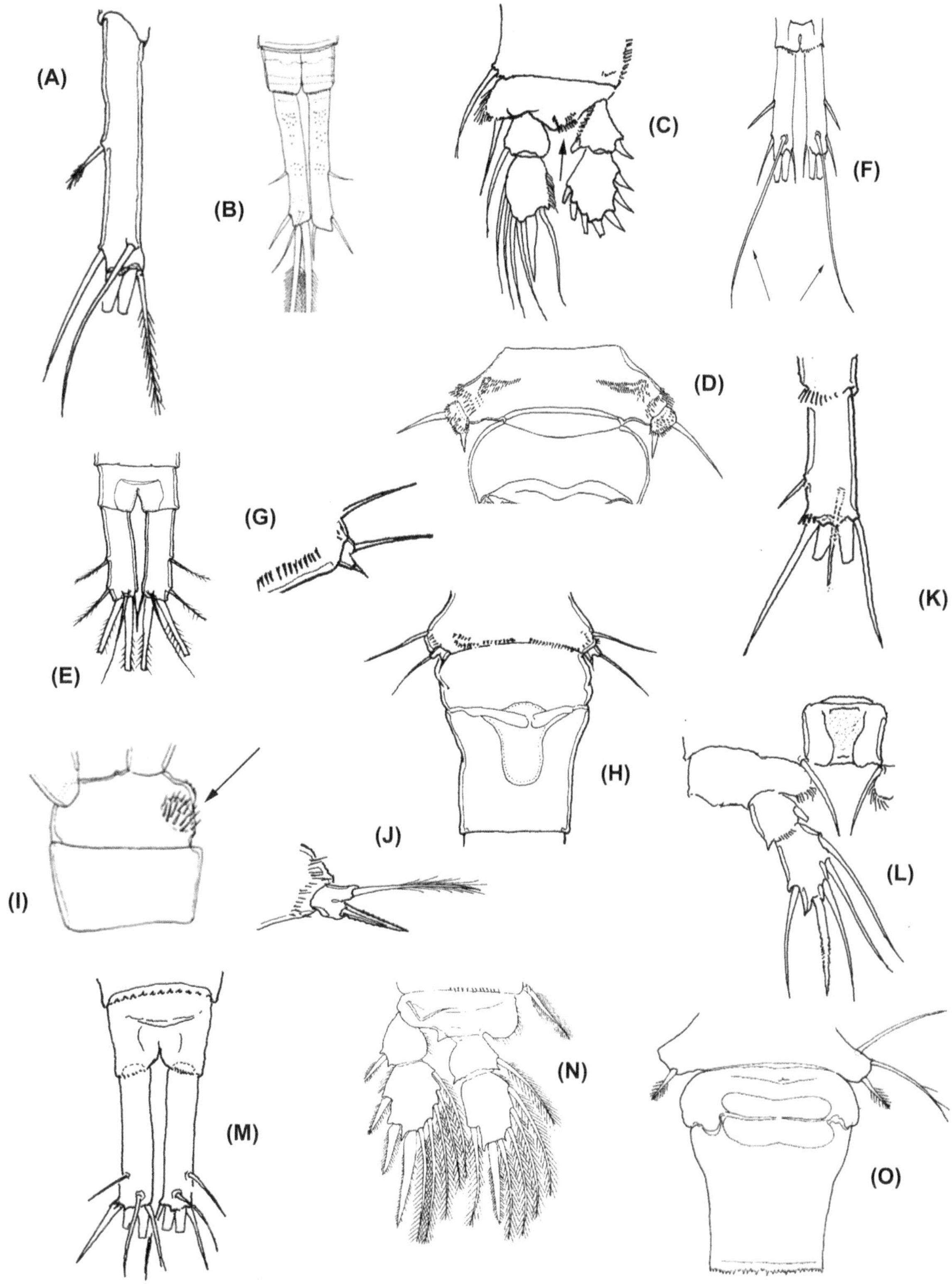

FIGURE 21.41 Morphology of Cyclopoida, Cyclopinae: (A) *Apocyclops dimorphus* caudal ramus dorsal view; (B) *Apocyclops dengizicus* caudal rami ventral view; (C) *A. dengizicus* leg 1 posterior; (D) *Apocyclops spartinus* fifth pedigerous somite and fifth leg ventral view; (E) *Apocyclops panamensis* anal somite and caudal rami dorsal view; (F) *Apocyclops distans* anal somite and caudal rami dorsal view showing dorsal caudal setae (arrows); (G) *A. panamensis* fifth leg; (H) *A. panamensis* fifth pedigerous somite and genital double somite ventral view; (I) *Apocyclops procerus* leg 4 coxa and basis showing anterior patch of spinulae (arrow); (J) *A. procerus* fifth leg; (K) *Yansacyclops neotropicalis* caudal ramus ventral view; (L) *Y. neotropicalis* leg 4 intercoxal sclerite basipodite and endopod; (M) *Yansacyclops ferrari* anal somite and caudal rami dorsal view; (N) *Y. ferrarii* leg 4 endopodal and exopodal rami; (O) *Y. ferrarii* fifth pedigerous somite and genital double-somite ventral view. *(B and I) After Coelho Botelho, (1999), (N) after Reid (1988).*

2(1) Antennule with 17 articles; leg 4 endopod medial spine length 1.4× lateral spine (Fig. 21.42 B) .... *Metacyclops grandis* (Kiefer, 1935)
[Uruguay]

2′ Antennule with 13 articles, Cyclopinae (Fig. 21.42 D); leg 4 endopod medial spine length 2× lateral spine; fifth leg as in Fig. 21.42 C .... *Metacyclops tredecimus* (Lowndes, 1934)
[Venezuela, Paraguay, Argentina]

3(1) Leg 4 second endopodal article with one apical spine (Fig. 21.42 F, H) 4

3′ Leg 4 second endopodal article with two apical spinae or one spine and one seta (Fig. 21.42 Y) 7

4(3) Caudal ramus medial most terminal seta (VI) length $\sim$ 1.3−1.8× lateral most terminal seta (III); urosomite posterior borders smooth 5

4′ Caudal ramus medial most terminal seta (VI) length $\leq$ lateral most terminal seta (III) 6

5(4) Caudal ramus medial most terminal seta (VI) length $\sim$ 1.8× lateral most terminal seta (III); leg 4 endopod apical spine stout, thick, shorter than article (Fig. 21.42 K, arrow) .... *Metacyclops curtispinosus* Dussart, 1984
[Venezuela]

5′ Caudal ramus medial most terminal seta (VI) length = 1.5−1.6× lateral most terminal seta (III); leg 4 endopod apical spine narrow, longer than article (Fig. 21.42 L, arrow); fifth leg free article cylindrical (Fig. 21.42 M) .... *Metacyclops agnitus* Herbst, 1988
[Barbados]

6(4) Genital double-somite and urosomites with posterior borders denticulate (Fig. 21.42 E, arrow); fifth leg apical spine as long as article (Fig. 21.42 G) .... *Metacyclops denticulatus* Dussart & Frutos, 1985
[Argentina]

6′ Urosomites with posterior borders smooth (Fig. 21.42 J); fifth leg apical spine length $<$ article (Fig. 21.42 I) .... *Metacyclops deserticus* Mercado-Salas & Suárez-Morales, 2013
[Mexico]

7(3) Leg 4 second endopodal article medial spine length 1.5−2.4× $>$ lateral spine or seta 8

7′ Leg 4 second endopodal article apical elements subequal (Fig. 21.42 N, arrow) .... *Metacyclops subaequalis* Dussart, 1984
[Venezuela]

8(7) Caudal rami length/width ratio $\sim$ 3.8−5.0 9

8′ Caudal rami length/width ratio $\sim$ 1.5−3.0 11

9(8) Fifth leg free article width subequal to length, or wider 1.6× length; caudal ramus medial most and lateral most terminal setae not subequal in length 10

9′ Fifth leg free article length $\sim$ 1.5× width (Fig. 21.42 O); caudal ramus medial most and lateral most terminal setae subequal in length; medial most terminal seta length 2.6−3.3× ramus .... *Metacyclops leptopus* (Kiefer, 1927)
[Venezuela, Colombia, Peru, Bolivia]

10(9) Fifth leg free article length subequal to width, apical seta 1.5× article, $>$ 2× adjacent spine (Fig. 21.42 R); caudal ramus medial most terminal seta (VI) length 1.8−2.0× ramus and 0.7× lateral most terminal seta (III); medial median terminal seta (V) not thickened proximally (Fig. 21.42 P); leg 4 endopod apical spinae as in Fig. 21.42 Q .... *Metacyclops mendocinus* (Wierzejski, 1892)
[Cuba, Haiti, Nicaragua, Venezuela, Colombia, Ecuador, Brazil, Peru, Bolivia, Chile, Paraguay, Argentina, Uruguay]

10′ Fifth leg free article $\sim$1.6× broader than long, apical seta slightly longer than spine (Fig. 21.42 S); caudal ramus medial most terminal seta (VI) length 3.4× ramus, and 1.3× lateral most terminal seta (III); medial median terminal seta (V) thickened proximally (Fig. 21.42 T, arrow) .... *Metacyclops mutatus* Herbst, 1988
[Aruba]

11(8) Antennule extending beyond 2nd prosomite posterior margin 12

11′ Antennule reaching cephalosome posterior margin, or shorter 13

12(11) Caudal rami length/width ratio = 1.5; leg 4 second endopodal article with stout medial spine and very slender, short lateral seta; antennule of 12 articles .... *Metacyclops hartmanni* Herbst, 1960
[Nicaragua]

12′ Caudal rami length/width ratio = 2.6; leg 4 second endopodal article length 2× width, with 2 stout spinae; antennule of 11 articles .... *Metacyclops rudis* Plesa, 1981
[Cuba]

13(11) Leg 4 intercoxal sclerite distal margin with triangular projections or spinae extending past sclerite margin 14

13′ Leg 4 intercoxal sclerite distal margin straight or with rounded but unornamented projections (Fig. 21.42 Z) (sclerite surface smooth or with hairs or spinulae) 16

14(13) Leg 4 intercoxal sclerite distal margin with two large triangular projections, sclerite otherwise smooth 15

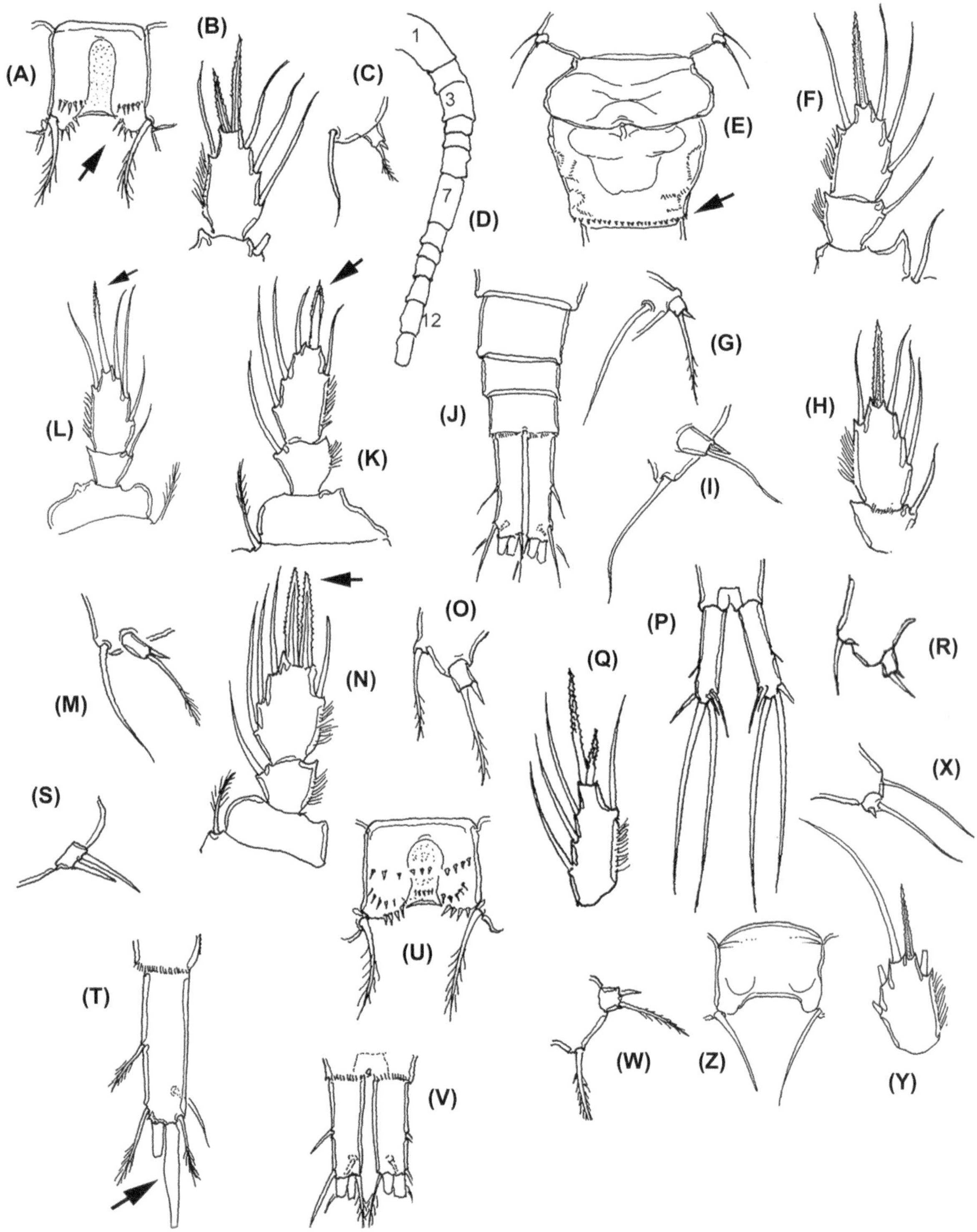

FIGURE 21.42 Morphology of Cyclopoida, Cyclopinae: (A) *Metacyclops grandis* leg 4 intercoxal sclerite showing spinule ornamentation (arrow); (B) *M. grandis* leg 4 second endopodal segment; (C) *Metacyclops tredecimus* fifth leg; (D) *M. tredecimus* antennule (armature omitted); (E) *Metacyclops denticulatus* fifth pedigerous and genital double-somite ventral view showing denticulate posterior margin (arrow); (F) *M. denticulatus* leg 4 endopod; (G) *M. denticulatus* fifth leg; (H) *Metacyclops deserticus* leg 4 endopod; (I) *M. deserticus* fifth leg; (J) *M. deserticus* urosome dorsal view; (K) *Metacyclops curtispinosus* leg 4 endopod showing short apical spine (arrow); (L) *Metacyclops agnitus* leg 4 endopod showing long apical spine (arrow); (M) *M. agnitus* fifth leg; (N) *Metacyclops subaequalis* leg 4 showing subequal apical spinae (arrow); (O) *Metacyclops leptopus* fifth leg; (P) *Metacyclops mendocinus* caudal rami dorsal view; (Q) *M. mendocinus* leg 4 second endopodal segment; (R) *M. mendocinus* fifth leg; (S) *Metacyclops mutatus* fifth leg; (T) *M. mutatus* caudal ramus showing thickened medial median terminal seta (arrow); (U) *Metacyclops botosaneanui* leg 4 intercoxal sclerite; (V) *M. botosaneanui* caudal rami ventral view; (W) *M. botosaneanui* fifth leg; (X) *Metacyclops cushae* fifth leg; (Y) *M. cushae* leg 4 second endopodal segment; (Z) *M. cushae* leg 4 intercoxal sclerite.

14′ Leg 4 intercoxal sclerite distal margin with four spinae on each side, and two spinule rows on surface (Fig. 21.42 U–W) .... *Metacyclops botosaneanui* Pesce, 1985
[Bonaire]

15(14) Leg 4 second endopodal article medial spine length 1.6× lateral spine or seta; length ratio of leg 4 medial spine/article = 0.5 .... *Metacyclops dentatus* (Plesa, 1981)
[Cuba]

15′ Leg 4 second endopodal article medial spine length 2.6× lateral spine or seta (Fig. 21.43 A, arrow); length ratio of leg 4 medial spine/article = 1.2 (Fig. 21.43 A), leg 5 as in Fig. 21.43 B .... *Metacyclops geltrudeae* Galassi & Pesce, 1994
[Venezuela]

16(13) Leg 4 second endopodal article length/width ratio = 2.4–2.8 17

16′ Leg 4 second endopodal article length/width ratio = 1.5–2.0 20

17(16) Caudal ramus lateral median terminal seta (IV) length 1.2–2.0× medial median terminal seta (V); leg 1 intercoxal sclerite lacking ornamentation 18

17′ Caudal ramus lateral median terminal seta (IV) length 0.7× medial median terminal seta (V) (Fig. 21.43 C, arrows); leg 1 intercoxal sclerite with two spinulae on each side (Fig. 21.43 D, arrow) .... *Metacyclops laticornis* (Lowndes, 1934)
[Colombia, Brazil, Paraguay. Probably Argentina]

18(17) Leg 4 intercoxal sclerite posterior margin straight, spinule row not reaching sclerite distal margin 19

18′ Leg 4 intercoxal sclerite posterior margin with rounded protrusions on each side, anterior surface smooth; leg 4 endopod second article with lateral spine length 0.5× medial spine (Fig. 21.43 E); fifth leg seta of terminal article shorter than genital somite anterior quarter (Fig. 21.43 F, arrow, G) .... *Metacyclops dianae* Pesce, 1985
[Haiti]

19(18) Fifth leg seta of terminal article reaching genital double-somite posterior third (Fig. 21.43 H, arrow, I) .... *Metacyclops brauni* Herbst, 1962
[Brazil]

19′ Fifth leg seta of terminal article length ~ 2× adjacent spine, not reaching genital double-somite proximal third (Fig. 21.43 J) .... *Metacyclops janstocki* Herbst, 1990
[Antigua]

20(16) Leg 4 second endopodal article length/width ratio = 1.8–3.3; fifth leg apical spine as long as free article; female antennule with 11 or 12 articles 21

20′ Leg 4 second endopodal article length/width ratio = 1.6–1.8; fifth leg apical spine clearly longer or shorter than free article; antennule with 11 articles 23

21(20) Antennule with 11 articles; caudal ramus length/width ratio = 3.0–3.3; second antenna exopodal seta short, not reaching distal article of appendage 22

21′ Antennule with 12 articles; caudal ramus length/width ratio = 2.5 (Fig. 21.43 N); caudal ramus medial most terminal seta (VI) length 1.4× lateral most terminal seta (III); caudal ramus lateral margin anterior third with two spinules (Fig. 21.43 P, arrow); second antenna exopodal seta reaching beyond appendage distal article (Fig. 21.43 O, arrow). .... *Metacyclops hirsutus* Rocha, 1994
[Brazil]

22(21) Caudal ramus length/width ratio = 3.3 (Fig. 21.43 K); caudal ramus medial most terminal seta (VI) length 0.9× lateral most terminal seta (III); caudal ramus lateral margin anterior third with three spinulae (Fig. 21.43 L, arrow), fifth leg apical spine longer than free article (Fig. 21.43 M) .... *Metacyclops campestris* Reid, 1987
[Brazil]

22′ Caudal ramus length/width ratio = 3.0; caudal ramus medial most terminal seta (VI) length 1.25× lateral most terminal seta (III); caudal ramus lateral margin anterior third without spinulae; fifth leg apical spine as long as free article (Fig. 21.42 X–Z) .... *Metacyclops cushae* Reid, 1991
[Brazil, Nearctic: Southern United States]

23(20) Caudal rami lateral margin anterior third with one spinule (Fig. 21.43 R, arrows); fifth leg apical spine smooth, shorter than free article (Fig. 21.43 Q) .... *Metacyclops paludicola* (Herbst, 1959)
[Brazil]

23′ Caudal rami lateral margin anterior third with three spinulae (Fig. 21.43 T); fifth leg apical spine serrate, length nearly 2× free article (Fig. 21.43 S) .... *Metacyclops oraemaris* Rocha, 1994
[Brazil]

Cyclopoida: Cyclopidae: Cyclopinae: *Muscocyclops*: Species

Based on Rocha & Bjornberg (1987, 1988).

1 Anal operculum with different features than below 2

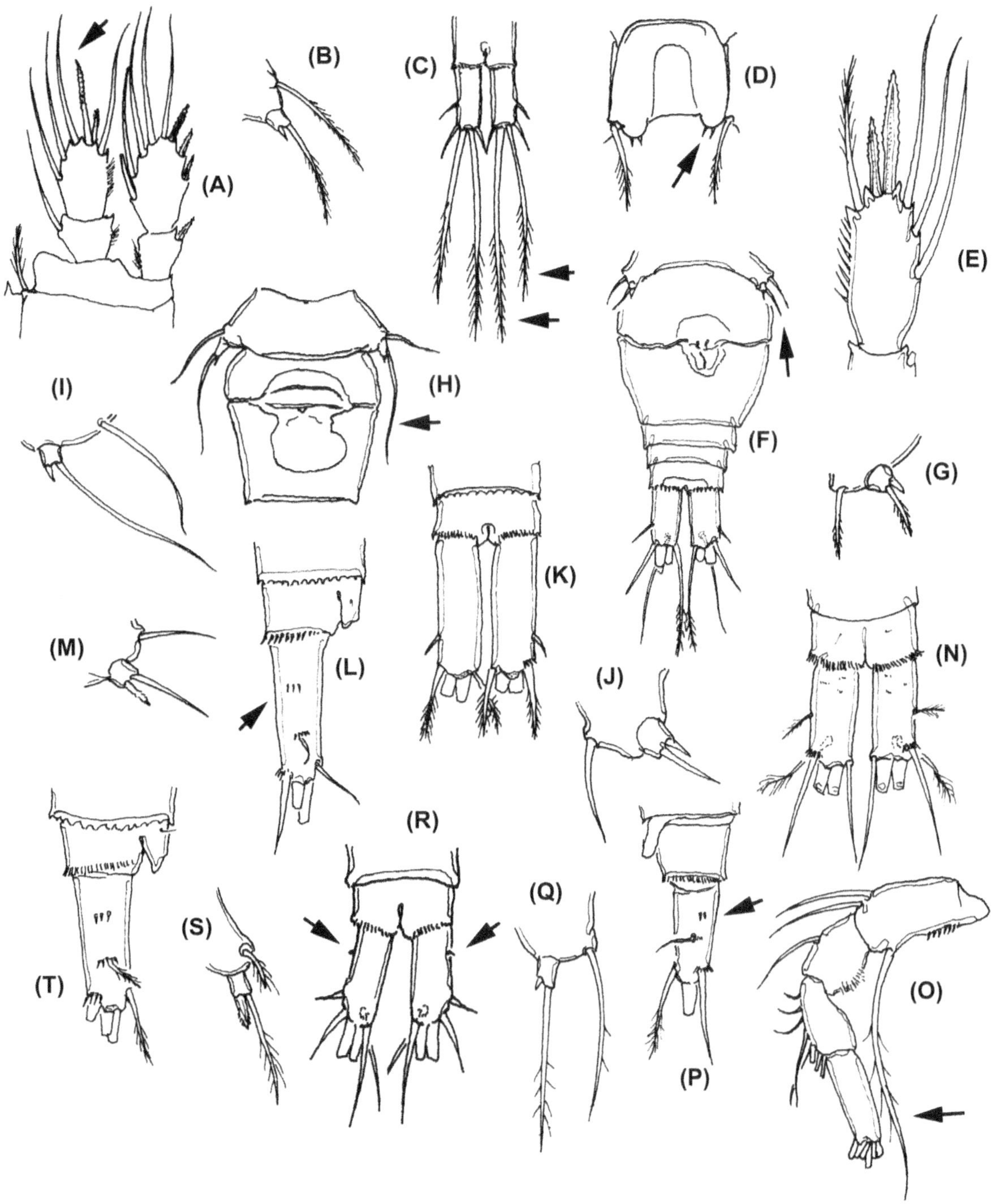

FIGURE 21.43 Morphology of Cyclopoida, Cyclopinae: (A) *Metacyclops geltrudeae* leg 4 showing long medial spine (arrow); (B) *M. geltrudeae* fifth leg; (C) *Metacyclops laticornis* anal somite and caudal rami showing length of lateral median and medial median terminal setae (arrows); (D) *M. laticornis* leg 1 intercoxal sclerite with spinae (arrow); (E) *Metacyclops dianae* leg 4 second endopodal segment; (F) *M. dianae* urosome ventral view showing short apical seta of fifth leg (arrow); (G) *M. dianae* fifth leg; (H) *Metacyclops brauni* fifth pedigerous and genital double-somite ventral view showing apical seta of fifth leg (arrow); (I) *M. brauni* fifth leg; (J) *Metacyclops janstocki* fifth leg; (K) *Metacyclops campestris* caudal rami ventral view; (L) *M. campestris* caudal ramus lateral view showing row of 3 spinulae (arrow); (M) *M. campestris* fifth leg; (N) *Metacyclops hirsutus* caudal rami ventral view; (O) *M. hirsutus* second antenna showing long exopodal seta (arrow); (P) *M. hirsutus* caudal ramus lateral view showing row of 2 spinulae (arrow); (Q) *Metacyclops paludicola* fifth leg; (R) *M. paludicola* caudal rami ventral view showing 1 spinule on lateral margin (arrows); (S) *Metacyclops oraemaris* fifth leg; (T) *M. oraemaris* caudal ramus lateral view.

1′ Anal operculum rounded, fringed with finely crenulate hyaline membrane; fifth leg free article apical setae plumose, subequal (Fig. 21.40 AA) .... *Muscocyclops operculatus* Chappuis, 1917
[Suriname, Brazil]

2(1) Anal operculum quadrate (Fig. 21.40 BB, arrow), posterior margin with 11 or 12 hyaline teeth (Fig. 21.40 BB); fifth leg free article with two stout apical setae, lateral seta length 2× medial (Fig. 21.40 CC); dorsal caudal seta length 1.2× ramus .... *Muscocyclops therasiae* Reid, 1987
[Brazil]

2′ Anal operculum convex, flanked by 2 large hyaline acute processes (Fig. 21.40 DD); fifth leg free article with apical spine length ~0.3× plumose lateral seta; dorsal caudal seta length 2.3× ramus .... *Muscocyclops bidentatus* Reid, 1987
[Brazil]

Cyclopoida: Cyclopidae: Cyclopinae: *Bryocyclops*: Species

For more information, consult Rocha & Bjornberg (1988).

1 Legs 1–4 second endopodal article setal formula: 5, 5, 5, 4; anal operculum with different features than below; pedigerous somites 2–4 without special ornamentations 2

1′ Legs 1–4 second endopodal article setal formula: 4, 5, 5, 4; anal operculum subtriangular, uniformly dentate; pedigerous somites 2 and 3 dorsal surface with transverse crenulate processes (Fig. 21.44 A, arrows, B) .... *Bryocyclops campaneri* Rocha & Björnberg, 1987
[Brazil]

2(1) Leg 4 second endopodal article midapical seta length ~ 4× adjacent apical seta (Fig. 21.44 E, arrow); operculum irregularly dentate (Fig. 21.44 D) .... *Bryocyclops caroli* Björnberg, 1985
[Brazil]

2′ Leg 4 second endopodal article midapical seta not as long as above; anal operculum acuminate with free margins smooth (Fig. 21.44 C) .... *Bryocyclops anninae* (Menzel, 1926)
[Puerto Rico]

Cyclopoida: Cyclopidae: Cyclopinae: *Yansacyclops*: Species

1 Caudal rami length/width ratio = 2.5 (Fig. 21.41 K); leg 4 second endopodal article with five setal elements (Fig. 21.41 L) .... *Yansacyclops neotropicalis* Dussart, 1984
[Venezuela]

1′ Caudal rami length/width ratio = 5.0 (Fig. 21.41 M); leg 4 second endopodal article with six setal elements (Fig. 21.41 N, O) .... *Yansacyclops ferrarii* (Reid, 1988)
[Brazil]

Cyclopoida: Cyclopidae: Cyclopinae: *Mesocyclops*: Species

Based on Suárez-Morales & Gutiérrez-Aguirre (2001) and Gutiérrez-Aguirre et al. (2006).

1 Leg 1 basipodite medial surface without seta 2

1′ Leg 1 basipodite medial surface with seta, or spiniform seta (Fig. 21.45 A, arrow) 6

2(1) Fifth pedigerous somite ventrally glabrous; leg 4 intercoxal sclerite with high, acute projections (Fig. 21.45 B, arrows) 3

2′ Fifth pedigerous somite ventrally pilose; leg 4 intercoxal sclerite with low, nonacute projections (Fig. 21.45 C) 4

3(2) Leg 4 basipodite medial margin pilose .... *Mesocyclops pescei* Petkovski, 1986
[Mexico, Bahamas]

3′ Leg 4 basipodite medial margin smooth .... *Mesocyclops pehpeiensis* Hu, 1943
[Mexico: Southeast. Cuba. Introduced. Oriental]

4(2) Maxillule palp medially without spinae 5

4′ Maxillule palp medially with spinae (Fig. 21.45 D, arrow) .... *Mesocyclops ogunnus* Onabamiro, 1957
[Cayman Islands, Brazil. Introduced. Afrotropical]

5(4) Caudal ramus smooth medially; female second antenna basipodite with row of large spinules (Fig. 21.45 E, arrow); male second antenna basipodite with row of tiny spinulae (Fig. 21.45 F, arrow) near insertion of setae opposite exopodal seta .... *Mesocyclops thermocyclopoides* Harada, 1931
[Introduced Central America, Brazil, Oriental]

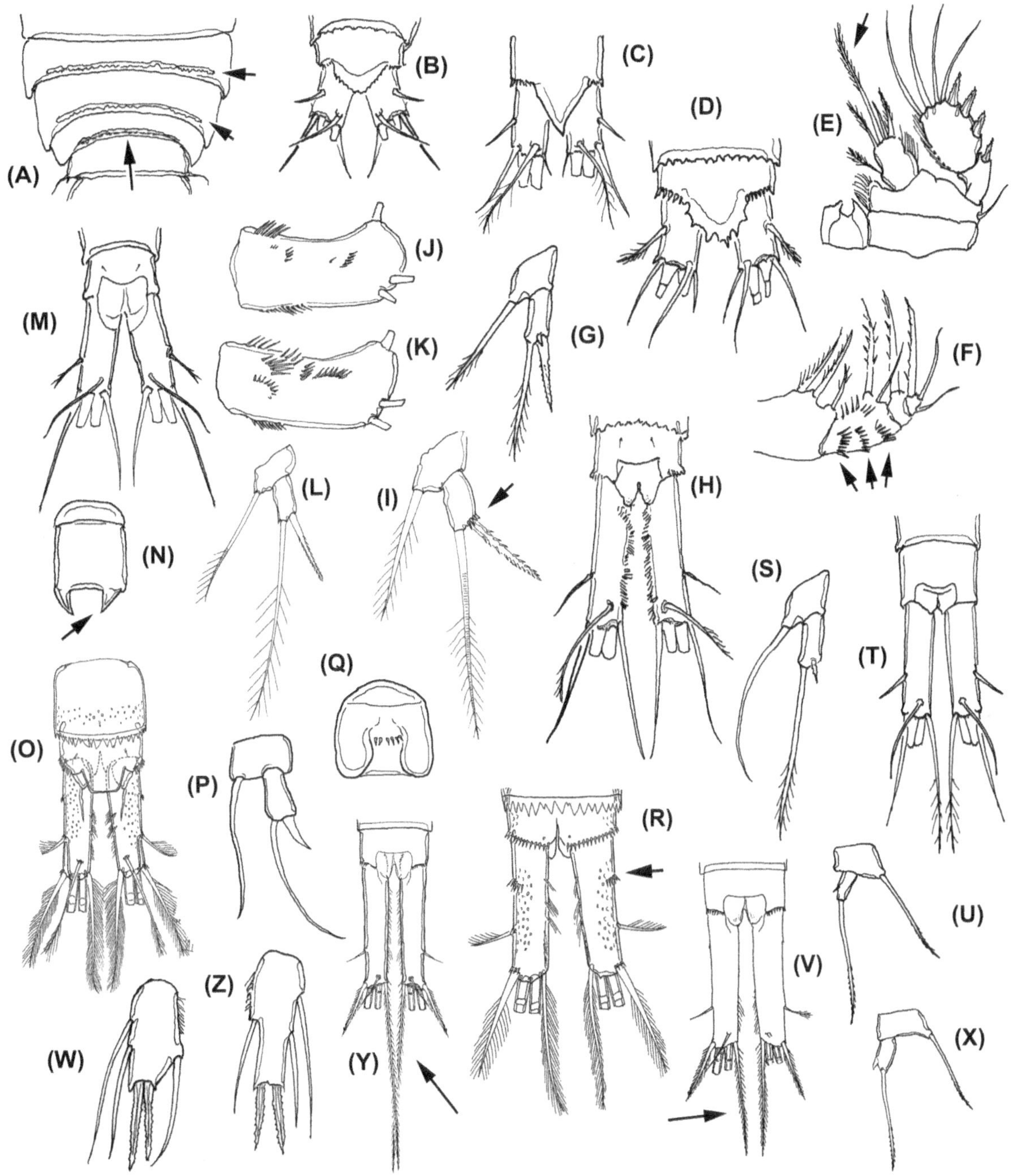

FIGURE 21.44 Morphology of Cyclopoida, Cyclopinae: (A) *Bryocyclops campaneri* pedigerous somites 2-4 showing dorsal crenulate processes (arrows); (B) *B. campaneri* anal somite and caudal rami, dorsal view; (C) *Bryocyclops anninae* anal somite and caudal rami dorsal view; (D) *Bryocyclops caroli* anal somite and caudal rami dorsal view; (E) *B. caroli* leg 4 showing long midapical seta (arrow); (F) *Diacyclops uruguayensis* maxilliped showing 3 rows of spinules (arrows); (G) *D. uruguayensis* fifth leg; (H) *D. uruguayensis* caudal rami dorsal view; (I) *Diacyclops chakan* fifth leg showing spinules at insertion of apical spine (arrow); (J) *D. chakan* second antenna basipodite caudal view; (K) *Diacyclops puuc* second antenna basipodite caudal view; (L) *D. puuc* fifth leg; (M) *Diacyclops bernardi* anal somite and caudal rami dorsal view; (N) *D. bernardi* leg 4 intercoxal sclerite showing acute processes (arrow); (O) *Diacyclops pilosus* anal somite and caudal rami dorsal view; (P) *Diacyclops andinus* fifth leg; (Q) *D. andinus* leg 4 intercoxal sclerite; (R) *Dryocyclops ecabensis* anal somite and caudal rami showing row of large spinules on proximal margin (arrow); (S) *Dryocyclops michaelseni* fifth leg; (T) *D. michaelseni* anal somite and caudal rami dorsal view; (U) *Megacyclops latipes* fifth leg; (V) *M. latipes* anal somite and caudal rami showing short terminal seta VI (arrow); (W) *M. latipes* leg 4 third endopodal segment; (X) *Megacyclops viridis* fifth leg; (Y) *M. viridis* caudal rami dorsal view showing elongate terminal seta VI (arrow); (Z) *M. viridis* leg 4 third endopodal segment.

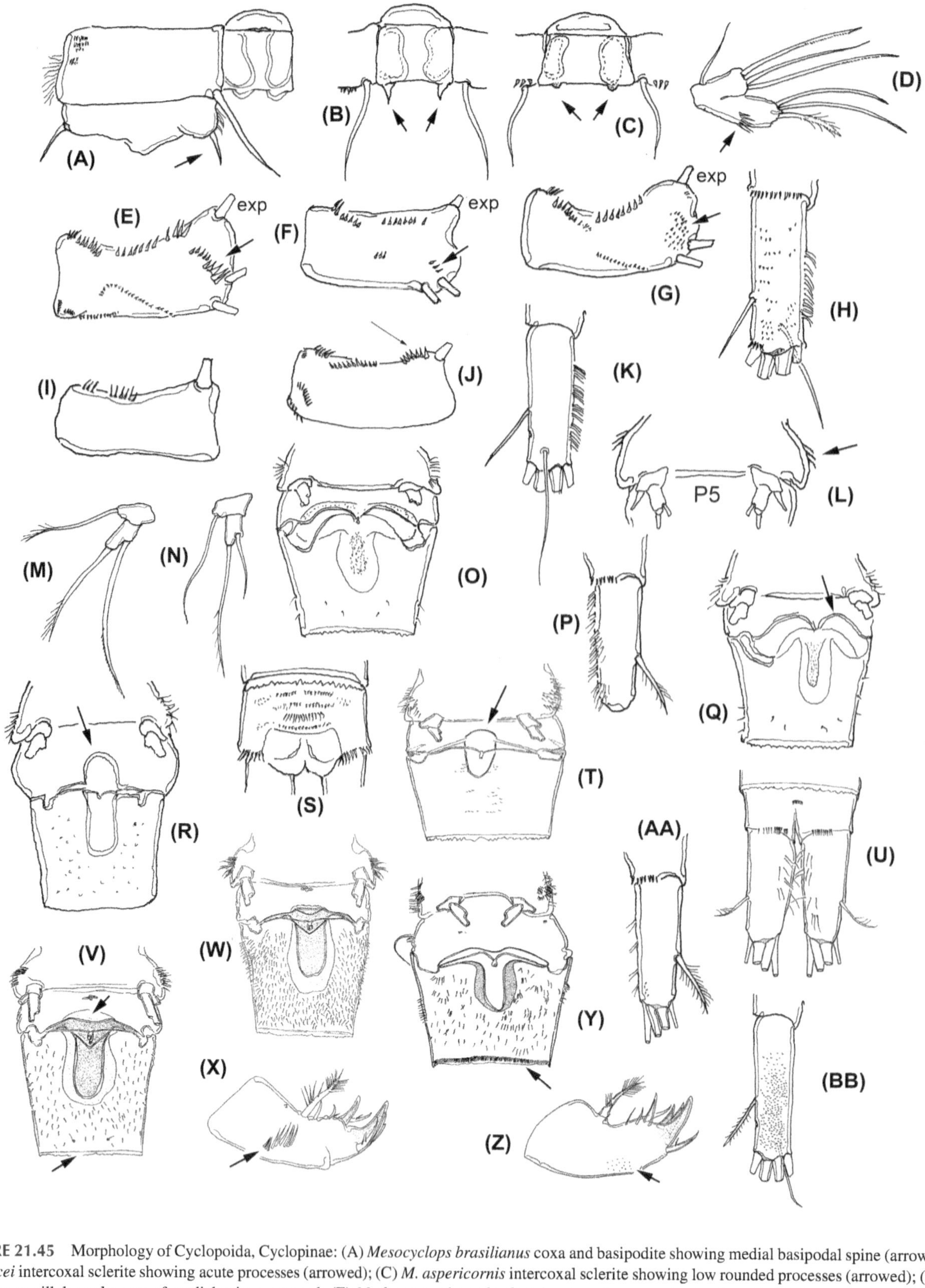

FIGURE 21.45 Morphology of Cyclopoida, Cyclopinae: (A) *Mesocyclops brasilianus* coxa and basipodite showing medial basipodal spine (arrow); (B) *M. pescei* intercoxal sclerite showing acute processes (arrowed); (C) *M. aspericornis* intercoxal sclerite showing low rounded processes (arrowed); (D) *M. ogunnus* maxillular palp, row of medial spinae arrowed; (E) *M. thermocyclopoides* female, second antenna basipodite; (F) *M. thermocyclopoides* male, second antenna basipodite; (G) *M. aspericornis* second antenna basipodite; (H) *M. aspericornis* caudal ramus ventral view; (I) *M. reidae* second antenna basipodite; (J) *M. longisetus* s.str. antenna basipodite; (K) *M. edax* caudal ramus dorsal view; (L) *M. reidae* fifth pedigerous somite showing ventrolateral spinae (arrow), ventral view; (M) *Mesocyclops chaci* leg 5; (N) *M. yutsil* fifth leg; (O) *M. longisetus* s.str. genital double-somite ventral view; (P) *M. longisetus* var. *curvatus* caudal ramus dorsal view; (Q) *M. longisetus* var. *curvatus* genital double-somite ventral view; (R) *M. ellipticus* fifth pedigerous and genital double-somite showing anterior margin of seminal receptacle (arrow), ventral view; (S) *M. ellipticus* anal somite dorsal view; (T) *M. intermedius* genital double somite showing anterior margin of seminal receptacle (arrow), ventral view; (U) *M. intermedius* caudal rami ventral view; (V) *M. brasilianus* fifth pedigerous and genital double-somite showing anterior margin of seminal receptacle (arrow), ventral view; (W) *M. meridianus* fifth pedigerous and genital double-somite ventral view; (X) *M. meridianus* maxillule showing setae (arrow) posterior; (Y) *M. meridionalis* fifth pedigerous and genital double-somite showing setae on posterior margin ventral view; (Z) *M. evadomingoi* maxillule showing scales (arrow) posterior; (AA) *M. pseudomeridianus* caudal ramus ventral view; (BB) *M. meridianus* caudal ramus ventral view.

5′ Caudal ramus pilose medially (Fig. 21.45 H); both sexes second antenna basipodite with dense group of tiny spinulae (Fig. 21.45 G, arrow) near insertion of setae opposite exopodal seta .... *Mesocyclops aspericornis* Daday, 1906
[Mexico, Caribbean Islands, Colombia, Brazil, Argentina. Afrotropica, Oriental. Palaearctic]

6(1) Leg 4 intercoxal sclerite with pointed projections 7

6′ Leg 4 intercoxal sclerite smooth 17

7(6) Second antenna basipodite exopodal seta insertion without spinulae (Fig. 21.45 I) 8

7′ Second antenna basipodite exopodal seta insertion with spinulae (Fig. 21.45 J, arrow) 11

8(7) Caudal rami smooth medially 9

8′ Caudal rami pilose (male and female) (Fig. 21.45 K) .... *Mesocyclops edax* (Forbes, 1891)
[Caribbean Islands, Nicaragua, Venezuela]

9(8) Fifth pedigerous somite ventrolateral surface smooth 10

9′ Fifth pedigerous somite with ventrolateral spinae (Fig. 21.45 L, arrow), caudal rami medially without setae .... *Mesocyclops reidae* Petkovski, 1986
[Mexico, Cuba, Caribbean Islands, Colombia]

10(9) Fifth leg apical article medial spine longer than lateral seta (male and female) (Fig. 21.45 M) .... *Mesocyclops chaci* Fiers, 1996
[Mexico: Yucatan]

10′ Fifth leg apical article medial spine shorter than lateral seta (Fig. 21.45 N) .... *Mesocyclops yutsil* Reid, 1996
[Mexico: Yucatan]

11(7) Leg 4 intercoxal sclerite with low, rounded projections 12

11′ Leg 4 intercoxal sclerite with high, acute projections 15

12(11) Seminal receptacle anterior margin concave, with wide lateral arms 13

12′ Seminal receptacle anterior margin convex, with thin lateral arms .... *Mesocyclops annulatus* Wierzejski, 1892
[Peru, Bolivia, Chile, Paraguay, Argentina, Uruguay]

13(12) Seminal receptacle lateral arms strongly curved (Fig. 21.45 Q, arrow); male antennule articles 15 and 16 medial surface with one or two spinule rows 14

13′ Seminal receptacle lateral arms not strongly curved (Fig. 21.45 O); male antennule article 16 medial surface with more than two spinule rows; caudal rami length/width ratio = 3.06 (female) and 2.93 (male); female caudal rami insertion of lateral most terminal seta (III) without adjacent spinulae; second antenna basipodite exopodal seta insertion with strong spinulae (Fig. 21.45 J) .... *Mesocyclops longisetus* s.str. (Thiébaud, 1912)
[Mexico, Guadeloupe, Guatemala, Venezuela, Colombia, Brazil, Bolivia, Paraguay, Argentina]

14(13) Leg 4 third endopodal article length/width ratio = 3.3–3.8 (female) and 3.14–3.46 (male); caudal rami length/width ratio = 3.79 (3.5–4.2) (female) and 3.27 (3.13–3.30) (male); female caudal rami insertion of lateral most terminal caudal seta (III) with spinulae .... *Mesocyclops araucanus* Löffler, 1961
[Chile]

14′ Female leg 4 third endopodal article length/width ratio = 2.6–3.2; female caudal rami length/width ratio = 2.6–3.2 (Fig. 21.45 P, Q) .... *Mesocyclops longisetus* var. *curvatus* Dussart, 1987
[Mexico, Guadeloupe, Honduras, Panama, Nearctic]

15(11) Seminal receptacle anterior margin weakly convex (Fig. 21.45 T, arrow); anal somite smooth dorsally 16

15′ Seminal receptacle anterior margin strongly convex (Fig. 21.45 R, arrow); anal somite dorsally with spinule rows and short setae (Fig. 21.45 S) .... *Mesocyclops ellipticus* Kiefer, 1936
[Haiti, Venezuela, Brazil]

16(15) Female anal somite posterior margin spinule row not dorsoventrally continuous; caudal ramus with few scattered pilose setae (Fig. 21.45 U) .... *Mesocyclops intermedius* Pesce, 1985
[Bonaire]

16′ Female anal somite posterior margin spinule row dorsoventrally continuous; caudal ramus hirsute .... *Mesocyclops paranaensis* Dussart & Frutos, 1986
[Argentina]

17(6) Female seminal receptacle anterior margin concave (Fig. 21.45 V, arrow) 18

17′ Female seminal receptacle anterior margin convex or straight (Fig. 21.45 W) 19

18(17) Female urosomites and genital double-somite posterior margins with row of setae (Fig. 21.45 Y, arrow) .... *Mesocyclops meridionalis* Dussart & Frutos, 1986
[Argentina]

18′ Female urosomites posterior margins without row of short setae (Fig. 21.45 V, arrow) .... *Mesocyclops brasilianus* Kiefer, 1933
[Venezuela, Nicaragua, Colombia, Brazil, Peru]

19(17) Maxillule praecoxal posterior surface with long setae (Fig. 21.45 X, arrow) 20

19′ Maxillule praecoxal posterior surface with small scales (male and female) (Fig. 21.45 Z, arrow) .... *Mesocyclops evadomingoi* Gutiérrez-Aguirre, 2001
[Mexico: Tabasco]

20(19) Female caudal rami medial margin pilose, ventrodistal surface with spinae (Fig. 21.45 AA) .... *Mesocyclops pseudomeridianus* Defaye & Dussart, 1989
[French Guiana]

20′ Female caudal ramus with medial margin smooth, ventral surface with scales (Fig. 21.45 BB) .... *Mesocyclops meridianus* Kiefer, 1926
[Bonaire, Venezuela, Colombia, Paraguay, Uruguay]

Cyclopoida: Cyclopidae: Cyclopinae: *Thermocyclops*: Species

Consult Reid (1989) and Silva & Matsumura-Tundisi (2005) for more details.

1 Caudal rami length > 3.5× width (Fig. 21.46 A) 2

1′ Caudal rami length 2.0–3.2× width 4

2(1) Caudal rami length 3.5–4.5× width 3

2′ Caudal rami length 5.0–5.1× width; leg 4 third endopodal article with medial spine length 2.5× lateral spine .... *Thermocyclops tenuis longifurcatus* Pesce, 1985
[Bonaire]

3(2) Leg 4 third endopodal article medial spine length 2.0–2.3× lateral spine (Fig. 21.46 B); dorsal caudal seta length 1.2× caudal ramus; leg 4 intercoxal sclerite with small rounded processes (Fig. 21.46 C, arrow) .... *Thermocyclops tenuis* (Marsh, 1909)
[Mexico, Cuba, Colombia, Brazil, Paraguay, Argentina. Nearctic: United States]

3′ Leg 4 third endopodal article medial spine as long as lateral spine (Fig. 21.46 N); dorsal caudal seta length 1.8–2.0× caudal ramus; fifth leg as in Fig. 21.46 M .... *Thermocyclops brehmi* (Kiefer, 1927)
[Venezuela, Argentina, Uruguay]

4(1) Leg 4 third endopodal article medial spine longer than or as long as lateral spine; intercoxal sclerite with or without projections 5

4′ Leg 4 third endopodal article medial spine length 0.6–0.8× lateral spine (Fig. 21.46 E, arrow); leg 4 intercoxal sclerite with subtriangular or rounded medial projections, unornamented; seminal receptacle with convex anterior margin (Fig. 21.46 D) .... *Thermocyclops inversus* (Kiefer, 1936)
[Mexico, Central America, Caribbean Islands, Brazil]

5(4) Leg 4 third endopodal article medial spine length > lateral spine; different combination of characters on leg 4 intercoxal sclerite than below 6

5′ Leg 4 third endopodal article medial spine length = lateral spine; leg 4 intercoxal sclerite without projections, only two small spinae on distal margin .... *Thermocyclops hastatus antillanus* Herbst, 1986
[Guadeloupe]

6(5) Leg 4 intercoxal sclerite with high rounded medial projections armed with distal row of spinulae or with two large spiny protrusions; leg 4 third endopodal article medial spine length 1.8–2.6× lateral spine 7

6′ Leg 4 intercoxal sclerite with smooth, low rounded projections; leg 4 third endopodal article medial spine length 3.5–5.1× lateral spine 8

7(6) Leg 4 third endopodal article medial spine length 1.8–2.3× lateral spine (Fig. 21.46 H); intercoxal sclerite with two protrusions ornamented with 3–4 spinae (Fig. 21.46 F, arrow); caudal rami length/width ratio ~2.0 (Fig. 21.46 G) .... *Thermocyclops crassus* (Fischer, 1853)
[Mexico, Costa Rica, Nearctic: United States]

7′ Leg 4 third endopodal article medial spine length 2.0–2.6× lateral spine (Fig. 21.46 P); intercoxal sclerite with two large, narrow protrusions ornamented with strong spiniform processes (Fig. 21.46 O); caudal rami length/width ratio ~2.5 .... *Thermocyclops decipiens* (Kiefer, 1929)
[Caribbean Islands, Costa Rica, Venezuela, Colombia, Brazil]

8(6) Leg 4 third endopodal article medial spine length 5.0× lateral spine (Fig. 21.46 J); seminal receptacle with medial depression on proximal margin (Fig. 21.46 I, arrow) .... *Thermocyclops minutus* (Lowndes, 1934)
[Venezuela, Brazil, Paraguay, Argentina]

8′ Leg 4 third endopodal article medial spine length 3.5× lateral spine (Fig. 21.46 L); seminal receptacle with convex anterior margin (Fig. 21.46 K, arrow) .... *Thermocyclops iguapensis* Silva & Matsumura-Tundisi, 2005
[Brazil]

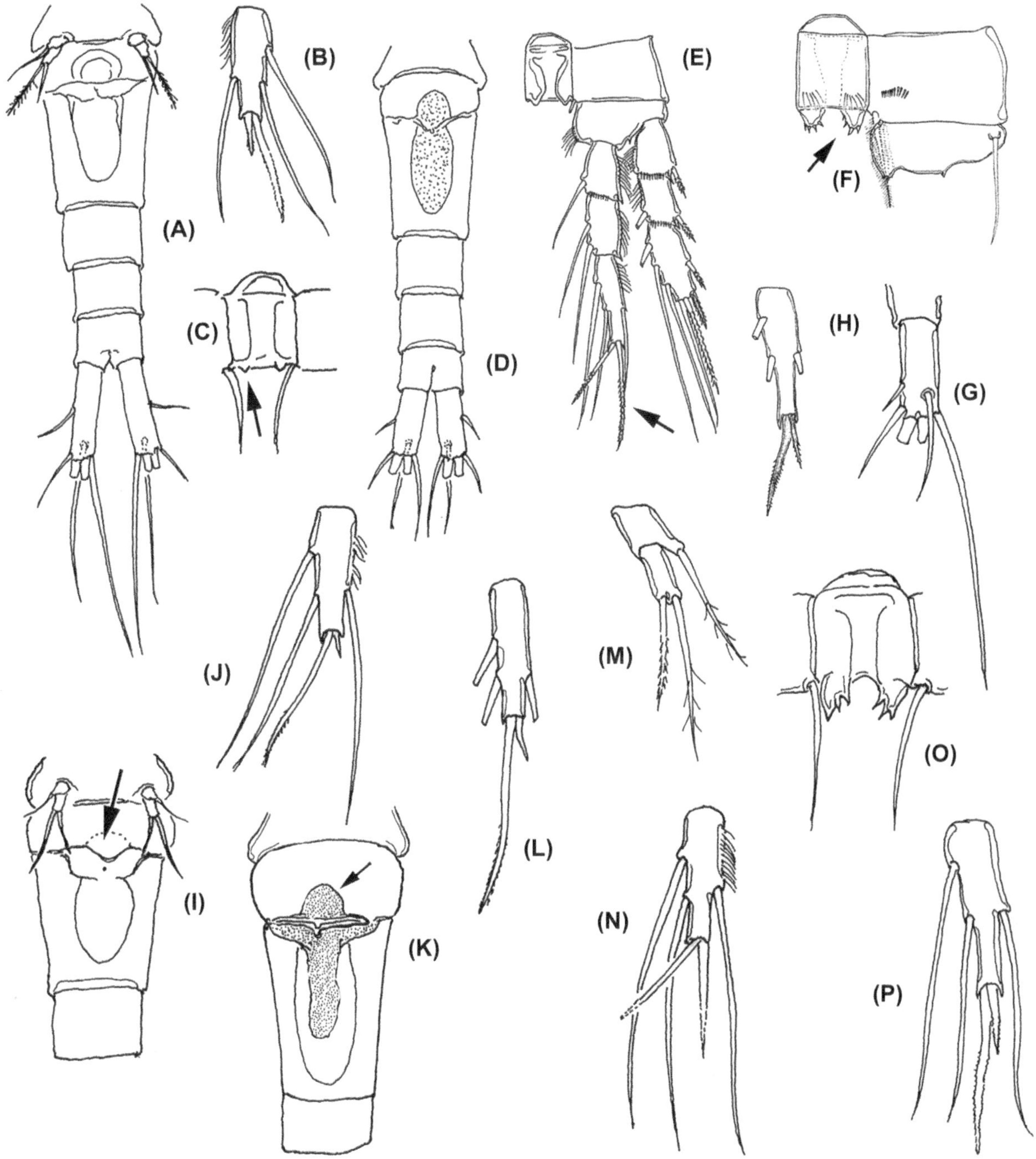

FIGURE 21.46 Morphology of Cyclopoida, Cyclopinae: (A) *Thermocyclops tenuis* urosome ventral view; (B) *T. tenuis* leg 4 third endopodal segment; (C) *T. tenuis* leg 4 intercoxal sclerite showing small rounded expansions (arrow); (D) *Thermocyclops inversus* urosome ventral view; (E) *T. inversus* leg 4 showing lateral spine on third endopodal segment (arrow); (F) *Thermocyclops crassus* leg 4 intercoxal sclerite, coxa, and basipodite showing spinous processes of intercoxal sclerite (arrow); (G) *T. crassus* caudal ramus dorsal view; (H) *T. crassus* leg 4 third endopodal segment; (I) *Thermocyclops minutus* urosome ventral view showing medially depressed anterior genital canal (arrow); (J) *T. minutus* leg 4 third endopodal segment; (K) *Thermocyclops iguapensis* genital double-somite ventral view showing convex anterior genital canal (arrow); (L) *T. iguapensis* leg 4 third endopodal segment; (M) *Thermocyclops brehmi* fifth leg; (N) *T. brehmi* leg 4 third endopodal segment; (O) *Thermocyclops decipiens* leg 4 intercoxal sclerite; (P) *T. decipiens* leg 4 third endopodal segment.

Cyclopoida: Cyclopidae: Cyclopinae: *Megacyclops*: Species

1 Leg 4 third endopodal article with setae not reaching distal margin of apical spinae (Fig. 21.44 Z); fifth leg terminal article apical spine barely reaching insertion of distal seta (Fig. 21.44 X); medial most terminal seta (VI) almost twice as long as caudal ramus (Fig. 21.44 Y, arrow) .... *Megacyclops viridis* (Jurine, 1820)
[Cosmopolitan. Mexico, Brazil, Uruguay, Argentina]

1′ Leg 4 third endopodal article with setae reaching distal margin of apical spinae (Fig. 21.44 W); fifth leg terminal article apical spine not reaching insertion of distal seta (Fig. 21.44 U); medial most terminal seta (VI) shorter than ramus (Fig. 21.44 V, arrow) .... *Megacyclops latipes* (Lowndes, 1927)
[Mexico]

Cyclopoida: Cyclopidae: Cyclopinae: *Diacyclops*: Species

Karanovic et al. (2014), Locascio de Mitrovich & Menu Marque (2001) are useful references for *Diacyclops* species.

1 Female antennules with 11 or 17 articles; legs 1–4 with 3-articulated rami; legs 1–4 exopodites spine formula: 2, 3, 3, 3; caudal rami length 2.5–4.8× width, medial margin smooth, pilose or with rows of spinulae on lateral margin; leg 4 intercoxal sclerite with low rounded projections or without projections 2

1′ Female antennules with 16 (sometimes 17) articles; legs 1–4 with 3-articulated rami; legs 1–4 spine formula: 2, 3, 3, 3; caudal rami length 2.6× width, smooth margins (Fig. 21.44 M); leg 4 intercoxal sclerite with high, sharp, triangular projections (Fig. 21.44 N) .... *Diacyclops bernardi* (Petkovski, 1986)
[Mexico, Colombia. Nearctic: Southern US]

2(1) Antennule with 11 or 12 articles 3

2′ Antennule with 17 articles 8

3(2) Caudal ramus length/width ratio = 3.3–4.8; fifth leg medial spine length < or as long as article 4

3′ Caudal ramus length/width ratio = 2.3–2.6; fifth leg medial spine length 2× article 7

4(3) Caudal ramus length/width ratio > 4.0; fifth leg medial spine length > 0.5× article 5

4′ Caudal ramus length/width ratio = 3.3; fifth leg medial spine length < 0.5× article (Fig. 21.44 S, T) .... *Diacyclops michaelseni* (Mrázek, 1901)
[Chile, Argentina] Note: *Diacyclops skottsbergi* (Lindberg, 1949), a presumed South American synonym, could be valid, but it is known from a single male only.

5(4) Caudal ramus medial and ventral surfaces ornamented with rows of slender spinae; caudal ramus length 4.7–4.8× width 6

5 Caudal ramus medial and ventral surfaces smooth; caudal ramus length = 4.0× width .... *Diacyclops joycei* Karanovic, Gibson, Hawes & Andersen, 2014
[Antarctica]

6(5) Dorsal caudal seta (VII) length 1.2× caudal ramus .... *Diacyclops walkeri* Karanovic, Gibson, Hawes & Andersen, 2014
[Antarctica]

6′ Dorsal caudal seta (VII) length 0.8× caudal ramus .... *Diacyclops kaupi* Karanovic, Gibson, Hawes & Andersen, 2014
[Antarctica]

7(3) Females total length = 0.8 mm; second antenna basipodite with dense spinulation pattern on caudal surface; fifth leg terminal article without spinules at insertion of medial spine (Fig. 21.44 K, L) .... *Diacyclops puuc* Fiers, 1996
[Mexico: Yucatan]

7′ Females total length < 0.7 mm; second antenna basipodite with reduced spinulation pattern caudal surface; fifth leg terminal article with spinules at insertion of medial spine (Fig. 21.44 I, J, arrow) .... *Diacyclops chakan* Fiers & Reid, 1996
[Mexico: Yucatan]

8(2) Caudal rami length 4× width, medial margin smooth or with rows of pilose setae 9

8′ Caudal rami length 3.3–3.7× width, medial margin with three or five rows of pilose setae 10

9(8) Caudal rami medial margin smooth; anal operculum with smooth rounded margin; fifth leg terminal article with subapical spine as long as or slightly longer than article (Fig. 21.44 P); leg 4 intercoxal sclerite with row of six spinulae on medial surface (Fig. 21.44 Q) .... *Diacyclops andinus* Locascio de Mitrovich & Menu-Marque, 2001
[Chile, Argentina]

9′ Caudal rami medial margin with four rows of pilose setae: three transverse, one longitudinal (Fig. 21.44 R); anal operculum from light to strongly projected, arched; leg 4 intercoxal sclerite with row of three to five spinules at base of each distal projection .... *Diacyclops ecabensis* Fiers, Ghenne & Suárez-Morales, 2000
[Mexico: Yucatan]

10(8) Pedigerous somites 2–4 posterior margins smooth; caudal rami medial margin with rows of hairy or spiniform setae; lateral margin without row of spinulae 11

10′ Pedigerous somites 2–4 posterior margins crenulated; caudal rami medial margin with three rows of hairy setae; lateral margin with row of spinulae on proximal third; anal operculum projected, subquadrate (Fig. 21.44 O) .... *Diacyclops pilosus* Fiers, Ghenne & Suárez-Morales, 2000
[Mexico: Yucatan. Brazil: Ceará]

11(10) Caudal rami medial margin with rows of slender short hairlike setae (Fig. 21.44 H); maxilliped second article with three rows of spinules (arrowed) along caudal surface (Fig. 21.44 F) .... *Diacyclops uruguayensis* (Kiefer, 1935)
[Brazil, Uruguay]

11′ Caudal rami medial margin with rows of short spinules; maxilliped second article with two rows of spinules along caudal surface .... *Diacyclops hispidus* Reid, 1988
[Colombia]

Cyclopoida: Cyclopidae: Cyclopinae: *Acanthocyclops*: Species

See also Reid (1985b), Einsle (1996), Reid & Suárez-Morales (1999), Fiers et al. (2000), Mirabdullayeb & Defaye (2002, 2004) and Mercado et al. (2006) for additional characters and illustrations.

1 Females with less than 17 antennulary articles (typically 11–12) 2

1′ Females always with 17 antennulary articles 3

2(1) Leg 4 third endopodal article lateral spine length 2× medial spine; legs 1–4 exopodites spine formula: 2, 3, 3, 3; legs 3 and 4 third endopodal article subdistal setae unmodified .... *Acanthocyclops exilis* (Coker, 1934)
[Mexico: Central] Note: Possibly refers to *Acanthocyclops dodsoni* Mercado-Salas & Suárez-Morales, 2006.

2′ Leg 4 third endopodal article lateral and medial spinae subequal in length; legs 1–4 exopodites spine formula: 3, 4, 4, 4; legs 3 and 4 third endopodal article subdistal setae modified, serrate, or with marginal teeth (Fig. 21.47 A, arrows, B) .... *Acanthocyclops dodsoni* Mercado-Salas & Suárez-Morales, 2006
[Mexico]

3(1) Caudal rami length > 3.5× (and up to 6×) width; dorsal caudal seta (VII) shorter than caudal ramus; legs 3 and 4 endopodal articles subdistal setae unmodified 4

3′ Caudal rami length ~ 3× width; dorsal caudal seta (VII) longer than the half-length or as long as the caudal ramus; medial surface smooth, or hairy 6

4(3) Caudal rami length/width ratio > 4.9; female caudal rami dorsal and ventral surfaces smooth; dorsal caudal seta (VII) length ~ 0.3–0.5x ramus 5

4′ Caudal rami length/width ratio = 4.3–4.9; female caudal rami dorsal and ventral surfaces uniformly covered with tiny spinulae; dorsal caudal seta (VII) length 0.5× ramus (Fig. 21.47 C, arrow) .... *Acanthocyclops eduardoi* Mercado-Salas & Álvarez-Silva, 2013
[Mexico]

5(4) Genital double-somite proximal lateral margin with chitinized angular process (arrows, Fig. 21.47 D); medial most terminal seta (VI) length ~ 0.5× caudal ramus; second antenna basipodite without group of spinulae near insertion of exopodal seta; caudal rami smooth .... *Acanthocyclops vernalis* (Fischer, 1853)
[Neotropical records of this species could represent undescribed forms of this species complex]

5′ Genital double-somite proximal outer margin rounded, not angled (Fig. 21.47 E); medial most terminal seta (VI) length > 0.5× caudal ramus (Fig. 21.47 F); second antenna basipodite with group of spinulae near insertion of exopodal seta (Fig. 21.47 G, arrow) .... *Acanthocyclops robustus* (G.O. Sars, 1863) s.str.
[Neotropical records of this nominal species could refer to undescribed forms of this complex: see Mirabdullayev & Defaye (2002, 2004)]

6(4) Cauda rami medial surface smooth 7

6′ Caudal rami medial surface hirsute 8

7(6) Second antenna basipodite frontal and caudal surfaces with rows of short, thin spinulae (Fig. 21.47 H); caudal ramus medial median terminal seta (V) thickened, stout; dorsal caudal seta length ≈ ramus (Fig. 21.47 I, arrow); legs 3 and 4 third endopodal articles with modified setae; legs 1–4 exopodites spine formula: 3, 4, 4, 4; fifth leg free article apical spine barely reaching distal margin of article (Fig. 21.47 H–K) .... *Acanthocyclops smithae* Reid & Suárez-Morales, 1999
[Cuba? Honduras, Mexico]

7′ Second antenna basipodite frontal and caudal surfaces with rows of thick spinae (Fig. 21.47 L, arrows); caudal ramus medial median terminal seta (V) not particularly thick or stout (Fig. 21.47 M, N); dorsal caudal seta (VII) length < ramus (Fig. 21.47 O arrow); legs 3 and 4 third endopodal articles with unmodified setae; legs 1–4 exopodites spine formula: 2, 3, 3, 3 .... *Acanthocyclops rebecae* Fiers, Ghenne & Suárez-Morales, 2000
[Mexico: Yucatan]

8(6) Caudal rami length/width ratio = 2.8; caudal rami dorsal surface with irregular patches of spinulae ventral surface with symmetrical patches; medial margin with short hairlike setulae; dorsal caudal seta (VII) length 0.63× ramus (Fig. 21.47 P); legs 1–4 exopodites spine formula: 3, 4, 4, 3 .... *Acanthocyclops marceloi* Mercado-Salas & Suárez-Morales, 2009
[Mexico]

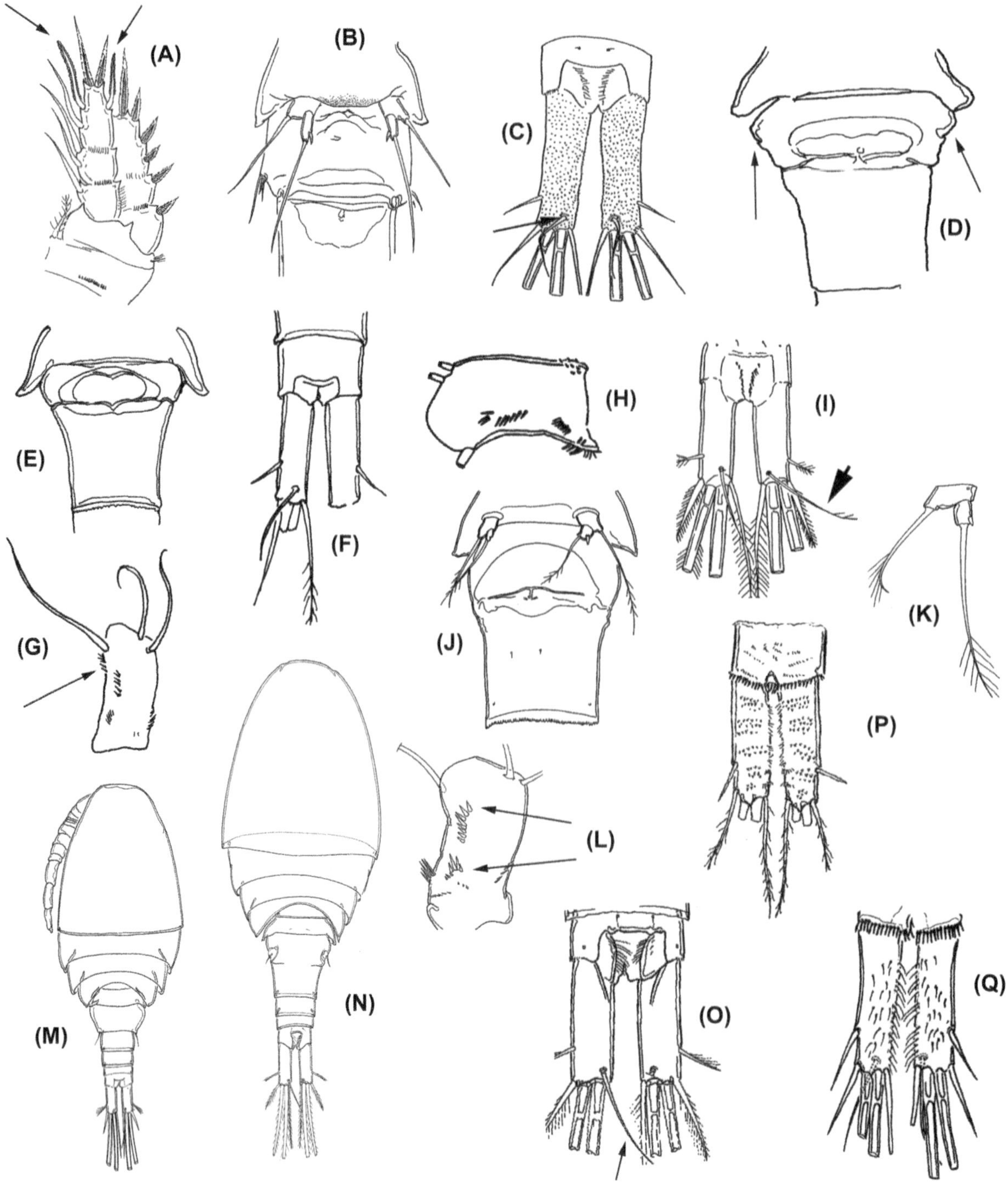

FIGURE 21.47 Morphology of Cyclopoida, Cyclopinae: (A) *Acanthocyclops dodsoni* leg 4 showing modified setae in third exopodal segment (arrows); (B) *A. dodsoni* fifth pedigerous and genital double-somite ventral view; (C) *Acanthocyclops eduardoi* caudal rami dorsal view showing dorsal caudal seta (arrow); (D) *Acanthocyclops vernalis* fifth pedigerous and genital double-somite ventral view showing angular, anterolateral margins (arrow); (E) *Acanthocyclops robustus* fifth pedigerous and genital double-somite ventral view; (F) *A. robustus* anal somite and caudal rami dorsal view; (G) *A. robustus* second antenna basipodite showing group of spinulae near insertion of exopodal seta (arrow); (H) *Acanthocyclops smithae* second antenna basipodite; (I) *A. smithae* caudal rami dorsal view showing caudal seta VII (arrow); (J) *A. smithae* fifth pedigerous and genital double-somite ventral view; (K) *A. smithae* fifth leg; (L) *Acanthocyclops rebecae* second antenna basipodite showing rows of thick spinae (arrows); (M) *A. rebecae* male habitus dorsal view; (N) *A. rebecae* female habitus dorsal view; (O) *A. rebecae* anal somite and caudal rami dorsal view showing caudal seta VII (arrow); (P) *Acanthocyclops marceloi* anal somite and caudal rami ventral view; (Q) *Acanthocyclops caesariatus* anal somite and caudal rami ventral view.

8′ Caudal rami length/width ratio = 3.8; caudal rami medial margin, dorsal and ventral surfaces with scattered hairlike elements; caudal rami dorsal caudal seta (VII) length 0.75× ramus; legs 1–4 exopodites spine formula: 3, 4, 4, 4 (Fig. 21.47 Q) .... *Acanthocyclops caesariatus* Mercado-Salas & Suárez-Morales, 2009 [Mexico]

Cyclopidae: Eucyclopinae

Eduardo Suárez-Morales

El Colegio de la Frontera Sur, Chetumal, Quintana Roo, México

Nancy F. Mercado-Salas

Centrum für Naturkunde—CeNak/Center of Natural History Hamburg, Universität Hamburg—Zoological Museum, Martin-Luther-King-Platz 3, 20146 Hamburg, Germany

Senckenberg am Meer Wilhelmshaven, German Centre for Marine Biodiversity Research, Südstrand 44, 26382, Wilhelmshaven

INTRODUCTION

The cyclopoid copepod subfamily Eucyclopinae is roughly defined by the presence of three elements (setae and/or spinae) on the P5 distal segment. In recent years, additional characters distinguishing the Eucyclopinae from other cyclopid subfamilies have strengthened the distinction of this group (Reid, 1985b; Dussart & Defaye, 2006; Mercado-Salas et al., 2016). The works by Karaytug & Boxshall (1998, 1999), Alekseev et al. (2006), Karanovic (2006), Karanovic & Tang (2009), Alekseev & Defaye (2011), and Mercado-Salas et al. (2016) have contributed to a clearer definition of some eucyclopine genera. Currently, the subfamily contains 185 species belonging to 11 genera (*Macrocyclops, Eucyclops, Paracyclops, Ectocyclops, Tropocyclops, Afrocyclops, Thaumasiocyclops, Ochridacyclops, Homocyclops, Australoeucyclops*, and *Austriocyclops*) distributed worldwide.

Many revisions and new species descriptions of Neotropical Eucyclopinae have been assessed during the past four decades (Reid, 1985b, Karaytug & Boxshall, 1998; Mercado-Salas & Suárez-Morales, 2009, 2014a, 2014b; Mercado-Salas et al. 2012, 2016). Among them are the benchmark key to the South American freshwater cyclopids made by Reid (1985b), the revision of the genus *Paracyclops* by Karaytug & Boxshall (1998) and the works on the speciose genus *Eucyclops* made by Mercado-Salas at al. (2012, 2016).

Overall, the keys to the species of Eucyclopinae provided in this document combine the evaluation of the traditionally used characters with those that have recently explored to separate closely related species, especially in the genera *Eucyclops* and *Paracyclops*. These microcharacters include the ornamentation of the second antenna (caudal and frontal surfaces), which involves a special nomenclature of each row by its position, from N1-N22 where N stands for the spinulae—setulae group number (Fig. 21.49 A, B) and the spinule rows on the intercoxal sclerites of P1-P4 (caudal surface) (Fig. 21.49 C—F). The nomenclature of the ornamentation of second antenna and P1-P4 follows Alekseev et al. (2006), Alekseev & Defaye (2012), and Mercado-Salas et al. (2016). It is expected that this new approach will allow an efficient identification process, leading to reliable identifications of Eucyclopinae, a genera of freshwater free-living cyclopoid copepods whose taxonomy is complex and highly dynamic, particularly in the Neotropical region.

KEY TO EUCYCLOPINAE

Cyclopoida: Cyclopidae: Eucyclopinae: Genera

1 P5 with one or two articles, distinct from somite, distal segment with three elements; caudal ramus length at least 2× width 2

1′ P5 represented by a narrow plate, completely fused to fifth thoracic somite, with two strong inner spinae and one outer seta; body flattened dorsoventrally, caudal ramus length ~1.8–2.0× width and ornamented with diagonal rows of spinae .... ***Ectocyclops* [p. 770]**

2(1) P5 with one article in males and females; apical article bearing three elements 3

2′ P5 with two articles in males and females; apical article bearing three elements (Fig. 21.48 E–F); body robust (Fig. 21.52 A) .... ***Macrocyclops* [p. 770]**

3(2) First antenna ♀ with 12 or less articles 4

3′ First antenna ♀ with 17 articles; body and urosome robust (Fig. 21.48 K) .... *Homocyclops ater* (Herrick, 1882) [Widespread in the Neotropics, from Mexico to Brazil]

4(3) Caudal ramus ♀ without longitudinal row of spinulae on outer margin 5

4′ Caudal ramus ♀ with longitudinal row of spinulae on outer margin, inner margin usually smooth (Fig. 21.49 G–J) .... ***Eucyclops* [p. 770]**

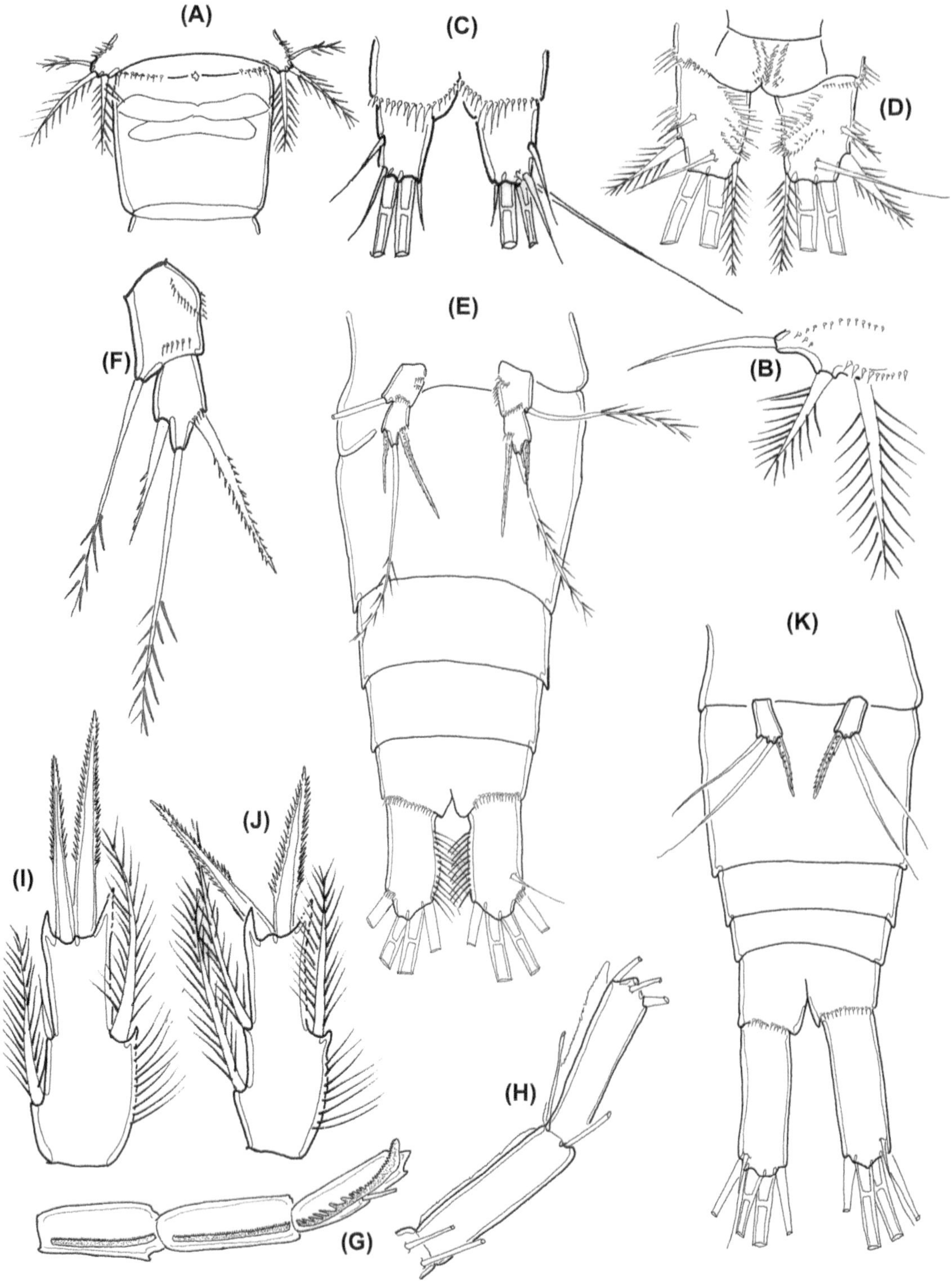

FIGURE 21.48 Representative structures of Eucyclopinae copepods: (A) *Ectocyclops herbsti*, genital somite; (B) *Ectocyclops herbsti*, P5; (C) *Ectocyclops strenskei*, caudal rami; (D) *E. herbsti*, caudal rami; (E) *Macrocyclops fuscus*, urosome; (F) *Macrocyclops albidus* s.str.; (G) *M. fuscus*, A1 (first antenna)—segments 15–17; (H) *Macrocyclops albidus* s.str.—segments 16–17; (I) *M albidus* s.str., P4 enp3 (endopodite 3); (J) *Macrocyclops albidus principalis*, P4 endopodite 3; (K) *Homocyclops ater*, urosome. *Figures (A), (B), (D), (F), (H–J) Redrawn from Dussart (1984a); (C) From Reid (1985b).*

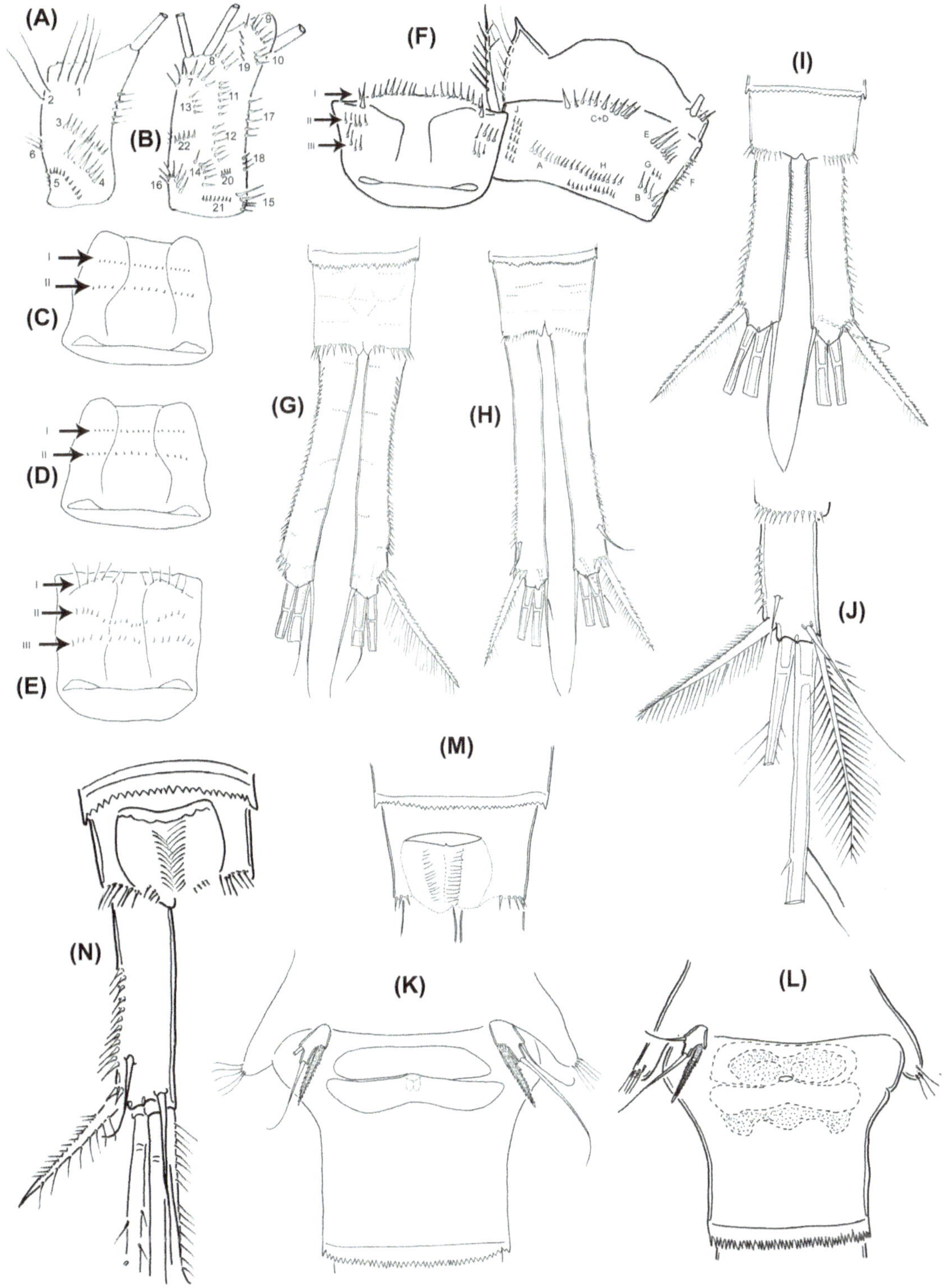

FIGURE 21.49 Representative structures of Eucyclopinae copepods. (A) *Eucyclops* A2 (second antenna) basis, frontal; (B) *Eucyclops* A2 basis, caudal; (C) *Eucyclops* P1 intercoxal sclerite, caudal; (D) *Eucyclops* P2 intercoxal sclerite, caudal; (E) *Eucyclops* P3 intercoxal sclerite, caudal; (F) *Eucyclops* P4 coxa and intercoxal sclerite, caudal; (G) *Eucyclops mittmanni,* caudal rami; (H) *Eucyclops neumani,* caudal rami; (I) *Eucyclops subciliatus,* caudal rami; (J) *Eucyclops siolii,* caudal rami; (K) *Eucyclops defayeae,* genital double-somite and seminal receptacle; (L) *Eucyclops angeli,* genital somite and seminal receptacle; (M) *Eucyclops torresphilipi,* anal operculum; (N) *Eucyclops tziscao,* anal operculum. *(J) Redrawn from Herbst (1962).*

5(4) Caudal ramus with row of spinulae transverse or oblique on dorsal surface at base of lateral caudal seta (II) (Figs. 21.51 A–D, D); first antenna ♀ last two articles no more than 2.0× longer than wide; first antenna ♂ first article usually with modified seta (Fig. 21.50 K) .. .. ***Paracyclops* [p. 778]**

5′ Caudal ramus without row of spinulae on dorsal surface at base of lateral caudal seta (II); first antenna ♀ last two articles more than 3.0× longer than wide; first antenna ♂ first article without modified seta modified seta; body small and slender (Fig. 21.52 D) .... ***Tropocyclops* [p. 779]**

Cyclopoida: Cyclopidae: Eucyclopinae: *Ectocyclops*: Species

1 Caudal ramus inner margin smooth (Fig. 21.48 C) 2

1′ Caudal ramus inner margin heavily ornamented with rows of spinulae/setulae on dorsal and ventral surface (Fig. 21.48 D) 3

2(1) Caudal ramus ♂ short, length 1.0× width; outermost caudal seta (III) as long as innermost caudal seta (VI) .... *Ectocyclops bromeliacola* Kiefer, 1935
[Brazil, Venezuela]. Female unknown

2′ Caudal ramus in both, ♂♀, length at least 1.6× width; outermost caudal seta (III) 2.0 times longer than innermost caudal seta (VI) (Fig. 21.48 C); first antenna ♀ 8-segmented .... *Ectocyclops strenskei* Herbst, 1959
[Brazil, species found in bromeliads]

3(1) Caudal ramus ♀relatively short, length ~ 1.3–1.5× width, dorsal caudal seta (VII) length > than ramus (Fig. 21.48 D) 4

3′ Caudal ramus length 2× width; insertion of outer and inner setae of leg 5 with patches of spinulae, innermost spiniform seta as long as or slightly longer than middle spiniform seta .... *Ectocyclops phaleratus* (Koch, 1838)
[Mexico, Cuba, Haiti, Hispaniola, Jamaica, Guatemala, Costa Rica, Panama, Venezuela, Colombia, Brazil, Argentina]

4(3) Dorsal caudal seta length 1.5× as ramus Fig. 21.48 B; P5 ♀ with three elements not reaching posterior margin of genital double somite (Fig. 21.48 A); P6 ♂ elements equally long .... *Ectocyclops herbsti* Dussart, 1984
[Venezuela, French Guiana, Brazil, Argentina] Note: possible synonym: *Ectocyclops hirsutus* Kiefer, 1930 (Dussart & Defaye, 2006).

4′ Dorsal caudal seta length 1.1–1.2× as ramus, P5 ♀ innermost seta slender, ~ 1.5× longer than medial seta; first antenna ♂ 14-segmented .... *Ectocyclops rubescens* Brady, 1904
[Mexico, Guatemala, Guadeloupe, Venezuela, Colombia, Brazil, Paraguay]

Cyclopoida: Cyclopidae: Eucyclopinae: *Macrocyclops*: Species

1 Inner margin of caudal rami smooth (Fig. 21.52 C); first antenna last article with finely serrated hyaline membrane (Figs. 21.48 H, 52 B) 2

1′ Inner margin of caudal rami thickly ornamented with setulae (Fig. 21.48 E); first antenna last article with coarsely serrate hyaline membrane on proximal half only (Fig. 21.48 G) .... *Macrocyclops fuscus* (Jurine, 1820)
[Brazil, Florida, Guatemala, Mexico]

2(1) P4 endopodite 3 with inner seta remarkably short, not reaching the apical margin of article (Fig. 21.48 I), P5 as in Fig. 21.48 F, first antenna hyaline membrane smooth or finely serrated (Figs. 21.48 H, 52 B) .... *Macrocyclops albidus* (Jurine, 1820) s.str.
[Florida, Mexico, Venezuela, Bahamas, Cuba, Guatemala, Haiti, Honduras, Nicaragua, Panama, Puerto Rico, Salvador, Bonaire, Cayman Islands, Jamaica, Turks and Caicos Islands, Costa Rica, Ecuador, Colombia, Brazil, Chile, Paraguay, Argentina, Uruguay]

2′ P4 endopodite 3 with inner seta as long as proximal seta, reaching beyond distal margin of article (Fig. 21.48 J) .... *Macrocyclops albidus principalis* Herbst, 1962
[Venezuela, Colombia, Brazil]

Cyclopoida: Cyclopidae: Eucyclopinae: *Eucyclops*: Species

1 Caudal ramus more than 5.8× longer than wide 2

1′ Caudal ramus less than 5.8× longer than wide 6

2(1) Caudal ramus with row of spinulae along outer margin (Fig. 21.49 G) 3

2′ Caudal ramus with row of spinulae only near insertion of lateral caudal seta II (Fig. 21.49 H) 4

3(2) Frontal surface of second antenna basis with N1 and N2 fused (Fig. 21.53 B, caudal surface not bearing N19 (Fig. 21.53 B); anal operculum slightly serrated (Fig. 21.53 E); body as in Fig. 21.53 A .... *Eucyclops elegans* (Herrick, 1884)
[Florida, Cuba, Mexico, Brazil, French Guiana, Argentina]

This species shows variations between North and South American populations, each eprobably representing a different taxon (Mercado-Salas et al., 2016). The South American specimens could belong to the synonymized nominal species *Eucyclops solitarius* described by Herbst (1959) from Brazil. Records of *Eucyclops speratus* in the Americas probably represent either *Eucyclops elegans* or *Eucyclops mittmanni*.

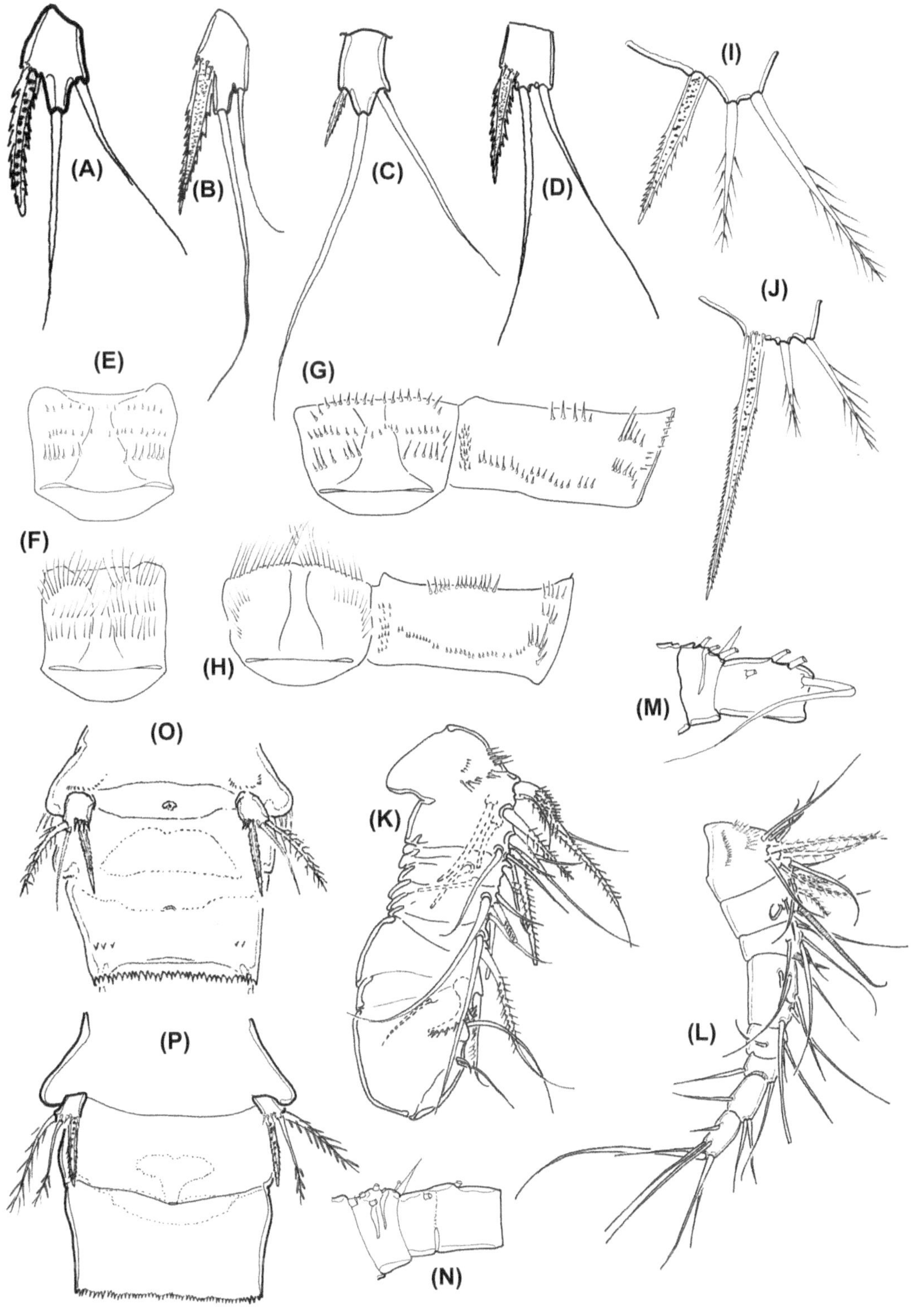

FIGURE 21.50 Representative structures of Eucyclopinae copepods: (A) *Eucyclops silvestri,* P5; (B) *Eucyclops defayeae,* P5; (C) *Eucyclops siolii* P5; (D) *Eucyclops cuatrocienegas,* P5; (E) *E. cuatrocienegas* P3 intercoxal sclerite, caudal; (F) *Eucyclops estherae* P3 intercoxal sclerite, caudal; (G) *E. cuatrocienegas* P4 coxa and intercoxal sclerite, caudal; (H) *E. estherae* P4coxa and intercoxal sclerite, caudal; (I) *Eucyclops bondi* P6, male; (J) *Eucyclops alticola* P6, male; (K) *Paracyclops hirsutus* A1 (first antenna), male; (L) *Paracyclops poppei* A1, female; (M) *Paracyclops bromelicola* A1—segments 3–4, female; (N) *Paracyclops punctatus* A1—segments 3–4, female; (O) *P. hirsutus* genital somite and seminal receptacle; (P) *Paracyclops hardingi* somite genital and seminal receptacle. *Redrawn: (A) from Kiefer (1936a); (C) from Herbst (1962); (I) from Kiefer (1957); (M), (N) and (P) from Karaytug (1999).*

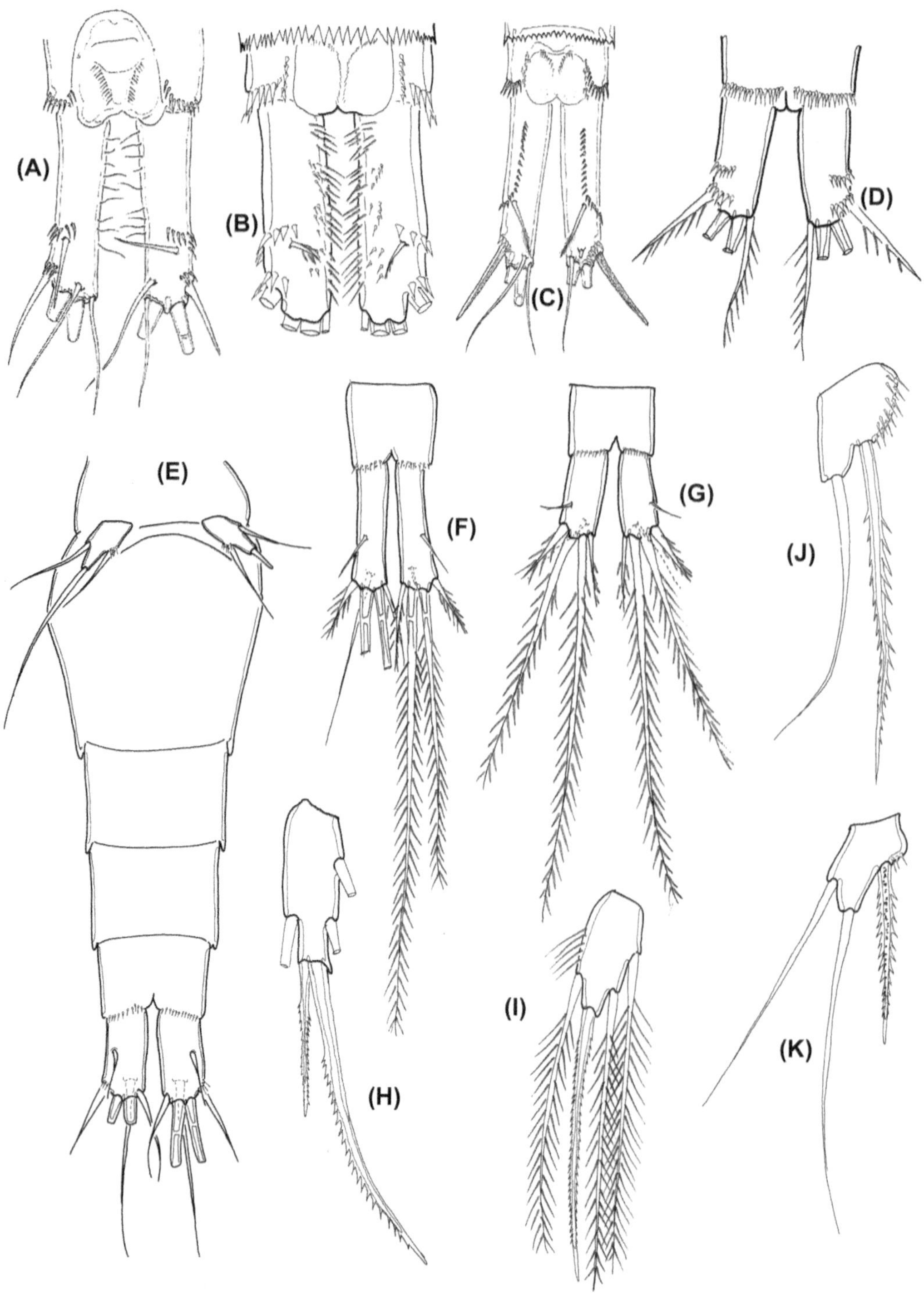

FIGURE 21.51 Representative structures of Eucyclopinae copepods: (A) *Paracyclops hirsutus* caudal rami; (B) *Paracyclops carectum* caudal rami; (C) *Paracyclops poppei* caudal rami; (D) *Paracyclops hardingi* caudal rami; (E) *Tropocyclops prasinus mexicanus*, urosome; (F) *Tropocyclops prasinus longispina*, caudal rami; (G) *T. prasinus* s.str., caudal rami; (H) *Tropocyclops schubarti* P4 enp3 (endopodite 3); (I) *Tropocyclops nananae* P4 enp3; (J) *Tropocyclops jamaicensis* P5; (K) *Tropocyclops prasinus altoandinus* P5. *Redrawn: (B) from Karaytug (1999); (D) from Reid (1985b); (F) and (G) from Kiefer (1931); (H) and (I) from Reid (1991a); (J) from Reid & Janetzki (1996); (K) from Gaviria (1994).*

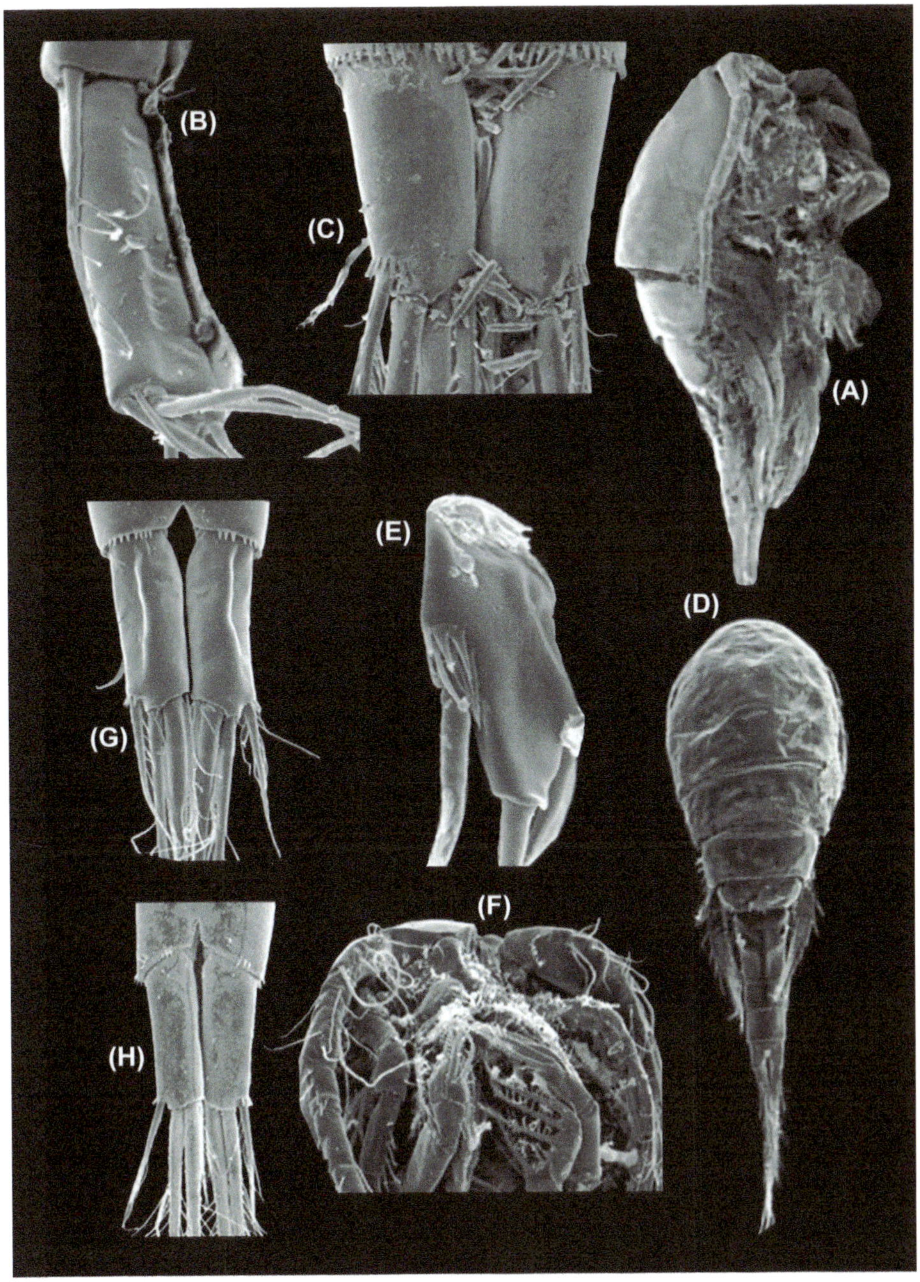

FIGURE 21.52 Representative structures of Eucyclopinae copepods: (A) *Macrocyclops albidus* s.str. habitus, lateral; (B) *M. albidus* s.str. A1 (first antenna)—segment 17; (C) *M. albidus* s.str. caudal rami; (D) *Tropocyclops prasinus aztequei* habitus, dorsal; (E) *T. prasinus* s.str. P5; (F) *T. prasinus aztequei* mouth parts; (G) *T. prasinus* s.str. caudal rami; (H) *T. prasinus aztequei,* caudal rami.

3′ Frontal surface of second antenna basis with N1 and N2 clearly separated, caudal surface bearing N19 (Fig. 21.54 G); anal operculum rounded and smooth; body as in Fig. 21.54 F .... *Eucyclops mittmanni* Mercado-Salas & Suárez-Morales, 2016 [Mexico]

4(1) P5 with long inner spine, at least 2.5× as long as the total length of segment 5

4′ P5 with a remarkably short inner spine, as long as/or shorter than total length of segment .... *Eucyclops macrurus* (G.O. Sars, 1863) [Mexico]

5(4) Length: width caudal rami between 7.0–8.0× (Fig. 21.49 H), first antenna ornamented with rows of small pits in segments 1–8; A2 frontal surface of basis with N1 bearing spinulae; anal operculum rounded but with small medial gap (Fig. 21.49 M); P5 medial seta as long or slightly shorter than inner spine (0.8–1.0) .... *Eucyclops neumani* (Pesta, 1927)

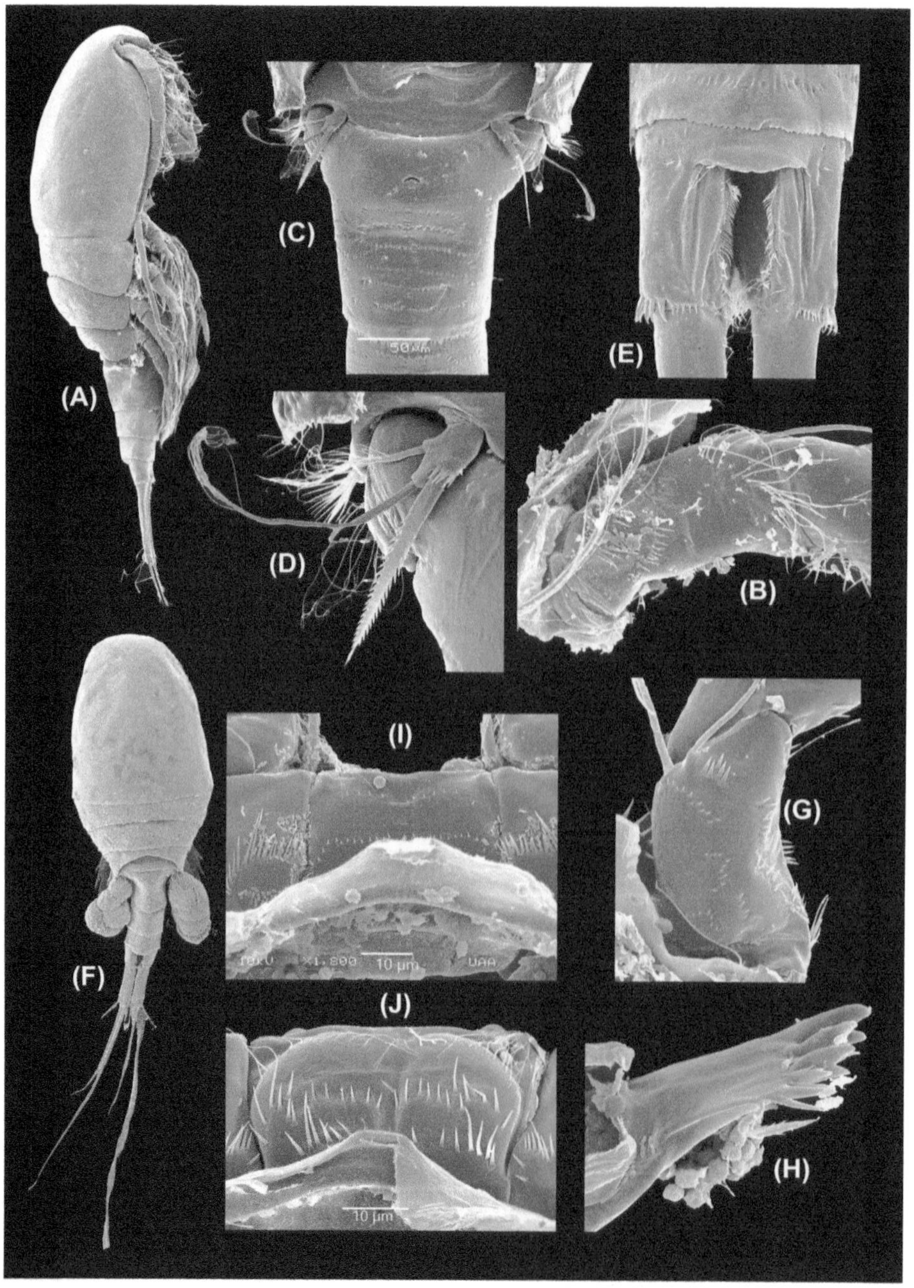

FIGURE 21.53 Representative structures of Eucyclopinae copepods: (A) *Eucyclops elegans* habitus, lateral; (B) *E. elegans* A2 (second antenna) basis, caudal; (C) *E. elegans* genital somite; (D) *E. elegans* P5; (E) *E. elegans* anal operculum; (F) *Eucyclops prionophorus* habitus, dorsal; (G) *E. prionophorus* A2 basis, caudal; (H) *E. prionophorus*, Md; (I) *E. prionophorus* P1 intercoxal sclerite, caudal; (J) *E. prionophorus* P3 intercoxal sclerite, caudal.

[Venezuela, Brazil, Peru, Bolivia, Chile, Argentina, Uruguay]

5′ Length: width caudal rami between 5.8–6.4×; first antenna without particular ornamentation; second antenna frontal surface of basis with N1 bearing setulae; anal operculum rounded and smooth; P5 with medial seta at least 1.3× longer than inner spine .... *Eucyclops titicacae* Kiefer, 1957
[Venezuela, Colombia, Peru]

6(1) Caudal ramus length less than 2.5× width 7

6′ Caudal ramus length between 3.1 and 5.5× width 9

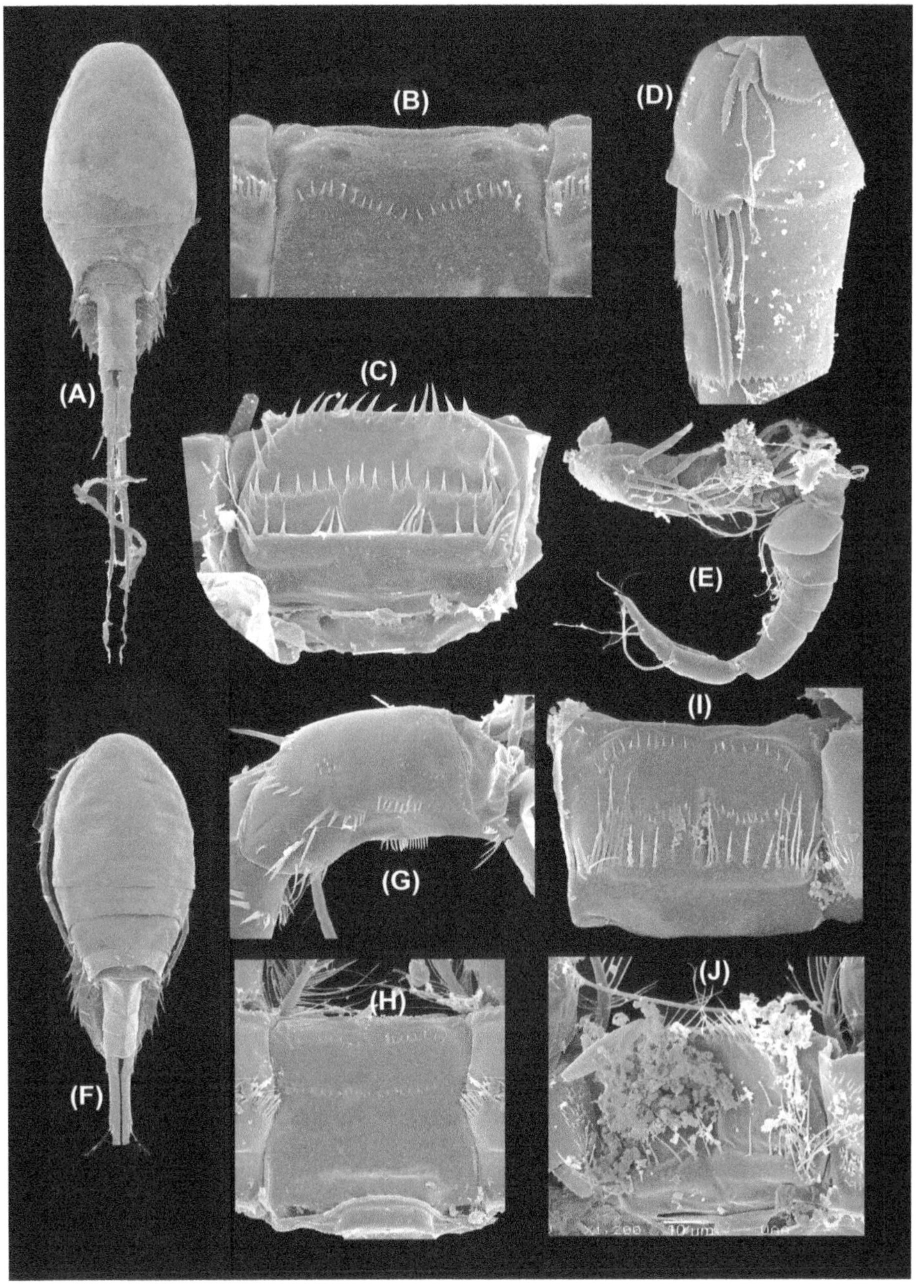

FIGURE 21.54 Representative structures of Eucyclopinae copepods. (A) *Eucyclops defayeae* habitus, dorsal; (B) *E. defayeae* P1 intercoxal sclerite, caudal; (C) *E. defayeae* P4 intercoxal sclerite, caudal; (D) *E. defayeae* P5 and P6, male; (E) *E. defayeae* A1 (first antenna), male; (F) *Eucyclops mittmanni* habitus, dorsal; (G) *E. mittmanni* A2 basis, caudal; (H) *E. mittmanni* P1 intercoxal sclerite, caudal; (I) *E. mittmanni* P3 intercoxal sclerite, caudal; (J) *E. mittmanni* P4 intercoxal sclerite, caudal.

7(6) P5 with long inner spine, at least 2.0× longer as segment (Fig. 21.49 L); dorsal caudal seta as long as or slightly shorter than caudal ramus length (0.8–1.1×) (Fig. 21.49 N) 8

7′ P5 with short inner spine, less than 0.8× as long as segment (Fig. 21.50 C); dorsal caudal seta longer than caudal ramus length (~1.3×) (Fig. 21.49 J) .... *Eucyclops siolii* Herbst, 1962
[Brazil]

8(7) P4 endopodite 3 being 1.8–2.0× longer than wide; P4 intercoxal sclerite with row I bearing small and slender spinulae; P5 outer seta as long as/or slightly longer than inner spine. .... *Eucyclops breviramatus* Löffler, 1963
[Ecuador]

8′ P4 endopodite 3 being 1.4–1.5× longer than wide; P4 intercoxal sclerite with row I bearing long and strongly chitinized spinulae; P5 outer seta short, ~0.5–0.7× as long as inner spine (Fig. 21.49 L) .... *Eucyclops angeli* Gutiérrez-Aguirre & Cervantes-Martínez, 2013 [Mexico]

9(6) Caudal ramus with row of spinulae extending not more than half of outer margin (Fig. 21.49 N) 10

9′ Caudal ramus with row of spinulae extending along more than half or outer margin (~60%––80%) 13

10(9) Caudal ramus inner margin ornamented with small setulae/spinulae (Fig. 21.49 I) 11

10′ Caudal ramus with inner margin naked (Fig. 21.49 N) 12

11(10) Caudal ramus ornamented along entire inner margin (Fig. 21.49 I); P4 endopodite 3 being 1.8× longer than wide; P5 with inner spine 0.6× as long as medial and outer setae .... *Eucyclops subciliatus* Dussart, 1984 [Brazil, Argentina]

11′ Caudal ramus inner margin ornamented only on proximal 1/5; P4 endopodite 3 being 2.0–2.2× longer than wide; P5 with inner spine slightly shorter than medial and outer setae, ~ 0.9× as long as other setae .... *Eucyclops neotropicus* Kiefer, 1936 [Venezuela, Brazil]

12(10) First antenna short, not reaching posterior margin of cephalothorax; P4 endopodite 3 being 1.6× longer than wide; P5 outer seta slightly longer than inner spine (1.1) and longer than medial seta (~1.4×); caudal ramus outer margin with row of 4–6 spinulae distributed near insertion of lateral caudal seta II; P6 ♂ with long strong inner spine, ~ 3× longer than medial seta and 2.2× longer than outer seta (Fig. 21.50 J) .... *Eucyclops alticola* Kiefer, 1957 [Colombia]

12′ First antenna reaching beyond posterior margin of cephalothorax; P4 endopodite 3 being 2.0× longer than wide; P5 outer seta slightly shorter than inner spine (0.8) and shorter than medial seta (~0.6×); caudal ramus outer margin with row of 4–10 spinulae covering 19%–43% of outer margin; P6 ♂ with long strong inner spine but not remarkably longer than other two elements, ~ 1.3× longer than medial seta and 1.8× longer than outer seta .... *Eucyclops delachauxi* Kiefer, 1925 [Colombia, Peru]

13(9) P6 ♂ with inner spine long and strong, always reaching beyond the posterior margin of succeeding somite (Fig. 21.54 D); inner spine longer than medial and outer setae 14

13′ P6 ♂ with inner spine remarkably short, shorter than succeeding somite; inner spine as long or slightly shorter than medial seta and shorter than outer seta (Fig. 21.50 I) .... *Eucyclops bondi* Kiefer, 1934 [Florida Cuba, Dominican Republic, Guatemala, Haiti, Trinidad, Colombia]

14(13) P5 inner spine as long or slightly longer than segment (not longer than 1.3×) 15

14′ P5 inner spine longer than segment, usually more than 2.0× longer 17

15(14) Second antenna frontal surface of basis with rows N1 and N2 (Fig. 21.53 B), P5 inner spine slender; dorsal caudal seta (VII) ~ 0.3× the length of the caudal 16

15′ Second antenna frontal surface of basis without rows N1 and N2, P5 inner spine strong and wide; dorsal caudal seta (VII) as long as/or slightly shorter than caudal ramus .... *Eucyclops conrowae* Reid, 1992 [Florida]

16(15) Caudal ramus 4.5–5.0× longer than wide; second antenna with rows N6, N16, and N17 present but without row N13; P5 with outer seta shorter than medial seta (~0.7×) .... *Eucyclops pectinifer* (Cragin, 1883) [Florida, Brazil, French Guiana, Argentina]

16′ Caudal ramus 3.7–4.0× longer than wide; second antenna with rows N6, N13, N14 present but without rows N16 and N17; P3, P4 intercoxal sclerite as in Fig. 21.50 E, G, respectively; P5 with outer seta as long as medial seta (Fig. 21.50 D) .... *Eucyclops cuatrocienegas* Suárez-Morales & Walsh, 2009 [Mexico]

17(14) P5 outer seta always longer than inner spine (Fig. 21.50 A) 18

17′ P5 outer seta as long as/or shorter than inner spine (Figs. 21.49 K, 50 B) 22

18(17) First antenna last segment not elongated, about the same size or slightly longer than preceding segment 19

18′ First antenna last segment remarkably elongated, more than 1.5× longer than previous segment .... *Eucyclops silvestri* (Brian, 1927) [Venezuela, Brazil, Chile, Argentina, Uruguay]

19(18) Caudal ramus with outermost caudal seta III as long as/or shorter than innermost caudal seta VI 20

19′ Caudal ramus with outermost caudal seta III longer than innermost caudal seta VI .... *Eucyclops ensifer* Kiefer, 1936 [Chile, Cuba, Nicaragua, Brazil, Peru, Argentina (Tierra de Fuego)]

20(19) P3 and P4 with modified setae on endopodite 3 and exopodite 3; P3 and P4 intercoxal sclerites with rows I–III bearing setulae and spinulae (Fig. 21.50 E, G); anal operculum rounded, serrate (Fig. 21.49 N) 21

20′ P3 and P4 without modified setae on endopodite 3 and exopodite 3; P3 and P4 intercoxal sclerites with rows I–III bearing setulae (Fig. 21.50 F, H); anal operculum rounded and smooth .... *Eucyclops chilensis* Suárez-Morales & Walsh, 2009 [Chile, Argentina]

21(20) Second antenna frontal surface of basis with N1 bearing setulae; P1 intercoxal sclerite with rows I and II present bearing tiny spinulae (Fig. 21.53 I); P4 intercoxal sclerite with rows I—III bearing strong spinulae, all rows distributed along sclerite width .... *Eucyclops tziscao* Mercado-Salas, 2013
[Mexico]

21′ Second antenna frontal surface of basis with N1 bearing spinulae; P1 intercoxal sclerite only with row I present, bearing tiny spinulae (Fig. 21.54 B); P4 intercoxal sclerite (caudal) with rows I bearing strong spinulae distributed width the sclerite; rows II and III bearing setulae only on lateral margins of sclerite .... *Eucyclops leptacanthus* Kiefer, 1956
[Mexico, Costa Rica, Venezuela, Brazil]

22(17) Second antenna frontal surface of basis with N1 and N2 (*serrulatus* group); P1 exopodite 3 bearing always five setae 23

22′ Second antenna frontal surface of basis without N1 and N2 (not belonging to *serrulatus* group); P1 exopodite 3 bearing three or five setae .... *Eucyclops pseudoensifer* Dussart, 1984
[Venezuela, Colombia]

23(22) Second antenna frontal surface of basis with N1 and N2 bearing setulae 24

23′ Second antenna frontal surface of basis with N1 bearing setulae or spinulae and N2 bearing spinulae 27

24(23) Second antenna frontal surface of basis with N18 present, caudal surface with rows N8 and N10 present 25

24′ Second antenna frontal surface of basis with N18 absent, caudal surface with rows N8 and N10 absent .... *Eucyclops serrulatus* (Fisher, 1851)
[Mexico, Cuba, Aruba, Hispaniola, Costa Rica, Guatemala, Honduras, Panama, Haiti, Venezuela, Colombia, Ecuador, Brazil, Peru, Chile, Bolivia, Argentina]. Records of *E. serrulatus* in the Americas continent possibly represent one of the species described by Mercado-Salas et al. (2016).

25(24) P4 intercoxal sclerite (caudal) with row I bearing setulae; P3 intercoxal sclerite (caudal) with rows I—III bearing setulae (Fig. 21.50 F) 26

25′ P4 intercoxal sclerite (caudal) with row I bearing spinulae (Fig. 21.54 C); P3 intercoxal sclerite (caudal) with rows I—III bearing spinulae, genital somite as in Fig. 21.49 K, P5 as in Fig. 21.50 B; body as in Fig. 21.54 A .... *Eucyclops defayeae* Mercado-Salas & Suárez-Morales, 2016
[Mexico]

26(25) P1 and P2 intercoxal sclerites (caudal) with rows I and II present, both bearing short setulae; innermost caudal seta (VI) as long as/or shorter than outermost caudal seta (III); P5 medial seta less than 1.5× longer than outer seta .... *Eucyclops wixarica* Mercado-Salas & Suárez-Morales, 2016
[Mexico]

26′ P1 and P2 intercoxal sclerites (caudal) only with row II present, bearing long setulae; innermost caudal seta (VI) more than 1.5× longer than outermost caudal seta (III); P3, P4 intercoxal sclerites as in Fig. 21.50 F, H, respectively; P5 medial seta more than 3.0× longer than outer seta .... *Eucyclops estherae* Mercado-Salas & Suárez-Morales, 2016
[Mexico]

27(23) Second antenna frontal surface of basis with N1 bearing long setulae or spinulae; P4 intercoxal sclerite (caudal) with row I bearing spinulae; P3 intercoxal sclerite with rows I—III bearing spinulae or a combination of rows with setulae 28

27′ Second antenna frontal surface of basis with N1 bearing long and strong spinulae; P4 intercoxal sclerite (caudal) with row I bearing setulae; P3 intercoxal sclerite with rows I—III bearing long setulae .... *Eucyclops chihuahuensis* Suárez-Morales & Walsh, 2009
[Mexico]

28(27) P1 intercoxal sclerite (caudal) with rows I and II present, bearing small spinulae 29

28′ P1 intercoxal sclerite (caudal) smooth or only with row II present, bearing small spinulae 31

29(28) P2 intercoxal sclerite (caudal) with rows I and II present, bearing small spinulae; P5 free segment elongated, more than 1.5× as long as wide 30

29′ P2 intercoxal sclerite (caudal) only with row II present, bearing small spinulae; P5 free segment not elongated 1.0—1.1× as long as wide .... *Eucyclops ishidai* Mercado-Salas & Suárez-Morales, 2016
[Mexico]

30(29) Second antenna caudal surface of basis with rows N6 and N13 present; P5 free segment ~ 2.1 × longer than wide; seminal receptacle with the typical shape of the *serrulatus* group (Fig. 21.49 K); caudal ramus with dorsal caudal seta (VII) less than 0.6× longer than outermost caudal seta III; total body length more than 800 μm .... *Eucyclops festivus* Lindberg, 1955
[Mexico]

30′ Second antenna caudal surface of basis with rows N6 and N13 absent (Fig. 21.53 G); P5 free segment ~ 1.5 × longer than wide; seminal receptacle with posterior lobe margin sinuous (Fig. 21.49 L); caudal ramus with dorsal caudal seta (VII) as long as/or slightly shorter than outermost caudal seta III; total body length not more than 700 μm (Fig. 21.53 F) .... *Eucyclops prionophorus* Kiefer, 1931
[Mexico, Uruguay, Paraguay, Argentina]

31(28) P4 coxa with row J divided into three groups of small spinulae; seminal receptacle with the typical shape of the *serrulatus* group (Fig. 21.49 K), with posterior lobe slightly expanded; anal operculum smooth and rounded but with a medial gap (Fig. 21.49 M) .... *Eucyclops torresphilipi* Suárez-Morales, 2004
[Mexico]

31′ P4 coxa with row J not divided into groups; seminal receptacle seminal receptacle with posterior lobe margin sinuous (Fig. 21.49 L); anal operculum smooth and rounded without medial .... *Eucyclops alekseevi* Mercado-Salas & Suárez-Morales, 2016 [Mexico]

Cyclopoida: Cyclopidae: Eucyclopinae: *Paracyclops*: Species

1 First antenna ♀ 8-articulated (Fig. 21.50 L) 2

1 First antenna ♀ 11-articulated .... *Paracyclops canadensis* (Willey, 1934) [Southern US Note: Records of *Paracyclops affinis* in the Americas correspond to *P. canadensis* (Karaytug & Boxshall, 1998, 1999)]

2(1) Caudal ramus with short diagonal or transverse row of spinae anterior to anterolateral caudal seta II (Fig. 21.51 A, B, D) 3

2 Caudal ramus with longitudinal row of spinulae along dorsal surface of ramus (Fig. 21.51 C) ...... *Paracyclops poppei* (Rehberg, 1880) [United States: Florida, Mexico, Guadeloupe, Brazil, Paraguay]

3(2) Second article of first antenna not elongate; terminal accessory caudal seta VI not more than 1.5× longer than caudal ramus 4

3 Second article of first antenna elongate; terminal accessory caudal seta VI ~ twice longer than caudal ramus .... *Paracyclops reidae* Karaytug & Boxshall, 1998 [Trinidad]

4(3) Inner margin of caudal rami ornamented (Fig. 21.51 A, B) 5

4′ Inner margin of caudal rami lightly or not ornamented (Fig. 21.51 C, D) 6

5(4) Caudal rami with strong spinulae along inner margin (Fig. 21.51 B); ornamentation of urosome and caudal rami similar in females and males; caudal ramus being 3.2× longer than wide .... *Paracyclops carectum* Reid, 1987 [Brazil]

5′ Caudal rami with scattered setulae along inner margin (Fig. 21.51 A); male urosome, caudal and ventral surfaces of caudal ramus with stronger ornamentation pattern than females; caudal rami more than 4.0× longer than wide .... *Paracyclops hirsutus* Mercado-Salas & Suárez-Morales 2009 [Mexico]

6(4) P3 endopodite 3 with four setae; P4 exopodite 3 with five setae; first and second articles of first antenna completely separated 7

6′ P3 endopodite 3 with three setae; P4 exopodite 3 with four setae; first and second articles of first antenna partially fused .... *Paracyclops pilosus* Dussart, 1983 [Venezuela, French Guiana]

7(6) Genital double-somite without ridge on ventral surface; seminal receptacle divided into broad anterior and posterior lobes 8

7′ Genital double-somite with surface ridge extending to either side of copulatory pore on ventral surface; seminal receptacle with a small conical anterior lobe and a broad posterior lobe (Fig. 21.50 P) .... *Paracyclops hardingi* Karaytug & Boxshall, 1998 [Ecuador, Peru]

8(7) Caudal surface of A2 basis with spinule row near insertion of two inner setae 9

8′ Caudal surface of A2 basis without spinule row near insertion of two inner setae 10

9(8) Ventral surface of caudal rami without cuticular depressions; caudal rami length 3.5–6.0× width; seminal receptacle divided into broad anterior and posterior lobes, posterior lobe weakly sinuous .... *Paracyclops fimbriatus* (Fischer, 1853) [Mexico, Brazil. Note: restricted to Europe and Asia (Karaytug, 1999; Karaytug & Boxshall, 1998), American records probably represent other species]

9′ Ventral surface of caudal rami with cuticular depressions; caudal rami length < 3.5× width; seminal receptacle divided into narrow anterior and broad posterior lobes .... *Paracyclops chiltoni* (Thomson, 1882) [Florida, Mexico, Cuba, Brazil, Chile, Argentina]

10(8) Second article of first antenna without suture line; outer seta of P5 length > inner spine 11

10′ Second article of first antenna with partial or complete suture line; outer seta of P5 length = inner spine .... *Paracyclops novenarius* Reid, 1987 [Colombia]

11(10) Only third article of first antenna with partial suture lines (Fig. 21.50 M); urosome and caudal rami lacking cuticular pits 12

11′ Third and fourth articles of first antenna with partial suture lines (Fig. 21.50 N); urosome and caudal rami distinctly ornamented with cuticular pits .... *Paracyclops punctatus* Karaytug & Boxshall, 1998 [Brazil, found in bromeliads]

12(11) Terminal accessory caudal seta VI length > posterolateral caudal seta III .... *Paracyclops bromeliacola* Karaytug & Boxshall, 1998 [Brazil, found in bromeliads]

12′ Terminal accessory caudal seta VI length < posterolateral caudal seta III .... *Paracyclops rochai* Karaytug & Boxshall, 1998 [Brazil]

Cyclopoida: Cyclopidae: Eucyclopinae: *Tropocyclops*: Species

1 P1-P4 exopodite 3 spine formula 2,3,3,3 or 3,3,3,3 2

1′ P1-P4 exopodite 3 spine formula 3,4,3,3 or 3,4,4,3 3

2(1) Spine formula P1–P4 exopodite 3: 2,3,3,3, dorsal caudal seta VII long, more than twice as long as caudal ramus; P5 inner spine as long as article .... *Tropocyclops pseudoparvus* Dussart & Fernando, 1986 [Costa Rica]

2′ P1-P4 exopodite 3 spine formula 3,3,3,3; P5 with inner spine slightly longer than free article of leg, middle and outer setae subequal in length, both reaching beyond posterior margin of genital double-somite .... *Tropocyclops federensis* Reid, 1991 [Brazil]

3(1) P1-P4 exopodite 3 spine formula 3,4,3,3 4

3′ P1-P4 exopodite 3 spine formula 3,4,4,3 9

4(3) P5 bearing three elements—one inner spine, one medial seta, and one inner seta (Fig. 21.51 E, K); P4 endopodite 1 with inner seta; P4 endopodite 3 not bended inward 5

4′ P5 bearing two elements—one long strong inner spine and one medial seta (Fig. 21.51 J); P4 endopodite 1 without inner seta; P4 endopodite 3 strongly bent inward .... *Tropocyclops jamaicensis* Reid & Janetzki, 1996 [Jamaica, in bromeliads]

5(4) P4 endopodite 3 with inner apical spine length 2.3–2.5× than outer apical spine 6

5′ P4 endopodite 3 with inner spine length 4.0–5.0× than outer apical spine 7

6(5) Caudal ramus length ~ 2× width .... *Tropocyclops confinis* (Kiefer, 1930) [Brazil]

6′ Caudal ramus length 2.3–2.4× width; P5 inner spine and outer seta equally long .... *Tropocyclops confinis longifurca* Dussart, 1983 [Antilles, Guadeloupe]

7(5) Caudal ramus length 2.5–3.0× width 8

7′ Caudal ramus length 2.0× width; P5 inner spine length 0.7× outer seta .... *Tropocyclops parvus* (Kiefer, 1931) [Guatemala]. Note: Considered a synonym of *Tropocyclops tenellus* (G.O. Sars, 1909) (Dussart & Defaye, 2006); however, *T. parvus* was described from Guatemala, whereas *T. tenellus* is known only from Africa. Both species can be easily separated because in *T. tenellus* the caudal ramus is 3.0× longer than wide, whereas in *T. parvus*, this value is not more than 2.0 (Kiefer, 1931)]

8(7) Terminal caudal seta (VI) and posterolateral caudal seta (III) subequal in length; P4 endopodite 3 lacking outer apical spine or when present ~ 0.2× shorter than inner apical spine (Fig. 21.51 I), P5 middle and outer setae long, but not reaching posterior margin of genital double-somite) .... *Tropocyclops nananae* Reid, 1991 [Brazil]

8′ Terminal accessory caudal seta (VI) length 1.8× posterolateral caudal seta (III); P4 endopodite 3 with outer apical spine always present (Fig. 21.51 H), being 0.25× shorter than inner apical spine .... *Tropocyclops rarus* Dussart, 1984 [French Guiana]

9(3) Caudal ramus 3.3–4.1× longer than broad (Fig. 21.51 F) 10

9′ Caudal ramus 3.0× longer than wide or shorter (Fig. 21.51 G) 11

10(9) P4 endopodite 3 with apical spine length ~ 1.5–2.1× the length of P4 endopodite 3 .... *Tropocyclops extensus* Kiefer, 1931 [Haiti, Cayman Islands, Mexico]

10′ P4 endopodite 3 with apical spine length ~ 2.3–2.5× the length of P4 endopodite 3 .... *Tropocyclops extensus* f. *longispina* Kiefer, 1931 [Mexico]

11(9) Dorsal caudal seta (VII) inserted between the insertion of lateral caudal seta (II) and distal margin of caudal ramus; P4 intercoxal sclerite naked or ornamented with short setulae or spinulae on margin 12

11′ Dorsal caudal seta (VII) inserted at level or slightly distal to insertion of lateral caudal seta (II); P4 intercoxal sclerite with long hairs on margin .... *Tropocyclops schubarti* s.str. Kiefer, 1935 [Brazil, in bromeliads]

12(11) Caudal ramus length 1.2–1.5× width; P4 intercoxal sclerite smooth 13

12′ Caudal ramus length 1.7–3.0× width; P4 intercoxal sclerite ornamented with short setulae or spinulae on margin (*Tropocyclops prasinus* complex) 14

13(12) Caudal ramus length 1.5× width; P4 endopodite 3 with inner apical spine length 2.0× the length of P4 endopodite 3 .... *Tropocyclops schubarti dispar* Herbst, 1962 [Brazil]

13′ Caudal ramus short, length 1.2× width; P4 endopodite 3 with inner apical spine length 2.4× the length of P4 endopodite 3 .... *Tropocyclops piscinalis* Dussart, 1984
[Brazil]

14(12) Dorsal caudal seta VII as long or shorter than caudal ramus 15

14′ Dorsal caudal seta VII length at least 1.3× caudal ramus 17

15(14) Caudal ramus length > 2.6× width 16

15′ Caudal ramus length 2.0–2.6× width (Figs. 21.51 G, 52 G), P5 as in Fig. 21.52 E .... *Tropocyclops prasinus* s.str. (Fisher, 1860)
[Mexico, Cuba, Hispaniola, Guatemala, Honduras, Guadeloupe, Panama, Brazil, Colombia, Chile, Argentina: Malvinas (Falkland) Island]

16(15) Caudal ramus parallel, length ~ 2.6–3.1× width (Fig. 21.52 D, H); P4 endopodite 3 length 2.5–3.0× width; P4 intercoxal sclerite ornamented with short, robust spinulae .... *Tropocyclops prasinus aztequei* Lindberg, 1955
[Mexico. Cayman Islands]

16′ Caudal ramus divergent, length ~3.0× width; P4 endopodite 3 length 2.2× width; P4 intercoxal sclerite naked .... *Tropocyclops prasinus divergens* Dussart, 1982
[Guadeloupe]

17(14) Dorsal caudal seta VII length 0.3–1.9× caudal ramus 18

17′ Dorsal caudal seta VII length > 2.0× caudal ramus (Fig. 21.51 E) .... *Tropocyclops prasinus mexicanus* Kiefer, 1938
[Florida, Mexico, Cuba, Jamaica, Guatemala, Brazil]

18(17) Caudal ramus length ~ 1.7–2.0× width; P4 endopodite 3 not extraordinarily long, < 3.2× width 19

18′ Caudal ramus length ~ 2.1–2.3× width; P4 endopodite 3 extraordinarily long, ~3.6× width .... *Tropocyclops prasinus altoandinus* Gaviria, 1994
[Colombia]

19(18) P4 endopodite 3 inner apical spine length 2.5–2.9× outer apical spine length; P4 endopodite 3 inner spine length 2.3–3.1× endopodite 3 length .... *Tropocyclops prasinus meridionalis* Kiefer, 1931
[New Caledonia, French Guiana, Brazil, Peru, Chile, Paraguay, Argentina: Malvinas (Falkland) Islands, Uruguay]

19′ P4 endopodite 3 inner apical spine length > 2.9× outer apical spine length; P4 endopodite 3 inner spine length < 2.2× endopodite 3 total length .... *Tropocyclops prasinus peruviana* Lindberg, 1955
[Peru] Note: Synonymized with *Tropocyclops prasinus meridionalis* by Herbst (1959).

Order Poecilostomatoida: Family Ergasilidae

Taisa M. Marques
Laboratório de Ecologia Molecular e Parasitologia Evolutiva, Universidade Federal do Paraná, Curitiba, Paraná, Brazil

Eduardo Suárez-Morales
El Colegio de la Frontera Sur, Chetumal, Quintana Roo, México

Yanis Cruz-Quintana and Ana María Santana-Piñeros
Grupo de Investigación en Sanidad Acuícola, Inocuidad y Salud Ambiental. Escuela de Acuicultura y Pesquería, Facultad de Ciencias Veterinarias, Universidad Técnica de Manabí, Portoviejo, Manabí, Ecuador

INTRODUCTION

Ergasilid copepods are parasitic on a wide variety of teleosts. The group is readily distinguished by its modified second antenna, which is modified as a prehensile organ that is used to attach the parasite to the host (Boxshall & Defaye, 2008). These parasites can also be a frequent element in the zooplankton because males, immature females, and juveniles occur as free-swimming forms (Amado & Rocha, 2001). After mating, adult females seek out and infect hosts to complete their life cycle. Attached females actively feed on the host epithelial tissues (Malta & Varella, 1996). In fish cultures, a high rate of infestation is usual, thus causing severe economic damage and mortality. Ergasilids are distributed mainly on the estuarine and coastal systems of all continents excluding Antarctica and occur in all salinity regimes, although the majority are found in freshwaters (Boxshall & Defaye, 2008). According to Boxshall & Defaye (2008) and Luque et al. (2013), the highest known diversity of the family is found in the Neotropical Region, with more than 75 species recorded in freshwater. The family contains around 28 genera of which 12 are currently known in the Neotropical Region. The genus *Ergasilus* is the most speciose and widespread among the Ergasilidae; it is known to contain more than 180 species worldwide. More than 20 species of *Ergasilus* are known from South American freshwater teleosts (Luque & Tavares, 2007; Luque et al., 2013; Marques, 2014; Marques et al., 2015; Muriel-Hoyos et al., 2015). In Mexico and Central America, records of ergasilids in

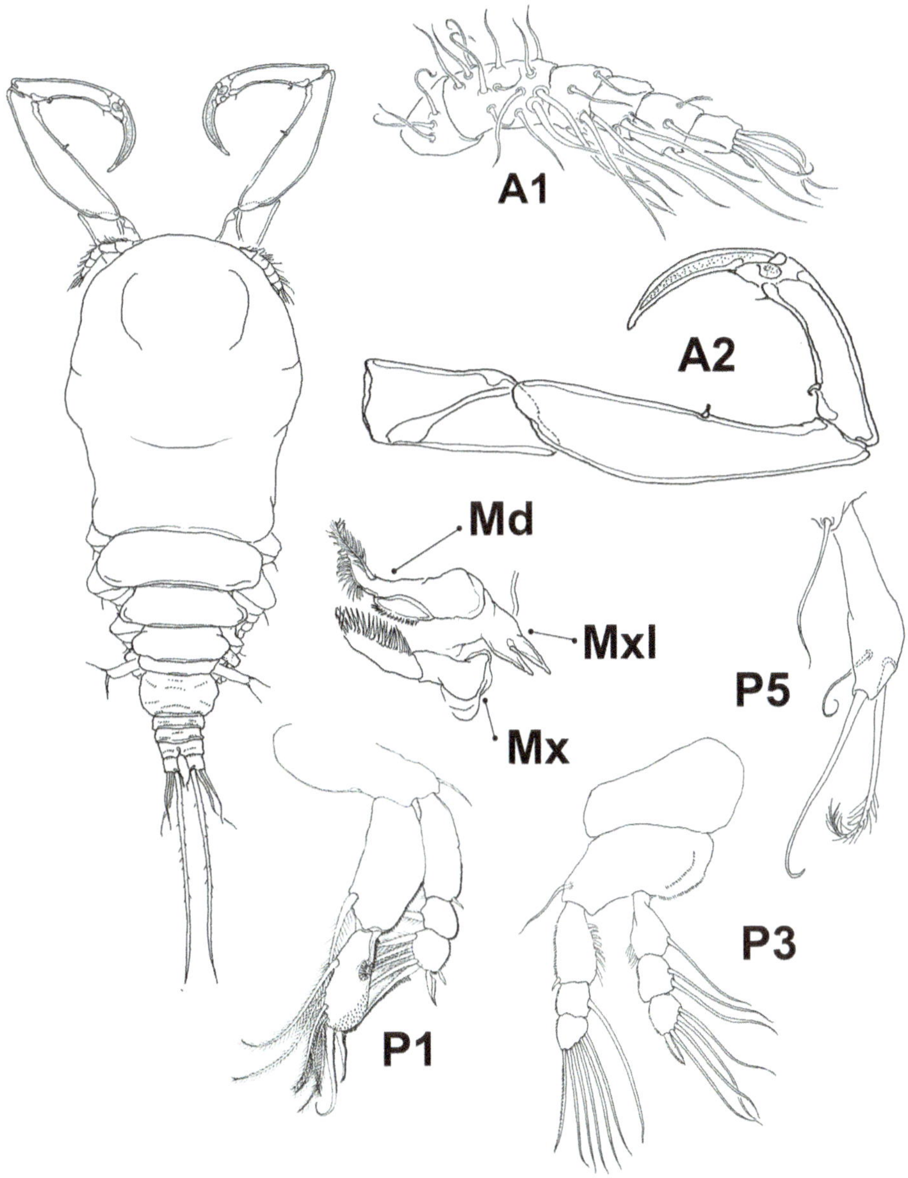

FIGURE 21.55 General morphology of *Ergasilus* (Poecilostomatoida, Ergasilidae). Abbreviations: *A1,* first antenna (antennule); *A2,* second antenna (antenna); *Md,* mandible; *Mx1,* first maxilla (maxillule); *Mx,* second maxilla (maxilla); *P1,* leg 1; *P3,* leg 3; *P5,* leg 5.

freshwater fishes are relatively scarce, largely limited to only three species (Jiménez-García & Suárez-Morales, 2017). The general morphology of *Ergasilus* is presented in Fig. 21.55. The taxonomy of this diverse genus relies on the structure of legs 1–4, particularly on their articulation and ornamentation patterns but also on the proportions of the cephalothorax, the armature and proportions of the second antenna (A2), and details of the mouthparts, including the mandible (Md), first maxillae (Mxl), and second maxillae (Mx). The present key includes only those genera and species of ergasilids that have been recorded from Neotropical freshwater fish.

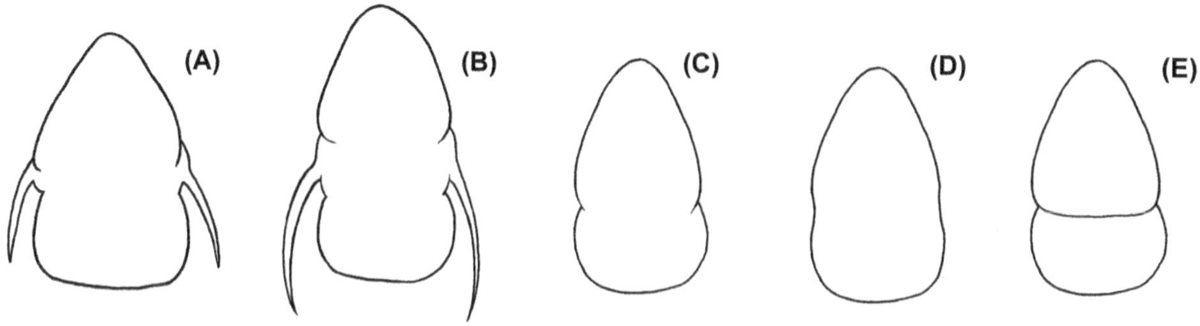

FIGURE 21.56 Morphology of the posterolateral corners of cephalosome in Ergasilidae: (A) posterolateral corners bearing retrostylets not exceeding the posterior margin; (B) retrostylets reaching beyond posterior margin; (C), (D), and (E) posterolateral corners of cephalosome rounded.

KEYS TO ERGASILIDAE

Poecilostomatoida: Ergasilidae: Genera

1 Leg 4 present 2

1′ Leg 4 absent........ ***Brasergasilus* [p. 784]**

2(1) Leg 4 biramous 3

2′ Leg 4 represented by one seta 16

3(2) Posterolateral corners of cephalosome rounded (Fig. 21.56 C, D) 4

3′ Posterolateral corners of cephalosome bearing retrostylets (Fig. 21.56 A, B) 12

4(3) First antenna 6-articulated 5

4′ First antenna 5-articulated 7

5(4) Leg 4 endopod articulate 6

5′ Leg 4 endopod 1-articulated, Leg 1 exopod 2 with large thumblike spine (Fig. 21.58 E) .... *Neoergasilus japonicus* (Harada, 1930) [Mexico: Introduced]

6(5) Leg 4 ENP2-articulated .... *Pindapixara tarira* Malta, 1995 [Brazil: Rondônia, Guaporé River]

6′ Leg 4 endopod 3-articulated .... ***Ergasilus* [p. 785]**

7(4) Second antenna prehensile (Fig. 21.57 G, H, K)........ 8

7′ Second antenna chelate (Fig. 21.57 J, L)........ ***Prehendorastrus* [p. 789]**

8(7) Preoral neck absent 9

8′ Preoral neck present, of variable length (Fig. 21.59 E) .... *Therodamas elongatus* (Thatcher, 1986) [Brazil: Amazon River]

9(8) Second antenna articles naked 10

9′ Second antenna articles covered by cuticular membrane .... ***Acusicola* [p. 789]**

10(9) Leg 1 endopod lacking setae or spinae 11

10′ Leg 1 endopod with setae and spinae (Fig. 21.58 D); claw with a subterminal barb (Fig. 21.57 D) .... *Gauchergasilus euripedese* (Montú, 1980) [Brazil: Rio Grande do Sul, Patos Lagoon]

11(10) Leg 1 endopod long, slender (Fig. 21.58 C) .... ***Miracetyma* [p. 790]**

11′ Leg 1 endopod ending in three small points (Fig. 21.58 B) .... *Amplexibranchius bryconis* Thatcher & Paredes, 1985 [Brazil: Amazonas, Amazon River]

12(3) Second antenna tipped with two claws (Fig. 21.57 E) 13

12′ Second antenna tipped with one claw 14

13(12) First antenna 6-articulated; rostral spine absent; retrostylets present on cephalosome (Fig. 21.56 B) .... ***Gamidactylus* [p. 790]**

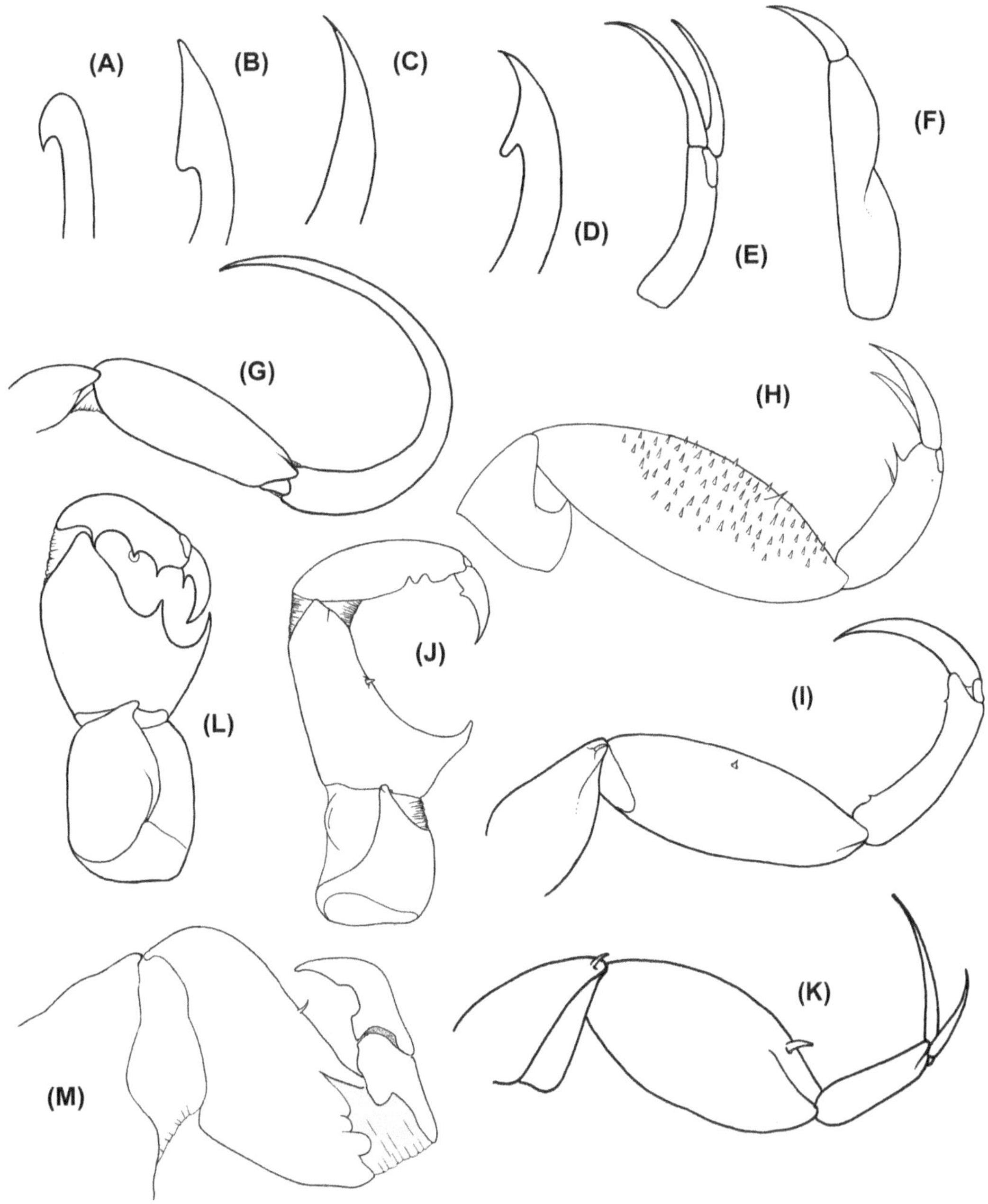

FIGURE 21.57 Second antenna morphology of Ergasilidae: (A) claw with curved distal enlargement; (B) claw with terminal barb; (C) claw ending in simple acute tip; (D) claw with a subterminal barb; (E) tipped with two claws; (F) claw very small; (G), (H), (I), and (K) prehensile; (J) and (L) chelate; (M) second antenna endopod 2 remarkably short.

13′ First antenna 5-articulated; rostral spine present (Fig. 21.59 A); retrostylets double, with outer margin flat and inner rounded process (Fig. 21.59 C) .... *Gamispatulus schizodontis* Thatcher & Boeger, 1984
[Brazil: Amazonas, Marchantaria Island]

14(12) First antenna 5-articulated 15

14′ First antenna 6-articulated .... *Vaigamus retrobarbatus* Thatcher & Robertson, 1984
[Brazil: Amazonas, Marchantaria Island]

15(14) Leg 4 endopod biarticulated; rostral spine absent; leg 4 exopod unarticulated; second antenna endopod 1 with several spinae on inner margin (Fig. 21.57 H) .... *Gamispinus diabolicus* Thatcher & Boeger, 1984
[Brazil: Paraná, Paraná River]

15′ Leg 4 endopod 3-articulated; rostral spine present; leg 4 exopod 2-articulated; second antenna endopod 1 smooth, lacking spinae .... *Pseudovaigamus spinicephalus* (Thatcher & Robertson, 1984)
[Brazil: Pará, Tocantins River]

16(2) Claw short; genital somite cylindrical, extremely long (Fig. 21.59 D) .... *Urogasilus brasiliensis* Rosim, Boxshall & Ceccarelli, 2013
[Brazil: Mato Grosso]

16′ Claw elongate; genital somite not remarkably elongate .... *Rhinergasilus piranhus* Boeger & Thatcher, 1988
[Brazil: Amazonas, Marchantaria Island, Catalão Lake]

Poecilostomatoida: Ergasilidae: *Brasergasilus*: Species

1 Claw with terminal barb (Fig. 21.57 B) or distally curved 2

1′ Claw ending in simple acute tip (Fig. 21.57 C) 4

2(1) Claw with terminal barb (Fig. 21.57 B) 3

2′ Claw with curved distal enlargement (Fig. 21.57 A) .... *Brasergasilus bifurcatus* Santos, Thatcher & Brasil-Sato, 2007
[Brazil: Três Marias Reservoir, São Francisco River]

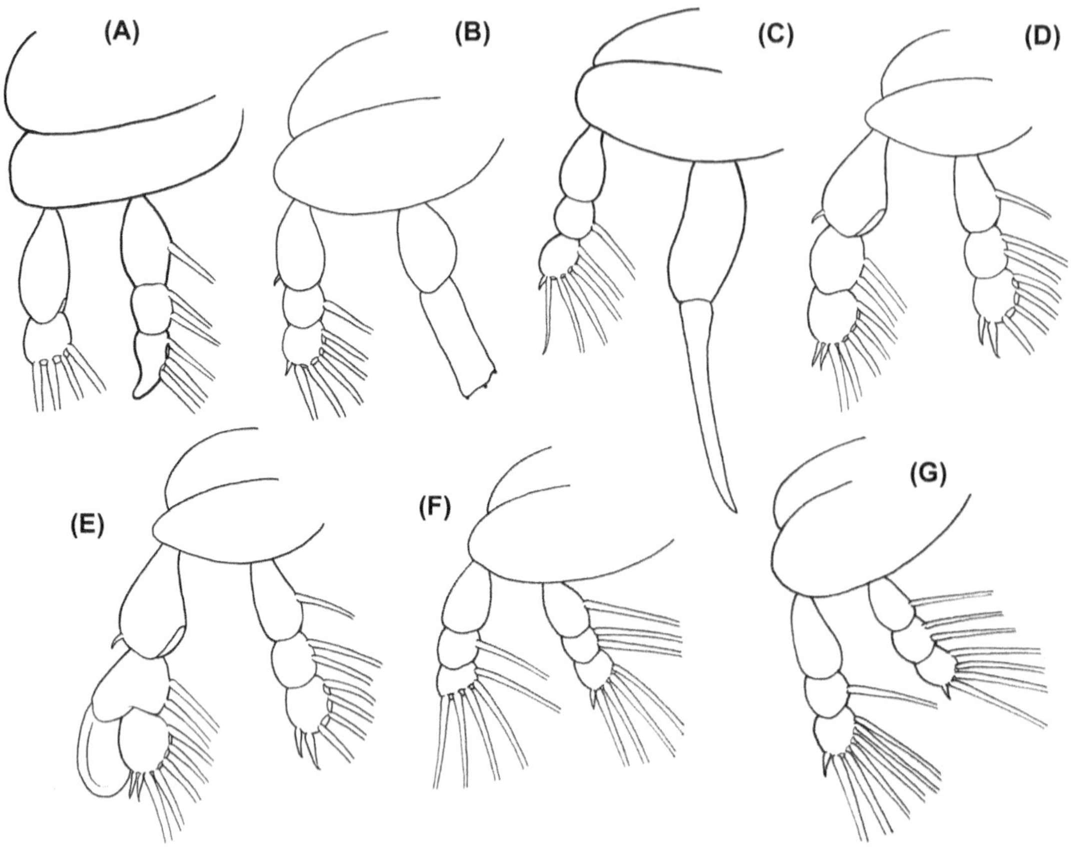

FIGURE 21.58 General morphology of legs 1 and 4 of Ergasilidae: (A) leg 4 exopod 2-articulated; (B) leg 1 endopod ending in three small points; (C) leg 1 endopod long, slender; (D) leg 1 endopod with setae and spinae; (E) leg 1 exopod 2 with large thumblike spine; (F) leg 3 exopod 3 armed with five setae, lacking spine; (G) leg 3 exopod 3 armed with one spine and six setae.

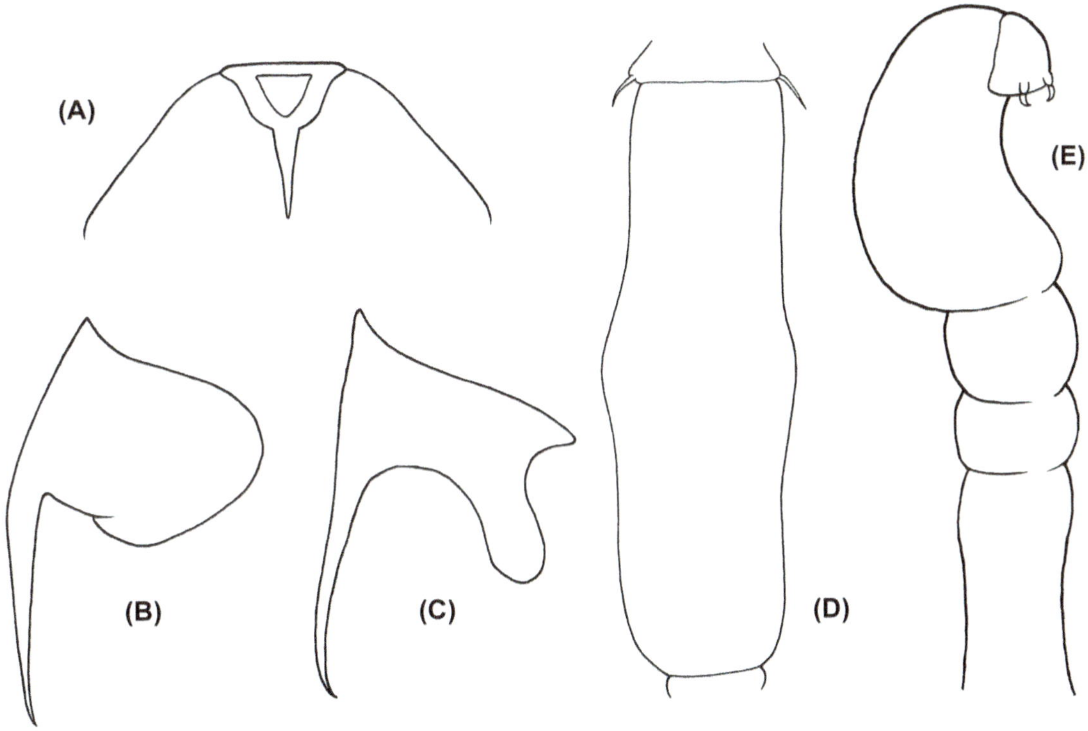

FIGURE 21.59 Ergasilidae morphology: (A) rostral spine present; (B) single retrostylet present on cephalosome; (C) double retrostylet with outer margin flat and inner rounded process; (D) genital somite cylindrical, extremely long; (E) preoral neck present, of variable length.

3(2) Leg 3 exopod 3 armed with one spine and six setae (Fig. 21.58 G) .... *Brasergasilus mamorensis* Varella & Malta, 2001
[Brazil: Rondônia, Mamoré River]

3′ Leg 3 exopod 3 armed with five setae, lacking spine (Fig. 21.58 F) .... *Brasergasilus guaporensis* Malta, 1995
[Brazil: Rondônia, Guaporé River]

4(1) Leg 3 exopod 3 armed with six setae, lacking spine; cephalothorax less than twice as long as wide 5

4′ Leg 3 exopod 3 armed with one spine and six setae (Fig. 21 4 G); cephalothorax twice as long as wide .... *Brasergasilus oranus* Thatcher & Boeger, 1985
[Brazil]

5(4) Second antenna endopod 1 with sensilla .... *Brasergasilus jaraquensis* Thatcher & Boeger, 1983
[Brazil: Manaus, Amazon River]

5′ Second antenna endopod 1 lacking sensilla .... *Brasergasilus anodus* Thatcher & Boeger, 1983
[Brazil: Pará, Tocantins River]

Poecilostomatoida: Ergasilidae: *Ergasilus*: Species

1 Third exopod article of leg 1 with one pectinate seta and four plumose setae 2

1′ Third exopod article of leg 1 lacking pectinate seta, with five plumose setae... 17

2(1) Two distal spinae on leg1 exopod 3 with spinulae or smooth 3

2′ Two distal spinae on leg 1 exopod 3 with membrane on outer margin; leg 1 endopod 2 with rosettelike array blunt of spinae .... *Ergasilus trygonophilus* Domingues & Marques, 2010
[Brazil: Pará, Marajó Bay]

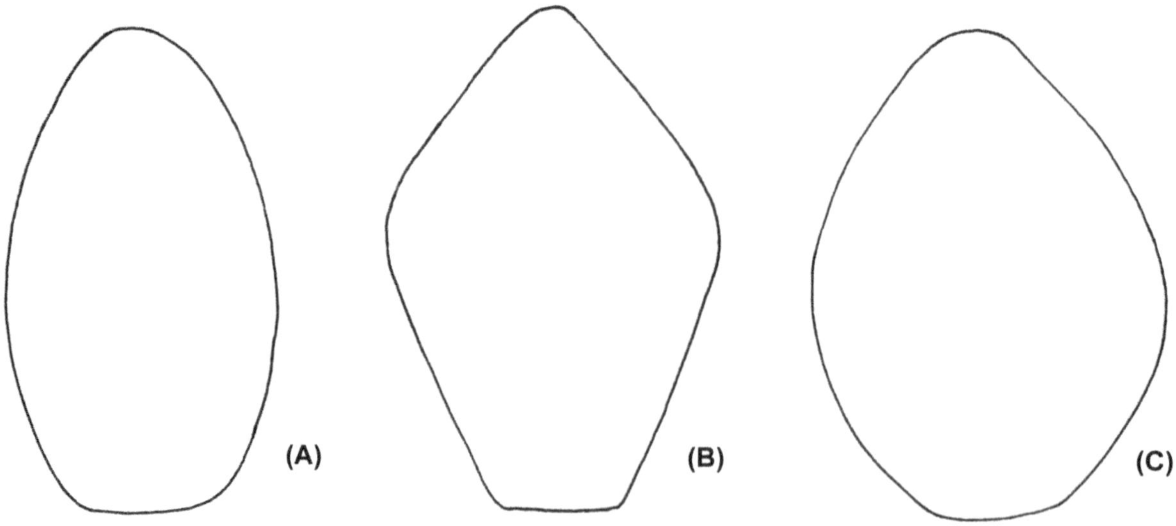

FIGURE 21.60 General morphology of cephalothorax in Ergasilidae: (A) oblong; (B) rhomboidal; (C) globose.

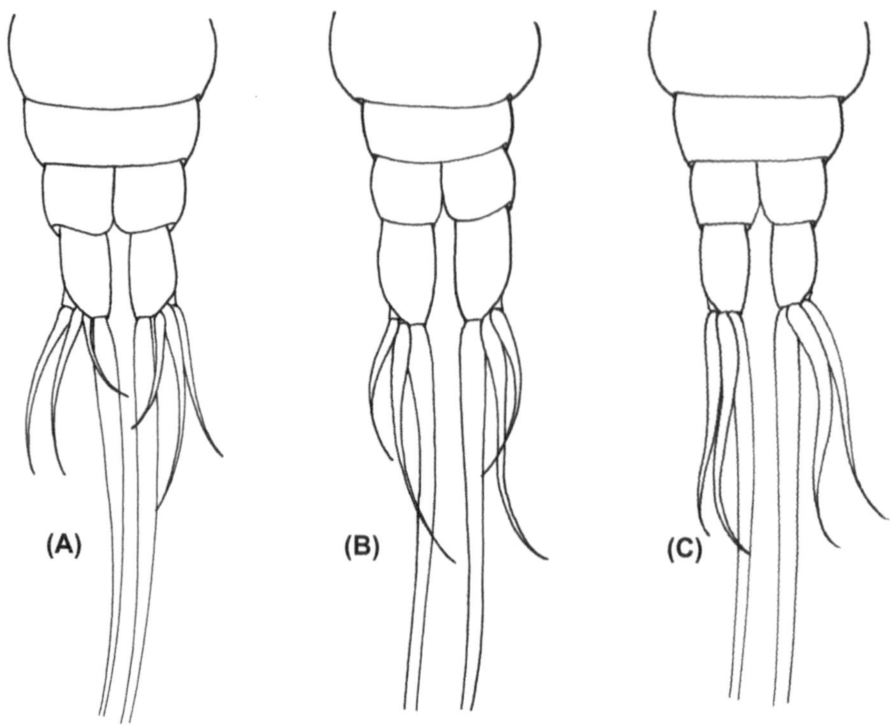

FIGURE 21.61 General morphology of caudal rami in *Acusicola* species (Poecilostomatoida, Ergasilidae): (A) caudal rami with one long seta, two median, and one shorter; (B) caudal rami with one long, one median, and one short seta; (C) caudal rami with one long and two small setae.

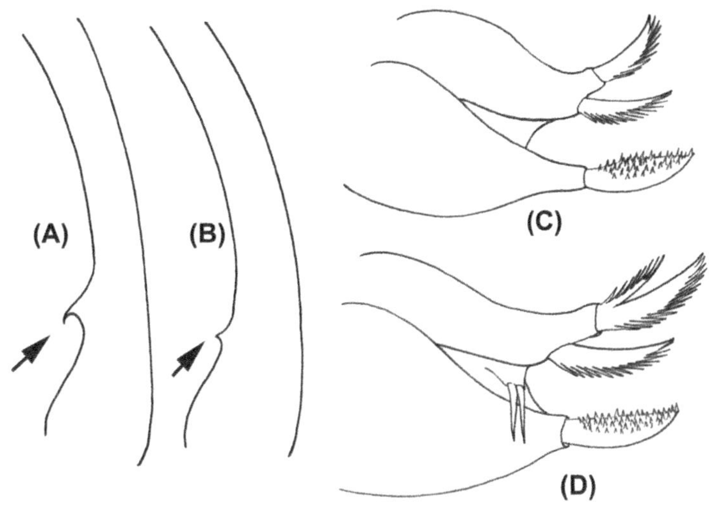

FIGURE 21.62 Proximal sensillum in the cephalosoma and mouth parts of *Ergasilus* (Poecilostomatoida, Ergasilidae): (A) proximal sensillum strongly developed; (B) proximal sensillum weakly developed; (C) first maxillae without setal elements and mandible with anterior blade absent; (D) first maxillae with two setal elements and mandible with anterior, median, and posterior blades.

3(2) Two distal spinae on leg1 exopod 3 ornamented with spinulae 4

3′ Two distal spinae on leg 1 exopod 3 smooth 11

4(3) Cephalothorax and first pedigerous somite completely or partially fused 5

4 Cephalothorax and first pedigerous somite unfused 9

5(4) Cephalothorax and first pedigerous somite partially fused 6

5′ Cephalothorax and first pedigerous somite completely fused ..... *Ergasilus chelangulatus* Thatcher & Brasil-Sato, 2008 [Brazil: Minas Gerais, São Francisco River]

6(5) Cephalothorax length ~ 0.5 body length; proximal sensillum short, weakly developed (Fig. 21.62 B) 7

6′ Cephalothorax length >0.5 body length; proximal sensillum strongly developed (Fig. 21.62 A) 8

7(6) Second antenna exopod 1 length/width ratio > 3.0; sensillum on second antenna endopod 1 inserted distally; internal margin of second antenna endopod 1 with spinae, endopod 2 curved, shorter than endopod 1 ..... *Ergasilus thatcheri* Engers, Boeger & Brandão, 2000 [Brazil: Rio Grande do Sul, Capané Dam]

7′ Second antenna endopod 1 length/width ratio < 3.0; sensillum on second antenna endopod 1 inserted at midlength of article; internal margin of second antenna endopod 1 lacking spine, endopod 2 not curved, also shorter than endopod 1 ..... *Ergasilus hypophthalmi* Boeger, Martins & Thatcher, 1993 [Brazil: Amazonas, Catalão Lake]

8(6) Sensillum on second antenna endopod 1 inserted distally; internal margin of second antenna endopod 1 with spinae ..... *Ergasilus salmini* Thatcher & Brasil-Sato, 2008 [Brazil: Minas Gerais, São Francisco River]

8′ Sensillum on second antenna endopod 1 inserted at midlength of article; internal margin of second antenna endopod 1 lacking spinae ..... *Ergasilus callophysus* Thatcher & Boeger, 1984 [Brazil: Amazonas, Amazon River]

9(4) Cephalothorax length ~0.5 body length; second antenna endopod 1 length/width ratio < 3.0; second antenna endopod 2 shorter than endopod 1; spinae on leg1 endopod 2 straight 10

9′ Cephalothorax > 0.5 body length; second antenna exopod 1 length/width ratio > 3.0; second antenna endopod 2 curved, shorter than endopod 1; spinae on leg 1 endopod 2 curved ..... *Ergasilus bryconis* Thatcher, 1981 [Brazil: Amazonas, Manaus]

10(9) Second antenna coxa with short distal seta; second antenna endopod 1 with one sensillum; two setal elements on first maxillae basis (Fig. 21.62 C) present; endopod 2 of legs 2 and 3 with two setae. ..... *Ergasilus turucuyus* Malta & Varella, 1996 [Brazil]

10′ Second antenna coxa lacking distal setae; second antenna endopod 1 with two sensilla; first maxillae smooth, setal elements absent or inconspicuous; endopod 2 of legs 2 and 3 with one seta .... *Ergasilus curticrus* Muriel-Hoyos, Santana-Piñeros, Cruz-Quintana & Suárez-Morales, 2015 [Colombia: Vichada River, Orinoco Basin River]

11(3) Cephalothorax length $\leq$ 0.5 body length 12

11′ Cephalothorax length > 0.5 body length; aesthetascs present on first antenna articles 5 and 6 13

12(11) Cephalothorax ~ 50% body length; aesthetascs absent or inconspicuous on first antenna articles 5 and 6 .... *Ergasilus hydrolycus* Thatcher, Boeger & Robertson, 1984 [Brazil: Amazonas, Marchantaria Island]

12′ Cephalothorax < 0.5 body length; aesthetascs present on first antenna articles 5 and 6 .... *Ergasilus urupaensis* Malta, 1995 [Brazil: Rondônia]

13(11) Second antenna endopod 1 length/width ratio > 3.0 14

13′ Second antenna ENP1 length/width ratio < 3.0 15

14(13) Cephalothorax and first pedigerous somite partially fused; second antenna coxa lacking seta; second antenna endopod 2 curved, shorter than endopod 1; endopod 2 with two sensilla .... *Ergasilus triangularis* Malta, 1996 [Brazil]

14′ Cephalothorax and first pedigerous somite not fused; second antenna coxa with short distal seta; second antenna endopod 2 shorter than endopod 1; endopod 2 with one sensillum .... *Ergasilus jaraquensis* Thatcher & Robertson, 1982 [Brazil]

15(13) Sensillum on second antenna endopod 1 inserted near midlength of article; endopod 2 with one sensillum; spinae on leg 1 endopod 2 straight; abdominal articles with spinae on entire posterior margin 16

15′ Sensillum on second antenna endopod 1 inserted distally; endopod 2 with two sensilla; spinae on leg 1 endopod 2 curved; abdominal articles lacking spinae .... *Ergasilus yumaricus* Malta &Varella, 1995 [Brazil]

16(15) Second antenna endopod 2 curved, shorter than endopod 1; mandible with anterior, median, and posterior blades; two setal elements on first maxillae present (Fig. 21.62 D); leg 1 exopod 2 lacking seta .... *Ergasilus lacusauratus* Marques, Boeger & Brasil-Sato, 2015 [Brazil]

16′ Second antenna endopod 2 shorter than endopod 1; mandible with median and posterior blades only, anterior blade absent (Fig. 21.62 C); two setal elements on first maxillae absent or inconspicuous; leg 1 exopod 2 with one seta .... *Ergasilus holobryconis* Malta & Varella, 1986 [Brazil]

17(1) Cephalothorax and first pedigerous somite completely or partially fused 18

17′ Cephalothorax and first pedigerous somite completely separated 21

18(17) Cephalothorax and first pedigerous somite completely fused 19

18′ Cephalothorax and first pedigerous somite partially fused, suture incomplete middorsally .... *Ergasilus leporinidis* Thatcher, 1981 [Brazil: Manaus, Amazonas]

19(18) Second antenna endopod 2 curved with one or two sensilla 20

19′ Second antenna endopod 2 remarkably short (Fig. 21.57 M), lacking sensillum .... *Ergasilus iheringi* Tidd, 1942 [Brazil]

20(19) Second antenna endopod 2 shorter than endopod 1 and with one sensillum .... *Ergasilus colomesus* Thatcher & Boeger, 1983 [Brazil]

20′ Second antenna endopod 2 as long as first article, with two sensilla .... *Ergasilus argulus* Cressey, 1970 [Colombia]

21(17) Second antenna endopod 1 length/width ratio $\leq$ 3.0 22

21′ Length/width ratio of antennary endopod 1 over > 3.0 23

22(21) Second antenna endopod 1 length/width ratio < 3.0 .... *Ergasilus coatiarus* Araujo & Varella, 1998 [Brazil: Amazonas, Marchantaria Island, Amazon-Solimões River]

22′ Second antenna endopod 1 length/width ratio ~ 3.0 .... *Ergasilus cerastes* Roberts, 1969 [Mexico. Introduced]

23(21) Second antenna coxa with short distal seta; second antenna endopod 2 curved, shorter than endopod 1; leg 1 exopod 3 with two distal smooth spinae .... *Ergasilus sinefalcatus* Marques, Boeger & Brasil-Sato, 2015
[Brazil]

23′ Second antenna coxa lacking seta; second antenna endopod 2 curved, as long as endopod 1; leg 1 exopod 3 with two unipinnate distal spinae .... *Ergasilus pitalicus* Thatcher, 1984
[Colombia]

Poecilostomatoida: Ergasilidae: *Prehendorastrus*: Species

1 Internal margin of second antenna endopod 2 with a large tooth (Fig. 21.57 J); cephalothorax rhomboidal (Fig. 21.60 B) .... *Prehendorastrus monodontus* Boeger & Thatcher, 1990
[Brazil: Manaus, Rio Negro]

1′ Internal margin of second antenna endopod 2 with two large teeth (Fig. 21.57 L); cephalothorax oblong .... *Prehendorastrus bidentatus* Boeger & Thatcher, 1990
[Brazil: Manaus, Rio Negro]

Poecilostomatoida: Ergasilidae: *Acusicola*: Species

1 Caudal rami with one long seta 2

1′ Caudal rami with two long, one median, and one short setae 8

2(1) Caudal rami with one long seta and two shorter seta 3

2′ Caudal rami with one long seta, two median, and one shorter (Fig. 21.61 A) 4

3(2) Caudal rami with one long, one median, and one short seta (Fig. 21.61 B) 6

3′ Caudal rami with one long and two small setae (Fig. 21.61 C) .... *Acusicola spinulosa* Amado & Rocha, 1996
[Brazil: Amazonas, José Açu Lake]

4(2) Cephalothorax oblong (Fig. 21.60 A) 5

4′ Cephalothorax globose (Fig. 21.60 C) .... *Acusicola joturicola* El-Rashidy & Boxshall, 1999
[Panama: Chirriqui Lagoon]

5(4) Cephalothorax and first pedigerous somite completely fused; anal somite with pair of sensilla on dorsal surface; coxobasis with sensory peg at inner distal corners .... *Acusicola spinuloderma* El-Rashidy & Boxshall, 1999
[Honduras: Río Tulito]

5′ Cephalothorax and first pedigerous somite partially fused; anal somite lacking sensilla; coxobasis lacking sensory peg .... *Acusicola mazatlanensis* El-Rashidy & Boxshall, 1999
[Mexico: Mazatlán]

6(3) Abdominal somites lacking ornamentation; second antenna endopod 1 with one sensillum 7

6′ Abdominal somites with spinulae on posteroventral margins; second antenna endopod 1 lacking ornamentation .... *Acusicola tucunarense* Thatcher, 1984
[Brazil: Amazonas]

7(6) Caudal rami with rows of spinulae on ventral surface .... *Acusicola lycengraulidis* Thatcher & Boeger, 1993
[Brazil: Amazonas, Manaus]

7′ Caudal rami lacking ornamentation .... *Acusicola pellonidis* Thatcher & Boeger, 1983
[Brazil: Manaus, Amazon River]

8(1) Cephalothorax and first pedigerous somite completely or partially fused 9

8′ Cephalothorax and first pedigerous somite not fused; ventral surface of genital double-somite lacking ornamentation .... *Acusicola rogeri* Amado & Rocha, 1996
[Guatemala]

9(8) Cephalothorax and first pedigerous somite partially fused 10

9′ Cephalothorax and first pedigerous somite completely fused, quadrangular; cephalosome with anterior projection; ventral surface of genital double-somite ornamented with curved rows and patches of acute spinulae .... *Acusicola rotunda* Amado & Rocha, 1996
[Brazil]

10(9) Ventral surface of genital double-somite with rows of spinulae on posterior margin and posterolateral corners .... *Acusicola cunula* Cressey, 1970
[Brazil: Amazonas, Porto do Moz]

10′ Cephalothorax shape oblong; cephalosome lacking projection; ventral surface of genital double-somite with rows of spinulae on posterior margin and posterolateral corners .... *Acusicola paracunula* Amado & Rocha, 1996
[Brazil]

Poecilostomatoida: Ergasilidae: *Miracetyma*: Species

1 First maxillae smooth 2

1′ First maxillae with one medial seta .... *Miracetyma etimaruya* Malta 1993
[Brazil: Jiparaná River, Amazon]

2(1) Cephalothorax oblong; cuticle extensions distally in external and internal margins of second antenna endopod 1 .... *Miracetyma kawa* Malta 1993
[Brazil: Mamoré River, Amazon]

2′ Cephalothorax globose (Fig. 21.60 C); cuticle extensions in external and internal margins of second antenna endopod 1 .... *Miracetyma piraya* Malta 1993
[Brazil: Mamoré River, Amazon]

Poecilostomatoida: Ergasilidae: *Gamidactylus*: Species

1 Retrostylets not exceeding the posterior margin of cephalothorax (Fig. 21.56 A) 2

1′ Retrostylets reaching beyond posterior margin of cephalothorax (Fig. 21.56 B) 3

2(1) Spine on second antennary article present .... *Gamidactylus bryconis* Varella, 1995
[Brazil: Rondônia, Jamari River; Mamoré River]

2′ Spine on second antennary article absent .... *Gamidactylus hoplius* Varella & Varella, 1996
[Brazil]

3(1) Spine on second antennal article present; exopod 1 of legs 1–3 with one spine on first article (Figs. 21.3 M, 4 D) .... *Gamidactylus jaraquensis* Thatcher & Boeger, 1984
[Brazil]

3′ Spine on second antennal article absent; exopod 1 of legs 1 - 3 lacking spinae on first article .... *Gamidactylus piranhus* Thatcher, Santos & Brasil-Sato, 2008
[Brazil]

REFERENCES

Adamowicz, S.J., S. Menu-Marque, S.A. Halse, J.C. Topan, T.S. Zemlak, P.D.N. Hebert & J.D.S. Witt. 2010. The evolutionary diversification of the Centropagidae (Crustacea, Calanoida): A history of habitat shifts. Molecular Phylogenetics and Evolution 55: 418–430.

Alekseev, V. & D. Defaye. 2011. Taxonomic differentiation and world geographical distribution of the *Eucyclops serrulatus* group (Copepoda, Cyclopidae, Eucyclopinae). Pages 41–72 *in*: D. Defaye, E. Suárez-Morales & J.C.von Vaupel Klein (eds.), Studies on freshwater Copepoda: a volume in honour of Bernard Dussart. Koninklijke Brill NV, Leiden.

Alekseev, V., H.J. Dumont, J. Pensaert, D. Baribwegure & J.R. Vanfleteren. 2006. A redescription of *Eucyclops serrulatus* (Fischer, 1885) (Crustacea: Copepoda: Cyclopoida) and some related taxa, with a phylogeny of the *E. serrulatus*-group. Zoologica Scripta 35:123–147.

Amado, M.A.P.M. & C.E.F. Rocha. 2001. Useful characters in identifying copepods of the genus *Ergasilus* from plankton, with the description of male and female of *E. sergipensis* n. sp. Hydrobiologia 450: 149–157.

Andrade, E.R. & G.O. Brandorff. 1975. Uma nova espécie de Diaptomidae *Diaptomus negrensis* das águas perto de Manaus. Acta Amazonica 5: 97–103.

Bayly, I.A.E. 1992a. The non-marine Centropagidae (Copepoda: Calanoida) of the world. Guides to the identification of the Microinvertebrates of the Continental Waters of the World. SPB Academic Publishing, The Hague.

Bayly, I.A.E. 1992b. Fusion of the genera *Boeckella* and *Pseudoboeckella* (Copepoda) and revision of their species from South America and Sub-antarctic Islands. Revista Chilena de Historia Natural 65: 17–63.

Bodin, P. 1968. Copépodes harpacticoides des étages bathyal et abyssal du Golfe de Gascogne. Mémoirs du Muséum National d'Histoire Naturelle, Sér. A 55(1): 1–107.

Borutzky, E.V. 1952. Fauna of U.S.S.R. Crustacea. Freshwater Harpacticoida. Volume 3, No. 4. Izdatel'stvo Akademii Nauk SSSR, Moskva-Leningrad. 424 pp. (In Russian, English translation by Israel Program for Scientific Translations, Jerusalem, 1964, 396 pp).

Bowman, T.E. 1986. Freshwater calanoid copepods from the West Indies. Syllogeus 58: 237-246.

Boxshall, G. 2008. Harpacticoida. *In*: T.C. Walter & G. Boxshall. 2019. World of Copepods database. Accessed through: World Register of Marine Species at http://www.marinespecies.org/aphia.php?p=taxdetails&id=1102 on 2019-07-20.

Boxshall, G.A. & D. Defaye. 2008. Global diversity of copepods (Crustacea: Copepoda) in freshwater. Hydrobiologia 595: 195–207.

Boxshall, G.A. & S.H. Halsey. 2004. An Introduction to Copepod Diversity. The Ray Society, London, UK.

Boxshall, G.A. & D. Jaume. 2000. Making waves: the repeated colonization of fresh water by copepod crustaceans. Advances in Ecological Research 31: 61–79.

Brandorff, G.O. 1973. Neue freilebende calanoide Copepoden (Crustaceae) aus dem Amazonasgebiet. Amazoniana 4: 205-218.

Brandorff, G.O. 1976. The geographic distribution of the Diaptomidae in South America (Crustacea, Copepoda). Revista Brasileira de Biologia 36: 613–627.

Brandorff, G.O. 1978. Ein neuer Diaptomidae (Crustaceae, Copepoda) "*Diaptomus*" *ohlei* aus dem brasilianischen Amazonasgebiet. Archives für Hydrobiology 82: 295-298.

Brandorff, G.O. 2012. Distribution of some Calanoida (Crustacea: Copepoda) from the Yucatán Peninsula, Belize and Guatemala. Revista de Biología Tropical 60: 187–202.

Brehm, V. 1935. Mitteilungen von den Forschungsreisen Prof. Rahms. Mitteilung II. Gibt es in der Chilenischen Region Diaptomiden '*Diaptomus diabolicus*' nov. spec. Zoologischer Anzeiger 112: 9–13.

Brehm, V.V. 1957. Sobre los copépodos hallados por el Prof. Birabén en la Argentina. 4a. Comunicación. Neotropica 3: 53–64.

Bruno, M.C. & V. Cottarelli. 1999. Harpacticoids from groundwaters in the Philipines: *Parastenocaris mangyans*, new species, *Epactophanes philippinus*, new species, and redescription of *Phyllogntahopus bassoti* (Copepoda). Journal of Crustacean Biology 19: 510–529.

Bruno, M.C., K.J. Cunningham & S.A. Perry. 2003a. Copepod communities from surface and ground waters in the Everglades, South Florida. Southeastern Naturalist 2: 523–546.

Bruno, M.C., J.W. Reid & S.A. Perry. 2000. New records of copepods from Everglades National Park (Florida): description of two new species of *Elaphoidella* (Harpacticoida, Canthocamptidae), and supplementary description of *Diacyclops nearcticus* Kiefer (Cyclopoida, Cyclopidae). Crustaceana 73: 1171–1204.

Bruno, M.C., J.W. Reid & S.A. Perry. 2002a. New records of harpacticoid copepods from Everglades National Park (Florida, U. S. A.): description of *Nitokra evergladensis*, new species (Ameiridae), supplementary description of *Attheyella americana*, and redescription of *Bryocamptus newyorkensis* (Canthocamptidae). Journal of Crustacean Biology 22: 834–854.

Bruno, M.C, J.W. Reid & S.A. Perry. 2003b. Checklist and identification key for free-living copepod crustaceans in freshwater habitats of Everglades National Park and adjacent areas, Florida, U.S.A., with notes on their ecology. Florida Scientist 66: 23–42.

Bruno, M.C, J.W. Reid & S.A. Perry. 2005. A list and identification key for the freshwater, free-living copepods of Florida (U.S.A.). Journal of Crustacean Biology 25: 384–400.

Bruno, M.C., G. Sagnotti & S.A. Perry. 2002b. Planktonic copepods from temporary in Everglades National Park (Florida, U.S.A.). Hydrobiologia 485: 1–18.

Cicchino, G., E.N. Santos Silva & B. Robertson. 2001. A new species of Notodiaptomus Kiefer, 1936 (Copepoda, Diaptomidae) from the Amazon and Orinoco river basins. Hydrobiologia 453/454: 539–548.

Coelho Botelho, M.C. 1999. Revisão do gênero *Apocyclops* Lindberg, 1942 (Copepoda: Cyclopoida) Ph. D. Thesis. Instituto de Biociências. Universidade de São Paulo, São Paulo, Brazil.

Corgosinho, P.H.C., M. Hołyńska, F. Marrone, L.J. de Oliveira Geraldes-Primeiro, E.N. dos Santos-Silva, G. Perbiche-Neves & C. Lopez. 2019. An annotated checklist of freshwater Copepoda (Crustacea, Hexanauplia) from continental Ecuador and the Galapagos Archipelago. Zookeys 871: 55–77.

Corgosinho, P.H.C. & P. Martínez Arbizu. 2005. Two new interstitial species of *Remaneicaris* Jakobi 1972 from the Ribeirão do Ouro River, Minas Gerais, Brazil, with a redefinition of the genus (Crustacea, Copepoda, Harpacticoida, Parastenocarididae). Senckenbergiana biologica 85: 147–162.

Corgosinho, P.H.C., P. Martínez Arbizu & E.N. Dos Santos-Silva. 2007a. Redescription of *Remaneicaris ignotus* (Dussart, 1983), a Parastenocarididae (Copepoda, Harpacticoida) with an unusual set of plesiomorphic characters. Invertebrate Zoology 4: 31–44.

Corgosinho, P H.C., P. Martínez Arbizu & E.N. Dos Santos-Silva. 2007b. Three new species of *Remaneicaris* Jakobi, 1972 (Copepoda, Harpacticoida, Parastenocarididae) from the Ribeirao do Ouro River, Minas Gerais, Brazil, with some remarks on the groundpattern of the Parastenocarididae. Zootaxa 1437: 1–28.

Corgosinho, P.H.C., P. Martínez Arbizu & E.N. Dos Santos-Silva. 2010a. Three new interstitial species of *Remaneicaris* Jakobi, 1972 (Copepoda: Harpacticoida: Parastenocarididae) from Southern and Southeastern Brazil. Invertebrate Zoology 7: 1–28.

Corgosinho, P.H.C., P. Martínez Arbizu & E.N. Dos Santos-Silva. 2010b. Revision of *Brasilibathynellocaris* Jakobi, 1972 (Copepoda: Harpacticoida: Parastenocarididae) with redefinition of the genus. Zoological Journal of the Linnean Society 159: 527–566.

Corgosinho, P.H.C., P. Martínez Arbizu & D. Previattelli. 2012b. Establishment of a new genus for *Parastenocaris itica* (Copepoda, Harpacticoida) from El Salvador, Central America, with discussion of the *Parastenocaris fontinalis* and *P. proserpina* groups. Iheringia, Sér. Zoologia 102: 401–411.

Corgosinho, P.H.C., P. Martínez Arbizu & J.W. Reid. 2008. Revision of the genus *Murunducaris* (Copepoda: Harpacticoida: Parastenocarididae), with descriptions of two new species from South America. Journal of Crustacean Biology 28: 700–720.

Corgosinho, P.H.C., Y. Ranga Reddy & P. Martínez Arbizu. 2012a. Revision of the genus *Siolicaris* Jakobi, 1972, with redescriptions of *S. sioli* (Noodt, 1963) and *S. jakobi* (Noodt, 1963) from South America, and *S. sandhya* (Ranga Reddi, 2001) comb. nov. from India (Copepoda, Harpacticoida, Parastenocarididae). Zootaxa 3493: 49–71.

Corgosinho, P.H.C., N.V. Schizas, D. Previattelli, C.E.F. da Rocha & E.N. dos Santos-Silva. 2017. A new genus of Parastenocarididae (Copepoda, Harpacticoida) from the Tocantins River basin (Goiás, Brazil), and a phylogenetic analysis of the Parastenocaridinae. Zoosystematics and Evolution 93(1): 167–187.

Defaye, D. & B.H. Dussart, 1988. Compléments à la fauna des Crustacés copépodes des eaux intérieurs de Guyane Française. Revue d'hydrobiologie tropicale 21: 109–125.

Defaye, D. & B. Dussart. 2011. World directory of Crustacea Copepoda of inland waters. III-Harpacticoida, IV-Gelyelloida. Backhuys Publishers.

Dussart, B. 1967. Les copèpodes des eaux continentales d'Europe occidentale. Tome I: Calanoïdes et Harpacticoïdes. Editions N. Boubée & Cie, Paris, France.

Dussart, B.H. 1984a. Some Crustacea Copepoda from Venezuela. Hydrobiologia 113: 25–67.

Dussart, B.H. 1984b. Sur quelques copépodes d'Amérique du Sud, IV. Revista Brasileira de Biologia 44: 255–265.

Dussart, B.H. 1985. Sur quelques copépodes d'Amérique du Sud, V. Diaptomidae. Archiv für Hydrobiologie 103: 201–215.

Dussart, B.H. & D. Defaye. 1995. Introduction to the Copepoda. Guides to the Identification of the Microinvertebrates of the Continental Waters of the World. Volume 7. SPB Academic Publishing, Amsterdam.

Dussart, B.H. & D. Defaye. 2001. Introduction to the Copepoda *in*: H.J.F. Dumont (ed.), Guides to the Identification of the Microinvertebrates of the Continental Waters of the World, Volume 16. SPB Academic Publishing, The Hague. 344pp.

Dussart, B.H. & D. Defaye. 2006. World Directory of Crustacea Copepoda of Inland Waters II- Cyclopiformes. Backhuys Publishers. The Netherlands. i–ii, 1–354.

Dussart, B.H. & S.M. Frutos. 1985. Sur quelques Copépodes d'Argentine. Revue d' Hydrobiologie Tropicale 18: 305–314.

Dussart, B.H. & S.M. Frutos. 1986. Sur quelques Copépodes d'Argentine. 2. Copépodes du Paraná Medio. Revue d' Hydrobiologie Tropicale 19: 241–262.

Einsle, U., 1996. Copepoda: Cyclopoida. Genera *Cyclops, Megacyclops, Acanthocyclops*. *In*: H.J.F. Dumont (ed.), Guides to the identification of the Microinvertebrates of the Continental Waters of the World, Volume 10. SPB Academic Publishing, The Hague. 83 pp.

Elías-Gutiérrez, M. & E. Suárez-Morales. 1998. Redescription of *Microdiaptomus cokeri* (Crustacea: Copepoda: Diaptomidae) from caves in central Mexico, with the description of a new diaptomid family. Proceedings of the Biological Society of Washington 111: 199–208.

Elías-Gutiérrez, M., E. Suárez Morales, M.A. Gutiérrez-Aguirre, M. Silva Briano, J.G. Granados & T. Garfias. 2008. Cladocera y Copepoda de las aguas continentales de México. Guía ilustrada. UNAM, ECOSUR, México.

Fiers, F. 2006. The generic concept of *Allocyclops* Kiefer, 1932: (Copepoda: Cyclopoida: Cyclopidae) an alternative view. Journal of Natural History 46: 175–247.

Fiers, F., V. Ghenne & E. Suárez-Morales. 2000. New species of continental copepods (Crustacea, Cyclopoida) from the Yucatan Peninsula, Mexico. Studies on the Neotropical Fauna and Environment 35: 209–251.

Fuentes-Reinés, J.M. & E. Suárez-Morales. 2014. Annotated checklist and new records of Harpacticoida (Copepoda) from a coastal system of northern Colombia, South America. Crustaceana 87: 212–255.

Fuentes-Reinés, J.M. & E. Zoppi de Roa. 2013. Harpacticoid coppods from Ciénaga Grande de Santa Marta, Colombia. Métodos en Ecología y Sistemática 8:5–28.

Galassi, D.M.P. & P. De Laurentiis. 2004. Towards a revisión of the genus *Parastenocaris* Kessler, 1913: establishment of *Simplicaris* gen. nov. from groundwaters in central Italy and review of the *P. brevipes*-group (Copepoda, Harpacticoida, Parastenocaridiidae). Zoological Journal of the Linnean Society 140: 417–436.

Galassi, D.M.P., P. De Laurentiis & B. Fiasca. 2011. Systematics of the Phyllognathopodidae (Copepoda, Harpacticoida): re-examination of *Phyllognathopus viguieri* (maupas, 1892) and *Parbatocamptus jochenmartensi* Dumont and Maas, 1988, proposal of a new genus for *Phyllognathopus bassoti* Rouch, 1972, and description of a new species of *Phyllognathopus*. Zookeys 104: 1–65.

Galassi, D.M.P., R. Huys & J.W. Reid. 2009. Diversity, ecology and evolution of groundwater copepods. Freshwater Biology 54: 691–708.

Galassi, D.P. & G.L. Pesce. 1992. The genus *Hesperocyclops* Herbst: an update, and description of *H. venezuelanus* n.sp. from Venezuela (Crustacea Copepoda: Cyclopidae). Stygologia 7: 219–224.

Gaviria, S. 1993. Zwei Canthocamptidae (Copepoda, Harpacticoida) aus kolumbianischen Andengewässern. Annalen des Naturhistorischen Museums in Wien 94/95: 361–375.

Gaviria, S. 1994. Los copépodos (Arthropoda, Crustacea) de vida libre de las aguas continentales de Colombia. Revista de la Academia Colombiana de Ciencias Exactas, Físicas y Naturales 19: 127–134.

Gaviria, S. & N. Aranguren. 2007. Especies de vida libre de la subclase Copepoda (Arthropoda, Crustacea) en aguas continentales de Colombia. Biota Colombiana 8: 53–68.

Gaviria, S. & D. Defaye. 2012. A new species of *Attheyella* (*Canthosella*) from Colombia and redescription of *Attheyella* (*Delachauxiella*) *freyi* (Copepoda: Harpacticoida: Canthocamptidae). Zootaxa 3179: 1–38.

Gaviria, S. & D. Defaye. 2015. Description of *Elaphoidella paramuna* n. sp. (Canthocamptidae), a new harpacticoid copepod from Colombia. Crustaceana 88: 1003–1029.

Gaviria, S., D. Defaye & P.H.C. Corgosinho. 2017. Two new parastenocaridid genera (Copepoda, Harpacticoida) from South America, including a revisión of Noodt's *columbiensis*-group and the redescription of *Parastenocaris tapajosensis* Noodt, 1963. Crustaceana 90(5): 535–570.

Gómez S. & J.M. Gee. 2009. On four new species of *Cletocamptus* Shmankevich, 1875 (Copepoda: Harpacticoida) from inland waters of Argentina. Journal of Natural History 43: 2853–2910.

Gómez, S., R. Gerber & J.M. Fuentes-Reinés. 2017. Redescription of *Cletocamptus albuquerquensis* and *C. dominicanus* (Harpacticoida: Canthocamptidae *incertae sedis*), and description of two new species from the US Virgin Islands and Bonaire. Zootaxa 4272: 301–359.

Gómez, S. & F.N. Morales-Serna. 2014. Updated checklist of published and unpublished records of harpacticoid copepods (Crustacea: Copepoda: Harpacticoida) from Mexico. Proceedings of the Biological Society of Washington 127: 99–121.

Granados-Ramírez, J. & Suárez-Morales, E. 2003. A new *Hesperodiaptomus* Light (Copepoda, Calanoida, Diaptomidae) from Mexico with comments on the distribution of the genus. Journal of Plankton Research 25:1383–1395.

Grimaldo-Ortega, D., M. Camacho, M. Elías-Gutiérrez & J. Ciros-Pérez. 1998. Additions to Mexican freshwater copepods with the description of the female of *Leptodiaptomus mexicanus* (Marsh). Journal of Marine Systems 15: 381–390.

Gutiérrez-Aguirre, M.A. & A. Cervantes-Martínez. 2013. Diversity of freshwater copepods (Maxillopoda: Copepoda: Calanoida, Cyclopoida) from Chiapas, Mexico with a description of *Mastigodiaptomus suarezmoralesi* sp. nov. Journal of Natural History 47: 479–498.

Gutiérrez-Aguirre, M.A. & A. Cervantes-Martínez. 2016a. A new species of *Mastigodiaptomus* Light, 1939 from Mexico, with notes of species diversity of the genus (Copepoda, Calanoida, Diaptomidae). ZooKeys 637: 61–79.

Gutiérrez-Aguirre, M.A. & A. Cervantes-Martínez. 2016b. Taxonomic evaluation of eleven species of *Microcyclops* Claus, 1893 (Copepoda, Cyclopoida) and description of *Microcyclops inarmatus* sp. n. from America. ZooKeys 603: 33–69.

Gutiérrez-Aguirre, M.A., A. Cervantes-Martínez & M. Elías-Gutiérrez. 2014. An example of how barcodes can clarify cryptic species: the case of the calanoid copepod *Mastigodiaptomus albuquerquensis* (Herrick). PLoS One 9(1): e85019.

Gutiérrez-Aguirre, M.A., E. Suárez-Morales, A. Cervantes-Martínez, M. Elías-Gutiérrez & D. Previatelli. 2006. The neotropical species of *Mesocyclops* (Copepoda, Cyclopoida): an upgraded identification key and comments on selected taxa. Journal of Natural History 40: 549–570.

Hamond, R. 1987. Non-marine harpacticoid copepods of Australia. I. Canthocamptidae of the genus *Canthocamptus* Westwood s. lat. and *Fibulacamptus*, gen. nov., and including the description of a related new species of *Canthocamptus* from New Caledonia. Invertebrate Taxonomy 1: 1023–1247

Herbst, H.V. 1959. Brasilianische Süsswassercyclopoiden (Crustacea Copepoda). Gewässer und Abwässer 24: 49–73.

Herbst, H.V. 1962. Crustacea aus dem Amazonasgebiet, gesammelt von Professor Dr. H. Sioli und Dr. R. Braun. Litorale und substratgebundene Cyclopoida Gnathostoma (Copepoda). Crustaceana 3: 259–278.

Herbst, H.V. 1988. Zwei neue *Metacyclops* (Crustacea Copepoda) von den Westindischen Inseln Barbados und Aruba: *M. agnitus* n. sp. und *M. mutatus* n. sp., sowie ein Bestimmungsschlüssel für das genus. Bijdragen tot de Dierkunde 58: 137–154.

Hribar, L.J. & J.W. Reid. 2008. New records of copepods (Crustacea) from the Florida Keys. Southeastern Naturalist 7: 219–228.

Huys, R. 1996. Superornatiremidae fam. Nov. (Copepoda: Harpacticoida): An enigmatic family from North Atlantic anchihaline caves. Scientia Marina 60(4): 497–542.

Huys, R. 2009. Unresolved cases of type fixation, synonymy and homonymy in harpacticoid copepod nomenclature (Crustacea: Copepoda). Zootaxa 2183: 1–99.

Huys, R. 2016. Harpacticoid copepods –their symbiotic associations and biogenic substrata: a review. Zootaxa 4174: 448–729.

Huys, R. & G. Boxshall. 1991. Copepod Evolution. The Ray Society, The Natural History Museum, London, UK.

Huys, R., J.M. Gee, C.G. Moore & R. Hamond. 1996. Marine and brackish water harpacticoid copepods. Part 1. *In*: R.S.K. Barnes & J.H. Crothers (eds.). Synopses of the British Fauna (New Series). The Linnean Society of London, and The Estuarine and Coastal Sciences Association. 352 pp..

Iepure, S. & D. Defaye. 2003. Redescription of *Teratocyclops cubensis* Plesa, 1981 (Copepoda, Cyclopoida, Cyclopidae) from Cuba. Crustaceana 76: 591–604.

Jiménez-García, M.I. & E. Suárez-Morales. 2017. Complementary description of *Ergasilus arthrosis* Roberts, 1969 (Copepoda: Cyclopoida: Ergasilidae), a new parasite of cichlid teleosts in southeast Mexico. Systematic Parasitology 94: 81–90.

Karanovic, T. 2006. Subterranean copepods (Crustacea, Copepoda) from the Pilbara region in Western Australia. Records of the Western Australian Museum Supplement 70: 1–239.

Karanovic, T., J.A.E. Gibson, I. Hawes, D.T. Andersen & M.I. Stevens. 2014. *Diacyclops* (Copepoda: Cyclopoida) in Continental Antarctica, including three new species. Antarctic Science 26: 250–260.

Karanovic, T. & Y. Ranga Reddy. 2004. First record of *Phyllognathopus bassoti* Rouch, 1972 from India, with remarks on the family Phyllognathopodidae Gurney, 1932 (Crustacea, Copepoda, Harpacticoida). Annales de Limnologie-Internatoional Journal of Limnology 40: 121–132.

Karanovic, T. & D. Tang. 2009. A new species of the copepod genus *Australoeucyclops* (Crustacea: Cyclopoida: Eucyclopinae) from Western Australia shows the role of aridity in habitat shift and colonization of ground water. Records of the Western Australian Museum Supplement 25: 247–263.

Karaytug, S. 1999. Copepoda: Cyclopoida. Genera Paracyclops, Orchidacyclops and Key to Eucyclopinae. Guides to the identification of the Microinvertebrates of the Continental Waters of the World. (ed.) H.J.F. Dumont, Volume 14. Backhuys Publishers, Leiden.

Karaytug, S. & G.A. Boxshall. 1998. Partial revision of *Paracyclops* Claus, 1893 (Copepoda, Cyclopoida, Cyclopidae) with descriptions of four new species. Bulletin of the Natural History Museum London (Zoology) 64: 111–205.

Karaytug, S. & G.A. Boxshall. 1999. Antennules of the male of *Paracyclops* (Copepoda): functional significance and their importance in systematics. Journal of Crustacean Biology 19: 371– 379.

Khodami, S., J. Vaun McArthur, L. Blanco-Bercial & P. Martínez Arbizu. 2017. Molecular phylogeny and revisión of copepod orders (Crustacea: Copepoda). Scientific Reports 7: 9164.

Kiefer, F. 1931. Die Untergattung *Tropocyclops* der Gattung *Eucyclops* (Copepoda, Cyclopoida). Zeitschrift für wissenschafliche Zoologie 138: 487–514.

Kiefer, F. 1934. Neue Ruderfuβkrebse von der Insel Haiti. Zoologischer Anzeiger 108: 227–233.

Kiefer, F. 1936a. Freilebende Süβ- und Salzwassercopepoden von der Insel Haiti. Mit einer Revision der Gattung *Halicyclops* Norman. Archiv für Hidrobiologie 30: 263–317.

Kiefer, F. 1936b. Über die Systematik der südamerikanischen Diaptomiden (Crustacea Copepoda). Zoologischer Anzeiger 116(7/8): 194–200.

Kiefer, F. 1939. Ein neuer *Arctodiaptomus* (Crust. Cop.) aus der Türkei. Zoologischer Anzeiger 126(1–2): 24–29.

Kiefer, F. 1954. Drei neue Ruderfusskrebse (Crust. Cop) aus Venezuela. Zoologischer Anzeiger 153(7–8): 170–174.

Kiefer, F. 1957. Freilebende Ruderfuβkrebse (Crustacea Copepoda) des Titicacasees. Verhandlungen Zoologische Staatssammlung München 4: 125–150.

Kitching, R.L. 2000. Food Webs and Container Habitats: The Natural History and Ecology of Phytotelmata. Cambridge University Press.

Lang, K. 1934. Marine Harpacticiden von der Campbell-Insel und einigen anderen südlichen Inseln. Acta Universitatis Lundensis, New Series, Avd. 2, 30(14): 1–56.

Lang, K. 1948. Monographie der Harpacticiden. Volumes I–II. Håkan Ohlsson, Lund. 1682 pp..

Lee, C.E. & B.W. Frost. 2002. Morphological stasis in the *Eurytemora affinis* species complex (Copepoda: Temoridae). Hydrobiologia 480: 111–128.

Locascio de Mitrovich, C. & S. Menu-Marque. 1997. Redescription and distribution of *Boeckella gibbosa* (Brehm, 1935) (Copepoda, Calanoida, Centropagidae). Hydrobiologia 354: 157–163.

Locascio de Mitrovich, C. & S. Menu-Marque. 2001. A new *Diacyclops* (Copepoda, Cyclopoida, Cyclopidae) from northwestern Argentina. Hydrobiologia 453: 533–538.

Lowndes, A.G. 1934. Reports of an expeditions to Brazil and Paraguay in 1926-7 supported by the Trustees of the Percy Sladen Memorial Fund and the executive Committee of the Carniegie Trust of Scotland. Journal of the Linnean Society of London Zoology 39: 83–131.

Luque, J.L. & L.E.R. Tavares. 2007. Checklist of Copepoda associated with fishes from Brazil. Zootaxa 1579: 1–39.

Luque, J.L., F.M. Vieira, R.M. Takemoto, G.C. Pavanelli & J.C. Eiras. 2013. Checklist of Crustacea parasitizing fishes from Brazil. Check List 9: 1449–1470.

Malta, J.C.O. & A.M.B. Varella. 1996. *Ergasilus turucuyus* sp. n. (Copepoda: Ergasilidae) das brânquias de *Acestrorhyncus falcatus* (Bloch, 1794) e *A. flacirostris* (Cuvier, 1819) (Characiformes: Characidae) da Amazônia Brasileira. Acta Amazonica 26: 69–76.

Marques, M.T. 2014. Taxonomic Revision and Morphological Phylogenetic Analysis of known species of *Ergasilus* (Crustacea: Poecilostomatoida, Ergasilidae) Parasites of Freshwater Neotropical Fishes. Master Thesis, Universidad Federal do Paraná. 104 pp..

Marques, M.T., W.A. Boeger & M.C. Brasil-Sato. 2015. Two new species of *Ergasilus* Nordmann, 1832 (Copepoda: Ergasilidae) and a redescription of *Ergasilus salmini* Thatcher & Brazil-Sato, 2008 from *Salminus brasiliensis* Cuvier and *S. franciscanus* Lima & Britsky (Teleostei: Characidae) in Brazil. Systematic Parasitology 90: 81–89.

Matsumura-Tundisi, T. 1986. Latitudinal distribution of Calanoida copepods in freshwater aquatic systems of Brasil. Revista Brasileira de Biologia 46: 527–553.

Menu-Marque, S. & E. Balseiro. 2000. *Boeckella antiqua* n.sp. (Copepoda, Calanoida, Centropagidae) from Patagonia. Hydrobiologia 429: 1–7.

Menu-Marque, S., J.J. Morrone & C. L. Mitrovich. 2000. Distributional patterns of the South American species of *Boeckella* (Copepoda: Centropagidae): a track analysis. Journal of Crustacean Biology 20: 262–272.

Menu-Marque, S. & L. Zúñiga 1994. *Boeckella diamantina* n. sp. (Calanoida, Centropagidae) from a high Andean lake in Mendoza, Argentina. Hydrobiologia 292/293: 81–87.

Mercado, N., E. Suárez-Morales & M. Silva-Briano. 2006. A new *Acanthocyclops* Kiefer 1927 (Copepoda, Cyclopoida) from Central Mexico with comments on the distribution of the genus in Middle America. International Review of Hydrobiology 91: 148–163.

Mercado-Salas, N.F., C. Pozo, J.J. Morrone & E. Suárez-Morales. 2012. Distribution patterns of the American species of the freshwater genus *Eucyclops* (Copepoda: Cyclopoida). Journal of Crustacean Biology 32: 457–464.

Mercado-Salas, N.F. & E. Suárez-Morales. 2009. A new species and illustrated records of *Paracyclops* Claus, 1893 (Copepoda: Cyclopoida: Cyclopidae) from Mexico. Journal of Natural History. 43: 2789–2808.

Mercado-Salas, N.F. & E. Suárez-Morales. 2014a. Morphological variation of the freshwater copepod *Eucyclops elegans* (Herrick, 1884) in the Americas and comments on records of *Eucyclops conrowae* Reid, 1992. Journal of Natural History 48: 2007–2026.

Mercado-Salas, N F. & E. Suárez-Morales. 2014b. On Kiefer's American *Eucyclops* (Copepoda: Eucyclopinae): *E. delachauxi, E. prionophorus, E. bondi* and *E. leptacanthus*, redescriptions and comments on historical records. Zookeys 402: 1–41.

Mercado-Salas, N.F., E. Suárez-Morales & M. Silva-Briano. 2016. Taxonomic revision of the Mexican *Eucyclops* (Copepoda: Cyclopoida) with comments on the biogeography of the genus. Journal of Natural History 50: 25–147.

Mirabdullayev, I.M. & D. Defaye, 2002. On the taxonomy of the *Acanthocyclops robustus* species complex (Copepoda, Cyclopidae). 1. *Acanthocyclops robustus* (G.O. Sars, 1863) and *Acanthocyclops trajani* n. sp. Selevinia 1–4: 7–20.

Mirabdullayev, I.M. & D. Defaye. 2004. On the taxonomy of the *Acanthocyclops robustus* species complex (Copepoda, Cyclopidae): *Acanthocyclops brevispinosus* and *A. einslei* sp. n. Vestnik Zoologii 38: 27–37.

Montiel-Martínez, A., J. Ciros-Pérez, E. Ortega-Mayagoitia & M. Elías-Gutiérrez. 2008. Morphological, ecological, reproductive and molecular evidence for *Leptodiaptomus garciai* (Osorio-Tafall 1942) as a valid endemic species. Journal of Plankton Research 30: 1079–1093.

Mrázek, A. 1901. Süsswasser Copepoden. Hamburger Mgalhaeusische Sammelreise, S. 1: 29.

Muriel-Hoyos, F., A.M. Santana-Piñeros, Y. Cruz-Quintana & E. Suárez-Morales. 2015. A new species of *Ergasilus* (Copepoda: Cyclopoida: Ergasilidae) from *Bryconops giacopinii* (Characidae) in the Vichada River Basin, Colombia. Systematic Parasitology 92: 291–299.

Osorio-Tafall, B. 1941. *Diaptomus cuauhtemoci* nov. sp. de la mesa central de México (Copepoda, Diaptomidae). Ciencia 2: 296–298.

Osorio-Tafall, B. 1942. *Diaptomus (Microdiaptomus) cokeri* nuevo subgénero y especie de diaptómido de las cuevas de la región de Valles (San Luis Potosí). Ciencia 3: 206–210.

Paggi, J.C. 2011. A new species of the rare genus *Idiodiaptomus* Kiefer, 1936 (Copepoda, Calanoida, Diaptomidae) from northeastern Argentina. Crustaceana Monographs 16: 1570–7024.

Perbiche-Neves, G., G.A. Boxshall, D. Previattelli, D.A. Naliato, M. Pie, C.E.F. Rocha & M.G. Nogueira. 2015. Regulation of the abundance and turnover of copepod species by temperature, turbidity and habitat type in a large river basin. Austral Ecology 40: 718–725.

Perbiche-Neves, G., G.A. Boxshall, J.C. Paggi, C.E.F. Rocha, D. Previattelli & M.G. Nogueira. 2013. Two new species of Diaptomidae (Crustacea: Copepoda: Calanoida) from the Neotropical Region (Paraná River). Journal of Natural History 47(5–12): 449–477.

Perbiche-Neves, G., M.G. Nogueira & C.E.F. Rocha. 2014a. Trends in planktonic copepod diversity in reservoirs and lotic stretches in a large river basin in South America. Marine and Freshwater Research 65: 727–737.

Perbiche-Neves, G., D. Previattelli & M.G. Nogueira. 2011. Record of *Argyrodiaptomus bergi* (Crustacea: Copepoda: Calanoida) after 36 years and first record in Brazil. Zoologia 28: 551–557.

Perbiche-Neves, G., D. Previattelli, M.R. Pie, A. Duran, E. Suárez-Morales, G.A. Boxshall, M.G. Nogueira & C.E.F. Rocha. 2014b. Historical biogeography of the Neotropical Diaptomidae (Crustacea: Copepoda). Frontiers in Zoology 11: 36.

Perbiche-Neves, G., V.S.S. Saito, D. Previattelli, C.E.F. Rocha & M.G. Nogueira. 2016. Cyclopoid copepods as bioindicators of eutrophication in reservoirs: Do patterns hold for large spatial extents? Ecological Indicators 70: 340–347.

Petkovski, T.K. & A. Brancelj. 1988. *Elaphoidella serbica* n. sp., ein neuer Harpacticoida aus Ostserbien (Crustacea, Copepoda). Fragmenta Balcanica, Musei Macedonici Scientiarum Naturalium, Skopje 14 (6/299): 47–58.

Pinto-da-Rocha, R. 1994. Invertebrados cavernicolas da porçao meridional da Provincia Espeleologica do Vale do Ribeira, sul do Brasil. Revista Brasileira de Zoologia 10: 229–255.

Pinto-da-Rocha, R. 1995. Sinopse da fauna cavernícola do Brasil (1907–1994). Papéis Avulsos de Zoologia, Museu de Zoologia da Universidade de São Paulo 39: 61–173.

Por, F.D. 1986. A reevaluation of the family Cletodidae Sars, Lang (Copepoda, Harpacticoida). *In* G. Schriever, H.K. Schminke & C.T. Shih (eds.). Proceedings of the Second International Conference on

Copepoda, Ottawa, Canada, 13–17 August, 1984. Syllogeus Volume 58: 420–425.

Previattelli, D. 2006. Sistemática do gênero Argyrodiaptomus Brehm, 1933 (Crustacea, Copepoda, Diaptomidae). Masters Thesis. Universidade Federal do Amazonas – Instituto Nacional de Pesquisas da Amazônia, Manaus, Brasil.

Previattelli, D., G. Perbiche-Neves & E.N. Santos-Silva. 2013. New Diaptomidae records (Crustacea: Copepoda: Calanoida: Diaptomidae) in the Neotropical region. Check List 9: 700–713.

Rayner, N.A. 1999. Copepoda: Calanoida (Diaptomidae: Paradiaptominae). Guides to the identification of the microinvertebrates of the continental waters of the world. Backhuys Publishers, Leiden. 122 pp..

Reeves, W.K. & J.W. Reid. 2002. New records and a checklist of continental free-living Copepoda (Crustacea) from Panama. Zootaxa 62: 1–8.

Reid, J.W. 1982. *Forficatocaris schadeni*, a new copepod (Harpacticoida) from central Brazil, with keys to the species of the genus. Journal of Crustacean Biology 2: 578–587.

Reid, J.W. 1984. Semiterrestrial meiofauna inhabiting a wet campo in central Brazil, with special reference to the Copepoda (Crustacea). Hydrobiologia 118: 95-111.

Reid, J.W. 1985a. Calanoid copepods (Diaptomidae) from coastal lakes, State of Rio de Janeiro, Brazil. Proceedings of the Biological Society of Washington 98: 574–590.

Reid, J.W. 1985b. Chave de identificação e lista de referencias bibliográficas para as especies continentais sulamericanas de vida livre da ordem Cyclopoida (Crustacea, Copepoda). Boletim de Zoologia, Universidad de São Paulo 9: 17–143.

Reid, J.W. 1987a. Some cyclopoid and harpacticoid copepods from Colombia, including descriptions of three new species. Proceedings of the Biological Society of Washington 100: 262–271.

Reid, J.W. 1987b. The cyclopoid copepods of a wet campo marsh in central Brazil. Hydrobiologia 153: 121–138.

Reid, J.W. 1988. Cyclopoid and harpacticoid copepods (Crustacea) from Mexico, Guatemala, and Colombia. Transactions of the American Microscopical Society 107: 190–202.

Reid, J.W. 1989. The distribution of species of the genus *Thermocyclops* (Copepoda, Cyclopoida) in the western hemisphere, with description of *T. parvus*, new species. Hydrobiologia 175: 149–174.

Reid, J.W. 1990. Continental and coastal free-living Copepoda (Crustacea) of Mexico, Central America and the Caribbean region. Pages 175–213 *in*: D. Navarro & J.G. Robinson (eds.), Diversidad Biológica en la Reserva de la Biósfera de Sian Ka'an, Quintana Roo, México. Centro de Investigaciones de Quintana Roo and Program of Studies in Tropical Conservation. University of Florida, Chetumal.

Reid, J. W. 1991a. Some species of *Tropocyclops* (Crustacea, Copepoda) from Brazil, with a key to the American species. Bijdragen tot de Dierkunde 61: 3–15.

Reid, J.W. 1991b. The neotropical genus *Potamocaris* Dussart (Copepoda: Harpacticoida: Parastenocarididae), with descriptions of two new species. Journal of Crustacean Biology 11: 463–472.

Reid, J.W. 1992. Copepoda (Crustacea) from fresh waters of the Florida Everglades, U.S.A., with a description of *Eucyclops conrowae* n. sp. Transactions of the American Microscopical Society 111: 229–254.

Reid, J.W. 1993a. *Fimbricyclops jimhensoni*, new genus, new species (Copepoda: Cyclopoida: Cyclopidae) from bromeliads in Puerto Rico. Journal of Crustacean Biology 13: 383–392.

Reid, J.W. 1993b. The harpacticoid and cyclopoid copepod fauna in the Cerrado Region of Central Brazil. 1. Species composition, habitats, and zoogeography. Acta Limnologica Brasiliensia 6: 56–68.

Reid, J.W. 1994. Latitudinal diversity patterns of continental benthic copepod species assemblages in the Americas. Hydrobiologia 292: 342–249.

Reid, J.W. 1995. Redescription of *Parastenocaris brevipes* Kessler and description of a new species of *Parastenocaris* (Copepoda: Harpacticoida: Parastenocarididae) from the U.S.A. Canadian Journal of Zoology 73: 173–187.

Reid, J.W. 1998. Maxillopoda-Copepoda, Harpacticoida. Pages 75–127 *in*: P.S. Young (ed.), Catalogue of Crustacea of Brazil. Museu Nacional Série livrvons, Rio de Janeiro: 6.

Reid, J.W. 2016. Subclass Copepoda. Pages 524–573 *in*: J.H. Thorp & D.C. Rogers (eds.), Volume II: Keys to Nearctic Fauna. Thorp and Covich's Freshwater Invertebrates, 4th edition. Academic Press, Elsevier, London.

Reid, J.W., R. Hamilton IV & R.M. Duffield. 2002. First confirmed New World record of *Apocyclops dengizicus* (Lepeshkin), with a key to the species of *Apocyclops* in North America and the Caribbean region (Crustacea: Copepoda: Cyclopidae). Jeffersoniana 10: 1–25.

Reid, J.W. & L.J. Hribar. 2006. Records of some Copepoda (Crustacea) from the Florida Keys. Proceedings of the Academy of Natural Sciences of Philadelphia 155: 1–7.

Reid, J.W. & W. Janetzki. 1996. Colonization of Jamaican Bromeliads by *Tropocyclops jamaicensis* n. sp. (Crustacea: Copepoda: Cyclopoida). Invertebrate Biology 115: 305–320.

Reid, J.W. & E. Suárez-Morales. 1999. A new, Neotropical species of *Acanthocyclops* (Copepoda: Cyclopoida: Cyclopidae). Beaufortia 49: 37–45.

Reid, J.W. & C.E. Williamson. 2010. Copepoda. Pages 829–899 *in*: J.H. Thorpe & A.P. Covich (eds.), Ecology and Classification of North American Freshwater Invertebrates, 3rd edition. Academic Press, New York.

Rocha, C.E.F. 1994. New species of *Metacyclops* (Copepoda, Cyclopidae) from Brazil, with remarks on *M. campestris*. Zoologica Scripta 23: 133–146.

Rocha, C.E.F. 1998. New morphological characters useful for the taxonomy of the genus *Microcyclops* (Copepoda, Cyclopoida). Journal of Marine Systems 15: 425–431.

Rocha, C.E.F. & M.H.G. Bjornberg. 1987. Copepods of the Juréia Ecological Reserve, State of São Paulo, Brazil. II. The genera *Hesperocyclops, Muscocyclops*, and *Bryocyclops* (Cyclopoida, Cyclopidae). Hydrobiologia 153: 97–107.

Rocha, C.E. F. & M.H.G. Bjornberg. 1988. *Allocyclops silvaticus* sp. n. (Copepoda, Cyclopoida, Cyclopidae), the first representative of the genus in South America. Hydrobiologia 167/168: 445–448.

Rocha, C.E.F. & M.C.J. Botelho. 1998. Maxillopoda-Copepoda. Cyclopoida *in*: P.S. Young (ed.), Catalogue of crustaceans of Brazil. Museu Nacional, Série Livros, Rio de Janeiro: 6.

Rocha, C.E.F., I. Claret Torres & P. Maia-Barbosa. 1998a. *Haplocyclops torresi* n.sp. and *Potamocaris estevesi* Reid, 1991 from Brazil, with a proposal for revalidation of the genus *Haplocyclops* Kiefer, 1952 (Copepoda). Beaufortia 48: 1–15.

Rocha, C.E. F., T.M. Iliffe, J.W. Reid & E. Suárez-Morales. 1998b. A new species of *Halicyclops* (Copepoda, Cyclopoida, Cyclopidae) from cenotes of the Yucatan peninsula, Mexico, with an identification key for the species of the genus from the Caribbean region and adjacent areas. Sarsia 83: 387–399.

Romero, A. 2009. Cave biology, life in darkness. Cambridge University Press. 291 pp..

Samanez, I. & D. López. 2014. Distribución geográfica de *Boeckella* y *Neoboeckella* (Calanoida: Centropagidae) en el Perú. Revista Peruana de Biología 21: 223–228.

Santos-Silva, E.N. 2008. Calanoid of the families Diaptomidae, Pseudodiaptomidae, and Centropagidae from Brasil. Biologia Geral e Experimental 8: 3–67.

Santos-Silva, E.N., G.A. Boxshall & C.E.F. Rocha. 2013. Revisão das espécies do "complexo *nordestinus*" (Wright, 1935) de *Notodiaptomus* Kiefer, 1936 (Copepoda: Calanoida: Diaptomidae). Biologia Geral e Experimental 13: 1–115.

Santos-Silva, E.N., M. Elías-Gutiérrez & M. Silva-Briano. 1996. Redescription and distribution of *Mastigodiaptomus montezumae* (Copepoda, Calanoida, Diaptomidae) in Mexico. Hydrobiologia 328: 207–213.

Santos-Silva, E.N., B.A. Robertson, J.W. Reid & E.R. Hardy. 1989. Atlas de copépodos planctônicos, Calanoida e Cyclopoida (Crustacea), da Amazônia Brasileira. I. Represa de Curuá-Una, Pará. Revista Brasileira de Zoologia 6: 725–758.

Sars, G.O. 1901. Contributions to the knowledge of the freshwater Entomostraca of South America. Part 11. Copepoda- Ostracoda. Archiv for Mathematik og Naturvidenskab 24(1): 1-52, pl. I–YIII.

Selden, P.A., R. Huys, M.H. Stephenson, A.P. Heward & P.N. Taylor. 2010. Crustaceans from bitumen clast in Carboniferous glacial diamictite extend fossil record of copepods. Nature Communications 1(50): 1–6.

Sewell, R.B.S. 1949. The littoral and semi parasitic Cyclopoida, the Monstrilloida and Notodelphyoida. The John Murray Expedition 1933–34 Scientific Reports 9: 17–199.

Shen, C.J., A.Y. Tai & Y.Z. Song. 1979. Fauna Sinica. Crustacea. Freshwater Copepoda. Harpacticoida:164–300. Science Press, Peking, China.

Silva, W.M. 2008. Diversity and distribution of the free-living freshwater Cyclopoida (Copepoda: Crustacea) in the Neotropics. Brazilian Journal of Biology 68: 1099–1106.

Silva, W.M. & T. Matsumura-Tundisi. 2005. Taxonomy, ecology and geographic distribution of the species of the genus *Thermocyclops* Kiefer, 1927 (Copepoda, Cyclopoida) in São Paulo State, Brazil, with description of a new species. Brazilian Journal of Biology 65: 533–540.

Silva-Briano, M. & E. Suárez-Morales. 2010. Illustrated record of the freshwater copepod *Leptodiaptomus dodsoni* (Calanoida, Diaptomidae) from central Mexico with comments on the distribution of the genus. Biota Neotropica 10: 409–414.

Suárez-Morales, E. 2003. Historical biogeography and distribution of the freshwater calanoid copepods (Crustacea: Copepoda) of the Yucatan Peninsula, Mexico. Journal of Biogeography 30: 1851–1859.

Suárez-Morales, E. 2015. Chapter 29. Class Maxillopoda. Pages 709–755 *in*: J.H. Thorp & D.C. Rogers (eds.), Thorp & Covich's Freshwater Invertebrates, 4th edition. Volume I: Ecology and General Biology. New York, Academic Press.

Suárez-Morales, E., M. De Troch & F. Fiers. 2006. A checklist of the marine Harpacticoida (Copepoda) of the Caribbean Sea. Zootaxa 1285:1–19.

Suárez-Morales, E. & M.A. Gutiérrez-Aguirre. 2001. Morfología y taxonomía de los *Mesocyclops* (Crustacea: Copepoda: Cycopoida) de México. El Colegio de la Frontera Sur (ECOSUR) and Consejo Nacional de Ciencia y Tecnología (CONACYT). Ciudad de México, Mexico.

Suárez-Morales, E., M.A. Gutiérrez-Aguirre & E.J. Walsh. 2010. Freshwater Copepoda (Crustacea) from the Chihuahuan desert with comments on biogeography. The Southwestern Naturalist 55: 525–531.

Suárez-Morales E. & N.A. Rayner. 2004. New records for the South African diaptomid fauna with a complementary description of *Paradiaptomus lamellatus* Sars (Copepoda, Diaptomidae). Journal of Natural History 38: 2803–2819.

Suárez-Morales, E. & J.W. Reid. 1998. An updated list of the free-living freshwater copepods (Crustacea) of Mexico. The Southwestern Naturalist 43:256–265.

Suárez Morales, E., J.W. Reid & M. Elías-Gutiérrez. 2005. Diversity and distributional patterns of Neotropical freshwater copepods (Calanoida: Diaptomidae). International Review of Hydrobiology 90: 71–83.

Suárez-Morales, E., J.W. Reid, T.M. Iliffe & F. Fiers. 1996. Catálogo de los copépodos (Crustacea) continentales de la Península de Yucatán, México. CONABIO/ECOSUR, Mexico City.

Suárez-Morales, E., M. Silva-Briano & M. Elías-Gutiérrez. 2000. Redescription and taxonomic validity of *Leptodiaptomus cuauhtemoci* (Osorio-Tafall, 1941) (Copepoda, Calanoida), with notes on its known distribution. Journal of Limnology 59: 5–14.

Suárez-Morales, E., R.J. Wasserman & T. Dalu. 2015. A new species of *Lovenula* Schmeil (Copepoda, Calanoida, Diaptomidae) from the Eastern Cape Province of South Africa. Crustaceana 88: 324–342.

Ueda, H. & J.W. Reid. 2003. Copepoda: Cyclopoida, Genera *Mesocyclops* and *Thermocyclops*. Guides to the Identification of the Microinvertebrates of the Continental Waters of the World. Backhuys Publishers. Leiden, The Netherlands.

Walter, T.C. 1989. Review of the new world species of *Pseudodiaptomus* (Copepoda: Calanoida), with a key to the species. Bulletin of Marine Science 43: 590–628.

Walter, T.C. & G. Boxshall. 2019. World of Copepods database. Accessed at http://www.marinespecies.org/copepoda on 2019-07-20. doi:10.14284/356.

Watling, L. 1989. A classification system for crustacean setae based on the homology concept. Pages 15–26 *in*: Felgenhauer, B. E., Watling, L. & Thisdtle, A. B. (eds.). Functional morphology of feeding and grooming in Crustacea. Balkema, Netherlands.

Wells, J.B.J. 2007. An annotated checklist and keys to the species of Copepoda Harpacticoida (Crustacea). Zootaxa 1568: 1–872.

Wilson, M.S. 1941. New species and distribution records of diaptomid copepods from the Marsh collection in the United States National Museum. Journal of the Washington Academy of Sciences 31: 509-515.

Wright, S. 1927. A revision of the South American species of *Diaptomus*. Transactions of the American Microscopical Society 46: 73–121.

Wright, S. 1935. Three new species of *Diaptomus* from Northeast Brazil. Annais da Academia Brasileira de Ciências 7:213-231.

Chapter 22

Class Branchiura

Eduardo Suárez-Morales
El Colegio de la Frontera Sur (ECOSUR), Chetumal, Mexico

CLASS BRANCHIURA

Chapter Outline

INTRODUCTION

Branchiurans are crustacean ectoparasites on ectoparasites teleosts and dwell mainly in freshwaters but occur frequently also in brackish and marine waters. These parasites are also found on some amphibians including the salamanders *Pseudobranchus axanthus* (Netting & Goin, 1942) from Florida and *Ambystoma dumerilii* (Dugès, 1870) from Mexico, and tadpoles of *Rana* species (Poly, 2003). There is a single family Argulidae, which contains about 155 nominal species belonging to 4 valid genera: *Argulus, Chonopeltis, Dipteropeltis*, and *Dolops* (Poly, 2008; Møller, 2009), each with a distinctive body structure (Fig. 22.1). The earliest descriptions of these crustaceans date back to the 10th century in China, and they did not escape the observations by Linné who described *A. foliaceus* (L., 1758) (as *Monoculus foliaceus*), but the knowledge about different aspects of this group have relied on a few widespread species only, thus leaving most argulids poorly studied (Møller, 2009).

Branchiurans are highly efficient parasites; they feed by piercing the skin of their fish hosts, injecting a toxin, and drawing off blood. Heavy infestations cause increased mortality of teleost hosts; infested individuals may stop feeding. Secondary bacterial infections are also another cause of mortality related to the infestation of branchiurans (Suárez-Morales, 2015). Argulids can move freely on their host skin, but they can also maintain their position using their first maxillae, modified either as suction discs (i.e., *Argulus*, *Dipteropeltis*, *Chonopeltis*) or as hooks (*Dolops*) (Møller et al., 2008).

Earliest classifications of this group included them among the Branchiopoda or with the Copepoda, as a separate suborder (Krøyer, 1863; Thorell, 1864). Branchiurans were separated as an independent group equal to the Copepoda based on the free nature of the maxillae during its ontogenetic development. Studies of higher level relationships link the Branchiura with either the Pentastomida or Ostracoda. Based on the sperm morphology several authors supported an alliance between the Pentastomida and the Branchiura, as sister groups (Martin & Davis, 2001), but the question is still open because of the recently hypothesized position of the Pentastomida far from the Crustacea (Møller, 2009). They are currently recognized as a separate subclass of the Class Ichthyostraca, superclass Oligostraca (WoRMS, 2019).

LIMITATIONS

Argulid taxonomy relies on a relatively reduced set of characters that includes the shape of the carapace, the shape and relative size of the carapace lobes, the size of the abdomen, the shape and arrangement of the respiratory areas, the number of sclerites in the sucker rods, and the structure of the maxillary teeth, among others. These characters are relatively easily observable (i.e., without dissection of appendages, using a stereomicroscope) in most species and their combination is unique for each. There are very detailed recent species descriptions and redescriptions (Poly, 2005; Møller & Olesen, 2010; Silva-Souza et al., 2011; Neethling et al., 2014; Aguiar et al., 2017) that add to

Thorp and Covich's Freshwater Invertebrates. https://doi.org/10.1016/B978-0-12-804225-0.00022-8

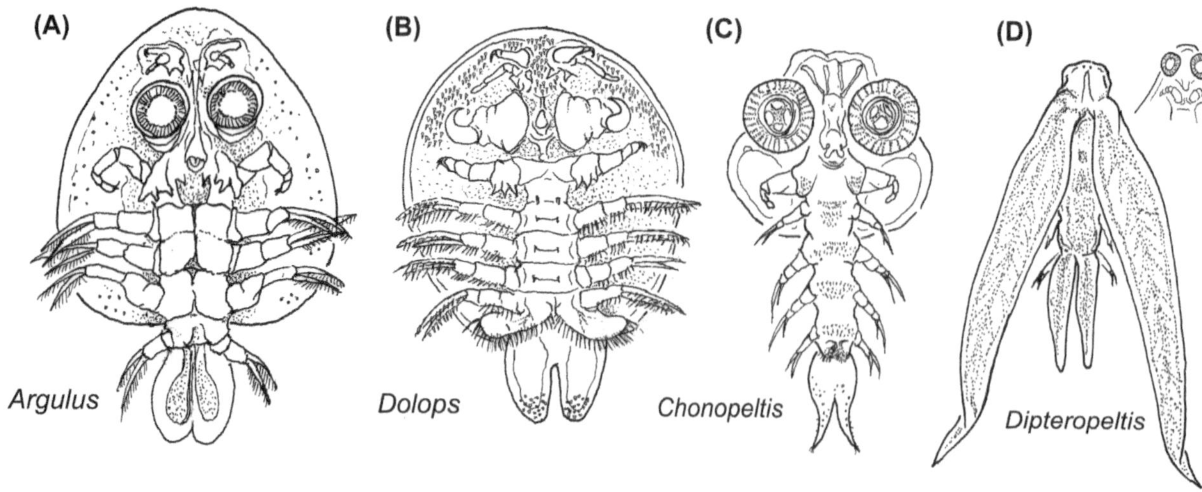

FIGURE 22.1 General view of the different body structures known in the Argulidae: (A) *Argulus* ventral view (approximately 1.7 mm long); (B) *Dolops* sp. ventral view; (C) *Chonopeltis* sp. ventral view; (D) *Dipteropeltis* sp. dorsal view with ventral view of suckers.

the knowledge of the group biodiversity and morphology but also provide a rich set of characters for further comparative and phylogenetic analyses. The key presented herein was designed to be applicable to either sex. It is intended to include all the species known from the Neotropical region including Mexico, an effort not previously attempted. The key is based on published regional identification keys and descriptions and illustrations provided by different authors (Bouvier, 1899a, 1899b; Wilson, 1902; Ringuelet, 1943, 1948; Weibezahn & Cobo, 1964; Cressey, 1972; Lemos de Castro, 1986; Thatcher, 2006) that were adapted where necessary. New species and records from Mexico (Pineda et al., 1995; Suárez-Morales et al., 1998; Poly, 2003, 2005), Bolivia (Mamani et al., 2004), Brazil (Malta & Santos-Silva, 1986; Malta & Varella, 2000; Neethling et al., 2014) not considered in recent accounts of the Neotropical argulids are included in this key. The recently described Cuban species *Argulus poeyi* (Fernández & Corrada, 2015), based on male specimens, was not included in the key because it was found on marine fish only and several important characters were not fully described or depicted; it has a distinctive carapace shape (Fernández & Corrada, 2015).

TERMINOLOGY AND MORPHOLOGY

Argulids are characterized by their dorsoventrally flattened bodies and the possession of **carapace** lobes which may or may not cover their four biramous thoracic legs. Their body size ranges between 4 and 15 mm, but most are usually 3–5 mm long. Specimens of *Dipteropeltis hirundo* (Calman, 1912) and *Dolops longicauda* (Heller, 1857) can reach up to 30 mm in length (Møller, 2009). They have a pair of compound eyes on the anterior part of the carapace, four pairs of swimming legs, and an unsegmented abdomen (Poly, 2008) (Figs. 22.1–3). The carapace coverage and shape shows distinctive variations at the genus level. Argulids have **cephalic shields** of different shapes which extend into the carapace lobes (Fig. 22.3) and can cover the legs. A slight indentation is present where the cephalic shield joins the carapace lobes (Figs. 22.1, 4). In *Argulus* and *Dolops* the carapace has two pairs of ventral respiratory areas located on each lateral lobe of the carapace.

The cephalothorax bears five pairs of appendages (Fig. 22.2) anterior to the first swimming leg. The first and second appendages are the **first antennae** and the **second antennae**, respectively. These differ among species in their structure and armature. The second antennae are inserted close together within a slight cuticular depression in front of each compound eye on the anteroventral surface of the shield. Posterior to these appendages is a **postantennal spine** (Fig. 22.2). In *Argulus*, the second antennae may have three or four articles each armed with distinct spines and processes (Tam et al., 2005) (Fig. 22.2). A second antenna may be absent in a few species (Overstreet et al., 1992).

The third pair of appendages is represented by the **first maxillae** (maxillula) which are usually modified as radially symmetrical suction discs (*Argulus*, *Chonopeltis*, *Dipteropeltis*) or as strong hooks (*Dolops* and *Argulus chinensis*; Ku & Yang, 1955) (Overstreet et al., 1992). **Suckers** are sclerotized, muscular organs (Fig. 22.2); their outermost

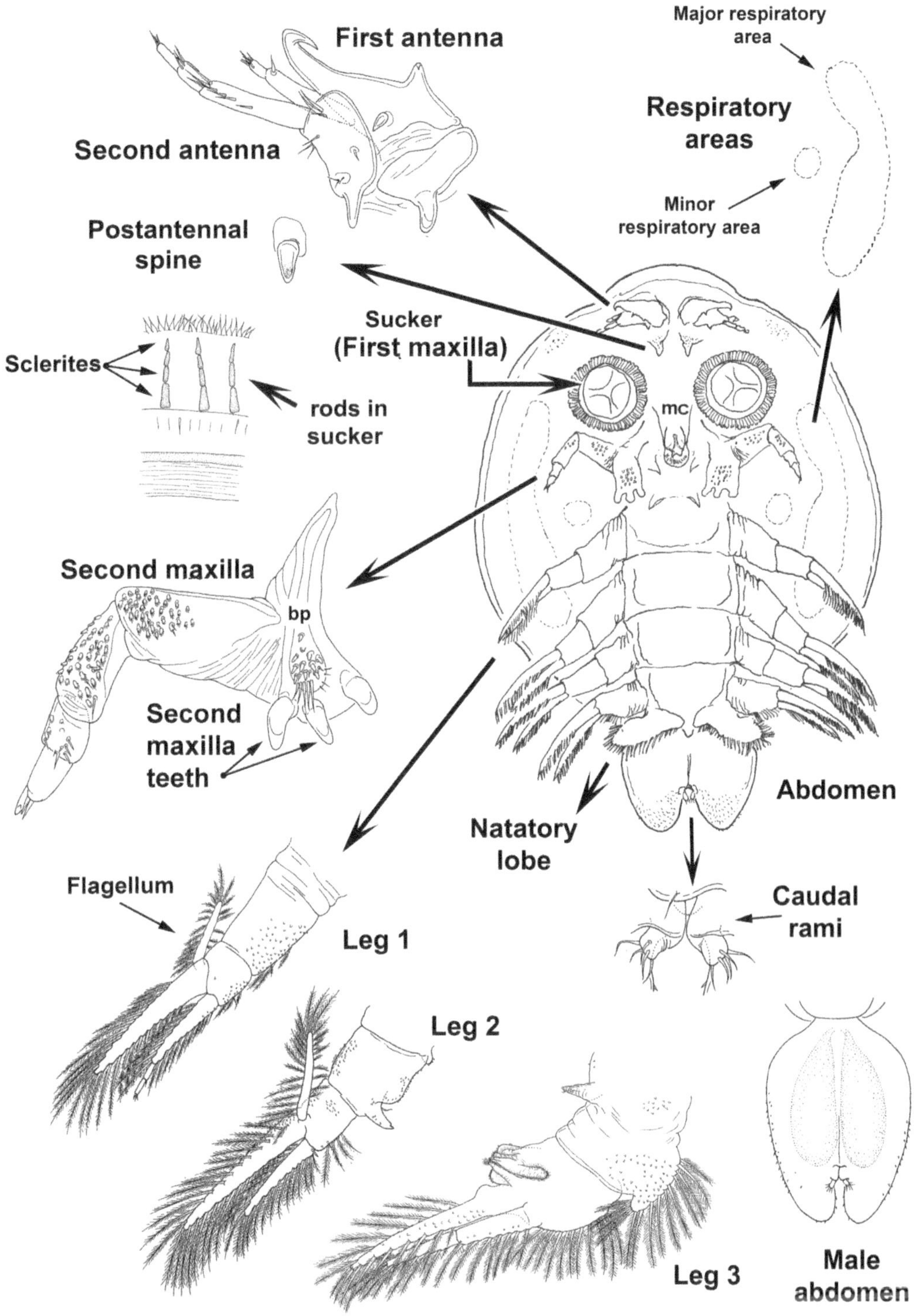

FIGURE 22.2 Generalized morphology of the body and appendages of argulid branchiurans: Ventral view of *Argulus* sp. (body length = 1.7 mm) showing general structure and basic appendage morphology. Abbreviations: *bp*, basipod; *mc*, mouth cone.

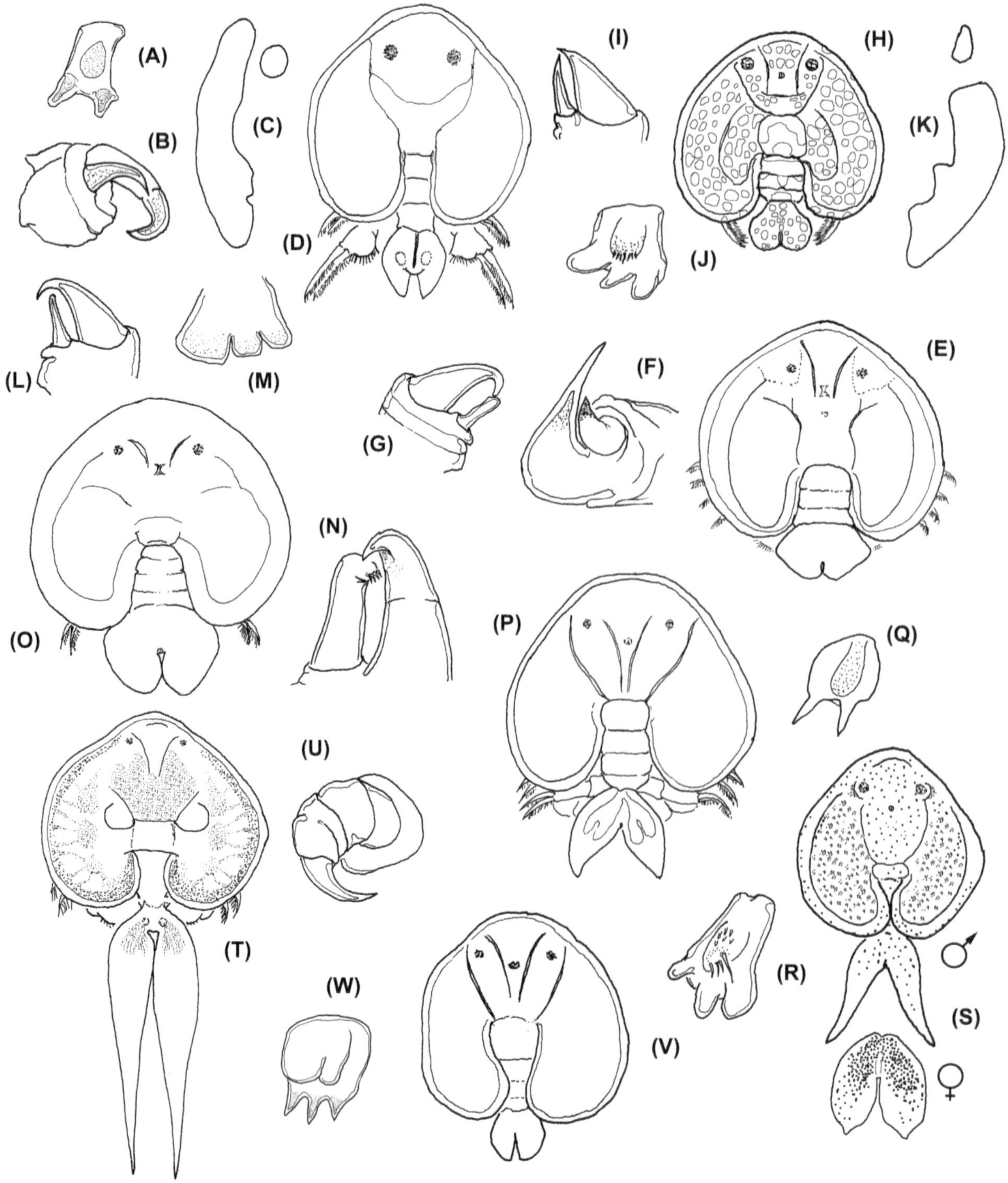

FIGURE 22.3 Structures of species of *Dolops* included in the key: (A) *D. bidentata* second maxilla teeth; (B) *D. bidentata* first maxilla terminal hook; (C) *D. bidentata* respiratory areas; (D) *D. bidentata* dorsal view of body; (E) *D. reperta* dorsal view of body; (F) *D. reperta* first maxilla hook; (G) *D. kollari* first maxilla hook; (H) *D. discoidalis* dorsal view of body; (I) *D. discoidalis* first maxilla hooks; (J) *D. discoidalis* second maxilla teeth; (K) *D. discoidalis* respiratory areas; (L) *D. striata* first maxilla hook; (M) *D. striata* second maxilla teeth; (N) *D. intermedia* first maxilla hook; (O) *D. intermedia* dorsal view of body; (P) *D. nana* dorsal view of body; (Q) *D. nana* second maxilla teeth; (R) *D. geayi* second maxilla teeth; (S) *D. geayi* dorsal view of body (male) and female abdomen; (T) *D. longicauda* dorsal view of body; (U) *D. longicauda* first maxilla hook; (V) *D. carvalhoi* dorsal view of body; (W) *D. carvalhoi* second maxilla teeth. *Figures redrawn: (A–D) from Silva-Souza et al. (2011); (E), (U) from Wilson (1902); (F), (H–J), (L), (M), (S), (T) from Bouvier (1899b); (G) from Thiele (1904); (K), (R) from Ringuelet (1943); (N), (O) from Silva (1978); (P), (Q) from Lemos de Castro (1950); V, W from Lemos de Castro (1949).*

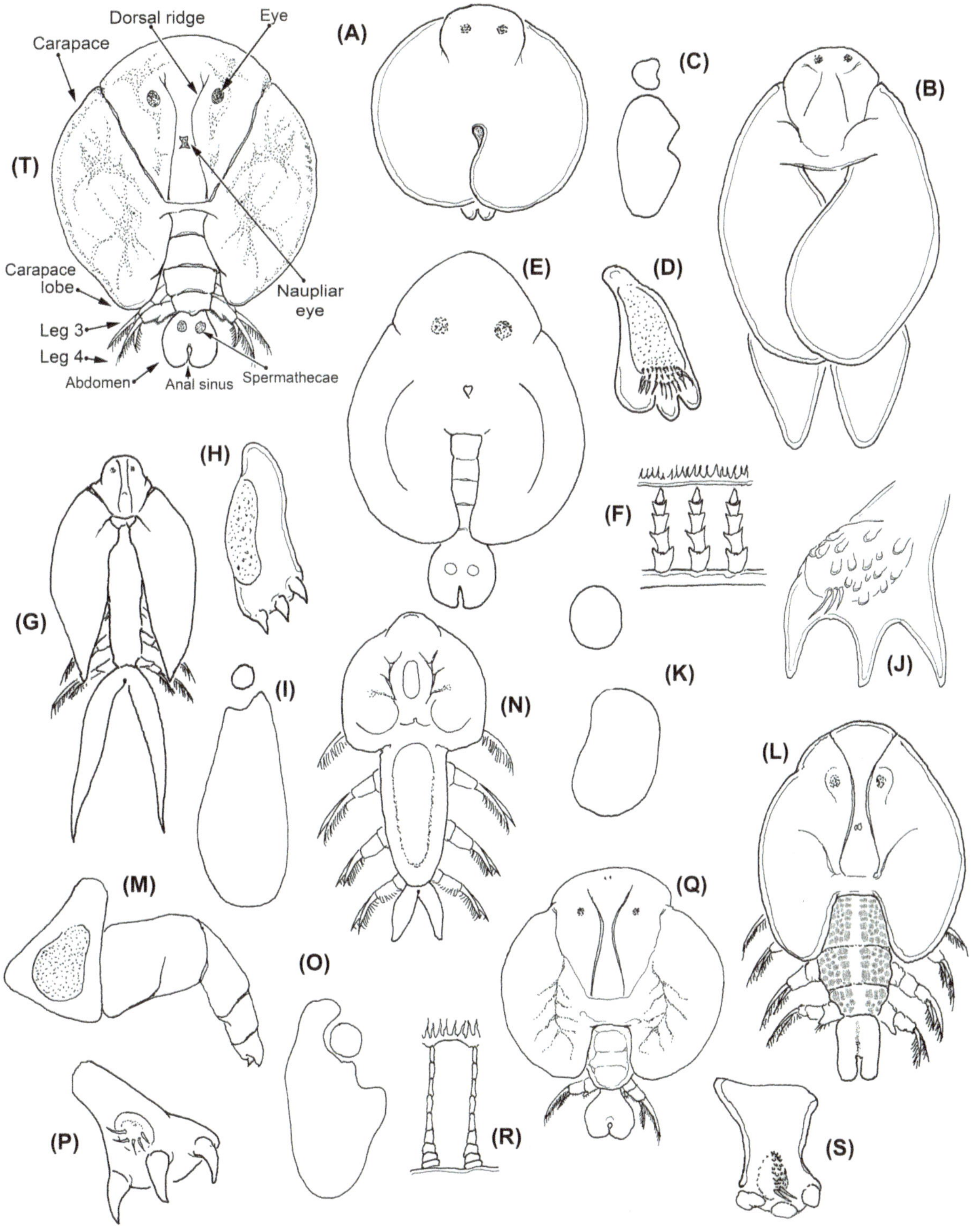

FIGURE 22.4 Structures of species of *Argulus* included in the key: (A) *A. nattereri* dorsal view of body; (B) *A. amazonicus* dorsal view of body; (C) *A. amazonicus* respiratory areas; (D) *A. rhamdiae* second maxilla teeth; (E) *A. rhamdiae* dorsal view of body; (F) *A. rhamdiae* sclerites of rods; (G) *A. paranensis* dorsal view of body; (H) *A. paranensis* second maxilla teeth; (I) *A. paranensis* respiratory areas; (J) *A. yucatanus* second maxilla teeth; (K) *A. yucatanus* respiratory areas; (L) *A. ichesi* dorsal view of body; (M) *A. multicolor* second maxilla; (N) *A. elongatus* dorsal view of body; (O) *A. mexicanus* respiratory areas; (P) *A. mexicanus* maxillary teeth; (Q) *A. salmini* dorsal view of body; (R) *A. salmini* sclerites in rods; (S) *A. salmini* maxillary teeth; (T) dorsal view of hypothetical argulid showing structures of carapace and abdomen. *Figures redrawn: (A), (N) from Wilson (1902); (B), (C) from Malta & Santos-Silva (1986); (D–F) from Wilson (1936); (G–I), (Q–S) from Ringuelet (1943); (J), (K) from Poly (2005); (L) from Bouvier (1910), (O), (P) from Pineda et al. (1995).*

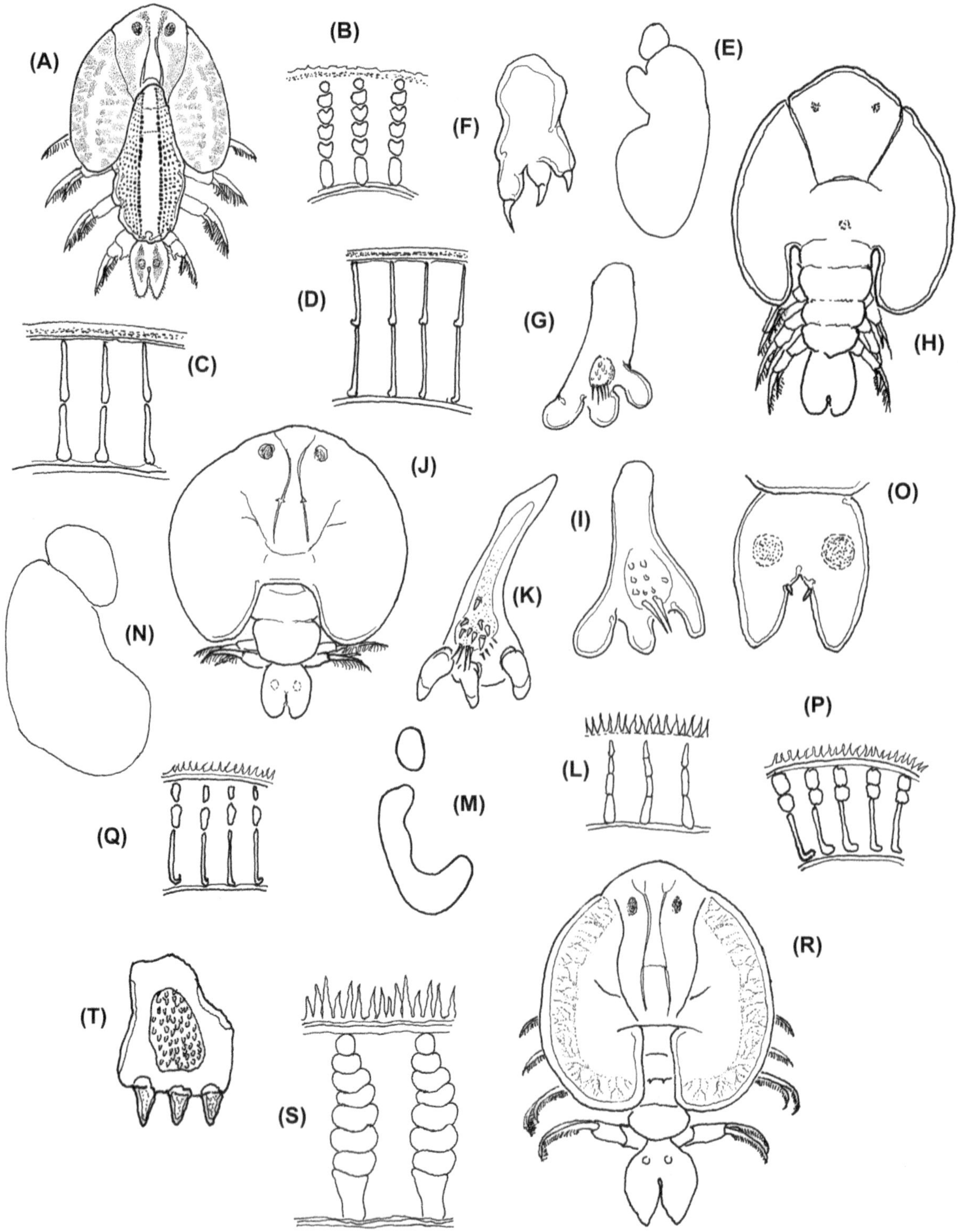

FIGURE 22.5 Structures of species of *Argulus* included in the key: (A) *A. chicomendesi* dorsal view of body; (B) *A. chicomendesi* sclerites in sucker rods; (C) *A. pestifer* sclerites in sucker rods; (D) *A. spinulosus* sclerites in rods; (E) *A. spinulosus* respiratory area; (F) *A. spinulosus* maxillary teeth; (G) *A. violaceus* second maxilla teeth; (H) *A. chromidis* dorsal view of body; (I) *A. cubensis* second maxilla teeth; (J) *A. flavescens* dorsal view of body; (K) *A. flavescens* second maxilla teeth; (L) *A. flavescens* sclerites in sucker rods; (M) *A. flavescens* respiratory areas; (N) *A. ambystoma* respiratory areas; (O) *A. carteri* abdominal papillae; (P) *A. carteri* sclerites in rods; (Q) *A. patagonicus* sclerites in sucker rods; (R) *A. ernsti* dorsal view of body; (S) *A. ernsti* sclerites in sucker rods; (T) *A. ernsti* second maxilla teeth. *Figures redrawn: (A), (B) from Malta & Varella (2000); (C), (G), (Q) from Ringuelet (1943); (D–F) from Silva (1980), (H) from Wilson (1902); (I) from Poly (2005); (J–M) from Suárez-Morales et al. (1998), (N) from Poly (2003); (O), (P) from Cunnington (1931); (R–T) from Weibezahn & Cobo (1964).*

circumference is supported by a ring of complex **sclerites** or sucker **rods** arranged on a flexible stalk. The number of rods in the sucker and the number of sclerites in each rod are also variable and has taxonomic value (Poly, 2003). The anterior and posterior supporting rods are usually regularly arranged, composed of rows of variously shaped sclerites fitting into one another, usually diminishing in size toward the outer rim. The basal sclerites are usually elongated and fusiform; some fusion among sclerites can occur.

The **mandibles** are contained in the cylindrical proboscis or mouth cone (Fig. 22.2, mc), situated medially between the suckers. The mouth opening and the mandibular gnathal processes are situated at the tip of the mouth cone. The **stylet** or **preoral spine** consists of a tapering spine contained within an eversible sheath; it is located midway between the first maxillae and it can have spines or scales.

The fifth pair of appendages is the **second maxilla** (or **maxilla**); they are inserted posterior to the suckers. These are uniramous structures of five articles (Fig. 22.2), characterized by the shape and armature of the robust and complex basal plate which is armed with different kinds of dentiform or digitiform processes. The plate usually has three such teeth but some species have two or none; the number of setae and scales of the basal segment are additional taxonomically valuable characters to examine in this appendage. The surface of the other second maxillary articles (2–5) can also bear patches of cuticular scales of different types and sizes (Overstreet et al., 1992; Tam et al., 2005). The distal fifth article usually ends in a reduced pair of claws. There is also a pair of accessory thoracic spines situated on either side of midline, and also a pair of postmaxillary spines inserted on the anterior end of the first thoracic segment (Tam et al., 2005).

The subcylindrical thorax consists of four somites, each having a pair of biramous swimming legs with a basally expanded coxal lobe. The legs of males bear the accessory copulatory structures which vary morphologically among the species. The posterior tagma is the **abdomen**, it has a distinctive shape in each species and terminates in paired abdominal lobes separated by a median anal cleft or sinus; adjacent to the anal opening is a pair of **uropods** or **caudal rami** (Overstreet et al., 1992; Boxshall, 2005) (Fig. 22.2).

The four swimming legs or **thoracopods** are adapted for swimming. Both rami (endopodal and exopodal) can bear rows of long, stout pinnate setae along the dorsal and ventral margins that function as oars. The first and second pairs of legs bear three articles each, and the two remaining pairs have two. In many species a flagellum is present; this is a posteriorly directed projection of the exopod armed with rows of pinnate setae used for grooming (Overstreet et al., 1992) (Fig. 22.2).

In male argulids the basal articles of legs 2–4 bear accessory copulatory structures which are species specific in shape and position. In most argulids, the most conspicuous difference between males and females is the presence of eggs in the thoracic region of the female and their absence in males. Additional differences are that the male abdomen contains elongated testes, which are whitish, full of sperm in live specimens, and dark in preserved specimens. The female abdomen bears the spermathecae which are always dark in color; the spermathecal spines occur anterior to the natatory lobes. Females can be readily recognized by the presence of the dark, rounded, or oblong spermathecae, located in the proximal part of the abdomen.

Most species of *Argulus* are transparent or opaque, and some have brown tones. Pigmentation of parasitic crustaceans is from either carotenoids or melanoid compounds. *Argulus multicolor* (Stekhoven, 1937) shows a unique color pattern, with dark spots of different sizes on the dorsal and ventral abdominal surfaces and a peripheral black stripe on the carapace framing a yellowish area. Some species of *Chonopeltis* acquire a pigmentation pattern that is similar to that of the host skin or have dark pigment bands overlaying the ovaries (van As & van As, 1999) probably as a protection from ultraviolet radiation for gonadic tissue.

MATERIAL PREPARATION AND PRESERVATION

Individuals are obtained from routine parasitological examination of fish hosts; they are attached either to the body surface or to the mouth and gill chamber. Many species are frequently found in zooplankton samples of fresh and brackish water habitats. The specimens can be fixed and preserved in cold 100% ethanol (if molecular analysis is required) or in a 10% solution of formalin, but ethanol is best for long-term storage. If genetic studies are required, 96% ethanol should be used for preservation. For microscopic examination specimens can be temporarily mounted in lactic acid or glycerine; this treatment clears the specimen to ease the observation of appendage details. The respiratory areas are usually easy to observe under reflected light but when they are not clearly seen, they can be stained with dilute Chlorazol black in both preserved and live specimens (Overstreet et al., 1992). Because of their rigid body structure, most argulids can be processed for SEM examination without the usual problems of shrinking or bending related to dehydration.

KEYS TO ARGULIDAE

Argulidae: Genera

1 First maxilla forming conspicuous rounded ventral suckers (Fig. 22.1) 2

1′ First maxilla forming strong claws or hamulate appendages (Fig. 22.1 B), suckers absent .... ***Dolops* [p. 804]**

2(1) First antenna and/or second antennae absent or with incomplete sutures; ventral preoral spine absent; second maxillae basal plates smooth (Fig. 22.1 C) .... ***Dipteropeltis* [p. 806]**

2′ First antenna well-developed, ornamented with stout hooks; ventral preoral spine present, caudal rami present; cephalic shield lateral lobes rounded, relatively short; second maxillae basal plates with teeth (Figs. 22.1 A, 2) .... ***Argulus* [p. 804]**

Argulidae: *Dolops*: Species

1 Carapace ventral surface with spines; first maxilla terminal hooks with long, conspicuous adjacent appendage (Fig. 22.3 B) 2

1′ Carapace ventral surface without spines; first maxilla terminal hooks without long adjacent appendage, or if present it is small 7

2(1) Second maxilla with three teeth 3

2′ Second maxilla with two teeth (Fig. 22.3 A); major respiratory area with lateral minor area (Fig. 22.3 A–D) .... *Dolops bidentata* (Bouvier, 1899)
[French Guiana, Brazil]

3(2) Carapace lateral areas with submarginal spines arranged in a single row, or absent 4

3′ Carapace lateral areas with submarginal spines arranged in more than one row, irregularly distributed; abdominal lobes length subequal to width (Fig. 22.3 E); first maxilla hooks as in Fig. 22.3 F .... *Dolops reperta* (Bouvier, 1899)
[French Guiana]

4(3) Carapace width $\geq$ length, discoid 5

4′ Carapace width $<$ length; first maxilla hooks as in Fig. 22.3 G .... *Dolops kollari* (Heller, 1857)
[Brazil]

5(4) Carapace lateral areas with posterior margin lacking submarginal row of ventral spines; respiratory areas not as below; second maxilla teeth unequal, external tooth of different size than remaining two 6

5′ Carapace lateral areas posterior margin with submarginal row of ventral spines; major respiratory area with two notches (Fig. 22.3 K); carapace discoid (Fig. 22.3 H); first maxilla acute (Fig. 22.3 I); second maxilla teeth rounded, subequal (Fig. 22.3 J), with row of small spinules .... *Dolops discoidalis* (Bouvier, 1899)
[Venezuela, Brazil, Bolivia]

6(5) First maxilla appendage blunt, smooth; second maxilla teeth with external tooth size $\sim$ 2× remaining two (Fig. 22.3 M) .... *Dolops striata* (Bouvier, 1899)
[Venezuela, French Guyana, Brazil. Bolivia. Paraguay, Argentina]

6′ First maxilla appendage with spinules (Fig. 22.3 N); second maxilla teeth with external tooth length $<$ remaining two; carapace widely rounded (Fig. 22.3 O) .... *Dolops intermedia* (da Silva, 1978)
[Brazil]

7(1) Second maxilla with three teeth 8

7′ Second maxilla with two teeth, both acute (Fig. 22.3 Q), carapace as in Fig. 22.3 P .... *Dolops nana* Lemos de Castro, 1950
[Brazil]

8(7) Carapace rounded; first three swimming legs with flagellum 9

8′ Carapace elliptical, abdomen acuminate, longer in male than in female (Fig. 22.3 S); first two swimming legs with flagellum; second maxilla teeth robust, flat, one smaller than other two (Fig. 22.3 R) .... *Dolops geayi* (Bouvier, 1897)
[Venezuela, French Guyana, Paraguay, Argentina]

9(8) Abdomen length $>$ carapace; abdominal lobes, lanceolate, acuminate (Fig. 22.3 T), first maxilla hook robust, acute, curved (Fig. 22.3 U) .... *Dolops longicauda* (Heller, 1857)
[Ecuador, Brazil, Paraguay, Argentina, Uruguay]

9′ Abdomen length $\sim$¼x carapace; abdominal lobes not acuminate; carapace without ventral spines (Fig. 22.3 V, W) .... *Dolops carvalhoi* Lemos de Castro, 1949
[Brazil, Bolivia]

Argulidae: *Argulus*: Species

Key based on Ringuelet (1943, 1948), Lemos de Castro (1949, 1950), and Thatcher (2006).

1 Carapace covering part of the abdomen and all four pairs of legs 2

1′ Carapace not covering abdomen and all four pairs of legs 5

2(1) Carapace length > width; second maxilla teeth different from below 3

2′ Carapace clearly circular; carapace posterior sinus narrow and deep; second maxilla teeth long, narrow (Fig. 22.4 A) .... *Argulus nattereri* (Heller, 1857)
[Brazil: Pantanal. Argentina]

3(2) Second maxilla teeth rounded or triangular 4

3′ Second maxilla teeth truncate .... *Argulus juparanaensis* Lemos de Castro, 1950
[Brazil, Bolivia]

4(3) Second maxilla teeth triangular; carapace lobes remarkably wide, right overlapping left at level of second swimming leg (Fig. 22.4 B); rods with six sclerites; respiratory areas as in Fig. 22.4 C .... *Argulus amazonicus* (Malta & Santos-Silva, 1986)
[Brazil: Amazon]

4′ Second maxilla short, rounded (Fig. 22.4 D); carapace lobes not as above; carapace length > width, tapering distally (Fig. 22.4 E); sucker rods with five sclerites (Fig. 22.4 F) .... *Argulus rhamdiae* (Wilson, 1936)
[Mexico: Yucatan]

5(1) First two pairs of swimming legs without flagellum 6

5′ First two pairs of swimming legs with flagellum 8

6(5) Carapace lobes rounded, not reaching fourth leg pair; abdomen length 0.5–0.3× carapace; second maxilla teeth spatulate or digitiform 7

6′ Carapace lobes acuminate, reaching abdomen proximal margin; cephalic area tapering distally; abdomen length > 0.5× carapace (Fig. 22.4 G); second maxilla with three spiniform, acute teeth (Fig. 22.4 H); minor respiratory areas small (Fig. 22.4 I) .... *Argulus paranensis* (Ringuelet, 1943)
[Argentina]

7(6) Abdomen length ~0.3× carapace; second maxilla teeth digitiform, basal plate with field of cuticular scales and row of few setae (Fig. 22.4 J); respiratory areas circular or ovoid, smaller one located anteriorly to and widely separated from larger (Fig. 22.4 K) .... *Argulus yucatanus* (Poly, 2005)
[Mexico: Yucatan]

7′ Abdomen length <0.5× carapace (Fig. 22.4 L); second maxilla teeth spatulate; respiratory areas not as above .... *Argulus ichesi* (Bouvier, 1910)
[Argentina]

8(5) Second maxilla with 2–3 teeth or spines; sucker rods with two or more sclerites 9

8′ Second maxilla lacking teeth or major spines (Fig. 22.4 M); sucker rods with single short sclerite alternating with a pattern of transverse, curved short lines .... *Argulus multicolor* (Stekhoven, 1937)
[Venezuela, Brazil]

9(8) Second maxilla with three well-developed teeth; carapace always covering at least first two swimming legs 10

9′ Second maxilla teeth reduced, only two short spines 22

10(9) Sucker rods with 6–16 short, free sclerites; first antenna anterior hook straight, distally blunt 11

10′ Sucker rods with five or less sclerites; first antenna anterior hook curved, distally acute 14

11(10) Minor respiratory area located anteriorly to major area 12

11′ Minor respiratory area partially surrounded by major area, latter area with two notches (Fig. 22.4 O); second maxilla teeth hamulate, basal plate with row of setae (Fig. 22.4 P) .... *Argulus mexicanus* (Pineda, Páramo & Del Río, 1995)
[Mexico: Tabasco]

12(11) Carapace covering third swimming leg (Figs. 22.4 Q, 5 R); sucker rods with nine sclerites 13

12′ Carapace covering only second swimming leg (Fig. 22.5 A); sucker rods with six sclerites (Fig. 22.5 B) .... *Argulus chicomendesi* Malta & Varella, 2000
[Brazil: Amazon, Pantanal]

13(12) Abdomen short, with broadly rounded lobes (Fig. 22.4 Q); sclerites slender, narrowing distally (Fig. 22.4 R); second maxilla teeth short, rounded, basal plate with field of spinules and few setae (Fig. 22.4 S) .... *Argulus salminei* (Krøyer, 1863)
[Brazil, Paraguay. Argentina]

13′ Abdomen with lobes longer than above, tapering distally (Fig. 22.5 R); rods with stacks of 8–10 subspherical sclerites (Fig. 22.5 S); second maxilla with three robust spiniform teeth: basal plate with field of small cuticular scales (Fig. 22.5 T) .... *Argulus ernsti* (Weibzahn & Cobo, 1964)
[Venezuela]

14(10) Sucker rods with two sclerites (Fig. 22.5 C, D) 15

14′ Sucker rods with 3–5 sclerites 17

15(14) Preoral spine robust 16

15′ Preoral spine small, inconspicuous, sclerites as in Fig. 22.5 C .... *Argulus pestifer* (Ringuelet, 1948)
[Brazil, Bolivia, Argentina]

16(15) Sucker rod sclerites elongate, narrow (Fig. 22.5 D); major respiratory area medially constricted, with derived secondary inner rounded process (Fig. 22.5 E); second maxilla teeth hamulate (Fig. 22.5 F) .... *Argulus spinulosus* (Silva, 1980)
[Brazil]

16′ Sucker rods with distal sclerites short, oblong; major respiratory area medially constricted, lacking derived inner process; second maxilla teeth lobulated (Fig. 22.5 G) .... *Argulus violaceus* Thomsen, 1925
[Argentina, Uruguay]

17(14) Character combination not as below 18

17′ Carapace tapering anteriorly, lobes separated from thorax by deep sinus reaching basis of leg 1 (Fig. 22.5 H); sucker rods with 3–5 sclerites .... *Argulus chromidis* Krøyer, 1863
[Nicaragua: Lake Nicaragua]

18(17) Character combination not as below 19

18′ Carapace not tapering anteriorly, lobes with relatively short sinus, not reaching basis of leg 1; sucker rods with four sclerites; second maxilla with three digitiform teeth, basal plate with few cuticular scales and two stout setae (Fig. 22.5 I) .... *Argulus cubensis* Wilson, 1935
[Cuba]

19(18) Carapace ventral surface with spines; sucker rods with 3–4 sclerites 20

19′ Carapace ventral surface without spines, carapace wide, nearly circular (Fig. 22.5 J); sucker rods with four sclerites (Fig. 22.5 L); respiratory areas as in Fig. 22.5 M; second maxilla teeth rounded, curved; basal plate with small scales, spinules, and few setae (Fig. 22.5 K) .... *Argulus flavescens* Wilson, 1916
[Mexico: Quintana Roo]

20(19) Respiratory areas not as below 21

20′ Respiratory areas with a smaller, oblong area anterior and attached to larger posterior area (Fig. 22.5 N), outlined with melanophores .... *Argulus ambystoma* (Poly, 2003)
[Mexico: Michoacán]

21(20) Abdominal sinus with papillae on lateral margins (Fig. 22.4 O); second maxilla teeth acute; rods with three sclerites (Fig. 22.5 P) .... *Argulus carteri* Cunnington, 1931
[Brazil, Paraguay]

21′ Abdominal sinus with papillae on vertex; second maxilla teeth digitiform; rods with three sclerites (Fig. 22.5 Q) .... *Argulus patagonicus* Ringuelet, 1943
[Argentina]

22(9) Carapace remarkably short, rounded, revealing legs 2–4 (Fig. 22.4 N); sucker rods with 10–11 short sclerites .... *Argulus elongatus* Heller, 1857
[Brazil]

22′ Carapace not as short; abdomen length ~0.5–0.7× carapace length, with rounded tips pyriform, carapace revealing legs 3 and 4; sucker rods with 8–12 sclerites .... *Argulus chiropteroideus* Aguiar, Rosim, Santos, Luque, Ceccarelli, Adriano & Tavares, 2017
[Brazil: Pantanal]

Argulidae: *Dipteropeltis*: Species

1 Head shield quadrate, unfolded, carapace lobes elliptical; abdomen with apices lanceolate, caudal rami absent .... *Dipteropeltis campanaformis* Neethling, Malta & Avenant-Oldewage, 2014
[Brazil]

1′ Head shield rounded, folded, carapace lobes straight; abdomen apices blunt, caudal rami present .... *Dipteropeltis hirundo* Calman, 1912
[Venezuela, Brazil, Bolivia, Argentina]

REFERENCES

Aguiar, J.C., D. F. Rosim, S.M. C. Santos, J.L. Luque, P.S. Ceccarelli, E.A. Adriano & L.E.R. Tavares. 2017. A new species of *Argulus* (Crustacea, Branchiura, Argulidae) from the skin of the catfish, with new records of branchiurans from wild fish in the Brazilian Pantanal wetland. Zootaxa 4320: 447–469.7

Bouvier, E.L. 1899a. Sur les "Argulides" du genre Gyropeltis, recueilles récemment par M. Geay dans la Guyane. Bulletin du Muséum d'Histoire Naturelle 5: 39-41.

Bouvier, E.L. 1899b. Les crustacés parasites du genre *Dolops* Audouin. Bulletin de la Societé Philomathique de Paris 8(10)2–3:53–81; 9, I: 12–40.

Bouvier, E.L. 1910. Un Argulidae nouveau de l' Argentine. Bulletin Museum Histoire Naturelle Paris 2: 92–95.

Boxshall, G.A. 2005. Crustacean parasites: Branchiura (fish lice). Pages 145–147 *in*: K. Rhode (ed.), Marine Parasitology. CSIRO Publication, Australia.

Cressey, R.F. 1972. The genus *Argulus* (Crustacea: Branchiura) of the United States. Biota of Freshwater Ecosystems, U.S. Environmental Protection Agency Identification Manual 2. U.S. Government Printing Office, Washington, DC, USA, viii + 14 p.

Cunnington, W.A. 1931. Reports of an expedition to Brazil and Paraguay in 1926–27, supported by the Trustees of the Percy Sladen Memorial Fund and the Executive Committee of the Carnegie Trust of Scotland. Argulidae. Journal of the Linnean Society of London 37: 259–265.

Fernández, R.A. & R.I. Corrada. 2015. Nueva especie de *Argulus* (Branchiura: Argulidae), parásito de *Chromis cyanea* (Pisces: Actinopterygii), para aguas cubanas. Serie Oceanológica 14: 38–45.

Krøyer, H. 1863. Bidrag till Kundskab om Snylterkrebsene. Naturhistorisk Tidsskrift, 3die Raekke, II, 75–426, pls. I–XVIII.

Lemos de Castro, A. 1949. Contribuição ao conhecimento dos crustáceos argulídeos do Brasil (Branchiura: Argulidae), com descrição de uma nova espécie. Boletim do Museu Nacional Río de Janeiro, Zoologia 93: 1–7.

Lemos de Castro, A. 1950. Contribuição ao conhecimento dos crustáceos Argulídeos do Brasil. II. Descrição de duas espécies novas. Anais da Academia Brasileira de Ciências 22: 245–252.

Lemos de Castro, A. 1986. Branchiura. Manual de identificação de invertebrados límnicos. Brasília, Ministério de Ciência e Tecnologia, 23p.

Malta, J.C.O. & E.N. Santos-Silva. 1986. *Argulus amazonicus* n.sp., crustáceo parasito de peixes da Amazônia Brasileira (Branchiura: Argulidae). Amazoniana 9: 485–492.

Malta, J.C.O. & A. Varella. 2000. *Argulus chicomendesi* sp. n. (Crustacea: Argulidae) parasita de peixes da Amazônia brasileira. Acta Amazonica 30: 481–498.

Mamani, M., C. Hamel & P.A. van Damme. 2004. Ectoparasites (Crustacea: Branchiura) of *Pseudoplatystoma fasciatum* (surubí) and *P. tigrinum* (chuncuina) in Bolivian white-water floodplains. Ecología en Bolivia 39: 9–20.

Martin, J.W. & G.E. Davis. 2001. An updated classification of the recent Crustacea. Natural History Museum of Los Angeles County Science Series 39: 1–124.

Møller, O.S. 2009. Branchiura (Crustacea) – survey of historical literature and taxonomy. Arthropod Systematics & Phylogeny 67: 41–55.

Møller, O.S. & J. Olesen. 2010. The little known *Dipteropeltis hirundo* Calman, 1912 (Crustacea, Branchiura): SEM investigations of paratype material in light of recent phylogenetic analyses. Experimental Parasitology 125: 30–41.

Møller, O.S., J. Olesen, A. Avenant-Oldewage, P.F. Thomsen & H. Glenner 2008. First maxillae suction discs in Branchiura (Crustacea): Development and evolution in light of the first molecular phylogeny of Branchiura, Pentastomida, and other "Maxillopoda". Arthropod Structure & Development 37: 338–346.

Neethling, L.A.M., J.C. Malta & A. Avenant-Oldewage. 2014. Additional morphological information on *Dipteropeltis hirundo* Calman, 1912, and a description of *Dipteropeltis campanaformis* n. sp. (Crustacea: Branchiura) from two characiform benthopelagic fish hosts from two Northern rivers of the Brazilian Amazon. Zootaxa 3755: 179–193.

Overstreet, R.M., I. Dykova & E. Hawkins. 1992. Chapter 8: Branchiura. Pages 385–413 *in*: F.W. Harrison & A.G. Humes (eds.), Microscopical Anatomy of Invertebrates. Crustacea. Wiley Liss Inc. New York.

Pineda, R., S. Páramo & R. Del Río. 1995. A new species of the genus *Argulus* (Crustacea: Branchiura) parasitic on *Atractosteus tropicus* (Pisces: Lepisosteidae) from Tabasco, Mexico. Systematic Parasitology 30: 199–206.

Poly, W.J. 2003. *Argulus ambystoma*, a new species parasitic on the salamander *Ambystoma dumerilii* from México (Crustacea: Branchiura: Argulidae). Ohio Journal of Science 103: 52–61.

Poly, W.J. 2005. *Argulus yucatanus* n. sp. (Crustacea: Branchiura) parasitic on *Cichlasoma urophthalmus* from Yucatan, Mexico. Gulf and Caribbean Research 17: 1–13.

Poly, W.J. 2008. Global diversity of fishlice (Crustacea: Branchiura: Argulidae) in freshwater. Hydrobiologia 595: 209–212.

Ringuelet, R. 1943. Revisión de los Argúlidos Argentinos (Crustacea, Branchiura) con el catálogo de las especies Neotropicales. Revista del Museo de La Plata 3: 43–99.

Ringuelet, R. 1948. Argúlidos del Museo de la Plata. Revista Museo de La Plata 5(33): 81–301.

Silva, N.M.M. 1978. Uma nova espécie de crustáceo argulídeo do Rio Grande do Sul, Brasil (Branchiura, Argulidae). Iheringia, Série Zoologia 52: 3–29.

Silva, N.M.M. 1980. *Argulus spinulosus* sp. n. (Branchiura, Argulidae), em peixes de água doce do Rio Grande do Sul, Brasil. Iheringia, Série Zoologia 56: 15–23.

Silva-Souza, A.T., V.D. Abdallah, R.K. de Azevedo, F.A. da Silva & J.L. Luque. 2011. Expanded description of *Dolops bidentata* (Bouvier, 1899) (Branchiura: Argulidae) based on specimens collected on *Pygocentrus nattereri* Kner, 1858 (Characiformes) from Poconé Wetland, MT, Brazil. Brazilian Journal of Biology 71: 1–9.

Suárez-Morales, E. 2015. Chapter 29. Class Maxillopoda. Pages 709–755 *in*: J.H. Thorp & D.C. Rogers (eds.), Thorp & Covich's Freshwater Invertebrates, 4th Edition. Volume I: Ecology and General Biology. Academic Press, New York.

Suárez-Morales, I.-H. Kim & I. Castellanos. 1998. A new geographic and host record for *Argulus flavescens* Wilson, 1916 (Crustacea, Arguloida) from Southeastern Mexico. Bulletin of Marine Science 62: 315–318.

Tam, Q., A. Avenant-Oldewage & E.H. Williams. 2005. An ultrastructural investigation of *Argulus personatus* Cunnington, 1913 (Crustacea: Branchiura) from Lake Tanganyika, northern Zambia. African Zoology 40: 301–308.

Thatcher, V.E. 2006. Branchiura. Pages 390–415 *in*: J. Adis, J. R. Arias, G. Rueda-Delgado & K.M. Wantzen (eds.), Aquatic biodiversity in Latin America. Vol. 1. Amazon fish parasites (2nd ed), Pensoft Publishers, Bulgaria.

Thiele, J. 1904. Beiträge zur Morphologie der Arguliden. Mitteilungen aus dem Zoologischen Museum in Berlin 2(4): 14–19.

Thorell, M.T. 1864. Om tvenne Europeiske Argulider; jemte anmärkningar om Argulidernas morfologi och systematiska ställning, samt en öfversigt af de för närvarande kända arterna af denna familj. Öfversigt af Kongl. Vetenskaps-Akademiens Förhandlingar, 1864: 7–72, pls. II-IV.

van As, L.L. & J.G. van As. 1999. Aspects of the morphology and a review of the taxonomic status of three species of the genus *Chonopeltis* (Crustacea: Branchiura) from the Orange-Vaal and South West Cape river systems, South Africa. Folia Parasitologica 46: 221–228.

Weibezahn, F.H. & T. Cobo. 1964. Seis argúlidos (Crustacea, Branchiura) parásitos de peces dulceacuícolas en Venezuela, con descripción de una nueva especie del género *Argulus*. Acta Biologica Venezuelica 4: 119–144.

Wilson, C.B. 1902. North American parasitic copepods of the family Argulidae, with a bibliography of the group and a systematic review of all known species. Proceedings of the United States National Museum 25(1302):635–742.

Wilson, C.B. 1936. Copepods from the cenotes and caves of the Yucatan Peninsula, with notes on cladocerans. Carnegie Institution of Washington, Publication 457:77–88.

WoRMS. 2019. Oligostraca. Accessed at: http://www.marinespecies.org/aphia.php?p=taxdetails&id=845957 on 2019-03-28.

Chapter 23

Phylum Arthropoda: Crustacea: Malacostraca

D. Christopher Rogers, Célio Magalhães, Marcela Peralta, Felipe Bezerra Ribeiro, Georgina Bond-Buckup, W. Wayne Price, Jürgen Guerrero-Kommritz, Fernando L. Mantelatto, Alessandra Bueno, Ana Isabel Camacho, Exequiel R. González, Carlos G. Jara, Manuel Pedraza, Carlos Pedraza-Lara, Erich Rudolph Latorre and Sandro Santos

Chapter Outline

PHYLUM ARTHROPODA: CRUSTACEA: MALACOSTRACA

Thorp and Covich's Freshwater Invertebrates. https://doi.org/10.1016/B978-0-12-804225-0.00023-X

Introduction to Malacostraca

D. Christopher Rogers
Kansas Biological Survey, and The Biodiversity Institute, University of Kansas, Lawrence, Kansas, United States

INTRODUCTION

Malacostraca is the largest crustacean class, with about 30,000 marine, freshwater, and terrestrial species worldwide. Common features of this class are four body regions: the head (five segments), pereon (eight segments), the pleon (six to seven segments), and the urosome (one to three segments), each with jointed appendages. The head and thorax are sometimes fused into a cephalothorax, as in the Decapoda. The pereopods are variously modified for feeding, defense, and locomotion. Pleonal appendages (pleopods) may be modified for swimming, reproduction, respiration, and/or care of eggs and young. Eleven orders occur in the Neotropical region.

LIMITATIONS

Much of the Neotropical region has not been surveyed, thus new, undescribed taxa are always a potential. In addition, new invasive species appear frequently. Many species appear in the pet trade or in aquaculture that have the potential to become invasive in the Neotropics. Many marine species are carried into the Panama Canal by ships, both on the vessel exterior and in ballast. Strictly marine taxa are not dealt with here, as species from all over the world frequently appear and then are never seen in the Canal again. Thus, it would be fruitless to try to provide a key to all such taxa. Only those taxa that are apparently established in the Panama Canal have been included.

KEY TO MALACOSTRACA

Malacostraca: Orders

1 Carapace not covering all thoracic segments or absent (Figs. 23.1, 2, 18, 48, 61) 2

1′ Carapace large, obvious (Figs. 23.64, 101, 110, 123, 186) 8

2(1) Thoracopods biramous (Fig. 23.6) 3

2′ Thoracopods uniramous 6

3(2) Carapace absent (Figs. 23.2, 18) 4

3′ Carapace present, covering one or more thoracic segments at least partially (Fig. 23.1) 5

4(3) Second antennae uniramous; uropods forming a "tail fan" (Fig. 23.9) .... **Anaspidacea [p. 819]**

4′ Second antennae biramous; uropods never forming a "tail fan" (Fig. 23.18). .... **Bathynellacea [p. 824]**

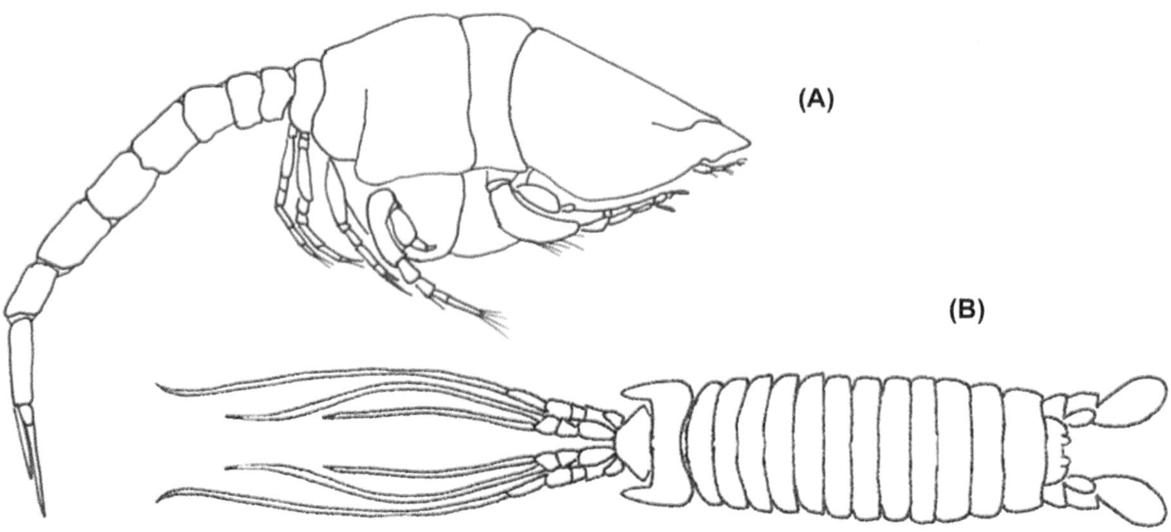

FIGURE 23.1 Representative Malacostraca minor orders: (A) Cumacea, *Claudicuma platense*; (B) Spelaeogriphaceae, *Potiicoara brasiliensis*.

5(3) Thorax subequal in diameter to abdomen; scaphocerite small, but present; eyes absent, eyestalk present; troglobitic (Fig. 23.1 B). .... Spelaeogriphaceae, Spelaeogriphidae one species: *Potiicoara brasiliensis* Pires, 1987 [Brazil]

5′ Thorax diameter at least 4× > abdomen diameter; scaphoceite absent; eyes fused medially (Fig. 23.1 A). .... Cumacea, Nannastacidae one species: *Claudicuma platense* Roccatagliata, 1981 [Argentina]

6(2) Telson and abdominal segments fused into a pleotelson subequal in width to body (Fig. 23.61) 7

6′ Telson projecting from abdomen, telson small, narrower than body (Fig. 23.48) .... **Amphipoda [p. 835]**

7(6) Body dorsoventrally compressed, first pereopods chelate; brackish .... **Tanaidacea [p. 860]**

7′ Body laterally compressed or subcylindrical, first pereopods as gnathopods or simple .... **Isopoda [p. 864]**

8(1) Compound eyes large, pedunculate, or eyestalks reduced to flat plate (Figs. 23.64, 100, 104, 110, 187, 193); pleopods developed and present 9

8′ Eyes absent, pleopods reduced on first abdominal segments, absent from remaining segments, eight pairs of thoracopods, first pair modified as a maxilliped. .... Thermosbanacea, Monodellidae, one genus: *Tethysbaena* [Caribbean Islands: many species]

9(8) Seven pairs of thoracopods (Fig. 23.186) 10

9′ Five pairs of walking legs, including chelae (Figs. 23.64, 101, 110, 123). .... **Decapoda [p. 873]**

10(9) Eyes normally developed (with stalk and cornea) or without cornea; uropodal endopod with statocyst (Fig. 187); uropodal protopod with no ventral medial lobe or process .... **Mysida [p. 974]**

10′ Eyes reduced to flat ocular plate with no visual elements; uropodal endopod lacking statocyst (Fig. 193); uropodal protopod with a relatively short ventral medial lobe or a very long process, bearing spiniform setae .... **Stygiomysida [p. 985]**

Anaspidacea

Marcela Peralta
Instituto de Invertebrados, Fundación M. Lillo, San Miguel de Tucumán, Argentina

INTRODUCTION

Anaspidacea is a basal and relictual lineage of Crustacea Eumalacostraca. Coineau & Camacho (2013), among others, included two orders in Syncarida, Anaspidacea, and Bathynellacea. Noodt (1965) proposed the creation of a third order, Stygocaridacea with only one family, Stygocarididae. However, Schminke (1978), based on comparative studies in primitive Eucarida, inserted Stygocarididae among Anaspidacea and dismissed Stygocaridacea. In the current phylogenetic context, Syncarida has no general consensus as a monophyletic taxon. In the phylogenetic reconstructions of higher Crustacea groups, Anaspidacea, Bathynellacea, and the entirely fossil Paleocaridacea, fall at the base of the Eumalacostraca. Currently, Anaspidacea includes five families: Anaspididae, Koonungidae, Psammaspididae, Stygocarididae, and Patagonaspididae, with 12 genera and 28 living epigean or hypogean species. The fossil record extends back to the Triassic. Anaspididae is possibly one of the best studied groups of small eumalacostracans including ecological, anatomical, and physiological studies, and they are considered living fossils (Schram & Hessler, 1984).

Anaspidaceans are confined to the Southern Hemisphere and considered a Gondwanan lineage, with records from continental Australia, Tasmania, New Zealand, and southern South America (central and southern Chile and western and southern Argentina). There are no records from Africa.

Two families are in the Neotropical region: Stygocarididae and Patagonaspididae. The latter, with only one species, is exclusively found in Patagonia, Argentina. Stygocarididae, the most differentiated and diverse anaspidacean family, is the only one with disjunct biogeographic distribution, with records from Australia, New Zealand, and South America. Eight living Stygocarididae species of three genera (*Oncostygocaris, Parastygocaris*, and *Stygocaris*) are known from southern South America, all with micro distributions.

Anaspidaceans inhabit exclusively freshwater environments, and recent species can live in a wide range of habitats. Epigean species (not known in Neotropical region) are found in streams, lakes, swamps, freshwater crayfishes' burrows, while stygobitic species inhabit groundwater of caves, cenotes (sinkholes), aquifers, and saturated interstitial spaces below the stream bed (hyporheic zone). Probably the primary habitat of hypogean Anaspidacea would be the interstitial spaces of deep phreatic water in ancient underground geological strata. All Neotropical anaspidacean species are stygobitic. It is also noteworthy that one of these species, *Stygocaris gomezmillasi* (Noodt, 1963),

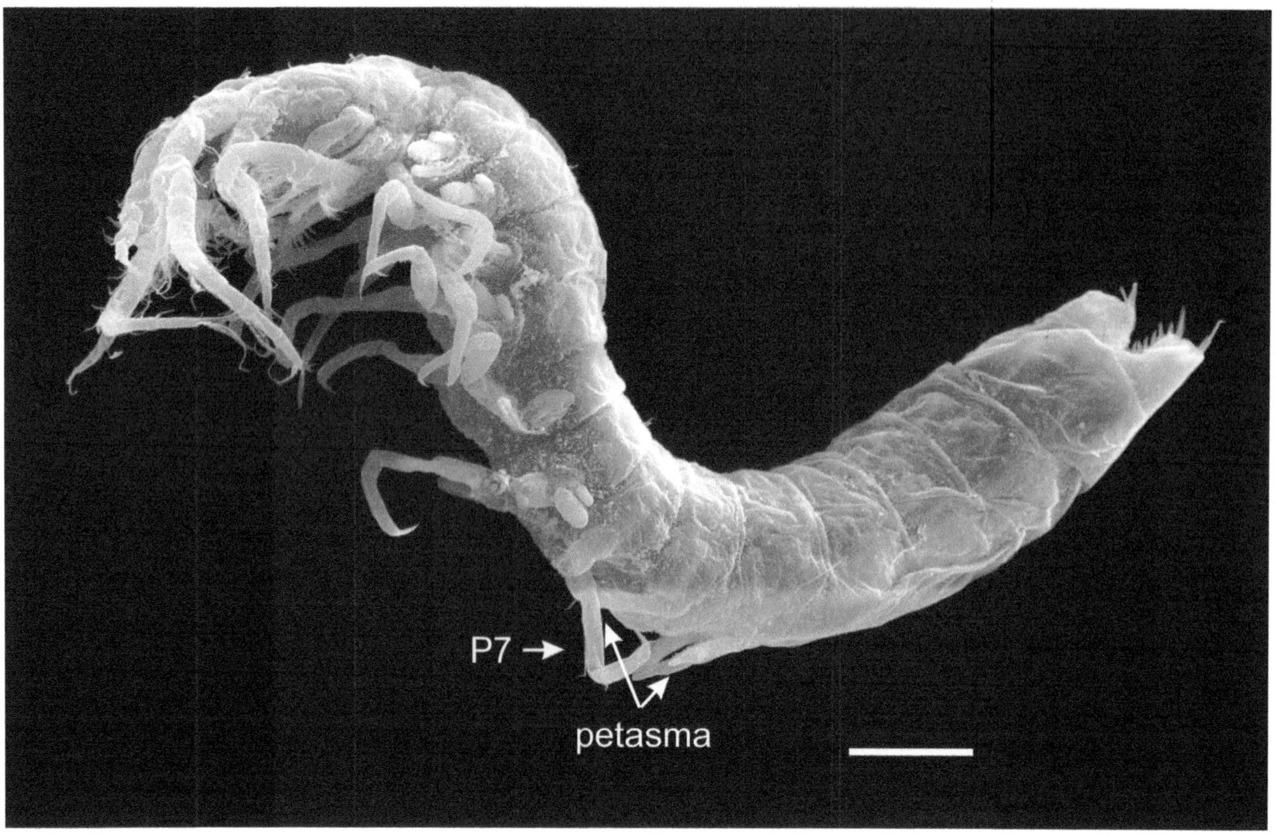

FIGURE 23.2 Anaspidacea; microphotograph of Stygocarididae, *Parastygocaris* sp. male (uropods broken rami). Abbreviation: *P7*, ambulatory pereopod 7. Scale bar = 100 μm.

PHYLUM ARTHROPODA: CRUSTACEA: MALACOSTRACA

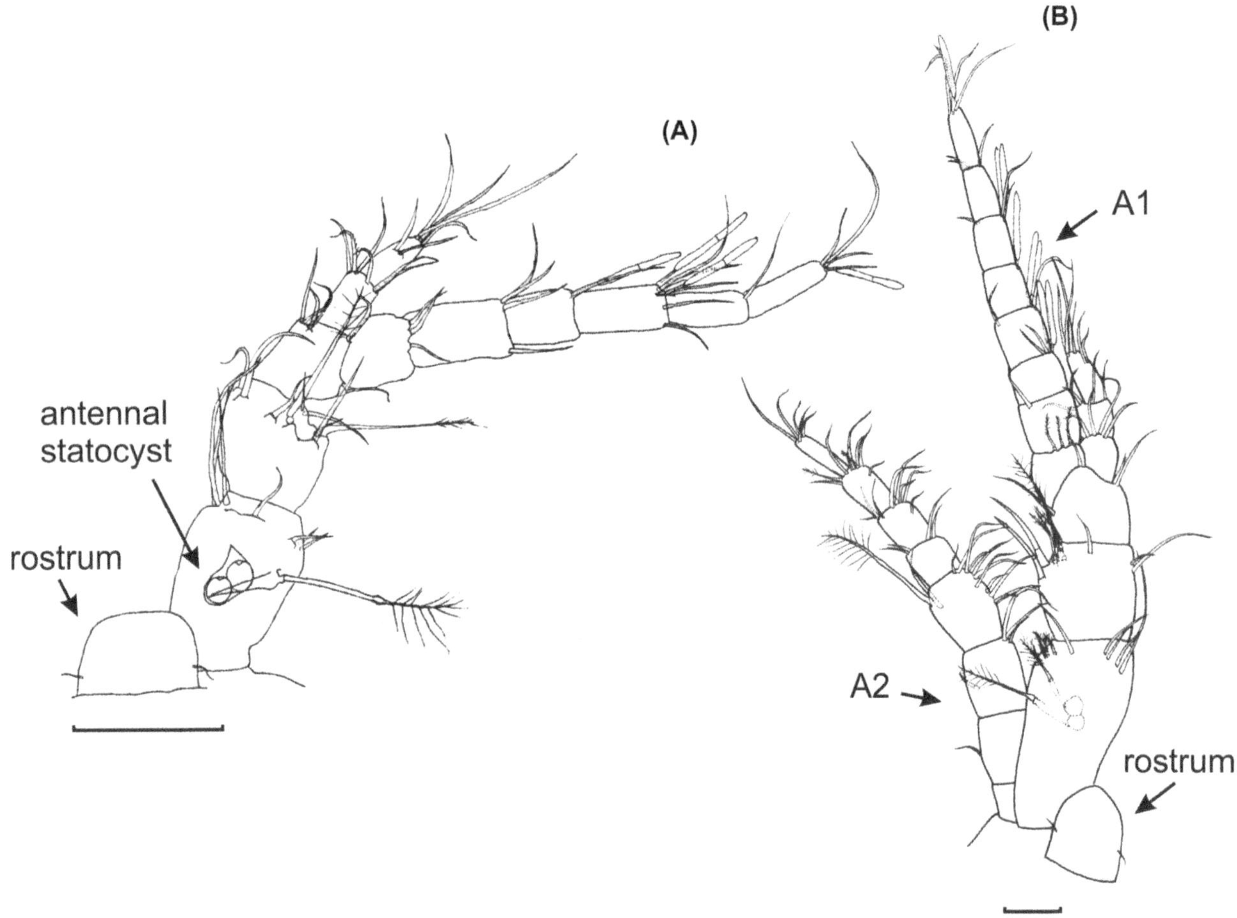

FIGURE 23.3 Rostrum and antennae of Anaspidacea: (A) rostrum and first antenna of *Stygocaris hugofernandezi* after Grosso & Peralta (1999); (B) rostrum, first and second antenna of *Parastygocaris* sp. Abbreviations: *A1*, first antenna; *A2*, second antenna. Scale bars = 0.05 mm.

inhabits underground waters near the sea opening, raising questions about the degree of salinity sensitivity. Bathynellacean species and *S. gomezmillasi* coexist in the hyporheic zone of a stream in Quebrada de Córdoba, Central Chile, a site that is approximately 2,000 m upstream from where the river enters the Pacific Ocean. In both crustacean and Hydracarina fauna from that hyporheic locality, a marine influence is indicated (Noodt, 1963a). However, there are no detailed salinity measurements that conclusively identify these species as living in brackish environments.

All Anaspidacea species have separate sexes, though mating has never been observed. Metamorphosis takes place inside the egg, the hatchlings emerge from the egg as juveniles (Schminke, 2014). Stygocarididae are adapted to a subterranean interstitial mode of life and show characters that are likely of neotenic origin (Schminke, 1978). The species of Stygocarididae are characterized by a transparent and vermiform body, absence of eyes, reduction and miniaturization of some appendages (particularly on the abdomen), and small sizes (rarely > 2 mm). They have no tail fan, antennal exopodite, or mandibular palp. They are not strong swimmers, instead they mostly walk through the microspace between sediment grains and explore the interstitial environment with their antennae. Their diet includes diatoms, algae, and the biofilm of bacteria on sand grains. Other biological aspects of stygobitic anaspidacean species are not well-known.

LIMITATIONS

Anaspidacean identification and comparative morphology studies (e.g., Grosso & Peralta, 1997; Noodt, 1963b; Peralta, 2014) require examination of buccal appendages. Patagonaspididae are monotypic, medium-sized (7 mm), eyeless, and with reduced appendages, particularly pleopods are vestigial. The Stygocarididae, the smallest anaspidaceans (1–4 mm), are interstitial with a marked simplification of the appendages. Small sizes make appendages dissection difficult.

We agree with Camacho & Valdecasas (2008) that full knowledge of anaspidacean diversity in the Neotropical region has not been achieved, because most hypogean fauna of the region is poorly known and its biodiversity is

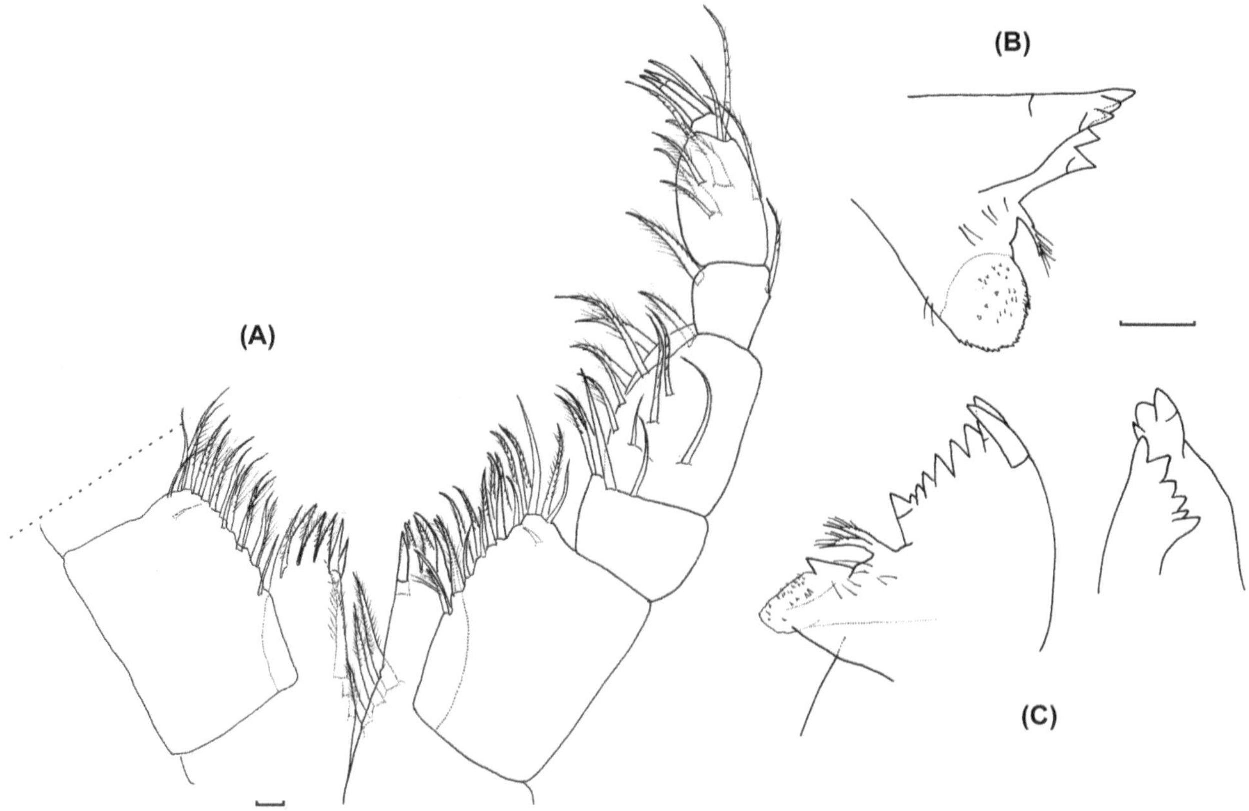

FIGURE 23.4 Maxilliped and mandibles of Anaspidacea: (A) Maxilliped of *Parastygocaris luisgrossoi*; (B) and (C) mandibles of *Parastygocaris* sp., (B) right Md, (C) left Md. Scale bars = 0.01 mm.

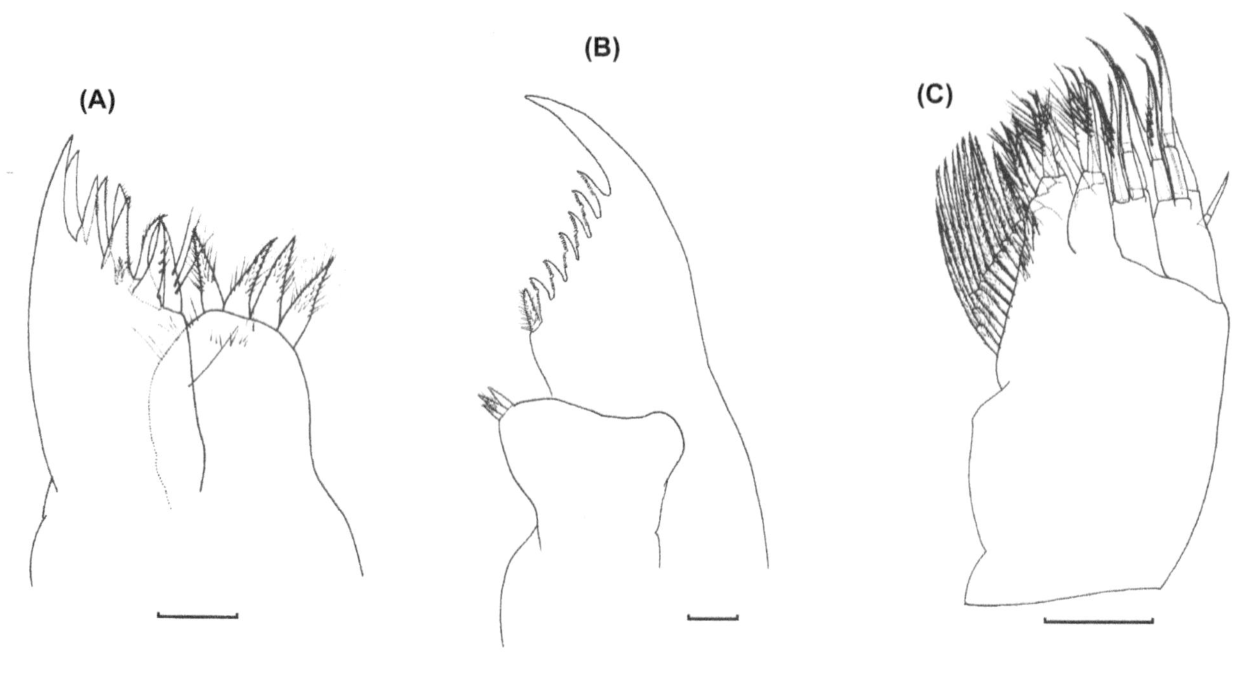

FIGURE 23.5 Maxillae of Anaspidacea: (A) Maxilla 1 of *Parastygocaris* sp.; (B) maxilla 1 of *Patagonaspides sandroruffoi*; (C) maxilla 2 of *Parastygocaris andina*. Scale bars: (A) = 0.01 mm; (B) and (C) = 0.05 mm.

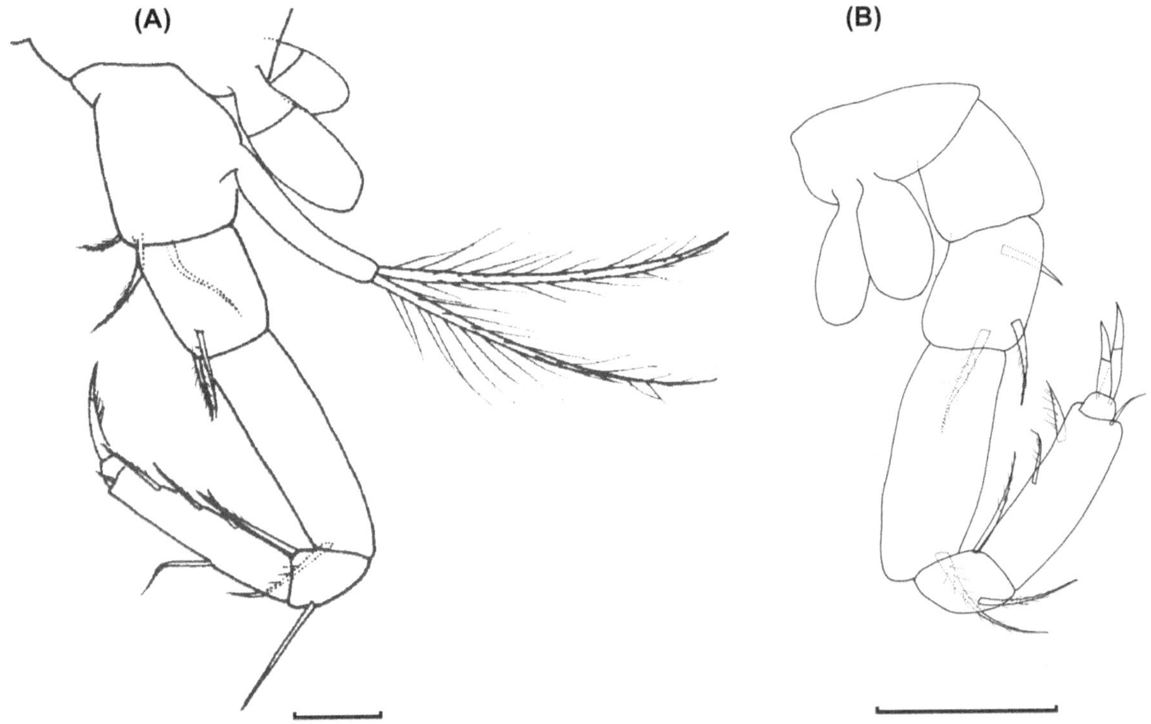

FIGURE 23.6 Pereopods of Anaspidacea: (A) Pereiopod 5 of *Parastygocaris andina*; (B) pereiopod 1 of *Stygocaris hugofernandezi*. Scale bars = 0.05 mm. *Redraw from Grosso & Peralta (1999).*

probably underestimated. In the Australian region, there are species such as *Allanaspides hickmani* (Swain, Wilson & Ong, 1971) that colonize crayfish burrows, and they are on Tasmanian threatened species list. There is no record of Anaspidacea in crayfish burrows in South America, and their conservation state is unknown.

The available information of anaspidacean biology is of mixed quantity and quality. Much of the information about life cycle, anatomy, habits, behavior, feeding mechanism, and evolution comes from *Anaspides tasmaniae* (Thompson, 1893), an iconic epigean species from southeastern Australia that has been studied as an evolutionary curiosity because of their similarities with ancient fossils. There is much to be discovered about the Neotropical species. Because it is a relict lineage, each new species leads to a reexamination of the status and the paleobiogeography of the group. The phylogenetic relationships among species have been partially explored by several authors. However, in order to interpret Anaspidacea biodiversity, it is still necessary to develop a comprehensive phylogenetic analysis to species level and to reexamine some incompletely described species.

TERMINOLOGY AND MORPHOLOGY

Anaspidaceans have elongated bodies with the basic eumalacostracan plan, with thoracomere 1 fused to the cephalon as a **cephalothorax**; in general the rest of thoracic and abdominal somites are separate and of similar size (Fig. 23.2). The species lack a true carapace and the cephalothorax has an anterior projection, the **rostrum**, which is a delicate plate of variable morphology and development. The rostrum of Stygocarididae is a thin lamellar structure, either quadrate or linguiform, entire, occasionally notched or deeply cleft (Fig. 23.3). Patagonaspididae species have a small articulated bilobated rostrum. The eyes are pedunculate or sessile, only developed in nonstygobiontic forms.

The cephalic appendages include biramous **first antennae** (= A1), **second antennae** (= A2), mouthparts, and one pair of **maxillipeds** (Mxp = first thoracopod). A1 consists of a 3-article **peduncle**, a **primary flagellum** (outer flagellum = OF), and an **accessory flagellum** (inner flagellum = IF) (Figs. 23.3, 10); an **antennal statocyst** (Fig. 23.3 A) is present on peduncle first article, which consists of mechanosensory setae (with dilated distal end) that respond to changes in spatial orientation. The statocyst cavity is opened by a dorsal transverse groove. The first antenna primary flagellum is the longest, with a number of articles between 6 and 130. In general, interstitial Neotropical species have short A1 (up to 20 OF articles). In most cases, there is sexual dimorphism in number of articles of the first antennae, development of antennular complex organs in males, or variations in chaetotaxy between both sexes. The first antenna accessory flagellum has between 2 and 20 articles. The second antenna is formed by

PHYLUM ARTHROPODA: CRUSTACEA: MALACOSTRACA

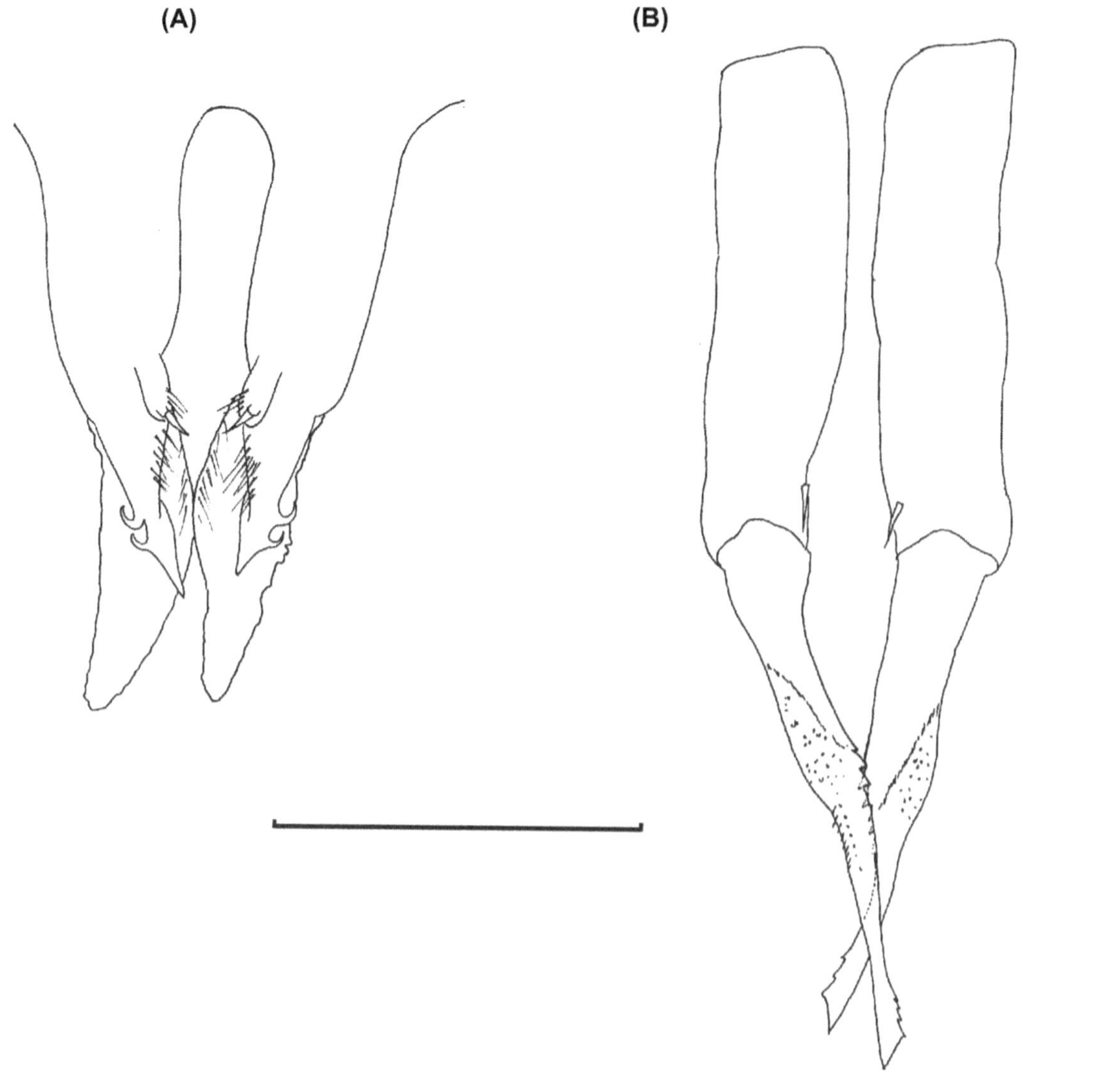

FIGURE 23.7 Petasma of *Parastygocaris schminkei* (Anaspidacea): (A) Pleopod 1; and (B) pleopod 2. Scale bar = 0.05 mm. *Redrawn from Grosso & Peralta (1997).*

a 4-article peduncle plus a variable number of flagellar articles. Second antenna is uniramous except in Anaspididae and some Psammaspididae species that have the exopodite developed as a scaphocerite.

The mouthparts have structural variations between families. The **maxilla 1** (= Mx1) consists of two endites, the medial one globoid and shorter (Fig. 23.5 A, B), in Stygocarididae and Patagonaspididae the maxilla 1 palp is absent. The **maxilla 2** (= Mx2) (following Noodt, 1970) in Stygocarididae has four successive endites along the transverse axis, all with terminal setae (Fig. 23.5 C); in Patagonaspididae the maxilla 2 has three distal endites. The maxillipeds (Fig. 23.4 A) are more robust and setose than the rest of the thoracopods and are directed anteriorly, covering mouthparts. In most maxillipeds, the basis has an internal expansion called "**pseudoendite**" (Coineau & Camacho, 2013). Each thoracic somite with a pair of ambulatory 7-articulate **thoracopods**, uni- or biramous.

The ambulatory **pereopods** (= P 1–7) are the walking legs (Fig. 23.6); the pereiopod 7 endopodite is flexed in the opposite direction as the anterior ones, with the dactylus directed anteriorly; pereiopods 1–5 generally with a cylindrical exopodite and accessory respiratory organs of varying morphology (only absent in some Stygocarididae); pereiopod 6 with or without an exopodite; on pereiopod 1–5 or 1–6 are one or two **branchial vesicular epipodites** which function as flattened gills; pereiopod 7 has no exopodite or epipodite.

Pleopods 1–5 variable, uni- or biramous, vestigial or absent except for an always developed petasma (Figs. 23.2, 7), complex copulatory structures consisting of male pleopods 1–2 endopodites modified, which are directed anteriorly against the sternal surface between the last pair

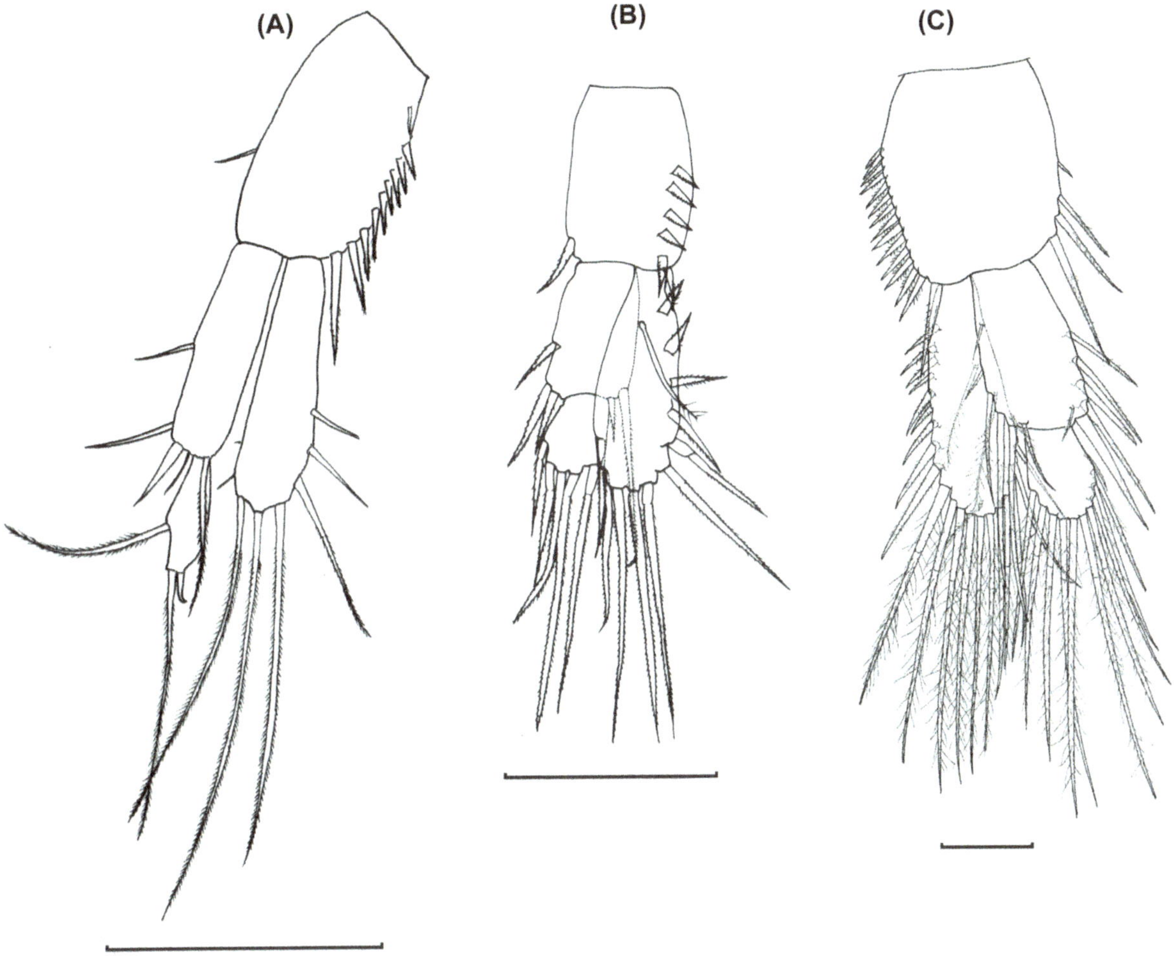

FIGURE 23.8 Uropods of Anaspidacea: (A) *Oncostygocaris patagonica* after Noodt (1963b); (B) *Stygocaris hugofernandezi* after Grosso & Peralta (1999); (C) *Parastygocaris andina*. Scale bars = 0.1 mm.

of pereopods. Females have no oostegites and generally have spermatheca. **Uropods** are biramous of variable morphology, rami uni- or biarticulate (Figs. 23.8, 9). The **telson** is generally entire with distal end not divided and variably straight, rounded or slightly pointed, and lined with setae. Only in Patagonaspididae is the telson bilobed with setose quadrangular lobes separated (Fig. 23.9 A). The Stygocarididae telson is a truncated cone, with a short anal plate with a lateral lobe (= caudal lobe) on either side of the anal opening, with two or three robust setae in each lobe (Fig. 23.9 B). Most Anaspidacea have the anal opening ventral, only in Stygocarididae is it terminal.

MATERIAL PREPARATION AND PRESERVATION

The specimens are collected using a 100-micron mesh net. The samples are fixed in 4% formaldehyde or 80% ethyl alcohol. Also 95% ethyl alcohol must be used on material for DNA analysis. In the laboratory, specimens may be transferred into glycerol for dissections and preservation. Accurate identification at generic or specific level requires careful dissection with a tungsten needle under stereoscope microscope. The appendages can be stained with Black Clorazol if necessary.

KEY TO ANASPIDACEA

Anaspidacea: Families

1 Body length ~7 mm (excluding antennae); telson flat, lamellar with two quadrangular lobes widely separated (Fig. 23.9 A); A1 accessory flagellum with six or seven articles; maxilla 1 outer endite with a row of seven teeth and a large terminal tooth (Fig. 23.5B); pereopods 1–5 with unarticulate exopodite, taeniform, with eight terminal setae .... Patagonaspididae, one species: *Patagonaspides sandroruffoi* Grosso & Peralta, 2002 [Argentina: Río Negro, Allén]

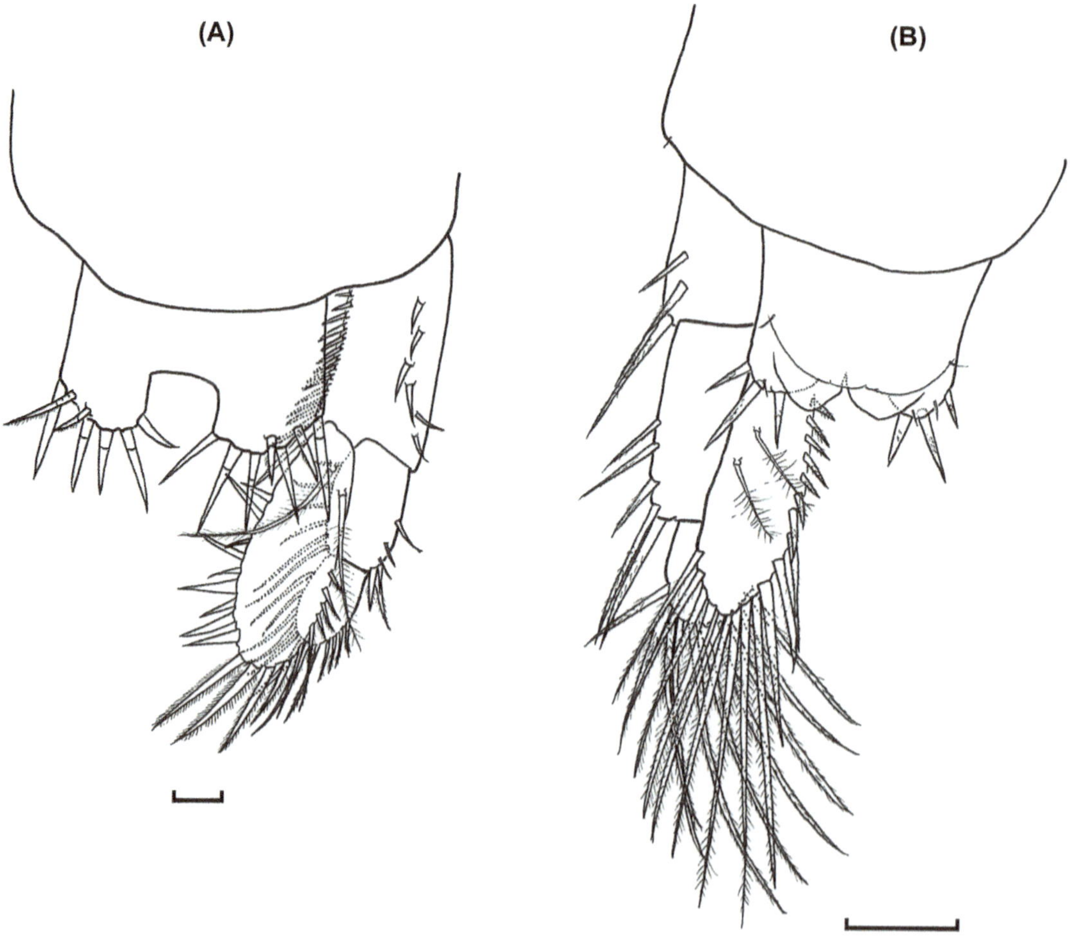

FIGURE 23.9 Telson and uropod of Anaspidacea: (A) *Patagonaspides sandroruffoi*; (B) *Parastygocaris* sp. Scale bars = 0.1 mm.

1′ Body length 1–4 mm; telson a truncated cone, reduced to a short anal plate with lateral lobes on both sides (Fig. 23.9 in B); A1 accessory flagellum with two or three articles; maxilla 1 outer endite with a double row of setae, all similar size (Fig. 23.5 A); pereopods 1–5 exopodite generally absent or unarticulate, cylindrical with two terminal setae (Fig. 23.6 A, B) .... **Stygocarididae [p. 820]**

Anaspidacea: Stygocarididae: Genera

1 Pereopods 1–5 without exopodite (Fig. 23.6 B) 2

1′ Pereopods 1–5 with unarticlulate exopodite (Fig. 23.6 A) .... ***Parastygocaris* [p. 820]**

2(1) Rostrum deeply cleft; uropod exopodite longer than endopodite, distal article with one long seta and one short, apically hamulate (Fig. 23.8 A) .... *Oncostygocaris patagonica* (Noodt, 1963)
[Chile: Aysén]

2′ Rostrum entire, linguiform (Fig. 23.3 B); uropod rami length subequal, lined with long setae (Fig. 23.8 B) .... ***Stygocaris* [p. 822]**

Anaspidacea: Stygocarididae: *Parastygocaris*: Species

1 A1 primary flagellum shorter than below; maxilla 2 medial endite medial row with less than 12 aligned setae; uropod protopod medial row with less than 13 setae 2

1 A1 primary flagellum with more than 13 articles (Fig. 23.10 A); maxilla 2 medial endite bears a medial row of 12 aligned setae (Fig. 23.5 C); uropod protopod medial row with 13–15 setae (Fig. 23.8 C) .... *Parastygocaris andina* Noodt, 1963
[Argentina: Mendoza]

2(1) A1 primary flagellum with 9 or 10 articles; uropod protopod medial row with seven or eight setae, endopodite lined with at least 13 setae 3

2′ A1 primary flagellum with eight or less articles; uropod protopod and endopodite with fewer setae than above 4

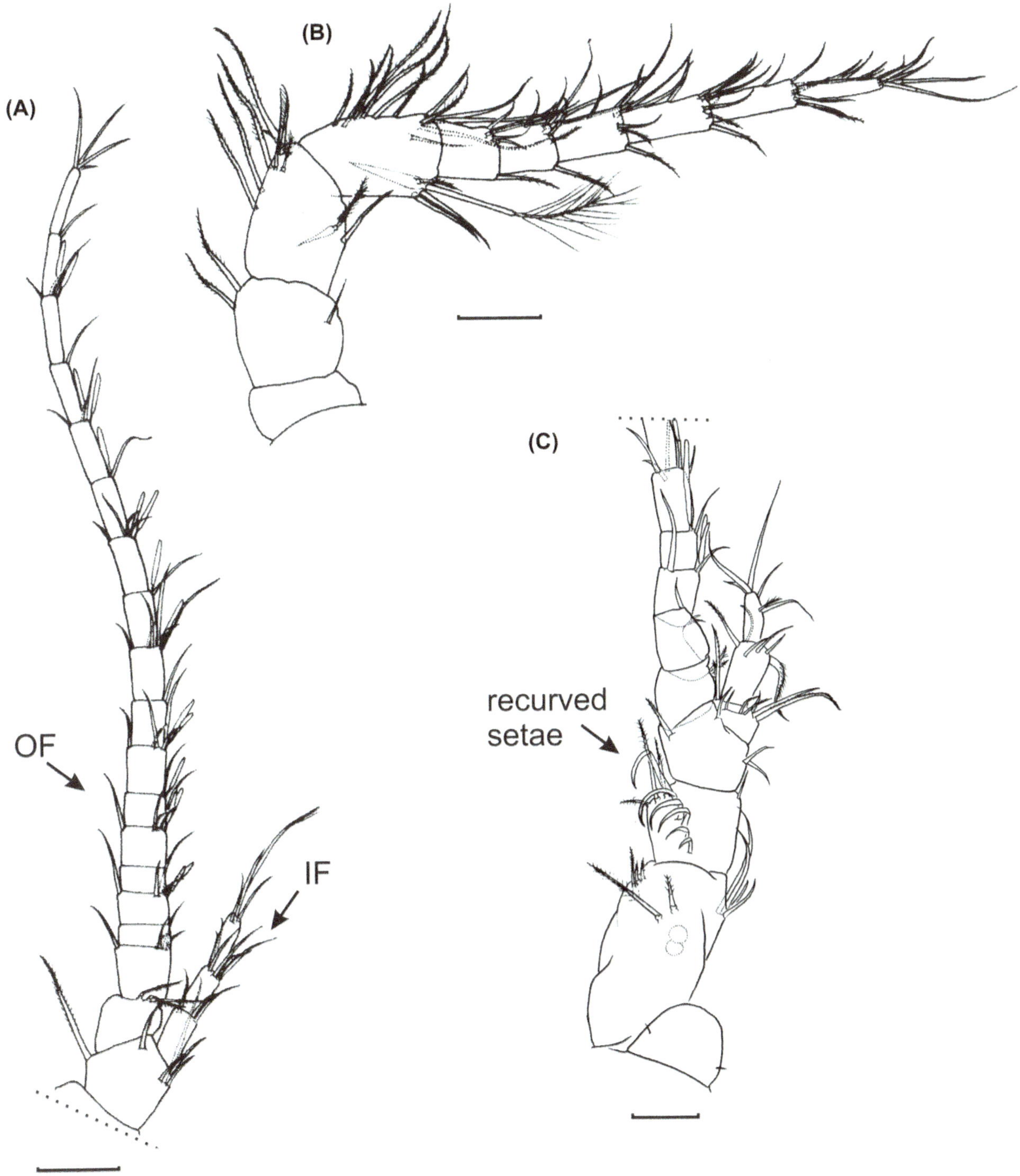

FIGURE 23.10 First antenna and second antenna of Anaspidacea: (A) first antenna third segment of peduncle, primary and accessory flagellum of *Parastygocaris andina* female; (B) second antenna of *Parastygocaris andina*; (C) first antenna peduncle of *Parastygocaris luisgrossoi* male. Abbreviations: *IF*, accessory flagellum; *OF*, primary flagellum. Scale bars = 0.05 mm.

3(2) Male A1 peduncle, second article dorsal surface with oblique row of recurved setae (Fig. 23.10 C); A2 with nine articles; maxilliped coxal endite with six distal and three medial setae (Fig. 23.4 A) .... *Parastygocaris luisgrossoi* Peralta, 2014
[Argentina: San Luis]

3′ Male A1 peduncle without recurved dorsal setae; A2 with eight articles; maxilliped coxal endite with eight distal and five medial setae .... *Parastygocaris goerssi* Noodt, 1963
[Argentina: San Luis]

4(2) A1 accessory flagellum not sexually dimorphic, with three articles; male pleopod 1 basipodite with distomedial prolongation (Fig. 23.7 A). .... *Parastygocaris schminkei* Grosso & Peralta, 1997
[Argentina: La Rioja]

4′ A1 accessory flagellum sexually dimorphic, with three articles in male and two in female; pleopod 1 basipodite without distomedial prolongation. .... *Parastygocaris clapsi* Grosso & Peralta, 1997
[Argentina: La Rioja]

Anaspidacea: Stygocarididae: *Stygocaris*: Species

1 Male A1 primary flagellum with eight articles; accessory flagellum with three articles, the second deeply excavated (only in male); uropod with unarticulate exopodite, distal third with constriction .... *Stygocaris gomezmillasi* Noodt, 1963
[Chile: San Antonio, Quebrada de Córdoba]

1′ Male A1 primary flagellum with six articles; accessory flagellum with two articles, not sexually dimorphic; uropod with biarticulate exopodite (Fig. 23.8 B) .... *Stygocaris hugofernandezi* Grosso & Peralta, 1999
[Argentina: La Rioja, Río Miranda]

REFERENCES

Camacho, A.I. & A.G. Valdecasas. 2008. Global diversity of syncarids (Syncarida; Crustacea) in freshwater. Hydrobiologia 595: 257–266.

Coineau, N. & A.I. Camacho. 2013. Sous-Classe des Eumalacostracés (Eumalacostraca Grobben, 1892) Super-Ordre des Syncarides (Syncarida Packard, 1885). Pages 897–954 *in*: J.C. Von Vaupel Klein, M. Charmantier-Daures & F.R. Schram (eds.), The Crustacea. Traité de Zoologie. P.P. Grassé. Vol. 4, Part A. Brill. Leiden, Boston.

Grosso, L.E. & M. Peralta. 1997. Hallazgo de un nuevo Stygocarididae en una antigua área de endemismo. *Parastygocaris schminkei* n. sp. (Crustacea, Syncarida). Physis B 54 (126–127): 21–26.

Grosso, L.E. & M. Peralta. 1999. Primera especie de *Stygocaris* (Crustacea, Syncarida) descubierta al Este de Los Andes. *S. hugofernandezi* n. sp. Physis B57 (132–133): 39–44.

Noodt, W. 1963a. Estudios sobre crustáceos de aguas subterráneas. III.- Crustacea Syncarida de Chile Central. Investigaciones Zoológicas Chilenas 10: 151–167.

Noodt, W. 1963b. Anaspidacea (Crustacea, Syncarida) in der südlichen Neotropis. Verhandlungen der Deutschen Zoologischen Gesellschaft in Wien 1962: 568–578.

Noodt, W. 1965. Natürliches Systemund Biogeographie der Syncarida (Crustacea, Malacostraca). Gewässerund Abwässer 37/38: 77–186.

Noodt, W. 1970. Zur Eidonomie der Stygocaridacea, einer Gruppe interstitieller Syncarida (Malacostraca). Crustaceana 19: 227–244.

Peralta, M. 2014. A new stygobitic species of Stygocarididae (Crustacea: Anaspidacea) from South America. Zootaxa 3760: 396–408.

Schminke, H.K. 1978. Die phylogenetische Stellung der Stygocarididae (Crustacea, Syncarida) —unterbesonderer Berücksichtigung morphologischer Ähnlich keiten mit Larven formen der Eucarida. Zeitschriftfür Zoologische Systematik und Evolutionsforschung 16: 225–239.

Schminke, H.K. 2014. Syncarida. Pages 190–193 *in:* J.W. Martin, J. Olesen and T. Høeg (eds.), Atlas of Crustacean Larvae. Johns Hopkins University Press, Baltimore.

Schram, F.R. & R.H. Hessler. 1984. Anaspidid Syncarida. Pages 192–195 *in*: N. Eldredge & S.M. Stanley (eds.), Living Fossils. Springer-Verlag. New York.

Bathynellacea

Marcela Peralta

Instituto de Invertebrados, Fundación M. Lillo, San Miguel de Tucumán, Argentina

Ana Isabel Camacho

Museo Nacional de Ciencias Naturales (CSIC), Dpto. Biodiversidad y Biología Evolutiva, Madrid, Spain

INTRODUCTION

Bathynellacea is an interesting group of stygobiotic crustaceans, which are presumably a primitive Eumalacostraca lineage with a long period of evolution but without a fossil record. Their distribution patterns indicate ancient Paleozoic roots in Pangea and pre-Pangea (Schram, 2008). The Bathynellacea is reported worldwide with records from all continents (except Antarctica). Most genera are endemic and a few have a disjunct distribution.

Traditionally, the Bathynellacea and Anaspidacea belong to superorder Syncarida. Based on comparative morphological and phylogenetic analysis (i.e., Schram & Hof, 1998; Camacho et al., 2002) it was shown that the Syncarida are not monophyletic. Serban (1970) erected the superorder Podophallocarida to contain Bathynellacea. Camacho & Valdecasas (2008) demonstrated that more studies are needed to resolve the validity of Podophallocarida.

Bathynellacea currently comprises about 298 species, and 79 genera worldwide (Camacho, 2015). This order is comprised of two families; Bathynellidae and Parabathynellidae. Some authors (Ranga Reddy & Totakura, 2012; Camacho et al., 2016) recognize a third family, Leptobathynellidae, although Schminke (1973) treated this group as Parabathynellidae. We recognize three valid families Bathynellidae, Parabathynellidae, and Leptobathynellidae.

Parabathynellidae is the most diverse and widely distributed of the families with about 200 species. All Neotropical records of the 10 species and seven genera are limited to South America, with the southernmost record for the order being *Atopobathynella valdiviana* from Valdivia, Chile, at 39°S. The Neotropical endemic genera are *Noodtibathynella* and the monotypic *Psalidobathynella*. Both *Atopobathynella* and *Chilibathynella* have disjunct distributions in eastern South America, Australia, and India. Similarly, *Cteniobathynella* (South America, Afrotropical region, Israel) and *Thermobathynella* (South America and Africa) are also broadly distributed. *Hexabathynella* has a nearly cosmopolitan distribution with 23 species from the Nearctic, Palearctic, Australian, Afrotropical, and Neotropical regions (Camacho & Valdecasas, 2008).

Leptobathynellidae is known only from the Southern Hemisphere, with 18 species in seven genera; four of which (with 7 species) inhabit in Neotropical region: *Parvulobathynella*; *Odontobathynella* (monotypic); *Leptobathynella*; and *Brasilibathynella* (monotypic). There is an undescribed *Brasilibathynella* recently found in India, thus it will soon no longer be considered a Neotropical endemic genus.

Bathynellidae is globally distributed. Three genera and four species of Bathynellidae occur in the Neotropics, with one genus (*Nannobathynella*) distributed across other biogeographic regions. Monotypic *Austrobathynella* (*A. patagonica* Delamare Deboutteville, 1960) from Lake Moreno, Patagonia, Argentina, is phylogenetically related to *Transvaalthynella* and *Transkeithynella*, both endemic to South Africa (Serban & Coineau, 1975). The third Neotropical genus, *Bathynella*, includes two species: *B. cautinensis* Noodt, 1965 (Chile) and *B. grossei* Noodt, 1965 (Chile and Argentina), which may represent a separate genus (Serban, 2000).

In general, these stygobionts represent ancient lineages of marine or freshwater crustaceans, which have long been isolated. Bathynellaceans generally live interstitially in aquifers, wells, hyporheic zones, lake psammon, and caves. Only a few species are found in oligo- to polyhaline waters (i.e., *Thermobathynella amyxi* Siewing, 1956).

LIMITATIONS

Although little is known about bathynellacean biology, two recently published reviews on Syncarida (Coineau & Camacho, 2013) and Parabathynellidae (Schminke & Cho, 2013) provide information on their habitat, behavior, locomotion, feeding, diet, and reproductive biology.

Bathynellacean adults are tiny pedomorphic malacostracans (Schminke, 1981). Their small size makes dissection of appendages difficult. Taxonomic identification requires detailed examination of the male thoracopod VIII and the mouth parts.

The Bathynellacea fauna is still imperfectly known and remains to be explored in most parts of the world (Schminke, 1982; Camacho et al., 2016). The Neotropics have been poorly sampled (Camacho & Valdecasas, 2008). This limited knowledge is due to the difficulty of accessing the interstitial underground habitats in which they live. The majority of species are treated as endemic not just because they have been collected from one locality, but because of the incompleteness of surveys (Camacho & Valdecasas, 2008).

The Bathynellacea are believed to have high levels of cryptic species (Camacho et al., 2011; Guzik et al., 2008), a situation that makes taxonomic studies of the group more intricate. Furthermore, there are few experts in the group and of stygobiont Crustacea in general. Globally, Bathynellidae is the least known group and *Bathynella* is in need of revision, as it is likely that it comprises several genera (Camacho, 2006; Coineau & Camacho, 2013).

TERMINOLOGY AND MORPHOLOGY

The Bathynellacea are pedomorphic, and tiny, rarely larger than 2 mm. The first thoracic somite is not fused with the head; therefore, the thorax consists of eight somites. There are five free abdominal somites and a pleotelson with furcal rami. The boundary between the thorax and abdomen is not very clear.

A carapace, eyes, and statocysts are absent. **Antenna I** (= A1) has a peduncle of three articles (the second article sometimes sexually dimorphic) and two flagella. **Antenna II** (= A2) is generally uniramous, the **endopod** with variable number of articles (up to five) or absent, and the **exopod** well developed only in Bathynellidae.

The **labrum** is variable, with the distal end concave, convex, entire, truncate, bilobed, glabrous (only in Bathynellidae), dentate, or spinulose. The **paragnaths** are developed only in Bathynellidae and some Parabathynellidae. The **mandibles** (= Md) are symmetrical, without a lacinia mobilis, and with the **palp** with one to three articles. In the Bathynellidae the palp is prehensile and armed with two strong setae (sometimes sexually dimorphic). The **maxilla I** (= Mx I) has two endites, the proximal one small, and both endites with distal setae. The **maxilla II** (= Mx II) has two to four articles, all with setae, the articles arrangement on a vertical axis, which is accentuated in Parabathynellidae. In Leptobathynellidae maxilla II is prehensile.

The thorax has eight pairs of **thoracopods (Th I–VIII)**, the first seven are locomotory appendages, usually biramous, and the eighth is transformed into complex copulatory organ (gonopod) in males, and is vestigial (or absent) in females. The locomotory thoracopods have an epipodite (sometimes absent in thoracopod I and/or II) with a respiratory function. In Parabathynellidae the exopod consists of 1 to 12 articles (usually 2) and an endopod of 4 articles. The Bathynellidae exopod is always of 1 article with the endopod of 3 or 4 articles; whereas in Leptobathynellidae the exopod is of 2 articles (always longer than the endopod) and the endopod is usually of 4 articles (sometimes with 3 in thoracopod I).

The abdomen is usually a little longer than the thorax, with the first two segments sometimes bearing vestigial **pleopods**. **Uropods** are positioned laterally with respect to the furca, and each consists of a one article sympod, with an exopod and endopod each of one article. Some genera have the exopod or endopod falcate through fusion with a terminal seta. The furcal rami are quadrangular or elongate, and with spines. The anus is located in the terminal portion of pleotelson and is covered by anal operculum in some Parabathynellidae.

MATERIAL PREPARATION AND PRESERVATION

The specimens are collected using a 100-microns mesh net and the samples are fixed in 4% formaldehyde or 80%–95% ethyl alcohol for molecular analysis. The tiny specimens may be transferred into glycerol for dissection and preservation. Accurate taxonomic identification at generic or specific level requires careful dissection with a tungsten needle under a microscope. Permanent slides are prepared with glycerine jelly stained with methylene blue and paraffin (Perina & Camacho, 2016).

KEY TO BATHYNELLACEA

Bathynellacea: Families

1 Antenna II directed posteriorly, exopod rudimentary or absent; mandible length/width = 0.8/1.5 (high/long), palp nonprehensile, of one or two articles; thoracopod exopod of more than one article; pleotelson without dorsal lateral seta 2

1′ Antenna II directed anteriorly, with exopod well developed (Fig. 23.11 A); labrum with distal free edge smooth (Fig. 23.12 A); mandible length/width =3 (very long), prehensile palp of one to three articles (Fig. 23.13 A); maxilla II of four articles, not prehensile (Fig. 23.14 A); thoracopod exopod of one article; pleotelson with dorsal lateral seta (Fig. 23.15) .... **Bathynellidae [p. 828]**

2(1) Antenna II without exopod (Fig. 23.11 B); labrum distal free edge dentate (Fig. 23.12 B); mandible length/width = 0.8 (high), palp of one article (Fig. 23.13 B); maxilla II not prehensile (Fig. 23.14 B); thoracopod exopod multiarticulate; pleotelson with ventrolateral seta (Figs. 23.16) .... **Parabathynellidae [p. 828]**

2′ Antenna II with rudimentary exopod (Fig. 23.11 C); labrum distal free edge with few spines (Fig. 23.12 C); mandible length/width = 1.5 (long), palp of one or two articles (Fig. 23.13 C); maxilla II prehensile (Fig. 23.14 C); thoracopod exopod of two articles; pleotelson without lateral seta .... **Leptobathynellidae [p. 829]**

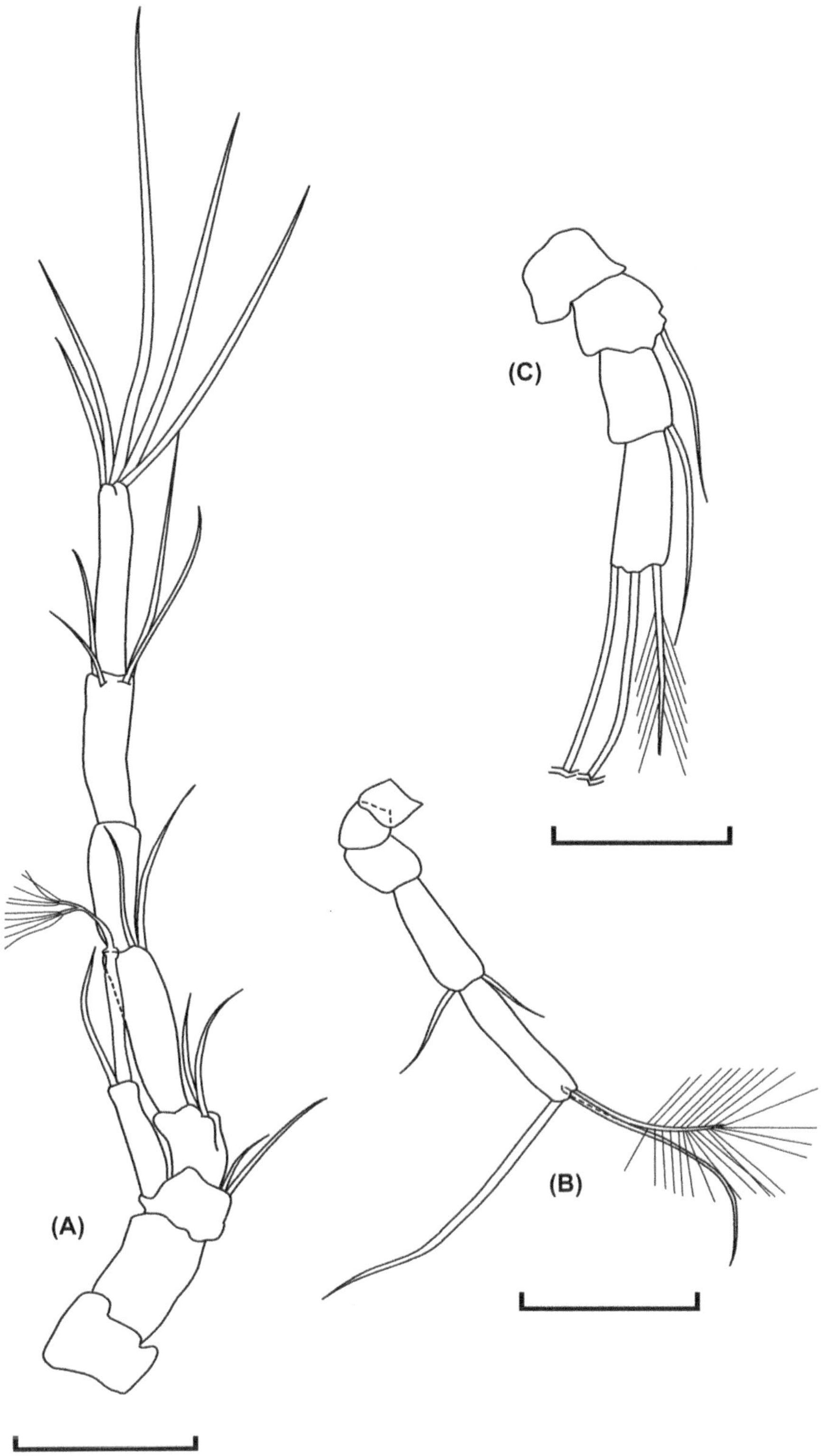

FIGURE 23.11 Antenna II of Bathynellacea: (A) Bathynellidae; (B) Parabathynellidae, *Hexabathynella szidati*; and (C) Leptobathynellidae, *Odontobathynella amazonica*. Scale bar (A) and (B) = 0.05 mm (C) = 0.02 mm. *(A) Original figures drawn by AIC; (B) after Peralta (2006), and (C) redrawn by AIC after Delamare Deboutteville & Serban (1979).*

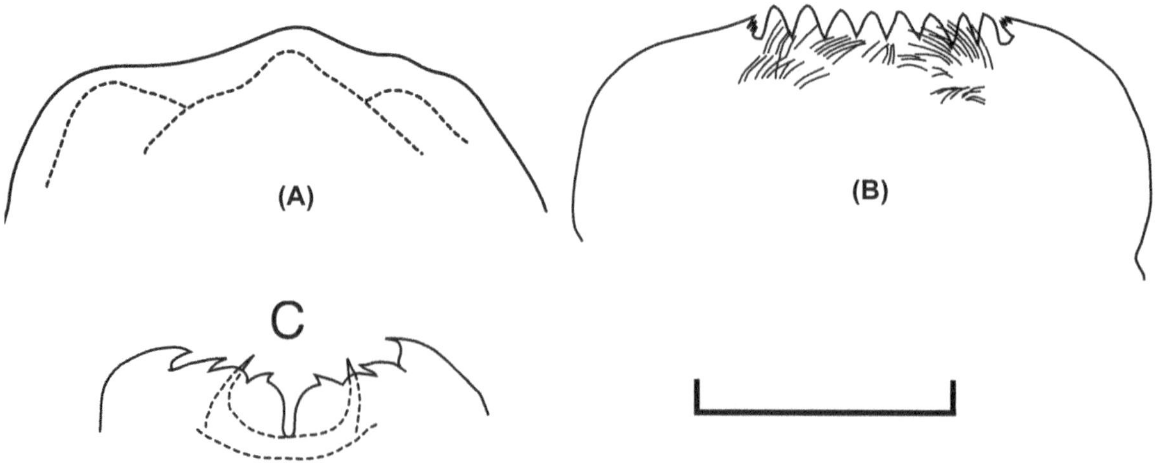

FIGURE 23.12 Labrum of Bathynellacea: (A) Bathynellidae; (B) Parabathynellidae, *Hexabathynella szidati*; and (C) Leptobathynellidae, *Odontobathynella amazonica*. Scale bar = 0.02 mm. *(A) Original figure drawn by AIC; (B) after Peralta (2006); and (C) redrawn by AIC after Delamare Deboutteville & Serban (1979).*

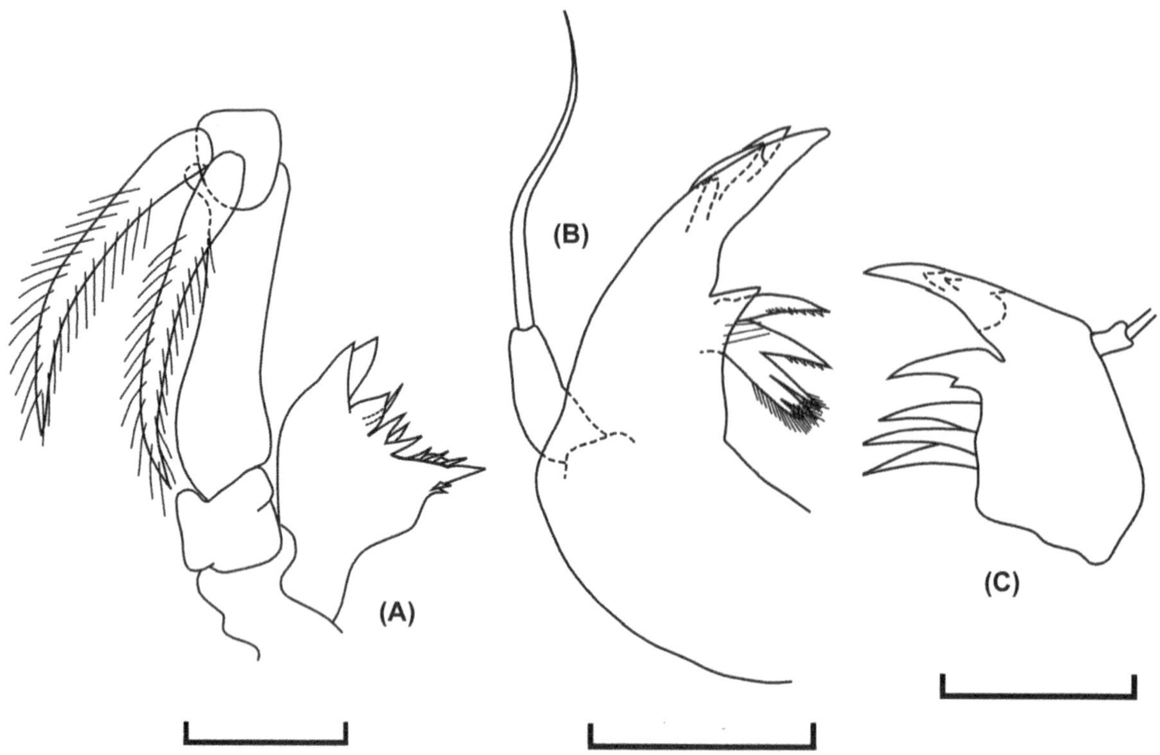

FIGURE 23.13 Mandibles of Bathynellacea: (A) Bathynellidae; (B) Parabathynellidae, *Hexabathynella szidati*; and (C) Leptobathynellidae, *Odontobathynella amazonica*. Scale bars = 0.02 mm. *(A) Original figure drawn by AIC; (B) after Peralta (2006); (C) redrawn by AIC after Delamare Deboutteville & Serban (1979).*

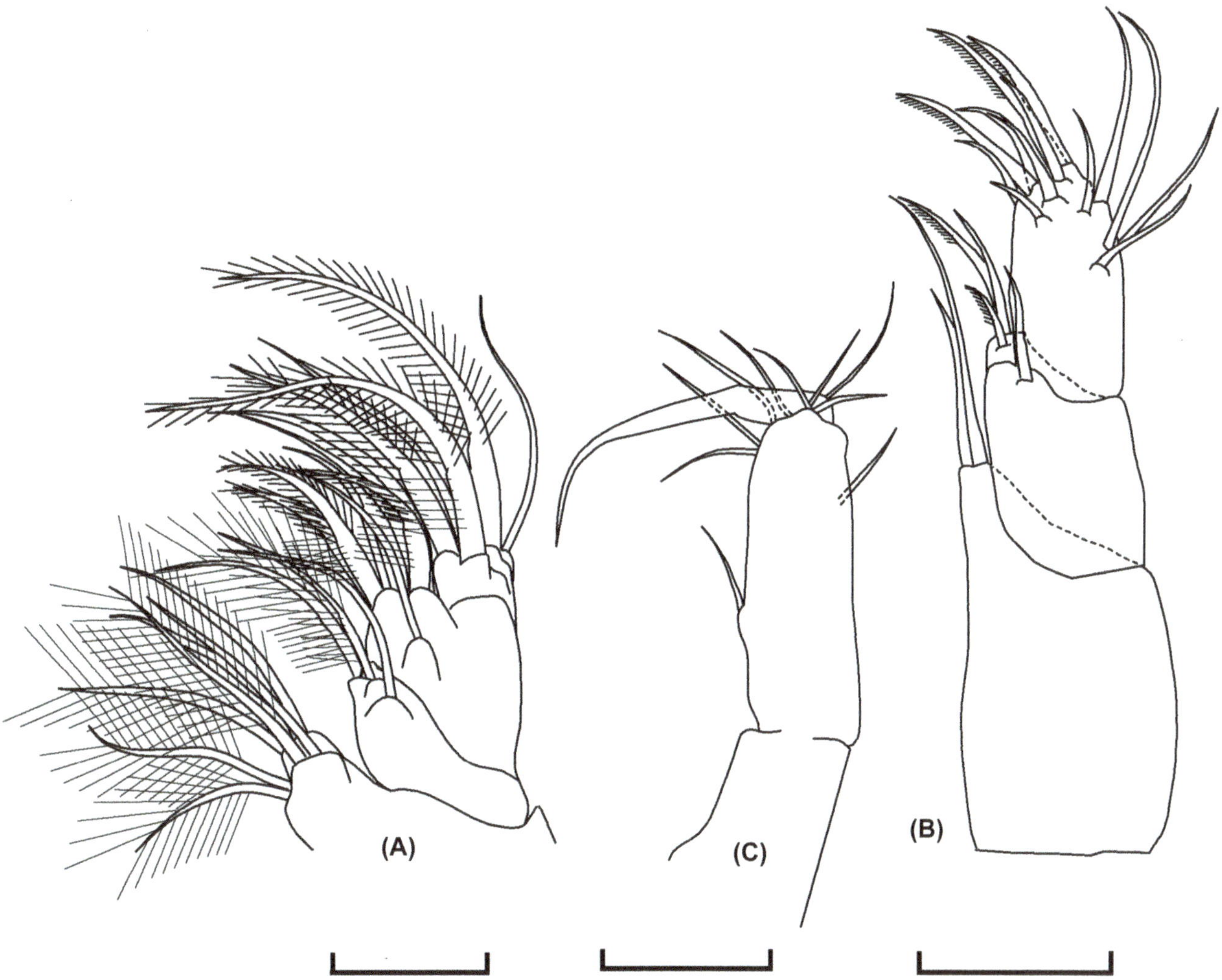

FIGURE 23.14 Maxilla II of Bathynellacea: (A) Bathynellidae; (B) Parabathynellidae, *Hexabathynella szidati*; and (C) Leptobathynellidae, *Odontobathynella amazonica*. Scale bars = 0.02 mm. *(A) Original figure drawn by AIC; (B) after Peralta (2006); redrawn by AIC after Delamare Deboutteville & Serban (1979).*

FIGURE 23.15 Photograph of a Neotropical Bathynellidae (Bathynellacea). Scale bar = 50 μm.

FIGURE 23.16 Photograph of a Neotropical Parabathynellidae (Bathynellacea). Scale bar = 1 mm.

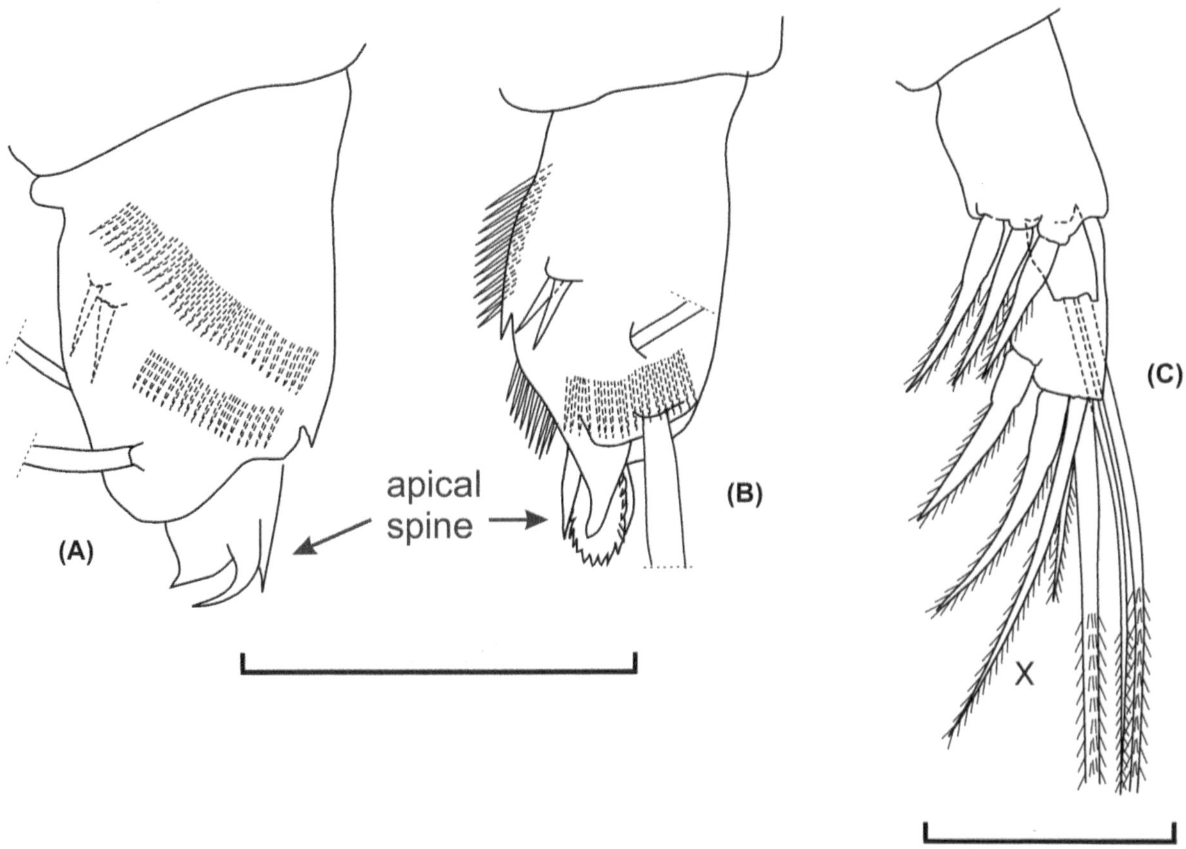

FIGURE 23.17 Furcal rami of Bathynellacea: (A) External side and (B) dorsal view of Leptobathynellidae, *Odontobathynella amazonica*; (C) uropod endopod of Bathynellidae, *Austrobathynella patagonica* with special seta ("X" of Serban). Scale bars: (A) and (B) = 0.02 mm, (C) = 0.05 mm. *(A) and (B) Redrawn after Delamare Deboutteville & Serban (1979); (C) redrawn after Delamare Deboutteville & Roland (1963).*

Bathynellacea: Bathynellidae: Genera

1 Thoracopod I–VII endopod with three or four articles; male thoracopod VIII exopod like exopod of thoracopod VII; female thoracopod VIII of one or two articles; uropod sympod with at least four spines; uropod endopod with three to four claws 2

1′ Thoracopod I–VII endopod with four articles; male thoracopod VIII exopod different than thoracopod VII; female thoracopod VIII of two articles; uropod sympod with at most four spines; uropod endopod with two claws and one special seta ("X" of Serban) (Fig. 23.17 C) .... *Austrobathynella patagonica* Delamare Deboutteville, 1960
[Argentina: Neuquén, Río Negro]

2(1) Thoracopod I to VII endopod of four articles; female thoracopod VIII of one article .... *Bathynella*
[Chile: Temuco; Aysén; Valdivia; Villa Rica Lake]

2′ Thoracopod I to VII endopod of three articles; female thoracopod VIII of two articles. .... *Nannobathynella marcusi* Noodt, 1969
[Brazil: São Paulo]

Bathynellacea: Parabathynellidae: Genera

1 Seven pairs of ambulatory thoracopods; antenna I with six or seven articles; antenna II with one to six articles 2

1′ Six pairs of ambulatory thoracopods; antenna I with six articles; antenna II with five articles; antenna I longer than antenna II (Figs. 23.11 B, 18) .... *Hexabathynella*
[Argentina: Córdoba, San Luis. Brazil: Cataratas do Iguaçu; Paraná]

2(1) Antenna I with six articles; antenna II with one or five articles 3

2′ Antenna I with seven articles; antenna II with four or six articles 5

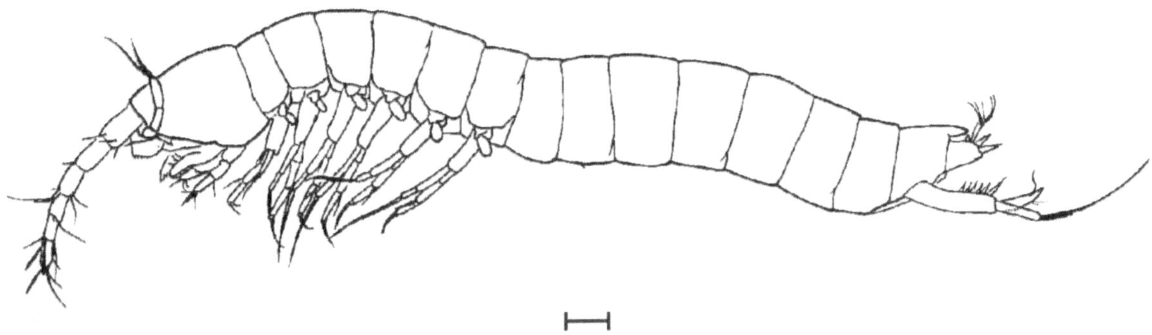

FIGURE 23.18 *Hexabathynella szidati* (Bathynellacea): Habitus female. Scale bar = 0.05 mm. *After Peralta (2006).*

3(2) Antenna II of five articles; maxilla II of three articles; thoracopod I–VII exopod of two articles; furcal rami with three spines 4

3′ Antenna II of one article; labrum heterodont; maxilla II of four articles; thoracopod I–VII exopod of one article; furcal rami with more than three spines. .... *Atopobathynella valdiviana* (Noodt, 1965)
[Chile: Valdivia]

4(3) Antenna I length > antenna II; labrum concave; uropod exopod with two terminal setae. .... *Cteniobathynella*
[Brazil: Mina Gerais; São Paulo]

4′ Antenna I subequal to antenna II; labrum flat; uropod exopod falcate .... *Thermobathynella amyxi* Siewing, 1956
[Brazil: Pará]

5(2) Antenna II of four articles; labrum flat; thoracopod I–VII exopod of two or four articles 6

5′ Antenna II of four or six articles; thoracopod I–VII exopod of one article .... *Chilibathynella clandestina* Noodt, 1963
[Chile: San Antonio]

6(5) Thoracopod I–VII exopod with four articles; uropod exopod with two distal setae; uropod endopod distally acute; furcal rami with three spines .... *Noodtibathynella*
[Colombia: Santander. Brazil: Pará. Argentina: Córdoba]

6′ Thoracopod I–VII exopod of two, three, or four articles; uropod exopod with three distal setae; uropod endopod with two short terminal spines opposing each other; furcal rami with more than three spines .... *Psalidobathynella stocki* Schminke, 1979
[Venezuela: Falcón]

Bathynellacea: Leptobathynellidae: Genera

1 Thoracopod I–VII endopod of four articles; thoracopod I–VII exopod of two articles; maxilla II with two or three articles 2

1′ Thoracopod I endopod with three articles; thoracopod II–VII endopod with four articles; thoracopod I–VII exopod of two articles; labrum free margin with spinules; maxilla II of two articles; uropod endopod with separated spines; furcal rami conical each with an apical spine and two medial small spines .... *Brasilibathynella florianopolis* Jakobi, 1958
[Brazil: Santa Catarina]

2(1) Uropod endopod with one distal spine; furcal rami with three spines 3

2′ Uropod endopod without spines, but rather similar in size to exopod, and dilated with three setae; furcal rami with apical spine of complex structure (Fig. 23.17 A, B); labrum cleft medially area, and dentate both lobes .... *Odontobathynella amazonica* Delamare Deboutteville & Serban, 1979
[Brazil: Amazonas]

3(2) Maxilla II of two articles; labrum smooth with lobulations and sometimes with spinulae; uropod endopod with separated medial spine .... *Leptobathynella*
[Brazil: Rio de Janeiro, São Paulo. Paraguay: Ypacaraí. Argentina: Córdoba]

3′ Maxilla II of two or three articles; labrum with lobes bearing spinulae; furcal rami rounded with three spines .... *Parvulobathynella*
[Brazil: Rio de Janeiro. Chile: Atacama, Choapa, San Antonio. Paraguay: Ypacaraí]

REFERENCES

Camacho, A.I. 2006. An annotated checklist of the Syncarida (Crustacea, Malacostraca) of the world. Zootaxa 1374: 1–54.

Camacho, A.I. 2015. Clase: Malacostraca. Orden Bathynellacea. Manual. Revista electrónica IDE@-SEA: 1–17.

Camacho, A.I., A. Brancelj, B.A. Dorda, A. Casado & I. Rey. 2016. New Parabathynellidae species in Africa: the first bathynellids from Chad and an assay of their phylogenetic position in the order Bathynellacea (Crustacea: Malacostraca) based on 18S sequences. Journal of Natural History 50: 2691–2726.

Camacho, A.I., B.A. Dorda & I. Rey. 2011. Identifying cryptic speciation across groudwater populations: first COI sequences of Bathynellidae (Crustacea, Syncarida). Graellsia 67: 7–12.

Camacho, A.I., I. Rey, B.A. Dorda, A. Machordom & A.G. Valdecasas. 2002. A note on the systematic position of the Bathynellacea (Crustacea, Malacostraca) using molecular evidence. Contribution to Zoology 71: 123–129.

Camacho, A.I. & A.G. Valdecasas. 2008. Global diversity of syncarids (Syncarida; Crustacea) in freshwater. Hydrobiologia 595: 257–266.

Coineau, N. & A.I. Camacho. 2013. Sous-Classe des Eumalacostracés (Eumalacostraca Grobben, 1892) Super-Ordre des Syncarides (Syncarida Packard, 1885). Pages 897–954 *in*: J.C. Von Vaupel Klein, M. Charmantier-Daures and F.R. Schram (eds.), The Crustacea. Traité de Zoologie. P. P. Grassé. Vol. 4, Part A. Brill. Leiden, Boston.

Delamare Deboutteville C. & C. Roland. 1963. Syncarides. *Austrobathynella patagonica* Delamare. Pages 55–62. *in*: C. Delamare-Deboutteville & E. Rapoport (eds), Biologie de l'Amérique Australe, II. Centre National de la Recherche Scientifique, París.

Delamare Deboutteville C. & E. Serban. 1979. *Odontobathynella amazonica* n. g. n. sp. Leptobathynellidae nouveau du Brésil (Malacostraca, Bathynellacea). Bulletin de la Societé de Sciences Naturelles de Tunisie 14: 5–20.

Guzik, M.T., K.M. Abrams, S.J.B. Cooper, W.F. Humphreys, J.-L. Cho & A. Austin. 2008. Phylogeography of the ancient Parabathynellidae (Crustacea: Bathynellacea) from the Yilgarn region of Western Australia. Invertebrate Systematics 22: 205–216.

Peralta, M. 2006. Redescripción de *Hexabathynella szidati* (Crustacea: Bathynellacea) de aguas subterráneas de Argentina. Segunda cita geográfica. Acta Zoológica Lilloana 50: 15–23.

Perina, G. & A.I Camacho. 2016. Permanent slides for morphological studies of small crustaceans: Serban's method and its variation applied on Bathynellacea (Malacostraca). Crustaceana 89: 1161–1173.

Ranga Reddy, Y. & V.R. Totakura. 2012. *Indobathynella prehensilis* n.gen, n.sp. an aberrant species of Bathynellacea (Eumalacostraca) from India. Journal of Crustacean Biology 32: 281–293.

Schminke, H.K. 1973. Evolution, System und Verbreitungsgeschichte der Familie Parabathynellidae (Bathynellacea, Malacostraca). Akademieder Wissenschaften und der Literaturzur Mainz, Mathematisch-naturwissenschaftliche Klasse. Mikrofauna des Meeresbodens 24: 1–192.

Schminke, H.K. 1981. Adaptations of Bathynellacea (Crustacea, Syncarida) to life in the interstitial ("Zoea theory"). Internationale Revue der gesamten Hydrobiologie und Hydrographie 66: 578–637.

Schminke, H.K. 1982. Syncarida. Pages 233–237 *in*: S. P. Parker (ed.), Synopsis and classification of living organisms Volumen 2. McGraw-Hill, New York.

Schminke H. K. & J-L Cho. 2013. Biology and Ecology of Parabathynellidae (Crustacea, Bathynellacea). A review. Crustaceana 86: 1266–1273.

Schram, F.R. 2008. Does biogeography have a future in a globalized world with globalized faunas? Contributions to Zoology 77: 127–133.

Schram, F.R. & C.H.J. Hof. 1998. Fossils and the interrelationships of major crustacean groups. Pages 233–302 *in*: G.D. Edgecombe (ed.), Arthropod Fossils and Phylogeny. Columbia University Press, New York.

Serban, E. 1970. A propos du genre *Bathynella* Vejdovsky (Crustacea, Syncarida). Pages 265–273 *in*: Livre du centenaire Emile G. Racovitza, Ed. Acad., Bucarest.

Serban, E. 2000. *Uenobathynella* n.g., *Parauenobathynella* n.g., *Morimotobathynella* n.g., *Nihobathynella* n.g. et *Paradoxibathynella* n.g., Bathynellinae du Japon (Bathynellidae, Bathynellacea, Podophallocarida). Travaux de l'Institut de Spéologie Émile Racovitza 36: 3–61.

Serban, E. & N. Coineau. 1975. Sur les Bathynellidae (Podophallocarida Bathynellacea) d'Afrique du Sud. Les genres *Transvaalthynella* nov. et *Transkeithynella* nov. Annales de Spéléologie 30: 137–165.

Amphipoda

Exequiel R. González
Universidad Santo Tomás, Santiago, Chile

Marcela Peralta
Instituto de Invertebrados, Fundación M. Lillo, San Miguel de Tucumán, Argentina

Alessandra Bueno
Universidade Federal de Lavras, Lavras, Brazil

INTRODUCTION

Amphipoda has traditionally included four suborders: Gammaridea, Caprellidea, Hyperiidea, and Ingolfiellidea. Lowry & Myers (2003, 2013) treated the higher classification with a morphological-based view, and erected the new suborder Senticaudata in 2013, which comprises 95 families formerly in Gammaridea and Corophiidea, the latter recategorized now as infraorder Corophiida (Senticaudata). Lowry & Myers (2013) defined Senticaudata having: (a) antenna 1 without a complex callynophore; (b) male antenna 2 without brush setae or elongate flagellum; and; (c) uropods 1 and 2 with apical robust setae. The suborder includes the infraorders: Carangoliopsida, Talitrida, Hadziida, Corophiida, Bogidiellida, and Gammarida. Only Carangoliopsida is absent from the Neotropics. The classification proposed by Lowry & Myers (2013) has been widely accepted.

Most Neotropical freshwater amphipods are included in Senticaudata, with others in Ingolfiellidea, plus the families Seborgiidae, Pardaliscidae, Lysianassidae, and Pseudingolfiellidae, which still have an uncertain taxonomic position. Recently De Smet (2015) considered Pseudingolfiellidae part of Senticaudata. We recognize 156 species, 47 genera in 20 families of freshwater or slightly brackish water amphipods from inland waters of Neotropical region (Väinölä et al., 2008; Lowry & Myers, 2013).

The Mexican and both Central and South American fauna have different origins. Amphipod paleobiogeographic history demonstrates that many geological and climatic changes have led to repeated continental invasions from the sea over time. These events have led to a heterogeneous faunal assemblage.

The family diversity is highest in the hypogean continental waters. The subterranean diversity is unusually high in some areas; for example, the island of Hispaniola (Greater Antilles) has at least 20 species (12 genera) (Stock, 1985a, 1985b; Jaume & Wagner, 1998; Jaume & Christenson, 2001).

LIMITATIONS

Neotropical inland Amphipoda are poorly known, particularly at species level. Many genera cannot be identified to species level due to the large numbers of undescribed species. Although there are several keys and reviews for some groups of freshwater amphipods (e.g., Peralta & Grosso, 2009), there are still large gaps of information, with most keys present in very specialized literature. The regional exploration of different Neotropical biotopes has been sparse and poorly inventoried, especially the subterranean, where freshwater or anchialine amphipods are among the most abundant and taxonomically diverse organisms (Holsinger, 1994). Overall hypogean environments require additional sampling efforts especially in phreatic and other subterranean aquatic habitats. These habitats are difficult or impossible for researchers to access, so the fauna is poorly understood and its biodiversity underestimated (Fernández, 2004). For example, Brazil, a country with remarkable subterranean diversity, has had probably less than 20% of the mapped caves sampled by biologists (Fišer et al., 2013).

Fresh estuarine aquatic habitats are also poorly sampled. The South American Atlantic coast includes three major river basins (Orinoco, Amazonas, and Río de la Plata) each with complex estuarine ecosystems that are very diverse biologically but poorly sampled.

We have included here only the hypogean saline species reported from slightly brackish water and those that are known simultaneously both in fresh and "oligohaline" waters. Some Neotropical species distributed in "oligohaline" groundwater are presumably of freshwater origin (Noodt, 1959). It is also important to point out that the genus *Hyalella* is diverse and complex, with low levels of morphological differentiation. Recent studies with genetic material suggest that some *Hyalella* "species" may actually be many closely related species. The *Hyalella* key included in this chapter is only valid for species that have been recently reviewed or described by contemporary specialists using modern methods. To use the key, do not go below genus level if your specimen does not exactly match the provided character states in each couplet.

TERMINOLOGY AND MORPHOLOGY

Amphipods are generally small to medium-sized peracarid crustaceans. Adults generally range from 7 to 12 mm. Amphipods are unique in possessing basomedial coxal gills; the first four limb pairs are directed anteriorly, and the last three pairs posteriorly; and the abdomen bears three anterior pairs of multiarticulate pleopods and three posterior pairs of uropods (Figs. 23.19, 20).

Other important terms are defined below.

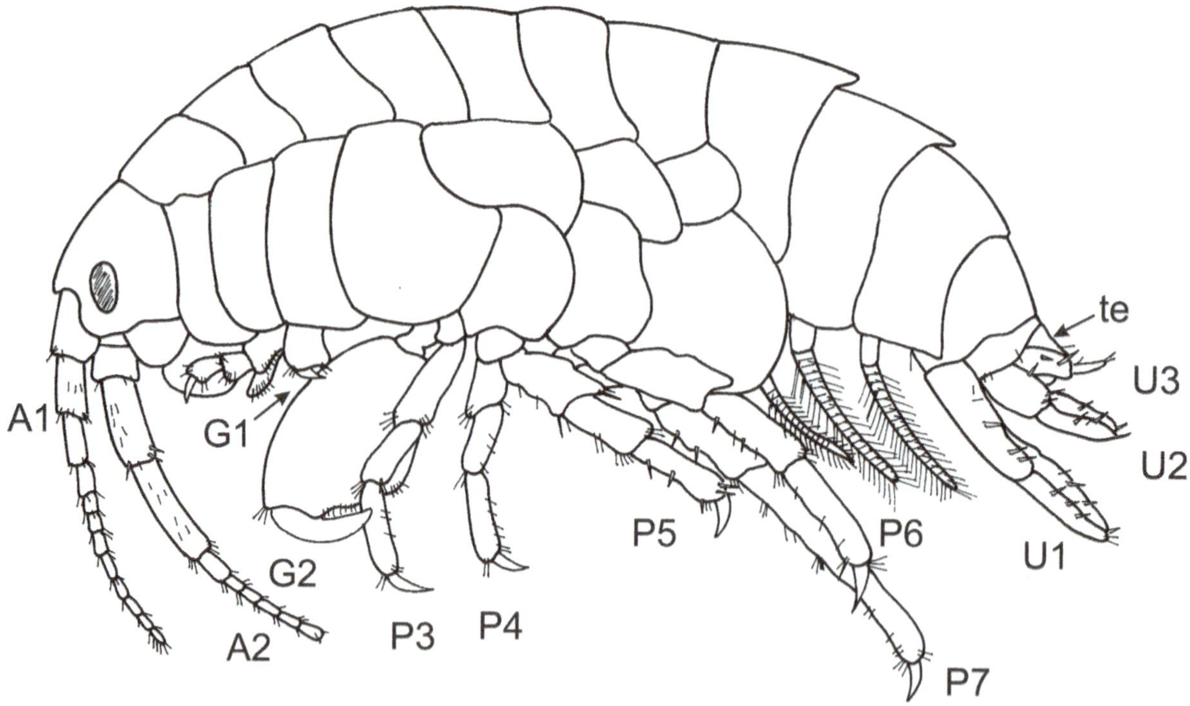

FIGURE 23.19 Hyalellidae: male of *Hyalella* sp. Abbreviations: *A1*, antenna 1; *A2*, antenna 2; *G1*, gnathopod 1; *G2*, gnathopod 2; *P*, pereopod; *U*, uropod; *te*, telson. Body size 5.6 mm.

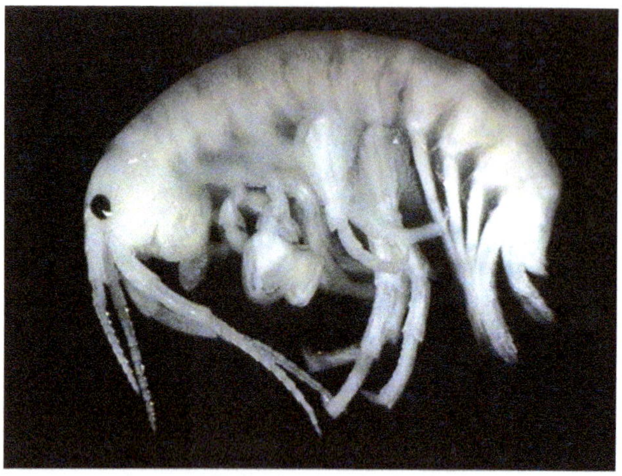

FIGURE 23.20 Hyalellidae: male of *Hyalella* sp. (body size 7.5 mm).

- **A-setae**: The mandible palp can show different groups of setae in its terminal article (A to E). The A-setae or A-group are located on the outer surface of the article; it may consist of a number of transverse subgroups. The B-group is located along the dorsal margin of the segment; the C-group is located on the inner surface at a slight distance from the ventral margin of the article; the D-group is placed in a kind of comblike row along the ventral margin of the article; the E-group consists of longer setae, implanted in terminal position.
- **Accessory flagellum**: The secondary ramus of antenna 1, often absent or vestigial, and attached medially to peduncle article 3.
- **Calceolus** (pl. calceoli): A small globular, linguiform or galeaform, articulate complex of sensory receptor found on antenna 2, or antenna 1 and 2, in some gammaridean amphipods. Lincoln & Hurley (1981) defined nine types of calceoli depending on the family.
- **Calceoli gammarid type 1**: The gammarid calceolus represents one of the simplest configurations. The proximal element forms a weakly concave crescentic plate closely applied along its inner margin to the distal element. The distal element usually has well-defined and variable transverse bands. Gammarid calceoli are usually confined to males, are typically few in number, and there is only one calceolus on each flagellar article (Fig. 23.21).
- **Carpus**: Article 5 of a thoracic appendage (gnathopod, pereopod).
- **Carpus palm**: Posterior surface or margin of gnathopod article 5 (Ingolfiellidea) on which articles 6 (propodus) and 7 (dactylus) close for the purpose of prehension.

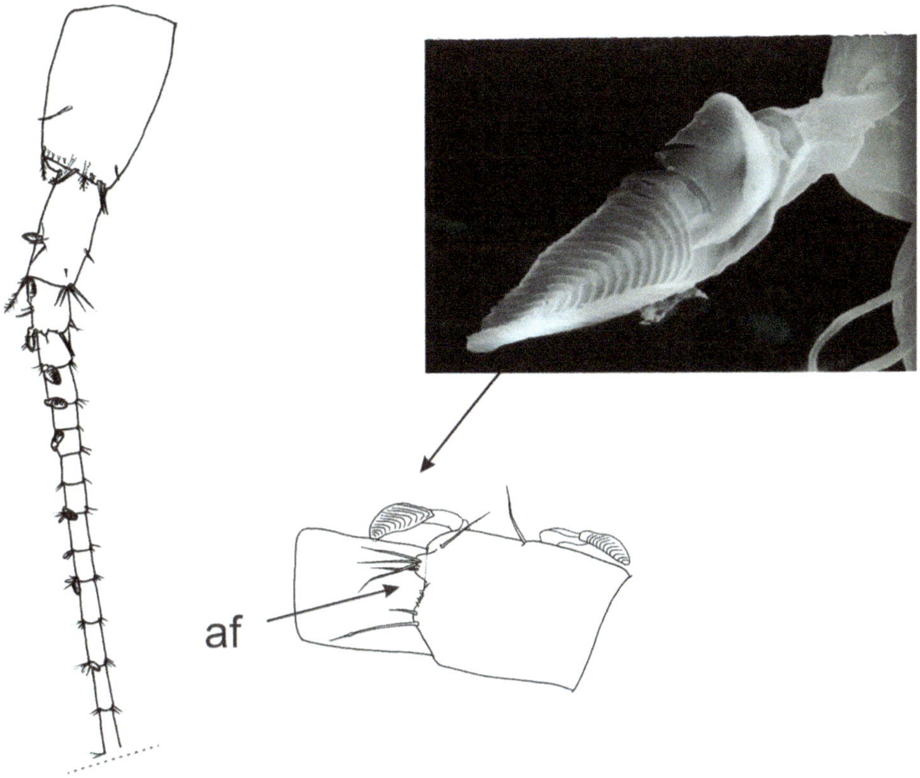

FIGURE 23.21 Paraleptamphopidae: Antenna 1 with detail of accessory flagellum (*af*) and microphotography of calceoli of *Rudolphia macrodactylus* (body size 5.8 mm).

- **Coxa, coxal plate**: Terms used synonymously. Article 1 of a pereonal appendage, expanded into a lateral lamella. Terms for other articles of the appendages such as basis, ischium, merus, carpus, propodus, and dactylus are frequently used; in some cases the articles are simply numbered.
- **Coxal gills**: Respiratory structures arising from inner proximal margin of coxal segment of pereopods 2–7.
- **Dactylus**: Terminal article of pereopods (article 7) or maxillipeds (articles 3 or 4).
- **Endopod**: Inner ramus of biramous appendages (see exopod).
- **Exopod**: Outer ramus of biramous appendage (see endopod).
- **Gnathopod**: One member of the first two pairs of free thoracic appendages; these appendages differ in function and usually in appearance from following pereopods.
- **Incisor**: The anterior apical part of the mandible usually formed into a toothed chewing edge or untoothed chopping plate.
- **Ischial endite**: For mouthparts, outer plate, especially for maxilla 1, maxilla 2, and maxilliped.
- **Lenticular organs** (Hertzog organs): Distinctive, semitransparent, "windows" of cuticle on the side of the articles of pereopods and pleopods.
- **Lower lip** (labium): A fleshy complex posterior to the mandibles, always composed of at least one pair of lobes (outer), often with a midproximal pair of inner lobes; the lateroproximal ends of outer lobes are often attenuated as alae and are denoted as mandibular lobes.
- **Maxilla 1**: A pair of cephalic appendages posterior to the lower lip; for taxonomic purposes only three parts of each member are named, the medial lobe (plate) or basal endite usually bearing marginal setae, the lateral and larger lobe (plate) or ischial endite bearing terminal spines, and, attached to the ischial endite, a palp usually composed of two articles but occasionally absent.
- **Maxilla 2**: A pair of cephalic appendages posterior to maxilla 1; for taxonomic purposes each member recognized as a pair of lobes (plates) medial or basal endite and lateral or ischial endite, usually strongly setose.
- **Maxillipeds**: The most posterior pair of "cephalic" appendages, representing the primitive first thoracic segment now amalgamated with the head but in amphipod taxonomy not included in the sequential

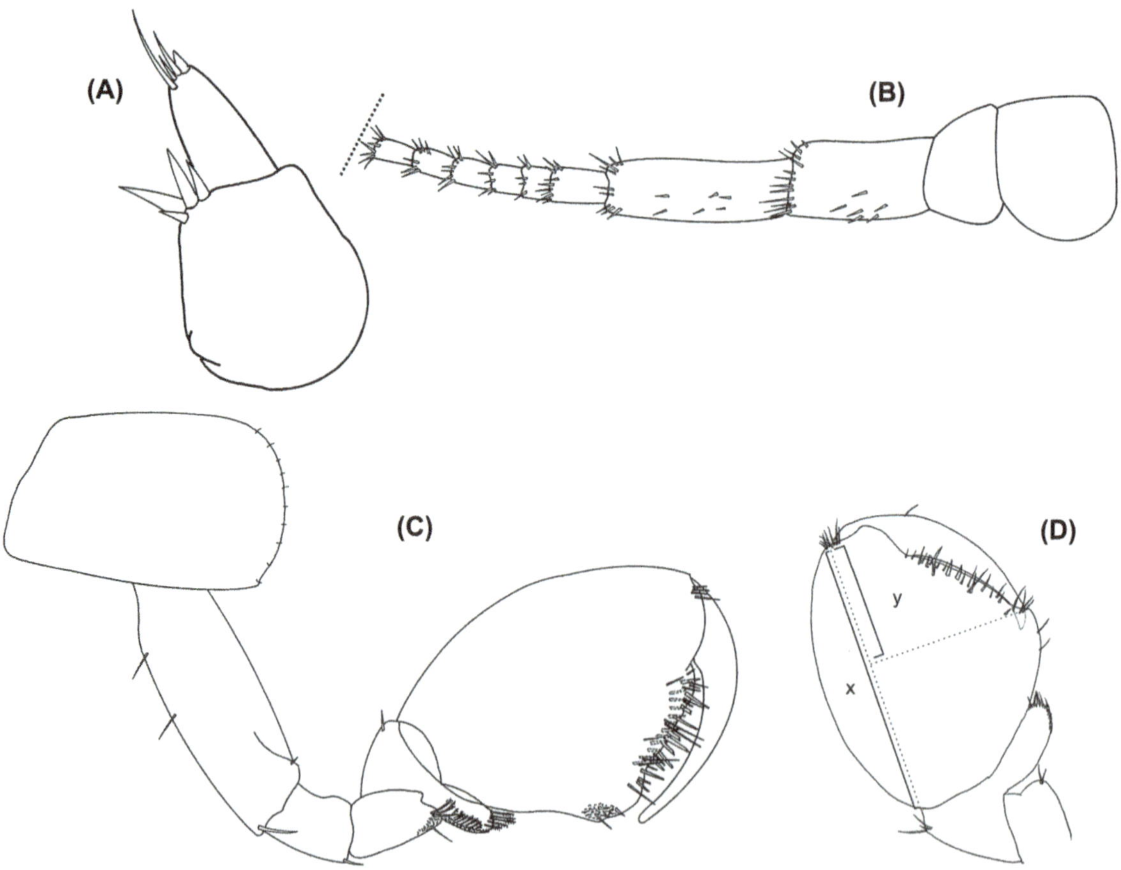

FIGURE 23.22 Hyalellidae: (A) Hyalellidae, uropod 3 of *Hyalella* sp.; (B) antenna 1 of *Hyalella sp.*; (C) Hyalellidae, gnathopod 2 male of *Hyalella gracilicornis* (body size 4.3 mm) (D) palm of gnathopod 2 male *Hyalella sp.* (*x*, longest axis of gnathopod; *y*, is the distal line of "x", it is determined by the intersection of line "x" with a perpendicular line drawn from the proximal palmar angle, see "palmar index" in glossary) (body size 7.4 mm). *Redrawn (C) from González & Watling (2003a).*

numbering of thoracic appendages; for taxonomic purposes recognized as a pair of basally amalgamated appendages, each member composed of a proximal plate (inner or basal endite), a distal (outer or ischial endite) plate, and a palp of four articles, rarely reduced to three or two articles or absent.

- **Merus**: Article 4 of a thoracic appendage (gnathopod, pereopod).
- **Molar**: A process of the mandible, located on the midmedial margin; when typically developed it is a medium-sized, subcylindrical body with a surface of ridges and teeth used for grinding (triturative). It evolves in two directions, becoming larger, less triturative, and finally smooth and pubescent; becoming smaller, less triturative and finally smooth or pubescent or spiny and finally disappearing.
- **Oostegites**: Flattened plates arising from inner proximal margin of coxa of pereopods two to five in female; forming brood chamber for retention of eggs and newly hatched young.
- **Outer plate**: Used for mouthparts, ischial endite.
- **Palm**: A posterior surface or margin of article 6 of a gnathopod or pereopod on which article 7 (dactylus) closes for the purpose of prehension; usually recognizable because of expansion of article 6 or by occurrence of special spines or ornamentation and usually with a proximal defining limit marked by a change in marginal slope or occurrence of special spines.
- **Palmar index**: Used to describe the shape of propodus in gnathopods 1 or 2. For its determination a straight line ("x") should be drawn between the upper distal angle of propodus with its proximal angle; line "y" is the distal line of "x," it is determined by the intersection of line "x" with a perpendicular line drawn from the proximal palmar angle; the ratio "y"/"x" is the palmar index (PI) which increases with the palmar inclination (Fig. 23.22 D).
- **Palp**: Terminal articles of a buccal appendage, in Amphipoda occurring only on mandibles, first maxillae, and maxillipeds as the stenopodous terminal articles distal to the expanded outer plates or main body.

- **Pappose setae**: Wide range of size, in all cases, long setules come off the shaft at all angles with a tendency to taper down to two rows toward the tips, which is pointed and simple. The apex bears a subterminal pore. Sometimes pappose setae bear a dense concentration of setules, others are very sparsely setulated. In all cases, the setules are present along the entire length of the shaft.
- **Pereon**: Thoracic division of body, 7-segmented carrying uniramous appendages (pereopods).
- **Pereonite**: A segment of the pereon.
- **Pereopod**: A walking graspy, standing or feeding appendage attached to a pereonite; normally composed of seven articles, including coxa; in Amphipoda the two first pairs are often termed gnathopods, and only the last five of thoracic legs are called pereopods (Fig. 23.19).
- **Pleon**: Three-segmented abdominal body division bearing biramous appendages.
- **Pleonite**: A segment of the pleon.
- **Propodus**: Article 6 of a thoracic appendage, especially used to denote the palmar article of a gnathopod.
- **Scale** (in relation to ornamentation of the body): A nonarticulated cuticular extension of which the base is generally very wide relative to its length; microscopic secondary features may arm the outer margin; this term should not be used to describe very small extensions of setal shafts.
- **Scale, scaled (in relation to antennae)**: Terms applied to the accessory flagellum when forming a small lamella immovably fused to article 3 of antenna l; and also to the uropod 3 endopod when it is strongly reduced and platelike.
- **Sternal gills**: Structures developed from midventral body wall of pereon segments.
- **Triturative**: Descriptive of the rasplike surface of a mandibular molar, composed of teeth, ridges, and cusps.
- **Urosome**: Posterior three segments of abdomen, carrying normally biramous uropods

MATERIAL PREPARATION AND PRESERVATION

Specimens once collected should be gently rinsed in freshwater, before preservation in 70% ethyl alcohol (ethanol) to remove flocculant material. Traditionally 10% formalin solution has been used for fixation. Specimens preserved this way should be transferred within a few hours, at most a few days, to fresh water, and then preserved in 70% ethyl alcohol (ethanol) for long-term storage. In some cases, it is advisable to preserve specimens in 95% ethanol in a volume of 3:1 alcohol to specimen(s), without any contact with formalin. Specimens thus preserved are of the highest quality and are available for future genetic studies. Accurate identification at generic or specific level requires dissection or manipulation of the specimens under a stereoscope and compound microscope. Usually a detailed examination and careful dissection of the appendages from one side of the body is necessary. These appendages can best be studied by preparation of temporary or permanent slide mounts. Barnard & Karaman (1991) give information on dissection and preparation of material for microscopic analysis.

KEY TO AMPHIPODA

Amphipoda: Families

1	First antenna without accessory flagellum (Fig. 23.22 B)	2
1′	First antenna with a well-developed or minute accessory flagellum (Fig. 23.21)	6
2(1)	Gnathopod 2 not sexually dimorphic	3
2′	Gnathopod 2 sexually dimorphic (Fig. 23.19)	4
3(2)	Pereopod coxa 4 without posterioventral lobe	**Hadziidae** (in part) **[p. 839]**
3′	Pereopod coxa 4 with a small or well-developed posterioventral lobe	**Bogidiellidae** (in part) **[p. 844]**
4(2)	Antenna 1 length > antenna 2 peduncle	5
4′	Antenna 1 length < antenna 2 peduncle	Talitridae, one species: *Orchestia scutigerula* (Dana, 1852) [Malvinas (Falkland) Islands]
5(4)	Uropod 3 uniramous; mandible without palp; maxilliped palp, article two length subequal to article one (Fig. 23.22 A)	Hyalellidae, one genus: ***Hyalella*** **[p. 849]**
5′	Uropod 3 uniramous or biramous; mandible with palp; maxilliped palp, article two length > article one (Fig. 23.23 A, B)	**Corophiidae [p. 856]**
6(1)	Accessory flagellum length < main flagellum three proximal articles	7
6′	Accessory flagellum > main flagellum three proximal articles	14

Couplet	Character	Result
7(6)	Uropod 3 uniramous	8
7′	Uropod 3 biramous	10
8(7)	Coxal gills present in segments 2–6 or 3–7	9
8′	Coxal gills present in segments 2–5 (Fig. 23.23 C, D)	Seborgiidae, one species: *Seborgia potiguar* Fišer, Zagmajster & Ferreira, 2013 [Brazil: Rio Grande do Norte]
9(8)	Sternal gills present (Fig. 23.24 A)	Pseudingolfiellidae, one species *Pseudingolfiella chilensis* (Noodt, 1959) [Chile: San Antonio]
9′	Sternal gills absent	Metacrangonyctidae, one genus: ***Metacrangonyx* [p. 856]**
10(7)	Eyes present	11
10′	Eyes absent	13
11(10)	Uropod 3 endopod subequal to exopod (Fig. 23.25 A)	12
11′	Uropod 3 endopod length < exopod (Fig. 23.24 B, C)	**Crangoweckeliidae [p. 856]**
12(11)	Eyes poorly developed (Fig. 23.25 B)	Paraleptamphopidae, one species: *Rudolphia macrodactylus* Grosso & Peralta, 2009 [Chile: Osorno]
12′	Eyes rounded	Phreatogammaridae, one species: *Ruffia patagonica* Bréhier, Vonk & Jaume, 2010 [Chile: Magallanes]
13(10)	Uropod 1 peduncle without basofacial robust seta (Fig. 23.26 A)	**Falklandellidae [p. 857]**
13′	Uropod 1 peduncle with basofacial robust seta (Fig. 23.26 B)	**Hadziidae** (in part) **[p. 839]**
14(6)	Antenna 2 calceoli present (at least in males)	15

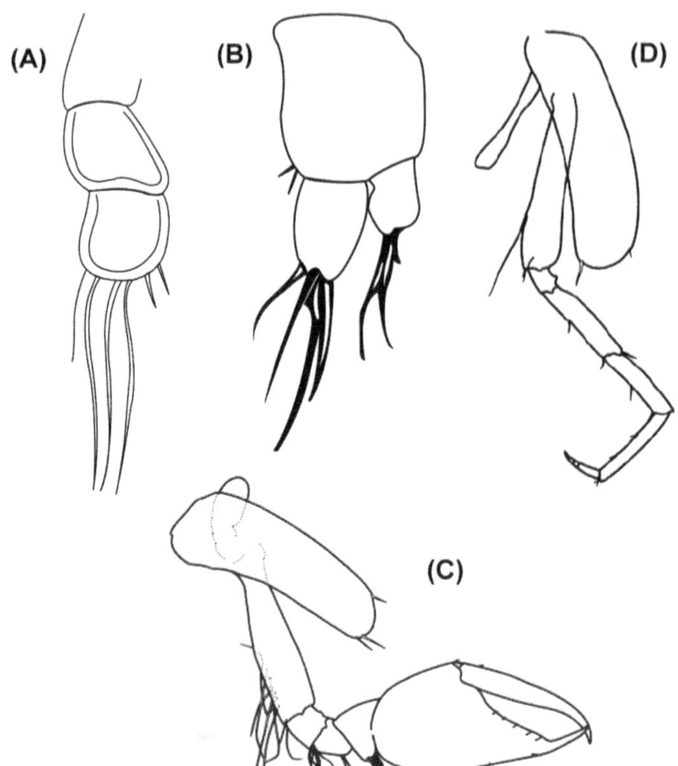

FIGURE 23.23 Corophiidae and Seborgiidae: (A) Corophiidae, uropod 3 of *Americorophium rioplatense* (body size 3–4 mm); (B) Corophiidae, uropod 3 of *Paracorophium hartmannorum* (body size 3.8 mm); (C) Seborgiidae, gnathopod 2 of *Seborgia potiguar* (body size 1.5 mm); (D) Seborgiidae, pereopod 3 of *S. potiguar* (body size 1.5 mm). *Redrawn (A) from Giambiagi (1929), (B) from González (1986), (C) and (D) from Fišer et al. (2013).*

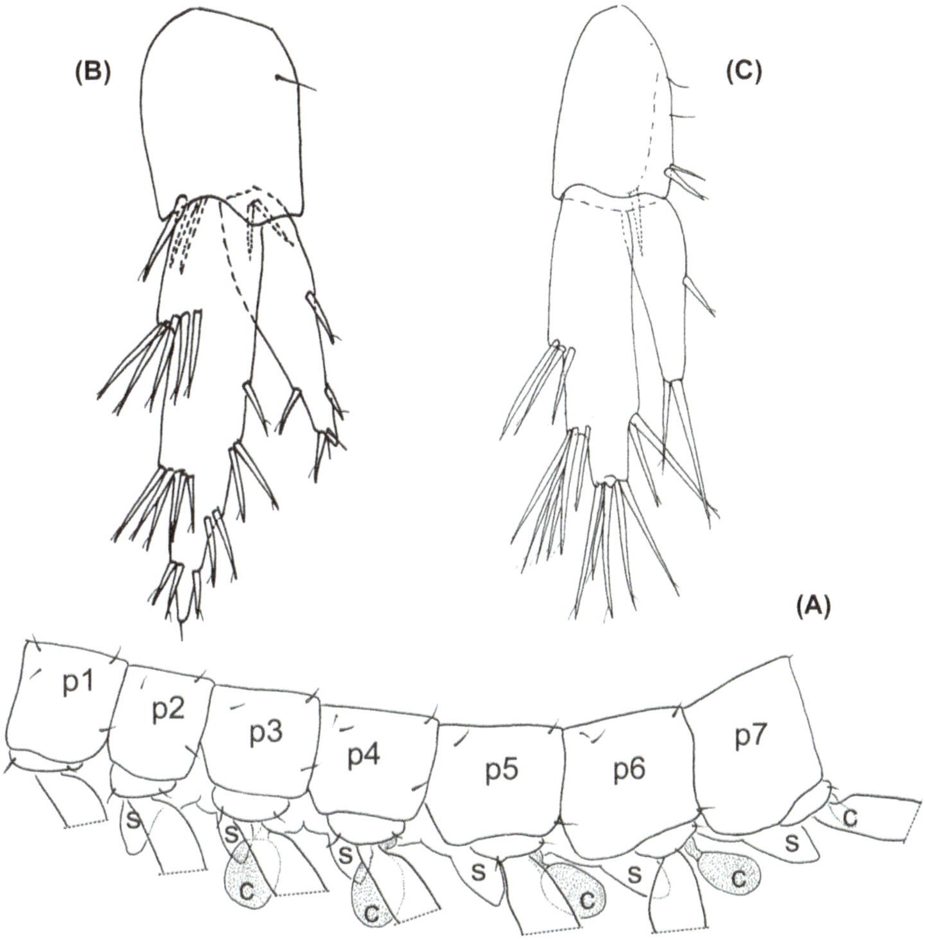

FIGURE 23.24 Pseudingolfiellidae and Crangoweckeliidae: (A) Pseudingolfiellidae, gills of *Pseudingolfiella* (c, coxal gills; s, sternal gills; p1-7, pereopds 1–7) (body size 2.15 mm); (B) Crangoweckeliidae, uropod 3 of *Crangoweckelia mixta* (body size 5–6 mm); (C) Crangoweckeliidae, uropod 3 of *Crangoweckelia spinicauda* (body size 7.0 mm). *Redrawn from: (B) and (C) from Stock (1985b).*

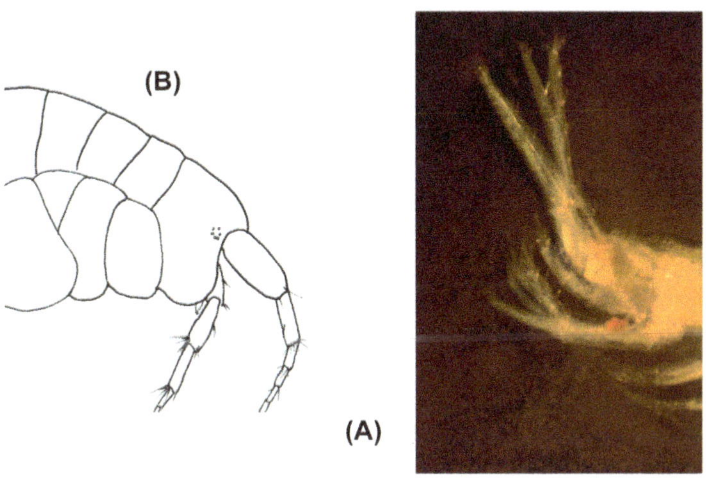

FIGURE 23.25 Paraleptamphopidae and Phreatogammaridae: (A) Paraleptamphopidae, uropod 3 of *Rudolphia macrodactylus* (body size 5.8 mm); (B) Paraleptamphopidae, head female of *Rudolphia macrodactylus* (body size 6.2 mm).

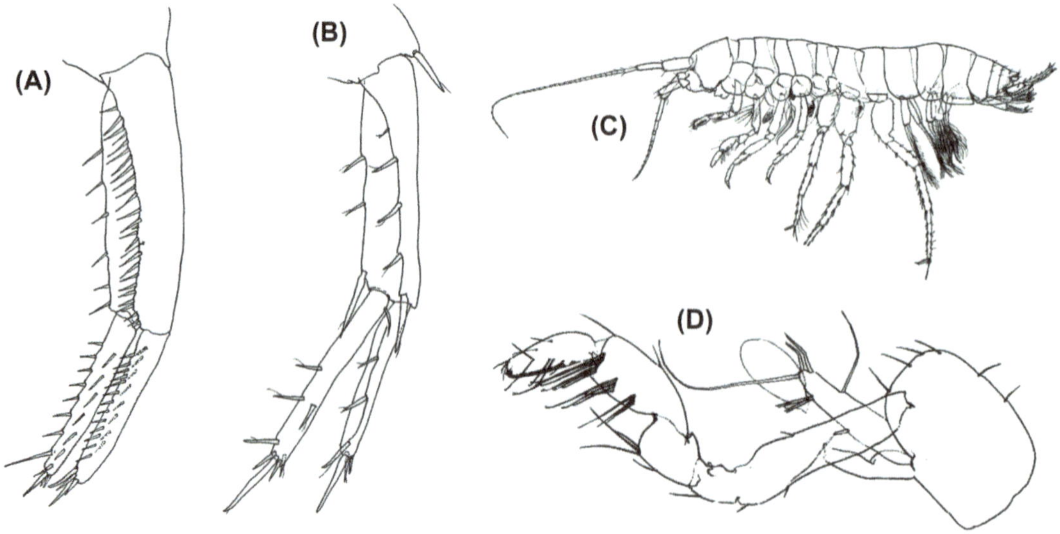

FIGURE 23.26 Falklandellidae, Hadziidae and Mesogammaridae: (A) Falklandellidae, uropod 1 of *Falklandella obtusa* (body size 6.5–7.5 mm); (B) Hadziidae, uropod 1 of *Bahadzia latipalpus* (body size 3.5–4.5 mm); (C) Mesogammaridae, body of *Potiberaba porakuara* (body size 3.5 mm); (D) Mesogammaridae, gnathopod 2 of *Potiberaba porakuara* (body size 3.5 mm). *Redrawn (A) from Stock & Platvoet (1991), (B) from Stock (1985b), (C) and (D) from Fišer et al. (2013).*

14′ Antenna 2 calceoli absent 16

15(14) Coxal gills present on pereopods 2–6; uropod 3 endopod length 0.40× exopod .... Gammaridae, one species: *Gammarus minus* Say, 1818 [Greater Antilles: Cuba]

15′ Coxal gills present on pereopods 2–7; uropod 3 endopod length 0.3× exopod .... Anisogammaridae, one species: *Eogammarus confervicolus* (Stimpson, 1856) [México: Baja California. Nearctic]

16(14) Gnathopod 2 normal, not reduced 17

16′ Gnathopod 2 dactylus remarkably reduced, attached to palm (Fig. 23.26 C, D) .... Mesogammaridae, one species: *Potiberaba porakuara* Fišer, Zagmajster & Ferreira, 2013 [Brazil: Rio Grande do Norte]

17(16) Mandible molar with some kind of triturative structure 18

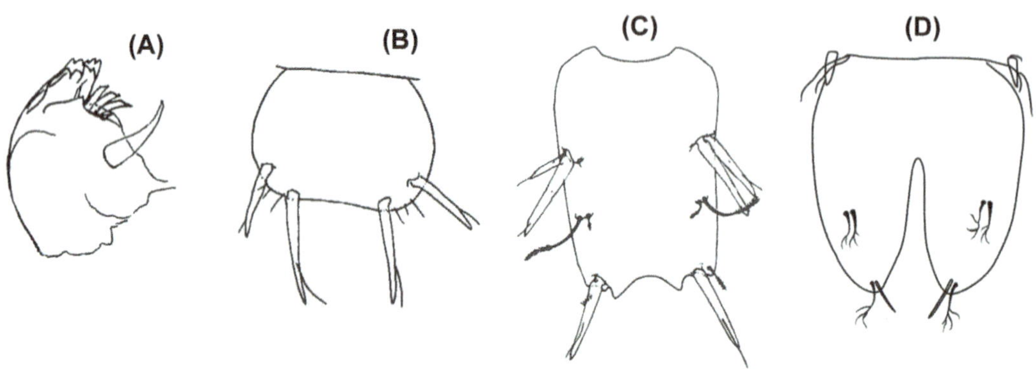

FIGURE 23.27 Ingolfiellidae, Bogidiellidae and Lysianassidae: (A) Ingolfiellidae, mandible of *Ingolfiella uspallatae* (body size 2.5 mm); (B) Bogidiellidae, telson of *Stygogidiella bredini* (body size 7 mm); (C) Bogidiellidae, telson of *Mesochthongidiella tucumanensis* (body size 3.2 mm); (D) Lysianassidae, telson of *Ottenwalderia kymbalion* (body size 12.1 mm). *Redrawn (A) from Noodt (1965), (B) from Stock (1981), (C) from Grosso & Fernández (1985), (D) from Jaume & Wagner (1998).*

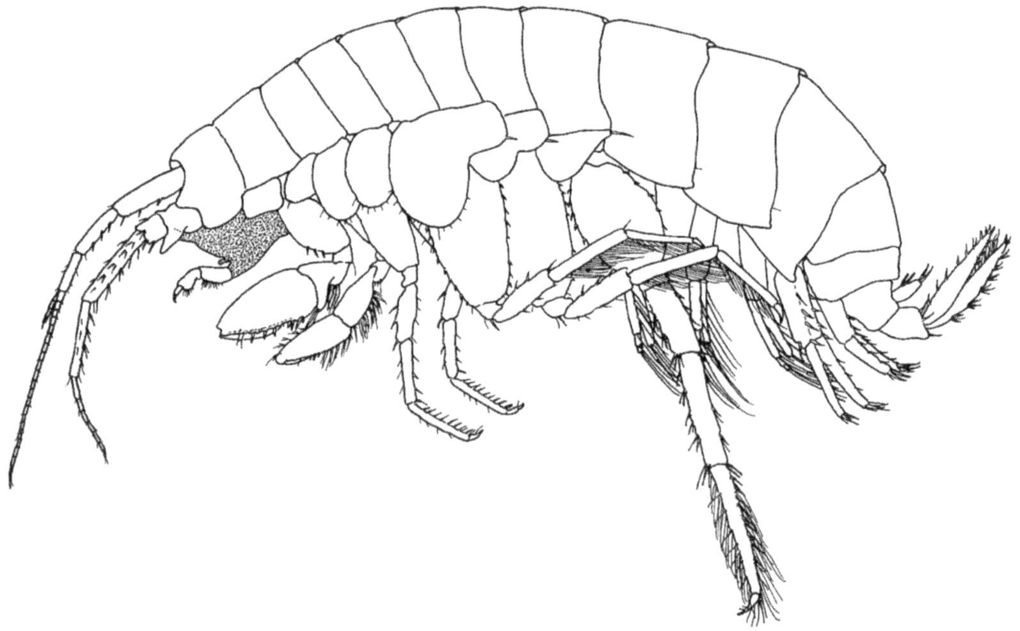

FIGURE 23.28 Artesiidae: body of *Spelaeogammarus spinilacertus* (body size 8.1 mm). *Redrawn from Koenemann & Holsinger (2000).*

17′ Mandible molar nontriturative (Fig. 23.27 A) 21

18(17) Telson entire to notched (Fig. 23.27 B, C) 19

18′ Telson cleft (Fig. 23.27 D) 20

19(18) Pereopod 5 length subequal to pereopod 6 (Fig. 23.28) .... Artesiidae, one genus: ***Spelaeogammarus* [p. 857]**

19′ Pereopod 5 length shorter than pereopod 6 .... **Bogidiellidae** (in part) **[p. 844]**

20(18) Antenna 1 subequal in length with antenna 2 .... Lysianassidae, one species: *Ottenwalderia kymbalion* Jaume & Wagner, 1998 [Greater Antilles: Dominican Republic]

20′ Antenna 1 longer than antenna 2 .... Maeridae, one species: *Quadrivisio lutzi* (Shoemaker, 1933) [Guyana: Georgetown]

21(17) Telson entire to notched without robust setae (Fig. 23.29 A, B). .... **Ingolfiellidae [p. 858]**

21′ Telson cleft with robust setae (Fig. 23.29 C) .... Pardaliscidae, one species: *Spelaeonicippe provo* Stock & Vermeulen, 1982 [Lesser Antilles: Isles Turks and Caicos]

Amphipoda: Hadziidae: Genera

1 Uropod 3 exopod with two articles (Fig. 23.30 A) 2

1′ Uropod 3 exopod with one article (Fig. 23.30 B) 3

2(1) Uropod 3 endopod wider and longer than exopod, with long and simple setae; telson rami much longer than wide, coxal gills not stalked .... ***Bahadzia* [p. 841]**

2′ Uropod 3 endopod narrower and shorter than exopod, with short simple or plumose setae; telson rami not much longer than wide, coxal gills stalked (Figs. 23.31 A, B) .... ***Metaniphargus* [p. 842]**

3(1) Uropod 3, rami similar in length; mandibular palp present 4

3′ Uropod 3, endopod much shorter than exopod; mandibular palp 1 article present or absent 9

4(3) Mandibular palp one to four articles 5

4′ Mandibular palp absent 8

5(4) Mandibular palp two to three articles 6

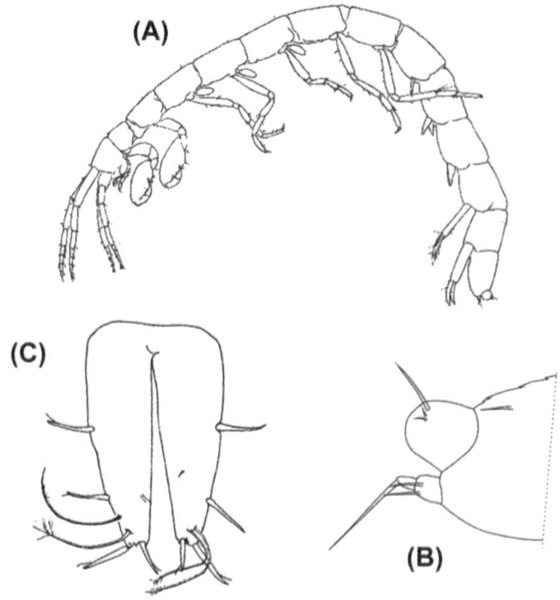

FIGURE 23.29 Ingolfiellidae and Pardaliscidae: (A) Ingolfiellidae, body of *Ingolfiella* sp. (body size 6.8 mm); (B) Ingolfiellidae, telson and uropod 3 of *Ingolfiella* sp., lateral view; (C) Pardaliscidae, telson of *Spelaeonicippe provo* (body size 3–4 mm). *Redrawn (C) from Stock & Vermeulen (1982).*

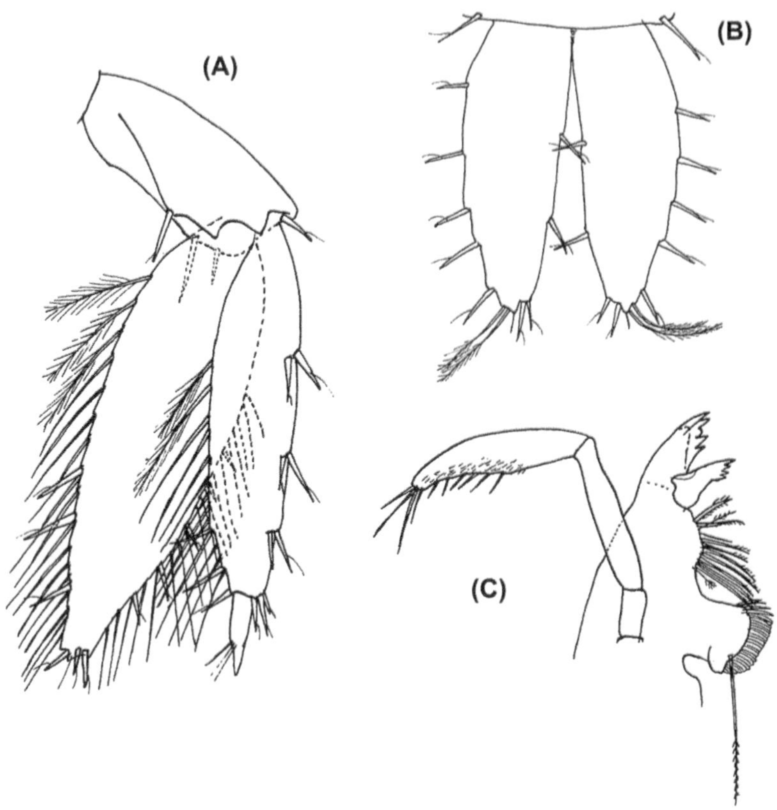

FIGURE 23.30 Hadziidae, *Bahadzia*: (A) uropod 3 and (B) telson of *B. latipalpus*body size 3.5–4.5 mm; (C) left mandible of *B. bozanici* (body size 6 mm). *Redrawn (A), (B) and (C) from Stock (1985b).*

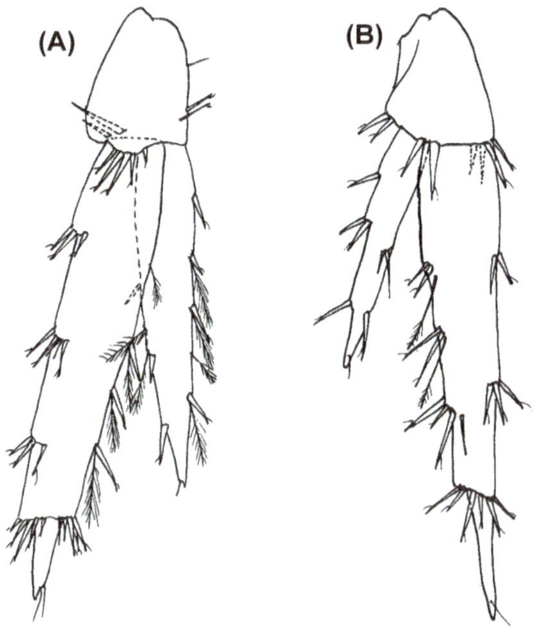

FIGURE 23.31 Hadziidae, *Metaniphargus*: (A) uropod 3 male of *M. plumicauda* (body size 4 mm); (B) uropod 3 male *M. crenatus* (body size 3.5–4 mm. *Redrawn (A) and (B) from Stock (1985b).*

5′ Mandibular palp one article; antenna 1 accessory flagellum four articles; lower lip inner lobes present .... *Weckelia caeca* (Weckel, 1907)
[Greater Antilles: Cuba]

6(5) Mandibular palp three articles; lower lip inner lobes present (vestigial or well-developed) 7

6′ Mandibular palp two to three articles; antenna 1 accessory flagellum three articles; lower lip inner lobes absent .... *Zombiweckelia parvipalpus* Stock, 1985
[Greater Antilles: Haiti]

7(6) Antenna 1 accessory flagellum two articles; lower lip inner lobes vestigial .... *Alloweckelia gurnei* Holsinger & Peck, 1968
[Puerto Rico: Bayaney]

7′ Antenna 1 accessory flagellum four articles; lower lip inner lobes present .... *Paraweckelia silvai* Shoemaker, 1959
[Greater Antilles: Cuba]

8(4) Coxa 4 not excavated posteriorly; antenna 1 accessory flagellum three articles; lower lip inner lobes present, but small (Fig. 23.32) .... *Tuluweckelia cernua* Holsinger, 1990
[México: Quintana Roo]

8′ Coxa 4 excavated posteriorly; antenna 1 accessory flagellum three articles; lower lip inner lobes present .... ***Mayaweckelia* [p. 844]**

9(3) Mandibular palp absent; antenna 1 accessory flagellum three articles; lower lip inner lobes present .... *Apoweckelia serrata* Stock, 1985
[Greater Antilles: Haiti]

9′ Mandibular palp one article; antenna 1 accessory flagellum four articles; lower lip inner lobes absent .... *Radaweckelia brevicauda* Stock, 1985
[Greater Antilles: Haiti]

Amphipoda: Hadziidae: *Bahadzia*: Species

1 Pereopods 5–7 dactylus posterior margin with subdistal spiniform process; mandibular palp without A-setae or A group-setae (see glossary) on third article (Fig. 23.30 C) 2

1′ Pereopods 5–7 dactylus normal, without spiniform process; mandibular palp article 3 bearing A-setae or A group-setae .... *Bahadzia bozanici* Holsinger 1992
[México: Cozumel Island]

2(1) Antenna 1 very long (59 to 87-articulate); pereopod 6 considerably longer than 7 .... *Bahadzia jaraguensis* Jaume & Wagner, 1998
[Greater Antilles: Dominican Republic]

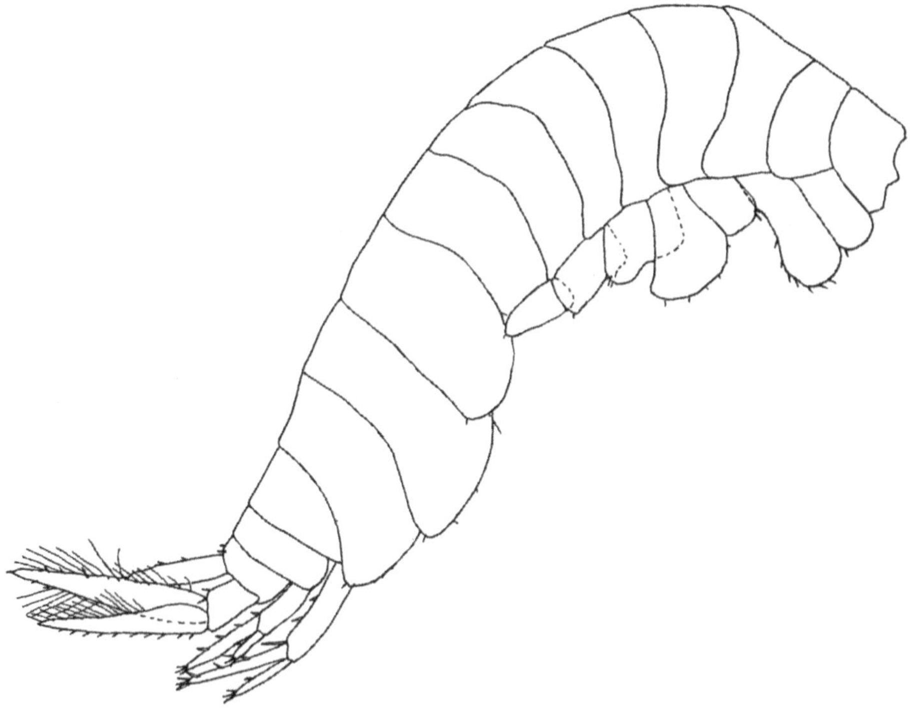

FIGURE 23.32 Hadziidae, *Tuluweckelia cernua*, habitus female (body size 6.8 mm). *Redrawn from Holsinger (1990).*

2′	Antenna 1 normal (about 29-articulate); pereopod 6 slightly shorter than 7 [Greater Antilles: Haiti]	*Bahadzia latipalpus* Stock, 1985

Amphipoda: Hadziidae: *Metaniphargus*: Species

1	Uropod 3 endopod with mostly simple setae, length < 66% exopod article one (Fig. 23.31 B)	2
1′	Uropod 3 endopod with one or more plumose setae, ramus length subequal to exopod article one (Fig. 23.31 A)	14
2(1)	Coxae 1–4 anterior margin not crenulate	3
2′	Coxae 1–4 anterior margin crenulate (Fig. 23.33 A) [Greater Antilles: Haiti]	*Metaniphargus crenatus* Stock, 1985
3(2)	Female gnathopod 2 palm armed with mostly long simple setae	4
3′	Female gnathopod 2 palm armed with simple setae [Lesser Antilles: Virgin Islands]	*Metaniphargus beatty* Shoemaker, 1942
4(3)	Pereopods 5–7 basis distoposterior corner neither produced nor overhanging	5
4′	Pereopods 5–7 basis distoposterior corner produced and overhanging [Lesser Antilles: Barbuda]	*Metaniphargus nicholsoni* Shoemaker, 1959
5(4)	Uropod 3 exopod distal article much longer than proximal article setae	6
5′	Uropod 3 exopod distal article length subequal to proximal article setae [Venezuela: Falcon]	*Metaniphargus venezolanus* Stock & Botosaneanu, 1983
6(5)	Mandible palp third article less elongate, less narrow	7
6′	Mandible palp third article very elongate and narrow [Lesser Antilles: St. Martin and Anguilla]	*Metaniphargus palpator* Stock, 1977
7(6)	Male pleopod 3 exopod with modest or strong lobe	9
7′	Male pleopod 3 exopod without lobe	8

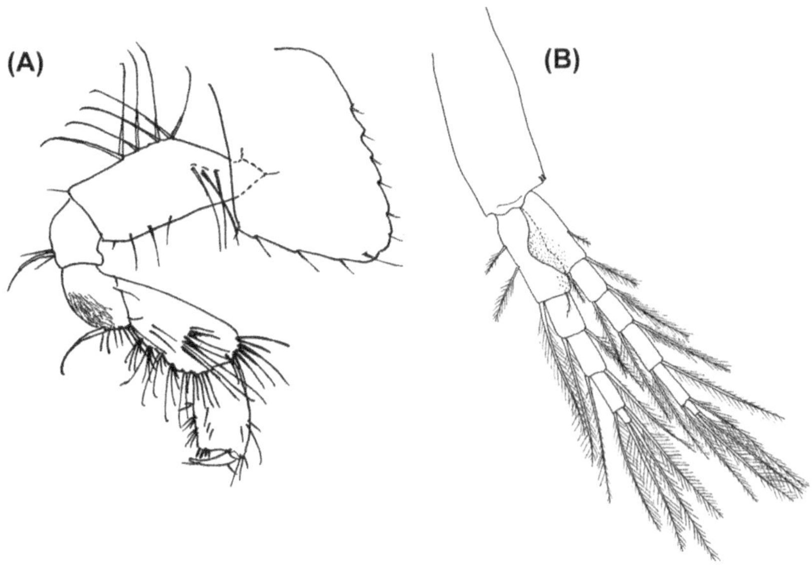

FIGURE 23.33 Hadziidae, *Metaniphargus*: (A) gnathopod 1 male *M. crenatus* (body size 3.5–4 mm); (B) pleopod 3 male *M. chaetodactylus*. *Redrawn (A) and (B) from Stock (1985b).*

8(7) Male pleopod 3 endopod mediolateral margin with a slight tumidity; mandibular palp article 3 straight .... *Metaniphargus hyporheicus* Stock, 1983
[Greater Antilles: Jamaica]

8′ Male pleopod 3 endopod without irregularities; mandibular palp article 3 ventral margin angular *Metaniphargus haitianus* Stock, 1985
[Greater Antilles: Haiti]

9(7) Male pleopod 3 exopod robust lateral lobe (Fig. 23.33 B) 10

9′ Male pleopod 3 exopod with weak lateral lobe; antenna 1 peduncle article two very slender and narrow, twice as long as article one .... *Metaniphargus bousfieldi* Stock, 1977
[Greater Antilles: Puerto Rico]

10(9) Mandibular palp article three length = article two 11

10′ Mandibular palp article three length ~2× article two 12

11(10) Uropod 3 endopod length > 0.5× article one exopod .... *Metaniphargus longipes* Stock, 1977
[Lesser Antilles: Aruba]

11′ Uropod 3 endopod length 0.5× article one exopod .... *Metaniphargus curasavicus* Stephensen, 1933
[Lesser Antilles: Curaçao]

12(10) Telson lobes with medial and terminal setae, pereopod 5–7 dactylus with lateral margin setae 13

12′ Telson lobes with only two terminal setae .... *Metaniphargus longipalpus* Stock, 1985
[Greater Antilles: Haiti]

13(11) Dactylus of pereopod 5–7 long and slender; uropod 3 endopod length < 0.5× exopod article one .... *Metaniphargus chaetodactylus* Stock, 1985
[Greater Antilles: Haiti]

13′ Dactylus of pereopod 5–7 short and robust; uropod 3 endopod length > 0.5× exopod article one .... *Metaniphargus pedunculatus* Stock, 1985
[Greater Antilles: Haiti]

14(1) Male and female uropod 3 endopod with mainly short or medium sized plumose setae (Fig. 23.31 A) 15

14′ Male and female uropod 3 endopod with one long, plumose setae .... *Metaniphargus craterensis* Stock, 1983
[Greater Antilles: Haiti]

15(14) Coxal gills on pereopods 2–5 .... *Metaniphargus plumicauda* Stock, 1985
[Greater Antilles: Haiti]

15′ Coxal gills on pereopods 2–6 .... *Metaniphargus jamaicae* (Holsinger, 1974)
[Greater Antilles: Jamaica]

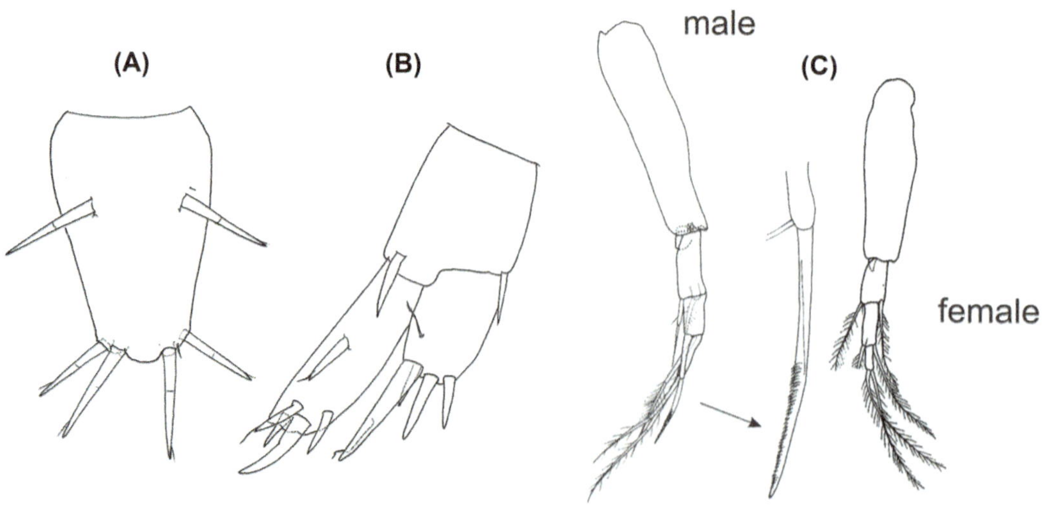

FIGURE 23.34 Bogidiellidae, (A) telson of *Patagongidiella danieli* (body size 6.5 mm), (B) uropod 2 male of *Argentinogidiella horcomollensis* (body size 2.5–5.2 mm); (C) pleopod 1 male and female (sexually dimorphous) of *Mexigidiella sbordonii* (body size 5–7 mm). *Redrawn (B) from Ruffo & Vigna Taglianti (1973).*

Amphipoda: Hadziidae: *Mayaweckelia*: Species

1 Antenna 1 length 0.5× body; gnathopod 2 propodus, maximum width 2.5× length; coxa 4 shallow, broader distally than proximally, margin with four rather short setae; pereopods 5–7 dactyls dorsal margin distal half without row of fine setae; pereopod 6 slightly longer than pereopod 7; telson as wide as long .... *Mayaweckelia yucatanensis* Holsinger, 1977 [México: Yucatan]

1′ Antenna 1 length subequal to slightly longer than body, distal 10–12 flagellar segments with aesthetascs (1 each); gnathopod 2 propodus, maximum width > 3× length; pereopod 6 length ≤ 1.15× pereopod 7; pereopods 5–7 dactyl upper margin, distal half with row of fine setae (pubescence), lower margin with two sets of setae; telson longer than wide .... *Mayaweckelia cenoticola* Holsinger, 1977 [México: Yucatan]

Amphipoda: Bogidiellidae: Genera

1 Pereonites 2–5 or 4–6 with large mediosternal processes (=sternal humps) (23.35 A) 2

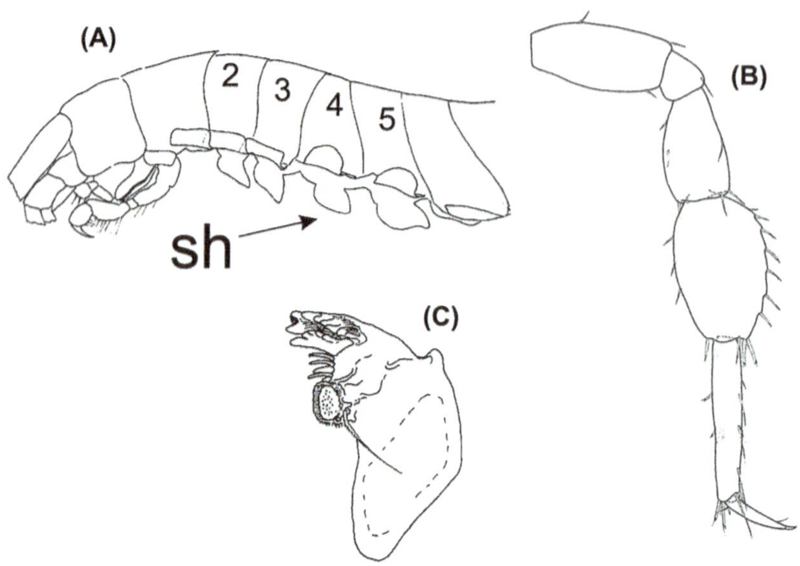

FIGURE 23.35 Bogidiellidae, (A) pereonites of *Grossogidiella mauryi* (sh, sternal humps) (body size 4.8–7.2 mm); (B) pereopod 7 male of *Orchestigidiella orchestipes* (body size 3.5–4 mm); (C) left mandible of Megagidiella azul (body size 16.2 mm). *Redrawn (B) from Ruffo & Vigna Taglianti (1977) and (C) from Koenemann & Holsinger (1999).*

1′ Pereonites without mediosternal process 3

2(1) Male pleopod 2 exopod without a special seta inserted laterally on second article; male uropod 1 peduncle without hypertrophied facial seta .... *Grossogidiella mauryi* (Grosso & Fernández, 1993)
[Argentina: Neuquén]

2′ Male uropod 1, peduncle with modified hypertrophied facial robust seta; male pleopod 2, exopod second article with modified distolateral seta .... ***Patagongidiella*** **[p. 845]**

3(1) Mandible with palp 4

3′ Mandible without palp (Fig. 23.35 C) .... *Megagidiella azul* Koenemann & Holsinger, 1999
[Brazil: Mato Grosso do Sul]

4(3) Male pleopod 2 exopod article 2 with setae unmodified 5

4′ Male pleopod 2 exopod with one seta flattened, concave with teeth, cilia, or hamulus 9

5(4) Maxilla 1 palp with two articles; antenna 1, accessory flagellum length ≥ primary flagellum first article 6

5′ Maxilla 1, palp with one article; antenna 1, accessory flagellum length < primary flagellum first article .... *Eobogidiella purmamarcensis* (Grosso & Ringuelet, 1979)
[Argentina: Jujuy]

6(5) Antenna 1, accessory flagellum with two or three articles; uropod 1 sexually dimorphic, male with exopod shortened and modified and/or with rasplike setae on rami modified 7

6′ Antenna 1, accessory flagellum with two articles; usually uropod 1 without sexual dimorphism .... ***Bogidiella*** **[p. 845]**

7(6) Gnathopod 1 basis posterior margin with three or five setae; antenna 1 accessory flagellum with two articles 8

7′ Gnathopod 1 basis posterior margin with one seta .... ***Arganogidiella*** **[p. 846]**

8(7) Pereopod 7 sexually dimorphic (carpus in male enlarged) (Fig. 23.35 B). .... *Orchestigidiella orchestipes* (Ruffo & Vigna Taglianti, 1977)
[México: Chiapas]

8′ Pereopod 7 without sexual dimorphism .... ***Guagidiella*** **[p. 847]**

9(4) Telson width ≥ length; uropod 2 without sexual dimorphism (Fig. 23.27 B) 10

9′ Telson length > width; usually male uropod 2 endopod fused with basipodite or shortened (Fig. 23.27 C) 11

10(9) Uropods 1–3 not sexually dimorphous 13

10′ Uropod 1 sexually dimorphic, female rami mucronate (male uropod normal) .... *Antillogidiella martini* (Stock, 1978)
[Lesser Antilles: St. Martin]

11(9) Pleopod 1 not sexually dimorphous, basipodite without ingrowth on the inner margin 12

11′ Pleopod 1 sexually dimorphic, endopod with only one article, reduced, exopod with three articles, male with the first two articles dilated with modified setae, second article external seta reduced and curved with the medial plumose part inserted at right angles before tip; pleopod 1, basipodite medial margin with a wide ingrowth. .... *Mesochthongidiella tucumanensis* (Grosso & Fernández, 1985)
[Argentina: Tucumán]

12(11) Male uropod 2 with endopod fused with basipodite or endopod very short .... ***Dycticogidiella*** **[p. 847]**

12′ Male uropod 2 rami subequal or endopod slightly shortened (Fig. 23.34 B) .... ***Argentinogidiella*** **[p. 848]**

13(10) Pleopod 1 not sexually dimorphic .... ***Stygogidiella*** **[p. 848]**

13′ Male pleopod 1 with differentiated seta (stiff, articulate, pointed, and serrate (Fig. 23.34 C) .... ***Mexigidiella*** **[p. 848]**

Amphipoda: Bogidiellidae: *Patagongidiella*: Species

1 Telson distal margin convex; antenna 1 primary flagellum with 16 articles (Fig. 23.34 A) .... *Patagongidiella danieli* Grosso & Fernández, 1993
[Argentina: Neuquén]

1′ Telson distal margin slightly emarginated; antenna 1 primary flagellum with 12 articles .... *Patagongidiella wefkoi* Pérez-Schultheiss, 2013
[Chile: Osorno]

Amphipoda: Bogidiellidae: *Bogidiella*: Species

1 Pleopods 1–2 without endopod 2

1′ Pleopods 1–2 uniarticulate or endopod vestigial 4

2(1) Gnathopod 1 and 2 propodi morphology dissimilar (Fig. 23.36 C, D) 3

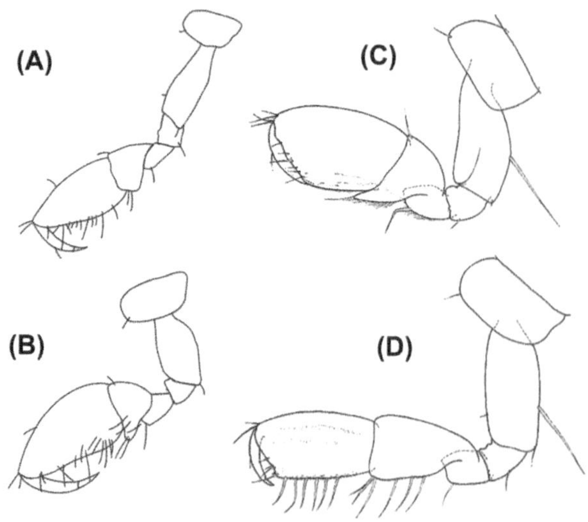

FIGURE 23.36 Bogidiellidae, *Bogidiella*: (A) gnathopod 1 and (B) gnathopod 2 of *B. neotropica* (body size 3 mm); (C) gnathopod 1 and (D) gnathopod 2 female of *B. michaelae* (body size 2.7 mm). *Redrawn (A) and (B) from Ruffo (1952), (C) and (D) from Ruffo & Vigna Taglianti (1977).*

2′ Gnathopod 1 and 2 propodi morphology similar, palms very oblique (Fig. 23.36 A, B) .... *Bogidiella neotropica* Ruffo, 1952
[Venezuela: Calabozo. Brazil: Amazonia]

3(2) Gnathopod 2 propodus subrectangular (palmar index: 0.17), palmar margin lightly concave; uropod 3 peduncle short, globose (Fig. 23.36 C, D). .... *Bogidiella michaelae* Ruffo & Vigna Taglianti, 1977
[Mexico: Oaxaca]

3′ Gnathopod 2 propodus suboval (palmar index: 0.42/0.44), palmar margin rectilineous; uropod 3 peduncle subrectangular .... *Bogidiella prionura* Stock, 1985
[Greater Antilles: Haiti]

4(1) Telson length > width 5

4′ Telson width > length 6

5(4) Antenna 1 accessory flagellum distal article elongate, shorter than basal; gnathopod 1 propodus palmar margin with rounded dentiform projection; maxilla 1 palp longer than lobes .... *Bogidiella gammariformis* Sket, 1985
[Ecuador: Archidona]

5′ Antenna 1 accessory flagellum distal article vestigial, globose; gnathopod 1 propodus, palmar margin without any projection; maxilla1 palp length equal to ischial endite (= lateral lobe) .... *Bogidiella cooki* Grosso & Ringuelet, 1979
[Argentina: Jujuy]

6(4) Total body length ~ 4 mm; antenna 1 length > 0.5× body, primary flagellum 18 articles; male gnathopod 1 pyriform (palmar index = 0.58); uropods 1–2 with marginal and apical setae (Fig. 23.37 A) .... *Bogidiella niphargoides* Ruffo & Vigna Taglianti, 1977
[Mexico: Oaxaca]

6′ Total body length ~2.2–2.6 mm; antenna 1 < 0.5× body, primary flagellum nine articles; male gnathopod 1 subtrapezoidal (palmar index = 0.48); uropods 1–2 without marginal setae (Fig. 23.37 B) .... *Bogidiella vomeroi* Ruffo & Vigna Taglianti, 1977
[Mexico: Chiapas]

Amphipoda: Bogidiellidae: *Arganogidiella*: Species

1 Antenna 1, accessory flagellum with two articles; male gnathopod 2, carpus with three posterior setae; Hertzog organ (= lenticular organs) present .... *Arganogidiella arganoides* (G. Karaman, 1982)
[Mexico: Etla]

1′ Antenna 1, accessory flagellum with three articles; male gnathopod 2, carpus with two posterior setae; Hertzog organ (= lenticular organs) absent .... *Arganogidiella arganoi* (Ruffo & Vigna Taglianti, 1973)
[Mexico: Veracuz]

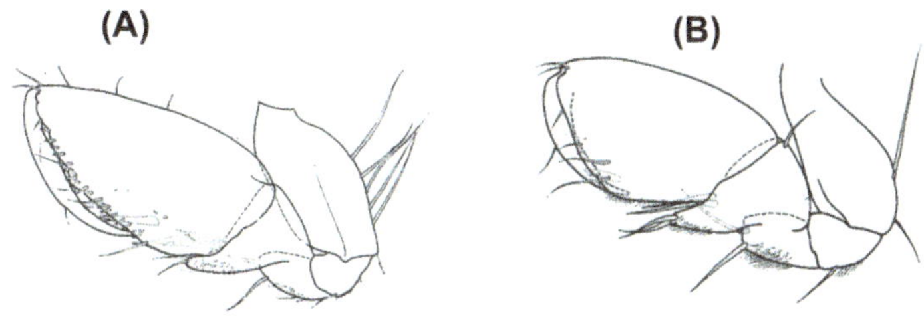

FIGURE 23.37 Bogidiellidae, *Bogidiella:* (A) gnathopod 1 male of *B. niphargoides* (body size 4.2 mm); (B) gnathopod 1 male of *B. vomeroi* (body size 2.2–2.3 mm). *Redrawn (A) and (B) from Ruffo & Vigna Taglianti (1977).*

Amphipoda: Bogidiellidae: *Guagidiella*: Species

1 Pereopod 3–7 dactylus elongate, ~ 0.5× protopod; male uropod exopod conical, reduced, bent at right angle to endopod, medially with one distinctive curved seta, longer than exopod; endopod lateral side subcylindrical (Fig. 23.38 A) .... *Guagidiella pasquinii* (Ruffo & Vigna Taglianti, 1977) [Guatemala: Huehuetenango]

1′ Pereopod 3–7 dactylus short, stout; male uropod 1 exopod reduced to an irregular scale, elongate, distally setaform, medially with a long, curved seta; endopod laterally concave (Fig. 23.38 B) .... *Guagidiella holsingeri* (Ruffo & Vigna Taglianti, 1973) [Guatemala: Alta Verapaz]

Amphipoda: Bogidiellidae: *Dycticogidiella:* Species

1 Male uropod 2 with endopod and basipodite fused, one endopod distal seta stout and helicoidal with fine and tiny setulae; telson distal edge scarcely sinuous .... *Dycticogidiella talampayensis* (Grosso & Claps, 1985) [Argentina: La Rioja]

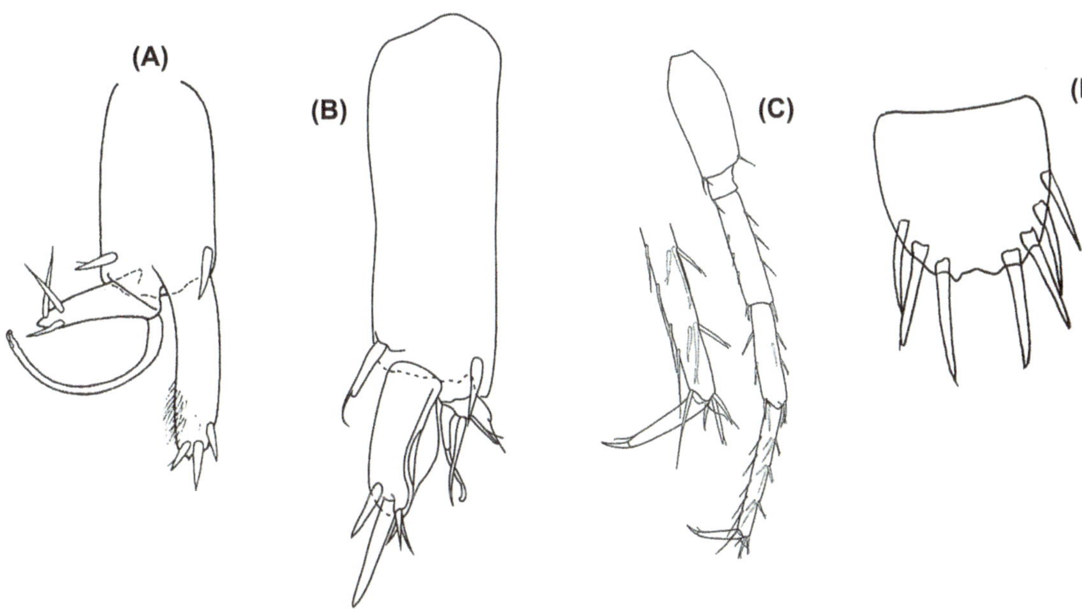

FIGURE 23.38 Bogidiellidae, *Guagidiella* and *Mexigidiella*: (A) uropod 1 male of *G. pasquinii* (body size 2.5 mm); (B) uropod 1 male of *G. holsingeri* (body size 4.5 mm); (C) pereopod 7 of *M. chitalensis* (body size 3–4.6 mm); (D) telson of *M. tabascensis* (body size 2.83–4.69 mm). *Redrawn (A) and (C) from Ruffo & Vigna Taglianti (1977), (B) from Ruffo & Vigna-Taglianti (1973), (D) from Villalobos (1961).*

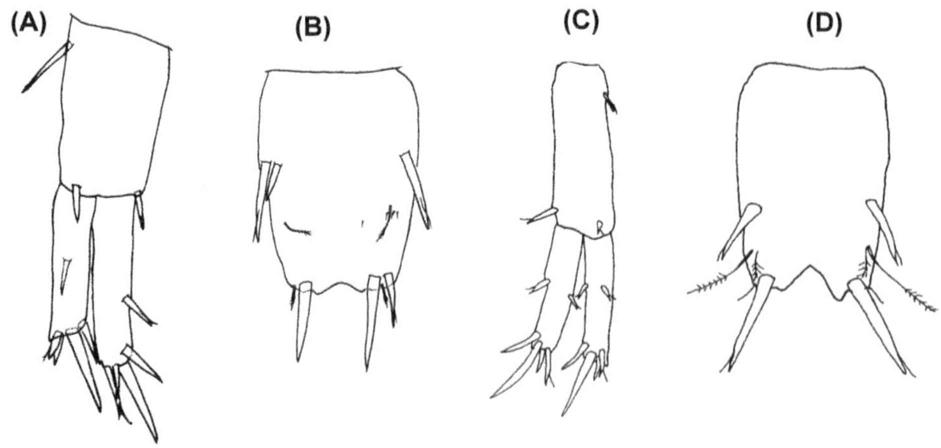

FIGURE 23.39 Bogidiellidae, *Argentinogidiella*: (A) uropod 1 male and (B) telson of *A. horcomollensis* (body size 2.5–5.2 mm); (C) uropod 1 male of *A. lavillai* (body size 4.3 mm); (D) telson of *A. lavillai* (body size 4.3 mm). *Redrawn (C) and (D) from Grosso & Claps (1984).*

1′ Male uropod 2 with basipodite shortened, endopod one article, endopod medially curved, with two distal setae elongate and robust; telson distal edge arcuately excavate .... *Dycticogidiella ringueleti* (Grosso & Fernández, 1988) [Argentina: Tucumán]

Amphipoda: Bogidiellidae: *Argentinogidiella*: Species

1 Uropod 1 rami, each with only a single lateral seta; telson distal edge sinuous with central depression and two shallow lateral depressions (Fig. 23.39 A, B) .... *Argentinogidiella horcomollensis* (Grosso & Fernández, 1988) [Argentina: Tucumán]

1′ Uropod 1 rami, each with two lateral margin setae; telson distal edge angulately excavate (Fig. 23.39 C, D) .... *Argentinogidiella lavillai* (Grosso & Claps, 1984) [Argentina: Jujuy]

Amphipoda: Bogidiellidae: *Stygogidiella*: Species

1 Antenna 1 accessory flagellum with three articles; pereopods 3 and 4, propodus distally with two sigmoid setae (Fig. 23.40 A) 2

1′ Antenna 1 accessory flagellum with two articles; pereopods 3 and 4, propodus distally without sigmoid setae .... *Stygogidiella perla* (Stock, 1981) [Venezuela: Isla de Margarita]

2(1) Gnathopods 1 and 2 propodus posteriolateral margin with few setae, gnathopod 1 with five setae and gnathopod 2 with six setae; pleopod endopod digitiform, length > width, length 0.5× exopod first article (Fig. 23.40 B) .... *Stygogidiella virginalis* (Stock, 1981) [Virgin Island: Tortola and St. John]

2′ Gnathopods 1 and 2 propodus posteriolateral margin with many setae, 8– 16 on gnathopod 1 and 6–13 on gnathopod 2; pleopod endopod scaliform, length < 0.5× exopod first article (Fig. 23.40 C) .... *Stygogidiella bredini* (Shoemaker, 1959) [Lesser Antilles: Barbuda]

Amphipoda: Bogidiellidae: *Mexigidiella*: Species

1 Pereopod 7 dactylus anterior margin with two setae 2

1′ Pereopod 7 dactylus anterior margin with one seta (Fig. 23.38 C) 3

2(1) Telson length = width, one distal seta and two to three marginal setae; antenna 1 accessory flagellum of two articles (Fig. 23.38 D) .... *Mexigidiella tabascensis* (Villalobos, 1961) [Mexico: Tabasco]

2′ Telson width > length, with one distal seta and one marginal seta; antenna 1 accessory flagellum triarticulate .... *Mexigidiella hamatula* (Stock, 1985) [Greater Antilles: Haiti]

3(1) Pereopod 7 with robust propodus; oostegites on pereopods 3–5 (Fig. 23.41 A) 4

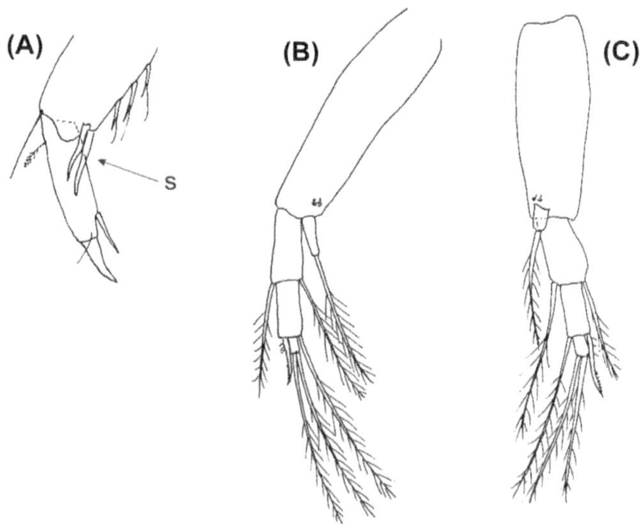

FIGURE 23.40 Bogidiellidae, *Stygogidiella*: (A) distal part of pereopod 3 male of *S. bredini* (s, sigmoid setae) (body size 5.8 mm); (B) pleopod 2 male of *S. virginalis* (body size 3.5–4 mm); (C) pleopod 2 male of *S. bredini* (body size 5.8 mm). *Redrawn (A), (B) and (C) from Stock (1981).*

3′ Pereopod 7 with slender propodus; oostegites on pereopods 2–5 (Fig. 23.38 C) .... *Mexigidiella chitalensis* (Karaman G., 1981)
[Mexico: Chiapas]

4(3) Gnathopod 2 propodus pyriform, palm length = posterior margin (Fig. 23.41 B) .... *Mexigidiella mexicana* (Karaman G., 1981)
[Mexico: Chiapas]

4′ Gnathopod 2 propodus trapezoidal, palm length < posterior margin (Fig. 23.41 C) ....
.... *Mexigidiella sbordonii* (Ruffo & Vigna Taglianti, 1973)
[Mexico: Chiapas]

Amphipoda: Hyalellidae: *Hyalella*: Species

This key is only to species described using modern standards. The user is greatly cautioned not to go beyond genus level if the material does not exactly match the provided character states in each couplet. Even then great caution must be exercised as many undescribed species are known to exist

1 Body with dorsal or lateral processes or mucronations (carinae, round ridges, teeth) (Fig. 23.42) 2

1′ Body dorsally smooth (Fig. 23.43) 19

2(1) Body with only dorsal processes 3

2′ Body with dorsal and lateral processes (teeth) (Fig. 23.44) 16

3(2) Some pereonites and pleonites with anterior or dorsoposterior carina 4

3′ Some pereonites and pleonites with rounded elevations 18

4(3) Pereon 1 with dorsoanterior carinae; pereonites 1–7 and pleonites 1–3 with dorsoposterior carina 5

4′ Some pereonites or pleonites with dorsoposterior carina (Fig. 23.45) 6

5(4) Pereonites 1–6 with long, anteriorly curved dorsal carinae, peronite 1 carina overlies head; coxae 1–4 distally rounded (Fig. 23.46) ...
.... *Hyalella lucifugax* (Faxon, 1876)
[Bolivia: Lake Titicaca]

5′ Pereonites 1–6 with short and not anteriorly curved dorsal; coxae 1–4 distally acuminate. .... *Hyalella longipes* (Faxon, 1876)
[Bolivia: Lake Titicaca]

6(4) Pleonites 1 and 2 only with dorsoposterior carinae 7

6′ Pereonites 5–7, or 7 only, and some pleonites with dorsoposterior carinae 9

7(6) Uropod 1 endopod medial margin without curved setae; sternal gills in segments 3–7 8

7′ Uropod 1 endopod medial margin with curved setae; sternal gills in segments 2–7 .... *Hyalella kaingang* Araujo & Cardoso, 2013
[Brazil: Rio Grande do Sul]

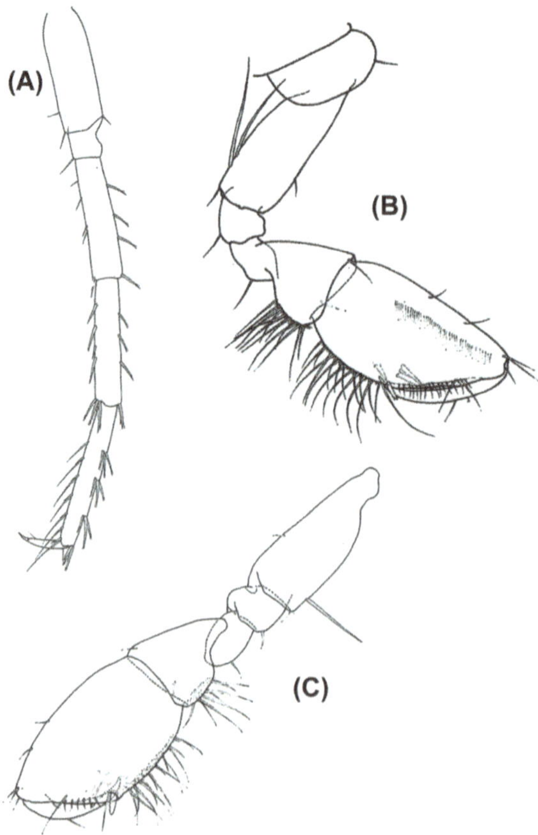

FIGURE 23.41 Bogidiellidae, *Mexigidiella*: (A) pereopod 7 and (B) gnathopod 2 male of *M. mexicana* (body size 5–6 mm); (C) gnathopod 2 male of *M. sbordonii* (body size 7 mm). *Redrawn (A) and (B) from Ruffo & Vigna Taglianti (1977), (C) from Ruffo & Vigna Taglianti (1973).*

FIGURE 23.42 Hyalellidae, *Hyalella armata*, dorsal habitus (body size 9.5 mm).

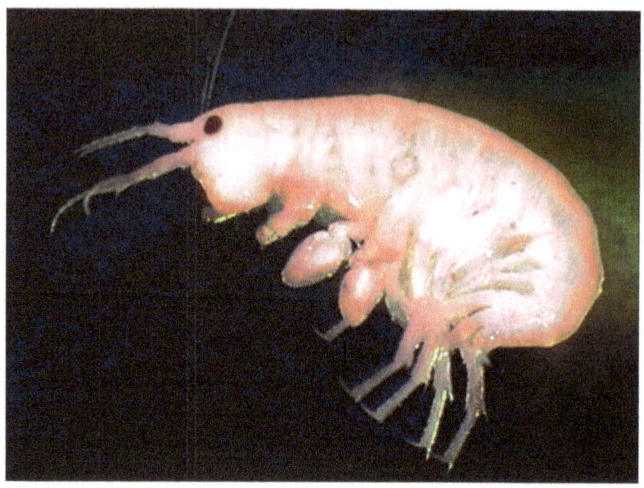

FIGURE 23.43 Hyalellidae, *Hyalella simplex*, lateral habitus (body size 7.8 mm).

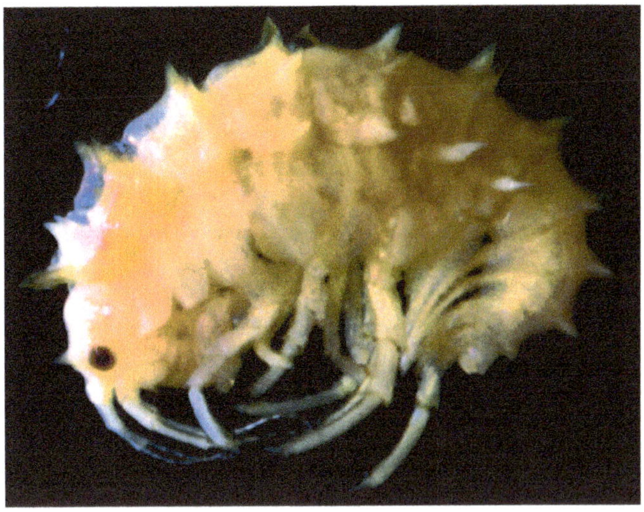

FIGURE 23.44 Hyalellidae, *Hyalella echinus* lateral habitus (body size 8.8 mm).

FIGURE 23.45 Hyalellidae, *Hyalella latimanus* lateral habitus (body size 7.1 mm).

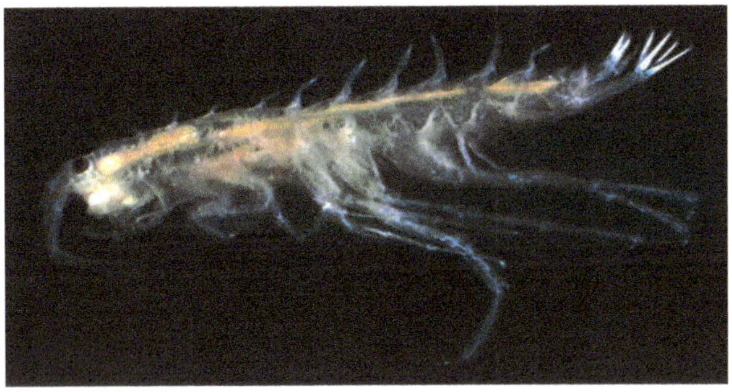

FIGURE 23.46 Hyalellidae, *Hyalella lucifugax* lateral habitus (body size 11 mm).

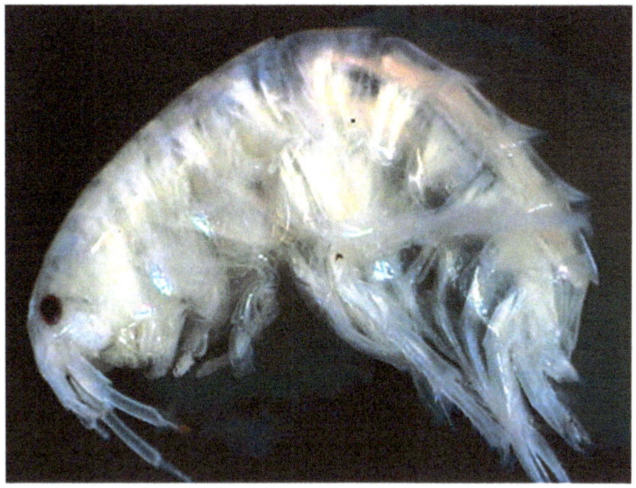

FIGURE 23.47 Hyalellidae, *Hyalella montforti* lateral habitus (body size 11.4 mm).

8(7) Eyes present 9

8′ Eyes absent .... *Hyalella cenotensis* Marrón-Becerra, Hermoso-Salazar & Solís-Weiss, 2014
[México: Quintana Roo]

9(8) Telson width > length, margin rounded .... *Hyalella maya* Marrón- Becerra, Hermoso-Salazar & Solís-Weiss, 2018
[México: Quintana Roo]

9′ Telson width = length, margin apically pointed .... *Hyalella azteca* (Saussure, 1858)
[México: Vera Cruz]

10(8) Pereonite 7 only, and some pleonites, with dorsoposterior carinae 11

10′ Pereonites 5−7 and pleonites 1−3 with dorsoposterior carinae .... *Hyalella neveulemairei* Chevreux, 1904
[Bolivia: Lake Titicaca]

11(10) Pereonite 7 and pleonites 1−2 with only dorsoposterior carinae 12

11′ Pereonite 7 and pleonites 1−3 with only dorsoposterior carinae (Fig. 23.47) 13

12(11) Mandible incisor untoothed; telson apically truncate, with two wide set, long, simple setae ....
.... *Hyalella nefrens* González & Watling, 2003
[Bolivia: Lake Titicaca]

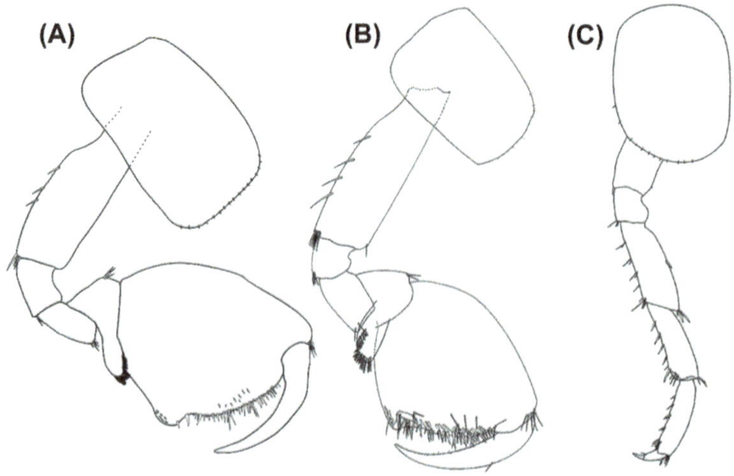

FIGURE 23.48 Hyalellidae, *Hyalella:* (A) gnathopod 2 male *H. pleoacuta* (body size 8.5 mm); (B) gnathopod 2 male *H. latimanus* (body size 7.1 mm); (C) pereopod 4 *H. spelaea* (body size 4.4 mm). *Redrawn (B) from González & Watling (2003b), (C) from Cardoso et al. (2011).*

12′ Mandible incisor toothed; telson apically acute, with two close set, long, simple setae .... *Hyalella pseudoazteca* González & Watling, 2003
[Brazil: Rio Grande do Sul]

13(11) Male gnathopod 2 propodus with palm oblique (Fig. 23.48 A) 14

13′ Male gnathopod 2 propodus with palm transverse (Fig. 23.48 B) 15

14(13) Male gnathopod 2 palm not distinguishable from posterior margin of propodus, dactylus nearly reaching posterior margin proximal end .... *Hyalella longipalma* (Faxon, 1876)
[Bolivia: Lake Titicaca]

14′ Male gnathopod 2 propodus with palm distinguishable from posterior margin of propodus dactylus reaches midway along posterior margin .... *Hyalella pleoacuta* González, Bond-Buckup & Araujo, 2006
[Brazil: Rio Grande do Sul]

15(13) Male uropod 1 endopod medially with curved setae; sternal gills on segments 2−7 .... *Hyalella montforti* Chevreux, 1907
[Bolivia: Lake Titicaca]

15′ Male uropod 1 endopod medially without curved setae; sternal gills on segments 3−7 .... *Hyalella latimanus* (Faxon, 1876)
[Bolivia: Lake Titicaca]

16(2) Pereon with two rows of dorsal processes (teeth) 17

16′ Pereon with three rows of dorsal processes (teeth) .... *Hyalella crawfordi* Coleman & González, 2006
[Bolivia: Lake Titicaca]

17(16) Pleon epimera posteriolateral margins with teeth .... *Hyalella gauthieri* Coleman & González, 2006
[Bolivia: Lake Titicaca]

17′ Pleon epimera posteriolateral margins without teeth (Fig. 23.44) .... *Hyalella echinus* (Faxon, 1876)
[Bolivia: Lake Titicaca]

18(3) Total body length 3× coxa 4 apical process .... *Hyalella armata* (Faxon, 1876)
[Bolivia: Lake Titicaca]

18′ Total body length 2× coxa 4 apical process .... *Hyalella longispina* González & Coleman, 2002
[Peru: Lake Titicaca]

19(1) Eyes absent or reduced 20

19′ Eyes present and pigmented 28

20(19) Sternal gills present on segments 3-7 20

20′ Sternal gills present on segments 2−7 23

21(20) Antenna 1 ≤ antenna 2 22

21′ Antenna 1 < antenna 2 .... *Hyalella caeca* Pereira, 1989
[Brazil: São Paulo]

22(21) Gnathopod 1 carpus posterior lobe with row of serrate setae ... *Hyalella imbya* Rodrigues & Bueno, 2012
[Brazil: Rio Grande do Sul]

22′ Gnathopod 1 carpus posterior lobe with a row of simple setae ... *Hyalella epikarstica* Rodrigues, Bueno & Ferreira, 2014
[Brazil: São Paulo]

23(20) Male uropod 1 endopod with curved setae ... 24

23′ Male uropod 1 endopod without curved setae ... 25

24(23) Male gnathopod 2 propodus posterior margin with comb-scales. ... *Hyalella veredae* Cardoso & Bueno, 2014
[Brazil: Minas Gerais]

24′ Male gnathopod 2 propodus posterior margin without comb-scales ... *Hyalella formosa* Cardoso & Araujo, 2014
[Brazil: Parana]

25(23) Coxa 4 excaved posteriorly ... 26

25′ Coxa 4 not excaved (Fig. 23.48 C) ... *Hyalella spelaea* Bueno & Cardoso, 2011
[Brazil: São Paulo]

26(25) Gnathopod 1 propodus length < 2× width ... 27

26′ Gnathopod 1 propodus length > 2× than width ... *Hyalella anophtalma* Ruffo, 1957
[Venezuela: Cueva de Rio Gueque]

27(26) Uropod 3 peduncle length > uropod 2 peduncle ... *Hyalella rionegrina* Grosso & Peralta, 1999
[Argentina: Río Negro]

27′ Uropod 3 peduncle length 0.5× uropod 2 ... *Hyalella troglofugia* Bastos-Pereira, Oliveira & Ferreira, 2018
[Brazil: Minas Gerais]

28(19) Uropod 3 ramus dorsal margin with setae ... 29

28′ Uropod 3 ramus dorsal margin without setae ... 31

29(28) Uropod 3 ramus length < peduncle ... 30

29′ Uropod 3 ramus length subequal to peduncle ... *Hyalella georginae* Streck & Castiglioni, 2017
[Brazil: Rio Grande do Sul]

30(29) Male gnathopod 2 propodus pyriform ... *Hyalella cuprea* (Faxon, 1876)
[Bolivia: Lake Titicaca]

30′ Male gnathopod 2 propodus ovate ... *Hyalella dielaii* Pereira, 2004
[Brazil: São Paulo]

31(28) Male uropod 1 endopod with curved setae (Fig. 23.49 A) ... 32

31′ Male uropod 1 endopod without curved setae ... 46

32(31) Male gnathopod 2 dactylus shorter than propodus palm ... 33

32′ Male gnathopod 2 dactylus as long as propodus palm ... 35

33(32) Telson width > length ... 34

33′ Telson width < length (Fig. 23.49 B) ... *Hyalella xakriaba* Bueno & Araujo, 2013
[Brazil: Minas Gerais]

34(33) Telson lateral margin with plumose setae ... *Hyalella cajasi* Alonso & Jaume, 2017
[Ecuador: Azuay]

34′ Telson lateral margin without plumose setae ... *Hyalella brasiliensis* Bousfield, 1996
[Brazil: Parana]

35(32) Uropod 3 peduncle apically with three to five conate setae ... 36

35′ Uropod 3 peduncle apically with < six conate setae ... 43

36(35) Sternal gills present on segments 2–7 ... 37

36′ Sternal gills present on segments 3–7 ... 39

37(36) Telson apical margin with slender setae ... 38

37′ Telson apical margin with robust setae ... *Hyalella carstica* Bastos-Pereira & Bueno, 2012
[Brazil: Minas Gerais]

38(37) Telson length > width ... *Hyalella curvispina* Shoemaker, 1942
[Uruguay: Montevideo]

38′ Telson length < width ... *Hyalella falklandensis* Bousfield, 1996
[Malvinas (Falkland) Islands: Sabruno Ditch]

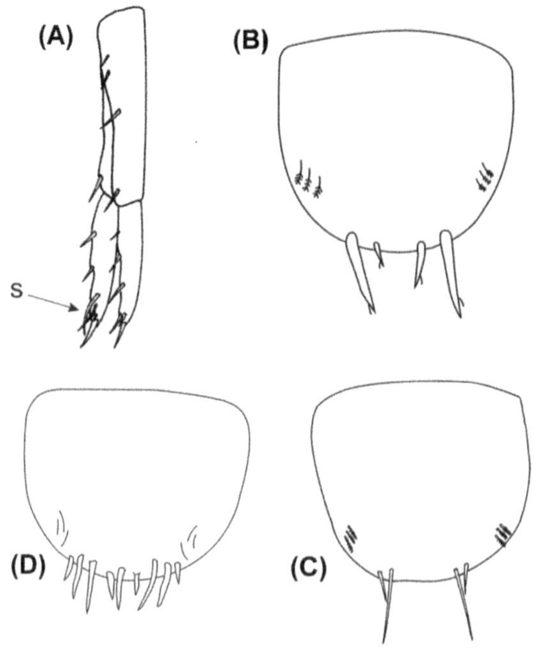

FIGURE 23.49 Hyalellidae, *Hyalella:* (A) uropod 1 male *H. kochi* (s, curved setae) (body size 6.9 mm); (B) telson *H. xakriaba* (body size 6.4 mm); (C) telson *H. costera* (body size 7.1 mm); (D) telson *H. montenegrinae* (body size 9.1–10.5 mm). *Redrawn (A) and (C) from González & Watling (2001), (B) from Bueno et al. (2013), (D) from Bond-Buckup & Araujo (1998).*

39(36)	Uropod 3 ramus length < peduncle	40
39′	Uropod 3 ramus length > peduncle	41
40(39)	Uropod 3 ramus, basal width ~ 2× apex [Chile: Bofedal de Arabilla]	*Hyalella kochi* González & Watling, 2001
40′	Uropod 3 ramus, basal width > 2× apex [Chile: Punta Arenas]	*Hyalella simplex* Schellenberg, 1943
41(39)	Telson width = length	42
41′	Telson length > width [Chile: Chiloe]	*Hyalella chiloensis* González & Watling, 2001
42(41)	Telson apical margin with slender setae (Fig. 23.49 C) [Chile: Antofagasta]	*Hyalella costera* González & Watling, 2001
42′	Telson apical margin with robust setae (Fig. 23.49 D) [Brazil: Rio Grande do Sul]	*Hyalella montenegrinae* Bond-Buckup & Araujo, 1998
43(36)	Uropod 3 peduncle length > width (rectangular)	44
43′	Uropod 3 peduncle length = width (quadrate) [Brazil: Rio Grande do Sul]	*Hyalella castroi* González, Bond-Buckup & Araujo, 2006
44(43)	Gnathopod 1 propodus rectangular	45
44′	Gnathopod 1 propodus quadrangular [Argentina: Buenos Aires]	*Hyalella pampeana* Cavalier, 1968
45(44)	Maxilla 1, maxilla 2 and maxilliped surface with several setulae [Argentina: Buenos Aires]	*Hyalella bonariensis* Bond-Buckup, Araujo & Santos, 2008
45′	Maxilla 1, maxilla 2 and maxilliped surface without several setulae [Brazil: Rio Grande do Sul]	*Hyalella gauchensis* Streck & Castiglioni, 2017
46(31)	Gnathopod 1 propodus length > 2× width (rectangular)	47
46′	Gnathopod 1 propodus length < 2× width (quadrangular)	50

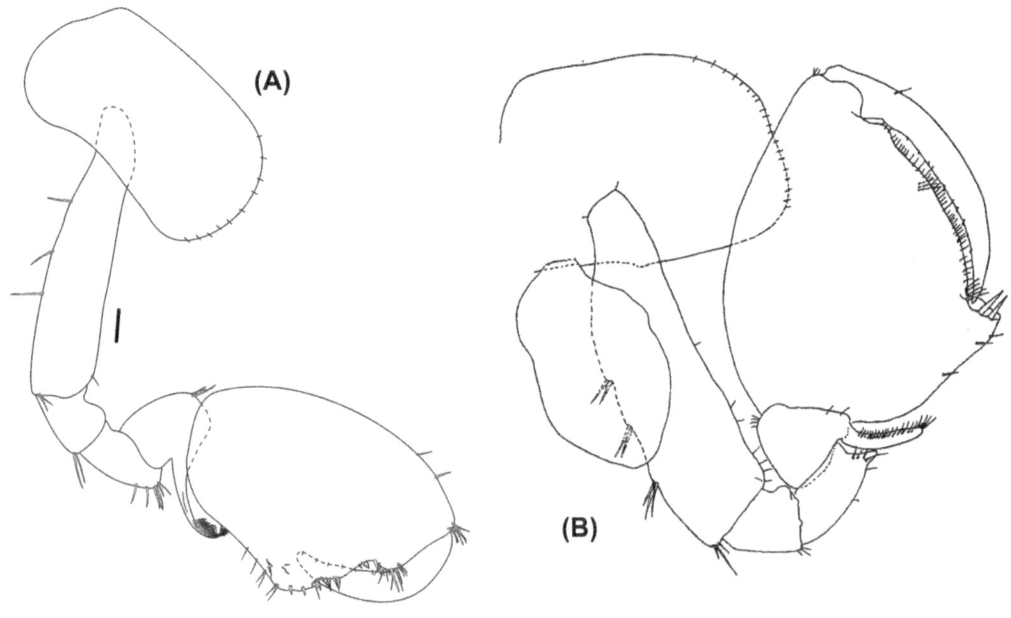

FIGURE 23.50 Hyalellidae, *Hyalella:* (A) gnathopod 2 male *H. minensis* (body size 6.9 mm); (B) gnathopod 2 male *H. neonoma* (body size 11 mm). *Redrawn (A) from Bastos-Pereira & Bueno (2013), (B) from Stock & Platvoet (1991).*

47(46) Gnathopod 1 propodus medial surface without comb-scales 48

47′ Gnathopod 1 propodus medial surface with comb-scales .... *Hyalella caribbeana* Bousfield, 1996
[West Indies: Grand Terre Guadeloupe]

48 (47) Gnathopod 1 propodus medial surface with pappose setae 49

48(42) Gnathopod 1 propodus medial surface with serrate setae .... *Hyalella montana* Rodrigues, Senna, Quadra & Bueno, 2017
[Brazil: Minas Gerais]

49(48) Gnathopod 1 propodus medial surface with five pappose setae .... *Hyalella longistila* (Faxon, 1876)
[Brazil: Rio de Janeiro]

49′ Gnathopod 1 propodus medial surface with ten pappose setae .... *Hyalella warmingi* Stebbing, 1899
[Brazil: Minas Gerais]

50(46) Male gnathopod 2 propodus with palm transverse 51

50′ Male gnathopod 2 propodus with palm oblique 52

51(50) Male gnathopod 2 propodus margin irregular with few robust, short setae (Fig. 23.22 C) .... *Hyalella gracilicornis* (Faxon, 1876)
[Brazil: Rio de Janeiro]

51′ Male gnathopod 2 propodus margin convex with several robust, short setae. .... *Hyalella squamosa* Mateus & Mateus, 1990
[Guadalupe: Grande Terra]

52(50) Male gnathopod 2 propodus triangular or rectangular 53

52′ Male gnathopod 2 propodus ovate 57

53(52) Male gnathopod 2 propodus rectangular (Fig. 23.50 A) 54

53′ Male gnathopod 2 propodus triangular (Fig. 23.50 B) 55

54(53) Male gnathopod 2 propodus margin with several fine short setae. .... *Hyalella minensis* Bastos-Pereira & Bueno, 2013
[Brazil: Minas Gerais]

54′ Male gnathopod 2 propodus margin with few robust short setae .... *Hyalella meinerti* Stebbing, 1899
[Venezuela: Laguna de Espino]

55(53) Male gnathopod 2 propodus with palm margin straight 56

55′ Male gnathopod 2 propodus with palm margin convex .... *Hyalella neonoma* Stock & Platvoet, 1991
[Malvinas (Falkland) Islands: Mount pleasant Road]

56(55) Dactylus with several endal setae .... *Hyalella pauperocavae* González & Watling, 2003
[Peru: Huancayo]

56′ Dactylus without endal setae .... *Hyalella franciscae* González & Watling, 2003
[Chile: Torres del Paine]

57(52) Sternal gills present on segments 2–7 58

57′ Sternal gills present on segments 3–7 60

58(57) Uropod 3 ramus length = peduncle 59

58′ Uropod 3 ramus length < peduncle .... *Hyalella tiwanaku* González & Watling, 2003
[Bolivia: Lake Titicaca]

59(58) Uropod 3 peduncle quadrate .... *Hyalella quindioensis* González & Watling, 2003
[Colombia: Quindio]

59′ Uropod 3 peduncle slender .... *Hyalella misionensis* Colla & Cesar, 2015
[Argentina: Misiones]

60(57) Uropod 3 peduncle distally with four or more robust setae 61

60′ Uropod 3 peduncle distally with two to three robust setae .... *Hyalella paramoensis* Andres, 1988
[Colombia: Paramo de Chisaca]

61(60) Telson width > length 62

61′ Telson width = length .... *Hyalella patagonica* Cunnigham, 1871
[Chile: Rio de los Ciervos]

62(61) Telson apically rounded .... *Hyalella faxoni* Stebbing, 1903
[Costa Rica: Volcan Reventado]

62′ Telson apically quadrate .... *Hyalella fossamancinii* Cavalieri, 1959
[Argentina: San Juan]

Amphipoda: Corophiidae: Genera and Species

1 Male gnathopod 2 merus enlarged, free along carpus anterioventral margin, both articles with elongated setal fringes; antenna 2 peduncle fourth article without distal process; epimera 2 ventral border with setae; epimera 3, posterior corner rounded; uropod 3 biramous (Fig. 23.23 B) .... *Paracorophium hartmannorum* Andres, 1975
[Chile: Concepción]

1′ Male gnathopod 2 merus not enlarged, both articles with elongated setal fringes; antenna 2 peduncle fourth article with bidentate distal process; epimera 2 ventral border without setae; epimera 3, posterior corner obtuse; uropod 3 uniramous (Fig. 23.23 A) .... *Americorophium rioplatense* (Giambiagi, 1929)
[Argentina: Río de la Plata]

Amphipoda: Metacrangonyctidae: *Metacrangonyx*: Species

1 Female gnathopod 2 long, slender, carpus length ~ 0.80 basis, length 3.9× width, propodus length 0.92 carpus, palmar index: 0.28, palm margin oblique; uropod 2 exopod with three setae along each margin; uropod 3, exopod longer than peduncle, medial margin with two setae .... *Metacrangonyx samanensis* Jaume & Christensen, 2001
[Greater Antilles: Dominican Republic]

1′ Gnathopod 2 in females, propodus length 0.94× carpus length, palmar index: 0.44, margin oblique, slightly convex; uropod 2, exopod with four to five lateral setae, uropod 3, exopod length 0.87× peduncle .... *Metacrangonyx dominicanus* Jaume & Christensen, 2001
[Greater Antilles: Dominican Republic]

Amphipoda: Crangoweckeliidae: Genera

1 Coxal plate 4 posteriorly emarginate; pereopods 5–7 basis elongate to oval, posterior margin convex to straight; maxilla 2 medial lobe with diagonal row of setae (Fig. 23.51 A) .... ***Crangoweckelia* [p. 857]**

1′ Coxal plate 4 not emarginate; pereopods 5–7 basis very elongate and narrow, nonlobate, posterior margin concave; maxilla 2 medial lobe without diagonal row of setae (Fig. 23.51 B) .... *Pintaweckelia grandis* Stock, 1985
[Greater Antilles: Haiti]

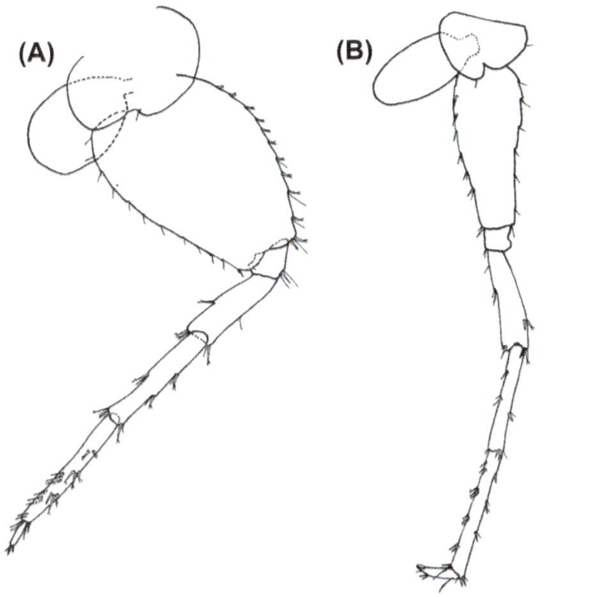

FIGURE 23.51 Crangoweckeliidae: (A) pereopod 5 of *Crangoweckelia mixta* (body size 5–6 mm); (B) pereopod 5 of *Pintaweckelia grandis* (body size 5–13 mm). *Redrawn (A) and (B) from Stock (1985b).*

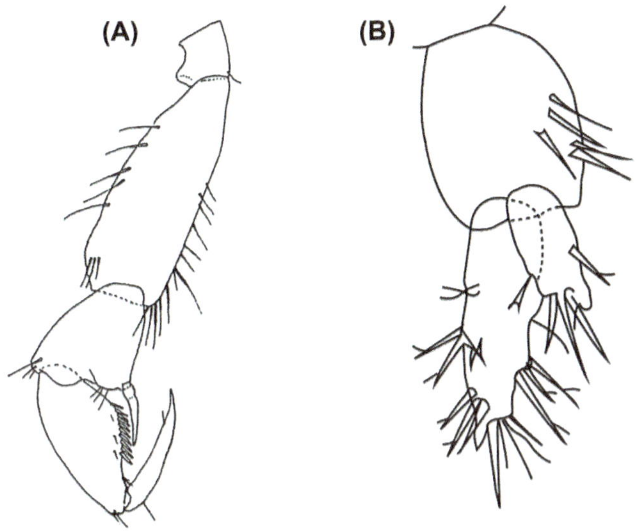

FIGURE 23.52 Falklandellidae: (A) pereopod 3 male of *Falklandella obtusa* (body size up to 5.7 mm); (B) uropod 3 of *Falklandella obtusa* (body size 5.7 mm). *Redrawn (A) and (B) from Stock & Platvoet (1991).*

Amphipoda: Crangoweckeliidae: *Crangoweckelia*: Species

1 Telson sensorial setae shorter than the two distal setae; uropod 3, exopod with short setae between medial seta and ramus apex (Fig. 23.24 B) .... *Crangoweckelia mixta* Stock, 1985
[Greater Antilles: Haiti]

1′ Telson sensorial setae longer than the single distal seta; uropod 3, exopod without short setae between medial seta and ramus apex, very long setae at exopod apex (Fig. 23.24 C) .... *Crangoweckelia spinicauda* Stock, 1985
[Greater Antilles: Haiti]

Amphipoda: Falklandellidae: Genera and Species

1 Gnathopod 1 larger than gnathopod 2; antennae calceoli absent or present; uropod 3 endopod reduced; telson emarginate, wider than long (Fig. 23.52 A) 2

1′ Gnathopod 1 smaller than gnathopod 2; antennae calceoli present; uropod 3 endopod not reduced; telson slightly cleft, longer than wide .... *Osornodella gabrielae* Pérez-Schultheiss, 2013
[Osorno: Chile]

2(1) Pereopod 7 with coxal gills; pereopod 3 not prehensile .... *Praefalklandella cuspidata* (Schellenberg, 1930)
[Malvinas (Falkland) Island]

2′ Pereopod 7 with coxal gills absent; pereopod 3 prehensile (Fig. 23.52 B) .... *Falklandella obtusa* Schellenberg, 1930
[Malvinas (Falkland) Islands]

Amphipoda: Artesiidae: *Spelaeogammarus:* Species

1 Antenna 1 accessory flagellum of four articles; gnathopod 1 propodus slightly larger than basis (Fig. 23.28) 2

1′ Antenna 1 accessory flagellum of five or more articles; gnathopod 1 propodus much larger than basis 4

2(1) Coxal plate 6 bearing one seta; maxilla 1 ischial endite with six multicuspidate setae and one plumose seta; maxilliped basal endite with four plumose setae 3

2′ Coxal plates 5 and 6 bearing one stout seta and about 20 smaller setae; antenna 1 accessory flagellum with four articles; maxilla 1 ischial endite with seven multicuspidate setae; maxilliped basal endite without plumose setae .... *Spelaeogammarus bahiensis* da Silva Brum, 1973

[Brazil: Bahia]

3(2) Gnathopod 1 basis anterior margin with five to nine short setae; coxal plate 5 bearing one stout setae and 17–18 short setae .... *Spelaeogammarus trajanoae* Koenemann & Holsinger, 2000
[Brazil: Bahia]

3′ Gnathopod 1 basis anterior margin with two to four long setae and one long and stout seta; coxal plate 5 bearing one stout setae and about nine short setae (Fig. 23.28) .... *Spelaeogammarus spinilacertus* Koenemann & Holsinger, 2000
[Brazil: Bahia]

4(1) Antenna 1 accessory flagellum with five articles; coxa 5 with posterior lobe round 5

4′ Antenna 1 accessory flagellum with six articles; coxa 5 with posterior lobe slightly concave .... *Spelaeogammarus titan* Senna, Andrade, Castelo-Branco & Ferreira, 2014
[Brazil: Bahia]

5(4) Gnathopod 1 basis posterior margin with 15–17 simple setae; uropod 3 endite with more than 18 dorsal bifid setae .... *Spelaeogammarus sanctus* Bastos-Pereira & Ferreira, 2015
[Brazil: Bahia]

5′ Gnathopod 1 basis posterior margin with 20 simple setae; uropod 3 endite with eight dorsal bifid setae .... *Spelaeogammarus santanensis* Koenemann & Holsinger, 2000
[Brazil: Bahia]

Amphipoda: Ingolfiellidae: Genera

1 Gnathopod 1, carpus with a finger-shaped process on the inner side, with one distal seta .... *Yacana ventania* Rodriguez, Armendáriz & Rodrígues Capítulo, 2017
[Argentina: Buenos Aires]

1′ Gnathopod 1, carpus without process on the inner side .... ***Ingolfiella*** **[p. 858]**

Amphipoda: Ingolfiellidae: *Ingolfiella*: Species

1 Pleopods 1–3 digitiform or triangular, at least pleopods 2 and 3 asymmetrical 2

1′ Pleopods 1–3 elongate spatulate, each pleopod symmetrical, maxilla 1 palp with one seta .... *Ingolfiella uspallatae* Noodt, 1965
[Argentina: Mendoza]

2(1) Pleopods triangular; gnathopod 2 carpus rounded to rectangular or ovate; some posterior pereopods dactylus with bifid claws 3

2′ Male pleopod 1 digitiform; gnathopod 2 carpus strongly triangular (i.e., palm more or less transverse), not sexually dimorphic; posterior pereopods dactylus lacks bifid claws; maxilla 1 palp with three distal setae .... *Ingolfiella manni* Noodt, 1961
[Chile: Antofagasta]

3(2) Male gnathopod 2 carpus palm with serrated teeth three to six well spaced (see glossary) 4

3′ Male gnathopod 2 carpus palm with serrated teeth three to six somewhat close together 5

4(3) Pereopod 3–7 dactylus narrower than propodus; maxilliped ischial endite slender almost digitiform, truncated and distally armed with one seta .... *Ingolfiella margaritae* Stock, 1979
[Venezuela: Isla de Margarita]

4′ Pereopod 3–7 dactylus not much narrower than propodus; maxilliped ischial endite digitiform, armed with two small setules .... *Ingolfiella putealis* Stock, 1976
[Lesser Antilles: Bonaire Islands]

5(3) Gnathopod 2, carpus palm with teeth three to six closely packed, similar and very small, antenna 2 flagellum length 0.5× peduncle .... *Ingolfiella tabularis* Stock, 1977
[Lesser Antiles: Curacao and Aruba Island]

5′ Gnathopod 2, carpus palm with teeth three to six somewhat close, tooth three larger than four to six; antenna 2 flagellum length 0.3× peduncle .... *Ingolfiella fontinalis* Stock, 1977
[Lesser Antilles: Bonaire Islands]

REFERENCES

Barnard, J.L. & G.S. Karaman. 1991. The families and genera of marine gammaridean Amphipoda (except marine gammaroids). Records of the Australian Museum, Supplement 13: 1–866.

Bastos-Pereira, R & A.A.P. Bueno. 2013. New species of freshwater Amphipod (Dogielinotidae, *Hyalella*) from Southeastern Brazil. Nauplius 21: 79–87.

Bond-Buckup, G., & P.B. Araujo. 1998. *Hyalella montenegrinae* sp. n., um Amphipoda de águas continentais do sul do Brasil (Crustacea, Peracarida, Hyalellidae). Nauplius 6: 53–59.

Bueno, A.A.P, P.B. Araujo, G.M. Cardoso, K.M. Gomes & G. Bond-Buckup. 2013. Two new species of *Hyalella* (Amphipoda, Dogielinotidae) from Brazil. Crustaceana, 86: 802–819.

Cardoso, G.M., A.A. Bueno & R.L. Ferreira. 2011. A new troglobiotic species of *Hyalella* (Crustacea, Amphipoda, Dogielinotidae) from Southeastern Brazil. Nauplius 19: 17–26.

De Smet, W.H. 2015. Description of *Pseudingolfiella possessionis* n. sp. (Crustacea, Amphipoda) from sub-Antarctic Île de La Possession, Crozet archipelago: the second freshwater amphipod known from the Antarctic biome, a human introduction of Gondwanan ancestry? Zootaxa 3941: 221–238.

Fernández, H. 2004. Does wrong procedence assignment lead to underestimates in groundwater biodiversity? Ecología Austral 14: 77–81.

Fišer, C., M. Zagmajster & R.L. Ferreira. 2013. Two new Amphipod families recorded in South America shed light on an old biogeographical enigma. Systematics and Biodiversity 11: 117–139.

Giambiagi, D. 1929. Un Nuevo anfípodo de agua dulce del género *Corophium*. Anales del Museo Nacional de Historia Natural de Buenos Aires, 34: 137–143.

González, E. 1986. A new record of *Paracorophium hartmannorum* Andres, 1975, from the Chilean coast, with a description of the adult (Amphipoda: Corophidae). Proceedings of the Biological Society of Washington 99: 21–28.

González E.R. & L. Watling. 2001. Three new species of *Hyalella* from Chile. Hydrobiologia, 464: 175–199.

González, E.R. & L. Watling. 2003a. A new species of *Hyalella* from Brazil (Crustacea: Amphipoda: Hyalellidae), with redescriptions of three other species in the genus. Journal of Natural History 37: 2045–2076.

González, E.R. & L. Watling. 2003b. Two new species of *Hyalella* from Lake Titicaca, and redescriptions of four others in the genus (Crustacea: Amphipoda). Hydrobiologia 497: 181–204.

González, E.R., G. Bond-Buckup & P.B. Araujo. 2006. Two new species of *Hyalella* from Southern Brazil (Amphipoda: Hyalellidae) with a taxonomic key. Journal of Crustacean Biology 26: 355–365.

Grosso, L.E. & G. Claps. 1984. Tercer Bogidiellidae (Crustacea, Amphipoda) de la cuenca del Río Grande (Jujuy, Argentina). Neotropica 30: 223–231.

Grosso, L.E. H.R Fernández. 1985. Una nueva *Bogidiella* (Amphipoda, Bogidiellidae) hiporreica de la provincia de Tucumán (Argentina). Neotropica 31: 201–209.

Grosso, L.E. & H. Fernández. 1993. Nuevo género cavernícola austral de Bogidiellidae; *Patagongidiella* n. g. del Noroeste Patagónico (Neuquén, Argentina). Bollettino del Museo Civico di Storia Naturale di Verona 17, 1990 (1993): 357–372.

Holsinger, J.R. 1990. *Tuluweckelia cernua*, a new genus and species of stygobiont amphipod crustacean (Hadziidae) from anchialine caves on the Yucatan Peninsula in Mexico. Beaufortia 41: 97–107.

Holsinger, J.R. 1994. Pattern and process in the biogeography of subterranean amphipods. Hydrobiologia 287: 131–145.

Jaume, D. & K. Christenson. 2001. Amphi-Atlantic distribution of the subterranean amphipod family Metacrangonyctidae (Gammaridea). Contributions to Zoology 70: 99–125.

Jaume, D. & H.P. Wagner. 1998. New cave-dwelling amphipods (Lysianassidae, Hadziidae) from the Dominican Republic (Hispaniola). Contributions to Zoology 68: 37–66.

Koenemann, S. & J.R. Holsinger. 1999. *Megagidiella azul*, a new genus and species of cavernicolous amphipod crustacean (Bogidiellidae) from Brazil, with remarks on its biogeographic and phylogenetic relationships. Proceedings of the Biological Society of Washington 112: 572–580.

Koenemann, S. & J.R. Holsinger. 2000. Revision of the subterranean amphipod genus *Spelaeogammarus* (Bogidiellidae) from Brazil, including descriptions of three new species and considerations of their phylogeny and biogeography. Proceedings of the Biological Society of Washington 113: 104–123.

Lincoln, R.J. & D.E. Hurley. 1981. The calceolus, a sensory structure of gammaridean amphipods (Amphipoda: Gammaridea). Bulletin of the British Museum (Natural History). 40: 103–116.

Lowry, J.K. & A.A. Myers. 2003. A Phylogeny and a new classification of the Corophiidea Leach, 1814 (Amphipoda). Journal of Crustacean Biology 23: 443–485.

Lowry, J.K. & A.A. Myers. 2013. A Phylogeny and Classification of the Senticaudata subord. nov. (Crustacea: Amphipoda). Zootaxa 3610: 1–80.

Noodt, W. 1959. Estudios sobre crustáceos chilenos de aguas subterráneas. I. *Ingolfiella chilensis* n. sp. de la playa marina de Chile Central (Crustacea, Amphipoda). Investigaciones Zoológicas Chilenas 5: 199–205.

Noodt, W. 1965. Interstitielle Amphipoden der konvergenten Gattungen *Ingolfiella* Hansen und *Pseudingolfiella* n. gen. aus Südamerika. Crustaceana 9: 17–30.

Peralta, M. & L.E. Grosso. 2009. Crustacea Syncarida, Amphipoda y Decapoda. Pages 469–495 *in*: E. Domínguez & H.R. Fernández (eds.). Macroinvertebrados Bentónicos de Sudamérica. Fundación Miguel Lillo, Tucumán, Argentina.

Ruffo, S. 1952. *Bogidiella neotropica* n. sp., nuovo Anfipodo dell' Amazonia. Schweizerische Zeitschrift für Hydrologie 14: 129–134.

Ruffo, S. & A. Vigna Taglianti. 1973. Three new subterranean *Bogidiella* from Mexico and Guatemala (Crustacea, Amphipoda). Subterranean Fauna of Mexico II. Quaderni Accademia Nazionale del Lincei 171: 105–133.

Ruffo, S. & A. Vigna Taglianti. 1977. Secondo contributo alla conoscenza del genere *Bogidiella* in Messico e Guatemala (Crustacea, Amphipoda, Gammaridae). Subterranean Fauna of Mexico III. Quaderni Accademia Nazionale del Lincei 171: 125–172.

Stock, J.H. 1981.The taxonomy and zoogeography of the family Bogidiellidae (Crustacea, Amphipoda), with emphasis on the West Indian taxa. Bijdragen tot de Dierkunde 51: 345–374.

Stock, J.H. 1985a. Bogidiellidae (Amphipoda) from Haiti and some general rules on the occurrence of crustacea Malacostraca in Inland Ground-waters of the West Indies. Stygologia 1: 208–223.

Stock, J.H. 1985b. Stygobiont amphipod crustaceans of the hadzioid group from Haiti. Bijdragen tot de Dierkunde 55: 331–426.

Stock, J.H. & D. Platvoet. 1991. The freshwater Amphipoda of the Falkland Islands. Journal of Natural History 25: 1469–1491.

Stock, J.H. & J.J. Vermeulen. 1982. A representative of the mainly abyssal family Pardaliscidae (Crustacea, Amphipoda) in cave waters of the Caicos Islands. Bijdragen tot de Dierkunde, 52: 3–12.

Väinölä, R., J.D.S. Witt, M. Grabowski, J.H. Bradbury, K. Jazdzewski & B. Sket. 2008. Global diversity of amphipods (Amphipoda; Crustacea) in freshwater. Hydrobiologia 595: 241–255.

Villalobos, F. A. 1961. Un anfipodo cavernícola nuevo de México: *Bogidiella tabascensis* n. sp. Anales del Instituto de Biología de México, ser. Zoología 31: 317–334.

Tanaidacea

Jürgen Guerrero-Kommritz
Fundabas Foundation, Bogotá, Colombia

INTRODUCTION

The Neotropical euryhaline and freshwater tolerant tanaidaceans consist of a small number of known benthic species that usually inhabit salt marshes, mangrove areas, bays and coastal rivers, and caves. No species have been found far away from the ocean. Like many other arthropod species, tanaids are usually overlooked and more specific searches for these taxa are needed.

These small (~3 mm long), free-living crustaceans lack planktonic larvae, and juveniles emerge as mancas from the female's brood pouch. Like other Peracarida, the geographical dispersion is limited because of the absence of a planktonic stage. Identification of species is based on females. Males are rare and the sexual dimorphism is very common, making the identification of males very difficult. Life cycles and special adaptations of freshwater Tanaidacea in the Neotropics are mostly unknown.

LIMITATIONS

Only three Neotropical freshwater species are known: *Sinelobus stanfordi* (Richardson, 1901), *Leptochelia dubia* (Koryer, 1842), and *Teleotanais gerlachi* Lang, 1956 (Lang, 1956; Gardiner, 1975; Sieg & Heard, 1983; Larsen & Hanskencht, 2004; Jaume & Boxshall, 2008).

The three recognized species that inhabit freshwater in the Neotropics are euryhaline and freshwater tolerant. Building a key with these three specimens is not possible at the moment. These reports correspond to species complexes. For that reason, it is necessary to refer the species with its author.

TERMINOLOGY AND MORPHOLOGY

Tanaids are easily recognized and separated from other peracarids. In some ways, they resemble little lobsters. The body is dorsoventrally depressed, and the most remarkable characteristic is the true **chelas**, on the first modified walking legs. The body is divided into a **cephalothorax**, a **pereon** with six segments, a **pleon** with five segments, and a **pleotelson**. Each segment has a pair of legs. The pleotelson has a pair of modified legs, the uropods. The cephalothorax has two antennae: the first pair (first antenna or antennule) is large; and the second pair (second antenna or antenna) is smaller. Next to them are the sessile compound eyes. Many species produce tubes that resemble long socks of sand and dirt with silk produced by their spinning glands. All known tanaids are bottom dwellers, no pelagic or terrestial tanaids have been found yet. Larsen (2005) provided a detailed account of external morphology.

There are three suborders: (1) the Apseudomorpha, which has antennular accessory flagellum as well a flagellum on the first true walking leg; (2) the Tanaidomorpha which lacks these flagella; and (3) the Neotanaidomorpha a deep-sea group with long uropods, huge chelas, and slender cephalothorax. Neotropical euryhaline and freshwater tolerant tanaidaceans belong to the Tanaidamorpha.

MATERIAL PREPARATION AND PRESERVATION

Taniads are difficult to see in the field, and it is often better to collect a portion of the likely habitat and sort them in the laboratory. The substrate should be fixed in 10% formalin, washed, and then stored in ethanol 70%. Further sorting is done afterward, and tanaids are separated from the other fauna and stored in vials. A strong dissecting microscope (60–80×) is needed to examine most tanaids. Field data (collecting site and date) should always be stored with the specimens and should be recorded with permanent ink or pencil.

Dissection is usually done in ethanol or glycerin with watchmaker forceps or chemically sharpened tungsten needles (Larsen & Hansknecht, 2004). Temporary mounts can be prepared with glycerin on slides with coverslips.

FRESHWATER TANAIDACEA

The three recognized species that inhabit freshwater in the Neotropics are euryhaline and freshwater tolerant.

Sinelobus stanfordi (Richardson, 1901) (Tanaididae) (Fig. 23.53) has a very wide distribution and is most probably a species complex. It can be recognized by the absence of an ischium on all legs, a very strong sexual dimorphism, males with a very large chela, uropods of four segments lacking exopod. Peropod six with a clawlike unguis.

Leptochelia dubia (Koryer, 1842) (Leptocheliidae) (Fig. 23.54) is a polymorphic species and most probably a species complex with a worldwide distribution; it was found on saltwater, brackish-, and freshwater. It can be recognized by presence of ischium on all legs, uropods long of six or seven segments, endopod of five or six segments, exopod of one segment, sexual dimorphism present, first antenna and chelas in males very large.

Teleotanais gerlachi Lang, 1956 (Teleotanaidae) (Fig. 23.55) has a very wide distribution and is known from salt, brackish, and freshwater habitats. It can be recognized by strong body pigmentation with darker regions on the body. Peropods with ischium, uropods short of three segments, endo- and exopod about the same length.

PHYLUM ARTHROPODA: CRUSTACEA: MALACOSTRACA

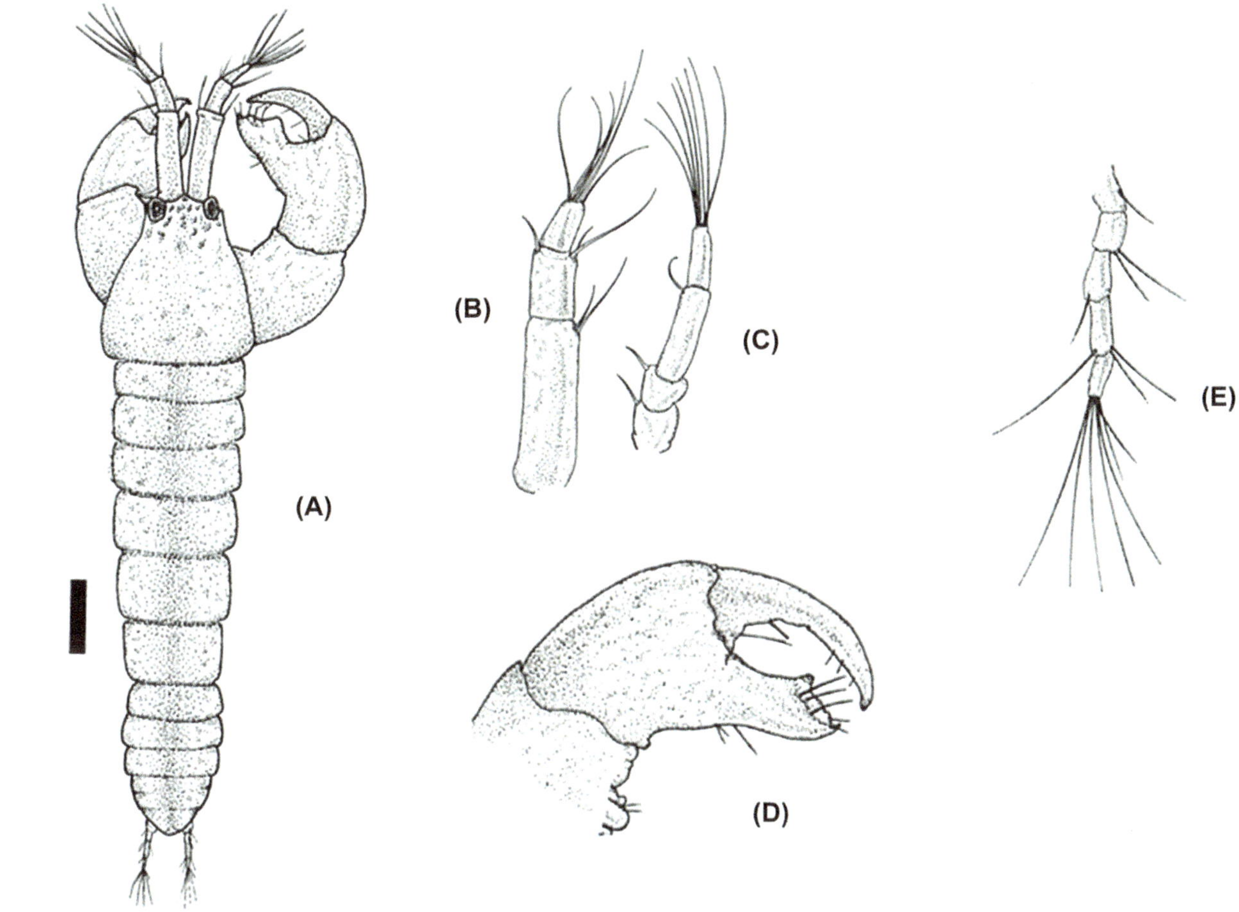

FIGURE 23.53 *Sinelobus stanfordi* (Richardson, 1901) (Tanaidacea): (A) dorsal view of male; (B) first antenna; (C) second antenna; (D) chela; (E) uropod. Scale bar = 0.2 mm. *Modified from Suarez-Morales et al. (2004).*

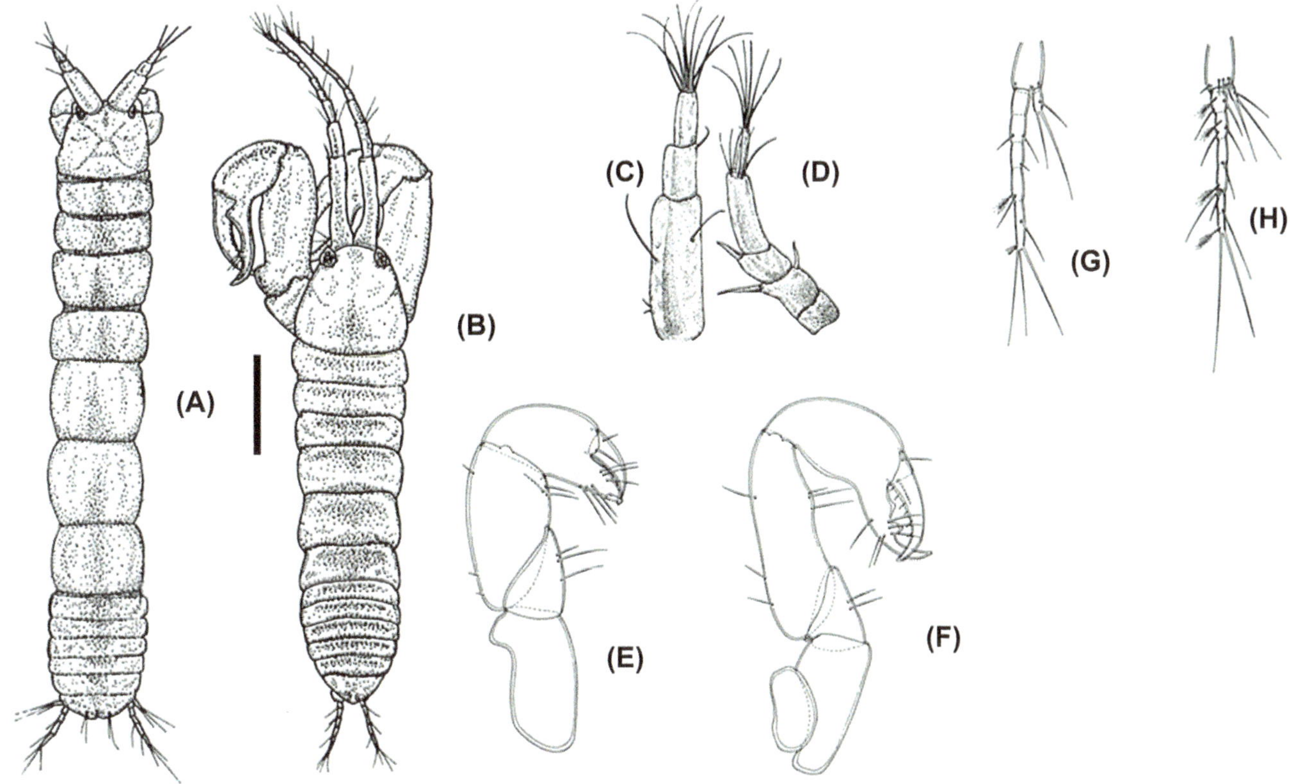

FIGURE 23.54 *Leptochelia dubia* (Koryer, 1842) (Tanaidacea): (A) dorsal view of female; (B) dorsal view of male; (C) first antenna; (D) second antenna; (E) cheliped female; (F) cheliped of male; (G) uropod of female; (H) uropod of male. Scale bar = 0.5 mm. *(A), (B), (D) Modified from Suarez-Morales et al. (2004), (E–H) from Sieg (1973).*

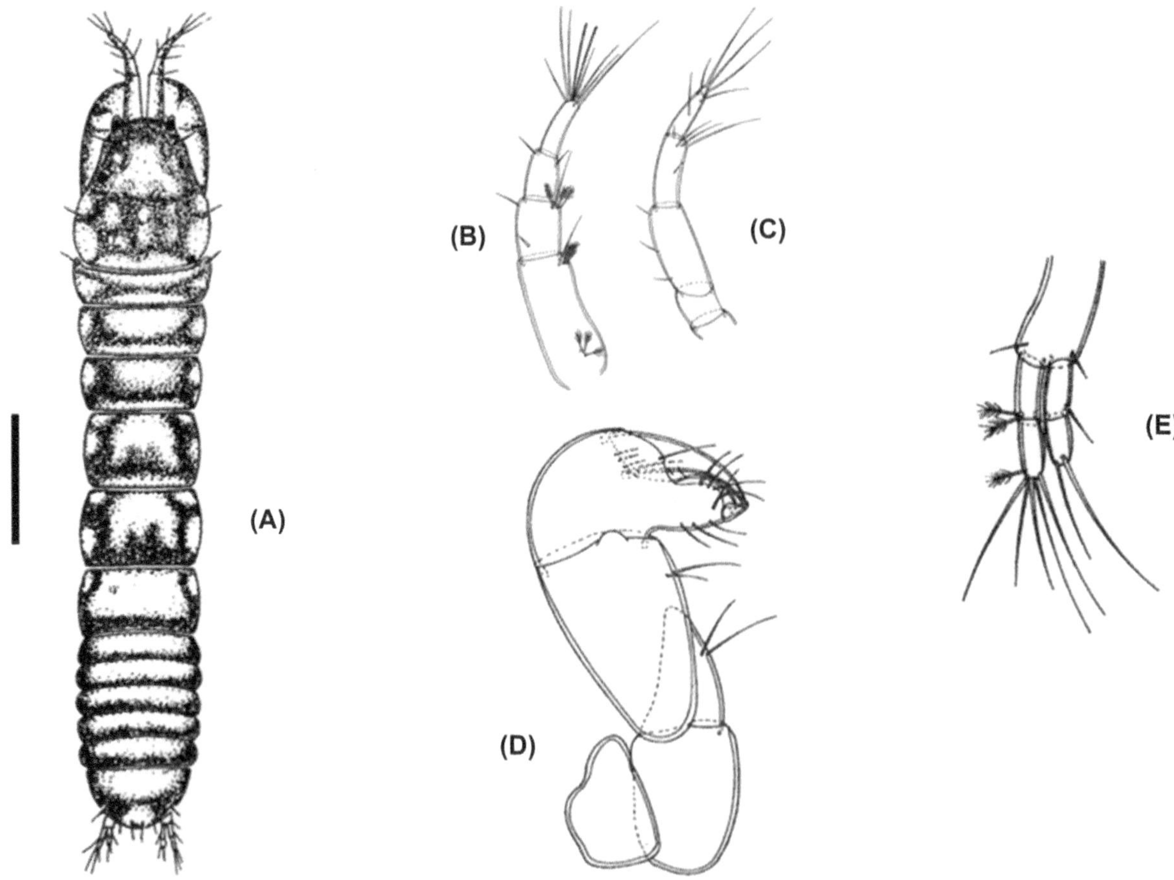

FIGURE 23.55 *Teleotanais gerlachi* Lang, 1956 (Tanaidacea): (A) dorsal view of female; (B) first antenna; (C) second antenna; (D) cheliped; (E) uropod. Scale bar = 0.5 mm. *Modified from Sieg (1973).*

REFERENCES

Gardiner, L.F. 1975. A fresh and brackish water tanaidacean, *Tanais stanfordi* Richardson, 1901, from a hypersaline lake in the Galapagos Archipelago, with a report on West Indian species. Crustaceana 29: 127–140.

Jaume, D. & G.A. Boxshall. 2008. Global diversity of cumaceans & tanaidaceans (Crustacea: Cumacea & Tanaidacea) in Freshwater. Hydrobiologia 595: 225–230.

Lang, K. 1956. Tanaidacea aus Brasilien gesammelt von Professor Dr. A. Remane und Dr. S. Gerlach. Kieler Meeresforchungen 12 (2): 249–260.

Larsen, K. 2005. Deep-sea Tanaidacea (Peracarida) from the Gulf of Mexico Crustaceana Monographs 5: 1–381.

Larsen, K. & T. Hansknecht. 2004. A new genus and species of freshwater tanaidacean, *Pseudohalmyrapseudes aquadulcis* (Apseudomorpha: Parapseudidae), from Northern Territory, Australia. Journal of Crustacean Biology 24: 567–575.

Suarez-Morales, E., R. Heard, S. Garcia-Madrigal, J.J. Oliva-Rivera & E. Escobar-Briones. 2004. Catálogo de los Tanaidaceos (Crustacea: Peracarida) del Caribe Mexicano. CONACYT, SEMARNAT y el colegio de la frontera sur Chetumal QR.121p.

Sieg, J. 1973. Ein Beitrag zum natürlichen System der Dikonophora (Crustacea, Tanaidacea) Dissertation Thesis Naturwissenschaftliche Fakultät der Christian Albrechts Universität Kiel Tafel 1–114.

Sieg, J. & R.W. Heard. 1983. Tanaidacea of the Gulf of Mexico. III. On the occurrence of *Teleotanais gerlachi* Lang, 1956 (Nototanaidae) in the Eastern Gulf. Gulf Research Reports 7: 267–271.

Isopoda

D. Christopher Rogers
Kansas Biological Survey, and The Biodiversity Institute, University of Kansas, Lawrence, KS, United States

INTRODUCTION

Neotropical freshwater isopods are typically inhabitants of streams of all sizes, ponds, lakes, and rivers where they may be found on the substratum, under stones and other debris, or at times crawling about vegetation. Perhaps over half of described species are obligatory groundwater inhabitants, where they occur in caves or other habitats where subterranean waters flow through unconsolidated deposits (e.g., gravel, sand, soil).

LIMITATIONS

Wilson (2008) reports 109 described Neotropical species in 42 genera, with the vast majority being among the Cymothoida + Sphaeromaridea. Most taxa globally are regional or localized endemics. This coupled with the fact that so little of the Neotropical region has been explored for freshwater isopods requires that the keys here be necessarily circumscribed to higher taxonomic levels. It is expected that many new taxa will be discovered in the Neotropical region, especially in groundwater habitats, as more areas are explored.

TERMINOLOGY AND MORPHOLOGY

Generalized body regions are presented in Fig. 23.56 A. The smallest of the Asellota are < 2 mm long, although most are somewhat longer. Epigean species are frequently mottled brown in appearance, with their subterranean counterparts a whitish color. Likewise, eyes are present in the surface species and vary from being reduced to completely absent in subterranean species. The body is comprised of three main sections. The head contains two pairs of antennae and the mouthparts, called (in sequence) the **mandibles** (sometimes with palps), two pairs of **maxillae**, and the **maxillipeds**. The thorax is comprised of seven **pereonites**, each with a pair of **pereopods**. The first pereopod pair may be modified for grasping with the penultimate article (the **propodus**) enlarged and opposed by the final article (the **dactylus**). The abdomen is comprised of five primary **pleonites**, which may be free or variously fused together, all with depending appendages called **pleopods**. Pleonite VI is fused with the telson into a single structure called a **pleotelson**. In the predominant asellids the second pleopod of the male is modified as a sperm reservoir and transfer mechanism. The apex of the male second pleopod endopod possesses a sperm transfer tube, called the **cannula**, which has a unique construction in each species, and is frequently surrounded by accessory processes. The identification of most asellid species relies on the examination of this structure. The terminal appendages are the **uropods**, which vary in their form and place of attachment to the pleon.

MATERIAL PREPARATION AND PRESERVATION

Isopods should be placed in 70% ethyl alcohol; however, 80%–95% may be required if many specimens are taken, thereby preventing dilution from the animal's high body water content. The ethanol must be replaced after 24–48 h for proper preservation. Specimens are best examined in 70% ethanol and viewed under low power magnification for family, or sometimes, genus level determination. Species identification frequently requires dissection of at least the male second pleopod, which can be accomplished by manipulating the specimen with fine forceps and probes constructed of insect minuten pins (minuten nadeln) or tungsten wire. By repeatedly piercing the exoskeleton in the sutures between the ventral pleonites, the tiny second pleopod can be teased away from the specimen.

The simplest procedure is to transfer the appendage into a drop of glycerin on a microscope slide. It is usually a good idea to look at the positioning of the pleopod in the glycerin to ensure than nothing needs to be adjusted before placing the coverslip. The pleopod is now ready for examination, which depending on the size of the specimen, may require 400–1000× for evaluation of the tip structures of the male second pleopod endopod. After examination the pleopods should be removed from the slide and placed in a microvial within the vial that accompanies the dissected specimen.

ACKNOWLEDGMENT

Niel Bruce provided an excellent review of this section, and I am grateful for his comments and discussion.

KEY TO ISOPODA

The Neotropical freshwater Isopoda are very poorly known, and we strongly suspect that only a fraction of the actual isopod diversity has been described. Wilson (2008) estimated that worldwide freshwater isopod diversity was only about 37% described. Therefore, we strongly suggest to the users that they exercise caution and not force a given specimen into one of the taxa presented. Rather, we encourage the user to use this key as a starting point, to direct the user to the original literature to make a proper determination.

Isopoda: Suborders

1 Uropods styliform, inserted terminally on pleotelson; body proportions variable (Fig. 23.56 B, E–G) 2

1′ Uropods lamellar, inserted at pleotelson anterolateral angles; body length $< 4\times$ width (Fig. 23.56 C, D) 4

2(1) One, two, or three free (unfused) pleonites; body proportions variable 3

2′ Five free pleonites; body length $< 5\times$ width; pigment and eyes absent .... Oniscidea, one family: **Calabozoidae [p. 864]**

3(2) Pleotelson domed; body length $\geq 6\times$ width; subterranean .... **Microcerberidea [p. 864]**

3′ Pleotelson never domed; body proportions variable .... **Asellota [p. 864]**

4(1) Uropods articulating ventrolaterally; pleon with four or five free pleonites plus pleotelson (Fig. 23.56 C, G) .... **Cymothoida [p. 866]**

4′ Uropods, particularly the peduncle/protopod, articulating anteriolaterally (Fig. 23.56 D); pleon dorsally with four or fewer free pleonites plus the pleotelson .... Sphaeromatidea, one family: **Sphaeromatidae [p. 871]**

Isopoda: Oniscidea: Calabozoidae: Genera and Species

1 Uropods lateral (Fig. 23.56 E); stygobiont .... *Calabozoa pellucida* Van Lieshout, 1983 [Venezuela]

1′ Uropods subterminal (Fig. 23.56 F); stygobiont .... *Pongycarcinia xiphidiourus* Messana, Baratti & Benvenuti, 2002 [Brazil]

Isopoda: Microcerberidea: Families

1 Body in dorsal aspect of typical isopod form; pereonites projecting laterally covering all or most of pereopods .... Atlantasellidae, one genus: ***Atlantasellus* [p. 864]**

1′ Body in dorsal aspect vermiform, elongate, subcylindrical; pereonites not projecting laterally over limbs .... **Microcerberidae [p. 864]**

Isopoda: Microcerberidea: Atlantasellidae: *Atlantasellus*: Species

1 Cephalothorax anterior margin acute, posteriolateral angles triangular; stygobiont .... *Atlantasellus cavernicolus* Sket, 1979 [Bermuda]

1′ Cephalothorax anterior margin convex, posteriolateral angles elongate; stygobiont .... *Atlantasellus dominicanus* Jaume, 2001 [Dominican Republic]

Isopoda: Microcerberidea: Microcerberidae: Genera and Species

1 Cephalothorax anterior margin acute; pleopod I present; hypogean .... *Mexicerberus troglodytes* Schultz, 1974 [Mexico]

1′ Cephalothorax anterior margin rounded; pleopod I absent; hypogean .... *Yvesia striata* Coineau & Botosaneanu, 1973 [Cuba]

Isopoda: Asellota: Families

1 Antenna I inserted anteriorly 2

1′ Antenna I inserted anteriodorsally .... Protojaniridae, one species: *Cuyojanira riojana* Grosso, 1992 [Caves. Argentina: La Rioja. Chile?]

2(1) One or zero free pleonites visible dorsally 3

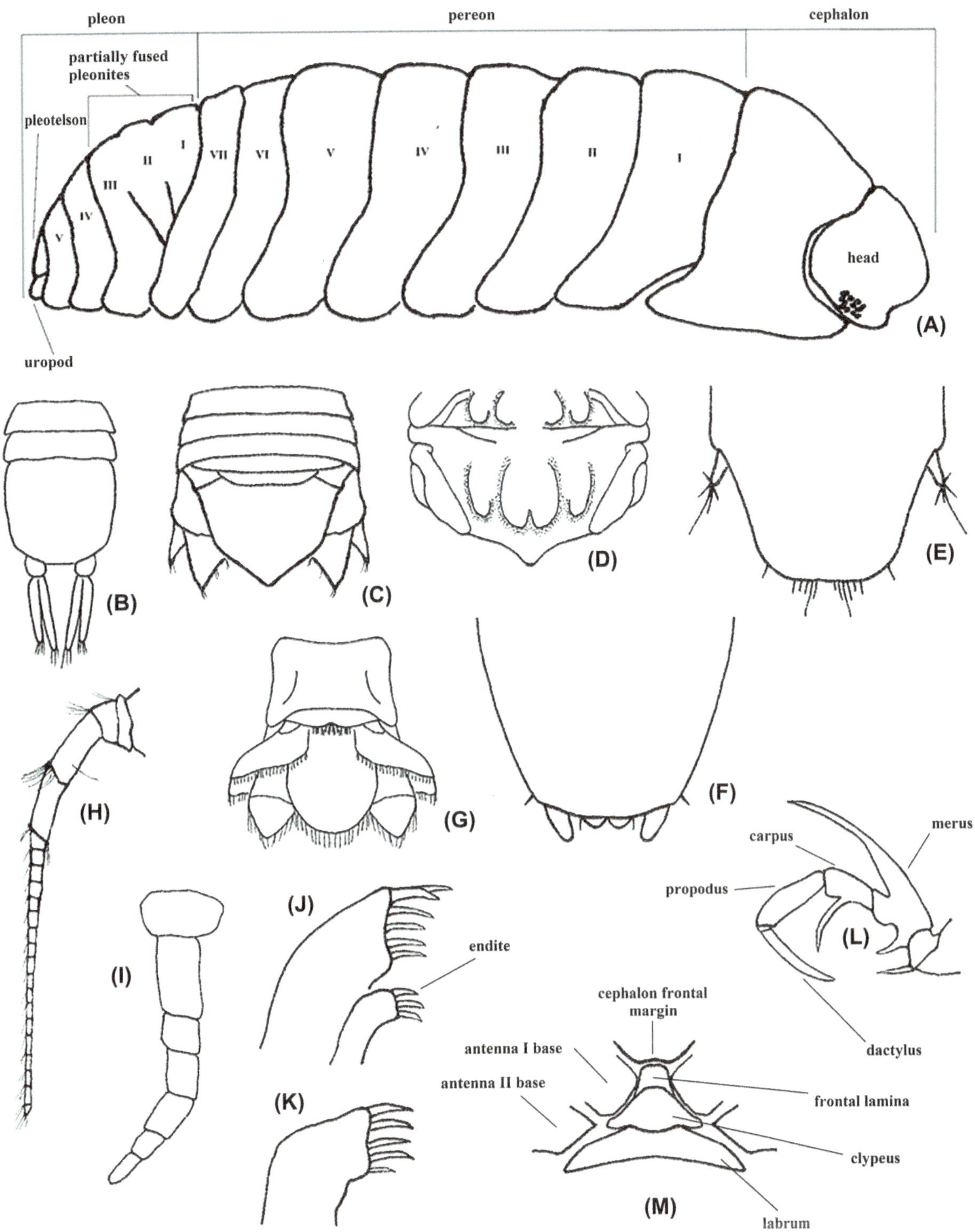

FIGURE 23.56 Isopoda: (A) generalized external dorsal morphology; (B) *Asellus* sp. pleotelson; (C) *Speocirolana thermydronis* pleotelson; (D) *Exosphaeroides quadricosta* pleotelson; (E) *Calabozoa pelucida* pleotelson; (F) *Pongycarcinia xiphidiourus* pleotelson; (G) Anthuridae generalised pleotelson; (H) *Bahalana caicosana* antenna II; (I) *Braga* sp. antenna II; (J) *Zulialana coalescens* maxilliped II; (K) *Excorallana* sp. maxilliped II; (L) *Bahalana bowmani* pereopod II; (M) generalized *Speocirolana* ventral view of the head.

2′ Two free pleonites .... Asellidae, one genus: *Caecidotea*
[Widespread, many undescribed species]

3(2) Pleotelson lateral margins parallel; body elongate, length $\geq 6\times$ body width .... Microparasellidae, one genus: ***Microcharon* [p. 866]**

3′ Pleotelson posteriolateral margins converging; uropods shorter than pleotelson; body broad in dorsal view, length less than four times width .... Janiridae, one species; *Heterias exul* (Müller, 1892)
[Brazil, Chile, Argentina. Afrotropical. Australian]

Isopoda: Asellota: Microparasellidae: *Microcharon*: Species

1 Rostrum acute; antenna II with eight flagellomeres; uropod peduncle length = pleotelson; hypogean .... *Microcharon phreaticus* Coineau & Botosaneanu, 1973
[Cuba]

1′ Rostrum rounded; antenna II with 12 flagellomeres; uropod peduncle length < pleotelson; interstitial, hypogean .... *Microcharon galapagoensis* Coineau & Schmidt, 1979
[Galápagos Islands]

Isopoda: Cymothoida: Superfamilies

1 Bilaterally symmetrical, or if asymmetrical, predatory or parasitic on fish 2

1′ Females bilaterally asymmetrical; males symbionts living on female's body; obligatory parasites of crustaceans .... Bopyroidea

2(1) Uropod exopod not curving over pleotelson dorsum; maxilliped coxae freely articulate; body length four times width or less .... **Cymothooidea [p. 866]**

2′ Uropod exopod curving over pleotelson dorsum (Fig. 23.56 G); maxilliped coxae fused to head; body elongate, length at least five times width .... Anthuridea, one family: **Anthuridae [p. 871]**

Isopoda: Cymothoida: Cymothooidea: Genera

1 Antennae with peduncle and flagellum distinctly different (Fig. 23.56 H); maxilliped palp with one or five palpomeres; predators (sometimes attacking fish many times larger than themselves, so may be mistaken for parasites) 2

1′ Antennae with peduncle and flagellum not distinctly different (Fig. 23.56 I); maxillipeds reduced to a palp with only two palpomeres; fish parasites .... **Cymothidae [p. 866]**
[Widespread]

2(1) Maxillipeds with endite (Fig. 23.56 J) .... **Cirolanidae [p. 867]**

2′ Maxilliped without endite (Fig. 23.56 K) .... Corallanidae, one genus: ***Excorallana* [p. 871]**

Isopoda: Cymothoida: Cymothidae: Genera

More freshwater cymothids occur in South America than anywhere else on Earth. These animals are all fish parasites, and are greatly in need of revision, with many poorly described species. This key is modified from Thatcher (2006).

1 Pereonite I with coxal plates fused to pereonite 2

1′ Pereonite I with coxal plate free .... *Anphira*

2(1) Pereopods I–VI prehensile with dactyls simple; pereopod VII ambulatory 3

2′ Pereopods I–VII prehensile with dactyls hamulate 4

3(2) Pleon segments obvious and separate from pleotelson .... *Artystone*

3′ Pleon and pleotelson fused .... *Riggia*

4(2) Pereonite I anterior margin not sinuate; pereonites II–VII posterior margins not projecting 5

4′ Pereonite I anterior margin with trisinuate; pereonites II–VII posterior margins acute and projecting .... *Nerocila*

5(4) Cephalon anterior margin without protrusions 6

5′ Cephalon anterior margin with three rounded protrusions .... *Asotana*

6(5) Pleopods all bilaminar 7

6′ Pleopods II–V with three to four lamina .... *Braga*

7(6) Pleotelson width ≤ pleonite I width 8

7′ Pleotelson width > pleonite I width, pleotelson length at midline > 1.8× basal width ..... *Paracymothoa*

8(7) Pleotelson width < pleonite I width 9

8′ Pleotelson width subequal to pleonite I width ..... *Telotha*

9(8) Pereon last segment not or slightly embracing pleon first segment, never reaching second 10

9′ Pereon last segment embracing pleon first segment and just coming into contact with pleon segment two ..... *Vanamea symmetrica* (Van Name, 1925)

10(9) Pleotelson rounded or subquadrate, never triangular 11

10′ Pleotelson triangular ..... *Isonebula*

11(10) Pleotelson without a central anteriodorsal carina ..... *Livoneca redmanii* Leach, 1818

11′ Pleotelson with central anteriodorsal carina ..... *Philostomella cigarra* Szidat & Schubart, 1960

Isopoda: Cymothoida: Cirolanidae: Genera

Many of the genera and species presented here are predominately marine, but occur in inland anchialine caves, and are included because they are "inland" aquatic taxa. *Conilera stygia* Packard, 1900 reported from México is a nomen dubium and is not included.

1 Seven pairs of pereopods 2

1′ Six pairs of pereopods (pair seven absent); anchialine ..... *Dodecalana yagerae* Carpenter, 1994 [Bahamas: Exumas, Grand Bahama, Sweeting's Cay]

2(1) Pleon with pleonites coalesced and/or fused; able to conglobate (roll into a ball) 3

2′ Pleon with three or more unfused pleonites; able to conglobate or not 4

3(2) Uropod endopod oval; pleonites I—III indicated by lateral sutures, fused medially, remaining pleonites fused completely; rostrum absent; second antenna flagellum with 30+ articles ..... *Zulialana coalescens* Botosaneanu & Viloria, 1993 [Venezuela]

3′ Uropod endopod lamellar, rounded; pleonites I and II, sometimes III free, remaining pleonites fused; rostrum present or absent; second antennal flagellum with 16 or fewer articles ..... ***Sphaerolana* [p. 868]** [Mexico]

4(2) Pleon with three pleonites with lateral margins free 5

4′ Pleon with four or five pleonites with lateral margins free 7

5(4) Pleonites I—III with lateral margins free, pleonite III embracing pleonites IV and V 6

5′ Pleonites I—III with lateral margins free, pleonite III embracing pleonite IV, pleonite V fused to pleotelson, apparent only as an incomplete medial suture line near the pleotelson anterior margin; pereopods I and II weakly haptorial ..... *Mexilana saluposi* Bowman, 1975 [Mexico: San Luis Potosi]

6(5) Antenna I reaching posteriorly to pereonite VII; unable to conglomerate ..... ***Speocirolana* [p. 868]** [Northern Mexico]

6′ Antenna I not or just reaching pereonite II; animal able to conglobate ..... *Creaseriella anops* (Creaser, 1936) [Yucatan Peninsula]

7(4) No or pereopods I—III haptorial (dactylus longer than propodus); antenna I with peduncle articles I, II, and III separate 8

7′ All pereopods haptorial; antenna I with peduncle articles I and II coalesced, giving the appearance of only two articles in the peduncle ..... ***Haptolana* [p. 869]**

8(7) Pereopods with meri not produced posteriodistally 9

8′ Pereopod I with merus produced posteriodistally; pereopods II and III with meri produced anteriodistally (Fig. 23.56 L) ..... ***Bahalana* [p. 869]** [Caribbean Islands. El Salvador]

9(8) Pereonite I anterior margin straight or smoothly curved, never embracing cephalon posterior margin 10

9′ Pereonite I anterior margin recessed to embrace cephalon posterior margin 11

10(9) Antenna I longer than antenna II; antennae reaching at most to pereonite V; pleotelson distal margin semicircular, wider than pleonites; pleonal epimeral plates not carinate; uropods shorter than pleotelson, nearly concealed in dorsal view; uropod endopod subdiscoid, length subequal to width; uropod endopod elongate suboval, lateral margins subparallel ..... *Exumalana reptans* Botosaneanu & Iliffe, 2003 [Bahamas]

10′ Antenna I shorter than antenna II; antenna II reaching at least to pleonite II; pleotelson subquadrate, narrower than pleonites; pleonal epimeral plates with perpendicular carinae; uropods length subequal to pleotelson, not concealed in dorsal view; uropod endopod broadly triangular, distal margin straight, length ~ 1.2× width; uropod endopod elongate hastate, lateral margins subparallel ..... *Metacirolana mayana* (Bowman, 1987) [Yucatan Peninsula]

PHYLUM ARTHROPODA: CRUSTACEA: MALACOSTRACA

11(9) At least first pereopod pair haptorial (dactyl longer than propodus), typically I—III haptorial; pleonite III not expanded ventrally, without a projection 12

11′ No pereopods haptorial; pleonite III expanded ventrally, forming a robust, ventrally directed projection .... *Lucayalana troglexuma* (Botosaneanu & Iliffe, 1997) [Bahamas]

12(11) Cephalon width > length, pentagonal or not, without lateral corners and rostrum projecting; pereonite I recessed < 20% its length to receive cephalon; pereonite I in dorsal view lateral margins evenly rounded 13

12′ Cephalon pentagonal, length / width subequal, with lateral corners and rostrum projecting; pereonite I recessed nearly 50% its length to receive cephalon; pereonite I in dorsal view lateral margin with a projecting angle at midpoint .... *Yucatalana robustispina* Botosaneanu & Iliffe, 1999 [Yucatan]

13(12) Clypeus flattened ventrally, not projecting .... ***Cirolana* [p. 870]**

13′ Clypeus projecting .... ***Arubolana* [p. 870]**

Isopoda: Cymothoida: Cirolanidae: *Sphaerolana*: Species

Hypogean.

1 Pleotelson apically rounded; two pleonites visible dorsally 2

1′ Telson apically acute; three pleonites visible dorsally .... *Sphaerolana karenae* Rodríguez-Almaraz & Bowman, 1995 [Mexico: Nuevo Leon]

2(1) Uropod peduncle with single marginal nub representing endopod; pleopod I exopod oval .... *Sphaerolana interstitialis* Cole & Minkley, 1970 [Mexico: Coahuila]

2′ Uropod peduncle with two marginal nubs representing endopod and exopod; pleopod I exopod medial margin straight .... *Sphaerolana affinis* Cole & Minkley, 1970 [Mexico: Coahuila]

Isopoda: Cymothoida: Cirolanidae: *Speocirolana*: Species

Stygobionts from Mexico and adjacent USA. This key is modified from Schotte (2002).

1 Pleotelson not acute 2

1′ Pleotelson acute distally .... *Speocirolana thermydronis* Cole & Minckley, 1966 [Mexico: Coahuila]

2(1) Pleotelson rounded distally 3

2′ Pleotelson truncated distally 6

3(2) Uropodal basis (protopod) not extending to telson distal margin 4

3′ Uropodal basis extending beyond pleotelson distal margin 5

4(3) Male uropodal endopod clavate, length > exopod .... *Speocirolana pubens* Bowman, 1982 [Mexico: San Luis Potosi]

4′ Male uropodal endopod not clavate, length subequal to exopod .... *Speocirolana prima* Schotte, 2002 [Mexico: Tamulipas]

5(3) Uropodal exopod acute, length = endopod; second antenna extending no further than pereonite VI, and bearing 40 or less flagellomeres .... *Speocirolana pelaezi* Bolivar & Pieltain, 1950 [Mexico: San Luis Potosi]

5′ Uropodal exopod lanceolate, length < endopod; second antenna extending to pereonite VII, and bearing 50 or more flagellomeres .... *Speocirolana xilitla* Alvarez & Villalobos, 2008 [Mexico: San Luis Potosi]

6(2) Uropodal exopod not clavate 7

6′ Uropodal exopod clavate, longer than endopod .... *Speocirolana fustiura* Botosaneanu & Iliffe, 1999 [Mexico: Nuevo León]

7(6) Pleotelson posterior margin width < 0.5× basal width 8

7′ Pleotelson posterior margin width > 0.5× basal width 10

8(7) Frontal lamina not narrow, not bent at one-third of length 9

8′ Frontal lamina narrow, elongated, bent at nearly 90° at one-third of length .... *Speocirolana lapenita* Botosaneanu & Iliffe, 1999 [Mexico: Tamulipas]

9(8) Uropod rami extending beyond pleotelson posterior margin .... *Speocirolana guerrai* Contreras-Balderas & Purata-Velarde, 1982 [Mexico: Nuevo León]

9′ Uropod exopod subequal in length to pleotelson, endopod distinctly shorter than pleotelson .... *Speocirolana zumbadora* Botosaneanu, Iliffe, & Hendrickson, 1998 [Mexico: Coahuila]

10(7) Frontal lamina (Fig. 23.56 M) flat, pentagonal; second antennae not extending beyond pleotelson middle 11

10′ Frontal lamina curved dorsally; second antennae nearly reaching pleotelson posterior margin .... *Speocirolana disparicornis* Botosaneanu & Iliffe, 1999 [Mexico: Tamulipas]

11(10) Frontal lamina length 1.5× width .... *Speocirolana endeca* Bowman, 1982 [Mexico: Tamulipas]

11′ Frontal lamina length 2.5× width .... *Speocirolana bolivari* Rioja, 1953 [Mexico: San Luis Potosi]

Isopoda: Cymothoida: Cirolanidae: *Haptolana*: Species

Hypogean. This genus also occurs in Africa and Australia.

1 Pleotelson apical margin rounded 2

1′ Pleotelson subquadrate, posterior margin distinctly transverse .... *Haptolana trichostoma* Bowman, 1966 [Cuba]

2(1) Second antenna peduncle of four articles, the two basal most coalescent; pereopods IV–VII propodi anterior width ~ 0.2× length, dactyli length ~ 0.70× propodi length .... *Haptolana belizana* Botosanenau & Iliffe, 1997 [Belize]

2′ Second antenna peduncle of five articles; pereopods IV–VII propodi anterior width ~ 0.5× length, dactyli length ~ 0.45× propodi length .... *Haptolana bowmani* Botosaneanu & Iliffe, 1997 [Yucatan]

Isopoda: Cymothoida: Cirolanidae: *Bahalana*: Species

1 Pereonite I anterior margin recessed to embrace cephalon posterior margin 2

1′ Pereonite I anterior margin straight or smoothly curved, never embracing cephalon posterior margin .... *Bahalana exumina* Botosaneanu & Iliffe, 2002 [Bahamas: Exuma Island]

2(1) Pleopods III–IV endopods subquadrate, with setae on distal margin 3

2′ Pleopods III–IV endopods rounded, lacking setae; maxillipedal endite with one coupling hook .... *Bahalana cardiopus* Notenboom, 1981 [Bahamas: Acklins Island, Mayaguana Island]

3(2) Pleopod V distal margin setose 4

3′ Pleopod V distal margin setose only on distomedial corner .... *Bahalana geracei* Carpenter, 1981 [Bahamas: San Salvador Island]

4(3) Pereon I anterior recess for receiving cephalon rectangular 5

4′ Pereon I anterior recess for receiving cephalon evenly curved .... *Bahalana bowmani* Ortiz, Lalana & Perez, 1997 [Cuba]

5(4) Males only: pene length ≤ 3× basal width .... *Bahalana caicosana* Botosaneanu & Iliffe, 2003 [Turks and Caicos Islands]

5′ Males only: pene length > 7× basal width .... *Bahalana abacoana* Botosaneau & Iliffe, 2006 [Bahamas: Abaco]

Isopoda: Cymothoida: Cirolanidae: *Cirolana*: Species

Most species are stygobionts.

1 Eyes present, brackish water, estuaries 2

1′ Eyes absent 7

2(1) At least some dorsal surfaces with tubercles 3

2′ Dorsal surfaces smooth 4

3(2) Cephalon with two fused tubercles; antenna II reaching to pereonite II or III; frontal lamina anteriorly rounded, projecting, length = width .... *Cirolana browni* Van Name, 1936 [Caribbean coasts and islands]

3′ Cephalon with three tubercles in male, female cephalon without tubercles; antenna II reaching to pereonite V; frontal lamina narrow, length > width, pentagonal, anteriorly acute, not projecting .... *Cirolana jonesi* (Kensley, 1987) [Belize]

4(2) Antenna I flagellum with nine or more articles; pleotelson subtriangular 5

4′ Antenna I flagellum with seven articles; pleotelson broadly rounded distally .... *Cirolana adriani* Ortiz & Cházaro, 2015 [Mexico; Cozumel Island]

5(4) Frontal lamina width ~ 0.5 length; Pacific drainages 6

5′ Frontal lamina width ≤ 0.25 length .... *Cirolana magna* (Ortiz, Lalana & Perez, 1997) [Cuba]

6(5) Antenna II flagellum with 20+ articles; frontal lamina anteriorly rounded .... *Cirolana aleci* (Brusca, Wetzer & France, 1995) [Panama]

6′ Antenna II flagellum with ≤ 19 articles; frontal lamina rectangular. .... *Cirolana sinu* (Kensley & Schotte, 1994) [Colombia]

7(1) Pleopods III–V entire 8

7′ Pleopods III–V bifurcate, deeply forked .... *Cirolana pleocissa* (Botosaneanu & Iliffe, 1997) [Jamaica]

8(7) Pleotelson distal margin with ten or more robust setae 9

8′ Pleotelson distal margin with eight or fewer robust setae 10

9(8) Cephalon (head region) length $\sim 0.6\times$ width, width $\sim 0.5\times$ pereonite I width; antenna II reaching posteriorly to pereonite VI or VII .... *Cirolana cubensis* Hay, 1903 [Cuba]

9′ Cephalon (head region) length $\sim 0.3\times$ width, width $\sim 0.7\times$ pereonite I width; antenna II reaching posteriorly to pereonite V .... *Cirolana acanthura* Notenboom, 1981 [Haiti]

10(8) Antenna II flagellum with <25 articles 11

10′ Antenna II flagellum with 30 + articles .... *Cirolana yunca* (Botosaneanu & Iliffe, 2000) [Yucatan]

11(10) Pleotelson distal margin with four robust setae 12

11′ Pleotelson distal margin with eight robust setae .... *Cirolana crenata* (Bowman & Franz, 1982) [Grand Cayman Islands]

12(11) Frontal lamina ovate; antenna I peduncle with articles I and II coalescent; body length 4–6 mm .... v .... *Cirolana yucatana* Botosaneanu & Iliffe, 2000 [Yucatan]

12′ Frontal lamina length > width, anteriorly expanded, rounded; antenna I peduncle with articles I and II separate; body length > 6 mm .... *Cirolana radicicola* Notenboom, 1981 [Haiti]

Isopoda: Cymothoida: Cirolanidae: *Arubalana*: Species

1 Eyes entirely absent 2

1′ Eyes present, but extremely minute; interstitial in rivers .... *Arubolana parvioculata* Notenboom, 1984 [Jamaica]

2(1) Free pereonites with lateral margin acute; pleotelson subtruncate 3

2′ Free pereonite I with lateral margin rounded, remaining free pereonites with lateral margin acute; pleotelson subcircular .... *Arubolana rotunditelson* (Botosaneanu & Stock, 2010)
[Dominican Republic]

3(2) Pleotelson posterior margin crenulate, but not emarginate .... *Arubolana aruboides* Bowman & Iliffe, 1983
[Bermuda]

3′ Pleotelson posterior margin emarginate .... *Arubolana imula* Botosaneanu & Stock, 1979
[Aruba]

Isopoda: Cymothoida: Cymothooidea: Corallanidae: *Excorallana*: Species

1 Male and female without cephalic horns .... *Excorallana berbicensis* Boone, 1918
[Guyana]

1′ Males with three cephalic horns, females with horns reduced or absent; undescribed species, depicted in photographs by Thatcher (2006) .... *Excorallana sp.*
[Brazil]

Isopoda: Cymothoida: Anthuridae: Genera

1 Pleon in dorsal view with only last pleonite not completely fused; eyes usually present; uropod exopod adpressed to telson dorsum, exopod width subequal to endopod width .... ***Cyathura* [p. 871]**
[Cuba]

1′ Pleonites in dorsal view free; eyes absent; uropod exopod not adpressed to telson dorsum, exopod width $< 0.5\times$ endopod; stygobionts; a dozen described species, at least three with males unknown, others with incomplete descriptions .... *Stygocyathura*
[Mexico. Antilles, Curaçao, Venezuela]

Isopoda: Cymothoida: Anthuridae: *Cyathura*: Species

1 Male copulatory stylets simple; female maxilliped endite present; brackish .... *Cyathura cubana* Negoescu, 1979
[Cuba]

1′ Male copulatory stylet medial subapical margin bearing a hamulate process, lateral subapical margin bearing lobe with short spinulae; female maxilliped endite absent; fresh/brackish .... *Cyathura odaliscae* Heard & Vargas, 2015
[Pacific Costa Rica]

Isopoda: Sphaeromatidea: Sphaeromatidae: Genera and Species

1 Dorsal surface without tubercles; uropod endopod acute (occurs in hot springs in southernmost Nearctic, and may be found in northern México) .... *Thermosphaeroma*
[Nearctic]

1′ Peronites with large, transverse ridge; pleotelson and last pleonites each with four large tubercles (Fig. 23.56 D); with uropod endopod apically rounded .... *Exosphaeroides quadricosta* Kensley, 2003
[Easter Island]

REFERENCES

Schotte, M. 2002. *Speocirolana prima*, a new species from Tamaulipas, Mexico with a key to known species of the genus (Crustacea: Isopoda: Cirolanidae). Proceedings of the Biological Society of Washington 115: 628–635.

Thatcher, V.E. 2006. Aquatic Biodiversity in Latin America, Volume I: Amazon Fish Parasites (2nd edition). Pensoft Publishing, Moscow.

Wilson, G.D.F. 2008. Global diversity of isopod crustacean (Crustacea; Isopoda) in freshwater. Hydrobiologia 595: 231–240.

Decapoda

D. Christopher Rogers
Kansas Biological Survey, and The Biodiversity Institute, University of Kansas, Lawrence, KS, United States

INTRODUCTION

The freshwater and terrestrial decapods of the Neotropical region are treated. All terrestrial species, especially younger individuals, may be often found in and around freshwater. Several invasive species have been introduced through the aquarium trade, aquaculture, and ship ballast.

LIMITATIONS

In general, sexually mature individuals (most frequently males) are required for species level identifications. In almost all crayfish and crabs fully mature males are required for any meaningful identification. Most females and juvenile males cannot be identified with reasonable accuracy. Identifying caridean shrimp is very difficult, and only mature specimens, and frequently only males, can be identified, and size is not a reliable indication of maturity. Many groups and genera need to be revised and new taxa are still being discovered and described particularly from cave systems.

However, the greatest limitation is that many shrimp, crab, and crayfish are introduced to new areas, sometimes as escaped or released bait, as aquarium specimens, or from aquaculture. Furthermore, many marine species are carried into the Panama Canal by ships, both on the exterior and in ballast. Strictly marine taxa are not dealt with here, as species from all over the world frequently appear and then are never seen in the Canal again. Thus, it would be fruitless to try to provide a key to all such taxa. Only those taxa that are apparently established in the Panama Canal have been included.

TERMINOLOGY AND MORPHOLOGY

Decapods are in part defined by their name: Decapoda, or "ten legs." The first or second pair of pereopods (walking limbs) may be chelate (modified as pinchers), or modified for filter-feeding, by possessing elongated **setal fans**.

The **carapace** is the dorsal shield that covers the animal. Anteriorly, the carapace may project between the eyes as a **rostrum** in shrimp and crayfish, or as the **front** in crabs. In shrimp, the presence, number, and position of spiniform **teeth** on the rostrum are important, as are presence or absence of specific spines on the carapace surface (Figs. 23.59, 78, 93, 106, 115): the **supraorbital spine** (above and just posterior to the eye), the **suborbital spine** (just below the eye), the **hepatic spine** (on the side of the carapace, posterior to the eye, and at least the length of the eye stalk away from the eye), the **branchostegial spine** (between the eye and the ventral margin of the carapace, set back posteriorly from the anterior carapace margin), and the **pterygostomial spine** (at the anteriolateral corner of the carapace). The aforementioned carapace anteriolateral corner is the **pterygostomial angle**, and may be rounded, produced, or acute, with or without a spine.

On the dorsal surface of the crayfish carapace, behind the medial arc of the cervical groove, are paired longitudinal, parallel lines (called the **branchiocardic grooves**) that arc toward each other medially. The region between them is the **areola**, which may be obliterated if the lines come together. Generally the areola is open and contains **punctations**, which are very small, shallow, usually circular depressions. The length and width of the areola, and sometimes the number of punctations are diagnostic characters.

Decapods bear five pairs of **pereopods** (ambulatory limbs; Figs. 23.96 A–F, 100), with the first and sometimes the second pair modified as **chelae** (claws). Each pereopod is composed of several podomeres, which are (from basalmost to distalmost) **coxa**, **basis**, **ischium**, **merus**, **carpus**, **propodus,** and **dactylus**. The coxae and ischia may bear diagnostic structures in the form of **bosses** or **hooks**. A **chela** claw is composed of the propodus and the dactylus. The propodus is comprised of the broad **palm** and a distal extension called the **pollex** (or fixed "finger"). The dactylus is the "movable finger" which moves against the pollex. The opposing margins of the dactylus and pollex may have defining characters in terms of gape or tubercle arrangement.

The chelae are important in species diagnoses; however, decapods commonly lose and regenerate them. Chelae characteristics used in this key apply only to non-regenerated chelae. Regenerated chelae differ from non-regenerated chelae in that they possess tubercles of the same size along the opposable margins of the dactyl and pollex, as well as by having a much shorter palm region.

Atyid shrimps are characterized by modified chelae of the first two pairs of pereopods. These bear thick bands of setae along the dactyl and propodus distal margins (Figs. 23.96 E, F, 100), thereby composing a filtering or scraping apparatus when the chela is open. The carpus in both pairs of pereopods is modified to receive the propodus.

The limb may have an elongate structure from the top of the ischum called an **exopod** (Fig. 23.96 D). A gill may grow from the coxa called a **podobranch**. If the gill grows from the body at the point of the limb attachment, then it is called an **arthrobranch**. If the gill grows from the body above the point of limb attachment it is called a **pleurobranch**.

The abdomen is posterior to the carapace and may be flat folded beneath the body (crabs) or projecting posteriorly (shrimp, crayfish). The **pleopods** depend from the ventral surface of the abdomen. The abdomen terminates in a **telson**, which has a **uropod** on either side. The pleopods and the uropods are generally biramal, with a basal **peduncle**, which bears two branches: the lateral **exopod** and the medial **endopod**. Male shrimp frequently have the endopod of the first and second pleopod further divided into a lateral lamella, an **appendix masculina** and a lateral **appendix interna**.

Dendrobranchiate shrimp are primarily identified by the form of the male sexual structures. The male first pleopods have the endopods greatly enlarged and coupled into a structure called a **petasma**. In juveniles and females the first pleopod is unmodified and resembles the remaining pairs. However, in juvenile males the endopod is shifted more distally than in the remaining pleopods. In adult females the posterior sternites are modified to form a **thelycum** for reproduction.

At the end of the abdomen is the terminal telson, flanked by two limb structures called **uropods**. The uropod has an oblique suture across the distal third or so of the exopod, called a **diaresis**.

The primary identification characters for many species are on the male **gonopods**; these are the first one or two pleopod pairs which have been modified for sperm transfer. As the external morphology of freshwater crabs is relatively conservative, the keys place greater emphasis on the taxonomically important gonopod I (first pleopod) and gonopod II (second pleopod) of mature male individuals.

MATERIAL PREPARATION AND PRESERVATION

Decapods are best preserved initially in 80%−90% ethyl alcohol, replaced after 24−48 h with 70% ethanol. This will probably need to be changed again after 6 months. The gonopod is typically dissected out and examined separately. Generally, the gonopod may be temporarily mounted on an insect pin for easy manipulation under the microscope. The gonopod is then generally kept in a smaller vial within the jar holding the larger specimen.

KEY TO DECAPODA

Decapoda: Infraorders

1	Abdomen projecting posteriorly (Figs. 23.57, 94); shrimp and crayfish	2
1′	Abdomen folded beneath body, flattened, pressed to thorax (Fig. 23.104); crabs	4
2(1)	Abdomen compressed laterally (Fig. 23.94); shrimp	3
2′	Abdomen compressed dorsoventrally (Fig. 23.57); crayfish	**Astacidea [p. 876]**
3(2)	Abdominal pleuron II not overlapping pleura I and III (Fig. 23.94 B)	**Dendrobranchiata [p. 902]**
3′	Abdominal pleuron II overlapping pleura I and III (Fig. 23.94 A, C, D)	**Caridea [p. 905]**
4(1)	Pereopod V greatly reduced, much smaller than preceeding pairs (Fig. 23.104).	Anomura, Aeglidae, one genus; ***Aegla* [p. 922]**
4′	Pereopod V serially homologous with preceeding pairs	**Brachyura [p. 934]**

Decapoda: Astacidea

Felipe Bezerra Ribeiro
Programa de Pós-Graduação em Biologia Animal, Departamento de Zoologia (Laboratório de Carcinologia), Instituto de Biociências, Universidade Federal do Rio Grande do Sul, Porto Alegre, Rio Grande do Sul, Brasil

Erich Rudolph Latorre
Departamento de Ciencias Biológicas y Biodiversidad, Universidad de Los Lagos, Osorno, Chile

Carlos Pedraza-Lara
Licenciatura en Ciencia Forense, Facultad de Medicina, Universidad Nacional Autónoma de México, Ciudad Universitaria, Ciudad de México, México

INTRODUCTION

The freshwater crayfish comprise about 668 species distributed worldwide, except in Antarctica and continental Africa (Toon et al., 2010; Crandall & De Grave, 2017). They are found in streams, lakes, burrows, caves, and swamp forests (Buckup & Rossi, 1980; Ribeiro et al., 2016, 2017; Huber et al., 2018). They are nocturnal, being mainly omnivorous and playing a key role in freshwater trophic webs (Reynolds et al., 2013). Most crayfish species are gonochoric and show sexual dimorphism. However, some species may have different modes of reproduction, such as hermaphroditism, intersexuality, and parthenogenesis (Yazicioglu et al., 2016). They have direct development with yolky-rich eggs and eclosion of juveniles. Females demonstrate posthatching brood care (Vogt, 2013). They are used as a human food source and can be a model organism for a variety of scientific studies.

Taxonomically, freshwater crayfishes are grouped in two superfamilies: Astacoidea and Parastacoidea (Crandall & De Grave, 2017). The former is composed of three families, Astacidae with 4 genera and 20 species; Cambaridae with 14 genera and about 444 species; and Cambaroididae with 1 genus with 6 species. The latter is composed of a single family, Parastacidae, which encompass 15 genera and 198 species (Crandall & De Grave, 2017; Huber et al., 2018; Miranda et al., 2018).

In the Neotropics the cambarids comprise 53 species belonging to the genera *Procambarus* and *Cambarellus* (Álvarez & Villalobos, 2015). Parastacids are represented by 19 species in *Samastacus* (1), *Parastacus* (14), and *Virilastacus* (4). They are distributed in southern Brazil (states of Rio Grande do Sul and Santa Catarina), Uruguay, Argentina, and Chile (Buckup & Rossi, 1980, 1993; Rudolph, 2010, 2015; Ribeiro & Araujo, 2017).

LIMITATIONS

Older identifications and descriptions of Neotropical freshwater crayfish were based on few morphological characters (Poeppig, 1835; von Martens, 1869; Philippi, 1882; Faxon, 1898, 1914; Buckup & Rossi, 1980). These characters included only general body shape, carapace, chelipeds, pleon, and tailfan. In addition, the absence of keys with good characters led to several misidentifications. Furthermore, the presence of supernumerary gonopores (intersexuality) in all *Parastacus* makes sex identification difficult (Rudolph & Almeida, 2000).

Recent contributions use integrative taxonomy and molecular tools with new characters and drawings (Ribeiro et al., 2017; Miranda et al., 2018). There is evidence of cryptic species and the inclusion of genetic data is very important (Miranda et al., 2018). However, *Parastacus* needs an extensive revision (F.B. Ribeiro, personal communication).

Other difficulties may arise in *Samastacus spinifrons* (Philippi, 1882), in which the occurrence of two morphotypes is documented. Specimens collected in streams have a robust body with relatively short and thick chelipeds, short rostrum, small eyes, and inconspicuous carapace ornamentation. On the other hand, specimens from lakes have a slender body, with elongated rostrum, large eyes, and the carapace ornamentation has conspicuous spines and tubercles (Rudolph et al., 2016).

Few reference collections for freshwater crayfish occur in South America. The most representative collection for the group is located in the Zoology Department of the Universidade Federal do Rio Grande do Sul, municipality of Porto Alegre, state of Rio Grande do Sul, southern Brazil.

TERMINOLOGY AND MORPHOLOGY

The crayfish body is divided in two regions: a **cephalothorax** (CT), covered by a **carapace**, and a segmented **pleon** (PL) (Fig. 23.57). The cephalothorax is divided in two regions (cephalic and thoracic) by the **cervical groove** (CG). The cephalic region is composed of six segments (*sensu* Brusca et al., 2016) the protocerebral segment being the first one, which bears the **pedunculate eyes** (Ey). The five subsequent segments bears five pairs of appendages: **first antennae** (At) and the **second antennae** (An) (preoral appendages); **mandibles** (Ma), **first maxillae** (Ml), and the **second maxillae** (Mx) (postoral appendages). In ventral view, the cephalic region is formed by a shieldlike **epistome** (Ep). The thoracic region is composed by eight segments, each with one pair of appendages. The first three pairs correspond to the **maxillipeds** (Mp1 to Mp3) and the posterior five pairs to the **pereopods** (P1 to P5). The first three pairs of

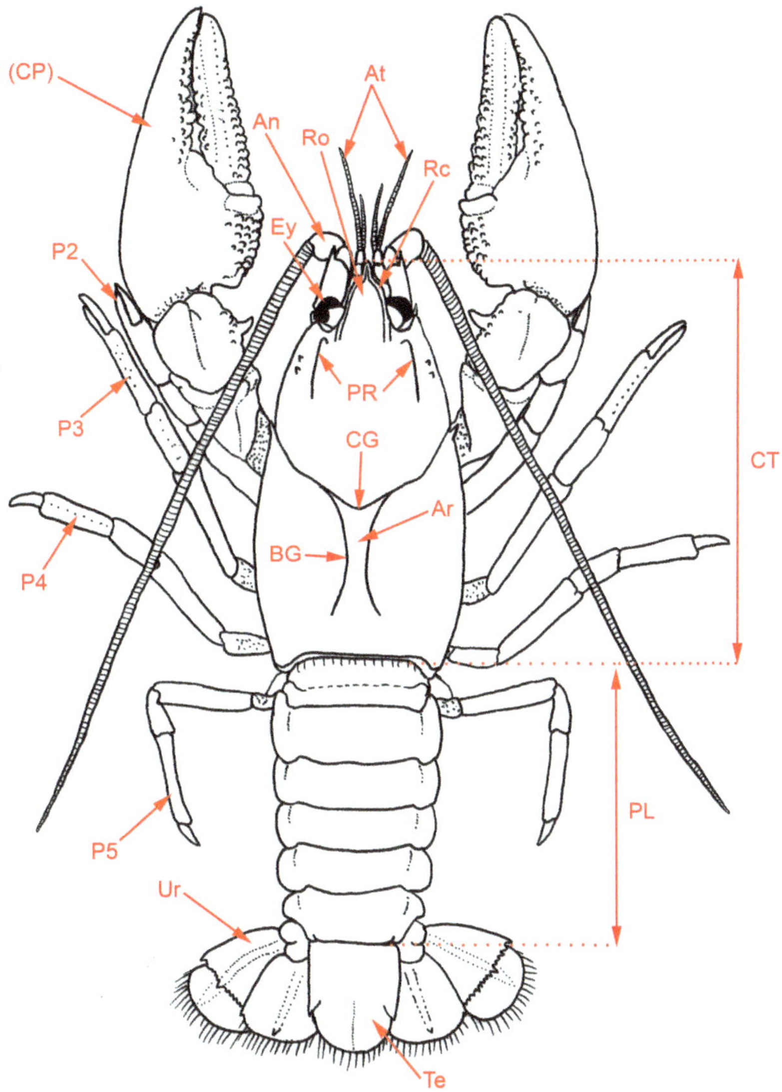

FIGURE 23.57 Crayfish general morphology. Abbreviation: *An*, antenna; *Ar*, areola; *At*, antennula; *BG*, branchiocardiac groove; *CG*, cervical groove; *CP*, cheliped; *CT*, cephalothorax; *Ey*, eye; *PL*, pleon; *P1–P5*, pereiopods 1–5; *POC*, postorbital carina; *Rc*, rostral carina; *Ro*, rostrum; *Te*, telson; *Ur*, uropod. *Illustration made from Loughman et al. (2017).*

pereopods are chelate, the first pair (P1) being the largest and called **chelipeds** (Cp). The anterior margin of the carapace has an anterior projection of variable shape and size, the **rostrum** (Ro). The rostrum margins bear the **rostral carinae** (Rc). Just below the rostrum, there are prominences in the carapace in both sides, the **post orbital carina** (POC). Beyond the cervical groove, two **branchiocardiac grooves** (BG) delimit a narrow portion of the carapace center, the **areola** (Ar). Laterally, the carapace extends toward the base of the appendages, forming the branchial chamber, which protects the set of branchiae (gills) under the carapace. These carapace lateral regions may be ornamented with spines or tubercles of variable size.

The cephalothorax ventral region between the appendages has a central lamina, the thoracic sternum or set of thoracic sternites. The gonopores of females open in the coxae of P3 and those of the males in the coxae of P5.

The **pleon** (PL) is composed of six somites plus the **telson** (Te). Each somite is enclosed by sclerites: a wide dorsal plate (**tergum**) and a narrow ventral plate (**sternum**) both connected on either side by a lateral plate (**pleuron**), whose ventral extensions in somites two to six form the **pleura** (Ple). Each somite bears a pair of **pleopods** (Pd), except in Parastacoidea, in which the first pair of pleopods is lacking. In Astacoidea, the first pair of pleopods is vestigial in females, and pleopods 2–5 are used for hatching eggs. In males, the first two pairs of pleopods are modified for sperm transfer (**gonopods**). Crayfish in the Cambaridae can only be properly identified by the form of the male gonopod. Furthermore, male cambarids have two sexual phases: **form I** and **form II**. Sexually active males are in the form I stage, where the gonopod has all necessary characters. However, form II males are like juveniles, in that the gonopodal characters are reduced and greatly obscured. Form II males are usually found in winter seasons.

The last pair of pleonal appendages are flattened and biramous, called **uropods** (Ur). They form the tail fan together with the telson, which is used as a propellent organ in rapid escape responses (tail-flipping).

Sexual dimorphism is usually observed in mature individuals. Females generally have a wider pleon and males have larger chelipeds. Additionally, cambarid males can bear small hooks on the coxa and ischium of some pereopods for use during copulation; females bear a spermatheca, the "annulus ventralis," between the coxa of P4 and P5. For more details regarding the external morphology and diagnostic characters of freshwater crayfishes see Cumberlidge et al. (2015) and Rogers (2016), respectively.

MATERIAL PREPARATION AND PRESERVATION

Freshwater crayfish are best preserved long term in 70% ethanol. For molecular studies, it is recommended to remove an appendage (pereopod or pleopod) and preserve it in 95% ethanol. They can be previously fixed in an 8% buffered formalin solution for a relatively short period of time. Specimens may also be encased in epoxy resin.

To preserve crayfish symbionts (e.g., protozoans, temnocephalids, branchiobdelids), specimens should be fixed in AFA (70% ethanol, 37% formalin, glacial acetic acid) for 24 h and then deposited in 70% ethanol. Finally, for additional information about the sampling and processing methods for crayfishes, see Crandall (2016).

KEY TO ASTACIDEA

Decapoda: Astacidea: Families

The keys for North American Astacidea are largely based on Hobbs (1972a, 1972b). Records follow Crandall & De Grave (2017) and Álvarez & Villalobos (2015). Most illustrations were redrawn from Hobbs (1972a, 1972b), Álvarez & Villalobos (2015), López-Mejía et al. (2004; 2005) and Pedraza-Lara & Doadrio (2015). Keys for South American Astacidea are based on Buckup & Rossi (1980), Hobbs (1991), Rudolph & Crandall (2005; 2007; 2012), Ribeiro et al. (2016; 2017), Ribeiro & Araujo (2017), Huber et al. (2018) and Miranda et al. (2018).

1 Sternum narrow, carinate; male pleopods I absent and II unmodified; female lacking pleopod I and annulus ventralis .... **Parastacidae [p. 876]**

1′ Sternum triangular, broadest between posterior pereopod pairs; mature males with pleopods I and II modified; pleopod I modified as a gonopod for sperm transfer, and some ischia bearing hooks; females with annulus ventralis .... **Cambaridae [p. 884]**

Decapoda: Astacidea: Parastacidae: Genera

1 Cheliped dactyl moving subhorizontally or obliquely, male genitalia articulated, slender, tubular phallic papilla, or complex uncalcified papilla (Fig. 23.58 A, B, D, E) 2

1′ Dactylus of chelipeds moving subvertically, male genitalia, a fixed and slightly elevated ventromesial ridge bearing a small noncalcified papilla (Fig. 23.58 C, F) .... ***Parastacus* [p. 876]**

2(1) Postorbital carinae well developed (Fig. 23.59 B) 3

2′ Postorbital carinae obsolete (Fig. 23.59 A) .... ***Virilastacus* [p. 884]**

3(2) Chelipeds with ventrolateral margin calcified .... *Samastacus*; one species, *Samastacus spinifrons* (Philippi, 1882) [Chile: Central-southern. Argentina: Southwest Neuquén, northeast Río Negro]

3′ Chelipeds with ventrolateral margin with noncalcified area .... ***Cherax* [p. 884]**

Decapoda: Astacidea: Parastacidae: *Parastacus*: Species

1 Telson with lateral spines 2

1′ Telson without lateral spines, maxilliped III exopod flagellum surpassing ischium distal margin (Fig. 23.60) .... *Parastacus nicoleti* (Philippi, 1882)
[Chile: Cautín, Valdivia, Osorno and Llanquihue provinces, Coastal Cordillera]

2(1) Postorbital carina proximal edge ending in a strong spine; uropod protopod proximal lobe bearing a spine; areola delimited by prominent carinae (Fig. 23.61 A, C) 3

2′ Postorbital carina proximal edge not ending in spine; uropod protopod proximal lobe lacking spine; areola not delimited by carinae (Fig. 23.61 B, D) 4

3(2) Cheliped merus, carpus, and propodus medially with tufts of long setae; maxilliped III ischium ventral margin with longitudinal spine row (Fig. 23.62 A, C) .... *Parastacus varicosus* Faxon, 1898
[Brazil: Rio Grande do Sul, Santa Catarina. Argentina: Entre Ríos. Uruguay: Maldonado, Montevideo, Rocha]

3′ Cheliped merus, carpus, and propodus of medially glabrous; maxilliped ischium ventral margin with longitudinal tubercle row (Fig. 23.62 B, D). .... *Parastacus saffordi* Faxon, 1898
[Brazil: Rio Grande do Sul, Santa Catarina. Uruguay: Montevideo, Maldonado]

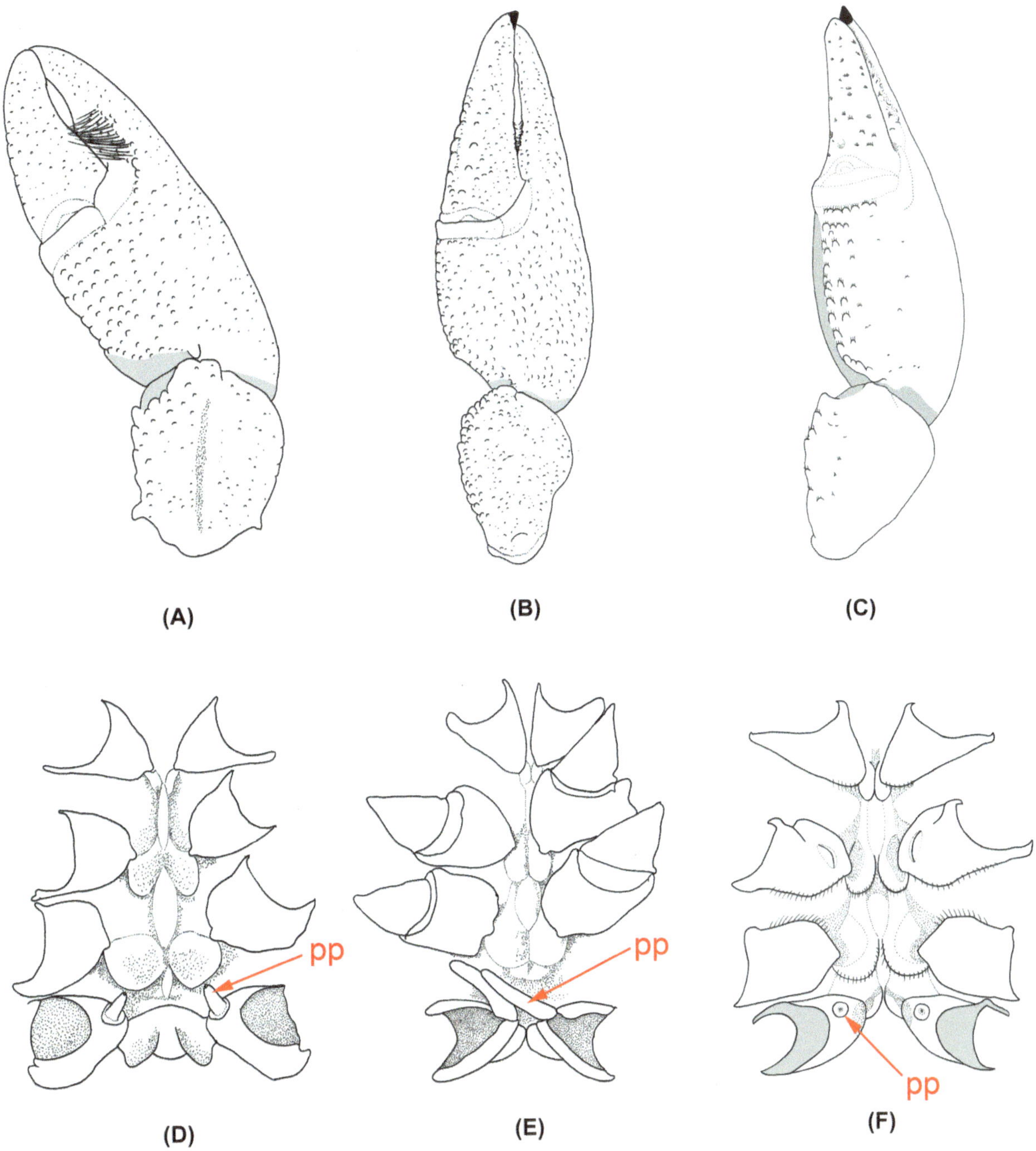

FIGURE 23.58 Astacidea: Cheliped and sternites in *Samastacus* (A, D), *Virilastacus* (B, E) and *Parastacus* (C, F). Abbreviation: *pp*, phallic papilla. *(A), (D) and (E) modified from Hobbs (1991); (B) modified from Rudolph & Crandall (2012); (F) modified from Ribeiro et al. (2017).*

4(2) Telson distal margin rounded (Fig. 23.63 A) 5

4′ Telson distal margin acute (Fig. 23.63 B) .... *Parastacus buckupi* Huber, Ribeiro & Araujo, 2018
[Brazil: Rio Grande do Sul]

5(4) Cheliped carpus distally glabrous; epistome anteromedian lobe hexagonal or pentagonal 6

5′ Cheliped carpus distally with tufts of long and simple setae; epistome anteromedian lobe heptagonal (Fig. 23.64) .... *Parastacus pilicarpus* Huber, Ribeiro & Araujo, 2018
[Brazil: Santa Catarina]

6(5) Cheliped dactyl and pollex cutting edges obscured by long setal tufts (Fig. 23.65 A) 7

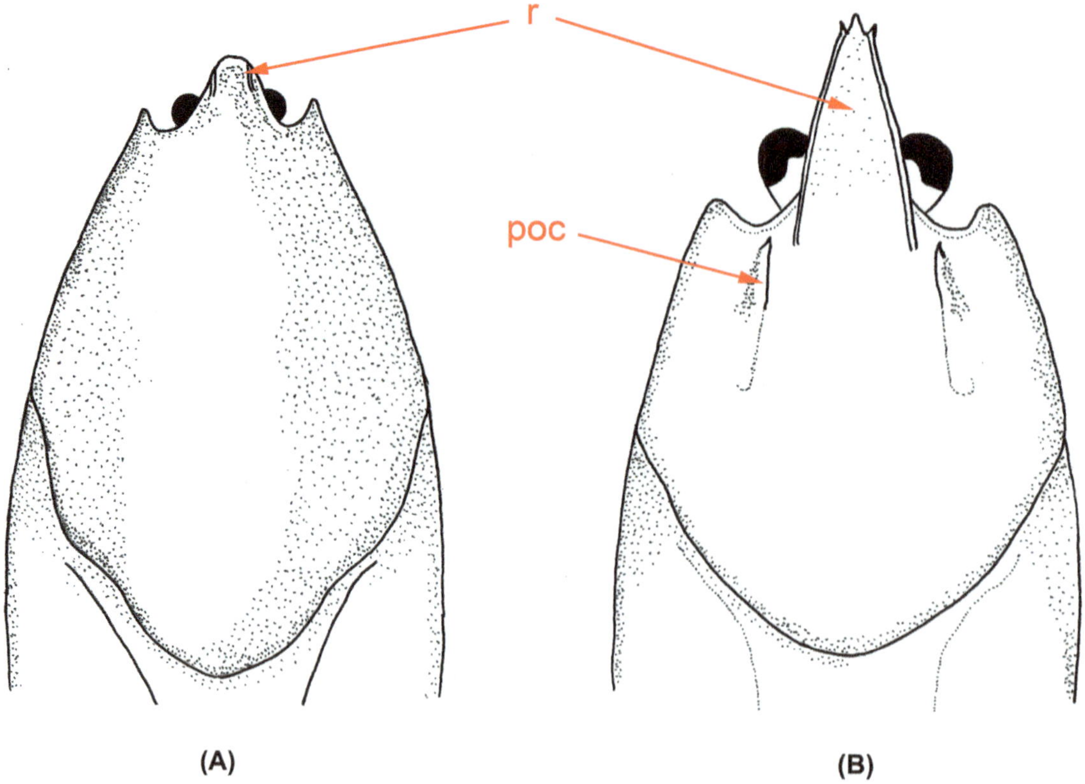

FIGURE 23.59 Astacidea: Dorsal view of cephalothorax in *Virilastacus* (A) and *Samastacus* (B). Abbreviations: *poc*, post orbital carinae; *r*, rostrum. *(A) Modified from Rudolph & Crandall (2012).*

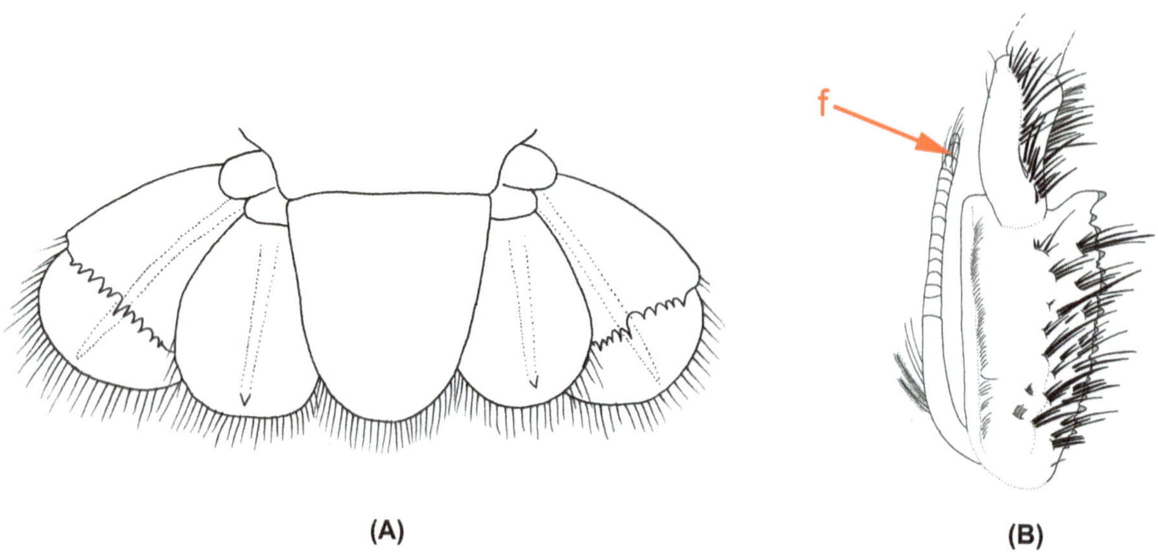

FIGURE 23.60 Astacidea: Dorsal view of tail-fan and ventral view of third maxilliped in *Parastacus nicoleti*. Abbreviation: *f*, flagellum. *(B) Modified from Ribeiro & Araujo (2017).*

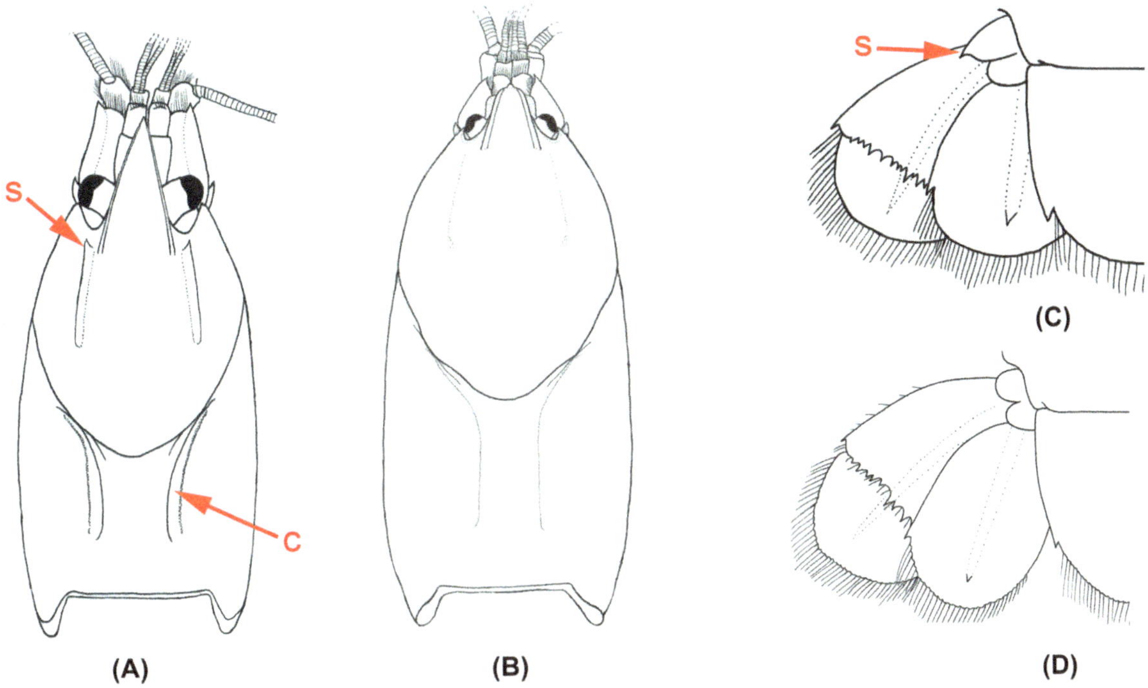

FIGURE 23.61 Astacidea: Dorsal view of cephalothorax and tailfan in *Parastacus* spp. Abbreviations: *c*, carina; *s*, spine. *(B) Modified from Huber et al. (2018).*

6′ Cheliped dactyl and pollex cutting edges clearly visible (Fig. 23.65 B) 8

7(6) Postorbital carinae prominent (Fig. 23.66 A) .... *Parastacus pilimanus* (von Martens, 1869) [Brazil: Rio Grande do Sul. Argentina: Catamarca, Corrientes, Entre Ríos, Santa Fé. Uruguay: Federación, Flores Rivera, Rocha, Cerro Largo]

7′ Postorbital carinae obsolete (Fig. 23.66 B) .... *Parastacus laevigatus* Buckup & Rossi, 1980 [Brazil: Santa Catarina]

8(6) Cervical groove V-shaped (Fig. 23.67 A) 9

8′ Cervical groove U-shaped (Fig. 23.67 B) .... *Parastacus caeruleodactylus* Ribeiro and Araujo in Ribeiro et al., 2016 [Brazil: Rio Grande do Sul]

9(8) Chelipeds with globose palm 10

9′ Chelipeds with flattened palm 11

10(8) Rostral carinae reaching postorbital carinae proximal edge (Fig. 23.68 A) .... *Parastacus defossus* Faxon, 1898 [Brazil: Rio Grande do Sul. Uruguay: Montevideo]

10′ Rostral carinae surpassing postorbital carinae proximal edge (Fig. 23.68 B) .... *Parastacus pugnax* (Poeppig, 1835) [Chile: from the Aconcagua River to the Toltén River]

11(9) Telson subrectangular (Fig. 23.69 A) 12

11′ Telson subtriangular (Fig. 23.69 B) .... *Parastacus fluviatilis* Ribeiro and Buckup in Ribeiro et al., 2016 [Brazil: Rio Grande do Sul]

12(11) Cheliped palm dorsal surface with tubercles irregularly distributed 13

12′ Cheliped palm dorsal surface with three well-defined tubercle rows (Fig. 23.70) .... *Parastacus tuerkayi* Ribeiro, Huber and Araujo in Ribeiro et al., 2017 [Brazil: Santa Catarina]

13(12) Cheliped palm medial margin with medial area tuberculate (Fig. 23.71 A) .... *Parastacus promatensis* Fontoura & Conter, 2008 [Brazil: Rio Grande do Sul]

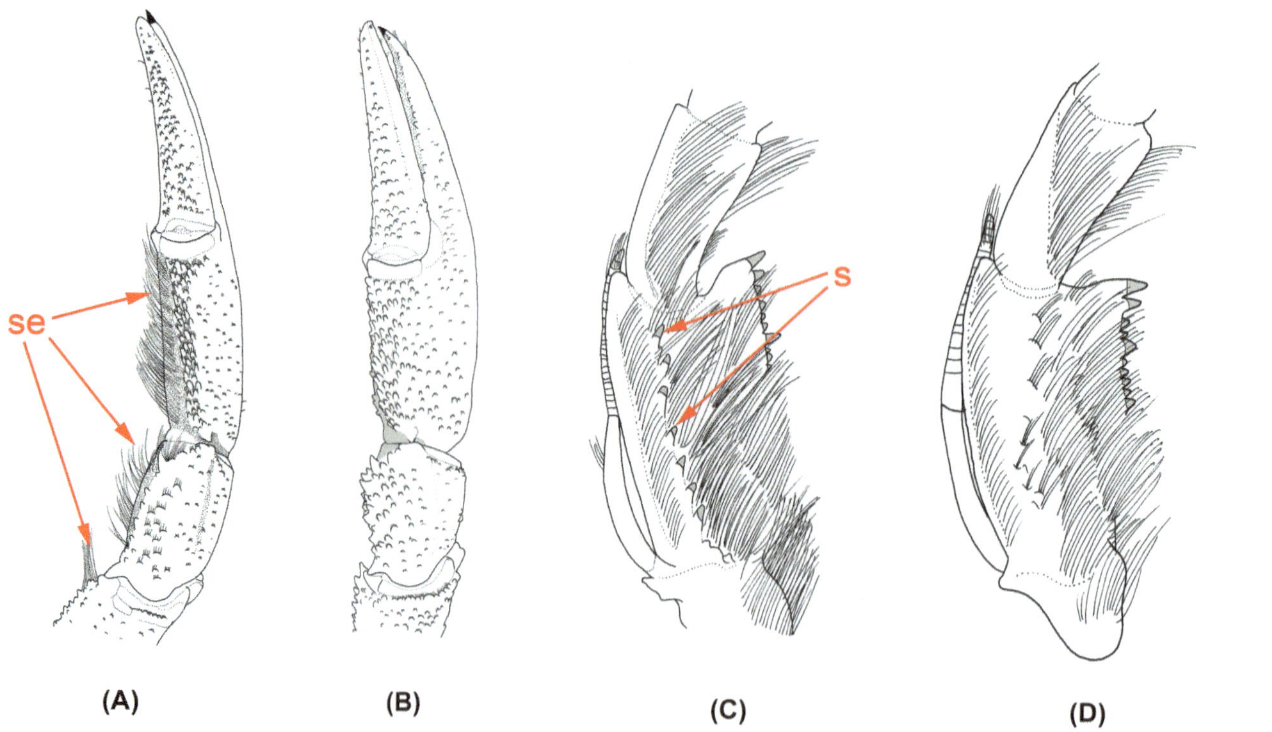

FIGURE 23.62 Astacidea: Dorsal view of chelipeds and ventral view of third maxilliped in *Parastacus varicosus* (A, C) and *P. saffordi* (B, D). Abbreviations: *s*, spine; *se*, setae.

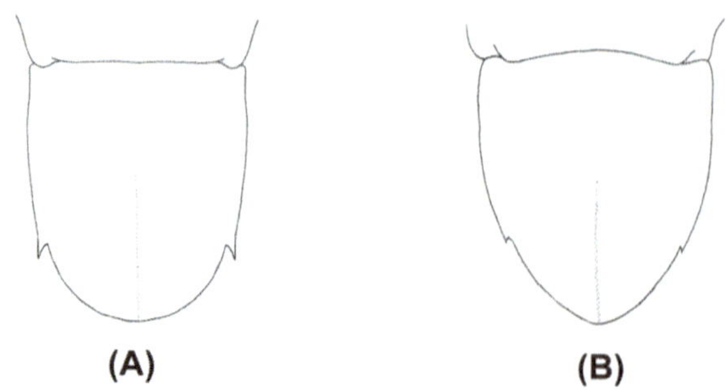

FIGURE 23.63 Astacidea: Dorsal view of telson in *Parastacus* spp. (A) and *Parastacus buckupi* (B). *Modified from Huber et al. (2018).*

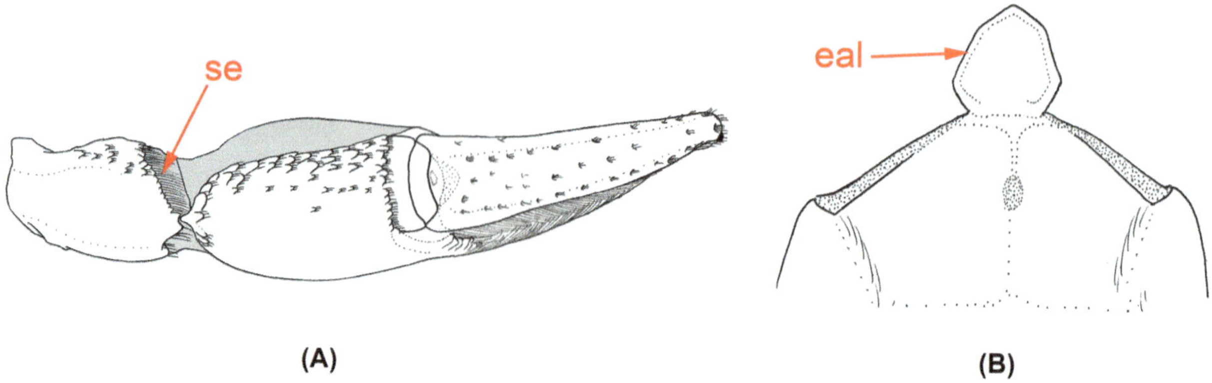

FIGURE 23.64 Astacidea: Cheliped (A) and epistome (B) in *Parastacus pilicarpus*. Abbreviations: *eal*, epistome anterior lobe; *se*, setae. *Modified from Huber et al. (2018).*

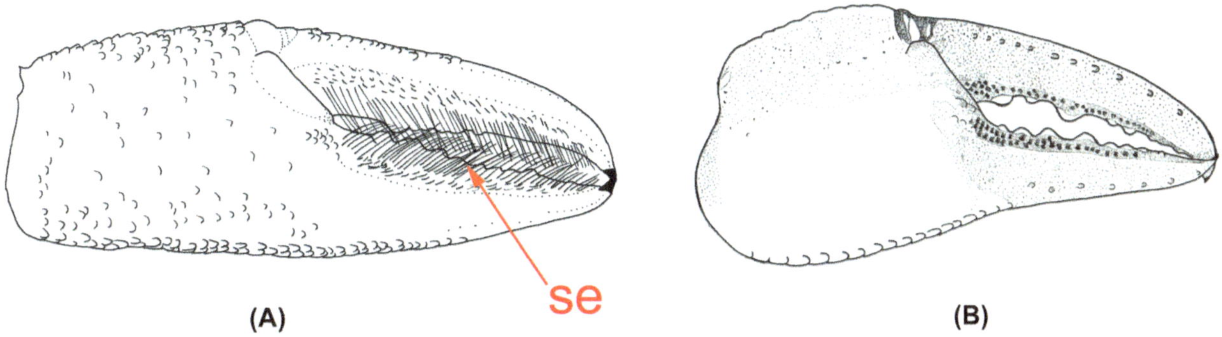

FIGURE 23.65 Astacidea: Cheliped propodus and dactylus in *Parastacus* spp. Abbreviations: *se*, setae. *(B) modified from Ribeiro et al. (2016).*

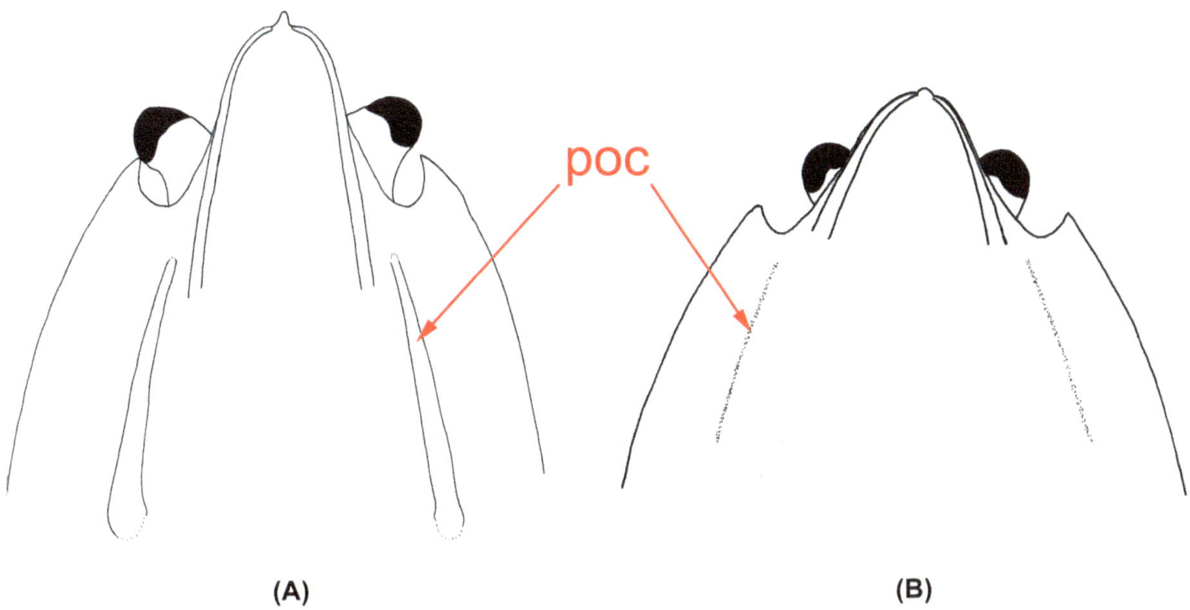

FIGURE 23.66 Astacidea: Cephalothorax anterior portion in *Parastacus pilimanus* (A) and *P. laevigatus* (B). Abbreviation: *poc*, postorbital carina.

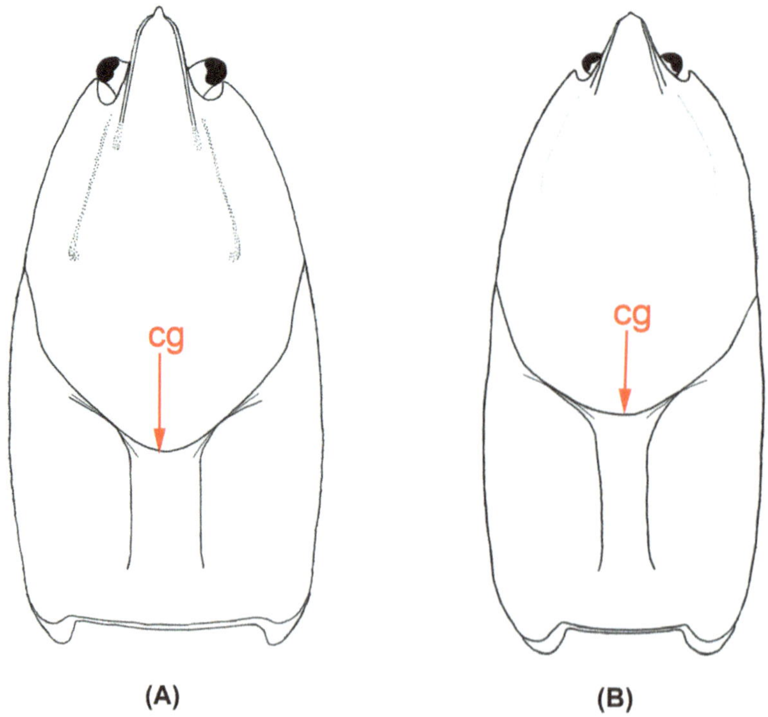

FIGURE 23.67 Astacidea: Cephalothorax in *Parastacus* spp. (A) and *Parastacus caeruleodactylus* (B). Abbreviation: *cg*, cervical groove. *(B) Modified from Ribeiro et al. (2016).*

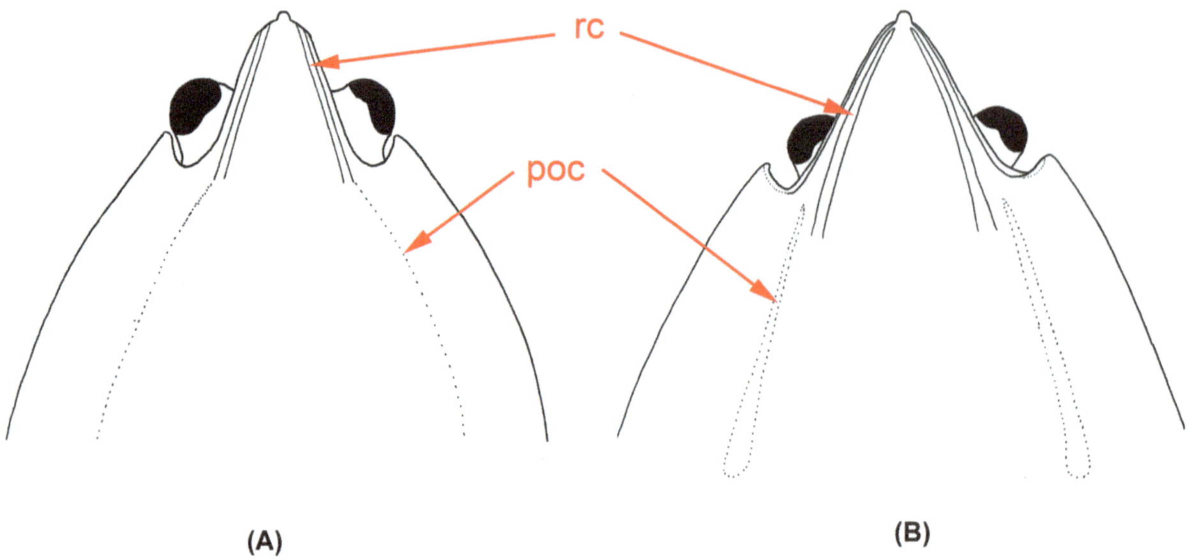

FIGURE 23.68 Astacidea: Dorsal view of chelipeds and ventral view of third maxilliped in *Parastacus defossus* (A) and *P. pugnax* (B). Abbreviations: *poc*, postorbital carina; *rc*, rostral carina.

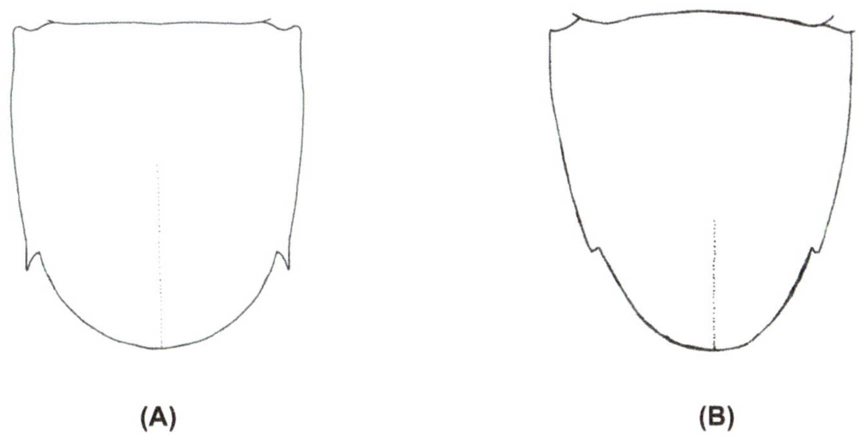

FIGURE 23.69 Astacidea: Dorsal view of telson in *Parastacus* spp. (A) and *Parastacus fluviatilis* (B).

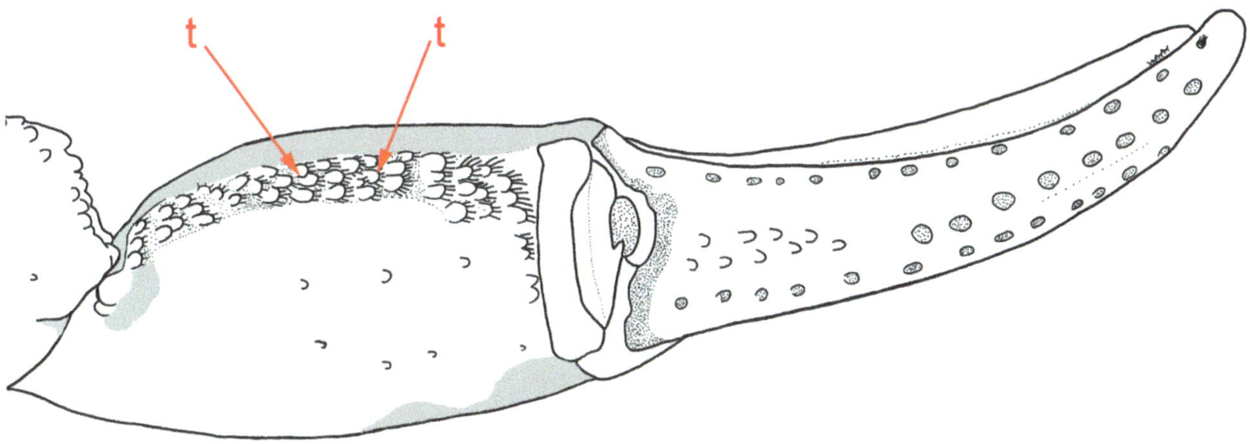

FIGURE 23.70 Astacidea: Cheliped propodus and dactylus in *Parastacus tuerkayi*. Abbreviation: *t*, tubercle. *Modified from Ribeiro et al. (2017).*

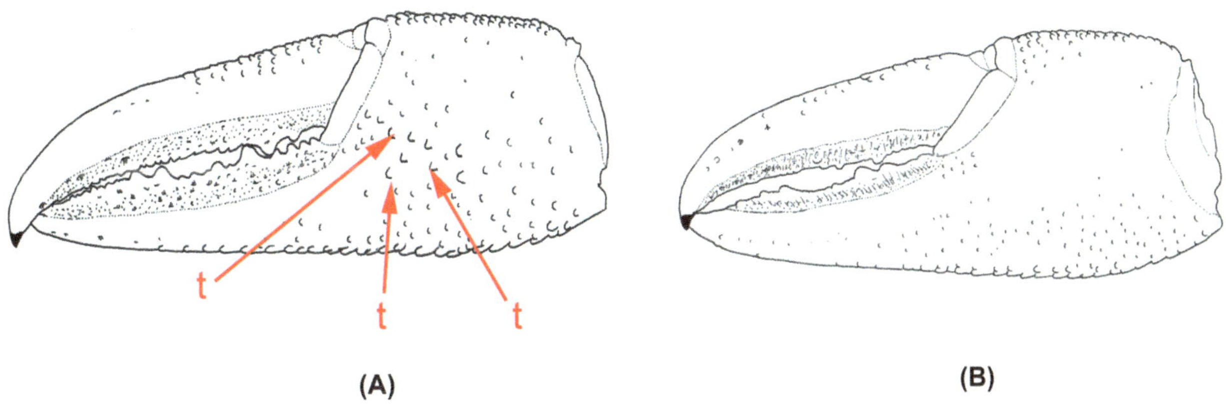

FIGURE 23.71 Astacidea: Inner margin of cheliped propodus and dactylus in *Parastacus promatensis* (A) and *P. brasiliensis* (B). Abbreviation: *t*, tubercles.

13′ Cheliped palm medial margin smooth (Fig. 23.71 B) .... *Parastacus brasiliensis* (von Martens, 1869)
[Brazil: Rio Grande do Sul]

Decapoda: Astacidea: Parastacidae: *Virilastacus*: Species

1 Cheliped propodus opposable margin with $>$ 10 teeth; pereopod IV coxae close set; male or female gonopores present; pereopod V coxae without male cuticle partition 2

1′ Cheliped propodus opposable margin with $<$ 10 teeth; pereopod IV coxae widely separated; supernumerary gonopores; pereopod V coxae with male cuticle partition (Fig. 23.72) .... *Virilastacus rucapihuelensis* Rudolph & Crandall, 2005
[Chile: Osorno province, Cordillera de la Costa]

2(1) Rostrum carinae not surpassing rostrum middle portion; cheliped propodus with pilous opposable margin 3

2′ Rostrum carinae reaching rostral basis; cheliped propodus with opposable margin glabrous (Fig. 23.73) .... *Virilastacus araucanius* (Faxon, 1914)
[Chile: coastal strip from Concepción to Maicolpué]

3(2) Cheliped dactylus moving in subhorizontal plane, propodus opposable margin dorsally and ventrally pilose (Fig. 23.74) .... *Virilastacus retamali* Rudolph & Crandall, 2007
[Chile: Osorno and Llanquihue provinces, Cordillera de la Costa]

3′ Cheliped dactylus moving in an oblique plane, propodus opposable margin dorsally and ventroproximally pilose; cervical groove V-shaped (Fig. 23.75) .... *Virilastacus jarai* Rudolph & Crandall, 2012
[Chile: Los Ángeles]

Decapoda: Astacidea: Parastacidae: *Cherax*: Species

1 Carapace with four obvious dorsal ridges, two extending posteriorly from the rostrum posterior corners, two extending posteriorly from the orbital spines; areola (area between branchiocardic grooves) length $\leq 3\times$ width .... *Cherax quadricarinatus* (von Martens, 1868)
[Australia. Invasive in Mexico: Morelos, San Luis Potosí, Tamaulipas. Common in aquaculture]

1′ Carapace with two obvious dorsal ridges, extending posteriorly from the rostrum posterior corners; areola length $> 4\times$ width .... *Cherax destructor* Clark, 1936
[Australia. Invasive in Neotropics. Common in aquaculture]

Decapoda: Astacidea: Cambaridae: Genera

Only form I males can be identified with this key.

1 Form I male pereopods II and III ischia each bearing a prominent medial projection (hook); Cambarellinae (Fig. 23.76 A) .... ***Cambarellus* [p. 884]**

1′ Form I male pereopods II ischia never bearing a prominent medial projection (hook), although other pereopods may; Cambarinae (Fig. 23.76 B–D) .... ***Procambarus* [p. 886]**

Decapoda: Astacidea: Cambaridae: *Cambarellus*: Species

Key modified from Hobbs (1972a), Villalobos-Figueroa & Hobbs (1981), Hobbs & Lodge (2010) and Pedraza-Lara & Doadrio (2015).

1 First pleopod mesial projection with a longitudinal groove (Fig. 23.77 A–F) 2

1′ First pleopod mesial projection subsetiform, without a longitudinal groove (Fig. 23.77 G) .... *Cambarellus alvarezi* Villalobos, 1952
[Mexico: El Potosí, Nuevo León, Mexico, reported as extinct in the wild]

2(1) Rostral acumen length $\geq$ distance between tips of marginal spines (Fig. 23.78 A–C) 3

2′ Rostral acumen length $<$ distance between tips of marginal spines 5

3(1) Rostral acumen length $\geq 2\times$ basal width (Fig. 23.78 B, C) 4

3′ Rostral acumen $\leq 1.2\times$ distance between tips of marginal spines (Fig. 23.78 A) .... *Cambarellus patzcuarensis* Villalobos, 1943
[Mexico: Michoacán]

4(3) Rostral acumen length $2\times$ basal width (Fig. 23.78 B). .... *Cambarellus chapalanus* (Faxon, 1898)
[Mexico: Jalisco, Michoacán]

4′ Rostral acumen $\geq 3\times$ basal width (Fig. 23.78 C). .... *Cambarellus prolixus* Villalobos & Hobbs, 1981
[Mexico: Jalisco, Michoacán]

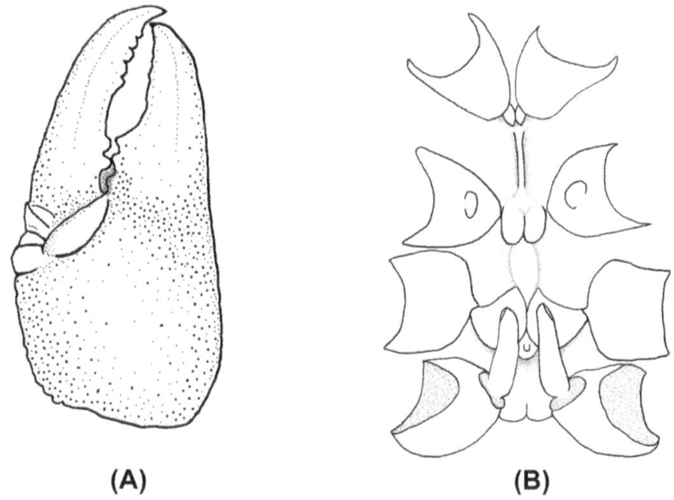

FIGURE 23.72 Astacidea: *Virilastacus rucapihuelensis*: (A) Cheliped propodus and dactylus; (B) sternites. *(A) modified from Rudolph & Crandall (2005).*

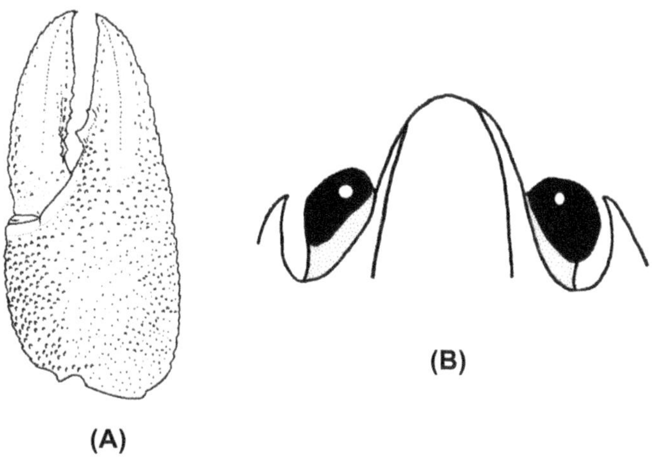

FIGURE 23.73 Astacidea: *Virilastacus araucanius*: (A) Chelipeds; (B) dorsal view of rostrum propodus and dactylus. *(B) modified from Hobbs (1989).*

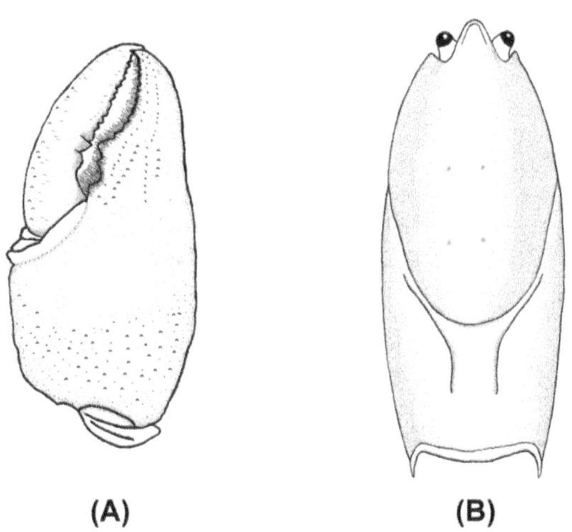

FIGURE 23.74 Astacidea: *Virilastacus retamali*: (A) Cheliped propodus and dactylus; (B) dorsal view of cephalotorax. *Modified from Rudolph & Crandall (2007).*

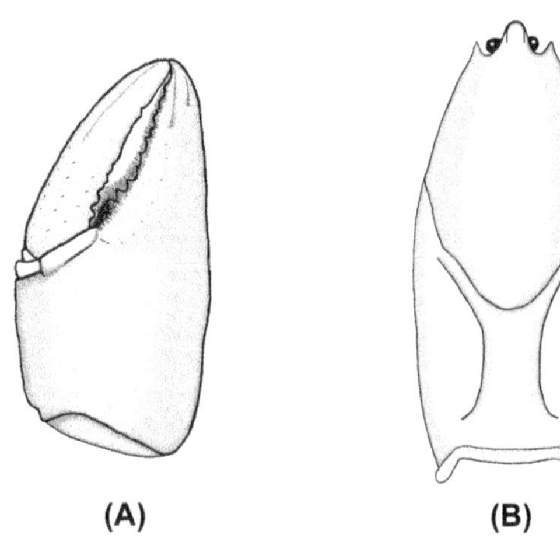

FIGURE 23.75 Astacidea: *Virilastacus jarai*: (A), Cheliped propodus and dactylus; (B), Dorsal view of cephalotorax. *(A) and (B) modified from Rudolph & Crandall (2012).*

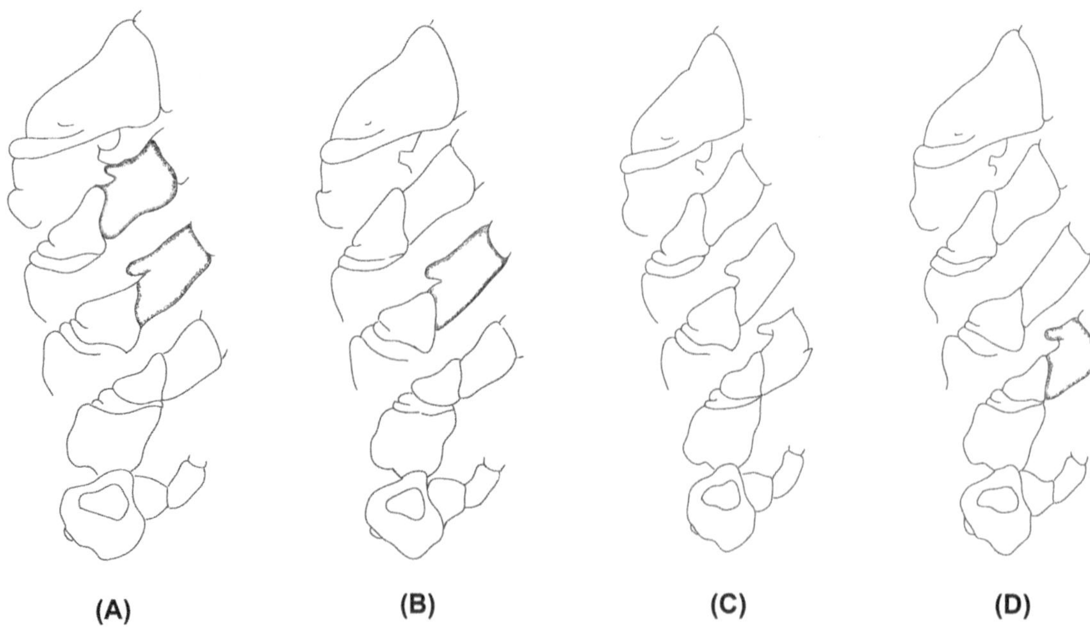

FIGURE 23.76 Astacidea: Ventral view of basal portions of left pereiopods of Cambaridae with ischia bearing hooks shaded.

5(2) Male pleopod I mesial process length subequal to other terminal elements or not extending beyond them posteriorly (Fig. 23.77 A–F) 6

5′ Male pleopod I mesial process length > other terminal elements (Fig. 23.77 H). .... *Cambarellus zacapuensis* Pedraza-Lara & Doadrio, 2015
[Mexico: Michoacán]

6(5) Chela dactyl and pollex directed distally (Fig. 23.79 A) 7

6′ Chela dactyl and pollex gently curved medially; chela lateral margin strongly convex (Fig. 23.79 B) .... *Cambarellus zempoalensis* Villalobos, 1943
[Mexico: Estado de Mexico, Hidalgo, Michoacán, Morelos, Querétaro and probably introduced in Puebla]

7(6) Male pleopod I central projection length ≥ 0.2× pleopod length (Fig. 23.77 B, C) 8

7′ Male pleopod I central projection length < 0.2× pleopod length (Fig. 23.77 F) .... *Cambarellus montezumae* (Saussure, 1857)
[Mexico: Valley of Mexico]

8(7) Areola length < 3.5× areola width (like Fig. 23.78 E) 9

8′ Areola length ≥ 3.5× areola width (Fig. 23.78 D) .... *Cambarellus occidentalis* (Faxon, 1898)
[Mexico: Jalisco, Sinaloa]

9(8) Cheliped carpus and merus with spines .... *Cambarellus chihuahuae* Hobbs, 1980
[Mexico: Chihuahua]

9′ Cheliped carpus and merus inerm .... *Cambarellus areolatus* (Faxon, 1885)
[Mexico: Coahuila, reported as extinct in the wild]

Decapoda: Astacidea: Cambaridae: *Procambarus*: Subgenera

Only form I males can be identified with this key. These keys are largely based on Hobbs (1972a, 1972b), Hobbs (1973), Hobbs et al. (1977), Fetzner (2015), and Villalobos Figueroa (1955). Drawings included below have been redrawn from their original sources Hobbs (1941, 1972a, 1972b), Álvarez et al. (2007), López et al. (2003), López-Mejía et al. (2004, 2005), Mejía-Ortiz et al. (2003), Villalobos Figueroa & Hobbs (1974), Rojas et al. (1999, 2000).

1′ Carapace with two or more cervical spines (Fig. 23.80 A) 2

1 Carapace with or without one cervical spine (Fig. 23.80 B) 3

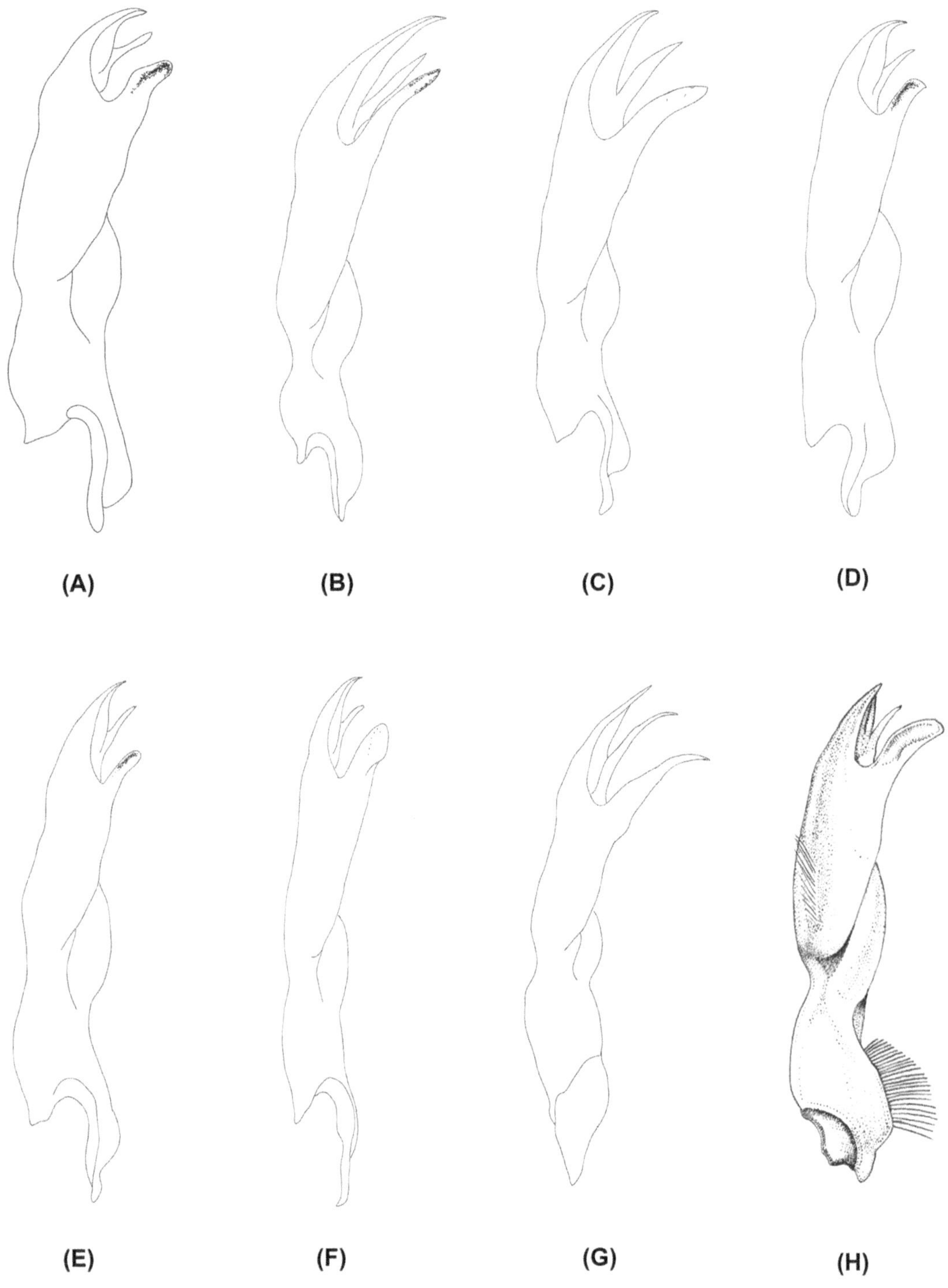

FIGURE 23.77 Astacidea: Mesial view of left first pleopods. (A) *Cambarellus chapalanus*; (B) *C. occidentalis*; (C) *C. areolatus*; (D) *C. patzcuarensis*; (E) *C. zempoalensis*; (F) *C. montezumae*; (G) *C. alvarezi*; (H) *C. zacapuensis*.

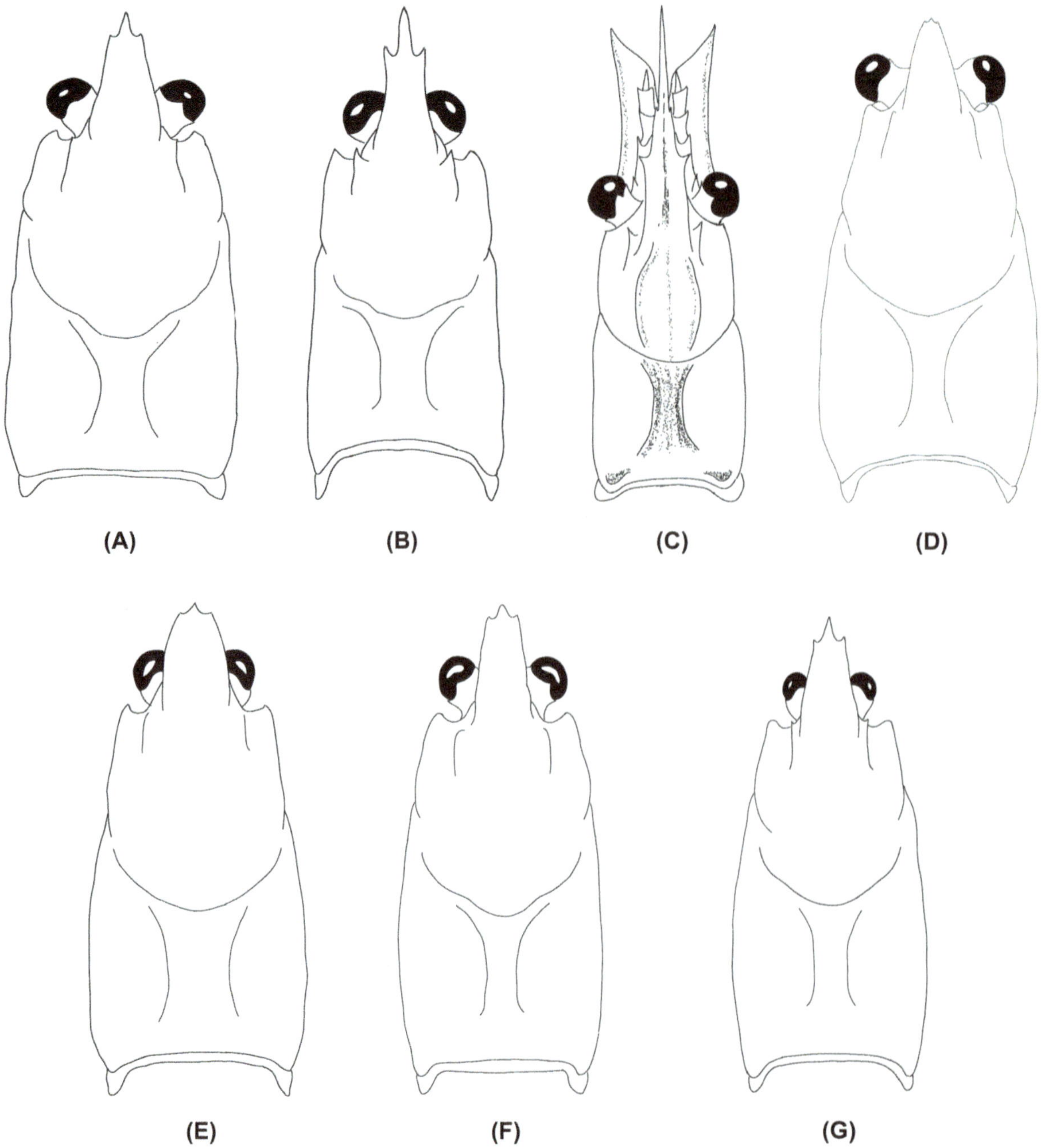

FIGURE 23.78 Astacidea: Dorsal view of carapaces. (A) *Cambarellus patzcuarensis*; (B) *C. chapalanus*; (C) *C. prolixus*; (D) *C. occidentalis*; (E) *C. areolatus*; (F) *C. zempoalensis*; (G) *C. montezumae*.

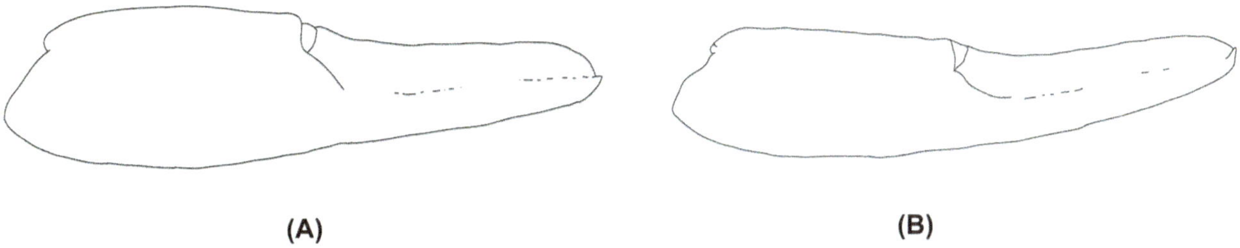

FIGURE 23.79 Astacidea: Dorsal view of chelae. (A) *Cambarellus montezumae*; (B) *C. zempoalensis*.

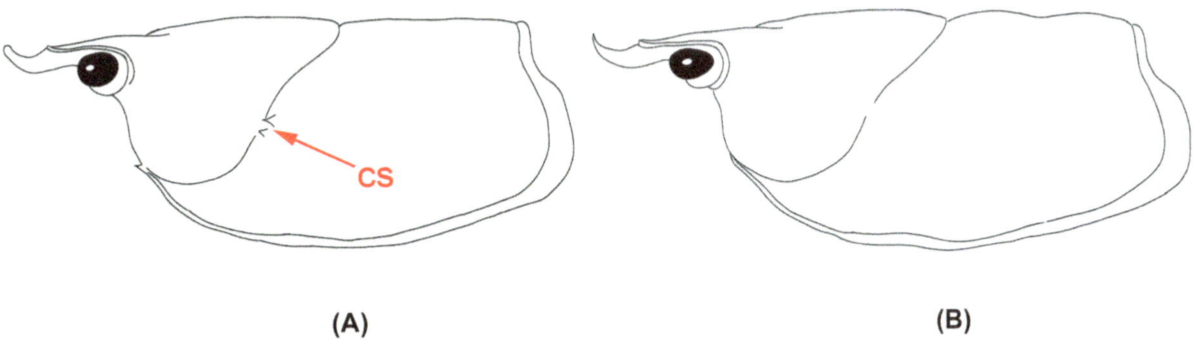

FIGURE 23.80 Astacidea: Lateral view of carapaces showing presence (A) and absence (B) of cervical spines. Abbreviation: *cs*, cervical spines.

2(1) First pair of pleopods assymetrical (Fig. 23.81 A) ..... *Pennides*; one species *Procambarus* (*Pennides*) *roberti* Villalobos & Hobbs, 1974 [Mexico: San Luis Potosí]

2′ First pair of pleopods symmetrical (Fig. 23.81 B) ..... ***Austrocambarus* (in part) [p. 891]**

3(1) Gonopod anterior surface with prominent angular or subangular shoulder situated far proximal to base of terminal elements (Fig. 23.87) 4

3′ Gonopod anterior surface with or without prominent angular or subangular shoulder; if present, situated immediately proximal to base of terminal elements (Fig. 23.82 D), or, if situated more proximally, somewhat rounded and anterior process never broadly rounded (Fig. 23.89 A) 5

4(2) Pereopod III and IV ischia bearing hooks (Fig. 23.76 C) ..... ***Scapulicambarus* [p. 895]**

4′ Pereopod III ischia bearing hooks (Fig. 23.76 B) ..... ***Austrocambarus* (in part) [p. 891]**

5(3) Only pereopod IV ischia bearing hook (Fig. 23.76 D) 6

5′ Pereopod III or III and IV bearing hooks (Fig. 23.76 B, C) 7

6(5) Gonopod central projection falciform, curved posteriorly (Fig. 23.82 A) ..... ***Paracambarus* [p. 895]**

6′ Gonopod central projection never falciform, directed distally (Fig. 23.82 B) ..... ***Villalobosus* [p. 895]**

7(5) Pereopod III ischia bearing hooks (Fig. 23.76 B) 8

7′ Pereopod III and IV ischia bearing hooks (Fig. 23.76 C) ..... ***Ortmannicus* [p. 896]**

8(7) Gonopod anterior process without a curved plate across anteriodistal extremity 9

8′ Gonopod cephalic process forming curved plate across anteriodistal extremity (Fig. 23.82 C) ..... *Mexicambarus*; one species: *Procambarus* (*Mexicambarus*) *bouvieri* (Ortmann 1909) [Mexico: Michoacán]

9(8) Rostrum with median carina; gonopod proximomedian lobe absent (Fig. 23.82 D) ..... *Procambarus*; one species: *Procambarus* (*Procambarus*) *digueti* (Bouvier, 1897) [Mexico: Michoacán]

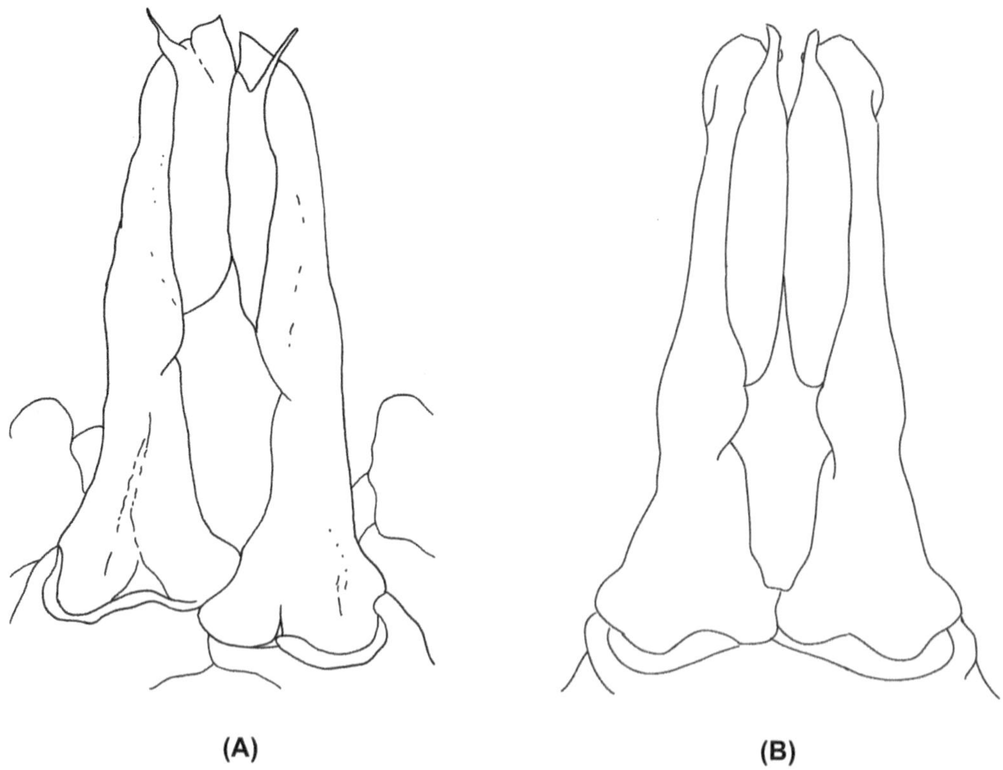

FIGURE 23.81 Astacidea: Ventral view of first pleopods. (A) Asymmetrical, *Procambarus roberti*; (B) Symmetrical, *P. llamasi*.

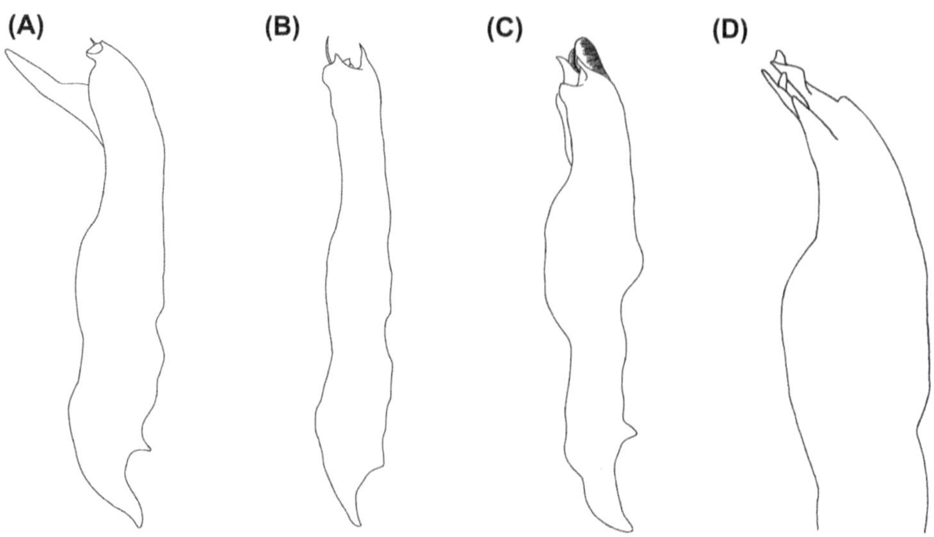

FIGURE 23.82 Astacidea: Lateral view of left first pleopods. (A) *P. paradoxus*; (B) *P. riojai*; (C) *P. bouvieri*; (D) *P. digueti*.

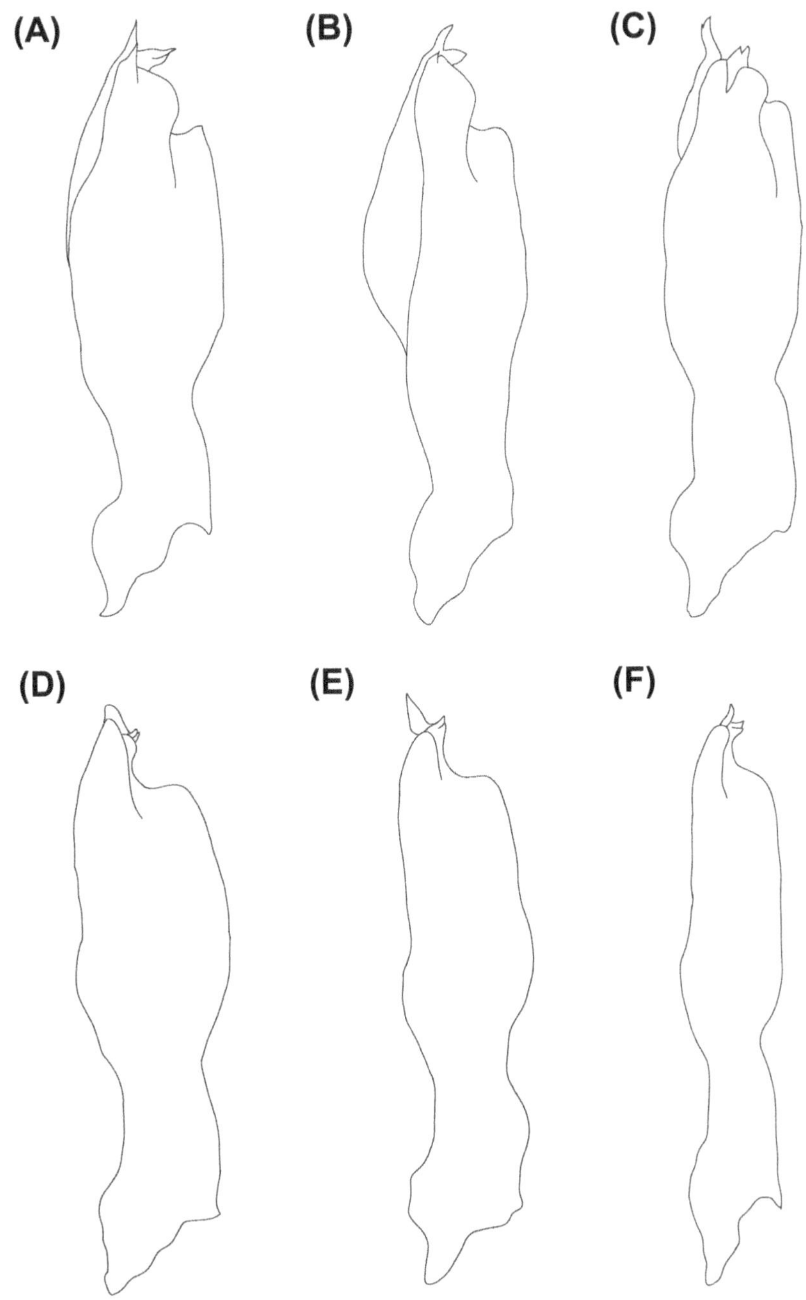

FIGURE 23.83 Astacidea: Lateral view of left first pleopods. (A) *Procambarus cubensis rivalis*; (B) *P. cubensis cubensis*; (C) *P. atkinsoni*; (D) *P. williamsoni*; (E) *P. mexicanus*; (F) *P. ruthveni*.

9′ Rostrum without median carina; gonopod proximomedian lobe very strongly developed .... *Girardiella* one species; *Procambarus* (*Girardiella*) *regiomontanus* Villalobos 1954 [Mexico, Nuevo León]

Decapoda: Astacidea: Cambaridae: *Procambarus* (*Austrocambarus*): Species

1 Gonopod with prominent bulbous enlargement between level of shoulder and base of central projection (Fig. 23.83 A–C) 2

1′ Gonopod lacking bulbous enlargement (slightly developed in *P. williamsoni*) between level of shoulder and base of central projection (Fig. 23.83 D–G) 5

2(1) Pigmented, eyes well developed 3

2′ Albinistic, eyes reduced but with small pigmented area .... *Procambarus (Austrocambarus) niveus* Hobbs & Villalobos, 1964
[Cuba: Pinar del Río]

3(2) Areola length usually < 4× breadth, seldom constituting as much as 26% of total carapace length; gonopod mesial part expanded little if any more than lateral part (Fig. 23.83 A, C) 4

3′ Areola length nearly always > 4× width, constituting at least 26% of total carapace length; gonopod mesial part expanded posteriorly more than lateral part (Fig. 23.83 B) .... *Procambarus (Austrocambarus) cubensis cubensis* (Erichson, 1846)
[Cuba]

4(3) Gonopod anterior surface shoulder rounded or tuberculiform; shaft middle part with anterior and posterior margins subparallel (Fig. 23.83 C) .... *Procambarus (Austrocambarus) atkinsoni* (Ortmann, 1913)
[Cuba: Isla de Pinos]

4′ Gonopod anterior surface shoulder angular; shaft middle part with anterior and posterior margins convex (Fig. 23.83 A) .... *Procambarus (Austrocambarus) cubensis rivalis* (Faxon, 1912)
[Cuba: Pinar del Río]

5(1) Carapace with two or more cervical spines (Fig. 23.80 A) 6

5′ Carapace with or without one cervical spine (Fig. 23.80 B) 11

6(5) Chela not conspicuously pubescent (Fig. 23.84 A) 7

6′ Chela dactyl and pollex conspicuously pubescent (Fig. 23.84 B) 8

7(6) Carapace hepatic region anterior margin with ≤ 1 spines (s) between cervical groove (cg) and suborbital angle (sa) (Fig. 23.85 A) .... *Procambarus (Austrocambarus) williamsoni* (Ortmann, 1905)
[Guatemala: Izabal. Honduras]

7′ Carapace hepatic region anterior margin with >1 spine between cervical groove and suborbital angle (Fig. 23.85 B). .... *Procambarus (Austrocambarus) maya* Álvarez, López-Mejía & Villalobos 2007
[Mexico: Quintana Roo]

8(6) Chela with conspicuous long pubescence on palm, dactyl, and pollex (Fig. 23.84 B) 9

8′ Chela with conspicuous long pubescence on dactyl and pollex (Fig. 23.84 C). .... *Procambarus (Austrocambarus) llamasi* Villalobos, 1954
[Mexico: Veracruz, Tabasco, Campeche, Yucatán. Guatemala]

9(8) Gonopod mesial process not as described below 10

9′ Gonopod mesial process folded over lateral shoulder .... *Procambarus (Austrocambarus) catemacoensis* Rojas, Álvarez & Villalobos 2000
[Mexico: Veracruz]

10(9) Carapace hepatic region anterior margin with three or four spines between cervical groove and suborbital angle (Fig. 23.85 B); pereopod III ischium hook apex not extending proximally of basioischial articulation (bi) (Fig. 23.86 B) .... *Procambarus (Austrocambarus) acanthophorus* Villalobos, 1948
[Mexico: Veracruz, Oaxaca]

10′ Carapace hepatic region anterior margin with only one spine between cervical groove and suborbital angle (Fig. 23.85 A); pereopod III ischium hook apex extending proximally of basioischial articulation (Fig. 23.86 A) .... *Procambarus (Austrocambarus) pilosimanus* (Ortmann, 1906)
[Mexico: Chiapas, Quintana Roo. Belize. Guatemala]

11(5) Pigmented; eyes well developed 12

11′ Albinistic; eyes reduced although a small pigment spot can be present 20

12(11) Areola length ≥ 9× breadth, with one or two punctations across narrowest part (Fig. 23.87 B–D) 13

12′ Areola length < 7× breadth, with three or four punctations across narrowest part (Fig. 23.87 A) 19

13(12) Gonopod central projection posteriodistal margin evenly contoured (Fig. 23.88 B); eyes not conspicuously large (Fig. 23.87 C) 14

13′ Gonopod central projection posteriodistal margin with small subacute projection near midlength (sp) (Fig. 23.88 A); eyes conspicuously large (Fig. 23.87 B) .... *Procambarus (Austrocambarus) vazquezae* Villalobos, 1954
[Mexico: Veracruz]

14(11) Gonopod anterior surface shoulder distal margin sloping proximally from base (Fig. 23.83 F) 15

14′ Gonopod anterior surface shoulder distal margin perpendicular to shaft or concave proximally (Fig. 23.83 E) 16

15(14) Pereopod III ischia hook extending proximally beyond basioischial articulation (Fig. 23.86 A) .... *Procambarus (Austrocambarus) zapoapensis* Villalobos, 1954
[Mexico: Veracruz]

15′ Pereopod III ischia hook never extending proximally beyond basioischial articulation (Figs. 23.83 F, 86 B) .... *Procambarus (Austrocambarus) ruthveni* (Pearse, 1911)
[Mexico: Veracruz]

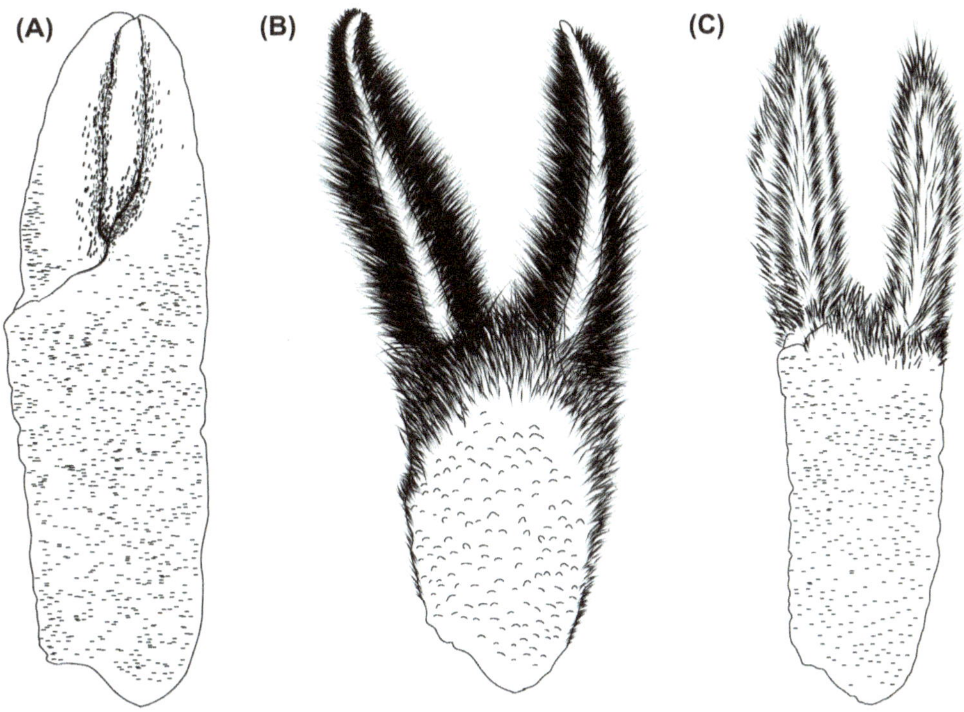

FIGURE 23.84 Astacidea: Dorsal view of chelae. (A) *Procambarus williamsoni*; (B) *P. pilosimanus*; (C) *P. llamasi*.

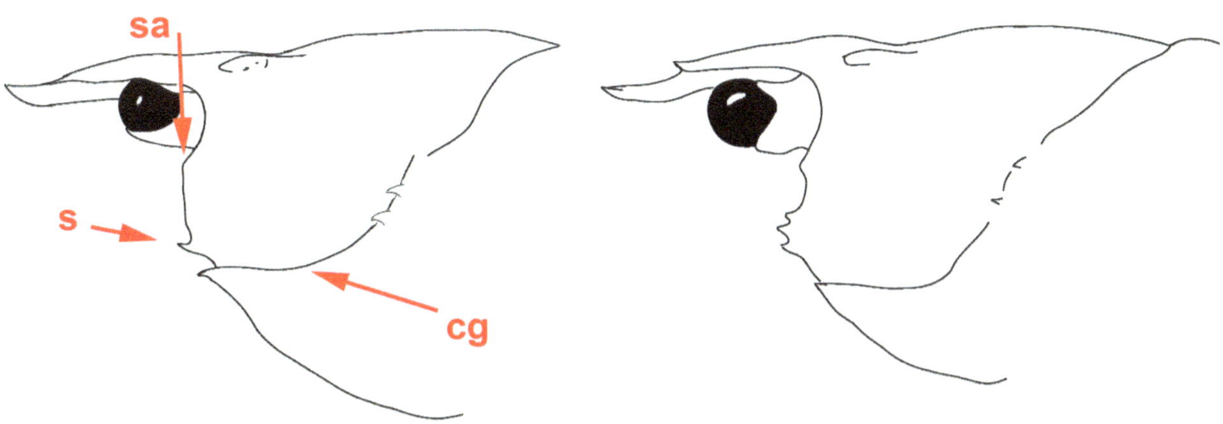

FIGURE 23.85 Astacidea: Lateral view of carapaces. (A) *Procambarus acanthophorus*; (B) *P. pilosimanus*. Abbreviation: *cg*, cervical groove; *s*, spine; *sa*, suborbital angle.

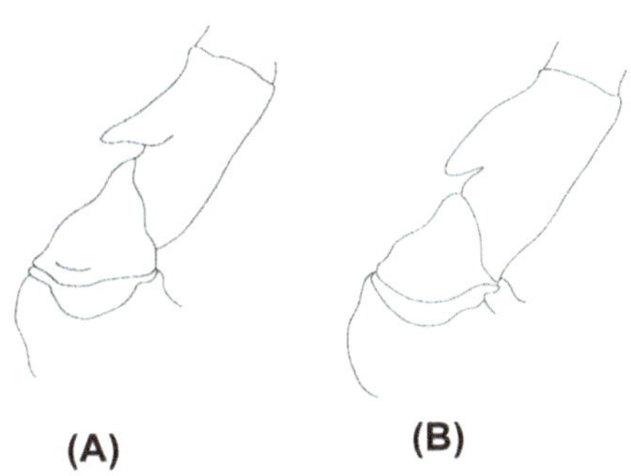

FIGURE 23.86 Astacidea: Caudoventral view of basal portions of third pereiopods. (A) *Procambarus pilosimanus*; (B) *P. acanthophorus*.

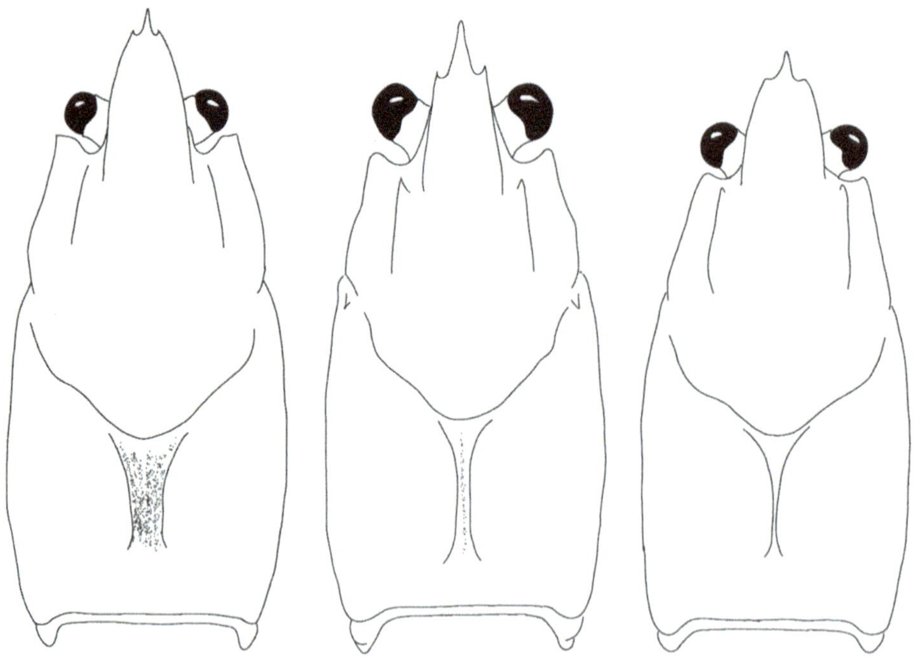

FIGURE 23.87 Astacidea: Dorsal view of carapaces. (A) *Procambarus mirandai*; (B) *P. vazquezae*; (C) *P. veracruzanus*.

16(14) Areola with utmost one punctation in narrowest part (like Fig. 23.87 C) 17

16′ Areola with at least two punctations across narrowest part (Fig. 23.87 B) 18

17(16) Rostral margins bearing spines at acumen base (Fig. 23.87 C) .... *Procambarus* (*Austrocambarus*) *veracruzanus* Villalobos, 1954
[Mexico: Veracruz]

17′ Rostral margins not bearing spines at acumen base .... *Procambarus* (*Austrocambarus*) *olmecorum* Hobbs, 1987
[Mexico: Veracruz]

18(16) Gonopod mesial process acute (Fig. 23.83 E) .... *Procambarus* (*Austrocambarus*) *mexicanus* (Erichson, 1846)
[Mexico: Veracruz]

18′ Gonopod mesial process lamellate .... *Procambarus* (*Austrocambarus*) *citlaltepetl* Rojas, Álvarez & Villalobos, 1999
[Mexico: Veracruz]

19(12) Gonopod mesial process hemispherical .... *Procambarus* (*Austrocambarus*) *mirandai* Villalobos, 1954
[Mexico: Chiapas]

19′ Gonopod mesial process acute .... *Procambarus* (*Austrocambarus*) *sbordonii* Hobbs, 1978
[Mexico: Chiapas]

20(11) Gonopod mesial process not strongly flattened; annulus ventralis gently rounded posterioventrally 21

20′ Gonopod mesial process somewhat flattened; annulus ventralis protruding prominently posterioventrally .... *Procambarus* (*Austrocambarus*) *rodriguezi* Hobbs, 1943
[Mexico: Veracruz]

21(20) Scaphocerite length $\geq 2\times$ width 22

21′ Scaphocerite length $< 2\times$ width. .... *Procambarus* (*Austrocambarus*) *cavernicola* Mejía-Ortíz, Hartnoll & Viccon-Pale, 2003

22(21) Areola constituting 35.8 to 37.8% of carapace total length; cervical spine absent .... *Procambarus* (*Austrocambarus*) *oaxacae oaxacae* Hobbs, 1973

22′ Areola constituting 32.9 to 35.6% of carapace total length; cervical spine present .... *Procambarus* (*Austrocambarus*) *oaxacae reddelli* Hobbs, 1973

Decapoda: Astacidea: Cambaridae: *Procambarus* (*Scapulicambarus*): Species

1 Gonopod conspicuously setose, cephalic process acute, broad, lamellate, with corneous margin partly surrounding central projection (Fig. 23.92 A); carapace dorsal surface with deep punctations .... *Procambarus* (*Scapulicambarus*) *strenthi* Hobbs, 1977
[Mexico: San Luis Potosí]

1′ Gonopod not conspicuously setose, cephalic process lobiform and directed posteriodistad; carapace dorsal surface with punctations normal (Fig. 23.92 B) .... *Procambarus* (*Scapulicambarus*) *clarkii* (Girard, 1852)
[Mexico: Chihuahua, Coahuila and Nuevo León. Introduced in Mexico and Brazil]

Decapoda: Astacidea: Cambaridae: *Procambarus* (*Paracambarus*): Species

1 Gonopod mesial process larger than central projection and extending posteriodistad much beyond it (Fig. 23.89 A) .... *Procambarus* (*Paracambarus*) *paradoxus* (Ortmann, 1906)
[Mexico: Puebla]

1′ Gonopod mesial process smaller than central projection and not extending so far distad as central projection (Fig. 23.89 B) .... *Procambarus* (*Paracambarus*) *ortmanni* (Villalobos, 1949)
[Mexico: Puebla]

Decapoda: Astacidea: Cambaridae: *Procambarus* (*Villalobosus*): Species

1 Gonopod mesial process extending distinctly farther distally than other terminal elements (Fig. 23.91 A–D) 2

1′ Gonopod mesial process extending little, if any, farther distally than other terminal elements (Fig. 23.91 E–M) 5

2(1) Gonopod mesial process length $\geq 1.6\times$ other terminal elements (Fig. 23.91 B–D) 3

2′ Gonopod mesial process length $\leq 1.5\times$ other terminal elements (Fig. 23.91 A) .... *Procambarus* (*Villalobosus*) *hoffmanni* (Villalobos, 1944)
[Mexico: Puebla, Veracruz]

3(2) Gonopod mesial process flattened or tapering (Fig. 23.91 C, D) 4

3′ Gonopod mesial process distally broadly rounded, anterior surface convex (Fig. 23.91 B) .... *Procambarus* (*Villalobosus*) *xochitlanae* Hobbs, 1975
[Mexico: Puebla]

4(2) Gonopod mesial process somewhat flattened, at least caudal margin sinuous (Fig. 23.91 C); areola with ≤ 2 punctations across narrowest part .... *Procambarus* (*Villalobosus*) *tlapacoyanensis* (Villalobos, 1947)
[Mexico: Veracruz]

4′ Gonopod mesial process conical and nearly straight (Fig. 23.91 D); areola with more than two punctations across narrowest part .... *Procambarus* (*Villalobosus*) *teziutlanensis* (Villalobos, 1947)
[Mexico: Puebla, Veracruz]

5(1) Gonopod with mesial and cephalic processes conspicuous, acute, subequal in length, projecting distally beyond other terminal elements (mesial process may reach little farther distally than cephalic process) (Fig. 23.91 E, F) 6

5′ Gonopod with only one conspicuous spiniform or subspiniform mesial process, or, if cephalic process somewhat spiniform, never extending distally beyond other terminal elements (Fig. 23.91 G–M) 7

6(5) Gonopod mesial and cephalic processes subequal in length, both projecting distally beyond other terminal elements, caudal process a small angular crest (Fig. 23.91 E) .... *Procambarus* (*Villalobosus*) *riojai* (Villalobos, 1944)
[Mexico: Hidalgo, Puebla]

6′ Gonopod mesial process reaching little farther distally than cephalic process, caudal process reduced, tuberculiform (Fig. 23.91 F) .... *Procambarus* (*Villalobosus*) *cuetzalanae* Hobbs, 1982
[Mexico: Puebla]

7(5) Gonopod cephalic process well developed, slender, never lamelliform (Fig. 23.91 G, L, M) 8

7′ Gonopod cephalic process either forming curved crest like lamella or reduced, never extending distally to level mesial process apex (Fig. 23.91 H–K) 10

8(7) Gonopod cephalic process emerging from caudal process, central projection triangular, divided in two processes 9

8′ Gonopod cephalic process situated anteriomesially, and extending distally to or slightly beyond level of tip of mesial process (Fig. 23.91 G) .... *Procambarus* (*Villalobosus*) *erichsoni* Villalobos, 1950
[Mexico: Hidalgo]

9(8) Gonopod mesial process flattened, directed anteriorly; central projection first process corneous and longer than the rest of terminal elements (Fig. 23.91 L) .... *Procambarus* (*Villalobosus*) *achilli* López, Mejía & Álvarez, 2003
[Mexico: Hidalgo]

9′ Gonopod mesial process acute, directed distolaterally, central projection not triangular and formed by an obscurely subdivided lamellar posterior process with a strong fold on its posteriolateral margin (Fig. 23.91 M) .... *Procambarus* (*Villalobosus*) *chacalli* López-Mejía, Álvarez & Mejía-Ortíz, 2004 [Mexico: Veracruz]

10(7) Curved crestlike plate on cephalodistal surface of pleopod not flared from base, extending distally (Fig. 23.91 I–K) 11

10′ Curved crestlike plate on cephalodistal surface of first pleopod flared from base (Fig. 23.91 H) .... *Procambarus* (*Villalobosus*) *zihuateutlensis* Villalobos, 1950 [Mexico: Puebla]

11(10) Gonopod central projection masked in lateral view by crestlike plate; cephalic process small and situated posteriomesial to crestlike plate (Figs. 23.91 I, J) .... *Procambarus* (*Villalobosus*) *contrerasi* (Creaser, 1931) [Mexico: Puebla]

11′ Gonopod central projection clearly evident in lateral view; cephalic process forming crestlike plate (Fig. 23.91 K) .... *Procambarus* (*Villalobosus*) *hortonhobbsi* Villalobos, 1950 [Mexico: Puebla]

Decapoda: Astacidea: Cambaridae: *Procambarus* (*Ortmannicus*): Species

1 Body pigmented, eyes well developed and pigmented 2

1′ Albinistic, eyes without pigment or with small pigment spot, troglobitic .... *Procambarus* (*Ortmannicus*) *xilitlae* Hobbs & Grubbs, 1982 [Mexico: San Luis Potosí]

2(1) Gonopod subapical setae, if present, never born on distinct knob 3

2′ Gonopod subapical setae born on anteriodistal knob (Fig. 23.92 A) .... *Procambarus* (*Ortmannicus*) *cuevachicae* (Hobbs, 1941) [Mexico: San Luis Potosí]

3(2) Gonopod caudal process prominent, arising from posteriolateral surface; caudal knob never well defined (Fig. 23.92 B) 4

3′ Gonopod caudal process rarely prominent; cephalic process arising from central projection lateral posterior side (see Fig. 23.92 C) ... 6

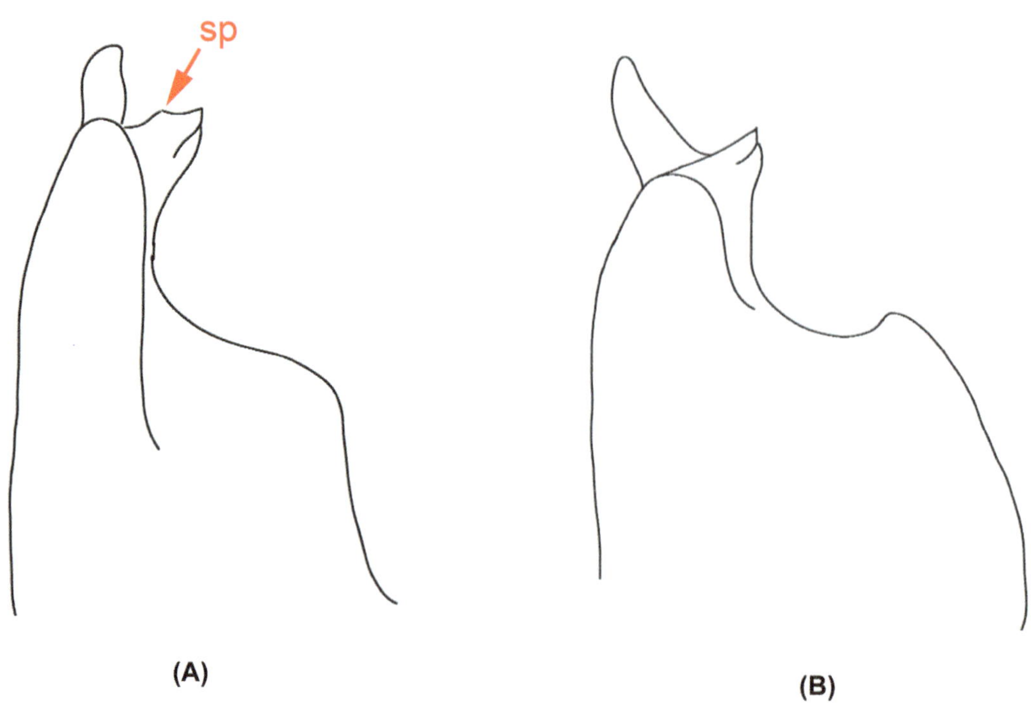

FIGURE 23.88 Astacidea: Lateral view of distal portion of first pleopods. (A) *Procambarus vazquezae*; (B) *P. mexicanus*. Abbreviation: *sp*, subacute projection.

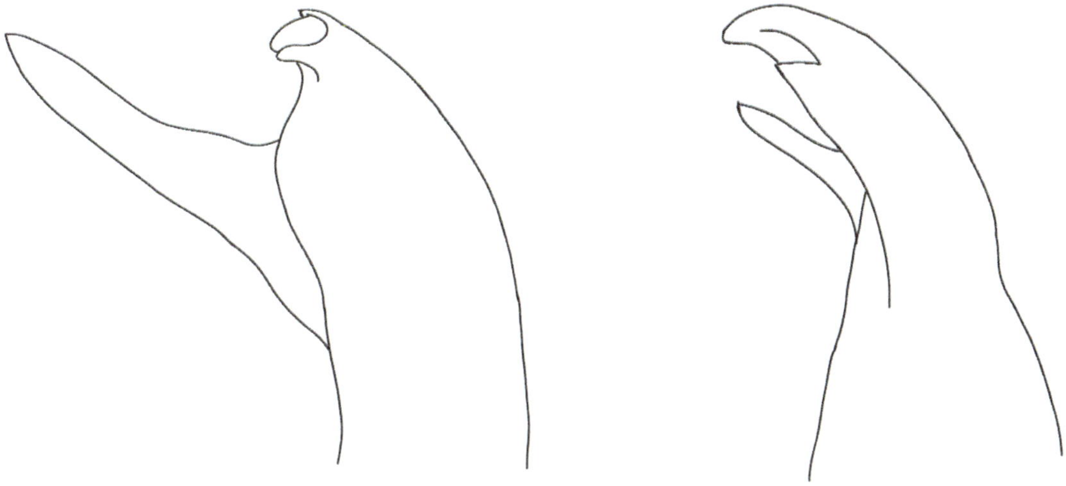

(A) (B)

FIGURE 23.89 Astacidea: Lateral view of left first pleopods. (A) *Procambarus paradoxus*; (B) *P. ortmanni*.

(A) (B)

FIGURE 23.90 Astacidea: Lateral view of left first pleopods. (A) *Procambarus strenthi*; (B) *P. clarkii*.

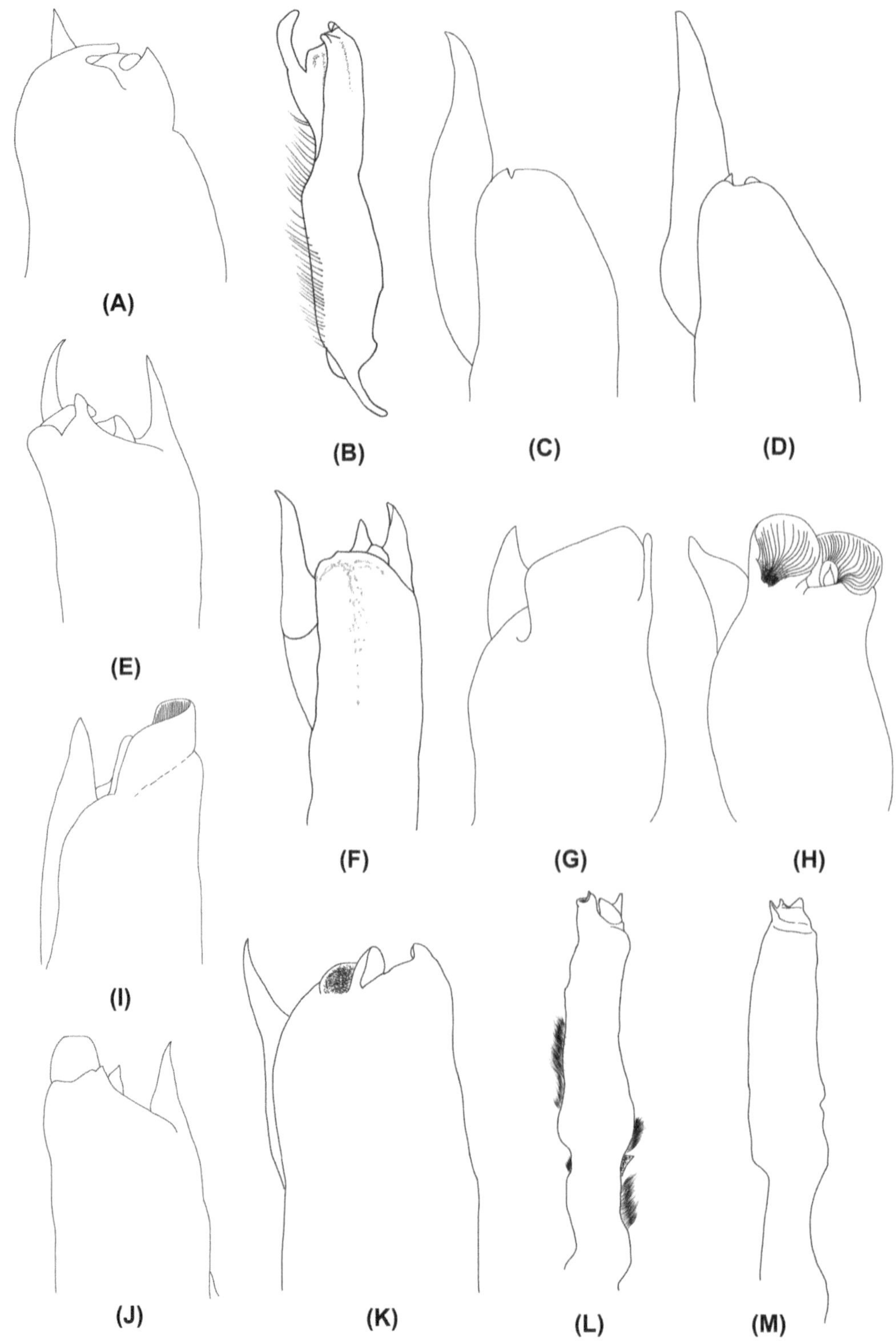

FIGURE 23.91 Astacidea: A–G, Lateral view of distal portion of left first pleopods. H, Mesial view of the same; L–M, lateral view of first pleopods. (A) *Procambarus hoffmanni*; (B) *P. xochitlanae*; (C) *P. tlapacoyanensis*; (D) *P. teziutlanensis*; (E) *P. riojai*; (F) *P. cuetzalanae*; (G) *P. erichsoni*; (H) *P. zihuateutlensis*; (I)–(J) *P. contrerasi*; (K) *P. hortonhobbsi*; (L) *P. achilli*; (M) *P. chacalli.*

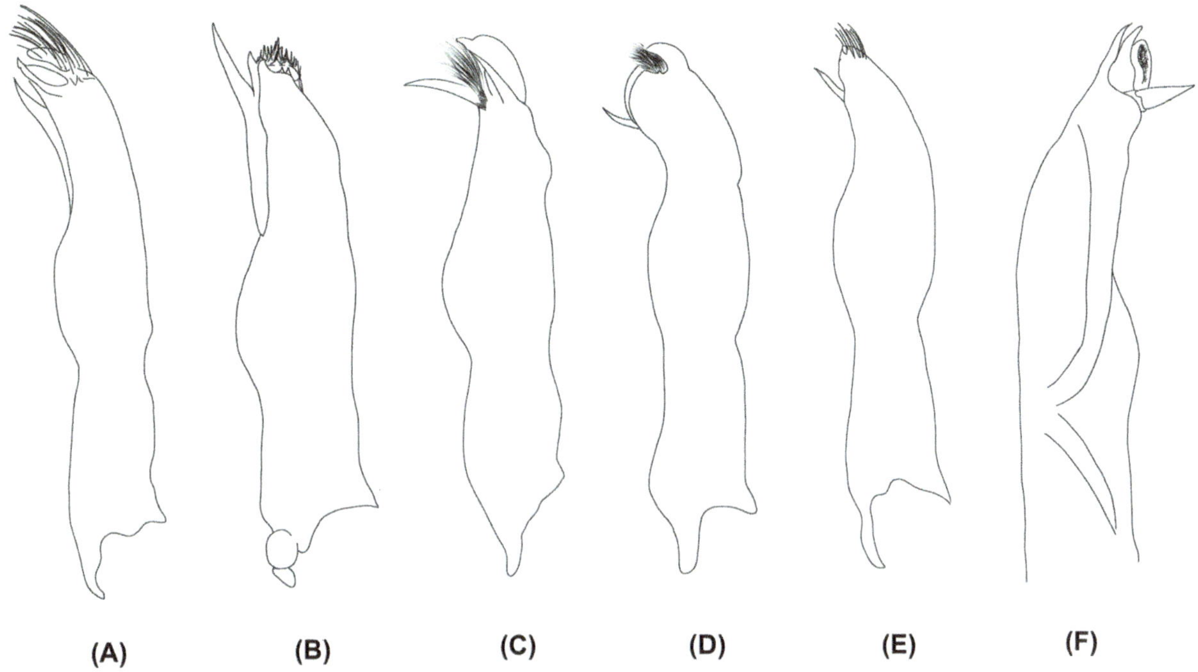

FIGURE 23.92 Astacidea: A–E, lateral view of left first pleopods. (A) *Procambarus cuevachicae*; (B) *P. villalobosi*; (C) P. *toltecae*; (D) *P. gonopodocristatus*; (E) *P. caballeroi*; (F) mesial view of first pleopod of *P. hidalgoensis*.

4(3) Gonopod cephalic process absent, or situated anteriorly or anteriomedially to central projection (Fig. 23.92 D, E) 5

4′ Gonopod cephalic process situated medially and directed distally (Fig. 23.92 B) .... *Procambarus* (*Ortmannicus*) *villalobosi* Hobbs, 1969
[Mexico: San Luis Potosí]

5(4) Gonopod caudal process forming long lamellar arc along distal posteriolateral surface (Fig. 23.92 D) .... *Procambarus* (*Ortmannicus*) *gonopodocristatus* Villalobos, 1958
[Mexico: Veracruz]

5′ Gonopod caudal process acute (Fig. 23.92 E) .... *Procambarus* (*Ortmannicus*) *caballeroi* Villalobos, 1944
[Mexico: Puebla]

6(3) Gonopod cephalic process situated lateral to central projection (Fig. 23.92 C) .... *Procambarus* (*Ortmannicus*) *toltecae* Hobbs, 1943
[Mexico: San Luis Potosí]

6′ Gonopod cephalic process originating on caudal surface (Fig. 23.92 F) .... *Procambarus* (*Ortmannicus*) *hidalgoensis* López-Mejía, Álvarez & Mejía-Ortíz, 2005
[Mexico: Hidalgo]

REFERENCES

Álvarez, F., M. López-Mejía & J.L. Villalobos. 2007. A new species of crayfish (Crustacea: Decapoda: Cambaridae) from a salt marsh in Quintana Roo, Mexico. Proceedings of the Biological Society of Washington 120: 311–319.

Álvarez, F. & J.L. Villalobos. 2015. The crayfish of middle America. *In*: T. Kawai, Z. Faulkes & G. Scholtz (eds.), Freshwater crayfish: A global overview, CRC Press.

Brusca, R.C., W. Moore & S.M. Shuster. 2016. Invertebrates. 3rd edition. Sinauer Associates, Oxford University Press.

Buckup, L. & A. Rossi. 1980. O Gênero *Parastacus* no Brasil (Crustacea, Decapoda, Parastacidade). Revista Brasileira de Biologia 40: 663–681.

Buckup, L. & A. Rossi. 1993. Os Parastacidae do espaço meridional andino (Crustacea, Astacidea). Revista Brasileira de Biologia 53: 167–176.

Crandall, K.A. 2016. Collecting and processing freshwater crayfishes. Journal of Crustacean Biology 36: 761–766.

Crandall, K.A. & S. De Grave. 2017. An updated classification of the freshwater crayfishes (Decapoda: Astacidea) of the world, with a complete species list. Journal of Crustacean Biology 39: 615–653.

Cumberlidge, N., H.H. Hobbs & D.M. Lodge. 2015. Chapter 32. Class Malacostraca, Order Decapoda. Pages 797–837 *in*: J.H. Thorp &

PHYLUM ARTHROPODA: CRUSTACEA: MALACOSTRACA

D.C. Rogers, Keys to Nearctic Fauna, Thorp and Covich's Freshwater Invertebrates, Vol. I, 4th Edition. Academic Press, New York.

Faxon, W. 1898. Observations on the Astacidae in the United States National Museum and in the Museum of Comparative Zoology, with descriptions of new species. Proceedings of the United States National Museum 20: 642–694.

Faxon, W. 1914. Notes on the crayfishes in the United States National Museum and the Museum of Comparative Zoology with descriptions of new species and subspecies to which is appended a catalogue of the known species and subspecies. Memoirs of the Museum of Comparative Zoology at Harvard College 40: 351–427.

Fetzner Jr, J.W. 2015. Crustacea: Malacostraca: Decapoda: Astacidea: Cambaridae: *Procambarus*: Subgenera. Pages 617–642 *in*: J.H. Thorp & D.C. Rogers (eds.), Keys to Nearctic Fauna, Thorp and Covich's Freshwater Invertebrates, Vol. II, 4th Edition. Academic Press, New York.

Hobbs, H.H. Jr. 1941. A new crayfish from San Luis Potosi, Mexico. Zoologica, Scientific Contributions of the New York Zoological Society 26: 1–4.

Hobbs, H.H. Jr. 1972a. Biota of Freshwater Ecosystems, Identification Manual no. 9, Crayfishes (Astacidae) of North and Middle America. EPA project # 18050 ELD.

Hobbs, H.H. Jr. 1972b. Crayfishes (Astacidae) of North and Middle America. Biota of freshwater Ecosystems. U.S. Environmental Protection Agency, Water Pollution Control Research Service Identification Manual 9: 1–173.

Hobbs, H.H. Jr. 1973. Three new troglobitic decapod crustaceans from Oaxaca, Mexico. Bulletin of the Association of Mexican Cave Studies 5: 25–38.

Hobbs, H.H. Jr. 1989. An illustrated checklist of the American crayfishes (Decapoda: Astacidae: Cambaridae, and Parastacidae). Smithsonian Contributions to Zoology 480: 1–236.

Hobbs, H.H. Jr. 1991. A new generic assignment for a South American crayfish (Decapoda, Parastacidae) with revised diagnoses of the South American genera and comments on the parastacid mandible. Proceeding of the Biological Society of Washington 104: 800–811.

Hobbs, H.H. Jr, H.H. Hobbs III & M.A. Daniel. 1977. A review of the troglobitic decapod crustaceans of the Americas. Smithsonian Contributions to Zoology 244: 1–154.

Hobbs, H.H. III & D.M. Lodge. 2010. Decapoda. Pages 901–967 *in*: J.H. Thorp & A.P. Covich (eds.), Ecology and Classification of North American Freshwater Invertebrates, 3rd edition. Academic Press, San Diego.

Huber, A.F., F.B. Ribeiro & P.B. Araujo. 2018. New endemic species of freshwater crayfish *Parastacus* Huxley, 1879 (Crustacea, Decapoda, Parastacidae) from the Atlantic forest in southern Brazil. Nauplius 26: e2018015.

López, M., L.M. Mejía & F. Álvarez. 2003. *Procambarus (Villalobosus) achilli* (Decapoda, Cambaridae): A new species of crayfish from Mexico. Crustaceana 76: 523–531.

López-Mejía, M., F. Álvarez & L.M. Mejía-Ortiz. 2004. A new species of *Procambarus* (Crustacea: Decapoda: Cambaridae) from Veracruz, Mexico. Proceedings of the Biological Society of Washington 117: 169–175.

López-Mejía, M., Álvarez, F. & L.M. Mejía-Ortiz. 2005. *Procambarus (Ortmannicus) hidalgoensis* (Crustacea: Decapoda: Cambaridae), a new species of crayfish from Mexico. Proceedings of the Biological Society of Washington 118: 558–565.

Loughman, Z.J., S.M. Henkanaththegedara, J.M. Fetzner Jr. & R.F. Thoma. 2017. A case of Appalachian endemism: Revision of the *Cambarus robustus* complex (Decapoda: Cambaridae) in the Kentucky and Licking River basins of Kentucky, USA, with the description of three new species. Zootaxa 4269: 460–494.

Martens, E. von. 1869. Sudbrasilische Suss und Brackwasser-Crustacean nach den Sammlundgen des Dr. Reinh. Hensel Archiv für Naturgeschichte 35: 1–37.

Mejía-Ortiz, L.M., Hartnoll, R.G. & J.A. Viccon-Pale. 2003. A new stygobitic crayfish from Mexico, *Procambarus cavernicola* (Decapoda: Cambaridae), with a review of cave-dwelling crayfishes in Mexico. Journal of Crustacean Biology 23: 391–401.

Miranda, I., K.M. Gomes, F.B. Ribeiro, P.B. Araujo, C. Souty-Grosset & C.D. Schubart. 2018. Molecular systematics reveals multiple lineages and cryptic speciation in the freshwater crayfish *Parastacus brasiliensis* (von Martens, 1869) (Crustacea, Decapoda, Parastacidae). Invertebrate Systematics 32: 1265–1281.

Pedraza-Lara, C. & I. Doadrio. 2015. A new species of dwarf crayfish (Decapoda: Cambaridae) from central México, as supported by morphological and genetic evidence. Zootaxa 3963: 583–594.

Philippi, R.A. 1882. Zoología chilena. Sobre los *Astacus*. Anales de la Universidad de Chile 61: 624–628.

Poeppig, E. 1835. Reise in Chile, Peru und auf dem Amazonas strome, während der Jahre 1827–1832, 1 (i-xviii), 466 pp.

Reynolds, J., C. Souty-Grosset & A. Richardson. 2013. Ecological roles of crayfish in freshwater and terrestrial habitats. Freshwater Crayfish 19: 197–218.

Ribeiro, F.B. & P.B. Araujo. 2017. Designation of a neotype for *Parastacus nicoleti* (Philippi, 1882) (Crustacea: Decapoda: Parastacidae). Zootaxa 4338: 393–400.

Ribeiro, F.B., L. Buckup, K.M. Gomes & P.B. Araujo. 2016. Two new species of South American freshwater crayfish genus *Parastacus* Huxley, 1879 (Crustacea: Decapoda: Parastacidae). Zootaxa 4158: 301–324.

Ribeiro, F.B., A.F. Huber, D. Schubart & P.B. Araujo. 2017. A new species of *Parastacus* Huxleys, 1879 (Crustacea, Decapoda, Parastacidae) from a swamp forest in southern Brazil. Nauplius 25: e2017008.

Rogers, D.C. 2016. Chapter 16. Phylum Arthropoda, Order Decapoda. Pages 605–11 *in*: J.H. Thorp & D.C. Rogers, Keys to Nearctic Fauna, Thorp and Covich's Freshwater Invertebrates, Vol. II, 4th Edition. Academic Press, New York.

Rojas, Y., F. Álvarez & J.L. Villalobos. 1999. A new species of crayfish of the genus *Procambarus* (Crustacea: Decapoda: Cambaridae) from Veracruz, Mexico. Proceedings of the Biological Society of Washington 112: 396–404.

Rojas, Y., F. Álvarez & J.L. Villalobos. 2000. A new crayfish (Crustacea: Decapoda: Cambaridae) from Lake Catemaco, Veracruz, Mexico. Proceedings of the Biological Society of Washington 113: 792–798.

Rudolph, E.H. 2010. Sobre la distribución geográfica de las especies chilenas de Parastacidae (Crustacea: Decapoda: Astacidea). Boletín de Biodiversidad de Chile 3: 32–46.

Rudolph, E.H. 2015. Current state of knowledge on *Virilastacus* species (Crustacea, Decapoda, Parastacidae). Latin American Journal of Aquatic Research 43: 807–818.

Rudolph, E. & A. Almeida. 2000. On the sexuality of South American Parastacidae (Crustacea: Decapoda). Invertebrate Reproduction and Development 37: 249–257.

Rudolph, E., N. Colihueque & M. Yañez. 2016. Morphological and genetic analysis in morphologically divergent river and lake specimens of the freshwater crayfish *Samastacus spinifrons* (Philippi, 1882) (Decapoda, Parastacidae). Crustaceana 89: 877–899.

Rudolph, E.H. & K, Crandall. 2005. A new species of burrowing crayfish *Virilastacus rucapihuelensis* (Crustacea: Decapoda: Parastacidae) from southern Chile. Proceedings of the Biological Society of Washington 118: 765–776.

Rudolph, E.H. & K. Crandall. 2007. A new species of burrowing crayfish *Virilastacus retamali* (Decapoda: Parastacidae) from the southern Chile peatland. Journal of Crustacean Biology 27: 502–512.

Rudolph, E.H. & K. Crandall. 2012. A new species of burrowing crayfish, *Virilastacus jarai* (Crustacea, Decapoda, Parastacidae) from central-southern Chile. Proceedings of the Biological Society of Washington 125: 258–275.

Toon, A., M. Pérez-Losada, C.E Schweitzer, R.M. Feldmann, M. Carlson & K. Crandall. 2010. Gondwanan radiation of the Southern Hemisphere crayfishes (Decapoda: Parastacidae): evidence from fossils and molecules. Journal of Biogeography 37: 2275–2290.

Villalobos, A. 1955. Cambarinos de la fauna Mexicana (Crustacea Decapoda). Doctoral dissertation. Universidad Nacional Autónoma de México, Facultad de Ciencias.

Villalobos Figueroa, A. & H.H. Hobbs Jr. 1974. Three new crustaceans from La Media Luna, San Luis Potosí, Mexico. Smithsonian Contributions to Zoology 174: 1–18.

Villalobos-Figueroa, A. & H.H. Hobbs Jr. 1981. A new dwarf crayfish from the Pacific versant of Mexico (Decapoda: Cambaridae). Proceedings of the Biological Society of Washington 94: 492–502.

Vogt, G. 2013. Abbreviation of larval development and extension of brood care as key features of the evolution of freshwater Decapoda. Biological Reviews 88: 81–116.

Yazicioglu, B., J. Reynolds & P. Kozák. 2016. Different aspects of reproduction strategies in crayfish: A review. Knowledge and Management of Aquatic Ecosystems 417: 1–15.

Decapoda: Dendrobranchiata

D. Christopher Rogers

Kansas Biological Survey, and The Biodiversity Institute, University of Kansas, Lawrence, KS, United States

KEY TO DENDROBRANCHIATA

Decapoda: Dendrobranchiata: Families

1 Rostrum extending beyond eyes (Fig. 23.93 A–C) .... **Penaeidae [p. 902]**

1′ Rostrum extending at most to middle of eye stalk (Figs. 23.94 B, 95 B) .... Sergestidae, one genus: ***Acetes* [p. 902]**

Decapoda: Dendrobranchiata: Penaeidae: Genera

1 Rostrum length $\leq$ scaphocerite, at least one or more ventral teeth typically present; antenna I dorsal flagellum shorter than peduncle; pereopods IV–V length $<$ pereopod III; pereopod II basis with a single spine (Fig. 23.93 B, C) 2

1′ Rostrum length $>$ scaphocerite, ventral teeth absent; antenna I dorsal flagellum length $\sim 6\times$ peduncle length; pereopods IV–V length $>$ pereopod III; pereopod II basis inerm; brackish water (Fig. 23.93 A) .... *Xiphopenaeus kroyeri* (Heller, 1862)
[Eastern coastal Neotropics from Mexico to Brazil. Caribbean Islands. Nearctic]

2(1) Rostral lateral grooves extending posteriorly nearly to carapace margin; brackish water, estuaries (Fig. 23.93 B) .... ***Farfantepenaeus* [p. 902]**

2′ Rostral lateral grooves extending posteriorly to just after posteriormost (gastric) dorsal tooth; brackish water (Fig. 23.93 C) .... *Litopenaeus schmitti* (Burkenroad, 1936)
[Eastern coastal Neoptropics from Honduruas to Brazil. Caribbean Islands]

Decapoda: Dendrobranchiata: Penaeidae: *Farfantepenaeus*: Species

1 Thelycum present; pleopod I unmodified (females) (Fig. 23.93 E) 2

1′ Petasma present (males) (Fig. 23.93 D) 4

2(1) Thelycum lateral membranous covers open at midline 3

2′ Thelycum lateral membranous covers meeting and closed .... *Farfantepenaeus brasiliensis* (Latreille, 1817)
[Caribbean and Atlantic islands and coasts south to Brazil. Nearctic]

3(2) Thelycum anteriomedial carina not anteriorly divided .... *Farfantepenaeus notialis* (Pérez-Farfante, 1967)
[Caribbean and Atlantic islands and coasts south to Brazil. Nearctic]

3′ Thelycum anteriomedial carina anteriorly bifurcated .... *Farfantepenaeus subtilis* (Pérez-Farfante, 1967)
[Caribbean and Atlantic islands and coasts south to Brazil. Nearctic]

4(1) Petasma rigid marginal band apex not projecting freely 5

4′ Petasma rigid marginal band apex projecting freely at distomedial end .... *Farfantepenaeus brasiliensis* (Latreille, 1817)
[Caribbean and Atlantic islands and coasts south to Brazil. Nearctic]

5(4) Rostral lateral grooves width $\geq 0.75\times$ medial ridge width; petasma distal margin with 2–12 spines .... *Farfantepenaeus notialis* (Pérez-Farfante, 1967)
[Caribbean and Atlantic islands and coasts south to Brazil. Nearctic]

5′ Rostral lateral grooves width $< 0.75\times$ medial ridge width; petasma distal margin inerm. .... *Farfantepenaeus subtilis* (Pérez-Farfante, 1967)
[Caribbean and Atlantic islands and coasts south to Brazil. Nearctic]

Decapoda: Dendrobranchiata: Penaeidae: Sergestidae: *Acetes*: Species

Modified from Vereshchaka et al. (2016).

1 Rostrum with one or two dorsal teeth 2

1′ Rostrum without dorsal teeth; estuaries .... *Acetes binghami* Burkenroad, 1934
[Pacific coast from Panama to Ecuador]

2(1) Rostrum with a single dorsal tooth 3

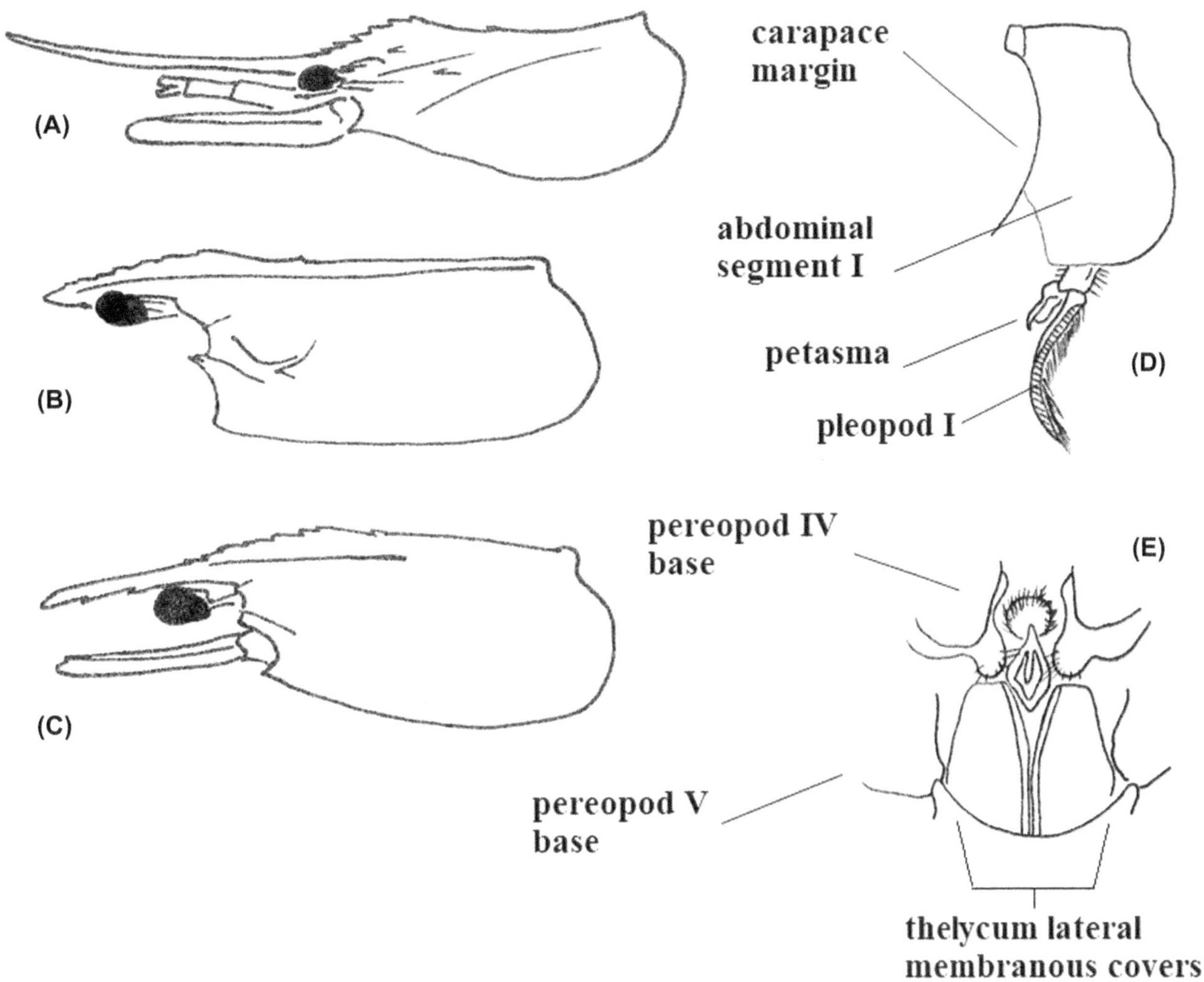

FIGURE 23.93 Dendrobranchiata: (A) *Xiphopennaeus kroyeri*, carapace, eye, antenna I peduncle and scaphocerite, left lateral view; (B) *Farfantepenaeus* sp, carapace and eye, left lateral view; (C) *Litopenaeus schmitti*, carapace, eye, and scaphocerite, left lateral view; (D) generalized dendrobranchiate shrimp male abdominal segment I, left lateral view; (E) generalized dendrobranchiate shrimp female abdominal segment I, ventral view.

2′ Rostrum with two dorsal teeth 4

3(2) Pereopod IV present but reduced, with three or five articles; estuaries .... *Acetes petrunkevitchi* (Burkenroad, 1945) [Brazil, Argentina, Uruguay]

3′ Pereopod IV absent; estuaries .... *Acetes americanus* (Ortmann, 1893) [Eastern coasts from México and Caribbean Islands to Brazil. Nearctic]

4(2) Female pereopod III coxa with small blunt tooth; male petasma without capitulum; male antenna I peduncle with proximal article shorter than second and third combined; rivers, freshwater .... *Acetes marinus* (Omori, 1975) [French Guaiana, Guyana, Suriname, Brazil]

4′ Female pereopod III coxa with large acute tooth; male petasma with rudimentary capitulum; male antenna I peduncle with proximal article longer than second and third combined; rivers, freshwater .... *Acetes paraguayensis* (Hansen, 1919) [Venezuela, Suriname, Brazil, Peru, Paraguay, Argentina]

REFERENCE

Vereshchaka, A.L., A.A. Lunina & J. Olesen. 2016. Phylogeny and classification of the shrimp genera *Acetes, Peisos*, and *Sicyonella* (Sergestidae: Crustacea: Decapoda). Zoological Journal of the Linnean Society 177: 353–377.

Decapoda: Caridea

Fernando L. Mantelatto

Laboratório de Bioecologia e Sistemática de Crustáceos (LBSC), Departamento de Biologia, Faculdade de Filosofia, Ciências e Letras de Ribeirão Preto (FFCLRP), Universidade de São Paulo (USP), São Paulo, SP, Brazil

Célio Magalhães

Instituto Nacional de Pesquisas da Amazônia (INPA), Manaus, AM, Brazil

D. Christopher Rogers

Kansas Biological Survey, and The Biodiversity Institute, University of Kansas, Lawrence, KS United States

INTRODUCTION

Among the extant diversity of caridean shrimp of more than 3,500 species (De Grave & Fransen, 2011), about one-fifth (~800 species) live in freshwater and related inland waters (De Grave et al., 2015). Particularities in the larval development (as abbreviation or complete suppression of larval stages), reproductive features (as extended of brood care), and physiological adaptations are among key adaptations to freshwater (Anger, 2016; Vogt, 2016). To the best of our knowledge, no established definition of true freshwater shrimp is available. So, by analogy, we consider all shrimps that present the combination of these previous characteristics and adults inhabiting a freshwater habitat in most part of their life cycle, either if some larval phases depend on marine or estuarine conditions for development. Neotropical freshwater shrimp are known from six families: Alpheidae, Atyidae, Euryrhynchidae, Hippolytidae, Palaemonidae, and Xiphocarididae.

For **Terminology and Morphology** see the section for Decapoda above (**p. 872**), and for **Material Preparation and Preservation** see the section for Decapoda above (**p. 873**).

LIMITATIONS

Freshwater shrimp taxonomy has changed greatly in the last decades, mostly promoted by the increase of surveys,

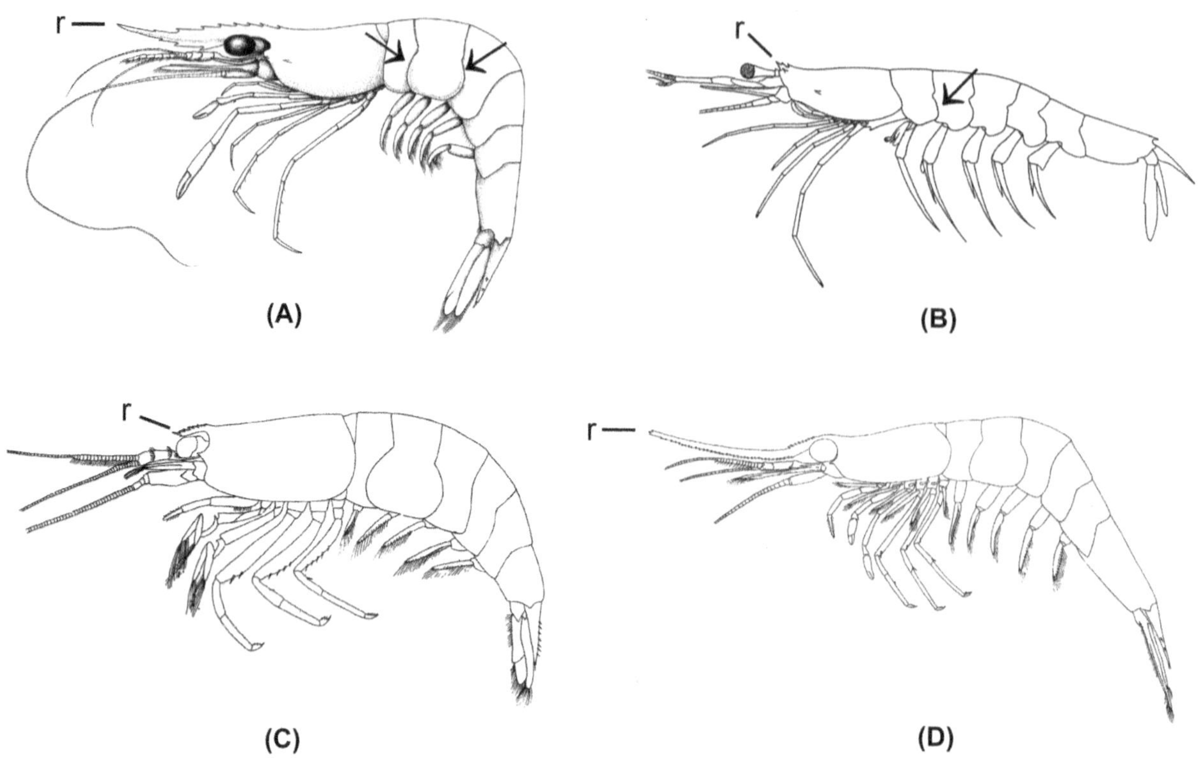

FIGURE 23.94 General morphology of decapod shrimp (Dendrobranchiata and Caridea), lateral view of: (A) Palaemonidae, *Pseudopalaemon nigramnis*; (B) Sergestidae, *Acetes paraguayensis*; (C) Atyidae, *Micratya poeyi*; (D) Atyidae, *Xiphocaris elongata*. Abbreviation: *r*, rostrum; *arrow*, pleonal somite 2. *Illustrations modified: (A) from Kensley & Walker (1982); (B) from Omori (1975); (C and D) from Pereira & García (1995).*

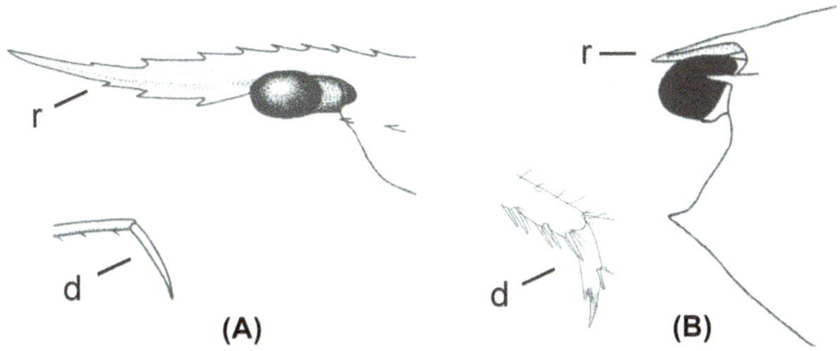

FIGURE 23.95 Lateral view of anterior part of the body and dactylus of Caridea spp.: (A) *Pseudopalaemon nigramnis*; (B) *Euryrhynchus amazoniensis*. Abbreviations: *d*, dactylus; *r*, rostrum. *Illustrations modified: (A) from Kensley & Walker (1982); (B) from Pachelle & Tavares (2018).*

cooperative projects in different regions, and through molecular techniques and morphological integrative studies. However, limitations still exist. There are strong indications that freshwater shrimp diversity is greater than currently known, and an increase in aquatic fauna surveys in some poorly explored regions may reveal additional diversity. Morphological variability within and between populations is responsible for many of the taxonomic inconsistencies among its representatives (García-Dávila & Magalhães, 2004; Pileggi & Mantelatto, 2012; Mantelatto et al., 2016). Part of this variability may be related to the presence of cryptic species not yet described, to ontogenetic development, and/or the presence of morphotypes (see Mantelatto et al., 2016 for review). The high degree of morphological conservatism and coexistence in same regions hinder the separation of some groups and aggravates this problem. Furthermore, the monophyly of many groups has not been tested, although recent work demonstrates the need for more study (De Grave & Ashelby, 2013; Carvalho et al., 2017, 2020).

The limited regional representation without a morphologically contextualized phylogenetic structure has led to mistakes in species identifications in the genus *Macrobrachium* (Pileggi & Mantelatto, 2010; 2012). In *Macrobrachium* only fully mature males can be identified and many characters used are variable (Holthuis, 1952). Similarly, some atyid genera can only be identified to species level using mature males.

KEY TO CARIDEA

Decapoda: Caridea: Families

1 Pereopod II carpus multiarticulate, of at least five articles (Fig. 23.96 A, B) 2

1′ Pereopod II carpus not subdivided (Figs. 23.94 A, 96 C) 4

2(1) Pereopod II carpus with > 25 articles; anchialine (Fig. 23.96 B) 3

2′ Pereopod II carpus with five articles (Fig. 23.96 A); eyes reduced, beneath carapace; cornea absent .... Alpheidae, one species: *Potamalpheops stygicola* (Hobbs, 1973) [Mexico]

3(2) Rostrum inerm .... Hippolytidae, one genus: ***Calliasmata*** **[p. 907]**

3′ Rostrum with dorsal and ventral teeth .... **Barbouriidae [p. 907]**

4(1) Pereopods I and II chelate, chelae without apical dense setal tufts (Fig. 23.96 C); pereopods always without exopods 5

4′ Pereopods I and II chelate, chelae with apical dense setal tufts (Figs. 23.94 C, D, 96 E, F) except *Xiphocaris*; pereopods either lacking exopods or exopods present (Fig. 23.96 D) on at least pereopod I and II .... **Atyidae [p. 907]**

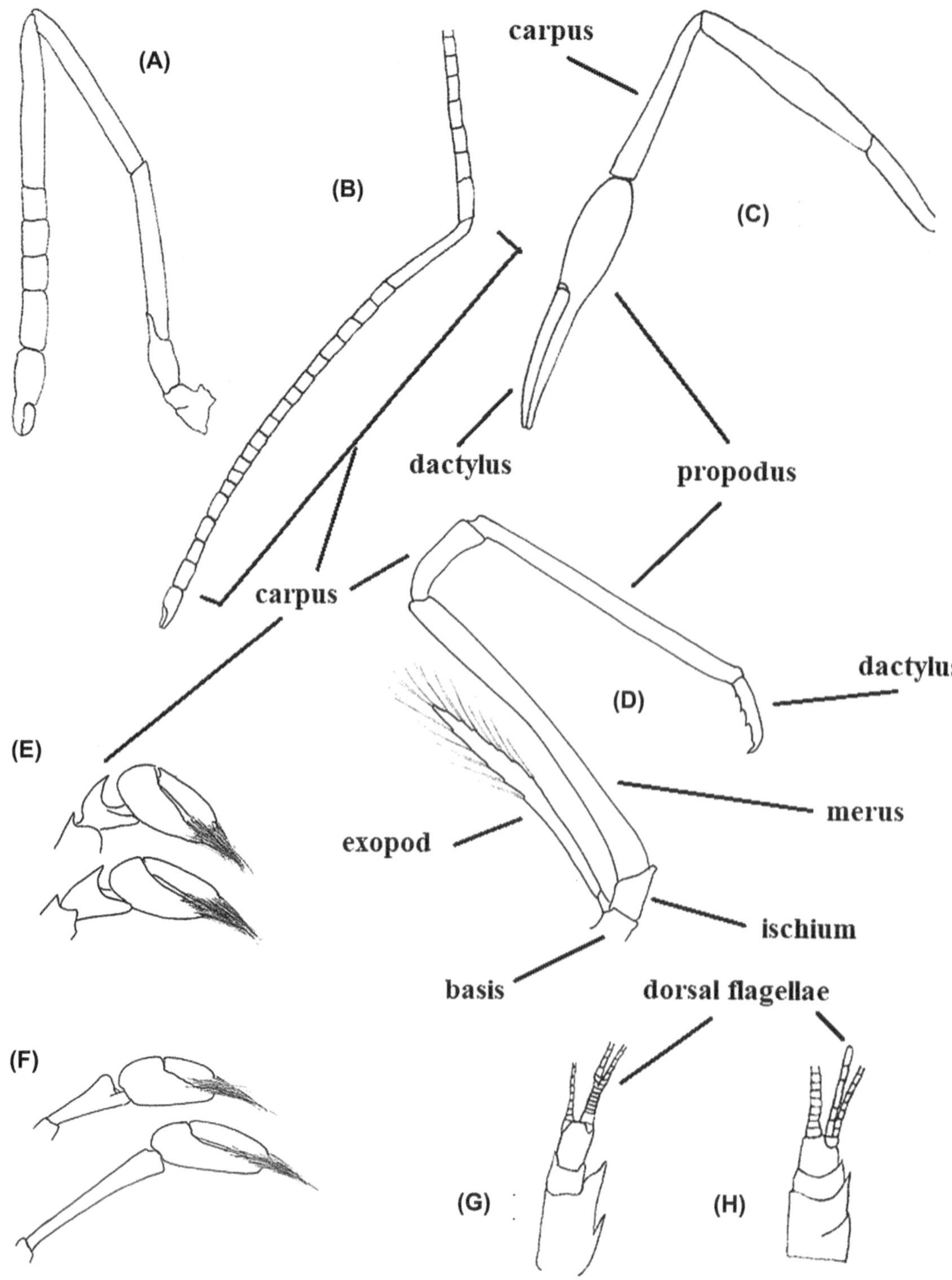

FIGURE 23.96 Caridea morphology: (A) *Potamalpheops stygicola* pereopod II; (B) *Calliasmata* sp. pereopod II; (C) *Palaemon* sp. pereopod II; (D) generalised caridean pereopod II; (E) *Atya* sp. carpi and chelae of pereopods I and II; (F) *Jonga serrei* carpi and chelae of pereopods I and II; (G) *Macrobrachium* sp. antenna I peduncle and proximal portion of flagella; (H) *Euryrhynchus* sp. antenna I peduncle and proximal portion of flagella.

5(4) Antenna I with dorsal flagella rami free (Fig. 23.96 H); males lacking appendix masculina, females lacking appendix interna .... Euryrhynchidae, one genus: ***Euryrhynchus*** **[p. 912]**

5′ Antenna I with dorsal flagella rami fused basally (Fig. 23.96 G); adult males with appendix masculina, adult females with appendix interna .... **Palaemonidae [p. 912]**

Decapoda: Caridea: Hippolytidae: *Calliasmata*: Species

1 Carapace and integument with setulae; antenna I stylocerite apex angular .... *Calliasmata nohochi* Escobar-Briones, Camacho, & Alcocer, 1997 [Mexico]

1′ Carapace and integument glaborous; antenna I stylocerite apex rounded .... *Calliasmata rimolii* Chace, 1975 [Dominican Republic]

Decapoda: Caridea: Barbouriidae: Genera

1 Rostrum length ~2.5× depth, reaching to or beyond antenna I peduncle second article base, with three dorsal (two postorbital) and 1–6 ventral teeth; cornea breadth > stalk; abdominal pleura I–III unarmed, pleura IV–VI posterioventral corner spiniform; anchialine .... *Parhippolyte sterreri* (Hart & Manning, 1981) [Bermuda, Bahamas]

1′ Rostrum ~5× depth, extending slightly beyond end of antenna I peduncle basal article, with four to seven dorsal (three postorbital) and one to four ventral teeth; cornea narrower than stalk; abdominal pleura I–IV rounded, pleura V and VI with posteroventral corner spiniform; anchialine .... ***Barbouria*** **[p. 907]**

Decapoda: Caridea: Barbouriidae: *Barbouria*: Species

1 Rostrum with five ventral teeth not reaching antenna I second article distal margin .... *Barbouria cubensis* (Von Martens, 1872) [Bermuda, Bahamas Islands, Cayman Islands, Cuba]

1′ Rostrum with three to nine ventral teeth extending to antenna I second article distal margin .... *Barbouria yanezi* Mejía, Zarza & López, 2008 [Mexico]

Decapoda: Caridea: Atyidae: Genera

The genus *Potimirim* has been examined using molecular techniques, demonstrating the probable presence of cryptic species (Torati & Mantelatto, 2012).

1 Pereopods all lacking exopods; eyes present and functional; body pigmented 2

1′ At least pereopod I and II with exopods present 6

2(1) Rostrum compressed dorsoventrally or laterally, if laterally, then without dorsomedian row of teeth; pereopod I with a basal arthrobranch; nine pairs of branchiae present (Fig. 23.98 B–D) 3

2′ Rostrum compressed laterally and bearing a dorsomedial row of teeth; pereopod I without a basal arthrobranch; eight pairs of branchiae present (Fig. 23.98 A) .... ***Micratya*** **[p. 910]**

3(2) Pereopod I and II with carpi subequal, each carpus length subequal to width, embracing propodus base (Figs. 23.96 E, 100 B) 4

3′ Pereopod I and II with carpus length ≥2× width; pereopod I carpus smaller than pereopod II carpus (Figs. 23.96 F, 100 A) 5

4(3) Maxilliped III terminating in a single spine, sometimes concealed by surrounding setae; male pleopod I tapering distally .... *Archaeatya chacei* Villalobos, 1960 [Western Costa Rica, Cocos Island]

4′ Maxilliped III terminating in stout setae, never with a terminal spine; male pleopod I explanate distally .... ***Atya*** **[p. 910]**

5(3) Orbit margined in acute denticulae; appendix masculina slender, apically acute (Fig. 23.97 B) .... *Jonga serrei* (Bouvier, 1909) [Mexico Caribbean Islands, Trinidad and Tobago, Venezuela]

5′ Orbit inerm; appendix masculina lamellar, margined in spinulae; in need of revision (Fig. 23.97 C) .... *Potimirim* [Mexico. Greater and Lesser Antilles. Honduras. El Salvador. Nicaragua. Costa Rica. Panama. Colombia. Venezuela. Brazil]

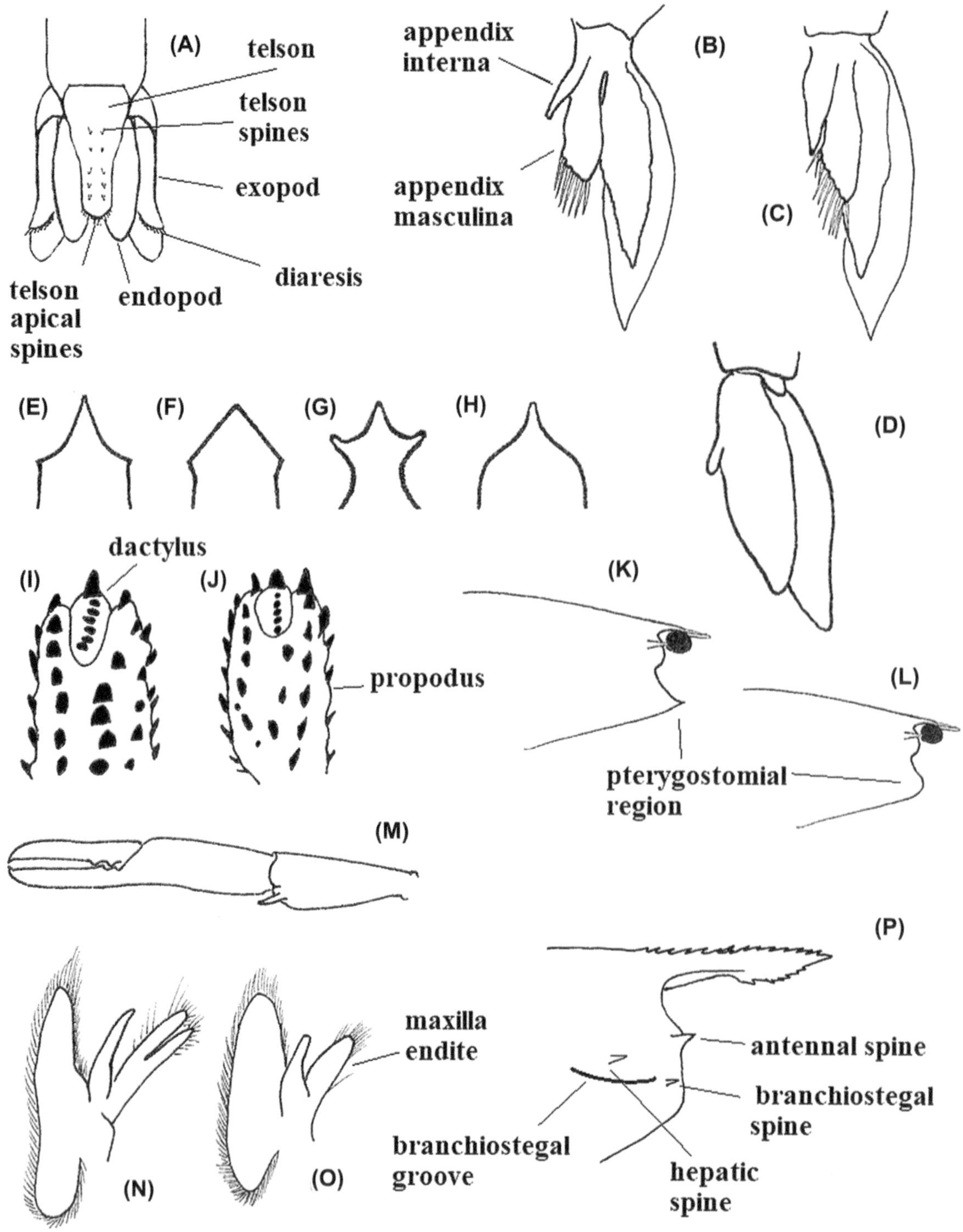

FIGURE 23.97 Caridea morphology: (A) generalized caridean shrimp telson; (B) *Micratya poeyi* male pleopod II; (C) *Micratya cooki* male pleopod II; (D) *Euryrhynchus taruman* male pleopod II; (E–H) *Atya* sp. variation in rostral form in dorsal view; (I) *Atya scabra* pereopod III distal portion of propodus and dactylus; (J) *Atya margaritacea* pereopod III distal portion of propodus and dactylus; (K) *Euryrhynchus amazoniensis* carapace, right lateral view; (L) *Euryrhynchus taruman* carapace, right lateral view; (M) *Euryrhynchus pemoni* pereopod II; (N) *Troglocubanus* sp. maxillae; (O) *Troglomexicanus* sp. maxillae; (P) generalized *Palaemonid* carapace, anterior portion, right lateral view.

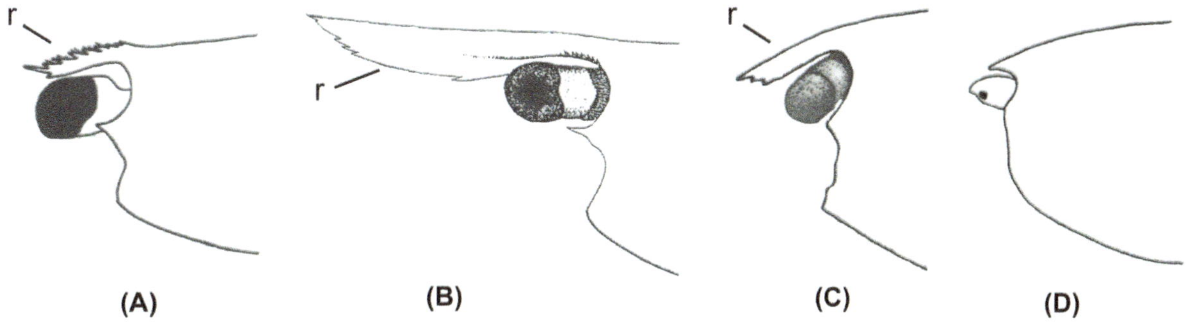

FIGURE 23.98 Lateral view of anterior part of the body of Caridea spp.: (A) *Micratya poeyi*; (B) *Jonga serrei*; (C) *Potimirim glabra*; (D) *Typhlatya monae*. Abbreviation: *r*, rostrum. *Illustrations modified: (A) from Pereira & García (1995); (B) from Hart (1961); (C) from Campos (2014); (D) from Chace (1954).*

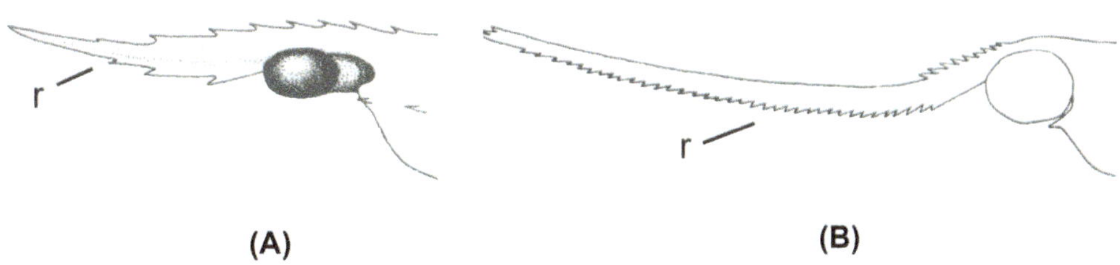

FIGURE 23.99 Lateral view of anterior part of the body of Caridea spp.: (A) *Pseudopalaemon nigramnis*; (B) *Xiphocaris elongata*. Abbreviation: *r*, rostrum. *Illustrations modified: (A) from Kensley & Walker (1982); (B) from Pereira & García (1995).*

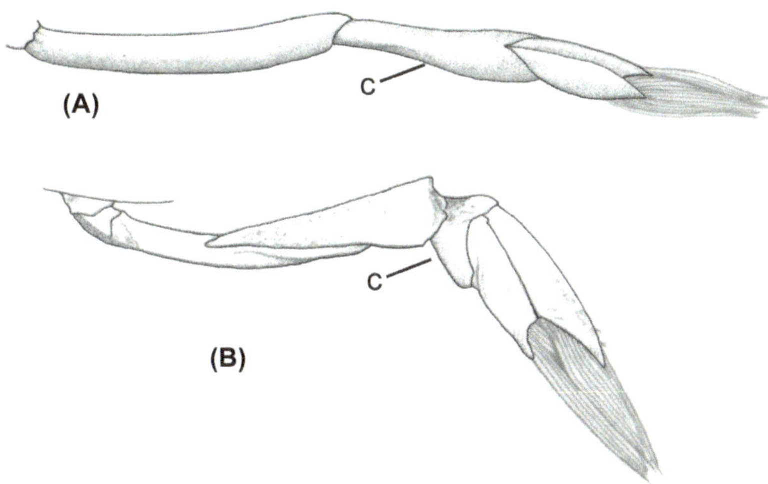

FIGURE 23.100 Second pereopod of Caridea spp.: (A) *Potimirim glabra*; (B) *Atya scabra*. Abbreviation: *c*, carpus. *Illustrations modified from Campos (2014).*

6(1) Pereopod I and II only with exopods present; eyes reduced, cornea reduced to a distolateral spot or absent; body unpigmented; carapace entirely without spines (Fig. 23.98 D) .... ***Typhlatya* [p. 911]**

6′ All pereopods with exopods; pereopods I and II chelate, chelae without apical dense setal tufts (Figs. 23.96 D, 94 D) .... *Xiphocaris elongata* Guérin-Méneville, 1855 [Caribbean Islands, Venezuela]

Decapoda: Caridea: Atyidae: *Micratya*: Species

1 Uropod diaresis (Fig. 23.97 A) with 20–23 movable spines; telson distal margin with 14–16 distal spines; appendix masculina with spines near distal; appendix interna apex not reaching the spinulated part of appendix masculina (Fig. 23.97 B) .... *Micratya poeyi* (Guérin-Méneville, 1855)
[Caribbean Islands. Costa Rica. Nicaragua. Panama. Venezuela]

1′ Uropod diaresis with 16–18 movable spines; telson distal margin with 10–12 distal spines; appendix masculina spines extending more proximally; appendix interna apex reaches spinulated part of appendix masculina (Fig. 23.97 C) .... *Micratya cooki* Karge, Page & Klotz, 2013
[Puerto Rico]

Decapoda: Caridea: Atyidae: *Atya*: Species

Modified from Hobbs & Hart (1982).

1 Rostrum with medial row of spines; carapace rough with spines or tubercles 2

1′ Rostrum lacking medial spine row 3

2(1) Rostral medial spines separated by a distance ≥ their basal width; carapace with prominent ridges bearing spines .... *Atya crassa* (Smith, 1871)
[Mexico: Pacific Basin. Nicaragua. Panama. Costa Rica. Colombia. Ecuador]

2′ Rostral medial spines separated by a distance <0.5× their basal width; carapace densely granulate dorsally .... *Atya limnetes* Holthuis, 1986
[Colombia]

3(1) Rostrum lateral margins in dorsal view an acute or right angle bend or with lateral projections (Fig. 23.97 E–G); pereopod III dactyl flexor surface with tubercles arranged in a small cluster or in a single row 4

3′ Rostrum margins without subangular bend or if present, then bend obtuse or with angles rounded (Fig. 23.97 H); pereopod III dactyl flexor surface with tubercles never arranged in a small cluster or in a single row 7

4(3) Preanal carina entire 5

4′ Preanal carina bifid in lateral view .... *Atya abelei* Felgenhauer & Martin, 1983
[Panama]

5(4) Pereopod III propodus flexor surface with few to many heavily sclerotized, elevated or squamous tubercles, if tubercles form linear series, then mesial row not quite contiguous (Fig. 23.97 J); abdominal segment II pleuron without sclerotized denticles 6

5′ Pereopod III propodus flexor surface with some heavily cornified squamous tubercles in longitudinal rows, those in mesial row subcontiguous (Fig. 23.97 I); abdominal segment II pleuron often with sclerotized denticles .... *Atya scabra* (Leach, 1816)
[Mexico to Brazil. Caribbean Islands. Afrotropical]

6(5) Carapace strongly sculptured; pereopod III propodus flexor surface with conspicuous tufts of long plumose setae and few squamous tubercles; pereopod III dactylus flexor surface with single or cluster of tubercles just proximal to sclerotized apex; abdominal segment V sternum with medial curved spiniform projection .... *Atya gabonensis* Giebel, 1875
[Venezuela to Brazil. Afrotropical]

6′ Carapace weakly sculptured; pereopod III propodus flexor surface without conspicuous setal tufts, with prominent squamous tubercles; pereopod III dactylus flexor surface with single row of tubercles; abdominal segment V sternum with small median tubercle .... *Atya margaritacea* A. Milne-Edwards, 1864
[Mexico: Baja California, to Peru]

7(3) Rostrum margins subparallel to strongly convergent, apex reaching or almost reaching antenna I second peduncle article base 8

7′ Rostrum margins strongly convex laterally, apex not nearly reaching antenna I peduncle second article distal margin .... *Atya brachyrhinus* Hobbs & Hart, 1982
[Barbados]

8(7) Rostral margins with strong subangular bend; abdominal pleura III–V ventral margin with or without sclerotized spinulae 9

8′ Rostral margins tapering distally; abdominal pleura III–V ventral margin without sclerotized spinulae 10

9(7) Pereopod III merus lateral surface with cornified tubercles in well-defined rows; carapace cephalolateral surface convex from anterior to posterior; rostrum directed anterioventrally, acumen length ≤ basal portion in dorsal view .... *Atya innocous* (Herbst, 1792)
[Caribbean Islands. Costa Rica. Nicaragua. Panama]

9′ Pereopod III merus lateral surface without tubercles or if present small, weakly cornified, never in well-defined rows; carapace cephalolateral surface straight; rostrum directed anteriorly, acumen length > basal portion in dorsal view .... *Atya ortmannioides* Villalobos, 1956
[Mexico: Pacific Basin]

10(8) Pterygostomian angle produced in prominent spine; antenna I peduncle article II dorsal surface with scattered spines; pereopod III merus dorsally and laterally without tubercles; preanal carina with vestigial spine; telson dorsal surface with rows of five to seven spines .... *Atya lanipes* Holthuis, 1963
[Caribbean Islands]

10′ Pterygostomian angle weak or obsolete; antenna I peduncle article II dorsal surface with lateral row of ~ six spines, plus one to three situated more mesially; pereopod III merus dorsally and laterally with distinct tubercles; preanal carina produced in well-developed caudally directed spine; telson dorsal surface rows of seven to nine spines .... *Atya dressleri* Abele, 1975
[Costa Rica. Panama]

Decapoda: Caridea: Atyidae: *Typhlatya*: Species

Modified from Alvarez et al. (2005). More species are expected to be discovered.

1 Eyes pigmented 2

1′ Eyes without pigment 6

2(1) Pereopods with ischium and merus free 3

2′ Pereopods with ischium and merus fused 5

3(2) Pereopod V exopod reduced, not reaching ischiomeral articulation 4

3′ Pereopod V exopod reaching beyond ischiomeral articulation .... *Typhlatya garciai* (Chace, 1942)
[Cuba: Oriente Province]

4(3) Eyes entire; uropod lateral, ramus lateral, margin without submarginal setae .... *Typhlatya monae* (Chace, 1954)
[Caribbean Islands]

4′ Eyes bilobed; uropod lateral ramus lateral margin with submarginal setae .... *Typhlatya utilaensis* Alvarez, Iliffe & Villalobos, 2005
[Honduras: Utila]

5(2) Pereopods III–V with ischium and merus fused .... *Typhlatya kakuki* Alvarez, Iliffe & Villalobos, 2005
[Bermuda]

5′ All pereopods with ischium and merus fused .... *Typhlatya iliffe* Hart & Manning, 1981
[Bermuda]

6(5) Rostrum extending beyond eye distal margin 7

6′ Rostrum not or barely reaching eye distal margin 10

7(6) Rostrum reaching antenna I first article 8

7′ Rostrum reaching antenna I second article 9

8(7) All pereiopods with ischium and merus free .... *Typhlatya campechae* Hobbs & Hobbs, 1976
[Mexico: Campeche]

8′ Pereopods I and II with ischium and merus fused, pereopods III–V with ischium and merus free .... *Typhlatya consobrina* Botasaneanu & Holthuis, 1970
[Cuba: Camagüey Province]

9(7) Pereopod V exopod length < basipodite length .... *Typhlatya pearsei* Creaser, 1936
[Mexico: Yucatan]

9′ Pereopod V exopod not reduced, reaching ischiomeral articulation .... *Typhlatya elena* Juarrero, 1994
[Cuba: Matanzas Province]

10(6) Abdominal somite VI length <4× somite V length 11

10′ Abdominal somite VI length 4× somite V length .... *Typhlatya garciadebrasi* Juarrero de Varona & Ortiz, 2000
[Cuba: Matanzas Province]

11(10) Pereopod II carpus/propodus ratio < 2.5 12

11′ Pereopod II carpus/propodus ratio > 2.5 .... *Typhlatya mitchelli* Hobbs & Hobbs, 1976
[Mexico: Yucatan]

12(11) Pereopods III–V with ischium and merus fused 13

12′ All pereopods with ischium and merus free .... *Typhlatya taina* Estrada & Gomez, 1987
[Cuba: Havanna]

13(12) Abdominal somite II pleura with ventral margin bilobed .... *Typhlatya galapagensis* Monod & Cals, 1970
[Ecuador: Galápagos Islands]

13′ Abdominal somite II pleura with ventral margin entire .... *Typhlatya dzilamensis* Alvarez, Iliffe & Villalobos, 2005
[Mexico: Yucatan]

Decapoda: Caridea: Euryrhynchidae: *Euryrhynchus*: Species

Modified from the complete revision of Pachelle & Tavares (2018).

1 Pereopod II merus distoventral angles inerm, smooth; pereopod III dactylus dorsal margin with one mesial cuspidate seta; pereopod IV propodus distodorsal angle usually without cuspidate setae 2

1′ Pereopod II merus distoventral angles with one or two spines (Fig. 23.97 M); pereopod III dactylus dorsal margin with pair of cuspidate setae; pereopod IV propodus distodorsal angle usually with cuspidate seta 5

2(1) Pereopod II carpus distomesial angle with spine; pereopod V dactylus dorsal margin with pair of cuspidate setae; gonopod stout, twisted, not overreaching exopod; appendix interna (if present) length < endopod length; pleopods IV and V endopod length ≥ 0.5× exopod 3

2′ Pereopod II carpus distomesial angle inerm, smooth; pereopod V dactylus dorsal margin with one lateral cuspidate seta; gonopod straight, overreaching exopod; appendix interna length ≥ endopod; pleopods IV and V endopod length < 0.5× exopod .... *Euryrhynchus wrzesniowskii* Miers, 1878
[Brazil. French Guiana. Guyana. Suriname]

3(2) Male pleopod II with appendix interna only (Fig. 23.97 D) 4

3′ Pleopod II and III (rarely IV) with appendix interna in both sexes .... *Euryrhynchus tuyuka* Pachelle & Tavares, 2018
[Brazil]

4(3) Carapace pterygostomial region produced anteriorly into acute, triangular projection (Fig. 23.97 K); large individuals with pereopod II carpus in dorsal view length 2× width .... *Euryrhynchus amazoniensis* Tiefenbacher, 1978
[Venezuela. Colombia. Brazil. Peru]

4′ Carapace pterygostomial region rounded or broadly angular, not produced (Fig. 23.97 L); large individuals with pereopod II carpus length ~2.5–3× width .... *Euryrhynchus taruman* Pachelle & Tavares, 2018
[Brazil]

5(1) Pereopod II merus with one spine on each distoventral angle; female pleopods III–V with appendix interna 6

5′ Pereopod II merus with single distomesial spine; female pleopods III–V without appendix interna .... *Euryrhynchus pemoni* Pereira, 1985
[French Guiana. Guyana. Venezuela]

6(5) Pereopod II without fine granules, carpus distomesial angle with spine; pereon sternite IV with transverse ridge produced medially into low, broad lobe .... *Euryrhynchus burchelli* Calman, 1907
[Brazil]

6′ Pereopod II carpus and propodus surface finely granulate, carpus distomesial angle inerm; pereon sternite IV with transverse ridge produced medially into prominent, triangular lobe .... *Euryrhynchus tomasi* De Grave, 2007
[Brazil. French Guiana]

Decapoda: Caridea: Palaemonidae: Genera

Modified from Holthuis's (1952) historical work. The genus *Macrobrachium* is the largest and most diverse genus of freshwater shrimp in the Neotropical region. It is also the most morphologically difficult group. Determinations to species level are only possible using fully mature males, and even then, these animals are highly variable both within and between species. Females, immatures, and juveniles of both sexes tend to be morphologically nondescript and generalized, and specimens may be fully sexually mature and still retain juvenile characteristics. The level of morphological variability and overlap, the need for revision in this group, and large number of known undescribed species prevent us from providing a key to *Macrobrachium* species.

1 Hepatic spine present, brachiostegal spine absent (Fig. 23.97 P) 2

1′ Hepatic spine absent, brachiostegal spine absent or present 4

2(1) Mandible palp present (Fig. 23.101 A) 3

2′ Mandible palp absent (Fig. 23.101 B) .... ***Pseudopalaemon* [p. 914]**

3(1) Pereopod V propodus without transverse setal row; mandible palp with two palpomeres; cornea without facets or pigment; troglobitic .... *Neopalaemon nahuatlus* Hobbs, 1973
[Mexico]

3′ Pereopod V propodus with one or more transverse setal rows; mandible palp with three palpomeres; eyes with or without corneal facets and pigment; epigean and troglobitic forms .... *Macrobrachium*
[Widely distributed in the Neotropics]

4(1) Branchiostegal spine present (Fig. 23.97 P) 5

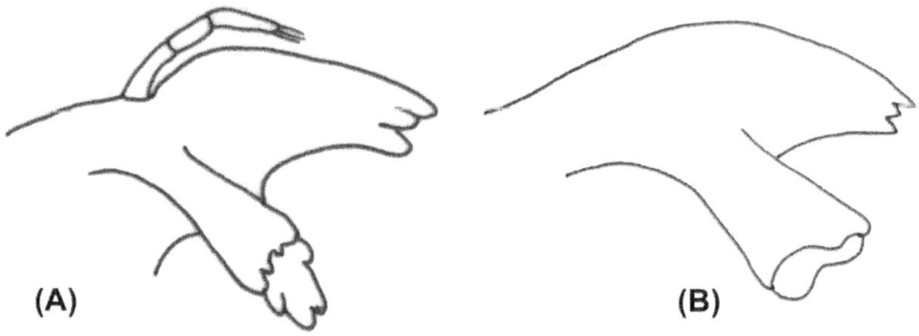

FIGURE 23.101 Mandible of Caridea spp.: (A) *Macrobrachium inpa*; (B) *Pseudopalaemon funchiae*. *Illustrations modified: (A) from Kensley & Walker (1982); (B) from García-Dávila & Magalhães (2004).*

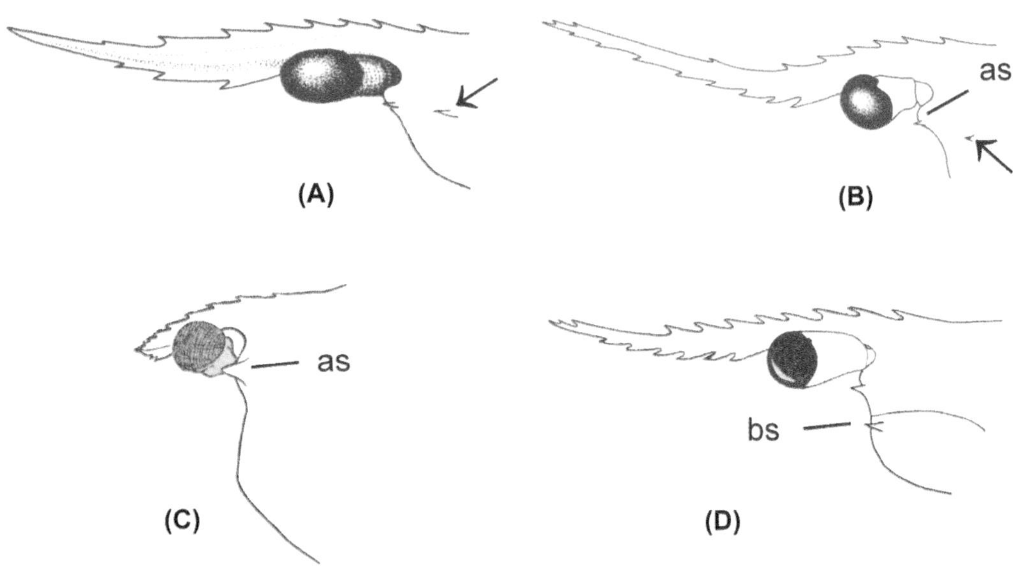

FIGURE 23.102 Lateral view of Caridea spp.: (A) *Pseudopalaemon nigramnis*; (B) *Macrobrachium amazonicum*; (C) *Cryphiops brasiliensis*; (D) *Palaemon carteri*. Abbreviation: *as*, antennal spine; *bs*, branchiostegal spine; *arrow*, hepatic spine. *Illustrations modified: (A, B) from Kensley & Walker (1982); (C) from Gomes Corrêa (1973); (D) from Holthuis (1959).*

4′	Brachiostegal spine absent	7
5(4)	Branchiostegal groove absent	6
5′	Branchiostegal groove present, distinct (Fig. 23.97 P); eyes with or without corneal facets and pigment; epigean and troglobitic forms pereopods I–III with dactyli length ≤ propodus	***Palaemon* [p. 914]**
6(5)	Pereopod III dactylus length ~ 0.2× propodus; eyes without pigment; troglobitic [Mexico]	*Creaseria morleyi* (Creaser, 1936)
6′	Pereopod III dactylus length ≥ propodus + carpus; eyes normal; estuariune [Brazil. French Guyana. Guyana. Suriname]	*Nematopalaemon schmitti* Holthuis, 1950
7(4)	Mandible palp absent (Fig. 23.101 B); pereopods gracile; cornea unfaceted, without pigment; troglobitic	8
7′	Mandible palp present, with three palpomeres (Fig. 23.102 C); pereopod II robust tuberculate; troglobitic or not	***Cryphiops* [p. 915]**
8(7)	Maxilla endite bilobate (Fig. 23.97 N); maxilliped I epipodite entire	***Troglocubanus* [p. 916]**
8′	Maxilla endite entire (Fig. 23.97 O); maxilliped I epipodite bilobed	***Troglomexicanus* [p. 916]**

Decapoda: Caridea: Palaemonidae: *Pseudopalaemon*: Species

Adapted and modified from Kensley & Walker (1982).

1 Rostrum length $\leq 1.2\times$ carapace length 2

1′ Rostrum length $\geq 1.5\times$ carapace length .... *Pseudopalaemon amazonensis* Ramos-Porto, 1979
[Brazil Venezuela]

2(1) One or more postorbital rostral spines 3

2′ No postorbital rostral spines .... *Pseudopalaemon bouvieri* Sollaud, 1911
[Brazil. Argentina. Uruguay]

3(2) One postorbital rostral spine 4

3′ Two or more postorbital rostral spines 5

4(3) Rostrum with seven to ten dorsal teeth; cornea dorsoventrally depressed .... *Pseudopalaemon nigramnis* Kensley & Walker, 1982
[Brazil]

4′ Rostrum with 11 to 13 dorsal teeth; cornea subspherical .... *Pseudopalaemon funchiae* García-Dávila & Magalhães, 2004
[Peru]

5(3) Both telson dorsal spine pairs situated in telson distal half 6

5′ Telson with anterior pair of dorsal spines situated at telson midlength .... *Pseudopalaemon chryseus* Kensley & Walker, 1982
[Brazil Venezuela]

6(5) Rostrum lightly shorter than scaphocerite; pereopod II dactyl and pollex length subequal to palm .... *Pseudopalaemon gouldingi* Kensley & Walker, 1982
[Brazil. Venezuela]

6′ Rostrum length subequal to scaphocerite; pereopod II dactyl and pollex length $0.8\times$ palm .... *Pseudopalaemon iquitoensis* García-Dávila & Magalhães, 2004
[Peru]

Decapoda: Caridea: Palaemonidae: *Palaemon*: Species

1 Rostrum length $\geq 1.5\times$ scaphocerite length, apex typically directed upward 2

1′ Rostrum length $\leq 1.2\times$ scaphocerite length 4

2(1) Telson gradually tapering to acute apex, no distinct posterior margin present; telson posterior spines surpassed by apex 3

2′ Telson apex with distinct posterior margin with central acute projection, which is overreached by central pair of posterior spines .... *Palaemon pandaliformis* (Stimpson, 1971)
[Caribbean Islands. Atlantic Drainages from Guatemala to Brazil]

3(2) Rostrum ventral margin with 9 to 12 teeth; maxilliped III distal article length subequal to penultimate article; pereopod II more robust than pereopod I, dactyl and pollex length subequal to palm, carpus length $> 2\times$ chela .... *Palaemon gracilis* (Smith, 1871)
[Pacific drainages from Nicaragua to Panama]

3′ Rostrum ventral margin with 11 to 16 teeth; maxilliped III distal article length $0.7\times$ penultimate article; pereopod II similar to pereopod I, dactyl and pollex length $<$ palm, carpus length $2\times$ chela .... *Palaemon hancocki* Holthuis, 1950
[Colombia. Ecuador]

4(1) Pereopod II carpus length $<$ chela 5

4′ Pereopod II carpus length $>$ chela 8

5(4) Pereopod II carpus length $0.7\times$ chela length 6

5′ Pereopod II carpus length subequal to palm length 7

6(5) Abdominal segment VI length = telston .... *Palaemon peruanus* Hothuis, 1950
[Peru]

6′ Abdominal segment VI length $<$ telston .... *Palaemon hiltoni* (Schmitt, 1916)
[Pacific drainages in Mexico. Nearctic]

7(6) Rostrum dorsal margin with 11–13 teeth, straight, though midrib curved; pereopod III reaching with part of propodus beyond scaphocerite; marine, but survives in portions of Panama Canal .... *Palaemon schmitti* Holthuis, 1950
[Panama]

7′ Rostrum dorsal margin concave, with 8 to 11 teeth, apex gradually curved upwards; pereopod III just reaching to or failing to reach scaphocerite apex .... *Palaemon vulgaris* Say, 1818
[Mexico. Nearctic]

8(4) Branchiostegal spine posterior to carapace anterior margin 9

8′ Branchiostegal spine placed on carapace anterior margin 12

9(8) Branchiostegal spine set back $\geq 2\times$ its length from carapace margin; rostrum ventral margin with two or three teeth 10

9′ Branchiostegal spine apex overreaching carapace margin; rostrum ventral margin with four to seven teeth 11

10(9) Telson basal width $\geq 2\times$ length .... *Palaemon ivonicus* Holthuis, 1950
[Brazil. Bolivia]

10′ Telson basal width $\sim 1.5\times$ length .... *Palaemon mercedae* Pereira, 1986
[Venezuela. Brazil]

11(9) Scaphocerite anteriolateral projection distal margin not or nearly reaching second article dorsodistal margin; scaphocerite anteriolateral spine usually reaching middle of anteriolateral projection; rostrum ventrally with three to seven teeth (usually > four) .... *Palaemon carteri* (Gordon, 1935)
[Venezuela. Brazil. French Guiana. Guyana. Suriname]

11′ Scaphocerite anteriolateral projection distal margin reaching or overreaching second article dorsodistal margin; scaphocerite anteriolateral spine not reaching middle of anteriolateral projection; rostrum ventrally with one to five teeth (usually $\leq$ four) .... *Palaemon yuna* Carvalho, Magalhães & Mantelatto, 2014
[Brazil. Peru. Bolivia]

12(8) Telson posterior dorsolateral spine pair situated midway between anterior pair and telson posterior margin 13

12′ Telson posterior dorsolateral spine pair situated at or adjacent to telson posterior margin or posterior spine pair absent .... *Palaemon mexicanus* (Strenth, 1976)
[Mexico: San Luis Potosí]

13(12) Pereopod II merus length = chela length 14

13′ Pereopod II merus length $\leq 08\times$ chela length .... *Palaemon suttkusi* (Smalley, 1963)
[Mexico: Coahuila]

14(13) Rostrum greatest depth $\geq 0.25\times$ rostrum length distance of orbit 15

14′ Rostrum greatest depth $\geq 0.15\times$ rostrum length distance of orbit .... *Palaemon hobbsi* (Strenth, 1994)
[Mexico: Tamaulipas]

15(14) Abdominal segment V with pleural distolateral margin angulate, projecting posteriorly .... *Palaemon lindsayi* (Villalobos Figueroa & Hobbs, 1974)
[Mexico: San Luis Potosí]

15′ Abdominal segment V with pleural distolateral margin vertical, straight, not projecting posteriorly .... *Palaemon argentinus* (Nobili, 1901)
[Brazil. Argentina. Uruguay]

Decapoda: Caridea: Palaemonidae: *Cryphiops*: Species

1 Male pereopod II pair slender, spiny, subequal in size; pereopods III–V propodus posterior margin with two longitudinal rows of spines; telson posterior margin triangular, acute or rounded, distal spines projecting beyond distal margin; rostrum length $\geq$ antenna I peduncle article II distal margin; freshwater 2

1′ Male pereopod II pair robust, spiny, unequal in size; pereopods III–V propodus spined, with two spine rows distally; telson posterior margin rounded or truncated, distal spines not exceeding posterior margin; rostrum reaching antenna I peduncle article I distal margin; fresh and brackish water .... *Cryphiops caementarius* (Molina, 1782)
[Peru. Chile]

2(1) Body lacking pigment; cornea pigment reduced to a small point; stygobionts 3

2′ Body with pigments; cornea with pigment 4

3(2) Pereopod I reaching pereopod II chela; pereopod II covered in spinulae, chelae with dactyl and pollex length $\leq$ palm length .... *Cryphiops luscus* (Holthuis, 1973)
[Mexico: Chiapas]

3′ Pereopod I not quite reaching pereopod II carpus distal margin; pereopod II covered in setal tufts, chelae with dactyl and pollex length $1.28\times$ palm length .... *Cryphiops sbordonii* Baldari, Mejía-Ortiz & López-Mejía, 2010
[Mexico: Chiapas]

4(2) Pereopod II chela palm length/width ratio > 3.25; adult male lower anteriolateral portion smooth; branchiostegal suture not terminating in a lobe 5

4′ Pereopod II chela palm length/width ratio < 3.0; adult male carapace lower anteriolateral portion with small and rounded spinulae; branchiostegal suture terminates in a shallow oval lobe below antennal spine .... *Cryphiops brasiliensis* (Gomes Corrêa, 1973)
[Brazil]

5(4) Appendix masculina not reaching endopod distal margin; appendix masculina/endopod length ratio 0.81–0.88; pereopod II carpus a little longer than palm .... *Cryphiops perspicax* (Holthuis, 1977)
[Mexico: Chiapas]

5′ Appendix masculina reaching endopod distal margin; appendix masculina/endopod length ratio 0.92–1.0; pereopod II carpus shorter than the palm. .... *Cryphiops villalobosi* Villalobos Hiriart, Nates Rodriguez & Cantú Diaz Barriga, 1989
[Mexico: Chiapas]

Decapoda: Caridea: Palaemonidae: *Troglocubanus*: Species

Adapted from Hobbs et al. (1977). All species are troglobitic.

1 Rostrum length < scaphocerite, six or fewer dorsal teeth; antennal spine marginal or absent 2

1′ Rostrum length subequal to scaphocerite, with six to eight dorsal teeth; antennal spine premarginal .... *Troglocubanus eigenmanni* (Hay, 1903)
[Cuba]

2(1) Rostrum apex not overreaching antenna I peduncle 3

2′ Rostrum apex overreaching antenna I peduncle .... *Troglocubanus calcis* (Rathbun, 1912)
[Cuba]

3(2) Rostrum with one or zero teeth 4

3′ Rostrum with two or three dorsal teeth .... *Troglocubanus gibarensis* (Chace, 1943)
[Cuba]

5(4) Antennal spine present .... *Troglocubanus jamaicensis* Holthuis, 1963
[Jamaica]

5′ Antennal spine absent .... *Troglocubanus inermis* (Chace, 1943)
[Cuba]

Decapoda: Caridea: Palaemonidae: *Troglomexicanus*: Species

Modified from Villalobos et al. (1999).

1 Telson length < uropod exopod 2

1′ Telson length subequal to uropod exopod .... *Troglomexicanus perezfarfante* Villalobos, 1971
[Mexico: San Luis Potosi]

2(1) Telson posterior margin rounded .... *Troglomexicanus tamaulipensis* Villalobos, Alvarez & Iliffe, 1999
[Mexico: Tamaulipas]

2′ Telson posterior margin truncate .... *Troglomexicanus huastecae* Villalobos, Alvarez & Iliffe, 1999
[Mexico: San Luis Potosi]

REFERENCES

Alvarez, F., T.M. Iliffe & J.L. Villalobos. 2005. New species of the genus *Typhlatya* (Decapoda: Atyidae) from anchialine caves in Mexico, the Bahamas, and Honduras. Journal of Crustacean Biology 25: 81–94.

Anger, K. 2016. Chapter 5: Adaptation to life in fresh water by decapod crustaceans: Evolutionary challenges in the early life-history stages. Pages 127–168 *in*: T. Kawai & N. Cumberlidge (eds.), A Global Overview of the Conservation of Freshwater Decapod Crustaceans. Springer, Cham, Switzerland.

Campos, M.R. 2014. Crustáceos decápodos de agua dulce de Colombia. Instituto de Ciencias Naturales. Biblioteca José JerónimoTriana, 27, 27, 692 pp.

Carvalho, F.L., S. De Grave & F.L. Mantelatto. 2017. An integrative approach to the evolution of shrimps of the genus *Palaemon* (Decapoda, Palaemonidae). Zoologica Scripta 46: 473–485.

Carvalho, F.L., C. Magalhães & F.L. Mantelatto. 2020. A molecular and morphological approach on the taxonomic status of the Brazilian species of *Palaemon* (Decapoda, Palaemonidae). Zoologica Scripta 49: 101–116.

Chace, F.A. Jr. 1954. Two new subterranean shrimps (Decapoda: Caridea) from Florida and the West indies, with a revised key to the American species. Journal of the Washington Academy of Sciences 44: 318–324.

De Grave, S. & C.W. Ashelby. 2013. A re-appraisal of the systematic status of selected genera in Palaemoninae (Crustacea: Decapoda: Palaemonidae). Zootaxa 3734: 331–344.

De Grave, S. & C.H.J.M. Fransen. 2011. Carideorum catalogus: The recent species of the dendrobranchiate, stenopodidean, procarididean and caridean shrimps (Crustacea: Decapoda). Zoologische Mededelingen 89: 195–589.

De Grave, S., K.G. Smith, N.A. Deler, D.J. Allen, F. Alvarez, A. Anker, Y. Cai, S.F. Carrizo, W. Klotz, F.L. Mantelatto, T.J. Page, J-Y Shy, J.L. Villalobos & D. Wowor. 2015. Dead shrimp blues: a global assessment of extinction risk in freshwater shrimps (Crustacea: Decapoda: Caridea). PLoS One 10: e0120198.

García-Dávila, C.R. & C. Magalhães. 2004. Revisão taxonômica dos camarões de água doce (Crustacea: Decapoda: Palaemonidae, Sergestidae) da Amazônia peruana. Acta Amazonica 33: 663–686 [2003].

Gomes Corrêa, M.M. 1973. Descrição de uma nova espécie do gênero *Cryphiops* (Decapoda: Natantia: Palaemonidae). Revista Brasileira de Biologia 33: 169–173.

Hart, C.W. Jr. 1961. *Jonga*, a new genus of freshwater atyid shrimp (Decapoda, Atyidae). Notulae Naturae 342: 1–3.

Hobbs, H.H. Jr. & C.W.J. Jr. Hart. 1982. The shrimp genus *Atya* (Decapoda: Atyidae). Smithsonian Contributions to Zoology 364: 1–143.

Hobbs, H.H. Jr., H.H. Hobbs III & M.A. Danielet. 1977. A Review of the Troglobitic Decapod Crustaceans of the Americas. Smithsonian Contributions to Zoology 224: 1–183.

Holthuis, L. 1952. A general revision of the Palaemonidae (Crustacea Decapoda Natantia) of the Americas. II. The subfamily Palaemoninae. Occasional Papers of the Allan Hancock Foundation 2.

Holthuis, L.B. 1959. The Crustacea Decapoda of Suriname (Dutch Guiana). Zoologische Verhandelingen 44: 1–296.

Kensley, B. & I. Walker. 1982. Palaemonid shrimps from the Amazon Basin, Brazil (Crustacea: Decapoda: Natantia). Smithsonian Contributions to Zoology 362.

Mantelatto, F.L. L.G. Pileggi, C. Magalhães, F.L. Carvalho, S.S. Rocha, E. C. Mossolin, N. Rossi & S.L.S. Bueno. 2016. Chapter 20: Avaliação dos camarões Palemonídeos (Decapoda: Palaemonidae). Pages 252–267 *in*: M.A.A. Pinheiro & H. Boos (Org.). Livro Vermelho dos Crustáceos do Brasil: Avaliação 2010–2014. Porto Alegre, RS: Sociedade Brasileira de Carcinologia - SBC, 466 pp.

Omori, M. 1975. The systemactics, biogeography, and fishery of epipelagic shrimps of the genus *Acetes* (Crustacea, Decapoda, Sergestidae). Bulletin of the Ocean Research Institute, University of Tokyo 7: 54–59.

Pachelle, P.P.G. & M. Tavares. 2018. The freshwater shrimp family Euryrhynchidae Holthuis, 1950 (Crustacea: Decapoda: Caridea) revisited, with a taxonomic revision of the genus *Euryrhynchus* Miers, 1878. Zootaxa 4380: 1–110.

Pereira, G. & J. García. 1995. Sobre la presencia de los crustaceos *Micratya poeyi, Xiphocaris elongata* (Decapoda, Atyidae y Xiphocarididae) y *Moina macrocopa macrocopa* (Cladocera, Moinidae) en Venezuela. Acta Biologica Venezuelica 15: 89–95.

Pileggi, L.G. & F.L. Mantelatto. 2010. Molecular phylogeny of the freshwater prawn genus *Macrobrachium* (Decapoda, Palaemonidae), with emphasis on the relationships among selected American species. Invertebrate Systematics 24: 194–208.

Pileggi, L.A. & F.L. Mantelatto. 2012. Taxonomic revision of doubtful Brazilian freshwater shrimp species of the genus *Macrobrachium* (Decapoda, Palaemonidae). Iheringia, Série Zoologia 102: 426–437.

Torati L.S. & F.L. Mantelatto. 2012. Ontogenetic and evolutionary change of external morphology of the neotropical shrimp *Potimirim* (Holthuis, 1954) explained by a molecular phylogeny of the genus. Journal of Crustacean Biology 32: 625–640.

Villalobos, J.L., F. Alvarez, & T.M. Iliffe. 1999. New species of troglobitic shrimps from Mexico, with the description of *Troglomexicanus*, new genus (Decapoda: Palaemonidae). Journal of Crustacean Biology 19: 111–122.

Vogt, G. 2016. Chapter 6: Direct development and posthatching brood care as key features of the evolution of freshwater Decapoda and challenges for conservation. Pages 169–198 *in*: T. Kawai & N. Cumberlidge (eds.), A Global Overview of the Conservation of Freshwater Decapod Crustaceans. Springer, Cham, Switzerland.

Decapoda: Anomura: Aeglidae

Georgina Bond-Buckup
Independent Researcher, Retired from the Universidade Federal do Rio Grande do Sul, Porto Alegre, RS, Brazil

Sandro Santos
Universidade Federal de Santa Maria, Santa Maria, RS, Brazil

Carlos G. Jara
Universidad Austral de Chile, Valdivia, Chile

INTRODUCTION

Aeglidae (infraorder Anomura) comprises only one extant genus, *Aegla* Leach, 1820, which inhabits continental waters (Fig. 23.103), and two marine fossil species, *Haumuriaegla glaessneri* Feldmann, 1984 and *Protaegla miniscula* Feldmann, Vega, Applegate & Bishop, 1998.

Presently, *Aegla* comprises 87 species (Bueno et al., 2017; Santos et al., 2017), although there are doubts about the existence of *Aegla intermedia* Girard, 1855. It was described based upon specimens collected by the "U. S. N. Astronomical Expedition," from tributaries of Maypu River, near to Santiago, Chile. Despite the effort of various researchers, the species has never been found again, and the type series has disappeared (Bond-Buckup & Buckup, 1994). The group is distributed from the Rio Grande Basin, on the border of São Paulo and Minas Gerais states, Brazil, to Duque de York Island, South River Basin, Chile (Bond-Buckup et al., 2008).

Several aeglids are under some level of threat, mainly due to the deterioration of environmental quality. The presence of some species in restricted areas, such as headwaters, some with few records, highlights the threat level to their conservation (Santos et al., 2017).

LIMITATIONS

Nearly a quarter of species were described in the last 20 years, and thus they are not cited in the most recent review (Bond-Buckup & Buckup, 1994), and the identification keys at that time. Here we present identification keys for all species, except *Aegla charon* Bueno & Moraes, 2017, *Aegla chilota* Jara, Pérez-Losada & Crandall, 2018, and *Aegla okora* Páez & Teixeira, 2018 (Bueno et al., 2017; Jara et al. 2018; Paez et al., 2018, respectively), all of them described during the revising of this book. The task of developing keys is difficult, as the increasing number of species makes it more and more difficult to identify the aeglids morphologically. The group has a very conservative morphology, and the taxonomic characters used may present plasticity that is sometimes difficult to delimit.

TERMINOLOGY AND MORPHOLOGY

The external general morphology was accurately described by Martin & Abele (1988), based on adult specimens of *Aegla platensis* and *Aegla uruguayana*. Here we only provide terms used in the keys

- Abdominal pleura II dorsum: (Fig. 23.110 B) smooth as in *A. jujuyana*, *A. pomerana*; carinate as in *A. rostrata*, *A. singularis*
- Abdominal pleura II dorsolateral angle: (Fig. 23.110 B) blunt as in *A. intercalata*, *A. oblata*; acute (with scale/tubercle or spine) as in *A. rostrata, A. longirostri*
- Areola shape: (Fig. 23.104 A) dimensional relationship between AL (areola length) and AW (areola width): quadrate/subquadrate: AL/AW = less than 1.69 as in *A. leachi*; *A. carinata*; rectangular/subrectangular: AL/AW = more than 1.70 as in *A. rostrata*, *A. pomerana*
- Cephalothorax dorsomedial carina (extended from behind epigastric prominences up to areola's rear margin): present as in *A. carinata*, *A. denticulata*; absent as in *A. spinipalma*, *A. serrana*, *A. jujuyana*
- Cheliped carpal ridges or crests, in dorsal view (one or two ridges) (Fig. 23.112): with two ridges as in *A. parana*, *A. denticulata*; with one ridge as in *A. lancinhas*, *A. leachi*
- Cheliped lateral margin proximal lobe: (Fig. 23.112) present (sometimes rudimentary) as in *A. scamosa*, *A. rostrata*; absent as in *A. grisella*, and *A. serrana*
- Cheliped size in the adult male: equal (same size) as in *A. humahuaca*, *A. d. lacustris*; unequal (different size) as in *A. neuquensis*, *A. lata*
- Dorsodistal lobe of palmar crest or predactylar lobe: (Fig. 23.112) present as in *A. affinis*, *A. intercalata*; absent s in *A. loyolai*
- Epibranchial lobe ornament: (Fig. 23.107 A) absent (apical portion of lobe blunt and smooth) as in *A. serrana*; tuberculate (apical portion of lobe with tubercle and scales) as in *A. itacolomiensis*, *A. violacea*; spiny (apical portion of lobe elongated with spine) as in *A. denticulata*, *A. jarai*
- Eyestalk: normal (Fig. 23.105) (eyestalks and corneae with normal size, as in most species found in rivers and lakes); reduced (eyestalks and corneae of small size, slender and with few ommatidia), as occurs in troglobitic species such as *A. cavernicola*, *A. microphthalma*, and *A. leptochela*
- Front relative width: (Fig. 23.104) ratio PCW/FW (precervical width distance between left and right epibranchial margins; frontal width distance between carapace anterolateral angles): wide: 1.79 to 1.89; narrow: more than 1.90
- Ornaments: scales, tubercles, and spines occur in all stages of development
- Palmar crest (Fig. 23.113): dorsal border of the propodus: obsolescent or subrectangular narrow as *A. inermis*;

PHYLUM ARTHROPODA: CRUSTACEA: MALACOSTRACA

FIGURE 23.103 Photograph of the dorsal view of *Aegla longirostri*. Scale bar = 1 cm. *Photograph of Sandro Santos.*

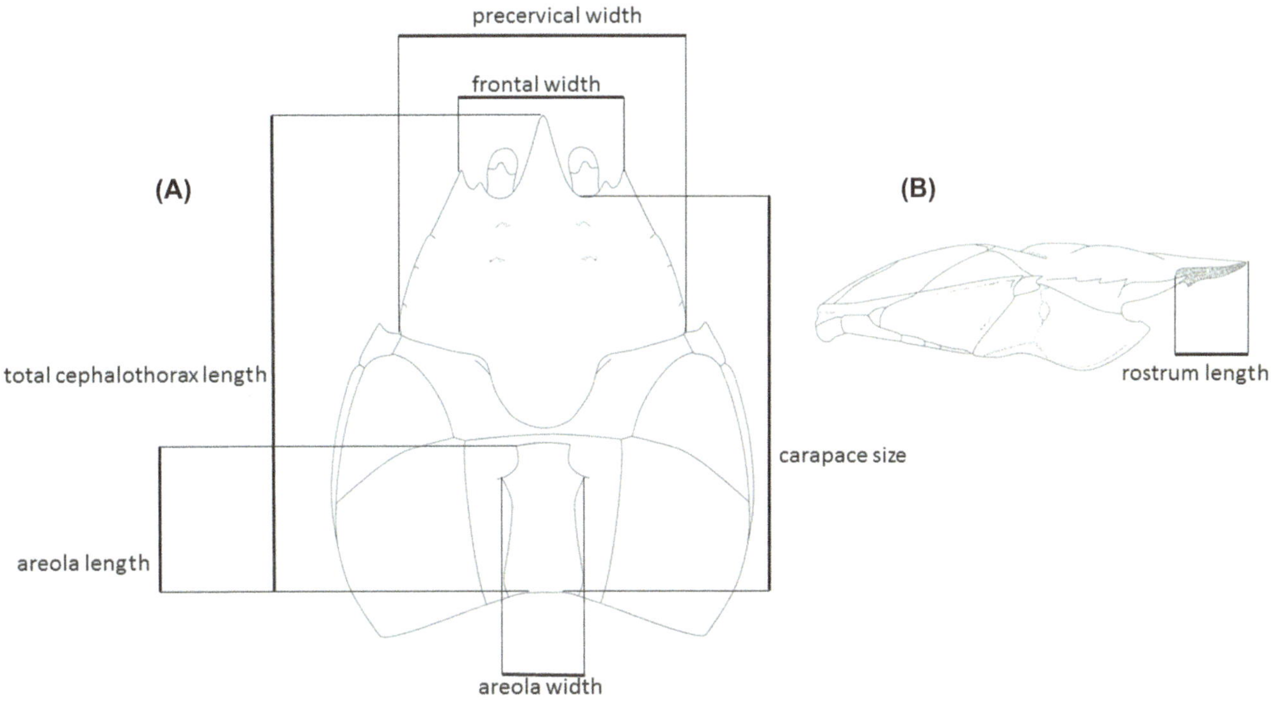

FIGURE 23.104 Measurements of aeglid carapace: (A) dorsal; (B) lateral.

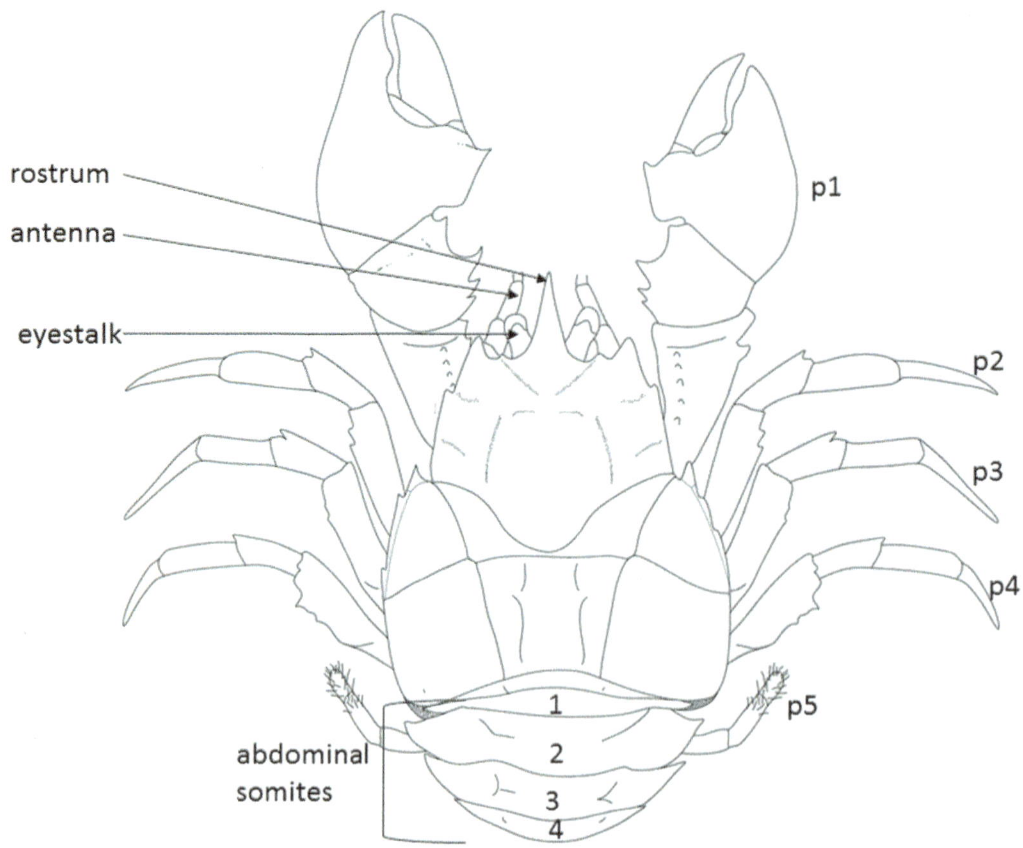

FIGURE 23.105 Diagrammatic view of aeglid carapace.Abbreviation: *p*: pereiopod.

subrectangular as *A. parana*; shape of heart inflated as *A. paulensis*; shape of heart depressed as *A. castro*

- Pereopod II merus dorsal margin: with scales and setae as in *A. loyolai*; smooth as in *A. renana*
- Pereopod II merus ventral margin: with tubercles/scales/or spines as in *A. expansa, A. jarai*; smooth as in *A. georginae, A. ludwigi*
- Posterior branchial area lateral margins: (Fig. 23.106) not recurved as in *A. papudo*; recurved as in *A. parana*
- Rostral carina extension: (Fig. 23.108) entire (carina extended all along rostral dorsum) as in *A. manni*; partial (carina extended scarcely on proximal third) as in *A. lancinhas*
- Rostral carina scales: with one row as in *A. brevipalma*; with two rows as in *A. bahamondei*, with multiple rows as in *A. carinata, A. manuinflata*
- Rostrum ventral portion height (in lateral view): (Fig. 23.107 B) the extent of protrusion is compared with the height of rostrum above, being: shallow, if protrusion of subrostral proces is equal or less than the height of rostrum above as in *A. spectabilis*; deep, if protrusion of ventral rostral process is more than the height of rostrum above as in *A. pomerana, A. papudo*
- Rostral shape in dorsal view: (Fig. 23.109) triangular (broad based gradually narrowing toward the rostral apex) as in *A. humahuaca, A. franciscana, A. denticulata*; linguiform (roughly triangular but lateral borders subparallel along central part of rostrum then converging onto rostral tip); dorsum flattened and carina evanescent as in *A. septentrionalis, A. papudo*, and *A. ligulata*; styliform: rostrum narrow, elongate, sharply tipped as in *A. uruguayana, A. spectabilis*, and *A. carinata*
- Rostral shape in lateral view: (Fig. 23.114) from epigastric prominence to rostral tip: straight as in *A. plana, A. hueicollensis*; recurved as in *A. manni, A spinipalma*; deflected as in *A. grisella, A. concepcionensis*

PHYLUM ARTHROPODA: CRUSTACEA: MALACOSTRACA

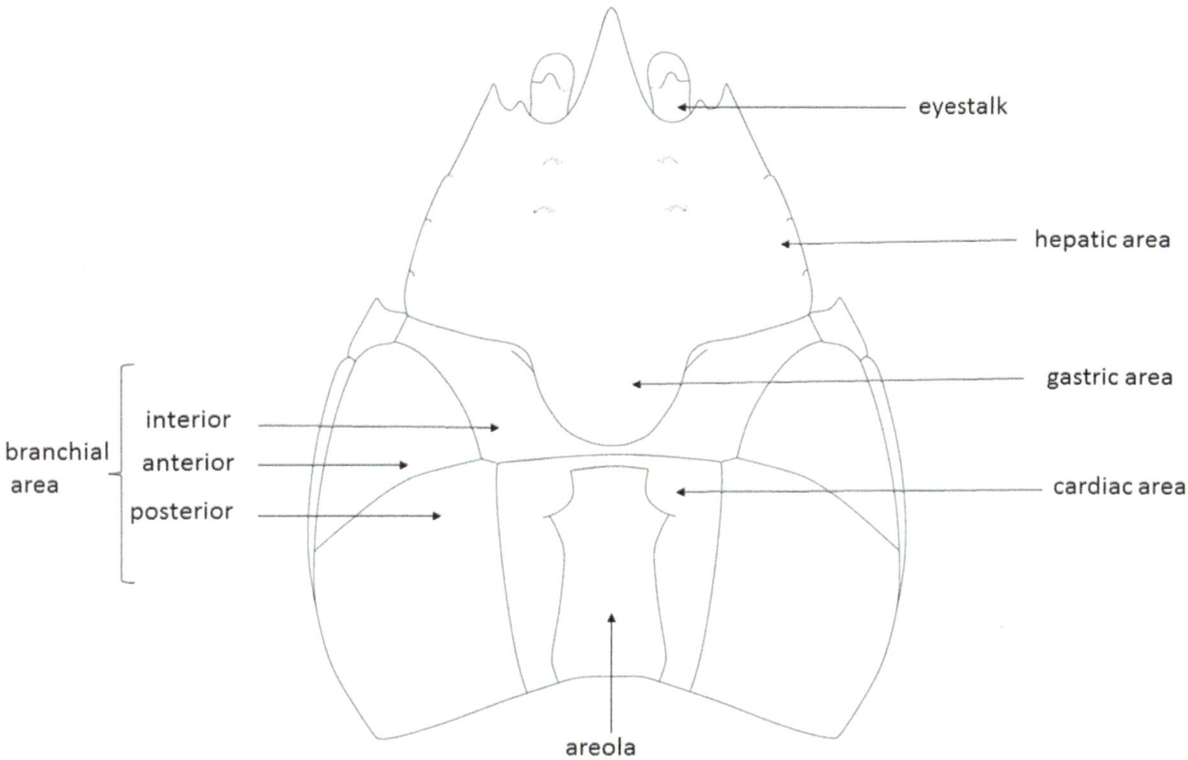

FIGURE 23.106 Dorsal view of the areas of the carapace of an aeglid.

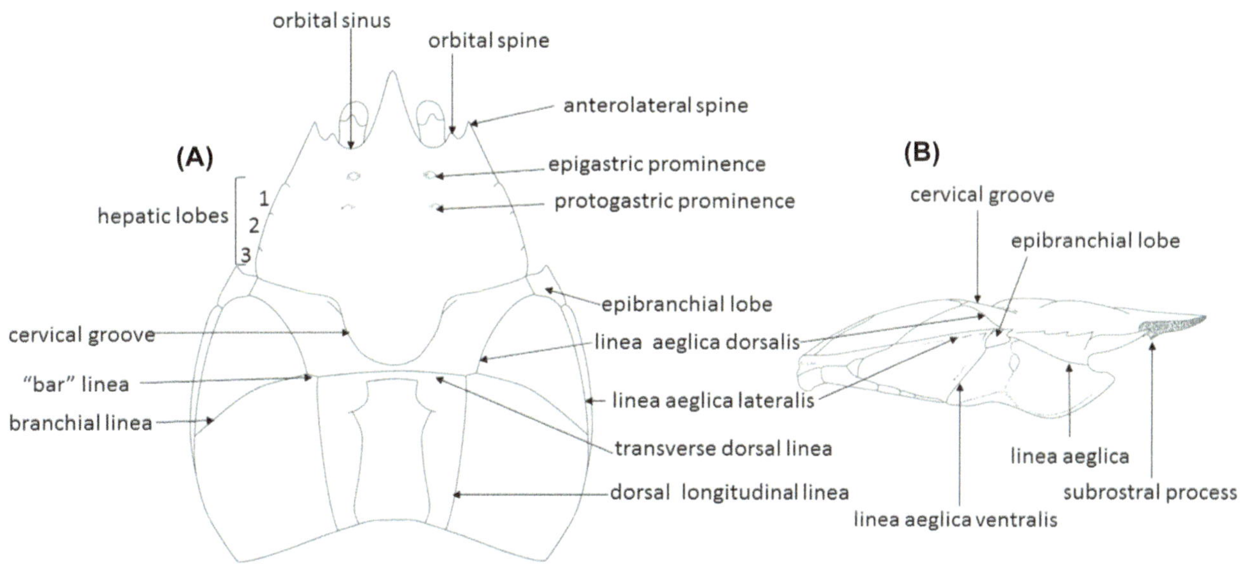

FIGURE 23.107 Dorsal view of the carapace of an aeglid showing characters and *lineae*: (A) dorsal view; (B) lateral view.

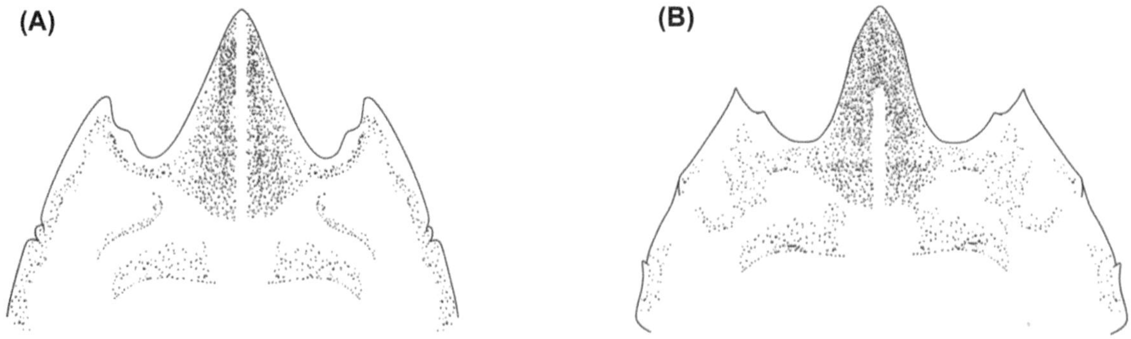

FIGURE 23.108 Anterior part of aeglid cephalothorax showing rostrum: (A) carina extended all along rostral dorsum: *A. manni;* (B) carina absent in the distal third: *A. lancinhas. After Bond-Buckup & Buckup (1994); Santos et al. (2015).*

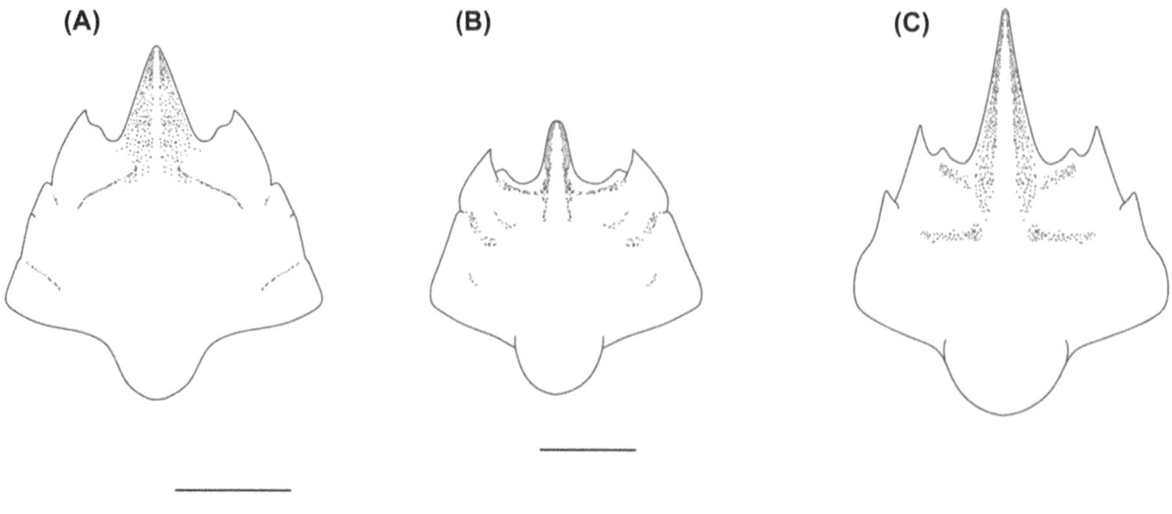

FIGURE 23.109 Rostral shape in dorsal view of aeglid: (A) triangular: *A. humahuaca*; (B) linguiform: *A. ligulata*; (C) styliform: *A. spectabilis*. Scale bars = 5 mm. *After Bond-Buckup & Buckup (1994).*

- Rostrum relative length: (Fig. 23.104 A, B) ratio CL/RL (carapace length: total cephalothorax length from the rostral apex to the carapace posterior margin midpoint; rostral length from rostral apex to the midpoint of a line drawn between orbital sinus deepest point: long = less than 3.9; medium = 4.0 to 4.9; short = more than 5.0
- Thoracic sternum IV ornamentation (spine, tubercles, or scale) (Fig. 23.111) absent as in *A. parva, A. spinosa*; present as in *A. carinata* and *A. platensis*

KEY TO AEGLIDAE

Decapoda: Anomura: Aeglidae: *Aegla*: Species

1	Specimens from the Pacific drainages of Chile and Argentina	2
1′	Specimens from the Atlantic drainages of northeastern Argentina, Bolivia, Brazil, Paraguay, and Uruguay	22
2(1)	Rostrum in dorsal view triangular	3

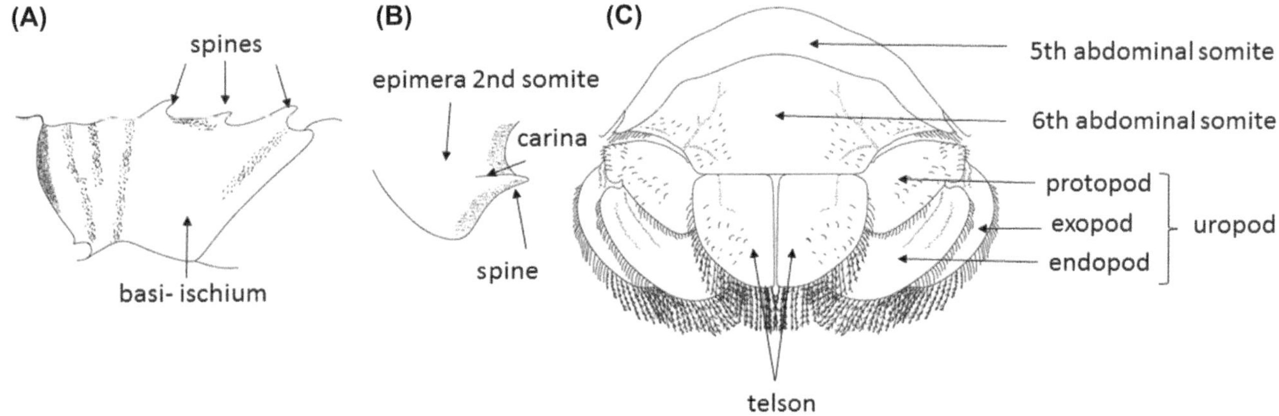

FIGURE 23.110 Ornaments and tailfan of aeglids: (A) ornaments on the ventral margin of basi-ischium of cheliped; (B) ornaments of latero dorsal pleura of second abdominal somite; (C) abdominal somites and tailfan (telson and uropod).

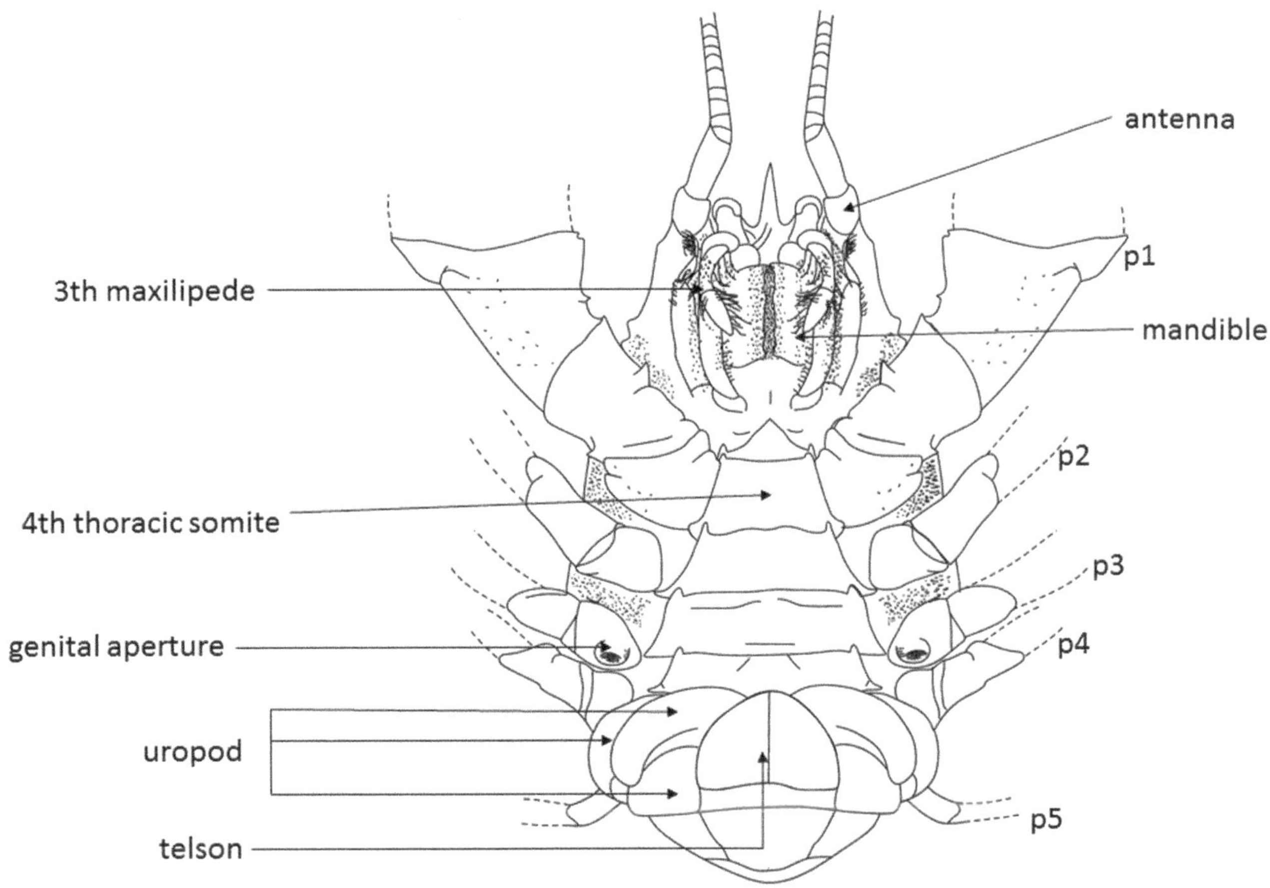

FIGURE 23.111 Ventral diagrammatic view of aeglid. Abbreviation: *p*: pereiopod.

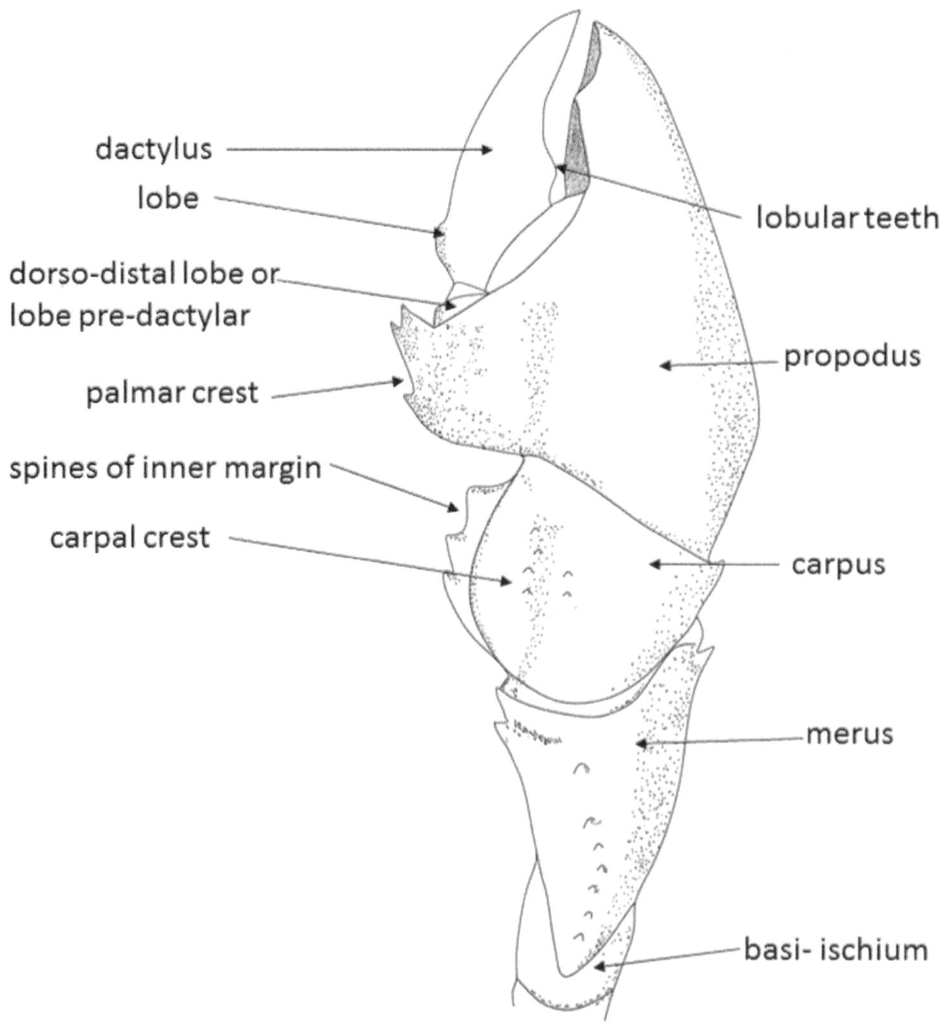

FIGURE 23.112 Schematic view of cheliped of aeglid.

2′	Rostrum in dorsal view linguiform or styliform	9
3(2)	Rostrum ratio length medium or short	4
3′	Rostrum ratio length long *Aegla denticulata denticulata* Nicolet, 1849 [Chile: Araucanía; Bio Bio; Los Lagos; Los Ríos; Chiloé]	
4(3)	Cephalothorax ratio length medium	5
4′	Cephalothorax ratio length short	8
5(4)	Abdominal pleura II smooth dorsally	6
5′	Abdominal pleura II carinate dorsally	7
6(5)	Rostrum in lateral view straight; posterior branchial lateral area margin shape not recurved; front relative width wide, abdominal pleura II dorsolateral angle unornamented *Aegla alacalufi* Jara & López, 1981 [Chile: Los Lagos; Chilóe, Palena, Aysén territory]	
6′	Rostrum in lateral view deflected; posterior branchial lateral area margin shape recurved; front relative width narrow; abdominal pleura II dorsolateral angle ornamented *Aegla abtao* Schmitt, 1942 [Chile: Araucanía; Los Lagos; Los Ríos; Chiloé]	
7(5)	Rostral carina scales in two rows; cephalothorax dorsal median line without carina; anterior branchial area lateral margin without lobes; adult male chelipeds unequal *Aegla bahamondei* Jara, 1982 [Chile: Bio Bio]	

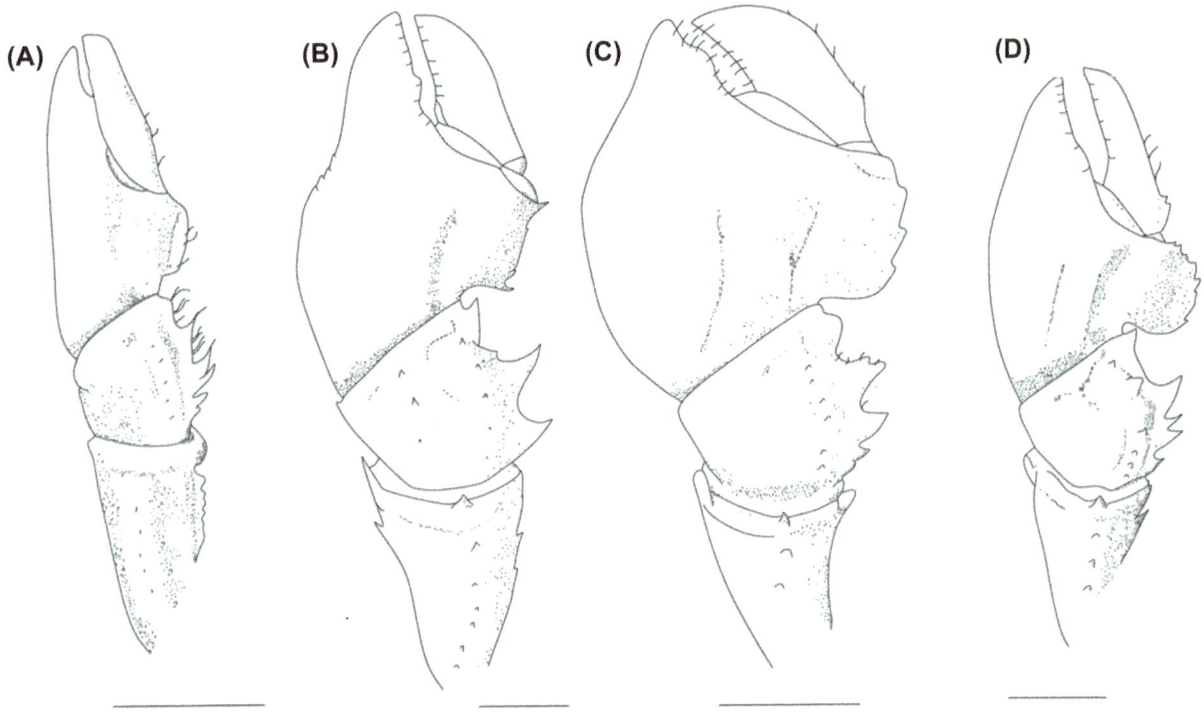

FIGURE 23.113 Cheliped shape of aeglid: (A) subrectangular narrow, *A. leptochela*; (B) subrectangular, *A. parana*; (C) shape of heart inflated, *A. paulensis*; (D) shape of heart depressed, *A. castro*. Scale bars = 5 mm. *After Bond-Buckup & Buckup (1994).*

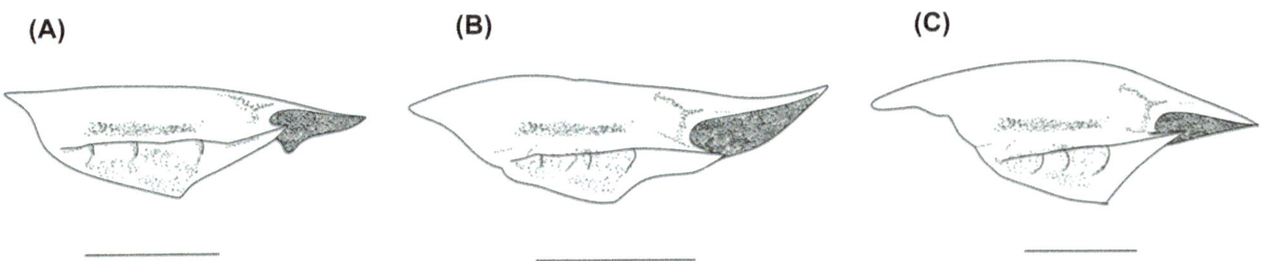

FIGURE 23.114 Rostral shape in lateral view of aeglid: (A) straight, *A. plana*; (B) recurved, *A. denticulata*; (C) deflected, *A. grisella*. Scale bars = 5 mm. *After Bond-Buckup & Buckup (1994).*

7′	Rostral carina scales in three or more rows; cephalothorax dorsal median line with carina; anterior branchial area lateral margin with lobes; adult male chelipeds equal .... *Aegla denticulata lacustris* Jara, 1989 [Chile: Los Lagos]
8(4)	Rostrum in lateral view straight; pereiopod II merus ventral margin with scales, tubercles, or spines; rostral carina extension only basal third; areola quadrate/subquadrate .... *Aegla hueicollensis* Jara & Palacios, 1999 [Chile: Los Ríos]
8′	Rostrum in lateral view recurved; pereiopod II merus ventral margin inerm; rostral carina extension entire; areola rectangular/sub-rectangular .... *Aegla manni* Jara, 1980 [Chile: Los Lagos; Los Ríos]
9(2)	Rostrum in dorsal view linguiform 10

9′ Rostrum in dorsal view styliform 15

10(9) Pereiopod II merus ventral margin with scales, tubercles, or spines 11

10′ Pereiopod II merus ventral margin inerm 13

11(10) Rostrum relative length medium; abdominal pleura II dorsally smooth; cheliped dactylus lateral margin with proximal lobe; dorsodistal lobe of palmar crest present 12

11′ Rostrum relative length short; abdominal pleura II dorsally carinate; cheliped dactylus lateral margin without lobe; chela palmar crest lacking dorsodistal lobe .... *Aegla talcahuano* Schmitt, 1942
[Chile: O'Higgins; Maule]

12(11) Areola quadrate/subquadrate; abdominal pleura II with dorsolateral ornamentation; rostrum in lateral view deflected; tooth or epibranchial lobe separate .... *Aegla concepcionensis* Schmitt, 1942
[Chile: Concepcion]

12′ Areola rectangular/subrectangular; abdominal pleura II without dorsolateral ornamentation; rostrum in lateral view recurved; tooth or epibranchial lobe inclusive .... *Aegla papudo* Schmitt, 1942
[Chile: Valparaiso]

13(10) Rostrum relative length medium; pereiopod II merus dorsal margin without plumose setae 14

13′ Rostrum relative length short; pereiopod II merus dorsal margin with plumose setae .... *Aegla laevis* (Latreille, 1818)
[Chile: Metropolitan region]

14(13) Rostral carina scales in two rows; cephalothorax medially without a carina; posterior branchial area lateral margin recurved; cheliped carpal ridges in dorsal view with two ridges .... *Aegla araucaniensis* Jara, 1980
[Chile: Los Lagos; Los Ríos; Chilóe]

14′ Rostral carina scales in multiple rows; cephalothorax medially with a carina; posterior branchial area lateral margin not recurved; cheliped carpal ridges in dorsal view with one ridge .... *Aegla expansa* Jara, 1992
[Chile: Bío Bío]

15(9) Rostrum in lateral view curved or deflected 16

15′ Rostrum in lateral view straight 17

16(15) Rostrum in lateral view recurved .... *Aegla neuquensis* Schmitt, 1942
[Chile: Aysen. Argentina: Buenos Aires; Chubut; Neuquén; Río Negro]

16′ Rostrum in lateral view deflected .... *Aegla affinis* Schmitt, 1942
[Chile: Maule. Argentina: Mendoza; Neuquén]

17(15) Areola quadrate/subquadrate; cheliped ischium ventral surface medial margin unornamented. 18

17′ Areola rectangular/subrectangular; cheliped ischium ventral surface medial margin ornamented 20

18(17) Anterior branchial area lateral margin with lobes; pereiopod II merus dorsal margin with plumose setae 19

18′ Anterior branchial area lateral margin without lobes; pereiopod II merus dorsal margin without plumose setae .... *Aegla chochol* Jara & Palacios, 1999
[Chile: Araucanía]

19(18) Rostrum relative length long; front relative width narrow; cheliped carpal ridges in dorsal view with two ridges .... *Aegla spectabilis* Jara, 1986
[Chile: Araucania]

19′ Rostrum relative length medium; front relative width narrow; cheliped carpal ridges in dorsal view with one ridge .... *Aegla occidentalis* Jara, Pérez-Losada & Crandall, 2003
[Chile: Bio Bio]

20′ Rostral carina scales in two or more rows 21

20(17) Rostral carina scales in one row .... *Aegla riolimayana* Schmitt, 1942
[Chile: Los Lagos; Los Ríos. Argentina: Neuquén; Río Negro]

21(20) Rostral carina scales in two rows .... *Aegla pewenchae* Jara, 1994
[Chile: Araucanía; Bío Bío; Maule; O'Higgins]

21′ Rostral carina scale in three or more rows .... *Aegla rostrata* Jara, 1977
[Chile: Araucania; Los Ríos]

22(1) Rostrum in lateral view recurved or deflected 23

22′ Rostrum in lateral view straight 46

23(22) Rostrum in lateral view recurved 24

23′ Rostrum in lateral view deflected 33

24(23) Rostrum in dorsal view linguiform or styliform 25

24′ Rostrum in dorsal view triangular 28

25(24) Rostrum in dorsal view linguiform 26

25′ Rostrum in dorsal view styliform 32

26(25) Front relative width narrow; abdominal pleura II laterodorsal angle unornamented; rostral carina extension entire; cheliped dactyl lateral margin proximal end without lobe 27

26′ Front relative width wide; abdominal pleura II laterodorsal angle ornamented; rostral carina extension to distal third; cheliped dactyl lateral margin proximal end with lobe .... *Aegla franciscana* Buckup & Rossi, 1977
[Brazil: Rio Grande do Sul]

27(26) Thoracic sternum IV unornamented; protogastric lobe present .... *Aegla renana* Bond-Buckup & Santos, 2010
[Brazil: Rio Grande do Sul]

27′ Thoracic sternum IV ornamented; protogastric lobe absent .... *Aegla septentrionalis* Bond-Buckup & Buckup, 1994
[Bolivia: Tarija; Tupiza. Argentina: Jujuy; Salta]

28(24) Front relative width wide 29

28′ Front relative width narrow 30

29(28) Rostral carina extension entire; rostral carina scales in more than two rows; extraorbital sinus absent; abdominal pleura II laterodorsal angle ornamented .... *Aegla rosanae* Campos Jr., 1998
[Brazil: São Paulo]

29′ Rostral carina extension to distal third; rostral carina scales in two rows; extraorbital sinus present (sometimes obsolete); abdominal pleura II laterodorsal angle unornamented .... *Aegla pomerana* Bond-Buckup & Buckup, 2010
[Brazil: Santa Catarina]

30(28) Rostral carina extension to distal third; abdominal pleura II laterodorsal angle unornamented 31

30′ Rostral carina extension entire; abdominal pleura II laterodorsal angle ornamented .... *Aegla castro* Schmitt, 1942
[Brazil: Paraná]

31(30) Areola quadrate/subquadrate; thoracic sternum IV ornamented; cheliped dactyl lateral margin proximal end without lobe; eyestalk reduced .... *Aegla cavernicola* Türkay, 1972
[Brazil: São Paulo]

31′ Areola rectangular/subrectangular; thoracic sternum IV unornamented; cheliped dactyl lateral margin proximal end with lobe; eyestalk normal .... *Aegla odebrechtii* Müller, 1876
[Brazil: Santa Catarina]

32(25) Rostral carina scales in two rows; front relative width narrow; abdominal pleura II laterodorsal angle unornamented; cheliped dactyl lateral margin proximal end with lobe .... *Aegla neuquensis* Schmitt, 1942
[Chile: Aysen. Argentina: Buenos Aires; Chubut; Neuquén; Río Negro]

32′ Rostral carina scales in three more rows; front relative width wide; abdominal pleura II laterodorsal angle ornamented; cheliped dactyl lateral margin proximal end without lobe .... *Aegla spinipalma* Bond-Buckup & Buckup, 1994
[Brazil: Rio Grande do Sul]

33(23) Rostral carina scales in two or more rows 34

33′ Rostral carina scales in one row 42

34(33) Rostral carina scales in two rows 35

34′ Rostral carina scales in three or more rows 39

35(34) Front relative width front relative width wide 36

35′ Front relative width narrow 37

36(35) Epibranchial lobe unornamented; cheliped dactyl lateral margin proximal end without lobe; cheliped palmar crest dorsodistal lobe absent; pereiopod II merus ventral margin with scales, tubercles, or spines .... *Aegla grisella* Bond-Buckup & Buckup, 1994
[Brazil: Rio Grande do Sul]

36′ Epibranchial lobe ornamented; cheliped dactyl lateral margin proximal end with lobe; cheliped palmar crest dorsodistal lobe present; pereiopod II merus ventral margin inerm .... *Aegla ludwigi* Santos & Jara, 2013
[Brazil: Rio Grande do Sul]

37(35) Rostral carina extending rostrum length; rostrum in dorsal view triangular; epibranchial lobe ornamented; thoracic sternum IV unornamented 38

37′ Rostral carina extending scarcely to distal third; rostrum in dorsal view linguiform; epibranchial lobe unornamented; thoracic sternum IV ornamented .... *Aegla obstipa* Bond-Buckup & Buckup, 1994
[Brazil: Rio Grande do Sul]

38(37) Areola quadrate/subquadrate; extraorbital sinus absent; cheliped dactyl lateral margin proximal end with lobe .... *Aegla leachi* Bond-Buckup & Buckup, 2012
[Brazil: Santa Catarina]

PHYLUM ARTHROPODA: CRUSTACEA: MALACOSTRACA

38′ Areola rectangular/subrectangular; extraorbital sinus present or obsolete; cheliped dactyl lateral margin proximal end without lobe ..... *Aegla saltensis* Bond-Buckup & Jara, 2010
[Argentina: Salta]

39(34) Front relative width wide 40

39′ Front relative width narrow 44

40(39) Rostral carina extension entire; epibranchial lobe ornamented; cheliped dactyl lateral margin proximal end without lobe; pereiopod II merus ventral margin with scales, tubercles, or spines 41

40′ Rostral carina extension scarcely to distal third; epibranchial lobe ornament absent; cheliped dactyl lateral margin proximal end with lobe; pereiopod II merus ventral margin inerm ..... *Aegla paulensis* Schmitt, 1942
[Brazil: São Paulo]

41(40) Areola quadrate/subquadrate; abdominal pleura II laterodorsal angle ornamented; thoracic sternum IV ornamented ..... *Aegla manuinflata* Bond-Buckup & Santos, 2009
[Brazil: Rio Grande do Sul]

41′ Areola rectangular/subrectangular; abdominal pleura II laterodorsal angle unornamented; thoracic sternum IV unornamented ..... *Aegla inermis* Bond-Buckup & Buckup, 1994
[Brazil: Rio Grande do Sul]

42(33) Rostrum in lateral view with ventral portion height greater than dorsal portion; chela palmar crest dorsodistal lobe present; pereiopod II merus dorsal margin without plumose setae 43

42′ Rostrum in lateral view with ventral portion height less than dorsal portion; chela palmar crest dorsodistal lobe absent; pereiopod II merus dorsal margin with plumose setae ..... *Aegla jujuyana* Schmitt, 1942
[Argentina: Jujuy; Salta]

43(42) Areola quadrate/subquadrate; front relative width wide; abdominal pleura II laterodorsal angle ornamented; thoracic sternum IV unornamented ..... *Aegla brevipalma* Bond-Buckup & Santos, 2012
[Brazil: Santa Catarina]

43′ Areola rectangular/subrectangular; front relative width narrow; abdominal pleura II laterodorsal angle unornamented; thoracic sternum IV ornamented ..... *Aegla strinatii* Türkay, 1972
[Brazil: São Paulo]

44(39) Rostral carina extension entire; rostrum triangular in dorsal view, in lateral view ventral portion height less than dorsal portion; cheliped ischium ventral surface medial margin ornamented 45

44′ Rostral carina extension partial, carina obscure in distal third; rostrum linguiform in dorsal view, in lateral view ventral portion higher than dorsal portion; cheliped ischium ventral surface medial margin not ornamented ..... *Aegla affinis* Schmitt, 1942
[Chile: Maule. Argentina: Mendoza; Neuquén]

45(44) Abdominal pleura II laterodorsal angle and thoracic sternum IV unornamented; chela palmar crest dorsodistal lobe absent; protogastric lobe present ..... *Aegla jundiai* Moraes, Tavares & Bueno, 2016
[Brazil: São Paulo]

45′ Abdominal pleura II laterodorsal angle and thoracic sternum IV ornamented; chela palmar crest dorsodistal lobe present; protogastric lobe absent ..... *Aegla humahuaca* Schmitt, 1942
[Argentina: Catamarca Jujuy, Salta, Santiago del Estero, Tucumán]

46(22) Rostrum in dorsal view linguiform or styliform 47

46′ Rostrum in dorsal view triangular 59

47(46) Rostrum in dorsal view linguiform 48

47′ Rostrum in dorsal view styliform 49

48(47) Rostrum in lateral view with ventral portion height greater than dorsal portion; extraorbital sinus present, sometimes obsolete; front relative width wide; epibranchial lobe ornamented ..... *Aegla ligulata* Bond-Buckup & Buckup, 1994
[Brazil: Rio Grande do Sul]

48′ Rostrum in lateral view with ventral portion height less than dorsal portion; extraorbital sinus absent; front relative width narrow; epibranchial lobe unornamented ..... *Aegla japi* Moraes, Tavares & Bueno, 2016
[Brazil: São Paulo]

49(47) Rostral carina scales in two or more rows 50

49′ Rostral carina scales in one row ..... *Aegla riolimayana* Schmitt, 1942
[Chile: Los Lagos; Los Ríos. Argentina: Neuquén, Río Negro]

50(49) Rostral carina scales in two rows 51

50′ Rostral carina scales in three or more rows 55

51(50) Cheliped dactyl lateral margin proximal end with lobe 52

51′ Cheliped dactyl lateral margin proximal end without lobe 53

52(51) Front relative width wide; chela palmar crest dorsodistal lobe absent; areola quadrate/subquadrate .... *Aegla loyolai* Bond-Buckup & Santos, 2015
[Brazil: Paraná]

52′ Front relative width narrow; cheliped palmar crest dorsodistal lobe present; areola rectangular/subrectangular .... *Aegla spinosa* Bond-Buckup & Buckup, 1994
[Brazil: Santa Catarina]

53(51) Rostrum in lateral view with ventral portion height greater than dorsal portion 54

53′ Rostrum in lateral view with ventral portion height less than dorsal portion .... *Aegla sanlorenzo* Schmitt, 1942
[Argentina: Jujuy, Salta]

54(53) Areola quadrate/subquadrate; thoracic sternum IV unornamented; pereiopod II merus dorsal margin with plumose setae, ventral margin inerm .... *Aegla meloi* Bond-Buckup & Santos, 2015
[Brazil: Paraná]

54′ Areola rectangular/subrectangular; thoracic sternum IV ornamented; pereiopod II merus dorsal margin without plumose setae, ventral margin with scales, tubercles, or spines .... *Aegla parana* Schmitt, 1942
[Brazil: Paraná]

55(50) Cheliped dactyl lateral margin proximal end with lobe 56

55′ Cheliped dactyl lateral margin proximal end without lobe 57

56(55) Pereiopod II merus ventral margin with scales, tubercles, or spines; cephalothorax median carina present .... *Aegla prado* Schmitt, 1942
[Brazil: Rio Grande do Sul. Uruguay: Canelones; Maldonado; Montevideo; Rocha; Taquarembó]

56′ Pereiopod II merus ventral margin inerm; cephalothorax without median carina .... *Aegla singularis* Ringuelet, 1948
[Brazil: Rio Grande do Sul. Argentina: Misiones]

57(55) Cephalothorax without median carina; areola rectangular/subrectangular; cheliped ischium ventromedial margin ornamented 58

57′ Cephalothorax with median carina; areola quadrate/subquadrate; cheliped ischium ventromedial margin unornamented .... *Aegla carinata* Bond-Buckup & Loureiro, 2014
[Uruguay: Rivera]

58(57) Front relative width wide; thoracic sternum IV ornamented; pereiopod II merus ventral margin inerm .... *Aegla uruguayana* Schmitt, 1942
[Brazil: Rio Grande do Sul. Argentina: Buenos Aires; Córdoba; Entre Ríos; Mendoza; Salta; Santa Fé; San Luis. Uruguay: Canelones; Colonia; Maldonado; Paysandu; Rocha; Soriano]

58′ Front relative width narrow; thoracic sternum IV unornamented; pereiopod II merus ventral margin with scales, tubercles or spines .... *Aegla longirostri* Bond-Buckup & Buckup, 1994
[Brazil: Rio Grande do Sul]

59(46) Abdominal pleura II dorsolateral angle unornamented 60

59′ Abdominal pleura II dorsolateral angle ornamented 74

60(59) Pereiopod II merus ventral margin with scales, tubercles, or spines 61

60′ Pereiopod II merus ventral margin inerm 69

61(60) Front relative width wide 62

61′ Front relative width narrow 65

62(61) Thoracic sternum IV unornamented; protogastric lobe absent 63

62′ Thoracic sternum IV ornamented; protogastric lobe present 64

63(62) Cheliped palmar crest obsolescent; cheliped ischium ventromedial margin with one spine and three to four elevations .... *Aegla inconspicua* Bond-Buckup & Buckup, 1994
[Brazil: Rio Grande do Sul]

63′ Cheliped palmar crest subrectangular; cheliped ischium ventromedial margin with one spine and tubercles .... *Aegla plana* Buckup & Rossi, 1977
[Brazil: Rio Grande do Sul]

64(62) Rostral carina scales in two rows; cheliped carpal ridges in dorsal view in two ridges; rostral carina extension scarcely distal third .... *Aegla leptodactyla* Buckup & Rossi, 1977
[Brazil: Rio Grande do Sul]

64′ Rostral carina scales in three or more rows; cheliped carpal ridges in dorsal view in one ridge; rostral carina extension entire .... *Aegla jarai* Bond-Buckup & Buckup, 1994
[Brazil: Santa Catarina]

65(61) Epibranchial lobe unornamented 66

65′ Epibranchial lobe ornament 67

66(65) Areola quadrate/subquadrate; rostral carina scales in more than two rows; thoracic sternum IV ornamented; cheliped dactyl lateral margin proximal end with lobe .... *Aegla marginata* Bond-Buckup & Buckup, 1994
[Brazil: Paraná; São Paulo]

66′ Areola rectangular/subrectangular; rostral carina scales in two rows; thoracic sternum IV unornamented; cheliped dactyl lateral margin proximal end without lobe .... *Aegla serrana* Buckup & Rossi, 1977
[Brazil: Rio Grande do Sul]

67(65) Thoracic sternum IV ornamented; cheliped palmar crest with dorsodistal lobe; chela carpal ridges in dorsal view single 68

67′ Thoracic sternum IV unornamented; cheliped palmar crest without dorsodistal lobe; chela carpal ridges in dorsal view double .... *Aegla itacolomiensis* Bond-Buckup & Buckup, 1994
[Brazil: Rio Grande do Sul]

68(67) Cheliped dactyl lateral margin proximal end with lobe; abdominal pleura II dorsally carinate .... *Aegla lata* Bond-Buckup & Buckup, 1994
[Brazil: Paraná]

68′ Cheliped dactyl lateral margin proximal end without lobe; abdominal pleura II dorsally smooth .... *Aegla perobae* Hebling & Rodrigues, 1977
[Brazil: São Paulo]

69(60) Pereiopod II merus dorsal margin without plumose setae 70

69′ Pereiopod II merus dorsal margin with plumose setae 72

70(69) Rostrum in lateral view with ventral portion height greater than dorsal portion; epibranchial lobe ornamented 71

70′ Rostrum in lateral view with ventral portion height less than dorsal portion; epibranchial lobe unornamented .... *Aegla franca* Schmitt, 1942
[Brazil: São Paulo]

71(70) Cheliped dactyl lateral margin proximal end with lobe; chela palmar crest with dorsodistal lobe; rostral carina extension entire; protogastric lobe absent .... *Aegla muelleri* Bond-Buckup & Buckup, 2010
[Brazil: Santa Catarina]

71′ Cheliped dactyl lateral margin proximal end without lobe; chela palmar crest without dorsodistal lobe; rostral carina extension scarcely to distal third; protogastric lobe present .... *Aegla lancinhas* Bond-Buckup & Buckup, 2015
[Brazil: Paraná]

72(69) Areola rectangular/subrectangular; extraorbital sinus absent 73

72′ Areola quadrate/subquadrate; extraorbital sinus present, sometimes obsolete .... *Aegla jaragua* Moraes, Tavares & Bueno, 2016
[Brazil: São Paulo]

73(72) Cheliped dactyl lateral margin proximal end with lobe; cheliped palmar crest with dorsodistal lobe; rostral carina scales in two rows; rostral carina extension entire .... *Aegla oblata* Bond-Buckup & Santos, 2012
[Brazil: Santa Catarina]

73′ Cheliped dactyl lateral margin proximal end without lobe; cheliped palmar crest without dorsodistal lobe; rostral carina scales in three or more rows; rostral carina extension scarcely to distal third .... *Aegla vanini* Moraes, Tavares & Bueno, 2016
[Brazil: São Paulo]

74(59) Rostral carina scales in two or more rows 75

74′ Rostral carina scales in one row .... *Aegla microphthalma* Bond-Buckup & Buckup, 1994
[Brazil: São Paulo]

75(74) Rostral carina scales in two rows 76

75′ Rostral carina scales in more than two rows 80

76(75) Rostrum in lateral view with ventral portion height greater than dorsal portion 77

76′ Rostrum in lateral view with ventral portion height less than dorsal portion 78

77(76) Front relative width wide; thoracic sternum IV unornamented; pereiopod II merus dorsal margin with plumose setae, ventral margin inerm .... *Aegla georginae* Santos & Jara, 2013
[Brazil: Rio Grande do Sul]

77′ Front relative width narrow; thoracic sternum IV ornamented; pereiopod II merus dorsal margin without plumose setae, ventral margin with scales, tubercles, or spines .... *Aegla scamosa* Ringuelet, 1948
[Argentina: Mendoza; San Juan]

78(76) Areola quadrate/subquadrate; cheliped dactyl lateral margin proximal end with lobe 79

78′ Areola rectangular/subrectangular; cheliped dactyl lateral margin proximal end without lobe .... *Aegla parva* Bond-Buckup & Buckup, 1994
[Brazil: Paraná, Santa Catarina]

79(78) Front relative width wide; epibranchial lobe and thoracic sternum IV unornamented; eyestalk normal .... *Aegla violacea* Bond-Buckup & Buckup, 1994
[Brazil: Rio Grande do Sul]

79′ Front relative width narrow; epibranchial lobe and thoracic sternum IV ornamented; eyestalk reduced .... *Aegla leptochela* Bond-Buckup & Buckup, 1994
[Brazil: São Paulo]

80(75) Thoracic sternum IV unornamented 81

80′ Thoracic sternum IV ornamented 83

81(80) Cheliped dactyl lateral margin proximal end with lobe; pereiopod II merus dorsal margin without plumose setae; protogastric lobe present 82

81′ Cheliped dactyl lateral margin proximal end without lobe; pereiopod II merus dorsal margin with plumose setae; protogastric lobe absent .... *Aegla ringueleti* Bond-Buckup & Buckup, 1994
[Argentina: Salta]

82(81) Front relative width wide; abdominal pleura II dorsum smooth; cheliped carpal ridges double in dorsal view .... *Aegla camargoi* Buckup & Rossi, 1977
[Brazil: Rio Grande do Sul]

82′ Front relative width narrow; abdominal pleura II dorsum carinate; cheliped carpal ridges single in dorsal view .... *Aegla rossiana* Bond-Buckup & Buckup, 1994
[Brazil: Rio Grande do Sul; Santa Catarina]

83(80) Rostrum in lateral view with ventral portion height greater than dorsal portion; front relative width narrow; cheliped palmar crest dorsodistal lobe present, ischium ventromedial margin ornamented 84

83′ Rostrum in lateral view with ventral portion height less than dorsal portion; front relative width wide; cheliped palmar crest absent; chela palmar crest dorsodistal lobe absent, ischium ventromedial margin unornamented .... *Aegla platensis* Schmitt, 1942
[Brazil: Rio Grande do Sul; Santa Catarina. Paraguay: Canindeyú. Argentina: Buenos Aires; Catamarca; Missiones; Tucumán. Uruguay: Canelones; Colonia]

84(83) Epibranchial lobe unornamented; abdominal pleura II dorsally carinate; cheliped dactyl lateral margin proximal end with lobe .... *Aegla schmitt* Hobbs III, 1979
[Brazil: Paraná; São Paulo]

84′ Epibranchial lobe ornamented; abdominal pleura II dorsally smooth; cheliped dactyl lateral margin proximal end without lobe .... *Aegla intercalata* Bond-Buckup & Buckup, 1994
[Argentina: Catamarca; Tucumán]

REFERENCES

Bond-Buckup, G. & L. Buckup. 1994. A família Aeglidae (Crustacea, Decapoda, Anomura). Arquivos de Zoologia 32:159–346.

Bond-Buckup, G., C.G. Jara, M. Pérez-Losada, L. Buckup & K.A. Crandall. 2008. Global diversity of crabs (Aeglidae: Anomura: Decapoda) in freshwater. Hydrobiologia 595: 267–273.

Bueno, S.L.S., A.L. Camargo & J.C. Moraes. 2017. A new species of stygobitic aeglid from lentic subterranean waters in southeastern Brazil, with an unusual morphological trait: short pleopods in adult males. Nauplius 25: 1–18, e201700021.

Jara, C. G., M. Pérez-Losada, & K. A. Crandall. 2018. *Aegla chilota*, new species of anomuran freshwater crab from Chiloé Island, western Patagonia. Nauplius 26: e2018029.

Martin, J.W. & L.G. Abele. 1988. External morphology of genus *Aegla* (Crustacea: Anomura: Aeglidae). Smithsonian Contributions to Zoology 453: 1–52.

Páez, F.P., I.C. Marçal, L. Souza-Shibatta, R.A. Gregati, S.H. Sofia & G.M. Teixeira. 2018. A new species of *Aegla* Leach, 1820 (Crustacea, Anomura) from the Iguaçu River basin, Brazil. Zootaxa 4527: 335–346.

Santos, S., G. Bond-Buckup, L. Buckup, M.L. Bartholomei-Santos, M. Pérez-Losada, C.G. Jara & K.A. Crandall. 2015. Three new species of Aeglidae (*Aegla* Leach, 1820) from Paraná State, Brazil. Journal of Crustacean Biology 35: 839–849.

Santos, S., G. Bond-Buckup, A.S. Gonçalves, M.L. Bartholomei-Santos, L. Buckup & C.G. Jara. 2017. Diversity and conservation status of *Aegla* spp. (Anomura, Aeglidae): an update. Nauplius 25: 1–14, e2017011.

Santos, S., G. Bond-Buckup, A.S. Gonçalves, M.L. Bartholomei-Santos, L. Buckup & C.G. Jara. 2017. Diversity and conservation status of *Aegla* spp. (Anomura, Aeglidae): an update. Nauplius 25: 1-14, e2017011.

Decapoda: Brachyura

Célio Magalhães
Instituto Nacional de Pesquisas da Amazônia, Manaus, AM, Brazil

Manuel Pedraza
Programa de Pós-Graduação em Sistemática, Taxonomia Animal e Biodiversidade, Museu de Zoologia da Universidade de São Paulo, São Paulo, SP, Brazil

D. Christopher Rogers
Kansas Biological Survey, and The Biodiversity Institute, University of Kansas, Lawrence, KS, United States

INTRODUCTION

Two nontaxonomic freshwater crab groups are recognized in the Neotropical region. The primary (or true) freshwater crabs occur exclusively in freshwater habitats, and have adopted aquatic, semiterrestrial, or terrestrial modes of life, never inhabiting brackish or marine environments, and all reproduce by direct development (never with larval stages) (Yeo et al., 2008; Cumberlidge & Ng, 2009; Cumberlidge, 2016). The Pseudothelphusidae and Trichodactylidae constitute this group in the Neotropics. The secondary freshwater crabs refer to those species belonging to originally marine groups that either occur sporadically or permanently in freshwater environments (Cumberlidge, 2016). The primary freshwater crabs form the largest number of species, representing the Pseudothelphusidae (34 genera, 264 species) and the less diverse Trichodactylidae (15 genera, 47 species) (Cumberlidge et al., 2014).

The primary freshwater crabs are, in general, distributed from Mexico to Argentina, also occurring in the Greater and Lesser Antilles. The distribution of each family, however, is not coincident along the whole region and can vary geographically, altitudinally, and ecologically; maps and details about the distribution of both families are presented in Cumberlidge et al. (2014) and Magalhães et al. (2016).

The most accepted classificatory systems for the Pseudothelphusidae (Pretzmann, 1972; Rodríguez, 1982; Ng et al., 2008) divide this family into two subfamilies: Epilobocerini and Pseudothelphusinae, which include the most diverse and derived groups, comprising five tribes: Hypolobocerini (9 genera), Kingsleyini (14 genera), Potamocarcinini (9 genera), Pseudothelphusini (5 genera), and Strengerianini (4 genera). The generic system adopted herein mainly follows Rodríguez (1982, 1992) and Ng et al. (2008), but considers five genera as "incertae sedis" due to some issues raised in recent studies.

The generic system used here for the Trichodactylidae follows Magalhães & Türkay (1996a) and is based on morphological characters of the internal skeleton, the male pleopods, and the abdominal segmentation. In this system, the family is divided into two subfamilies: Trichodactylinae (3 genera) and Dilocarcininae, which in turn is split into two tribes, Dilocarcinini (6 genera) and Valdiviini (6 genera).

LIMITATIONS

The construction of comprehensive keys to Neotropical primary freshwater crabs poses some difficulties, particularly regarding the Pseudothelphusidae. Rodríguez (1982) published the last complete taxonomic revision of the family. Rodríguez (1992) amended his own 1982 proposal of generic arrangement for the Hypolobocerini and Potamocarcinini and recommended keeping that of Pretzmann (1972). Since then, the group has been the target of many studies with descriptions of new taxa, establishment of new synonyms, new combinations, range extensions, etc., that have greatly increased the taxonomic complexity of the group and made this classification somewhat inadequate to properly accommodate the current known diversity. This means that the family is in need of a modern, extensive systematic review. Such a review, however, seems to be increasingly difficult to accomplish because of the high diversity of the group, its wide distribution, and restrictions on collecting, exchanging, and examining material. In addition, many taxa are very poorly described and illustrated, several species are only known from a single adult specimen, and many nomenclatural issues and taxonomic questions are still pending resolution. It is vital to stress the importance of thorough descriptions and detailed illustrations of the male gonopod I as current pseudothelphusid systematics relies almost exclusively on the morphology of this structure.

Consequently, no attempt to update the keys in Rodríguez' (1982) monograph has been made since then, although several valuable keys were presented in more restricted contexts. Regional faunas from Brazil, Colombia, Costa Rica, and Peru were covered, respectively, in the keys offered by Magalhães (2003), Campos (2005, 2014), Magalhães et al. (2015), and Rodríguez & Suárez (2004). Key to species were also provided in the scope of a restricted taxonomic revision of a few taxa: the tribe Pseudothelphusini, by Villalobos Hiriart (2005); and the genera *Typhlopseudothelphusa*, by Hobbs (1986); *Fredius*, by Rodríguez & Campos (1998); *Phrygiopilus*, by Rodríguez (2001); the Colombian *Hypolobocera*, by Campos (2003); *Tehuana*, by Villalobos & Álvarez (2003); *Neostrengeria*, by Campos & Pedraza (2008); *Phallangothelphusa*, by Cardona & Campos (2012); and *Ptychophallus*, by Magalhães et al. (2015).

The somewhat unstable systematic position of some genera also precludes the composition of comprehensive keys for some groups. In a recent molecular phylogenetic

PHYLUM ARTHROPODA: CRUSTACEA: MALACOSTRACA

study of the Pseudothelphusidae that included at least a few representatives of all subfamilies and tribes, Poettinger et al. (2016) found that the genera *Achlidon*, *Allacanthos*, and *Ptychophallus* form a highly supported monophyletic clade, each of them positioned outside the tribes in which they were originally assigned. The authors, however, did not propose a new suprageneric rank to include these genera in the classification. Similarly, a phylogenetic analysis based on morphological data made by Villalobos & Álvarez (2010) showed that the Pseudothelphusini was a paraphyletic group, and proposed that *Ehecatusa* Ng & Low, 2010 (= *Epithelphusa* Rodríguez & Smalley in Smalley, 1970) was excluded from the subfamily and accommodated into two different genera (at least one yet to be established), although they did not indicate where to position them. Furthermore, two other species were considered to represent new genera within the subfamily. As they have not been established yet, they are herein referred as "*Pseudothelphusa*" in the keys. Rodríguez & Hobbs (1989), reviewing the structure of the Potamocarcinini, excluded *Spirocarcinus garthi* Pretzmann, 1971 from the tribe and tentatively moved it to Kingsleyini. Magalhães et al. (2013) discussed the inadequacies of such rearrangement, but were inconclusive to establish its position even within the subfamilies. These genera are herein provisionally included in the keys of the tribes in which they were originally classified according to Rodríguez (1982).

The current generic system of the much less diverse Trichodactylidae is more stable, but the taxonomic status of a few species is still disputable. Therefore, the keys presented in this chapter do not include *Trichodactylus crassus* A. Milne-Edwards, 1869, *Valdivia haraldi* Bott, 1969, and *V. serrata harttii* (Rathbun, 1905), although they were listed in the classification of Magalhães & Türkay (1996a) and Ng et al. (2008). In addition, new molecular-based studies may prove that some taxa might represent a complex of species, as it was recently demonstrated for *Trichodactylus fluviatilis* Latreille, 1828 (Souza-Carvalho et al., 2017).

Taking into account such constraints, the keys to the Pseudothelphusidae especially should be used with caution. It is strongly advisable to doublecheck any identification with the primary taxonomic literature.

TERMINOLOGY AND MORPHOLOGY

In most freshwater crabs, the somatic characters are very conservative and have little taxonomic value. The shape and dentition of the frontal and lateral carapace margins, the degree of arching of the carapace dorsal surface, the abdominal segmentation, the overall shape of the thoracic sternum, and elements of the endophragmal system (usually deduced from the sternal furrows) are useful in some groups. These characters are explained in detail in Magalhães & Türkay (1996a) and schematically illustrated in Magalhães (2003). The maxilliped III is used, particularly for suprageneric groups.

Crab morphological identification relies almost exclusively on the adult male pleopods, which are modified into **gonopods**. Only the first two pairs of gonopods are present in males, but it is the first pair that bears a multitude of features that usually provide consistent taxonomic characters. The flagelliform second pair is much less differentiated but can sometimes be useful.

The **gonopod I** is especially important. The structures and processes are named in terms of their position (proximal, median, subdistal, distal, terminal, apical) and orientation (mesial, lateral, posterior, anterior) along the gonopod. Although the left gonopod I is usually dissected for examination under stereoscope, it is advisable to leave the right gonopod I attached to the body for checking the orientation of its surfaces "in loco," being "mesial" the surface of the gonopod directed toward the middle of the body, "lateral" toward the side, "posterior" toward the ventral side, and "anterior" toward the dorsal side. These structures and process provide a number of (usually) consistent characters that vary greatly in shape and number, especially in the distal half of the gonopod I; the main ones are the "marginal suture," "marginal process," "mesial process," "lateral process," "apical process," and "apical spine field."

The terminology herein used in each couplet of the keys is indicated in the caption of the corresponding illustrations wherever needed. All illustrations are schematic.

MATERIAL PREPARATION AND PRESERVATION

The male gonopod I is vital for identifying most specimens. The position of gonopod I is generally defined with the specimen in dorsal view. When the specimen is turned upside-down and seen from a ventral perspective, the abdomen is opened to expose the sternoabdominal cavity and the gonopods; in this case, the left gonopod I actually lies on the right side. For the assessment of the gonopod morphology, the left gonopod I is traditionally most commonly used for the identification of pseudothelphusid crabs, whereas the right gonopod I is employed in identifying trichodactylid crabs.

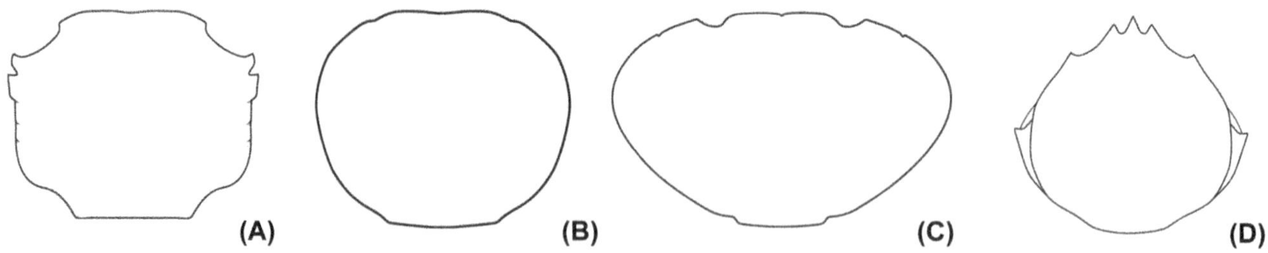

FIGURE 23.115 Carapace outline of brachyuran crabs: (A) Sesarmidae, *Sesarma fossarum*; (B) Trichodactylidae, *Trichodactylus fluviatilis*; (C) Pseudothelphusidae, *Fredius fittkaui*; and (D) Hymenosomatidae, *Neorhynchoplax kempi*.

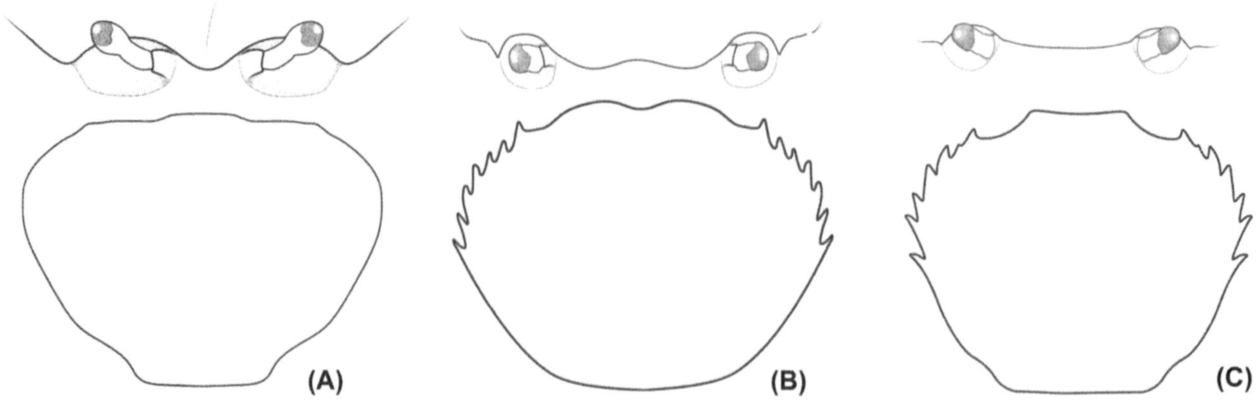

FIGURE 23.116 Frontal margin and carapace outline of brachyuran crabs: (A) *Ucides cordatus*; (B) *Dilocarcinus pagei*; (C) *Valdivia serrata*.

KEY TO BRACHYURA

Many marine species are carried into the Panama Canal by ships, both on the exterior and in ballast. Strictly marine taxa are not dealt with here, as species from all over the world may appear and then are never seen in the Canal again. Thus, it would be fruitless to try to provide a key to all such taxa. Only those taxa that are apparently established in the Panama Canal have been included.

Decapoda: Brachyura: Families

1 Carapace in dorsal view ovate to quadrangular, rostrum absent (Figs. 23.115 A–C, 116) 2

1′ Carapace in dorsal view subcircular, bearing a rostrum (Fig. 23.115 D) .... Hymenosomatidae, one species: *Neorhynchoplax kempi* (Chopra & Das, 1930) [Invasive; Panama Canal. Oriental]

2(1) Combined orbit lengths ≤ half the carapace front margin width 3

2′ Combined orbit lengths > half the carapace front margin width (ghost and fiddler crabs) .... **Ocypodidae [p. 935]**

3(2) Maxilliped III carpus articulating at or near merus distal margin middle or distolateral angle, occasionally partially or completely concealed by expanded merus; adult carapace length at midline ≥ 0.75× maximum width, or carapace anterolateral margin inerm 4

3′ Maxilliped III carpus articulating at or near merus distomesial angle, never concealed by expanded merus; adult carapace length at midline ≤ 0.67× maximum width, or if greater than 0.75×, carapace anterolateral margin with 8–11 teeth 7

4(3) Maxilliped III merus without three longitudinal sulci 5

4′ Maxilliped III merus with three longitudinal sulci .... **Glyptograpsidae [p. 937]**

5(4) Carapace lateral margins dentate, carinate, or angulate 6

5′ Adult carapace anteriolaterally tumescent, lateral margins broadly rounded; mostly terrestrial .... **Gecarcinidae [p. 938]**

6(5) Maxilliped III merus and ischium with an oblique, setose ridge; pterygostomian region with reticulated pattern of evenly and densely spaced short, stiff setae .... **Sesarmidae [p. 938]**

6′ Maxilliped III merus and ischium without clear, oblique, setose ridge; pterygostomian region with scattered setae, not set in a reticulated pattern .... **Grapsidae [p. 941]**

7(3) Pereopod V dactylus compressed, densely margined in natatory setae 8

7′ Pereopod V dactylus not compressed, without dense setal margins; limbs II–V ventrally generally with spines or scattered setae 9

8(7) Pereopod V dactylus broadly ovate .... Portunidae, one genus: ***Callinectes* [p. 941]**

8′ Pereopod V dactylus arcuate .... **Trichodactylidae [p. 942]**

9(7) Carapace anteriolateral margins inerm, or with row of strong denticles, or with seven or more teeth; freshwater .... **Pseudothelphusidae [p. 956]**

9′ Carapace anteriolateral margins with three to five large teeth; brackish water, estuaries .... **Panopeidae [p. 970]**

Decapoda: Brachyura: Ocypodidae: Genera

This key is based on the excellent work of Shih et al. (2016). It must be stressed that the only fiddler crab species included in the keys below are those that occur in freshwater, brackish water, or inland saline aquatic habitats. Fiddler crab diversity is very high, and other species may be encountered where there is a greater marine influence.

1 Pereopods II and III separated basally by a recess densely fringed with setae; male chela greatly unequal (fiddler crabs) 2

1′ Pereopods II and III not separated basally by a hirsuite recess .... ***Ucides* [p. 935]**

2(1) Orbital floor without tubercle at medial end adjacent to antenna I; two genera, very difficult to separate with many overlapping characteristics; we strongly recommend that material be run through the keys for each of the following genera 3

2′ Orbital floor with tubercle at medial end adjacent to antenna I .... ***Uca* [p. 935]**

3(2) Carapace anterolateral margin short; carapace with zero to two posterolateral striae; pleonites free or somites four to six partly or completely fused; major pollex without ventral carina .... ***Leptuca* [p. 935]**

3′ Carapace anterolateral long, curving into dorsolateral surface; carapace with two posterolateral striae; all pleonites free; major pollex may possess ventral carina .... ***Minuca* [p. 937]**

Decapoda: Brachyura: Ocypodidae: *Ucides*: Species

1 Chela palm ventral margin convex; male carapace width 1.25× length; fresh and brackish water, mangroves .... *Ucides cordatus* (Linnaeus, 1763)
[Atlantic and Caribbean coasts and islands, south to Brazil. Nearctic]

1′ Chela palm ventral margin straight; male carapace width 1.50× length; brackish water, mangroves .... *Ucides occidentalis* (Ortmann, 1897)
[Pacific coasts from Mexico to Peru]

Decapoda: Brachyura: Ocypodidae: *Uca*: Species

Key based on Crane (1975).

1 Pollex without pits or punctae, width $\geq$ dactyl; salt marshes, river mouths 2

1′ Pollex with distinct pits or punctae, width $<$ dactyl; estuaries, mangroves, coastal freshwater streams and rivers .... *Uca ornata* (Smith, 1870)
[Pacific coasts from El Salvador to Peru]

2(1) Carapace anteriolateral and dorsolateral margins granulate nearly to posterior margins .... *Uca princeps* (Smith, 1870)
[Pacific coast from México to Peru]

2′ Carapace anteriolateral margins indistinct, inerm or utmost weakly serrate; carapace dorsolateral margins absent in male, irregularly granulate in female .... *Uca monilifera* Rathbun, 1914
[Mexico]

Decapoda: Brachyura: Ocypodidae: *Leptuca*: Species

Key based on Crane (1975). This genus is very difficult to separate from *Minuca* (Shih et al., 2016). The range of variation can cause apparent overlap in the genus level characteristics. Thus, the user is strongly urged that material be tried through both keys to better establish its identity.

1 Pacific coastal taxa 2

1′ Atlantic and Caribbean coasts and islands 12

2(1) Male small chela widest portion of gape ≤ adjacent dactyl width 3

2′ Male small chela widest portion of gape ≥ adjacent dactyl width; tidally influenced streams .... *Leptuca latimanus* (Rathbun, 1893)
[Mexico south to Ecuador]

3(2) Male small chela with dactyl and propodus cutting edge serrations weak, never in contact 4

3′ Male small chela with dactyl and propodus cutting edge serrations robust, in distal half either in contact or nearly so 6

4(3) Carapace anteriolateral margins robust, divergent, extending to or nearly to ventral margin; estuaries 5

4 Carapace anteriolateral margins obscure, extending at most one-fourth of carapace length; streams ...... *Leptuca limicola* (Crane, 1941)
[Costa Rica, El Salvador, Panama]

5(4) Carapace anteriolateral angles acute, produced; female carapace glaborous .... *Leptuca crenulata* (Lockington, 1877)
[Mexico. Nearctic]

5′ Carapace anteriolateral angles subquadrate; female carapace posteriorly with a pair of tomentose patches............ *Leptuca coloradensis* (Rathbun, 1893)
[Mexico]

6(3) Carapace anteriolateral margins robust, obvious 7

6′ Carapace anteriolateral margins obscure to absent; freshwater streams .... *Leptuca pygmaea* (Crane, 1941)
[Costa Rica]

7(6) Suborbital crenulations absent or obscure, never increasing in size distally 8

7′ Suborbital crenulations well developed in lateral half, increasing in size distally 9

8(7) Pereopod II–II merus with dorsal margin slightly convex; male major chela dactyl cutting edge with two large tubercles; river mouths .... *Leptuca saltitanta* (Crane, 1941)
[Costa Rica, El Salvador, Panama]

8′ Pereopod II–II merus with dorsal margin flat; male major chela dactyl cutting edge with zero or one large tubercle; mangroves, estuaries .... *Leptuca tenuipedis* (Crane, 1941)
[Costa Rica to Peru]

9(7) Carapace dorsally glaborous, or with pilosity in small patches in and near central "H" depression 10

9′ Carapace dorsally pilose, especially females and immatures, may be worn off, and persists only in grooves and sulci; mangrove estuaries, streams .... *Leptuca umbratila* Crane, 1941
[Costa Rica, El Salvador, Panama]

10(9) Carapace dorsally without pilosities, at most some setae in grooves, female with posteriodorsal pilosities; rivers, streams 11

10′ Carapace dorsally with curved pilosities near centre of disc, six in male, two in female; estuaries .... *Leptuca tallanica* (von Hagen, 1968)
[Ecuador. Peru]

11(10) Male major chela with dactyl and pollex length 1.5–2× palm length; female carapace with posteriodorsal pilosities only .... *Leptuca beebei* (Crane, 1941)
[El Salvador to Peru]

11′ Male major chela with dactyl and pollex 3–7× palm length; female carapace with posteriodorsal and posteriolateral pilosities .... *Leptuca festae* (Nobili, 1902)
[El Salvador to Ecuador]

12(1) Carapace anterior width ≤ 3.5× width between eyestalks; freshwater sloughs and streams 13

12′ Carapace anterior width < 4.5× width between eyestalks; mangrove estuaries, streams .... *Leptuca thayeri* (Rathbun, 1900)
[Caribbean Islands. Brazil. Nearctic]

13(12) Carapace anteriodorsal surface granulate, remaining surface pilose (unless abraded) .... *Leptuca speciosa* (Ives, 1891)
[Cuba. Jamaica. Atlantic Mexico. Nearctic]

13′ Carapace anteriodorsal surface smooth, remaining surface glaborous or with pilose patches in grooves and sulci .... *Leptuca spinicarpa* (Rathbun, 1900)
[Atlantic Mexico. Nearctic]

Decapoda: Brachyura: Ocypodidae: *Minuca*: Species

Key based on Crane (1975). This genus is very difficult to separate from *Leptuca* (Shih et al. 2016). The range of variation can cause apparent overlap in the genus-level characteristics. Thus, the user is strongly urged that material be tried through both keys to better establish its identity.

1 Pacific coast taxa 2

1′ Atlantic and Caribbean coasts and islands 6

2(1) Orbits slightly oblique to straight; male major chela medial surface with oblique tuberculate ridge (maybe obscure) 3

2′ Orbits oblique; male major chela medial surface without oblique tuberculate ridge; rivers, streams, estuaries .... *Minuca zacae* (Crane, 1941)

[Mexico to Costa Rica]

3(2) Male major chela palm medial surface with robust oblique tuberculate ridge; both sexes carapace glaborous 4

3′ Male major chela palm medial surface with robust oblique tuberculate ridge obscure; both sexes carapace with scattered pilosities (often worn off); shaded or partly shaded stream and rivers, mangroves .... *Minuca ecuadoriensis* (Maccagno, 1928)

[Mexico to Peru]

4(3) Male major chela palm lateral surface with arcuate beaded ridge from dorsal margin, directed ventrally; palmar proximal ridge at dactyl base parallel to adjacent groove; pollex cutting edge distal half without large tubercles; stream mouths, mangroves 5

4′ Male major chela palm lateral surface without arcuate beaded ridge from dorsal margin; palmar proximal ridge at dactyl base diverging from adjacent groove; pollex cutting edge distal half with several large tubercles; freshwater rivers, streams, wet forest .... *Minuca brevifrons* (Stimpson, 1860)

[Mexico to Panama]

5(4) Pereopods II–V carpus and manus glabrous .... *Minuca galapagensis* (Rathbun, 1902)

[Colombia to Peru (Chile?), Galapágos Islands]

5′ Pereopods II–V carpus and manus hirsute .... *Minuca herradurensis* (Bott, 1954)

[El Salvador to Panama]

6(1) Male major chela palm medial surface with robust oblique tuberculate ridge; female pereopods II–V hirsute; both sexes carapace glaborous or with scattered setae in grooves or disc 7

6′ Male major chela palm medial surface without robust oblique tuberculate ridge; female pereopods II–V glaborous; both sexes carapace with pilosities in marbelled pattern; shaded or partly shaded stream and rivers, mangroves .... *Minuca vocator* (Herbst, 1804)

[Coasts and islands from Belize and Puerto Rico to Trinidad and Brazil]

7(6) Pereopods II–V ventrally glaborous 8

7′ Pereopods II–V carpus and manus ventrally hirsute; rivers, estuaries .... *Minuca mordax* (Smith, 1870)

[Guatemala to Brazil]

8(7) Male major chela palm proximal ridge at dactyl base paralleling adjacent groove, or diverging only apically 9

8′ Male major chela palm proximal ridge at dactyl base diverging from adjacent groove; estuaries, stream and river mouths, mangroves .... *Minuca burgersi* (Holthuis, 1967)

[Atlantic and Caribbean coasts and islands, south to Brazil. Nearctic]

9(8) Male major chela palm disc with flattened tubercules; estuaries .... *Minuca virens* (Salmon & Atsaides, 1968)

[Northern coastal Mexico. Nearctic]

9′ Male major chela palm disc finely granulate to smooth; coastal streams, rivers, deltas, mudflats .... *Minuca rapax* (Smith, 1870)

[Atlantic and Caribbean coasts and islands, south to Brazil. Nearctic]

Decapoda: Brachyura: Glyptograpsidae: Genera

Coastal riverine habitats.

1 Major chela propodus proximolateral margin elongated, extending beyond carpus articulation .... *Platychirograpsus spectabilis* de Man, 1896

[Atlantic and Caribbean coasts and islands, south to Brazil. Nearctic]

1′ Major chela propodus not elongated .... ***Glyptograpsus*** **[p. 937]**

Decapoda: Brachyura: Glyptograpsidae: *Glyptograpsus*: Species

1 Major chela dorsomedial surface with two parallel rows of rounded tubercles, meeting proximally .... *Glyptograpsus jamaicensis* (Benedict, 1802)

[Jamaica]

1′ Major chela dorsomedial surface with scattered, flattened, round tubercles, dorsomedial margin with a single row of small denticles .... *Glyptograpsus impressus* Smith, 1870

[Atlantic and Caribbean drainages Mexico to Panama]

Decapoda: Brachyura: Gecarcinidae: Genera

Primarily terrestrial, these crabs may occasionally be found in and around streams, rivers, or wetlands

1 Carapace anterior margin between eyestalks > 0.5 carapace width .... ***Cardisoma*** **[p. 938]**

1′ Carapace anterior margin between eyestalks < 0.5 carapace width .... ***Gecarcinus*** **[p. 938]**

Decapoda: Brachyura: Gecarcinidae: *Cardisoma*: Species

1 Pacific coasts; legs hirsute; male pleopod I with two slender projections .... *Cardisoma crassum* Smith, 1870
[Mexico to Peru (Chile?)]

1′ Atlantic and Caribbean coasts and islands; legs with short sparse setae; male pleopod I apically broad, rounded .... *Cardisoma guanhumi* Latrielle, 1825
[Bahamas. Mexico to Brazil. Caribbean Islands. Nearctic]

Decapoda: Brachyura: Gecarcinidae: *Gecarcinus*: Species

Modified from Rathbun (1918).

1 Maxilliped III merus medial margin entire 2

1′ Maxilliped III merus medial margin concave 3

2(1) Maxilliped III merus length > width, covering epistome and antenna I cavities; pereopods II–V dactyls each with six rows of spines .... *Gecarcinus ruricola* (Linnaeus, 1758)
[Gulf of Mexico coasts and Caribbean Islands. Nearctic]

2′ Maxilliped III merus length subequal to width, not covering epistome; pereopods II–V dactyls each with four or five rows of spines .... *Gecarcinus quadratus* Saussure, 1853
[Pacific Coast from Mexico to Colombia]

3(1) Maxilliped III merus medial margin angulate, sometimes continued into merus as a closed fissure 4

3′ Maxilliped III merus medial margin concavity shallow, arcuate .... *Gecarcinus lateralis* (Freminville, 1835)
[Caribbean Islands to Brazil. Bermuda. Ascension Island. Afrotropical]

4(3) Maxilliped III covering epistome and antenna I cavity .... *Gecarcinus lagostoma* Milne Edwards, 1836
[Gulf of Mexico coasts and Caribbean Islands. Nearctic]

4′ Maxilliped III not covering epistome .... *Gecarcinus platanus* Stimpson, 1808
[Pacific coastal Mexico]

Decapoda: Brachyura: Sesarmidae: Genera

1 Carapace dorsally convex 2

1′ Carapace flat; in phytotelmata .... *Metopaulias depressus* (Rathbun, 1896)
[Jamaica]

2(1) Carapace anteriolateral margin with or without tooth posterior to orbital angle, if tooth present carapace lateral margins subparallel or diverging posteriorly, rarely converging slightly 3

2′ Carapace anteriolateral margin without tooth posterior to orbital angle; carapace with lateral margins strongly converging posteriorly .... ***Aratus*** **[p. 938]**

3(2) Chela palm dorsal margin carinate, carina composed of a single row of acute tubercles .... ***Sesarma*** **[p. 939]**

3′ Chela palm dorsal margin without distinct carina, but bearing scattered granulae .... ***Armases*** **[p. 940]**

Decapoda: Brachyura: Sesarmidae: *Aratus*: Species

1 Eastern Neotropics; estuaries, coastal streams, lagoons, mangroves .... *Aratus pisonii* (H. Milne Edwards, 1837)
[Atlantic and Caribbean coasts and Islands south to Brazil. Nearctic]

1′ Western Neotropics; estuaries, coastal streams, lagoons, mangroves .... *Aratus pacificus* Thiercelin & Schubart, 2014
[Pacific coasts from Nicaragua to Peru]

Decapoda: Brachyura: Sesarmidae: *Sesarma*: Species

Adapted in part from Abele (1992)

1 Pacific coasts 2

1′ Atlantic and Caribbean coasts and islands 5

2(1) Carapace with anteriolateral tooth posterior to orbital angle 3

2′ Carapace without anteriolateral tooth posterior to orbital angle; mangroves ..... *Sesarma rubinofforum* Abele, 1973
[Costra Rica. Panama]

3(2) Carapace dorsally granulate; adult carapace breadth > 15cm 4

3′ Carapace dorsally not granulate; adult carapace breadth < 13cm; mangroves ..... *Sesarma rhizophorae* Rathbun, 1906
[Costa Rica to Peru]

4(3) Carapace anterior margin in dorsal view, between eyes (frontal region) sinuate with anteriormost projecting portions between carapace, dorsomedial longitudinal groove and first submedial groove; brackish water, mangroves ..... *Sesarma sulcatum* Smith, 1870
[Mexico to Colombia]

4′ Carapace frontal region sinuate with anteriormost projecting portions in line with carapace first submedial groove; semiterrestrial, brackish, and freshwater ..... *Sesarma aequatoriale* Ortmann, 1894
[Mexico to Ecuador]

5(1) Carapace with anteriolateral tooth posterior to orbital angle; pereopods II–V meri length $> 2\times$ greatest width 6

5′ Carapace with anteriolateral ridge posterior to orbital angle; pereopods II–V meri length $< 2\times$ greatest width; estuaries, coastal mud banks ..... *Sesarma rectum* Randall, 1840
[Trinidad & Tobago. Grenada. Brazil. Guyana. Suriname]

6(5) Carapace regions well defined, disc granulate at least in some areas, pubescence present or absent 7

6′ Carapace regions obscure, disc nearly smooth, often with fine pubescence; mangroves ..... *Sesarma curacaoense* De Man, 1892
[Caribbean Islands. Mexico to Brazil. Nearctic]

7(6) Pereopods II–V meri length $\geq 3\times$ greatest width; Jamaican endemics 8

7′ Pereopods II–V meri length $\sim 2\times$ greatest width; deep rivers and estuaries ..... *Sesarma crassipes* Cano, 1889
[Costa Rica to Brazil]

8(7) Eyes normal, reaching nearly to orbital angles 9

8′ Eyes reduced, reaching only half way to orbital angles; caves ..... *Sesarma verleyi* Rathbun, 1914
[Jamaica]

9(8) Pereopod III length $\leq 2.5\times$ carapace mid-dorsal length, merus length $\sim 3\times$ merus width 10

9′ Pereopod III length $\sim 3\times$ carapace mid-dorsal length, merus length $\sim 4\times$ merus width; terrestrial to 650m asl ..... *Sesarma cookei* Hartnoll, 1971
[Jamaica]

10(9) Chela palm dorsal margin with ridge interrupted 11

10′ Chela palm dorsal margin with ridge continuous 13

11(10) Cornea width subequal to eyestalk width 12

11′ Cornea width $\sim 0.75\times$ eyestalk width; caves ..... *Sesarma windsor* Türkay & Diesel, 1994
[Jamaica]

12(11) Chela carpus in dorsal view with anteriolateral corner at nearly right angle; mountain freshwater rivers and streams ..... *Sesarma bidentatum* Benedict, 1892
[Jamaica]

12′ Chela carpus in dorsal view with anteriolateral corner broadly rounded; mountain freshwater streams ..... *Sesarma ayatum* Schubart, Reimer, & Diesel, 1988
[Jamaica]

13(11) Carapace with scattered setal tufts 14

13′ Carapace mostly glaborous with a few scattered setae in the branchial regions (posteriolateral portions); freshwater ..... *Sesarma merides* Schubart & Koller, 2005
[Jamaica]

14(13) Pereopods IV with dactyl shorter than propodus, pereopod III dactyl subequal to propodus 15

14′ Pereopods III and IV with dactyli longer than propodi; terrestrial, to 900 m asl ..... *Sesarma jarvisi* Rathbun, 1914
[Jamaica]

15(14) Chela dactyl dorsally smooth 16

15′ Chela dactyl dorsally with longitudinal row of tubercles; freshwater ..... *Sesarma fossarum* Schubart, Reimer, Diesel & Türkay, 1988
[Jamaica]

16(15) Ratio of pereopod IV length to carapace mid-dorsal length > 2; freshwater ..... *Sesarma dolphinum* Reimer, Schubart & Diesel, 1988
[Jamaica]

16′ Ratio of pereopod IV length to carapace mid-dorsal length < 2; freshwater ..... *Sesarma abeokuta* Scubart & Santl, 2014
[Jamaica]

Decapoda: Brachyura: Sesarmidae: *Armases*: Species

Adapted in part from Abele (1992).

1 Pacific coasts 2

1′ Atlantic and Caribbean coasts and islands 5

2(1) Carapace mid-dorsal length $\leq$ width; fresh, brackish, and terrestrial 3

2′ Carapace mid-dorsal length $>$ width; freshwater ..... *Armases angustum* Smith, 1870
[Pacific coast from El Salvador to Ecuador, and associated islands]

3(1) Carapace mid-dorsal length $<$ width 4

3′ Carapace mid-dorsal length $=$ width; brackish, terrestrial ..... *Armases occidentale* Smith, 1870
[Pacific coast from El Salvador to Ecuador]

4(3) Pereopod V dactylus with lateral and ventral margins with black spines ..... *Armases magdalenense* (Rathbun, 1918)
[Mexico]

4′ Pereopod V dactylus ventral margins only with black spines; mangroves ..... *Armases gorei* (Abele, 1981)
[Peru]

5(1) Male chela dactyl basal width $< 1.5\times$ length, subequal or narrower than propodus 6

5′ Male chela dactyl basal width $\geq 1.5\times$ length, wider than propodus; brackish to freshwater ..... *Armases benedicti* (Rathbun, 1897)
[Venezuela to Suriname. Nearctic (?)]

6(5) Pereopod IV and V with black spines at least ventrally 7

6′ Pereopod IV and V without black spines; mangroves ..... *Armases rubripes* (Rathbun, 1897)
[Atlantic and Caribbean coasts and Islands from Nicaragua to Uruguay]

7(6) Carapace front in anterior view with lateral margins subparallel 8

7′ Carapace front in anterior view with lateral margins diverging 9

8(7) Carapace front in dorsal view with shallow central emagination, $\sim 0.3\times$ width of front; muddy river and stream banks ..... *Armases americanum* (Saussure, 1858)
[Atlantic and Caribbean coasts of México. Caribbean Islands. Nearctic]

8′ Carapace front in dorsal view with deep central emagination, $\leq 0.25\times$ width of front; freshwater to 350m asl ..... *Armases roberti* (H. Milne-Edwards, 1853)
[Caribbean Islands]

9(7) Gonopod (denuded) apically subrectangular, directed distolaterally 10

9′ Gonopod (denuded) apex small, centralized; terrestrial ..... *Armases ricordi* (H. Milne-Edwards, 1853)
[Bahamas. Bermuda. Caribbean Islands. México to Suriname. Nearctic]

10(9) Pereopod IV dactylus with lateral and ventral margins with black spines 11

10′ Pereopod IV dactylus ventral margins only with black spines; mudflats ..... *Armases miersii* (Rathbun, 1897)
[Bahamas. Caribbean Islands. Nearctic]

11(10) Pereopod IV merus length $\sim 3.0\times$ width; gonopod blending distally into apical portion without interruption; phytotelmata, sinkholes (?) ..... *Armases angustipes* (Dana, 1852)
[Bahamas. Caribbean Islands. Mexico to Brazil. Trinidad]

11′ Pereopod IV merus length $\sim 2.6\times$ width; gonopod with small convex portion proximal to apical portion; fresh, brackish, terrestrial, swamps, intertidal ..... *Armases cinereum* (Bosc, 1802)
[Mexico. Nearctic]

Decapoda: Brachyura: Grapsidae: Genera

1 Antenna I excluded from orbit; estuaries, river mouths .... *Goniopsis cruentata* (Latreille, 1802)
[Bermuda south to Brazil. Afrotropics]

1′ Antenna I included in orbit; coastal river banks, mangroves .... *Pachygrapsus gracilis* (De Saussure, 1858)
[Bermuda south to Brazil. Afrotropics. Nearctic]

Decapoda: Brachyura: Portunidae: *Callinectes*: Species

Based on Rathbun (1930), Chace & Hobbs (1969), and Williams (1974). Only mature adults maybe identified here.

1 Carapace front in dorsal view, with four teeth between medial orbital teeth, or with two large teeth, length greater than basal width, bearing a pair of rudimentary mesial teeth 2

1′ Carapace front in dorsal view between medial orbital teeth with two triangular teeth, length subequal to basal width, with or without a pair of rudimentary mesial teeth; estuaries and tidally influenced freshwater .... *Callinectes sapidus* Rathbun, 1895
[Atlantic and Caribbean coasts and islands south to Uruguay. Nearctic. Invasive elsewhere]

2(1) Carapace front mesial pair of teeth length $>0.5\times$ lateral pair of teeth 3

2′ Carapace front mesial pair of teeth length $\leq 0.5\times$ lateral pair of teeth 7

3(2) Carapace four frontal teeth subequal in length 4

3′ Carapace mesial frontal teeth obviously shorter than lateral pair; estuaries, tidally influenced freshwaters .... *Callinectes exasperatus* (Gerstaecker, 1856)
[Atlantic and Caribbean coasts and Islands, south to Brazil. Nearctic]

4(3) Carapace frontal teeth with one or both pairs triangular 5

4′ Carapace frontal teeth with both pairs lobiform; estuaries .... *Callinectes toxotes* Ordway, 1863
[Pacific coasts from México to Chile]

5(4) Carapace frontal teeth apically rounded, lateral pair more broadly triangular than mesial pair, with mesial side slope more oblique than lateral side slope 6

5′ Carapace frontal teeth apically acute, lateral pair usually broader than mesial pair; estuaries .... *Callinectes rathbunae* Contreras, 1930
[Gulf of Mexico. Nearctic]

6(5) Carapace frontal teeth with submedial pair directed anteriorly, with anterior margins< posterior margins; mature female abdominal segment VI distal margin broadly triangular; estuaries, tidally influenced freshwaters .... *Callinectes bocourti* A. Milne Edwards, 1879
[Atlantic and Caribbean coasts and Islands, south to Brazil. Nearctic]

6′ Carapace frontal teeth with submedial pair directed laterally, apices acuminate and margins shouldered at least anteriorly; mature female abdominal segment VI distal margin semiellipsoid; estuaries, tidally influenced freshwaters .... *Callinectes maracaiboensis* Taissoun, 1969
[Venezuela]

7(2) Carapace remarkably smooth, granule lines visible, epibranchial line variably prominent, remaining barely perceptible to touch 8

7′ Carapace with scattered granulae, granule lines visible and raised 9

8(7) Carapace frontal teeth mesial pair vestigial .... *Callinectes bellicosus* (Stimpson, 1859)
[Pacific Mexico. Nearctic]

8′ Carapace frontal teeth mesial pair small but clearly present; tidally influenced rivers, marine .... *Callinectes similis* Williams, 1966
[Gulf of Mexico. Nearctic]

9(7) Carapace finely granulate; distalmost one or two anteriolateral teeth curved anteriorly, remaining teeth with shoulders 10

9′ Carapace coarsely granulate; anteriolateral teeth except first two curved anteriorly, shoulders, brackish water, mangroves, marine .... *Callinectes marginatus* (A. Milne Edwards, 1861)
[Atlantic and Caribbean coasts and Islands, south to Brazil. Afrotropics. Nearctic]

10(9) Carapace frontal teeth mesial pair small but clearly present, length $\leq 0.5\times$ lateral pair 11

10′ Carapace frontal teeth mesial pair absent or vestigial; fresh to marine .... *Callinectes ornatus* Ordway, 1863
[Atlantic and Caribbean coasts and Islands, south to Brazil. Nearctic]

11(10) Mature male gonopod I with subterminal dorsal setae 12

PHYLUM ARTHROPODA: CRUSTACEA: MALACOSTRACA

11′ Mature male gonopod I with subterminal dorsal spines; brackish water .... *Callinectes affinis* Fausto-Filho, 1980 [Brazil]

12(11) Mature male gonopod I with > four subterminal dorsal setae in a row .... *Callinectes arcuatus* Ordway, 1863 [Pacific coasts and islands south to Peru. Galapágos Islands. Nearctic]

12′ Mature male gonopod I with ≤ four subterminal dorsal setae, often inconspicuous or missing; estuaries and marine .... *Callinectes danae* Smith, 1869 [Atlantic and Caribbean coasts and Islands, south to Brazil. Nearctic]

Decapoda: Brachyura: Trichodactylidae: Subfamilies

Modified from Rodríguez (1992), Magalhães & Türkay (1996a, 1996b, 1996c; 2008a, 2008b; 2012), Magalhães (2003).

1 Carapace dorsally without postgastric pits (Fig. 23.117 A); abdominal somites each separated by distinct, shallow sulcus (sometimes marked only by distinct pigmentation); gonopod I straight (except in *Avotrichodactylus constrictus*), marginal suture following basal main axis along mesial side, not twisted ventrally or dorsally in distal half, distal opening usually wide, distinctly V-shaped proximally (except in *Avotrichodactylus*) .... **Trichodactylinae [p. 942]**

1′ Carapace dorsally with postgastric pits (Fig. 23.117 B, arrow); abdomen with at least somites III–V fused or obsolete; gonopod I nearly straight to curved, marginal suture slightly to greatly twisted towards ventral or dorsal sides in distal half, distal opening usually narrow, V-shaped, or rimate .... **Dilocarcininae [p. 945]**

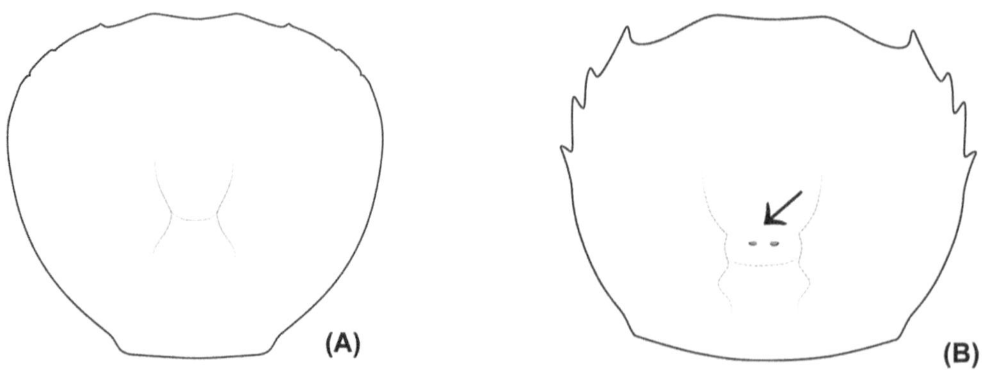

FIGURE 23.117 Carapace outline of Trichodactylidae: (A) *Trichodactylus fluviatilis*; (B) *Sylviocarcinus australis*. *Arrow*, carapace postgastric pit.

Decapoda: Brachyura: Trichodactylidae: Trichodactylinae: Genera

1 Gonopod II length < gonopod I (Fig. 23.118 A) 2

1′ Gonopod II length > gonopod I (Fig. 23.118 B) .... ***Trichodactylus* [p. 942]**

2(1) Gonopod I distal portion gradually tapering, conical, slightly to strongly curved ventrolaterally, distal opening subcircular and very narrow (Fig. 23.119 A, B). .... ***Avotrichodactylus* [p. 943]**

2′ Gonopod I distal portion subcylindrial, stout, straight, distal opening wide; proximally V-shaped (Figs. 23.118 A, 119 C) .... ***Rodriguezia* [p. 943]**

Decapoda: Brachyura: Trichodactylidae: Trichodactylinae: *Trichodactylus*: Species

1 Carapace lateral margin (Fig. 23.120 A–D) inerm or with one to three, large or small (sometimes vestigial) teeth 2

1′ Carapace lateral margin (Fig. 23.120 E–H) with five small teeth (often two anteriormost or the two posteriormost teeth reduced, sometimes both) 6

2(1) Carapace anteriolateral margin with three large (sometimes reduced), usually sharp teeth (Fig. 23.120 C, D) 3

2′ Carapace anteriolateral margin inerm or with one or two (rarely three) reduced teeth or notches (Fig. 23.120 A, B) 5

3(2) Carapace dorsal region uneven, usually with distinct transversal median crest; anteriolateral margin usually with three large, sharp teeth subequally spaced (sometimes posteriormost is vestigial) (Fig. 23.120 C) 4

PHYLUM ARTHROPODA: CRUSTACEA: MALACOSTRACA

3′ Carapace dorsal region smooth; anteriolateral margin usually with three small, sharp teeth, the anterior two closer set than the second and third, which are vestigial) (Fig. 23.120 D) .... *Trichodactylus dentatus* H. Milne Edwards, 1853

[Brazil: Paraná, Rio de Janeiro, Santa Catarina, São Paulo]

4(3) Telson luniform, proximal width < abdominal somite VI, lateral margins discontinuous (Fig. 23.121 A, arrow); medial orbital angle with robust curved spine directed mesially .... *Trichodactylus kensleyi* Rodríguez, 1992

[Brazil: Rio Grande do Sul, Santa Catarina. Paraguay. Argentina: Misiones]

4′ Telson subtriangular, proximal width equal to abdominal somite VI, lateral margins continuous (Fig. 23.121 B, arrow); medial orbital angle pyriform, prominent, acute, or rounded .... *Trichodactylus petropolitanus* (Göldi, 1886)

[Brazil: Minas Gerais, Paraná, Rio de Janeiro, Santa Catarina, São Paulo]

5(2) Carapace nearly flat, anteriolateral margin with two small, widely spaced, sharp teeth (sometimes reduced or vestigial) (Fig. 23.120 B); gonopod I apex distal margin mesial side angled, distal opening subquadrate (Fig. 23.122 A) .... *Trichodactylus faxoni* Rathbun, 1905

[Colombia: Amazonas. Brazil: Amapá, Amazonas. Peru: Huánuco, Loreto]

5′ Carapace nearly flat to strongly convex, anteriolateral margin usually smooth or with one or two notches or small, usually blunt, close set teeth, posterior one sometimes vestigial (Fig. 23.120 A); gonopod I apex distal margin rounded, distal opening subtriangular (Fig. 23.122 B) .... *Trichodactylus fluviatilis* Latreille, 1828

[Brazil: Alagoas, Bahia, Espírito Santo, Minas Gerais, Paraná, Pernambuco, Rio de Janeiro, Rio Grande do Sul, Santa Catarina, São Paulo, Sergipe]

6(1) Carapace lateral margin with five acute, regularly spaced, teeth 7

6′ Carapace lateral margin with ≤ five teeth, but one to three small teeth irregularly spaced 8

7(6) Gonopod I apex asymmetrical, distal margin projected to mesodorsally (Fig. 23.123 A) .... *Trichodactylus panoplus* (von Martens, 1869)

[Brazil: Rio Grande do Sul, Santa Catarina?, São Paulo?. Argentina: Buenos Aires, Corrientes. Uruguay]

7′ Gonopod I apex symmetrical, distal margin nearly straight (Fig. 23.123 B) .... *Trichodactylus quinquedentatus* Rathbun, 1893

[Nicaragua. Colombia: Atlántico, Bolivar, Cesar, Córdoba, Magdalena, Meta]

8(6) Carapace lateral margin with one median acute tooth and one or two posterolateral, vestigial teeth, or posteriolateral margin smooth (Figs. 23.124 A, B) 9

8′ Carapace lateral margin with one median and two posteriolateral acute teeth (Fig. 23.124 C) .... *Trichodactylus borellianus* Nobili, 1896

[Brazil: Mato Grosso, Mato Grosso do Sul, Pará. Paraguay. Argentina: Corrientes, Formosa, Misiones, Santa Fe]

9(8) Gonopod I apex symmetrical, distal margin truncate and concave (Fig. 23.125 A) .... *Trichodactylus ehrhardti* Bott, 1969

[Brazil: Amazonas, Pará]

9′ Gonopod I apex asymmetrical, with distal margin convex (Fig. 23.125 B) .... *Trichodactylus parvus* Moreira, 1912

[Brazil: Mato Grosso]

Decapoda: Brachyura: Trichodactylidae: Trichodactylinae: *Avotrichodactylus*: Species

1 Gonopod I distal portion strongly curved ventrolaterally (Fig. 23.119 A); carapace usually with two or three anterolateral teeth .... *Avotrichodactylus constrictus* (Pearse, 1911)

[Mexico: Chiapas, Oaxaca, Tabasco, Veracruz]

1′ Gonopod I distal portion slightly curved ventrolaterally (Fig. 23.119 B); carapace usually with four or five anterolateral teeth .... *Avotrichodactylus oaxensis* Rodríguez, 1992

[Mexico: Oaxaca, Veracruz]

Decapoda: Brachyura: Trichodactylidae: Trichodactylinae: *Rodriguezia*: Species

1 Ocular peduncle reduced, cornea absent; gonopod II mesial margin subdistal portion bearing minute spines (Fig. 23.126 B); troglobitic 2

1′ Ocular peduncle developed, cornea present; gonopod II mesial margin subdistal portion smooth (Fig. 23.126 A); epigean .... *Rodriguezia villalobosi* (Rodríguez & Manrique, 1967)

[Mexico: Chiapas]

2(1) Pereopods length > carapace breadth .... *Rodriguezia mensabak* (Cottarelli & Argano, 1977)

[Mexico: Chiapas]

2′ Pereopods length < carapace breadth .... *Rodriguezia adani* Alvarez & Villalobos, 2018

[Mexico: Tabasco]

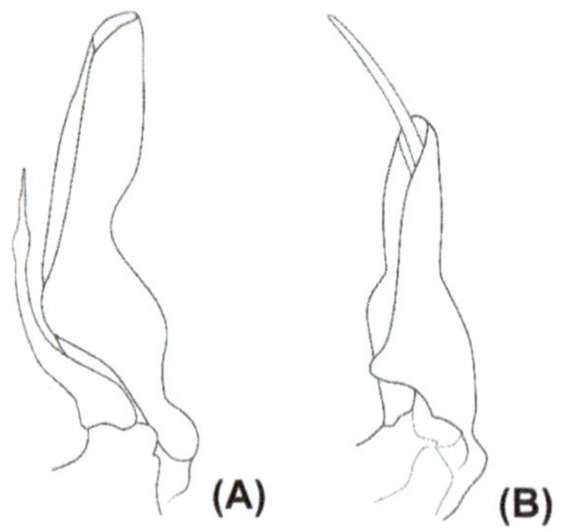

FIGURE 23.118 Gonopods I and II of Trichodactylinae: (A) *Rodriguezia villalobosi*; (B) *Trichodactylus fluviatilis*.

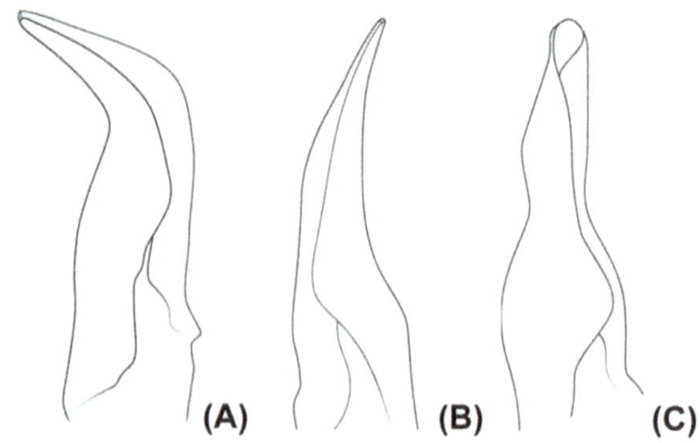

FIGURE 23.119 Gonopod I of Trichodactylinae: (A) *Avotrichodactylus constrictus*; (B) *A. oaxensis*; (C) *Rodriguezia villalobosi*.

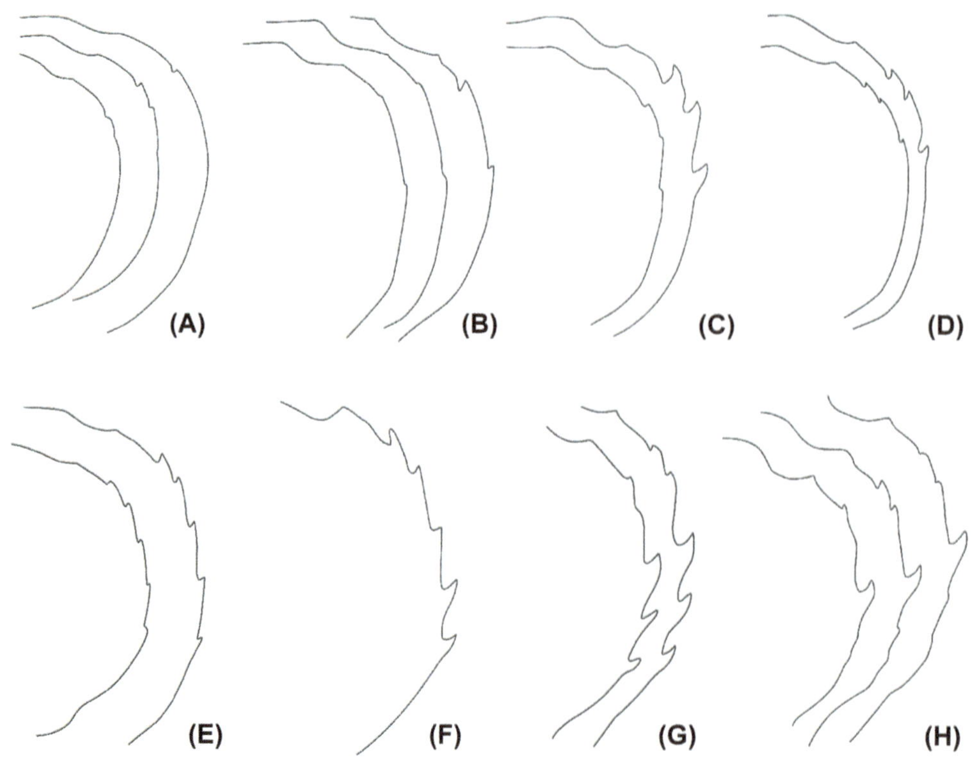

FIGURE 23.120 Lateral margin of carapace of *Trichodactylus* spp., with some common variations: (A) *T. fluviatilis*; (B) *T. faxoni*; (C) *T. petropolitanus*; (D) *T. dentatus*; (E) *T. quinquedentatus*; (F) *T. panoplus*; (G) *T. borellianus*; (H) *T. parvus*.

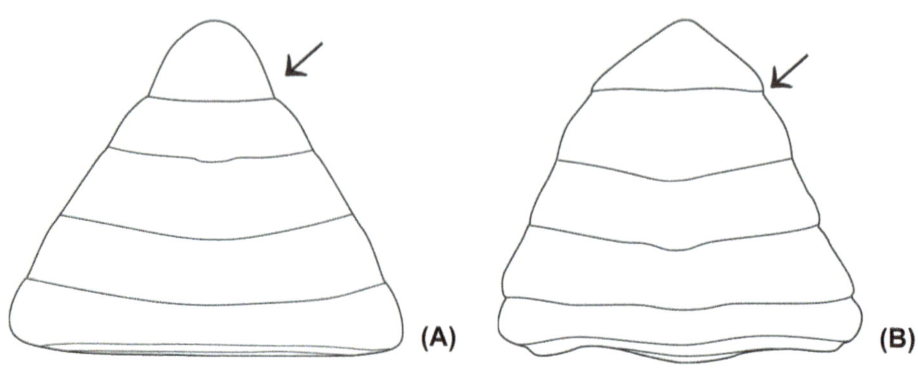

FIGURE 23.121 Abdomen outline of *Trichodactylus* spp.: (A) *T. kensleyi*; (B) *T. petropolitanus*. *Arrow*, lateral margin.

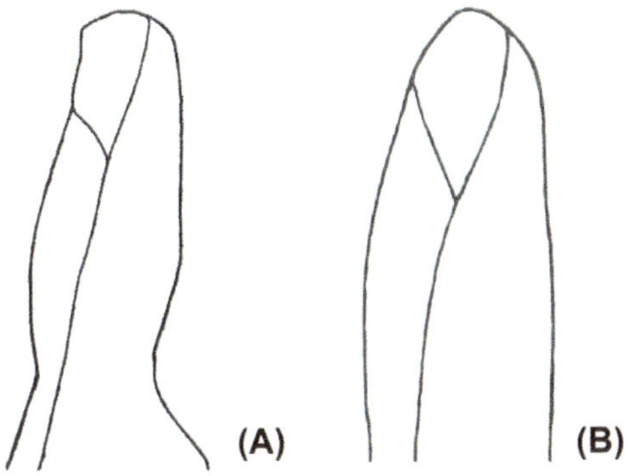

FIGURE 23.122 Gonopod I (distal part) of *Trichodactylus* spp.: (A) *T. faxoni*; (B) *T. fluviatilis*.

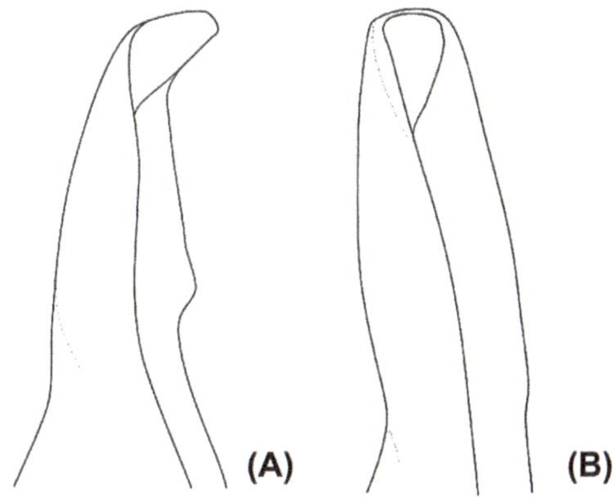

FIGURE 23.123 Gonopod I (distal part) of *Trichodactylus* spp.: (A) *T. panoplus*; (B) *T. quinquedentatus*.

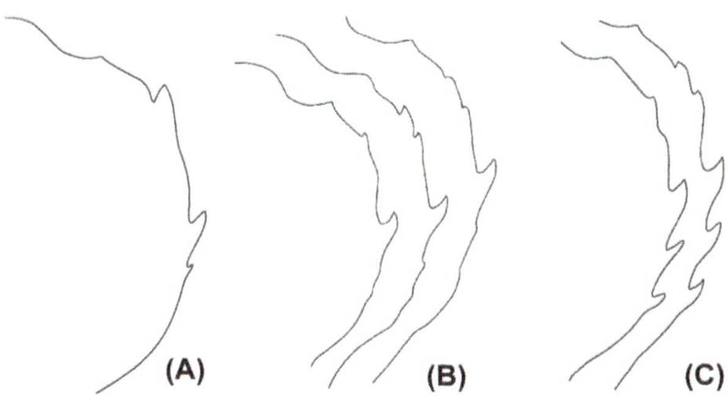

FIGURE 23.124 Lateral margin of carapace of *Trichodactylus* spp.: (A) *T. ehrhardti*; (B) *T. parvus*; (C) *T. borellianus*.

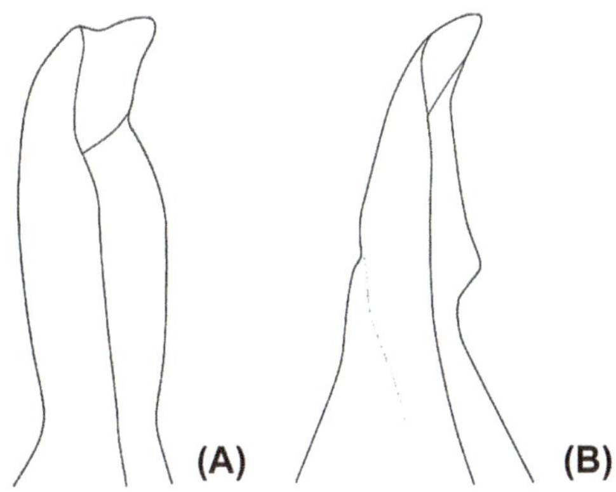

FIGURE 23.125 Gonopod I (distal part) of *Trichodactylus* spp.: (A) *T. ehrhardti*; (B) *T. parvus*.

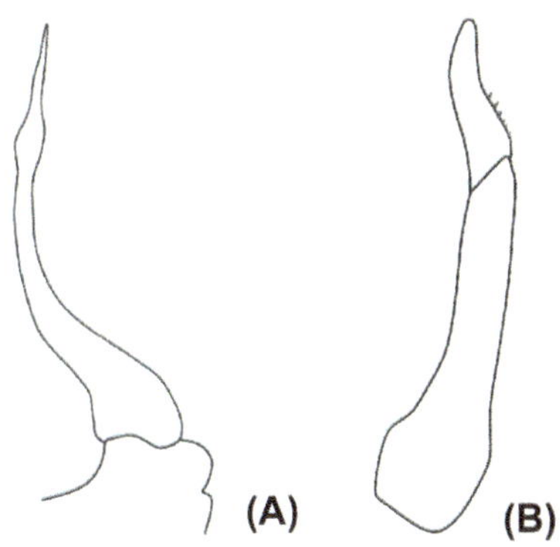

FIGURE 23.126 Gonopod II of *Rodriguezia* spp.: (A) *R. villalobosi*; (B) *R. mensabak*.

Decapoda: Brachyura: Trichodactylidae: Dilocarcininae: Tribes

1 Thoracic sternum with medial line present on somites V–VIII (Fig. 23.127 A, arrow); carapace orbicular .... **Dilocarcinini [p. 946]**

1′ Thoracic sternum with medial line present on somites VII–VIII (Fig. 23.127 B, arrow); carapace subhexagonal ..... **Valdiviini [p. 947]**

Decapoda: Brachyura: Trichodactylidae: Dilocarcininae: Dilocarcinini: Genera

1 Abdominal somites III–VI fused (Fig. 23.128 A, B) 2

1′ Abdominal segments free, in some females, segmentation sometimes only marked by pigmentation (Fig. 23.128 C) .... ***Goyazana* [p. 946]**

2(1) Gonopod I marginal suture near apex (Fig. 23.129 A, B, arrow) twisted ventrolaterally; apex flattened, distal portion with strong torsion 3

2′ Gonopod I marginal suture near apex straight (Fig. 23.129 D–E, arrow) or twisted mesodorsally (Fig. 23.129 C, arrow); apex subcylindrical, distal portion without torsion 4

3(2) Male abdomen narrowly triangular, lateral margins concave; gonopod I ventral margin (Fig. 23.128 A, arrow) continuous, subterminal field of spines more or less continuous .... ***Dilocarcinus* [p. 946]**

3′ Male abdomen broadly triangular, lateral margins convex; gonopod I ventral margin (Fig. 23.130 B, arrow) medially discontinuous, bulging proximally, subterminal spines in two discontinuous patches .... ***Fredilocarcinus* [p. 946]**

4(2) Gonopod I distal opening simple, V-shaped or rimate 5

4′ Gonopod I distal opening (Fig. 23.131 C) circular to broadly oval, its margins rolled .... ***Poppiana* [p. 947]**

5(4) Gonopod I distal portion tapering, straight or slightly bent mesodorsally, caudal margin (Fig. 23.132 A, arrow) continuous, distal opening (Fig. 23.131 A) V-shaped .... ***Moreirocarcinus* [p. 947]**

5′ Gonopod I distal portion subcylindrical, slightly or strongly bent ventrally, caudal margin (Fig. 23.132 B, arrow) with medially discontinuous, bulging proximally, distal opening (Fig. 23.131 B) reduced, rimate .... ***Bottiella* [p. 947]**

Decapoda: Brachyura: Trichodactylidae: Dilocarcininae: Dilocarcinini: *Goyazana*: Species

1 Male abdomen triangular, lateral margins straight or slightly concave; gonopod I slender, caudal margin (Fig. 23.133 A, arrow) gently sinuous, proximal portion caudal margin slightly convex, apical field of spines weakly developed .... *Goyazana castelnaui* (H. Milne Edwards, 1853)
[Brazil: Bahia, Goiás, Maranhão, Mato Grosso, Mato Grosso do Sul, Pará, Pernambuco, Piauí, São Paulo, Sergipe, Tocantins]

1′ Male abdomen subtriangular, broadly rounded, lateral margins convex; gonopod I stout, caudal margin medially discontinuous (Fig. 23.133 B, arrow), proximal portion caudal margin straight, subrectangular, apical field of spines well developed .... *Goyazana rotundicauda* Magalhães & Türkay, 1996
[Ecuador: Pastaza. Brazil: Amazonas. Peru: Loreto]

Decapoda: Brachyura: Trichodactylidae: Dilocarcininae: Dilocarcinini: *Dilocarcinus*: Species

1 Abdominal somite III anterior margin without transversal carinae (Fig. 23.134 A, arrow) 2

1′ Abdominal somite III anterior margin with elevated transversal carinae (Fig. 23.134 B, arrow) .... *Dilocarcinus pagei* Stimpson, 1861
[Colombia: Amazonas. Brazil: Acre, Amapá, Amazonas, Mato Grosso, Mato Grosso do Sul, Pará, Paraná, Rondônia, São Paulo. Peru: Loreto, Madre de Díos, Ucayali. Bolivia: Beni, Cochabamba, Santa Cruz. Paraguay. Argentina: Buenos Aires, Chaco, Corrientes, Entre Ríos, Formosa, Santa Fe]

2(1) Gonopod I distally slightly to strongly curved laterally, apex (Fig. 23.135 A, arrow) compressed, length > subdistal lobe (Fig. 23.135 A, sl), distal opening directed laterally .... *Dilocarcinus septemdentatus* (Herbst, 1783)
[Suriname. French Guiana. Brazil: Acre, Amapá, Amazonas, Goiás, Maranhão, Pará. Peru: Loreto. Argentina: Santa Fe]

2′ Gonopod I distal portion straight, apex (Fig. 23.135 B, arrow) straight, length < subdistal lobe (Fig. 23.135 B, sl), distal opening directed apically .... *Dilocarcinus truncatus* Rodríguez, 1992
[Bolivia: Beni]

Decapoda: Brachyura: Trichodactylidae: Dilocarcininae: Dilocarcinini: *Fredilocarcinus*: Species

1 Gonopod I apex flattened, relatively large, marginal suture apically (Fig. 23.136 A, B, arrow) twisted ventrolaterally 2

1′ Gonopod I apex slender, tapering, marginal suture apically (Fig. 23.136 C, arrow) twisted ventrally .... *Fredilocarcinus apyratii* Magalhães & Türkay, 1996
[Brazil: Acre]

2(1) Gonopod I subdistal lobe (Fig. 23.136 A, sl) developed, bulging, apex distally acute, length < subdistal lobe .... *Fredilocarcinus musmuschiae* Pretzmann & Mayta, 1980
[Peru: Huánuco, Loreto]

2′ Gonopod I subdistal lobe (Fig. 23.136 B, sl) rudimentary, weakly bulging, apex rounded, length > subdistal lobe .... *Fredilocarcinus raddai* Pretzmann, 1978
[Colombia: Amazonas; Peru: Loreto]

Decapoda: Brachyura: Trichodactylidae: Dilocarcininae: Dilocarcinini: *Poppiana*: Species

1 Carapace frontal margin unarmed, smooth (Fig. 23.137 A) 2

1′ Carapace frontal margin with several acute spines (Fig. 23.137 B) .... *Poppiana dentata* (Randall, 1840)
[Nicaragua. Venezuela: Apure, Aragua, Anzoátegui, Barinas, Bolívar, Carabobo, Cojedes, Delta Amacuro, Distrito Federal, Falcón, Guárico, Miranda, Monagas, Portuguesa, Sucre, Táchira, Yaracuy, Zulia. Trinidad. Colombia: Arauca, Bolívar, Casanare, César, Córdoba, La Guajira, Santander. Guyana. Suriname. French Guiana. Brazil: Amazonas, Roraima]

2(1) Gonopod I ventral margin with weakly developed, nearly indistinct subdistal lobe (Fig. 23.138 A, arrow) .... *Poppiana argentiniana* (Rathbun, 1905)
[Bolivia: Pando, Beni, La Paz. Brazil: Amazonas, Mato Grosso, Pará; Paraguay. Argentina: Buenos Aires, Chaco, Corrientes, Formosa, Santa Fe]

2′ Gonopod I ventral margin with robust, bulbiform, subdistal lobe (Fig. 23.138 B, arrow). .... *Poppiana bulbifer* (Rodríguez, 1992)
[Brazil: Amazonas; Peru: Madre de Dios]

Decapoda: Brachyura: Trichodactylidae: Dilocarcininae: Dilocarcinini: *Moreirocarcinus*: Species

1 Carapace, lateral margin with ≤ five teeth (Fig. 23.139 A, B) 2

1′ Carapace, lateral margin with six to eight teeth (Fig. 23.139 C). .... *Moreirocarcinus laevifrons* (Moreira, 1901)
[Venezuela: Amazonas. Colombia: Amazonas, Guainía. Brazil: Amazonas]

2(1) Male abdomen lateral margins nearly straight (Fig. 23.140 A); gonopod I subterminal spine fields well developed, with a conspicuous mesial patch .... *Moreirocarcinus emarginatus* (H. Milne-Edwards, 1853)
[Venezuela: Amazonas. Colombia: Amazonas, Caquetá, Casanare, Cundinamarca, Guaviare, Meta, Putumayo, Vaupés. Ecuador: Napo. Brazil: Amazonas. Peru: Loreto]

2′ Male abdomen lateral margins convex (Fig. 23.140 B); gonopod I subterminal spine fields poorly developed, without mesial patch .... *Moreirocarcinus chacei* (Pretzmann, 1968)
[Ecuador: Morona-Santiago, Napo, Pastaza]

Decapoda: Brachyura: Trichodactylidae: Dilocarcininae: Dilocarcinini: *Bottiella*: Species

1 Gonopod I apex slightly asymmetric, mesial margin greatly expanded subterminally, apical process (Fig. 23.141 A, B, arrow) subterminal, not overreaching apex distal margin 2

1′ Gonopod I apex strongly asymmetric, mesial margin little expanded subterminally, apical process (Fig. 23.141 C, arrow) terminal, overreaching apex distal margin. .... *Bottiella cucutensis* (Pretzmann, 1968)
[Venezuela: Mérida, Zulia. Colombia: Norte de Santander]

2(1) Gonopod I distal portion nearly straight or curved ~25° ventrolaterally, dorsomedial side subterminal longitudinal ridge robust (Fig. 23.141 A, lr), visible in ventromesdal view .... *Bottiella medemi* (Smalley & Rodríguez, 1972)
[Colombia: Córdoba]

2′ Gonopod I distal portion curved ~70°–90° ventrolaterally, dorsomedial side subterminal longitudinal ridge obscure or absent (Fig. 23.141 B) .... *Bottiella niceforei* (Schmitt & Pretzmann, 1968)
[Venezuela: Zulia. Colombia: Norte de Santander]

Decapoda: Brachyura: Trichodactylidae: Dilocarcininae: Valdiviini: Genera

1 Gonopod I ventral margin slightly to prominently broadly rounded or with subrectangular proximal lobe (Fig. 23.142 E, F, arrow) 2

1′ Gonopod I ventral margin without proximal lobe or proximal lobe slightly convex (Fig. 23.142 A–D, arrow) 3

2(1) Gonopod I ventral margin with prominent, broadly rounded proximal lobe (Fig. 23.142 E, arrow), and rounded subdistal lobe (Fig. 23.142 E, dl). .... *Melocarcinus meekei* (Pretzmann, 1968)

[Panama]

2 Gonopod I ventral margin with subrectangular proximal lobe (Fig. 23.142 F, arrow), subdistal lobe absent or weakly developed .... ***Valdivia* [p. 952]**

3(1) Gonopod I distal portion strongly curved ventrolaterally (Fig. 23.142 A, B) 4

3′ Gonopod I distal portion straight or slightly curved dorsomedially (Fig. 23.142 C, D) 5

4(3) Gonopod I distal portion margins asymmetrical, ventrolaterally bearing subdistal lobe (Fig. 23.142 A, dl) .... *Forsteriana venezuelensis* (Rathbun, 1905)

[Venezuela: Apure, Aragua, Bolívar, Cojedes, Guárico, Monagas. Colombia: Arauca, Caquetá, Casanare, Cundinamarca, Meta]

4′ Gonopod I distal portion margins subparallel, subdistal lobe absent (Fig. 23.142 B) .... *Rotundovaldivia latidens* (A. Milne-Edwards, 1869)

[Ecuador: Morona-Santiago. Brazil: Amazonas. Peru: Amazonas, Loreto, Pasco]

5(3) Gonopod I distal portion slightly bent dorsomedially, ventrally with slight to robust subdistal lobe (Fig. 23.142 D, dl); spine field well developed, distally in coalescent patches .... ***Sylviocarcinus* [p. 952]**

5′ Gonopod I distal portion bent ventrolaterally, ventrally without subdistal lobe (Fig. 23.142 C); spine field weakly developed in discontinuous patches .... ***Zilchiopsis* [p. 953]**

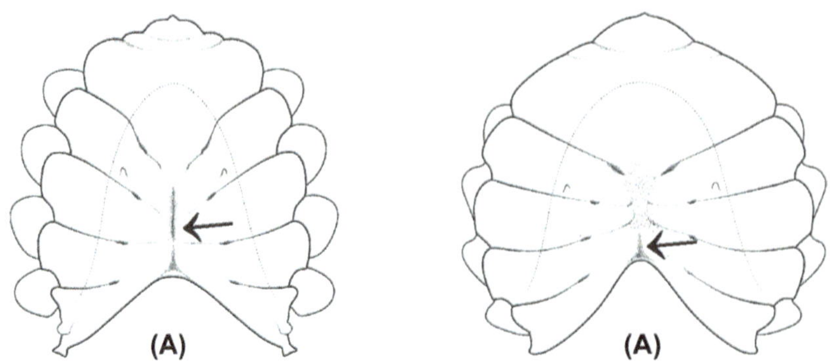

FIGURE 23.127 Thoracic sternum outline of Dilocarcininae: (A) Dilocarcinini, *Dilocarcinus pagei*; (B) Valdiviini, *Sylviocarcinus pictus*. *Arrow*, medial line of thoracic sternum.

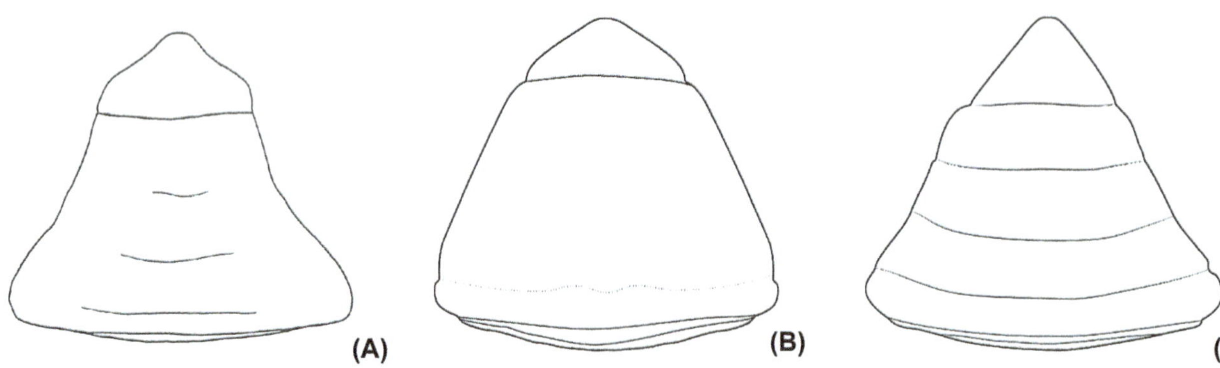

FIGURE 23.128 Abdomen outline: (A) *Dilocarcinus septemdentatus*; (B) *Poppiana dentate*; (C) *Goyazana castelnaui*.

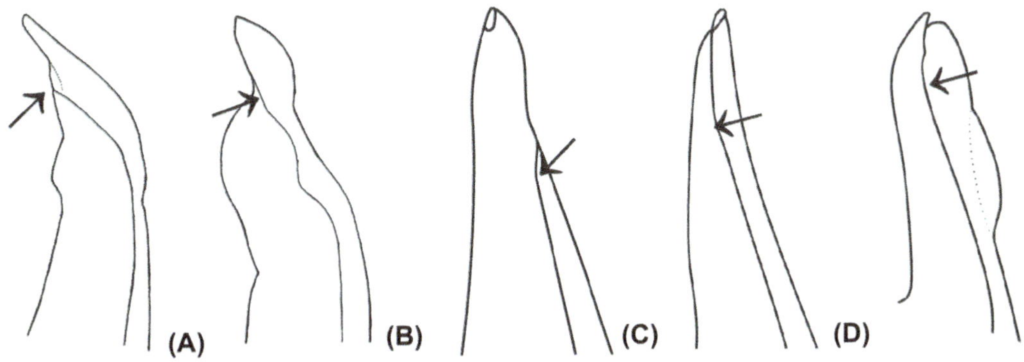

FIGURE 23.129 Gonopod 1 (distal part): (A) *Dilocarcinus septemdentatus*; (B) *Fredilocarcinus musmuschiae*; (C) *Poppiana dentata*; (D) *Moreirocarcinus emarginatus*; (E) *Bottiella medemi*. *Arrow*, marginal suture.

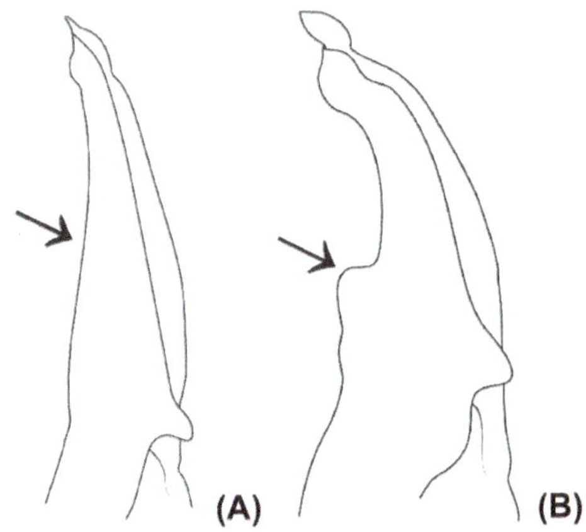

FIGURE 23.130 Gonopod I of Dilocarcinini: (A) *Dilocarcinus septemdentatus*; (B) *Fredilocarcinus musmuschiae*. *Arrow*, ventral margin.

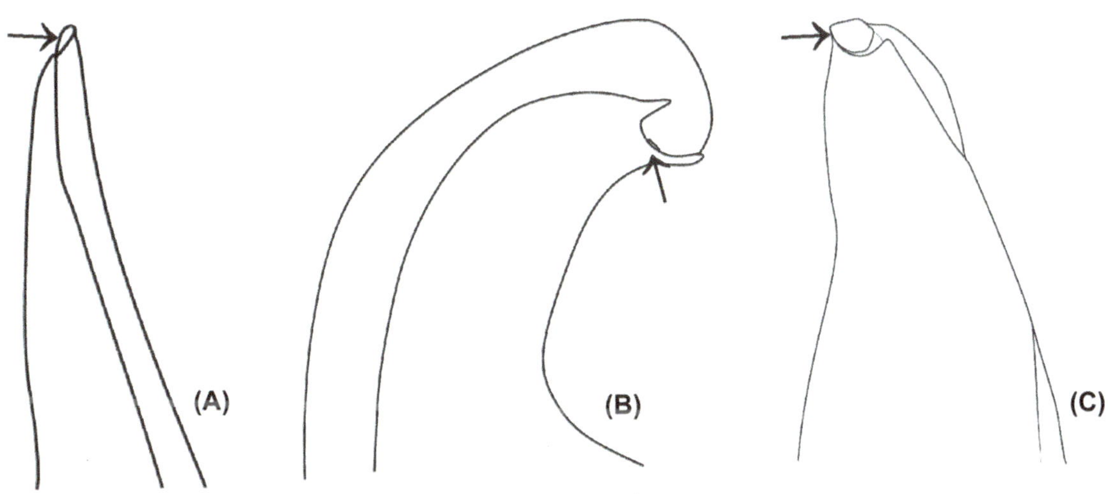

FIGURE 23.131 Gonopod I (distal part) of Dilocarcinini, showing the distal opening: (A) *Moreirocarcinus chacei*; (B) *Bottiella cucutensis*; (C) *Poppiana argentiniana*. *Arrow*, distal opening.

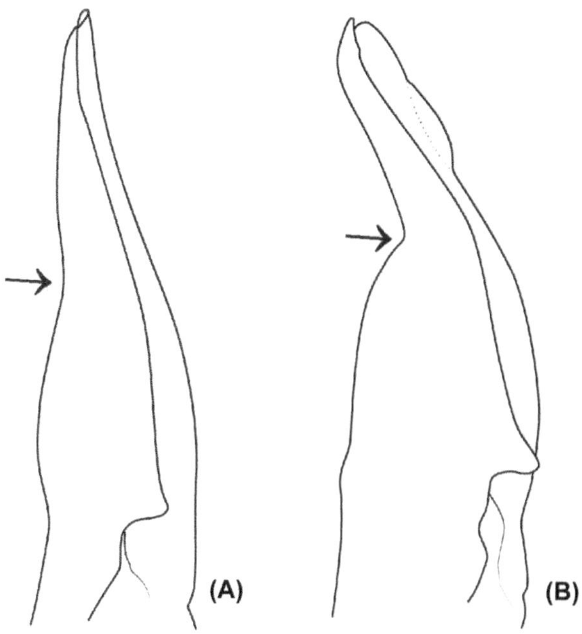

FIGURE 23.132 Gonopod I of Dilocarcinini: (A) *Moreirocarcinus chacei*; (B) *Bottiella medemi*. *Arrow*, caudal margin.

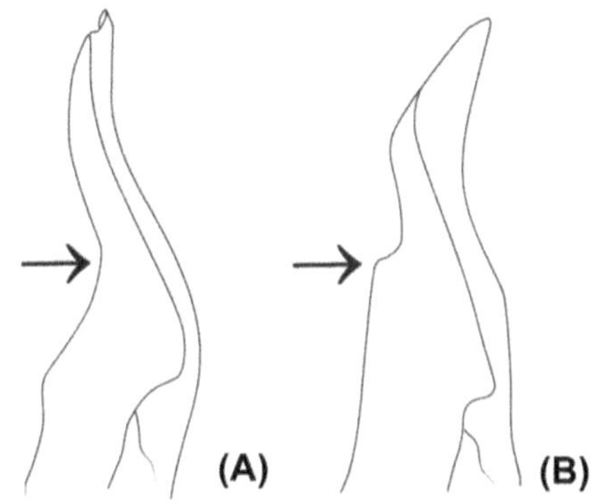

FIGURE 23.133 Gonopod I of *Goyazana* spp.: (A) *G. castelnaui*; (B) *G. rotundicauda*. *Arrow*, caudal margin.

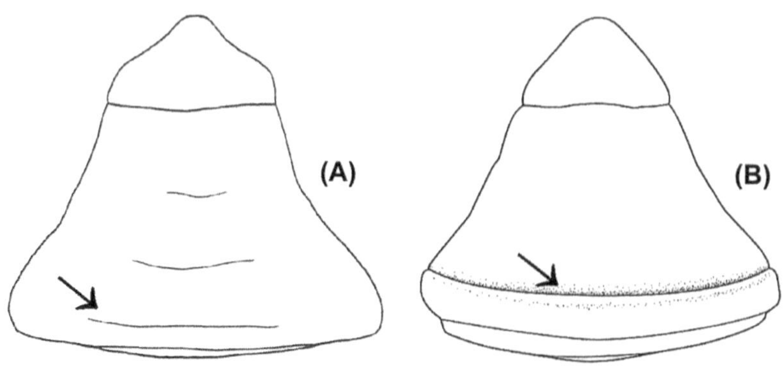

FIGURE 23.134 Abdomen outline of *Dilocarcinus* spp.: (A) *D. septemdentatus*; (B) *D. pagei*. *Arrow*, anterior margin of abdominal somite III.

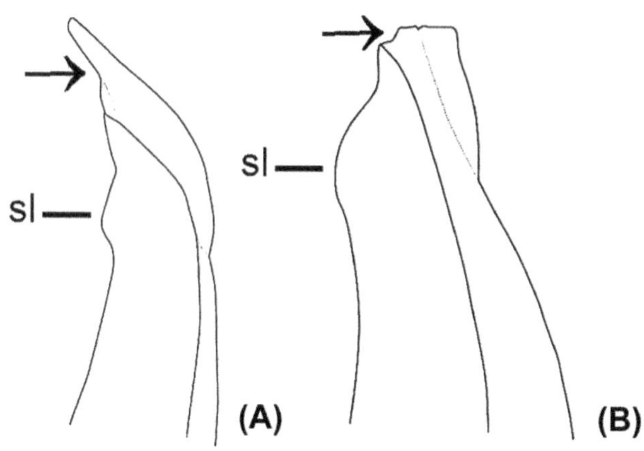

FIGURE 23.135 Gonopod I (distal part) of *Dilocarcinus* spp.: (A) *D. septemdentatus*; (B) *D. truncatus*. Abbreviation: *sl*, subdistal lobe; *arrow*, apex.

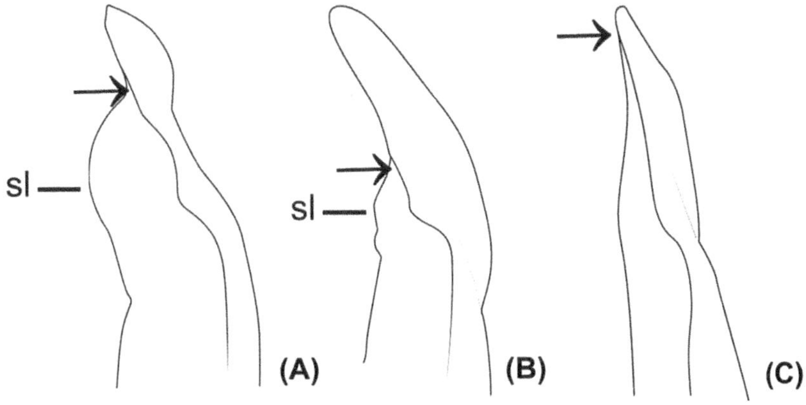

FIGURE 23.136 Gonopod 1 (distal part) of *Fredilocarcinus* spp.: (A) *F. musmuschiae*; (B) *F. raddai*; (C) *F. apyratii*. Abbreviation: *sl*, subdistal lobe; *arrow*, marginal suture.

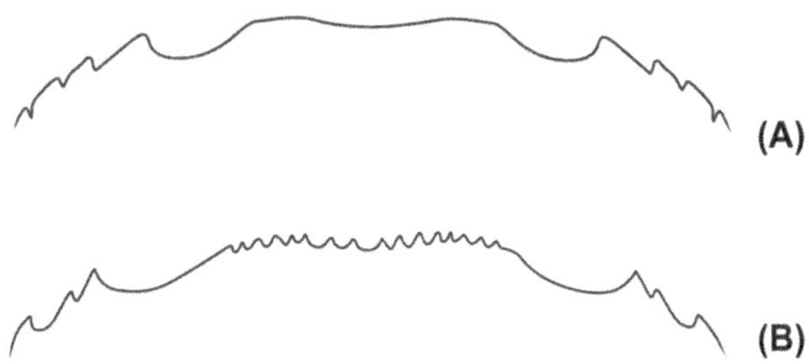

FIGURE 23.137 Frontal margin of *Poppiana* spp.: (A) *P. argentiniana*; (B) *P. dentata*.

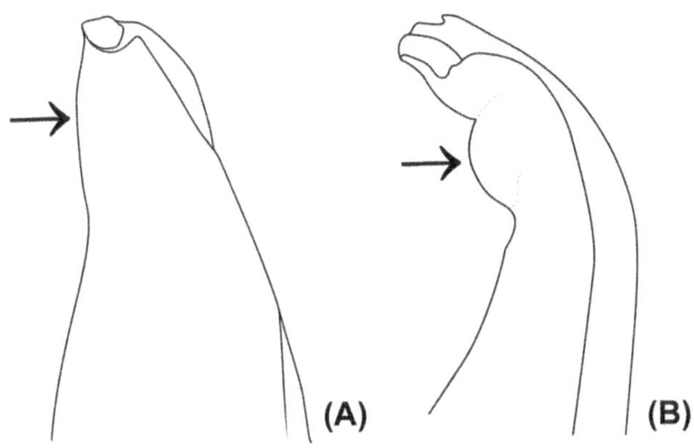

FIGURE 23.138 Gonopod I (distal part) of *Poppiana* spp.: (A) *P. argentiniana*; (B) *P. bulbifer*. *Arrow*, subdistal lobe.

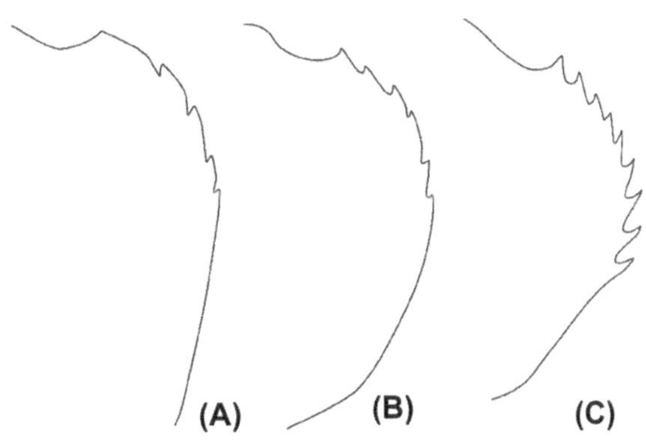

FIGURE 23.139 Lateral margin of carapace of *Moreirocarcinus* spp.: (A) *Moreirocarcinus emarginatus*; (B) *M. chacei*; (C) *M. laevifrons*.

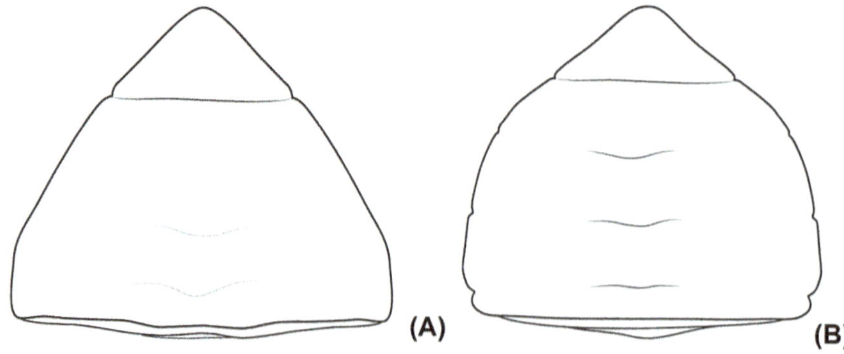

FIGURE 23.140 Abdomen outline of *Moreirocarcinus* spp.: (A) *M. emarginatus*; (B) *M. chacei*.

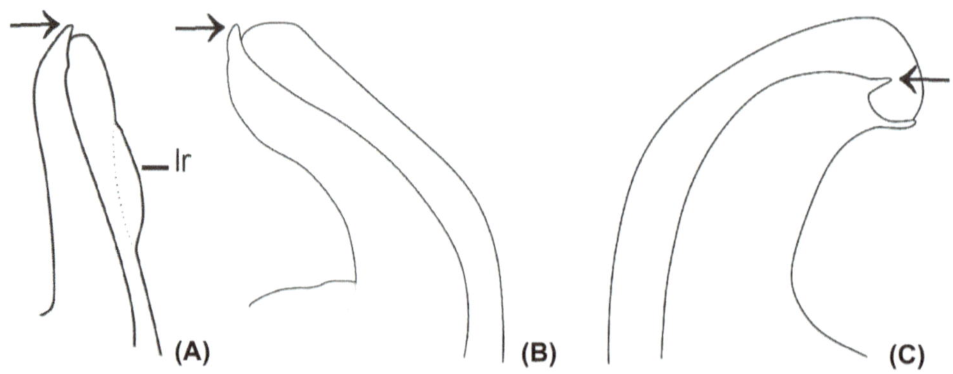

FIGURE 23.141 Gonopod I (distal part) *Bottiella* spp.: (A) *Bottiella medemi*; (B) *B. niceforei*; (C) *B. cucutensis*. Abbreviation: *lr*, longitudinal ridge; *arrow*, apical process.

Decapoda: Brachyura: Trichodactylidae: Dilocarcininae: Valdiviini: *Valdivia*: Species

1 Gonopod I distal portion bent ventrolaterally, lateral side smooth, subdistal lobe absent (Fig. 23.143 A, B, arrow) 2

1′ Gonopod I distal portion bent dorsomedially, lateral side with poorly developed subdistal lobe (Fig. 23.143 C, arrow) .... *Valdivia novemdentata* (Pretzmann, 1968)
[Brazil: Amazonas, Rondônia. Peru: Loreto, Madre de Dios]

2(1) Gonopod I apex symmetric or slightly asymmetric, lateral margin (Fig. 23.144 A, lm) nearly straight or rounded, length subequal to slightly longer than mesial margin (Fig. 23.144 A, mm) .... *Valdivia serrata* White, 1847
[Venezuela: Amazonas, Bolívar, Delta Amacuro. Colombia: Amazonas, Arauca, Caquetá, Guaviare, Meta, Putumayo, Vichada. Guyana. Suriname. Ecuador: Napo. Brazil: Acre, Amapá, Amazonas, Maranhão, Pará, Rondônia, Roraima. Peru: Loreto, Madre de Díos. Bolivia: Pando]

2′ Gonopod I apex distinctly asymmetric, lateral margin (Fig. 23.144 B, lm) pointed, length > mesial (Fig. 23.144 B, mm) .... *Valdivia camerani* (Nobili, 1896)
[Brazil: Mato Grosso, Mato Grosso do Sul. Bolivia: Beni. Paraguay. Argentina: Santa Fe]

Decapoda: Brachyura: Trichodactylidae: Dilocarcininae: Valdiviini: *Sylviocarcinus*: Species

1 Abdominal somites III–V fused, somite VI free (Fig. 23.145, arrow) 2

1′ Abdominal somites III–VI fused (Fig. 23.145 B) 4

2(1) Frontal margin inerm (Fig. 23.146 B); gonopod I ventrally with subdistal lobe absent or weak (Fig. 23.147 A–D, arrow) 3

2′ Frontal margin with rounded granules or spines (Fig. 23.146 A); gonopod I ventrally with subdistal lobe robust (Fig. 23.147 E, arrow). .... *Sylviocarcinus devillei* H. Milne Edwards, 1853
[Colombia: Amazonas, Caquetá. Ecuador: Napo, Pastaza. Brazil: Acre, Amapá, Amazonas, Goiás, Mato Grosso, Pará. Peru: Amazonas, Junín, Loreto, Madre de Díos, San Martin. Bolivia: Pando]

3(2) Gonopod I distal portion subcylindrical, ventral subdistal lobe absent (Fig. 23.147 A) .... *Sylviocarcinus maldonadoensis* (Pretzmann, 1978)

[Brazil: Amazonas, Pará. Peru: Loreto, Madre de Díos. Bolivia: Pando]

3′ Gonopod I distal portion subconical, ventral subdistal lobe weak (Fig. 23.147 B) .... *Sylviocarcinus piriformis* (Pretzmann, 1968)

[Venezuela: Falcón, Lara, Mérida, Trujillo, Zulia. Colombia: Antioquia, Bolívar, Boyacá, Caldas, Cesar, Chocó, Córdoba, Cundinamarca, La Guajira, Magdalena, Norte de Santander, Tolima]

4(1) Gonopod I subdistal lobe robust, median spine patch (Fig. 23.148 A, arrow) well developed .... *Sylviocarcinus pictus* (H. Milne Edwards, 1853)

[Venezuela: Bolívar. Colombia: Amazonas. Brazil: Amapá, Amazonas, Maranhão, Pará, Piauí, Roraima, Rondonia. Guyana. French Guiana. Peru: Loreto. Bolivia: Cochabamba. Argentina: Santa Fe]

4′ Gonopod I subdistal lobe weak, median spine patch (Fig. 23.148 B, arrow) weak .... *Sylviocarcinus australis* Magalhães & Türkay, 1996

[Brazil: Mato Grosso, Mato Grosso do Sul, Rondônia. Paraguay. Argentina: Misiones, Santa Fe]

Decapoda: Brachyura: Trichodactylidae: Dilocarcininae: Valdiviini: *Zilchiopsis*: Species

1 Carapace subhexagonal, dorsally uneven (regions distinct), lateral margins with three to four distinct, rounded, acute, or evanescent teeth (Fig. 23.149 A, B); gonopod I apex (Fig. 23.150 A, B, arrow) long, slender, length $\geq 0.5\times$ subterminal spine field (Fig. 23.150 A, B, fss) 2

1′ Carapace ellipsoid, dorsally smooth (regions indistinct), lateral margins with very faint indications of teeth (Fig. 23.149 C); gonopod I apex (Fig. 23.150 C, arrow) $<0.5\times$ subterminal spine field (Fig. 23.150 C, fss) .... *Zilchiopsis cryptoda* (Ortmann, 1893)

[Peru: Amazonas, Loreto]

2(1) Gonopod I apex (Fig. 23.150 A, arrow) length $\sim 0.5\times$ subterminal spine field (Fig. 23.150 A, fss) .... *Zilchiopsis oronensis* (Pretzmann, 1968)

[Brazil: Amazonas, Mato Grosso, Pará, Rondônia. Bolivia: Beni. Paraguay. Argentina: Chaco?, Formosa?, Santa Fe]

2′ Gonopod I apex (Fig. 23.150 B, arrow) long and slender, length always $>0.5\times$ subterminal spine field (Fig. 23.150 B, fss) .... *Zilchiopsis collastinensis* (Pretzmann, 1968)

[Bolivia?: Cochabamba?. Argentina: Buenos Aires, Chaco, Corrientes, Entre Ríos, Formosa, Santa Fe, Santiago del Estero, Tucumán]

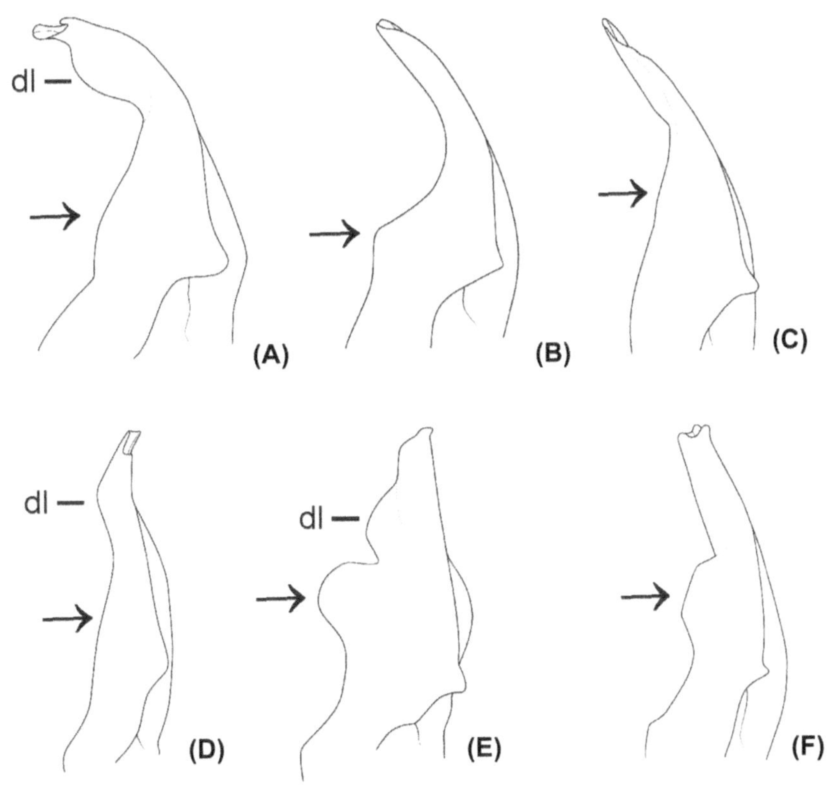

FIGURE 23.142 Gonopod I of Valdiviini: (A) *Forsteriana venezeuelensis*; (B) *Rotundovaldivia latidens*; (C) *Zilchiopsis oronensis*; (D) *Sylviocarcinus pictus*; (E) *Melocarcinus meekei*; (F) *Valdivia serrata*. Abbreviation: *dl*, subdistal lobe; *arrow*, proximal lobe.

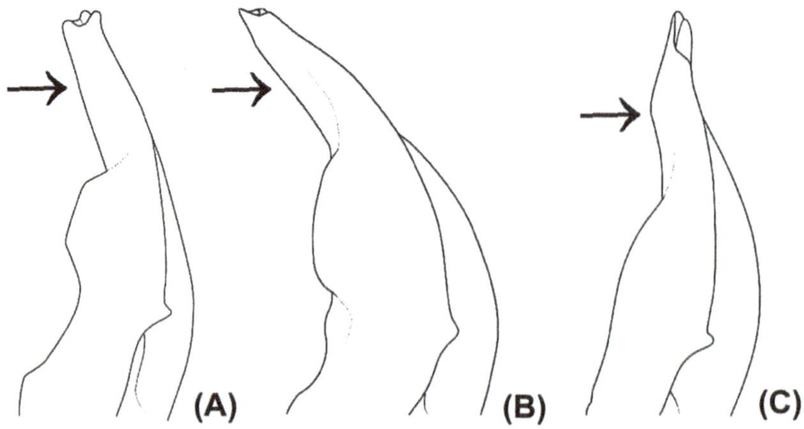

FIGURE 23.143 Gonopod I of *Valdivia* spp.: (A) *V. serrata*; (B) *V. camerani*; (C) *V. novemdentata*. *Arrow*, subdistal lobe.

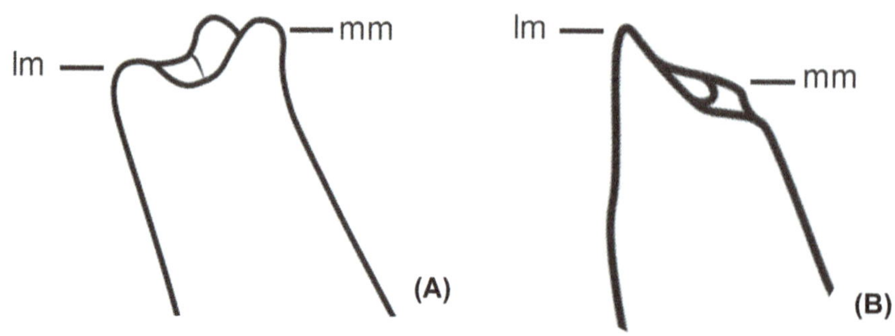

FIGURE 23.144 Gonopod I (apex) of *Valdivia* spp.: (A) *V. serrate*; (B) *V. camerani*. *lm*, lateral margin; *mm*, mesial margin.

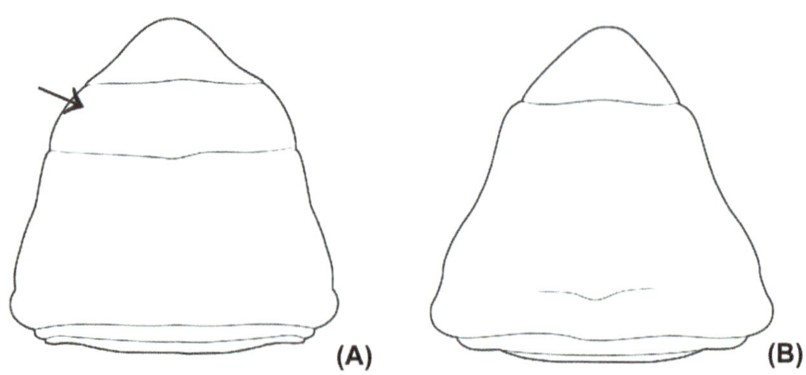

FIGURE 23.145 Abdomen outline of *Sylviocarcinus* spp.: (A) *S. devillei*; (B) *S. australis*. *Arrow*, abdominal somite VI.

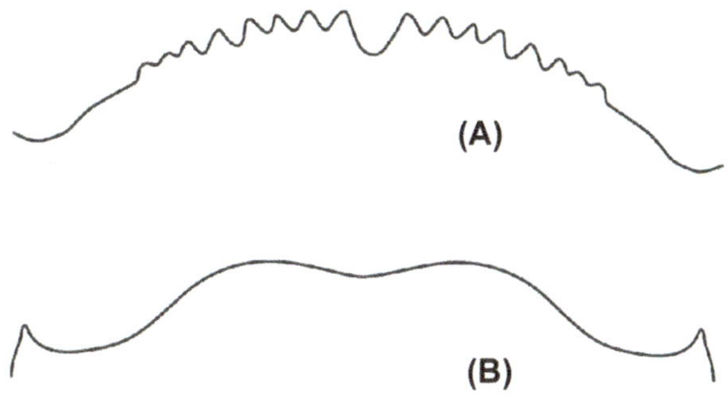

FIGURE 23.146 Frontal margin of *Sylviocarcinus* spp.: (A) *S. australis*; (B) *S. devillei*.

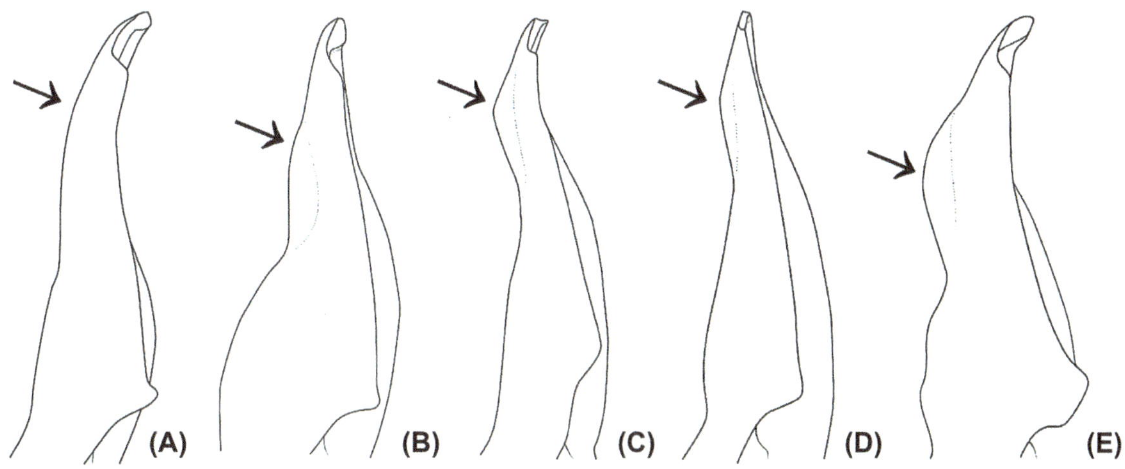

FIGURE 23.147 Gonopod I of *Sylviocarcinus* spp.: (A) *S. maldonadoensis*; (B) *S. piriformis*; (C) *S. pictus*; (D) *S. australis*; (E) *S. devillei*. Arrow, subdistal lobe.

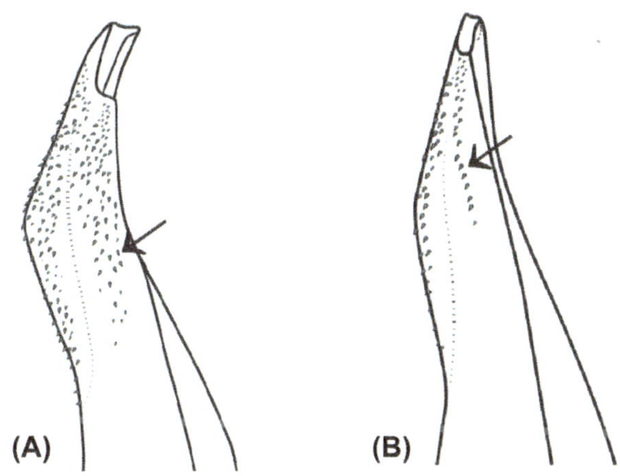

FIGURE 23.148 Gonopod I (distal part) of *Sylviocarcinus* spp.: (A) *S. pictus*; (B) *S. australis*. Arrow, median patch of the field of spines.

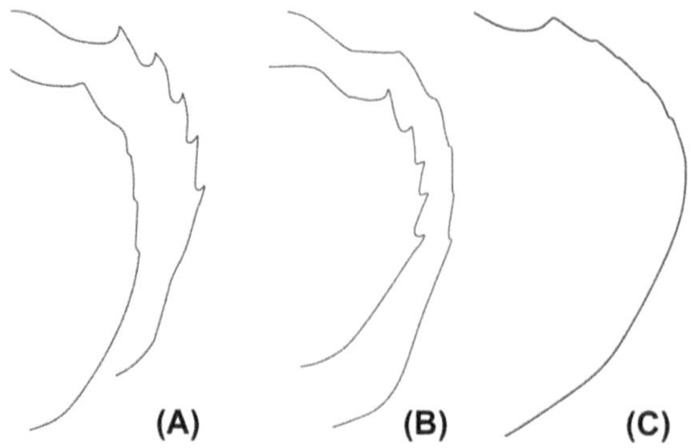

FIGURE 23.149 Lateral margin of carapace of *Zilchiopsis* spp.: (A) *Z. oronensis*; (B) *Z. collastinensis*; (C) *Z. cryptoda*.

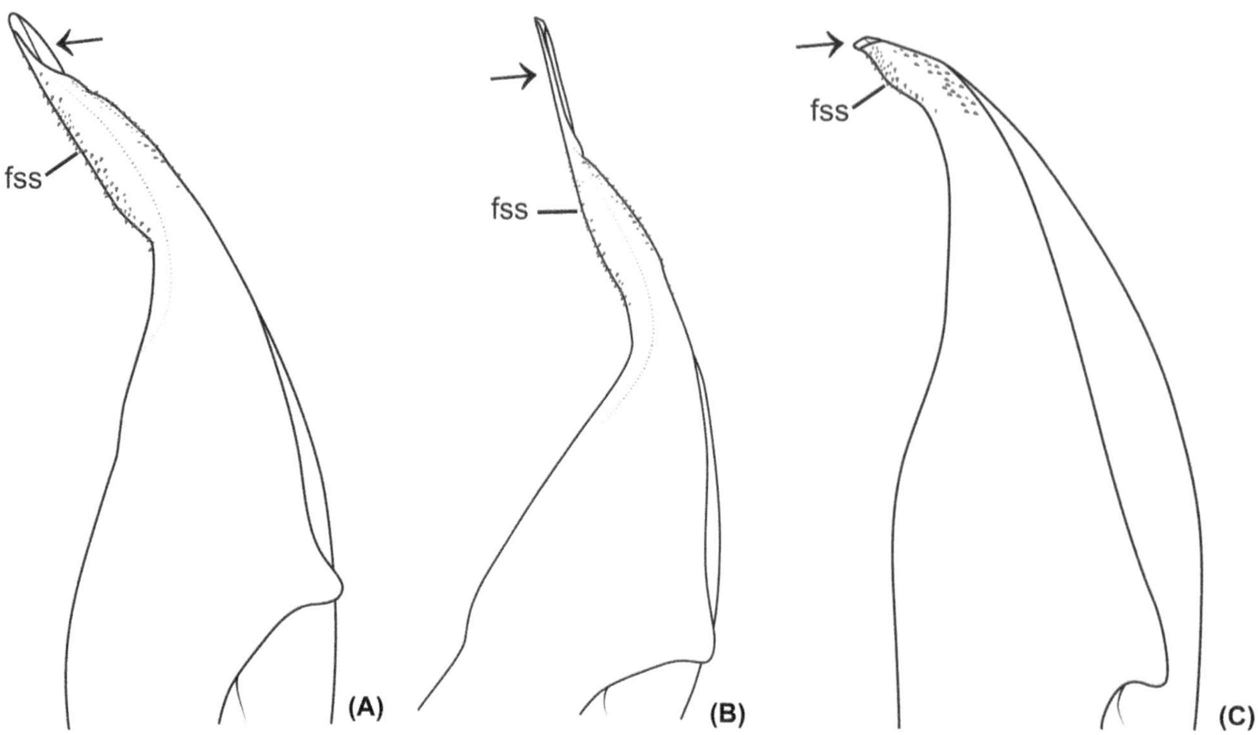

FIGURE 23.150 Gonopod I of *Zilchiopsis* spp.: (A) *Z. oronensis*; (B) *Z. collastinensis*; (C) *Z. cryptoda*. Abbreviation: *fss*, field of the subterminal spine; *arrow*, GI apex.

Decapoda: Brachyura: Pseudothelphusidae: Subfamilies

This key is modified from Pretzmann (1972) and Rodriguez (1982).

1	Maxilliped III exopod distinctly overreaching ischium (Fig. 23.151 A)	Epilobocerinae, one genus: ***Epilobocera*** **[p. 956]**
1′	Maxilliped III exopod slightly overreaching or not reaching distal margin of ischium (Figs. 23.151 B, C)	**Pseudothelphusinae [p. 957]**

Decapoda: Brachyura: Pseudothelphusidae: *Epilobocera*: Species

This genus needs to be revised. Key modified and updated from Rodriguez (1982).

1	Maxilliped III merus rectangular	2

1′ Maxilliped III merus rounded 3

2(1) Maxilliped III without palp .... *Epilobocera haytensis* Rathbun, 1893
[Hispaniola]

2′ Maxilliped III with palp .... *Epilobocera sinuatifrons* (A. Milne-Edwards, 1866)
[Puerto Rico. St. Croix]

3(1) Pereopods V merus length $\geq$ carapace medial length; caves 4

3′ Pereopods V merus length $<$ carapace medial length 5

4(3) Chela carpus with one spine .... *Epilobocera gertraudae* Pretzmann, 1965
[Cuba]

4′ Chela carpus with two to five spines .... *Epilobocera diazbeltrani* Capolongo, 2005
[Cuba]

5(3) Carapace anteriolateral margin with teeth or at least papillae; adult carapace medial length $>$ 40mm 6

5′ Carapace anteriolateral margin smooth; adult carapace medial length $<$ 30mm ....
.... *Epilobocera wetherbeei* Rodríguez & Williams, 1995
[Dominican Republic]

6(5) Front declivious, margin entirely visible in dorsal view 7

6′ Front delimited dorsally by transverse crest or ridge, generally concealing part of true frontal margin in dorsal view ....
.... *Epilobocera armata* Smith, 1870
[Cuba]

7(6) Maxilliped III exopod width $<$ 0.4× ischium width 8

7′ Maxilliped III exopod width $>$ 0.5× ischium width .... *Epilobocera capolongoi* Pretzmann, 2000
[Cuba]

8(7) Maxilliped III merus distal margin slightly concave 9

8′ Maxilliped III merus distal margin convex 10

9(6) Suborbital margin smooth .... *Epilobocera cubensis* Stimpson, 1860
[Cuba]

9′ Suborbital margin granulate .... *Epilobocera najasensis* Capolongo & Pretzmann, 2002
[Cuba]

10(8) Maxilliped III exopod extending to merus distal margin 11

10′ Maxilliped III exopod extending well beyond merus distal margin .... *Epilobocera synoecia* Capolongo & Pretzmann, 2002
[Cuba]

11(10) Carapace front in anterior view straight .... *Epilobocera placensis* Capolongo & Pretzmann, 2002
[Cuba]

11′ Carapace front in anterior view curved .... *Epilobocera gilmanii* (Smith, 1870)
[Cuba: Isla de Pinos]

Decapoda: Brachyura: Pseudothelphusidae: Pseudothelphusinae: Tribes

1 Gonopod I apical spine field (Fig. 23.152 A, B, arrow) transverse, usually situated distally 2

1′ Gonopod I apical spine field (Fig. 23.152 C, arrow) longitudinal, usually situated laterally 4

2(1) Gonopod I apex without papillae near opening, distolateral margin (Fig. 23.153 A, lm) without lobes or crests 3

2′ Gonopod I apex with two papillae (Fig. 23.153 C, arrows) near opening, distolateral margin enlarged (Fig. 23.153 B, lm), usually with one or two lobes and a crest .... **Hypolobocerini [p. 959]**

3(2) Gonopod I marginal suture straight, without torsion; marginal process (Fig. 23.154 A, mp) free along entire length, overreaching apex or not; anterior side distally (Fig. 23.154 B, cs) usually with at least one robust subtriangular tooth .... **Potamocarcinini [p. 960]**

3′ Gonopod I marginal suture twisted mesially; marginal process fused with caudal plate forming an anteriorly directed caudomarginal uni-, bi-, or trilobed process, never overreaching apex (Fig. 23.154 C); anterior side distally inerm (Fig. 23.154 D, cs). ....
.... **Pseudothelphusini [p. 963]**

4(1) Maxilliped III exopod length $>$ 0.8× ischium (Fig. 23.155 A); efferent branchial chamber aperture partially closed or closed by spine at jugal angle (Fig. 23.155 C, D) (except in *Phallangothelphusa*); gonopod I without an apical or distal projection, basis flattened dorsoventrally .... **Strengerianini [p. 964]**

4′ Maxilliped III exopod length $<$ 0.5× ischium (Fig. 23.155 B); efferent branchial chamber aperture open, usually subquadrate (Fig. 23.155 E); gonopod I with apical complex extending beyond distal opening, basis enlarged dorsoventrally .... **Kingsleyini [p. 964]**

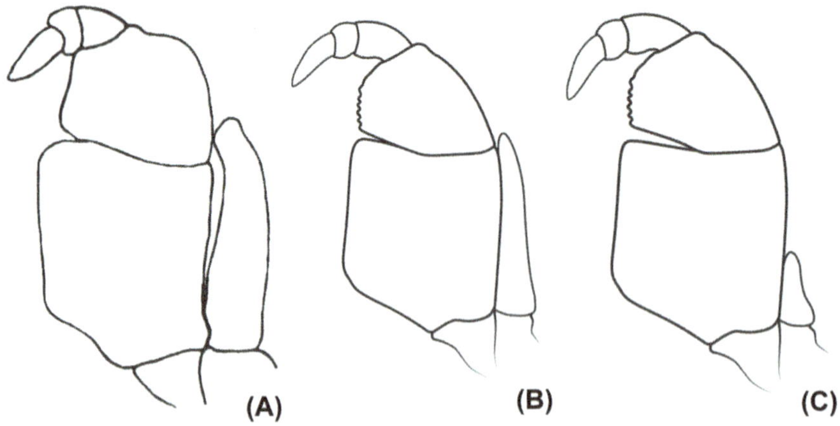

FIGURE 23.151 Maxilliped III of Pseudothelphusidae: (A) Epilobocerinae, *Epilobocera cubensis*; (B) Strengerianini, *Strengeriana bolivariensis*; (C) Kingsleyini, *Kingsleya latifrons*.

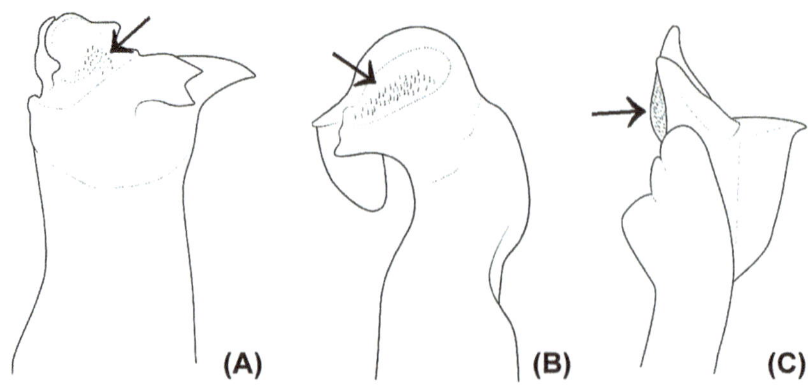

FIGURE 23.152 Gonopod I (distal part) of Pseudothelphusinae: (A) *Potamocarcinus magnus* (laterocephalic view); (B) *Pseudothelphusa leiophrys* (laterocephalic view); (C) *Kingsleya latifrons* (caudal view). *Arrow*, field of apical spines.

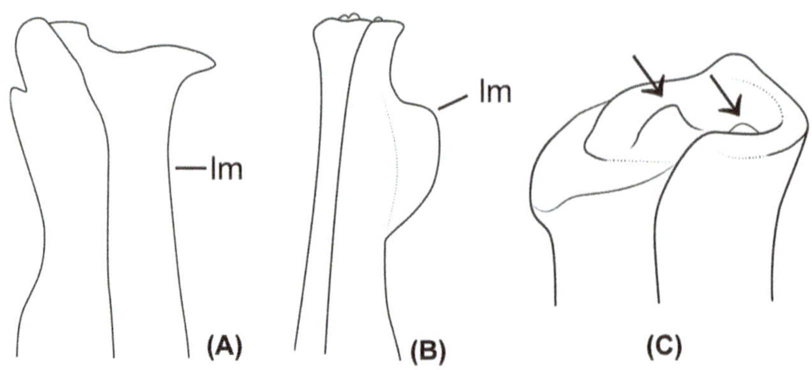

FIGURE 23.153 Gonopod I of Pseudothelphusidae: (A) *Potamocarcinus magnus* (distal part, caudomesial view); (B) *Hypolobocera bouvieri* (distal part, mesial view); (C) *H. bouvieri* (apex, apical view). Abbreviation: *lm*, lateral margin; *arrow*, apex with papillae.

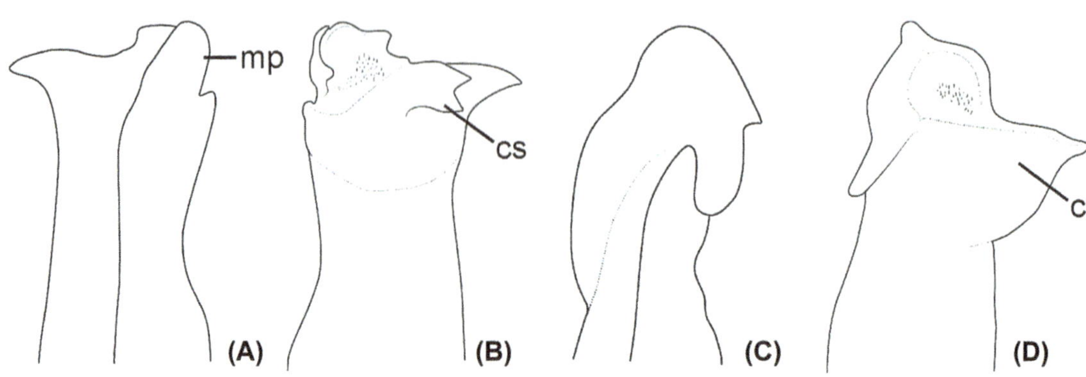

FIGURE 23.154 Gonopod I, distal part of Pseudothelphusidae: *Potamocarcinus magnus*: (A) caudomesial view; (B) laterocephalic view and *Pseudothelphusa americana*: (C) caudomesial view; (D) cephalic view. Abbreviations: *cs*, cephalic side; *mp*, marginal process.

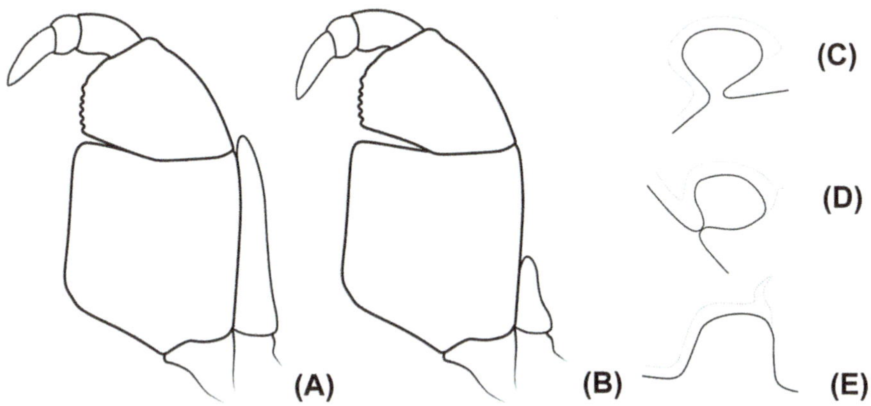

FIGURE 23.155 Maxilliped III of Strengerianini and Kingsleyini: (A) *Strengeriana bolivariensis* and (B) *Kingsleya latifrons*; aperture of efferent branchial chamber of (C) *Martiana clausa*, (D) *Strengeriana furhmanni* and (E) *Kingsleya gustavoi*.

Decapoda: Brachyura: Pseudothelphusidae: Pseudothelphusinae: Hypolobocerini: Genera

The following key is modified from Rodríguez (1982), Rodríguez & von Sternberg (1998), and Campos (2005).

1 Gonopod I cephalic surface bearing subdistal processes (Fig. 23.156) 2

1′ Gonopod I cephalic surface devoid of subdistal processes (Fig. 23.157) 9

2(1) Gonopod I lateral lobe, if present, length < gonopod distal half (Fig. 23.156 A–E) 3

2′ Gonopod I lateral lobe (Fig. 23.156 F, lo) prominent, broad, semicircular, length subequal to gonopod distal half .... *Spirothelphusa verticalis* (Rathbun, 1893) [Mexico]

3(2) Gonopod I apical spine field (Fig. 23.156 A–C, fas) defined, surrounded by distal crests 4

3′ Gonopod I apical spine field diffuse (Fig. 23.156 D–E, fas), not totally surrounded by distal crests 8

4(3) Gonopod I mesial surface apically smooth; maxilliped III merus (Fig. 23.158, arrow) width:length ratio < 1.4 5

4′ Gonopod I mesial surface apically with numerous tubercles (Fig. 23.156 A, mt); maxilliped III merus (Fig. 23.158 B, arrow) width:length ratio > 1.8 .... *Lobithelphusa mexicana* Rodríguez, 1982 [Mexico]

5(4) Gonopod I apex laterally rounded (Fig. 23.156 C, le) 6

5′ Gonopod I apex laterally with an acute spine (Fig. 23.156 B, le) .... *Colombiathelphusa culmarcuata* Campos & Magalhães, 2014 [Colombia]

6(5) Gonopod I lateral and anterior surfaces apically smooth 7

6′ Gonopod I lateral and anterior surfaces apically with setal patches (Fig. 23.156 C, arrow) .... ***Elsalvadoria* [p. 960]**

7(6) Gonopod I slender, apex elongated and slim, straight; lateral surface simple; posterior surface distal portion smooth .... ***Achlidon* [p. 960]** [Costa Rica]

7′ Gonopod I stout, apex oblong, bent anteriorly; lateral surface with lateral process large and usually bilobed; posterior surface distal portion with longitudinal ridge .... *Ptychophallus* [Costa Rica, Panama]

8(3) Gonopod I with one cephalic process short, blunt (Fig. 23.156 D, cp), marginal process enlarged medially as large, serrulate lobe, mesial process bifid (Fig. 23.156 D, me). .... *Camptophallus botti* Smalley, 1965 [Nicaragua]

8′ Gonopod I with three robust spiniform anterior processes (Fig. 23.156 E, cp), marginal process rounded, smooth, not enlarged medially, mesial process simple (Fig. 23.156 E, me) .... *Eidocamptophallus chacei* Rodríguez & Hobbs, 1989 [Colombia]

9(1) Gonopod I lateral lobe (Figs. 23.157 A–C, arrow) simple 10

9′ Gonopod I lateral lobe (Fig. 23.157 D, arrow) divided into two subparallel, longitudinal accessory lobes .... *Neostrengeria* [Colombia]

10(9) Gonopod I apex in apical view (Figs. 23.159 A—C) rounded, oval, subtriangular or moderately elongate, posterior end rounded; lateral lobe (Figs. 23.157 B, C, arrow) defined, although sometimes reduced 11

10′ Gonopod I apex in apical view (Fig. 23.159 D) narrow, slender, very elongate, posterior end digitiform; lateral lobe in posterior view (Fig. 23.157 A, arrow) gradually widening distally, appearing elongate, subtriangular ..... *Moristchus*
[Colombia. Ecuador. Peru]

11(10) Gonopod I lateral lobe densely spinulate (Fig. 23.157 B, arrow); maxilliped III exopod length usually > 0.45 ischium length ..... *Lindacatalina*
[Colombia. Ecuador. Peru]

11′ Gonopod I lateral lobe smooth or with scattered spinulae and short setae (Fig. 23.157 C, arrow); maxilliped III exopod length usually < 0.45 ischium ..... *Hypolobocera*
[Panama. Venezuela. Colombia. Ecuador. Peru]

Decapoda: Brachyura: Pseudothelphusidae: Pseudothelphusinae: Hypolobocerini: *Elsalvadoria*: Species

1 Gonopod I cephalic process distally unilobed or faintly bilobed, lateral process enlarged, rounded ..... *Elsalvadoria tomhaasi* Bott, 1970
[Guatemala]

1′ Gonopod I cephalic process distinctly bilobed, lateral process conical, acute ..... *Elsalvadoria zurstrasseni* (Bott, 1956)
[El Salvador]

Decapoda: Brachyura: Pseudothelphusidae: Pseudothelphusinae: Hypolobocerini: *Achlidon*: Species

The following key is modified from Magalhães et al. (2015).

1 Gonopod I in posteriomedial view curved anteriorly, marginal process rounded not overreaching mesial crest apex, mesial process weakly developed, rounded ..... *Achlidon agrestis* (Rathbun, 1898)
[Costa Rica]

1′ Gonopod I in posteriomedial view curved anteriolaterally, marginal process triangular overreaching mesial crest apex, mesial process strongly developed, reniform ..... *Achlidon puntarenas* (Hobbs, 1991)
[Costa Rica]

Decapoda: Brachyura: Pseudothelphusidae: Pseudothelphusinae: Potamocarcinini: Genera

Key developed from Pretzmann (1972), Rodríguez (1982, 2001), Rodríguez & Hobbs (1989), and Villalobos & Álvarez (2013).

1 Gonopod I anterior surface smooth 2

1′ Gonopod I anterior surface with obvious patch of subapical spinulae ..... *Allacanthos*
[Costa Rica]

2(1) Gonopod I apex with mesial and lateral borders ending nearly symmetrically (Fig. 23.160 A, B, arrows) 3

2′ Gonopod I apex with mesial and lateral borders ending assymmetrically (Fig. 23.160 C, D, arrows). 4

3(2) Gonopod I apex flattened, distal margins subparallel, forming medial triangular projection (Fig. 23.160 A, arrows), apical spine field narrow, rimate, medial process rather narrow, subtriangular (Fig. 23.160 A, me) ..... *Odontothelphusa*

3′ Gonopod I apex widened, distal margins not parallel, medially rounded (Fig. 23.160 B, arrows), apical spine field subcircular or oval (Fig. 23.160 B, fas), medial process wide, broadly subtriangular or subrectangular (Fig. 23.160 B, me) ..... *Raddaus*
[Mexico. Guatemala. Belize. Honduras. El Salvador. Nicaragua]

4(2) Gonopod I apex medial surface (Figs. 23.161 B, C, arrow) ending straight or produced terminally into supraapical process with length and breadth < proximal gonopod portion 5

4′ Gonopod I apex medial surface (Fig. 23.161 A, arrow) produced terminally into oval supraapical process with length $\geq 0.3\times$ proximal gonopod portion, width $\geq$ proximal gonopod portion ..... *Phrygiopilus*
[Mexico. Guatemala]

5(4) Gonopod I apical spine field (Fig. 23.160 C—E, fas) narrow, terminal, directed apically 6

5′ Gonopod I apical spine field (Fig. 23.160 F, fas) narrow to wide, subterminal, directed anteriorly, laterally or posteriorly 8

6(5) Gonopod I distal half arcing laterally (Fig. 23.160 D, E), mesial process (Fig. 23.160 D, E, me) narrowly subtriangular, usually falciform, length $<$ to slightly $>$ distal gonopod portion width 7

6′ Gonopod I straight (Fig. 23.160 C), mesial process (Fig. 23.160 C, me) broadly subtriangular, length $<$ distal gonopod portion width .... ***Villalobosius*** **[p. 962]**

[Mexico]

7(6) Eyes unpigmented; maxilliped III exopod length 0.1−0.7× ischium; gonopod I mesial process (Fig. 23.160 E, me) subdistal; stygobitic species .... *Typhlopseudothelphusa*

[Mexico. Guatemala]

7′ Eyes pigmented; maxilliped III exopod length 0.8−0.7× ischium; gonopod I mesial process (Fig. 23.160 D, me) distal; epigean species .... *Zilchia*

[Mexico. Guatemala. Belize. Honduras. El Salvador]

8(5) Gonopod I in posterior view straight or distal part slightly arced, medial surface distal margin with medial lobe (Fig. 23.161 B, C, ml) as an apical projection smooth, granulate, or spinose 9

8′ Gonopod I in posterior view curved, medial surface distal margin with medial lobe (Fig. 23.161 D, ml) as apical projection with spinulae .... *Tzotzilthelphusa villarosalensis* Villalobos & Álvarez, 2013

[Mexico]

9(8) Gonopod I medial surface distal margin medial lobe (Fig. 23.161 B, ml) with linguiform apical projection with margin granulate or spinose .... ***Sylvathelphusa*** **[p. 963]**

[Mexico]

9′ Gonopod I medial surface distal margin medial lobe (Fig. 23.161 C, ml) poorly to moderately developed, with apical projection narrow, subtriangular, subrectangular, or rounded and margin inerm .... *Potamocarcinus*

[Mexico. Guatemala. Belize. Honduras. El Salvador. Costa Rica. Panama. Colombia]

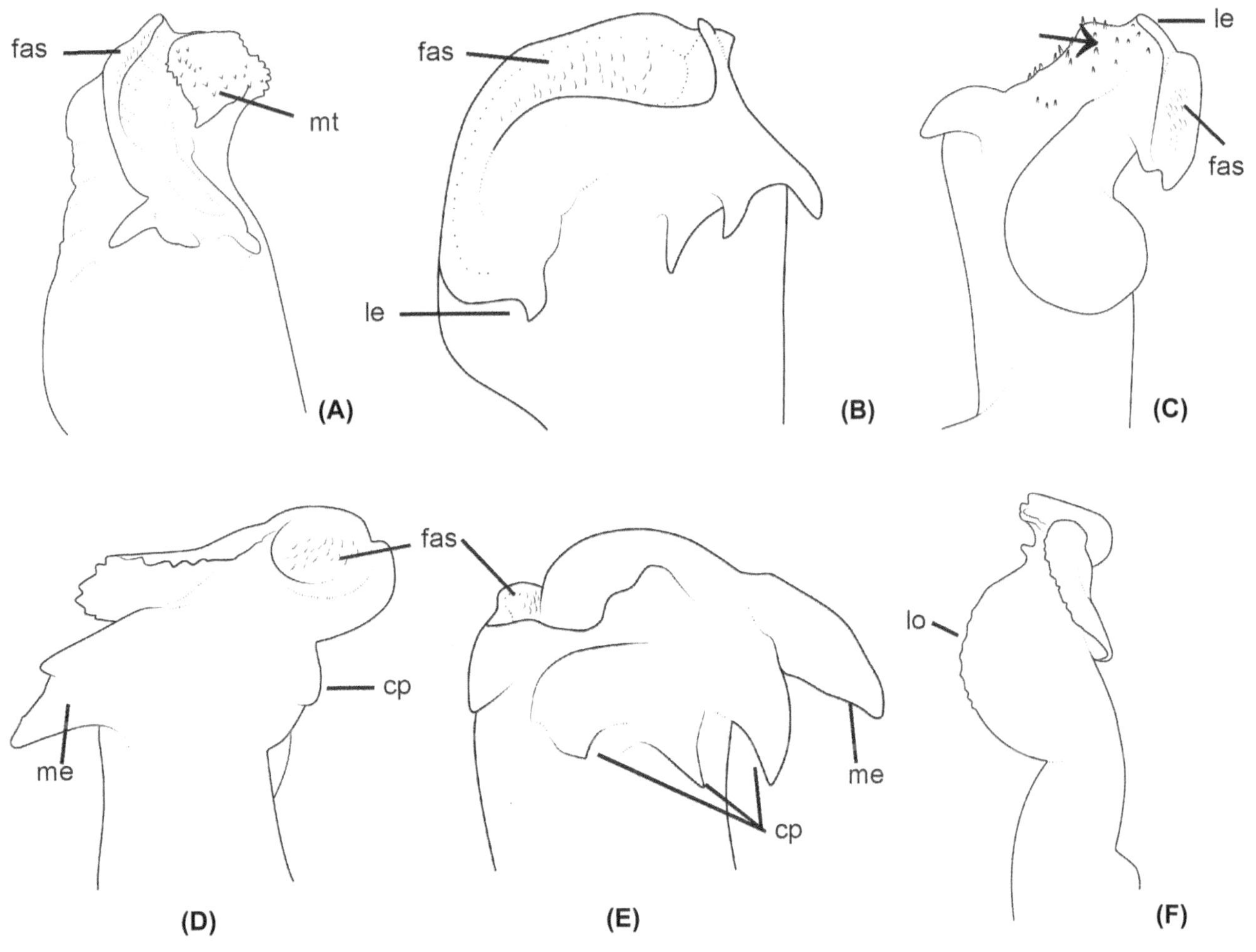

FIGURE 23.156 Gonopod I of Hypolobocerini: (A) *Lobithelphusa mexicana* (cephalic view); (B) *Colombiathelphusa culmarcuata* (distocephalic view); (C) *Elsalvadoria tomhaasi* (cephalic view); (D) *Camptophallus botti* (cephalic view); (E) *Eidocamptophallus chacei* (laterocephalic view); (F) *Spirothelphusa verticalis* (cephalic view). Abbreviations: *cp*, cephalic process; *fas*, field of apical spines; *le*, lateral end; *lo*, lateral lobe; *me*, mesial process; *mt*, tubercles in the mesial surface.

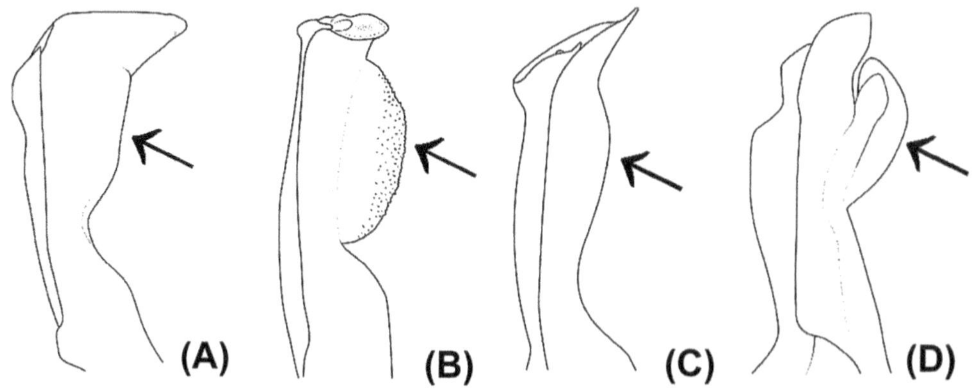

FIGURE 23.157 Gonopod I of Hypolobocerini: (A) *Moritschus henrici*; (B) *Lindacatalina latipenis*; (C) *Hypolobocera rathbuni*; (D) *Neostrengeria guenteri*. *Arrow*, lateral lobe.

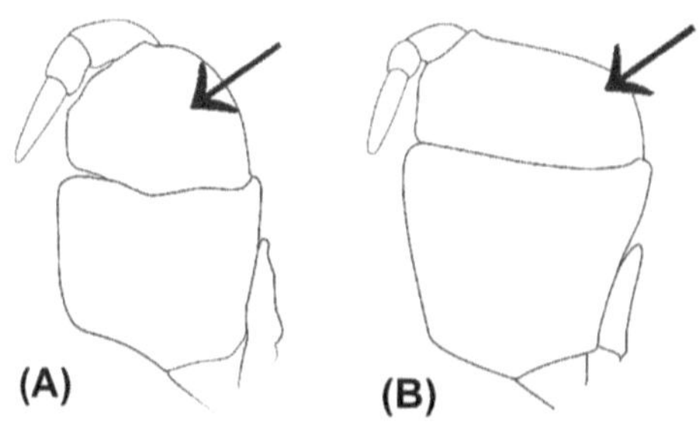

FIGURE 23.158 Maxilliped III of Hypolobocerini: (A) *Colombiathelphusa culmarcuata*; (B) *Lobithelphusa mexicana*. *Arrow*, merus.

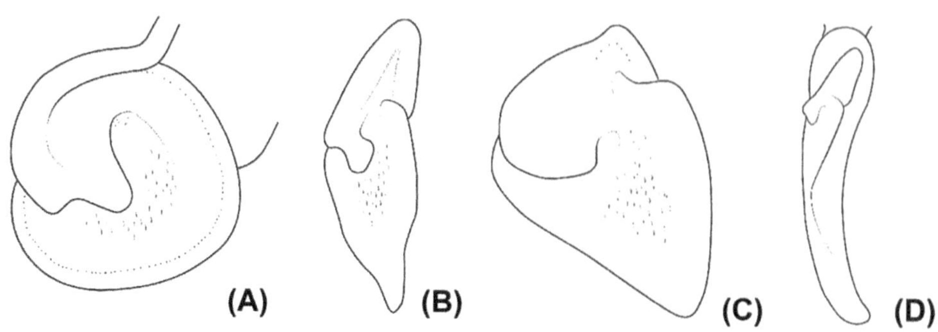

FIGURE 23.159 Gonopod I (apical view) of Hypolobocerini: (A) *Lindacatalina latipensis*; (B) *Hypolobocera rathbuni*; (C) *H. exuca*; (D) *Moritschus henrici*.

Decapoda: Brachyura: Pseudothelphusidae: Pseudothelphusinae: Potamocarcinini: *Villalobosius*: Species

1 Gonopod I marginal process subtriangular and overreaching apex, mesial process directed distomedially .... *Villalobosius leptomelus* Rodríguez & Hobbs, 1989 [Mexico]

1′ Gonopod I marginal process subquadrate reaching apex, mesial process directed laterally .... *Villalobosius lopezformenti* Alvarez & Villalobos, 1991 [Mexico]

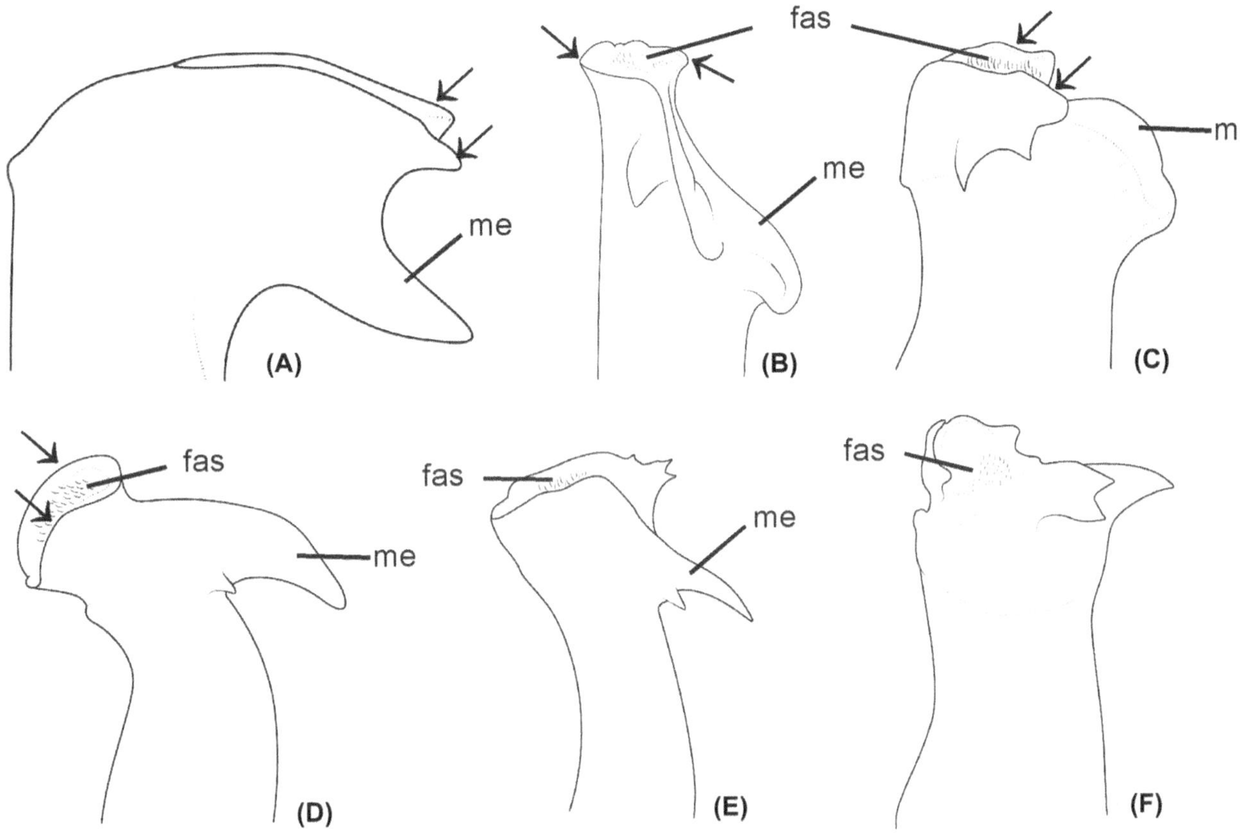

FIGURE 23.160 Gonopod I of Potamocarcinini: (A) *Odontothelphusa maxillipes* (lateral view); (B) *Raddaus bocourti* (mesial view); (C) *Villalobosius lopezformenti* (lateral view); (D) *Zilchia falcata* (cephalic view); (E) *Typhlopseudothelphusa juberthiei* (cephalic view) and (F) *Potamocarcinus magnus* (laterocephalic view). Abbreviations: *fas*, field of apical spines, *me*, mesial process.

Decapoda: Brachyura: Pseudothelphusidae: Pseudothelphusinae: Potamocarcinini: *Sylvathelphusa*: Species

1 Gonopod I medial surface distal margin with acute, robust spines; medial surface distal margin thin, serrate; mesial process directed laterally .... *Sylvathelphusa kalebi* Villalobos & Álvarez, 2013
[Mexico]

1′ Gonopod I medial surface distal margin broad, rounded, with shallow longitudinal grooves and scattered subacute granules; medial surface distal margin broad, rounded, with few granulae; mesial process directed anteriorly .... *Sylvathelphusa cavernicola* Villalobos & Álvarez, 2013
[Mexico]

Decapoda: Brachyura: Pseudothelphusidae: Pseudothelphusinae: Pseudothelphusini: Genera

Key is modified from Rodriguez (1982) and Villalobos Hiriart (2005). Two taxa presented in quotation marks represent undescribed genera.

1 Gonopod I lateral apical process absent 2

1′ Gonopod I lateral apical process present .... **Ehecatusa [p. 964]**
[Mexico]

2(1) Gonopod I apex smooth or bearing spicules in distal crest of caudomarginal and lateral surfaces, posteriomarginal and lateral crests bearing spiculae; posteriomarginal caudomarginal projection (Fig. 23.162 A–E, arrow) ending in one, two, or three lobes, the distalmost absent or an acute, triangular, or rounded tooth 3

2′ Gonopod I apex bearing spines; posteriomarginal projection unilobed, ending in a series of spines or teeth (Fig. 23.162 F, arrow) .... *Smalleyus tricristatus* Alvarez, 1989
[Mexico]

3(2) Gonopod I posteriomarginal projection bi- or trilobed (Fig. 23.162 B–E, arrow); mesial process a triangular or subrectangular plate, laterally with a long, acute spine, or two or more teeth or spines, mesially with a small subtriangular expansion not surpassing the distal crests 4

3′ Gonopod I posteriomarginal projection unilobed, securiform (Fig. 23.162 A, arrow); mesial process an irregular plate, laterally with two teeth, mesially as a circular expansion surpassing distal crests .... "*Pseudothelphusa*" *galloi* Álvarez & Villalobos, 1990
[Mexico]

4(3) Gonopod I distomedial prominence absent; posteriomarginal projection, proximal lobe medially smooth or with low, obscure crest; lateral surface constricted in distal third 5

4′ Gonopod I distomedial prominence (Fig. 23.162 B, md) conical or compressed longitudinally; posteriomarginal projection proximal lobe medially with obvious crest; lateral surface constricted at midpoint .... *Tehuana*
[Mexico]

5(4) Gonopod I posteriomarginal projection bilobed (Fig. 23.162 C, arrow), distal lobe robust with anterior end rounded, proximal and distal lobes separated by deep invagination; proximal lobe width > length, distal margin rounded. .... "*Pseudothelphusa*" *sulcifrons* (Rathbun, 1898)
[Mexico]

5′ Gonopod I posteriomarginal projection bi- or trilobed (Fig. 23.162 D, E, arrow), distal lobe poorly developed, subtriangular, proximal and distal lobes separated by "V" or "U"-shaped notch; proximal lobe usually length > width, distal margin angled, acute, or rounded .... *Pseudothelphusa*
[Mexico]

Decapoda: Brachyura: Pseudothelphusidae: Pseudothelphusinae: Pseudothelphusini: *Ehecatusa*: Species

1 Gonopod I apicolateral process separated from mesial process .... *Ehecatusa chiapensis* (Rodríguez & Smalley, 1972)
[Mexico]

1′ Gonopod I apicolateral process present .... *Ehecatusa mixtepensis* (Rodríguez & Smalley, 1972)
[Mexico]

Decapoda: Brachyura: Pseudothelphusidae: Pseudothelphusinae: Strengerianini: Genera

Key is modified from Campos (2005).

1 Efferent branchial channel aperture (Fig. 23.163 A–D) partially closed or closed by spine at jugal angle and epistome lateral lobe extension 2

1′ Efferent branchial channel aperture open (Fig. 23.163 E) .... *Phallangothelphusa*
[Colombia]

2(1) Maxilliped III exopod (Fig. 23.164 A, arrow) length < ischium 3

2′ Maxilliped III exopod (Fig. 23.164 B, arrow) length > ischium .... *Strengeriana*
[Colombia]

3(2) Gonopod I apex with one process (Fig. 23.165 A, arrow) .... *Martiana clausa* (Rathbun, 1915)
[Colombia]

3′ Gonopod I apex with two processes (Fig. 23.165 B, arrow) .... *Chaceus*
[Colombia]

Decapoda: Brachyura: Pseudothelphusidae: Pseudothelphusinae: Kingsleyini: Genera

1 Gonopod I marginal suture straight along medial surface, marginal process shorter than apex 2

1′ Gonopod I marginal suture distally twisted toward anterior surface, marginal process reaching apex .... *Spirocarcinus garthi* (Pretzmann, 1971)
[Panama]

2(1) Gonopod I terminal portion straight or nearly straight (Fig. 23.166 A) 3

2′ Gonopod I terminal portion deflected ~90 degrees (Fig. 23.166 B) .... *Kunziana irengis* (Pretzmann, 1971)
[Guyana]

3(2) Gonopod I apical spine field (Fig. 23.167 A, arrow) longitudinal or oblique, not or partially exposed in distal view; gonopod I distal margins of anterior and posterior surfaces fused as apical plate 4

3′ Gonopod I apical spine field (Fig. 23.167 B, arrow) transverse, exposed in distal view; gonopod I distal margins of anterior and posterior surfaces not fused .... *Guinotia dentata* (Latreille, 1825)
[Lesser Antilles: Guadaloupe, Dominica, Martinique, Saint Lucia, Saint Vincent]

4(3) Gonopod I mesial process (Fig. 23.168 A, arrow) subtriangular or subquadrate, ≤ apical plate 5

4′ Gonopod I mesial process spherical (Fig. 23.168 B, arrow), encompassing most of apical portion, >2× apical plate .... *Oedothelphusa orientalis* Rodríguez, 1980
[Venezuela]

5(4) Gonopod I subapical anteriolateral portion nearly flat (Fig. 23.169 A); gonopod II apex abruptly tapering 6

5′ Gonopod I subapical anteriolateral portion tumid (Fig. 23.169 B, arrow); gonopod II apex gradually tapering .... *Fredius*
[Venezuela. Colombia. Guyana. Suriname. French Guiana. Brazil. Peru]

6(5) Gonopod I spermatic channel opening delimited by proximal and distal lobe crests; apical plate proximal lobe not carinate (Fig. 23.170 A) 7

6′ Gonopod I spermatic channel opening delimited by proximal lobe crest; apical plate proximal lobe with transverse carina (Fig. 23.170 B, arrow) .... ***Melothelphusa* [p. 970]**
[Brazil]

7(6) Gonopod I apical plate with robust distal lobe, subequal to proximal lobe (Fig. 23.171 A, arrow) 8

7′ Gonopod I apical plate with distal lobe (Fig. 23.171 B, arrow) ≥ 5× proximal lobe. .... *Brasiliothelphusa tapajoense* Magalhães & Türkay, 1986
[Brazil]

8(7) Gonopod I mesial process (Fig. 23.172 A, arrow) obscure, fused to apical plate distal margin 9

8′ Gonopod I mesial process (Fig. 23.172 B, arrow) robust, distinct from apical plate 10

9(8) Gonopod I mesial process (Fig. 23.173 A, arrow) distal margin dentate .... *Prionothelphusa eliasi* Rodríguez, 1980
[Venezuela. Colombia. Brazil]

9′ Gonopod I mesial process distal margin inerm (Fig. 23.173 B, arrow) .... *Orthothelphusa*
[Venezuela. Colombia]

10(8) Gonopod I intermediate process between marginal process and apical plate absent (Fig. 23.174 A); major cheliped pollex without proximal rounded tubercle 11

10′ Gonopod I intermediate process present (Fig. 23.174 B, arrow) between marginal process and apical plate; major cheliped pollex proximally with rounded tubercle (except in *R. trujillensis*) .... *Rodriguezus*
[Venezuela. Trinidad and Tobago]

11(10) Gonopod I penian foramen with subcircular lobe; marginal process lamelliform (Fig. 23.175 A, arrow) 12

11′ Gonopod I penian foramen with subtriangular lobe; marginal process enlarged (Fig. 23.175 B, arrow) 15

12(11) Gonopod I apical plate proximal and distal lobes separated by notch (Fig. 23.176 A, arrow) 13

12′ Gonopod I apical plate proximal and distal lobes fused (Fig. 23.176 B, arrow) 14

13(12) Maxilliped III exopod (Fig. 23.177 A, arrow) length 0.5× ischium .... ***Eudaniela* [p. 970]**

13′ Maxilliped III exopod (Fig. 23.177 B, arrow) length 0.25× ischium .... *Microthelphusa*
[Venezuela. Trinidad. Guyana. Suriname. Brazil]

14(12) Gonopod I marginal process (Fig. 23.178 A, arrow) truncate, distally fused to distal lobe crest .... *Neopseudothelphusa fossor* (Rathbun, 1898)
[Venezuela]

14′ Gonopod I marginal process (Fig. 23.178 B, arrow) acute, free .... *Culterthelphusa simoni* (Rathbun, 1905)
[Venezuela]

15(11) Gonopod I marginal and lateral sutures fused at distal third, followed by strong twist toward lateral side; apical plate depressed .... *Michaelthelphusa tuerkayi* Magalhães, 2017
[Brazil]

15′ Gonopod I marginal and lateral sutures not fused; apical plate compressed .... *Kingsleya*
[Venezuela. Guyana. Suriname. French Guiana. Brazil]

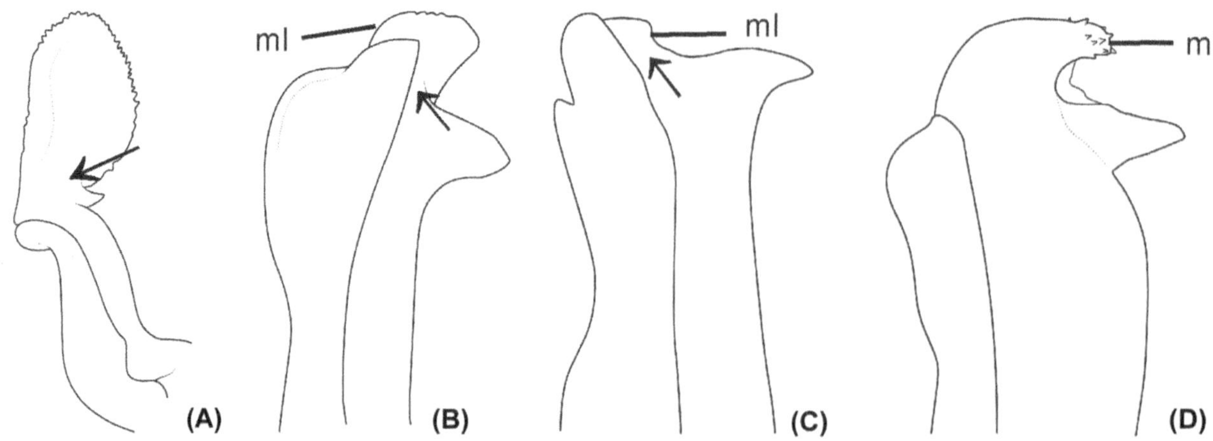

FIGURE 23.161 Gonopod I of Potamocarcinini: (A) *Phrygiopilus acanthophallus* (caudomesial view); (B) *Sylvathelphusa kalebi* (mesial view); (C) *Potamocarcinus magnus* (caudomesial view); (D) *Tzotzilthelphusa villarosalensis* (mesial view). Abbreviations: *ml*, mesial lobe; *arrow*, mesial surface.

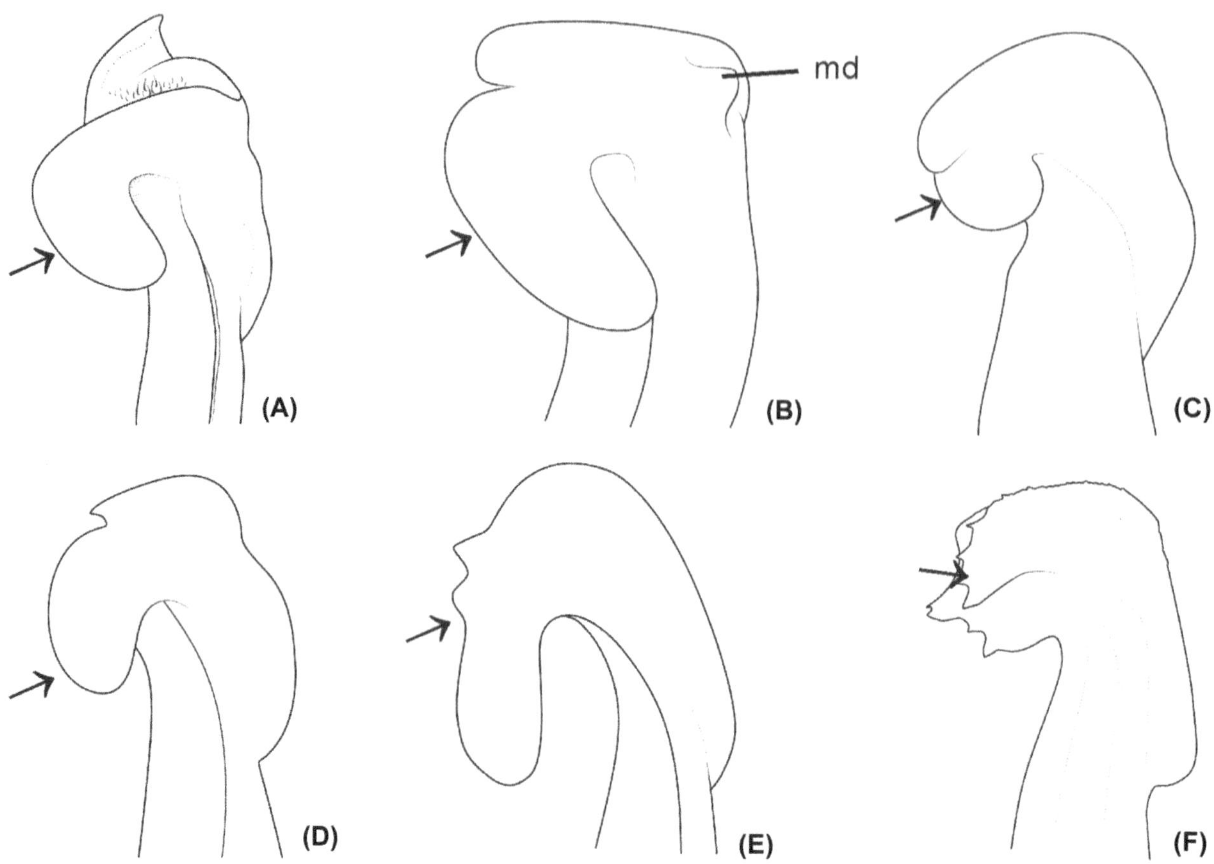

FIGURE 23.162 Gonopod I (mesial view) of Pseudothelphusini: (A) Genus 1: "*Pseudothelphusa*"*galloi*; (B) *Tehuana veracruzana*; (C) Genus 2: "*Pseudothelphusa*" *sulcifrons*; (D) *Pseudothelphusa parabelliana*; (E) *Pseudothelphusa pecki*; (F) *Smalleyus tricristatus*. Abbreviation: *md*, mesodistal prominence; *arrow*, caudal-marginal projection.

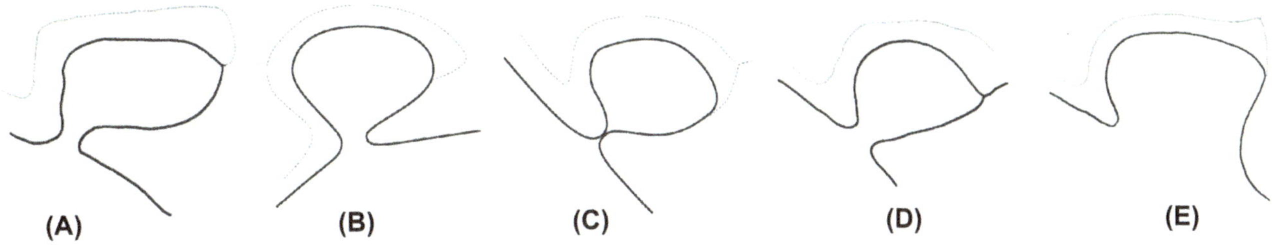

FIGURE 23.163 Aperture of efferent branchial channel of Strengerianini: (A) *Chaceus pearsei*; (B) *Martiniana clausa*; (C) *Strengeriana fuhrmanni*; (D) *Strengeriana restrepoi*; (E) *Phallangothelphusa dispar*.

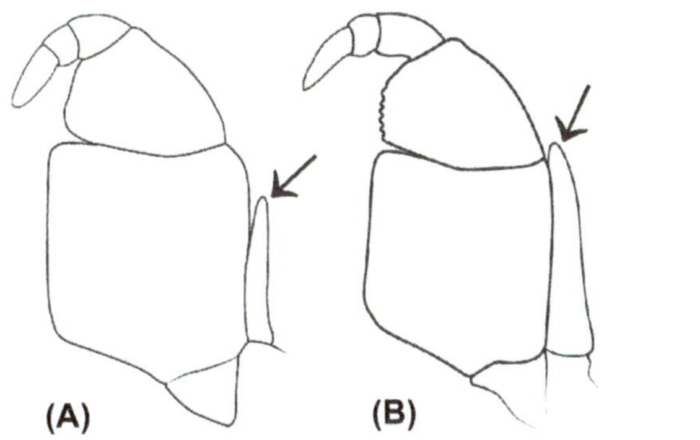

FIGURE 23.164 Maxilliped III of Strengerianini: (A) *Chaceus nasutus*; (B) *Strengeriana bolivariensis*. *Arrow*, exopod of maxilliped III.

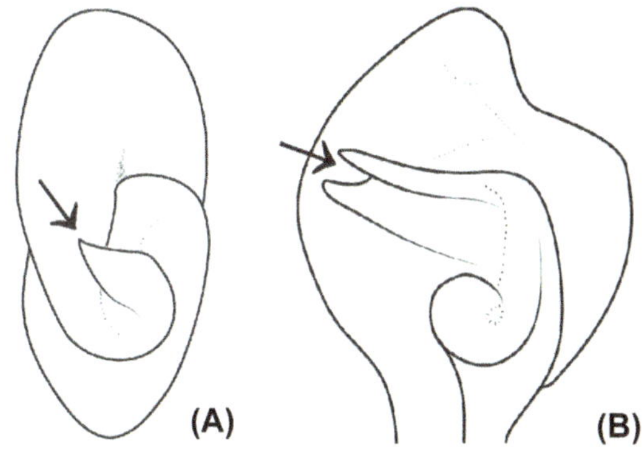

FIGURE 23.165 Gonopod I (apical view) of Strengerianini: (A) *Martiana clausa*; (B) *Chaceus davidi*. *Arrow*, apical process.

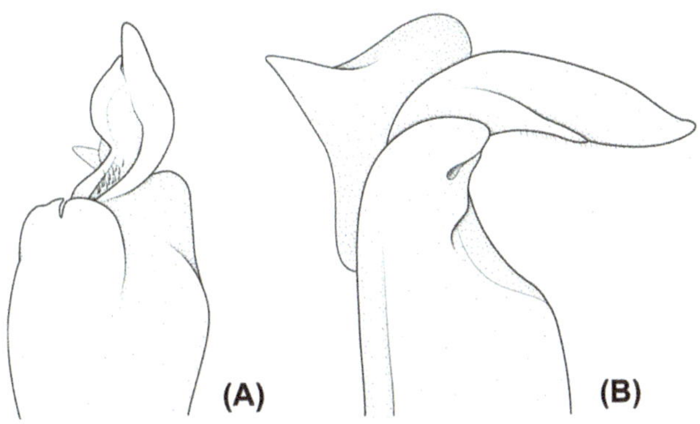

FIGURE 23.166 Gonopod I (distal part, lateral view) of Kingsleyini: (A) *Kingsleya latifrons*; (B) *Kunziana irengis*.

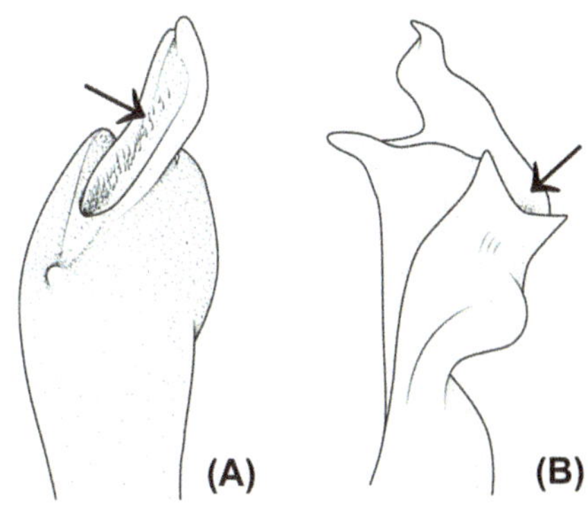

FIGURE 23.167 Gonopod I (distal part) of Kingsleyini: (A) *Eudaniela casanarensis* (lateral view); (B) *Guinotia dentata* (caudal view). *Arrow*, field of apical spines.

PHYLUM ARTHROPODA: CRUSTACEA: MALACOSTRACA

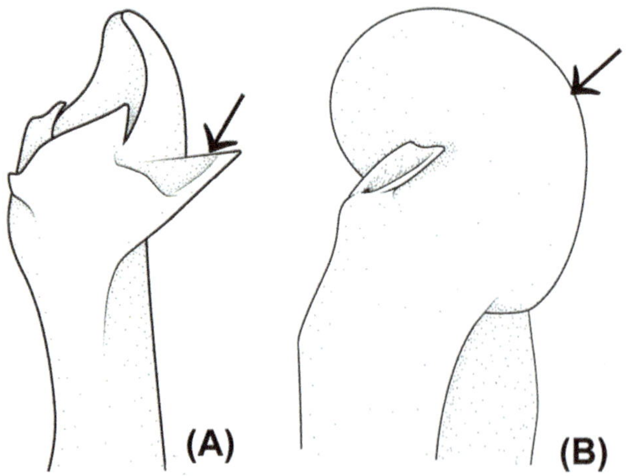

FIGURE 23.168 Gonopod I (distal part) of Kingsleyini: (A) *Rodriguezus iturbei* (cephalic view); (B) *Oedothelphusa orientalis* (lateral view). Arrow, mesial process.

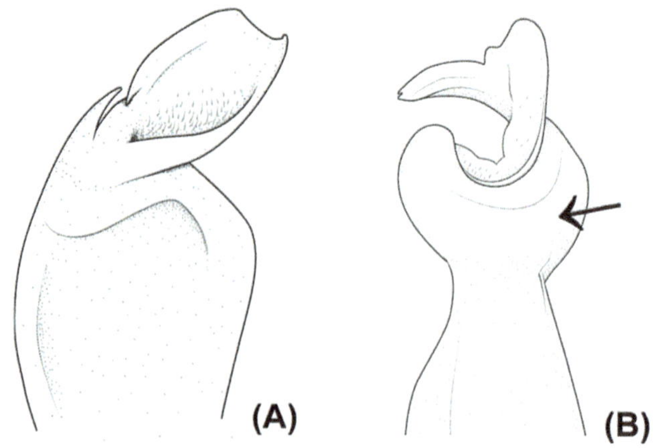

FIGURE 23.169 Gonopod I (distal part, lateral view) of Kingsleyini: (A) *Neopseudothelphusa simony*; (B) *Fredius fittkaui*. *Arrow*, subapical portion: latero-cephalic surface.

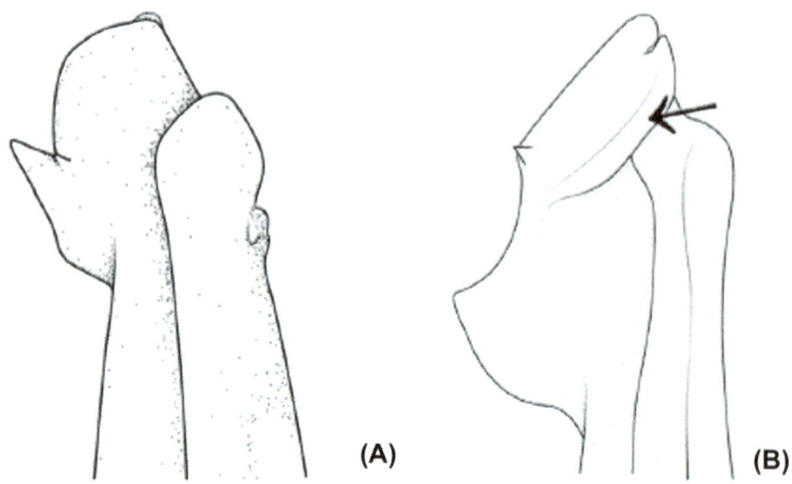

FIGURE 23.170 Gonopod I (distal part, caudal view) of Kingsleyini: (A) *Eudaniela casanarensis*; (B) *Melothelphusa dardanelosensis*. *Arrow*, keellike process.

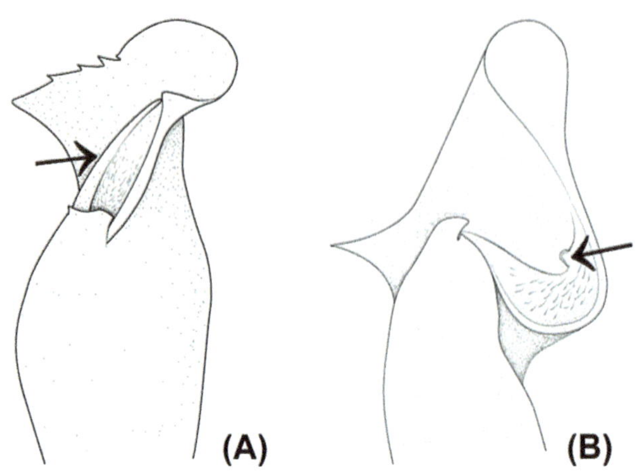

FIGURE 23.171 Gonopod I (distal part, lateral view) of Kingsleyini: (A) *Prionothelphusa eliasi*; (B) *Brasiliothelphusa tapajoense*. *Arrow*, proximal lobe of the apical plate.

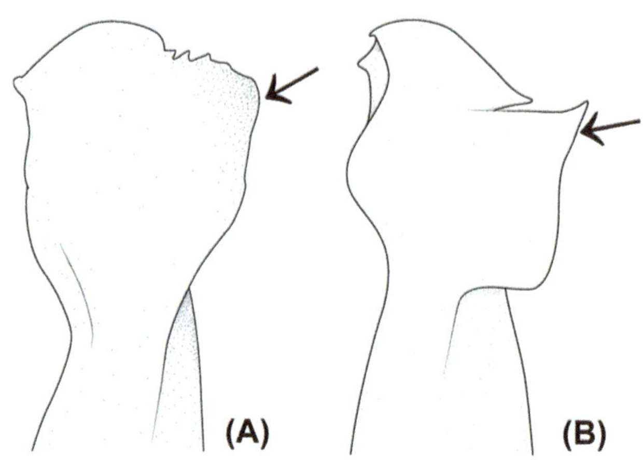

FIGURE 23.172 Gonopod I (distal part, cephalic view) of Kingsleyini: (A) *Prionothelphusa eliasi*; (B) *Microthelphusa bolivari*. *Arrow*, mesial process.

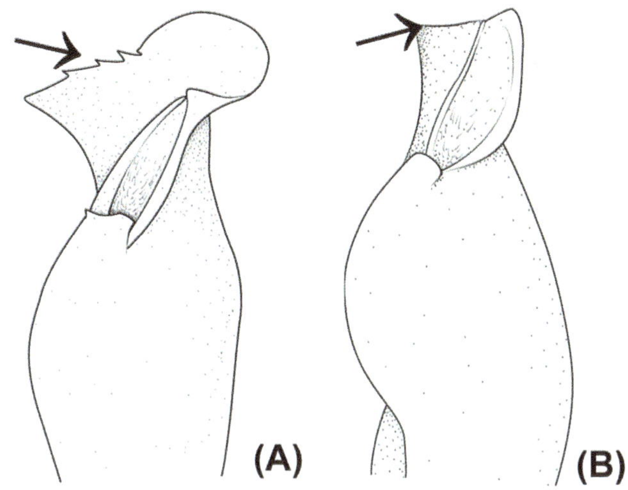

FIGURE 23.173 Gonopod I (distal part, lateral view) of Kingsleyini: (A) *Prionothelphusa eliasi*; (B) *Orthothelphusa venezuelensis*. *Arrow*, distal margin of mesial process.

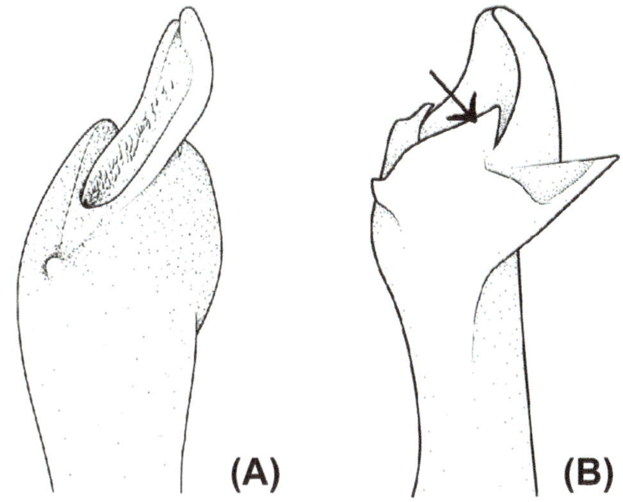

FIGURE 23.174 Gonopod I (distal part) of Kingsleyini: (A) *Eudaniela casanarensis* (lateral view); (B) *Rodriguezus iturbei* (cephalic view). *Arrow*, intermediate process.

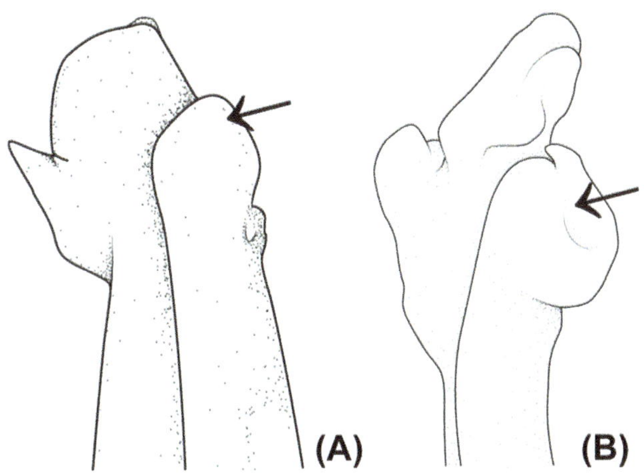

FIGURE 23.175 Gonopod I (distal part, caudal view) of Kingsleyini: (A) *Eudaniela casanarensis*; (B) *Kingsleya ytupora*. *Arrow*, marginal process.

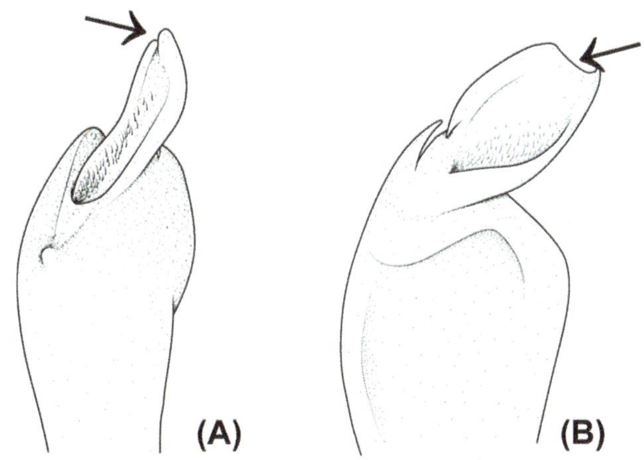

FIGURE 23.176 Gonopod I (distal part, lateral view) of Kingsleyini: (A) *Eudaniela casanarensis*; (B) *Culterthelphusa simoni*. *Arrow*, lobes of the apical plate.

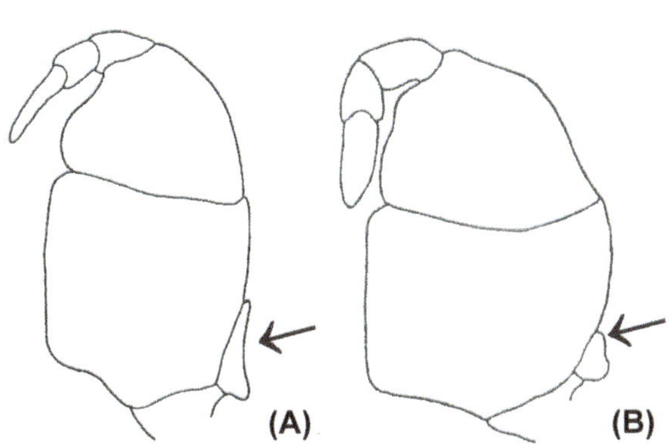

FIGURE 23.177 Maxilliped III of Kingsleyini: (A) *Eudaniela casanarensis*; (B) *Microthelphusa rodriguezi*.

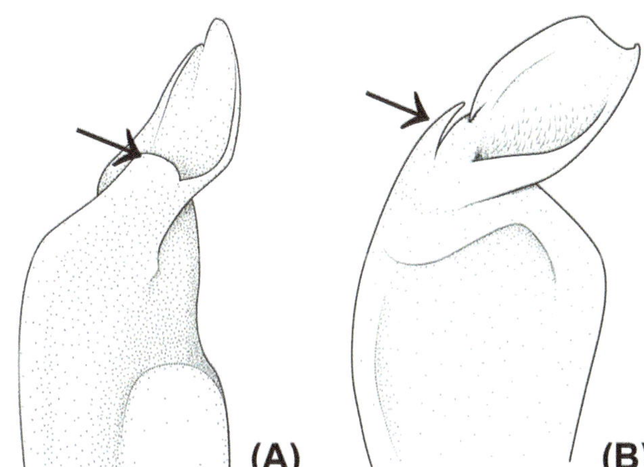

FIGURE 23.178 Gonopod I (distal part, lateral view) of Kingsleyini: (A) *Prionothelphusa eliasi*; (B) *Brasiliothelphusa tapajoense*. *Arrow*, proximal lobe of the apical plate.

Decapoda: Brachyura: Pseudothelphusidae: Pseudothelphusinae: Kingsleyini: *Melothelphusa*: Species

1	Gonopod I apical plate mesial side devoid of spines, mesial process subquadrate .... *Melothelphusa apiaka* Pedraza, Tavares & Magalhães, 2016 [Brazil]	
1′	Gonopod I apical plate mesial side with triangular spine, mesial process subtriangular .... *Melothelphusa dardanelosensis* (Magalhães & Türkay, 2010) [Brazil]	

Decapoda: Brachyura: Pseudothelphusidae: Pseudothelphusinae: Kingsleyini: *Eudaniela*: Species

1	Gonopod I apical plate mesial margin rounded in ventral view .... *Eudaniela pestai* (Pretzmann, 1965) [Venezuela]	
1′	Gonopod I apical plate mesial margin rounded in ventral view .... *Eudaniela casanarensis* (Campos, 2001) [Colombia]	

Decapoda: Brachyura: Panopeidae: Genera and Species

Following Chace & Hobbs (1969).

1	Chelae with pollex and dactyl whitish	2
1′	Chelae with pollex and dactyl dark .... *Panopeus herbstii* H. Milne-Edwards, 1834 [Atlantic and Caribbean coasts and islands south to Uruguay. Nearctic]	
2(1)	Carapace widest posterior lateral teeth apices .... *Eurytium limosum* (Say, 1818) [Atlantic and Caribbean coasts and islands south to Brasil. Nearctic]	
2′	Carapace widest between tips of posterior lateral teeth .... *Rhithropanopeus harissii* (Gould, 1841) [Nearctic. Atlantic México. Invasive elsewhere, including Panama Canal Zone and Brazil]	

REFERENCES

Abele, L. 1992. A review of the Grapsoid crab genus *Sesarma* (Crustacea: Decapoda: Grapsidae) in America, with the description of a new genus. Smithsonian Contributions to Zoology, 527.

Campos, M.R. 2003. A review of the freshwater crabs of the genus *Hypolobocera* Ortmann, 1897 (Crustacea: Decapoda: Brachyura: Pseudothelphusidae), from Colombia. Proceedings of the Biological Society of Washington 116: 754–802.

Campos, M.R. 2005. Freshwater crabs from Colombia. A taxonomic and distributional study. Academia Colombiana de Ciencias Exactas Físicas y Naturales. Col. Jorge Álvarez Lleras 24: 1–363.

Campos, M.R. 2014. Crustáceos decápodos de agua dulce de Colombia. Biblioteca José Jerónimo Triana, Instituto de Ciencias Naturales 27: 1–692.

Campos, M.R. & M. Pedraza. 2008. Two new species of freshwater crab of the genus *Neostrengeria* Pretzmann, 1965, from Colombia (Crustacea: Decapoda: Pseudothelphusidae), with an updated key of the species of the genus. Caldasia, 30: 457–468.

Cardona, L. & M.R. Campos. 2012. A new freshwater crab of the genus *Phallangothelphusa* Pretzmann, 1965, from Colombia (Crustacea: Decapoda: Pseudothelphusidae). Zootaxa, 3515: 83–88.

Chace, F.A., Jr. & H.H. Hobbs, Jr. 1969. The freshwater and terrestrial decapod crustaceans of the West Indies with special reference to Dominica. Bulletin of the United States National Museum 292: 1–258.

Crane, J. 1975. Fiddler Crabs of the World, Ocypodidae: Genus *Uca*. Princeton University Press.

Cumberlidge, N. 2016. Chapter 1: Global diversity and conservation of freshwater crabs (Crustacea: Decapoda: Brachyra). Pages 1–22 in: T. Kawai and N. Cumberlidge (eds.), A Global Overview of the Conservation of Freshwater Decapod Crustaceans. Springer, Cham, Switzerland.

Cumberlidge, N., F. Álvarez & J.L. Villalobos. 2014. Results of the global conservation assessment of the freshwater crabs (Brachyura, Pseudothelphusidae and Trichodactylidae): The Neotropical region, with an update on diversity. Zookeys 457: 133–157.

Cumberlidge, N. & P.K.L. Ng. 2009. Systematics, evolution, and biogeography of freshwater crabs. Pages 491–508 *in*: J.W. Martin, K.A. Crandall, and D.L. Felder (eds.), Decapod Crustacean Phylogenetics. Crustacean Issues, vol. 18. Koenemann, S. (series ed.). CRC Press, Taylor & Francis Group, Boca Raton, London, New York.

Hobbs Jr., H.H. 1986. A new troglobitic crab (Crustacea: Decapoda: Pseudothelphusidae) from Belize. Texas Memorial Museum, Speleological Monographs 1: 1–4.

Magalhães, C. 2003. Brachyura: Pseudothelphusidae e Trichodactylidae. Pages 143–297 *in*: G.A.S. Melo (Org.), Manual de identificação dos crustáceos decápodos de água doce brasileiros. São Paulo, Edições Loyola.

Magalhães, C., M.R. Campos, P.A. Collins & F.L. Mantelatto. 2016. Chapter 10: Diversity, distribution and conservation of freshwater crabs and shrimps in South America. Pages 303322 *in*: T. Kawai and N. Cumberlidge (eds.), A Global Overview of the Conservation of Freshwater Decapod Crustaceans. Springer, Cham, Switzerland.

Magalhães, C., M.R. Campos & M. Türkay. 2013. Freshwater crabs from eastern Panamá: a new species of *Potamocarcinus* H. Milne Edwards, 1853, and new records of two little-known species (Crustacea: Decapoda: Pseudothelphusidae, Trichodactylidae). Zootaxa 3702: 348–356.

Magalhães, C. & M. Türkay. 1996a. Taxonomy of the Neotropical freshwater crab family Trichodactylidae I. The generic system with description of some new genera (Crustacea: Decapoda: Brachyura). Senckenbergiana biologica 75(1/2): 63–95.

Magalhães, C. & M. Türkay. 1996b. Taxonomy of the Neotropical freshwater crab family Trichodactylidae. II. The genera *Forsteria, Melocarcinus, Sylviocarcinus*, and *Zilchiopsis* (Crustacea: Decapoda: Brachyura). Senckenbergiana biologica 75: 97–130.

Magalhães, C. & M. Türkay. 1996c. Taxonomy of the Neotropical freshwater crab family Trichodactylidae. III. The genera *Fredilocarcinus* and *Goyazana* (Crustacea: Decapoda: Brachyura). Senckenbergiana biologica 75: 131–142.

Magalhães, C. & M. Türkay. 2008a. Taxonomy of the Neotropical freshwater crab family Trichodactylidae. IV. The genera *Dilocarcinus* and *Poppiana* (Crustacea, Decapoda, Trichodactylidae). Senckenbergiana biologica 88: 185–215.

Magalhães, C. & M. Türkay. 2008b. Taxonomy of the neotropical freshwater crab family Trichodactylidae. V. The genera *Bottiella* and *Rotundovaldivia* (Crustacea, Decapoda, Trichodactylidae). Senckenbergiana biologica 88: 217–230.

Magalhães, C. & M. Türkay. 2012. Taxonomy of the neotropical freshwater crab family Trichodactylidae. VI. The genera *Avotrichodactylus* and *Rodriguezia* (Decapoda: Brachyura: Trichodactylidae). Nauplius 20: 27–40.

Magalhães, C., I.S. Wehrtmann, L.R. Lara & F.L. Mantelatto. 2015. Freshwater crabs from Costa Rica, with a taxonomic revision of the genus *Ptychophallus* Smalley, 1964 (Crustacea: Decapoda: Pseudothelphusidae). Zootaxa 3905: 301–344.

Ng, P.K.L., D. Guinot & P.J.F. Davie. 2008. Systema Brachyurorum: Part I. An annotated checklist of extant brachyuran crabs of the world. The Raffles Bulletin of Zoology, Supplement 17: 1–286.

Pretzmann, G. 1972. Die Pseudothelphusidae (Crustacea Brachyura). Zoologica 120: 1–184.

Poettinger, T.S., S. Klaus & C. D. Schubart. 2016. Phylogenetic relationships among Neotropical freshwater crabs of the Pseudothelphusidae (Decapoda, Brachyura) and the taxonomic position of *Ptychophallus* and related genera. Crustaceana 89: 1717–1728.

Rathbun, M.J. 1918. The Grapsoid Crabs of America.United States National Museum Bulletin 97.

Rathbun, M.J. 1930. The Cancroid Crabs of America of the families Euryalidae, Portunidae, Atelecyclidae, Cancridae, and Xanthidae. United States National Museum Bulletin 152.

Rodríguez, G. 1982. Les crabes d'eau douce d'Amerique. Famille des Pseudothelphusidae. Editions ORSTOM, Paris. 224 pp. (Collection Faune Tropicale 22).

Rodríguez. G. 1992. The Freshwater Crabs of America. Family Trichodactylidae and Supplement to the Family Pseudothelphusidae. Editions ORSTOM, Paris. 189 pp. (Collection Faune Tropicale 31).

Rodríguez, G. 2001. New species and records of pseudothelphusid crabs (Crustacea: Brachyura) from Central America in the Museum of Natural History of Tulane University. Proceedings of the Biological Society of Washington 114: 435–443.

Rodríguez, G. & M.R. Campos. 1998. A cladistic revision of the genus Fredius (Crustacea: Decapoda: Pseudothelphusidae) and its significance to the biogeography of the Guianan lowlands of South Amaerica. Journal of Natural History 32: 763–775.

Rodríguez, G. & H.H. Hobbs Jr. 1989. A new cavernicolous crab, *Zilchia falcata*, from Guatemala, with notes on the genera of the Potamocarcinini (Crustacea Decapoda, Pseudothelphusidae). Bulletin du Muséum national d'Histoire naturelle, Section A, Zoologie, Biologie et Ecologie Animales, Paris, 4e série 11: 183–192.

Rodríguez, G. & H. Suárez. 2004. A revision of the freshwater crabs of the family Pseudothelphusidae (Decapoda: Brachyura) from Peru with notes on the southern limits of the family. Amazoniana 18: 11–28.

Rodríguez, G. & R. von Sternberg. 1998. A revision of the freshwater crabs of the family Pseudothelphusidae (Decapoda: Brachyura) from Ecuador. Proceedings of the Biological Society of Washignton 111: 110–139.

Shih, H.-T., P.K.L. Ng, P.J.F. Davie, C.D. Schubart, M. Türkay, R. Naderloo, D. Jones & M.-Y. Liu. 2016. Systematics of the family Ocypodidae Rafinesque, 1815 (Crustacea: Brachyura), based on phylogenetic relationships, with a reorganization of subfamily rankings and a review of the taxonomic status of *Uca* Leach, 1814, sensu lato and its subgenera. Raffles Bulletin of Zoology 64: 139–175.

Souza-Carvalho, E.A., C. Magalhães &. F.L. Mantelatto. 2017. Molecular phylogeny of the *Trichodactylus fluviatilis* Latreille, 1828 (Brachyura, Trichodactylidae) species complex. Journal of Crustacean Biology, 37: 187–194.

Villalobos, J.L. & F. Alvarez. 2003. Two new species of freshwater crabs of the genus *Tehuana* (Brachyura: Pseudothelphusidae) from Southern Mexico. Journal of Crustacean Biology 23: 223–229.

Villalobos, J.L. & F. Alvarez. 2010. Phylogenetic analysis of the Mexican freshwater crabs of the tribe Pseudothelphusini (Decapoda: Brachyura: Pseudothelphusidae). Zoological Journal of the Linnean Society 160: 457–481.

Villalobos, J.L. & F. Álvarez. 2013. Two new genera and three new species of freshwater crabs (Crustacea: Pseudothelphusidae: Potamocarcinini) from Chiapas, Mexico. Zootaxa 3599: 457–470.

Villalobos Hiriart, J.S. 2005. Sistemática de los cangrejos de agua dulce de México, tribu Pseudothelphusini Ortmann, 1897 (Crustacea: Decapoda: Brachyura: Pseudothelphusidae). Análisis filogenético, biogeográfico y descripción de especies nuevas. Doctorate Thesis, Instituto de Biología, Universidad Nacional Autónoma de México. 446 pp.

Williams, A.B. 1974. The swimming crabs of the genus *Callinectes* (Decapoda: Portunidae). Fishery Bulletin 72: 685–798.

Yeo, D.C.J., P.K.L. Ng, N. Cumberlidge, C. Magalhães, S.R. Daniels & M.R. Campos. 2008. Global diversity of crabs (Crustacea: Decapoda: Brachyura) in freshwater. Hydrobiologia 595: 275–286.

Mysida

W. Wayne Price
Department of Biology, University of Tampa, Tampa, FL, United States

INTRODUCTION

A total of 21 species of Mysida (mysids) or "opossum shrimp" have been reported in Neotropical freshwater habitats. Members are represented by three groups: subterranean or hypogean Tethyan relicts (6 species), surface water or epigenean fauna (13 species), and euryhaline estuarine fauna (2 species).

All hypogean Neotropical mysids belong to the genus *Antromysis* and are inhabitants of groundwater where they are found in freshwater and anchialine caves and crab burrows in Mexico, Central America, Cuba, and Jamaica. All species have reduced, separate eyes and are blind with the exception of *A. anophelinae* Tattersall, 1951, which has reduced, fused eyes that have retained some pigmentation. This species lives in mangrove areas along the west coast of Costa Rica in deep burrows of the large land crab *Cardisoma crassum* (Tattersall, 1951). The distribution of *Antromysis* in the Neotropics as well as the worldwide distribution of members of the Order Stygiomysida and *Troglomysis vjetrenicensis* Stammer, 1933 (Mysida) suggest a Tethyan origin for these groups of subterranean species (Porter et al., 2008). Boxshall & Jaume (2000) hypothesized that the expansion and dispersal of marine ancestors occurred in the Tethys Sea resulting in widespread genera prior to the tectonic events that led to the closure of this sea and the opening of the Atlantic. These events resulted in the uplift and stranding of ancestral populations in subterranean habitats around the Tethyan coastline. Modern distributions may have resulted from the isolation and subsequent speciation of these populations.

The 13 surface or epigean species of Mysida belong to two genera, *Parvimysis* and *Surinamysis,* which are distributed in Brazil (10), Suriname (2), and Venezuela (1) in South America. Nine of the 10 species in Brazil are located in the Amazon Basin, making it the world's second most species-rich area for freshwater mysids, exceeded only by the 24 species reported from the Pontocaspian (Wittmann, 2018). Three of the South American species, *P. almyra* (Bowman, 1964), *S. americana* (Tattersall, 1951), and *S. aestuaria* Wittmann, 2017, occupy habitats ranging from brackish and freshwater ditches and streams to open water near the coast. The remaining 10 species are found in freshwater streams, rivers, and lakes farther inland. Both genera are closely related to *Antromysis* (Bamber & Henderson, 1990), and, at one time, were designated as subgenera within that genus (Bowman, 1977, 1980). This led Porter et al. (2008) to hypothesize that these three genera may have a similar Tethyan distributional pattern.

Two species of mysids are primarily allied to estuarine environments, but can tolerate freshwater. *Americamysis almyra* and *Taphromysis bowmani* (Băcescu, 1961) (Fig. 23.179) occur in bays, but can also inhabit upper reaches of estuaries and their freshwater tributaries. The Neotropical distributions of both species are restricted to Florida.

LIMITATIONS

Identification of Neotropical mysids in this chapter is based mainly on variations in the morphology of the eye, antenna 1, antennal scale, telson, and uropod, characters that exhibit little sexual dimorphism. Whenever possible, mature adult specimens should be examined because many characters used in the key are applicable mostly or only to adults. This is especially important for meristic characters such as the number of spiniform setae on the lateral and posterior margins of telsons and uropodal endopods. Setae may be added with successive molts as the organism matures. In addition, the length:width ratios of telsons and antennal scales as well as the degree of male pleopod dimorphism may change with growth. Characters that separate immature from mature Mysida as well as females and males are presented in the next section.

Most mature Neotropical Mysida are not difficult to identify to genus regardless of sex, although having male *Parvimysis* and *Surinamysis* make it easier to separate these two genera as well as species within each genus. Although the six species of the hypogean genus *Antromysis* may be tentatively separated based on eye and telsonal morphology, the morphological variation for at least half of these species is underestimated since their descriptions are based on examinations of five or fewer individuals. However, since the endemicity of *Antromysis* appears to be pronounced, range and habitat information may be helpful in identifying members of this subterranean genus.

TERMINOLOGY AND MORPHOLOGY

Neotropical adult Mysida range in length from less than 3 to 11 mm and have a carapace that covers most or all of the thoracic somites. **Antenna 1** consists of a peduncle of three articles followed by two flagella (except for *Surinamysis* males) and is often more robust in males than females. Males have a setose conical process at the distal end of article 3 of the peduncle; this is the male lobe (**appendix masculina**). This lobe is well developed in most mature Neotropical mysids, but is reduced in some *Parvimysis* species. In addition, males of *Surinamysis* have a third flagellum, termed accessory flagellum extending from article 3 of the peduncle (Wittmann, 2017). The biramous antenna 2 has an endopod with a 3-articulated peduncle and a flagellum. The exopod or **antennal scale** is in the form of a flat plate that is usually longer than the antennal peduncle (except *Antromysis*). Epigean species have well-developed, stalked, moveable eyes with pigmented corneas; however, subterranean species (*Antromysis*) have reduced eyes that may be fused or separate and have little to no pigmentation.

Neotropical Mysida have eight pairs of biramous thoracic appendages (**thoracopods**) with the anterior two pairs of **endopods** modified for feeding and often termed **maxillipeds**. All **exopods** are well developed and used for swimming. Mature females have a well-developed **marsupium** or brood pouch, with or without eggs/larvae, that is comprised of usually two to three paired plates (**oostegites**). These structures are attached to posterior thoracopods and extend between them on the ventral side of the thorax. Each oostegite of *Parvimysis* species has a circular discoidal structure on its inner wall to which one egg or larva is attached. This "oostegital discoid" is unique to this genus within the Mysida (Wittmann, 2018). Most mature males have a pair of tubular **penes** extending from the medial bases of the eighth thoracic appendages where the gonopores are located. The gonopores are usually flanked by setose lobes that form a closing apparatus (Wittmann et al., 2014).

A pair of **pleopods** is present on each of the first five abdominal somites and exhibits marked sexual dimorphism in all Neotropical species. Female pleopods are uniramous simple plates, but male pleopods are either all biramous with pleopods 2–5 well developed (*Americamysis*), or pleopods 1, 2, 3, and 5 are uniramous plates and pleopod 4 is biramous with an elongated exopod (*Antromysis*, *Parvimysis*, *Surinamysis*, *Taphromysis*). The modified pleopod 4 may be important in the taxonomic diagnosis of genera and species.

Abdominal somite 6 bears the telson as well as the last abdominal appendages, the uropods. The telson is the terminal somite (rather than appendage) of the body and thus the terms anterior-posterior rather than proximal-distal apply. The apex or posterior margin of the telson of Neotropical mysids may be convex, truncate, emarginate, or cleft; the cleft/emargination may be unarmed or armed with spinulae. The lateral margins of the telson may be unarmed or armed partially or completely with spiniform setae. Each uropod comprises a uniarticulate protopod and two branches, the exopod and endopod. All Neotropical mysids have a statocyst located at the base of each uropodal endopod; this equilibrium organ appears as a large vesicle containing a statolith. A spiniform seta may extend distal to the statocyst.

MATERIAL PREPARATION AND PRESERVATION

Mysids can be preserved in the field in a 5% buffered formalin solution or 70%–95% ethanol. If formalin is used, specimens should be transferred to 70% ethanol within seven to ten days. For long-term storage, 70% ethanol with 1% glycerine is a good medium because flexibility of specimens is maintained and desiccation is prevented.

To identify mysids, manipulate specimens in ethanol or water. Although dissection is often not necessary for larger specimens that can be identified using a dissecting microscope, smaller specimens may require dissection, clearing, and viewing with a compound microscope. Examine smaller specimens <10 mm using a depression slide by placing the specimen in the depression with liquid, and covering with a cover slip. By rolling the specimen under the cover slip, you can obtain the desired orientations. If clearing is preferred, place in the depression a few drops of an ethanol/glycerine mixture (8 parts of 70% ethanol: 2 parts glycerine), wait a few minutes for the alcohol to evaporate leaving mostly glycerine, and then place a cover slip over the depression. Any dissections should be performed in this mixture and body parts should be transferred with glycerine to the flat part of the depression slide or to another slide and covered with small round or square cover slip. To protect delicate or thick parts from too much compression, support the cover slip with small pieces of broken cover slips. Add pure glycerine as necessary. These "temporary" glycerine mounts partially clear specimens or parts and last for an extended period of time. If glycerine does not provide sufficient clearing, use lactic acid for a temporary mount. Place specimens in a 50% aqueous solution and then full strength lactic acid. However, do not store specimens or parts in acid as they will continue to clear and may disarticulate.

Staining specimens in methylene blue or chlorazol black E may facilitate the examination of certain external morphological structures, especially setae. Both stains are prepared as 1% solutions by weight in distilled water or 70% ethanol. Add a few drops of the methylene blue solution to diluted or full strength lactic acid. To use chlorazol black E, add the stain to lactic acid or glycerine to make about a 5% solution. Since chlorazol black E is chitin specific, leaving specimens too long in the stain solution will turn them very dark.

Although specimens or parts can be mounted permanently, the mounted parts cannot be manipulated for three-dimensional observation. I favor permanent storage in an ethanol-glycerine solution in glass or plastic vials with small specimens or parts placed in polyethylene embedding capsules with hinged caps, containers used in electron microscopy specimen preparation. Several capsules may be placed in one vial or larger container. For permanent slides, a variety of mounting media, such as glycerine jelly, lactophenol, and Canada balsam, may be used. Some of these substances must be sealed by ringing the cover slip with waterproof polyurethane, Canada balsam, or clear nail polish; slides should be stored horizontally.

KEY TO MYSIDA

Mysida: Mysidae: Genera

1 Eyes, normally developed, eyestalks separate, cornea pigmented; telson, length (excluding apical setae) 1.1–1.7× width at broadest interval, lateral margin with 4–20 or more spiniform setae (Figs. 23.180 A, C, D, 181 G–P, 182 A–D, 183 H–L, 184 A–D) 2

1′ Eyes, eyestalks separate without pigmentation or fused into ocular plate with narrow band of pigmentation along outer edge; telson, length 0.9–1.1× width at broadest interval, lateral margin with 0–4 spiniform setae (Figs. 23.181 A–F, 183 E–G) .... ***Antromysis* [p. 975]**

2(1) Telson, posterior margin cleft, emarginate, or truncate, lateral margin with 4–11 spiniform setae (Figs. 23.180 D, 181 H–P, 182 A–D) 3

2′ Telson, posterior margin convex, lateral margin with at least 13 spiniform setae (Fig. 23.181 G) .... *Americamysis almyra* (Bowman, 1964)
[USA: Florida. Nearctic]

3(2) Telson, posterior margin cleft, emarginate or truncate, lined with 0–25 spinulae or spiniform setae, lateral margin with four to eight spiniform setae; uropodal endopod without spiniform seta on medial margin near distal edge of statocyst (Figs. 23.180 D, 181 I–P, 182 A–D, F) 4

3′ Telson, posterior margin cleft, lined with 15–20 spinulae, lateral margin with 7–11 spiniform setae; uropodal endopod with one spiniform seta on medial margin near distal edge of statocyst (Figs. 23.181 H, 182 E) .... *Taphromysis bowmani* Băcescu, 1961
[USA: Florida. Nearctic]

4(3) Telson, posterior margin emarginate or truncate, lined with 0–10 spinulae or spiniform setae, emargination length ≤ 0.05× telson length; antennular peduncle, article 3 distally with two flagella in males and females (Figs. 23.181 I–P, 183 H–L) .... ***Parvimysis* [p. 979]**

4′ Telson, posterior margin cleft, lined with 12–25 spinulae, cleft length 0.08–0.20× telson length; antennular peduncle, article 3 distally with three flagella in males and two in females (23.180 D, 182 A–D, 184 D, E) .... ***Surinamysis* [p. 980]**

FIGURE 23.179 Mysida: (A) *Taphromysis bowmani*; (B) *Americamysis almyra*. *Photographs courtesy of McLaughlin et al. (2005).*

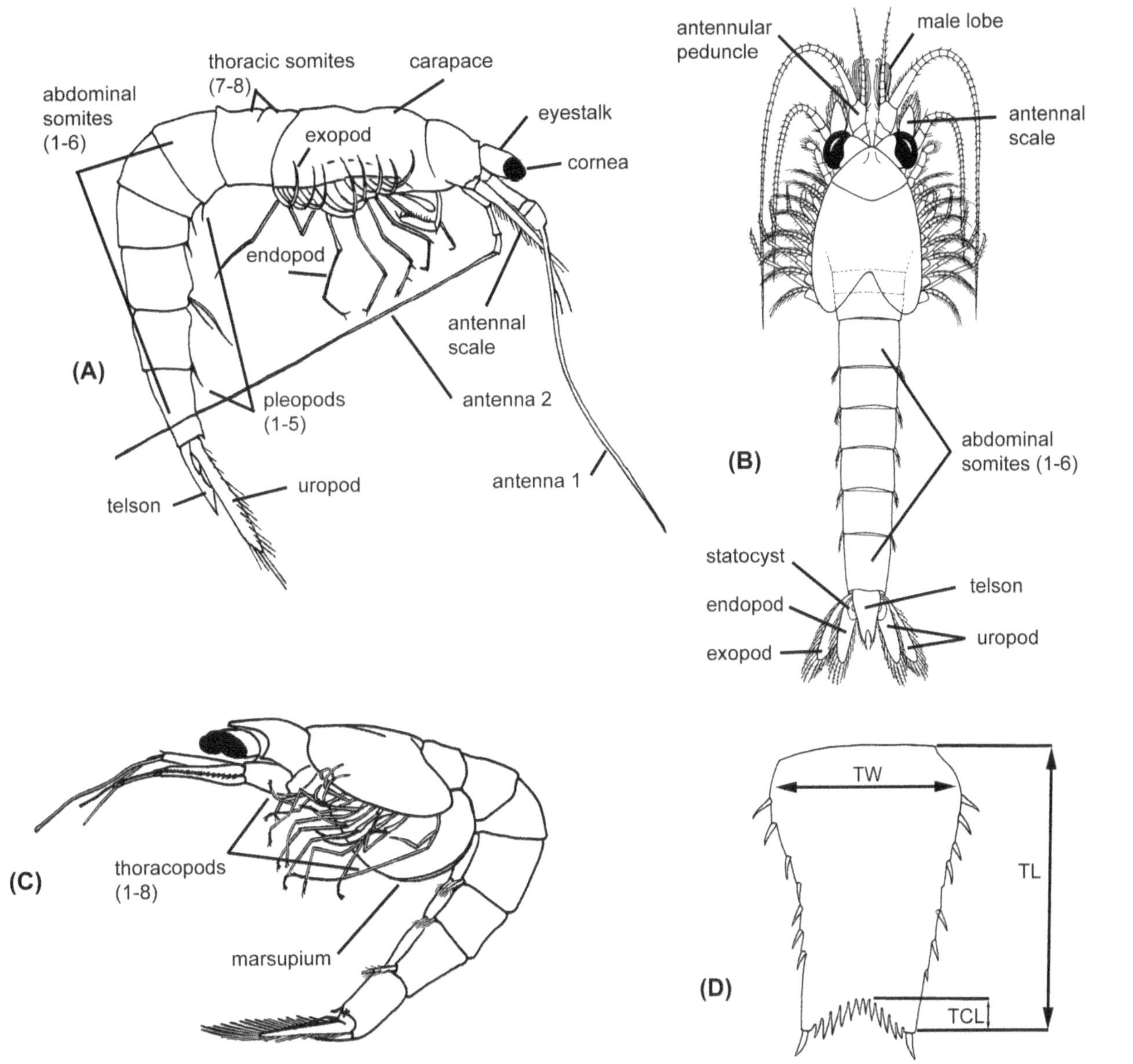

FIGURE 23.180 Mysida morphology: (A) *Surinamysis robertsonae*, male, lateral; (B) Morphology of a typical male mysid, dorsal; (C) *Parvimysis pisciscibus*, ovigerous female, lateral; (D) *Surinamysis merista*, telson. Abbreviations: *TCL*, telson cleft length; *TL*, telson length; *TW*, telson width at broadest interval. *(A) from Bamber & Henderson (1990); (B) modified from Smith (1964); (C) modified from Henderson & Bamber (1983); (D) from Bowman (1980).*

Mysida: Mysidae: *Antromysis*: Species

1 Eyes, eyestalks separated with no pigmentation; telson, lateral margins basally with no spiniform setae (Figs. 23.181 B–F, 183 F, G) 2

1′ Eyes, eyestalks fused proximally to form ocular plate, pigmentation forming narrow band along edge of ocular plate; telson, lateral margins basally with one pair of spiniform setae (Figs. 23.181 A, 183 E) .... *Antromysis anophelinae* Tattersall, 1951
[Costa Rica: Puntarenas]

2(1) Telson, posterior end with 0–6 spiniform setae, no marginal setae (Fig. 23.181 C–F) 3

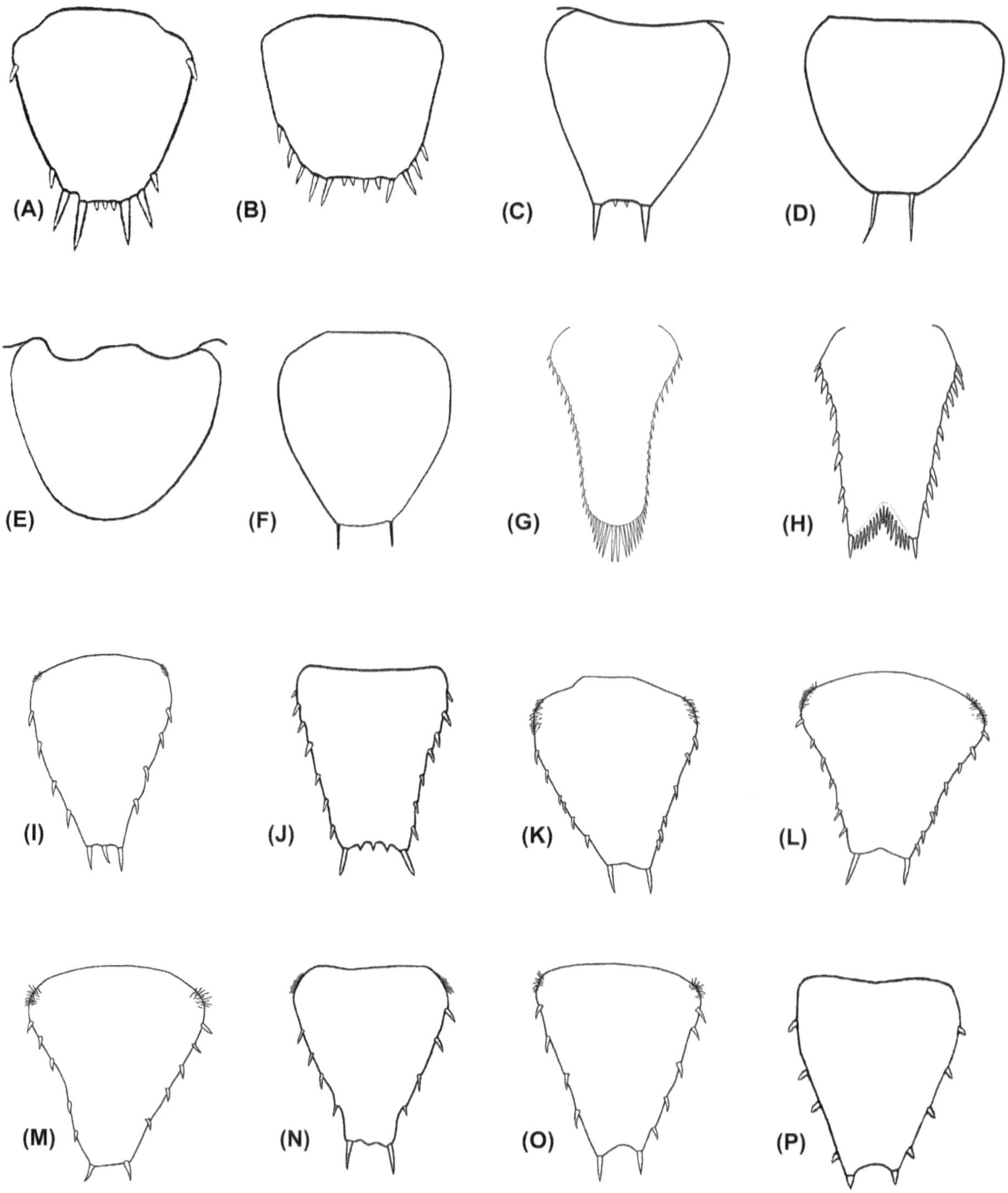

FIGURE 23.181 Mysida telson morphology: (A) *Antromysis anophelinae*; (B) *A. reddelli*; (C) *A. cenotensis*; (D) *A. peckorum*; (E) *A. juberthiei*; (F) *A. cubanica*; (G) *Americamysis almyra*; (H) *Taphromysis bowmani*; (I) *Parvimysis tridens*; (J) *P. almyra*; (K) *P. macrops*; (L) *P. fluviatilis*; (M) *P. lacustris*; (N) *P. amazonica*; (O) *P. fittkaui*; (P) *P. pisciscibus*. *(A) From Tattersall (1951); (B–D) from Bowman (1977); (E) modified from Băcescu & Orghidan (1977); (F) modified from Băcescu & Orghidan (1971); (G, H) from Brattegard (1969); (I, K–O) modified from Wittmann (2018); (J) from Brattegard (1977); (P) modified from Henderson & Bamber (1983).*

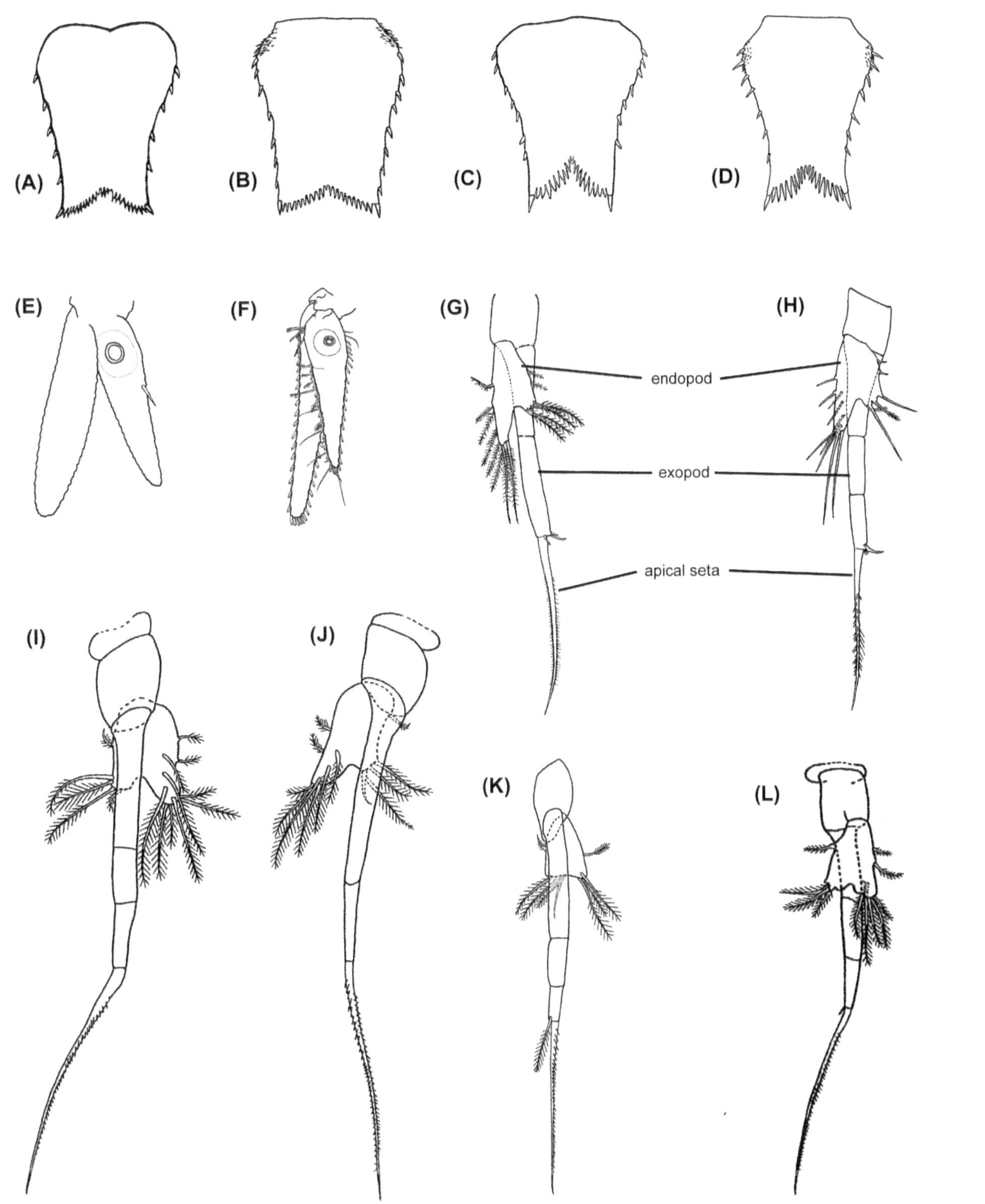

FIGURE 23.182 Mysida morphology. Telson: (A) *Surinamysis robertsonae*; (B) *S. rionegrensis*; (C) *S. americana*; (D) *S. aestuaria*. Uropod: (E) *Taphromysis bowmani*; (F) *Surinamysis merista*. Male pleopod 4: (G) S. *americana*; (H) *S. merista*; (I) *Parvimysis macrops*; (J) *P. fluviatilis*; (K) *P. lacustris*; (L) *P. amazonica*. *(A) From Bamber & Henderson (1990); (B, D) modified from Wittmann (2017); (C, G) from Brattegard (1977); (E) from Brattegard (1969); (F, H) from Bowman (1980); (I–L) modified from Wittmann (2018).*

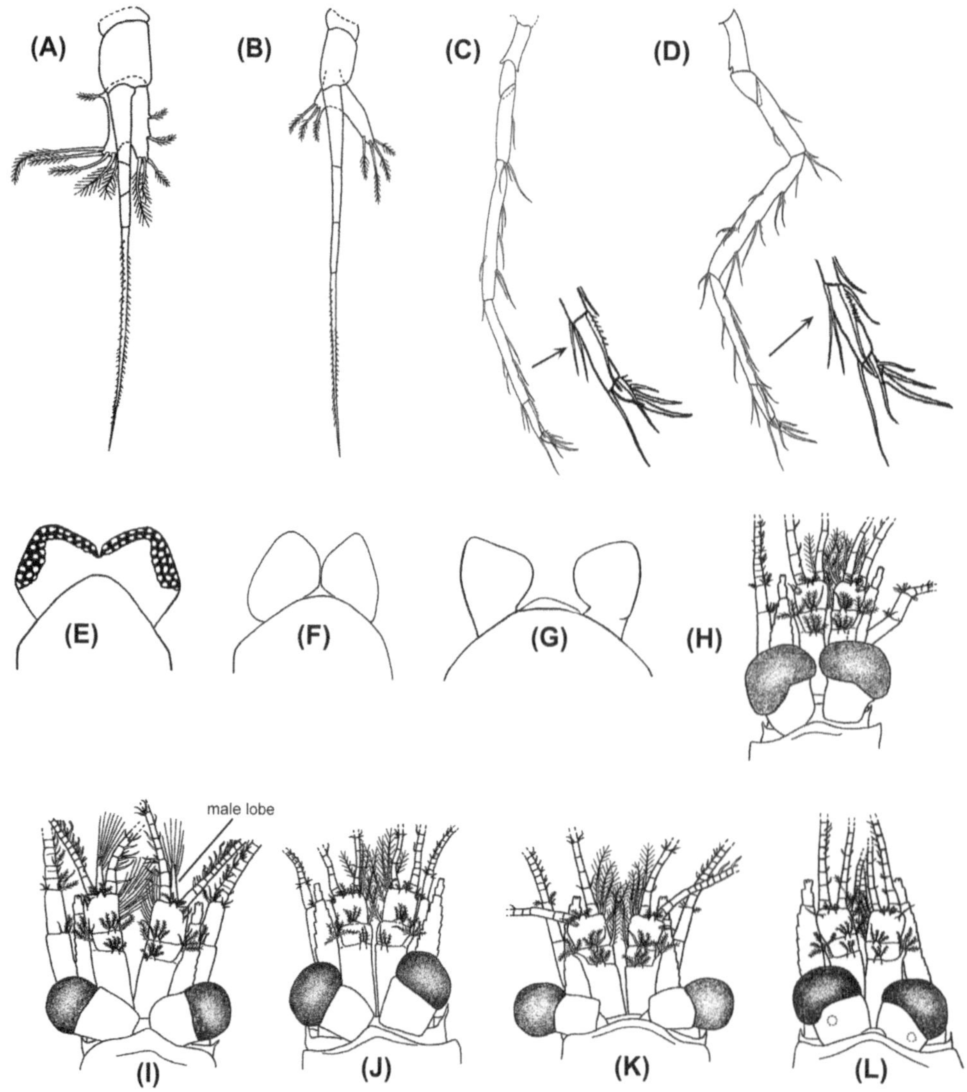

FIGURE 23.183 Mysida morphology. Male pleopod 4: (A) *Parvimysis fittkaui*; (B) *P. pisciscibus*. Thoracic endopod 8: (C) *P. pisciscibus*; (D) *P. amazonica*. Anterior end: (E) *Antromysis anophelinae*; (F) *A. cenotensis*; (G) *A. cubanica*; (H) *Parvimysis macrops*, female; (I) *P. lacustris*, male; (J) *P. amazonica*, female; (K) *P. fittkaui*, female; (L) *P. pisciscibus*, female. *(A–D, H–L) Modified from Wittmann (2018); (E) from Tattersall (1951); (F) from Bowman (1977); (G) modified from Băcescu & Orghidan (1971).*

2′ Telson, 12–14 spiniform setae, posterior end with four short medial setae bordered by four to five longer setae decreasing in length anteriorly and extending along 1/4–1/3 of lateral margins (Fig. 23.181 B) .... *Antromysis reddelli* Bowman, 1977
[Mexico: Oaxaca]

3(2) Eyes, eyestalks close together, medial margins almost in contact (Fig. 23.183 F) 4

3′ Eyes, eyestalks widely separated (Fig. 23.183 G) 5

4(3) Telson, length (excluding apical setae) slightly greater than width at broadest interval, posterior end with a strong spiniform seta on each corner, one to three, usually two, medial short spiniform setae, posterior end width 0.3–0.5× width at broadest interval (Fig. 23.181 C) .... *Antromysis cenotensis* Creaser, 1936
[Mexico: Yucatan]

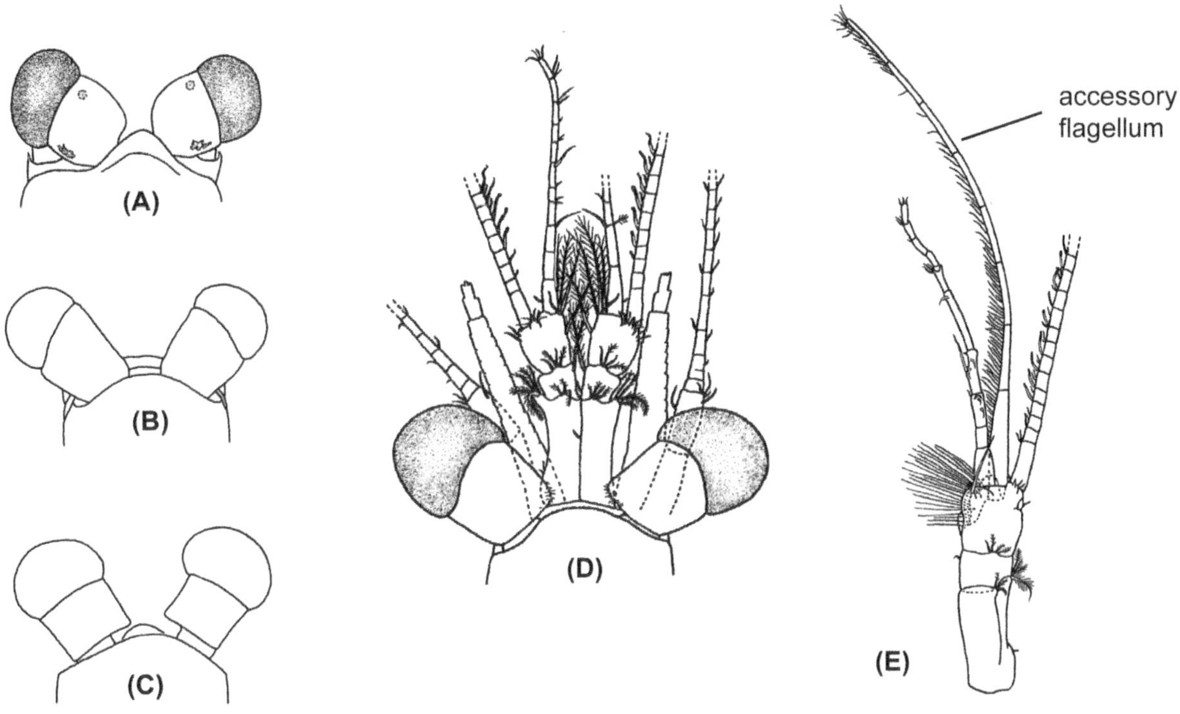

FIGURE 23.184 Mysida morphology. Anterior end: (A) *Parvimysis fluviatilis*; (B) *Surinamysis americana*; (C) *S. merista*; (D) *S. aestuaria*, female. Antenna 1, male: (E) *S. aestuaria*. *(A) Modified from Wittmann (2018); (B) from Brattegard (1977); (C) from Bowman (1980); (D) and (E) from Wittmann (2017).*

4′ Telson, width at broadest interval slightly wider than length, posterior end with a strong spiniform seta on each corner, 0–2, usually 0, medial short spiniform setae, posterior end width ~ 0.2× width at broadest interval (Fig. 23.181 D) .... *Antromysis peckorum* Bowman, 1977
[Jamaica: Clarendon]

5(3) Telson, posterior end with no setae (Fig. 23.181 E). .... *Antromysis juberthiei* Bacescu & Orghidan, 1977
[Cuba: Isla de la Juventud]

5′ Telson, posterior end with one to three, usually two setae (Figs. 23.181 F). .... *Antromysis cubanica* Bacescu & Orghidan, 1971
[Cuba: Isla de la Juventud, Sancti Spiritus]

Mysida: Mysidae: *Parvimysis*: Species

Modified from Wittmann (2018).

1 Telson, posterior margin armed with two apical lobes, each with one short, spiniform seta, 1–10 spinulae or spiniform setae between lobes (Fig. 23.181 I, J) 2

1′ Telson, posterior margin armed with two apical lobes, each with one short, spiniform seta, no spinulae or spiniform setae between lobes (Fig. 23.181 K–P) 3

2(1) Telson, posterior margin with one median spiniform seta between the two apical lobes, anterior maximum width ~ 3.8× as wide as width at posterior end (Fig. 23.181 I) .... *Parvimysis tridens* Wittmann, 2018
[Brazil: Amazonas]

2′ Telson, posterior margin with 3–10 spinulae between the two apical lobes, anterior maximum width 1.8–2.1× as wide as width at posterior end (Fig. 23.181 J) .... *Parvimysis almyra* Brattegard, 1977
[Suriname: Commewijne, Wanica]

3(1) Cornea diameter $< 2.5\times$ length of article 3 of antennular peduncle; fourth male pleopod, exopod 2 or 3-articulated, terminal article $\leq$ subterminal article length (Figs. 23.182 J–L, 183 A, B, I–L, 184 A) 4

3′ Cornea large, maximum diameter $>2.5\times$ length of article 3 of antennular peduncle; male pleopod 4, exopod 3-articulated, article 3 longer than article 2; telson, lateral margin with six spiniform setae (Figs. 23.181 K, 182 I, 183 H) .... *Parvimysis macrops* Wittmann, 2018
[Brazil: Amazonas]

4(3) Telson, lateral margin with three to five spiniform setae; male pleopod 4, exopod 3-articulated (Figs. 23.181 M–P, 182 K, L, 183 A, B) 5

4′ Telson, lateral margin with six to eight spiniform setae; male pleopod 4, exopod 2-articulated (Figs. 23.181 L, 182 J) .... *Parvimysis fluviatilis* Wittmann, 2018
[Brazil: Amazonas]

5(4) Thoracopod 8, endopod with oblique articulation between carpus and propodus; male pleopod 4, exopod ending in one large modified seta accompanied by one shorter (minute to well-developed) seta (Figs. 23.182 K, L, 183 D) 6

5′ Thoracopod 8, endopod with transverse articulation between carpus and propodus; male pleopod 4, exopod ending in one large modified seta with no accompanying setae (Fig. 23.183 A–C) 7

6(5) Antenna 1, peduncle extending to slightly less than end of antennal scale; male pleopod 4, exopod, article 1 > combined length of articles 2 and 3, article 3 ending in one large modified seta accompanied by a shorter plumose seta (Figs. 23.182 K, 183 I) .... *Parvimysis lacustris* Wittmann, 2018
[Brazil: Amazonas]

6′ Antenna 1, peduncle extending to or slightly beyond article 1 of antennal scale; male pleopod 4, exopod, article 1 < combined length of articles 2 and 3, article 3 ending in one large modified seta accompanied by a minute simple seta (Figs. 23.182 L, 183 J) .... *Parvimysis amazonica* Wittmann, 2018
[Brazil: Amazonas, Roraima]

7(5) Antenna 1, peduncle extending slightly beyond article 1 of antennal scale; male pleopod 4, exopod, terminal large modified seta much longer than combined lengths of three articles of exopod (Fig. 23.183 A, K) .... *Parvimysis fittkaui* Wittmann, 2018
[Brazil: Amazonas, Roraima]

7′ Antenna 1, peduncle not extending to article 2 of antennal scale; male pleopod 4, exopod, terminal large modified seta $\leq$ exopod length (Fig. 23.183 B, L). .... *Parvimysis pisciscibus* Henderson & Bamber, 1983
[Brazil: Amazonas, Roraima]

Mysida: Mysidae: *Surinamysis*: Species

Modified from Wittmann (2017, 2018)

1 Telson, apical cleft ~ 0.08–0.10× telson length (excluding apical setae) (Fig. 23.182 A, B) 2

1′ Telson, apical cleft ~0.12–0.20× telson length (Figs. 23.180 D, 182 C, D) 3

2(1) Telson, lateral margins armed with four to five spiniform setae (Figs. 23.180 A, 182 A) .... *Surinamysis robertsonae* Bamber & Henderson, 1990
[Brazil: Amazonas, Roraima]

2′ Telson, lateral margins armed with seven to eight spiniform setae (Figs. 23.182 B) .... *Surinamysis rionegrensis* Wittmann, 2017
[Brazil: Amazonas]

3(1) Cornea length 0.45–0.50× eye length; male pleopod 4, exopod 3-articulated (Fig. 23.182 H, 184 C, D) 4

3′ Cornea length 0.30–0.38× eye length; male pleopod 4, exopod 2-articulated (Figs. 23.182 G, 184 B) .... *Surinamysis americana* (Tattersall, 1951)
[Suriname: Commewijne, Paramaribo]

4(3) Telson, apical cleft 0.15–0.20× telson length (excluding apical setae), length 1.3× width at broadest interval, apical cleft armed with 15–19 spinulae (23.182 D) .... *Surinamysis aestuaria* Wittmann, 2017
[Brazil: Para]

4′ Telson, apical cleft ~ 0.12× telson length, length ~ 1.5× width at broadest interval, apical cleft armed with 12–16 spinulae (Fig. 23.180 D) .... *Surinamysis merista* (Bowman, 1980)
[Venezuela: Monagas]

REFERENCES

Băcescu, M. & T.N. Orghidan. 1971. *Antromysis cubanica* n.sp. et *Spelaeomysis nuniezi* n.sp., Mysis cavernicoles nouvelles de Cuba. Revue Roumaine de Biologie. Serie de Zoologie 16: 225–231.

Băcescu, M., T.N. Orghidan. 1977. New contribution to the knowledge of troglobian mysids of Cuba: *Antromysis juberthiei* n.sp. Pages 263–265 *in*: T. Orghidan, A. Nuñez Jiménez, V. Decou, S. Negra, & N. Vina Bayes (eds.), Résultats des Expéditions Biospéléologiques Cubano-Roumaines à Cuba, Vol. 2. Editura Academiei Republicii Socialiste România, Bucuresti.

Bamber R.N. & P.A. Henderson. 1990. A new freshwater mysid from the Amazon, with a reassessment of *Surinamysis* Bowman (Crustacea: Mysidacea). Zoological Journal of the Linnean Society 100: 393–401.

Bowman, T. 1977. A review of the genus *Antromysis* (Crustacea: Mysidacea), including new species from Jamaica and Oaxaca, Mexico, and a redescription and new records for *A. cenotensis*. Association for Mexican Cave Studies, Bulletin 6: 27–38.

Bowman, T. 1980. *Antromysis* (*Surinamysis*) *merista*, a new freshwater mysid from Venezuela (Crustacea: Mysidacea). Proceedings of the Biological Society of Washington 93: 208–215.

Boxshall, G.A. & D. Jaume. 2000. Discoveries of cave misophrioids (Crustacea: Copepoda) shed new light on the origin of anchialine faunas. Zoologischer Anzeiger 239: 1–19.

Brattegard, T. 1969. Marine biological investigations in the Bahamas. 10. Mysidacea from shallow water in the Bahamas and Southern Florida. Part I. Sarsia 39: 17–106.

Brattegard, T. 1977. Three species of Mysidacea (Crustacea) from Surinam. Zoologische Mededelingen 50: 282–293.

Henderson, P.A. & R.N. Bamber. 1983. A new species of mysid (Crustacea: Mysidacea) from the Amazon Basin. Journal of Natural History 17: 139–143.

McLaughlin, P.A., D.K. Camp, M.V. Angel, E.L. Bousfield, P. Brunei, R.C. Brusca, D. Cadien, A.C. Cohen, K. Conlan, L.G. Eldredge, D.L. Felder, J.W. Goy, T. Haney, B. Hann, R.W. Heard, E.A. Hendrycks, H.H. Hobbs III, J.R. Holsinger, B. Kensley, D.R. Laubitz, S.E. LeCroy, R. Lemaitre, R.F. Maddocks, J.W. Martin, P. Mikkelsen, E. Nelson, W.A. Newman, R.M. Overstreet, W.J. Poly, W.W. Price, J.W Reid, A. Robertson, D.C. Rogers, A. Ross, M. Schotte, F.R. Schram, C. Shih, L. Watling, G.D.F. Wilson & D.D. Turgeon. 2005. Common and Scientific Names of Aquatic Invertebrates from the United States and Canada: Crustaceans. American Fisheries Society Porter, M., K. Meland & W. Price. 2008. Global diversity of mysids (Crustacea-Mysida) in freshwater. Hydrobiologia 595: 213–218.

Porter, M., K. Meland & W. Price. 2008. Global diversity of mysids (Crustacea-Mysida) in freshwater. Hydrobiologia 595: 213–218.

Smith, R.I. 1964. Keys to the Marine Invertebrates of the Woods Hole Region. Systematics-Ecology Program, Contribution No. 11, Marine Biological Laboratory, Woods Hole, MA. 208 pp.

Tattersall, W. 1951. A review of the Mysidacea of the United States National Museum. Smithsonian Institution, United States National Museum Bulletin 201: 1–292.

Wittmann, K.J. 2017. The genus *Surinamysis* (Mysida, Mysidae, Diamysini) from Amazonia and the coast of Brazil, with descriptions of two new species. Crustaceana 90: 359–380.

Wittmann, K.J. 2018. Six new freshwater species of *Parvimysis*, with notes on breeding biology, statocyst composition, and a key to the Mysidae (Mysida) of Amazonia. Crustaceana 91: 537–576.

Wittmann, K.J., A.P. Ariani & J-P Lagardere. 2014. Orders Lophogastrida Boas, 1883, Stygiomysida Tchindonova, 1981, and Mysida Boas, 1883 (also known collectively as Mysidacea). Pages 189–396 *in*: J.C von Vaupel Klein, M. Charmantier-Daures & F.R. Schram (eds.). Treatise on Zoology- Anatomy, Taxonomy, Biology, The Crustacea vol. 4, Part B. Brill, Leiden, Netherlands.

Stygiomysida

W. Wayne Price
Department of Biology, University of Tampa, Tampa, FL, United States

INTRODUCTION

The Stygiomysida (stygiomysids), a small group of subterranean mysidaceans, comprises two monogeneric families: Lepidomysidae (*Spelaeomysis*, 9 species) and Stygiomysidae (*Stygiomysis*, 7 species) (Mees & Meland, 2012). Most occur in coastal fresh to brackish groundwater environments such as caves, wells, and the freshwater layer in anchihaline habitats in the tropics and subtropics. The majority of stygiomysids are found in the Caribbean and Mediterranean and surrounding areas, a distributional pattern suggesting a Tethyan origin for this group (see Order Mysida-Introduction for details) (Porter et al., 2008; Wittmann et al., 2014).

Of the 16 known species, 9 are found in Neotropical fresh to brackish water habitats, mostly in the Caribbean and Mexico. All Neotropical species are blind and live in wells, caves, and anchihaline habitats, except for *Spelaeomysis cardisomae*, which has a few ommatidia on reduced corneas (Fig. 23.185). This species has been reported from pools of brackish water at the bottom of burrows of two species of land crabs, *Cardisoma crassipes* from Peru and *C. guanhumi* from San Andrés, Isla Providencia (Bowman, 1973) and Belize (present study) in the western Caribbean. Additional specimens were collected from caves on San Andrés (Bowman, 1986) and from the upper reaches of estuaries and their freshwater tributaries from both coasts in south Florida. *Spelaeomysis cardisomae* may be an interstitial species that tunnels through groundwater habitats into crab burrows, caves, and estuaries (Bowman, 1986). The geographic distribution of this species is the most widespread (and interesting) of the Neotropical stygiomysids. Of the 8 remaining species, 6 have distributions restricted to one Caribbean island or one area in Mexico. Only *Stygiomysis holthuisi* (five Caribbean islands; Mexico) and *S. clarkei* (Turks and Caicos; Bahamas) are more widespread.

LIMITATIONS

Identification of Neotropical stygiomysids is based mainly on variations in the morphology of the carapace, eye, antennal scale, telson, and uropod, characters that exhibit little sexual dimorphism. These and other basic morphological characters included in the key are presented in Fig. 23.186. Whenever possible, mature adult specimens should be examined because many characters used in the key are from adults as opposed to juveniles and immatures. This is especially important for meristic characters such as the number and size of setae on the lateral and posterior margins of telsons and uropodal protopods. Setae may be added with successive molts as the organism matures. In addition, the length:width ratios of telsons may change with growth. Unfortunately, it may be difficult to separate immature stygiomysids from adults and even adult females from males since little sexual dimorphism exists. Some characters that separate these developmental stages are presented in the next section.

Additional problems with identification of stygiomysids result from the small number of specimens examined for some species and the limited areas collected. Many more collections are needed before the morphological variability and geographic distribution of the species in the Stygiomysida can be described adequately.

TERMINOLOGY AND MORPHOLOGY

Neotropical adult stygiomysids vary in length from less than 5 to more than 20 mm. The morphology of this group is similar to the Mysida (see above), but varies in several characters that are useful in separating the two orders. All

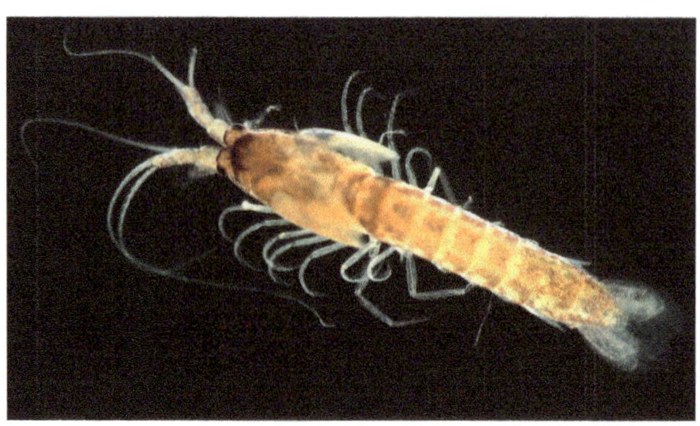

FIGURE 23.185 Stygiomysida, *Spelaeomysis cardisomae. Courtesy of Ernst Peebles.*

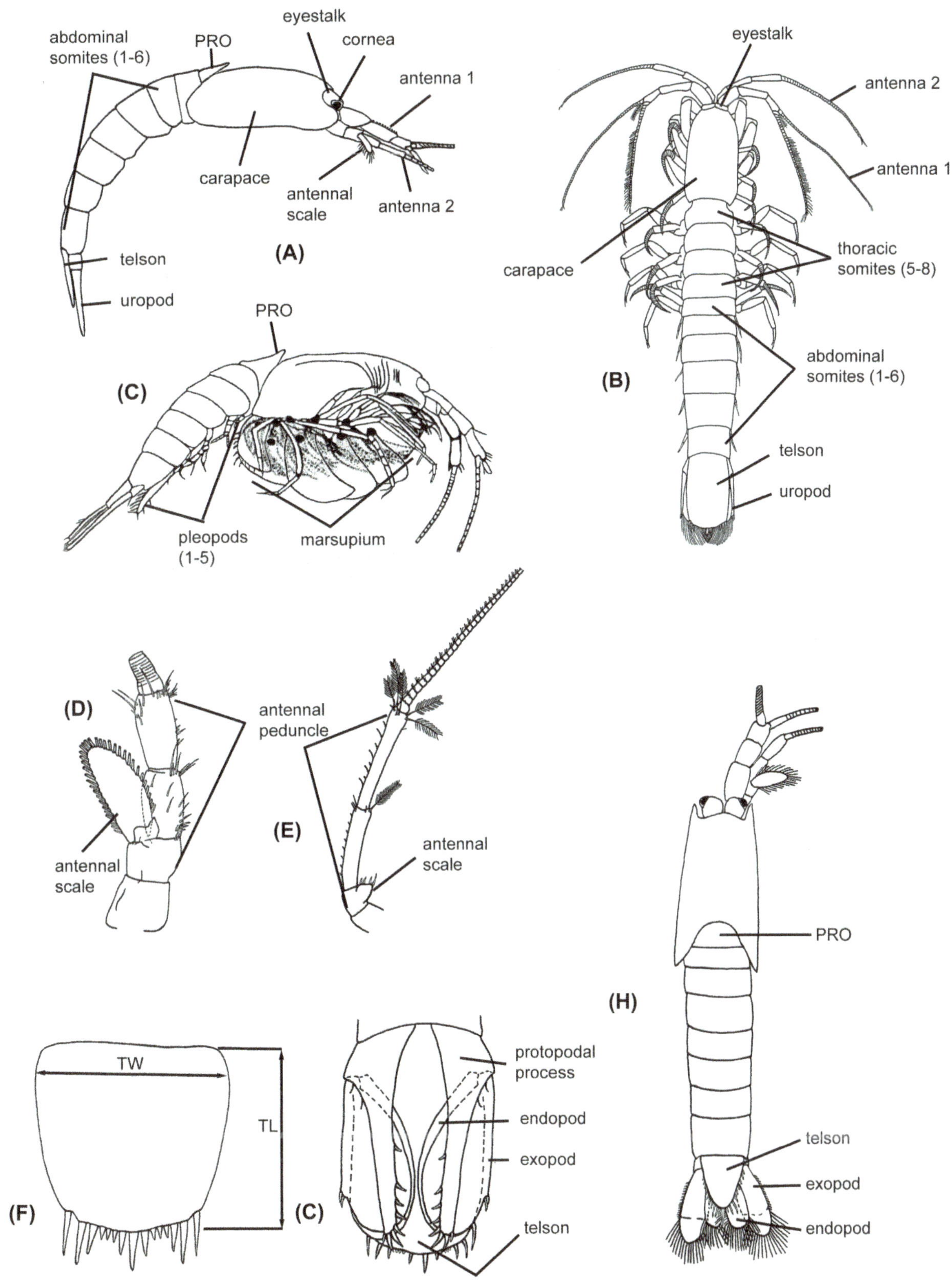

FIGURE 23.186 Stygiomysida morphology: (A) *Spelaeomysis cardisomae*, male, lateral; (B) *Stygiomysis cokei*, dorsal; (C) *Spelaeomysis nuniezi*, ovigerous female, lateral; (D) *S. cardisomae*, antenna 2; (E) *Stygiomysis cokei*, antenna 2; (F) *Stygiomysis clarkei*, telson; (G) *Stygiomysis cokei*, telson and uropods (setae removed); (H) *Spelaeomysis cardisomae*, male, dorsal. Abbreviations: *PRO*, thoracic somite 7 projection; *TL*, telson length; *TW*, telson width at broadest interval. *(A), (D) and (H) from Bowman (1973); (B), (E) and (G) from Kallmeyer & Carpenter (1996); (C) from Ortiz et al. (2005); (F) from Bowman et al. (1984).*

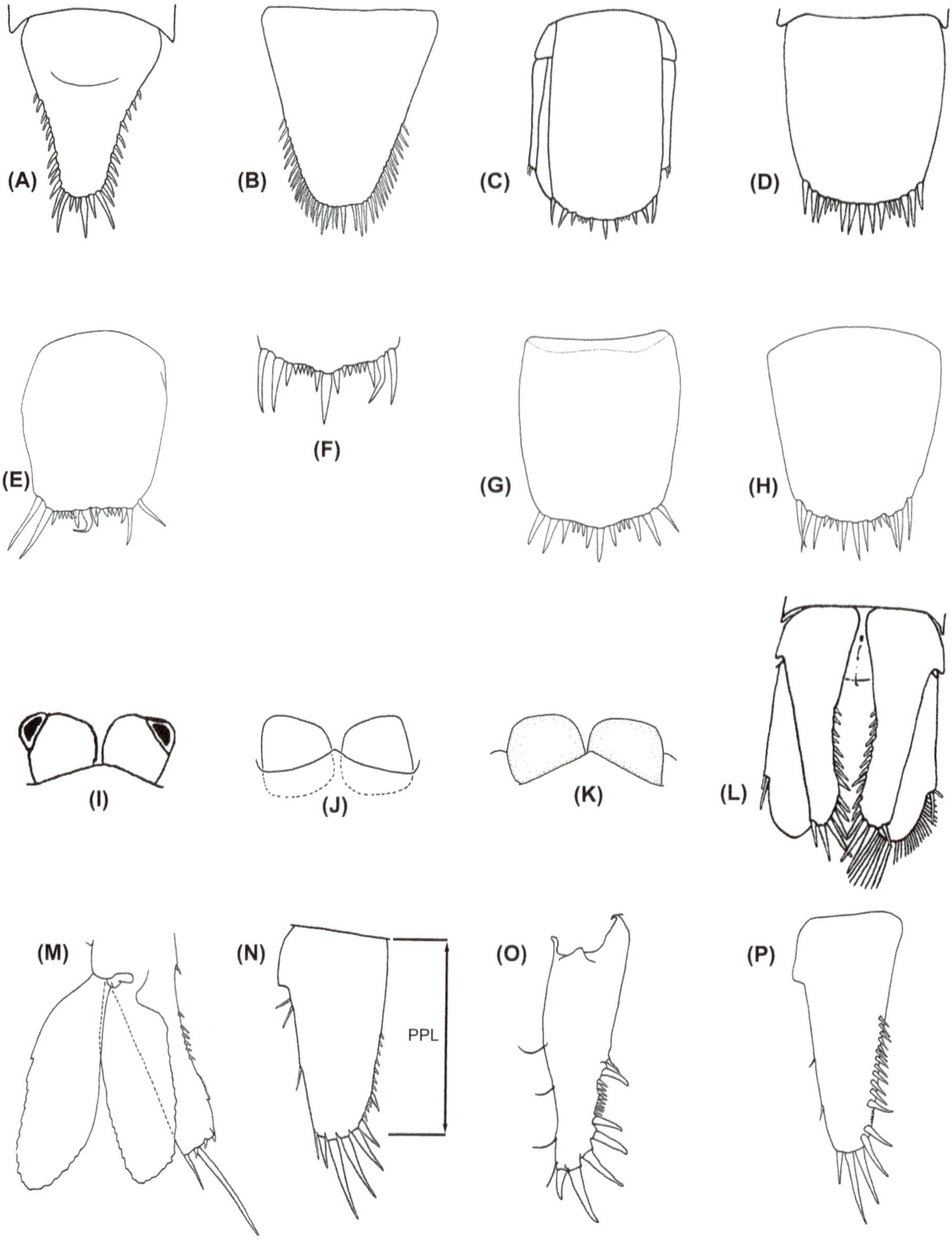

FIGURE 23.187 Stygiomysida telson morphology: (A) *Spelaeomysis nuniezi*; (B) *S. olivae*; (C) *Stygiomysis cokei*; (D) *S. major*; (E) *S. ibarrae*; (F) *S. ibarrae*, posterior end, close-up; (G) *S. aemete*; (H) *S. holthuisi*. Eyes: (I) *Spelaeomysis cardisomae*; (J) *S. nuniezi*; (K) *S. olivae*. Uropod(s): (L) *Stygiomysis major*, ventral; (M) *S. ibarrae*, dorsal. Protopodal process of uropod: (N) *S. clarkei*. Abbreviations: PPL, protopodal process length; (O) *S. aemete*; (P) *S. holthuisi*. *(A) and (J) from Gordon & Botosaneanu (1985); (B), (I) and (K) from Bowman (1973); (C) modified from Kallmeyer & Carpenter (1996); (D) and (L) from Bowman (1976); (E), (F) and (M) from Ortiz et al. (1996); (G) and (O) from Wagner (1992); (H), (N) and (P) from Bowman et al. (1984).*

stygiomysids exhibit some degree of eye reduction in which the cornea is diminished to a few or no ommatidia.

The marsupium of mature female stygiomysids is composed of four to seven pairs of oostegites. Members of the Stygiomysidae have four pairs on thoracopods 3–6 and Lepidomysidae has seven pairs on thoracopods 2–8. In contrast to most Mysida, mature male stygiomysids have no penes, but their gonopores are bordered by an anterior setose lobe and a posterior bare lobe (Wittmann, 2013; Wittmann et al., 2014; Meland et al., 2015).

The **pleopods** of stygiomysids are biramous, reduced in both sexes and consist of a uniarticulate protopod, a uniarticulate endopod, and a multiarticulate (2–4) exopod. Sexual dimorphism is exhibited by the pleopod 2 exopod in which males have fewer articles but more setae than females in the species for which both sexes have been examined. Abdominal somites 3–5 have transverse membranous lamellae extending backwards from their posterior sternal margins (Wittmann et al., 2014; Meland et al., 2015).

The **telson** of stygiomysids is entire with spiniform setae on the apex and, in some species, along the lateral margins. The **uropodal** protopod is produced on the ventral surface into a posteriorly directed prolongation, the **protopodal process**, which is armed with a series of apical and medial spiniform setae and spinulae. This process is distinct in both families, but more robust and longer in Stygiomysidae than Lepidomysidae. The uropodal endopod lacks a statocyst, and the exopod of the Lepidomysidae, but not the Stygiomysidae, has a distal suture (Wittmann et al., 2014; Meland et al., 2015).

Although the Lepidomysidae and the Stygiomysidae are closely related, several morphological characteristics, in addition to the ones mentioned above, separate them. In the Lepidomysidae, the carapace extends at least to thoracic somite 7, the antennal scale is small, but normally developed, only thoracic endopods 1 and 2 are normally modified as maxillipeds, and a dorsal scaliform projection from thoracic somite 7 overlaps with or fits into the posteriomedial margin of the carapace. Members of the Stygiomysidae have a reduced carapace that does not extend beyond thoracic somite 4, the antennal scale is reduced to a small plate with few setae, thoracic endopods 1–4 are modified as maxillipeds, and there is no anterior projection from thoracic somite 7 (Meland & Willassen, 2007; Wittmann et al., 2014; Meland et al., 2015)

MATERIAL PREPARATION AND PRESERVATION

Stygiomysids are prepared and preserved as for mysids (see Order Mysida, p. 973).

KEY TO STYGIOMYSIDA

Stygiomysida: Families

1 Antennal scale small, but normally developed, setose all around; carapace reaching somite 7 or beyond; thoracic somite 7 with dorsal scaliform projection overlapping or fitting into carapace posteriomedial margin from behind (Fig. 23.186 A, C, D, H). .... Lepidomysidae, one genus: ***Spelaeomysis*** **[p. 985]**

1′ Antennal scale reduced to small plate, with five or fewer setae; carapace reduced, covering only first four thoracic somites; thoracic somite 7 without dorsal scaliform projection (Fig. 23.186 B, E) .... Stygiomysidae, one genus: ***Stygiomysis*** **[p. 985]**

Stygiomysida: Lepidomysidae: *Spelaeomysis*: Species

1 Eyestalks separate, without ommatidia (Fig. 23.187 J, K) 2

1′ Eyestalks separate, with a few distolateral ommatidia (Fig. 23.186 A, D, H, 187 I). .... *Spelaeomysis cardisomae* Bowman, 1973
[USA: Florida. Belize: Belize. Colombia: Isla Providencia, San Andrés. Peru: Tumbes]

2(1) Telson, length (excluding apical setae) ~ 1.5× width at broadest interval, armed with 20–25 spiniform setae on apex and posterior 60% of somewhat concave lateral margins (Figs. 23.186 C, 187 A, J) .... *Spelaeomysis nuniezi* Băcescu & Orghidan, 1971
[Cuba: Sancti Spiritus]

2′ Telson, length (excluding apical setae) ~ 1.1–1.2× width at broadest interval, armed with about 40 spiniform setae on apex and posterior half or less of straight lateral margins (Fig. 23.187 B, K) .... *Spelaeomysis olivae* Bowman, 1973
[Mexico: Oaxaca]

Stygiomysida: Stygiomysidae: *Stygiomysis*: Species

1 Telson, length (excluding apical setae) < 1.3× width at broadest interval; uropod, protopodal process with at least 11 apical and medial spiniform setae, including one or more spinulae (Figs. 23.186 F, 187 D–H, L–P) 2

1′ Telson, length (excluding apical setae) > 1.6× width at broadest interval; uropod, protopodal process with seven to eight fairly large apical and medial spiniform setae, no small spinulae (Figs. 25.186 B, E, G, 187 C) .... *Stygiomysis cokei* Kallmeyer & Carpenter, 1996
[Mexico: Quintana Roo, Yucatan]

2(1) Telson, posterior margin armed with median group of three spiniform setae, outer pair subequal and shorter than medial single seta (Figs. 23.186 F, 187 E–H) 3

2′ Telson, posterior margin armed with median group of three to five subequal spiniform setae (Fig. 23.187 D, L) .... *Stygiomysis major* Bowman, 1976
[Jamaica: Clarendon]

3(2) Telson, posterior margin armed with median group of three spiniform setae, outer pair 0.6–0.8× length of medial single seta; uropod, protopodal process, five strong apical or apical and subapical spiniform setae, shortest seta 0.3–0.5× length of longest (Figs. 23.186 F, 187 G, H, N–P) 4

3′ Telson, posterior margin armed with median group of three spiniform setae, outer pair 0.4–0.5× length of medial single seta; uropod, protopodal process, five apical spiniform setae, shortest seta <0.1× length of longest (Figs. 23.186 C, 187 E, F, M) .... *Stygiomysis ibarrae* Ortiz, Lalana, & Pérez, 1996
[Cuba: Matanzas]

4(3) Telson, length (excluding apical setae) 1.1–1.2× width at broadest interval; uropod, protopodal process, five large apical and subapical spiniform setae, large subapical setae extending proximally 0.1–0.2× length of process (excluding apical setae) (Fig. 23.187 G, H, O, P) 5

4′ Telson, length (excluding apical setae) 0.9–1.0× width at broadest interval; uropod, protopodal process, five large apical spiniform setae (Figs. 23.186 F, 187 N) .... *Stygiomysis clarkei* Bowman, Iliffe & Yager, 1984
[Bahamas: East Grand Bahama Island. Turks and Caicos: Middle Caicos. Providenciales]

5(4) Uropod, protopodal process, five large apical and subapical spiniform setae followed medially by nine small spinulae, then two larger spiniform setae; length to 15 mm (Fig. 23.187 G, O) .... *Stygiomysis aemete* Wagner, 1992
[Dominican Republic: Pedernales]

5′ Uropod, protopodal process, five large apical and subapical spiniform setae followed medially by a series of two to eight small spinulae, then a row of 4–12 large and smaller spiniform setae; length 3.5–10 mm (Fig. 23.187 H, P) .... *Stygiomysis holthuisi* Gordan, 1958
[Bahamas: East Grand Bahama Island. Mexico: Quintana Roo, Yucatan. Puerto Rico: Mayagüez. Anguilla. St. Martin]

REFERENCES

Bowman, T. 1973. Two new American species of *Spelaeomysis* (Crustacea: Mysidacea) from a Mexican cave and land crab burrows. Association for Mexican Cave Studies, Bulletin 5: 13–20.

Bowman, T. 1976. *Stygiomysis major*, a new troglobitic mysid from Jamaica, and extension of the range of *S. holthuisi* to Puerto Rico (Crustacea: Stygiomysidae). International Journal of Speleology 8: 365–373.

Bowman, T. 1986. Mysidacea. Pages 405–409 *in*: L. Botosaneanu (ed.), Stygiofauna Mundi. E. J. Brill, Leiden, Netherlands.

Bowman, T.E., T.M. Iliffe & J. Yager. 1984. New records of the troglobitic mysid genus *Stygiomysis: S. clarkei*, new species, from the Caicos Islands, and *S. holthuisi* (Gordon) from Grand Bahama Island (Crustacea: Mysidacea). Proceedings of the Biological Society of Washington 97: 637–644.

Gordon, I. & L. Botosaneanu. 1985. On a cave-dwelling mysid from Cuba, of the genus *Spelaeomysis* (Mysidacea, Lepidomysinae). Crustaceana 49: 139–149.

Kallmeyer, D.E. & J.H. Carpenter. 1996. *Stygiomysis cokei*, new species, a troglobitic mysid from Quintana Roo, Mexico (Mysidacea: Stygiomysidae). Journal of Crustacean Biology 16: 418–427.

Mees, J. & K. Meland. 2012 onwards. World List of Lophogastrida, Stygiomysida and Mysida. http://www.marinespecies.org/mysidacea accessed at on 2019-10-25. doi:10.14284/369.

Meland, K., J. Mees, M. Porter & K. J Wittmann. 2015. Taxonomic review of the Orders Mysida and Stygiomysida (Crustacea, Peracarida). PloS One 10: 1–28.

Meland, K. & E. Willassen. 2007. The disunity of "Mysidacea" (Crustacea). Molecular Phylogenetics and Evolution 44: 1083–1104.

Ortiz, M., A. García-Debrás, A. Pérez & R. Lalana. 2005. Estudio del marsupio y las larvas de *Spelaeomysis nuniezi* Băcescu y Orghidan, 1971. Revista Biologia 19: 83–84.

Ortiz, M., R. Lalana &. A. Perez. 1996. El primer registro del género *Stygiomysis* (Crustacea, Mysidacea) en la Isla de Cuba y description de una especie nueva. Revista de Investigaciones Marinas 17: 107–115.

Porter, M., K. Meland & W. Price. 2008. Global diversity of mysids (Crustacea-Mysida) in freshwater. Hydrobiologia 595: 213–218.

Wagner, H.P. 1992. *Stygiomysis aemete* n. sp., a new subterranean mysid (Crustacea, Mysidacea, Stygiomysidae) from the Dominican Republic, Hispaniola. Bijdragen tot de Dierkunde 62: 71–79.

Wittmann, K. J. 2013. Comparative morphology of the external male genitalia in Lophogastrida, Stygiomysida, and Mysida (Crustacea, Eumalacostraca). Zoomorphology 132: 389–401.

Wittmann, K.J., A.P. Ariani & J-P Lagardere. 2014. Orders Lophogastrida Boas, 1883, Stygiomysida Tchindonova, 1981, and Mysida Boas, 1883 (also known collectively as Mysidacea) 189–396 *in*: J.C. von Vaupel Klein, M. Charmantier-Daures, & F.R. Schram (eds.), Treatise on Zoology- Anatomy, Taxonomy, Biology, The Crustacea: 4, Part B. Brill. Leiden, Netherlands.

Taxonomic Index

'*Note*: Page numbers followed by "f" indicate figures and "t" indicate tables.'

B

C

D

E

F

I

M

N

Q

R

S

U